P9-CLI-710

Molecular Biology

Fifth Edition

Robert F. Weaver
University of Kansas

The McGraw·Hill Companies

MOLECULAR BIOLOGY, FIFTH EDITION

Published by McGraw-Hill, a business unit of The McGraw-Hill Companies, Inc., 1221 Avenue of the Americas, New York, NY 10020.

This book is printed on acid-free paper.

1 2 3 4 5 6 7 8 9 0 QDB /QDB 1 0 9 8 7 6 5 4 3 2 1

ISBN 978-0-07-352532-7
MHID 0-07-352532-4

Vice President & Editor-in-Chief: *Marty Lange*
Vice President EDP/Central Publishing Services: *Kimberly Meriwether David*
Publisher: *Janice Roerig-Blong*
Executive Marketing Manager: *Patrick E. Reidy*
Project Manager: *Robin A. Reed*
Design Coordinator: *Brenda A. Rolwes*
Cover Designer: *Studio Montage, St. Louis, Missouri*
Lead Photo Research Coordinator: *Carrie K. Burger*
Cover Image: © *Getty Images RF*
Buyer: *Sandy Ludovissy*
Media Project Manager: *Balaji Sundararaman*
Compositor: *Aptara®, Inc.*
Typeface: *10/12 Sabon*
Printer: *Quad/Graphics*

Library of Congress Cataloging-in-Publication Data

Weaver, Robert Franklin, 1942-
Molecular biology / Robert F. Weaver.—5th ed.
p. cm.
ISBN 978–0–07–352532–7 (hardcover : alk. paper)
1. Molecular biology. I. Title.
QH506.W43 2011
572.8—dc22

2010051759

www.mhhe.com

To Camilla and Nora

ABOUT THE AUTHOR

(*Source:* Ashvini C. Ganesh)

Rob Weaver was born in Topeka, Kansas, and grew up in Arlington, Virginia. He received his bachelor's degree in chemistry from the College of Wooster in Wooster, Ohio, in 1964. He earned his Ph.D. in biochemistry at Duke University in 1969, then spent two years doing postdoctoral research at the University of California, San Francisco, where he studied the structure of eukaryotic RNA polymerases with William J. Rutter.

He joined the faculty of the University of Kansas as an assistant professor of biochemistry in 1971, was promoted to associate professor, and then to full professor in 1981. In 1984, he became chair of the Department of Biochemistry, and served in that capacity until he was named Associate Dean of the College of Liberal Arts and Sciences in 1995.

Prof. Weaver is the divisional dean for the science and mathematics departments within the College, which includes supervising 10 different departments and programs. As a professor of molecular biosciences, he teaches courses in introductory molecular biology and the molecular biology of cancer. In his research laboratory, undergraduates and graduate students have participated in research on the molecular biology of a baculovirus that infects caterpillars.

Prof. Weaver is the author of many scientific papers resulting from research funded by the National Institutes of Health, the National Science Foundation, and the American Cancer Society. He has also coauthored two genetics textbooks and has written two articles on molecular biology in the *National Geographic Magazine.* He has spent two years performing research in European laboratories as an American Cancer Society Research Scholar, one year in Zurich, Switzerland, and one year in Oxford, England.

BRIEF CONTENTS

C O N T E N T S

PREFACE

One of my most exciting educational experiences was my introductory molecular biology course in graduate school. My professor used no textbook, but assigned us readings directly from the scientific literature. It was challenging, but I found it immensely satisfying to meet the challenge and understand, not only the conclusions, but how the evidence supported those conclusions.

When I started teaching my own molecular biology course, I adopted this same approach, but tried to reduce the challenge to a level more appropriate for undergraduate students. I did this by narrowing the focus to the most important experiments in each article, and explaining those carefully in class. I used hand-drawn cartoons and photocopies of the figures as illustrations.

This approach worked well, and the students enjoyed it, but I really wanted a textbook that presented the concepts of molecular biology, along with experiments that led to those concepts. I wanted clear explanations that showed students the relationship between the experiments and the concepts. So, I finally decided that the best way to get such a book would be to write it myself. I had already coauthored a successful introductory genetics text in which I took an experimental approach—as much as possible with a book at that level. That gave me the courage to try writing an entire book by myself and to treat the subject as an adventure in discovery.

Organization

The book begins with a four-chapter sequence that should be a review for most students. Chapter 1 is a brief history of genetics. Chapter 2 discusses the structure and chemical properties of DNA. Chapter 3 is an overview of gene expression, and Chapter 4 deals with the nuts and bolts of gene cloning. All these are topics that the great majority of molecular biology students have already learned in an introductory genetics course. Still, students of molecular biology need to have a grasp of these concepts and may need to refresh their understanding of them. I do not deal specifically with these chapters in class; instead, I suggest students consult them if they need more work on these topics. These chapters are written at a more basic level than the rest of the book.

Chapter 5 describes a number of common techniques used by molecular biologists. It would not have been possible to include all the techniques described in this book in one chapter, so I tried to include the most common or, in a few cases, valuable techniques that are not mentioned elsewhere in the book. When I teach this course, I do not present Chapter 5 as such. Instead, I refer students to it when we first encounter a technique in a later chapter. I do it that way to avoid boring my students with technique after technique. I also realize that the concepts behind some of these techniques are rather sophisticated, and the students' appreciation of them is much deeper after they've acquired more experience in molecular biology.

Chapters 6–9 describe transcription in bacteria. Chapter 6 introduces the basic transcription apparatus, including promoters, terminators, and RNA polymerase, and shows how transcripts are initiated, elongated, and terminated. Chapter 7 describes the control of transcription in three different operons, then Chapter 8 shows how bacteria and their phages control transcription of many genes at a time, often by providing alternative sigma factors. Chapter 9 discusses the interaction between bacterial DNA-binding proteins, mostly helix-turn-helix proteins, and their DNA targets.

Chapters 10–13 present control of transcription in eukaryotes. Chapter 10 deals with the three eukaryotic RNA polymerases and the promoters they recognize. Chapter 11 introduces the general transcription factors that collaborate with the three RNA polymerases and points out the unifying theme of the TATA-box-binding protein, which participates in transcription by all three polymerases. Chapter 12 explains the functions of gene-specific transcription factors, or activators. This chapter also illustrates the structures of several representative activators and shows how they interact with their DNA targets. Chapter 13 describes the structure of eukaryotic chromatin and shows how activators and silencers can interact with coactivators and corepressors to modify histones, and thereby to activate or repress transcription.

Chapters 14–16 introduce some of the posttranscriptional events that occur in eukaryotes. Chapter 14 deals with RNA splicing. Chapter 15 describes capping and polyadenylation, and Chapter 16 introduces a collection of fascinating "other posttranscriptional events," including rRNA and tRNA processing, *trans*-splicing, and RNA editing. This chapter also discusses four kinds of posttranscriptional control of gene expression: (1) RNA interference; (2) modulating mRNA stability (using the transferrin receptor mRNA as the prime example); (3) control by microRNAs, and (4) control of transposons in germ cells by Piwi-interacting RNAs (piRNAs).

Chapters 17–19 describe the translation process in both bacteria and eukaryotes. Chapter 17 deals with initiation of translation, including the control of translation at the initiation step. Chapter 18 shows how polypeptides are elongated, with the emphasis on elongation in bacteria. Chapter 19 provides details on the structure and function of two of the key players in translation: ribosomes and tRNA.

Chapters 20–23 describe the mechanisms of DNA replication, recombination, and translocation. Chapter 20 introduces the basic mechanisms of DNA replication and repair, and some of the proteins (including the DNA polymerases) involved in replication. Chapter 21 provides details of the initiation, elongation, and termination steps in DNA replication in bacteria and eukaryotes. Chapters 22 and 23 describe DNA rearrangements that occur naturally in cells. Chapter 22 discusses homologous recombination and Chapter 23 deals with translocation.

Chapters 24 and 25 present concepts of genomics, proteomics, and bioinformatics. Chapter 24 begins with an old-fashioned positional cloning story involving the Huntington disease gene and contrasts this lengthy and heroic quest with the relative ease of performing positional cloning with the human genome (and other genomes). Chapter 25 deals with functional genomics (transcriptomics), proteomics, and bioinformatics.

New to the Fifth Edition

The most obvious change in the fifth edition is the splitting of old Chapter 24 (Genomics, Proteomics, and Bioinformatics) in two. This chapter was already the longest in the book, and the field it represents is growing explosively, so a split was inevitable. The new Chapter 24 deals with classical genomics: the sequencing and comparison of genomes. New material in Chapter 24 includes an analysis of the similarity between the human and chimpanzee genomes, and a look at the even closer similarity between the human and Neanderthal genomes, including recent evidence for interbreeding between humans and Neanderthals. It also includes an update on the new field of synthetic biology, made possible by genomic work on microorganisms, and contains a report of the recent success by Craig Venter and colleagues in creating a living *Mycoplasma* cell with a synthetic genome.

Chapter 25 deals with fields allied with Genomics: Functional Genomics, Proteomics, and Bioinformatics. New material in Chapter 25 includes new applications of the ChIP-chip and ChIP-seq techniques—the latter using next-generation DNA sequencing; collision-induced dissociation mass spectrometry, which can be used to sequence proteins; and the use of isotope-coded affinity tags (ICATs) and stable isotope labeling by amino acids (SILAC) to make mass spectrometry (MS) quantitative. Quantitative MS in turn enables comparative proteomics, in which the concentrations of large numbers of proteins can be compared between species.

All but the introductory chapters of this fifth edition have been updated. Here are a few highlights:

- Chapter 5: Introduces high-throughput (next generation) DNA sequencing techniques. These have revolutionized the field of genomics. Chromatin immunoprecipitation (ChIP) and the yeast two-hybrid assay have been moved to Chapter 5, in light of their broad applicabilities. A treatment of the energies of the β-electrons from ^{3}H, ^{14}C, ^{35}S, and ^{32}P has been added, and the fluorography technique, which captures information from the lower-energy emissions, is discussed.
- Chapter 6: Adds a discussion of FRET-ALEX (FRET with alternating laser excitation), along with a description of how this technique has been used to support (1) the stochastic release model of the σ-cycle and (2) the scrunching hypothesis to explain abortive transcription. This chapter also updates the structure of the bacterial elongation complex, including a discussion of a two-state model for nucleotide addition.
- Chapter 7: Introduces the riboswitch in the mRNA from the *glmS* gene of *B. subtilis,* in which the end product of the gene turns expression of the gene off by stimulating the mRNA to destroy itself. This chapter also introduces a hammerhead ribozyme as a possible mammalian riboswitch that may operate by a similar mechanism.
- Chapter 8: Introduces the concepts of anti-σ-factors and anti-anti-σ-factors as controllers of transcription during sporulation in *B. subtilis*.
- Chapter 9: Emphasizes the dynamic nature of protein structure, and points out that a given crystal structure represents just one of a range of different possible protein conformations.
- Chapter 10: Presents a new study by Roger Kornberg's group that identifies the RNA polymerase II trigger loop as a key determinant in transcription specificity, along with a discussion of how the enzyme distinguishes between ribonuncleotides and deoxyribonucleotides. This chapter also introduces the concepts of core promoter and proximal promoter, where the core promoter contains any combination of TFIIB recognition element, TATA box, initiator, downstream promoter element, downstream core element, and motif ten element, and the proximal promoter contains upstream promoter elements.
- Chapter 11: Introduces the concept of core TAFs—those associated with class II preinitiation complexes from a wide variety of eukaryotes, and introduces the new nomenclature (TAF1–TAF13), which replaces the old, confusing nomenclature that was

based on molecular masses (e.g., $TAF_{II}250$). This chapter also describes an experiment that shows the importance of TFIIB in setting the start site of transcription. It also shows that a similar mechanism applies in the archaea, which use a TFIIB homolog known as transcription factor B.

- Chapter 12: Introduces the technique of chromosome conformation capture (3C) and shows how it can be used to detect DNA looping between an enhancer and a promoter. This chapter also introduces the concept of imprinting during gametogenesis, and explains the role of methylation in imprinting, particularly methylation of the imprinting control region of the mouse *Igf2/H19* locus. It also introduces the concept of transcription factories, where transcription of multiple genes occurs. Finally, this chapter refines and updates the concept of the enhanceosome.
- Chapter 13: Presents a new table showing all the ways histones can be modified in vivo; brings back the solenoid, alongside the two-start helix, as a candidate for the 30-nm fiber structure; and presents evidence that chromatin adopts one or the other structure, depending on its nucleosome repeat length. This chapter also introduces the concept of specific histone methylations as markers for transcription initiation and elongation, and shows how this information can be used to infer that RNA polymerase II is poised between initiation and elongation on many human protein-encoding genes. It also emphasizes the importance of histone modifications in affecting not only histone–DNA interactions, but also nucleosome–nucleosome interactions and recruitment of histone-modifying and chromatin-remodeling proteins. Finally, this chapter shows how PARP1 (poly[ADP-ribose] polymerase-1) can facilitate nucleosome loss from chromatin by poly(ADP-ribosyl)ating itself.
- Chapter 14: Introduces the exon junction complex (EJC), which is added to mRNAs during splicing in the nucleus, and shows how the EJC can stimulate transcription by facilitating the association of mRNAs with ribosomes. This chapter also introduces exon and intron definition modes of splicing and shows how they can be distinguished experimentally. This test has revealed that higher eukaryotes primarily use exon definition and lower eukaryotes primarily use intron definition.
- Chapter 15: Demonstrates that a subunit of CPSF (CPSF-73) is responsible for cutting a pre-mRNA at a polyadenylation signal. It also shows that serine 7, in addition to serines 2 and 5 in the repeating heptad in the CTD of the largest RNA polymerase subunit, can be phosphorylated, and shows that this serine 7 phosphorylation controls the expression of certain genes (e.g., the U2 snRNA gene) by controlling the 3′-end processing of their mRNAs.
- Chapter 16: Identifies a single enzyme, tRNA 3′ processing endoribonuclease, as the agent that cleaves excess nucleotides from the 3′-end of a eukaryotic tRNA precursor; points out the overwhelming prevalence of trans-splicing in *C. elegans;* presents a new model for removal of the passenger strand of a double-stranded siRNA—cleavage of the passenger strand by Ago2; introduces Piwi-interacting RNAs (piRNAs) and presents the ping-pong model by which they are assumed to amplify themselves and inactivate transposons in germ cells; introduces plant RNA polymerases IV and V, and describes their roles in gene silencing. This chapter also greatly expands the coverage of miRNAs, and points out that hundreds of miRNAs control thousands of plant and animal genes, and that mutations in miRNA genes typically have very deleterious effects. Chapter 16 also updates the biogenesis of miRNAs, introducing two pathways to miRNA production: the Drosha and mirtron pathways. Finally, this chapter introduces P-bodies, which are involved in mRNA decay and translational repression.
- Chapter 17: Updates the section on eukaryotic viral internal ribosome entry sequences (IRESs). Some viruses cleave eIF4G, leaving a remnant called p100. Poliovirus IRESs bind to p100 and thereby gain access to ribosomes, but hepatitis C virus IRESs bind directly to eIF3, while hepatitis A virus IRESs bind even more directly to ribosomes. This chapter also refines the model describing how the cleavage of eIF4G affects mammalian host mRNA translation. Different cell types respond differently to this cleavage. Finally, this chapter introduces the concept of the pioneer round of translation, and points out that different initiation factors are used in the pioneer round than in all subsequent rounds.
- Chapter 18: Introduces the concept of superwobble, which holds that a single tRNA with a U in its wobble position can recognize codons ending in any of the four bases, and presents evidence that superwobble works. This chapter also introduces the hybrid P/I state as the initial ribosomal binding state for fMet-$tRNA_f^{Met}$. In this state, the anticodon is in the P site, but the fMet and acceptor stem are in an "initiator" site between the P site and the E site. This chapter also describes no-go decay, which degrades mRNA containing a stalled ribosome, and introduces the concept of codon bias to explain inefficiency of translation. Finally, this chapter explains how the slowing of translation by rare codons can influence protein folding both negatively and positively.

- Chapter 19: Includes a new section based on recent crystal structures of the ribosome in complex with various elongation factors. One of these structures involves aminoacyl-tRNA and EF-Tu, and has shown that the tRNA is bent by about 30 degrees in forming an A/T complex. This bend is important in fidelity of translation, and also facilitates the GTP hydrolysis that permits EF-Tu to leave the ribosome. Another crystal structure involves EF-G–GDP and shows the ribosome in the post-translocation E/E, P/P state, as opposed to the spontaneously achieved pre-translocation P/E, A/P hybrid state. This chapter also provides links to two excellent new movies describing the elongation process and an overview of translation initiation, elongation, and termination. Finally, this chapter describes crystal structures that illustrate the functions of two critical parts of RF1 and RF2 in stop codon recognition and cleavage of polypeptides from their tRNAs.
- Chapter 20: Introduces the controversial proposal, with evidence, that DNA replication in *E. coli* is discontinuous on both strands. This chapter also introduces ACL1, a chromatin remodeler recruited via its macrodomain to sites of double-strand breaks by poly(ADP-ribose) formed at these sites by poly(ADP-ribose) polymerase 1 (PARP-1).
- Chapter 21: Presents a co-crystal structure of a β dimer bound to a primed DNA template, showing that the β clamp really does encircle the DNA, but that the DNA runs through the circle at an angle of 20 degrees with respect to the horizontal. This chapter also includes a corrected and updated Figure 21.17 (model of the polIII* subassembly) to show a single γ-subunit and the two τ-subunits joined to the core polymerases through their flexible C-terminal domains. This section also clarifies that the γ- and τ-subunits are products of the same gene, but the former lacks the C-terminal domain of the latter. This chapter also introduces the complex of telomere-binding proteins known as shelterin, and focuses on the six shelterin proteins of mammals and their roles in protecting telomeres, and in preventing inappropriate repair and cell cycle arrest in response to normal chromosome ends.
- Chapter 22: Adds a new figure (Figure 22.3) to show how different nicking patterns to resolve the Holliday junction in the RecBCD pathway lead to different recombination products (crossover or noncrossover recombinants).
- Chapter 23: Reports that piRNAs targeting P element transposons are likely to be the transposition suppressors in the P-M system. Similarly, piRNAs appear to play the suppressor role in the I-R transposon system.

Supplements

For the Student www.mhhe.com/weaver5e
The text website for *Molecular Biology* is a great place to review chapter material and to enhance your study routine. Here you will have access to:

- digital image files
- animation quizzes
- PowerPoint lecture outlines
- answers to end-of-chapter
- questions
- web links.

For the Instructor www.mhhe.com/weaver5e
The *Molecular Biology* website offers a wealth of teaching and learning aids for instructors and students. Instructors will appreciate:

- Test bank questions and software options with EZ Test Online, desktop version or Word docs.

- Answers to end-of-chapter questions
- Lecture outline PowerPoint files
- Image PowerPoint files
- McGraw-Hill Presentation Center

McGraw-Hill Presentation Center
Build instructional materials wherever, whenever, and however you want! Presentation Center is an online digital library containing assets such as photos, artwork, PowerPoints, animations, and other media types that can be used to create customized lectures, visually enhanced tests and quizzes, compelling course websites, or attractive printed support materials.

Options

You're in charge of your course, so why not be in control of the content of your textbook? At McGraw-Hill Custom Publishing, we can help you create the ideal text—the one you've always imagined. Quickly. Easily. With more than 20 years of experience in custom publishing, we're experts. But at McGraw Hill, we're also innovators, leading the way with new methods and means for creating simplified value added custom textbooks.

eBooks

Going green . . . it's on everybody's minds these days. It's not only about saving trees; it's also about saving money. Available for a greatly reduced price, McGraw-Hill eBooks are an eco-friendly and cost-savings alternative to the traditional print textbook. So, you do some good for the environment . . . and you do some good for your wallet. Visit www.mhhe.com/ebooks for details.

ACKNOWLEDGMENTS

In writing this book, I have been aided immeasurably by the advice of many editors and reviewers. They have contributed greatly to the accuracy and readability of the book, but they cannot be held accountable for any remaining errors or ambiguities. For those, I take full responsibility. I would like to thank the following people for their help.

Fifth Edition Reviewers

Aimee Bernard
University of Colorado–Denver

Brian Freeman
University of Illinois, Urbana–Champaign

Dennis Bogyo
Valdosta State University

Donna Hazelwood
Dakota State University

Margaret Ritchey
Centre College

Nemat Kayhani
University of Florida

Nicole Bournias-Bardiabasis
California State University, San Bernardino

Ruhul Kuddus
Utah Valley University

Tao Weitao
University of Texas at San Antonio

Fourth Edition Reviewers

Dr. David Asch
Youngstown State University

Christine E. Bezotte
Elmira College

Mark Bolyard
Southern Illinois University, Edwardsville

Diane Caporale
University of Wisconsin, Stevens Point

Jianguo Chen
Claflin University

Chi-Lien Cheng
Department of Biological Sciences, University of Iowa

Mary Ellard-Ivey
Pacific Lutheran University

Olukemi Fadayomi
Ferris State University, Big Rapids, Michigan

Charles Giardina
University of Connecticut

Eli V. Hestermann
Furman University

Dr. Dorothy Hutter
Monmouth University

Cheryl Ingram-Smith
Clemson University

Dr. Cynthia Keler
Delaware Valley College

Jack Kennell
Saint Louis University

Charles H. Mallery
College of Arts and Sciences, University of Miami

Jon L. Milhon
Azusa Pacific University

Hao Nguyen
California State University, Sacramento

Thomas Peterson
Iowa State University

Ed Stellwag
East Carolina University

Katherine M. Walstrom
New College of Florida

Cornelius A. Watson
Roosevelt University

Fadi Zaher
Gateway Technical College

Third Edition Reviewers

David Asch
Youngstown State University

Gerard Barcak
University of Maryland School of Medicine

Bonnie Baxter
Hobart & William Smith Colleges

André Bédard
McMaster University

Felix Breden
Simon Fraser University

Laura Bull
UCSF Liver Center Laboratory

James Ellis
Developmental Biology Program, Hospital for Sick Children, Toronto, Ontario

Robert Helling
The University of Michigan

David Hinkle
University of Rochester

Robert Leamnson
University of Massachusetts at Dartmouth

David Mullin
Tulane University

Marie Pizzorno
Bucknell University

Michael Reagan
College of St. Benedict/ St. John's University

Rodney Scott
Wheaton College

Second Edition Reviewers

Mark Bolyard
Southern Illinois University

M. Suzanne Bradshaw
University of Cincinnati

Anne Britt
University of California, Davis

Robert Brunner
University of California, Berkeley

Caroline J. Decker
Washington State University

Jeffery DeJong
University of Texas, Dallas

Stephen J. D'Surney
University of Mississippi

John S. Graham
Bowling Green State University

Ann Grens
Indiana University

Ulla M. Hansen
Boston University

Laszlo Hanzely
Northern Illinois University

Robert B. Helling
University of Michigan

Martinez J. Hewlett
University of Arizona

David C. Hinkle
University of Rochester

Barbara C. Hoopes
Colgate University

Richard B. Imberski
University of Maryland

Cheryl Ingram-Smith
Pennsylvania State University

Alan Kelly
University of Oregon

Robert N. Leamnson
University of Massachusetts, Dartmouth

Karen A. Malatesta
Princeton University

Robert P. Metzger
San Diego State University

David A. Mullin
Tulane University

Brian K. Murray
Brigham Young University

Michael A. Palladino
Monmouth University

James G. Patton
Vanderbilt University

Martha Peterson
University of Kentucky

Marie Pizzorno
Bucknell University

Florence Schmieg
University of Delaware

Zhaomin Yang
Auburn University

First Edition Reviewers

Kevin L. Anderson
Mississippi State University

Rodney P. Anderson
Ohio Northern University

Prakash H. Bhuta
Eastern Washington University

Dennis Bogyo
Valdosta State University

Richard Crawford
Trinity College

Christopher A. Cullis
Case Western Reserve University

Beth De Stasio
Lawrence University

R. Paul Evans
Brigham Young University

Edward R. Fliss
Missouri Baptist College

Michael A. Goldman
San Francisco State University

Robert Gregerson
Lyon College

Eileen Gregory
Rollins College

Barbara A. Hamkalo
University of California, Irvine

Mark L. Hammond
Campbell University

Terry L. Helser
State University of New York, Oneonta

Carolyn Herman
Southwestern College

Andrew S. Hopkins
Alverno College

Carolyn Jones
Vincennes University

Teh-Hui Kao
Pennsylvania State University

Mary Evelyn B. Kelley
Wayne State University

Harry van Keulen
Cleveland State University

Leo Kretzner
University of South Dakota

Charles J. Kunert
Concordia University

Robert N. Leamnson
University of Massachusetts, Dartmouth

James D. Liberatos
Louisiana Tech University

Cran Lucas
Louisiana State University

James J. McGivern
Gannon University

James E. Miller
Delaware Valley College

Robert V. Miller
Oklahoma State University

George S. Mourad
Indiana University-Purdue University

David A. Mullin
Tulane University

James R. Pierce
Texas A&M University, Kingsville

Joel B. Piperberg
Millersville University

John E. Rebers
Northern Michigan University

Florence Schmieg
University of Delaware

Brian R. Shmaefsky
Kingwood College

Paul Keith Small
Eureka College

David J. Stanton
Saginaw Valley State University

Francis X. Steiner
Hillsdale College

Amy Cheng Vollmer
Swarthmore College

Dan Weeks
University of Iowa

David B. Wing
New Mexico Institute of Mining & Technology

GUIDE TO EXPERIMENTAL TECHNIQUES IN MOLECULAR BIOLOGY

CHAPTER 1

A Brief History

Garden pea flowers. Flower color (purple or white) was one of the traits Mendel studied in his classic examination of inheritance in the pea plant. *© Shape'n'colour/Alamy, RF.*

What is molecular biology? The term has more than one definition. Some define it very broadly as the attempt to understand biological phenomena in molecular terms. But this definition makes molecular biology difficult to distinguish from another well-known discipline, biochemistry. Another definition is more restrictive and therefore more useful: the study of gene structure and function at the molecular level. This attempt to explain genes and their activities in molecular terms is the subject matter of this book.

Molecular biology grew out of the disciplines of genetics and biochemistry. In this chapter we will review the major early developments in the history of this hybrid discipline, beginning with the earliest genetic experiments performed by Gregor Mendel in the mid-nineteenth century.

In Chapters 2 and 3 we will add more substance to this brief outline. By definition, the early work on genes cannot be considered molecular biology, or even molecular genetics, because early geneticists did not know the molecular nature of genes. Instead, we call it transmission genetics because it deals with the transmission of traits from parental organisms to their offspring. In fact, the chemical composition of genes was not known until 1944. At that point, it became possible to study genes as molecules, and the discipline of molecular biology began.

1.1 Transmission Genetics

In 1865, Gregor Mendel (Figure 1.1) published his findings on the inheritance of seven different traits in the garden pea. Before Mendel's research, scientists thought inheritance occurred through a blending of each trait of the parents in the offspring. Mendel concluded instead that inheritance is particulate. That is, each parent contributes particles, or genetic units, to the offspring. We now call these particles **genes.** Furthermore, by carefully counting the number of progeny plants having a given **phenotype,** or observable characteristic (e.g., yellow seeds, white flowers), Mendel was able to make some important generalizations. The word *phenotype,* by the way, comes from the same Greek root as *phenomenon,* meaning *appearance.* Thus, a tall pea plant exhibits the tall phenotype, or appearance. *Phenotype* can also refer to the whole set of observable characteristics of an organism.

Figure 1.1 Gregor Mendel. (*Source:* © Pixtal/age Fotostock RF.)

Mendel's Laws of Inheritance

Mendel saw that a gene can exist in different forms called **alleles.** For example, the pea can have either yellow or green seeds. One allele of the gene for seed color gives rise to yellow seeds, the other to green. Moreover, one allele can be **dominant** over the other, **recessive,** allele. Mendel demonstrated that the allele for yellow seeds was dominant when he mated a green-seeded pea with a yellow-seeded pea. All of the progeny in the first filial generation (F_1) had yellow seeds. However, when these F_1 yellow peas were allowed to self-fertilize, some green-seeded peas reappeared. The ratio of yellow to green seeds in the second filial generation (F_2) was very close to 3:1.

The term *filial* comes from the Latin: *filius,* meaning son; *filia,* meaning daughter. Therefore, the first filial generation (F_1) contains the offspring (sons and daughters) of the original parents. The second filial generation (F_2) is the offspring of the F_1 individuals.

Mendel concluded that the allele for green seeds must have been preserved in the F_1 generation, even though it did not affect the seed color of those peas. His explanation was that each parent plant carried two copies of the gene; that is, the parents were **diploid,** at least for the characteristics he was studying. According to this concept, **homozygotes** have two copies of the same allele, either two alleles for yellow seeds or two alleles for green seeds. **Heterozygotes** have one copy of each allele. The two parents in the first mating were homozygotes; the resulting F_1 peas were all heterozygotes. Further, Mendel reasoned that sex cells contain only one copy of the gene; that is, they are **haploid.** Homozygotes can therefore produce sex cells, or **gametes,** that have only one allele, but heterozygotes can produce gametes having either allele.

This is what happened in the matings of yellow with green peas: The yellow parent contributed a gamete with a gene for yellow seeds; the green parent, a gamete with a gene for green seeds. Therefore, all the F_1 peas got one allele for yellow seeds and one allele for green seeds. They had not lost the allele for green seeds at all, but because yellow is dominant, all the seeds were yellow. However, when these heterozygous peas were self-fertilized, they produced gametes containing alleles for yellow and green color in equal numbers, and this allowed the green phenotype to reappear.

Here is how that happened. Assume that we have two sacks, each containing equal numbers of green and yellow marbles. If we take one marble at a time out of one sack and pair it with a marble from the other sack, we will wind up with the following results: one-quarter of the pairs will be yellow/yellow; one-quarter will be green/green; and the remaining one-half will be yellow/green. The alleles for yellow and green peas work the same way.

Recalling that yellow is dominant, you can see that only one-quarter of the progeny (the green/green ones) will be green. The other three-quarters will be *yellow* because they have at least one allele for yellow seeds. Hence, the ratio of yellow to green peas in the second (F_2) generation is 3:1.

Mendel also found that the genes for the seven different characteristics he chose to study operate independently of one another. Therefore, combinations of alleles of two different genes (e.g., yellow or green peas with round or wrinkled seeds, where yellow and round are dominant and green and wrinkled are recessive) gave ratios of 9:3:3:1 for yellow/round, yellow/wrinkled, green/round, and green/wrinkled, respectively. Inheritance that follows the simple laws that Mendel discovered can be called **Mendelian inheritance.**

SUMMARY Genes can exist in several different forms, or alleles. One allele can be dominant over another, so heterozygotes having two different alleles of one gene will generally exhibit the characteristic dictated by the dominant allele. The recessive allele is not lost; it can still exert its influence when paired with another recessive allele in a homozygote.

The Chromosome Theory of Inheritance

Other scientists either did not know about or uniformly ignored the implications of Mendel's work until 1900 when three botanists, who had arrived at similar conclusions independently, rediscovered it. After 1900, most geneticists accepted the particulate nature of genes, and the field of genetics began to blossom. One factor that made it easier for geneticists to accept Mendel's ideas was a growing understanding of the nature of chromosomes, which had begun in the latter half of the nineteenth century. Mendel had predicted that gametes would contain only one allele of each gene instead of two. If chromosomes carry the genes, their numbers should also be reduced by half in the gametes—and they are. Chromosomes therefore appeared to be the discrete physical entities that carry the genes.

This notion that chromosomes carry genes is the **chromosome theory of inheritance.** It was a crucial new step in genetic thinking. No longer were genes disembodied factors; now they were observable objects in the cell nucleus. Some geneticists, particularly Thomas Hunt Morgan (Figure 1.2), remained skeptical of this idea. Ironically, in 1910 Morgan himself provided the first definitive evidence for the chromosome theory.

Morgan worked with the fruit fly *(Drosophila melanogaster),* which was in many respects a much more

Figure 1.2 Thomas Hunt Morgan. (*Source:* National Library of Medicine.)

convenient organism than the garden pea for genetic studies because of its small size, short generation time, and large number of offspring. When he mated red-eyed flies (dominant) with white-eyed flies (recessive), most, but not all, of the F_1 progeny were red-eyed. Furthermore, when Morgan mated the red-eyed males of the F_1 generation with their red-eyed sisters, they produced about one-quarter white-eyed males, but no white-eyed females. In other words, the eye color phenotype was **sex-linked.** It was transmitted along with sex in these experiments. How could this be?

We now realize that sex and eye color are transmitted together because the genes governing these characteristics are located on the same chromosome—the X chromosome. (Most chromosomes, called **autosomes,** occur in pairs in a given individual, but the X chromosome is an example of a **sex chromosome,** of which the female fly has two copies and the male has one.) However, Morgan was reluctant to draw this conclusion until he observed the same sex linkage with two more phenotypes, miniature wing and yellow body, also in 1910. That was enough to convince him of the validity of the chromosome theory of inheritance.

Before we leave this topic, let us make two crucial points. First, every gene has its place, or **locus,** on a chromosome. Figure 1.3 depicts a hypothetical chromosome and the positions of three of its genes, called *A, B,* and *C.* Second, diploid organisms such as human beings normally have two copies of all chromosomes (except sex chromosomes). That means that they have two copies of most genes, and that these copies can be the same alleles, in which case the organism is **homozygous,** or different

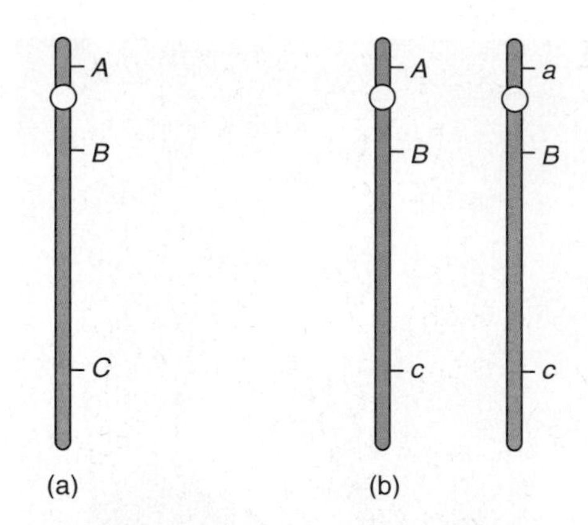

Figure 1.3 Location of genes on chromosomes. (a) A schematic diagram of a chromosome, indicating the positions of three genes: *A, B,* and *C.* **(b)** A schematic diagram of a diploid pair of chromosomes, indicating the positions of the three genes—*A, B,* and *C*—on each, and the genotype (*A* or *a; B* or *b;* and *C* or *c*) at each locus.

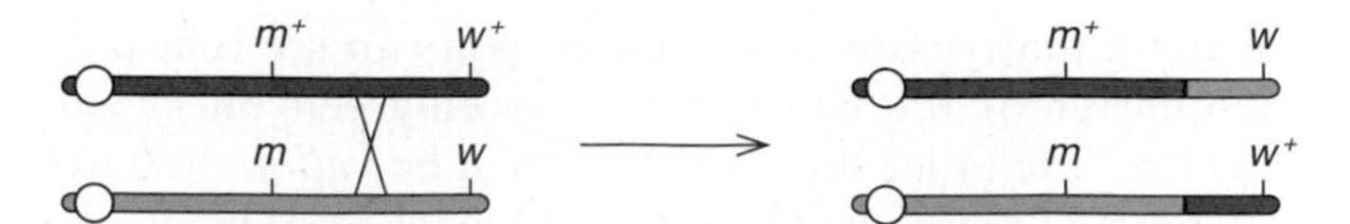

Figure 1.4 Recombination in *Drosophila.* The two X chromosomes of the female are shown schematically. One of them (red) carries two wild-type genes: (m^+), which results in normal wings, and (w^+), which gives red eyes. The other (blue) carries two mutant genes: *miniature* (*m*) and *white* (*w*). During egg formation, a recombination, or crossing over, indicated by the crossed lines, occurs between these two genes on the two chromosomes. The result is two recombinant chromosomes with mixtures of the two parental alleles. One is $m^+ \, w$, the other is $m \, w^+$.

alleles, in which case it is **heterozygous.** For example, Figure 1.3b shows a diploid pair of chromosomes with different alleles at one locus *(Aa)* and the same alleles at the other two loci (*BB* and *cc*). The **genotype,** or allelic constitution, of this organism with respect to these three genes, is *AaBBcc*. Because this organism has two different alleles (*A* and *a*) in its two chromosomes at the *A* locus, it is heterozygous at that locus (Greek: *hetero,* meaning different). Since it has the same, dominant *B* allele in both chromosomes at the *B* locus, it is homozygous dominant at that locus (Greek: *homo,* meaning same). And because it has the same, recessive *c* allele in both chromosomes at the *C* locus, it is homozygous recessive there. Finally, because the *A* allele is dominant over the *a* allele, the phenotype of this organism would be the dominant phenotype at the *A* and *B* loci and the recessive phenotype at the *C* locus.

This discussion of varying phenotypes in *Drosophila* gives us an opportunity to introduce another important genetic concept: **wild-type** versus **mutant.** The wild-type phenotype is the most common, or at least the generally accepted standard, phenotype of an organism. To avoid the mistaken impression that a wild organism is automatically a wild-type, some geneticists prefer the term **standard type.** In *Drosophila,* red eyes and full-size wings are wild-type. Mutations in the *white* and *miniature* genes result in mutant flies with white eyes and miniature wings, respectively. Mutant alleles are usually recessive, as in these two examples, but not always.

Genetic Recombination and Mapping

It is easy to understand that genes on separate chromosomes behave independently in genetic experiments, and that genes on the same chromosome—like the genes for miniature wing *(miniature)* and white eye *(white)*—behave as if they are linked. However, genes on the same chromosome usually do not show perfect **genetic linkage.** In fact, Morgan discovered this phenomenon when he examined the behavior of the sex-linked genes he had found. For example, although *white* and *miniature* are both on the X chromosome, they remain linked in offspring only 65.5% of the time. The other offspring have a new combination of alleles not seen in the parents and are therefore called **recombinants.**

How are these recombinants produced? The answer was already apparent by 1910, because microscopic examination of chromosomes during meiosis (gamete formation) had shown crossing over between **homologous chromosomes** (chromosomes carrying the same genes, or alleles of the same genes). This resulted in the exchange of genes between the two homologous chromosomes. In the previous example, during formation of eggs in the female, an X chromosome bearing the *white* and *miniature* alleles experienced crossing over with a chromosome bearing the red eye and normal wing alleles (Figure 1.4). Because the crossing-over event occurred between these two genes, it brought together the *white* and normal wing alleles on one chromosome and the red (normal eye) and *miniature* alleles on the other. Because it produced a new combination of alleles, we call this process **recombination.**

Morgan assumed that genes are arranged in a linear fashion on chromosomes, like beads on a string. This, together with his awareness of recombination, led him to propose that the farther apart two genes are on a chromosome, the more likely they are to recombine. This makes sense because there is simply more room between widely spaced genes for crossing over to occur. A.H. Sturtevant extended this hypothesis to predict that a mathematical relationship exists between the distance separating two genes on a chromosome and the frequency of recombination between these two genes. Sturtevant collected data on recombination in the fruit fly that supported his hypothesis. This established the rationale for **genetic mapping** techniques still in use today. Simply stated, if two loci recombine with a frequency of 1%, we say that they are separated by a map distance of one **centimorgan** (named for Morgan

Figure 1.5 Barbara McClintock. (*Source:* Bettmann Archive/Corbis.)

Figure 1.6 Friedrich Miescher. (*Source:* National Library of Medicine.)

himself). By the 1930s, other investigators found that the same rules applied to other **eukaryotes** (nucleus-containing organisms), including the mold *Neurospora,* the garden pea, maize (corn), and even human beings. These rules also apply to **prokaryotes,** organisms in which the genetic material is not confined to a nuclear compartment.

Physical Evidence for Recombination

Barbara McClintock (Figure 1.5) and Harriet Creighton provided a direct physical demonstration of recombination in 1931. By examining maize chromosomes microscopically, they could detect recombinations between two easily identifiable features of a particular chromosome (a knob at one end and a long extension at the other). Furthermore, whenever this physical recombination occurred, they could also detect recombination genetically. Thus, they established a direct relationship between a region of a chromosome and a gene. Shortly after McClintock and Creighton performed this work on maize, Curt Stern observed the same phenomenon in *Drosophila.* So recombination could be detected both physically and genetically in animals as well as plants. McClintock later performed even more notable work when she discovered transposons, moveable genetic elements (Chapter 23), in maize.

SUMMARY The chromosome theory of inheritance holds that genes are arranged in linear fashion on chromosomes. The reason that certain traits tend to be inherited together is that the genes governing these traits are on the same chromosome. However, recombination between two homologous chromosomes during meiosis can scramble the parental alleles to give nonparental combinations. The farther apart two genes are on a chromosome the more likely such recombination between them will be.

1.2 Molecular Genetics

The studies just discussed tell us important things about the transmission of genes and even about how to map genes on chromosomes, but they do not tell us what genes are made of or how they work. This has been the province of molecular genetics, which also happens to have its roots in Mendel's era.

The Discovery of DNA

In 1869, Friedrich Miescher (Figure 1.6) discovered in the cell nucleus a mixture of compounds that he called nuclein. The major component of nuclein is **deoxyribonucleic acid (DNA).** By the end of the nineteenth century, chemists had learned the general structure of DNA and of a related compound, **ribonucleic acid (RNA).** Both are long polymers—chains of small compounds called nucleotides. Each nucleotide is composed of a sugar, a phosphate group, and a base. The chain is formed by linking the sugars to one another through their phosphate groups.

The Composition of Genes By the time the chromosome theory of inheritance was generally accepted, geneticists agreed that the chromosome must be composed of a polymer of some kind. This would agree with its role as a string of genes. But which polymer is it? Essentially, the choices were three: DNA, RNA, and **protein.** Protein was the other major component of Miescher's nuclein; its chain is composed of links called **amino acids.** The amino acids in protein are joined by **peptide bonds,** so a single protein chain is called a **polypeptide.**

Oswald Avery (Figure 1.7) and his colleagues demonstrated in 1944 that DNA is the right choice (Chapter 2). These investigators built on an experiment performed earlier by Frederick Griffith in which he transferred a genetic trait from one strain of bacteria to another. The trait was virulence, the ability to cause a lethal infection,

Figure 1.7 Oswald Avery. (*Source:* National Academy of Sciences.)

(a)

(b)

Figure 1.8 (a) George Beadle; (b) E. L. Tatum. (*Source:* (*a, b*) AP/Wide World Photos.)

and it could be transferred simply by mixing dead virulent cells with live avirulent (nonlethal) cells. It was very likely that the substance that caused the transformation from avirulence to virulence in the recipient cells was the gene for virulence, because the recipient cells passed this trait on to their progeny.

What remained was to learn the chemical nature of the transforming agent in the dead virulent cells. Avery and his coworkers did this by applying a number of chemical and biochemical tests to the transforming agent, showing that it had the characteristics of DNA, not of RNA or protein.

The Relationship Between Genes and Proteins

The other major question in molecular genetics is this: How do genes work? To lay the groundwork for the answer to this question, we have to backtrack again, this time to 1902. That was the year Archibald Garrod noticed that the human disease alcaptonuria seemed to behave as a Mendelian recessive trait. It was likely, therefore, that the disease was caused by a defective, or mutant, gene. Moreover, the main symptom of the disease was the accumulation of a black pigment in the patient's urine, which Garrod believed derived from the abnormal buildup of an intermediate compound in a biochemical pathway.

By this time, biochemists had shown that all living things carry out countless chemical reactions and that these reactions are accelerated, or catalyzed, by proteins called **enzymes.** Many of these reactions take place in sequence, so that one chemical product becomes the starting material, or substrate, for the next reaction. Such sequences of reactions are called **pathways,** and the products or substrates within a pathway are called **intermediates.** Garrod postulated that an intermediate accumulated to abnormally high levels in alcaptonuria because the enzyme that would normally convert this intermediate to the next was defective. Putting this idea together with the finding that alcaptonuria behaved genetically as a Mendelian recessive trait, Garrod suggested that a defective gene gives rise to a defective enzyme. To put it another way: A gene is responsible for the production of an enzyme.

Garrod's conclusion was based in part on conjecture; he did not really know that a defective enzyme was involved in alcaptonuria. It was left for George Beadle and E. L. Tatum (Figure 1.8) to prove the relationship between genes and enzymes. They did this using the mold *Neurospora* as their experimental system. *Neurospora* has an enormous advantage over the human being as the subject of genetic experiments. By using *Neurospora,* scientists are not limited to the mutations that nature provides, but can use **mutagens** to introduce mutations into genes and then observe the effects of these mutations on biochemical pathways. Beadle and Tatum found many instances where they could create *Neurospora* mutants and then pin the defect down to a single step in a biochemical pathway, and therefore to a single enzyme (see Chapter 3). They did this by adding the intermediate that would normally be made by the defective enzyme and showing that it restored normal growth. By circumventing the blockade, they discovered where it was. In these same cases, their genetic experiments showed that a single gene was involved. Therefore, a defective gene gives a defective (or absent) enzyme. In other words, a gene seemed to be responsible for making one enzyme. This was the one-gene/one-enzyme hypothesis. This hypothesis was actually not quite right for at least three reasons: (1) An enzyme can be composed of

Figure 1.9 James Watson (left) and Francis Crick. (*Source:* © A. Barrington Brown/Photo Researchers, Inc.)

(a)

(b)

Figure 1.10 (a) Rosalind Franklin; (b) Maurice Wilkins. (*Sources:* (*a*) From *The Double Helix* by James D. Watson, 1968, Atheneum Press, NY. © Cold Spring Harbor Laboratory Archives. (*b*) Courtesy Professor M. H. F. Wilkins, Biophysics Dept., King's College, London.)

more than one polypeptide chain, whereas a gene has the information for making only one polypeptide chain. (2) Many genes contain the information for making polypeptides that are not enzymes. (3) As we will see, the end products of some genes are not polypeptides, but RNAs. A modern restatement of the hypothesis would be: Most genes contain the information for making one polypeptide. This hypothesis is correct for prokaryotes and lower eukaryotes, but must be qualified for higher eukaryotes, such as humans, where a gene can give rise to different polypeptides through an *alternative splicing* mechanism we will discuss in Chapter 14.

(a)

(b)

Figure 1.11 (a) Matthew Meselson; (b) Franklin Stahl. (*Sources:* (*a*) Courtesy Dr. Matthew Meselson. (*b*) Cold Spring Harbor Laboratory Archives.)

Activities of Genes

Let us now return to the question at hand: How do genes work? This is really more than one question because genes do more than one thing. First, they are replicated faithfully; second, they direct the production of RNAs and proteins; third, they accumulate mutations and so allow evolution. Let us look briefly at each of these activities.

How Genes Are Replicated First of all, how is DNA replicated faithfully? To answer that question, we need to know the overall structure of the DNA molecule as it is found in the chromosome. James Watson and Francis Crick (Figure 1.9) provided the answer in 1953 by building models based on chemical and physical data that had been gathered in other laboratories, primarily x-ray diffraction data collected by Rosalind Franklin and Maurice Wilkins (Figure 1.10).

Watson and Crick proposed that DNA is a **double helix**—two DNA strands wound around each other. More important, the bases of each strand are on the inside of the helix, and a base on one strand pairs with one on the other in a very specific way. DNA has only four different bases: adenine, guanine, cytosine, and thymine, which we abbreviate A, G, C, and T. Wherever we find an A in one strand, we always find a T in the other; wherever we find a G in one strand, we always find a C in the other. In a sense, then, the two strands are complementary. If we know the base sequence of one, we automatically know the sequence of the other. This complementarity is what allows DNA to be replicated faithfully. The two strands come apart, and enzymes build new partners for them using the old strands as templates and following the Watson–Crick base-pairing rules (A with T, G with C). This is called **semiconservative replication** because one strand of the parental double helix is conserved in each of the daughter double helices. In 1958, Matthew Meselson and Franklin Stahl (Figure 1.11)

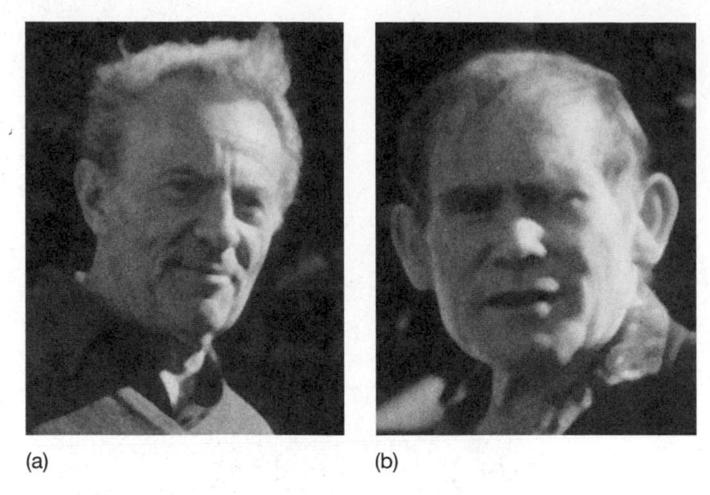

(a) (b)

Figure 1.12 (a) François Jacob; (b) Sydney Brenner. (*Source:* (*a, b*) Cold Spring Harbor Laboratory Archives.)

Figure 1.13 Gobind Khorana (left) and Marshall Nirenberg. (*Source:* Corbis/Bettmann Archive.)

proved that DNA replication in bacteria follows the semiconservative pathway (see Chapter 20).

How Genes Direct the Production of Polypeptides Gene expression is the process by which a cell makes a gene product (an RNA or a polypeptide). Two steps, called **transcription** and **translation,** are required to make a polypeptide from the instructions in a DNA gene. In the transcription step, an enzyme called RNA polymerase makes a copy of one of the DNA strands; this copy is not DNA, but its close cousin RNA. In the translation step, this RNA (**messenger RNA,** or **mRNA**) carries the genetic instructions to the cell's protein factories, called **ribosomes.** The ribosomes "read" the **genetic code** in the mRNA and put together a protein according to its instructions.

Actually, the ribosomes already contain molecules of RNA, called **ribosomal RNA (rRNA).** Francis Crick originally thought that this RNA residing in the ribosomes carried the message from the gene. According to this theory, each ribosome would be capable of making only one kind of protein—the one encoded in its rRNA. François Jacob and Sydney Brenner (Figure 1.12) had another idea: The ribosomes are nonspecific translation machines that can make an unlimited number of different proteins, according to the instructions in the mRNAs that visit the ribosomes. Experiment has shown that this idea is correct (Chapter 3).

What is the nature of this genetic code? Marshall Nirenberg and Gobind Khorana (Figure 1.13), working independently with different approaches, cracked the code in the early 1960s (Chapter 18). They found that 3 bases constitute a code word, called a **codon,** that stands for one amino acid. Out of the 64 possible 3-base codons, 61 specify amino acids; the other three are stop signals.

The ribosomes scan a messenger RNA 3 bases at a time and bring in the corresponding amino acids to link to the growing polypeptide chain. When they reach a stop signal, they release the completed polypeptide.

How Genes Accumulate Mutations Genes change in a number of ways. The simplest is a change of one base to another. For example, if a certain codon in a gene is GAG (for the amino acid called glutamate), a change to GTG converts it to a codon for another amino acid, valine. The protein that results from this mutated gene will have a valine where it ought to have a glutamate. This may be one change out of hundreds of amino acids, but it can have profound effects. In fact, this specific change has occurred in the gene for one of the human blood proteins and is responsible for the genetic disorder we call sickle cell disease.

Genes can suffer more profound changes, such as deletions or insertions of large pieces of DNA. Segments of DNA can even move from one locus to another. The more drastic the change, the more likely that the gene or genes involved will be totally inactivated.

Gene Cloning Since the 1970s, geneticists have learned to isolate genes, place them in new organisms, and reproduce them by a set of techniques collectively known as **gene cloning.** Cloned genes not only give molecular biologists plenty of raw materials for their studies, they also can be induced to yield their protein products. Some of these, such as human insulin or blood clotting factors, can be very useful. Cloned genes can also be transplanted to plants and animals, including humans.

These transplanted genes can alter the characteristics of the recipient organisms, so they may provide powerful tools for agriculture and for intervening in human genetic diseases. We will examine gene cloning in detail in Chapter 4.

SUMMARY All cellular genes are made of DNA arranged in a double helix. This structure explains how genes engage in their three main activities: replication, carrying information, and collecting mutations. The complementary nature of the two DNA strands in a gene allows them to be replicated faithfully by separating and serving as templates for the assembly of two new complementary strands. The sequence of nucleotides in a gene is a genetic code that carries the information for making an RNA. Most of these are messenger RNAs that carry the information to protein-synthesizing ribosomes. The end result is a new polypeptide chain made according to the gene's instructions. A change in the sequence of bases constitutes a mutation, which can change the sequence of amino acids in the gene's polypeptide product. Genes can be cloned, allowing molecular biologists to harvest abundant supplies of their products.

1.3 The Three Domains of Life

In the early part of the twentieth century, scientists divided all life into two kingdoms: animal and plant. Bacteria were considered plants, which is why we still refer to the bacteria in our guts as intestinal "flora." But after the middle of the century, this classification system was abandoned in favor of a five-kingdom system that included bacteria, fungi, and protists, in addition to plants and animals.

Then in the late 1970s, Carl Woese (Figure 1.14) performed sequencing studies on the ribosomal RNA genes of many different organisms and reached a startling conclusion: A class of organisms that had been classified as bacteria have rRNA genes that are more similar to those of eukaryotes than they are to those of classical bacteria like *E. coli.* Thus, Woese named these organisms *archaebacteria,* to distinguish them from true bacteria, or *eubacteria.* However, as more and more molecular evidence accumulated, it became clear that the archaebacteria, despite a superficial resemblance, are not really bacteria. They represent a distinct domain of life, so Woese changed their name to archaea. Now we recognize three **domains of life: bacteria, eukaryota,** and **archaea.** Like bacteria, archaea are **prokaryotes**—organisms without nuclei—but their molecular biology is actually more like that of eukaryotes than that of bacteria.

Figure 1.14 Carl Woese. (*Source:* Courtesy U. of Ill at Urbana Champaign.)

The archaea live in the most inhospitable regions of the earth. Some of them are **thermophiles** ("heat-lovers") that live in seemingly unbearably hot zones at temperatures above 100°C near deep-ocean geothermal vents or in hot springs such as those in Yellowstone National Park. Others are **halophiles** (halogen-lovers) that can tolerate very high salt concentrations that would dessicate and kill other forms of life. Still others are **methanogens** ("methane-producers") that inhabit environments such as a cow's stomach, which explains why cows are such a good source of methane.

In this book, we will deal mostly with the first two domains, because they are the best studied. However, we will encounter some interesting aspects of the molecular biology of the archaea throughout this book, including details of their transcription in Chapter 11. And in Chapter 24, we will learn that an archaeon, *Methanococcus jannaschii,* was among the first organisms to have its genome sequenced.

SUMMARY All living things are grouped into three domains: bacteria, eukaryota, and archaea. Although the archaea resemble the bacteria physically, some aspects of their molecular biology are more similar to those of eukaryota.

This concludes our brief chronology of molecular biology. Table 1.1 reviews some of the milestones. Although it is a very young discipline, it has an exceptionally rich history, and molecular biologists are now adding new knowledge at an explosive rate. Indeed, the pace of discovery in molecular biology, and the power of its techniques, has led many commentators to call it a revolution. Because some of the most important changes in medicine and agriculture over the next few decades are likely to depend on the manipulation of genes by molecular biologists, this revolution will touch everyone's life in one way or another. Thus, you are

Table 1.1 Molecular Biology Time Line

Year	Investigator(s)	Contribution
1859	Charles Darwin	Published *On the Origin of Species*
1865	Gregor Mendel	Advanced the principles of segregation and independent assortment
1869	Friedrich Miescher	Discovered DNA
1900	Hugo de Vries, Carl Correns, Erich von Tschermak	Rediscovered Mendel's principles
1902	Archibald Garrod	First suggested a genetic cause for a human disease
1902	Walter Sutton, Theodor Boveri	Proposed the chromosome theory
1910, 1916	Thomas Hunt Morgan, Calvin Bridges	Demonstrated that genes are on chromosomes
1913	A.H. Sturtevant	Constructed a genetic map
1927	H.J. Muller	Induced mutation by x-rays
1931	Harriet Creighton, Barbara McClintock	Obtained physical evidence for recombination
1941	George Beadle, E.L. Tatum	Proposed the one-gene/one-enzyme hypothesis
1944	Oswald Avery, Colin McLeod, Maclyn McCarty	Identified DNA as the material genes are made of
1953	James Watson, Francis Crick, Rosalind Franklin, Maurice Wilkins	Determined the structure of DNA
1958	Matthew Meselson, Franklin Stahl	Demonstrated the semiconservative replication of DNA
1961	Sydney Brenner, François Jacob, Matthew Meselson	Discovered messenger RNA
1966	Marshall Nirenberg, Gobind Khorana	Finished unraveling the genetic code
1970	Hamilton Smith	Discovered restriction enzymes that cut DNA at specific sites, which made cutting and pasting DNA easy, thus facilitating DNA cloning
1972	Paul Berg	Made the first recombinant DNA in vitro
1973	Herb Boyer, Stanley Cohen	First used a plasmid to clone DNA
1977	Frederick Sanger	Worked out methods to determine the sequence of bases in DNA and determined the base sequence of an entire viral genome (ϕX174)
1977	Phillip Sharp, Richard Roberts, and others	Discovered interruptions (introns) in genes
1993	Victor Ambros and colleagues	Discovered that a cellular microRNA can decrease gene expression by base-pairing to an mRNA
1995	Craig Venter, Hamilton Smith	Determined the base sequences of the genomes of two bacteria: *Haemophilus influenzae* and *Mycoplasma genitalium,* the first genomes of free-living organisms to be sequenced
1996	Many investigators	Determined the base sequence of the genome of brewer's yeast, *Saccharomyces cerevisiae,* the first eukaryotic genome to be sequenced
1997	Ian Wilmut and colleagues	Cloned a sheep (Dolly) from an adult sheep udder cell
1998	Andrew Fire and colleagues	Discovered that RNAi works by degrading mRNAs containing the same sequence as an invading double-stranded RNA
2003	Many investigators	Reported a finished sequence of the human genome
2005	Many investigators	Reported the rough draft of the genome of the chimpanzee, our closest relative
2007	Craig Venter and colleagues	Used traditional sequencing to obtain the first sequence of an individual human (Craig Venter).
2008	Jian Wang and colleagues	Used "next generation" sequencing to obtain the first sequence of an Asian (Han Chinese) human.
2008	David Bentley and colleagues	Used single molecule sequencing to obtain the first sequence of an African (Nigerian) human.

embarking on a study of a subject that is not only fascinating and elegant, but one that has practical importance as well. F. H. Westheimer, professor emeritus of chemistry at Harvard University, put it well: "The greatest intellectual revolution of the last 40 years may have taken place in biology. Can anyone be considered educated today who does not understand a little about molecular biology?" Happily, after this course you should understand more than a little.

SUMMARY

Genes can exist in several different forms called alleles. A recessive allele can be masked by a dominant one in a heterozygote, but it does not disappear. It can be expressed again in a homozygote bearing two recessive alleles.

Genes exist in a linear array on chromosomes. Therefore, traits governed by genes that lie on the same chromosome can be inherited together. However, recombination between homologous chromosomes occurs during meiosis, so that gametes bearing nonparental combinations of alleles can be produced. The farther apart two genes lie on a chromosome, the more likely such recombination between them will be.

Most genes are made of double-stranded DNA arranged in a double helix. One strand is the complement of the other, which means that faithful gene replication requires that the two strands separate and acquire complementary partners. The linear sequence of bases in a typical gene carries the information for making a protein.

The process of making a gene product is called gene expression. It occurs in two steps: transcription and translation. In the transcription step, RNA polymerase makes a messenger RNA, which is a copy of the information in the gene. In the translation step, ribosomes "read" the mRNA and make a protein according to its instructions. Thus, a change (mutation) in a gene's sequence may cause a corresponding change in the protein product.

All living things are grouped into three domains: bacteria, eukaryota, and archaea. The archaea resemble bacteria physically, but their molecular biology more closely resembles that of eukaryota.

SUGGESTED READINGS

Creighton, H.B., and B. McClintock. 1931. A correlation of cytological and genetical crossing-over in *Zea mays*. *Proceedings of the National Academy of Sciences* 17:492–97.

Mirsky, A.E. 1968. The discovery of DNA. *Scientific American* 218 (June):78–88.

Morgan, T.H. 1910. Sex-limited inheritance in *Drosophila*. *Science* 32:120–22.

Sturtevant, A.H. 1913. The linear arrangement of six sex-linked factors in *Drosophila*, as shown by their mode of association. *Journal of Experimental Zoology* 14:43–59.

CHAPTER 2

The Molecular Nature of Genes

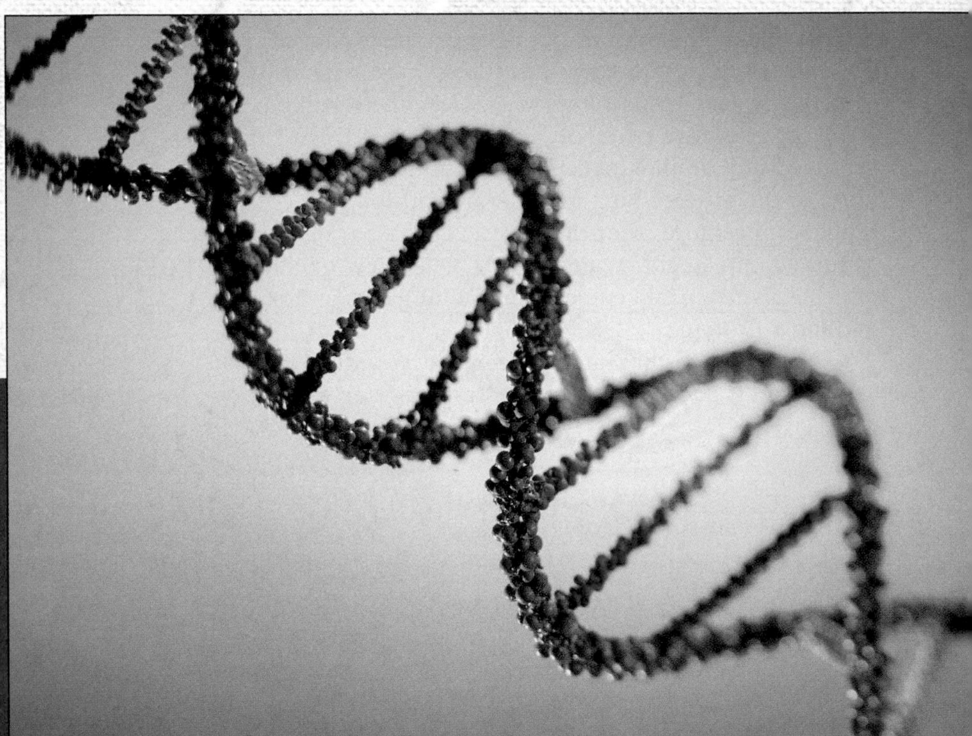

Computer model of the DNA double helix.
© Comstock Images/Jupiter RF.

Before we begin to study in detail the structure and activities of genes, and the experimental evidence underlying those concepts, we need a fuller outline of the adventure that lies before us. Thus, in this chapter and in Chapter 3, we will flesh out the brief history of molecular biology presented in Chapter 1. In this chapter we will begin this task by considering the behavior of genes as molecules.

2.1 The Nature of Genetic Material

The studies that eventually revealed the chemistry of genes began in Tübingen, Germany, in 1869. There, Friedrich Miescher isolated nuclei from pus cells (white blood cells) in waste surgical bandages. He found that these nuclei contained a novel phosphorus-bearing substance that he named *nuclein*. Nuclein is mostly **chromatin,** which is a complex of **deoxyribonucleic acid (DNA)** and chromosomal proteins.

By the end of the nineteenth century, both DNA and **ribonucleic acid (RNA)** had been separated from the protein that clings to them in the cell. This allowed more detailed chemical analysis of these **nucleic acids.** (Notice that the term *nucleic acid* and its derivatives, *DNA* and *RNA*, come directly from Miescher's term *nuclein*.) By the beginning of the 1930s, P. Levene, W. Jacobs, and others had demonstrated that RNA is composed of a sugar (ribose) plus four nitrogen-containing bases, and that DNA contains a different sugar (deoxyribose) plus four bases. They discovered that each base is coupled with a sugar–phosphate to form a nucleotide. We will return to the chemical structures of DNA and RNA later in this chapter. First, let us examine the evidence that genes are made of DNA.

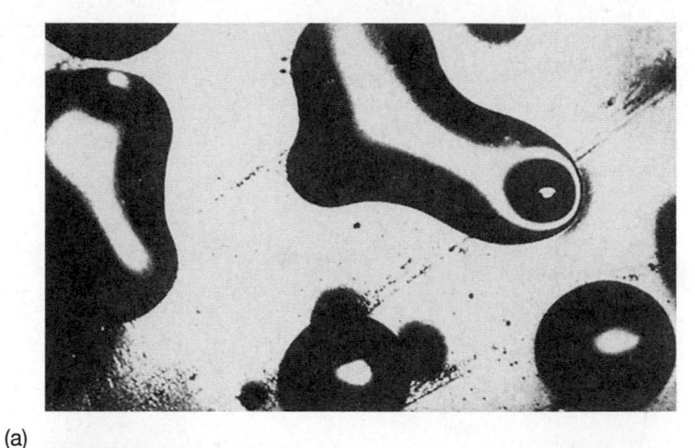
(a)

(b)

Figure 2.1 Variants of *Streptococcus pneumoniae:* (a) The large, glossy colonies contain smooth (S) virulent bacteria; **(b)** the small, mottled colonies are composed of rough (R) avirulent bacteria. (*Source:* (*a, b*) Harriet Ephrussi-Taylor.)

Transformation in Bacteria

Frederick Griffith laid the foundation for the identification of DNA as the genetic material in 1928 with his experiments on **transformation** in the bacterium pneumococcus, now known as *Streptococcus pneumoniae*. The wild-type organism is a spherical cell surrounded by a mucous coat called a capsule. The cells form large, glistening colonies, characterized as smooth (S) (Figure 2.1a). These cells are **virulent,** that is, capable of causing lethal infections upon injection into mice. A certain mutant strain of *S. pneumoniae* has lost the ability to form a capsule. As a result, it grows as small, rough (R) colonies (Figure 2.1b). More importantly, it is **avirulent;** because it has no protective coat, it is engulfed by the host's white blood cells before it can proliferate enough to do any damage.

The key finding of Griffith's work was that heat-killed virulent colonies of *S. pneumoniae* could **transform** avirulent cells to virulent ones. Neither the heat-killed virulent bacteria nor the live avirulent ones by themselves could cause a lethal infection. Together, however, they were deadly. Somehow the virulent trait passed from the dead cells to the live, avirulent ones. This transformation phenomenon is illustrated in Figure 2.2. Transformation was not transient; the ability to make a capsule and therefore to kill host animals, once conferred on the avirulent bacteria, was passed to their descendants as a heritable trait. In other words, the avirulent cells somehow gained the gene for virulence during transformation. This meant that the transforming substance in the heat-killed bacteria was probably the gene for virulence itself. The missing piece of the puzzle was the chemical nature of the transforming substance.

DNA: The Transforming Material Oswald Avery, Colin MacLeod, and Maclyn McCarty supplied the missing piece in 1944. They used a transformation test similar to the one that Griffith had introduced, and they took pains to define the chemical nature of the transforming substance from virulent cells. First, they removed the protein from the extract with organic solvents and found that the extract still transformed. Next, they subjected it to digestion with various enzymes. Trypsin and chymotrypsin, which destroy protein, had no effect on transformation. Neither did ribonuclease, which degrades RNA. These experiments ruled out protein or RNA as the transforming material. On the other hand, Avery and his coworkers found that the enzyme deoxyribonuclease (DNase), which breaks down DNA, destroyed the transforming ability of the virulent cell extract. These results suggested that the transforming substance was DNA.

Direct physical-chemical analysis supported the hypothesis that the purified transforming substance was DNA. The analytical tools Avery and his colleagues used were the following:

1. *Ultracentrifugation* They spun the transforming substance in an ultracentrifuge (a very high-speed

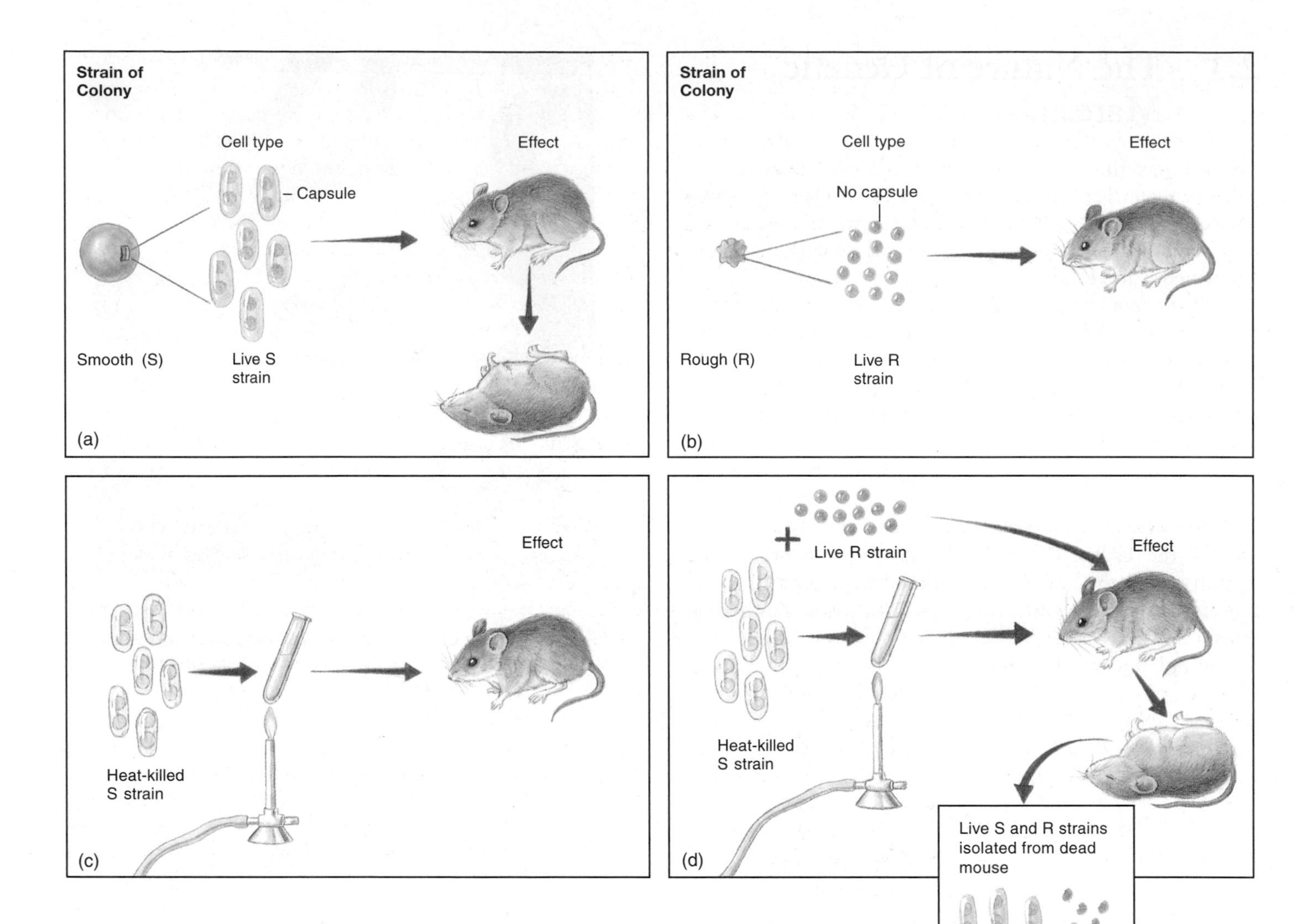

Figure 2.2 Griffith's transformation experiments. (a) Virulent strain S *S. pneumoniae* bacteria kill their host; **(b)** avirulent strain R bacteria cannot infect successfully, so the mouse survives; **(c)** strain S bacteria that are heat-killed can no longer infect; **(d)** a mixture of strain R and heat-killed strain S bacteria kills the mouse. The killed virulent (S) bacteria have transformed the avirulent (R) bacteria to virulent (S).

centrifuge) to estimate its size. The material with transforming activity sedimented rapidly (moved rapidly toward the bottom of the centrifuge tube), suggesting a very high molecular weight, characteristic of DNA.

2. *Electrophoresis* They placed the transforming substance in an electric field to see how rapidly it moved. The transforming activity had a relatively high mobility, also characteristic of DNA because of its high charge-to-mass ratio.
3. *Ultraviolet Absorption Spectrophotometry* They placed a solution of the transforming substance in a spectrophotometer to see what kind of ultraviolet (UV) light it absorbed most strongly. Its absorption spectrum matched that of DNA. That is, the light it absorbed most strongly had a wavelength of about 260 nanometers (nm), in contrast to protein, which absorbs maximally at 280 nm.
4. *Elementary Chemical Analysis* This yielded an average nitrogen-to-phosphorus ratio of 1.67, about what one would expect for DNA, which is rich in both elements, but vastly lower than the value expected for protein, which is rich in nitrogen but poor in phosphorus. Even a slight protein contamination would have raised the nitrogen-to-phosphorus ratio.

Further Confirmation These findings should have settled the issue of the nature of the gene, but they had little immediate effect. The mistaken notion, from early chemical analyses, that DNA was a monotonous repeat of a four-nucleotide sequence, such as ACTG-ACTG-ACTG, and so on, persuaded many geneticists that it could not be the genetic material. Furthermore, controversy persisted about possible protein contamination in the transforming material, whether transformation could be accomplished with other genes besides those governing R and S, and even whether bacterial genes were like the genes of higher organisms.

Yet, by 1953, when James Watson and Francis Crick published the double-helical model of DNA structure, most geneticists agreed that genes were made of DNA. What had changed? For one thing, Erwin Chargaff had shown in 1950 that the bases were not really found in equal proportions in DNA, as previous evidence had suggested, and that the base composition of DNA varied from one species to another. In fact, this is exactly what one would expect for genes, which also vary from one species to another. Furthermore, Rollin Hotchkiss had refined and extended Avery's findings. He purified the transforming substance to the point where it contained only 0.02% protein and showed that it could still change the genetic characteristics of bacterial cells. He went on to show that such highly purified DNA could transfer genetic traits other than R and S.

Finally, in 1952, A.D. Hershey and Martha Chase performed another experiment that added to the weight of evidence that genes were made of DNA. This experiment involved a **bacteriophage** (bacterial virus) called T2 that infects the bacterium *Escherichia coli* (Figure 2.3).

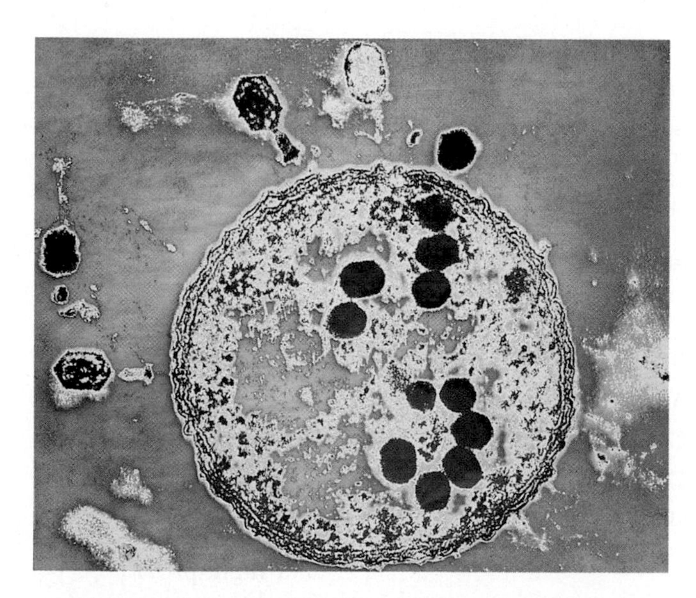

Figure 2.3 A false color transmission electron micrograph of T2 phages infecting an *E. coli* cell. Phage particles at left and top appear ready to inject their DNA into the host cell. Another T2 phage has already infected the cell, however, and progeny phage particles are being assembled. The progeny phage heads are readily discernible as dark polygons inside the host cell. (*Source:* © Lee Simon/Photo Researchers, Inc.)

(The term *bacteriophage* is usually shortened to ***phage***.) During infection, the phage genes enter the host cell and direct the synthesis of new phage particles. The phage is composed of protein and DNA only. The question is this: Do the genes reside in the protein or in the DNA? The Hershey–Chase experiment answered this question by showing that, on infection, most of the DNA entered the bacterium, along with only a little protein. The bulk of the protein stayed on the outside (Figure 2.4). Because DNA was the major component that got into the host cells, it likely contained the genes. Of course, this conclusion was not unequivocal; the small amount of protein that entered along with the DNA could conceivably have carried the genes. But taken together with the work that had gone before, this study helped convince geneticists that DNA, and not protein, is the genetic material.

The Hershey–Chase experiment depended on radioactive labels on the DNA and protein—a different label for each. The labels used were phosphorus-32 (^{32}P) for DNA and sulfur-35 (^{35}S) for protein. These choices make sense, considering that DNA is rich in phosphorus but phage protein has none, and that protein contains sulfur but DNA does not.

Hershey and Chase allowed the labeled phages to attach by their tails to bacteria and inject their genes into their hosts. Then they removed the empty phage coats by mixing vigorously in a blender. Because they knew that the genes must go into the cell, their question was: What went in, the ^{32}P-labeled DNA or the ^{35}S-labeled protein? As we have seen, it was the DNA. In general, then, genes are made of DNA. On the other hand, as we will see later in this chapter, other experiments showed that some viral genes consist of RNA.

SUMMARY Physical-chemical experiments involving bacteria and a bacteriophage showed that their genes are made of DNA.

The Chemical Nature of Polynucleotides

By the mid-1940s, biochemists knew the fundamental chemical structures of DNA and RNA. When they broke DNA into its component parts, they found these constituents to be nitrogenous **bases, phosphoric acid,** and the sugar **deoxyribose** (hence the name *deoxyribonucleic acid*). Similarly, RNA yielded bases and phosphoric acid, plus a different sugar, **ribose.** The four bases found in DNA are **adenine** (A), **cytosine** (C), **guanine** (G), and **thymine** (T). RNA contains the same bases, except that **uracil** (U) replaces thymine. The structures of these bases,

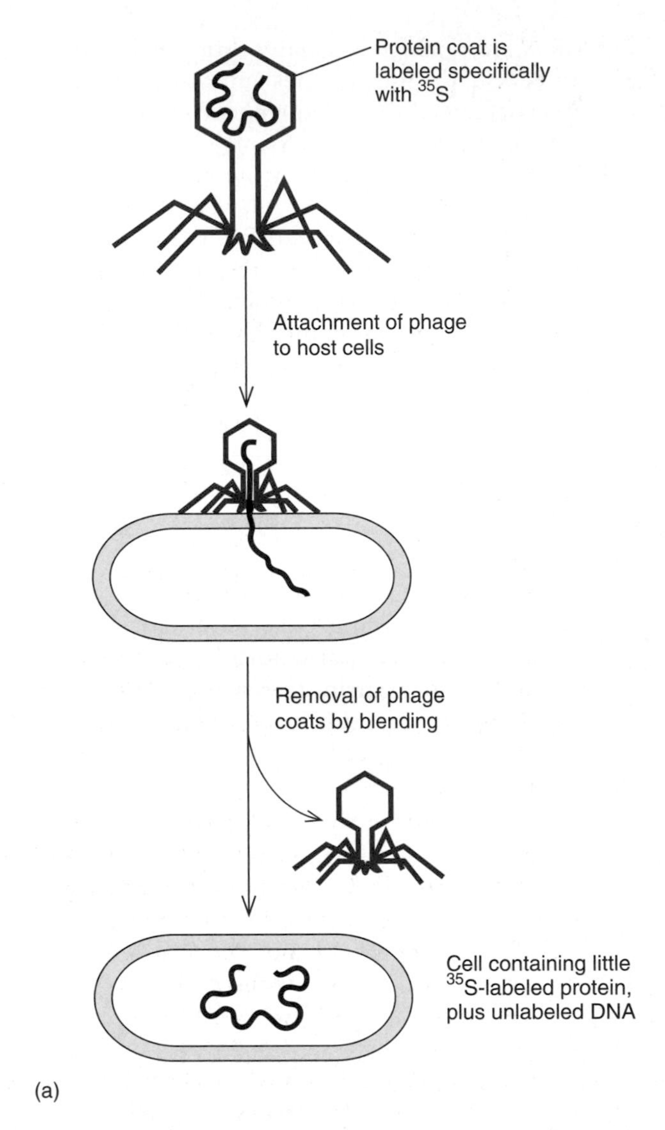

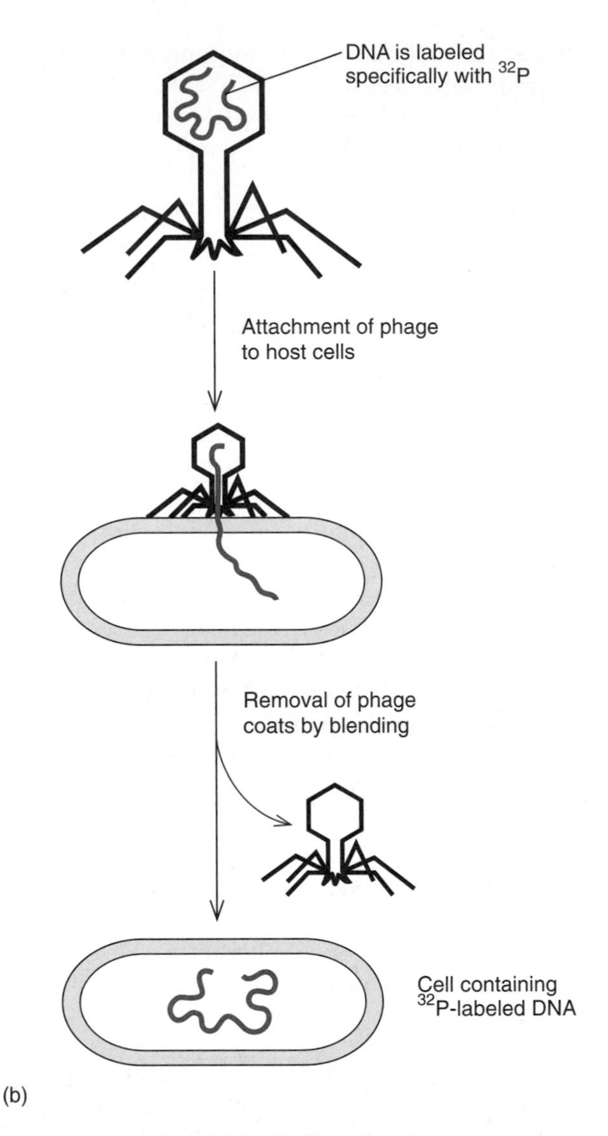

Figure 2.4 The Hershey–Chase experiment. Phage T2 contains genes that allow it to replicate in *E. coli.* Because the phage is composed of DNA and protein only, its genes must be made of one of these substances. To discover which, Hershey and Chase performed a two-part experiment. In the first part **(a),** they labeled the phage protein with ^{35}S (red), leaving the DNA unlabeled (black). In the second part **(b),** they labeled the phage DNA with ^{32}P (red), leaving the protein unlabeled (black). Since the phage genes must enter the cell, the experimenters reasoned that the type of label found in the infected cells would indicate the nature of the genes. Most of the labeled protein remained on the outside and was stripped off the cells by use of a blender (a), whereas most of the labeled DNA entered the infected cells (b). The conclusion was that the genes of this phage are made of DNA.

shown in Figure 2.5, reveal that adenine and guanine are related to the parent molecule, purine. Therefore, we refer to these compounds as **purines.** The other bases resemble pyrimidine, so they are called **pyrimidines.** These structures constitute the alphabet of genetics.

Figure 2.6 depicts the structures of the sugars found in nucleic acids. Notice that they differ in only one place. Where ribose contains a hydroxyl (OH) group in the 2-position, deoxyribose lacks the oxygen and simply has a hydrogen (H), represented by the vertical line. Hence the name *deoxyribose.* The bases and sugars in RNA and DNA are joined together into units called **nucleosides** (Figure 2.7). The names of the nucleosides derive from the corresponding bases:

Base	*Nucleoside (RNA)*	*Deoxynucleoside (DNA)*
Adenine	Adenosine	Deoxyadenosine
Guanine	Guanosine	Deoxyguanosine
Cytosine	Cytidine	Deoxycytidine
Uracil	Uridine	Not usually found
Thymine	Not usually found	(Deoxy)thymidine

Because thymine is not usually found in RNA, the "deoxy" designation for its nucleoside is frequently assumed, and the deoxynucleoside is simply called **thymidine.** The numbering of the carbon atoms in the sugars of the nucleosides (see Figure 2.7) is important. Note that the ordinary numbers are used in the bases, so

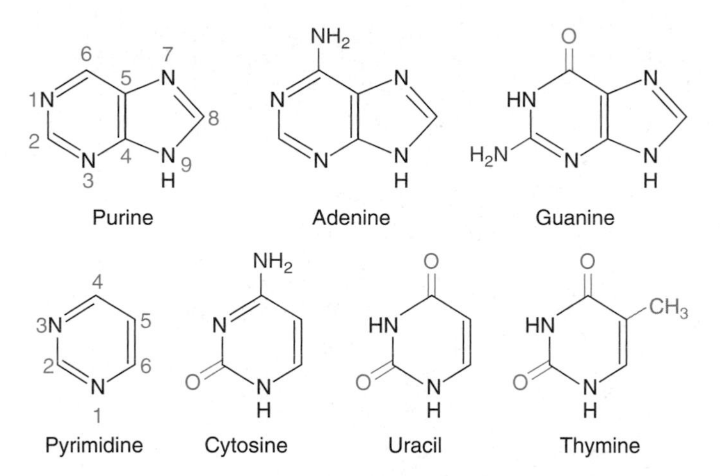

Figure 2.5 The bases of DNA and RNA. The parent bases, purine and pyrimidine, on the left, are not found in DNA and RNA. They are shown for comparison with the other five bases.

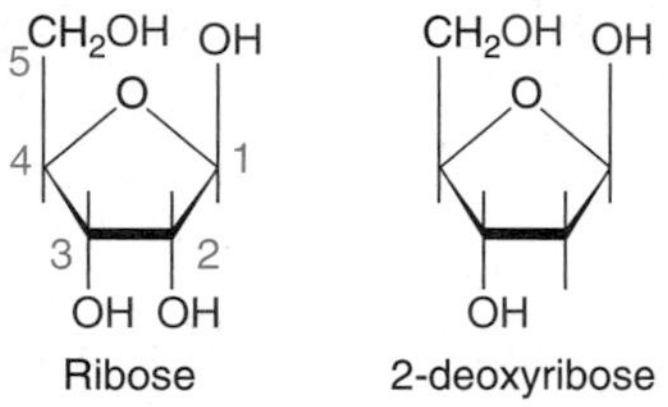

Figure 2.6 The sugars of nucleic acids. Note the OH in the 2-position of ribose and its absence in deoxyribose.

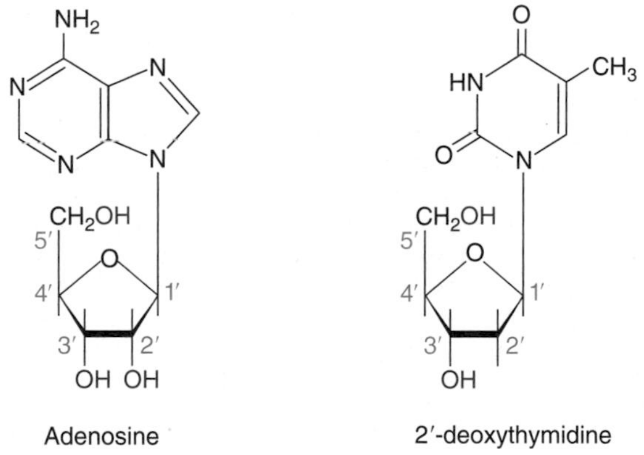

Figure 2.7 Two examples of nucleosides.

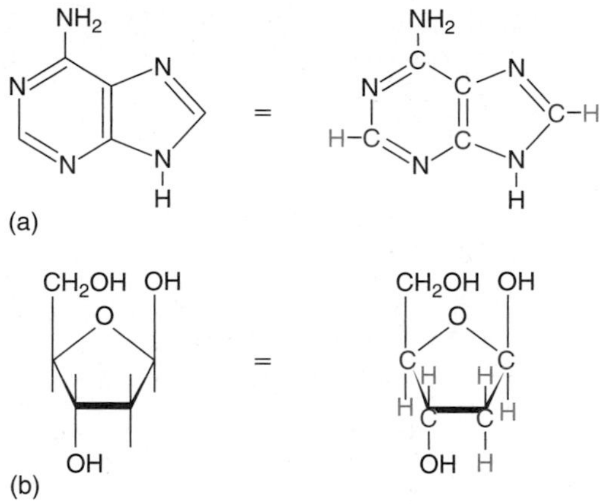

Figure 2.8 The structures of (a) adenine and (b) deoxyribose. Note that the structures on the left do not designate most or all of the carbons and some of the hydrogens. These designations are included in the structures on the right, in red and blue, respectively.

the carbons in the sugars are called by primed numbers. Thus, for example, the base is linked to the 1′-position of the sugar, the 2′-position is deoxy in deoxynucleosides, and the sugars are linked together in DNA and RNA through their 3′- and 5′-positions.

The structures in Figure 2.5 were drawn using an organic chemistry shorthand that leaves out certain atoms for simplicity's sake. Figures 2.6 and 2.7 use a slightly different convention, in which a straight line with a free end denotes a C–H bond with a hydrogen atom at the end. Figure 2.8 shows the structures of adenine and deoxyribose, first in shorthand, then with every atom included.

The subunits of DNA and RNA are **nucleotides,** which are nucleosides with a phosphate group attached through a phosphoester bond (Figure 2.9). An ester is an organic compound formed from an alcohol (bearing a hydroxyl group) and an acid. In the case of a nucleotide, the alcohol group is the 5′-hydroxyl group of the sugar, and the acid is phosphoric acid, which is why we call the ester a *phosphoester.* Figure 2.9 also shows the structure of one of the four DNA precursors, deoxyadenosine-5′-triphosphate (dATP). When synthesis of DNA takes place, two phosphate groups are removed from dATP, leaving deoxyadenosine-5′-monophosphate (dAMP). The other three nucleotides in DNA (dCMP, dGMP, and dTMP) have analogous structures and names.

We will discuss the synthesis of DNA in detail in Chapters 20 and 21. For now, notice the structure of the bonds that join nucleotides together in DNA and RNA (Figure 2.10). These are called **phosphodiester bonds** because they involve phosphoric acid linked to *two* sugars: one through a sugar 5′-group, the other through a sugar 3′-group. You will notice that the bases have been rotated in this picture, relative to their positions in previous figures. This more closely resembles their geometry in DNA or RNA. Note also that this **trinucleotide,** or string of three nucleotides, has polarity: The top of the molecule bears a free 5′-phosphate group, so it is called the **5′-end.** The bottom, with a free 3′-hydroxyl group, is called the **3′-end.**

Figure 2.11 introduces a shorthand way of representing a nucleotide or a DNA chain. This notation presents the deoxyribose sugar as a vertical line, with the base joined to the 1′-position at the top and the phosphodiester links to neighboring nucleotides through the 3′-(middle) and 5′-(bottom) positions.

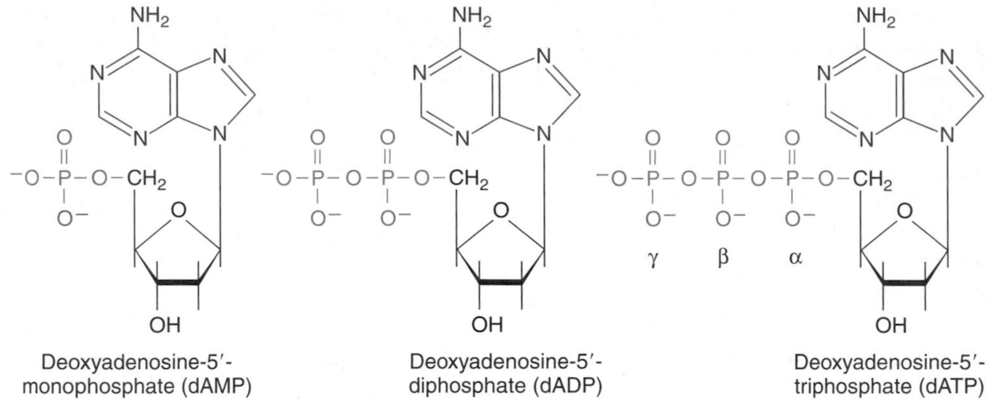

Deoxyadenosine-5′-monophosphate (dAMP)

Deoxyadenosine-5′-diphosphate (dADP)

Deoxyadenosine-5′-triphosphate (dATP)

Figure 2.9 Three nucleotides. The 5′-nucleotides of deoxyadenosine are formed by phosphorylating the 5′-hydroxyl group. The addition of one phosphate results in deoxyadenosine-5′-monophosphate (dAMP). One more phosphate yields deoxyadenosine-5′-diphosphate (dADP). Three phosphates (designated α, β, γ) give deoxyadenosine-5′-triphosphate (dATP).

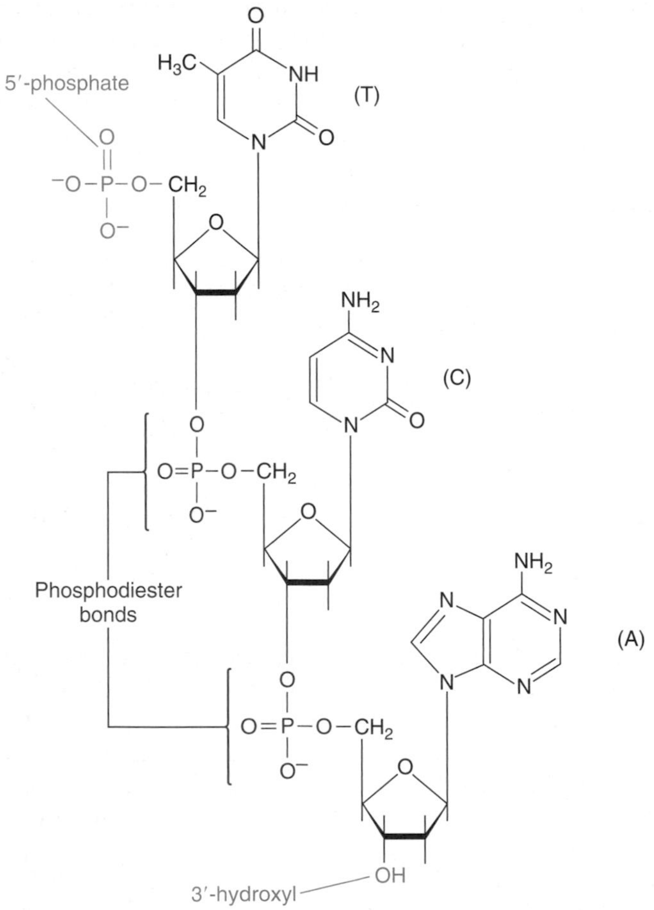

Figure 2.10 A trinucleotide. This little piece of DNA contains only three nucleotides linked together by phosphodiester bonds (red) between the 5′- and 3′-hydroxyl groups of the sugars. The 5′-end of this DNA is at the top, where a free 5′-phosphate group (blue) is located; the 3′-end is at the bottom, where a free 3′-hydroxyl group (also blue) appears. The sequence of this DNA could be read as 5′pdTpdCpdA3′. This would usually be simplified to TCA.

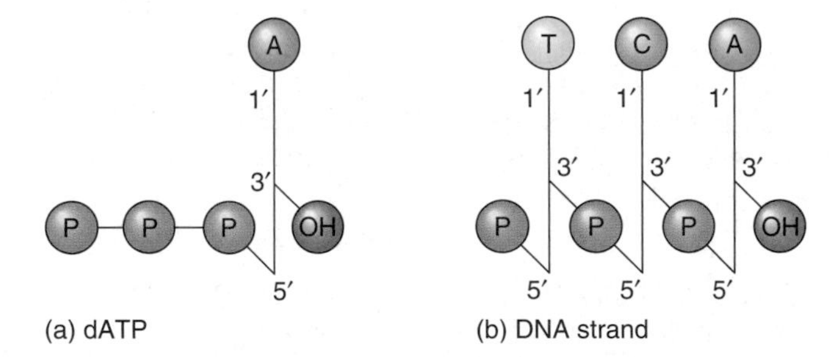

Figure 2.11 Shorthand DNA notation. (a) The nucleotide dATP. This illustration highlights four features of this DNA building block: (1) The deoxyribose sugar is represented by the vertical black line. (2) At the top, attached to the 1′-position of the sugar is the base, adenine (green). (3) In the middle, at the 3′-position of the sugar is a hydroxyl group (OH, orange). (4) At the bottom, attached to the 5′-position of the sugar is a triphosphate group (purple). **(b)** A short DNA strand. The same trinucleotide (TCA) illustrated in Figure 2.10 is shown here in shorthand. Note the 5′-phosphate and the phosphodiester bonds (purple), and the 3′-hydroxyl group (orange). According to convention, this little piece of DNA is written 5′ to 3′ left to right.

SUMMARY DNA and RNA are chain-like molecules composed of subunits called nucleotides. The nucleotides contain a base linked to the 1′-position of a sugar (ribose in RNA or deoxyribose in DNA) and a phosphate group. The phosphate joins the sugars in a DNA or RNA chain through their 5′- and 3′-hydroxyl groups by phosphodiester bonds.

2.2 DNA Structure

All the facts about DNA and RNA just mentioned were known by the end of the 1940s. By that time it was also becoming clear that DNA was the genetic material and that it therefore stood at the very center of the study of life. Yet the three-dimensional structure of DNA was unknown. For these reasons, several researchers dedicated themselves to finding this structure.

Experimental Background

One of the scientists interested in DNA structure was Linus Pauling, a theoretical chemist at the California Institute of Technology. He was already famous for his studies on chemical bonding and for his elucidation of the α-helix; an important feature of protein structure. Indeed, the α-helix, held together by hydrogen bonds, laid the intellectual groundwork for the double-helix model of DNA proposed by Watson and Crick. Another group trying to find the structure of DNA included Maurice Wilkins, Rosalind Franklin, and their colleagues at King's College in London. They were using x-ray diffraction to analyze the three-dimensional structure of DNA. Finally, James Watson and Francis Crick entered the race. Watson, still in his early twenties, but already holding a Ph.D. degree from Indiana University, had come to the Cavendish Laboratories in Cambridge, England, to learn about DNA. There he met Crick, a physicist who at age 35 was retraining as a molecular biologist. Watson and Crick performed no experiments themselves. Their tactic was to use other groups' data to build a DNA model.

Erwin Chargaff was another very important contributor. We have already seen how his 1950 paper helped identify DNA as the genetic material, but the paper contained another piece of information that was even more significant. Chargaff's studies of the base compositions of DNAs from various sources revealed that the content of purines was always roughly equal to the content of pyrimidines. Furthermore, the amounts of adenine and thymine were always roughly equal, as were the amounts of guanine and cytosine. These findings, known as Chargaff's rules, provided a valuable foundation for Watson and Crick's model. Table 2.1 presents Chargaff's data. You will notice some deviation from the rules due to incomplete recovery of some of the bases, but the overall pattern is clear.

Perhaps the most crucial piece of the puzzle came from an x-ray diffraction picture of DNA taken by Franklin in 1952—a picture that Wilkins shared with James Watson in London on January 30, 1953. The x-ray technique worked as follows: The experimenter made a very concentrated, viscous solution of DNA, then reached in with a needle and pulled out a fiber. This was not a single molecule, but a whole batch of DNA molecules, forced into side-by-side alignment by the pulling action. Given the right relative humidity in the surrounding air, this fiber was enough like a crystal that it diffracted x-rays in an interpretable way. In fact, the x-ray diffraction pattern in Franklin's picture (Figure 2.12) was so simple—a series of spots arranged in an X shape—that it indicated that the DNA structure itself must be very simple. By contrast, a complex, irregular molecule like a protein gives a complex x-ray diffraction pattern with many spots, rather like a surface peppered by a shotgun blast. Because DNA is very large, it can be simple only if it has a regular, repeating structure. And the simplest repeating shape that a long, thin molecule can assume is a corkscrew, or helix.

The Double Helix

Franklin's x-ray work strongly suggested that DNA was a helix. Not only that, it gave some important information about the size and shape of the helix. In particular, the spacing between adjacent bands in an arm of the X is inversely related to the overall repeat distance in the helix, 33.2 angstroms (33.2 Å), and the spacing from the top of the X to the bottom is inversely related to the spacing (3.32 Å) between the repeated elements (**base pairs**) in the helix. (See Chapter 9 for information on how Bragg's law explains these inverse relationships.) However, even though the Franklin picture told much about

Table 2.1 Composition of DNA in Moles of Base per Mole of Phosphate

	Human							Bovine				
	Sperm		Thymus	Liver Carcinoma	Yeast		Avian Tubercle Bacilli	Thymus			Spleen	
	#1	#2			#1	#2		#1	#2	#3	#1	#2
A:	0.29	0.27	0.28	0.27	0.24	0.30	0.12	0.26	0.28	0.30	0.25	0.26
T:	0.31	0.30	0.28	0.27	0.25	0.29	0.11	0.25	0.24	0.25	0.24	0.24
G:	0.18	0.17	0.19	0.18	0.14	0.18	0.28	0.21	0.24	0.22	0.20	0.21
C:	0.18	0.18	0.16	0.15	0.13	0.15	0.26	0.16	0.18	0.17	0.15	0.17
Recovery:	0.96	0.92	0.91	0.87	0.76	0.92	0.77	0.88	0.94	0.94	0.84	0.88

Source: E. Chargaff "Chemical Specificity of Nucleic Acids and Mechanism of Their Enzymatic Degradation," *Experientia* 6:206, 1950.

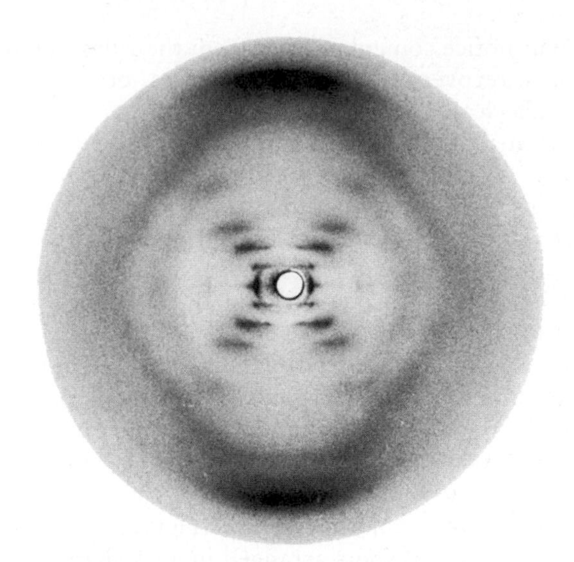

Figure 2.12 Franklin's x-ray picture of DNA. The regularity of this pattern indicated that DNA is a helix. The spacing between the bands at the top and bottom of the X gave the spacing between elements of the helix (base pairs) as 3.32 Å. The spacing between neighboring bands in the pattern gave the overall repeat of the helix (the length of one helical turn) as 33.2 Å. (*Source:* Courtesy Professor M.H.F. Wilkins, Biophysics Dept., King's College, London.)

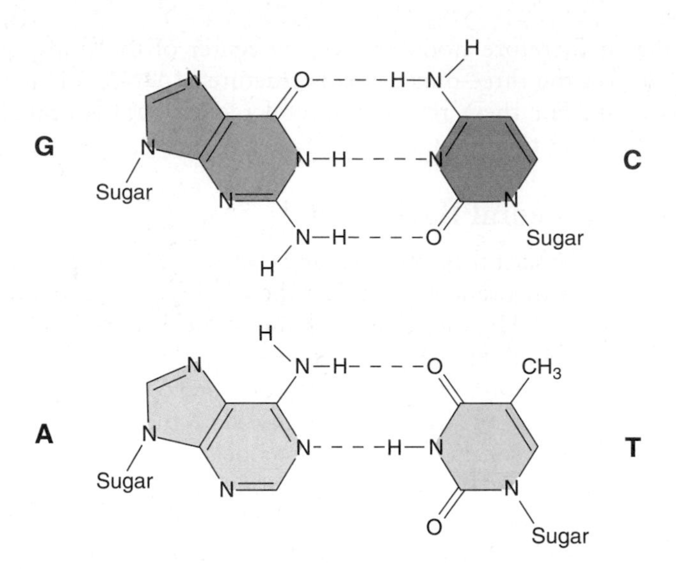

Figure 2.13 The base pairs of DNA. A guanine–cytosine pair (G–C), held together by three hydrogen bonds (dashed lines), has almost exactly the same shape as an adenine–thymine pair (A–T), held together by two hydrogen bonds.

DNA, it presented a paradox: DNA was a helix with a regular, repeating structure, but for DNA to serve its genetic function, it must have an *irregular* sequence of bases.

Watson and Crick saw a way to resolve this contradiction and satisfy Chargaff's rules at the same time: DNA must be a **double helix** with its sugar–phosphate backbones on the outside and its bases on the inside. Moreover, the bases must be paired, with a purine in one strand always across from a pyrimidine in the other. This way the helix would be uniform; it would not have bulges where two large purines were paired or constrictions where two small pyrimidines were paired. Watson has joked about the reason he seized on a double helix: "I had decided to build two-chain models. Francis would have to agree. Even though he was a physicist, he knew that important biological objects come in pairs."

But Chargaff's rules went further than this. They decreed that the amounts of adenine and thymine were equal and so were the amounts of guanine and cytosine. This fit very neatly with Watson and Crick's observation that an adenine–thymine base pair held together by hydrogen bonds has almost exactly the same shape as a guanine–cytosine base pair (Figure 2.13). So Watson and Crick postulated that adenine must always pair with thymine, and guanine with cytosine. This way, the double-stranded DNA will be uniform, composed of very similarly shaped base pairs, regardless of the unpredictable sequence of either DNA strand by itself. This was their crucial insight, and the key to the structure of DNA.

The double helix, often likened to a twisted ladder, is presented in three ways in Figure 2.14. The curving sides of the ladder represent the sugar–phosphate backbones of the two DNA strands; the rungs are the base pairs. The spacing between base pairs is 3.32 Å, and the overall helix repeat distance is about 33.2 Å, meaning that there are about 10 base pairs (**bp**) per turn of the helix. (One angstrom [Å] is one ten-billionth of a meter or one-tenth of a nanometer [nm].) The arrows indicate that the two strands are **antiparallel.** If one has $5' \rightarrow 3'$ polarity from top to bottom, the other must have $3' \rightarrow 5'$ polarity from top to bottom. In solution, DNA has a structure very similar to the one just described, but the helix contains about 10.4 bp per turn.

Watson and Crick published the outline of their model in the journal *Nature,* back-to-back with papers by Wilkins and Franklin and their coworkers showing the x-ray data. The Watson–Crick paper is a classic of simplicity—only 900 words, barely over a page long. It was published very rapidly, less than a month after it was submitted. Actually, Crick wanted to spell out the biological implications of the model, but Watson was uncomfortable doing that. They compromised on a sentence that is one of the greatest understatements in scientific literature: "It has not escaped our notice that the specific base pairing we have proposed immediately suggests a possible copying mechanism for the genetic material."

As this provocative sentence indicates, Watson and Crick's model does indeed suggest a copying mechanism

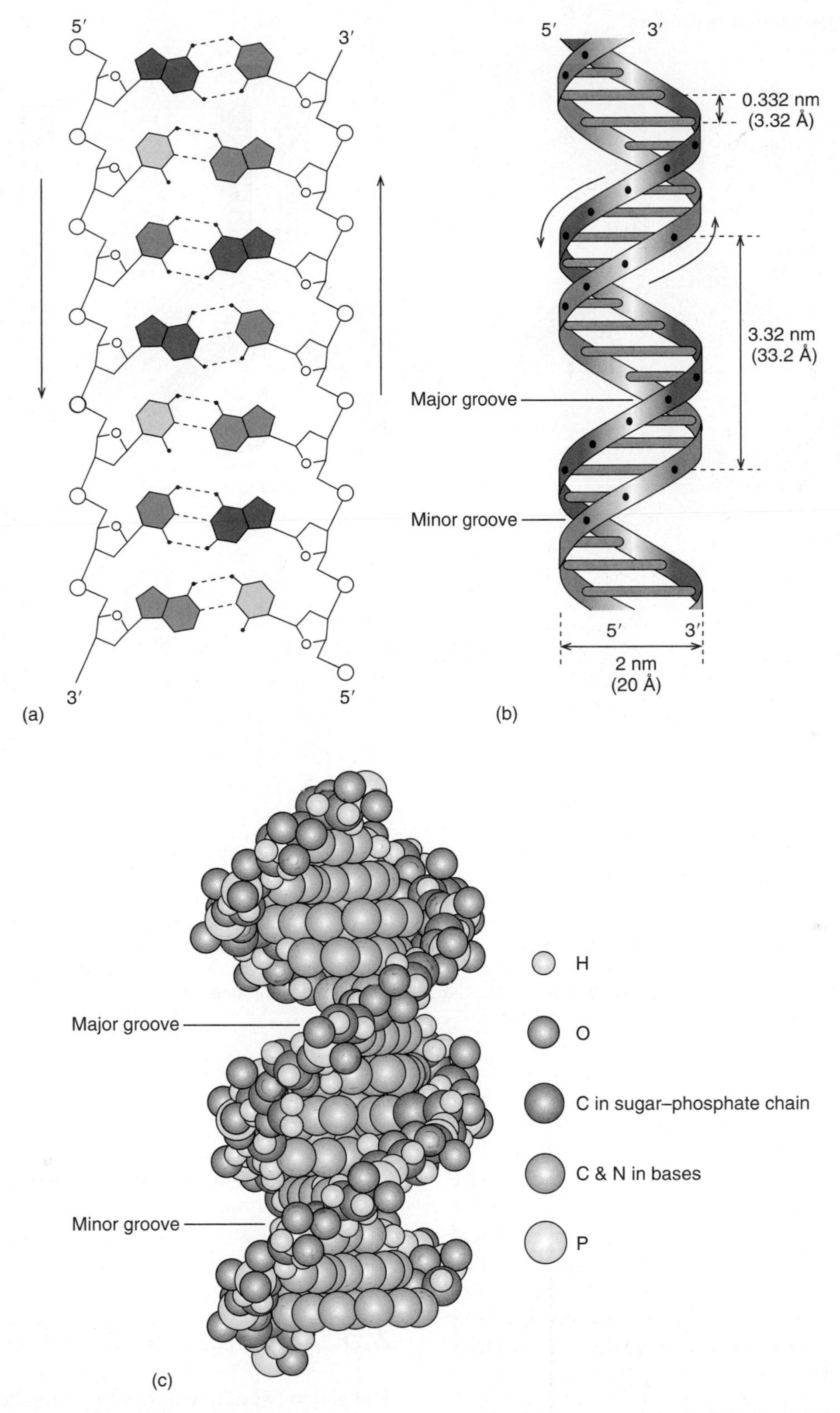

Figure 2.14 Three models of DNA structure. **(a)** The helix is straightened out to show the base pairing in the middle. Each type of base is represented by a different color, with the sugar–phosphate backbones in black. Note the three hydrogen bonds in the G–C pairs and the two in the A–T pairs. The vertical arrows beside each strand point in the 5′→3′ direction and indicate the antiparallel nature of the two DNA strands. The left strand runs 5′→3′, top to bottom; the right strand runs 5′→3′, bottom to top. The deoxyribose rings (white pentagons with O representing oxygen) also show that the two strands have opposite orientations: The rings in the right strand are inverted relative to those in the left strand. **(b)** The DNA double helix is presented as a twisted ladder whose sides represent the sugar–phosphate backbones of the two strands and whose rungs represent base pairs. The curved arrows beside the two strands indicate the 5′→3′ orientation of each strand, further illustrating that the two strands are antiparallel. **(c)** A space-filling model. The sugar–phosphate backbones appear as strings of dark gray, red, light gray, and yellow spheres, whereas the base pairs are rendered as horizontal flat plates composed of blue spheres. Note the major and minor grooves in the helices depicted in parts (b) and (c).

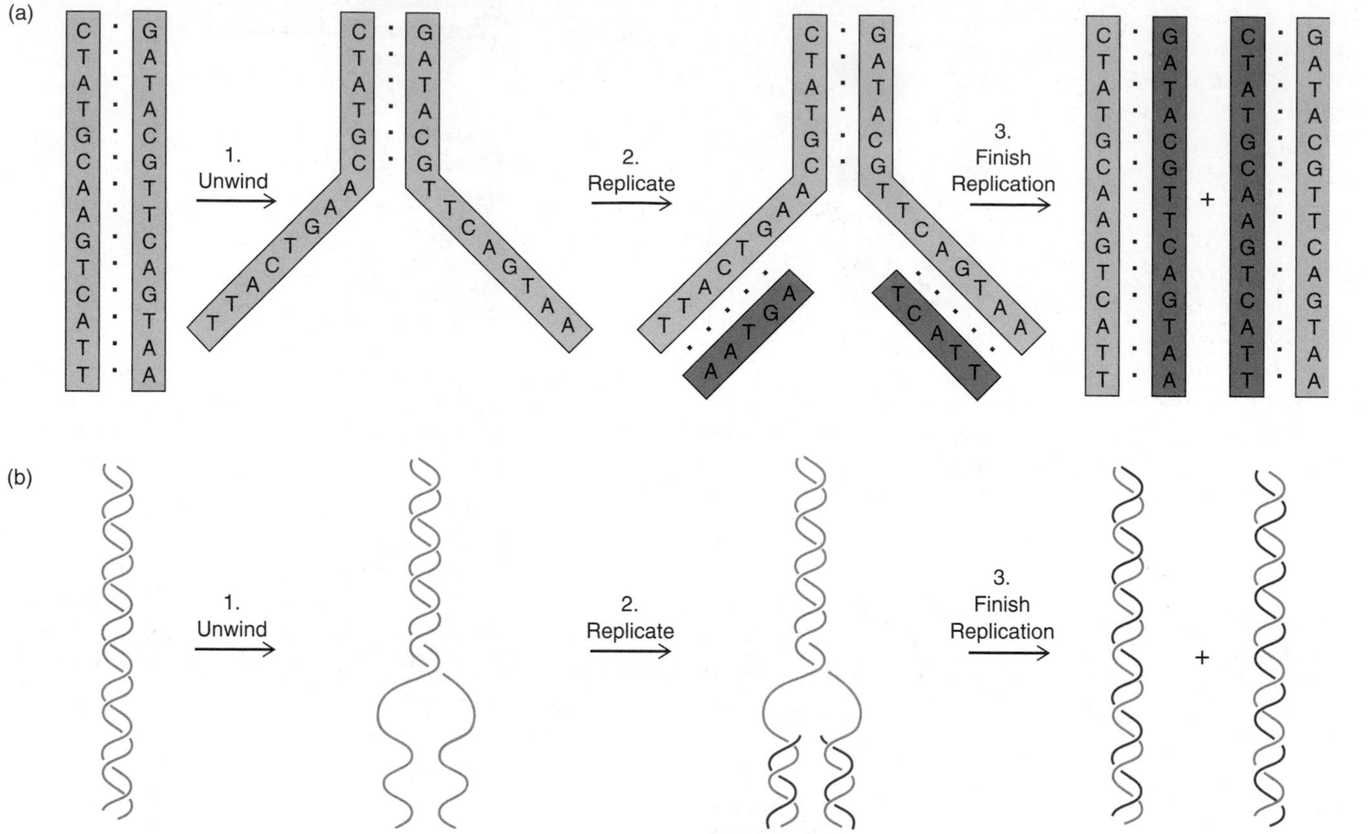

Figure 2.15 Replication of DNA. **(a)** For simplicity, the two parental DNA strands (blue) are represented as parallel lines. Step 1: During replication these parental strands separate, or unwind. Step 2: New strands (pink) are built with bases complementary to those of the separated parental strands. Step 3: Replication is finished, with the parental strands totally separated and the new strands completed. The end result is two double-stranded DNA duplexes identical to the original. Therefore, each daughter duplex gets one parental strand (blue) and one new strand (pink). Because only one parental strand is conserved in each of the daughter duplexes, this mechanism of replication is called semiconservative. **(b)** A more realistic portrayal of the same process. Here the strands are shown in a double helix instead of as parallel lines. Notice again that two daughter duplexes are generated, each with one parental strand (blue) and one new strand (pink).

for DNA. Because one strand is the **complement** of the other, the two strands can be separated, and each can then serve as the template for building a new partner. Figure 2.15 shows schematically how this is accomplished. Notice how this mechanism, known as semiconservative replication, ensures that the two daughter DNA duplexes will be exactly the same as the parent, preserving the integrity of the genes as cells divide. In 1958, Matthew Meselson and Franklin Stahl demonstrated that this really is how DNA replicates (Chapter 20).

SUMMARY The DNA molecule is a double helix, with sugar–phosphate backbones on the outside and base pairs on the inside. The bases pair in a specific way: adenine (A) with thymine (T), and guanine (G) with cytosine (C). The replication of DNA is semiconservative, with each strand serving as the template for building a complementary partner.

2.3 Genes Made of RNA

The genetic system explored by Hershey and Chase was a phage, a bacterial virus. A virus particle by itself is essentially just a package of genes. It has no life of its own, no metabolic activity; it is inert. But when the virus infects a host cell, it seems to come to life. Suddenly the host cell begins making viral proteins. Then the viral genes are replicated and the newly made genes, together with viral coat proteins, assemble into progeny virus particles. Because of their behavior as inert particles outside, but life-like agents inside their hosts, viruses resist classification. Some scientists refer to them as "living things" or even "organisms." Others prefer a label that, although more cumbersome, is also more descriptive of a virus's less-than-living status: infectious agent.

All true organisms and some viruses contain genes made of DNA. But other viruses, including several phages, plant and animal viruses (e.g., HIV, the AIDS virus), have RNA genes. Sometimes viral RNA genes are double-stranded, but usually they are single-stranded.

We have already encountered one famous example of the use of viruses in molecular biology research. We will see many more in subsequent chapters. In fact, without viruses, the field of molecular biology would be immeasurably poorer.

SUMMARY Certain viruses contain genes made of RNA instead of DNA.

2.4 Physical Chemistry of Nucleic Acids

DNA and RNA molecules can assume several different structures. Let us examine these and the behavior of DNA under conditions that encourage the two strands to separate and then come together again.

A Variety of DNA Structures

The structure for DNA proposed by Watson and Crick (see Figure 2.14) represents the sodium salt of DNA in a fiber produced at very high relative humidity (92%). This is called the **B form** of DNA. Although it is probably close to the conformation of most DNA in the cell, it is not the only conformation available to double-stranded nucleic acids. If we reduce the relative humidity surrounding the DNA fiber to 75%, the sodium salt of DNA assumes the **A form** (Figure 2.16a). This differs from the B form (Figure 2.16b) in several respects. Most obviously, the plane of a base pair is no longer roughly perpendicular to the helical axis, but tilts 20 degrees away from horizontal.

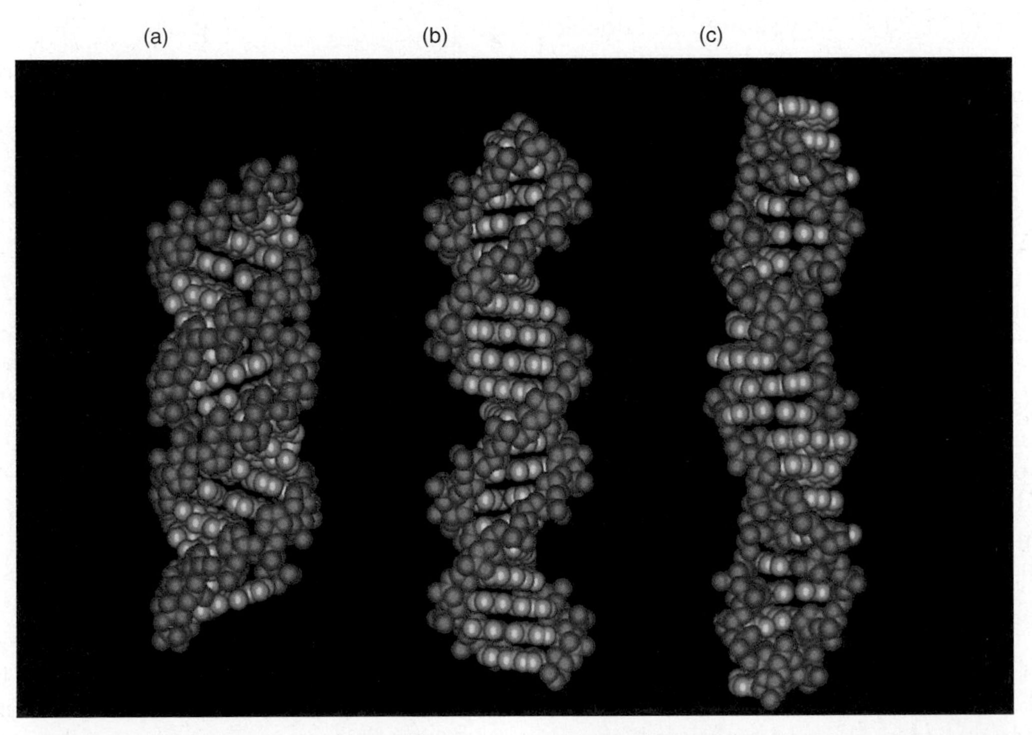

Figure 2.16 Computer graphic models of A-, B-, and Z-DNA. **(a)** A-DNA. Note the base pairs (blue), whose tilt up from right to left is especially apparent in the major grooves at the top and near the bottom. Note also the right-handed helix traced by the sugar–phosphate backbone (red). **(b)** B-DNA. Note the familiar right-handed helix, with roughly horizontal base pairs. **(c)** Z-DNA. Note the left-handed helix. All these DNAs are depicted with the same number of base pairs, emphasizing the differences in compactness of the three DNA forms. (*Source:* Courtesy Fusao Takusagawa.)

Table 2.2 Forms of DNA

Form	Pitch Å	Residues per Turn	Inclination of Base Pair from Horizontal (degrees)
A	24.6	10.7	+19
B	33.2	~10	−1.2
Z	45.6	12	−9

Also, the A helix packs in 10.7 bp per helical turn instead of the 10 found in the B form crystal structure, and each turn occurs in only 24.6 instead of 33.2 Å. This means that the pitch, or distance required for one complete turn of the helix, is only 24.6 instead of 33.2 Å, as in B-DNA. A hybrid polynucleotide containing one DNA and one RNA strand assumes the A form in solution, as does a double-stranded RNA. Table 2.2 presents these helical parameters for A and B form DNA, and for a left-handed Z-form of DNA, discussed in the next paragraph.

Both the A and B form DNA structures are right-handed: The helix turns clockwise away from you whether you look at it from the top or the bottom. Alexander Rich and his colleagues discovered in 1979 that DNA does not always have to be right-handed. They showed that double-stranded DNA containing strands of alternating purines and pyrimidines (e.g., poly[dG-dC] · poly[dG-dC]):

—GCGCGCGC—
—CGCGCGCG—

can exist in an extended left-handed helical form. Because of the zigzag look of this DNA's backbone when viewed from the side, it is often called **Z-DNA.** Figure 2.16c presents a picture of Z-DNA. The helical parameters of this structure are given in Table 2.2. Although Rich discovered Z-DNA in studies of model compounds like poly[dG-dC] · poly[dG-dC], this structure seems to be more than just a laboratory curiosity. Evidence suggests that living cells contain a small proportion of Z-DNA. Moreover, Keji Zhao and colleagues discovered in 2001 that activation of at least one gene requires that a regulatory sequence switch to the Z-DNA form.

SUMMARY In the cell, DNA may exist in the common B form, with base pairs horizontal. A small fraction of the DNA may assume an extended left-handed helical form called Z-DNA (at least in eukaryotes). An RNA–DNA hybrid assumes a third helical shape, called the A form, with base pairs tilted away from the horizontal.

Separating the Two Strands of a DNA Double Helix Although the ratios of G to C and A to T in an organism's DNA are fixed, the GC content (percentage of G + C) can vary considerably from one DNA to another. Table 2.3 lists the GC contents of DNAs from several organisms and viruses. The values range from 22–73%, and these differences are reflected in differences in the physical properties of DNA.

When a DNA solution is heated enough, the noncovalent forces that hold the two strands together weaken and finally break. When this happens, the two strands come apart in a process known as **DNA denaturation,** or **DNA melting.** The temperature at which the DNA strands are half denatured is called the melting temperature, or T_m. Figure 2.17 contains a melting curve for DNA from *Streptococcus pneumoniae.* The amount of strand separation, or melting, is measured by the absorbance of the DNA solution at 260 nm. Nucleic acids absorb light at this

Table 2.3 Relative G + C Contents of Various DNAs

Sources of DNA	Percent (G + C)
Dictyostelium (slime mold)	22
Streptococcus pyogenes	34
Vaccinia virus	36
Bacillus cereus	37
B. megaterium	38
Haemophilus influenzae	39
Saccharomyces cerevisiae	39
Calf thymus	40
Rat liver	40
Bull sperm	41
Streptococcus pneumoniae	42
Wheat germ	43
Chicken liver	43
Mouse spleen	44
Salmon sperm	44
B. subtilis	44
T1 bacteriophage	46
Escherichia coli	51
T7 bacteriophage	51
T3 bacteriophage	53
Neurospora crassa	54
Pseudomonas aeruginosa	68
Sarcina lutea	72
Micrococcus lysodeikticus	72
Herpes simplex virus	72
Mycobacterium phlei	73

Source: From Davidson, *The Biochemistry of the Nucleic Acids,* 8th ed. revised by Adams et al., Lippencott.

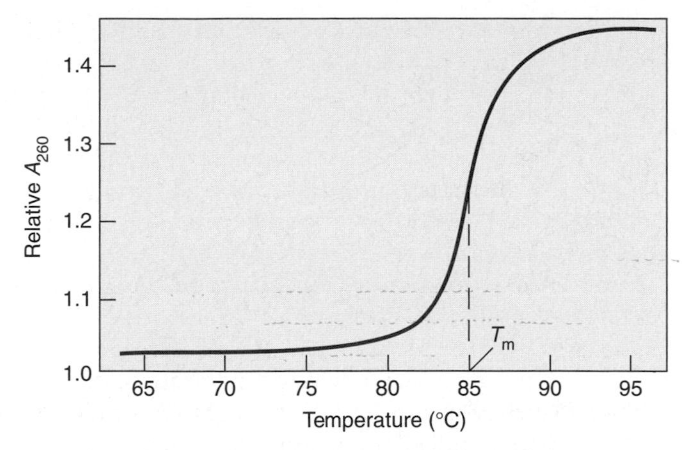

Figure 2.17 Melting curve of *Streptococcus pneumoniae* DNA. The DNA was heated, and its melting was measured by the increase in absorbance at 260 nm. The point at which the melting is half complete is the melting temperature, or T_m. The T_m for this DNA under these conditions is about 85°C. (Adapted from P. Doty, *The Harvey Lectures* 55:121, 1961.)

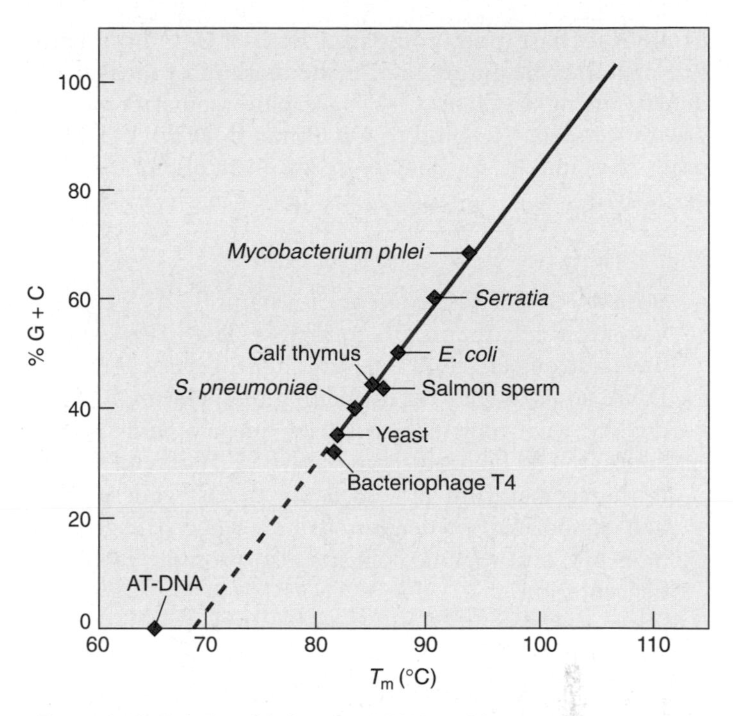

Figure 2.18 Relationship between DNA melting temperature and GC content. AT-DNA refers to synthetic DNAs composed exclusively of A and T (GC content = 0). (Adapted from P. Doty, *The Harvey Lectures* 55:121, 1961.)

wavelength because of the electronic structure in their bases, but when two strands of DNA come together, the close proximity of the bases in the two strands quenches some of this absorbance. When the two strands separate, this quenching disappears and the absorbance rises 30–40%. This is called the **hyperchromic shift.** The precipitous rise in the curve shows that the strands hold fast until the temperature approaches the T_m and then rapidly let go.

The GC content of a DNA has a significant effect on its T_m. In fact, as Figure 2.18 shows, the higher a DNA's GC content, the higher its T_m. Why should this be? Recall that one of the forces holding the two strands of DNA together is hydrogen bonding. Remember also that G–C pairs form three hydrogen bonds, whereas A–T pairs have only two. It stands to reason, then, that two strands of DNA rich in G and C will hold to each other more tightly than those of AT-rich DNA. Consider two pairs of embracing centipedes. One pair has 200 legs each, the other 300. Naturally the latter pair will be harder to separate.

Heating is not the only way to denature DNA. Organic solvents such as dimethyl sulfoxide and formamide, or high pH, disrupt the hydrogen bonding between DNA strands and promote denaturation. Lowering the salt concentration of the DNA solution also aids denaturation by removing the ions that shield the negative charges on the two strands from each other. At very low ionic strength, the mutually repulsive forces of these negative charges are strong enough to denature the DNA at a relatively low temperature.

The GC content of a DNA also affects its density. Figure 2.19 shows a direct, linear relationship between GC content and density, as measured by density gradient centrifugation in a CsCl solution (see Chapter 20). Part of the reason for this dependence of density on base composition seems to be real: the larger molar volume

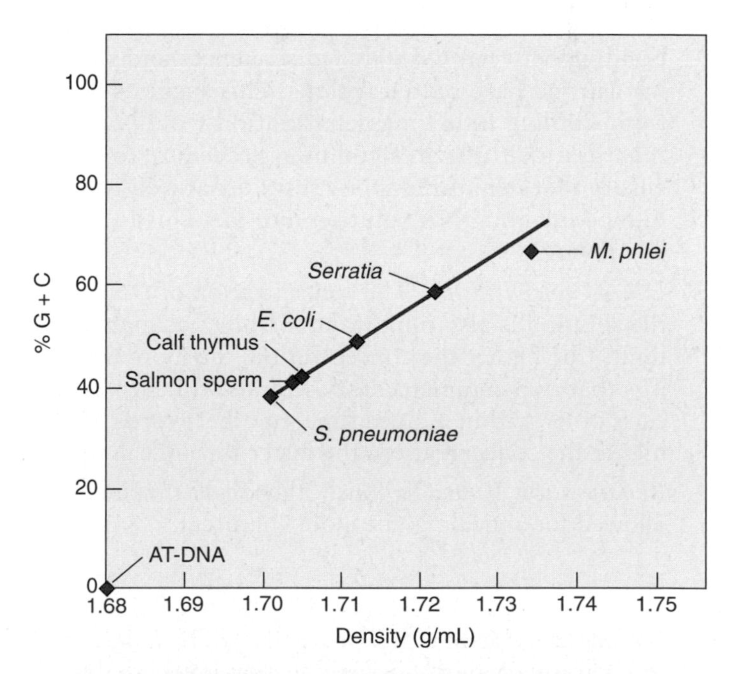

Figure 2.19 Relationship between the GC contents and densities of DNAs from various sources. AT-DNA is a synthetic DNA that is pure A + T; its GC content is therefore zero. (Adapted from P. Doty, *The Harvey Lectures* 55:121, 1961.)

of an A–T base pair, compared with a G–C base pair. But part may be an artifact of the method of measuring density using CsCl: A G–C base pair seems to have a greater tendency to bind to CsCl than does an A–T base pair. This makes its density seem even higher than it actually is.

SUMMARY The GC content of a natural DNA can vary from less than 25% to almost 75%. This can have a strong effect on the physical properties of the DNA, in particular on its melting temperature and density, each of which increases linearly with GC content. The melting temperature (T_m) of a DNA is the temperature at which the two strands are half-dissociated, or denatured. Low ionic strength, high pH, and organic solvents also promote DNA denaturation.

Reuniting the Separated DNA Strands Once the two strands of DNA separate, they can, under the proper conditions, come back together again. This is called **annealing** or **renaturation.** Several factors contribute to renaturation efficiency. Here are three of the most important:

1. *Temperature* The best temperature for renaturation of a DNA is about 25°C below its T_m. This temperature is low enough that it does not promote denaturation, but high enough to allow rapid diffusion of DNA molecules and to weaken the transient bonding between mismatched sequences and short intrastrand base-paired regions. This suggests that rapid cooling following denaturation would prevent renaturation. Indeed, a common procedure to ensure that denatured DNA stays denatured is to plunge the hot DNA solution into ice. This is called *quenching*.
2. *DNA Concentration* The concentration of DNA in the solution is also important. Within reasonable limits, the higher the concentration, the more likely it is that two complementary strands will encounter each other within a given time. In other words, the higher the concentration, the faster the annealing.
3. *Renaturation Time* Obviously, the longer the time allowed for annealing, the more will occur.

SUMMARY Separated DNA strands can be induced to renature, or anneal. Several factors influence annealing; among them are (1) temperature, (2) DNA concentration, and (3) time.

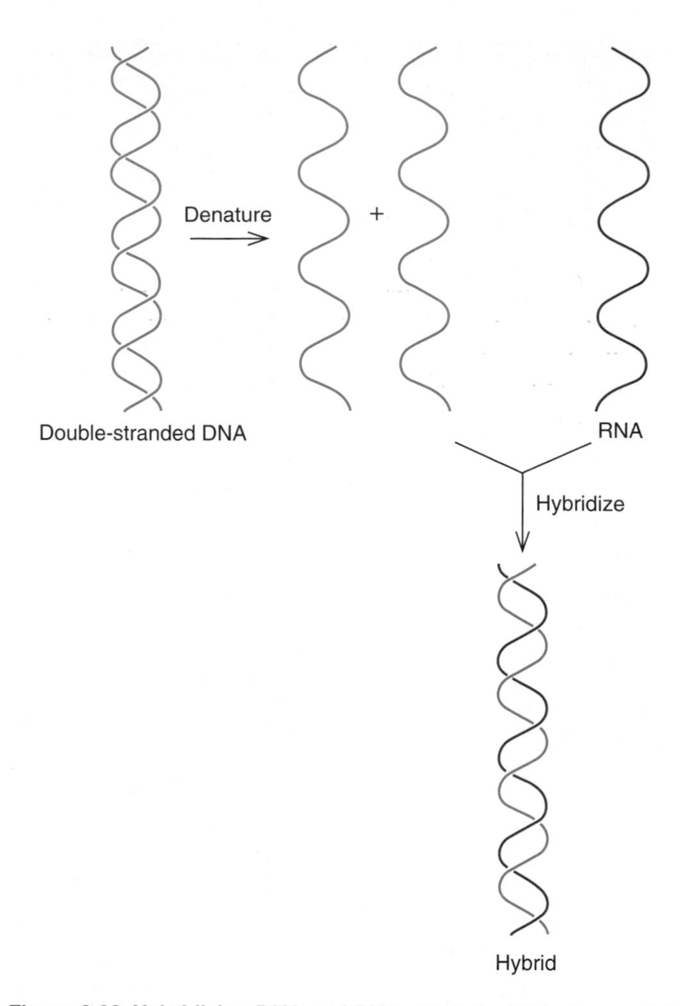

Figure 2.20 Hybridizing DNA and RNA. First, the DNA at upper left is denatured to separate the two DNA strands (blue). Then the DNA strands are mixed with a strand of RNA (red) that is complementary to one of the DNA strands. This hybridization reaction is carried out at a relatively high temperature, which favors RNA–DNA hybridization over DNA–DNA duplex formation. This hybrid has one DNA strand (blue) and one RNA strand (red).

Hybridization of Two Different Polynucleotide Chains So far, we have dealt only with two separated DNA strands simply getting back together again, but other possibilities exist. Consider, for example, a strand of DNA and a strand of RNA getting together to form a double helix. This could happen if one separated the two strands of a gene, and placed it together with an RNA strand complementary to one of the DNA strands (Figure 2.20). We would not refer to this as annealing; instead, we would call it **hybridization** because we are putting together a **hybrid** of two different nucleic acids. The two chains do not have to be as different as DNA and RNA. If we put together two different strands of DNA having complementary, or nearly complementary, sequences we could still call it hybridization—as long as the strands are of different origin. The difference between the two complementary strands may be very subtle; for example, one may be radioactive and the other not. As we will see later in this book, hybridization is an extremely valuable technique. In fact, it would be difficult to overestimate the importance of hybridization to molecular biology.

DNAs of Various Sizes and Shapes

Table 2.4 shows the sizes of the haploid genomes of several organisms and viruses. The sizes are expressed three ways: molecular weight, number of base pairs, and length. These are all related, of course. We already know how to convert number of base pairs to length, because about 10.4 bp occur per helical turn, which is 33.2 Å long. To convert base pairs to molecular weight, we simply need to multiply by 660, which is the approximate molecular weight of one average nucleotide pair.

How do we measure these sizes? For small DNAs, this is fairly easy. For example, consider phage PM2 DNA, which contains a double-stranded, circular DNA. How do we know it is circular? The most straightforward way to find out is simply by looking at it. We can do this using an electron microscope, but first we have to treat the DNA so that it stops electrons and will show up in a micrograph just as bones stop x-rays and therefore show up in an x-ray picture. The most common way of doing this is by *shadowing* the DNA with a heavy metal such as platinum. One places the DNA on an electron microscope grid and bombards it with minute droplets of metal from a shallow angle. This makes the metal pile up beside the DNA like snow behind a fence. One rotates the DNA on the grid so it becomes shadowed all around. Now the metal will stop the electrons in the electron microscope and make the DNA appear as light strings against a darker background. Printing reverses this image to give a picture such as Figure 2.21, which is an electron micrograph of PM2 DNA in two forms: an open circle (lower left) and a **supercoil** (upper right), in which the DNA coils around itself rather like a twisted rubber band. We can also use pictures like these to measure the length of the DNA. This is more accurate if we include a standard DNA of known length in the same picture.

The size of a DNA can also be estimated by gel electrophoresis, a topic we will discuss in Chapter 5.

SUMMARY Natural DNAs come in sizes ranging from several kilobases to thousands of megabases. The size of a small DNA can be estimated by electron microscopy. This technique can also reveal whether a DNA is circular or linear, and whether it is supercoiled.

Table 2.4 Sizes of Various DNAs

Source	Molecular Weight	Base Pairs (bp)	Length
Viruses and Mitochondria:			
SV40 (mammalian tumor virus)	3.5×10^6	5226	1.7 μm
Bacteriophage φX174 (double-stranded form)	3.2×10^6	5386	1.8 μm
Bacteriophage λ	3.3×10^7	4.85×10^4	13 μm
Bacteriophage T2 or T4	1.3×10^8	2×10^5	50 μm
Human mitochondria	9.5×10^6	16,596	5 μm
Bacteria:			
Haemophilus influenzae	1.2×10^9	1.83×10^6	620 μm
Escherichia coli	3.1×10^9	4.64×10^6	1.6 mm
Salmonella typhimurium	8×10^9	1.1×10^7	3.8 mm
Eukaryotes (content per haploid nucleus):			
Saccharomyces cerevisiae (yeast)	7.9×10^9	1.2×10^7	4.1 mm
Neurospora crassa (pink bread mold)	$\approx 1.9 \times 10^{10}$	$\approx 2.7 \times 10^7$	≈9.2 mm
Drosophila melanogaster (fruit fly)	$\approx 1.2 \times 10^{11}$	$\approx 1.8 \times 10^8$	≈6.0 cm
Mus musculus (mouse)	$\approx 1.5 \times 10^{12}$	$\approx 2.2 \times 10^9$	≈750 cm
Homo sapiens (human)	$\approx 2.3 \times 10^{12}$	$\approx 3.2 \times 10^9$	≈1.1 m
Zea mays (corn, or maize)	$\approx 4.4 \times 10^{12}$	$\approx 6.6 \times 10^9$	≈2.2 m
Rana pipiens (frog)	$\approx 1.4 \times 10^{13}$	$\approx 2.3 \times 10^{10}$	≈7.7 m
Lilium longiflorum (lily)	$\approx 2 \times 10^{14}$	$\approx 3 \times 10^{11}$	≈100 m

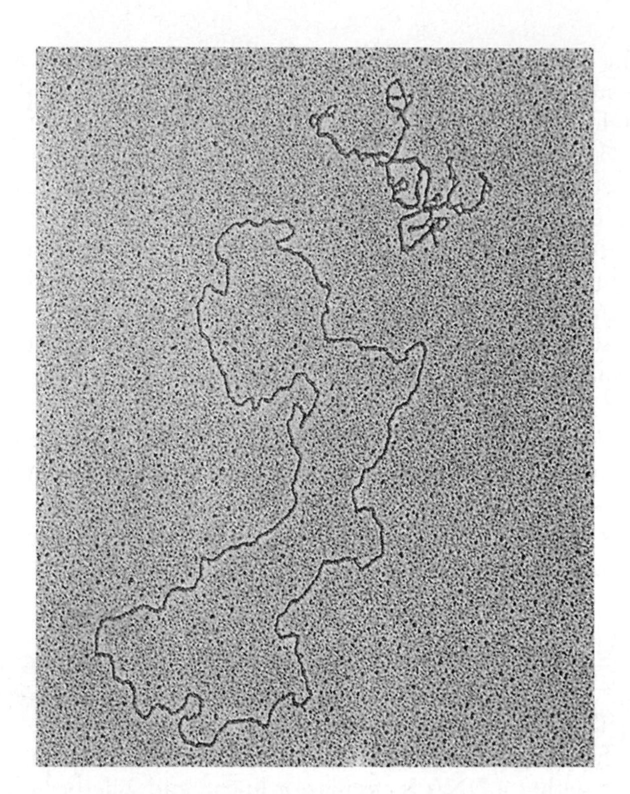

Figure 2.21 Electron micrograph of phage PM2 DNA. The open circular form is shown on the lower left and the supercoiled form is shown at the upper right. (*Source:* © Jack Griffith.)

The Relationship Between DNA Size and Genetic Capacity How many genes are in a given DNA? It is impossible to tell just from the size of the DNA, because we do not know how much of a given DNA is devoted to genes and how much is space between genes, or even intervening sequences within genes. We can, however, estimate an upper limit on the number of genes a DNA can hold. We start with the assumption that the genes we are discussing here are those that encode proteins. In Chapter 3 and other chapters, we will see that many genes simply encode RNAs, but we are ignoring them here. We also assume that an average protein has a molecular mass of about 40,000 D. How many amino acids does this represent? The molecular masses of amino acids vary, but they average about 110 D. To simplify our calculation, let us assume that the average is 110. That means our average protein contains 40,000/110, or about 364 amino acids. Because each amino acid requires 3 bp of DNA to code for it, a protein containing 364 amino acids needs a gene of about 1092 bp.

Consider a few of the DNAs listed in Table 2.4. The *E. coli* chromosome contains 4.6×10^6 bp, so it could encode about 4200 average proteins. Phage λ, which infects *E. coli,* has only 4.85×10^4 bp, so it can code for only about 44 proteins. One of the smallest double-stranded DNAs on the list, belonging to the phage ϕX174, has a mere 5375 bp. In principle, that is only enough to code for about five proteins, but the phage squeezes in some extra information by overlapping its genes.

DNA Content and the C-Value Paradox You would probably predict that complex organisms such as vertebrates need more genes than simple organisms like yeast. Therefore, they should have higher **C-values,** or DNA content per haploid cell. In general, your prediction would be right; mouse and human haploid cells contain more than 100 times more DNA than yeast haploid cells. Furthermore, yeast cells have about five times more DNA than *E. coli* cells, which are even simpler. However, this correspondence between an organism's physical complexity and the DNA content of its cells is not perfect. Consider, for example, the frog. Intuitively, you would not suspect that an amphibian would have a higher C-value than a human, yet the frog has seven times more DNA per cell. Even more dramatic is the fact that the lily has 100 times more DNA per cell than a human.

This perplexing situation is called the **C-value paradox.** It becomes even more difficult to explain when we look at organisms within a group. For example, some amphibian species have C-values 100 times higher than those of others, and the C-values of flowering plants vary even more widely. Does this mean that one kind of higher plant has 100 times more genes than another? That is simply unbelievable. It would raise questions about what all those extra genes are good for and why we do not notice tremendous differences in physical complexity among these organisms. The more plausible explanation of the C-value paradox is that organisms with extraordinarily high C-values simply have a great deal of extra, noncoding DNA. The function, if any, of this extra DNA is still mysterious.

In fact, even mammals have much more DNA than they need for genes. Applying our simple rule (dividing the number of base pairs by 1090) to the human genome yields an estimate of about 3 million for the maximum number of genes, which is far too high.

In fact, the finished version of the human genome suggests that there are only about 20–25,000 genes. This means that human cells contain more than 100 times more DNA than they apparently need. Much of this extra DNA is found in intervening sequences within eukaryotic genes (Chapter 14). The rest is in noncoding regions outside of genes.

SUMMARY There is a rough correlation between the DNA content and the number of genes in a cell or virus. However, this correlation breaks down in several cases of closely related organisms where the DNA content per haploid cell (C-value) varies widely. This C-value paradox is probably explained, not by extra genes, but by extra noncoding DNA in some organisms.

SUMMARY

Genes of all true organisms are made of DNA; certain viruses have genes made of RNA. DNA and RNA are chain-like molecules composed of subunits called nucleotides. DNA has a double-helical structure with sugar–phosphate backbones on the outside and base pairs on the inside. The bases pair in a specific way: adenine (A) with thymine (T) and guanine (G) with cytosine (C). When DNA replicates, the parental strands separate; each then serves as the template for making a new, complementary strand.

The G + C content of a natural DNA can vary from 22–73%, and this can have a strong effect on the physical properties of DNA, particularly its melting temperature. The melting temperature (T_m) of a DNA is the temperature at which the two strands are half-dissociated, or denatured. Separated DNA strands can be induced to renature, or anneal. Complementary strands of polynucleotides (either RNA or DNA) from different sources can form a double helix in a process called hybridization. Natural DNAs vary widely in length. The size of a small DNA can be estimated by electron microscopy.

A rough correlation occurs between the DNA content and the number of genes in a cell or virus. However, this correlation does not hold in several cases of closely related organisms in which the DNA content per haploid cell (C-value) varies widely. This C-value paradox is probably explained by extra noncoding DNA in some organisms.

REVIEW QUESTIONS

1. Compare and contrast the experimental approaches used by Avery and colleagues, and by Hershey and Chase, to demonstrate that DNA is the genetic material.
2. Draw the general structure of a deoxynucleoside monophosphate. Show the sugar structure in detail and indicate the positions of attachment of the base and the phosphate. Also indicate the deoxy position.
3. Draw the structure of a phosphodiester bond linking two nucleotides. Show enough of the two sugars that the sugar positions involved in the phosphodiester bond are clear.
4. Which DNA purine forms three H bonds with its partner in the other DNA strand? Which forms two H bonds? Which DNA pyrimidine forms three H bonds with its partner? Which forms two H bonds?
5. The following drawings are the outlines of two DNA base pairs, with the bases identified as *a, b, c,* and *d*. What are the real identities of these bases?

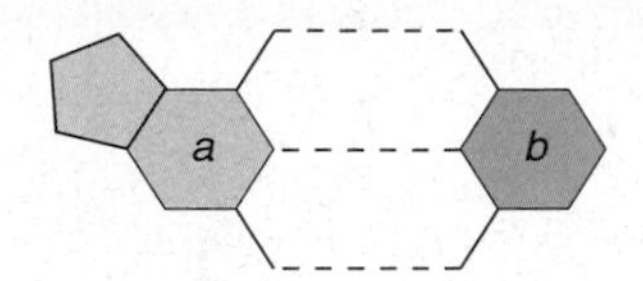

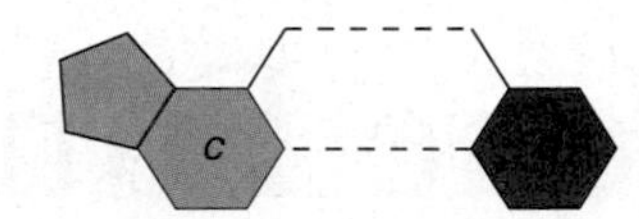

6. Draw a typical DNA melting curve. Label the axes and point out the melting temperature.
7. Use a graph to illustrate the relationship between the GC content of a DNA and its melting temperature. What is the explanation for this relationship?
8. Use a drawing to illustrate the principle of nucleic acid hybridization.

ANALYTICAL QUESTIONS

1. The double-stranded DNA genome of human herpes simplex virus 1 has a molecular mass of about 1.0×10^5 kD. (a) How many base pairs does this virus contain? (b) How many full double-helical turns does this DNA contain? (c) How long is this DNA in microns?
2. How many proteins of average size could be encoded in a virus with a DNA genome having 12,000 bp, assuming no overlap of genes?

SUGGESTED READINGS

Adams, R.L.P., R.H. Burdon, A.M. Campbell, and R.M.S. Smellie, eds. 1976. *Davidson's The Biochemistry of the Nucleic Acids,* 8th ed. The structure of DNA, chapter 5. New York: Academic Press.

Avery, O.T., C.M. McLeod, and M. McCarty. 1944. Studies on the chemical nature of the substance-inducing transformation of pneumococcal types. *Journal of Experimental Medicine* 79:137–58.

Chargaff, E. 1950. Chemical specificity of the nucleic acids and their enzymatic degradation. *Experientia* 6:201–9.

Dickerson, R.E. 1983. The DNA helix and how it reads. *Scientific American* 249 (December): 94–111.

Hershey, A.D., and M. Chase. 1952. Independent functions of viral protein and nucleic acid in growth of bacteriophage. *Journal of General Physiology* 36:39–56.

Watson, J.D., and F.H.C. Crick. 1953. Genetical implications of the structure of deoxyribonucleic acid. *Nature* 171:964–67.

Watson, J.D., and F.H.C. Crick. 1953. Molecular structure of the nucleic acids: A structure for deoxyribose nucleic acid. *Nature* 171:737–38.

CHAPTER 3

An Introduction to Gene Function

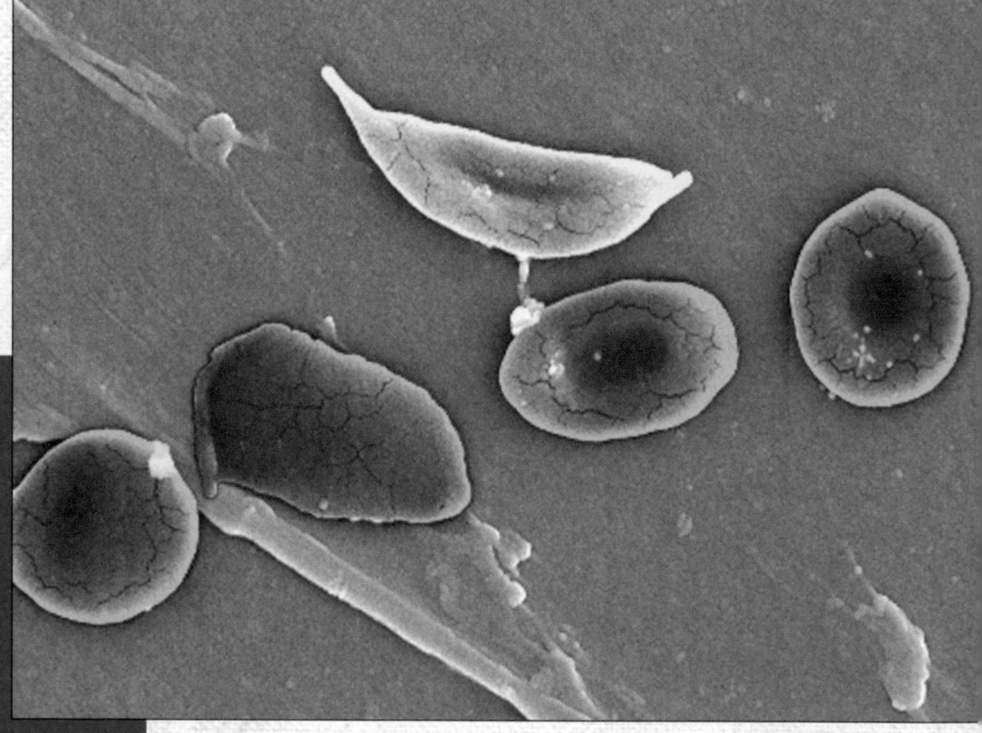

Red blood cells from a sickle cell disease patient showing one obviously sickled cell (top center). *© Courtesy Centers for Disease Control and Prevention.*

As we learned in Chapter 1, a gene participates in three major activities:

1. A gene is a repository of information. That is, it holds the information for making one of the key molecules of life, an RNA. The sequence of bases in the RNA depends directly on the sequence of bases in the gene. Most of these RNAs, in turn, serve as templates for making other critical cellular molecules, proteins. The production of RNAs and proteins from a DNA blueprint is called gene expression. Chapters 6–19 deal with various aspects of gene expression.
2. A gene can be replicated. This duplication is very faithful, so the genetic information can be passed essentially unchanged from generation to generation. We will discuss gene replication in Chapters 20–21.

3. A gene can accept occasional changes, or mutations. This allows organisms to evolve. Sometimes, these changes involve recombination, exchange of DNA between chromosomes or sites within a chromosome. A subset of recombination events involve pieces of DNA (transposable elements) that move from one place to another in the genome. We will deal with recombination and transposable elements in Chapters 22 and 23.

Chapter 3 outlines the three activities of genes and provides some background information that will be useful in our deeper explorations in subsequent chapters.

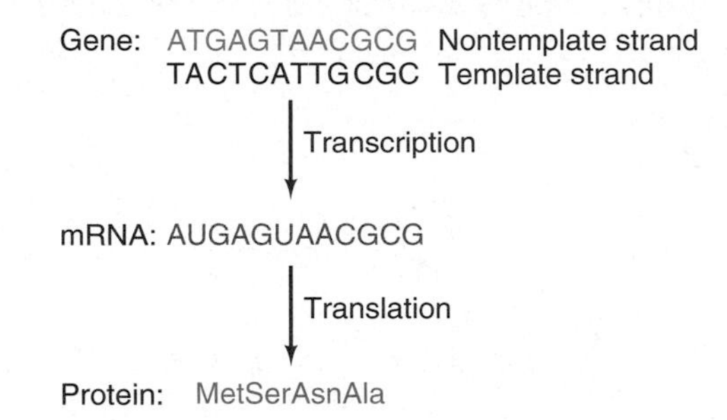

Figure 3.1 Outline of gene expression. In the first step, transcription, the template strand (black) is transcribed into mRNA. Note that the nontemplate strand (blue) of the DNA has the same sequence (except for the T–U change) as the mRNA (red). In the second step, the mRNA is translated into protein (green). This little "gene" is only 12 bp long and codes for only four amino acids (a tetrapeptide). Real genes are much larger.

3.1 Storing Information

Let us begin by examining the gene expression process, starting with a brief overview, followed by an introduction to protein structure and an outline of the two steps in gene expression.

Overview of Gene Expression

As we have seen, producing a protein from information in a DNA gene is a two-step process. The first step is synthesis of an RNA that is complementary to one of the strands of DNA. This is called **transcription.** In the second step, called **translation,** the information in the RNA is used to make a polypeptide. Such an informational RNA is called a **messenger RNA (mRNA)** to denote the fact that it carries information—like a message—from a gene to the cell's protein factories.

Like DNA and RNA, proteins are polymers—long, chain-like molecules. The monomers, or links, in the protein chain are called amino acids. DNA and protein have this informational relationship: Three nucleotides in the DNA gene stand for one amino acid in a protein.

Figure 3.1 summarizes the process of expressing a protein-encoding gene and introduces the nomenclature we apply to the strands of DNA. Notice that the mRNA has the same sequence (except that U's substitute for T's) as the top strand (blue) of the DNA. An mRNA holds the information for making a polypeptide, so we say it "codes for" a polypeptide, or "encodes" a polypeptide. (*Note:* It is redundant to say "encodes for" a polypeptide.) In this case, the mRNA codes for the following string of amino acids: methionine-serine-asparagine-alanine, which is abbreviated Met-Ser-Asn-Ala. We can see that the codeword (or **codon**) for methionine in this mRNA is the triplet AUG; similarly, the codons for serine, asparagine, and alanine are AGU, AAC, and GCG, respectively.

Because the bottom DNA strand is complementary to the mRNA, we know that it served as the template for making the mRNA. Thus, we call the bottom strand the **template strand,** or the transcribed strand. For the same reason, the top strand is the **nontemplate strand,** or the nontranscribed strand. Because the top strand in our example has essentially the same coding properties as the corresponding mRNA, many geneticists call it the *coding strand.* The opposite strand would therefore be the *anticoding strand.* Also, since the top strand has the same sense as the mRNA, this same system of nomenclature refers to this top strand as the *sense strand,* and to the bottom strand as the *antisense strand.* However, many other geneticists use the "coding strand" and "sense strand" conventions in exactly the opposite way. From now on, to avoid confusion, we will use the unambiguous terms *template strand* and *nontemplate strand.*

Protein Structure

Because we are seeking to understand gene expression, and because proteins are the final products of most genes, let us take a brief look at the nature of proteins. Proteins, like nucleic acids, are chain-like polymers of small subunits. In the case of DNA and RNA, the links in the chain are nucleotides. The chain links of proteins are **amino acids.** Whereas DNA contains only four different nucleotides, proteins contain 20 different amino acids. The structures of these compounds are shown in Figure 3.2. Each amino acid has an amino group (NH_3^+), a carboxyl group (COO^-), a hydrogen atom (H), and a side chain. The only difference between any two amino acids is in their different side chains. Thus, it is the arrangement of amino acids, with their distinct side chains, that gives each protein its unique character. The amino acids join together in proteins via **peptide bonds,** as shown in Figure 3.3. This gives rise to the name **polypeptide** for a chain of

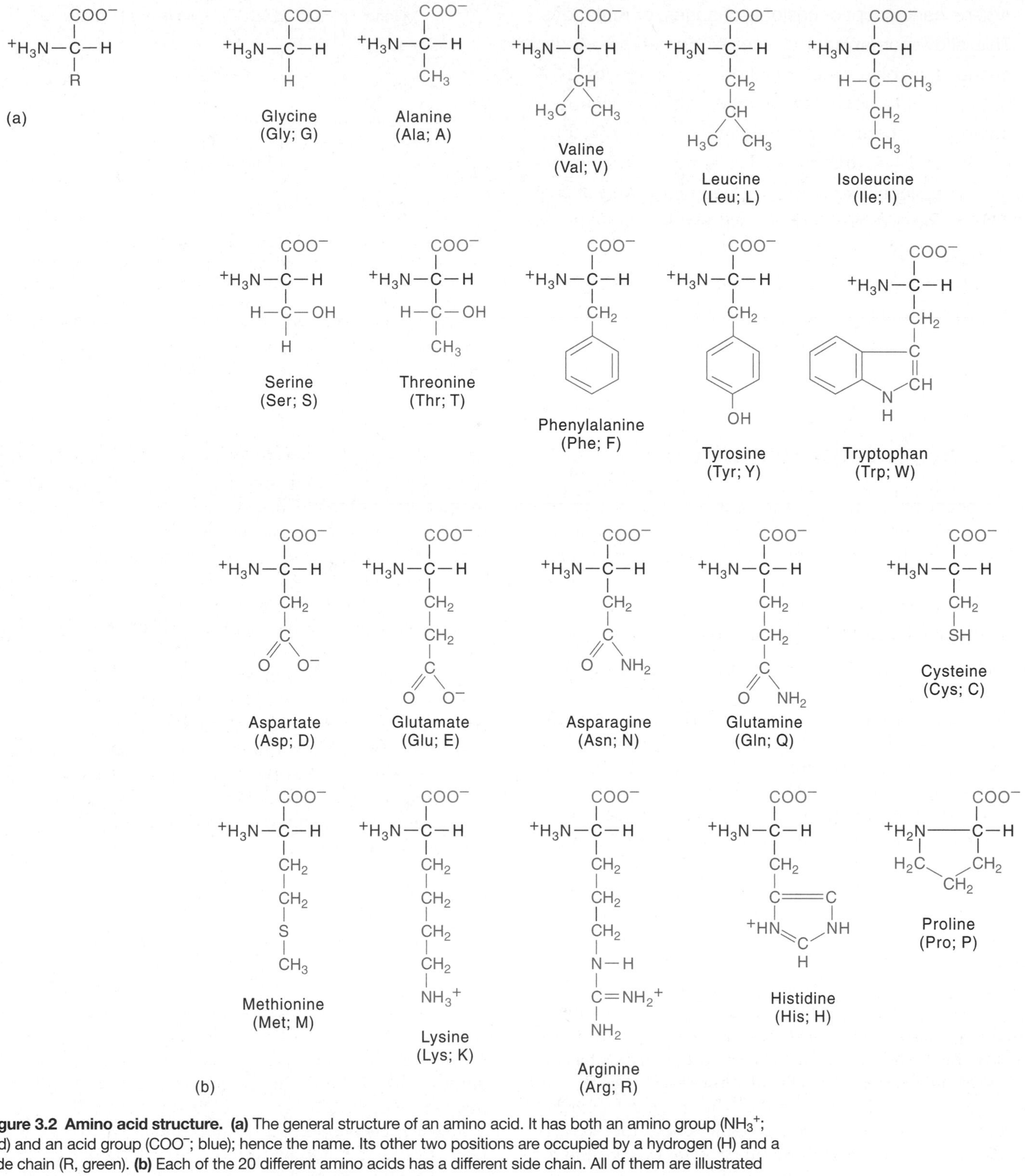

(a)

Glycine (Gly; G)
Alanine (Ala; A)
Valine (Val; V)
Leucine (Leu; L)
Isoleucine (Ile; I)
Serine (Ser; S)
Threonine (Thr; T)
Phenylalanine (Phe; F)
Tyrosine (Tyr; Y)
Tryptophan (Trp; W)
Aspartate (Asp; D)
Glutamate (Glu; E)
Asparagine (Asn; N)
Glutamine (Gln; Q)
Cysteine (Cys; C)
Methionine (Met; M)
Lysine (Lys; K)
Arginine (Arg; R)
Histidine (His; H)
Proline (Pro; P)

(b)

Figure 3.2 Amino acid structure. (a) The general structure of an amino acid. It has both an amino group (NH_3^+; red) and an acid group (COO^-; blue); hence the name. Its other two positions are occupied by a hydrogen (H) and a side chain (R, green). **(b)** Each of the 20 different amino acids has a different side chain. All of them are illustrated here. Three-letter and one-letter abbreviations are in parentheses.

amino acids. A protein can be composed of one or more polypeptides.

A polypeptide chain has polarity, just as the DNA chain does. The *dipeptide* (two amino acids linked together) shown on the right in Figure 3.3 has a free amino group at its left end. This is the **amino terminus,** or **N-terminus.** It also has a free carboxyl group at its right end, which is the **carboxyl terminus,** or **C-terminus.**

The linear order of amino acids constitutes a protein's **primary structure.** The way these amino acids interact

$$^{+}H_3N-\underset{R}{\overset{H}{C}}-\overset{O}{\overset{\|}{C}}-O^{-} + {}^{+}H_3N-\underset{R}{\overset{H}{C}}-\overset{O}{\overset{\|}{C}}-O^{-} \longrightarrow {}^{+}H_3N-\underset{R}{\overset{H}{C}}-\overset{O}{\overset{\|}{C}}-\underset{H}{N}-\underset{R}{\overset{H}{C}}-\overset{O}{\overset{\|}{C}}-O^{-} + H_2O$$

Peptide bond

Figure 3.3 Formation of a peptide bond. Two amino acids with side chains R and R′ combine through the acid group of the first and the amino group of the second to form a dipeptide, two amino acids linked by a peptide bond. One molecule of water also forms as a by-product.

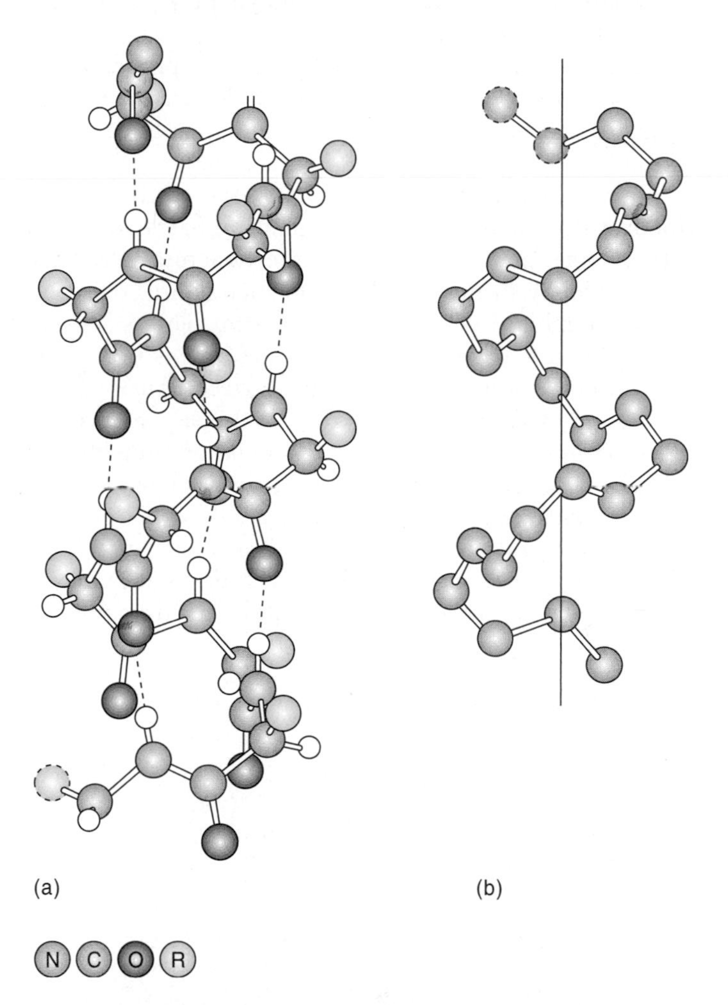

Figure 3.4 An example of protein secondary structure: The α-helix. (a) The positions of the amino acids in the helix are shown, with the helical backbone in gray and blue. The dashed lines represent hydrogen bonds between hydrogen and oxygen atoms on nearby amino acids. The small white circles represent hydrogen atoms. **(b)** A simplified rendition of the α-helix, showing only the atoms in the helical backbone.

with their neighbors gives a protein its **secondary structure.** The *α-helix* is a common form of secondary structure. It results from hydrogen bonding among near-neighbor amino acids, as shown in Figure 3.4. Another common secondary structure found in proteins is the β-pleated sheet (Figure 3.5). This involves extended protein chains, packed side by side, that interact by hydrogen bonding. The packing of the chains next to each other creates the

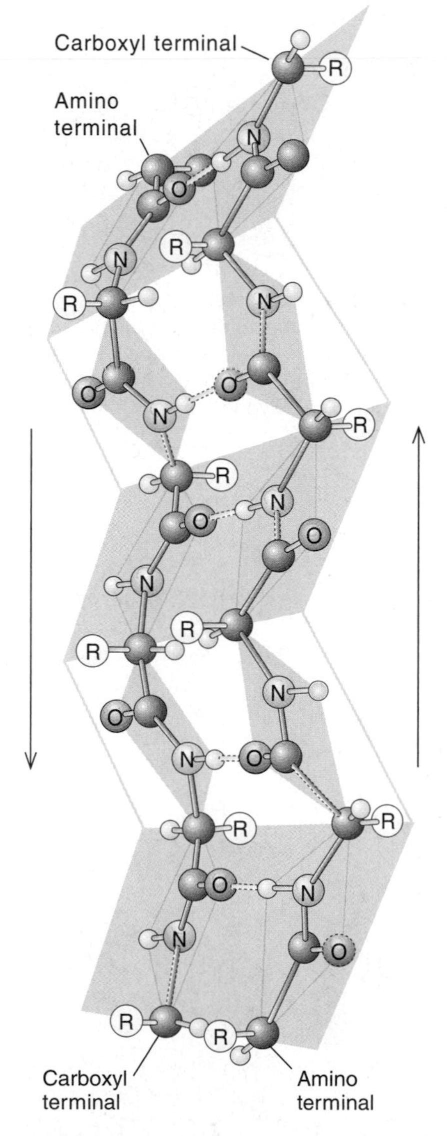

Figure 3.5 An antiparallel β-sheet. Two polypeptide chains are arranged side by side, with hydrogen bonds (dashed lines) between them. The green and white planes show that the β-sheet is pleated. The chains are antiparallel in that the amino terminus of one and the carboxyl terminus of the other are at the top. The arrows indicate that the two β-strands run from amino to carboxyl terminal in opposite directions. Parallel β-sheets, in which the β-strands run in the same direction, also exist.

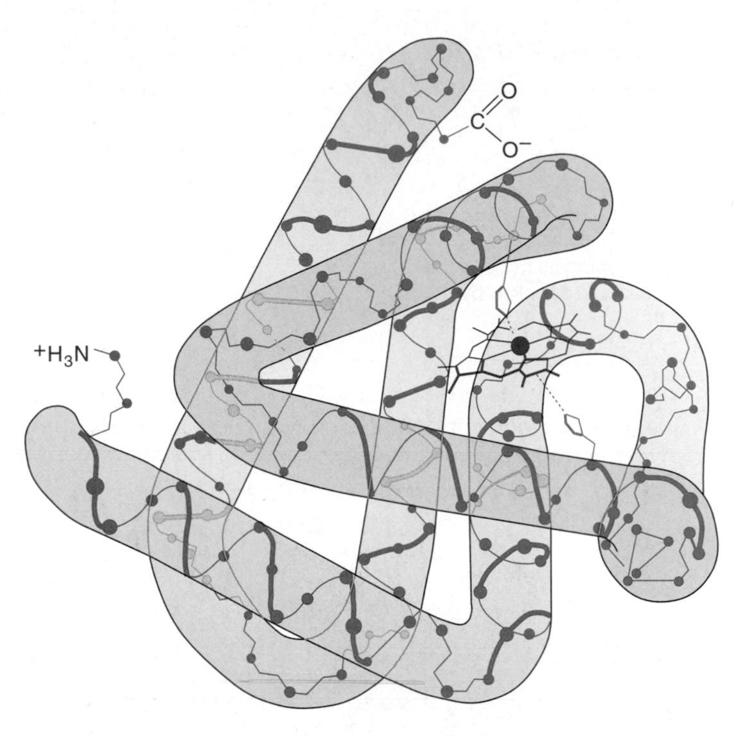

Figure 3.6 Tertiary structure of myoglobin. The several α-helical regions of this protein are represented by turquoise corkscrews. The overall molecule seems to resemble a sausage, twisted into a roughly spherical or globular shape. The heme group is shown in red, bound to two histidines (turquoise polygons) in the protein.

sheet appearance. Silk is a protein very rich in β-pleated sheets. A third example of secondary structure is simply a turn. Such turns connect the α-helices and β-pleated sheet elements in a protein.

The total three-dimensional shape of a polypeptide is its **tertiary structure.** Figure 3.6 illustrates how the protein myoglobin folds up into its tertiary structure. Elements of secondary structure are apparent, especially the several α-helices of the molecule. Note the overall roughly spherical shape of myoglobin. Most polypeptides take this form, which we call *globular.*

Figure 3.7 is a different representation of protein structure called a ribbon model. This model depicts the tertiary structure of an enzyme known as guanidinoacetate methyltransferase (GAMT). Here we can clearly see three types of secondary structure: α-helices, represented by helical ribbons; β-pleated sheets, represented by flat arrows laid side by side; and turns between the structural elements, represented by strings. The ball and stick figures represent two small molecules bound to the protein. This is a stereo diagram that you can view in three dimensions with a stereo viewer, or by using the "magic eye" technique.

Both myoglobin and GAMT are composed of a single, more or less globular, structure, but other proteins can contain more than one compact structural region. Each of these regions is called a **domain.** Antibodies (the proteins that white blood cells make to repel invaders) provide a good example of domains. Each of the four polypeptides in the IgG-type antibody contains globular domains, as

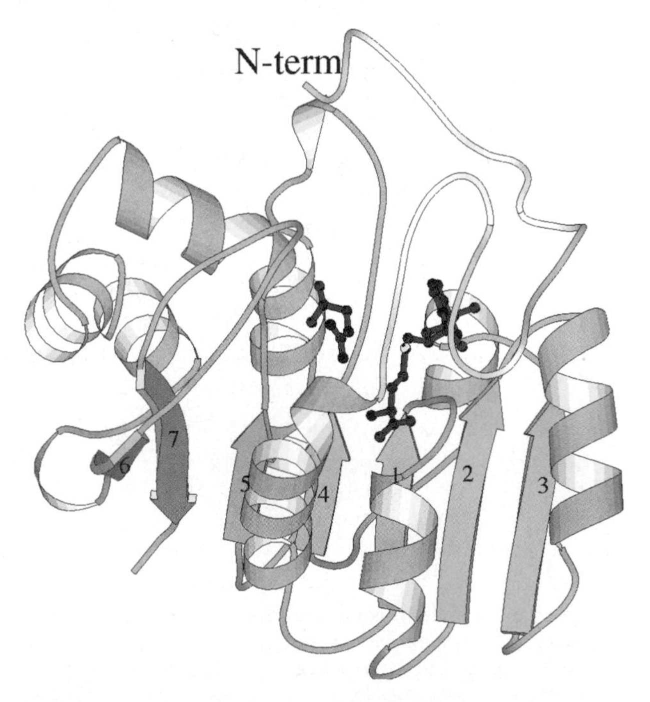

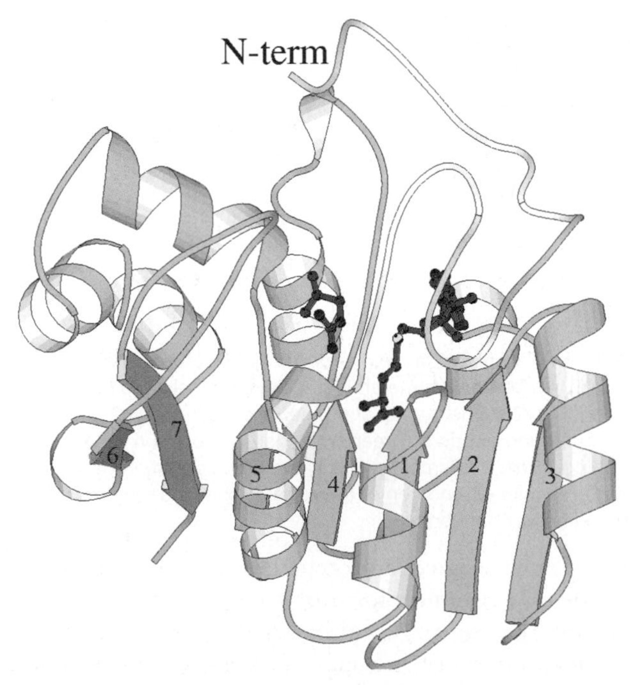

Figure 3.7 Tertiary structure of guanidinoacetate methyltransferase (GAMT). Secondary structure elements, including α-helices (coiled ribbons), β-pleated sheets (numbered flat arrows), and turns (strings) are apparent. The two bound molecules (ball and stick figures) are guanidinoacetate (left) and S-adenosylhomocysteine (right). Guanidinoacetate is one of the substrates of the enzyme and S-adenosylhomocysteine is a product inhibitor. (*Source:* Reprinted with permission from Fusao Takusagawa, University of Kansas.)

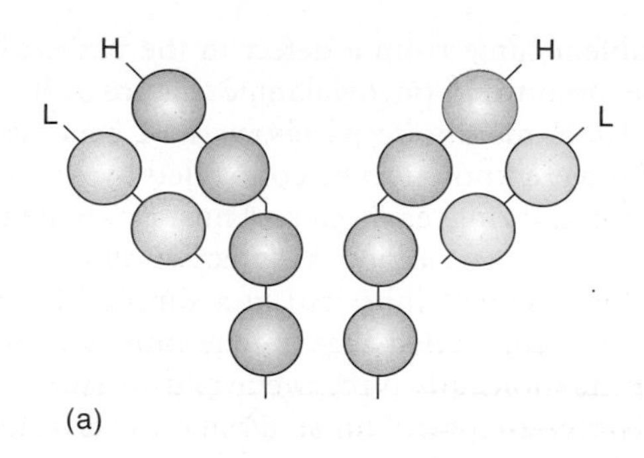

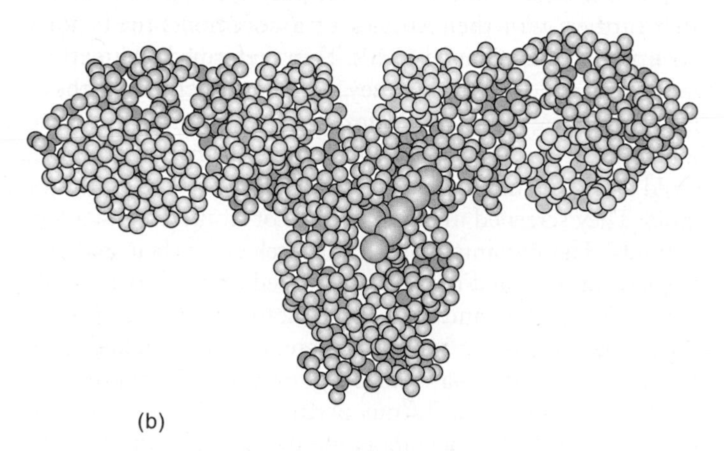

Figure 3.8 The globular domains of an immunoglobulin. **(a)** Schematic diagram, showing the four polypeptides that constitute the immunoglobulin: two light chains (L) and two heavy chains (H). The light chains each contain two globular regions, and the heavy chains have four globular domains apiece. **(b)** Space-filling model of an immunoglobulin. The colors correspond to those in part **(a).** Thus, the two H chains are in peach and blue; the L chains are in green and yellow. A complex sugar attached to the protein is shown in gray. Note the globular domains in each of the polypeptides. Also note how the four polypeptides fit together to form the quaternary structure of the protein.

shown in Figure 3.8. When we study protein–DNA binding in Chapter 9, we will see that domains can contain common structural–functional **motifs.** For example, a finger-shaped motif called a zinc finger is involved in DNA binding. Figure 3.8 also illustrates the highest level of protein structure—**quaternary structure**—which is the way two or more individual polypeptides fit together in a complex protein. It has long been assumed that a protein's amino acid sequence determines all of its higher levels of structure, much as the linear sequence of letters in this book determines word, sentence, and paragraph structure. However, this analogy is an oversimplification. Most proteins cannot fold properly by themselves outside their normal cellular environment. Some cellular factors besides the protein itself seem to be required in these cases, and folding often must occur during synthesis of a polypeptide.

What forces hold a protein in its proper shape? Some of these are covalent bonds, but most are noncovalent. The principal covalent bonds within and between polypeptides are disulfide (S–S) bonds between cysteines. The noncovalent bonds are primarily hydrophobic and hydrogen bonds. Predictably, hydrophobic amino acids cluster together in the interior of a polypeptide, or at the interface between polypeptides, so they can avoid contact with water (*hydrophobic,* meaning water-fearing). Hydrophobic interactions play a major role in tertiary and quaternary structures of proteins.

SUMMARY Proteins are polymers of amino acids linked through peptide bonds. The sequence of amino acids in a polypeptide (primary structure) gives rise to that molecule's local shape (secondary structure), overall shape (tertiary structure), and interaction with other polypeptides (quaternary structure).

Protein Function

Why are proteins so important? Some proteins provide the structure that helps give cells integrity and shape. Others serve as *hormones* to carry signals from one cell to another. For example, the pancreas secretes the hormone insulin that signals liver and muscle cells to take up the sugar glucose from the blood. Proteins can also bind and carry substances. The protein hemoglobin carries oxygen from the lungs to remote areas of the body; myoglobin stores oxygen in muscle tissue until it is used. Proteins also control the activities of genes, as we will see many times in this book. And proteins serve as enzymes that catalyze the hundreds of chemical reactions necessary for life. Thus, different proteins give different cells their distinctive functions: A pancreas islet cell makes insulin, while a red blood cell makes hemoglobin. Similarly, different organisms make different proteins: Birds make feather proteins, and mammals make hair proteins, for example. While this is part of what sets one organism apart from another, these differences are often more subtle than you would expect, as we will see in Chapters 24 and 25.

The Relationship Between Genes and Proteins Our knowledge of the gene–protein link dates back as far as 1902, when a physician named Archibald Garrod noticed that a human disease, *alcaptonuria,* behaved as if it were caused by a single recessive gene. Fortunately, Mendel's work had been rediscovered 2 years earlier and provided the theoretical background for Garrod's observation. Patients with alcaptonuria excrete copious amounts of *homogentisic acid,* which has the startling effect of coloring their urine black. Garrod reasoned that the abnormal buildup of this compound resulted from a defective metabolic pathway. Somehow, a blockage somewhere in the pathway was causing the intermediate, homogentisic acid, to accumulate to abnormally high levels, much as a dam causes water to accumulate behind it. Several years later, Garrod proposed

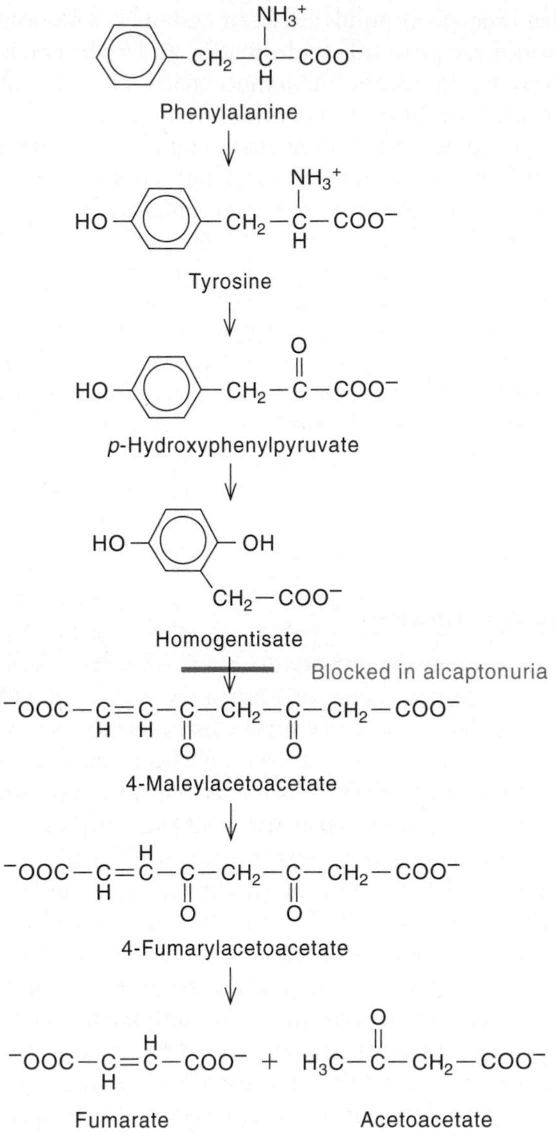

Phenylalanine

Tyrosine

p-Hydroxyphenylpyruvate

Homogentisate

Blocked in alcaptonuria

4-Maleylacetoacetate

4-Fumarylacetoacetate

Fumarate + Acetoacetate

Figure 3.9 Pathway of phenylalanine breakdown. Alcaptonuria patients are defective in the enzyme that converts homogentisate to 4-maleylacetoacetate.

that the problem came from a defect in the pathway that degrades the amino acid phenylalanine (Figure 3.9).

By that time, metabolic pathways had been studied for years and were known to be controlled by enzymes—one enzyme catalyzing each step. Thus, it seemed that alcaptonuria patients carried a defective enzyme. And because the disease was inherited in a simple Mendelian fashion, Garrod concluded that a gene must control the enzyme's production. When that gene is defective, it gives rise to a defective enzyme. This suggested the crucial conceptual link between genes and proteins.

George Beadle and E. L. Tatum carried this argument a step further with their studies of a common bread mold, ***Neurospora crassa,*** in the 1940s. They performed their experiments as follows: First, they bombarded the peritheca (spore-forming parts) of *Neurospora* with x-rays to cause mutations. Then, they collected the spores from the irradiated mold and germinated them separately to give pure strains of mold. They screened many thousands of strains to find a few mutants. The mutants revealed themselves by their inability to grow on minimal medium composed only of sugar, salts, inorganic nitrogen, and the vitamin biotin. Wild-type *Neurospora* grows readily on such a medium; the mutants had to be fed something extra—a vitamin, for example—to survive.

Next, Beadle and Tatum performed biochemical and genetic analyses on their mutants. By carefully adding substances, one at a time, to the mutant cultures, they pinpointed the biochemical defect. For example, the last step in the synthesis of the vitamin pantothenate involves putting together the two halves of the molecule: pantoate and β-alanine (Figure 3.10). One "pantothenateless" mutant would grow on pantothenate, but not on the two halves of the vitamin. This demonstrated that the last step (step 3) in the biochemical pathway leading to pantothenate was blocked, so the enzyme that carries out that step must have been defective.

The genetic analysis was just as straightforward. *Neurospora* is an *ascomycete,* in which nuclei of two different mating types fuse and undergo meiosis to give eight haploid *ascospores,* borne in a fruiting body called an *ascus.*

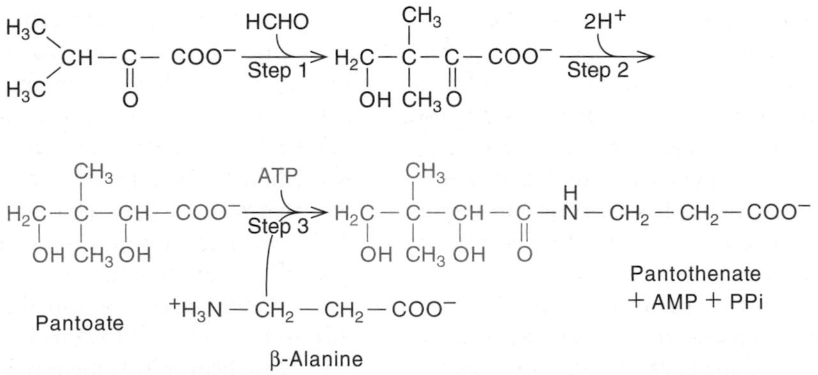

HCHO, Step 1; $2H^+$, Step 2; ATP, Step 3

Pantoate + β-Alanine → Pantothenate + AMP + PPi

Figure 3.10 Pathway of pantothenate synthesis. The last step (step 3), formation of pantothenate from the two half-molecules, pantoate (blue) and β-alanine (red), was blocked in one of Beadle and Tatum's mutants. The enzyme that carries out this step must have been defective.

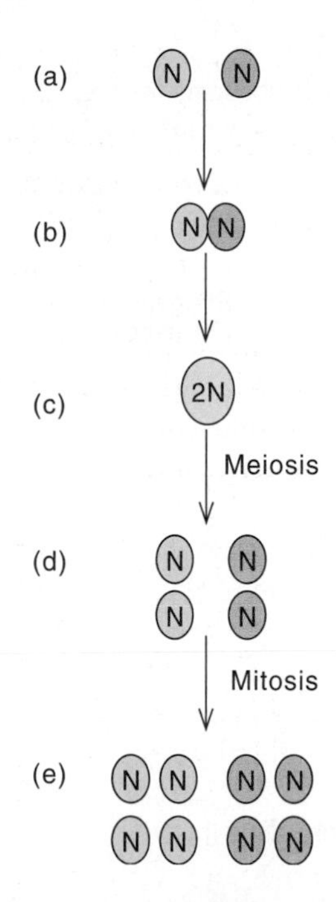

Figure 3.11 Sporulation in the mold *Neurospora crassa.* **(a)** Two haploid nuclei, one wild-type (yellow) and one mutant (blue), have come together in the immature fruiting body of the mold. **(b)** The two nuclei begin to fuse. **(c)** Fusion is complete, and a diploid nucleus (green) has formed. One haploid set of chromosomes is from the wild-type nucleus, and one set is from the mutant nucleus. **(d)** Meiosis occurs, producing four haploid nuclei. If the mutant phenotype is controlled by one gene, two of these nuclei (blue) should have the mutant allele and two (yellow) should have the wild-type allele. **(e)** Finally, mitosis occurs, producing eight haploid nuclei, each of which will go to one ascospore. Four of these nuclei (blue) should have the mutant allele and four (yellow) should have the wild-type allele. If the mutant phenotype is controlled by more than one gene, the results will be more complex.

Therefore, a mutant can be crossed with a wild-type strain of the opposite mating type to give eight spores (Figure 3.11). If the mutant phenotype results from a mutation in a single gene, then four of the eight spores should be mutant and four should be wild-type. Beadle and Tatum collected the spores, germinated them separately, and checked the phenotypes of the resulting molds. Sure enough, they found that four of the eight spores gave rise to mutant molds, demonstrating that the mutant phenotype was controlled by a single gene. This happened over and over again, leading these investigators to the conclusion that each enzyme in a biochemical pathway is controlled by one gene.

Subsequent work has shown that many enzymes contain more than one polypeptide chain and that each polypeptide is usually encoded in one gene. This is the **one-gene/one-polypeptide hypothesis.** As noted in Chapter 1, this hypothesis needs to be modified to account for, among other things, genes, such as the tRNA and rRNA genes, that simply encode RNAs. For decades, one assumed that the number of such genes was small—considerably less than 100. But the twenty-first century has seen explosive growth in the discovery of non-coding RNAs, which now number in the thousands in humans alone. Some of these RNAs may not have any function, and so would not satisfy everyone's definition of true gene products, but many others have demonstrable and important functions. Thus, the very definition of the word "gene" has become more complex and debatable. We now recognize overlapping genes, genes-within-genes, and fragmented genes, as well as more exotic possibilities. We will discuss these complications later in the book. For the remainder of this chapter, we will consider expression of "traditional" genes—those that encode proteins.

SUMMARY Most genes contain the information for making one polypeptide.

Discovery of Messenger RNA

The concept of a messenger RNA carrying information from gene to ribosome developed in stages during the years following the publication of Watson and Crick's DNA model. In 1958, Crick himself proposed that RNA serves as an intermediate carrier of genetic information. He based his hypothesis in part on the fact that the DNA resides in the nucleus of eukaryotic cells, whereas proteins are made in the cytoplasm. This means that something must carry the information from one place to the other. Crick noted that ribosomes contain RNA and suggested that this ribosomal RNA (rRNA) is the information bearer. But rRNA is an integral part of ribosomes; it cannot escape. Therefore, Crick's hypothesis implied that each ribosome, with its own rRNA, would produce the same kind of protein over and over.

François Jacob and colleagues proposed an alternative hypothesis calling for nonspecialized ribosomes that translate unstable RNAs called **messengers.** The messengers are independent RNAs that bring genetic information from the genes to the ribosomes. In 1961, Jacob, along with Sydney Brenner and Matthew Meselson, published their proof of the messenger hypothesis. This study used the same bacteriophage (T2) that Hershey and Chase had employed almost a decade earlier to show that genes were made of DNA (Chapter 2). The premise of the experiments was this: When phage T2 infects *E. coli,* it subverts its host from making bacterial proteins to making phage proteins. If Crick's hypothesis were correct, this switch to phage protein synthesis should be accompanied by the production of new ribosomes equipped with phage-specific RNAs.

To distinguish new ribosomes from old, these investigators labeled the ribosomes in uninfected cells with heavy isotopes of nitrogen (^{15}N) and carbon (^{13}C). This made "old" ribosomes heavy. Then they infected these cells with phage T2 and simultaneously transferred them to medium containing light nitrogen (^{14}N) and carbon (^{12}C). Any "new" ribosomes made after phage infection would therefore be light and would separate from the old, heavy ribosomes during density

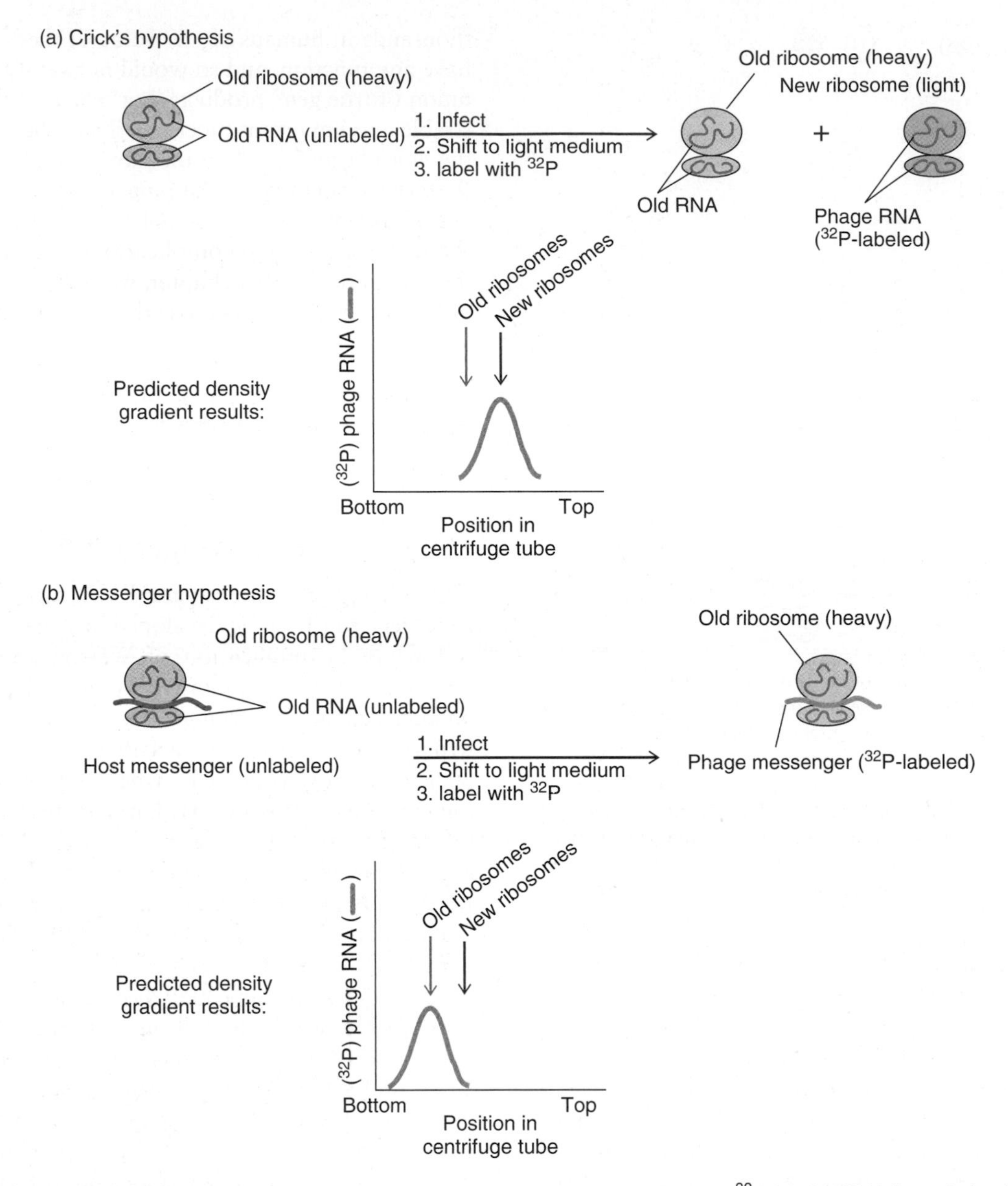

Figure 3.12 Experimental test of the messenger hypothesis. Heavy *E. coli* ribosomes were made by labeling the bacterial cells with heavy isotopes of carbon and nitrogen. The bacteria were then infected with phage T2 and simultaneously shifted to "light" medium containing the normal isotopes of carbon and nitrogen, plus some ^{32}P to make the phage RNA radioactive. **(a)** Crick had proposed that ribosomal RNA carried the message for making proteins. If this were so, then whole new ribosomes with phage-specific ribosomal RNA would have been made after phage infection. In that case, the new ^{32}P-labeled RNA (green) should have moved together with the new, light ribosomes (pink). **(b)** Jacob and colleagues had proposed that a messenger RNA carried genetic information to the ribosomes. According to this hypothesis, phage infection would cause the synthesis of new, phage-specific messenger RNAs that would be ^{32}P-labeled (green). These would associate with old, heavy ribosomes (blue). The radioactive label would therefore move together with the old, heavy ribosomes in the density gradient. This was indeed what happened.

gradient centrifugation. Brenner and colleagues also labeled the infected cells with ^{32}P to tag any phage RNA as it was made. Then they asked this question: Was the radioactively labeled phage RNA associated with new or old ribosomes?

Figure 3.12 shows that the phage RNA was found on old ribosomes whose rRNA was made before infection even began. Clearly, this old rRNA could not carry phage genetic information; by extension, it was very unlikely that it could carry host genetic information, either. Thus, the ribosomes are constant. The nature of the polypeptides they make depends on the mRNA that associates with them. This relationship resembles that of a DVD player and DVD. The nature of the movie (polypeptide) depends on the DVD (mRNA), not the player (ribosome).

Other workers had already identified a better candidate for the messenger: a class of unstable RNAs that associate transiently with ribosomes. Interestingly enough, in phage T2-infected cells, this RNA had a base composition very

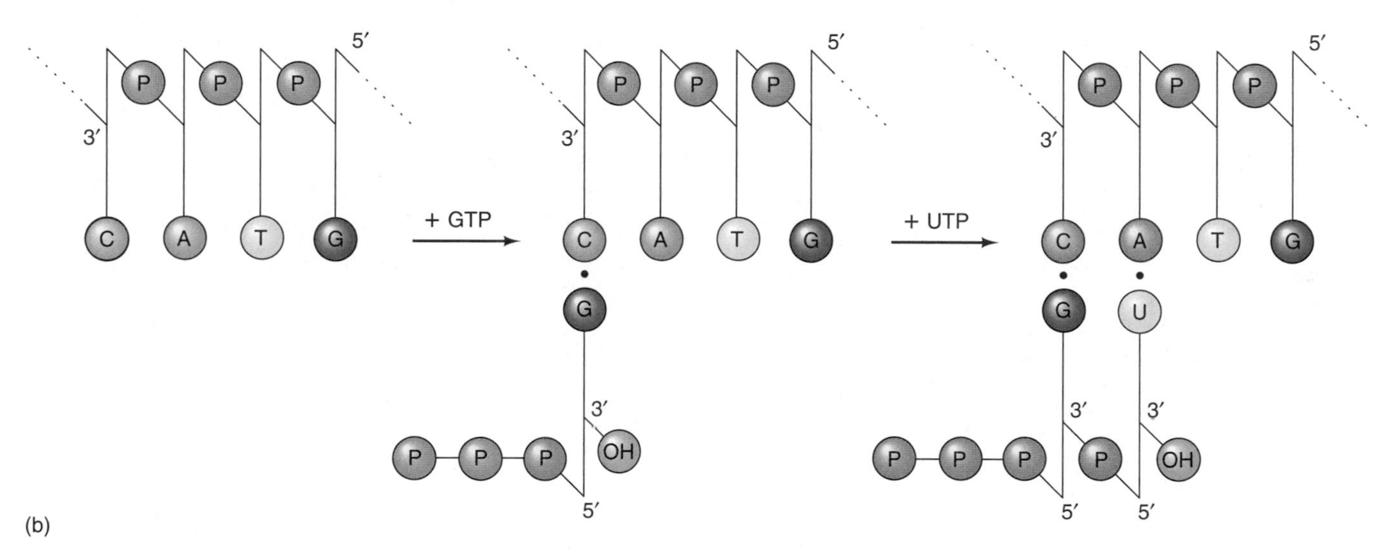

Figure 3.13 Making RNA. (a) Phosphodiester bond formation in RNA synthesis. ATP and GTP are joined together to form a dinucleotide. Note that the phosphorus atom closest to the guanosine is retained in the phosphodiester bond. The other two phosphates are removed as a by-product called pyrophosphate. **(b)** Synthesis of RNA on a DNA template. The DNA template at top contains the sequence 3′-dC-dA-dT-dG-5′ and extends in both directions, as indicated by the dashed lines. To start the RNA synthesis, GTP forms a base pair with the dC nucleotide in the DNA template. Next, UTP provides a uridine nucleotide, which forms a base pair with the dA nucleotide in the DNA template and forms a phosphodiester bond with the GTP. This produces the dinucleotide GU. In the same way, a new nucleotide joins the growing RNA chain at each step until transcription is complete. The pyrophosphate by-product is not shown.

similar to that of phage DNA—and quite different from that of bacterial DNA and RNA. This is exactly what we would expect of phage **messenger RNA (mRNA)**, and that is exactly what it is. On the other hand, host mRNA, unlike host rRNA, has a base composition similar to that of host DNA. This lends further weight to the hypothesis that mRNA, not rRNA, is the informational molecule.

SUMMARY Messenger RNAs carry the genetic information from the genes to the ribosomes, which synthesize polypeptides.

Transcription

As you might expect, transcription follows the same base-pairing rules as DNA replication: T, G, C, and A in the DNA pair with A, C, G, and U, respectively, in the RNA product. (Notice that uracil appears in RNA in place of thymine in DNA.) This base-pairing pattern ensures that an RNA transcript is a faithful copy of the gene (Figure 3.13).

Of course, highly directed chemical reactions such as transcription do not happen at significant rates by themselves—they are enzyme-catalyzed. The enzyme that directs transcription is called **RNA polymerase.** Figure 3.14 presents a schematic diagram of *E. coli* RNA polymerase at work. Transcription has three phases: initiation, elongation, and termination. The following is an outline of these three steps in bacteria:

1. *Initiation* First, the enzyme recognizes a region called a **promoter,** which lies just "upstream" of the gene. The polymerase binds tightly to the promoter and causes localized melting, or separation, of the two DNA strands within the promoter. At least 12 bp are melted. Next, the polymerase starts building the RNA chain. The substrates, or building blocks, it uses for

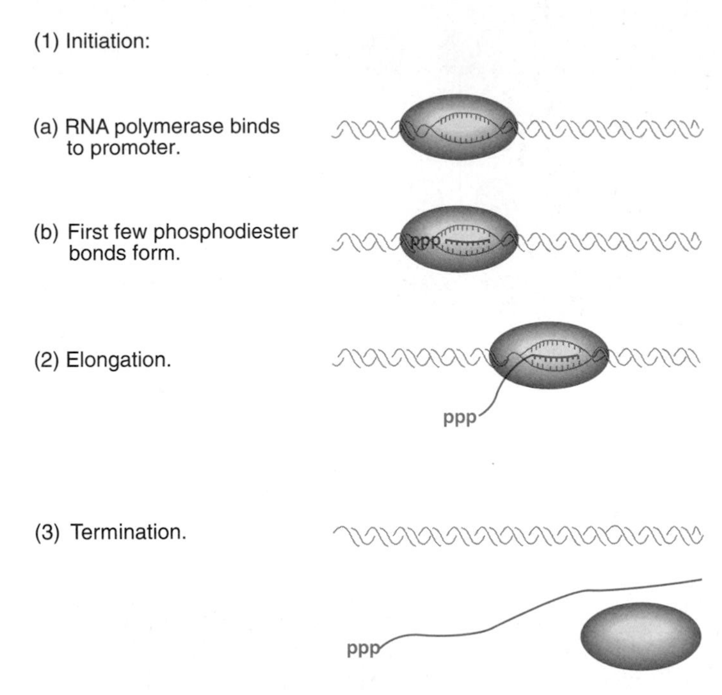

Figure 3.14 Transcription. (1a) In the first stage of initiation, RNA polymerase (red) binds tightly to the promoter and "melts" a short stretch of DNA. **(1b)** In the second stage of initiation, the polymerase joins the first few nucleotides of the nascent RNA (blue) through phosphodiester bonds. The first nucleotide retains its triphosphate group (ppp). **(2)** During elongation, the melted bubble of DNA moves with the polymerase, allowing the enzyme to "read" the bases of the DNA template strand and make complementary RNA. **(3)** Termination occurs when the polymerase reaches a termination signal, causing the RNA and the polymerase to fall off the DNA template.

this job are the four **ribonucleoside triphosphates:** ATP, GTP, CTP, and UTP. The first, or initiating, substrate is usually a purine nucleotide. After the first nucleotide is in place, the polymerase joins a second nucleotide to the first, forming the initial phosphodiester bond in the RNA chain. Several nucleotides may be joined before the polymerase leaves the promoter and elongation begins.

2. *Elongation* During the elongation phase of transcription, RNA polymerase directs the sequential binding of ribonucleotides to the growing RNA chain in the $5' \rightarrow 3'$ direction (from the $5'$-end toward the $3'$-end of the RNA). As it does so, it moves along the DNA **template,** and the "bubble" of melted DNA moves with it. This melted region exposes the bases of the template DNA one by one so they can pair with the bases of the incoming ribonucleotides. As soon as the transcription machinery passes, the two DNA strands wind around each other again, re-forming the double helix. This points to two fundamental differences between transcription and DNA replication: (a) RNA polymerase makes only one RNA strand during transcription, which means that it copies only one DNA strand in a given gene. (However, the opposite strand may be transcribed in another gene.) Transcription is therefore said to be **asymmetrical.** This contrasts with semiconservative DNA replication, in which both DNA strands are copied. (b) In transcription, DNA melting is limited and transient. Only enough strand separation occurs to allow the polymerase to "read" the DNA template strand. However, during replication, the two parental DNA strands separate permanently.
3. *Termination* Just as promoters serve as initiation signals for transcription, other regions at the ends of genes, called **terminators,** signal termination. These work in conjunction with RNA polymerase to loosen the association between RNA product and DNA template. The result is that the RNA dissociates from the RNA polymerase and DNA, thereby stopping transcription.

A final, important note about conventions: RNA sequences are usually written $5'$ to $3'$, left to right. This feels natural to a molecular biologist because RNA is made in a $5'$-to-$3'$ direction, and, as we will see, mRNA is also translated $5'$ to $3'$. Thus, because ribosomes read the message $5'$ to $3'$, it is appropriate to write it $5'$ to $3'$ so that we can read it like a sentence.

Genes are also usually written so that their transcription proceeds in a left-to-right direction. This "flow" of transcription from one end to the other gives rise to the term *upstream,* which refers to the DNA close to the start of transcription (near the left end when the gene is written conventionally). Thus, we can describe most promoters as lying just upstream of their respective genes. By the same convention, we say that genes generally lie *downstream* of their promoters. Genes are also conventionally written with their nontemplate strands on top.

SUMMARY Transcription takes place in three stages: initiation, elongation, and termination. Initiation involves binding RNA polymerase to the promoter, local melting, and forming the first few phosphodiester bonds; during elongation, the RNA polymerase links together ribonucleotides in the $5' \rightarrow 3'$ direction to make the rest of the RNA. Finally, in termination, the polymerase and RNA product dissociate from the DNA template.

Translation

The mechanism of translation is also complex and fascinating. The details of translation will concern us in later chapters; for now, let us look briefly at two substances that play key roles in translation: ribosomes and transfer RNA.

Ribosomes: Protein-Synthesizing Machines Figure 3.15 shows the approximate shapes of the *E. coli* **ribosome** and its two subunits: the 50S and 30S subunits. The numbers

(a)

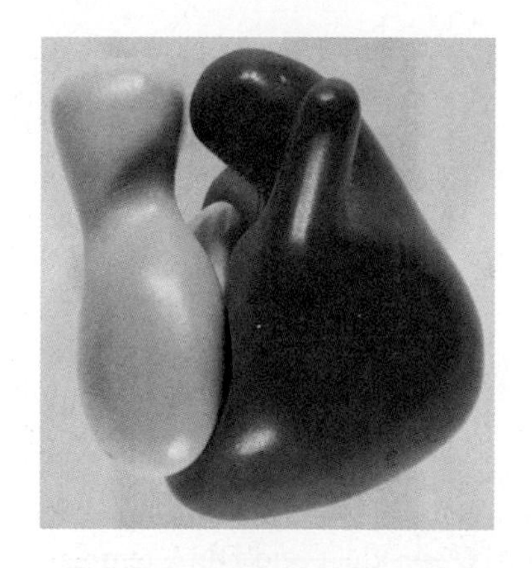

(b)

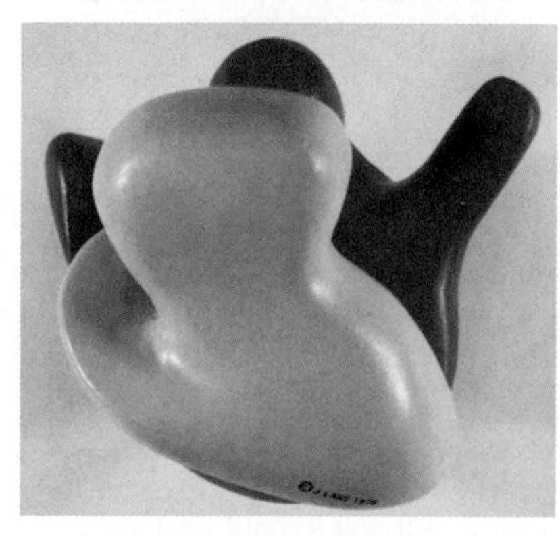

Figure 3.15 *E. coli* ribosome structure. (a) The 70S ribosome is shown from the "side" with the 30S particle (yellow) and the 50S particle (red) fitting together. **(b)** The 70S ribosome is shown rotated 90 degrees relative to the view in part **(a).** The 30S particle (yellow) is in front, with the 50S particle (red) behind. (*Source:* Lake, J. Ribosome structure determined by electron microcopy of *Escherichia coli* small subunits, large subunits, and monomeric ribosomes. *J. Mol. Biol.* 105 (1976), p. 155, fig. 14, by permission of Academic Press.)

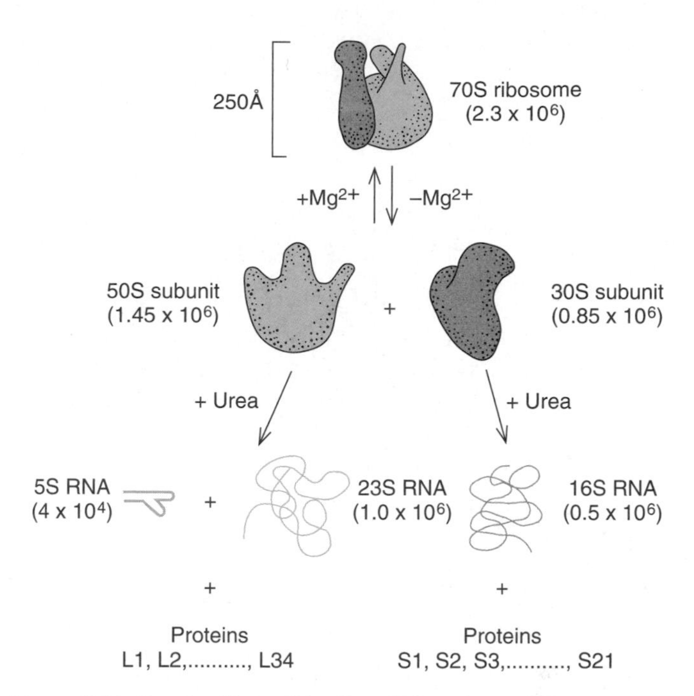

Figure 3.16 Composition of the *E. coli* ribosome. The arrows at the top denote the dissociation of the 70S ribosome into its two subunits when magnesium ions are withdrawn. The lower arrows show the dissociation of each subunit into RNA and protein components in response to the protein denaturant, urea. The masses (M_r, in daltons) of the ribosome and its components are given in parentheses.

50S and 30S refer to the **sedimentation coefficients** of the two subunits. These coefficients are a measure of the speed with which the particles sediment through a solution when spun in an ultracentrifuge. The 50S subunit, with a larger sedimentation coefficient, migrates more rapidly to the bottom of the centrifuge tube under the influence of a centrifugal force. The coefficients are functions of the mass and shape of the particles. Heavy particles sediment more rapidly than light ones; spherical particles migrate faster than extended or flattened ones—just as a skydiver falls more rapidly in a tuck position than with arms and legs extended. The 50S subunit is actually about twice as massive as the 30S. Together, the 50S and 30S subunits compose a **70S** ribosome. Notice that the numbers do not add up. This is because the sedimentation coefficients are not proportional to the particle mass; in fact, they are roughly proportional to the two-thirds power of the particle mass.

Each ribosomal subunit contains RNA and protein. The 30S subunit includes one molecule of **ribosomal RNA (rRNA)** with a sedimentation coefficient of 16S, plus 21 ribosomal proteins. The 50S subunit is composed of 2 rRNAs (23S + 5S) and 34 proteins (Figure 3.16). All these ribosomal proteins are of course gene products themselves. Thus, a ribosome is produced by dozens of different genes. Eukaryotic ribosomes are even more complex, with one more rRNA and more proteins.

Note that rRNAs participate in protein synthesis but do not code for proteins. Transcription is the only step in expression of the genes for rRNAs, aside from some trimming of the transcripts. No translation of these RNAs occurs.

SUMMARY Ribosomes are the cell's protein factories. Bacteria contain 70S ribosomes with two subunits, called 50S and 30S. Each of these contains ribosomal RNA and many proteins.

Transfer RNA: The Adapter Molecule The transcription mechanism was easy for molecular biologists to predict. RNA resembles DNA so closely that it follows the same base-pairing rules. By following these rules, RNA polymerase produces replicas of the genes it transcribes. But what rules govern the ribosome's translation of mRNA to protein? This is a true translation problem. A nucleic acid language must be translated to a protein language.

Francis Crick suggested the answer to this problem in a 1958 paper before much experimental evidence was available

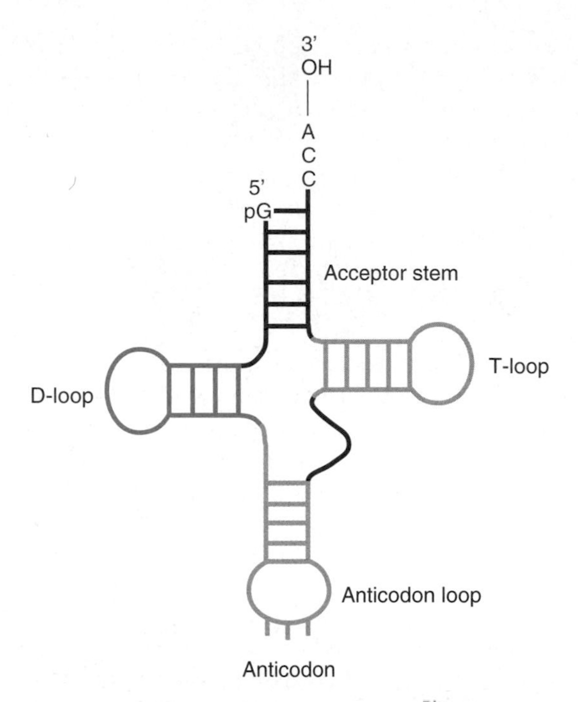

Figure 3.17 Cloverleaf structure of yeast tRNA^{Phe}. At top is the acceptor stem (red), where the amino acid binds to the 3′-terminal adenosine. At left is the dihydro U loop (D-loop, blue), which contains at least one dihydrouracil base. At bottom is the anticodon loop (green), containing the anticodon. The T-loop (right, gray) contains the virtually invariant sequence TψC. Each loop is defined by a base-paired stem of the same color.

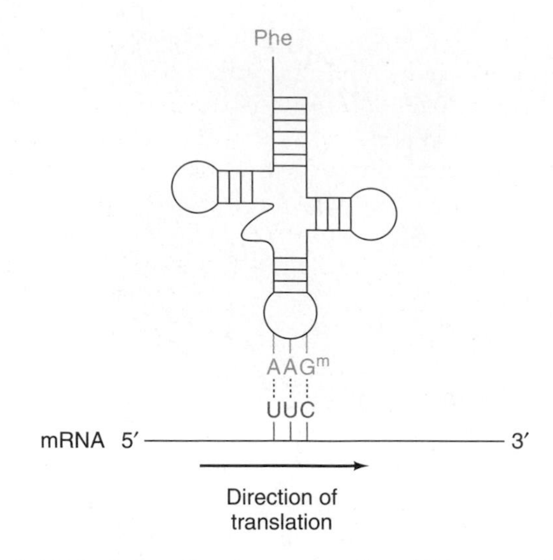

Figure 3.18 Codon–anticodon recognition. The recognition between a codon in an mRNA and a corresponding anticodon in a tRNA obeys essentially the same Watson–Crick rules as apply to other polynucelotides. Here, a 3′AAG^{m}5′ anticodon (blue) on a tRNA^{Phe} is recognizing a 5′UUC3′ codon (red) for phenylalanine in an mRNA. The G^m denotes a methylated G, which base-pairs like an ordinary G. Notice that the tRNA is pictured backwards (3′→5′) relative to normal convention, which is 5′→3′, left to right. That was done to put its anticodon in the proper orientation (3′→5′, left to right) to base-pair with the codon, shown conventionally reading 5′→3′, left to right. Remember that the two strands of DNA are antiparallel; this applies to any double-stranded polynucleotide, including one as small as the 3-bp codon–anticodon pair.

to back it up. What is needed, Crick reasoned, is some kind of adapter molecule that can recognize the nucleotides in the RNA language as well as the amino acids in the protein language. He was right. He even noted that a type of small RNA of unknown function might play the adapter role. Again, he guessed right. Of course he made some bad guesses in this paper as well, but even they were important. By their very creativity, Crick's ideas stimulated the research (some from Crick's own laboratory) that led to solutions to the puzzle of translation.

The adapter molecule in translation is indeed a small RNA that recognizes both RNA and amino acids; it is called **transfer RNA (tRNA).** Figure 3.17 shows a schematic diagram of a tRNA that recognizes the amino acid phenylalanine (Phe). In Chapter 19 we will discuss the structure and function of tRNA in detail. For the present, the *cloverleaf* model, though it bears scant resemblance to the real shape of tRNA, will serve to point out the fact that the molecule has two "business ends." One end (the top of the model) attaches to an amino acid. Because this is a tRNA specific for phenylalanine (tRNA^{Phe}), only phenylalanine will attach. An enzyme called *phenylalanine-tRNA synthetase* catalyzes this reaction. The generic name for such enzymes is **aminoacyl-tRNA synthetase.**

The other end (the bottom of the model) contains a 3-bp sequence that pairs with a complementary 3-bp sequence in an mRNA. Such a triplet in mRNA is called a **codon;** naturally enough, its complement in a tRNA is called an **anticodon.** The codon in question here has attracted the anticodon of a tRNA bearing a phenylalanine. That means that this codon tells the ribosome to insert one phenylalanine into the growing polypeptide. The recognition between codon and anticodon, mediated by the ribosome, obeys the same Watson–Crick rules as any other double-stranded polynucleotide, at least in the case of the first two base pairs. The third pair is allowed somewhat more freedom, as we will see in Chapter 18.

It is apparent from Figure 3.18 that UUC is a codon for phenylalanine. This implies that the **genetic code** contains three-letter words, as indeed it does. We can predict the number of possible 3-bp codons as follows: The number of permutations of 4 different bases taken 3 at a time is 4^3, which is 64. But only 20 amino acids exist. Are some codons not used? Actually, three of the possible codons (UAG, UAA, and UGA) code for termination; that is, they tell the ribosome to stop. All of the other codons specify amino acids. This means that most amino acids have more than one codon; the genetic code is therefore said to be **degenerate.** Chapter 18 presents a fuller description of the code and how it was broken.

SUMMARY Two important sites on tRNAs allow them to recognize both amino acids and nucleic acids. One site binds covalently to an amino acid. The other site contains an anticodon that base-pairs with a 3-bp codon in mRNA. The tRNAs are therefore capable of serving the adapter role postulated by Crick and are the key to the mechanism of translation.

Initiation of Protein Synthesis We have just seen that three codons terminate translation. A codon (AUG) also usually initiates translation. The mechanisms of these two processes are markedly different. As we will see in Chapter 18, the three termination codons interact with protein factors, whereas the initiation codon interacts with a special aminoacyl-tRNA. In eukaryotes this is methionyl-tRNA (a tRNA with methionine attached); in bacteria it is a derivative called *N*-formylmethionyl-tRNA. This is just methionyl tRNA with a formyl group attached to the amino group of methionine.

We find AUG codons not only at the beginning of mRNAs, but also in the middle of messages. When they are at the beginning, AUGs serve as initiation codons, but when they are in the middle, they simply code for methionine. The difference is context. Bacterial messages have a special sequence, called a **Shine–Dalgarno sequence,** named for its discoverers, just upstream of the initiating AUG. The Shine–Dalgarno sequence attracts ribosomes to the nearby AUG so translation can begin. Eukaryotes, by contrast, do not have Shine–Dalgarno sequences. Instead, their mRNAs have a special methylated nucleotide called a cap at their 5′ ends. A cap-binding protein known as **eIF4E** binds to the cap and then helps attract ribosomes. We will discuss these phenomena in greater detail in Chapter 17.

SUMMARY AUG is usually the initiating codon. It is distinguished from internal AUGs by a Shine–Dalgarno ribosome-binding sequence near the beginning of bacterial mRNAs, and by a cap structure at the 5′ end of eukaryotic mRNAs.

Translation Elongation At the end of the initiation phase of translation, the initiating aminoacyl-tRNA is bound to a site on the ribosome called the **P site.** For elongation to occur, the ribosome needs to add amino acids one at a time to the initiating amino acid. We will examine this process in detail in Chapter 18. For the moment, let us consider a simple overview of the elongation process in *E. coli* (Figure 3.19). Elongation begins with the binding of the second aminoacyl-tRNA to another site on the ribosome called the **A site.** This process requires an **elongation factor** called **EF-Tu,** where EF stands for "elongation factor," and energy provided by GTP.

Next, a peptide bond must form between the two amino acids. The large ribosomal subunit contains an enzyme known as **peptidyl transferase,** which forms a peptide bond between the amino acid or peptide in the P site (formylmethionine [fMet] in this case) and the amino acid part of the aminoacyl tRNA in the A site. The result is a dipeptidyl-tRNA in the A site. The dipeptide is composed of fMet plus the second amino acid, which is still bound to its tRNA. The large ribosomal RNA contains the peptidyl transferase active center.

The third step in elongation, **translocation,** involves the movement of the mRNA one codon's length through the ribosome. This maneuver transfers the dipeptidyl-tRNA from the A site to the P site and moves the deacetylated tRNA from the P site to another site, the **E site,** which provides an exit from the ribosome. Translocation requires another elongation factor called **EF-G** and GTP.

SUMMARY Translation elongation involves three steps: (1) transfer of an aminoacyl-tRNA to the A site; (2) formation of a peptide bond between the amino acid in the P site and the aminoacyl-tRNA in the A site; and (3) translocation of the mRNA one codon's length through the ribosome, bringing the newly formed peptidyl-tRNA to the P site.

Termination of Translation and mRNA Structure Three different codons (UAG, UAA, and UGA) cause termination of translation. Protein factors called **release factors** recognize these **termination codons** (or **stop codons**) and cause translation to stop, with release of the polypeptide chain. The initiation codon at one end, and the termination codon at the other end of a coding region of a gene identify an **open reading frame (ORF).** It is called "open" because it contains no internal termination codons to interrupt the translation of the corresponding mRNA. The "reading frame" part of the name refers to the way the ribosome can read the mRNA in three different ways, or "frames," depending on where it starts.

Figure 3.20 illustrates the reading frame concept. This minigene (shorter than any gene you would expect to find) contains a start codon (ATG) and a stop codon (TAG). (Remember that these DNA codons will be transcribed to mRNA with the corresponding codons AUG and UAG.) In between (and including these codons) we have a short open reading frame that can be translated to yield a tetrapeptide (a peptide containing four amino acids): fMet-Gly-Tyr-Arg. In principle, translation could also begin four nucleotides upstream at another AUG, but notice that translation would be in another **reading frame,** so the codons would be different: AUG, CAU, GGG, AUA, UAG. Translation in this second reading frame would therefore produce another tetrapeptide: fMet-His-Gly-Ile. The third reading frame has no initiation

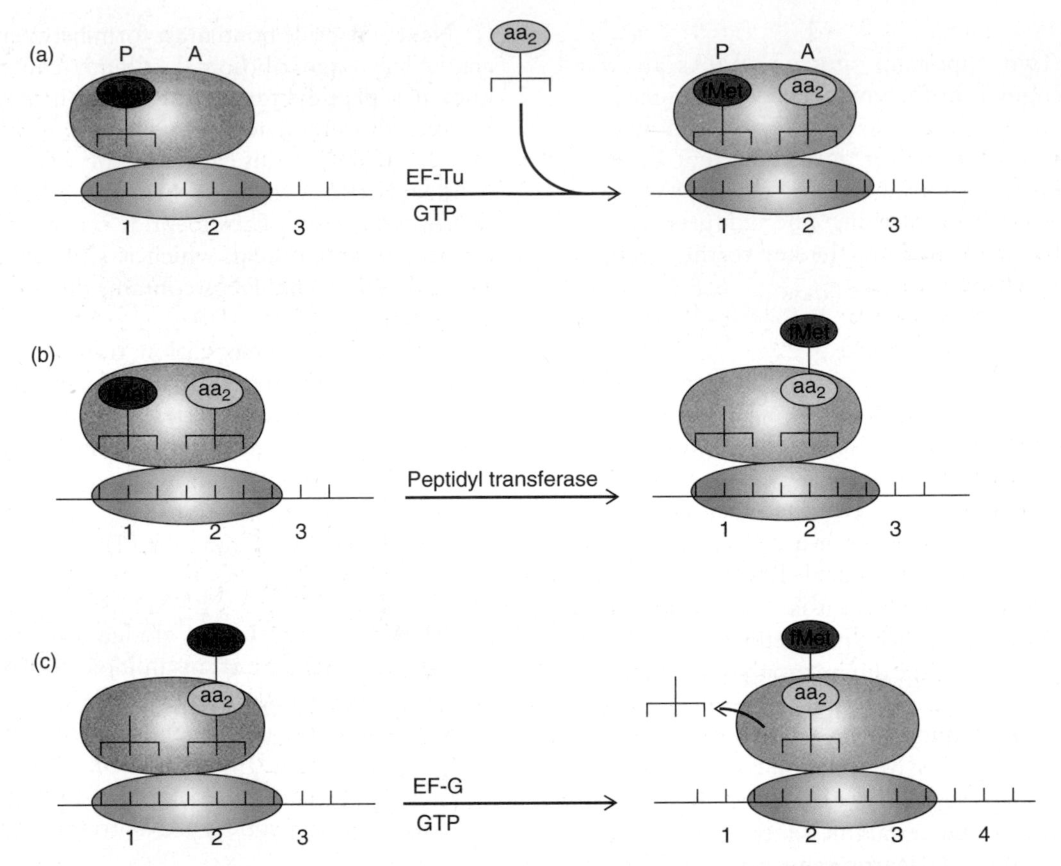

Figure 3.19 Summary of translation elongation. (a) EF-Tu, with help from GTP, transfers the second aminoacyl-tRNA to the A site. (The P and A sites are conventionally represented on the left and right halves of the ribosome, as indicated at top.) **(b)** Peptidyl transferase, an integral part of the large rRNA in the 50S subunit, forms a peptide bond between fMet and the second aminoacyl-tRNA. This creates a dipeptidyl-tRNA in the A site. **(c)** EF-G, with help from GTP, translocates the mRNA one codon's length through the ribosome. This brings codon 2, along with the peptidyl-tRNA, to the P site, and codon 3 to the A site. It also moves the deacylated tRNA out of the P site into the E site (not shown), from which it is ejected. The A site is now ready to accept another aminoacyl-tRNA to begin another round of elongation.

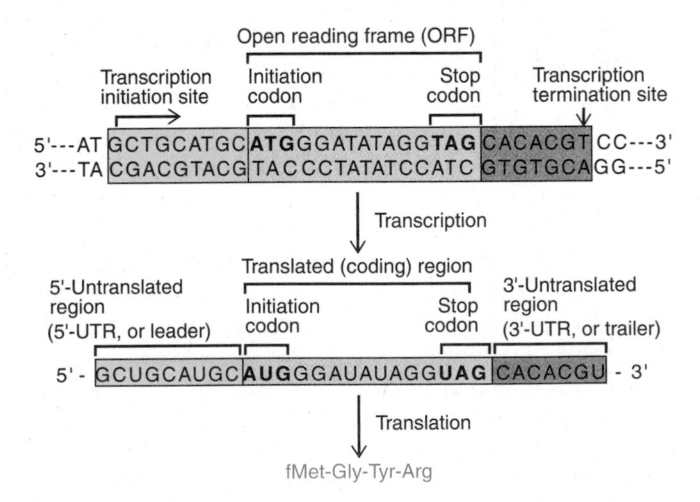

Figure 3.20 Simplified gene and mRNA structure. At top is a simplified gene that begins with a transcription initiation site and ends with a transcription termination site. In between are the translation initiation codon and the stop codon, which define an open reading frame that can be translated to yield a polypeptide (a very short polypeptide with only four amino acids, in this case). The gene is transcribed to give an mRNA with a coding region that begins with the initiation codon and ends with the termination codon. This is the RNA equivalent of the open reading frame in the gene. The material upstream of the initiation codon in the mRNA is the leader, or 5′-untranslated region. The material downstream of the termination codon in the mRNA is the trailer, or 3′-untranslated region. Note that this gene has another open reading frame that begins four bases farther upstream, and it codes for another tetrapeptide. Notice also that this alternative reading frame is shifted 1 bp to the left relative to the other.

codon. A natural mRNA may also have more than one open reading frame, but the largest is usually the one that is used.

Figure 3.20 also shows that transcription and translation in this gene do not start and stop at the same places. Transcription begins with the first G and translation begins 9 bp downstream at the start codon (AUG). Thus, the mRNA produced from this gene has a 9-bp **leader,** which is also called the **5′-untranslated region,** or **5′-UTR.** Similarly, a **trailer** is present at the end of the mRNA between the stop codon and the transcription termination site. The trailer is also called the **3′-untranslated region,** or **3′-UTR.** In a eukaryotic gene, the transcription termination site would probably be farther downstream, but the mRNA would be cleaved downstream of the translation stop codon and a string of A's [poly(A)] would be added to the 3′-end of the mRNA. In that case, the trailer would be the stretch of RNA between the stop codon and the poly(A).

SUMMARY Translation terminates at a stop codon (UAG, UAA, or UGA). The genetic material including a translation initiation codon, a coding region, and a termination codon, is called an open reading frame. The piece of an mRNA between its 5′-end and the initiation codon is called a leader or 5′-UTR. The part between the 3′-end [or the poly(A)] and the termination codon is called a trailer or 3′-UTR.

3.2 Replication

A second characteristic of genes is that they replicate faithfully. The Watson–Crick model for DNA replication (introduced in Chapter 2) assumes that as new strands of DNA are made, they follow the usual base-pairing rules of A with T and G with C. This is essential because the DNA-replicating machinery must be capable of discerning a good pair from a bad one, and the Watson–Crick base pairs give the best fit. The model also presupposes that the two parental strands separate and that each then serves as a template for a new progeny strand. This is called **semiconservative replication** because each daughter double helix has one parental strand and one new strand (Figure 3.21a). In other words, one of the parental strands is "conserved" in each daughter double helix. This is not the only possibility. Another potential mechanism (Figure 3.21b) is **conservative replication,** in which the two parental strands stay together and somehow produce another daughter helix with two completely new strands. Yet another possibility is **dispersive replication,** in which the DNA becomes fragmented so that new and old DNA regions coexist in the same strand after replication (Figure 3.21c). As mentioned in Chapter 1, Matthew Meselson and Franklin Stahl proved that DNA really does replicate by a semiconservative mechanism. Chapter 20 will present this experimental evidence.

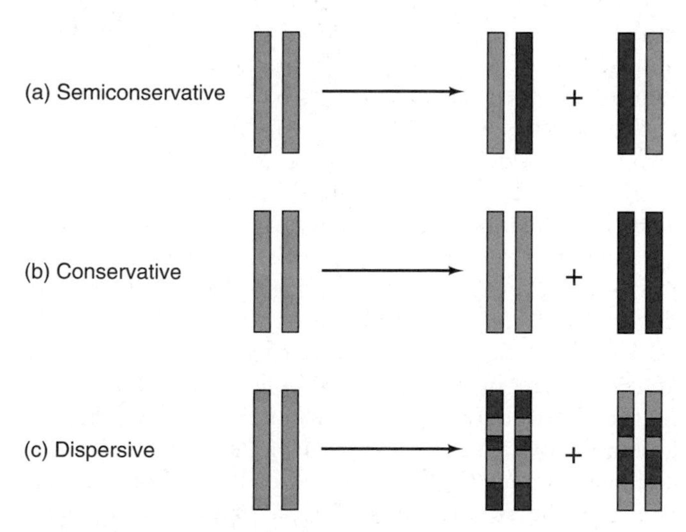

Figure 3.21 Three hypotheses for DNA replication. **(a)** Semiconservative replication (see also Figure 2.15) gives two daughter duplex DNAs, each of which contains one old strand (blue-green) and one new strand (red). **(b)** Conservative replication yields two daughter duplexes, one of which has two old strands (blue-green) and one of which has two new strands (red). **(c)** Dispersive replication gives two daughter duplexes, each of which contains strands that are a mixture of old and new DNA.

3.3 Mutations

A third characteristic of genes is that they accumulate changes, or mutations. By this process, life itself can change, because mutation is essential for evolution. Now that we know most genes are strings of nucleotides that code for polypeptides, which in turn are strings of amino acids, it is easy to see the consequences of changes in DNA. If a nucleotide in a gene changes, it is likely that a corresponding change will occur in an amino acid in that gene's protein product. Sometimes, because of the degeneracy of the genetic code, a nucleotide change will not affect the protein. For example, changing the codon AAA to AAG is a mutation, but it would probably not be detected because both AAA and AAG code for the same amino acid: lysine. Such innocuous alterations are called **silent mutations.** More often, a changed nucleotide in a gene results in an altered amino acid in the protein. This may be harmless if the amino acid change is **conservative** (e.g., a leucine changed to an isoleucine). But if the new amino acid is much different from the old one, the change frequently impairs or destroys the function of the protein.

Sickle Cell Disease

An excellent example of a disease caused by a defective gene is **sickle cell disease,** a true genetic disorder. People who are homozygous for this condition have normal-looking red blood cells when their blood is rich in oxygen. The shape of normal cells is a *biconcave disc;* that is, the disc is concave viewed from both the top and bottom. However, when these people exercise, or otherwise deplete the oxygen in their blood, their red blood cells change dramatically to a sickle, or crescent, shape. This has dire consequences. The sickle cells cannot fit through tiny capillaries, so they clog and rupture them, starving parts of the body for blood and causing internal bleeding and pain. Furthermore, the sickle cells are so fragile that they burst, leaving the patient anemic. Without medical attention, patients undergoing a sickling crisis are in mortal danger.

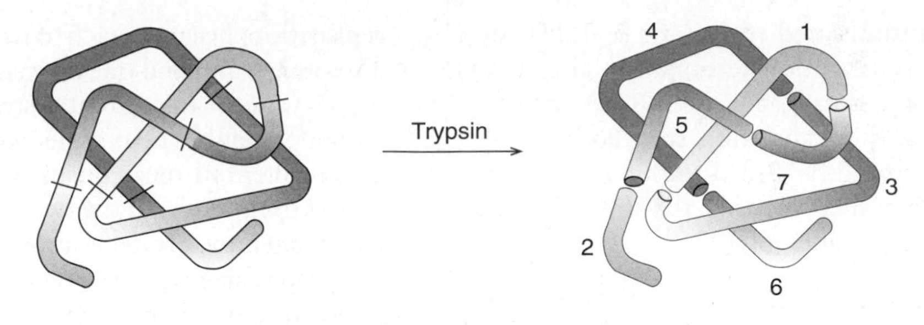

(a) Cutting protein to peptides

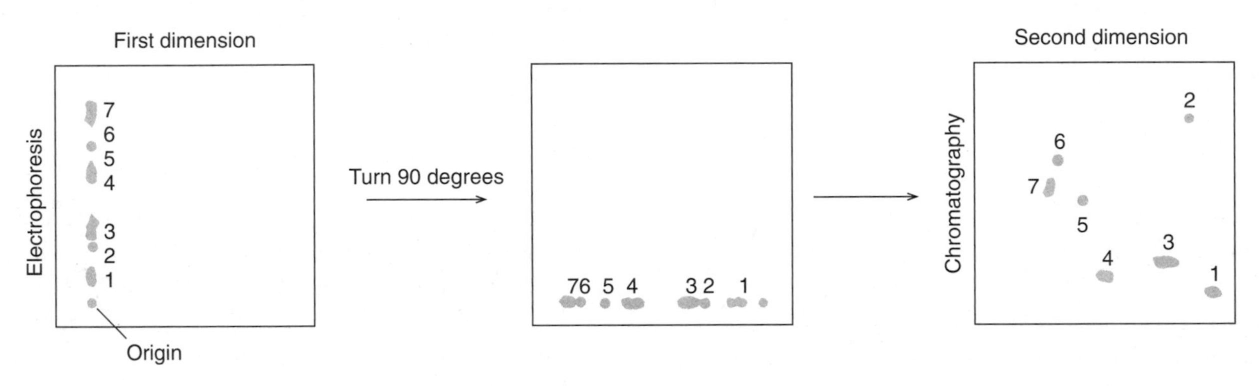

(b) Two-dimensional separation of peptides

Figure 3.22 Fingerprinting a protein. (a) A hypothetical protein, with six trypsin-sensitive sites indicated by slashes. After digestion with trypsin, seven peptides are released. **(b)** These tryptic peptides separate partially during electrophoresis in the first dimension, then fully after the paper is turned 90 degrees and chromatographed in the second dimension with another solvent.

What causes this sickling of red blood cells? The problem is in **hemoglobin,** the red, oxygen-carrying protein in the red blood cells. Normal hemoglobin remains soluble under ordinary physiological conditions, but the hemoglobin in sickle cells precipitates when the blood oxygen level falls, forming long, fibrous aggregates that distort the blood cells into the sickle shape.

What is the difference between normal hemoglobin (HbA) and sickle cell hemoglobin (HbS)? Vernon Ingram answered this question in 1957 by determining the amino acid sequences of parts of the two proteins using a process that was invented by Frederick Sanger and is known as **protein sequencing.** Ingram focused on the β-globins of the two proteins. β-globin is one of the two different polypeptide chains found in the tetrameric (four-chain) hemoglobin protein. First, Ingram cut the two polypeptides into pieces with an enzyme that breaks selected peptide bonds. These pieces, called *peptides,* can be separated by a two-dimensional method called **fingerprinting** (Figure 3.22). The peptides are separated in the first dimension by paper electrophoresis. Then the paper is turned 90 degrees and the peptides are subjected to paper chromatography to separate them still farther in the second dimension. The peptides usually appear as spots on the paper. Different proteins, because of their different amino acid compositions, give different patterns of spots. These patterns are aptly named **fingerprints.**

When Ingram compared the fingerprints of HbA and HbS, he found that all the spots matched except for one (Figure 3.23). This spot had a different mobility in the HbS fingerprint than in the normal HbA fingerprint, which indicated that it had an altered amino acid composition. Ingram checked the amino acid sequences of the two peptides in these spots. He found that they were the amino-terminal peptides located at the very beginning of both proteins. And he found that they differed in only one amino acid. The glutamate in the sixth position of HbA becomes a valine in HbS (Figure 3.24). This is the only difference in the two proteins, yet it is enough to cause a profound distortion of the protein's behavior.

Knowing the genetic code (Chapter 18), we can ask: What change in the β-globin gene caused the change Ingram detected in its protein product? The two codons for glutamate (Glu) are GAA and GAG; two of the four codons for valine (Val) are GUA and GUG. If the glutamate codon in the HbA gene is GAG, a single base change to GTG would alter the mRNA to GUG, and the amino acid inserted into HbS would be valine instead of glutamate. A similar argument can be made for a GAA→GTA change. Notice that, by convention, we are presenting the DNA strand that has the same sense as the mRNA (the nontemplate strand). Actually, the opposite strand (the template strand), reading CAC, is transcribed to give a GUG

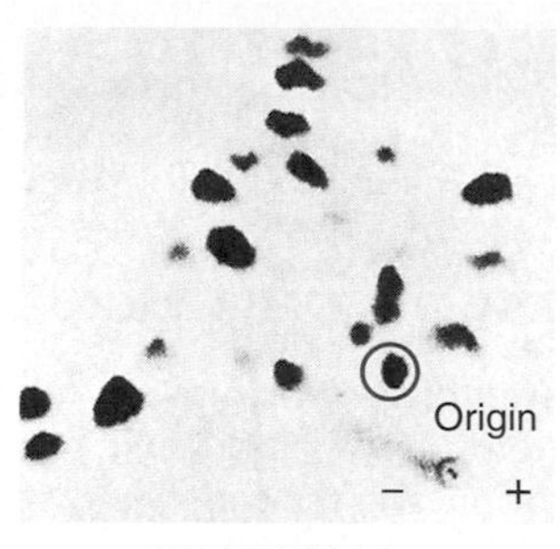

Hemoglobin A

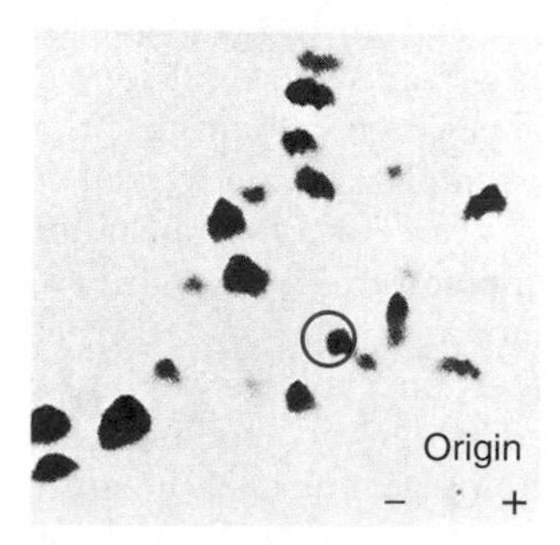

Hemoglobin S

Figure 3.23 Fingerprints of hemoglobin A and hemoglobin S. The fingerprints are identical except for one peptide (circled), which shifts up and to the left in hemoglobin S. (*Source:*Dr. Corrado Baglioni.)

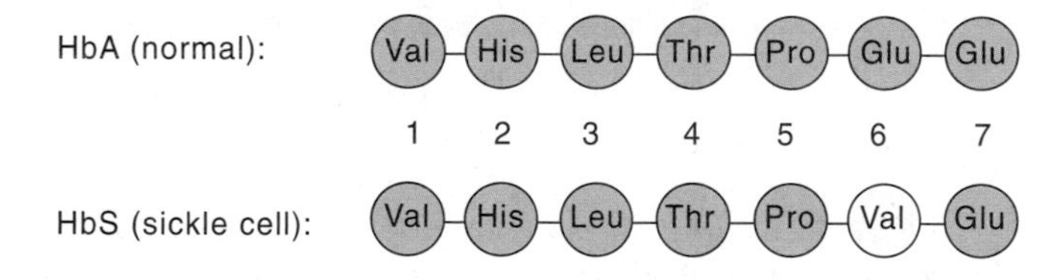

Figure 3.24 Sequences of amino-terminal peptides from normal and sickle cell β-globin. The numbers indicate the positions of the corresponding amino acids in the mature protein. The only difference is in position 6, where a valine (Val) in HbS replaces a glutamate (Glu) in HbA.

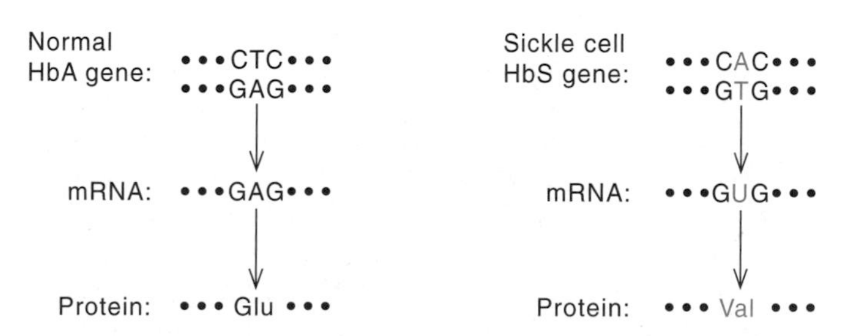

Figure 3.25 The sickle cell mutation and its consequences. The GAG in the sixth codon of the nontemplate strand of the normal gene changes to GTG. This leads to a change from GAG to GUG in the sixth codon of the β-globin mRNA of the sickle cells. This, in turn, results in insertion of a valine in the sixth amino acid position of sickle cell β-globin, where a glutamate ought to go.

sequence in the mRNA. Figure 3.25 presents a summary of the mutation and its consequences. We can see how changing the blueprint does indeed change the product.

Sickle cell disease is a very common problem among people of central African descent. Why has this deleterious mutation spread so successfully through the population? The answer seems to be that although the homozygous condition can be lethal, heterozygotes have little if any difficulty because their normal allele makes enough product to keep their blood cells from sickling. Moreover, heterozygotes are at an advantage in central Africa, where malaria is rampant, because HbS helps protect against replication of the malarial parasite when it tries to infect their blood cells.

SUMMARY Sickle cell disease is a human genetic disorder. It results from a single base change in the gene for β-globin. The altered base causes insertion of the wrong amino acid into one position of the β-globin protein. This altered protein results in distortion of red blood cells under low-oxygen conditions. This disease illustrates a fundamental genetic concept: a change in a gene can cause a corresponding change in the protein product of that gene.

SUMMARY

The three main activities of genes are information storing, replication, and accumulating mutations. Proteins, or polypeptides, are polymers of amino acids linked through peptide bonds. Most genes contain the information for making one polypeptide and are expressed in a two-step process: transcription or synthesis of an mRNA copy of the gene, followed by translation of this message to protein. Translation takes place on structures called ribosomes, the cell's protein factories. Translation also requires adapter molecules called transfer RNAs (tRNAs) that can recognize both the genetic code in mRNA and the amino acids the mRNA encodes.

Translation elongation involves three steps: (1) transfer of an aminoacyl-tRNA to the A site; (2) formation of a peptide bond between the amino acid in the P site and the aminoacyl-tRNA in the A site; and (3) translocation of the mRNA one codon's length through the ribosome, bringing the newly formed peptidyl-tRNA to the P site. Translation terminates at a stop codon (UAG, UAA, or UGA). A region of RNA or DNA including a translation initiation codon, a coding region, and a termination codon, is called an open reading frame. The piece of an mRNA between its 5′-end and the initiation codon is called a leader or 5′-UTR. The part between the 3′-end [or the poly(A)] and the termination codon is called a trailer or 3′-UTR.

DNA replicates in a semiconservative manner: When the parental strands separate, each serves as the template for making a new, complementary strand. A change, or mutation, in a gene frequently causes a change at a corresponding position in the polypeptide product. Sickle cell disease is an example of the deleterious effect of such mutations.

REVIEW QUESTIONS

1. Draw the general structure of an amino acid.
2. Draw the structure of a peptide bond.
3. Use a rough diagram to compare the structures of a protein α-helix and an antiparallel β-sheet. For simplicity, show only the backbone atoms of the protein.
4. What do we mean by primary, secondary, tertiary, and quaternary structures of proteins?
5. What was Garrod's insight into the relationship between genes and proteins, based on the disease alcaptonuria?
6. Describe Beadle and Tatum's experimental approach to demonstrating the relationship between genes and proteins.
7. What are the two main steps in gene expression?
8. Describe and give the results of the experiment of Jacob and colleagues that demonstrated the existence of mRNA.
9. What are the three steps in transcription? With a diagram, illustrate each one.
10. What ribosomal RNAs are present in *E. coli* ribosomes? To which ribosomal subunit does each rRNA belong?
11. Draw a diagram of the cloverleaf structure of a tRNA. Point out the site to which the amino acid attaches and the site of the anticodon.
12. How does a tRNA serve as an adapter between the 3-bp codons in mRNA and the amino acids in protein?
13. Explain how a single base change in a gene could lead to premature termination of translation of the mRNA from that gene.
14. Explain how a single base deletion in the middle of a gene would change the reading frame of that gene.
15. Explain how a single base change in a gene can lead to a single amino acid change in that gene's polypeptide product. Illustrate with an example.

ANALYTICAL QUESTIONS

1. Here is the sequence of a portion of a bacterial gene:

   5′GTATCGTATGCATGCATCGTGAC3′
   3′CATAGCATACGTACGTAGCACTG5′

   The template strand is on the bottom. (a) Assuming that transcription starts with the first T in the template strand, and continues to the end, what would be the sequence of the mRNA derived from this fragment? (b) Find the initiation codon in this mRNA. (c) Would there be an effect on translation of changing the first G in the template strand to a C? If so, what effect? (d) Would there be an effect on translation of changing the second T in the template strand to a G? If so, what effect? (e) Would there be an effect on translation of changing the last T in the template strand to a C? If so, what effect? (Hint: You do not need to know the genetic code to answer these questions; you just need to know the nature of initiation and termination codons given in this chapter.)
2. You are performing genetic experiments on *Neurospora crassa,* similar to the ones Beadle and Tatum did. You isolate one pantothenateless mutant that cannot synthesize pantothenate unless you supply it with pantoate. What step in the pantothenate pathway is blocked?

SUGGESTED READINGS

Beadle, G.W., and E.L. Tatum. 1941. Genetic control of biochemical reactions in *Neurospora. Proceedings of the National Academy of Sciences* 27:499–506.

Brenner, S., F. Jacob, and M. Meselson. 1961. An unstable intermediate carrying information from genes to ribosomes for protein synthesis. *Nature* 190:576–81.

Crick, F.H.C. 1958. On protein synthesis. *Symposium of the Society for Experimental Biology* 12:138–63.

Meselson, M., and F.W. Stahl. 1958. The replication of DNA in *Escherichia coli. Proceedings of the National Academy of Sciences* 44:671–82.

CHAPTER 4

Molecular Cloning Methods

Close-up of bacteria in a Petri dish. Bacteria, especially *E. coli,* are favorite organisms in which to clone genes. *© Glowimages/Getty RF.*

Now that we have reviewed the fundamentals of gene structure and function, we are ready to start a more detailed study of molecular biology. The main focus of our investigation will be the experiments that molecular biologists have performed to elucidate the structure and function of genes. For this reason, we need to pause at this point to discuss some of the major experimental techniques of molecular biology. Because it would be impractical to talk about them all at this early stage, we will deal with the common ones in the next two chapters and introduce the others as needed throughout the book. We will begin in this chapter with the technique that revolutionized the discipline, gene cloning.

Imagine that you are a geneticist in the year 1972. You want to investigate the function of eukaryotic genes at the molecular level. In particular, you are curious about the molecular structure and function of the human growth hormone (hGH) gene. What is the base sequence of this gene? What does its promoter look like? How does RNA polymerase interact with this gene? What changes occur in this gene to cause conditions like hypopituitary dwarfism?

These questions cannot be answered unless you can purify enough of the gene to study—probably about a milligram's worth. A milligram does not sound like much, but it is an overwhelming amount when you imagine purifying it from whole human DNA. Consider that the DNA involved in one hGH gene is much less than one part per million in the human genome. And even if you could collect that much material somehow, you would not know how to separate the one gene you are interested in from all the rest of the DNA. In short, you would be stuck.

Gene cloning neatly solves these problems. By linking eukaryotic genes to small bacterial or phage DNAs and inserting these recombinant molecules into bacterial hosts, one can produce large quantities of these genes in pure form. In this chapter we will see how to clone genes in bacteria and in eukaryotes.

4.1 Gene Cloning

One product of any cloning experiment is a **clone,** a group of identical cells or organisms. We know that some plants can be cloned simply by taking cuttings (Greek: *klon,* meaning twig), and that others can be cloned by growing whole plants from single cells collected from one plant. Even vertebrates can be cloned. John Gurdon produced clones of identical frogs by transplanting nuclei from a single frog embryo to many enucleate eggs, and a sheep named Dolly was cloned in Scotland in 1997 using an enucleate egg and a nucleus from an adult sheep mammary gland. Identical twins constitute a natural clone.

The usual procedure in a gene cloning experiment is to place a foreign gene into bacterial cells, separate individual cells, and grow colonies from each of them. All the cells in each colony are identical and will contain the foreign gene. Thus, as long as we ensure that the foreign gene can replicate, we can clone the gene by cloning its bacterial host. Stanley Cohen, Herbert Boyer, and their colleagues performed the first cloning experiment in 1973.

The Role of Restriction Endonucleases

Cohen and Boyer's elegant plan depended on invaluable enzymes called **restriction endonucleases.** Stewart Linn and Werner Arber discovered restriction endonucleases in *E. coli* in the late 1960s. These enzymes get their name from the fact that they prevent invasion by foreign DNA, such as viral DNA, by cutting it up. Thus, they "restrict" the host range of the virus. Furthermore, they cut at sites within the foreign DNA, rather than chewing it away at the ends, so we call them endonucleases (Greek: *endo,* meaning within) rather than exonucleases (Greek: *exo,* meaning outside). Linn and Arber hoped that their enzymes would cut DNA at specific sites, giving them finely honed molecular knives with which to slice DNA. Unfortunately, these particular enzymes did not fulfill that hope.

However, an enzyme from *Haemophilus influenzae* strain R_d, discovered by Hamilton Smith, did show specificity in cutting DNA. This enzyme is called *Hin*dII (pronounced Hin-dee-two). Restriction enzymes derive the first three letters of their names from the Latin name of the microorganism that produces them. The first letter is the first letter of the genus and the next two letters are the first two letters of the species (hence: *Haemophilus influenzae* yields *Hin*). In addition, the strain designation is sometimes included; in this case, the "d" from R_d is used. Finally, if the strain of microorganism produces just one restriction enzyme, the name ends with the Roman numeral I. If more than one enzyme is produced, the others are numbered II, III, and so on.

*Hin*dII recognizes this sequence:

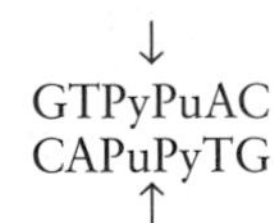

and cuts both DNA strands at the points shown by the arrows. Py stands for either of the pyrimidines (T or C), and Pu stands for either purine (A or G). Wherever this sequence occurs, and *only* when this sequence occurs, *Hin*dII will make a cut. Happily for molecular biologists, *Hin*dII turned out to be only one of hundreds of restriction enzymes, each with its own specific recognition sequence. Table 4.1 lists the sources and recognition sequences for several popular restriction enzymes. Note that some of these enzymes recognize 4-bp sequences instead of the more common 6-bp sequences. As a result, they cut much more frequently. This is because a given sequence of 4 bp will occur about once in every $4^4 = 256$ bp, whereas a sequence of 6 bp will occur only about once in every $4^6 = 4096$ bp. Thus, a 6-bp cutter will yield DNA fragments of average length about

Table 4.1 Recognition Sequences and Cutting Sites of Selected Restriction Endonucleases

Enzyme	Recognition Sequence*
*Alu*I	AG↓CT
*Bam*HI	G↓GATCC
*Bgl*II	A↓GATCT
*Cla*I	AT↓CGAT
*Eco*RI	G↓AATTC
*Hae*III	GG↓CC
*Hin*dII	GTPy↓PuAC
*Hin*dIII	A↓AGCTT
*Hpa*II	C↓CGG
*Kpn*I	GGTAC↓C
*Mbo*I	↓GATC
*Pst*I	CTGCA↓G
*Pvu*I	CGAT↓CG
*Sal*I	G↓TCGAC
*Sma*I	CCC↓GGG
*Xma*I	C↓CCGGG
*Not*I	GC↓GGCCGC

*Only one DNA strand, written 5′→3′ left to right is presented, but restriction endonucleases actually cut double-stranded DNA as illustrated in the text for *Eco*RI. The cutting site for each enzyme is represented by an arrow.

4000 bp, or 4 **kilobases (4 kb).** Some restriction enzymes, such as *Not*I, recognize 8-bp sequences, so they cut much less frequently (once in $4^8 \approx 65{,}000$ bp); they are therefore called **rare cutters.** In fact, *Not*I cuts even less frequently than you would expect in mammalian DNA, because its recognition sequence includes two copies of the rare dinucleotide CG. Notice also that the recognition sequences for *Sma*I and *Xma*I are identical, although the cutting sites within these sequences are different. We call such enzymes that recognize different sites in identical sequences **heteroschizomers** (Greek: *hetero,* meaning different; *schizo,* meaning split) or **neoschizomers** (Greek: *neo,* meaning new). We call enzymes that cut at the same site in the same sequence **isoschizomers** (Greek: *iso,* meaning equal).

The main advantage of restriction enzymes is their ability to cut DNA strands reproducibly in the same places. This property is the basis of many techniques used to analyze genes and their expression. But this is not the only advantage. Many restriction enzymes make staggered cuts in the two DNA strands (they are the ones with off-center cutting sites in Table 4.1), leaving single-stranded overhangs, or **sticky ends,** that can base-pair together briefly. This makes it easier to stitch two different DNA molecules together, as we will see. Note, for example, the complementarity between the ends created by *Eco*RI (pronounced Eeko R-1 or Echo R-1):

```
  ↓
5′---GAATTC---3′         ---G3′          5′AATTC---
3′---CTTAAG---5′    →    ---CTTAA5′   +        3′G---
         ↑
```

Note also that *Eco*RI produces 4-base overhangs that protrude from the 5′-ends of the fragments. *Pst*I cuts at the 3′-ends of its recognition sequence, so it leaves 3′-overhangs. *Sma*I cuts in the middle of its sequence, so it produces blunt ends with no overhangs.

Restriction enzymes can make staggered cuts because the sequences they recognize usually display twofold symmetry. That is, they are identical after rotating them 180 degrees. For example, imagine inverting the *Eco*RI recognition sequence just described:

```
     ↓
5′---GAATTC---3′
3′---CTTAAG---5′
         ↑
```

You can see it will still look the same after the inversion. In a way, these sequences read the same forward and backward. Thus, *Eco*RI cuts between the G and the A in the top strand (on the left), and between the G and the A in the bottom strand (on the right), as shown by the vertical arrows.

Sequences with twofold symmetry are also called **palindromes.** In ordinary language, palindromes are sentences that read the same forward and backward. Examples are Napoleon's lament: "Able was I ere I saw Elba," or a wart remedy: "Straw? No, too stupid a fad; I put soot on warts," or a statement of preference in Italian food: "Go hang a salami! I'm a lasagna hog." DNA palindromes also read the same forward and backward, but you have to be careful to read the same sense (5′→3′) in both directions. This means that you read the top strand left to right and the bottom strand right to left.

One final question about restriction enzymes: If they can cut up invading viral DNA, why do they not destroy the host cell's own DNA? The answer is this: Almost all restriction endonucleases are paired with methylases that recognize and methylate the same DNA sites. The two enzymes—the restriction endonuclease and the methylase—are collectively called a **restriction–modification system,** or an **R-M system.** After methylation, DNA sites are protected against most restriction endonucleases so the methylated DNA can persist unharmed in the host cell. But what about DNA replication? Doesn't that create newly replicated DNA strands that are unmethylated, and therefore vulnerable to cleavage? Figure 4.1 explains how DNA continues to be protected during replication. Every time the cellular DNA replicates, one strand of the

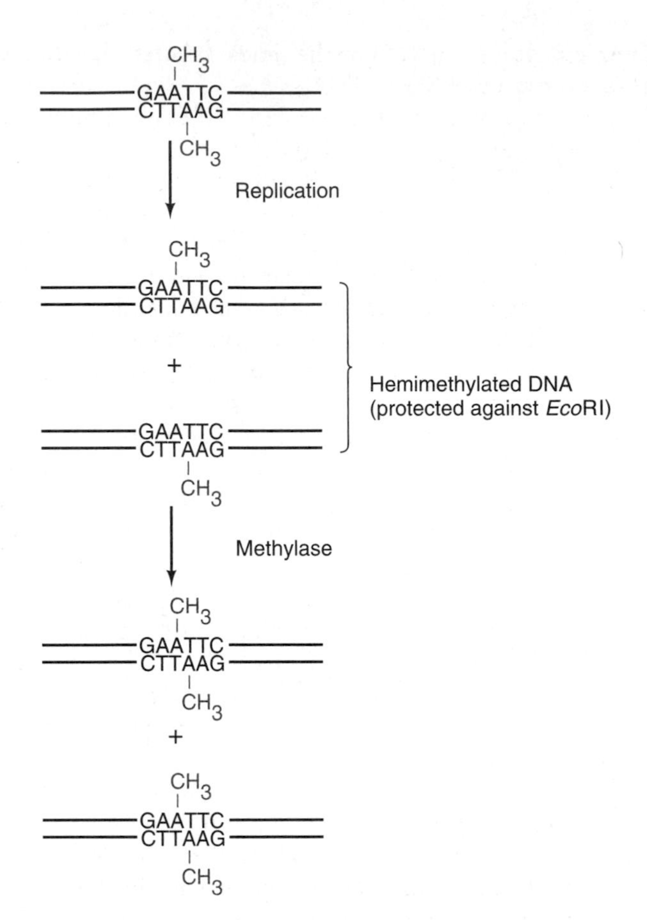

Figure 4.1 Maintaining restriction endonuclease resistance after DNA replication. We begin with an *Eco*RI site that is methylated (red) on both strands. After replication, the parental strand of each daughter DNA duplex remains methylated, but the newly made strand of each duplex has not been methylated yet. The one methylated strand in these hemimethylated DNAs is enough to protect both strands against cleavage by *Eco*RI. Soon, the methylase recognizes the unmethylated strand in each *Eco*RI site and methylates it, regenerating the fully methylated DNA.

daughter duplex will be a newly made strand and will be unmethylated. But the other will be a parental strand and therefore be methylated. This half-methylation (hemimethylation) is enough to protect the DNA duplex against cleavage by the great majority of restriction endonucleases, so the methylase has time to find the site and methylate the other strand yielding fully methylated DNA.

Cohen and Boyer took advantage of the sticky ends created by a restriction enzyme in their cloning experiment (Figure 4.2). They cut two different DNAs with the same restriction enzyme, *Eco*RI. Both DNAs were **plasmids,** small, circular DNAs that are independent of the host chromosome. The first, called pSC101, carried a gene that conferred resistance to the antibiotic tetracycline; the other, RSF1010, conferred resistance to both streptomycin and sulfonamide. Both plasmids had just one *Eco*RI **restriction site,** or cutting site for *Eco*RI. Therefore, when *Eco*RI cut these circular DNAs, it converted them to linear molecules and left them with the same sticky ends. These sticky ends then base-paired with each other, at least briefly. Of course, some of this base-pairing involved sticky ends on the same DNA, which simply closed up the circle again. But some base-pairing of sticky ends brought the two different DNAs together. Finally, **DNA ligase** completed the task of joining the two DNAs covalently. DNA ligase is an enzyme that forms covalent bonds between the ends of DNA strands.

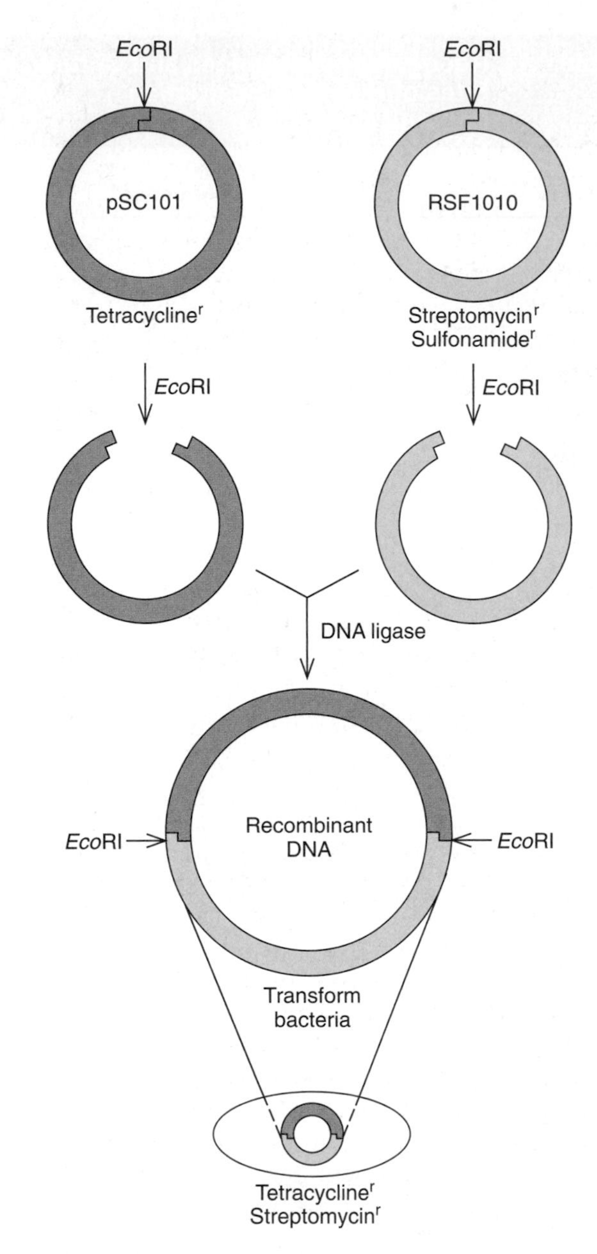

Figure 4.2 The first cloning experiment involving a recombinant DNA assembled in vitro. Boyer and Cohen cut two plasmids, pSC101 and RSF1010, with the same restriction endonuclease, *Eco*RI. This gave the two linear DNAs the same sticky ends, which were then linked in vitro using DNA ligase. The investigators reintroduced the recombinant DNA into *E. coli* cells by transformation and selected clones that were resistant to both tetracycline and streptomycin. These clones were therefore harboring the recombinant plasmid.

The desired result was a **recombinant DNA,** two previously separate pieces of DNA linked together. This new, recombinant plasmid was probably outnumbered by the two parental plasmids that had been cut and then religated, but it was easy to detect. When introduced into bacterial cells, it conferred resistance to both tetracycline, a property of pSC101, and to streptomycin, a property of RSF1010. Recombinant DNAs abound in nature, but this one differs from most of the others in that it was not created naturally in a cell. Instead, molecular biologists put it together in a test tube.

SUMMARY Restriction endonucleases recognize specific sequences in DNA molecules and make cuts in both strands. This allows very specific cutting of DNAs. Also, because the cuts in the two strands are frequently staggered, restriction enzymes can create sticky ends that help link together two DNAs to form a recombinant DNA in vitro.

Vectors

Both plasmids in the Cohen and Boyer experiment are capable of replicating in *E. coli.* Thus, both can serve as carriers to allow replication of recombinant DNAs. All gene cloning experiments require such carriers, which we call **vectors,** but a typical experiment involves only one vector, plus a piece of foreign DNA that depends on the vector for its replication. The foreign DNA has no **origin of replication,** the site where DNA replication begins, so it cannot replicate unless it is placed in a vector that does have an origin of replication. Since the mid-1970s, many vectors have been developed; these fall into two major classes: plasmids and phages. Regardless of the nature of the vector, the recombinant DNA must be introduced into bacterial cells by transformation (Chapter 2). The traditional way to do this is to incubate the cells in a concentrated calcium salt solution to make their membranes leaky, then mix these permeable cells with the DNA to allow the DNA entrance to the leaky cells. Alternatively, one can use high voltage to drive the DNA into cells—a process called **electroporation.**

Plasmids as Vectors In the early years of the cloning era, Boyer and his colleagues developed a set of very popular vectors known as the pBR plasmid series. Nowadays, one can choose from many plasmid cloning vectors besides the pBR plasmids. One useful, though somewhat dated, class of plasmids is the **pUC** series. These plasmids are based on pBR322, from which about 40% of the DNA has been deleted. Furthermore, the pUC vectors have many restriction sites clustered into one small area called a **multiple cloning site (MCS).** The pUC vectors contain an ampicillin resistance gene to allow selection for bacteria that have received a copy of the vector. Moreover, they have genetic elements that provide a convenient way of screening for clones that have recombinant DNAs.

The multiple cloning sites of the pUC vectors lie within a DNA sequence (called *lacZ′*) coding for the amino terminal portion (the α-peptide) of the enzyme β-galactosidase. The host bacteria used with the pUC vectors carry a gene fragment that encodes the carboxyl portion of β-galactosidase (the ω-peptide). By themselves, the β-galactosidase fragments made by these partial genes have no activity. But they can complement each other in vivo by so-called α-complementation. In other words, the two partial gene products can associate to form an active enzyme. Thus, when pUC18 by itself transforms a bacterial cell carrying the partial β-galactosidase gene, active β-galactosidase is produced. If these clones are plated on medium containing a β-galactosidase indicator, colonies with the pUC plasmid will turn color. The indicator X-gal, for instance, is a synthetic, colorless galactoside; when β-galactosidase cleaves X-gal, it releases galactose plus an indigo dye that stains the bacterial colony blue.

On the other hand, interrupting the plasmid's partial β-galactosidase gene by placing an insert into the multiple cloning site usually inactivates the gene. It can no longer make a product that complements the host cell's β-galactosidase fragment, so the X-gal remains colorless. Thus, it is a simple matter to pick the clones with inserts. They are the white ones; all the rest are blue. Notice that this is a one-step process. One looks simultaneously for a clone that (1) grows on ampicillin and (2) is white in the presence of X-gal. The multiple cloning sites have been carefully constructed to preserve the reading frame of β-galactosidase. Thus, even though the gene is interrupted by 18 codons, a functional protein still results. But further interruption by large inserts is usually enough to destroy the gene's function.

Even with the color screen, cloning into pUC can give false-positives, that is, white colonies without inserts. This can happen if the vector's ends are "nibbled" slightly by nucleases before ligation to the insert. Then, if these slightly degraded vectors simply close up during the ligation step, chances are that the *lacZ′* gene has been changed enough that white colonies will result. This underscores the importance of using clean DNA and enzymes that are free of nuclease activity.

This phenomenon of a vector religating with itself can be a greater problem when we use vectors that do not have a color screen, because then it is more difficult to distinguish colonies with inserts from those without. Even with pUC and related vectors, we would like to minimize vector religation. A good way to do this is to treat the vector with alkaline phosphatase, which removes the 5′-phosphates necessary for ligation. Without

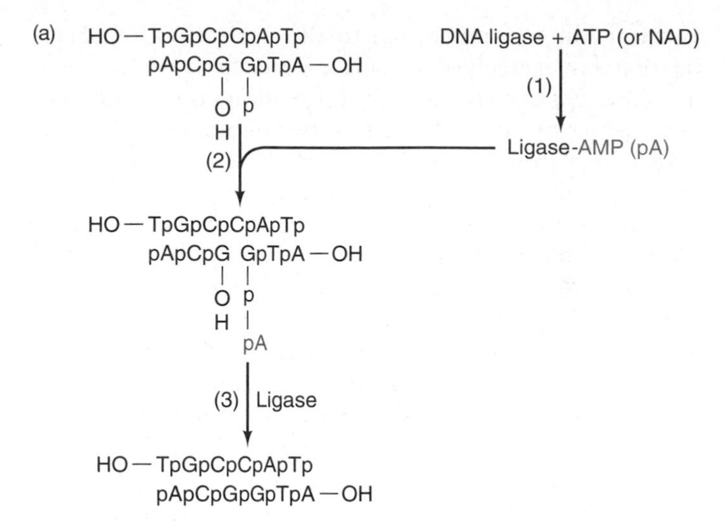

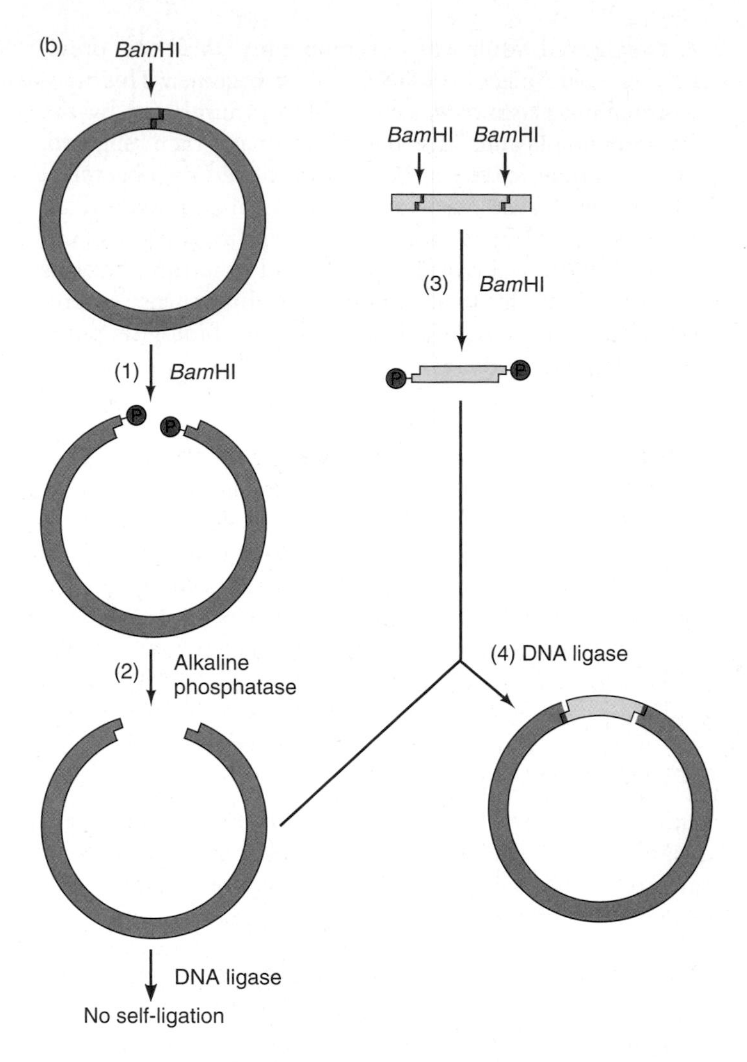

Figure 4.3 Joining of vector to insert. (a) Mechanism of DNA ligase. Step 1: DNA ligase reacts with an AMP donor—either ATP or NAD (nicotinamide adenine dinucleotide), depending on the type of ligase. This produces an activated enzyme (ligase-AMP). Step 2: The activated enzyme donates the AMP (blue) to the free 5′-phosphate (red) at the nick in the lower strand of the DNA duplex, creating a high-energy diphosphate group on one side of the nick. Step 3: With energy provided by cleavage of the bond between the phosphate groups, a new phosphodiester bond (red) is created, sealing the nick in the DNA. This reaction can occur in both DNA strands, so two independent DNAs can be joined together by DNA ligase. **(b)** Alkaline phosphatase prevents vector religation. Step 1: Cut the vector (blue, top left) with *Bam*HI. This produces sticky ends with 5′-phosphates (red). Step 2: Remove the phosphates with alkaline phosphatase, making it impossible for the vector to religate with itself. Step 3: Also cut the insert (yellow, upper right) with *Bam*HI, producing sticky ends with phosphates that are not removed. Step 4: Finally, ligate the vector and insert together. The phosphates on the insert allow two phosphodiester bonds to form (red), but leave two unformed bonds, or nicks. These are completed once the DNA is in the transformed bacterial cell.

these phosphates, the vector cannot ligate to itself, but can still ligate to the insert that retains its 5′-phosphates. Figure 4.3b illustrates this process. Notice that, because only the insert has phosphates, two nicks (unformed phosphodiester bonds) remain in the ligated product. These are not a problem; they will be sealed by DNA ligase in vivo once the ligated DNA has made its way into a bacterial cell.

The multiple cloning site also allows one to cut it with two different restriction enzymes (say, *Eco*RI and *Bam*HI) and then to clone a piece of DNA with one *Eco*RI end and one *Bam*HI end. This is called **directional cloning,** because the insert DNA is placed into the vector in only one orientation. (The *Eco*RI and *Bam*HI ends of the insert have to match their counterparts in the vector.) Knowing the orientation of an insert has certain benefits, which we will explore later in this chapter. Directional cloning also has the advantage of preventing the vector from simply religating by itself because its two restriction sites are incompatible. Even more convenient vectors than these are now available. We will discuss some of them later in this chapter.

SUMMARY Among the first generations of plasmid cloning vectors were pBR322 and the pUC plasmids. The latter have an ampicillin resistance gene and a multiple cloning site that interrupts a partial β-galactosidase gene. One screens for ampicillin-resistant clones that do not make active β-galactosidase and therefore do not turn the indicator, X-gal, blue. The multiple cloning site also makes it convenient to carry out directional cloning into two different restriction sites.

Phages as Vectors Bacteriophages are natural vectors that transduce bacterial DNA from one cell to another. It was only natural, then, to engineer phages to do the same thing for *all* kinds of DNA. Phage vectors have a natural advantage over plasmids: They infect cells much more efficiently than plasmids transform cells, so the yield of clones with phage vectors is usually higher. With phage vectors, clones are not colonies of cells, but **plaques** formed when a phage clears out a hole in a lawn of bacteria. Each plaque derives from a single phage that infects a cell, producing progeny phages that burst out of the cell, killing it and infecting surrounding cells. This process continues until a visible patch, or plaque, of dead cells appears. Because all the phages in the plaque derive from one original phage, they are all genetically identical—a clone.

λ Phage Vectors Fred Blattner and his colleagues constructed the first phage vectors by modifying the well-known λ phage (Chapter 8). They took out the region in the middle of the phage DNA, but retained the genes needed for phage replication. The missing phage genes could then be replaced with foreign DNA. Blattner named these vectors **Charon phages** after Charon, the boatman on the river Styx in classical mythology. Just as Charon carried souls to the underworld, the Charon phages carry foreign DNA into bacterial cells. Charon the boatman is pronounced "Karen," but Charon the phage is often pronounced "Sharon." A more general term for λ vectors such as Charon 4 is **replacement vectors** because λ DNA is removed and replaced with foreign DNA.

One clear advantage of the λ phages over plasmid vectors is that they can accommodate much more foreign DNA. For example, Charon 4 can accept up to about 20 kb of DNA, a limit imposed by the capacity of the λ phage head. By contrast, traditional plasmid vectors with inserts that large replicate poorly. When would one need such high capacity? A common use for λ replacement vectors is in constructing **genomic libraries.** Suppose we wanted to clone the entire human genome. This would obviously require a great many clones, but the larger the insert in each clone, the fewer total clones would be needed. In fact, such genomic libraries have been constructed for the human genome and for genomes of a variety of other organisms, and λ replacement vectors have been popular vectors for this purpose.

Aside from their high capacity, some of the λ vectors have the advantage of a minimum size requirement for their inserts. Figure 4.4 illustrates the reason for this requirement: To get the Charon 4 vector ready to accept an insert, it can be cut with *Eco*RI. This cuts at three sites near the middle of the phage DNA, yielding two "arms" and two "stuffer" fragments. Next, the arms are purified by gel electrophoresis or ultracentrifugation and the stuffers are discarded. The final step is to ligate the arms to the insert, which then takes the place of the discarded stuffers.

At first glance, it may appear that the two arms could simply ligate together without accepting an insert. Indeed, this happens, but it does not produce a clone, because the two arms constitute too little DNA and will not be packaged into a phage. The packaging is done in vitro when the recombinant DNA is mixed with all the components needed to put together a phage particle. Nowadays one can buy the purified λ arms, as well as the packaging extract in cloning kits. The extract has rather stringent requirements as to the size of DNA it will package. It must have at least 12 kb of DNA in addition to λ arms, but no more than 20 kb.

Because each clone has at least 12 kb of foreign DNA, the library does not waste space on clones that contain insignificant amounts of DNA. This is an important consideration because, even at 12–20 kb per clone, the library needs at least half a million clones to ensure that each human gene is represented at least once. It would be much more difficult to make a human genomic library in pBR322 or a pUC vector because bacteria selectively take up and reproduce small plasmids. Therefore, most of the clones would contain inserts of a few thousand, or even just a few hundred base pairs. Such a library would have to contain many millions of clones to be complete.

Because *Eco*RI produces fragments with an average size of about 4 kb, but the vector will not accept any inserts smaller than 12 kb, the DNA cannot be completely cut with *Eco*RI, or most of the fragments will be too small to clone. Furthermore, *Eco*RI, and most other restriction enzymes, cut in the middle of most eukaryotic genes one or more times, so a complete digest would contain only fragments of most genes. One can minimize these problems by performing an incomplete digestion with *Eco*RI (using a low concentration of enzyme or a short reaction time, or both). If the enzyme cuts only about every fourth or fifth site, the average length of the resulting fragments will be about 16–20 kb, just the size the vector will accept and big enough to include the entirety of most eukaryotic genes. If we want a more random set of fragments, we can also use mechanical means such as ultrasound instead of a restriction endonuclease to shear the DNA to an appropriate size for cloning.

A genomic library is very handy. Once it is established, one can search for any gene of interest. The only problem is that no catalog exists for such a library to help find particular clones, so some kind of probe is needed to show which clone contains the gene of interest. An ideal probe would be a labeled nucleic acid whose sequence matches that of the gene of interest. One would then carry out a **plaque hybridization** procedure in which the DNA from each of the thousands of λ phages from the library is hybridized to the labeled probe. The plaque with the DNA that forms a labeled hybrid is the right one.

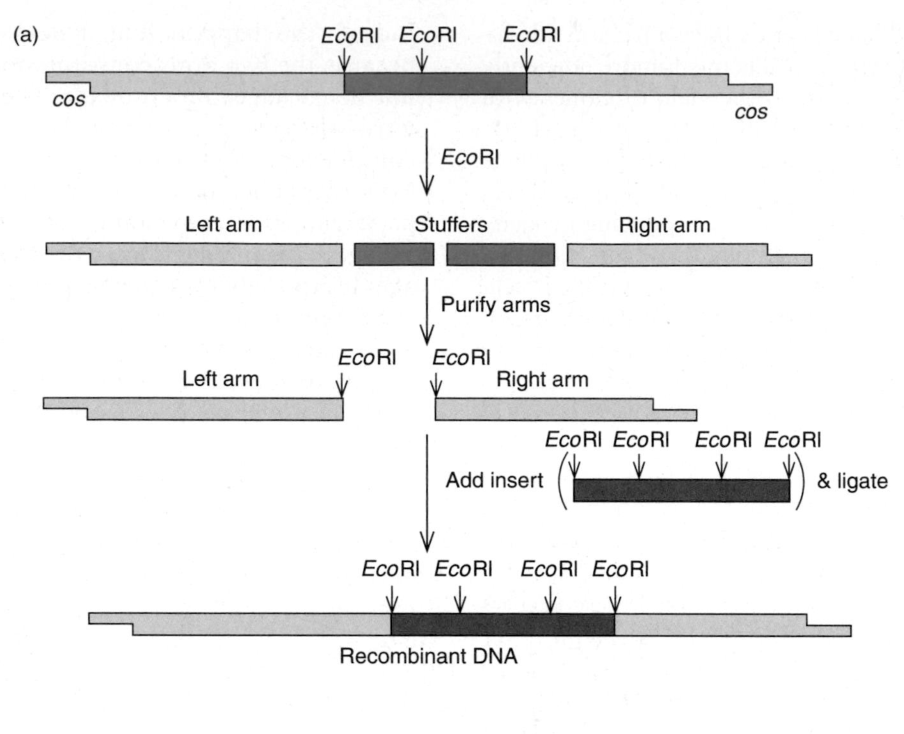

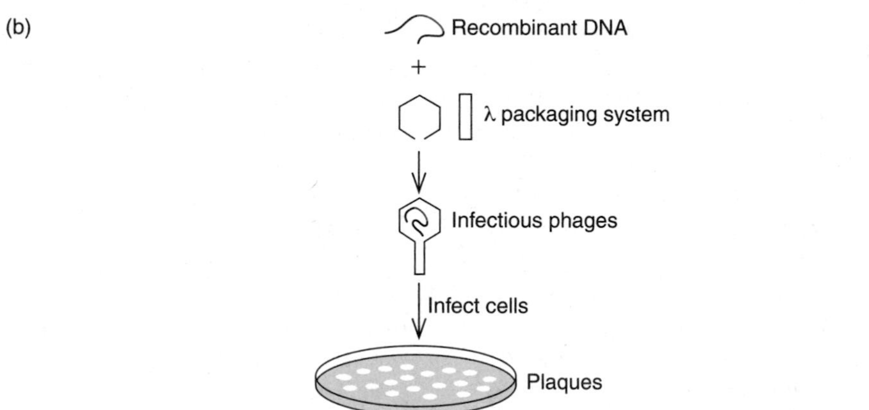

Figure 4.4 Cloning in Charon 4. (a) Forming the recombinant DNA. Cut the vector (yellow and blue) with *Eco*RI to remove the stuffer fragments (blue) and save the arms. Next, ligate partially digested insert DNA (red) to the arms. The extensions of the ends are 12-base cohesive ends (*cos* sites), whose size is exaggerated here. **(b)** Packaging and cloning the recombinant DNA. Mix the recombinant DNA from part (a) with an in vitro packaging extract that contains λ phage head and tail components and all other factors needed to package the recombinant DNA into functional phage particles. Finally, plate these particles on *E. coli* and collect the plaques that form.

We have encountered hybridization before in Chapter 2, and we will discuss it again in Chapter 5. Figure 4.5 shows how plaque hybridization works. Thousands of plaques are grown on each of several Petri dishes (only a few plaques are shown here for simplicity). Next, a filter made of a DNA-binding material such as **nitrocellulose** or coated nylon is touched to the surface of the Petri dish. This transfers some of the phage DNA from each plaque to the filter. The DNA is then denatured with alkali and hybridized to the labeled probe. Before the probe is added, the filter is saturated with a nonspecific DNA or protein to prevent nonspecific binding of the probe. When the probe encounters complementary DNA, which should be only the DNA from the clone of interest, it will hybridize, labeling that DNA spot. This labeled spot is then detected with x-ray film. The black spot on the film shows where to look on the original Petri dish for the plaque containing the gene of interest. In practice, the original plate may be so crowded with plaques that it is impossible to pick out the right one, so several plaques can be picked from that area, replated at a much lower phage density, and the hybridization process can be repeated to find the positive clone.

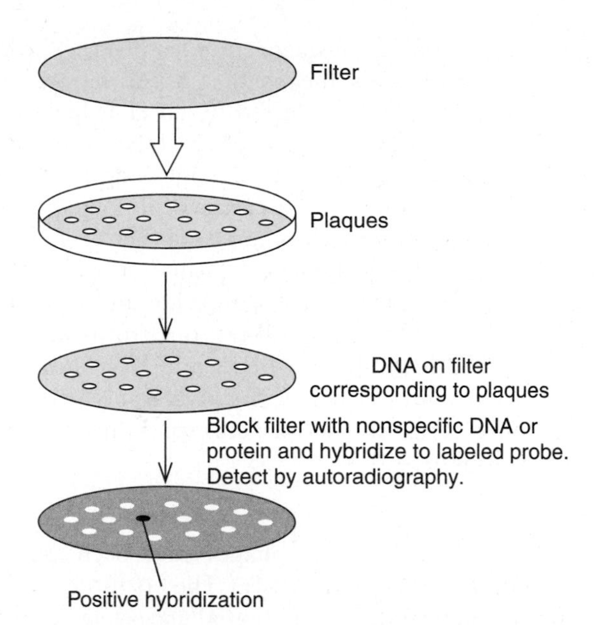

Figure 4.5 Selection of positive genomic clones by plaque hybridization. First, touch a nitrocellulose or similar filter to the surface of the dish containing the Charon 4 plaques from Figure 4.4. Phage DNA released naturally from each plaque sticks to the filter. Next, denature the DNA with alkali and hybridize the filter to a labeled probe for the gene under study, then use x-ray film to reveal the position of the label. Cloned DNA from one plaque near the center of the filter has hybridized, as shown by the dark spot on the film.

We have introduced λ phage vectors as agents for genomic cloning. But other types of λ vectors are very useful for making another kind of library—a cDNA library—as we will learn later in this chapter.

Cosmids Another vector designed especially for cloning large DNA fragments is called a **cosmid.** Cosmids behave both as plasmids and as phages. They contain the *cos* sites, or cohesive ends, of λ phage DNA, which allow the DNA to be packaged into λ phage heads (hence the "cos" part of the name "cosmid"). They also contain a plasmid origin of replication, so they can replicate as plasmids in bacteria (hence the "mid" part of the name).

Because almost the entire λ genome, except for the *cos* sites, has been removed from the cosmids, they have room for large inserts (40–50 kb). Once these inserts are in place, the recombinant cosmids are packaged into phage particles in vitro. These particles cannot replicate as phages because they have almost no phage DNA, but they are infectious, so they carry their recombinant DNA into bacterial cells. Once inside, the DNA can replicate as a plasmid because it has a plasmid origin of replication.

M13 Phage Vectors Another phage used as a cloning vector is the filamentous (long, thin, filament-like) phage M13. Joachim Messing and his coworkers endowed the phage DNA with the same β-galactosidase gene fragment and multiple cloning sites found in the pUC family of vectors. In fact, the M13 vectors were engineered first; then the useful cloning sites were simply transferred to the pUC plasmids.

What is the advantage of the M13 vectors? The main factor is that the genome of this phage is a single-stranded DNA, so DNA fragments cloned into this vector can be recovered in single-stranded form. As we will see later in this chapter, single-stranded DNA can be an aid to site-directed mutagenesis, by which we can introduce specific, premeditated alterations into a gene.

Figure 4.6 illustrates how to clone a double-stranded piece of DNA into M13 and harvest a single-stranded

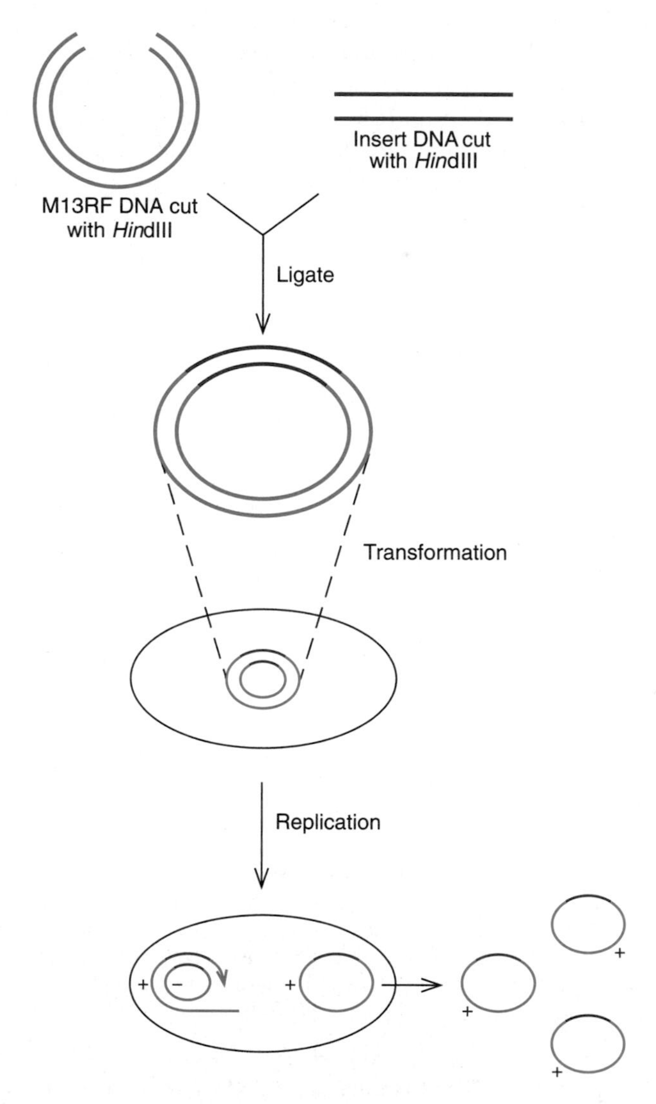

Figure 4.6 Obtaining single-stranded DNA by cloning in M13 phage. Foreign DNA (red), cut with *Hin*dIII, is inserted into the *Hin*dIII site of the double-stranded phage DNA. The resulting recombinant DNA is used to transform *E. coli* cells, whereupon the DNA replicates, producing many single-stranded product DNAs. The product DNAs are called positive (+) strands, by convention. The template DNA is therefore the negative (−) strand.

DNA product. The DNA in the phage particle itself is single-stranded, but after infecting an *E. coli* cell, the DNA is converted to a double-stranded replicative form (RF). This double-stranded replicative form of the phage DNA is used for cloning. After it is cut with one or two restriction enzymes at its multiple cloning site, foreign DNA with compatible ends can be inserted. This recombinant DNA is then used to transform host cells, giving rise to progeny phages that bear single-stranded recombinant DNA. The phage particles, containing phage DNA, are secreted from the transformed cells and can be collected from the growth medium.

Phagemids Another class of vectors that produce single-stranded DNA has also been developed. These are like the cosmids in that they have characteristics of both phages and plasmids; thus, they are called **phagemids.** One popular variety (Figure 4.7) goes by the trade name pBluescript (pBS). Like the pUC vectors, pBluescript has a multiple cloning site inserted into the *lacZ'* gene, so clones with inserts can be distinguished by white versus blue staining with X-gal. This vector also has the origin of replication of the single-stranded phage f1, which is related to M13. This means that a cell harboring a recombinant phagemid, if infected by an f1 helper phage that supplies the single-stranded phage DNA replication machinery, will produce and package single-stranded phagemid DNA. A final useful feature of this class of vectors is that the multiple cloning site is flanked by two different phage RNA polymerase promoters. For example, pBS has a T3 promoter on one side and a T7 promoter on the other. This allows one to isolate the double-stranded recombinant phagemid DNA and transcribe it in vitro with either of the phage polymerases to produce pure RNA transcripts corresponding to either strand of the insert.

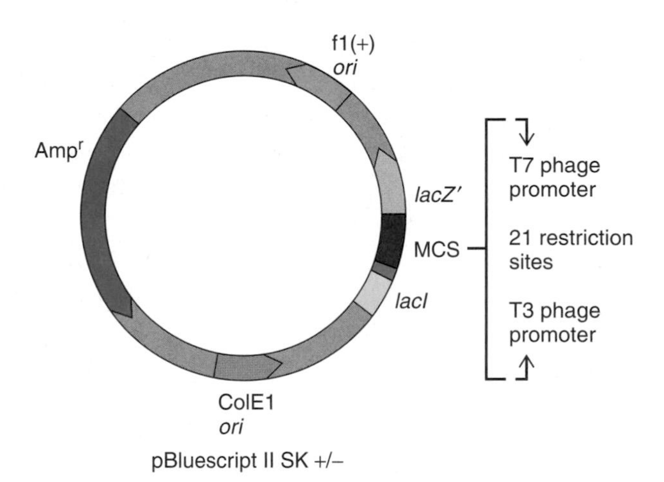

Figure 4.7 The pBluescript vector. This plasmid is based on pBR322 and has that vector's ampicillin resistance gene (green) and origin of replication (purple). In addition, it has the phage f1 origin of replication (orange). Thus, if the cell is infected by an f1 helper phage to provide the replication machinery, single-stranded copies of the vector can be packaged into progeny phage particles. The multiple cloning site (MCS, red) contains 21 unique restriction sites situated between two phage RNA polymerase promoters (T7 and T3). Thus, any DNA insert can be transcribed in vitro to yield an RNA copy of either strand, depending on which phage RNA polymerase is provided. The MCS is embedded in an *E. coli lacZ'* gene (blue), so the uncut plasmid will produce the β-galactosidase N-terminal fragment when an inducer such as isopropylthiogalactoside (IPTG) is added to counteract the repressor made by the *lacI* gene (yellow). Thus, clones bearing the uncut vector will turn blue when the indicator X-gal is added. By contrast, clones bearing recombinant plasmids with inserts in the MCS will have an interrupted *lacZ'* gene, so no functional β-galactosidase is made. Thus, these clones remain white.

SUMMARY Two kinds of phages have been especially popular as cloning vectors. The first of these is λ, from which certain nonessential genes have been removed to make room for inserts. Some of these engineered phages can accommodate inserts up to 20 kb, which makes them useful for building genomic libraries, in which it is important to have large pieces of genomic DNA in each clone. Cosmids can accept even larger inserts—up to 50 kb—making them a favorite choice for genomic libraries. The second major class of phage vectors consists of the M13 phages. These vectors have the convenience of a multiple cloning site and the further advantage of producing single-stranded recombinant DNA, which can be used for DNA sequencing and for site-directed mutagenesis. Plasmids called phagemids have also been engineered to produce single-stranded DNA in the presence of helper phages.

Eukaryotic Vectors and Very High Capacity Vectors Several very useful vectors have been designed for cloning genes into eukaryotic cells. Later in this chapter, we will consider some vectors that are designed to yield the protein products of genes in eukaryotes. We will also introduce vectors based on the Ti plasmid of *Agrobacterium tumefaciens* that can carry genes into plant cells. In Chapter 24 we will discuss vectors known as yeast artificial chromosomes (YACs) and bacterial artificial chromosomes (BACs) designed for cloning huge pieces of DNA (up to hundreds of thousands of base pairs).

Identifying a Specific Clone with a Specific Probe

We have already mentioned the need for a probe to identify a desired clone among the thousands of irrelevant ones. What sort of probe could be employed? Two different kinds are widely used: polynucleotides (or oligonucleotides) and antibodies. Both are molecules able to bind very specifically to other molecules. We will discuss polynucleotide probes here and antibody probes later in this chapter.

Polynucleotide Probes To probe for the gene you want, you might use the homologous gene from another organism if someone has already cloned it. You would hope the two genes have enough similarity in sequence that one would hybridize to the other. This hope is usually fulfilled. However, you generally have to lower the **stringency** of the hybridization conditions so that the hybridization reaction can tolerate some mismatches in base sequence between the probe and the cloned gene.

Researchers use several means to control stringency. High temperature, high organic solvent concentration, and low salt concentration all tend to promote the separation of the two strands in a DNA double helix. You can therefore adjust these conditions until only perfectly matched DNA strands will form a duplex; this is high stringency. By relaxing these conditions (lowering the temperature, for example), you lower the stringency until DNA strands with a few mismatches can hybridize.

Without homologous DNA from another organism, what could you use? There is still a way out if you know at least part of the sequence of the protein product of the gene. We faced a problem just like this in our lab when we cloned the gene for a plant toxin known as ricin. Fortunately, the entire amino acid sequences of both polypeptides of ricin were known. That meant we could examine the amino acid sequence and, using the genetic code, deduce a set of nucleotide sequences that would code for these amino acids. Then we could construct these nucleotide sequences chemically and use these synthetic probes to find the ricin gene by hybridization. The probes in this kind of procedure are strings of several nucleotides, so they are called **oligonucleotides.** Why did we have to use more than one oligonucleotide to probe for the ricin gene? The genetic code is degenerate, which means that most amino acids are encoded by more than one triplet codon. Thus, we had to consider several different nucleotide sequences for most amino acids.

Fortunately, we were spared some inconvenience because one of the polypeptides of ricin includes this amino acid sequence: Trp-Met-Phe-Lys-Asn-Glu. The first two amino acids in this sequence have only one codon each, and the next three only two each. The sixth gives us two extra bases because the degeneracy occurs only in the third base. Thus, we had to make only eight 17-base oligonucleotides *(17-mers)* to be sure of getting the exact coding sequence for this string of amino acids. This degenerate sequence can be expressed as follows:

		U	G	U	
UGG	AUG	UUC	AAA	AAC	GA
Trp	Met	Phe	Lys	Asn	Glu

Using this mixture of eight 17-mers (UGGAUGUUCAAAAACGA, UGGAUGUUUAAAAACGA, etc.), we quickly identified several ricin-specific clones. Nowadays, so many genomes have been sequenced that we already know the sequences of many genes. Probes with these exact sequences can therefore be synthesized.

Solved Problem

Problem

Here is the amino acid sequence of part of a hypothetical protein whose gene you want to clone:

Arg-Leu-Met-Glu-Trp-Ile-Cys-Pro-Met-Leu

a. What sequence of five amino acids would give a 17-mer probe (including two bases from the next codon) with the least degeneracy?

b. How many different 17-mers would you have to synthesize to be sure your probe matches the corresponding sequence in your cloned gene perfectly?

c. If you started your probe two codons to the right of the optimal one (the one you chose in part *a*), how many different 17-mers would you have to make?

Solution

a. Begin by consulting the genetic code (Chapter 18) to determine the coding degeneracy of each amino acid in the sequence. This yields

6 6 1 2 1 3 2 4 1 6
Arg-Leu-Met-Glu-Trp-Ile-Cys-Pro-Met-Leu

where the numbers above the amino acids represent the coding degeneracy for each. In other words, arginine has six codons, leucine six, methionine one, and so on. Now the task is to find the contiguous set of five codons with the lowest degeneracy. A quick inspection shows that Met-Glu-Trp-Ile-Cys works best.

b. To find how many different 17-mers you would have to prepare, multiply the degeneracies at all positions within the region covered by your probe. For the five amino acids you have chosen, this is $1 \times 2 \times 1 \times 3 \times 2 = 12$. Note that you can use the first two bases (CC) in the proline (Pro) codons without encountering any degeneracy because the fourfold degeneracy in coding for proline all occurs in the third base in the codon (CCU, CCA, CCC, CCG). Thus, your probe can be 17 bases long, instead of the 15 bases you get from the codons for the five amino acids selected.

c. If you had started two amino acids farther to the right, starting with Trp, the degeneracy would have been $1 \times 3 \times 2 \times 4 \times 1 = 24$, so you would have had to prepare 24 different probes instead of just 12. ■

SUMMARY Specific clones can be identified using polynucleotide probes that bind to the gene itself. Knowing the amino acid sequence of a gene product, one can design a set of oligonucleotides that encode part of this amino acid sequence. This can be one of the quickest and most accurate means of identifying a particular clone.

cDNA Cloning

A **cDNA** (short for **complementary DNA** or **copy DNA**) is a DNA copy of an RNA, usually an mRNA. Sometimes we want to make a **cDNA library,** a set of clones representing as many as possible of the mRNAs in a given cell type at a given time. Such libraries can contain tens of thousands of different clones. Other times, we want to make one particular cDNA—a clone containing a DNA copy of just one mRNA. The technique we use depends in part on which of these goals we wish to achieve.

Figure 4.8 illustrates one simple, yet effective method for making a cDNA library. The central part of any cDNA cloning procedure is synthesis of the cDNA from an mRNA template using **reverse transcriptase** (RNA-dependent DNA polymerase). Reverse transcriptase is like any other DNA-synthesizing enzyme in that it cannot initiate DNA synthesis without a primer. To get around this problem, we take advantage of the poly(A) tail at the 3′-end of most eukaryotic mRNAs and use oligo(dT) as the primer. The oligo(dT) is complementary to poly(A), so it binds to the poly(A) at the 3′-end of the mRNA and primes DNA synthesis, using the mRNA as the template.

After the mRNA has been copied, yielding a single-stranded DNA (the "first strand"), the mRNA is partially degraded with **ribonuclease H (RNase H).** This enzyme degrades the RNA strand of an RNA–DNA hybrid—just what we need to begin to digest the RNA base-paired to the first-strand cDNA. The remaining RNA fragments serve as primers for making the "second strand," using the first as the template. This phase of the process depends on a phenomenon called **nick translation,** which is illustrated in Figure 4.9. The net result is a double-stranded cDNA with a small fragment of RNA at the 5′-end of the second strand.

The essence of nick translation is the simultaneous removal of DNA ahead of a **nick** (a single-stranded DNA break) and synthesis of DNA behind the nick, rather like a road paving machine that tears up old pavement at its front end and lays down new pavement at its back end. The net result is to move, or "translate," the nick in the 5′→3′ direction. The enzyme usually used for nick translation is *E. coli* DNA polymerase I, which has a 5′→3′ exonuclease activity that allows the enzyme to degrade DNA ahead of the nick as it moves along.

The next task is to ligate the cDNA to a vector. This was easy with pieces of genomic DNA cleaved with restriction

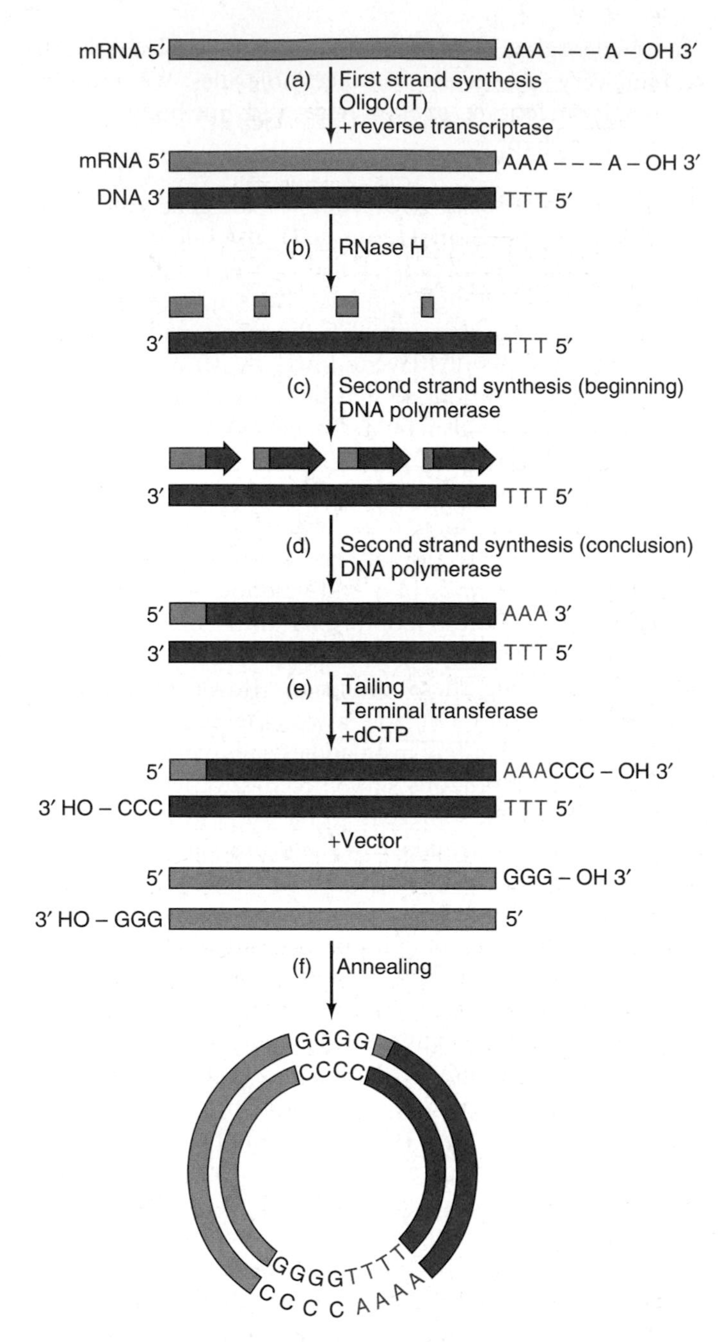

Figure 4.8 Making a cDNA library. (a) Use oligo(dT) as a primer and reverse transcriptase to copy the mRNA (blue), producing a cDNA (red) that is hybridized to the mRNA template. **(b)** Use RNase H to partially digest the mRNA, yielding a set of RNA primers base-paired to the first-strand cDNA. **(c)** Use *E. coli* DNA polymerase I to build second-strand cDNAs on the RNA primers. **(d)** The second-strand cDNA growing from the leftmost primer (blue) has been extended all the way to the 3′-end of the oligo(dA) corresponding to the oligo(dT) primer on the first-strand cDNA. **(e)** To place sticky ends on the double-stranded cDNA, add oligo(dC) with terminal transferase. **(f)** Anneal the oligo(dC) ends of the cDNA to complementary oligo(dG) ends of a suitable vector (purple). The recombinant DNA can then be used to transform bacterial cells. Enzymes in these cells remove remaining nicks and replace any remaining RNA with DNA.

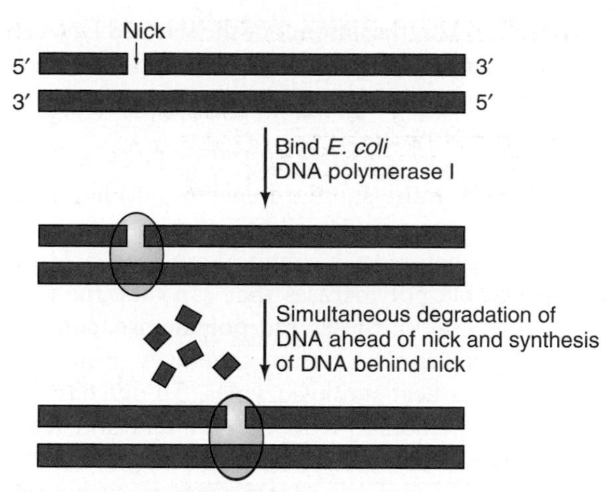

Figure 4.9 Nick translation. This illustration is a generic example with double-stranded DNA, but the same principles apply to an RNA–DNA hybrid. Beginning with a double-stranded DNA with a nick in the top strand, *E. coli* DNA polymerase I binds to this nick and begins elongating the DNA fragment on the top left in the 5′→3′ direction (left to right). At the same time, the 5′→3′ exonuclease activity degrades the DNA fragment to its right to make room for the growing fragment behind it. The small red rectangles represent nucleotides released by exonuclease digestion of the DNA.

enzymes, but cDNAs have no sticky ends. It is true that blunt ends can be ligated together, even though the process is relatively inefficient. However, to get the efficient ligation afforded by sticky ends, one can create sticky ends (oligo[dC] in this case) on the cDNA, using an enzyme called **terminal deoxynucleotidyl transferase (TdT)** or simply **terminal transferase** and one of the deoxyribonucleoside triphosphates. In this case, dCTP was used. The enzyme adds dCMPs, one at a time, to the 3′-ends of the cDNA. In the same way, oligo(dG) ends can be added to a vector. Annealing the oligo(dC) ends of the cDNA to the oligo(dG) ends of the vector brings the vector and cDNA together in a recombinant DNA that can be used directly for transformation. The base pairing between the oligonucleotide tails is strong enough that no ligation is required before transformation. The DNA ligase inside the transformed cells finally performs the ligation, and DNA polymerase I removes any remaining RNA and replaces it with DNA.

What kind of vector should be used to ligate to a cDNA or cDNAs? Several choices are available, depending on the method used to detect positive clones (those that bear the desired cDNA). A plasmid or phagemid vector such as pUC or pBS can be used; if so, positive clones are usually identified by **colony hybridization** with a labeled DNA probe. This procedure is analogous to the plaque hybridization described previously. Or one can use a λ phage, such as λgt11, as a vector. This vector places the cloned cDNA under the control of a *lac* promoter, so that transcription and translation of the cloned gene can occur. One can then use an antibody to screen directly for the protein product of the correct gene. We will describe this procedure in more detail later in this chapter. Alternatively, a polynucleotide probe can be used to hybridize to the recombinant phage DNA.

Rapid Amplification of cDNA Ends

Very frequently, a cDNA is not full-length, possibly because the reverse transcriptase, for whatever reason, did not make it all the way to the end of the mRNA. This does not mean one has to be satisfied with an incomplete cDNA, however. Fortunately, one can fill in the missing pieces of a cDNA, using a procedure called **rapid amplification of cDNA ends (RACE)**. Figure 4.10 illustrates the technique (5′-RACE) for filling in the 5′-end of a cDNA (the usual

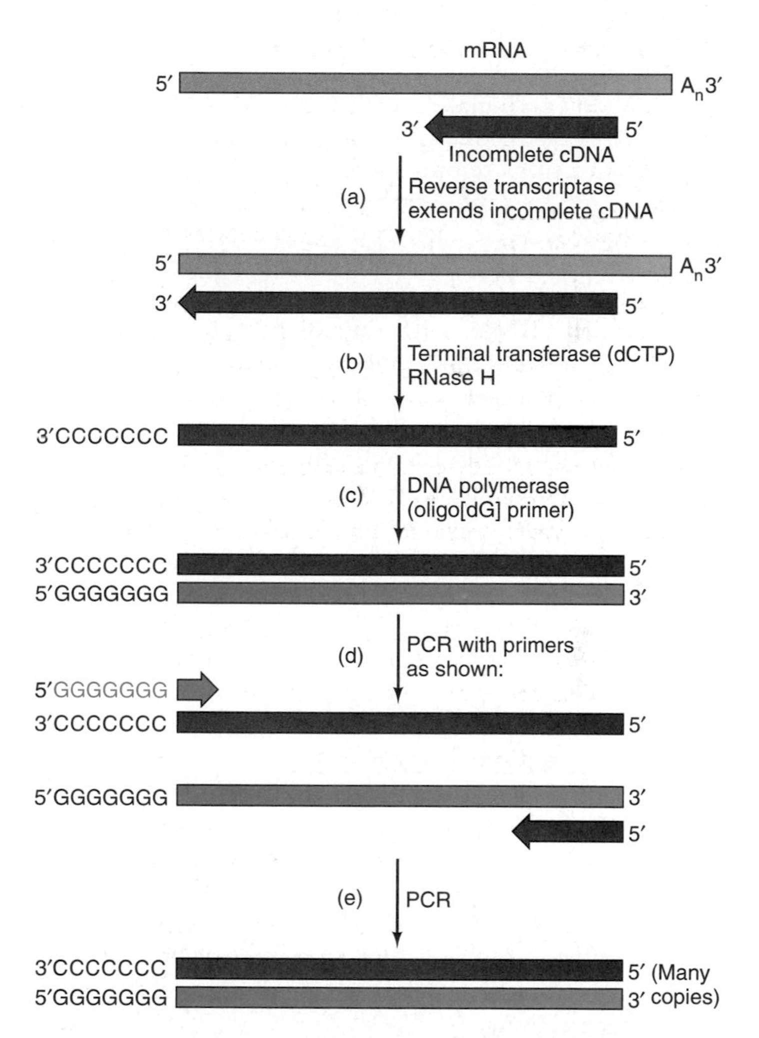

Figure 4.10 RACE procedure to fill in the 59-end of a cDNA. **(a)** Hybridize an incomplete cDNA (red), or an oligonucleotide segment of a cDNA to mRNA (green), and use reverse transcriptase to extend the cDNA to the 5′-end of the mRNA. **(b)** Use terminal transferase and dCTP to add C residues to the 3′-end of the extended cDNA; also, use RNase H to degrade the mRNA. **(c)** Use an oligo(dG) primer and DNA polymerase to synthesize a second strand of cDNA (blue). **(d)** and **(e)** Perform PCR with oligo(dG) as the forward primer and an oligonucleotide that hybridizes to the 3′-end of the cDNA as the reverse primer. The product is a cDNA that has been extended to the 5′-end of the mRNA. A similar procedure (3′-RACE) can be used to extend the cDNA in the 3′-direction. In that case, there is no need to tail the 3′-end of the cDNA with terminal transferase because the mRNA already contains poly(A); thus, the reverse primer would be oligo(dT).

problem), but an analogous technique (3′-RACE) can be used to fill in a missing 3′-end of a cDNA.

A 5′-RACE procedure begins with an RNA preparation containing the mRNA of interest and a partial cDNA whose 5′-end is missing. An incomplete strand of the cDNA can be annealed to the mRNA and then reverse transcriptase can be used to copy the rest of the mRNA. Then the completed cDNA can be tailed with oligo(dC) (for example), using terminal transferase and dCTP. Next, oligo(dG) is used to prime second-strand synthesis. This step produces a double-stranded cDNA that can be amplified by PCR, using oligo(dG) and a 3′-specific oligonucleotide as primers.

SUMMARY To make a cDNA library, one can synthesize cDNAs one strand at a time, using mRNAs from a cell as templates for the first strands and these first strands as templates for the second strands. Reverse transcriptase generates the first strands and *E. coli* DNA polymerase I generates the second strands. One can endow the double-stranded cDNAs with oligonucleotide tails that base-pair with complementary tails on a cloning vector, then use these recombinant DNAs to transform bacteria. Particular clones can be detected by colony hybridization with radioactive DNA probes, or with antibodies if an expression vector such as λgt11 is used. Incomplete cDNA can be filled in by 5′- or 3′-RACE.

4.2 The Polymerase Chain Reaction

We have now seen how to clone fragments of DNA generated by cleavage with restriction endonucleases, or by physical shearing of DNA, and we have examined a classical technique for cloning cDNAs. But a newer technique, called **polymerase chain reaction (PCR)**, can also yield a DNA fragment for cloning and is especially useful for cloning cDNAs.

Standard PCR

PCR was invented by Kary Mullis and his colleagues in the 1980s. As Figure 4.11 explains, this technique uses the enzyme DNA polymerase to make a copy of a selected region of DNA. Mullis and colleagues chose the part (X) of the DNA they wanted to amplify by putting in short pieces of DNA (primers) that hybridized to DNA sequences on each side of X and caused initiation (priming) of DNA synthesis through X. The copies of both strands of X, as well as the original DNA strands, then serve as templates for the next round of synthesis. In this way, the amount of the selected DNA region doubles over and over with each cycle—up to millions of times the starting amount—until enough is present to be seen by gel electrophoresis.

Originally, workers had to add fresh DNA polymerase at every round because standard enzymes do not stand up to the high temperatures (over 90°C) needed to separate the strands of DNA before each round of replication. However, special heat-stable polymerases that can take the heat are now available. One of these, **Taq polymerase,** comes from *Thermus aquaticus,* a bacterium that lives in hot springs and therefore has heat-stable enzymes. All one has to do is mix the Taq polymerase with the primers and template DNA in a test tube, seal the tube, then place it in a **thermal cycler.** The thermal cycler is programmed to cycle over and over again among three different temperatures: first a high temperature (about 95°C) to separate the DNA strands; then a relatively low temperature (about 50°C) to allow the primers to anneal to the template DNA strands; then a medium temperature (about 72°C) to allow DNA synthesis. Each cycle takes as little as a few minutes, and it usually takes fewer than 20 cycles to produce as much amplified DNA as one needs. PCR is such a powerful amplifying device that it has even helped spawn science fiction stories such as *Jurassic Park* (see Box 4.1).

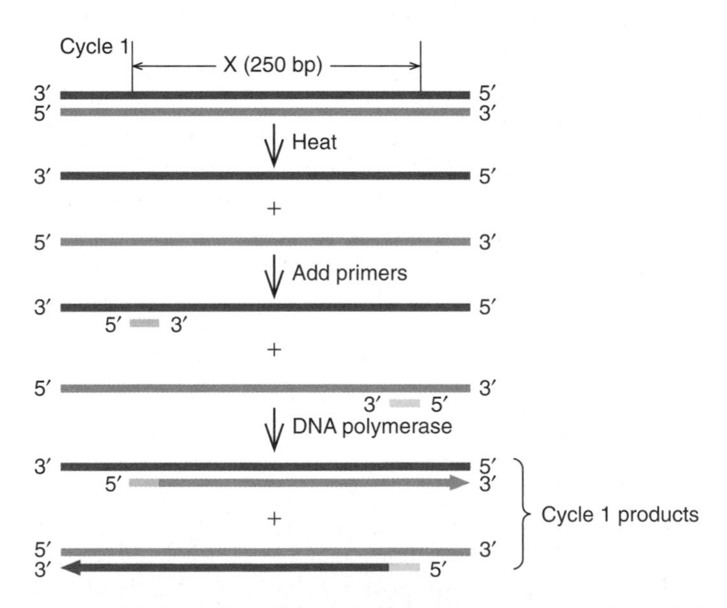

Figure 4.11 Amplifying DNA by the polymerase chain reaction. Start with a DNA duplex (top) and heat it to separate its two strands (red and blue). Then add short, single-stranded DNA primers (purple and yellow) complementary to sequences on either side of the region (X, 250 bp) to be amplified. The primers hybridize to the appropriate sites on the separated DNA strands; now a special heat-stable DNA polymerase uses these primers to start synthesis of complementary DNA strands. The arrows represent newly made DNA in which replication has stopped at the tip of the arrowhead. At the end of cycle 1, two DNA duplexes are present, including the region to be amplified, whereas we started with only one. The 5′→3′ polarities of all DNA strands and primers are indicated. The same principles apply in every cycle thereafter.

BOX 4.1

Jurassic Park: More than a Fantasy?

In Michael Crichton's book *Jurassic Park,* and in the movie of the same name, a scientist and an entrepreneur collaborate in a fantastic endeavor: to generate living dinosaurs. Their strategy is to isolate dinosaur DNA, but not directly from dinosaur remains, from which DNA would be impossible to get. Instead they find Jurassic-period blood-sucking insects that had feasted on dinosaur blood and had then become mired in tree sap, which had turned to amber, entombing and preserving the insects. They reason that, because blood contains blood cells that have DNA, the insect gut contents should contain this dinosaur DNA. The next step is to use PCR to amplify the dinosaur DNA, piece the fragments together, place them in an egg, and *voila!* A dinosaur is hatched.

This scenario sounds preposterous, and indeed certain practical problems keep it totally in the realm of science fiction. But it is striking that some parts of the story are already in the scientific literature. In June 1993, the same month that *Jurassic Park* opened in movie theaters, a paper appeared in the journal *Nature* describing the apparent PCR amplification and sequencing of part of a gene from an extinct weevil trapped in Lebanese amber for 120–135 million years. That takes us back to the Cretaceous period, not quite as ancient as the Jurassic, but a time when plenty of dinosaurs were still around. If this work were valid, it would indeed be possible to find preserved, blood-sucking insects with dinosaur DNA in their guts. Furthermore, it would be possible that this DNA would still be intact enough that it could serve as a template for PCR amplification. After all, the PCR technique is powerful enough to start with a single molecule of DNA and amplify it to any degree we wish.

So what stands in the way of making dinosaurs? Leaving aside the uncharted territory of creating a vertebrate animal from naked DNA, we have to consider first the simple limitations of the PCR process itself. One of these is the present limit to the size of a DNA fragment that we can amplify by PCR: up to 40 kb. That is probably on the order of one-hundred thousandth the size of the whole dinosaur genome, which means that we would ultimately have to piece together at least a hundred thousand PCR fragments to reconstitute the whole genome. And that assumes that we know enough about the sequence of the dinosaur DNA, at the start, to make PCR primers for all of those fragments.

But what if we worked out a way to make PCR go much farther than 40,000 bp? What if PCR became so powerful that we could amplify whole chromosomes, up to hundreds of millions of base pairs at a time? Then we would run up against the fact that DNA is an inherently unstable molecule, and no full-length chromosomes would be expected to survive for millions of years, even in an insect embalmed in amber. PCR can amplify relatively short stretches because the primers need to find only one molecule that is unbroken over that short stretch. But finding a whole unbroken chromosome, or even an unbroken megabase-size stretch of DNA, appears to be impossible.

These considerations have generated considerable uncertainty about the few published examples of amplifying ancient DNA by PCR. Many scientists argue that it is simply not credible that a molecule as fragile as DNA can last for millions of years. They believe that dinosaur DNA would long ago have decomposed into nucleotides and be utterly useless as a template for PCR amplification. Indeed, this appears to be true for all ancient DNA more than about 100,000 years old.

On the other hand, the PCR procedure has amplified some kind of DNA from the ancient insect samples. If it is not ancient insect DNA, what is it? This brings us to the second limitation of the PCR method, which is also its great advantage: its exquisite sensitivity. As we have seen, PCR can amplify a single molecule of DNA, which is fine if that is the DNA we want to amplify. It can, however, also seize on tiny quantities—even single molecules—of contaminating DNAs in our sample and amplify them instead of the DNA we want.

For this reason, the workers who examined the Cretaceous weevil DNA did all their PCR amplification and sequencing on that DNA before they even began work on modern insect DNA, to which they compared the weevil sequences. That way, they minimized the worry that they were amplifying trace contaminants of modern insect DNA left over from previous experiments, when they thought they were amplifying DNA from the extinct weevil. But DNA is everywhere, especially in a molecular biology lab, and eliminating every last molecule is agonizingly difficult.

Furthermore, dinosaur DNA in the gut of an insect would be heavily contaminated with insect DNA, not to mention DNA from intestinal bacteria. And who is to say the insect fed on only one type of dinosaur before it died in the tree sap? If it fed on two, the PCR procedure would probably amplify both their DNAs together, and there would be no way to separate them.

In other words, some of the tools to create a real Jurassic Park are already in hand, but, as exciting as it is to imagine seeing a living dinosaur, the practical problems make it seem impossible.

On a far more realistic level, the PCR technique is already allowing us to compare the sequences of genes from extinct organisms with those of their present-day relatives. And this is spawning an exciting new field, which botanist Michael Clegg calls "molecular paleontology."

SUMMARY PCR amplifies a region of DNA between two predetermined sites. Oligonucleotides complementary to these sites serve as primers for synthesis of copies of the DNA between the sites. Each cycle of PCR doubles the number of copies of the amplified DNA until a large quantity has been made.

Using Reverse Transcriptase PCR (RT-PCR) in cDNA Cloning

If one wants to clone a cDNA from just one mRNA whose sequence is known, one can use a type of PCR called **reverse transcriptase PCR (RT-PCR)** as illustrated in Figure 4.12. The main difference between this procedure and the PCR method described earlier in this chapter is that this one starts with an mRNA instead of a double-stranded DNA. Thus, one begins by converting the mRNA to DNA. As usual, this RNA→DNA step can be done with reverse transcriptase and a reverse primer: One reverse transcribes the mRNA to make a single-stranded DNA, then uses a forward primer to convert the single-stranded DNA to double-stranded. Then one can use standard PCR to amplify the cDNA until enough is available for cloning. One can even add restriction sites to the ends of the cDNA by using primers that contain these sites. In this example, a *Bam*HI site is present on one primer and a *Hin*dIII site is present on the other (placed a few nucleotides from the ends to allow the restriction enzymes to cut efficiently). Thus, the PCR product is a cDNA with these two restriction sites at its two ends. Cutting the PCR product with these two restriction enzymes creates sticky ends that can be ligated into the vector of choice. Having two *different* sticky ends allows directional cloning, so the cDNA will have only one of two possible orientations in the vector. This is very useful when a cDNA is cloned into an expression vector, because the cDNA must be in the same orientation as the promoter that drives transcription of the cDNA. A caveat is necessary, however: One must make sure that the cDNA itself has neither of the restriction sites that have been added to its ends. If it does, the restriction enzymes will cut within the cDNA, as well as at the ends, and the products will be useless.

SUMMARY RT-PCR can be used to generate a cDNA from a single type of mRNA, but the sequence of the mRNA must be known so the primers for the PCR step can be designed. Restriction site sequences can be placed on the PCR primers, so these sites appear at the ends of the cDNA. This makes it easy to cleave them and then to ligate the cDNA into a vector.

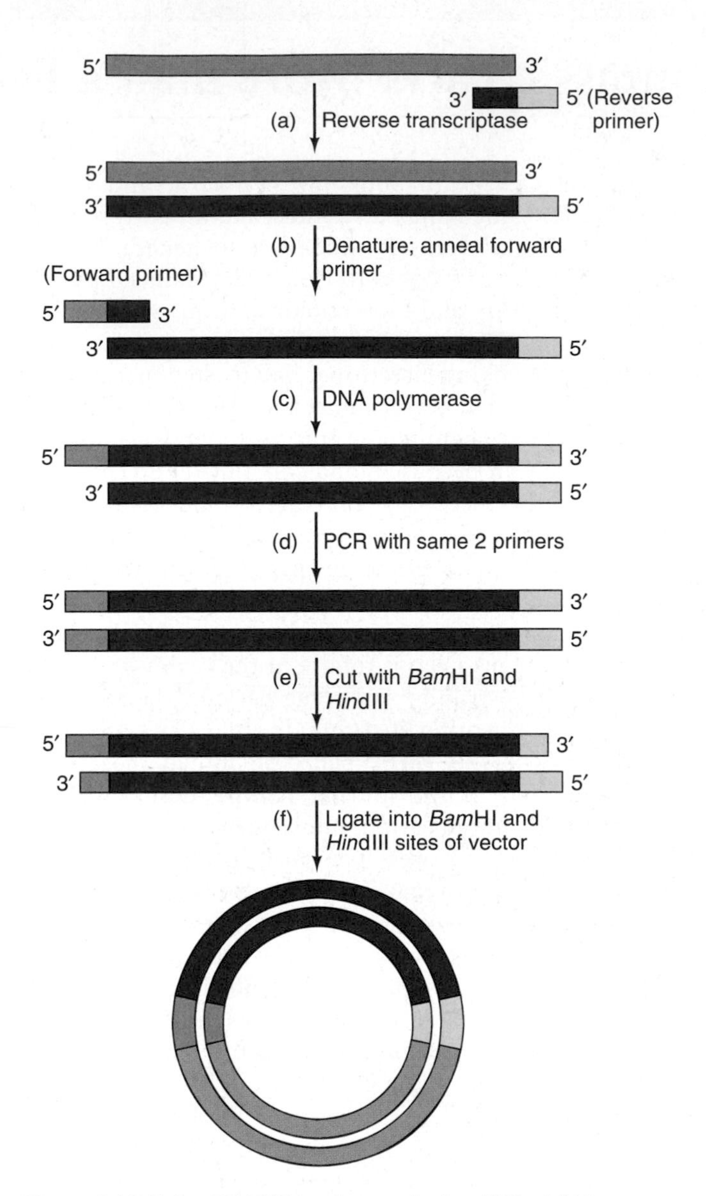

Figure 4.12 Using RT-PCR to clone a single cDNA. (a) Use a reverse primer (red) with a *Hin*dIII site (yellow) at its 5′-end to start first-strand cDNA synthesis, with reverse transcriptase to catalyze the reaction. **(b)** Denature the mRNA–cDNA hybrid and anneal a forward primer (red) with a *Bam*HI site (green) at its 5′-end. **(c)** This forward primer initiates second-strand cDNA synthesis, with DNA polymerase catalyzing the reaction. **(d)** Continue PCR with the same two primers to amplify the double-stranded cDNA. **(e)** Cut the cDNA with *Bam*HI and *Hin*dIII to generate sticky ends. **(f)** Ligate the cDNA to the *Bam*HI and *Hin*dIII sites of a suitable vector (purple). Finally, transform cells with the recombinant cDNA to produce a clone.

Real-Time PCR

Real-time PCR is a way of quantifying the amplification of a DNA as it occurs—that is, in real time. Figure 4.13 illustrates the basis of one real-time PCR method. After the two DNA strands are separated, they are annealed, not only to the forward and reverse primers, but also to a fluorescent-tagged oligonucleotide that is complementary

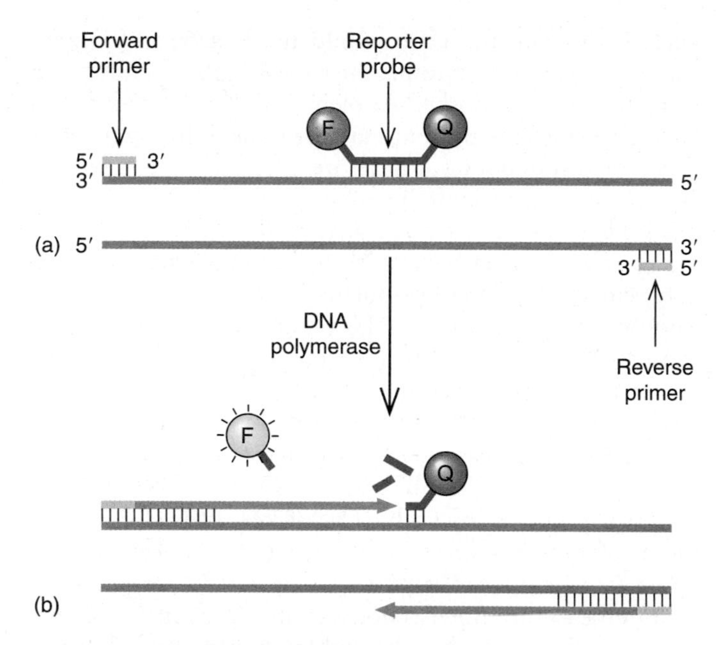

Figure 4.13 Real-time PCR. (a) The forward and reverse primers (purple) are annealed to the two separated DNA strands (blue), and a reporter probe (red) is annealed to the top DNA strand. The reporter probe has a fluorescent tag (gray) at its 5′-end and a fluorescence quenching tag (brown) at its 3′-end. **(b)** DNA polymerase has extended the primers, with the new DNA depicted in green. To make way for replicating the top strand, the DNA polymerase has also degraded part of the reporter probe. This separates the fluorescent tag from the quenching tag, and allows the fluorescent tag to exhibit its normal fluorescence (yellow). The more DNA strands are replicated, the more fluorescence will be observed.

to part of one of the DNA strands and serves as a **reporter probe.** The reporter probe has a fluorescent tag (F) at its 5′-end, and a fluorescence quenching tag (Q) at its 3′-end.

During the PCR polymerization step, the DNA polymerase extends the forward primer and then encounters the reporter probe. When that happens, the polymerase begins degrading the reporter probe so it can make new DNA in that region. As the reporter probe is degraded, the fluorescent tag is separated from the quenching tag, so its fluorescence increases dramatically. The whole process takes place inside a fluorimeter that measures the fluorescence of the tag, which in turn measures the progress of the PCR reaction. Enough reporter probe is present to anneal to each newly-made DNA strand, so fluorescence increases with each round of amplification.

It is unfortunate that "real-time" and "reverse transcriptase" can both be abbreviated "RT." Thus, when you see "RT-PCR" in the scientific literature, you need to see it in context to know which kind of PCR is being used. One can even do real-time reverse transcriptase PCR, starting with an RNA instead of double-stranded DNA. One way to abbreviate that method is **"real-time RT-PCR."**

SUMMARY Real-time PCR keeps track of the progress of PCR by monitoring the degradation of a reporter probe hybridized to the strand complementary to the forward primer. As this probe is degraded, a fluorescent tag is separated from a quenching tag, so fluorescence increases, and this increase can be measured in real time in a fluorimeter.

4.3 Methods of Expressing Cloned Genes

Why would we want to clone a gene? An obvious reason, suggested at the beginning of this chapter, is that cloning allows us to produce large quantities of particular DNA sequences so we can study them in detail. Thus, the gene itself can be a valuable product of gene cloning. Another goal of gene cloning is to make a large quantity of the gene's product, either for investigative purposes or for profit.

If the goal is to use bacteria to produce the protein product of a cloned eukaryotic gene—especially a higher eukaryotic gene—a cDNA will probably work better than a gene cut directly out of the genome. That is because most higher eukaryotic genes contain interruptions called introns (Chapter 14) that bacteria cannot deal with. Eukaryotic cells usually transcribe these interruptions, forming a pre-mRNA, and then cut them out and stitch the remaining parts (exons) of the pre-mRNA together to form the mature mRNA. Thus, a cDNA, which is a copy of an mRNA, already has its introns removed and can be expressed correctly in a bacterial cell.

Expression Vectors

The vectors we have examined so far are meant to be used primarily in the first stage of cloning—when one first puts a foreign DNA into a bacterium and gets it to replicate. By and large, they work well for that purpose, growing readily in *E. coli* and producing high yields of recombinant DNA. Some of them even work as **expression vectors** that can yield the protein products of the cloned genes. For example, the pUC and pBS vectors place inserted DNA under the control of the *lac* promoter, which lies upstream of the multiple cloning site. If an inserted DNA happens to be in the same reading frame as the *lacZ′* gene it interrupts, a **fusion protein** will result. It will have a partial β-galactosidase protein sequence at its amino end and another protein sequence, encoded in the inserted DNA, at its carboxyl end (Figure 4.14).

However, if one is interested in high-level expression of a cloned gene, specialized expression vectors usually work better. Bacterial expression vectors typically have two elements that are required for active gene expression: a strong promoter and a ribosome binding site near an initiating AUG codon (ATG in the DNA).

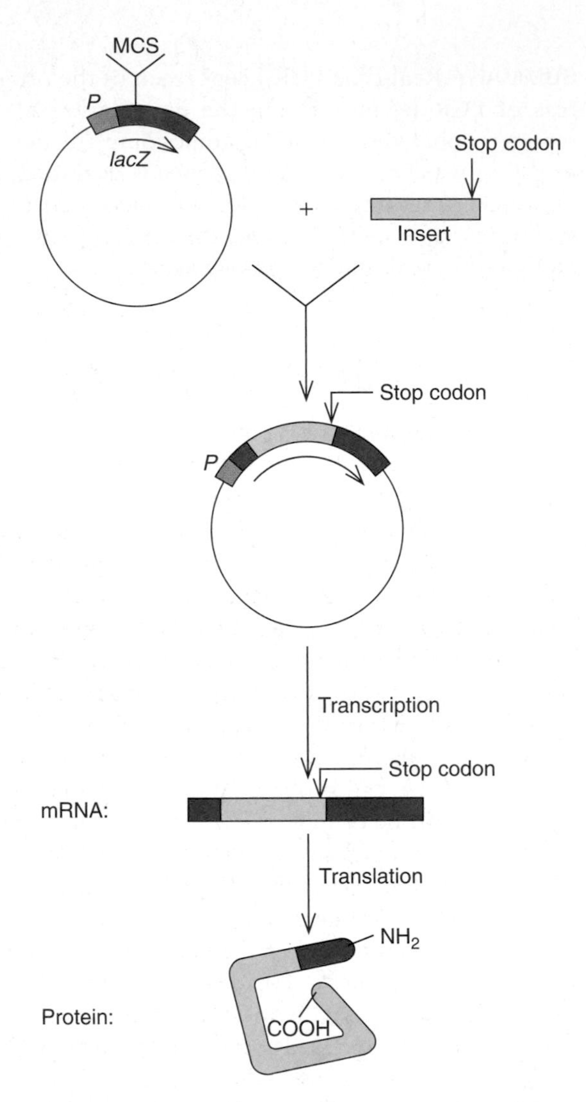

Figure 4.14 Producing a fusion protein by cloning in a pUC plasmid. Insert foreign DNA (yellow) into the multiple cloning site (MCS); transcription from the *lac* promoter (purple) gives a hybrid mRNA beginning with a few *lacZ′* codons, changing to insert sequence, then back to *lacZ′* (red). This mRNA is translated to a fusion protein containing a few β-galactosidase amino acids at the beginning (amino end), followed by the insert amino acids for the remainder of the protein. Because the insert contains a translation stop codon, the remaining *lacZ′* codons are not translated.

Inducible Expression Vectors The main function of an expression vector is to yield the product of a gene—usually, the more product the better. Therefore, expression vectors are ordinarily equipped with very strong promoters; the rationale is that the more mRNA that is produced, the more protein product will be made.

It is usually advantageous to keep a cloned gene repressed until it is time to express it. One reason is that eukaryotic proteins produced in large quantities in bacteria can be toxic. Even if these proteins are not actually toxic, they can build up to such great levels that they interfere with bacterial growth. In either case, if the cloned gene were allowed to remain turned on constantly, the bacteria bearing the gene would never grow to a great enough concentration to produce meaningful quantities of protein product. Another problem with high expression in bacteria is that the protein may form insoluble aggregates called **inclusion bodies.** Therefore, it is helpful to keep the cloned gene turned off by placing it downstream of an inducible promoter that can be turned off.

The *lac* promoter is inducible to a certain extent, presumably remaining off until stimulated by the synthetic inducer isopropylthiogalactoside (IPTG). However, the repression caused by the *lac* repressor is incomplete (leaky), and some expression of the cloned gene will be observed even in the absence of inducer. One way around this problem is to express a gene in a plasmid or phagemid that carries its own *lacI* (repressor) gene, as pBS does (see Figure 4.7). The excess repressor produced by such a vector keeps the cloned gene turned off until it is time to induce it with IPTG. (For a review of the *lac* operon, see Chapter 7.)

But the *lac* promoter is not very strong, so many vectors have been designed with a hybrid ***trc* promoter,** which combines the strength of the *trp* (tryptophan operon) promoter with the inducibility of the *lac* promoter. The *trp* promoter is much stronger than the *lac* promoter because of its –35 box (Chapter 6). Accordingly, molecular biologists have combined the –35 box of the *trp* promoter with the –10 box of the *lac* promoter, plus the *lac* operator (Chapter 7). The –35 box of the *trp* promoter makes the hybrid promoter strong, and the *lac* operator makes it inducible by IPTG.

A promoter from the *ara* (arabinose) operon, P_{BAD}, allows fine control of transcription. This promoter is inducible by the sugar arabinose (Chapter 7), so no transcription occurs in the absence of arabinose, but more and more transcription occurs as more and more arabinose is added to the medium. Figure 4.15 illustrates this phenomenon in an experiment in which the green fluorescent protein (GFP) gene was cloned in a P_{BAD} vector and expression was induced with increasing concentrations of arabinose.

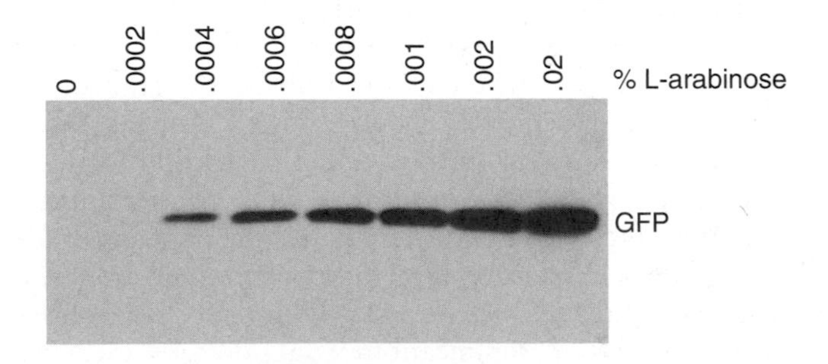

Figure 4.15 Using a P_{BAD} vector. The green fluorescent protein (GFP) gene was cloned into a vector under control of the P_{BAD} promoter and promoter activity was induced with increasing concentrations of arabinose. GFP production was monitored by electrophoresing extracts from cells induced with the arabinose concentrations given at top, blotting the proteins to a membrane, and detecting GFP with an anti-GFP antibody (immunoblotting, Chapter 5). (*Source:* Copyright 2003 Invitrogen Corporation. All Rights Reserved. Used with permission.)

No GFP appeared in the absence of arabinose, but concentrations of arabinose 0.0004% and above yielded increasing quantities of the protein.

Another strategy is to use a tightly controlled promoter such as the lambda (λ) phage promoter P_L. Expression vectors with this promoter–operator system are cloned into host cells bearing a temperature-sensitive λ repressor gene (*cI857*). As long as the temperature of these cells is kept relatively low (32°C), the repressor functions, and no expression takes place. However, when the temperature is raised to the nonpermissive level (42°C), the temperature-sensitive repressor can no longer function and the cloned gene is derepressed.

A popular method of ensuring tight control, as well as high-level induced expression, is to place the gene to be expressed in a plasmid under control of a T7 phage promoter. Then this plasmid is placed in a cell that contains a tightly regulated gene for T7 RNA polymerase. For example, the T7 RNA polymerase gene may be under control of a modified *lac* promoter in a cell that also carries the gene for the *lac* repressor. Thus, the T7 polymerase gene is strongly repressed unless the *lac* inducer is present. As long as no T7 polymerase is present, transcription of the gene of interest cannot take place because the T7 promoter has an absolute requirement for its own polymerase. But as soon as a *lac* inducer is added, the cell begins to make T7 polymerase, which transcribes the gene of interest. And because many molecules of T7 polymerase are made, the gene is turned on to a very high level and abundant amounts of protein product are made.

SUMMARY Expression vectors are designed to yield the protein product of a cloned gene, usually in the greatest amount possible. To optimize expression, these vectors include strong bacterial or phage promoters and bacterial ribosome binding sites that would be missing on cloned eukaryotic genes. Most cloning vectors are inducible, which avoids premature overproduction of a foreign product that could poison the bacterial host cells.

Expression Vectors That Produce Fusion Proteins Most expression vectors produce fusion proteins. This might at first seem a disadvantage because the natural product of the inserted gene is not made. However, the extra amino acids on the fusion protein can be a great help in purifying the protein product.

Consider the oligohistidine expression vectors, one of which has the trade name pTrcHis (Figure 4.16). These have a short sequence just upstream of the multiple cloning site that encodes a stretch of six histidines. Thus, a protein expressed in such a vector will be a fusion protein with six histidines at its amino end. Why would one want to attach six histidines to a protein? Oligohistidine regions like this have a high affinity for divalent metal ions like nickel (Ni^{2+}), so proteins that have such regions can be purified using nickel **affinity chromatography.** The beauty of this method is its simplicity and speed. After the bacteria have made the fusion protein, one simply lyses them, adds the crude bacterial extract to a nickel affinity column, washes out all unbound proteins, then releases the fusion protein with histidine or a histidine analog called imidazole. This procedure allows one to harvest essentially pure fusion protein in only one step. This is possible because very few if any natural proteins have oligohistidine regions, so the fusion protein is essentially the only one that binds to the column.

What if the oligohistidine tag interferes with the protein's activity? The designers of these vectors have thoughtfully provided a way to remove it. Just before the multiple cloning site is a coding region for a stretch of amino acids recognized by the enzyme enterokinase (a protease, not really a kinase at all). So enterokinase can be used to cleave the fusion protein into two parts: the oligohistidine tag and the protein of interest. The site recognized by enterokinase is very rare, and the chance that it exists in any given protein is insignificant. Thus, the rest of the protein should not be chopped up as its oligohistidine tag is removed. The enterokinase-cleaved protein can be run through the nickel column once more to separate the oligohistidine fragments from the protein of interest.

Lambda (λ) phages have also served as the basis for expression vectors; one designed specifically for this purpose is λgt11. This phage (Figure 4.17) contains the *lac* control region followed by the *lacZ* gene. The cloning sites are located within the *lacZ* gene, so products of a gene inserted correctly into this vector will be fusion proteins with a leader of β-galactosidase.

The expression vector λgt11 has been a popular vehicle for making and screening cDNA libraries. In the examples of screening presented earlier, the proper DNA sequence was detected by probing with a labeled oligonucleotide or polynucleotide. By contrast, λgt11 allows one to screen a group of clones directly for the expression of the right protein. The main ingredients required for this procedure are a cDNA library in λgt11 and an antiserum directed against the protein of interest.

Figure 4.18 shows how this works. Lambda phages with various cDNA inserts are plated, and the proteins released by each clone are blotted onto a support such as nitrocellulose. Once the proteins from each plaque have been transferred to nitrocellulose, they can be probed with antiserum. Next, antibody bound to protein from a particular plaque can be detected, using labeled protein A from *Staphylococcus aureus*. This protein binds tightly to antibody and labels the corresponding spot on the nitrocellulose. This label can be detected by autoradiography or by phosphorimaging (Chapter 5), then the corresponding plaque

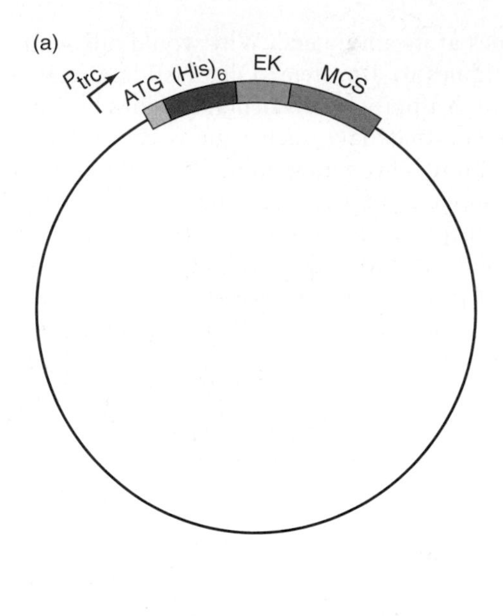

Figure 4.16 Using an oligohistidine expression vector. (a) Map of a generic oligohistidine vector. Just after the ATG initiation codon (green) lies a coding region (red) encoding six histidines in a row [$(His)_6$]. This is followed by a region (orange) encoding a recognition site for the proteolytic enzyme enterokinase (EK). Finally, the vector has a multiple cloning site (MCS, blue). Usually, the vector comes in three forms with the MCS sites in each of the three reading frames. One can select the vector that puts the gene in the right reading frame relative to the oligohistidine. **(b)** Using the vector. 1. Insert the gene of interest (yellow) into the vector in frame with the oligohistidine coding region (red) and transform bacterial cells with the recombinant vector. The cells produce the fusion protein (red and yellow), along with other, bacterial proteins (green). 2. Lyse the cells, releasing the mixture of proteins. 3. Pour the cell lysate through a nickel affinity chromatography column, which binds the fusion protein but not the other proteins. 4. Release the fusion protein from the column with histidine or with imidazole, a histidine analogue, which competes with the oligohistidine for binding to the nickel. 5. Cleave the fusion protein with enterokinase. 6. Pass the cleaved protein through the nickel column once more to separate the oligohistidine from the desired protein.

can be picked from the master plate. Note that a fusion protein is detected, not the protein of interest by itself. Furthermore, it does not matter if a whole cDNA has been cloned or not. The antiserum is a mixture of antibodies that will react with several different parts of the protein, so even a partial gene will do, as long as its coding region is cloned in the same orientation and reading frame as the β-galactosidase coding region.

Even partial cDNAs are valuable because they can be completed by RACE, as we saw earlier in this chapter. The β-galactosidase tag on the fusion proteins helps to stabilize them in the bacterial cell, and can even make them easy to

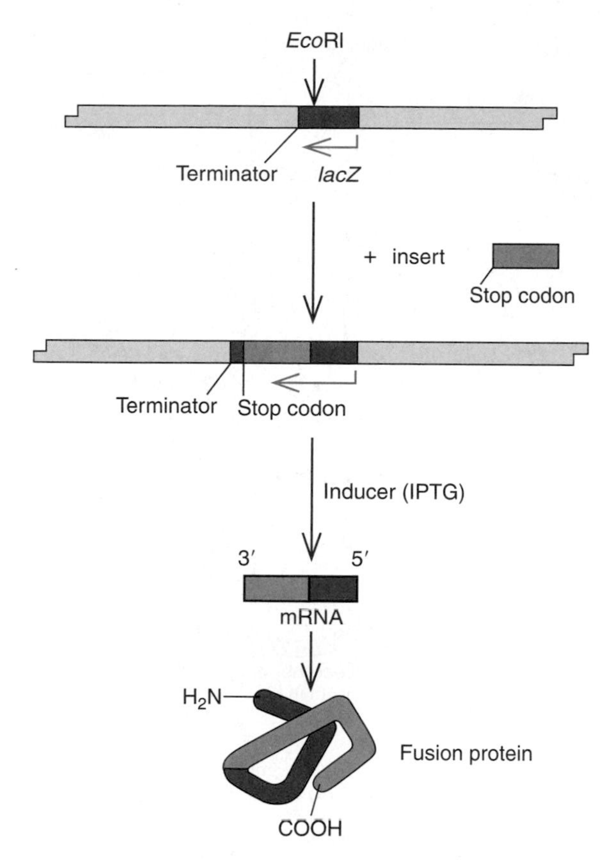

Figure 4.17 Synthesizing a fusion protein in λgt11. The gene to be expressed (green) is inserted into the *Eco*RI site near the end of the *lacZ* coding region (red) just upstream of the transcription terminator. Thus, on induction of the *lacZ* gene by IPTG, a fused mRNA results, containing the inserted coding region just downstream of the bulk of the coding region of β-galactosidase. This mRNA is translated by the host cell to yield a fusion protein.

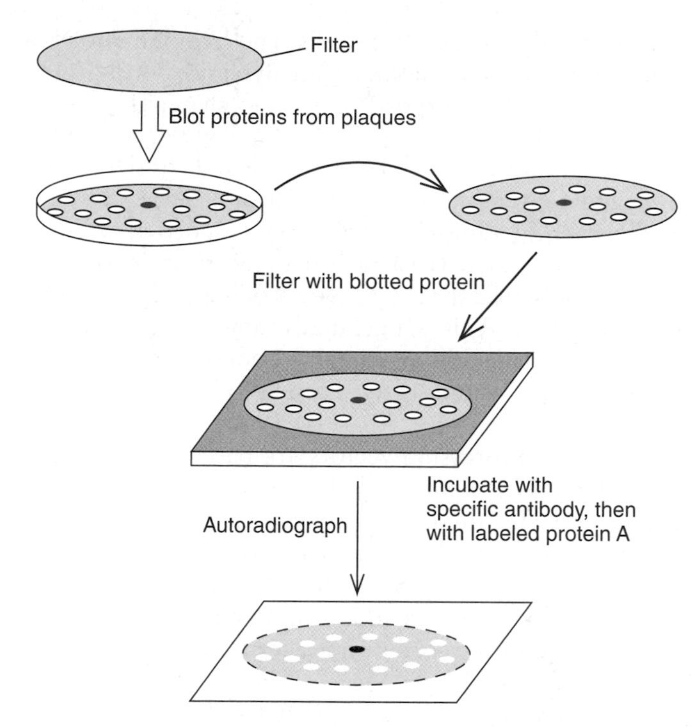

Figure 4.18 Detecting positive λgt11 clones by antibody screening. A filter is used to blot proteins from phage plaques on a Petri dish. One of the clones (red) has produced a plaque containing a fusion protein including β-galactosidase and a part of the protein of interest. The filter with its blotted proteins is incubated with an antibody directed against the protein of interest, then with labeled *Staphylococcus* protein A, which binds to most antibodies. It will therefore bind only to the antibody–antigen complexes at the spot corresponding to the positive clone. A dark spot on the film placed in contact with the filter reveals the location of the positive clone.

purify by affinity chromatography on a column containing an anti-β-galactosidase antibody.

SUMMARY Expression vectors frequently produce fusion proteins, with one part of the protein coming from coding sequences in the vector and the other part from sequences in the cloned gene itself. Many fusion proteins have the great advantage of being simple to isolate by affinity chromatography. The λgt11 vector produces fusion proteins that can be detected in plaques with a specific antiserum.

Eukaryotic Expression Systems Eukaryotic genes are not really "at home" in bacterial cells, even when they are expressed under the control of their bacterial vectors. One reason is that *E. coli* cells sometimes recognize the protein products of cloned eukaryotic genes as outsiders and destroy them. Another is that bacteria do not carry out the same kinds of posttranslational modifications as eukaryotes do. For example, a protein that would ordinarily be coupled to sugars in a eukaryotic cell will be expressed as a naked protein when cloned in bacteria. This can affect a protein's activity or stability, or at least its response to antibodies. A more serious problem is that the interior of a bacterial cell is not as conducive to proper folding of eukaryotic proteins as the interior of a eukaryotic cell. Frequently, the result is improperly folded, inactive products of cloned genes. This means that one can frequently express a cloned gene at a stupendously high level in bacteria, but the product forms highly insoluble, inactive granules called inclusion bodies. These are of no use unless one can get the protein to refold and regain its activity. Fortunately, it is frequently possible to renature the proteins from inclusion bodies. In that case, the inclusion bodies are an advantage because they can be separated from almost all other proteins by simple centrifugation.

To avoid the potential incompatibility between a cloned gene and its host, the gene can be expressed in a eukaryotic cell. In such cases, the initial cloning is usually done in *E. coli*, using a **shuttle vector** that can replicate in both bacterial and eukaryotic cells. The recombinant DNA is then transferred to the eukaryote of choice. One eukaryote suited for this purpose is yeast. It shares the advantages of rapid growth and ease of culture with bacteria, yet it is a eukaryote and thus it carries out some of the protein folding and glycosylation (adding sugars) characteristic of a eukaryote. In addition, by splicing a cloned gene to the coding region for a yeast export signal peptide, one can usually ensure that the gene product will be secreted to the growth medium. This is a great advantage in purifying the protein. The yeast cells are simply removed in a centrifuge, leaving relatively pure secreted gene product behind in the medium.

The yeast expression vectors are based on a plasmid, called the *2-micron plasmid,* that normally inhabits yeast cells. It provides the origin of replication needed by any vector that must replicate in yeast. Yeast–bacterial shuttle vectors also contain the pBR322 origin of replication, so they can also replicate in *E. coli.* In addition, of course, a yeast expression vector must contain a strong yeast promoter.

Another eukaryotic vector that has been remarkably successful is derived from the **baculovirus** that infects the caterpillar known as the alfalfa looper. Viruses in this class have a rather large circular DNA genome, approximately 130 kb in length. The major viral structural protein, polyhedrin, is made in copious quantities in infected cells. In fact, it has been estimated that when a caterpillar dies of a baculovirus infection, up to 10% of the dry mass of the dead insect is this one protein. This huge mass of protein indicates that the polyhedrin gene must be very active, and indeed it is—apparently due to its powerful promoter. Max Summers and his colleagues, and Lois Miller and her colleagues first developed successful vectors using the polyhedrin promoter in 1983 and 1984, respectively. Since then, many other baculovirus vectors have been constructed using this and other viral promoters.

At their best, baculovirus vectors can produce up to half a gram per liter of protein from a cloned gene—a large amount indeed. Figure 4.19 shows how a typical baculovirus expression system works. First, the gene of interest is cloned in one of the vectors. In this example, let us consider a vector with the polyhedrin promoter. (The polyhedrin coding region has been deleted from the vector. This does not inhibit virus replication because polyhedrin is not required for transmission of the virus from cell to cell in culture.) Most such vectors have a unique *Bam*HI site directly downstream of the promoter, so they can be cut with *Bam*HI and a fragment with *Bam*HI-compatible ends can be inserted into the vector, placing the cloned gene under the control of the polyhedrin promoter. Next the recombinant plasmid (vector plus insert) is mixed with wild-type viral DNA that has been cleaved so as to remove a gene essential for viral replication, along with the polyhedrin gene. Cultured insect cells are then transfected with this mixture.

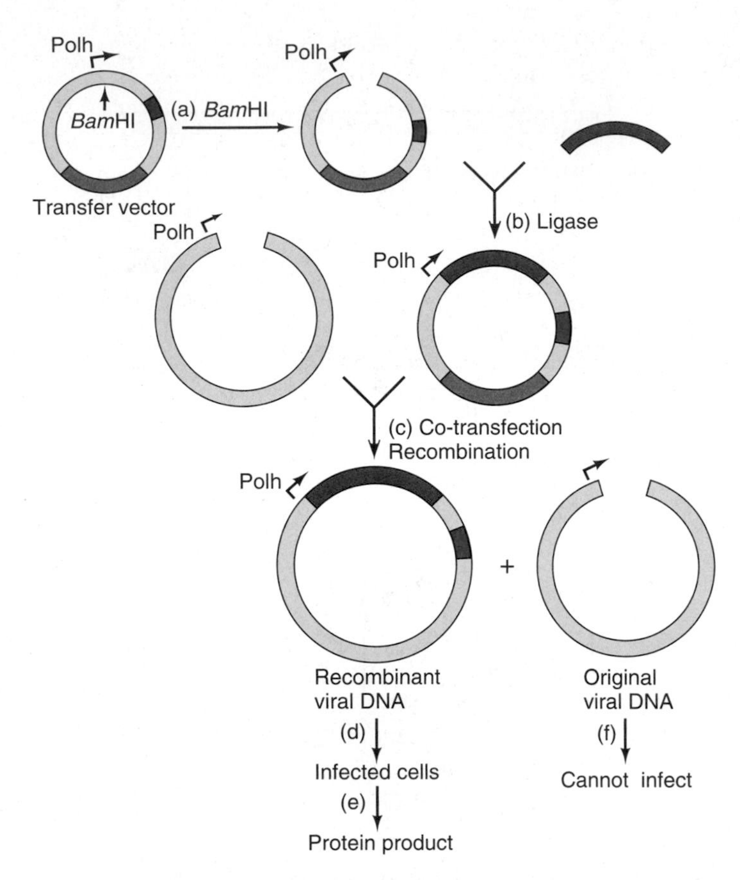

Figure 4.19 Expressing a gene in a baculovirus. First, insert the gene to be expressed (red) into a baculovirus transfer vector. In this case, the vector contains the powerful polyhedrin promoter (Polh), flanked by the DNA sequences (yellow) that normally surround the polyhedrin gene, including a gene (green) that is essential for virus replication; the polyhedrin coding region itself is missing from this transfer vector. Bacterial vector sequences are in blue. Just downstream of the promoter is a *Bam*HI restriction site, which can be used to open up the vector (step **a**) so it can accept the foreign gene (red) by ligation (step **b**). In step **c,** mix the recombinant transfer vector with linear viral DNA that has been cut so as to remove the essential gene. Transfect insect cells with the two DNAs together. This process is known as co-transfection. The two DNAs are not drawn to scale; the viral DNA is actually almost 15 times the size of the vector. Inside the cell, the two DNAs recombine by a double crossover that inserts the gene to be expressed, along with the essential gene, into the viral DNA. The result is a recombinant virus DNA that has the gene of interest under the control of the polyhedrin promoter. Finally, in steps **d** and **e,** infect cells with the recombinant virus and collect the protein product these cells make. Notice that the original viral DNA is linear and it is missing the essential gene, so it cannot infect cells **(f).** This lack of infectivity selects automatically for recombinant viruses; they are the only ones that can infect cells.

Because the vector has extensive homology with the regions flanking the polyhedrin gene, recombination can occur within the transfected cells. This transfers the cloned gene into the viral DNA, still under the control of the polyhedrin promoter. Now this recombinant virus can be used to infect cells and the protein of interest can be harvested after these cells enter the very late phase of infection, during which the polyhedrin promoter is most active. What about the nonrecombinant viral DNA that enters the transfected cells along with the recombinant vector? It cannot give rise to infectious virus because it lacks an essential gene that can only be supplied by the vector.

Notice the use of the term *transfected* with eukaryotic cells instead of *transformed,* which we use with bacteria. We make this distinction because *transformation* has another meaning in eukaryotes: the conversion of a normal cell to a cancer-like cell. To avoid confusion with this phenomenon, we use **transfection** to denote introducing new DNA into a eukaryotic cell.

Transfection in animal cells is conveniently carried out in at least two ways: (1) Cells can be mixed with DNA in a phosphate buffer, then a solution of a calcium salt can be added to form a precipitate of $Ca_3(PO_4)_2$. The cells take up the calcium phosphate crystals, which also include some DNA. (2) The DNA can be mixed with lipid, which forms **liposomes,** small vesicles that include some DNA solution inside. These DNA-bearing liposomes then fuse with the cell membranes, delivering their DNA into the cells. Plant cells are commonly transfected by a **biolistic** method in which small metal pellets are coated with DNA and literally shot into cells.

SUMMARY Foreign genes can be expressed in eukaryotic cells, and these eukaryotic systems have some advantages over their prokaryotic counterparts for producing eukaryotic proteins. Two of the most important advantages are (1) Eukaryotic proteins made in eukaryotic cells tend to be folded properly, so they are soluble, rather than aggregated into insoluble inclusion bodies. (2) Eukaryotic proteins made in eukaryotic cells are modified (phosphorylated, glycosylated, etc.) in a eukaryotic manner.

Other Eukaryotic Vectors

Some well-known eukaryotic vectors serve purposes other than expressing foreign genes. For example, yeast artificial chromosomes (YACs), bacterial artificial chromosomes (BACs), and P1 phage artificial chromosomes (PACs) are capable of accepting huge chunks of foreign DNA and therefore find use in large sequencing programs such as the human genome project, where big pieces of cloned DNA are especially valuable. We will discuss the artificial chromosomes in Chapter 24 in the context of genomics. Another important eukaryotic vector is the Ti plasmid, which can transport foreign genes into plant cells and ensure their replication there.

Using the Ti Plasmid to Transfer Genes to Plants

Genes can also be introduced into plants, using vectors that can replicate in plant cells. The common bacterial vectors do not serve this purpose because plant cells cannot recognize their bacterial promoters and replication origins. Instead, a plasmid containing so-called **T-DNA** can be used. This is a piece of DNA from a plasmid known as **Ti** (tumor-inducing).

The Ti plasmid inhabits the bacterium *Agrobacterium tumefaciens,* which causes tumors called **crown galls** (Figure 4.20) in dicotyledonous plants. When this bacterium infects a plant, it transfers its Ti plasmid to the host cells, whereupon the T-DNA integrates into the plant DNA, causing the abnormal proliferation of plant cells that gives rise to a crown gall. This is advantageous for the invading bacterium, because the T-DNA has genes directing the synthesis of unusual organic acids called **opines.** These opines are worthless to the plant, but the bacterium has enzymes that can break down opines so they can serve as an exclusive energy source for the bacterium.

The T-DNA genes coding for the enzymes that make opines (e.g., mannopine synthetase) have strong promoters. Plant molecular biologists take advantage of them by putting T-DNA into small plasmids, then placing foreign genes under the control of one of these promoters. Figure 4.21 outlines the process used to transfer a foreign gene to a tobacco plant, producing a **transgenic plant.** One punches out a small disk (7 mm or so in diameter) from a tobacco leaf and places it in a dish with nutrient medium. Under these conditions, tobacco tissue will grow around the edge of the disk. Next, one adds *Agrobacterium* cells containing the foreign gene cloned into a Ti plasmid; these bacteria infect the growing tobacco cells and introduce the cloned gene.

When the tobacco tissue grows roots around the edge, those roots are transplanted to medium that encourages shoots to form. These plantlets give rise to full-sized tobacco plants whose cells contain the foreign gene. This gene can confer new properties on the plant, such as pesticide resistance, drought resistance, or disease resistance.

One of the most celebrated successes so far in plant genetic engineering has been the development of the "Flavr Savr" tomato. Calgene geneticists provided this plant with an antisense copy of a gene that contributes to fruit softening during ripening. The RNA product of this antisense gene is complementary to the normal mRNA, so it hybridizes to the mRNA and blocks expression of the gene. This allows tomatoes to ripen without softening as much, so they can ripen naturally on the vine instead of being picked green and ripened artificially.

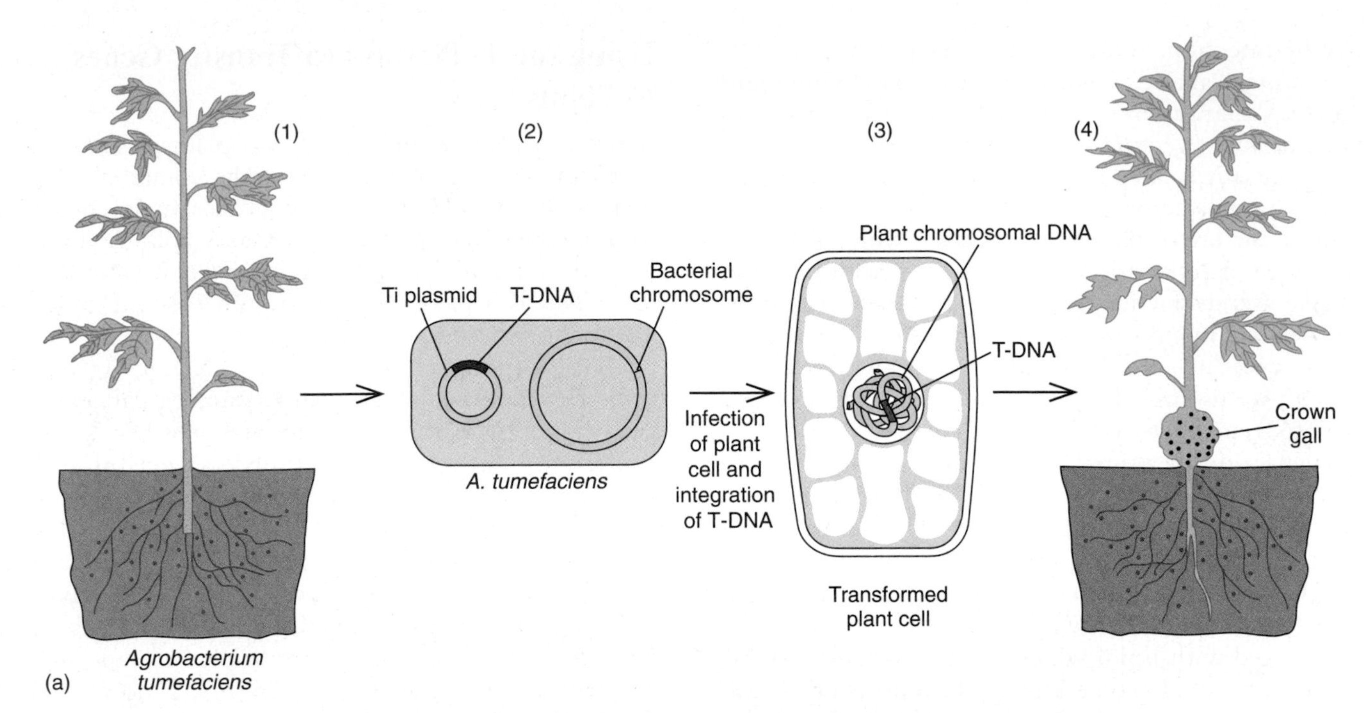

Figure 4.20 Crown gall tumors. **(a)** Formation of a crown gall 1. *Agrobacterium* cells enter a wound in the plant, usually at the crown, or the junction of root and stem. 2. The *Agrobacterium* contains a Ti plasmid in addition to the much larger bacterial chromosome. The Ti plasmid has a segment (the T-DNA, red) that promotes tumor formation in infected plants. 3. The bacterium contributes its Ti plasmid to the plant cell, and the T-DNA from the Ti plasmid integrates into the plant's chromosomal DNA. 4. The genes in the T-DNA direct the formation of a crown gall, which nourishes the invading bacteria. **(b)** Photograph of a crown gall tumor generated by cutting off the top of a tobacco plant and inoculating with *Agrobacterium.* This crown gall tumor is a teratoma, which generates normal as well as tumorous tissues. (*Source: (b)* Dr. Robert Turgeon and Dr. B. Gillian Turgeon, Cornell University.)

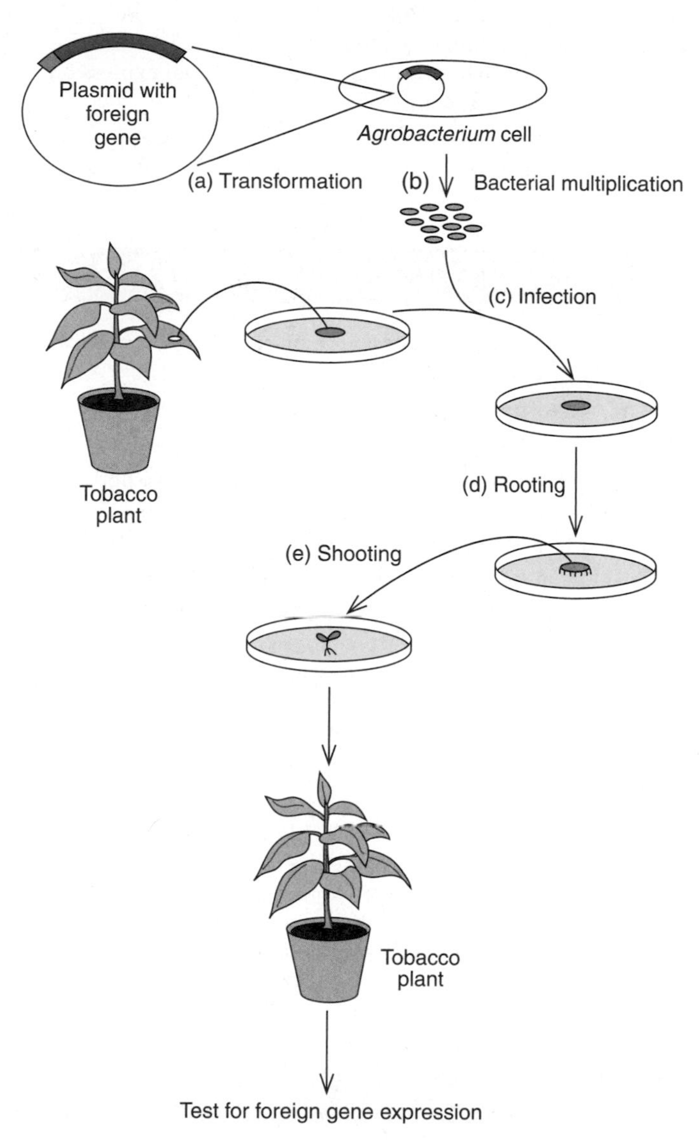

Figure 4.21 Using a T-DNA plasmid to introduce a gene into tobacco plants. (a) A plasmid is constructed with a foreign gene (red) under the control of the mannopine synthetase promoter (blue). This plasmid is used to transform *Agrobacterium* cells. **(b)** The transformed bacterial cells divide repeatedly. **(c)** A disk of tobacco leaf tissue is removed and incubated in nutrient medium, along with the transformed *Agrobacterium* cells. These cells infect the tobacco tissue, transferring the plasmid bearing the cloned foreign gene, which integrates into the plant genome. **(d)** The disk of tobacco tissue sends out roots into the surrounding medium. **(e)** One of these roots is transplanted to another kind of medium, where it forms a shoot. This plantlet grows into a transgenic tobacco plant that can be tested for expression of the transplanted gene.

Other plant molecular biologists have made additional strides, including the following: (1) conferring herbicide resistance on plants; (2) conferring virus resistance on tobacco plants by inserting a gene for the viral coat protein; (3) endowing corn and cotton plants with a bacterial pesticide; and (4) inserting the gene for firefly luciferase into tobacco plants—this experiment has no practical value, but it does have the arresting effect of making the plant glow in the dark.

SUMMARY Molecular biologists can transfer cloned genes to plants, creating transgenic organisms with altered characteristics, using a plant vector such as the Ti plasmid.

SUMMARY

To clone a gene, one must insert it into a vector that can carry the gene into a host cell and ensure that it will replicate there. The insertion is usually carried out by cutting the vector and the DNA to be inserted with the same restriction endonucleases to endow them with the same "sticky ends." Vectors for cloning in bacteria come in two major types: plasmids and phages.

Among the plasmid cloning vectors are pBR322 and the pUC plasmids. Screening is convenient with the pUC plasmids and pBS phagemids. These vectors have an ampicillin resistance gene and a multiple cloning site that interrupts a partial β-galactosidase gene whose product is easily detected with a color test. The desired clones are ampicillin-resistant and do not make active β-galactosidase.

Two kinds of phages have been especially popular as cloning vectors. The first is λ (lambda), which has had certain nonessential genes removed to make room for inserts. In some of these engineered phages, inserts up to 20 kb in length can be accommodated. Cosmids, a cross between phage and plasmid vectors, can accept inserts as large as 50 kb. This makes these vectors very useful for building genomic libraries. The second major class of phage vectors is the M13 phages. These vectors have the convenience of a multiple cloning region and the further advantage of producing single-stranded recombinant DNA, which can be used for DNA sequencing and for site-directed mutagenesis. Plasmids called phagemids have an origin of replication for a single-stranded DNA phage, so they can produce single-stranded copies of themselves.

Expression vectors are designed to yield the protein product of a cloned gene, usually in the greatest amount possible. To optimize expression, bacterial expression vectors provide strong bacterial promoters and bacterial ribosome binding sites that would be missing from cloned eukaryotic genes. Most cloning vectors are inducible, to avoid premature overproduction of a foreign product that could poison the bacterial host cells. Expression vectors frequently produce fusion proteins, which can often be isolated quickly and easily. Eukaryotic expression systems have the advantages that the protein products are usually soluble, and these products are modified in a eukaryotic manner.

Cloned genes can also be transferred to plants, using a plant vector such as the Ti plasmid. This procedure can alter the plants' characteristics.

REVIEW QUESTIONS

1. Consulting Table 4.1, determine the length and the nature (5′ or 3′) of the overhang (if any) created by the following restriction endonucleases:
   a. *Alu*I
   b. *Bgl*II
   c. *Cla*I
   d. *Kpn*I
   e. *Mbo*I
   f. *Pvu*I
   g. *Not*I
2. Why does one need to attach DNAs to vectors to clone them?
3. Describe the process of cloning a DNA fragment into the *Bam*HI and *Pst*I sites of the vector pUC18. How would you screen for clones that contain an insert?
4. Describe the process of cloning a DNA fragment into the *Eco*RI site of the Charon 4 vector.
5. You want to clone a 1-kb cDNA. Which vectors discussed in this chapter would be appropriate to use? Which would be inappropriate? Why?
6. You want to make a genomic library with DNA fragments averaging about 45 kb in length. Which vector discussed in this chapter would be most appropriate to use? Why?
7. You want to make a library with DNA fragments averaging over 100 kb in length. Which vectors discussed in this chapter would be most appropriate to use? Why?
8. You have constructed a cDNA library in a phagemid vector. Describe how you would screen the library for a particular gene of interest. Describe methods using oligonucleotide and antibody probes.
9. How would you obtain single-stranded cloned DNAs from an M13 phage vector? From a phagemid vector?
10. Diagram a method for creating a cDNA library.
11. Diagram the process of nick translation.
12. Outline the polymerase chain reaction (PCR) method for amplifying a given stretch of DNA.
13. What is the difference between reverse transcriptase PCR (RT-PCR) and standard PCR? For what purpose would you use RT-PCR?
14. Describe the use of a vector that produces fusion proteins with oligohistidine at one end. Show the protein purification scheme to illustrate the advantage of the oligohistidine tag.
15. What is the difference between a λ insertion vector such as λgt11 and a λ replacement vector? What is the advantage of each?
16. Describe the use of a baculovirus system for expressing a cloned gene. What advantages over a bacterial expression system does the baculovirus system offer?
17. What kind of vector would you use to insert a transgene into a plant such as tobacco? Diagram the process you would use.

ANALYTICAL QUESTIONS

1. Here is the amino acid sequence of part of a hypothetical gene you want to clone:

   Pro-Arg-Tyr-Met-Cys-Trp-Ile-Leu-Met-Ser

   a. What sequence of five amino acids would give a 14-mer probe with the least degeneracy for probing a library to find your gene of interest? Notice that you do not use the last base in the fifth codon because of its degeneracy.
   b. How many different 14-mers would you have to make in order to be sure that your probe matches the corresponding sequence in your cloned gene perfectly?
   c. If you started your probe one amino acid to the left of the one you chose in (a), how many different 14-mers would you have to make? Use the genetic code to determine degeneracy.
2. You are cloning the genome of a new DNA virus into pUC18. You plate out your transformants on ampicillin plates containing X-gal and pick one blue colony and one white colony. When you check the size of the inserts in each plasmid (blue and white), you are surprised to find that the plasmid from the blue colony contains a very small insert of approximately 60 bp, while the plasmid from the white colony does not appear to contain any insert at all. Explain these results.

SUGGESTED READINGS

Capecchi, N.R. 1994. Targeted gene replacement. *Scientific American* 270 (March):52–59.

Chilton, M.-D. 1983. A vector for introducing new genes into plants. *Scientific American* 248 (June):50–59.

Cohen, S. 1975. The manipulation of genes. *Scientific American* 233 (July):24–33.

Cohen, S., A. Chang, H. Boyer, and R. Helling. 1973. Construction of biologically functional bacterial plasmids in vitro. *Proceedings of the National Academy of Sciences* 70:3240–44.

Gasser, C.S., and R.T. Fraley. 1992. Transgenic crops. *Scientific American* 266 (June):62–69.

Gilbert, W., and L. Villa-Komaroff. 1980. Useful proteins from recombinant bacteria. *Scientific American* 242 (April):74–94.

Nathans, D., and H.O. Smith. 1975. Restriction endonucleases in the analysis and restructuring of DNA molecules. *Annual Review of Biochemistry* 44:273–93.

Sambrook, J., and D. Russell. 2001. *Molecular Cloning: A Laboratory Manual,* 3rd ed. Plainview, NY: Cold Spring Harbor Laboratory Press.

Watson, J.D., J. Tooze, and D.T. Kurtz. 1983. *Recombinant DNA: A Short Course.* New York: W.H. Freeman.

CHAPTER 5

Molecular Tools for Studying Genes and Gene Activity

Two scientists examine an autoradiograph of an electrophoretic gel.

In this chapter we will describe the most popular techniques that molecular biologists use to investigate the structure and function of genes. Most of these start with cloned genes. Many use gel electrophoresis. Many also use labeled tracers, and many rely on nucleic acid hybridization. We have already examined gene cloning techniques. Let us continue by briefly considering three other mainstays of molecular biology research: molecular separations including gel electrophoresis; labeled tracers; and hybridization.

5.1 Molecular Separations

Gel Electrophoresis

It is very often necessary in molecular biology research to separate proteins or nucleic acids from each other. For example, we may need to purify a particular enzyme from a crude cellular extract in order to use it or to study its properties. Or we may want to purify a particular RNA or DNA molecule that has been produced or modified in an enzymatic reaction, or we may simply want to separate a series of RNAs or DNA fragments from each other. We will describe here some of the most common techniques used in such molecular separations, including gel electrophoresis of both nucleic acids and proteins, ion exchange chromatography, and gel filtration chromatography.

Gel electrophoresis can be used to separate different nucleic acid or protein species. We will begin by considering DNA gel electrophoresis. In this technique one makes an agarose gel with slots in it, as shown in Figure 5.1. The slots are formed by pouring a hot (liquid) agarose solution into a shallow box equipped with a removable "comb" with teeth that point downward into the agarose. Once the agarose has gelled, the comb is removed, leaving rectangular holes, or slots, in the gel. One puts a little DNA in a slot and runs an electric current through the gel at neutral pH. The DNA is negatively charged because of the phosphates in its backbone, so it migrates toward the positive pole (the *anode*) at the end of the gel. The secret of the gel's ability to separate DNAs of different sizes lies in friction. Small DNA molecules experience little frictional drag from solvent and gel molecules, so they migrate rapidly. Large DNAs, by contrast, encounter correspondingly more friction, so their mobility is lower. The result is that the electric current will distribute the DNA fragments according to their sizes: the largest near the top, the smallest near the bottom. Finally, the DNA is stained with a fluorescent dye and the gel is examined under ultraviolet illumination. Figure 5.2 depicts the results of such analysis on fragments of phage DNA of known size. The mobilities of these fragments are plotted versus the log of their molecular weights (or number of base pairs). Any unknown DNA can be electrophoresed in parallel with the standard fragments, and its size can be estimated if it falls within the range of the standards. For example, a DNA with a mobility of 20 mm in Figure 5.2 would contain about 910 bp. The same principles apply to electrophoresing RNAs of various sizes.

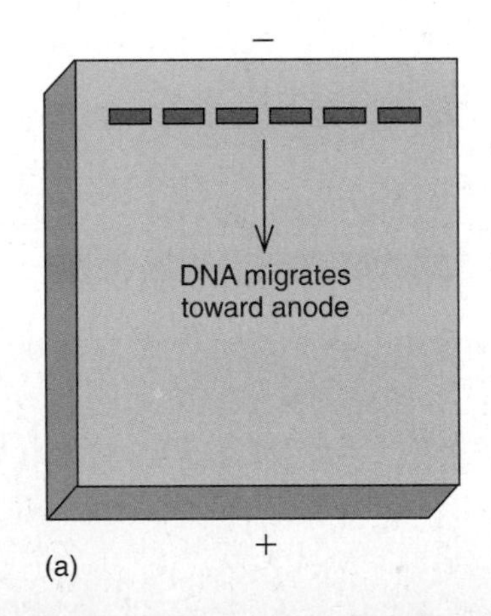

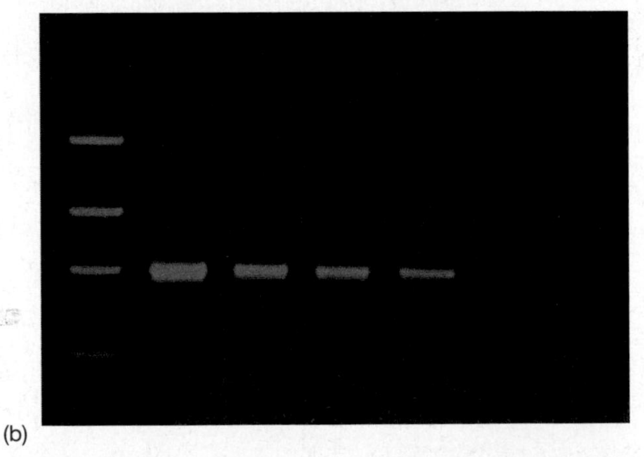

Figure 5.1 DNA gel electrophoresis. (a) Scheme of the method: This is a horizontal gel made of agarose (a substance derived from seaweed, and the main component of agar). The agarose melts at high temperature, then gels as it cools. A "comb" is inserted into the molten agarose; after the gel cools, the comb is removed, leaving slots, or wells (orange). The DNA is then placed in the wells, and an electric current is run through the gel. Because the DNA is an acid, it is negatively charged at neutral pH and electrophoreses, or migrates, toward the positive pole, or anode. **(b)** A photograph of a gel after electrophoresis showing the DNA fragments as bright bands. DNA binds to a dye that fluoresces orange under ultraviolet light, but the bands appear pink in this photograph.
(*Source:* (*b*) Reproduced with permission from Life Technologies, Inc.)

Solved Problem

Problem 1

Following is a graph showing the results of a gel electrophoresis experiment on double-stranded DNA fragments having sizes between 0.3 and 1.2 kb.

On the basis of this graph, answer the following questions:

a. What is the size of a fragment that migrated 16 mm in this experiment?

b. How far would a 0.5-kb fragment migrate in this experiment?

Solution

a. Draw a vertical dashed line from the 16-mm point on the x axis up to the experimental line. From the point where that vertical line intersects the experimental line, draw a horizontal dashed line to the

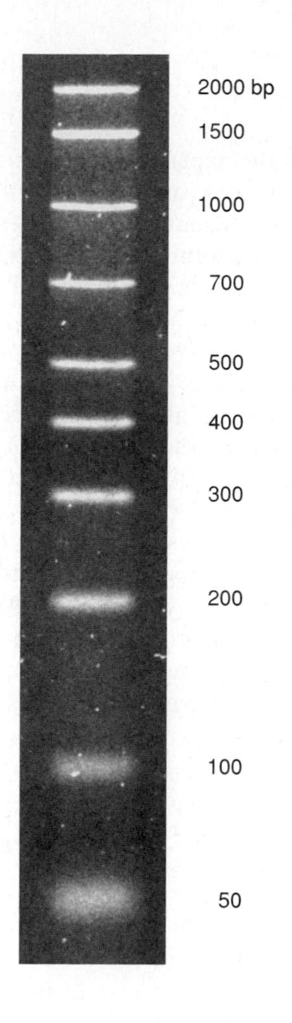

(a)

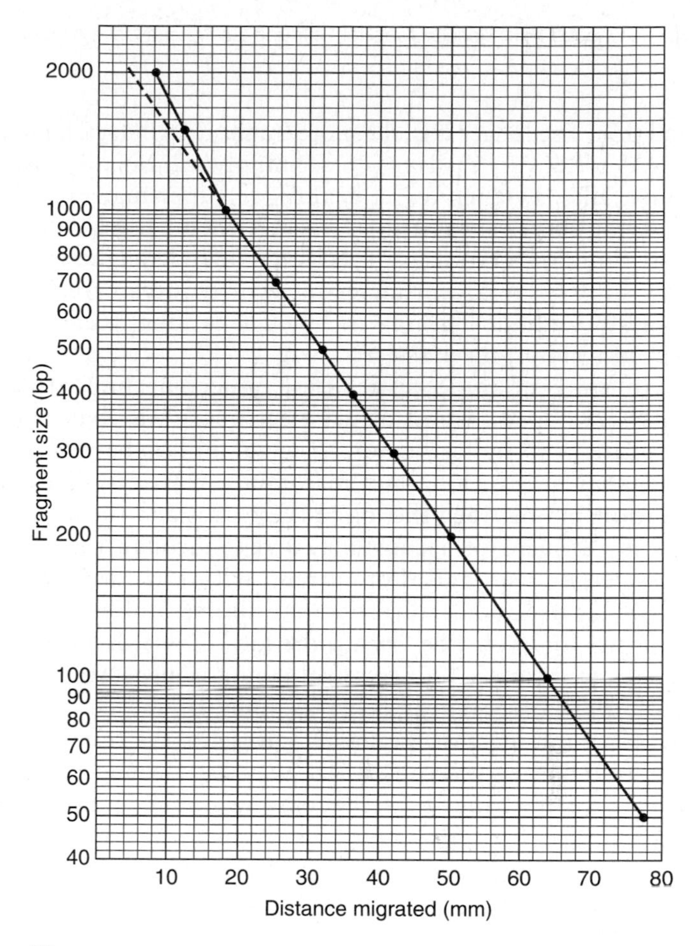

(b)

Figure 5.2 Analysis of DNA fragment size by gel electrophoresis. (a) Photograph of a stained gel of commercially prepared fragments after electrophoresis. The bands that would be orange in a color photo show up white in a black-and-white photo taken with an orange filter. The sizes of the fragments (in bp) are given at right. Note that this photo has been enlarged somewhat, so the mobilities of the bands appear a little higher than they really were. **(b)** Graph of the migration of the DNA fragments versus their sizes in base pairs. The vertical axis is logarithmic rather than linear, because the electrophoretic mobility (migration rate) of a DNA fragment is inversely proportional to the log of its size. However, notice the departure from this proportionality at large fragment sizes, represented by the difference between the solid line (actual results) and the dashed line (theoretical behavior). This suggests the limitations of conventional electrophoresis for measuring the sizes of very large DNAs. (*Source:* (a) Courtesy Bio-Rad Laboratories.)

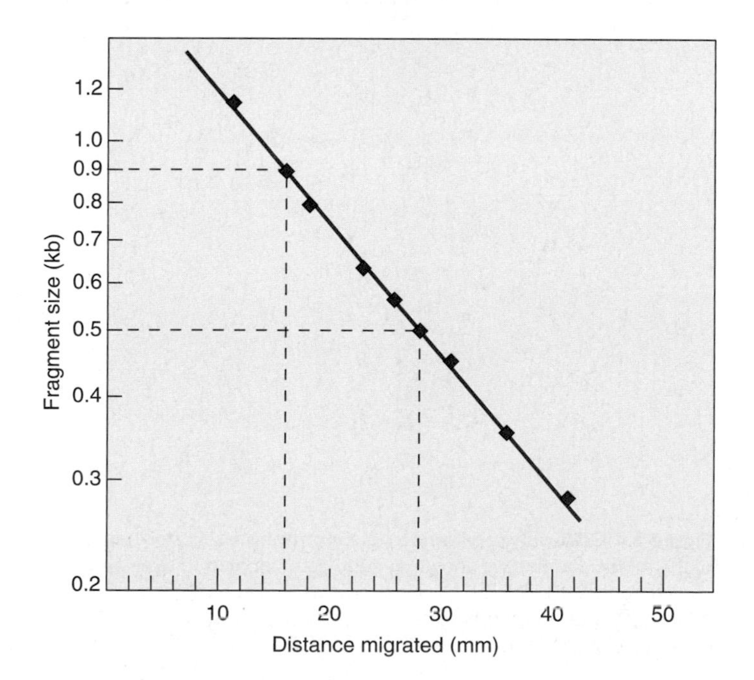

y axis. This line intersects the *y* axis at the 0.9-kb point. This shows that fragments that migrate 16 mm in this experiment are 0.9 kb (or 900 bp) long.

b. Draw a horizontal dashed line from the 0.5-kb point on the *y* axis across to the experimental line. From the point where that horizontal line intersects the experimental line, draw a vertical dashed line down to the *x* axis. This line intersects the *x* axis at the 28-mm point. This shows that 0.5-kb fragments migrate 28 mm in this experiment. ■

Determining the size of a large DNA by gel electrophoresis requires special techniques. One reason is that the relationship between the log of a DNA's size and its electrophoretic mobility deviates strongly from linearity if the DNA is very large. A hint of this deviation is apparent at the top left of Figure 5.2b. Another reason is that double-stranded DNA is a relatively rigid rod—very long and thin. The longer it is, the more fragile it is. In fact, large DNAs break very easily; even seemingly mild manipulations, like swirling in a beaker or pipetting, create shearing forces sufficient to fracture them. To visualize this, think of DNA as a piece of uncooked spaghetti. If it is short—say a centimeter or two—you can treat it roughly without harming it, but if it is long, breakage becomes almost inevitable.

In spite of these difficulties, molecular biologists have developed a kind of gel electrophoresis that can separate DNA molecules up to several million base pairs (**megabases, Mb**) long and maintain a relatively linear relationship between the log of their sizes and their mobilities. Instead of a constant current through the gel, this method uses pulses of current, with relatively long pulses in the forward direction and shorter pulses in the opposite, or even sideways, direction. This **pulsed-field gel electrophoresis (PFGE)** is valuable for measuring the sizes of DNAs even as large as some of the chromosomes found in yeast. Figure 5.3 presents the results of pulsed-field gel electrophoresis on yeast chromosomes. The 16 visible bands represent chromosomes containing 0.2–2.2 Mb.

Electrophoresis is also often applied to proteins, in which case the gel is usually made of polyacrylamide. We therefore call it **polyacrylamide gel electrophoresis,** or **PAGE.** To determine the polypeptide makeup of a complex protein, the experimenter must treat the protein so that the polypeptides, or subunits, will electrophorese independently. This is usually done by treating the protein with a detergent (**sodium dodecyl sulfate,** or **SDS**) to **denature** the subunits so they no longer bind to one another. The SDS has two added advantages: (1) It coats all the polypeptides with negative charges, so they all electrophorese toward the anode. (2) It masks the natural charges of the subunits, so they all electrophorese according to their molecular masses and not by their native charges. Small polypeptides fit easily through the pores in the gel, so they migrate rapidly. Larger polypeptides migrate more slowly. Researchers also usually employ a reducing agent to break covalent bonds between subunits.

Figure 5.4 shows the results of **SDS-PAGE** on a series of polypeptides, each of which is attached to a dye so they can be seen during electrophoresis. Ordinarily, the polypeptides would all be stained after electrophoresis with a dye such as Coomassie Blue.

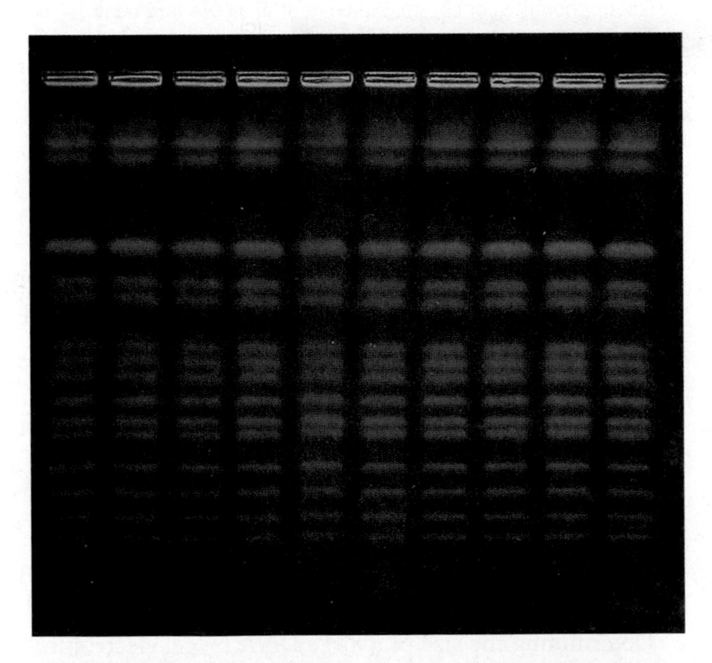

Figure 5.3 Pulsed-field gel electrophoresis of yeast chromosomes. Identical samples of yeast chromosomes were electrophoresed in 10 parallel lanes and stained with ethidium bromide. The bands represent chromosomes having sizes ranging from 0.2 Mb (at bottom) to 2.2 Mb (at top). Original gel is about 13 cm wide by 12.5 cm long. (*Source*: Courtesy Bio-Rad Laboratories/CHEF-DR(R)II pulsed-field electrophoresis systems.)

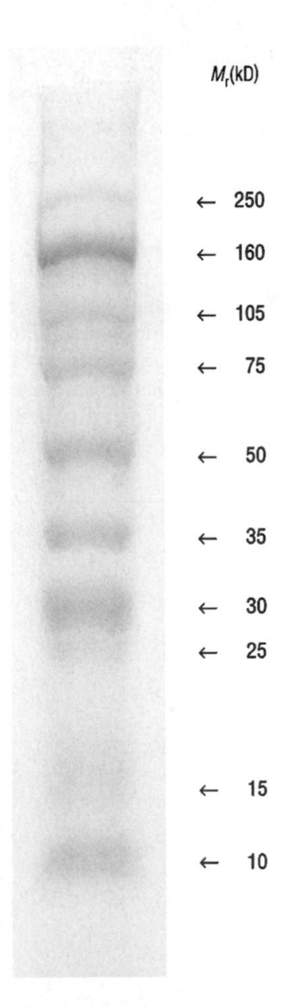

Figure 5.4 SDS-polyacrylamide gel electrophoresis. Polypeptides of the molecular masses shown at right were coupled to dyes and subjected to SDS-PAGE. The dyes allow us to see each polypeptide during and after electrophoresis. (*Source:* Courtesy of Amersham Pharmacia Biotech.)

SUMMARY DNAs, RNAs, and proteins of various masses can be separated by gel electrophoresis. The most common gel used in nucleic acid electrophoresis is agarose, but polyacrylamide is usually used in protein electrophoresis. SDS-PAGE is used to separate polypeptides according to their masses.

Two-Dimensional Gel Electrophoresis

SDS-PAGE gives very good resolution of polypeptides, but sometimes a mixture of polypeptides is so complex that we need an even better method to resolve them all. For example, we may want to separate all of the thousands of polypeptides present at a given time in a given cell type. This is very commonly done now as part of a subfield of molecular biology known as proteomics, which we will discuss in Chapter 24.

To improve on the resolving power of a one-dimensional SDS-PAGE procedure, molecular biologists have developed two-dimensional methods. In one simple method, described in Chapter 19, one can simply run nondenaturing gel electrophoresis (no SDS) in one dimension at one pH and one polyacrylamide gel concentration, then in a second dimension at a second pH and a second polyacrylamide concentration. Proteins will electrophorese at different rates at different pH values because their net charges change with pH. They will also behave differently at different polyacrylamide concentrations according to their sizes. But individual polypeptides cannot be analyzed by this method because the lack of detergent makes it impossible to separate the polypeptides that make up a complex protein.

An even more powerful method is commonly known as **two-dimensional gel electrophoresis,** even though it involves a bit more than the name implies. In the first step, the mixture of proteins is electrophoresed through a narrow tube gel containing molecules called ampholytes that set up a pH gradient from one end of the tube to the other. A negatively charged molecule will electrophorese toward the anode until it reaches its **isoelectric point,** the pH at which it has no net charge. Without net charge, it is no longer drawn toward the anode, or the cathode, for that matter, so it stops. This step is called **isoelectric focusing** because it focuses proteins at their isoelectric points in the gel.

In the second step, the gel is removed from the tube and placed at the top of a slab gel for ordinary SDS-PAGE. Now the proteins that have been partially resolved by isoelectric focusing are further resolved according to their sizes by SDS-PAGE. Figure 5.5 presents two-dimensional gel electrophoresis separations of *E. coli* proteins grown in the presence and absence of benzoic acid. Proteins from the

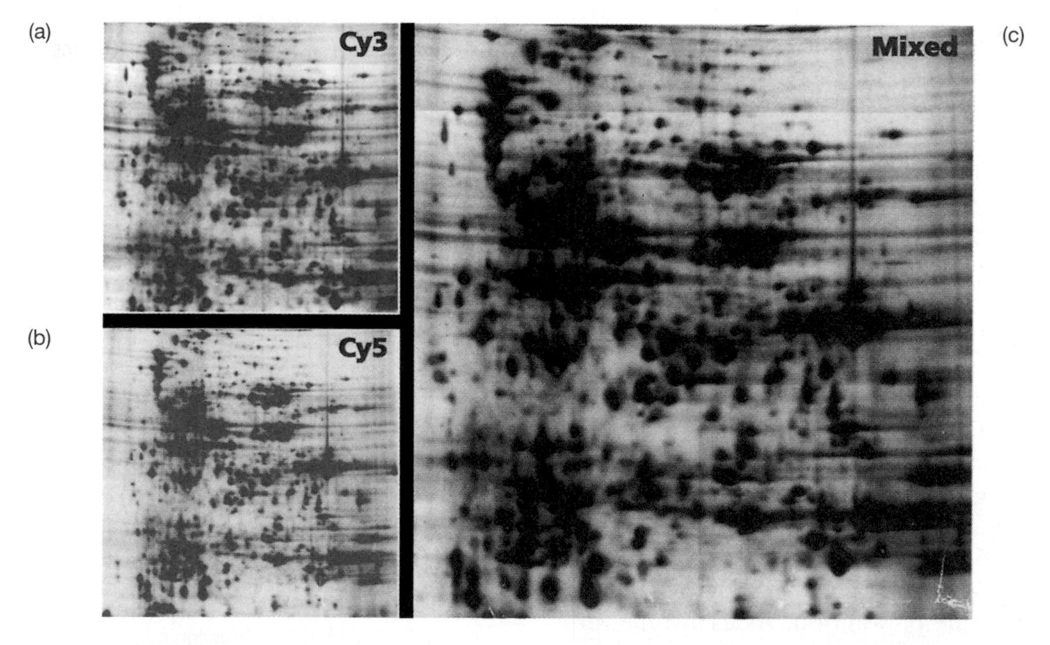

Figure 5.5 Two-dimensional gel electrophoresis. In this experiment, the investigators grew *E. coli* cells in the presence or absence of benzoic acid. Then they stained a lysate of the cells grown in the absence of benzoic acid with the red fluorescent dye Cy3, so the proteins from that lysate would fluoresce red. They stained a lysate of the cells grown in the presence of benzoic acid with the blue fluorescent dye Cy5, so those proteins would fluoresce blue. Finally, they performed two-dimensional gel electrophoresis on **(a)** the proteins from cells grown in the absence of benzoic acid, **(b)** on the proteins grown in the presence of benzoic acid, and **(c)** on a mixture of the two sets of proteins. In panel (c), the proteins that accumulate only in the absence of benzoic acid fluoresce red, those that accumulate only in the presence of benzoic acid fluoresce blue, and those that accumulate under both conditions fluoresce both red and blue, and so appear purple or black. (*Source:* Courtesy of Amersham Pharmacia Biotech.)

cells grown without benzoic acid were stained with the red fluorescent dye Cy3, and proteins from the cells grown with benzoic acid were stained with the blue fluorescent dye Cy5. Two-dimensional gel electrophoresis of these two sets of proteins, separately and together allows us to see which proteins are prevalent in the presence or absence of benzoic acid, and which are prevalent under both conditions.

SUMMARY High-resolution separation of polypeptides can be achieved by two-dimensional gel electrophoresis, which uses isoelectric focusing in the first dimension and SDS-PAGE in the second.

Ion-Exchange Chromatography

Chromatography is a term that originally referred to the pattern one sees after separating colored substances on paper (**paper chromatography**). Nowadays, many different types of chromatography exist for separating biological substances. **Ion-exchange chromatography** uses a resin to separate substances according to their charges. For example, DEAE-Sephadex chromatography uses an ion-exchange resin that contains positively charged diethylaminoethyl (DEAE) groups. These positive charges attract negatively charged substances, including proteins. The greater the negative charge, the tighter the binding.

In Chapter 10, we will see an example of DEAE-Sephadex chromatography in which the experimenters separated three forms of an enzyme called RNA polymerase. They made a slurry of DEAE-Sephadex and poured it into a column. After the resin had packed down, they loaded the sample, a crude cellular extract containing the RNA polymerases. Finally, they **eluted,** or removed, the substances that had bound to the resin in the column by passing a solution of gradually increasing ionic strength (or salt concentration) through the column. The purpose of this salt gradient was to use the negative ions in the salt solution to compete with the proteins for ionic binding sites on the resin, thus removing the proteins one by one. This is why we call it ion-exchange chromatography.

As the ionic strength of the elution buffer increases, samples of solution flowing through the column are collected using a fraction collector. This device works by positioning test tubes, one at a time, beneath the column to collect a given volume of solution. As each tube finishes collecting its fraction of the solution, it moves aside and a new tube moves into position to collect its fraction. Finally, each fraction is assayed (tested) to determine how much of the substance of interest it contains. If the substance is an enzyme, the fractions are assayed for that particular enzyme activity. It is also useful to measure the ionic strength of each fraction to determine what salt concentration is necessary to elute each of the enzymes of interest.

One can also use a negatively charged resin to separate positively charged substances, including proteins. For example, phosphocellulose is commonly used to separate proteins by cation-exchange chromatography. Note that it is not essential for a protein to have a net positive charge to bind to a cation-exchange resin like phosphocellulose. Most proteins have a net negative charge, yet they can still bind to a cation exchange resin if they have a significant center of positive charge. Figure 5.6 depicts the results of a hypothetical ion-exchange chromatography experiment in which two forms of an enzyme are separated.

SUMMARY Ion-exchange chromatography can be used to separate substances, including proteins, according to their charges. Positively charged resins like DEAE-Sephadex are used for anion-exchange chromatography, and negatively charged resins like phosphocellulose are used for cation-exchange chromatography.

Gel Filtration Chromatography

Standard biochemical separations of proteins usually require more than one step, and, because valuable protein is lost at each step, it is important to minimize the number of these steps. One way to do this is to design a strategy that enables each step to take advantage of a different property of the protein of interest. Thus, if anion-exchange chromatography is the first step and cation-exchange chromatography is the second, a third step that separates proteins on some other basis besides charge is needed. Protein size is an obvious next choice.

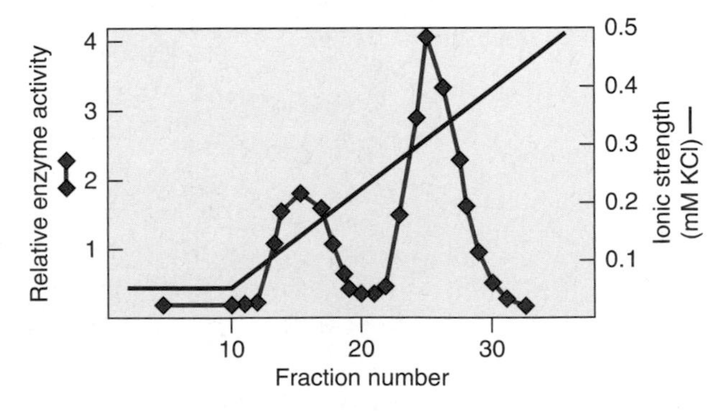

Figure 5.6 Ion-exchange chromatography. Begin by loading a cell extract containing two different forms of an enzyme onto an ion-exchange column. Then pass a buffer of increasing ionic strength through the column and collect fractions (32 fractions in this case). Assay each fraction for enzyme activity (red) and ionic strength (blue), and plot the data as shown. The two forms of the enzyme are clearly separated by this procedure.

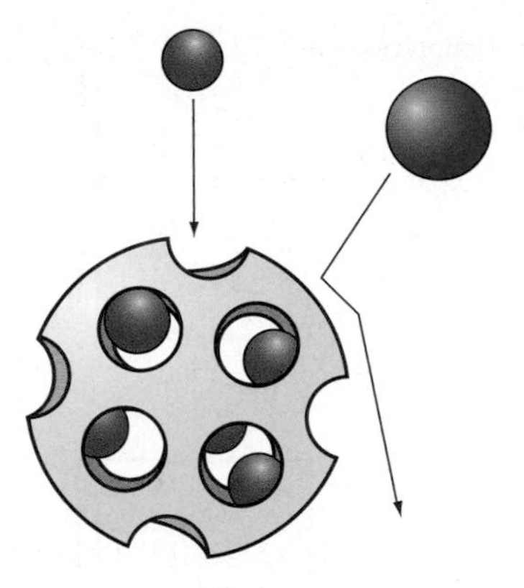
(a)

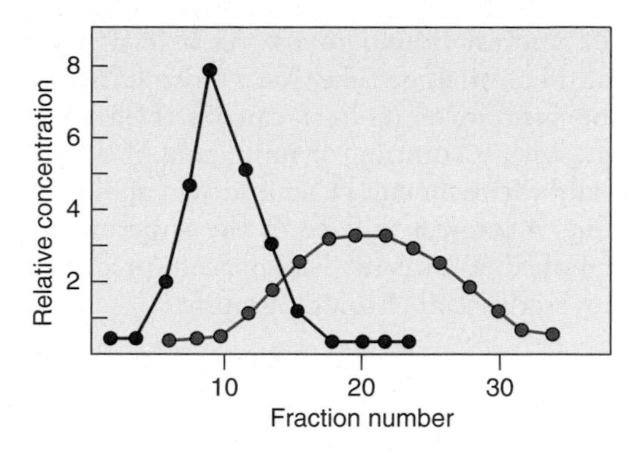

(b)

Figure 5.7 Gel filtration chromatography. (a) Principle of the method. A resin bead is schematically represented as a "whiffle ball" (yellow). Large molecules (blue) cannot fit into the beads, so they are confined to the relatively small buffer volume outside the beads. Thus, they emerge quickly from the column. Small molecules (red), by contrast, can fit into the beads and so have a large buffer volume available to them. Accordingly, they take a longer time to emerge from the column. **(b)** Experimental results. Add a mixture of large and small molecules from panel (a) to the column, and elute them by passing buffer through the column. Collect fractions and assay each for concentration of the large (blue) and small (red) molecules. As expected, the large molecules emerge earlier than the small ones.

Gel filtration chromatography is one method that separates molecules based on their physical dimensions. Gel filtration resins such as Sephadex are porous beads of various sizes that can be likened to "whiffle balls," hollow plastic balls with holes in them. Imagine a column filled with tiny whiffle balls. When one passes a solution containing different size molecules through this column, the small molecules will easily enter the holes in the whiffle balls (the pores in the beads) and therefore flow through the column slowly. On the other hand, large molecules will not be able to enter any of the beads and will flow more quickly through the column. They emerge with the so-called **void volume**—the volume of buffer surrounding the beads, but not included in the beads. Intermediate-size molecules will enter some beads and not others and so will have an intermediate mobility. Thus, large molecules will emerge first from the column, and small molecules will emerge last. Many different resins with different size pores are available for separating different size molecules. Figure 5.7 illustrates this method.

SUMMARY Gel filtration chromatography uses columns filled with porous resins that let in smaller substances, but exclude larger ones. Thus, the smaller substances are slowed in their journey through the column, but larger substances travel relatively rapidly through the column.

Affinity Chromatography

One of the most powerful separation techniques is **affinity chromatography,** in which the resin contains a substance (an *affinity reagent*) to which the molecule of interest has strong and specific affinity. For example, the resin may be coupled to an antibody that recognizes a specific protein, or it may contain an unreactive analog of an enzyme's substrate. In the latter case, the enzyme will bind strongly to the analog, but will not metabolize it. After virtually all the contaminating proteins have flowed through the column because they have no (or weak) affinity for the affinity reagent, the molecule of interest can be eluted from the column using a solution of a substance that competes with binding between the molecule of interest and the affinity reagent. For example, a solution of the enzyme analog could be used. In this case, the analog in solution will compete with the analog on the resin for binding to the enzyme and the enzyme will elute from the column.

The power of affinity chromatography lies in the specificity of binding between the affinity reagent on the resin and the molecule to be purified. Indeed, it is possible to design an affinity chromatography procedure to purify a protein in a single step because that protein is the only one in the cell that will bind to the affinity reagent. In Chapter 4 we saw a good example: the use of a nickel column to purify a protein tagged with oligohistidine. Because all of the other proteins in the cell are natural and are therefore not tagged with oligohistidine, the tagged protein is the only one that will stick to the affinity reagent, nickel. In that case, one could elute the protein from the column with a nickel solution, but that would yield a protein-nickel complex, rather than a pure protein. So investigators use a histidine analog, imidazole, which also disrupts binding between the affinity reagent and the protein of interest—by binding to the nickel on the column.

When the molecule to be purified (e.g., an oligohistidine-tagged protein) is the only one that binds to the affinity resin, column chromatography is not even needed. Instead, the investigator can simply mix the resin with a cell extract, spin down the resin in a centrifuge, throw away the remaining solution (the **supernatant**), leaving the protein of interest bound to the resin in a pellet at the bottom of the centrifuge tube. After rinsing the pellet with buffer, the protein of interest can be released from the resin (e.g., with a solution of imidazole, if a nickel resin is used), and the resin can be spun down again. This time, the protein of interest will be in the supernatant, which can be removed and saved. This procedure is simpler and faster than traditional chromatography.

SUMMARY Affinity chromatography is a powerful purification technique that exploits an affinity reagent with strong and specific affinity for a molecule of interest. That molecule binds to a column coupled to the affinity reagent but all or most other molecules flow through without binding. Then the molecule of interest can be eluted from the column with a solution of a substance that disrupts the specific binding.

5.2 Labeled Tracers

Until recently, "labeled" has been virtually synonymous with "radioactive" because radioactive tracers have been available for decades, and they are easy to detect. Radioactive tracers allow vanishingly small quantities of substances to be detected. This is important in molecular biology because the substances we are trying to detect in a typical experiment are present in very tiny amounts. Let us assume, for example, that we are attempting to measure the appearance of an RNA product in a transcription reaction. We may have to detect RNA quantities of less than a picogram (pg; only one trillionth of a gram, or 10^{-12} g). Direct measurement of such tiny quantities by UV light absorption or by staining with dyes is not possible because of the limited sensitivities of these methods. On the other hand, if the RNA is radioactive we can measure small amounts of it easily because of the great sensitivity of the equipment used to detect radioactivity. Let us now consider the favorite techniques molecular biologists use to detect radioactive tracers: autoradiography, phosphorimaging, and liquid scintillation counting.

Autoradiography

Autoradiography is a means of detecting radioactive compounds with a photographic emulsion. The form of emulsion favored by molecular biologists is a piece of x-ray film. Figure 5.8 presents an example in which the investigator electrophoreses some radioactive DNA fragments on a gel and then places the gel in contact with the x-ray film and leaves it in the dark for a few hours, or even days. The radioactive emissions from the bands of DNA expose the film, just as visible light would. Thus, when the

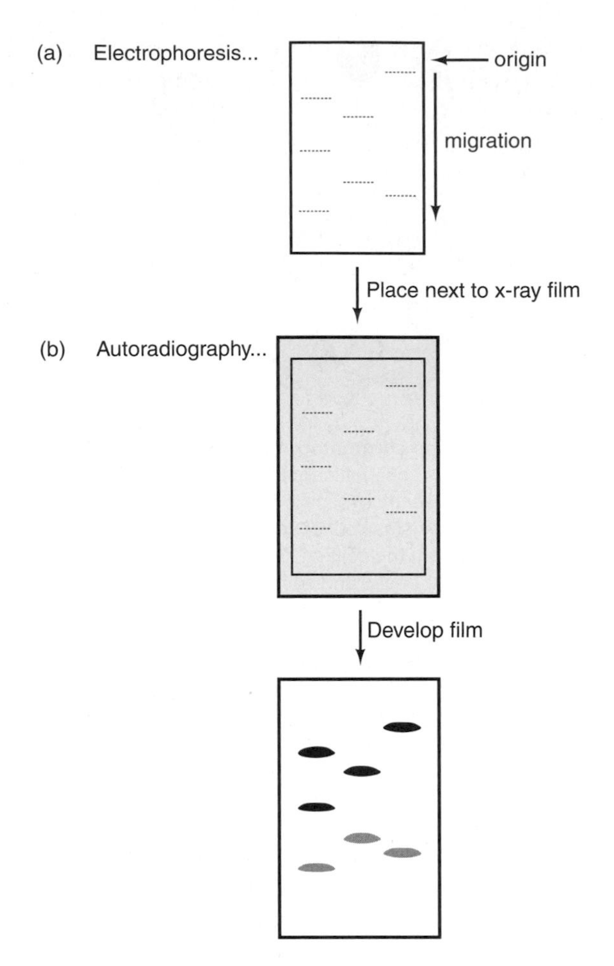

Figure 5.8 Autoradiography. (a) Gel electrophoresis. Electrophorese radioactive DNA fragments in three parallel lanes on a gel, either agarose or polyacrylamide, depending on the sizes of the fragments. At this point the DNA bands are invisible, but their positions are indicated here with dotted lines. **(b)** Autoradiography. Place a piece of x-ray film in contact with the gel and leave it for several hours, or even days if the DNA fragments are only weakly radioactive. Finally, develop the film to see where the radioactivity has exposed the film. This shows the locations of the DNA bands on the gel. In this case, the large, slowly migrating bands are the most radioactive, so the bands on the autoradiograph that correspond to them are the darkest.

film is developed, dark bands appear, corresponding to the DNA bands on the gel. In effect, the DNA bands take a picture of themselves, which is why we call this technique *auto*radiography.

To enhance the sensitivity of autoradiography, at least with ^{32}P, one can use an **intensifying screen.** This is a screen coated with a compound that fluoresces when it is excited by β electrons at low temperature. (β electrons are the radioactive emissions from the common radioisotopes used in molecular biology: ^{3}H, ^{14}C, ^{35}S, and ^{32}P.) Thus, one can put a radioactive gel (or other medium) on one side of a photographic film and the intensifying screen on the other. Some β electrons expose the film directly, but others pass right through the film and would be lost without the screen.

When these high-energy electrons strike the screen, they cause fluorescence, which is detected by the film.

An intensifying screen works well with ^{32}P β electrons because they have high energy and therefore can pass easily through an x-ray film. The β electrons emitted by ^{14}C and ^{35}S are about 10-fold less energetic, and so barely make it out of a gel, let alone through an x-ray film. Tritium (^{3}H) β electrons are about 10-fold weaker still, and so cannot reach the x-ray film in significant numbers. For these lower energy radioisotopes, **fluorography** provides a way to enhance the image. In this technique, the experimenter soaks the gel in a **fluor,** a compound that fluoresces when it is impacted by a β electron, even one from ^{3}H. Because the fluor disperses throughout the gel, there are always fluor molecules very close to the radioactive nuclei, so even weak β electrons will excite them and give rise to light. This light then exposes the x-ray film.

What if the goal is to measure the exact amount of radioactivity in a fragment of DNA? One can get a rough estimate by looking at the intensity of a band on an autoradiograph, and an even better estimate by scanning the autoradiograph with a **densitometer.** This instrument passes a beam of light through a sample—an autoradiograph in this case—and measures the absorbance of that light by the sample. If the band is very dark, it will absorb most of the light, and the densitometer records a large peak of absorbance (Figure 5.9). If the band is faint, most of the light passes through, and the densitometer records only a minor peak of absorbance. By measuring the area under each peak, one can get an estimate of the radioactivity in each band. This is still an indirect measure of radioactivity, however. To get a really accurate reading of the radioactivity in each band, one can scan the gel with a phosphorimager, or subject the DNA to liquid scintillation counting.

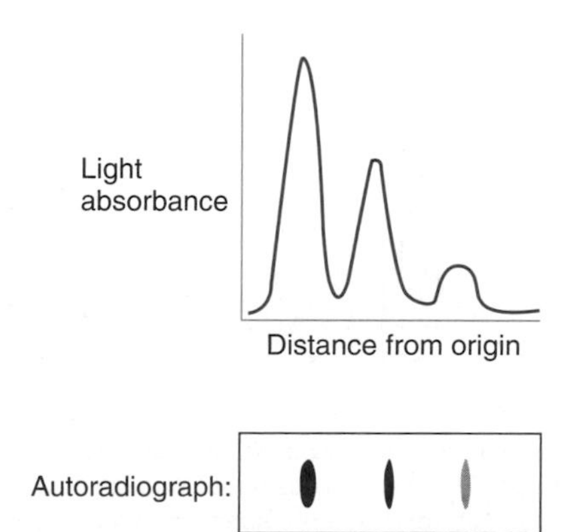

Figure 5.9 Densitometry. An autoradiograph is pictured beneath a densitometer scan of the same film. Notice that the areas under the three peaks of the scan are proportional to the darkness of the corresponding bands on the autoradiograph.

Phosphorimaging

The technique of **phosphorimaging** has several advantages over standard autoradiography, but the most important is that it is much more accurate in quantifying the amount of radioactivity in a substance. This is because its response to radioactivity is far more linear than that of an x-ray film. With standard autoradiography, a band with 50,000 radioactive **disintegrations per minute (dpm)** may look no darker than one with 10,000 dpm because the emulsion in the film is already saturated at 10,000 dpm. But the **phosphorimager** detects radioactive emissions and analyzes them electronically, so the difference between 10,000 dpm and 50,000 dpm would be obvious. Here is how this technique works: One starts with a radioactive sample—a blot with RNA bands that have hybridized with a labeled probe, for example. This sample is placed in contact with a phosphorimager plate, which absorbs β electrons. These electrons excite molecules on the plate, and these molecules remain in an excited state until the phosphorimager scans the plate with a laser. At that point, the β electron energy trapped by the plate is released and monitored by a computerized detector. The computer converts the energy it detects to an image such as the one in Figure 5.10. This is a false color image, in which the different colors represent different degrees of radioactivity, from the lowest (yellow) to the highest (black).

Figure 5.10 False color phosphorimager scan of an RNA blot. After hybridizing a radioactive probe to an RNA blot and washing away unhybridized probe, the blot was exposed to a phosphorimager plate. The plate collected energy from β electrons from the radioactive probe bound to the RNA bands, then gave up this energy when scanned with a laser. A computer converted this energy into an image in which the colors correspond to radiation intensity according to the following color scale: yellow (lowest) $<$ purple $<$ magenta $<$ light blue $<$ green $<$ dark blue $<$ black (highest). (*Source:* © Jay Freis/Image Bank/Getty.)

Liquid Scintillation Counting

Liquid scintillation counting uses the radioactive emissions from a sample to create photons of visible light that a photomultiplier tube can detect. To do this, one places the radioactive sample (a band cut out of a gel, for example), into a vial with **scintillation fluid.** This fluid contains a fluor, which, in effect, converts the invisible radioactivity into visible light, just as it does in the fluorography technique discussed earlier in this chapter. A liquid scintillation counter is an instrument that lowers the vial into a dark chamber with a photomultiplier tube. There, the tube detects the light resulting from the radioactive emissions exciting the fluor. The instrument counts these bursts of light, or **scintillations,** and records them as **counts per minute (cpm).** This is not the same as disintegrations per minute because the scintillation counter is not 100% efficient. One common radioisotope used by molecular biologists is ^{32}P. The β electrons emitted by this isotope are so energetic that they create photons even without a fluor, so a liquid scintillation counter can count them directly, though at a lower efficiency than with scintillation fluid.

SUMMARY Detection of the tiny quantities of substances used in molecular biology experiments generally requires the use of labeled tracers. If the tracer is radioactive one can detect it by autoradiography, using x-ray film or a phosphorimager, or by liquid scintillation counting.

Nonradioactive Tracers

As we pointed out earlier in this section, the enormous advantage of radioactive tracers is their sensitivity, but now nonradioactive tracers rival the sensitivity of their radioactive forebears. This can be a significant advantage because radioactive substances pose a potential health hazard and must be handled very carefully. Furthermore, radioactive tracers create radioactive waste, and disposal of such waste is increasingly difficult and expensive. How can a nonradioactive tracer compete with the sensitivity of a radioactive one? The answer is, by using the multiplier effect of an enzyme. An enzyme is coupled to a probe that detects the molecule of interest, so the enzyme will produce many molecules of product, thus amplifying the signal. This works especially well if the product of the enzyme is chemiluminescent (light-emitting, like the tail of a firefly), because each molecule emits many photons, amplifying the signal again. Figure 5.11 shows the principle behind one such tracer method. The light can be detected by autoradiography with x-ray film, or by a phosphorimager.

To avoid the expense of a phosphorimager or x-ray film, one can use enzyme substrates that change color instead of becoming chemiluminescent. These **chromogenic substrates** produce colored bands corresponding to the location of the enzyme and, therefore, to the location of the molecule of interest. The intensity of the color is directly related to the amount of that molecule, so this is also a quantitative method.

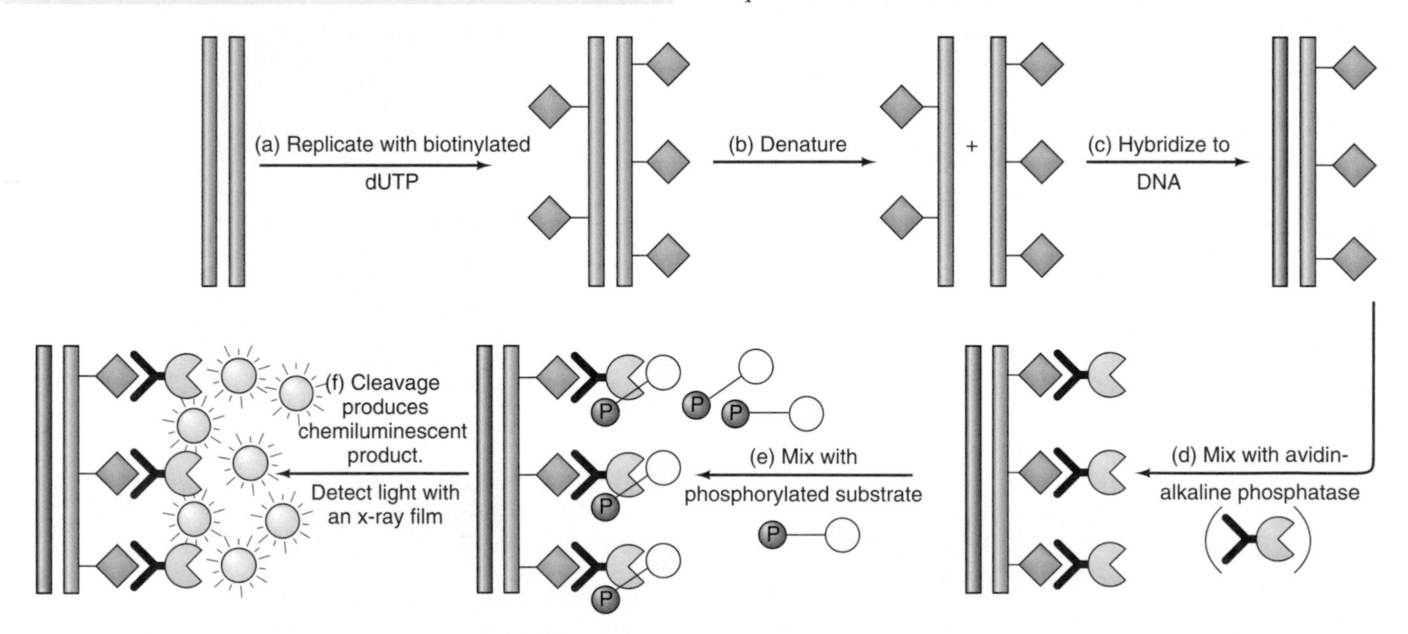

Figure 5.11 Detecting nucleic acids with a nonradioactive probe. This sort of technique is usually indirect; detecting a nucleic acid of interest by hybridization to a labeled probe that can in turn be detected by virtue of its ability to produce a colored or light-emitting substance. In this example, the following steps are executed. **(a)** Replicate the probe DNA in the presence of dUTP that is tagged with the vitamin biotin (blue). This generates biotinylated probe DNA. **(b)** Denature this probe and **(c)** hybridize it to the DNA to be detected (pink). **(d)** Mix the hybrids with a bifunctional reagent containing both avidin and the enzyme alkaline phosphatase (green). The avidin binds tightly and specifically to the biotin in the probe DNA. **(e)** Add a phosphorylated compound that will become chemiluminescent as soon as its phosphate group is removed. **(f)** The alkaline phosphatase enzymes attached to the probe cleave the phosphates from these substrate molecules, rendering them chemiluminescent (light-emitting). The light emitted from the chemiluminescent substrate can be detected with an x-ray film.

SUMMARY Some very sensitive nonradioactive labeled tracers are now available. Those that employ chemiluminescence can be detected by autoradiography or by phosphorimaging, just as if they were radioactive. Those that produce colored products can be detected directly, by observing the appearance of colored spots.

5.3 Using Nucleic Acid Hybridization

The phenomenon of hybridization—the ability of one single-stranded nucleic acid to form a double helix with another single strand of complementary base sequence—is one of the backbones of modern molecular biology. We have already encountered plaque and colony hybridization in Chapter 4. Here we will illustrate several further examples of hybridization techniques.

Southern Blots: Identifying Specific DNA Fragments

Many eukaryotic genes are parts of families of closely related genes. How would one determine the number of family members in a particular gene family? If a member of that gene family—even a partial cDNA—has been cloned, one can estimate this number.

One begins by using a restriction enzyme to cut genomic DNA isolated from the organism. It is best to use a restriction enzyme such as *Eco*RI or *Hin*dIII that recognizes a 6-bp cutting site. These enzymes will produce thousands of fragments of genomic DNA, with an average size of about 4000 bp. Next, these fragments are electrophoresed on an agarose gel (Figure 5.12). The result, if the bands are visualized by staining, will be a blurred streak of thousands of bands, none distinguishable from the others (although Figure 5.12, for simplicity's sake shows just a few bands). Eventually, a labeled probe will be hybridized to these bands to see how many of them contain coding sequences for the gene of interest. First, however, the bands are transferred to a medium on which hybridization is more convenient.

Edward Southern was the pioneer of this technique; he transferred, or blotted, DNA fragments from an agarose gel to nitrocellulose by diffusion, as depicted in Figure 5.12. This process has been called **Southern blotting** ever since. Nowadays, blotting is frequently done by electrophoresing the DNA bands out of the gel and onto the blot. Before blotting, the DNA fragments are denatured with alkali so that the resulting single-stranded DNA can bind to the nitrocellulose, forming the Southern blot. Media superior to nitrocellulose are now available; some use nylon

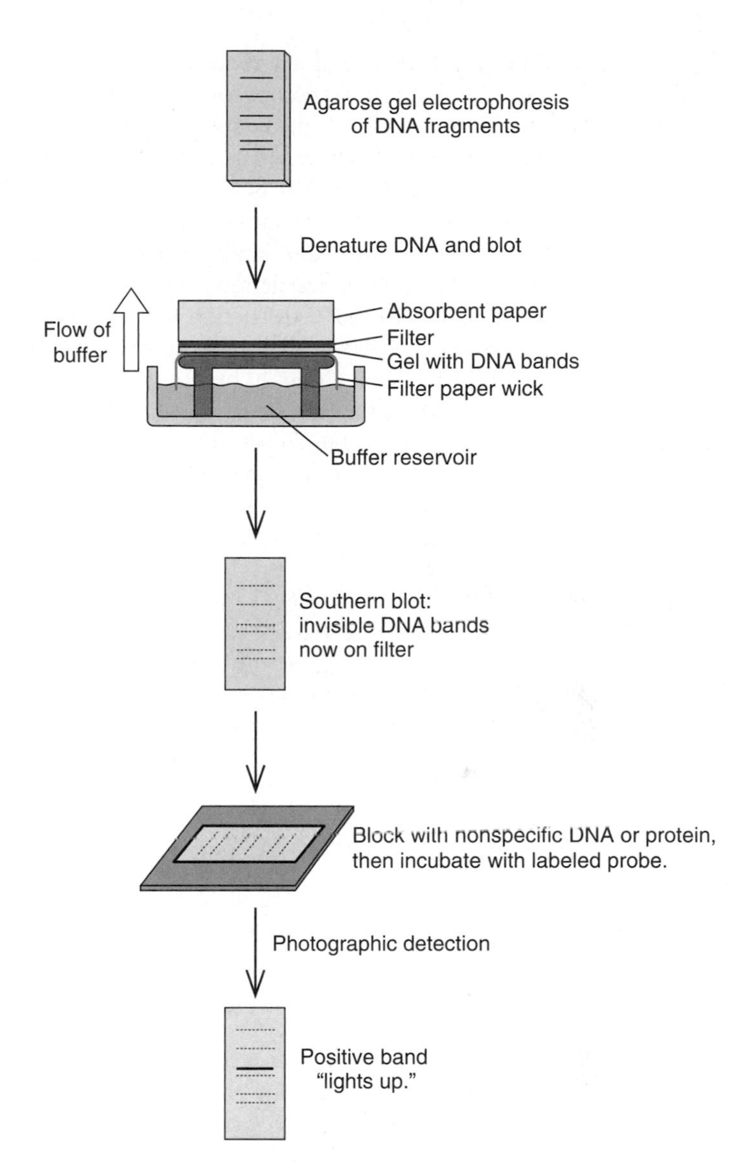

Figure 5.12 Southern blotting. First, electrophorese DNA fragments in an agarose gel. Next, denature the DNA with base and transfer the single-stranded DNA fragments from the gel (yellow) to a sheet of nitrocellulose or another DNA-binding material (red). One can do this in two ways: by diffusion, as shown here, in which buffer passes through the gel, carrying the DNA with it, or by electrophoresis (not shown). Next, hybridize the blot to a labeled probe and detect the labeled bands by autoradiography or phosphorimaging.

supports that are far more flexible than nitrocellulose. Next, the cloned DNA is labeled by adding DNA polymerase to it in the presence of labeled DNA precursors. Then this labeled probe is denatured and hybridized to the Southern blot. Wherever the probe encounters a complementary DNA sequence, it hybridizes, forming a labeled band corresponding to the fragment of DNA containing the gene of interest. Finally, these bands are visualized by autoradiography with x-ray film or by phosphorimaging.

If only one band is seen, the interpretation is relatively easy; probably only one gene has a sequence matching the cDNA probe. Alternatively, a gene (e.g., a histone or ribosomal

RNA gene) could be repeated over and over again in tandem, with a single restriction site in each copy of the gene. This would yield a single very dark band. If multiple bands are seen, multiple genes are probably present, but it is difficult to tell exactly how many. One gene can give more than one band if it contains one or more cutting sites for the restriction enzyme used. One can minimize this problem by using a short probe, such as a 100–200-bp restriction fragment of the cDNA, for example. Chances are, a restriction enzyme that cuts on average only every 4000 bp will not cut within the 100–200-bp region of the genes that hybridize to such a probe. If multiple bands are still obtained with a short probe, they probably represent a gene family whose members' sequences are similar or identical in the region that hybridizes to the probe.

SUMMARY Labeled DNA (or RNA) probes can be used to hybridize to DNAs of the same, or very similar, sequence on a Southern blot. The number of bands that hybridize to a short probe gives an estimate of the number of closely related genes in an organism.

DNA Fingerprinting and DNA Typing

Southern blots are not just a research tool. They are widely used in forensic laboratories to identify individuals who have left blood or other DNA-containing material at the scenes of crimes. Such **DNA typing** has its roots in a discovery by Alec Jeffreys and his colleagues in 1985. These workers were investigating a DNA fragment from the gene for a human blood protein, α-globin, when they discovered that this fragment contained a sequence of bases repeated several times. This kind of repeated DNA is called a **minisatellite.** More interestingly, they found similar minisatellite sequences in other places in the human genome, again repeated several times. This simple finding turned out to have far-reaching consequences, because individuals differ in the pattern of repeats of the basic sequence. In fact, they differ enough that two individuals have only a remote chance of having exactly the same pattern. That means that these patterns are like fingerprints; indeed, they are called **DNA fingerprints.**

A DNA fingerprint is really just a Southern blot. To make one, investigators first cut the DNA under study with a restriction enzyme such as *Hae*III. Jeffreys chose this enzyme because the repeated sequence he had found did not contain a *Hae*III recognition site. That means that *Hae*III will cut on either side of the minisatellite regions, but not inside, as shown in Figure 5.13a. In this case, the DNA has three sets of repeated regions, containing four, three, and two repeats, respectively. Thus, three different-size fragments bearing these repeated regions will be produced.

Next, the fragments are electrophoresed, denatured, and blotted. The blot is then probed with a labeled minisatellite DNA, and the labeled bands are detected with x-ray film, or by phosphorimaging. In this case, three labeled bands occur, so three dark bands will appear on the film (Figure 5.13d).

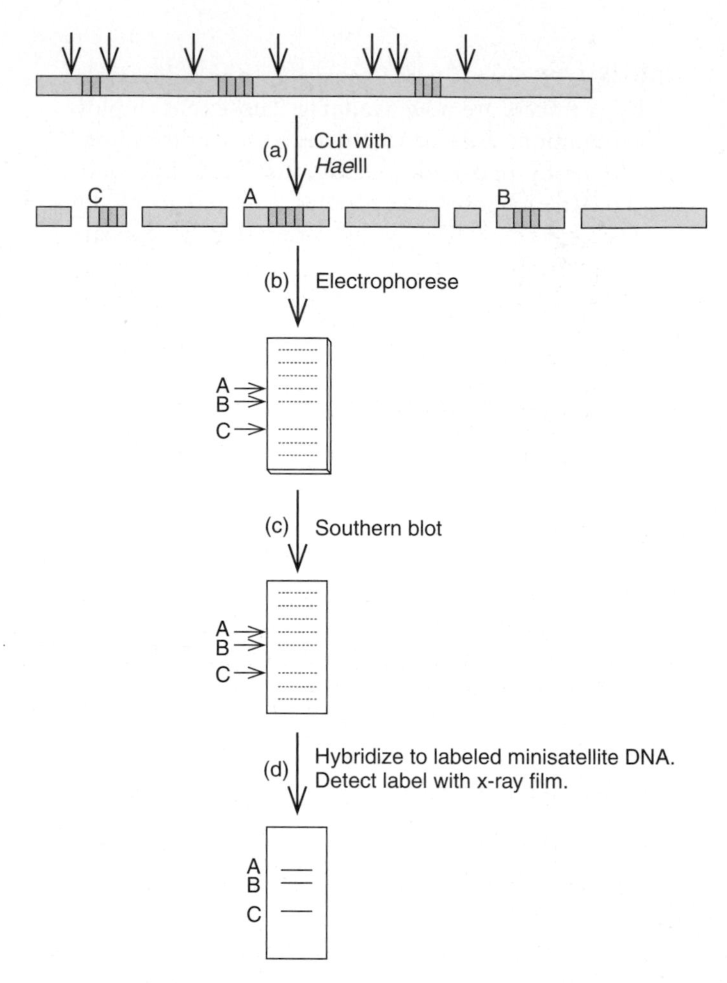

Figure 5.13 DNA fingerprinting. (a) First, cut the DNA with a restriction enzyme. In this case, the enzyme *Hae*III cuts the DNA in seven places (short arrows), generating eight fragments. Only three of these fragments (labeled A, B, and C according to size) contain the minisatellites, represented by blue boxes. The other fragments (yellow) contain unrelated DNA sequences. **(b)** Electrophorese the fragments from part (a), which separates them according to their sizes. All eight fragments are present in the electrophoresis gel, but they remain invisible. The positions of all the fragments, including the three (A, B, and C) with minisatellites are indicated by dotted lines. **(c)** Denature the DNA fragments and Southern blot them. **(d)** Hybridize the DNA fragments on the Southern blot to a labeled DNA with several copies of the minisatellite. This probe will bind to the three fragments containing the minisatellites, but with no others. Finally, use x-ray film or phosphorimaging to detect the three labeled bands.

Real animals have a much more complex genome than the simple piece of DNA in this example, so they will have many more than three fragments that contain a minisatellite sequence that will react with the probe. Figure 5.14 shows an example of the DNA fingerprints of several unrelated people and a set of monozygotic twins. As we have already mentioned, this is such a complex pattern of fragments that

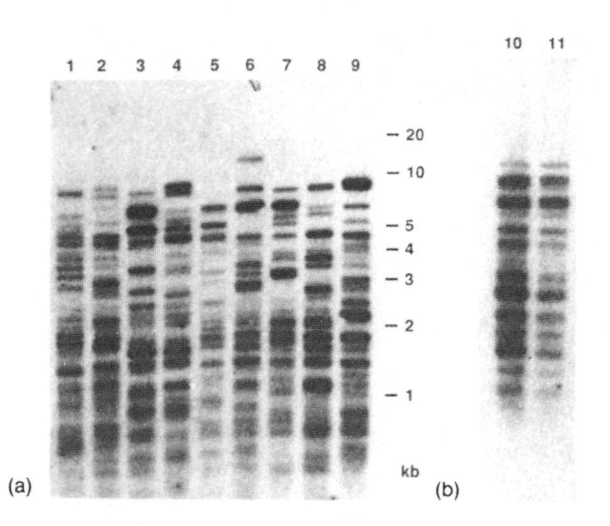

Figure 5.14 DNA fingerprint. (a) The nine lanes contain DNA from nine unrelated white subjects. Note that no two patterns are identical, especially at the upper end. **(b)** The two lanes contain DNA from monozygotic twins, so the patterns are identical (although there is more DNA in lane 10 than in lane 11). (*Source:* G. Vassart et al., A sequence in M13 phage detects hypervariable minisatellites in human and animal DNA. *Science* 235 (6 Feb 1987) p. 683, f. 1. © AAAS.)

the patterns for two individuals are extremely unlikely to be identical, unless they come from monozygotic twins. This complexity makes DNA fingerprinting a very powerful identification technique.

Forensic Uses of DNA Fingerprinting and DNA Typing

A valuable feature of DNA fingerprinting is the fact that, although almost all individuals have different patterns, parts of the pattern (sets of bands) are inherited in a Mendelian fashion. Thus, fingerprints can be used to establish parentage. An immigration case in England illustrates the power of this technique. A Ghanaian boy born in England had moved to Ghana to live with his father. When he wanted to return to England to be with his mother, British authorities questioned whether he was a son or a nephew of the woman. Information from blood group genes was equivocal, but DNA fingerprinting of the boy demonstrated that he was indeed her son.

In addition to testing parentage, DNA fingerprinting has the potential to identify criminals. This is because a person's DNA fingerprint is, in principle, unique, just like a traditional fingerprint. Thus, if a criminal leaves some of his cells (blood, semen, or hair, for example) at the scene of a crime, the DNA from these cells can identify him. As Figure 5.14 showed, however, DNA fingerprints are very complex. They contain dozens of bands, some of which smear together, which can make them hard to interpret.

To solve this problem, forensic scientists have developed probes that hybridize to a single DNA locus that varies from one individual to another, rather than to a whole set of DNA loci as in a classical DNA fingerprint. Each probe now gives much simpler patterns, containing only one or a few bands. This is an example of a restriction fragment length polymorphism (RFLP) disussed in detail in Chapter 24. RFLPs occur because the pattern of restriction fragment sizes at a given locus varies from one person to another. Of course, each probe by itself is not as powerful an identification tool as a whole DNA fingerprint with its multitude of bands, but a panel of four or five probes can give enough different bands to be definitive. We sometimes still call such analysis DNA fingerprinting, but a better, more inclusive term is *DNA typing.*

One early, dramatic case of DNA typing involved a man who murdered a man and woman as they slept in a pickup truck, then about forty minutes later went back and raped the woman. This act not only compounded the crime, it also provided forensic scientists with the means to convict the perpetrator. They obtained DNA from the sperm cells in the semen he had left behind, typed it, and showed that the pattern matched that of the suspect's DNA. This evidence helped convince the jury to convict the defendant. Figure 5.15 presents an example of DNA typing that was used to identify another rape suspect. The pattern from the suspect clearly matches that from the sperm DNA. This is the result from only one probe. The others also gave patterns that matched the sperm DNA.

One advantage of DNA typing is its extreme sensitivity. Only a few drops of blood or semen are sufficient to perform a test. However, sometimes forensic scientists have even less to go on—a hair pulled out by the victim, for example. Although the hair by itself may not be enough for DNA typing, it can be useful if it is accompanied by hair follicle cells. Selected segments of DNA from these cells can be amplified by PCR and typed.

In spite of its potential accuracy, DNA typing has sometimes been effectively challenged in court, most famously in the O.J. Simpson trial in Los Angeles in 1995. Defense lawyers have focused on two problems with DNA typing: First, it is tricky and must be performed very carefully to give meaningful results. Second, there has been controversy about the statistics used in analyzing the data. This second question revolves around the use of the product rule in deciding whether the DNA typing result uniquely identifies a suspect. Let us say that a given probe detects a given allele (a set of bands in this case) in one in a hundred people in the general population. Thus, the chance of a match with a given person with this probe is one in a hundred, or 10^{-2}. If we use five probes, and all five alleles match the suspect, we might conclude that the chances of such a match are the product of the chances of a match with each individual probe, or $(10^{-2})^5$ or 10^{-10}. Because fewer than 10^{10} (10 billion) people are now on earth, this would mean this DNA typing would statistically eliminate everyone but the suspect. Prosecutors have used a more conservative estimate that takes into account the fact that members of some ethnic groups have higher probabilities of matches

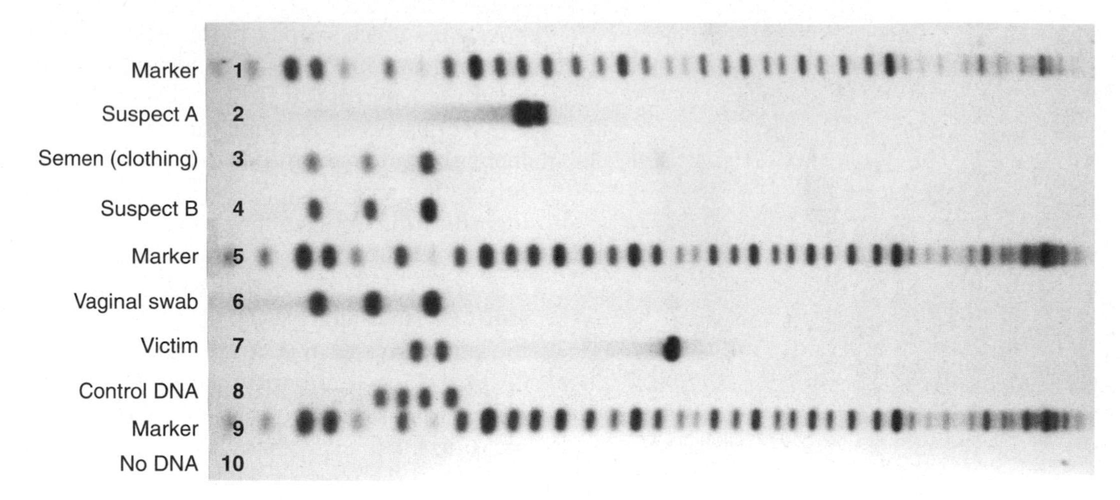

Figure 5.15 Use of DNA typing to help identify a rapist. Two suspects have been accused of attacking and raping a young woman, and DNA analyses have been performed on various samples from the suspects and the woman. Lanes 1, 5, and 9 contain marker DNAs. Lane 2 contains DNA from the blood cells of suspect A. Lane 3 contains DNA from a semen sample found on the woman's clothing. Lane 4 contains DNA from the blood cells of suspect B. Lane 6 contains DNA obtained by swabbing the woman's vaginal canal. (Too little of the victim's own DNA was present to detect.) Lane 7 contains DNA from the woman's blood cells. Lane 8 contains a control DNA. Lane 10 is a control containing no DNA. Partly on the basis of this evidence, suspect B was found guilty of the crime. Note how his DNA fragments in lane 4 match the DNA fragments from the semen in lane 3 and the vaginal swab in lane 6. (*Source:* Courtesy Lifecodes Corporation, Stamford, CT.)

with certain probes. Still, probabilities greater than one in a million are frequently achieved, and they can be quite persuasive in court. Of course, DNA typing can do more than identify criminals. It can just as surely eliminate a suspect (see suspect A in Figure 5.15).

SUMMARY Modern DNA typing uses a battery of DNA probes to detect variable sites in individual animals, including humans. As a forensic tool, DNA typing can be used to test parentage, to identify criminals, or to remove innocent people from suspicion.

In Situ Hybridization: Locating Genes in Chromosomes

This chapter has illustrated the use of probes to identify the band on a Southern blot that contains a gene of interest. Labeled probes can also be used to hybridize to chromosomes and thereby reveal which chromosome has the gene of interest. The strategy of such **in situ hybridization** is to spread the chromosomes from a cell and partially denature the DNA to create single-stranded regions that can hybridize to a labeled probe. One can use x-ray film to detect the label in the spread after it is stained and probed. The stain allows one to visualize and identify the chromosomes, and the darkening of the photographic emulsion locates the labeled probe, and therefore the gene to which it hybridized.

Other means of labeling the probe are also available. Figure 5.16 shows the localization of the muscle glycogen

Figure 5.16 Using a fluorescent probe to find a gene in a chromosome by in situ hybridization. A DNA probe specific for the human muscle glycogen phosphorylase gene was coupled to dinitrophenol. A human chromosome spread was then partially denatured to expose single-stranded regions that can hybridize to the probe. The sites where the DNP-labeled probe hybridized were detected indirectly as follows: A rabbit anti-DNP antibody was bound to the DNP on the probe; then a goat antirabbit antibody, coupled with fluorescein isothiocyanate (FITC), which emits yellow fluorescent light, was bound to the rabbit antibody. The chromosomal sites where the probe hybridized show up as bright yellow fluorescent spots against a red background that arises from staining the chromosomes with the fluorescent dye propidium iodide. This analysis identifies chromosome 11 as the site of the glycogen phosphorylase gene. (*Source:* Courtesy Dr. David Ward, *Science* 247 (5 Jan 1990) cover. © AAAS.)

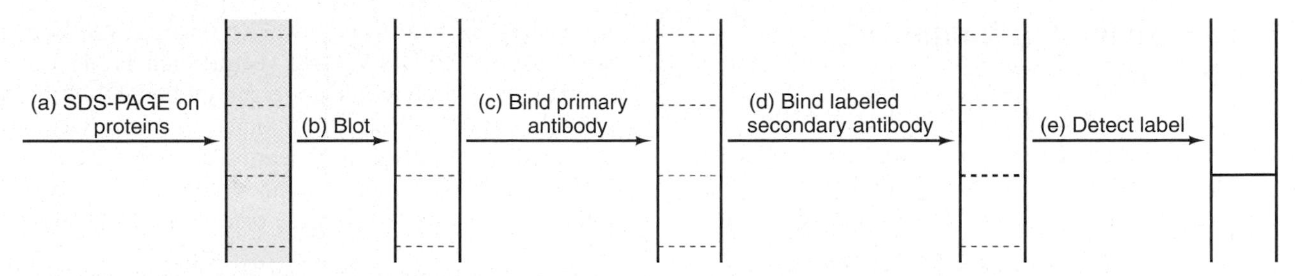

Figure 5.17 Immunoblotting (Western blotting). (a) An immunoblot begins with separation of a mixture of proteins by SDS-PAGE. **(b)** Next, the separated proteins, represented by dotted lines, are blotted to a membrane. **(c)** The blot is probed with a primary antibody specific for a protein of interest on the blot. Here, the antibody has reacted with one of the protein bands (red), but the reaction is undetectable so far. **(d)** A labeled secondary antibody (or protein A) is used to detect the primary antibody, and therefore the protein of interest. Here, the presence of the secondary antibody attached to the primary antibody is denoted by the change in color of the band from red to purple, but this reaction is also undetectable so far. **(e)** Finally, the labeled band is detected—using an x-ray film or a phosphorimager if the label is radioactive. If the label is nonradioactive, it can be detected as described in Figure 5.11.

phosphorylase gene to human chromosome 11 using a DNA probe labeled with dinitrophenol, which can be detected with a fluorescent antibody. The chromosomes are counterstained with propidium iodide, so they will fluoresce red. Against this background, the yellow fluorescence of the antibody probe on chromosome 11 is easy to see. This technique is known as **fluorescence in situ hybridization (FISH).**

SUMMARY One can hybridize labeled probes to whole chromosomes to locate genes or other specific DNA sequences. This type of procedure is called in situ hybridization; if the probe is fluorescently labeled, the technique is called fluorescence in situ hybridization (FISH).

Immunoblots (Western Blots)

Immunoblots (also known as **Western blots,** keeping to the Southern nomenclature system), although they do not use hybridization, follow the same experimental pattern as Southern blots: The investigator electrophoreses molecules and then blots these molecules to a membrane where they can be identified readily. However, immunoblots involve electrophoresis of proteins instead of nucleic acids. We have seen that DNAs on Southern blots are detected by hybridization to labeled oligonucleotide or polynucleotide probes. But hybridization is appropriate only for nucleic acids, so how are the blotted proteins detected? Instead of a nucleic acid, one uses an antibody (or antiserum) specific for a particular protein. That antibody binds to the target protein on the blot. Then a labeled secondary antibody (for example, a goat antibody that recognizes all rabbit antibodies in the IgG class), or a labeled IgG-binding protein such as Staphylococcal protein A, can be used to label the band with the target protein, by binding to the antibody already attached there. (The fact that antibodies are products of the immune system gives rise to the term "immunoblot.") Immunoblots can tell us whether or not a particular protein is present in a mixture, and can also give at least a rough idea of the quantity of that protein.

Why bother with a secondary antibody or protein A; why not just use a labeled primary antibody? The main reason is that this would require individually labeling every different antibody used to probe a series of immunoblots. It is much simpler and cheaper to use unlabeled primary antibody, and buy a stock of labeled secondary antibody or protein A that can bind to and detect any primary antibody. Figure 5.17 illustrates the process of making and probing an immunoblot for a particular protein.

SUMMARY Proteins can be detected and quantified in complex mixtures using immunoblots (or Western blots). Proteins are electrophoresed, then blotted to a membrane and the proteins on the blot are probed with specific antibodies that can be detected with labeled secondary antibodies or protein A.

5.4 DNA Sequencing and Physical Mapping

In 1975, Frederick Sanger and his colleagues, and Alan Maxam and Walter Gilbert developed two different methods for determining the exact base sequence of a cloned piece of DNA. These spectacular breakthroughs revolutionized molecular biology and won the 1980 Nobel prize in chemistry for Gilbert and Sanger. They have allowed molecular biologists to determine the sequences of thousands of genes and many whole genomes, including the human genome. Modern DNA sequencing derives from the Sanger method, so that is the one we will describe here.

The Sanger Chain-Termination Sequencing Method

The original method of sequencing a piece of DNA by the Sanger method (Figure 5.18) is presented here to explain the principles. In practice, it is rarely done manually this way anymore. In the next section we will see how the technique has been automated. The original method began with cloning the DNA into a vector, such as M13 phage or a phagemid, that would give the cloned DNA in single-stranded form. These days, one can start with double-stranded DNA and simply heat it to create single-stranded DNAs for sequencing. To the single-stranded DNA one hybridizes an oligonucleotide primer about 20 bases long.

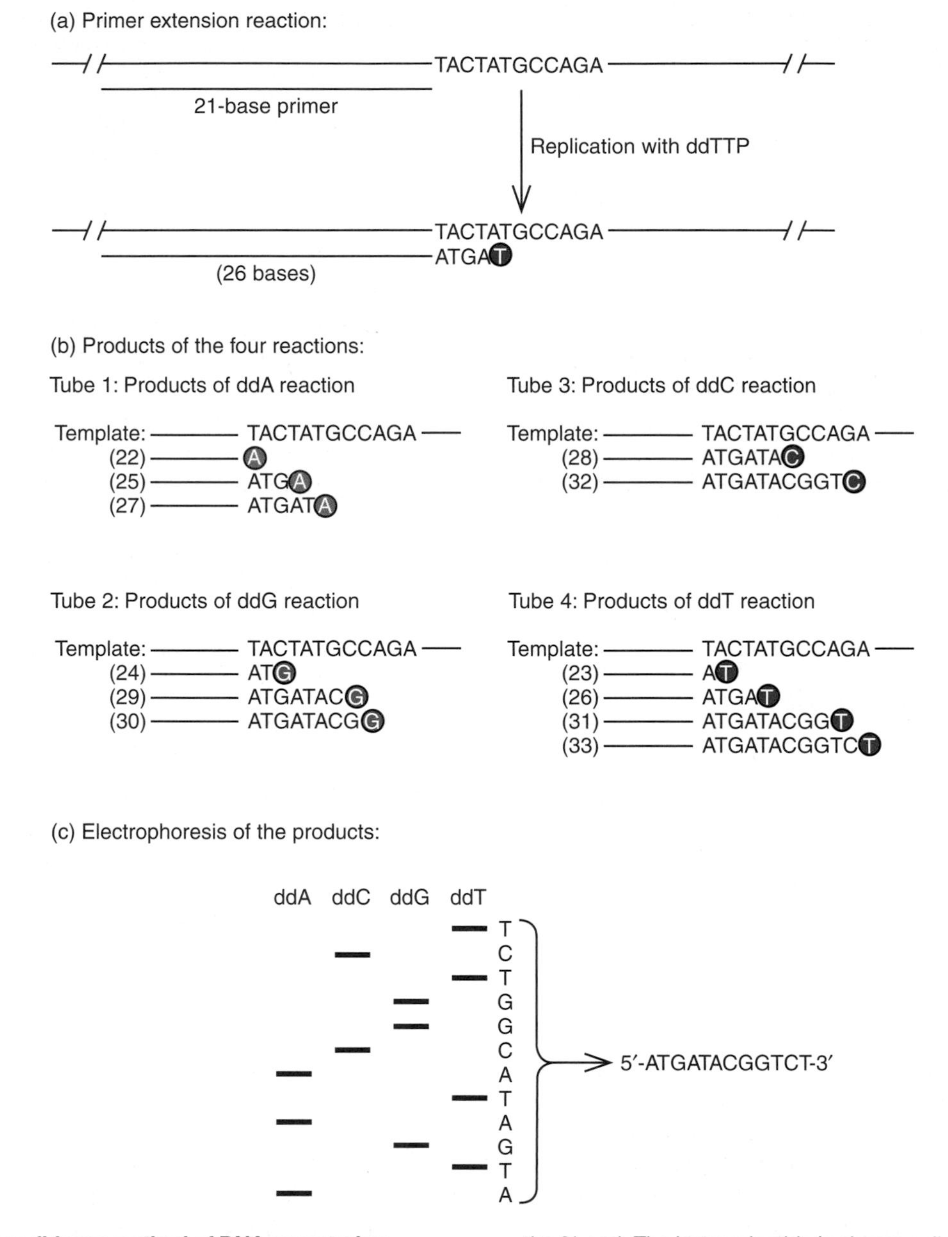

Figure 5.18 The Sanger dideoxy method of DNA sequencing. **(a)** The primer extension (replication) reaction. A primer, 21 nt long in this case, is hybridized to the single-stranded DNA to be sequenced, then mixed with the Klenow fragment of DNA polymerase and dNTPs to allow replication. One dideoxy NTP is included to terminate replication after certain bases; in this case, ddTTP is used, and it has caused termination at the second position where dTTP was called for. **(b)** Products of the four reactions. In each case, the template strand is shown at the top, with the various products underneath. Each product begins with the 21-nt primer and has one or more nucleotides added to the 3′-end. The last nucleotide is always a dideoxy nucleotide (color) that terminated the chain. The total length of each product is given in parentheses at the left end of the fragment. Thus, fragments ranging from 22 to 33 nt long are produced. **(c)** Electrophoresis of the products. The products of the four reactions are loaded into parallel lanes of a high-resolution electrophoresis gel and electrophoresed to separate them according to size. By starting at the bottom and finding the shortest fragment (22 nt in the A lane), then the next shortest (23 nt in the T lane), and so forth, one can read the sequence of the product DNA. Of course, this is the complement of the template strand.

This synthetic primer is designed to hybridize to a sequence adjacent to the multiple cloning site of the vector and is oriented with its 3′-end pointing toward the insert in the multiple cloning site.

Extending the primer using the Klenow fragment of DNA polymerase (Chapter 20) produces DNA complementary to the insert. The trick to Sanger's method is to carry out such DNA synthesis reactions in four separate tubes and to include in each tube a different chain terminator. The chain terminator is a **dideoxy nucleotide** such as **dideoxy ATP (ddATP).** Not only is this terminator 2′-deoxy, like a normal DNA precursor, it is 3′-deoxy as well. Thus, it cannot form a phosphodiester bond because it lacks the necessary 3′-hydroxyl group. That is why we call it a chain terminator; whenever a dideoxy nucleotide is incorporated into a growing DNA chain, DNA synthesis stops.

Dideoxy nucleotides by themselves do not permit any DNA synthesis at all, so an excess of normal deoxy nucleotides must be used, with just enough dideoxy nucleotide to stop DNA strand extension once in a while at random. This random arrest of DNA growth means that some strands will terminate early, others later. Each tube contains a different dideoxy nucleotide: ddATP in tube 1, so chain termination will occur with A's; ddCTP in tube 2, so chain termination will occur with C's; and so forth. Radioactive dATP is also included in all the tubes so the DNA products will be radioactive.

The result is a series of fragments of different lengths in each tube. In tube 1, all the fragments end in A; in tube 2, all end in C; in tube 3, all end in G; and in tube 4, all end in T. Next, all four reaction mixtures are electrophoresed in parallel lanes in a high-resolution polyacrylamide gel under denaturing conditions, so all DNAs are single-stranded. Finally, autoradiography is performed to visualize the DNA fragments, which appear as horizontal bands on an x-ray film.

Figure 5.18c shows a schematic of the sequencing film. To begin reading the sequence, start at the bottom and find the first band. In this case, it is in the A lane, so you know that this short fragment ends in A. Now move to the next longer fragment, one step up on the film; the gel electrophoresis has such good resolution that it can separate fragments differing by only one base in length, at least until the fragments become much longer than this. And the next fragment, one base longer than the first, is found in the T lane, so it must end in T. Thus, so far you have found the sequence AT. Simply continue reading the sequence in this way as you work up the film. The sequence is shown, reading bottom to top, at the right of the drawing. At first you will be reading just the sequence of part of the multiple cloning site of the vector. However, before very long, the DNA chains will extend into the insert—and unknown territory. An experienced sequencer can continue to read sequence from one film for hundreds of bases.

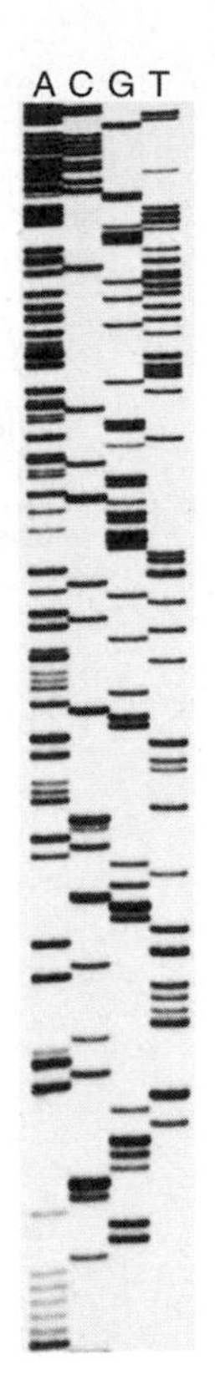

Figure 5.19 A typical sequencing film. The sequence begins CAAAAAACGG. You can probably read the rest of the sequence to the top of the film. (*Source:* Courtesy Life Technologies, Inc., Gaithersburg, MD.)

Figure 5.19 shows a typical sequencing film. The shortest band (at the very bottom) is in the C lane. After that, a series of six bands occurs in the A lane. So the sequence begins CAAAAAA. It is easy to read many more bases on this film; try it yourself.

Automated DNA Sequencing

The "manual" sequencing technique just described is powerful, but it is still relatively slow. If one is to sequence a really large amount of DNA, such as the 3 billion base pairs found in the human genome, then rapid, automated sequencing methods are required. Indeed, automated DNA sequencing has been in use for many years. Figure 5.20a describes one such technique, again based on Sanger's chain-termination method. This procedure uses dideoxy nucleotides, just as in the manual method, with one important exception. The primers, or, more commonly, the dideoxy nucleotides used in each of the four reactions are tagged with a different fluorescent molecule, so the products from each tube will emit a different color fluorescence when excited by light.

After the extension reactions and chain termination are complete, all four reactions are mixed and electrophoresed together in the same lane on a gel in a short, thin column (Figure 5.20b). Near the bottom of the gel is an

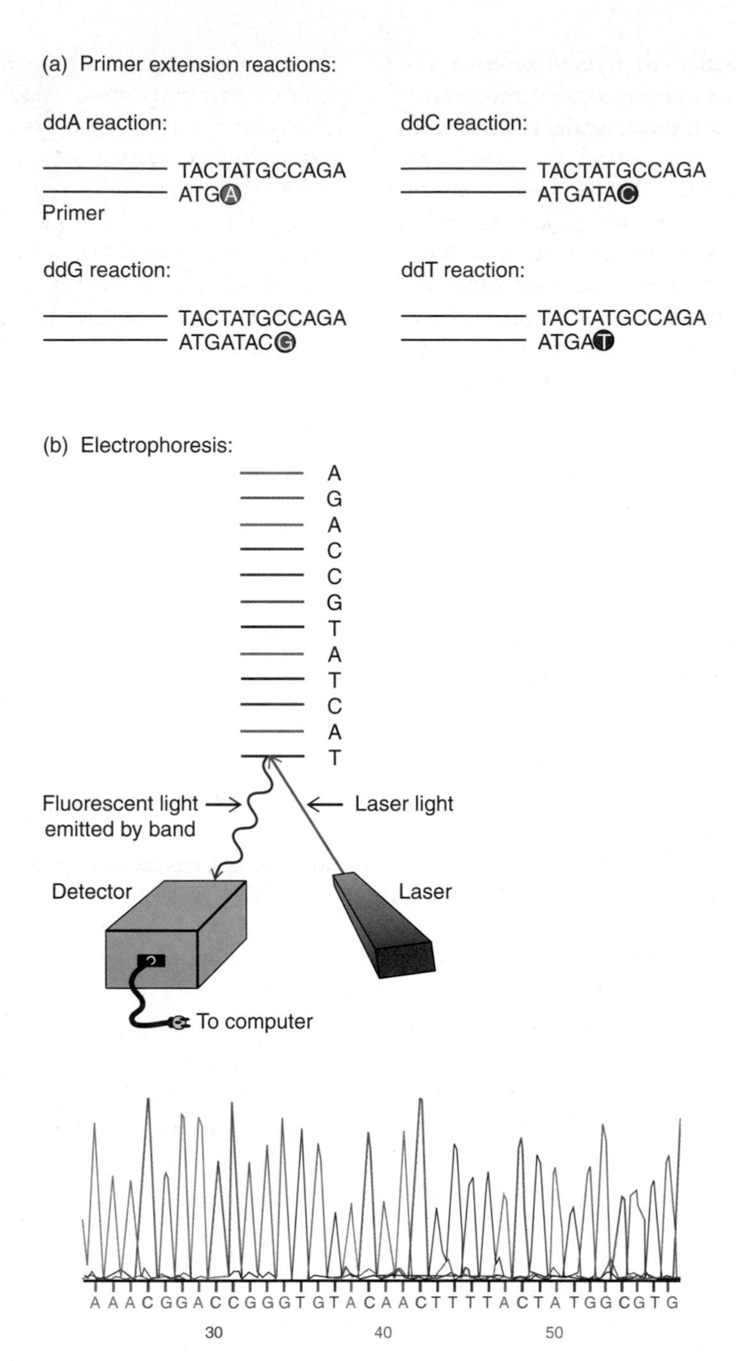

Figure 5.20 Automated DNA sequencing. **(a)** The primer extension reactions are run in the same way as in the manual method, except that the dideoxy nucleotides in each reaction are labeled with a different fluorescent molecule that emits light of a distinct color. Only one product is shown for each reaction, but all possible products are actually produced, just as in manual sequencing. **(b)** Electrophoresis and detection of bands. The various primer extension reaction products separate according to size on gel electrophoresis. The bands are color-coded according to the termination reaction that produced them (e.g., green for oligonucleotides ending in ddA, blue for those ending in ddC, and so forth). A laser scanner excites the fluorescent tag on each band as it passes by, and a detector analyzes the color of the resulting emitted light. This information is converted to a sequence of bases and stored by a computer. **(c)** Sample printout of an automated DNA sequencing experiment. Each colored peak is a plot of the fluorescence intensity of a band as it passes through the laser beam. The colors of these peaks, and those of the bands in part (b) and the tags in part (a), were chosen for convenience. They may not correspond to the actual colors of the fluorescent light.

analyzer that excites the fluorescent oligonucleotides with a laser beam as they pass by. Then the color of the fluorescent light emitted from each oligonucleotide is detected electronically. This information then passes to a computer, which has been programmed to convert the color information to a base sequence. If it "sees" blue, for example, this might mean that this oligonucleotide came from the dideoxy C reaction, and therefore ends in C (actually a ddC). Green may indicate A; orange, G; and red, T. The computer gives a printout of the profile of each passing fluorescent band, color-coded for each base (Figure 5.20c), and stores the sequence of these bases in its memory for later use.

Nowadays, automated sequencers (**sequenators**) may simply print out the sequence or send it directly to a computer for analysis. Large genome projects use many sequenators with 96, or even 384, columns apiece, running simultaneously to obtain millions or even billions of bases of sequence (Chapter 24). One 384-column sequenator can produce 200,000 nt of sequence in one three-hour run.

SUMMARY The Sanger DNA sequencing method uses dideoxy nucleotides to terminate DNA synthesis, yielding a series of DNA fragments whose sizes can be measured by electrophoresis. The last base in each of these fragments is known, because we know which dideoxy nucleotide was used to terminate each reaction. Therefore, ordering these fragments by size—each fragment one (known) base longer than the next—tells us the base sequence of the DNA. Automated sequenators make this process very efficient.

High-Throughput Sequencing

Once an organism's genome sequence is known, very rapid sequencing techniques can be applied to sequence the genome of another member of the same species. These **high-throughput DNA sequencing** techniques (also called **next-generation sequencing**) typically produce relatively short **reads,** or contiguous sequences obtained from a single run of the sequencing apparatus. Whereas Sanger sequencing typically produces reads more than 500 bases long, high-throughput sequencing typically produces reads in the 25–35-base or 200–300-base range, depending on the specific method. These relatively short snippets of sequence make finding overlaps among reads difficult, but that is not a problem if a reference sequence is already available, as it can serve as a guide for piecing the reads together.

In the late 1990s, one such high-throughput method, called **pyrosequencing,** was reported. This technique has the great advantages of speed and accuracy, and it does not require electrophoresis. With refinements introduced by 2005, a company known as 454 Life Sciences launched a commercial automated sequencer that could read 20 million base pairs per 4.5-h run.

The idea behind pyrosequencing is to allow DNA polymerase (usually the Klenow fragment of DNA polymerase I; Chapter 20) to replicate the DNA to be sequenced and follow the incorporation of each nucleotide in real time. Each nucleotide incorporation event results in the release of pyrophosphate (PPi), and that can be measured quantitatively by coupling it to the generation of light according to the following sequence of reactions:

$$1)\ \text{Growing DNA fragment } (\text{dNMP}_n) + \text{dNTP} \xrightarrow{\text{DNA polymerase}} \text{dNMP}_{n+1} + \text{PPi}$$

$$2)\ \text{PPi} + \text{adenosine phosphosulfate} \xrightarrow{\text{ATP sulfurylase}} \text{ATP} + \text{sulfate}$$

$$3)\ \text{ATP} + \text{luciferin} + O_2 \xrightarrow{\text{Luciferase}} \text{AMP} + \text{PPi} + \text{oxyluciferin} + CO_2 + \text{light.}$$

The pyrosequencing system is automated, so the apparatus feeds the DNA polymerase each of the four deoxynucleotides in turn. For example, it could supply them in the order dA, dG, dC, then dT. In a solid-state system, the DNA and DNA polymerase are tethered to a solid support, such as a resin bead, and the reagents, including each dNTP, are quickly washed away after allowing time for each dNMP to be incorporated. If a dAMP is incorporated, it liberates PPi, which results in a burst of light that is detected and quantified by the apparatus as a peak. If two dAMPs in a row are incorporated, the peak of light will be twice as high. This linearity persists in strings of up to eight dAMPs in a row. After that, the ratio of light intensity to number of nucleotides incorporated levels off, and analysis becomes more difficult. If, on the other hand, dAMP is not incorporated, only a small peak, perhaps due to contamination of the dATP reagent by another nucleotide, will be seen.

In a liquid system, the DNA and DNA polymerase are in solution, not tethered to a bead, so there must be a system to remove each dNTP before the next one is added. That is typically accomplished by the enzyme apyrase, which carries out a two-step degradation of dNTPs:

$$\text{dNTP} \xrightarrow{\text{Apyrase}} \text{dNDP} \xrightarrow{\text{Apyrase}} \text{dNMP.}$$

This removal of the dNTP allows dNTPs to be added in very rapid succession without washing in between.

The light produced by each deoxynucleotide incorporation stimulates a charge-coupled device (CCD) camera,

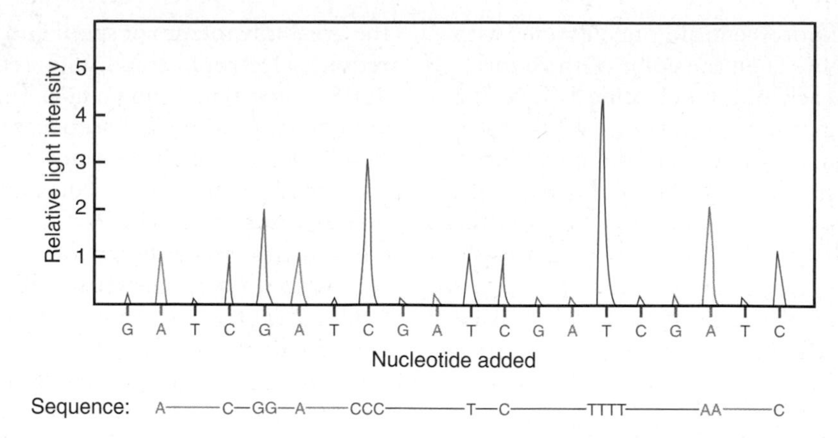

Figure 5.21 A hypothetical pyrogram. The light produced from the addition of each dNTP in a pyrosequencing run is recorded as a peak. Nucleotides that are not incorporated generate only a small amount of light. Incorporation of a single nucleotide yields a relative light intensity of 1. Incorporation of two, three, or four nucleotides of the same kind in a row generate relative light intensities of 2, 3, or 4, respectively. Thus, the sequence of bases added to this growing oligonucleotide can be determined and is presented at bottom: ACGGACCCTCTTTTAAC

which sends the signal to a computer, which produces a **pyrogram,** as illustrated in Figure 5.21. It is easy to see from the peak height the difference in incorporation of one, two, three, or four nucleotides of the same kind in a row. It is also easy to distinguish between incorporation of a nucleotide and nonincorporation, which gives only a small blip. The computer converts the series of peaks into a sequence.

One drawback of the pyrosequencing technique is that each read on a given piece of DNA can currently go only about 200–300 nt before the sequence accuracy is unacceptably degraded. In the liquid version of the procedure, this degradation comes from dilution of the sample by repeated additions of reagents, and buildup of inhibitory products, as well as the fact that some chains inevitably get ahead of the majority, and some fall behind. With increasing chain length, these asynchronous chain elongations build up to the point that the pyrogram is difficult to interpret. In the solid-state version, the first two problems don't arise, because of the washing step before each nucleotide addition, but the last one still limits accuracy in long reads. The inability of pyrosequencing to perform long reads prevents its use in sequencing new, large genomes because repetitive DNAs with repeats longer than about 250 nt do not have unique regions that would allow the short reads to be ordered properly.

On the other hand, the speed and economy of pyrosequencing make it a powerful tool for resequencing known genomes. For example, it works well for sequencing parts of an individual's genes to detect mutations that can cause disease. In fact, in cases like this, nucleotides can be added in the known, normal sequence, speeding up the process. A mutation is then readily detected by the failure of the normal nucleotide to be incorporated at a particular position. Pyrosequencing is also very useful in a method called ChIPSeq (Chapter 24), which can be used to locate binding sites for transcription factors.

Each pyrosequencing run is inherently fast, but the factor that gives the technique its great advantage in speed is the ability to perform many runs in parallel. For example, 96 different runs can be carried out simultaneously in a 96-well microtiter plate. The light from each well can be focused onto the chip of a CCD camera, so the camera can keep track of all 96 reactions simultaneously. The whole process is automated, so it requires very little human attention.

Another high-throughput method, developed by the Illumina company, starts by attaching short pieces of DNA to a solid surface, amplifying each DNA in a tiny patch on the surface, then sequencing the patches together by extending them one nucleotide at a time using fluorescent chain-terminating nucleotides. After each cycle of nucleotide addition, in which all four chain-terminating nucleotides are provided, the surface is scanned by a CCD camera attached to a microscope to detect the color of the fluorescent tag added to each patch. That color reveals the identity of the nucleotide just added. The fluorescent tags and chain-terminating groups (3′-azidomethyl groups) are easily removed chemically, so the process can be repeated over and over until the whole piece of DNA (averaging about 35 nt long) is sequenced. So many patches of DNA can be analyzed simultaneously that 1–2 billion base pairs can be sequenced in one 72-hour run of the sequencer. Figure 5.22 shows a representation of the colored patches the camera would see in a field with a very low density of patches. Overlapping patches would confuse the analysis and so are automatically discarded.

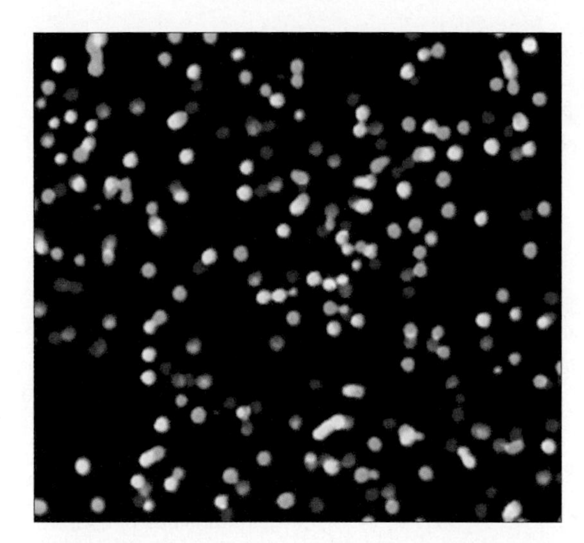

Figure 5.22 Image of clusters of growing DNA chains in an Illumina Genome Analyzer (GA1). The camera actually uses four filters to detect each color individually, so all colors would not really reach the camera at the same time. This is a simulated image in which the patches in each of the four images have been colored artificially and combined, so it approximates what the eye would see at one point during the sequencing process. Patches that overlap are discarded because they would give confusing results. (*Source:* Reprinted by permission from Macmillan Publishers Ltd: ***Nature,*** 456, 53–59, 6 November 2008. Bentley et al, Accurate whole human genome sequencing using reversible terminator chemistry. © 2008.)

SUMMARY High-throughput sequencing allows very rapid sequencing of genomes if the genome of one member of the species has already been sequenced. In pyrosequencing, nucleotides are added one by one, and the incorporation of a nucleotide is detected by the release of pyrophosphate, which leads through a chain of reactions to a flash of light. Many reactions can be carried out simultaneously in automated sequencing machines. Another method, developed by the Illumina company, uses short pieces of DNA amplified in tiny, closely spaced patches on a support surface. These DNA pieces are sequenced by adding fluorescent, chain-terminating nucleotides, the color of whose fluorescence reveals their identity. The colors are visualized with a microscope fitted with a CCD camera. After each round of DNA elongation, the fluorescent and chain-terminating groups are removed and the process is repeated to obtain the whole fragment's sequence.

Restriction Mapping

Before sequencing a large stretch of DNA, some preliminary mapping is usually done to locate landmarks on the DNA molecule. These are not genes, but small regions of the DNA—cutting sites for restriction enzymes, for example. A map based on such physical characteristics is called, naturally enough, a **physical map.** (If restriction sites are the only markers involved, we can also call it a **restriction map.**)

To introduce the idea of restriction mapping, let us consider the simple example illustrated in Figure 5.23. We start with a *Hin*dIII fragment 1.6 kb (1600 bp) long (Figure 5.23a). When this fragment is cut with another restriction enzyme (*Bam*HI), two fragments are generated, 1.2 and 0.4 kb long. The sizes of these fragments can be measured by electrophoresis, as pictured in Figure 5.23a. The sizes reveal that *Bam*HI cuts 0.4 kb from one end of the 1.6-kb *Hin*dIII fragment, and 1.2 kb from the other.

Now suppose the 1.6-kb *Hin*dIII fragment is cloned into the *Hin*dIII site of a hypothetical plasmid vector, as illustrated in Figure 5.23b. Because this is not directional cloning, the fragment will insert into the vector in either of the two possible orientations: with the *Bam*HI site on the right (left side of Figure 5.23), or with the *Bam*HI site on the left (right side of the Figure 5.23). How can you determine which orientation exists in a given clone? To answer this question, locate a restriction site asymmetrically situated in the vector, relative to the *Hin*dIII cloning site. In this case, an *Eco*RI site is only 0.3 kb from the *Hin*dIII site. This means that if you cut the cloned DNA pictured on the left with *Bam*HI and *Eco*RI, you will generate two fragments: 3.6 and 0.7 kb long. On the other hand, if you cut the DNA pictured on the right with the same two enzymes, you will generate two fragments: 2.8 and 1.5 kb in size. You can distinguish between these two possibilities easily by electrophoresing the fragments to measure their sizes, as shown at the bottom of Figure 5.23. Usually, DNA is prepared from several different clones, each of them is cut with the two enzymes, and the fragments are electrophoresed side by side with one lane reserved for marker fragments of known sizes. On average, half of the clones will have one orientation, and the other half will have the opposite orientation.

These examples are relatively simple, but we use the same kind of logic to solve much more complex mapping problems. Sometimes it helps to label (radioactively or nonradioactively) one restriction fragment and hybridize it to a Southern blot of fragments made with another restriction enzyme to help sort out the relationships among fragments. For example, consider the linear DNA in Figure 5.24. We might be able to figure out the order of restriction sites without the use of hybridization, but it is not simple. Consider the information we get from just a few hybridizations. If we Southern blot the *Eco*RI fragments and hybridize them to the labeled *Bam*HI-A fragment, for example, the *Eco*RI-A and *Eco*RI-C fragments will become labeled. This demonstrates that *Bam*HI-A overlaps these two *Eco*RI fragments. If we hybridize the blot to the *Bam*HI-B fragment, the *Eco*RI-A and *Eco*RI-D fragments become labeled. Thus, *Bam*HI-B overlaps *Eco*RI-A and *Eco*RI-D. Ultimately, we will discover that no other *Bam*HI fragments besides A and B hybridize to *Eco*RI-A, so *Bam*HI-A and *Bam*HI-B must be adjacent. Using this kind of approach, we can piece together the physical map of the whole 30-kb fragment.

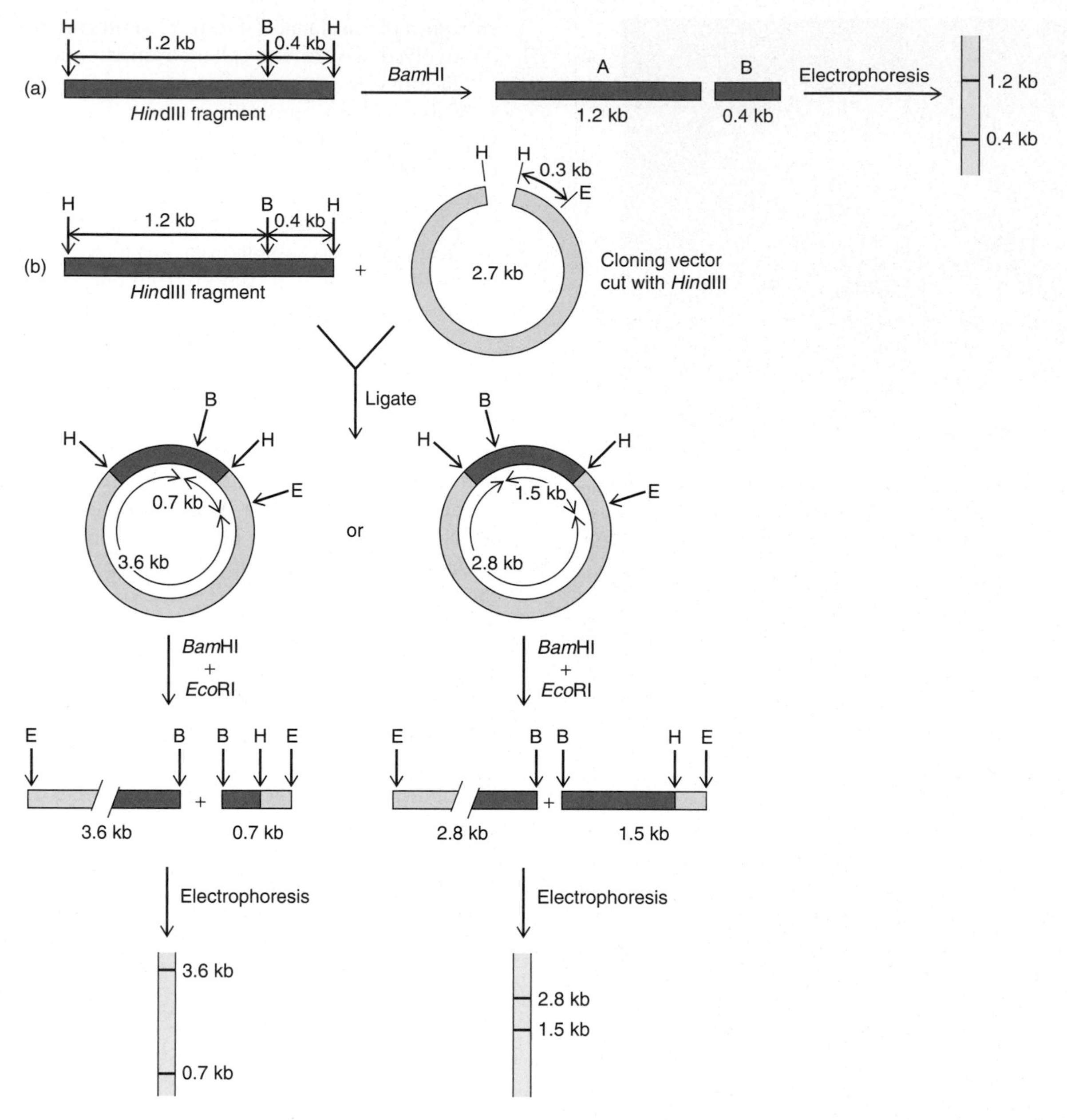

Figure 5.23 A simple restriction mapping experiment. **(a)** Determining the position of a *Bam*HI site. A 1.6-kb *Hin*dIII fragment is cut by *Bam*HI to yield two subfragments. The sizes of these fragments are determined by electrophoresis to be 1.2 kb and 0.4 kb, demonstrating that *Bam*HI cuts once, 1.2 kb from one end of the *Hin*dIII fragment and 0.4 kb from the other end. **(b)** Determining the orientation of the *Hin*dIII fragment in a cloning vector. The 1.6-kb *Hin*dIII fragment can be inserted into the *Hin*dIII site of a cloning vector, in either of two ways: (1) with the *Bam*HI site near an *Eco*RI site in the vector or (2) with the *Bam*HI site remote from an *Eco*RI site in the vector. To determine which, cleave the DNA with both *Bam*HI and *Eco*RI and electrophorese the products to measure their sizes. A short fragment (0.7 kb) shows that the two sites are close together (left). On the other hand, a long fragment (1.5 kb) shows that the two sites are far apart (right).

SUMMARY A physical map tells us about the spatial arrangement of physical "landmarks," such as restriction sites, on a DNA molecule. One important strategy in restriction mapping (mapping of restriction sites) is to cut the DNA in question with two or more restriction enzymes in separate reactions, measure the sizes of the resulting fragments, then cut each with another restriction enzyme and measure the sizes of the subfragments by gel electrophoresis. These sizes allow us to locate at least some of the recognition sites relative to the others. We can improve this process considerably by Southern blotting some of the fragments and then hybridizing these fragments to labeled fragments generated by another restriction enzyme. This strategy reveals overlaps between individual restriction fragments.

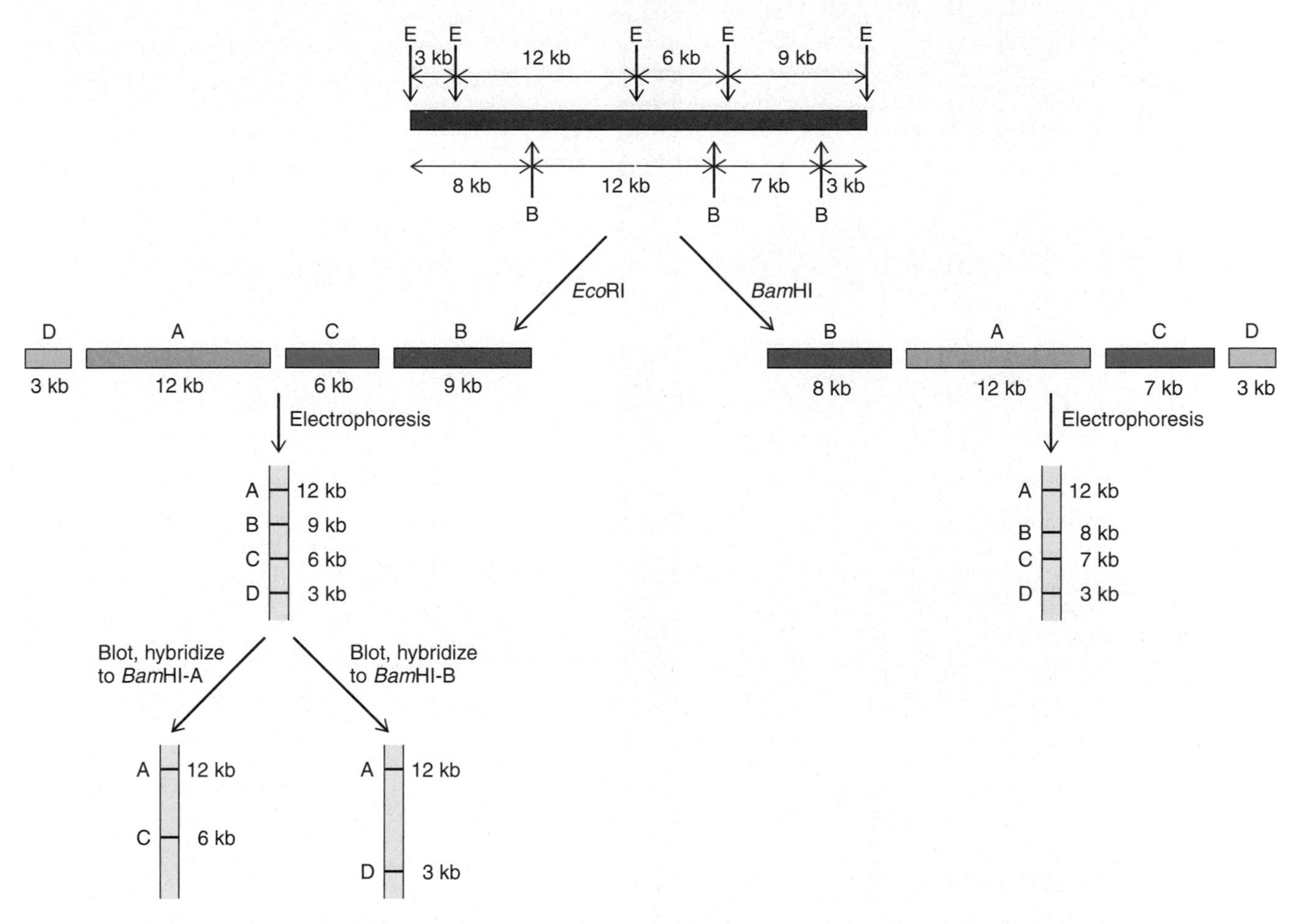

Figure 5.24 Using Southern blots in restriction mapping. A 30-kb fragment is being mapped. It is cut three times each by *Eco*RI (E) and *Bam*HI (B). To aid in the mapping, first cut with *Eco*RI, and electrophorese the four resulting fragments (*Eco*RI-A, -B, -C, and -D); next, Southern blot the fragments and hybridize them to labeled, cloned *Bam*HI-A and -B fragments. The results, shown at lower left, demonstrate that the *Bam*HI-A fragment overlaps *Eco*RI-A and -C, and the *Bam*HI-B fragment overlaps *Eco*RI-A and -D. This kind of information, coupled with digestion of *Eco*RI fragments by *Bam*HI (and vice versa), allows the whole restriction map to be pieced together.

5.5 Protein Engineering with Cloned Genes: Site-Directed Mutagenesis

Traditionally, protein biochemists relied on chemical methods to alter certain amino acids in the proteins they studied, so they could observe the effects of these changes on protein activities. But chemicals are rather crude tools for manipulating proteins; it is difficult to be sure that only one amino acid, or even one kind of amino acid, has been altered. Cloned genes make this sort of investigation much more precise, allowing us to perform microsurgery on a protein. By changing specific bases in a gene, we also change amino acids at corresponding sites in the protein product. Then we can observe the effects of those changes on the protein's function.

Let us suppose that we have a cloned gene in which we want to change a single codon. In particular, the gene codes for a sequence of amino acids that includes a tyrosine. The amino acid tyrosine contains a phenolic group:

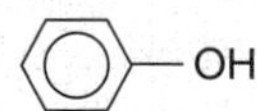

To investigate the importance of this phenolic group, we can change the tyrosine codon to a phenylalanine codon. Phenylalanine is just like tyrosine except that it lacks the phenolic group; instead, it has a simple phenyl group:

If the tyrosine phenolic group is important to a protein's activity, replacing it with phenylalanine's phenyl group should diminish that activity.

In this example, let us assume that we want to change the DNA codon TAC (Tyr) to TTC (Phe). How do we perform such **site-directed mutagenesis?** A popular technique, depicted in Figure 5.25, relies on PCR (Chapter 4). We begin with a cloned gene containing a tyrosine codon (TAC) that we want to change to a phenylalanine codon (TTC).

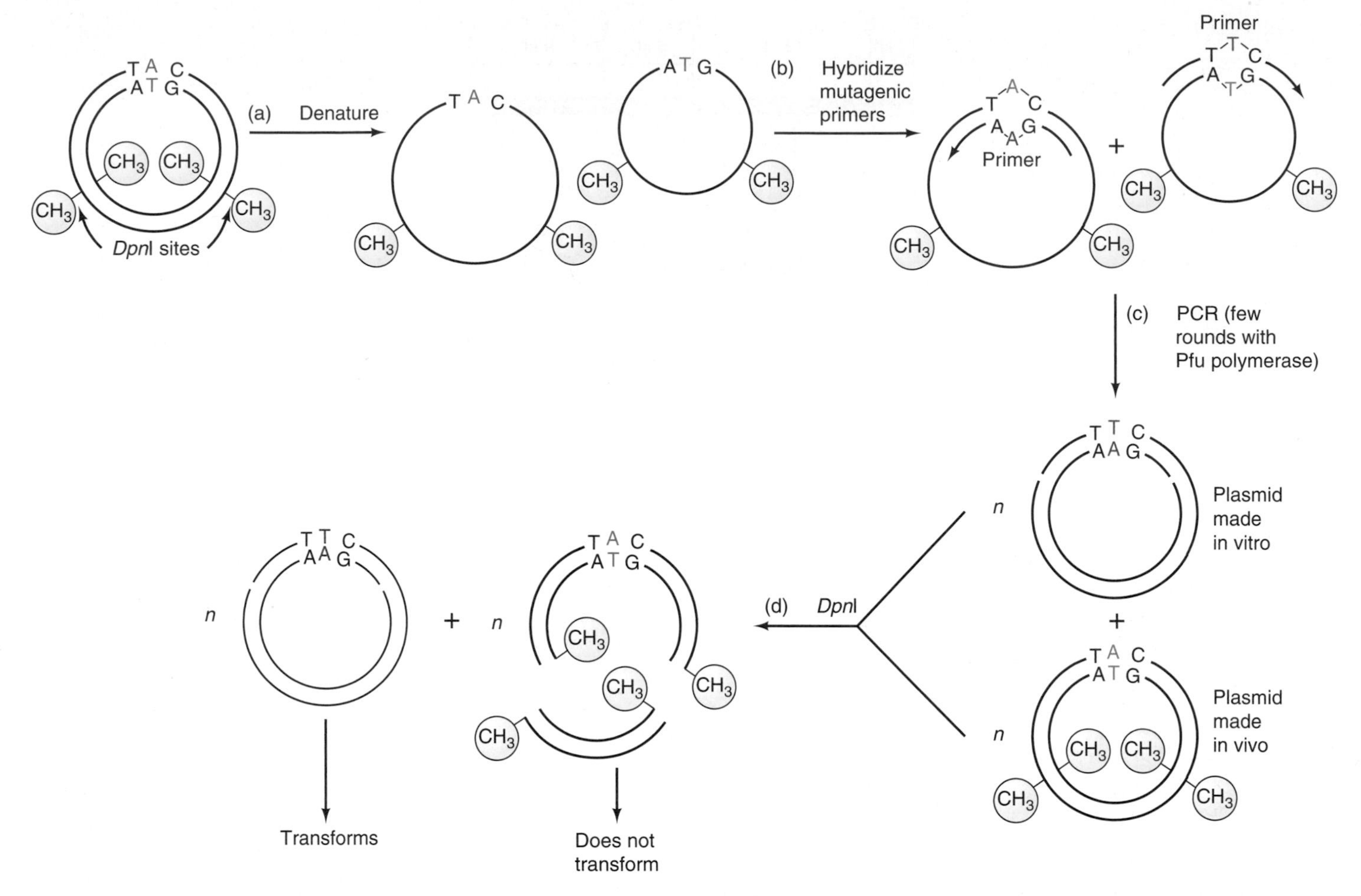

Figure 5.25 PCR-based site-directed mutagenesis. Begin with a plasmid containing a gene with a TAC tyrosine codon that is to be altered to a TTC phenylalanine codon. Thus, the A–T pair (blue) in the original must be changed to a T–A pair. This plasmid was isolated from a normal strain of *E. coli* that methylates the A's of GATC sequences (*Dpn*I sites). The methyl groups are indicated in yellow. **(a)** Heat the plasmid to separate its strands. The strands of the original plasmid are intertwined, so they don't completely separate. They are shown here separating completely for simplicity's sake. **(b)** Hybridize mutagenic primers that contain the TTC codon, or its reverse complement, GAA, to the single-stranded DNA. The altered base in each primer is indicated in red. **(c)** Perform a few rounds of PCR (about eight) with the mutagenic primers to amplify the plasmid with its altered codon. Use a faithful, heat-stable DNA polymerase, such as Pfu polymerase, to minimize mistakes in copying the plasmid. **(d)** Treat the DNA in the PCR reaction with *Dpn*I to digest the methylated wild-type DNA. Because the PCR product was made in vitro, it is not methylated and is not cut. Finally, transform *E. coli* cells with the treated DNA. In principle, only the mutated DNA survives to transform. Check this by sequencing the plasmid DNA from several clones.

The CH_3 symbols indicate that this DNA, like DNAs isolated from most strains of *E. coli,* is methylated on 5′-GATC-3′ sequences. This methylated sequence happens to be the recognition site for the restriction enzyme *Dpn*I, which will come into play later in this procedure. Two methylated *Dpn*I sites are shown, even though many more are usually present because GATC occurs about once every 250 bp in a random sequence of DNA.

The first step is to denature the DNA by heating. The second step is to hybridize mutagenic primers to the DNA. One of these primers is a 25-base oligonucleotide (a 25-mer) with the following sequence:

3′-CGAGTCTGCCAA<u>A</u>GCATGTATAGTA-5′

This primer was designed to have the same sequence as a piece of the gene's nontemplate strand, except that the central triplet has been changed from A<u>T</u>G to A<u>A</u>G, with the altered base underlined. The other primer is the complementary 25-mer. Both primers incorporate the altered base to change the codon we are targeting. The third step is to use a few rounds of PCR with these primers to amplify the DNA, and incorporate the change we want to make. We deliberately use just a few rounds of PCR to minimize other mutations that might creep in by accident during DNA replication. For the same reason, we use a very faithful DNA polymerase called Pfu polymerase. This enzyme is purified from archaea called *Pyrococcus furiosus* (Latin: *furious fireball*), which live in the boiling hot water surrounding undersea thermal vents. It has the ability to "proofread" the DNA it synthesizes, so it makes relatively few mistakes. A similar enzyme from another hyperthermophilic (extreme heat-loving) archeon is called vent polymerase.

Once the mutated DNA is made, we must either separate it from the remaining wild-type DNA or destroy the latter. This is where the methylation of the wild-type DNA comes in handy. *Dpn*I will cut only GATC sites that are methylated. Because the wild-type DNA is methylated, but the mutated DNA, which was made in vitro, is not, only the wild-type DNA will be cut. Once cut, it is no longer capable of transforming *E. coli* cells, so the mutated DNA is the only species that yields clones. We can check the sequence of DNA from several clones to make sure it is the mutated sequence and not the original, wild-type sequence. Usually, it is mutated.

SUMMARY Using cloned genes, we can introduce changes at will, thus altering the amino acid sequences of the protein products. The mutagenized DNA can be made with double-stranded DNA, two complementary mutagenic primers, and PCR. Simply digesting the PCR product with *Dpn*I removes almost all of the wild-type DNA, so cells can be transformed primarily with mutagenized DNA.

5.6 Mapping and Quantifying Transcripts

One recurring theme in molecular biology has been mapping transcripts (locating their starting and stopping points) and quantifying them (measuring how much of a transcript exists at a certain time). Molecular biologists use a variety of techniques to map and quantify transcripts, and we will encounter several in this book.

You might think that the simplest way of finding out how much transcript is made at a given time would be to label the transcript by allowing it to incorporate labeled nucleotides in vivo or in vitro, then to electrophorese it and detect the transcript as a band on the electrophoretic gel by autoradiography. In fact, this has been done for certain transcripts, both in vivo and in vitro. However, it works in vivo only if the transcript in question is quite abundant and easy to separate from other RNAs by electrophoresis. Transfer RNA and 5S ribosomal RNA satisfy both these conditions and their synthesis has been traced in vivo by simple electrophoresis (Chapter 10). This direct method succeeds in vitro only if the transcript has a clear-cut terminator, so a discrete species of RNA is made, rather than a continuum of species with different 3′-ends that would produce an unintelligible smear, rather than a sharp band. Again, in some instances this is true, most notably in the case of prokaryotic transcripts, but eukaryotic examples are rare. Thus, we frequently need to turn to other, less direct, but more specific methods. Several popular techniques are available for mapping the 5′-ends of transcripts, and one of these also locates the 3′-end. Some of them can also tell how much of a given transcript is in a cell at a given time. These methods rely on the power of nucleic acid hybridization to detect just one kind of RNA among thousands.

Northern Blots

Suppose you have cloned a cDNA (a DNA copy of an RNA) and want to know how actively the corresponding gene (gene X) is expressed in a number of different tissues of organism Y. You could answer that question in several ways, but the method we describe here will also tell the size of the mRNA the gene produces.

You would begin by collecting RNA from several tissues of the organism in question. Then you electrophorese these RNAs in an agarose gel and blot them to a suitable support. Because a similar blot of DNA is called a Southern blot, it was natural to name a blot of RNA a **Northern blot.**

Next, you hybridize the Northern blot to a labeled cDNA probe. Wherever an mRNA complementary to the probe exists on the blot, hybridization will occur, resulting in a labeled band that you can detect with x-ray film. If you run marker RNAs of known size next to the unknown RNAs, you can tell the sizes of the RNA bands that "light up" when hybridized to the probe.

Furthermore, the Northern blot tells you how abundant the gene X transcript is. The more RNA the band contains, the more probe it will bind and the darker the band will be on the film. You can quantify this darkness by measuring the amount of light it absorbs in a densitometer. Or you can quantify the amount of label in the band directly by phosphorimaging. Figure 5.26 shows a Northern blot of RNA from eight different rat tissues, hybridized to a probe for a gene encoding G3PDH (glyceraldehyde-3-phosphate dehydrogenase), which is involved in sugar metabolism. Clearly, transcripts of this gene are most abundant in the heart and skeletal muscle, and least abundant in the lung.

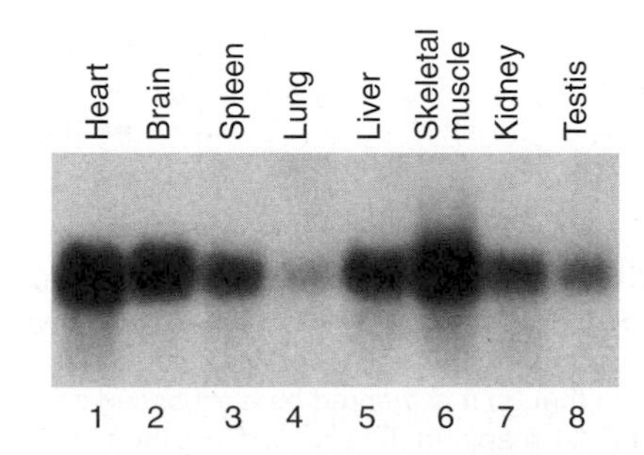

Figure 5.26 A Northern blot. Cytoplasmic mRNA was isolated from the rat tissues indicated at the top, then equal amounts of RNA from each tissue were electrophoresed and Northern blotted. The RNAs on the blot were hybridized to a labeled probe for the rat glyceraldehyde-3-phosphate dehydrogenase (G3PDH) gene, and the blot was then exposed to x-ray film. The bands represent the G3PDH mRNA, and their intensities are indicative of the amounts of this mRNA in each tissue. (*Source:* Courtesy Clontech.)

SUMMARY A Northern blot is similar to a Southern blot, but it contains electrophoretically separated RNAs instead of DNAs. The RNAs on the blot can be detected by hybridizing them to a labeled probe. The intensities of the bands reveal the relative amounts of specific RNA in each.

S1 Mapping

S1 mapping is used to locate the 5′- or 3′-ends of RNAs and to quantify the amount of a given RNA in cells at a given time. The principle behind this method is to label a single-stranded DNA probe that can hybridize only to the transcript of interest. The probe must span the sequence where the transcript starts or ends. After hybridizing the probe to the transcript, one applies **S1 nuclease,** which degrades only single-stranded DNA and RNA; double-stranded DNAs, RNAs, and hybrids are protected from S1 nuclease degradation. Thus, because the transcript forms a double-stranded RNA–DNA hybrid with the DNA probe, it protects part of the probe from degradation. The size of this part can be measured by gel electrophoresis, and the extent of protection tells where the transcript starts or ends. Figure 5.27 shows in detail how S1 mapping can be used to find the transcription start site. First, the DNA probe is labeled at its 5′-end with ^{32}P-phosphate. The 5′-end of a DNA strand usually already contains a nonradioactive phosphate, so this phosphate is removed with an enzyme called alkaline phosphatase before the labeled phosphate is added. Then the enzyme polynucleotide kinase is used to transfer the ^{32}P-phosphate group from [γ-^{32}P]ATP to the 5′-hydroxyl group at the beginning of the DNA strand.

In this example, a *Bam*HI fragment has been labeled on both ends, which would yield two labeled single-stranded probes. However, this would needlessly confuse the analysis, so the label on the left end must be removed. That task is accomplished here by recutting the DNA with another restriction enzyme, *Sal*I, then using gel electrophoresis to separate the short, left-hand fragment from the long fragment that will produce the probe. Now the double-stranded DNA is labeled

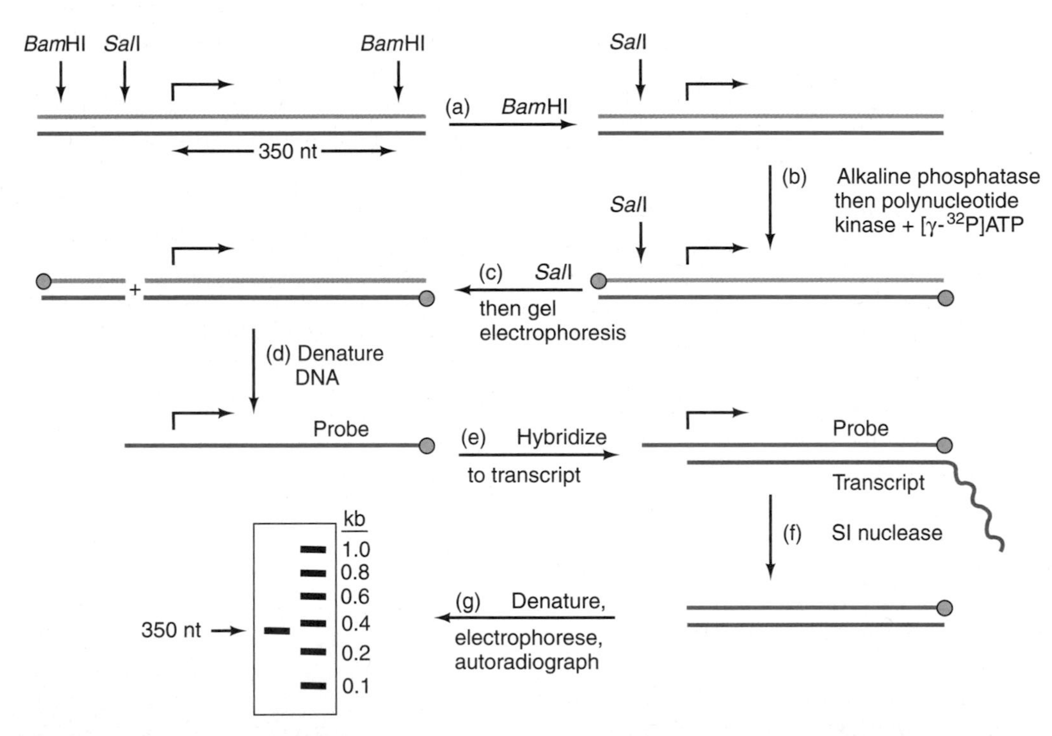

Figure 5.27 S1 mapping the 5′-end of a transcript. Begin with a cloned piece of double-stranded DNA with several known restriction sites. In this case, the exact position of the transcription start site is not known, even though it is marked here (↱) based on what will be learned from the S1 mapping. It is known that the transcription start site is flanked by two *Bam*HI sites, and a single *Sal*I site occurs upstream of the start site. In step **(a)** cut with *Bam*HI to produce the *Bam*HI fragment shown at upper right. In step **(b)** remove the unlabeled phosphates on this fragment's 5′-hydroxyl groups, then label these 5′-ends with polynucleotide kinase and [γ-^{32}P]ATP. The orange circles denote the labeled ends. In step **(c)** cut with *Sal*I and separate the two resulting fragments by electrophoresis. This removes the label from the left end of the double-stranded DNA. In step **(d)** denature the DNA to generate a single-stranded probe that can hybridize with the transcript (red) in step **(e).** In step **(f)** treat the hybrid with S1 nuclease. This digests the single-stranded DNA on the left and the single-stranded RNA on the right of the hybrid from step (e), but leaves the hybrid intact. In step **(g)** denature the remaining hybrid and electrophorese the protected piece of the probe to see how long it is. DNA fragments of known length are included as markers in a separate lane. The length of the protected probe indicates the position of the transcription start site. In this case, it is 350 bp upstream of the labeled *Bam*HI site in the probe.

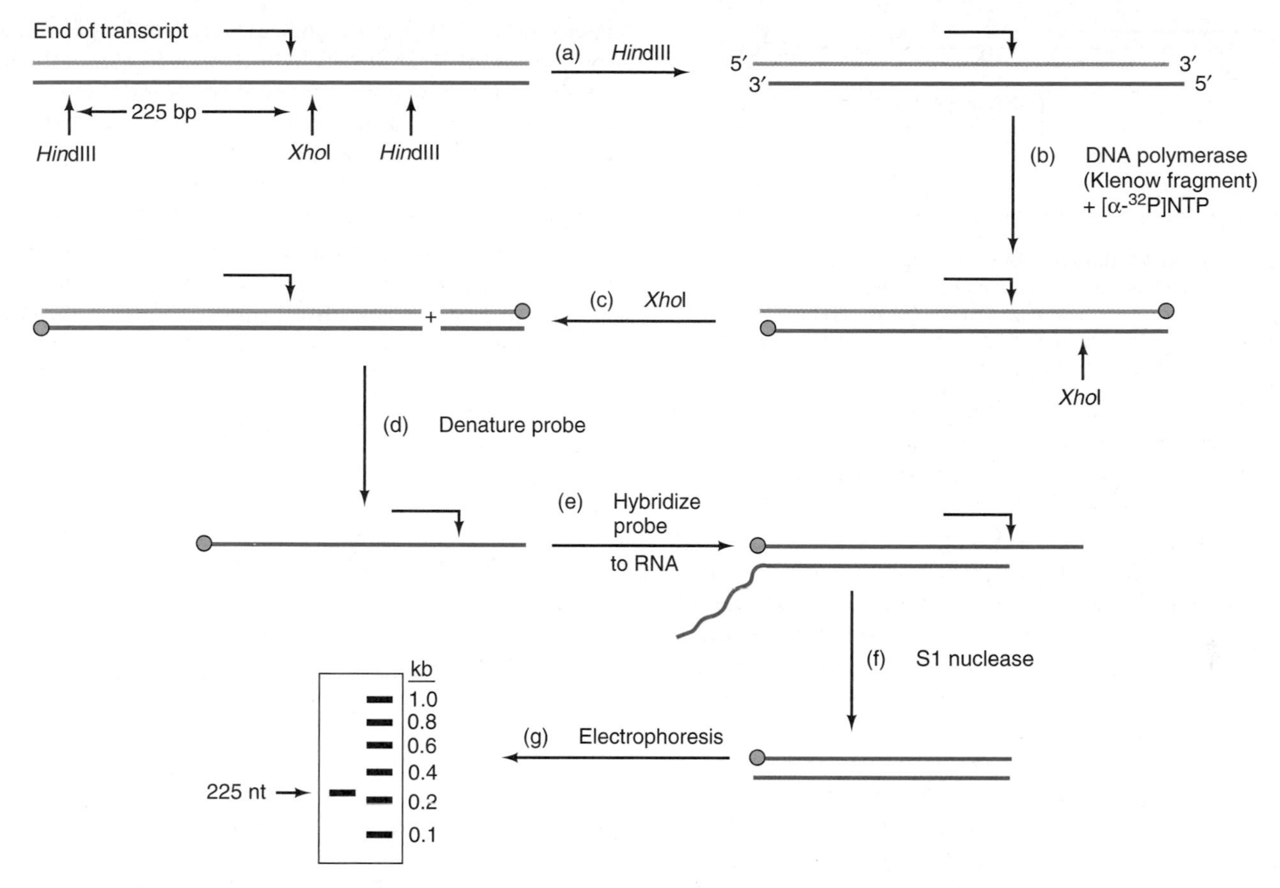

Figure 5.28 S1 mapping the 3′-end of a transcript. The principle is the same as in 5′-end mapping except that a different means of labeling the probe—at its 3′-end instead of its 5′-end—is used (detailed in Figure 5.29). In step **(a)** cut with *Hin*dIII, then in step **(b)** label the 3′-ends of the resulting fragment. The orange circles denote these labeled ends. In step **(c)** cut with *Xho*I and purify the left-hand labeled fragment by gel electrophoresis. In step **(d)** denature the probe and hybridize it to RNA (red) in step **(e).** In step **(f)** remove the unprotected region of the probe (and of the RNA) with S1 nuclease. Finally, in step **(g)** electrophorese the labeled protected probe to determine its size. In this case it is 225 nt long, which indicates that the 3′-end of the transcript lies 225 bp downstream of the labeled *Hin*dIII site on the left-hand end of the probe.

on only one end, and it can be denatured to yield a labeled single-stranded probe. Next, the probe DNA is hybridized to a mixture of cellular RNAs that contains the transcript of interest. Hybridization between the probe and the complementary transcript will leave a tail of single-stranded DNA on the left, and single-stranded RNA on the right. Next, S1 nuclease is used. This enzyme specifically degrades single-stranded DNA or RNA, but leaves double-stranded polynucleotides, including RNA–DNA hybrids, intact. Thus, the part of the DNA probe, including the terminal label, that is hybridized to the transcript will be protected. Finally, the length of the protected part of the probe is determined by high-resolution gel electrophoresis alongside marker DNA fragments of known length. Because the location of the right-hand end of the probe (the labeled *Bam*HI site) is known exactly, the length of the protected probe automatically tells the location of the left-hand end, which is the transcription start site. In this case, the protected probe is 350 nt long, so the transcription start site is 350 bp upstream of the labeled *Bam*HI site.

One can also use S1 mapping to locate the 3′-end of a transcript. It is necessary to hybridize a 3′-end-labeled probe to the transcript, as shown in Figure 5.28. All other aspects of the assay are the same as for 5′-end mapping. 3′-end-labeling is different from 5′-labeling because polynucleotide kinase will not phosphorylate 3′-hydroxyl groups on nucleic acids. One way to label 3′-ends is to perform **end-filling,** as shown in Figure 5.29. When a DNA is cut with a restriction enzyme that leaves a recessed 3′-end, that recessed end can be extended in vitro until it is flush with the 5′-end. If labeled nucleotides are included in this end-filling reaction, the 3′-ends of the DNA will become labeled.

S1 mapping can be used not only to map the ends of a transcript, but to tell the transcript concentration. Assuming that the probe is in excess, the intensity of the band on the autoradiograph is proportional to the concentration of the transcript that protected the probe. The more transcript, the more protection of the labeled probe, so the more intense the band on the autoradiograph. Thus, once it is known which band corresponds to the transcript of interest, its intensity can be used to measure the transcript concentration.

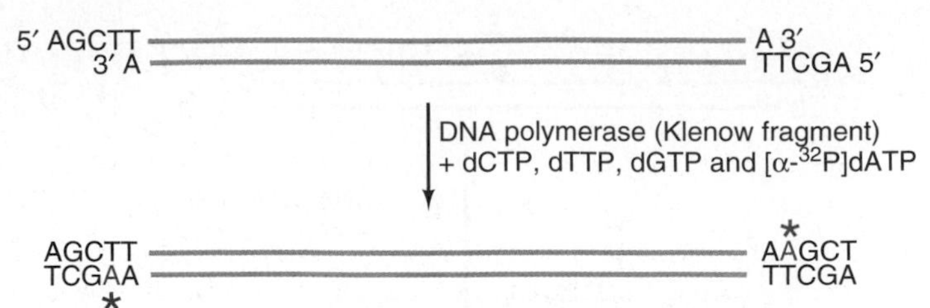

Figure 5.29 3′-end-labeling a DNA by end-filling. The DNA fragment at the top has been created by cutting with *Hin*dIII, which leaves 5′-overhangs at each end, as shown. These can be filled in with a fragment of DNA polymerase called the Klenow fragment (Chapter 20). This enzyme fragment has an advantage over the whole DNA polymerase in that it lacks the normal 5′→3′ exonuclease activity, which could degrade the 5′-overhangs before they could be filled in. The end-filling reaction is run with all four nucleotides, one of which (dATP) is labeled, so the DNA end becomes labeled. If more labeling is desired, more than one labeled nucleotide can be used.

One important variation on the S1 mapping theme is **RNase mapping (RNase protection assay).** This procedure is analogous to S1 mapping and can yield the same information about the 5′- and 3′-ends and concentration of a specific transcript. The probe in this method, however, is made of RNA and can therefore be degraded with RNase instead of S1 nuclease. This technique is very popular, partly because of the relative ease of preparing RNA probes (**riboprobes**) by transcribing recombinant plasmids or phagemids in vitro with purified phage RNA polymerases. Another advantage of using riboprobes is that they can be labeled to very high specific activity by including a labeled nucleotide in the in vitro transcription reaction, yielding a uniformly-labeled, rather than an end-labeled probe. The higher the specific activity of the probe, the more sensitive it is in detecting tiny quantities of transcripts.

SUMMARY In S1 mapping, a labeled DNA probe is used to detect the 5′- or 3′-end of a transcript. Hybridization of the probe to the transcript protects a portion of the probe from digestion by S1 nuclease, which specifically degrades single-stranded polynucleotides. The length of the section of probe protected by the transcript locates the end of the transcript, relative to the known location of an end of the probe. Because the amount of probe protected by the transcript is proportional to the concentration of transcript, S1 mapping can also be used as a quantitative method. RNase mapping is a variation on S1 mapping that uses an RNA probe and RNase instead of a DNA probe and S1 nuclease.

Primer Extension

S1 mapping has been used in some classic experiments we will introduce in later chapters, and it is the best method for mapping the 3′-end of a transcript, but it has some drawbacks. One is that the S1 nuclease tends to "nibble" a bit on the ends of the RNA–DNA hybrid, or even within the hybrid where A–T-rich regions can melt transiently. On the other hand, sometimes the S1 nuclease will not digest the single-stranded regions completely, so the transcript appears to be a little longer than it really is. These can be serious problems if we need to map the end of a transcript with one-nucleotide accuracy. But another method, called **primer extension,** can tell the 5′-end (but not the 3′-end) to the exact nucleotide.

Figure 5.30 shows how primer extension works. The first step, transcription, generally occurs naturally in vivo.

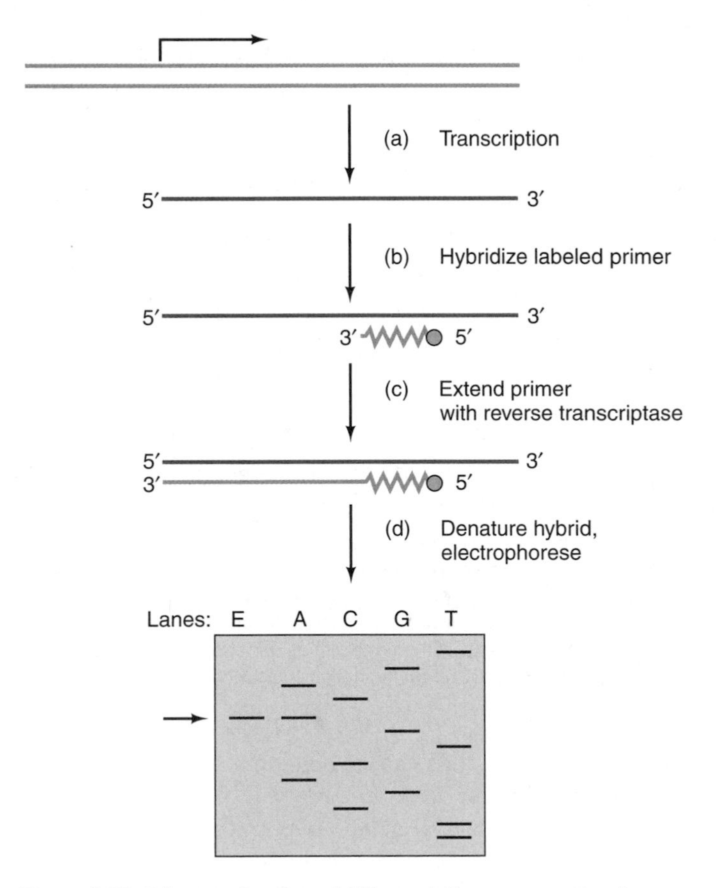

Figure 5.30 Primer extension. (a) Transcription occurs naturally within the cell, so begin by harvesting cellular RNA. **(b)** Knowing the sequence of at least part of the transcript, synthesize and label a DNA oligonucleotide that is complementary to a region not too far from the suspected 5′-end, then hybridize this oligonucleotide to the transcript. It should hybridize specifically to this transcript and to no others. **(c)** Use reverse transcriptase to extend the primer by synthesizing DNA complementary to the transcript, up to its 5′-end. If the primer itself is not labeled, or if it is desirable to introduce extra label into the extended primer, labeled nucleotides can be included in this step. **(d)** Denature the hybrid and electrophorese the labeled, extended primer (experimental lane, E). In separate lanes (lanes A, C, G, and T) run sequencing reactions, performed with the same primer and a DNA from the transcribed region, as markers. In principle, this can indicate the transcription start site to the exact base. In this case, the extended primer (arrow) coelectrophoreses with a DNA fragment in the sequencing A lane. Because the same primer was used in the primer extension reaction and in all the sequencing reactions, this shows that the 5′-end of this transcript corresponds to the middle A (underlined) in the sequence TTCGACTGACAGT.

One simply harvests cellular RNA containing the transcript whose 5′-end is to be mapped and whose sequence is known. Next, one hybridizes a labeled oligonucleotide (the primer) of approximately 18 nt to the cellular RNA. Notice that the specificity of this method derives from the complementarity between the primer and the transcript, just as the specificity of S1 mapping comes from the complementarity between the probe and the transcript. In principle, this primer (or an S1 probe) will be able to pick out the transcript to be mapped from a sea of other, unrelated RNAs.

Next, one uses reverse transcriptase to extend the oligonucleotide primer to the 5′-end of the transcript. As presented in Chapter 4, reverse transcriptase is an enzyme that performs the reverse of the transcription reaction; that is, it makes a DNA copy of an RNA template. Hence, it is perfectly suited for the job we are asking it to do: making a DNA copy of the RNA to be mapped. Once this primer extension reaction is complete, one can denature the RNA–DNA hybrid and electrophorese the labeled DNA along with markers on a high-resolution gel such as the ones used in DNA sequencing. In fact, it is convenient to use the same primer used during primer extension to do a set of sequencing reactions with a cloned DNA template. One can then use the products of these sequencing reactions as markers. In the example illustrated here, the product comigrates with a band in the A lane, indicating that the 5′-end of the transcript corresponds to the second A (underlined) in the sequence TTCGACTGACAGT. This is a very accurate determination of the transcription start site.

Just as with S1 mapping, primer extension can also give an estimate of the concentration of a given transcript. The higher the concentration of transcript, the more molecules of labeled primer will hybridize and therefore the more labeled reverse transcripts will be made. The more labeled reverse transcripts, the darker the band on the autoradiograph of the electrophoretic gel.

SUMMARY Using primer extension one can locate the 5′-end of a transcript by hybridizing an oligonucleotide primer to the RNA of interest, extending the primer with reverse transcriptase to the 5′-end of the transcript, and electrophoresing the reverse transcript to determine its size. The intensity of the signal obtained by this method is a measure of the concentration of the transcript.

Run-Off Transcription and G-Less Cassette Transcription

Suppose you want to mutate a gene's promoter and observe the effects of the mutations on the accuracy and efficiency of transcription. You would need a convenient assay that would tell you two things: (1) whether transcription is accurate (i.e., it initiates in the right place, which you have already mapped in previous primer extension or other experiments); and (2) how much of this accurate transcription occurred. You could use S1 mapping or primer extension, but they are relatively complicated. A simpler method, called **run-off transcription,** will give answers more rapidly.

Figure 5.31 illustrates the principle of run-off transcription. You start with a DNA fragment containing the gene you want to transcribe, then cut it with a restriction enzyme in the middle of the transcribed region. Next, you transcribe this truncated gene fragment in vitro with labeled nucleotides so the transcript becomes labeled.

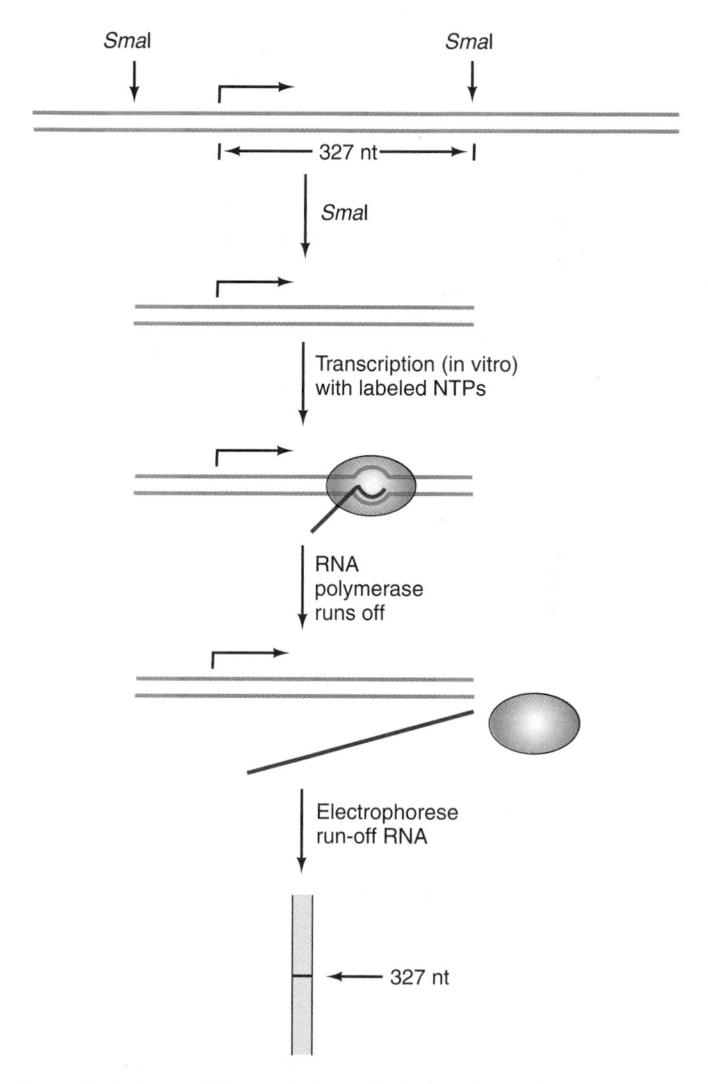

Figure 5.31 Run-off transcription. Begin by cutting the cloned gene, whose transcription is to be measured, with a restriction enzyme. Then transcribe this truncated gene in vitro. When the RNA polymerase (orange) reaches the end of the shortened gene, it falls off and releases the run-off transcript (red). The size of the run-off transcript (327 nt in this case) can be measured by gel electrophoresis and corresponds to the distance between the start of transcription and the known restriction site at the 3′-end of the shortened gene (a *Sma*I site in this case). The more actively this gene is transcribed, the stronger the 327-nt signal.

Because you have cut the gene in the middle, the polymerase reaches the end of the fragment and simply "runs off." Hence the name of this method. Now you can measure the length of the run-off transcript. Because you know precisely the location of the restriction site at the 3′-end of the truncated gene (a *Sma*I site in this case), the length of the run-off transcript (327 nt in this case) confirms that the start of transcription is 327 bp upstream of the *Sma*I site.

Notice that S1 mapping and primer extension are well suited to mapping transcripts made in vivo; by contrast, run-off transcription relies on transcription in vitro. Thus, it will work only with genes that are accurately transcribed in vitro and cannot give information about cellular transcript concentrations. However, it is a good method for measuring the rate of in vitro transcription. The more transcript is made, the more intense will be the run-off transcription signal. Indeed, run-off transcription is most useful as a quantitative method. After you have identified the physiological transcription start site by S1 mapping or primer extension you can use run-off transcription in vitro.

A variation on the run-off theme of quantifying accurate transcription in vitro is the **G-less cassette** assay (Figure 5.32). Here, instead of cutting the gene, a G-less cassette, or stretch of nucleotides lacking guanines in the nontemplate strand, is inserted just downstream of the promoter. This template is transcribed in vitro with CTP, ATP, and UTP, one of which is labeled, but no GTP. Transcription will stop at the end of the cassette where the first G is required, yielding an aborted transcript of predictable size (based on the size of the G-less cassette, which is usually a few hundred base pairs long). These transcripts are electrophoresed, and the gel is autoradiographed to measure the transcription activity. The stronger the promoter, the more of these aborted transcripts will be produced, and the stronger the corresponding band on the autoradiograph will be.

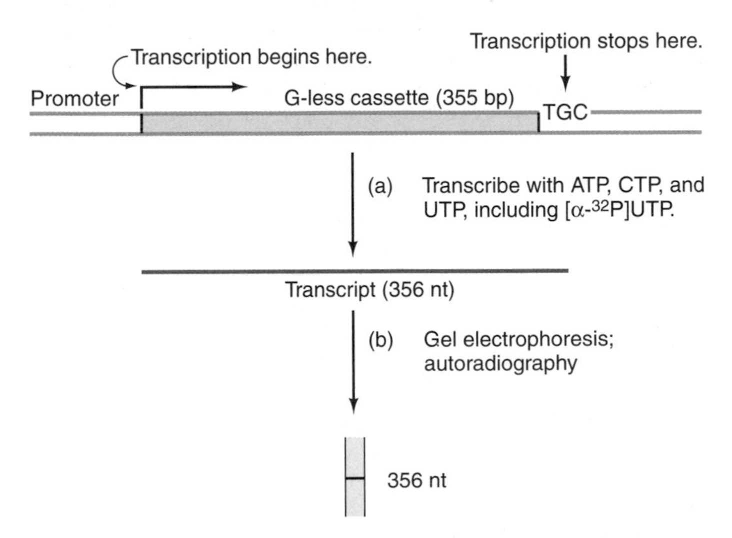

Figure 5.32 G-less cassette assay. (a) Transcribe a template with a G-less cassette (pink) inserted downstream of the promoter in vitro in the absence of GTP. This cassette is 355 bp long, contains no G's in the nontemplate strand, and is followed by the sequence TGC, so transcription stops just before the G, producing a transcript 356 nt long. **(b)** Electrophorese the labeled transcript and autoradiograph the gel and measure the intensity of the signal, which indicates how actively the cassette was transcribed.

SUMMARY Run-off transcription is a means of checking the efficiency and accuracy of in vitro transcription. A gene is truncated in the middle and transcribed in vitro in the presence of labeled nucleotides. The RNA polymerase runs off the end and releases an incomplete transcript. The size of this run-off transcript locates the transcription start site, and the amount of this transcript reflects the efficiency of transcription. In G-less cassette transcription, a promoter is fused to a double-stranded DNA cassette lacking G's in the nontemplate strand, then the construct is transcribed in vitro in the absence of GTP. Transcription aborts at the end of the cassette, yielding a predictable size band on gel electrophoresis.

5.7 Measuring Transcription Rates in Vivo

Primer extension, S1 mapping, and Northern blotting are useful for determining the concentrations of specific transcripts in a cell at a given time, but they do not necessarily tell us the rates of synthesis of the transcripts. That is because the transcript concentration depends not only on its rate of synthesis, but also on its rate of degradation. To measure transcription rates, we can employ other methods, including nuclear run-on transcription and reporter gene expression.

Nuclear Run-On Transcription

The idea of this assay (Figure 5.33a) is to isolate nuclei from cells, then allow them to extend in vitro the transcripts they had already started in vivo. This continuing transcription in isolated nuclei is called **run-on transcription** because the RNA polymerase that has already initiated transcription in vivo simply "runs on" or continues to elongate the same RNA chains. The run-on reaction is usually done in the presence of labeled nucleotides so the products will be labeled. Because initiation of new RNA chains in isolated nuclei does not generally occur, one can be fairly confident that any transcription observed in the isolated nuclei is simply a continuation of transcription that was already occurring in vivo. Therefore, the transcripts obtained in a run-on reaction should reveal not only transcription rates but also give an idea about which genes are transcribed in vivo. To eliminate the possibility of

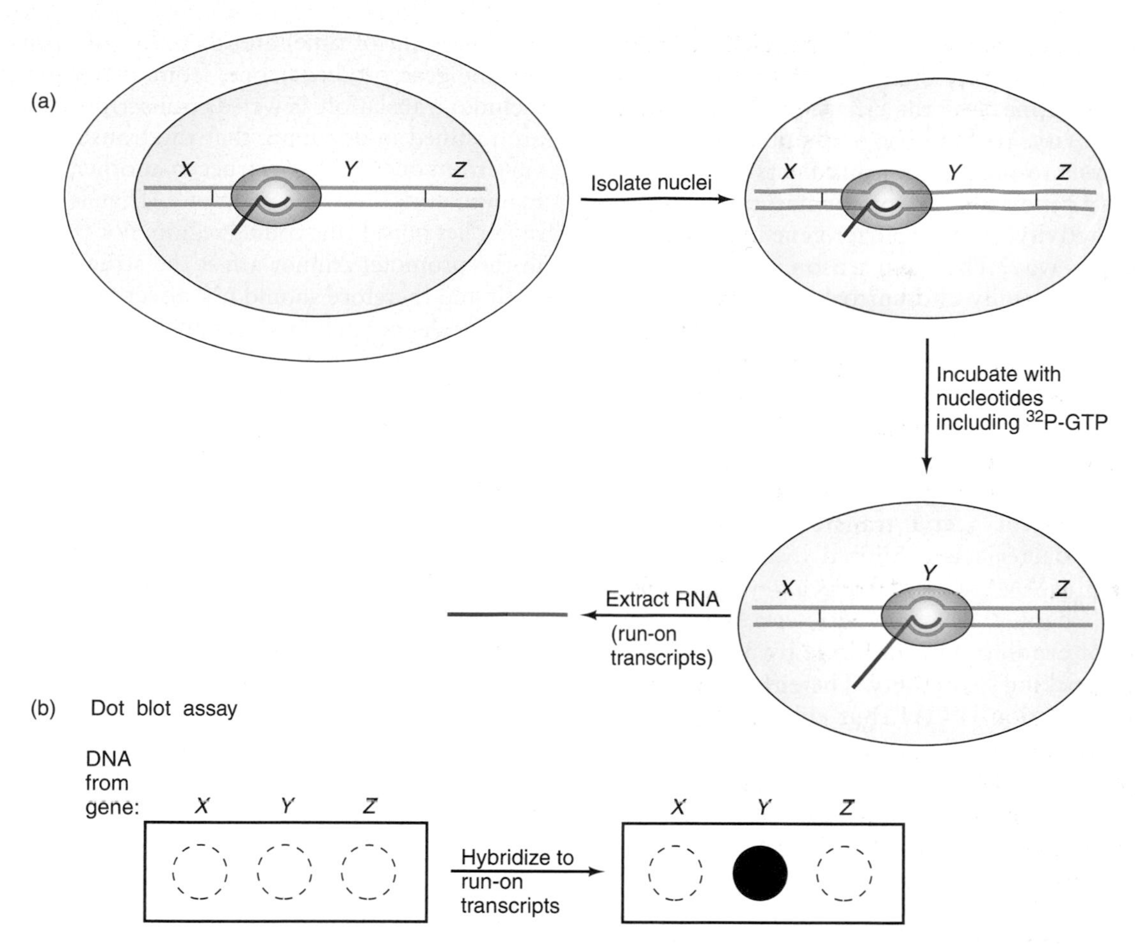

Figure 5.33 Nuclear run-on transcription. (a) The run-on reaction. Start with cells that are in the process of transcribing the *Y* gene, but not the *X* or *Z* genes. The RNA polymerase (orange) is making a transcript (blue) of the *Y* gene. Isolate nuclei from these cells and incubate them with nucleotides so transcription can continue (run-on). Also include a labeled nucleotide in the run-on reaction so the transcripts become labeled (red). Finally, extract the labeled run-on transcripts. **(b)** Dot blot assay. Spot single-stranded DNA from genes *X*, *Y*, and *Z* on nitrocellulose or another suitable medium, and hybridize the blot to the labeled run-on transcripts. Because gene *Y* was transcribed in the run-on reaction, its transcript will be labeled, and the gene *Y* spot becomes labeled. The more active the transcription of gene *Y*, the more intense the labeling will be. On the other hand, because genes *X* and *Z* were not active, no labeled *X* and *Z* transcripts were made, so the *X* and *Z* spots remain unlabeled.

initiation of new RNA chains in vitro, one can add heparin, an anionic polysaccharide that binds to any free RNA polymerase and prevents reinitiation.

Once labeled run-on transcripts have been produced, they must be identified. Because few if any of them are complete transcripts, their sizes will not be meaningful. The easiest way to perform the identification is by dot blotting (see Figure 5.33b). Samples of known, denatured DNAs are spotted on a filter and this **dot blot** is hybridized to the labeled run-on RNA. The RNA is identified by the DNA to which it hybridizes. Furthermore, the relative activity of a given gene is proportional to the degree of hybridization to the DNA from that gene. The conditions in the run-on reaction can also be manipulated, and the effects on the products can be measured. For example, inhibitors of certain RNA polymerases can be included to see if they inhibit transcription of a certain gene. If so, the RNA polymerase responsible for transcribing that gene can be identified.

SUMMARY Nuclear run-on transcription is a way of ascertaining which genes are active in a given cell by allowing transcription of these genes to continue in isolated nuclei. Specific transcripts can be identified by their hybridization to known DNAs on dot blots. The run-on assay can also be used to determine the effects of assay conditions on nuclear transcription.

Reporter Gene Transcription

Another way to measure transcription in vivo is to place a surrogate **reporter gene** under control of a specific promoter, and then measure the accumulation of the product of this reporter gene. For example, imagine that you want to examine the structure of a eukaryotic promoter. One way to do this is to make mutations in the DNA region that contains the promoter, then introduce the mutated DNA into cells and measure the effects of the mutations on promoter activity. You can use S1 mapping or primer extension analysis to do this measurement, but you can also substitute a reporter gene for the natural gene, and then assay the activity of the reporter gene product.

Why do it this way? The main reason is that reporter genes have been carefully chosen to have products that are very convenient to assay—more convenient than S1 mapping or primer extension. One of the most popular reporter genes is *lacZ,* whose product, β-galactosidase, can be measured using chromogenic substrates such as X-gal, which turns blue on cleavage. Another widely used reporter gene is the bacterial gene *(cat)* encoding the enzyme **chloramphenicol acetyl transferase (CAT).** The growth of most bacteria is inhibited by the antibiotic chloramphenicol (CAM), which blocks a key step in protein synthesis (Chapter 18). Some bacteria have developed a means of evading this antibiotic by acetylating it and therefore blocking its activity. The enzyme that carries out this acetylation is CAT. But eukaryotes are not susceptible to this antibiotic, so they have no need for CAT. Thus, the background level of CAT activity in eukaryotic cells is zero. That means that one can introduce a *cat* gene into these cells, under control of a eukaryotic promoter, and any CAT activity observed is due to the introduced gene.

How could you measure CAT activity in cells that have been transfected with the *cat* gene? In one of the most popular methods, an extract of the transfected cells is mixed with radioactive chloramphenicol and an acetyl donor (acetyl-CoA). Then thin-layer chromatography is used to separate the chloramphenicol from its acetylated products. The greater the concentrations of these products, the higher the CAT activity in the cell extract, and therefore the higher the promoter activity. Figure 5.34 outlines this procedure.

(Thin layer chromatography uses a thin layer of adsorbent material, such as silica gel, attached to a plastic backing. One places substances to be separated in spots near the bottom of the thin layer plate, then places the plate into a chamber with a shallow pool of solvent in the bottom. As the solvent wicks upward through the thin layer, substances move upward as well, but their mobilities depend on their relative affinities for the adsorbent material and the solvent.)

Another standard reporter gene is the **luciferase** gene from firefly lanterns. The enzyme luciferase, mixed with ATP and luciferin, converts the luciferin to a chemiluminescent compound that emits light. That is the secret of the firefly's lantern, and it also makes a convenient reporter because the light can be detected easily with x-ray film, or even in a scintillation counter.

In the experiments described here, we are assuming that the amount of reporter gene product is a reasonable measure of transcription rate (the number of RNA chain initiations per unit of time) and therefore of promoter activity. But the gene products come from a two-step process that includes translation as well as transcription. Ordinarily, we are justified in assuming that the translation rates do not vary from one DNA construct to another, as long as we are manipulating only the promoter. That is because the promoter lies outside the coding region. For this reason changes in the promoter cannot affect the structure of the mRNA itself and therefore should not affect translation. However, one can deliberately make changes in the region of a gene that will be transcribed into mRNA and then use a reporter gene to measure the effects of these changes on translation. Thus, depending on where the changes to a gene are made, a reporter gene can detect alterations in either transcription or translation rates.

SUMMARY To measure the activity of a promoter, one can link it to a reporter gene, such as the genes encoding β-galactosidase, CAT, or luciferase, and let the easily assayed reporter gene products indicate the activity of the promoter. One can also use reporter genes to detect changes in translational efficiency after altering regions of a gene that affect translation.

Measuring Protein Accumulation in Vivo

Gene activity can also be measured by monitoring the accumulation of the ultimate products of genes—proteins. This is commonly done in two ways, **immunoblotting (Western blotting)**, which we discussed earlier in this chapter, and immunoprecipitation.

Immunoprecipitation begins with labeling proteins in a cell by growing the cells with a labeled amino acid, typically [^{35}S] methionine. Then the labeled cells are homogenized and a particular labeled protein is bound to a specific antibody or antiserum directed against that protein. The antibody-protein complex is precipitated with a secondary antibody or protein A coupled to resin beads that can be sedimented in a low-speed centrifuge, or coupled to magnetic beads that can be isolated magnetically. Then the precipitated protein is released from the antibody, electrophoresed, and detected by autoradiography. Note that the antibody and other reagents will also be present in the precipitate, but will not be detected because they are not labeled. The more label in the protein band, the more that protein has accumulated in vivo.

SUMMARY Gene expression can be quantified by measuring the accumulation of the protein products of genes. Immunoblotting and immunoprecipitation are the favorite ways of accomplishing this task.

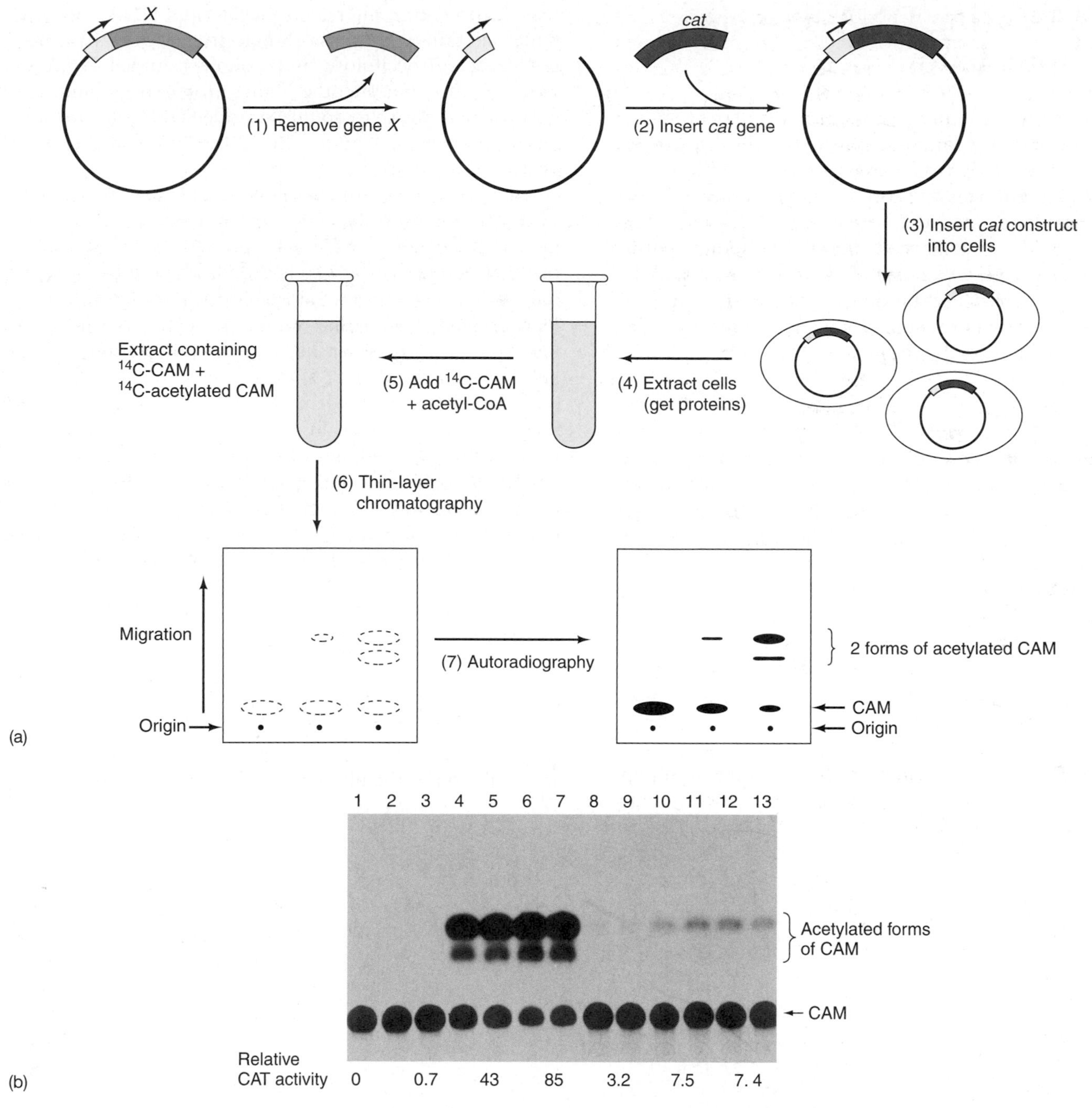

Figure 5.34 Using a reporter gene. (a) Outline of the method. Step 1: Start with a plasmid containing gene *X,* (blue) under control of its own promoter (yellow) and use restriction enzymes to remove the coding region of gene *X.* Step 2: Insert the bacterial *cat* gene under control of the *X* gene's promoter. Step 3: Place this construct into eukaryotic cells. Step 4: After a period of time, make an extract of the cells that contains soluble cellular proteins. Step 5: To begin the CAT assay, add ^{14}C-CAM and the acetyl donor acetyl-CoA. Step 6: Perform thin-layer chromatography to separate acetylated and unacetylated species of CAM. Step 7: Finally, subject the thin layer to autoradiography to visualize CAM and its acetylated derivatives. Here CAM is seen near the bottom of the autoradiograph and two acetylated forms of CAM, with higher mobility, are seen near the top. **(b)** Actual experimental results. Again, the parent CAM is near the bottom, and two acetylated forms of CAM are nearer the top. The experimenters scraped these radioactive species off the thin-layer plate and subjected them to liquid scintillation counting, yielding the CAT activity values reported at the bottom (averages of duplicate lanes). Lane 1 is a negative control with no cell extract. *Abbreviations:* CAM = chloramphenicol; CAT = chloramphenicol acetyl transferase. (*Source:* (*b*) Qin, Liu, and Weaver. Studies on the control region of the p10 gene of the *Autographa californica* nuclear polyhedrosis virus. *J. General Virology* 70 (1989) f. 2, p. 1276. (Society for General Microbiology, Reading, England.)

5.8 Assaying DNA–Protein Interactions

Another of the recurring themes of molecular biology is the study of DNA–protein interactions. We have already discussed RNA polymerase–promoter interactions, and we will encounter many more examples. Therefore, we need methods to quantify these interactions and to determine exactly what part of the DNA interacts with a given protein. We will consider here two methods for detecting protein–DNA binding and three examples of methods for showing which DNA bases interact with a protein.

Filter Binding

Nitrocellulose membrane filters have been used for decades to filter–sterilize solutions. Part of the folklore of molecular biology is that someone discovered by accident that DNA can bind to such nitrocellulose filters because they lost their DNA preparation that way. Whether this story is true or not is unimportant. What is important is that nitrocellulose filters can indeed bind DNA, but only under certain conditions. Single-stranded DNA binds readily to nitrocellulose, but double-stranded DNA by itself does not. On the other hand, protein does bind, and if a protein is bound to double-stranded DNA, the protein–DNA complex will bind. This is the basis of the assay portrayed in Figure 5.35.

In Figure 5.35a, labeled double-stranded DNA is poured through a nitrocellulose filter. The amount of label in the filtrate (the material that passes through the filter) and in the filter-bound material is measured, which shows that all the labeled material has passed through the filter into the filtrate. This confirms that double-stranded DNA does not bind to nitrocellulose. In Figure 5.35b, a solution of a labeled protein is filtered, showing that all the protein is bound to the filter. This demonstrates that proteins bind by themselves to the filter. In Figure 5.35c, double-stranded DNA is again labeled, but this time it is mixed with a protein to which it binds. Because the protein binds to the filter, the protein–DNA complex will also bind, and the radioactivity is found bound to the filter, rather than in the filtrate. Thus, filter binding is a direct measure of DNA–protein interaction.

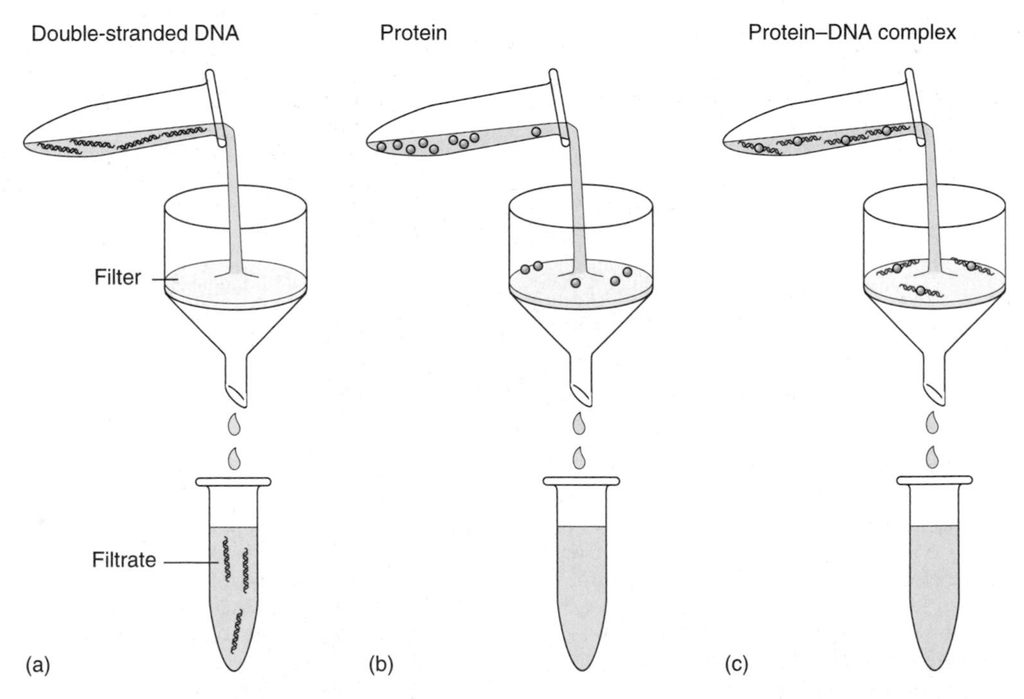

Figure 5.35 Nitrocellulose filter-binding assay. (a) Double-stranded DNA. End-label double-stranded DNA (red), and pass it through a nitrocellulose filter. Then monitor the radioactivity on the filter and in the filtrate by liquid scintillation counting. None of the radioactivity sticks to the filter, indicating that double-stranded DNA does not bind to nitrocellulose. Single-stranded DNA, on the other hand, binds tightly. **(b)** Protein. Label a protein (green), and filter it through nitrocellulose. The protein binds to the nitrocellulose. **(c)** Double-stranded DNA–protein complex. Mix an end-labeled double-stranded DNA (red) with an unlabeled protein (green) to which it binds to form a DNA–protein complex. Then filter the complex through nitrocellulose. The labeled DNA now binds to the filter because of its association with the protein. Thus, double-stranded DNA–protein complexes bind to nitrocellulose, providing a convenient assay for association between DNA and protein.

SUMMARY Filter binding as a means of measuring DNA–protein interaction is based on the fact that double-stranded DNA will not bind by itself to a nitrocellulose filter or similar medium, but a protein–DNA complex will. Thus, one can label a double-stranded DNA, mix it with a protein, and assay protein–DNA binding by measuring the amount of label retained by the filter.

Gel Mobility Shift

Another method for detecting DNA–protein interaction relies on the fact that a small DNA has a much higher mobility in gel electrophoresis than the same DNA does when it is bound to a protein. Thus, one can label a short, double-stranded DNA fragment, then mix it with a protein, and electrophorese the complex. Then one subjects the gel to autoradiography to detect the labeled species. Figure 5.36 shows the electrophoretic mobilities of three different species. Lane 1 contains naked DNA, which has a very high mobility because of its small size. Recall from earlier in this chapter that DNA electropherograms are conventionally depicted with their origins at the top, so high-mobility DNAs are found near the bottom, as shown here. Lane 2 contains the same DNA bound to a protein, and its mobility is greatly reduced. This is the origin of the name for this technique: **gel mobility shift assay** or **electrophoretic mobility shift assay (EMSA).** Lane 3 depicts the behavior of the same DNA bound to two proteins. The mobility is reduced still further because of the greater mass of protein clinging to the DNA. This is called a **supershift.** The protein could be another DNA-binding protein, or a second protein that binds to the first one. It can even be an antibody that specifically binds to the first protein.

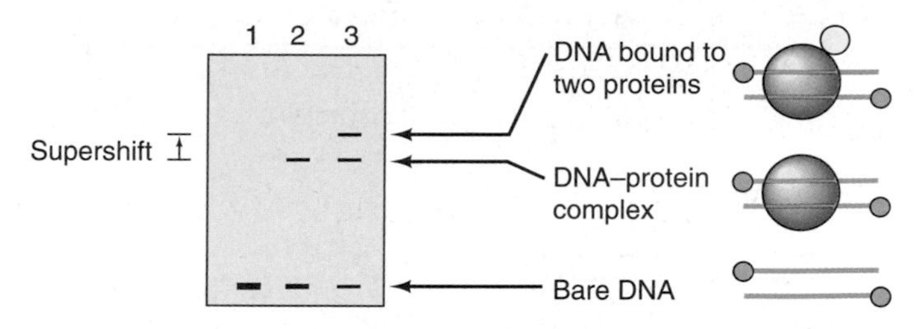

Figure 5.36 Gel mobility shift assay. Subject pure, labeled DNA or DNA–protein complexes to gel electrophoresis, then autoradiograph the gel to detect the DNAs and complexes. Lane 1 shows the high mobility of bare DNA. Lane 2 shows the mobility shift that occurs on binding a protein (red) to the DNA. Lane 3 shows the supershift caused by binding a second protein (yellow) to the DNA–protein complex. The orange dots at the ends of the DNAs represent terminal labels.

SUMMARY A gel mobility shift assay detects interaction between a protein and DNA by the reduction of the electrophoretic mobility of a small DNA that occurs on binding to a protein.

DNase Footprinting

Footprinting is a method for detecting protein–DNA interactions that can tell where the target site lies on the DNA and even which bases are involved in protein binding. Several methods are available, but three are very popular: DNase, dimethylsulfate (DMS), and hydroxyl radical footprinting. **DNase footprinting** (Figure 5.37) relies on the fact that a protein, by binding to DNA, covers the binding site and so protects it from attack by DNase. In this sense, it leaves its "footprint" on the DNA. The first step in a footprinting experiment is to end-label the DNA. Either strand can be labeled, but only one strand per experiment. Next, the protein (yellow in the figure) is bound to the DNA. Then the DNA–protein complex is treated with DNase I under mild conditions (very little DNase), so that an average of only one cut occurs per DNA molecule. Next, the protein is removed from the DNA, the DNA strands are separated, and the resulting fragments are electrophoresed on a high-resolution polyacrylamide gel alongside size markers (not shown). Of course, fragments will arise from the other end of the DNA as well, but they will not be detected because they are unlabeled. A control with DNA alone (no protein) is always included, and more than one protein concentration is usually used so the gradual disappearance of the bands in the footprint region reveals that protection of the DNA depends on the concentration of added protein. The footprint represents the region of DNA protected by the protein, and therefore tells where the protein binds.

DMS Footprinting and Other Footprinting Methods

DNase footprinting gives a good idea of the location of the binding site for the protein, but DNase is a macromolecule and is therefore a rather blunt instrument for probing the fine details of the binding site. That is, gaps may occur in the interaction between protein and DNA that DNase would not fit into and therefore would not detect. Moreover, DNA-binding proteins frequently perturb the DNA within the binding region, distorting the double helix. These perturbations are interesting, but are not generally detected by DNase footprinting because the protein keeps the DNase away. More detailed footprinting requires a smaller molecule that can fit into the nooks and crannies of the DNA–protein complex and reveal more of the subtleties of the interaction. A favorite tool for this job is the methylating agent **dimethyl sulfate (DMS).**

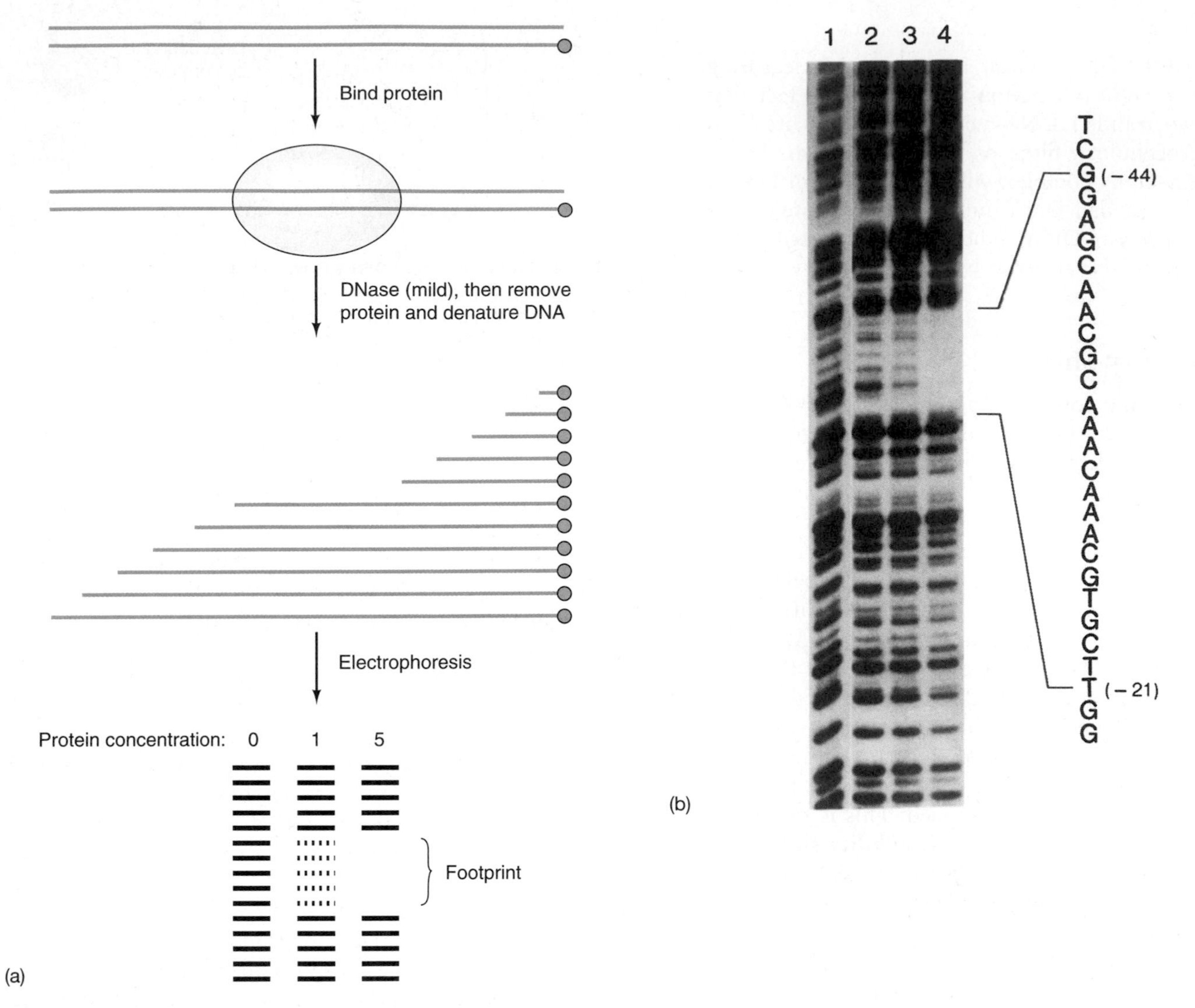

Figure 5.37 DNase footprinting. **(a)** Outline of method. Begin with a double-stranded DNA, labeled at one end (orange). Next, bind a protein to the DNA. Next, digest the DNA–protein complex under mild conditions with DNase I, so as to introduce approximately one break per DNA molecule. Next, remove the protein and denature the DNA, yielding the end-labeled fragments shown at center. Notice that the DNase cut the DNA at regular intervals except where the protein bound and protected the DNA. Finally, electrophorese the labeled fragments, and perform autoradiography to detect them. The three lanes represent DNA that was bound to 0, 1, and 5 units of protein. The lane with no protein shows a regular ladder of fragments. The lane with one unit of protein shows some protection, and the lane with five units of protein shows complete protection in the middle. This protected area is called the footprint; it shows where the protein binds to the DNA. Sequencing reactions performed on the same DNA in parallel lanes are usually included. These serve as size markers that show exactly where the protein bound. **(b)** Actual experimental results. Lanes 1–4 contained DNA bound to 0, 10, 18, and 90 pmol of protein, respectively (1 pmol = 10^{-12} mol). The DNA sequence was obtained previously by standard dideoxy sequencing. (*Source:* (*b*) Ho et al., Bacteriophage lambda protein cII binds promoters on the opposite face of the DNA helix from RNA polymerase. *Nature* 304 (25 Aug 1983) p. 705, f. 3, © Macmillan Magazines Ltd.)

Figure 5.38 illustrates DMS footprinting, which starts in the same way as DNase footprinting, with end-labeling the DNA and binding the protein. Then the DNA–protein complex is methylated with DMS, using a mild treatment so that on average only one methylation event occurs per DNA molecule. Next, the protein is dislodged, and the DNA is treated with piperidine, which removes methylated purines, creating apurinic sites (deoxyriboses without bases), then breaks the DNA at these apurinic sites. Finally, the DNA fragments are electrophoresed, and the gel is autoradiographed to detect the labeled DNA bands. Each band ends next to a nucleotide that was methylated and thus unprotected by the protein. In this example, three bands progressively disappear as more and more protein is added. But one band actually becomes more prominent at high protein concentration. This suggests that binding the protein distorts

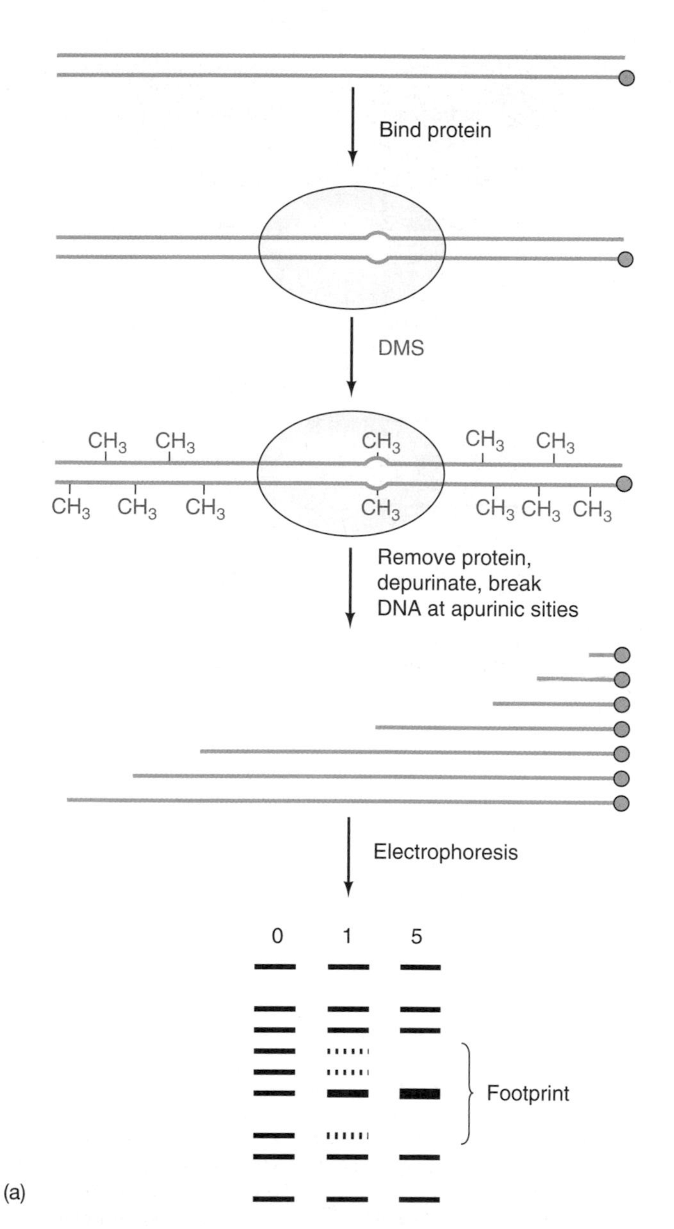

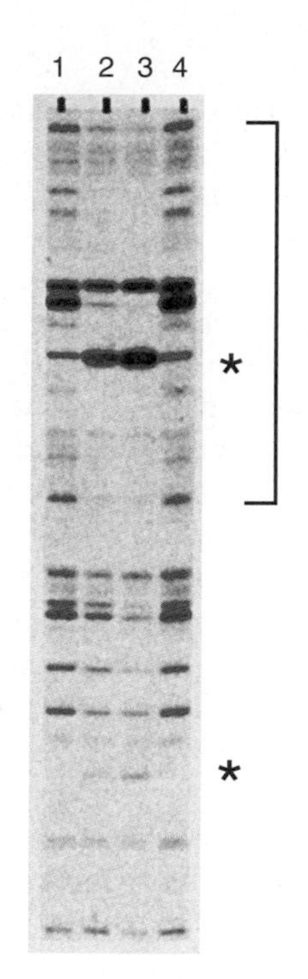

Figure 5.38 DMS footprinting. (a) Outline of the method. As in DNase footprinting, start with an end-labeled DNA, then bind a protein (yellow) to it. In this case, the protein causes some tendency of the DNA duplex to melt in one region, represented by the small "bubble." Next, methylate the DNA with DMS. This adds methyl groups (CH_3, red) to certain bases in the DNA. Do this under mild conditions so that, on average, only one methylated base occurs per DNA molecule (even though all seven methylations are shown together on one strand for convenience here). Next, use piperidine to remove methylated purines from the DNA, then to break the DNA at these apurinic sites. This yields the labeled DNA fragments depicted at center. Electrophorese these fragments and autoradiograph the gel to give the results shown at bottom. Notice that three sites are protected against methylation by the protein, but one site is actually made more sensitive to methylation (darker band). This is because of the opening up of the double helix that occurs in this position when the protein binds. **(b)** Actual experimental results. Lanes 1 and 4 have no added protein, whereas lanes 2 and 3 have increasing concentrations of a protein that binds to this region of the DNA. The bracket indicates a pronounced footprint region. The asterisks denote bases made *more* susceptible to methylation by protein binding. (*Source:* (*b*) Learned et al., Human rRNA transcription is modulated by the coordinate binding of two factors to an upstream control element. *Cell* 45 (20 June 1986) p. 849, f. 2a. Reprinted by permission of Elsevier Science.)

the DNA double helix such that it makes the base corresponding to this band more vulnerable to methylation.

In addition to DNase and DMS, other reagents are commonly used to footprint protein–DNA complexes by breaking DNA except where it is protected by bound proteins. For example, organometallic complexes containing copper or iron act by generating **hydroxyl radicals** that attack and break DNA strands.

SUMMARY Footprinting is a means of finding the target DNA sequence, or binding site, of a DNA-binding protein. DNase footprinting is performed by binding the protein to its end-labeled DNA target, then attacking the DNA–protein complex with DNase. When the resulting DNA fragments are electrophoresed, the protein binding site shows up as a gap, or "footprint" in the pattern where the protein protected the DNA from degradation. DMS footprinting follows a similar principle, except that the DNA methylating agent DMS, instead of DNase, is used to attack the DNA– protein complex. The DNA is then broken at the methylated sites. Unmethylated (or hypermethylated) sites show up on electrophoresis of the labeled DNA fragments and demonstrate where the protein bound to the DNA. Hydroxyl radical footprinting uses copper- or iron-containing organometallic complexes to generate hydroxyl radicals that break DNA strands.

Chromatin Immunoprecipitation (ChIP)

Chromatin immunoprecipitation (**ChIP**) is a way of discovering whether a given protein is bound to a given gene in chromatin—the DNA–protein complex that is the natural state of DNA in a living cell (Chapter 13). Figure 5.39 illustrates the method. One starts with chromatin isolated from cells and adds formaldehyde to form covalent bonds between DNA and any proteins bound to it. Then one shears the chromatin by sonication to produce short, double-stranded DNA fragments cross-linked to proteins. Next, one makes cell extracts and immunoprecipitates the protein–DNA complexes with antibodies directed against a protein of interest, as described earlier in this chapter. This precipitates that specific protein, and the DNA to which it binds. To see if that DNA contains the gene of interest, one performs PCR (Chapter 4) on the immunoprecipitate with primers designed to amplify that gene. If the gene is present, a DNA fragment of predictable size will result and be detectable as a band after gel electrophoresis.

SUMMARY Chromatin immunoprecipitation detects a specific protein–DNA interaction in chromatin in vivo. It uses an antibody to precipitate a particular protein in complex with DNA, and PCR to determine whether the protein binds near a particular gene.

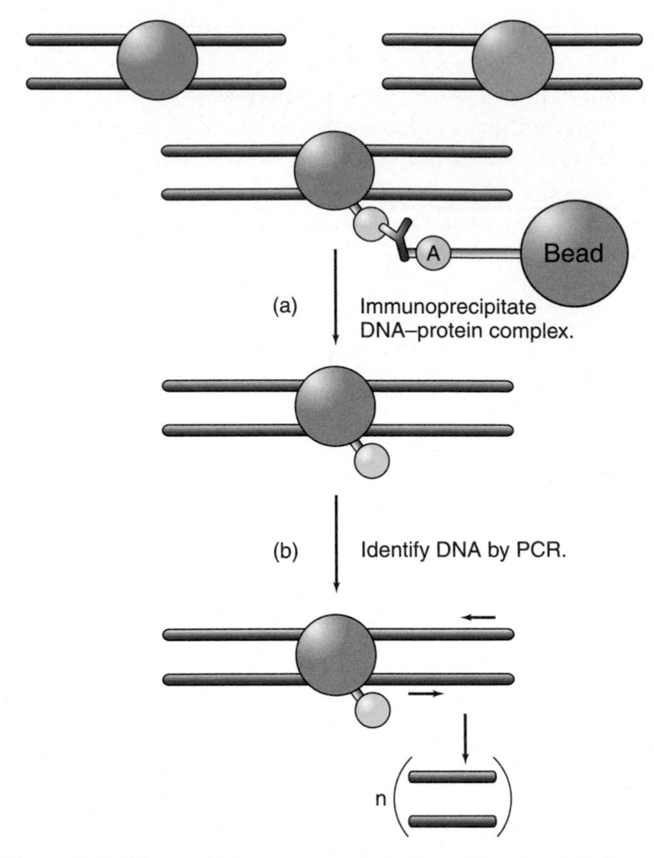

Figure 5.39 Chromatin immunoprecipitation. The chromatin has already been cross-linked with formaldehyde and sheared into short pieces. **(a)** The immunoprecipitation step. An antibody (red) has bound to an epitope (yellow) attached to a protein of interest (purple), which in turn is bound to a specific site on a double-stranded DNA (blue). The antibody is bound to staphylococcal protein A (or G), which is coupled to a large bead that can be easily purified by centrifugation. The bead can even be magnetic, so the immune complexes can be drawn to the bottom of a tube with a magnet. The antibody does not bind to the other proteins (green and orange) to which the epitope is not attached. **(b)** Identifying the DNA in the immunoprecipitate. Primers specific for the DNA of interest are used in a PCR reaction to amplify a portion of the DNA. Production of a DNA fragment of the correct predicted size indicates that the protein did indeed bind to the DNA of interest. (The primers do not amplify the exact sequence to which the protein binds,but an adjacent portion of the gene of interest.)

5.9 Assaying Protein–Protein Interactions

Protein–protein interactions are also extremely important in molecular biology, and there are a number of ways to assay them. Immunoprecipitation, which we discussed earlier in this chapter, is one way: If an antibody directed against a particular protein (X) precipitates both proteins X and Y together, but has no affinity for protein Y on its own, it is very likely that protein Y associates with protein X.

Another popular method is called a **yeast two-hybrid assay.** Figure 5.40 describes a generic version of this very sensitive technique, which is designed to demonstrate binding—even transient binding—between two proteins. The yeast two-hybrid assay takes advantage of two facts, discussed in Chapter 12: (1) that transcription activators typically have a DNA-binding domain and a transcription-activating domain; and (2) that these two domains have self-contained activities. To assay for binding between two proteins, X and Y, one can arrange for yeast cells to express these two proteins as fusion proteins, pictured in Figure 5.40b. Protein X is fused to a DNA-binding domain, and protein Y is fused to a transcription-activating domain. Now if proteins X and Y interact, that brings the DNA-binding and transcription-activating domains together, and activates transcription of a reporter gene (typically *lacZ*).

One can even use the yeast two-hybrid system to fish for unknown proteins that interact with a known protein (Z). In a screen such as Figure 5.40c, one would prepare a library of cDNAs linked to the coding region for a transcription-activating domain and express these hybrid genes, along with a gene encoding the DNA-binding domain—Z hybrid gene, in yeast cells. In practice, each yeast cell would make a different fusion protein (AD–A, AD–B, AD–C, etc.), along with the BD–Z fusion protein, but they are all pictured here together for simplicity. We can see that AD–D binds to BD–Z and activates transcription, but none of the other fusion proteins can do this because they cannot interact with BD–Z. Once clones that activate transcription are found, the plasmid bearing the AD–D hybrid gene is isolated and the D portion is sequenced to find out what it codes for. Because the yeast two-hybrid assay is indirect, it is subject to artifacts.

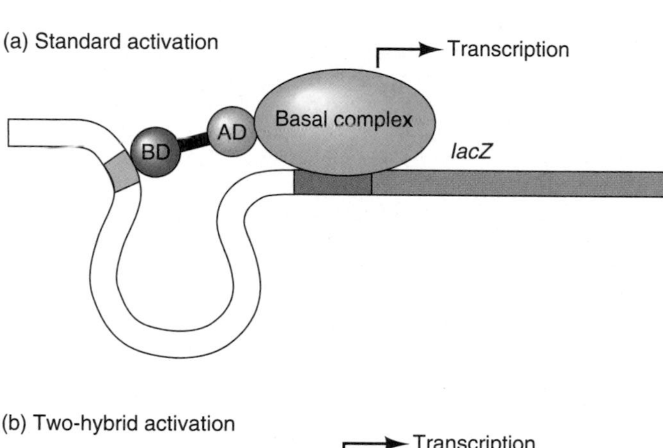

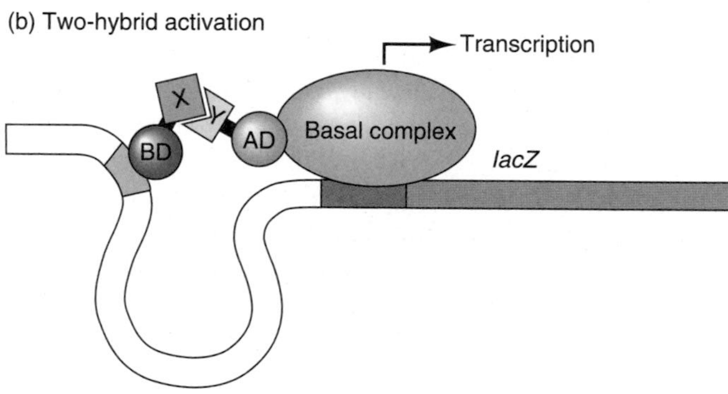

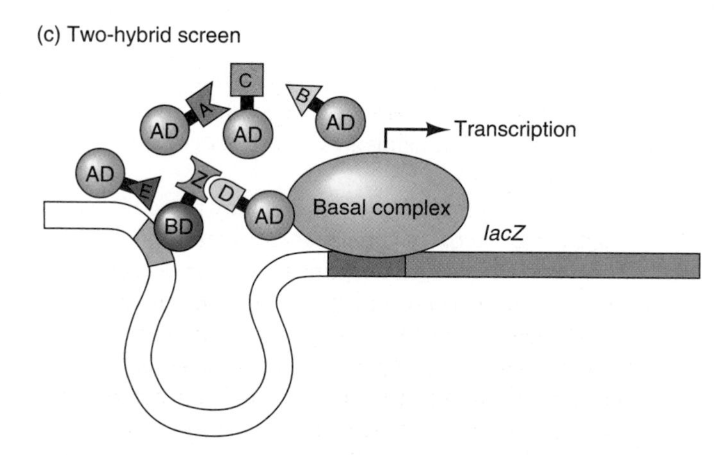

Figure 5.40 Principle of the yeast two-hybrid assay. (a) Standard model of transcription activation. The DNA-binding domain (BD, red) of an activator binds to an enhancer (pink), and the activating domain (AD green) interacts with the basal complex (orange), recruiting it to the promoter (brown). This stimulates transcription. **(b)** Two-hybrid assay for protein–protein interaction. Using gene cloning techniques (Chapter 4), link the gene for one protein (X, turquoise) to the part of a gene encoding a DNA-binding domain to encode one hybrid protein; link the gene for another protein (Y, yellow) to the part of a gene encoding a transcription-activating domain to encode a second hybrid protein. When plasmids encoding these two hybrid proteins are introduced into yeast cells bearing the appropriate promoter, enhancer, and reporter gene (*lacZ*, purple, in this case), the two hybrid proteins can get together as shown to serve as an activator. Activated transcription produces abundant reporter gene product, which can be detected with a colorimetric assay, using X-gal, for example. One hybrid protein contributes a DNA-binding domain, and the other contributes a transcription-activating domain. The two parts of the activator are held together by the interaction between proteins X and Y. If X and Y interact, and X-gal is used in the assay for the reporter gene product, the yeast cells will turn blue. If X and Y do not interact, no activator will form, and no activation of the reporter gene will occur. In this case, the yeast cells will remain white in the presence of X-gal. The GAL4 DNA-binding domain and transcription-activating domain are traditionally used in this assay, but other possibilities exist. **(c)** Two-hybrid screen for a protein that interacts with protein Z. Yeast cells are transformed with two plasmids: one encoding a DNA-binding domain (red) coupled to a "bait" protein (Z, turquoise). The other is a set of plasmids containing many cDNAs coupled to the coding region for a transcription-activating domain. Each of these encodes a fusion protein containing the activating domain (green) fused to an unknown cDNA product (the "prey"). Each yeast cell is transformed with just one of these prey-encoding plasmids, but several of their products are shown together here for convenience. One prey protein (D, yellow) interacts with the bait protein, Z. This brings together the DNA-binding domain and the transcription-activating domain so they can activate the reporter gene. Now the experimenter can purify the prey plasmid from this positive clone and thereby get an idea about the nature of the prey protein.

Thus, the protein–protein interactions suggested by such an assay should be verified with a direct assay, such as immunoprecipitation.

SUMMARY Protein–protein interactions can be detected in a number of ways, including immunoprecipitation and yeast two-hybrid assay. In the latter technique, three plasmids are introduced into yeast cells. One encodes a hybrid protein composed of protein X and a DNA-binding domain. The second encodes a hybrid protein composed of protein Y and a transcription-activating domain. The third has a promoter-enhancer region linked to a reporter gene such as *lacZ*. The enhancer interacts with the DNA-binding domain linked to protein X. If proteins X and Y interact, they bring together the two parts of a transcription activator that can activate the reporter gene, giving a product that can catalyze a colorimetric reaction. If X-gal is used, for example, the yeast cells will turn blue.

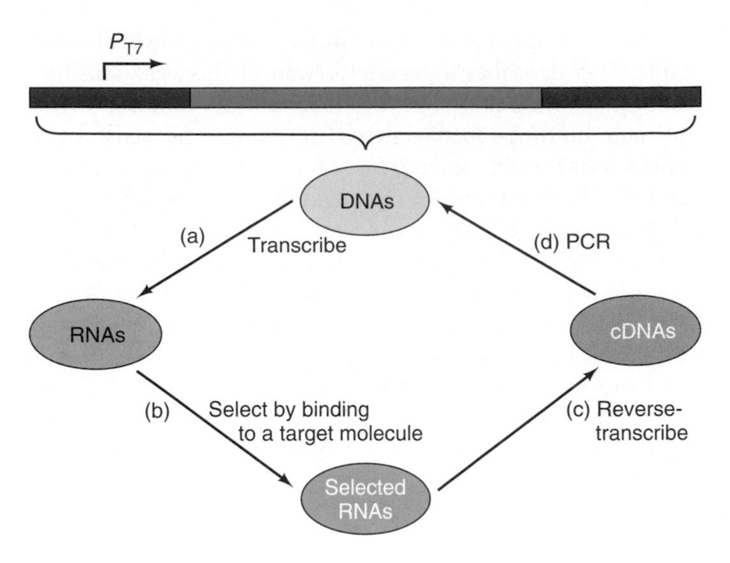

Figure 5.41 SELEX. Start with a large collection of DNAs (top) that have a random sequence (blue) flanked by constant sequences (red). **(a)** Transcribe the DNA pool to produce a pool of RNAs that also contain a random sequence flanked by constant sequences. **(b)** Select for aptamers by affinity chromatography with the target molecule. **(c)** Reverse-transcribe the selected RNAs to produce a pool of cDNAs. **(d)** Amplify the cDNAs by PCR, using primers complementary to the constant regions at the ends of the DNAs. This cycle is repeated several times to enrich the aptamers in the pool.

5.10 Finding RNA Sequences That Interact with Other Molecules

SELEX

SELEX (systematic evolution of ligands by exponential enrichment) is a method that was originally developed to discover short RNA sequences (**aptamers**) that bind to particular molecules. Figure 5.41 illustrates the classical SELEX procedure. One starts with a pool of PCR-amplified synthetic DNAs that have constant end regions (red), but random central regions (blue) that can potentially encode over 10^{15} different RNA sequences. In the first step, these DNAs are transcribed in vitro, using the phage T7 RNA polymerase, which recognizes the T7 promoter in the upstream constant region of every DNA in the pool. In the next step, the aptamers are selected by affinity chromatography (this chapter), using a resin with the target molecule immobilized. The selected RNAs bind to the resin and then can be released with a solution containing the target molecule. These selected RNAs are then reverse-transcribed to yield double-stranded DNA, which is then subjected to PCR, using primers specific for the DNAs' constant ends.

One round of SELEX yields a population of molecules only partially enriched in aptamers, so the process is repeated several more times to produce a highly enriched population of aptamers. SELEX has been extensively exploited to find the RNA sequences that are contacted by proteins. It is extremely powerful in that it finds a few aptamers among an astronomically high number of starting RNA sequences.

Functional SELEX

Functional SELEX is similar to classical SELEX in that it finds a few "needles" (RNA sequences) in a "haystack" of starting sequences. But instead of finding aptamers that bind to other molecules, it finds RNA sequences that carry out, or make possible, some function. With simple binding, selection is easy; it just requires affinity chromatography. But selection based on function is trickier and requires creativity in designing the selection step. For instance, the first functional SELEX procedures detected a ribozyme (an RNA with enzymatic activity), and this ribozyme activity altered the RNA itself to allow it to be amplified. One simple example is a ribozyme that can add an olignucleotide to its own end. This activity allowed the investigators to supply an oligonucleotide of defined sequence to the ribozyme, which then added this tag to itself. Once tagged, the ribozyme becomes subject to amplification using a PCR primer complementary to the tag.

A pool of random RNA sequences may not contain any RNAs with high activity. But that problem can be overcome by carrying out the amplification step under mutagenizing conditions, such that many variants of the mildly active sequences are created. Some of these will probably have greater activity than the original. After several rounds of selection and mutagenesis, RNAs with very strong enzymatic activity can be produced.

SUMMARY SELEX is a method that allows one to find RNA sequences that interact with other molecules, including proteins. RNAs that interact with a target molecule are selected by affinity chromatography, then converted to double-stranded DNAs and amplified by PCR. After several rounds of this procedure, the RNAs are highly enriched for sequences that bind to the target molecule. Functional SELEX is a variation on this theme in which the desired function somehow alters the RNA so it can be amplified. If the desired function is enzymatic, mutagenesis can be introduced into the amplification step to produce variants with higher activity.

5.11 Knockouts and Transgenics

Most of the techniques we have discussed in Chapter 5 are designed to probe the structures and activities of genes. But these frequently leave a big question about the role of the gene being studied: What purpose does the gene play in the life of the organism? We can answer this question best by seeing what happens when we create deliberate deletions or additions of genes to a living organism. We now have techniques for targeted disruption of genes in several organisms. For example, we can disrupt genes in mice, and when we do, we call the products **knockout mice.** We can also add foreign genes, or **transgenes,** to organisms. For example, adding a transgene to mice creates **transgenic mice.** Let us examine each of these techniques.

Knockout Mice

Figure 5.42 explains one way to begin the process of creating a knockout mouse. We start with cloned DNA containing the mouse gene we want to knock out. We interrupt this gene with another gene that confers resistance to the antibiotic neomycin. Elsewhere in the cloned DNA, outside the target gene, we introduce a thymidine kinase *(tk)* gene. Later, these extra genes will enable us to weed out those clones in which targeted disruption did not occur.

Next, we mix the engineered mouse DNA with **embryonic stem cells (ES cells)** from an embryonic brown mouse. By definition, these ES cells can differentiate into any kind of mouse cell. In a small percentage of these cells, the interrupted gene will find its way into the nucleus, and homologous recombination will occur between the altered gene and the resident, intact gene. This recombination places the altered gene into the mouse genome and removes the *tk* gene. Unfortunately, such recombination events are relatively rare, so many stem cells will experience no recombination and therefore will suffer no interruption of their resident gene. Still other cells will experience nonspecific recombination, in which the interrupted gene will insert randomly into the genome without replacing the intact gene.

The problem now is to eliminate the cells in which homologous recombination did not occur. This is where the extra genes we introduced earlier come in handy. Cells in which no recombination took place will have no neomycin-resistance gene. Thus, we can eliminate them by growing the cells in medium containing the neomycin derivative G418. Cells that experienced nonspecific recombination will have incorporated the *tk* gene, along with the interrupted gene, into their genome. We can kill these cells with gangcyclovir, a drug that is lethal to tk^+ cells. (The stem cells we used are tk^-.) Treatment with these two drugs leaves us with engineered cells that have undergone homologous recombination and are therefore heterozygous for an interruption in the target gene.

Our next task is to introduce this interrupted gene into a whole mouse (Figure 5.43). We do this by injecting our engineered cells into a mouse blastocyst that is destined to develop into a black mouse. Because the ES cells can differentiate into any kind of mouse cell, they act like the normal blastocyst cells, cooperating to form an embryo that can be placed into a surrogate mother, which eventually gives birth to a chimeric mouse. We can recognize this mouse as a chimera by its patchy coat; the black zones come from the original black embryo, and the brown zones result from the transplanted engineered cells.

To get a mouse that is a true heterozygote instead of a chimera, we allow the chimera to mature, then mate it with a black mouse. Because brown (agouti) is dominant, some of the progeny should be brown. In fact, all of the offspring resulting from gametes derived from the engineered stem cells should be brown. But only half of these brown mice will carry the interrupted gene because the engineered stem cells were heterozygous for the knockout. Southern blots showed that two of the brown mice in our example carry the interrupted gene. We mate these and look for progeny that are homozygous for the knockout by Examining their DNA. In our example, one of the mice from this mating is a knockout, and now our job is to observe its phenotype. Frequently, as here, the phenotype is not obvious. (It's there; can you see it?) But obvious or not, it can be very instructive.

In other cases, the knockout is lethal and the affected mouse fetuses die before birth. Still other knockouts have intermediate effects. For example, consider the tumor suppressor gene called *p53*. Humans with defects in this gene are highly susceptible to certain cancers. Mice with their *p53* gene knocked out develop normally but are afflicted with tumors at an early age.

Transgenic Mice

Molecular biologists use two popular methods to generate transgenic mice. In the first, they simply inject a cloned

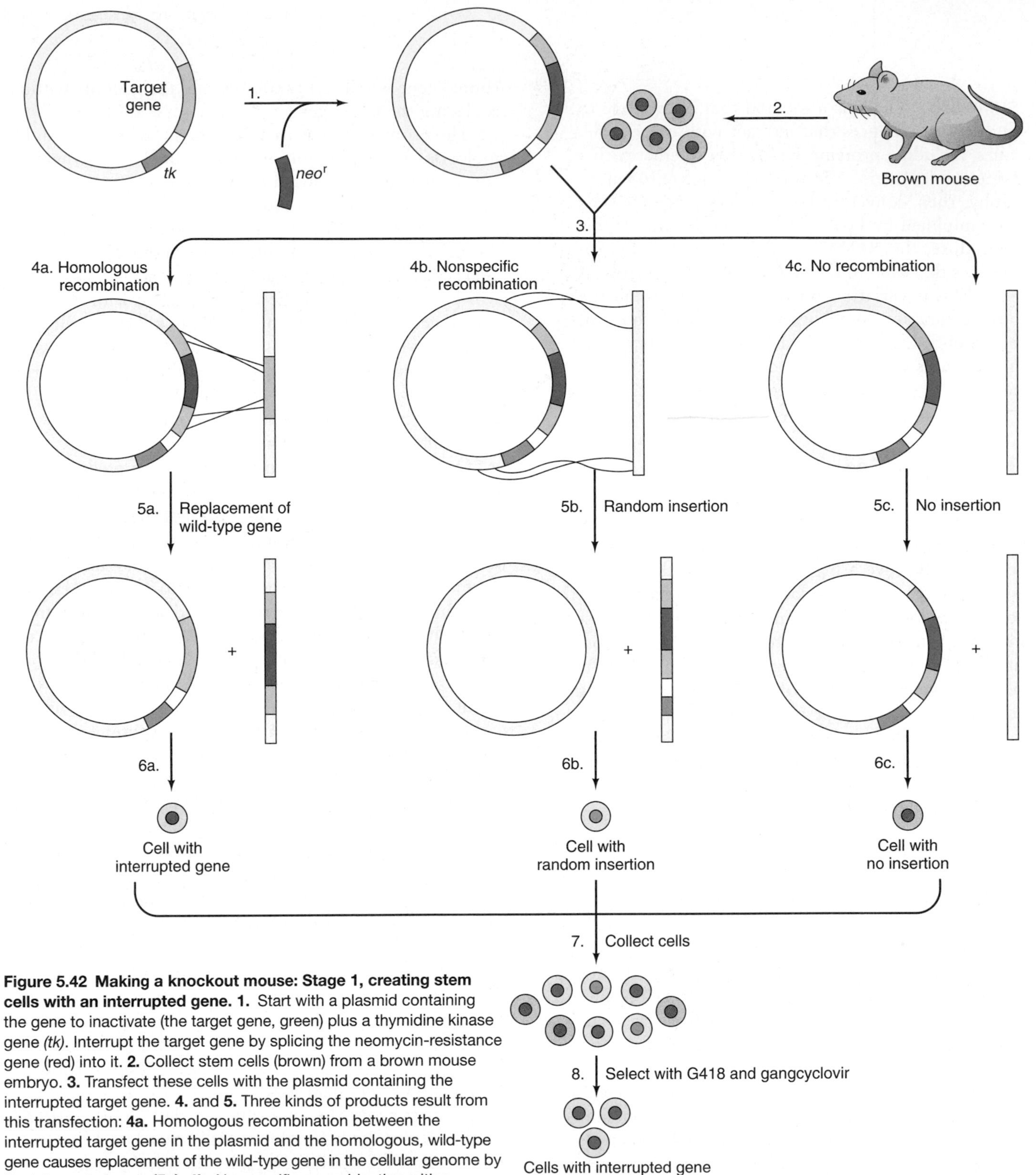

Figure 5.42 Making a knockout mouse: Stage 1, creating stem cells with an interrupted gene. 1. Start with a plasmid containing the gene to inactivate (the target gene, green) plus a thymidine kinase gene *(tk)*. Interrupt the target gene by splicing the neomycin-resistance gene (red) into it. **2.** Collect stem cells (brown) from a brown mouse embryo. **3.** Transfect these cells with the plasmid containing the interrupted target gene. **4.** and **5.** Three kinds of products result from this transfection: **4a.** Homologous recombination between the interrupted target gene in the plasmid and the homologous, wild-type gene causes replacement of the wild-type gene in the cellular genome by the interrupted gene **(5a). 4b.** Nonspecific recombination with a nonhomologous sequence in the cellular genome results in random insertion of the interrupted target gene plus the *tk* gene into the cellular genome **(5b). 4c.** When no recombination occurs, the interrupted target gene is not integrated into the cellular genome at all **(5c). 6.** The cells resulting from these three events are color-coded as indicated: Homologous recombination yields a cell (red) with an interrupted target gene **(6a)**; nonspecific recombination yields a cell (blue) with the interrupted target gene and the *tk* gene inserted at random **(6b);** no recombination yields a cell (brown) with no integration of the interrupted gene **(6c). 7.** Collect the transfected cells, containing all three types (red, blue, and brown). **8.** Grow the cells in medium containing the neomycin analog G418 and the drug gangcyclovir. The G418 kills all cells without a neomycin-resistance gene, namely those cells (brown) that did not experience a recombination event. The gangcyclovir kills all cells that have a *tk* gene, namely those cells (blue) that experienced a nonspecific recombination. This leaves only the cells (red) that experienced homologous recombination and therefore have an interrupted target gene.

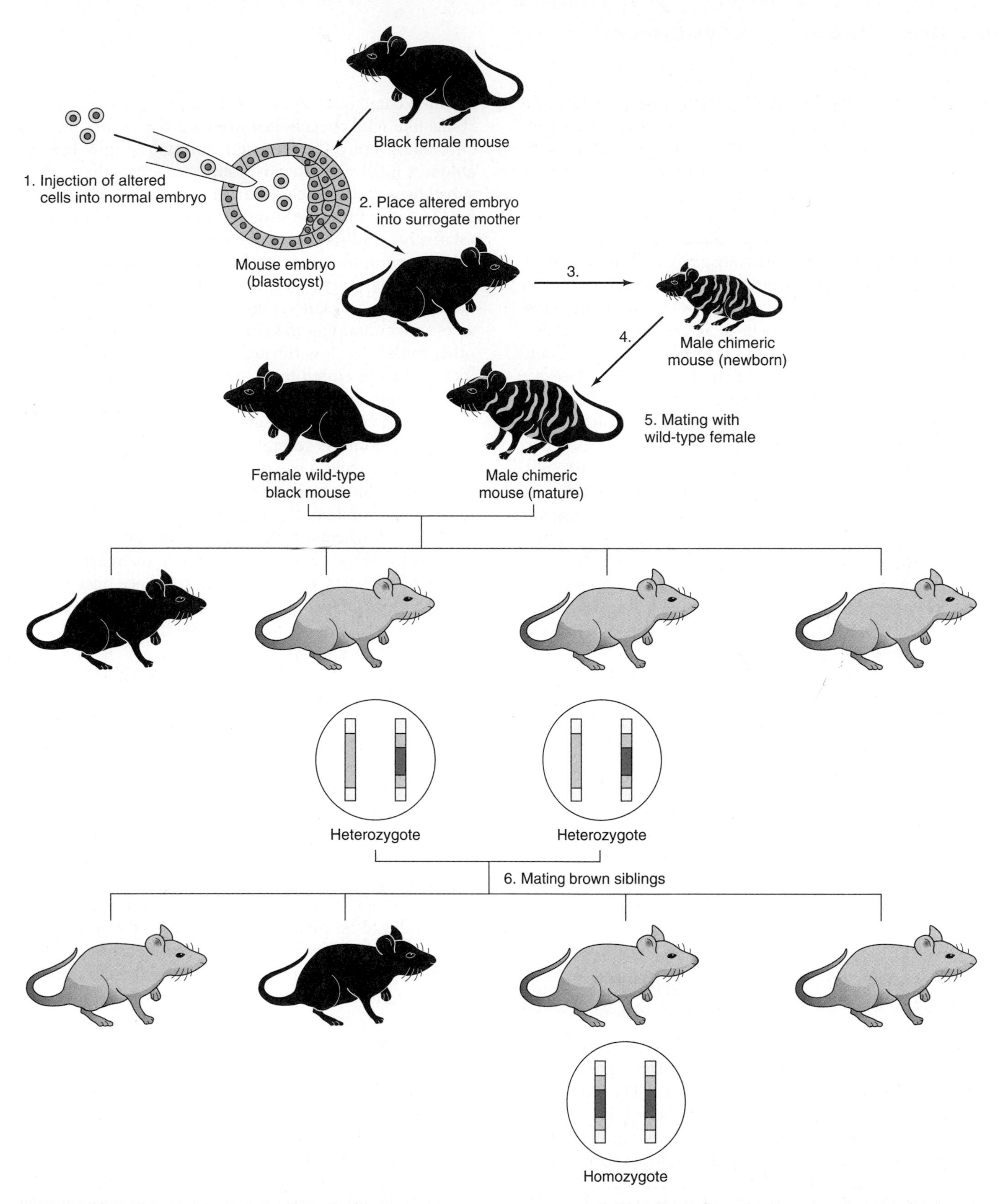

Figure 5.43 Making a knockout mouse: Stage 2, placing the interrupted gene in the animal. (1) Inject the cells with the interrupted gene (see Figure 5.42) into a blastocyst-stage embryo from black parent mice. **(2)** Transplant this mixed embryo to the uterus of a surrogate mother. **(3)** The surrogate mother gives birth to a chimeric mouse, which one can identify by its black and brown coat. (Recall that the altered cells came from an agouti [brown] mouse, and they were placed into an embryo from a black mouse.) **(4)** Allow the chimeric mouse (a male) to mature. **(5)** Mate it with a wild-type black female. Discard any black offspring, which must have derived from the wild-type blastocyst; only brown mice could have derived from the transplanted cells. **(6)** Select a brown brother and sister pair, both of which show evidence of an interrupted target gene (by Southern blot analysis), and mate them. Again, examine the DNA of the brown progeny by Southern blotting. This time, one animal that is homozygous for the interrupted target gene is found. This is the knockout mouse. Now observe this animal to determine the effects of knocking out the target gene.

foreign gene into the sperm pronucleus just after fertilization of a mouse egg, before the sperm and egg nuclei have fused. This allows the foreign DNA to insert itself into the embryonic cell DNA, often as strings of tandemly repeated genes. This insertion occurs very early in embryonic development, but even if one or two embryonic cell divisions have already taken place, some cells in the resulting adult organism will not contain the transgene, and the organism will be a chimera. Thus, the next step is to breed the chimera with a wild-type mouse and select pups that have the transgene. The fact that they have it at all means that they derived from a sperm or an egg that had the transgene, and therefore they have it in every cell in their bodies. These are true transgenic mice. Notice that the transgene they carry can come from any organism—even another mouse.

The second method is to inject the foreign DNA into mouse embryonic stem cells, creating transgenic ES cells. As mentioned in the previous section, these ES cells can behave like normal embryonic cells. Thus, if the transgenic ES cells are mixed with early normal mouse embryos, they will begin differentiating, along with the normal embryonic cells, producing a chimera, some of whose cells contain the transgene, and some that do not. From here on, the second method is just like the first: breed the chimera with a wild-type mouse and select true transgenic pups, with the transgene in all their cells.

SUMMARY To probe the role of a gene, molecular biologists can perform targeted disruption of the corresponding gene in a mouse, and then look for the effects of that mutation on the "knockout mouse." One can also create a transgenic mouse that carries a gene from another organism, and observe the effect of that transgene on the mouse. These techniques can be used with many other organisms besides mice.

SUMMARY

Methods of purifying proteins and nucleic acids are crucial in molecular biology. DNAs, RNAs, and proteins of various sizes can be separated by gel electrophoresis. The most common gel used in nucleic acid electrophoresis is agarose, and polyacrylamide is usually used in protein electrophoresis. Sodium dodecyl sulfate polyacrylamide gel electrophoresis (SDS-PAGE) is used to separate polypeptides according to their sizes. High-resolution separation of polypeptides can be achieved by two-dimensional gel electrophoresis, which uses isoelectric focusing in the first dimension and SDS-PAGE in the second.

Ion-exchange chromatography can be used to separate substances, including proteins, according to their charges. Positively charged resins like DEAE-Sephadex are used for anion-exchange chromatography, and negatively charged resins like phosphocellulose are used for cation-exchange chromatography. Gel filtration chromatography uses columns filled with porous resins that let smaller substances in, but exclude larger substances. Thus, the smaller substances are slowed, but larger substances travel relatively rapidly through the column. Affinity chromatography is a powerful purification technique that exploits an affinity reagent with strong and specific affinity for a molecule of interest. That molecule binds to a column containing the affinity reagent, but all or most other molecules flow through. Then the molecule of interest can be eluted from the column with a substance that disrupts the specific binding.

Detection of the tiny quantities of substances in molecular biology experiments generally requires labeled tracers. If the tracer is radioactive it can be detected by autoradiography, using x-ray film or a phosphorimager, or by liquid scintillation counting. Nonradioactive labeled tracers can produce light (chemiluminescence) or colored spots.

Labeled DNA (or RNA) probes can be hybridized to DNAs of the same, or very similar, sequence on a Southern blot. Modern DNA typing uses Southern blots and a battery of DNA probes to detect variable sites in individual animals, including humans.

Labeled probes can be hydridized to whole chromosomes to locate genes or other specific DNA sequences. This is called in situ hybridization or, if the probe is fluorescently labeled, fluorescence in situ hybridization (FISH). Proteins can be detected and quantified in complex mixtures using immunoblots (or Western blots). Proteins are electrophoresed, then blotted to a membrane and the proteins on the blot are probed with specific antibodies that can be detected with labeled secondary antibodies or protein A.

The Sanger DNA sequencing method uses dideoxy nucleotides to terminate DNA synthesis, yielding a series of DNA fragments whose sizes can be measured by electrophoresis. The last base in each of these fragments is known because we know which dideoxy nucleotide was used to terminate each reaction. Therefore, ordering these fragments by size—each fragment one (known) base longer than the next—tells us the base sequence of the DNA. Automated DNA sequencing speeds this process up, and high throughput sequencing, by running many reactions simultaneously, achieves even greater speed.

A physical map depicts the spatial arrangement of physical "landmarks," such as restriction sites, on a DNA molecule. Overlaps can be detected by Southern blotting some of the fragments and then hybridizing these fragments to labeled fragments generated by another restriction enzyme.

Using cloned genes, one can introduce changes conveniently by site-directed mutagenesis, thus altering the amino acid sequences of the protein products.

A Northern blot is similar to a Southern blot, but it contains electrophoretically separated RNAs instead of DNAs. The RNAs on the blot can be detected by hybridizing them to a labeled probe. The intensities of the bands reveal the relative amounts of specific RNA in each, and the positions of the bands indicate the lengths of the respective RNAs.

In S1 mapping, a labeled DNA probe is used to detect the 5′- or 3′-end of a transcript. Hybridization of the probe to the transcript protects a portion of the probe from digestion by S1 nuclease. The length of the section of probe protected by the transcript locates the end of the transcript, relative to the known location of an end of the probe. Because the amount of probe protected by the transcript is proportional to the concentration of transcript, S1 mapping can also be used as a quantitative method. RNase mapping is a variation on S1 mapping that uses an RNA probe and RNase instead of a DNA probe and S1 nuclease.

Using primer extension one can locate the 5′-end of a transcript by hybridizing an oligonucleotide primer to the RNA of interest, extending the primer with reverse transcriptase to the 5′-end of the transcript, and electrophoresing the reverse transcript to determine its size. The intensity of the signal obtained by this method is a measure of the concentration of the transcript.

Run-off transcription is a means of checking the efficiency and accuracy of in vitro transcription. One truncates a gene in the middle and transcribes it in vitro in the presence of labeled nucleotides. The RNA polymerase runs off the end and releases an incomplete transcript. The size of this run-off transcript locates the transcription start site, and the amount of this transcript reflects the efficiency of transcription. G-less cassette transcription also produces a shortened transcript of predictable size, but does so by placing a G-less cassette just downstream of a promoter and transcribing this construct in the absence of GTP.

Nuclear run-on transcription is a way of ascertaining which genes are active in a given cell by allowing transcription of these genes to continue in isolated nuclei. Specific transcripts can be identified by their hybridization to known DNAs on dot blots. One can also use the run-on assay to determine the effects of assay conditions on nuclear transcription.

To measure the activity of a promoter, one can link it to a reporter gene, such as the genes encoding β-galactosidase, CAT, or luciferase, and let the easily assayed reporter gene products tell us indirectly the activity of the promoter. One can also use reporter genes to detect changes in translational efficiency after altering regions of a gene that affect translation.

Gene expression can be quantified by measuring the accumulation of the protein products of genes by immunoblotting or immunoprecipitation.

Filter binding as a means of measuring DNA–protein interaction is based on the fact that double-stranded DNA will not bind by itself to a nitrocellulose filter, or similar medium, but a protein–DNA complex will. Thus, one can label a double-stranded DNA, mix it with a protein, and assay protein–DNA binding by measuring the amount of label retained by the filter. A gel mobility shift assay detects interaction between a protein and DNA by the reduction of the electrophoretic mobility of a small DNA that occurs when the DNA binds to a protein.

Footprinting is a means of finding the target DNA sequence, or binding site, of a DNA-binding protein. We perform DNase footprinting by binding the protein to its DNA target, then digesting the DNA–protein complex with DNase. When we electrophorese the resulting DNA fragments, the protein binding site shows up as a gap, or "footprint," in the pattern where the protein protected the DNA from degradation. DMS footprinting follows a similar principle, except that we use the DNA methylating agent DMS, instead of DNase, to attack the DNA–protein complex. Unmethylated (or hypermethylated) sites show up on electrophoresis and demonstrate where the protein is bound to the DNA. Hydroxyl radical footprinting uses organometallic complexes to generate hydroxyl radicals that break DNA strands.

Chromatin immunoprecipitation detects a specific protein–DNA interaction in chromatin in vivo. It uses an antibody to precipitate a particular protein in complex with DNA, and PCR to determine whether the protein binds near a particular gene.

Protein–protein interactions can be detected in a number of ways, including immunoprecipitation and yeast two-hybrid assay. In the latter technique, three plasmids are introduced into yeast cells. One encodes a hybrid protein composed of protein X and a DNA-binding domain. The second encodes a hybrid protein composed of protein Y and a transcription-activating domain. The third has a promoter-enhancer region linked to a reporter gene such as *lacZ*. The enhancer interacts with the DNA-binding domain linked to protein X. If proteins X and Y interact, they bring together the two parts of a transcription activator that can activate the reporter gene, giving a product that can catalyze a colorimetric reaction. If X-gal is used, for example, the yeast cells will turn blue.

SELEX is a method that allows one to find RNA sequences that interact with other molecules, including proteins. RNAs that interact with a target molecule are selected by affinity chromatography, then converted to double-stranded DNAs and amplified by PCR. After several rounds of this procedure, the RNAs are highly enriched for sequences that bind to the target molecule. Functional SELEX is a variation on this theme in which the desired function somehow alters the RNA so it can be amplified. If the desired function is enzymatic, mutagenesis can be introduced into the amplification step to produce variants with higher activity.

To probe the role of a gene, one can perform targeted disruption of the corresponding gene in a mouse, and then look for the effects of that mutation on the "knockout mouse." One can also create a transgenic mouse that carries a gene from another organism, and observe the effect of that transgene on the mouse.

REVIEW QUESTIONS

1. Use a drawing to illustrate the principle of DNA gel electrophoresis. Indicate roughly the comparative electrophoretic mobilities of DNAs with 150, 600, and 1200 bp.
2. What is SDS? What are its functions in SDS-PAGE?
3. Compare and contrast SDS-PAGE and modern two-dimensional gel electrophoresis of proteins.
4. Describe the principle of ion-exchange chromatography. Use a graph to illustrate the separation of three different proteins by this method.
5. Describe the principle of gel filtration chromatography. Use a graph to illustrate the separation of three different proteins by this method. Indicate on the graph the largest and smallest of these proteins.
6. Compare and contrast the principles of autoradiography and phosphorimaging. Which method provides more quantitative information?
7. Describe a nonradioactive method for detecting a particular nucleic acid fragment in an electrophoretic gel.
8. Diagram the process of Southern blotting and probing to detect a DNA of interest. Compare and contrast this procedure with Northern blotting.
9. Describe a DNA fingerprinting method using a minisatellite probe. Compare this method with a modern forensic DNA typing method using probes to detect single variable DNA loci.
10. What kinds of information can we obtain from a Northern blot?
11. Describe fluorescence in situ hybridization (FISH). When would you use this method, rather than Southern blotting?
12. Draw a diagram of an imaginary Sanger sequencing autoradiograph, and provide the corresponding DNA sequence.
13. Show how a manual DNA sequencing method can be automated.
14. Show how to use restriction mapping to determine the orientation of a restriction fragment ligated into a restriction site in a vector. Use fragment sizes different from those in the text.
15. Explain the principle of site-directed mutagenesis, then describe a method to carry out this process.
16. Compare and contrast the S1 mapping and primer extension methods for mapping the 5′-end of an mRNA. Which of these methods can be used to map the 3′-end of an mRNA. Why would the other method not work?
17. Describe the run-off transcription method. Why does this method not work with in vivo transcripts, as S1 mapping and primer extension do?
18. How would you label the 5′-ends of a double-stranded DNA? The 3′-ends?
19. Describe a nuclear run-on assay, and show how it differs from a run-off assay.
20. How does a dot blot differ from a Southern blot?
21. Describe the use of a reporter gene to measure the strength of a promoter.
22. Describe a filter-binding assay to measure binding between a DNA and a protein.
23. Compare and contrast the gel mobility shift and DNase footprinting methods of assaying specific DNA–protein interactions. What information does DNase footprinting provide that gel mobility shift does not?
24. Compare and contrast DMS and DNase footprinting. Why is the former method more precise than the latter?
25. Describe a ChIP assay to detect binding between protein X and gene Y. Show sample positive results.
26. Describe a yeast two-hybrid assay for interaction between two known proteins.
27. Describe a yeast two-hybrid screen for finding an unknown protein that interacts with a known protein.
28. Describe a method for creating a knockout mouse. Explain the importance of the thymidine kinase and neomycin-resistance genes in this procedure. What information can a knockout mouse provide?
29. Describe a procedure to produce a transgenic mouse.

ANALYTICAL QUESTIONS

1. You have electrophoresed some DNA fragments on an agarose gel and obtain the results shown in Figure 5.2. (a) What is the size of a fragment that migrated 25 mm? (b) How far did the 200 bp fragment migrate?
2. Design a Southern blot experiment to check a chimeric mouse's DNA for insertion of the neomycin-resistance gene. You may assume any array of restriction sites you wish in the target gene and in the neo^r gene. Show sample results for a successful and an unsuccessful insertion.
3. In a DNase footprinting experiment, either the template or nontemplate strand can be end-labeled. In Figure 5.37a, the template strand is labeled. Which strand is labeled in Figure 5.37b? How do you know?
4. Invent a pyrogram with 12 peaks and write the corresponding DNA sequence.

SUGGESTED READINGS

Galas, D.J. and A. Schmitz. 1978. DNase footprinting: A simple method for the detection of protein–DNA binding specificity. *Nucleic Acids Research* 5:3157–70.

Lichter, P. 1990. High resolution mapping of human chromosome 11 by in situ hybridization with cosmid clones. *Science* 247:64–69.

Sambrook, J., and D.W. Russell. 2001. *Molecular Cloning: A Laboratory Manual,* 3rd ed. Plainview, NY: Cold Spring Harbor Laboratory Press.

CHAPTER 6

The Mechanism of Transcription in Bacteria

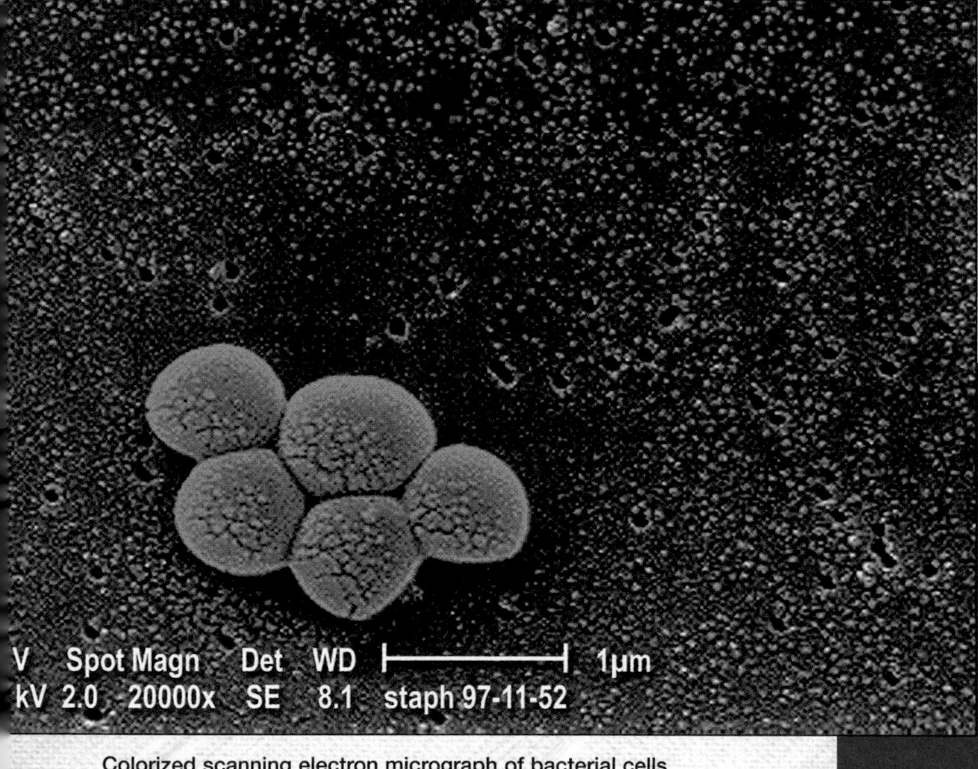

Colorized scanning electron micrograph of bacterial cells *(Staphylococcus aureus).* *Centers for Disease Control and Prevention*

In Chapter 3 we learned that transcription is the first step in gene expression. Indeed, transcription is a vital control point in the expression of many genes. Chapters 6–9 will examine in detail the mechanism of transcription and its control in bacteria. In Chapter 6 we will focus on the basic mechanism of transcription. We will begin with RNA polymerase, the enzyme that catalyzes transcription. We will also look at the interaction between RNA polymerase and DNA. This interaction begins when an RNA polymerase docks at a promoter (a specific polymerase binding site next to a gene), continues as the polymerase elongates the RNA chain, and ends when the polymerase reaches a terminator, or stopping point, and releases the finished transcript.

6.1 RNA Polymerase Structure

As early as 1960–1961, RNA polymerases were discovered in animals, plants, and bacteria. And, as you might anticipate, the bacterial enzyme was the first to be studied in great detail. By 1969, the polypeptides that make up the *E. coli* RNA polymerase had been identified by SDS polyacrylamide gel electrophoresis (SDS-PAGE) as described in Chapter 5.

Figure 6.1, lane 1, presents the results of an SDS-PAGE separation of the subunits of the *E. coli* RNA polymerase by Richard Burgess, Andrew Travers, and their colleagues. This enzyme preparation contained two very large subunits: beta (β) and beta-prime (β′), with molecular masses of 150 and 160 kD, respectively. These two subunits were not well separated in this experiment, but they were clearly distinguished in subsequent studies. The other RNA polymerase subunits visible on this gel are called **sigma** (σ) and alpha (α), with molecular masses of 70 and 40 kD, respectively. Another subunit, omega (ω), with a molecular mass of 10 kDa is not detectable here, but was clearly visible in urea gel electrophoresis experiments performed on this same enzyme preparation. In contrast to the other subunits, the ω-subunit is not required for cell viability, nor for enzyme activity in vitro. It seems to play a role, though not a vital one, in enzyme assembly. The polypeptide marked with an asterisk was a contaminant. Thus, the subunit content of an **RNA polymerase holoenzyme** is β′, β, σ, α_2, ω; in other words, two molecules of α and one of all the others are present.

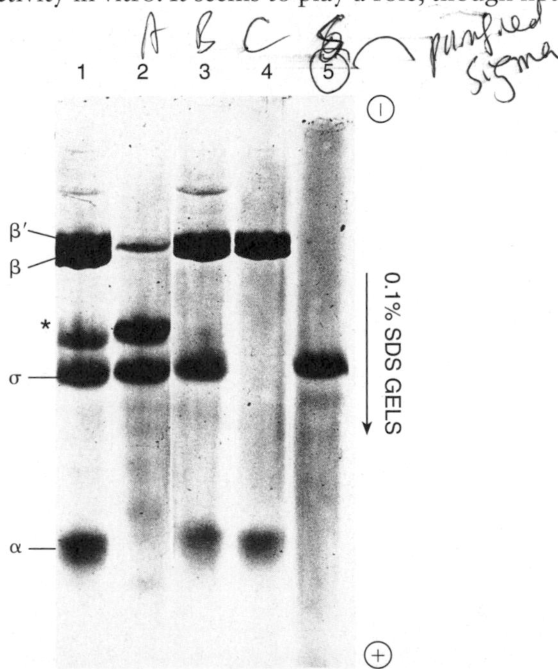

Figure 6.1 Separation of σ-factor from core *E. coli* RNA polymerase by phosphocellulose chromatography. Burgess, Travers, and colleagues subjected RNA polymerase holoenzyme to phosphocellulose chromatography, which yielded three peaks of protein: A, B, and C. Then they performed SDS-PAGE on the holoenzyme (lane 1), peaks A, B, and C (lanes 2–4, respectively), and purified σ (lane 5). Peak A contained σ, along with some contaminants (the most prominent of which is marked with an asterisk), B contained the holoenzyme, and C contained the functional core polymerase (subunits α, β, and β′). (*Source:* Burgess et al., "Factor Stimulating Transcription by RNA Polymerase." *Nature* 221 (4 January 1969) p. 44, fig. 3. © Macmillan Magazines Ltd.

Table 6.1 Ability of Core and Holoenzyme to Transcribe DNAs

	Relative Transcription Activity	
DNA Template	**Core**	**Holoenzyme**
T4 (native, intact)	0.5	33.0
Calf thymus (native, nicked)	14.2	32.8

When Burgess, Travers, and colleagues subjected the RNA polymerase holoenzyme to cation exchange chromatography (Chapter 5) using a phosphocellulose resin, they detected three peaks of protein, which they labeled A, B, and C. When they performed SDS-PAGE analysis of each of these peaks, they discovered that they had separated the σ-subunit from the remainder of the enzyme, called the **core polymerase.** Figure 6.1, lane 2 shows the composition of peak A, which contained the σ-subunit, along with a prominent contaminating polypeptide and perhaps a bit of β′. Lane 3 shows the polypeptides in peak B, which contained the holoenzyme. Lane 4 shows the composition of peak C, containing the core polymerase, which clearly lacks the σ-subunit. Further purification of the σ-subunit yielded the preparation in lane 5, which was free of most contamination.

Next, the investigators tested the RNA polymerase activities of the two separated components of the enzyme: the core polymerase and the σ-factor. Table 6.1 shows that this separation had caused a profound change in the enzyme's activity. Whereas the holoenzyme could transcribe intact phage T4 DNA in vitro quite actively, the core enzyme had little ability to do this. On the other hand, core polymerase retained its basic RNA polymerizing function because it could still transcribe highly nicked templates (DNAs with single-stranded breaks) very well. (As we will see, transcription of nicked DNA is a laboratory artifact and has no biological significance.)

Sigma (σ) as a Specificity Factor

Adding σ back to the core reconstituted the enzyme's ability to transcribe unnicked T4 DNA. Even more significantly, Ekkehard Bautz and colleagues showed that the holoenzyme transcribed only a certain class of T4 genes (called immediate early genes), but the core showed no such specificity.

Not only is the core enzyme indiscriminate about the T4 genes it transcribes, it also transcribes both DNA strands. Bautz and colleagues demonstrated this by hybridizing the labeled product of the holoenzyme or the core enzyme to authentic T4 phage RNA and then checking for RNase resistance. That is, they attempted to get the two RNAs to base-pair together and form an RNase-resistant double-stranded RNA. Because authentic T4 RNA is made **asymmetrically** (only one DNA strand in any given region is copied), it should not hybridize to T4 RNA made properly in vitro because this RNA is also made asymmetrically and is therefore identical, not complementary, to the authentic RNA. Bautz and associates did indeed observe this behavior with RNA made in vitro by the holoenzyme. However, if the RNA is made symmetrically in vitro, up to half of it will be complementary to the in vivo RNA and will be able to hybridize to it and thereby become resistant to RNase. In fact, Bautz and associates found that about 30% of the labeled RNA made by the core polymerase in vitro became RNase-resistant after hybridization to authentic T4 RNA. Thus, the core enzyme acts in an unnatural way by transcribing both DNA strands.

Clearly, depriving the holoenzyme of its σ-subunit leaves a core enzyme with basic RNA synthesizing capability, but lacking specificity. Adding σ back restores specificity. In fact, σ was named only after this characteristic came to light, and the σ, or Greek letter s, was chosen to stand for "specificity."

SUMMARY The key player in the transcription process is RNA polymerase. The *E. coli* enzyme is composed of a core, which contains the basic transcription machinery, and a σ-factor, which directs the core to transcribe specific genes.

6.2 Promoters

In the T4 DNA transcription experiments presented in Table 6.1, why was core RNA polymerase still capable of transcribing nicked DNA, but not intact DNA? Nicks and gaps in DNA provide ideal initiation sites for RNA polymerase, even core polymerase, but this kind of initiation is necessarily nonspecific. Few nicks or gaps occurred on the intact T4 DNA, so the core polymerase encountered only a few such artificial initiation sites and transcribed this DNA only weakly. On the other hand, when σ was present, the holoenzyme could recognize the authentic RNA polymerase binding sites on the T4 DNA and begin transcription there. These polymerase binding sites are called **promoters.** Transcription that begins at promoters in vitro is specific and mimics the initiation that would occur in vivo. Thus, σ operates by directing the polymerase to initiate at specific promoter sequences. In this section, we will examine the interaction of bacterial polymerase with promoters, and the structures of these promoters.

Binding of RNA Polymerase to Promoters

How does σ change the way the core polymerase behaves toward promoters? David Hinkle and Michael Chamberlin used nitrocellulose filter-binding studies (Chapter 5) to help answer this question. To measure how tightly holoenzyme and core enzyme bind to DNA, they isolated these enzymes from *E. coli* and bound them to ^{3}H-labeled T7 phage DNA, whose early promoters are recognized by the *E. coli* polymerase. Then they added a great excess of unlabeled T7 DNA, so that any polymerase that dissociated from a labeled DNA had a much higher chance of rebinding to an unlabeled DNA than to a labeled one. After varying lengths of time, they passed the mixture through nitrocellulose filters. The labeled DNA would bind to the filter only if it was still bound to polymerase. Thus, this assay measured the dissociation rate of the polymerase–DNA complex. As the last (and presumably tightest bound) polymerase dissociated from the labeled DNA, that DNA would no longer bind to the filter, so the filter would become less radioactive.

Figure 6.2 shows the results of this experiment. Obviously, the polymerase holoenzyme binds much more tightly to the T7 DNA than does the core enzyme. In fact, the holoenzyme dissociates with a half time ($t_{1/2}$) of 30–60 h, which lies far beyond the timescale of Figure 6.2. This means that after 30–60 h, only half of the complex had

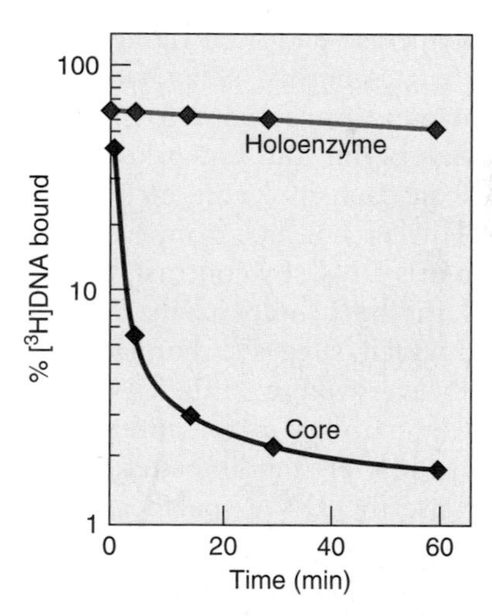

Figure 6.2 Sigma stimulates tight binding between RNA polymerase and promoter. Hinkle and Chamberlin allowed ^{3}H-labeled T7 DNA to bind to *E. coli* core polymerase (blue) or holoenzyme (red). Next, they added an excess of unlabeled T7 DNA, so that any polymerase that dissociated from the labeled DNA would be likely to rebind to unlabeled DNA. They filtered the mixtures through nitrocellulose at various times to monitor the dissociation of the labeled T7 DNA–polymerase complexes. (As the last polymerase dissociates from the labeled DNA, the DNA will no longer bind to the filter, which loses radioactivity.) The much slower dissociation rate of the holoenzyme (red) relative to the core polymerase (blue) shows much tighter binding between T7 DNA and holoenzyme. (*Source:* Adapted from Hinckle, D.C. and Chamberlin, M.J., "Studies of the Binding of *Escherichia coli* RNA Polymerase to DNA," *Journal of Molecular Biology,* Vol. 70, 157–85, 1972.)

dissociated, which indicates very tight binding indeed. By contrast, the core polymerase dissociated with a $t_{1/2}$ of less than a minute, so it bound much less tightly than the holoenzyme did. Thus, the σ-factor can promote tight binding, at least to certain DNA sites.

In a separate experiment, Hinkle and Chamberlin switched the procedure around, first binding polymerase to unlabeled DNA, then adding excess labeled DNA, and finally filtering the mixture at various times through nitrocellulose. This procedure measured the dissociation of the first (and loosest bound) polymerase, because a newly dissociated polymerase would be available to bind to the free labeled DNA and thereby cause it to bind to the filter. This assay revealed that the holoenzyme, as well as the core, had loose binding sites on the DNA.

Thus, the holoenzyme finds two kinds of binding sites on T7 DNA: tight binding sites and loose ones. On the other hand, the core polymerase is capable of binding only loosely to the DNA. Because Bautz and coworkers had already shown that the holoenzyme, but not the core, can recognize promoters, it follows that the tight binding sites are probably promoters, and the loose binding sites represent the rest of the DNA. Chamberlin and colleagues also showed that the tight complexes between holoenzyme and T7 DNA could initiate transcription immediately on addition of nucleotides, which reinforces the conclusion that the tight binding sites are indeed promoters. If the polymerase had been tightly bound to sites remote from the promoters, a lag would have occurred while the polymerases searched for initiation sites. Furthermore, Chamberlin and coworkers titrated the tight binding sites on each molecule of T7 DNA and found only eight. This is not far from the number of early promoters on this DNA. By contrast, the number of loose binding sites for both holoenzyme and core enzyme is about 1300, which suggests that these loose sites are found virtually everywhere on the DNA and are therefore nonspecific. The inability of the core polymerase to bind to the tight (promoter) binding sites accounts for its inability to transcribe DNA specifically, which requires binding at promoters.

Hinkle and Chamberlin also tested the effect of temperature on binding of holoenzyme to T7 DNA and found a striking enhancement of tight binding at elevated temperature. Figure 6.3 shows a significantly higher dissociation rate at 25° than at 37°C, and a much higher dissociation rate at 15°C. Because high temperature promotes DNA melting (strand separation, Chapter 2) this finding is consistent with the notion that tight binding involves local melting of the DNA. We will see direct evidence for this hypothesis later in this chapter.

Hinkle and Chamberlin summarized these and other findings with the following hypothesis for polymerase–DNA interaction (Figure 6.4): RNA polymerase holoenzyme binds loosely to DNA at first. It either binds initially

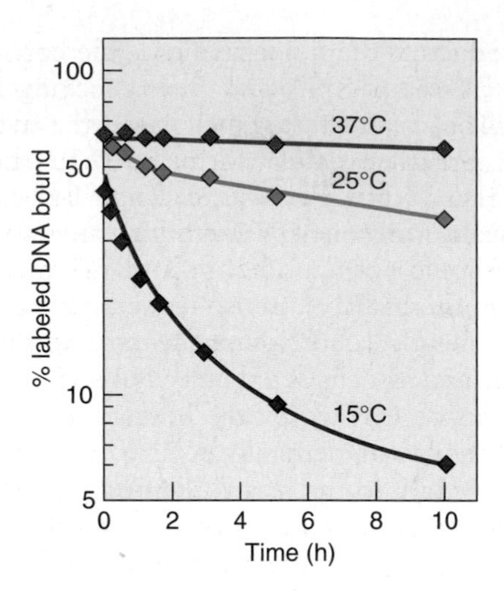

Figure 6.3 The effect of temperature on the dissociation of the polymerase–T7 DNA complex. Hinkle and Chamberlin formed complexes between *E. coli* RNA polymerase holoenzyme and ^{3}H-labeled T7 DNA at three different temperatures: 37°C (red), 25°C (green), and 15°C (blue). Then they added excess unlabeled T7 DNA to compete with any polymerase that dissociated; they removed samples at various times and passed them through a nitrocellulose filter to monitor dissociation of the complex. The complex formed at 37°C was more stable than that formed at 25°C, which was much more stable than that formed at 15°C. Thus, higher temperature favors tighter binding between RNA polymerase holoenzyme and T7 DNA. (*Source:* Adapted from Hinckle, D.C. and Chamberlin, M.J., "Studies of the Binding of *Escherichia coli* RNA Polymerase to DNA," *Journal of Molecular Biology,* Vol. 70, 157–85, 1972.)

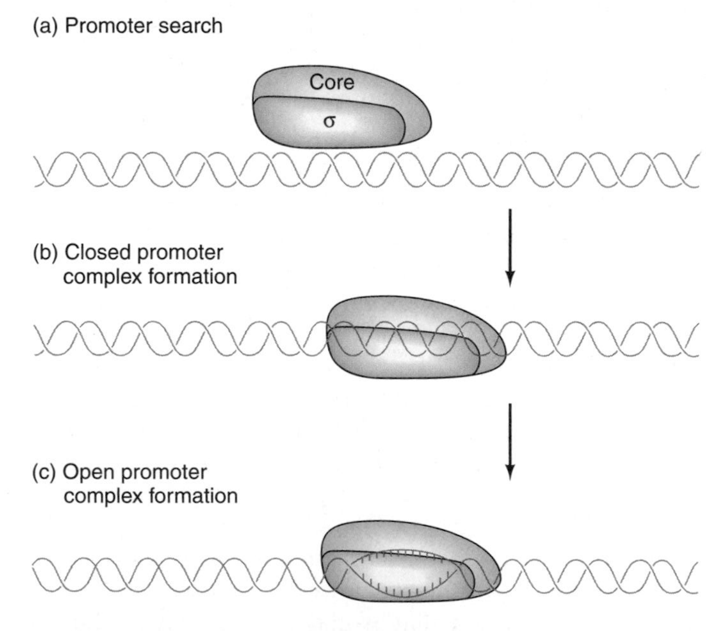

Figure 6.4 RNA polymerase/promoter binding. (a) The holoenzyme binds and rebinds loosely to the DNA, searching for a promoter. **(b)** The holoenzyme has found a promoter and has bound loosely, forming a closed promoter complex. **(c)** The holoenzyme has bound tightly, melting a local region of DNA and forming an open promoter complex.

at a promoter or scans along the DNA until it finds one. The complex with holoenzyme loosely bound at the promoter is called a **closed promoter complex** because the DNA remains in closed double-stranded form. Then the holoenzyme can melt a short region of the DNA at the promoter to form an **open promoter complex** in which the polymerase is bound tightly to the DNA. This is called an open promoter complex because the DNA has to open up to form it.

It is this conversion of a loosely bound polymerase in a closed promoter complex to the tightly bound polymerase in the open promoter complex that requires σ, and this is also what allows transcription to begin. We can now appreciate how σ fulfills its role in determining specificity of transcription: It selects the promoters to which RNA polymerase will bind tightly. The genes adjacent to these promoters will then be transcribed.

SUMMARY The σ-factor allows initiation of transcription by causing the RNA polymerase holoenzyme to bind tightly to a promoter. This tight binding depends on local melting of the DNA to form an open promoter complex and is stimulated by σ. The σ-factor can therefore select which genes will be transcribed.

Promoter Structure

What is the special nature of a bacterial promoter that attracts RNA polymerase? David Pribnow compared several *E. coli* and phage promoters and discerned a region they held in common: a sequence of 6 or 7 bp centered approximately 10 bp upstream of the start of transcription. This was originally dubbed the "Pribnow box," but is now usually called the **−10 box.** Mark Ptashne and colleagues noticed another short sequence centered approximately 35 bp upstream of the transcription start site; it is known as the **−35 box.** Thousands of promoters have now been examined and a typical, or **consensus sequence** for each of these boxes has emerged (Figure 6.5).

These so-called consensus sequences represent probabilities. The capital letters in Figure 6.5 denote bases that have a high probability of being found in the given position. The lowercase letters correspond to bases that are usually found in the given position, but at a lower frequency than those denoted by capital letters. The probabilities are such that one rarely finds −10 or −35 boxes that match the consensus sequences perfectly. However, when such perfect matches are found, they tend to occur in very strong promoters that initiate transcription unusually actively. In fact, mutations that destroy matches with the consensus sequences tend to be **down mutations.** That is, they make the promoter weaker, resulting in less transcription. Mutations that make the promoter sequences more like the consensus sequences usually make the promoters stronger; these are called **up mutations.** The spacing between promoter elements is also important, and deletions or insertions that move the −10 and −35 boxes unnaturally close together or far apart are deleterious. In Chapter 10 we will see that eukaryotic promoters have their own consensus sequences, one of which resembles the −10 box quite closely.

In addition to the −10 and −35 boxes, which we can call **core promoter elements,** some very strong promoters have an additional element farther upstream called an **UP element.** *E. coli* cells have seven genes (***rrn* genes**) that encode rRNAs. Under rapid growth conditions, when rRNAs are required in abundance, these seven genes by themselves account for the majority of the transcription occurring in the cell. Obviously, the promoters driving these genes are extraordinarily powerful, and their UP elements are part of the explanation. Figure 6.6 shows the structure of one of these promoters, the *rrnB* P1 promoter. Upstream of the core promoter (blue), there is an UP element (red) between positions −40 and −60. We know that the UP element is a true promoter element because it stimulates transcription of the *rrnB* P1 gene by a factor of 30 in the presence of RNA polymerase alone. Because it is recognized by the polymerase itself, we conclude that it is a promoter element.

This promoter is also associated with three so-called Fis sites between positions −60 and −150, which are binding sites for the transcription-activator protein Fis. The Fis sites, because they do not bind to RNA polymerase itself, are not classical promoter elements, but instead are members of another class of transcription-activating DNA elements called **enhancers.** We will discuss bacterial enhancers in greater detail in Chapter 9.

The *E. coli rrn* promoters are also regulated by a pair of small molecules: the initiating NTP (the iNTP) and an **alarmone,** guanosine 5′-diphosphate 3′-diphosphate (**ppGpp**). An abundance of iNTP indicates that the concentration of

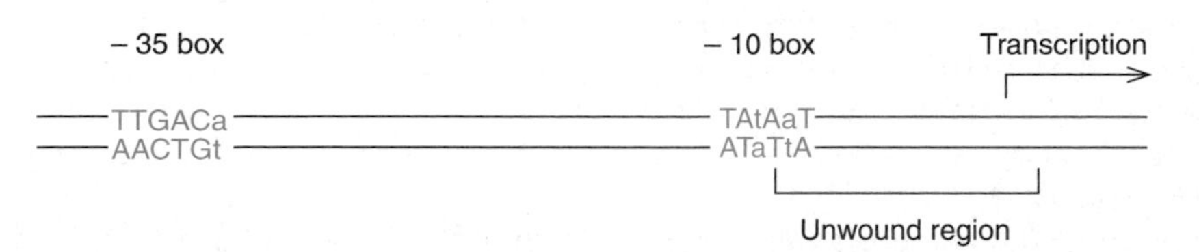

Figure 6.5 A bacterial promoter. The positions of −10 and −35 boxes and the unwound region are shown relative to the start of transcription for a typical *E. coli* promoter. Capital letters denote bases found in those positions in more than 50% of promoters examined; lower-case letters denote bases found in those positions in 50% or fewer of promoters examined.

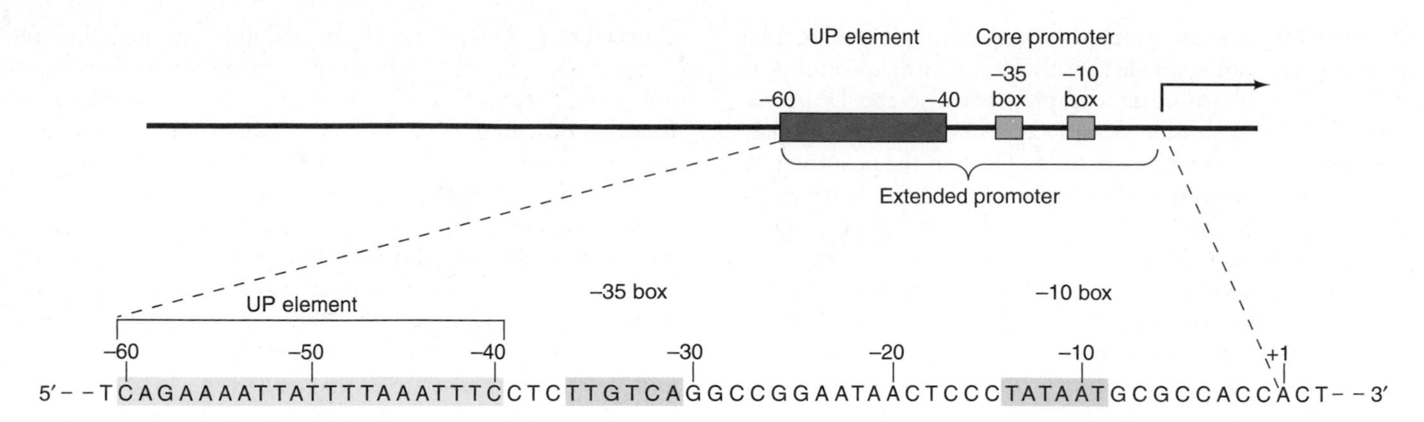

Figure 6.6 The *rrnB* P1 promoter. The core promoter elements (−10 and −35 boxes, blue) and the UP element (red) are shown schematically above, and with their complete base sequences (nontemplate strand) below, with the same color coding. (*Source:* Adapted from Ross et al., "A third recognition element in bacterial promoters: DNA binding by the alpha subunit of RNA polymerase." *Science* 262:1407, 1993.)

nucleotides is high, and therefore it is appropriate to synthesize plenty of rRNA. Accordingly, iNTP stabilizes the open promoter complex, stimulating transcription.

On the other hand, when cells are starved for amino acids, protein synthesis cannot occur readily and the need for ribosomes (and rRNA) decreases. Ribosomes sense the lack of amino acids when uncharged tRNAs bind to the ribosomal site where aminoacyl-tRNAs would normally bind. Under these conditions, a ribosome-associated protein called RelA receives the "alarm" and produces the "alarmone" ppGpp, which destabilizes open promoter complexes whose lifetimes are normally short, thus inhibiting transcription.

The protein **DskA** also plays an important role. It binds to RNA polymerase and reduces the lifetimes of the *rrn* open promoters to a level at which they are responsive to changes in iNTP and ppGpp concentrations. Thus, DskA is required for the regulation of *rrn* transcription by these two small molecules. Indeed, *rrn* transcription is insensitive to iNTP and ppGpp in mutants lacking DskA.

SUMMARY Bacterial promoters contain two regions centered approximately at −10 and −35 bp upstream of the transcription start site. In *E. coli,* these bear a greater or lesser resemblance to two consensus sequences: TATAAT and TTGACA, respectively. In general, the more closely regions within a promoter resemble these consensus sequences, the stronger that promoter will be. Some extraordinarily strong promoters contain an extra element (an UP element) upstream of the core promoter. This makes these promoters even more attractive to RNA polymerase. Transcription from the *rrn* promoters responds positively to increases in the concentration of iNTP, and negatively to the alarmone ppGpp.

6.3 Transcription Initiation

Until 1980, it was a common assumption that transcription initiation ended when RNA polymerase formed the first phosphodiester bond, joining the first two nucleotides in the growing RNA chain. Then, Agamemnon Carpousis and Jay Gralla reported that initiation is more complex than that. They incubated *E. coli* RNA polymerase with DNA bearing a mutant *E. coli lac* promoter known as the *lac* UV5 promoter. Along with the polymerase and DNA, they included heparin, a negatively charged polysaccharide that competes with DNA in binding tightly to free RNA polymerase. The heparin prevented any reassociation between DNA and polymerase released at the end of a cycle of transcription. These workers also included labeled ATP in their assay to label the RNA products. Then they subjected the products to gel electrophoresis to measure their sizes. They found several very small oligonucleotides, ranging in size from dimers to hexamers (2–6 nt long), as shown in Figure 6.7. The sequences of these oligonucleotides matched the sequence of the beginning of the expected transcript from the *lac* promoter. Moreover, when Carpousis and Gralla measured the amounts of these oligonucleotides and compared them to the number of RNA polymerases, they found many oligonucleotides per polymerase. Because the heparin in the assay prevented free polymerase from reassociating with the DNA, this result implied that the polymerase was making many small, **abortive transcripts** without ever leaving the promoter. Other investigators have since verified this result and have found abortive transcripts up to 9 or 10 nt in size.

Thus, we see that transcription initiation is more complex than first supposed. It is now commonly represented in four steps, as depicted in Figure 6.8: (1) formation of a closed promoter complex; (2) conversion of the closed promoter complex to an open promoter complex; (3) polymerizing the first few nucleotides (up to 10) while the polymerase remains at the promoter, in an **initial transcribing complex;**

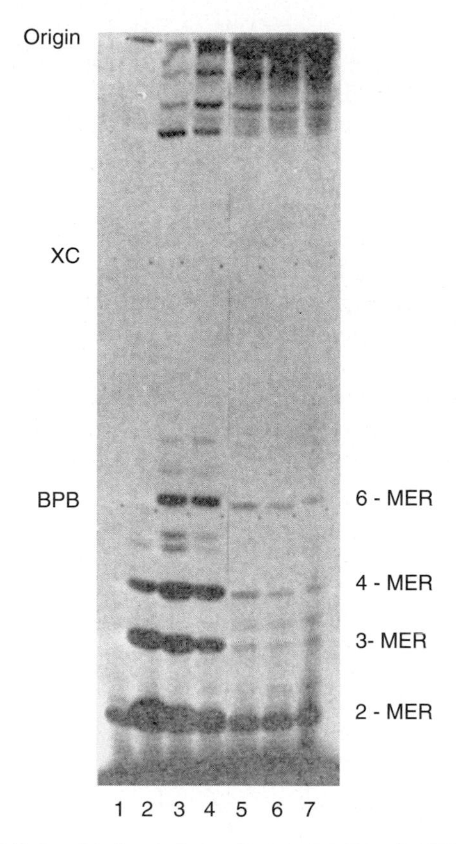

Figure 6.7 Synthesis of short oligonucleotides by RNA polymerase bound to a promoter. Carpousis and Gralla allowed *E. coli* RNA polymerase to synthesize ^{32}P-labeled RNA in vitro using a DNA containing the *lac* UV5 promoter, heparin to bind any free RNA polymerase, [^{32}P]ATP, and various concentrations of the other three nucleotides (CTP, GTP, and UTP). They electrophoresed the products on a polyacrylamide gel and visualized the oligonucleotides by autoradiography. Lane 1 is a control with no DNA; lane 2, ATP only; lanes 3–7; ATP with concentrations of CTP, GTP, and UTP increasing by twofold in each lane, from 25 μM in lane 3 to 400 μM in lane 7. The positions of 2-mers through 6-mers are indicated at right. The positions of two marker dyes (bromophenol blue [BPB] and xylene cyanol [XC]) are indicated at left. The apparent dimer in lane 1, with no DNA, is an artifact caused by a contaminant in the labeled ATP. The same artifact can be presumed to contribute to the bands in lanes 2–7. (*Source:* Carpousis A.J. and Gralla J.D. Cycling of ribonucleic acid polymerase to produce oligonucleotides during initiation in vitro at the *lac* UV5 promoter. *Biochemistry* 19 (8 Jul 1980) p. 3249, f. 2, © American Chemical Society.)

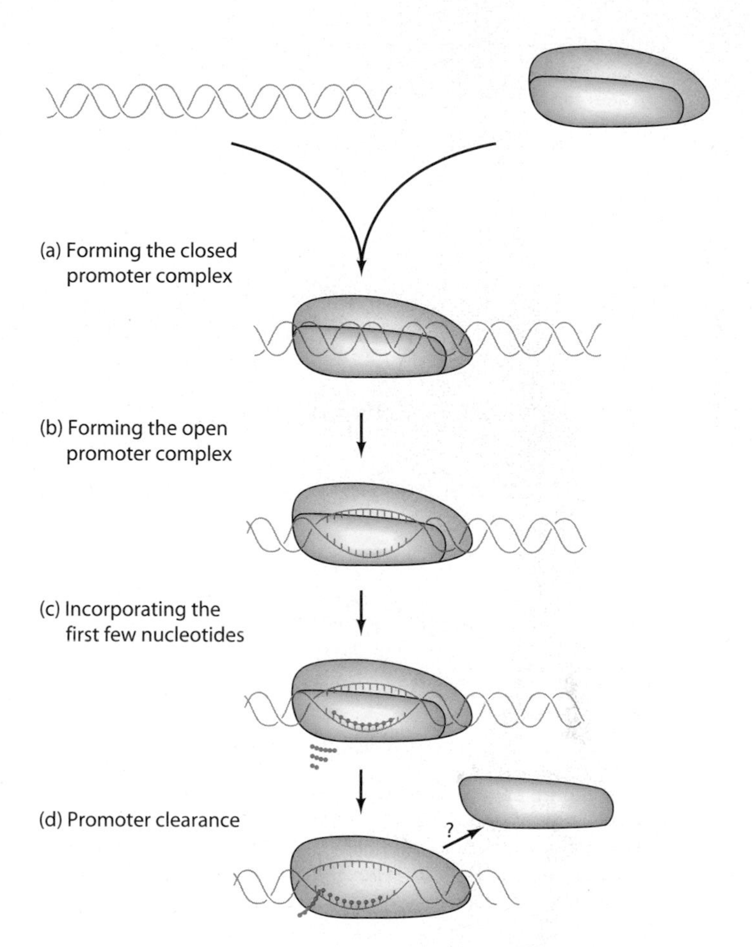

Figure 6.8 Stages of transcription initiation. (a) RNA polymerase binds to DNA in a closed promoter complex. **(b)** The σ-factor stimulates the polymerase to convert the closed promoter complex to an open promoter complex. **(c)** The polymerase incorporates the first 9 or 10 nt into the nascent RNA. Some abortive transcripts are pictured at left. **(d)** The polymerase clears the promoter and begins the elongation phase. The σ-factor may be lost at this point or later, during elongation.

and (4) **promoter clearance,** in which the transcript becomes long enough to form a stable hybrid with the template strand. This helps to stabilize the transcription complex, and the polymerase changes to its elongation conformation and moves away from the promoter. In this section, we will examine the initiation process in more detail.

Sigma Stimulates Transcription Initiation

Because σ directs tight binding of RNA polymerase to promoters, it places the enzyme in a position to initiate transcription—at the beginning of a gene. Therefore, we would expect σ to stimulate initiation of transcription. To test this, Travers and Burgess took advantage of the fact that the first nucleotide incorporated into an RNA retains all three of its phosphates (α, β, and γ), whereas all other nucleotides retain only their α-phosphate (Chapter 3). These investigators incubated polymerase core in the presence of increasing amounts of σ in two separate sets of reactions. In some reactions, the labeled nucleotide was [^{14}C]ATP, which is incorporated throughout the RNA and therefore measures elongation, as well as initiation, of RNA chains. In the other reactions, the labeled nucleotide was [γ-^{32}P]ATP or [γ-^{32}P]GTP, whose label should be incorporated only into the first position of the RNA, and therefore is a measure of transcription initiation. (They used ATP and GTP because transcription usually starts with a purine nucleotide—more often ATP than GTP.) The results in Figure 6.9 show that σ stimulated the incorporation of both ^{14}C- and γ-^{32}P-labeled nucleotides, which suggests that σ enhanced both initiation and elongation.

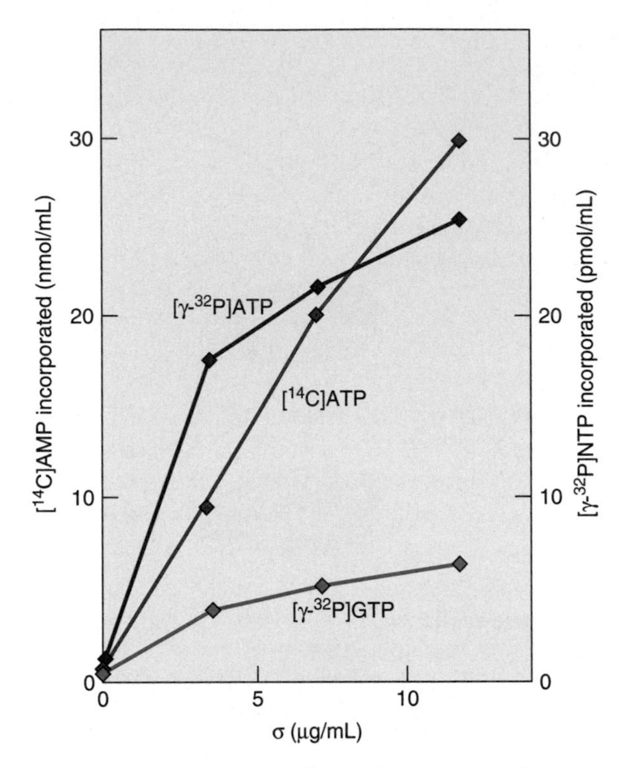

Figure 6.9 Sigma seems to stimulate both initiation and elongation. Travers and Burgess transcribed T4 DNA in vitro with *E. coli* RNA polymerase core plus increasing amounts of σ. In separate reactions, they included [^{14}C]ATP (red), [γ-^{32}P]ATP (blue), or [γ-^{32}P]GTP (green) in the reaction mix. The incorporation of the [^{14}C]ATP measured bulk RNA synthesis, or elongation; the incorporation of the γ-^{32}P-labeled nucleotides measured initiation. Because all three curves rise with increasing σ concentration, this experiment makes it appear that σ stimulates both elongation and initiation. (*Source:* Adapted from Travers, A.A. and R.R. Burgess, "Cyclic re-use of the RNA polymerase sigma factor." *Nature* 222:537–40, 1969.)

However, initiation is the rate-limiting step in transcription (it takes longer to get a new RNA chain started than to extend one). Thus, σ could appear to stimulate elongation by stimulating initiation and thereby providing more initiated chains for core polymerase to elongate.

Travers and Burgess proved that is the case by demonstrating that σ really does not accelerate the rate of RNA chain growth. To do this, they held the number of RNA chains constant and showed that under those conditions σ did not affect the length of the RNA chains. They held the number of RNA chains constant by allowing a certain amount of initiation to occur, then blocking any further chain initiation with the antibiotic **rifampicin,** which blocks bacterial transcription initiation, but not elongation. Then they used ultracentrifugation to measure the length of RNAs made in the presence or absence of σ. They found that σ made no difference in the lengths of the RNAs. If it really had stimulated the rate of elongation, it would have made the RNAs longer. Therefore, σ does not stimulate elongation, and the apparent stimulation in the previous experiment was simply an indirect effect of enhanced initiation.

SUMMARY Sigma stimulates initiation, but not elongation, of transcription.

Reuse of σ

In the same 1969 paper, Travers and Burgess demonstrated that σ can be recycled. The key to this experiment was to run the transcription reaction at low ionic strength, which prevents RNA polymerase core from dissociating from the DNA template at the end of a gene. This caused transcription initiation (as measured by the incorporation of γ-^{32}P-labeled purine nucleotides into RNA) to slow to a stop, as depicted in Figure 6.10 (red line). Then, when they added

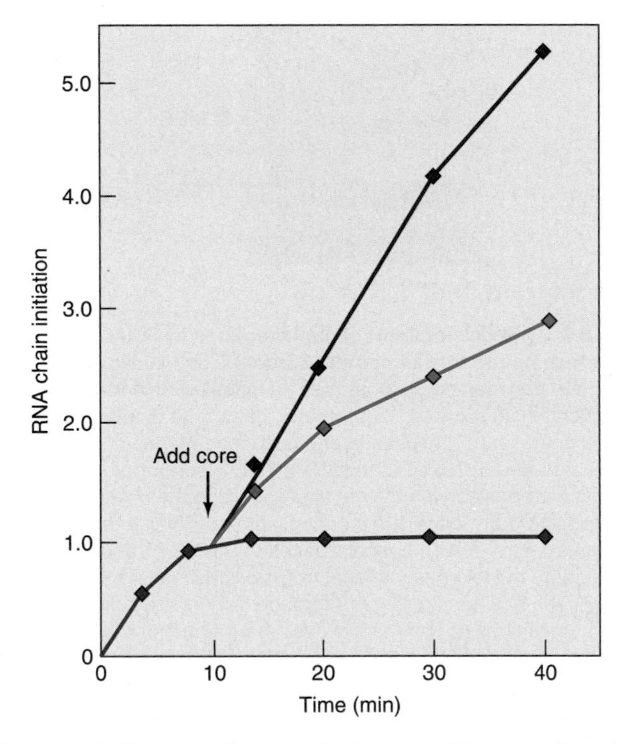

Figure 6.10 Sigma can be reused. Travers and Burgess allowed RNA polymerase holoenzyme to initiate and elongate RNA chains on a T4 DNA template at low ionic strength, so the polymerases could not dissociate from the template to start new RNA chains. The red curve shows the initiation of RNA chains, measured by [γ-^{32}P]ATP and [γ-^{32}P]GTP incorporation, under these conditions. After 10 min (arrow), when most chain initiation had ceased, the investigators added new, rifampicin-resistant core polymerase in the presence (green) or absence (blue) of rifampicin. The immediate rise of both curves showed that addition of core polymerase can restart RNA synthesis, which implied that the new core associated with σ that had been associated with the original core. In other words, the σ was recycled. The fact that transcription occurred even in the presence of rifampicin showed that the new core, which was from rifampicin-resistant cells, together with the old σ, which was from rifampicin-sensitive cells, could carry out rifampicin-resistant transcription. Thus, the core, not the σ, determines rifampicin resistance or sensitivity. (*Source:* Adapted from Travers, A.A. and R.R. Burgess, "Cyclic re-use of the RNA polymerase sigma factor." *Nature* 222:537–40, 1969.)

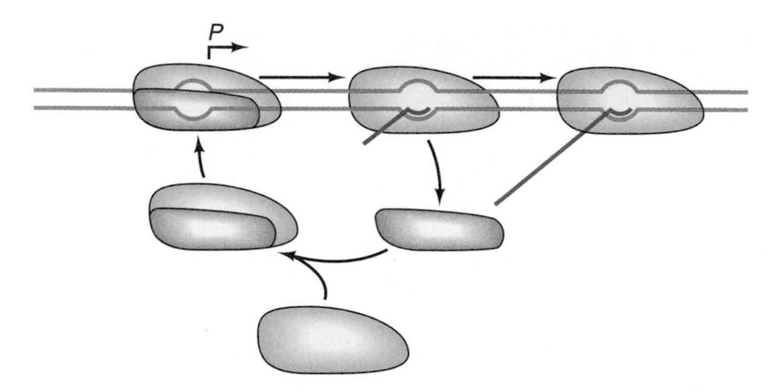

Figure 6.11 The σ cycle. RNA polymerase binds to the promoter at left, causing local melting of the DNA. As the polymerase moves to the right, elongating the RNA, the σ-factor dissociates and joins with a new core polymerase (lower left) to initiate another RNA chain.

new core polymerase, these investigators showed that transcription began anew (blue line). This meant that the new core was associating with σ that had been released from the original holoenzyme. In a separate experiment, they demonstrated that the new transcription could occur on a different kind of DNA added along with the new core polymerase. This supported the conclusion that σ had been released from the original core and was associating with a new core on a new DNA template. Accordingly, Travers and Burgess proposed that σ cycles from one core to another, as shown in Figure 6.11. They dubbed this the **"σ cycle."**

Figure 6.10 contains still another piece of valuable information. When Travers and Burgess added rifampicin, along with the core polymerase, which came from a rifampicin-resistant mutant, transcription still occurred (green line). Because the σ was from the original, rifampicin-sensitive polymerase, the rifampicin resistance in the renewed transcription must have been conferred by the newly added core. The fact that less initiation occurred in the presence of rifampicin probably means that the rifampicin-resistant core is still somewhat sensitive. We might have expected the σ-factor, not the core, to determine rifampicin sensitivity or resistance because rifampicin blocks initiation, and σ is the acknowledged initiation factor. Nevertheless, the core is the key to rifampicin sensitivity, and experiments to be presented later in this chapter will provide some clarification of why this is so.

SUMMARY At some point after σ has participated in initiation, it appears to dissociate from the core polymerase, leaving the core to carry out elongation. Furthermore, σ can be reused by different core polymerases, and the core, not σ, governs rifampicin sensitivity or resistance.

The Stochastic σ-Cycle Model

The σ-cycle model that arose from Travers and Burgess's experiments called for the dissociation of σ from core as the polymerase undergoes promoter clearance and switches from initiation to elongation mode. This has come to be known as the **obligate release** version of the σ-cycle model. Although this model has held sway for over 30 years and has considerable experimental support, it does not fit all the data at hand. For example, Jeffrey Roberts and colleagues demonstrated in 1996 that σ is involved in pausing at position +16/+17 downstream of the late promoter ($P_{R'}$) in λ phage. This implies that σ is still attached to core polymerase at position +16/+17, well after promoter clearance has occurred.

Based on this and other evidence, an alternative view of the σ-cycle was proposed: the **stochastic release model.** ("Stochastic" means "random"; Greek: *stochos,* meaning guess.) This hypothesis holds that σ is indeed released from the core polymerase, but there is no discrete point during transcription at which this release is required; rather, it is released randomly. As we will see, the preponderance of evidence now favors the stochastic release model.

Richard Ebright and coworkers noted in 2001 that all of the evidence favoring the obligate release model relies on harsh separation techniques, such as electrophoresis or chromatography. These could strip σ off of core if σ is weakly bound to core during elongation and, thus, make it appear that σ had dissociated from core during promoter clearance. These workers also noted that previous work had generally failed to distinguish between active and inactive RNA polymerases. This is a real concern because a significant fraction of RNA polymerase molecules in any population is not competent to switch from initiation to elongation mode.

To test the obligate release hypothesis, Ebright and coworkers used a technique, **fluorescence resonance energy transfer (FRET),** that allows the position of σ relative to a site on the DNA to be measured without using separation techniques that might themselves displace σ from core. The FRET technique relies on the fact that two fluorescent molecules close to each other will engage in transfer of resonance energy, and the efficiency of this energy transfer (FRET efficiency) will decrease rapidly as the two molecules move apart.

Ebright and coworkers measured FRET with fluorescent molecules (**fluorescence probes**) on both σ and DNA. The probe on σ serves as the fluorescence donor, and the probe on the DNA serves as the fluorescence acceptor. Sometimes the probe on the DNA was at the 5′, or upstream end (trailing-edge FRET), which allowed the investigators to observe the drop in FRET as the polymerase moved away from the promoter and the 5′-end of the DNA. In other experiments, the probe on the DNA was at the 3′-, or downstream end (leading-edge FRET), which allowed the investigators to observe the increase in FRET as the polymerase moved toward the downstream end. Figure 6.12 illustrates the strategies of trailing-edge and leading-edge FRET.

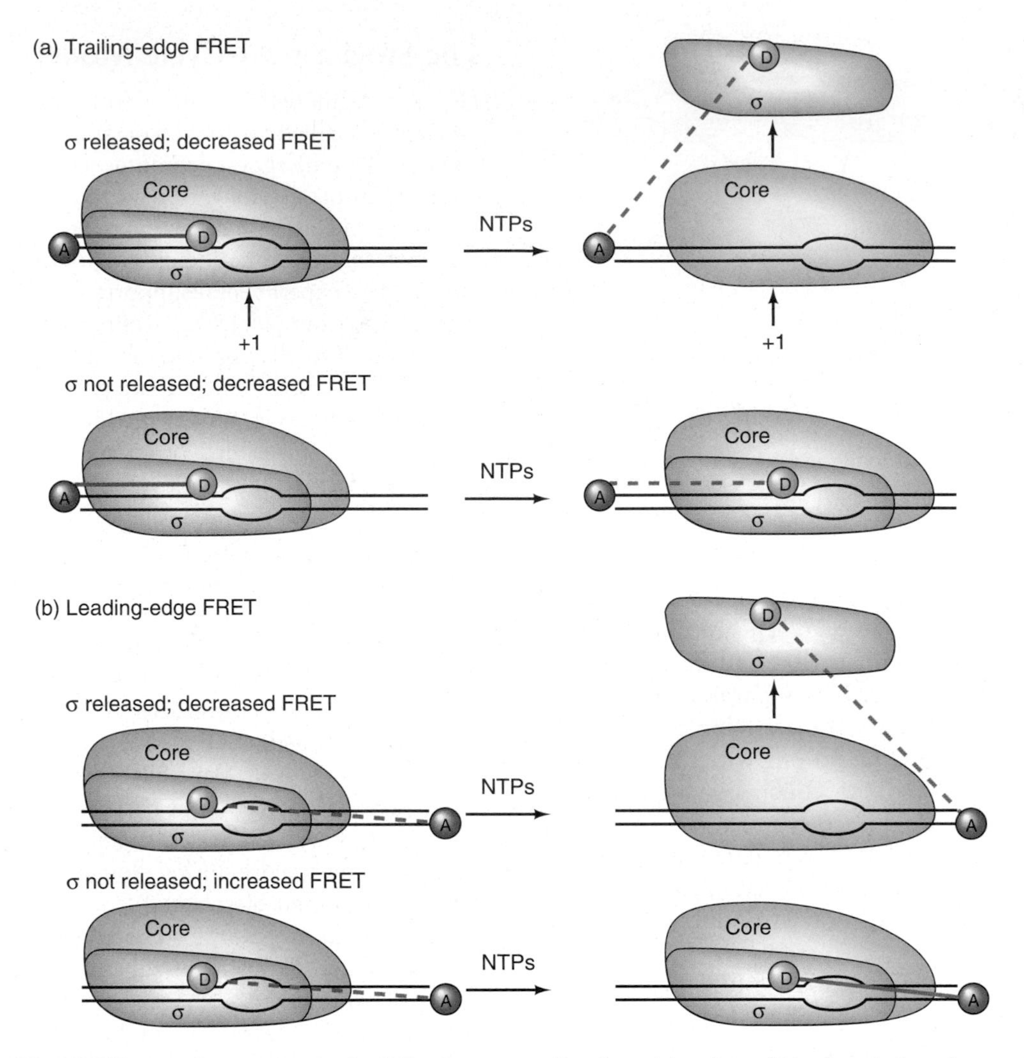

Figure 6.12 Rationale of FRET assay for σ movement relative to DNA. (a) Trailing-edge FRET. A fluorescence donor (D, green) is attached to the single cysteine residue in a σ^{70} mutant that had been engineered to eliminate all but one cysteine. A fluorescence acceptor (A, red) is attached to the 5′-end of the DNA. FRET efficiency is high (solid purple line) in the open promoter complex (RP_o) because the two probes are close together. On addition of 3 of the 4 nucleotides, the polymerase moves to a position downstream at which the fourth nucleotide (CTP) is required. This is at least position +11, so promoter clearance occurs. FRET efficiency decreases (dashed purple line) regardless of whether σ dissociates from the core, because the two probes grow farther apart in either case. If σ does not dissociate, it would travel with the core downstream during elongation, taking it farther from the probe at the 5′-end of the DNA. If σ dissociates, it would be found at random positions in solution, but, on average, it would be much farther away from the core than it was in the open promoter complex before transcription began. **(b)** Leading-edge FRET. Again a fluorescence donor is attached to σ^{70}, but this time, the fluorescence acceptor is attached to the 3′-end of the DNA. FRET efficiency is low (dashed purple line) in the open promoter complex because the two probes are far apart. On the addition of nucleotides, the polymerase undergoes promoter clearance and elongates to a downstream position as in (a). Now FRET can distinguish between the two hypotheses. If σ dissociates from core, FRET should decrease (dashed purple line), as it did in panel (a). On the other hand, if σ remains bound to core, the two probes will grow closer together as the polymerase moves downstream, and FRET efficiency will increase (solid purple line).

The trailing-edge FRET strategy does not distinguish between one model in which σ dissociates from the core, and a second model in which σ does not dissociate, after promoter clearance. In both cases, the donor probe on σ gets farther away from the acceptor probe at the upstream end of the DNA after promoter clearance and the FRET efficiency therefore decreases. Indeed, Figure 6.13a shows that the FRET efficiency does decrease with time when the probe on the DNA is at the upstream end.

On the other hand, the leading-edge strategy can distinguish between the two models (Figure 6.12b). If σ dissociates from the core, then FRET efficiency should decrease, just as it did in the trailing-edge experiment. But if σ is not released from the core, it should move closer to the probe at the downstream end of the DNA with time, and FRET efficiency should increase. Figure 6.13b shows that FRET efficiency did indeed increase, which supports the hypothesis that σ remains with the core after promoter clearance. In fact, the

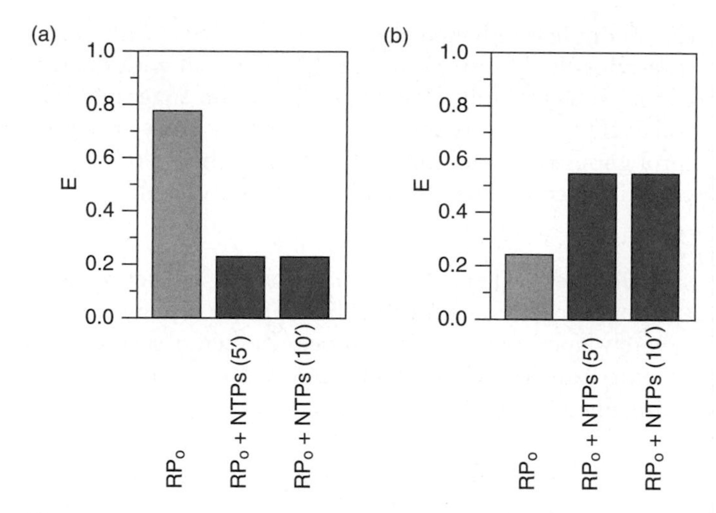

Figure 6.13 FRET analysis of σ-core association after promoter clearance. Ebright and coworkers performed FRET analysis as described in Figure 6.12. **(a)** Trailing-edge FRET results; **(b)** leading-edge FRET results. Blue bars, FRET efficiency (E) of open promoter complex (RP_O); red bars, FRET efficiency after 5 and 10 min, respectively, in the presence of the three nucleotides that allow the polymerase to move 11 bp downstream of the promoter.

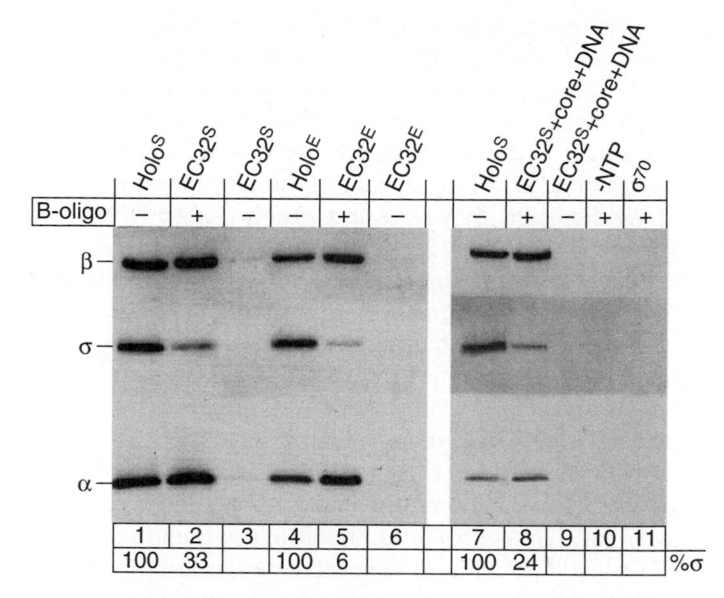

Figure 6.14 Measuring σ associated with transcription elongation complexes. Bar-Nahum and Nudler purified elongation complexes stalled at position +32 from stationary cells ($EC32^S$ complexes) or from exponentially growing cells ($EC32^E$ complexes), released the proteins from the nascent RNAs with nuclease, and subjected the proteins to SDS-PAGE, followed by immunoblotting. The nature of the complex and the presence or absence of an oligonucletide on the beads used to purify the complexes is denoted at the top. Lanes 8 and 9 are controls in which excess amounts of core and DNA were added to $EC32^S$ complexes prior to binding to the oligonucleotide beads. The purpose was to rule out σ attachment to beads due to nonspecific binding between σ and core or DNA. (*Source:* Reprinted from *Cell* v. 106, Bar-Nahum and Nudler, p. 444, © 2001, with permission from Elsevier Science.)

magnitude of the FRET efficiency increase suggests that 100% of the complexes after promoter clearance still retained their σ-factor.

Ebright and coworkers performed the experiments in Figure 6.13a and b in a polyacrylamide gel as follows. They formed open promoter complexes in solution, then added heparin to bind to any uncomplexed polymerase. Then they subjected the complexes to nondenaturing electrophoresis in a polyacrylamide gel. They located the complexes in the gel, sliced the gel and removed the slice containing the complexes, placed that gel slice in a container called a cuvette that fits into the fluorescence-measuring instrument (a fluorometer), added transcription buffer, and measured FRET efficiency on RP_O. Then they added three nucleotides to allow the polymerase to move downstream, and measured FRET efficiency on the elongation complex. This in-gel assay has the advantage of measuring FRET efficiency only on active complexes, because gel electrophoresis removes inactive (closed promoter) complexes. To eliminate the possibility that electrophoresis introduced an artifact of some kind, Ebright and coworkers performed the same experiments in solution and obtained very similar results.

In 2001, Bar-Nahum and Nudler also presented evidence for retention of σ. They formed complexes between holoenzyme and a DNA containing one promoter, then added three out of four nucleotides to allow the polymerase to move to position +32. Then they purified this elongation complex (called EC32) rapidly and gently by annealing the upstream end of the elongating RNA to a complementary oligonucleotide attached to resin beads. This allowed the beads, along with the complexes, to be purified quickly by low-speed centrifugation. Only elongation complexes are purified this way, because they are the only ones with a nascent RNA that can bind to the complementary oligonucleotide.

Finally, Bar-Nahum and Nudler released the complexes from the beads with nuclease, subjected the proteins to SDS-PAGE, and performed an immunoblot (Chapter 5) to identify the proteins associated with the complexes. Figure 6.14 shows that the purified EC32 complexes contained at least some σ. Quantification showed that complexes isolated from stationary phase cells contained 33 ± 2% of the full complement of σ per complex, and complexes isolated from exponential phase cells contained 6 ± 1% of the full complement of σ per complex. This is considerably less than the 100% observed by Ebright and coworkers and suggests relatively weak binding between σ and core in elongation complexes. Nevertheless, even these amounts of complexes that retain σ could aid considerably in reinitiation of transcription, because the association of core with σ is the rate-limiting step in transcription initiation.

Although the results of Bar-Nahum and Nudler, and those of Ebright and colleagues appear to rule out the obligate release model, and may seem to argue against the σ-cycle in general, they are actually consistent with the stochastic release version of the σ-cycle, which calls for σ release at multiple points throughout transcription. Bar-Nahum and Nudler collected elongation complexes after only 32-nt of transcription, which could be too early in transcription to see complete σ release. And, while it is true that Ebright and colleagues did not observe significant σ dissociation after 50 nt of transcription in the experiments we have discussed, they were unwittingly using a DNA template (the *E. coli lac*UV5 promoter) that contributed to this phenomenon. This promoter contains a second −10-like box just downstream of the transcription start site. It has recently been learned that this sequence causes pausing that depends on σ, and indeed appears to aid in σ retention. When this second −10-like box was mutated, the FRET signal decreased, and σ dissociation increased more than 4-fold. Furthermore, when they performed their original experiments with fluorescent labels on σ and core, rather than σ and DNA, Ebright and colleagues found that their FRET signal did decrease with increasing transcript length. All of these findings suggest that some σ was dissociating from core during the transcription process, and that the DNA sequence can influence the rate of such dissociation.

To probe further the validity of the σ-cycle hypothesis, Ebright and colleagues used leading and trailing edge single-molecule FRET analysis with alternating-laser excitation (single-molecule FRET ALEX). For leading edge FRET, they tagged the leading edge of σ with the donor fluorophore and a downstream DNA site with the acceptor. For trailing edge FRET, they tagged the trailing edge of σ with the donor and an upstream DNA site with the acceptor fluorophore. They measured both fluorescence efficiency and "stoichiometry," or the presence of one or both of the fluorophores (donor and acceptor) in a small (femtoliter [10^{-15} L] scale) excitation volume, which should have at most one copy of the elongation complex at any given time. They switched rapidly between exciting the donor and acceptor fluorophore, such that each would be excited multiple times during the approximately 1 ms transit time through the excitation volume. Furthermore, they stalled the elongation complex at various points (nascent RNAs 11, 14, and 50 nt long) by coupling the *E. coli lac*UV5 promoter to various G-less cassettes (Chapter 5) and leaving out CTP in the transcription reaction. By measuring both fluorescence efficiency and stoichiometry for the same elongation complex, they could tell: (1) how far transcription had progressed (by the fluorescence efficiency, which grows weaker in trailing edge FRET, and stronger in leading edge FRET, as transcription progresses); and (2) whether or not σ had dissociated from core (by the stoichiometry, which should be approximately 0.5 for holoenzyme, but nearer 0 for core alone and 1.0 for σ alone).

These studies confirmed that σ did indeed remain associated with the great majority (about 90%) of elongation complexes that had achieved promoter clearance (with transcripts 11 nt long). Again, this finding argued strongly against the obligate release model. But they also showed that about half of halted elongation complexes with longer transcripts had lost their σ-factors, in accord with the stochastic release model. Finally, their results suggested that some elongation complexes may retain their σ-factors throughout the transcription process. If that is true, these elongation complexes are avoiding the σ cycle altogether.

SUMMARY The σ-factor appears to be released from the core polymerase, but not usually immediately upon promoter clearance. Rather, σ seems to exit from the elongation complex in a stochastic manner during the elongation process.

Local DNA Melting at the Promoter

Chamberlin's studies on RNA polymerase–promoter interactions showed that such complexes were much more stable at elevated temperature. This suggested that local melting of DNA occurs on tight binding to polymerase, because high temperature would tend to stabilize melted DNA. Furthermore, such DNA melting is essential because it exposes bases of the template strand so they can base-pair with bases on incoming nucleotides.

Tao-shih Hsieh and James Wang provided more direct evidence for local DNA melting in 1978. They bound *E. coli* RNA polymerase to a restriction fragment containing three phage T7 early promoters and measured the hyperchromic shift (Chapter 2) caused by such binding. This increase in the DNA's absorbance of 260-nm light is not only indicative of DNA strand separation, its magnitude is directly related to the number of base pairs that are opened. Knowing the number of RNA polymerase holoenzymes bound to their DNA, Hsieh and Wang calculated that each polymerase caused a separation of about 10 bp.

In 1979, Ulrich Siebenlist, identified the base pairs that RNA polymerase melted in a T7 phage early promoter. Figure 6.15 shows the strategy of his experiment. First he end-labeled the promoter DNA, then added RNA polymerase to form an open promoter complex. As we have seen, this involves local DNA melting, and when the strands separate, the N_1 of adenine—normally involved in hydrogen bonding to a T in the opposite strand—becomes susceptible to attack by certain chemical agents. In this case, Siebenlist methylated the exposed adenines with dimethyl sulfate (DMS). Then, when he removed the RNA polymerase and the melted region closed up again, the methyl groups prevented proper base-pairing between these N_1-methyl-adenines and the thymines in the opposite strand and thus preserved at least some of the single-stranded character of the formerly melted region. Next, he treated the DNA with S1 nuclease, which specifically cuts single-stranded DNA. This enzyme should therefore cut wherever an adenine had been in a melted region of the

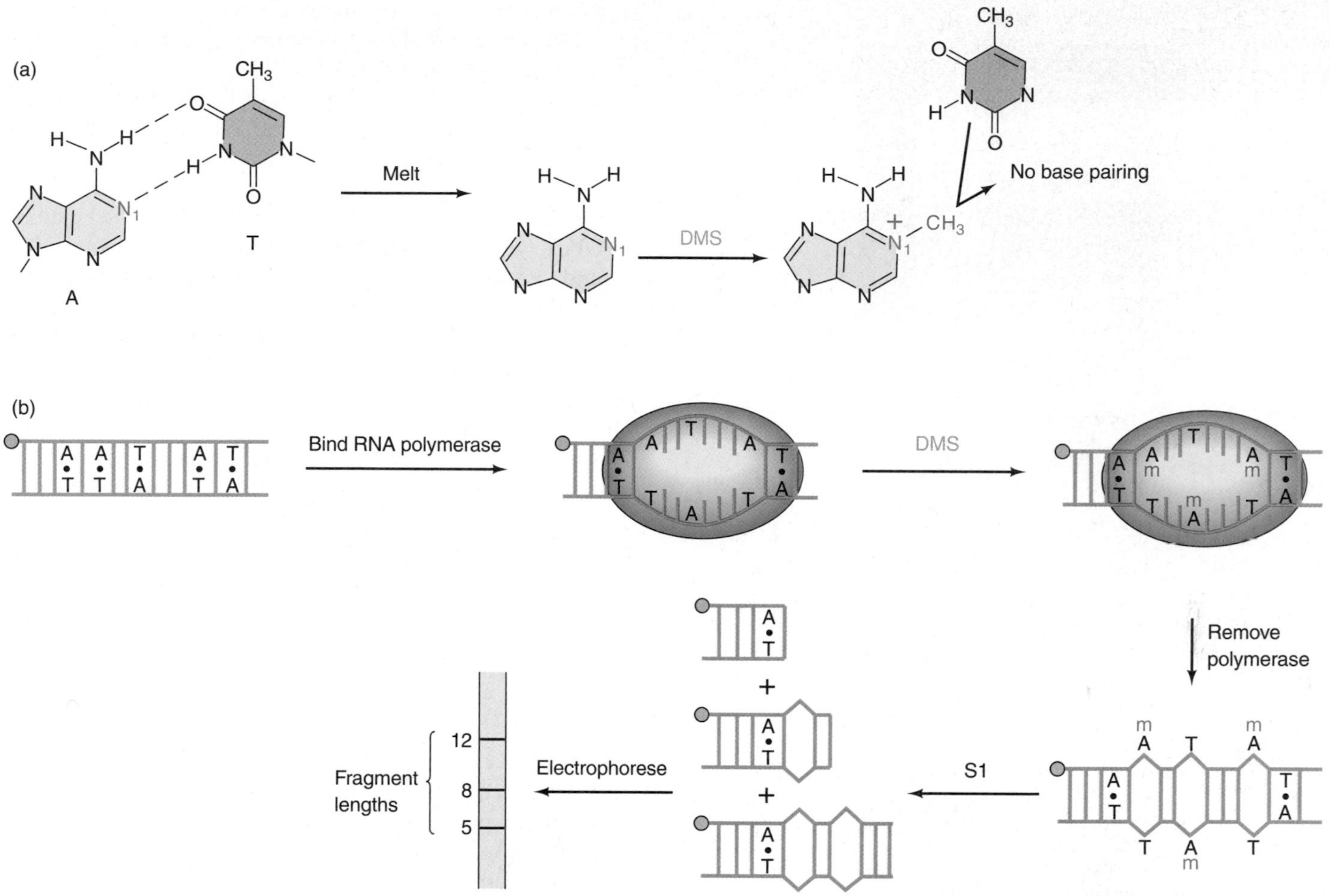

Figure 6.15 Locating the region of a T7 phage early promoter melted by RNA polymerase. (a) When adenine is base-paired with thymine (left) the N_1 nitrogen of adenine is hidden in the middle of the double helix and is therefore protected from methylation. On melting (right), the adenine and thymine separate; this opens the adenine up to attack by dimethyl sulfate (DMS, blue), and the N_1 nitrogen is methylated. Once this occurs, the methyl-adenine can no longer base-pair with its thymine partner. **(b)** A hypothetical promoter region containing five A–T base pairs is end-labeled (orange), then RNA polymerase (red) is bound, which causes local melting of the promoter DNA. The three newly exposed adenines are methylated with dimethyl sulfate (DMS). Then, when the polymerase is removed, the A–T base pairs cannot reform because of the interfering methyl groups (m, blue). Now S1 nuclease can cut the DNA at each of the unformed base pairs because these are local single-stranded regions. Very mild cutting conditions are used so that only about one cut per molecule occurs. Otherwise, only the shortest product would be seen. The resulting fragments are denatured and electrophoresed to determine their sizes. These sizes tell how far the melted DNA region was from the labeled DNA end.

promoter and had become methylated. In principle, this should produce a series of end-labeled fragments, each one terminating at an adenine in the melted region. Finally, Siebenlist electrophoresed the labeled DNA fragments to determine their precise lengths. Then, knowing these lengths and the exact position of the labeled end, he could calculate accurately the position of the melted region.

Figure 6.16 shows the results. Instead of the expected neat set of fragments, we see a blur of several fragments extending from position +3 to −9. The reason for the blur seems to be that each of the multiple methylations in the melted region introduced a positive charge and therefore weakened base pairing so much that few strong base pairs could re-form; the whole melted region retained at least partially single-stranded character and therefore remained open to cutting by S1 nuclease. The length of the melted region detected by this experiment is 12 bp, roughly in agreement with Hsieh and Wang's estimate, although this may be an underestimate because the next base pairs on either side are G–C pairs whose participation in the melted region would not have been detected. This is because neither guanines nor cytosines are readily methylated under the conditions used in this experiment. It is also satisfying that the melted region is just at the place where RNA polymerase begins transcribing.

The experiments of Hsieh and Wang, and of Siebenlist, as well as other early experiments, measured the DNA melting in a simple binary complex between polymerase and DNA. None of these experiments examined the size

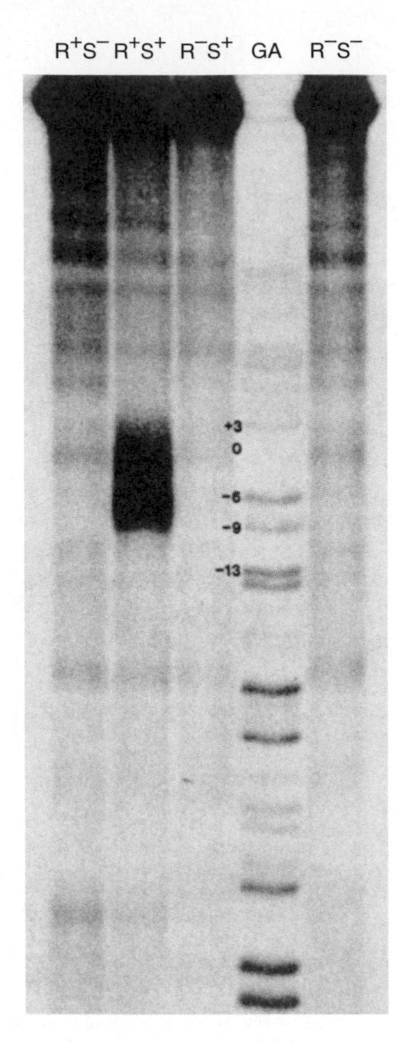

Figure 6.16 RNA polymerase melts the DNA in the −9 to +3 region of the T7 A3 promoter. Siebenlist performed a methylation-S1 assay as described in Figure 6.15. Lane R^+S^+ shows the results when both RNA polymerase (R) and S1 nuclease (S) were used. The other lanes were controls in which Siebenlist left out either RNA polymerase, or S1 nuclease, or both. The partial sequencing lane (GA) served as a set of markers and allowed him to locate the melted region approximately between positions −9 and +3. (*Source:* Siebenlist. RNA polymerase unwinds an 11-base pair segment of a phage T7 promoter. *Nature* 279 (14 June 1979) p. 652, f. 2, © Macmillan Magazines Ltd.)

of a DNA bubble in complexes in which initiation or elongation of RNA chains was actually taking place. Thus, in 1982, Howard Gamper and John Hearst set out to estimate the number of base pairs melted by polymerases, not only in binary complexes, but also in actively transcribing complexes that also contained RNA (ternary complexes). They used SV40 DNA, which happens to have one promoter site recognized by the *E. coli* RNA polymerase. They bound RNA polymerase to the SV40 DNA at either 5°C or 37°C in the absence of nucleotides to form binary complexes, or in the presence of nucleotides to form ternary complexes. Under the conditions of the experiment, each polymerase initiated only once, and no polymerase terminated transcription, so all polymerases remained complexed to the DNA. This allowed an accurate assessment of the number of polymerases bound to the DNA.

After binding a known number of *E. coli* RNA polymerases to the DNA, Gamper and Hearst relaxed any supercoils that had formed with a crude extract from human cells, then removed the polymerases from the relaxed DNA (Figure 6.17a). The removal of the protein left melted regions of DNA, which meant that the whole DNA was underwound. Because the DNA was still a covalently closed circle, this underwinding introduced strain into the circle that was relieved by forming supercoils (Chapters 2 and 20). The higher the superhelical content, the greater the double helix unwinding that has been caused by the polymerase. The superhelical content of a DNA can be measured by gel electrophoresis because the more superhelical turns a DNA contains, the faster it will migrate in an electrophoretic gel.

Figure 6.17b is a plot of the change in the superhelicity as a function of the number of active polymerases per genome at 37°C. A linear relationship existed between these two variables, and one polymerase caused about 1.6 superhelical turns, which means that each polymerase unwound 1.6 turns of the DNA double helix. If a double helical turn contains 10.5 bp, then each polymerase melted about 17 bp ($1.6 \times 10.5 = 16.8$). A similar calculation of the data from the 5°C experiment yielded a value of 18 bp melted by one polymerase. From these data, Gamper and Hearst concluded that a polymerase binds at the promoter, melts 17 ± 1 bp of DNA to form a **transcription bubble,** and a bubble of this size moves with the polymerase as it transcribes the DNA. Subsequent experimental and theoretical work has suggested that the size of the transcription bubble actually increases and decreases within a range of approximately 11–16 nt, according to conditions, including the base sequence within the bubble. Larger bubbles can form, but their abundance decreases exponentially with size because of the energy required to melt more base pairs.

SUMMARY On binding to a promoter, RNA polymerase causes melting that has been estimated at 10–17 bp in the vicinity of the transcription start site. This transcription bubble moves with the polymerase, exposing the template strand so it can be transcribed.

Promoter Clearance

RNA polymerases cannot work if they do not recognize promoters, so they have evolved to recognize and bind strongly to them. But that poses a challenge when it comes time for promoter clearance: Somehow those strong bonds between polymerase and promoter must be broken in order for the polymerase to leave the promoter and enter the elongation phase. How can we explain that phenomenon?

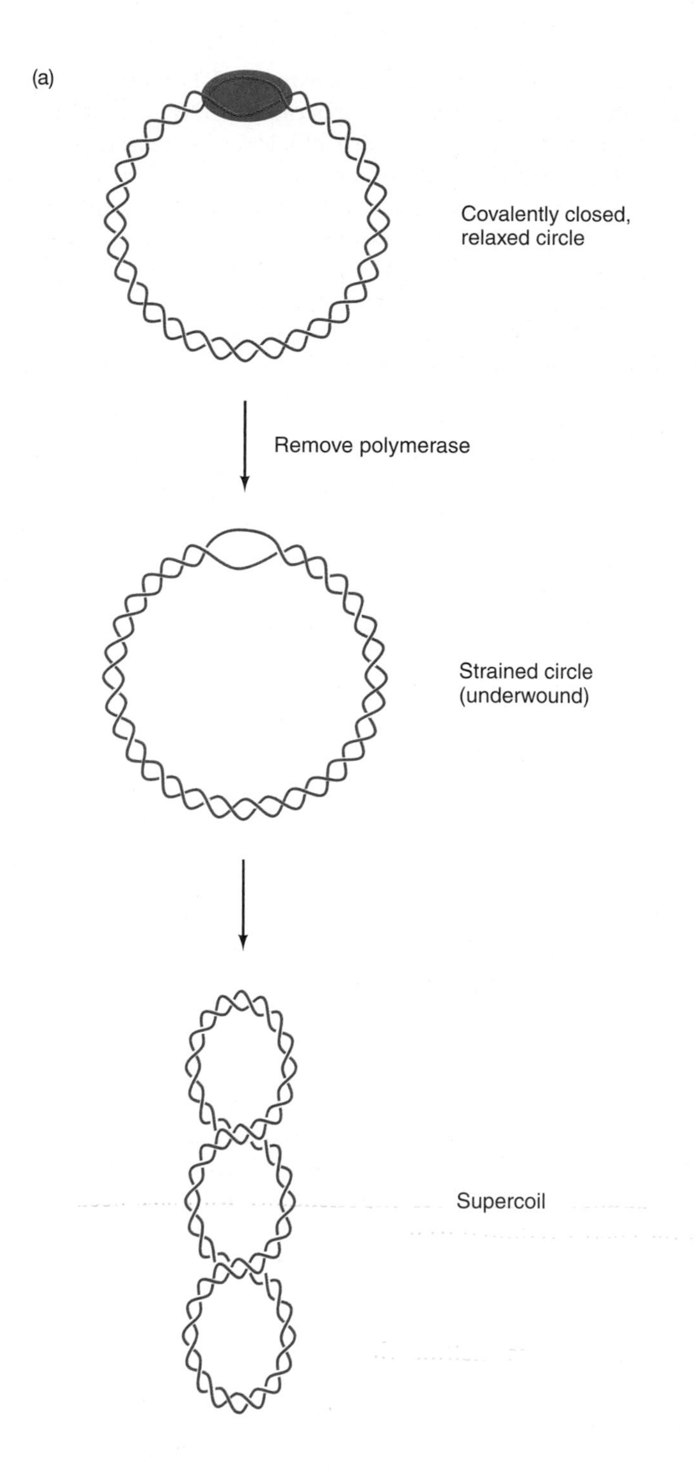

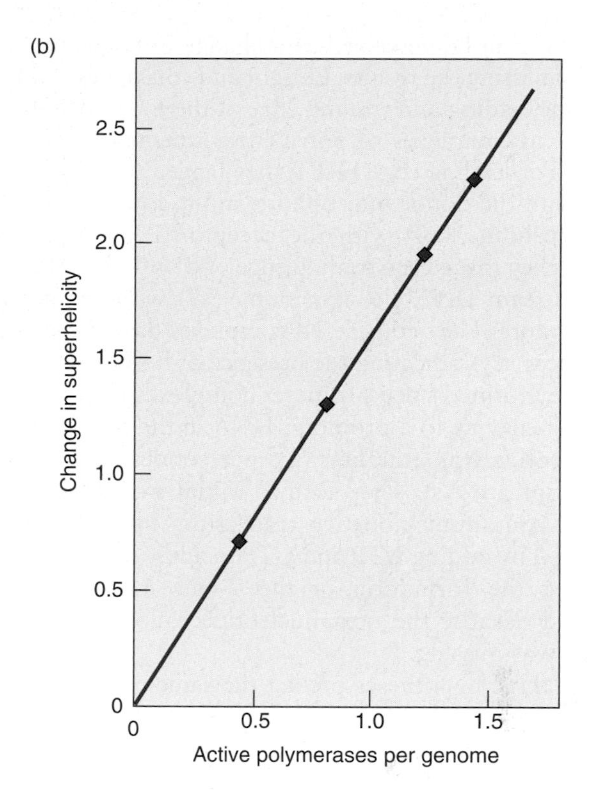

Figure 6.17 Measuring the melting of DNA by polymerase binding. **(a)** Principle of the experiment. Gamper and Hearst added *E. coli* RNA polymerase (red) to SV40 DNA, then relaxed any supercoils with a nicking-closing extract to produce the complexes shown at top. Then they removed the polymerase, leaving the DNAs strained (middle) because of the region that had been melted by the polymerase. This strain was quickly relieved by forming supercoils (bottom). The greater the superhelicity, the greater the unwinding caused by the polymerase. **(b)** Experimental results. Gamper and Hearst plotted the change in superhelicity of DNA as a function of the number of polymerases added. The plot was a straight line with a slope of 1.6 (1.6 superhelical turns introduced per polymerase).

Several hypotheses have been proposed, including the idea that the energy released by forming a short transcript (up to 10 nt long) is stored in a distorted polymerase or DNA, and the release of that energy in turn allows promoter clearance. However this process works, it is clearly not perfect, as it fails more often than not, giving rise to abortive transcripts.

The polymerase cannot move enough downstream to make a 10-nt transcript without doing one of three things: moving briefly downstream and then snapping back to the starting position (transient excursion); stretching itself by leaving its trailing edge in place while moving its leading edge downstream (inchworming); or compressing the DNA without moving itself (**scrunching**). In 2006, Richard Ebright and colleagues applied two single-molecule strategies to show that scrunching appears to be the correct answer.

The first set of experiments used single-molecule FRET as described earlier in this chapter, but with a twist known as "FRET analysis with alternating-laser excitation" (**FRET-ALEX**). This adaptation can correct for the fact that the spectrum of a donor fluorophore depends on its exact protein environment, which can change during an experiment because proteins are dynamic molecules. This change in spectrum can be perceived as a change in fluorescence energy, confusing the results. Ebright and colleagues examined both the leading and trailing edge of the *E. coli* RNA polymerase in complexes of polymerase attached to promoter DNA. For leading edge FRET, they tagged the leading edge of σ with the donor fluorophore and a downstream DNA site (position +20) with the acceptor. For trailing edge FRET, they tagged the trailing edge of σ with the donor and an upstream DNA site (position −39) with the acceptor fluorophore. They considered complexes only if they had a stoichiometry indicating the presence of both fluorophores.

They formed open promoter complexes (RP_o) by binding holoenzyme to a promoter DNA in the presence of the dinucleotide ApA (the first two nucleotides in the nascent transcript are A's). They formed initial transcribing complexes containing abortive transcripts up to 7 nt long ($RP_{itc\leq7}$) by adding UTP and GTP in addition to ApA. This allowed the formation of the 7-mer AAUUGUG, but stopped because the next nucleotide called for was ATP, which was missing.

All three hypotheses predict the same result with leading edge FRET ALEX: All three should yield a decreased separation between the fluorophores, as illustrated in Figure 6.18a. Indeed, a comparison of RP_o and $RP_{itc\leq7}$ showed an increase in FRET efficiency as the polymerase formed abortive transcripts up to 7 nt long, and therefore a decreased distance between fluorophores.

To begin to distinguish among the hypotheses, Ebright and colleagues performed trailing edge FRET ALEX (Figure 6.18b). Both the inchworming and scrunching models predict no change in the position of the trailing edge of the polymerase during abortive transcript production. But the transient excursion model predicts that the polymerase moves downstream in producing abortive transcripts and therefore $RP_{itc\leq7}$ complexes should show a decrease in FRET efficiency relative to RP_o complexes. In fact, Ebright and colleagues observed no difference in FRET efficiency, ruling out the transient excursion model.

To distinguish between the inchworming and scrunching models, Ebright and colleagues placed the donor fluorophore on the leading edge of σ and the acceptor fluorophore on the DNA spacer between the −10 and −35 boxes of the promoter (Figure 6.18c). If the polymerase stretches, as the inchworming model predicts, the separation between fluorophores should increase, and the fluorescence efficiency should fall. On the other hand, the scrunching model predicts that downstream DNA is drawn into the enzyme, which should not change the separation between fluorophores. Indeed, the fluorescence efficiency did not change, supporting the scrunching model.

To check this result, Ebright and colleagues tested directly for the scrunching of DNA. They placed the donor fluorophore at DNA position −15, and the acceptor fluorophore in the downstream DNA, at position +15. If the polymerase really does pull downstream DNA into itself, the distance between fluorophores on the DNA should decrease. Indeed, the fluorescence efficiency increased, supporting the scrunching hypothesis.

Thus, it may be the scrunched DNA that stores the energy expended in abortive transcript formation, rather like a spring, and enables the RNA polymerase finally to break away from the promoter and shift to the elongation phase. In another study, Ebright, Terence Strick, and colleagues used single-molecule DNA nanomanipulation to show that DNA scrunching indeed accompanies, and is probably required for, promoter clearance.

In this method, Ebright, Strick, and colleagues tethered a magnetic bead to one end of a piece of DNA, and a glass surface to the other (Figure 6.19). They made the DNA stick straight up from the glass surface by placing a pair of magnets above the magnetic bead. By rotating the magnets, they could rotate the DNA, introducing either positive or negative supercoils, depending on the direction of rotation. Then they added RNA polymerase, which bound to a promoter in the DNA. By adding different subsets of nucleotides, they could form either RP_o, $RP_{itc\leq4}$, $RP_{itc\leq8}$, or an elongation complex (RP_e). (With this promoter, addition of ATP and UTP leads to an abortive transcript up to 4 nt long, and addition of ATP, UTP, and CTP produces an abortive transcript up to 8 nt long.)

If scrunching occurs during abortive transcription, then the DNA will experience an extra unwinding, which causes a compensating loss of negative supercoiling, or gain of positive supercoiling. Every unwinding of one helical turn (about 10 bp) leads to loss of one negative, or gain of one positive, supercoil. The change in supercoiling can be measured as shown in Figure 6.19. Gain of one positive supercoil should decrease the apparent length (l) of the DNA (the distance between the bead and the glass surface) by 56 nm. Similarly, loss of one negative supercoil should increase l by 56 nm. Such changes in the position of the magnetic bead can be readily observed in real time by videomicroscopy, yielding estimates of DNA unwinding with a resolution of about 1 bp.

Ebright, Strick, and colleagues observed the expected change in l upon converting RP_o to $RP_{itc\leq4}$ and $RP_{itc\leq8}$. Thus, unwinding of DNA accompanies formation of abortive transcripts, and the degree of unwinding depends on the length of the abortive transcript made. In particular, formation of abortive transcripts 4 and 8 nt long led to unwinding of 2 and 6 nt, respectively. This is consistent with the hypothesis that the active center of RNA polymerase can polymerize two nucleotides without moving relative to the DNA, but further RNA synthesis requires scrunching.

Does scrunching also accompany promoter clearance? To find out, Ebright, Strick, and colleagues looked at individual complexes over time: from the addition of polymerase and all four nucleotides until termination at a

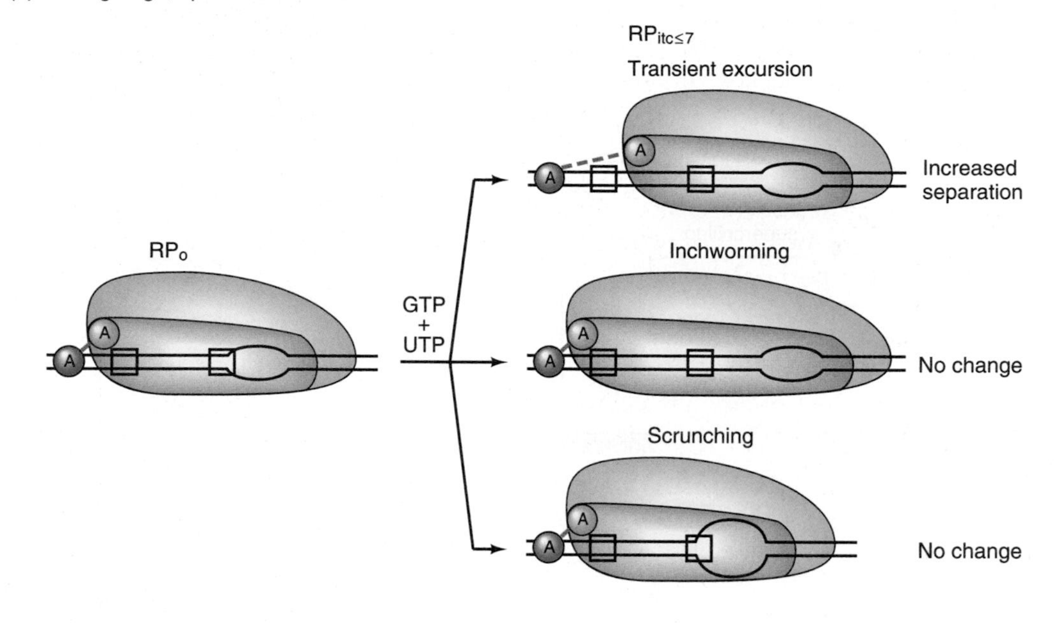

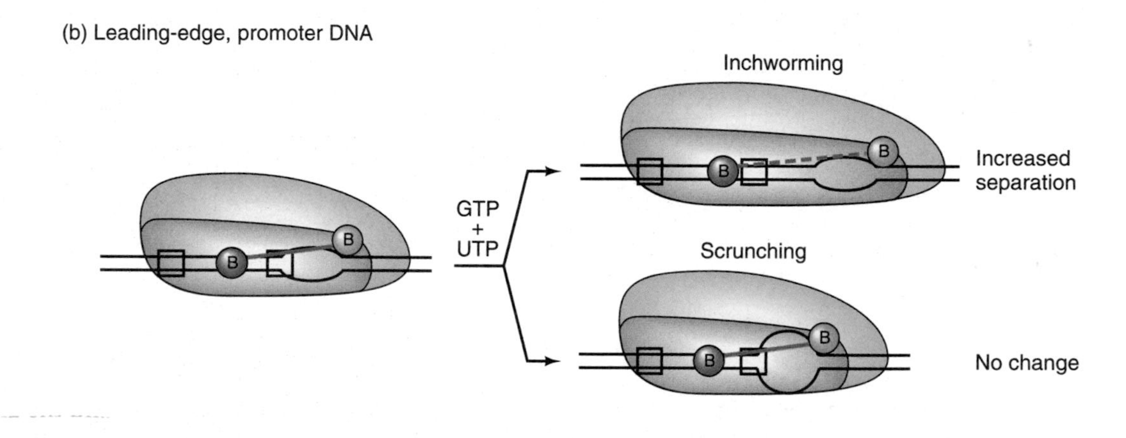

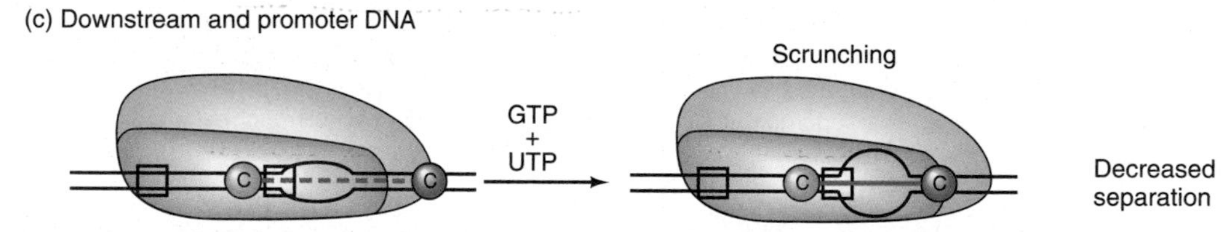

Figure 6.18 Evidence for DNA scrunching during abortive transcription. Ebright and colleagues used single-molecule FRET ALEX to distinguish among three hypotheses for the mechanism of abortive transcription: transient excursion, inchworming, and scrunching. They compared the average efficiency of single-molecule FRET of RP_o and $RP_{itc\leq7}$ complexes of *E. coli* RNA polymerase with promoter DNA. The latter complexes contained abortive transcripts up to 7 nt in length and were created by allowing transcription in the presence of the primer ApA plus UTP and GTP. ATP is required in the eighth position, limiting the abortive transcripts to 7 nt. The position of the donor fluorophore is denoted in green, and the acceptor fluorophore in red, throughout. High-efficiency FRET, indicating short distance between fluorophores, is denoted by a solid purple line throughout. Lower-efficiency FRET, indicating a greater distance between fluorophores, is denoted by a dashed purple line throughout. The three experiments depicted in panels **(a)–(c)** are described in the text. The boxes represent the −10 and −35 boxes of the promoter.

terminator either 100 or 400 bp downstream of the promoter. In fact, since reinitiation could occur, the investigators could look at multiple rounds of transcription on each DNA. They found a four-phase pattern that repeated over and over with each round. Considering a positively supercoiled DNA: First, the superhelicity increased, reflecting the DNA unwinding that occurs during RP_o formation. Second, the superhelicity increased still further, relecting the

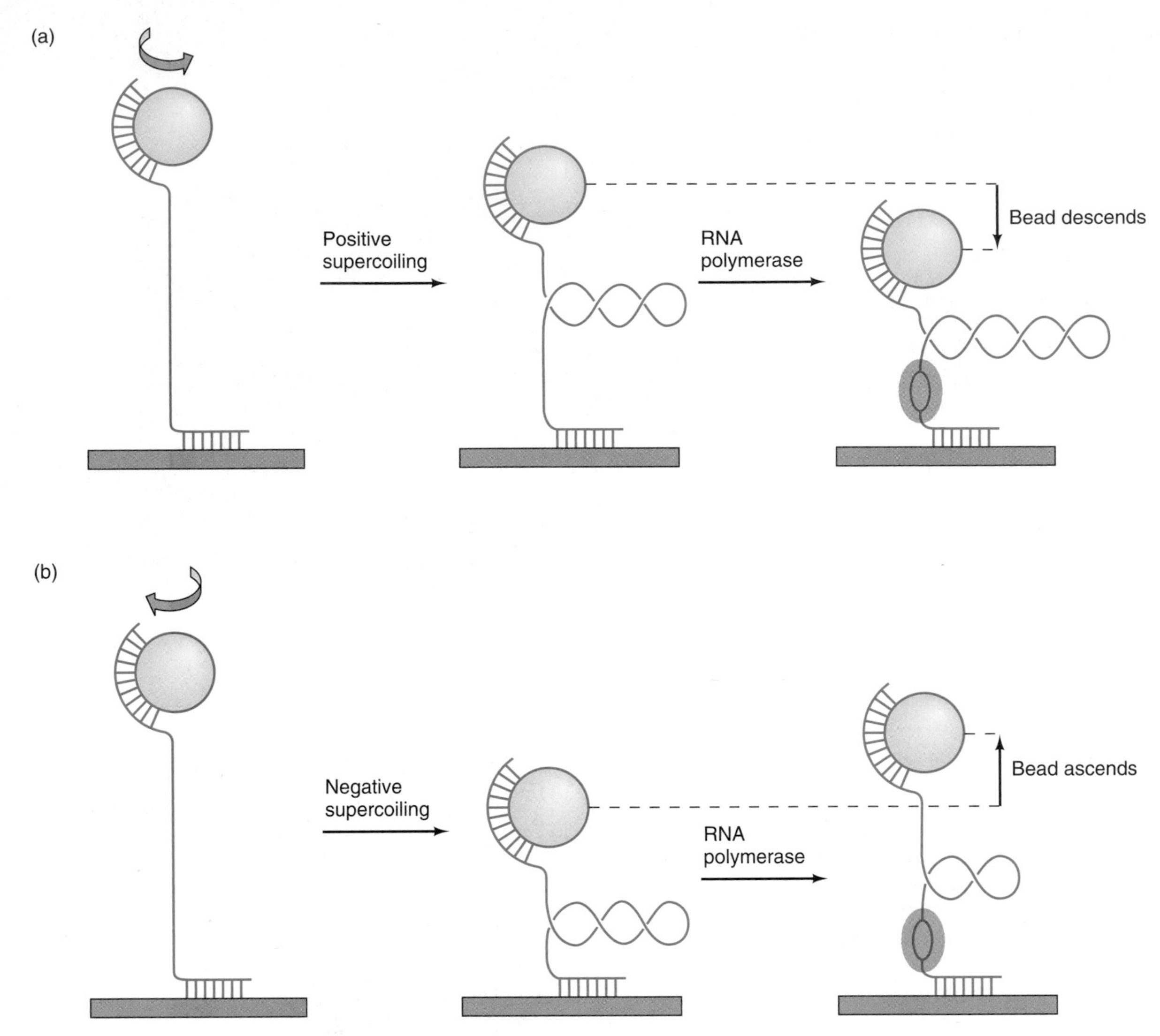

Figure 6.19 Basis of single-molecule nanomanipulation procedure. One end of a promoter-containing piece of DNA is tethered to a magnetic bead (yellow), and the other end is tethered to a glass surface (blue). A pair of magnets at the top extend the DNA vertically, and introduce a rightward **(a)** or leftward **(b)** twist to the bead, and therefore to the DNA. Every full turn of the bead introduces one superhelical turn into the DNA. The supercoiling is positive in (a) and negative in (b). When RNA polymerase (pink) is added to the DNA, it binds to the promoter and unwinds about one double-helical turn of DNA, which adds one positive supercoil (a), which drags the magnetic bead down about 56 nm for every such supercoil. Similarly, unwinding of promoter DNA by the polymerase subtracts one negative supercoil (b). These changes in bead position are detected by videomicroscopy.

scrunching that occurs during RP_{itc} formation. Third, the superhelicity decreased, reflecting the reversal of scrunching during promoter clearance and RP_e formation. Finally, the superhelicity decreased back to the original level, reflecting the loss of RNA polymerase at termination. The amount of scrunching observed in these experiments was 9 ± 2 bp, which is within experimental error of the amount expected: Promoter clearance at this promoter was known to occur upon formation of an 11-nt transcript, 9 nt of which should require 9 bp of DNA scrunching, and 2 nt of which the polymerase can synthesize without scrunching.

Eighty percent of the transcription cycles studied had detectable scrunches. But 20% of the cycles were predicted to have scrunches that lasted less than 1 s, and 1 s was the limit of resolution in these experiments. So this 20% of cycles probably also had scrunches. The authors concluded that approximately 100% of all the transcription cycles involve scrunching, which suggests that scrunching is required for promoter clearance.

E. coli RNA polymerase was used in all these studies, but the similarity among RNA polymerases, the strength of binding between polymerases and promoters, and the necessity to break that binding to start productive transcription, all suggest that scrunching could be a general phenomenon, and could be universally required for promoter clearance.

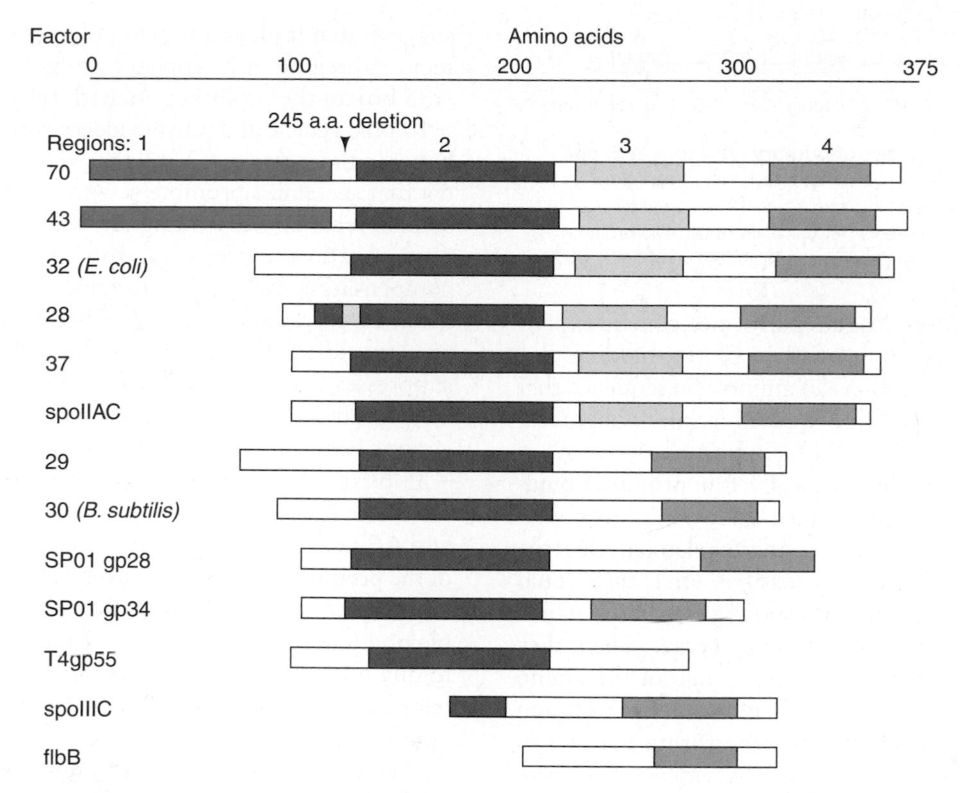

Figure 6.20 Homologous regions in various *E. coli* and *B. subtilis* σ-factors. The σ proteins are represented as horizontal bars, with homologous regions aligned vertically. Only the top two, the primary σ-factors of *E. coli* and *B. subtilis,* respectively, contain the first homologous region. Also, σ^{70} contains a sequence of 245 amino acids between regions 1 and 2 that is missing in σ^{43}. This is marked above the σ^{70} bar. Lighter shading denotes an area that is conserved only in some of the proteins.

SUMMARY The *E. coli* RNA polymerase achieves abortive transcription by scrunching: drawing downstream DNA into the polymerase without actually moving and losing its grip on promoter DNA. The scrunched DNA could store enough energy to allow the polymerase to break its bonds to the promoter and begin productive transcription.

Structure and Function of σ

By the late 1980s, the genes encoding a variety of σ-factors from various bacteria had been cloned and sequenced. As we will see in Chapter 8, each bacterium has a primary σ-factor that transcribes its vegetative genes—those required for everyday growth. For example, the primary σ in *E. coli* is called σ^{70}, and the primary σ in *B. subtilis* is σ^{43}. These proteins are named for their molecular masses, 70 and 43 kD, respectively, and they are also called σ^{A} because of their primary nature. In addition, bacteria have alternative σ-factors that transcribe specialized genes (heat shock genes, sporulation genes, and so forth). In 1988, Helmann and Chamberlin reviewed the literature on all these factors and analyzed the striking similarities in amino acid sequence among them, which are clustered in four regions (regions 1–4, see Figure 6.20). The conservation of sequence in these regions suggests that they are important in the function of σ, and in fact they are all involved in binding to core and positively or negatively, in binding to DNA. Helmann and Chamberlin proposed the following functions for each region.

Region 1 This region is found only in the primary σ's (σ^{70} and σ^{43}). Its role appears to be to prevent σ from binding by itself to DNA. We will see later in this chapter that a fragment of σ is capable of DNA binding, but region 1 prevents the whole polypeptide from doing that. This is important because free σ binding to promoters could inhibit holoenzyme binding and thereby inhibit transcription.

Region 2 This region is found in all σ-factors and is the most highly conserved σ region. It can be subdivided into four parts, 2.1–2.4 (Figure 6.21).

We have good evidence that region 2.4 is responsible for a crucial σ activity, recognition of the promoter's −10 box. First of all, if σ region 2.4 does recognize the −10 box, then σ's with similar specificities should have similar regions 2.4. This is demonstrable; σ^{43} of *B. subtilis* and σ^{70} of *E. coli* recognize identical promoter sequences, including −10 boxes. Indeed, these two σ's are interchangeable. And the regions 2.4 of these two σ's are 95% identical.

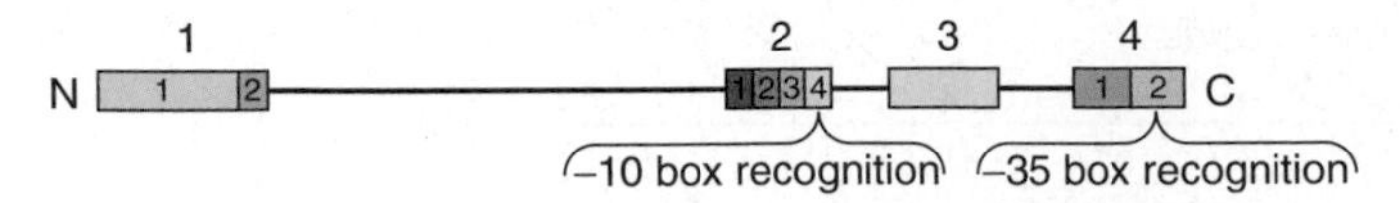

Figure 6.21 Summary of regions of primary structure in *E. coli* σ^{70}. The four conserved regions are indicated, with subregions delineated in regions 1, 2, and 4. (*Source:* Adapted from Dombroski, A.J., et al., "Polypeptides containing highly conserved regions of the transcription initiation factor σ^{70} exhibit specificity of binding to promoter DNA." *Cell* 70:501–12, 1992.)

Richard Losick and colleagues performed genetic experiments that also link region 2.4 with −10 box binding. Region 2.4 of the σ-factor contains an amino acid sequence that suggests it can form an α-helix. We will learn in Chapter 9 that an α-helix is a favorite DNA-binding motif, which is consistent with a role for this part of the σ in promoter binding. Losick and colleagues reasoned as follows: If this potential α-helix is really a −10 box-recognition element, then the following experiment should be possible. First, they could make a single base change in a promoter's −10 box, which destroys its ability to bind to RNA polymerase. Then, they could make a compensating mutation in one of the amino acids in region 2.4 of the σ-factor. If the σ-factor mutation can suppress the promoter mutation, restoring binding to the mutated promoter, it provides strong evidence that there really is a relationship between the −10 box and region 2.4 of the σ. So Losick and colleagues caused a G→A transition in the −10 box of the *B. subtilis spo*VG promoter, which prevented binding between the promoter and RNA polymerase. Then they caused a Thr → Ile mutation at amino acid 100 in region 2.4 of σ^H, which normally recognizes the *spo*VG promoter. This σ mutation restored the ability of the polymerase to recognize the mutant promoter.

Region 3 We will see later in this chapter that region 3 is involved in both core and DNA binding.

Region 4 Like region 2, region 4 can be subdivided into subregions. Also like region 2, region 4 seems to play a key role in promoter recognition. Subregion 4.2 contains a helix-turn-helix DNA-binding domain (Chapter 9), which suggests that it plays a role in polymerase–DNA binding. In fact, subregion 4.2 appears to govern binding to the −35 box of the promoter. As with the σ region 2.4 and the −10 box, genetic and other evidence supports the relationship between the σ region 4.2 and the −35 box. Again, we see that σ's that recognize promoters with similar −35 boxes have similar regions 4.2. And again, we observe suppression of mutations in the promoter (this time in the −35 box) by compensating mutations in region 4.2 of the σ-factor. For instance, Miriam Susskind and her colleagues showed that an Arg→His mutation in position 588 of the *E. coli* σ^{70} suppresses G→A or G→C mutations in the −35 box of the *lac* promoter. Figure 6.22 summarizes this and other interactions between regions 2.4 and 4.2 of σ and the −10 and −35 boxes, respectively, of bacterial promoters.

These results all suggest the importance of σ regions 2.4 and 4.2 in binding to the −10 and −35 boxes, respectively, of the promoter. The σ-factor even has putative DNA-binding domains in strategic places. But we are left with the perplexing fact that σ by itself does not bind to promoters, or to any other region of DNA. Only when it is bound to the core can σ bind to promoters. How do we resolve this apparent paradox?

Carol Gross and her colleagues suggested that regions 2.4 and 4.2 of σ are capable of binding to promoter regions on their own, but other domains in σ interfere with this binding. In fact, we now know that region 1.1 prevents σ from binding to DNA in the absence of core. Gross and colleagues further suggested that when σ associates with core it changes conformation, unmasking its DNA-binding domains, so it can bind to promoters. To test this hypothesis, these workers made fusion proteins (Chapter 4) containing glutathione-*S*-transferase (GST) and fragments of the *E. coli* σ-factor (region 2.4, or 4.2, or both). (These fusion proteins are easy to purify because of the affinity of GST for glutathione.) Then they showed that a fusion protein containing region 2.4 could bind to a DNA fragment containing a −10 box, but not a −35 box. Furthermore, a fusion protein containing region 4.2 could bind to a DNA fragment containing a −35 box, but not a −10 box.

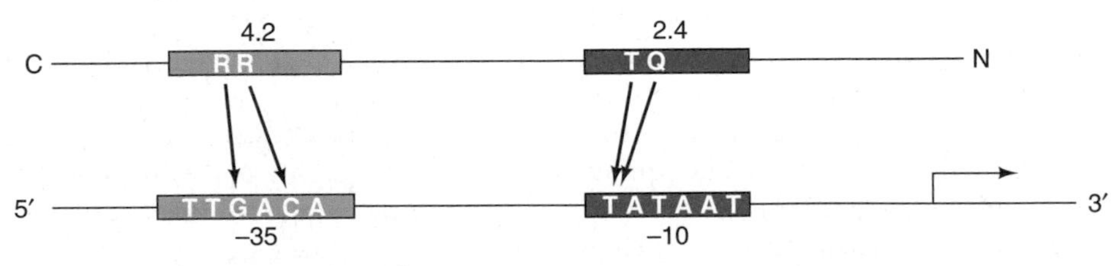

Figure 6.22 Specific interactions between σ regions and promoter regions. Arrows denote interactions revealed by mutation suppression experiments involving σ^{70}. The letters in the upper bar, representing the σ^{70} protein show the amino acid mutated and the arrows point to bases in the promoter that the respective amino acids in σ^{70} appear to contact. The two R's in σ^{70} region 4.2 represent arginines 584 and 588 (the 584th and 588th amino acids in the protein), and these amino acids contact a C and a G, respectively, in the −35 box of the promoter. The Q and T in the σ^{70} 2.4 region represent glutamine 437 and threonine 440, respectively, both of which contact a T in the −10 box of the promoter. Notice that the linear structure of the σ-factor (top) is written with the C-terminus at left, to match the promoter written conventionally, 5′→3′ left to right (bottom). (*Source:* Adapted from Dombroski, A.J., et al., "Polypeptides containing highly conserved regions of transcription initiation factor σ^{70} exhibit specificity of binding to promoter DNA." *Cell* 70:501–12, 1992.)

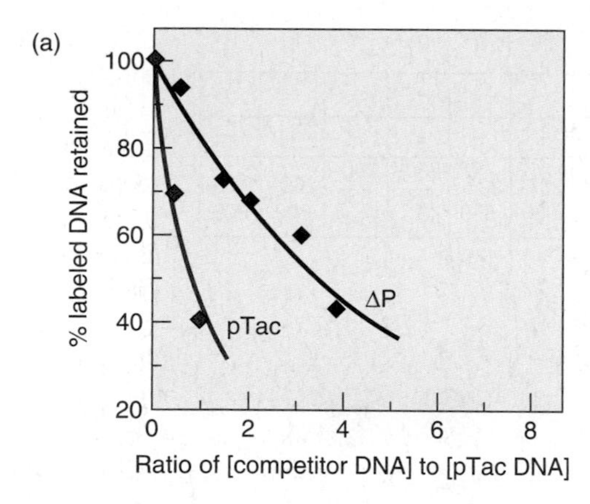

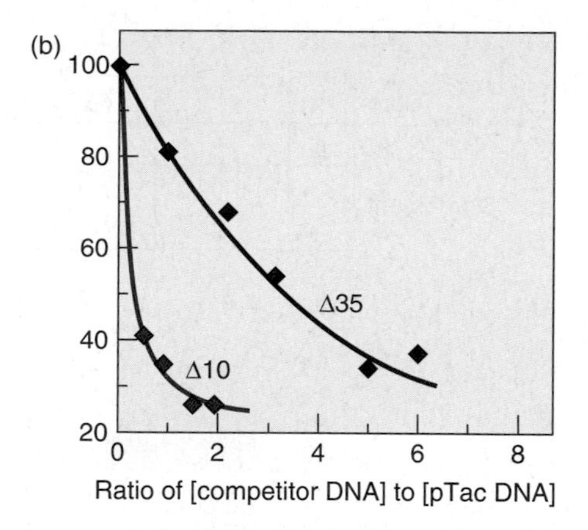

Figure 6.23 Analysis of binding between σ region 4.2 and the promoter −35 box. (a) Recognition of the promoter. Gross and colleagues measured binding between a σ fragment-GST fusion protein and a labeled DNA fragment (pTac) containing the *tac* promoter. The σ fragment in this experiment contained only the 108 amino acids at the C-terminus of the *E. coli* σ, which includes region 4, but not region 2. Gross and coworkers measured binding of the labeled DNA–protein complex to nitrocellulose filters in the presence of competitor DNA containing the *tac* promoter (pTac), or lacking the *tac* promoter (ΔP). Because pTac DNA competes much better than ΔP DNA, they concluded that the fusion protein with region 4 can bind to the *tac* promoter. **(b)** Recognition of the −35 region. Gross and colleagues repeated the experiment but used two different competitor DNAs: One (Δ10) had a *tac* promoter with a 6-bp deletion in the −10 box; the other (Δ35) had a *tac* promoter with a 6-bp deletion in the −35 box. Because deleting the −35 box makes the competitor no better than a DNA with no *tac* promoter at all and removing the −10 box had no effect, it appears that the σ fragment with region 4 binds to the −35 box, but not to the −10 box. (*Source:* Adapted from Dombroski, A.J., et al., "Polypeptides containing highly conserved regions of transcription initiation factor σ^{70} exhibit specificity of binding to promoter DNA." *Cell* 70:501–12, 1992.)

To measure the binding between fusion proteins and promoter elements, Gross and coworkers used a nitrocellulose filter-binding assay. They labeled the target DNA containing one or both promoter elements from the composite *tac* promoter. The *tac* promoter has the −10 box of the *lac* promoter and the −35 box of the *trp* promoter. Then they added a fusion protein to the labeled target DNA in the presence of excess unlabeled competitor DNA and measured the formation of a labeled DNA–protein complex by nitrocellulose binding.

Figure 6.23a shows the results of an experiment in which Gross and colleagues bound a labeled *tac* promoter to a GST–σ-region 4 fusion protein. Because σ-region 4 contains a putative −35 box-binding domain, we expect this fusion protein to bind to DNA containing the *tac* promoter more strongly than to DNA lacking the *tac* promoter. Figure 6.23a demonstrates this is just what happened. Unlabeled DNA containing the *tac* promoter was an excellent competitor, whereas unlabeled DNA missing the *tac* promoter competed relatively weakly. Thus, the GST–σ region 4 protein binds weakly to nonspecific DNA, but strongly to *tac* promoter-containing DNA, as we expect.

Figure 6.23b shows that the binding between the GST–σ region 4 proteins and the promoter involves the −35 box, but not the −10 box. As we can see, a competitor from which the −35 box was deleted competed no better than nonspecific DNA, but a competitor from which the −10 box was deleted competed very well because it still contained the −35 box. Thus, σ region 4 can bind specifically to the −35 box, but not to the −10 box. Similar experiments with a GST–σ region 2 fusion protein showed that this protein can bind specifically to the −10 box, but not the −35 box.

We have seen that the polymerase holoenzyme can recognize promoters and form an open promoter complex by melting a short region of the DNA, approximately between positions −11 and +1. We suspect that σ plays a big role in this process, but we know that σ cannot form an open promoter complex on its own. One feature of open complex formation is binding of polymerase to the nontemplate strand in the −10 region of the promoter. Again, σ cannot do this on its own so, presumably, some part of the core enzyme is required to help σ with this task. Gross and colleagues have posed the question: What part of the core enzyme is required to unmask the part of σ that binds to the nontemplate strand in the −10 region of the promoter?

To answer this question, Gross and colleagues focused on the β′ subunit, which had already been shown to collaborate with σ in binding to the nontemplate strand in the −10 region. They cloned different segments of the β′ subunit, then tested these, together with σ, for ability to bind to radiolabeled single-stranded oligonucleotides corresponding to the template and nontemplate strands in the −10 region of a promoter. They incubated the β′ segments, along with σ, with the labeled DNAs, then subjected the complexes to UV irradiation to cross-link σ to the DNA. Then they performed SDS-PAGE on the cross-linked complexes. If the β′ fragment induced binding between σ and the DNA, then σ would be cross-linked to the labeled DNA and the SDS-PAGE band corresponding to σ would become labeled.

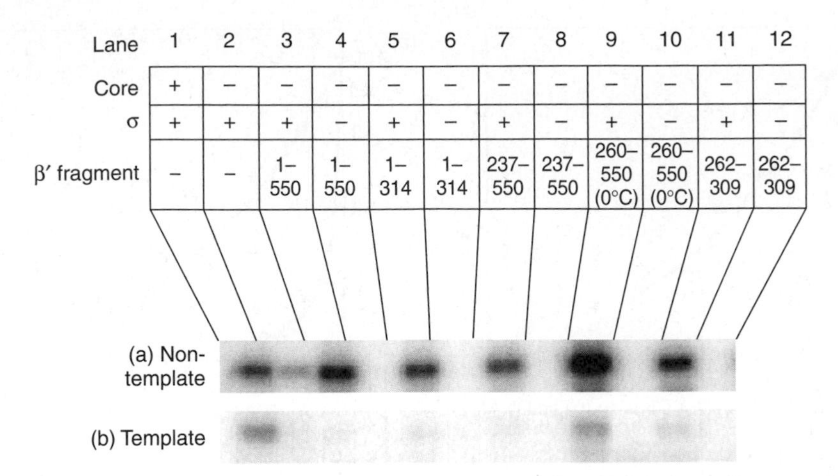

Figure 6.24 Induction of σ binding to the −10 region of a promoter. Gross and colleagues mixed σ plus various fragments of β′, as indicated at top, with labeled oligonucleotides representing either the nontemplate or template stand in the −10 region of the promoter. Then they UV-irradiated the complexes to cross-link any σ-subunit bound to the DNA, subjected the complexes to SDS-PAGE, and performed autoradiography to detect σ bound to labeled DNA. Lane 1 is a positive control with whole core instead of a β′ fragment; lane 2 is a control with no β′ fragment; and all the other even-numbered lanes are negative controls with no protein. The experiments in lanes 9 and 10 were performed at 0°C; all other experiments were performed at room temperature. The autoradiography results are shown for experiments with **(a)** the nontemplate strand and **(b)** the template strand. (*Source:* Reprinted from *Cell* v. 105, Young et al., p. 940 © 2001, with permission from Elsevier Science.)

Figure 6.24 shows that the fragment of β′ containing amino acids 1–550 caused binding between σ and the nontemplate strand DNA (but not the template strand), whereas σ by itself showed little binding. Next, Gross and colleagues used smaller fragments of the 1–550 region to pinpoint the part of β′ that was inducing the binding. All of the fragments illustrated in Figure 6.24 could induce binding, although the 260–550 fragment would work only at low temperature. Strikingly, the very small 262–309 fragment, with only 48 amino acids, could stimulate binding very actively, even at room temperature. Mutations in three amino acids in this region (R275, E295, and A302) were already known to interfere with σ binding to promoters. Accordingly, Gross and colleagues tested these mutations for interference with σ binding to the nontemplate strand in the −10 region. In every case, these mutations caused highly significant interference.

SUMMARY Comparison of the sequences of different σ genes reveals four regions of similarity among a wide variety of σ-factors. Subregions 2.4 and 4.2 are involved in promoter −10 box and −35 box recognition, respectively. The σ-factor by itself cannot bind to DNA, but interaction with core unmasks a DNA-binding region of σ. In particular, the region between amino acids 262 and 309 of β′ stimulates σ binding to the nontemplate strand in the −10 region of the promoter.

The Role of the α-Subunit in UP Element Recognition

As we learned earlier in this chapter, RNA polymerase itself can recognize an upstream promoter element called an UP element. We know that the σ-factor recognizes the core promoter elements, but which polymerase subunit is responsible for recognizing the UP element? Based on the following evidence, it appears to be the α-subunit of the core polymerase.

Richard Gourse and colleagues made *E. coli* strains with mutations in the α-subunit and found that some of these were incapable of responding to the UP element—they gave no more transcription from promoters with UP elements than from those without UP elements. To measure transcription, they placed a wild-type form of the very strong *rrnB* P1 promoter, or a mutant form that was missing its UP element, about 170 bp upstream of an *rrnB* P1 transcription terminator in a cloning vector. They transcribed these constructs with three different RNA polymerases, all of which had been reconstituted from purified subunits: (1) wild-type polymerase with a normal α-subunit; (2) α-235, a polymerase whose α-subunit was missing 94 amino acids from its C-terminus; and (3) R265C, a polymerase whose α-subunit contained a cysteine (C) in place of the normal arginine (R) at position 265. They included a labeled nucleotide to label the RNA, then subjected this RNA to gel electrophoresis, and finally performed autoradiography to visualize the RNA products.

Figure 6.25a depicts the results with wild-type polymerase. The wild-type promoter (lanes 1 and 2) allowed a great deal more transcription than the same promoter with vector DNA substituted for its UP element (lanes 3 and 4), or having its UP element deleted (lanes 5 and 6). Figure 6.25b shows the same experiment with the polymerase with 94

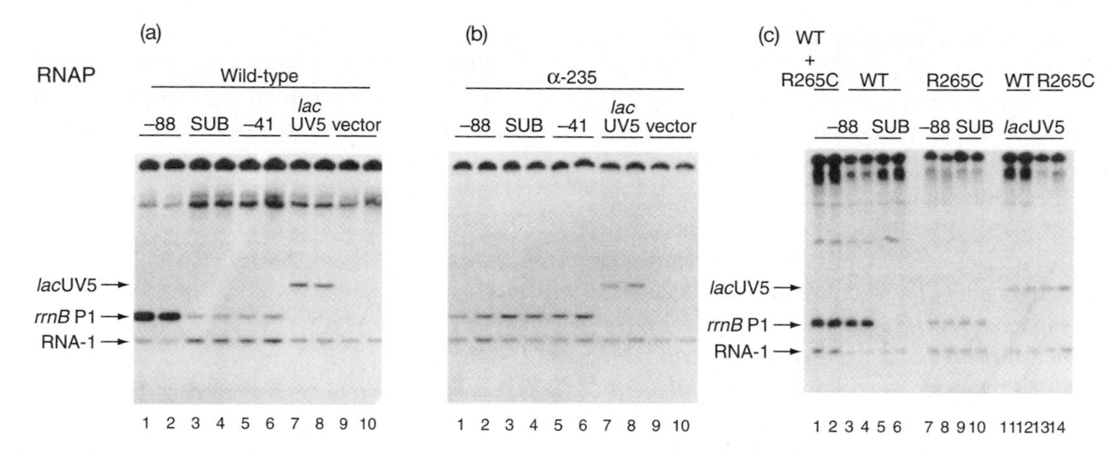

Figure 6.25 Importance of the α-subunit of RNA polymerase in UP element recognition. Gourse and colleagues performed in vitro transcription on plasmids containing the promoters indicated at top. They placed the promoters between 100 and 200 nt upstream of a transcription terminator to produce a transcript of defined size. After the reaction, they subjected the labeled transcripts to gel electrophoresis and detected them by autoradiography. The promoters were as follows: −88 contained wild-type sequence throughout the region between positions −88 and +1; SUB contained an irrelevant sequence instead of the UP element between positions −59 and −41; −41 lacked the UP element upstream of position −41 and had vector sequence instead; *lac*UV5 is a *lac* promoter without an UP element; vector indicates a plasmid with no promoter inserted. The positions of transcripts from the *rrnB* P1 and *lac*UV5 promoters, as well as an RNA (RNA-1) transcribed from the plasmid's origin of replication, are indicated at left. RNAP at top indicates the RNA polymerase used, as follows: **(a)** Wild-type polymerase used throughout. **(b)** α-235 polymerase (missing 94 C-terminal amino acids of the α-subunit) used throughout. **(c)** Wild-type (WT) polymerase or R265C polymerase (with cysteine substituted for arginine 265) used, as indicated. (*Source:* Ross et al., A third recognition element in bacterial promoters: DNA binding by the alpha subunit of RNA polymerase. *Science* 262 (26 Nov 1993) f. 2, p. 1408. © AAAS.)

C-terminal amino acids missing from its α-subunit. We see that this polymerase is just as active as the wild-type polymerase in transcribing a gene with a core promoter (compare panels a and b, lanes 3–6). However, in contrast to the wild-type enzyme, this mutant polymerase did not distinguish between promoters with and without an UP element (compare lanes 1 and 2 with lanes 3–6). The UP element provided no benefit at all. Thus, it appears that the C-terminal portion of the α-subunit enables the polymerase to respond to an UP element.

Figure 6.25c demonstrates that the polymerase with a cysteine in place of an arginine at position 265 of the α-subunit (R265C) does not respond to the UP element (lanes 7–10 all show modest transcription). Thus, this single amino acid change appears to destroy the ability of the α-subunit to recognize the UP element. This phenomenon was not an artifact caused by an inhibitor in the R265C polymerase preparation because a mixture of R265C and the wild-type polymerase still responded to the UP element (lanes 1–4 all show strong transcription).

To test the hypothesis that the α-subunit actually contacts the UP element, Gourse and coworkers performed DNase footprinting experiments (Chapter 5) with DNA containing the *rrnB* P1 promoter and either wild-type or mutant RNA polymerase. They found that the wild-type polymerase made a footprint in the core promoter and the UP element, but that the mutant polymerase lacking the C-terminal domain of the α-subunit made a footprint in the core promoter only (data not shown). This indicates that the α-subunit C-terminal domain is required for interaction between polymerase and UP elements. Further evidence for this hypothesis came from an experiment in which Gourse and coworkers used purified α-subunit dimers to footprint the UP element of the *rrnB* P1 promoter. Figure 6.26 shows the results—a clear footprint in the UP element caused by the α-subunit dimer all by itself.

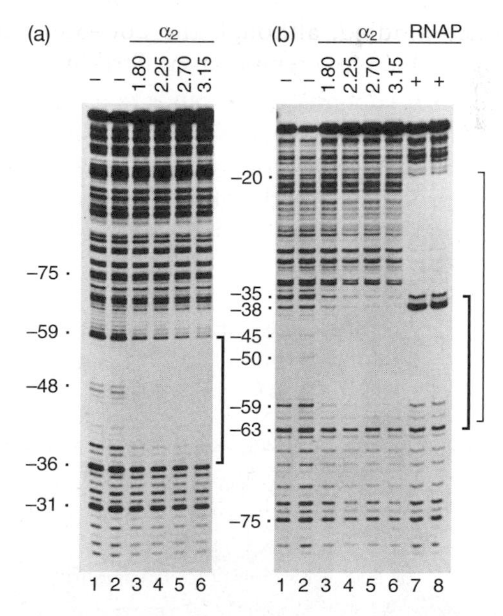

Figure 6.26 Footprinting the UP element with pure α-subunit. Gourse and colleagues performed DNase footprinting with end-labeled template strand **(a)** or nontemplate strand **(b)** from the *rrnB* P1 promoter. They used the amounts listed at top (in micrograms) of purified α-dimers, or 10 nM RNA polymerase holoenzyme (RNAP). The bold brackets indicate the footprints in the UP element caused by the α-subunit, and the thin bracket indicates the footprint caused by the holoenzyme. (*Source:* Ross et al., A third recognition element in bacterial promoter: DNA binding by the α-subunit of RNA ploymerase. *Science* 262 (26 Nov 1993) f. 5, p. 1408. © AAAS.)

Richard Gourse, Richard Ebright, and their colleagues used **limited proteolysis** analysis to show that the α-subunit N-terminal and C-terminal domains (the α-NTD and α-CTD, respectively) fold independently to form two domains that are tethered together by a flexible linker. A protein **domain** is a part of a protein that folds independently to form a defined structure. Because of their folding, domains tend to resist proteolysis, so limited digestion with a proteolytic enzyme will attack unstructured elements between domains and leave the domains themselves alone. When Gourse and Ebright and collaborators performed limited proteolysis on the *E. coli* RNA polymerase α-subunit, they released a polypeptide of about 28 kD, and three polypeptides of about 8 kD. The sequences of the ends of these products showed that the 28-kD polypeptide contained amino acids 8–241, whereas the three small polypeptides contained amino acids 242–329, 245–329, and 249–329. This suggested that the α-subunit folds into two domains: a large N-terminal domain encompassing (approximately) amino acids 8–241, and a small C-terminal domain including (approximately) amino acids 249–329.

Furthermore, these two domains appear to be joined by an unstructured linker that can be cleaved in at least three places by the protease used in this experiment (Glu-C). This linker seems at first glance to include amino acids 242–248. Because Glu-C requires three unstructured amino acids on either side of the bond that is cleaved, however, the linker is longer than it appears at first. In fact, it must be at least 13 amino acids long (residues 239–251).

These experiments suggest a model such as the one presented in Figure 6.27. RNA polymerase binds to a core promoter via its σ-factor, with no help from the C-terminal domains of its α-subunits, but it binds to a promoter with an UP element using σ plus the α-subunit C-terminal domains. This allows very strong interaction between polymerase and promoter and therefore produces a high level of transcription.

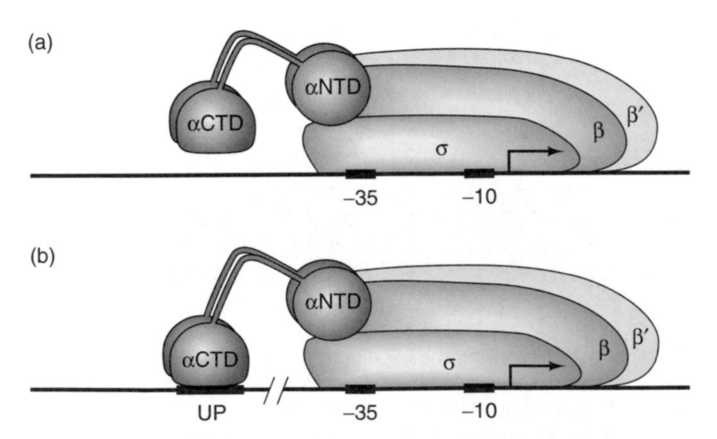

Figure 6.27 Model for the function of the C-terminal domain (CTD) of the polymerase α-subunit. (a) In a core promoter, the α-CTDs are not used, but **(b)** in a promoter with an UP element, the α-CTDs contact the UP element. Notice that two α-subunits are depicted: one behind the other.

SUMMARY The RNA polymerase α-subunit has an independently folded C-terminal domain that can recognize and bind to a promoter's UP element. This allows very tight binding between polymerase and promoter.

6.4 Elongation

After initiation of transcription is accomplished, the core continues to elongate the RNA, adding one nucleotide after another to the growing RNA chain. In this section we will explore this elongation process.

Core Polymerase Functions in Elongation

So far we have been focusing on the role of σ because of the importance of this factor in determining the specificity of initiation. However, the core polymerase contains the RNA synthesizing machinery, so the core is the central player in elongation. In this section we will see evidence that the β- and β′-subunits are involved in phosphodiester bond formation, that these subunits also participate in DNA binding, and that the α-subunit has several activities, including assembly of the core polymerase.

The Role of β in Phosphodiester Bond Formation Walter Zillig was the first to investigate the individual core subunits, in 1970. He began by separating the *E. coli* core polymerase into its three component polypeptides and then combining them again to reconstitute an active enzyme. The separation procedure worked as follows: Alfred Heil and Zillig electrophoresed the core enzyme on cellulose acetate in the presence of urea. Like SDS, urea is a denaturing agent that can separate the individual polypeptides in a complex protein. Unlike SDS, however, urea is a mild denaturant that is relatively easy to remove. Thus, it is easier to renature a urea-denatured polypeptide than an SDS-denatured one. After electrophoresis was complete, Heil and Zillig cut out the strips of cellulose acetate containing the polymerase subunits and spun them in a centrifuge to drive the buffer, along with the protein, out of the cellulose acetate. This gave them all three separated polypeptides, which they electrophoresed individually to demonstrate their purity (Figure 6.28).

Once they had separated the subunits, they recombined them to form active enzyme, a process that worked best in the presence of σ. Using this separation–reconstitution system, Heil and Zillig could mix and match the components from different sources to answer questions about their functions. For example, recall that the core polymerase determines sensitivity or resistance to the antibiotic rifampicin, and that rifampicin blocks transcription initiation.

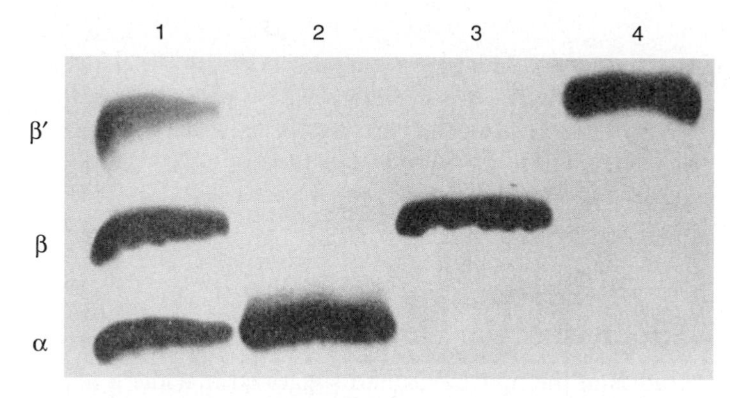

Figure 6.28 Purification of the individual subunits of *E. coli* RNA polymerase. Heil and Zillig subjected the *E. coli* core polymerase to urea gel electrophoresis on cellulose acetate, then collected the separated polypeptides. Lane 1, core polymerase after electrophoresis; lane 2, purified α; lane 3, purified β; lane 4, purified β′. (*Source:* Heil, A. and Zillig, W. Reconstitution of bacterial DNA-dependent RNA-polymerase from isolated subunits as a tool for the elucidation of the role of the subunits in transcription. *FEBS Letters* 11 (Dec 1970) p. 166, f. 1.)

Separation and reconstitution of the core allowed Heil and Zillig to ask which core subunit confers this antibiotic sensitivity or resistance. When they recombined the α-, β′-, and σ-subunits from a rifampicin-sensitive bacterium with the β-subunit from a rifampicin-resistant bacterium, the resulting polymerase was antibiotic-resistant (Figure 6.29). Conversely, when the β-subunit came from an antibiotic-sensitive bacterium, the reconstituted enzyme was antibiotic-sensitive, regardless of the origin of the other subunits. Thus, the β-subunit is obviously the determinant of rifampicin sensitivity or resistance.

Another antibiotic, known as *streptolydigin,* blocks RNA chain elongation. By the same separation and reconstitution strategy used for rifampicin, Heil and Zillig showed that the β-subunit also governed streptolydigin resistance or sensitivity. At first this seems paradoxical. How can the same core subunit be involved in both initiation and elongation? The answer, which we will discuss in detail later in this chapter, is that rifampicin actually blocks *early* elongation, preventing the RNA from growing more than 2–3 nucleotides long. Thus, strictly speaking, it blocks initiation, because initiation is not complete until the RNA is up to 10 nucleotides long, but its effect is really on the elongation that is part of initiation.

In 1987, M. A. Grachev and colleagues provided more evidence for the notion that β plays a role in elongation, using a technique called **affinity labeling.** The idea behind this technique is to label an enzyme with a derivative of a normal substrate that can be cross-linked to protein. In this way, one can use the affinity reagent to seek out and then tag the active site of the enzyme. Finally, one can dissociate the enzyme to see which subunit the tag is attached to. Grachev and coworkers used 14 different affinity reagents, all ATP or GTP analogs. One of these, which was the first in the series, and therefore called I, has the structure shown in Figure 6.30a. When it was added to RNA polymerase, it went to the active site, as an ATP that is initiating transcription would normally do, and then formed a covalent bond with an amino group at the active site according to the reaction in Figure 6.30b.

In principle, these investigators could have labeled the affinity reagent itself and proceeded from there. However, they recognized a pitfall in that simple strategy: The affinity reagent could bind to other amino groups on the enzyme surface in addition to the one(s) in the active site. To circumvent this problem, they used an unlabeled affinity reagent, followed by a radioactive nucleotide ([α-^{32}P]UTP or CTP) that would form a phosphodiester bond with the affinity reagent in the active site and therefore label that site and no others on the enzyme. Finally, they dissociated the labeled enzyme and subjected the subunits to SDS-PAGE.

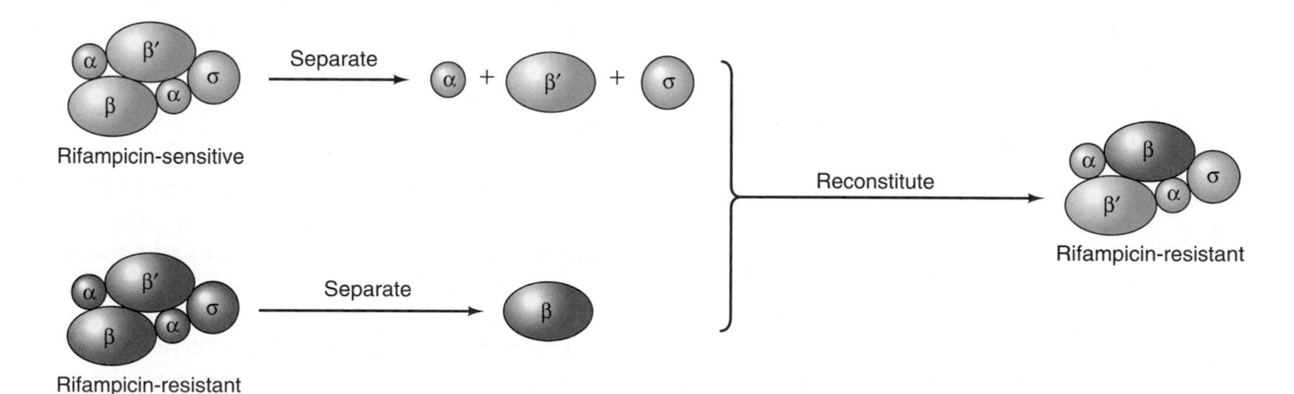

Figure 6.29 Separation and reconstitution of RNA polymerase to locate the determinant of antibiotic resistance. Start with RNA polymerases from rifampicin-sensitive and -resistant *E. coli* cells, separate them into their component polypeptides, and recombine them in various combinations to reconstitute the active enzyme. In this case, the α-, β′-, and σ-subunits came from the rifampicin-sensitive polymerase (blue), and the β-subunit came from the antibiotic-resistant enzyme (red). The reconstituted polymerase is rifampicin-resistant, which shows that the β-subunit determines sensitivity or resistance to this antibiotic.

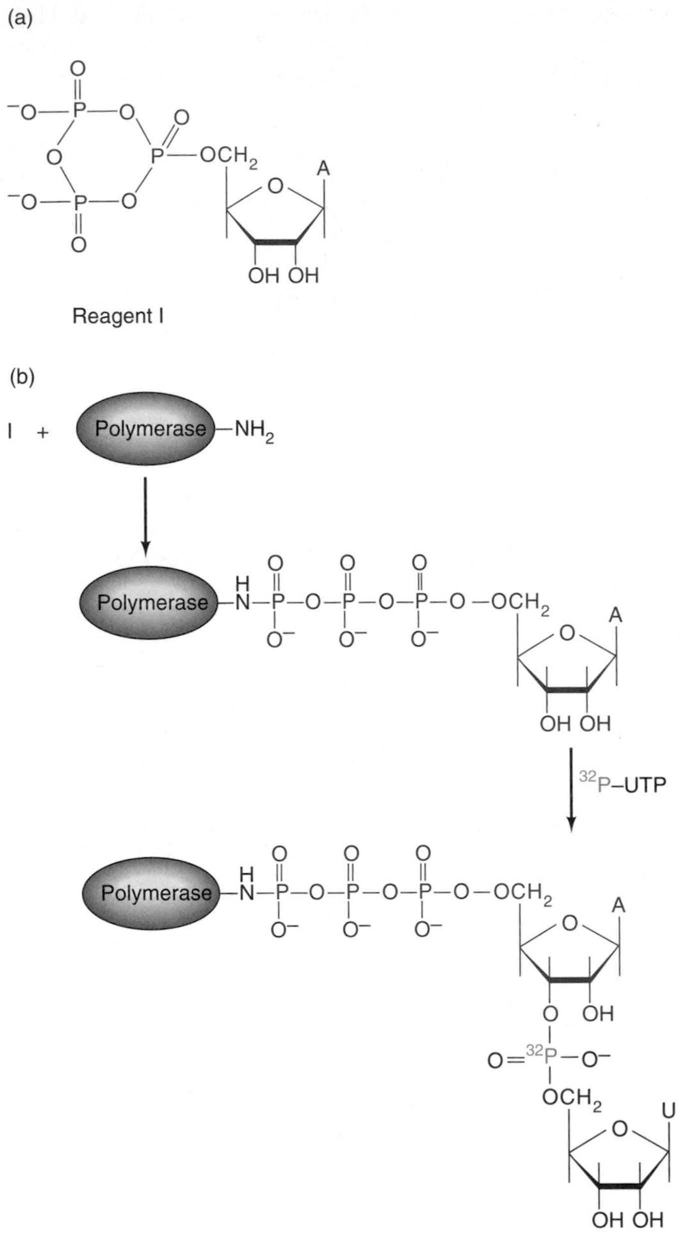

Figure 6.30 Affinity labeling RNA polymerase at its active site. **(a)** Structure of one of the affinity reagents (I), an ATP analog. **(b)** The affinity-labeling reactions. First, add reagent I to RNA polymerase. The reagent binds covalently to amino groups at the active site (and perhaps elsewhere). Next, add radioactive UTP, which forms a phosphodiester bond (blue) with the enzyme-bound reagent I. This reaction should occur only at the active site, so only that site becomes radioactively labeled.

The results are presented in Figure 6.31. Obviously, the β-subunit is the only core subunit labeled by any of the affinity reagents, suggesting that this subunit is at or very near the site where phosphodiester bond formation occurs. In some cases, we also see some labeling of σ, suggesting that it too may lie near the catalytic center.

SUMMARY The core subunit β lies near the active site of the RNA polymerase where phosphodiester bonds are formed. The σ-factor may also be near the nucleotide-binding site, at least during the initiation phase.

Structure of the Elongation Complex

Studies in the mid-1990s had suggested that the β and β′ subunits are involved in DNA binding. In this section, we will see how well these predictions have been borne out by structural studies. We will also consider the topology of elongation: How does the polymerase deal with the problems of unwinding and rewinding its template, and of moving along its twisted (helical) template without twisting its RNA product around the template?

The RNA–DNA Hybrid Up to this point we have been assuming that the RNA product forms an RNA–DNA hybrid with the DNA template strand for a few bases before peeling off and exiting from the polymerase. But the length of this hybrid has been controversial, with estimates ranging from 3–12 bp, and some investigators even doubted whether it existed. But Nudler and Goldfarb and their colleagues applied a transcript walking technique, together with RNA–DNA cross-linking, to prove that an RNA–DNA hybrid really does occur within the elongation complex, and that this hybrid is 8–9 bp long.

The transcript walking technique works like this: Nudler and colleagues used gene cloning techniques described in Chapter 4 to engineer an RNA polymerase with six extra histidines at the C-terminus of the β–subunit. This string of histidines, because of its affinity for divalent metals such as nickel, allowed them to tether the polymerase to a nickel resin so they could change substrates rapidly by washing the resin, with the polymerase stably attached, and then adding fresh reagents. Accordingly, by adding a subset of nucleotides (e.g., ATP, CTP, and GTP, but no UTP), they could "walk" the polymerase to a particular position on the template (where the first UTP is required, in the present case). Then they could wash away the first set of nucleotides and add a second subset to walk the polymerase to a defined position further downstream.

These workers incorporated a UMP derivative (U•) at either position 21 or 45 with respect to the 5′-end of a ^{32}P-labeled nascent RNA. U• is normally unreactive, but in the presence of $NaBH_4$ it becomes capable of cross-linking to a base-paired base, as shown in Figure 6.32a. Actually, U• can reach to a purine adjacent to the base-paired A in the DNA strand, but this experiment was

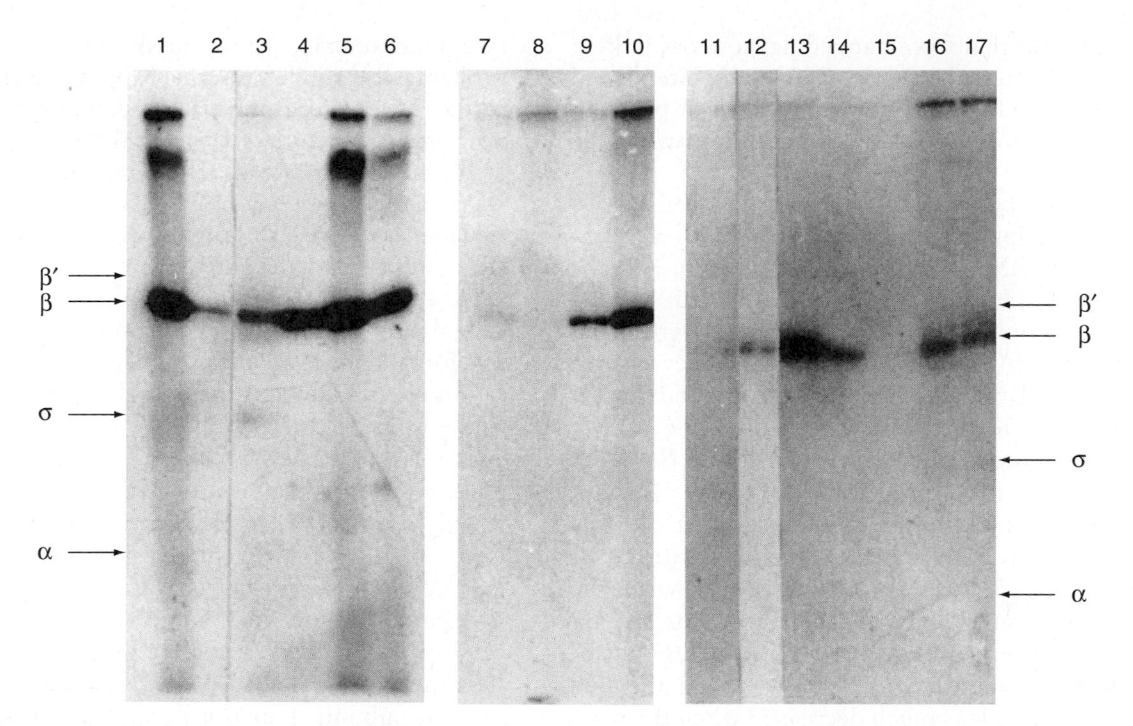

Figure 6.31 The β-subunit is at or near the active site where phosphodiester bonds are formed. Grachev and colleagues labeled the active site of *E. coli* RNA polymerase as described in Figure 6.30, then separated the polymerase subunits by electrophoresis to identify the subunits that compose the active site. Each lane represents labeling with a different nucleotide-affinity reagent plus radioactive UTP, except lanes 5 and 6, which resulted from using the same affinity reagent, but either radioactive UTP (lane 5) or CTP (lane 6). The autoradiograph of the separated subunits demonstrates labeling of the β-subunit with most of the reagents. In a few cases, σ was also faintly labeled. Thus, the β-subunit appears to be at or near the phosphodiester bond-forming active site. (*Source:* Grachev et al., Studies on the functional topography of *Escherichia coli* RNA polymerase. *European Journal of Biochemistry* 163 (16 Dec 1987) p. 117, f. 2.)

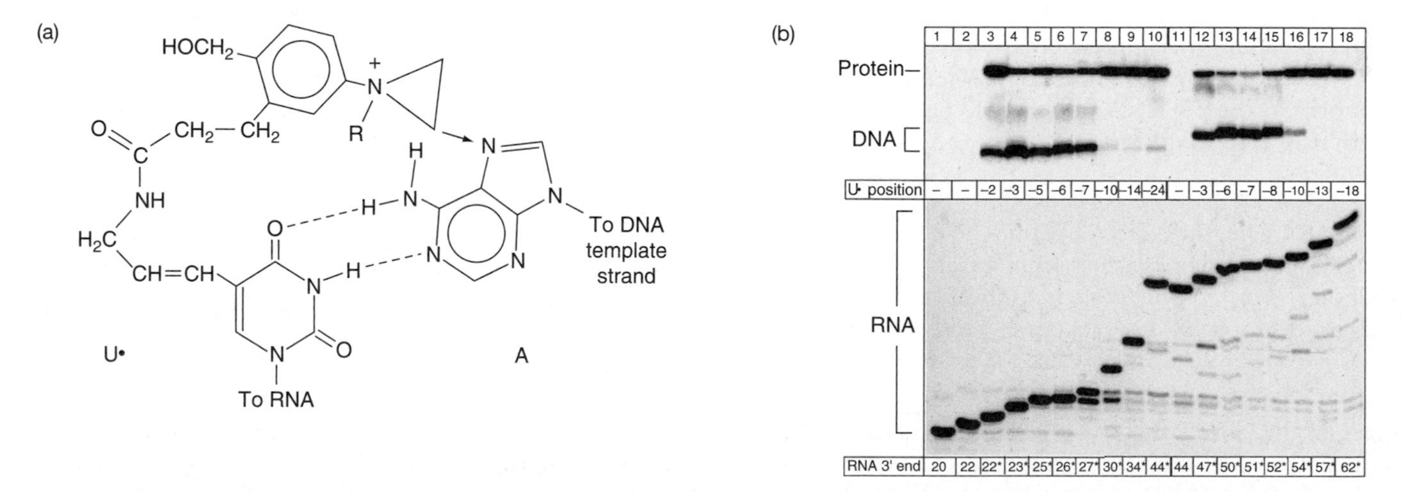

Figure 6.32 RNA–DNA and RNA–protein cross-linking in elongation complexes. (a) Structure of the cross-linking reagent U• base-paired with an A in the DNA template strand. The reagent is in position to form a covalent bond with the DNA as shown by the arrow. **(b)** Results of cross-linking. Nudler, Goldfarb, and colleagues incorporated U• at position 21 or 45 of a [^{32}P]nascent RNA in an elongation complex. Then they walked the U• to various positions between −2 and −24 with respect to the 3′-end (position −1) of the nascent RNA. Then they cross-linked the RNA to the DNA template (or the protein in the RNA polymerase). They then electrophoresed the DNA and protein in one gel (top) and the free RNA transcripts in another (bottom) and autoradiographed the gels. Lanes 1, 2, and 11 are negative controls in which the RNA contained no U•. Lanes 3−10 contained products from reactions in which the U• was in position 21; lanes 12–18 contained products from reactions in which the U• was in position 45 of the nascent RNA. Asterisks at bottom denote the presence of U• in the RNA. Cross-linking to DNA was prevalent only when U• was between positions −2 and −8. (*Sources:* (a) Reprinted from *Cell* 89, Nudler, E. et al. The RNA-DNA hybrid maintains the register of transcription by preventing backtracking of RNA polymerase fig.1, p. 34 © 1997 from Elsevier (b) Nudler, E. et al. The RNA–DNA hybrid maintains the register of transcription by preventing backtracking of RNA polymerase. *Cell* 89 (1997) f. 1, p. 34. Reprinted by permission of Elsevier Science.)

designed to prevent that from happening. So cross-linking could occur only to an A in the DNA template strand that was base-paired to the U• base in the RNA product. If no base-pairing occurred, no cross-linking would be possible.

Nudler, Goldfarb, and their colleagues walked the U• base in the transcript to various positions with respect to the 3′-end of the RNA, beginning with position −2 (the nucleotide next to the 3′-end, which is numbered −1) and extending to position −44. Then they tried to cross-link the RNA to the DNA template strand. Finally, they electrophoresed both the DNA and protein in one gel, and just the RNA in another. Note that the RNA will always be labeled, but the DNA or protein will be labeled only if the RNA has been cross-linked to them.

Figure 6.32b shows the results. The DNA was strongly labeled if the U• base was in position −2 through position −8, but only weakly labeled when the U• base was in position −10 and beyond. Thus, the U• base was base-paired to its A partner in the DNA template strand only when it was in position −2 through −8, but base-pairing was much decreased when the reactive base was in position −10. So the RNA–DNA hybrid extends from position −1 to position −8, or perhaps −9, but no farther. (The nucleotide at the very 3′-end of the RNA, at position −1, must be base-paired to the template to be incorporated correctly.) This conclusion was reinforced by the protein labeling results. Protein in the RNA polymerase became more strongly labeled when the U• was not within the hybrid region (positions −1 through −8). This presumably reflects the fact that the reactive group was more accessible to the protein when it was not base-paired to the DNA template. More recent work on the T7 RNA polymerase has indicated a hybrid that is 8 bp long.

SUMMARY The RNA–DNA hybrid within the *E. coli* elongation complex extends from position −1 to position −8 or −9 with respect to the 3′-end of the nascent RNA. The T7 hybrid appears to be 8 bp long.

Structure of the Core Polymerase To get the clearest picture of the structure of the elongation complex, we need to know the structure of the core polymerase. X-ray crystallography would give the best resolution, but it requires three-dimensional crystals and, so far, no one has succeeded in preparing three-dimensional crystals of the *E. coli* polymerase. However, in 1999 Seth Darst and colleagues crystallized the core polymerase from another bacterium, *Thermus aquaticus,* and obtained a crystal structure to a resolution of 3.3 Å. This structure is very similar in overall shape to the lower-resolution structure of the *E. coli* core polymerase obtained by electron microscopy of two-dimensional crystals, so the detailed structures are probably also similar. In other words, the crystal structure of the *T. aquaticus* polymerase is our best window right now on the structure of a bacterial polymerase. As we look at this and other crystal structures throughout this book, we need to remember a principle we will discuss more fully in Chapters 9 and 10: Proteins do not have just one static structure. Instead, they are dynamic molecules that can assume a wide range of conformations. The one we trap in a crystal may not be the one (or more than one) that the active form of the protein assumes in vivo.

Figure 6.33 depicts the overall shape of the enzyme in three different orientations. We notice first of all that it resembles an open crab claw. The four subunits (β, β′, and two α) are shown in different colors so we can distinguish them. This coloring reveals that half of the claw is composed primarily of the β-subunit, and the other half is composed primarily of the β′-subunit. The two α-subunits lie at the "hinge" of the claw, with one of them (αI, yellow) associated with the β-subunit, and the other (αII, green) associated with the β′-subunit. The small ω-subunit is at the bottom, wrapped around the C-terminus of β′.

Figure 6.34 shows the **catalytic center** of the core polymerase. We see that the enzyme contains a channel, about 27 Å wide, between the two parts of the claw, and the template DNA presumably lies in this channel. The catalytic center of the enzyme is marked by the Mg^{2+} ion, represented here by a pink sphere. Three pieces of evidence place the Mg^{2+} at the catalytic center. First, an invariant string of amino acids (NADFDGD) occurs in the β′-subunit from all bacteria examined so far, and it contains three aspartate residues (D) suspected of chelating a Mg^{2+} ion. Second, mutations in any of these Asp residues are lethal. They create an enzyme that can form an open-promoter complex at a promoter, but is devoid of catalytic activity. Thus, these Asp residues are essential for catalytic activity, but not for tight binding to DNA. Finally, as Figure 6.34 demonstrates, the crystal structure of the *T. aquaticus* core polymerase shows that the side chains of the three Asp residues (red) are indeed coordinated to a Mg^{2+} ion. Thus, the three Asp residues and a Mg^{2+} ion are at the catalytic center of the enzyme.

Figure 6.34 also identifies a rifampicin-binding site in the part of the β-subunit that forms the ceiling of the channel through the enzyme. The amino acids whose alterations cause rifampicin resistance are tagged with purple dots. Clearly, these amino acids are tightly clustered in the three-dimensional structure, presumably at the site of rifampicin binding. We also know that rifampicin allows RNA synthesis to begin, but blocks elongation of the RNA chain beyond just a few nucleotides. On the other hand, the antibiotic has no effect on elongation once promoter clearance has occurred.

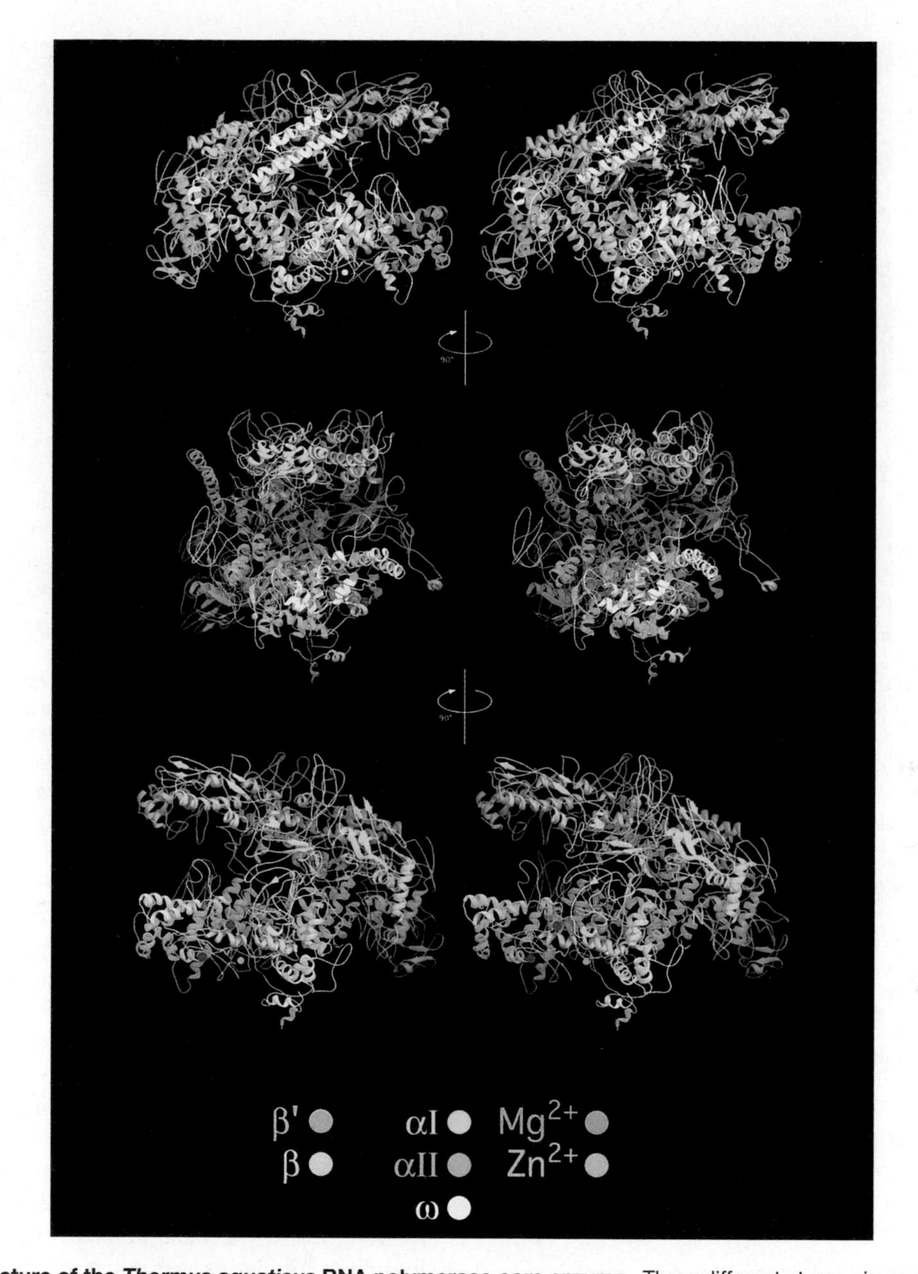

Figure 6.33 Crystal structure of the *Thermus aquaticus* RNA polymerase core enzyme. Three different stereo views are shown, differing by 90-degree rotations. The subunits and metal ions in the enzyme are color-coded as indicated at the bottom. The metal ions are depicted as small colored spheres. The larger red dots denote unstructured regions of the β- and β′-subunits that are missing from these diagrams.
(*Source:* Zhang, G. et al., Crystal structure of *Thermus aquaticus* core RNA polymerase at 3.3 Å resolution. *Cell* 98 (1999) 811–24. Reprinted by permission of Elsevier Science.)

How can we interpret the location of the rifampicin-binding site in terms of the antibiotic's activity? One hypothesis is that rifampicin bound in the channel blocks the exit through which the growing RNA should pass, and thus prevents growth of a short RNA. Once an RNA reaches a certain length, it might block access to the rifampicin-binding site, or at least prevent effective binding of the antibiotic.

Darst and colleagues validated this hypothesis by determining the crystal structure of the *T. aquaticus* polymerase core complexed with rifampicin. The antibiotic lies in the predicted site in such a way that it would block the exit of the elongating transcript when the RNA reaches a length of 2 or 3 nt.

SUMMARY X-ray crystallography on the *Thermus aquaticus* RNA polymerase core has revealed an enzyme shaped like a crab claw designed to grasp DNA. A channel through the enzyme includes the catalytic center (a Mg^{2+} ion coordinated by three Asp residues), and the rifampicin-binding site.

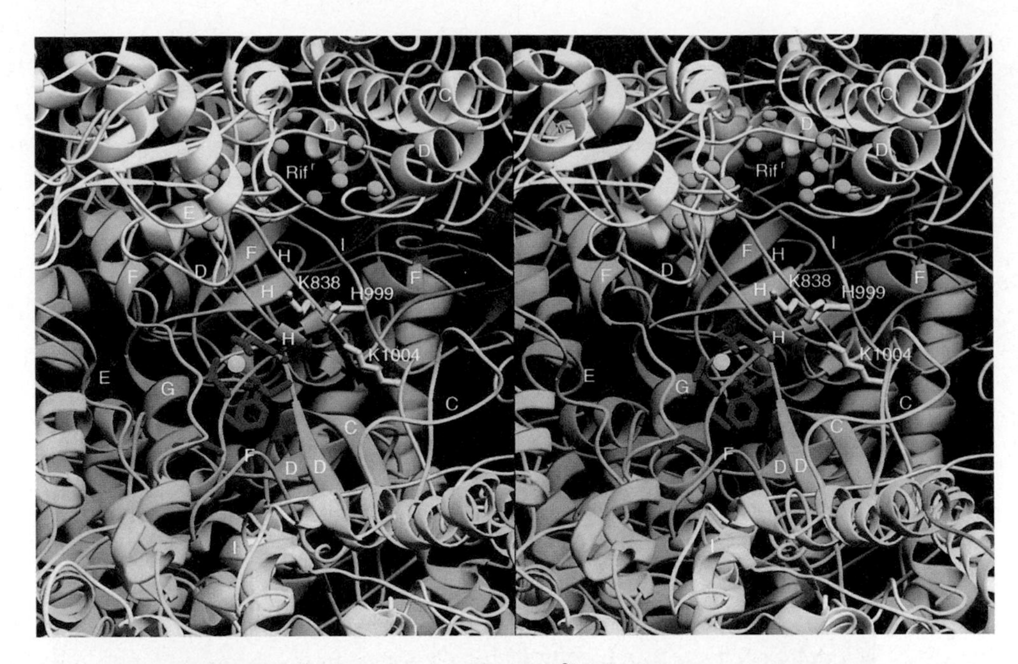

Figure 6.34 Stereo view of the catalytic center of the core polymerase. The Mg^{2+} ion is shown as a pink sphere, coordinated by three aspartate side chains (red) in this stereo image. The amino acids involved in rifampicin resistance are denoted by purple spheres at the top of the channel, surrounding the presumed rifampicin-binding site, or Rif pocket, labeled Rif^r. The colors of the polymerase subunits are as in Figure 6.33 (β′, pink; β, turquoise; α's yellow and green). Note that the two panels of this figure are the two halves of the stereo image. (*Source:* Zhang G. et al., "Crystal structure of *Thermus aquaticus* core RNA polymerase at 3.3 Å resolution." *Cell* 98 (1999) 811–24. Reprinted by permission of Elsevier and Green Science.)

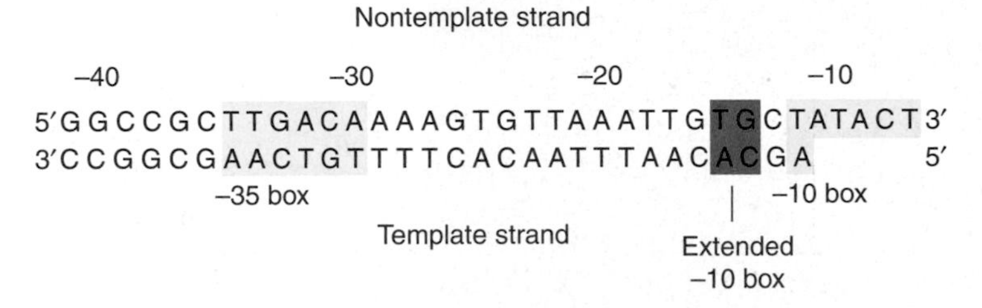

Figure 6.35 Structure of the DNA used to form the RF complex. The −10 and −35 boxes are shaded yellow, and an extended −10 element is shaded red. Bases −11 through −7 are in single-stranded form, as they would be in an open promoter complex.

Structure of the Holoenzyme–DNA Complex To generate a homogeneous holoenzyme–DNA complex, Darst and colleagues bound the *T. aquaticus* holoenzyme to the "fork-junction" DNA pictured in Figure 6.35. This DNA is mostly double-stranded, including the −35 box, but has a single-stranded projection on the nontemplate strand in the −10 box region, beginning at position −11. This simulates the character of the promoter in the open promoter complex, and locks the complex into a form (RF, where F stands for "fork junction") resembling RP_O.

Figure 6.36a shows an overall view of the holoenzyme–promoter complex. The first thing to notice is that the DNA stretches across the top of the polymerase in this view—where the σ-subunit is located. In fact, all of the specific DNA–protein interactions involve σ, not the core. Considering the importance of σ in initiation, that is not surprising.

Looking more closely (Figure 6.36b) we can see that the structure corroborates several features already inferred from biochemical and genetic experiments. First of all, as we saw earlier in this chapter, σ region 2.4 is implicated in recognizing the −10 box of the promoter. In particular, mutations in Gln 437 and Thr 440 of *E. coli* σ^{70} can suppress mutations in position −12 of the promoter, suggesting an interaction between these two amino acids and the base at position −12 (recall Figure 6.22). Gln 437 and Thr 440 in *E. coli* σ^{70} correspond to Gln 260 and Asn 263 of *T. aquaticus* σ^{A}, so we would expect these two amino acids to be close to the base at position −12 in the promoter. Figure 6.36b bears out part of this prediction. Gln 260 (Q260, green) is indeed close enough to contact base −12. Asn 263 (N263, also colored green) is too far away to make contact in this structure, but a minor movement, which could easily occur in vivo, would bring it close enough.

Three highly conserved aromatic residues in *E. coli* σ^{70} (corresponding to Phe 248 (F248), Tyr 253 (Y253), and Trp 256 (W256) of *T. aquaticus* σ^{A}) have been implicated in promoter melting. These amino acids presumably bind the nontemplate strand in the −10 box in the open promoter complex. These amino acids (colored yellow-green in Figure 6.36b) are indeed in position to interact with the single-stranded nontemplate strand in the RF complex. In fact, Trp 256 is neatly positioned to stack with base pair 12, which is the last base pair before the melted region of the −10 box. In this way, Trp 256 would substitute for a base pair in position −11 and help melt that base pair.

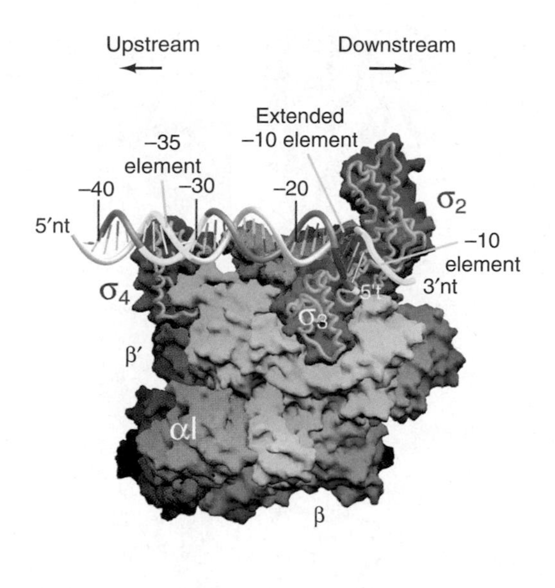

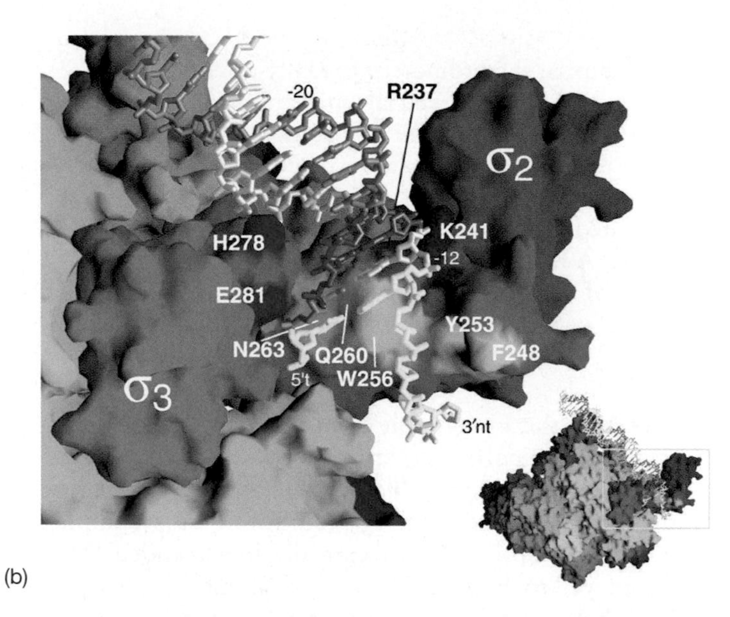

Figure 6.36 Structure of the RF complex. (a) The whole complex. The various subunits are color coded as follows: β, turquoise; β′, brown; α, gray; regions of σ (σ_2–σ_4), tan and orange (σ_1 is not included in this crystal structure). The DNA is shown as a twisted ladder. The surface of σ is rendered partially transparent to reveal the path of the α-carbon backbone. **(b)** Contacts between the holoenzyme and downstream DNA. The σ_2 and σ_3 domains are colored as in (a), except for residues that have been implicated by genetic studies in downstream promoter binding. These are: extended −10 box recognition, red; −10 box recognition, green; −10 box melting and nontemplate strand binding, yellow-green; and invariant basic residues implicated in DNA binding, blue. The −10 box DNA is yellow and the extended −10 box DNA is red. The 3′-end of the nontemplate strand is denoted 3′nt. Specific amino acid side chains that are important in DNA binding are labeled. The box in the small structure at lower right shows the position of the magnified structure within the RF complex. *(Source:* Murakami et al., *Science* 296: (a), p. 1287; (b), p. 1288. Copyright 2002 by the AAAS.)

Two invariant basic residues in σ regions 2.2 and 2.3 (Arg 237 [R237] and Lys 241 [K241]) are known to participate in DNA binding. Figure 6.36b shows why: These two residues (colored blue in the figure) are well positioned to bind to the acidic DNA backbone by electrostatic interaction. These interactions are probably not sequence-specific.

Previous studies implicated region 3 of σ in DNA binding, in particular binding to the extended (upstream) −10 box. Specifically, Glu 281 (E281) was found to be important in recognizing the extended −10 box, while His 278 (H278) was implicated in more general DNA-binding in this region. The structure in Figure 6.36b is consistent with those findings: Both Glu 281 and His 278 (red shading on σ region 3) are exposed on an α-helix, and face the major groove of the extended −10 box (red DNA). Glu 281 is probably close enough to contact a thymine at position −13, and His 278 is close enough to the extended −10 box that it could interact nonspecifically with the phosphodiester bond linking the nontemplate strand residues −17 and −18.

We saw earlier in this chapter that specific residues in σ region 4.2 are instrumental in binding to the −35 box of the promoter. But, surprisingly, the RF structure does not confirm these findings. In particular, the −35 box seems about 6 Å out of position relative to $\sigma_{4.2}$, and the DNA is straight instead of bending to make the necessary interactions. Because the evidence for these −35 box–$\sigma_{4.2}$ interactions is so strong, Darst and colleagues needed to explain why their crystal structure does not allow them. They concluded that the −35 box DNA in the RF structure is pushed out of its normal position relative to $\sigma_{4.2}$ by crystal packing forces—a reminder that the shape a molecule or a complex assumes in a crystal is not necessarily the same as its shape in vivo, and indeed that proteins are dynamic molecules that can change shape as they do their jobs.

The studies of Darst and colleagues, and others, have revealed only one Mg^{2+} ion at the active site. But all DNA and RNA polymerases are thought to use a mechanism that requires two Mg^{2+} ions. In accord with this mechanism, Dmitry Vassylyev and colleagues have determined the crystal structure of the *T. thermophilus* polymerase at 2.6 Å resolution. Their asymmetric crystals contained two polymerases, one with one Mg^{2+} ion, and one with two. The latter is probably the form of the enzyme that takes part in RNA synthesis. The two Mg^{2+} ions are held by the same three aspartate side chains that hold the single Mg^{2+} ion, in a network involving several nearby water molecules.

SUMMARY The crystal structure of a *Thermus aquaticus* holoenzyme–DNA complex mimicking an open promoter complex reveals several things. First, the DNA is bound mainly to the σ-subunit, which makes all the important interactions with the promoter DNA. Second, the predicted interactions between amino acids in region 2.4 of σ and the

−10 box of the promoter are really possible. Third, three highly conserved aromatic amino acids are predicted to participate in promoter melting, and they really are in a position to do so. Fourth, two invariant basic amino acids in σ are predicted to participate in DNA binding and they are in a position to do so. A higher resolution crystal structure reveals a form of the polymerase that has two Mg^{2+} ions, in accord with the probable mechanism of catalysis.

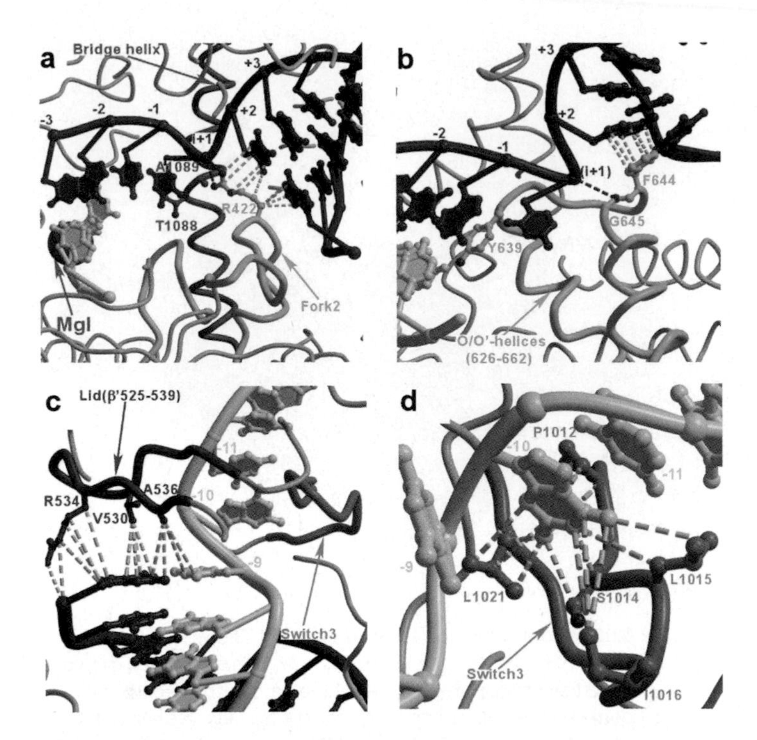

Figure 6.37 Strand separation in the DNA template and in the RNA–DNA hybrid. (a) Downstream DNA strand separation in the *T. thermophilus* polymerase. Note the interactions between R422 (green) and the template nucleotide phosphate and the +2 base pair. In all panels, polar interactions are in dark blue, and van der Waals interactions are in blue-green dashed lines. **(b)** Downstream DNA strand separation in the T7 enzyme. Note the interactions between F644 (green) and the template nucleotide phosphate and the +2 base pair. **(c)** RNA–DNA hybrid strand separation in the *T. thermophilus* enzyme. Note the stacking of three amino acids in the β′ lid (blue) and the −9 base pair, and the interaction of the first displaced RNA base (−10, light green) with the pocket in the β switch 3 loop (orange). **(d)** Detail of interactions between the first displaced RNA base (−10) and five amino acids in the β switch 3 loop (orange). *Source:* Reprinted by permission from Macmillan Publishers Ltd: ***Nature,*** 448, 157–162, 20 June 2007. Vassylyev et al, Structural basis for transcription elongation by bacterial RNA polymerase. © 2007.

Structure of the Elongation Complex In 2007, Dmitry Vassylyev and colleagues presented the x-ray crystal structure of the *Thermus thermophilus* RNA polymerase elongation complex at 2.5Å resolution. This complex contained 14 bp of downstream double-stranded DNA that had yet to be melted by the polymerase, 9 bp of RNA–DNA hybrid, and 7 nt of RNA product in the RNA exit channel. Several important observations came from this work.

First, a valine residue in the β′ subunit inserts into the minor groove of the downstream DNA. This could have two important consequences: It could prevent the DNA from slipping backward or forward in the enzyme; and it could induce the screw-like motion of the DNA through the enzyme, which we will examine later in this chapter. (Consider a screw being driven through a threaded hole in a piece of metal. The metal threads, because of their position between the threads of the screw, require the screw to turn in order to penetrate or withdraw.) There are analogous residues in the single-subunit phage T7 RNA polymerase (Chapter 8), and in the multi-subunit yeast enzyme (Chapter 10) that probably play the same role as the valine residue in the *T. thermophilus* β′ subunit.

Second, as Figure 6.37a shows, the downstream DNA is double-stranded up to and including the +2 base pair, where +1 is the position at which the new nucleotide is added. This means that only one base pair (at position +1) is melted and available for base-pairing with an incoming nucleotide, so only one nucleotide at a time can bind specifically to the complex. Figure 6.37a also demonstrates that one amino acid in the β subunit is situated in a key position right at the site where nucleotides are added to the growing RNA chain. This is arginine 422 of the β fork 2 loop. It makes a hydrogen bond with the phosphate of the +1 template nucleotide, and van der Waals interactions with both bases of the +2 base pair. In the T7 polymerase elongation complex, phenylalanine 644 is in a similar position (Figure 6.37b). The proximity of these amino acids to the active site, and their interactions with key nucleotides there, suggests that they play a role in molding the active site for accurate substrate recognition. If this is so, then mutations in these amino acids should decrease the accuracy of transcription. Indeed, changing phenylalanine 644 (or glycine 645) of the T7 polymerase to alanine does decrease fidelity. At the time this work appeared, the effect of mutations in arginine 422 of the bacterial enzyme had not been checked.

Third, in agreement with previous biochemical work, the enzyme can accommodate nine base pairs of RNA–DNA hybrid. Furthermore, at the end of this hybrid, a series of amino acids of the β′ lid (valine 530, arginine 534, and alanine 536) stack on base pair −9, stabilizing it, and limiting any further base-pairing (Figure 6.37c). These interactions therefore appear to play a role in strand separation at the end of the RNA–DNA hybrid. A variety of experiments have shown the hybrid to vary between 8–10 bp in length, and the β′ lid appears to be flexible enough to handle that kind of variability. But other forces are at work in limiting the length of the hybrid. One is the tendency of the two DNA strands to reanneal. Another is the trapping of the first displaced RNA base (−10) in a hydrophobic pocket of a β loop known as switch 3 (Figure 6.37c). Five amino acids in this pocket make van der Waals interactions with the displaced RNA base (Figure 6.37d), stabilizing the displacement.

Fourth, the RNA product in the exit channel is twisted into the shape it would assume as one-half of an A-form double-stranded RNA. Thus, it is ready to form a hairpin that will cause pausing, or even termination of transcription (see later in this chapter and Chapter 8). Because RNA in hairpin form was not used in this structural study, we cannot see exactly how a hairpin would fit into the exit channel. However, Vassylyev and colleagues modeled the fit of an RNA hairpin in the exit channel, and showed that such a fit can be accomplished with only minor alterations of the protein structure. Indeed, the RNA hairpin could fit with the core enzyme in much the same way as the σ-factor fits with the core in the initiation complex.

In a separate study, Vassylyev and colleagues examined the structure of the elongation complex including an unhydrolyzable substrate analog, adenosine-5′-[(α, β)-methyleno]-triphosphate (AMPcPP), which has a methylene (CH2) group instead of an oxygen between the α- and β- phosphates of ATP. Since this is the bond that is normally broken when the substrate is added to the growing RNA chain, the substrate analog binds to the catalytic site and remains there unaltered. These investigators also looked at the elongation complex structure with AMPcPP and with and without the elongation inhibitor streptolydigin. This comparison yielded interesting information about how the substrate associates with the enzyme in a two-step process.

In the absence of streptolydigin, the so-called trigger loop (residues 1221–1266 of the β′ subunit) is fully folded into two α-helices with a short loop in between. (Figure 6.38a). This brings the substrate into the active site in a productive way, with two metal ions (Mg^{2+}, in this case) close enough together to collaborate in forming the phosphodiester bond that will incorporate the new substrate into the growing RNA chain. Studies of many RNA and DNA polymerases (see Chapter 10) have shown that two metal ions participate in phosphodiester bond formation. One of these is permanently held in the active site, and the other shuttles in, bound to the β- and (γ-phosphates of the NTP substrate. Once the substrate is added to the growing RNA, the second metal ion leaves, bound to the by-product, inorganic pyrophosphate (which comes from the β- and (γ-phosphates of the substrate).

In the presence of streptolydigin, by contrast, the antibiotic forces a change in the trigger loop conformation: The two α-helices unwind somewhat to form a larger loop in between. This in turn forces a change in the way the substrate binds to the active site: The base and sugar of the substrate bind in much the same way, but the triphosphate part extends a bit farther away from the active site, taking

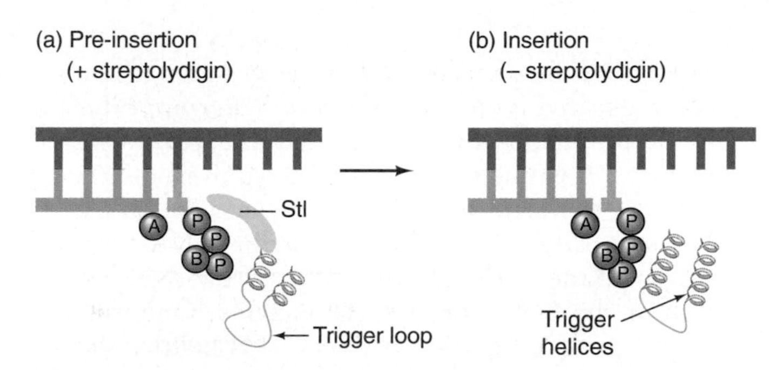

Figure 6.38 A two-step model for nucleotide insertion during RNA synthesis. (a) Pre-insertion state. This is presumably a natural first step in vivo, but it is stabilized by the antibiotic streptolydigin in vitro. Here, streptolydigin (yellow) is forcing the trigger loop out of its normal position close to the active site, which in turn allows the incoming nucleotide (orange with purple triphosphate) to extend its triphosphate moiety away from the active site (exaggerated in this illustration). Because the second metal (metal B) essential for catalysis is complexed to the β- and γ-phosphates of the incoming nucleotide, this places metal B too far away from metal A to participate in catalysis. **(b)** Insertion state. No streptolydigin is present, so the trigger loop can fold into trigger helices that lie closer to the active site, allowing the triphosphates of the incoming nucleotide, and their complexed metal B, to approach closer to metal A at the active site. This arrangement allows the two metal ions to collaborate in nucleotide insertion into the growing RNA chain.

with it one of the metal ions required for catalysis (Figure 6.38b). This makes catalysis impossible and explains how streptolydigin blocks transcription elongation.

Vassylyev and colleagues concluded that the two states of the elongation complex revealed by streptolydigin correspond to two natural states: a **preinsertion state** (seen in the presence of the antibiotic) and an **insertion state** (seen in the absence of the antibiotic). Presumably, the substrate normally binds first in the preinsertion state (Figure 6.38b), and this allows the enzyme to examine it for correct base-pairing and for the correct sugar (ribose vs. deoxyribose) before it switches to the insertion state (Figure 6.38a), where it can be examined again for correct base-pairing with the template base. Thus, the two-state model helps to explain the fidelity of transcription.

The great similarity in structure of the active site among RNA polymerases from all kingdoms of life suggests that all should use the same mechanism of substrate addition, including the two-state model described here. However, as we will see in Chapter 10, investigators of the yeast RNA polymerase have described a two-state model that includes an "entry state" that differs radically from the preinsertion state described here. The substrate in the "entry site" is essentially upside down with respect to the substrate in the insertion state. Clearly, in such a position, it cannot be checked for proper fit with the template base. Vassylyev and colleagues do not dispute the existence of the entry site, but postulate that, if it exists, it must represent a third state of the entering substrate, which must precede the preinsertion state.

SUMMARY Structural studies of the elongation complex involving the *Thermus thermophilus* RNA polymerase have revealed the following features: A valine residue in the β′ subunit inserts into the minor groove of the downstream DNA. In this position, it could prevent the DNA from slipping, and it could induce the screw-like motion of the DNA through the enzyme. Only one base-pair of DNA (at position +1) is melted and available for base-pairing with an incoming nucleotide, so only one nucleotide at a time can bind specifically to the complex. Several forces limit the length of the RNA–DNA hybrid. One of these is the length of the cavity in the enzyme that accommodates the hybrid. Another is a hydrophobic pocket in the enzyme at the end of the cavity that traps the first RNA base displaced from the hybrid. The RNA product in the exit channel assumes the shape of one-half of a double-stranded RNA. Thus, it can readily form a hairpin to cause pausing, or even termination of transcription. Structural studies of the enzyme with an inactive substrate analog and the antibiotic streptolydigin have identified a preinsertion state for the substrate that is catalytically inactive, but could provide for checking that the substrate is the correct one.

Topology of Elongation Does the core, moving along the DNA template, maintain the local melted region created during initiation? Common sense tells us that it does because this would help the RNA polymerase "read" the bases of the template strand and therefore insert the correct bases into the transcript. Experimental evidence also demonstrates that this is so. Jean-Marie Saucier and James Wang added nucleotides to an open promoter complex, allowing the polymerase to move down the DNA as it began elongating an RNA chain, and found that the same degree of melting persisted. Furthermore, the crystal structure of the polymerase–DNA complex shows clearly that the two DNA strands feed through separate channels in the holoenzyme, and we assume that this situation persists with the core polymerase during elongation.

The static nature of the transcription models presented in Chapter 6 is somewhat misleading. If we could see transcription as a dynamic process, we would observe the DNA double helix opening up in front of the moving "bubble" of melted DNA and closing up again behind. In theory, RNA polymerase could accomplish this process in two ways, and Figure 6.39 presents both of them. One way would be for the polymerase and the growing RNA to rotate around and around the DNA template, following the natural twist of the double-helical DNA, as transcription progressed (Figure 6.39a). This would not twist the DNA at all, but it would require considerable energy to make the polymerase gyrate that much, and it would leave the transcript hopelessly twisted around the DNA template, with no known enzyme to untwist it.

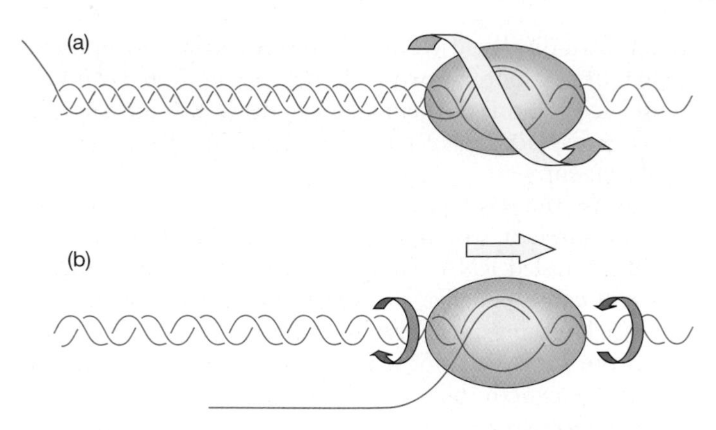

Figure 6.39 Two hypotheses of the topology of transcription of double-stranded DNA. (a) The RNA polymerase (pink) moves around and around the double helix, as indicated by the yellow arrow. This avoids straining the DNA, but it wraps the RNA product (red) around the DNA template. **(b)** The polymerase moves in a straight line, as indicated by the yellow arrow. This avoids twisting the RNA product (red) around the DNA, but it forces the DNA ahead of the moving polymerase to untwist and the DNA behind the polymerase to twist back up again. These two twists, represented by the green arrows, introduce strain into the DNA template that must be relieved by topoisomerases.

The other possibility is that the polymerase moves in a straight line, with the template DNA rotating in one direction ahead of it to unwind, and rotating in the opposite direction behind it to wind up again (Figure 6.39b). But this kind of rotating of the DNA introduces strain. To visualize this, think of unwinding a coiled telephone cord, or actually try it if you have one available. You can feel (or imagine) the resistance you encounter as the cord becomes more and more untwisted, and you can appreciate that you would also encounter resistance if you tried to wind the cord more tightly than its natural state. It is true that the rewinding of DNA at one end of the melted region creates an opposite and compensating twist for the unwinding at the other. But the polymerase in between keeps this compensation from reaching across the melted region, and the long span of DNA around the circular chromosome insulates the two ends of the melted region from each other the long way around.

So if this second mechanism of elongation is valid, we have to explain how the strain of unwinding the DNA is relaxed. As we will see in Chapter 20 when we discuss DNA replication, a class of enzymes called **topoisomerases** can introduce transient breaks into DNA strands and so relax this kind of strain. We will see that strain due to twisting a double-helical DNA causes the helix to tangle up like a twisted rubber band. This process is called **supercoiling,** and the supercoiled DNA is called a **supercoil** or **superhelix.** Unwinding due to the advancing polymerase causes a compensating overwinding ahead of the unwound region. (Compensating overwinding is what makes it difficult to unwind a coiled telephone cord.) The supercoiling due to overwinding is by convention called positive. Thus, positive supercoils build up in front of the advancing polymerase. Conversely, negative supercoils form behind the polymerase. One line of evidence that directly supports this model of transcription comes from studies with topoisomerase mutants that cannot relax supercoils. If the mutant cannot relax positive supercoils, these build up in DNA that is being transcribed. On the other hand, negative supercoils accumulate during transcription in topoisomerase mutants that cannot relax that kind of superhelix.

SUMMARY Elongation of transcription involves the polymerization of nucleotides as the RNA polymerase travels along the template DNA. As it moves, the polymerase maintains a short melted region of template DNA. This requires that the DNA unwind ahead of the advancing polymerase and close up again behind it. This process introduces strain into the template DNA that is relaxed by topoisomerases.

Pausing and Proofreading The process of elongation is far from uniform. Instead, the polymerase repeatedly pauses, and in some cases backtracks, while elongating an RNA chain. Under in vitro conditions of 21°C and 1 mM NTPs, pauses in bacterial systems have been found to be very brief: generally only 1–6 sec. But repeated short pauses significantly slow the overall rate of transcription. Pausing is physiologically important for at least two reasons: First, it allows translation, an inherently slower process, to keep pace with transcription. This is important for phenomena such as attenuation (Chapter 7), and aborting transcription if translation fails. The second important aspect of pausing is that it is the first step in termination of transcription, as we will see later in this chapter.

Sometimes the polymerase even backtracks by reversing its direction and thereby extruding the 3′-end of the growing transcript out of the active site of the enzyme. This is more than just an exaggerated pause. For one thing, it tends to last much longer: 20 sec, up to irreversible arrest. For another, it occurs only under special conditions: when nucleotide concentrations are severely reduced, or when the polymerase has added the wrong nucleotide to the growing RNA chain. In the latter case, backtracking is part of a proofreading process in which auxiliary proteins known as **GreA** and **GreB** stimulate an inherent RNase activity of the polymerase to cleave off the end of the growing RNA, removing the misincorporated nucleotide, and allowing transcription to resume. GreA produces only short RNA end fragments 2–3 nt long, and can prevent, but not reverse transcription arrest. GreB can produce RNA end fragments up to 18 nt long, and can reverse arrested transcription. We will discuss the analogous proofreading mechanism in eukaryotes in greater detail in Chapter 11.

One complication to this proofreading model is that the auxiliary proteins are dispensable in vivo. And yet one would predict that mRNA proofreading would be important for life. In 2006, Nicolay Zenkin and colleagues suggested a resolution to this apparent paradox: The nascent RNA itself appears to participate in its own proofreading.

Zenkin and colleagues simulated an elongation complex by mixing RNA polymerase with a piece of single-stranded DNA and an RNA that was either perfectly complementary to the DNA or had a mismatched base at its 3′-end. When they added Mg^{2+}, they observed that the mismatched RNA lost a dinucleotide from its 3′-end, including the mismatched nucleotide and the penultimate (next-to-last) nucleotide. This proofreading did not occur with the perfectly matched RNA. The fact that two nucleotides were lost suggests that the polymerase had backtracked one nucleotide in the mismatched complex. And this in turn suggested a chemical basis for the RNA-assisted proofreading: In the backtracked complex, the mismatched nucleotide, because it is not base-paired to the template DNA, is flexible enough to bend back and contact metal II, holding it at the active site of the enzyme. This would be expected to enhance phosphodiester bond cleavage, because metal II is presumably involved in the enzyme's RNase activity. In addition, the mismatched nucleotide can orient a water molecule to make it a better nucleophile in attacking the phosphodiester bond that links the terminal dinucleotide to the rest of the RNA. Both of these considerations help to explain why the mismatched RNA can stimulate its own cleavage, while a perfectly matched RNA cannot.

SUMMARY RNA polymerase frequently pauses, or even backtracks, during elongation. Pausing allows ribosomes to keep pace with the RNA polymerase, and it is also the first step in termination. Backtracking aids proofreading by extruding the 3′-end of the RNA out of the polymerase, where misincorporated nucleotides can be removed by an inherent nuclease activity of the polymerase, stimulated by auxiliary factors. Even without these factors, the polymerase can carry out proofreading: The mismatched nucleotide at the end of a nascent RNA plays a role in this process by contacting two key elements at the active site: metal II and a water molecule.

6.5 Termination of Transcription

When the polymerase reaches a **terminator** at the end of a gene it falls off the template, releasing the RNA. *E. coli* cells contain about equal numbers of two kinds of terminators. The first kind, known as **intrinsic terminators,** function with the RNA polymerase by itself without help from other proteins. The second kind depend on an auxiliary factor called **rho** (ρ). Naturally, these are called rho-dependent terminators. Let us consider the mechanisms of termination employed by these two systems, beginning with the simpler, intrinsic terminators.

Rho-Independent Termination

Rho-independent, or intrinsic, termination depends on terminators consisting of two elements: an inverted repeat followed immediately by a T-rich region in the nontemplate strand of the gene. The model of termination we will present later in this section depends on a "hairpin" structure in the RNA transcript of the inverted repeat. Before we get to the model, we should understand how an inverted repeat predisposes a transcript to form a hairpin.

Inverted Repeats and Hairpins Consider this inverted repeat:

5′-TACGAAGTTCGTA-3′
•
3′-ATGCTTCAAGCAT-5′

Such a sequence is symmetrical around its center, indicated by the dot; it would read the same if rotated 180 degrees in the plane of the paper, and if we always read the strand that runs 5′→3′ left to right. Now observe that a transcript of this sequence

UACGAAGUUCGUA

is self-complementary around its center (the underlined G). That means that the self-complementary bases can pair to form a hairpin as follows:

U • A
A • U
C • G
G • C
A • U
A U
G

The A and the U at the apex of the hairpin cannot form a base pair because of the physical constraints of the turn in the RNA.

The Structure of an Intrinsic Terminator The *E. coli trp* operon (Chapter 7) contains a DNA sequence called an attenuator that causes premature termination of transcription. The *trp* attenuator contains the two elements (an inverted repeat and a string of T's in the nontemplate DNA strand) suspected to be vital parts of an intrinsic terminator, so Peggy Farnham and Terry Platt used attenuation as an experimental model for normal termination.

The inverted repeat in the *trp* attenuator is not perfect, but 8 bp are still possible, and 7 of these are strong G–C pairs, held together by three hydrogen bonds. The hairpin looks like this:

A • U
G • C
C • G
C • G
C • G
G • C
C • G
C • G > A
U U
A A

Notice that a small loop occurs at the end of this hairpin because of the U–U and A–A combinations that cannot base-pair. Furthermore, one A on the right side of the stem has to be "looped out" to allow 8 bp instead of just 7. Still, the hairpin should form and be relatively stable.

Farnham and Platt reasoned as follows: As the T-rich region of the attenuator is transcribed, eight A–U base pairs would form between the A's in the DNA template strand and the U's in the RNA product. They also knew that rU–dA base pairs are exceptionally weak; they have a melting temperature 20°C lower than even rU–rA or dT–rA pairs. This led the investigators to propose that the polymerase paused at the terminator, and then the weakness of the rU–dA base pairs allowed the RNA to dissociate from the template, terminating transcription.

What data support this model? If the hairpin and string of rU–dA base pairs in the *trp* attenuator are really important, we would predict that any alteration in the base sequence that would disrupt either one would be deleterious to attenuation. Farnham and Platt devised the following in vitro assay for attenuation (Figure 6.40): They started with a *Hpa*II restriction fragment containing the *trp* attenuator and transcribed it in vitro. If attenuation works, and transcription terminates at the attenuator, a short (140-nt) transcript should be the result. On the other hand, if transcription fails to terminate at the attenuator, it will continue to the end of the fragment, yielding a run-off transcript 260 nt in length. These two transcripts are easily distinguished by electrophoresis.

When these investigators altered the string of eight T's in the nontemplate strand of the terminator to the sequence TTTTGCAA, creating the mutant they called *trp a*1419,

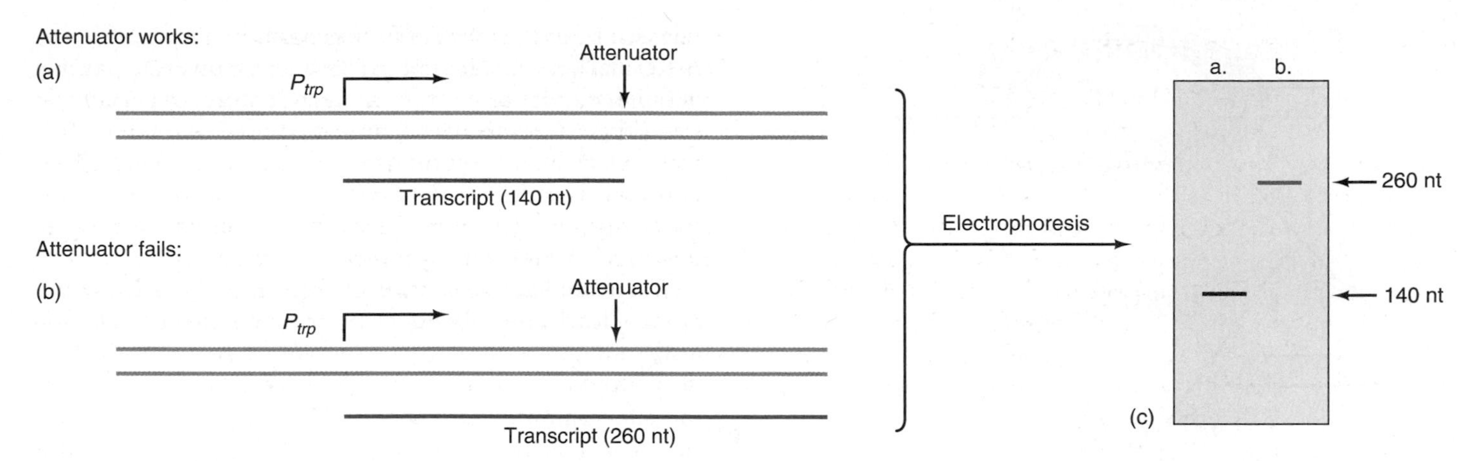

Figure 6.40 An assay for attenuation. (a) When the DNA fragment containing the *trp* promoter and attenuator is transcribed under conditions in which the attenuator works, transcription stops in the attenuator, and a 140-nt transcript (red) results. **(b)** When the same DNA fragment is transcribed under conditions that cause the attenuator to fail, a run-off transcript of 260 nt (green) is the result. **(c)** The transcripts from the two different reactions can be distinguished easily by electrophoresis. Using this assay, one can tell whether the attenuator works under a variety of conditions.

attenuation was weakened. This is consistent with the hypothesis that the weak rU–dA pairs are important in termination, because half of them would be replaced by stronger base pairs in this mutant.

Moreover, this mutation could be overridden by substituting the nucleotide iodo-CTP (I-CTP) for normal CTP in the in vitro reaction. The most likely explanation is that base-pairing between G and iodo-C is stronger than between G and ordinary C. Thus, the GC-rich hairpin should be stabilized by I-CMP, and this effect counteracts the loss of weak base pairs in the region following the hairpin. On the other hand, IMP (inosine monophosphate, a GMP analog) should weaken base-pairing in the hairpin because I–C pairs, with only two hydrogen bonds holding them together, are weaker than G–C pairs with three. Sure enough, substituting ITP for GTP in the transcription reaction weakened termination at the attenuator. Thus, all of these effects are consistent with the hypothesis that the hairpin and string of U's in the transcript are important for termination. However, they do not identify the roles that these RNA elements play in pausing and termination.

SUMMARY Using the *trp* attenuator as a model terminator, Farnham and Platt showed that intrinsic terminators have two important features: (1) an inverted repeat that allows a hairpin to form at the end of the transcript; (2) a string of T's in the nontemplate strand that results in a string of weak rU–dA base pairs holding the transcript to the template strand.

A Model for Termination Several hypotheses have been proposed for the roles of the hairpin and string of rU–dA base pairs in the mechanism of termination. Two important clues help narrow the field of hypotheses. First, hairpins are found to destabilize elongation complexes that are stalled artificially (not at strings of rU–dA pairs). Second, terminators in which half of the inverted repeat is missing still stall at the strings of rU–dA pairs, even though no hairpin can form. This leads to the following general hypothesis: The rU–dA pairs cause the polymerase to pause, allowing the hairpin to form and destabilize the already weak rU–dA pairs that are holding the DNA template and RNA product together. This destabilization results in dissociation of the RNA from its template, terminating transcription.

W. S. Yarnell and Jeffrey Roberts proposed a variation on this hypothesis in 1999, as illustrated in Figure 6.41. This model calls for the withdrawal of the RNA from the active site of the polymerase that has stalled at a terminator—either because the newly formed hairpin helps pull it out or because the polymerase moves downstream without elongating the RNA, thus leaving the RNA behind. To test their hypothesis, Yarnell and Roberts used a DNA template that contained two mutant terminators (ΔtR2 and Δt82) downstream of a strong promoter. These terminators had a T-rich region in the nontemplate strand, but only half of an inverted repeat, so hairpins could not form. To compensate for the hairpin, these workers added an oligonucleotide that was complementary to the remaining half of the inverted repeat. They reasoned that the oligonucleotide would base-pair to the transcript and restore the function of the hairpin.

To test this concept, they attached magnetic beads to the template, so it could be easily removed from the mixture magnetically. Then they used *E. coli* RNA polymerase to synthesize labeled RNAs in vitro in the presence and absence of the appropriate oligonucleotides. Finally, they removed the template magnetically to form

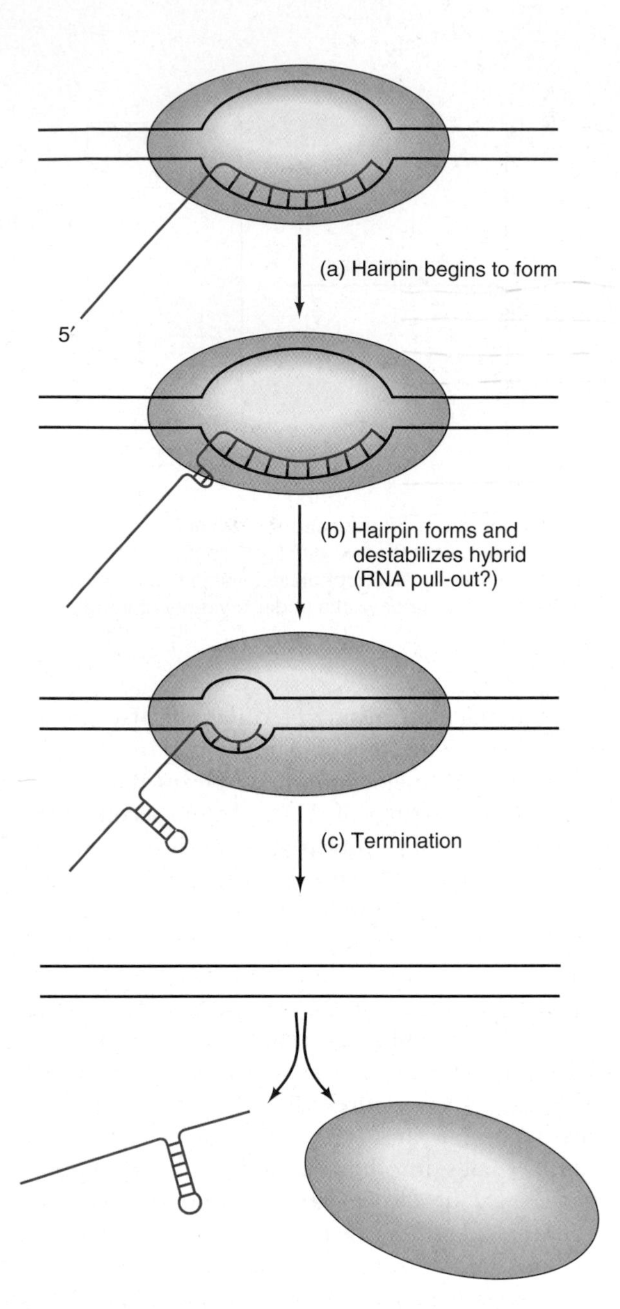

Figure 6.41 A model for rho-independent, or intrinsic termination. **(a)** The polymerase has paused at a string of weak rU–dA base pairs, and a hairpin has started to form just upstream of these base pairs. **(b)** As the hairpin forms, it further destabilizes the RNA–DNA hybrid. This destabilization could take several forms: The formation of the hairpin could physically pull the RNA out of the polymerase, allowing the transcription bubble to collapse; conversely, it could cause the transcription bubble to collapse, expelling the RNA from the hybrid. **(c)** The RNA product and polymerase dissociate completely from the DNA template, terminating transcription.

a pellet and electrophoresed the material in the pellet and the supernatant and detected the RNA species by autoradiography.

Figure 6.42 shows the results. In lanes 1–6, no oligonucleotides were used, so little incomplete RNA was released into the supernatant (see faint bands at ΔtR2 and Δt82 markers in lanes 1, 3, and 5). However, pausing definitely did occur at both terminators, especially at short times (see stronger bands in lanes 2, 4, and 6). This was a clear indication that the hairpin is not required for pausing, though it is required for efficient release of the transcript. In lanes 7–9, Yarnell and Roberts included an oligonucleotide (t19) complementary to the remaining, downstream half of the inverted repeat in the ΔtR2 terminator. Clearly, this oligonucleotide stimulated termination at the mutant terminator, as the autoradiograph shows a dark band corresponding to a labeled RNA released into the supernatant. This labeled RNA is exactly the same size as an RNA released by the wild-type terminator would be. Similar results, though less dramatic, were obtained with an oligonucleotide (t18) that is complementary to the downstream half of the inverted repeat in the Δt82 terminator.

To test further the importance of base-pairing between the oligonucleotide and the half-inverted repeat, these workers mutated one base in the t19 oligonucleotide to yield an oligonucleotide called t19H1. Lane 13 shows that this change caused a dramatic reduction in termination at ΔtR2. Then they made a compensating mutation in ΔtR2 and tested t19H1 again. Lane 14 shows that this restored strong termination at ΔtR2. This template also contained the wild-type t82 terminator, so abundant termination also occurred there. Lanes 15 and 16 are negative controls in which no t19H1 oligonucleotide was present, and, as expected, very little termination occurred at the ΔtR2 terminator.

Together, these results show that the hairpin itself is not required for termination. All that is needed is something to base-pair with the downstream half of the inverted repeat to destabilize the RNA–DNA hybrid. Furthermore, the T-rich region is not required if transcription can be slowed to a crawl artificially. Yarnell and Roberts advanced the polymerase to a site that had neither an inverted repeat nor a T-rich region and made sure it paused there by washing away the nucleotides. Then they added an oligonucleotide that hybridized upstream of the artificial pause site. Under these conditions, they observed release of the nascent RNA.

Termination is also stimulated by a protein called **NusA,** which appears to promote hairpin formation in the terminator. The essence of this model, presented in 2001 by Ivan Gusarov and Evgeny Nudler, is that the upstream half of the hairpin binds to part of the core polymerase called the **upstream binding site (UBS).** This protein–RNA binding slows down hairpin formation and so makes termination less likely. But NusA loosens the association between the RNA and the UBS, thereby stimulating hairpin formation. This makes termination more likely. In Chapter 8, we will discuss NusA and its mode of action in more detail and see evidence for the model mentioned here.

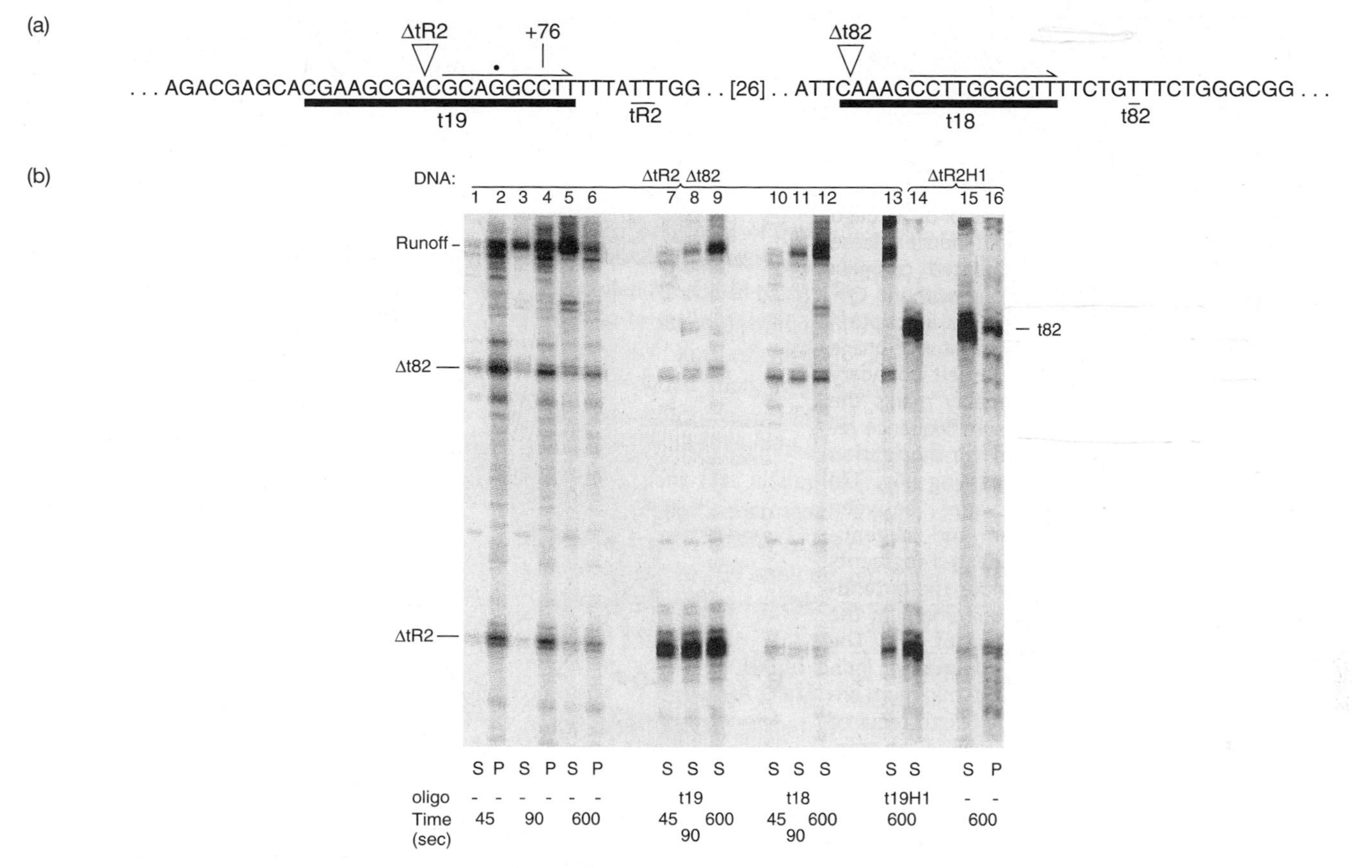

Figure 6.42 Release of transcripts from elongation complexes by oligonucleotides complementary to mutant terminators. **(a)** Scheme of the template used in these experiments. The template contained two mutant terminators, ΔtR2, and Δt82, situated as shown, downstream of a strong promoter. The normal termination sites for these two terminators are labeled with thin underlines. The black bars denote regions complementary to the oligonucleotides used (t19 and t18). The rightward arrows denote the half inverted repeats remaining in the mutant terminators. The dot indicates the site of a base altered in the t19HI oligonucleotide and of a compensating mutation in the DNA template in certain of the experiments. The template was attached to a magnetic bead so it could be removed from solution easily by centrifugation. **(b)** Experimental results. Yarnell and Roberts synthesized labeled RNA in the presence of the template in panel **(a)** and; no oligonucleotide (lanes 1–6 and 15–16), the t19 oligonucleotide (lanes 7–9), the t18 oligonucleotide (lanes 10–12); and the t19HI oligonucleotide (lanes 13–14). They allowed transcription for the times given at bottom, then removed the template and any RNA attached to it by centrifugation. They electrophoresed the labeled RNA in the pellet (P) or supernatant (S), as indicated at bottom, and autoradiographed the gel. The positions of run-off transcripts, and of transcripts that terminated at the ΔtR2 and Δt82 terminators, are indicated at left. (*Source:* (a–b) Yarnell, W.S. and Roberts, J.W. Mechanism of intrinsic transcription termination and antitermination. *Science* 284 (23 April 1999) 611–12. © AAAS.)

SUMMARY The essence of a bacterial terminator is twofold: (1) base-pairing of something to the transcript to destabilize the RNA–DNA hybrid; and (2) something that causes transcription to pause. A normal intrinsic terminator satisfies the first condition by causing a hairpin to form in the transcript, and the second by causing a string of U's to be incorporated just downstream of the hairpin.

Rho-Dependent Termination

Jeffrey Roberts discovered rho as a protein that caused an apparent depression of the ability of RNA polymerase to transcribe certain phage DNAs in vitro. This depression is simply the result of termination. Whenever rho causes a termination event, the polymerase has to reinitiate to begin transcribing again. And, because initiation is a time-consuming event, less net transcription can occur. To establish that rho is really a termination factor, Roberts performed the following experiments.

Rho Affects Chain Elongation, But Not Initiation Just as Travers and Burgess used [γ-^{32}P]ATP and [^{14}C]ATP to measure transcription initiation and total RNA synthesis, respectively, Roberts used [γ-^{32}P]GTP and [^{3}H]UTP for the same purposes. He carried out in vitro transcription reactions with these two labeled nucleotides in the presence of increasing concentrations of rho. Figure 6.43 shows the results. We see that rho had little effect on initiation; if anything, the rate of initiation went up. But rho caused a

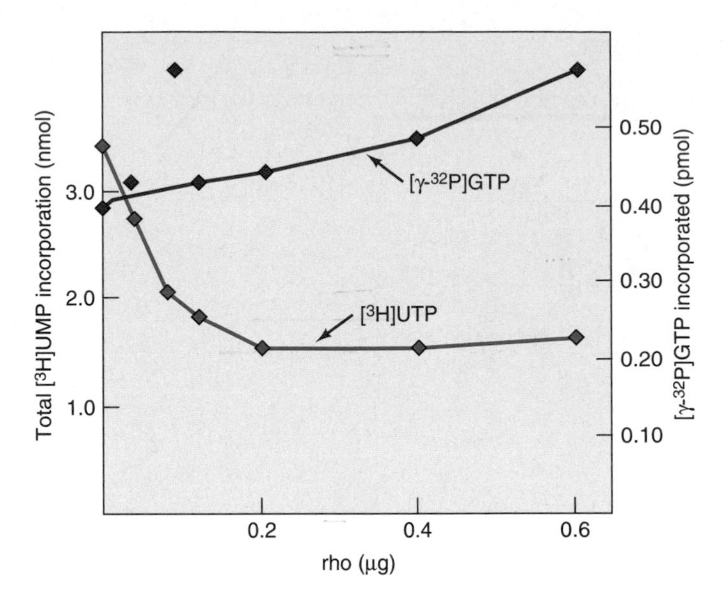

Figure 6.43 Rho decreases the net rate of RNA synthesis. Roberts allowed *E. coli* RNA polymerase to transcribe λ phage DNA in the presence of increasing concentrations of rho. He used [γ-^{32}P]GTP to measure initiation (red) and [^{3}H]UTP to measure elongation (green). Rho depressed the elongation rate, but not initiation. (*Source:* Adapted from Roberts, J.W. Termination factor for RNA synthesis, *Nature* 224:1168–74, 1969.)

significant decrease in total RNA synthesis. This is consistent with the notion that rho terminates transcription, thus forcing time-consuming reinitiation. This hypothesis predicts that rho would cause shorter transcripts to be made.

Rho Causes Production of Shorter Transcripts It is relatively easy to measure the size of RNA transcripts by gel electrophoresis or, in 1969, when Roberts performed his experiments, by ultracentrifugation. But just finding short transcripts would not have been enough to conclude that rho was causing termination. It could just as easily have been an RNase that chopped up longer transcripts into small pieces.

To exclude the possibility that rho was simply acting as a nuclease, Roberts first made ^{3}H-labeled λ RNA in the absence of rho, then added these relatively large pieces of RNA to new reactions carried out in the presence of rho, in which [^{14}C]UTP was the labeled RNA precursor. Finally, he measured the sizes of the ^{14}C- and ^{3}H-labeled λ RNAs by ultracentrifugation. Figure 6.44 presents the results. The solid curves show no difference in the size of the preformed ^{3}H-labeled RNA even when it had been incubated with rho in the second reaction. Rho therefore shows no RNase activity. However, the ^{14}C-labeled RNA made in the presence of rho (red line in Figure 6.44b) is obviously much smaller than the preformed RNA made without rho. Thus, rho is causing the synthesis of much smaller RNAs. Again, this is consistent with the role of rho in terminating transcription. Without rho, the transcripts grew to abnormally large size.

Rho Releases Transcripts from the DNA Template Finally, Roberts used ultracentrifugation to analyze the sedimentation properties of the RNA products made in the presence and absence of rho. The transcripts made without rho (Figure 6.45a) cosedimented with the DNA template, indicating that they had not been released from their association with the DNA. By contrast, the transcripts made in the presence of rho (Figure 6.45b) sedimented at a much lower rate, independent of the DNA. Thus, rho seems to release RNA transcripts from the DNA template. In fact, rho (the Greek letter ρ) was chosen to stand for "release."

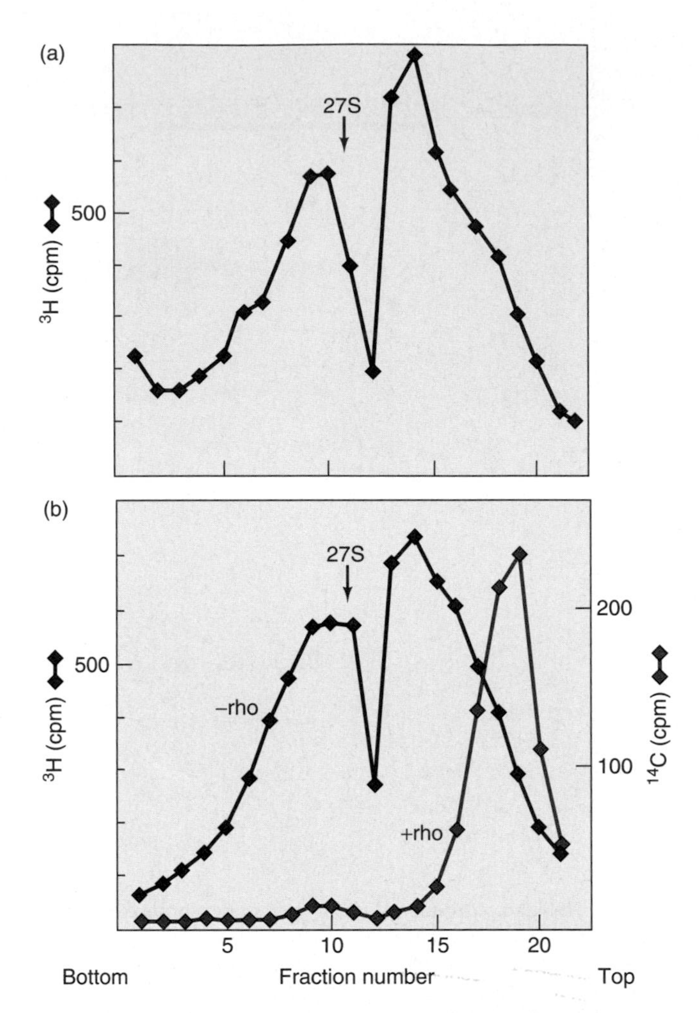

Figure 6.44 Rho reduces the size of the RNA product. (a) Roberts allowed *E. coli* RNA polymerase to transcribe λ DNA in the absence of rho. He included [^{3}H]UTP in the reaction to label the RNA. Finally, he used ultracentrifugation to separate the transcripts by size. He collected fractions from the bottom of the centrifuge tube, so low-numbered fractions, at left, contained the largest RNAs. **(b)** Roberts used *E. coli* RNA polymerase to transcribe λ DNA in the presence of rho. He also included [^{14}C]ATP to label the transcripts, plus the ^{3}H-labeled RNA from panel **(a).** Again, he ultracentrifuged the transcripts to separate them by size. The ^{14}C-labeled transcripts (red) made in the presence of rho were found near the top of the gradient (at right), indicating that they were relatively small. On the other hand, the ^{3}H-labeled transcripts (blue) from the reaction lacking rho were relatively large and the same size as they were originally. Thus, rho has no effect on the size of previously made transcripts, but it reduces the size of the transcripts made in its presence. (*Source:* Adapted from Roberts, J.W. Termination factor for RNA synthesis, *Nature* 224:1168–74, 1969.)

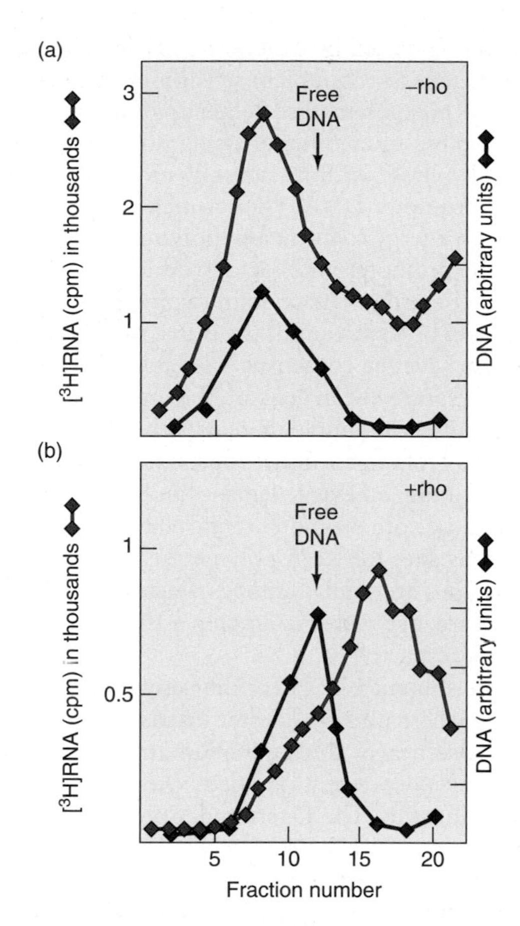

Figure 6.45 Rho releases the RNA product from the DNA template. Roberts transcribed λ DNA under the same conditions as in Figure 6.44, in the **(a)** absence or **(b)** presence of rho. Then he subjected the ^{3}H-labeled product (red) to ultracentrifugation to see whether the product was associated with the DNA template (blue). (a) The RNA made in the absence of rho sedimented together with the template in a complex that was larger than free DNA. (b) The RNA made in the presence of rho sedimented independently of DNA at a position corresponding to relatively small molecules. Thus, transcription with rho releases transcripts from the DNA template. (*Source:* Adapted from Roberts, J.W. Termination factor for RNA synthesis, *Nature* 224:1168–74, 1969.)

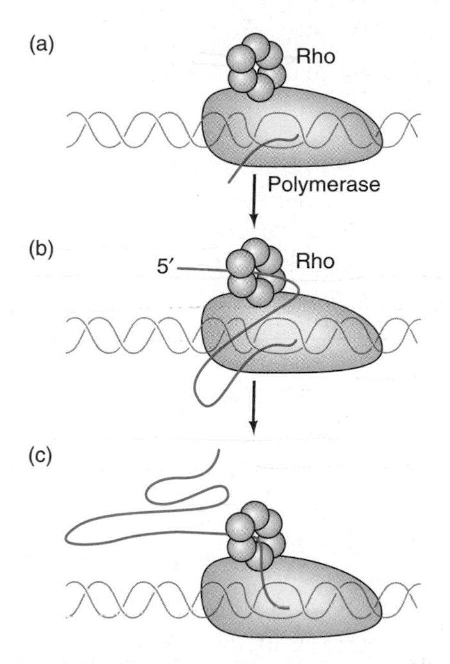

Figure 6.46 A model of rho-dependent termination. (a) Rho (blue) has joined the elongation complex by binding directly to RNA polymerase. The end of the nascent transcript (green) has just emerged from the polymerase. **(b)** The transcript has lengthened and has bound to rho via a rho loading site, forming an RNA loop. Rho can now feed the transcript through its central cavity. **(c)** The polymerase has paused at a terminator. By continuously feeding the transcript through itself, rho has tightened the RNA loop and irreversibly trapped the elongation complex. Rho has also begun to dissociate the RNA–DNA hybrid, which will lead to transcript release.

The Mechanism of Rho How does rho do its job? It has been known for some time that rho is able to bind to RNA at a so-called **rho loading site,** or **rho utilization (*rut*) site,** and has ATPase activity that can provide the energy to propel it along an RNA chain. Accordingly, a model has arisen that calls for rho to bind to a nascent RNA, and follow the polymerase by moving along the RNA chain in the $5'\rightarrow3'$ direction. This chase continues until the polymerase stalls in the terminator region just after making the RNA hairpin. Then rho can catch up and release the transcript. In support of this hypothesis, Terry Platt and colleagues showed in 1987 that rho has RNA–DNA helicase activity that can unwind an RNA–DNA hybrid. Thus, when rho encounters the polymerase stalled at the terminator, it can unwind the RNA–DNA hybrid within the transcription bubble, releasing the RNA and terminating transcription.

Evgeny Nudler and colleagues presented evidence in 2010 that this attractive hypothesis is probably wrong. These workers used their transcription walking method, as described earlier in this chapter, using His_6-tagged rho coupled to nickel beads. They found that elongation complexes (ECs) with RNA products only 11 nt long were retained by the beads. Because an 11-nt RNA is completely contained within RNA polymerase, this behavior means that the association between rho and the EC must involve the polymerase, not the RNA. Thus, if rho binds directly to the polymerase, it does not need to bind to the nascent RNA first and chase the polymerase until it catches up.

Furthermore, the EC tethered to the rho-nickel beads could be walked along the DNA template without dissociating, proving that the association between rho and the EC is stable. And the complex could terminate normally at rho-dependent terminators, showing that the rho that is bound to the polymerase is capable of sponsoring termination.

If rho is already bound to the polymerase at an early stage in transcription, how does its affinity for RNA come into play in termination? Nudler and colleagues proposed the model in Figure 6.46. First, rho binds to the polymerase when the transcript is still very short. When the transcript grows longer, and includes a rho loading site, the RNA binds to rho. X-ray crystallography studies have shown that rho is a hexamer of identical subunits arranged in the shape of a lock washer—an open circle with slightly offset ends. This presumably allows the growing RNA to enter the hole in the middle of the hexamer, forming an RNA loop. As transcription progresses, rho continues to feed the RNA product through itself, progressively tightening the RNA loop. Ultimately, when the polymerase encounters a termination signal, it pauses, allowing the RNA loop to tighten so much that further transcription cannot occur. This creates a "trapped" elongation complex. Finally, rho could invade the RNA–DNA hybrid within the polymerase and cause termination in one of two ways: It could use its RNA–DNA helicase activity to unwind the hybrid, or it could unwind the hybrid by physically disrupting it.

SUMMARY Rho-dependent terminators consist of an inverted repeat, which can cause a hairpin to form in the transcript, but no string of T's. Rho binds to the RNA polymerase in an elongation complex. When the RNA transcript has grown long enough, rho binds to it via a rho loading site, forming an RNA loop between the polymerase and rho. Rho continues to feed the growing transcript through itself until the polymerase pauses at a terminator. This pause allows rho to tighten the RNA loop and trap the elongation complex. Rho then dissociates the RNA–DNA hybrid, terminating transcription.

SUMMARY

The catalytic agent in the transcription process is RNA polymerase. The *E. coli* enzyme is composed of a core, which contains the basic transcription machinery, and a σ-factor, which directs the core to transcribe specific genes. The σ-factor allows initiation of transcription by causing the RNA polymerase holoenzyme to bind tightly to a promoter. This σ-dependent tight binding requires local melting of 10–17 bp of the DNA in the vicinity of the transcription start site to form an open promoter complex. Thus, by directing the holoenzyme to bind only to certain promoters, a σ-factor can select which genes will be transcribed. The initiation process continues until 9 or 10 nt have been incorporated into the RNA, the core changes to an elongation-specific conformation, leaves the promoter, and carries on with the elongation process. The σ-factor appears to be released from the core polymerase, but not usually immediately upon promoter clearance. Rather, σ seems to exit from the elongation complex in a stochastic manner during the elongation process. The σ-factor can be reused by different core polymerases. The core, not σ, governs rifampicin sensitivity or resistance. The *E. coli* RNA polymerase achieves abortive transcription by scrunching: drawing downstream DNA into the polymerase without actually moving and losing its grip on promoter DNA. The scrunched DNA could store enough energy to allow the polymerase to break its bonds to the promoter and begin productive transcription.

Prokaryotic promoters contain two regions centered at −10 and −35 bp upstream of the transcription start site. In *E. coli,* these have the consensus sequences TATAAT and TTGACA, respectively. In general, the more closely regions within a promoter resemble these consensus sequences, the stronger that promoter will be. Some extraordinarily strong promoters contain an extra element (an UP element) upstream of the core promoter. This makes these promoters better binding sites for RNA polymerase.

Four regions are similar among σ-factors, and subregions 2.4 and 4.2 are involved in promoter −10 box and −35 box recognition, respectively.

The core subunit β lies near the active site of the RNA polymerase where phosphodiester bonds are formed. The σ-factor is also nearby during the initiation phase. The α-subunit has independently folded N-terminal and C-terminal domains. The C-terminal domain can recognize and bind to a promoter's UP element. This allows very tight binding between polymerase and promoter.

Elongation of transcription involves the polymerization of nucleotides as the RNA polymerase core travels along the template DNA. As it moves, the polymerase maintains a short melted region of template DNA. This transcription bubble is 11-16 bases long and contains an RNA–DNA hybrid about 9 bp long. The movement of the transcription bubble requires that the DNA unwind ahead of the advancing polymerase and close up again behind it. This process introduces strain into the template DNA that is relaxed by topoisomerases.

The crystal structure of the *T. aquaticus* RNA polymerase core is shaped like a crab claw. The catalytic center, containing a Mg^{2+} ion coordinated by three Asp residues, lies in a channel that conducts DNA through the enzyme.

The crystal structure of a *T. aquaticus* holoenzyme–DNA complex mimicking an open promoter complex allows the following conclusions. (1) The DNA is bound mainly to the σ-subunit. (2) The predicted interactions between amino acids in region 2.4 of σ and the −10 box of the promoter are really possible. (3) Three highly conserved aromatic amino acids that are predicted to participate in promoter melting are really in a position to do so. (4) Two invariant basic amino acids in σ that are predicted to participate in DNA binding are in proper position to do so. A higher resolution crystal structure reveals a form of the polymerase that has two Mg^{2+} ions, in accord with the probable mechanism of catalysis.

Structural studies of the elongation complex involving the *Thermus thermophilus* RNA polymerase revealed that: A valine residue in the β' subunit inserts into the minor groove of the downstream DNA; thus, it could prevent the DNA from slipping, and it could induce the screw-like motion of the DNA through the enzyme. Only one base pair of DNA (at position +1) is melted and available for base-pairing with an incoming nucleotide, so only one nucleotide at a time can bind specifically to the complex. Several forces limit the length of the RNA–DNA hybrid, including the length of the cavity in the enzyme that accommodates the hybrid and a hydrophobic pocket in the enzyme at the end of the cavity that traps the first RNA base displaced from the hybrid. The RNA product in the exit channel assumes the shape of one-half of a double-stranded RNA. Thus, it can readily form a hairpin to cause pausing, or even termination of transcription. Structural studies of the enzyme with an inactive substrate analog and the antibiotic streptolydigin have identified a preinsertion state for the substrate that is catalytically inactive, but could provide for checking that the substrate is the correct one.

Intrinsic terminators have two important elements: (1) an inverted repeat that allows a hairpin to form at the end of the transcript to destabilize the RNA–DNA hybrid; (2) a string of T's in the nontemplate strand that results in a string of weak rU–dA base pairs holding the transcript to the template. Together, these elements cause the polymerase to pause and the transcript to be released. Rho-dependent terminators consist of an inverted repeat, which can cause a hairpin to form in the transcript, but no string of T's. Rho binds to the RNA polymerase in an elongation complex. When the RNA transcript has grown long enough, rho binds to it via a rho loading site, forming an RNA loop between the polymerase and rho. Rho continues to feed the growing transcript through itself until the polymerase pauses at a terminator. This pause allows rho to tighten the RNA loop and trap the elongation complex. Rho then dissociates the RNA–DNA hybrid, terminating transcription.

REVIEW QUESTIONS

1. Explain the following findings: (1) Core RNA polymerase transcribes intact T4 phage DNA only weakly, whereas holoenzyme transcribes this template very well; but (2) core polymerase can transcribe calf thymus DNA about as well as the holoenzyme can.
2. How did Bautz and colleagues show that the holoenzyme transcribes phage T4 DNA asymmetrically, but the core transcribes this DNA symmetrically?
3. Describe an experiment to measure the dissociation rate of the tightest complex between a protein and a DNA. Show sample results of weak and tight binding. How do these results relate to the binding of core polymerase and holoenzyme to DNA that contains promoters?
4. What effect does temperature have on the dissociation rate of polymerase–promoter complexes? What does this suggest about the nature of the complex?
5. Diagram the difference between a closed and an open promoter complex.
6. Diagram a typical prokaryotic promoter, and a promoter with an UP element. Exact sequences are not necessary.
7. Describe and give the results of an experiment that demonstrates the formation of abortive transcripts by *E. coli* RNA polymerase.
8. Diagram the four-step transcription initiation process in *E. coli.*
9. Describe and show the results of an experiment that measures the effects of σ on transcription initiation and elongation rates.
10. How can you show that σ does not really accelerate the rate of transcription elongation?
11. What final conclusion can you draw from the experiments in the previous two questions?
12. Describe and show the results of an experiment that demonstrates the reuse of σ. On the same graph, show the results of an experiment that shows that the core polymerase determines resistance to rifampicin.
13. Draw a diagram of the "σ-cycle," assuming σ dissociates from core during elongation.
14. Describe and show the results of a fluorescence resonance energy transfer (FRET) experiment that suggests that σ does not dissociate from the core polymerase during elongation.
15. In the σ-cycle, what is obligate release and what is stochastic release? Which is the favored hypothesis?
16. Propose three hypotheses for the mechanism of abortive transcription in *E. coli.* Describe and give the results of a FRET experiment that supports one of these hypotheses.
17. Describe and show the results of an experiment that shows which base pairs are melted when RNA polymerase binds to a promoter. Explain how this procedure works.
18. Describe and show the results of an experiment that gives an estimate of the number of base pairs melted during transcription by *E. coli* RNA polymerase.
19. What regions of the σ-factor are thought to be involved in recognizing (1) the −10 box of the promoter and (2) the −35 box of the promoter? Without naming specific residues, describe the genetic evidence for these conclusions.
20. Describe a binding assay that provides biochemical evidence for interaction between σ-region 4.2 and the −35 box of the promoter.
21. Cite evidence to support the hypothesis that the α-subunit of *E. coli* RNA polymerase is involved in recognizing a promoter UP element.

22. Describe how limited proteolysis can be used to define the domains of a protein such as the α-subunit of *E. coli* RNA polymerase.
23. Describe an experiment to determine which polymerase subunit is responsible for rifampicin and streptolydigin resistance or sensitivity.
24. Describe and give the results of an experiment that shows that the β-subunit of *E. coli* RNA polymerase is near the active site that forms phosphodiester bonds.
25. Describe an RNA–DNA cross-linking experiment that demonstrates the existence of an RNA–DNA hybrid at least 8 bp long within the transcription elongation complex.
26. Draw a rough sketch of the structure of a bacterial RNA polymerase core based on x-ray crystallography. Point out the positions of the subunits of the enzyme, the catalytic center, and the rifampicin-binding site. Based on this structure, propose a mechanism for inhibition of transcription by rifampicin.
27. Based on the crystal structure of the *E. coli* elongation complex, what factors limit the length of the RNA–DNA hybrid?
28. Based on the crystal structures of the *E. coli* elongation complex with and without the antibiotic streptolydigin, propose a mechanism for the antibiotic.
29. Draw a rough sketch of the crystal structure of the holoenzyme–DNA complex in the open promoter form. Focus on the interaction between the holoenzyme and DNA. What enzyme subunit plays the biggest role in DNA binding?
30. Sigma regions 2.4 and 4.2 are known to interact with the −10 and −35 boxes of the promoter, respectively. What parts of this model are confirmed by the crystal structure of the holoenzyme–DNA complex? Provide explanations for the parts that are not confirmed.
31. Present two models for the way the RNA polymerase can maintain the bubble of melted DNA as it moves along the DNA template. Which of these models is favored by the evidence? Cite the evidence in a sentence or two.
32. What are the two important elements of an intrinsic transcription terminator? How do we know they are important? (Cite evidence.)
33. Present evidence that a hairpin is not required for pausing at an intrinsic terminator.
34. Present evidence that base-pairing (of something) with the RNA upstream of a pause site is required for intrinsic termination.
35. What does a rho-dependent terminator look like? What role is rho thought to play in such a terminator?
36. How can you show that rho causes a decrease in net RNA synthesis, but no decrease in chain initiation? Describe and show the results of an experiment.
37. Describe and show the results of an experiment that demonstrates the production of shorter transcripts in the presence of rho. This experiment should also show that rho does not simply act as a nuclease.
38. Describe and show the results of an experiment that demonstrates that rho releases transcripts from the DNA template.

ANALYTICAL QUESTIONS

1. Draw the structure of an RNA hairpin with a 10-bp stem and a 5-nt loop. Make up a sequence that will form such a structure. Show the sequence in the linear as well as the hairpin form.
2. An *E. coli* promoter recognized by the RNA polymerase holoenzyme containing σ^{70} has a −10 box with the following sequence in the nontemplate strand: 5′-CATAGT-3′. (a) Would a C→T mutation in the first position likely be an up or a down mutation? (b) Would a T→A mutation in the last position likely be an up or down mutation? Explain your answers.
3. You are carrying out experiments to study transcription termination in an *E coli* gene. You sequence the 3′-end of the gene and get the following results:

   5′ – CGAAGCGCCG**ATTGC**CGGCGCTTTTTTTTT -3′
   3′ – GCTTCGCGGC**TAACG**GCCGCGAAAAAAAAA -5′

   You then create mutant genes with this sequence changed to the following (top, or nontemplate strand, 5′→3′):

   Mutant A: CGAAACTAAG**ATTGC**AGCAGTTTTTTTTT

   Mutant B: CGAAGCGCCG**TAGCA**CGGCGCTTTTTTTTT

   Mutant C: CGAAGCGCCG**ATTGC**CGGCGCTTACGGCCC

   You put each of the mutant genes into an assay that measures termination and get the following results:

Mutant Gene Tested	Without Rho	With Rho
Wild-type gene	100% termination	100% termination
Mutant A	40% termination	40% termination
Mutant B	95% termination	95% termination
Mutant C	20% termination	80% termination

   a. Draw the structure of the RNA molecule that results from transcription of the wild-type sequence above.
   b. Explain these experimental results as completely as possible.
4. Examine the sequences below and determine the consensus sequence.

   TAGGACT – TCGCAGA – AAGCTTG – TACCAAG – TTCCTCG

SUGGESTED READINGS

General References and Reviews

Busby, S. and R.H. Ebright. 1994. Promoter structure, promoter recognition, and transcription activation in prokaryotes. *Cell* 79:743–46.

Cramer, P. 2007. Extending the message. *Nature* 448:142–43.

Epshtein, V., D. Dutta, J. Wade, and E. Nudler. 2010. An allosteric mechanism of Rho-dependent transcription termination. *Nature* 463:245–50.

Geiduschek, E.P. 1997. Paths to activation of transcription. *Science* 275:1614–16.

Helmann, J.D. and M.J. Chamberlin. 1988. Structure and function of bacterial sigma factors. *Annual Review of Biochemistry* 57:839–72.

Landick, R. 1999. Shifting RNA polymerase into overdrive. *Science* 284:598–99.

Landick, R. and J.W. Roberts. 1996. The shrewd grasp of RNA polymerase. *Science* 273:202–3.

Mooney, R.A., S.A. Darst, and R. Landick. 2005. Sigma and RNA polymerase: An on-again, off-again relationship? *Molecular Cell* 20:335–46.

Richardson, J.P. 1996. Structural organization of transcription termination factor rho. *Journal of Biological Chemistry* 271:1251–54.

Roberts, J.W. 2006. RNA polymerase, a scrunching machine. *Science* 314:1097–98.

Young, B.A., T.M. Gruber, and C.A. Gross. 2002. Views of transcription initiation. *Cell* 109:417–20.

Research Articles

Bar-Nahum, G. and E. Nudler. 2001. Isolation and characterization of σ^{70}-retaining transcription elongation complexes from *E. coli. Cell* 106:443–51.

Bautz, E.K.F., F.A. Bautz, and J.J. Dunn. 1969. *E. coli* σ factor: A positive control element in phage T4 development. *Nature* 223:1022–24.

Blatter, E.E., W. Ross, H. Tang, R.L. Gourse, and R.H. Ebright. 1994. Domain organization of RNA polymerase α subunit: C-terminal 85 amino acids constitute a domain capable of dimerization and DNA binding. *Cell* 78:889–96.

Brennan, C.A., A.J. Dombroski, and T. Platt. 1987. Transcription termination factor rho is an RNA–DNA helicase. *Cell* 48:945–52.

Burgess, R.R., A.A. Travers, J.J. Dunn, and E.K.F. Bautz. 1969. Factor stimulating transcription by RNA polymerase. *Nature* 221:43–46.

Campbell, E.A., N. Korzheva, A. Mustaev, K. Murakami, S. Nair, A. Goldfarb, and S.A. Darst. 2001. Structural mechanism for rifampicin inhibition of bacterial RNA polymerase. *Cell* 104:901–12.

Carpousis, A.J. and J.D. Gralla. 1980. Cycling of ribonucleic acid polymerase to produce oligonucleotides during initiation in vitro at the *lac* UV5 promoter. *Biochemistry* 19:3245–53.

Dombroski, A.J., W.A. Walter, M.T. Record, Jr., D.A. Siegele, and C.A. Gross. (1992). Polypeptides containing highly conserved regions of transcription initiation factor σ^{70} exhibit specificity of binding to promoter DNA. *Cell* 70:501–12.

Farnham, P.J. and T. Platt. 1980. A model for transcription termination suggested by studies on the *trp* attenuator in vitro using base analogs. *Cell* 20:739–48.

Grachev, M.A., T.I. Kolocheva, E.A. Lukhtanov, and A.A. Mustaev. 1987. Studies on the functional topography of *Escherichia coli* RNA polymerase: Highly selective affinity labelling of initiating substrates. *European Journal of Biochemistry* 163:113–21.

Hayward, R.S., K. Igarashi, and A. Ishihama. 1991. Functional specialization within the α-subunit of *Escherichia coli* RNA polymerase. *Journal of Molecular Biology* 221:23–29.

Heil, A. and W. Zillig. 1970. Reconstitution of bacterial DNA-dependent RNA polymerase from isolated subunits as a tool for the elucidation of the role of the subunits in transcription. *FEBS Letters* 11:165–71.

Hinkle, D.C. and M.J. Chamberlin. 1972. Studies on the binding of *Escherichia coli* RNA polymerase to DNA: I. The role of sigma subunit in site selection. *Journal of Molecular Biology* 70:157–85.

Hsieh, T. -s. and J.C. Wang. 1978. Physicochemical studies on interactions between DNA and RNA polymerase: Ultraviolet absorbance measurements. *Nucleic Acids Research* 5:3337–45.

Kapanidis, A.N., E. Margeat, S. O. Ho, E. Kortkhonjia, S. Weiss, and R.H. Ebright. 2006. Initial transcription by RNA polymerase proceeds through a DNA-scrunching mechanism. *Science* 314:1144–47.

Malhotra, A., E. Severinova, and S.A. Darst. 1996. Crystal structure of a σ^{70} subunit fragment from *E. coli* RNA polymerase. *Cell* 87:127–36.

Mukhopadhyay, J., A.N. Kapanidis, V. Mekler, E. Kortkhonjia, Y.W. Ebright, and R.H. Ebright. 2001. Translocation of σ^{70} with RNA polymerase during transcription: Fluorescence resonance energy transfer assay for movement relative to DNA. *Cell* 106:453–63.

Murakami, K.S., S. Masuda, E.A. Campbell, O. Muzzin, and S.A. Darst. 2002. Structural basis of transcription initiation: An RNA polymerase holoenzyme-DNA complex. *Science* 296:1285–90.

Nudler, E., A. Mustaev, E. Lukhtanov, and A. Goldfarb. 1997. The RNA–DNA hybrid maintains the register of transcription by preventing backtracking of RNA polymerase. *Cell* 89:33–41.

Paul, B.J., M.M. Barker, W. Ross, D.A. Schneider, C. Webb, J.W. Foster, and R.L. Gourse. 2004. DskA. A critical component of the transcription initiation machinery that potentiates the regulation of rRNA promoters by ppGpp and the initiating NTP. *Cell* 118:311–22.

Revyakin, A., C. Liu, R.H. Ebright, and T.R. Strick. 2006. Abortive initiation and productive initiation by RNA polymerase involve DNA scrunching. *Science* 314:1139–43.

Roberts, J.W. 1969. Termination factor for RNA synthesis. *Nature* 224:1168–74.

Ross, W., K.K. Gosink, J. Salomon, K. Igarashi, C. Zou, A. Ishihama, K. Severinov, and R.L. Gourse. 1993. A third recognition element in bacterial promoters: DNA binding by the α subunit of RNA polymerase. *Science* 262:1407–13.

Saucier, J. -M. and J.C. Wang. 1972. Angular alteration of the DNA helix by *E. coli* RNA polymerase. *Nature New Biology* 239:167–70.

Sidorenkov, I., N. Komissarova, and M. Kashlev. 1998. Crucial role of the RNA:DNA hybrid in the processivity of transcription. *Molecular Cell* 2:55–64.

Siebenlist, U. 1979. RNA polymerase unwinds an 11-base pair segment of a phage T7 promoter. *Nature* 279:651–52.

Toulokhonov, I., I. Artsimovitch, and R. Landick. 2001. Allosteric control of RNA polymerase by a site that contacts nascent RNA hairpins. *Science* 292:730–33.

Travers, A.A. and R.R. Burgess. 1969. Cyclic re-use of the RNA polymerase sigma factor. *Nature* 222:537–40.

Vassylyev, D.G., S.-i Sekine, O. Laptenko, J. Lee, M.N. Vassylyeva, S. Borukhov, and S. Yokoyama. 2002. Crystal structure of bacterial RNA polymerase holoenzyme at 2.6 Å resolution. *Nature* 417:712–19.

Vassylyev, D.G., M.N. Vassylyeva, A. Perederina, T.H. Tahirov, and I. Artsimovitch. 2007. Structural basis for transcription elongation by bacterial RNA polymerase. *Nature* 448:157–62.

Vassylyev, D.G., M.N. Vassylyeva, J. Zhang, M. Palangat, and I. Artsimovitch. 2007. Structural basis for substrate loading in bacterial RNA polymerase. *Nature* 448:163–68.

Yarnell, W.S. and J.W. Roberts. 1999. Mechanism of intrinsic transcription termination and antitermination. *Science* 284:611–15.

Young, B.A., L.C. Anthony, T.M. Gruber, T.M. Arthur, E. Heyduk, C.Z. Lu, M.M. Sharp, T. Heyduk, R.R. Burgess, and C.A. Gross. 2001. A coiled-coil from the RNA polymerase β′ subunit allosterically induces selective nontemplate strand binding by σ^{70}. *Cell* 105:935–44.

Zhang, G., E.A. Campbell, L. Minakhin, C. Richter, K. Severinov, and S.A. Darst. 1999. Crystal structure of *Thermus aquaticus* core RNA polymerase at 3.3 Å resolution. *Cell* 98:811–24.

Zhang, G. and S.A. Darst. 1998. Structure of the *Escherichia coli* RNA polymerase α subunit amino terminal domain. *Science* 281:262–66.

CHAPTER 7

Operons: Fine Control of Bacterial Transcription

X-ray crystal structure of the *lac* repressor tetramer bound to two operator fragments. *Lewis et al, Crystal structure of the lactose operon repressor and its complexes with DNA and inducer. Science 271 (1 Mar 1996), f. 6, p. 1251. © AAAS*

The *E. coli* genome contains over 3000 genes. Some of these are active all the time because their products are in constant demand. But some of them are turned off most of the time because their products are rarely needed. For example, the enzymes required for the metabolism of the sugar arabinose would be useful only when arabinose is present and when the organism's favorite energy source, glucose, is absent. Such conditions are not common, so the genes encoding these enzymes are usually turned off. Why doesn't the cell just leave all its genes on all the time, so the right enzymes are always there to take care of any eventuality? The reason is that gene expression is an expensive process. It takes a lot of energy to produce RNA and protein. In fact, if all of an *E. coli* cell's genes were turned on all the time, production of RNAs and proteins would drain the cell of so much energy that it could not compete with more efficient organisms. Thus, control of gene expression is essential to life. In this chapter we will explore one strategy bacteria employ to control the expression of their genes: by grouping functionally related genes together so they can be regulated together easily. Such a group of contiguous, coordinately controlled genes is called an *operon.*

7.1 The *lac* Operon

The first operon to be discovered has become the prime example of the operon concept. It contains three genes that code for the proteins that allow *E. coli* cells to use the sugar **lactose,** hence the name ***lac* operon.** Consider a flask of *E. coli* cells growing on a medium containing the sugars **glucose** and **lactose** (Figure 7.1). The cells exhaust the glucose and stop growing. Can they adjust to the new nutrient source? For a short time it appears that they cannot; but then, after a lag period of about an hour, growth resumes. During the lag, the cells have been turning on the *lac* operon and beginning to accumulate the enzymes they need to metabolize lactose. The growth curve in Figure 7.1 is called "diauxic" from the Latin *auxilium,* meaning help, because the two sugars help the bacteria grow.

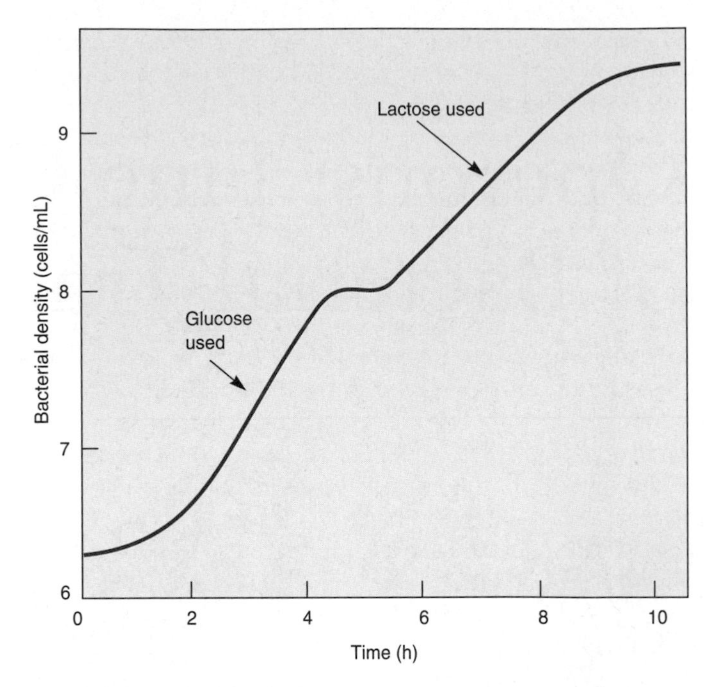

Figure 7.1 Diauxic growth. *E. coli* cells are grown on a medium containing both glucose and lactose, and the bacterial density (number of cells/mL) is plotted versus time in hours. The cells grow rapidly on glucose until that sugar is exhausted, then growth levels off while the cells induce the enzymes needed to metabolize lactose. As those enzymes appear, growth resumes.

What are these enzymes? First, the bacteria need an enzyme to transport the lactose into the cells. The name of this enzyme is **galactoside permease.** Next, the cells need an enzyme to break the lactose down into its two component sugars: galactose and glucose. Figure 7.2 shows this reaction. Because lactose is composed of two simple sugars, we call it a *disaccharide.* These six-carbon sugars, galactose and glucose, are joined together by a linkage called a β-galactosidic bond. Lactose is therefore called a β-galactoside, and the enzyme that cuts it in half is called **β-galactosidase.** The genes for these two enzymes, galactoside permease and β-galactosidase, are found side by side in the *lac* operon, along with another structural gene—for **galactoside transacetylase**—whose function in lactose metabolism is still unclear.

The three genes coding for enzymes that carry out lactose metabolism are grouped together in the following order: β-galactosidase ***(lacZ)***, galactoside permease ***(lacY)***, galactoside transacetylase ***(lacA)***. They are all transcribed together to produce one messenger RNA, called a **polycistronic message,** starting from a single promoter. Thus, they can all be controlled together simply by controlling that promoter. The term *polycistronic* comes from **cistron,** which is a synonym for *gene.* Therefore, a polycistronic message is simply a message with information from more than one gene. Each cistron in the mRNA has its own ribosome binding site, so each cistron can be translated by separate ribosomes that bind independently of each other.

As mentioned at the beginning of this chapter, the *lac* operon (like many other operons) is tightly controlled.

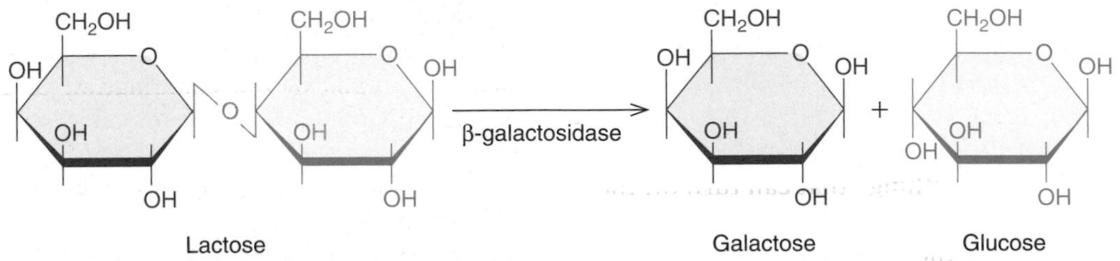

Figure 7.2 The β-galactosidase reaction. The enzyme breaks the β-galactosidic bond (gray) between the two sugars, galactose (pink) and glucose (blue), that compose lactose.

In fact, two types of control are operating. First is **negative control,** which is like the brake of a car: You need to release the brake for the car to move. The "brake" in negative control is a protein called the ***lac* repressor,** which keeps the operon turned off (or repressed) as long as lactose is absent. That is economical; it would be wasteful for the cell to produce enzymes that use an absent sugar.

If negative control is like the brake of a car, **positive control** is like the accelerator pedal. In the case of the *lac* operon, removing the repressor from the operator (releasing the brake) is not enough to activate the operon. An additional positive factor called an **activator** is needed. We will see that the activator responds to low glucose levels by stimulating transcription of the *lac* operon, but high glucose levels keep the concentration of the activator low, so transcription of the operon cannot be stimulated. The advantage of this positive control system is that it keeps the operon turned nearly off when the level of glucose is high. If there were no way to respond to glucose levels, the presence of lactose alone would suffice to activate the operon. But that is inappropriate when glucose is still available, because *E. coli* cells metabolize glucose more easily than lactose; it would therefore be wasteful for them to activate the *lac* operon in the presence of glucose.

SUMMARY Lactose metabolism in *E. coli* is carried out by two enzymes, with possible involvement by a third. The genes for all three enzymes are clustered together and transcribed together from one promoter, yielding a polycistronic message. These three genes, linked in function, are therefore also linked in expression. They are turned off and on together. Negative control keeps the *lac* operon repressed in the absence of lactose, and positive control keeps the operon relatively inactive in the presence of glucose, even when lactose is present.

Negative Control of the *lac* Operon

Figure 7.3 illustrates one aspect of *lac* operon regulation: the classical version of negative control. We will see later in this chapter and in Chapter 9 that this classical view is oversimplified, but it is a useful way to begin consideration of the operon concept. The term "negative control" implies that the operon is turned on unless something intervenes to stop it. The "something" that can turn off the *lac* operon is the ***lac* repressor.** This repressor, the product of a regulatory gene called the ***lacI* gene** shown at the extreme left in Figure 7.3, is a tetramer of four identical polypeptides; it binds to the **operator** just to the right of the promoter. When the repressor is bound to the operator, the operon is **repressed.** That is because the operator and promoter are contiguous, and when the repressor occupies the operator, it appears to prevent RNA polymerase from binding to the promoter and transcribing the operon. Because its genes are not transcribed, the operon is off, or repressed.

The *lac* operon is repressed as long as no lactose is available. On the other hand, when all the glucose is gone and lactose is present, a mechanism should exist for removing the repressor so the operon can be derepressed to take advantage of the new nutrient. How does this mechanism work? The repressor is a so-called **allosteric protein:** one in which the binding of one molecule to the protein changes the shape of a remote site on the protein and alters its interaction with a second molecule (Greek: *allos,* meaning other + *stereos,* meaning shape). The first molecule in this case is called the **inducer** of the *lac* operon because it binds to the repressor, causing the protein to change to a conformation that favors dissociation from the operator (the second molecule), thus inducing the operon (Figure 7.3b).

What is the nature of this inducer? It is actually an alternative form of lactose called **allolactose** (again, Greek: *allos,* meaning other). When β-galactosidase cleaves lactose to galactose plus glucose, it rearranges a small fraction of the lactose to allolactose. Figure 7.4 shows that allolactose is just galactose linked to glucose in a different way than in lactose. (In lactose, the linkage is through a β-1,4 bond; in allolactose, the linkage is β-1,6.)

You may be asking yourself: How can lactose be metabolized to allolactose if no permease is present to get it into the cell and no β-galactosidase exists to perform the metabolizing because the *lac* operon is repressed? The answer is that repression is somewhat leaky, and a low basal level of the *lac* operon products is always present. This is enough to get the ball rolling by producing a little inducer. It does not take much inducer to do the job, because only about 10 tetramers of repressor are present per cell. Furthermore, the derepression of the operon will snowball as more and more operon products are available to produce more and more inducer.

Discovery of the Operon

The development of the operon concept by François Jacob and Jacques Monod and their colleagues was one of the classic triumphs of the combination of genetic and biochemical analysis. The story begins in 1940, when Monod began studying the inducibility of lactose metabolism in *E. coli.* Monod learned that an important feature of lactose metabolism was β-galactosidase, and that this enzyme was inducible by lactose and by other galactosides. Furthermore, he and Melvin Cohn had used an anti-β-galactosidase antibody to detect β-galactosidase protein, and they

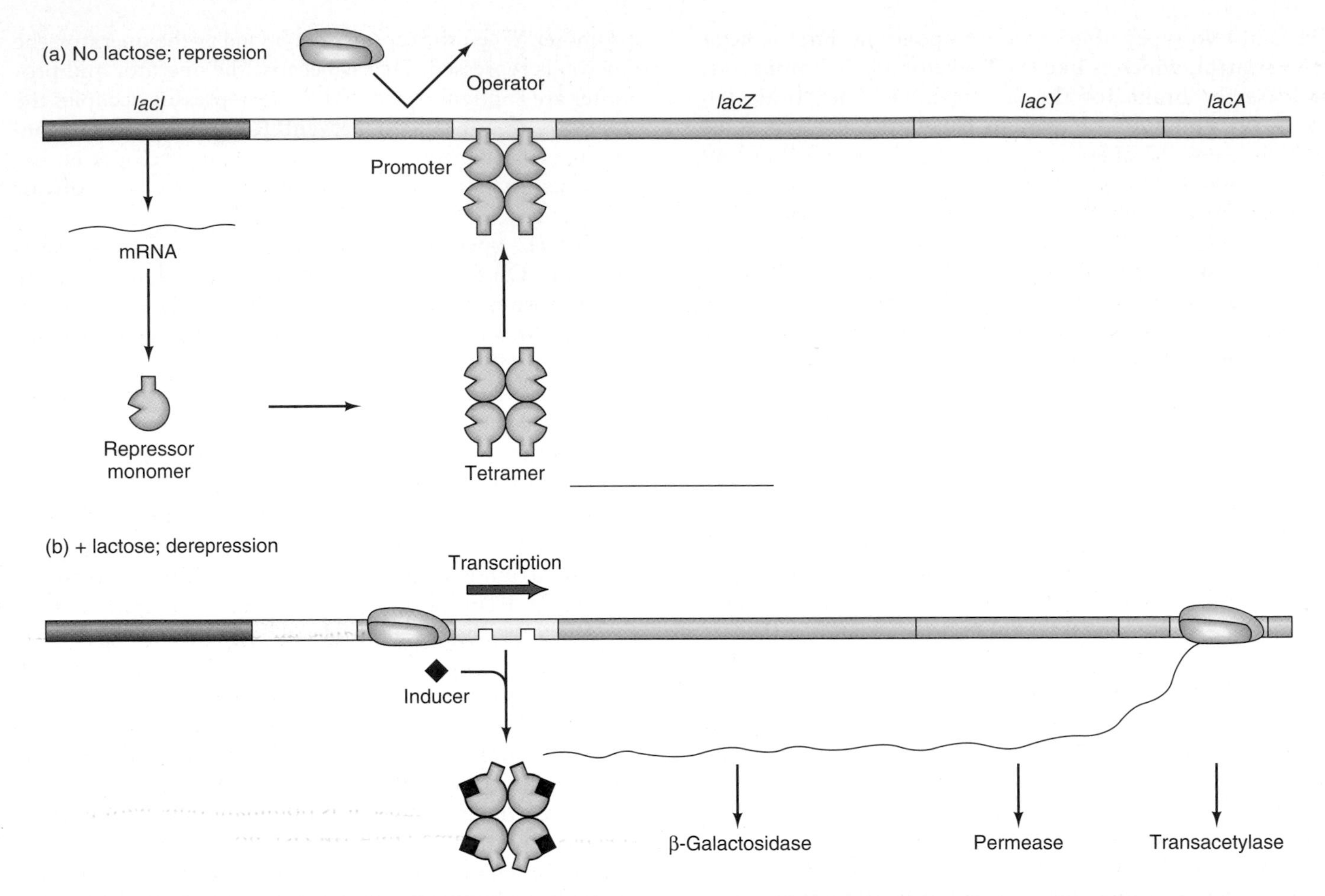

Figure 7.3 Negative control of the *lac* operon. (a) No lactose; repression. The *lacI* gene produces repressor (green), which binds to the operator and blocks RNA polymerase from transcribing the *lac* genes. **(b)** Presence of lactose, derepression. The inducer (black) binds to repressor, changing it to a form (bottom) that no longer binds well to the operator. This removes the repressor from the operator, allowing RNA polymerase to transcribe the structural genes. This produces a polycistronic mRNA that is translated to yield β-galactosidase, permease, and transacetylase.

showed that the amount of this protein increased on induction. Because more gene product appeared in response to lactose, the β-galactosidase gene itself was apparently being induced.

To complicate matters, certain mutants (originally called "cryptic mutants") were found that could make β-galactosidase but still could not grow on lactose. What was missing in these mutants? To answer this question, Monod and his coworkers added a radioactive galactoside to wild-type and mutant bacteria. They found that uninduced wild-type cells did not take up the galactoside, and neither did the mutants, even if they were induced. Induced

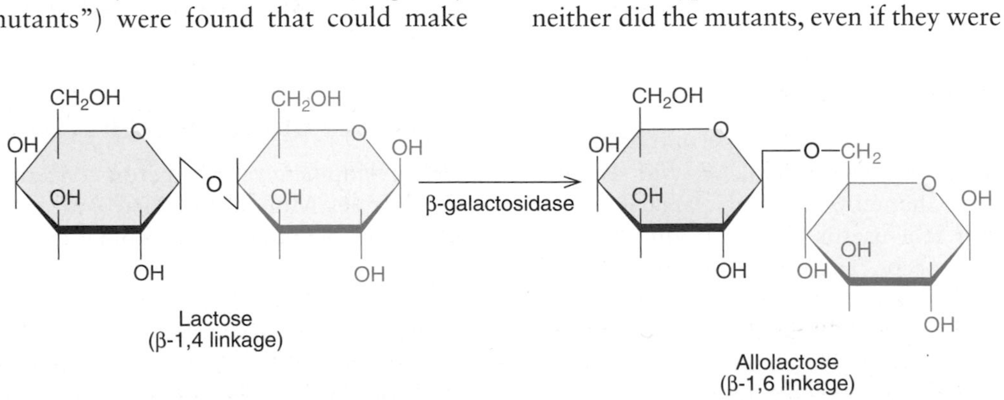

Figure 7.4 Conversion of lactose to allolactose. A side reaction carried out by β-galactosidase rearranges lactose to the inducer, allolactose. Note the change in the galactosidic bond from β-1,4 to β-1,6.

wild-type cells did accumulate the galactoside. This revealed two things: First, a substance (galactoside permease) is induced along with β-galactosidase in wild-type cells and is responsible for transporting galactosides into the cells; second, the mutants seem to have a defective gene (Y^-) for this substance (Table 7.1).

Monod named this substance galactoside permease, and then endured criticism from his colleagues for naming a protein before it had been isolated. He later remarked, "This attitude reminded me of that of two traditional English gentlemen who, even if they know each other well by name and by reputation, will not speak to each other before having been formally introduced." In their efforts to purify galactoside permease, Monod and his colleagues identified another protein, galactoside transacetylase, which is induced along with β-galactosidase and galactoside permease.

Thus, by the late 1950s, Monod knew that three enzyme activities (and therefore presumably three genes) were induced together by galactosides. He had also found some mutants, called **constitutive mutants,** that needed no induction. They produced the three gene products all the time. Monod realized that further progress would be greatly accelerated by genetic analysis, so he teamed up with François Jacob, who was working just down the hall at the Pasteur Institute.

In collaboration with Arthur Pardee, Jacob and Monod created **merodiploids** (partial diploid bacteria) carrying both the wild-type (inducible) and constitutive alleles. The inducible allele proved to be dominant, demonstrating that wild-type cells produce some substance that keeps the *lac* genes turned off unless they are induced. Because this substance turned off the genes from the constitutive as well as the inducible parent, it made the merodiploids inducible. Of course, this substance is the *lac* repressor. The constitutive mutants had a defect in the gene *(lacI)* for this repressor. These mutants are therefore *lacI*$^-$ (Figure 7.5a).

The existence of a repressor required that some specific DNA sequence exists to which the repressor would bind. Jacob and Monod called this the operator. The specificity of this interaction suggested that it should be subject to genetic mutation; that is, some mutations in the operator should abolish its interaction with the repressor. These would also be constitutive mutations, so how can they be distinguished from constitutive mutations in the repressor gene?

Jacob and Monod realized that they could make this distinction by determining whether the mutation was dominant or recessive. Because the repressor gene produces a repressor protein that can diffuse throughout the cell, it can bind to both operators in a merodiploid. We call such a gene ***trans*-acting** because it can act on loci on both DNA molecules in the merodiploid (Latin: *trans,* meaning across). A mutation in one of the repressor genes will still leave the other repressor gene undamaged, so its wild-type product can still diffuse to both operators and turn them off. In other words, both *lac* operons in the merodiploid would still be repressible. Thus, such a mutation should be recessive (Figure 7.5a), and we have already observed that it is.

Table 7.1 Effect of Cryptic Mutant (*lacY*$^-$) on Accumulation of Galactoside

Genotype	Inducer	Accumulation of Galactoside
Z^+Y^+	–	–
Z^+Y^+	+	+
Z^+Y^- (cryptic)	–	–
Z^+Y^- (cryptic)	+	–

On the other hand, because an operator controls only the operon on the same DNA molecule, we call it ***cis*-acting** (Latin: *cis,* meaning here). Thus, a mutation in one of the operators in a merodiploid should render the operon on that DNA molecule unrepressable, but should not affect the operon on the other DNA molecule. We call such a mutation ***cis*-dominant** because it is dominant only with respect to genes on the same DNA (***in cis***), not on the other DNA in the merodiploid (***in trans***). Jacob and Monod did indeed find such *cis*-dominant mutations, and they proved the existence of the operator. These mutations are called O^c, for **operator constitutive.**

What about mutations in the repressor gene that render the repressor unable to respond to inducer? Such mutations should make the *lac* operon uninducible and should be dominant both *in cis* and *in trans* because the mutant repressor will remain bound to both operators even in the presence of inducer or of wild-type repressor (Figure 7.5c). Monod and his colleagues found two such mutants, and Suzanne Bourgeois later found many others. These are named I^s to distinguish them from constitutive repressor mutants (I^-), which make a repressor that cannot recognize the operator.

Both of the common kinds of constitutive mutants (I^- and O^c) affected all three of the *lac* genes (*Z, Y,* and *A*) in the same way. The genes had already been mapped and were found to be adjacent on the chromosome. These findings strongly suggested that the operator lay near these three genes.

We now recognize yet another class of repressor mutants, those that are constitutive and dominant (I^{-d}). This kind of mutant gene (Figure 7.5d) makes a defective product that can still form tetramers with wild-type repressor monomers. However, the defective monomers spoil the activity of the whole tetramer so it cannot bind to the operator. Hence the dominant nature of this mutation. These mutations are not just *cis*-dominant because the "spoiled" repressors cannot bind to either operator in a merodiploid. This kind of "spoiler" mutation is widespread in nature, and it is called by the generic name **dominant-negative.**

Thus, Jacob and Monod, by skillful genetic analysis, were able to develop the operon concept. They predicted the existence of two key control elements: the repressor gene and the operator. Deletion mutations revealed a third element (the promoter) that was necessary for expression of all three *lac* genes. Furthermore, they could conclude that all three *lac* genes (*lacZ, Y,* and *A*) were clustered into a single control unit: the *lac* operon. Subsequent biochemical studies have amply confirmed Jacob and Monod's beautiful hypothesis.

SUMMARY Negative control of the *lac* operon occurs as follows: The operon is turned off as long as the repressor binds to the operator, because the repressor keeps RNA polymerase from transcribing the three *lac* genes. When the supply of glucose is exhausted and lactose is available, the few molecules of *lac* operon enzymes produce a few molecules of allolactose from the lactose. The allolactose acts as an inducer by binding to the repressor and causing a conformational shift that encourages dissociation from the operator. With the repressor removed, RNA polymerase is free to transcribe the three *lac* genes. A combination of genetic and biochemical experiments revealed the two key elements of negative control of the *lac* operon: the operator and the repressor.

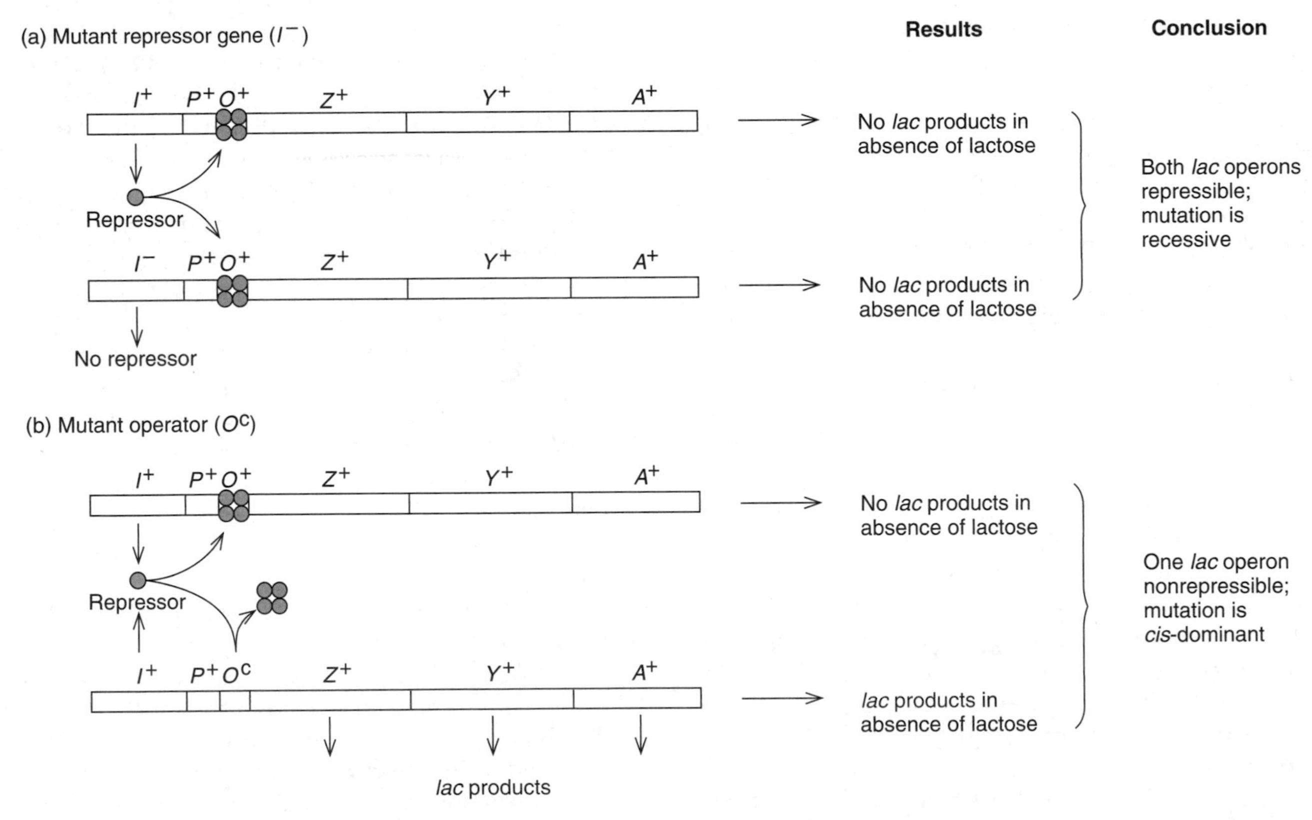

Figure 7.5 Effects of regulatory mutations in the *lac* operon in merodiploids. Jacob, Monod, and others created merodiploid *E. coli* strains as described in panels (a)–(d) and tested them for *lac* products in the presence and absence of lactose. **(a)** This merodiploid has one wild-type operon (top) and one operon (bottom) with a mutation in the repressor gene (I^-). The wild-type repressor gene (I^+) makes enough normal repressor (green) to repress both operons, so the I^- mutation is recessive. **(b)** This merodiploid has one wild-type operon (top) and one operon (bottom) with a mutation in the operator (O^c) that makes it defective in binding repressor (green). The wild-type operon remains repressible, but the mutant operon is not; it makes *lac* products even in the absence of lactose. Because only the operon connected to the mutant operator is affected, this mutation is *cis*-dominant.

(continued)

Repressor–Operator Interactions

After the pioneering work of Jacob and Monod, Walter Gilbert and Benno Müller-Hill succeeded in partially purifying the *lac* repressor. This work is all the more impressive, considering that it was done in the 1960s, before the advent of modern gene cloning. Gilbert and Müller-Hill's challenge was to purify a protein (the *lac* repressor) that is present in very tiny quantities in the cell, without an easy assay to identify the protein. The most sensitive assay available to them was binding a labeled synthetic inducer (isopropylthiogalactoside, or IPTG) to the repressor. But, with a crude extract of wild-type cells, the repressor was in such low concentration that this assay could not detect it. To get around this problem, Gilbert and Müller-Hill used a mutant *E. coli* strain with a repressor mutation (*lacI*^{t}) that causes the repressor to bind IPTG more tightly than normal. This tight binding allowed the mutant repressor to bind enough inducer that the protein could be detected even in very impure extracts. Because they could detect the protein, Gilbert and Müller-Hill could purify it.

Melvin Cohn and his colleagues used repressor purified by this technique in operator-binding studies. To assay repressor–operator binding, Cohn and colleagues used the nitrocellulose filter-binding assay we discussed in Chapters 5 and 6. If repressor–operator interaction worked normally, we would expect it to be blocked by inducer. Indeed, Figure 7.6 shows a typical saturation curve for repressor–operator binding in the absence of inducer, but no binding in the presence of the synthetic inducer, IPTG. In another binding experiment (Figure 7.7), Cohn and coworkers showed that DNA containing the constitutive mutant operator (*lacO*^{c}) required a higher concentration of repressor to achieve full binding than did the wild-type operator. This was an important

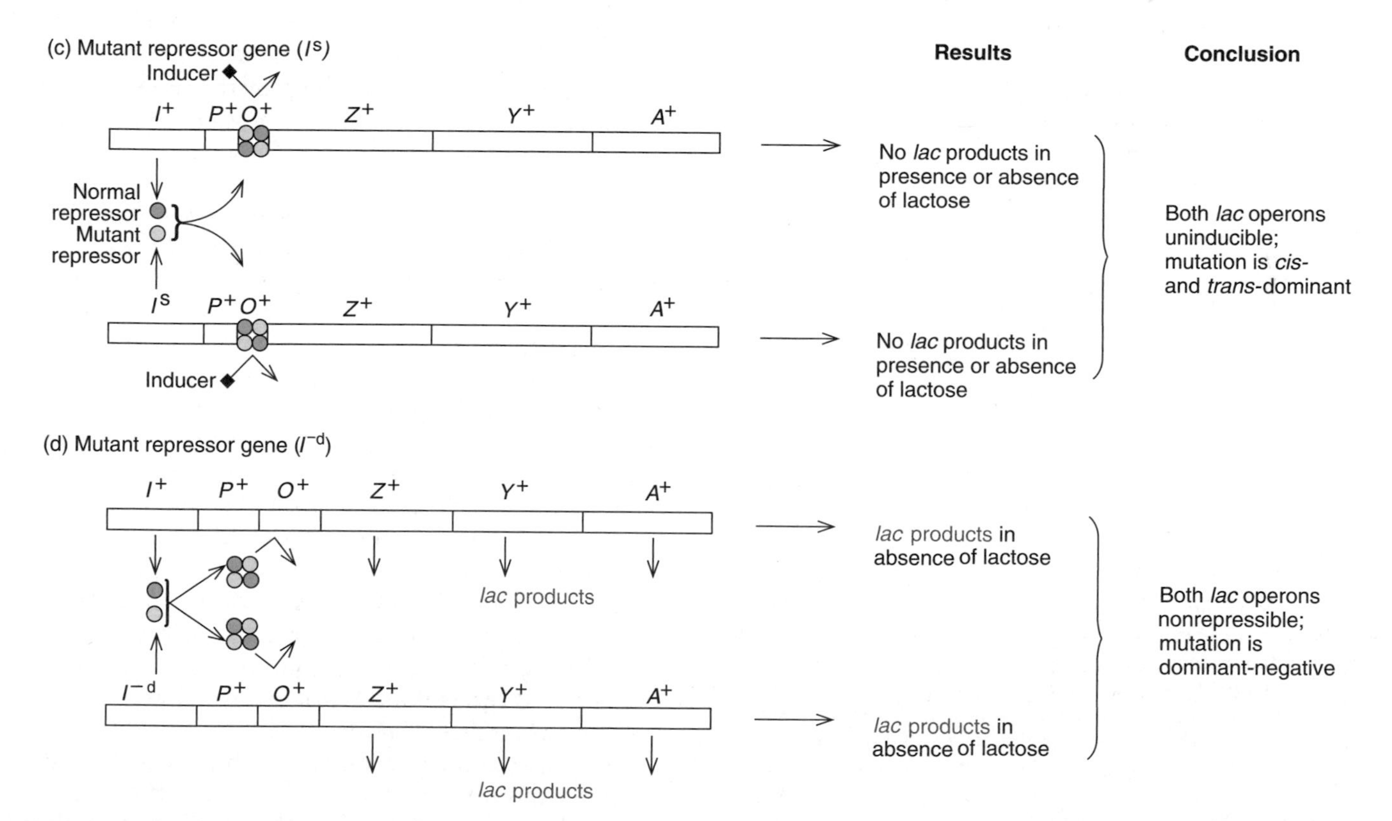

Figure 7.5 *(continued)* (c) This merodiploid has one wild-type operon (top) and one operon (bottom) with a mutant repressor gene (*Is*) whose product (yellow) cannot bind inducer. The mutant repressor therefore binds irreversibly to both operators and renders both operons uninducible. This mutation is therefore dominant. Notice that these repressor tetramers containing some mutant and some wild-type subunits behave as mutant proteins. That is, they remain bound to the operator even in the presence of inducer. **(d)** This merodiploid has one wild-type operon (top) and one operon (bottom) with a mutant repressor gene (I^{-d}) whose product (yellow) cannot bind to the *lac* operator. Moreover, mixtures (heterotetramers) composed of both wild-type and mutant repressor monomers still cannot bind to the operator. Thus, because the operon remains derepressed even in the absence of lactose, this mutation is dominant. Furthermore, because the mutant protein poisons the activity of the wild-type protein, we call the mutation dominant-negative.

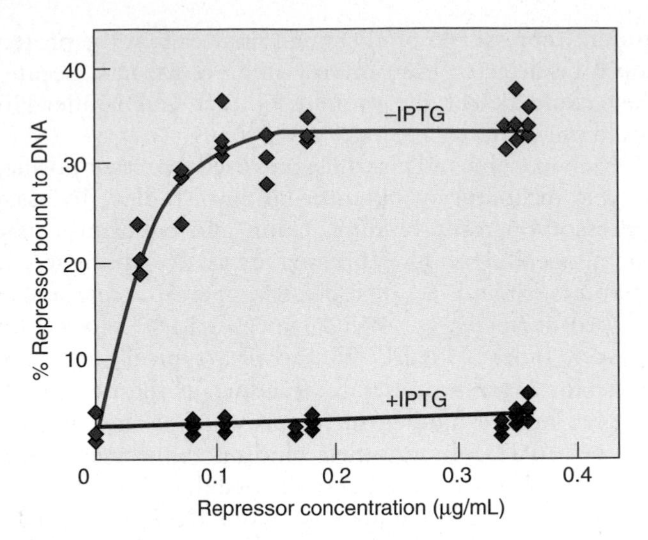

Figure 7.6 Assaying the binding between *lac* operator and *lac* repressor. Cohn and colleagues labeled *lacO*-containing DNA with ^{32}P and added increasing amounts of *lac* repressor. They assayed binding between repressor and operator by measuring the radioactivity attached to nitrocellulose. Only labeled DNA bound to repressor would attach to nitrocellulose. Red: repressor bound in the absence of the inducer IPTG. Blue: repressor bound in the presence of 1 mM IPTG, which prevents repressor–operator binding. (*Source:* Adapted from Riggs, A.D., et al.,1968. DNA binding of the *lac* repressor, *Journal of Molecular Biology,* Vol. 34: 366.)

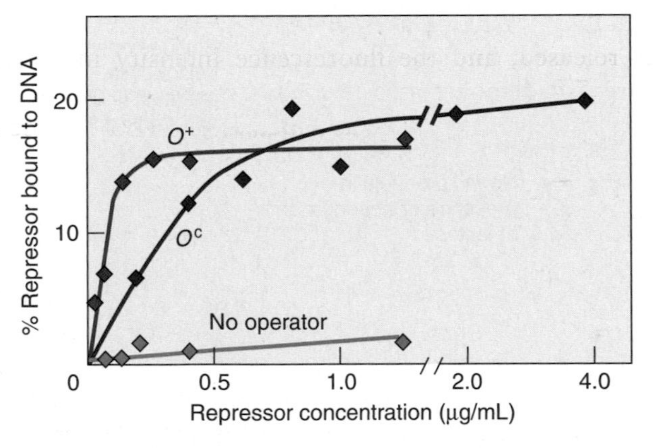

Figure 7.7 The O^c *lac* operator binds repressor with lower affinity than does the wild-type operator. Cohn and colleagues performed a *lac* operator–repressor binding assay as described in Figure 7.6, using three different DNAs as follows: red, DNA containing a wild-type operator (O^+); blue, DNA containing an operator-constitutive mutation (O^c) that binds repressor with a lower affinity; green, control, λϕ80 DNA, which does not have a *lac* operator. (*Source:* Adapted from Riggs, A.D., et al. 1968. DNA binding of the *lac* repressor. *Journal of Molecular Biology,* Vol. 34: 366.)

demonstration: What Jacob and Monod had defined genetically as the operator really was the binding site for repressor. If it were not, then mutating it should not have affected repressor binding.

SUMMARY Cohn and colleagues demonstrated with a filter-binding assay that *lac* repressor binds to *lac* operator. Furthermore, this experiment showed that a genetically defined constitutive *lac* operator has lower than normal affinity for the *lac* repressor, demonstrating that the sites defined genetically and biochemically as the operator are one and the same.

The Mechanism of Repression

For years it was assumed that the *lac* repressor acted by denying RNA polymerase access to the promoter, in spite of the fact that Ira Pastan and his colleagues had shown as early as 1971 that RNA polymerase could bind tightly to the *lac* promoter, even in the presence of repressor. Pastan's experimental plan was to incubate polymerase with DNA containing the *lac* operator in the presence of repressor, then to add inducer (IPTG) and rifampicin together. As we will see later in this chapter, rifampicin will inhibit transcription unless an open promoter complex has already formed. (Recall from Chapter 6 that an open promoter complex is one in which the RNA polymerase has caused local DNA melting at the promoter and is tightly bound there.) In this case, transcription did occur, showing that the *lac* repressor had not prevented the formation of an open promoter complex. Thus, these results suggested that the repressor does not block access by RNA polymerase to the *lac* promoter. Susan Straney and Donald Crothers reinforced this view in 1987 by showing that polymerase and repressor can bind together to the *lac* promoter.

If we accept that RNA polymerase can bind tightly to the promoter, even with repressor occupying the operator, how do we explain repression? Straney and Crothers suggested that repressor blocks the formation of an open promoter complex, but that would be hard to reconcile with the rifampicin resistance of the complex observed by Pastan. Barbara Krummel and Michael Chamberlin proposed an alternative explanation: Repressor blocks the transition from the initial transcribing complex state (Chapter 6) to the elongation state. In other words, repressor traps the polymerase in a nonproductive state in which it spins its wheels making abortive transcripts without ever achieving promoter clearance.

Jookyung Lee and Alex Goldfarb provided some evidence for this idea. First, they used a run-off transcription assay (Chapter 5) to show that RNA polymerase is already engaged on the DNA template, even in the presence of repressor. The experimental plan was as follows: First, they incubated repressor with a 123-bp DNA fragment containing the *lac* control region plus the beginning of the *lacZ* gene. After allowing 10 min for the repressor to bind to the operator, they added polymerase. Then they added heparin—a polyanion that binds to any RNA polymerase that is free or loosely bound to DNA and keeps it from binding to DNA. They also added all the remaining components of the RNA polymerase reaction except CTP. Finally, they added labeled CTP with or without the inducer IPTG. The question is this: Will a run-off transcript be made? If so, the RNA polymerase has formed a heparin-resistant (open) complex with the promoter even in the presence of the repressor. In fact, as Figure 7.8 shows, the run-off transcript did appear, just as if repressor had not been present. Thus, under these conditions in vitro, repressor does not seem to inhibit tight binding between polymerase and the *lac* promoter.

Figure 7.8 RNA polymerase forms an open promoter complex with the *lac* promoter even in the presence of *lac* repressor in vitro. Lee and Goldfarb incubated a DNA fragment containing the *lac* UV5 promoter with (lanes 2 and 3) or without (lane 1) *lac* repressor (LacR). After repressor–operator binding had occurred, they added RNA polymerase. After allowing 20 min for open promoter complexes to form, they added heparin to block any further complex formation, along with all the other reaction components except CTP. Finally, after 5 more minutes, they added [α-^{32}P]CTP alone or with the inducer IPTG. They allowed 10 more minutes for RNA synthesis and then electrophoresed the transcripts. Lane 3 shows that transcription occurred even when repressor bound to the DNA before polymerase could. Thus, repressor did not prevent polymerase from binding and forming an open promoter complex. (*Source:* Lee J., and Goldfarb A., *lac* repressor acts by modifying the initial transcribing complex so that it cannot leave the promoter. *Cell* 66 (23 Aug 1991) f. 1, p. 794. Reprinted by permission of Elsevier Science.)

If it does not inhibit transcription of the *lac* operon by blocking access to the promoter, how would the *lac* repressor function? Lee and Goldfarb noted the appearance of shortened abortive transcripts (Chapter 6), only about 6 nt long, in the presence of repressor. Without repressor, the abortive transcripts reached a length of 9 nt. The fact that *any* transcripts—even short ones—were made in the presence of repressor reinforced the conclusion that, at least under these conditions, RNA polymerase really can bind to the *lac* promoter in the presence of repressor. This experiment also suggested that repressor may limit *lac* operon transcription by locking the polymerase into a nonproductive state in which it can make only abortive transcripts. Thus, extended transcription cannot get started.

One problem with the studies of Lee and Goldfarb and the others just cited is that they were performed in vitro under rather nonphysiological conditions. For example, the concentrations of the proteins (RNA polymerase and repressor) were much higher than they would be in vivo. To deal with such problems, Thomas Record and colleagues performed kinetic studies in vitro under conditions likely to prevail in vivo. They formed RNA polymerase/*lac* promoter complexes, then measured the rate of abortive transcript synthesis by these complexes alone, or after addition of either heparin or *lac* repressor. They measured transcription by using a UTP analog with a fluorescent tag on the γ-phosphate (*pppU). When UMP was incorporated into RNA, tagged pyrophosphate (*pp) was released, and the fluorescence intensity increased. Figure 7.9 demonstrates that the rate of abortive transcript synthesis continued at a high level in the absence of competitor, but rapidly leveled off in the presence of either heparin or repressor.

Record and colleagues explained these results as follows: The polymerase–promoter complex is in equilibrium with free polymerase and promoter. Moreover, in the absence of competitor (curve 1), the polymerases that dissociate go right back to the promoter and continue making abortive transcripts. However, both heparin (curve 2) and repressor (curve 3) prevent such reassociation. Heparin does so by binding to the polymerase and preventing its association with DNA. But the repressor presumably does so by binding to the operator adjacent to the promoter and blocking access to the promoter by RNA polymerase. Thus, these data support the old hypothesis of a competition between polymerase and repressor.

We have seen that the story of the *lac* repressor mechanism has had many twists and turns. Have we seen the last twist? The latest results suggest that the original, competition hypothesis is correct, but we may not have heard the end of the story yet.

Another complicating factor in repression of the *lac* operon is the presence of not one, but three operators: one *major*

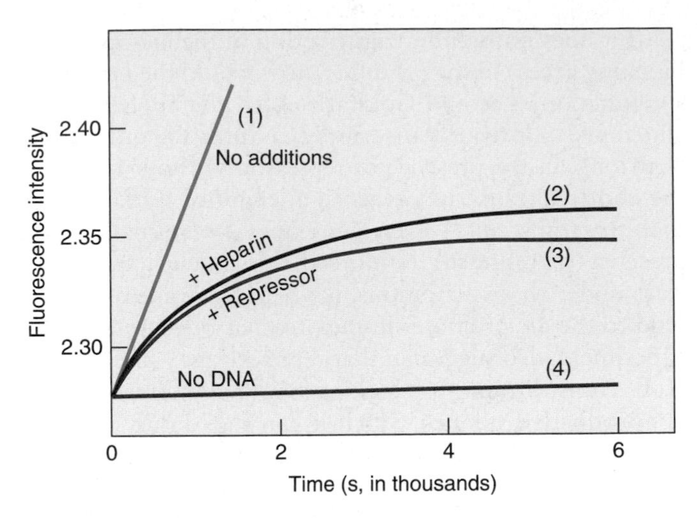

Figure 7.9 Effect of *lac* repressor on dissociation of RNA polymerase from the *lac* promoter. Record and colleagues made complexes between RNA polymerase and DNA containing the *lac* promoter–operator region. Then they allowed the complexes to synthesize abortive transcripts in the presence of a UTP analog fluorescently labeled in the γ-phosphate. As the polymerase incorporates UMP from this analog into transcripts, the labeled pyrophosphate released increases in fluorescence intensity. The experiments were run with no addition (curve 1, green), with heparin to block reinitiation by RNA polymerase that dissociates from the DNA (curve 2, blue), and with a low concentration of *lac* repressor (curve 3, red). A control experiment was run with no DNA (curve 4, purple). The repressor inhibited reinitiation of abortive transcription as well as heparin, suggesting that it blocks dissociated RNA polymerase from reassociating with the promoter. (*Source:* Adapted from Schlax, P.J., Capp, M.W., and M.T. Record, Jr. Inhibition of transcription initiation by *lac* repressor, *Journal of Molecular Biology* 245: 331–50.)

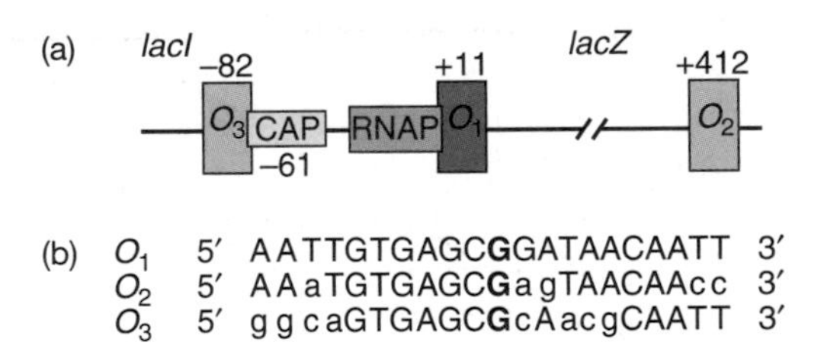

Figure 7.10 The three *lac* operators. (a) Map of the *lac* control region. The major operator (O_1) is shown in red; the two auxiliary operators are shown in pink. The CAP and RNA polymerase binding sites are in yellow and blue, respectively. CAP is a positive regulator of the *lac* operon discussed in the next section of this chapter. **(b)** Sequences of the three operators. The sequences are aligned, with the central G of each in boldface. Sites at which the auxiliary operator sequences differ from the major operator are lower case in the O_2 and O_3 sequences.

operator near the transcription start site and two *auxiliary operators* (one upstream and one downstream). Figure 7.10 shows the spatial arrangement of these operators, the classical (major) operator O_1, centered at position +11, the downstream auxiliary operator O_2, centered at position +412, and the upstream auxiliary operator O_3, centered at position −82. We have already discussed the classical operator, and the role investigators have traditionally ascribed to it alone. But Müller-Hill and others have more recently investigated the auxiliary operators and have discovered that they are not just trivial copies of the major operator. Instead, they play a significant role in repression. Müller-Hill and colleagues demonstrated this role by showing that removal of either of the auxiliary operators decreased repression only slightly, but removal of both auxiliary operators decreased repression about 50-fold. Figure 7.11 outlines the results of these experiments and shows that all three operators together repress transcription 1300-fold, two operators together repress from 440- to 700-fold, but the classical operator by itself represses only 18-fold.

In 1996, Mitchell Lewis and coworkers provided a structural basis for this cooperativity among operators. They determined the crystal structure of the *lac* repressor and its complexes with 21-bp DNA fragments containing operator sequences. Figure 7.12 summarizes their findings. We can see that the two dimers in a repressor tetramer are independent DNA-binding entities that interact with the major groove of the DNA. It is also clear that the two dimers within the tetramer are bound to separate operator sequences. It is easy to imagine these two operators as part of a single long piece of DNA.

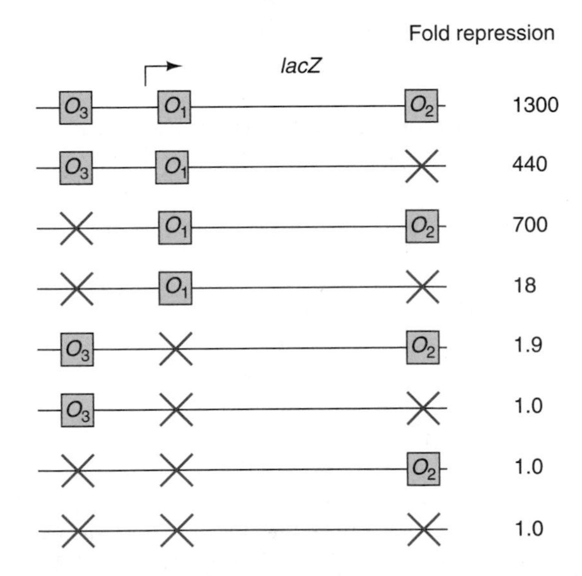

Figure 7.11 Effects of mutations in the three *lac* operators. Müller-Hill and colleagues placed wild-type and mutated *lac* operon fragments on λ phage DNA and allowed these DNAs to lysogenize *E. coli* cells (Chapter 8). This introduced these *lac* fragments, containing the three operators, the *lac* promoter, and the *lacZ* gene, into the cellular genome. The cell contained no other *lacZ* gene, but it had a wild-type *lacI* gene. Then Müller-Hill and coworkers assayed for β-galactosidase produced in the presence and absence of the inducer IPTG. The ratio of activity in the presence and absence of inducer is the repression given at right. For example, the repression observed with all three operators was 1300-fold. λ Ewt 123 (top) was wild-type in all three operators (green). All the other phages had one or more operators deleted (red X). *Source:* Adapted from Oehler, S., E.R. Eismann, H. Krämer, and B. Müller-Hill. 1990. The three operators of the *lac* operon cooperate in repression. *The EMBO Journal* 9:973–79.

(a)

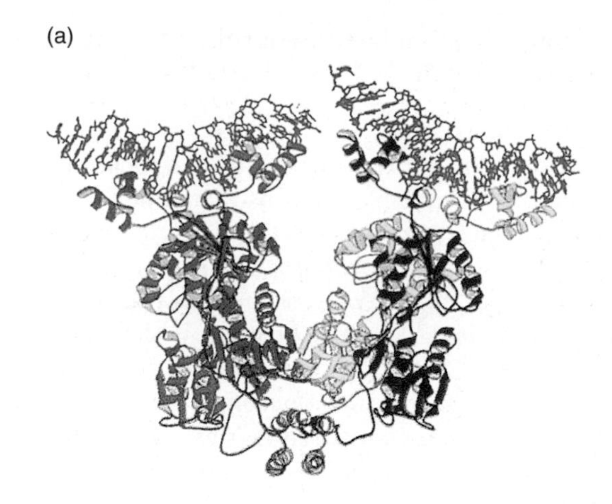

(b)

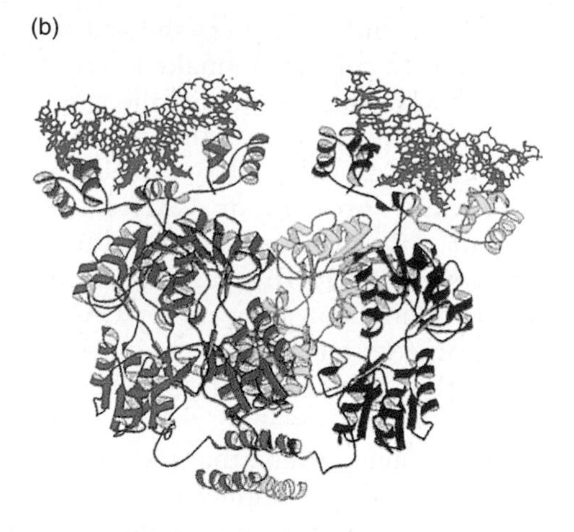

Figure 7.12 Structure of the *lac* repressor tetramer bound to two operator fragments. Lewis, Lu, and colleagues performed x-ray crystallography on *lac* repressor bound to 21-bp DNA fragments containing the major *lac* operator sequence. The structure presents the four repressor monomers in pink, green, yellow, and red, and the DNA fragments in blue. Two repressor dimers interact with each other at bottom to form tetramers. Each of the dimers contains two DNA-binding domains that can be seen interacting with the DNA major grooves at top. The structure shows clearly that the two dimers can bind independently to separate *lac* operators. Panels **(a)** and **(b)** are "front" and "side" views of the same structure. (*Source:* Lewis et al., Crystal structure of the lactose operon processor and its complexes with DNA and inducer. *Science* 271 (1 Mar 1996), f. 6, p. 1251. © AAAS.)

SUMMARY Two competing hypotheses seek to explain the mechanism of repression of the *lac* operon. One is that the RNA polymerase can bind to the *lac* promoter in the presence of the repressor, but the repressor inhibits the transition from abortive transcription to processive transcription. The other is that the repressor, by binding to the operator, blocks access by the polymerase to the adjacent promoter. The latest evidence supports the latter hypothesis. In addition to the classical (major) *lac* operator adjacent to the promoter, two auxiliary *lac* operators exist: one each upstream and downstream. All three operators are required for optimum repression, two work reasonably well, but the classical operator by itself produces only a modest amount of repression.

Positive Control of the *lac* Operon

As we learned earlier in this chapter, *E. coli* cells keep the *lac* operon in a relatively inactive state as long as glucose is present. This selection in favor of glucose metabolism and against use of other energy sources has long been attributed to the influence of some breakdown product, or *catabolite,* of glucose. It is therefore known as **catabolite repression.**

The ideal positive controller of the *lac* operon would be a substance that sensed the lack of glucose and responded by activating the *lac* promoter so that RNA polymerase could bind and transcribe the *lac* genes (assuming, of course, that lactose is present and the repressor is therefore not bound to the operator). One substance that responds to glucose concentration is a nucleotide called **cyclic-AMP (cAMP)** (Figure 7.13). As the level of glucose drops, the concentration of cAMP rises.

Catabolite Activator Protein Ira Pastan and his colleagues demonstrated that cAMP, added to bacteria, could overcome catabolite repression of the *lac* operon and a number of other operons, including the *gal* and *ara* operons. The latter two govern the metabolism of the sugars galactose and arabinose, respectively. In other words, cAMP rendered these genes active, even in the presence of glucose. This finding implicated cAMP strongly in the positive control of the *lac* operon. Does this mean that cAMP is the positive effector? Not exactly. The positive controller of the *lac* operon is a complex composed of two parts: cAMP and a protein factor.

Cyclic-AMP

Figure 7.13 Cyclic-AMP. Note the cyclic 5′-3′ phosphodiester bond (blue).

Geoffrey Zubay and coworkers showed that a crude cell-free extract of *E. coli* would make β-galactosidase if supplied with cAMP. This finding led the way to the discovery of a protein in the extract that was necessary for the stimulation by cAMP. Zubay called this protein **catabolite activator protein,** or **CAP.** Later, Pastan's group found the same protein and named it **cyclic-AMP receptor protein,** or **CRP.** To avoid confusion, we will refer to this protein from now on as CAP, regardless of whose experiments we are discussing. However, the gene encoding this protein has been given the official name *crp*.

Pastan and colleagues found that the dissociation constant for the CAP–cAMP complex was $1–2 \times 10^{-6}$ M. However, they also isolated a mutant whose CAP bound about 10 times less tightly to cAMP. If CAP–cAMP really is important to positive control of the *lac* operon, we would expect reduced production of β-galactosidase by a cAMP-supplemented cell-free extract of these mutant cells. Figure 7.14 shows that this is indeed the case. To make the point even more strongly, Pastan showed that β-galactosidase synthesis by this mutant extract (plus cAMP) could be stimulated about threefold by the addition of wild-type CAP.

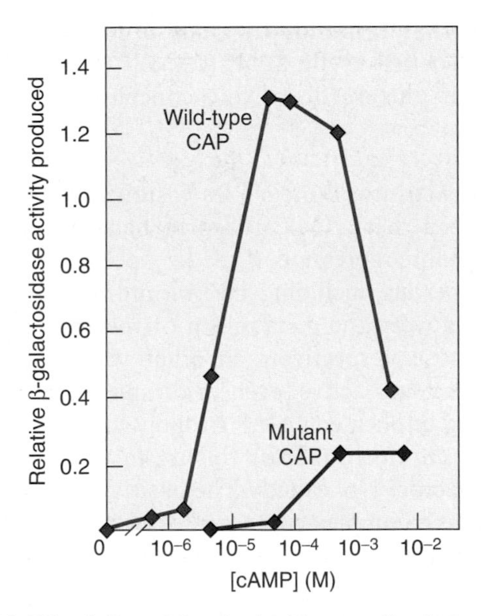

Figure 7.14 Stimulation of β-galactosidase synthesis by cAMP with wild-type and mutant CAP. Pastan and colleagues stimulated cell-free bacterial extracts to make β-galactosidase in the presence of increasing concentrations of cAMP with a wild-type extract (red), or an extract from mutant cells that have a CAP with reduced affinity for cAMP (blue). This mutant extract made much less β-galactosidase, which is what we expect if the CAP–cAMP complex is important in *lac* operon transcription. Too much cAMP obviously interfered with β-galactosidase synthesis in the wild-type extract. This is not surprising because cAMP has many effects, and some may indirectly inhibit some step in expression of the *lacZ* gene in vitro. (*Source:* Adapted from Emmer, M., et al., Cyclic AMP receptor protein of *E. coli:* Its role in the synthesis of inducible enzymes, *Proceedings of the National Academy of Sciences* 66(2): 480–487, June 1970.)

SUMMARY Positive control of the *lac* operon, and certain other inducible operons that code for sugar-metabolizing enzymes, is mediated by a factor called catabolite activator protein (CAP), which, in conjunction with cyclic-AMP, stimulates transcription. Because cyclic-AMP concentration is depressed by glucose, this sugar prevents stimulation of transcription. Thus, the *lac* operon is activated only when glucose concentration is low and therefore a need arises to metabolize an alternative energy source.

The Mechanism of CAP Action

How do CAP and cAMP stimulate *lac* transcription? Zubay and colleagues discovered a class of *lac* mutants in which CAP and cAMP could not stimulate *lac* transcription. These mutations mapped to the *lac* promoter, suggesting that the binding site for the CAP–cAMP complex lies in the promoter. Later molecular biological work, which we will discuss shortly, has shown that the CAP–cAMP binding site (the **activator-binding site**) lies just upstream of the promoter. Pastan and colleagues went on to show that this binding of CAP and cAMP to the activator site helps RNA polymerase to form an open promoter complex. The role of cAMP is to change the shape of CAP to increase its affinity for the activator-binding site.

Figure 7.15 shows how this experiment worked. First, Pastan and colleagues allowed RNA polymerase to bind to the *lac* promoter in the presence or absence of CAP and cAMP. Then they challenged the promoter complex by adding nucleotides and rifampicin simultaneously to see if an open promoter complex had formed. If not, transcription should be rifampicin-sensitive because the DNA melting step takes so much time that it would allow the antibiotic to inhibit the polymerase before initiation could occur. However, if it was an open promoter complex, it would be primed to polymerize nucleotides. Because nucleotides reach the polymerase before the antibiotic, the polymerase has time to initiate transcription. Once it has initiated an RNA chain, the polymerase becomes resistant to rifampicin until it completes that RNA chain. In fact, Pastan and colleagues found that when the polymerase– promoter complex formed in the absence of CAP and cAMP it was still rifampicin-sensitive. Thus, it had not formed an open promoter complex. On the other hand, when CAP and cAMP were present when polymerase associated with the promoter, a rifampicin-resistant open promoter complex formed.

Figure 7.15b presents a dimer of CAP–cAMP at the activator site on the left and polymerase at the promoter

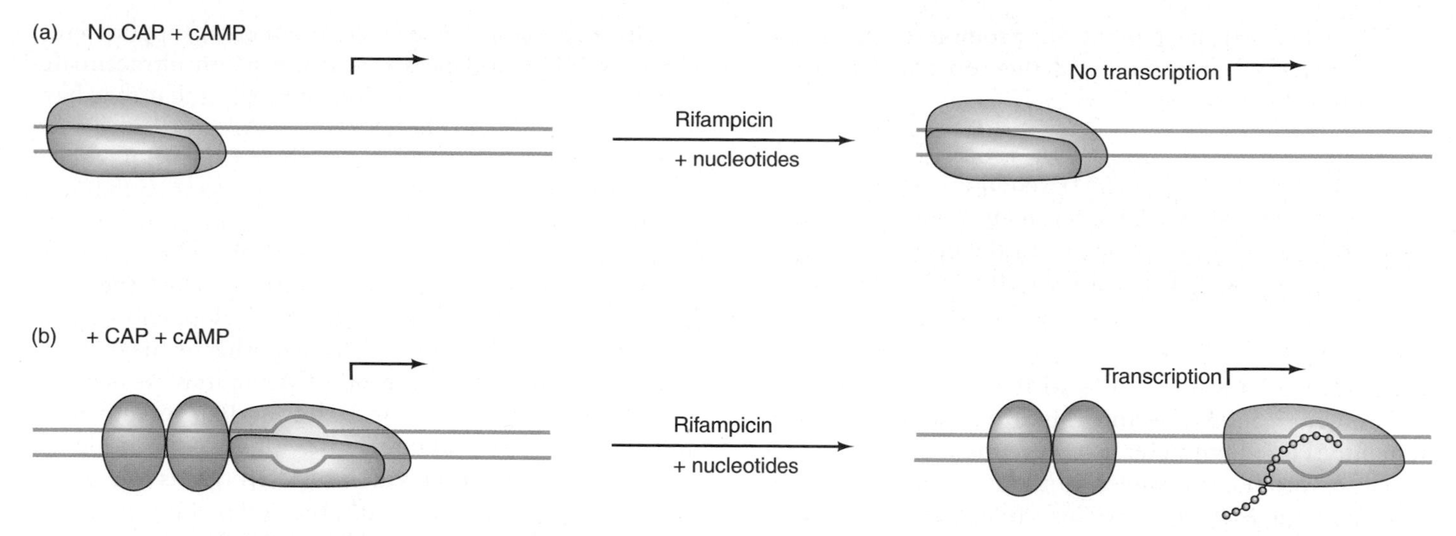

Figure 7.15 CAP plus cAMP allow formation of an open promoter complex. (a) When RNA polymerase binds to DNA containing the *lac* promoter without CAP, it binds randomly and weakly to the DNA. This binding is susceptible to inhibition when rifampicin is added along with nucleotides, so no transcription occurs. **(b)** When RNA polymerase binds to the *lac* promoter in the presence of CAP and cAMP (purple), it forms an open promoter complex. This is not susceptible to inhibition when rifampicin and nucleotides are added at the same time because the open promoter complex is ready to polymerize the nucleotides, which reach the polymerase active site before the antibiotic. Once the first few phosphodiester bonds form, the polymerase is resistant to rifampicin inhibition until it reinitiates. Thus, transcription occurs under these conditions, demonstrating that CAP and cAMP facilitate formation of an open promoter complex. The RNA is shown as a green chain.

on the right. How do we know that is the proper order? The first indication came from genetic experiments. Mutations to the left of the promoter prevent stimulation of transcription by CAP and cAMP, but still allow a low level of transcription. An example is a deletion called L1, whose position is shown in Figure 7.16. Because this deletion completely obliterates positive control of the *lac* operon by CAP and cAMP, the CAP-binding site must lie at least partially within the deleted region. On the other hand, since the L1 deletion has no effect on CAP-independent transcription, it has not encroached on the promoter, where RNA polymerase binds. Therefore, the right-hand end of this deletion serves as a rough dividing line between the activator-binding site and the promoter.

The CAP-binding sites in the *lac, gal,* and *ara* operons all contain the sequence TGTGA. The conservation of this sequence suggests that it is an important part of the CAP-binding site, and we also have direct evidence for this notion. For example, footprinting studies show that binding of the CAP–cAMP complex protects the G's in this sequence against methylation by dimethyl sulfate, suggesting that the CAP–cAMP complex binds tightly enough to these G's that it hides them from the methylating agent.

The *lac* operon, and other operons activated by CAP and cAMP, have remarkably weak promoters. Their −35 boxes are particularly unlike the consensus sequences; in fact, they are scarcely recognizable. This situation is actually not surprising. If the *lac* operon had a strong promoter, RNA polymerase could form open promoter complexes readily without help from CAP and cAMP, and it would therefore be active even in the presence of glucose. Thus, this promoter has to be weak to be dependent on CAP and

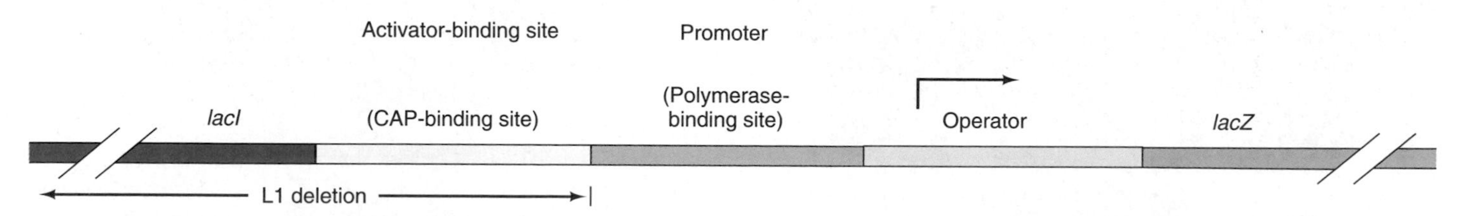

Figure 7.16 The *lac* control region. The activator–promoter region, just upstream of the operator, contains the activator-binding site, or CAP-binding site, on the left (yellow) and the promoter, or polymerase-binding site, on the right (pink). These sites have been defined by footprinting experiments and by genetic analysis. An example of the latter approach is the L1 deletion, whose right-hand end is shown. The L1 mutant shows basal transcription of the *lac* operon, but no stimulation by CAP and cAMP. Thus, it still has the promoter, but lacks the activator-binding site.

cAMP. In fact, strong mutant *lac* promoters are known (e.g., the *lac*UV5 promoter) and they do not depend on CAP and cAMP.

SUMMARY The CAP–cAMP complex stimulates transcription of the *lac* operon by binding to an activator-binding site adjacent to the promoter and helping RNA polymerase bind to the promoter.

Recruitment How does CAP–cAMP recruit polymerase to the promoter? Such **recruitment** has two steps: (1) Formation of the closed promoter complex, and (2) conversion of the closed promoter complex to the open promoter complex. William McClure and his colleagues summarized these two steps in the following equation:

$$R + P \underset{K_B}{\rightleftarrows} RP_c \underset{k_2}{\rightarrow} RP_o$$

where R is RNA polymerase, P is the promoter, RP_c is the closed promoter complex, and RP_o is the open promoter complex. McClure and coworkers devised kinetic methods of distinguishing between the two steps and determined that CAP–cAMP acts directly to stimulate the first step by increasing K_B. CAP–cAMP has little if any effect on k_2, so the second step is not accelerated. Nevertheless, by increasing the rate of formation of the closed promoter complex, CAP–cAMP provides more raw material (closed promoter complex) for conversion to the open promoter complex. Thus, the net effect of CAP–cAMP is to increase the rate of open promoter complex formation.

How does binding CAP–cAMP to the activator-binding site facilitate binding of polymerase to the promoter? One long-standing hypothesis is that CAP and RNA polymerase actually touch as they bind to their respective DNA target sites and therefore they bind cooperatively.

This hypothesis has much experimental support. First, CAP and RNA polymerase cosediment on ultracentrifugation in the presence of cAMP, suggesting that they have an affinity for each other. Second, CAP and RNA polymerase, when both are bound to their DNA sites, can be chemically cross-linked to each other, suggesting that they are in close proximity. Third, DNase footprinting experiments (Chapter 5) show that the CAP–cAMP footprint lies adjacent to the polymerase footprint. Thus, the DNA binding sites for these two proteins are close enough that the proteins could interact with each other as they bind to their DNA sites. Fourth, several CAP mutations decrease activation without affecting DNA binding (or bending), and some of these mutations alter amino acids in the region of CAP (**activation region I [ARI]**) that is thought to interact with polymerase. Fifth, the polymerase site that is presumed to interact with ARI on CAP is the carboxyl terminal domain of the α-subunit (the αCTD), and deletion of the αCTD prevents activation by CAP–cAMP.

Sixth, Richard Ebright and colleagues performed x-ray crystallography in 2002 on a complex of DNA, CAP– cAMP, and the αCTD of RNA polymerase. They showed that the ARI site on CAP and the αCTD do indeed touch in the crystal structure, although the interface between the two proteins is not large. They arranged for the αCTD to bind on its own to the complex by changing the sequences flanking the CAP-binding site to A–T-rich sequences (5′-AAAAAA-3′) that are attractive to the αCTD. Figure 7.17a presents the crystal structure they determined. One molecule of αCTD (αCTD^{DNA}) binds to DNA alone; the other molecule ($\alpha CTD^{CAP,DNA}$) binds to both DNA and CAP. The latter αCTD clearly contacts the part of CAP identified as ARI, and detailed analysis of the structure showed exactly which amino acids in each protein were involved in the interaction. The fact that only one monomer of αCTD binds to a monomer of CAP reflects the situation in vivo; the other monomer of αCTD does not contact CAP either in the crystal structure or in vivo.

(a)

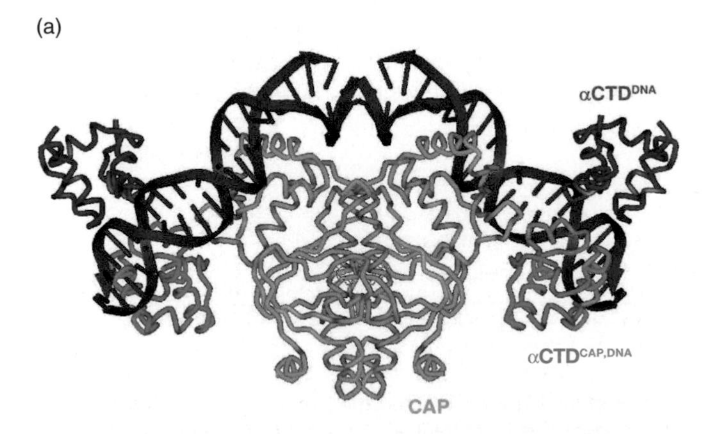

(b)

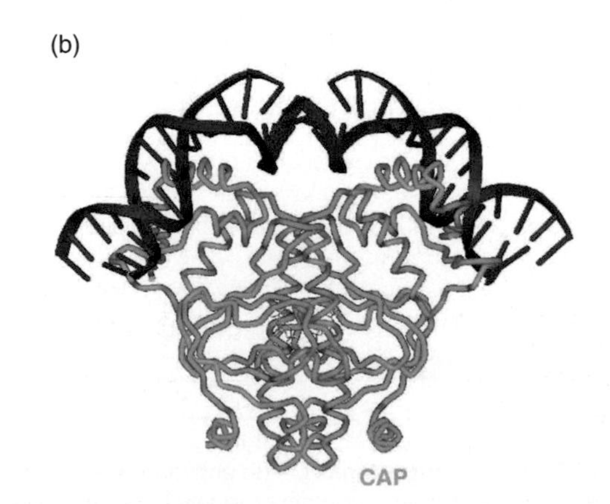

Figure 7.17 Crystal structures of the CAP–cAMP–αCTD–DNA complex and the CAP–cAMP–DNA complex. (a) The CAP–cAMP–αCTD–DNA complex. DNA is in red, CAP is in cyan, with cAMP represented by thin red lines, αCTD^{DNA} is in dark green, and $\alpha CTD^{CAP,DNA}$ is in light green. **(b)** CAP–cAMP–DNA complex. Same colors as in panel (a). (*Source:* Benoff et al., *Science* 297 © 2002 by the AAAS.)

Another thing to notice about Figure 7.17a is that binding of CAP–cAMP to its DNA target bends the DNA considerably—about 100 degrees. This bend had been noticed before in the crystal structure of the CAP–cAMP–DNA complex in the absence of αCTD, determined by Thomas Steitz and colleagues in 1991, and can be seen again in an equivalent crystal structure determined in this study (Figure 7.17b). It is interesting that the structure of the DNA and CAP in the CAP–cAMP–DNA complex and in the CAP–cAMP–DNA–αCTD complex are superimposable. This means that the αCTD did not perturb the structure.

The DNA bend observed in the crystallography studies had been detected as early as 1984 by Hen-Ming Wu and Donald Crothers, using electrophoresis (Figure 7.18). When a piece of DNA is bent, it migrates more slowly during electrophoresis. Furthermore, as Figure 7.18b and c illustrate, the closer the bend is to the middle of the DNA, the more slowly the DNA electrophoreses. Wu and Crothers took advantage of this phenomenon by preparing DNA fragments of the *lac* operon, all the same length, with the CAP-binding site located at different positions in each. Next, they bound CAP–cAMP to each fragment and electrophoresed the DNA–protein complexes. If CAP binding really did bend the DNA, then the different fragments should have migrated at different rates. If the DNA did not bend, they all should have migrated at the same rate. Figure 7.18d demonstrates that the fragments really did migrate at different rates. Moreover, the more pronounced the DNA bend, the greater the difference in electrophoretic rates should be. In other words, the shape of the curve in Figure 7.18 should give us an estimate of the degree of bending of DNA by CAP–cAMP. In fact the bending seems to be about 90 degrees, which agrees reasonably well with

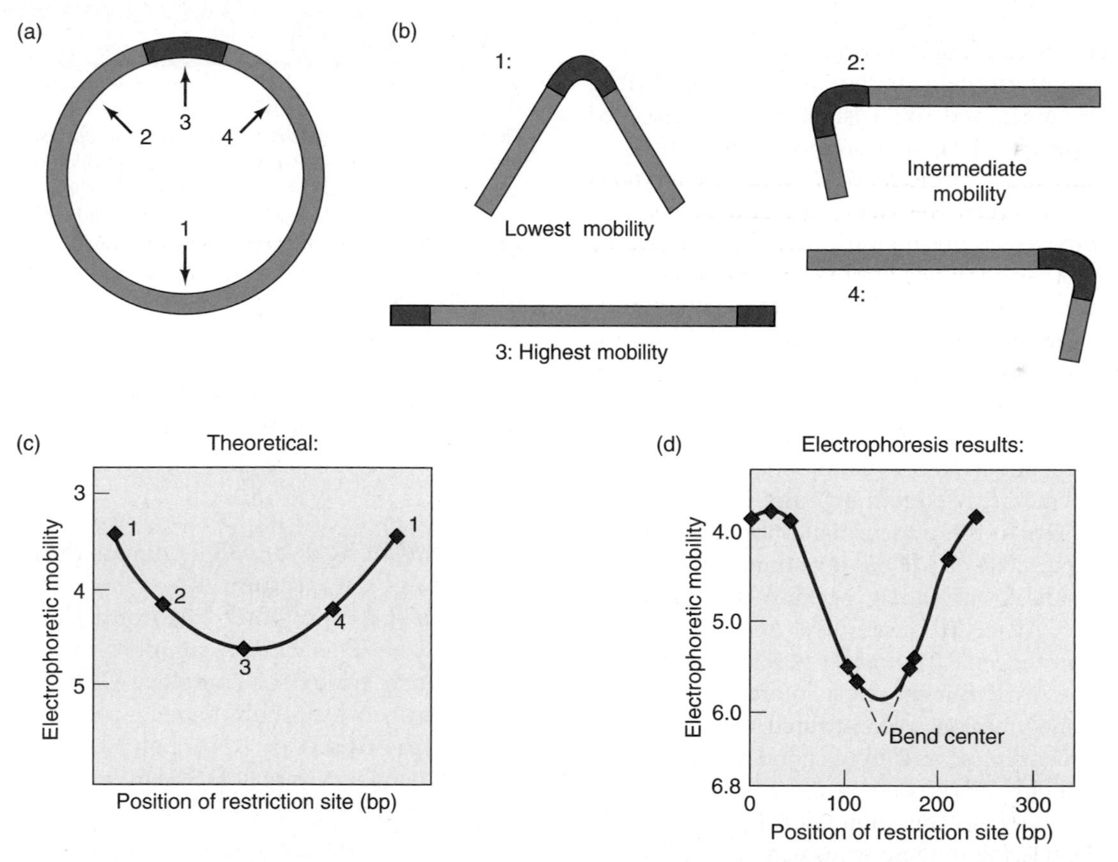

Figure 7.18 Electrophoresis of CAP–cAMP–promoter complexes. **(a)** Map of a hypothetical DNA circle, showing a protein-binding site at center (red), and cutting sites for four different restriction enzymes (arrows). **(b)** Results of cutting DNA in panel (a) with each restriction enzyme, then adding a DNA-binding protein, which bends DNA. Restriction enzyme 1 cuts across from the binding site, leaving it in the middle; restriction enzymes 2 and 4 place the binding site off center; and restriction enzyme 3 cuts within the binding site, allowing little if any bending of the DNA. **(c)** Theoretical curve showing the relationship between electrophoretic mobility and bent DNA, with the bend at various sites along the DNA. Note that the mobility is lowest when the bend is closest to the middle of the DNA fragment (at either end of the curve). Note also that mobility increases in the downward direction on the *y* axis. **(d)** Actual electrophoresis results with CAP–cAMP and DNA fragments containing the *lac* promoter at various points in the fragment, depending on which restriction enzyme was used to cut the DNA. The symmetrical curve allowed Wu and Crothers to extrapolate to a bend center that corresponds to the CAP–cAMP-binding site in the *lac* promoter. (*Source:* Wu, H.M., and D.M. Crothers, The locus of sequence-directed and protein-induced DNA bending. *Nature* 308:511, 1984.)

Table 7.2 Activation of *lac* P1 Transcription by CAP–cAMP

	Transcripts (cpm)				P1/UV5 (%)		
	−cAMP–CAP		+cAMP–CAP				
Enzyme	**P1**	**UV5**	**P1**	**UV5**	**−cAMP–CAP**	**+cAMP–CAP**	**Activation (fold)**
α-WT	46	797	625	748	5.8	83.6	14.4
α-256	53	766	62	723	6.9	8.6	1.2
α-235	51	760	45	643	6.7	7.0	1.0

the 100 degrees determined later by x-ray crystallography. This bending is presumably necessary for optimal interaction among the proteins and DNA in the complex.

All of the studies we have cited point to the importance of protein–protein interaction between CAP and RNA polymerase—the αCTD of polymerase, in particular. This hypothesis predicts that mutations that remove the αCTD should prevent transcription stimulation by CAP–cAMP. In fact, Kazuhiko Igarashi and Akira Ishihama have provided such genetic evidence for the importance of the αCTD of RNA polymerase in activation by CAP–cAMP. They transcribed cloned *lac* operons in vitro with RNA polymerases reconstituted from separated subunits. All the subunits were wild-type, except in some experiments, in which the α-subunit was a truncated version lacking the CTD. One of the truncated α-subunits ended at amino acid 256 (of the normal 329 amino acids); the other ended at amino acid 235. Table 7.2 shows the results of run-off transcription (Chapter 5) from a CAP–cAMP-dependent *lac* promoter (P1) and a CAP–cAMP-independent *lac* promoter (*lac*UV5) with reconstituted polymerases containing the wild-type or truncated α-subunits in the presence and absence of CAP–cAMP. As expected, CAP–cAMP did not stimulate transcription from the *lac*UV5 promoter because it is a strong promoter that is CAP–cAMP-insensitive. Also as expected, transcription from the *lac* P1 promoter was stimulated over 14-fold by CAP–cAMP. But the most interesting behavior was that of the polymerases reconstituted with truncated α-subunits. These enzymes were just as good as wild-type in transcribing from either promoter in the absence of CAP–cAMP, but they could not be stimulated by CAP–cAMP. Thus, the αCTD, missing in these truncated enzymes, is not necessary for reconstitution of an active RNA polymerase, but it is necessary for stimulation by CAP–cAMP.

Figure 7.19 illustrates the hypothesis of activation we have been discussing, in which the CAP–cAMP dimer binds to its activator site and simultaneously binds to the carboxyl-terminal domain of the polymerase α-subunit (αCTD), facilitating binding of polymerase to the promoter. This would be the functional equivalent of the αCTD binding to an UP element in the DNA (Chapter 6), thereby enhancing polymerase binding to the promoter.

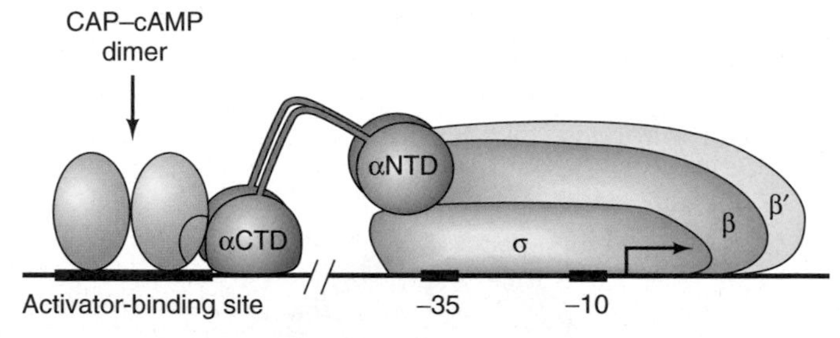

Figure 7.19 Hypothesis for CAP–cAMP activation of *lac* transcription. The CAP–cAMP dimer (purple) binds to its target site on the DNA, and the αCTD (red) interacts with a specific site on the CAP protein (brown). This strengthens binding between polymerase and promoter. (*Source:* Adapted from Busby, S. and R.H. Ebright, Promoter structure, promoter recognition, and transcription activation in prokaryotes, *Cell* 79:742, 1994.)

CAP stimulates transcription at over 100 promoters, and it is just one of a growing number of bacterial transcription activators. We will examine more examples in Chapter 9.

SUMMARY CAP–cAMP binding to the *lac* activator-binding site recruits RNA polymerase to the adjacent *lac* promoter to form a closed promoter complex. This closed complex then converts to an open promoter complex. CAP–cAMP causes recruitment through protein–protein interaction with the αCTD of RNA polymerase. CAP–cAMP also bends its target DNA by about 100 degrees when it binds.

7.2 The *ara* Operon

We have already mentioned that the *ara* operon of *E. coli*, which codes for the enzymes required to metabolize the sugar arabinose, is another catabolite-repressible operon. It has several interesting features to compare with the *lac* operon. First, two *ara* operators exist: *ara*O_1 and *ara*O_2. The former regulates transcription of a control gene called *ara*C.

The other operator is located far upstream of the promoter it controls (P_{BAD}), between positions −265 and −294, yet it still governs transcription. Second, the CAP-binding site is about 200 bp upstream of the *ara* promoter, yet CAP can still stimulate transcription. Third, the operon has another system of negative regulation, mediated by the AraC protein.

The *ara* Operon Repression Loop

How can *ara*O_2 control transcription from a promoter over 250 bp downstream? The most reasonable explanation is that the DNA in between these remote sites (the operator and the promoter) loops out as illustrated in Figure 7.20a. Indeed, we have good evidence that DNA looping is occurring. Robert Lobell and Robert Schleif found that if they inserted DNA fragments containing an integral number of double-helical turns (multiples of 10.5 bp) between the operator and the promoter, the operator still functioned. However, if the inserts contained a nonintegral number of helical turns (e.g., 5 or 15 bp), the operator did not function. This is consistent with the general notion that a double-stranded DNA can loop out and bring two protein-binding sites together as long as these sites are located on the same face of the double helix. However, the DNA cannot twist through the 180 degrees required to bring binding sites on opposite faces around to the same face so they can interact with each other through looping (see Figure 7.20). In this respect, DNA resembles a piece of stiff coat hanger wire: It can be bent relatively easily, but it resists twisting.

The simple model in Figure 7.20 assumes that proteins bind first to the two remote binding sites, then these proteins interact to cause the DNA looping. However, Lobell and Schleif found that the situation is more subtle than that. In fact, the *ara* control protein (**AraC**), which acts as both a positive and a negative regulator, has three binding sites, as illustrated in Figure 7.21a. In addition to the far upstream site, *ara*O_2, AraC can bind to *ara*O_1, located between positions −106 and −144, and to *araI*, which really includes two half-sites: *ara*I_1 (−56 to −78) and *ara*I_2 (−35 to −51), each of which can bind one monomer of AraC. The *ara* operon is also known as the *araCBAD* operon, for its four genes, *araA–D*. Three of these genes, *araB, A,* and *D,* encode the arabinose metabolizing enzymes; they are transcribed rightward from the promoter *ara*P_{BAD}. The other gene, *araC,* encodes the control protein AraC and is transcribed leftward from the *ara*P_C promoter.

In the absence of arabinose, when no *araBAD* products are needed, AraC exerts negative control, binding to *ara*O_2 and *ara*I_1, looping out the DNA in between and repressing the operon (Figure 7.21b). On the other hand, when arabinose is present, it apparently changes the conformation of AraC so that it no longer binds to *ara*O_2, but occupies *ara*I_1 and *ara*I_2 instead. This breaks the repression loop, and the operon is derepressed (Figure 7.21c). As in the *lac* operon, however, derepression isn't the whole story. Positive control mediated by CAP and cAMP also occurs, and Figure 7.21c shows this complex attached to its binding site upstream of the *araBAD* promoter. DNA looping presumably explains how binding of CAP–cAMP at a site remote from the *araBAD* promoter can control transcription. The looping would allow CAP to contact the polymerase and thereby stimulate its binding to the promoter.

Evidence for the *ara* Operon Repression Loop

What is the evidence for the looping model of *ara* operon repression? First, Lobell and Schleif used electrophoresis to show that AraC can cause loop formation in the absence

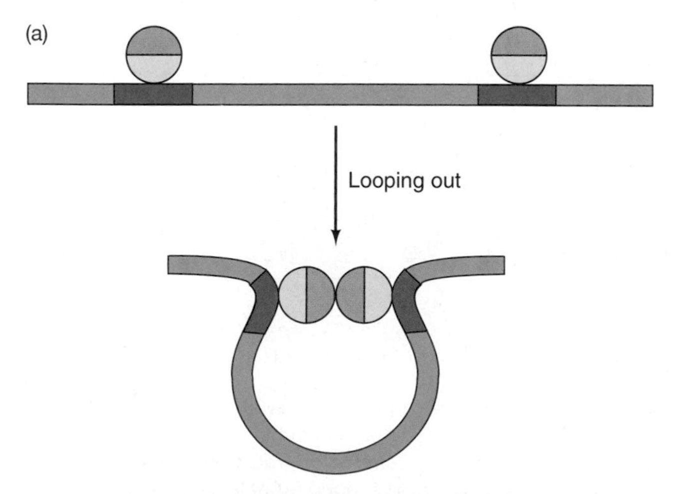

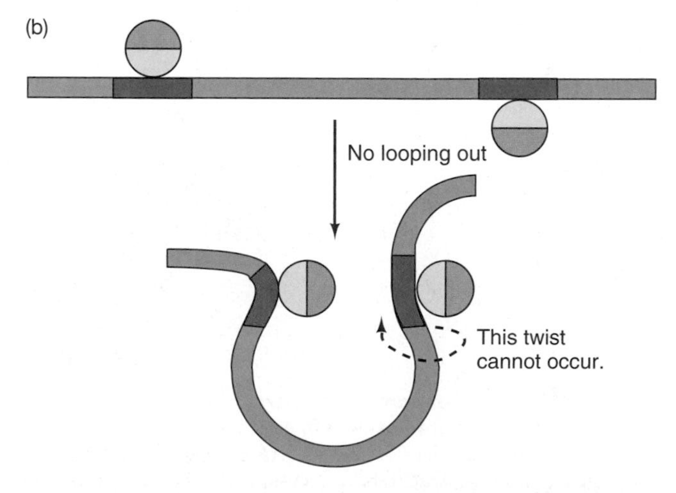

Figure 7.20 Proteins must bind to the same face of the DNA to interact by looping out the DNA. (a) Two proteins with DNA-binding domains (yellow) and protein–protein interaction domains (blue) bind to sites (red) on the same face of the DNA double helix. These proteins can interact because the intervening DNA can loop out without twisting. **(b)** Two proteins bind to sites on opposite sides of the DNA duplex. These proteins cannot interact because the DNA is not flexible enough to perform the twist needed to bring the protein interaction sites together.

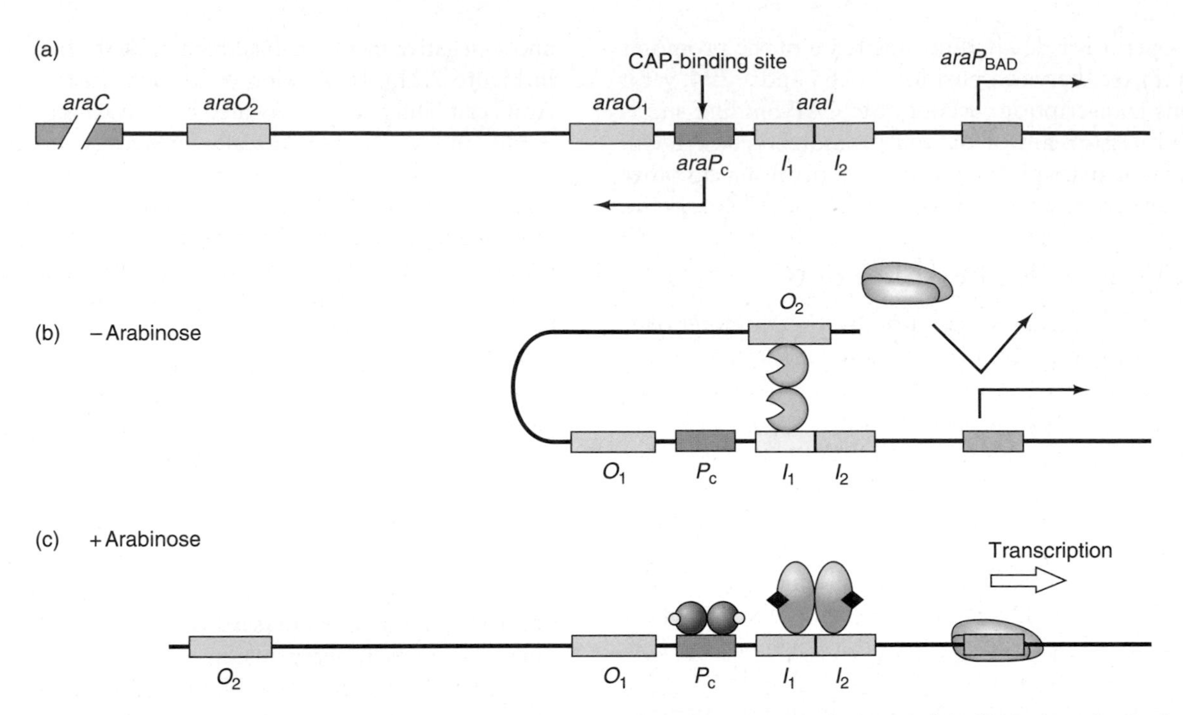

Figure 7.21 Control of the *ara* operon. (a) Map of the *ara* control region. There are four AraC-binding sites (*ara*O_1, *ara*O_2, *araI*$_1$, and *araI*$_2$), which all lie upstream of the *ara* promoter, *araP*$_{BAD}$. The *araP*$_c$ promoter drives leftward transcription of the *araC* gene at far left. **(b)** Negative control. In the absence of arabinose, monomers of AraC (green) bind to O_2 and I_1, bending the DNA and blocking access to the promoter by RNA polymerase (red and blue). **(c)** Positive control. Arabinose (black) binds to AraC, changing its shape so it prefers to bind as a dimer to I_1 and I_2 and not to O_2. This opens up the promoter (pink) to binding by RNA polymerase. If glucose is absent, the CAP–cAMP complex (purple and yellow) is in high enough concentration to occupy the CAP-binding site, which stimulates polymerase binding to the promoter. Now active transcription can occur.

of arabinose. Instead of the entire *E. coli* DNA, they used a small (404-bp) supercoiled circle of DNA, called a *minicircle,* that contained the *ara*O_2 and *araI* sites, 160 bp apart. They then added AraC and measured looping by taking advantage of the fact that looped supercoiled DNAs have a higher electrophoretic mobility than the same DNAs that are unlooped. Figure 7.22 shows one such assay. Comparing lanes 1 and 2, we can see that the addition of AraC causes the appearance of a new, high-mobility band that corresponds to the looped minicircle.

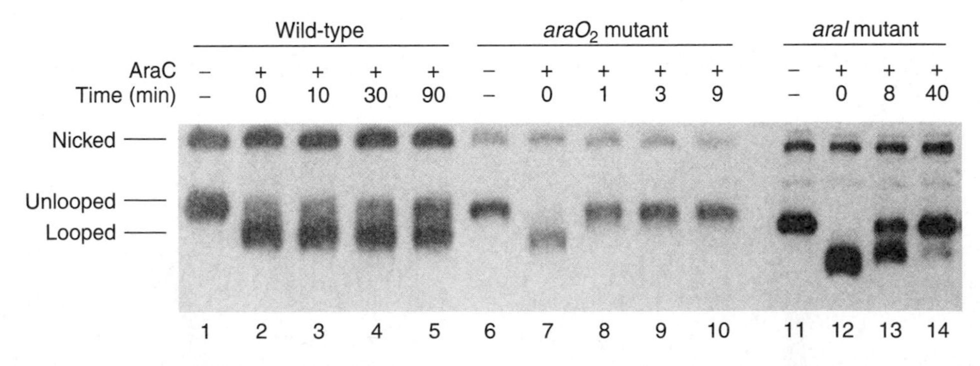

Figure 7.22 Effects of mutations in *ara*O_2 and *araI* on the stability of looped complexes with AraC. Lobell and Schleif prepared labeled minicircles (small DNA circles) containing either wild-type or mutant AraC binding sites, as indicated at top. Then they added AraC to form a complex with the labeled DNA. Next they added an excess of unlabeled DNA containing an *araI* site as a competitor, for various lengths of time. Finally they electrophoresed the protein–DNA complexes to see whether they were still in looped or unlooped form. The looped DNA was more supercoiled than the unlooped DNA, so it migrated faster. The wild-type DNA remained in a looped complex even after 90 min in the presence of the competitor. By contrast, dissociation of AraC from the mutant DNAs, and therefore loss of the looped complex, occurred much faster. It lasted less than 1 min with the *ara*O_2 mutant DNA and was half gone in less than 10 min with the *araI* mutant DNA. (*Source:* Lobell, R.B. and Schleif, R.F., DNA looping and unlooping by AraC protein. *Science* 250 (1990), f. 2, p. 529. © AAAS.)

This experiment also shows that the stability of the loop depends on binding of AraC to both *ara*O_2 and *araI*. Lobell and Schleif made looped complexes with a wild-type minicircle, with a minicircle containing a mutant *ara*O_2 site, and with a minicircle containing mutations in both *araI* sites. They then added an excess of unlabeled wild-type minicircles and observed the decay of each of the looped complexes. Lanes 3–5 show only about 50% conversion of the looped to unlooped wild-type minicircle in 90 min. Thus, the half-time of dissociation of the wild-type looped complex is about 100 min. In contrast, the *ara*O_2 mutant minicircle's conversion from looped to unlooped took less than 1 min (compare lanes 7 and 8). The *araI* mutant's half-time of loop breakage is also short—less than 10 min. Thus, both *ara*O_2 and *araI* are involved in looping by AraC because mutations in either one greatly weaken the DNA loop.

Next, Lobell and Schleif demonstrated that arabinose breaks the repression loop. They did this by showing that arabinose added to looped minicircles immediately before electrophoresis eliminates the band corresponding to the looped DNA. Figure 7.23 illustrates this phenomenon. In a separate experiment, Lobell and Schleif showed that a broken loop could re-form if arabinose was removed. They used arabinose to prevent looping, then diluted the DNA into buffer containing excess competitor DNA, either with or without arabinose. The buffer with arabinose maintained the broken loop, but the buffer without arabinose diluted the sugar to such an extent that the loop could re-form.

What happens to the AraC monomer bound to *ara*O_2 when the loop opens up? Apparently it binds to *ara*I_2. To demonstrate this, Lobell and Schleif first showed by **methylation interference** that AraC contacts *ara*I_1, but not *ara*I_2, in the looped state. The strategy was to partially methylate the minicircle DNA, bind AraC to loop the DNA, separate looped from unlooped DNA by electrophoresis, and then break the looped and unlooped DNAs at their methylated sites. Because methylation at important sites blocks looping, those sites that are important for looping will be unmethylated in the looped DNA, but methylated in the unlooped DNA. Indeed, two *ara*I_1 bases were heavily methylated in the unlooped DNA, but only lightly methylated in the looped DNA. In contrast, no *ara*I_2 bases showed this behavior. Thus, it appears that AraC does not contact *ara*I_2 in the looped state.

Lobell and Schleif confirmed this conclusion by showing that mutations in *ara*I_2 have no effect on AraC binding in the looped state, but have a strong effect on binding in the unlooped state. We infer that *ara*I_2 is necessary for AraC binding in the unlooped state and is therefore contacted by AraC under these conditions.

These data suggest the model of AraC–DNA interaction that was depicted in Figure 7.21b and c. A dimer of AraC causes looping by simultaneously interacting with *ara*I_1 and *ara*O_2. Arabinose breaks the loop by changing the conformation of AraC so the protein loses its affinity for *ara*O_2 and binds instead to *ara*I_2.

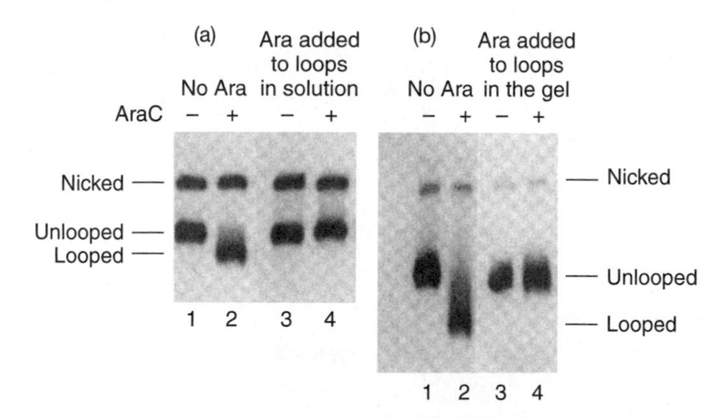

Figure 7.23 Arabinose breaks the loop between *ara*O_2 and *araI*. **(a)** Lobell and Schleif added arabinose to preformed loops before electrophoresis. In the absence of arabinose, AraC formed a DNA loop (lane 2). In the presence of arabinose, the loop formed with AraC was broken (lane 4). **(b)** This time the investigators added arabinose to the gel after electrophoresis started. Again, in the absence of arabinose, looping occurred (lane 2). However, in the presence of arabinose, the loop was broken (lane 4). The designation Ara at top refers to arabinose. (*Source:* Lobell R.B., and Schleif R.F., DNA looping and unlooping by AraC protein. *Science* 250 (1990), f. 4, p. 530. © AAAS.)

Autoregulation of *araC*

So far, we have only briefly mentioned a role for *ara*O_1. It does not take part in repression of *araBAD* transcription; instead it allows AraC to regulate its own synthesis. Figure 7.24 shows the relative positions of *araC*, P_c, and *ara*O_1. The *araC* gene is transcribed from P_c in the leftward direction, which puts *ara*O_1 in a position to control this transcription. As the level of AraC rises, it binds to *ara*O_1 and inhibits leftward transcription, thus preventing an accumulation of too much repressor. This kind of mechanism, where a protein controls its own synthesis, is called **autoregulation.**

SUMMARY The *ara* operon is controlled by the AraC protein. AraC represses the operon by looping out the DNA between two sites, *ara*O_2 and *ara*I_1, that are 210 bp apart. Arabinose can derepress the operon by causing AraC to loosen its attachment to *ara*O_2 and to bind to *ara*I_2 instead. This breaks the loop and allows transcription of the operon. CAP and cAMP further stimulate transcription by binding to a site upstream of *araI*. AraC controls its own synthesis by binding to *ara*O_1 and preventing leftward transcription of the *araC* gene.

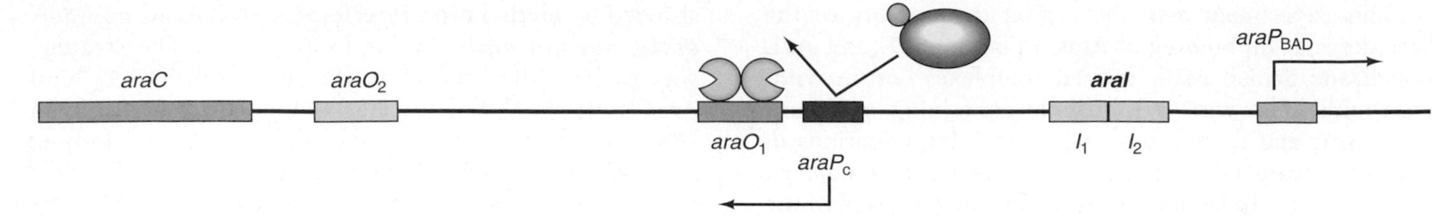

Figure 7.24 Autoregulation of *araC*. AraC (green) binds to $araO_1$ and prevents transcription leftward from P_c through the *araC* gene. This can presumably happen whether or not arabinose is bound to AraC, that is, with the control region either unlooped or looped.

7.3 The *trp* Operon

The *E. coli* **trp** (pronounced "trip") **operon** contains the genes for the enzymes that the bacterium needs to make the amino acid tryptophan. Like the *lac* operon, it is subject to negative control by a repressor. However, there is a fundamental difference. The *lac* operon codes for **catabolic** enzymes—those that break down a substance. Such operons tend to be turned on by the presence of that substance, lactose in this case. The *trp* operon, on the other hand, codes for **anabolic** enzymes—those that build up a substance. Such operons are generally turned off by that substance. When the tryptophan concentration is high, the products of the *trp* operon are not needed any longer, and we would expect the *trp* operon to be repressed. That is what happens. The *trp* operon also exhibits an extra level of control, called attenuation, not seen in the *lac* operon.

Tryptophan's Role in Negative Control of the *trp* Operon

Figure 7.25 shows an outline of the structure of the *trp* operon. Five genes code for the polypeptides in the enzymes that convert a tryptophan precursor, chorismic acid, to tryptophan. In the *lac* operon, the promoter and operator precede the genes, and the same is true in the *trp* operon. However, the *trp* operator lies wholly within the *trp* promoter, whereas the two loci are merely adjacent in the *lac* operon.

In the negative control of the *lac* operon, the cell senses the presence of lactose by the appearance of tiny amounts of its rearranged product, allolactose. In effect, this inducer causes the repressor to fall off the *lac* operator and derepresses the operon. In the case of the *trp* operon, a plentiful supply of tryptophan means that the cell does not need to spend any more energy making this amino acid. In other words, a high tryptophan concentration is a signal to turn off the operon.

How does the cell sense the presence of tryptophan? In essence, tryptophan helps the *trp* repressor bind to its operator. Here is how that occurs: In the absence of tryptophan, no *trp* repressor exists—only an inactive protein called the **aporepressor.** When the aporepressor binds tryptophan, it changes to a conformation with a much higher affinity for the *trp* operator (Figure 7.25b). This is another allosteric

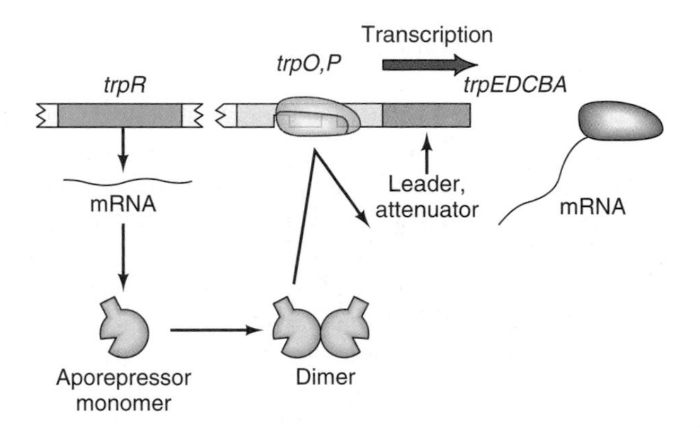

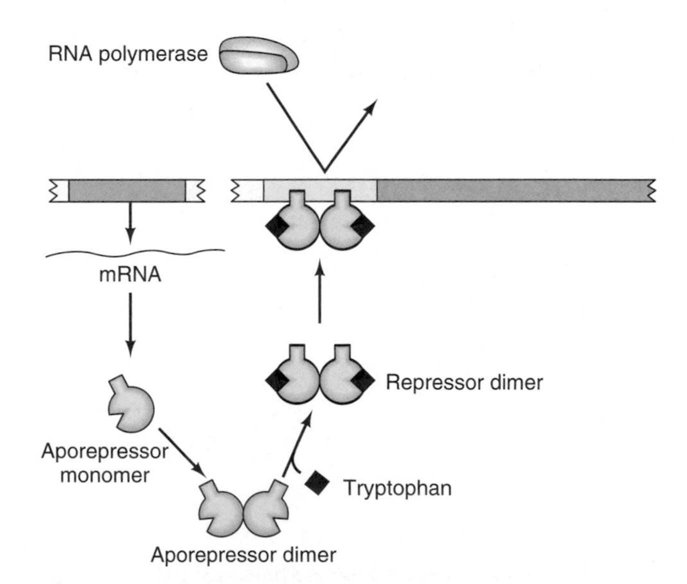

Figure 7.25 Negative control of the *trp* operon. (a) Derepression. RNA polymerase (red and blue) binds to the *trp* promoter and begins transcribing the *trp* genes (*trpE, D, C, B,* and *A*). Without tryptophan, the aporepressor (green) cannot bind to the operator. **(b)** Repression. Tryptophan, the corepressor (black), binds to the inactive aporepressor, changing it to repressor, with the proper shape for binding successfully to the *trp* operator. This prevents RNA polymerase from binding to the promoter, so no transcription occurs.

transition like the one we encountered in our discussion of the *lac* repressor. The combination of aporepressor plus tryptophan is the ***trp* repressor;** therefore, tryptophan is called a **corepressor.** When the cellular concentration of tryptophan is high, plenty of corepressor is available to bind and form the active *trp* repressor. Thus, the operon is repressed. When the tryptophan level in the cell falls, the amino acid dissociates from the aporepressor, causing it to shift back to the inactive conformation; the repressor–operator complex is thus broken, and the operon is derepressed. In Chapter 9, we will examine the nature of the conformational shift in the aporepressor that occurs on binding tryptophan and see why this is so important in operator binding.

SUMMARY The negative control of the *trp* operon is, in a sense, the mirror image of the negative control of the *lac* operon. The *lac* operon responds to an inducer that causes the repressor to dissociate from the operator, derepressing the operon. The *trp* operon responds to a repressor that includes a corepressor, tryptophan, which signals the cell that it has made enough of this amino acid. The corepressor binds to the aporepressor, changing its conformation so it can bind better to the *trp* operator, thereby repressing the operon.

Control of the *trp* Operon by Attenuation

In addition to the standard, negative control scheme we have just described, the *trp* operon employs another mechanism of control called **attenuation.** Why is this extra control needed? The answer probably lies in the fact that repression of the *trp* operon is weak—much weaker, for example, than that of the *lac* operon. Thus, considerable transcription of the *trp* operon can occur even in the presence of repressor. In fact, in attenuator mutants where only repression can operate, the fully repressed level of transcription is only 70-fold lower than the fully derepressed level. The attenuation system permits another 10-fold control over the operon's activity. Thus, the combination of repression and attenuation controls the operon over a 700-fold range, from fully inactive to fully active: (70-fold [repression] $\times$ 10-fold [attenuation] $=$ 700-fold). This is valuable because synthesis of tryptophan requires considerable energy.

Here is how attenuation works. Figure 7.25 lists two loci, the ***trp* leader** and the ***trp* attenuator,** in between the operator and the first gene, *trpE*. Figure 7.26 gives a closer view of the leader–attenuator, whose purpose is to attenuate, or weaken, transcription of the operon when tryptophan is relatively abundant. The attenuator operates by causing premature termination of transcription. In other words, transcription that gets started, even though the tryptophan concentration is high, stands a 90% chance of terminating in the attenuator region.

(a) Low tryptophan: transcription of *trp* structural genes

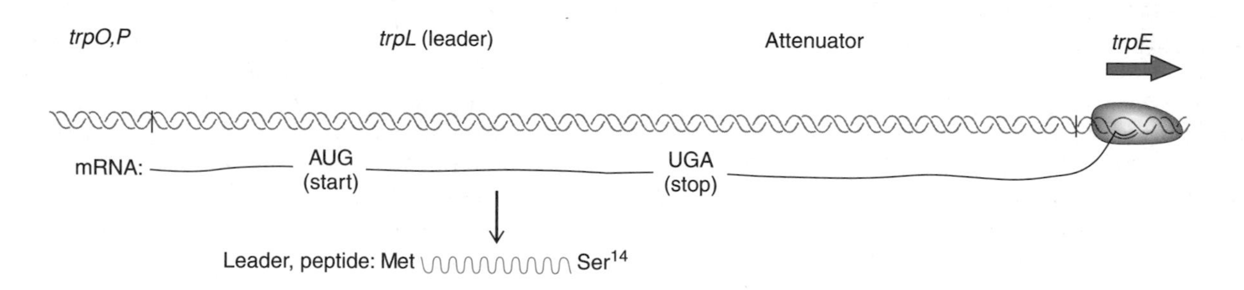

(b) High tryptophan: attenuation, premature termination

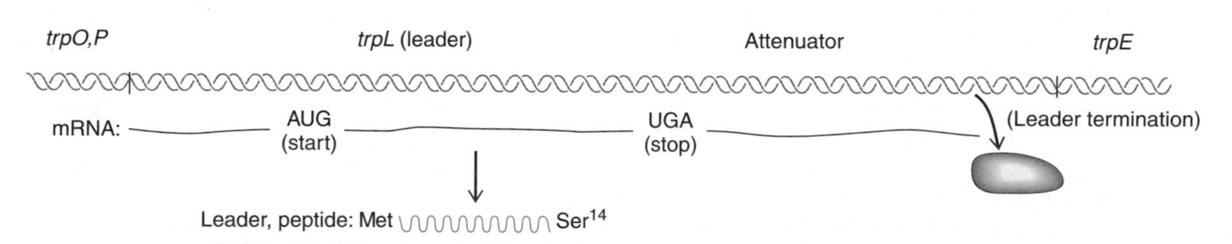

Figure 7.26 Attenuation in the *trp* operon. (a) Under low tryptophan conditions, the RNA polymerase (red) reads through the attenuator, so the structural genes are transcribed. **(b)** In the presence of high tryptophan, the attenuator causes premature termination of transcription, so the *trp* genes are not transcribed.

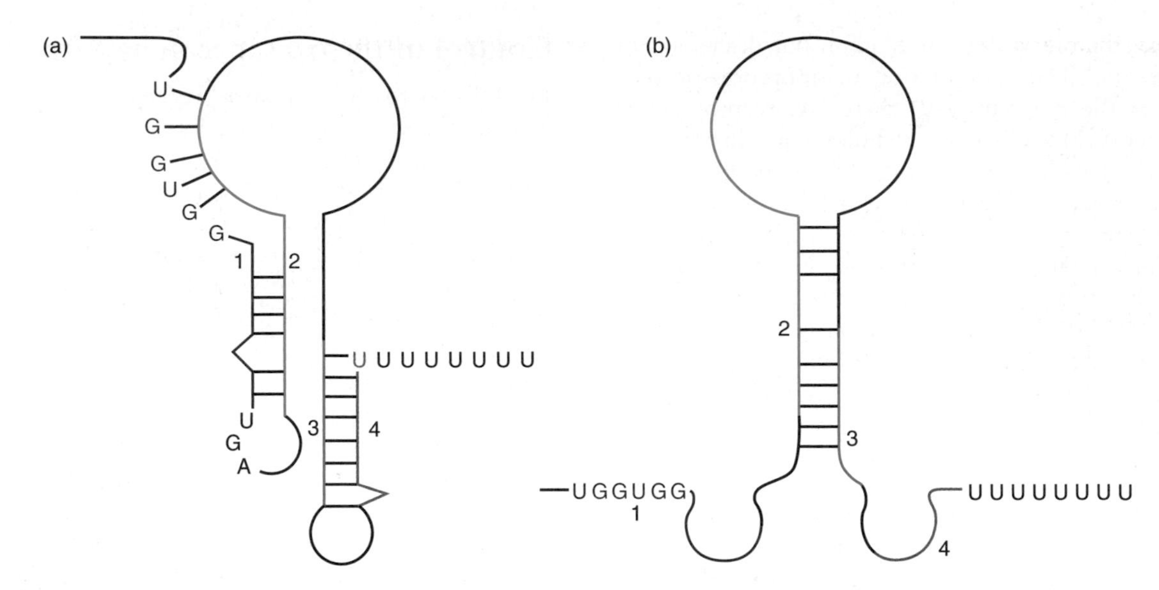

Figure 7.27 Two structures available to the leader–attenuator transcript. **(a)** The more stable structure, with two hairpin loops. **(b)** The less stable structure, containing only one hairpin loop. The curved shape of the RNA at the bottom is not meant to suggest a shape for the molecule—it is drawn this way simply to save space. The base-paired segments (1–4) in (a) are colored, and these same regions are colored the same way in (b) so they can be recognized.

The reason for this premature termination is that the attenuator contains a transcription stop signal (terminator): an inverted repeat followed by a string of eight A–T pairs in a row. Because of the inverted repeat, the transcript of this region would tend to engage in intramolecular base pairing, forming a "hairpin". As we learned in Chapter 6, a hairpin followed by a string of U's in a transcript destabilizes the binding between the transcript and the DNA and thus causes termination.

SUMMARY Attenuation imposes an extra level of control on an operon, over and above the repressor–operator system. It operates by causing premature termination of transcription of the operon when the operon's products are abundant.

Defeating Attenuation

When tryptophan is scarce, the *trp* operon must be activated, and that means that the cell must somehow override attenuation. Charles Yanofsky proposed this hypothesis: Something preventing the hairpin from forming would destroy the termination signal, so attenuation would break down and transcription would proceed. A look at Figure 7.27a reveals not just one potential hairpin near the end of the leader transcript, but two. However, the terminator includes only the second hairpin, which is adjacent to the string of U's in the transcript. Furthermore, the two-hairpin arrangement is not the only one available; another, containing only one hairpin, is shown in Figure 7.27b. Note that this alternative hairpin contains elements from each of the two hairpins in the first structure. Figure 7.27 illustrates this concept by labeling the sides of the original two hairpins 1, 2, 3, and 4. If the first of the original hairpins involves elements 1 and 2 and the second involves 3 and 4, then the alternative hairpin in the second structure involves 2 and 3. This means that the formation of the alternative hairpin (Figure 7.27b) precludes formation of the other two hairpins, including the one adjacent to the string of U's, which is a necessary part of the terminator (Figure 7.27a).

The two-hairpin structure involves more base pairs than the alternative, one-hairpin structure; therefore, it is more stable. So why should the less stable structure ever form? A clue comes from the base sequence of the leader region shown in Figure 7.28. One very striking feature of this sequence is that two codons for tryptophan (UGG) occur in a row in element 1 of the first potential hairpin. This

Met Lys Ala Ile Phe Val Leu Lys Gly Trp Trp Arg Thr Ser Stop
pppA---AUGAAAGCAAUUUUCGUACUGAAAGGUUGGUGGCGCACUUCCUGA

Figure 7.28 Sequence of the leader. The sequence of part of the leader transcript is presented, along with the leader peptide it encodes. Note the two Trp codons in tandem (blue).

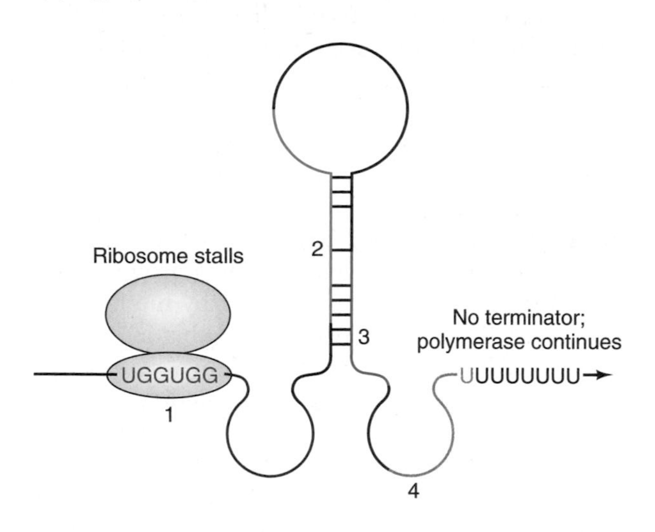

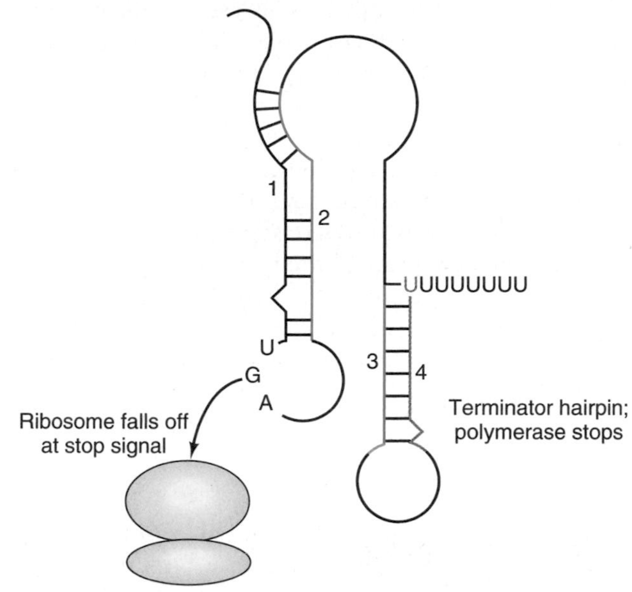

Figure 7.29 Overriding attenuation. (a) Under conditions of tryptophan starvation, the ribosome (yellow) stalls at the Trp codons and prevents element 1 (red) from pairing with element 2 (blue). This forces the one-hairpin structure, which lacks a terminator, to form, so no attenuation should take place. **(b)** Under conditions of tryptophan abundance, the ribosome reads through the two tryptophan codons and falls off at the translation stop signal (UGA), so it cannot interfere with base pairing in the leader transcript. The more stable, two-hairpin structure forms; this structure contains a terminator, so attenuation occurs.

may not seem unusual, but tryptophan (Trp) is a rare amino acid in most proteins; it is found on average only once in every 100 amino acids. So the chance of finding two Trp codons in a row *anywhere* is quite small, and the fact that they are found in the *trp* operon is very suspicious.

In bacteria, transcription and translation occur simultaneously. Thus, as soon as the *trp* leader region is transcribed, ribosomes begin translating this emerging mRNA. Think about what would happen to a ribosome trying to translate the *trp* leader under conditions of tryptophan starvation (Figure 7.29a). Tryptophan is in short supply, and here are two demands in a row for that very amino acid. In all likelihood, the ribosome will not be able to satisfy those demands immediately, so it will pause at one of the Trp codons. And where does that put the stalled ribosome? Right on element 1, which should be participating in formation of the first hairpin. The bulky ribosome clinging to this RNA site effectively prevents its pairing with element 2, which frees 2 to pair with 3, forming the one-hairpin alternative structure. Because the second hairpin (elements 3 and 4) cannot form, transcription does not terminate and attenuation has been defeated. This is desirable, of course, because when tryptophan is scarce, the *trp* operon should be transcribed.

Notice that this mechanism involves a coupling of transcription and translation, where the latter affects the former. It would not work in eukaryotes, where transcription and translation take place in separate compartments. It also depends on transcription and translation occurring at about the same rate. If RNA polymerase outran the ribosome, it might pass through the attenuator region before the ribosome had a chance to stall at the Trp codons.

You may be wondering how the polycistronic mRNA made from the *trp* operon can be translated if ribosomes are stalled in the leader at the very beginning. The answer is that each of the genes represented on the mRNA has its own translation start signal (AUG). Ribosomes recognize each of these independently, so translation of the *trp* leader does not affect translation of the *trp* genes.

On the other hand, consider a ribosome translating the leader transcript under conditions of abundant tryptophan (Figure 7.29b). Now the dual Trp codons present no barrier to translation, so the ribosome continues through element 1 until it reaches the stop signal (UGA) between elements 1 and 2 and falls off. With no ribosome to interfere, the two hairpins can form, completing the transcription termination signal that halts transcription before it reaches the *trp* genes. Thus, the attenuation system responds to the presence of adequate tryptophan and prevents wasteful synthesis of enzymes to make still more tryptophan.

Other *E. coli* operons besides *trp* use the attenuation mechanism. The most dramatic known use of consecutive codons to stall a ribosome occurs in the *E. coli* histidine *(his)* operon, in which the leader region contains seven histidine codons in a row!

SUMMARY Attenuation operates in the *E. coli trp* operon as long as tryptophan is plentiful. When the supply of this amino acid is restricted, ribosomes stall at the tandem tryptophan codons in the *trp* leader. Because the *trp* leader is being synthesized just as stalling occurs, the stalled ribosome will influence the way this RNA folds. In particular, it prevents the formation of a hairpin, which is part of the transcription termination signal that causes attenuation. Therefore, when tryptophan is scarce, attenuation is defeated and the operon remains active. This means that the control exerted by attenuation responds to tryptophan levels, just as repression does.

7.4 Riboswitches

We have just seen an example of controlling gene expression by manipulating the structure of the 5′-untranslated region (UTR) of an mRNA (the *trp* mRNA of *E. coli*). In this case, a macromolecular assembly (the ribosome) senses the concentration of a small molecule (tryptophan) and binds to the *trp* 5′-UTR, altering its shape, thereby controlling its continued transcription. So this is an example of a group of macromolecules mediating the effect of a small molecule (or *ligand*) on gene expression.

We also have a growing number of examples of small molecules acting *directly* on mRNAs (usually on their 5′-UTRs) to control their expression. The regions of these mRNAs that are capable of altering their structures to control gene expression in response to ligand binding are called **riboswitches.** Riboswitches are responsible for 2–3% of gene expression control in bacteria, and they are also found in archaea, fungi, and plants. Later in this section we will learn of a possible example in animals.

The region of a riboswitch that binds to the ligand is called an **aptamer.** Aptamers were first discovered by scientists studying evolution in a test tube, who exploited rapidly replicating RNAs to select for short RNA sequences that bind tightly and specifically to ligands. As the RNAs replicate, they make mistakes, producing new RNA sequences, and those that bind best to a particular ligand are selected. Experimenters found many such aptamers in these in vitro experiments and wondered why living things did not take advantage of them. Now we know that they do.

A classic example of a riboswitch is the *ribD* operon in *B. subtilis*. This operon controls the synthesis and transport of the vitamin riboflavin and one of its products, flavin mononucleotide (FMN). Bacterial *rib* operons contain a conserved element in their 5′-UTRs known as the ***RFN* element.** Mutations in this region abolish normal control of the *ribD* operon by FMN, which led to the hypothesis that this *RFN* element interacts with a protein that responds to FMN or, perhaps, with FMN itself.

To test the hypothesis that the *RFN* element is an aptamer that binds directly to FMN, Ronald Breaker and colleagues used a technique called **in-line probing.** This method relies on the fact that efficient hydrolysis (breakage) of a phosphodiester bond in RNA needs a 180-degree ("in-line") arrangement among the attacking nucleophile (water), the phosphorus atom in the phosphodiester bond, and the leaving hydroxyl group at the end of one of the RNA fragments created by the hydrolysis. Unstructured RNA can easily assume this in-line conformation, but RNA that is constrained by secondary structure (intramolecular base pairing) or by binding to a ligand cannot. Thus, spontaneous cleavage of linear, unstructured RNA will occur much more readily than will cleavage of a structured RNA with lots of base pairing or with a ligand bound to it.

Thus, Breaker and colleagues incubated a labeled RNA fragment containing the *RFN* element in the presence and absence of FMN. Figure 7.30a shows that the patterns of spontaneous hydrolysis of the RNA were different in the presence and absence of FMN, suggesting that FMN binds directly to the RNA and causes it to shift its conformation. This is what we would expect of an aptamer bound to its ligand.

In particular, Breaker and colleagues found that FMN binding rendered certain phosphodiester bonds less susceptible to cleavage, whereas others retained their normal susceptibility (Figure 7.30b). Furthermore, the changes in susceptibility were half-maximal at an FMN concentration of only 5 nM. This indicates high affinity between the RNA and its ligand.

The patterns of decreased susceptibility to cleavage in the presence of FMN suggested the two alternative conformations of the *RFN* element depicted in Figure 7.30c. In the absence of FMN, the element should form an antiterminator, with the hairpin remote from the string of six U's. But FMN would cause the conformation of the element to shift such that it forms a terminator, blocking expression of the operon. This makes sense because, with abundant FMN, there is no need to express the *ribD* operon, so the proposed attenuation by FMN would save the cell energy.

To test this hypothesis, Breaker and colleagues performed an in vitro transcription assay with a cloned DNA template containing both the *RFN* element and the proposed terminator. They found that transcription terminated about 10% of the time at the terminator even in the absence of FMN, but FMN raised the frequency of termination to 30%. They mapped the termination site with a run-off transcription assay (Chapter 5) and showed that transcription terminated right at the end of the string of U's. Next, they used a mutant version of the DNA template that encoded fewer than six U's in the putative terminator. In this case, FMN caused no change

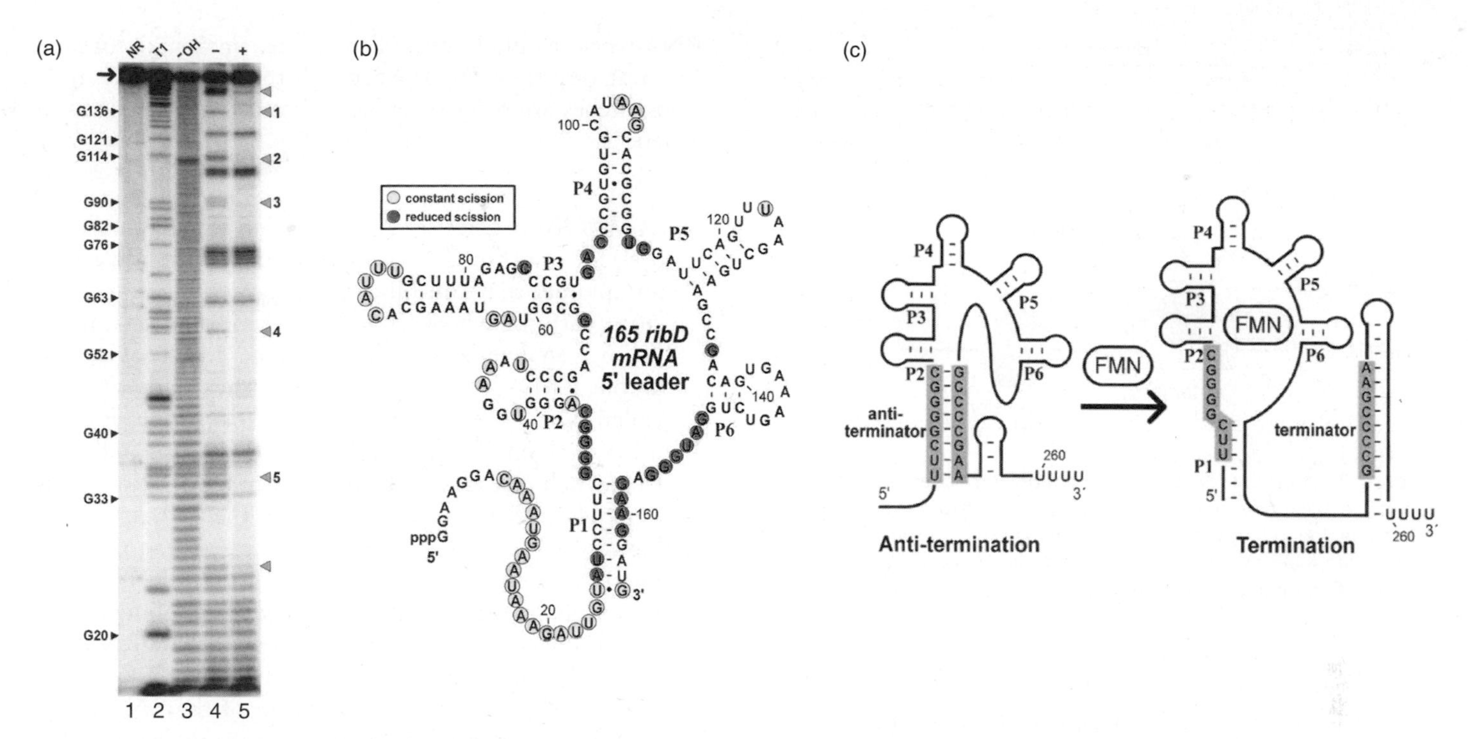

Figure 7.30 Results of in-line probing of *RFN* element and model for the action of the *ribD* riboswitch. (a) Gel electrophoresis results of in-line probing. Lane 1, no RNA; lane 2, RNA cut with RNase T1; lane 3, RNA cut with base; lanes 4 and 5, RNAs subjected to spontaneous cleavage in the absence (−) and presence (+) of FMN for 40 h at 25°C. Arrows at right denote regions of the RNA that became less susceptible to cleavage in the presence of FMN. **(b)** Sequence of part of the 5′-UTR of the *B. subtilis ribD* mRNA, showing the internucleotide linkages that became less susceptible to spontaneous cleavage upon FMN binding (red), and those that showed constant susceptibility (yellow). The secondary structure of the element is based on comparisons of sequences of many *RFN* elements. **(c)** Proposed change in structure of the riboswitch upon FMN binding. In the absence of FMN, base pairing between the two yellow regions forces the riboswitch to assume an antiterminator conformation, with the hairpin remote from the string of U's. Conversely, binding of FMN to the growing mRNA allows the GCCCCGAA sequence to base-pair with another part of the riboswitch, creating a terminator that stops transcription. (*Source: (a-c)* © 2002 National Academy of Science. Proceedings of the National Academy of Sciences, vol. 99, no. 25, December 10, 2002, pp. 15908–15913 "An mRNA structure that controls gene expression by binding FMN," Chalamish, and Ronald R. Breaker, fig.1, p. 15909 & fig. 3, p. 15911.)

in the frequency of termination, presumably because the shorter string of U's considerably lowered the efficiency of the terminator, even with FMN. Thus, with the wild-type gene, FMN really does appear to force more of the growing transcripts to form terminators that halt transcription.

Breaker and colleagues discovered another riboswitch in a conserved region in the 5′-untranslated region (5′-UTR) of the *glmS* gene of *Bacillus subtilis* and at least 17 other Gram-positive bacteria. This gene encodes an enzyme known as glutamine-fructose-6-phosphate amidotransferase, whose product is the sugar glucosamine-6-phosphate (GlcN6P). Breaker and colleagues found that the riboswitch in the 5′-UTR of the *glmS* mRNA is a ribozyme (an RNase) that can cleave the mRNA molecule itself. It does this at a low rate when concentrations of GlcN6P are low. However, when the concentration of GlcN6P rises, the sugar binds to the riboswitch in the mRNA and changes its conformation to make it a much better RNase (about 1000-fold better). This RNase destroys the mRNA, so less of the enzyme is made, so the GlcN6P concentration falls.

This riboswitch mechanism may not be confined to bacteria. In 2008, Harry Noller, William Scott, and colleagues discovered a very active **hammerhead ribozyme** in the 3′-UTRs of rodent C-type lectin type II (*Clec2*) mRNAs. Hammerhead ribozymes are so named because their secondary structure loosely resembles a hammer, with three base-paired stems constituting the "handle," "head," and "claw" of the hammer. At the junction of these three stems is a highly conserved group of 17 nucleotides that make up the RNase and the cleavage site, which lies at the bottom of the hammerhead where it joins the handle. Presumably, the hammerhead ribozyme in the *Clec2* mRNA responds to some cellular cue by cleaving itself and thus reducing *Clec2* gene expression, but it is not yet known what that cue is.

We will see another example of a riboswitch in Chapter 17, when we study the control of translation. We will learn that a ligand can bind to a riboswitch in an mRNA's 5′-UTR, and can control translation of that mRNA by changing the conformation of the 5′-UTR to hide the ribosome-binding site.

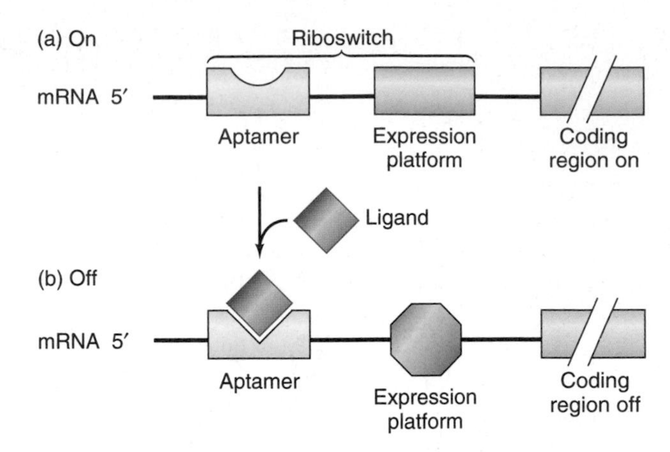

Figure 7.31 A model for riboswitch action. (a) Absence of the ligand. Gene expression is turned on. **(b)** Presence of the ligand. The ligand has bound to the aptamer in the riboswitch, causing a change in the conformation of the riboswitch, including the expression platform. This turns gene expression off.

These examples of riboswitches both operate by depressing gene expression: one at the transcriptional level, and one at the translational level. Indeed, all riboswitches studied to date work that way, although there is no reason why a riboswitch could not work by stimulating gene expression. These examples, among others, also lead to a general model for riboswitches (Figure 7.31). They are regions in the 5′-UTRs of mRNAs that contain two modules: an aptamer and another module, which Breaker and colleagues call an *expression platform*. The expression platform can be a terminator, a ribosome-binding site, or another RNA element that affects gene expression. By binding to its aptamer and changing the conformation of the riboswitch, a ligand can affect an expression platform, and thereby control gene expression.

Note that a riboswitch is another example of allosteric control, that is, one in which a ligand causes a conformational change in a large molecule that in turn affects the ability of the large molecule to interact with something else. We encountered an allosteric mechanism earlier in this chapter in the context of the *lac* operon, where a ligand (allolactose) bound to a protein (*lac* repressor) and interfered with its ability to bind to the *lac* operator. In fact, many examples of allosteric control are known, but up until recently they all involved allosteric proteins. Riboswitches work similarly, except that the large molecule is an RNA, rather than a protein.

Finally, riboswitches may provide a window on the "RNA world," a hypothetical era early in the evolution of life, in which proteins and DNA had not yet evolved. In this world, genes were made of RNA, not DNA, and enzymes were made of RNA, not protein. (We will see modern examples of catalytic RNAs in Chapters 14, 17, and 19.) Without proteins to control their genes, life forms in the RNA world would have had to rely on small molecules interacting directly with their genes. If this hypothesis is true, riboswitches are relics of one of the most ancient forms of genetic control.

SUMMARY A riboswitch is a region, usually in the 5′-UTR of an mRNA, that contains two modules: an aptamer that can bind a ligand, and an expression platform whose change in conformation can cause a change in expression of the gene. For example, FMN can bind to an aptamer in a riboswitch called the *RFN* element in the 5′-UTR of the *ribD* mRNA. Upon binding FMN, the base pairing in the riboswitch changes to create a terminator that attenuates transcription. This saves the cell energy because FMN is one of the products of the *ribD* operon. In another example, the *glmS* mRNA of *B. subtilis* contains a riboswitch that responds to the product of the enzyme encoded by the mRNA. When this product builds up, it binds to the riboswitch, changing the conformation of the RNA to stimulate an inherent RNase activity in the RNA so it cleaves itself.

SUMMARY

Lactose metabolism in *E. coli* is carried out by two proteins, β-galactosidase and galactoside permease. The genes for these two, and one additional enzyme, are clustered together and transcribed together from one promoter, yielding a polycistronic message. These functionally related genes are therefore controlled together.

Control of the *lac* operon occurs by both positive and negative control mechanisms. Negative control appears to occur as follows: The operon is turned off as long as repressor binds to the operator, because the repressor prevents RNA polymerase from binding to the promoter to transcribe the three *lac* genes. When the supply of glucose is exhausted and lactose is available, the few molecules of *lac* operon enzymes produce a few molecules of allolactose from the lactose. The allolactose acts as an inducer by binding to the repressor and causing a conformational shift that encourages dissociation from the operator. With the repressor removed, RNA polymerase is free to transcribe the three *lac* genes. A combination of genetic and biochemical experiments revealed the two key elements of negative control of the *lac* operon: the operator and the repressor. DNA sequencing revealed the presence of two auxiliary *lac* operators: one upstream, and one downstream of the major operator. All three are required for optimal repression.

Positive control of the *lac* operon, and certain other inducible operons that code for sugar-metabolizing enzymes, is mediated by a factor called catabolite activator protein (CAP), which, in conjunction with cyclic-AMP (cAMP), stimulates transcription. Because cAMP concentration is depressed by glucose, this sugar prevents positive control from operating. Thus, the *lac* operon is activated only when glucose concentration is low and a corresponding need arises to metabolize an alternative energy source. The CAP–cAMP complex stimulates expression of the *lac* operon by binding to an activator site adjacent to the promoter. CAP–cAMP binding helps RNA polymerase form an open promoter complex. It does this by recruiting polymerase to form a closed promoter complex, which then converts to an open promoter complex. Recruitment of polymerase occurs through protein–protein interactions between CAP and the αCTD of RNA polymerase.

The *ara* operon is controlled by the AraC protein. AraC represses the operon by looping out the DNA between two sites, $araO_2$ and $araI_1$, that are 210 bp apart. Arabinose can induce the operon by causing AraC to loosen its attachment to $araO_2$ and to bind to $araI_1$ and $araI_2$ instead. This breaks the loop and allows transcription of the operon. CAP and cAMP further stimulate transcription by binding to a site upstream of *araI*. AraC controls its own synthesis by binding to $araO_1$ and preventing leftward transcription of the *araC* gene.

The *trp* operon responds to a repressor that includes a corepressor, tryptophan, which signals the cell that it has made enough of this amino acid. The corepressor binds to the aporepressor, changing its conformation so it can bind better to the *trp* operator, thereby repressing the operon.

Attenuation operates in the *E. coli trp* operon as long as tryptophan is plentiful. When the supply of this amino acid is restricted, ribosomes stall at the tandem tryptophan codons in the *trp* leader. Because the *trp* leader is being synthesized just as this is taking place, the stalled ribosome will influence the way this RNA folds. In particular, it prevents the formation of a hairpin, which is part of the transcription termination signal that causes attenuation. When tryptophan is scarce, attenuation is therefore defeated and the operon remains active. This means that the control exerted by attenuation responds to tryptophan levels, just as repression does.

A riboswitch is a region in the 5′-UTR of an mRNA that contains two modules: an aptamer that can bind a ligand, and an expression platform whose change in conformation can cause a change in expression of the gene. For example, FMN can bind to an aptamer in a riboswitch called the *RFN* element in the 5′-UTR of the *ribD* mRNA. Upon binding FMN, the base pairing in the riboswitch changes to create a terminator that attenuates transcription.

REVIEW QUESTIONS

1. Draw a growth curve of *E. coli* cells growing on a mixture of glucose and lactose. What is happening in each part of the curve?
2. Draw diagrams of the *lac* operon that illustrate (a) negative control and (b) positive control.
3. What are the functions of β-galactosidase and galactoside permease?
4. Why are negative and positive control of the *lac* operon important to the energy efficiency of *E. coli* cells?
5. Describe and give the results of an experiment that shows that the *lac* operator is the site of repressor binding.
6. Describe and give the results of an experiment that shows that RNA polymerase can bind to the *lac* promoter, even if repressor is already bound at the operator.
7. Describe and give the results of an experiment that shows that *lac* repressor prevents RNA polymerase from binding to the *lac* promoter.
8. How do we know that all three *lac* operators are required for full repression? What are the relative effects of removing each or both of the auxiliary operators?
9. Describe and give the results of an experiment that shows the relative levels of stimulation of β-galactosidase synthesis by cAMP, using wild-type and mutant extracts, in which the mutation reduces the affinity of CAP for cAMP.
10. Present a hypothesis for activation of *lac* transcription by CAP–cAMP. Include the C-terminal domain of the polymerase α-subunit (the αCTD) in the hypothesis. What evidence supports this hypothesis?
11. Describe and give the results of an electrophoresis experiment that shows that binding of CAP–cAMP bends the *lac* promoter region.
12. What other data support DNA bending in response to CAP–cAMP binding?
13. Explain the fact that insertion of an integral number of DNA helical turns (multiples of 10.5 bp) between the $araO_2$ and *araI* sites in the *araBAD* operon permits repression by AraC, but insertion of a nonintegral number of helical turns prevents repression. Illustrate this phenomenon with diagrams.
14. Use a diagram to illustrate how arabinose can relieve repression of the *araBAD* operon. Show where AraC is located (a) in the absence of arabinose, and (b) in the presence of arabinose.
15. Describe and give the results of an experiment that shows that arabinose can break the repression loop formed by AraC.
16. Describe and give the results of an experiment that shows that both $araO_2$ and *araI* are involved in forming the repression loop.
17. Briefly outline evidence that shows that $araI_2$ is important in binding AraC when the DNA is in the unlooped, but not the looped, form.

18. Present a model to explain negative control of the *trp* operon in *E. coli*.
19. Present a model to explain attenuation in the *trp* operon in *E. coli*.
20. Why does translation of the *trp* leader region not simply continue into the *trp* structural genes (*trpE*, etc.) in *E. coli*?
21. How is *trp* attenuation overridden in *E. coli* when tryptophan is scarce?
22. What is a riboswitch? Illustrate with an example.
23. Describe what is meant by "in-line probing."

ANALYTICAL QUESTIONS

1. The table below gives the genotypes (with respect to the *lac* operon) of several partial diploid *E. coli* strains. Fill in the phenotypes, using a "+" for β-galactosidase synthesis and "−" for no β-galactosidase synthesis. Glucose is absent in all cases. Give a brief explanation of your reasoning.

	Phenotype for β-galactosidase Production	
Genotype	*No Inducer*	*Inducer*
a. $I^+O^+Z^+/I^+O^+Z^+$		
b. $I^+O^+Z^-/I^+O^+Z^+$		
c. $I^-O^+Z^+/I^+O^+Z^+$		
d. $I^sO^+Z^+/I^+O^+Z^+$		
e. $I^+O^cZ^+/I^+O^+Z^+$		
f. $I^+O^cZ^-/I^+O^+Z^+$		
g. $I^sO^cZ^+/I^+O^+Z^+$		

2. (*a*) In the genotype listed in the following table, the letters *A*, *B*, and *C* correspond to the *lacI*, and *lacO*, *lacZ* loci, though not necessarily in that order. From the mutant phenotypes exhibited by the first three genotypes listed in the table, deduce the identities of *A*, *B*, and *C* as they correspond to the three loci of the *lac* operon. The minus superscripts (e.g., A^-) can refer to the following aberrant functions: Z^-, O^c, or I^-.
   (*b*) Determine the genotypes, in conventional *lac* operon genetic notation, of the partial diploid strains shown in lines 4 and 5 of the table. Here, I^+, I^-, and I^s are all possible.

	Phenotype for β-galactosidase Production	
Genotype	*No Inducer*	*Inducer*
1. $A^+B^+C^-$	+	+
2. $A^-B^+C^+$	+	+
3. $A^+B^+C^-/A^+B^+C^+$	+	+
4. $A^-B^+C^+/A^+B^+C^+$	−	−
5. $A^-B^+C^+/A^+B^+C^+$	−	+

3. Consider *E. coli* cells, each having one of the following mutations:
   a. a mutant *lac* operator (the O^c locus) that cannot bind repressor
   b. a mutant *lac* repressor (the I^- gene product) that cannot bind to the *lac* operator
   c. a mutant *lac* repressor (the I^s gene product) that cannot bind to allolactose
   d. a mutant *lac* promoter region that cannot bind CAP plus cAMP

   What effect would each mutation have on the function of the *lac* operon (assuming no glucose is present)?
4. You are studying a new operon in *E. coli* involved in phenylalanine biosynthesis.
   a. How would you predict this operon is regulated (inducible or repressible by phenylalanine, positive or negative)? Why?
   b. You sequence the operon and discover that it contains a short open reading frame near the 5′-end of the operon that contains several codons for phenylalanine. What prediction would you make about this leader sequence and the peptide that it encodes?
   c. What would happen if the sequence of this leader were changed so that the phenylalanine codons (UUU, UUU) were changed to leucine codons (UUA, UUG)?
   d. What is this kind of regulation called and would it work in a eukaryotic cell? Why or why not?
5. You suspect that the mRNA from gene X of *E. coli* contains an aptamer that binds to a small molecule, Y. Describe an experiment to test this hypothesis.
6. The *aim* operon includes sequences A, B, C, and D. Mutations in these sequences have the following effects, where a plus sign (+) indicates that a functional enzyme is produced and a minus sign (−) indicates that a functional enzyme is not produced. X is a metabolite.

	X present		X absent	
Mutation in sequence:	*Enzyme 1*	*Enzyme 2*	*Enzyme 1*	*Enzyme 2*
A	−	−	−	−
B	+	+	+	+
C	+	−	−	−
D	−	+	−	−
Wild-Type	+	+	−	−

   a. Do the structural gene products from the *aim* operon participate in an anabolic or catabolic process?
   b. Is the repressor protein associated with the *aim* operon produced in an initially active or inactive form?
   c. What does sequence D encode?
   d. What does sequence B encode?
   e. What is sequence A?

SUGGESTED READINGS

General References and Reviews

Beckwith, J.R. and D. Zipser, eds. 1970. *The Lactose Operon.* Plainview, NY: Cold Spring Harbor Laboratory Press.

Corwin, H.O. and J.B. Jenkins. *Conceptual Foundations of Genetics: Selected Readings.* 1976. Boston: Houghton Mifflin Co.

Jacob, F. 1966. Genetics of the bacterial cell (Nobel lecture). *Science* 152:1470–78.

Matthews, K.S. 1996. The whole lactose repressor. *Science* 271:1245–46.

Miller, J.H. and W.S. Reznikoff, eds. 1978. *The Operon.* Plainview, NY: Cold Spring Harbor Laboratory Press.

Monod, J. 1966. From enzymatic adaptation to allosteric transitions (Nobel lecture). *Science* 154:475–83.

Ptashne, M. 1989. How gene activators work. *Scientific American* 260 (January):24–31.

Ptashne, M. and W. Gilbert. 1970. Genetic repressors. *Scientific American* 222 (June):36–44.

Vitreschak, A.G., D.A. Rodionov, A.A. Mironov, and M.S. Gelfand. 2004. Riboswitches: The oldest mechanism for the regulation of gene expression? *Trends in Genetics* 20:44–50.

Winkler, W.C. and R.R. Breaker. 2003. Genetic control by metabolite-binding riboswitches. *Chembiochem* 4:1024–32.

Research Articles

Adhya, S. and S. Garges. 1990. Positive control. *Journal of Biological Chemistry* 265:10797–800.

Benoff, B., H. Yang, C.L. Lawson, G. Parkinson, J. Liu, E. Blatter, Y.W. Ebright, H.M. Berman, and R.H. Ebright. 2002. Structural basis of transcription activation: The CAP–αCTD–DNA complex. *Science* 297:1562–66.

Busby, S. and R.H. Ebright. 1994. Promoter structure, promoter recognition, and transcription activation in prokaryotes. *Cell* 79:743–46.

Chen, B., B. deCrombrugge, W.B. Anderson, M.E. Gottesman, I. Pastan, and R.L. Perlman. 1971. On the mechanism of action of *lac* repressor. *Nature New Biology* 233:67–70.

Chen, Y., Y.W. Ebright, and R.H. Ebright. 1994. Identification of the target of a transcription activator protein by protein–protein photocrosslinking. *Science* 265:90–92.

Emmer, M., B. deCrombrugge, I. Pastan, and R. Perlman. 1970. Cyclic-AMP receptor protein of *E. coli:* Its role in the synthesis of inducible enzymes. *Proceedings of the National Academy of Sciences USA* 66:480–87.

Gilbert, W. and B. Müller-Hill. 1966. Isolation of the *lac* repressor. *Proceedings of the National Academy of Sciences USA* 56:1891–98.

Igarashi, K. and A. Ishihama. 1991. Bipartite functional map of the *E. coli* RNA polymerase α subunit: Involvement of the C-terminal region in transcription activation by cAMP–CRP. *Cell* 65:1015–22.

Jacob, F. and J. Monod. 1961. Genetic regulatory mechanisms in the synthesis of proteins. *Journal of Molecular Biology* 3:318–56.

Krummel, B. and M.J. Chamberlin. 1989. RNA chain initiation by *Escherichia coli* RNA polymerase. Structural transitions of the enzyme in the early ternary complexes. *Biochemistry* 28:7829–42.

Lee, J. and A. Goldfarb. 1991. *Lac* repressor acts by modifying the initial transcribing complex so that it cannot leave the promoter. *Cell* 66:793–98.

Lewis, M., G. Chang, N.C. Horton, M.A. Kercher, H.C. Pace, M.A. Schumacher, R.G. Brennan, and P. Lu. 1996. Crystal structure of the lactose operon repressor and its complexes with DNA and inducer. *Science* 271:1247–54.

Lobell, R.B. and R.F. Schleif. 1991. DNA looping and unlooping by AraC protein. *Science* 250:528–32.

Malan, T.P. and W.R. McClure. 1984. Dual promoter control of the *Escherichia coli* lactose operon. *Cell* 39:173–80.

Oehler, S., E.R. Eismann, H. Krämer, and B. Müller-Hill. 1990. The three operators of the *lac* operon cooperate in repression. *The EMBO Journal* 9:973–79.

Riggs, A.D., S. Bourgeois, R.F. Newby, and M. Cohn. 1968. DNA binding of the *lac* repressor. *Journal of Molecular Biology* 34:365–68.

Schlax, P.J., M.W. Capp, and M.T. Record, Jr. 1995. Inhibition of transcription initiation by *lac* repressor. *Journal of Molecular Biology* 245:331–50.

Schultz, S.C., G.C. Shields, and T.A. Steitz. 1991. Crystal structure of a CAP–DNA complex: The DNA is bent by 90 degrees. *Science* 253:1001–7.

Straney, S. and D.M. Crothers. 1987. *Lac* repressor is a transient gene-activating protein. *Cell* 51:699–707.

Winkler, W.C., S. Cohen-Chalamish, and R.R. Breaker. 2002. An mRNA structure that controls gene expression by binding FMN. *Proceedings of the National Academy of Sciences, USA* 99:15908–13.

Wu, H.-M. and D.M. Crothers. 1984. The locus of sequence-directed and protein-induced DNA bending. *Nature* 308:509–13.

Yanofsky, C. 1981. Attenuation in the control of expression of bacterial operons. *Nature* 289:751–58.

Zubay, G., D. Schwartz, and J. Beckwith. 1970. Mechanism of activation of catabolite-sensitive genes: A positive control system. *Proceedings of the National Academy of Sciences USA* 66:104–10.

CHAPTER 8

Major Shifts in Bacterial Transcription

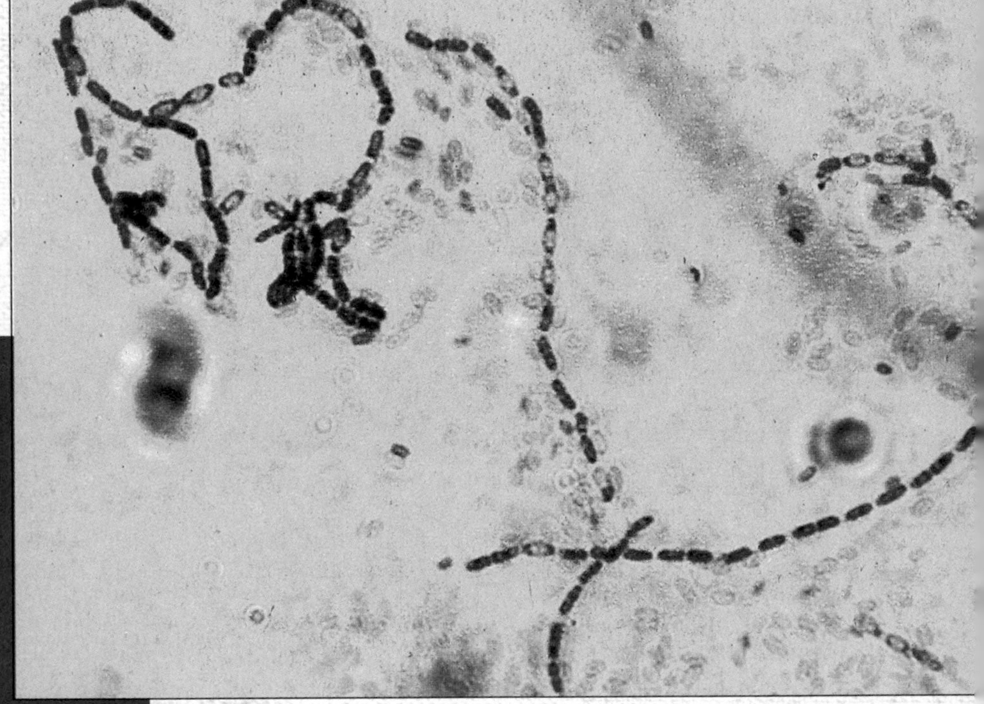

Chains of *Bacillus* bacterial cells. © Steven P. Lynch

In Chapter 7, we discussed the ways in which bacteria control the transcription of a very limited number of genes at a time. For example, when the *lac* operon is switched on, only three genes are activated. At other times in a bacterial cell's life more radical shifts in gene expression take place. These shifts require more fundamental changes in the transcription machinery than are possible in the operon model. In this chapter, we will examine three mechanisms of major shifts in transcription: σ-factor switching; RNA polymerase switching; and antitermination. We will use the λ phage to illustrate the antitermination mechanism, and also discuss the genetic switch used by λ phage to change from one kind of infection strategy to another.

8.1 Sigma Factor Switching

When a phage infects a bacterium, it usually subverts the host's transcription machinery to its own use. In the process, it establishes a time-dependent, or temporal, program of transcription. In other words, the early phage genes are transcribed first, then the later genes. By the time phage T4 infection of *E. coli* reaches its late phase, essentially no more transcription of host genes takes place—only transcription of phage genes. This massive shift in specificity would be hard to explain by the operon mechanisms described in Chapter 7. Instead, it is engineered by a fundamental change in the transcription machinery—a change in RNA polymerase itself.

Another profound change in gene expression occurs during sporulation in bacteria such as *Bacillus subtilis.* Here, genes that are needed in the vegetative phase of growth are turned off, and other, sporulation-specific genes are turned on. Again, this switch is accomplished by changes in RNA polymerase. Bacteria also experience stresses such as starvation, heat shock, and lack of nitrogen, and they also respond to these by shifting their patterns of transcription.

Thus, bacteria respond to changes in their environment by global changes in transcription, and these changes in transcription are accomplished by changes in RNA polymerase. Most often, these are changes in the σ-factor.

Phage Infection

What part of RNA polymerase would be the logical candidate to change the specificity of the enzyme? In Chapter 6, we learned that σ is the key factor in determining specificity of phage T4 DNA transcription in vitro, so σ is the most reasonable answer to our question, and experiments have confirmed that σ is the correct answer. However, these experiments were not done first with the *E. coli* T4 system, but with *B. subtilis* and its phages, especially phage SPO1.

SPO1, like T4, has a large DNA genome. It has a temporal program of transcription as follows: In the first 5 min or so of infection, the early genes are expressed; next, the middle genes turn on (about 5–10 min after infection); from about the 10-min point until the end of infection, the late genes switch on. Because the phage has a large number of genes, it is not surprising that it uses a fairly elaborate mechanism to control this temporal program. Janice Pero and her colleagues were the leaders in developing the model illustrated in Figure 8.1.

The host RNA polymerase holoenzyme handles transcription of early SPO1 genes, which is analogous to the T4 model, where the earliest genes are transcribed by the host holoenzyme (Chapter 6). This arrangement is necessary because the phage does not carry its own

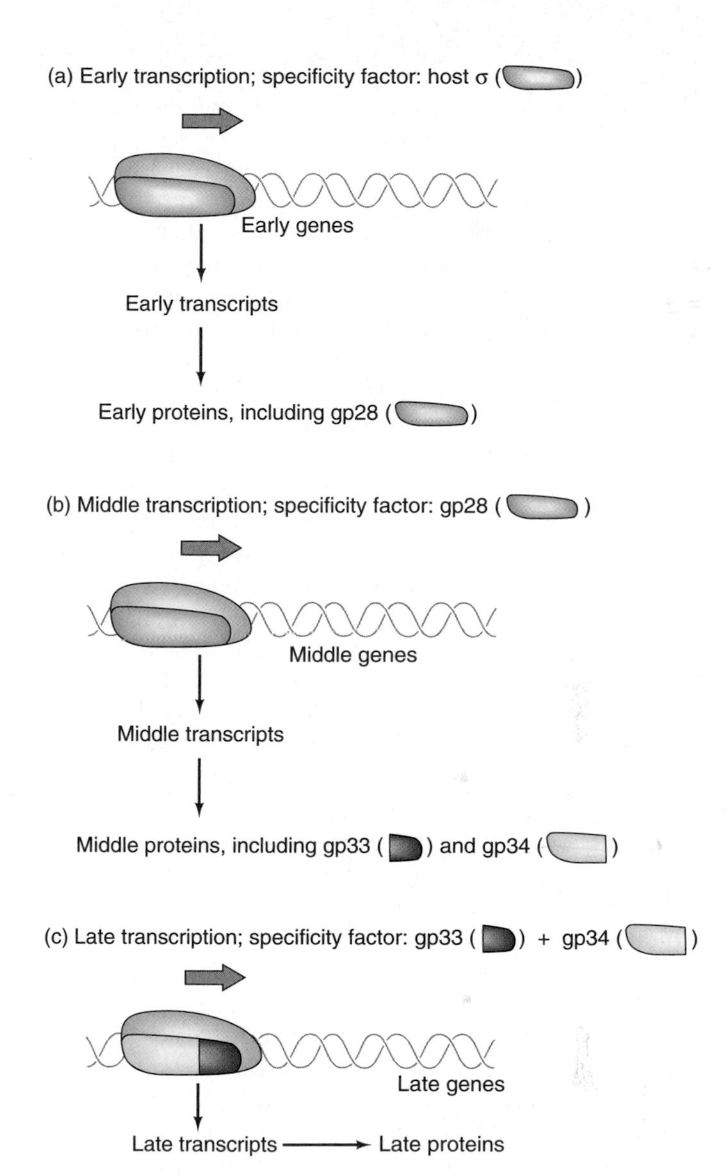

Figure 8.1 Temporal control of transcription in phage SPO1-infected *B. subtilis.* **(a)** Early transcription is directed by the host RNA polymerase holoenzyme, including the host σ-factor (blue); one of the early phage proteins is gp28 (green), a new σ-factor. **(b)** Middle transcription is directed by gp28, in conjunction with the host core polymerase (red); two middle phage proteins are gp33 and gp34 (purple and yellow, respectively); together, these constitute yet another σ-factor. **(c)** Late transcription depends on the host core polymerase plus gp33 and gp34.

RNA polymerase. When the phage first infects the cell, the host holoenzyme is therefore the only RNA polymerase available. The *B. subtilis* holoenzyme closely resembles the *E. coli* enzyme. Its core consists of two large (β and β′), two small (α), and one very small (ω) polypeptides; its primary σ-factor has a molecular mass of 43,000 kD, somewhat smaller than *E. coli*'s primary σ (70,000 kD). In addition, the polymerase includes a δ-subunit with a molecular mass of about 20,000 kD.

This subunit helps to prevent binding to nonpromoter regions, a function performed by the *E. coli* σ-factor but not by the smaller *B. subtilis* σ-factor.

One of the genes transcribed in the early phase of SPO1 infection is called gene 28. Its product, **gp28,** associates with the host core polymerase, displacing the host σ (σ^{43}). With this new, phage-encoded polypeptide in place, the RNA polymerase changes specificity. It begins transcribing the phage middle genes instead of the phage early genes and host genes. In other words, gp28 is a novel σ-factor that accomplishes two things: It diverts the host's polymerase from transcribing host genes, and it switches from early to middle phage transcription.

The switch from middle to late transcription occurs in much the same way, except that two polypeptides team up to bind to the polymerase core and change its specificity. These are **gp33** and **gp34,** the products of two phage middle genes (genes 33 and 34, respectively). These proteins constitute a σ-factor that can replace gp28 and direct the altered polymerase to transcribe the phage late genes in preference to the middle genes. Note that the polypeptides of the host core polymerase remain constant throughout this process; it is the progressive substitution of σ-factors that changes the specificity of the enzyme and thereby directs the transcription program. Of course, the changes in transcription specificity also depend on the fact that the early, middle, and late genes have promoters with different sequences. That is how they can be recognized by different σ-factors.

One striking aspect of this process is that the different σ-factors vary quite a bit in size. In particular, host σ, gp28, gp33, and gp34 have molecular masses of 43,000, 26,000, 13,000, and 24,000 kD, respectively. Yet they are capable of associating with the core enzyme and performing a σ-like role. (Of course, gp33 and gp34 must combine forces to play this role.) In fact, even the *E. coli* σ, with a molecular mass of 70,000 kDa, can complement the *B. subtilis* core in vitro. The core polymerase apparently has a versatile σ-binding site.

How do we know that the σ-switching model is valid? Two lines of evidence, genetic and biochemical, support it. First, genetic studies have shown that mutations in gene 28 prevent the early-to-middle switch, just as we would predict if the gene 28 product is the σ-factor that turns on the middle genes. Similarly, mutations in either gene 33 or 34 prevent the middle-to-late switch, again in accord with the model.

Pero and colleagues performed the biochemical studies. First, they purified RNA polymerase from SPO1-infected cells. This purification scheme included a phosphocellulose chromatography step, which separated three forms of the polymerase. The first of the separated polymerases, enzyme A, contains the host core polymerase, including δ, plus all the phage-encoded factors.

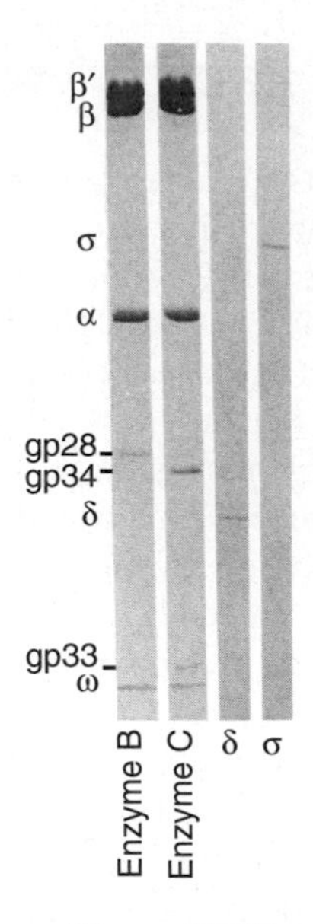

Figure 8.2 Subunit compositions of RNA polymerases in SPO1 phage-infected *B. subtilis* cells. Polymerases were separated by chromatography and subjected to SDS-PAGE to display their subunits plus. Enzyme B (first lane) contains the core subunits (β′, β, α, and ω), as well as gp28. Enzyme C (second lane) contains the core subunits plus gp34 and gp33. The last two lanes contain separated δ- and σ-subunits, respectively. (*Source:* Pero J., R. Tjian, J. Nelson, and R. Losick. In vitro transcription of a late class of phage SPO1 genes. *Nature 257* (18 Sept 1975): f. 1, p. 249 © Macmillan Magazines Ltd.)

The other two polymerases, B and C, were missing δ, but B contained gp28 and C contained gp33 and gp34. Figure 8.2 presents the subunit compositions of these latter two enzymes, determined by SDS-PAGE. Without δ, these two enzymes were incapable of specific transcription because they could not distinguish clearly between promoter and nonpromoter regions of DNA. However, when Pero and colleagues added δ back and assayed transcription specificity, they found that B was specific for the delayed early phage genes, and C was specific for the late genes.

SUMMARY Transcription of phage SPO1 genes in infected *B. subtilis* cells proceeds according to a temporal program in which early genes are transcribed first, then middle genes, and finally late genes. This switching is directed by a set of phage-encoded σ-factors that associate with the host core RNA polymerase and change its specificity of promoter recognition from early to middle to late. The host σ is specific for the phage early genes; the phage gp28 protein switches the specificity to the middle genes; and the phage gp33 and gp34 proteins switch to late specificity.

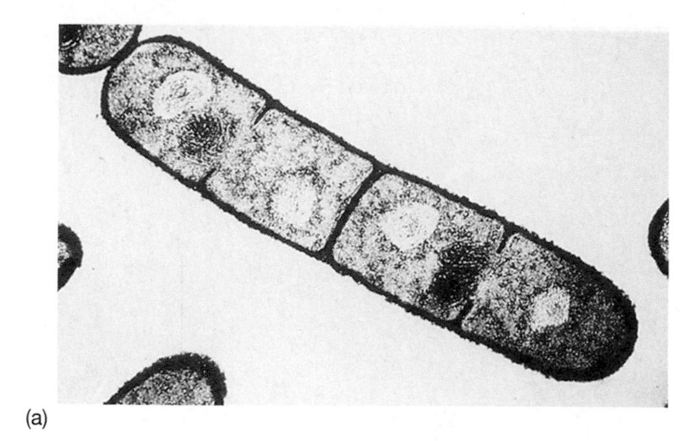

(a)

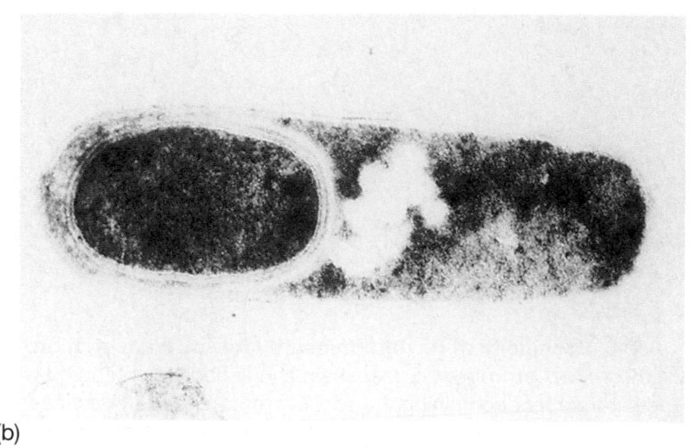

(b)

Figure 8.3 Two developmental fates of *B. subtilis* cells. **(a)** *B. subtilis* vegetative cells dividing and **(b)** sporulation, with an endospore developing at the left end, and the mother cell at the right and surrounding the endospore. (*Source:* Courtesy Dr. Kenneth Bott.)

Sporulation

We have already seen how phage SPO1 changes the specificity of its host's RNA polymerase by replacing its σ-factor. In the following section, we will show that the same kind of mechanism applies to changes in gene expression in the host itself during the process of **sporulation.** *B. subtilis* can exist indefinitely in the **vegetative,** or growth, state, as long as nutrients are available and other conditions are appropriate for growth. But, under starvation or other adverse conditions, this organism forms **endospores—**tough, dormant bodies that can survive for years until favorable conditions return (Figure 8.3).

Sporulation begins with the formation of a polar septum between daughter cells. Unlike a vegetative septum that divides the cell equally, the polar septum forms toward one end, dividing the cell into two unequal parts. The smaller part (on the left in Figure 8.3), is the forespore, which develops into a mature endospore. The larger part is the mother cell, which surrounds the endospore.

Gene expression must change during sporulation; cells as different in morphology and metabolism as vegetative and sporulating cells must contain at least some different gene products. In fact, when *B. subtilis* cells sporulate, they activate a whole new set of sporulation-specific genes. The switch from the vegetative to the sporulating state is accomplished by a complex σ-switching scheme that turns off transcription of some vegetative genes and turns on sporulation-specific transcription.

As you might anticipate, more than one new σ-factor is involved in sporulation. In fact, several participate: σ^F, σ^E, σ^H, σ^C, and σ^K each play a role, in addition to the vegetative σ^A. Each σ-factor recognizes a different class of promoter. For example, the vegetative σ^A recognizes promoters that are very similar to the promoters recognized by the major *E. coli* σ-factor, with a −10 box that looks something like TATAAT and a −35 box having the consensus sequence TTGACA. By contrast, the sporulation-specific factors recognize quite different sequences. The σ^F-factor appears first in the sporulation process, in the forespore. It activates transcription of about 16 genes, including the genes that encode the other sporulation-specific σ-factors. In particular, it activates *spoIIR,* which in turn activates the gene encoding σ^E in the mother cell. Together, σ^F and σ^E put the forespore and mother cell, respectively, on an irreversible path to sporulation.

To illustrate the techniques used to demonstrate that these are authentic σ-factors, let us consider some work by Richard Losick and his colleagues on one of them, σ^E. First, they showed that this σ-factor confers specificity for a known sporulation gene. To do this, they used polymerases containing either σ^E or σ^A to transcribe a plasmid containing a piece of *B. subtilis* DNA in vitro in the presence of labeled nucleotides. The *B. subtilis* DNA (Figure 8.4) contained promoters for both vegetative and sporulation genes. The vegetative promoter lay in a restriction fragment 3050 bp long, and the sporulation promoter was in a 770-bp restriction fragment. Losick and coworkers then hybridized the labeled RNA products to Southern blots (Chapter 5) of the template

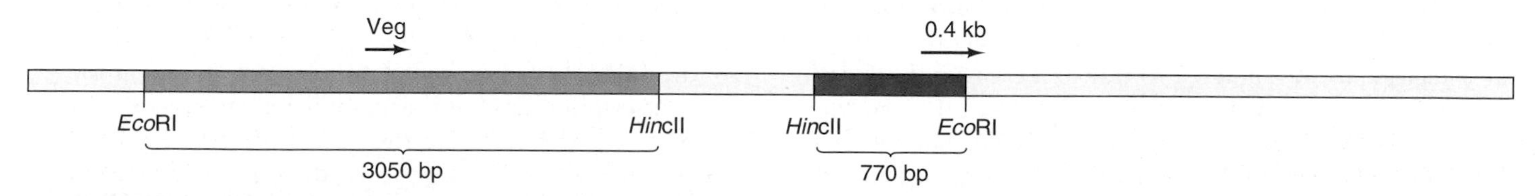

Figure 8.4 Map of part of plasmid p213. This DNA region contains two promoters: a vegetative promoter (Veg) and a sporulation promoter (0.4 kb). The former is located on a 3050-bp *Eco*RI–*Hinc*II fragment (blue); the latter is on a 770-bp fragment (red). (*Source:* Adapted from Haldenwang W.G., N. Lang, and R. Losick, A sporulation-induced sigma-like regulatory protein from *B. subtilis*. *Cell* 23:616, 1981.)

Figure 8.5 Specificities of σ^A and σ^E. Losick and colleagues transcribed plasmid p213 in vitro with RNA polymerase containing σ^A (lane 1) or σ^E (lane 2). Next they hybridized the labeled transcripts to Southern blots containing *Eco*RI–*Hin*cII fragments of the plasmid. As shown in Figure 8.4, this plasmid has a vegetative promoter in a 3050-bp *Eco*RI–*Hin*cII fragment, and a sporulation promoter in a 770-bp fragment. Thus, transcripts of the vegetative gene hybridized to the 3050-bp fragment, while transcripts of the sporulation gene hybridized to the 770-bp fragment. The autoradiograph in the figure shows that the σ^A-enzyme transcribed only the vegetative gene, but the σ^E-enzyme transcribed both the vegetative and sporulation genes. (*Source:* Haldenwang W.G., N. Lang, and R. Losick, A sporulation-induced sigma-like regulatory protein from *B. subtilis*. *Cell* 23 (Feb 1981), f. 4, p. 618. Reprinted by permission of Elsevier Science.)

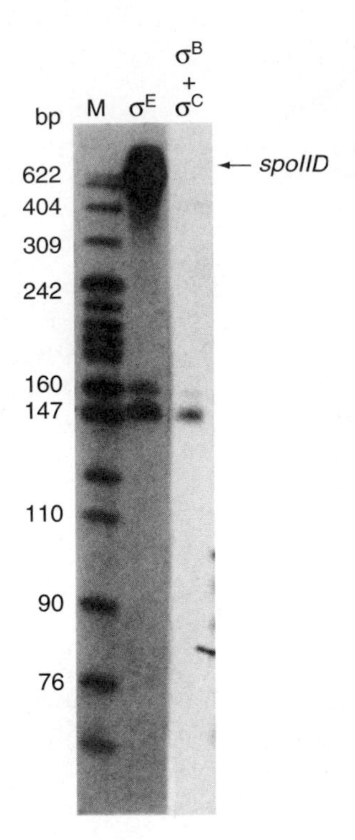

Figure 8.6 Specificity of σ^E determined by run-off transcription from the *spoIID* promoter. Sonenshein and associates prepared a restriction fragment containing the *spoIID* promoter and transcribed it in vitro with *B. subtilis* core RNA polymerase plus σ^E (middle lane) or σ^B plus σ^C (right lane). Lane M contained marker DNA fragments whose sizes in base pairs are indicated at left. The arrow at the right indicates the position of the expected run-off transcript from the *spoIID* promoter (about 700 nt). Only the enzyme containing σ^E made this transcript. (*Source:* Rong S., M.S. Rosenkrantz, and A.L. Sonenshein, Transcriptional control of the *Bacillus subtilis spoIID* gene. *Journal of Bacteriology* 165, no. 3 (1986) f. 7, p. 777, by permission of American Society for Microbiology.)

DNA. This procedure revealed the specificities of the σ-factors: If the vegetative gene was transcribed in vitro, the resulting labeled RNA would hybridize to a 3050-bp band on the Southern blot of the template DNA. On the other hand, if the sporulation gene was transcribed in vitro, the labeled RNA product would hybridize to the 770-bp band. Figure 8.5 shows that when the polymerase contained σ^A, the transcript hybridized only to the vegetative band (3050 bp). By contrast, when the polymerase contained σ^E, the transcript hybridized to both vegetative and sporulation bands (3050 and 770 bp). Apparently σ^E has some ability to recognize vegetative promoters; however, its main affinity seems to be for sporulation promoters—at least those of the type contained in the 770-bp DNA fragment.

The nature of the sporulation gene contained in the 770-bp fragment was not known, so Abraham Sonenshein and colleagues set out to show that σ^E could transcribe a well-characterized sporulation gene. They chose the *spoIID* gene, which was known to be required for sporulation and had been cloned. They used polymerases containing three different σ-factors, σ^B, σ^C, and σ^E, to transcribe a truncated fragment of the gene so as to produce a run-off transcript (Chapter 5). Previous S1 mapping with RNA made in vivo had identified the natural transcription start site. Because the truncation in the *spoIID* gene occurred 700 bp downstream of this start site, transcription from the correct start site in vitro produced a 700-nt run-off transcript. As Figure 8.6 shows, only σ^E could produce this transcript; neither of the other σ-factors could direct the RNA polymerase to recognize the *spoIID* promoter. A similar experiment showed that σ^A could not recognize this promoter, either.

Losick and his colleagues established that σ^E is itself the product of a sporulation gene, originally called *spoIIG*. Predictably, mutations in this gene block sporulation at an early stage. Without a σ-factor to recognize sporulation genes such as *spoIID*, these genes cannot be expressed, and therefore sporulation cannot occur.

SUMMARY When the bacterium *B. subtilis* sporulates, a whole new set of sporulation-specific genes is turned on, and many, but not all, vegetative genes are turned off. This switch takes place largely at the transcription level. It is accomplished by several new σ-factors that displace the vegetative σ-factor from the core RNA polymerase and direct transcription of sporulation genes instead of vegetative genes. Each σ-factor has its own preferred promoter sequence.

Genes with Multiple Promoters

The story of *B. subtilis* sporulation is an appropriate introduction to our next topic, multiple promoters, because sporulation genes provided some of the first examples of this phenomenon. Some genes must be expressed during two or more phases of sporulation, when different σ-factors predominate. Therefore, these genes have multiple promoters recognized by the different σ-factors.

One of the sporulation genes with two promoters is *spoVG,* which is transcribed by both $E\sigma^B$ and $E\sigma^E$ (holoenzymes bearing σ^B or σ^E, respectively). Losick and colleagues achieved a partial separation of these holoenzymes by DNA-cellulose chromatography of RNA polymerases from sporulating cells. Then they performed run-off transcription of a cloned, truncated *spoVG* gene with fractions from the peak of polymerase activity. The fractions on the leading edge of the peak produced primarily a 110-nt run-off transcript. On the other hand, fractions from the trailing edge of the peak made predominately a 120-nt run-off transcript. The fraction in the middle made both run-off transcripts.

These workers succeeded in completely separating the two polymerase activities, using another round of DNA-cellulose chromatography. One set of fractions, containing σ^E, synthesized only the 110-nt run-off transcript. Furthermore, the ability to make this transcript paralleled the content of σ^E in the enzyme preparation, suggesting that σ^E was responsible for this transcription activity. To reinforce the point, Losick's group purified σ^E using gel electrophoresis, combined it with core polymerase, and showed that it made only the 110-nt run-off transcript (Figure 8.7). This same experiment also established that σ^B plus core polymerase made only the 120-nt run-off transcript.

These experiments demonstrated that the *spoVG* gene can be transcribed by both $E\sigma^B$ and $E\sigma^E$, and that these two

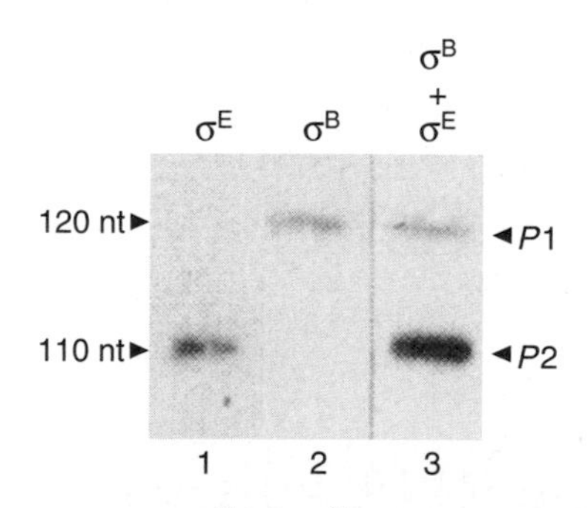

Figure 8.7 Specificities of σ^B and σ^E. Losick and colleagues purified the σ-factors σ^B and σ^E by gel electrophoresis and tested them with core polymerase using a run-off transcription assay. Lane 1, containing σ^E, caused initiation selectively at the downstream promoter (*P2*). Lane 2, containing σ^B, caused initiation selectively at the upstream promoter (*P1*). Lane 3, containing both σ-factors, caused initiation at both promoters. (*Source:* Adapted from Johnson W.C., C.P. Moran, Jr., and R. Losick, Two RNA polymerase sigma factors from *Bacillus subtilis* discriminate between overlapping promoters for a developmentally regulated gene. *Nature* 302 (28 Apr 1983), f. 4, p. 803. © Macmillan Magazines Ltd.)

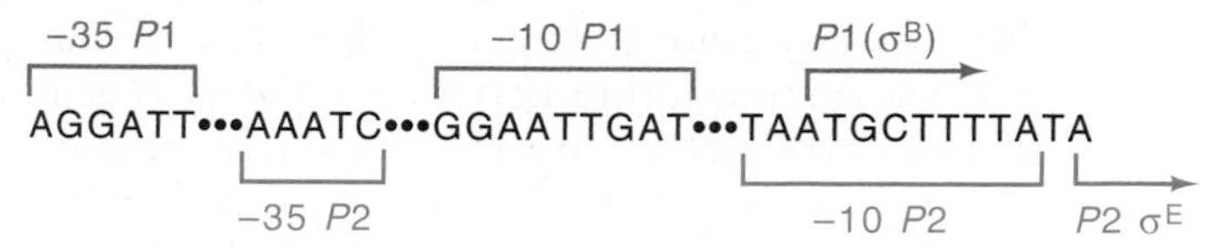

Figure 8.8 Overlapping promoters in *B. subtilis spoVG.* *P1* denotes the upstream promoter, recognized by σ^B; the start of transcription and −10 and −35 boxes for this promoter are indicated in red above the sequence. *P2* denotes the downstream promoter, recognized by σ^E, the start of transcription and −10 and −35 boxes for this promoter are indicated in blue below the sequence.

enzymes have transcription start sites that lie 10 bp apart, as shown in Figure 8.8. Knowing the locations of these start sites, we can count the appropriate number of base pairs upstream and find the −10 and −35 boxes of the promoters recognized by each of these σ-factors (also shown in Figure 8.8). Comparing many −10 and −35 boxes recognized by the same σ-factor allowed the identification of consensus sequences, such as those reported in Chapter 6.

SUMMARY Some prokaryotic genes must be transcribed under conditions where two different σ-factors are active. These genes are equipped with two different promoters, each recognized by one of the two σ-factors. This ensures their expression no matter which factor is present and allows for differential control under different conditions.

Other σ Switches

When cells experience an increase in temperature, or a variety of other environmental insults, they mount a defense called the **heat shock response** to minimize damage. They start producing proteins called **molecular chaperones** that bind to proteins partially unfolded by heating and help them fold properly again. They also produce proteases that degrade proteins that are so badly unfolded that they cannot be refolded, even with the help of chaperones. Collectively, the genes encoding the proteins that help cells survive heat shock are called **heat shock genes.**

Almost immediately after *E. coli* cells are heated from their normal growth temperature (37°C) to a higher temperature (42°C), normal transcription ceases, or at least decreases, and the synthesis of 17 new, heat shock transcripts begins. These transcripts encode the molecular chaperones and proteases that help the cell survive heat shock. This shift in transcription requires the product of the *rpoH* gene, which encodes a σ-factor with a molecular mass of 32 kD. Hence this factor is called σ^{32}, but it is also known as σ^H, where the H stands for heat shock. In 1984, Grossman and coworkers demonstrated that σ^H really is a σ-factor. They did this by combining σ^H with core polymerase and showing that this mixture could transcribe a variety of heat shock genes in vitro from their natural transcription start sites.

The heat shock response begins in less than 1 min, which is not enough time for transcription of the *rpoH* gene and translation of the mRNA to yield a significant amount of new σ-factor. Instead, two other processes explain the rapid accumulation of σ^H. First, the protein itself becomes stabilized at elevated temperatures. This phenomenon can be explained as follows: Under normal growth conditions, σ^H is destabilized by binding to heat shock proteins, which cause its destruction. But when the temperature rises, many other proteins become unfolded and that causes the heat shock proteins to leave σ^H alone and attempt to save or degrade these other, unfolded proteins.

The second effect of high temperature on σ^H concentration operates at the translation level: High temperature causes melting of secondary structure in the 5′-untranslated region of the *rpoH* mRNA, rendering the mRNA more accessible to ribosomes. Miyo Morita and colleagues tested this hypothesis with mutations in the suspected critical secondary structure region. They found that the melting temperatures of the secondary structures in wild-type and mutant mRNAs correlated with the inducibility of σ^H synthesis. We will discuss this mechanism in more detail in Chapter 17.

During nitrogen starvation, another σ-factor (σ^{54}, or σ^N) directs transcription of genes that encode proteins responsible for nitrogen metabolism. In addition, although gram-negative bacteria such as *E. coli* do not sporulate, they do become relatively resistant to stresses, such as extreme pH or starvation. The genes that confer stress resistance are switched on in stationary phase (nonproliferating) *E. coli* cells by an RNA polymerase bearing the alternative σ-factor, σ^S or σ^{38}. These are all examples of a fundamental coping mechanism: Bacteria tend to deal with changes in their environment with global changes in transcription mediated by shifts in σ-factors.

SUMMARY The heat shock response in *E. coli* is governed by an alternative σ-factor, σ^{32} (σ^H) which displaces σ^{70} (σ^A) and directs the RNA polymerase to the heat shock gene promoters. The accumulation of σ^H in response to high temperature is due to stabilization of σ^H and enhanced translation of the mRNA encoding σ^H. The responses to low nitrogen and other stresses, such as starvation, depend on genes recognized by σ^{54} (σ^N) and σ^{38} (σ^S), respectively.

Anti-σ-Factors

In addition to the σ replacement mechanisms we have just discussed, bacterial cells have evolved ways of controlling transcription using **anti-σ-factors.** These proteins do not compete with a σ-factor for binding to a core polymerase. Instead, they bind directly to a σ-factor and inhibit its function. One example of such an anti-σ is the product of the *E. coli rsd* **gene**. The name of the gene derives from its product's ability to regulate (inhibit) the activity of the major vegetative σ, σ^{70}, the product of the *rpoD* gene. Thus, *rsd* stands for "regulator of sigma D.

As long as *E. coli* cells are growing rapidly, most genes are transcribed by $E\sigma^{70}$, and no *rsd* product, **Rsd,** is made. However, as we have just seen, when cells are stressed by such insults as loss of nutrients, high osmolarity, or high temperature, they stop growing and enter the stationary phase. At this point, a new set of stress genes is activated by the new σ-factor, σ^S, which accounts for about one-third of the total amount of RNA polymerase in the cell. This means that about two-thirds of the σ present in the cell is still σ^{70}; nevertheless, expression of genes transcribed by $E\sigma^{70}$ has fallen by over 10-fold. These observations suggest that something else besides relative availability of σ-factors is influencing gene expression, and that extra factor appears to be Rsd, which is made as cells enter stationary phase, then binds to σ^{70} and prevents its association with the core polymerase. Thus, anti-σs can supplement the σ replacement mechanisms by inhibiting the activity of one σ in favor of another.

Some anti-σs are subject to control by **anti-anti-σ-factors.** In sporulating *B. subtilis,* for example, the anti-σ-factor SpoIIAB binds to and inhibits the activities of two σ-factors required at the onset of sporulation, σ^F and σ^G. But another protein, SpoIIAA, binds to complexes of SpoIIAB plus σ^F or σ^G and releases the σ-factors, thus counteracting the effect of the anti-σ-factor. Amazingly enough, the anti-σ-factor SpoIIAB can also act as an **anti-anti-anti-σ-factor** by phosphorylating and inactivating SpoIIAA.

SUMMARY Many σ-factors are controlled by anti-σ-factors that bind to a specific σ and block its binding to the core polymerase. Some of these anti-σ-factors are even controlled by anti-anti-σ-factors that bind to the complexes between a σ and an anti-σ-factor and release the σ-factor. In at least one case, an anti-σ-factor is also an anti-anti-anti-σ-factor that phosphorylates and inactivates the cognate anti-anti-σ-factor.

8.2 The RNA Polymerase Encoded in Phage T7

Phage T7 belongs to a class of relatively simple *E. coli* phages that also includes T3 and ϕII. These have a considerably smaller genome than SPO1 and, therefore, many fewer genes. In these phages we distinguish three phases of transcription: an early phase called class I, and two late phases called classes II and III. One of the five class I genes (gene 1) is necessary for class II and class III gene expression. When it is mutated, only the class I genes are transcribed. Having just learned the SPO1 story, you may be expecting to hear that gene 1 codes for a σ-factor directing the host RNA polymerase to transcribe the late phage genes. In fact, this was the conclusion reached by some workers on T7 transcription, but it was erroneous.

The gene 1 product is actually not a σ-factor but a phage-specific RNA polymerase contained in one polypeptide. This polymerase, as you might expect, transcribes the T7 phage class II and III genes specifically, leaving the class I genes completely alone. Indeed, this enzyme is unusually specific; it will transcribe the class II and III genes of phage T7 and virtually no other natural template. The switching mechanism in this phage is thus quite simple (Figure 8.9). When the phage DNA enters the host cell, the *E. coli* holoenzyme transcribes the five class I genes, including gene 1. The gene 1 product—the phage-specific RNA polymerase—then transcribes the phage class II and class III genes.

A similar polymerase has been isolated from phage T3. It is specific for T3, rather than T7 genes. In fact, T7 and T3 promoters have been engineered into cloning vectors such as pBluescript (Chapter 4). These DNAs can be transcribed in vitro by one phage polymerase or the other to produce strand-specific RNA.

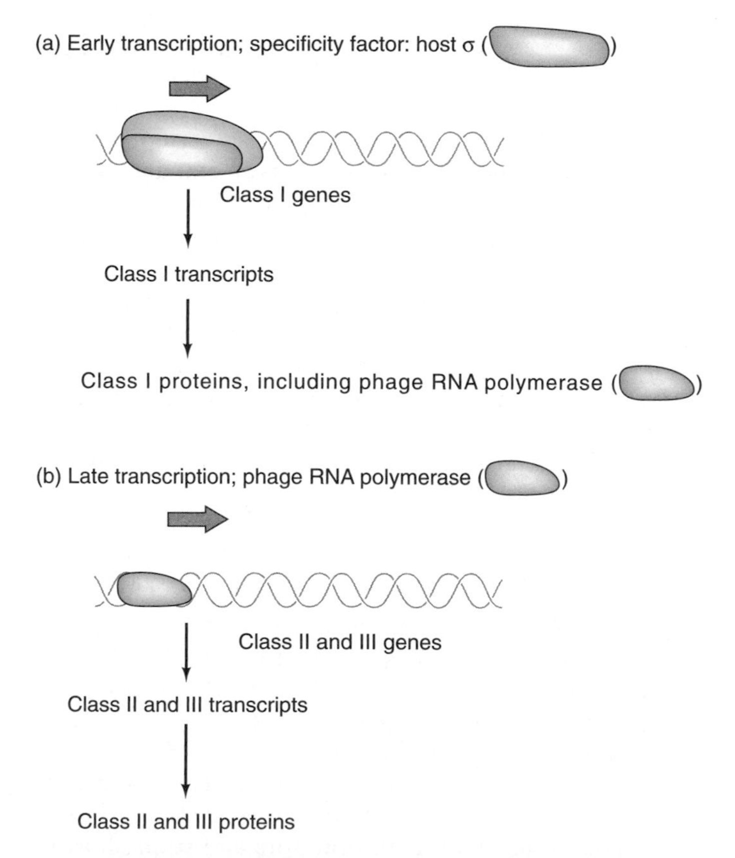

Figure 8.9 Temporal control of transcription in phage T7-infected *E. coli*. **(a)** Early (class I) transcription depends on the host RNA polymerase holoenzyme, including the host σ-factor (blue); one of the early phage proteins is the T7 RNA polymerase (green). **(b)** Late (class II and III) transcription depends on the T7 RNA polymerase.

SUMMARY Phage T7, instead of coding for a new σ-factor to change the host polymerase's specificity from early to late, encodes a new RNA polymerase with absolute specificity for the late phage genes. This polymerase, composed of a single polypeptide, is a product of one of the earliest phage genes, gene 1. The temporal program in the infection by this phage is simple. The host polymerase transcribes the early (class I) genes, one of whose products is the phage polymerase, which then transcribes the late (class II and class III) genes.

8.3 Infection of *E. coli* by Phage λ

Many of the phages we have studied so far (T2, T4, T7, and SPO1, for example) are **virulent** phages. When they replicate, they kill their host by **lysing** it, or breaking it open. On the other hand, **lambda (λ)** is a **temperate** phage; when it infects an *E. coli* cell, it does not necessarily kill it. In this respect, λ is more versatile than many phages; it can follow two paths of reproduction (Figure 8.10). The first is the **lytic** mode, in which infection progresses just as it would with a virulent phage. It begins with phage DNA entering the host cell and then serving as the template for transcription by host RNA polymerase. Phage mRNAs are translated to yield phage proteins, the phage DNA replicates, and progeny phages assemble from these DNA and protein components. The infection ends when the host cell lyses to release the progeny phages.

In the **lysogenic** mode, something quite different happens. The phage DNA enters the cell, and its early genes are transcribed and translated, just as in a lytic infection. But then a 27-kD phage protein (the **λ repressor,** or CI) appears and binds to the two phage operator regions, ultimately shutting down transcription of all genes except for ***cI*** (pronounced "c-one," not "c-eye"), the gene for the λ repressor itself. Under these conditions, with only one phage gene active, it is easy to see why no progeny phages can be produced. Furthermore, when lysogeny is established, the phage DNA integrates into the host genome. A bacterium harboring this integrated phage DNA is called a **lysogen.** The integrated DNA is called a **prophage.** The lysogenic state can exist indefinitely and should not be considered a disadvantage for the phage, because the phage DNA in the lysogen replicates right along with the host DNA. In this way, the phage genome multiplies without the necessity of making phage particles; thus, it gets a "free ride." Under

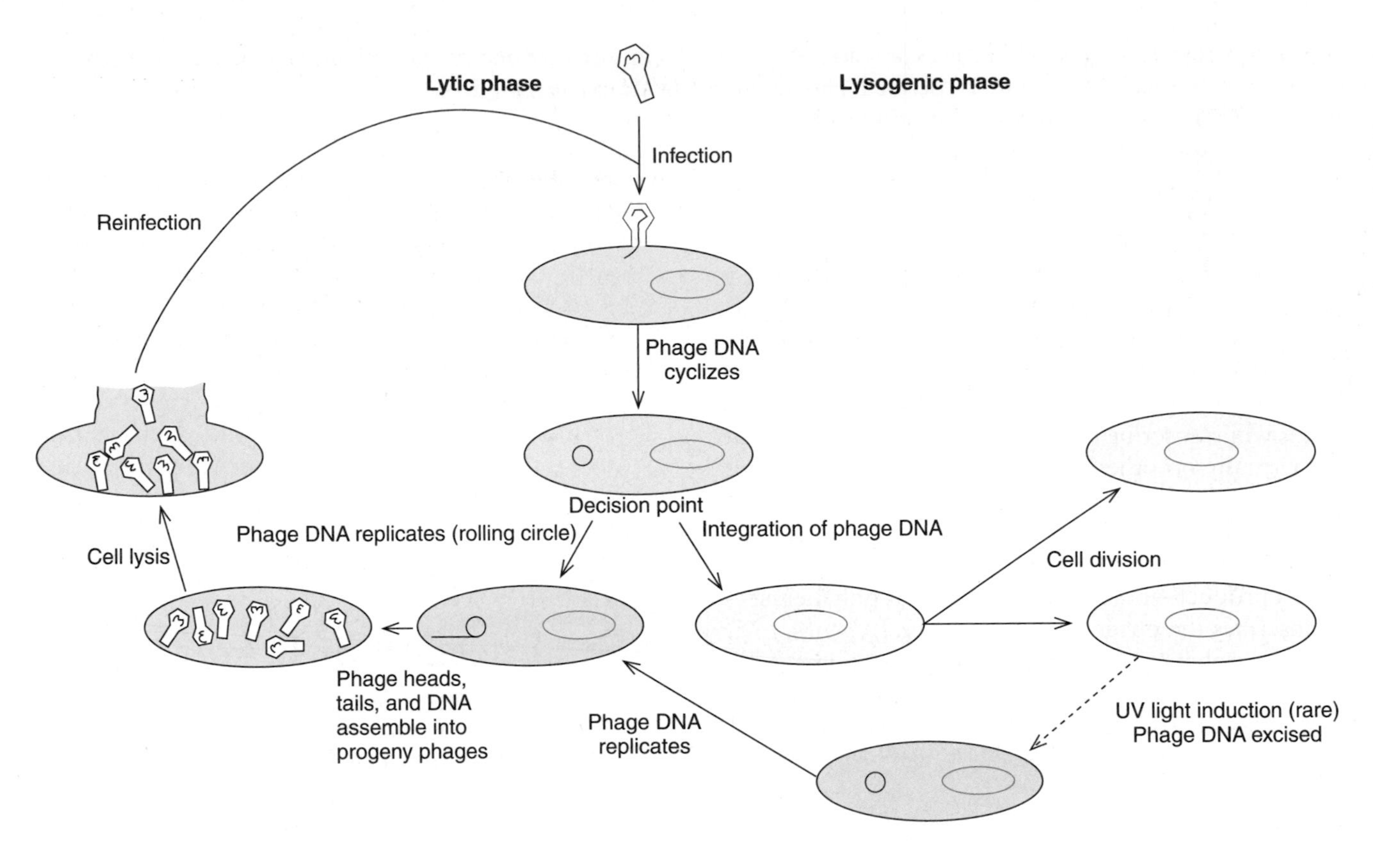

Figure 8.10 Lytic versus lysogenic infection by phage λ. Blue cells are in the lytic phase; yellow cells are in the lysogenic phase; green cells are uncommitted.

certain conditions, such as when the lysogen encounters mutagenic chemicals or radiation, lysogeny can be broken and the phage enters the lytic phase.

SUMMARY Phage λ can replicate in either of two ways: lytic or lysogenic. In the lytic mode, almost all of the phage genes are transcribed and translated, and the phage DNA is replicated, leading to production of progeny phages and lysis of the host cells. In the lysogenic mode, the λ DNA is incorporated into the host genome; after that occurs, only one gene is expressed. The product of this gene, the λ repressor, prevents transcription of all the rest of the phage genes. However, the incorporated phage DNA (the prophage) still replicates, because it has become part of the host DNA.

Lytic Reproduction of Phage λ

The lytic reproduction cycle of phage λ resembles that of the virulent phages we have studied in that it contains three phases of transcription, called **immediate early, delayed early,** and **late.** These three classes of genes are sequentially arranged on the phage DNA, which helps explain how they are regulated, as we will see. Figure 8.11 shows the λ genetic map in two forms: linear, as the DNA exists in the phage particles, and circular, the shape the DNA assumes shortly after infection begins. The circularization is made possible by 12-base overhangs, or "sticky" ends, at either end of the linear genome. These cohesive ends go by the name **cos.** Note that cyclization brings together all the late genes, which had been separated at the two ends of the linear genome.

As usual, the program of gene expression in this phage is controlled by transcriptional switches, but λ uses a switch we have not seen before: **antitermination.** Figure 8.12 outlines this scheme. Of course, the host RNA polymerase holoenzyme transcribes the immediate early genes first. There are only two of these genes, *cro* and *N*, which lie immediately downstream of the rightward and leftward promoters, P_R and P_L, respectively. At this stage in the lytic cycle, no repressor is bound to the operators that govern these promoters (O_R and O_L, respectively), so transcription proceeds unimpeded. When the polymerase reaches the ends of the immediate early genes, it encounters terminators and stops short of the delayed early genes.

The products of both immediate early genes are crucial to further expression of the λ program. The ***cro*** gene product is a repressor that blocks transcription of the λ repressor

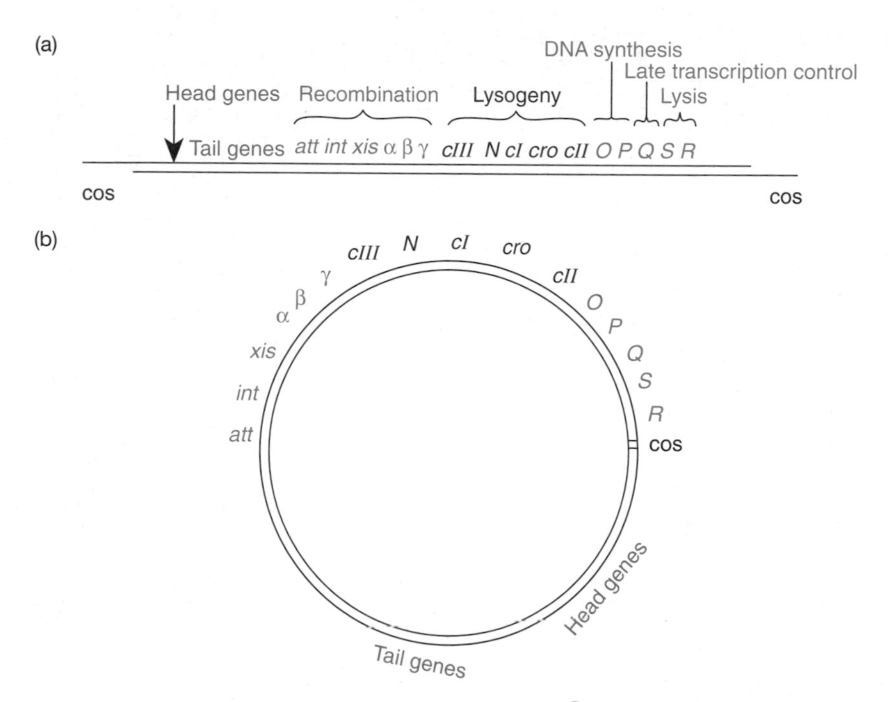

Figure 8.11 Genetic map of phage λ. (a) The map is shown in linear form, as the DNA exists in the phage particles; the cohesive ends (cos) are at the ends of the map. The genes are grouped primarily according to function. **(b)** The map is shown in circular form, as it exists in the host cell during a lytic infection after annealing of the cohesive ends.

gene, *cI,* and therefore prevents synthesis of λ repressor protein. This is necessary for expression of the other phage genes, which would be blocked by λ repressor. The *N* gene product, **N,** is an **antiterminator** that permits RNA polymerase to ignore the terminators at the ends of the immediate early genes and continue transcribing into the delayed early genes. When this happens, the delayed early phase begins. Note that the same promoters (P_R and P_L) are used for both immediate early and delayed early transcription. The switch does not involve a new σ-factor or RNA polymerase that recognizes new promoters and starts new transcripts, as we have seen with other phages; instead, it involves an extension of transcripts controlled by the same promoters.

The delayed early genes are important in continuing the lytic cycle and, as we will see in the next section, in establishing lysogeny. Genes *O* and *P* code for proteins that are necessary for phage DNA replication, a key part of lytic growth. The *Q* gene product (**Q**) is another antiterminator, which permits transcription of the late genes.

The late genes are all transcribed in the rightward (clockwise) direction, but not from P_R. The late promoter, $P_{R'}$, lies just downstream of *Q*. Transcription from this promoter terminates after only 194 bases, unless Q intervenes to prevent termination. The *N* gene product cannot substitute for Q; it is specific for antitermination after *cro* and *N*. The late genes code for the proteins that make up the phage head and tail, and for proteins that lyse the host cell so the progeny phages can escape.

SUMMARY The immediate early/delayed early/late transcriptional switching in the lytic cycle of phage λ is controlled by antiterminators. One of the two immediate early genes is *cro,* which codes for a repressor of the *cI* gene that allows the lytic cycle to continue. The other, *N,* codes for an antiterminator, N, that overrides the terminators after the *N* and *cro* genes. Transcription then continues into the delayed early genes. One of the delayed early genes, *Q,* codes for another antiterminator (Q) that permits transcription of the late genes from the late promoter, $P_{R'}$, to continue without premature termination.

Antitermination How do N and Q perform their antitermination functions? They appear to use two different mechanisms. Let us first consider antitermination by N. Figure 8.13 presents an outline of the process. Panel (a) shows the genetic sites surrounding the *N* gene. On the right is the leftward promoter P_L and its operator O_L. This is where leftward transcription begins. Downstream (left) of the *N* gene is a transcription terminator, where transcription ends in the absence of the *N* gene product (N). Panel (b) shows what happens in the absence of N. The RNA polymerase (pink) begins transcription at P_L and transcribes *N* before reaching the terminator and falling off the DNA, releasing the *N* mRNA. Now that *N* has

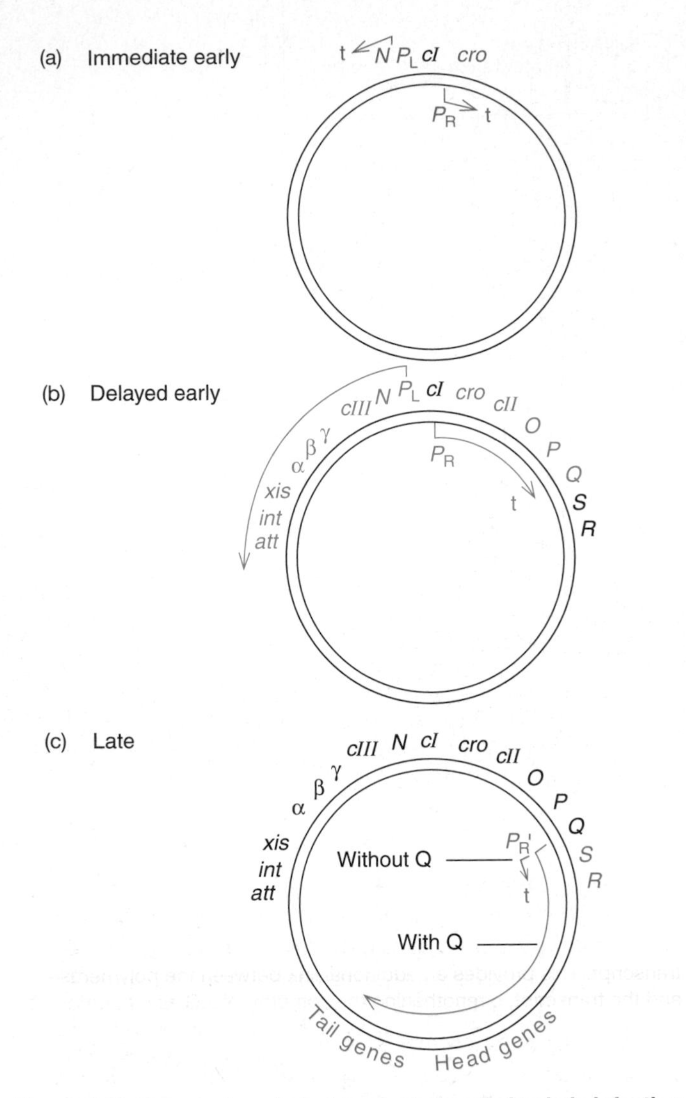

Figure 8.12 Temporal control of transcription during lytic infection by phage λ. (a) Immediate early transcription (red) starts at the rightward and leftward promoters (P_R and P_L, respectively) that flank the repressor gene (*cI*); transcription stops at the rho-dependent terminators (t) after the *N* and *cro* genes. **(b)** Delayed early transcription (blue) begins at the same promoters, but bypasses the terminators by virtue of the *N* gene product, N, which is an antiterminator. **(c)** Late transcription (green) begins at a new promoter ($P_{R'}$); it would stop short at the terminator (t) without the *Q* gene product, Q, another antiterminator. Note that *O* and *P* are protein-encoding delayed early genes, not operator and promoter.

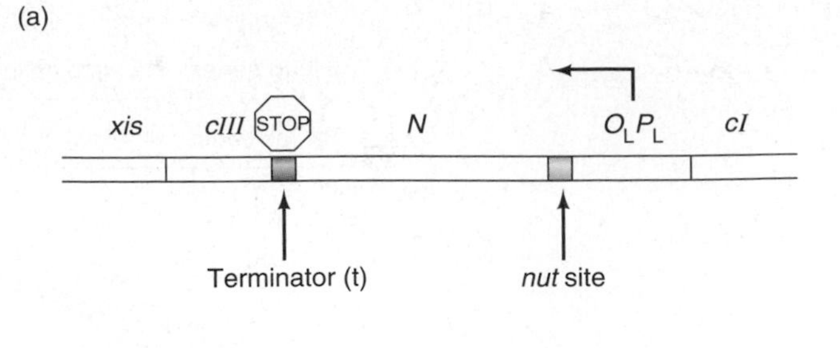

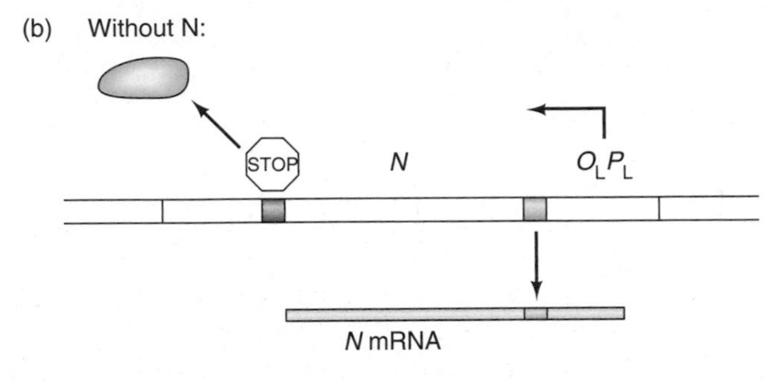

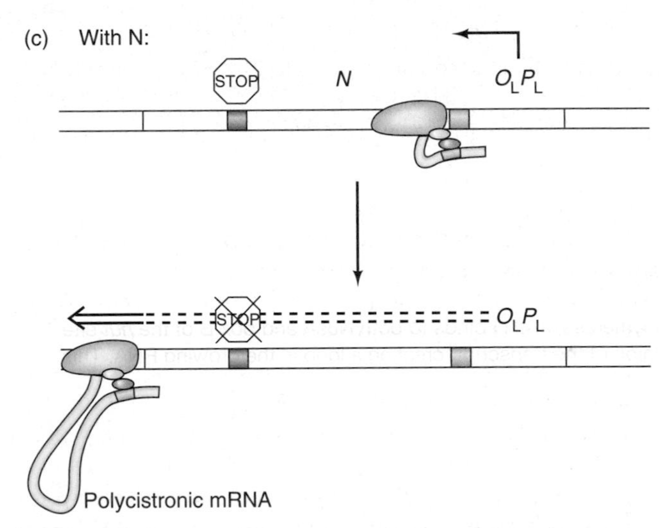

Figure 8.13 Effect of N on leftward transcription. (a) Map of *N* region of λ genome. The genes surrounding *N* are depicted, along with the leftward promoter (P_L) and operator (O_L), the terminator (red), and the *nut* site (green). **(b)** Transcription in the absence of N. RNA polymerase (pink) begins transcribing leftward at P_L and stops at the terminator at the end of *N*. The *N* mRNA is the only product of this transcription. **(c)** Transcription in the presence of N. N (purple) binds to the *nut* region of the transcript, and also to NusA (yellow), which, along with other proteins not shown, has bound to RNA polymerase. This complex of proteins alters the polymerase so it can read through the terminator and continue into the delayed early genes.

been transcribed, the N protein appears, and panel (c) shows what happens next. The N protein (purple) binds to the *transcript* of the **N utilization site** (***nut*** **site,** green) and interacts with a complex of host proteins (yellow) bound to the RNA polymerase. This somehow alters the polymerase, turning it into a "juggernaut" that ignores the terminator and keeps on transcribing into the delayed early genes. The same mechanism applies to rightward transcription from P_R, because a site just to the right of *cro* allows the polymerase to ignore the terminator and enter the delayed early genes beyond *cro*.

How do we know that host proteins are involved in antitermination? Genetic studies have shown that mutations in four host genes interfere with antitermination. These genes encode the proteins **NusA,** NusB, NusG, and the ribosomal S10 protein. It may seem surprising that host proteins cooperate in a process that leads to host cell death, but this is just one of many examples in which a virus harnesses a cellular process for its own benefit. In this case, the cellular process served by the S10 protein is obvious:

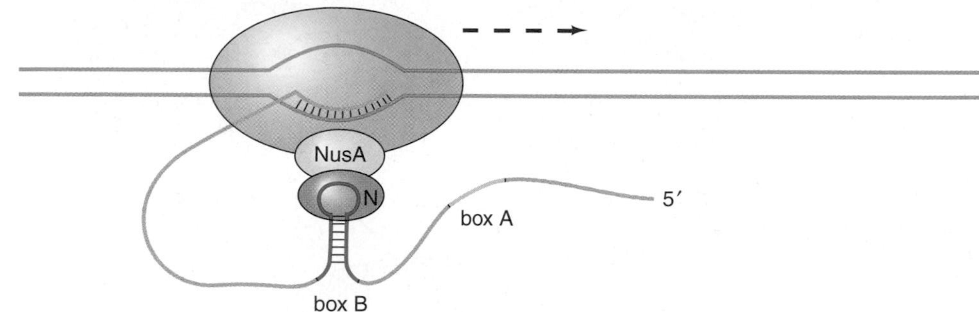

(b) Strong, processive complex

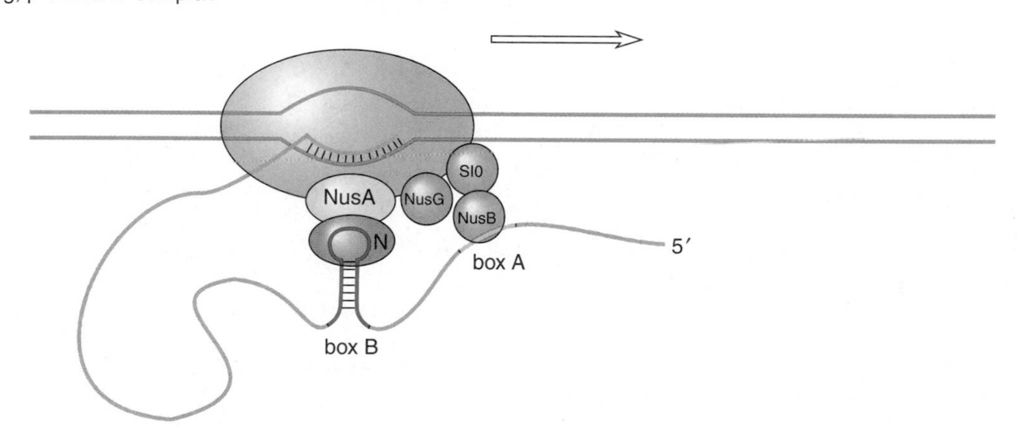

Figure 8.14 Protein complexes involved in N-directed antitermination. (a) Weak, nonprocessive complex. NusA binds to polymerase, and N binds to both NusA and box B of the *nut* site region of the transcript, creating a loop in the growing RNA. This complex is relatively weak and can cause antitermination only at terminators near the *nut* site (dashed arrow). These conditions exist only in vitro. **(b)** Strong, processive complex. NusA tethers N and box B to the polymerase, as in panel (a); in addition, S10 binds to polymerase, and NusB binds to box A of the *nut* site region of the transcript. This provides an additional link between the polymerase and the transcript, strengthening the complex. NusG also contributes to the strength of the complex. This complex is processive and can cause antitermination thousands of base pairs downstream in vivo (open arrow).

protein synthesis. But the Nus proteins also have cellular roles. They allow antitermination in the seven *rrn* operons that encode ribosomal RNAs, as well as some tRNAs. In addition, NusA actually stimulates termination, as we will see.

In vitro studies have shown that two proteins, N and NusA, can cause antitermination if the terminator is close enough to the *nut* site. Figure 8.14a shows the protein complex involved in this short-range antitermination and illustrates the fact that N does not bind by itself to RNA polymerase. It binds to NusA, which in turn binds to polymerase. This figure also introduces the two parts of the *nut* site, known as *box A* and *box B*. Box A is highly conserved among *nut* sites, but box B varies quite a bit from one *nut* site to another. The transcript of box B contains an inverted repeat, which presumably forms a stem-loop, as shown in Figure 8.14.

Antitermination in vivo is not likely to use this simple scheme because it occurs at terminators that are at least hundreds of base pairs downstream of the corresponding *nut* sites. We call this kind of natural antitermination **processive** because the antitermination factors remain associated with the polymerase as it moves a great distance along the DNA. Such processive antitermination requires more than just N and NusA. It also requires the other three host proteins: NusB, NusG, and S10. These proteins presumably help to stabilize the antitermination complex so it persists until it reaches the terminator. Figure 8.14b depicts this stable complex that includes all five antitermination proteins.

Perhaps the most unexpected feature of the antitermination complex depicted in Figure 8.14 is the interaction of the complex with the *transcript* of the *nut* site, rather than with the *nut* site itself. How do we know this is what happens? One line of evidence is that the region of N that is essential for *nut* recognition is an arginine-rich domain that resembles an RNA-binding domain. Asis Das provided more direct evidence, using a gel mobility shift assay to demonstrate binding between N and an RNA fragment containing box B. Furthermore, when N and NusA have both bound to the complex, they partially protect box B, but not box A, from RNase attack. Only when all five proteins have bound is box A protected from RNase. This is consistent with the model in Figure 8.14.

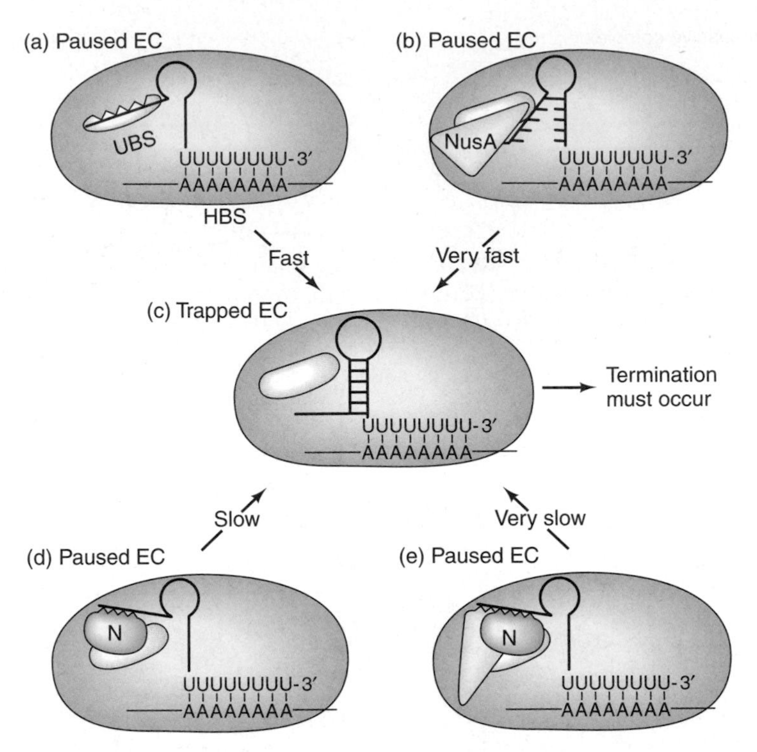

Figure 8.15 Model for the function of NusA and N in intrinsic termination. (a) The upstream half of the potential terminator hairpin is bound to the upstream binding site (UBS) in a pocket in the core polymerase. Nevertheless, the protein–RNA bonds (pink "sawteeth") can break, and hairpin formation can occur rapidly to yield the trapped elongation complex (EC) that is committed to terminate **(c). (b)** NusA helps break the bonds between the UBS and the upstream half of the potential hairpin. This facilitates hairpin formation and therefore stimulates termination. **(d)** N binds to the upstream half of the potential terminator hairpin (protein–RNA bonds represented by green "sawteeth") and slows down hairpin formation. This makes termination less likely. **(e)** N not only binds to the upstream half of the potential hairpin, it also facilitates NusA binding at an adjacent position where it can also bind the upstream half of the potential hairpin (yellow "sawteeth"). This slows hairpin formation even further and renders termination even less likely. (*Source:* Adapted from Gusarov I. and E. Nudler. 2001. Control of intrinsic transcription termination by N and NusA: The basic mechanisms. *Cell* 107:444.)

How do we know the RNA loops as shown in Figure 8.14? We don't know for sure, but the easiest way to imagine a signal to the polymerase that persists from the time N binds to the *nut* site transcript until the polymerase reaches the terminator is to envision N maintaining its association with both the polymerase and the RNA. This would require that the RNA form a loop as shown. In accord with this hypothesis, Jack Greenblatt and colleagues have isolated mutants with alterations in the gene encoding the RNA polymerase β-subunit that interfere with N-mediated antitermination. These mutants also fail to protect the *nut* site transcript during transcription in vitro. This suggests that an association exists between the RNA polymerase, N, and the *nut* site transcript during transcription. Again, this is easiest to imagine if the RNA between the *nut* site transcript and the polymerase forms a loop.

How does N prevent termination? One hypothesis was that it limited the pausing by RNA polymerase that is essential for termination. But Ivan Gusarov and Evgeny Nudler demonstrated in 2001 that N does not affect pausing enough to have a significant effect on termination. They went on to show that N binds to the RNA that is destined to form the upstream part of the terminator hairpin and thereby slows down hairpin formation. Without the hairpin, termination cannot occur. This is reminiscent of the mechanism of overriding transcription attenuation (termination) in the *trp* operon, which involves a ribosome stalled on the upstream part of one of the attenuator hairpins (Chapter 7).

Figure 8.15 presents the model of Gusarov and Nudler (2001) (part of which we already discussed in Chapter 6), which also shows a role for **NusA** in termination. When the **elongation complex (EC)** synthesizes the string of U's, it pauses after incorporating the seventh nucleotide in the string. This places the upstream portion of the potential hairpin in position to bind to an **upstream binding site (UBS)** of the RNA polymerase. The pause lasts for only about 2 s, so the hairpin must form within this time period or the polymerase will move on without terminating. If the hairpin does form, it traps the elongation complex in a state that is bound to terminate. NusA acts by weakening the binding between the upstream part of the potential hairpin and the UBS, thereby encouraging the hairpin to form before the end of the pause. This stimulates termination.

The model in Figure 8.15 also calls for binding of N to the upstream part of the potential hairpin, which blocks hairpin formation. Moreover, once N has bound to the RNA, it also binds to NusA. In this position, NusA also binds to the upstream part of the potential hairpin. With both N and NusA bound to the RNA, it forms a hairpin only very slowly, so the polymerase moves on without terminating.

What is the evidence for this scheme? The key part of the model is the interaction between N (and N + NusA) and the upstream part of the RNA hairpin. Gusarov and Nudler demonstrated that these interactions really take place using a protein–RNA cross-linking technique. By walking, or elongating, a $[^{32}P]$RNA one to a few nucleotides at a time (Chapter 6) in the presence or absence of N and NusA, they introduced **4-thioU (sU)** into position +45 of the RNA. Then, by walking further, they created RNAs that had reached positions +50, +54, +58, +62, +68, and +75. This placed the sU at positions −6, −10, −14, −18, −24, or −31, with respect to the 3′-end of the RNA.

The 4-thiouracil base is photoreactive and, upon UV irradiation, will cross-link to any tightly bound protein (within about 1 Å of the base). So Gusarov and Nudler UV-irradiated the complexes with sU in various positions relative to the 3′-end of the nascent RNA. Then they subjected the complexes to SDS-PAGE and autoradiography to detect cross-linking between the RNA and N, NusA, and the core polymerase (α, β, and β′). If RNA became cross-linked to a protein, that protein would become radio-labeled, and the protein band would appear dark in the autoradiograph.

Figure 8.16a shows the results. Both N and NusA cross-linked to the sU when it was between positions −18 and −24 relative to the 3′-end of the nascent RNA (lanes 6, 7, 12, and 13). This is where the upstream part of the hairpin lies. Furthermore, when N and NusA were present together, NusA bound more strongly to the RNA, and its binding extended to the −31 region. The fact that N and NusA bind to the upstream half of the hairpin at the time of termination suggests that these two proteins are in a position to control termination by controlling whether the hairpin forms. A mutant form of N, N^{RRR}, with a mutated RNA-binding RRR motif, could bind to the RNA hairpin as well as wild-type N, suggesting that this motif is not required for binding to the hairpin.

To test the hypothesis that N and NusA control hairpin formation, Gusarov and Nudler prepared a different elongation complex, again by walking, but this time they walked the labeled RNA to a mutated terminator (T7-tR2^{mut2}), in which two U's in the oligo(U) region at the end of the RNA are changed to G's. This change delays hairpin formation and allows study of the elongation complex just before termination. This time, Gusarov and Nudler placed another photoreactive nucleotide, 6-thioG **(sG)**, in either of two positions, −14, or −24. Again, they performed the walking in the presence of either N or NusA, or both. And again, they UV-irradiated the complexes and then subjected them to SDS-PAGE and autoradiography to detect cross-linking of RNA to proteins.

Figure 8.16b shows that no cross-linking of RNA to either N or NusA occurred when the sG was in the −14 position (the downstream part of the hairpin), but it did occur when the sG was in the −24 position (the upstream part of the hairpin). Thus, both N and NusA appear to contact the RNA that forms the upstream half of the hairpin. Furthermore, NusA binding to the RNA decreased binding to the core polymerase (to the α-subunit and to the β- and β′-subunits, compare lanes 5 and 6). By contrast, N binding to the RNA did not decrease binding to the core polymerase (compare lanes 5 and 7). Similarly, when N and NusA bound together to the RNA, no decrease in RNA binding to the core was observed (compare lanes 5 and 8). These results suggested that NusA interferes with binding of the upstream part of the hairpin to the polymerase, and that N counters that interference and restores binding between polymerase and the hairpin.

Robert Landick and colleagues performed similar cross-linking experiments, but they used 5-iodoU as the cross-linking reagent, and they placed it at position −11, which is in the loop of a pause hairpin, which causes transcription pausing, but not termination. These workers found that NusA caused a strong RNA cross-link to β to be replaced by a weaker cross-link to NusA. Moreover, the link between the RNA hairpin loop and the β-subunit of RNA polymerase was to a region of β called the **flap-tip helix.** Furthermore, Landick and colleagues showed that removal of a few amino acids from the flap-tip helix abolished stimulation of pausing by NusA. Thus, the flap-tip helix is required for NusA activity. Because the flap is connected directly to the part of β at the active site, Landick and colleagues proposed an allosteric mechanism: The interaction between the pause hairpin loop and the flap-tip helix changes the conformation of the active site enough that elongation becomes more difficult, so the polymerase pauses. NusA presumably facilitates this process.

If the Gusarov–Nudler model in Figure 8.15 is correct, then placing a large loop of RNA between the upstream and downstream parts of the hairpin should interfere with the activities of both N and NusA. That is because N and NusA bind to the UBS in such a way as to interact with the RNA that comes just before the downstream part of the hairpin. Ordinarily, this is the upstream part of the hairpin, but in this case it is the beginning of the large tR2^{loop}. Thus, N and NusA are not in position to influence the formation of the hairpin, and these proteins should therefore have little effect on termination. Gusarov and Nudler tested this hypothesis using the tR2^{loop} terminator. As predicted, neither N nor NusA had much effect on termination.

The control of late λ transcription also uses an antitermination mechanism, but with significant differences

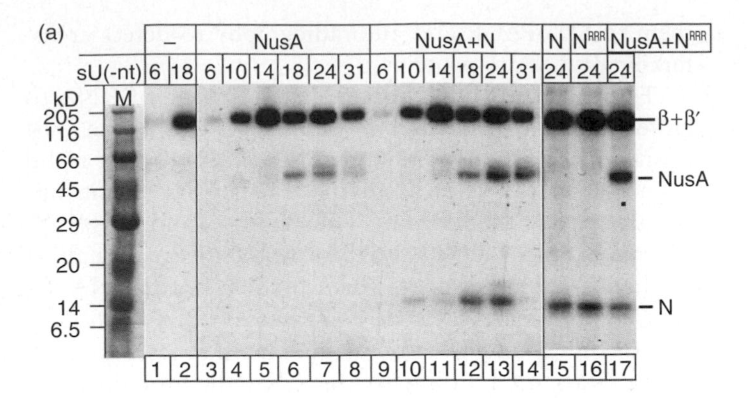

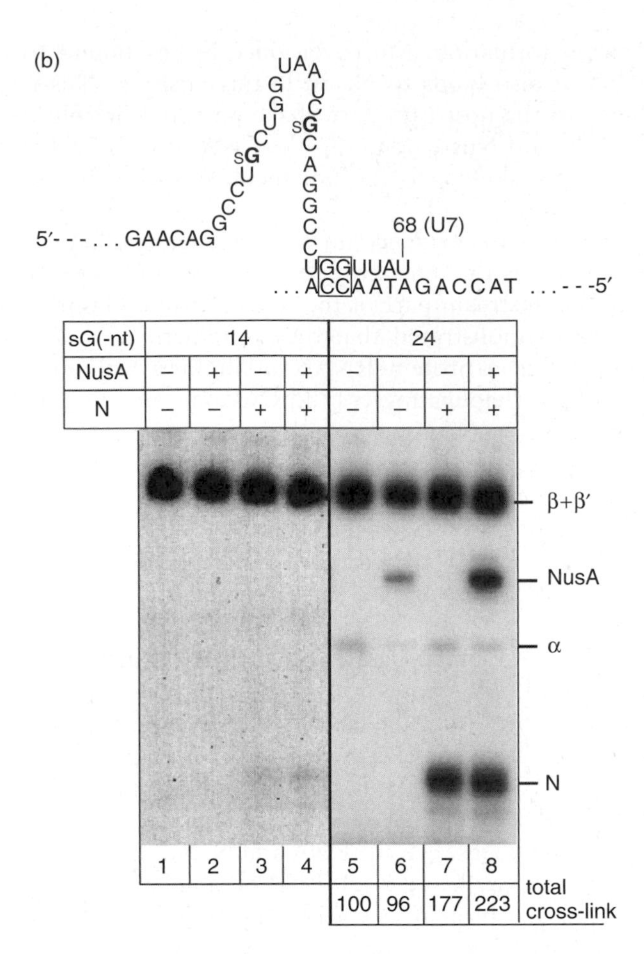

Figure 8.16 Demonstration of protein–RNA contacts in the paused EC, with and without NusA and N. (a) Cross-linking of NusA and N to the upstream half of the RNA hairpin at the terminator. Gusarov and Nudler labeled a nascent terminator-containing transcript with ^{32}P and incorporated a photoreactive nucleotide (4-thio-UMP) into position +45 by walking the elongation complex (see Chapter 6). Then they walked the complex to positions that placed the 4-thio-UMP at positions −6, −10, −14, −18, −24, and −31, as indicated at top, in the presence of NusA, NusA + N, or N, also as indicated at top. Then they exposed the complexes to UV light to cross-link the 4-thio-UMP to any protein tightly bound to the RNA in that region. Then they subjected the complexes to SDS-PAGE and autoradiography to identify proteins covalently linked to the RNA. The positions of N, NusA, and RNA polymerase β + β′ subunits are indicated at right. Lane M contains markers. N^{RRR} denotes a mutant N with the RRR RNA-binding motif mutated. N and NusA both bound to RNA in the −18 to −24 region, which includes the upstream half of the hairpin. **(b)** Effect of N and NusA on interaction of polymerase core with the hairpin. Gusarov and Nudler performed an experiment similar to the one in panel (a), but used 6-thio-G (sG) as a cross-linking agent, and used a mutant terminator ($tR2^{mut2}$) to slow down termination. The position of the sG (–14 or −24 relative to the 3′-end of the RNA) is indicated at top, as is the presence of NusA, N, or both. The positions of the proteins β + β′, NusA, α, and N are indicated at right. The cartoon at top illustrates the locations of sG in the −14 and −24 positions. The boxed base pairs are the ones that are altered in the $tR2^{mut2}$ mutant terminator. NusA caused a decrease in the binding of the upstream half of the hairpin (position −24) to the core polymerase (β + β′ and α), but N appeared to cause an increase in RNA binding to the core polymerase, and N plus NusA caused a greater increase (see boxes at bottom labeled "total cross-link"). (*Source:* Reprinted from *Cell* v. 107, Gusarov and Nudler, p. 443 © 2001, with permission from Elsevier Science.)

from the N-antitermination system. Figure 8.17 shows that a **Q utilization *(qut)* site** overlaps the late promoter ($P_{R'}$). This *qut* site also overlaps a pause site at 16–17 bp downstream of the transcription initiation site. In contrast to the N system, Q binds directly to the *qut* site, not to its transcript.

In the absence of Q, RNA polymerase pauses for several minutes at this site just after transcribing the pause signal. After it finally leaves the pause site, RNA polymerase transcribes to the terminator, where it aborts late transcription. On the other hand, if Q is present, it recognizes the paused complex and binds to the *qut* site. Q then binds to the polymerase and alters it in such a way that it resumes transcription and ignores the terminator, continuing on into the late genes. The Q-altered polymerase appears to inhibit RNA hairpin formation immediately behind the polymerase, thereby inhibiting terminator activity. Q can cause antitermination by itself in the λ late control region, but NusA makes the process more efficient.

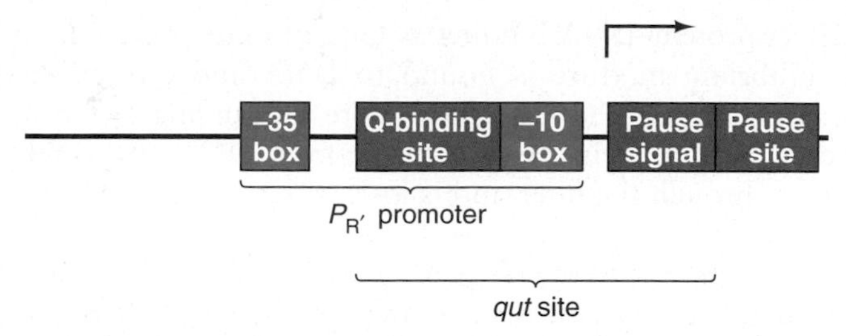

Figure 8.17 Map of the $P_{R'}$ region of the λ genome. The $P_{R'}$ promoter comprises the −10 and −35 boxes. The *qut* site overlaps the promoter and includes the Q-binding site upstream of the −10 box, the pause signal downstream of the transcription start site, and the pause site at positions +16 and +17. (*Source:* Adapted from *Nature* 364:403, 1993.)

SUMMARY Five proteins (N, NusA, NusB, NusG, and S10) collaborate in antitermination at the λ immediate early terminators. NusA and S10 bind to RNA polymerase, and N and NusB bind to the box B and box A regions, respectively, of the *nut* site in the growing transcript. N and NusB bind to NusA and S10, respectively, probably tethering the transcript to the polymerase. This alters the polymerase so it reads through the terminators at the ends of the immediate early genes. NusA stimulates termination at an intrinsic terminator by interfering with the binding between the upstream part of the terminator hairpin and the core polymerase, thereby facilitating the formation of the hairpin in the nascent RNA. The hairpin traps the complex in a form that is irreversibly committed to terminate. N interferes with this process by binding to the upstream part of the terminator hairpin, preventing hairpin formation. N even helps NusA bind farther upstream to the nascent RNA, thereby inhibiting hairpin formation still more. Without hairpin formation, termination cannot occur. Antitermination in the λ late region requires Q, which binds to the Q-binding region of the *qut* site as RNA polymerase is stalled just downstream of the late promoter. Subsequent binding of Q to the polymerase appears to alter the enzyme so it can ignore the terminator and transcribe the late genes.

Establishing Lysogeny

We have mentioned that the delayed early genes are required not only for the lytic cycle but for establishing lysogeny. The delayed early genes help establish lysogeny in two ways: (1) Some of the delayed early gene products are needed for integration of the phage DNA into the host genome, a prerequisite for lysogeny. (2) The products of the *cII* and *cIII* genes allow transcription of the *cI* gene and therefore production of the λ repressor, the central component in lysogeny.

Two promoters control the *cI* gene (Figure 8.18): P_{RM} and P_{RE}. P_{RM} stands for "promoter for repressor maintenance." This is the promoter that is used *during* lysogeny to ensure a continuing supply of repressor to maintain the lysogenic state. It has the peculiar property of requiring its own product—repressor—for activity. We will discuss the basis for this requirement; however, we can see immediately one important implication. This promoter cannot be used to establish lysogeny, because at the start of infection no repressor is present to activate it. Instead, the other promoter, P_{RE}, is used. P_{RE} stands for "promoter for repressor establishment." P_{RE} lies to the right of both P_R and *cro*. It directs transcription leftward through *cro* and then through *cI*. Thus, P_{RE} allows *cI* expression before any repressor is available.

Of course, the natural direction of transcription of *cro* is rightward from P_R, so the leftward transcription from P_{RE}

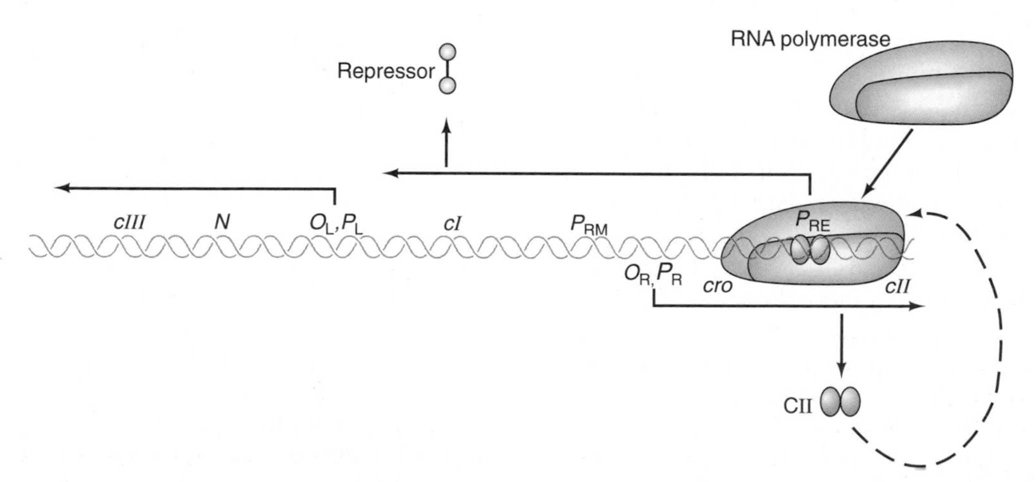

Figure 8.18 Establishing lysogeny. Delayed early transcription from P_R gives *cII* mRNA that is translated to CII (purple). CII allows RNA polymerase (red and blue) to bind to P_{RE} and transcribe the *cI* gene, yielding repressor (green).

gives an RNA product that is an **antisense** transcript of *cro,* as well as a "sense" transcript of *cI.* The *cI* part of the RNA can be translated to give repressor, but the anti-sense *cro* part cannot be translated. This antisense RNA also contributes to the establishment of lysogeny because the *cro* antisense transcript binds to *cro* mRNA and interferes with its translation. Because *cro* works against lysogeny, blocking its action promotes lysogeny. Similarly, CII stimulates transcription from a leftward promoter ($P_{\text{anti-Q}}$) within *Q*. This "backwards" transcription produces *Q* antisense RNA that blocks production of Q. Because Q is required for late transcription in the lytic phase, obstructing its synthesis favors the alternative pathway—lysogeny.

P_{RE} has some interesting requirements of its own. It has −10 and −35 boxes with no clear resemblance to the consensus sequences ordinarily recognized by *E. coli* RNA polymerase. Indeed, it cannot be transcribed by this polymerase alone in vitro. However, the *cII* gene product, CII, helps RNA polymerase bind to this unusual promoter sequence.

Hiroyuki Shimatake and Martin Rosenberg demonstrated this activity of CII in vitro. Using a filter-binding assay, they showed that neither RNA polymerase nor CII protein alone could bind to a fragment of DNA containing P_{RE}, and therefore could not cause this DNA to bind to the nitrocellulose filter. On the other hand, CII protein plus RNA polymerase together could bind the DNA to the filter. Thus, CII must be stimulating RNA polymerase binding to P_{RE}. Furthermore, this binding is specific. CII could stimulate polymerase to bind only to P_{RE} and to one other promoter, P_I. The latter is the promoter for the *int* gene, which is also necessary for establishing lysogeny and which also requires CII. The *int* gene is involved in integration of the λ DNA into the host genome.

We saw in Chapter 7 that CAP–cAMP stimulates RNA polymerase binding to the *lac* promoter by protein–protein interaction. Mark Ptashne and his colleagues have provided evidence that CII may work in a similar way. They used DNase footprinting (Chapter 5) to map the binding site for CII in the DNA fragment containing P_{RE} and found that CII binds between positions −21 and −44 of the promoter. Of course, this includes the unrecognizable −35 box of the promoter and raises the question: How can two proteins (CII and RNA polymerase) bind to the same site at the same time? The probable answer comes from Ptashne and coworkers' DMS footprinting experiments (Chapter 5) with CII, which showed that the bases that appear to be contacted by CII (e.g., G's at positions −26 and −36) are on the opposite side of the helix from those thought to be in contact with RNA polymerase (e.g., G at position −41). In other words, these two proteins seem to bind to opposite sides of the DNA helix, and so can bind cooperatively instead of competitively.

Why could CII footprint λ promoter DNA by itself, whereas it could not bind to λ promoter DNA in a filter-binding assay? The former is a more sensitive assay. It will detect protein–DNA binding as long as some protein in an equilibrium mixture is bound to DNA and can protect it. By contrast, the latter is a more demanding assay. As soon as a protein dissociates from a DNA, the DNA flows through the filter and is lost, with no opportunity to rebind.

The product of the *cIII* gene, CIII, is also instrumental in establishing lysogeny, but its effect is less direct. It retards destruction of CII by cellular proteases. Thus, the delayed early products CII and CIII cooperate to make possible the establishment of lysogeny by activating P_{RE} and P_I.

SUMMARY Phage λ establishes lysogeny by causing production of enough repressor to bind to the early operators and prevent further early RNA synthesis. The promoter used for establishment of lysogeny is P_{RE}, which lies to the right of P_R and *cro*. Transcription from this promoter goes leftward through the *cI* gene. The products of the delayed early genes *cII* and *cIII* also participate in this process: CII, by directly stimulating polymerase binding to P_{RE} and P_I; CIII, by slowing degradation of CII.

Autoregulation of the *cI* Gene During Lysogeny

Once the λ repressor appears, it binds as a dimer to the λ operators, O_R and O_L. This has a double-barreled effect, both tending toward lysogeny. First, the repressor turns off further early transcription, thus interrupting the lytic cycle. The turnoff of *cro* is especially important to lysogeny because the *cro* product (**Cro**) acts to counter repressor activity, as we will see. The second effect of repressor is that it stimulates its own synthesis by activating P_{RM}.

Figure 8.19 illustrates how this self-activation works. The key to this phenomenon is the fact that both O_R and O_L are subdivided into three parts, each of which can bind repressor. The O_R region is more interesting because it controls leftward transcription of *cI,* as well as rightward transcription of *cro*. The three binding sites in the O_R region are called O_R1, O_R2, and O_R3. Their affinities for repressor are quite different: Repressor binds most tightly to O_R1, and less tightly to O_R2 and O_R3. However, binding of repressor to O_R1 and O_R2 is cooperative. This means that as soon as repressor dimer binds to its "favorite" site, O_R1, it facilitates binding of another repressor dimer to O_R2. No cooperative binding to O_R3 normally occurs.

Repressor protein is a dimer of two identical subunits, each of which is represented by a dumbbell shape in Figure 8.19. This shape indicates that each subunit has two domains, one at each end of the molecule. The two domains

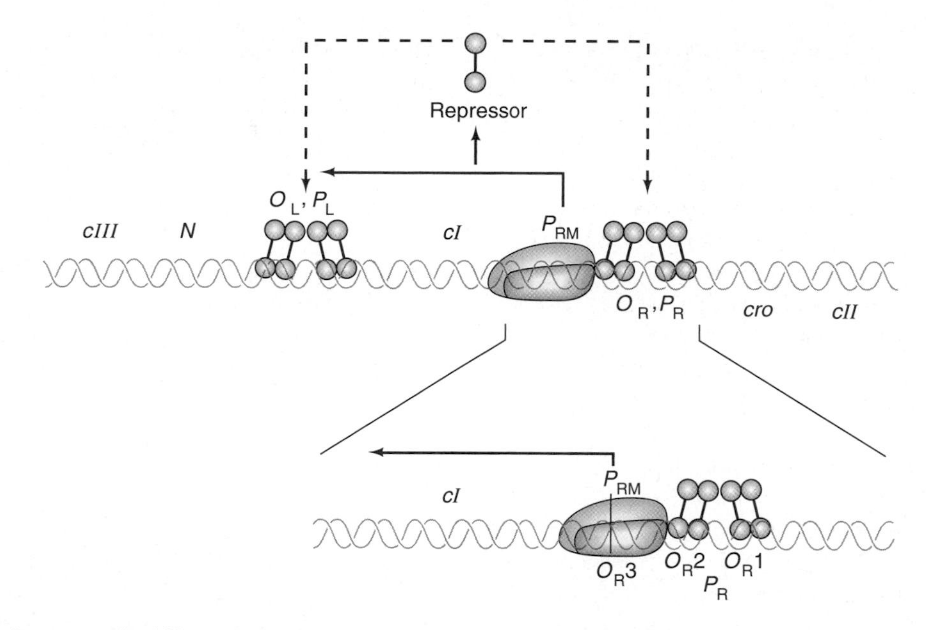

Figure 8.19 Maintaining lysogeny. (top) Transcription (from P_{RM}) and translation of the *cI* mRNA yields a continuous supply of repressor, which binds to O_R and O_L and prevents transcription of any genes aside from *cI.* (bottom) Detail of control region. Repressor (green) forms dimers and binds cooperatively to O_R1 and 2. The protein–protein contact between repressor on O_R2 and RNA polymerase (red and blue) allows polymerase to bind to P_{RM} and transcribe *cI.*

have distinct roles: The amino–terminal domain is the DNA-binding end of the molecule; the carboxyl–terminal domain is the site for the repressor–repressor interaction that makes dimerization and cooperative binding possible. Once repressor dimers have bound to both O_R1 and O_R2, the repressor occupying O_R2 lies very close to the binding site for RNA polymerase at P_{RM}. So close, in fact, that the two proteins touch each other. Far from being a hindrance, as you might expect, this protein–protein contact strengthens the binding of RNA polymerase to a very weak promoter, much as CII facilitates binding of RNA polymerase to P_{RE}.

With repressors bound to O_R1 and O_R2, no more transcription from P_{RE} can occur because the repressors block *cII* and *cIII* transcription, and the products of these genes, needed for transcription from P_{RE}, break down very rapidly. However, the disappearance of CII and CIII is not usually a problem, because lysogeny is already established and a small supply of repressor is all that is required to maintain it. That small supply of repressor can be provided as long as O_R3 is left open, because RNA polymerase can transcribe *cI* freely from P_{RM}. Also, repressor bound to O_R1 and O_R2 blocks *cro* transcription by interfering with polymerase binding to P_R.

It is conceivable that the concentration of repressor could build up to such a level that it would even fill its weakest binding site, O_R3. In that case, all *cI* transcription would cease because even P_{RM} would be blocked. This halt in *cI* transcription would allow the repressor level to drop, at which time repressor would dissociate first from O_R3, allowing *cI* transcription to begin anew. This mechanism would allow repressor to keep its own concentration from rising too high.

Ptashne and colleagues demonstrated with one in vitro experiment three of the preceding assertions: (1) the λ repressor at low concentration can stimulate transcription of its own gene; (2) the repressor at high concentration can inhibit transcription of its own gene; and (3) the repressor can inhibit *cro* transcription. The experiment was a modified run-off assay that used a low concentration of UTP so transcription frequently paused before it ran off the end of the DNA template. This produced so-called "stutter" transcripts resulting from the pauses. The template was a 790-bp *Hae*III restriction fragment that included promoters for both the *cI* and *cro* genes (Figure 8.20). Ptashne called the *cro* gene by another name, *tof,* in this paper.

The experimenters added RNA polymerase, along with increasing amounts of repressor, to the template and observed the rate of production of the 300-nt run-off

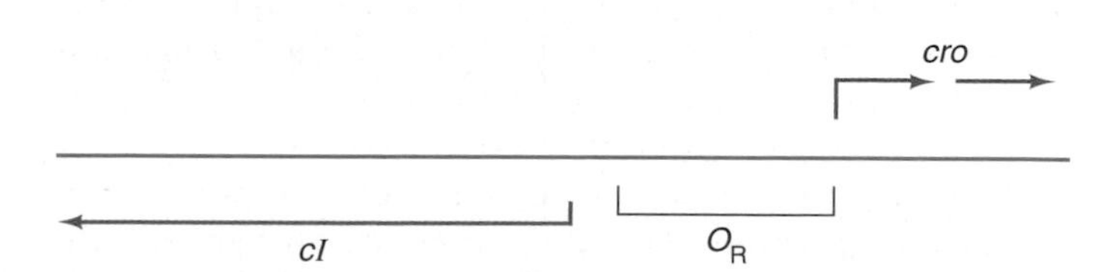

Figure 8.20 Map of the DNA fragment used to assay transcription from *cI* and *cro* promoters. The red arrows denote the in vitro *cI* and *cro* transcripts, some of which ("stutter transcripts") terminate prematurely. (*Source:* Adapted from Meyer B.J., D.G. Kleid, and M. Ptashne. *Proceedings of the National Academy of Sciences* 72:4787, 1975.)

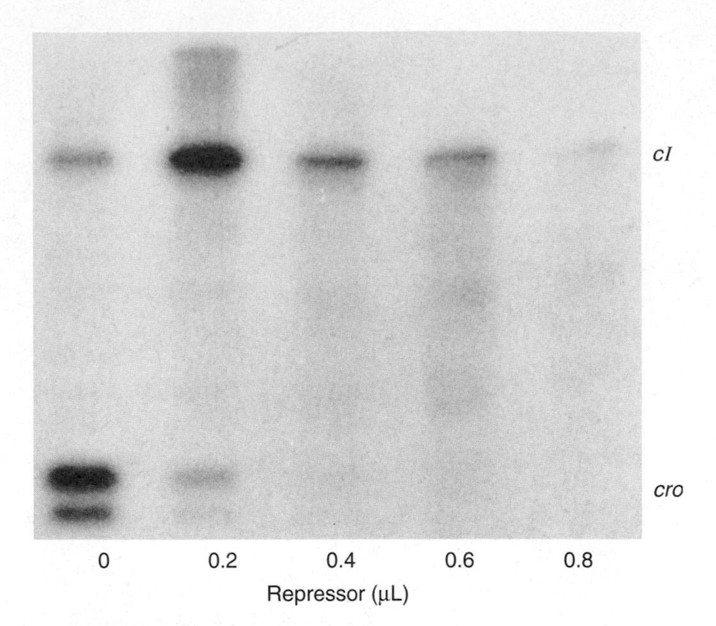

Figure 8.21 Analysis of the effect of λ repressor on *cI* and *cro* transcription in vitro. Ptashne and colleagues performed run-off transcription (which also produced "stutter" transcripts) using the DNA template depicted in Figure 8.20. They included increasing concentrations of repressor as shown at bottom. Electrophoresis separated the *cI* and *cro* stutter transcripts, which are identified at right. The repressor clearly inhibited *cro* transcription, but it greatly stimulated *cI* transcription at low concentration, then inhibited *cI* transcription at high concentration. (*Source:* Meyer B.J., D.G. Kleid, and M. Ptashne. Repressor turns off transcription of its own gene. *Proceedings of the National Academy of Sciences* 72 (Dec 1975), f. 5, p. 4788.)

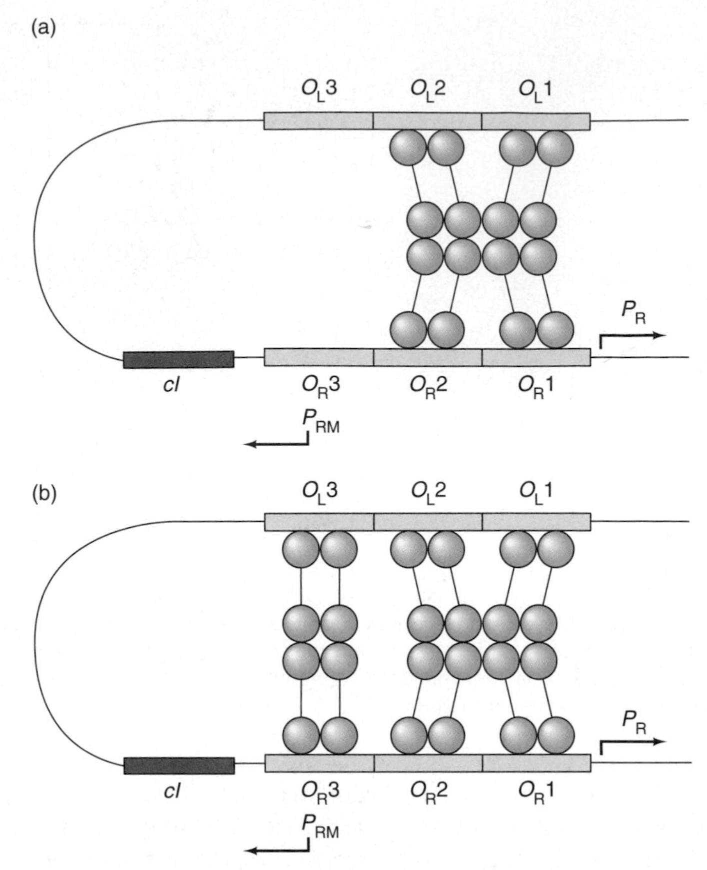

Figure 8.22 Model for involvement of O_L in repression of P_R and P_{RM}. (a) Repression of P_R. A repressor octamer binds cooperatively to O_R1, O_R2, O_L1, and O_L2, looping out the DNA in between the two operators. This involvement of O_L is not necessary for repression of P_R, but it sets the stage for repression of P_{RM}. **(b)** Repression of P_{RM}. With the octamer of repressor formed, a tetramer of repressor can bind cooperatively to O_R3 and O_L3. This causes effective repression of P_{RM} that would not be possible with just a dimer of repressor bound to O_R3.

transcript from the *cI* gene and the two approximately 110-nt stutter transcripts from the *cro* gene. Figure 8.21 shows the results. At low repressor concentration we can already see some inhibition of *cro* transcription, and we also notice a clear stimulation of *cI* transcription. At higher repressor concentration, *cro* transcription ceases entirely, and at even higher concentration, *cI* transcription is severely inhibited.

There is a significant problem with the in vitro experiment we just discussed: It included concentrations of repressor that are higher than one would expect to find in a real lysogen in vivo. In fact, when Ptashne and colleagues used physiological levels of repressor in later experiments, the inhibition of *cI* transcription from the P_{RM} promoter was only 5–20%; to get even 50% inhibition, Ptashne and colleagues had to add a concentration of repressor 15 times higher than normally found in a lysogen.

In light of these later results, we are left wondering whether λ repressor can really repress its own gene at all. And if not, we wonder about the need for O_R3 and O_L3. If they are not needed for either positive or negative autoregulation, why are they there? Ian Dodd and colleagues investigated this question by testing the levels of expression from P_{RM} and found that repressor at the levels found in lysogens really can repress transcription from P_{RM}, but only if O_L3 is present. (The reason Ptashne and colleagues did not observe strong repression of P_{RM} at physiological repressor concentration in their earlier experiments was because their constructs did not include O_L.) Furthermore, mutations in O_R3 that prevent repression of transcription from P_{RM} produce abnormally high levels of repressor, and are defective in switching from the lysogenic to the lytic phase (see later in this chapter).

Dodd and colleagues explained these data on the basis of DNA looping that is known to be possible between O_R 1 and 2 and O_L 1 and 2, and is predicted to involve a repressor octamer, as illustrated in Figure 8.22. When the loop forms, it would open the way for another tetramer of repressor to bind to O_R3 and O_L3, and that would repress transcription from P_{RM}.

SUMMARY The promoter that is used to maintain lysogeny is P_{RM}. It comes into play after transcription from P_{RE} makes possible the burst of repressor synthesis that establishes lysogeny. This repressor binds to O_R1 and O_R2 cooperatively, but leaves O_R3 open. RNA polymerase binds to P_{RM}, which overlaps O_R3 in such a way that it contacts the repressor bound to O_R2. This protein–protein interaction is required for this promoter to work efficiently. High levels of repressor can repress transcription from P_{RM}, and this process may involve interaction of repressor dimers bound to O_R1, O_R2, and O_R3, with repressor dimers bound to O_L1, O_L2, and O_L3, via DNA looping.

RNA Polymerase/Repressor Interaction How do we know that λ repressor/RNA polymerase interaction is essential for stimulation at P_{RM}? In 1994, Miriam Susskind and colleagues performed a genetic experiment that provided strong support for this hypothesis. They started with a mutant λ phage in which a key aspartate in the λ repressor had been changed to an asparagine. This mutant belongs to a class called *pc* for positive control; the mutant repressor is able to bind to the λ operators and repress transcription from P_R and P_L, but it is not able to stimulate transcription from P_{RM} (i.e., positive control of *cI* does not function). Susskind and coworkers reasoned as follows: If direct interaction between repressor and polymerase is necessary for efficient transcription from P_{RM}, then a mutant with a compensating amino acid change in a subunit of RNA polymerase should be able to restore the interaction with the mutant repressor, and therefore restore active transcription from P_{RM}. Figure 8.23 illustrates this concept, which is known as **intergenic suppression,** because a mutation in one gene suppresses a mutation in another.

The search for such intergenic suppressor mutants could be extremely tedious if each one had to be tested separately for the desired activity. It is much more feasible to use a **selection** (Chapter 4) to eliminate any wild-type polymerase genes, or those with irrelevant mutations, and just keep those with the desired mutations. Susskind and colleagues used such a selection, which is described in Figure 8.24. They used *E. coli* cells with two prophages. One was a λ prophage bearing a *cI* gene with a *pc* mutation indicated by the black X; expression of this *cI* gene was driven by a weak version of the *lac* promoter. The other was a P22 prophage bearing the kanamycin-resistance gene under control of the P_{RM} promoter. Susskind and colleagues grew these cells in the presence of the antibiotic kanamycin, so the cells' survival depended on expression of the kanamycin-resistance gene. With the mutated repressor and the RNA polymerase provided by the cell, these cells could not survive because they could not activate transcription from P_{RM}.

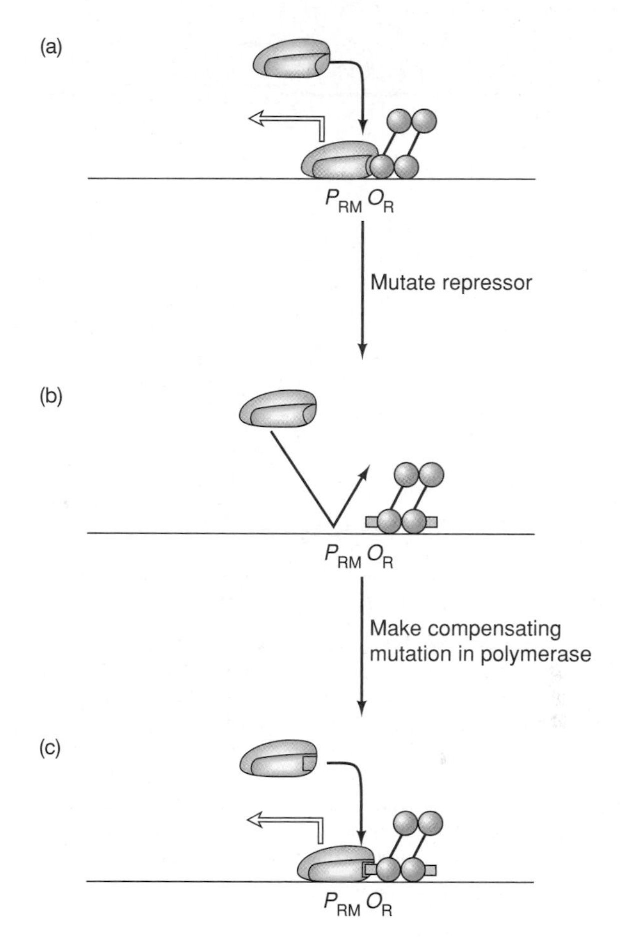

Figure 8.23 Principle of intergenic suppression to detect interaction between λ repressor and RNA polymerase. (a) With wild-type repressor and polymerase, the two proteins interact closely, which stimulates polymerase binding and transcription from P_{RM}. **(b)** The repressor gene has been mutated, yielding repressor with an altered amino acid (yellow). This prevents binding to polymerase. **(c)** The gene for one polymerase subunit has been mutated, yielding polymerase with an altered amino acid (represented by the square cavity) that restores binding to the mutated repressor. Because polymerase and repressor can now interact, transcription from P_{RM} is restored.

Next, the investigators transformed the cells with plasmids bearing wild-type and mutant versions of a gene *(rpoD)* encoding the RNA polymerase σ-subunit. If the *rpoD* gene was wild-type or contained an irrelevant mutation, the product (σ) could not interact with the mutant repressor, so the cells would not grow in kanamycin. On the other hand, if they contained a suppressor mutation, the σ-factor could join with core polymerase subunits provided by the cell to form a mutated polymerase that could interact with the mutated repressor. This interaction would allow activation of transcription from P_{RM}, and the cells could therefore grow in kanamycin. In principle, these cells with a suppressor mutation in *rpoD* were the *only* cells that could grow, so they were easy to find.

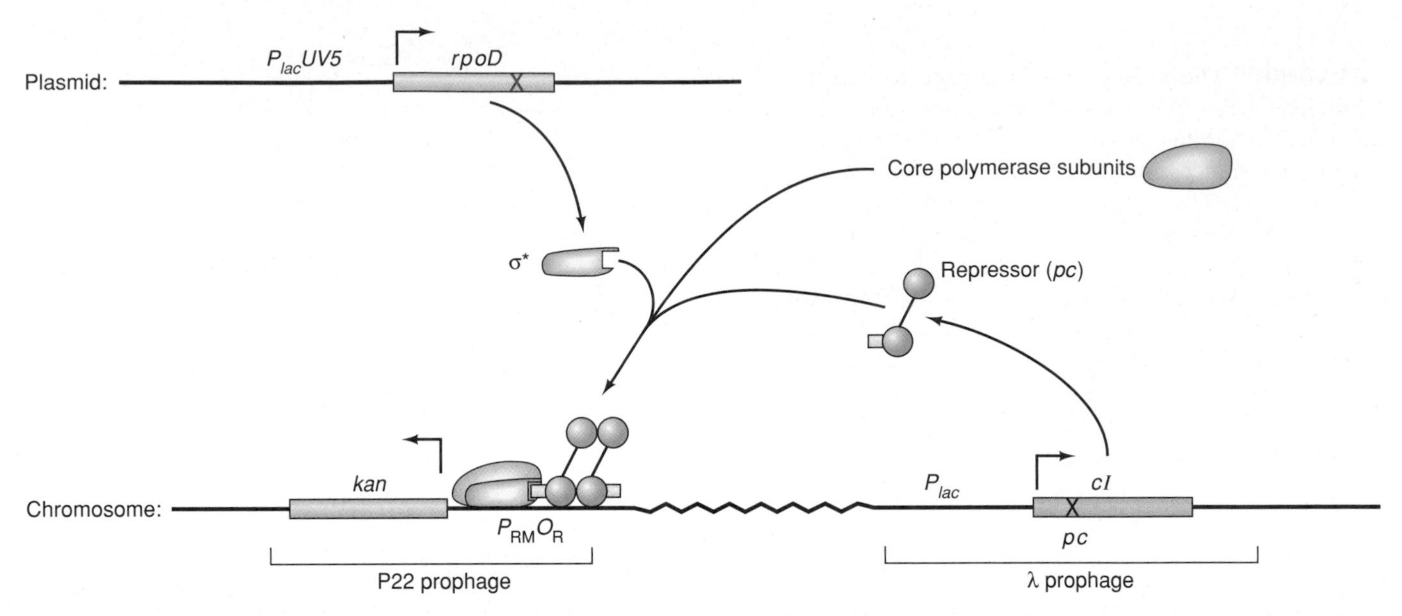

Figure 8.24 Selection for intergenic suppressor of λ *cI pc* mutation. Susskind and colleagues used bacteria with the chromosome illustrated (in small part) at bottom. The chromosome included two prophages: (1) a P22 prophage with a kanamycin-resistance gene (orange) driven by a λ P_{RM} promoter with adjacent λ O_R; (2) a λ prophage containing the λ *cI* gene (light green) driven by a weak *lac* promoter. Into these bacteria, Susskind and colleagues placed plasmids bearing mutagenized *rpoD* (σ-factor) genes (light blue) driven by the *lac UV5* promoter. Then they challenged the transformed cells with medium containing kanamycin. Cells transformed with a wild-type *rpoD* gene, or with *rpoD* genes bearing irrelevant mutations, could not grow in kanamycin. However, cells transformed with *rpoD* genes having a mutation (red X) that compensated for the mutation (black X) in the *cI* gene could grow. This mutation suppression is illustrated by the interaction between the mutant σ-factor (σ*, blue) and the mutant repressor (green), which permits transcription of the kanamycin-resistance gene from P_{RM}.

Susskind and associates double-checked their mutated *rpoD* genes by introducing them into cells just like the ones used for selection, but bearing a *lacZ* gene instead of a kanamycin-resistance gene under control of P_{RM}. They assayed for production of the *lacZ* product, β-galactosidase, which is a measure of transcription from P_{RM}, and therefore of interaction between polymerase and repressor. As expected, cells with the mutated repressor that were transformed with the mutated *rpoD* gene gave as many units of β-galactosidase (120) as cells with both wild-type *cI* and *rpoD* genes (100 units). By contrast, cells with the mutated repressor transformed with the wild-type *rpoD* gene gave only 18.5 units of β-galactosidase. When Susskind and colleagues sequenced the *rpoD* genes of all eight suppressor mutants they isolated, they found the same mutation, which resulted in the change of an arginine at position 596 in the σ-factor to a histidine. This is in region 4 of the σ-factor, which is the region that recognizes the −35 box.

These data strongly support the hypothesis that polymerase–repressor interactions are essential for activation of transcription from P_{RM}. They also demonstrate that the activating interaction between the two proteins involves the σ-factor, and not the α-subunit, as one might have expected. This provides an example of activation via σ, illustrated in Figure 8.25. Promoters subject to such activation have weak −35 boxes, which are poorly recognized by σ. The activator site, by overlapping the −35 box, places the activator (λ repressor, in this case) in position to interact with region 4 of the σ-factor—in effect substituting for the weakly recognized −35 box.

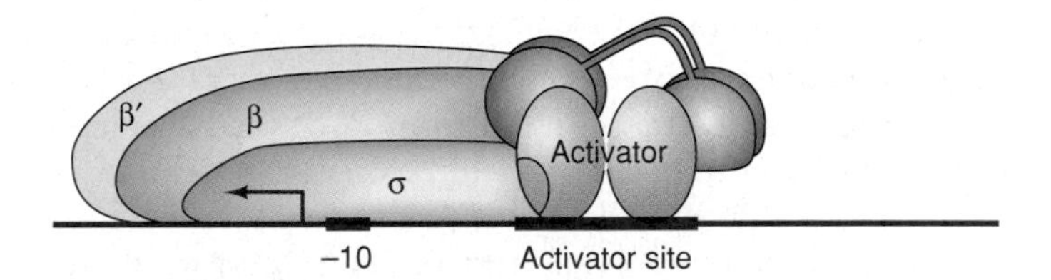

Figure 8.25 Activation by contacting σ. The activator (e.g., λ repressor) binds to an activator site that overlaps the weak −35 box of the promoter. This allows interaction between the activator and region 4 of σ, which would otherwise bind weakly, if at all, to the −35 box. This allows the polymerase to bind tightly to a very weak promoter and therefore to transcribe the adjacent gene successfully.
(*Source:* Adapted from Busby S. and R.H. Ebright, Promoter structure, promoter recognition, and transcription activation in prokaryotes. *Cell* 79:743, 1994.)

We saw in Chapter 7 that CAP–cAMP recruits RNA polymerase to the *lac* promoter by stimulating formation of the closed promoter complex. But Diane Hawley and William McClure showed that the λ repressor does not affect this step at P_{RM}. Instead, it stimulates the second step in recruitment: helping to convert the closed promoter complex at P_{RM} to the open promoter complex.

SUMMARY Intergenic suppressor mutation studies show that the crucial interaction between repressor and RNA polymerase involves region 4 of the σ-subunit of the polymerase. This polypeptide binds near the weak −35 box of P_{RM}, which places the σ-region 4 close to the repressor bound to O_R2. Thus, the repressor can interact with the σ-factor, helping to compensate for the weak promoter. In this way, O_R2 serves as an activator site, and λ repressor is an activator of transcription from P_{RM}. It stimulates conversion of the closed promoter complex to the open promoter complex.

Determining the Fate of a λ Infection: Lysis or Lysogeny

What determines whether a given cell infected by λ will enter the lytic cycle or lysogeny? The balance between these two fates is delicate, and we usually cannot predict the actual path taken in a given cell. Support for this assertion comes from a study of the appearance of *E. coli* cells infected by λ phage. When a few phage particles are sprinkled on a lawn of bacteria in a Petri dish, they infect the cells. If a lytic infection takes place, the progeny phages spread to neighboring cells and infect them. After a few hours, we can see a circular hole in the bacterial lawn caused by the death of lytically infected cells. This hole is called a **plaque.** If the infection were 100% lytic, the plaque would be clear, because all the host cells would be killed. But λ plaques are not usually clear. Instead, they are turbid, indicating the presence of live lysogens. This means that even in the local environment of a plaque, some infected cells suffer the lytic cycle, and others are lysogenized.

Let us digress for a moment to ask this question: Why are the lysogens not infected lytically by one of the multitude of phages in the plaque? The answer is that if a new phage DNA enters a lysogen, plenty of repressor is present in the cell to bind to the new phage DNA and prevent its expression. Therefore, we can say that the lysogen is **immune** to **superinfection** by a phage with the same control region, or **immunity region,** as that of the prophage.

Now let us return to the main problem. We have seen that some cells in a plaque can be lytically infected, whereas others are lysogenized. The cells within a plaque are all genetically identical, and so are the phages, so the choice of fate is not genetic. Instead, it seems to represent a race between the products of two genes: *cI* and *cro*. This race is rerun in each infected cell, and the winner determines the pathway of the infection in that cell. If *cI* prevails, lysogeny will be established; if *cro* wins, the infection will be lytic. We can already appreciate the basics of this argument: If the *cI* gene manages to produce enough repressor, this protein will bind to O_R and O_L, prevent further transcription of the early genes, and

(a) *cI* wins, lysogeny

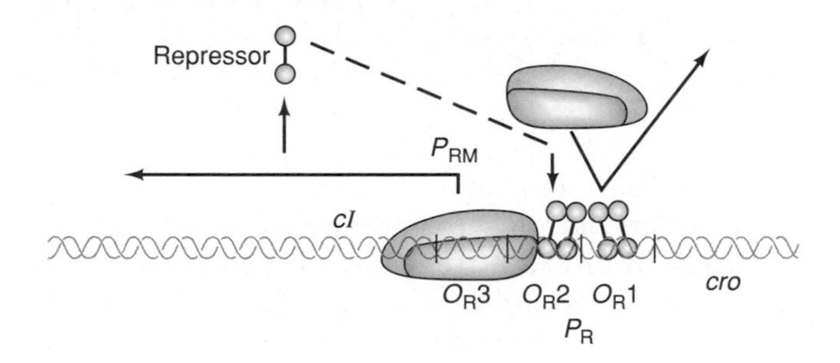

(b) *cro* wins, lytic cycle

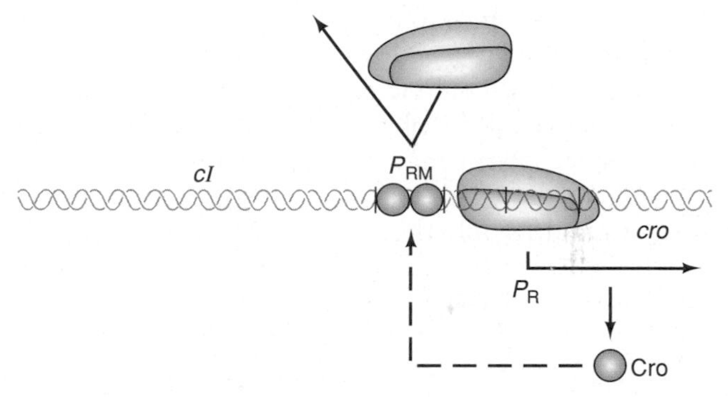

Figure 8.26 The battle between *cI* and *cro*. (a) *cI* wins. Enough repressor (green) is made by transcription of the *cI* gene from P_{RM} (and P_{RE}) that it blocks polymerase (red and blue) from binding to P_R and therefore blocks *cro* transcription. Lysogeny results. **(b)** *cro* wins. Enough Cro (purple) is made by transcription from P_R that it blocks polymerase from binding to P_{RM} and therefore blocks *cI* transcription. The lytic cycle results.

thereby prevent expression of the late genes that would cause progeny phage production and lysis. On the other hand, if enough Cro is made, this protein can prevent *cI* transcription and thereby block lysogeny (Figure 8.26).

The key to Cro's ability to block *cI* transcription is the nature of its affinity for the λ operators: Cro binds to both O_R and O_L, as does repressor, but its order of binding to the three-part operators is exactly opposite to that of repressor. Instead of binding in the order 1, 2, 3, as repressor does, Cro binds first to O_R3. As soon as that happens, *cI* transcription from P_{RM} stops, because O_R3 overlaps P_{RM}. In other words, Cro acts as a repressor. Furthermore, when Cro levels build up to the point that Cro fills up all the rightward and leftward operators, it prevents transcription of all the early genes from P_R and P_L, including *cII* and *cIII*. Without the products of these genes, P_{RE} cannot function; so all repressor synthesis ceases. Lytic infection is then ensured. Cro's turning off of early transcription is also required for lytic growth. Continued production of delayed early proteins late in infection aborts the lytic cycle.

But what determines whether *cI* or *cro* wins the race? Surely it is more than just a flip of the coin. Actually, the most important factor seems to be the concentration of the *cII* gene product, CII. The higher the CII concentration within the cell, the more likely lysogeny becomes. This fits with what we have already learned about CII—it activates P_{RE} and thereby helps turn on the lysogenic program. We have also seen that the activation of P_{RE} by CII works against the lytic program by producing *cro* antisense RNA that can inhibit translation of *cro* sense RNA.

And what controls the concentration of CII? We have seen that CIII protects CII against cellular proteases, but high protease concentrations can overwhelm CIII, destroy CII, and ensure that the infection will be lytic. Such high protease concentrations occur under good environmental conditions—rich medium, for example. By contrast, protease levels are depressed under starvation conditions. Thus, starvation tends to favor lysogeny, whereas rich medium favors the lytic pathway. This is advantageous for the phage because the lytic pathway requires considerable energy to make all the phage DNA, RNAs, and proteins, and this much energy may not be available during starvation. In comparison, lysogeny is cheap; after it is established, it requires only the synthesis of a little repressor.

SUMMARY Whether a given cell is lytically or lysogenically infected by phage λ depends on the outcome of a race between the products of the *cI* and *cro* genes. The *cI* gene codes for repressor, which blocks O_R1, O_R2, O_L1, and O_L2, turning off all early transcription, including transcription of the *cro* gene. This leads to lysogeny. On the other hand, the *cro* gene codes for Cro, which blocks O_R3 (and O_L3), turning off *cI* transcription. This leads to lytic infection. Whichever gene product appears first in high enough concentration to block its competitor's synthesis wins the race and determines the cell's fate. The winner of this race is determined by the CII concentration, which is determined by the cellular protease concentration, which is in turn determined by environmental factors such as the richness of the medium.

Lysogen Induction

We mentioned that a lysogen can be induced by treatment with mutagenic chemicals or radiation. The mechanism of this induction is as follows: *E. coli* cells respond to DNA-damaging environmental insults, such as mutagens or radiation, by inducing a set of genes whose collective activity is called the **SOS response;** one of these genes, in fact the most important, is ***recA***. The *recA* product (**RecA**) participates in recombination repair of DNA damage (Chapter 20), which explains part of its usefulness to the SOS response, but environmental insults also induce a new activity in the RecA protein. It becomes a coprotease that stimulates

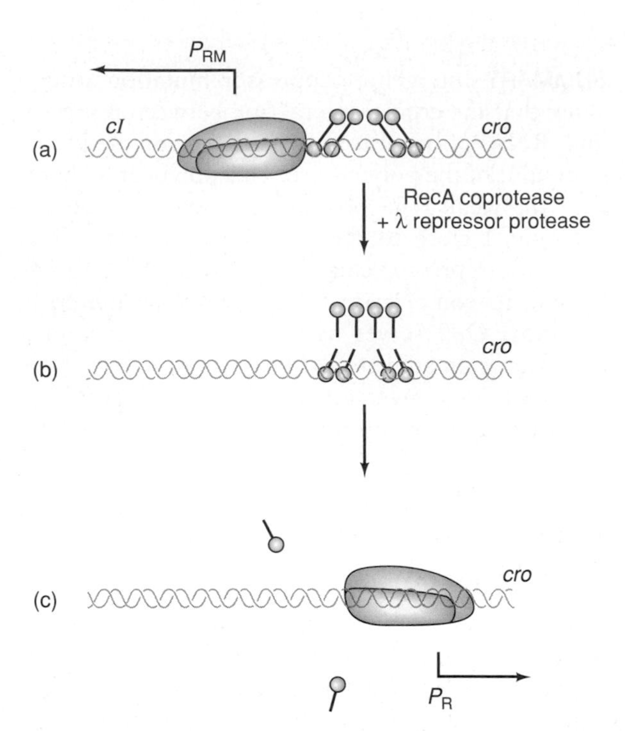

Figure 8.27 Inducing the λ prophage. (a) Lysogeny. Repressor (green) is bound to O_R (and O_L) and *cI* is being actively transcribed from the P_{RM} promoter. **(b)** The RecA coprotease (activated by ultraviolet light or other mutagenic influence) unmasks a protease activity in the repressor, so it can cleave itself. **(c)** The severed repressor falls off the operator, allowing polymerase (red and blue) to bind to P_R and transcribe *cro*. Lysogeny is broken.

a latent protease, or protein-cleaving activity, in the λ repressor. This protease then cleaves the repressor in half and releases it from the operators, as shown in Figure 8.27. As soon as that happens, transcription begins from P_R and P_L. One of the first genes transcribed is *cro*, whose product shuts down any further transcription of the repressor gene. Lysogeny is broken and lytic phage replication begins.

Surely λ would not have evolved with a repressor that responds to RecA by chopping itself in half unless it provided an advantage to the phage. That advantage seems to be this: The SOS response signals that the lysogen is under some kind of DNA-damaging attack. It is expedient under those circumstances for the prophage to get out by inducing the lytic cycle, rather like rats deserting a sinking ship.

SUMMARY When a lysogen suffers DNA damage, it induces the SOS response. The initial event in this response is the appearance of a coprotease activity in the RecA protein. This causes the repressors to cut themselves in half, removing them from the λ operators and inducing the lytic cycle. In this way, progeny λ phages can escape the potentially lethal damage that is occurring in their host.

SUMMARY

Bacteria experience many different major shifts in transcription pattern (e.g., during phage infection or sporulation), and several mechanisms have evolved to effect these shifts. For example, transcription of phage SPO1 genes in infected *B. subtilis* cells proceeds according to a temporal program in which early genes are transcribed first, then middle genes, and finally late genes. This switching is directed by a set of phage-encoded σ-factors that associate with the host core RNA polymerase and change its specificity from early to middle to late. The host σ is specific for the phage early genes, the phage gp28 switches the specificity to the middle genes; and the phage gp33 and gp34 switch to late specificity.

When the bacterium *B. subtilis* sporulates, a whole new set of sporulation-specific genes turns on, and many, but not all, vegetative genes turn off. This switch takes place largely at the transcription level. It is accomplished by several new σ-factors that displace the vegetative σ-factor from the core RNA polymerase and direct transcription of sporulation genes instead of vegetative genes. Each σ-factor has its own preferred promoter sequence.

Some prokaryotic genes must be transcribed under conditions where two different σ-factors are active. These genes are equipped with two different promoters, each recognized by one of the two σ-factors. This ensures their expression no matter which factor is present and allows for differential control under different conditions. The heat shock response, as well as the response to low nitrogen and starvation stress in *E. coli* are governed by alternative σ-factors, σ^{32} (σ^{H}), σ^{54} (σ^{N}), and σ^{34} (σ^{S}), respectively, which displace σ^{70} (σ^{A}) and direct the RNA polymerase to alternative promoters. Many σ-factors are controlled by anti-σ-factors that bind to a specific σ and block its binding to the core polymerase. Some of these anti-σ-factors are even controlled by anti-anti-σ-factors that bind to the complexes between a σ and an anti-σ-factor and release the σ-factor. In at least one case, an anti-σ-factor is also an anti-anti-anti-σ-factor that phosphorylates and inactivates the cognate anti-anti-σ-factor.

Phage T7, instead of coding for a new σ-factor to change the host polymerase's specificity from early to late, encodes a new RNA polymerase with absolute specificity for the later phage genes. This polymerase, composed of a single polypeptide, is a product of one of the early phage genes, gene 1. The temporal program in the infection by this phage is simple. The host polymerase transcribes the early (class I) genes, one of whose products is the phage polymerase, which then transcribes the late genes (classes II and III).

The immediate early/delayed early/late transcriptional switching in the lytic cycle of phage λ is controlled by antiterminators. One of the two immediate early genes is *cro,* which codes for a repressor of *cI* that allows the lytic cycle to continue. The other, *N,* codes for an antiterminator, N, that overrides the terminators located after the *N* and *cro* genes. Transcription then continues into the delayed early genes. One of the delayed early genes, *Q,* codes for another antiterminator (Q) that permits transcription of the late genes from the late promoter, $P_{R'}$, to continue without premature termination.

Five proteins (N, NusA, NusB, NusG, and S10) collaborate in antitermination at the λ immediate early terminators. NusA and S10 bind to RNA polymerase, and N and NusB bind to the box B and box A regions, respectively, of the *nut* site in the growing transcript. N and NusB bind to NusA and S10, respectively, probably tethering the transcript to the polymerase. This alters the polymerase so it reads through the terminators at the ends of the immediate early genes. NusA stimulates termination at an intrinsic terminator by interfering with the binding between the upstream part of the RNA hairpin and the core polymerase, thereby facilitating the formation of the hairpin in the nascent RNA. The hairpin traps the complex in a form that is irreversibly committed to terminate. N interferes with this termination process by binding to the upstream part of the RNA hairpin, preventing hairpin formation. N even helps NusA bind farther upstream on the nascent RNA, further inhibiting hairpin formation. Without hairpin formation, termination cannot occur. Antitermination in the λ late region requires Q, which binds to the Q-binding region of the *qut* site as RNA polymerase is stalled just downstream of the late promoter. Subsequent binding of Q to the polymerase appears to alter the enzyme so it can ignore the terminator and transcribe the late genes.

Phage λ establishes lysogeny by causing production of enough repressor to bind to the early operators and prevent further early RNA synthesis. The promoter used for establishment of lysogeny is P_{RE}, which lies to the right of P_R and *cro*. Transcription from this promoter goes leftward through the *cI* gene. The products of the delayed early genes *cII* and *cIII* also participate in this process: CII, by directly stimulating polymerase binding to P_{RE}; CIII, by slowing degradation of CII.

The promoter that is used to maintain lysogeny is P_{RM}. It comes into play after transcription from P_{RE} makes possible the burst of repressor synthesis that establishes lysogeny. This repressor binds to O_R1 and O_R2 cooperatively, but leaves O_R3 open. RNA polymerase binds to P_{RM}, which overlaps O_R3 in such a way that it just touches the repressor bound to O_R2. This protein–protein interaction is required for this promoter to work efficiently.

The crucial interaction between the λ repressor and RNA polymerase involves region 4 of the σ-subunit of the polymerase. This polypeptide binds near the weak

−35 box of P_{RM}, which places the σ-region 4 close to the repressor bound to O_R2. Thus, the repressor can interact with the σ-factor, attracting RNA polymerase to the weak promoter. In this way, O_R2 serves as an activator site, and λ repressor is an activator of transcription from P_{RM}.

Whether a given cell is lytically or lysogenically infected by phage λ depends on the outcome of a race between the products of the *cI* and *cro* genes. The *cI* gene codes for repressor, which blocks O_R1, O_R2, O_L1, and O_L2, turning off all early transcription, including transcription of the *cro* gene. This leads to lysogeny. On the other hand, the *cro* gene codes for Cro, which blocks O_R3 (and O_L3), turning off *cI* transcription. This leads to lytic infection. Whichever gene product appears first in high enough concentration to block its competitor's synthesis wins the race and determines the cell's fate. The winner of this race is determined by the CII concentration, which is determined by the cellular protease concentration, which is in turn determined by environmental factors such as the richness of the medium.

When a λ lysogen suffers DNA damage, it induces the SOS response. The initial event in this response is the appearance of a coprotease activity in the RecA protein. This causes the repressors to cut themselves in half, removing them from the λ operators and inducing the lytic cycle. In this way, progeny λ phages can escape the potentially lethal damage that is occurring in their host.

REVIEW QUESTIONS

1. Present a model to explain how the phage SPO1 controls its transcription program.
2. Summarize the evidence to support your answer to question 1.
3. Describe and give the results of an experiment that shows that the *B. subtilis* σ^E recognizes the sporulation-specific 0.4-kb promoter, whereas σ^A recognizes a vegetative promoter.
4. Summarize the mechanism *B. subtilis* cells use to alter their transcription program during sporulation.
5. Describe and give the results of an experiment that shows that the *B. subtilis* σ^E recognizes the *spoIID* promoter, but other σ-factors do not.
6. Present an explanation for the rapid response of *E. coli* to heat shock.
7. Present a model to explain how the phage T7 controls its transcription program.
8. How does λ phage switch from immediate early to delayed early to late transcription during lytic infection?
9. Present a model for N-directed antitermination in λ phage-infected *E. coli* cells without giving details of the antitermination mechanism. What proteins are involved?
10. Present a model to explain the effects of N and NusA in intrinsic termination and in antitermination.
11. Describe and present the results of an experiment that shows that N and NusA control hairpin formation at an intrinsic terminator.
12. The P_{RE} promoter is barely recognizable and is not in itself attractive to RNA polymerase. How can the *cI* gene be transcribed from this promoter?
13. How can CII and RNA polymerase bind to the same region of DNA at the same time?
14. Describe and give the results of an experiment that shows that λ repressor regulates its own synthesis both positively and negatively. What does this same experiment show about the effect of repressor on *cro* transcription?
15. Diagram the principle of an intergenic suppression assay to detect interaction between two proteins.
16. Describe and give the results of an intergenic suppression experiment that shows interaction between the λ repressor and the σ-subunit of the *E. coli* RNA polymerase.
17. Present a model to explain the struggle between *cI* and *cro* for lysogenic or lytic infection of *E. coli* by λ phage. What tips the balance one way or the other?
18. Present a model to explain the induction of a λ lysogen by mutagenic insults.

ANALYTICAL QUESTIONS

1. You are studying a gene that you suspect has two different promoters, recognized by two different σ-factors. Design an experiment to test your hypothesis.
2. What would happen if you infected *E. coli* cells lysogenized by λ phage with the identical strain of λ? Would you get superinfection? Why or why not?
3. You repeat the experiment in Question 2 with a different strain of λ having operator sequences significantly different from those of the lysogen. What results would you expect? Why?
4. You infect two, genetically different strains of *E. coli* with wild-type λ phage. You select lysogens from each strain. When you irradiate these lysogens with UV light you get lytic infection from one strain, but nothing from the other. Explain these results.
5. What might be mutated in a strain of λ phage that always produces 100% lytic infections?
6. What might be mutated in a strain of λ phage that always produces 100% lysogenic infections?
7. You are working with a strain of λ phage with an inactivated *N* gene. Would you expect this phage to produce lytic infections, lysogenic infections, both, or neither. Why?

SUGGESTED READINGS

General References and Reviews

Busby, S. and R.H. Ebright. 1994. Promoter structure, promoter recognition, and transcription activation in prokaryotes. *Cell* 79:743–46.

Goodrich, J.A. and W.R. McClure. 1991. Competing promoters in prokaryotic transcription. *Trends in Biochemical Sciences* 15:394–97.

Gralla, J.D. 1991. Transcriptional control—lessons from an *E. coli* data base. *Cell* 66:415–18.

Greenblatt, J., J.R. Nodwell, and S.W. Mason. 1993. Transcriptional antitermination. *Nature* 364:401–6.

Helmann, J.D. and M.J. Chamberlin. 1988. Structure and function of bacterial sigma factors. *Annual Review of Biochemistry* 57:839–72.

Ptashne, M. 1992. *A Genetic Switch.* Cambridge, MA: Cell Press.

Research Articles

Dodd, I.B., A.J. Perkins, D. Tsemitsidis, and J.B. Egan. 2001. Octamerization of λ CI repressor is needed for effective repression of P_{RM} and efficient switching from lysogeny. *Genes and Development* 15:3013–21.

Gusarov, I. and E. Nudler. 1999. The mechanism of intrinsic transcription termination. *Molecular Cell* 3:495–504.

Gusarov, I. and E. Nudler. 2001. Control of intrinsic transcription termination by N and NusA: The basic mechanisms. *Cell* 107:437–49.

Haldenwang, W.G., N. Lang, and R. Losick. 1981. A sporulation-induced sigma-like regulatory protein from *B. subtilis. Cell* 23:615–24.

Hawley, D.K. and W.R. McClure. 1982. Mechanism of activation of transcription initiation from the λ P_{RM} promoter. *Journal of Molecular Biology* 157:493–525.

Ho, Y.-S., D.L. Wulff, and M. Rosenberg. 1983. Bacteriophage λ protein cII binds promoters on the opposite face of the DNA helix from RNA polymerase. *Nature* 304:703–8.

Johnson, W.C., C.P. Moran, Jr., and R. Losick. 1983. Two RNA polymerase sigma factors from *Bacillus subtilis* discriminate between overlapping promoters for a developmentally regulated gene. *Nature* 302:800–4.

Li, M., H. Moyle, and M.M. Susskind. 1994. Target of the transcriptional activation function of phage λ cI protein. *Science* 263:75–77.

Meyer, B.J., D.G. Kleid, and M. Ptashne. 1975. λ repressor turns off transcription of its own gene. *Proceedings of the National Academy of Science, USA* 72:4785–89.

Pero, J., R. Tjian, J. Nelson, and R. Losick. 1975. In vitro transcription of a late class of phage SPO1 genes. *Nature* 257:248–51.

Rong, S., M.S. Rosenkrantz, and A.L. Sonenshein. 1986. Transcriptional control of the *B. subtilis spoIID* gene. *Journal of Bacteriology* 165:771–79.

Stragier, P., B. Kunkel, L. Kroos, and R. Losick. 1989. Chromosomal rearrangement generating a composite gene for a developmental transcription factor. *Science* 243:507–12.

Toulokhonov, I., I. Artsimovitch, and R. Landick. 2001. Allosteric control of RNA polymerase by a site that contacts nascent RNA hairpins. *Science* 292:730–33.

CHAPTER 9

DNA–Protein Interactions in Bacteria

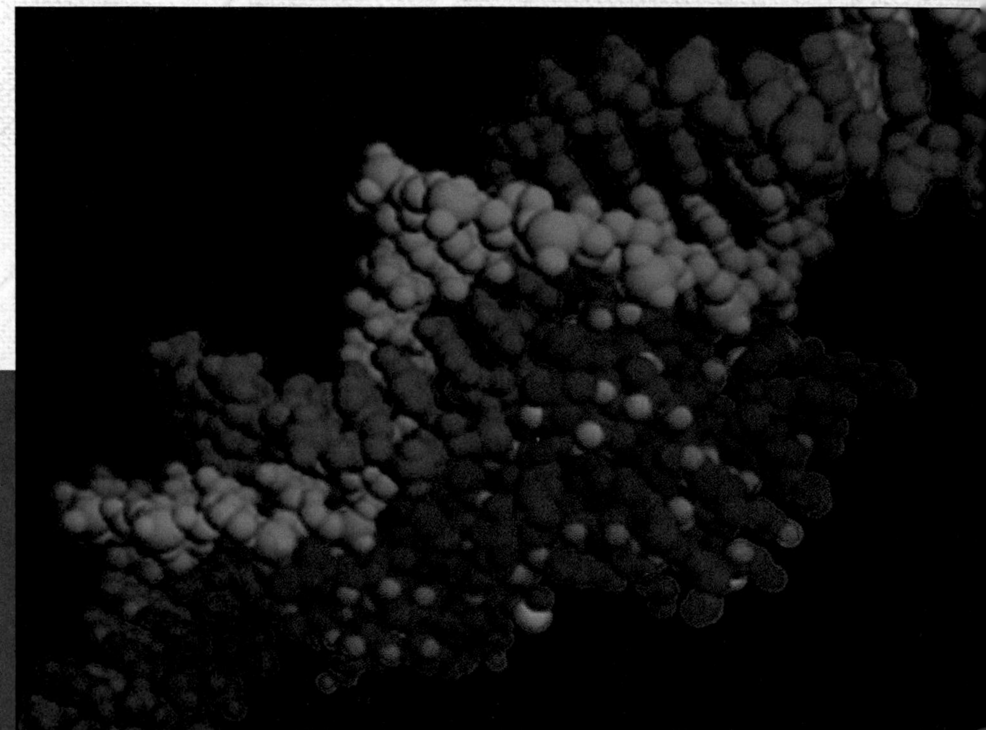

Gene regulation: Computer model of Cro protein bound to DNA.
© Ken Eward/SS/Photo Researchers, Inc.

In Chapters 7 and 8 we discussed several proteins that bind tightly to specific sites on DNA. These include RNA polymerase, *lac* repressor, CAP, *trp* repressor, λ repressor, and Cro. All of these have been studied in detail, and all can locate and bind to one particular short DNA sequence among a vast excess of unrelated sequences. How do these proteins accomplish such specific binding—akin to finding a needle in a haystack? The latter five proteins have a similar structural motif: two α-helices connected by a short protein "turn." This **helix-turn-helix motif** (Figure 9.1a) allows the second helix (the **recognition helix**) to fit snugly into the major groove of the target DNA site (Figure 9.1b). We will see that the configuration of this fit varies considerably from one protein to another, but all the proteins fit their DNA binding sites like a key in a lock. In this chapter we will explore several

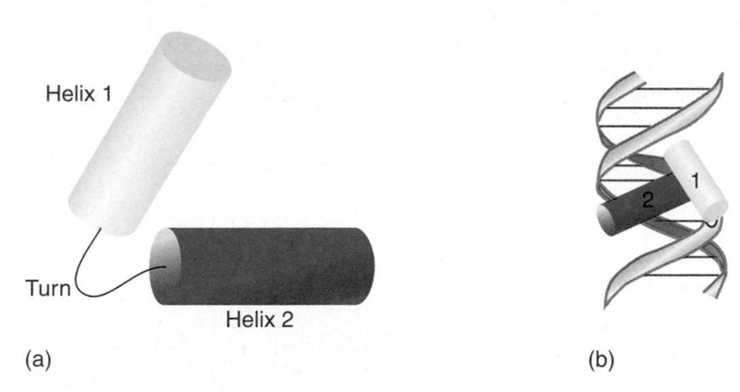

Figure 9.1 The helix-turn-helix motif as a DNA-binding element. **(a)** The helix-turn-helix motif of the λ repressor. **(b)** The fit of the helix-turn-helix motif of one repressor monomer with the λ operator. Helix 2 of the motif (red) lies in the major groove of its DNA target; some of the amino acids on the back of this helix (away from the viewer) are available to make contacts with the DNA.

well-studied examples of specific DNA–protein interactions that occur in prokaryotic cells to see what makes them so specific. In Chapter 12 we will consider several other DNA-binding motifs that occur in eukaryotes.

9.1 The λ Family of Repressors

The repressors of λ and similar phages have recognition helices that lie in the major groove of the appropriate operator as shown in Figure 9.2. The specificity of this binding depends on certain amino acids in the recognition helices

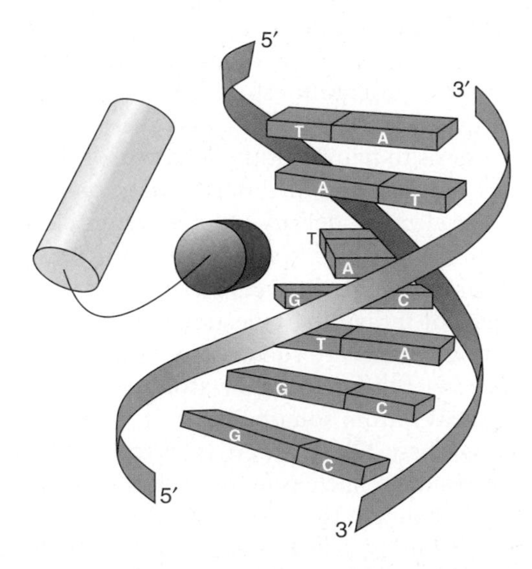

Figure 9.2 Schematic representation of the fit between the recognition helix of a λ repressor monomer and the major groove of the operator region of the DNA. The recognition helix is represented by a red cylinder that lies in the major groove in a position to facilitate hydrogen bonding with the edges of base pairs in the DNA. (*Source:* Adapted from Jordan, S.R. and C.O. Pabo. Structure of the lambda complex at 2.5 Å resolution. Details of the repressor–operator interaction. *Science* 242:896, 1988.)

that make specific contact with functional groups of certain bases protruding into the major DNA groove, and with phosphate groups in the DNA backbone. Other proteins with helix-turn-helix motifs are not able to bind as well at that same site because they do not have the correct amino acids in their recognition helices. We would like to know which are the important amino acids in these interactions.

Probing Binding Specificity by Site-Directed Mutagenesis

Mark Ptashne and his colleagues have provided part of the answer to the specificity question, using repressors from two λ-like phages, **434** and **P22,** and their respective operators. These two phages have very similar molecular genetics, but they have different immunity regions: They make different repressors that recognize different operators. Both repressors resemble the λ repressor in that they contain helix-turn-helix motifs. However, because they recognize operators with different base sequences, we would expect them to have different amino acids in their respective recognition helices, especially those amino acids that are strategically located to contact the bases in the DNA major groove.

Using x-ray diffraction analysis (Box 9.1) of operator–repressor complexes, Stephen Harrison and Ptashne identified the face of the recognition helix of the 434 phage repressor that contacts the bases in the major groove of its operator. By analogy, they could make a similar prediction for the P22 repressor. Figure 9.3 schematically illustrates the amino acids in each repressor that are most likely to be involved in operator binding.

If these are really the important amino acids, one ought to be able to change only these amino acids and thereby alter the specificity of the repressor. In particular, one should be able to employ such changes to alter the 434 repressor so that it recognizes the P22 operator instead of its own. This is exactly what Robin Wharton and Ptashne did. They started with a cloned gene for the 434 repressor and, using mutagenesis techniques similar to those described in Chapter 5, systematically altered the codons for five amino acids in the 434 recognition helix to codons for the five corresponding amino acids in the P22 recognition helix.

Next, they expressed the altered gene in bacteria and tested the product for ability to bind to 434 and P22 operators, both in vivo and in vitro. The in vivo assay was to check for immunity. Recall that an *E. coli* cell lysogenized by λ phage is immune to superinfection by λ because the excess λ repressor in the lysogen immediately binds to the superinfecting λ DNA and prevents its expression (Chapter 8). Phages 434 and P22 are λ-like (lambdoid) phages, but they differ in their immunity regions, the control regions that include the repressor genes and the operators. Thus, a 434 lysogen is immune to superinfection by 434, but not by P22. The 434 repressor cannot bind to the

B O X 9.1

X-Ray Crystallography

This book contains many examples of structures of DNA-binding proteins obtained by the method of **x-ray diffraction analysis,** also called **x-ray crystallography.** This box provides an introduction to this very powerful technique.

X-rays are electromagnetic radiation, just like light rays, but with much shorter wavelengths so they are much more energetic. Thus, it is not surprising that the principle of x-ray diffraction analysis is in some ways similar to the principle of light microscopy. Figure B9.1 illustrates this similarity. In light microscopy (Figure B9.1), visible light is scattered by an object; then a lens collects the light rays and focuses them to create an image of the object.

In x-ray diffraction, x-rays are scattered by an object (a crystal). But here we encounter a major problem: No lens is capable of focusing x-rays, so one must use a relatively indirect method to create the image. That method is based on the following considerations: When x-rays interact with an electron cloud around an atom, the x-rays scatter in every direction. However, because x-ray beams interact with multiple atoms, most of the scattered x-rays cancel one another due to their wave nature. But x-rays scattered to certain specific directions are amplified in a phenomenon called *diffraction.* Bragg's law, $2d \sin \theta = \lambda$, describes the relationship between the angle (θ) of diffraction and spacing (d) of scattering planes. As you can see in Figure B9.2, x-ray 2 travels $2 \times d \sin \theta$ longer than x-ray 1. Thus, if the wavelength (λ) of x-ray 2 is equal to $2\ d \sin \theta$, the resultant rays from the scattered x-ray 1 and x-ray 2 have the same phase and are therefore amplified. On the other hand,

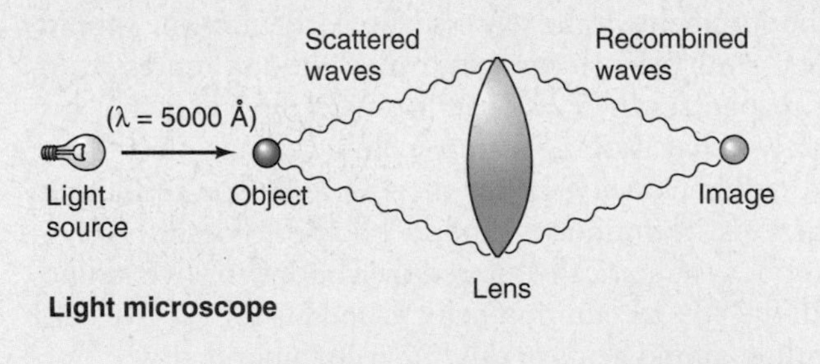

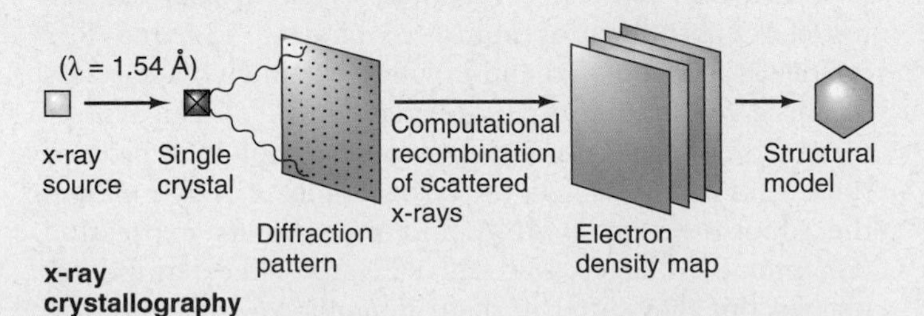

Figure B9.1 Schematic diagram of the procedures followed for image reconstruction in light microscopy (top) and x-ray crystallography (bottom).

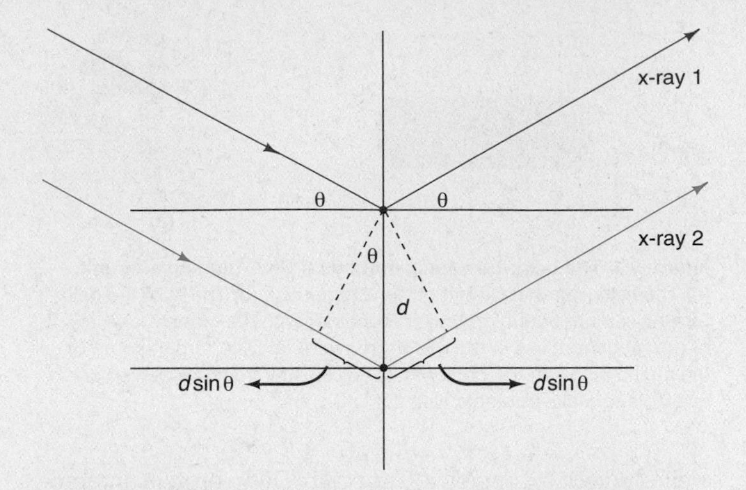

Figure B9.2 Reflection of two x-rays from parallel planes of a crystal. The two x-rays (1 and 2) strike the planes at angle θ and are reflected at the same angle. The planes are separated by distance d. The extra distance traveled by x-ray 2 is $2\ d \sin \theta$.

the resultant rays are diminished if λ is not equal to $2\ d \sin \theta$. The diffracted x-rays are recorded as spots on a collecting device (a detector) placed in the path of the x-rays. This device can be as simple as a sheet of x-ray film, but nowadays much more efficient electronic detectors are available. Figure B9.3 shows a diffraction pattern of a simple protein, lysozyme. Even though the protein is relatively simple (only 129 amino acids), the pattern of spots is complex. To obtain the protein structure in three dimensions, one must rotate the crystal and record diffraction patterns in many different orientations.

The next task is to use the arrays of spots in the diffraction patterns to figure out the structure of the molecule that caused the diffraction. Unfortunately, one cannot reconstruct the **electron-density map** (electron cloud distribution) from the arrays of spots in the diffraction patterns, because information about the physical parameters, called *phase angles,* of individual reflections are not included in the diffraction pattern. To solve this problem, crystallographers make 3–10 different heavy-atom derivative crystals by soaking heavy atom solutions (Hg, Pt, U, etc.) into protein crystals. These heavy atoms tend to bind to reactive amino acid residues, such as cysteine, histidine, and aspartate, without changing the protein structure.

This procedure is called multiple isomorphous replacement (MIR). The phase angles of individual reflections are determined by comparing the diffraction patterns from the native and heavy-atom derivative crystals. Once the phase angles are obtained, the diffraction pattern is mathematically converted to an electron-density map of the diffracting molecule. Then the electron-density map can be used to infer the

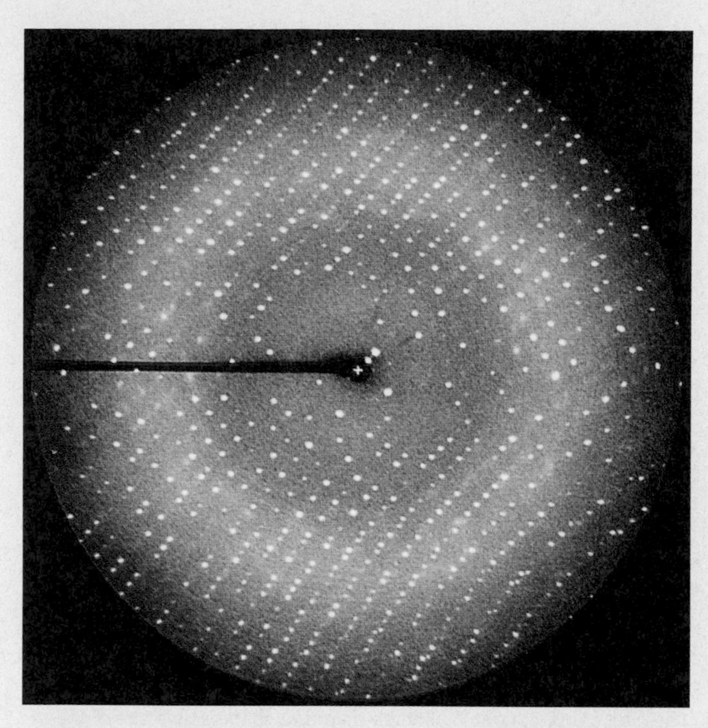

Figure B9.3 Sample diffraction pattern of a crystal of the protein lysozyme. The dark line from the left is the shadow of the arm that holds the beam stop, which protects the detector from the x-ray beam. The location of the crystal is marked by the (+) at the center. (*Source:* Courtesy of Fusao Takusagawa.)

structure of the diffracting molecule. Using the diffracted rays to create an image of the diffracting object is analogous to using a lens. But this is not accomplished physically, as a lens would; it is done mathematically. Figure B9.4 shows the electron-density map of part of the structure of lysozyme, surrounding a stick diagram representing the molecular structure inferred from the map. Figure B9.5 shows three different representations of the whole lysozyme molecule deduced from the electron-density map of the whole molecule.

Why are single crystals used in x-ray diffraction analysis? It is clearly impractical to place a single molecule of a protein in the path of the x-rays; even if it could be done, the diffraction power from a single molecule would be too weak to detect. Therefore, many molecules of protein are placed in the x-ray beam so the signal will be strong enough to detect. Why not just use a protein powder or a solution of protein? The problem with this approach is that the molecules in a powder or solution are randomly oriented, so x-rays diffracted by such a sample would not have an interpretable pattern.

The solution to the problem is to use a crystal of protein. A crystal is composed of many small repeating units (**unit cells**) that are three-dimensionally arranged in a regular

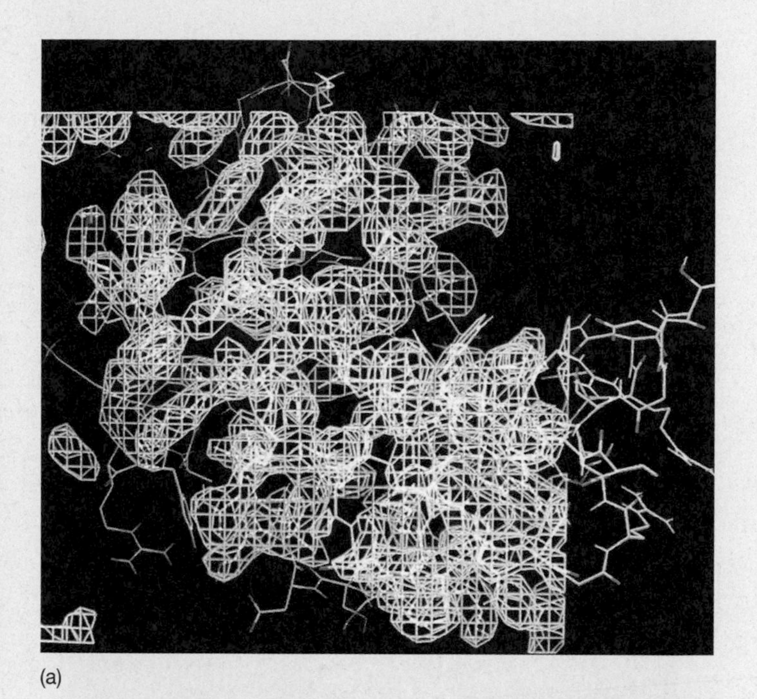

(a)

(b)

Figure B9.4 Electron-density map of part of the lysozyme molecule. (a) Low magnification, showing the electron density map of most of the molecule. The blue cages correspond to regions of high electron density. They surround a stick model of the molecule (red, yellow, and blue) inferred from the pattern of electron density. **(b)** High magnification, showing the center of the map in panel (a). The resolution of this structure was 2.4 Å so the individual atoms were not resolved. But this resolution is good enough to identify the unique shape of each amino acid. (*Source:* Courtesy Fusao Takusagawa.)

continued

BOX 9.1

X-Ray Crystallography *(continued)*

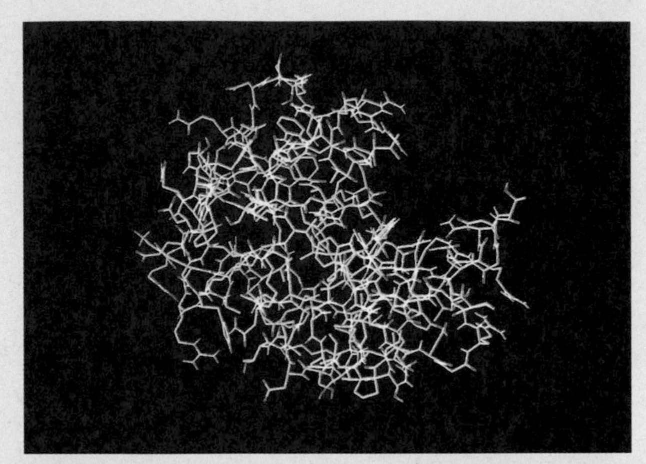
(a)

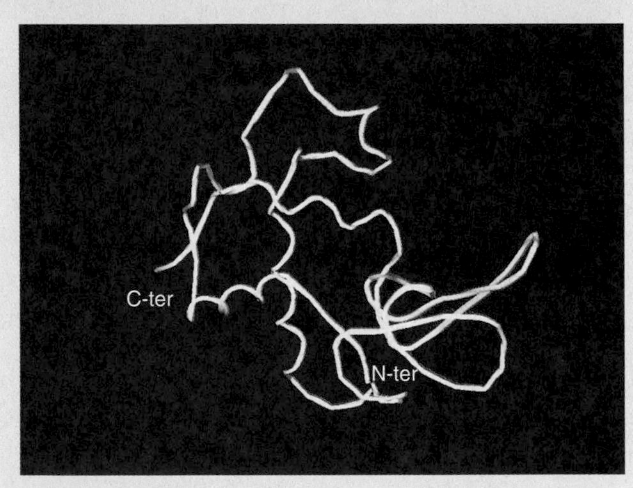

(b)

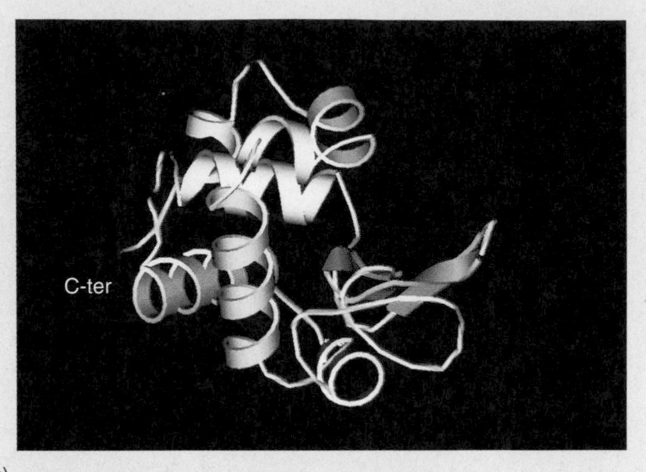

(c)

Figure B9.5 Three representations of the structure of lysozyme calculated from electron density maps such as those in Figure B9.4. **(a)** Stick diagram as in Figure B9.4a. **(b)** String diagram with α-helices in green, β-sheets in magenta, and random coils in blue. The N-terminus and C-terminus of the protein are marked N-ter and C-ter, respectively. **(c)** Ribbon diagram with same color coding as in panel (b). The helical nature of the α-helices is obvious in this diagram. The cleft at upper right in all three diagrams is the active site of the enzyme. (*Source:* Courtesy Fusao Takusagawa.)

way. A unit cell of a protein contains several protein molecules that are usually related by special symmetries. Thus, diffractions by all the molecules in a unit cell in the crystal are the same, and they reinforce one another. To be useful for x-ray diffraction, the smallest dimension of a protein crystal should be at least 0.1 mm. A cubic crystal of this size contains more than 10^{12} molecules (assuming that one protein molecule occupies a $50 \times 50 \times 50$ Å space). Figure B9.6 presents a photograph of crystals of lysozyme suitable for x-ray diffraction analysis. Protein crystals contain not only pure protein but also a large amount of solvent (30–70% of their weight). Thus, their environment in the crystal resembles that in solution, and their three-dimensional structure in the crystal should therefore be close to their structure in solution. In general, then, we can be confident that the protein structures determined by x-ray crystallography are close to their structures in the cell. In fact, most enzyme crystals retain their enzymatic activities.

Why not just use visible light rays to see the structures of proteins and avoid all the trouble involved with x-rays? The problem with this approach lies in **resolution**—the ability to distinguish separate parts of the molecule. The ultimate goal in analyzing the structure of a molecule is to distinguish each atom, so the exact spatial relationship of all the atoms in the molecule is apparent. But atoms have dimensions on the order of angstroms (1 Å = 10^{-10} m), and the maximum resolving power of radiation is one-third of its wavelength ($0.6\lambda/2 \sin \theta$). So we need radiation with a very short wavelength (measured in angstroms) to resolve the atoms in a

P22 operators and therefore cannot prevent superinfection by the P22 phage. The reverse is also true: A P22 lysogen is immune to superinfection by P22, but not by 434.

Instead of creating lysogens, Wharton and Ptashne transformed *E. coli* cells with a plasmid encoding the recombinant 434 repressor, then asked whether the recombinant 434 repressor (with its recognition helix altered to be like the P22 recognition helix) still had its original binding specificity. If so, cells producing the recombinant repressor should have been immune to 434 infection. On

Figure B9.6 Crystals of lysozyme. The photograph was taken using polarizing filters to produce the color in the crystals. The actual size of these crystals is approximately 0.5 × 0.5 × 0.5 mm. (*Source:* Courtesy Fusao Takusagawa.)

protein. But visible light has wavelengths averaging about 500 nm (5000 Å). Thus, it is clearly impossible to resolve atoms with visible light. By contrast, x-rays have wavelengths of one to a few angstroms. For example, the characteristic x-rays emitted by excited copper atoms have a wavelength of 1.54 Å, which is ideal for high-resolution x-ray diffraction analysis of proteins.

In this chapter we will see protein structures at various levels of resolution. What is the reason for these differences in resolution? A protein crystal in which the protein molecules are relatively well ordered gives many diffraction spots far from the incident beam, that is, from the center of the detector. These spots are produced by x-rays with large diffraction angles (θ, see Figure B9.2). An electron-density map calculated from these diffraction spots from a relatively ordered crystal gives a high-resolution image of the diffracting molecule. On the other hand, a protein crystal whose molecules are relatively poorly arranged gives diffraction spots only near the center of the detector, resulting from x-rays with small diffraction angles. Such data produce a relatively low-resolution image of the molecule.

This relationship between resolution and diffraction angle is another consequence of Bragg's law $2d \sin \theta = \lambda$. Rearranging Bragg's equation, we find $d = \lambda/2 \sin \theta$. So we see that d, the distance between structural elements in the protein, is inversely related to sin θ. Therefore, the larger the distance between structural elements in the crystal, the smaller the angle of diffraction and the closer to the middle of the pattern the diffracted ray will fall. This is just another way of saying that low-resolution structure (with large distances between elements) gives rise to the pattern of spots near the middle of the diffraction pattern. By the same argument, high-resolution structure gives rise to spots near the periphery of the pattern because they diffract the x-rays at a large angle. When crystallographers can make crystals that are good enough to give this kind of high resolution, they can build a detailed model of the structure of the protein.

The proteins we are considering in this chapter are DNA-binding proteins. In many cases, investigators have prepared cocrystals of the protein and a double-stranded DNA fragment containing the target sequence recognized by the protein. These can reveal not only the shapes of the protein and DNA in the protein–DNA complex, but also the atoms that are involved in the protein–DNA interaction.

It is important to note that x-ray crystallography captures but one conformation of a molecule or collection of molecules. But proteins generally do not have just one possible conformation. They are dynamic molecules in constant motion and are presumably continuously sampling a range of different conformations. The particular conformation revealed by x-ray crystallography depends on the ligands that co-crystallize with the protein, and on the conditions used during crystallization.

Furthermore, a protein by itself may have a preferred conformation that seems incompatible with binding to a ligand, but its dynamic motions lead to other conformations that do permit ligand binding. For example, Max Perutz noted many years ago that the x-ray crystal structure of hemoglobin was not compatible with binding to its ligand, oxygen. Yet hemoglobin obviously does bind oxygen, and it does so by changing its shape enough to accommodate the ligand. Similarly, a DNA-binding protein by itself may prefer a conformation that cannot admit the DNA, but dynamic motions lead to another conformation that can bind the DNA, and the DNA traps the protein in that conformation.

the other hand, if the binding specificity had changed, the cells producing the recombinant repressor should have been immune to P22 infection. Actually, 434 and P22 do not infect *E. coli* cells, so the investigators used recombinant λ phages with the 434 and P22 immunity regions (λ_{imm}434 and λ_{imm}P22, respectively) in these tests. They found that the cells producing the altered 434 repressor *were* immune to infection by the λ phage with the P22 immunity region, but not to infection by the λ phage with the 434 immunity region.

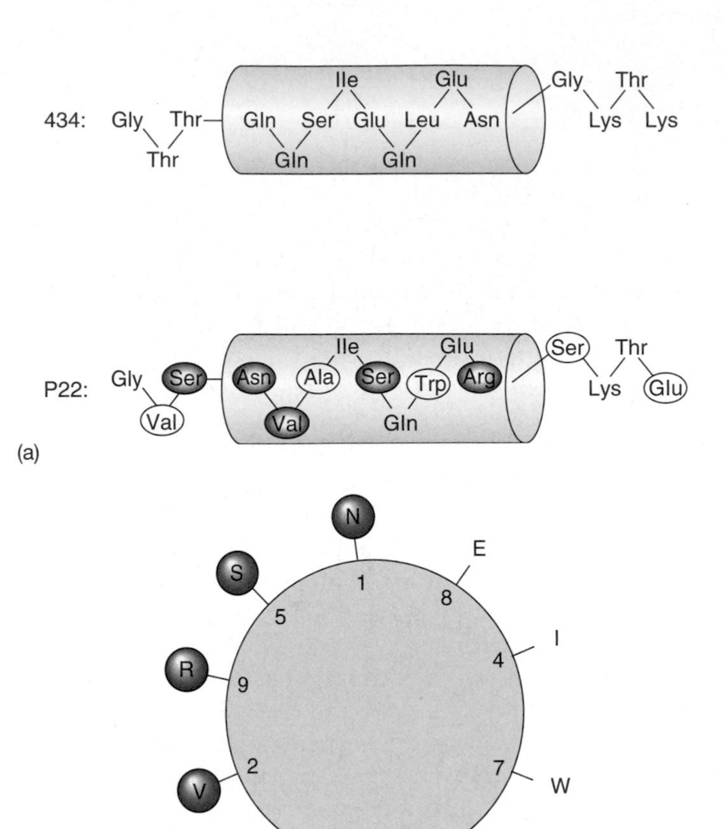

Figure 9.3 The recognition helices of two λ-like phage repressors. (a) Key amino acids in the recognition helices of two repressors. The amino acid sequences of the recognition helices of the 434 and P22 repressors are shown, along with a few amino acids on either side. Amino acids that differ between these two proteins are circled in the P22 diagram; these are more likely to contribute to differences in specificity. Furthermore, the amino acids on the side of the helix that faces the DNA are most likely to be involved in DNA binding. These, along with one amino acid in the turn just before the helix (red), were changed to alter the binding specificity of the protein. **(b)** The recognition helix of the P22 repressor viewed on end. The numbers represent the positions of the amino acids in the protein chain. The left-hand side of the helix faces toward the DNA, so the amino acids on that side are more likely to be important in binding. Those that differ from amino acids in corresponding positions in the 434 repressor are circled in red. (*Source: (b)* Adapted from Wharton, R.P. and M. Ptashne, Changing the binding specificity of a repressor by redesigning an alpha-helix. *Nature* 316:602, 1985.)

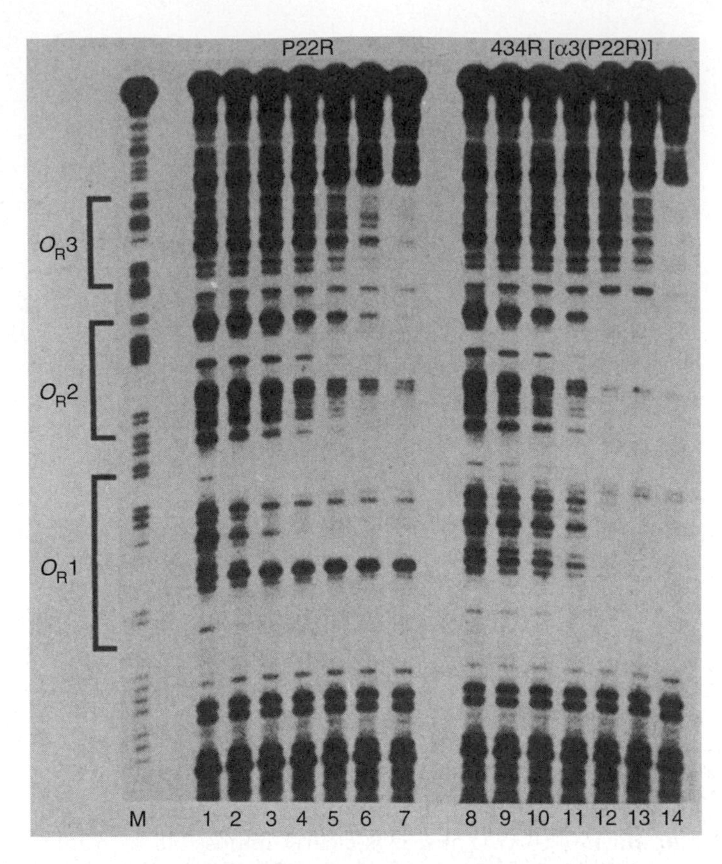

Figure 9.4 DNase footprinting with the recombinant 434 repressor. Wharton and Ptashne performed DNase footprinting with end-labeled P22 phage O_R and either P22 repressor (P22R, lanes 1–7) or the 434 repressor with five amino acids in the recognition helix (α-helix 3) changed to match those in the phage P22 recognition helix (434R[α3(P22R)], lanes 8–14). The two sets of lanes contained increasing concentrations of the respective repressors (0 M in lanes 1 and 8, and ranging from 7.6×10^{-10} M to 1.1×10^{-8} M in lanes 2–7 and from 5.2×10^{-9} M to 5.6×10^{-7} M in lanes 8–14). The marker lane (M) contained the A + G reaction from a sequencing procedure. The positions of all three rightward operators are indicated with brackets at left. (*Source:* Wharton, R.P. and M. Ptashne, Changing the binding specificity of a repressor by redesigning an alpha-helix. *Nature* 316 (15 Aug 1985), f. 3, p. 603. © Macmillan Magazines Ltd.)

To check these results, Wharton and Ptashne measured DNA binding in vitro by DNase footprinting (Chapter 5). They found that the purified recombinant repressor could make a "footprint" in the P22 operator, just as the P22 repressor can (Figure 9.4). In control experiments (not shown) they demonstrated that the recombinant repressor could no longer make a footprint in the 434 operator. Thus, the binding specificity really had been altered by these five amino acid changes. In further experiments, Ptashne and colleagues showed that the first four of these amino acids were necessary and sufficient for either binding activity. That is, if the repressor had TQQE (threonine, glutamine, glutamine, glutamate) in its recognition helix, it would bind to the 434 operator. On the other hand, if it had SNVS (serine, asparagine, valine, serine), it would bind to the P22 operator.

What if Wharton and Ptashne had not tried to *change* the specificity of the repressor, but just to *eliminate* it? They could have identified the amino acids in the repressor that were probably important to specificity, then changed them to other amino acids chosen at random and shown that this recombinant 434 repressor could no longer bind to its operator. If that is all they had done, they could have said that the results were consistent with the hypothesis that the altered amino acids are directly involved in binding. But an alternative explanation would remain: These amino acids could simply be important to the overall three-dimensional shape of the repressor protein, and changing them changed this shape and therefore indirectly prevented binding. By contrast, changing specificity by changing amino acids is

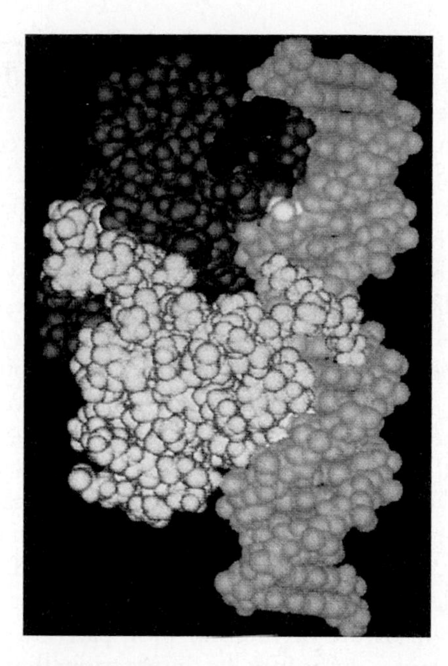

Figure 9.5 Computer model of the λ repressor dimer binding to λ operator (O_R2). The DNA double helix (light blue) is at right. The two monomers of the repressor are in dark blue and yellow. The helix-turn-helix motif of the upper monomer (dark red and blue) is inserted into the major groove of the DNA. The arm of the lower monomer reaches around to embrace the DNA. (*Source:* Hochschild, A., N. Irwin, and M. Ptashne, Repressor structure and the mechanism of positive control. *Cell* 32 (1983) p. 322. Reprinted by permission of Elsevier Science. Photo by Richard Feldman.)

strong evidence for the direct involvement of these amino acids in binding.

In a related x-ray crystallographic study, Ptashne and coworkers showed that the λ repressor has an amino-terminal arm not found in the repressors of the 434 and P22 phages. This arm contributes to the repressor's binding to the λ operator by embracing the operator. Figure 9.5 shows a computer model of a dimer of λ repressor interacting with λ operator. In the repressor monomer at the top, the helix-turn-helix motif is visible projecting into the major groove of the DNA. At the bottom, we can see the arm of the other repressor monomer reaching around to embrace the DNA.

Cro also uses a helix-turn-helix DNA binding motif and binds to the same operators as the λ repressor, but it has the exact opposite affinity for the three different operators in a set (Chapter 8). That is, it binds first to O_R3 and last to O_R1, rather than vice versa. Therefore, by changing amino acids in the recognition helices, one ought to be able to identify the amino acids that give Cro and the λ repressor their different binding specificities. Ptashne and his coworkers accomplished this task and found that amino acids 5 and 6 in the recognition helices are especially important, as is the amino-terminal arm in the λ repressor. When these workers altered base pairs in the operators, they discovered that the base pairs critical to discriminating between O_R1 and O_R3 are at position 3, to which Cro is more sensitive, and at positions 5 and 8, which are selective for repressor binding.

SUMMARY The repressors of the λ-like phages have recognition helices that fit sideways into the major groove of the operator DNA. Certain amino acids on the DNA side of the recognition helix make specific contact with bases in the operator, and these contacts determine the specificity of the protein–DNA interactions. In fact, changing these amino acids can change the specificity of the repressor. The λ repressor itself has an extra motif not found in the other repressors, an amino-terminal arm that aids binding by embracing the DNA. The λ repressor and Cro share affinity for the same operators, but they have microspecificities for O_R1 or O_R3, determined by interactions between different amino acids in the recognition helices of the two proteins and different base pairs in the two operators.

High-Resolution Analysis of λ Repressor–Operator Interactions

Steven Jordan and Carl Pabo wished to visualize the λ repressor–operator interaction at higher resolution than previous studies allowed. They were able to achieve a resolution of 2.5 Å by making excellent cocrystals of a repressor fragment and an operator fragment. The repressor fragment encompassed residues 1–92, which included all of the DNA-binding domain of the protein. The operator fragment (Figure 9.6) was 20 bp long and contained one complete site to which the repressor dimer attached. That is, it had two half-sites, each of which bound to a repressor monomer. Such use of partial molecules is a common trick employed by x-ray crystallographers to make better crystals than they can obtain with whole proteins or whole DNAs. In this case, because the primary goal was to elucidate the structure of the interface between the repressor and the operator, the protein and DNA fragments were probably just as useful as the whole protein and DNA because they contained the elements of interest.

General Structural Features Figure 9.2, used at the beginning of this chapter to illustrate the fit between λ repressor and operator, is based on the high-resolution model from the

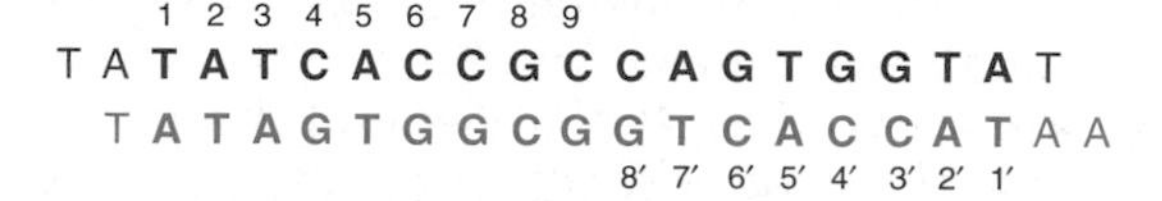

Figure 9.6 The operator fragment used to prepare operator–repressor cocrystals. This 20-mer contains the two λ O_L1 half-sites, each of which binds a monomer of repressor. The half-sites are included within the 17-bp region in boldface; each half-site contains 8 bp, separated by a G–C pair in the middle (9). The half-site on the left has a consensus sequence; that on the right deviates somewhat from the consensus. The base pairs of the consensus half-site are numbered 1–8; those in the other half-site are numbered 1′–8′.

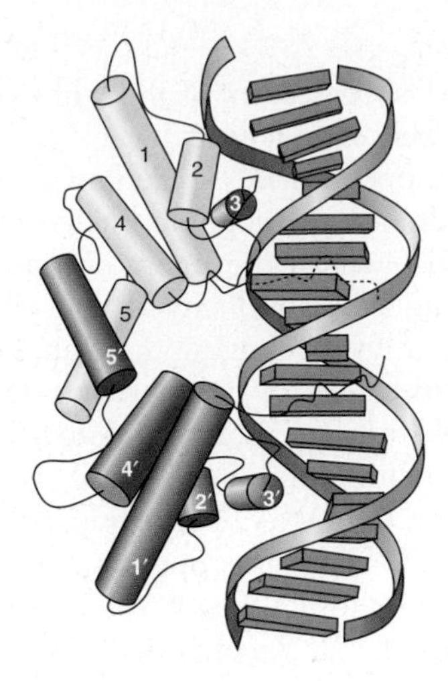

Figure 9.7 Geometry of the λ repressor–operator complex. The DNA (blue) is bound to the repressor dimer, whose monomers are depicted in yellow and purple. The recognition helix of each monomer is shown in red and labeled 3 and 3′. (*Source:* Adapted from Jordan, S.R. and C.O. Pabo, Structure of the lambda complex at 2.5 Å resolution. Details of the repressor–operator interaction. *Science* 242:895, 1988.)

Jordan and Pabo analysis we are now considering. Figure 9.7, a more detailed representation of the same model, reveals several general aspects of the protein–DNA interaction. First of all, of course, we can see the recognition helices (3 and 3′, red) of each repressor monomer nestled into the DNA major grooves in the two half-sites. We can also see how helices 5 and 5′ approach each other to hold the two monomers together in the repressor dimer. Finally, note that the DNA is similar in shape to the standard B-form of DNA. We can see a bit of bending of the DNA, especially at the two ends of the DNA fragment, as it curves around the repressor dimer, but the rest of the helix is relatively straight.

Interactions with Bases Figure 9.8 shows the details of the interactions between amino acids in a repressor monomer and bases in one operator half-site. The crucial amino acids participating in these interactions are glutamine 33 (Gln 33), glutamine 44 (Gln 44), serine 45 (Ser 45), lysine 4 (Lys 4), and asparagine 55 (Asn 55). Figure 9.8a is a stereo view of the interactions, where α-helices 2 and 3 are represented by bold lines. The recognition helix (3) is almost perpendicular to the plane of the paper, so the helical polypeptide backbone looks like a bumpy circle. The key amino acid side chains are shown making hydrogen bonds (dashed lines) to the DNA and to one another.

Figure 9.8b is a schematic diagram of the same amino acid/DNA interactions. It is perhaps easier to see the hydrogen bonds in this diagram. We see that three of the important bonds to DNA bases come from amino acids in the recognition helix. In particular, Gln 44 makes two hydrogen bonds to adenine-2, and Ser 45 makes one hydrogen bond to guanine-4. Figure 9.8c depicts these hydrogen bonds in detail and also clarifies a point made in parts (a) and (b) of the figure: Gln 44 also makes a hydrogen bond to Gln 33, which in turn is hydrogen-bonded to the phosphate preceding base pair number 2. This is an example of a **hydrogen bond network,** which involves three or more entities (e.g., amino acids, bases, or DNA backbone). The participation of Gln 33 is critical. By bridging between the DNA backbone and Gln 44, it positions Gln 44 and the rest of the recognition helix to interact optimally with the operator. Thus, even though Gln 33 resides at the beginning of helix 2, rather than on the recognition helix, it plays an important role in protein–DNA binding. To underscore the importance of this glutamine, we note that it also appears in the same position in the 434 phage repressor and plays the same role in interactions with the 434 operator, which we will examine later in this chapter.

Serine 45 also makes an important hydrogen bond with a base pair, the guanine of base pair number 4. In addition, the methylene (CH_2) group of this serine approaches the methyl group of the thymine of base pair number 5 and participates in a hydrophobic interaction that probably also includes the methyl group of Ala 49. Such hydrophobic interactions involve nonpolar groups like methyl and methylene, which tend to come together to escape the polar environment of the water solvent, much as oil droplets coalesce to minimize their contact with water. Indeed, hydrophobic literally means "water-fearing."

The other hydrogen bonds with base pairs involve two other amino acids that are not part of the recognition helix. In fact, these amino acids are not part of any helix: Asn 55 lies in the linker between helices 3 and 4, and Lys 4 is on the arm that reaches around the DNA. Here again we see an example of a hydrogen bond network, not only between amino acid and base, but between two amino acids. Figure 9.8c makes it particularly clear that these two amino acids each form hydrogen bonds to the guanine of base pair number 6, and also to each other. Such networks add considerably to the stability of the whole complex.

Amino Acid/DNA Backbone Interactions We have already seen one example of an amino acid (Gln 33) that forms a hydrogen bond with the DNA backbone (the phosphate between base pairs 1 and 2). However, this is only one of five such interactions in each half-site. Figure 9.9 portrays these interactions in the consensus half-site, which involve five different amino acids, only one of which (Asn 52) is in the recognition helix. The dashed lines represent hydrogen bonds from the NH groups of the peptide backbone, rather than from the amino acid side chains.

One of these hydrogen bonds, involving the peptide NH at Gln 33, is particularly interesting because of an electrostatic contribution of helix 2 as a whole. To appreciate this, recall from Chapter 3 that all the C=O bonds in a protein α-helix point in one direction. Because each of these bonds is polar, with a partial negative charge on the oxygen and a partial positive charge on the carbon, the whole α-helix has

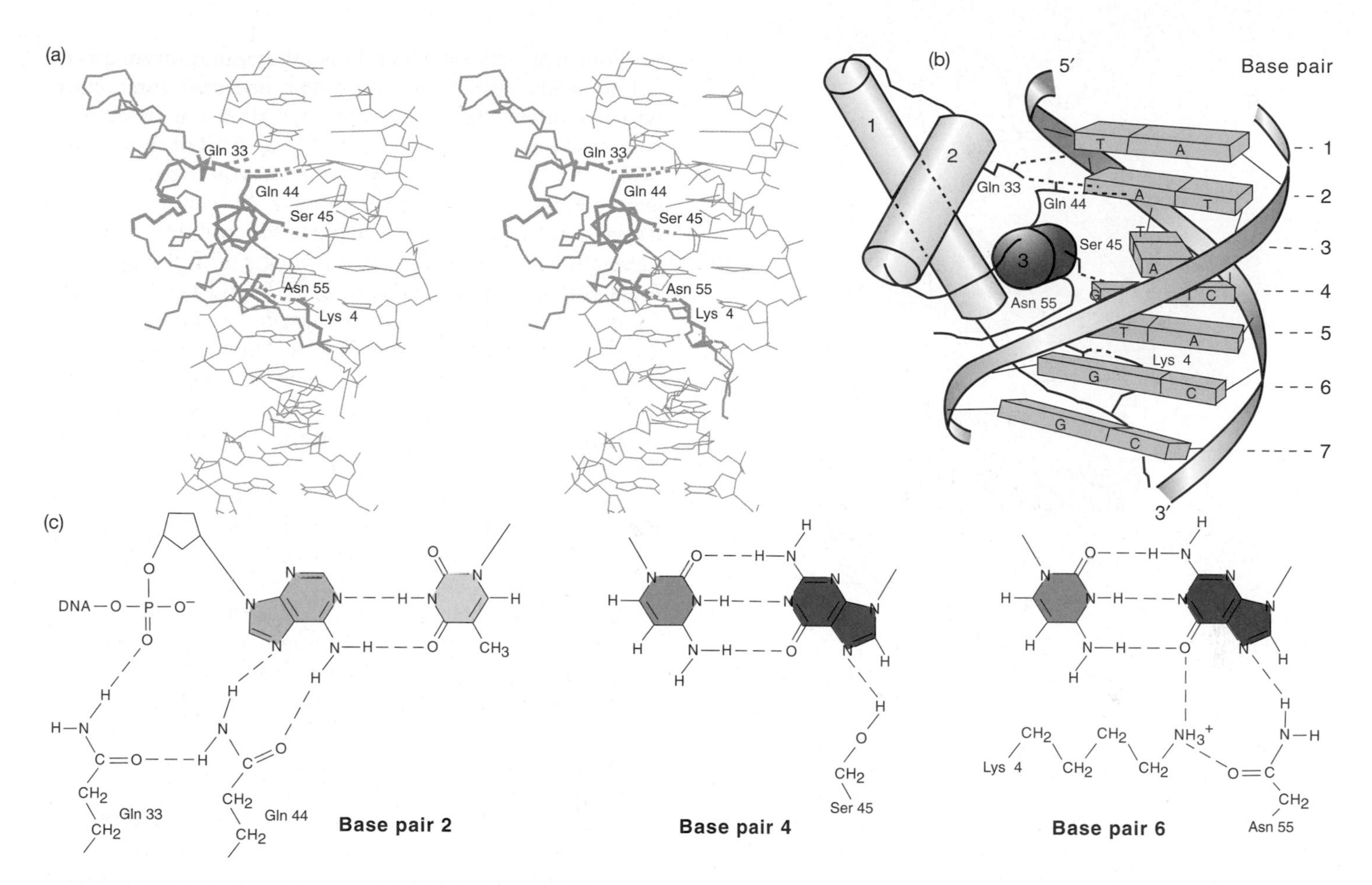

Figure 9.8 Hydrogen bonds between λ repressor and base pairs in the major groove of the operator. (a) Stereo diagram of the complex, with the DNA double helix on the right and the amino terminal part of the repressor monomer on the left. α-Helices 2 and 3 are rendered in bold lines, with the recognition helix almost perpendicular to the plane of the paper. Hydrogen bonds are represented by dashed lines. **(b)** Schematic diagram of the hydrogen bonds shown in panel (a). Only the important amino acid side chains are shown. The base pairs are numbered at right. **(c)** Details of the hydrogen bonds. Structures of the key amino acid side chains and bases are shown, along with the hydrogen bonds in which they participate. (*Source:* From Jordan, S.R. and C.O. Pabo, Structure of the lambda complex at 2.5 Å resolution: Details of the repressor-operator interactions. *Science* 242:896, 1988. Copyright © 1988 AAAS. Reprinted with permission from AAAS.)

a considerable polarity, with practically a full net positive charge at the amino terminus of the helix. This end of the helix will therefore have a natural affinity for the negatively charged DNA backbone. Now look again at Figure 9.9 and notice that the amino end of helix 2, where Gln 33 is located, points directly at the DNA backbone. This maximizes the electrostatic attraction between the positively charged amino end of the α-helix and the negatively charged DNA and stabilizes the hydrogen bond between the peptide NH of Gln 33 and the phosphate group in the DNA backbone.

Other interactions involve hydrogen bonds between amino acid side chains and DNA backbone phosphates. For example, Lys 19 and Asn 52 both form hydrogen bonds with phosphate P_B. The amino group of Lys 26 carries a full positive charge. Although it may be too far away from the DNA backbone to interact directly with a phosphate, it may contribute to the general affinity between protein and DNA. The large number of amino acid/DNA phosphate contacts suggests that these interactions play a major role in the stabilization of the protein–DNA complex. Figure 9.9 also shows the position of the side chain of Met 42. It probably forms a hydrophobic interaction with three carbon atoms on the deoxyribose between P_C and P_D.

Confirmation of Biochemical and Genetic Data Before the detailed structure of the repressor–operator complex was known, we already had predictions from biochemical and genetic experiments about the importance of certain repressor amino acids and operator bases. In almost all cases, the structure confirms these predictions.

First, ethylation of certain operator phosphates interfered with repressor binding. Hydroxyl radical footprinting had also implicated these phosphates in repressor binding. Now we see that these same phosphates (five per half-site) make important contacts with repressor amino acids in the cocrystal.

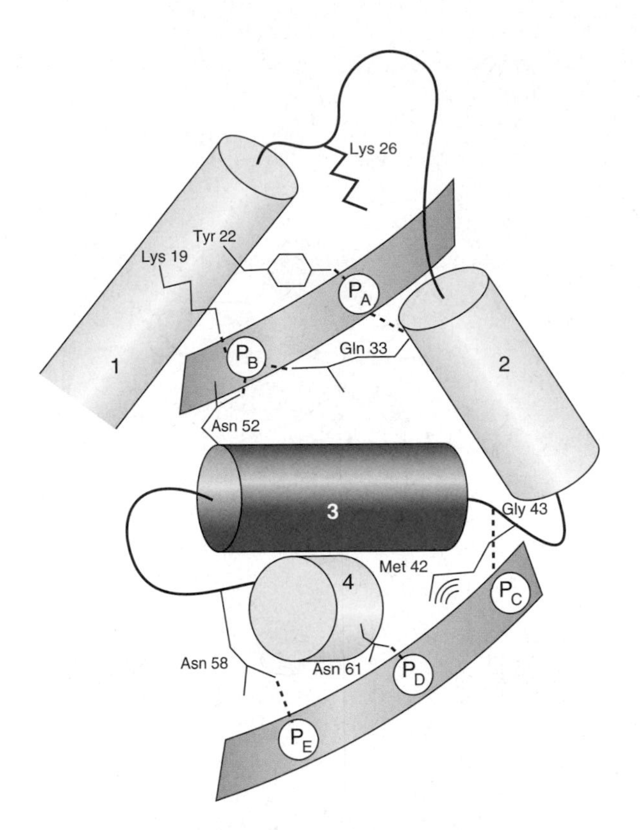

Figure 9.9 Amino acid/DNA backbone interactions. α-Helices 1–4 of the λ repressor are shown, along with the phosphates (P_A–P_E) that are involved in hydrogen bonds with the protein. This diagram is perpendicular to that in Figure 9.8. The side chains of the important amino acids are shown. The two dashed lines denote hydrogen bonds between peptide NH groups and phosphates. Concentric arcs denote a hydrophobic interaction. (*Source:* Adapted from Jordan S.R. and C.O. Pabo, Structure of the lambda complex at 2.5 Å resolution: Details of the repressor–operator interactions. *Science* 242:897, 1988.)

Second, methylation protection experiments had predicted that certain guanines in the major groove would be in close contact with repressor. The crystal structure now shows that all of these are indeed involved in repressor binding. One major-groove guanine actually became more sensitive to methylation on repressor binding, and this guanine (G8′, Figure 9.6) is now seen to have an unusual conformation in the cocrystal. Base pair 8′ is twisted more than any other on its horizontal axis, and the spacing between this base pair and the next is the widest. This unusual conformation could open guanine 8′ up to attack by the methylating agent DMS. Also, adenines were not protected from methylation in previous experiments. This makes sense because adenines are methylated on N3, which resides in the minor groove. Because no contacts between repressor and operator occur in the minor groove, repressor cannot protect adenines from methylation.

Third, DNA sequence data had shown that the A–T base pair at position 2 and the G–C base pair at position 4 (Figure 9.8) were conserved in all 12 half-sites of the operators O_R and O_L. The crystal structure shows why these base pairs are so well conserved: They are involved in important contacts with the repressor.

Fourth, genetic data had shown that mutations in certain amino acids destabilized repressor–operator interaction, whereas other changes in repressor amino acids actually enhanced binding to the operator. Almost all of these mutations can be explained by the cocrystal structure. For example, mutations in Lys 4 and Tyr 22 were particularly damaging, and we now see (Figures 9.8 and 9.9) that both these amino acids make strong contacts with the operator: Lys 4 with guanine-6 (and with Asn 55) and Tyr 22 with P_A. As an example of a mutation with a positive effect, consider the substitution of lysine for Glu 34. This amino acid is not implicated by the crystal structure in any important bonds to the operator, but a lysine in this position could rotate so as to form a salt bridge with the phosphate before P_A (Figure 9.9) and thus enhance protein–DNA binding. This salt bridge would involve the positively charged ε-amino group of the lysine and the negatively charged phosphate.

SUMMARY The cocrystal structure of a λ repressor fragment with an operator fragment shows many details about how the protein and DNA interact. The most important contacts occur in the major groove, where amino acids make hydrogen bonds with DNA bases and with the DNA backbone. Some of these hydrogen bonds are stabilized by hydrogen-bond networks involving two amino acids and two or more sites on the DNA. The structure derived from the cocrystal is in almost complete agreement with previous biochemical and genetic data.

High-Resolution Analysis of Phage 434 Repressor–Operator Interactions

Harrison, Ptashne, and coworkers used x-ray crystallography to perform a detailed analysis of the interaction between phage 434 repressor and operator. As in the λ cocrystal structure, the crystals they used for this analysis were not composed of full-length repressor and operator, but fragments of each that contained the interaction sites. As a substitute for the repressor, they used a peptide containing the first 69 amino acids of the protein, including the helix-turn-helix DNA-binding motif. For the operator, they used a synthetic 14-bp DNA fragment that contains the repressor-binding site. These two fragments presumably bound together as the intact molecules would, and the complex could be crystallized relatively easily. We will focus here on concepts that were not clearly demonstrated by the λ repressor–operator studies.

Contacts with Base Pairs Figure 9.10 summarizes the contacts between the side chains of Gln 28, Gln 29, and Gln 33, all in the recognition helix (α3) of the 434 repressor. Starting at the bottom of the figure, note the two possible

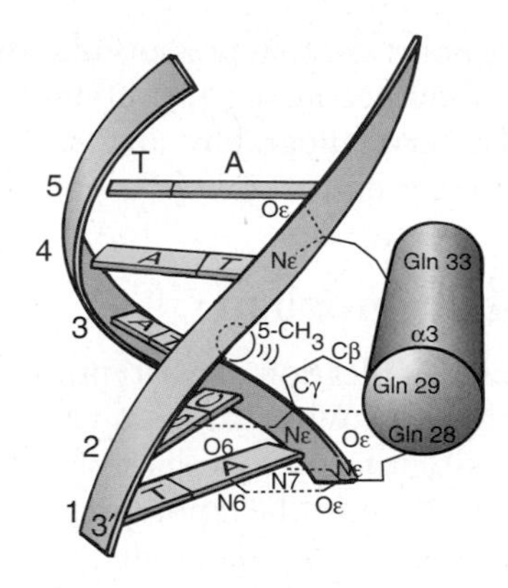

Figure 9.10 Detailed model of interaction between recognition helix amino acid side chains and one 434 operator half-site. Hydrogen bonds are represented by dashed lines. The van der Waals interaction between the Gln 29 side chain and the 5-methyl group of the thymine paired to adenine 3 is represented by concentric arcs. (*Source:* Adapted from Anderson, J. E., M. Ptashne, and S. C. Harrison, Structure of the repressor–operator complex of bacteriophage 434. *Nature* 326:850, 1987.)

hydrogen bonds (represented by dashed lines) between the Oε and Nε of Gln 28 and the N6 and N7 of adenine 1. Next, we see that a possible hydrogen bond between the Oε of Gln 29 and the protein backbone NH of the same amino acid points the Nε of this amino acid directly at the O6 of the guanine in base pair 2 of the operator, which would allow a hydrogen bond between this amino acid and base. Note also the potential van der Waals interactions (represented by concentric arcs) between Cβ and Cγ of Gln 29 and the 5-methyl group of the thymine in base pair 3. Such van der Waals interactions can be explained roughly as follows: Even though all the groups involved are nonpolar, at any given instant they have a very small dipole moment due to random fluctuations in their electron clouds. These small dipole moments can cause a corresponding opposite polarity in a very close neighbor. The result is an attraction between the neighboring groups.

SUMMARY X-ray crystallography of a phage 434 repressor-fragment/operator-fragment complex shows probable hydrogen bonding between three glutamine residues in the recognition helix and three base pairs in the repressor. It also reveals a potential van der Waals contact between one of these glutamines and a base in the operator.

Effects of DNA Conformation The contacts between the repressor and the DNA backbone require that the DNA double helix curve slightly. Indeed, higher-resolution crystallography studies by Harrison, Ptashne, and colleagues show that the DNA does curve this way in the DNA–protein complex (Figure 9.11); we do not know yet whether the DNA bend preexists in this DNA region or whether it is induced by repressor binding. In either case, the base sequence of the operator plays a role by facilitating this bending. That is, some DNA sequences are easier to bend in a given way than others, and the 434 operator sequence is optimal for the bend it must make to fit the repressor. We will discuss this general phenomenon in more detail later in this chapter.

Another notable feature of the conformation of the operator DNA is the compression of the DNA double helix between base pairs 7 and 8, which lie between the two half-sites of the operator. This compression amounts to an overwinding of 3 degrees between base pairs 7 and 8, or 39 degrees, compared with the normal 36 degrees helical twist between base pairs. Notice the narrowness of the minor groove at center right in Figure 9.11b, compared to Figure 9.11a. The major grooves on either side are wider than normal, due to a compensating underwinding of that DNA. Again, the base sequence at this point is optimal for assuming this conformation.

SUMMARY The x-ray crystallography analysis of the partial phage 434 repressor–operator complex shows that the DNA deviates significantly from its normal regular shape. It bends somewhat to accommodate the necessary base/amino acid contacts. Moreover, the central part of the helix, between the two half-sites, is wound extra tightly, and the outer parts are wound more loosely than normal. The base sequence of the operator facilitates these departures from normal DNA shape.

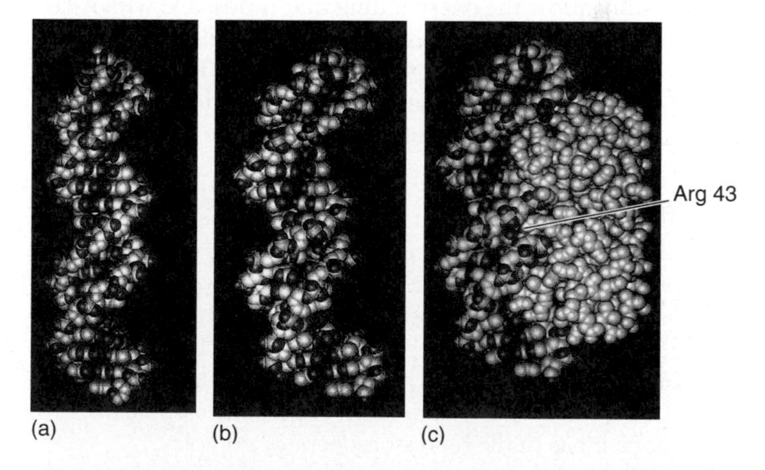

Figure 9.11 Space-filling computer model of distorted DNA in the 434 repressor–operator complex. (a) Standard B-DNA. **(b)** Shape of the operator-containing 20-mer in the repressor–operator complex with the protein removed. Note the overall curvature, and the narrowness of the minor groove at center right. **(c)** The repressor–operator complex, with the repressor in orange. Notice how the DNA conforms to the shape of the protein to promote intimate contact between the two. The side chain of Arg 43 can be seen projecting into the minor groove of the DNA near the center of the model. (*Source:* Aggarwal et al., Recognition of a DNA operator by the repressor of phage 434: A view at high resolution. *Science* 242 (11 Nov 1988) f. 3b, f. 3c, p. 902. © AAAS.)

Genetic Tests of the Model If the apparent contacts we have seen between repressor and operator are important, mutations that change these amino acids or bases should reduce or abolish DNA–protein binding. Alternatively, we might be able to mutate the operator so it does not fit the repressor, then make a compensating mutation in the repressor that restores binding. Also, if the unusual shape assumed by the operator is important, mutations that prevent it from taking that shape should reduce or abolish repressor binding. As we will see, all those conditions have been fulfilled.

To demonstrate the importance of the interaction between Gln 28 and A1, Ptashne and colleagues changed A1 to a T. This destroyed binding between repressor and operator, as we would expect. However, this mutation could be suppressed by a mutation at position 28 of the repressor from Gln to Ala. Figure 9.10 reveals the probable explanation: The two hydrogen bonds between Gln 28 and A1 can be replaced by a van der Waals contact between the methyl groups on Ala 28 and T1. The importance of this contact is underscored by the replacement of T1 with a uracil, which does not have a methyl group, or 5-methylcytosine (5MeC), which does. The U-substituted operator does not bind the repressor with Ala 28, but the 5MeC-substituted operator does. Thus, the methyl group is vital to interactions between the mutant operator and mutant repressor, as predicted on the basis of the van der Waals contact.

We strongly suspect that the overwinding of the DNA between base pairs 7 and 8 is important in repressor–operator interaction. If so, substituting G–C or C–G base pairs for the A–T and T–A pairs at positions 6–9 should decrease repressor–operator binding, because G–C pairs do not readily allow the overwinding that is possible with A–T pairs. As expected, repressor did not bind well to operators with G–C or C–G base pairs in this region. This failure to bind well did not prove that overwinding exists, but it was consistent with the overwinding hypothesis.

SUMMARY The contacts between the phage 434 repressor and operator predicted by x-ray crystallography can be confirmed by genetic analysis. When amino acids or bases predicted to be involved in interaction are altered, repressor–operator binding is inhibited. Furthermore, binding is also inhibited when the DNA is mutated so it cannot as readily assume the shape it has in the repressor–operator complex.

9.2 The *trp* Repressor

The *trp* repressor is another protein that uses a helix-turn-helix DNA-binding motif. However, recall from Chapter 7 that the aporepressor (the protein without the tryptophan corepressor) is not active. Paul Sigler and colleagues used x-ray crystallography of *trp* repressor and aporepressor to point out the subtle but important difference that tryptophan makes. The crystallography also sheds light on the way the *trp* repressor interacts with its operator.

The Role of Tryptophan

Here is a graphic indication that tryptophan affects the shape of the repressor: When you add tryptophan to crystals of aporepressor, the crystals shatter! When the tryptophan wedges itself into the aporepressor to form the repressor, it changes the shape of the protein enough to break the lattice forces holding the crystal together.

This raises an obvious question: What moves when free tryptophan binds to the aporepressor? To understand the answer, it helps to visualize the repressor as illustrated in Figure 9.12. The protein is actually a dimer of identical subunits, but these subunits fit together to form a three-domain structure. The central domain, or "platform," comprises the A, B, C, and F helices of each monomer, which are grouped together on the right, away from the DNA. The other two domains, found on the left close to the DNA, are the D and E helices of each monomer.

Now back to our question: What moves when we add tryptophan? The platform apparently remains stationary, whereas the other two domains tilt, as shown in Figure 9.12. The recognition helix in each monomer is helix E, and we can see an obvious shift in its position when tryptophan binds. In the top monomer, it shifts from a somewhat downward orientation to a position in which it points directly into the major groove of the operator. In this position, it is ideally situated to make contact with (or "read") the DNA, as we will see.

Sigler refers to these DNA-reading motifs as *reading heads,* likening them to the heads in the hard drive of a computer. In a computer, the reading heads can assume two positions: engaged and reading the drive, or disengaged and away from the drive. The *trp* repressor works the same way. When tryptophan is present, it inserts itself between the platform and each reading head, as illustrated in Figure 9.12, and forces the reading heads into the best position (transparent helices D and E) for fitting into the major groove of the operator. On the other hand, when tryptophan dissociates from the aporepressor, the gap it leaves allows the reading heads to fall back toward the central platform and out of position to fit with the operator (gray helices D and E).

Figure 9.13a shows a closer view of the environment of the tryptophan in the repressor. It is a hydrophobic pocket that is occupied by the side chain of a hydrophobic amino acid (sometimes tryptophan) in almost all comparable helix-turn-helix proteins, including the λ repressor, Cro, and CAP. However, in these other proteins the hydrophobic amino acid is actually part of the protein chain, not a free amino acid, as in the *trp* repressor. Sigler likened the arrangement of the tryptophan between Arg 84 and Arg 54

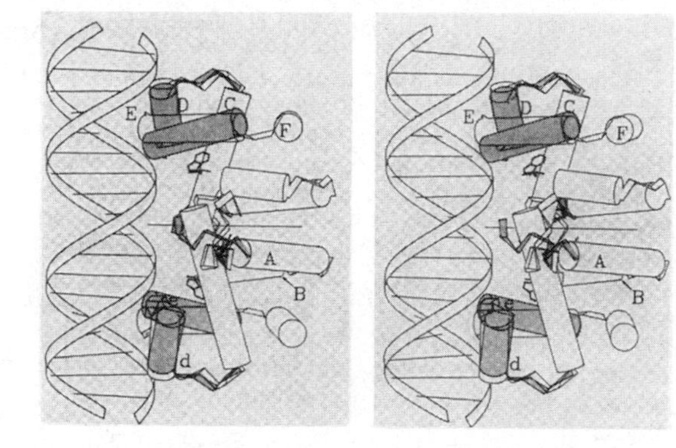

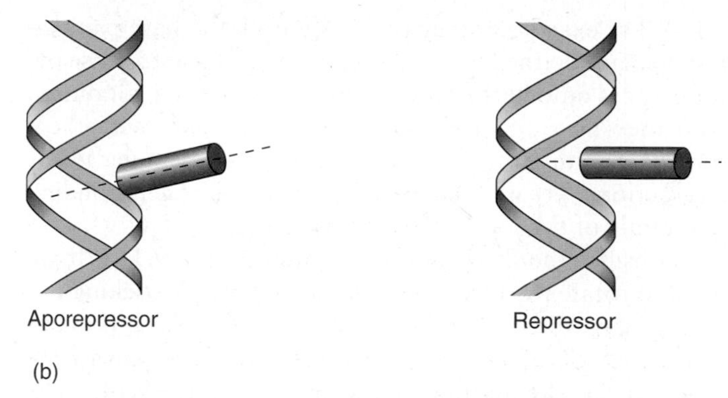

Figure 9.12 Comparison of the fit of *trp* repressor and aporepressor with *trp* operator. (a) Stereo diagram. The helix-turn-helix motifs of both monomers are shown in the positions they assume in the repressor (transparent) and aporepressor (dark). The position of tryptophan in the repressor is shown (black polygons). Note that the recognition helix (helix E) in the aporepressor falls back out of ideal position for inserting into the major groove of the operator DNA. The two almost identical drawings constitute a stereo presentation that allows you to view this picture in three dimensions. To get this 3-D effect, use a stereo viewer, or alternatively, hold the picture 1–2 ft in front of you and let your eyes relax as they would when you are staring into the distance or viewing a "magic eye" picture. After a few seconds, the two images should fuse into one in the center, which appears in three dimensions. This stereo view gives a better appreciation for the fit of the recognition helix and the major groove of the DNA, but if you cannot get the 3-D effect, just look at one of the two pictures. **(b)** Simplified (nonstereo) diagram comparing the positions of the recognition helix (red) of the aporepressor (left) and the repressor (right) with respect to the DNA major groove. Notice that the recognition helix of the repressor points directly into the major groove, whereas that of the aporepressor points more downward. The dashed line emphasizes the angle of the recognition helix in each drawing.

to a salami sandwich, in which the flat tryptophan is the salami. When it is removed, as in Figure 9.13b, the two arginines come together as the pieces of bread would when you remove the salami from a sandwich. This model has implications for the rest of the molecule, because Arg 54 is on the surface of the central platform of the repressor dimer, and Arg 84 is on the facing surface of the reading head. Thus, inserting the tryptophan between these two arginines pushes the reading head away from the platform and points it toward the major groove of the operator, as we saw in Figure 9.12.

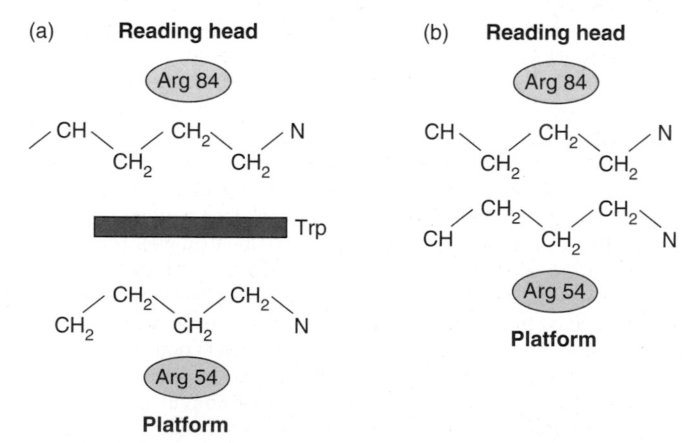

Figure 9.13 Tryptophan-binding site in the *trp* repressor. (a) Environment surrounding the tryptophan (Trp) in the *trp* repressor. Notice the positions of Arg 84 above and Arg 54 below the tryptophan side chain (red). **(b)** The same region in the aporepressor, without tryptophan. Notice that the Arg side chains have moved together to fill the gap left by the absent tryptophan.

SUMMARY The *trp* repressor requires tryptophan to force the recognition helices of the repressor dimer into the proper position for interacting with the *trp* operator.

9.3 General Considerations on Protein–DNA Interactions

What contributes to the specificity of binding between a protein and a specific stretch of DNA? The examples we have seen so far suggest two answers: (1) specific interactions between bases and amino acids; and (2) the ability of the DNA to assume a certain shape, which also depends on the DNA's base sequence (a phenomenon Sigler calls "indirect readout"). These two possibilities are clearly not mutually exclusive, and both apply to many of the same protein–DNA interactions.

Hydrogen Bonding Capabilities of the Four Different Base Pairs

We have seen that different DNA-binding proteins depend to varying extents on contacts with the bases in the DNA. To the extent that they "read" the sequence of bases, one can ask, What exactly do they read? After all, the base pairs do not open up, so the DNA-binding proteins have to sense the differences among the bases in their base-paired condition. And they have to make base-specific contacts with these base pairs, either through hydrogen bonds or van der Waals interactions. Let us examine further the hydrogen-bonding potentials of the four different base pairs.

Consider the DNA double helix in Figure 9.14a. If we were to rotate the DNA 90 degrees so that it is sticking out of the page directly at us, we would be looking straight down the helical axis. Now consider one base pair of the DNA in this orientation, as pictured in Figure 9.14b. The major groove is on top, and the minor groove is below. A DNA-binding protein can approach either of these grooves to interact with the base pair. As it does so, it "sees" four possible contours in each groove, depending on whether the base pair is a T–A, A–T, C–G, or G–C pair.

Figure 9.14c presents two of these contours from both the major and minor groove perspectives. At the very bottom we see line diagrams (Figure 9.14d) that summarize what the protein encounters in both grooves for an A–T and a G–C base pair. Hydrogen bond acceptors (oxygen and nitrogen atoms) are denoted "Acc," and hydrogen bond donors (hydrogen atoms) are denoted "Don." The major and minor grooves lie above and below the horizontal lines, respectively. The lengths of the vertical lines represent the relative distances that the donor or acceptor atoms project away from the helical axis toward the outside of the DNA groove. We can see that the A–T and G–C base pairs present very different profiles to the outside world, especially in the major groove. The difference between a pyrimidine–purine pair and the purine–pyrimidine pairs shown here would be even more pronounced.

These hydrogen-bonding profiles assume direct interactions between base pairs and amino acids. However, other possibilities exist. There is indirect readout, in which amino acids "read" the shape of the DNA backbone, either by direct hydrogen bonding or by forming salt bridges. Amino acids and bases can also interact indirectly through hydrogen bonds to an intervening water molecule, but these "indirect interactions" are no less specific than direct ones.

SUMMARY The four different base pairs present four different hydrogen-bonding profiles to amino acids approaching either the major or minor DNA groove.

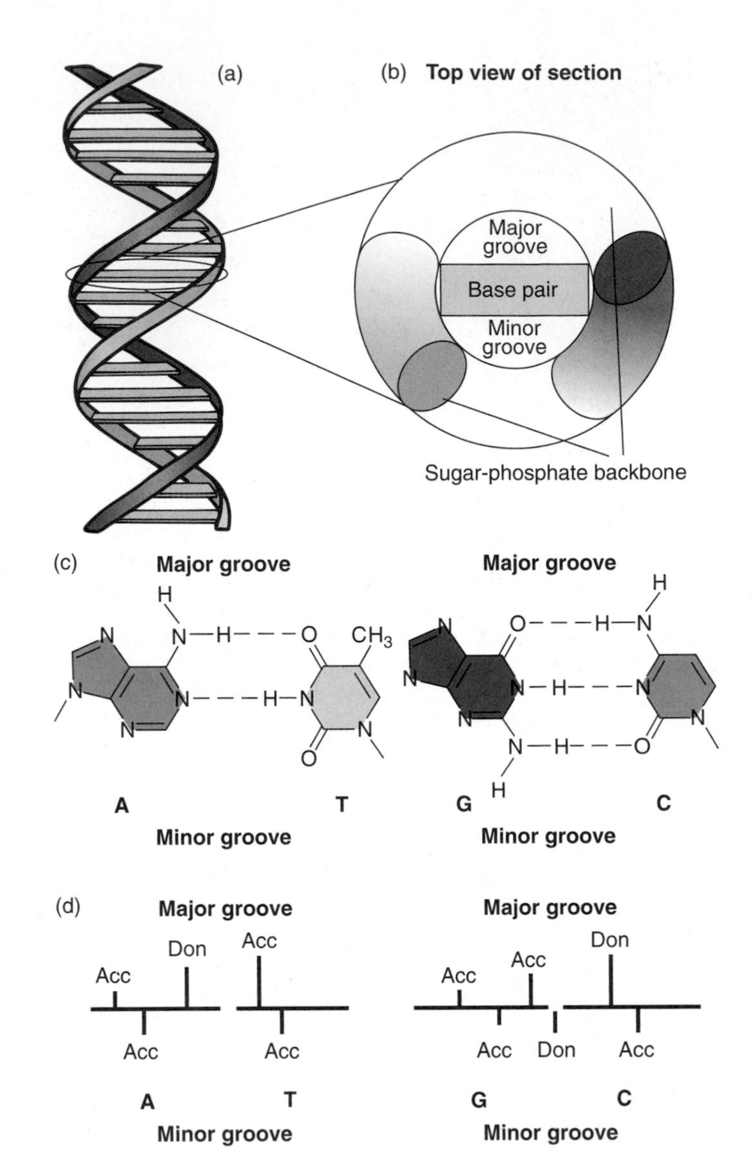

Figure 9.14 Appearance of base pairs in the major and minor grooves of DNA. **(a)** Standard B-form DNA, with the two backbones in red and blue, and the base pairs in yellow. **(b)** Same DNA molecule seen from the top. Notice the wider opening to the major groove (top), compared with the minor groove (bottom). **(c)** Structural formulas of the two base pairs. Again, the major groove is on top, and the minor groove on the bottom. **(d)** Line diagrams showing the positions of hydrogen bond acceptors (Acc) and donors (Don) in the major and minor grooves. For example, reading left to right, the major groove of the T–A pair has an acceptor (the N-7 in the ring of the adenine), then a donor (the NH_2 of the adenine), then an acceptor (the C=O of the thymine). The relative horizontal positions of these groups are indicated by the point of intersection with the vertical lines. The relative vertical positions are indicated by the lengths of the vertical lines. The two base pairs present different patterns of donors and acceptors in both major and minor grooves, so they are perceived differently by proteins approaching from the outside. By inverting these diagrams left-to-right, you can see that T–A and C–G pairs would present still different patterns. (*Source:* Adapted from R. Schleif, DNA binding by proteins. *Science* 241:1182–3, 1988.)

The Importance of Multimeric DNA-Binding Proteins

Robert Schleif noted that the target sites for DNA-binding proteins are usually symmetric, or repeated, so they can interact with multimeric proteins—those composed of more than one subunit. Most DNA-binding proteins are dimers (some are even tetramers), and this greatly enhances the binding between DNA and protein because the two protein subunits bind cooperatively. Having one at the binding site automatically increases the concentration of the other. This boost in concentration is important because DNA-binding proteins are generally present in the cell in very small quantities.

Another way of looking at the advantage of dimeric DNA-binding proteins uses the concept of *entropy*. Entropy can be considered a measure of disorder in the universe. It probably does not come as a surprise to you to learn that entropy, or disorder, naturally tends to increase with time. Think of what happens to the disorder of your room, for example. The disorder increases with time until you expend energy to straighten it up. Thus, it takes energy to push things in the opposite of the natural direction—to create order out of disorder, or make the entropy of a system decrease.

A DNA–protein complex is more ordered than the same DNA and protein independent of each other, so bringing them together causes a decrease in entropy. Binding *two* protein subunits, independently of each other, causes twice the decrease in entropy. But if the two protein subunits are already stuck together in a dimer, orienting one relative to the DNA automatically orients the other, so the entropy change is much less than in independent binding, and therefore requires less energy. Looking at it from the standpoint of the DNA–protein complex, releasing the dimer from the DNA does not provide the same entropy gain as releasing two independently bound proteins would, so the protein and DNA stick together more tightly.

SUMMARY Multimeric DNA-binding proteins have an inherently higher affinity for binding sites on DNA than do multiple monomeric proteins that bind independently of one another.

9.4 DNA-Binding Proteins: Action at a Distance

So far, we have dealt primarily with DNA-binding proteins that govern events that occur very nearby. For example, the *lac* repressor bound to its operator interferes with the activity of RNA polymerase at an adjacent DNA site; or λ repressor stimulates RNA polymerase binding at an adjacent site. However, numerous examples exist in which DNA-binding proteins can influence interactions at remote sites in the DNA. We will see that this phenomenon is common in eukaryotes, but several prokaryotic examples occur as well.

The *gal* Operon

In 1983, S. Adhya and colleagues reported the unexpected finding that the *E. coli gal* operon, which codes for enzymes

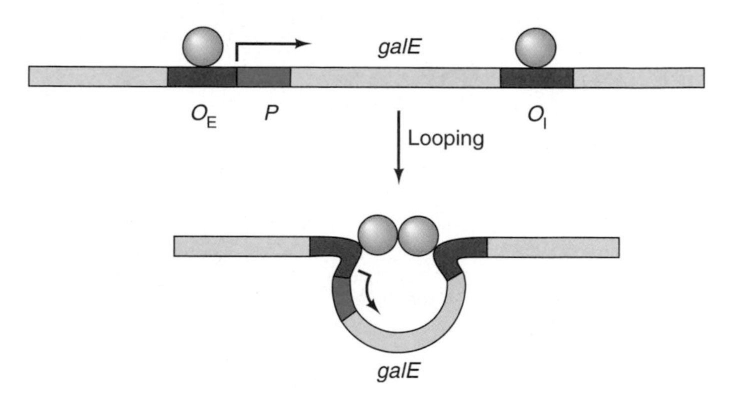

Figure 9.15 Repression of the *gal* operon. The *gal* operon has two operators (red): one external (O_E), adjacent to the promoter (green), and one internal (O_I), within the *galE* gene (yellow). Repressor molecules (blue) bind to both operators and appear to interact by looping out the intervening DNA (bottom).

needed to metabolize the sugar galactose, has two distinct operators, about 97 bp apart. One is located where you would expect to find an operator, adjacent to the *gal* promoter. This one is called O_E, for "external" operator. The other is called O_I, for "internal" operator and is located within the first structural gene, *galE*. The downstream operator was discovered by genetic means: O^c mutations were found that mapped to the *galE* gene instead of to O_E. One way to explain the function of two separated operators is by assuming that they both bind to repressors, and the repressors interact by **looping out** the intervening DNA, as pictured in Figure 9.15. We have already seen examples of this kind of repression by looping out in our discussion of the *lac* and *ara* operons in Chapter 7.

Duplicated λ Operators

The brief discussion of the *gal* operon just presented strongly suggests that proteins interact over a distance of almost 100 bp, but provided no direct evidence for this contention. Ptashne and colleagues used an artificial system to obtain such evidence. The system was the familiar λ operator–repressor combination, but it was artificial in that the experimenters took the normally adjacent operators and separated them to varying extents. We have seen that repressor dimers normally bind cooperatively to O_R1 and O_R2 when these operators are adjacent. The question is this: Do repressor dimers still bind cooperatively to the operators when they are separated? The answer is that they do, as long as the operators lie on the same face of the DNA double helix. This finding supports the hypothesis that repressors bound to separated *gal* operators probably interact by DNA looping.

Ptashne and coworkers used two lines of evidence to show cooperative binding to the separated λ promoters: DNase footprinting and electron microscopy. If we

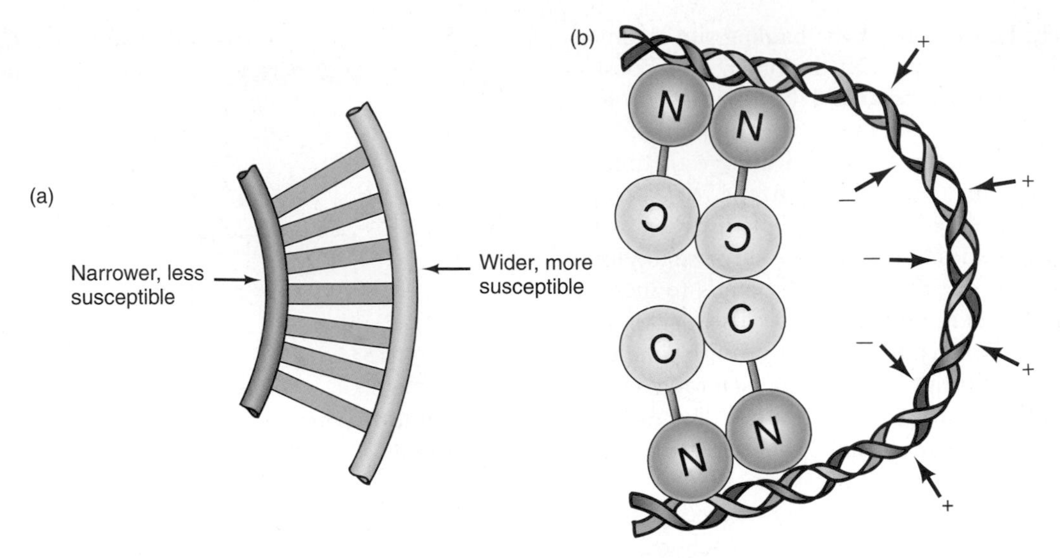

Figure 9.16 Effect of DNA looping on DNase susceptibility. **(a)** Simplified schematic diagram. The double helix is depicted as a railroad track to simplify the picture. The backbones are in red and blue, and the base pairs are in orange. As the DNA bends, the strand on the inside of the bend is compressed, restricting access to DNase. By the same token, the strand on the outside is stretched, making it easier for DNase to attack. **(b)** In a real helix each strand alternates being on the inside and the outside of the bend. Here, two dimers of a DNA-binding protein (λ repressor in this example) are interacting at separated sites, looping out the DNA in between. This stretches the DNA on the outside of the loop, opening it up to DNase I attack (indicated by + signs). Conversely, looping compresses the DNA on the inside of the loop, obstructing access to DNase I (indicated by the – signs). The result is an alternating pattern of higher and lower sensitivity to DNase in the looped region. Only one strand (red) is considered here, but the same argument applies to the other. (*Source:* (*b*) Adapted from Hochschild A. and M. Ptashne, Cooperative binding of lambda repressors to sites separated by integral turns of the DNA helix. *Cell* 44:685, 1986.)

DNase-footprint two proteins that bind independently to remote DNA sites, we see two separate footprints. However, if we footprint two proteins that bind cooperatively to remote DNA sites through DNA looping, we see two separate footprints just as in the previous example, but this time we also see something interesting in between that does not occur when the proteins bind independently. This extra feature is a repeating pattern of insensitivity, then hypersensitivity to DNase. The reason for this pattern is explained in Figure 9.16. When the DNA loops out, the bend in the DNA compresses the base pairs on the inside of the loop, so they are relatively protected from DNase. On the other hand, the base pairs on the outside of the loop are spread apart more than normal, so they become extra sensitive to DNase. This pattern repeats over and over as we go around and around the double helix.

Using this assay for cooperativity, Ptashne and colleagues performed DNase footprinting on repressor bound to DNAs in which the two operators were separated by an integral or nonintegral number of double-helical turns. Figure 9.17a shows an example of cooperative binding, when the two operators were separated by 63 bp—almost exactly six double-helical turns. We can see the repeating pattern of lower and higher DNase sensitivity in between the two binding sites. By contrast, Figure 9.17b presents an example of noncooperative binding, in which the two operators were separated by 58 bp—just 5.5 double-helical turns. Here we see no evidence of a repeating pattern of DNase sensitivity between the two binding sites.

Electron microscopy experiments enabled Ptashne and coworkers to look directly at repressor–operator complexes with integral and nonintegral numbers of double-helical turns between the operators to see if the DNA in the former case really loops out. As Figure 9.18 shows, it does loop out. It is clear when such looping out is occurring, because the DNA is drastically bent. By contrast, Ptashne and colleagues almost never observed bent DNA when the two operators were separated by a nonintegral number of double-helical turns. Thus, as expected, these DNAs have a hard time looping out. These experiments demonstrate clearly that proteins binding to DNA sites separated by an integral number of double-helical turns can bind cooperatively by looping out the DNA in between.

SUMMARY When λ operators are separated by an integral number of double-helical turns, the DNA in between can loop out to allow cooperative binding. When the operators are separated by a nonintegral number of double-helical turns, the proteins have to bind to opposite faces of the DNA double helix, so no cooperative binding can take place.

Enhancers

Enhancers are nonpromoter DNA elements that bind protein factors and stimulate transcription. By definition, they

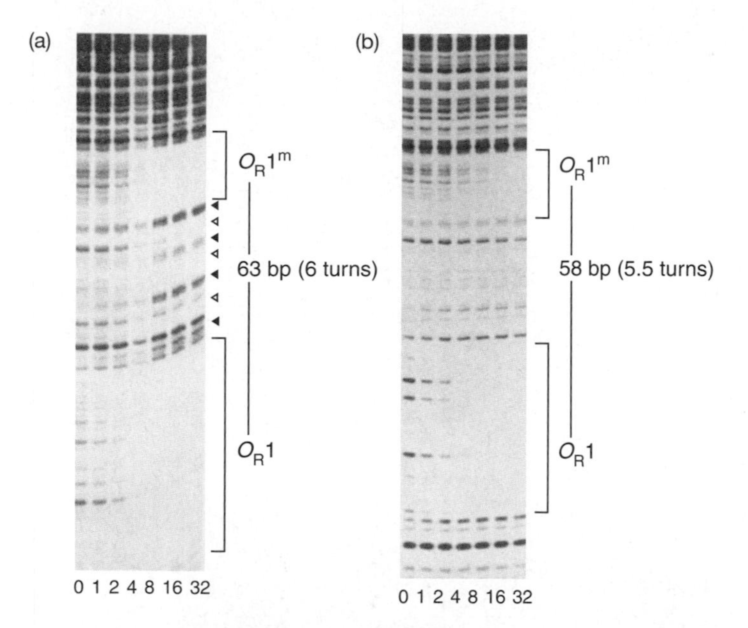

Figure 9.17 DNase footprints of dual operator sites. **(a)** Cooperative binding. The operators are almost exactly six double-helical turns apart (63 bp), and an alternating pattern of enhanced and reduced cleavage by DNase I appears between the two footprints when increasing amounts of repressor are added. The enhanced cleavage sites are denoted by filled arrowheads, the reduced cleavage sites by open arrowheads. This suggests looping of DNA between the two operators on repressor binding. **(b)** Noncooperative binding. The operators are separated by a nonintegral number of double-helical turns (58 bp, or 5.5 turns). No alternating pattern of DNase susceptibility appears on repressor binding, so the repressors bind at the two operators independently, without DNA looping. In both (a) and (b), the number at the bottom of each lane gives the amount of repressor monomer added, where 1 corresponds to 13.5 nM repressor monomer in the assay, 2 corresponds to 27 nM repressor monomer, and so on. (*Source:* Adapted from Hochschild, A. and M. Ptashne, Cooperative binding of lambda repressors to sites separated by integral turns of the DNA helix. *Cell* 44 (14 Mar 1986) f. 3a&4, p. 683.)

can act at a distance. Such elements have been recognized in eukaryotes since 1981, and we will discuss them at length in Chapter 12. More recently, enhancers have also been found in prokaryotes. In 1989, Popham and coworkers described an enhancer that aids in the transcription of genes recognized by an auxiliary σ-factor in *E. coli:* σ^{54}. We encountered this factor in Chapter 8; it is the σ-factor, also known as σ^N, that comes into play under nitrogen starvation conditions to transcribe the *glnA* gene from an alternative promoter.

The σ^{54} factor is defective. DNase footprinting experiments demonstrate that it can cause the $E\sigma^{54}$ holoenzyme to bind stably to the *glnA* promoter, but it cannot do one of the important things normal σ-factors do: direct the formation of an open promoter complex. Popham and coworkers assayed this function in two ways: heparin resistance and DNA methylation. When polymerase forms an open promoter complex, it is bound very tightly to DNA. Adding heparin as a DNA competitor does not inhibit the polymerase. On the other hand, when polymerase forms a closed promoter complex, it is relatively loosely bound and will dissociate at a much higher rate. Thus, it is subject to inhibition by an excess of the competitor heparin. Furthermore, when polymerase forms an open promoter complex, it exposes the cytosines in the melted DNA to methylation by DMS. Because no melting occurs in the closed promoter complex, no methylation takes place.

By both these criteria—heparin sensitivity and resistance to methylation—$E\sigma^{54}$ fails to form an open promoter complex. Instead, another protein, NtrC (the product of the *ntrC* gene), binds to the enhancer and helps $E\sigma^{54}$ form an open promoter complex. The energy for the DNA melting comes from the hydrolysis of ATP, performed by an ATPase domain of NtrC.

How does the enhancer interact with the promoter? The evidence strongly suggests that DNA looping is involved. One clue is that the enhancer has to be at least 70 bp away from the promoter to perform its function. This would allow enough room for the DNA between the promoter and enhancer to loop out. Moreover, the enhancer can still function even if it and the promoter are on separate DNA molecules, as long as the two molecules are linked in a catenane, as shown in Figure 9.19. This would still allow the enhancer and promoter to interact as they would during looping, but it precludes any mechanism (e.g., altering the degree of supercoiling or sliding proteins along the DNA) that requires the two elements to be on the same DNA molecule. We will discuss this phenomenon in more detail in Chapter 12. Finally, and perhaps most tellingly, we can actually observe the predicted DNA loops between NtrC bound to the enhancer and the σ^{54} holoenzyme bound to the promoter. Figure 9.20 shows the results of electron microscopy experiments performed by Sydney Kustu, Harrison Echols, and colleagues with cloned DNA containing the enhancer–*glnA* region. These workers inserted 350 bp of DNA between the enhancer and promoter to make the loops easier to see. The polymerase holoenzyme stains more darkly than NtrC in most of these electron micrographs, so we can distinguish the two proteins at the bases of the loops, just as we would predict if the two proteins interact by looping out the DNA in between. The loops were just the right size to account for the length of DNA between the enhancer and promoter.

Phage T4 provides an example of an unusual, mobile enhancer that is not defined by a set base sequence. Transcription of the late genes of T4 depends on DNA replication; no late transcription occurs until the phage DNA begins to replicate. One reason for this linkage between late transcription and DNA replication is that the late phage σ-factor (σ^{55}), like σ^{54} of *E. coli,* is defective. It cannot function without an enhancer. But the late T4 enhancer is not a fixed DNA sequence like the NtrC-binding site. Instead, it is the DNA replicating fork. The enhancer-binding protein, encoded by phage genes 44, 45, and 62, is part of the phage DNA replicating machinery. Thus, this protein migrates along with the

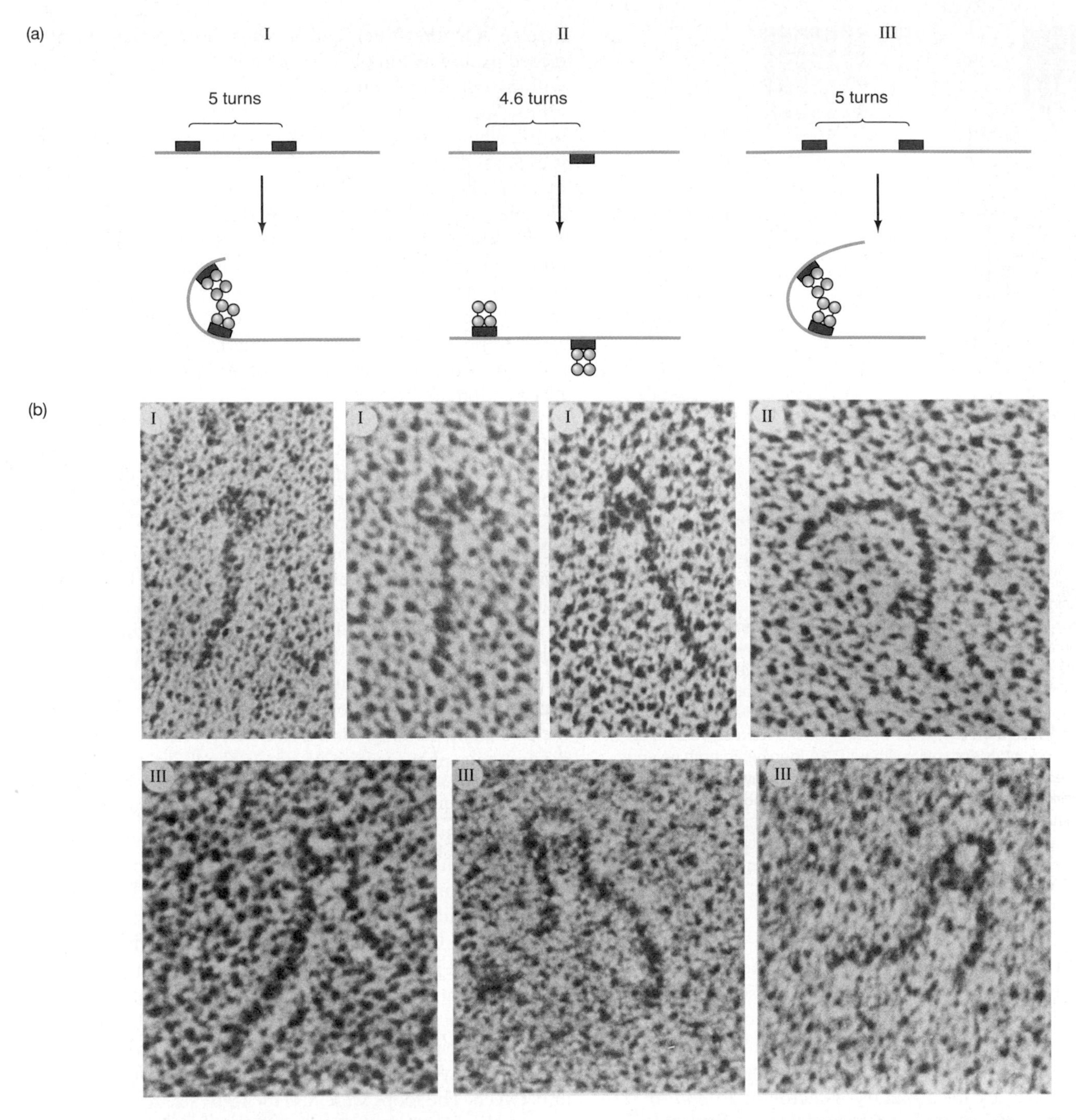

Figure 9.18 Electron microscopy of λ repressor bound to dual operators. (a) Arrangement of dual operators in three DNA molecules. In I, the two operators are five helical turns apart near the end of the DNA; in II, they are 4.6 turns apart near the end; and in III they are five turns apart near the middle. The arrows in each case point to a diagram of the expected shape of the loop due to cooperative binding of repressor to the two operators. In II, no loop should form because the two operators are not separated by an integral number of helical turns and are consequently on opposite sides of the DNA duplex. **(b)** Electron micrographs of the protein–DNA complexes. The DNA types [I, II, or III from panel (a) used in the complexes are given at the upper left of each picture. The complexes really do have the shapes predicted in panel (a). (*Source:* (a) Griffith et al., DNA loops induced by cooperative binding of lambda repressor. *Nature* 322 (21 Aug 1986) f. 2, p. 751. © Macmillan Magazines Ltd.)

replicating fork, which keeps it in contact with the moving enhancer.

One can mimic the replicating fork in vitro with a simple nick in the DNA, but the polarity of the nick is important: It works as an enhancer only if it is in the nontemplate strand. This suggests that the T4 late enhancer probably does not act by DNA looping because polarity does not matter in looping. Furthermore, unlike typical enhancers

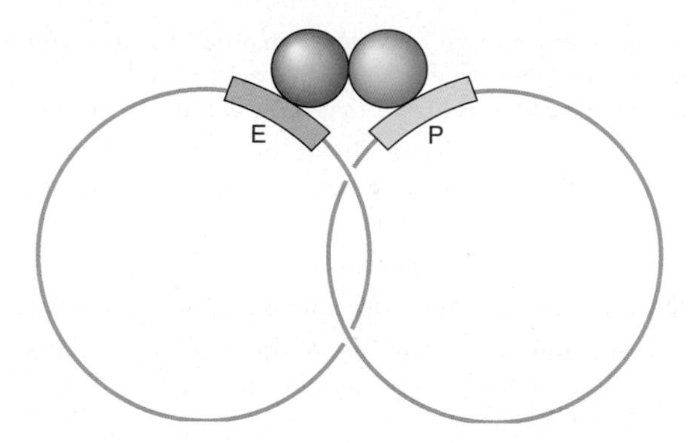

Figure 9.19 Interaction between two sites on separate but linked DNA molecules. An enhancer (E, pink) and a promoter (P, light green) lie on two separate DNA molecules that are topologically linked in a catenane (intertwined circles). Thus, even though the circles are distinct, the enhancer and promoter cannot ever be far apart, so interactions between proteins that bind to them (red and green, respectively) are facilitated.

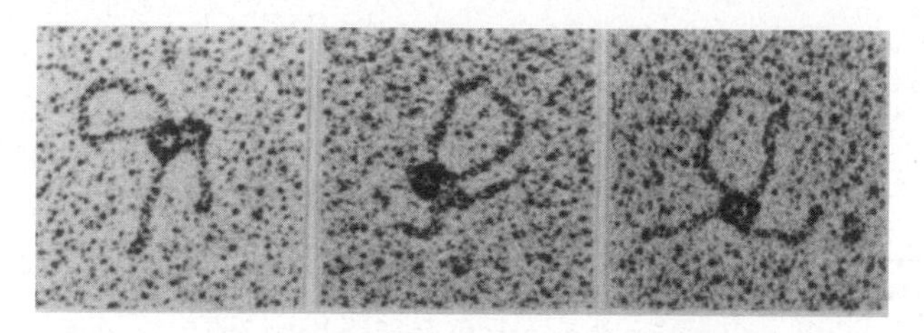

Figure 9.20 Looping the *glnA* promoter–enhancer region. Kustu, Echols, and colleagues moved the *glnA* promoter and enhancer apart by inserting a 350-bp DNA segment between them, then allowed the NtrC protein to bind to the enhancer, and RNA polymerase to bind to the promoter. When the two proteins interacted, they looped out the DNA in between, as shown in these electron micrographs. (*Source:* Su, W., S. Porter, S. Kustu, and H. Echols, DNA-looping and enhancer activity: Association between DNA-bound NtrC activator and RNA polymerase at the bacterial *glnA* promoter. *Proceedings of the National Academy of Sciences USA 87* (July 1990) f. 4, p. 5507.)

such as the *glnA* enhancer, the T4 late enhancer must be on the same DNA molecule as the promoters it controls. It does not function *in trans* as part of a catenane. This argues against a looping mechanism.

SUMMARY The *E. coli glnA* gene is an example of a prokaryotic gene that depends on an enhancer for its transcription. The enhancer binds the NtrC protein, which interacts with polymerase bound to the promoter at least 70 bp away. Hydrolysis of ATP by NtrC allows the formation of an open promoter complex so transcription can take place. The two proteins appear to interact by looping out the DNA in between. The phage T4 late enhancer is mobile; it is part of the phage DNA-replication apparatus. Because this enhancer must be on the same DNA molecule as the late promoters, it probably does not act by DNA looping.

SUMMARY

The repressors of the λ-like phages have recognition helices that fit sideways into the major groove of the operator DNA. Certain amino acids on the DNA side of the recognition helix make specific contact with bases in the operator, and these contacts determine the specificity of the protein–DNA interactions. Changing these amino acids can change the specificity of the repressor. The λ repressor and Cro protein share affinity for the same operators, but they have microspecificities for O_R1 or O_R3, determined by interactions between different amino acids in the recognition helices of the two proteins and base pairs in the different operators.

The cocrystal structure of a λ repressor fragment with an operator fragment shows many details about how the protein and DNA interact. The most important contacts occur in the major groove, where amino acids on the recognition helix, and other amino acids, make hydrogen bonds with the edges of DNA bases and with the DNA backbone. Some of these hydrogen bonds are stabilized by hydrogen bond networks involving two amino acids and two or more sites on the DNA. The structure derived from the cocrystal is in almost complete agreement with previous biochemical and genetic data.

X-ray crystallography of a phage 434 repressor-fragment/operator-fragment complex shows probable hydrogen bonding between amino acid residues in the recognition helix and base pairs in the repressor. It also reveals a potential van der Waals contact between an amino acid in the recognition helix and a base in the operator. The DNA in the complex deviates significantly from its normal regular shape. It bends somewhat to accommodate the necessary base/amino acid contacts. Moreover, the central part of the helix, between the two half-sites, is wound extra tightly, and the outer parts are wound more loosely than normal. The base sequence of the operator facilitates these departures from normal DNA shape.

The *trp* repressor requires tryptophan to force the recognition helices of the repressor dimer into the proper position for interacting with the *trp* operator.

A DNA-binding protein can interact with the major or minor groove of the DNA (or both). The four different base pairs present four different hydrogen-bonding profiles to amino acids approaching either the major or minor DNA groove, so a DNA-binding protein can recognize base pairs in the DNA even though the two strands do not separate.

Multimeric DNA-binding proteins have an inherently higher affinity for binding sites on DNA than do multiple monomeric proteins that bind independently of one another. The advantage of multimeric proteins is that they can bind cooperatively to DNA.

When λ operators are separated by an integral number of helical turns, the DNA in between can loop out to allow cooperative binding. When the operators are separated by a nonintegral number of helical turns, the proteins have to bind to opposite faces of the DNA double helix, so no cooperative binding can take place.

The *E. coli glnA* gene is an example of a bacterial gene that depends on an enhancer for its transcription. The enhancer binds the NtrC protein, which interacts with polymerase bound to the promoter at least 70 bp away. Hydrolysis of ATP by NtrC allows the formation of an open promoter complex so transcription can take place. The two proteins appear to interact by looping out the DNA in between. The phage T4 late enhancer is mobile; it is part of the phage DNA-replication apparatus. Because this enhancer must be on the same DNA molecule as the late promoters, it probably does not act by DNA looping.

REVIEW QUESTIONS

1. Draw a rough diagram of a helix-turn-helix domain interacting with a DNA double helix.
2. Describe and give the results of an experiment that shows which amino acids are important in binding between λ-like phage repressors and their operators. Present two methods of assaying the binding between the repressors and operators.
3. In general terms, what accounts for the different preferences of λ repressor and Cro for the three operator sites?
4. Glutamine and asparagine side chains tend to make what kind of bonds with DNA?
5. Methylene and methyl groups on amino acids tend to make what kind of bonds with DNA?
6. What is meant by the term *hydrogen bond network* in the context of protein–DNA interactions?
7. Draw a rough diagram of the "reading head" model to show the difference in position of the recognition helix of the *trp* repressor and aporepressor, with respect to the *trp* operator.
8. Draw a rough diagram of the "salami sandwich" model to explain how adding tryptophan to the *trp* aporepressor causes a shift in conformation of the protein.
9. In one sentence, contrast the orientations of the λ and *trp* repressors relative to their respective operators.
10. Explain the fact that protein oligomers (dimers or tetramers) bind more successfully to DNA than monomeric proteins do.
11. Use a diagram to explain the alternating pattern of resistance and elevated sensitivity to DNase in the DNA between two separated binding sites when two proteins bind cooperatively to these sites.
12. Describe and give the results of a DNase footprinting experiment that shows that λ repressor dimers bind cooperatively to two operators separated by an integral number of DNA double-helical turns, but noncooperatively to two operators separated by a nonintegral number of turns.
13. Describe and give the results of an electron microscopy experiment that shows the same thing as the experiment in the preceding question.
14. In what way is σ^{54} defective?
15. What substances supply the missing function to σ^{54}?
16. Describe and give the results of an experiment that shows that DNA looping is involved in the enhancement of the *E. coli glnA* locus.
17. In what ways is the enhancer for phage T4 σ^{55} different from the enhancer for the *E. coli* σ^{54}?

ANALYTICAL QUESTIONS

1. An asparagine in a DNA-binding protein makes an important hydrogen bond with a cytosine in the DNA. Changing this glutamine to alanine prevents formation of this hydrogen bond and blocks the DNA–protein interaction. Changing the cytosine to thymine restores binding to the mutant protein. Present a plausible hypothesis to explain these findings.
2. You have the following working hypothesis: To bind well to a DNA-binding protein, a DNA target site must twist less tightly and widen the narrow groove between base pairs 4 and 5. Suggest an experiment to test your hypothesis.
3. Draw a T–A base pair. Based on that structure, draw a line diagram indicating the relative positions of the hydrogen bond acceptor and donor groups in the major and minor grooves. Represent the horizontal axis of the base pair by two segments of a horizontal line, and the relative horizontal positions of the hydrogen bond donors and acceptors by vertical lines. Let the lengths of the vertical lines indicate the relative vertical positions of the acceptors and donors. What relevance does this diagram have for a protein that interacts with this base pair?

SUGGESTED READINGS

General References and Reviews

Geiduschek, E.P. 1997. Paths to activation of transcription. *Science* 275:1614–16.

Kustu, S., A.K. North, and D.S. Weiss. 1991. Prokaryotic transcriptional enhancers and enhancer-binding proteins. *Trends in Biochemical Sciences* 16:397–402.

Schleif, R. 1988. DNA binding by proteins. *Science* 241:1182–87.

Research Articles

Aggarwal, A.K., D.W. Rodgers, M. Drottar, M. Ptashne, and S.C. Harrison. 1988. Recognition of a DNA operator by the repressor of phage 434: A view at high resolution. *Science* 242:899–907.

Griffith, J., A. Hochschild, and M. Ptashne. 1986. DNA loops induced by cooperative binding of λ repressor. *Nature* 322:750–52.

Herendeen, D.R., G.A. Kassavetis, J. Barry, B.M. Alberts, and E.P. Geiduschek. 1990. Enhancement of bacteriophage T4 late transcription by components of the T4 DNA replication apparatus. *Science* 245:952–58.

Hochschild, A., J. Douhann III, and M. Ptashne. 1986. How λ repressor and λ cro distinguish between O_R1 and O_R3. *Cell* 47:807–16.

Hochschild, A. and M. Ptashne. 1986. Cooperative binding of λ repressors to sites separated by integral turns of the DNA helix. *Cell* 44:681–87.

Jordan, S.R. and C.O. Pabo. 1988. Structure of the lambda complex at 2.5 Å resolution: Details of the repressor–operator interactions. *Science* 242:893–99.

Popham, D.L., D. Szeto, J. Keener, and S. Kustu. 1989. Function of a bacterial activator protein that binds to transcriptional enhancers. *Science* 243:629–35.

Sauer, R.T., R.R. Yocum, R.F. Doolittle, M. Lewis, and C.O. Pabo. 1982. Homology among DNA-binding proteins suggests use of a conserved super-secondary structure. *Nature* 298:447–51.

Schevitz, R.W., Z. Otwinowski, A. Joachimiak, C.L. Lawson, and P. B. Sigler. 1985. The three-dimensional structure of *trp* repressor. *Nature* 317:782–86.

Su, W., S. Porter, S. Kustu, and H. Echols. 1990. DNA looping and enhancer activity: Association between DNA-bound NtrC activator and RNA polymerase at the bacterial *glnA* promoter. *Proceedings of the National Academy of Sciences USA* 87:5504–8.

Wharton, R.P. and M. Ptashne. 1985. Changing the binding specificity of a repressor by redesigning an α-helix. *Nature* 316:601–5.

Zhang, R.-g., A. Joachimiak, C.L. Lawson, R.W. Schevitz, Z. Otwinowski, and P.B. Sigler. 1987. The crystal structure of *trp* aporepressor at 1.8 Å shows how binding tryptophan enhances DNA affinity. *Nature* 327:591–97.

CHAPTER 10

Eukaryotic RNA Polymerases and Their Promoters

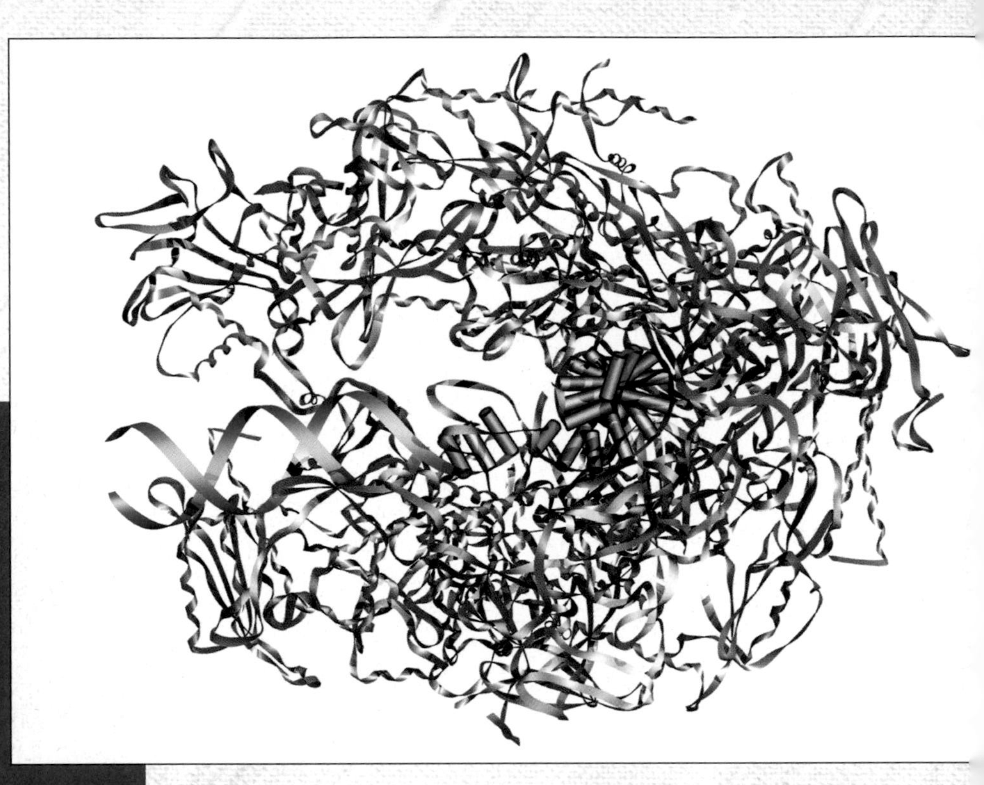

Computer-generated model of yeast Pol II Δ4/7 protein with RNA–DNA hybrid in the active site. *© David A. Bushnell, Kenneth D. Westover, and Roger D. Kornberg.*

In Chapter 6 we learned that bacteria have only one RNA polymerase, which makes all three of the familiar RNA types: mRNA, rRNA, and tRNA. True, the polymerase can switch σ-factors to meet the demands of a changing environment, but the core enzyme remains essentially the same. Quite a different situation prevails in the eukaryotes. In this chapter we will see that three distinct RNA polymerases occur in the nuclei of eukaryotic cells. Each of these is responsible for transcribing a separate set of genes, and each recognizes a different kind of promoter.

10.1 Multiple Forms of Eukaryotic RNA Polymerase

Several early studies suggested that at least two RNA polymerases operate in eukaryotic nuclei: one to transcribe the major ribosomal RNA genes (those coding for the 28S, 18S, and 5.8S rRNAs in vertebrates), and one or more to transcribe the rest of the nuclear genes.

To begin with, the ribosomal genes are different in several ways from other nuclear genes: (1) They have a different base composition from that of other nuclear genes. For example, rat rRNA genes have a GC content of 60%, but the rest of the DNA has a GC content of only 40%. (2) They are unusually repetitive; depending on the organism, each cell contains from several hundred to over 20,000 copies of the rRNA gene. (3) They are found in a different compartment—the nucleolus—than the rest of the nuclear genes. These and other considerations suggested that at least two RNA polymerases were operating in eukaryotic nuclei. One of these synthesized rRNA in the nucleolus, and the other synthesized other RNA in the nucleoplasm (the part of the nucleus outside the nucleolus).

Separation of the Three Nuclear Polymerases

Robert Roeder and William Rutter showed in 1969 that eukaryotes have not two, but three different RNA polymerases. Furthermore, these three enzymes have distinct roles in the cell. These workers separated the three enzymes by DEAE-Sephadex ion-exchange chromatography (Chapter 5).

They named the three peaks of polymerase activity in order of their emergence from the ion-exchange column: **RNA polymerase I, RNA polymerase II,** and **RNA polymerase III** (Figure 10.1). The three enzymes have different properties besides their different behaviors on DEAE-Sephadex chromatography. For example, they have different responses to ionic strength and divalent metals. More importantly, they have distinct roles in transcription: Each makes different kinds of RNA.

Roeder and Rutter next looked in purified nucleoli and nucleoplasm to see if these subnuclear compartments were enriched in the appropriate polymerases. Figure 10.2 shows that polymerase I is indeed located primarily in the nucleolus, and polymerases II and III are found in the nucleoplasm. This made it very likely that polymerase I is the rRNA-synthesizing enzyme, and that polymerases II and III make some other kinds of RNA.

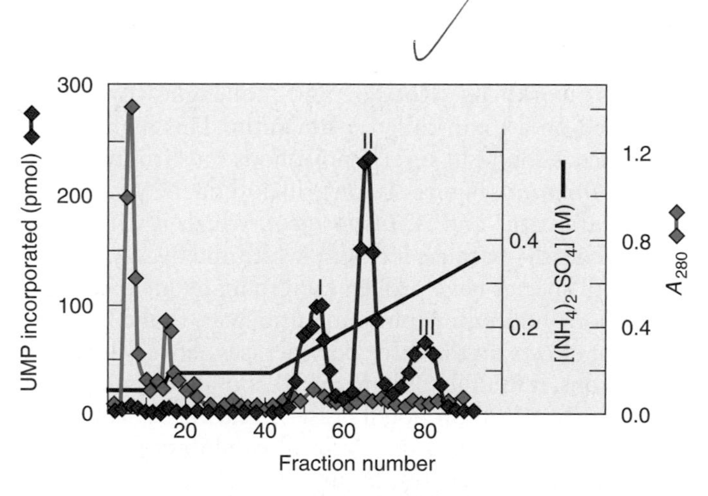

Figure 10.1 Separation of eukaryotic RNA polymerases. Roeder and Rutter subjected extracts from sea urchin embryos to DEAE-Sephadex chromatography. Green, protein measured by A_{280}; red, RNA polymerase activity measured by incorporation of labeled UMP into RNA; blue, ammonium sulfate concentration. (*Source:* Adapted from Roeder, R.G. and W.J. Rutter, Multiple forms of DNA dependent RNA polymerase in eukaryotic organisms. *Nature* 224:235, 1969.)

(a)

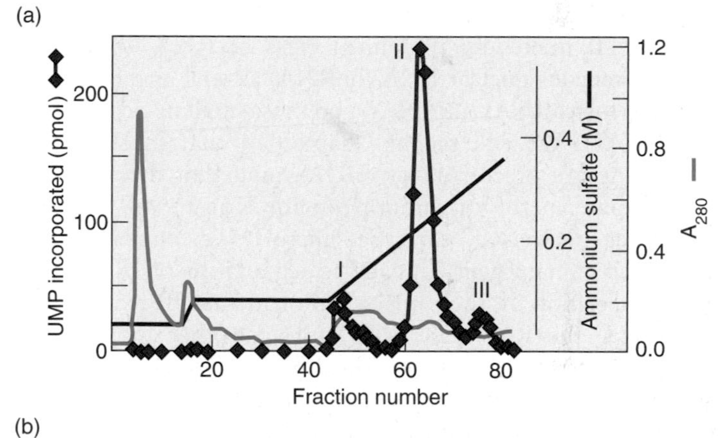

(b)

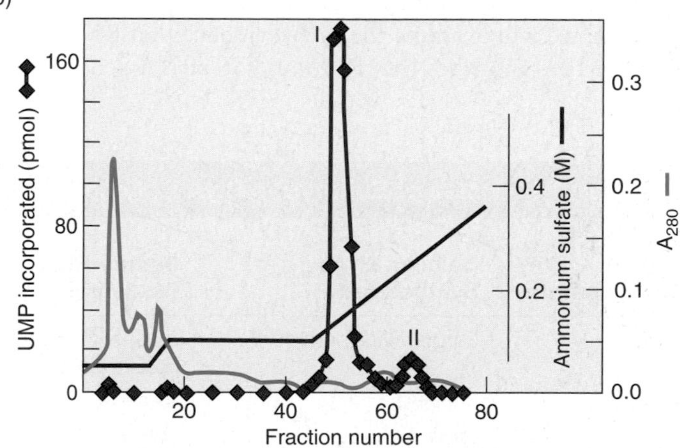

Figure 10.2 Cellular localization of the three rat liver RNA polymerases. Roeder and Rutter subjected the polymerases found in the nucleoplasmic fraction **(a)** or nucleolar fraction **(b)** of rat liver to DEAE-Sephadex chromatography as described in Figure 10.1. Colors have the same meanings as in Figure 10.1. (*Source:* Adapted from Roeder, R.G. and W.J. Rutter, Specific nucleolar and nucleoplasmic RNA polymerases, *Proceedings of the National Academy of Sciences* 65(3):675–82, March 1970.)

SUMMARY Eukaryotic nuclei contain three RNA polymerases that can be separated by ion-exchange chromatography. RNA polymerase I is found in the nucleolus; the other two polymerases (RNA polymerases II and III) are located in the nucleoplasm. The location of RNA polymerase I in the nucleolus suggests that it transcribes the rRNA genes.

The Roles of the Three RNA Polymerases

How do we know that the three RNA polymerases have different roles in transcription? The clearest evidence for these roles has come from studies in which the purified polymerases were shown to transcribe certain genes, but not others, in vitro. Such studies have demonstrated that the three RNA polymerases have the following specificities (Table 10.1): Polymerase I makes the large rRNA precursor. In mammals, this precursor has a sedimentation coefficient of 45S and is processed to the 28S, 18S, and 5.8S mature rRNAs. Polymerase II makes an ill-defined class of RNA known as **heterogeneous nuclear RNA (hnRNA)** as well as the precursors of **microRNAs (miRNAs)** and most **small nuclear RNAs (snRNAs).** We will see in Chapter 14 that most of the hnRNAs are precursors of mRNAs and that the snRNAs participate in the maturation of hnRNAs to mRNAs. In Chapter 16, we will learn that microRNAs control the expression of many genes by causing degradation of, or limiting the translation of, their mRNAs. Polymerase III makes precursors to the tRNAs, 5S rRNA, and some other small RNAs.

However, even before cloned genes and eukaryotic in vitro transcription systems were available, we had evidence to support most of these transcription assignments. In this section, we will examine the early evidence that RNA polymerase III transcribes the tRNA and 5S rRNA genes.

Table 10.1 Roles of Eukaryotic RNA Polymerases

RNA Polymerase	Cellular RNAs Synthesized	Mature RNA (Vertebrate)
I	Large rRNA precursor	28S, 18S, and 5.8S rRNAs
II	hnRNAs	mRNAs
	snRNAs	snRNAs
	miRNA precursors	miRNAs
III	5S rRNA precursor	5S rRNA
	tRNA precursors	tRNAs
	U6 snRNA (precursor?)	U6 snRNA
	7SL RNA (precursor?)	7SL RNA
	7SK RNA (precursor?)	7SK RNA

(a)

(b)

Figure 10.3 Alpha-amanitin. (a) *Amanita phalloides* ("the death cap"), one of the deadly poisonous mushrooms that produce α-amanitin. **(b)** Structure of α-amanitin. (*Source:* (a) Arora, D. *Mushrooms Demystified* 2e, 1986, Plate 50 (Ten Speed Press).)

This work, by Roeder and colleagues in 1974, depended on a toxin called **α-amanitin.** This highly toxic substance is found in several poisonous mushrooms of the genus *Amanita* (Figure 10.3a), including *A. phalloides,* "the death cap," and *A. bisporigera,* which is called "the angel of death" because it is pure white and deadly poisonous. Both species have proven fatal to many inexperienced mushroom hunters. Alpha-amanitin was found to have different effects on the three polymerases. At very low concentrations, it inhibits polymerase II completely while having no effect at all on polymerases I and III. At 1000-fold higher concentrations, the toxin also inhibits polymerase III from most eukaryotes (Figure 10.4).

The plan of the experiment was to incubate mouse cell nuclei in the presence of increasing concentrations of α-amanitin, then to electrophorese the transcripts to observe the effect of the toxin on the synthesis of small RNAs. Figure 10.5 reveals that high concentrations of α-amanitin inhibited the synthesis of both 5S rRNA and 4S tRNA

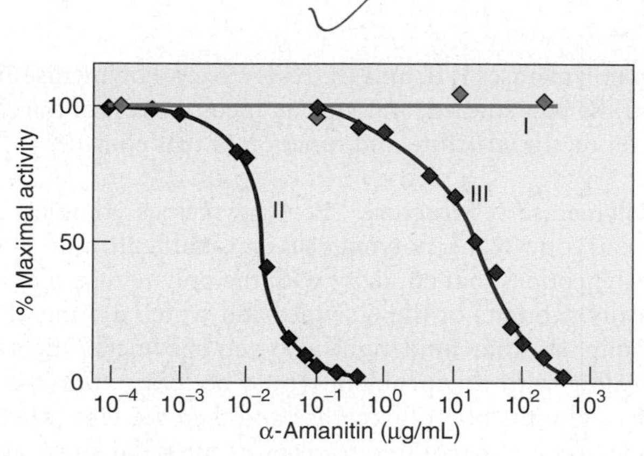

Figure 10.4 Sensitivity of purified RNA polymerases to α-amanitin. Weinmann and Roeder assayed RNA polymerases I (green), II (blue), and III (red) with increasing concentrations of α-amanitin. Polymerase II was 50% inhibited by about 0.02 μg/mL of the toxin, whereas polymerase III reached 50% inhibition only at about 20 μg/mL of toxin. Polymerase I retained full activity even at an α-amanitin concentration of 200 μg/mL. (*Source:* Adapted From R. Weinmann and R.G. Roeder, Role of DNA-dependent RNA polymerase III in the transcription of the tRNA and 5S RNA genes, *Proceedings of the National Academy of Sciences USA* 71(5):1790–4, May 1974.)

precursor. Moreover, this pattern of inhibition of 5S rRNA and tRNA precursor synthesis matched the pattern of inhibition of RNA polymerase III: They both were about half-inhibited at 10 μg/mL of α-amanitin. Therefore, these data support the hypothesis that RNA polymerase III makes these two kinds of RNA. (Actually, polymerase III synthesizes the 5S rRNA as a slightly larger precursor, but this experiment did not distinguish the precursor from the mature 5S rRNA.) Polymerase III also makes a variety of other small cellular and viral RNAs. These include **U6 snRNA,** a small RNA that participates in RNA splicing (Chapter 14); **7SL RNA,** a small RNA involved in signal peptide recognition in the synthesis of secreted proteins; 7SK RNA, a small nuclear RNA that binds and inhibits the class II transcription elongation factor P-TEFb, the adenovirus VA (virus-associated) RNAs; and the Epstein–Barr virus EBER2 RNA.

Similar experiments were performed to identify the genes transcribed by RNA polymerases I and II. But these studies were not as easy to interpret and they have been confirmed by much more definitive in vitro studies.

The sequencing of the first plant genome (*Arabidopsis thaliana,* or thale cress) in 2000 led to the discovery of two

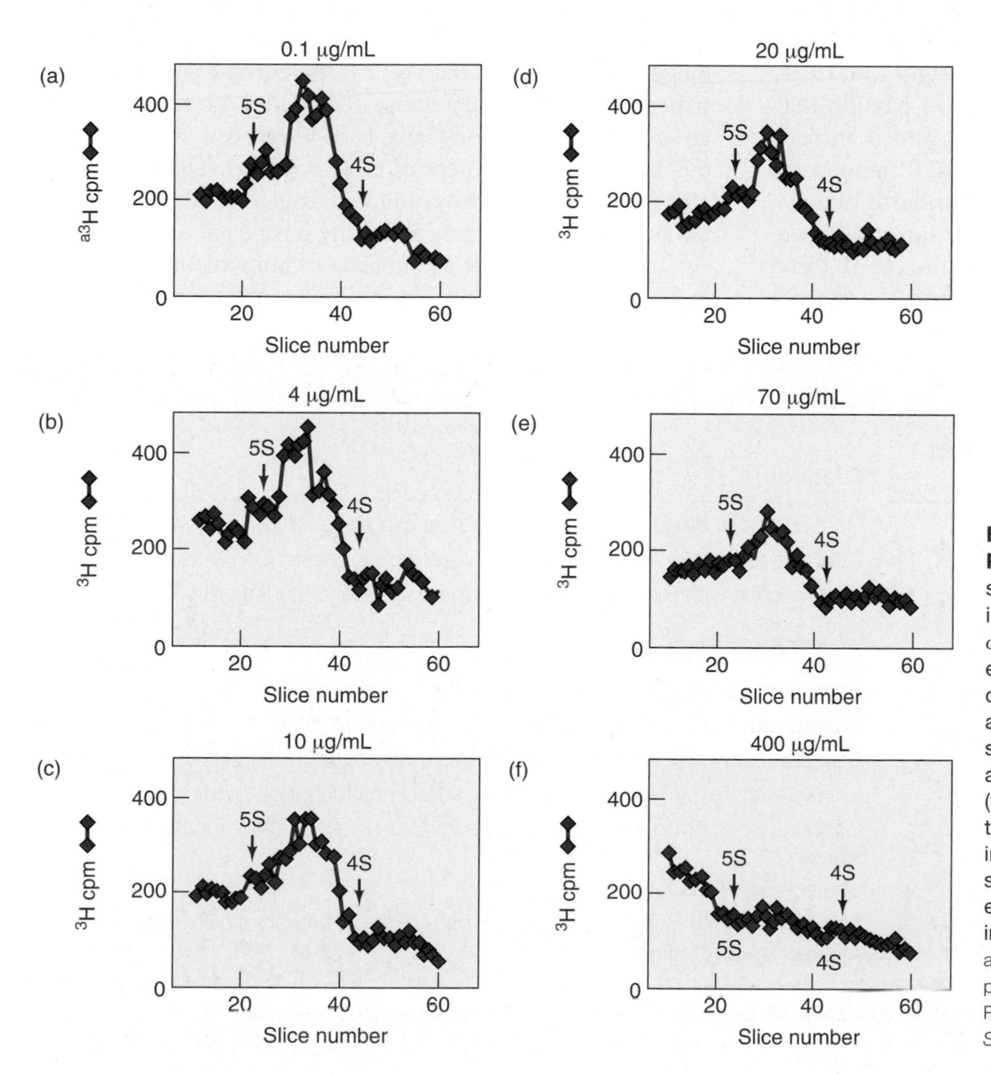

Figure 10.5 Effect of α-amanitin on small RNA synthesis. Weinmann and Roeder synthesized labeled RNA in isolated nuclei in the presence of increasing amounts of α-amanitin (concentration given at the top of each panel). The small labeled RNAs leaked out of the nuclei and were found in the supernatant after centrifugation. The researchers then subjected these RNAs to PAGE, sliced the gel, and determined the radioactivity in each slice (red). They also ran markers (5S rRNA and 4S tRNA) in adjacent lanes of the same gel. The inhibition of 5S rRNA and 4S tRNA precursor synthesis by α-amanitin closely parallels the effect of the toxin on polymerase III, determined in Figure 10.4. (*Source:* Adapted from R. Weinmann and R.G. Roeder, Role of DNA-dependent RNA polymerase III in the transcription of the tRNA and 5S RNA genes, *Proceedings of the National Academy of Sciences USA* 71(5):1790–4, May 1974.)

additional RNA polymerases in flowering plants: **RNA polymerase IV** and **RNA polymerase V.** These enzymes produce noncoding RNAs that are involved in a mechanism that silences genes. (Similar transcriptional tasks are performed by polymerase II in other eukaryotes, and indeed the largest subunits of both polymerases IV and V are evolutionarily related to the largest subunit of polymerase II.) We will discuss such gene silencing mechanisms in more detail in Chapter 16.

SUMMARY The three nuclear RNA polymerases have different roles in transcription. Polymerase I makes the large precursor to the rRNAs (5.8S, 18S, and 28S rRNAs in vertebrates). Polymerase II makes hnRNAs, which are precursors to mRNAs, miRNA precursors, and most of the snRNAs. Polymerase III makes the precursors to 5S rRNA, the tRNAs, and several other small cellular and viral RNAs.

RNA Polymerase Subunit Structures

The first subunit structures for a eukaryotic RNA polymerase (polymerase II) were reported independently by Pierre Chambon and Rutter and their colleagues in 1971, but they were incomplete. We should note in passing that Chambon named his three polymerases A, B, and C, instead of I, II, and III, respectively. However, the I, II, III nomenclature of Roeder and Rutter has become the standard. We now have very good structural information on all three polymerases from a variety of eukaryotes. The structures of all three polymerases are quite complex, with 14, 12, and 17 subunits in polymerases I, II, and III, respectively. Polymerase II is by far the best studied, and we will focus the rest of our discussion on the structure and function of that enzyme.

Polymerase II Structure For enzymes as complex as the eukaryotic RNA polymerases it is difficult to tell which polypeptides that copurify with the polymerase activity are really subunits of the enzymes and which are merely contaminants that bind tightly to the enzymes. One way of dealing with this problem would be to separate the putative subunits of a polymerase and then see which polypeptides are really required to reconstitute polymerase activity. Although this strategy worked beautifully for the prokaryotic polymerases, no one has yet been able to reconstitute a eukaryotic nuclear polymerase from its separate subunits. Thus, one must try a different tack.

Another way of approaching this problem is to find the genes for all the putative subunits of a polymerase, mutate them, and determine which are required for activity. This has been accomplished for one enzyme: polymerase II of baker's yeast, *Saccharomyces cerevisiae.* Several investigators used traditional methods to purify yeast polymerase II to homogeneity and identified 10 putative subunits. Later, some of the same scientists discovered two other subunits that had been hidden in the earlier analyses, so the current concept of the structure of yeast polymerase II includes 12 subunits. The genes for all 12 subunits have been sequenced, which tells us the amino acid sequences of their products. The genes have also been systematically mutated, and the effects of these mutations on polymerase II activity have been observed.

Table 10.2 lists the 12 subunits of human and yeast polymerase II, along with their molecular masses and some of

Table 10.2 Human and Yeast RNA Polymerase II Subunits

Subunit	Yeast Gene	Yeast Protein (kD)	Features
hRPB1	*RPB1*	192	Contains CTD; binds DNA; involved in start site selection; β′ ortholog
hRPB2	*RPB2*	139	Contains active site; involved in start site selection, elongation rate; β ortholog
hRPB3	*RPB3*	35	May function with Rpb11 as ortholog of the α dimer of prokaryotic RNA polymerase
hRPB4	*RPB4*	25	Subcomplex with Rpb7; involved in stress response
hRPB5	*RPB5*	25	Shared with Pol I, II, III; target for transcriptional activators
hRPB6	*RPB6*	18	Shared with Pol I, II, III; functions in assembly and stability
hRPB7	*RPB7*	19	Forms subcomplex with Rpb4 that preferentially binds during stationary phase
hRPB8	*RPB8*	17	Shared with Pol I, II, III; has oligonucleotide/oligosaccharide-binding domain
hRPB9	*RPB9*	14	Contains zinc ribbon motif that may be involved in elongation: functions in start site selection
hRPB10	*RPB10*	8	Shared with Pol I, II, III
hRPB11	*RPB11*	14	May function with Rpb3 as ortholog of the α dimer of prokaryotic RNA polymerase
hRPB12	*RPB12*	8	Shared with Pol I, II, III

Source: ANNUAL REVIEW OF GENETICS. Copyright © 2002 by ANNUAL REVIEWS. Reproduced with permission of ANNUAL REVIEWS in the format textbook via Copyright Clearance Center.

their characteristics. Each of these polypeptides is encoded in a single gene in the yeast and human genomes. The names of these polymerase subunits, Rpb1, and so on, derive from the names of the genes that encode them (*RPB1,* and so on). Note the echo of the Chambon nomenclature in the name *RPB,* which stands for RNA polymerase B (or II).

How do the structures of polymerases I and III compare with this polymerase II structure? First, all the polymerase structures are complex—even more so than the structures of the bacterial polymerases. Second, all the structures are similar in that each contains two large (greater than 100 kD) subunits, plus a variety of smaller subunits. In this respect, these structures resemble those of the prokaryotic core polymerases, which contain two high-molecular-mass subunits (β and β′) plus three low-molecular-mass subunits (two α's and an ω). In fact, as we will see later in this chapter, an evolutionary relationship is evident between three of the prokaryotic core polymerase subunits and three of the subunits of all of the eukaryotic polymerases. In other words, the three eukaryotic polymerases are related to the prokaryotic polymerase and to one another.

A third message from Table 10.2 is that the three yeast nuclear polymerases have several subunits in common. In fact, five such *common subunits* exist. In the polymerase II structure, these are called Rpb5, Rpb6, Rpb8, Rpb10, and Rpb12. These are identified on the right in Table 10.2.

Richard Young and his coworkers originally identified 10 polypeptides that are authentic polymerase II subunits, or at least tightly bound contaminants. The method they used is called **epitope tagging** (Figure 10.6), in which they attached a small foreign epitope to one of the yeast polymerase II subunits (Rpb3) by engineering its gene. Then they introduced this gene into yeast cells lacking a functional Rpb3 gene, labeled the cellular proteins with either ^{35}S or ^{32}P, and used an antibody directed against the foreign epitope to precipitate the whole enzyme. After immunoprecipitation, they separated the labeled polypeptides of the precipitated protein by SDS-PAGE and detected them by autoradiography. Figure 10.7a presents the results. This single-step purification method yielded essentially pure polymerase II with 10 apparent subunits. We can also see a few minor polypeptides, but they are equally visible in the control in which wild-type enzyme, with no epitope tag, was used. Therefore, they are not polymerase-associated. Figure 10.7b shows a later SDS-PAGE analysis of the same polymerase, performed by Roger Kornberg and colleagues, which distinguished 12 subunits. Rpb11 had coelectrophoresed with Rpb9, and Rpb12 had coelectrophoresed with Rpb10, so both Rpb11 and Rpb12 had been missed in the earlier experiments.

Because Young and colleagues already knew the amino acid compositions of all 10 original subunits, the relative labeling of each polypeptide with ^{35}S-methionine gave them a good estimate of the stoichiometries of subunits, which are listed in Table 10.3. Figure 10.7a also shows us that two

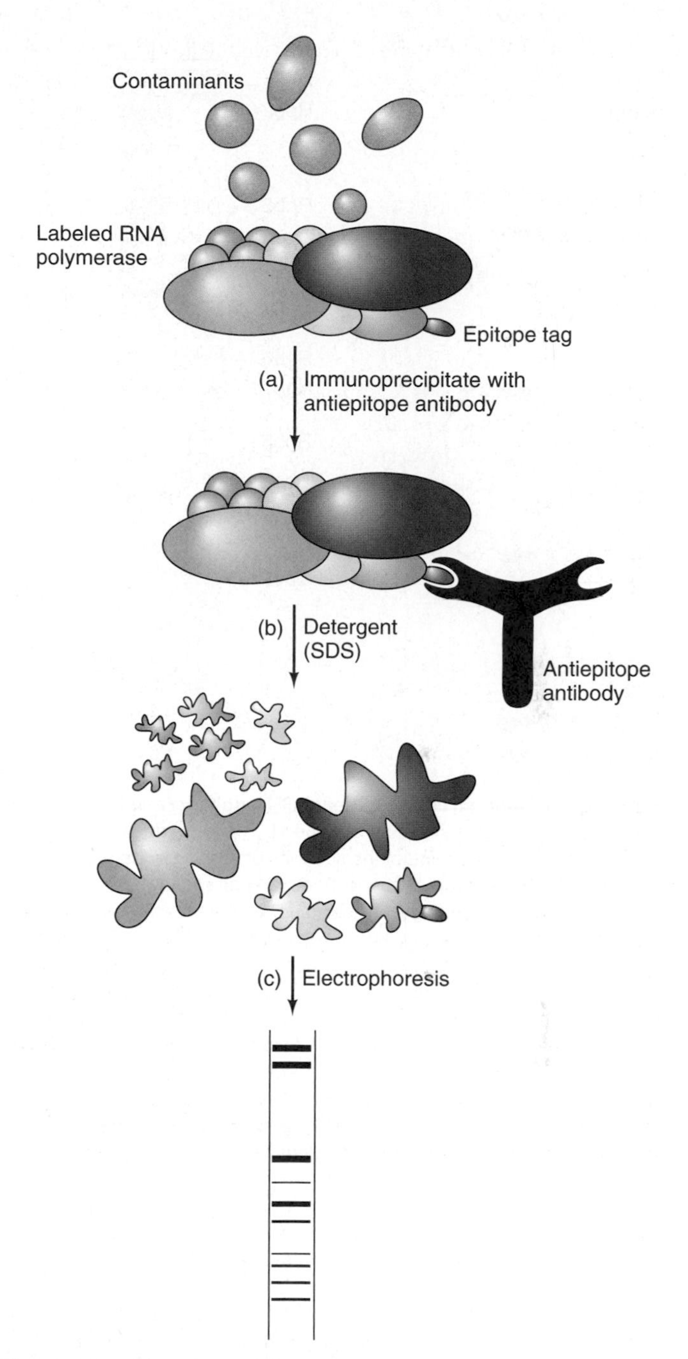

Figure 10.6 Principle of epitope tagging. An extra domain (an epitope tag, red) has been added genetically to one subunit (Rpb3) of the yeast RNA polymerase II. All the other subunits are normal, and assemble with the altered Rpb3 subunit to form an active polymerase. This polymerase has also been labeled by growing cells in labeled amino acids. **(a)** Add an antibody directed against the epitope tag, which immunoprecipitates the whole RNA polymerase, separating it from contaminating proteins (gray). This gives very pure polymerase in just one step. **(b)** Add the strong detergent SDS, which separates and denatures the subunits of the purified polymerase. **(c)** Electrophorese the denatured subunits of the polymerase to yield the electropherogram at bottom.

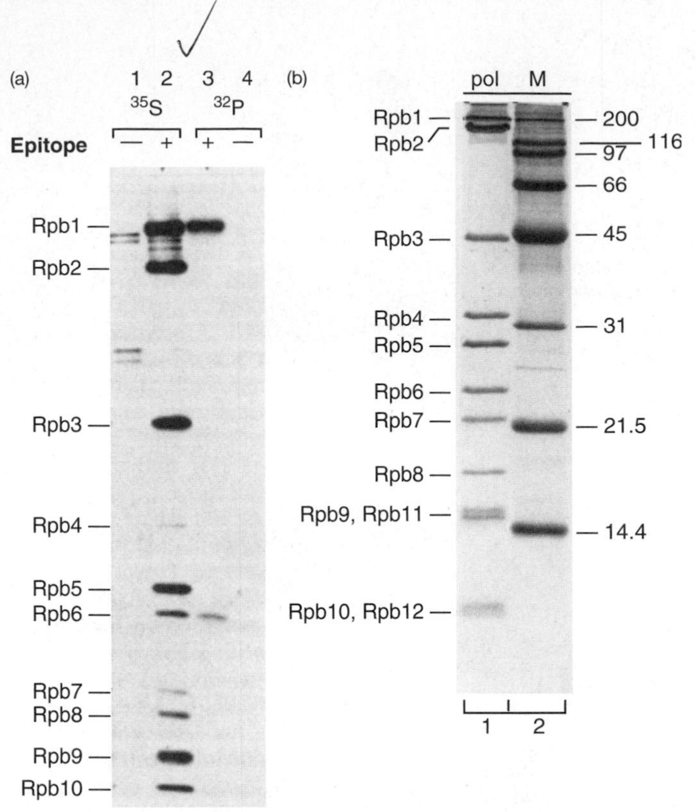

Figure 10.7 Subunit structure of yeast RNA polymerase II. **(a)** Apparent 10-subunit structure obtained by epitope tagging. Young and colleagues endowed one of the subunits of yeast polymerase II (Rpb3) with an extra group of amino acids (an epitope tag) by substituting a gene including the codons for this tag for the usual yeast *RPB3* gene. Then they labeled these engineered yeast cells with either [^{35}S]methionine to label all the polymerase subunits, or [γ-^{32}P]ATP to label the phosphorylated subunits only. They immunoprecipitated the labeled protein with an antibody directed against the epitope tag and electrophoresed the products. Lane 1, ^{35}S-labeled protein from wild-type yeast without the epitope tag; lane 2, ^{35}S-labeled protein from yeast having the epitope tag on Rpb3; lane 3, ^{32}P-labeled protein from yeast with the epitope tag; lane 4, ^{32}P-labeled protein from wild-type yeast. The polymerase II subunits are identified at left. **(b)** Apparent 12-subunit structure obtained by multistep purification including immunoprecipitation. Kornberg and colleagues immunoprecipitated yeast RNA polymerase II and subjected it to SDS-PAGE (lane 1), alongside molecular mass markers (lane 2). The marker molecular masses are given at right, and the polymerase II subunits are identified at left. Notice that Rpb9 and Rpb11 almost comigrate, as do Rpb10 and Rpb12. (*Sources:* (a) Kolodziej, P.A., N. Woychik, S.-M. Liao, and R. Young, RNA polymerase II subunit composition, stoichiometry, and phosphorylation, *Molecular and Cellular Biology* 10 (May 1990) p. 1917, f. 2. American Society for Microbiology. (b) Sayre, M.H., H. Tschochner, and R.D. Kornberg, Reconstitution of transcription with five purified initiation factors and RNA polymerase II from *Saccharomyces cerevisiae. Journal of Biological Chemistry.* 267 (15 Nov 1992) p. 23379, f. 3b. American Society for Biochemistry and Molecular Biology.)

polymerase II subunits are phosphorylated, because they were labeled by [γ-^{32}P]ATP. These phosphoproteins are subunits Rpb1 and Rpb6. Rpb2 is also phosphorylated, but at such a low level that Figure 10.7a does not show it.

Core Subunits These three polypeptides, Rpb1, Rpb2, and Rpb3, are all absolutely required for enzyme activity. They are homologous to the β'-, β-, and α-subunits, respectively, of *E. coli* RNA polymerase.

How about functional relationships? We have seen (Chapter 6) that the *E. coli* β'-subunit binds DNA, and so does Rpb1. Chapter 6 also showed that the *E. coli* β-subunit is at or near the nucleotide-joining active site of the enzyme. Using the same experimental design, André Sentenac and his colleagues have established that Rpb2 is also at or near the active site of RNA polymerase II. The functional similarity among the second largest subunits in all three nuclear RNA polymerases, as well as prokaryotic polymerases, is mirrored by structural similarities among these same subunits, as revealed by the sequences of their genes.

Although Rpb3 does not closely resemble the *E. coli* α-subunit, there is one 20-amino-acid region of great similarity. In addition, the two subunits are about the same size and have the same stoichiometry, two monomers per holoenzyme. Furthermore, the same kinds of polymerase assembly defects are seen in *RPB3* mutants as in *E. coli* α-subunit mutants. All of these factors suggest that Rpb3 and *E. coli* α are homologous.

Common Subunits Five subunits—Rpb5, Rpb6, Rpb8, Rpb10, and Rpb12—are found in all three yeast nuclear polymerases. We know little about the functions of these subunits, but the fact that they are found in all three polymerases suggests that they play roles fundamental to the transcription process.

SUMMARY The genes encoding all 12 RNA polymerase II subunits in yeast have been sequenced and subjected to mutation analysis. Three of the subunits resemble the core subunits of bacterial RNA polymerases in both structure and function, five are found in all three nuclear RNA polymerases, two are not required for activity, at least at 37°C, and two fall into none of these three categories. Two subunits, especially Rpb1, are heavily phosphorylated, and one is lightly phosphorylated.

Heterogeneity of the Rpb1 Subunit The very earliest studies on RNA polymerase II structure showed some heterogeneity in the largest subunit. Figure 10.8 illustrates this phenomenon in polymerase II from a mouse tumor called a plasmacytoma. We see three polypeptides near the top of the electrophoretic gel, labeled IIo, IIa, and IIb, that are present in smaller quantities than polypeptide IIc. These three polypeptides appear to be related to one another, and indeed two of them seem to derive from the other one. But which is the parent and which are the offspring? Sequencing of the yeast *RPB1* gene predicts a polypeptide product of 210 kD, so the IIa subunit, which has a molecular mass close to 210 kD, seems to be the parent.

Table 10.3 Yeast RNA Polymerase II Subunits

Subunit	SDS-PAGE Mobility (kD)	Protein Mass (kD)	Stoichiometry	Deletion Phenotype
Rpb1	220	190	1.1	Inviable
Rpb2	150	140	1.0	Inviable
Rpb3	45	35	2.1	Inviable
Rpb4	32	25	0.5	Conditional
Rpb5	27	25	2.0	Inviable
Rpb6	23	18	0.9	Inviable
Rpb7	17	19	0.5	Inviable
Rpb8	14	17	0.8	Inviable
Rpb9	13	14	2.0	Conditional
Rpb10	10	8.3	0.9	Inviable
Rpb11	13	14	1.0	Inviable
Rpb12	10	7.7	1.0	Inviable

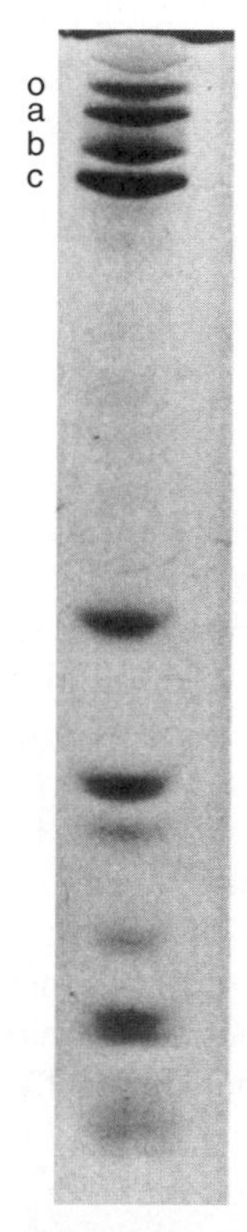

Figure 10.8 Partial subunit structure of mouse plasmacytoma RNA polymerase II. The largest subunits are identified by letter on the left, although these subunit designations are not the same as those applied to the yeast polymerase II (see Figure 10.7). Subunits o, a, and b are three forms of the largest subunit, corresponding to yeast Rpb1. Subunit c corresponds to yeast Rpb2. (*Source:* Sklar, V.E.F., L.B. Schwartz, and R.G. Roeder, Distinct molecular structures of nuclear class I, II, and III DNA-dependent RNA polymerases. *Proceedings of the National Academy of Sciences USA* 72 (Jan 1975) p. 350, f. 2C.)

Furthermore, amino acid sequencing has shown that the IIb subunit lacks a repeating string of seven amino acids (a *heptad*) with the following consensus sequence: Tyr-Ser-Pro-Thr-Ser-Pro-Ser. Because this sequence is found at the carboxyl terminus of the IIa subunit, it is called the **carboxyl-terminal domain,** or **CTD.** Antibodies against the CTD react readily with the IIa subunit, but not with IIb, reinforcing the conclusion that IIb lacks this domain. A likely explanation for this heterogeneity is that a proteolytic enzyme clips off the CTD, converting IIa to IIb. Because IIb has not been observed in vivo, this clipping seems to be an artifact that occurs during purification of the enzyme. In fact, the sequence of the CTD suggests that it will not fold into a compact structure; instead, it is probably extended and therefore highly accessible to proteolytic enzymes.

What about the IIo subunit? It appears bigger than IIa, so it cannot arise through proteolysis. Instead, it seems to be a phosphorylated version of IIa. Indeed, subunit IIo can be converted to IIa by incubating it with a phosphatase that removes the phosphate groups. Furthermore, serines 2, 5, and sometimes 7 in the heptad are found to be phosphorylated in the IIo subunit.

Can we account for the difference in apparent molecular mass between IIo and IIa simply on the basis of phosphate groups? Apparently not; even though mammalian polymerase II contains 52 repeats of the heptad, not enough phosphates are present, so we must devise another explanation for the low electrophoretic mobility of IIo. Perhaps phosphorylation of the CTD induces a conformational change in IIo that makes it electrophorese more slowly and therefore seem larger than it really is. But this conformational change would have to persist even in the denatured protein. Figure 10.9 shows the probable relationships among the subunits IIo, IIa, and IIb.

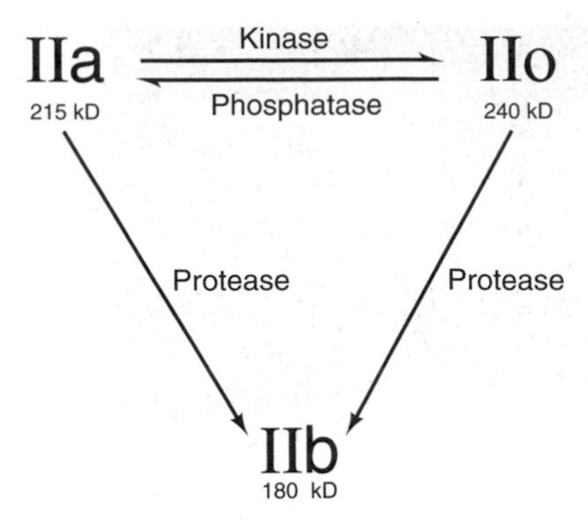

Figure 10.9 Proposed relationships among the different forms of the largest subunit of RNA polymerase II.

The fact that cells contain two forms of the Rpb1 subunit (IIo and IIa) implies that two different forms of RNA polymerase II exist, each of which contains one of these subunits. We call these **RNA polymerase IIO** and **RNA polymerase IIA,** respectively. The nonphysiological form of the enzyme, which contains subunit IIb, is called RNA polymerase IIB.

Do polymerases IIO and IIA have identical or distinct roles in the cell? The evidence strongly suggests that IIA (the unphosphorylated form of the enzyme) is the species that initially binds to the promoter, and that IIO (with its CTD phosphorylated) is the species that carries out elongation. Thus, phosphorylation of the CTD appears to accompany the transition from initiation to elongation. We will examine the evidence for this hypothesis, and refine it further, in Chapter 11.

SUMMARY Subunit IIa is the primary product of the *RPB1* gene in yeast. It can be converted to IIb in vitro by proteolytic removal of the carboxyl-terminal domain (CTD), which is essentially a heptapeptide repeated over and over. Subunit IIa can be converted to IIo by phosphorylating two serines in the repeating heptad that makes up the CTD. The enzyme (polymerase IIA) with the IIa subunit is the one that binds to the promoter; the enzyme (polymerase IIO) with the IIo subunit is the one involved in transcript elongation.

The Three-Dimensional Structure of RNA Polymerase II

The most powerful method for determining the shape of a protein, as we have seen in Chapter 9, is x-ray crystallography. This has been done with RNA polymerases from *Thermus aquaticus* and phage T7, but, until 1999, it was difficult to produce crystals of RNA polymerase II of high enough quality for x-ray crystallography studies. The problem lay in the heterogeneity of the polymerase caused by the loss of the Rpb4 and Rpb7 subunits from some of the enzymes. (Heterogeneous mixtures of proteins do not form crystals readily.) Roger Kornberg and colleagues solved this heterogeneity problem by using a mutant yeast polymerase (pol II Δ4/7) lacking Rbp4 (and therefore lacking Rpb7, because Rpb7 binds to Rpb4 and depends on the latter for binding to the rest of the enzyme). This polymerase is capable of transcription elongation, though not initiation at promoters. Thus, it should be adequate for modeling the elongation complex. It produced crystals that were good enough for x-ray crystallography leading to a model with up to 2.8 Å resolution in 2001.

Figure 10.10 presents a stereo view of this model of yeast RNA polymerase II. Each of the subunits is color-coded and their relative positions are illustrated in the small diagram at the upper right. The most prominent feature of the enzyme is the deep DNA-binding cleft, with the active site, containing a Mg^{2+} ion, at the base of the cleft. The opening of the cleft features a pair of jaws. The upper jaw is composed of part of Rpb1 plus Rpb9, and the lower jaw is composed of part of Rpb5.

Previous, lower resolution structural studies by Kornberg and colleagues had shown that the DNA template lay in the cleft in the enzyme. The newer structure strengthened this hypothesis by showing that the cleft is lined with basic amino acids, whereas almost the entire remainder of the surface of the enzyme is acidic. The basic residues in the cleft presumably help the enzyme bind to the acidic DNA template.

Structural studies of all single-subunit RNA and DNA polymerases had shown two metal ions at the active center, and a mechanism relying on both metal ions was therefore proposed. Thus, it came as a surprise to find only one Mg^{2+} ion in previous crystal structures of yeast polymerase II. However, the higher-resolution structure showed two Mg^{2+} ions, though the signal for one of them was weak. Kornberg and colleagues theorized that the strong metal signal corresponds to a strongly bound Mg^{2+} ion *(metal A)*, but the weak signal corresponds to a weakly bound Mg^{2+} ion *(metal B)* that may enter bound to the substrate nucleotide. Metal A is bound to three invariant aspartate residues (D481, D483, and D485 of Rpb1). Metal B is also surrounded by three acidic residues (D481 of Rpb1 and E836 and D837 of Rpb2), but they are too far away in the crystal structure to coordinate the metal. Nevertheless, during catalysis, they may move closer to metal B, coordinate it, and thereby create the proper conformation at the active center to accelerate the polymerase reaction.

SUMMARY The structure of yeast polymerase II (pol II Δ4/7) reveals a deep cleft that can accept a DNA template. The catalytic center, containing a Mg^{2+} ion, lies at the bottom of the cleft. A second Mg^{2+} ion is present in low concentration, and presumably enters the enzyme bound to each substrate nucleotide.

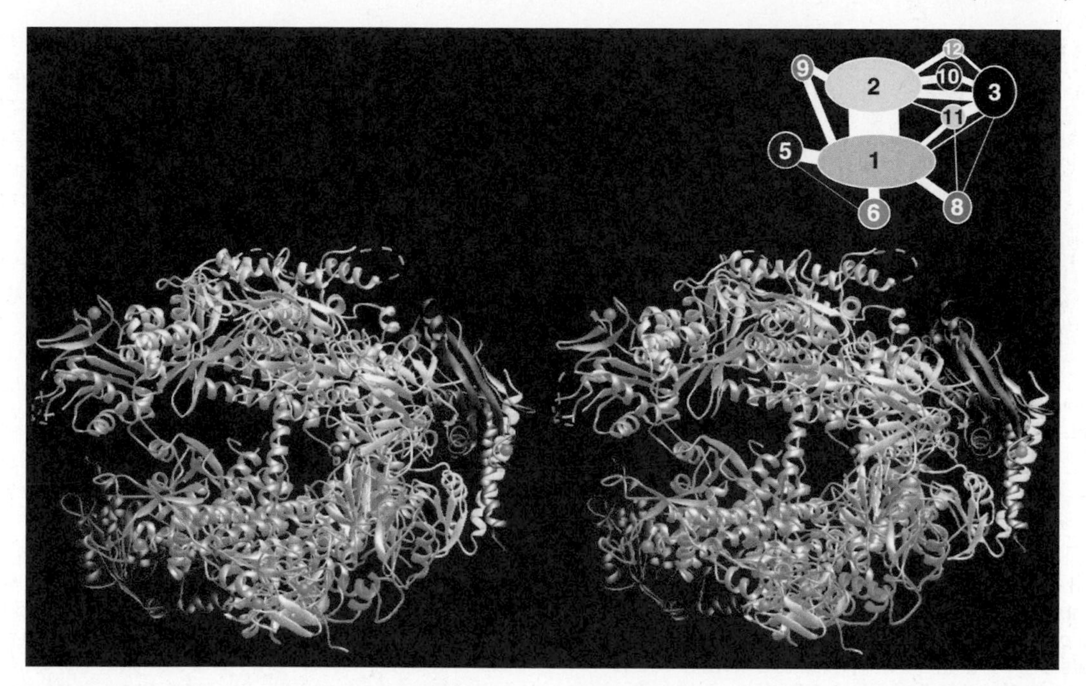

Figure 10.10 Crystal structure of yeast RNA polymerase II. The stereo view at bottom shows all 10 subunits of the enzyme (lacking Rpb4 and Rpb7), color-coded according to the small diagram at the upper right. The thickness of the white lines connecting the subunits in the small diagram indicate the extent of contact between the subunits. The metal ion at the active center in the stereo view is represented by a magenta sphere. Zn^{2+} ions are represented by blue spheres. (*Source:* Cramer, et al., *Science* 292: p. 1864.)

Three-Dimensional Structure of RNA Polymerase II in an Elongation Complex The previous section has shown the shape of yeast RNA polymerase II by itself. But Kornberg and colleagues have also determined the structure of yeast polymerase II bound to its DNA template and RNA product in an elongation complex. The resolution is not as high (3.3 Å) as in the structure of the polymerase by itself, but it still gives a wealth of information about the interaction between the enzyme and the DNA template and RNA product.

To induce polymerase II to initiate on its own without help from any transcription factors, Kornberg and colleagues used a DNA template with a 3′-single-stranded oligo[dC] tail, which allows polymerase II to initiate in the tail, 2–3nt from the beginning of the double-stranded region. The template was also designed to allow the polymerase to elongate the RNA to a 14-mer in the absence of UTP and then pause at the point where it needed the first UTP. This sequence of events created a homogeneous population of elongation complexes, contaminated with inactive polymerases that did not bind to DNA. The inactive enzymes were removed on a heparin column. Heparin is a polyanionic substance that can bind in the basic cleft of the polymerase if the cleft is not occupied by DNA. Thus, inactive enzymes bound to the heparin on the column, but the active elongation complexes passed through. These complexes could then be crystallized.

Figure 10.11a shows the crystal structure of the elongation complex, together with the crystal structure of the polymerase by itself. One of the most obvious differences, aside from the presence of the nucleic acids in the elongation complex, is the position of the **clamp.** In the polymerase itself, the clamp is open to allow access to the active site. But in the elongation complex, the clamp is closed over the DNA template and RNA product. This ensures that the enzyme will be **processive**—able to transcribe a whole gene without falling off and terminating transcription prematurely.

Figure 10.11b shows a closer view of the elongation complex, with part of the enzyme cut away to reveal the nucleic acids in the enzyme's cleft. Several features are apparent. We can see that the axis of the DNA–RNA hybrid (formed from the template DNA strand and the RNA product) lies at an angle with respect to the downstream DNA duplex that has yet to be transcribed. This turn is forced by the closing of the clamp and is facilitated by the single-stranded DNA between the RNA–DNA hybrid and the downstream DNA duplex. (Kornberg and colleagues' later crystal structure of a post-translocation complex showed that the RNA–DNA hybrid is actually 8 bp long.)

We can also see the catalytic Mg^{2+} ion at the active center—the point where a nucleotide has just been added to the growing RNA chain. This ion corresponds to metal A detected in the structure of polymerase itself. Finally, we can see a **bridge helix** that spans the cleft near the active center. We will discuss this bridge helix in more detail later in this section.

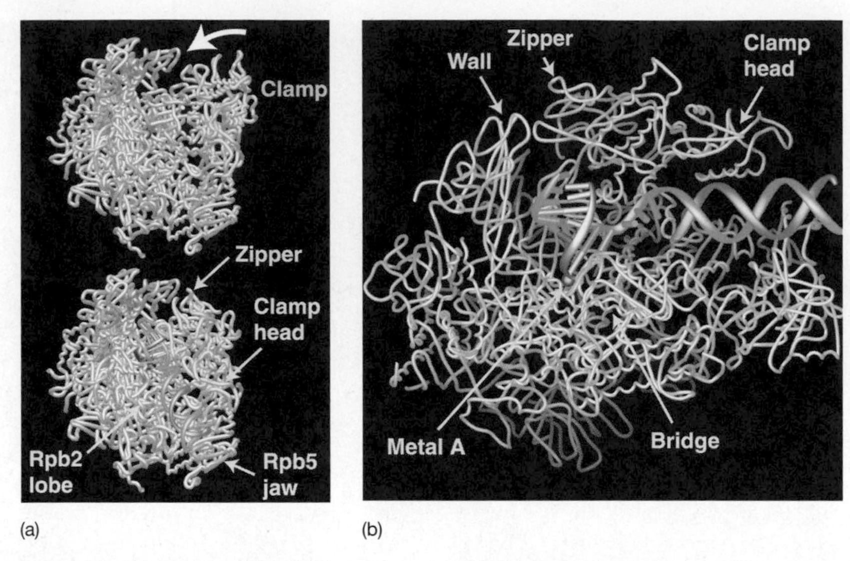

Figure 10.11 Crystal structure of the elongation complex. **(a)** Comparison of the crystal structures of the free polymerase II (top) and the elongation complex (bottom). The clamp is highlighted in yellow. The template DNA strand, the nontemplate DNA strand, and RNA product are highlighted in blue, green, and red, respectively. **(b)** Detailed view of the elongation complex. Color codes are the same as in panel (a). The active center metal is in magenta and the bridge helix is in green. (*Source:* Gnatt et al., *Science* 292: p. 1877.)

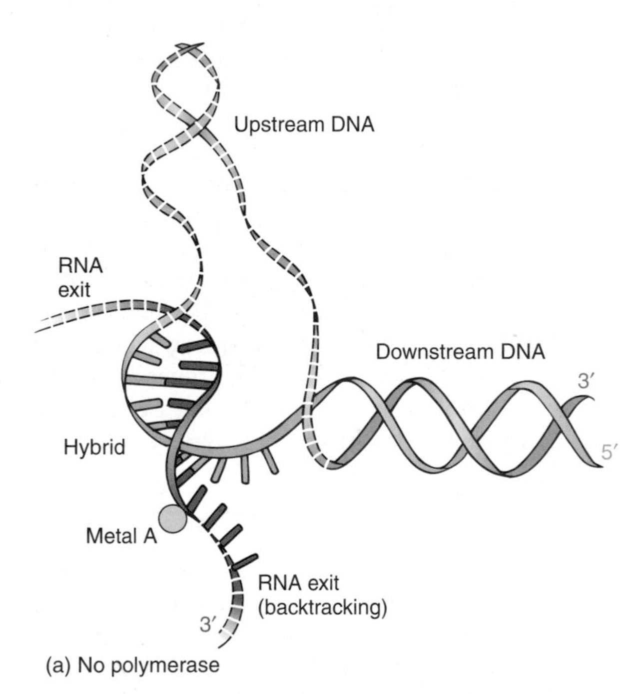

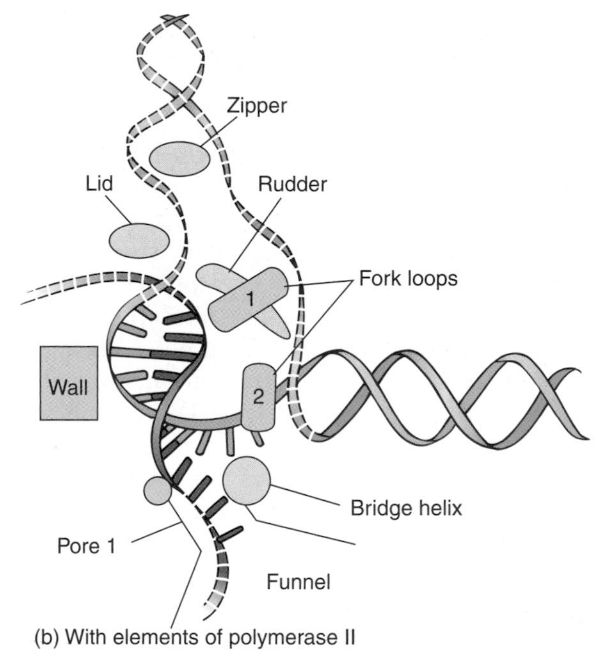

Figure 10.12 The transcription bubble. **(a)** Positions of the nucleic acids. The DNA template strand is in blue, the nontemplate strand in green, and the RNA in red. Solid lines correspond to nucleic acids represented in the crystal structure. Dashed lines show hypothetical paths for nucleic acids not represented in the crystal structure. **(b)** Nucleic acids plus key elements of RNA polymerase II. The nucleic acids from panel (a) are superimposed on critical elements of polymerase II: the protein loops extending from the clamp (the zipper, lid, and rudder); fork loops 1 and 2; the bridge helix; the funnel; pore 1; and the wall. (*Source:* Adapted from Gnatt, A.L., P. Cramer, J. Fu, D.A. Bushnell, and R.D. Kornberg, Structural basis of transcription: An RNA polymerase II elongation complex at 3.3 Å resolution. *Science* 292 (2001) p. 1879, f. 4.)

The Mg^{2+} ion in the elongation complex (metal A) is positioned so that it can bind to the phosphate linking nucleotides +1 and −1 (the last two nucleotides added to the growing RNA; Figure 10.12a). Metal B is missing from this complex, presumably because it has departed along with the pyrophosphate released from the last nucleotide added to the RNA. The nucleotide in position +1 lies just at the entrance to **pore 1** (Figure 10.12b), strongly suggesting that the nucleotides enter the active site through this pore. Indeed, there would not be room for them to enter any other way without significant rearrangements of the nucleic acids and proteins. Moreover, pore 1 is in perfect position for extrusion of the 3′-end of the RNA when the polymerase backtracks. Such backtracks occur when a nucleotide is misincorporated (recall Chapter 6), thus exposing the misincorporated nucleotide to removal by TFIIS (Chapter 11), which binds to the funnel at the other end of the pore 1.

Figure 10.12b also illustrates the probable roles of three loops, called the *lid, rudder,* and *zipper,* which extend from the clamp. These loops are in position to affect several important events, including formation and maintenance of the transcription bubble and dissociation of the RNA–DNA hybrid. If the RNA–DNA hybrid extended farther than 9 bp, the rudder would be in the way. Thus, the rudder may facilitate the dissociation of the hybrid.

Kornberg and colleagues noted that the bridge helix is straight in the elongation complex, but bent in the bacterial polymerase crystal structures. This bend occurs in the neighborhood of conserved residues corresponding to Thr 831 and Ala 832 and would interfere with nucleotide binding to the active site. This observation led these authors to speculate about the role of the bridge helix in **translocation** (the 1-nt steps of DNA template and RNA product through the polymerase), as illustrated in Figure 10.13. They suggest that the bridge helix oscillates between straight and bent conformations during the translocation step as follows: With the bridge helix in the straight state, the active site is open for addition of a nucleotide, so the nucleotide enters through pore 1 of the enzyme, just below the active site. The polymerase adds this new nucleotide to the growing RNA chain, filling the space between the 3′-end of the RNA and the straight bridge helix. Next, coincident with translocation, the bridge helix shifts to the bent state. When it shifts back to the straight state, it reopens the space at the 3′-end of the RNA, and the cycle is ready to repeat.

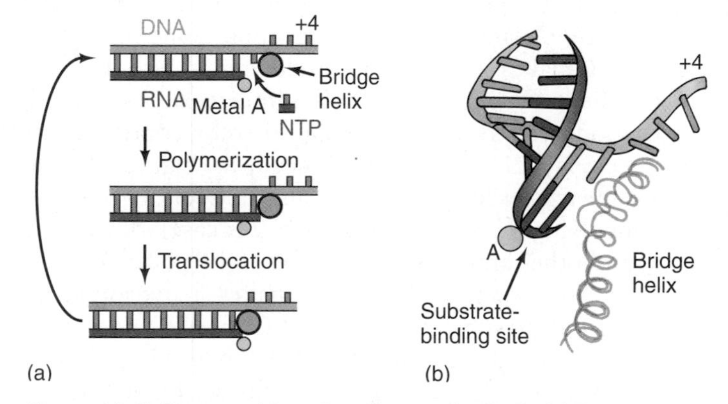

Figure 10.13 Proposed translocation mechanism. (a) The model. We begin with the bridge helix in the straight state (orange), leaving a gap for a nucleotide (NTP) to enter the active site, marked by the yellow circle (metal A). During the synthesis step, the nucleotide joins the growing RNA (red), filling the gap between the end of the RNA and the straight bridge helix. During the translocation step, the RNA–DNA hybrid moves one bp to the left, bringing a new template strand nucleotide into the active site. Simultaneously, the bridge helix bends (represented by the green dot), remaining close to the end of the RNA. When the bridge helix returns to the straight state (arrow at left), it reopens the active site so another nucleotide can enter. **(b)** The straight and bent states of the bridge helix. The straight state is represented by the orange helix, and the bent state by the green helix. Note that bending the bridge helix brings it very close to the end of the growing RNA. (*Source:* Adapted from Gnatt, A.L., P. Cramer, J. Fu, D.A. Bushnell, and R.D. Kornberg, Structural basis of transcription: An RNA polymerase II elongation complex at 3.3 Å resolution. *Science* 292 (2001) p.1880, F.6.)

Further support for this hypothesis comes from the crystal structure of the cocrystal of yeast RNA polymerase II and α-amanitin. The α-amanitin-binding site lies so close to the bridge helix that hydrogen bonds form between the two. Binding of α-amanitin to this site thus severely constrains the bending of the bridge helix necessary for translocation. This explains how α-amanitin can block RNA synthesis without blocking nucleotide entry or phosphodiester bond formation—it blocks translocation after a phosphodiester bond forms.

SUMMARY The crystal structure of a transcription elongation complex involving yeast RNA polymerase II (lacking Rpb 4/7) reveals that the clamp is indeed closed over the RNA–DNA hybrid in the enzyme's cleft, ensuring processivity of transcription. In addition, three loops of the clamp—the rudder, lid, and zipper—appear to play important roles in, respectively: initiating dissociation of the RNA–DNA hybrid, maintaining this dissociation, and maintaining dissociation of the template DNA. The active center of the enzyme lies at the end of pore 1, which appears to be the conduit for nucleotides to enter the enzyme and for extruded RNA to exit the enzyme during backtracking. A bridge helix lies adjacent to the active center, and flexing of this helix could play a role in translocation during transcription. Binding of α-amanitin to a site near this helix appears to block flexing of the helix, and therefore blocks translocation.

Structural Basis of Nucleotide Selection In 2004, Kornberg and colleagues published x-ray diffraction data on a posttranslocation complex. First, they bound RNA polymerase II to a set of synthetic oligonucleotides representing a partially double-stranded DNA template and a 10-nt RNA product terminated in 3′-deoxyadenosine, which, as we have just seen, prevents addition of any more nucleotides, and traps the polymerase in the posttranslocation state. Then they soaked crystals of this complex with either a nucleotide (UTP) that paired correctly with the next nucleotide in the DNA template strand, or a mismatched nucleotide, then obtained the crystal structures of the resulting complexes. The difference between the two structures was striking: The mismatched nucleotide lay in a site adjacent to the one occupied by the correct nucleotide, and it was inverted relative to the correct nucleotide (Figure 10.14).

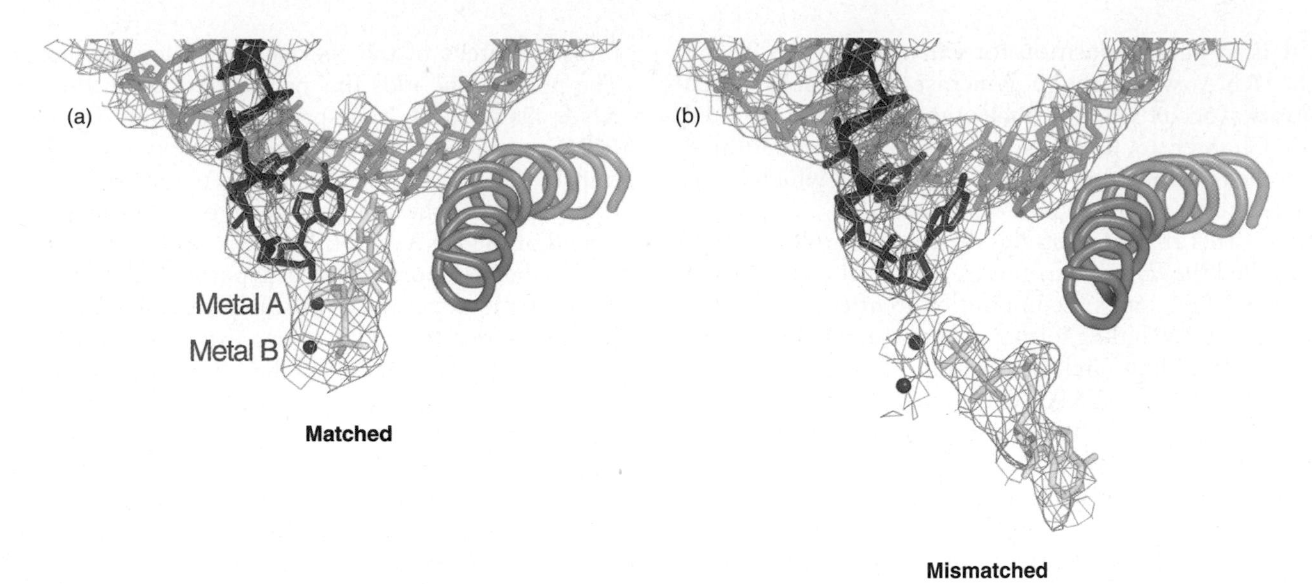

Figure 10.14 Matched (a) and mismatched (b) nucleotides in A and E sites, respectively. Metals A and B at the active site are labeled and represented by magenta spheres. DNA is in blue, RNA is in red, and the nucleotides in the A and E sites are in yellow. The green coil is the bridge helix of the RNA polymerase. (*Source:* Reprinted from Cell, Vol. 119, Kenneth D. Westover, David A. Bushnell and Roger D. Kornberg, "Structural Basis of Transcription: Nucleotide Selection by Rotation in the RNA Polymerase II Active Center," p. 481–489, Copyright 2004 with permission from Elsevier.

These data revealed two distinct nucleotide-binding sites at the active center of RNA polymerase II. The previously-known site, where phosphodiester bond formation, or nucleotide addition, occurs, had already been named the **A site, for "addition."** The second site, where nucleotides bind prior to entering the A site, had been predicted by Alexander Goldfarb and colleagues based on biochemical studies of the *E. coli* RNA polymerase; they had named this the **E site,** for "entry." The two sites overlap somewhat and Kornberg and colleagues noted that nucleotides, in moving through the nucleotide entry pore toward the A site, must pass through the E site.

The crystal structures also reinforced the case for two metal ions at the active site. One metal ion (metal A) is permanently attached to the enzyme, but the other (metal B) enters the enzyme attached to the incoming nucleotide (coordinated to the β- and γ-phosphates). In contrast to previous structures, the two metal ions had equivalent intensities in the latest structures. Thus, the mechanism of phosphodiester bond formation in RNA polymerases almost certainly relies on two metal ions at the active site.

The discovery of the E and A sites, though interesting, did not illuminate the mechanism by which the polymerase discriminates among the four ribonucleoside triphosphates, or how it excludes dNTPs. Then, in 2006, Kornberg and colleagues obtained the crystal structure of a very similar complex, but with GTP, rather than UTP, in the A site, opposite a C, rather than an A, in the template i+1 site. In this structure, and in a further refined version of their previous structure, they could see the **trigger loop,** a part of Rpb1 roughly encompassing residues 1070 to 1100, very near the substrate in the A site (Figure 10.15a). In both of these structures, the correct nucleotide occupied the A site. In 12 other crystal structures without the correct substrate in the A site, three alternative positions for the trigger loop were observed, all remote from the A site (Figure 10.15b).

Thus, only when the correct substrate nucleotide occupies the A site does the trigger loop come into play, and then it makes several important contacts with the substrate. These contacts presumably stabilize the substrate's association with the active site, and thereby contribute to the specificity of the enzyme. Indeed, as Figure 10.16a shows, the trigger loop is involved in a network of interactions involving the substrate (GTP in this case), the bridge helix, and other amino acids of Rpb1 and Rpb2 at the active site. For example, Leu 1081 makes a hydrophobic contact with the substrate base, and Gln 1078 engages in a hydrogen bond network with Rpb1-Asn 479 and the 3′-hydroxyl group of the substrate ribose. Indeed, there could even be a weak direct H-bond between this 3′-hydroxyl group and Gln 1078. In addition, His 1085 makes an H-bond or salt bridge to the β-phosphate of the substrate, and His 1085 is held in proper position by H-bonds to Asn 1082 and the Rpb2-Ser1019 backbone carbonyl group. Finally, Rpb1 Arg 446 (not part of the trigger loop) lies close to the 2′-hydroxyl group of the substrate ribose. Thus, this network of contacts recognizes all parts of the substrate nucleotide: the base, both hydroxyl groups of the sugar, and one of the phosphates.

Why is this network of contacts so important to nucleotide specificity? Presumably, the enzyme requires these contacts to create the proper environment for catalysis. Even more explicitly, the trigger loop His 1085

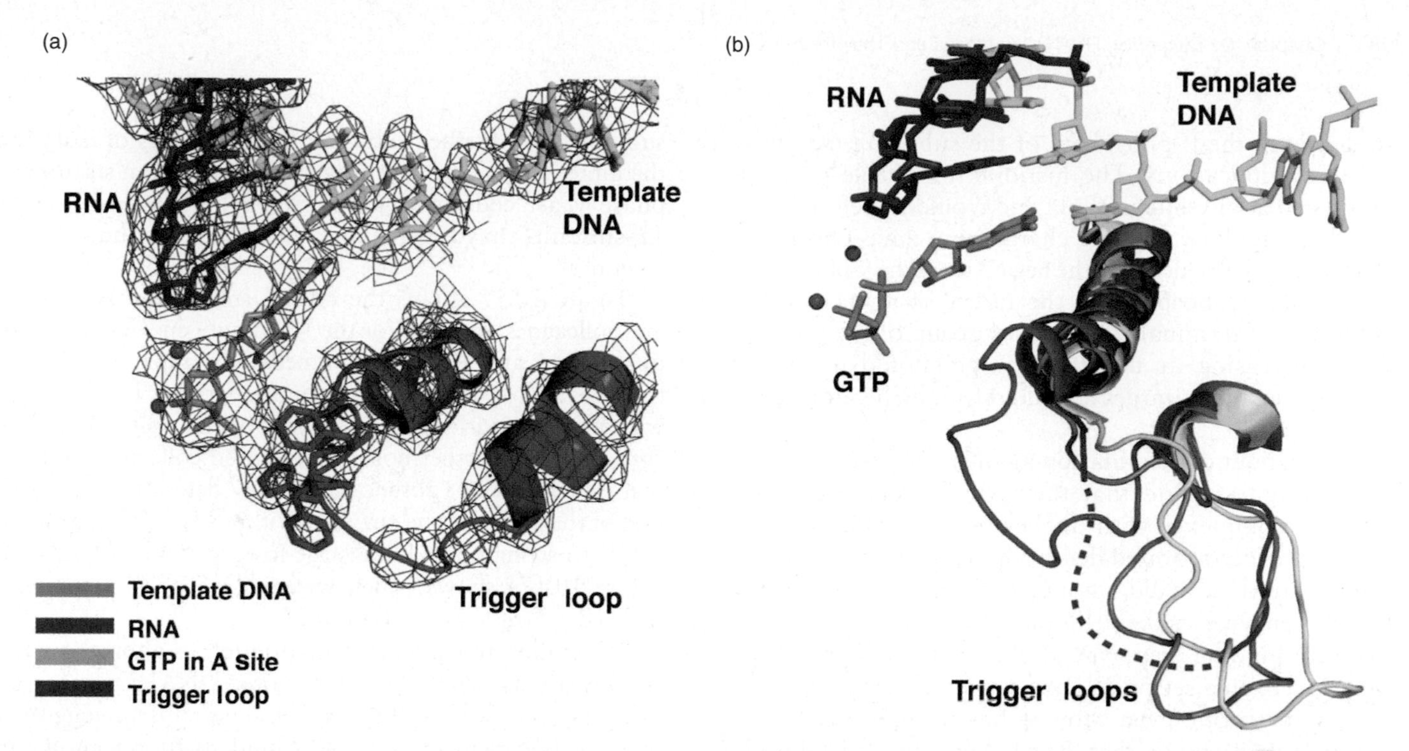

Figure 10.15 RNA polymerase II active site, including trigger loop. **(a)** The active site is shown with the proper NTP (GTP) in the A site. The electron densities are modeled with blue mesh. The trigger loop is in magenta, the GTP in orange, the RNA in red, and the template DNA strand in cyan. The Mg^{2+} ions are represented by magenta spheres. **(b)** Four different conformations for the trigger loop. Magenta, as in panel (a), with GTP in the A site at low Mg^{2+} concentration; red, ATP in the E site, low Mg^{2+}; blue, UTP in the E site, high Mg^{2+}; yellow, RNA polymerase II-TFIIS complex (see Chapter 11) with no nucleotide and high Mg^{2+}. (*Source:* Reprinted from CELL, Vol. 127, Wang et al, Structural Basis of Transcription: Role of the Trigger Loop in Substrate Specificity and Catalysis, Issue 5, 1 December 2006, pages 941–954, © 2006, with permission from Elsevier.)

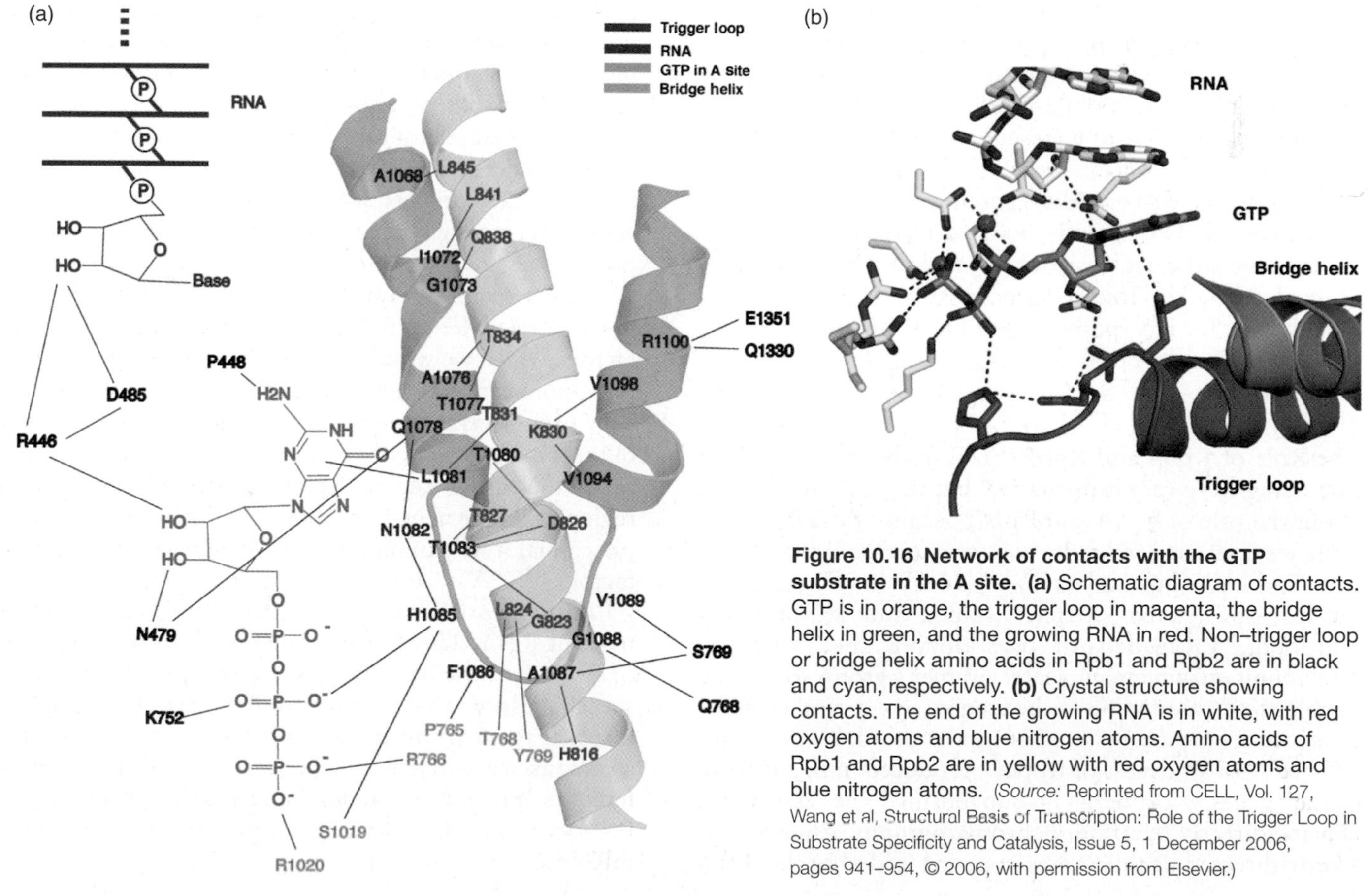

Figure 10.16 Network of contacts with the GTP substrate in the A site. **(a)** Schematic diagram of contacts. GTP is in orange, the trigger loop in magenta, the bridge helix in green, and the growing RNA in red. Non–trigger loop or bridge helix amino acids in Rpb1 and Rpb2 are in black and cyan, respectively. **(b)** Crystal structure showing contacts. The end of the growing RNA is in white, with red oxygen atoms and blue nitrogen atoms. Amino acids of Rpb1 and Rpb2 are in yellow with red oxygen atoms and blue nitrogen atoms. (*Source:* Reprinted from CELL, Vol. 127, Wang et al, Structural Basis of Transcription: Role of the Trigger Loop in Substrate Specificity and Catalysis, Issue 5, 1 December 2006, pages 941–954, © 2006, with permission from Elsevier.)

contact with the β-phosphate of the substrate may have catalytic implications. The histidine imidazole group is protonated at physiological pH and would therefore be expected to withdraw negative charge from the β-phosphate, which could in turn decrease the negativity of the γ-phosphate. Because the γ-phosphate is the target of a nucleophilic attack by the terminal 3′-hydroxyl group of the growing RNA, decreasing its negative charge should make it a better nucleophilic target and therefore help catalyze the reaction.

What about discrimination against dNTPs? Kornberg and colleagues found that they could prepare enzyme-substrate complexes with dNTPs in the A site, but that the enzyme incorporated deoxyribonucleotides at a much slower rate than it did ribonucleotides. They concluded that the enzyme makes this discrimination, not at the substrate binding step, but at the catalytic step. Moreover, the enzyme seems to have a way of removing a deoxyribonucleotide even after it has been incorporated. Figure 10.16a shows that Rpb1 Arg 446 and Glu 485 contact the 2′-hydroxyl group of the nucleotide that had been incorporated just before the new substrate bound. If this hydroxyl group is missing because a dNMP was incorporated by accident, these contacts can't be made, and the enzyme will presumably stall until the misincorporated dNMP can be removed.

SUMMARY In moving through the entry pore toward the active site of RNA polymerase II, an incoming nucleotide first encounters the E (entry) site, where it is inverted relative to its position in the A site, the active site where phosphodiester bonds are formed. Two metal ions (Mg^{2+} or Mn^{2+}) are present at the active site. One is permanently bound to the enzyme and one enters the active site complexed to the incoming nucleotide. The trigger loop of Rpb1 positions the substrate for incorporation and discriminates against improper nucleotides.

The Role of Rpb4 and Rpb7 The studies we have been discussing were very informative, but they told us nothing about the role of Rpb4 and Rpb7, because these two subunits were missing from the core polymerase II that Kornberg and colleagues crystallized. To fill in this gap, two groups, one led by Patrick Cramer, and the other by Kornberg, succeeded in crystallizing the complete, 12-subunit enzyme from yeast. Cramer's group solved the problem of producing a homogeneous population of 12-subunit enzyme by incubating the purified 10-subunit enzyme with an excess of Rbp4/7 produced in *E. coli* from cloned genes. Kornberg's group purified the 12-subunit enzyme directly by affinity chromatography, using an antibody directed against an epitope tag added to the Rpb4 subunit. They further enhanced their chances of isolating the intact enzyme by isolating the enzyme from stationary phase yeast cells, which contain a high proportion of 12-subunit enzyme, rather than the 10-subunit core enzyme.

Figure 10.17 shows the crystal structure that Cramer and colleagues obtained for the 12-subunit enzyme. The subunits Rbp4 and Rpb7 are immediately apparent because they stick out to the side of the enzyme, rather like a wedge, with its thin end lodged in the rest of the polymerase (the core enzyme). Furthermore, Cramer and colleagues noticed that the presence or absence of Rpb4/7 determines the position of the clamp of the enzyme. Without Rpb4/7, the clamp is free to swing open, but, as the inset at the lower right in Figure 10.17a shows, when wedge-like Rpb4/7 is present, the wedge forces the clamp shut.

What does this new information tell us about how the polymerase associates with promoter DNA? Cramer and colleagues, as well as Bushnell and Kornberg, suggested that the polymerase core could bind to the promoter in double-stranded form, the promoter could then melt, and then Rpb4/7 could bind and close the clamp over the template DNA strand, excluding the nontemplate strand from the active site. But these authors also point out that this simple model is contradicted by other evidence: First, RNA polymerases from other organisms have Rpb4/7 homologs that are not thought to dissociate from the core enzyme. Similarly, the crystal structure of the *E. coli* RNA polymerase holoenzyme, the form of the enzyme involved in initiation (Chapter 6), has a closed conformation that seems incapable of allowing access to double-stranded DNA. So both sets of authors proposed that the promoter DNA could bind to the outer surface of the enzyme and melt, and the template strand could then descend into the active site, with accompanying pronounced bending of the promoter DNA.

Both research groups also noted a potential strong influence of Rpb4/7 on interaction with general transcription factors, which we will discuss in Chapter 11. We know that RNA polymerase II cannot bind to promoter DNA without help from several general transcription factors, and some of these make direct contact with an area of the polymerase called the "dock" region. Rpb4/7 greatly extends the dock region, as shown in Figure 10.17b. Thus, Rpb4/7 could play a major role in binding the vital general transcription factors.

Further work has shown that Rpb7 can bind to a nascent RNA. This finding, together with the proximity of Rpb4/7 to the base of the CTD of Rpb1 has prompted the suggestion that it can bind the nascent RNA and direct it toward the CTD. This could be important because, as we will see in Chapters 14 and 15, the CTD harbors proteins that make essential modifications (splicing, capping, and polyadenylation) to nascent mRNAs.

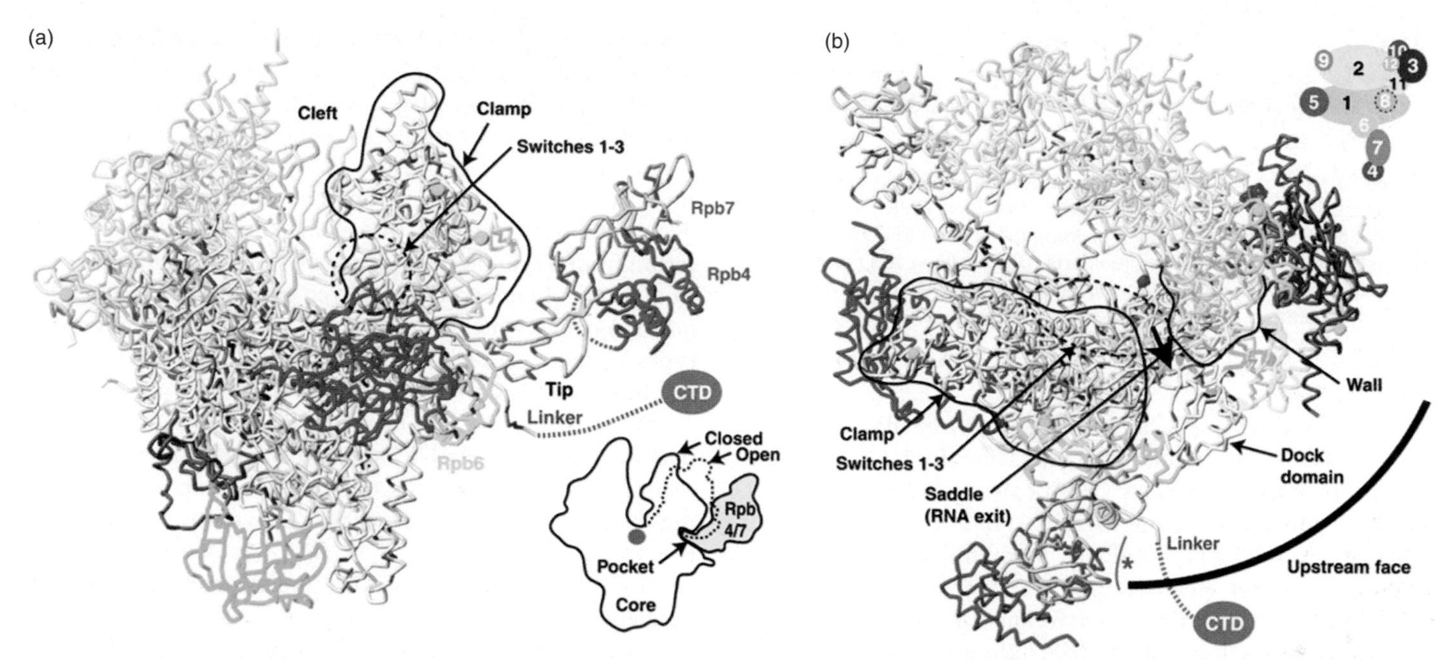

Figure 10.17 Crystal structure of the 12-subunit RNA polymerase II from yeast. **(a)** Structure showing the interaction between Rpb4/7 and the core polymerase. Rpb4 and Rpb7 are in magenta and blue, respectively, and are labeled. The clamp is outlined in solid black. The location of switches 1–3 is denoted by a dashed circle. Eight zinc ions are denoted by cyan spheres, and the magnesium ion at the active center at the base of the cleft (difficult to see in this panel) is represented by a pink sphere. The linker to the CTD of Rpb1 is denoted by a dashed line. The inset at lower right shows the closed and open positions of the clamp, and demonstrates that binding of Rpb4/7 is incompatible with the clamp's open position; that is, binding of Rpb4/7 wedges the clamp shut. **(b)** Another view of the structure, with the subunits color-coded as shown at upper right. This view emphasizes the effect of Rpb4/7 on extension of the dock domain of the enzyme. The solid circle segment at lower right represents a 25-bp radius, centered on the active site, which is the minimum distance between the TATA box and the transcription start site. The blue asterisk at lower center indicates a potential RNA-binding site on Rpb7. *(Source: (a-b)* © 2003 National Academy of Sciences Proceedings of the National Academy of Sciences, Vol. 100, no. 12, June 10, 2003, p. 6964–6968 "Architecture of initiation-competent 12-subunit RNA polymerase II," Karim-Jean Armache, Hubert Kettenberger, and Patrick Cramer, Fig. 2, p. 6966.

SUMMARY The structure of the 12-subunit RNA polymerase II reveals that, with Rpb4/7 in place, the clamp is forced shut. Because initiation occurs with the 12-subunit enzyme, with its clamp shut, it appears that the promoter DNA must melt before the template DNA strand can descend into the enzyme's active site. It also appears that Rpb4/7 extends the dock region of the polymerase, making it easier for certain general transcription factors to bind, thereby facilitating transcription initiation.

10.2 Promoters

We have seen that the three eukaryotic RNA polymerases have different structures and they transcribe different classes of genes. We would therefore expect that the three polymerases would recognize different promoters, and this expectation has been borne out. We will conclude this chapter by looking at the structures of the promoters recognized by all three polymerases.

Class II Promoters

We begin with the promoters recognized by RNA polymerase II (**class II promoters**) because these are the most complex and best studied. Class II promoters can be considered as having two parts: the **core promoter** and the **proximal promoter.** The core promoter attracts general transcription factors and RNA polymerase II at a basal level and sets the transcription start site and direction of transcription. It consists of elements lying within about 37 bp of the transcription start site, on either side. The proximal promoter helps attract general transcription factors and RNA polymerase and includes promoter elements that can extend from about 37 bp up to 250 bp upstream of the transcription start site. Elements of the proximal promoter are also sometimes called **upstream promoter elements.**

The core promoter is modular and can contain almost any combination of the following elements (Figure 10.18). The **TATA box** is centered at approximately position −28 (about −31 to −26) and has the consensus sequence TATA(A/T)AA(G/A); the **TFIIB recognition element (BRE)** lies just upstream of the TATA box (about

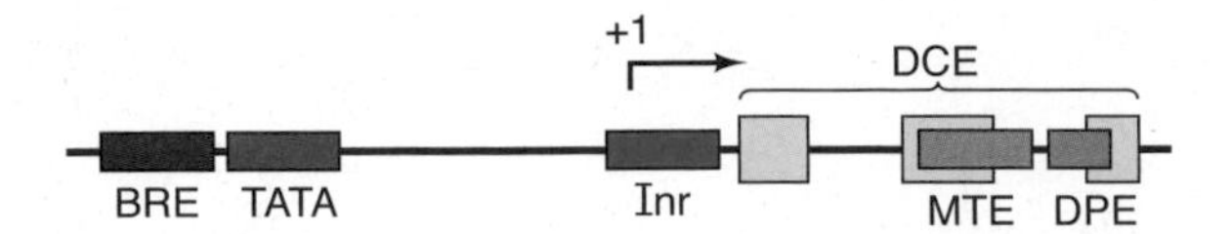

Figure 10.18 A generic class II core promoter. This core promoter contains up to six elements. These are, 5′ to 3′: the TFIIB-recognition element (BRE, purple); the TATA box (red); the initiator (green); the downstream core element, in three parts (DCE, yellow); the motif ten element (MTE, blue); and the downstream promoter element (DPE, orange). The exact locations of these promoter elements are given in the text.

position −37 to −32) and has the consensus sequence (G/C)(G/C)(G/A)CGCC; the **initiator (Inr)** is centered on the transcription start site (position −2 to +4) and has the consensus sequence GCA(G/T)T(T/C) in *Drosophila,* or PyPyAN(T/A)PyPy in mammals; the **downstream promoter element (DPE)** is centered on position +30 (+28 to +32); the **downstream core element (DCE)** has three parts located at approximately +6 to +12, +17 to +23, and +31 to +33, and these have the consensus sequences CTTC, CTGT, and AGC, respectively; and the **motif ten element (MTE)** lies approximately between positions +18 and +27.

The TATA Box By far the best-studied element in the many class II promoters is a sequence of bases with the consensus sequence TATAAA (in the nontemplate strand). The last A of this sequence usually lies 25 to 30 bp upstream of the transcription start site in higher eukaryotes. Its name, *TATA box,* derives from its first four bases. You may have noticed the close similarity between the eukaryotic TATA box and the prokaryotic −10 box. The major difference between the two is position with respect to the transcription start site: −25 to −30 versus −10. (TATA boxes in yeast [*Saccharomyces cerevisiae*] have a more variable location, from 30 to more than 300 bp upstream of their transcription start sites.)

As usual with consensus sequences, exceptions to the rule exist. Indeed, in this case they are plentiful. Sometimes G's and C's creep in, as in the TATA box of the rabbit β-globin gene, which starts with the sequence CATA. Frequently, no recognizable TATA box is evident at all. Such TATA-less promoters tend to be found in two classes of genes: (1) The first class comprises the **housekeeping genes** that are constitutively active in virtually all cells because they control common biochemical pathways, such as nucleotide synthesis, needed to sustain cellular life. Thus, we find TATA-less promoters in the cellular genes for adenine deaminase, thymidylate synthetase, and dihydrofolate reductase, all of which encode enzymes necessary for making nucleotides, and in the SV40 region encoding the viral late proteins. These genes sometimes have GC boxes that appear to compensate for the lack of a TATA box (Chapter 11). In *Drosophila,* only about 30% of class II promoters have recognizable TATA boxes, but many TATA-less promoters have DPEs that play the same role as a TATA box. (2) The second class of genes with TATA-less promoters are developmentally regulated genes such as the homeotic genes that control development of the fruit fly or genes that are active during development of the immune system in mammals. We will examine one such gene (the mouse terminal deoxynucleotidyltransferase [TdT] gene) later in this chapter. In general, **specialized genes** (sometimes called *luxury genes*), which encode proteins made only in certain types of cells (e.g., keratin in skin cells and hemoglobin in red blood cells), do have TATA boxes.

What is the function of the TATA box? That seems to depend on the gene. The first experiments to probe this question involved deleting the TATA box and then assaying the deleted DNA for promoter activity by transcription in vitro.

In 1981, Christophe Benoist and Pierre Chambon performed a deletion mutagenesis study of the SV40 early promoter. The assays they used for promoter activity were primer extension and S1 mapping. These techniques, described in Chapter 5, produce labeled DNA fragments whose lengths tell us where transcription starts and whose abundance tells us how active the promoter is. As Figure 10.19a shows, the P1A, AS, HS0, HS3, and HS4 mutants, which Benoist and Chambon had created by deleting progressively more of the DNA downstream of the TATA box, including the initiation site, simply shortened the S1 signal by an amount equal to the number of base pairs removed by the deletion. This result is consistent with a downstream shift in the transcription start site caused by the deletion. Such a shift is just what we would predict if the TATA box positions transcription initiation approximately 25 to 30 bp downstream of the last base of the TATA box. If this is so, what should be the consequences of deleting the TATA box altogether? The H2 deletion extends the H4 deletion through the TATA box and therefore provides the answer to our question: Lane 8 of Figure 10.19b shows that removing the TATA box caused transcription to initiate at a wide variety of sites, while not decreasing the efficiency of transcription. If anything, the darkness of the S1 signals suggests an increase in transcription. Thus, it appears that the TATA box is involved in positioning the start of transcription.

In further experiments, Benoist and Chambon reinforced this conclusion by systematically deleting DNA between the TATA box and the initiation site of the SV40 early gene and locating the start of transcription in the resulting shortened DNAs by S1 mapping. Transcription of the wild-type gene begins at three different guanosines, clustered 27–34 bp downstream of the first T of the TATA box. As Benoist and Chambon removed more and more of the DNA between the TATA box and these initiation sites, they noticed that transcription no longer initiated at these sites. Instead, transcription started at other bases, usually purines, that

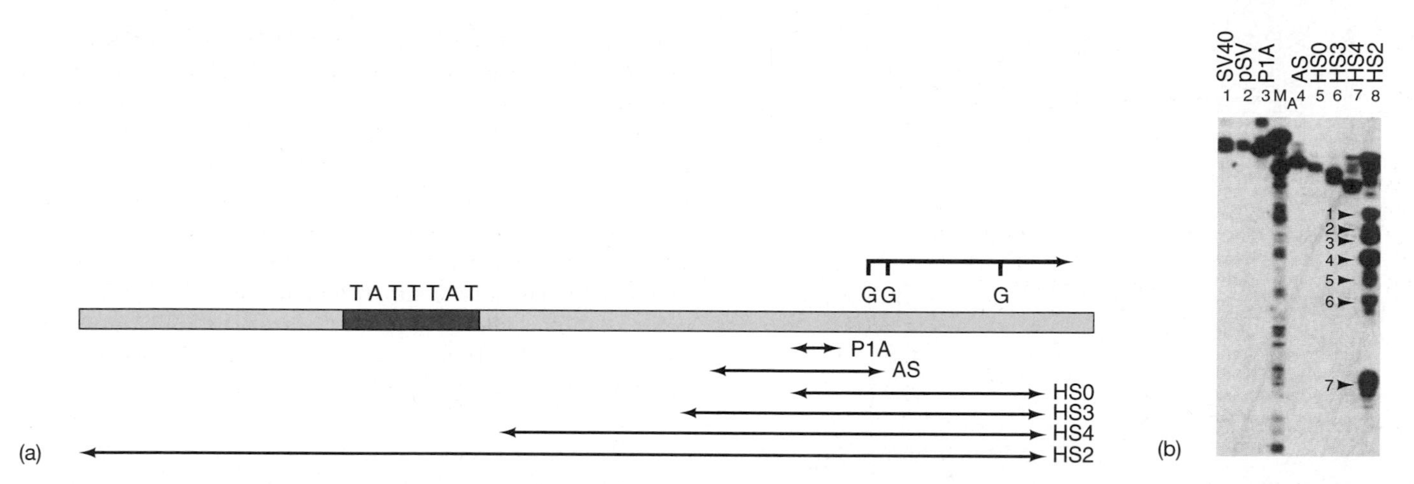

Figure 10.19 Effects of deletions in the SV40 early promoter. **(a)** Map of the deletions. The names of the mutants are given at the right of each arrow. The arrows indicate the extent of each deletion. The positions of the TATA box (TATTTAT, red) and the three transcription start sites (all G's) are given at top. **(b)** Locating the transcription start sites in the mutants. Benoist and Chambon transfected cells with either SV40 DNA, or a plasmid containing the wild-type SV40 early region (pSV1), or a derivative of pSV1 containing one of the mutated SV40 early promoters described in panel (a). They located the initiation site (or sites) by S1 mapping. The names of the mutants being tested are given at the top of each lane. The lane denoted M_A contained size markers. The numbers to the left of the bands in the HS2 lane denote novel transcription start sites not detected with the wild-type promoter or with any of the other mutants in this experiment. The heterogeneity in the transcription initiation sites was apparently due to the lack of a TATA box in this mutant. (*Source:* (*b*) Benoist C. and P. Chambon, In vivo sequence requirements of the SV40 early promoter region. *Nature* 290 (26 Mar 1981) p. 306, f. 3.)

were about 30 bp downstream of the first T of the TATA box. In other words, the distance between the TATA box and the transcription initiation sites remained constant, with little regard to the exact sequence at these initiation sites.

In this example, the TATA box appears to be important for locating the start of transcription, but not for regulating the efficiency of transcription. However, in some other promoters, removal of the TATA box impairs promoter function to such an extent that transcription, even from aberrant start sites, cannot be detected.

Steven McKnight and Robert Kingsbury provided an example with their studies of the herpes virus thymidine kinase (tk) promoter. They performed **linker scanning mutagenesis,** in which they systematically substituted a synthetic 10-bp linker for 10-bp sequences throughout the tk promoter. One of the results of this analysis was that mutations within the TATA box destroyed promoter activity (Figure 10.20). In the mutant with the lowest promoter activity (LS –29/–18), the normal sequence in the region of the TATA box had been changed from GCATATTA to CCGGATCC.

Thus, some class II promoters require the TATA box for function, but others need it only to position the transcription start site. And, as we have seen, some class II promoters, most notably the promoters of housekeeping genes, have no TATA box at all, and they still function quite well. How do we account for these differences? As we will see in Chapters 11 and 12, promoter activity depends on assembling a collection of transcription factors and RNA polymerase called a preinitiation complex. This complex forms at the transcription start site and launches the transcription process. In class II promoters, the TATA box serves as the site where this assembly of protein factors begins. The first protein to bind is TFIID, including the TATA-box-binding protein (TBP), which then attracts the other factors. But what about promoters that lack TATA boxes? These still require TBP, but because TBP has no TATA box to which it can bind, it depends on other proteins, which bind to other promoter elements, to hold it in place.

Initiators, Downstream Promoter Elements, and TFIIB Recognition Elements Some class II promoters have conserved sequences around their transcription start sites that are required for optimal transcription. These are called **initiators,** and mammalian initiators have the consensus sequence PyPyAN(T/A)PyPy, where Py stands for either pyrimidine (C or T), N stands for any base, and the underlined A is the transcription start point. *Drosophila* initiators have the consensus sequence TCA(G/T)T(T/C). The classic example of an initiator comes from the adenovirus major late promoter. This initiator, together with the TATA box, constitutes a core promoter that can drive transcription of any gene placed downstream of it, though at a very low level. This promoter is also susceptible to stimulation by upstream elements or enhancers connected to it.

Another example of a gene with an important initiator is the mammalian terminal deoxynucleotidyltransferase (TdT) gene, which is activated during development of B and T lymphocytes. Stephen Smale and David Baltimore studied the mouse TdT promoter and found that it contains no TATA box and no apparent upstream promoter elements, but it does contain an initiator. This initiator is sufficient to drive basal-level transcription of the gene from a single start

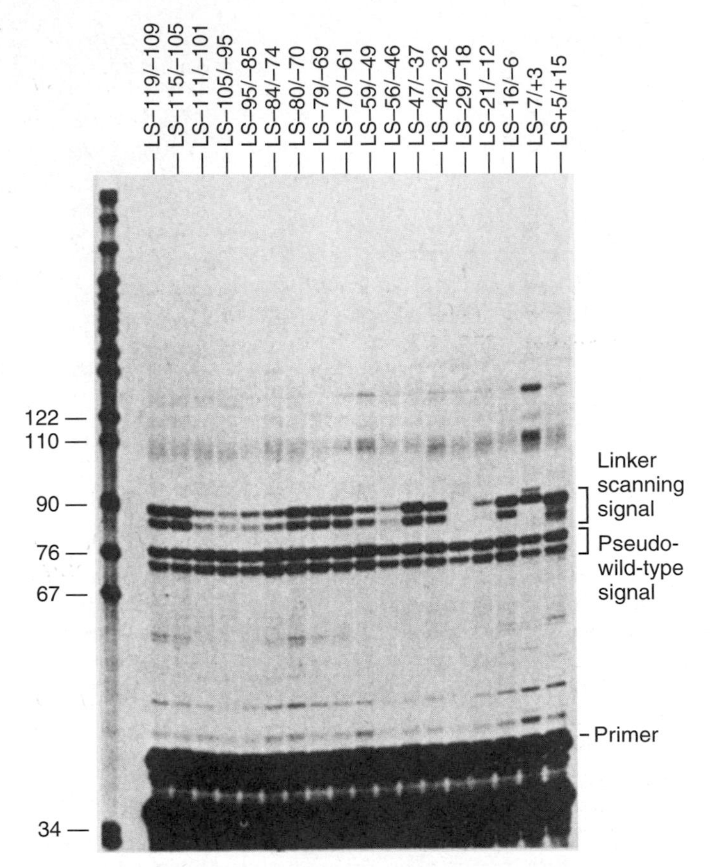

Figure 10.20 Effects of linker scanning mutations in the herpes virus tk promoter. McKnight and Kingsbury made linker scanning mutations throughout the tk promoter, then injected the mutated DNAs into frog oocytes, along with a pseudo-wild-type DNA (mutated at the +21 to +31 position). Transcription from this pseudo-wild-type promoter was just as active as that from the wild-type promoter, so this DNA served as an internal control. The investigators assayed for transcription from the test plasmid and from the control plasmid by primer extension analysis. Transcription from the control plasmid remained relatively constant, as expected, but transcription from the test plasmid varied considerably depending on the locus of the mutations. (*Source:* Adapted from McKnight, S.L. and R. Kingsbury, Transcriptional control signals of a eukaryotic protein-coding gene. *Science* 217 (23 July 1982) p. 322, f. 5.)

site located within the initiator sequence. Smale and Baltimore also found that a TATA box or the GC boxes from the SV40 promoter could greatly stimulate transcription starting at the initiator. Thus, this initiator alone constitutes a very simple, but functional, promoter whose efficiency can be enhanced by other promoter elements.

Downstream promoter elements are very common in *Drosophila*. In fact, in 2000 Alan Kutach and James Kadonaga reported the surprising discovery that DPEs are just as common in *Drosophila* as TATA boxes. These DPEs are found about 30 bp downstream of the transcription initiation site and include the consensus sequence G(A/T)CG. They can compensate for the loss of the TATA box from a promoter. Indeed, many naturally TATA-less promoters in *Drosophila* contain DPEs, which accounts for the abundance of DPEs in this organism. It is common to find a DPE coupled with an Inr in TATA-less *Drosophila* promoters. The similarity between the TATA box and the DPE extends to their ability to bind to a key general transcription factor known as TFIID (Chapter 11).

Another important general transcription factor is TFIIB, which binds to the promoter along with TFIID, RNA polymerase II, and other factors, to form a preinitiation complex that is competent to begin transcription. Some promoters have a DNA element just upstream of the TATA box that helps TFIIB to bind to the DNA. These are called TFIIB recognition elements (BREs).

SUMMARY Class II promoters may consist of a core promoter immediately surrounding the transcription start site, and a proximal promoter further upstream. The core promoter may contain up to six conserved elements: the TFIIB recognition element (BRE), the TATA box, the initiator (Inr), the downstream core element (DCE), the motif ten element (MTE), and the downstream promoter element (DPE). At least one of these elements is missing in most promoters. In fact, TATA-less promoters tend to have DPEs, at least in *Drosophila*. Promoters for highly expressed specialized genes tend to have TATA boxes, but promoters for housekeeping genes tend to lack them.

Proximal Promoter Elements McKnight and Kingsbury's linker scanning analysis of the herpes virus tk gene revealed other important promoter elements upstream of the TATA box. Figure 10.20 shows that mutations in the −47 to −61 and in the −80 to −105 regions caused significant loss of promoter activity. The nontemplate strands of these regions contain the sequences GGGCGG and CCGCCC, respectively. These are so-called **GC boxes,** which are found in a variety of promoters, usually upstream of the TATA box. Notice that the two GC boxes are in opposite orientations in their two locations in the herpes virus tk promoter.

Chambon and colleagues also found GC boxes in the SV40 early promoter, and not just two copies, but six. Furthermore, mutations in these elements significantly decreased promoter activity. For example, loss of one GC box decreased transcription to 66% of the wild-type level, and loss of a second GC box decreased transcription all the way down to 13% of the control level. We will see in Chapter 12 that a specific transcription factor called Sp1 binds to the GC boxes and stimulates transcription. Later in this chapter we will discuss DNA elements called enhancers that stimulate transcription, but differ from promoters in two important respects: They are position- and orientation-independent. The GC boxes are orientation-independent; they can be flipped 180 degrees and they still function (as occurs naturally in the herpes virus tk promoter). But the GC boxes do not have the position independence of classical enhancers, which can be moved as much as several kilobases away from a promoter, even downstream of a gene's coding region, and still function. If the GC boxes are moved more than a few dozen base pairs away from their own TATA box, they lose the ability to stimulate transcription. Thus, it is probably more proper to consider the GC boxes, at least in these two genes, as proximal promoter elements, rather than enhancers. On the other hand, the distinction is subtle and perhaps borders on semantic.

Another upstream element found in a wide variety of class II promoters is the so-called **CCAAT box** (pronounced "cat box"). In fact, the herpes virus tk promoter has a CCAAT box; the linker scanning study we have discussed failed to detect any loss of activity when this CCAAT box was mutated, but other investigations have clearly shown the importance of the CCAAT box in this and in many other promoters. Just as the GC box has its own transcription factor, so the CCAAT box must bind a transcription factor (the **CCAAT-binding transcription factor [CTF]**, among others) to exert its stimulatory influence.

SUMMARY Proximal promoter elements are usually found upstream of class II core promoters. They differ from the core promoter in that they bind to relatively gene-specific transcription factors. For example, GC boxes bind the transcription factor Sp1, while CCAAT boxes bind CTF. The proximal promoter elements, unlike the core promoter, can be orientation-independent, but they are relatively position-dependent, unlike classical enhancers.

Class I Promoters

What about the promoter recognized by RNA polymerase I? We can refer to this promoter in the singular because almost all species have only one kind of gene recognized by polymerase I: the rRNA precursor gene. The one known exception is the trypanosome, in which polymerase I transcribes two protein-encoding genes, in addition to the rRNA precursor gene. It is true that the rRNA precursor gene is present in hundreds of copies in each cell, but each copy is virtually the same as the others, and they all have the same promoter sequence. However, this sequence is quite variable from one species to another—more variable than those of the promoters recognized by polymerase II, which tend to have conserved elements, such as TATA boxes, in common.

Robert Tjian and colleagues used linker scanning mutagenesis to identify the important regions of the human rRNA promoter. Figure 10.21 shows the results of this analysis: The promoter has two critical regions in which mutations cause a great reduction in promoter strength. One of these, the **core element,** also known at the initiator (rINR), is located at the start of transcription, between positions −45 and +20. The other is the **upstream promoter element (UPE),** located between positions −156 and −107.

The presence of two promoter elements raises the question of the importance of the spacing between them. In this case, spacing is very important. Tjian and colleagues deleted or added DNA fragments of various lengths between the UPE and the core element of the human rRNA promoter. When they removed only 16 bp between the two promoter elements, the promoter

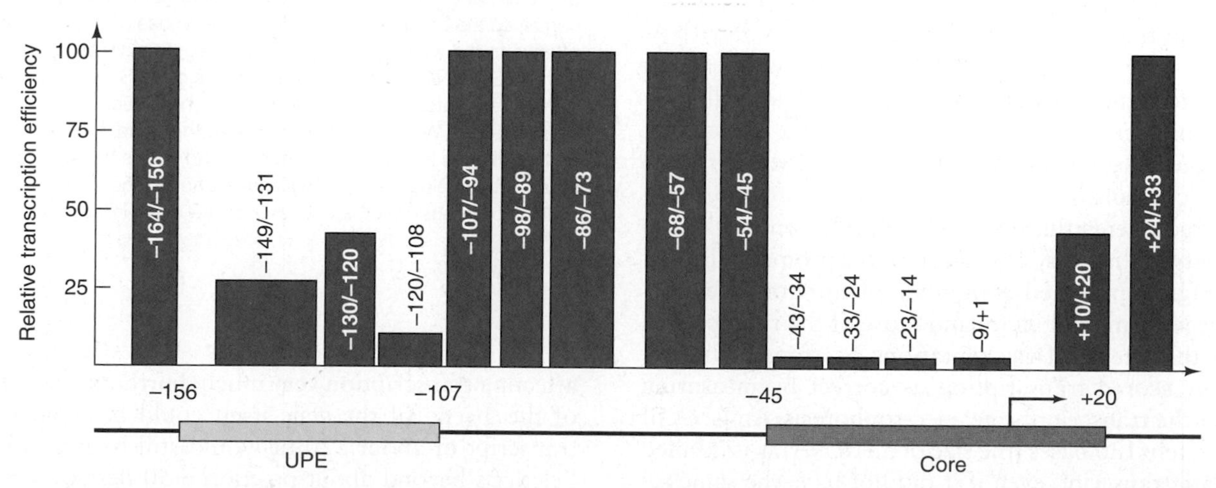

Figure 10.21 Two rRNA promoter elements. Tjian and colleagues used linker scanning to mutate short stretches of DNA throughout the 5′-flanking region of the human rRNA gene. They then tested these mutated DNAs for promoter activity using an in vitro transcription assay. The bar graph illustrates the results, which show that the promoter has two important regions: labeled UPE (upstream promoter element) and Core. The UPE is necessary for optimal transcription, but basal transcription is possible in its absence. On the other hand, the core element is absolutely required for any transcription to occur. (*Source:* Adapted from Learned, R.M., T.K. Learned, M.M. Haltiner, and R.T. Tjian, Human rRNA transcription is modulated by the coordinated binding of two factors to an upstream control element. *Cell* 45:848, 1986.)

strength dropped to 40% of wild-type; by the time they had deleted 44 bp, the promoter strength was only 10%. On the other hand, they could add 28 bp between the elements without affecting the promoter, but adding 49 bp reduced promoter strength by 70%. Thus, the promoter efficiency is more sensitive to deletions than to insertions between the two promoter elements.

SUMMARY Class I promoters are not well conserved in sequence from one species to another, but the general architecture of the promoter is well conserved. It consists of two elements, a core element surrounding the transcription start site, and an upstream promoter element (UPE) about 100 bp farther upstream. The spacing between these two elements is important.

Class III Promoters

As we have seen, RNA polymerase III transcribes a variety of genes that encode small RNAs. These include (1) the "classical" class III genes, including the 5S rRNA and tRNA genes, and the adenovirus VA RNA genes; and (2) some relatively recently discovered class III genes, including the U6 snRNA gene, the 7SL RNA gene, the 7SK RNA gene, and the Epstein–Barr virus EBER2 gene. The latter, "nonclassical" class III genes have promoters that resemble those found in class II genes. By contrast, the "classical" class III genes have promoters located entirely within the genes themselves.

Class III Genes with Internal Promoters Donald Brown and his colleagues performed the first analysis of a class III promoter, on the gene for the *Xenopus borealis* 5S rRNA. The results they obtained were astonishing. Whereas the promoters recognized by polymerases I and II, as well as by bacterial polymerases, are located mostly in the 5′-flanking region of the gene, the 5S rRNA promoter is located *within* the gene it controls.

The experiments that led to this conclusion worked as follows: First, to identify the 5′-end of the promoter, Brown and colleagues prepared a number of mutant 5S rRNA genes that were missing more and more of their 5′-end and observed the effects of the mutations on transcription in vitro. They scored transcription as correct by measuring the size of the transcript by gel electrophoresis. An RNA of approximately 120 bases (the size of 5S rRNA) was deemed an accurate transcript, even if it did not have the same sequence as real 5S rRNA. They had to allow for incorrect sequence in the transcript because they changed the internal sequence of the gene to disrupt the promoter.

The surprising result (Figure 10.22) was that the entire 5′-flanking region of the gene could be removed without affecting transcription very much. Furthermore, big chunks of the 5′-end of the gene itself could be removed, and a transcript of about 120 nt would still be made. However, deletions beyond about position +50 destroyed promoter function.

Using a similar approach, Brown and colleagues identified a sensitive region between bases 50 and 83 of the transcribed sequence that could not be encroached on without destroying promoter function. These are the apparent outer

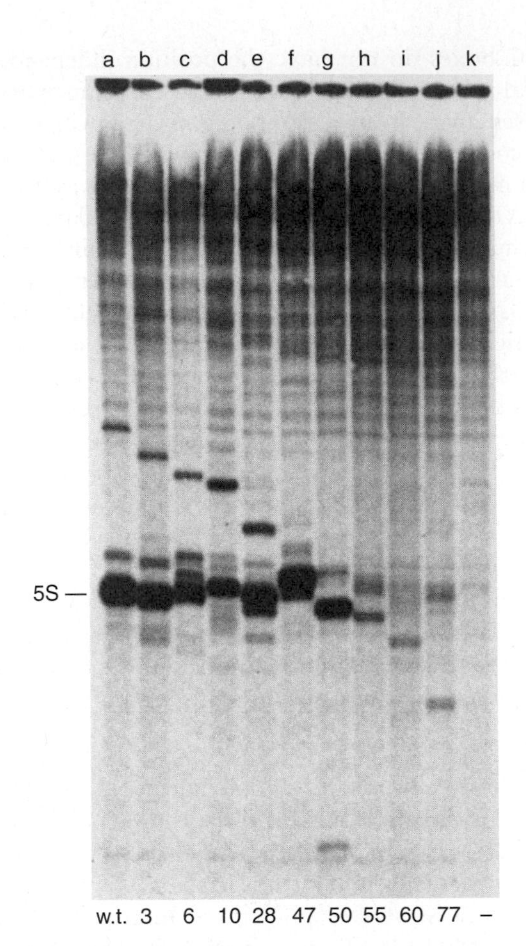

Figure 10.22 Effect of 5′-deletions on 5S rRNA gene transcription. Brown and colleagues prepared a series of deleted *Xenopus borealis* 5S rRNA genes with progressively more DNA deleted from the 5′-end of the gene itself. Then they transcribed these deleted genes in vitro in the presence of labeled substrate and electrophoresed the labeled products. DNA templates: lane a, undeleted positive control; lanes b–j, deleted genes with the position of the remaining 5′-end nucleotide denoted at bottom (e.g., lane b contained the product of a 5S rRNA gene whose 5′-end is at position +3 relative to the wild-type gene); lane k, negative control (pBR322 DNA with no 5S rRNA gene). Strong synthesis of a 5S-size RNA took place with all templates through lane g, in which deletion up to position +50 had occurred. With further deletion into the gene, this synthesis ceased. Lanes h–k also contained a band in this general area, but it is an artifact unrelated to 5S rRNA gene transcription. (*Source:* Sakonju, S., D.F. Bogenhagen, and D.D. Brown. A control region in the center of the 5S RNA gene directs specific initiation of transcription: I. The 5′ border of the region. *Cell* 19 (Jan 1980) p. 17, f. 4.)

boundaries of the internal promoter of the *Xenopus* 5S rRNA gene. Other experiments showed that it is possible to *add* chunks of DNA outside this region without harming the promoter. Roeder and colleagues later performed systematic mutagenesis of bases throughout the promoter region and identified three regions that could not be changed without greatly diminishing promoter function. These sensitive regions are called *box A,* the *intermediate element,* and *box C.* (No box B occurs because a box B had already been discovered in other class III genes, and it had no counterpart in the 5S rRNA promoter.) Figure 10.23a summarizes the results of these experiments on the 5S rRNA promoter. Similar experiments on the other two classical class III genes, the tRNA and VA RNA genes, showed that their promoters contain a *box A* and a *box B* (Figure 10.23b). The sequence of the box A is similar to that of the box A of the 5S rRNA gene. Furthermore, the space in between the two blocks can be altered somewhat without destroying promoter function. Such alteration does have limits, however; if one inserts too much DNA between the two promoter boxes, efficiency of transcription suffers.

Thus, we see that there are several kinds of class III promoters. The 5S rRNA genes are in a group by themselves, called *type I* (Figure 10.23a). Do not confuse this with "class I;" we are discussing only class III promoters here. The second group, *type II,* contains most class III promoters, which look like the tRNA and VA RNA promoters in Figure 10.23b. The third group, *type III,* contains the nonclassical promoters with control elements restricted to the 5′-flanking region of the gene. These, promoters are typified by the human 7SK RNA promoter and the human U6 RNA promoter (Figure 10.23c). By the way, the U6 RNA is a member of a group of small nuclear RNAs (snRNAs) that are key players in mRNA splicing, which we will discuss in Chapter 14. Finally, there are promoters that appear to be hybrids of types II and III, such as the human 7SL promoter. These have both internal and external elements that are important for promoter activity.

SUMMARY RNA polymerase III transcribes a set of short genes. The classical class III genes (types I and II) have promoters that lie wholly within the genes. The internal promoter of the type I class III gene (the 5S rRNA gene) is split into three regions: box A, a short intermediate element, and box C. The internal promoters of the type II genes (e.g., the tRNA genes) are split into two parts: box A and box B. The promoters of the nonclassical (type III) class III genes resemble those of class II genes.

Class III Genes with Class II-like Promoters After Brown and other investigators established the novel idea of internal promoters for class III genes, it was generally assumed that all class III genes worked this way. However, by the mid-1980s some exceptions were discovered. The **7SL RNA** is part of the signal recognition particle that recognizes a signal sequence in certain mRNAs and targets their translation to membranes such as the endoplasmic reticulum. In 1985, Elisabetta Ullu and Alan Weiner conducted in vitro transcription studies on wild-type and mutant 7SL RNA genes that showed that the 5′-flanking region was required for high-level transcription. Without this DNA region, transcription efficiency dropped by 50–100-fold. Ullu and Weiner concluded that the most important DNA element for transcription of this gene lies upstream of the gene. Nevertheless, the fact that transcription still occurred in mutant genes lacking the 5′-flanking region implies that these genes also contain a weak internal promoter. These data help explain why the hundreds

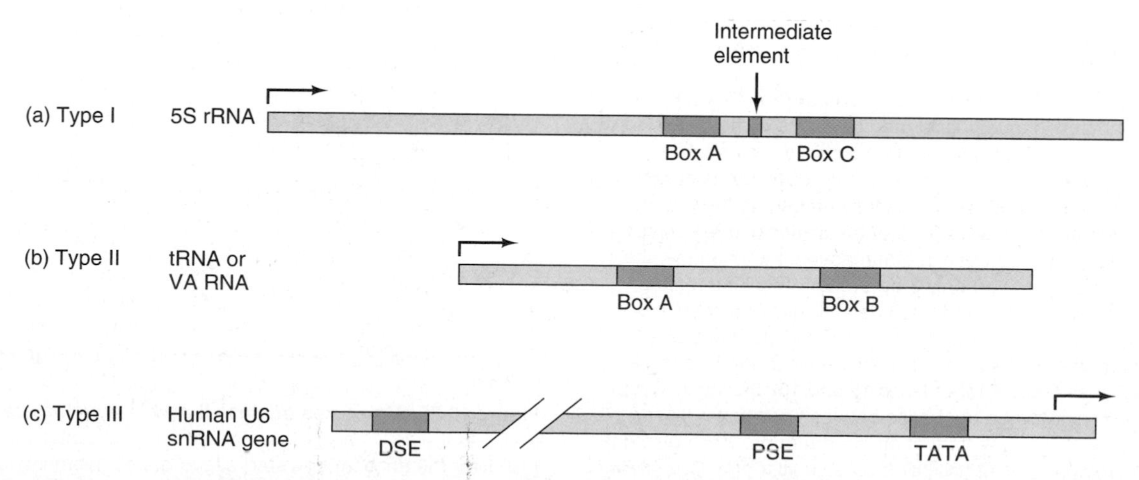

Figure 10.23 Promoters of some class III genes. The promoters of the 5S, tRNA and U6 RNA genes are depicted as groups of blue boxes within the genes they control. DSE and PSE are distal and proximal sequence elements, respectively.

of 7SL RNA pseudogenes (nonfunctional copies of the 7SL gene) in the human genome, as well as the related *Alu* sequences (remnants of transposons, Chapter 23), are relatively poorly transcribed in vivo: They lack the upstream element required for high-level transcription.

Marialuisa Melli and colleagues noticed that the 7SK RNA gene does not have internal sequences that resemble the classic class III promoter. On the other hand, the 7SK RNA gene does have a 5′-flanking region homologous to that of the 7SL RNA gene. On the basis of these observations, they proposed that this gene has a completely external promoter. To prove the point, they made successive deletions in the 5′-flanking region of the gene and tested them for ability to support transcription in vitro. Figure 10.24 shows that deletions up to position −37 still allowed production of high levels of 7SK RNA, but deletions downstream of this point were not tolerated. On the other hand, the coding region was not needed for transcription: In vitro transcription analysis of another batch of deletion mutants, this time with deletions within the coding region, showed that transcription still occurred, even when the whole coding region was removed. Thus, this gene lacks an internal promoter.

What is the nature of the promoter located in the region encompassing the 37 bp upstream of the start site? Interestingly enough, a TATA box resides in this region, and changing three of its bases (TAT→GCG) reduced transcription by 97%. Thus the TATA box is required for good promoter function. All this may make you wonder whether polymerase II, not polymerase III, really transcribes this gene after all. If that were the case, low concentrations of α-amanitin should inhibit transcription, but it takes high concentrations of this toxin to block 7SK RNA synthesis. In fact, the profile of inhibition of 7SK RNA synthesis by α-amanitin is exactly what we would expect if polymerase III, not polymerase II, is involved. By the way, the **7SK RNA** plays a role in controlling the phosphorylation of one serine (serine 2) in the repeating heptad of the CTD of Rpb1 of RNA polymerase II. We will see in Chapter 11 that this phosphorylation is required for the transition from transcription initiation to elongation.

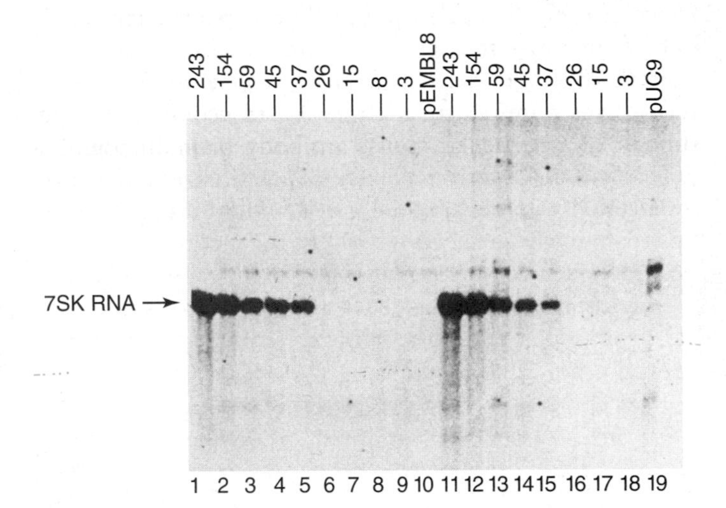

Figure 10.24 Effects of 5′-deletion mutations on the 7SK RNA promoter. Melli and colleagues performed deletions in the 5′-flanking region of the human 7SK RNA gene and transcribed the mutated genes in vitro. Then they electrophoresed the products to determine if 7SK RNA was still synthesized. The negative numbers at the top of each lane give the number of base pairs of the 5′-flanking region still remaining in the deleted gene used in that reaction. For example, the template used in lane 9 retained only 3 bp of the 5′-flanking region—up to position −3. Lanes 1–10 contained deleted genes cloned into the vector pEMBL8; lanes 11–19 contained genes cloned into pUC9. The cloning vectors themselves were transcribed in lanes 10 and 19. A comparison of lanes 5 and 6 (or of lanes 15 and 16) shows an abrupt drop in promoter activity when the bases between position −37 and −26 were removed. This suggests that an important promoter element lies in this 11-bp region. (*Source:* Murphy, S., C. DiLiegro, and M. Melli, The in vitro transcription of the 7SK RNA gene by RNA polymerase III is dependent only on the presence of an upstream promoter. *Cell* 51 (9) (1987) p. 82, f. 1b.)

Now we know that the other nonclassical class III genes, including the U6 RNA gene and the EBER2 gene, behave the same way. They are transcribed by polymerase III, but they have polymerase II-like promoters. In Chapter 11 we will see that this is not as strange as it seems at first because the TATA-binding protein (TBP) is involved in class III (and class I) transcription, in addition to its well-known role in class II gene transcription.

The small nuclear RNA (snRNA) genes present a fascinating comparison of class II and class III nonclassical promoters. In Chapter 14 we will learn that many eukaryotic mRNAs are synthesized as over-long precursors that need to have internal sections (introns) removed in a process called splicing. This pre-mRNA splicing requires several small nuclear RNAs (snRNAs). Most of these, including U1 and U2 snRNAs, are made by RNA polymerase II. But their promoters do not look like typical class II promoters. Instead, in humans, each promoter contains two elements (Figure 10.25a): a **proximal sequence element (PSE)**, which is essential, and a **distal sequence element (DSE)**, which confers greater efficiency.

One of the snRNAs, U6 snRNA, is made by RNA polymerase III. As usual with nonclassical class III promoters, the human U6 snRNA promoter (Figure 10.25b), with its TATA

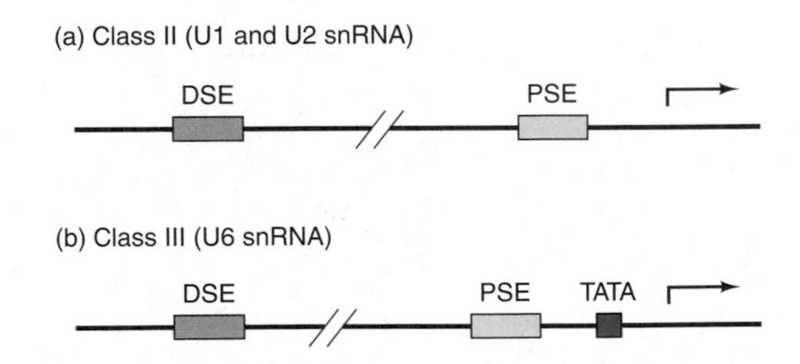

Figure 10.25 Structures of class II and III nonclassical promoters. **(a)** Class II: The U1 and U2 snRNA promoters contain an essential PSE near the transcription start site and a supplementary DSE further upstream. **(b)** Class III: The U6 snRNA promoter contains a TATA box in addition to the PSE and DSE.

box, looks more like a class II promoter. Paradoxically, removal of that TATA box converts the promoter from class III to class II. Similarly, adding a TATA box to a U1 or U2 snRNA promoter converts it from class II to class III. One might have predicted just the opposite. By contrast, in *Drosophila* and in sea urchins, some snRNA genes have TATA boxes and others do not, but other sequence elements, not the TATA boxes, determine whether the promoters are class II or class III.

SUMMARY At least one class III gene, the 7SL RNA gene, contains a weak internal promoter, as well as a sequence in the 5′-flanking region of the gene that is required for high-level transcription. Other nonclassical class III genes (e.g., 7SK, and U6 RNA genes) lack internal promoters altogether, and contain promoters that strongly resemble class II promoters in that they lie in the 5′-flanking region and contain TATA boxes. The U1 and U6 snRNA genes have nonclassical class II and III promoters, respectively. The U1 snRNA promoter has an essential proximal sequence element (PSE), and a distal sequence element (DSE) and is transcribed by polymerase II. The U6 snRNA promoter has a PSE, a DSE, and a TATA box, and is transcribed by polymerase III.

10.3 Enhancers and Silencers

Many eukaryotic genes, especially class II genes, are associated with *cis*-acting DNA elements that are not strictly part of the promoter, yet strongly influence transcription. As we learned in Chapter 9, **enhancers** are elements that stimulate transcription. **Silencers,** by contrast, depress transcription. We will discuss these elements briefly here and expand on their modes of action in Chapters 12 and 13.

Enhancers

Chambon and colleagues discovered the first enhancer in the 5′-flanking region of the SV40 early gene. This DNA region had been noticed before because it contains a conspicuous duplication of a 72-bp sequence, called the *72-bp repeat* (Figure 10.26). When Benoist and Chambon made deletion mutations in this region, they observed profoundly depressed transcription in vivo. This behavior suggested that the 72-bp repeats constituted another upstream promoter element. However, Paul Berg and his colleagues discovered that the 72-bp repeats still stimulated transcription even if they were inverted or moved all the way around to the opposite side of the circular SV40 genome, over 2 kb away from the promoter. The latter behavior, at least, is very un-promoter-like. Thus, such orientation- and position-independent DNA elements are called enhancers to distinguish them from promoter elements.

How do enhancers stimulate transcription? We will see in Chapter 12 that enhancers act through proteins that bind to them. These have several names: **transcription factors, enhancer-binding proteins,** or **activators.** These proteins appear to stimulate transcription by interacting with other proteins called **general transcription factors** at the promoter. This interaction promotes formation of a preinitiation complex, which is necessary for transcription. Thus, enhancers usually allow a gene to be induced (or sometimes repressed) by activators. We will discuss these interactions in much greater detail in Chapters 11 and 12 and we will see that activators frequently require help from other molecules (e.g., hormones and coactivator proteins) to exert their effects.

We frequently find enhancers upstream of the promoters they control, but this is by no means an absolute rule. In fact, as early as 1983 Susumo Tonegawa and his colleagues found an example of an enhancer *within* a gene. These investigators were studying a gene that encodes the larger subunit of a particular mouse antibody, or immunoglobulin, called γ_{2b}. They introduced this gene into mouse plasmacytoma cells that normally expressed antibody genes, but not this particular gene. To detect efficiency of expression of the transfected cells, they added a labeled amino acid to tag newly made proteins, then **immunoprecipitated** the labeled γ_{2b} protein (Chapter 5) with an antibody directed against γ_{2b}. Then they electrophoresed the immunoprecipitated proteins and detected them by autoradiography. The suspected enhancer lay in one of the gene's introns, a region within the gene that is transcribed, but is subsequently cut out of the transcript by a process called splicing (Chapter 14). Tonegawa and colleagues began by deleting two chunks of DNA from this suspected enhancer region, as shown in Figure 10.27a. Then they assayed for expression of the γ_{2b} gene in cells transfected by this mutated DNA. Figure 10.27b shows the results: The deletions within the intron, though they should have no effect on the protein product because they are in a noncoding region of

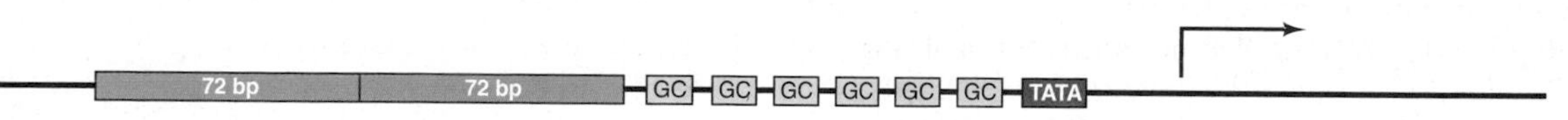

Figure 10.26 Structure of the SV40 virus early control region. As usual, an arrow with a right-angle bend denotes the transcription initiation site, although this is actually a cluster of three sites, as we saw in Figure 10.19. Upstream of the start site we have, in right-to-left order, the TATA box (red), six GC boxes (yellow), and the enhancer (72-bp repeats, blue).

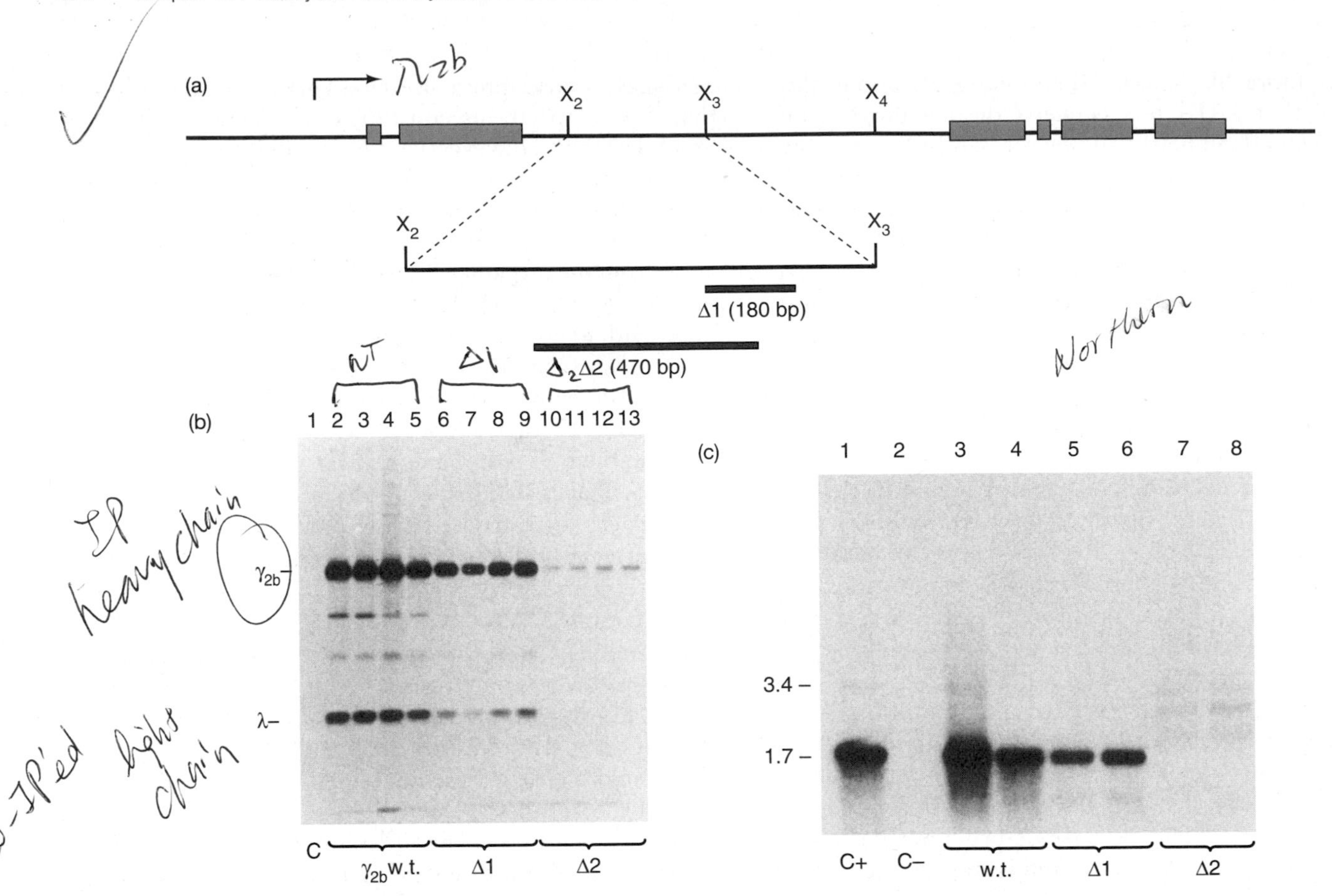

Figure 10.27 Effects of deletions in the immunoglobulin γ_{2b} H-chain enhancer. (a) Map of the cloned γ_{2b} gene. The blue boxes represent the exons of the gene, the parts that are included in the mRNA that comes from this gene. The lines in between boxes are introns, regions of the gene that are transcribed, but then cut out of the mRNA precursor as it is processed to the mature mRNA. X_2, X_3, and X_4 represent cutting sites for the restriction enzyme *Xba*I. Tonegawa and colleagues suspected an enhancer lay in the X_2–X_3 region, so they made deletions Δ1 and Δ2 as indicated by the red boxes. **(b)** Assay of expression of the γ_{2b} gene at the protein level. Tonegawa and colleagues transfected plasmacytoma cells with the wild-type gene (lanes 2–5), the gene with deletion Δ1 (lanes 6–9), or the gene with deletion Δ2 (lanes 10–13). Lane 1 was a control with untransfected plasmacytoma cells. After transfecting the cells, these investigators added a radioactive amino acid to label any newly made protein, then extracted the protein, immunoprecipitated the γ_{2b} protein, electrophoresed the precipitated protein, and detected the radioactive protein by fluorography (a modified version of autoradiography in which a compound called a fluor is added to the electrophoresis gel). Radioactive emissions excite this fluor to give off photons that are detected by x-ray film. The Δ1 deletion produced only a slight reduction in expression of the gene, but the Δ2 deletion gave a profound reduction. **(c)** Assay of transcription of the γ_{2b} gene. Tonegawa and colleagues electrophoresed and Northern blotted RNA from the following cells: lane 1 (positive control), untransfected plasmacytoma cells (MOPC 141) that expressed the γ_{2b} gene; lane 2 (negative control), untransfected plasmacytoma cells (J558L) that did not express the γ_{2b} gene; lanes 3 and 4, J558L cells transfected with the wild-type γ_{2b} gene; lanes 5 and 6, J558L cells transfected with the gene with the Δ1 deletion; lanes 7 and 8, J558L cells transfected with the gene with the Δ2 deletion. The Δ1 deletion decreased transcription somewhat, but the Δ2 deletion abolished transcription. (*Source:* (*b–c*) Gillies, S.D., S.L. Morrison, V.T. Oi, and S. Tonegawa, A tissue-specific transcription enhancer element is located in the major intron of a rearranged immunoglobulin heavy chain gene. *Cell* 33 (July 1983) p. 719, f. 2&3.)

the gene, caused a decrease in the amount of gene product made. This was especially pronounced in the case of the larger deletion (Δ2).

Is this effect due to decreased transcription, or some other cause? Tonegawa's group answered this question by performing Northern blots (Chapter 5) with RNA from cells transfected with normal and deleted γ_{2b} genes. These blots, shown in Figure 10.27c, again demonstrated a profound loss of function when the suspected enhancer was deleted. But is this really an enhancer? If so, one should be able to move it or invert it and it should retain its activity. Tonegawa and colleagues did this by first inverting the X_2–X_3 fragment, which contained the enhancer, as shown in position/orientation B of Figure 10.28a. Figure 10.28b shows that the enhancer still functioned. Next, they took fragment X_2–X_3 out of the intron and placed it upstream of the promoter (position/orientation C). It still worked. Then they inverted it in its new location (position/orientation D). Still it functioned. Thus, some region within the X_2–X_3 fragment behaved as an enhancer: It stimulated transcription from a nearby promoter, and it was position- and orientation-independent.

Finally, these workers compared the expression of this gene when it was transfected into two different types of mouse cells: plasmacytoma cells as before, and fibroblasts.

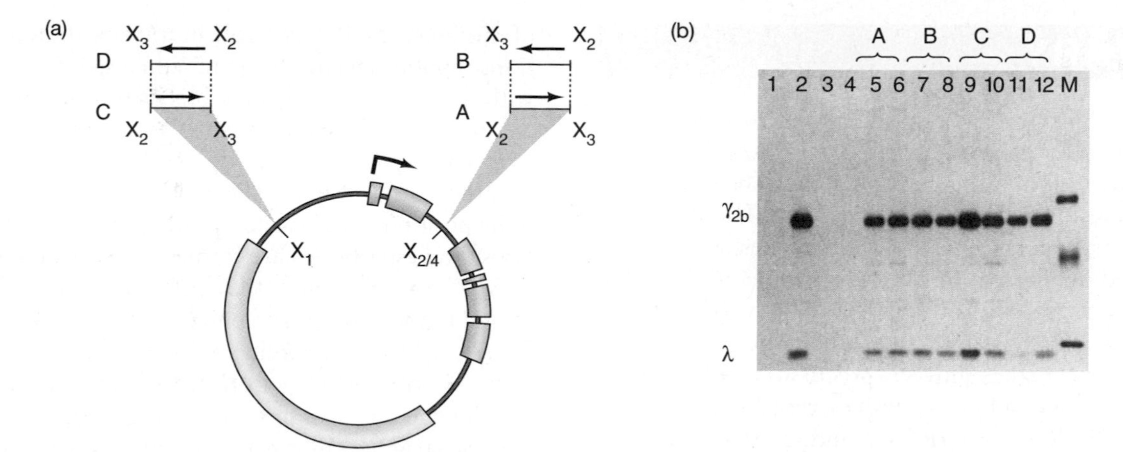

Figure 10.28 The enhancing element in the γ_{2b} gene is orientation- and position-independent. (a) Outline of the mutant plasmids. Tonegawa and colleagues removed the X_2–X_3 region of the parent plasmid containing the γ_{2b} gene (see Figure 10.27a). This deleted the enhancer. Then they reinserted the X_2–X_3 fragment (with the enhancer) in four different ways: plasmids A and B, the fragment was inserted back into the intron in its usual location in the forward (normal) orientation (A), or in the backward orientation (B); plasmids C and D, the fragment was inserted into another *Xba*I site (X_1) hundreds of base pairs upstream of the gene in the forward orientation (C), or in the backward orientation (D). **(b)** Experimental results. Tonegawa and colleagues tested all four plasmids from (a), as well as the parent, for efficiency of expression as in Figure 10.27b. All functioned equally well. Lane 1, untransfected J558L cells lacking the γ_{2b} gene. Lanes 2–12, J558L cells transfected with the following plasmids: lane 2, the parent plasmid with no deletions; lanes 3 and 4, the parent plasmid with the X_2–X_3 fragment deleted; lanes 5 and 6, plasmid A; lanes 7 and 8, plasmid B; lanes 9 and 10, plasmid C; lanes 11 and 12, plasmid D. Lane M contained protein size markers. (*Source:* (a) Adapted from Gillies, S.D., S.L. Morrison, V.T. Oi, and S. Tonegawa, A tissue-specific transcription enhancer element is located in the major intron of a rearranged immunoglobulin heavy chain gene. *Cell* 33 (July 1983) p. 721, f. 5.)

Expression was much more active in plasmacytoma cells. This is also consistent with enhancer behavior because fibroblasts do not make antibodies and therefore should not contain enhancer-binding proteins capable of activating the enhancer of an antibody gene. Thus, the antibody gene should not be expressed actively in such cells.

The finding that a gene is much more active in one cell type than in another leads to an extremely important point: All cells contain the same genes, but different cell types differ greatly from one another: A nerve cell, for example, is much different from a liver cell, in shape and function. What makes these cells differ so much? The proteins in the cells. And, as we have learned, the suite of proteins in each cell type is determined by the genes that are active in those cells. And what activates those genes? We now see that the activators are transcription factors that bind to enhancers. Thus, different cell types express different activators that turn on different genes that produce different proteins. We will expand on this vital theme in several chapters to follow.

Silencers

Enhancers are not the only DNA elements that can act at a distance to modulate transcription. Silencers also do this, but—as their name implies—they inhibit rather than stimulate transcription. The mating system *(MAT)* of yeast provides a good example. Yeast chromosome III contains three loci of very similar sequence: *MAT, HML,* and *HMR*. Though *MAT* is expressed, the other two loci are not, and silencers located at least 1 kb away seem to be responsible for this genetic inactivity. We know that something besides the inactive genes themselves is at fault, because active yeast genes can be substituted for *HML* or *HMR* and the transplanted genes become inactive. Thus, they seem to be responding to an external negative influence: a silencer. How do silencers work? The available data indicate that they cause the chromatin to coil up into a condensed, inaccessible, and therefore inactive form, thereby preventing transcription of neighboring genes. We will examine this process in more detail in Chapter 13.

Sometimes the same DNA element can have both enhancer and silencer activity, depending on the protein bound to it. For example, the thyroid hormone response element acts as a silencer when the thyroid hormone receptor binds to it without its ligand, thyroid hormone. But it acts as an enhancer when the thyroid hormone receptor binds along with thyroid hormone. We will revisit this concept in Chapter 12.

SUMMARY Enhancers and silencers are position- and orientation-independent DNA elements that stimulate or depress, respectively, the transcription of associated genes. They are also tissue-specific in that they rely on tissue-specific DNA-binding proteins for their activities. Sometimes a DNA element can act as either an enhancer or a silencer depending on what is bound to it.

SUMMARY

Eukaryotic nuclei contain three RNA polymerases that can be separated by ion-exchange chromatography. RNA polymerase I is found in the nucleolus; the other two polymerases are located in the nucleoplasm. The three nuclear RNA polymerases have different roles in transcription. Polymerase I makes a large precursor to the major rRNAs (5.8S, 18S, and 28S rRNAs in vertebrates). Polymerase II synthesizes hnRNAs, which are precursors to mRNAs. It also makes miRNA precursors and most small nuclear RNAs (snRNAs). Polymerase III makes the precursors to 5S rRNA, the tRNAs, and several other small cellular and viral RNAs.

The subunit structures of all three nuclear polymerases from several eukaryotes have been determined. All of these structures contain many subunits, including two large ones, with molecular masses greater than 100 kD. All eukaryotes seem to have at least some common subunits that are found in all three polymerases. The genes encoding all 12 RNA polymerase II subunits in yeast have been sequenced and subjected to mutation analysis. Three of the subunits resemble the core subunits of bacterial RNA polymerases in both structure and function, five are found in all three nuclear RNA polymerases, two are not required for activity, at least at normal temperatures, and two fall into none of these three categories.

Subunit IIa is the primary product of the RPB1 gene in yeast. It can be converted to IIb in vitro by proteolytic removal of the carboxyl-terminal domain (CTD), which is essentially a heptapeptide repeated over and over. Subunit IIa is converted in vivo to IIo by phosphorylating two serines within the CTD heptad. The enzyme (polymerase IIA) with the IIa subunit is the one that binds to the promoter; the enzyme (polymerase IIO) with the IIo subunit is the one involved in transcript elongation.

The structure of yeast pol II Δ4/7 reveals a deep cleft that can accept a DNA template. The catalytic center, containing a Mg^{2+} ion lies at the bottom of the cleft. A second Mg^{2+} ion is present in low concentration and presumably enters the enzyme bound to each substrate nucleotide.

The crystal structure of a transcription elongation complex involving yeast RNA polymerase II (lacking Rpb4/7) reveals that the clamp is indeed closed over the RNA–DNA hybrid in the enzyme's cleft, ensuring processivity of transcription. In addition, three loops of the clamp—the rudder, lid, and zipper—appear to play important roles in, respectively: initiating dissociation of the RNA–DNA hybrid, maintaining this dissociation, and maintaining dissociation of the template DNA. The active center of the enzyme lies at the end of pore 1, which appears to be the conduit for nucleotides to enter the enzyme and for extruded RNA to exit the enzyme during backtracking. A bridge helix lies adjacent to the active center, and flexing of this helix could play a role in translocation during transcription. The toxin α-amanitin appears to interfere with this flexing and thereby blocks translocation.

In moving through the entry pore toward the active site of RNA polymerase II, an incoming nucleotide first encounters the E (entry) site, where it is inverted relative to its position in the A site, the active site where phosphodiester bonds are formed. Two metal ions (Mg^{2+} or Mn^{2+}) are present at the active site. One is permanently bound to the enzyme and one enters the active site complexed to the incoming nucleotide. The trigger loop of Rpb1 positions the substrate for incorporation and discriminates against improper nucleotides.

The structure of the 12-subunit RNA polymerase II reveals that, with Rpb4/7 in place, the clamp is forced shut. Because initiation occurs with the 12-subunit enzyme, with its clamp shut, it appears that the promoter DNA must melt before the template DNA strand can descend into the enzyme's active site. It also appears that Rpb4/7 extends the dock region of the polymerase, making it easier for certain general transcription factors to bind, thereby facilitating transcription initiation.

Class II promoters may consist of a core promoter immediately surrounding the transcription start site, and a proximal promoter farther upstream. The core promoter may contain up to six conserved elements: the TFIIB recognition element (BRE), the TATA box, the initiator (Inr), the downstream core element (DCE), the motif ten element (MTE), and the downstream promoter element (DPE). At least one of these elements is missing in most promoters. Promoters for highly expressed specialized genes tend to have TATA boxes, but promoters for housekeeping genes tend to lack them.

Proximal promoter elements are usually found upstream of class II core promoters. They differ from the core promoter in that they bind to relatively gene-specific transcription factors. For example, GC boxes bind the transcription factor Sp1, while CCAAT boxes bind CTF. The proximal promoter elements, unlike the core promoter, can be orientation-independent, but they are relatively position-dependent, unlike classical enhancers.

Class I promoters are not well conserved in sequence from one species to another, but the general architecture of the promoter is well conserved. It consists of two elements: a core element surrounding the transcription start site, and an upstream promoter element (UPE) about 100 bp farther upstream. The spacing between these two elements is important.

RNA polymerase III transcribes a set of short genes. The classical class III genes (types I and II) have promoters that lie wholly within the genes. The internal promoter of the type I class III gene (the 5S rRNA gene) is split into three regions: box A, a short intermediate element, and box C. The internal promoters of the type II genes (e.g., the tRNA gene) are split into two parts: box A and box B.

Other class III genes called type III (e.g., 7SK, and U6 RNA genes) lack internal promoters altogether and contain promoters that strongly resemble class II promoters in that they lie in the 5′-flanking region and contain TATA boxes. The U1 and U6 snRNA genes have nonclassical class II and III promoters, respectively. The U1 snRNA promoter has an essential proximal sequence element (PSE) and a distal sequence element (DSE). The U6 snRNA promoter has a PSE, a DSE, and a TATA box.

Enhancers and silencers are position- and orientation-independent DNA elements that stimulate or depress, respectively, the transcription of associated genes. They are also tissue-specific in that they rely on tissue-specific DNA-binding proteins for their activities.

REVIEW QUESTIONS

1. Diagram the elution pattern of the eukaryotic nuclear RNA polymerases from DEAE-Sephadex chromatography. Show what you would expect if you assayed the same fractions in the presence of 1 μg/mL of α-amanitin.
2. Describe and give the results of an experiment that shows that polymerase I is located primarily in the nucleolus of the cell.
3. Describe and give the results of an experiment that shows that polymerase III makes tRNA and 5S rRNA.
4. How many subunits does yeast RNA polymerase II have? Which of these are "core" subunits? How many subunits are common to all three nuclear RNA polymerases?
5. Describe how epitope tagging can be used to purify polymerase II from yeast in one step.
6. Some preparations of polymerase II show three different forms of the largest subunit (RPB1). Give the names of these subunits and show their relative positions after SDS-PAGE. What are the differences among these subunits? Present evidence for these conclusions.
7. What is the structure of the CTD of RPB1?
8. Draw a rough diagram of the structure of yeast RNA polymerase II. Show where the DNA lies, and provide another piece of evidence that supports this location for DNA. Also, show the location of the active site.
9. How many Mg^{2+} ions are proposed to participate in catalysis at the active center of RNA polymerases? Why is one of these metal ions difficult to see in the crystal structure of yeast RNA polymerase II?
10. Cite evidence to support pore 1 as the likely exit point for RNA extrusion during polymerase II backtracking.
11. What is meant by the term "processive transcription?" What part of the polymerase II structure ensures processivity?
12. What is the probable function of the rudder of polymerase II?
13. What is the probable function of the bridge helix? What is the relationship of α-amanitin to this function?
14. What are the E site and A site of RNA polymerase II? What roles are they thought to play in nucleotide selection?
15. What role does the polymerase II trigger loop play in nucleotide selection? Illustrate with a schematic diagram of contacts to the base, sugar, and triphosphate.
16. What role does the Rpb4/7 complex play in opening or closing the clamp of RNA polymerase II? What evidence supports this role?
17. The 12-subunit RNA polymerase II interacts with promoter DNA. What implications does this have for the state of the promoter DNA with which the polymerase must interact?
18. Draw a diagram of a composite polymerase II promoter, showing all of the types of elements it could have.
19. What kinds of genes tend to have TATA boxes? What kinds of genes tend not to have them?
20. What is the probable relationship between TATA boxes and DPEs?
21. What are the two most likely effects of removing the TATA box from a class II promoter?
22. Describe the process of linker scanning. What kind of information does it give?
23. List two common proximal promoter elements of class II promoters. How do they differ from core promoter elements?
24. Diagram a typical class I promoter.
25. How were the elements of class I promoters discovered? Present experimental results.
26. Describe and give the results of an experiment that shows the importance of spacing between the elements of a class I promoter.
27. Compare and contrast (with diagrams) the classical and nonclassical class III promoters. Give an example of each.
28. Diagram the structures of the U1 and U6 snRNA promoters. Which RNA polymerase transcribes each? What is the effect of moving the TATA box from one of these promoters to the other? Why does this seem paradoxical?
29. Describe and give the results of an experiment that locates the 5′-border of the 5S rRNA gene's promoter.
30. Explain the fact that enhancer activity is tissue-specific.

ANALYTICAL QUESTIONS

1. Transcription of a class II gene starts at a guanosine 25 bp downstream of the last base of the TATA box. You delete 20 bp of DNA between this guanosine and the TATA box and transfect cells with this mutated DNA. Will transcription still start at the same guanosine? If not, where? How would you locate the transcription start site?
2. You suspect that a repeated sequence just upstream of a gene is acting as an enhancer. Describe and predict the results of an experiment you would run to test your hypothesis. Be sure your experiment shows that the sequence acts as an enhancer and not as a promoter element.
3. You are investigating a new class II promoter, but you can find no familiar sequences. Design an experiment to locate the promoter sequences, and show sample results.
4. Describe a primer extension assay you could use to define the 3′-end of the 5S rRNA promoter.

SUGGESTED READINGS

General References and Reviews

Corden, J.L. 1990. Tales of RNA polymerase II. *Trends in Biochemical Sciences* 15:383–87.

Klug, A. 2001. A marvelous machine for making messages. *Science* 292:1844–46.

Landick, R. 2004. Active-site dynamics in RNA polymerases. *Cell* 116:351–53.

Lee, T.I. and R.A. Young. 2000. Eukaryotic transcription. *Annual Review of Genetics* 34:77–137.

Paule, M.R. and R.J. White. 2000. Transcription by RNA polymerases I and III. *Nucleic Acids Research* 28:1283–98.

Sentenac, A. 1985. Eukaryotic RNA polymerases. *CRC Critical Reviews in Biochemistry* 18:31–90.

Woychik, N.A. and R.A. Young. 1990. RNA polymerase II: Subunit structure and function. *Trends in Biochemical Sciences* 15:347–51.

Research Articles

Benoist, C. and P. Chambon. 1981. In vivo sequence requirements of the SV40 early promoter region. *Nature* 290:304–10.

Bogenhagen, D.F., S. Sakonju, and D.D. Brown. 1980. A control region in the center of the 5S RNA gene directs specific initiation of transcription: II. The 3′ border of the region. *Cell* 19:27–35.

Bushnell, D.A., P. Cramer, and R.D. Kornberg. 2002. Structural basis of transcription: α-amanitin–RNA polymerase cocrystal at 2.8 Å resolution. *Proceedings of the National Academy of Sciences USA* 99:1218–22.

Cramer, P., D.A. Bushnell, and R.D. Kornberg. 2001. Structural basis of transcription: RNA polymerase II at 2.8 Ångstrom resolution. *Science* 292:1863–76.

Das, G., D. Henning, D. Wright, and R. Reddy. 1988. Upstream regulatory elements are necessary and sufficient for transcription of a U6 RNA gene by RNA polymerase III. *EMBO Journal* 7:503–12.

Gillies, S.D., S.L. Morrison, V.T. Oi, and S. Tonegawa. 1983. A tissue-specific transcription enhancer element is located in the major intron of a rearranged immunoglobulin heavy chain gene. *Cell* 33:717–28.

Gnatt, A.L., P. Cramer, J. Fu, D.A. Bushnell, and R.D. Kornberg. 2001. Structural basis of transcription: An RNA polymerase II elongation complex at 3.3 Å resolution. *Science* 292:1876–82.

Haltiner, M.M., S.T. Smale, and R. Tjian. 1986. Two distinct promoter elements in the human rRNA gene identified by linker scanning mutagenesis. *Molecular and Cellular Biology* 6:227–35.

Kolodziej, P.A., N. Woychik, S.-M. Liao, and R. Young. 1990. RNA polymerase II subunit composition, stoichiometry, and phosphorylation. *Molecular and Cellular Biology* 10:1915–20.

Kutach, A.K. and J.T. Kadonaga. 2000. The downstream promoter element DPE appears to be as widely used as the TATA box in *Drosophila* core promoters. *Molecular and Cellular Biology* 20:4754–64.

Learned, R.M., T.K. Learned, M.M. Haltiner, and R.T. Tjian. 1986. Human rRNA transcription is modulated by the coordinate binding of two factors to an upstream control element. *Cell* 45:847–57.

McKnight, S.L. and R. Kingsbury. 1982. Transcription control signals of a eukaryotic protein-coding gene. *Science* 217:316–24.

Murphy, S., C. Di Liegro, and M. Melli. 1987. The in vitro transcription of the 7SK RNA gene by RNA polymerase III is dependent only on the presence of an upstream promoter. *Cell* 51:81–87.

Pieler, T., J. Hamm, and R.G. Roeder. 1987. The 5S gene internal control region is composed of three distinct sequence elements, organized as two functional domains with variable spacing. *Cell* 48:91–100.

Roeder, R.G. and W.J. Rutter. 1969. Multiple forms of DNA-dependent RNA polymerase in eukaryotic organisms. *Nature* 224:234–37.

Roeder, R.G. and W.J. Rutter. 1970. Specific nucleolar and nucleoplasmic RNA polymerases. *Proceedings of the National Academy of Sciences USA* 65:675–82.

Sakonju, S., D.F. Bogenhagen, and D.D. Brown. 1980. A control region in the center of the 5S RNA gene directs initiation of transcription: I. The 5′ border of the region. *Cell* 19:13–25.

Sayre, M.H., H. Tschochner, and R.D. Kornberg. 1992. Reconstitution of transcription with five purified initiation factors and RNA polymerase II from *Saccharomyces cerevisiae*. *Journal of Biological Chemistry* 267:23376–82.

Sklar, V.E.F., L.B. Schwartz, and R.G. Roeder. 1975. Distinct molecular structures of nuclear class I, II, and III DNA-dependent RNA polymerases. *Proceedings of the National Academy of Sciences USA* 72:348–52.

Smale, S.T. and D. Baltimore. 1989. The "initiator" as a transcription control element. *Cell* 57:103–13.

Ullu, E. and A.M. Weiner. 1985. Upstream sequences modulate the internal promoter of the human 7SL RNA gene. *Nature* 318:371–74.

Wang, D., D.A. Bushnell, K.D. Westover, C.D. Kaplan, and R.D. Kornberg. 2006. Structural basis of transcription: Role of the trigger loop in substrate specificity and catalysis. *Cell* 127:941–954.

Weinman, R. and R.G. Roeder. 1974. "Role of DNA-dependent RNA polymerase III in the transcription of the tRNA and 5S rRNA genes." *Proceedings of the National Academy of Sciences USA* 71:1790–94.

Westover, K.D., D.A. Bushnell, and R.D. Kornberg. 2004. Structural basis of transcription: Separation of RNA from DNA by RNA polymerase II. *Science* 303:1014–16.

Westover, K.D., D.A. Bushnell, and R.D. Kornberg. 2004. Structural basis of transcription: Nucleotide selection by rotation in the RNA polymerase II active center. *Cell* 119:481–89.

Woychik, N.A., S.M. Liao, P.A. Kolodziej, and R.A. Young. 1990. Subunits shared by eukaryotic nuclear RNA polymerases. *Genes and Development* 4:313–23.

Woychik, N.A., et al. 1993. Yeast RNA polymerase II subunit RPB11 is related to a subunit shared by RNA polymerase I and III. *Gene Expression* 3:77–82.

CHAPTER 11

General Transcription Factors in Eukaryotes

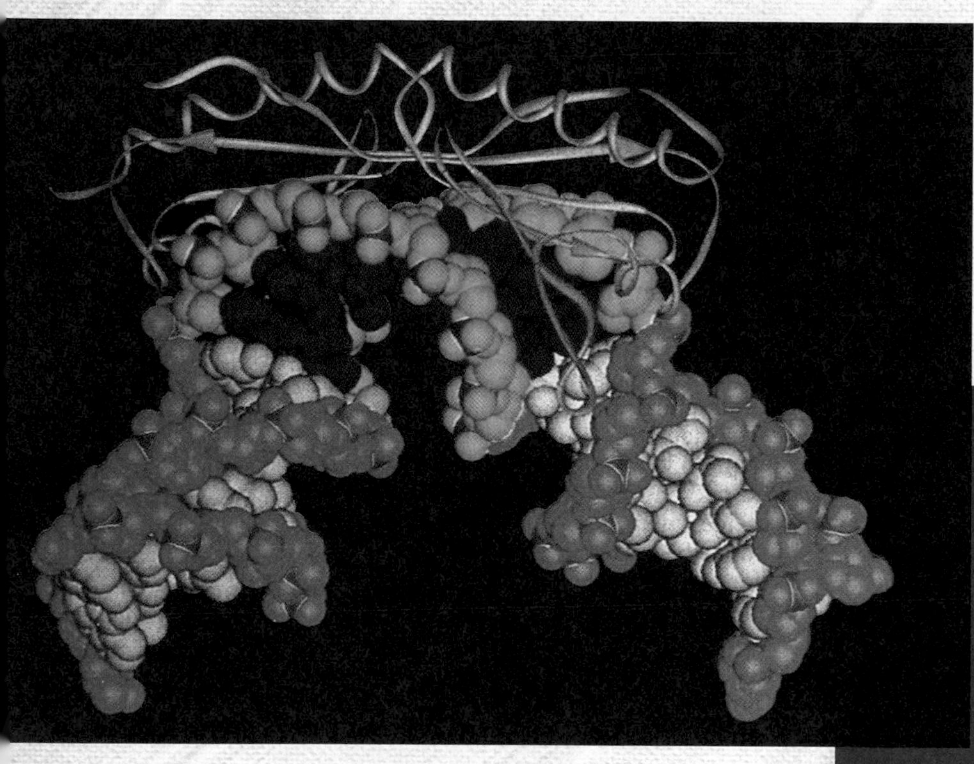

X-ray crystal structure of the TBP-TATA box complex. © Klug, A. Opening the gateway. *Nature* 365 (7 Oct 1993) p. 487, f. 2. © Macmillan Magazines Ltd.

Eukaryotic RNA polymerases, unlike their bacterial counterparts, are incapable of binding by themselves to their respective promoters. Instead, they rely on proteins called transcription factors to show them the way. Such factors are grouped into two classes: **general transcription factors** and gene-specific transcription factors (**activators**). Without activators, the general transcription factors can attract the RNA polymerases to their respective promoters, but only to a weak extent. Therefore, these factors can support only a basal level of transcription. Furthermore, general transcription factors and the three polymerases alone allow for only minimal transcription control, whereas activators help cells exert exquisitely fine control over transcription. Nevertheless, the task performed by the general transcription factors—getting the RNA polymerases together with their promoters—is not only vital, but also very complex because many polypeptides are required to do the job. In this chapter we will survey the general transcription factors that interact with all three RNA polymerases and their promoters.

11.1 Class II Factors

The general transcription factors combine with RNA polymerase to form a **preinitiation complex** that is competent to initiate transcription as soon as nucleotides are available. This tight binding involves formation of an open promoter complex in which the DNA at the transcription start site has melted to allow the polymerase to read it. We will begin with the assembly of preinitiation complexes involving polymerase II. Even though these are by far the most complex, they are also the best studied. Once we see how the class II general transcription factors work, the class I and III mechanisms will be relatively easy to understand.

The Class II Preinitiation Complex

The class II preinitiation complex contains polymerase II and six general transcription factors named **TFIIA, TFIIB, TFIID, TFIIE, TFIIF,** and **TFIIH.** Many studies have shown that the class II general transcription factors and RNA polymerase II bind in a specific order to the growing preinitiation complex, at least in vitro. In particular, Danny Reinberg, as well as Phillip Sharp and their colleagues, performed DNA gel mobility shift and DNase and hydroxyl radical footprinting experiments (Chapter 5) that defined most of the order of factor binding in building the class II preinitiation complex.

Figure 11.1a presents the results of a gel mobility shift assay performed by Danny Reinberg and Jack Greenblatt

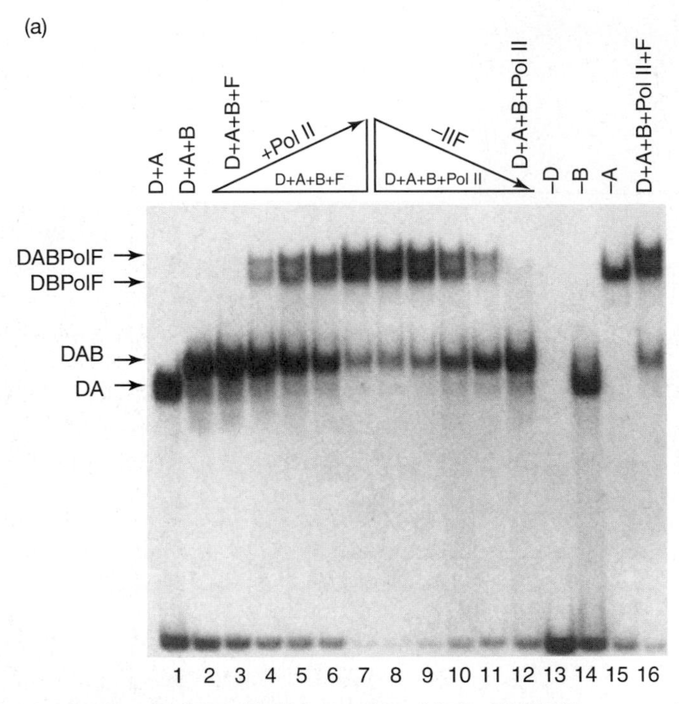

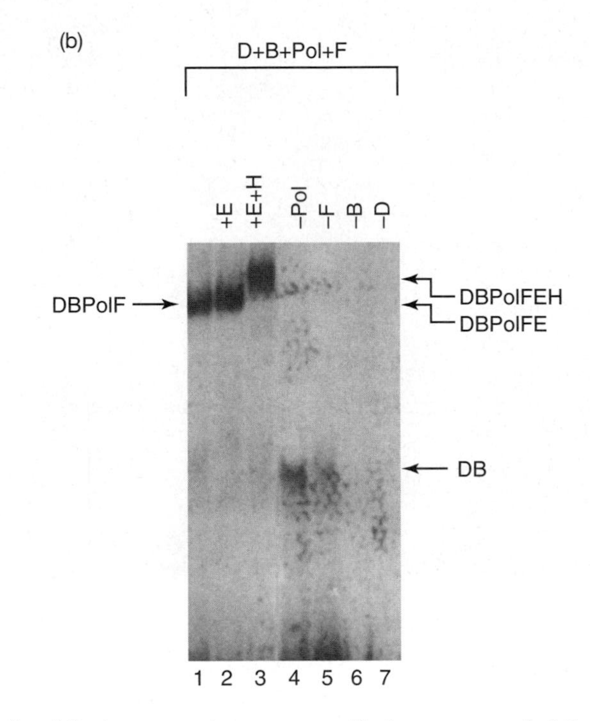

Figure 11.1 Building the preinitiation complex. (a) The DABPolF complex. Reinberg and colleagues performed gel mobility shift assays with TFIID, A, B, and F, and RNA polymerase II, along with labeled DNA containing the adenovirus major late promoter. Lane 1 shows the DA complex, formed with TFIID and A. Lane 2 demonstrates that adding TFIIB caused a new complex, DAB, to form. Lane 3 contained TFIID, A, B, and F, but it looks identical to lane 2. Thus, TFIIF did not seem to bind in the absence of polymerase II. Lanes 4–7 show what happened when the investigators added more and more polymerase II in addition to the four transcription factors: More and more of the large complexes, DABPolF and DBPolF, appeared. Lanes 8–11 contained less and less TFIIF, and we see less and less of the large complexes. Finally, lane 12 shows that essentially no DABPolF or DBPolF complexes formed when TFIIF was absent. Thus, TFIIF appears to bring polymerase II to the complex. The lanes on the right show what happened when Reinberg and colleagues left out one factor at a time. In lane 13, without TFIID, no complexes formed at all. Lane 14 shows that the DA complex, but no others, formed in the absence of TFIIB. Lane 15 demonstrates that DBPolF could still develop without TFIIA. Finally, all the large complexes appeared in the presence of all the factors (lane 16). **(b)** The DBPolFEH complex. Reinberg and colleagues started with the DBPolF complex (lacking TFIIA, lane 1) assembled on a labeled DNA containing the adenovirus major late promoter. Next, they added TFIIE, then TFIIH, in turn, and performed gel mobility shift assays. With each new transcription factor, the complex grew larger and its mobility decreased further. The mobilities of both complexes are indicated at right. Lanes 4–7 show again the result of leaving out various factors, denoted at the top of each lane. At best, only the DB complex forms. At worst, in the absence of TFIID, no complex at all forms. (*Sources:* (*a*) Flores, O., H. Lu, M. Killeen, J. Greenblatt, Z.F. Burton, and D. Reinberg, The small subunit of transcription factor IIF recruits RNA polymerase II into the preinitiation complex. *Proceedings of the National Academy of Sciences USA,* 88 (Nov 1991) p. 10001, f. 2a. (*b*) Cortes, P., O. Flores, and D. Reinberg. 1992. Factors involved in specific transcription by mammalian RNA polymerase II: Purification and analysis of transcription factor IIA and identification of transcription factor IIJ. *Molecular and Cellular Biology* 12: 413–21. American Society for Microbiology.)

and their colleagues using TFIIA, TFIID, TFIIB, and TFIIF, as well as RNA polymerase II. This experiment reveals the existence of four distinct complexes, which are labeled at the left of the figure. When the investigators added TFIID and A alone to DNA containing the adenovirus major late promoter, a *DA complex* formed (lane 1). When they added TFIIB in addition to D and A, a new, *DAB complex* formed (lane 2). The central part of the figure shows what happened when they added various concentrations of RNA polymerase II and TFIIF to the DAB complex. In lane 3, labeled D+A+B+F, all four of those factors were present, but RNA polymerase was missing. No difference was detectable between the complex formed with these four factors and the DAB complex. Thus, TFIIF does not seem to bind independently to DAB. But when the investigators added increasing amounts of polymerase (lanes 4–7), two new complexes appeared. These seem to include both polymerase and TFIIF, so the top complex is called the *DABPolF complex*. The other new complex *(DBPolF)* migrates somewhat faster because it is missing TFIIA, as we will see. After they had added enough polymerase to give a maximum amount of DABPolF, the investigators started decreasing the quantity of TFIIF (lanes 8–11). This reduction in TFIIF concentration decreased the yield of DABPolF, until, with no TFIIF but plenty of polymerase (lane 12), essentially no DABPolF (or DABPol) complexes formed. These data indicated that RNA polymerase and TFIIF are needed *together* to join the growing preinitiation complex.

Reinberg, Greenblatt, and colleagues assessed the order of addition of proteins by performing the same kind of mobility shift assays, but leaving out one or more factors at a time. In the most extreme example, lane 13, labeled −D, shows what happened when the investigators left out TFIID. No complexes formed, even with all the other factors present. This dependence on TFIID reinforced the hypothesis that TFIID is the first factor to bind; the binding of all the other factors depends on the presence of TFIID at the TATA box. Lane 14, marked −B, shows that TFIIB was needed to add polymerase and TFIIF. In the absence of TFIIB, only the DA complex could form. Lane 15, labeled −A, demonstrates that leaving out TFIIA made little difference. Thus, at least in vitro, TFIIA did not seem to be critical. Also, the fact that the band in this lane comigrated with the smaller of the two big complexes suggests that this smaller complex is DBPolF. Finally, the last lane contained all the proteins and displayed the large complexes as well as some residual DAB complex.

Reinberg and his coworkers extended this study in 1992 with TFIIE and H. Figure 11.1b demonstrates that they could start with the DBPolF complex and then add TFIIE and TFIIH in turn, producing a larger complex, with reduced mobility, with each added factor. The final preinitiation complex formed in this experiment was DBPolFEH. The last four lanes in this experiment show again that leaving out any of the early factors (polymerase II, TFIIF, TFIIB, or TFIID) prevents formation of the full preinitiation complex.

Thus, the order of addition of the general transcription factors (and RNA polymerase) to the preinitiation complex in vitro is as follows: TFIID (or TFIIA + TFIID), TFIIB, TFIIF + polymerase II, TFIIE, TFIIH. Now let us consider the question of where on the DNA each factor binds. Several groups, beginning with Sharp's, approached this question using footprinting. Figure 11.2 shows the results of a footprinting study on the DA and DAB complexes. Reinberg and colleagues used two different reagents to cut the protein–DNA complexes: 1,10-phenanthroline (OP)-copper ion complex, which creates hydroxyl radicals (lanes 1–4 in both panels), and DNase I (lanes 5–8 in both panels). Panel (a) depicts the data on the template strand, and panel (b) presents the results for the nontemplate strand. Panel (a), lanes 3 and 7 show that TFIID and A protect the TATA box. Lanes 3 and 7 in panel (b) show that the DA complex also protects the TATA box region on the nontemplate strand. Lanes 4 and 8 in panel (a) show no change in the template strand footprint after adding TFIIB to form the DAB complex. Essentially the same results were obtained with the nontemplate strand, but one subtle difference is apparent. As lane 8 shows, addition of TFIIB makes the DNA at position +10 even more sensitive to DNase. Thus, TFIIB does not seem to cover a significant expanse of DNA, but it does perturb the DNA structure enough to alter its susceptibility to DNase attack.

RNA polymerase II is a very big protein, so we would expect it to cover a large stretch of DNA and leave a big footprint. Figure 11.3 bears out this prediction. Whereas TFIID, A, and B protected the TATA box region (between positions −17 and −42) in the DAB complex, RNA polymerase II and TFIIF extended this protected region another 34 bases on the nontemplate strand, from position −17 to about position +17. Figure 11.4 summarizes what we have learned about the role of TFIIF in building the DABPolF complex. Polymerase II (red) and TFIIF (green) bind cooperatively, perhaps by forming a binary complex that joins the preformed DAB complex.

SUMMARY Transcription factors bind to class II promoters, including the adenovirus major late promoter, in the following order in vitro: (1) TFIID, apparently with help from TFIIA, binds to the TATA box, forming the DA complex. (2) TFIIB binds next. (3) TFIIF helps RNA polymerase bind to a region extending from at least position −34 to position +17. The remaining factors bind in this order: TFIIE and TFIIH, forming the DABPolFEH preinitiation complex. The participation of TFIIA seems to be optional in vitro.

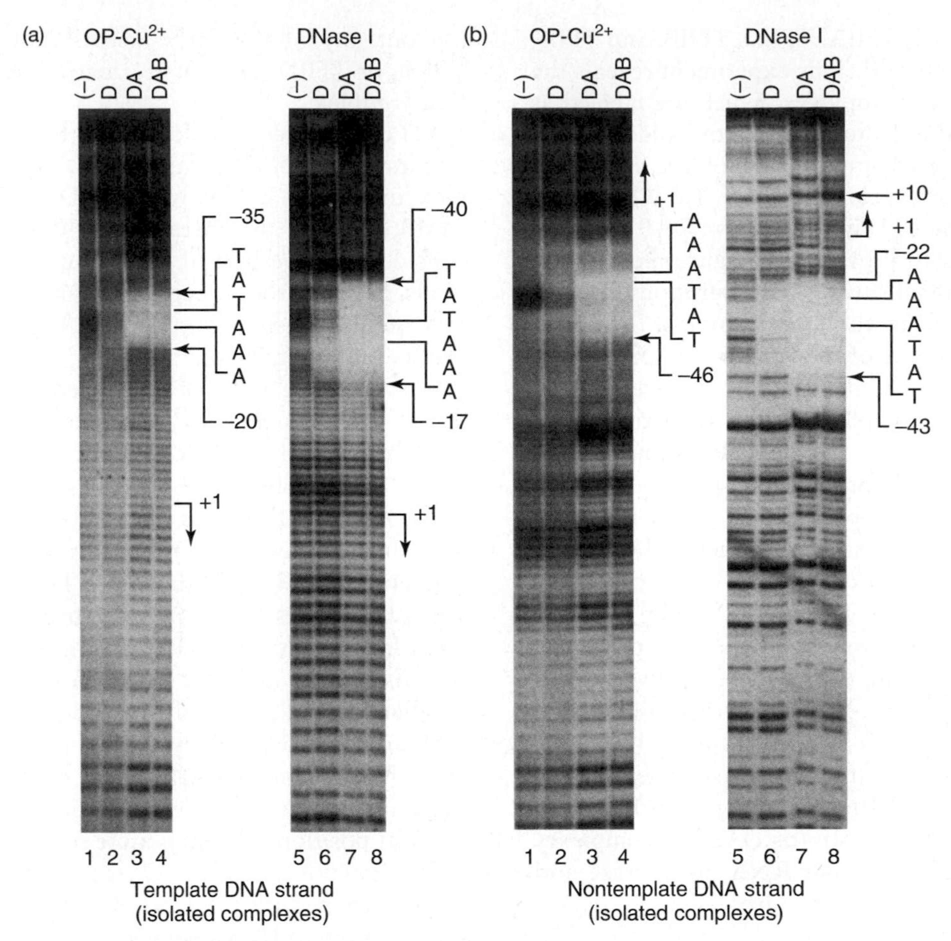

Figure 11.2 Footprinting the DA and DAB complexes. Reinberg and colleagues performed footprinting on the DA and DAB complexes with both DNase I (lanes 1–4) and another DNA strand breaker: a 1,10-phenanthroline-copper ion complex (OP-Cu^{2+}, lanes 5–8). **(a)** Footprinting on the template strand. The DA and DAB complexes formed on the TATA box (TATAAA, indicated at right, top to bottom). **(b)** Footprinting on the nontemplate strand. Again, the protected region in both the DA and DAB complexes was centered on the TATA box (TATAAA, indicated at right, bottom to top). The arrow near the top at right denotes a site of enhanced DNA cleavage at position +10. (*Source:* Adapted from Maldonado E., I. Ha, P. Cortes, L. Weiss, and D. Reinberg, Factors involved in specific transcription by mammalian RNA polymerase II: Role of transcription Factors IIA, IID, and IIB during formation of a transcription-competent complex. *Molecular and Cellular Biology* 10 (Dec 1990) p. 6344, f. 9. American Society for Microbiology.)

Structure and Function of TFIID

TFIID is a complex protein containing a **TATA-box-binding protein (TBP)** and 13 core **TBP-associated factors (TAFs,** or more specifically, **TAF_{II}s**). The subscript “II” was traditionally used when the context was unclear, because TBP also participates in transcription of class I and III genes and is associated with different TAFs (TAF_Is and TAF_{III}s) in class I and III preinitiation complexes, respectively. We will discuss the role of TBP and its TAFs in transcription from class I and III promoters later in this chapter. Let us first discuss the components of TFIID and their activities, beginning with TBP and concluding with the TAFs.

The TATA-Box-Binding Protein TBP, the first polypeptide in the TFIID complex to be characterized, is highly evolutionarily conserved: Organisms as disparate as yeast, fruit flies, plants, and humans have TATA-box-binding domains that are more than 80% identical in amino acid sequence. These domains encompass the carboxyl-terminal 180 amino acids of each protein and are very rich in basic amino acids. Another indication of evolutionary conservation is the fact that the yeast TBP functions well in a preinitiation complex in which all the other general transcription factors are mammalian.

Tjian’s group demonstrated the importance of the carboxyl-terminal 180 amino acids of TBP when they showed by DNase I footprinting that a truncated form of human TBP containing only the carboxyl-terminal 180 amino acids of a human recombinant TBP is enough to bind to the TATA box region of a promoter, just as the native TFIID would.

How does the TBP in TFIID bind to the TATA box? The original assumption was that it acts like most other DNA-binding proteins (Chapter 9) and makes specific contacts with the base pairs in the major groove of the TATA box DNA. However, this assumption proved to be wrong.

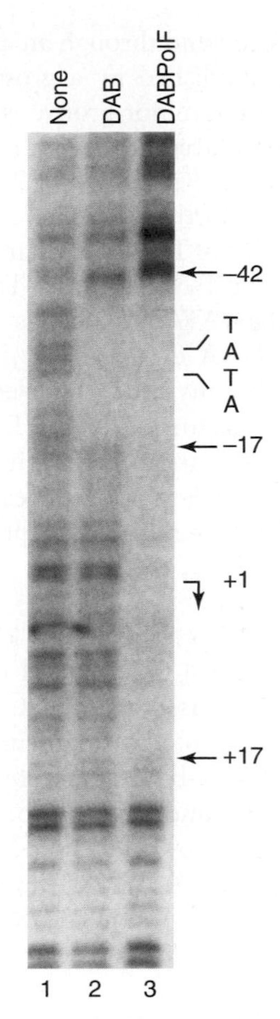

Figure 11.3 Footprinting the DABPolF complex. Reinberg and colleagues performed DNase footprinting with TFIID, A, and B (lane 2) and with TFIID, A, B, and F, and RNA polymerase II (lane 3). When RNA polymerase and TFIIF joined the complex, they caused a large extension of the footprint, to about position +17. This is consistent with the large size of RNA polymerase II. (*Source:* Flores O., H. Lu, M. Killeen, J. Greenblatt, Z.F. Burton, and D. Reinberg, The small subunit of transcription factor IIF recruits RNA polymerase II into the preinitiation complex. *Proceedings of the National Academy of Sciences USA* 88 (Nov 1991) p. 10001, f. 2b.)

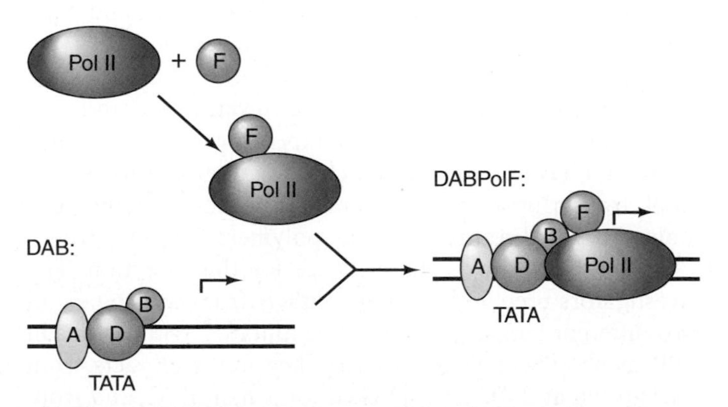

Figure 11.4 Model for formation of the DABPolF complex. TFIIF (green) binds to polymerase II (Pol II, red) and together they join the DAB complex. The result is the DABPolF complex. This model conveys the idea that polymerase II extends the DAB footprint in the downstream direction, and therefore binds to DNA downstream of the binding sites for TFIID, A, and B, which center on the TATA box.

Two research groups, headed by Diane Hawley and Robert Roeder, showed convincingly that the TBP in TFIID binds to the minor groove of the TATA box.

Barry Starr and Hawley changed all the bases of the TATA box, such that the major groove was changed, but the minor groove was not. This is possible because the hypoxanthine base in inosine (I) looks just like adenine (A) in the minor groove, but much different in the major groove (Figure 11.5a). Similarly, cytosine looks like thymine in the minor, but not the major, groove. Thus, Starr and Hawley made an adenovirus major late TATA box with all C's instead of T's, and all I's instead of A's (CICIIII instead of TATAAAA, Figure 11.5b). Then they measured TFIID binding to this CICI box and to the standard TATA box by a DNA mobility shift assay. As Figure 11.5c shows, the CICI box worked just as well as the TATA box, but a nonspecific

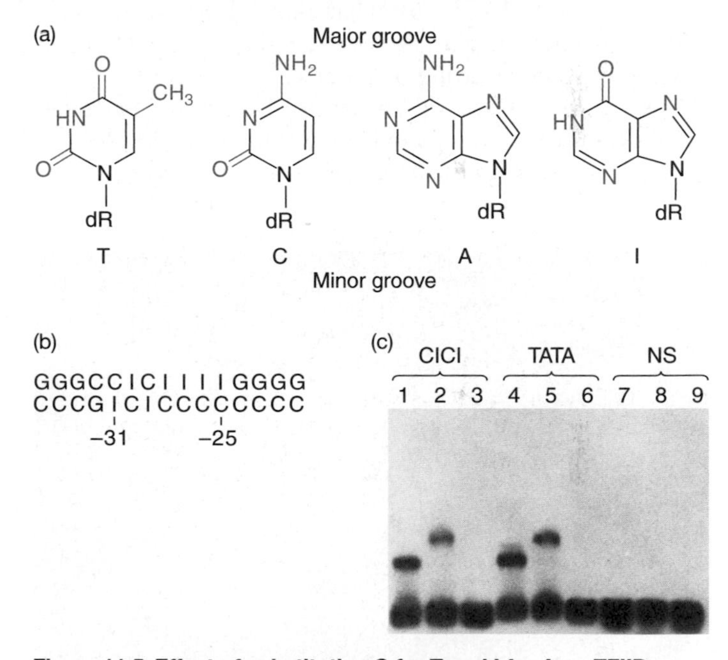

Figure 11.5 Effect of substituting C for T and I for A on TFIID binding to the TATA box. (a) Appearance of nucleosides as viewed from the major and minor grooves. Notice that thymidine and cytidine look identical from the minor groove (green, below), but quite different from the major groove (red, above). Similarly, adenosine and inosine look the same from the minor groove, but very different from the major groove. **(b)** Sequence of the adenovirus major late promoter (MLP) TATA box with C's substituted for T's and I's substituted for A's, yielding a CICI box. **(c)** Binding TBP to the CICI box. Starr and Hawley performed gel mobility shift assays using DNA fragments containing the MLP with a CICI box (lanes 1–3) or the normal TATA box (lanes 4–6), or a nonspecific DNA (NS) with no promoter elements (lanes 7–9). The first lane in each set (1, 4, and 7) contained yeast TBP; the second lane in each set (2, 5, and 8) contained human TBP; and the third lane in each set contained just buffer. The yeast and human TBPs gave rise to slightly different size protein–DNA complexes, but substituting a CICI box for the TATA box had little effect on the yield of the complexes. Thus, TBP binding to the TATA box was not significantly diminished by the substitutions. (*Source:* (*b–c*) Starr, D.B. and D.K. Hawley, TFIID binds in the minor groove of the TATA box. *Cell* 67 (20 Dec 1991) p. 1234, f. 2b. Reprinted by permission of Elsevier Science.)

DNA did not bind TFIID at all. Therefore, changing the bases in the TATA box did not affect TFIID binding as long as the minor groove was unaltered. This is strong evidence for binding of TFIID to the minor groove of the TATA box, and for no significant interaction in the major groove.

How does TFIID associate with the TATA box minor groove? Nam-Hai Chua, Roeder, and Stephen Burley and colleagues began to answer this question when they solved the crystal structure of the TBP of a plant, *Arabidopsis thaliana*. The structure they obtained was shaped like a saddle, complete with two "stirrups," which naturally suggested that TBP sits on DNA the way a saddle sits on a horse. The TBP structure has rough two-fold symmetry corresponding to the two sides of the saddle with their stirrups. Then, in 1993, Paul Sigler and colleagues and Stephen Burley and colleagues independently solved the crystal structure of TBP bound to a small synthetic piece of double-stranded DNA that contained a TATA box. That allowed them to see how TBP really interacts with the DNA, and it was not nearly as passive as a saddle sitting on a horse.

Figure 11.6 shows this structure. The curved undersurface of the saddle, instead of fitting neatly over the DNA, is roughly aligned with the long axis of the DNA, so its curvature forces the DNA to bend through an angle of 80 degrees. This bending is accomplished by a gross distortion in the DNA helix in which the minor groove is forced open. This opening is most pronounced at the first and last steps of the TATA box (between base pairs 1 and 2 and between base pairs 7 and 8). At each of those sites, two phenylalanine side chains from the stirrups of TBP intercalate, or insert, between base pairs, causing the DNA to kink. This distortion may help explain why the TATA sequence is so well conserved: The T–A step in a DNA double helix is relatively easy to distort, compared with any other dinucleotide step. This argument assumes that distortion of the TATA box is important to transcription initiation. Indeed, it is easy to imagine that peeling open the DNA minor groove aids the local DNA melting that is part of forming an open promoter complex.

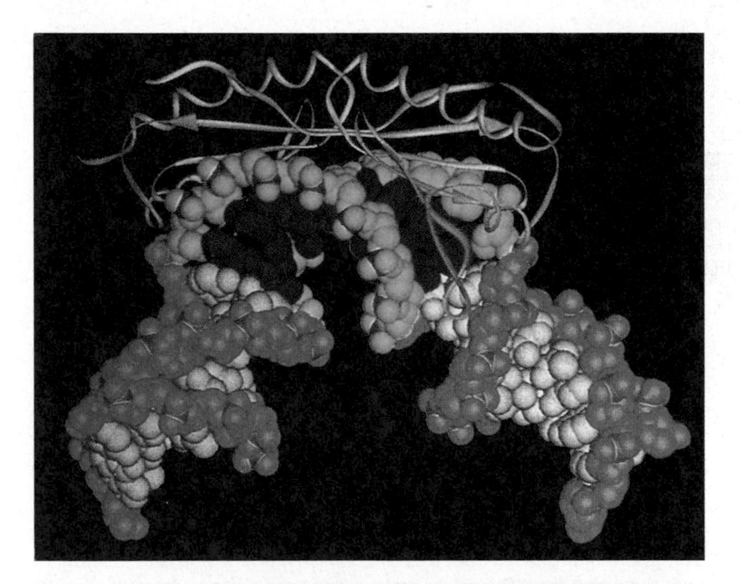

Figure 11.6 Structure of the TBP–TATA box complex. This diagram, based on Sigler and colleagues' crystal structure of the TBP–TATA box complex, shows the backbone of the TBP in olive at top. The long axis of the "saddle" is in the plane of the page. The DNA below the protein is in multiple colors. The backbones in the region that interacts with the protein are in orange, with the base pairs in red. Notice how the protein has opened up the narrow groove and almost straightened the helical twist in that region. One stirrup of the TBP is seen as an olive loop at right center, inserting into the minor groove. The other stirrup performs the same function, but it is out of view in back of the DNA. The two ends of the DNA, which do not interact with the TBP, are in blue and gray: blue for the backbones, and gray for the base pairs. The left end of the DNA sticks about 25 degrees out of the plane of the page, and the right end points inward by the same angle. The overall bend of about 80 degrees in the DNA, caused by TBP, is also apparent. (*Source:* Klug, A. Opening the gateway. *Nature* 365 (7 Oct 1993) p. 487, f. 2. © Macmillan Magazines Ltd.)

SUMMARY TFIID contains a 38-kD TATA-box-binding protein (TBP) plus several other polypeptides known as TBP-associated factors (TAFs). The C-terminal 180 amino acid fragment of the human TBP is the TATA-box-binding domain. The interaction between a TBP and a TATA box takes place in the DNA minor groove. The saddle-shaped TBP lines up with the DNA, and the underside of the saddle forces open the minor groove and bends the TATA box through an 80-degree curve angle.

The Versatility of TBP Molecular biology is full of wonderful surprises, and one of these is the versatility of TBP. This factor functions not only with polymerase II promoters that have a TATA box, but with TATA-less polymerase II promoters. Astonishingly, it also functions with TATA-less polymerase III promoters, *and* with TATA-less polymerase I promoters. In other words, TBP appears to be a universal eukaryotic transcription factor that operates at all promoters, regardless of their TATA content, and even regardless of the polymerase that recognizes them.

One indication of the widespread utility of TBP came from work by Ronald Reeder and Steven Hahn and colleagues on mutant yeasts with temperature-sensitive TBPs. We would have predicted that elevated temperature would block transcription by polymerase II in these mutants, but it also impaired transcription by polymerases I and III.

Figure 11.7 shows the evidence for this assertion. The investigators prepared cell-free extracts from wild-type and two different temperature-sensitive mutants, with lesions in TBP, as shown in Figure 11.7a. They made extracts from cells grown at 24°C and shocked for 1 h at 37°C, and from cells kept at the lower temperature. Then they added DNAs containing promoters recognized by all three polymerases and assayed transcription by S1 analysis. Figure 11.7b–e depicts the results. The heat shock had no effect on the wild-type extract, as expected (lanes 1 and 2). By contrast, the I143→N mutant extract could barely support transcription

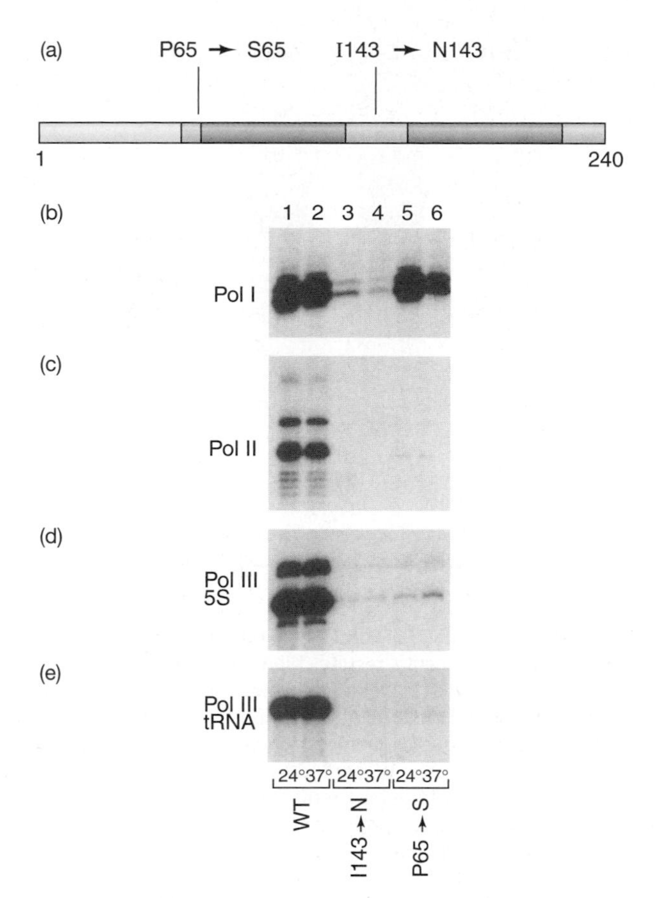

Figure 11.7 Effects of mutations in TBP on transcription by all three RNA polymerases. **(a)** Locations of the mutations. The blue and red regions indicate the conserved C-terminal domain of the TBP; red areas denote two repeated elements involved in DNA binding. The two mutations are: P65→S, in which proline 65 is changed to a serine; and I143→N, in which isoleucine 143 is changed to asparagine. **(b–e)** Effects of the mutations. Reeder and Hahn made extracts from wild-type or mutant yeasts, as indicated at bottom, and either heat-shocked them at 37°C or left them at 24°C, again as indicated at bottom. Then they tested these extracts by S1 analysis for ability to start transcription at promoters recognized by all three nuclear RNA polymerases: **(b)** the rRNA promoter (polymerase I); **(c)** the CYC1 promoter (polymerase II); **(d)** the 5S rRNA promoter (polymerase III); and **(e)** the tRNA promoter (also polymerase III). The I143→N extract was deficient in transcribing from all four promoters even when not heat-shocked. The P65→S extract was deficient in transcribing from polymerase II and III promoters, but could recognize the polymerase I promoter, even after heat shock. (*Source:* (a) Adapted from Schultz, M.C., R.H. Reeder, and S. Hahn. 1992. Variants of the TATA binding protein can distinguish subsets of RNA polymerase I, II, and III promoters. *Cell* 69:697–702.)

by any of the three polymerases, whether it was heat shocked or not (lanes 3 and 4). Clearly, the mutation in TBP was affecting not only polymerase II transcription, but transcription by the other two polymerases as well. The other mutant, P65→S, shows an interesting difference between the behavior of polymerase I and the other two polymerases. Whereas this mutant extract could barely support transcription of polymerase II and III genes, whether it had been heat shocked or not, it allowed wild-type levels of transcription by polymerase I if it was not heat-shocked, but heating reduced transcription by polymerase I by about twofold. Finally, wild-type TBP could restore transcription by all three polymerases in mutant extracts (data not shown).

Not only is TBP universally involved in eukaryotic transcription, it also seems to be involved in transcription in a whole different kingdom of organisms: the **archaea.** Archaea (formerly known as archaebacteria) are single-celled organisms that lack nuclei and usually live in extreme environments, such as hot springs or boiling hot deep ocean vents. They are as different from bacteria as they are from eukaryotes, and in several ways they resemble eukaryotes more than they do prokaryotes. In 1994, Stephen Jackson and colleagues reported that one of the archaea, *Pyrococcus woesei,* produces a protein that is structurally and functionally similar to eukaryotic TBP. This protein is presumably involved in recognizing the TATA boxes that frequently map to the 5′-flanking regions of archaeal genes. Moreover, a TFIIB-like protein has also been found in archaea. Thus, the transcription apparatus of the archaea bears at least some resemblance to that in eukaryotes, and suggests that the archaea and the eukaryotes diverged *after* their common ancestor diverged from the bacteria. This evolutionary scheme is also supported by the sequence of archaeal rRNA genes, which bear more resemblance to eukaryotic than to bacterial sequences.

SUMMARY Genetic studies have demonstrated that TBP mutant cell extracts are deficient, not only in transcription of class II genes, but also in transcription of class I and III genes. Thus, TBP is a universal transcription factor required by all three classes of genes. A similar factor has also been found in archaea.

The TBP-Associated Factors Many researchers have contributed to our knowledge of the TBP-associated factors (**TAFs**) in TFIIDs from several organisms. To identify TAFs from *Drosophila* cells, Tjian and his colleagues used an antibody specific for TBP to immunoprecipitate TFIID from a crude TFIID preparation. Then they treated the immunoprecipitate with 2.5 M urea to strip the TAFs off of the TBP–antibody precipitate and displayed the TAFs by SDS-PAGE. These and subsequent experiments have led to the identification of 13 TAFs associated with class II preinitiation complexes from a wide variety of organisms, from yeasts to humans.

These **core TAFs** were at first named according to their molecular masses, so the largest *Drosophila* TAF, with a molecular mass of 230 kD, was called $TAF_{II}230$, and the homologous human TAF was called $TAF_{II}250$. To avoid that kind of confusion, the core TAFs have been renamed according to their sizes, from largest to smallest, as TAF1 through TAF13. Thus, *Drosophila* $TAF_{II}230$, human $TAF_{II}250$, and fission yeast $TAF_{II}111$ are all now called TAF1. This nomenclature allows equivalent TAFs from different organisms to be compared easily because they have the same names, regardless of their exact sizes. Note that the subscript II has been deleted. The context of the discussion should prevent confusion with class I and III TAFs. Some organisms encode TAF paralogs (homologous proteins in the same organism that have descended from a common ancestor protein). For example, we now know that human $TAF_{II}130/135$ and $TAF_{II}105$ are paralogs, so they are named TAF4 and TAF4b to indicate their homology. Some organisms encode TAF-like proteins that are similar, but not homologous to one of the core TAFs. These are given the designation L (for -like), as in TAF5L in humans and *Drosophila*. Some organisms (yeast and human, at least) have extra, non-core TAFs (TAF14 in yeast, and TAF15 in humans) that have no obvious homologs in other organisms.

Investigators have discovered several functions of the TAFs, but two that have received considerable attention are interaction with the promoter and interaction with gene-specific transcription factors. Let us consider the evidence for each of these functions and, where possible, the specific TAFs involved in each.

We have already seen the importance of the TBP in binding to the TATA box. But footprinting studies have indicated that the TAFs attached to TBP extend the binding of TFIID well beyond the TATA box in some promoters. In particular, Tjian and coworkers showed in 1994 that TBP protected the 20 bp or so around the TATA box in some promoters, but that TFIID protected a region extending to position +35, well beyond the transcription start site. This suggested that the TAFs in TFIID were contacting the initiator and downstream elements in these promoters.

To investigate this phenomenon in more detail, Tjian's group tested the abilities of TBP and TFIID to transcribe DNAs bearing two different classes of promoters in vitro. The first class (the adenovirus E1B and E4 promoters) contained a TATA box, but no initiator or downsteam promoter element (DPE). The second class (the adenovirus major late [AdML] promoter and the *Drosophila* heat shock protein [*hsp70*] promoter) contained a TATA box, an initiator, and a DPE. Figure 11.8 depicts the structures of these promoters, as well as the results of the in vitro transcription experiments. We can see that TBP and TFIID sponsored transcription equally well from the promoters that contained only the TATA box (compare lanes 1 and 2 and lanes 3 and 4). But TFIID had a decided advantage in sponsoring transcription from the promoters that also had an initiatior and DPE (compare lanes 5 and 6 and lanes 7 and 8). Thus, TAFs apparently help TBP facilitate transcription from promoters with initiators and DPEs.

Which TAFs are responsible for recognizing the initiator and DPE? To find out, Tjian and colleagues performed a photo-cross-linking experiment with *Drosophila* TFIID and a radioactively labeled DNA fragment containing the *hsp70* promoter. They incorporated bromodeoxyuridine (BrdU) into the promoter-containing DNA, then allowed TFIID to

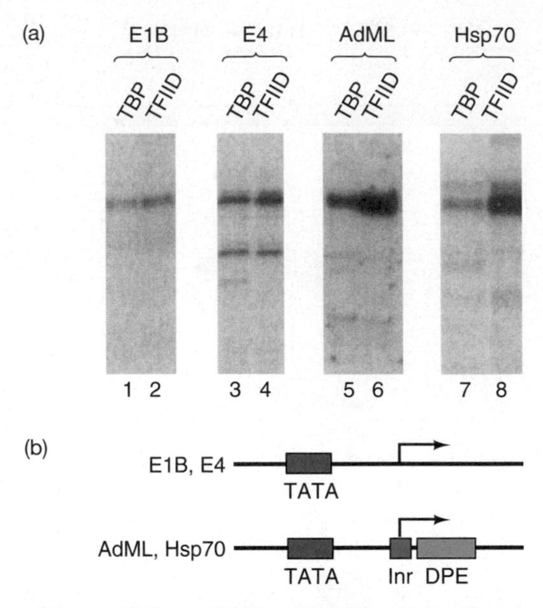

Figure 11.8 Activities of TBP and TFIID on four different promoters. (a) Experimental results. Tjian and colleagues tested a reconstituted *Drosophila* transcription system containing either TBP or TFIID (indicated at top) on templates bearing four different promoters (also as indicated at top). The promoters were of two types diagrammed in panel **(b).** The first type, represented by the adenovirus E1B and E4 promoters, contained a TATA box (red). The second type, represented by the adenovirus major late promoter (AdML) and the *Drosophila hsp70* promoter, contained a TATA box plus an initiator (Inr, green) and a DPE (blue). After transcription in vitro, Tjian and coworkers assayed the RNA products by primer extension (top). The autoradiographs show that TBP and TFIID fostered transcription equally well from the first type of promoter (TATA box only), but that TFIID worked much better than TBP in supporting transcription from the second type of promoter (TATA box plus Inr plus DPE). (*Source:* Verrijzer, C.P., J.-L. Chen, K. Yokomari, and R. Tijan, Promoter recognition by TAFs. *Cell* 81 (30 June 1995) p. 1116, f. 1. Reprinted with permission of Elsevier Science.)

bind to the promoter, then irradiated the complexes with UV light to cross-link the protein to the BrdU in the DNA. After washing away unbound protein, the investigators digested the DNA with nuclease to release the proteins, then subjected the labeled proteins to SDS-PAGE. Figure 11.9, lane 1, shows that two TAFs (TAF1 and TAF2) bound to the *hsp70* promoter and thereby became labeled. When TFIID was omitted (lane 2), no proteins became labeled. Following up on these findings, Tjian and coworkers reconstituted a ternary complex containing only TBP, TAF1, and TAF2 and tested it in the same photo-cross-linking assay. Lane 3 shows that this experiment also yielded labeled TAF1 and TAF2, and lane 4 shows that TBP did not become labeled when it was bound to the DNA by itself. We know that TBP binds to this TATA-box-containing DNA, but it does not become cross-linked to BrdU and therefore does not become labeled. Why not? Probably because this kind of photo-cross-linking works well only with proteins that bind in the major groove, and TBP binds in the minor groove of DNA.

To double-check the binding specificity of the ternary complex (TBP–TAF1–TAF2), Tjian and colleagues performed

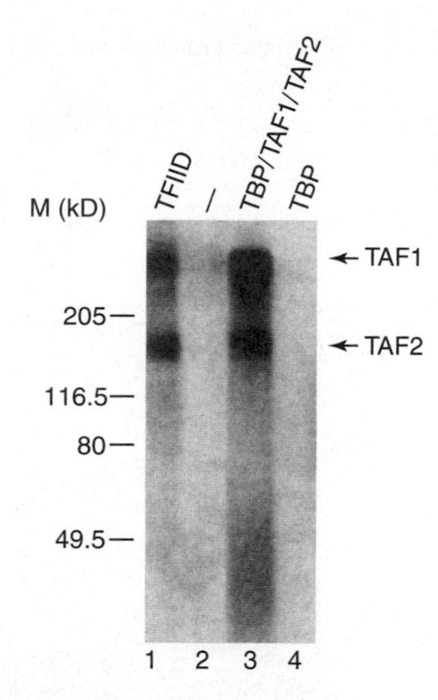

Figure 11.9 Identifying the TAFs that bind to the *hsp70* promoter. Tjian and colleagues photo-cross-linked TFIID to a ^{32}P-labeled template containing the *hsp70* promoter as follows: First, they bound the TFIID to the labeled template, which had also been substituted with the photosensitive nucleoside bromodeoxyuridine (BrdU). Next, these investigators irradiated the TFIID–DNA complex with UV light to form covalent bonds between the DNA and any proteins in close contact with the major groove of the DNA. Next, they digested the DNA with nuclease and subjected the proteins to SDS-PAGE. Lane 1 of the autoradiograph shows the results when TFIID was the input protein. TAF1 and TAF2 became labeled, implying that these two proteins had been in close contact with the labeled DNA's major groove. Lane 2 is a control with no TFIID. Lane 3 shows the results when a ternary complex containing TBP, TAF1, and TAF2 was the input protein. Again, the two TAFs became labeled, suggesting that they bound to the DNA. Lane 4 shows the results when TBP was the input protein. It did not become labeled, which was expected because it does not bind in the DNA major groove. (*Source:* Verrijzer, C.P., J.-L. Chen, K. Yokomari, and R. Tjian, *Cell* 81 (30 June 1995) p. 1117, f. 2a. Reprinted with permission of Elsevier Science.)

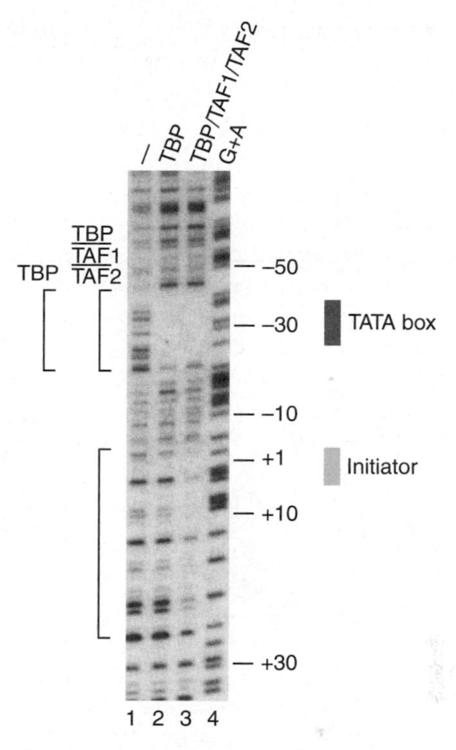

Figure 11.10 DNase footprinting the *hsp70* promoter with TBP and the ternary complex (TBP, TAF1, and TAF2). Lane 1, no protein; lane 2, TBP; lane 3, ternary complex. In both lanes 2 and 3, TFIIA was also added to stabilize the DNA–protein complexes, but separate experiments indicated that it did not affect the extent of the footprints. Lane 4 is a G+A sequencing lane used as a marker. The extents of the footprints caused by TBP and the ternary complex are indicated by brackets at left. The locations of the TATA box and initiator are indicated by boxes at right. (*Source:* Verrijzer, C.P., J.-L. Chen, K. Yokomori, and R. Tjian, *Cell* 81 (30 June 1995) p. 1117, f. 2c. Reprinted with permission of Elsevier Science.)

a DNase footprinting experiment with TBP or the ternary complex. Figure 11.10 shows that TBP caused a footprint only in the TATA box, whereas the ternary complex caused an additional footprint in the initiator and downstream sequences. This reinforced the hypothesis that the two TAFs bind at least to the initiator, and perhaps to the DPE.

Further experiments with binary complexes (TBP–TAF1 or TBP–TAF2) showed that these complexes were no better than TBP alone in recognizing initiators and DPEs. Thus, both TAFs seem to cooperate in enhancing binding to these promoter elements. Furthermore, the ternary complex (TBP–TAF1–TAF2) is almost as effective as TFIID in recognizing a synthetic promoter composed of the AdML TATA box and the TdT initiator. By contrast, neither binary complex functions any better than TBP in recognizing this promoter. These findings support the hypothesis that TAF1 and TAF2 cooperate in binding to the initiator alone, as well as to the initiator plus a DPE.

The TBP part of TFIID is of course important in recognizing the majority of the well-studied class II promoters, which contain TATA boxes (Figure 11.11a). But what about promoters that lack a TATA box? Even though these promoters cannot bind TBP directly, most still depend on this transcription factor for activity. The key to this apparent paradox is the fact that these TATA-less promoters contain other elements that ensure the binding of TBP. These other elements can be initiators and DPEs, to which TAF1 and TAF2 can bind and thereby secure the whole TFIID to the promoter (Figure 11.11b). Or they can be upstream elements that bind gene-specific transcription factors, which in turn interact with one or more TAFs to anchor TFIID to the promoter. For example, the activator Sp1 binds to proximal promoter elements (GC boxes) and also interacts with at least one TAF (TAF4). This bridging activity apparently helps TFIID bind to the promoter (Figure 11.11c).

The second major activity of the TAFs is to participate in the transcription stimulation provided by activators, some of which we will study in Chapter 12. Tjian and colleagues demonstrated in 1990 that TFIID is sufficient to

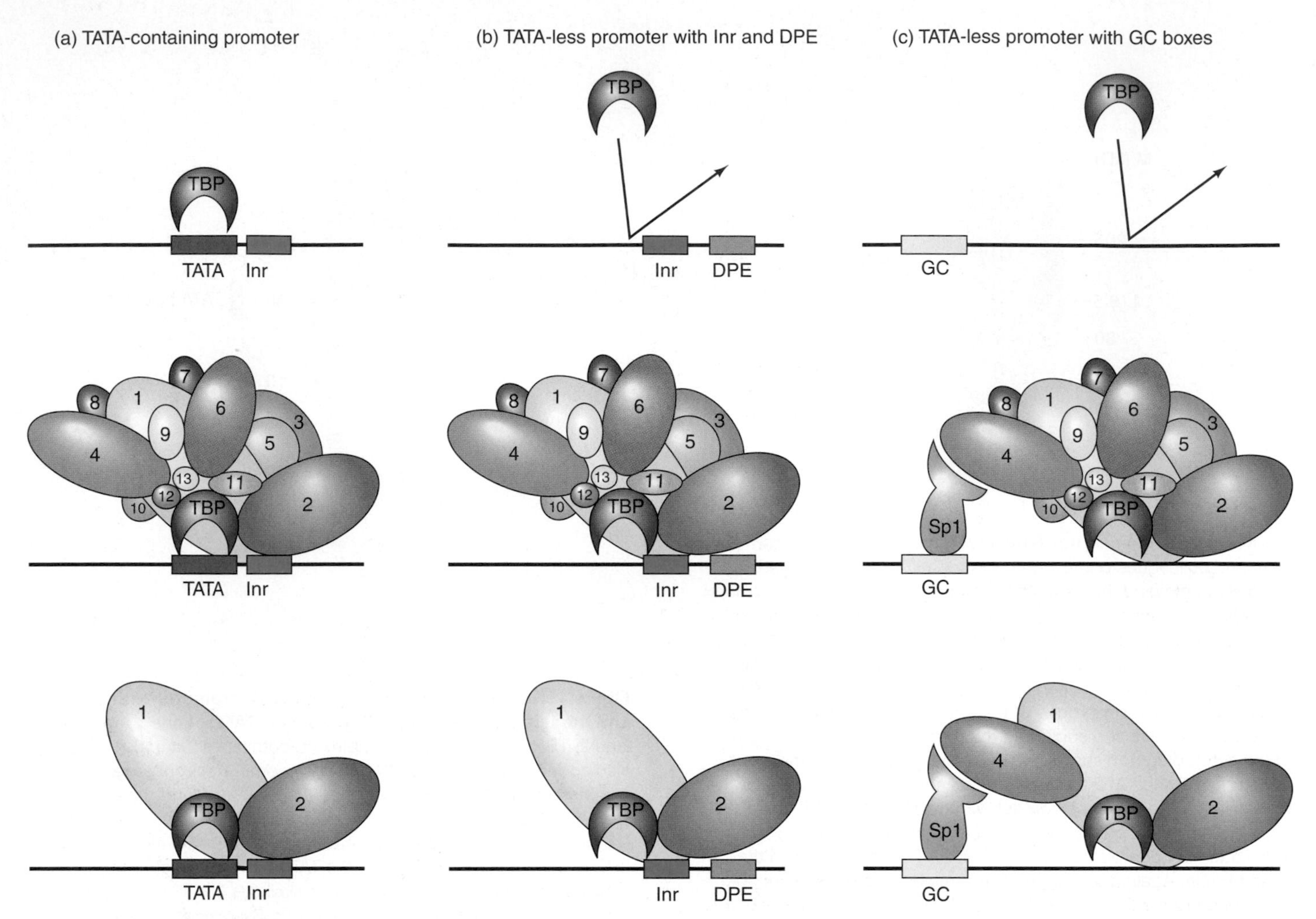

Figure 11.11 Model for the interaction between TBP and TATA-containing or TATA-less promoters. (a) TATA-containing promoter. TBP can bind by itself to the TATA box of this promoter (top). It can also bind in the company of all the TAFs in TFIID (middle). And it can bind with a subset of TAFs (bottom). **(b)** TATA-less promoter with initiator element and DPE. TBP cannot bind by itself to this promoter, which contains no TATA box (top). The whole TFIID is competent to bind to the TATA-less promoter through interactions between TAF1 (yellow) and TAF2 (brown, middle). TAF1 and TAF2 are sufficient to tether TBP to the initiator and DPE (bottom). **(c)** TATA-less promoter with GC boxes. TBP cannot bind to this promoter by itself (top). The whole TFIID can bind to this promoter through interactions with Sp1 bound at the GC boxes (middle). TAF1, TAF2, and TAF4 are sufficient to anchor TBP to the Sp1 bound to the GC boxes. (*Source:* Adapted from Goodrich, J.A., G. Cutter, and R. Tjian, Contacts in context: Promoter specificity and macromolecular interactions in transcription. *Cell* 84:826, 1996.)

participate in such stimulation by the factor Sp1, but TBP is not. These results suggest that some factors in TFIID are necessary for interaction with upstream-acting factors such as Sp1 and that these factors are missing from TBP. By definition, these factors are TAFs, and they are sometimes called **coactivators.**

We have seen that mixing TBP with subsets of TAFs can produce a complex with the ability to participate in transcription from certain promoters. For example, the TBP–TAF1–TAF2 complex functioned almost as well as the whole TFIID in recognizing a promoter composed of a TATA box and an initiator. Tjian and colleagues used a similar technique to discover which TAFs are involved in activation by Sp1. They found that activation by Sp1 in *Drosophila* or human extracts occurred only when TAF4 was present. Thus, TBP and TAF1 plus TAF2 were sufficient for basal transcription, but could not support activation by Sp1. Adding TAF4 in addition to the other two factors and TBP allowed Sp1 to activate transcription.

Tjian and colleagues also showed that Sp1 binds directly to TAF4, but not to TAF1 or TAF2. They built an affinity column containing GC boxes and Sp1 and tested it for the ability to retain the three TAFs. As predicted, only TAF4 was retained.

Using the same strategy, Tjian and colleagues demonstrated that another activator, NTF-1, binds to TAF2 and requires either TAF1 and TAF2 or TAF1 and TAF6 to activate transcription in vitro. Thus, different activators work with different combinations of TAFs to enhance transcription, and all of them seem to have TAF1 in common. This suggests that TAF1 serves as an assembly factor around which other TAFs can aggregate. These findings are compatible with the model in Figure 11.12: Each activator interacts with a particular subset of TAFs, so the holo-TFIID can

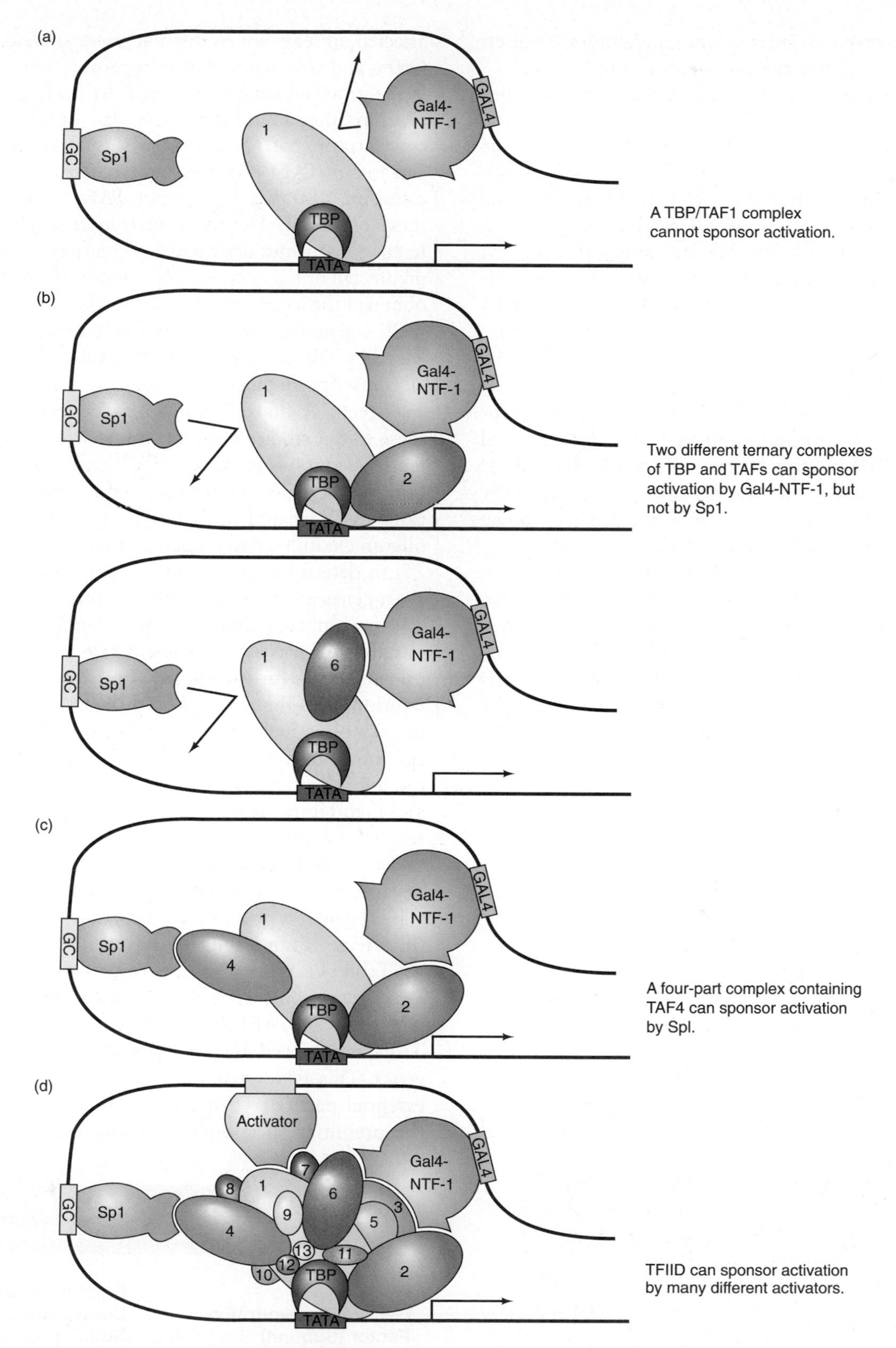

Figure 11.12 A model for transcription enhancement by activators. **(a)** TAF1 does not interact with either Sp1 or Gal4-NTF-1 (a hybrid activator with the transcription-activating domain of NTF-1), so no activation takes place. **(b)** Gal4-NTF-1 can interact with either TAF2 or TAF6 and activate transcription; Sp1 cannot interact with either of these TAFs or with TAF1 and does not activate transcription. **(c)** Gal4-NTF-1 interacts with TAF2 and Sp1 interacts with TAF4, so both factors activate transcription. **(d)** Holo-TFIID contains the complete assortment of TAFs, so it can respond to a wide variety of activators, represented here by Sp1, Gal4-NTF-1, and a generic activator (green) at top. (*Source:* Adapted from Chen, J.L., L.D. Attardi, C.P. Verrijzer, K. Yokomori, and R. Tjian, Assembly of recombinant TFIID reveals differential coactivator requirements for distinct transcriptional activators. *Cell* 79:101, 1994).

interact with several activators at once, magnifying their effect and producing strong enhancement of transcription.

In addition to their abilities to interact with promoter elements and activators, TAFs can have enzymatic activities. The best studied of these is TAF1, which has two known enzymatic activities. It is a histone acetyltransferase (HAT), which attaches acetyl groups to lysine residues of histones. Such acetylation is generally a transcription-activating event. We will study this process in greater detail in Chapter 13. TAF1 is also a protein kinase that can phosphorylate itself and TFIIF (and TFIIA and TFIIE, though to a lesser extent). These phosphorylation events may modulate the efficiency of assembly of the preinitiation complex.

Despite early indications that it was not required for preinitiation complex formation in vitro, TFIIA is essential for TBP (or TFIID) binding to promoters. Much evidence leads to this conclusion, but one experiment is particularly easy to describe: Mutations in either of the genes encoding the two subunits of TFIIA in yeast are lethal.

TFIIA not only stabilizes TBP-TATA box binding, it also stimulates TFIID-promoter binding by an *antirepression* mechanism, as follows: When TFIID is not bound to a promoter, the DNA-binding surface of TBP is covered by the N-terminal domain of TAF1, which inhibits TFIID binding to the promoter. But TFIIA can interfere with the interaction between the TAF1 N-terminal domain and the DNA-binding surface of TBP, freeing up TBP for binding to the promoter.

SUMMARY TFIID contains 13 TAFs, in addition to TBP. Most of these TAFs are evolutionarily conserved in the eukaryotes. The TAFs serve several functions, but two obvious ones are interacting with core promoter elements and interacting with activators. TAF1 and TAF2 help TFIID bind to the initiator and DPEs of promoters and therefore can enable TBP to bind to TATA-less promoters that contain such elements. TAF1 and TAF4 help TFIID interact with Sp1 that is bound to GC boxes upstream of the transcription start site. These TAFs therefore ensure that TBP can bind to TATA-less promoters that have GC boxes. Different combinations of TAFs are apparently required to respond to various activators, at least in higher eukaryotes. TAF1 also has two enzymatic activities. It is a histone acetyltransferase and a protein kinase.

Exceptions to the Universality of TAFs and TBP Genetic studies in yeast call into question the generality of the model in Figure 11.12. Michael Green and Kevin Struhl and their colleagues independently discovered that mutations in yeast TAF genes were lethal, but transcription activation was not affected, at least not in the first genes studied. For example, Green and colleagues made temperature-sensitive mutations in the gene encoding yeast TAF1. At the nonpermissive temperature, they found that there was a rapid decrease in the concentration of TAF1, and at least two other yeast TAFs. The loss of TAF1 apparently disrupted the TFIID enough to cause the degradation of other TAFs. However, in spite of these losses of TAFs, the in vivo transcription rates of five different yeast genes activated by a variety of activators were unaffected at the nonpermissive temperature. These workers obtained the same results with another mutant in which the TAF14 gene had been deleted. By contrast, when the genes encoding TBP or an RNA polymerase subunit were mutated, all transcription quickly ceased.

Green, Richard Young, and colleagues followed up these initial studies with a genome-wide analysis of the effects of mutations in two TAF genes, as well as several other yeast genes. They made temperature-sensitive mutations in TAF1 and in TAF9. Then they used high-density oligonucleotide arrays (such as those described in Chapter 25) to determine the extent of expression of each of 5460 yeast genes at an elevated temperature at which the mutant TAF was inactive and at a lower temperature at which the mutant TAF was active. These arrays contained oligonucleotides specific for each gene. Total yeast RNA can then be hybridized to these arrays, and the extent of hybridization to each oligonucleotide is a measure of the extent of expression of the corresponding gene. The investigators compared the hybridization of RNA to each oligonucleotide at low and high temperature and compared the response with the results of a similar analysis of a temperature-sensitive mutation in the largest subunit of RNA polymerase II (Rpb1). Because the latter mutation prevented transcription of *all* class II genes, it provided a baseline with which to compare the effects of mutations in other genes.

Table 11.1 presents the results of this analysis. It is striking that only 16% of the yeast genes analyzed were as dependent on TAF1 as they were on Rpb 1, indicating that TAF1 is required for transcription of only 16% of yeast genes. This is not what we would expect if the TAFs are essential parts of TFIID, and TFIID is an essential part of the preinitiation complexes formed at all class II genes.

Table 11.1 Whole Genome Analysis of Transcription Requirements in Yeast

General Transcription Factor (Subunit)	Fraction of Genes Dependent on Subunit Function (%)
TFIID (TAF1)	16
TFIID (TAF9)	67
TFIIE (Tfa1)	54
TFIIH (Kin28)	87

Indeed, TAF1, along with TBP, had been regarded as a keystone of TFIID, helping to assemble all the other TAFs in that factor, but this view is clearly not supported by the genome-wide expression analysis. Instead, TAF1 and its homolog in higher organisms appear to be required in the preinitiation complexes formed at only a subset of genes. In yeast, these genes tend to be ones governing progression through the cell cycle.

Mutation of the other yeast TAF (TAF9) had a more pronounced effect. Sixty-seven percent of the yeast genes analyzed were as dependent on this TAF as they were on Rpb1. But that does not mean that TFIID is required for transcription of all these genes, because TAF9 is also part of a transcription adapter complex known as **SAGA** (named for three classes of proteins it contains—SPTs, ADAs, and GCN5—and its enzymatic activity, histone acetyltransferase). Like TFIID, SAGA contains TBP, a number of TAFs, and histone acetyltransferase activity, and appears to mediate the effects of certain transcription activator proteins. So the effect of mutating TAF9 may be due to its role in SAGA or perhaps in other protein complexes yet to be discovered, rather than in TFIID.

Not only are some TAFs not universally required for transcription, the TFIIDs appear to be heterogenous in their TAF compositions. For example, TAF10 is found in only a fraction of human TFIIDs, and its presence correlates with responsiveness to estrogen.

Even more surprisingly, TBP is not universally found in preinitiation complexes in higher eukaryotes. The most celebrated example of an alternative TBP is **TRF1** (**TBP-related factor 1**) in *Drosophila melanogaster*. This protein is expressed in developing neural tissue, binds to TFIIA and TFIIB, and stimulates transcription just as TBP does, and it has its own group of TRF-associated factors called **nTAFs** (for neural TAFs). In 2000, Michael Holmes and Robert Tjian used primer extension analysis in vivo and in vitro to show that TRF1 stimulates transcription of the *Drosophilia tudor* gene. Furthermore, this analysis revealed that the *tudor* gene has two distinct promoters. The first is a downstream promoter with a TATA box recognized by a complex including TBP. The second promoter lies about 77 bp upstream of the first and has a TC box recognized by a complex including TRF1 (Figure 11.13). The TC box extends from position −22 to −33 with respect to the start of transcription and has the sequence ATTGCTTTTCTT in the nontemplate strand. It is protected by a complex of TRF1, TFIIA, and TFIIB in DNase footprinting experiments. However, none of these proteins alone make a footprint in this region, and neither does TBP, or TBP plus TFIIA and TFIIB.

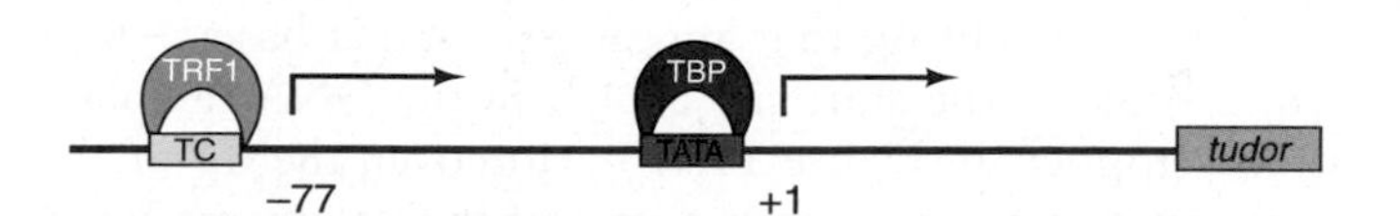

Figure 11.13 The *Drosophila tudor* control region. This gene has two promoters about 77 bp apart. The downstream promoter has a TATA box that attracts a preinitiation complex based on TBP. The upstream promoter has a TC box that attracts a preinitiation complex based on TRF1.

Thus, TRF appears to be a cell type-specific variant of TBP. The presence of alternative TBPs and TAFs raises the possibility that gene expression in higher eukaryotes could be controlled in part by the availability of the appropriate TBP and TAFs, as well as by the activator proteins we will study in Chapter 12. Indeed, the recognition of two different *tudor* promoters by two different TBPs is reminiscent of the recognition of two different prokaryotic promoters for the same gene by RNA polymerases bearing different σ-factors, as we saw in Chapter 8.

Actually, TRF appears to be unique to *Drosophila*. But another **TBP-like factor** (**TLF**) has been found in all multicellular animals investigated to date. TLF differs from TBP in lacking the pairs of phenylalanines that intercalate between base pairs in TATA boxes and help bend the DNA at the promoter. Accordingly, TLF appears not to bind to TATA boxes and may direct transcription at other, TATA-less promoters.

The central role of TBP in forming preinitiation complexes has been further challenged by the discovery of a **TBP-free TAF-containing complex** (**TFTC**) that is able to sponsor preinitiation complex formation without any help from TFIID or TBP. Structural studies by Patrick Schultz and colleagues have provided some insight into how TFTC can substitute for TFIID. They have performed electron microscopy and digital image analysis on both TFTC and TFIID and found that they have strikingly similar three-dimensional structures. Figure 11.14 shows three-dimensional models of the two protein complexes in three different orientations. The most obvious characteristics of both complexes is a groove large enough to accept a double-stranded DNA. In fact, it appears that the protein of both complexes would encircle the DNA and hold it like a clamp. The only major difference between the two complexes is the projection at the top of TFTC due to domain 5. TFIID lacks both the projection and domain 5.

In Chapter 10 we learned that many promoters in *Drosophila* lack a TATA box; instead, they have a DPE, usually coupled with an initiator element (Inr). We also learned that the DPE can attract TFIID through one or more of its TAFs. In 2000, James Kadonaga and colleagues also discovered a factor in *Drosophila* (dNC2) that is homologous to a factor from other organisms known as **NC2** (**negative cofactor 2**) or Dr1-Drap1. For simplicity's sake, we can refer to all such factors as NC2. Kadonaga and colleagues also made the interesting discovery that NC2 can discriminate between TATA box-containing promoters and DPE-containing promoters. In fact, NC2 stimulates transcription from DPE-containing promoters and represses transcription from TATA

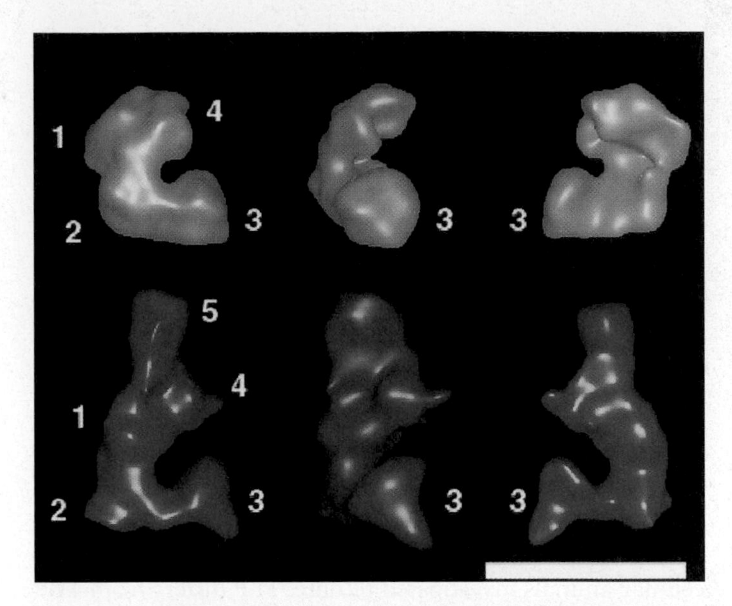

Figure 11.14 Three-dimensional models of TFIID and TFTC. Schultz and colleagues made negatively stained electron micrographs (see Chapter 19, for method) of TFIID and TFTC, then digitally combined images to arrive at an average. Then they tilted the grid in the microscope and analyzed the resulting micrographs to glean three-dimensional information for both proteins. The resulting models for TFIID (green) and TFTC (blue) are shown. (*Source:* Brand, M., C. Leurent, V. Mallouh, L. Tora, and P. Schuttz, Three-dimensional structures of the TAF_{II}-containing complexes TFDIID and TFTC. *Science* 286 (10 Dec 1999) f. 3, p. 2152. Copyright © AAAS.)

box-containing promoters. Thus, NC2 may be a focal point of gene regulation.

The crystal structure of an NC2–TATA box–TBP complex, determined by Stephen Burley and colleagues in 2001, shows how NC2 can inhibit transcription from TATA box-containing promoters. It binds to the underside of the DNA that has been bent by the saddle-shaped TBP. Once NC2 has bound to the promoter, one of its α-helices blocks TFIIB from joining the complex, and another part of NC2 interferes with TFIIA binding. Without TFIIA or TFIIB, the preinitiation complex cannot form and transcription cannot initiate.

SUMMARY The TAFs do not appear to be universally required for transcription of class II genes. Even TAF1 is not required for transcription of the great majority of yeast class II genes. Even TBP is not universally required. Some promoters in higher eukaryotes respond to an alternative protein such as TRF1 and not to TBP. Some promoters can be stimulated by a TBP-free TAF-containing complex (TFTC), rather than by TFIID. The general transcription factor NC2 stimulates transcription from DPE-containing promoters but represses transcription from TATA-containing promoters.

Structure and Function of TFIIB

Danny Reinberg and his coworkers cloned and expressed the gene for human TFIIB. This cloned TFIIB product can substitute for the authentic human protein in all in vitro assays, including response to activators such as Sp1. This suggests that TFIIB is a single-subunit factor (M_r = 35 kD) that requires no auxiliary polypeptides such as the TAFs. As we have already discovered, TFIIB is the third general transcription factor to join the preinitiation complex in vitro (after TFIID and A), or the second if TFIIA has not yet bound. It is essential for binding RNA polymerase because the polymerase–TFIIF complex will bind to the DAB complex, but not to the DA complex.

The position of TFIIB between TFIID and TFIIF/RNA polymerase II in the assembly of the preinitiation complex suggests that TFIIB is part of the measuring device that places RNA polymerase II in the proper position to initiate transcription. If so, TFIIB should have two domains: one to bind to each of these proteins. Indeed, TFIIB does have two domains: an N-terminal domain (**$TFIIB_N$**), and a C-terminal domain (**$TFIIB_C$**). Subsequent structural work in 2004 by Roger Kornberg and colleagues revealed that these two domains really do function to bridge between TFIID at the TATA box and RNA polymerase II so as to position the active center of the polymerase about 26–31 bp downstream of the TATA box, just where transcription should begin. In particular, this work showed that TBP, by bending the DNA at the TATA box, wraps the DNA around $TFIIB_C$, and that $TFIIB_N$ binds to a site on the polymerase that positions the enzyme correctly at the transcription initiation site.

Kornberg and colleagues crystallized a complex of RNA polymerase II and TFIIB from budding yeast *(Saccharomyces cerevisiae)*. Figure 11.15 shows two views of the structure of this complex, along with the positions of TBP and promoter DNA inferred from previous work. We can see the two domains of TFIIB in this complex. $TFIIB_C$ (magenta) appears to interact with TBP and DNA at the TATA box. Indeed, the DNA bent by TBP at the TATA box appears to wrap around $TFIIB_C$ and the polymerase. After the bend, the DNA extends straight toward $TFIIB_N$, which lies near the active site of the polymerase.

Previous studies had shown that mutations in $TFIIB_N$ altered the start site of transcription, and the present work provides a rationale for those findings. In particular, it was known that mutations in residues 62–66 cause changes in the initiation site. These amino acids lie on the side of a finger domain in $TFIIB_N$ that appears to contact bases −6 to −8, relative to the start site at +1, in the DNA template strand (top left in Figure 11.16). Moreover, the tip of the finger approaches the active center of the polymerase, and lies near the initiator region of the promoter (Chapter 10), which surrounds the transcription start site.

In the human TFIIB, the fingertip contains two basic residues (lysine), which could bind well to the DNA at the

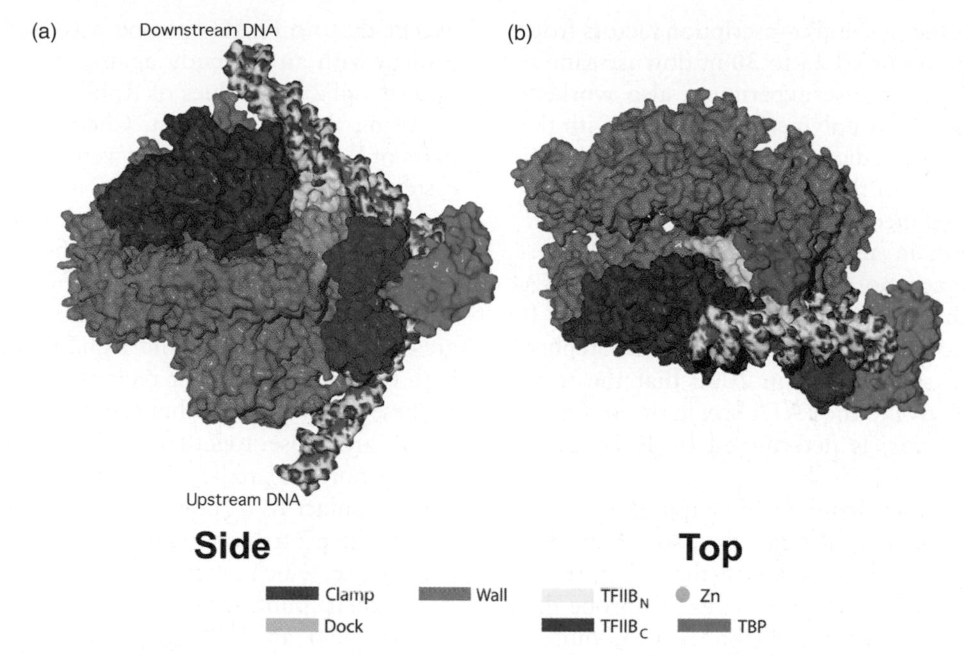

Figure 11.15 A model for the TFIIB–TBP–polymerase II-DNA structure. (a) and **(b)** show two different views of the structure, which Kornberg and colleagues inferred from separate structures of TFIIB$_C$–TBP–TATA box DNA and RNA polymerase II-TFIIB. The color key at bottom identifies TBP, the domains of TFIIB, and domains of the polymerase that interact with TFIIB. Other regions of the polymerase are in gray. The bent TATA box DNA, with 20-bp B-form DNA extensions, is in red, white, and blue. (*Source:* (a–b) Reprinted with permission from *Science,* Vol. 303, David A. Bushnell, Kenneth D. Westover, Ralph E. Davis, Roger D. Kornberg, "Structural Basis of Transcription: An RNA Polymerase II-TFIIB Cocrystal at 4.5 Angstroms" Fig. 3 c&d, p. 986. Copyright 2004, AAAS.)

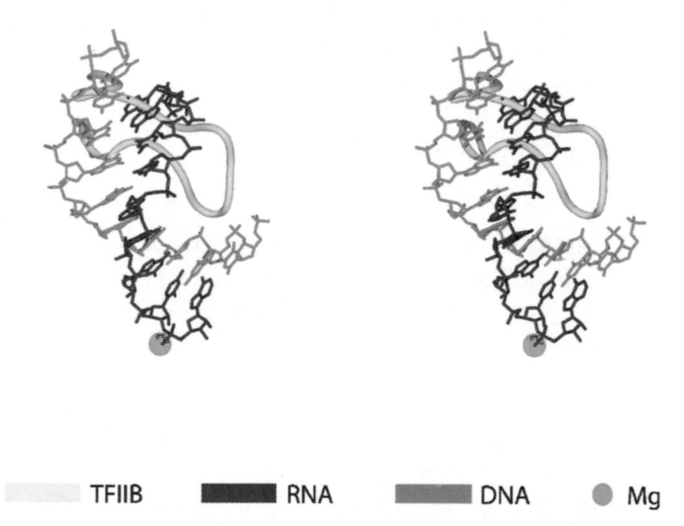

Figure 11.16 Stereo view of the interaction between the B finger of TFIIB$_N$, the DNA template strand, and the RNA product. The elements of the structure are identified by the color key at bottom. (*Source:* Reprinted with permission from *Science,* Vol 303, David A. Bushnell, Kenneth D. Westover, Ralph E. Davis, Roger D. Kornberg, "Structural Basis of Transcription: An RNA Polymerase II-TFIIB Cocrystal at 4.5 Angstroms" Fig. 4, p. 987. Copyright 2004, AAAS.)

initiator, thus positioning the start of transcription there. However, these two basic amino acids are replaced by acidic amino acids in yeast TFIIB, and initiator sequences do not exist in yeast promoters. These considerations may help explain why the human preinitiation complex can successfully position the start of transcription approximately 25–30 bp downstream of the TATA box, whereas transcription initiation is much more variable (40–120 bp downstream of the TATA box) in yeast.

Kornberg and colleagues concluded that TFIIB plays a dual role in positioning the transcription start site. First, it achieves coarse positioning by binding via its TFIIB$_C$ domain to TBP at the TATA box and binding to RNA polymerase via the finger and an adjacent zinc ribbon in the TFIIB$_N$ domain. In most eukaryotes, this places the polymerase in position to start about 25–30 bp downstream of the TATA box. Then, upon DNA unwinding, TFIIB achieves fine positioning by interacting with DNA at, and just upstream of, the initiator via the finger of TFIIB$_N$. Notice that TFIIB not only determines the start site of transcription, it also determines the *direction* of transcription. That is because its asymmetry of binding to the promoter—with its C-terminal domain upstream and its N-terminal domain downstream—establishes an asymmetry to the preinitiation complex, which in turn establishes the direction of transcription.

The importance of TFIIB and RNA polymerase II in establishing the transcription start site is underscored by the following experiment. In the budding yeast *Saccharomyces cerevisiae,* the start site is about 40 to 120 nt downstream of the TATA box, whereas in the fission yeast *Saccharomyces pombe,* it is about 25 to 30 nt downstream of the TATA box. However, when *S. pombe* TFIIB and RNA polymerase II were mixed with the other general transcription factors from *S. cerevisiae,* initiation occurred 25 to 30 nt downstream of the TATA box. And the reverse experiment also worked: *S. cerevisiae* TFIIB and RNA polymerase II, mixed with the other factors from *S. pombe,* dictated transcription initiation 40 to 120 nt downstream of the TATA box.

A similar measuring mechanism appears to apply to the archaea. Transcription in archaea requires a basal transcription apparatus composed of a multisubunit RNA polymerase, an arachaeal TBP, and **transcription factor B (TFB)**, which is homologous to eukaryotic TFIIB. Stephen Bell and Stephen Jackson showed in 2000 that the transcription start site, relative to the TATA box in the archaeon *Sulfolobus acidocaldarius,* is determined by RNA polymerase and TFB.

The model presented in Figure 11.15 is appealing, but it is cobbled together from partial structures, so we are left wondering how closely it corresponds to the structure we would see in an intact preinitiation complex. To probe this question, Hung-Ta Chen and Steven Hahn used a combination of photo-cross-linking and **hydroxyl radical probing** to map the interactions between domains of yeast TFIIB and domains of yeast RNA polymerase II.

Hydroxyl radical probing uses the following strategy: The experimenters introduce cysteine residues into one protein by site-directed mutagenesis (Chapter 5). To each cysteine in turn, they attach an iron-EDTA (ethylenediamine tetraacetate) complex known as Fe-BABE, which can generate hydroxyl radicals that can cleave protein chains within about 15 Å. After cleavage, the protein fragments can be displayed by gel electrophoresis and detected by Western blotting. This procedure identifies any regions of a second protein lying within 15 Å of a given cysteine on the first protein.

In their first experiment, Chen and Hahn changed several amino acids in the finger and linker regions of TFIIB to cysteines, which were then linked to Fe-BABE. After assembling preinitiation complexes with these modified TFIIB molecules, they activated hydroxyl radical formation to cleave proteins in close proximity to the cysteines in the finger and linker regions of TFIIB. To facilitate Western blotting, they attached an epitope (FLAG) to the end of either Rpb1 or Rpb2, so they could use anti-FLAG antibodies to probe their Western blots. Figure 11.17a–c shows the results of the Western blots probed with anti-FLAG antibody when the FLAG epitope was placed at the N- or C- terminus of Rpb2, or the C-terminus of Rpb1. The novel bands created by hydroxyl radical cleavage (not found in lanes with no substituted cysteines [wt] or no Fe-BABE [–]) are marked with brackets.

These bands contain protein fragments of known length, and we know that they include either the protein's N-terminus or C-terminus because they are detected by an anti-FLAG antibody, and the FLAG epitope is attached to a protein terminus. Thus, the cleavage sites could be mapped to locations on the known crystal structure of the protein. Figure 11.17d presents a similar experiment, except that no FLAG epitope was used, and the blot was probed with an antibody against a natural epitope in the N-terminal 200 residues of Rpbl.

Using this information, Chen and Hahn mapped the parts of Rpb1 and Rpb2 that were in close contact with the cysteine attached to the Fe-BABE in each case. Figure 11.17e and f depict the maps of cleavages caused by TFIIB variants with cysteines introduced into the finger and linker regions, respectively. Dark blue and light blue regions denote strong and moderate-to-weak cleavage, respectively. These are the regions of Rpb1 and Rpb2 that are in close contact with the finger and linker regions of TFIIB. The similarities of these maps suggests that the finger and linker regions of TFIIB are close together in the preinitiation complex. Furthermore, as predicted, this part of TFIIB ($TFIIB_N$) does indeed contact RNA polymerase II. In particular, it contacts sites in the protrusion, wall, clamp, and fork regions of the polymerase, which are near the active center.

In their photo-cross-linking experiments, Chen and Hahn linked an ^{125}I-tagged photo-cross-linking reagent called PEAS to the cysteines in the same TFIIB cysteine variants used in the hydroxyl radical probing. After assembling preinitiation complexes with these derivatized TFIIBs, they irradiated the complexes to form covalent cross-links, then observed the cross-links by SDS-PAGE and autoradiography to detect the ^{125}I tags. As expected, they found that the TFIIB finger and linker domains cross-linked to RNA polymerase II. However, they also discovered something unexpected: The TFIIB finger and linker domains also cross-linked to the largest subunit of TFIIF, placing this polypeptide close to the active center of polymerase II.

SUMMARY Structural studies on a TFIIB-polymerase II complex show that TFIIB binds to TBP at the TATA box via its C-terminal domain, and to polymerase II via its N-terminal domain. This bridging action effects a coarse positioning of the polymerase active center about 25–30 bp downstream of the TATA box. In mammals, a loop motif of the N-terminal domain of TFIIB effects a fine positioning of the start of transcription by interacting with the single-stranded template DNA strand very near the active center. Biochemical studies confirm that the TFIIB N-terminal domain (the finger and linker domains, in particular) lies close to the RNA polymerase II active center, and to the largest subunit of TFIIF, in the preinitiation complex.

Structure and Function of TFIIH

TFIIH is the last general transcription factor to join the preinitiation complex. It appears to play two major roles in transcription initiation; one of these is to phosphorylate

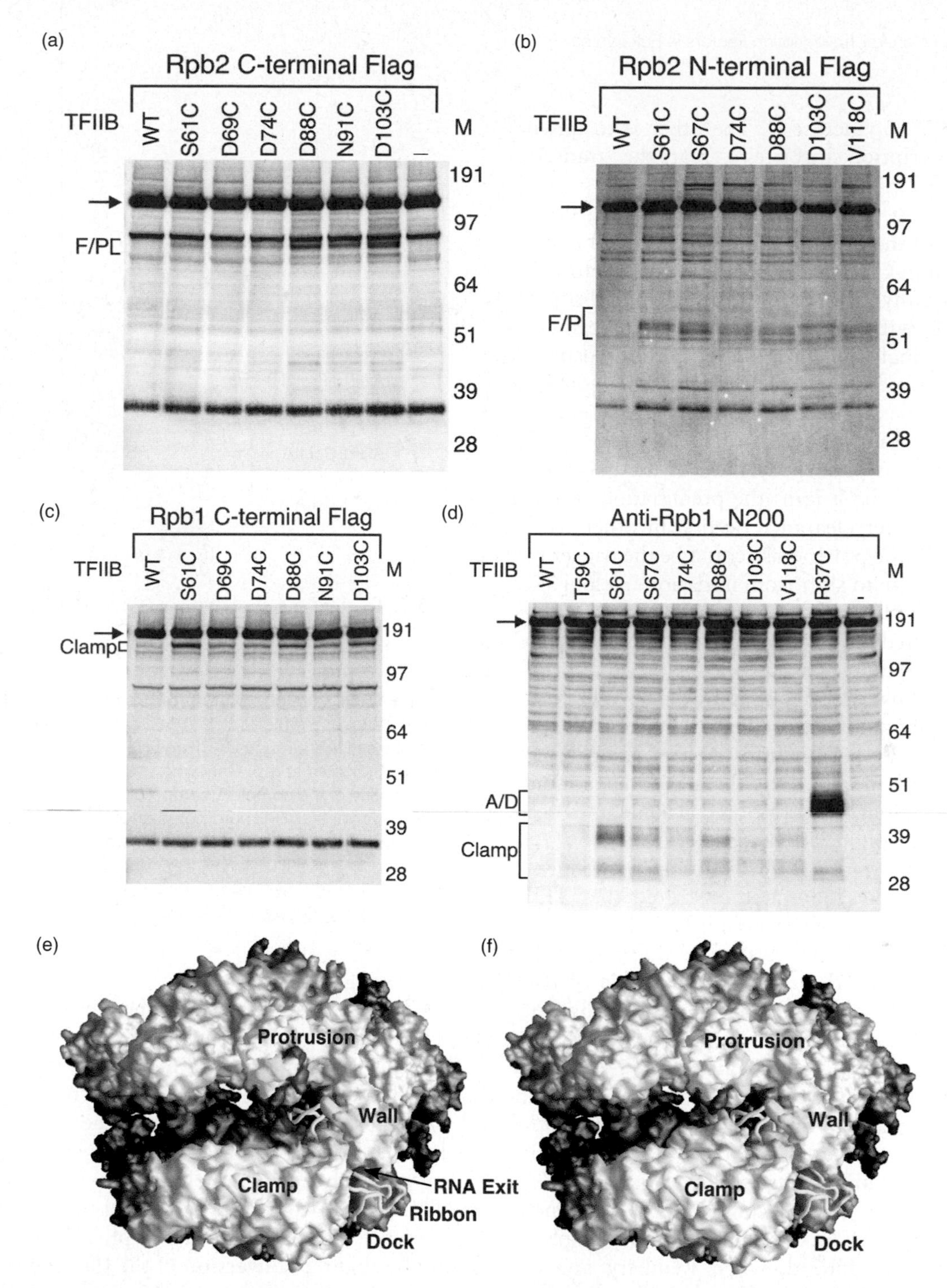

Figure 11.17 Mapping contacts between TFIIB and RNA polymerase II in the yeast preinitiation complex. (a–d) Chen and Hahn attached Fe-BABE hydroxyl-radical-generating reagents to cysteines that had been substituted for other amino acids (positions indicated at tops of lanes) in the finger and linker domains of TFIIB. Then they formed preinitiation complexes that included these substituted TFIIBs and RNA polymerases whose Rpb2 C-terminus (a) or N-terminus (b), or whose Rpbl C-terminus (c) had been tagged with the FLAG epitope, as indicated at the top of each gel. Then they activated hydroxyl radical formation to cleave proteins within about 15 Å of the cysteine in the TFIIB. Then they performed SDS-PAGE on the preinitiation complex proteins and protein fragments, and on proteins from complexes that did not contain substituted cysteines (wt), or did not contain TFIIB complexed with Fe-BABE ( ). They blotted the protein bands and visualized them by probing the blots with an anti-FLAG antibody (a–c) or with an antibody against a natural epitope in the terminal 200 amino acids of Rpbl. The novel bands (brackets) that do not appear in the control lanes (wt and −) represent polypeptide fragments generated by hydroxyl radical cleavage. The lengths of these fragments, compared to markers (M), together with the knowledge that they contain one of the ends of either Rpbl or 2, allows the cleavage site to be determined to within four amino acids on either side. The locations of these cleavage sites are identified beside each bracket: clamp; F/P (fork and protrusion); or A/D (active site and dock regions). **(e)** and **(f)** Mapping the cleavage sites to the known crystal structure of the yeast RNA polymerase II when TFIIB contained substituted cysteines in the finger domain (e), or the linker domain (f). Dark blue represents strong cleavages, and light blue represents weak to moderate cleavages. To take account of the error inherent in the method, the color was spread out over nine amino acids, centered on the apparent cleavage site. (*Source: (a–f)* Reprinted from *Cell,* Vol. 119, Hung-Ta Chen and Steven Hahn, "Mapping the Location of TFIIB within the RNA Polymerase II Transcription Preinitiation Complex: A Model for the Structure of the PIC," pp. 169–180, fig 2, p. 172. Copyright 2004 with permission from Elsevier.)

the CTD of RNA polymerase II. The other is to unwind DNA at the transcription start site to create the "transcription bubble."

Phosphorylation of the CTD of RNA Polymerase II As we have already seen in Chapter 10, RNA polymerase II exists in two physiologically meaningful forms: IIA (unphosphorylated) and IIO (with many phosphorylated serines in the carboxyl-terminal domain [CTD]). The unphosphorylated enzyme, polymerase IIA, is the form that joins the preinitiation complex. But the phosphorylated enzyme, polymerase IIO, carries out RNA chain elongation. This behavior suggests that phosphorylation of the polymerase occurs between the time it joins the preinitiation complex and the time promoter clearance occurs. In other words, phosphorylation of the polymerase could be the trigger that allows the polymerase to shift from initiation to elongation mode. This hypothesis receives support from the fact that the unphosphorylated CTD in polymerase IIA binds much more tightly to TBP than does the phosphorylated form in polymerase IIO. Thus, phosphorylation of the CTD could break the tether that binds the polymerase to the TBP at the promoter and thereby permit transcription elongation to begin. On the other hand, this hypothesis is damaged somewhat by the finding that transcription can sometimes occur in vitro without phosphorylation of the CTD.

Whatever the importance of CTD phosphorylation, Reinberg and his colleagues have demonstrated that TFIIH was a good candidate for the protein kinase that catalyzes this process. First, these workers showed that the purified transcription factors, by themselves, are capable of phosphorylating the CTD of polymerase II, converting polymerase IIA to IIO. The evidence, shown in Figure 11.18 came from a gel mobility shift assay. Lanes 1–6 demonstrate that adding ATP had no effect on the mobility of the DAB, DABPolF, or DABPolFE complexes. On the other hand, after TFIIH was added to form the DABPolFEH complex, ATP produced a change to lower mobility. What accounted for this change? One possibility is that one of the transcription factors in the complex had phosphorylated the polymerase. Indeed, when Reinberg and colleagues isolated the polymerase from the lower mobility complex, it proved to be the phosphorylated form, polymerase IIO. But polymerase IIA had been added to the complex in the first place, so one of the transcription factors had apparently performed the phosphorylation.

Next Reinberg and colleagues demonstrated directly that the TFIIH preparation phosphorylates polymerase IIA. To do this, they incubated purified polymerase IIA and TFIIH together with [γ-^{32}P]ATP under DNA-binding conditions. A small amount of polymerase phosphorylation occurred, as shown in Figure 11.19a. Thus, this TFIIH preparation by itself is capable of carrying out the phosphorylation. By contrast, all the other factors together caused no such phosphorylation. However, these factors

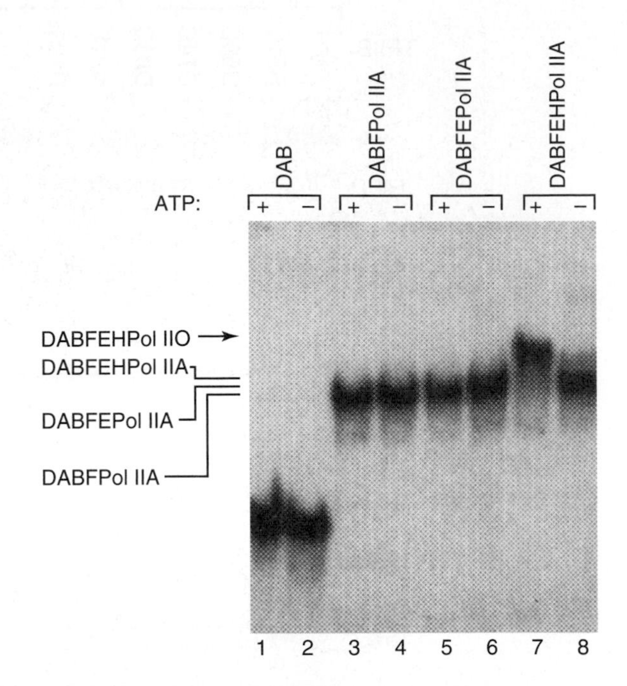

Figure 11.18 Phosphorylation of preinitiation complexes. Reinberg and colleagues performed gel mobility shift assays with preinitiation complexes DAB through DABPolFEH, in the presence and absence of ATP, as indicated at top. Only when TFIIH was present did ATP shift the mobility of the complex (compare lanes 7 and 8). The simplest explanation is that TFIIH promotes phosphorylation of the input polymerase (polymerase IIA) to polymerase IIO. (*Source:* Lu, H., I. Zawel, L. Fisher, J.M. Egly, and D. Reinberg, Human general transcription factor IIH phosphorylates the C-terminal domain of RNA polymerase II. *Nature* 358 (20 Aug 1992) p. 642, f. 1. Copyright © Macmillan Magazines Ltd.)

could greatly stimulate the phosphorylating capability of TFIIH. Lanes 6–9 show the results with TFIIH plus an increasing set of the other factors. As Reinberg and associates added each new factor, they noticed an increasing efficiency of phosphorylation of the polymerase and accumulation of polymerase IIO. Because the biggest increase in polymerase IIO labeling came with the addition of TFIIE, these workers performed a time-course study in the presence of TFIIH or TFIIH plus TFIIE. Figure 11.19b shows that the conversion of the IIa subunit to the IIo subunit was much more efficient when TFIIE was present. Figure 11.19c shows the same results graphically.

We know that the CTD of the polymerase IIa subunit is the site of the phosphorylation because polymerase IIB, which lacks the CTD, is not phosphorylated by the TFIIDBFEH complex, while polymerase IIA, and to a lesser extent, polymerase IIO, are phosphorylated (Figure 11.20a). Also, as we have seen, phosphorylation produces a polypeptide that coelectrophoreses with the IIo subunit, which does have a phosphorylated CTD. To demonstrate directly the phosphorylation of the CTD, Reinberg and colleagues cleaved the phosphorylated enzyme with chymotrypsin, which cuts off the CTD, and electrophoresed the products. The autoradiograph of the chymotrypsin

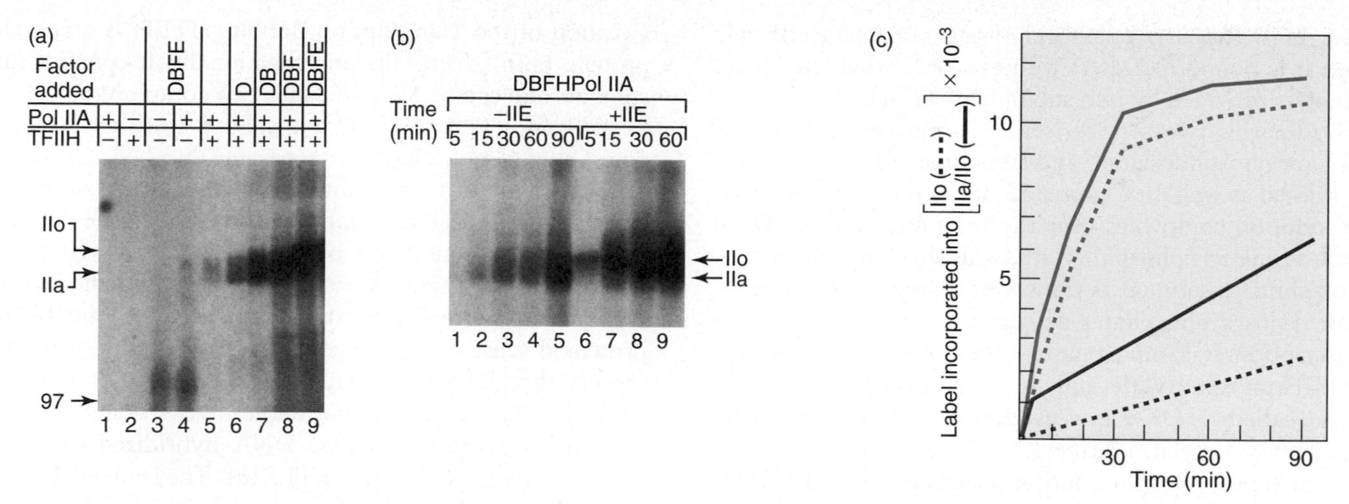

Figure 11.19 TFIIH phosphorylates RNA polymerase II. **(a)** Reinberg and colleagues incubated polymerase IIA (containing the hypophosphorylated subunit IIa) with various mixtures of transcription factors, as shown at top. They included [γ-^{32}P]ATP in all reactions to allow phosphorylation of the polymerase, then electrophoresed the proteins and performed autoradiography to visualize the phosphorylated polymerase. Lane 4 shows that TFIID, B, F, and E, were insufficient to cause phosphorylation. Lanes 5–9 demonstrate that TFIIH alone is sufficient to cause some polymerase phosphorylation, but that the other factors enhance the phosphorylation. TFIIE provides particularly strong stimulation of phosphorylation of the polymerase IIa subunit to IIo. **(b)** Time course of polymerase phosphorylation. Reinberg and colleagues performed the same assay for polymerase phosphorylation with TFIID, B, F, and H in the presence or absence of TFIIE, as indicated at top. They carried out the reactions for 60 or 90 min, sampling at various intermediate times, as shown at top. Arrows at right mark the positions of the two polymerase subunit forms. Note that polymerase phosphorylation is more rapid in the presence of TFIIE. **(c)** Graphic presentation of the data from panel (b). Green and red curves represent phosphorylation in the presence and absence, respectively, of TFIIE. Solid lines and dotted lines correspond to appearance of phosphorylated polymerase subunits IIa and IIo, or just IIo, respectively. (*Source:* Adapted from Lu, H., I. Zawel, L. Fisher, J.-M. Egly, and D. Reinberg, Human general transcription factor IIH phosphorylates the C-terminal domain of RNA polymerase II. *Nature* 358 (20 Aug 1992) p. 642, f. 2. Copyright © Macmillan Magazines Ltd.)

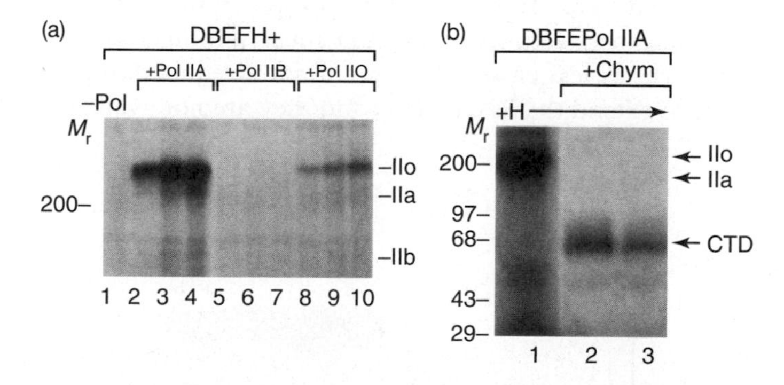

Figure 11.20 TFIIH phosphorylates the CTD of polymerase II. **(a)** Reinberg and colleagues phosphorylated increasing amounts of polymerases IIA, IIB, or IIO, as indicated at top, with TFIID, B, F, E, and H and radioactive ATP as described in Figure 11.19. Polymerase IIB, lacking the CTD, could not be phosphorylated. The unphosphorylated polymerase IIA was a much better phosphorylation substrate than IIO, as expected. **(b)** Purification of the phosphorylated CTD. Reinberg and colleagues cleaved the CTD from the phosphorylated polymerase IIa subunit with the protease chymotrypsin (Chym), electrophoresed the products, and visualized them by autoradiography. Lane 1, reaction products before chymotrypsin cleavage; lanes 2 and 3, reaction products after chymotrypsin cleavage. The position of the CTD had been identified in a separate experiment. (*Source:* Lu, H., L. Zawel, L. Fisher, J.-M. Egly, and D. Reinberg, Human general transcription factor IIH phosphorylates the C terminal domain of RNA polymerase II. *Nature* 358 (20 Aug 1992) p. 642, f. 3. Copyright © Macmillan Magazines Ltd.)

products (Figure 11.20b) shows a labeled CTD fragment, indicating that labeled phosphate has been incorporated into the CTD part of the large polymerase II subunit. The rest of the subunit was not labeled.

To prove that none of the subunits of RNA polymerase II was helping in the kinase reaction, Reinberg and coworkers cloned a chimeric gene that codes for the CTD as a fusion protein that also includes the DNA-binding domain from the transcription factor GAL4 and the enzyme glutathione-*S*-transferase. It appeared that TFIIH, all by itself, was capable of phosphorylating the CTD domain of this fusion protein. Thus, this TFIIH preparation had the appropriate kinase activity, even in the absence of other polymerase II subunits.

All of the experiments described so far were done under conditions in which the polymerase (or polymerase domain) was bound to DNA. Is this important? To find out, Reinberg's group tried the kinase assay with polymerase II in the presence of DNA that had a complete promoter, or merely the TATA box or the initiator regions of the promoter, or even no promoter at all. The result was that the TFIIH preparation performed the phosphorylation quite well in the presence of a TATA box, or an initiator, but did very poorly with a synthetic DNA (poly [dI-dC]) that contained neither.

Thus, TFIIH appears to phosphorylate polymerase II only when it is bound to DNA. We now know that the kinase activity is provided by two subunits of TFIIH.

Ordinarily, two serines (serine 2 and serine 5) of the CTD are phosphorylated, and sometimes serine 7 is phosphorylated as well. In Chapter 15, we will see evidence that transcription complexes near the promoter have CTDs in which serine 5 is phosphorylated, but that this phosphorylation shifts to serine 2 as transcription progresses. That is, serine 5 loses phosphates as serine 2 gains them during transcription. It is important to note that the protein kinase of TFIIH phosphorylates only serine 5 of the CTD. Another kinase, called **CTDK-1** in yeast and **CDK9 kinase** in metazoans, phosphorylates serine 2.

Sometimes, phosphorylation on serine 2 of the CTD is also lost during elongation, and that can cause pausing of the polymerase. In order for elongation to begin again, re-phosphorylation of serine 2 of the CTD must occur.

SUMMARY The preinitiation complex forms with the hypophosphorylated form of RNA polymerase II (IIA). Then, TFIIH phosphorylates serine 5 in the heptad repeat in the carboxyl-terminal domain (CTD) of the largest RNA polymerase II subunit, creating the phosphorylated form of the enzyme (IIO). TFIIE greatly stimulates this process in vitro. This phosphorylation is essential for initiation of transcription. During the shift from initiation to elongation, phosphorylation shifts from serine 5 to serine 2. If phosphorylation of serine 2 is also lost, the polymerase pauses until re-phosphorylation by a non-TFIIH kinase occurs.

Creation of the Transcription Bubble TFIIH is a complex protein, both structurally and functionally. It contains nine subunits and can be separated into two complexes: a protein kinase complex composed of four subunits, and a five-subunit core TFIIH complex with two separate DNA helicase/ATPase activities. One of these, contained in the largest subunit of TFIIH, is essential for viability: When its gene in yeast *(RAD25)* is mutated, the organism cannot survive. Satya Prakash and colleagues demonstrated that this helicase is essential for transcription. First they overproduced the **RAD25** protein in yeast cells, purified it almost to homogeneity, and showed that this product had helicase activity. For a helicase substrate, they used a partial duplex DNA composed of a ^{32}P-labeled synthetic 41-base DNA hybridized to single-stranded M13 DNA (Figure 11.21a). They mixed RAD25 with this substrate in the presence and absence of ATP and electrophoresed the products. Helicase activity released the short, labeled DNA from its much longer partner, so it had a much higher electrophoretic mobility and was found at the bottom of the gel. As Figure 11.21b demonstrates, RAD25 has an ATP-dependent helicase activity.

Next, Prakash and colleagues showed that transcription was temperature-sensitive in cells bearing a temperature-sensitive *RAD25* gene (*rad25*-ts$_{24}$). Figure 11.22 shows the results of an in vitro transcription assay using a G-less cassette (Chapter 5) as template. This template had a yeast TATA box upstream of a 400-bp region with no G's in the nontemplate strand. Transcription in the presence of ATP, CTP, and UTP (but no GTP) apparently initiated (or terminated) at two sites within this G-less region and gave rise to two transcripts, 375 and 350 nt in length, respectively. Transcription must terminate at the end of the G-less cassette because G's are required at that point to extend the RNA chain, and they are not available.

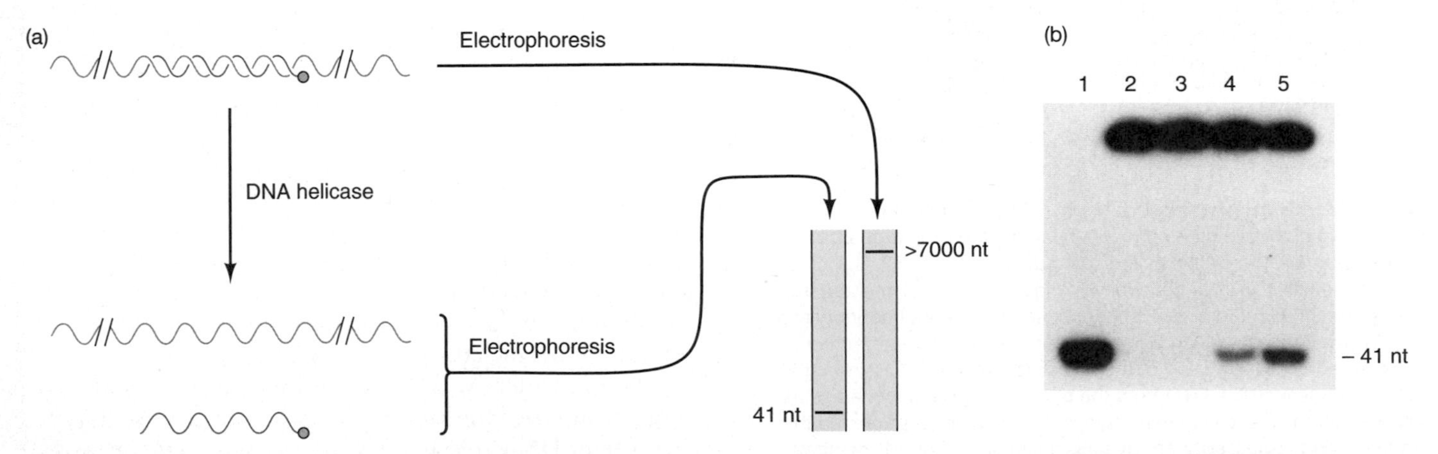

Figure 11.21 Helicase activity of TFIIH. **(a)** The helicase assay. The substrate consisted of a labeled 41-nt piece of DNA (red) hybridized to its complementary region in a much larger, unlabeled, single-stranded M13 phage DNA (blue). DNA helicase unwinds this short helix and releases the labeled 41-nt DNA from its larger partner. The short DNA is easily distinguished from the hybrid by electrophoresis. **(b)** Results of the helicase assay. Lane 1, heat-denatured substrate; lane 2, no protein; lane 3, 20 ng of RAD25 with no ATP; lane 4, 10 ng of RAD25 plus ATP; lane 5, 20 ng of RAD25 plus ATP. (*Source:* (*b*) Gudzer, S.N., P. Sung, V. Bailly, L. Prakash, and S. Prakash, RAD25 is a DNA helicase required for DNA repair and RNA polymerase II transcription. *Nature* 369 (16 June 1994) p. 579, f. 2c. Copyright © Macmillan Magazines Ltd.)

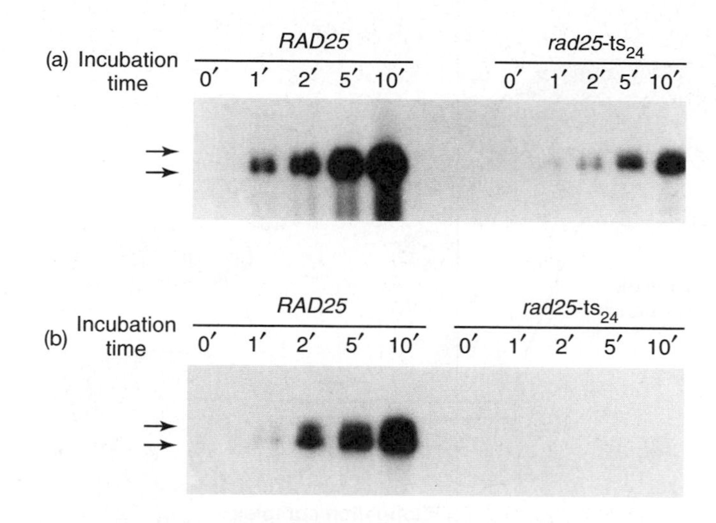

Figure 11.22 The TFIIH DNA helicase gene product (RAD25) is required for transcription in yeast: Prakash and colleagues tested extracts from wild-type (*RAD25*) and temperature-sensitive mutant (*rad25*-ts_{24}) cells for transcription of a G-less cassette template at the **(a)** permissive and **(b)** nonpermissive temperatures. After allowing transcription for 0–10 min in the presence of ATP, CTP, and UTP (but no GTP), with one ^{32}P-labeled nucleotide, they electrophoresed the labeled products and detected the bands by autoradiography. The origin of the extract (*RAD25* or *rad25*-ts_{24} cells), as well as the time of incubation in minutes, is given at top. Arrows at left denote the positions of the two G-less transcripts. We can see that transcription is temperature-sensitive when the TFIIH DNA helicase (RAD25) is temperature-sensitive.
(*Source:* Gudzer, S.N., P. Sung, V. Bailly, L. Prakash, and S. Prakash, RAD25 is a DNA helicase required for DNA repair and RNA polymerase II transcription. *Nature* 369 (16 June 1994) p. 580, f. 3 b–c. Copyright © Macmillan Magazines Ltd.)

(The shorter transcript may have come from premature termination within the G-less cassette, rather than from a different initiation site.) Panel (a) shows the results of transcription for 0–10 min at the permissive temperature (22°C). It is clear that the *rad25*-ts_{24} mutant extract gave weaker transcription than the wild-type *(RAD25)* extract even at low temperature. Panel (b) shows the results of transcription at the nonpermissive temperature (37°C). The elevated temperature completely inactivated transcription in the *rad25*-ts_{24} mutant extract. Thus, the *RAD25* product (the TFIIH DNA helicase) is required for transcription.

What step in transcription requires DNA helicase activity? The chain of evidence leading to the answer begins with the following consideration: Transcription of class II genes, unlike transcription of class I and III genes, requires ATP (or dATP) hydrolysis. Of course, the α-β-bonds of all four nucleotides, including ATP, are hydrolyzed during all transcription, but class II transcription requires hydrolysis of the β-γ-bond of ATP. The question arises: What step requires ATP hydrolysis? We would naturally be tempted to look at TFIIH for the answer to this question because it has two activities (CTD kinase and DNA helicase) that involve hydrolysis of ATP. The answer appears to be that the helicase activity of TFIIH is the ATP-requiring step. The main evidence in favor of this hypothesis is that GTP can substitute for ATP in CTD phosphorylation, but GTP cannot satisfy the ATP hydrolysis requirement for transcription. Thus, transcription requires ATP hydrolysis for some process besides CTD phosphorylation, and the best remaining candidate is DNA helicase.

Now let us return to the main question: What transcription step requires DNA helicase activity? The most likely answer is promoter clearance. In Chapter 6 we defined transcription initiation to include promoter clearance, but promoter clearance can also be considered a separate event that serves as the boundary between initiation and elongation. James Goodrich and Tjian asked this question: Are TFIIE and TFIIH required for initiation or for promoter clearance? To find the answer, they devised an assay that measures the production of abortive transcripts (trinucleotides). The appearance of abortive transcripts indicates that a productive transcription initiation complex has formed, including local DNA melting and synthesis of the first phosphodiester bond. Goodrich and Tjian found that TFIIE and TFIIH were not required for production of abortive transcripts, but TBP, TFIIB, TFIIF, and RNA polymerase II were required. Thus, TFIIE and TFIIH are not required for transcription initiation, at least up to the promoter clearance step. However, TFIIH is required for full DNA melting at promoters. If the largest subunit of human TFIIH is mutated, the DNA helicase of that subunit is defective, and the DNA at the promoter does not open completely. This could block promoter clearance, as explained later in this section.

These findings left open the possiblitity that TFIIE and TFIIH are required for either promoter clearance or RNA elongation, or both. To distinguish among these possibilities, Goodrich and Tjian assayed for elongation and measured the effect of TFIIE and TFIIH on that process. By leaving out the nucleotide required in the 17th position, but not before, they allowed transcription to initiate (without TFIIE and TFIIH) on a supercoiled template and proceed to the 16-nt stage. (They used a supercoiled template because transcription on such templates in vitro does not require TFIIE and TFIIH, nor does it require ATP.) Then they linearized the template by cutting it with a restriction enzyme and added ATP to allow transcription to continue in the presence or absence of TFIIE and TFIIH. They found that TFIIE and TFIIH made no difference in this elongation reaction. Thus, because TFIIE and TFIIH appear to have no effect on initiation or elongation, Goodrich and Tjian concluded that TFIIE and TFIIH are required in the promoter clearance step. Figure 11.23 summarizes these findings and more recent data discussed in the next paragraphs.

Tjian and others assumed that the DNA helicase activity of TFIIH acted directly on the DNA at the initiator to melt it. But cross-linking studies performed in 2000 by Tae-Kyung Kim, Richard Ebright, and Danny Reinberg showed that TFIIH (in particular, the subunit bearing the promoter-melting DNA helicase) forms cross-links with DNA between positions +3 and +25, and perhaps farther downstream. This site of interaction for TFIIH is downstream of

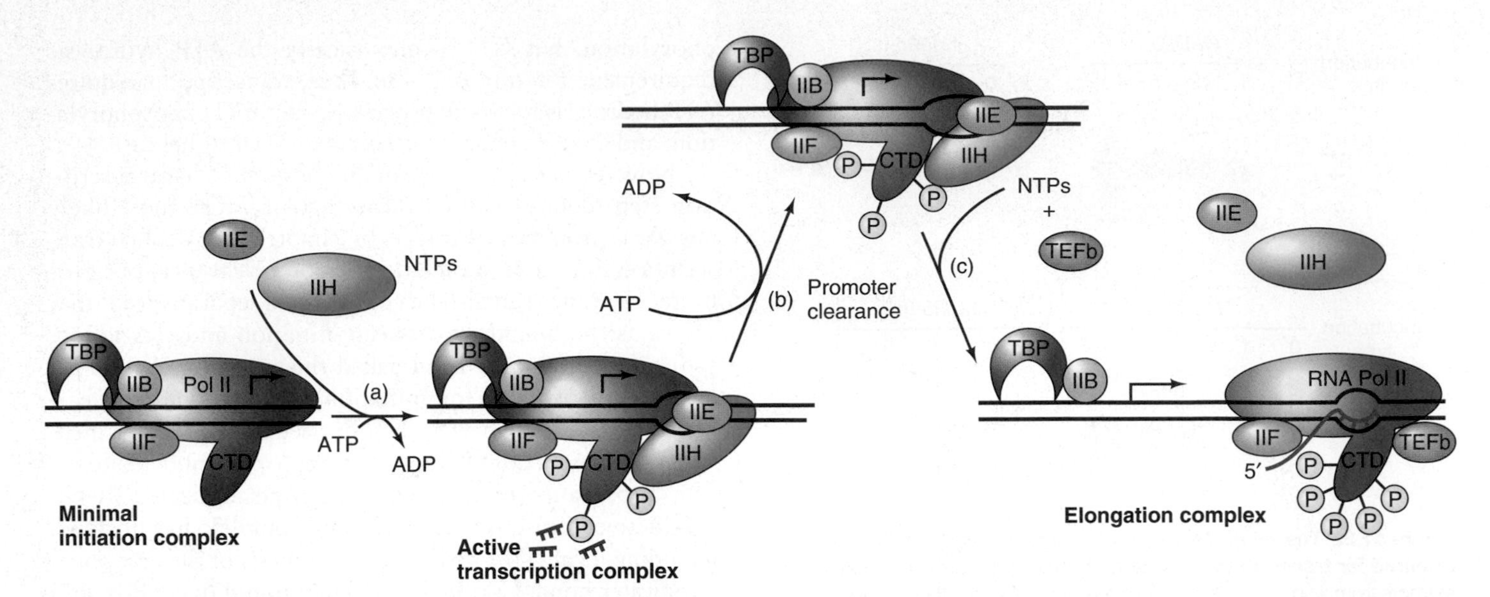

Figure 11.23 A model for the participation of general transcription factors in initiation, promoter clearance, and elongation. (a) TBP (or TFIID), along with TFIIB, TFIIF, and RNA polymerase II form a minimal initiation complex at the initiator. Addition of TFIIH, TFIIE, and ATP allows DNA melting at the initiator region and partial phosphorylation of the CTD of the largest subunit of RNA polymerase. These events allow production of abortive transcripts (magenta), but the polymerase stalls at position +10 to +12. **(b)** With energy provided by ATP, the DNA helicase of TFIIH causes further unwinding of the DNA, expanding the transcription bubble. This expansion releases the stalled polymerase and allows it to clear the promoter. **(c)** With further phosphorylation of the polymerase CTD by TEFb and with continuous addition of NTPs, the elongation complex continues elongating the RNA. TBP and TFIIB remain at the promoter. TFIIE and TFIIH are not needed for elongation and dissociate from the elongation complex. (*Source:* Adapted from Goodrich, J.A. and T. Tjian. 1994. Transcription factors IIE and IIH and ATP hydrolysis direct promoter clearance by RNA polymerase II. *Cell* 77:145–56.)

the site of the first transcription bubble (position −9 to +2). On the other hand, TFIIE cross-links to the transcription bubble region; TFIIB, TFIID, and TFIIF cross-link to the region upstream of the bubble; and RNA polymerase cross-links to the entire region encompassing all the other factors. These findings imply that the DNA helicase of TFIIH is not in contact with the first transcription bubble, and therefore cannot create the bubble by directly unwinding DNA there. Addition of ATP has no effect on the interactions upstream of the transcription bubble, but it does perturb the interactions within and downstream of the bubble.

We know from previous work that the helicase of TFIIH is responsible for creating the transcription bubble, but the cross-linking work described here indicates that it cannot directly unwind the DNA at the transcription bubble. So how does it create the bubble? Kim and associates suggested that it acts like a molecular "wrench" by untwisting the downstream DNA. Because TFIID and TFIIB (and perhaps other proteins) hold the DNA upstream of the bubble tightly, and this binding persists after addition of ATP, untwisting the downstream DNA would create strain in between and open up the DNA at the transcription bubble. This would allow the polymerase to initiate transcription and move 10–12 bp downstream. But previous work has shown that the polymerase stalls at that point unless it gets further help from TFIIH, which apparently twists the downstream DNA further to lengthen the transcription bubble, releasing the stalled polymerase to clear the promoter.

Figure 11.23 is drawn schematically so the effects of TFIIH on CTD phosphorylation and DNA unwinding are easy to see. But the real structure of the preinitiation complex is more complicated. Kornberg and colleagues modeled the positions of all the general transcription factors (except TFIIA) in the preinitiation complex, based on previous structural studies of TFIIE-polymerase II, TFIIF-polymerase II, and TFIIE-TFIIH complexes (Figure 11.24). The second-largest subunit of TFIIF (Tfg2) is homologous to the bacterial σ-factor, and lies at approximately the same position relative to the promoter as σ. In fact, two domains of Tfg2 that are homologous to domains 2 and 3 of *E. coli* σ-factor are labeled "2" and "3" in the figure. TFIIE lies about 25 bp downstream of the polymerase active center, in position to fulfill its role in recruiting TFIIH. And TFIIH is in position for its DNA helicase activity to act as a molecular wrench to open the promoter DNA, either directly, or indirectly by inducing negative supercoiling.

SUMMARY TFIIE and TFIIH are not essential for formation of an open promoter complex, or for elongation, but they are required for promoter clearance. TFIIH has a DNA helicase activity that is essential for transcription, presumably because it causes full melting of the DNA at the promoter and thereby facilitates promoter clearance.

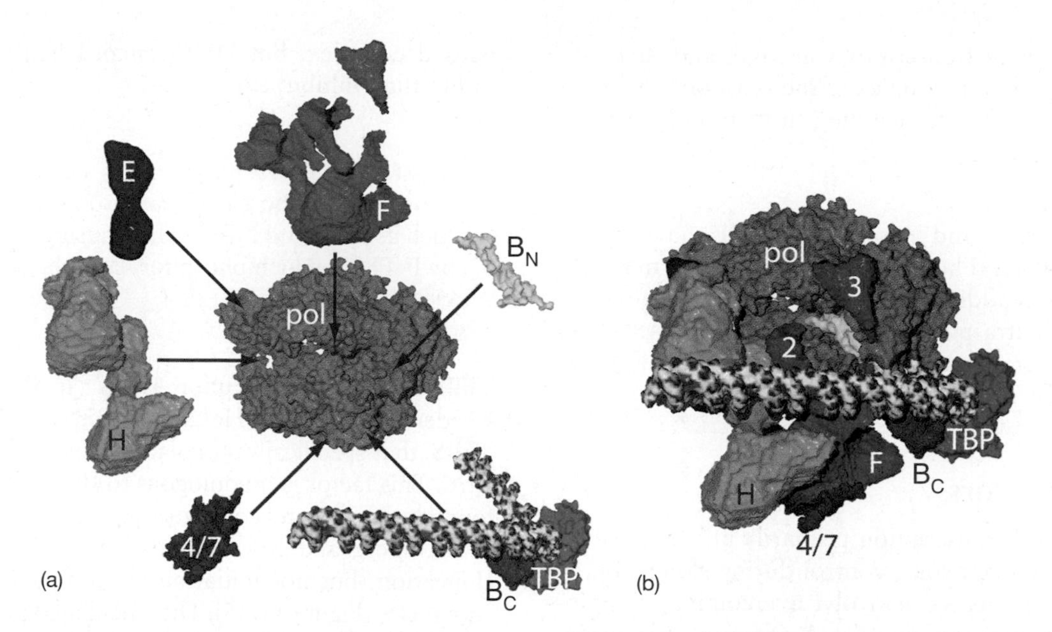

Figure 11.24 A model for the class II preinitiation complex. Kornberg and colleagues added previous structural information about the positions of promoter DNA, TFIIF, TFIIE, and TFIIH to their crystal structure of the TFIIB-RNA polymerase II complex to generate this composite model. **(a)** A blow-up to show the identities of all the components of the complex. The red component (4/7) represents Rpb4 and Rpb7, and pol (gray) denotes the rest of RNA polymerase II. B_N and B_C denote the N-terminal and C-terminal domains of TFIIB, respectively. The promoter DNA is represented by a red, white, and blue model, with a pronounced bend caused by binding of TBP. **(b)** Intact structure. Note that the transcription bubble has not yet formed. The direction of transcription is right to left. (*Source:* (*a–b*) Reprinted with permission from *Science,* Vol. 303, David A. Bushnell, Kenneth D. Westover, Ralph E. Davis, Roger D. Kornberg, "Structural Basis of Transcription: An RNA Polymerase II-TFIIB Cocrystal at 4.5 Angstroms" Fig. 6, p. 986. Copyright 2004, AAAS.)

The Mediator Complex and the RNA Polymerase II Holoenzyme

Another collection of proteins, known as **Mediator,** can also be considered a general transcription factor because it is part of most, if not all, class II preinitiation complexes. Unlike the other general transcription factors, Mediator is not required for initiation per se. But it is required for activated transcription, as we will see in Chapter 12. Mediator was first discovered in yeast, and found to contain about 20 polypeptides. A human Mediator was discovered later, and it is also a very large complex of over 20 polypeptides, only a minority of which have clear homology to those of yeast Mediator.

Our discussion so far has assumed that a preinitiation complex assembles at a class II promoter one protein at a time. This may indeed occur, but some evidence suggests that class II preinitiation complexes can assemble by binding a preformed **RNA polymerase II holoenzyme** to the promoter. The holoenzyme contains RNA polymerase, a subset of general transcription factors, and the Mediator complex.

Evidence for the holoenzyme concept came in 1994 with work from the laboratories of Roger Kornberg and Richard Young. Both groups isolated a complex protein from yeast cells, which contained RNA polymerase II and many other proteins. Kornberg and colleagues used immunoprecipitation with an antibody directed against one component of the holoenzyme to precipitate the whole complex. They recovered the subunits of RNA polymerase II, the subunits of TFIIF, and 17 other polypeptides. They could restore accurate transcription activity to this holoenzyme by adding TBP, TFIIB, E, and H. TFIIF was not required because it was already part of the holoenzyme.

Anthony Koleske and Young used a series of purification steps to isolate a holoenzyme from yeast that contained RNA polymerase II, TFIIB, TFIIF, and TFIIH. All this holenzyme needed for accurate transcription in vitro was TFIIE and TBP, so it contained more of the general transcription factors than the holoenzyme isolated by Kornberg and associates. Koleske and Young also identified some of the Mediator polypeptides in their holoenzyme and named them *SRB proteins* (*SRB2, SRB4, SRB5,* and *SRB6*).

The SRB proteins were discovered by Young and colleagues in a genetic screen whose logic went like this: Deletion of part of the CTD of the largest polymerase II subunit led to ineffective stimulation of transcription by the GAL4 protein, a transcription activator we will study in greater detail in Chapter 12. Young and coworkers then screened for mutants that could suppress this weak stimulation by GAL4. They identified several suppressor mutations in genes they named *SRB*s, for "suppressor of RNA polymerase B." We will discuss the probable basis for this suppression in Chapter 12. For now, it is enough to stress that these SRB proteins are required, at least in yeast, for optimal activation of transcription in vivo, and that they are part of the Mediator complex of the yeast polymerase II holoenzyme. Mammalian, including human, holoenzymes have also been isolated.

SUMMARY Yeast and mammalian cells have an RNA polymerase II holoenzyme that contains many polypeptides in addition to the subunits of the polymerase. The extra polypeptides include a subset of general transcription factors (not including TBP) and Mediator.

Elongation Factors

Eukaryotes control transcription primarily at the initiation step, but they also exert some control during elongation, at least in class II genes. This can involve overcoming transcription pausing or transcription arrest. A common characteristic of RNA polymerases is that they do not transcribe at a steady rate. Instead, they pause, sometimes for a long time, before resuming transcription. These pauses tend to occur at certain defined **pause sites,** because the DNA sequences at these sites destabilize the RNA–DNA hybrid and cause the polymerase to backtrack, probably extruding the free 3′-end of the nascent RNA into a pore in the enzyme, as we learned in Chapter 10. If the backtracking is limited to just a few nucleotides, the pause is relatively short, and the polymerase can resume transcribing on its own. On the other hand, if the backtracking goes too far, the polymerase cannot recover on its own, but needs help from an elongation factor. This more severe situation is termed a **transcription arrest** rather than a **transcription pause.**

Promoter Proximal Pausing Genome-wide analysis of the positions of RNA polymerase II on genes has shown that a sizable fraction of genes (perhaps 20–30%) contain polymerases paused at specific pause sites lying 20–50 bp downstream of the transcription start site. Some of the genes with such paused polymerases are those, such as the *Drosophila Hsp70* gene, that need to be activated quickly upon induction—in this case, by heat shock. These genes have polymerases poised to resume transcribing, as soon as they receive the signal to do so.

To understand this signal, it helps to understand how the polymerase became paused in the first place. Two protein factors are known to help stabilize RNA polymerase II in the paused state. These are **DRB sensitivity-inducing factor (DSIF)** and **negative elongation factor (NELF).** DSIF comprises two subunits, the elongation factors Spt4 and Spt5, which are found in eukaryotes from yeast to humans. NELF, on the other hand, is found in vertebrates, but not in all metazoans.

The signal to leave the paused state is delivered by **positive transcription elongation factor-b (P-TEFb).** This factor has a protein kinase that can phosphorylate polymerase II, DSIF, and NELF. Upon phosphorylation, NELF leaves the paused complex, but DSIF remains behind to stimulate, rather than inhibit, elongation.

SUMMARY RNA polymerases can be induced to pause at specific sites near promoters by proteins such as DSIF and NELF. This pausing can be reversed by P-TEFb, which phosphorylates the polymerase, as well as DSIF and NELF.

TFIIS Reverses Transcription Arrest In 1987, Reinberg and Roeder discovered a HeLa cell factor, which they named **TFIIS,** that specifically stimulates transcription elongation in vitro. This factor is homologous to IIS, which was originally found by Natori and colleagues in Ehrlich ascites tumor cells.

Reinberg and Roeder demonstrated that TFIIS affects elongation, but not initiation, by testing it on preinitiated complexes (Figure 11.25). They incubated polymerase II with a DNA template and nucleotides to allow initiation to occur, then added heparin (a polyanion that can bind to RNA polymerase as DNA would) to bind any free polymerase and block new initiation, then added either TFIIS or buffer and measured the rate of incorporation of labeled GMP into RNA. Figure 11.25 shows that TFIIS enhanced RNA synthesis considerably: the vertical dashed lines show that TFIIS

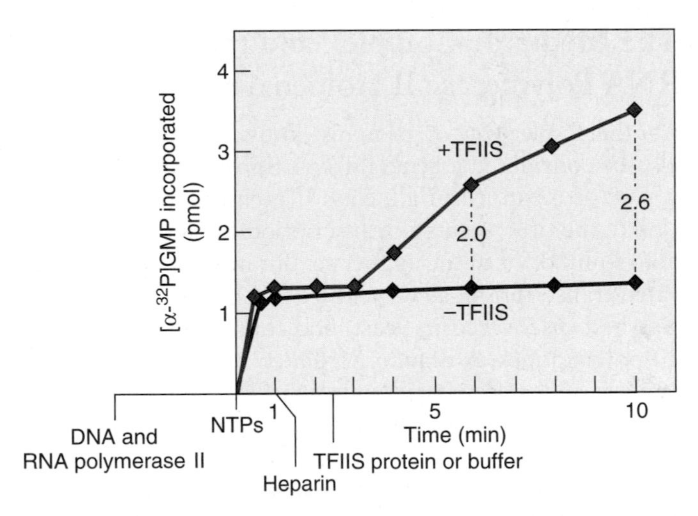

Figure 11.25 Effect of TFIIS on transcription elongation. Reinberg and Roeder formed elongation complexes as outlined in the time line at bottom. At time –3 min, they added DNA and RNA polymerase, then at time 0 they started the reaction by adding all four NTPs, one of which (GTP) was ^{32}P-labeled. At time +1 min, they added heparin to bind any free RNA polymerase, so all transcription complexes thereafter should be elongation complexes. Finally, at time +2.5 min, they added either TFIIS (red) or buffer (blue) as a negative control. They allowed labeled GMP incorporation to occur for various lengths of time, then took samples of the reaction mixture and measured the label incorporated into RNA. The dashed vertical lines indicate the fold stimulation of total RNA synthesis by TFIIS. (*Source:* Adapted from D. Reinberg and R.G. Roeder, Factors involved in specific transcription by mammalian RNA polymerase II. Transcription factor IIS stimulates elongation of RNA chains. *Journal of Biological Chemistry* 262:3333, 1987.)

stimulated GMP incorporation 2.0-fold by the 6-min mark, and 2.6-fold by the 10-min mark. Clearly, the *rate* of elongation increased even more dramatically—at least 10-fold.

It remained possible that TFIIS also stimulated transcription initiation. To investigate this possibility, Reinberg and Roeder repeated the experiment, but added TFIIS in the initial incubation, before they added heparin. If TFIIS really did stimulate initiation as well as elongation, then it should have produced a greater stimulation in this experiment than in the first. But the stimulations by TFIIS in the two experiments were almost identical. Thus, TFIIS appears to stimulate elongation only.

How does TFIIS enhance transcription elongation? Reinberg and Roeder performed an experiment that strongly suggested it does so by limiting transcription arrest.

One can detect pausing (or arresting) during in vitro transcription by electrophoresing the in vitro transcripts and finding discrete bands that are shorter than full-length transcripts. Reinberg and Roeder found that TFIIS minimized the appearance of these short transcripts, indicating that it minimized transcription arrest. Other workers have since confirmed this conclusion.

Daguang Wang and Diane Hawley demonstrated in 1993 that RNA polymerase II has an inherent, weak RNase activity that can be stimulated by TFIIS. This finding, and subsequent studies, led to a hypothesis to explain how TFIIS can restart arrested transcription (Figure 11.26). The arrested RNA polymerase has backtracked so far that the 3′-end of the nascent RNA is no longer in the enzyme's active site. Instead, it is extruded out through the pore and funnel that lead to the active site. With no 3′-terminal nucleotide to add to, the polymerase is stuck. So TFIIS activates the RNase activity in RNA polymerase II, which cleaves off the extruded part of the nascent RNA and creates a new 3′-terminus in the enzyme's active site.

How does TFIIS convert an enzyme that normally synthesizes RNA to one that breaks down RNA? Patrick Cramer and colleagues have obtained an x-ray crystal structure of an RNA polymerase II-TFIIS complex that sheds additional light on this question. Figure 11.27 shows a cutaway diagram of the complex, based on the crystal structure. TFIIS consists of three domains, including one that features a zinc ribbon. This zinc ribbon lies in the same pore and funnel of polymerase II as the extruded RNA. Just at the tip of the zinc ribbon are two acidic residues in very close proximity to metal A at the active site of the enzyme. In this position, the acidic side chains are ideally located to coordinate a second magnesium ion that would participate, along with the first, in ribonuclease activity.

Thus, TFIIS appears to change the activity of RNA polymerase, not by binding to the surface of the enzyme and effecting some conformational change within, but by getting right into the active site of the enzyme and actively participating in catalysis. This hypothesis receives strong support from the finding of a bacterial protein, called **GreB** in *E. coli,* that has the same function as TFIIS in restarting

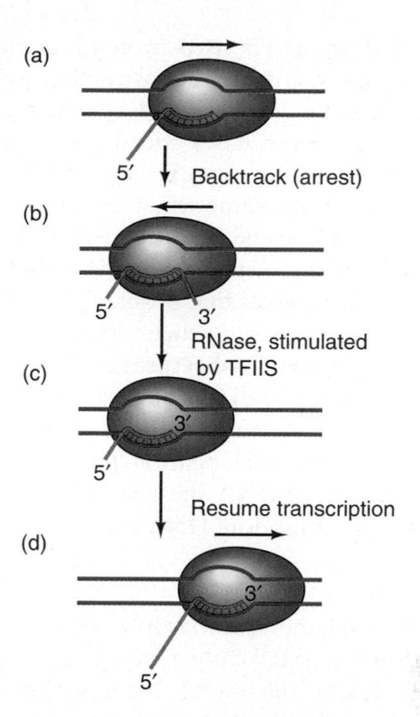

Figure 11.26 A model for reversal of transcription arrest by TFIIS. (a) RNA polymerase II, transcribing the DNA from left to right, has paused at a pause site. **(b)** The polymerase has backtracked to the left, extruding the 3′-end of the nascent RNA out of the enzyme's active site. This has caused a transcription arrest from which the polymerase cannot recover on its own. **(c)** A latent ribonuclease activity of the polymerase, stimulated by TFIIS, has cleaved off the extruded 3′-end of the nascent RNA. **(d)** With a free RNA 3′-end back in the active site, the polymerase can resume transcription.

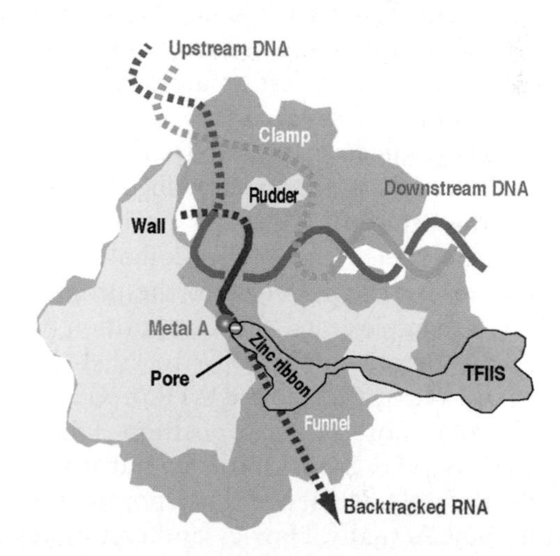

Figure 11.27 Cutaway view of the arrested yeast RNA polymerase II-TFIIS complex. The polymerase has backtracked, extruding the 3′-end of the nascent RNA (red) out of the enzyme's active site, into the pore and funnel. The zinc ribbon of TFIIS (orange) also lies in the pore and funnel, and its tip, containing two acidic residues, represented by the green circle and minus sign, approaches the metal A at the catalytic center of the polymerase, represented by the magenta circle. In this position, the two acidic residues can coordinate a second metal that collaborates with the first to constitute a ribonuclease activity that cleaves off the end of the extruded RNA. (*Source:* Reprinted from *Cell,* Vol 114, Conaway et al., "TFIIS and GreB: Two Like-Minded Transcription Elongation Factors with Sticky Fingers," fig. 1, pp. 272–274. Copyright 2003, with permission from Elsevier. Image courtesy of Joan Weliky Conaway and Patrick Cramer.)

arrested transcription. The two proteins are not **homologous**; that is, they share no sequence similarity, so they do not seem to have descended from a common evolutionary ancestor. However, GreB has a coiled-coil domain that extends into the exit channel for extruded RNA in the *E. coli* RNA polymerase in the same way the zinc ribbon in TFIIS does. Furthermore, located at the tip of the coiled-coil of GreB, adjacent to the metal ion at the polymerase active site, are two acidic residues that probably play the same role in ribonuclease catalysis as their counterparts in TFIIS appear to. This apparent convergent evolution of function argues for the validity of that proposed function.

It is interesting that an initiation factor (TFIIF) is also reported to play a role in elongation. It apparently does not limit arrests at defined DNA sites, as TFIIS does, but limits transient pausing at random DNA sites.

SUMMARY Polymerases that have backtracked and have become arrested can be rescued by TFIIS. This factor performs the rescue by inserting into the active site of RNA polymerase and stimulating an RNase that cleaves off the extruded 3′-end of the nascent RNA, which is causing transcription arrest. TFIIF also stimulates elongation, apparently by limiting transient pausing.

TFIIS Stimulates Proofreading of Transcripts Not only does TFIIS counteract pausing, it also contributes to proofreading of transcripts, presumably by a variation on the mechanism it uses to restart arrested transcription: stimulating an inherent RNase in the RNA polymerase to remove misincorporated nucleotides. Diane Hawley and her colleagues followed the procedure described in Figure 11.28a to measure the effect of TFIIS on proofreading. First, they isolated unlabeled elongation complexes that were paused at a variety of sites close to the promoter. Next, they walked the complexes to a defined position (Chapter 6) in the presence of radioactive UTP to label the RNA in the complexes. Next, they added ATP or GTP to extend the RNA by one more base, to position +43. The base that is called for at this position is A, but if G is all that is available, the polymerase will incorporate it, though at lower efficiency. Actually, Hawley and colleagues discovered that their ultrapure GTP contained a small amount of ATP, so AMP and GMP were incorporated in about equal quantities at position +43, even though ultrapure GTP was the only nucleotide they added. Next, they either cleaved the products with RNase T1, which cuts after G's, or chased with all four nucleotides to extend the labeled RNA to full length and then cut it with RNase T1. Finally, they subjected all RNase T1 products to electrophoresis and visualized the labeled products by autoradiography.

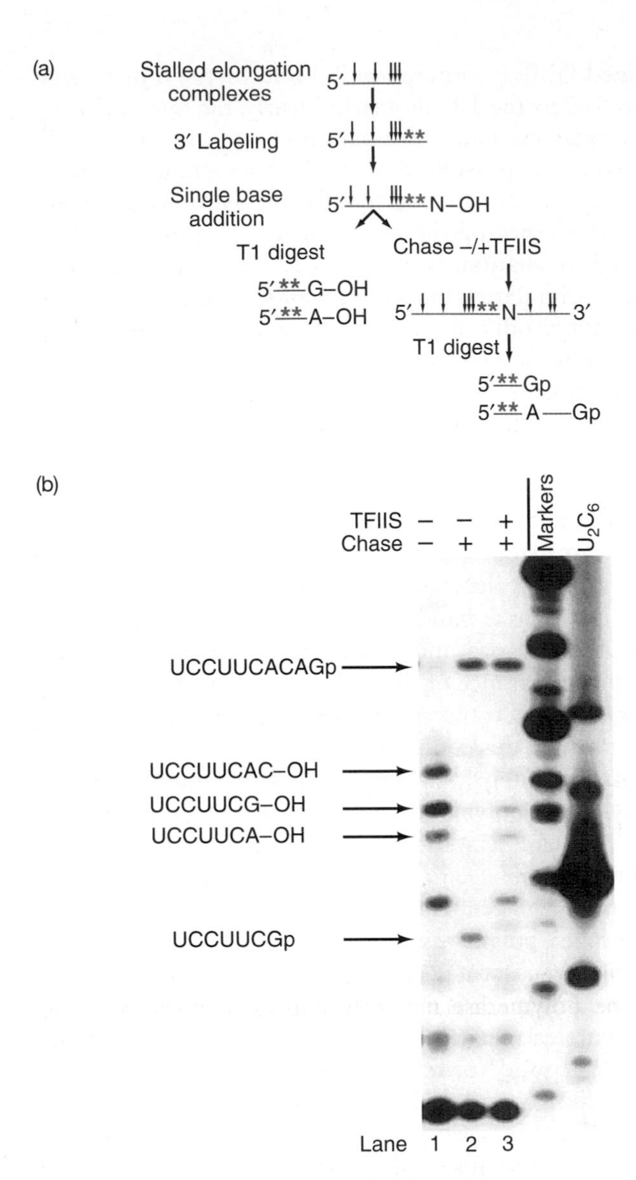

Figure 11.28 TFIIS stimulates proofreading by RNA polymerase II. **(a)** Experimental scheme. Hawley and colleagues started with short elongation complexes and 3′-end-labeled the short transcripts by walking the polymerase farther in the presence of [α–^{32}P]UTP. Then they added GTP to force misincorporation of G into position +43 where an A was called for. Then they digested the labeled transcripts with RNase T1 to measure the misincorporation of G (left), or chased the transcripts into full length with all four nucleotides, then cleaved the transcripts with RNase T1 to measure the loss of G from position +43 by proofreading. **(b)** Experimental results. Hawley and colleagues electrophoresed the RNase T1 products from part (a) and visualized them by autoradiography. Lane 1 contained unchased transcripts. The 7-mer resulting from misincorporation of G (UCCUUCG–OH), and the 7-mer (UCCUUCA) and 8-mer (UCCUUCAC) resulting from normal incorporation of A (or A and C) are indicated by arrows at left. Lanes 2 and 3 contained RNase T1 products of transcripts chased in the absence (lane 2) or presence (lane 3) of TFIIS. The 7-mer (UCCUUCGp) indicative of the misincorporated G that remained in the chased transcript is denoted by an arrow at left. The 10-mer (UCCUUCACAGp) indicative of incorporation of A in position +43, or G replaced by A at that position by proofreading, is also denoted by an arrow at left. TFIIS allowed removal of all detectable misincorporated G. (*Source:* (*b*) Thomas, M.J., A.A. Platas, and D.K. Hawley, Transcriptional fidelity and proofreading by RNA Polymerase II. *Cell* 93 (1998) f. 4, p. 631. Reprinted by permission of Elsevier Science.)

Simply cleaving the transcript with RNase T1 allowed Hawley and coworkers to measure the relative incorporations of AMP and GMP into position +43 because electrophoresis clearly separated the terminal 7-mers ending in A and G. Figure 11.28b, lane 1 shows the results of an experiment with no chasing. The 7-mer ending in G, the result of misincorporation of G, is about equally represented with the combination of a 7-mer ending in A, and an 8-mer ending in AC, which result from correct incorporation of A (or AC) from nucleotides contaminating the GTP substrate. Lanes 2 and 3 show the effects of chasing in the absence or presence, respectively, of TFIIS. The chased, full-length transcripts were cleaved with RNase T1, which yielded a 7-mer ending in Gp from a full-length transcript that still contained the misincorporated G, or a 10-mer ending in Gp from a full-length transcript in which proofreading had changed the misincorporated G to an A. When Hawley and colleagues did the chase in the absence of TFIIS, a significant amount of the misincorporated G remained in the RNA (see the band in lane 2 opposite the arrow indicating the 7-mer UCCUUCGp). However, most of the product appeared in the 10-mer (arrow labeled UCCUUCACAGp), which indicates that the polymerase was able to do some proofreading even without TFIIS. On the other hand, when they included TFIIS in the chase, Hawley and colleagues discovered that the 7-mer disappeared, and all of the labeled product was in the form of the 10-mer. Thus, TFIIS stimulates proofreading of the transcript.

The current model for proofreading (recall Figure 11.26) is that the polymerase not only pauses in response to a misincorporated nucleotide, it backtracks, extruding the 3′-end of the RNA out of the polymerase. This causes transcription to arrest. Then, TFIIS stimulates the latent RNase activity of the polymerase, which cuts off the extruded end of the RNA, including the misincorporated nucleotide, allowing the polymerase to resume transcribing.

Recall from Chapter 6 that the auxiliary factors that stimulate proofreading in bacteria are dispensable, but that the polymerase, with help from the mismatched end of a nascent RNA, can carry out proofreading in the absence of auxiliary factors. The strong conservation of the active site of RNA polymerases suggests that the same phenomenon will be observed in eukaryotic RNA polymerases, too. Indeed, this notion fits with the finding of Hawley and colleagues that polymerase II can carry out proofreading without any help from TFIIS.

SUMMARY TFIIS stimulates proofreading—the correction of misincorporated nucleotides—presumably by stimulating the RNase activity of the RNA polymerase, allowing it to cleave off a misincorporated nucleotide (with a few other nucleotides) and replace it with the correct one.

11.2 Class I Factors

The preinitiation complex that forms at rRNA promoters is much simpler than the polymerase II preinitiation complex we have just discussed. It involves polymerase I, of course, in addition to just two transcription factors. The first is a core-binding factor called **SL1** in humans, and **TIF-IB** in some other organisms; the second is a UPE-binding factor called **upstream-binding factor (UBF)** in mammals and **upstream activating factor (UAF)** in yeast. SL1 (or TIF-IB) is the core-binding factor. Along with RNA polymerase I, it is required for basal transcription activity. In fact, the core-binding factor is necessary to recruit polymerase I to the promoter. UBF (or UAF) is the factor that binds to the UPE. It is an **assembly factor** that helps the core-binding factor bind to the core promoter element. It does so by bending the DNA dramatically, so it can also be called an **architectural transcription factor** (Chapter 12). Humans and *Xenopus laevis* exhibit an almost absolute reliance on UBF for transcription of class I genes, whereas other organisms, including yeast, rats, and mice, can carry out some transcription without the help of the assembly factor. Still other organisms, such as the amoeba *Acanthamoeba castellanii,* show relatively little need for the assembly factor.

The Core-Binding Factor

Tjian and his colleagues discovered SL1 in 1985, when they separated a HeLa cell extract into two functional fractions. One fraction had RNA polymerase I activity, but no ability to initiate accurate transcription of a human rRNA gene in vitro. Another fraction had no polymerase activity of its own, but could direct the polymerase fraction to initiate accurately on a human rRNA template. Furthermore, this transcription factor, SL1, showed species specificity. That is, it could distinguish between the human and mouse rRNA promoter.

The experiments described so far used impure polymerase I and SL1. Further experiments with highly purified components revealed that human SL1 by itself cannot stimulate human polymerase I to bind to class I promoters and begin transcribing. It requires the UBF to assist its binding, as we will see in the next section.

Because human class I transcription works so poorly with the core-binding factor SL1 in the absence of UBF, the human system is not well suited to studies of the role of the core-binding factor in recruiting polymerase I to the promoter. On the other hand, *A. castellanii,* which exhibits little dependence on a UPE-binding protein, is a better choice because the effect of the core-binding factor can be studied by itself. Marvin Paule and Robert White exploited this system to show that the core-binding factor (TIF-IB) can recruit polymerase I to the promoter and stimulate initiation in the proper place. The actual DNA sequence where the polymerase binds appears to be irrelevant.

Paule and colleagues created mutant templates with various numbers of base pairs inserted or deleted between the TIF-IB-binding site and the normal transcription initiation site. This is reminiscent of the experiment performed by Benoist and Chambon with a class II promoter, reported in Chapter 10. In that experiment, deleting base pairs between the TATA box and the normal transcription initiation site did not alter the strength of transcription and did not change the transcription initiation site relative to the TATA box. In all cases, transcription began about 30 bp downstream of the TATA box.

With the class I promoter, Paule and colleagues reached a similar conclusion. They found that adding or subtracting up to 5 base pairs between the TIF-IB binding site and the normal transcription start site still allowed transcription to occur. Furthermore, the initiation site moved upstream or downstream according to the number of base pairs added or deleted (Figure 11.29). Adding or subtracting more than 5 bp blocked transcription activity (data not shown). Paule and colleagues concluded that TIF-IB contacts polymerase I and positions it for initiation a set number of base pairs downstream.

The exact base sequence contacted by the polymerase must not matter, because it is different in each mutant DNA. To confirm that the polymerase is contacting DNA in the same place relative to the TIF-IB-binding site in each mutant, Paule and colleagues performed DNase footprinting with a wild-type template and with each mutant template. The footprints were essentially indistinguishable, reinforcing the conclusion that the polymerase binds in the same spot regardless of the DNA sequence there. This is consistent with the hypothesis that TIF-IB binds to its DNA target and positions the polymerase I by direct protein–protein contact. The polymerase appears to contact the DNA because it extends the footprint caused by TIF-IB, but this contact appears to be nonspecific.

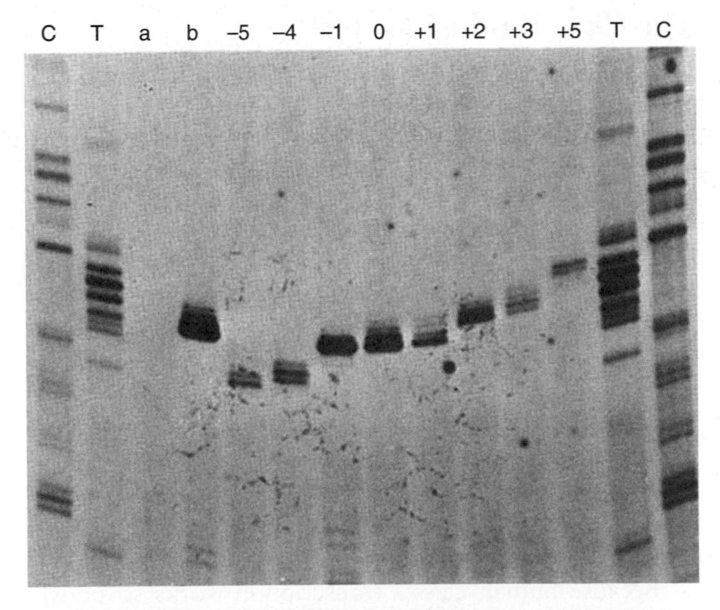

Figure 11.29 Effect of insertions and deletions on polymerase I transcription initiation site. Paule and colleagues made insertions and deletions of up to 5 bp, as indicated at top, between the TIF-IB binding site and the normal transcription start site in an *A. castellanii* rRNA promoter. Then they transcribed these templates in vitro and performed primer extension analysis (Chapter 5) with a ^{32}P-labeled 17-nt sequencing primer. They electrophoresed the labeled extended primers alongside C and T sequencing lanes using the same primer (lanes C and T). Lane a is a negative control run with vector DNA, but no rRNA promoter, lane b is a positive control containing a wild-type rRNA promoter. Lane 0 also contained the extended primer generated from the transcript of wild-type DNA with no deletion. (*Source:* Reprinted from *Cell* v. 50, Kownin et al., p. 695 © 2001, with permission from Elsevier Science.)

SUMMARY Class I promoters are recognized by two transcription factors, a core-binding factor and a UPE-binding factor. The human core-binding factor is called SL1; in some other organisms, such as *A. castellanii,* the homologous factor is known as TIF-IB. The core-binding factor is the fundamental transcription factor required to recruit RNA polymerase I. This factor also determines species specificity, at least in animals. The factor that binds the UPE is called UBF in mammals and most other organisms, but UAF in yeast. It is an assembly factor that helps the core-binding factor bind to the core promoter element. The degree of reliance on the UPE-binding factor varies considerably from one organism to another. In *A. castellanii,* TIF-IB alone suffices to recruit the RNA polymerase I and position it correctly for initiation of transcription.

The UPE-Binding Factor

Because human SL1 by itself did not appear to bind directly to the rRNA promoter, but a partially purified RNA polymerase I preparation did, Tjian and his coworkers began a search for DNA-binding proteins in the polymerase preparation. This led to the purification of human UBF in 1988. The factor as purified was composed of two polypeptides, of 97 and 94 kD. However, the 97-kD polypeptide alone is sufficient for UBF activity. When Tjian and colleagues performed footprint analysis with this highly purified UBF, they found that it had the same behavior as observed previously with partially purified polymerase I. That is, it gave the same footprint in the core element and a section of the UPE called *site A,* and SL1 intensified this footprint and extended it to a part of the UPE called *site B* (Figure 11.30). Thus, UBF, not polymerase I, was the agent that bound to the promoter in the previous experiments, and SL1 facilitates this binding. These studies did not reveal whether SL1 actually contacts the DNA in a complex with UBF, or whether it merely changes the conformation of UBF so it can contact a longer stretch of DNA that

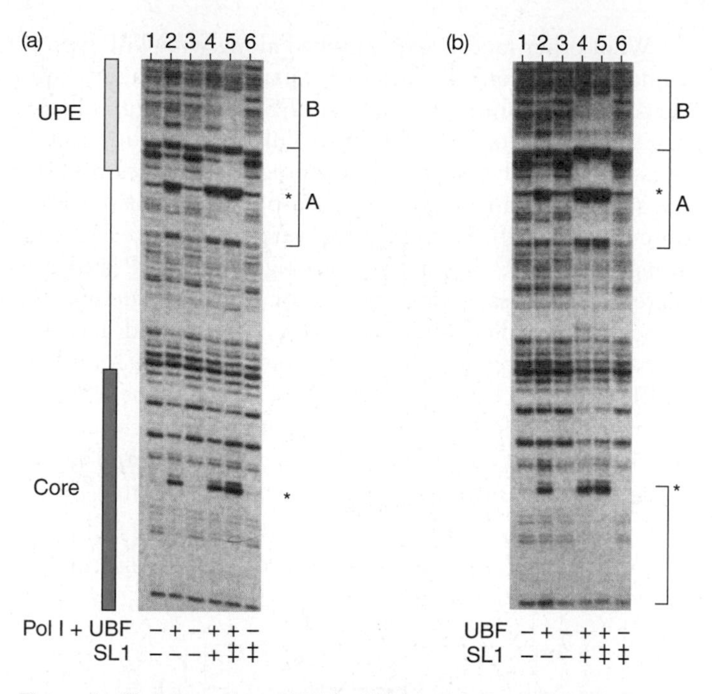

Figure 11.30 Interaction of UBF and SL1 with the rRNA promoter. Tjian and colleagues performed DNase footprinting with the human rRNA promoter and various combinations of **(a)** polymerase I + UBF and SL1 or **(b)** UBF and SL1. The proteins used in each lane are indicated at bottom. The positions of the UPE and core elements are shown at left, and the locations of the A and B sites are illustrated with brackets at right. Asterisks mark the positions of enhanced DNase sensitivity. SL1 caused no footprint on its own, but enhanced and extended the footprints of UBF in both the UPE and the core element. This enhancement is especially evident in the absence of polymerase I (panel b). (*Source:* Adapted from Bell S.P., R.M. Learned, H.-M. Jantzen, and R. Tjian, Functional cooperativity between transcription factors UBF1 and SL1 mediates human ribosomal RNA synthesis. *Science* 241 (2 Sept 1988) p. 1194, f. 3 a–b.)

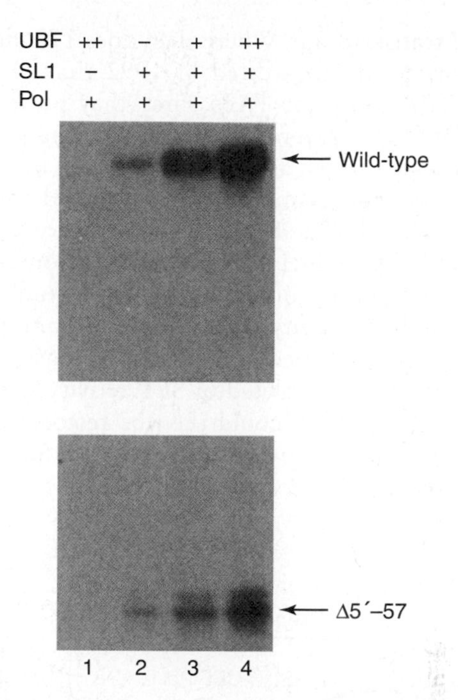

Figure 11.31 Activation of transcription from the rRNA promoter by UBF and SL1. Tjian and colleagues used an S1 assay to measure transcription from the human rRNA promoter in the presence of RNA polymerase I and various combinations of UBF and SL1, as indicated at top. The top panel shows transcription from the wild-type promoter; the bottom panel shows transcription from a mutant promoter (Δ5′–57) lacking UPE function. SL1 was required for at least basal activity, but UBF enhanced this activity on both templates. (*Source:* Bell S.P., R.M. Learned, H.-M. Jantzen, and R. Tjian, Functional cooperativity between transcription factors UBF1 and SL1 mediates human ribosomal RNA synthesis, *Science* 241 (2 Sept 1988) p. 1194, f. 4. Copyright © AAAS.)

extends into site B. Based on this and other data, we can conclude that SL1 cannot bind by itself, while UBF can. However, SL1 and UBF appear to bind cooperatively to give more extensive binding together than either could accomplish on its own.

Tjian and associates also found that UBF stimulates transcription of the rRNA gene in vitro. Figure 11.31 depicts the results of a transcription experiment using the wild-type human rRNA promoter and the mutant promoter (Δ5′–57) that lacks the UPE, and including various combinations of SL1 and UBF. Polymerase I was present in all reactions, and transcription efficiency was assayed by the S1 technique (Chapter 5). Lane 1 contained UBF, but no SL1, and showed no transcription of either template. This reaffirms that SL1 is absolutely required for transcription. Lane 2 had SL1, but no UBF, and showed a basal level of transcription. This demonstrates again that SL1 by itself is capable of stimulating basal transcription. Moreover, about as much transcription occurred on the mutant template that lacks the UPE as on the wild-type template. Thus, UBF is required for stimulation of transcription through the UPE. Lanes 3 and 4 contained both SL1 and an increasing amount of UBF. Significantly enhanced transcription occurred on both templates, but especially on the template containing the UPE. Tjian and colleagues concluded that UBF is a transcription factor that can stimulate transcription by binding to the UPE, but it can also exert an effect in the absence of the UPE, presumably by binding to the core element.

SUMMARY Human UBF is a transcription factor that stimulates transcription by polymerase I. It can activate the intact promoter, or the core element alone, and it mediates activation by the UPE. UBF and SL1 act synergistically to stimulate transcription.

Structure and Function of SL1

We have been discussing just two human factors, UBF and SL1, that are involved in transcription by polymerase I, and one of these, UBF, is probably just a single 97-kD polypeptide. But work presented earlier in this chapter showed that TATA box-binding protein (TBP) is essential for class I transcription. Where then does TBP fit in? Tjian and coworkers demonstrated in 1992 that SL1 is composed of TBP and three TAFs. First, they purified human (HeLa cell) SL1 by several different procedures. After each step, they used an S1 assay to locate SL1 activity. Then they assayed these same fractions for TBP by Western blotting. Figure 11.32 shows the striking correspondence they found between SL1 activity and TBP content.

If SL1 really does contain TBP, then it should be possible to inhibit SL1 activity with an anti-TBP antibody. Tjian and colleagues confirmed that this worked as predicted. A nuclear extract was depleted of SL1 activity with an anti-TBP antibody. Activity could then be restored by adding back SL1, but not just by adding back TBP. Something besides TBP must have been removed.

What other factors are removed along with TBP by immunoprecipitation? To find out, Tjian and colleagues subjected the immunoprecipitate to SDS-PAGE. Figure 11.33 depicts the results. In addition to TBP and antibody (IgG), we see three polypeptides, with molecular masses of 110, 63, and 48 kD (although the 48-kD polypeptide is partially obscured by TBP). Because these were immunoprecipitated along with TBP, they must bind tightly to TBP and are therefore TBP-associated factors, or **TAF_Is**, by definition. Hence, Tjian called them TAF_I110, TAF_I63, and TAF_I48. These are completely different from the TAFs found in TFIID (compare lanes 4 and 5). The TAFs could be stripped off of the TBP and antibody in the immunoprecipitate by treating the precipitate with 1 M guanidine-HCl and reprecipitating. The antibody and TBP remained together in the

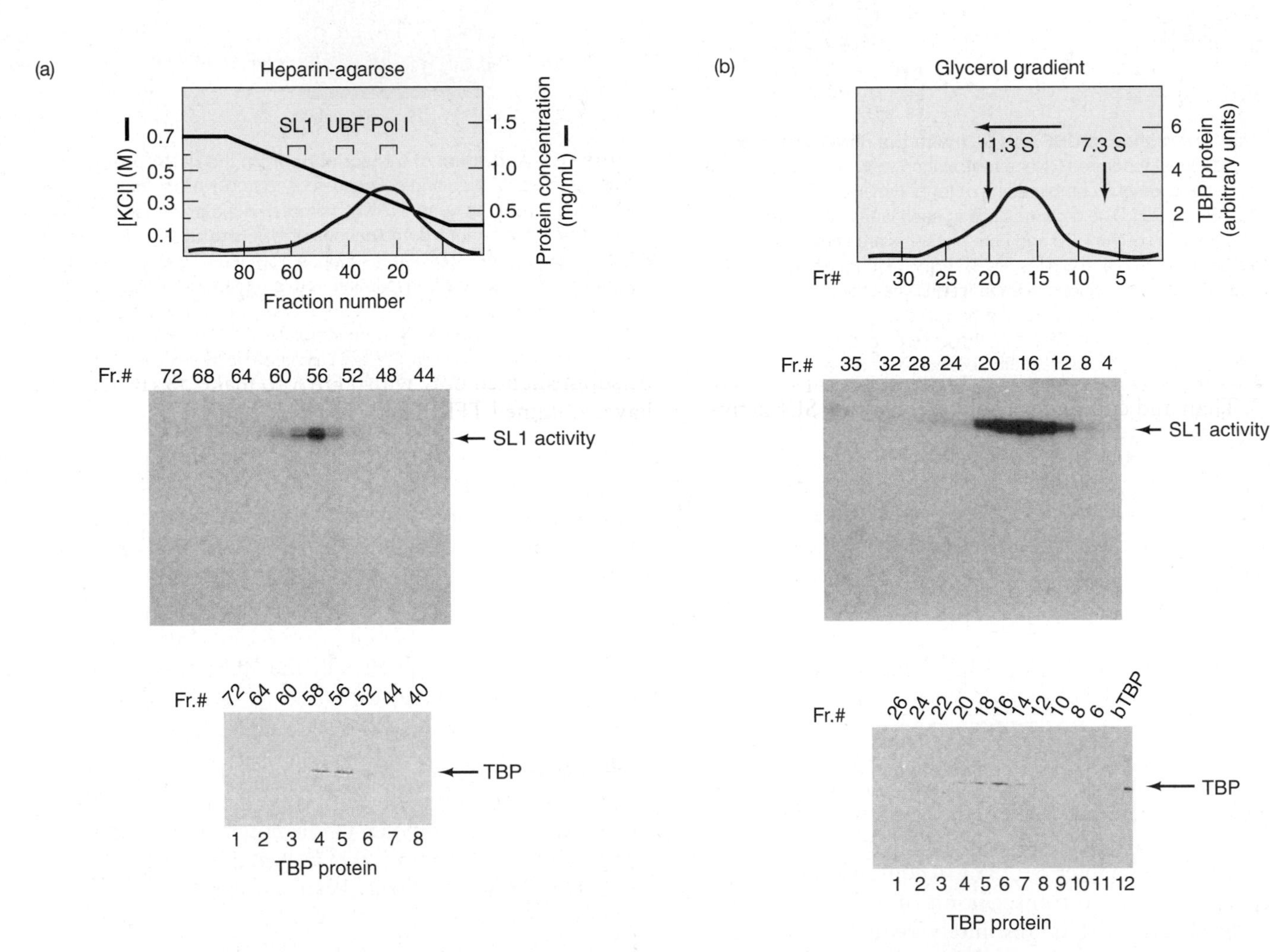

Figure 11.32 Co-purification of SL1 and TBP. (a) Heparin–agarose column chromatography (see Chapter 5 for column chromatography methods.) Top: Pattern of elution from the column of total protein (red) and salt concentration (blue), as well as three specific proteins (brackets). Middle: SL1 activity, measured by S1 protection analysis, in selected fractions. Bottom: TBP protein, detected by Western blotting, in selected fractions. Both SL1 and TBP were centered on fraction 56. **(b)** Glycerol gradient ultracentrifugation. Top: Sedimentation profile of TBP. Two other proteins, catalase and aldolase, with sedimentation coefficients of 11.3 S and 7.3 S, respectively, were run in a parallel centrifuge tube as markers. Middle and bottom panels, as in panel (a). Both SL1 and TBP sedimented to a position centered around fraction 16. (*Source:* Comai, L., N. Tanese, and R. Tjian, The TATA-binding protein and associated factors are integral components of the RNA polymerase I transcription factor, SL1. *Cell* 68 (6 Mar 1992) p. 968, f. 2a–b. Reprinted by permission of Elsevier Science.)

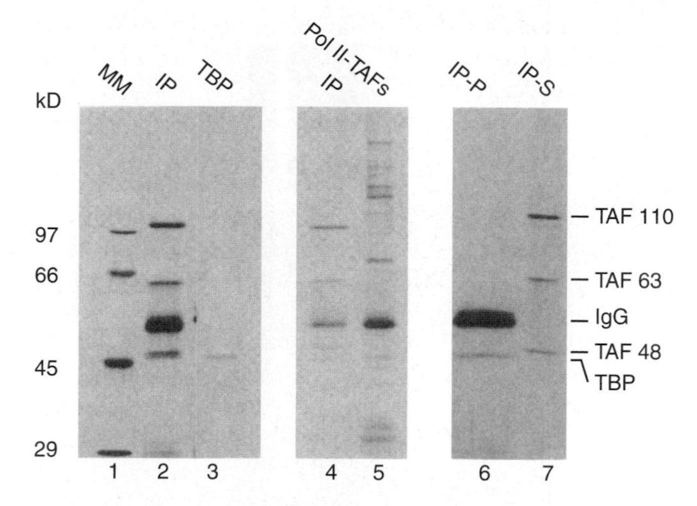

Figure 11.33 The TAFs in SL1. Tjian and colleagues immunoprecipitated SL1 with an anti-TBP antibody and subjected the polypeptides in the immunoprecipitate to SDS-PAGE. Lane 1, molecular weight markers; lane 2, immunoprecipitate (IP); lane 3, purified TBP for comparison; lane 4, another sample of immunoprecipitate; lane 5, TFIID TAFs (Pol II-TAFs) for comparison; lane 6, pellet after treating immunoprecipitate with 1 M guanidine–HCl and reprecipitating, showing TBP and antibody (IgG); lane 7, supernatant after treating immunoprecipitate with 1 M guanidine–HCl and reprecipitating, showing the three TAFs (labeled at right). (*Source:* Comai, L., N. Tanese, and R. Tjian, The TATA-binding protein and associated factors are integral components of the RNA polymerase I transcription factor, SL1. *Cell* 68 (6 Mar 1992) p. 971, f. 5. Reprinted by permission of Elsevier Science.)

precipitate (lane 6) and the TAFs stayed in the supernatant (lane 7). Tjian and colleagues could reconstitute SL1 activity by adding together purified TBP and the three TAFs, and this activity was species-specific, as one would expect. In later work, Tjian and coworkers showed that the TAF_Is and TAF_{II}s could compete with each other for binding to TBP. This finding suggested that binding of one set of TAFs to TBP is mutually exclusive of binding of the other set.

Thus, both polymerase I and polymerase II rely on transcription factors (SL1 and TFIID, respectively) composed of TBP and several TAFs. The TBP is identical in the two factors but the TAFs are completely different.

A unifying theme for all class I core-binding factors, except in yeast, is TBP. Yeast TBP binds to the core-binding factor, but not stably, the way other TBPs bind to their corresponding TAF_Is. The number and sizes of the TAF_Is we have discussed are typical of human cells. Other organisms have their own spectrum of TAF_Is.

SUMMARY Human-SL1 is composed of TBP and three TAFs: TAF_I110, TAF_I63, and TAF_I48. Fully functional and species-specific SL1 can be reconstituted from these purified components, and binding of TBP to the TAF_Is precludes binding to the TAF_{II}s. Other organisms have their own groups of TAF_Is.

11.3 Class III Factors

In 1980, Roeder and his colleagues discovered a factor that bound to the internal promoter of the 5S rRNA gene and stimulated its transcription. They named the factor **TFIIIA.** Since then, two other factors, TFIIIB and C, have been discovered. These two factors participate, not only in 5S rRNA gene transcription, but in all transcription by polymerase III.

Barry Honda and Robert Roeder demonstrated the importance of the TFIIIA factor in 5S rRNA gene transcription when they developed the first eukaryotic in vitro transcription system, from *Xenopus laevis,* and found that it could make no 5S rRNA unless they added TFIIIA. Donald Brown and colleagues went on to show that similar cell-free extracts provided with a 5S rRNA gene and a tRNA gene could make both 5S rRNA and tRNA simultaneously. Furthermore, an antibody against TFIIIA could effectively halt the production of 5S rRNA, but had no effect on tRNA synthesis (Figure 11.34). Thus, TFIIIA is required for transcription of the 5S rRNA genes, but not the tRNA genes.

If transcription of the tRNA genes does not require TFIIIA, what factors *are* involved? In 1982, Roeder and colleagues separated two new factors they called **TFIIIB** and **TFIIIC** and found that they are necessary and sufficient for transcription of the tRNA genes. We have subsequently learned that these two factors govern transcription of all classical polymerase III genes, including the 5S rRNA genes. That means that the original extracts that needed to be supplemented only with TFIIIA to make 5S rRNA must have contained TFIIIB and C.

SUMMARY Transcription of all classical class III genes requires TFIIIB and C, and transcription of the 5S rRNA genes requires these two plus TFIIIA.

TFIIIA

As the very first eukaryotic transcription factor to be discovered, TFIIIA received a considerable amount of attention. It was the first member of a large group of DNA-binding proteins that feature a so-called **zinc finger.** We will discuss the zinc finger proteins in detail in Chapter 12. Here, let us concentrate on the zinc fingers of TFIIIA. The essence of a zinc finger is a roughly finger-shaped protein domain containing four amino acids that bind a single zinc ion. In TFIIIA, and in other typical zinc finger proteins, these four amino acids are two cysteines, followed by two histidines. However, some other zinc finger-like proteins have four cysteines and no histidines. TFIIIA has nine zinc fingers in a row, and these appear to insert into the DNA major groove on either side of the internal promoter of the 5S rRNA gene. This allows specific amino acids to make contact with specific base pairs, forming a tight protein–DNA complex.

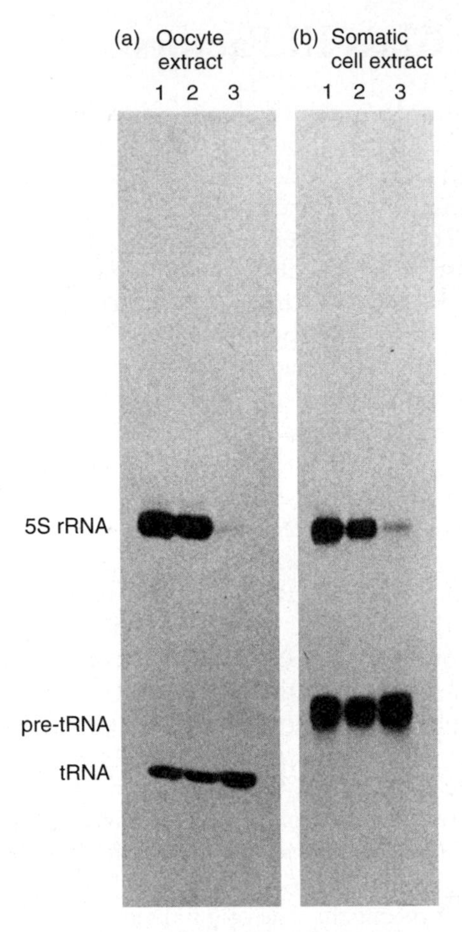

Figure 11.34 Effect of anti-TFIIIA antibody on transcription by polymerase III. Brown and colleagues added cloned 5S rRNA and tRNA genes to **(a)** an oocyte extract, or **(b)** a somatic cell extract in the presence of labeled nucleotide and: no antibody (lanes 1), an irrelevant antibody (lanes 2), or an anti-TFIIIA antibody (lanes 3). After transcription, these workers electrophoresed the labeled RNAs. The anti-TFIIIA antibody blocked 5S rRNA gene transcription in both extracts, but did not inhibit tRNA gene transcription in either extract. The oocyte extract could process the pre-tRNA product to the mature tRNA form, but the somatic cell extract could not. Nevertheless, transcription occurred in both cases. (*Source:* Pelham, H.B., W.M. Washington, and D.D. Brown, Related 5S rRNA transcription factors in *Xenopus* oocytes and somatic cells. *Proceedings of The National Academy of Sciences USA* 78 (Mar 1981) p. 1762, f. 3.)

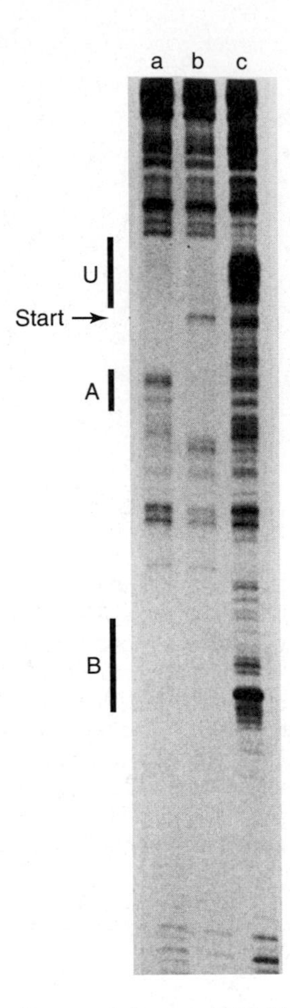

Figure 11.35 Effect of transcription on DNA binding between a tRNA gene and transcription factors. Geiduschek and colleagues performed DNase footprinting with a tRNA gene and an extract containing polymerase III, TFIIIB, and TFIIIC. Lane a contained transcription factors, but no nucleotides. Lane b had factors plus three of the four nucleotides (all but GTP), so transcription could progress for 17 nt, until GTP was needed. Lane c was a control with no added protein. The 17-bp migration of the polymerase in lane b relative to lane a caused a corresponding downstream shift in the footprint around the transcription start site, to a position extending upstream and downstream of the A box. On the other hand, the footprint in the region just upstream of the start of transcription remained unchanged. (*Source:* Kassavetis, G.A., D.L. Riggs, R. Negri, L.H. Nguyen, and E.P. Geiduschek, Transcription factor III B generates extended DNA interactions in RNA polymerase III transcription complexes on tRNA genes. *Molecular and Cellular Biology.* 9, no.171 (June 1989) p. 2555, f. 3. Copyright © 1989 American Society for Microbiology, Washington, DC. Reprinted with permission.)

TFIIIB and C

TFIIIB and C are both required for transcription of the classical polymerase III genes, and it is difficult to separate the discussion of these two factors because they depend on each other for their activities. Peter Geiduschek and coworkers established in 1989 that a crude transcription factor preparation bound both the internal promoter and an upstream region in a tRNA gene. Figure 11.35 contains DNase footprinting data that led to this conclusion. Lane c is the digestion pattern with no added protein, lane a is the result with factors and polymerase III, and lane b has all this plus three nucleoside triphosphates (ATP, CTP, and UTP), which allowed transcription for just 17 nt, until the first GTP was needed. Notice in lane a that the factors and polymerase strongly protected box B of the internal promoter and the upstream region (U) and weakly protected box A of the internal promoter. Lane b shows that the polymerase shifted downstream and a new region overlapping box A was protected. However, the protection of the upstream region persisted even after the polymerase moved away.

What accounts for the persistent binding to the upstream region? To find out, Geiduschek and colleagues partially purified TFIIIB and C and performed footprinting studies with these separated factors. Figure 11.36

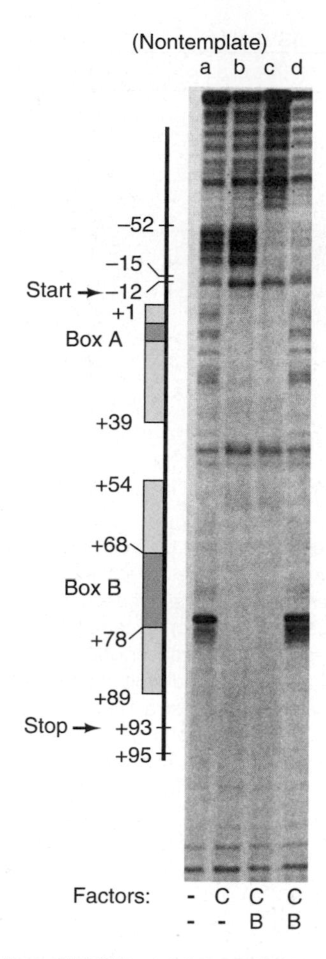

Figure 11.36 Binding of TFIIIB and C to a tRNA gene. Geiduschek and coworkers performed DNase footprinting with a labeled tRNA gene (all lanes), and combinations of purified TFIIIB and C. Lane a, negative control with no factors; lane b, TFIIIC only; lane c, TFIIIB plus TFIIIC; lane d, TFIIIB plus TFIIIC added, then heparin added to strip off any loosely bound protein. Note the added protection in the upstream region afforded by TFIIIB in addition to TFIIIC (lane c). Note also that this upstream protection provided by TFIIIB survives heparin treatment, but the protection of boxes A and B does not. Yellow boxes represent coding regions for mature tRNA. Boxes A and B within these regions are indicated in blue. (*Source:* From Kassavetis, G.A., D.L. Riggs, R. Negri, L.H. Nguyen, and E.P. Geiduschek, Transcription factor III B generates extended DNA interactions in RNA polymerase III transcription complexes on tRNA genes. *Molecular and Cellular Biology* 9:2558, 1989. Copyright © 1989 American Society for Microbiology, Washington, DC. Reprinted by permission.)

shows the results of one such experiment. Lane b, with TFIIIC alone, reveals that this factor protects the internal promoter, especially box B, but does not bind to the upstream region. When both factors are present, the upstream region is also protected (lane c). Similar DNase footprinting experiments made it clear that TFIIIB by itself does not bind to any of these regions. Its binding is totally dependent on TFIIIC. However, once TFIIIC has sponsored the binding of TFIIIB to the upstream region, TFIIIB appears to remain there, even after polymerase has moved on (recall Figure 11.35). Moreover, Figure 11.36, lane d,

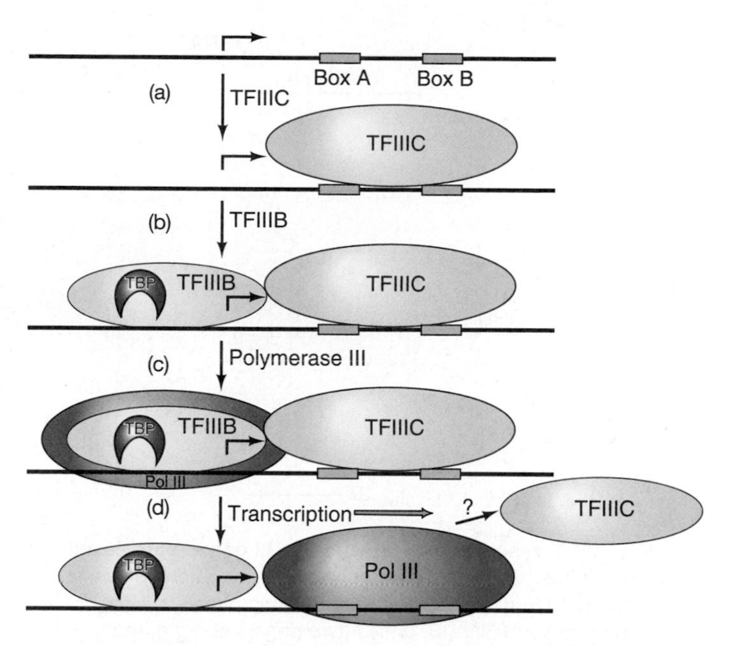

Figure 11.37 Hypothetical scheme for assembly of the preinitiation complex on a classical polymerase III promoter (tRNA), and start of transcription. (a) TFIIIC (light green) binds to the internal promoter's A and B blocks (green). **(b)** TFIIIC promotes binding of TFIIIB (yellow), with its TBP (blue), to the region upstream of the transcription start site. **(c)** TFIIIB promotes polymerase III (red) binding at the start site, ready to begin transcribing. **(d)** Transcription begins. As the polymerase moves to the right, making RNA (not shown), it may or may not remove TFIIIC from the internal promoter. But TFIIIB remains in place, ready to sponsor a new round of polymerase binding and transcription.

shows that TFIIIB binding persists even after heparin has stripped TFIIIC away from the internal promoter, as the upstream region is still protected from DNase, even though boxes A and B are not.

The evidence we have seen so far suggests the following model for involvement of transcription factors in polymerase III transcription (Figure 11.37): First, TFIIIC (or TFIIIA and C, in the case of the 5S rRNA genes) binds to the internal promoter; then these **assembly factors** allow TFIIIB to bind to the upstream region; then TFIIIB helps polymerase III bind at the transcription start site; finally, the polymerase transcribes the gene, perhaps removing TFIIIC (or A and C) in the process, but TFIIIB remains bound, so it can continue to promote further rounds of transcription.

Geiduschek and colleagues have provided further evidence to bolster this hypothesis. They bound TFIIIC and B to a tRNA gene (or TFIIIA, C, and B to a 5S rRNA gene), then removed (stripped) the assembly factors, TFIIIC (or A and C) with either heparin or high salt, then separated the remaining TFIIIB–DNA complex from the other factors. Finally, they demonstrated that this TFIIIB–DNA complex was still capable of supporting one round, or even multiple rounds, of transcription by polymerase III (Figure 11.38). How does TFIIIB remain so tightly bound to its DNA

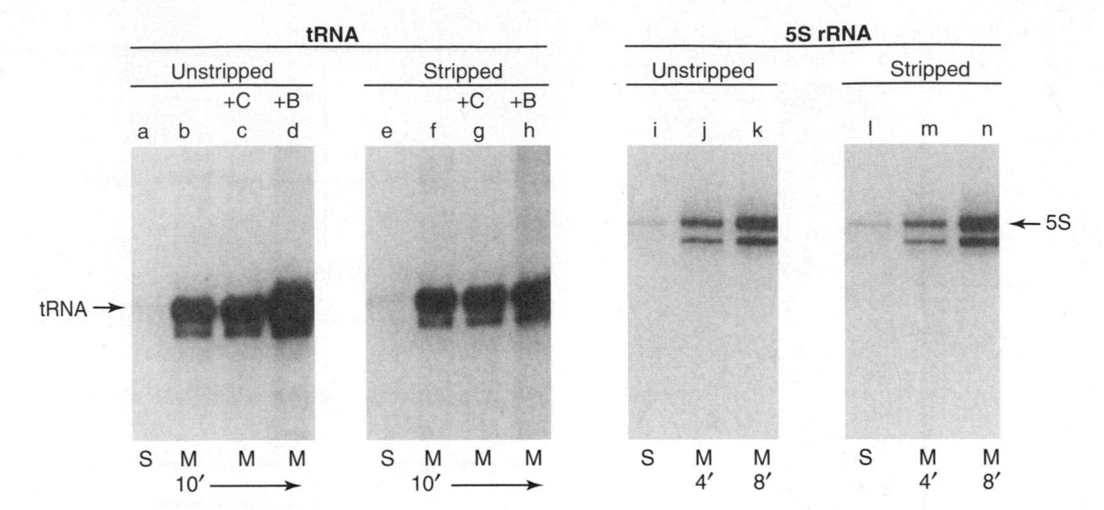

Figure 11.38 Transcription of polymerase III genes complexed only with TFIIIB. Geiduschek and coworkers made complexes containing a tRNA gene and TFIIIB and C (two panels at left), or a 5S rRNA gene and TFIIIA, B, and C (two panels at right), then stripped off TFIIIC with heparin (lanes e–h), or TFIIIA and C with a high ionic strength buffer (lanes l–n). They passed the stripped templates through gel filtration columns to remove any unbound factors, and demonstrated by gel mobility shift and DNase footprinting (not shown) that the purified complexes contained only TFIIIB bound to the upstream regions of the respective genes. Next, they tested these stripped complexes alongside unstripped complexes for ability to support single-round transcription (S; lanes a, e, i, and l), or multiple-round transcription (M; all other lanes) for the times indicated at bottom. (The single-round signals are faint, but visible.) They added extra TFIIIC in lanes c and g, and extra TFIIIB in lanes d and h as indicated at top. They confined transcription to a single round in lanes a, e, i, and l by including a relatively low concentration of heparin, which allowed elongation of RNA to be completed, but then bound up the released polymerase so it could not reinitiate. Notice that the stripped template, containing only TFIIIB, supported just as much transcription as the unstripped template in both single-round and multiple-round experiments, even when the experimenters added extra TFIIIC (compare lanes c and g, and lanes k and n). The only case in which the unstripped template performed better was in lane d, which was the result of adding extra TFIIIB. This presumably resulted from some remaining free TFIIIC that helped the extra TFIIIB bind, thus allowing more preinitiation complexes to form. (*Source:* Kassavetis, G.A., B.R. Brawn, L.H. Nguyen, and E.P. Geiduschek, *S. cerevisiae* TFIIIB is the transcription initiation factor proper of RNA polymerase III, while TFIIIA and TFIIIC are assembly factors. *Cell* 60 (26 Jan 1990) p. 237, f. 3. Reprinted by permission of Elsevier Science.)

target when it has no affinity for this DNA on its own? The answer may be that TFIIIC (or TFIIIA and TFIIIC) can cause a conformational shift in TFIIIB, revealing a site that can bind tenaciously to DNA.

TFIIIC is a remarkable protein. It can bind to both box A and box B of tRNA genes, as demonstrated by DNase footprinting and protein–DNA cross-linking studies. In some tRNA genes there is an intron between boxes A and B, and TFIIIC still manages to contact both promoter elements. How can it do that? It helps that TFIIIC is one of the largest and most complex of all the known transcription factors. The yeast TFIIIC contains six subunits with a combined molecular mass of about 600 kD. Furthermore, electron microscopic studies have shown that TFIIIC has a dumbbell shape with two globular regions separated by a stretchable linker region that allows the protein to span a surprisingly long distance.

In these studies, André Sentenac and colleagues bound yeast TFIIIC (which they called τ factor) to cloned tRNA genes having variable distances between their boxes A and B. Then they visualized the complexes by scanning transmission electron microscopy. Figure 11.39 shows the results: When the distance between boxes A and B was zero, TFIIIC appeared as a large blob on the DNA. However, with increasing distance between boxes A and B, TFIIIC appeared as two globular domains separated by a linker of increasing length between them. Thus, the combination of large size and stretchability allows TFIIIC to contact two widely separated promoter regions with its two globular domains.

SUMMARY Classical class III genes require two factors, TFIIIB and C, in order to form a preinitiation complex with the polymerase. The 5S rRNA genes also require TFIIIA. TFIIIC and A are assembly factors that bind to the internal promoter and help TFIIIB bind to a region just upstream of the transcription start site. TFIIIB then remains bound and can sponsor the initiation of repeated rounds of transcription. TFIIIC is a very large protein. The yeast protein has six subunits that are arranged into two globular regions joined through a flexible linker. The stretchability of this linker allows the protein to cover the long distance between boxes A and B of the internal promoter.

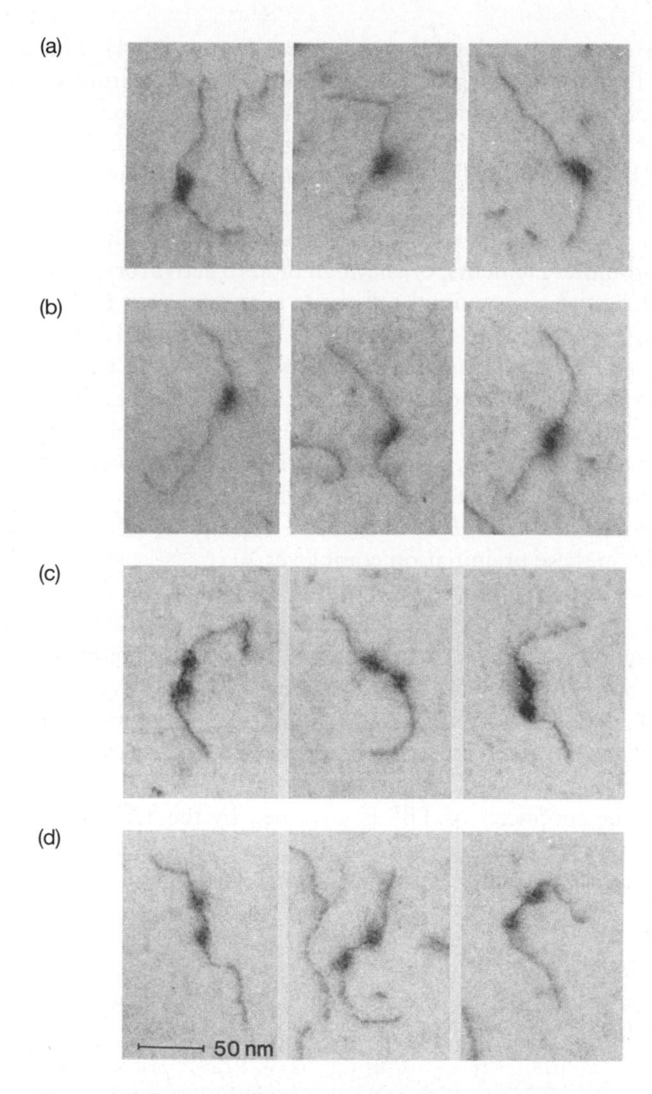

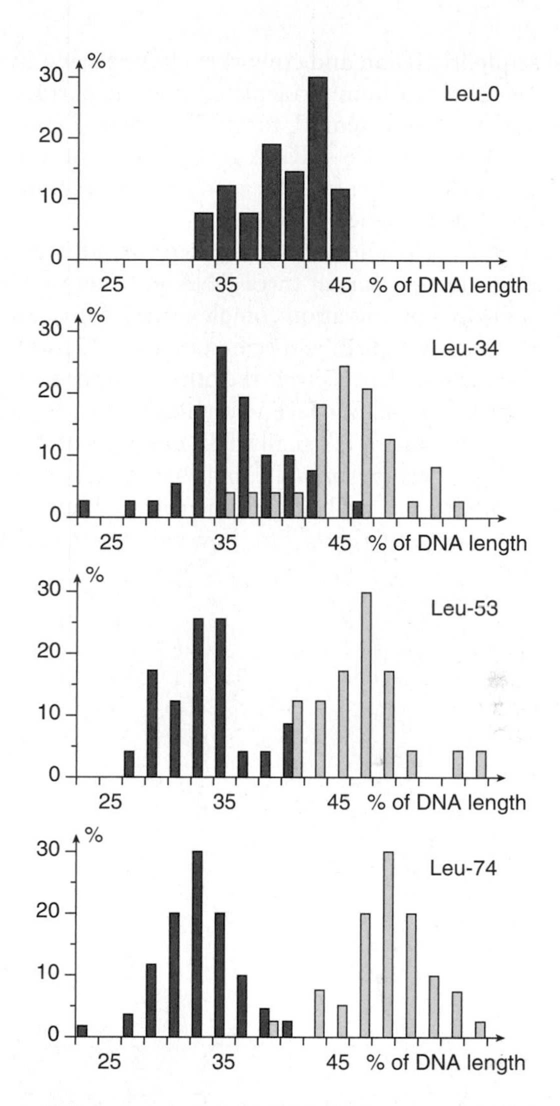

Figure 11.39 Yeast TFIIIC contains two globular domains connected by a flexible linker. Sentenac and colleagues bound yeast TFIIIC to cloned tRNA genes with variable distances between their boxes A and B. Next, they subjected the complexes to negative staining with uranyl acetate, then submitted them for scanning transmission electron microscopy. The distances between boxes A and B are given at right: **(a)** 0 bp; **(b)** 34 bp; **(c)** 53 bp; and **(d)** 74 bp, which is the wild-type distance. Three examples of micrographs with each DNA are presented at left. The histograms at right display the positions of the globular domains of TFIIIC on the DNA, determined from many different micrographs. The bars show the percentages of DNAs with globular domains at each location along the DNA. The red bars show the locations of the globular domain closest to the end of the DNA, and the yellow bars show the locations of the other globular domain. (*Source:* Schultz et al *EMBO Journal* 8: p. 3817 © 1989.)

The Role of TBP

If TFIIIC is necessary for TFIIIB binding in classical class III genes, what about nonclassical genes that have no boxes A or B to which TFIIIC can bind? What stimulates TFIIIB binding to these genes? Because the promoters of these genes have TATA boxes (Chapter 10), and we have already seen that TBP is required for their transcription, it makes sense to propose that the TBP binds to the TATA box and anchors TFIIIB to its upstream binding site.

But what about classical polymerase III genes? These have no TATA box, and yet we have seen that TBP is required for transcription of classical class III genes such as the tRNA and 5S rRNA genes in yeast and human cells. Where does TBP fit into this scheme? It has now become clear that TFIIIB contains TBP along with a small number of TAFs. In mammals, these TAFs are called Brf1 and Bdp1. Geiduschek and coworkers showed that TBP was present even in the purest preparations of TFIIIB. Further studies on yeast TFIIIB, including reconstitution from cloned components, have revealed that the factor is composed of three subunits: TBP and two TAF_{III}s. These two proteins have different names in different organisms. The yeast versions are called *B″* and *TFIIB-related factor,* or *BRF,* because of its homology to TFIIB.

Subsequently, Tjian and coworkers have shown by adding factors back to immunodepleted nuclear extracts that TRFI, not TBP, is essential for transcribing *Drosophila* tRNA, 5S rRNA and U6 snRNA genes. Thus, transcription by polymerase III in the fruit fly is another exception to the generality of dependence on TBP.

A unifying principle that emerges from the studies on transcription factors for all three RNA polymerases is that the assembly of a preinitiation complex starts with an assembly factor that recognizes a specific binding site in the promoter. This protein then recruits the other components of the preinitiation complex. For TATA-containing class II promoters, the assembly factor is usually TBP, and its binding site is the TATA box. This presumably applies to TATA-containing class III promoters as well, at least in yeast and human cells. We have already seen a model for how this process begins in TATA-containing class II promoters (Figure 11.4). Figure 11.40 shows, in highly schematic form, the nature of these preinitiation complexes for all kinds of TATA-less promoters. In class I promoters, the assembly factor is UBF, which binds to the UPE and then attracts the TBP-containing SL1 to the core element. TATA-less class II promoters can attract TBP in at least two ways. TAFs in TFIID can bind to core promoter elements, or they can bind to activators, such as Sp1 bound to proximal promoter elements, such as GC boxes. Both methods anchor TFIID to the TATA-less promoter. Classical class III promoters, at least in yeast and human cells, follow the same general scheme. TFIIIC, or in the case of the 5S rRNA genes, TFIIIA plus TFIIIC, play the role of assembly factor, binding to the internal promoter and attracting the TBP-containing TFIIIB to a site upstream of the start point. In *Drosophila* cells, TRFI appears to substitute for TBP in these preinitiation complexes.

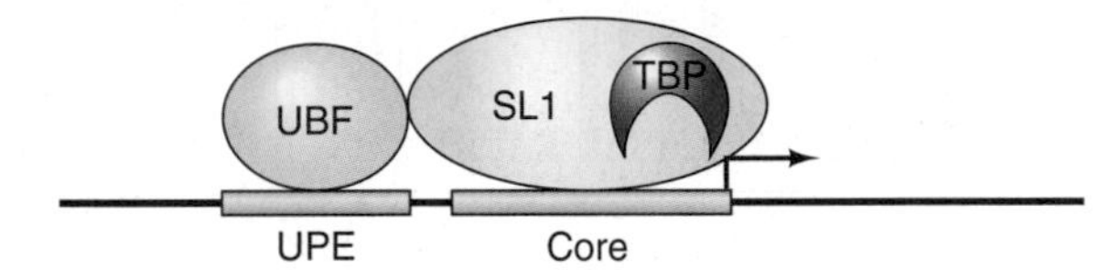

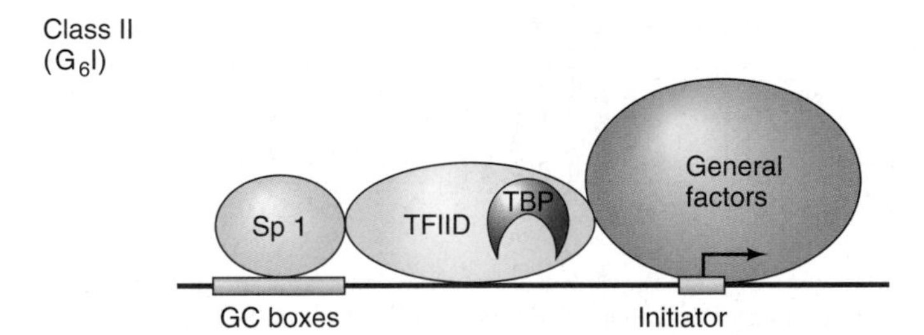

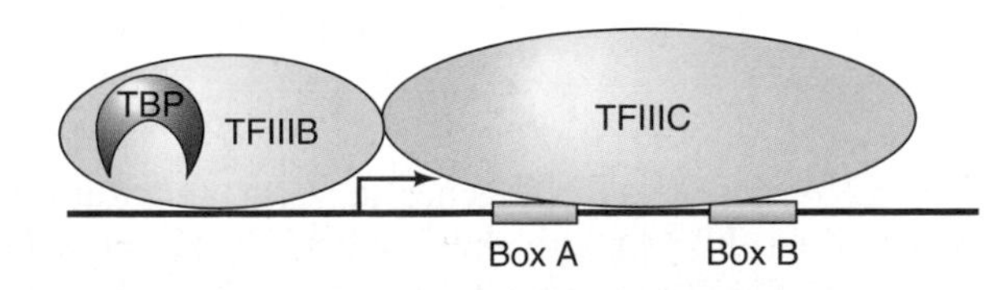

Figure 11.40 Model of preinitiation complexes on TATA-less promoters recognized by all three polymerases. In each case, an assembly factor (green) binds first (UBF, Sp1, and TFIIIC in class I, II, and III promoters, respectively). This in turn attracts another factor (yellow), which contains TBP (blue); this second factor is SL1, TFIID, or TFIIIB in class I, II, or III promoters, respectively. These complexes are sufficient to recruit polymerase for transcription of class I and III promoters, but in class II promoters more general factors (purple) besides polymerase II must bind before transcription can begin. (*Source:* Adapted from White, R.J. and S.P. Jackson, Mechanism of TATA-binding protein recruitment to a TATA-less class III promoter. *Cell* 71:1051, 1992.)

Just because TBP does not always bind first, we should not discount its importance in organizing the preinitiation complex on these TATA-less promoters. Once TBP binds, it helps bring the remaining factors, including RNA polymerase, to the complex. This is a second unifying principle: TBP plays an organizing role in preinitiation complexes on most types of eukaryotic promoters. A third unifying principle is that the specificity of TBP is governed by the TAFs with which it associates; thus, TBP affiliates with different TAFs when it binds to each of the various kinds of promoter.

SUMMARY The assembly of the preinitiation complex on each kind of eukaryotic promoter begins with the binding of an assembly factor to the promoter. With TATA-containing class II (and presumably class III) promoters, this factor is TBP, but other promoters have their own assembly factors. Even if TBP is not the first-bound assembly factor at a given promoter, it becomes part of the growing preinitiation complex on most known promoters and serves an organizing function in building the complex. The specificity of the TBP—which kind of promoter it will bind to—depends on its associated TAFs. TRFI substitutes for TBP, at least in some preinitiation complexes in *Drosophila* class III genes.

SUMMARY

Transcription factors bind to class II promoters in the following order in vitro: (1) TFIID, apparently with help from TFIIA, binds to the TATA box. (2) TFIIB binds next. (3) TFIIF helps RNA polymerase II bind. The remaining factors bind in this order: TFIIE and TFIIH, forming the DABPolFEH preinitiation complex. The participation of TFIIA seems to be optional in vitro.

TFIID contains a TATA-box-binding protein (TBP) plus 13 other polypeptides known as TBP-associated factors (TAFs). The C-terminal 180 amino acid fragment of the human TBP is the TATA-box-binding domain. The interaction between a TBP and a TATA box takes place in the DNA minor groove. The saddle-shaped TBP lines up with the DNA, and the underside of the saddle forces open the minor groove and bends the TATA box through an 80-degree angle. TBP is required for transciption of most members of all three classes of genes, not just class II genes.

Most of the TAFs are evolutionarily conserved in the eukaryotes. They serve several functions, but two obvious ones are interacting with core promoter elements and interacting with gene-specific transcription factors. TAF1 and TAF2 help TFIID bind to the initiator and DPEs of promoters and therefore can enable TBP to bind to certain TATA-less promoters that contain such elements. TAF1 and TAF4 help TFIID interact with Sp1 that is bound to GC boxes upstream of the transcription start site. These TAFs therefore ensure that TBP can bind to TATA-less promoters that have GC boxes. Different combinations of TAFs are apparently required to respond to various transcription activators, at least in higher eukaryotes. TAF1 also has two enzymatic activities. It is a histone acetyltransferase and a protein kinase. TFIID is not universally required, at least in higher eukaryotes. Some promoters in *Drosophila* require an alternative factor, TRF1, and some promoters require a TBP-free TAF-containing complex.

Structural studies on a TFIIB-polymerase II complex show that TFIIB binds to TBP at the TATA box via its C-terminal domain, and to polymerase II via its N-terminal domain. This bridging action effects a coarse positioning of the polymerase active center about 25–30 bp downstream of the TATA box. In mammals, a loop motif of the N-terminal domain of TFIIB effects a fine positioning of the start of transcription by interacting with the single-stranded template DNA strand very near the active center. Biochemical studies confirm that the TFIIB N-terminal domain (the finger and linker domains, in particular) lies close to the RNA polymerase II active center, and to the largest subunit of TFIIF, in the preinitiation complex.

The preinitiation complex forms with the hypophosphorylated form to RNA polymerase II (IIA). Then, a subunit of TFIIH phosphorylates serine 5 in the heptad repeat in the carboxyl-terminal domain (CTD) of the largest RNA polymerase II subunit, creating the phosphorylated form of the enzyme (IIO). TFIIE greatly stimulates this process in vitro. This phosphorylation is essential for initiation of transcription. During the shift from initiation to elongation, phosphorylation shifts from serine 5 to serine 2. If phosphorylation of serine 2 is also lost, the polymerase pauses until re-phosphorylation by a non-TFIIH kinase occurs.

TFIIE and TFIIH are not essential for formation of an open promoter complex, or for elongation, but they are required for promoter clearance. TFIIH has a DNA helicase activity that is essential for transcription, presumably because it facilitates promoter clearance by fully melting the DNA at the promoter.

RNA polymerases can be induced to pause at specific sites near promoters by proteins such as DSIF and NELF. This pausing can be reversed by P-TEFb, which phosphorylates the polymerase, as well as DSIF and NELF. Polymerases that have backtracked and have become arrested can be rescued by TFIIS. This factor inserts into the active site of the polymerase, stimulates an RNase activity inherent in the polymerase, which cleaves off the 3′-end of the nascent RNA, extruded during backtracking. This allows resumption of elongation. TFIIS also stimulates proofreading, presumably by stimulating the RNase activity of RNA polymerase II, allowing it to remove misincorporated nucleotides.

Yeast and mammalian cells have been shown to contain an RNA polymerase II holoenzyme with many polypeptides in addition to the subunits of the polymerase.

Class I promoters are recognized by two transcription factors, a core-binding factor and a UPE-binding factor. The human core-binding factor is called SL1; in some other organisms, such as *A. castellanii,* the homologous factor is known as TIF-IB. The core-binding factor is the fundamental transcription factor required to recruit RNA polymerase I. This factor also determines species specificity, at least in animals. The factor that binds the UPE is called UBF in mammals and most other organisms, but UAF in yeast. It is an assembly factor that helps the core-binding factor bind to the core promoter element. The degree of reliance on the UPE-binding factor varies considerably from one organism to another. In *A. castellanii,* TIF-IB alone suffices to recruit the RNA polymerase I and position it correctly for initiation of transcription. Human UBF is a transcription factor that stimulates transcription by polymerase I. It can activate the intact promoter, or the core element alone, and it mediates activation by the UCE. UBF and SL1 act synergistically to stimulate transcription.

Human SL1 is composed of TBP and three TAFs, TAF_I110, TAF_I63, and TAF_I48. Fully functional and species-specific SL1 can be reconstituted from these purified components, and binding of TBP to the TAF_Is precludes binding to the $TAF_{II}s$. Other organisms have their own groups of TAF_Is.

Classical class III genes require two factors, TFIIIB and C, to form a preinitiation complex with the polymerase. The 5S rRNA genes also require TFIIIA. TFIIIC and A are assembly factors that bind to the internal promoter and help TFIIIB bind to a region just upstream of the transcription start site. TFIIIB then remains bound and can sponsor the initiation of repeated rounds of transcription.

The assembly of the preinitiation complex on each kind of eukaryotic promoter begins with the binding of an assembly factor to the promoter. With TATA-containing class II (and presumably class III) promoters, this factor is usually TBP, but other promoters have their own assembly factors. Even if TBP is not the first-bound assembly factor at a given promoter, it becomes part of the growing preinitiation complex on most known promoters and serves an organizing function in building the complex. The specificity of the TBP—which kind of promoter it will bind to—depends on its associated TAFs, and there are TAFs specific for each of the promoter classes.

REVIEW QUESTIONS

1. List in order the proteins that assemble in vitro to form a class II preinitiation complex.
2. Describe and give the results of an experiment that shows that TFIID is the fundamental building block of the class II preinitiation complex.
3. Describe and give the results of an experiment that shows that TFIIF and polymerase II bind together, but neither can bind independently to the preinitiation complex.
4. Describe and give the results of an experiment that shows where TFIID binds.
5. Show the difference between the footprints caused by the DAB and the DABPolF complexes. What conclusion can you reach, based on this difference?
6. Present a hypothesis that explains the fact that substitution of dCs for dTs and dIs for dAs, in the TATA box (making a CICI box) has no effect on TFIID binding. Provide the rationale for your hypothesis.
7. What shape does TBP have? What is the geometry of interaction between TBP and the TATA box?
8. Describe and give the results of an experiment that shows TBP is required for transcription from all three classes of promoters.
9. Describe and give the results of an experiment that shows that a class II promoter is more active in vitro with TFIID than with TBP.
10. Describe and give the results of an experiment that identifies the TAFs that bind to a class II promoter containing a TATA box, an initiator, and a downstream promoter element.
11. Describe and give the results of a DNase footprinting experiment that shows how the footprint is expanded by TAF1 and TAF2 compared with TBP alone.
12. Draw a diagram of a model for the interaction of TBP (and other factors) with a TATA-less class II promoter.
13. Whole genome expression analysis indicates that yeast TAF1 is required for transcription of only 16% of yeast genes, and TAF9 is required for transcription of 67% of yeast genes. Provide a rationale for these results.
14. Present examples of class II preinitiation complexes with:
    a. An alternative TBP
    b. A missing TAF
    c. No TBP or TBP-like protein
15. What are the apparent roles of TFIIA and TFIIB in transcription?
16. Draw a rough sketch of the TBP–TFIIB–RNA polymerase II complex bound to DNA, showing the relative positions of the proteins. How do these positions correlate with the apparent roles of the proteins? Include an explanation of how TFIIB determines the direction of transcription.
17. Describe and give the results of an experiment that mapped the sites on Rpb1 and Rpb2 that are in close contact with the finger and linker regions of TFIIB.
18. Describe and give the results of an experiment that shows that TFIIH, but not the other general transcription factors, phosphorylates the IIA form of RNA polymerase II to the IIO form. In addition, include data that show that the other general transcription factors help TFIIH in this task.
19. Describe and give the results of an experiment that shows that TFIIH phosphorylates the CTD of polymerase II.
20. Describe an assay for DNA helicase and show how it can be used to demonstrate that TFIIH is associated with helicase activity.
21. Describe a G-less cassette transcription assay and show how it can be used to demonstrate that the RAD25 DNA helicase activity associated with TFIIH is required for transcription in vitro.
22. Draw a rough diagram of the class II preinitiation complex, showing the relative positions of the polymerase, the promoter DNA, TBP, and TFIIB, E, F, and H. Show the direction of transcription.
23. Describe and give the results of an experiment that shows that TFIIS stimulates transcription elongation by RNA polymerase II.
24. Present a model for reversal of transcription arrest by TFIIS. What part of TFIIS participates most directly? How?
25. Describe and give the results of an experiment that shows that TFIIS stimulates proofreading by RNA polymerase II.
26. What is the meaning of the term *RNA polymerase II holoenzyme?* How does the holoenzyme differ from the core polymerase II?
27. Describe and give the results of an experiment that shows the effect of adding or removing a few base pairs between the core element and the transcription start site in a class I promoter.
28. Which general transcription factor is the assembly factor in class I promoters? In other words, which binds first

and helps the other bind? Describe a DNase footprinting experiment you would perform to prove this, and show idealized results, not necessarily those that Tjian and colleagues actually obtained. Make sure your diagrams indicate an effect of both transcription factors on the footprints.

29. Describe and give the results of copurification and immunoprecipitation experiments that show that SL1 contains TBP.
30. Describe and give the results of an experiment that identified the TAFs in SL1.
31. How do we know that TFIIIA is necessary for transcription of 5S rRNA, but not tRNA, genes?
32. Geiduschek and colleagues performed DNase footprinting with polymerase III plus TFIIIB and C and a tRNA gene. Show the results they obtained with: No added protein; polymerase and factors; and polymerase, factors and three of the four NTPs. What can you conclude from these results?
33. The classical class III genes have internal promoters. Nevertheless, TFIIIB and C together cause a footprint in a region upstream of the gene's coding region. Draw a diagram of the binding of these two factors that explains these observations.
34. Draw a diagram of what happens to TFIIIB and C after polymerase III has begun transcribing a classical class III gene such as a tRNA gene. How does this explain how new polymerase III molecules can continue to transcribe the gene, even though factors may not remain bound to the internal promoter?
35. Describe and give the results of a DNase footprint experiment that shows that TFIIIB + C, but not TFIIIC alone, can protect a region upstream of the transcription start site in a tRNA gene. Show also what happens to the footprint when you strip off TFIIIC with heparin.
36. Describe and give the results of an experiment that shows the following: Once TFIIIB binds to a classical class III gene, it can support multiple rounds of transcription, even after TFIIIC (or C and A) are stripped off the promoter.
37. Describe and give the results of an experiment that demonstrates the flexibility of TFIIIC in binding to boxes A and B that are close together or far apart in a class III promoter.
38. Diagram the preinitiation complexes with all three classes of TATA-less promoters. Identify the assembly factors in each case.

ANALYTICAL QUESTIONS

1. You are studying a new class of eukaryotic promoters (class IV) recognized by a novel RNA polymerase IV. You discover two general transcription factors that are required for transcription from these promoters. Describe experiments you would perform to determine which, if any, is an assembly factor, and which is required to recruit the RNA polymerase to the promoter. Provide sample results of your experiments.
2. You discover that one of your novel class IV transcription factors contains TBP. Describe an experiment you would perform to identify the TAFs in this factor.
3. Some of the class IV promoters contain two DNA elements (boxes X and Y), others contain just one (box X). Describe experiments you would perform to identify the TAFs that bind to each of these two types of promoters.
4. You incubate cells with an inhibitor of the protein kinase activity of TFIIH and then perform in vitro transcription and DNase footprinting experiments. What step in transcription would you expect to see blocked? What kind of assay would reveal such a blockage? Would you still expect to see a footprint at the promoter? Why or why not? If so, how large would the footprint be, compared to the footprint in the absence of the inhibitor?
5. You know that protein X and protein Y interact, but you want to know whether a particular domain of protein X interacts with protein Y, and if so, where. Design a hydroxyl radical cleavage analysis experiment to answer this question.

SUGGESTED READINGS

General References and Reviews

Asturias, F.J. and J.L. Craighead. 2003. RNA polymerase II at initiation. *Proceedings of the National Academy of Sciences USA*. 100:6893–95.

Berk, A.J. 2000. TBP-like factors come into focus. *Cell* 103:5–8.

Buratowski, S. 1997. Multiple TATA-binding factors come back into style. *Cell* 91:13–15.

Burley, S.K. and R.G. Roeder. 1996. Biochemistry and structural biology of transcription factor IID (TFIID). *Annual Review of Biochemistry* 65:769–99.

Chao, D.M. and R.A. Young. 1996. Activation without a vital ingredient. *Nature* 383:119–20.

Conaway, R.C., S.E. Kong, and J.W. Conaway. 2003. TFIIS and GreB: Two like-minded transcription elongation factors with sticky fingers. *Cell* 114:272–74.

Goodrich, J.A., G. Cutler, and R. Tjian. 1996. Contacts in context: Promoter specificity and macromolecular interactions in transcription. *Cell* 84:825–30.

Grant, P. and J.L. Workman. 1998. A lesson in sharing? *Nature* 396:410–11.

Green, M.A. 1992. Transcriptional transgressions. *Nature* 357:364–65.

Hahn, S. 1998. The role of TAFs in RNA polymerase II transcription. *Cell* 95:579–82.

Hahn, S. 2004. Structure and mechanism of the RNA polymerase II transcription machinery. *Nature Structural & Molecular Biology* 11:394–403.

Klug, A. 1993. Opening the gateway. *Nature* 365:486–87.

Paule, M.R. and R.J. White. 2000. Transcription by RNA polymerases I and III. *Nucleic Acids Research* 28:1283–98.

Sharp, P.A. 1992. TATA-binding protein is a classless factor. *Cell* 68:819–21.

White, R.J. and S.P. Jackson. 1992. The TATA-binding protein: A central role in transcription by RNA polymerases I, II, and III. *Trends in Genetics* 8:284–88.

Research Articles

Armache, K.-J., H. Kettenberger, and P. Cramer. 2003. Architecture of initiation-competent 12-subunit RNA polymerase II. *Proceedings of the National Academy of Sciences USA*. 100:6964–68.

Bell, S.P., R.M. Learned, H.-M. Jantzen, and R. Tjian. 1988. Functional cooperativity between transcription factors UBF1 and SL1 mediates human ribosomal RNA synthesis. *Science* 241:1192–97.

Brand, M., C. Leurent, V. Mallouh, L. Tora, and P. Schultz. 1999. Three-dimensional structures of the TAF_{II}-containing complexes TFIID and TFTC. *Science* 286:2151–53.

Bushnell, D.A. and R.D. Kornberg. 2003. Complete, 12-subunit RNA polymerase II at 4.1-Å resolution: Implications for the initiation of transcription. *Proceedings of the National Academy of Sciences USA*. 100:6969–73.

Bushnell, D.A., K.D. Westover, R.E. Davis, and R.D. Kornberg. 2004. Stuctural basis of transcription: An RNA polymerase II-TFIIB cocrystal at 4.5 angstroms. *Science* 303:983–88.

Chen, H.-T. and S. Hahn. 2004. Mapping the location of TFIIB within the RNA polymerase II transcription preinitiation complex: A model for the structure of the PIC. *Cell* 119: 169–80.

Dynlacht, B.D., T. Hoey, and R. Tjian. (1991). Isolation of coactivators associated with the TATA-binding protein that mediate transcriptional activation. *Cell* 66:563–76.

Flores, O., H. Lu, M. Killeen, J. Greenblatt, Z.F. Burton, and D. Reinberg. 1991. The small subunit of transcription factor IIF recruits RNA polymerase II into the preinitiation complex. *Proceedings of the National Academy of Sciences USA* 88:9999–10003.

Flores, O., E. Maldonado, and D. Reinberg. 1989. Factors involved in specific transcription by mammalian RNA polymerase II: Factors IIE and IIF independently interact with RNA polymerase II. *Journal of Biological Chemistry* 264:8913–21.

Guzder, S.N., P. Sung, V. Bailly, L. Prakash, and S. Prakash. 1994. RAD25 is a DNA helicase required for DNA repair and RNA polymerase II transcription. *Nature* 369:578–81.

Hansen, S.K., S. Takada, R.H. Jacobson, J.T. Lis, and R. Tjian. 1997. Transcription properties of a cell type-specific TATA binding protein, TRF. *Cell* 91:71–83.

Holmes, M.C. and R. Tjian. 2000. Promoter-selective properties of the TBP-related factor TRF1. *Science* 288:867–70.

Holstege, F.C.P., E.G. Jennings, J.J. Wyrick, T.I. Lee, C.J. Hengartner, M.R. Green, T.R. Golub, E.S. Lander, and R.A. Young. 1998. Dissecting the regulatory circuitry of a eukaryotic genome. *Cell* 95:717–28.

Honda, B.M. and R.G. Roeder. 1980. Association of a 5S gene transcription factor with 5S RNA and altered levels of the factor during cell differentiation. *Cell* 22:119–26.

Kassavetis, G.A., B.R. Braun, L.H. Nguyen, and E.P. Geiduschek. 1990. *S. cerevisiae* TFIIIB is the transcription initiation factor proper of RNA polymerase III, while TFIIIA and TFIIIC are assembly factors. *Cell* 60:235–45.

Kassavetis, G.A., D.L. Riggs, R. Negri, L.H. Nguyen, and E.P. Geiduschek. 1989. Transcription factor IIIB generates extended DNA interactions in RNA polymerase III transcription complexes on tRNA genes. *Molecular and Cellular Biology* 9:2551–66.

Kettenberger, H., K.-J. Armache, and P. Cramer. 2003. Architecture of the RNA polymerase II-TFIIS complex and implications for mRNA cleavage. *Cell* 114:347–57.

Kim, J.L., D.B. Nikolov, and S.K. Burley. 1993. Co-crystal structure of a TBP recognizing the minor groove of a TATA element. *Nature* 365:520–27.

Kim, T.-K., R.H. Ebright, and D. Reinberg. 2000. Mechanism of ATP-dependent promoter melting by transcription factor IIH. *Science* 288:1418–21.

Kim, Y.J., S. Björklund, Y. Li, M.H. Sayre, and R.D. Kornberg. 1994. A multiprotein mediator of transcriptional activation and its interaction with the C-terminal repeat domain of RNA polymerase II. *Cell* 77:599–608.

Koleske, A.J. and R.A. Young. 1994. An RNA polymerase II holoenzyme responsive to activators. *Nature* 368:466–69.

Kownin, P., E. Bateman, and M.R. Paule. 1987. Eukaryotic RNA polymerase I promoter binding is directed by protein contacts with transcription initiation factor and is DNA sequence-independent. *Cell* 50:693–99.

Learned, R.M., S. Cordes, and R. Tjian. 1985. Purification and characterization of a transcription factor that confers promoter specificity to human RNA polymerase I. *Molecular and Cellular Biology* 5:1358–69.

Lobo, S.L., M. Tanaka, M.L. Sullivan, and N. Hernandez. 1992. A TBP complex essential for transcription from TATA-less but not TATA-containing RNA polymerase III promoters is part of the TFIIIB fraction. *Cell* 71:1029–40.

Lu, H., L. Zawel, L. Fisher, J.-M. Egly, and D. Reinberg. 1992. Human general transcription factor IIH phosphorylates the C-terminal domain of RNA polymerase II. *Nature* 358:641–45.

Maldonado, E., I. Ha, P. Cortes, L. Weis, and D. Reinberg. 1990. Factors involved in specific transcription by mammalian RNA polymerase II: Role of transcription factors IIA, IID, and IIB during formation of a transcription-competent complex. *Molecular and Cellular Biology* 10:6335–47.

Ossipow, V., J.-P. Tassan, E.I. Nigg, and U. Schibler. 1995. A mammalian RNA polymerase II holoenzyme containing all components required for promoter-specific transcription initiation. *Cell* 83:137–46.

Pelham, H.B., Wormington, W.M., and D.D. Brown. 1981. Related 5S rRNA transcription factors in *Xenopus* oocytes and somatic cells. *Proceeding of the National Academy of Sciences USA* 78:1760–64.

Pugh, B.F. and R. Tjian. 1991. Transcription from a TATA-less promoter requires a multisubunit TFIID complex. *Genes and Development* 5:1935–45.

Rowlands, T., P. Baumann, and S.P. Jackson. 1994. The TATA-binding protein: A general transcription factor in eukaryotes and archaebacteria. *Science* 264:1326–29.

Sauer, F., D.A. Wassarman, G.M. Rubin, and R. Tjian. 1996. TAF_{II}s mediate activation of transcription in the *Drosophila* embryo. *Cell* 87:1271–84.

Schultz, M.C., R.H. Roeder, and S. Hahn. 1992. Variants of the TATA-binding protein can distinguish subsets of RNA polymerase I, II, and III promoters. *Cell* 69:697–702.

Setzer, D.R. and D.D. Brown. 1985. Formation and stability of the 5S RNA transcription complex. *Journal of Biological Chemistry* 260:2483–92.

Shastry, B.S., S.-Y. Ng, and R.G. Roeder. 1982. Multiple factors involved in the transcription of class III genes in *Xenopus laevis. Journal of Biological Chemistry* 257:12979–86.

Starr, D.B. and D.K. Hawley. 1991. TFIID binds in the minor groove of the TATA box. *Cell* 67:1231–40.

Taggart, K.P., J.S. Fisher, and B.F. Pugh. 1992. The TATA-binding protein and associated factors are components of Pol III transcription factor TFIIIB. *Cell* 71:1051–28.

Takada, S., J.T. Lis, S. Zhou, and R. Tjian. 2000. A TRF1:BRF complex directs *Drosophila* RNA polymerase III transcription. *Cell* 101:459–69.

Tanese, N. 1991. Coactivators for a proline-rich activator purified from the multisubunit human TFIID complex. *Genes and Development* 5:2212–24.

Thomas, M.J., A.A. Platas, and D.K. Hawley. 1998. Transcriptional fidelity and proofreading by RNA polymerase II. *Cell* 93:627–37.

Verrijzer, C.P., J.-L. Chen, K. Yokomori, and R. Tjian. 1995. Binding of TAFs to core elements directs promoter selectivity by RNA polymerase II. *Cell* 81:1115–25.

Walker, S.S., J.C. Reese, L.M. Apone, and M.R. Green. 1996. Transcription activation in cells lacking TAF_{II}s. *Nature* 383:185–88.

Wieczorek, E., M. Brand, X. Jacq, and L. Tora. 1998. Function of TAF_{II}-containing complex without TBP in transcription by RNA polymerase II. *Nature* 393:187–91.

CHAPTER 12

Transcription Activators in Eukaryotes

Computer model of the transcription factor p53 interacting with its target DNA site. *Courtesy Nicola P. Pavletich, Sloan-Kettering Cancer Center, Science (15 July 1994) cover. Copyright © AAAS.*

In Chapters 10 and 11 we learned about the basic machinery involved in eukaryotic transcription: the three RNA polymerases, their promoters, and the general transcription factors that bring RNA polymerase and promoter together. However, it is clear that this is not the whole story. The general transcription factors by themselves dictate the starting point and direction of transcription, but they are capable of sponsoring only a very low level of transcription (basal level transcription). But transcription of active genes in cells rises above (frequently far above) the basal level. To provide the needed extra boost in transcription, eukaryotic cells have additional, gene-specific transcription factors **(activators)** that bind to DNA elements called enhancers (Chapter 10). The transcription activation provided by these activators also permits cells to control the expression of their genes.

In addition, eukaryotic DNA is complexed with protein in a structure called chromatin. Some chromatin, called heterochromatin, is highly condensed and inaccessible to RNA polymerases, so it cannot be transcribed. Other chromatin (euchromatin) still contains protein, but it is relatively extended. Much of this euchromatin, even though it is relatively open, contains genes that are not transcribed in a given cell because the appropriate activators are not available to turn them on. Instead, other proteins may hide the promoters from RNA polymerase and general transcription factors to ensure that they remain turned off. In this chapter, we will examine the activators that control eukaryotic genes. Then, in Chapter 13, we will look at the crucial relationship among activators, chromatin structure, and gene activity.

12.1 Categories of Activators

Activators can either stimulate or inhibit transcription by RNA polymerase II, and they have structures composed of at least two functional domains: a **DNA-binding domain** and a **transcription-activating domain.** Many also have a **dimerization domain** that allows the activators to bind to each other, forming homodimers (two identical monomers bound together), heterodimers (two different monomers bound together), or even higher multimers such as tetramers. Some even have binding sites for effector molecules like steroid hormones. Let us consider some examples of these three kinds of structural–functional domains, bearing in mind an important principle we discussed in Chapters 6 and 9: A protein does not have just one shape. Rather, it is a dynamic molecule that assumes many possible conformations. Some of these may be especially advantageous for binding to other molecules, such as a specific DNA sequence, and these conformations would be stabilized by binding to such DNA sequences. Thus, when we refer to the shape of a DNA-binding protein, or a domain within such a protein, we mean one of many possible shapes, which happens to fit particularly well with the DNA in question.

DNA-Binding Domains

A protein **domain** is an independently folded region of a protein. Each DNA-binding domain has a **DNA-binding motif,** which is the part of the domain that has a characteristic shape specialized for specific DNA binding. Most DNA-binding motifs fall into the following classes:

1. *Zinc-containing modules.* At least three kinds of zinc-containing modules act as DNA-binding motifs. These all use one or more zinc ions to create the proper shape so an α-helix within the motif can fit into the DNA major groove and make specific contacts there. These zinc-containing modules include:
   a. **Zinc fingers,** such as those found in TFIIIA and Sp1, two transcription factors we have already encountered.
   b. Zinc modules found in the glucocorticoid receptor and other members of this group of nuclear receptors.
   c. Modules containing two zinc ions and six cysteines, found in the yeast activator GAL4 and its relatives.
2. **Homeodomains (HDs).** These contain about 60 amino acids and resemble in structure and function the helix-turn-helix DNA-binding domains of prokaryotic proteins such as the λ phage repressor. HDs, found in a variety of activators, were originally identified in activators called homeobox proteins that regulate development in the fruit fly *Drosophila.*
3. **bZIP and bHLH motifs.** The CCAAT/enhancer-binding protein (C/EBP), the MyoD protein, and many other eukaryotic transcription factors have a highly basic DNA-binding motif linked to one or both of the protein dimerization motifs known as leucine zippers and helix-loop-helix (HLH) motifs. (By the way C/EBP is different from the CCAAT-binding transcription factor [CTF, Chapter 10]).

This list is certainly not exhaustive. In fact, several transcription factors have now been identified that do not fall into any of these categories.

Transcription-Activating Domains

Most activators have one of these domains, but some have more than one. So far, most of these domains fall into three classes, as follows:

1. *Acidic domains.* The yeast activator GAL4 typifies this group. It has a 49-amino-acid domain with 11 acidic amino acids.
2. *Glutamine-rich domains.* The activator Sp1 has two such domains, which are about 25% glutamine. One of these has 39 glutamines in a span of 143 amino acids. In addition, Sp1 has two other activating domains that do not fit into any of these three main categories.
3. *Proline-rich domains.* The activator CTF, for instance, has a domain of 84 amino acids, 19 of which are prolines.

Our descriptions of the transcription-activating domains are necessarily nebulous, because the domains themselves are rather ill-defined. The acidic domain, for example, has seemed to require nothing more than a preponderance of acidic residues to make it function, which led to the name "acid blob" to describe this presumably unstructured domain. On the other hand, Stephen Johnston and his colleagues have shown that the acidic activation domain of GAL4 tends to form a defined structure—a β-sheet—in slightly acidic solution. It is possible that the β-sheet also forms under the slightly basic conditions in vivo, but this is not yet clear. These workers also removed all six of the acidic amino acids in the GAL4 acidic domain and showed that it still retained 35% of its normal ability to activate transcription. Thus, not only is the structure of the acidic activating domain unclear, the importance of its acidic nature is even in doubt.

With such persistent uncertainty, it has been difficult to draw conclusions about how the structure and function of transcription-activating domains are related. On the other hand, some evidence suggests that the glutamine-rich activation domain of Spl operates by interacting with glutamine-rich domains of other transcription factors.

SUMMARY Eukaryotic activators are composed of at least two domains: a DNA-binding domain and a transcription-activating domain. DNA-binding domains contain motifs such as zinc modules, homeodomains, and bZIP or bHLH motifs. Transcription-activating domains can be acidic, glutamine-rich, or proline-rich.

12.2 Structures of the DNA-Binding Motifs of Activators

By contrast to the transcription-activating domains, most DNA-binding domains have well-defined structures, and x-ray crystallography studies have shown how these structures interact with their DNA targets. Furthermore, these same structural studies have frequently elucidated the dimerization domains responsible for interaction between protein monomers to form a functional dimer, or in some cases, a tetramer. This is crucial, because most classes of DNA-binding proteins are incapable of binding to DNA in monomer form; they must form at least dimers to function. Let us explore the structures of several classes of DNA-binding motifs and see how they mediate interaction with DNA. In the process we will discover the ways some of these proteins can dimerize.

Zinc Fingers

In 1985, Aaron Klug noticed a periodicity in the structure of the general transcription factor TFIIIA. This protein has nine repeats of a 30-residue element. Each element has two closely spaced cysteines followed 12 amino acids later by two closely spaced histidines. Furthermore, the protein is rich in zinc—enough for one zinc ion per repeat. This led Klug to predict that each zinc ion is complexed by the two cysteines and two histidines in each repeat unit to form a finger-shaped domain.

Finger Structure Michael Pique and Peter Wright used nuclear magnetic resonance spectroscopy to determine the structure in solution of one of the zinc fingers of the *Xenopus laevis* protein Xfin, an activator of certain class II promoters. Note that this structure, depicted in Figure 12.1, really is not very finger-shaped, unless it is a rather wide, stubby finger. It is also worth noting that this finger shape by itself does not confer any binding specificity, since there are many different finger proteins, all with the same shape fingers but each binding to its own unique DNA target sequence. Thus, it is the precise amino acid sequences of the fingers, or of neighboring parts of the protein, that determine the DNA sequence to which the protein can bind. In the Xfin finger, an α-helix (on the left in Figure 12.1) contains several basic amino acids—all on the side that seems to contact the DNA. These and other amino acids in the helix presumably determine the binding specificity of the protein.

Carl Pabo and his colleagues used x-ray crystallography to obtain the structure of the complex between DNA,

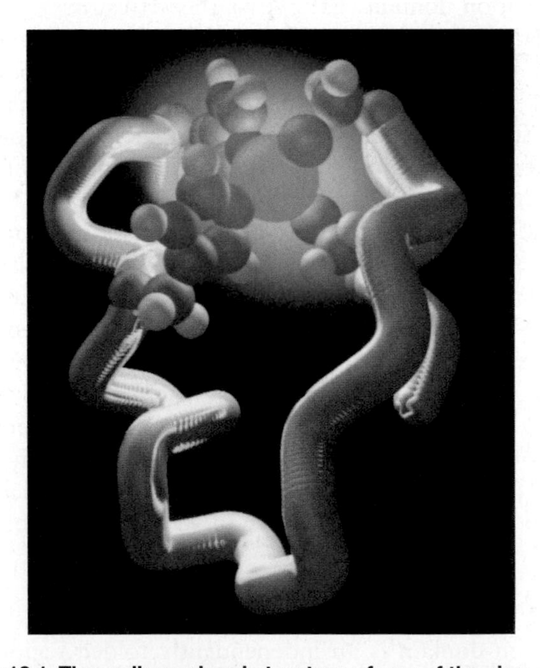

Figure 12.1 Three-dimensional structure of one of the zinc fingers of the *Xenopus* protein Xfin. The zinc is represented by the turquoise sphere at top center. The sulfurs of the two cysteines are represented by yellow-green spheres. The two histidines are represented by the blue-green structures at upper left. The backbone of the finger is represented by the purple tube. (*Source:* Pique, Michael and Peter E. Wright, Dept. of Molecular Biology, Scripps Clinic Research Institute, La Jolla, CA. (cover photo, *Science* 245 (11 Aug 1989).)

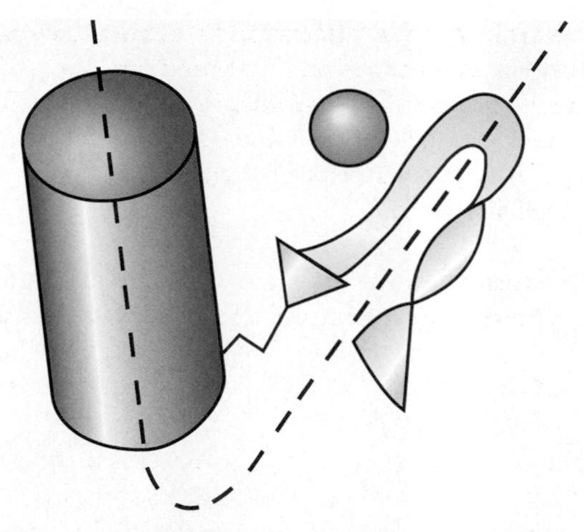

Figure 12.2 Schematic diagram of zinc finger 1 of the Zif268 protein. The right-hand side of the finger is an antiparallel β-sheet (yellow), and the left-hand side is an α-helix (red). Two cysteines in the β-sheet and two histidines in the α-helix coordinate the zinc ion in the middle (blue). The dashed line traces the outline of the "finger" shape. (*Source:* Adapted from Pavletich, N.P. and C.O. Pabo, Zinc finger–DNA recognition: Crystal structure of a Zif268–DNA complex at 2.1 Å. *Science* 252:812, 1991.)

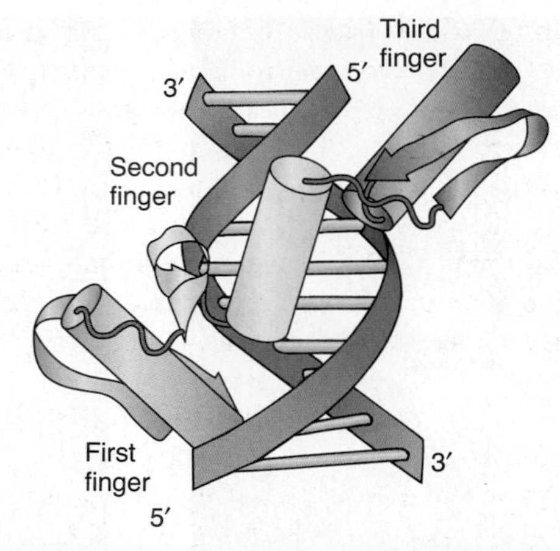

Figure 12.3 Arrangement of the three zinc fingers of Zif268 in a curved shape to fit into the major groove of DNA. As usual, the cylinders and ribbons stand for α-helices and β-sheets, respectively. (*Source:* Adapted from Pavletich, N.P. and C.O. Pabo, Zinc finger–DNA recognition: Crystal structure of a Zif268–DNA complex at 2.1 Å. *Science* 252:811, 1991.)

and a member of the TFIIIA class of zinc finger proteins—the mouse protein Zif268. This is a so-called *immediate early protein,* which means that it is one of the first genes to be activated when resting cells are stimulated to divide. The Zif268 protein has three adjacent zinc fingers that fit into the major groove of the DNA double helix. We will see the arrangement of these three fingers a little later in the chapter. For now, let us consider the three-dimensional structure of the fingers themselves. Figure 12.2 presents the structure of finger 1 as an example. The finger shape in this presentation is perhaps not obvious. Still, on close inspection we can see the finger contour, which is indicated by the dashed line. As in the Xfin zinc finger, the left side of each Zif268 finger is an α-helix. This is connected by a short loop at the bottom to the right side of the finger, a small antiparallel β-sheet. Do not confuse this β-sheet itself with the finger; it is only one half of it. The zinc ion (blue sphere) is in the middle, coordinated by two histidines in the α-helix and by two cysteines in the β-sheet. All three fingers have almost exactly the same shape.

Interaction with DNA How do the fingers interact with their DNA targets? Figure 12.3 shows all three Zif268 fingers lining up in the major groove of the DNA. In fact, the three fingers are arranged in a curve, or C-shape, which matches the curve of the DNA double helix. All the fingers approach the DNA from essentially the same angle, so the geometry of protein–DNA contact is very similar in each case. Binding between each finger and its DNA-binding site relies on direct amino acid–base interactions, between amino acids in the α-helix and bases in the major groove of the DNA. For more detailed descriptions of amino acid–base interactions, see Chapter 9.

Comparison with Other DNA-Binding Proteins One unifying theme emerging from studies of many, but not all, DNA-binding proteins is the utility of the α-helix in contacting the DNA major groove. We saw many examples of this with the prokaryotic helix-turn-helix domains (Chapter 9), and we will see several other eukaryotic examples. What about the β-sheet in Zif268? It seems to serve the same function as the first α-helix in a helix-turn-helix protein, namely to bind to the DNA backbone and help position the recognition helix for optimal interaction with the DNA major groove.

Zif268 also shows some differences from the helix-turn-helix proteins. Whereas the latter proteins have a single DNA-binding domain per monomer, the finger protein DNA-binding domains have a modular construction, with several fingers making contact with the DNA. This arrangement means that these proteins, in contrast to most DNA-binding proteins, do not need to form dimers or tetramers to bind to DNA. They already have multiple binding domains built in. Also, most of the protein–DNA contacts are with one DNA strand, rather than both, as in the case of the helix-turn-helix proteins. At least with this particular finger protein, most of the contacts are with bases, rather than the DNA backbone.

In 1991, Nikola Pavletich and Carl Pabo solved the structure of a cocrystal between DNA and a five-zinc-finger human protein called GLI. This provided an interesting contrast with the three-finger Zif268 protein. Again, the major groove is the site of finger–DNA contacts, but in this case one finger (finger 1) does not contact the DNA. Also, the overall geometries of the two finger–DNA complexes are similar, with the fingers wrapping around the DNA

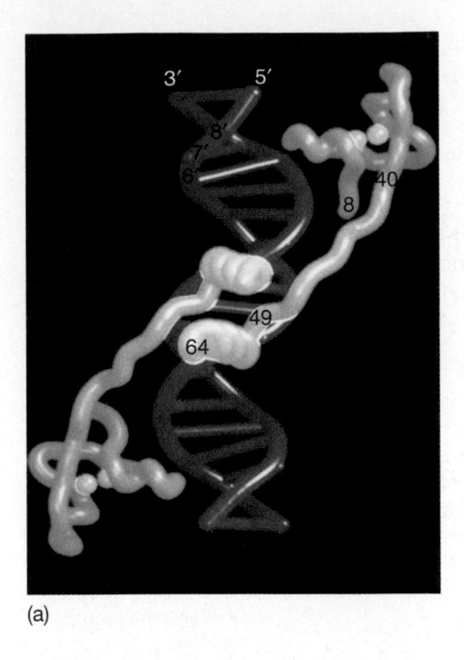

(a)

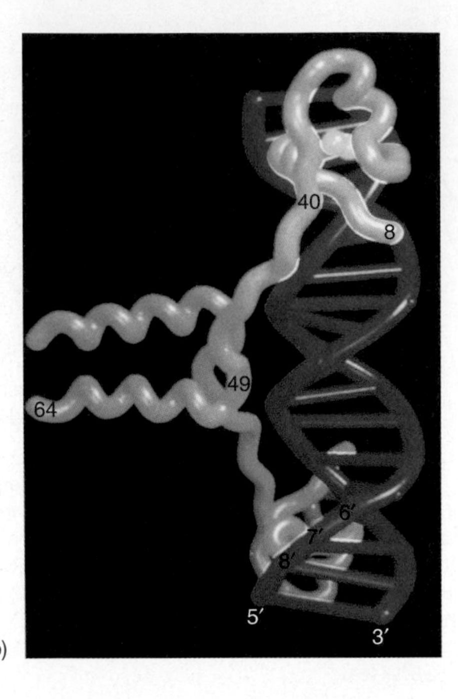

(b)

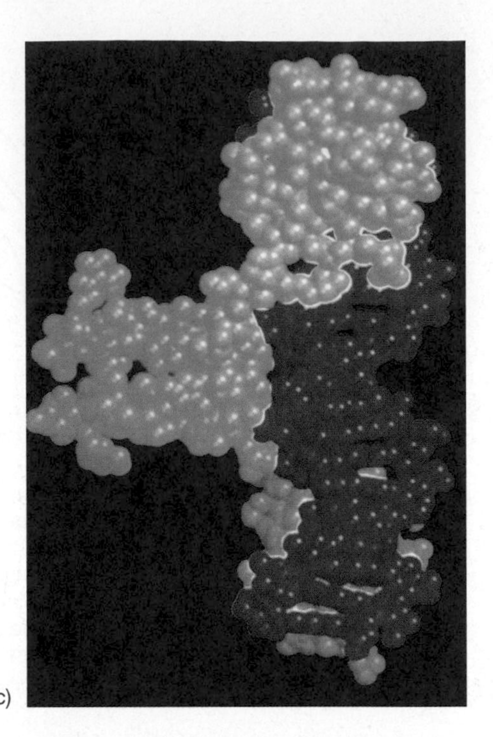

(c)

Figure 12.4 Three views of the GAL4–DNA complex. (a) The complex viewed approximately along its two-fold axis of symmetry. The DNA is in red, the protein is in blue, and the zinc ions are represented by yellow spheres. Amino acid residue numbers at the beginnings and ends of the three domains are given on the top monomer: The DNA recognition module extends from residue 8 to 40. The linker, from residue 41 to 49, and the dimerization domain, from residue 50 to 64. **(b)** The complex viewed approximately perpendicular to the view in panel (a). The dimerization elements appear roughly parallel to one another at left center. **(c)** Space-filling model of the complex in the same orientation as in panel (b). Notice that the recognition modules on the two GAL4 monomers make contact with opposite faces of the DNA. Notice also the neat fit between the coiled coil of the dimerization domain and the minor groove of the DNA helix. (*Source:* Marmorstein, R., M. Carey, M. Ptashne, and S.C. Harrison, DNA recognition by GAL4: Structure of a protein–DNA complex. *Nature* 356 (2 April 1992) p. 411, f. 3. Copyright © Macmillan Magazines Ltd.)

major groove, but no simple "code" of recognition between certain bases and amino acids exists.

SUMMARY Zinc fingers are composed of an antiparallel β-sheet, followed by an α-helix. The β-sheet contains two cysteines, and the α-helix two histidines, that are coordinated to a zinc ion. This coordination of amino acids to the metal helps form the finger-shaped structure. The specific recognition between the finger and its DNA target occurs in the major groove.

The GAL4 Protein

The **GAL4** protein is a yeast activator that controls a set of genes responsible for metabolism of galactose. Each of these GAL4-responsive genes contains a GAL4 target site (enhancer) upstream of the transcription start site. These target sites are called **upstream activating sequences,** or **UAS_Gs.** GAL4 binds to a UAS_G as a dimer. Its DNA-binding motif is located in the first 40 amino acids of the protein, and its dimerization motif is found in residues 50–94. The DNA-binding motif is similar to the zinc finger in that it contains zinc and cysteine residues, but its structure must be different: Each motif has six cysteines and no histidines, and the ratio of zinc ions to cysteines is 1:3.

Mark Ptashne and Stephen Harrison and their colleagues performed x-ray crystallography on cocrystals of the first 65 amino acids of GAL4 and a synthetic 17-bp piece of DNA. This revealed several important features of the protein–DNA complex, including the shape of the DNA-binding motif and how it interacts with its DNA target, and part of the dimerization motif in residues 50–64.

The DNA-Binding Motif Figure 12.4 depicts the structure of the GAL4 peptide dimer–DNA complex. One end of each monomer contains a DNA-binding motif containing six cysteines that complex two zinc ions (yellow spheres), forming a *bimetal thiolate cluster.* Each of these motifs also features a short α-helix that protrudes into the major groove of the DNA double helix, where its amino acid side chains can make specific interactions with the DNA bases and backbone. The other end of each monomer is an α-helix that serves a dimerization function that we will discuss later in this chapter.

The Dimerization Motif The GAL4 monomers also take advantage of α-helices in their dimerization, forming a parallel **coiled coil** as illustrated at left in Figure 12.4b and c. This figure also shows that the dimerizing α-helices point directly at the minor groove of the DNA. Finally, note in Figure 12.4 that the DNA recognition module and the dimerization module in each monomer are joined by an extended linker domain. We will see other examples of coiled coil dimerization motifs when we discuss bZIP and bHLH motifs later in this chapter.

SUMMARY The GAL4 protein is a member of the zinc-containing family of DNA-binding proteins, but it does not have zinc fingers. Instead, each GAL4 monomer contains a DNA-binding motif with six cysteines that coordinate two zinc ions in a bimetal thiolate cluster. The recognition module contains a short α-helix that protrudes into the DNA major groove and makes specific interactions there. The GAL4 monomer also contains an α-helical dimerization motif that forms a parallel coiled coil as it interacts with the α-helix on the other GAL4 monomer.

The Nuclear Receptors

A third class of zinc module is found in the **nuclear receptors.** These proteins interact with a variety of endocrine-signaling molecules (steroids and other hormones) that diffuse through the cell membrane. They form hormone-receptor complexes that function as activators by binding to enhancers, or **hormone response elements,** and stimulating transcription of their associated genes. Thus, these activators differ from the others we have studied in that they must bind to an effector (a hormone) in order to function as activators. This implies that they must have an extra important domain—a hormone-binding domain—and indeed they do.

Some of the hormones that work this way are the sex hormones (androgens and estrogens); progesterone, the hormone of pregnancy (and principal ingredient of common birth control pills); the glucocorticoids, such as cortisol; vitamin D, which regulates calcium metabolism; and thyroid hormone and retinoic acid, which regulate gene expression during development. Each hormone binds to its specific receptor, and together they activate their own set of genes.

The nuclear receptors have traditionally been divided into three classes. The **type I receptors** include the steroid hormone receptors, typified by the **glucocorticoid receptor.** In the absence of their hormone ligands, these receptors reside in the cytoplasm, coupled with another protein. When a type I receptor binds to its hormone ligand, it releases its protein partner and migrates to the nucleus, where it binds as a homodimer to its hormone response element. For example, the glucocorticoid receptor exists in the cytoplasm complexed with a partner known as heat shock protein 90 (Hsp90). When the receptor binds to its glucocorticoid ligand (Figure 12.5), it changes conformation, dissociates from Hsp90, and moves into the nucleus

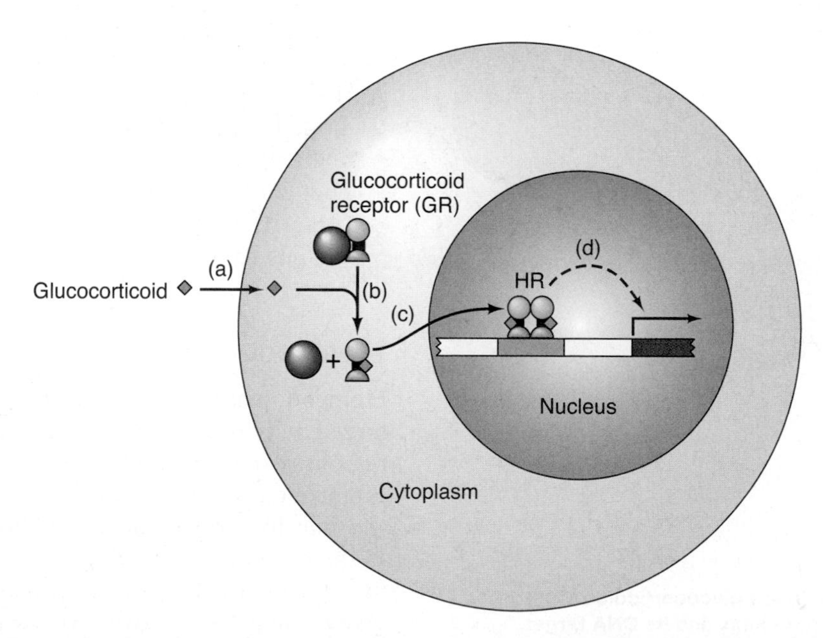

Figure 12.5 Glucocorticoid action. The glucocorticoid receptor (GR) exists in an inactive form in the cytoplasm complexed with heat shock protein 90 (Hsp90). **(a)** The glucocorticoid (blue diamond) diffuses across the cell membrane and enters the cytoplasm. **(b)** The glucocorticoid binds to its receptor (GR, red and green), which changes conformation and dissociates from Hsp90 (orange). **(c)** The hormone–receptor complex (HR) enters the nucleus, dimerizes with another HR, and binds to a hormone-response element, or enhancer (pink), upstream of a hormone-activated gene (brown). **(d)** Binding of the HR dimer to the enhancer activates (dashed arrow) the associated gene, so transcription occurs (bent arrow).

to activate genes controlled by enhancers called **glucocorticoid response elements (GREs)**.

Sigler and colleagues performed x-ray crystallography on cocrystals of the glucocorticoid receptor and an oligonucleotide containing two target half-sites.

The crystal structure revealed several aspects of the protein–DNA interaction: (1) The binding domain dimerizes, with each monomer making specific contacts with one target half-site. (2) Each binding motif is a zinc module that contains two zinc ions, rather than the one found in a classical zinc finger. (3) Each zinc ion is complexed to four cysteines to form a finger-like shape. (4) The amino-terminal finger in each binding domain engages in most of the interactions with the DNA target. Most of these interactions involve an α-helix. The crystal structure revealed several aspects of the protein-DNA interaction: Figure 12.6 illustrates the specific amino-acid–base associations between this recognition helix and the DNA target site. Some amino acids outside this helix also make contact with the DNA through its backbone phosphates.

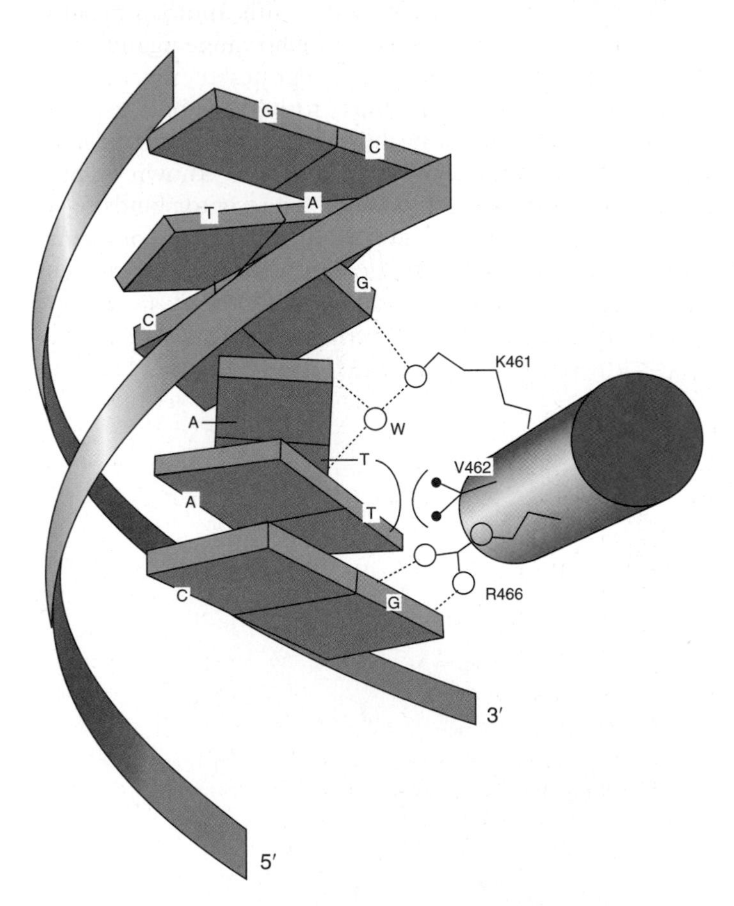

Figure 12.6 Association between the glucocorticoid receptor DNA-binding domain's recognition helix and its DNA target. The specific amino-acid–base interactions are shown. A water molecule (W) mediates some of the H-bonding between lysine 461 and the DNA. (*Source:* Adapted from Luisi, B.F., W.X. Xu, Z. Otwinowski, L.P. Freedman, K.R. Yamamoto, and P.B. Sigler, Crystallographic analysis of the interaction of the glucocorticoid receptor with DNA. *Nature* 352 (8 Aug 1991) p. 500, f. 4a. Copyright © Macmillan Magazines Ltd.)

The **type II receptors**, exemplified by the **thyroid hormone receptor**, stay in the nucleus, where they form dimers with another protein called **retinoic acid receptor X (RXR)**, whose ligand is 9-*cis* retinoic acid. These receptors bind to their target sites in both the presence and absence of their ligands. As we will see in Chapter 13, binding of these type II receptors in the absence of ligand can repress transcription, whereas binding of the receptors along with their ligands can stimulate transcription. Thus, the same protein can act as either an activator or a repressor, depending on environmental conditions.

The **type III receptors** are not as well understood. They are also known as "orphan receptors" because their ligands have not been identified. Perhaps further study will show that some or all of these type III receptors really belong with the type I or type II receptors.

Finally, note that all three classes of zinc-containing DNA-binding modules use a common motif—an α-helix—for most of the interactions with their DNA targets.

SUMMARY Type I nuclear receptors reside in the cytoplasm, bound to another protein. When these receptors bind to their hormone ligands, they release their cytoplasmic protein partners and move to the nucleus where they bind to enhancers, and thereby act as activators. The glucocorticoid receptor is representative of this group. It has a DNA-binding domain with two zinc-containing modules. One module contains most of the DNA-binding residues (in a recognition α-helix), and the other module provides the surface for protein–protein interaction to form a dimer. Type II nuclear receptors, e.g., thyroid hormone receptor, stay in the nucleus, bound to their target DNA sites. In the absence of their ligands they repress gene activity, but when they bind their ligands they activate transcription. Type III receptors are "orphan" receptors whose ligands have not been identified.

Homeodomains

Homeodomains are DNA-binding domains found in a large family of activators. Their name comes from the gene regions, called **homeoboxes**, in which they are encoded. Homeoboxes were first discovered in regulatory genes of the fruit fly *Drosophila,* called homeotic genes. Mutations in these genes cause strange transformations of body parts in the fruit fly. For example, a mutation called *Antennapedia* causes legs to grow where antennae would normally be (Figure 12.7).

Homeodomain proteins are members of the helix-turn-helix family of DNA-binding proteins (Chapter 9). Each homeodomain contains three α-helices; the second and third of these form the helix-turn-helix motif, with the third

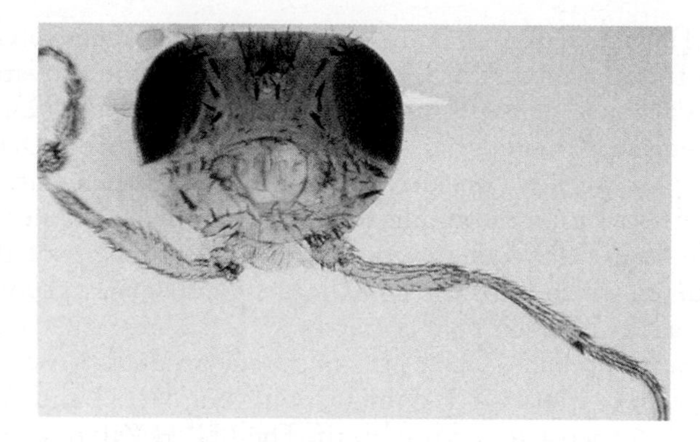

Figure 12.7 The *Antennapedia* phenotype. Legs appear on the head where antennae would normally be. (*Source:* Courtesy Walter J. Gehring, University of Basel, Switzerland.)

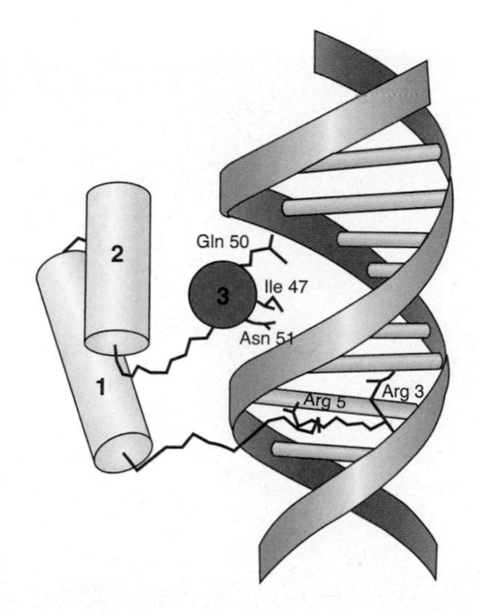

Figure 12.8 Representation of the homeodomain–DNA complex. Schematic model with the three helices numbered on the left, and a ribbon diagram of the DNA target on the right. The recognition helix (labeled 3, red) is shown on end, resting in the major groove of the DNA. The N-terminal arm is also shown, inserted into the DNA minor groove. Key amino acid side chains are shown interacting with DNA. (*Source:* Adapted from Kissinger, C.R., B. Liu, E. Martin-Blanco, T.B. Kornberg, and C.O. Pabo, Crystal structure of an engrailed homeodomain–DNA complex at 2.8 Å resolution: A framework for understanding homeodomain–DNA interactions. *Cell* 63 (2 November, 1990) p. 582. f. 5b.)

serving as the recognition helix. But most homeodomains have another element, not found in helix-turn-helix motifs: The N-terminus of the protein forms an arm that inserts into the the minor groove of the DNA. Figure 12.8 shows the interaction between a typical homeodomain, from the *Drosophila* homeotic gene *engrailed,* and its DNA target. This view of the protein–DNA complex comes from Thomas Kornberg's and Carl Pabo's x-ray diffraction analysis of cocrystals of the engrailed homeodomain and an oligonucleotide containing the engrailed binding site. Most homeodomain proteins have weak DNA-binding specificity on their own. As a result, they rely on other proteins to help them bind specifically and efficiently to their DNA targets.

SUMMARY The homeodomains in eukaryotic activators contain a DNA-binding motif that functions in much the same way as helix-turn-helix motifs in which a recognition helix fits into the DNA major groove and makes specific contacts there. In addition, the N-terminal arm nestles in the adjacent minor groove.

The bZIP and bHLH Domains

As with several of the other DNA-binding domains we have studied, the **bZIP** and **bHLH** domains combine two functions: DNA binding and dimerization. The *ZIP* and *HLH* parts of the names refer to the **leucine zipper** and **helix-loop-helix parts,** respectively, of the domains, which are the dimerization motifs. The *b* in the names refers to a basic region in each domain that forms the majority of the DNA-binding motif.

Let us consider the structures of these combined dimerization/DNA-binding domains, beginning with the bZIP domain. This domain actually consists of two polypeptides, each of which contains half of the zipper: an α-helix with leucine (or other hydrophobic amino acid) residues spaced seven amino acids apart, so they are all on one face of the helix. The spacing of the hydrophobic amino acids on one monomer puts them in position to interact with a similar string of amino acids on the other protein monomer. In this way, the two helices act like the two halves of a zipper.

To get a better idea of the structure of the zipper, Peter Kim and Tom Alber and their colleagues crystallized a synthetic peptide corresponding to the bZIP domain of GCN4, a yeast activator that regulates amino acid metabolism. The x-ray diffraction pattern shows that the dimerized bZIP domain assumes a parallel coiled coil structure (Figure 12.9). The α-helices are parallel in that their amino to carboxyl orientations are the same (left to right in panel b). Figure 12.9a, in which the coiled coil extends directly out at the reader, gives a good feel for the extent of supercoiling in the coiled coil. Notice the similarity between this and the coiled coil dimerization motif in GAL4 (see Figure 12.4).

This crystallographic study, which focused on the zipper in the absence of DNA, did not shed light on the mechanism of DNA binding. However, Kevin Struhl and Stephen Harrison and their colleagues performed x-ray crystallography on the bZIP domain of GCN4, bound to its DNA target. Figure 12.10 shows that the leucine zipper not only brings the two monomers together, it also places the two basic parts of the domain in position to grasp the DNA like a pair of forceps, or fireplace tongs, with the basic motifs fitting into the DNA major groove.

Harold Weintraub and Carl Pabo and colleagues solved the crystal structure of the bHLH domain of the activator **MyoD** bound to its DNA target. The structure (Figure 12.11) is remarkably similar to that of the bZIP domain–DNA complex we just considered. The helix-loop-helix part is the dimerization motif, but the long helix (helix 1) in each helix-loop-helix domain contains the basic region of the domain, which grips the DNA target via its major groove, just as the bZIP domain does.

Some proteins, such as the oncogene products Myc and Max, have **bHLH-ZIP domains** with both HLH and ZIP motifs adjacent to a basic motif. The bHLH-ZIP domains interact with DNA in a manner very similar to that employed by the bHLH domains. The main difference between bHLH and bHLH-ZIP domains is that the latter

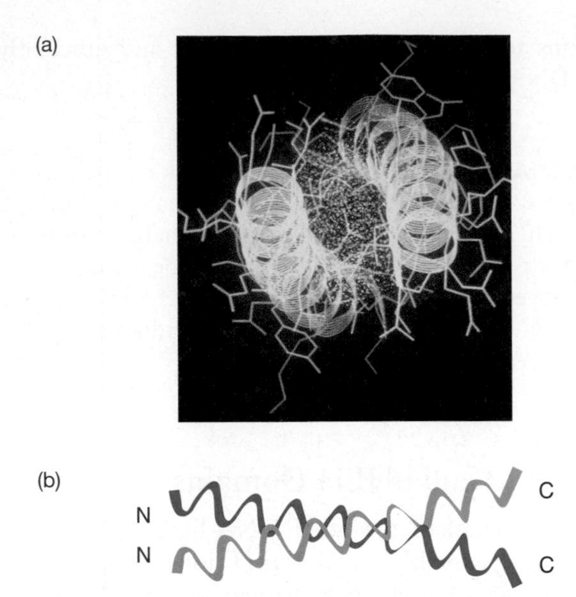

Figure 12.9 Structure of a leucine zipper. (a) Kim and Alber and colleagues crystallized a 33-amino-acid peptide containing the leucine zipper motif of the transcription factor GCN4. X-ray crystallography on this peptide yielded this view along the axis of the zipper with the coiled coil pointed out of the plane of the paper. **(b)** A side view of the coiled coil with the two α-helices colored red and blue. Notice that the amino ends of both peptides are on the left. Thus, this is a parallel coiled coil. (*Source:* (a) O'Shea, E.K., J.D. Klemm, P.S. Kim, and T. Alber, X-ray structure of the GCN4 leucine zipper, a two-stranded, parallel coiled coil. *Science* 254 (25 Oct 1991) p. 541, f. 3. Copyright © AAAS.)

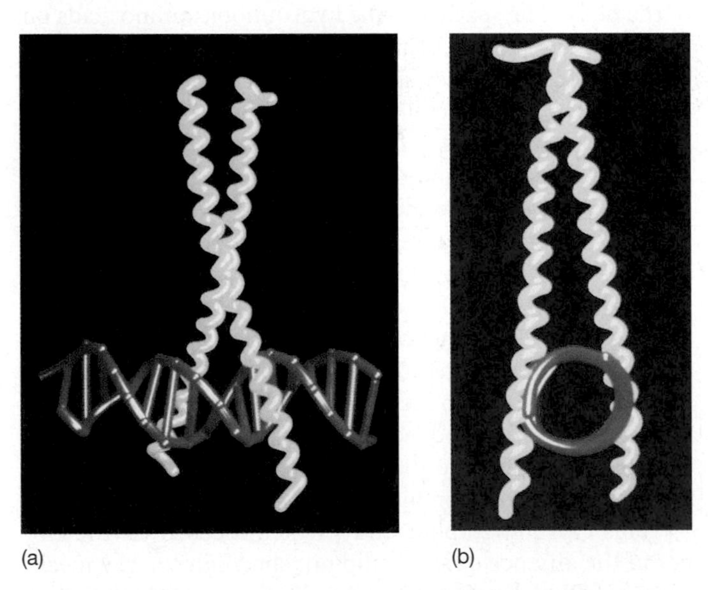

Figure 12.10 Crystal structure of the bZIP motif of GCN4 bound to its DNA target. The DNA (red) contains a target for the bZIP motif (yellow). Notice the coiled coil nature of the interaction between the protein monomers, and the tong-like appearance of the protein grasping the DNA. **(a)** Side view of DNA. **(b)** End view of DNA. (*Source:* Ellenberger, T.E., C.J. Brandl, K. Struhl, and S.C. Harrison, The GCN4 basic region leucine zipper binds DNA as a dimer of uninterrupted alpha helices: Crystal structure of the protein–DNA complex. *Cell* 71 (24 Dec 1992) p. 1227, f. 3a–b. Reprinted by permission of Elsevier Science.)

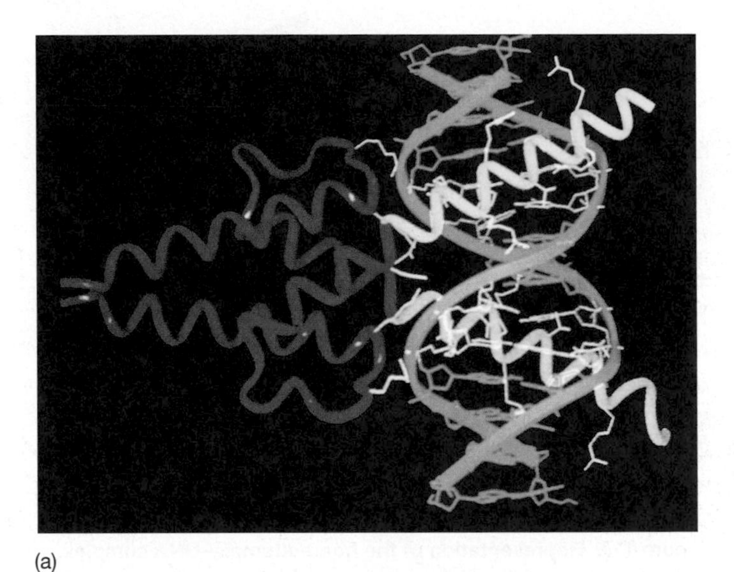

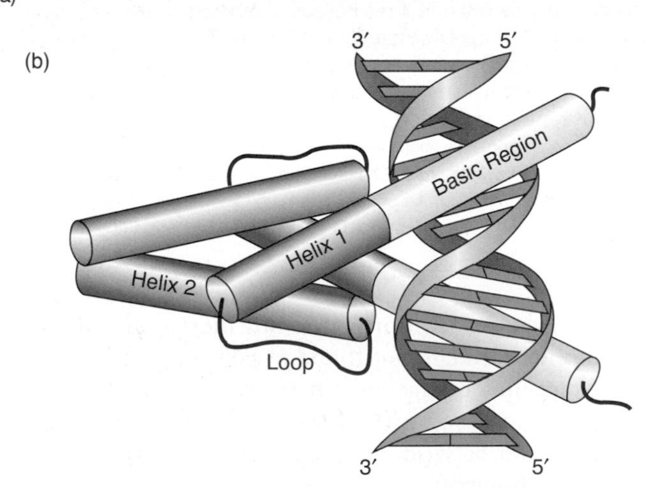

Figure 12.11 Crystal structure of the complex between the bHLH domain of MyoD and its DNA target. (a) Diagram with coiled ribbons representing α-helices. **(b)** Diagram with cylinders representing α-helices. (*Source:* Ma, P.C.M., M.A. Rould, H. Weintraub, and C.O. Palo, Crystal structure of MyoD bHLH domain-DNA complex: Perspectives on DNA recognition and implications for transcriptional activation. *Cell* 77 (6 May 1994) p. 453, f. 2a. Reprinted by permission of Elsevier Science.)

may require the extra interaction of the leucine zippers to ensure dimerization of the protein monomers.

SUMMARY The bZIP proteins dimerize through a leucine zipper, which puts the adjacent basic regions of each monomer in position to embrace the DNA target site like a pair of tongs. Similarly, the bHLH proteins dimerize through a helix-loop-helix motif, which allows the basic parts of each long helix to grasp the DNA target site, much as the bZIP proteins do. The bHLH and bHLH-ZIP domains bind to DNA in the same way, but the latter have extra dimerization potential due to their leucine zippers.

12.3 Independence of the Domains of Activators

We have now seen several examples of DNA-binding and transcription-activating domains in activators. These domains are separated physically on the proteins, they fold independently of each other to form distinct three-dimensional structures, and they operate independently of each other. Roger Brent and Mark Ptashne demonstrated this independence by creating a chimeric factor with the DNA-binding domain of one protein and the transcription-activating domain of the other. This hybrid protein functioned as an activator, with its specificity dictated by its DNA-binding domain.

Brent and Ptashne started with the genes for two proteins: GAL4 and LexA. We have already studied the DNA-binding and transcription-activating domains of GAL4; LexA is a prokaryotic repressor that binds to *lexA* operators and represses downstream genes in *E. coli* cells. It does not normally have a transcription-activating domain, because that is not its function. By cutting and recombining fragments of the two genes, Brent and Ptashne created a chimeric gene containing the coding regions for the transcription-activating domain of GAL4 and the DNA-binding domain of LexA. To assay the activity of the protein product of this gene, they introduced two plasmids into yeast cells. The first plasmid had the chimeric gene, which produced its hybrid product. The second contained a promoter responsive to GAL4 (either the *GAL1* or the *CYC1* promoter), linked to the *E. coli* β-galactosidase gene, which served as a reporter gene (Chapter 5). The more transcription from the GAL4-responsive promoter, the more β-galactosidase was produced. Therefore, by assaying for β-galactosidase, Brent and Ptashne could determine the transcription rate.

One more element was necessary to make this assay work: a binding site for the chimeric protein. The normal binding site for GAL4 is an upstream enhancer called UAS_G. However, this site would not be recognized by the chimeric protein, which has a LexA DNA-binding domain. To make the *GAL1* promoter responsive to activation, the investigators had to introduce a DNA target for the LexA DNA-binding domain. Therefore, they inserted a *lexA* operator in place of UAS_G. It is important to note that a *lexA* operator would not normally be found in a yeast cell; it was placed there just for the purpose of this experiment. Now the question is: Did the chimeric protein activate the *GAL1* gene?

The answer is yes, as Figure 12.12 demonstrates. The three test plasmids contained UAS_G, no target site, or the *lexA* operator. The activator was either LexA-GAL4, as we have discussed, or LexA (a negative control). With UAS_G present (Figure 12.12a), a great deal of β-galactosidase was made, regardless of which activator was present. This is because the yeast cells themselves make GAL4, which can activate via UAS_G. When no DNA target site was present (Figure 12.12b), no β-galactosidase could be made. Finally, when the *lexA* operator replaced UAS_G (Figure 12.12c), the LexA-GAL4 chimeric protein could activate β-galactosidase production over 500-fold. Thus, one can replace the

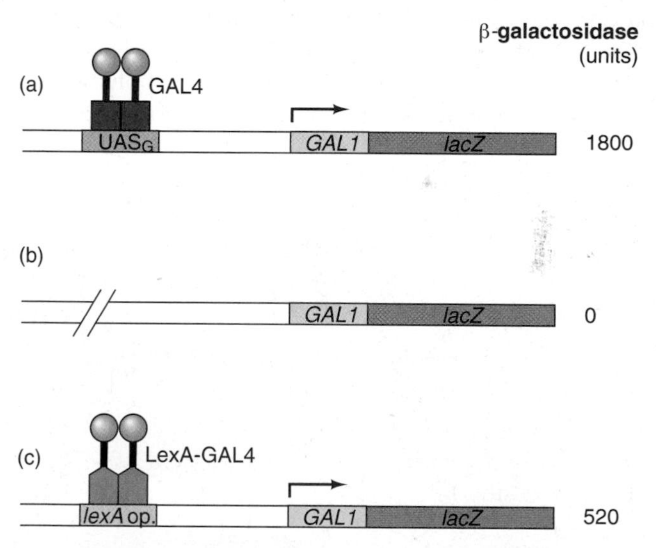

Figure 12.12 Activity of a chimeric transcription factor. Brent and Ptashne introduced two plasmids into yeast cells: (1) a plasmid encoding LexA-GAL4, a hybrid protein containing the transcription-activating domain of GAL4 (green) and the DNA-binding domain of LexA (blue); and (2) one of the test plasmid constructs shown in panels a–c. Each of the test plasmids had the *GAL1* promoter linked to a reporter gene (the *E. coli lacZ* gene). The chimeric protein LexA-GAL4 was used as the activator. The production of β-galactosidase (given at right) is a measure of promoter activity. **(a)** With a UAS_G element, transcription was very active and did not depend on the added transcription factor, because endogenous GAL4 could activate via UAS_G. **(b)** With no DNA target site, LexA-GAL4 could not activate, because it could not bind to the DNA near the *GAL1* promoter. **(c)** With the *lexA* operator, transcription was greatly stimulated by the LexA-GAL4 chimeric factor. The LexA DNA-binding domain could bind to the *lexA* operator, and the GAL4 transcription-activating domain could enhance transcription from the *GAL1* promoter.

DNA-binding domain of GAL4 with the DNA-binding domain of a completely unrelated protein, and produce a functional activator. This demonstrates that the transcription-activating and DNA-binding domains of GAL4 can operate quite independently.

SUMMARY The DNA-binding and transcription-activating domains of activator proteins are independent modules. We can make hybrid proteins with the DNA-binding domain of one protein and the transcription-activating domain of another, and show that the hybrid protein still functions as an activator.

12.4 Functions of Activators

In bacteria, the core RNA polymerase is incapable of initiating meaningful transcription, but the RNA polymerase holoenzyme can catalyze basal level transcription. Basal level transcription is frequently insufficient at weak promoters, so cells have activators to boost this basal transcription to higher levels by a process called **recruitment.** Recruitment leads to the tight binding of RNA polymerase holoenzyme to a promoter.

Eukaryotic activators also recruit RNA polymerase to promoters, but not as directly as prokaryotic activators. The eukaryotic activators stimulate binding of general transcription factors and RNA polymerase to a promoter. Figure 12.13 presents two hypotheses to explain this recruitment: (1) the general transcription factors cause a stepwise build-up of a preinitiation complex; or (2) the general transcription factors and other proteins are already bound to the polymerase in a complex called the **RNA polymerase II holoenzyme,** and the factors and polymerase are recruited together to the promoter. The truth may be a combination of the two hypotheses. In any event, it appears that direct contacts between general transcription factors and activators are necessary for recruitment. (However, as we will see later in this chapter, some activators require other proteins called coactivators to mediate the contact with the general transcription factors.) Which factors do the activators contact? The answer seems to be that many factors can be targets, but the one that was discovered first was TFIID.

Recruitment of TFIID

In 1990, Keith Stringer, James Ingles, and Jack Greenblatt performed a series of experiments to identify the factor that binds to the acidic transcription-activating domain of the herpesvirus transcription factor **VP16.** These workers expressed the VP16 transcription-activating domain as a fusion protein with the *Staphylococcus aureus* protein A, which binds tightly and specifically to immunoglobulin IgG.

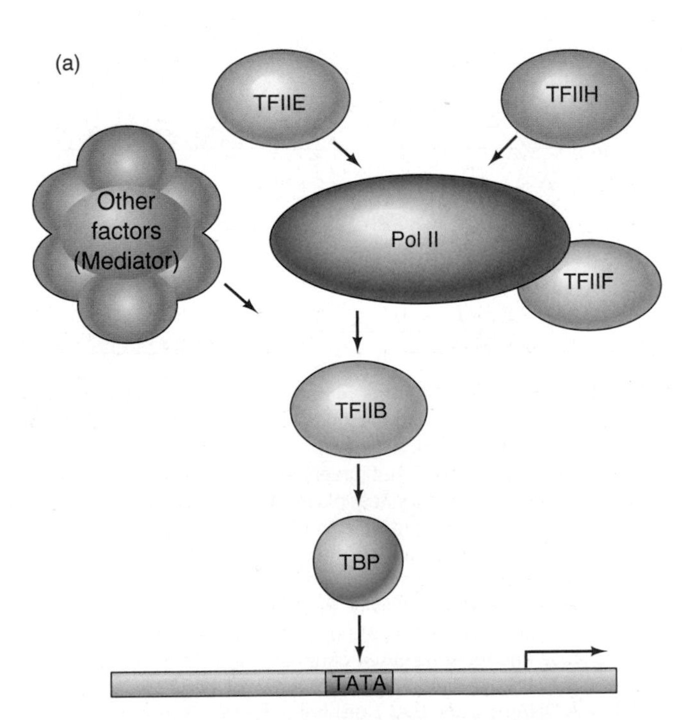

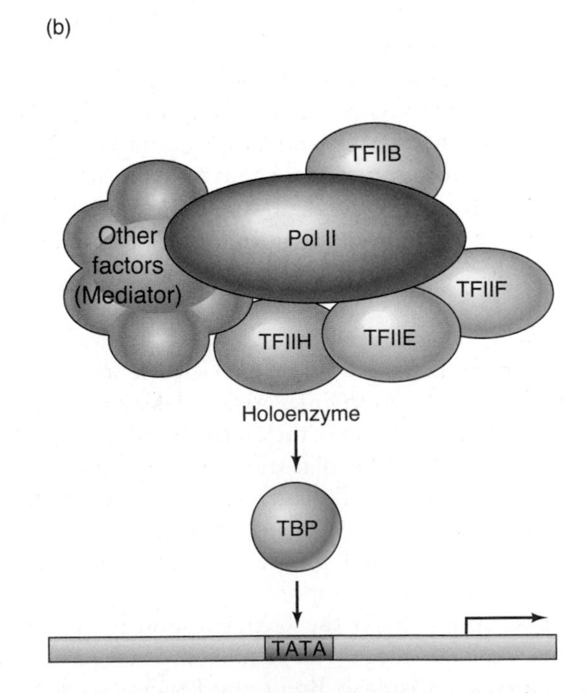

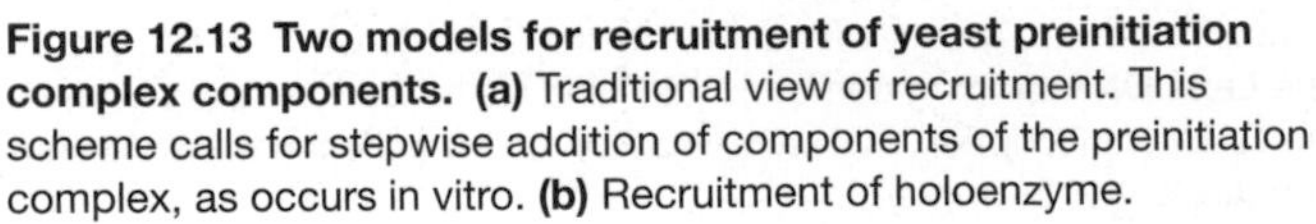
Figure 12.13 Two models for recruitment of yeast preinitiation complex components. (a) Traditional view of recruitment. This scheme calls for stepwise addition of components of the preinitiation complex, as occurs in vitro. **(b)** Recruitment of holoenzyme. Here, TBP binds first, then the holoenzyme binds to form the preinitiation complex. (*Source:* Adapted from Koleske, A.J. and R.A. Young, An RNA polymerase II holoenzyme responsive to activators. *Nature* 368:466, 1994.)

They immobilized the fusion protein (or protein A by itself) on an agarose IgG column and used these as affinity columns to "fish out" proteins that interact with the VP16-activating domain. To find out what proteins bind to the VP16-activating domain, they poured HeLa cell nuclear extracts through the columns containing either protein A by itself or the protein A/VP16-activating domain fusion protein. Then they used run-off transcription (Chapter 5) to assay various fractions for ability to transcribe the adenovirus major late locus accurately in vitro. They found that the flow-through from the protein A column still had abundant ability to support transcription, indicating no nonspecific binding of any essential factors to protein A. However, when they tested the flow-through from the protein A/VP16-activating domain column they found no transcription activity until they added back the proteins that bound to the column. Thus, some factor or factors essential for in vitro transcription bound to the VP16-activating domain.

Stringer and colleagues knew that TFIID was rate-limiting for transcription in their in vitro system, so they suspected that TFIID was the factor that bound to the affinity column. To find out, they depleted a nuclear extract of TFIID by heating it, then added back the material that bound to either the protein A column or the column containing the protein A/VP16-activating domain. Figure 12.14 shows that the material that bound to protein A by itself could not reconstitute the activity of a TFIID-depleted extract, but the material that bound to the protein A/VP16-activating domain could. This strongly suggested that TFIID binds to the VP16-activating domain.

To check this conclusion, Stringer and colleagues first showed that the material that bound to the VP16-activating domain column behaved just like TFIID on DEAE-cellulose ion-exchange chromatography. Then they assayed the material that bound to the VP16-activating domain column for the ability to substitute for TFIID in a template commitment experiment. In this experiment, they formed preinitiation complexes on one template, then added a second template to see whether it could also be transcribed. Under these experimental conditions, the commitment to transcribe the second template depended on TFIID. These workers found that the material that bound to the VP16-activating domain column could shift commitment to the second template, but the material that bound to the protein A column could not. These, and similar experiments performed with yeast nuclear extracts, provided convincing evidence that TFIID is the important target of the VP16 transcription-activating domain in this experimental system.

SUMMARY The acidic transcription-activating domain of the herpesvirus transcription factor VP16 binds to TFIID under affinity chromatography conditions.

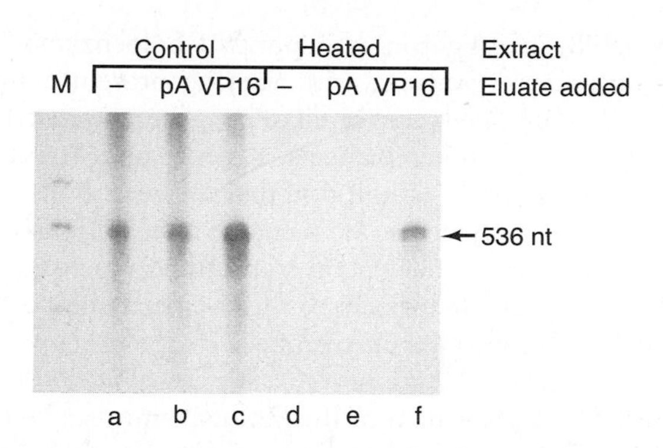

Figure 12.14 Evidence that an acidic activation domain binds TFIID. Stringer and colleagues fractionated a HeLa cell extract by affinity chromatography with a resin containing a fusion protein composed of protein A fused to the VP16-activating domain, or a resin containing just protein A. Then they eluted the proteins bound to each affinity column and tested them for ability to restore in vitro run-off transcription activity to an extract that had been heated to destroy TFIID specifically. Lanes a–c are controls in which the extract had not been heated. Because TFIID was still active, all lanes showed activity. Lanes d–f contained heated extract supplemented with: nothing (–), the eluate from the protein A column (pA), or the eluate from the column that contained the fusion protein composed of protein A and the transcription-activating domain of the VP16 protein (VP16). Only the eluate from the column containing the VP16 fusion protein could replace the missing TFIID and give an accurately initiated run-off transcript with the expected length (536 nt, denoted at right). Thus, TFIID must have bound to the VP16 transcription-activating domain in the affinity column. (*Source:* Stringer, K.F., C.J. Ingles, and J. Greenblatt, Direct and selective binding of an acidic transcriptional activation domain to the TATA-box factor TFIID. *Nature* 345 (1990) f. 2, p. 784. Copyright © Macmillan Magazines Ltd.)

Recruitment of the Holoenzyme

In Chapter 11 we learned that RNA polymerase II can be isolated from eukaryotic cells as a holoenzyme—a complex containing a subset of general transcription factors and other polypeptides. Much of our discussion so far has been based on the assumption that activators recruit general transcription factors one at a time to assemble the preinitiation complex. But it is also possible that activators recruit the holoenzyme as a unit, leaving only a few other proteins to be assembled at the promoter. In fact, there is good evidence that recruitment of the holoenzyme really does occur.

In 1994, Anthony Koleske and Richard Young isolated from yeast cells a holoenzyme that contained polymerase II, TFIIB, F, and H, and SRB2, 4, 5, and 6. They went on to demonstrate that this holoenzyme, when supplemented with TBP and TFIIE, could accurately transcribe a template bearing a *CYC1* promoter in vitro. Finally, they showed that the activator GAL4-VP16 could activate this transcription. Because the holoenzyme was provided intact, this last finding suggested that the activator recruited the intact holoenzyme to the promoter rather than building it up step by step on the promoter (recall Figure 12.13).

By 1998, investigators had purified holoenzymes from many different organisms, with varying protein compositions. Some contained most or all of the general transcription factors and many other proteins. Koleske and Young suggested the simplifying assumption that the yeast holoenzyme contains RNA polymerase II, a coactivator complex called Mediator, and all of the general transcription factors except TFIID and TFIIE. In principle, this holoenzyme could be recruited as a preformed unit, or piece by piece.

Evidence for Recruitment of the Holoenzyme as a Unit In 1995, Mark Ptashne and colleagues added another strong argument for the holoenzyme recruitment model. They reasoned as follows: If the holoenzyme is recuited as a unit, then interaction between *any* part of an activator (bound near a promoter) and *any* part of the holoenzyme should serve to recruit the holoenzyme to the promoter. This protein–protein interaction need not involve the normal transcription-activating domain of the activator, nor the activator's normal target on a general transcription factor. Instead, any contact between the activator and the holoenzyme should cause activation. On the other hand, if the preinitiation complex must be built up protein by protein, then an abnormal interaction between an activator and a seemingly unimportant member of the holoenzyme should not activate transcription.

Ptashne and colleagues took advantage of a chance observation to test these predictions. They had previously isolated a yeast mutant with a point mutation that changed a single amino acid in a holoenzyme protein (GAL11). They named this altered protein GAL11P (for potentiator) because it responded strongly to weak mutant versions of the activator GAL4. Using a combination of biochemical and genetic analysis, they found the source of the potentiation by GAL11P: The alteration in GAL11 caused this protein to bind to a region of the dimerization domain of GAL4, between amino acids 58 and 97. Because GAL11 (or GAL11P) is part of the holoenzyme, this novel association between GAL11P and GAL4 could recruit the holoenzyme to GAL4-responsive promoters, as illustrated in Figure 12.15. We call the association between GAL11P and GAL4 novel because the part of GAL11P involved is normally functionally inactive, and the part of GAL4 involved is in the dimerization domain, not the activation domain. It is highly unlikely that any association between these two protein regions occurs normally.

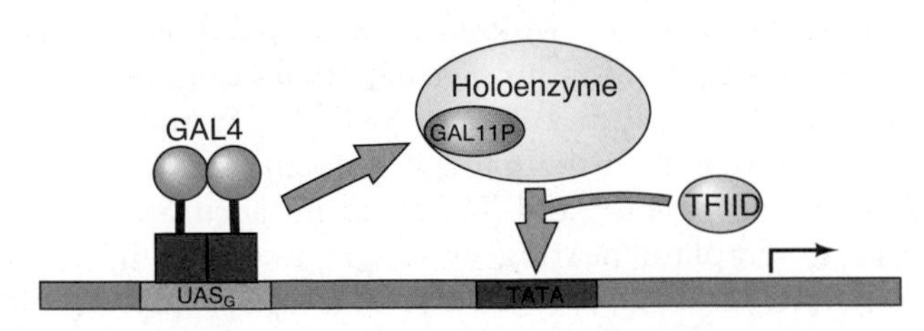

Figure 12.15 Model for recruitment of the GAL11P-containing holoenzyme by the dimerization domain of GAL4. The dimerization domain of GAL4 binds (orange arrow) to GAL11P (purple) in the holoenzyme. This causes the holoenzyme, along with TFIID, to bind to the promoter, activating the gene.

To test the hypothesis that the region of GAL4 between amino acids 58 and 97 is responsible for activation by GAL11P, Ptashne and colleagues performed the following experiment. Using gene-cloning techniques, they made a plasmid encoding a fusion protein containing the region between amino acids 58 and 97 of GAL4 and the LexA DNA-binding domain. They introduced this plasmid into yeast cells along with a plasmid encoding either GAL11 or GAL11P, and a plasmid bearing two binding sites for LexA upstream of a GAL1 promoter driving transcription of the *E. coli lacZ* reporter gene. Figure 12.16 summarizes this experiment and shows the results. The LexA-GAL4(58–97) protein is ineffective as an activator when wild-type GAL11 is in the holoenzyme (Figure 12.16a), but works well as an activator when GAL11P is in the holoenzyme (Figure 12.16b).

If activation is really due to interaction between LexA-GAL4(58–97) and GAL11P, we would predict that fusing the LexA DNA-binding domain to GAL11 would also cause activation, as illustrated in Figure 12.16c. In fact, this construct did cause activation, in accord with the hypothesis. Here, no novel interaction between LexA-GAL4 and GAL11P was required because LexA and GAL11 were already covalently joined.

The simplest explanation for these data is that activation, at least in this system, can operate by recruitment of the holoenzyme, rather than by recruitment of individual general transcription factors. It is possible, but not likely, that GAL11 is a special protein whose recruitment causes the stepwise assembly of a preinitiation complex. But it is much more likely that association between an activator and any component of the holoenzyme can recruit the holoenzyme and thereby cause activation. Ptashne and colleagues conceded that TFIID is an essential part of the preinitiation complex, but is apparently not part of the yeast holoenzyme. They proposed that TFIID might have bound to the promoter cooperatively with the holoenzyme in their experiments.

On the other hand, at least two lines of evidence suggest that the holoenzyme is not recruited as a whole. First, David Stillman and colleagues have performed kinetic studies of the binding of various factors to the *HO* promoter region in yeast. These studies showed that one part of the holoenzyme, Mediator, binds to the promoter earlier in G1 phase than does RNA polymerase II. Thus, the holoenzyme is certainly not binding as a complete unit, at least to this yeast promoter.

Second, Roger Kornberg and colleagues reasoned that, if the holenzyme binds as a unit to promoters, one should find all the components of the holoenzyme in roughly equal amounts in cells. They also knew that determining the concentrations of proteins in cells is tricky. One

(a) **WT cells**

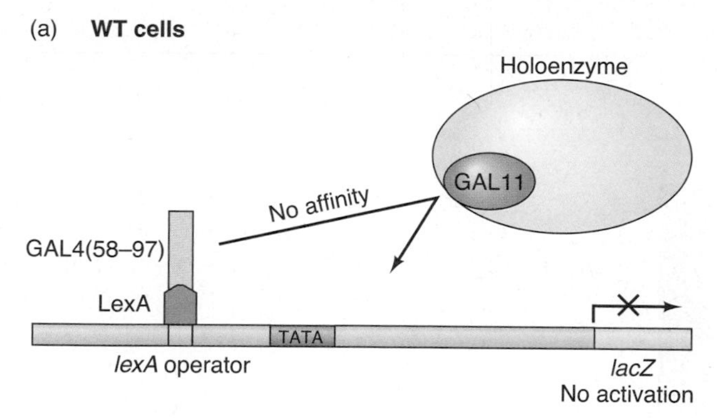

(b) ***GAL11P* cells**

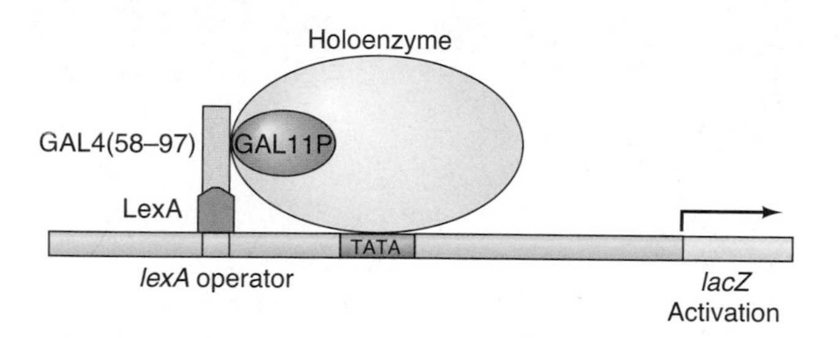

(c) ***gal11* cells**

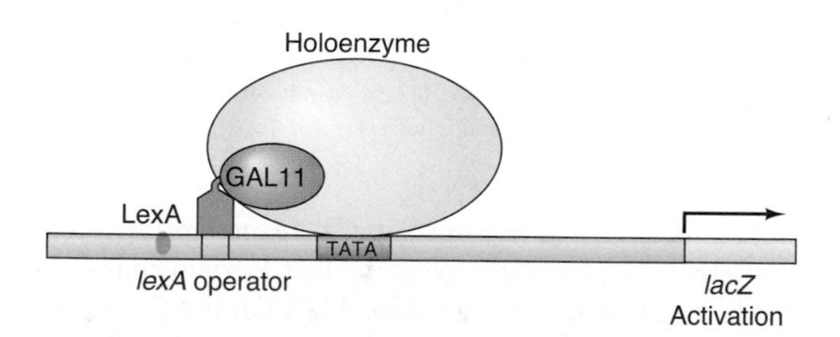

Figure 12.16 Activation by GAL11P and GAL11-LexA. Ptashne and colleagues transformed cells with a plasmid containing a *lexA* operator 50 bp upstream of a promoter driving transcription of a *lacZ* reporter gene, plus the following plasmids: **(a)** a plasmid encoding amino acids 58–97 of GAL4 coupled to the DNA-binding domain of LexA plus a plasmid encoding wild-type GAL11; **(b)** a plasmid encoding amino acids 58–97 of GAL4 coupled to the DNA-binding domain of LexA plus a plasmid encoding GAL11P; **(c)** a plasmid encoding GAL11 coupled to the DNA-binding domain of LexA. They assayed for production of the *lacZ* product, β-galactosidase. Results: **(a)** The GAL4(58–97) region did not interact with GAL11, so no activation occurred. **(b)** The GAL4(58–97) region bound to GAL11P, recruiting the holoenzyme to the promoter, so activation occurred. **(c)** The LexA-GAL11 fusion protein could bind to the *lexA* operator, recruiting the holoenzyme to the promoter, so activation occurred. (*Source:* Adapted from Barberis A., J. Pearlberg, N. Simkovich, S. Farrell, P. Resnagle, C. Bamdad, G. Sigal, and M. Ptashne, with a component of the polymerase II holoenzyme suffices for gene activation. *Cell* 81:365, 1995.)

cannot do it by measuring mRNA levels because of wide variation in posttranscriptional events such as mRNA degradation and nuclear export. Indeed, concentrations of mRNAs and their respective protein products can deviate from expected values by up to 20- or 30-fold. One can separate proteins by two-dimensional gel electrophoresis and determine their concentrations by mass spectrometry (Chapter 24), but that method is not sensitive enough for proteins, such as transcription factors, found in very low concentrations in vivo.

So Kornberg and colleagues chose a method that combines high sensitivity and great accuracy. They began by using gene cloning techniques to attach "TAP" tags to the genes encoding seven different components of the polymerase II holoenzyme. These included RNA polymerase II, Mediator, and five general transcription factors. The TAP tag contains a region from *Staphylococcus* protein A (Chapter 4) that binds to antibodies of the IgG class. Thus, Kornberg and colleagues could dot-blot cell extracts from the yeast strains carrying genes for TAP-tagged proteins, then probe the blots with an antiperoxidase antibody. The TAP tag on a protein on the blot bound to the antibody, which in turn bound to peroxidase added later, which in turn converted a peroxidase substrate to a chemiluminescent product that could be detected photographically (Chapter 5).

The intensities of the bands on the film corresponded to the concentration of TAP-tagged proteins on the blots. With serial dilutions of each extract, these band intensities could be converted to concentrations of each protein per cell by comparing them with the results of a blot of known amounts of a standard, GST-TAP. Figure 12.17 shows sample results. It is clear from the wild-type lane with no TAP-tagged proteins that the background of this method is essentially zero, which is important for accuracy of quantification. It is also clear that there is considerably more RNA polymerase II than Med8, one of the subunits of Mediator. Quantification (Figure 12.17b) showed five to six times as much Rpb3 as any of the subunits of Mediator or of TFIIH. Table 12.1 presents a quantification of the amounts of TFIIF, TFIIE, TFIIB, and TFIID, in addition to the proteins considered in Figure 12.17. Again, RNA polymerase was more abundant than any of the other factors, but the four other general transcription factors were more abundant than either Mediator or TFIIH.

Because all of the components of the holoenzyme are *not* found in roughly equal amounts, it is unlikely that the holoenzyme binds to most promoters as a unit. It is still possible, though, that it is recruited to some promoters as a unit.

SUMMARY Activation, at least in certain promoters in yeast, appears to function by recruitment of the holoenzyme, rather than by recruitment of individual components of the holoenzyme one at a time. However, other evidence suggests that recruitment of the holoenzyme as a unit is not common.

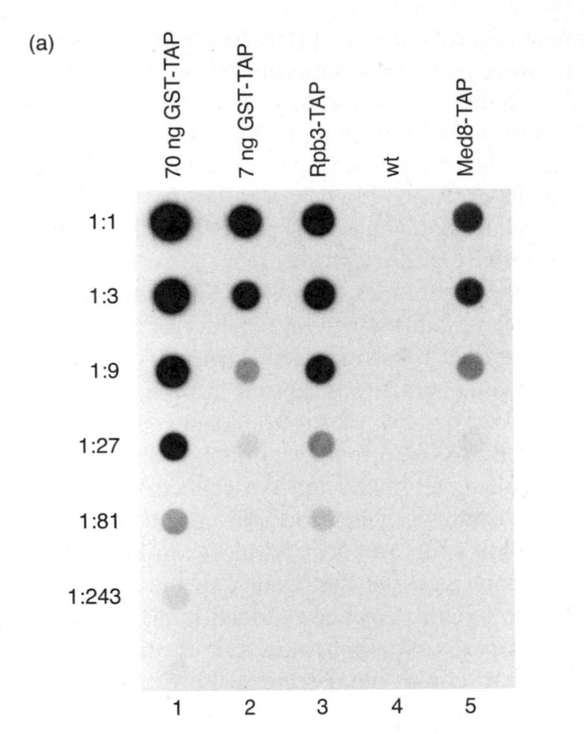

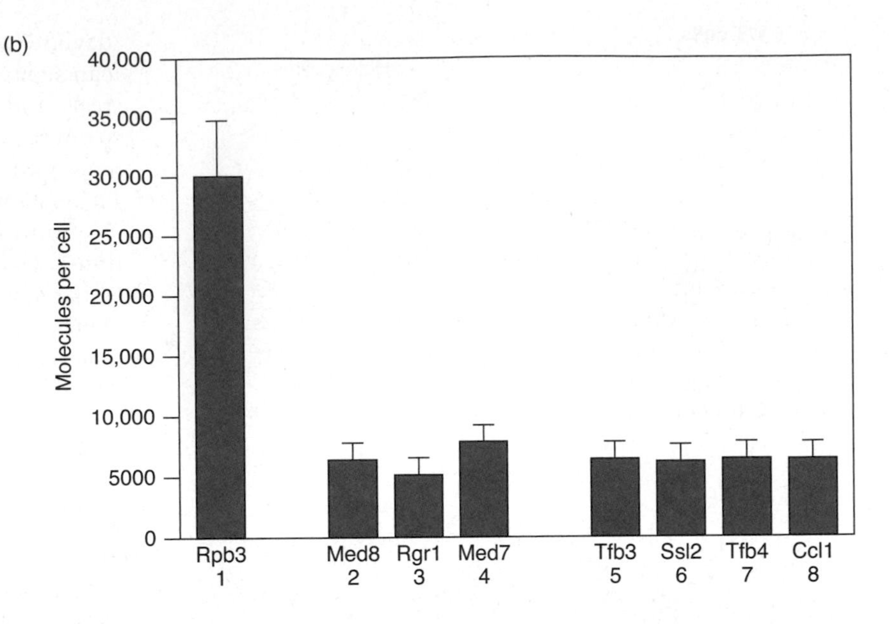

Figure 12.17 Determining the concentration of holoenzyme subunits by dot blotting. **(a)** Dot blot results. Kornberg and colleagues dot-blotted serial dilutions of extracts from cells bearing chimeric genes encoding holoenzyme subunits tagged with TAP sequences. The TAP sequences contained two *Staphylococcus* A protein sequences that bind to IgG immunoglobulins. The investigators reacted TAP sequences on the dot blot with an IgG immunoglobulin directed against peroxidase (rabbit antiperoxidase IgG). The IgG was in turn detected photographically with peroxidase and a substrate that becomes chemiluminescent on reaction with peroxidase. The dilutions are given at left. Columns 1 and 2 contained serial dilutions of two different amounts of GST-TAP, as given at top. Columns 3–5 contained serial dilutions of extracts from cells containing TAP-tagged Rpb3, wild-type cells with no TAP tags, and cells containing TAP-tagged Med8, respectively. **(b)** Cellular concentrations of Rpb3 (bar 1), three subunits of Mediator (bars 2–4), and four subunits of TFIIH (bars 5–8), determined by dot blotting. (*Source: Journal of Biological Chemistry* by Borggrefe et al. Copyright 2001 by Am. Soc. For Biochemistry & Molecular Biol. Reproduced with permission of Am. Soc. For Biochemistry & Molecular Biol. in the format Textbook via Copyright Clearance Center.)

Table 12.1 Number of Selected Protein Molecules per Yeast Cell

Protein	Copies per Cell
RNA polymerase II (Rpb3)	30,000
TFIIF (Tfg2)	24,000
TFIIE (Tfa2)	24,000
TFIIB (Sua7)	20,000
TFIID (TBP)	20,000
Mediator (Med8)	6000
TFIIH (Tfb3)	6000

Source: Borggrefe, T., R. Davis, A. Bareket-Samish, and R.D. Kornberg, Quantitation of the RNA polymerase II transcription machinery in yeast. *Journal of Biological Chemistry* 276 (2001): 47150–53, tll. Reprinted with permission.

12.5 Interaction Among Activators

We have seen several examples of crucial interactions among different types of transcription factors. Obviously, the general transcription factors must interact to form the preinitiation complex. But activators and general transcription factors also interact. For example, we have just learned that GAL4 and other activators interact with TFIID and other general transcription factor(s). In addition, activators usually interact with one another in activating a gene. This can occur in two ways: Individual factors can interact to form a protein dimer to facilitate binding to a single DNA target site. Alternatively, specific factors bound to different DNA target sites can collaborate in activating a gene.

Dimerization

We have already mentioned a number of different means of interaction between protein monomers in DNA-binding proteins. In Chapter 9 we discussed the helix-turn-helix proteins such as the λ repressor and observed that the interaction between the monomers of this protein place the recognition helices of the two monomers in just the right position to interact with two major grooves exactly one helical turn apart. The recognition helices are antiparallel to each other so they can recognize the two parts of a palindromic DNA target. Earlier in this chapter we discussed the coiled coil dimerization domains of the GAL4 protein and the similar leucine zippers of the bZIP proteins.

In Chapter 9 we discussed the advantage that a protein dimer has over a monomer in binding to DNA. This advantage can be summarized as follows: The affinity of binding between a protein and DNA varies with the square of the free energy of binding. Because the free energy depends on the number of protein–DNA contacts, doubling the contacts by using a protein dimer instead of a monomer quadruples the affinity between the protein and the DNA. This is significant because most activators have to operate at very low concentrations. The fact that the great majority of DNA-binding proteins are dimers is a testament to the advantage of this arrangement. We have seen that some activators, such as GAL4, form homodimers; others, such as the thyroid hormone receptor, form heterodimers.

SUMMARY Dimerization is a great advantage to an activator because it increases the affinity between the activator and its DNA target. Some activators form homodimers, but others function as heterodimers.

Action at a Distance

We have seen that both bacterial and eukaryotic enhancers can stimulate transcription, even though they are located some distance away from the promoters they control. How does this action at a distance occur? In Chapter 9 we learned that the evidence favors looping out of DNA in between the two remote sites to allow bacterial DNA-binding proteins to interact. We will see that this same scheme also seems to apply to eukaryotic enhancers.

Among the most reasonable hypotheses to explain the ability of enhancers to act at a distance are the following (Figure 12.18): (a) An activator binds to an enhancer and changes the topology, or shape, of the whole DNA duplex, perhaps by causing supercoiling. This in turn opens the promoter up to general transcription factors. (b) An activator binds to an enhancer and then slides along the DNA until it encounters the promoter, where it can activate transcription by virtue of its direct contact with the promoter DNA. (c) An activator binds to an enhancer and, by looping out DNA in between, interacts with proteins at the promoter, stimulating transcription. (d) An activator binds to an enhancer and a downstream segment of DNA to form a DNA loop. By enlarging this loop, the protein tracks toward the promoter. When it reaches the promoter, it interacts with proteins there to stimulate transcription.

Notice that the first two of these models demand that the two elements, enhancer and promoter, be on the same DNA molecule. A change in topology of one DNA molecule cannot influence transcription on a second, and an activator cannot bind to an enhancer on one DNA and slide onto a second molecule that contains the promoter. On the other hand, the third model simply requires that the enhancer and promoter be relatively near each other, not necessarily on the same molecule. This is because the essence of the looping model is not the looping itself, but the interaction between the proteins bound to remote sites. In principle, this would work just as well if the proteins were bound to two sites on different DNA molecules, as long as the molecules were tethered together somehow so they would not float apart and prevent interactions between the bound proteins. Figure 12.19 shows how this might happen.

Thus, if we could arrange to put an enhancer on one DNA molecule and a promoter on another, and get the two molecules to link together in a **catenane**, (circles linked as in a chain) we could test the hypotheses. If the enhancer still functioned, we could eliminate the first two. Marietta Dunaway and Peter Dröge did just that. They constructed a plasmid with the *Xenopus laevis* rRNA promoter plus an rRNA minigene on one side and the rRNA enhancer on the other, with the λ phage integration sites, *attP* and *attB*, in between. These are targets of site-specific recombination, so placing them on the same molecule and allowing recombination produces a catenane, as illustrated in Figure 12.19.

Finally, these workers injected combinations of plasmids into *Xenopus* oocytes and measured their transcription by quantitative S1 mapping. The injected plasmids were the catenane, the unrecombined plasmid containing both enhancer and promoter, or two separate plasmids, each containing either the enhancer or promoter. In quantitative S1 mapping, a reference plasmid is needed to correct for the variations among oocytes. In this case, the reference plasmid contained an rRNA minigene (called ψ52) with a 52-bp insert, whereas the rRNA minigenes of the test plasmids (called ψ40) all contained a 40-bp insert. Dunaway and Dröge included probes for both these minigenes in their assay, so we expect to see two signals, 12 nt apart, if both genes are transcribed. We are most interested in the *ratio* of these two signals, which tells us how well each test plasmid is transcribed relative to the reference plasmid, which should behave the same in each case.

Figure 12.20a shows the test plasmid results in the lanes marked "a" and the reference plasmid results in the lanes marked "b." The plasmids used to produce the transcripts in each lane are pictured in panel (b). Note that the same plasmids were used in both lane a and lane b of each set in panel (a). Only the probes were different. These were the results: Lanes 1 show that when the plasmid contained the promoter alone, the test plasmid signal was weaker than the reference plasmid signal. That is because the test probe was less radioactive than the reference probe. Lanes 2 demonstrate that the enhancer adjacent to the promoter (its normal position) greatly enhanced transcription in the test plasmid—its signal was much stronger than the reference plasmid signal. Lanes 3 show that the enhancer still worked, though not quite as well, when placed opposite the promoter on the plasmid. Lanes 4 are the most important. They show that the enhancer still worked when it was on a separate plasmid that formed a catenane with the

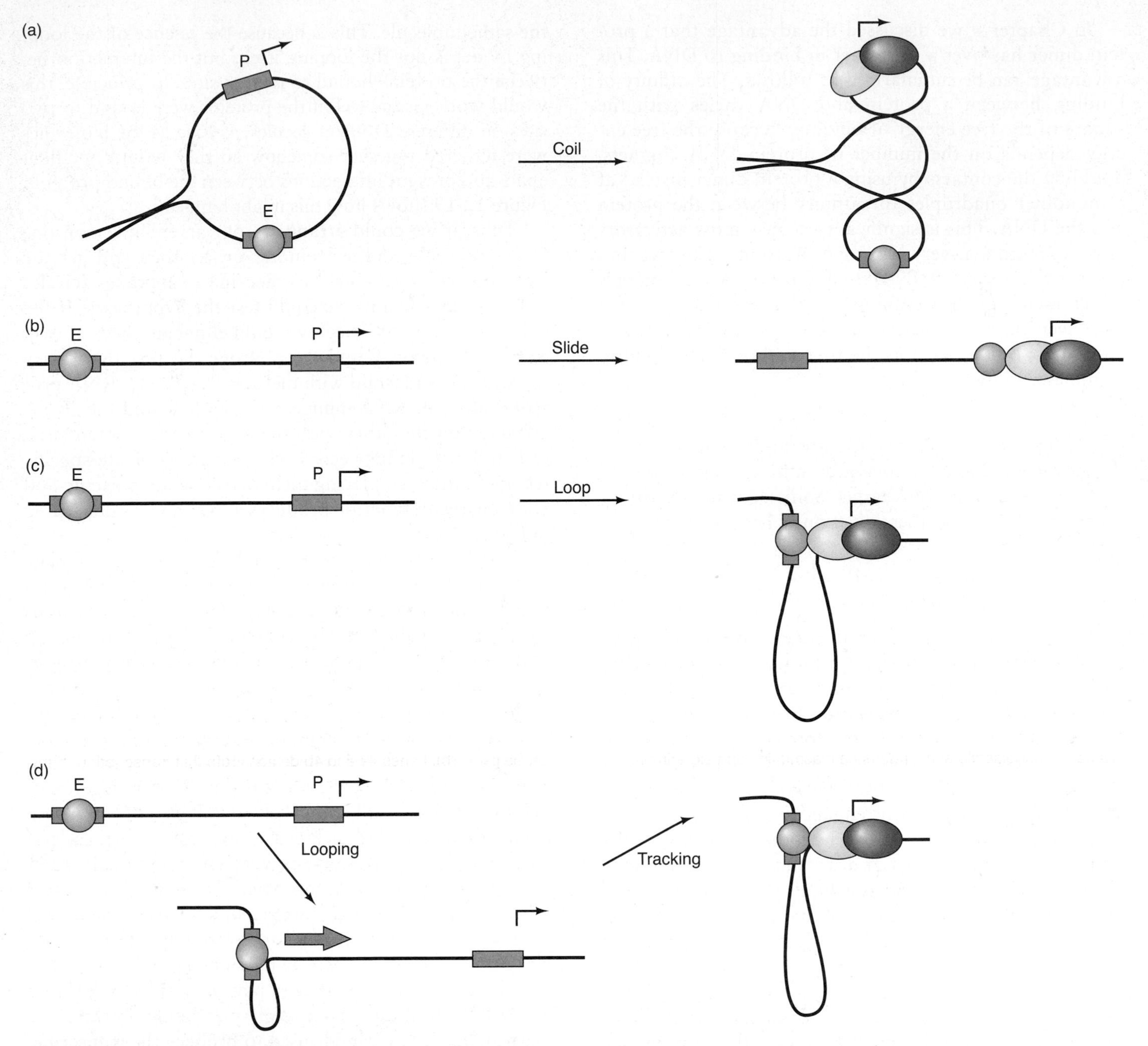

Figure 12.18 Four hypotheses of enhancer action. (a) Change in topology. The enhancer (E, blue) and promoter (P, orange) are both located on a loop of DNA. Binding of a gene-specific transcription factor (green) to the enhancer causes supercoiling that facilitates binding of general transcription factors (yellow) and polymerase (red) to the promoter. **(b)** Sliding. A transcription factor binds to the enhancer and slides down the DNA to the promoter, where it facilitates binding of general transcription factors and polymerase. **(c)** Looping. A transcription factor binds to the enhancer and, by looping out the DNA in between, binds to and facilitates the binding of general transcription factors and polymerase to the promoter. **(d)** Facilitated tracking. A transcription factor binds to the enhancer and causes a short DNA segment to loop out downstream. Increasing the size of this loop allows the factor to track along the DNA until it reaches the promoter, where it can facilitate the binding of general transcription factors and RNA polymerase.

plasmid containing the promoter. Lanes 5 verify that the enhancer did not work if it was on a separate plasmid *not* linked in a catenane with the promoter plasmid. Finally, lanes 6 show that the enhancement observed in lanes 4 was not due to a small amount of contamination by unrecombined plasmid. In lanes 6, the investigators added 5% of such a plasmid and observed no significant increase in the test plasmid signal.

These results lead to the following conclusion about enhancer function: The enhancer does not need to be on the same DNA with the promoter, but it does need to be able to approach the promoter, so the proteins bound to

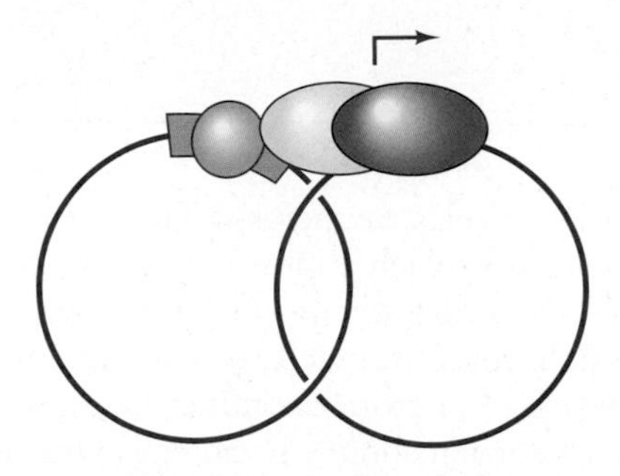

Figure 12.19 Interaction between enhancer and promoter on two plasmids linked in a catenane. Hypothetical interaction between an activator (green) bound to an enhancer (blue) on one plasmid, and general transcription factors (yellow) and RNA polymerase (red) bound to the promoter (not visible beneath the bent arrow) in the other plasmid of the catenane.

enhancer and promoter can interact. This is difficult to reconcile with models involving supercoiling or sliding (Figure 12.18a and b), but is consistent with the DNA looping and facilitated tracking models (Figure 12.18c and d). In the catenane, no looping or tracking is required because the enhancer and promoter are on different DNA molecules; instead, protein–protein interactions can occur without looping, as illustrated in Figure 12.19a.

If enhancer action requires DNA looping, then we should be able to observe it directly, using appropriate tools. A technique called **chromosome conformation capture (3C)** provides just such a tool. This method, illustrated in Figure 12.21, is designed to test whether two remote DNA regions, such as an enhancer and a promoter, are brought together—by interactions between DNA-binding

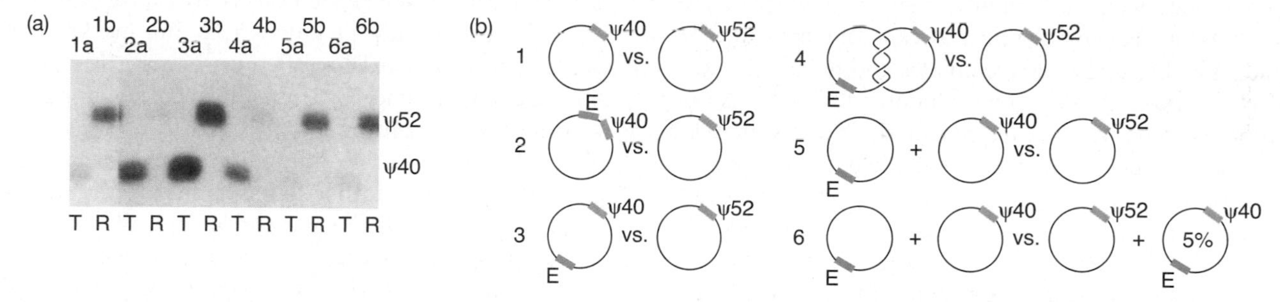

Figure 12.20 Results of the catenane experiment. Dunaway and Dröge injected mixtures of plasmids into *Xenopus* oocytes and measured transcription rates by quantitative S1 mapping. They injected a test plasmid and a reference plasmid in each experiment and assayed for transcription of each with separate probes. **(a)** Experimental results. The results of the test (T) and reference (R) assays are given in lanes a and b, respectively, of each experiment. The plasmids injected in each experiment are given in panel **(b).** For example, the plasmids used in the experiments in lanes 1a and 1b are labeled 1. The plasmids on the left, labeled Ψ40 (or Ψ40 plus another plasmid), are the test plasmids. The ones on the right, labeled Ψ52, are the reference plasmids. The 40 and 52 in these names denote the size inserts each has to distinguish it from the other. Both plasmids were injected and then assayed with the test probe (lane 1a) or the reference probe (lane 1b). Lanes 4a and 4b demonstrate that transcription of the catenane with the enhancer on one plasmid and the promoter on the other is enhanced relative to transcription of the plasmid containing just the promoter (lanes 1a and 1b). This is evident in the much higher ratio of the signals in lanes 4a and 4b relative to the ratio of the signals in lanes 1a and 1b. (*Source:* Adapted from Dunaway M. and P. Dröge, Transactivation of the *Xenopus* rRNA gene promoter by its enhancer. *Nature* 341 (19 Oct 1989) p. 658, f. 2a. Copyright © Macmillan Magazines Ltd.)

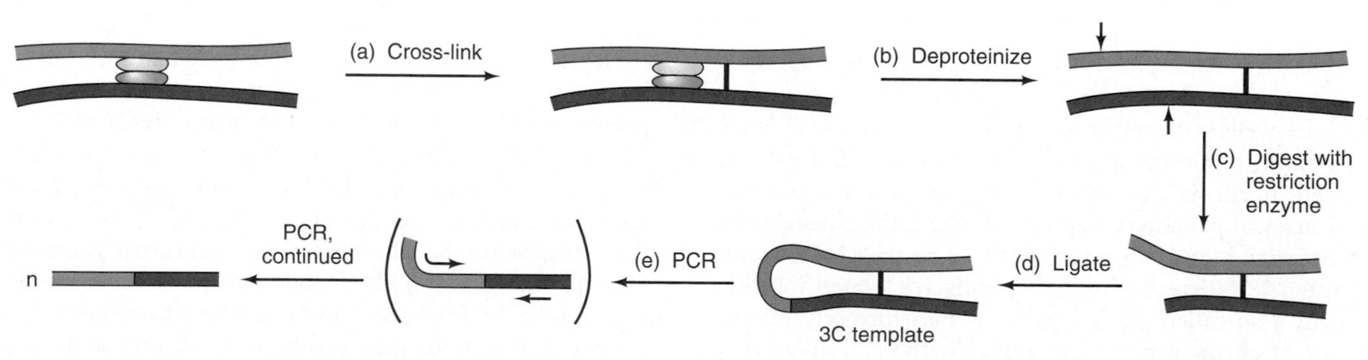

Figure 12.21 Chromatin conformation capture (3C). (a) Begin with chromatin in which you believe two sites are brought together by interaction between two DNA-binding proteins (green and yellow). The two segments of chromosome (red and blue) can be on separate chromosomes, or the same chromosome. Cross-link the two separate chromosome segments with formaldehyde. **(b)** Deproteinize the chromatin. **(c)** Digest the DNA with a restriction enzyme. Arrows show two restriction sites. **(d)** Ligate the nearby DNA ends under conditions (low DNA concentration) in which intramolecular ligation is favored. This yields the 3C template. **(e)** PCR on the 3C template with primers indicated by the short arrows yields a significant amount of PCR product, showing that the two chromosome segments represented by the primers are probably close together in this chromatin.

BOX 12.1

Genomic Imprinting

Because most eukaryotes are diploid organisms, you would probably predict that it doesn't matter which allele of any gene pair came from the mother and which came from the father. In most cases, you would be right, but there are important exceptions. The first evidence for one very important class of exceptions came from studies with mouse eggs just after fertilization, in which the maternal and paternal nuclei had not yet fused. At this stage, the maternal nucleus can be removed and replaced with a second paternal nucleus. Similarly, the paternal nucleus can be removed and replaced with a second maternal nucleus. In either case, the embryo will have chromosomes contributed by only one parent. In principle, that should not have made a big difference, because the parental mice were from an inbred strain in which all the individuals are genetically identical (except, of course, for the XY versus XX difference between males and females).

In fact, however, it made a tremendous difference. All of these embryos died during development, most at a very early stage. Those that made it the longest before dying showed an interesting difference, depending on whether their genes came from the mother or the father. Those with genes derived only from the mother had few abnormalities in the embryo itself, but had abnormal and stunted placentas and yolk sacs. Embryos with genes derived only from the father were small and poorly formed, but had relatively normal placentas and yolk sacs. How can we account for this difference if the genes contributed by the mother and father are identical? One explanation for this phenomenon is that the genes—that is, the base sequences of the genes—are identical, but they are somehow modified, or imprinted, differently in males and females.

Bruce Cattanach provided more evidence for imprinting with his studies on mice with fused chromosomes. For example, in some mice, chromosome 11 is fused, so it cannot separate during mitosis or meiosis. This means that some gametes produced by such a mouse will have two copies of chromosome 11, while some will have none. These mice made it possible for Cattanach to produce offspring with both chromosomes 11 from the father (using sperm with a double dose of chromosome 11 and eggs with no chromosome 11, or both from the mother (by reversing the procedure). Again, if the parental source of the chromosome did not matter, these offspring should have been normal. But they were not. In cases where both chromosomes came from the mother, the pups were abnormally small; if both chromosomes came from the father, the pups were giants.

Furthermore, these experiments demonstrated that the imprint is erased at each generation. That is, a runty male mouse whose chromosomes 11 came from his mother generally would produce normal-size offspring himself. The production of male gametes somehow erased the maternal imprint.

Genomic imprinting also occurs in humans, occasionally with tragic results. Inheritance of a deleted chromosome 15 from the father is associated with Prader-Willi syndrome, in which the patient is typically mentally impaired, short, and obese, because of an uncontrollable appetite. The lack of a particular part of the paternal copy of chromosome 15 is important because the gene associated with Prader-Willi syndrome is imprinted, and therefore inactivated, on the maternal chromosome 15. Thus, deletion of the paternal allele, and imprinting of the maternal allele, leaves no functioning copy of the gene. By contrast, inheritance of a deleted chromosome 15 from the mother is connected with Angelman syndrome, characterized by a large mouth and abnormally red cheeks, as well as by severe mental impairment, with inappropriate laughter and jerky movements. The lack of a particular part of the maternal

proteins, for example. First, chromatin with suspected DNA looping is fixed with formaldehyde to form covalent bonds between chromatin regions that are in close contact. (Chromatin is the natural state of DNA within a eukaryotic cell. It consists of DNA bound to an approximately equal mass of protein (Chapter 13). Next, the chromatin is deproteinized and digested with a restriction enzyme (Chapter 4). Next, the free DNA ends are ligated together to form a so-called 3C template. If two formerly remote regions of chromatin are in contact with each other, they will be ligated together in the 3C template, and PCR primers specific for these two regions will produce a relatively short PCR product. The more prevalent this product, the more often the two chromatin regions are in contact. This method can be used to detect either intra- or interchromosomal interactions.

Karl Pfeifer and colleagues exploited the 3C method to demonstrate interaction between an enhancer and a promoter. They focused on the mouse ***Igf2/H19* locus** (Figure 12.22a). The *Igf2* gene, driven by three promoters, spaced 2 kb apart, encodes IGF2 (**interferon-like growth factor 2**), and *H19* encodes a noncoding RNA. Interestingly, the *Igf2* gene on the male chromosome is turned on, but the homologous gene on the female chromosome is silenced. Conversely, the *H19* gene on the female chromosome is on, but the homologous gene on the male chromosome is off. This chromosome-specific behavior is explained by **imprinting,** which is established during gametogenesis by methylation of the **imprinting control region (*ICR*)**. Box 12.1 gives further insight into the biology of imprinting, and this locus in particular. Later in this chapter, we will learn more about the mechanism of imprinting.

copy of chromosome 15 is important because the gene, or genes, associated with Angelman syndrome are imprinted, and therefore inactivated, on the paternal chromosome. Thus, deletion of the maternal copies, and imprinting of the paternal copies, leaves no functioning copies of these genes.

How can the DNA be modified in a reversible way so the imprint can be erased? The evidence points to DNA methylation. First, experiments show that genes derived from males and females are methylated differently, and this methylation correlates with gene activity. In general, methylated genes are found in females, and the methylated genes are inactivated. (However, note that in the *Igf2* example in the main text, it is an insulator that gets methylated in male mice, and this allows *Igf2* expression, whereas the unmethylated insulator in females blocks *Igf2* expression.)

Furthermore, methylation can be reversed. Philip Leder and colleagues used transgenic mice (Chapter 5) to follow the methylated state of a transgene as it moves through gametogenesis (the production of sperm or eggs) and into the developing embryo. These experiments revealed that the methyl groups on the transgene are removed in the early stages of gametogenesis in both males and females. The developing egg then establishes the maternal methylation pattern before the oocyte is completely mature. In the male, some methylation occurs during sperm development, but this methylation pattern is further modified in the developing embryo. Thus, methylation has all the characteristics we expect in an imprinting mechanism: It occurs differently in male and female gametes; it is correlated with gene activity; and it is erased after each generation.

Do any benefits derive from genomic imprinting, or is it just another cause of genetic disorders? David Haig has cited an imprinting example that he believes has evolved in response to environmental demands: The insulin-like growth factor (IGF-2), and its receptor in the mouse. The growth factor tends to make baby mice bigger, but it must interact with its receptor (the type-1 IGF receptor) in order to do so. To complicate the problem, mice have an alternate receptor (a type-2 receptor) that binds IGF-2 but does not pass the growth-promoting signal along. Thus, expression of the *Igf2* gene in developing mice will produce bigger offspring, but expression of the type-2 receptor will sop up the IGF-2 and keep it away from the type-1 receptor, and therefore produce smaller offspring.

Haig points to an inherent biological conflict between the interests of the mother and those of the father of a baby mammal. If the benefits to the mother and father are viewed simply in terms of getting their own genes passed on to their offspring, then the father should favor large offspring, and the mother should favor small ones. The reason is that a large baby is more likely to survive and therefore perpetuate the father's genes. On the other hand, a large baby saps the mother's strength and leaves her fewer resources to provide to other offspring, which could be sired by a different father, but still would perpetuate her genes. This is a coldhearted way of looking at parenthood, but it is the sort of thing that can influence evolution.

Viewed in this context, it is very interesting that imprinting of male and female gametes in the mouse dictate that the *Igf2* gene provided by a mother mouse is repressed, while that provided by the father is active. On the other hand, the type-2 IGF receptor gene from the father is turned off, whereas that from the mother is active. Both of these phenomena fit with the premise that a male should favor large offspring and a female should favor small ones. We seem to have a battle of the sexes going on at the molecular level, but neither side is winning, because the strategies of each side are canceled by those of the other!

The *Igf2/H19* locus also contains two enhancers, one of which is active in endodermal cells, and the other in mesodermal cells. These enhancers can stimulate transcription of both the *Igf2* and *H19* genes. Notice that the ICR lies between the enhancers and the *Igf2* promoters, but not between the enhancers and the *H19* promoter. This location enables the ICR to function as an insulator to shield the *Igf2* promoters from the stimulatory effect of the enhancers, but only on the maternal chromosome. We will learn about insulator activity later in this chapter; for now, it is sufficient to know that the *Igf2* gene is active only on the paternal chromosome.

The imprinted nature of the *Igf2* locus allowed Pfeifer and colleagues to look at DNA looping between enhancers and promoters on active (paternal) and inactive (maternal) chromosomes in the same cells. If the looping model of enhancer action is correct, such looping would be observed only on the paternal chromosomes—and that is what happened.

To distinguish between maternal and paternal chromosomes in the 3C experiments, Pfeifer and colleagues bred mice that had *Igf2* loci from two different mouse species, as follows: They intercrossed FVB mice (*Mus domesticus*) with Cast7 mice, which are just like FVB mice, but have the distal part of chromosome 7, including the *Igf2* locus, derived from another mouse species (*Mus castaneus*). The *Igf2* loci of the two mouse species differ in several restriction sites, so cleavage with certain restriction enzymes yields different-size restriction fragments from DNAs of the two species. These variations are called **restriction fragment length polymorphisms (RFLPs,** Chapter 24), and can be used to determine whether a PCR product in a 3C experiment comes from the maternal or paternal chromosome.

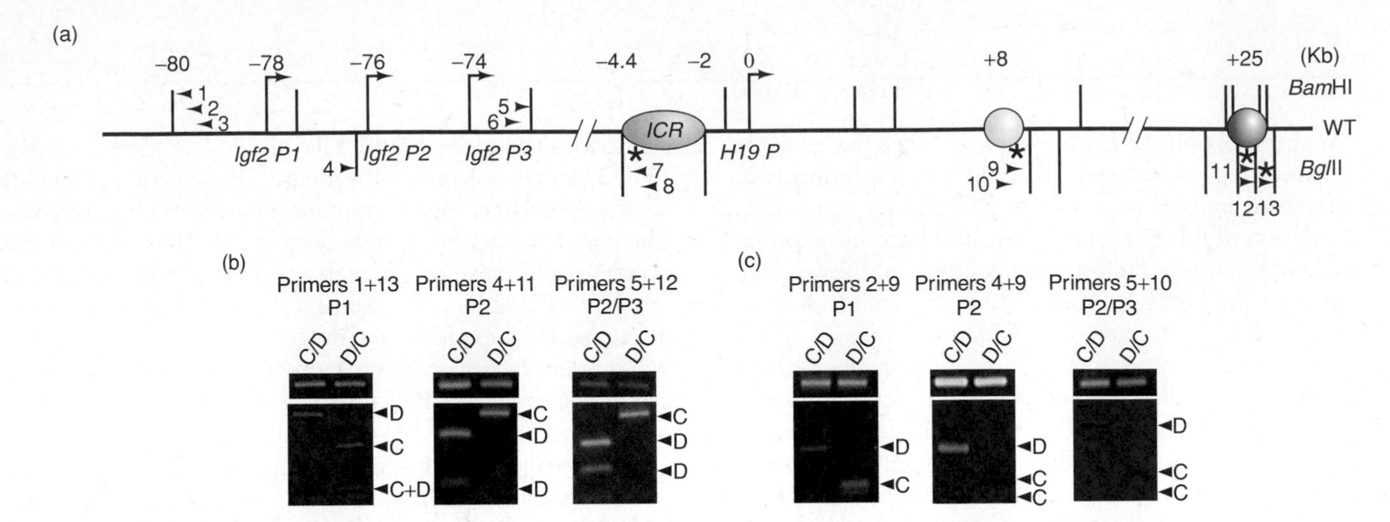

Figure 12.22 Association of chromatin elements in the mouse *Ifg2* locus. (a) Map of the wild-type locus. The whole locus is just over 100 kb long, as indicated at top. The three *Igf2* promoters are indicated near positions −78, −76, and −74, and the *H19* promoter is indicated at position 0. The ICR is in blue and the endodermal and mesodermal enhancers are in yellow and red, respectively. The vertical bars above and below the DNA represent *Bam*HI and *Bgl*II sites, respectively. Asterisks indicate *Bgl*II RFLPs that distinguish between *M. domesticus* and *M. castaneus* DNAs. Short arrows represent PCR primers used in the 3C analysis. Note that these primers always point toward the nearby restriction site. Thus, they are in position to create a short PCR product whenever two remote sections of DNA are cut with the corresponding restriction enzyme and then ligated together. **(b-c)** 3C analysis of long-range interactions in **(a)** mouse fetal muscle (mesodermal) cells and **(b)** fetal liver (endodermal) cells, respectively, using the indicated primers. The source of the embryo chromosomes (*M. domesticus* [Dl or *M. castaneus* [C]) is shown at top of each panel, with the maternal chromosome first. The upper panels in each case show the PCR product of the 3C analysis. The lower panels show the RFLP analysis on the PCR products. Arrowheads labeled C or D point to RFLP bands that are characteristic of *M. castaneus* or *M. domesticus,* respectively. C+D denotes an RFLP band resulting from comigration of bands from both mouse species. (*Source:* Yoon et al, Analysis of the H19*ICR*. *Molecular and Cellular Biology,* May 2007, pp. 3499–3510, Vol. 27, No. 9. Copyright © 2007 American Society for Microbiology.)

Figure 12.22b and c show the 3C results in fetal muscle (mesodermal) cells and fetal liver (endodermal) cells, respectively. The top part of each panel contains the 3C PCR product, and the bottom part contains the results of RFLP analysis to identify the maternal or paternal origin of each PCR product. The C/D and D/C designations at the top refer to the *M. castaneus* or *M. domesticus Igf2* locus, with the maternal allele always presented first. Thus, C/D mice had the *M. cataneus Igf2* locus on the maternal chromosome and the *M. domesticus Igf2* locus on the paternal chromosome. The C and D designations beside the gels show RFLP bands corresponding to *M. castaneus* and *M. domesticus,* respectively. Note that the 3C PCR products always derived from the paternal chromosome. For example, in the first lane in the first gel in Figure 12.22b, the paternal chromosome was from *M. domesticus,* and the RFLP analysis identified the PCR product as coming from *M. domesticus* (D). On the other hand, in the second lane in the first gel, the paternal chromosome was from *M. castaneus,* and the RFLP analysis showed that the PCR product came from *M. castaneus* (C). This demonstrated that the enhancer and promoters are brought together by DNA looping only on the paternal chromosome, where the *Igf2* gene is active.

Pfeifer and colleagues chose the primers to show linkages between each of the three *Igf2* promoters and the appropriate enhancer. Thus, in muscle cells, DNA looping brought each of the promoters (defined by primers 1, 4, and 5, respectively), close to the mesodermal enhancer (the one on the far right in Figure 12.22a, and defined by primers 11, 12, and 13). On the other hand, in liver cells, DNA looping brought the promoters and the endodermal enhancer (defined by primers 9 and 10) together. Thus, the 3C technique demonstrates that tissue-appropriate enhancers and promoters are brought together, presumably by DNA looping.

SUMMARY The essence of enhancer function—protein–protein interaction between activators bound to the enhancers, and general transcription factors and RNA polymerase bound to the promoter—seems in many cases to be mediated by looping out the DNA in between. This can also account for the effects of multiple enhancers on gene transcription, at least in theory. DNA looping could bring the activators bound to each enhancer close to the promoter where they could stimulate transcription, perhaps in a cooperative way.

Transcription Factories

The notion of DNA loops discussed in the previous section is consistent with the concept of **transcription factories**—discrete nuclear sites where transcription of multiple genes occurs: If two or more active genes on the same chromosome are clustered in the same transcription factory, this would naturally form DNA loops between them. Thus, the existence of transcription factories implies the existence of DNA loops in eukaryotic nuclei. During the 1990s, several research groups provided evidence for the existence of these transcription factories. This concept raises at least two interesting questions: (1) How many transcription factories exist in a nucleus? (2) How many polymerases are active in a transcription factory?

To count the number of transcription factories, Peter Cook and colleagues performed the following experiment in 1998. They labeled growing RNA chains in HeLa cells with bromouridine (BrU). They followed this BrU labeling in vivo by permeabilizing the cells and further labeling growing RNA chains in vitro with biotin-CTP. The labeled RNA could then be detected with primary antibodies against either BrU or biotin, and secondary antibodies or protein A labeled with gold particles. BrU labeling was detected with 9-nm gold particles, and biotin labeling was detected with 5-nm particles. Figure 12.23a shows the results of labeling with BrU at low magnification, and Figure 12.23b shows the results of labeling with both BrU and biotin at higher power. Note that transcription does not occur uniformly across the nucleus, but is concentrated into patches, most of which contain more than one growing RNA chain.

The purpose of the in vitro labeling with biotin is to control for migration of finished RNAs away from their site of synthesis. If RNAs do this in groups, these would appear just like transcription factories and the number of apparent factories would therefore be inflated. But labeling in vitro does not allow for RNA chains to be finished and leave their sites of synthesis, so in vitro-labeled RNAs (small gold particles) should represent real transcription factories. Cook and colleagues found a high level of correspondence between in vivo- and in vitro–labeled clusters, as long as the in vivo labeling times were kept short (2.5 min). That is, large gold particles were found in the same clusters with small gold particles about 85% of the time. With longer in vivo labeling times (10 min or more), many BrU-labeled clusters were not associated with biotin-labeled clusters, and were therefore probably not transcription factories.

Do the clusters really represent sites of transcription? If so, we would expect the number of particles to increase with time, as more polymerases initiate RNA chains. Figure 12.23c shows that the number of particles in clusters does indeed increase with time, while the number of single particles does not. Thus, transcription is associated with the clusters, not the single particles.

On average, Cook and colleagues found one cluster per μm^2: in their nuclear sections. Knowing the total nucleoplasmic volume, this allowed them to calculate that there are about 5500 nucleoplasmic transcription factories with active polymerases II and III per cell. Extending preinitiated RNA chains in vitro with labeled UTP in the presence and

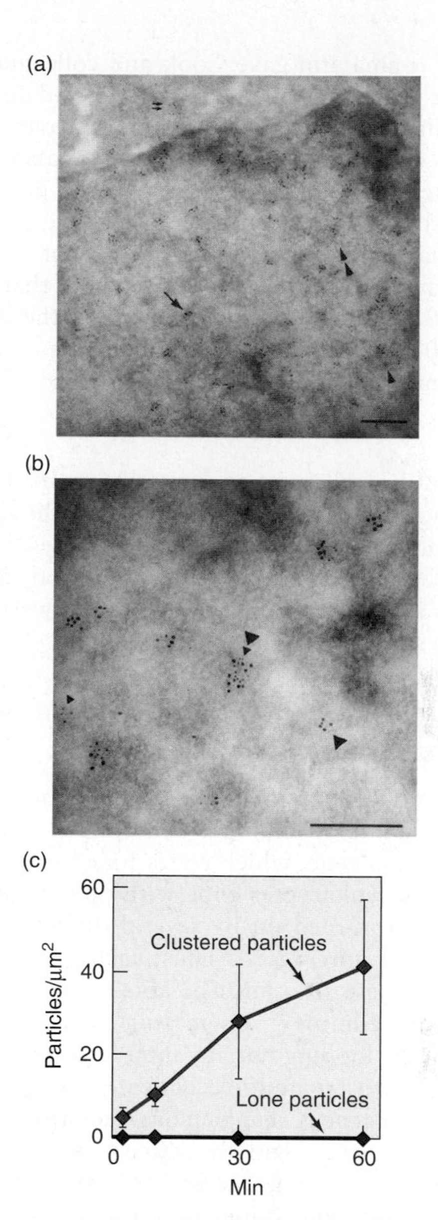

Figure 12.23 Detecting transcription factories. (a) Low-magnification view. Cook and colleagues labeled growing RNA chains in HeLa cells with BrU and detected the label by indirect immunostaining with 9-nm gold particles. They found most of the labeled RNA in clusters (arrow). Most of these clusters represent transcription factories, but some represent sites of RNA processing, or even mature RNAs in the cytoplasm (two small arrows). Weak label was found in interchromatin clusters (double arrowhead). No label was found in perichromatin clusters (single arrowhead). **(b)** High-magnification view. Cook and colleagues labeled nascent RNA with BrU in vivo and then extended these growing RNAs in vitro and labeled them with biotin-CTP. They detected BrU- and biotin-labeled RNAs by indirect immunostaining with 9-nm and 5-nm gold particles, respectively. They found most gold particles in clusters. Large and small arrowheads point to clusters with large and small gold particles, respectively. Most clusters contained both sizes of particles. **(c)** Clustered particles correspond to transcription sites. Cook and colleagues grew cells for various times in medium containing BrU, then detected BrU-RNA by immunostaining with 9-nm gold particles. (*Source:* Jackson et al, Numbers and Organization of RNA Polymerases, Nascent Transcripts, and Transcription Units in HeLa Nuclei. *Molecular Biology of the Cell* Vol. 9, 1523–1536, June 1998. Copyright © 1998 by The American Society for Cell Biology.)

absence of α-amanitin gave Cook and colleagues an estimate of the total amount of RNA synthesized during the in vitro labeling period. Knowing the approximate length each RNA chain would grow during the labeling period, these workers could estimate the number of growing RNA chains, and therefore the number of active polymerases. They calculated that each cell contained about 75,000 active RNA polymerases II and III. Thus, given that there are about 5500 transcription factories per cell, there are about 75,000/5500, or about 14 active polymerases II and III per transcription factory.

SUMMARY Transcription appears to be concentrated in transcription factories within the nucleus, where an average of about 14 polymerases II and III are active. The existence of transcription factories implies the existence of DNA loops between genes being transcribed in the same factory.

Complex Enhancers

Many genes have more than one activator-binding site, so they can respond to multiple stimuli. For example, the metallothionine gene, which codes for a protein that apparently helps eukaryotes cope with poisoning by heavy metals, can be turned on by several different agents, as illustrated in Figure 12.24. Thus, each of the activators that bind at these sites must be able to interact with the preinitiation complex assembling at the promoter, presumably by looping out any intervening DNA.

The finding that multiple activator-binding sites can control a given gene is changing our definition of the word "enhancer." It was originally defined as a nonpromoter DNA element that, together with at least one enhancer-binding protein, could stimulate transcription of a nearby gene. Thus, the control region of the metallothionine gene upstream of the TATA box in Figure 12.24 was considered to contain many enhancers. But the definition has evolved toward a concept that embraces an entire contiguous control region outside the promoter itself. Thus, the entire control region of the metallothionine gene can be considered an enhancer, and the BLE, for example, is only one element of the whole enhancer. Even using the newer definition, we can still say that some genes are controlled by multiple enhancers. For example, the *Drosophila yellow* and *white* genes considered later in this chapter are controlled by three enhancers—three clusters of contiguous binding sites for activators.

Enhancers that interact with many activators allow for very fine control over the expression of genes. Different combinations of activators produce different levels of expression of a given gene in different cells. In fact, the presence or absence of various enhancer elements near a gene reminds one of a binary code, where the presence is an "on" switch, and the absence is an "off" switch. Of course, the activators also have to be present to throw the switches. It may not be a simple additive arrangement, however, since multiple enhancer elements are known to act cooperatively.

Another metaphor that works well in describing the actions of multiple activators on multiple enhancer elements is a **combinatorial code.** The concentrations of all the activators in any given cell at a given time constitute the code. A gene can read the code if it has a battery of enhancer elements, each responsive to one or more of the activators. The result is an appropriate level of expression of the gene.

Eric Davidson and colleagues provided a beautiful example of multiple enhancer elements in the *Endo 16* gene of a sea urchin. This gene is active in the early embryo's vegetal plate—a group of cells that produces the endodermal tissues, including the gut. Davidson and colleagues began by testing DNA in the *Endo 16* 5′-flanking region for the ability to bind nuclear proteins. They found dozens of such regions, arranged into six modules, as illustrated in Figure 12.25.

How do we know that all these modules that bind nuclear proteins are actually involved in gene activation? Chiou-Hwa-Yuh and Davidson tested them by linking them alone and in combinations to the *cat* reporter gene (Chapter 5), reintroducing these constructs into sea urchin eggs, and observing the patterns of expression of the reporter gene in the resulting developing embryo. They found that the reporter gene was switched on in different parts of the embryo and at different times, depending on the exact combination of modules attached. Thus, the modules were responding to activators that were distributed nonuniformly in the developing embryo.

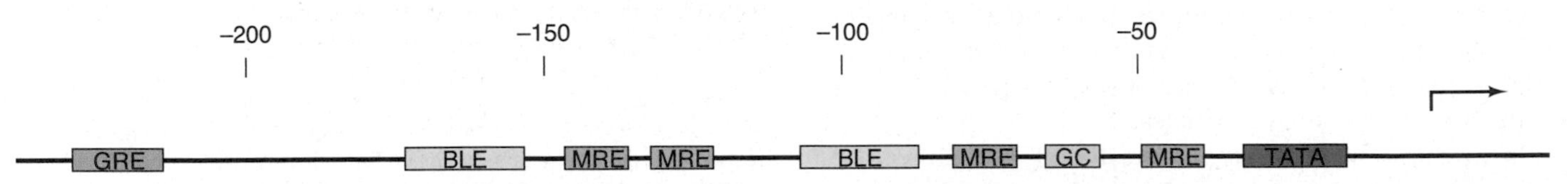

Figure 12.24 Control region of the human metallothionine gene. Upstream of the transcription start site at position +1 we find, in 3′–5′ order: the TATA box; a metal response element (MRE) that allows the gene to be stimulated in response to heavy metals; a GC box that responds to the activator Sp1; another MRE; a basal level enhancer (BLE) that responds to the activator AP-1; two more MREs; another BLE; and a glucocorticoid response element (GRE) that allows the gene to be stimulated by an activator composed of a glucocorticoid hormone and its nuclear receptor.

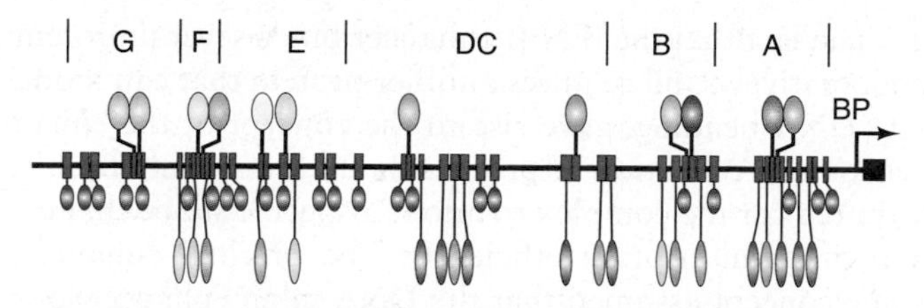

Figure 12.25 Modular arrangement of enhancers at the sea urchin *Endo 16* gene. The large colored ovals represent activators, and the small blue ovals represent architectural transcription factors, bound to enhancer elements (red boxes). The enhancers are arranged in clusters, or modules, as indicated by the regions labeled G, F, E, DC, B, and A. Long vertical lines denote restriction sites that define the modules. BP stands for "basal promoter." (*Source:* Adapted from Romano, L.A. and G.A. Wray, Conversation of *Endo 16* expression in sea urchins despite evolutionary divergence in both cis and trans-acting components of transcriptional regulation, *Development 130 (17): 4189, 2003.*)

Although all the elements may be able to function independently in vitro, the situation is more organized in vivo. Module A appears to be the only one that interacts directly with the basal transcription apparatus; all the other modules work through A. Some of the upstream modules (B and G) act synergistically through A to stimulate *Endo 16* transcription in endoderm cells. The other modules (DC, E, and F) act synergistically through A to block *Endo 16* transcription in nonendoderm cells (modules E and F play this role in ectoderm cells, and module DC plays this role in skeletogenic mesenchyme cells).

SUMMARY Complex enhancers enable a gene to respond differently to different combinations of activators. This arrangement gives cells exquisitely fine control over their genes in different tissues, or at different times in a developing organism.

Architectural Transcription Factors

The looping mechanism we have discussed for bringing together activators and general transcription factors is quite feasible for proteins bound to DNA elements that are separated by at least a few hundred base pairs because DNA is flexible enough to allow such bending. On the other hand, many enhancers are located much closer to the promoters they control, and that presents a problem: DNA looping over such short distances will not occur spontaneously, because short DNAs behave more like rigid rods than like flexible strings.

How then do activators and general transcription factors bound close together on a stretch of DNA interact to stimulate transcription? They can still approach each other if something else intervenes to bend the DNA more than the DNA itself would normally permit. We now have several examples of **architectural transcription factors** whose

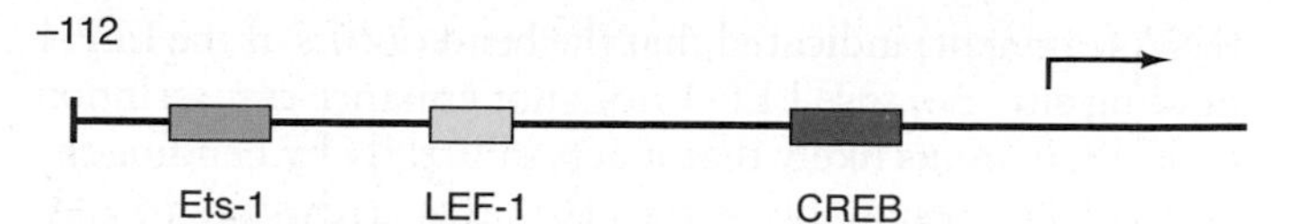

Figure 12.26 Control region of the human T-cell receptor α-chain (TCRα) gene. Within 112 bp upstream of the start of transcription lie three enhancer elements, which bind Ets-1, LEF-1, and CREB. These three enhancers are identified here by the transcription factors they bind, not by their own names.

sole (or main) purpose seems to be to change the shape of a DNA control region so that other proteins can interact successfully to stimulate transcription. Rudolf Grosschedl and his colleagues provided the first example of a eukaryotic architectural transcription factor. They used the human T-cell receptor α-chain (TCRα) gene control region, which contains three enhancers, binding sites for the activators Ets-1, LEF-1, and CREB within just 112 bp of the transcription start site (Figure 12.26).

LEF-1 is the lymphoid enhancer-binding factor, which binds to the middle enhancer pictured in Figure 12.26 and helps activate the TCRα gene. However, previous work by Grosschedl and others had shown that LEF-1 by itself cannot activate TCRα gene transcription. So what is its role? Grosschedl and coworkers established that it acts by binding primarily to the minor groove of the enhancer and bending the DNA by 130 degrees.

These workers demonstrated minor groove binding by two methods. First, they showed that methylating six enhancer adenines on N3 (in the minor groove) interfered with enhancer function. Then they substituted these six A–T pairs with I–C pairs, which look the same in the minor groove, but not the major groove, and found no loss of enhancer activity. This is the same strategy Stark and Hawley used to demonstrate that TBP binds to the minor groove of the TATA box (Chapter 11).

Next, using the same electrophoretic assay Wu and Crothers used to show that CAP bends *lac* operon DNA (Chapter 7), Grosschedl and coworkers showed that LEF-1 bends DNA. They placed the LEF-1 binding site at different positions on linear DNA fragments, bound LEF-1, and measured the electrophoretic mobilities. The mobility was greatly retarded when the binding site was in the middle of the fragment, suggesting significant bending.

They also showed that the DNA bending is due to a so-called **HMG domain** on LEF-1. **HMG proteins** are small nuclear proteins that have a high electrophoretic mobility (hence, high mobility group, or HMG). To show the importance of the HMG domain of LEF-1, these workers prepared a purified peptide containing just the HMG domain and showed that it caused the same degree of bending (130 degrees) as the full-length protein. Extrapolation of the mobility curve to the point of maximum mobility (where the bend-inducing element should be right at the end of the

DNA fragment) indicated that the bend occurs at the LEF-1 binding site. Because LEF-1 does not enhance transcription by itself, it seems likely that it acts indirectly by bending the DNA. This presumably allows the other activators to contact the basal transcription machinery at the promoter and thereby enhance transcription.

SUMMARY The activator LEF-1 binds to the minor groove of its DNA target through its HMG domain and induces strong bending in the DNA. LEF-1, an architectural transcription factor, does not enhance transcription by itself, but the bending it induces probably helps other activators bind and interact with other activators and the general transcription factors to stimulate transcription.

Enhanceosomes

We have discussed several examples of enhancers, ranging from modular and spread out (the sea urchin *Endo 16* enhancer) to compact (the TCRα enhancer). We saw that transcription of the *Endo 16* gene responds differently to different combinations of activators, which also means that the *Endo 16* gene can be activated by subsets of activators. But not all enhancers work that way. Tom Maniatis and colleagues have studied an enhancer at the other end of the continuum of enhancer size and complexity: the human interferon-β (IFN-β) enhancer, which responds to viral infection. This enhancer contains binding sites for only eight polypeptides: two from the heterodimer ATF-2/cJun; four from two copies each of the interferon response factors IRF-3 and IRF-7; and two from the heterodimer **nuclear factor kappa B (NFκB)**, whose two subunits are p50 and RelA. These proteins interact with proteins at the promoter through a coactivator known as CREB-binding protein (CBP), or its closely related cousin, p300.

In contrast to the *Endo 16* enhancer, the IFN-β enhancer works only when all of its activators are present at the same time in a cell. This is important because all of these activators activate many genes and are present in a wide variety of cells. Nevertheless, the IFN-β gene is strongly activated only when it is needed: when a cell is under attack by a virus. The requirement for all the activators at once explains this paradox, because all the activators are present together essentially only when cells are virus-infected.

Another protein that plays an important role in IFN-β activation is another member of the HMG family: **HMGA1a.** Unlike LEF-1, proteins of the HMGA1a type do not bend DNA. Instead they modulate the natural bending of A-T rich DNA regions. HMGA1a is essential for activation of the IFN-β gene, and its role is to ensure cooperative binding of the other activators to the enhancer.

The fact that the IFN-β enhancer binds several proteins cooperatively, and requires another protein that can modulate DNA bending, gave rise to the concept of the **enhanceosome,** a collection of proteins bound to an enhancer, all required for the complex to adopt a specific shape that can activate transcription efficiently. The original enhanceosome concept assumed that the DNA in an enhanceosome would be significantly bent, and that HMG proteins would play a role in such bending. However, we now know that HMGA1a does not bend DNA and, as we will soon see, it is not even part of the IFN-β enhanceosome, so the assumption of an enhanceosome with a strongly bent DNA rested on shaky ground.

Indeed, in 2007 Maniatis and colleagues assembled the crystal structure of the IFN-β enhanceosome (Figure 12.27) from two parts: The DNA-binding domains of IRF-3, IRF-7, and NFκB from one-half of the enhanceosome and a previously determined structure for the other half. They found that the DNA within the enhanceosome is essentially straight, experiencing only a gentle undulation. The IFN-β

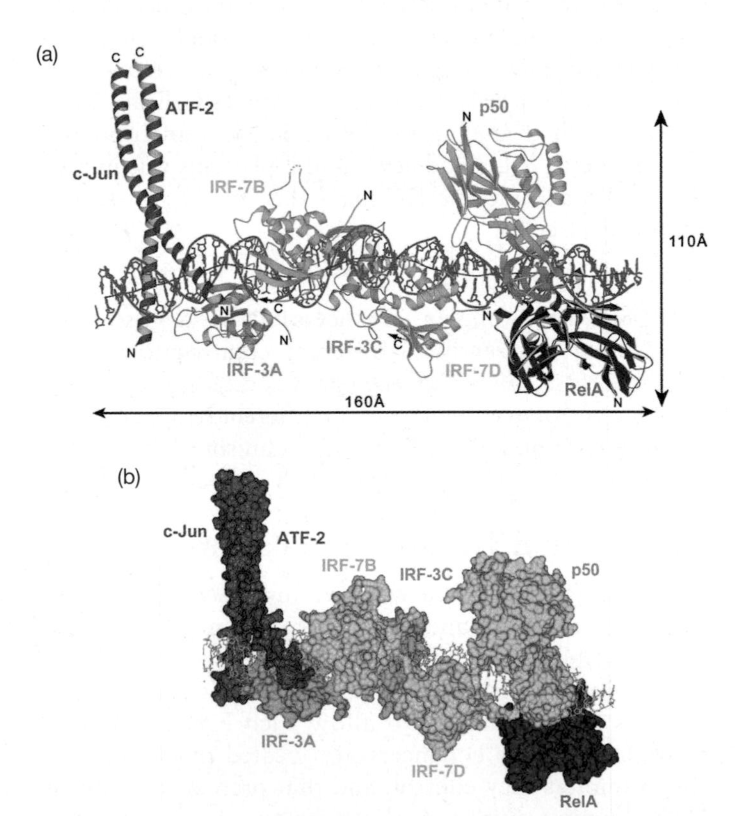

Figure 12.27 Model for the human IFN-β enhanceosome. **(a)** Ribbon diagram of the enhanceosome showing the gently undulating path of the DNA, whose local axis is traced by the dotted red line. The two IRF-3 molecules are designated -3A and -3C, and the two IRF-7 molecules are designated -7B and -7D. The overlapping binding sites for all the activators are shown on the DNA sequence below the diagram. **(b)** Molecular surface diagram of the enhanceosome in the same orientation as in panel (a). (*Source:* Reprinted from CELL, Vol. 129, Panne et al, An Atomic Model of the Interferon-β Enhanceosome, Issue 6, 15 June 2007, pages 1111–1123, © 2007, with permission from Elsevier.)

enhancer contains four binding sites for HMGA1a, but this protein is apparently not bound along with all the other activators. There is simply not room for it in the final enhancesome. But the crystal structure does emphasize the role of HMGA1a in cooperative binding of the other activators to the enhancer: It shows that, although the activators bind close together, they interact with each other to a remarkably small extent. Thus, HMGA1a presumably stimulates cooperativity by binding transiently to the DNA and other activators and helping them come together.

SUMMARY An enhanceosome is a nucleoprotein complex containing a collection of activators bound to an enhancer in such a way that stimulates transcription. The archetypical enhanceosome involves the IFN-β enhancer. Its structure involves eight polypeptides bound cooperatively to an essentially straight 55-bp stretch of DNA. HMGA1a is essential for this cooperative binding, but it is not part of the final enhanceosome.

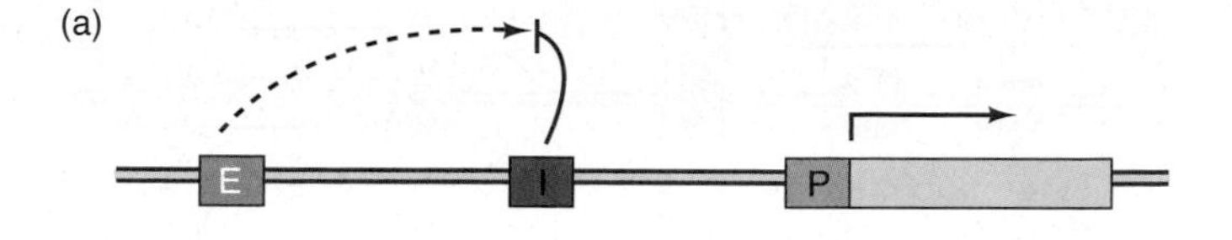

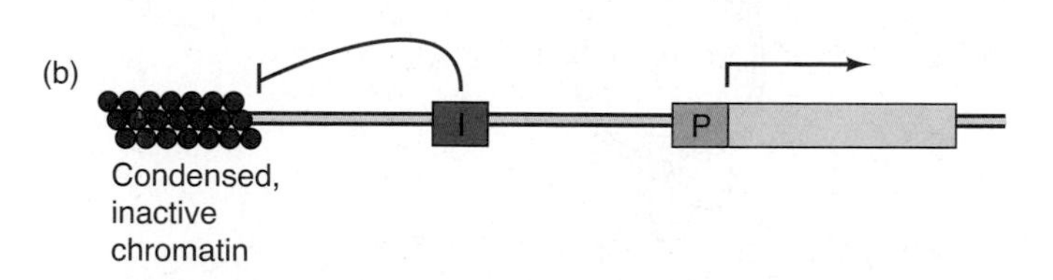

Figure 12.28 Insulator function. (a) Enhancer-blocking activity. The insulator between a promoter and an enhancer prevents the promoter from feeling the activating effect of the enhancer. **(b)** Barrier activity. The insulator between a promoter and condensed, repressive chromatin (induced by a silencer) prevents the promoter from feeling the repressive effect of the condensed chromatin (indeed, prevents the condensed chromatin from engulfing the promoter).

Insulators

We know that enhancers can act at a great distance from the promoters they activate. For example, the wing margin enhancer in the *Drosophila cut* locus is separated by 85 kb from the promoter. With a range that large, some enhancers will likely be close enough to other, unrelated genes to activate them as well. How does the cell prevent such inappropriate activation? Higher organisms, including at least *Drosophila* and mammals, use DNA elements called **insulators** to block activation of unrelated genes by nearby enhancers.

Gary Felsenfeld has defined an insulator as a "barrier to the influence of neighboring elements." An insulator that can protect a gene from activation by nearby enhancers is called an **enhancer blocking insulator.** On the other hand, an insulator that stops the encroachment of condensed chromatin into a target gene, thereby preventing gene silencing, is called a **barrier insulator.** Although many do, not all insulators have both blocking and barrier activities. Some are specialized for one activity or the other. The yeast elements that serve as barriers to the silencers at telomeres are prominent examples of insulators with only barrier activity.

How do insulators work? The details are not clear yet, but we do know that insulators define boundaries between DNA domains. Thus, an insulator abolishes activation if placed between an enhancer and a promoter. Similarly, an insulator abolishes repression if placed between a silencer and a silenced gene. It appears that the insulator creates a boundary between the domain of the gene and that of the enhancer (or silencer) so the gene can no longer feel the activating (or repressing) effects (Figure 12.28).

We also know that insulator function depends on protein binding. For example, certain *Drosophila* insulators contain the sequence GAGA and are known as **GAGA boxes.** These require the GAGA-binding protein **Trl** for insulator activity. Genetic experiments have shown that insulator activity can be abolished by mutations in either the GAGA box itself, or in the *trl* gene, which encodes Trl.

One can imagine many mechanisms for insulator function. We can easily eliminate one of these: a model in which the insulator induces a silenced, condensed chromatin domain upstream of the insulator. If that were the case, then a gene placed upstream of an insulator would always be silenced. But experiments with *Drosophila* have shown that such upstream genes are still potentially active and can be activated by their own enhancers.

Figure 12.29 illustrates two more models of insulator action. The first involves a signal that somehow moves progressively from the enhancer to the promoter, and the insulator blocks the progression of this signal. The second requires interaction between insulators on either side of an enhancer, which isolates the enhancer on a loop so it cannot interact with the promoter.

The first hypothesis is hard to reconcile with an experiment performed by J. Krebs and Dunaway similar in concept to the one by Dunaway and Dröge we discussed earlier in this chapter. In that earlier experiment (see Figure 12.20), Dunaway and Dröge placed a promoter and an enhancer on separate DNA circles linked in a catenane and showed that the enhancer still worked. In the later experiment, Krebs and Dunaway used the same catenane construct, but this time they surrounded either the enhancer or promoter with two *Drosophila* insulators: *scs* and *scs'*. They found that in both cases, the insulators blocked enhancer activity.

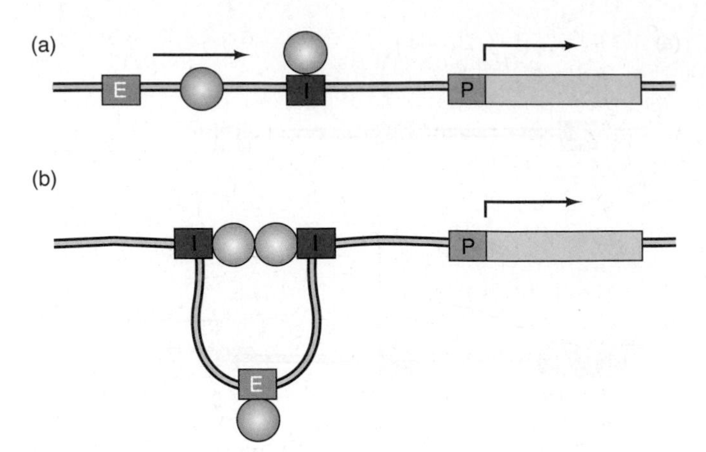

Figure 12.29 Two hypotheses for the mechanism of insulator activity. (a) Sliding model. An activator has bound to an enhancer and a stimulatory signal (green), perhaps the activator itself, is sliding along the DNA from the enhancer toward the promoter. But the insulator (red), perhaps with a protein or proteins attached, stands in the way and prevents the signal from reaching the promoter. **(b)** Looping model. Two insulators (red) flank an enhancer (blue). When proteins (purple) bind to these insulators, they interact with one another, isolating the enhancer on a loop so it cannot stimulate transcription from the nearby promoter (orange).

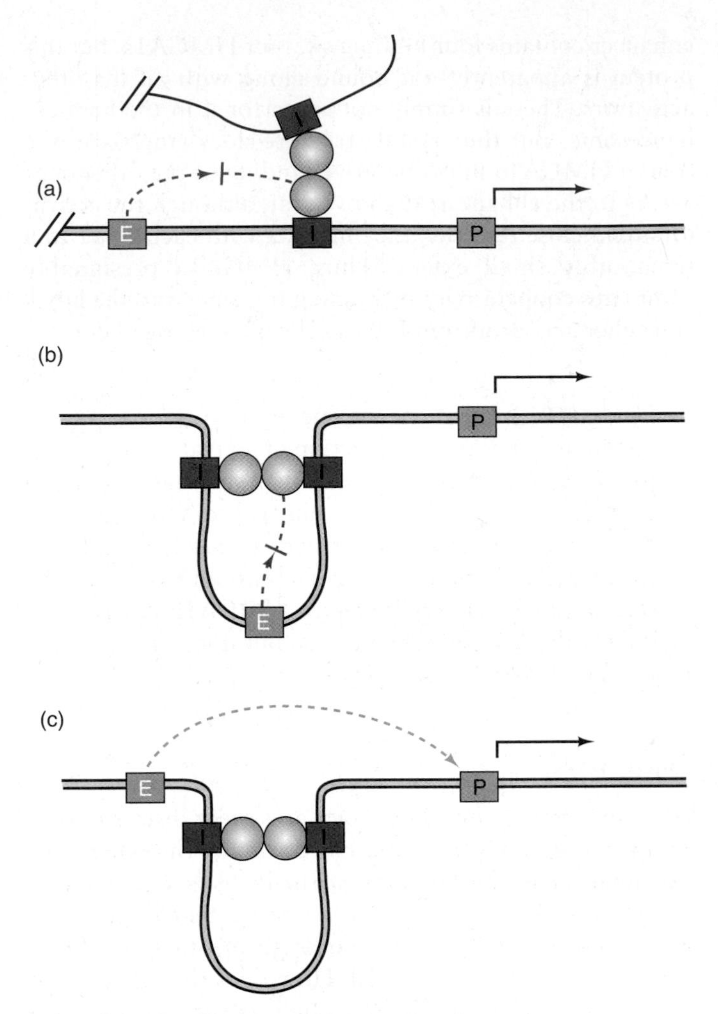

Figure 12.30 Model of multiple insulator action. (a) A single insulator (I, red) between an enhancer (E, blue) and a promoter (P, orange) binds to a protein(s) (purple) that interact with other protein(s), not necessarily of the same type, that are bound to another, remote insulator, also not necessarily of the same type. These protein–protein interactions isolate the enhancer from the promoter and block enhancement of transcription. **(b)** Two insulators flanking an enhancer bind to proteins that interact, looping the DNA and isolating the enhancer from the promoter. This prevents enhancement. **(c)** Two (or more) insulators between the enhancer and promoter bind to proteins that interact and loop out the DNA in between but do not isolate the enhancer from the promoter; in fact they bring the two elements closer together. Thus, the two insulators cancel each other out and do not block enhancement. The enhancer and promoter probably interact by DNA looping that is not illustrated here. (*Source:* Adapted from Cai, H.N. and P. Shen, Effects of *cis* arrangement of chromatin insulators on enhancer-blocking activity. *Science* 291 [2001] p. 495, (4.))

On the other hand, a single insulator in either circle had little effect on enhancement. Both experiments from Dunaway's group are incompatible with a signal propagating from the enhancer to the promoter unless the signal can jump from one DNA circle to another.

Arguments against the second hypothesis have been presented as well. Chief among them is the fact that some insulators work as single copies, so it is not apparent that there are two insulators flanking an enhancer. However, it is possible that the second insulator is present but not recognized in these experiments. It could attract novel proteins that can interact with the proteins that bind to the known insulator. Thus, the chromatin could be forced to loop in such a way as to prevent the enhancer from interacting with a promoter on one side, but not on the other.

Haini Cai and Ping Shen have performed experiments that support this hypothesis. When they placed a single copy of a known *Drosophila* insulator [su(Hw); (suppressor of *Hairy wing*)] between an enhancer and a promoter, they observed some insulator activity (a decrease in the effectiveness of the enhancer). However, when they placed two copies of the same insulator in the same place, they observed no insulator activity. Finally, when they placed single copies of the su(Hw) insulators on either side of the enhancer, they observed the most insulator activity of all. By the way, the Su(Hw) insulator is part of a retrotransposon (Chapter 23) known as *gypsy*. The insulator binds to a protein that is also known as Su(Hw).

Figure 12.30 illustrates Cai and Shen's interpretation of these results. Panel (a) shows what happened with the single insulator. It teamed up with an unknown insulator (I) somewhere upstream of the enhancer to block the action of the enhancer. Panel (b) shows what happened with an insulator on either side of the enhancer. Proteins bound to the insulators and caused the DNA to loop, isolating the enhancer in the loop in such a way that it could no longer interact with the promoter. In panel (c), the two adjacent insulators between enhancer and promoter bound proteins that interacted with each other, looping out the DNA in between, but did not interfere with enhancer activity. In fact, the looped DNA actually brought the enhancer closer to the promoter and presumably made the enhancer more effective.

At the same time in 2001, Vincenzo Pirrotta and coworkers reported work in which they performed the same kind of experiment with the su(Hw) insulator in single and multiple copies, but with different *Drosophila* promoters, and obtained the same results. Then they added a new wrinkle: two different genes in tandem, instead of just one, with three upstream enhancers, and one to three insulators in various positions. The two genes were *yellow* and *white,* which are responsible for dark body and wing color, and for red eye color, respectively, in adult flies. When the *yellow* gene is inactivated (or mutated) dark pigment fails to be made, and the body and wings are yellow instead of black. When the *white* gene is inactivated (or mutated), red eye pigment synthesis fails and the eyes of the fly are white.

Figure 12.31 illustrates the constructs Pirrotta and coworkers used, and the results they obtained. The first construct (EyeSYW) contained one copy of the insulator between the enhancers and the two genes. As Cai and Shen's model predicted, the insulator prevented activation of both genes by the enhancers. The second construct (EyeYSW) contained an insulator between the *yellow* and *white* genes. Again, predictably, the *yellow* gene was activated, but the *white* gene was not.

The third construct (EyeSYSW), in which two insulators flanked the *yellow* gene, is more interesting. This time, the *yellow* gene was not activated, but the *white* gene was. Again, Cai and Shen's model is compatible with these results: The two insulators flanking the *yellow* gene prevented its activation, but they constituted two insulators together between the enhancers and the *white* gene, so they cancelled each other and allowed activation of that gene. Thus, the interaction of the two insulators, while it cancelled their effect on the *white* gene, did not really inactivate them: They could still prevent inactivation of the *yellow* gene that lay between them. The fourth construct (EyeSYWS) contained two insulators flanking the *yellow* and *white* genes. Predictably, the insulators prevented activation of both genes.

Finally, the fifth construct (EyeSFSYSW) contained three insulators, two between the enhancers and the *yellow* gene, and one between the *yellow* and *white* genes. Because both genes were activated, we see that two *or more* copies of the insulator between an enhancer and a gene neutralizes the effect of the insulators. (There are two copies between the enhancers and the *yellow* gene, but three between the enhancers and the *white* gene.) We might have expected the two insulators upstream of the *yellow* gene to neutralize each other and allow activation of the *yellow* gene, but the single remaining insulator between the *yellow* and *white* genes might have been expected to block activation of the *white* gene. Instead, none of the three insulators had any effect, and both genes were activated. This experiment therefore revealed that the inactivation of two tandem insulators is not due to a simple, exclusive interaction between the two. Somehow, proteins bound to all three

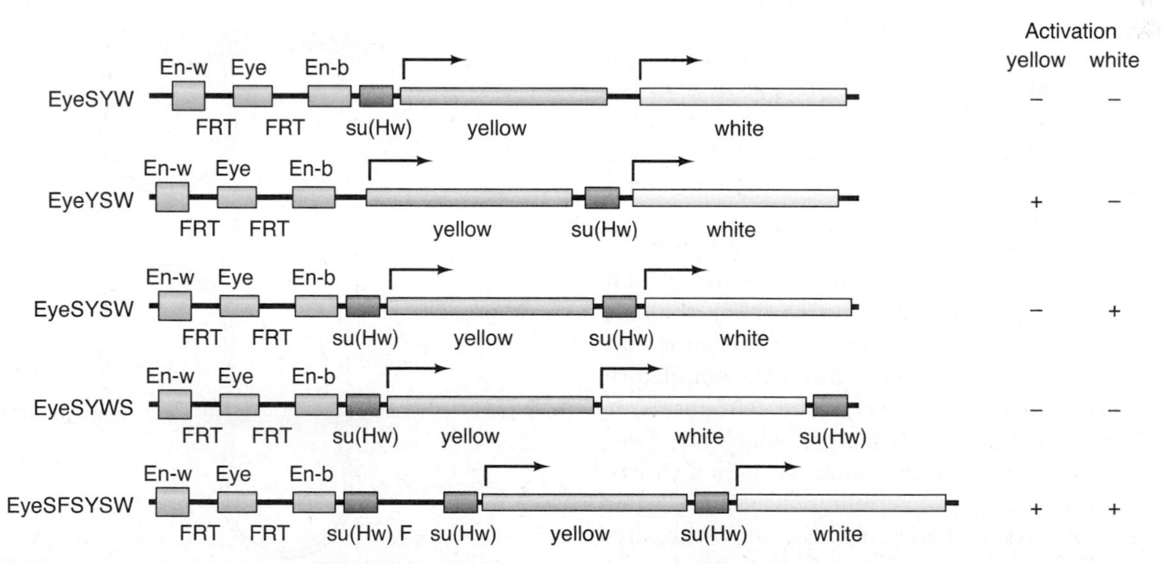

Figure 12.31 Effects of insulators on two tandem *Drosophila* genes. The structures of the constructs are given on the left, with the results (activation [+] or no activation [–] of the *yellow* and *white* genes) on the right. The names of the constructs all begin with Eye, which stands for the eye-specific enhancer found in the cluster of three enhancers (blue) upstream of both the *yellow* and *white* genes. The S, Y, and W in the names stand for the insulator [su(Hw), red], the *yellow* gene, and the *white* gene, respectively. The F in the last construct stands for a spacer fragment. The positions of the letters in the construct names indicate the positions of the corresponding elements in the constructs. Pirrotta and coworkers placed each construct into *Drosophila* embryos and observed the effects on body and wing color (*yellow* gene activity) and on eye color (*white* gene activity). (*Source:* Adapted from Muravyova, E., A. Golovnin, E. Gracheva, A. Parshikov, T. Belenkaya, V. Pirotta, and P. Georgiev, Loss of insulator activity by paired Su(Hw) Chromatin Insultators. *Science* 291 [2001] p. 497, f. 2.)

insulators appear to interact in such a way as to permit the enhancers upstream to do their job.

All of these results on enhancement and insulator action are easiest to explain on the basis of DNA looping, as illustrated in Figure 12.30. But looping is not the only possible explanation. Experimental evidence to date cannot rule out some kind of tracking mechanism (see Figure 12.18d) to explain enhancement. And proteins bound to the enhancer and tracking toward the promoter would be readily blocked by placing a single insulator between the enhancer and the promoter. How then can we explain the canceling effect of two or more insulators between the enhancer and the promoter? One way is to invoke **insulator bodies,** which are conglomerations of two or more insulators and their binding proteins that have been detected at the periphery of the nucleus. The formation of insulator bodies is thought to play a critical role in insulator activity, but we have no accepted hypothesis for how the insulator bodies play this role. In the absence of such a hypothesis, we cannot rule out the possibility that two or more insulators (lying between an enhancer and a promoter) and their binding proteins interact with each other in such a way as to prevent the association of the insulators with insulator bodies. And such interactions would thereby block insulator activity.

Another model for insulator activity, proposed by Pfeifer and colleagues, is that the insulator blocks association between enhancers and promoters by forming associations of its own with these chromosomal elements. Of course, it is not the DNA regions themselves, but the proteins bound to these DNA regions, that are interacting. As we learned earlier in this chapter, Pfeifer and colleagues showed that the *Igf2* enhancers and promoters are brought together by DNA looping when the gene is activated, but not when it is silenced. Furthermore, we learned that the maternal copy of the gene is silenced by imprinting (see Box 12.1), while the paternal copy remains active in fetal muscle and liver cells.

It was already known by 2007 that silencing of the maternal *Igf2* gene depended on the **imprinting control region** (***ICR***, refer back to Figure 12.22a). Furthermore, the *ICR* silences the maternal gene by acting as an insulator that shields the maternal *Igf2* promoters from the stimulatory effects of the two nearby enhancers. The *ICR* insulator binds to **CTCF** (CCCTC-binding factor), which is a common insulator-binding protein that interacts with a variety of insulators found throughout vertebrate genomes. Pfeifer and colleagues, and others, had previously shown that removal of the *ICR* from the maternal chromosome allowed expression of the maternal copy of the *Igf2* gene. Then, Pfeifer and colleagues demonstrated (by the same kind of 3C and RFLP analysis shown in Figure 12.22) that removal of the *ICR* from the maternal chromosome also allowed the maternal enhancers to associate with the *Igf2* promoters. This bolstered the hypothesis that physical association between enhancers and promoters is essential for enhancer activity, and the *ICR* insulator acts by blocking that essential association.

But how does the *ICR* insulator block association between the *Igf2* enhancers and promoters? Pfeifer and colleagues proposed that CTCF bound to the insulator interacts with the enhancers and promoters, or proteins bound to both, and prevents their interaction with each other (Figure 12.32). To test this hypothesis, they performed 3C and RFLP analysis on maternal and paternal chromosomes, with and without the insulator, and showed that indeed the insulator interacts with both enhancers and promoters, but only on the maternal chromosome, in which *Igf2* transcription is silenced.

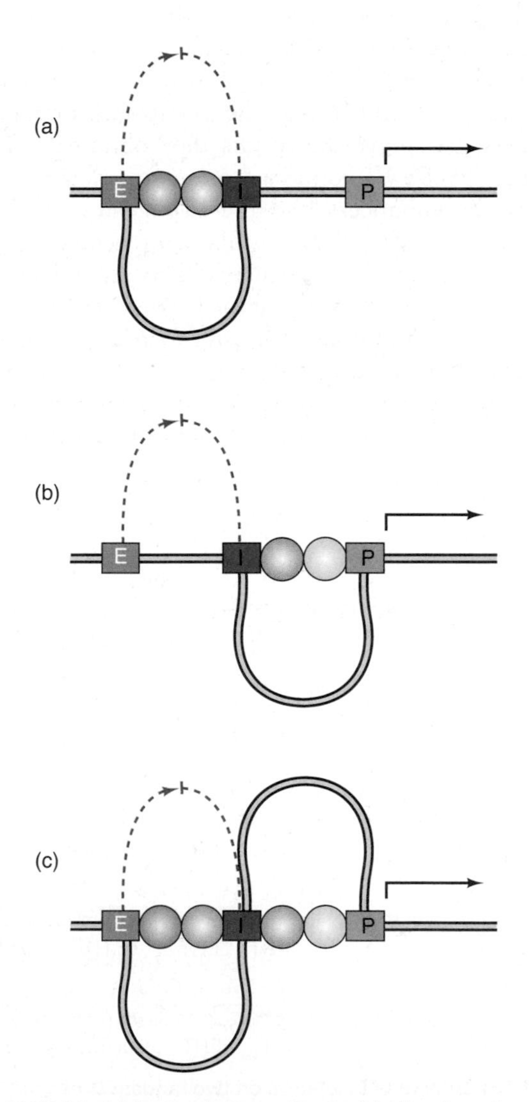

Figure 12.32 Model for insulator action by binding to enhancers and/or promoters. (a) The insulator binds to an enhancer (through proteins bound to both) and prevents its interaction with a promoter. **(b)** The insulator binds to a promoter (again through proteins) and prevents its interaction with an enhancer. **(c)** The insulator binds to both promoter and enhancer (through proteins) and prevents interaction between the promoter and enhancer.

Thus, in this system at least, insulator action appears to depend on the insulator's interacting with the enhancers and promoters in such a way that they cannot interact with each other. In some ways, this is an attractive hypothesis, but it has serious limitations as a general explanation for insulator action. First, insulators are position dependent. They block enhancer action only when placed between the enhancer and a promoter. In the present example, the *ICR* insulator blocks the enhancers from stimulating transcription from the *Igf2* promoters, but not from the *H19* promoter. It is not obvious why the position of the *ICR* insulator between the *Igf2* promoters and enhancers would cause it to interact only with those promoters, and not the *H19* promoter, which is much closer to the insulator. Second, insulators do not inactivate enhancers. While they block the action of an enhancer on one set of promoters (e.g., the *Igf2* promoters), they leave it free to stimulate transcription from another (e.g., the *H19* promoter). It is not clear how binding of the insulator to the *Igf2* enhancers and promoters would prevent their interaction with each other, and still allow them to interact productively with other chromosomal partners such as the *H19* promoter.

Finally, you may be wondering why the paternal copy of the *Igf2* gene is not affected by the insulator. The paternal *ICR* becomes methylated during and after spermiogenesis, so it cannot bind CTCF. Without the insulator-binding protein, the insulator cannot function, so the enhancers are allowed to stimulate transcription from the paternal *Igf2* promoters. Thus, methylation of the insulator is the functional equivalent of its removal.

Perhaps the best way to summarize our knowledge about the mechanism of insulator action is to acknowledge that there may not be a single mechanism. Some insulators may work one way, and others may have another mode of action.

SUMMARY Insulators are DNA elements that can shield genes from activation by enhancers (enhancer-blocking activity) or repression by silencers (barrier activity). Some insulators have both enhancer-blocking and barrier activities, but some have only one or the other. Insulators may do their job by working in pairs that bind proteins that can interact to form DNA loops. These loops would isolate enhancers and silencers so they can no longer stimulate or repress promoters. In this way, insulators may establish boundaries between DNA regions in a chromosome. Two or more insulators between an enhancer and a promoter cancel each other's effect, perhaps by binding proteins that interact with each other, thereby preventing the DNA looping that would isolate the enhancer from the promoter. Alternatively, the interaction between adjacent insulator-binding proteins could prevent the association of the insulators with insulator bodies, and this could block insulator activity. Insulators may also act as a barrier to a signal propagating along the chromosome from an enhancer or silencer. The nature of this signal is not defined, but it may be a sliding protein or a sliding (and growing) loop of chromatin. Finally, enhancer-blocking insulators may act by binding proteins that interact with proteins and/or DNA at enhancers and promoters, thereby preventing those enhancers and promoters from interacting with each other, which is essential for efficient transcription.

12.6 Regulation of Transcription Factors

Transcription factors regulate transcription both positively and negatively, but what regulates the regulators? We have already seen one example earlier in this chapter, and we will see several other examples in the last section of this chapter and in Chapter 13. They fall into the following categories:

- As we learned earlier in this chapter, binding between nuclear receptors (e.g., the glucocorticoid receptor) and their ligands (e.g., the glucocorticoids) can cause the receptors to dissociate from an inhibitory protein in the cytoplasm, translocate to the nucleus, and activate transcription.
- As we will see in Chapter 13, binding between nuclear receptors and their ligands can change the receptors from transcription repressors to activators.
- Phosphorylation of activators can allow them to interact with coactivators that in turn stimulate transcription.
- Ubiquitylation of transcription factors (attachment of the polypeptide ubiquitin to them) can mark them for destruction by proteolysis.
- Alternatively, ubiquitylation of transcription factors can stimulate their activity instead of marking them for destruction.
- Sumoylation of transcription factors (attachment of the polypeptide SUMO to them) can target them for incorporation into compartments of the nucleus where their activity cannot be expressed.
- Methylation of transcription factors can modulate their activity.
- Acetylation of transcription factors can modulate their activity.

Let us examine some of these regulation phenomena.

Coactivators

Some class II activators may be capable of recruiting the basal transcription complex all by themselves, possibly by contacting one or more general transcription factors or RNA polymerase. But many, if not most, cannot. Roger Kornberg and colleagues provided the first evidence that something else must be involved when they studied **activator interference,** or **squelching,** in 1989 and 1990. Squelching occurs when increasing the concentration of one activator inhibits the activity of another activator in an in vitro transcription experiment, presumably by competing for a scarce factor required by both activators. A reasonable candidate for such a limiting factor would be a general transcription factor, but Kornberg and coworkers discovered that adding very large quantities of the general transcription factors did not relieve squelching. This finding suggested that some other factor must be required by both activators.

What was this other factor? In 1990, Kornberg and colleagues partially purified a yeast protein that could relieve squelching. Then, in 1991, they purified this factor further and demonstrated directly that it had coactivator activity. That is, it could stimulate activated transcription, but not basal transcription in vitro. They called it **Mediator** because it appeared to mediate the effect of an activator. (We have already encountered Mediator in Chapter 11 in the context of the polymerase II holoenzyme.)

Kornberg and colleagues' assay for transcription used a G-less cassette (Chapter 5) driven by the yeast *CYC1* promoter and a GAL4-binding site. They added increasing concentrations of Mediator in the absence and presence of the activator GAL4-VP16, a chimeric activator with the DNA-binding domain of GAL4 and the transcription-activating domain of VP16. Figure 12.33 shows the results: Mediator had no effect on transcription in the absence of the activator (lanes 3–6), but it greatly stimulated transcription in the presence of the activator (lanes 7–10). A similar experiment with the yeast activator GCN4 yielded comparable results, showing that Mediator could cooperate with more than one activator having an acidic activation domain.

Mediator-like complexes have also been purified from higher eukaryotes, including humans. One such complex has been purified independently by two different groups and is therefore called by two different names: **SRB and MED-containing cofactor** (**SMCC**), and **thyroid-hormone-receptor-associated protein** (**TRAP**). SMCC/TRAP is the most complex of the known Mediator-like complexes in mammals, but there are others that seem to be structurally and functionally related to Mediator. One of these is CRSP, which we will discuss later in this section.

Further work has shown that Mediator and its homologs are ubiquitous participants at active class II promoters. Indeed, they are so widespread that they can be considered general transcription factors, rather than true coactivators. A typical **coactivator** is a protein that has no activator function of its own, but collaborates with one or more activators to stimulate the expression of a set of genes.

(a)

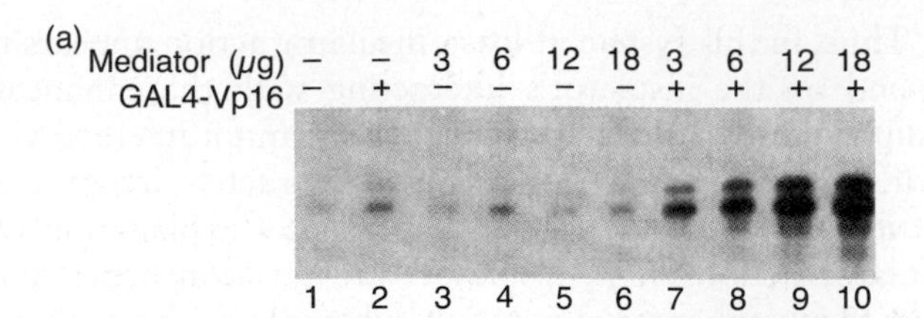

(b)

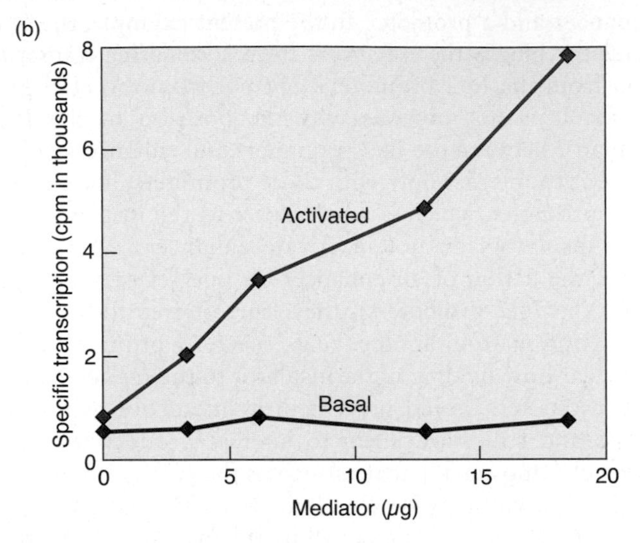

Figure 12.33 Discovery of Mediator. Kornberg and colleagues placed the yeast *CYC1* promoter downstream of a GAL4-binding site and upstream of a G-less cassette, so transcription of the G-less cassette depended on both the *CYC1* promoter and GAL4. Then they transcribed this construct in vitro in the absence of GTP and in the presence of the amounts of Mediator shown at the top of panel (a), and in the absence (–) or presence (+) of the activator GAL4-VP16 as indicated at the top of panel (a). They included a labeled nucleotide to label the products of the in vitro transcription reactions and electrophoresed the labeled RNAs. **(a)** Phosphorimager scan of the electropherogram. **(b)** Graphical presentation of the results in panel (a). Note that Mediator greatly stimulates transcription in the presence of the activator, but has no effect on unactivated (basal) transcription. (*Source:* Flanagan, P.M., R.J. Kelleher, 3rd, M.H. Sayre, H. Tschochner, and R.D. Kornberg, A mediator required for activation of RNA polymerase II transcription in vitro. *Nature* 350 (4 Apr 1991) f. 2, p. 437. Copyright © Macmillan Magazines Ltd.)

For example, in Chapter 7 we learned that cyclic-AMP (cAMP) stimulates transcription of bacterial operons by binding to an activator (CAP) and causing it to bind to activator target sites in the operon control regions. Cyclic-AMP also participates in transcription activation in eukaryotes, but it does so in a less direct way, through a series of steps called a **signal transduction pathway.** When the level of cAMP rises in a eukaryotic cell, it stimulates the activity of **protein kinase A** (**PKA**) and causes this enzyme to move into the cell nucleus. In the nucleus, PKA phosphorylates an activator called the **cAMP response element-binding protein** (**CREB**), which binds to the **cAMP response element** (**CRE**) and activates associated genes.

Because phosphorylation of CREB is necessary for activation of transcription, one would expect this phosphorylation to cause CREB to move into the nucleus or to bind more strongly to CREs, but neither of these things actually seems to happen—CREB localizes to the nucleus and binds to CREs very well even without being phosphorylated. How, then, does phosphorylation of CREB cause activation? The key to the answer appeared in 1993 with the discovery of the **CREB-binding protein (CBP).** CBP binds to CREB much more avidly after CREB has been phosphorylated by protein kinase A. Then, CBP can contact and recruit elements of the basal transcription apparatus, or it could recruit the holoenzyme as a unit. By coupling CREB to the transcription apparatus, CBP acts as a coactivator (Figure 12.34).

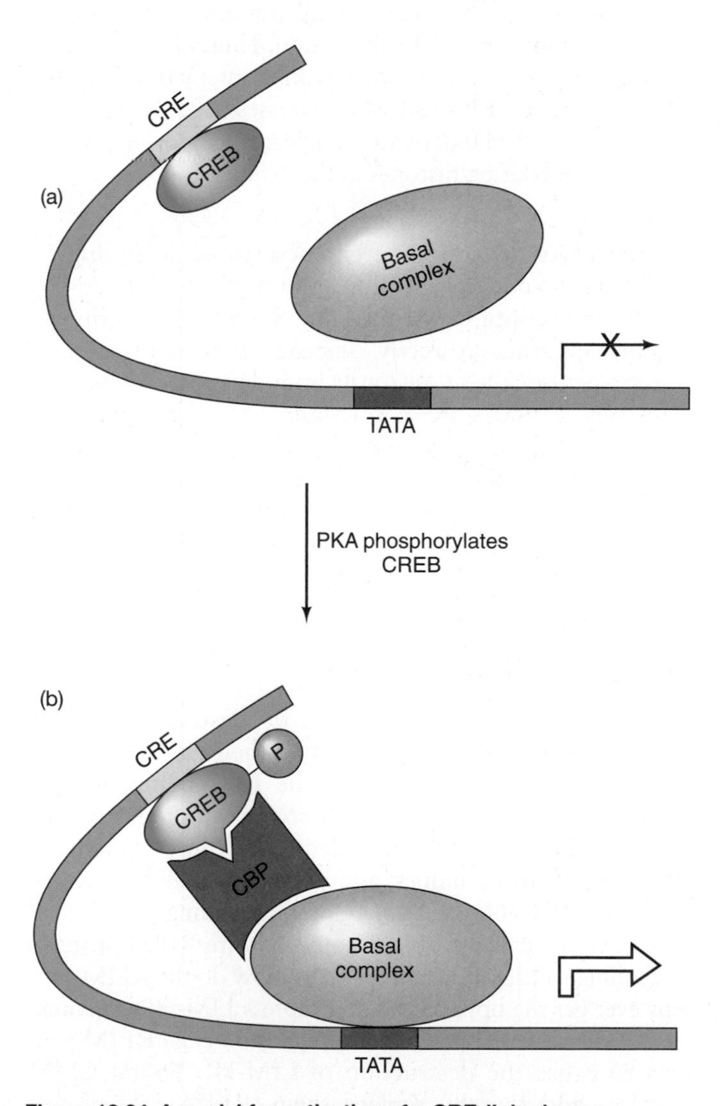

Figure 12.34 A model for activation of a CRE-linked gene. **(a)** Unphosphorylated CREB (turquoise) is bound to CRE, but the basal complex (RNA polymerase plus general transcription factors, orange) is not bound to the promoter in significant quantity and may not even have assembled yet. Thus, the gene is not activated. **(b)** PKA has phosphorylated CREB, which causes CREB to associate with CBP (red). CBP, in turn, associates with at least one component of the basal transcription complex, recruiting it to the promoter. Now transcription is activated.

Since 1993 when CBP was discovered, many coactivators, have been identified. In 1999, Tjian and colleagues isolated a coactivator required for activation of transcription in vitro by the transcription factor Sp1. When they purified this coactivator, which they called **cofactor required for Sp1 activation (CRSP),** they discovered that it had nine putative subunits. They separated these subunits by SDS-PAGE, transferred them to a nitrocellulose membrane, then cleaved each polypeptide with a protease to generate peptides that could be sequenced. The sequences revealed that some of the subunits of CRSP are unique, but many of them are identical, or at least homologous, to other known coactivators—subunits of the yeast Mediator, for example. Thus, different coactivators seem to be assembled by "mixing and matching" subunits from a variety of other coactivators. Mediator and CRSP also seem to share a mode of action in common. Both contact the CTD of RNA polymerase II. That interaction may explain how these coactivators help recruit the basal transcription complex.

The coactivator role of CBP is not limited to cAMP-responsive genes. It also serves as a coactivator in genes responsive to the nuclear receptors. This helps to explain why no one could detect direct interaction between the transcription-activation domains of the nuclear receptors and any of the general transcription factors. Part of the reason is that the nuclear receptors do not contact the basal transcription apparatus directly. Instead, CBP, or its homologue, **p300,** acts as a coactivator, helping to bring together the nuclear receptors and the basal transcription apparatus. But CBP does not perform this task alone. It collaborates with another family of coactivators called the **steroid receptor coactivator (SRC) family.** This group of proteins is also sometimes called the p160 family because of their molecular masses of 160 kD. The SRC family includes three groups of homologous proteins, **SRC-1, SRC-2,** and **SRC-3,** which interact with liganded (but not ligand-free) nuclear receptors. This interaction occurs between the nuclear receptor's activation domain and a so-called LXXLL box (where L stands for leucine and X stands for any amino acid) in the middle of the SRC protein chain. The SRC proteins also bind to CBP and can therefore help the nuclear receptors recruit CBP, which in turn recruits the basal transcription apparatus. The first SRC family member to be discovered was SRC-1 (Figure 12.35). It interacts with the ligand-bound forms of: progestin receptor; estrogen receptor; and thyroid hormone receptor. Not only does it bridge between nuclear receptors and CBP, it recruits a protein called **coactivator-associated arginine methyltransferase (CARM1),** which methylates proteins in the vicinity of the promoter, activating transcription. We will examine the role of CARM1 later in this section.

Still another important class of activators use CBP as a coactivator. A variety of growth factors and cellular stresses initiate a cascade of events (another signal transduction pathway) that results in the phosphorylation and activation

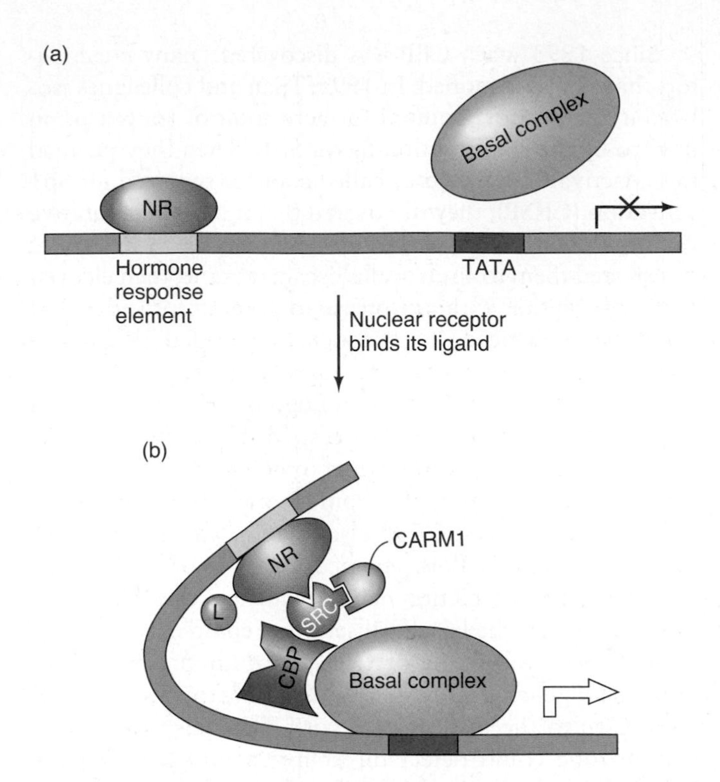

Figure 12.35 Models for activation of a nuclear receptor-activated gene. (a) A nuclear receptor (without its ligand) is bound to its hormone response element, but it cannot contact the basal transcription complex, so the linked gene is not activated. Depending on the type, the nuclear receptor could also be dissociated from its DNA target in the absence of its ligand. The nuclear receptor bound to its DNA target without its ligand may also actively inhibit transcription. **(b)** The nuclear receptor has bound to its ligand (purple) and is now able to interact with SRC (green), which in turn binds to CBP, which binds to at least one component of the basal transcription apparatus, recruiting it to the promoter and activating transcription. SRC also binds to CARM1 (torquoise), which methylates proteins near the promoter, further simulating transcription.

of a protein kinase called **mitogen-activated protein kinase (MAPK).** The activated MAPK enters the nucleus and phosphorylates activators such as Sap-1a and the Jun monomers in AP-1. These activators then use CBP to mediate activation of their target genes, which finally stimulate cell division.

Besides recruiting the basal transcription apparatus to the promoter, CBP plays another role in gene activation. CBP has a powerful histone acetyltransferase activity, which adds acetyl groups to histones. As we will see in Chapter 13, histones are general repressors of gene activity. Moreover, acetylation of histones causes them to loosen their grip on DNA and relax their repression of transcription. Thus, the association between activators and CBP at an enhancer brings the histone acetyltransferase to the enhancer, where it can acetylate histones and activate the nearby gene. We will discuss this phenomenon in greater detail in Chapter 13.

We have seen that CBP and p300 can serve as a coactivator for a variety of activators, including CREB and nuclear receptors. This means that the CREB and nuclear receptor pathways could potentially compete with each other for activation of different genes through the same coactivator. Ronald Evans and colleagues discovered that one way cells limit that competition is through methylation of CBP or p300. To simplify our discussion of this mechanism, we will refer to these proteins as CBP/p300.

Nuclear receptors attract not only CBP/p300, but several other proteins as well. One of these others is CARM1. The CARM1 activity methylates arginines on histones after they have been acetylated by CBP/p300, (Chapter 13) and this methylation has a transcription-activating effect. But CARM1 also methylates an arginine on CBP/p300 itself. The target arginine on CBP/p300 is in the so-called *KIX domain,* which is necessary for recruitment of CREB, but has no effect on the nuclear receptor-CBP/p300 interaction. Thus, CARM1 serves as a transcriptional switch. By blocking interaction between CBP/300 and CREB, CARM1 represses CREB-responsive genes, but CARM1 activates nuclear receptor-responsive genes by methylating histones in the vicinity.

SUMMARY Several different activators, including CREB, the nuclear receptors, and AP-1, do not activate transcription by contacting the basal transcription apparatus directly. Instead, they contact a coactivator called CBP (or its homolog p300), which in turn contacts the basal transcription apparatus and recruits it to promoters. CBP/p300 bound to nuclear receptor-response elements can also recruit CARM1, which methylates an arginine on CBP/p300 required to interact with CREB. This prevents activation of CREB-responsive genes.

Activator Ubiquitylation

Sometimes genes are inactivated by destruction of the activators that have been stimulating their activity. For example, transcription factors in the **LIM homeodomain (LIM-HD)** family associate with corepressors and coactivators. The coactivators are called **CLIM,** for "cofactor of LIM," among other names, and the corepressors are called **RLIM,** for "RING finger LIM domain-binding protein."

CLIM proteins are able to compete with RLIM proteins for binding to LIM-HD activators, so how do the RLIM proteins ever get the upper hand and repress LIM-HD-activated genes? The secret appears to lie in the ability of RLIM proteins to cause the destruction of LIM-HD-bound CLIM proteins, and thereby replace them. RLIM proteins set CLIM proteins up for destruction by binding to them and attaching several copies of a small protein called **ubiquitin** to lysine residues of the protein, creating what we call a **ubiquitylated protein.** Once the chain of ubiquitin molecules becomes long enough, it targets the ubiquitylated protein to a cytoplasmic structure called the **proteasome.** The proteasome is a collection of proteins with a combined sedimentation coefficient of 26S. It includes proteases that degrade any ubiquitylated protein brought to it.

The normal function of the ubiquitin-linked proteasome appears to be quality control. It is estimated that about 20% of cellular proteins are made incorrectly because of mistakes in transcription or translation. These aberrant proteins are potentially damaging to the cell, so they are tagged with ubiquitin and sent to the proteasome for degradation before they can cause any trouble. Other proteins that are made correctly can become denatured by stresses such as oxidation or heat. The cell has **chaperone proteins** that can unfold and then allow such denatured proteins to refold correctly. But sometimes the denaturation is so extensive that proper refolding is impossible. In such cases, the denatured proteins would be ubiquitylated and then destroyed by the proteasome.

It may seem surprising that ubiquitylation can also affect activators without causing their destruction. One example comes from the *MET* genes of yeast, which are required to produce the sulfur-containing amino acids methionine and cysteine. These genes are controlled by the concentration of the methyl donor *S*-adenosylmethionine, known as SAM or AdoMet (Chapter 15). When the concentration of SAM is low, the *MET* genes are stimulated by the activator Met4. However, when the concentration of SAM rises, Met4 is inactivated by a process that involves ubiquitylation. This seems to imply that Met4 is ubiquitylated and then destroyed by the proteasome. However, things are not that simple.

It is true that Met4 degradation can play a role in its inactivation, but under certain conditions (rich medium supplemented with methionine), Met4 remains stable despite being ubiquitylated. However, even though it is stable, ubiquitylated Met4 loses its ability to activate the *MET* genes. It can no longer bind properly to these genes, even though it is still able to bind and activate another class of genes called the *SAM* genes. Thus, ubiquitylation of Met4 can inactivate it directly, without causing its destruction. And this inactivation is selective. It affects the ability of Met4 to activate some genes, but not others.

Several studies have indicated that very strong transcription factors tend to be regulated by ubiquitylation and subsequent destruction by the proteasome. This allows a cell some flexibility in controlling gene expression because it provides a mechanism for quickly shutting off strong expression of genes driven by powerful activators. But, again, the picture is not quite as simple as just protein degradation. Some of these activators are actually activated by monoubiquitylation (tagging the protein with a single copy of ubiquitin). But polyubiquitylation of the same activator can mark it for destruction.

Recently, evidence has accumulated for another kind of involvement of the proteasome in transcription regulation. Proteins belonging to the 19S regulatory particle of the proteasome have been discovered in complexes with transcription factors at active promoters. Moreover, the **19S particle** can strongly stimulate transcription elongation in vitro. Also, a subset of proteins from the 19S particle can be recruited to promoters by the activator GAL4. These proteins include ATPases that are necessary for unfolding proteins prior to their degradation but not proteins involved in proteolysis itself. Thus, the activation effect of the 19S particle proteins appears to be independent of proteolysis. Joan Conaway and colleagues speculated that the proteasomal proteins stimulate transcription by at least partially unfolding transcription factors so that they can be remodeled in such a way that stimulates transcription initiation, or elongation, or both.

SUMMARY RLIM proteins, which are LIM-HD corepressors, can bind to LIM-HD coactivators such as CLIM proteins and ubiquitylate them. This marks the coactivators for destruction by the 26S proteasome and allows the RLIM corepressors to take their place. Ubiquitylation (especially monoubiquitylation) of some activators can have an activating effect, but polyubiquitylation marks these same proteins for destruction. Proteins from the 19S regulatory particle of the proteasome can stimulate transcription, perhaps by remodeling and thereby activating transcription factors.

Activator Sumoylation

Sumoylation is the addition of one or more copies of the 101-amino-acid polypeptide **SUMO** (**small ubiquitin-related modifier**) to lysine residues on a protein. This process is accomplished by a mechanism very similar to the one used in ubiquitylation, but the results are quite different. Instead of being destroyed, sumoylated activators appear to be targeted to a specific nuclear compartment that keeps them stable, but unable to reach their target genes.

For example, certain activators, including one called PML, for "promyelocytic leukemia," are normally sumoylated and sequestered in nuclear bodies called PML oncogenic domains (PODs). In promyelocytic leukemia cells, the PODs are disrupted, and the released transcription factors, including PML, presumably reach and activate their target genes, and this activation contributes to the leukemic state.

Another example involves the Wnt signal transduction pathway, which ends when an activator called β-catenin enters the nucleus and teams up with LEF-1, an architectural transcription factor that we discussed earlier in this chapter, to activate transcription of certain genes. LEF-1 is subject to sumoylation, which causes it to be sequestered in nuclear bodies. Without LEF-1, β-catenin cannot activate its target genes, and Wnt signaling is blocked. And, as we have already learned, LEF-1 is involved in activating other genes, such as the TCR-α gene, independent of Wnt signaling, and those activations are also blocked by LEF-1 sumoylation.

Another consideration is that LEF-1 can also partner with repressors, such as Groucho, and this repression is presumably also blocked by sumoylation of LEF-1.

SUMMARY Some activators can be sumoylated (coupled to a small protein called SUMO), which causes them to be sequestered in nuclear bodies, where they cannot carry out their transcription activation function.

Activator Acetylation

In Chapter 13 we will learn that basic proteins called histones associate with DNA and repress transcription. It has been known for a long time that these histones can be acetylated on lysine residues by enzymes called histone acetyltransferases (HATs), which decreases the histones' repressive activity. Recently, investigators have shown that HATs can also acetylate nonhistone activators and repressors, and this can have either positive or negative effects on the acetylated protein's activities.

The tumor suppressor protein p53 is an example of an activator whose acetylation stimulates its activity. The coactivator p300 has HAT activity that can acetylate p53. When this happens, the activity of p53 increases, resulting in stronger stimulation of transcription of this activator's target genes.

The HAT activity of p300 can also acetylate the repressor BCL6, and this acetylation inactivates the repressor. Thus, the result in this case, as with the activation of an activator, is stimulation of transcription.

SUMMARY Nonhistone activators and repressors can be acetylated by HATs, and this acetylation can have either positive or negative effects.

Signal Transduction Pathways

The phosphorylations of CREB, Jun, and β-catenin, mentioned in the preceding section, are all the results of signal transduction pathways. So signal transduction pathways play a major role in the control of transcription. Let us explore the concept of signal transduction further and examine some examples. Cells are surrounded by a semipermeable membrane that keeps the cell contents from escaping and provides some protection from noxious substances in the cell's environment. This barrier between the interior of a cell and its environment means that mechanisms had to evolve to allow cells to sense the conditions in their surroundings and to respond accordingly. Signal transduction pathways provide these mechanisms. Because the responses a cell makes to its environment usually require changes in gene expression, signal transduction pathways usually end with activation of a transcription factor that activates a gene or set of genes.

Figure 12.36 outlines three signal transduction pathways: the protein kinase A pathway; the Ras–Raf pathway;

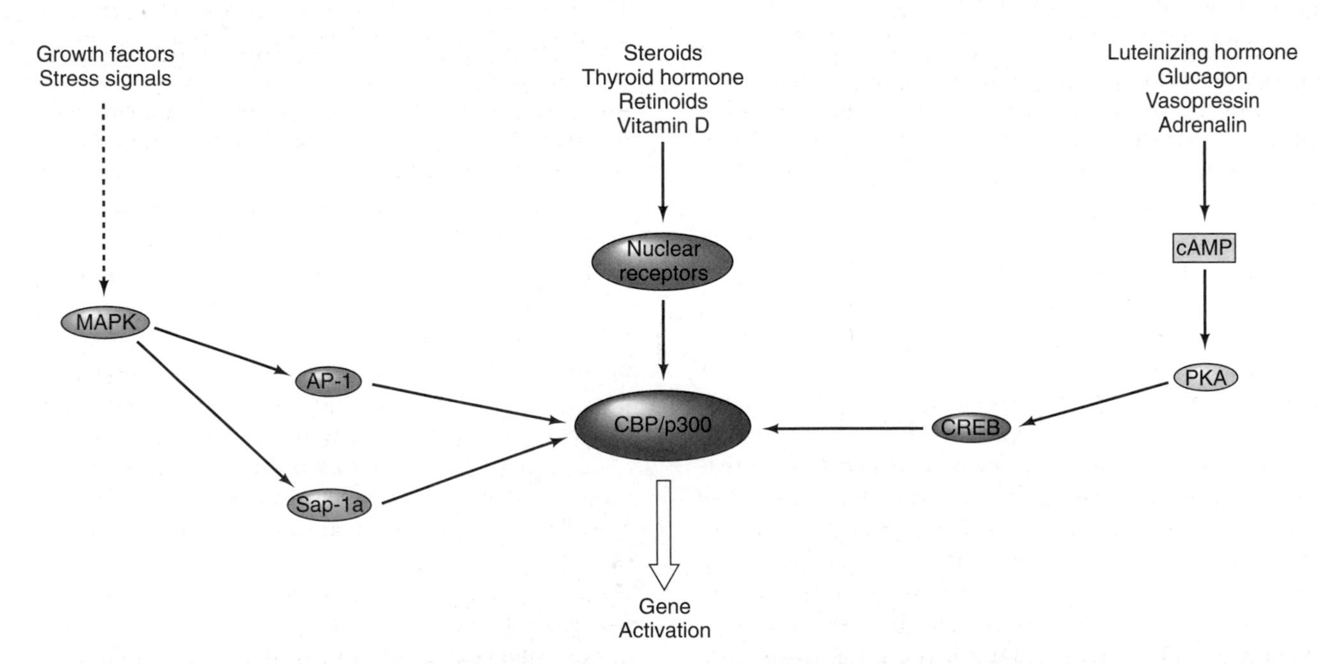

Figure 12.36 Multiple roles of CBP/p300. Three signal transduction pathways that use CBP/p300 to mediate transcription activation are shown converging on CBP/p300 (red) at center. The arrows between pathway members simply indicate position within the pathway (e.g., MAPK acts on AP-1), without indicating the nature of the action (e.g., phosphorylation). This scheme has also been simplified by omitting branches in the pathways. For example MAPK and PKA also phosphorylate nuclear receptors, although the importance of this phosphorylation is unclear. (*Source:* Adapted from Jankneht, R. and T. Hunter, Transcription: A growing coactivator network. *Nature* 383:23, 1996. Copyright © 1996.)

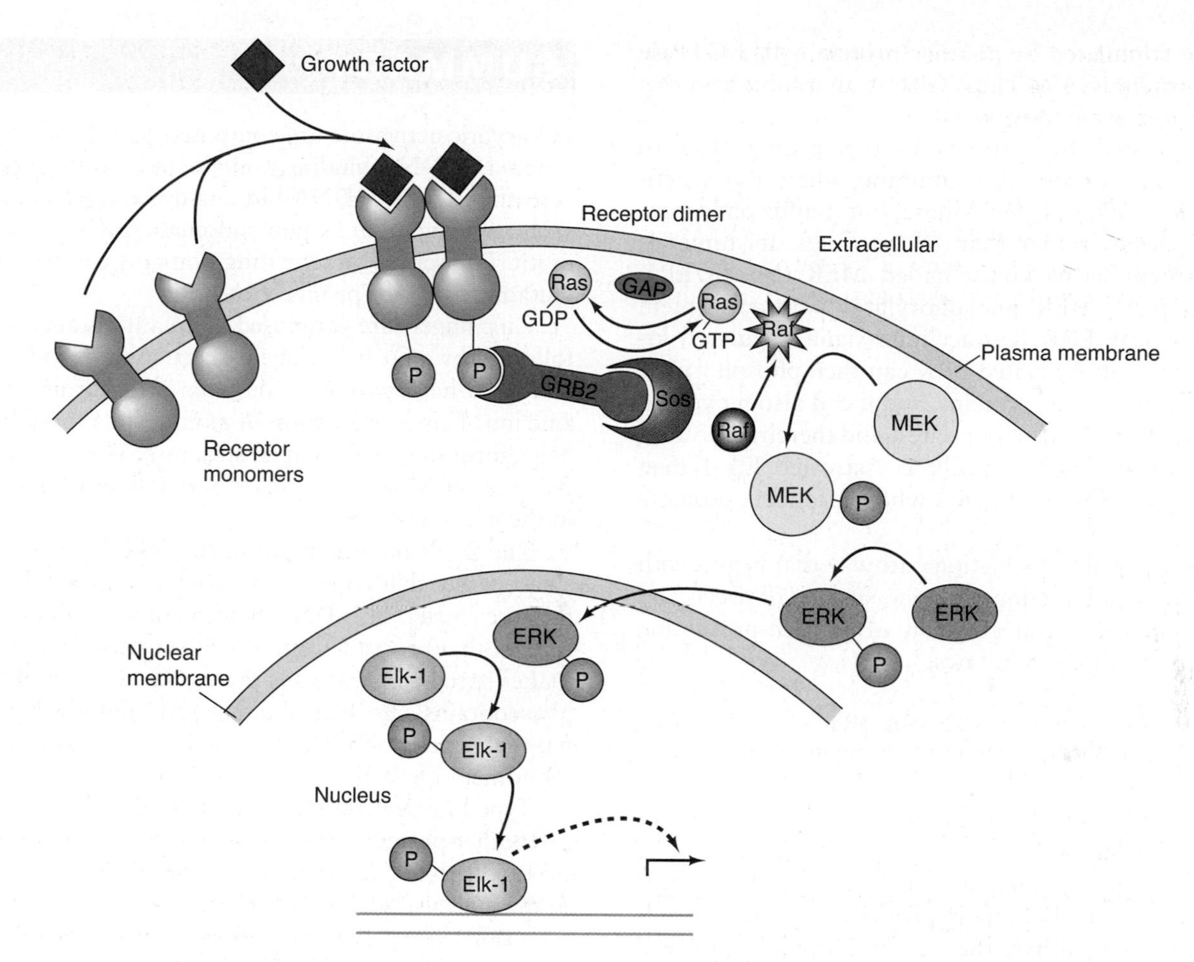

Figure 12.37 Signal transduction pathway involving Ras and Raf. Signal transduction begins (top) when a growth factor or other extracellular signaling molecule (red) binds to its receptor (blue). In this case, the receptor dimerizes on binding its ligand. The intracellular protein tyrosine kinase domain of each receptor monomer then phosphorylates its partner. The new phosphotyrosines can then be recognized by an adapter molecule called GRB2 (dark green), which in turn binds to the Ras exchanger Sos. Sos (gray) is activated to replace GDP on Ras with GTP, thus activating Ras (yellow). Ras delivers Raf (purple) to the cell membrane, where Raf becomes activated. The protein serine/threonine kinase domain of Raf is activated at the membrane, so it phosphorylates MAPK/ERK kinase (MEK, pale yellow), which phosphorylates extracellular-signal-regulated kinase (ERK, pink), which enters the nucleus and phosphorylates the transcription factor Elk-1 (light green). This activates Elk-1, which stimulates transcription of certain genes. The end result is more rapid cell division.

and the nuclear receptor pathway. The first two rely heavily on protein phosphorylation cascades to activate members of the pathway and ultimately to activate transcription. Let us explore the Ras–Raf pathway in more detail and see how aberrant members of the pathway can lead a cell to lose control over its growth and become a cancer cell.

Figure 12.37 presents a Ras–Raf pathway with mammalian names for the proteins. The same pathway operates in other organisms (famously in *Drosophila*) where the proteins have different names. The pathway begins when an extracellular agent, such as a growth factor, interacts with a receptor in the cell membrane. The agent (**epidermal growth factor [EGF]**, for example) binds to the extracellular domain of its receptor. This binding stimulates two adjacent receptors to come together to form a dimer, causing the intracellular domains, which have protein tyrosine kinase activity, to phosphorylate each other. Notice how the transmembrane receptor has transduced the signal across the cell membrane into the cell (Latin, *transducere,* meaning "to lead across"). Once the intracellular domains of the receptors are phosphorylated, the new phosphotyrosines attract adapter proteins such as **GRB2** (pronounced "grab two") that have specialized phosphotyrosine binding sites called **SH2 domains.** These are named for similar sites on an oncoprotein called pp60^{src}, which can transform cells from normal to tumor-like behavior; SH stands for "Src homology." GRB2 has another domain called **SH3** (also found in pp60^{src}) that attracts proteins with a particular kind of hydrophobic α-helix, such as **Sos.** Sos is a **Ras exchanger** that can replace GDP on the protein **Ras** with GTP, thereby activating the Ras protein. Ras contains an endogenous GTPase activity that can hydrolyze the GTP to GDP, inactivating the Ras protein. This GTPase activity is very weak by itself, but it can be strongly stimulated by another protein called **GTPase activator protein (GAP).** Thus, GAP is an inhibitor of this signal transduction pathway.

Once activated, Ras attracts another protein, **Raf,** to the inner surface of the cell membrane, where Raf is activated. Raf is another protein kinase, but it adds phosphate groups to serines rather than to tyrosines. Its target is another protein serine kinase called **MEK (MAPK/ERK kinase).** In turn, MEK phosphorylates another protein kinase known as **ERK (extracellular-signal-regulated kinase),** activating it. Activated ERK can then phosphorylate a variety of cytoplasmic proteins, and it can also move into the nucleus, where it phosphorylates, and thereby activates, several activators, including **Elk-1.** Activated Elk-1 then stimulates transcription of genes whose products promote cell division.

Thus, one signal transduction pathway that begins with a growth factor interacting with the surface of a cell and ends with enhanced transcription of growth-promoting genes, can be pictured as follows:

Growth factor→receptor→GRB2→Sos→Ras→Raf→MEK→
ERK→Elk-1→enhanced transcription→more cell division

It is not surprising that the genes encoding many of the carriers in this pathway are **proto-oncogenes,** whose mutation can lead to runaway cell growth and cancer. If these genes overproduce their products, or make products that are hyperactive, the whole pathway can speed up, leading to abnormally enhanced cell growth and, ultimately, to cancer.

Notice the amplifying power of this pathway. One molecule of EGF can lead to the activation of many molecules of Ras, each of which can activate many molecules of Raf. And, because Raf and the kinases that follow it in the pathway are all enzymes, each can activate many molecules of the next member of the pathway. By the end, one molecule of EGF can yield a great number of activated transcription factors, leading to a burst of new transcription. We should also note that this is only one pathway leading through Ras. In reality, the pathway branches at several points, rather like a web. This kind of interaction between members of different signal transduction pathways is called **cross talk.**

SUMMARY Signal transduction pathways usually begin with a signaling molecule that interacts with a receptor on the cell surface, which sends the signal into the cell, and frequently leads to altered gene expression. Many signal transduction pathways, including the Ras–Raf pathway, rely on protein phosphorylation to pass the signal from one protein to another. This amplifies the signal at each step.

SUMMARY

Eukaryotic activators are composed of at least two domains: a DNA-binding domain and a transcription-activating domain. DNA-binding domains include motifs such as a zinc module, homeodomain, bZIP, or bHLH motif. Transcription-activating domains can be acidic, glutamine-rich, or proline-rich.

Zinc fingers are composed of an antiparallel β-sheet, followed by an α-helix. The β-sheet contains two cysteines, and the α-helix two histidines, that are coordinated to a zinc ion. This coordination of amino acids to the metal helps form the finger-shaped structure. The specific recognition between the finger and its DNA target occurs in the major groove.

The DNA-binding motif of the GAL4 protein contains six cysteines that coordinate two zinc ions in a bimetal thiolate cluster. This DNA-binding motif contains a short α-helix that protrudes into the DNA major groove and makes specific interactions there. The GAL4 monomer also contains an α-helical dimerization motif that forms a parallel coiled coil with the α-helix on the other GAL4 monomer.

Type I nuclear receptors reside in the cytoplasm, bound to another protein. When they bind their hormone ligands, these receptors release their cytoplasmic partners, move to the nucleus, bind to enhancers, and thereby act as activators. The glucocorticoid receptor is representative of this group. It has a DNA-binding domain containing two zinc modules. One module contains most of the DNA-binding residues (in a recognition α-helix), and the other module provides the surface for protein–protein interaction to form a dimer. These zinc modules use four cysteine residues to complex the zinc ion, instead of two cysteines and two histidines as seen in classical zinc fingers.

The homeodomains in eukaryotic activators contain a DNA-binding motif that functions in much the same way as prokaryotic helix-turn-helix motifs, where a recognition helix fits into the DNA major groove and contacts specific residues there.

The bZIP proteins dimerize through a leucine zipper, which puts the adjacent basic regions of each monomer in position to embrace the DNA target site like a pair of tongs. Similarly, the bHLH proteins dimerize through a helix-loop-helix motif, which allows the basic parts of each long helix to grasp the DNA target site, much as the bZIP proteins do. The bHLH and bHLH-ZIP domains bind to DNA in the same way, but the latter have extra dimerization potential due to their leucine zippers.

The DNA-binding and transcription-activation domains of activator proteins are independent modules. Hybrid proteins with the DNA-binding domain of one protein and the transcription-activation domain of another still function as activators.

Activators function by contacting general transcription factors and stimulating the assembly of preinitiation complexes at promoters. For class II promoters, this assembly may occur by stepwise buildup of the general transcription factors and RNA polymerase II, as observed in vitro, or it may occur by recruitment of a large holoenzyme that includes RNA polymerase and most of the general transcription factors. Additional factors (perhaps just TBP or TFIID) may be recruited independently of the holoenzyme.

Dimerization is a great advantage to an activator because it increases the affinity between the activator and its DNA target. Some activators form homodimers, but others function better as heterodimers.

The essence of enhancer function—protein–protein interaction between activators bound to the enhancers, and general transcription factors and RNA polymerase bound to the promoter—seems in many cases to be mediated by looping out the DNA in between. At least in theory, this can also account for the effects of multiple enhancers on gene transcription. DNA looping could bring the activators bound to each enhancer close to the promoter where they could stimulate transcription, perhaps in a cooperative way.

Transcription appears to be concentrated in transcription factories within the nucleus, where an average of about 14 polymerases II and III are active. The existence of transcription factories implies the existence of DNA loops between genes being transcribed in the same factory.

Complex enhancers enable a gene to respond differently to different combinations of activators. This arrangement gives cells exquisitely fine control over their genes in different tissues, or at different times in a developing organism.

The architectural transcription factor LEF-1 binds to the minor groove of its DNA target through its HMG domain and induces strong bending in the DNA. LEF-1 does not enhance transcription by itself, but the bending it induces probably helps other activators bind and interact with still other activators and the general transcription factors to stimulate transcription.

An enhanceosome is a nucleoprotein complex containing a collection of activators bound to an enhancer so as to stimulate transcription. The archetypical enhanceosome involves the IFN-β enhancer. Its structure involves eight polypeptides bound cooperatively to an essentially straight 55-bp stretch of DNA. HMGA1a is essential for this cooperative binding, but it is not part of the final enhanceosome.

Insulators are DNA elements that can shield genes from activation by enhancers (enhancer-blocking activity) or repression by silencers (barrier activity). Some insulators have both enhancer blocking and barrier activities, but some have only one or the other. Insulators may do their job by working in pairs that bind proteins that can interact to form DNA loops. These loops would isolate enhancers and silencers so they can no longer stimulate or repress promoters. In this way, insulators may establish boundaries between DNA regions in a chromosome. Two or more insulators between an enhancer and a promoter cancel each other's effect, perhaps by binding proteins that interact with each other and thereby prevent the DNA looping that would isolate the enhancer from the promoter. Alternatively, the interaction between adjacent insulator-binding proteins could prevent the association of the insulators with insulator bodies, and this could block insulator activity.

Several different activators, including CREB, the nuclear receptors, and AP-1, do not activate transcription by contacting the basal transcription apparatus directly. Instead, upon being phosphorylated, they contact a coactivator called CBP (or its homolog p300), which in turn contacts the basal transcription apparatus and recruits it to promoters. CBP/p300 bound to nuclear receptor-response elements can also recruit CARM1, which methylates an arginine on CBP/p300 required to interact with CREB. This prevents activation of CREB-responsive genes.

Some activators and coactivators are controlled by ubiquitin-mediated destruction. The proteins are ubiquitylated, which marks them for destruction by the 26S proteasome. Ubiquitylation (especially monoubiquitylation) of some activators can have an activating effect, but polyubiquitylation marks these same proteins for destruction. Proteins from the 19S regulatory particle of the proteasome can stimulate transcription, perhaps by remodeling and thereby activating transcription factors.

Some activators can be sumoylated (coupled to a small, ubiquitin-like protein called SUMO), which causes them to be sequestered in nuclear bodies, where they cannot carry out their transcription activation function. Nonhistone activators and repressors can be acetylated by HATs, and this acetylation can have either positive or negative effects.

Signal transduction pathways usually begin with a signaling molecule that interacts with a receptor on the cell surface, which sends the signal into the cell, and frequently leads to altered gene expression. Many signal transduction pathways, including the Ras–Raf pathway, rely on protein phosphorylation to pass the signal from one protein to another. This enzymatic action amplifies the signal at each step. However, ubiquitylation and sumoylation of activators and other signal transduction pathway members can also play major roles in these pathways.

REVIEW QUESTIONS

1. List three different classes of DNA-binding domains found in eukaryotic transcription factors.
2. List three different classes of transcription activation domains in eukaryotic transcription factors.

3. Draw a diagram of a zinc finger. Point out the DNA-binding motif of the finger.
4. List one important similarity and three differences between a typical prokaryotic helix-turn-helix domain and the Zif268 zinc finger domain.
5. Draw a diagram of the dimer composed of two molecules of the N-terminal 65 amino acids of the GAL4 protein, interacting with DNA. Your diagram should show clearly the dimerization domains and the motifs in the two DNA-binding domains interacting with their DNA-binding sites. What metal ions and coordinating amino acids, and how many of each, are present in each DNA-binding domain?
6. In general terms, what is the function of a nuclear receptor?
7. Explain the difference between type I and II nuclear receptors and give an example of each.
8. What metal ions and coordinating amino acids, and how many of each, are present in each DNA-binding domain of a nuclear receptor? What part of the DNA-binding domain contacts the DNA bases?
9. What is the nature of the homeodomain? What other DNA-binding domain does it most resemble?
10. Draw a diagram of a leucine zipper seen from the end. How does this diagram illustrate the relationship between the structure and function of the leucine zipper?
11. Draw a diagram of a bZIP protein interacting with its DNA-binding site.
12. Describe and show the results of an experiment that illustrates the independence of the DNA-binding and transcription-activating domains of an activator.
13. Present two models of recruitment of the class II preinitiation complex, one involving a holoenzyme, the other not.
14. Describe and give the results of an experiment that shows that an acidic transcription-activating domain binds to TFIID.
15. Present evidence that favors the holoenzyme recruitment model.
16. Present two lines of evidence that argue against the holoenzyme recruitment model.
17. Why is a protein dimer (or tetramer) so much more effective than a monomer in DNA binding? Why is it important for a transcription activator to have a high affinity for specific sequences in DNA?
18. Present three models to explain how an enhancer can act on a promoter hundreds of base pairs away.
19. Describe and give the results of an experiment that shows the effect of isolating an enhancer on a separate circle of DNA intertwined with another circle of DNA that contains the promoter. Which model(s) of enhancer activity does this experiment favor? Why?
20. Describe how you would perform a hypothetical 3C experiment. Describe the results you would get, and give an interpretation.
21. What advantage do complex enhancers confer on a gene?
22. Describe how you would identify transcription factories in a cell nucleus. Why are both in vitro and in vivo transcription essential parts of the procedure? Why does the existence of transcription factories imply that chromatin loops occur in the nucleus?
23. LEF-1 is an activator of the human T-cell receptor α-chain, yet LEF-1 by itself does not activate this gene. How does LEF-1 act? Describe and show the results of an experiment that supports your answer.
24. Does LEF-1 bind in the major or minor groove of its DNA target? Present evidence to support your answer.
25. What do insulators do?
26. Diagram a model to explain the following results: (a) Having one insulator between an enhancer and a promoter partially blocks enhancer activity. (b) Having two insulators between an enhancer and a promoter does not block enhancer activity. (c) Having one insulator on either side of an enhancer strongly blocks enhancer activity.
27. What is the effect of three copies of an insulator between an enhancer and a promoter? How do you explain this phenomenon?
28. Present evidence for the hypothesis that an insulator blocks enhancement by interacting with nearby enhancers and promoters. What are the difficulties in generalizing this hypothesis to all insulators?
29. Describe and give the results of an experiment that shows the effects of Mediator.
30. Draw diagrams to illustrate the action of CBP as a coactivator of (a) phosphorylated CREB; (b) a nuclear receptor.
31. How do signal transduction pathways amplify their signals? Present an example.
32. Present a hypothesis to explain the negative effect of ubiquitin on transcription.
33. Present a hypothesis to explain the positive effect of proteasome proteins on transcription.

ANALYTICAL QUESTIONS

1. Design an experiment to show that TFIID binds directly to an acidic activating domain. Show sample positive results.
2. You are studying the human *BLU* gene, which is under the control of three enhancers. You suspect that the proteins that bind to these enhancers interact with each other to form an enhanceosome that is required for activation. What spacing among these enhancers is optimal for such interaction? What changes in this spacing could you introduce to test your hypothesis? What results would you expect?
3. Consider Figure 12.22a. What primers would you use in a 3C experiment to show association between the *ICR* insulator and each of the *Igf2* promoters *P1*, *P2*, and *P3*, on the maternal chromosome.
4. You are going to create a human activator (eA1) that controls a set of genes responsible for academic success. You aim to create an activator that includes the components deemed essential through the study of other activators.

What is the composition of your activator? In the process you also create two additional distinct activators (eA2, eA3). What experiments would you run to determine which activator works best? Suppose you wanted the activator to work in women students but not in men. How might you arrange that? What kind of activator would you have to design to make that work?

SUGGESTED READINGS

General References and Reviews

Bell, A.C. and G. Felsenfeld. 1999. Stopped at the border: boundaries and insulators. *Current Opinion in Genetics and Development* 9:191–98.

Blackwood, E.M. and J.T. Kadonaga. 1998. Going the distance: A current view of enhancer action. *Science* 281:60–63.

Carey, M. 1994. Simplifying the complex. *Nature* 368:402–3.

Conaway, R.C., C.S. Brower, and J.W. Conaway. 2002. Emerging roles of ubiquitin in transcription regulation. *Science* 296:1254–58.

Freiman, R.N. and R. Tjian. 2003. Regulating the regulators: Lysine modifications make their mark. *Cell* 112:11–17.

Goodrich, J.A., G. Cutler, and R. Tjian. 1996. Contacts in context: Promoter specificity and macromolecular interactions in transcription. *Cell* 84:825–30.

Hampsey, M. and D. Reinberg. 1999. RNA polymerase II as a control panel for multiple coactivator complexes. *Current Opinion in Genetics and Development* 9:132–39.

Janknecht, R. and T. Hunter. 1996. Transcription. A growing coactivator network. *Nature* 383:22–23.

Montminy, M. 1997. Something new to hang your hat on. *Nature* 387:654–55.

Myer, V.E. and R.A. Young. 1998. RNA polymerase II holoenzymes and subcomplexes. *Journal of Biological Chemistry* 273:27757–60.

Nordheim, A. 1994. CREB takes CBP to tango. *Nature* 370:177–78.

Ptashne, M. and A. Gann. 1997. Transcriptional activation by recruitment. *Nature* 386:569–77.

Roush, W. 1996. "Smart" genes use many cues to set cell fate. *Science* 272:652–53.

Sauer, F. and R. Tjian. 1997. Mechanisms of transcription activation: Differences and similarities between yeast, *Drosophila,* and man. *Current Opinion in Genetics and Development* 7:176–81.

Wallace, J.A. and G. Felsenfeld. 2007. We gather together: insulators and genome organization. *Current Opinion in Genetics and Development* 17:400–407.

Werner, M.H. and S.K. Burley. 1997. Architectural transcription factors: Proteins that remodel DNA. *Cell* 88:733–36.

West, A.G., M. Gaszner, and G. Felsenfeld. 2002. Insulators: many functions, many mechanisms. *Genes and Development* 16:271–88.

Research Articles

Barberis, A., J. Pearlberg, N. Simkovich, S. Farrell, P. Reinagle, C. Bamdad, G. Sigal, and M. Ptashne. 1995. Contact with a component of the polymerase II holoenzyme suffices for gene activation. *Cell* 81:359–68.

Borggrefe, T., R. Davis, A. Bareket-Samish, and R.D. Kornberg. 2001. Quantitation of the RNA polymerase II transcription machinery in yeast. *Journal of Biological Chemistry* 276:47150–53.

Brent, R. and M. Ptashne. 1985. A eukaryotic transcriptional activator bearing the DNA specificity of a prokaryotic repressor. *Cell* 43:729–36.

Cai, H.N. and P. Shen 2001. Effects of *cis* arrangement of chromatin insulators on enhancer-blocking activity. *Science* 291:493–95.

Dunaway, M. and P. Dröge. 1989. Transactivation of the *Xenopus* rRNA gene promoter by its enhancer. *Nature* 341:657–59.

Ellenberger, T.E., C.J. Brandl, K. Struhl, and S.C. Harrison. 1992. The GCN4 basic region leucine zipper binds DNA as a dimer of uninterrupted α helices: Crystal structure of the protein–DNA complex. *Cell* 71:1223–37.

Flanagan, P.M., R.J. Kelleher, 3rd, M.H. Sayre, H. Tschochner, and R.D. Kornberg. 1991. A mediator required for activation of RNA polymerase II transcription in vitro. *Nature* 350:436–38.

Geise, K., J. Cox, and R. Grosschedl. 1992. The HMG domain of lymphoid enhancer factor 1 bends DNA and facilitates assembly of functional nucleoprotein structures. *Cell* 69:185–95.

Jackson, D.A., F.J. Iborra, E.M.M. Manders, and P.R. Cook. 1998. Numbers and organization of RNA polymerases, nascent transcripts, and transcription units in HeLa nuclei. *Molecular Biology of the Cell* 9:1523–36.

Kissinger, C.R., B. Liu, E. Martin-Blanco, T.B. Kornberg, and C.O. Pabo. 1990. Crystal structure of an engrailed homeodomain–DNA complex at 2.8 Å resolution: A framework for understanding homeodomain–DNA interactions. *Cell* 63:579–90.

Koleske, A.J. and R.A. Young. 1994. An RNA polymerase II holoenzyme responsive to activators. *Nature* 368:466–69.

Krebs, J.E. and Dunaway, M. 1998. The scs and scs′ elements impart a *cis* requirement on enhancer–promoter interactions. *Molecular Cell* 1:301–08.

Lee, M.S. 1989. Three-dimensional solution structure of a single zinc finger DNA-binding domain. *Science* 245:635–37.

Leuther, K.K., J.M. Salmeron, and S.A. Johnston. 1993. Genetic evidence that an activation domain of GAL4 does not require acidity and may form a β sheet. *Cell* 72:575–85.

Luisi, B.F., W.X. Xu, Z. Otwinowski, L.P. Freedman, K.R. Yamamoto, and P.B. Sigler. 1991. Crystallographic analysis of the interaction of the glucocorticoid receptor with DNA. *Nature* 352:497–505.

Ma, P.C.M., M.A. Rould, H. Weintraub, and C.O. Pabo. 1994. Crystal structure of MyoD bHLH domain–DNA complex: Perspectives on DNA recognition and implications for transcription activation. *Cell* 77:451–59.

Marmorstein, R., M. Carey, M. Ptashne, and S.C. Harrison. 1992. DNA recognition by GAL4: Structure of a protein–DNA complex. *Nature* 356:408–14.

Muravyova, E., A. Golovnin, E. Gracheva, A. Parshikov, T. Belenkaya, V. Pirrotta, and P. Georgiev. 2001. Loss of insulator activity by paired Su(Hw) chromatin insulators. *Science* 291:495–98.

O'Shea, E.K., J.D. Klemm, P.S. Kim, and T. Alber. 1991. X-ray structure of the GCN4 leucine zipper, a two-stranded, parallel coiled coil. *Science* 254:539–44.

Panne, D., T. Maniatis, and S.C. Harrison. 2007. An atomic model of the interferon-β enhanceosome. *Cell* 129:1111–23.

Pavletich, N.P. and C.O. Pabo. 1991. Zinc finger–DNA recognition: Crystal structure of a Zif 268–DNA complex at 2.1 Å. *Science* 252:809–17.

Romano, L.A. and G.A. Wray. 2003. Conservation of *Endo 16* expression in sea urchins despite evolutionary divergence in both cis and trans-acting components of transcriptional regulation. *Development* 130:4187–99.

Ryu, L., L. Zhou, A.G. Ladurner, and R. Tjian. 1999. The transcriptional cofactor complex CRSP is required for activity of the enhancer-binding protein Sp1. *Nature* 397:446–50.

Stringer, K.F., C. J. Ingles, and J. Greenblatt. 1990. Direct and selective binding of an acidic transcriptional activation domain to the TATA-box factor TFIID. *Nature* 345:783–86.

Xu, W., H. Chen, K. Du, H. Asahara, M. Tini, B.M. Emerson, M. Montminy, and R.M Evans. 2001. A transcriptional switch mediated by cofactor methylation. *Science* 294:2507–11.

Yoon, Y.S., S. Jeong, Q. Rong, K.-Y. Park, J.H. Chung, and K. Pfeifer. 2007. Analysis of the *H19ICR* insulator. *Molecular and Cellular Biology* 27:3499–3510.

Yuh, C.-H., B. Hamid, and E.H. Davidson. 1998. Genomic *cis*-regulatory logic: Experimental and computational analysis of a sea urchin gene. *Science* 279:1896–1902.

Zhu, J. and M. Levine. 1999. A novel *cis*-regulatory element, the PTS, mediates an antiinsulator activity in the *Drosophila* embryo. *Cell* 99:567–75.

CHAPTER 13

Chromatin Structure and Its Effects on Transcription

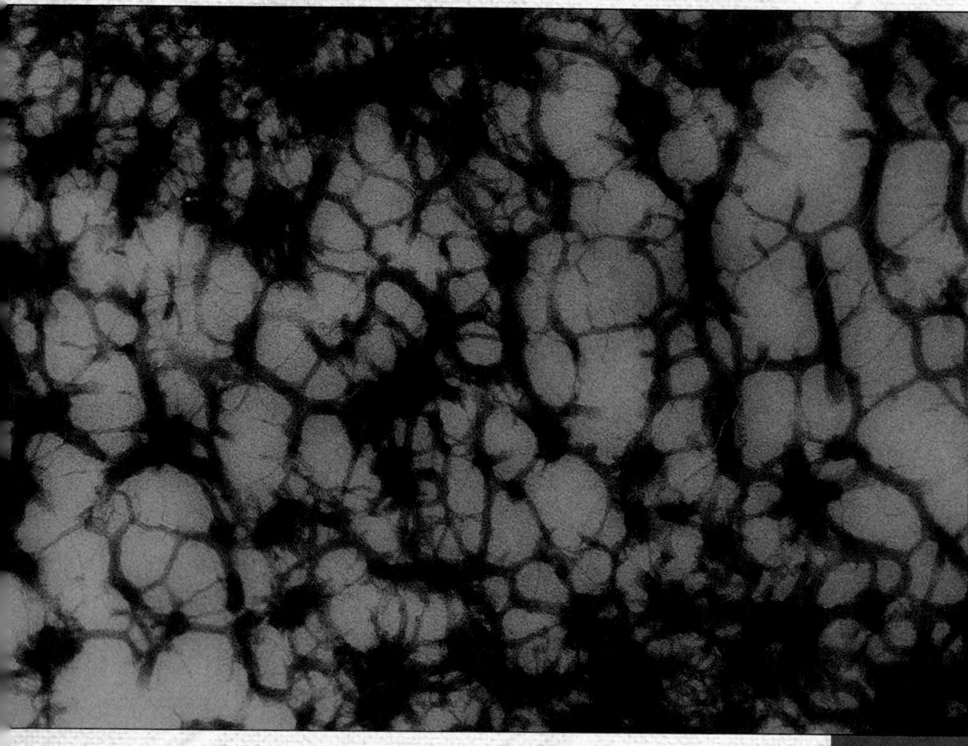

Chromatin in developing human spermatid (×300,000).

In our discussion of transcription of eukaryotic genes, we have so far been ignoring an important point: Eukaryotic genes do not exist naturally as naked DNA molecules, or even as DNA molecules bound only to transcription factors. Instead, they are complexed with an equal mass of other proteins to form a substance known as chromatin. As we will see, the chemical nature of chromatin is variable, and these variations play an enormous role in chromatin structure and in the control of gene expression.

13.1 Chromatin Structure

Chromatin is composed of DNA and proteins, mostly basic proteins called **histones** that help chromatin fold so it can pack into the tiny volume of a cell's nucleus. In this section we will examine the structure of histones, and the role they play in folding chromatin. In a later section we will look at the roles histones play in modifying the structure of chromatin and in controlling transcription.

Histones

Most eukaryotic cells contain five different kinds of **histones: H1, H2A, H2B, H3,** and **H4.** These are extremely abundant proteins; the mass of histones in eukaryotic nuclei is equal to the mass of DNA. They are also unusually basic—at least 20% of their amino acids are arginine or lysine—and have a pronounced positive charge at neutral pH. For this reason, they can be extracted from cells with strong acids, such as 1.5 N HCl—conditions that would destroy most proteins. Also because of their basic nature, the histones migrate toward the cathode during nondenaturing electrophoresis, unlike most other proteins, which are acidic and therefore move toward the anode. Most of the histones are also well conserved from one organism to another. The most extreme example of this is histone H4. Cow histone H4 differs from pea H4 in only two amino acids out of a total of 102, and these are conservative changes—one basic amino acid (lysine) substituted for another (arginine), and one hydrophobic amino acid (valine) substituted for another (isoleucine). In other words, in the more than one billion years since the cow and pea lines have diverged from a common ancestor, only two amino acids in histone H4 have changed. Histone H3 is also extremely well conserved; histones H2A and H2B are moderately well conserved; but histone H1 varies considerably among organisms. Table 13.1 lists some of the characteristics of histones.

Low-resolution gel electrophoresis of the histones gives the impression that each histone is a homogeneous species. However, higher resolution separations of the histones have revealed much greater variety. This variety stems from two sources: gene reiteration and posttranslational modification. The histone genes are not single-copy genes like most protein-encoding genes in eukaryotes. Instead, they are repeated many times: 10–20 times in the mouse, and about 100 times in *Drosophila.* Many of these copies are identical, but some are quite different. Histone H1 (the lysine-rich histone) shows the greatest variation, with at least six subspecies in the mouse. One H1 variant is called H1°. Birds, fish, amphibians, and reptiles have another lysine-rich histone that could be an extreme variant of H1, but it is so different from H1 that it is generally called by a distinct name, H5. Histone H4 shows the least variation; only two variant species have ever been reported, and these are rare. It is assumed that the variant species of a given histone all play essentially the same role, but each may influence the properties of chromatin somewhat differently.

The second cause of histone heterogeneity, posttranslational modification, is an exceedingly rich source of variation. The most common histone modification is acetylation, which can occur on N-terminal amino groups and on lysine ε-amino groups. Other modifications include lysine ε-amino methylation and phosphorylation, including serine and threonine O-phosphorylation. These and other histone modifications are summarized in Table 13.2. These modifications are dynamic processes, so modifying groups can be removed as well as added. These histone modifications influence chromatin structure and function, and play important roles in governing gene activity. We will discuss this phenomenon later in this chapter.

Table 13.1 General Properties of the Histones

Histone Type	Histone	Molecular Mass (M_r)
Core histones	H3	15,400
	H4	11,340
	H2A	14,000
	H2B	13,770
Linker histones	H1	21,500
	H1°	~21,500
	H5	21,500

Source: Adapted from Critical Reviews in Biochemistry and Molecular Biology, by Butler, P.J.C., 1983. Taylor & Francis Group. LLC., http://www.taylorandfrancis.com

Table 13.2 Histone Modifications

Modification	Amino Acids Modified
Acetylation (ac)	Lysine
Methylation (me)	Lysine (mono-, di-, or tri-me)
Methylation	Arginine (mono- or di-me[symmetric and asymmetric])
Phosphorylation	Serine and threonine
Ubiquitylation	Lysine
Sumoylation	Lysine
ADP ribosylation	Glutamate
Deimination	Arginine → Citrulline
Proline isomerization	Proline (cis → trans)

Nucleosomes

The length-to-width ratio of a typical human chromosome is more than 10 million to one. Such a long, thin molecule would tend to get tangled if it were not folded somehow. Another way of considering the folding problem is that the total length of human DNA, if stretched out, would be about 2 m, and this all has to fit into a nucleus only about 10 μm in diameter. In fact, if you laid all the DNA molecules in your body end to end, they would reach to the sun and back hundreds of times. Obviously, a great deal of DNA folding must occur in your body and in all other living things. We will see that eukaryotic chromatin is indeed folded in several ways. The first order of folding involves structures called **nucleosomes,** which have a core of histones, around which the DNA winds.

Maurice Wilkins showed as early as 1956 that x-ray diffraction patterns of DNA in intact nuclei exhibited sharp bands, indicating a repeating structure larger than the double helix itself. Subsequent x-ray diffraction work by Aaron Klug, Roger Kornberg, Francis Crick, and others showed a strong repeat at intervals of approximately 100 Å. This corresponds to a string of nucleosomes, which are about 110 Å in diameter. Kornberg found in 1974 that he could chemically cross-link histones H3 and H4, or histones H2A and H2B in solution. Moreover, he found that H3 and H4 exist as a tetramer $(H3–H4)_2$ in solution. He also noted that chromatin is composed of roughly equal masses of histones and DNA. In addition, the concentration of histone H1 is about half that of the other histones. This corresponds to one histone octamer (two molecules each of H2A, H2B, H3, and H4) plus one molecule of histone H1 per 200 bp of DNA. Finally, he reconstituted chromatin from H3–H4 tetramers, H2A–H2B oligomers, and DNA and found that this reconstituted chromatin produced the same x-ray diffraction pattern as natural chromatin. Several workers, including Gary Felsenfeld and L.A. Burgoyne, had already shown that chromatin cut with a variety of nucleases yielded DNA fragments about 200 bp long. Based on all these data, Kornberg proposed a repeating structure of chromatin composed of the histone octamer plus one molecule of histone H1 complexed with about 200 bp of DNA.

G.P. Georgiev and coworkers discovered that histone H1 is much easier than the other four histones to remove from chromatin. In 1975, Pierre Chambon and colleagues took advantage of this phenomenon to selectively remove histone H1 from chromatin with trypsin or high salt buffers, and found that this procedure yielded chromatin with a "beads-on-a-string" appearance (Figure 13.1a). They named the beads **nucleosomes.** Figure 13.1b shows some of the nucleosomes that Chambon and coworkers purified from chicken red blood cells, using micrococcal nuclease to cut the DNA string between the beads.

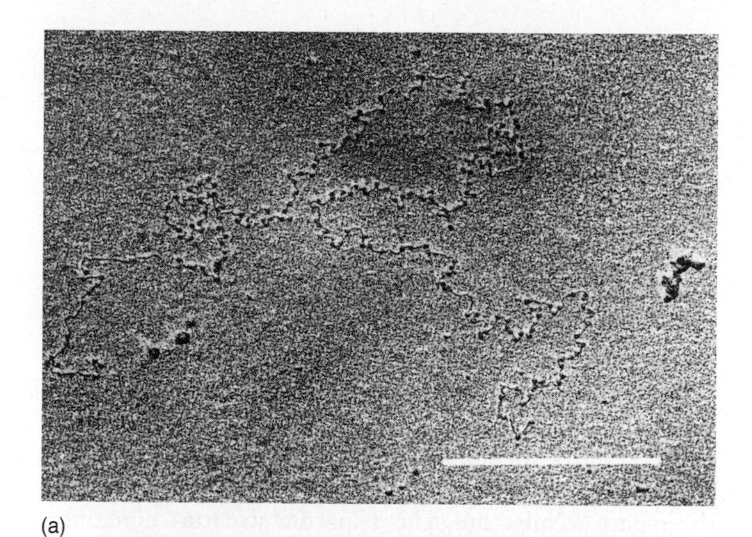

(a)

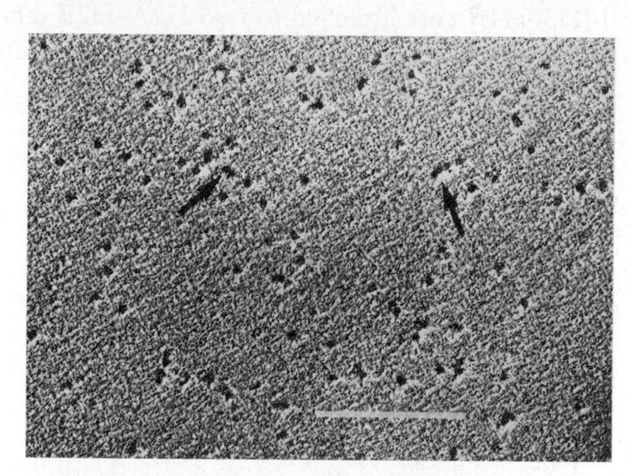

(b)

Figure 13.1 Early electron micrographs of nucleosomes. **(a)** Nucleosome strings. Chambon and colleagues used trypsin to remove histone H1 from chromatin isolated from chicken red blood cells, revealing a beads-on-a-string structure. The bar represents 500 nm. **(b)** Isolated nucleosomes. Chambon's group used micrococcal nuclease to cut between nucleosomes, then isolated these particles by ultracentrifugation. The arrows point to two representative nucleosomes. The bar represents 250 nm. (*Source:* Oudet P., M. Gross-Bellarard, and P. Chanaban, Electron microscopic and biochemical evidence that chromatin structure is a repeating unit. *Cell* 4 (1975), f. 4b & 5, pp. 286–87. Reprinted by permission of Elsevier Science.)

J.P. Baldwin and colleagues subjected chromatin to neutron-scattering analysis, which is similar to x-ray diffraction, but uses a beam of neutrons instead of x-rays. The pattern of scattering of the neutrons by the sample gives clues to the three-dimensional structure of the molecules in the sample. These investigators found a ring of scattered neutrons corresponding to a repeat distance of about 105 Å, which agreed with the x-ray diffraction analysis. Moreover, the overall pattern suggested that the protein and DNA occupied separate regions within the nucleosomes. Based on these data, Baldwin and coworkers proposed that the core histones (H2A, H2B,

H3, and H4) form a ball, with the DNA wrapped around the outside. Having the DNA on the outside also has the advantage that it minimizes the amount of bending the DNA would have to do. In fact, double-stranded DNA is such a stiff structure that it could not bend tightly enough to fit inside a nucleosome. These workers also placed histone H1 on the outside, in accord with its ease of removal from chromatin. In fact, H1 binds to the linker DNA between nucleosomes, which is why it is called a **linker histone.**

Several research groups have used x-ray crystallography to determine a structure for the histone octamer. According to the work of Evangelos Moudrianakis and his colleagues in 1991, the octamer takes on different shapes when viewed from different directions, but most viewpoints reveal a three-part architecture. This tripartite structure contains a central $(H3–H4)_2$ core attached to two H2A–H2B dimers, as shown in Figure 13.2a and b. The overall structure is shaped roughly like a disc, or hockey puck, that has been worn down to a wedge shape. Notice that this structure is consistent with Kornberg's data on the association between histones in solution and with the fact that the histone octamer dissociates into an $(H3–H4)_2$ tetramer and two H2A–H2B dimers.

Where does the DNA fit in? It was not possible to tell from these data, because the crystals did not include DNA. However, grooves on the surface of the proposed octamer defined a left-handed helical ramp that could provide a path for the DNA (Figure 13.2c). In 1997, Timothy Richmond and colleagues succeeded in crystallizing a nucleosomal core particle that did include DNA. The nucleosome, as originally defined, contained about 200 bp of DNA. This is the length of DNA released by subjecting chromatin to a mild nuclease treatment. However, exhaustive digestion with nuclease gives a **core nucleosome** with 146 bp of DNA and the histone octamer containing all four **core histones** (H2A, H2B, H3, and H4), but no histone H1, which is relatively easily removed because it binds to the

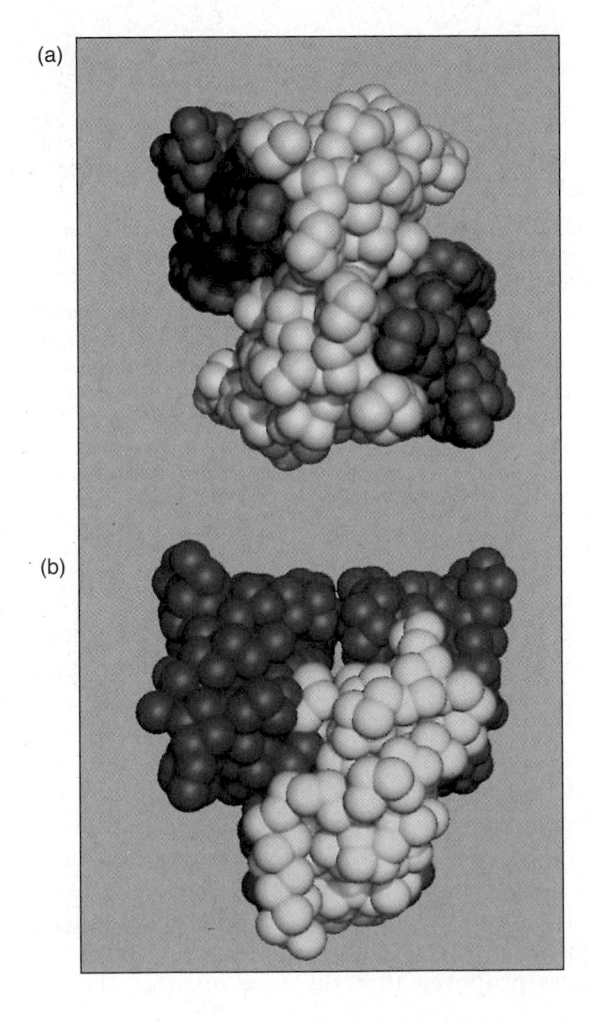

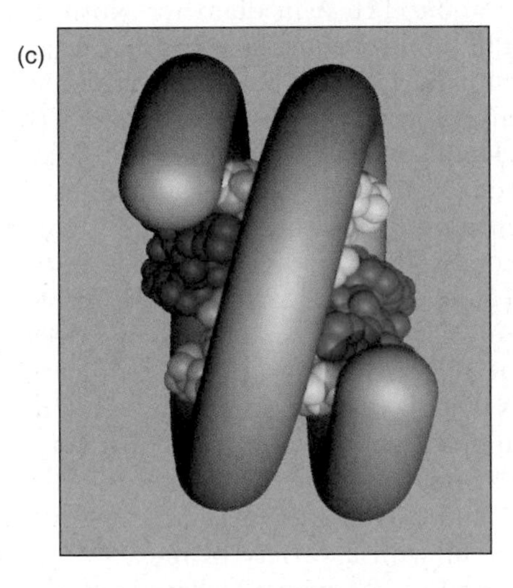

Figure 13.2 Two views of the histone octamer based on x-ray crystallography and a hypothetical path for the nucleosomal DNA. The H2A–H2B dimers are dark blue; the $(H3–H4)_2$ tetramer is light blue. The octamer in panel **(b)** is rotated 90 degrees downward relative to the octamer in panel **(a).** The thin edge of the wedge is pointing toward the viewer in panel (a) and downward in panel (b), where it is clear that the narrowing of the wedge occurs primarily in the H3–H4 tetramer. **(c)** Hypothetical path of the DNA around the histone octamer. The 20 Å-diameter DNA (blue-gray tube) nearly obscures the octamer, which is shown in the same orientation as in panel (a). (*Sources: (a–b)* Arents, A., R.W. Burlingame, B.-C. Wang, W.B. Love, and E.N. Moudrianakis, The nucleosomal core histone octamer at 3.1Å resolution: A tripartite protein assembly and a left-handed superhelix. *Proceedings of the National Academy of Sciences USA* 88 (Nov 1991), f. 3, p. 10150. (c) Arents, A. and E.N. Moudrianakis, Topography of the histone octamer surface: Repeating structural motifs utilized in the docking of nucleosomal DNA. *Proceedings of the National Academy of Sciences USA* 90 (Nov 1993), f. 3a, 1 & 4, pp. 10490–91. Copyright © National Academy of Sciences, USA.)

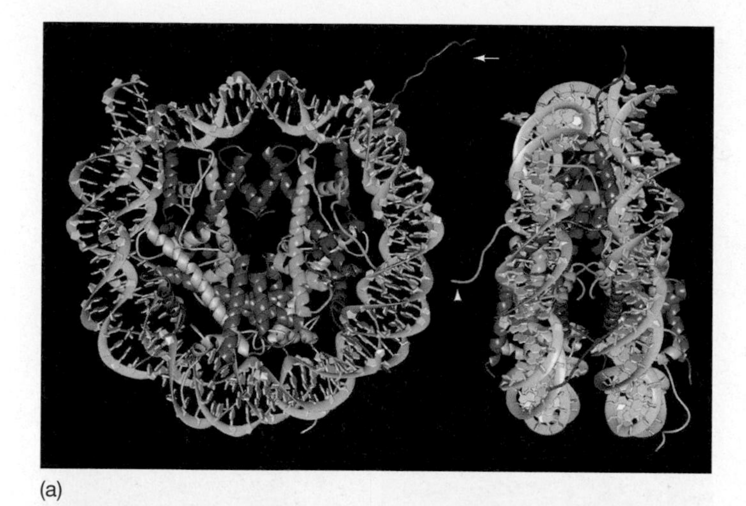

(a)

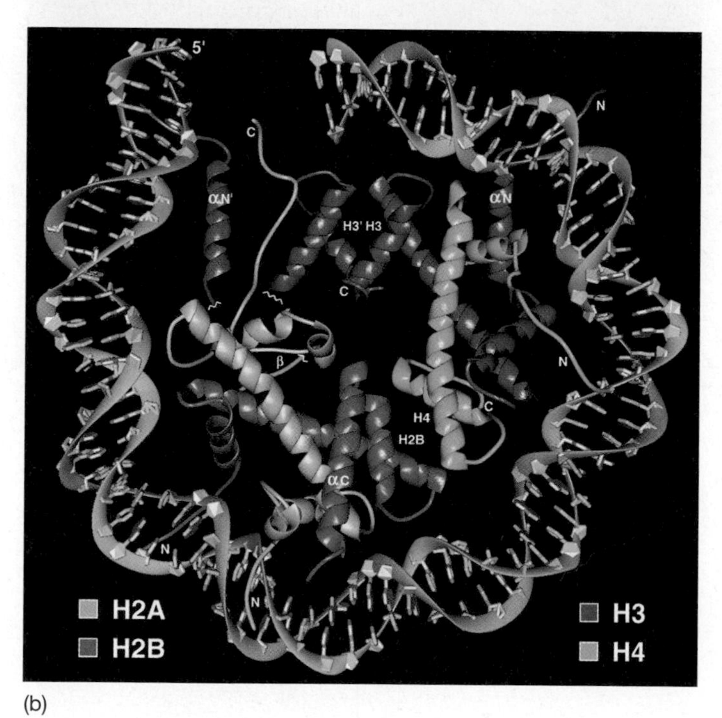

(b)

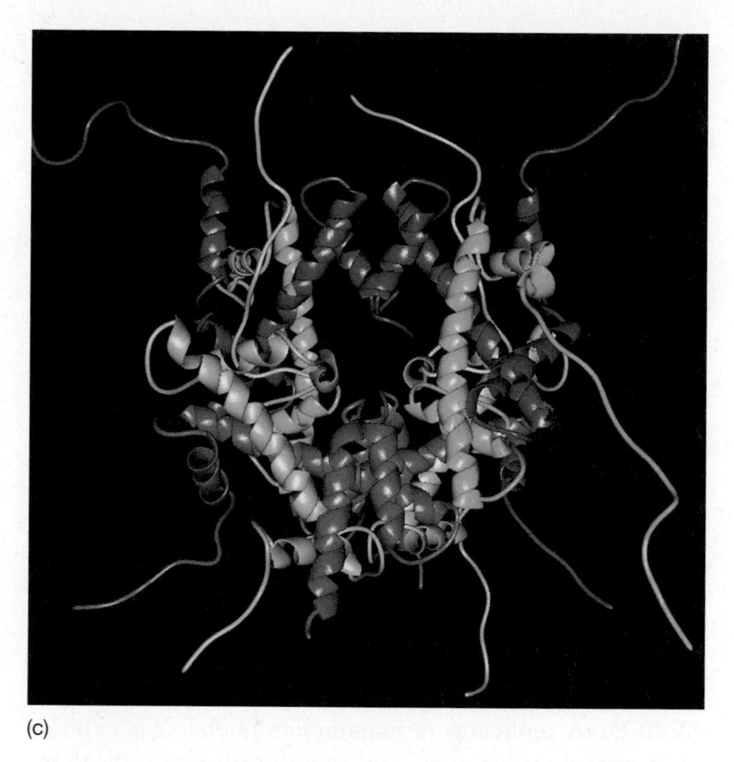

(c)

Figure 13.3 Crystal structure of a nucleosomal core particle. Richmond and colleagues crystallized a core particle composed of a 146-bp DNA and cloned core histones, then determined its crystal structure. **(a)** Two views of the core particle, seen face-on (left) and edge-on (right). The DNA on the outside is rendered in tan and green. The core histones are rendered as follows: H2A, yellow; H2B, red; H3, purple; and H4, green. Note the H3 tail (arrow) extending through a cleft between the minor grooves of the two adjacent turns of the DNA around the core particle. **(b)** Half of the core particle, showing 73 bp of DNA plus at least one molecule each of the core histones. **(c)** Core particle with DNA removed. (*Sources:* (*a–b*) Luger, K., A.W. Mäder, R.K. Richmond, D.F. Sargent, and T.J. Richmond, Crystal structure of the nucleosome core particle at 2.8 Å Resolution. *Nature* 389 (18 Sep 1997) f. 1, p. 252. Copyright © Macmillan Magazines Ltd. (*c*) Rhodes, D., Chromatin structure: The nucleosome core all wrapped up. *Nature* 389 (18 Sep 1997) f. 2, p. 233. Copyright © Macmillan Magazines Ltd.)

linker DNA outside the nucleosome, and this linker DNA is digested by the nuclease.

Figure 13.3 depicts the core nucleosome structure determined by Richmond and colleagues. We can see the DNA winding almost twice around the core histones. We can also see the H3–H4 tetramer near the top and the two H2A–H2B dimers near the bottom. This arrangement is particularly obvious on the right in panel a. The architecture of the histones themselves is interesting. All of the core histones contain the same fundamental **histone fold,** which consists of three α-helices linked by two loops. All of them also contain extended tails that make up about 28% of the mass of the core histones. Because the tails are relatively unstructured, the crystal structure does not include most of their length. The tails are especially evident with the DNA removed in panel c. The tails of H2B and H3 pass out of the core particle through a cleft formed from two adjacent DNA minor grooves (see the long purple tail at the top of the left part of panel a). One of the H4 tails is exposed to the side of the core particle (see the right part of panel a). This tail is rich in basic residues and can interact strongly with an acidic region of an H2A–H2B dimer in an adjacent nucleosome. Such interactions may play a role in nucleosome cross-linking, which we will discuss later in this chapter.

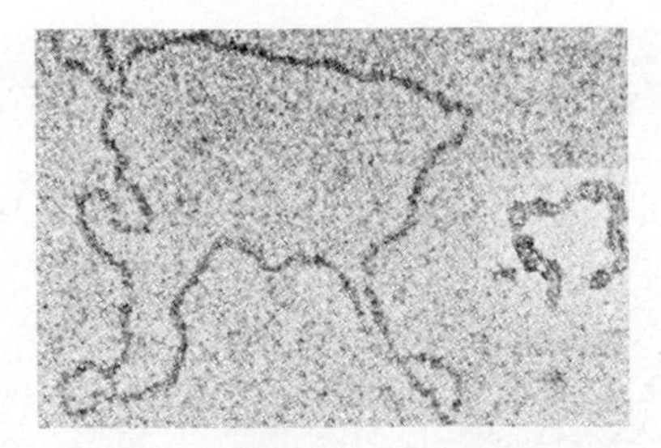

Figure 13.4 Condensation of DNA in nucleosomes. Deproteinized SV40 DNA is shown next to an SV40 minichromosome (inset, right) in electron micrographs enlarged to the same scale. The condensation of DNA afforded by nucleosome formation is apparent. (*Source:* Griffith, J., Chromatin structure: Deduced from a minichromosome. *Science* 187:1202 (28 March 1975). Copyright © AAAS.)

This and other models of the nucleosome indicate that the DNA winds about 1.65 times around the core, condensing the length of the DNA by a factor of 6 to 7. Jack Griffith also observed this magnitude of condensation in his 1975 study of the SV40 minichromosome. Because SV40 DNA replicates in mammalian nuclei, it is exposed to mammalian histones, and therefore forms typical nucleosomes. Figure 13.4 shows two views of the SV40 DNA. The main panel shows the DNA after all protein has been stripped off. The inset shows the minichromosome with all its protein—at the same scale. The reason the minichromosome looks so much smaller is that the DNA is condensed by winding around the histone cores in the nucleosomes.

SUMMARY Eukaryotic DNA combines with basic protein molecules called histones to form structures known as nucleosomes. These structures contain four pairs of core histones (H2A, H2B, H3, and H4) in a wedge-shaped disc, around which is wrapped a stretch of about 146 bp of DNA. Histone H1 is more easily removed from chromatin than the core histones and is not part of the core nucleosome.

The 30-nm Fiber

After the string of nucleosomes, the next order of chromatin folding produces a fiber about 30 nm in diameter. Until 2005, it had not been possible to crystallize any component of chromatin larger than the nucleosome core, so researchers had to rely on lower-resolution methods such as electron microscopy (EM) to investigate higher-order chromatin structure. Figure 13.5 depicts the results of an EM study that shows how the string of nucleosomes condenses to form the 30-nm fiber at increasing ionic strength. The degree of this condensation is another six- to sevenfold in

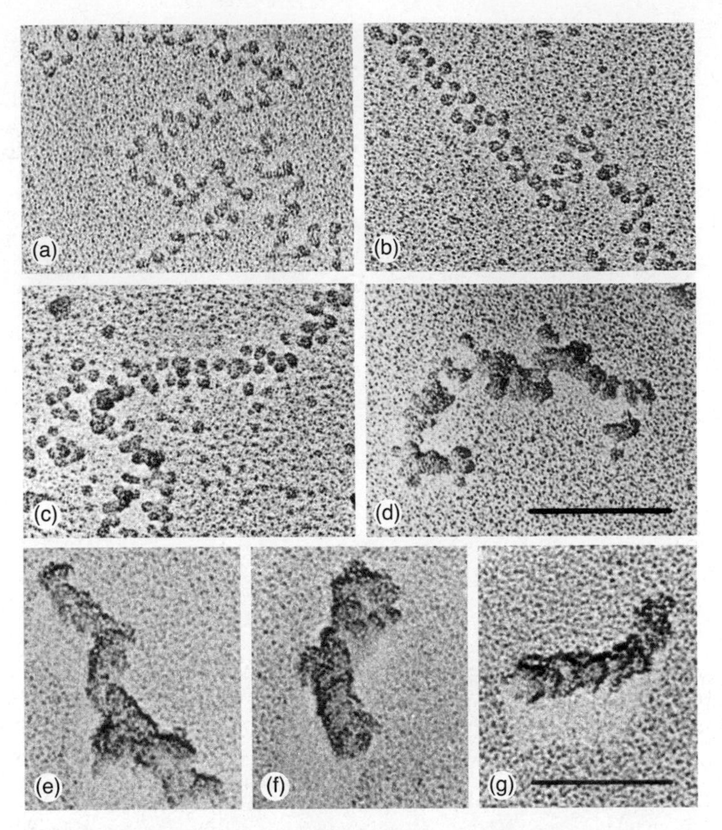

Figure 13.5 Condensation of chromatin on raising the ionic strength. Klug and colleagues subjected rat liver chromatin to buffers of increasing ionic strength, during fixation for electron microscopy. Panels **(a)–(c)** were at low ionic strength, panel **(d)** at moderate ionic strength, and panels **(e)–(g)** at high ionic strength. More specifically, the fixing conditions in each panel were the following, plus 0.2 mM EDTA in each case: (a) 1 mM triethylamine hydrochloride (TEACl); (b and c) 5 mM TEACl; (d) 40 mM NaCl, 5 mM TEACl; (e)–(g) 100 mM NaCl, 5 mM TEACl. The bars represent 100 nm. (*Source:* Thoma, F., T. Koller, and A. Klug, Involvement of histone H1 in the organization of the nucleosome and of the salt-dependent superstructures of chromatin. *Journal of Cell Biology* 83 (1979) f. 4, p. 408. Copyright © Rockefeller University Press.)

addition to the approximately six- to sevenfold condensation in the nucleosome itself.

What is the structure of the 30-nm fiber? This question has vexed molecular biologists for decades. In 1976, Aaron Klug and his colleagues, on the basis of electron microscopy and small angle x-ray scattering data, proposed a **solenoid** model (Figure 13.6), in which the nucleosomes were arranged in a hollow, compact helix (Greek: *solen* = pipe). But others, not convinced by the data behind the solenoid model, proposed various other schemes: a zigzag ribbon of nucleosomes; a superbead, with relatively disordered nucleosomes; an irregular, open helical arrangement of nucleosomes; and a two-start helix, in which the linker DNA between nucleosomes zigzags back and forth between two helical arrangements of stacked nucleosomes, such that one helix contains the odd-numbered nucleosomes and the other contains the even-numbered ones.

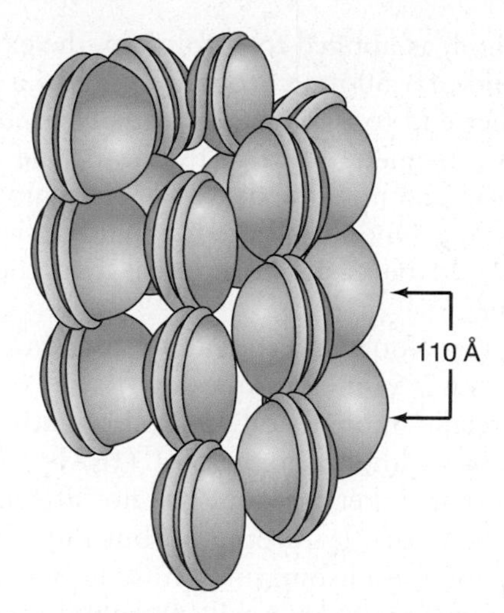

Figure 13.6 The solenoid model of chromatin folding. A string of nucleosomes coils into a hollow tube, or solenoid. Each nucleosome is represented by a blue cylinder with DNA (pink) coiled around it. For simplicity, the solenoid is drawn with six nucleosomes per turn and the nucleosomes parallel to the solenoid axis.
Source: Adapted from Widom, J. and A. Klug. Structure of the 300 Å chromatin filament: X-ray diffraction from oriented samples. Cell 43:210, 1985.

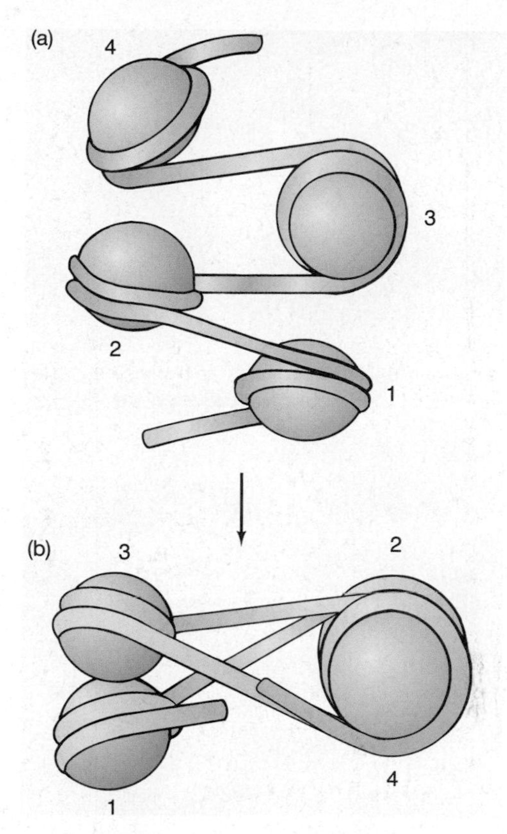

Figure 13.7 Structure of a tetranucleosome. (a) Diagrams of tetranucleosomes in two conformations. **(a)** A hypothetical conformation constrained only by the known degree of winding of DNA around the histone cores. **(b)** The conformation determined by x-ray crystallography. The nucleosomes form two stacks, and the linker DNA zigzags back and forth between nucleosomes in the two stacks. Consequently, consecutive nucleosomes are no longer nearest neighbors. Instead, alternate nucleosomes are nearest neighbors.
(*Source:* Adapted from Woodcock, C.L. *Nature Structural & Molecular Biology* 12, 2005, 1, p. 639.)

To resolve this long-standing controversy, higher-resolution structural data were needed. Finally, in 2005, Richmond and colleagues achieved a breakthrough by reporting the x-ray crystal structure of a tetranucleosome, or string of four nucleosomes. The resolution of this structure was not very high, only 9 Å, but it was good enough that the high resolution structure of an individual nucleosome could be incorporated. Figure 13.7 illustrates the structure of the tetranucleosome. Panel (a) of this figure starts with a string of nucleosomes, which is constrained only by the number of turns the DNA duplex makes around each nucleosome, and the length of the linker DNA between nucleosomes. One could wind the linker DNA in such a way as to stack the nucleosomes on top of each other. Or one could keep the zigzag arrangement and form two stacks, each containing every-other nucleosome, as shown in panel (b).

In fact, this zigzag arrangement is supported by the crystal structure of the tetranucleosome. This representation of the tetranucleosome structure is complex. The schematic in panel (a) helps interpret it, but it is best viewed in three dimensions. You can do this with a video, using this link:

http://www.nature.com/nature/journal/v436/n7047/suppinfo/nature03686.html

As the video runs, the structure rotates so you can see the connections among all the nucleosomes, which are represented by their DNA only, and appreciate the zigzag nature of the structure.

The zigzag structure has important implications for the overall structure of chromatin. It is incompatible with most of the previous suggestions, including the solenoid model. But it is consistent with the crossed-linker, two-start helix, in which each of the two stacks of nucleosomes forms a left-handed helix. The exact nature of this double helix of polynucleosomes is not clarified by the tetranucleosome structure, but Richmond and colleagues speculated as follows. First, they built a "direct" model by essentially stacking tetranucleosomes on top of each other. But this led to intolerable steric interference between neighboring tetranucleosomes, so the authors built an "idealized" model by equalizing the angles between each pair of nucleosomes in a stack. This procedure distorted the angles between nucleosomes seen in the tetranucleosome structure, but it avoided steric interference and generated a reasonable model, as illustrated in Figure 13.8. The two helices of polynucleosomes are apparent in this structure, and the zigzags of linker DNA can even be seen between some of the nucleosomes in the two helices.

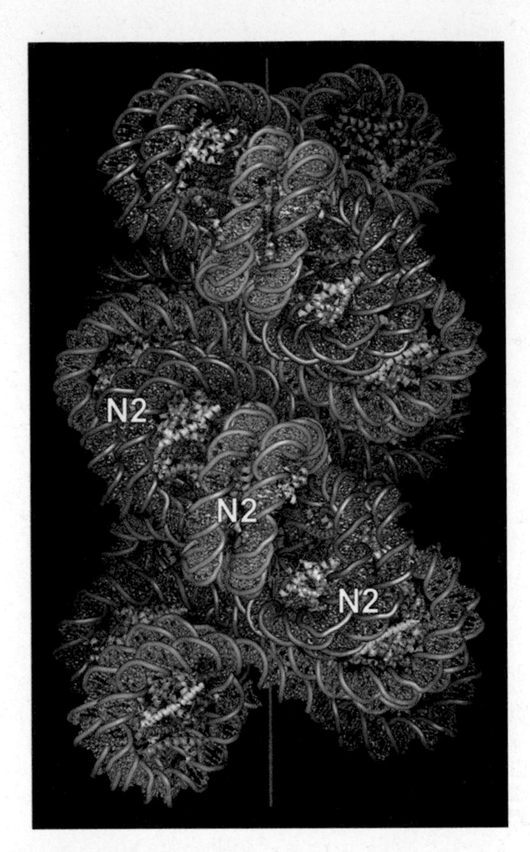

Figure 13.8 A model for the 30-nm fiber. Richmond and colleagues built this "idealized" model based on the tetranucleosome structure. It is arranged so that the dyad axis of each nucleosome (a line through the middle of the nucleosome, between the two coils of DNA) is perpendicular to the axis of the 30-nm fiber (gray vertical line). Also, the angles between any two adjacent nucleosomes are equal. (*Source:* Reprinted by permission from Macmillan Publishers Ltd: *Nature,* 436, 138–141, Thomas Schalch, Sylwia Duda, David F. Sargent and Timothy J. Richmond, "X-ray structure of a tetranucleosome and its implications for the chromatin fibre," fig. 3, p. 140, copyright 2005.)

Two of the models for the 30-nm fiber—the solenoid and the two-start double helix—have considerable experimental support, but which is the right one? In 2009, John van Noort and colleagues presented data that suggested that both models may be right. They proposed that the structure of the 30-nm fiber may depend on the exact nature of the chromatin, and in particular on the **nucleosome repeat length (NRL)**. This length of the DNA from the beginning of one nucleosome to the beginning of the next varies between about 165 bp and 212 bp in vivo, but most chromatin has an NRL of about 188 or 196. Chromatin of this type is generally transcriptionally inactive and associated with a linker histone such as H1. A smaller proportion of chromatin has an NRL of 167, tends to be transcriptionally active, and lacks a linker histone. Could it be that one type of chromatin forms one kind of 30-nm fiber, and the other type forms the other?

To answer this question, van Noort and colleagues used a technique called **single-molecule force spectroscopy.** In this method, as applied to chromatin, the experimenter links one end of a 30-nm chromatin fiber to a glass slide, and the other end to a magnetic bead. Then, by applying an attractive magnetic force to the bead, one can stretch the chromatin and note the degree of stretching produced by a given force. One would predict that the simple helical solenoid would be easier to stretch than the two-start double helix.

Indeed, van Noort and colleagues found that chromatin containing 25 nucleosomes with the longer NRL (197 bp) stretches more readily than chromatin containing 25 nucleosomes with the shorter NRL (167 bp). In addition, they found that linker histones did not affect the length or stretchability of the chromatin, but they did stabilize the folding of the chromatin. Thus, it is possible that most of the chromatin in a cell (presumably the inactive fraction) adopts a solenoid shape for its 30-nm fiber, while a minor fraction (at least potentially active) forms a 30-nm fiber according to the two-start double helical model. It is interesting in this regard that Richmond and colleagues, in forming their tetranucleosomes for x-ray crystallography, used an NRL of 167, and found a two-start double helical structure. Such chromatin has also been shown by van Noort and colleagues to conform to the two-start double helical model.

Some have even questioned whether the 30-nm fiber exists in vivo at all. It is well documented in vitro, but has never been visualized in intact nuclei. There are several ways to explain this inability to find the 30-nm fiber in vivo. First, as unlikely as this may seem, it may not exist in vivo. But there are other possibilities: It may exist, but is not seen because higher-order chromatin folding obscures it. Or it may simply be that our tools for visualizing chromatin in intact nuclei are not adequate to detect the 30-nm fiber.

SUMMARY A string of nucleosomes folds into a 30-nm fiber in vitro, and presumably also in vivo. structural studies suggest that the 30-nm chromatin fiber in the nucleus exists in at least two forms: inactive chromatin tends to have a high nucleosome repeat length (about 197 bp) and favors a solenoid folding structure. This kind of chromatin interacts with histone h1, which helps to stabilize its structure. Active chromatin tends to have a low nucleosome repeat length (about 167 bp) and folds according to the two-start double helical model.

Higher-Order Chromatin Folding

The 30-nm fiber probably accounts for most of the chromatin in a typical interphase nucleus, but further

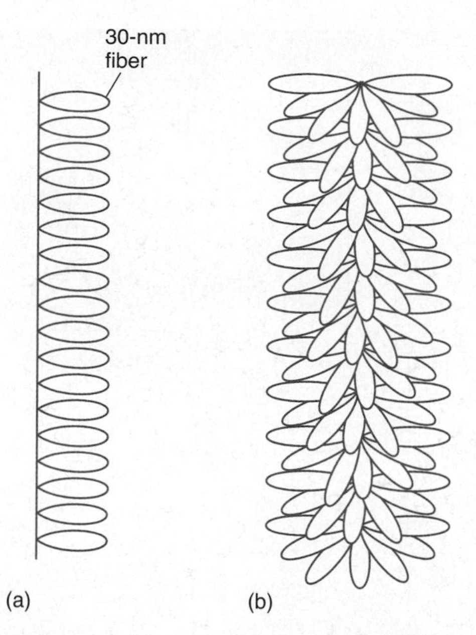

Figure 13.9 Radial loop models of chromatin folding. (a) This is only a partial model, showing some of the loops of chromatin attached to a central scaffold; of course, all the loops are composed of the same continuous 30-nm fiber. **(b)** A more complete model, showing how the loops are arranged in three dimensions around the central scaffold. (*Source:* Adapted from Marsden, M.P.F. and U.K. Laemmli, Metaphase chromosome structure: Evidence of a radial loop model. *Cell* 17:856, 1979.)

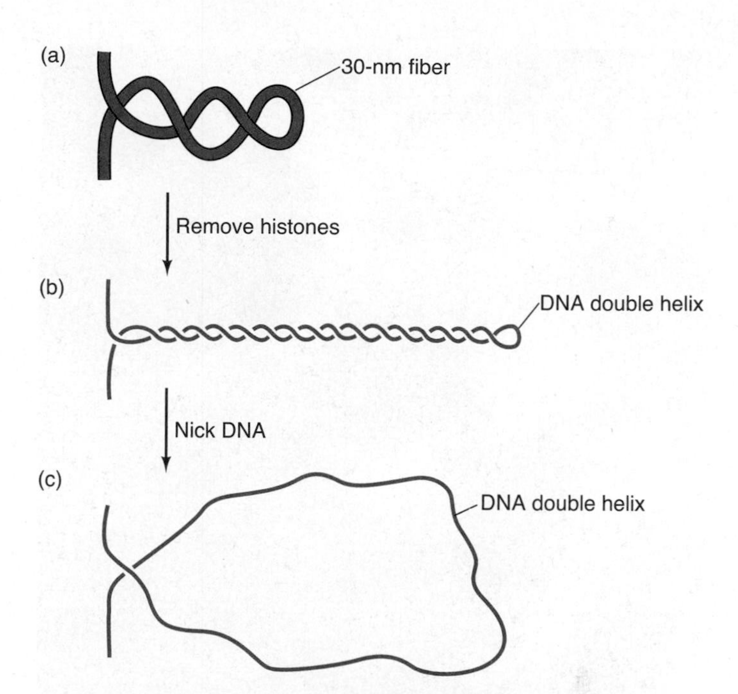

Figure 13.10 Relaxing supercoiling in chromatin loops. (a) A hypothetical chromatin loop composed of the 30-nm fiber, with some superhelical turns. **(b)** The chromatin loop with histones removed. Without histones, the nucleosomes and 30-nm fiber have disappeared, leaving a supercoiled DNA duplex. Note that the helical turns here are superhelices, not ordinary turns in a DNA double helix. **(c)** A relaxed chromatin loop. The DNA has been nicked to relax the superhelix. Now we see a relaxed DNA double helix that forms a loop. With each step from (a) to (c), the apparent length of the loop increases, but these increases are not drawn to scale.

orders of folding are clearly needed, especially in mitotic chromosomes, which have condensed so much that they become visible with a light microscope. The favorite model for the next order of condensation is a series of radial loops, as pictured in Figure 13.9. Cheeptip Benyajati and Abraham Worcel produced the first evidence in support of this model in 1976 when they subjected *Drosophila* chromatin to mild digestion with DNase I, then measured the sedimentation coefficients of the digested chromatin. They found that the coefficients decreased gradually with digestion, then reached a plateau value. Worcel had previously shown that the *E. coli* nucleoid (the DNA-containing complex) exhibited similar behavior, which was caused by the introduction of nicks into more and more superhelical loops of the bacterial DNA. As each loop was nicked once, it relaxed to an open circular form and slightly decreased the sedimentation coefficient of the whole complex. But eukaryotic chromosomes are linear, so how can the DNA in them be supercoiled? If the chromatin fiber is looped as it is in *E. coli* and held fast at the base of each loop, then each loop would be the functional equivalent of a circle and could be supercoiled. Indeed, the winding of DNA in the nucleosomes would provide the strain necessary for supercoiling. Figure 13.10 illustrates this concept and shows how relaxation of a supercoiled loop gives much less compact chromatin in that region, which would reduce the sedimentation coefficient.

How big are the loops? Worcel calculated that each loop in a *Drosophila* chromosome contains about 85 kb, but other investigators, working with vertebrate species and using a variety of techniques, have made estimates ranging from 35 to 83 kb.

The images of chromosomes in Figure 13.11 also support the loop idea. Figure 13.11a shows the edge of a human metaphase chromosome, with loops clearly visible. Figure 13.11b depicts a cross section of a swollen human chromosome in which the 30-nm fiber is preserved. Radial loops are clearly visible. Figure 13.11c shows part of a deproteinized human chromosome. Loops of DNA are anchored to a central scaffold in the skeleton of the chromosome. All these pictures strongly support the notion of a radially looped fiber in chromosomes.

SUMMARY Sedimentation and EM studies have revealed a radial loop structure in eukaryotic chromosomes. The 30-nm fiber seems to form loops between 35 and 85 kb long, anchored to the central matrix of the chromosome.

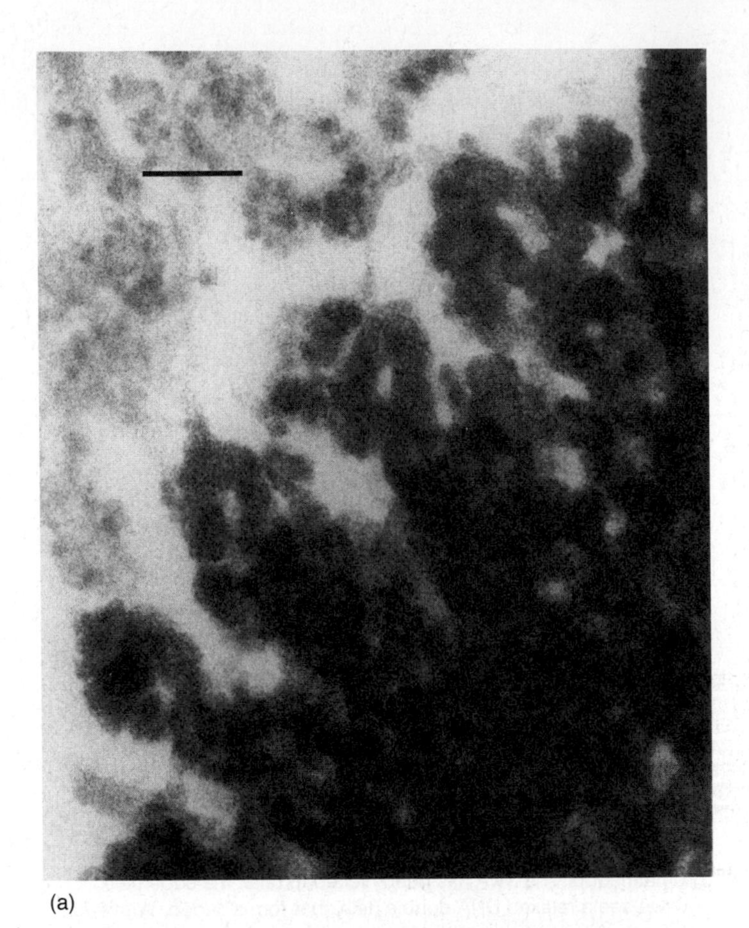
(a)

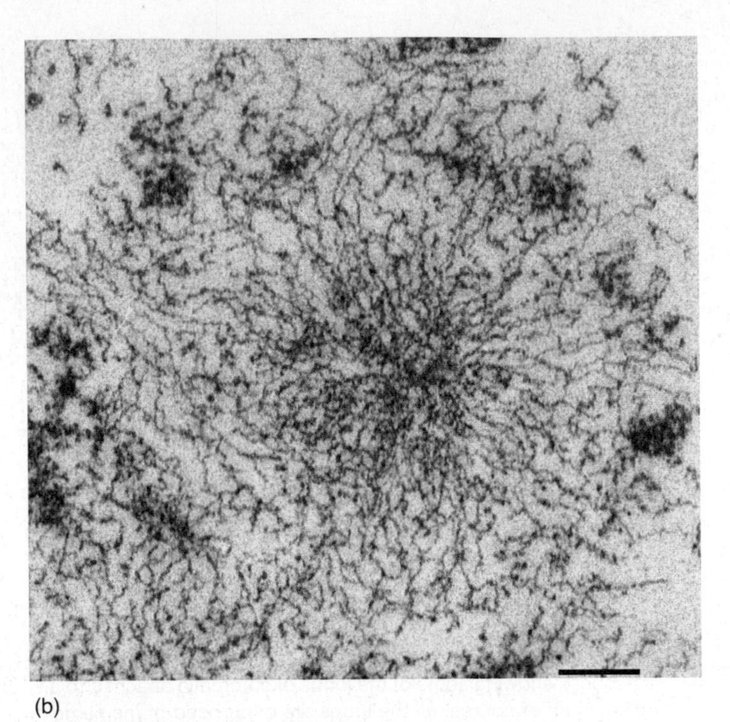
(b)

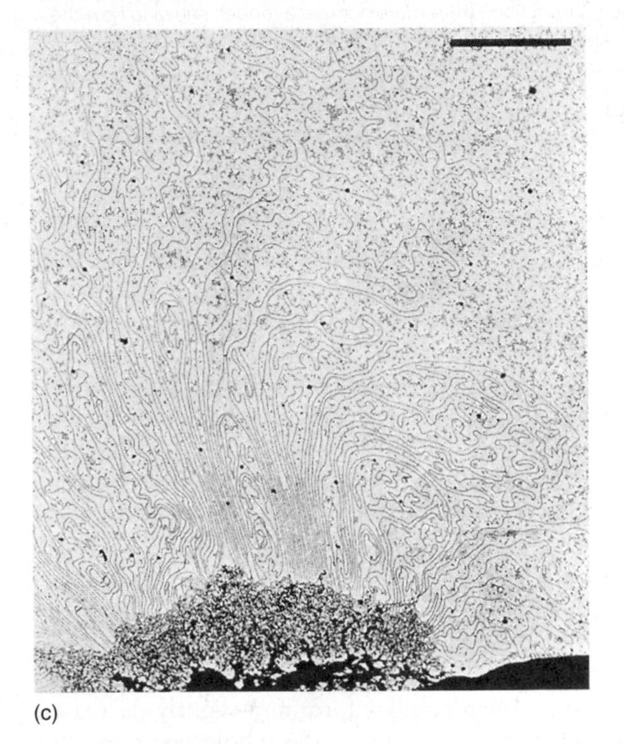
(c)

Figure 13.11 Three views of loops in human chromosomes. **(a)** Scanning transmission electron micrograph of the edge of a human chromosome isolated with hexylene glycol. Bar represents 100 nm. **(b)** Transmission electron micrograph of cross sections of human chromosomes swollen with EDTA. The chromatin fiber visible here is the 30-nm nucleosome fiber. Bar represents 200 nm. **(c)** Transmission electron micrograph of a deproteinized human chromosome showing DNA loops emanating from a central scaffold. Bar represents 2 μm (2000 nm). (*Sources:* (*a*) Marsden, M.P.F. and U.K. Laemmli, Metaphase chromosome structure: Evidence for a radial loop model. *Cell* 17 (Aug 1979) f. 5, p. 855. Reprinted by permission of Elsevier Science. (*b*) Marsden and Laemmli, *Cell* 17 (Aug 1979) f. 1, p. 851. Reprinted by permission of Elsevier Science. (*c*) Paulson, J.R. and U.K. Laemmli, The structure of histone-depleted metaphase chromosomes. *Cell* 12 (1977) f. 5, p. 823. Reprinted by permission of Elsevier Science.)

13.3 Chromatin Structure and Gene Activity

Enthusiasm for histones as important regulators of gene activity has been inconsistent. When it first became clear that histones could turn off transcription when added to DNA in vitro, molecular biologists got excited. Then, when the role of histones in chromatin structure was elucidated, most investigators tended to focus on this structural role and forget about histones as regulators of genetic activity. Histones were then viewed as mere scaffolding for the DNA. Now we have come full circle and molecular biologists are elucidating the regulatory functions of histones.

The Effects of Histones on Transcription of Class II Genes

In the 1980s, Donald Brown and his colleagues showed that the 5S rRNA genes (class III genes) of *Xenopus laevis* can be selectively repressed in vitro by addition of histone H1, and that this repression increased dramatically as the level of histone H1 reached one molecule per 200 bp of DNA, its natural level in chromatin. In the 1990s, James Kadonaga and his colleagues showed that the same principles concerning the interactions between histones and class III genes also apply to histones and class II genes.

Core Histones In 1991, Paul Laybourne and Kadonaga performed a detailed study to distinguish between the effects of the core histones and of histone H1 on transcription by RNA polymerase II in vitro. They found that the core histones (H2A, H2B, H3, and H4) formed core nucleosomes with cloned DNA and caused a mild repression (about fourfold) of genetic activity. Transcription factors had no effect on this repression. When they added histone H1, in addition to the core histones, the repression became much more profound: 25- to 100-fold. This repression *could* be blocked by activators. In this respect, these factors resembled the class III factors (presumably TFIIIA, B, and C), which could compete with histone H1 for the control region of the *Xenopus* 5S rRNA gene.

Laybourne and Kadonaga's experimental strategy was to reconstitute chromatin from plasmid DNA containing a well-defined cloned gene, and histones in the presence or absence of activators that were known to affect transcription of the cloned gene in question. They also added topoisomerase I to keep the DNA relaxed. Then they used a primer extension assay to test whether the reconstituted chromatin could be transcribed by a nuclear extract. In the first studies, these workers used only the core histones, not histone H1. They added a mass ratio of histones to DNA of 0.8 to 1.0, which is enough to form an average of one nucleosome per 200 bp of DNA.

Using such reconstituted chromatin that contained the *Drosophila Krüppel* gene, Laybourne and Kadonaga showed that a *Drosophila* nuclear extract could transcribe the *Krüppel* gene (Figure 13.12). However, core histones in quantities that produced nucleosomes at a density of one nucleosome per 200 bp, which is the physiological density, caused partial repression of transcription (down to 25% of the control value; compare lanes 2 and 5). Notice that the transcription start sites as detected by this method are quite heterogeneous in this gene, so we see a cluster of primer extension products.

The authors pointed to two possible explanations for the 75% repression observed with the core histones. First, the nucleosomes could slow the progress of all RNA polymerases by about 75%, but not stop any of them. Second, 75% of the polymerases could be blocked entirely by nucleosomes, but 25% of the promoters might have been left free of nucleosomes and thus could remain available to

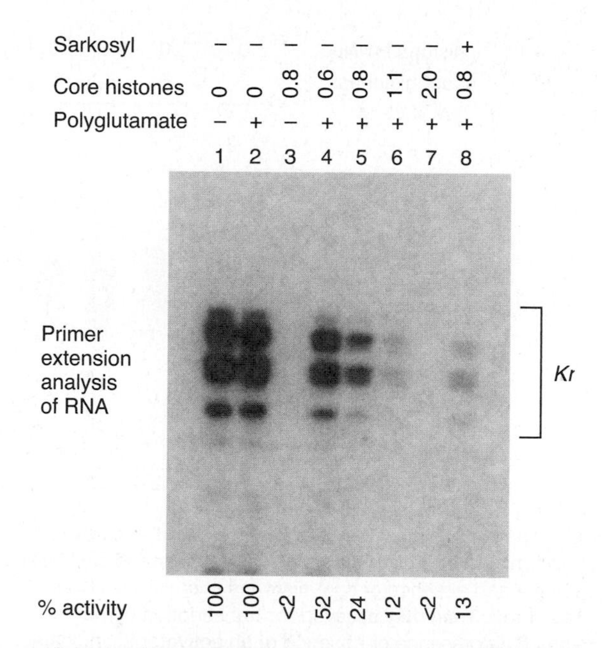

Figure 13.12 In vitro transcription of reconstituted chromatin. Laybourne and Kadonaga reconstituted chromatin with plasmid DNA containing the *Drosophila Krüppel* gene and core histones in varying ratios of protein to DNA, as indicated at top. Then they performed primer extension analysis to measure efficiency of transcription. Diverse signals corresponding to *Krüppel* gene transcription are indicated by the bracket at right. Lane 1, naked DNA; lane 2, naked DNA plus polyglutamate (used as a vehicle to help histones deposit onto DNA); lanes 3–7, chromatin at various core histone–DNA ratios; lane 8, sarkosyl was included to prevent reinitiation, so only one round of transcription occurred. Core histones can apparently inhibit transcription of the *Krüppel* gene in a dose-dependent manner. (*Source:* Laybourn, P.J. and J.T. Kadonaga, Role of nucleosomal cores and histone H1 in regulation of transcription by RNA polymerase II. *Science* 254 (11 Oct 1991) f. 2B, p. 239. Copyright © AAAS.)

RNA polymerase. A control experiment showed that the remaining 25% transcription could be eliminated by cutting the chromatin with a restriction enzyme that cleaves just downstream of the transcription start site. The fact that this site was available indicated that it was nucleosome-free. Thus, hypothesis 2 is the right one.

SUMMARY The core histones (H2A, H2B, H3, and H4) assemble nucleosome cores on naked DNA. Transcription of reconstituted chromatin with an average of one nucleosome core per 200 bp of DNA exhibits about 75% repression relative to naked DNA. The remaining 25% is due to promoter sites not covered by nucleosome cores.

Histone H1 Based on its suspected role as a nucleosome stabilizer, we would expect that histone H1 would add to the inhibition of transcription caused by the core histones in reconstituted chromatin. This is indeed the case, as Laybourne and Kadonaga demonstrated. They reconstituted

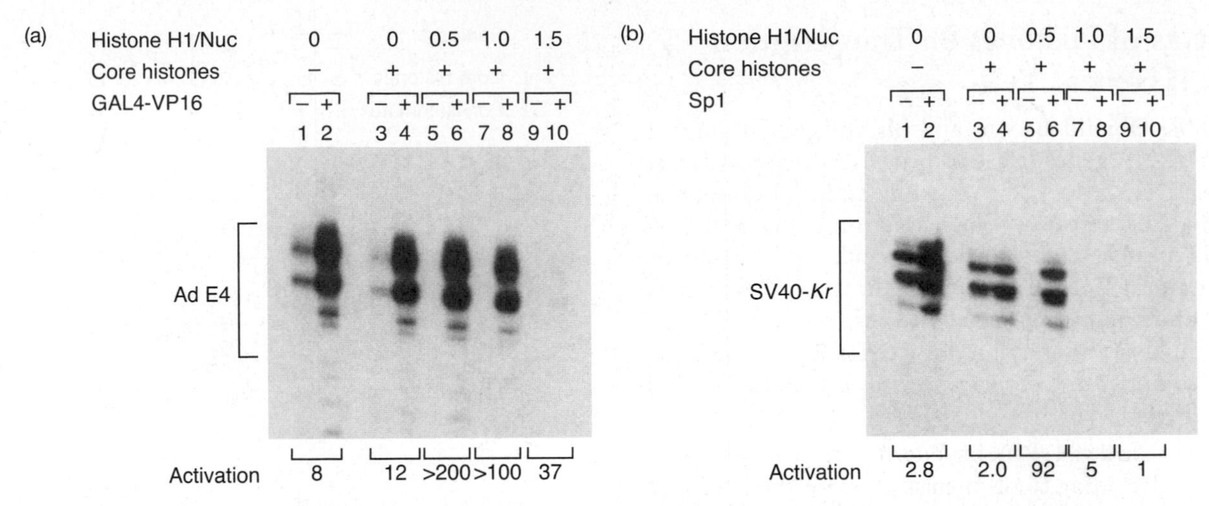

Figure 13.13 Competing effects of histones and activators on transcription. Laybourne and Kadonaga reconstituted chromatin in the presence and absence of core histones and histone H1 as indicated at top. Then they assayed for transcription by primer extension in the presence or absence of an activator as indicated. Apparent degrees of activation by each activator are given below each pair of lanes. The true activation by each activator is seen in lanes 1 and 2 of each panel, where naked DNA was the template. Any higher levels of apparent activation in the other lanes, where chromatin served as the template, were due to antirepression. **(a)** Effect of GAL4-VP16. Chromatin contained the adenovirus E4 promoter with five GAL4-binding sites. The signals corresponding to E4 transcription are indicated by the bracket at left. **(b)** Effect of Sp1. Chromatin contained the *Krüppel* minimal promoter plus the SV40 promoter GC boxes, which are responsive to Sp1. The signals corresponding to *Krüppel* transcription are indicated at left. (*Source:* Laybourn, P.J. and J.T. Kadonaga, Role of nucleosomal cores and histone H1 in regulation of transcription by RNA polymerase II. *Science* 254 (11 Oct 1991) f. 7, p. 243. Copyright © AAAS.)

chromatin with DNA containing two enhancer–promoter constructs: (1) pG_5E4 (five GAL4-binding sites coupled to the adenovirus E4 minimal promoter); and (2) pSV-Kr (six GC boxes from the SV40 early promoter coupled to the *Drosophila Krüppel* minimal promoter). In this experiment, they added not only the core histones, but histone H1 in various quantities, from 0 to 1.5 molecules per core nucleosome. Then they transcribed the reconstituted chromatin in vitro.

The odd lanes in Figure 13.13 show that increasing amounts of histone H1 caused a progressive loss of template activity, until transcription was barely detectable. However, at moderate histone H1 levels (0.5 molecules per core histone), activators could prevent much of the repression. For example, on chromatin reconstituted from the pG_5E4 plasmid, the hybrid activator GAL4-VP16, which interacts with GAL4-binding sites, caused a 200-fold greater template activity. Part of this (eightfold) is due to the stimulatory activity of the activator, observed even on naked DNA. The remaining 25-fold stimulation is apparently due to **antirepression,** the prevention of repression by histones. Similarly, when the reconstituted chromatin contained the pSV-Kr promoter, the activator Sp1, which binds to the GC boxes in the promoter, caused a 92-fold increase in template activity. Because true activation by Sp1 on naked DNA was only 2.8-fold, 33-fold of the 92-fold stimulation was antirepression. The true activation component is what we studied in Chapter 12, in which the experimenters used naked DNAs as the templates in their transcription assays.

These data are consistent with the model in Figure 13.14. Histone H1 can cause repression in the cases studied here by binding to the linker DNA between nucleosomes that happens to contain a transcription start site. Activators, represented by the green oval, can prevent this effect if added at the same time as histone H1. But these factors cannot reverse the effects of preformed nucleosome cores, even without histone H1. In other words, there is a sort of race between these activators and histone H1. If the activators get to the DNA first, they block the repressive action of histone H1. But if histone H1 reaches the DNA first, it stabilizes the nucleosomes and blocks activation. Other activators, represented by the purple oval, when confronted by a nucleosome blocking the promoter, can team up with chromatin-remodeling factors (see later in this chapter) to shoulder nucleosomes aside, at least if the nucleosomes are not stabilized by histone H1.

Kadonaga and colleagues have also studied another protein, called *GAGA factor,* which binds to several GA-rich sequences in the *Krüppel* promoter and to other *Drosophila* promoters. It has no transcription-stimulating activity of its own; in fact it slightly inhibits transcription. But GAGA factor prevents repression by histone H1 when added to DNA before the histone and can therefore cause a significant net increase in transcription rate. Thus, the GAGA factor seems to be a pure antirepressor, unlike the more typical activators we have been studying, which have both antirepression and transcription stimulation activities.

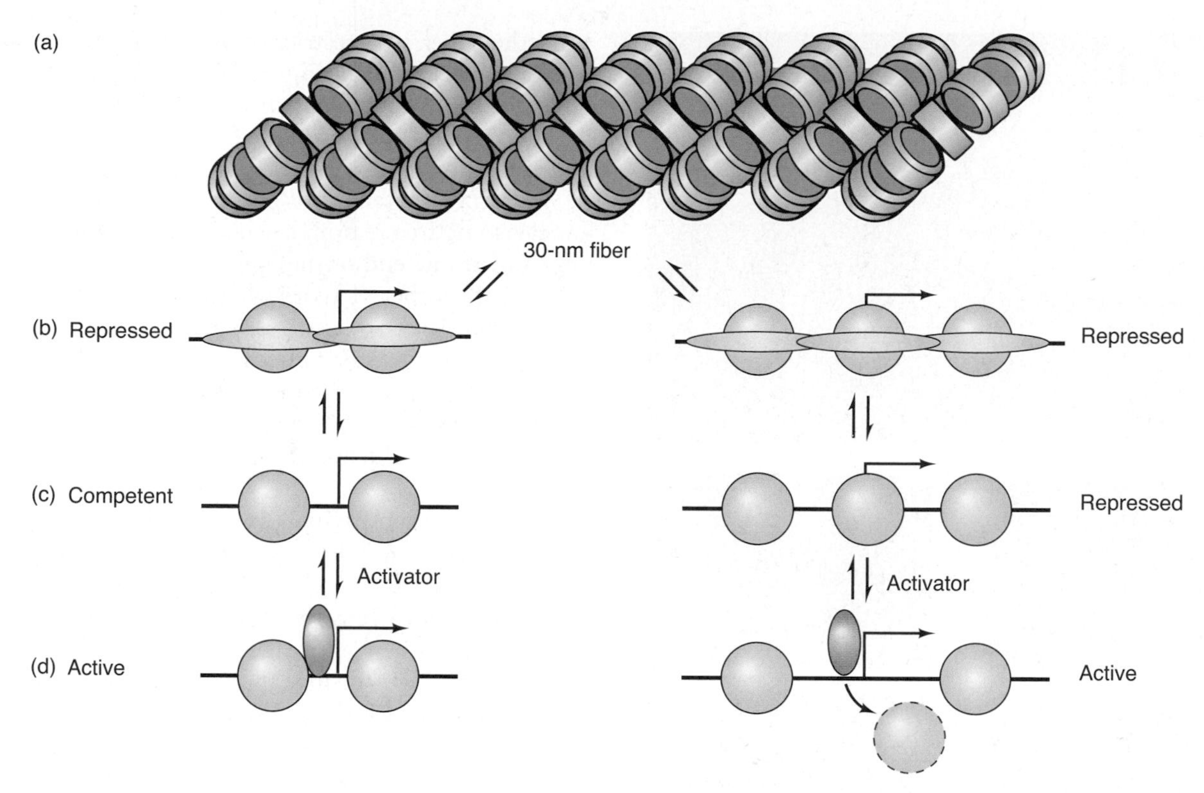

Figure 13.14 A model of transcriptional activation. (a) We start at the top with a 30-nm fiber. **(b)** The 30-nm fiber can open up to give two kinds of repressed chromatin. On the right, a stabilized nucleosome (blue) covers the promoter, keeping it repressed. On the left, no nucleosomes cover the promoter, but histone H1 (yellow) stabilizes nucleosomes flanking the promoter, so the gene is still repressed. **(c)** When we remove histone H1, we can get two chromatin states: On the left the promoter is uncovered, so the gene is competent to be transcribed. On the right, a nucleosome still covers the promoter, so it remains repressed. **(d)** Antirepression. If the gene's control region is not blocked by a nucleosome (left), the activator (green) can bind and, together with other factors, cause transcription initiation. If the gene's control region is blocked by one or more nucleosomes (right), the activator (purple), together with other factors, including chromatin-remodeling factors, can move the nucleosome aside (not necessarily removing it from the DNA, as shown here) and cause transcription to initiate. (*Source:* Adapted from Laybourn, P.J. and J.T. Kadonaga, Role of nucleosomal cores and histone H1 in regulation of transcription by polymerase II. *Science* 254:243, 1991.)

SUMMARY Histone H1 causes a further repression of template activity, in addition to that produced by core nucleosomes. This repression can be counteracted by transcription factors. Some, like Sp1 and GAL4, act as both antirepressors (preventing repression by histones) and as transcription activators. Others, like GAGA factor, are just antirepressors.

Nucleosome Positioning

The model of activation and antirepression in Figure 13.14 asserts that transcription factors can cause antirepression by removing nucleosomes that obscure a promoter or by preventing their binding to the promoter in the first place. Both these scenarios embody the idea of **nucleosome positioning**, in which activators force the nucleosomes to take up positions around, but not within, the promoter.

Nucleosome-Free Zones Several lines of evidence demonstrate nucleosome-free zones in the control regions of active genes. M. Yaniv and colleagues performed a particularly graphic experiment on the control region of SV40 virus DNA. SV40 DNA in an infected mammalian cell exists as a minichromosome, as described earlier in this chapter. Yaniv noticed that some actively transcribed SV40 minichromosomes have a conspicuous nucleosome-free zone late in infection (Figure 13.15). We would expect this nucleosome-free region to include at least one late promoter. In fact, the SV40 early and late promoters lie very close to each other, with the 72-bp repeat enhancer in between. Is this the nucleosome-free zone? The problem with a circular chromosome is that it has no beginning and no end, so we cannot tell what part of the circle we are looking at without a marker of some kind. Yaniv and colleagues used restriction sites as markers. A *Bgl*I restriction site occurs close to one end of the control region, and *Bam*HI and *Eco*RI sites occur on the other side of the circle, as illustrated in Figure 13.16a. Therefore, if the nucleosome-free region includes the control region, *Bgl*I will cut within that zone,

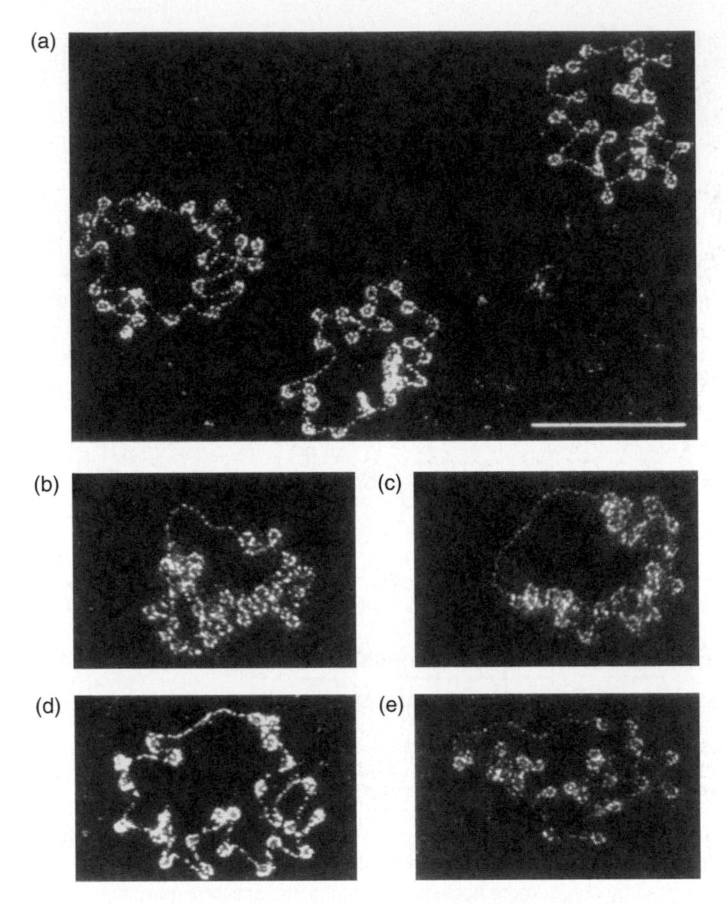

Figure 13.15 Nucleosome-free zones in SV40 minichromosomes. **(a)** Three examples of minichromosomes with no extensive nucleosome-free zones. **(b–e)** Four examples of SV40 minichromosomes with easily detectable nucleosome-free regions. The bar represents 100 nm. (*Source:* Saragosti, S., G. Moyne, and M. Yaniv, Absence of nucleosomes in a fraction of SV40 chromatin between the origin of replication and the region coding for the late leader RNA. *Cell* 20 (May 1980) f. 2, p. 67. Reprinted by permission of Elsevier Science.)

and the other two restriction enzymes will cut at remote sites, as illustrated in Figure 13.16b. Figure 13.17 shows that cutting with *Bam*HI or *Bgl*I produced exactly the expected results. Cutting with *Eco*RI (not shown) also fulfilled the prediction.

We can even tell that *Bgl*I cut asymmetrically within the nucleosome-free region, because it left a long nucleosome-free tail at one end of the linearized minichromosome, but not at the other. This is what we would expect if the nucleosome-free zone corresponds to one of the SV40 promoters, which are asymmetrically arranged relative to the *Bgl*I site. On the other hand, it is not what we would expect if the nucleosome-free zone corresponds to the viral origin of replication, which almost coincides with the *Bgl*I site.

DNase Hypersensitivity Another sign of a nucleosome-free DNA region is hypersensitivity to DNase. Chromatin regions that are actively transcribed are **DNase-sensitive** (≈10-fold more sensitive than bulk chromatin). But the control regions of active genes are **DNase-hypersensitive** (≈100-fold more sensitive than bulk chromatin). For example, the control region of SV40 DNA is DNase-hypersensitive, as we would expect. Yaniv demonstrated this by isolating chromatin from SV40 virus-infected monkey cells, mildly digesting this chromatin with DNase I, then purifying the SV40 DNA, cutting it with *Eco*RI, electrophoresing the fragments, Southern blotting, and probing the blot with radioactive SV40 DNA. Figure 13.16a shows that the *Eco*RI and *Bgl*I sites lie 67% (and 33%) apart on the circle. Therefore, if the nucleosome-free region near the *Bgl*I site is really DNase-hypersensitive, then DNase will cut there and *Eco*RI will cut at its unique site, yielding two fragments containing about 67% and 33% of the total SV40 genome.

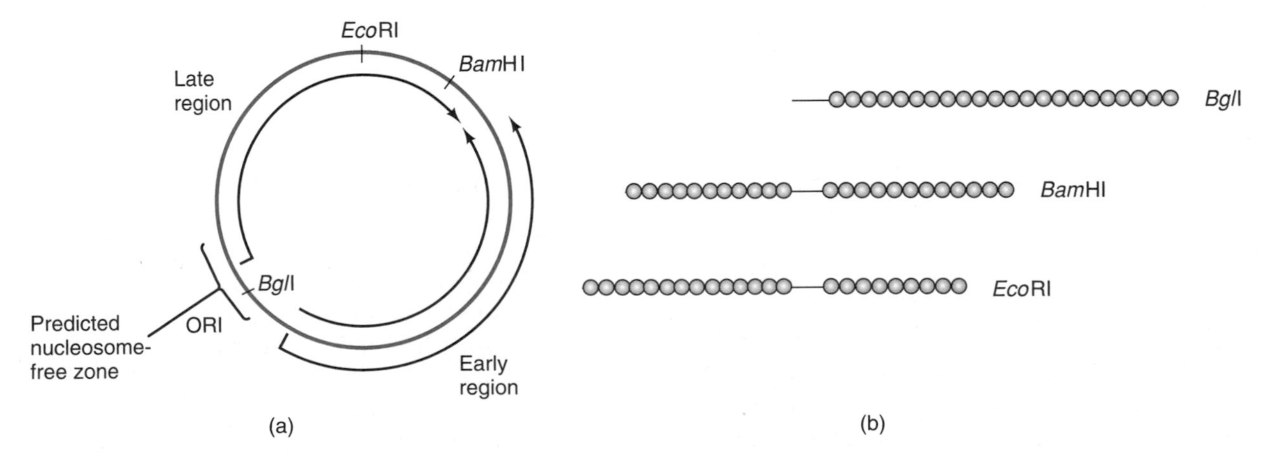

Figure 13.16 Experimental scheme to locate the nucleosome-free zone in the SV40 minichromosome. **(a)** Map of SV40 genome showing the cutting sites for three restriction enzymes *Bgl*I, *Bam*HI, and *Eco*RI. The control region surrounds the origin of replication (ORI), with the late control region on the clockwise side. **(b)** Expected results of cleavage of minichromosome from late infected cells with three restriction enzymes, assuming that the late control region is nucleosome-free. All three enzymes should cut once to linearize the minichromosome. *Bgl*I is predicted to cut near one end of the nucleosome-free zone and should therefore produce a minichromosome with a nucleosome-free zone at one end. *Bam*HI is predicted to cut at a site diametrically opposed to the nucleosome-free zone and should therefore produce a minichromosome with the zone in the middle. In the same way, *Eco*RI should yield a minichromosome with the zone somewhat asymmetrically located.

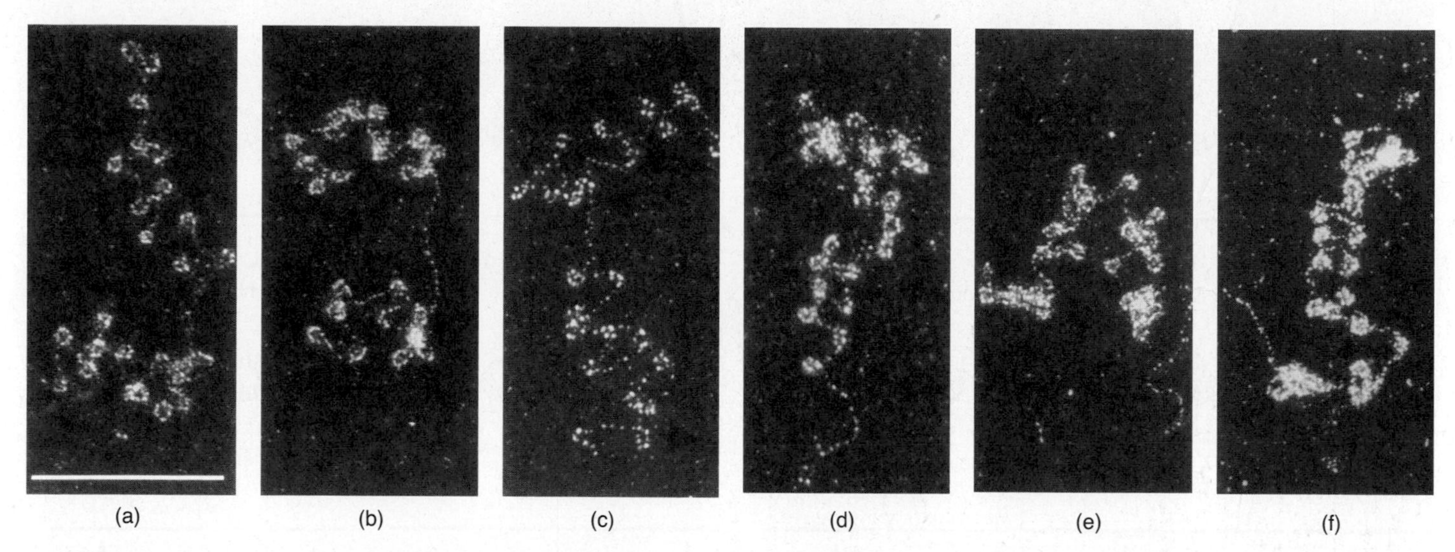

Figure 13.17 Locating the nucleosome-free zone on the SV40 minichromosome. Yaniv and colleagues cut SV40 minichromosomes from late infected cells with either *Bam*HI (panels **a–c**) or *Bgl*I (panels **d–f**). Just as predicted in Figure 13.16, *Bam*HI produced a centrally located nucleosome-free zone, and *Bgl*I yielded a nucleosome-free zone at the end of the minichromosome. The bar represents 100 nm. (*Source:* Saragosti, S., G. Moyne, and M. Yaniv, Absence of nucleosomes in a fraction of SV40 chromatin between the origin of replication and the region coding for the late leader RNA. *Cell* 20 (May 1980) f. 4, p. 69. Reprinted by permission of Elsevier Science.)

In fact, as Figure 13.18 demonstrates, experiments carried out 24 h, 34 h, and 44 h after virus infection all produced a large amount of the 67% product, and lesser amounts of the 33% product and shorter fragments. This suggests that DNase I is really cutting the chromatin in a relatively small region around the *Bgl*I site. Thus, the nucleosome-free region and the DNase-hypersensitive region coincide.

DNase hypersensitivity of the control regions of active genes is a general phenomenon. For example, the 5′-flanking region of the ε-globin gene in red blood cells is DNase-hypersensitive. In fact, the DNase hypersensitivity of the globin genes gives a good indication of the activity of those genes at any given time.

Figure 13.19 illustrates the principle involved in detecting a DNase-hypersensitive gene by Southern blotting. We see at the top of panels a and b the arrangement of nucleosomes on an active and an inactive gene, and the positions of two recognition sites for a restriction endonuclease (RE). If DNase I is used to lightly digest nuclei containing the inactive gene, nothing happens because no DNase-hypersensitive sites are present. On the other hand, if the same thing is done to nuclei containing the active gene, the DNase will attack the hypersensitive site near the promoter. Now the protein is removed from both DNAs, which are then cut with the RE. The restriction fragments are then electrophoresed, Southern blotted, and the blots are probed with a short gene-specific probe (green). DNA from the inactive chromatin will be intact, so the RE will generate a 13-kb fragment that will hybridize to the probe. But DNA from the active chromatin contains a DNase-hypersensitive

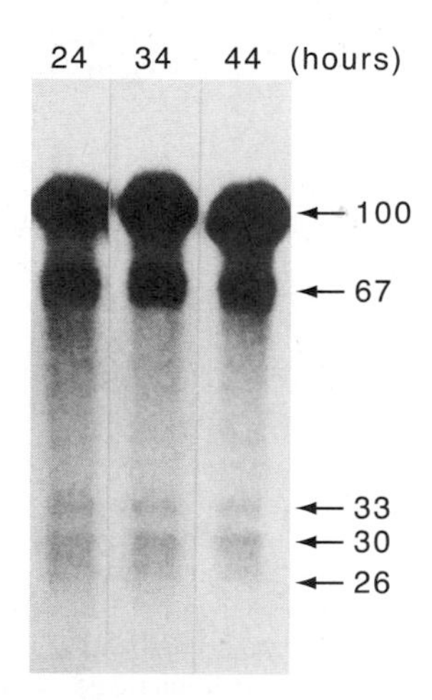

Figure 13.18 Locating the region of DNase hypersensitivity in the SV40 minichromosome. Yaniv and colleagues isolated nuclei from SV40 virus-infected monkey cells at 24, 34, and 44 h after infection and treated them with DNase I. Then they cleaved the treated minichromosomes with *Eco*RI and analyzed the DNA products by electrophoresis, Southern blotting, and probing with radioactive SV40 DNA. Because *Eco*RI cuts 33% of the way clockwise around the circle from the nucleosome-free zone, we would expect to see two fragments, corresponding to 33% and 67% of the whole length of the SV40 genome, assuming that the nucleosome-free zone and the DNase-hypersensitive region coincide. Actually, the 67% fragment is very prevalent, but the 33% fragment is partially degraded into smaller fragments. Thus, the DNase hypersensitive region does correspond to the nucleosome-free zone, which is large enough to produce a range of degradation products. (*Source:* Saragosti, S., G. Moyne, and M. Yaniv, Absence of nucleosomes in a fraction of SV40 chromatin between the origin of replication and the region coding for the late leader RNA. *Cell* 20 (May 1980) f. 7, p. 71. Reprinted by permission of Elsevier Science.)

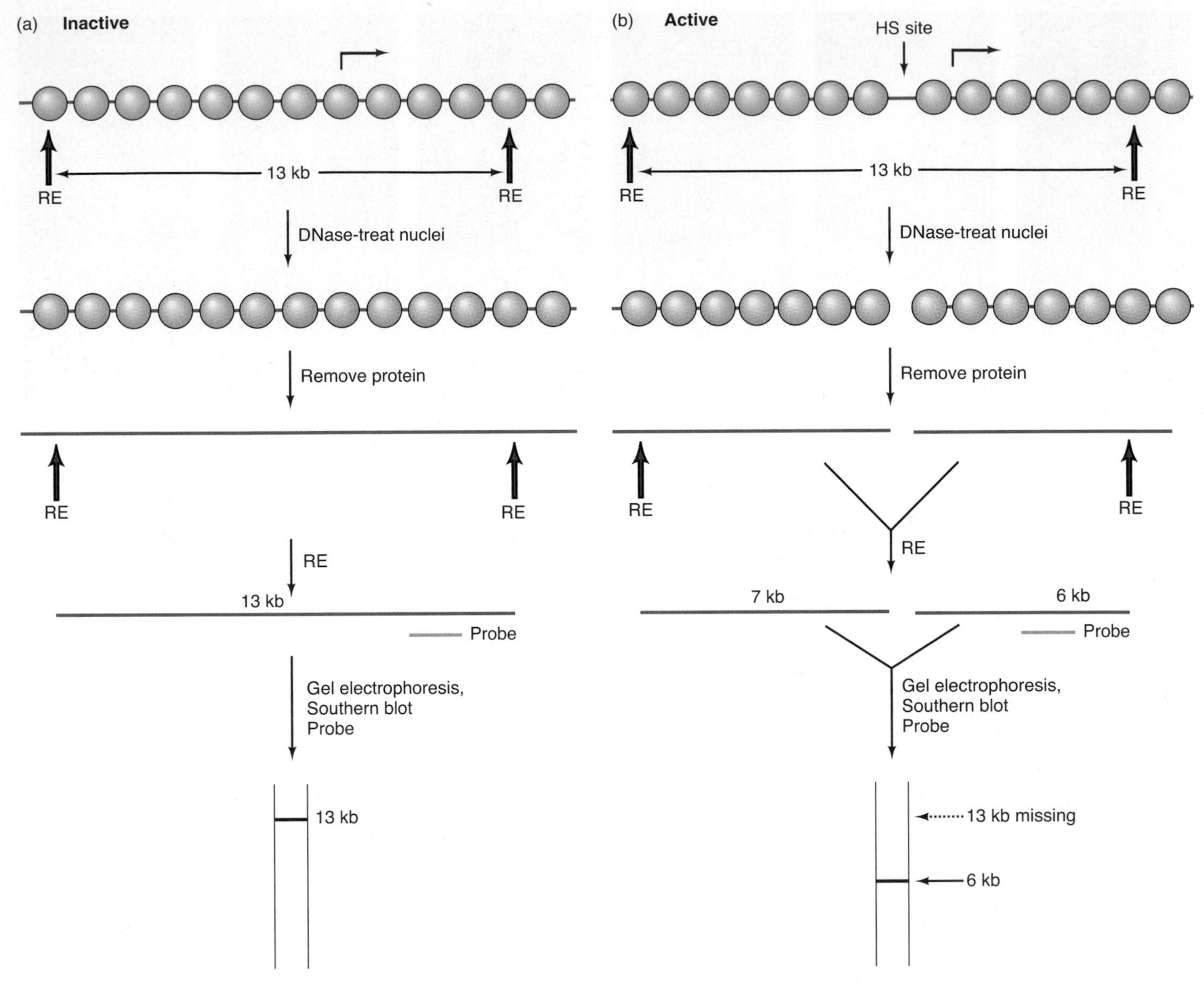

Figure 13.19 Experimental scheme for detecting DNase-hypersensitive regions. (a) Inactive gene, no DNase hypersensitivity. The gene and its control region are complexed with nucleosomes; therefore, no DNA will be degraded when nuclei containing this gene are subjected to mild treatment with DNase I. Next, isolate the DNA from these nuclei, removing all the protein, and digest with a restriction endonuclease (RE). This creates a DNA fragment 13 kb long that spans the gene's control region. Electrophorese the RE digestion products, Southern blot the fragments, and probe the blot with the gene-specific probe (green). This will "light up" the 13-kb fragment. **(b)** Active gene, DNase hypersensitivity. An active gene has one or more nucleosome-free zones that may correspond to a promoter, an enhancer, an insulator, or another control region. Thus, when nuclei containing this active gene are subjected to mild DNase I treatment, that hypersensitive site (HS site) will be digested, as shown. Next, isolate the DNA, remove protein, digest with a restriction endonuclease, electrophorese the fragments, blot, and probe as in panel (a). The 13-kb fragment has disappeared because of its cleavage by DNase, but a new fragment at 6 kb has appeared. The 7-kb fragment will not be detected because it does not hybridize to the probe. This experiment has revealed a DNase-hypersensitive site approximately 6 kb upstream of the downstream RE site. In practice, increasing concentrations of DNase are often used, which would cause a gradual decrease in the intensity of the 13-kb band as the 6-kb band increases in intensity.

site, so two fragments (6 kb and 7 kb) are generated by the combination of DNase I and RE. The 6-kb fragment will be detected by the probe, but the 7-kb fragment will not. And the 13-kb fragment will usually disappear with longer DNase I treatment.

Figure 13.20 shows the results of just such an experiment performed by Frank Grosveld and colleagues in 1987 on the human globin gene cluster, which contains five active globin genes in this order: 5′-ε-^{G}γ-^{A}γ-δ-β-3′. Grosveld and colleagues noted that when the β-globin gene is

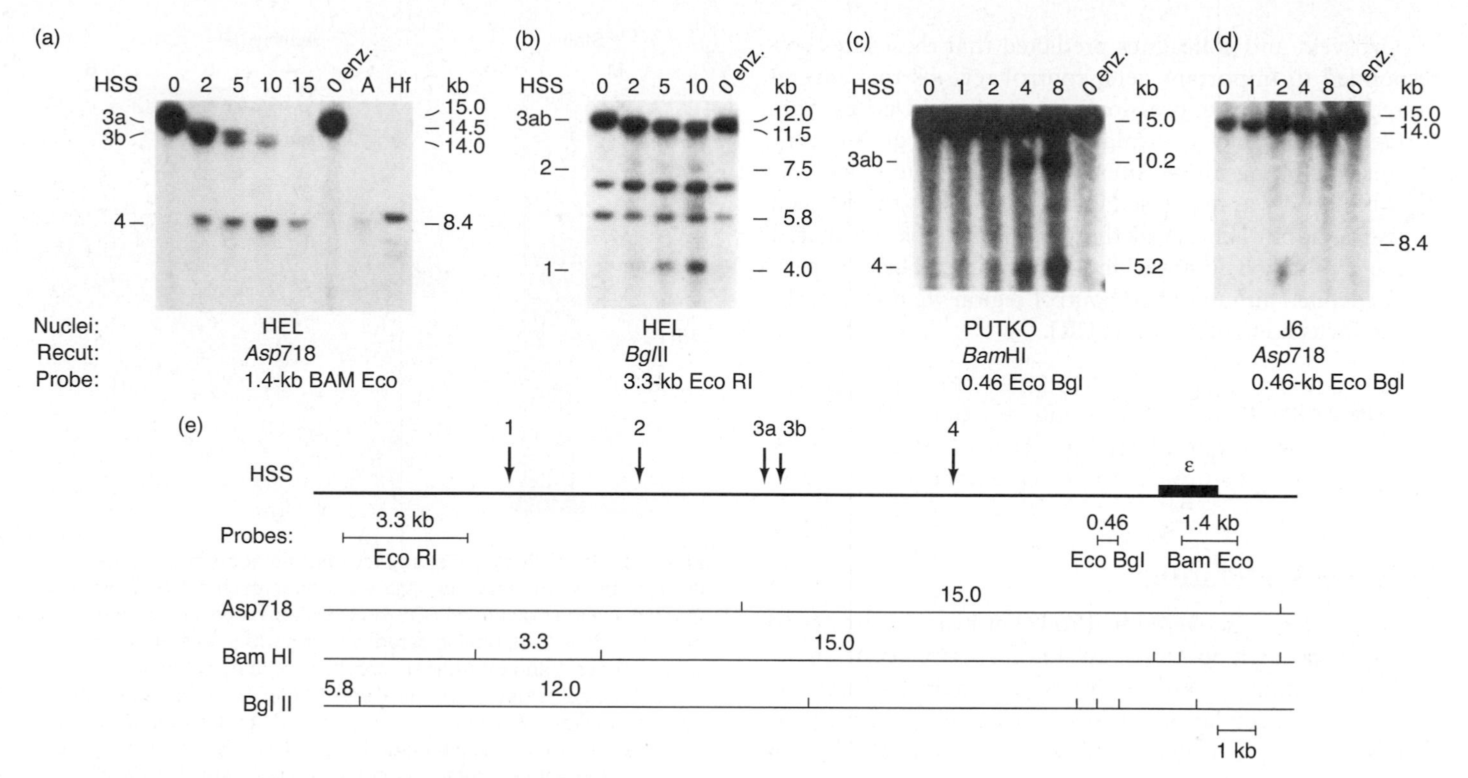

Figure 13.20 Mapping DNase-hypersensitive sites in the 5′-flanking region of the human globin gene. Panels **(a–d)** Grosveld and colleagues treated nuclei from HEL, PUTKO, or J6 cells, as indicated at bottom ("Nuclei:"), with a low concentration of DNase I for the times (in minutes) indicated at top, or with zero enzyme (0 enz.). Then they extracted DNA from the nuclei, deproteinized it with proteinase K, cleaved it with the restriction enzymes indicated at bottom ("Recut:"), electrophoresed the fragments, blotted them, and probed the blots with the probes indicated at bottom ("Probe:"). The fragments corresponding to cleavage at hypersensitive sites (HSS) 1, 2, 3a, 3b, and 4, are indicated at left. The lanes labeled A and Hf in panel (a) contained DNA cut with *Alu*I or *Hin*fI instead of DNase I. **(e)** Map of the 5′-flanking region of the human ε-globin locus, showing the positions of the three probes, and the restriction sites for the three restriction endonucleases used in panels (a–d). (*Source:* Reprinted from *Cell* v. 51, Grosveld et al., p. 976. © 1987, with permission from Elsevier Science.)

transferred by itself to transgenic mice (Chapter 5), it functions at best at only about 10% of its normal level. And when it was inserted into some chromosomal locations it functioned much better than in others. They reasoned that something outside the β-globin gene itself governs efficiency of expression. In fact, several sites contribute to this efficiency, and they are all DNase-hypersensitive.

Five of these sites (1, 2, 3a, 3b, and 4) lie upstream of the ε-globin locus, as shown in Figure 13.20e. Grosveld and colleagues assayed for DNase-hypersensitive sites as described previously in Figure 13.19. The positions of three different probes (Eco RI, Eco Bgl, and Bam Eco) are shown in panel (e). Grosveld and colleagues treated nuclei from two human cell lines that express the β-globin gene—erythroleukemia (HEL) cells, and another human erythroid cell line (PUTKO)—and a cell line that does not express the β-globin gene—human T cells (J6). The "0 enz." lane in each panel shows the results of treatment with no DNase I, and the other numbered lanes show the results of treatment with DNase I for increasing times.

Panel (a) shows the results with HEL cells, the restriction enzyme *Asp*718, and the 1.4-kb Bam Eco probe. DNase I cleavage was readily observed at sites 3a, 3b, and 4. To detect hypersensitive sites farther upstream of the gene, Grosveld and colleagues used the 3.3-kb Eco RI probe, as shown in panel (b). This time, cleavages at sites 1, 2, 3a, and 3b were observed, although cleavage at site 2 was delayed and relatively weak. The 5.8-kb band corresponds to the 5.8-kb fragment that reacts with the probe, as shown in panel (e). The 6.8-kb band came from nonspecific hybridization to an unrelated gene and could be eliminated by hybridization at higher stringency. Panel (c) shows the results with PUTKO cells, the restriction enzyme *Bam*HI and the 0.46 Eco Bgl probe. Cleavage at sites 3a, 3b, and 4 could be observed. Using the same kind of approach, Grosveld and colleagues detected another DNase-hypersensitive site downstream of the β-globin gene.

Finally, Grosveld and colleagues tested for DNase hypersensitivity in J6 T cells, which do not have active globin genes. As panel (d) shows, no DNase hypersensitivity was detected. This result supports the hypothesis that hypersensitivity corresponds to the presence of gene-specific factors that exclude nucleosomes from active genes, but not from inactive genes.

Grosveld and colleagues predicted that these sites corresponded to important gene control regions that are required for optimal expression of transplanted genes. Sure enough, when they transplanted the whole globin gene cluster, including these sites, into transgenic mice, the β-globin gene was expressed just as actively as the resident mouse β-globin gene. And the gene was active no matter where it inserted into the mouse genome. These experiments defined an important control region we now call the globin **locus control region** (**LCR**).

SUMMARY Active genes tend to have DNase-hypersensitive control regions. At least part of this hypersensitivity is due to the absence of nucleosomes.

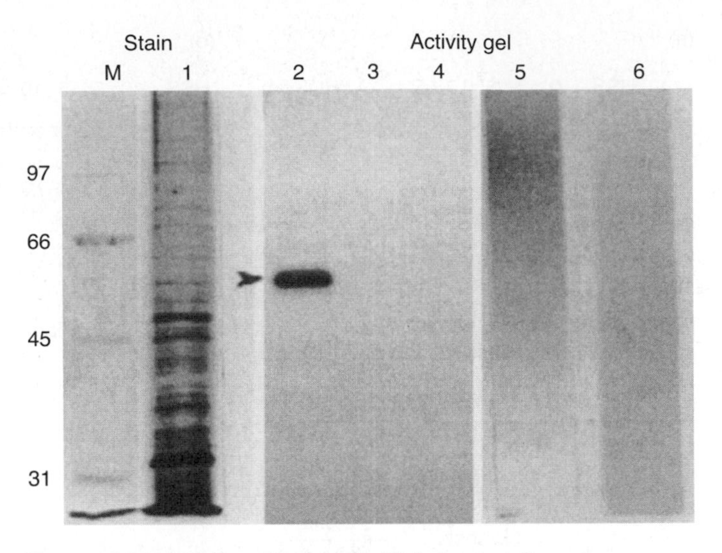

Figure 13.21 Activity gel assay for histone acetyltransferase (HAT) activity. Brownell and Allis electrophoresed a *Tetrahymena* macronuclear extract in an SDS-polyacrylamide gel containing histones (lanes 2–4), bovine serum albumin (BSA, lane 5), or no protein (lanes 1 and 6). After electrophoresis, they either silver-stained the gel to detect protein (lanes M and 1), or treated it with acetyl-CoA labeled in its acetyl group with ^{3}H to detect HAT activity. After washing to remove unreacted acetyl-CoA, they subjected the gel to fluorography to detect ^{3}H-acetyl groups. Lane 2 showed a clear band of ^{3}H-acetylated histones, which indicated the presence of HAT activity. Lanes 3 and 4 failed to show activity because the HAT in the nuclear extracts was inactivated by heating (lane 3) or by treatment with *N*-ethylmaleimide (lane 4) prior to electrophoresis. Lane 5, with BSA instead of histones, also showed no activity, as did lane 6, with no protein substrate. Lane M contained molecular mass marker proteins. (*Source:* Brownell, J.E. and C.D. Allis, An activity gel assay defects a single, catalytically active histone acetyltransferase subunit in *Tetrahymena* macronuclei. *Proceedings of the National Academy of Sciences USA* (July 1995) f. 1, p. 6365. Copyright © National Academy of Sciences, USA.)

Histone Acetylation

Vincent Allfrey discovered in 1964 that histones are found in both acetylated and unacetylated forms. Acetylation occurs on the amino groups on lysine side chains. Allfrey also showed that acetylation of histones correlates with gene activity. That is, unacetylated histones, added to DNA, tend to repress transcription, but acetylated histones are weaker repressors of transcription. These findings implied that enzymes in nuclei acetylate and deacetylate histones and thereby influence gene activity. To investigate this hypothesis, one needs to identify these enzymes, yet they remained elusive for over 30 years, in part because they are present in low quantities in cells.

Finally, in 1996, James Brownell and David Allis succeeded in identifying and purifying a **histone acetyltransferase** (**HAT**), an enzyme that transfers acetyl groups from a donor (acetyl-CoA) to core histones. These investigators used a creative strategy to isolate the enzyme: They started with *Tetrahymena* (ciliated protozoan) cells because this organism has histones that are heavily acetylated, which suggests that the cells contain relatively high concentrations of HAT. They prepared extracts from macronuclei (the large *Tetrahymena* nuclei that contain the active genes) and subjected them to gel electrophoresis in an SDS gel impregnated with histones. To detect HAT activity, they soaked the gel in a solution of acetyl-CoA with a radioactive label in the acetyl group. If the gel contained a band with HAT activity, the HAT would transfer labeled acetyl groups from acetyl-CoA to the histones. This would create a labeled band of acetylated histones in the gel at the position of the HAT activity. To detect the labeled histones, they washed away the unreacted acetyl-CoA, then subjected the gel to fluorography. Figure 13.21 shows the result: a band of HAT activity corresponding to a protein 55 kD in size. Accordingly, Brownell and Allis named this protein p55.

Allis and colleagues followed this initial identification of the HAT activity with a classic molecular cloning scheme to learn more about p55 and its gene. They began by purifying the HAT activity further, using standard biochemical techniques. Once they had purified the HAT activity essentially to homogeneity, they isolated enough of it to obtain a partial amino acid sequence. Using this sequence, they designed a set of degenerate oligonucleotides (Chapter 4) that coded for parts of the amino acid sequence and therefore hybridized to the macronuclear genomic DNA (or to cellular RNA). Using these oligonucleotides as primers, and total cellular RNA as template, they performed RT-PCR as explained in Chapter 4, then cloned the PCR products. They obtained the base sequences of some of the cloned PCR products and checked them to verify that the internal parts also coded for known HAT amino acid sequences. None of the PCR clones contained complete cDNAs, so these workers extended them in both the 5′- and 3′-directions, using rapid amplification of cDNA ends (RACE, Chapter 4). Finally, they obtained a cDNA clone that encoded the full 421-amino-acid p55 protein.

The amino acid sequence inferred from the base sequence of the p55 cDNA was very similar to the amino acid sequence of a yeast protein called Gcn5p. Gcn5p had been identified as a coactivator of acidic transcription activators such as Gcn4p, so the amino acid sequence similarity suggested that both p55 and Gcn5p are HATs that are involved in gene activation. To verify that Gcn5p has HAT activity, Allis and colleagues expressed its gene in *E. coli*, then subjected it and p55 to the SDS-PAGE activity gel assay. Both proteins showed clear HAT activity. Thus, at least one HAT (Gcn5p) has both HAT and transcription coactivator activities. It appears to play a direct role in gene activation by acetylating histones.

It is important to note that p55 and Gcn5p are type A HATs (**HAT A's**) that exist in the nucleus and are apparently involved in gene regulation. They acetylate the lysine-rich N-terminal tails of core histones. Fully acetylated histone H3 has acetyl groups on lysines 9, 14, and 18, and fully acetylated histone H4 has acetyl groups on lysines 5, 8, 12, and 16. Lysines 9 and 14 of histone H3 and Lysines 5, 8, and 16 of histone H4 are acetylated in active chromatin and deacetylated in inactive chromatin. Type B HATs (**HAT B's**) are found in the cytoplasm and acetylate newly synthesized histones H3 and H4 so they can be assembled properly into nucleosomes. The acetyl groups added by HAT B's are later removed in the nucleus by histone deacetylases. All known HAT A's, including p55 and Gcn5p, contain a **bromodomain,** while all known HAT B's lack a bromodomain. Bromodomains allow proteins to bind to acetylated lysines. This is useful to HAT A's, which must recognize partially acetylated histone tails and add acetyl groups to the other lysine residues. But HAT B's have no use for a bromodomain, because they must recognize newly synthesized core histones that are unacetylated.

Since Allis's group's initial discovery of p55, several coactivators besides Gcn5p have been found to have HAT A activity. Among these are CBP/p300 (Chapter 12) and TAF1 (Chapter 11). All three of these coactivators cooperate with activators to enhance transcription. The fact that they have HAT A activity suggests a mechanism for part of this transcription enhancement: By binding near the transcription start site, they could acetylate core histones in the nucleosomes in the neighborhood, neutralizing some of their positive charge and thereby loosening their hold on the DNA (and perhaps on neighboring nucleosomes). This would allow remodeling of the chromatin to make it more accessible to the transcription apparatus, thus stimulating transcription.

It is interesting in this context that TAF1 has a double bromodomain module capable of recognizing two neighboring acetylated lysines, such as we would find on partially acetylated core histones in inactive chromatin. Thus, another role of TAF1 may be to recognize partially acetylated histones in inactive chromatin and to usher its partners, TBP and the other TAFs, into such chromatin to begin the activation process. We will see evidence for this hypothesis later in this chapter.

SUMMARY Histone acetylation occurs in both the cytoplasm and nucleus. Cytoplasmic acetylation is carried out by a HAT B and prepares histones for incorporation into nucleosomes. The acetyl groups are later removed in the nucleus. Nuclear acetylation of core histone N-terminal tails is catalyzed by a HAT A and correlates with transcription activation. A variety of coactivators have HAT A activity, which may allow them to loosen the association of nucleosomes with a gene's control region. Acetylation of core histone tails also attracts bromodomain proteins such as TAF1, which are essential for transcription.

Histone Deacetylation

If core histone acetylation is a transcription-activating event, we would predict that core histone deacetylation would be a repressing event. In accord with this hypothesis, chromatin with underacetylated core histones is less transcriptionally active than average chromatin. Figure 13.22 outlines the apparent mechanism behind this repression: Known transcription repressors, such as nuclear receptors without their ligands, interact with corepressors, which in turn interact with histone deacetylases. These deacetylases then remove acetyl groups from the basic tails of core histones in nearby nucleosomes, tightening the grip of the histones on the DNA, thus stabilizing the nucleosomes and keeping transcription

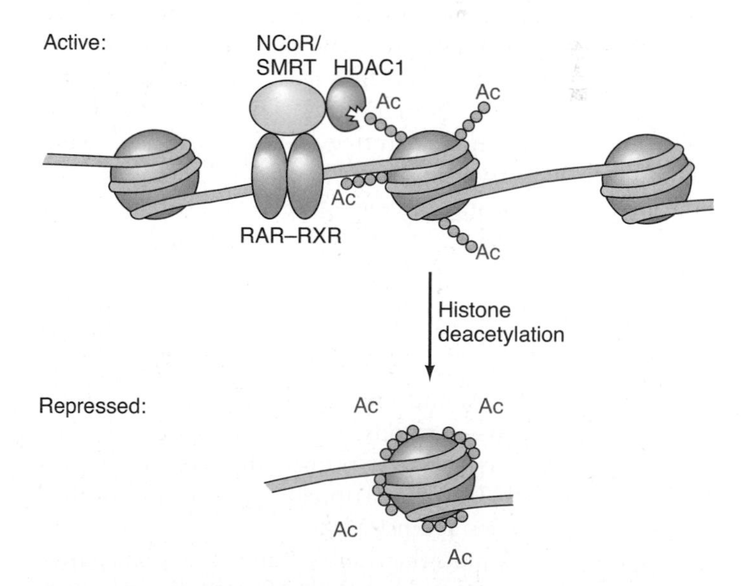

Figure 13.22 Model for participation of histone deacetylase in transcription repression. A heterodimer of retinoic acid receptor (RAR) and retinoic acid receptor X (RXR) binds to an enhancer (top). In the absence of the ligand, retinoic acid, the receptor dimer binds to the corepressor NcoR/SMRT, which binds to the histone deacetylase HDAC1. The deacetylase then removes acetyl groups (red) from the lysine side chains (gray) on core histones of nearby nucleosomes. This deacetylation allows the lysine side chains to associate more closely with DNA (bottom), stabilizing the nucleosomes, and thereby inhibiting transcription.

repressed. This repression can be considered **silencing,** although it is less severe than the silencing seen in heterochromatic regions of chromosomes, such as the ends, or telomeres. Some of the best studied corepressors are **SIN3** (yeast), **SIN3A** and **SIN3B** (mammals), and **NCoR/SMRT** (mammals). NCoR stands for "nuclear receptor corepressor" and SMRT stands for "silencing mediator for retinoid and thyroid hormone receptors." These proteins interact with unliganded retinoic acid receptor (RAR-RXR), a heterodimeric nuclear receptor.

How do we know a physical association exists among transcription factors, corepressors, and histone deacetylases? One way to answer this question has been to add epitope tags to one of the components, then to immunoprecipitate the whole complex with an antibody against the tag. For example, Robert Eisenman and coworkers used epitope tagging to demonstrate a ternary complex among a transcription factor **Mad-Max,** a mammalian Sin3 corepressor (SIN3A), and a histone deacetylase (HDAC2). Max is a transcription factor that can serve as an activator or a repressor, depending on its partner in the heterodimer. If it associates with Myc to form a Myc-Max dimer, it acts as a transcription activator. On the other hand, if it associates with Mad to form a Mad-Max dimer, it acts as a repressor.

Part of the repression caused by Mad-Max comes from histone deacetylation, which suggests some kind of interaction between a histone deacetylase and Mad. By analogy to the RAR-RXR–NCoR/SMRT–HDAC1 interaction illustrated in Figure 13.22, we might expect some corepressor like NCoR/SMRT to mediate this interaction between Mad and a histone deacetylase. To show that this interaction really does occur in vivo, and that it is mediated by a corepressor (SIN3A), Eisenman and coworkers used the following epitope-tagging strategy. They transfected mammalian cells with two plasmids. The first plasmid encoded epitope-tagged histone deacetylase (**HDAC2** tagged with a small peptide called the FLAG epitope [FLAG-HDAC2]). The second plasmid encoded Mad1, or a mutant Mad1 (Mad1Pro) having a proline substitution that blocked both interaction with SIN3A and repression of transcription. Then Eisenman and coworkers prepared extracts from these transfected cells and immunoprecipitated complexes using an anti-FLAG antibody. After electrophoresis, they blotted the proteins and first probed the blots with antibodies against SIN3A, then stripped the blots and probed them with antibodies against Mad1.

Figure 13.23 depicts the results. Lanes 1–3 are negative controls from cells containing a FLAG-encoding plasmid, rather than a FLAG-HDAC2-encoding plasmid. Immunoprecipitation of these lysates with an anti-FLAG antibody should not have precipitated HDAC2 or any proteins associated with it. Accordingly, no SIN3A or Mad1 were found in the blots. Lanes 4–6 contained extracts from cells transfected with a plasmid encoding FLAG-HDAC2, and plasmids encoding: no Mad1 (lane 4); Mad1 (lane 5); and Mad1Pro (lane 6). All three lanes contained SIN3A, which indicated that this protein coprecipitated with FLAG-HDAC2. However, only lane 5 contained Mad1. It was expected that lane 4 would not contain Mad1 because no Mad1 plasmid was provided. It is significant that lane 6 did not contain Mad1Pro, even though a plasmid encoding this protein was included in the transfection. Because Mad1Pro cannot bind to SIN3A, it would not be expected to coprecipitate with FLAG-HDAC2 unless it interacted directly with HDAC2. The fact that it did not coprecipitate supports the hypothesis that Mad1 must bind to SIN3A, and not to HDAC2. This is another way of saying that the corepressor SIN3A mediates the interaction between the transcription factor Mad1 and the histone deacetylase HDAC2. Lanes 7–9 show the results of simply electrophoresing whole-cell lysates without any immunoprecipitation. The two blots show that these lysates contained plenty of SIN3A and abundant Mad1, if a Mad1-encoding plasmid was given (lane 8), or Mad1Pro, if that was the protein present (lane 9). Thus, the lack of Mad1Pro in lane 6 could

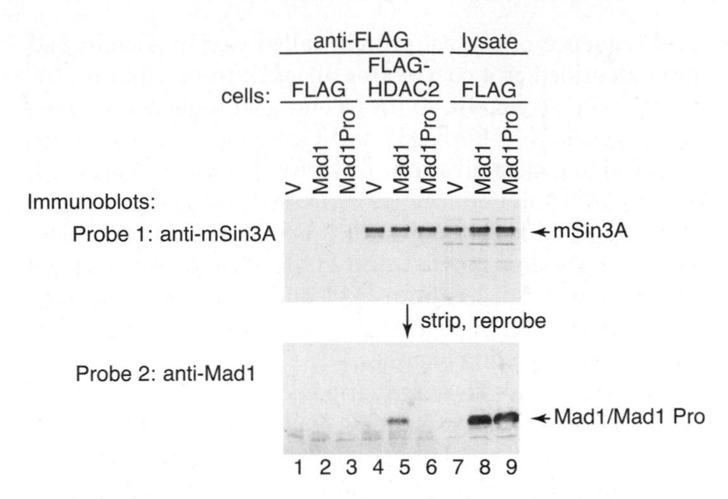

Figure 13.23 Evidence for a ternary complex involving HDAC2, SIN3A, and Mad1. Eisenman and coworkers transfected cells with a plasmid encoding either the FLAG epitope alone, or FLAG–HDAC2, as indicated at the top beside the designation "cells"; and a plasmid encoding either no Mad1 (V), Mad1, or Mad1Pro, also as indicated at top. They immunoprecipitated cell lysates with an anti-FLAG antibody (lanes 1–6, designated "anti-FLAG" at top) or just collected lysates (lanes 7–9, designated "lysate" at top) and electrophoresed the immunoprecipitates or lysates. After electrophoresis, they blotted the proteins to a membrane and probed the immunoblots, first with an anti-SIN3A antibody (top blot). Then, after stripping the first blot, they probed with an anti-Mad1 antibody that reacts with both Mad1 and Mad1Pro (bottom blot). Finally, they detected antibodies bound to proteins on the blot with a secondary antibody conjugated to horseradish peroxidase. They detected the presence of this enzyme with a substrate that becomes chemiluminescent on reaction with peroxidase. The positions of SIN3A and Mad1/Mad1Pro are indicated beside the blots at right. (*Source:* Laherty, C.D., W.-M. Yang, J.-M. Sun, J.R. Davie, E. Seto, and R.N. Eisenman, Histone deacetylases associated with the mSin3 co-repressor mediate Mad transcriptional repression. *Cell* 89 (2 May 1997) f. 3, p. 352. Reprinted by permission of Elsevier Science.)

not be explained by the failure of the plasmid encoding Mad1Pro to produce Mad1Pro protein.

We have now seen two examples of proteins that can be either activators or repressors, depending on other molecules bound to them. Some nuclear receptors behave this way depending on whether or not they are bound to their ligands. Max proteins behave this way depending on whether they are bound to Myc or Mad proteins. Figure 13.24 illustrates this phenomenon for a nuclear receptor, **thyroid hormone receptor (TR).** TR forms heterodimers with RXR and binds to the enhancer known as the **thyroid hormone response element (TRE).** In the absence of **thyroid hormone,** it serves as a repressor. Part of this repression is due to its interaction with NCoR, SIN3, and a histone deacetylase known as mRPD3, which deacetylates core histones in neighboring nucleosomes. This deacetylation stabilizes the nucleosomes and therefore represses transcription.

In the presence of thyroid hormone, the TR–RXR dimer serves as an activator. Part of the activation is due to binding to CBP/p300, P/CAF, and TAF1, all three of which are histone acetyltransferases that acetylate histones in neighboring nucleosomes. This acetylation destabilizes the nucleosomes and therefore stimulates transcription. Notice that the significant targets of the histone acetyltransferases and the histone deacetylases are core histones, not histone H1. Thus, the core histones, as well as H1, play important roles in nucleosome stabilization and destabilization.

Acetylation of core histone tails apparently does more than just inhibit binding of these tails to DNA. As we saw earlier in this chapter (see Figure 13.3), Timothy Richmond and colleagues' x-ray crystallography of core nucleosome particles revealed an interaction between histone H4 in one nucleosome core and the histone H2A–H2B dimer in the adjacent nucleosome core in the crystal lattice. In particular, the very basic region of the N-terminal tail of histone H4 (residues 16–25) interacts with an acidic pocket in the H2A–H2B dimer of the adjoining nucleosome. This interaction could help explain the cross-linking of nucleosomes that blocks access to transcription factors and therefore

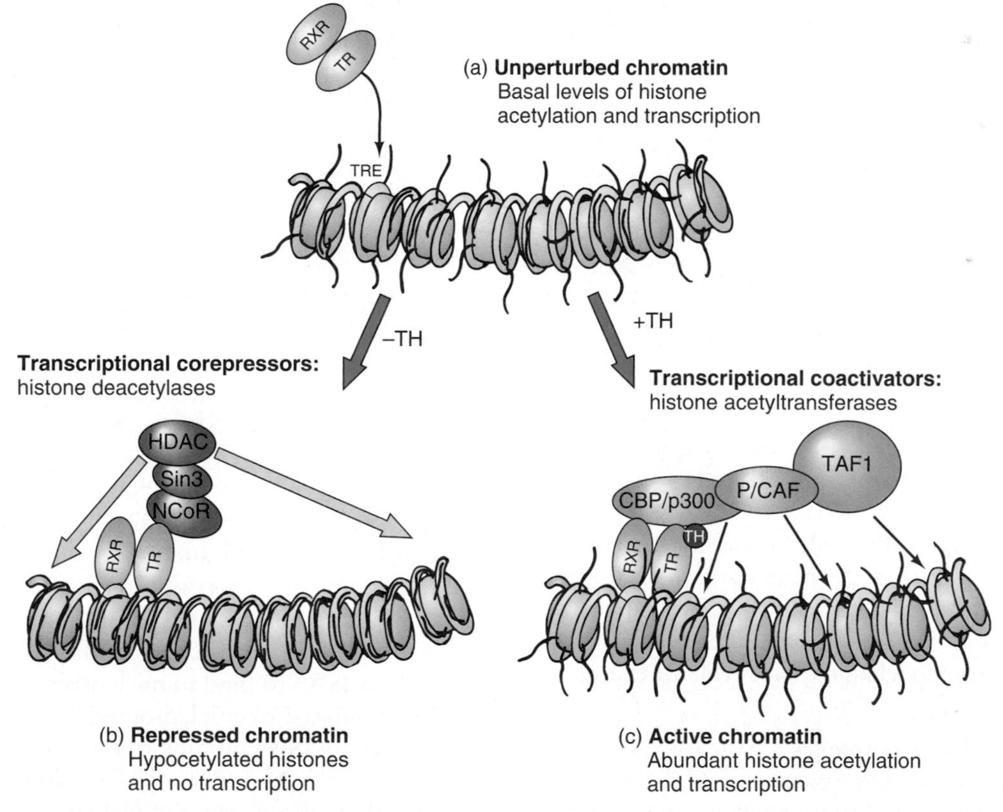

Figure 13.24 A model for activation and repression by the same nuclear receptor. (a) Unperturbed chromatin. No nuclear receptor (TR–RXR dimer) is bound to the thyroid hormone response element (TRE). Core histone tails are moderately acetylated. Transcription occurs at a basal level. **(b)** Repressed chromatin. The nuclear receptor is bound to the TRE in the absence of thyroid hormone (TH). The nuclear receptor interacts with either of the corepressors SIN3 and NCoR, which interact with a histone deacetylase (HDAC). The deacetylase cleaves acetyl groups off of the tails of core histones in surrounding nucleosomes, tightening the binding between histones and DNA, and between histones in neighboring nucleosomes, thereby helping to repress transcription. **(c)** Active chromatin. Thyroid hormone (purple) binds to the TR part of the nuclear receptor dimer, changing its conformation so it binds to one or more of the coactivators CBP/p300, P/CAF, and TAF1. These coactivators are all HAT A's that acetylate the tails of core histones in nearby nucleosomes, loosening the binding between histones and DNA and between histones on neighboring nucleosomes and helping to activate transcription. (*Source:* Adapted from Wolfe, A.P. 1997. Sinful repression. *Nature* 387:16–17.)

represses transcription. This hypothesis would also help explain why acetylating the tails of the core histones has an activating effect: Neutralizing the positive charge of the N-terminal tail of histone H4 by acetylation would help prevent nucleosome cross-linking and therefore help deter repression of transcription.

However, as mentioned in the previous section, simple charge neutralization is only part of the story. Acetylated lysines on core histone tails provide a docking site for bromodomain proteins such as TAF1, which are essential for transcription. In fact, as we will see in the next section, acetylation and other modifications of core histones may constitute a "histone code" that can be interpreted by other proteins that stimulate and repress transcription.

SUMMARY Transcription repressors such as unliganded nuclear receptors and Mad-Max bind to DNA sites and interact with corepressors such as NCoR/SMRT and SIN3, which in turn bind to histone deacetylases such as HDAC1 and 2. This assembly of ternary protein complexes brings the histone deacetylases close to nucleosomes in the neighborhood. The deacetylation of core histones allows the basic tails of the histones to bind strongly to DNA and to histones in neighboring nucleosomes, stabilizing and cross-linking the nucleosomes, and thereby inhibiting transcription. Deacetylation of core histones also removes binding sites for bromodomain proteins that are essential for transcription activation.

Chromatin Remodeling

Histone acetylation is frequently essential for gene derepression but it is not sufficient because it deals only with the tails of the core histones, which lie outside the nucleosome core. Acetylation of these core histone tails can disrupt nucleosome cross-linking, as we will see in the next section, but it leaves the nucleosomes intact. Something else is needed to "remodel" the nucleosome cores to permit access to transcription factors, and this remodeling requires ATP for energy.

Chromatin Remodeling Complexes At least four classes of protein complexes participate in this **chromatin remodeling**, and they are distinguished by their ATPase component, which harnesses the energy of ATP hydrolysis to the task of chromatin remodeling. These are the **SWI/SNF** family (pronounced "switch-sniff"), the **ISWI** ("imitation switch") family, the NuRD family, and the **INO80** family. All four classes of proteins alter the structure of nucleosome cores to make the DNA more accessible, not only to transcription activators, but also to nucleases and other proteins.

SWI/SNF complexes have been isolated from eukaryotic organisms ranging from yeast to human. They were originally identified in yeast, and found to regulate the *HO* endonuclease gene, which was responsible for mating type switching (hence the "SWI" part of the name). They also regulated the *SUC2* gene, which encodes invertase, the enzyme that begins the sucrose fermentation process. Thus, mutants with defects in the genes encoding the subunits of the complex were sucrose non-fermenters (hence the "SNF" part of the name). The SWI/SNF complexes all share an ATPase known as **BRG1** (or Brm in certain organisms). Gerald Crabtree and colleagues used an antibody to BRG1 to immunoprecipitate SWI/SNF complexes from several mammalian species, and found 9–12 **BRG1-associated factors (BAFs)** that co-precipitated with BRG1.

There are many similarities between mammalian and yeast BAFs, but some proteins distinct to each. In addition, mammalian BAFs are more diverse than their yeast counterparts. This could reflect the complexity of mammalian development relative to that of yeast, and different mammalian complexes could be devoted to different developmental processes.

One of the BAFs is called BAF 155 or BAF 170, depending on the species. It contains a so-called **SANT domain** ("SANT" is an acronym that refers to four proteins in which the domain is found). This domain has a sequence and three-dimensional structure that resembles that of the DNA-binding domain (DBD) of a transcription factor known as Myb. But some amino acid differences between SANT and the Myb DBD suggest that SANT does not bind DNA. In particular, the putative DNA-binding fold of the domain is lined with acidic residues, rather than basic ones, which is consistent with a role in binding histones, which are basic, and not DNA, which is acidic.

Members of the ISWI class of chromatin remodeling proteins also contain a SANT domain; in fact, they contain two. The first is a canonical SANT domain with a preponderance of acidic residues. The second has a net positive charge at neutral pH and could therefore be involved in DNA binding. This second domain is known as a SANT-like ISWI domain (**SLIDE**) to distinguish it from ordinary SANT domains. Both SANT and SLIDE domains are required for ISWI to bind to nucleosomes, and for its ATPase to be stimulated by nucleosomes. Thus, these domains appear to allow ISWI binding to nucleosomes and to transfer a stimulatory signal to the ATPase domain of ISWI, which then enables chromatin remodeling.

All these families of proteins may yield the nucleosome-free regions around enhancers and promoters that are characteristic of active genes. In fact, we would predict that a nucleosome-free enhancer would be an important early requirement for gene activation. Thus, it is not surprising that SWI/SNF appears to be one of the first coactivators to arrive on the scene when many yeast genes are activated.

SUMMARY Activation of many eukaryotic genes requires chromatin remodeling. Several different protein complexes carry out this remodeling, and all of them have an ATPase that harvests the energy from ATP hydrolysis to use for remodeling. The remodeling complexes are distinguished by their ATPase component, and two of the best-studied complexes are SWI/SNF and ISWI. The SWI/SNF complex in mammals has BRG1 as its ATPase, and 9–12 BRG1-associated factors (BAFs). One of the highly conserved BAFs is called BAF 155 or 170. It has a SANT domain that appears to be responsible for histone binding. This would help SWI/SNF bind to nucleosomes. Members of the ISWI class of remodeling complexes have a SANT domain, and another domain called SLIDE that appears to be involved in DNA binding.

The Mechanism of Chromatin Remodeling It is still not clear exactly what "remodeling" means. Sometimes it involves movement of nucleosomes away from their starting positions, opening up promoters to transcription factors. But remodeling does not necessarily involve simple sliding of nucleosomes. For example, remodeling can occur in chromatin in which nucleosomes are arrayed back-to-back through a promoter, and simply sliding them all in tandem would not open up significant amounts of DNA. Also, as we will see later in this chapter, remodeling sometimes involves a loosening of one or more nucleosomes so they can be moved aside by *other* proteins, such as TFIID. Perhaps the best provisional description of remodeling is that it mobilizes nucleosomes. That is, it allows nucleosomes to move by sliding or by other mechanisms. This movement can be caused by the remodeling complexes themselves, or by other proteins.

Furthermore, the effect of chromatin remodeling is not always activation of transcription; all known remodeling complexes sometimes collaborate in repression. Thus, remodeling of nucleosomes can make it easier to move them away from promoters, activating transcription. But remodeling can also make it easier to move nucleosomes into position to repress transcription. In fact, one of the subunits of the NuRD complex is a histone deacetylase, which can help repress transcription.

Robert Kingston and colleagues examined the nature of chromatin remodeling activity, focusing on the BRG1 subunit of SWI/SNF. They reasoned that one aspect of remodeling is making DNA more accessible, so they studied DNA accessibility as a measure of remodeling activity. They imagined two models for remodeling (Figure 13.25): Model 1 involves the formation of several different conformations of the nucleosomal DNA with respect to the core histones. Model 2 involves the formation of a single remodeled conformation. This would occur if the DNA simply peeled away from the core histones from the point of the DNA's entry to or exit from the nucleosome, as it does in uncatalyzed DNA exposure in mononucleosomes. Model 2 would also apply if the nucleosome simply slid along the DNA, as it does in heated nucleosomes in vitro.

Kingston and colleagues devised several ways to distinguish between the two models, all of which led to the conclusion that model 1 is correct, and remodeled chromatin exists in several different conformations. They started with a model nucleosome, which included a labeled 157-bp

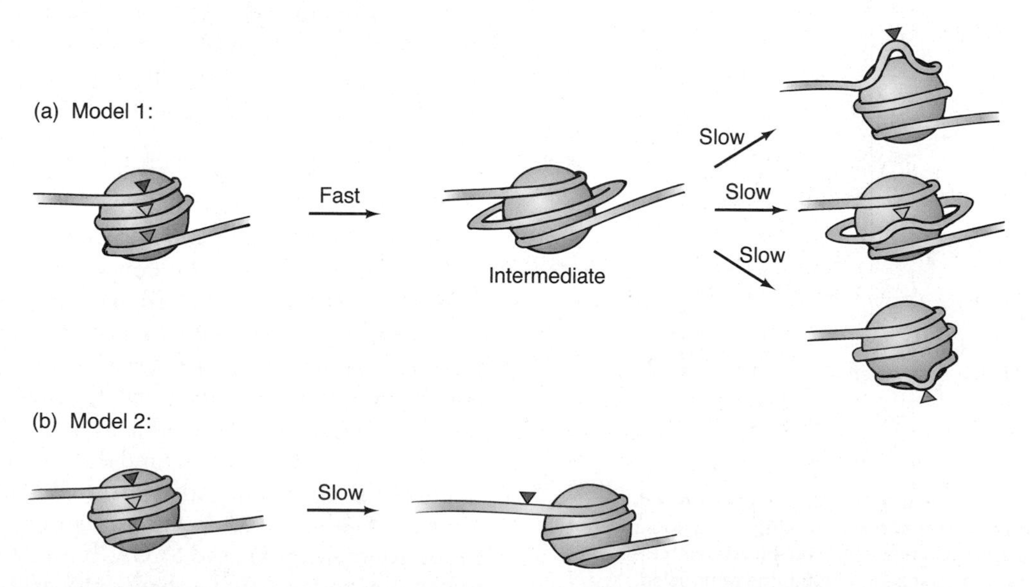

Figure 13.25 Two models for chromatin remodeling by SWI/SNF. **(a)** Model 1. This nucleosome contains three restriction sites, denoted by the colored triangles. In the first (fast) step, the nucleosome may generate an intermediate, which then converts in rate limiting steps to various remodeled conformations. Each of the three conformations illustrated here have opened up one of the restriction sites. **(b)** Model 2. Remodeling yields a single conformation, which, in this case, opens up one of the restriction sites.

DNA fragment that contained cleavage sites for three restriction enzymes, *Pst*I, *Spe*I, and *Xho*I. They reasoned that the two models made different predictions about the rates at which the three restriction sites would become available during remodeling.

Notice that the actual rates of cutting by the restriction enzymes are very fast, so they are not rate limiting. The change in chromatin conformation, which makes the restriction sites accessible, is relatively slow, so that is what limits the rate of cutting. Thus, model 1, in which different conformations are produced, predicts that the rates of cutting by the three enzymes will be different. That is because different conformations will have different accessibilities to the three enzymes, and these different conformations are reached at different rates. Model 2, which produces a single conformation, should yield accessibility to all three enzymes at the same rate, so they should all cut at the same rate.

Thus, Kingston and colleagues added BRG1 and ATP to their labeled model nucleosome and measured the rate of cleavage by each restriction enzyme during remodeling. Figure 13.26 shows that the rates differed by as much as a factor of 9, supporting model 1. Furthermore, the rate of cutting by DNase 1 was 10–20 times faster than the rate of cutting by *Pst*I, which also fits model 1, but not model 2. Finally, Kingston and colleagues repeated their experiments with whole SWI/SNF, instead of just BRG1, and obtained the same results. Thus, model 1 also describes remodeling carried out by intact SWI/SNF, and these experiments make clear that authentic, catalyzed chromatin remodeling is quite different from the simple alterations in chromatin that can occur in the absence of a catalyst.

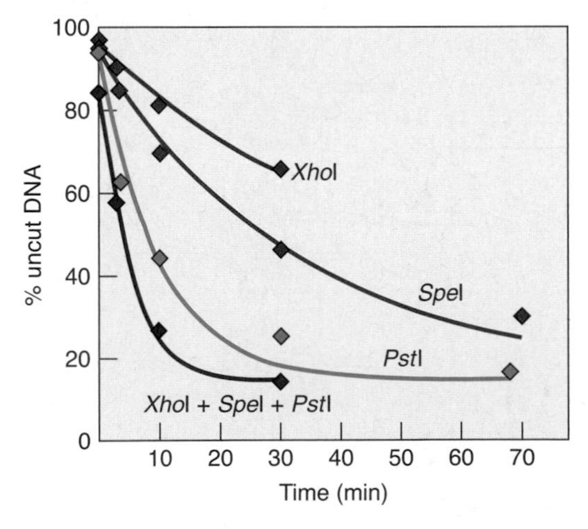

Figure 13.26 Restriction sites are revealed at different rates during BRG1-catalyzed chromatin remodeling. Kingston and colleagues incubated nucleosomes with labeled DNA with BRG1 and ATP for various times up to 70 min and tested the remodeled nucleosomes for susceptibility to cleavage by three restriction enzymes: *Xho*I, *Spe*I, and *Pst*I. They plotted uncut DNA, revealed by electrophoresis of deproteinized DNA, versus time. (*Source:* Adapted from Narlikar G.J. et al., *Molecular Cell* 8, 2001. f. 4A, p. 1224.)

SUMMARY The mechanism of chromatin remodeling is not understood in detail, but it does involve mobilization of nucleosomes, with loosening of the association between DNA and core histones. In contrast to uncatalyzed DNA exposure in nucleosomes, or simple sliding of nucleosomes along a stretch of DNA, catalyzed remodeling of nucleosomes involves the formation of distinct conformations of the nucleosomal DNA with respect to the core histones.

Remodeling in Yeast *HO* Gene Activation Kim Nasmyth and colleagues studied protein association with the *HO* gene of yeast, which plays a key role in switching the mating type. The expression of *HO* depends on a series of protein factors that appear at different phases of the cell cycle. Nasmyth and colleagues used a technique called chromatin immunoprecipitation (ChIP; Chapter 5) as follows: First, they fused DNA fragments encoding short regions (epitopes) of a protein (Myc) to the ends of genes encoding the proteins known to associate with the *HO* gene. This led to the production of fusion proteins with the Myc epitopes at their C-termini. Then they synchronized the yeast cells, so most of them went through the cell cycle together. They obtained cells in various phases of the cell cycle and added formaldehyde to form covalent bonds between DNA and any proteins bound to it. Then they sheared the chromatin by sonication to produce short, double-stranded DNA fragments cross-linked to proteins. Next, they made cell extracts and immunoprecipitated the protein–DNA complexes with antibodies directed against the Myc epitopes. Recall that the Myc epitopes were attached to the proteins known to associate with *HO*, so the immunoprecipitated protein–DNA complexes should contain both these fusion proteins and the *HO* gene. To verify that these complexes contained the *HO* gene, Nasmyth and colleagues performed PCR with *HO*-specific primers. The PCR product should be a band of predictable size if the *HO* gene is really present.

The experimental results showed that a protein known as Swi5 bound first to the control region of *HO*. Next, SWI/SNF bound, followed by the SAGA complex (Chapter 11), which contains the HAT Gcn5p, which then recruited the activator SBF. Other proteins, including general transcription factors and RNA polymerase II bound in turn after SBF. Both SWI/SNF and SAGA are absolutely required for activation of *HO*, and they could act in concert to remodel the chromatin around the *HO* promoter. For example, SWI/SNF could disrupt the core histones around the gene's control region, and SAGA, by acetylating the tails of the core histones, could enhance the disruption and possibly make it permanent. Other work strongly suggests that the factors do not have to act in the order presented here. At other promoters, they can act in many different orders and can help each other perform their functions. In the next section we will see an example of a gene that recruits a HAT *before* the SWI/SNF complex.

SUMMARY ChIP analysis can reveal the order of binding of factors to a gene during activation. As the yeast *HO* gene is activated, the first factor to bind is Swi5, followed by SWI/SNF and SAGA, which contains the HAT Gcn5p. Next, the general transcription factors and other proteins bind. Thus, chromatin remodeling is among the first steps in activation of this gene, but the order can be different in other genes.

Remodeling in the Human IFN-β Gene: The Histone Code We have seen that the core histone tails are subject to acetylation and deacetylation, which tend to activate, and deactivate transcription, respectively. But histone tails are subject to several other modifications, including methylation, phosphorylation, ubiquitylation, and sumoylation. Each of these modifications affects the transcription levels of nearby genes, which has given rise to the concept of a **histone code.** This concept, elaborated by Thomas Jenuwein and David Allis in 2001, holds that the combination of histone modifications on a given nucleosome near a gene's control region affects the efficiency of transcription of that gene. The histone code is an **epigenetic** code (not affecting the base sequence of DNA itself), which adds to the code written in the base sequence of the gene and its control region. Since 2001, many studies have supported the histone code hypothesis. Let us examine one such study, on the human interferon-β (IFN-β) gene.

Dimitris Thanos and colleagues have investigated a well-studied example of chromatin remodeling that occurs during the activation of the human IFN-β gene. When this gene is activated by viral infection, transcription activators bind to nucleosome-free regions near the promoter, forming an enhanceosome, as we learned in Chapter 12. The activators in the enhanceosome recruit factors that modify and remodel the chromatin around the transcription start site. In particular, one nucleosome is moved out of the way so transcription can initiate.

This process involves the following events: The activators recruit HATs, the SWI/SNF complex, and the general transcription factors. The HATs acetylate core histone tails in the nucleosome, which attracts the CBP–RNA polymerase II holoenzyme via one or more bromodomains in CBP. The SWI/SNF complex in the holoenzyme loosens the association between the nucleosome and the promoter DNA. Then, when TFIID binds to the TATA box and bends it, the remodeled nucleosome slides to a new location 36 bp downstream, allowing transcription initiation to occur.

Thanos and colleagues looked at the ordered acetylation of nucleosome core histones and found that acetylation of lysine 8 of histone H4 causes recruitment of the SWI/SNF complex, and acetylation of lysines 9 and 14 in histone H3 causes recruitment of TFIID.

These investigators began by looking at the time course of histone acetylation after Sendai virus infection of HeLa cells, using ChIP analysis. They immunoprecipitated cross-linked chromatin with antibodies against acetylated and phosphorylated histones H3 and H4. Figure 13.27a shows that chromatin bearing the IFN-β gene could be immunoprecipitated with antibodies against acetylated lysines 8 and 12 on histone H4, and with antibodies against acetylated lysines 9 and 14 and phosphorylated serine 10 on histone H3. But the same chromatin could not be immunoprecipitated with antibodies against acetylated lysines 5 and 16 on histone H4. Thus, the pattern of histone acetylation was not random. In a separate experiment, Thanos and colleagues showed that the antibodies against acetylated lysines 5 and 16 of histone H4 were capable of precipitating chromatin if these lysines really were acetylated.

Furthermore, the timing of histone modification varied from position to position. Thus, lysine 8 of histone H4 was acetylated from 3 to 8 h after virus infection, but lysine 12 of H4 was acetylated only at 6 h. Also, phosphorylation of serine 10 of histone H3 began at about 3 h after infection and peaked strongly at 6 h, whereas acetylation of lysine 14 of H3 began at about 6 h, and acetylation of lysine 9 of H3 began earlier and lasted until at least 19 h.

The timing of serine 10 phosphorylation and lysine 14 acetylation of histone H3 supported an earlier hypothesis that phosphorylation of serine 10 is necessary for lysine 14 acetylation. These results also revealed a perfect correspondence between the timing of acetylation of lysine 14 and the recruitment of TBP to the promoter. (Compare row 9 with row 10, showing immunoprecipitation with an antibody against TBP.) This finding is consistent with the hypothesis that acetylation of lysine 14 of H3 is required to recruit TBP to the promoter.

Thanos and colleagues performed similar experiments in vitro with chromatin reconstituted from histones expressed in bacteria and modified at selected sites in vitro. They found that the sites acetylated in vitro were the same ones acetylated in vivo. Furthermore, they performed the same experiments with extracts missing one or more HATs to see the effects on specific lysine acetylations. Figure 13.27b shows that extracts immunodepleted of the HAT GCN5/PCAF were defective in acetylating lysine 8 of histone H4. On the other hand, extracts immunodepleted of the HAT CBP/p300 or the SWI/SNF component BRG1/BRM could still acetylate lysine 8 of H4. A separate, control experiment demonstrated that depletion of GCN5/PCAF did not cause a depletion of CBP/p300, and vice versa. Thus, it appears that GCN5/PCAF is responsible for acetylating lysine 8 of histone H4, and a separate experiment (not shown) made the same case that this HAT is also responsible for acetylating lysine 14 of

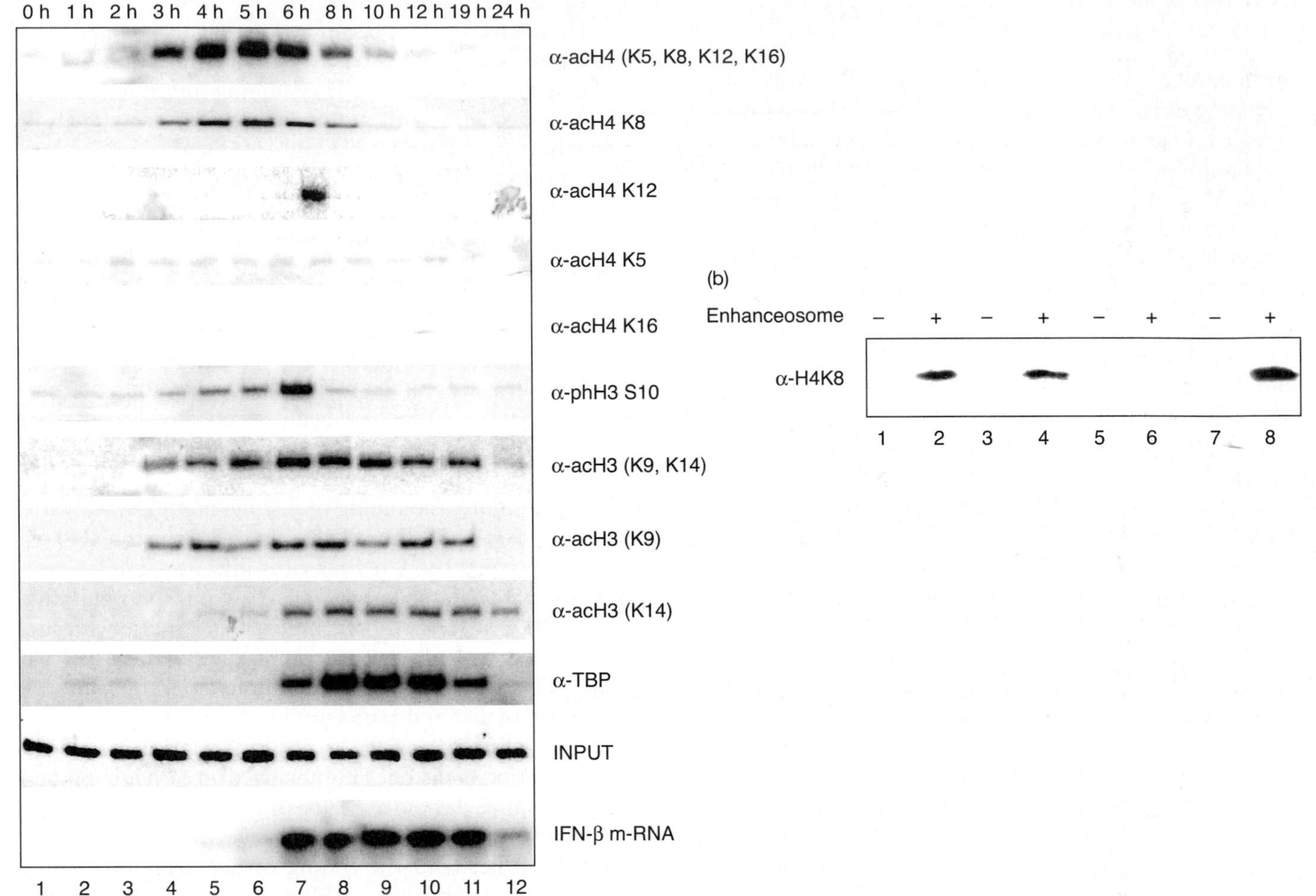

Figure 13.27 Timing of histone acetylation in chromatin at the IFN-β promoter after virus infection. (a) ChIP analysis. Thanos and colleagues performed ChIP with HeLa cell nuclear extracts at various times after infection with Sendai virus, using antibodies (indicated at right) directed against histone H4 acetylated on: lysine 8 (α-acH4 K8), lysine 12 (α-acH4 K12), lysine 5 (α-acH4 K5), or lysine 16 (α-acH4 K16), or all of these antibodies (α-acH4 [K5, K8, K12, K16]); or histone H3 phosphorylated on serine 10 (α-phH3 S10); or histone H3 acetylated on: lysine 9 (α-acH3 K9), or lysine 14 (α-acH3 K14), or both (α-acH3 [K9, K14]). They also performed ChIP with an antibody directed against TBP. Then they performed PCR on all the immunoprecipitated chromatins with primers specific for the IFN-β promoter. These PCR signals are presented, along with an RT-PCR signal that shows the abundance of IFN-β mRNA at the various times. The input lane shows the PCR signal using the input chromatin to show that roughly equal amounts of chromatin were used in each experiment. **(b)** Effects of immunodepletion of HATs on acetylation of lysine 8 of histone H4. Thanos and colleagues assembled the IFN-β enhanceosome on a biotinylated piece of DNA containing the IFN-β promoter and enhancers. Then they incubated the enhanceosome (even lanes) or buffer (odd lanes) with wild-type cell nuclear extracts (lanes 1 and 2), or nuclear extracts depleted of: CBP/p300 (lanes 3 and 4); GCN5/PCAF (lanes 5 and 6); or the SWI/SNF component BRG1/BRM. Then they electrophoresed the proteins, Western blotted the gels, and probed the blots with an antibody directed against histone H4 acetylated on lysine 8. (*Source:* Reprinted from *Cell* v. 111, Agalioti et al., p. 383. © 2002, with permission from Elsevier Science.)

histone H3. (Note that GCN5 is the human homolog of yeast Gcn5p.)

To investigate the effects of core histone tail acetylations on recruitment of SWI/SNF and TFIID, Thanos and colleagues reconstituted chromatin with the IFN-β promoter coupled to resin beads and core histones, then incubated the chromatin with nuclear extracts in the presence or absence of the acetyl donor acetyl-CoA, washed unbound proteins away, then disrupted the chromatin with SDS and subjected the released proteins to Western blotting and probed the blots with antibodies against a SWI/SNF component (BRG1) and a TFIID component (TAF1).

Figure 13.28a, shows that the chromatin bound only small amounts of BRG1 and TAF1 when it was not acetylated (lanes 1 and 2), but larger amounts of both proteins when it was acetylated (lanes 3 and 4). When chromatin was

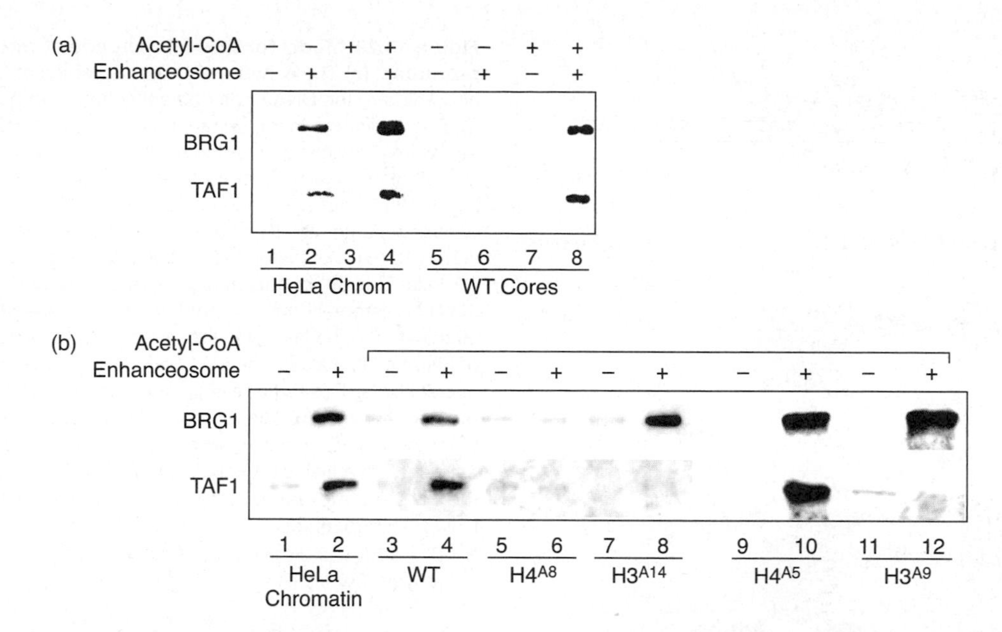

Figure 13.28 Recruitment of SWI/SNF and TFIID to IFN-β promoters: in the presence of wild-type and mutant core histones **(a)** Thanos and colleagues reconstituted chromatin on an IFN-β promoter attached to Dyna-beads, then incubated it with HeLa cell nuclear extracts, washed away unbound protein, then assayed for bound BRG1 and TAF1 by Western blotting and probing with antibodies against these two proteins. Each lane either contained the enhanceosome, or not, as indicated at top, and acetyl-CoA was included in the nuclear extract incubation to allow acetylation of histones, or not, also as indicated at top. Lanes 1–4 contained chromatin reconstituted from native HeLa cell chromatin. Lanes 5–8 contained chromatin reconstituted from recombinant wild-type core histones expressed in *E. coli* (WT Cores). **(b)** Conditions were as in panel (a) except that mutant core histones were used in some experiments, as indicated below each lane. Again the presence or absence of enhanceosomes was indicated at top, along with presence of acetyl-CoA, indicated by the bracket at top. Examples of mutant nomenclature: $H4^{A8}$ indicates a histone H4 in which lysine 8 has been changed to alanine. (*Source:* Reprinted from *Cell* v. 111, Agalioti et al., p. 386. © 2002, with permission from Elsevier.)

reconstituted with histones produced from cloned genes in *E. coli*, it bound no detectable BRG1 and TAF1 when it was not acetylated (lanes 5 and 6), but abundant quantities of both proteins when it was acetylated (lanes 7 and 8).

To investigate the role of acetylation of specific histone lysines, Thanos and colleagues reconstituted chromatin with mutant histones in which one lysine had been converted to an alanine. Figure 13.28b shows the results. Natural HeLa chromatin bound both BRG1 and TAF1 (lanes 1 and 2), as we have already seen in panel (a). Predictably, chromatin reconstituted with wild-type histones also bound the two proteins (lanes 3 and 4). But chromatin reconstituted with histone H4 lacking lysine 8 (which had been converted to alanine) failed to bind either BRG1 or TAF1 (lanes 5 and 6). This result can be explained by the failure of this mutant chromatin to recruit SWI/SNF (BRG1), which is required to recruit TFIID (TAF1).

When lysine 14 of histone H3 was changed to alanine, the reconstituted chromatin could recruit BRG1, but not TAF1 (lanes 7 and 8). The same behavior was observed when lysine 9 of histone H3 was changed to alanine (lanes 11 and 12). Thus, acetylation of lysines 9 and 14 appear to be required for TFIID recruitment, but not for SWI/SNF recruitment. In a control experiment, lysine 5 of histone H4 was changed to alanine. This lysine is known not to be acetylated on virus infection, so it is not surprising that its mutation had no effect on recruitment of either BRG1 or TAF1 (lanes 9 and 10).

Using the same method, Thanos and colleagues showed that substitution of lysine 12 of histone H3 with alanine did not affect recruitment of either TAF1 or BRG1. This lysine was acetylated in vivo, but only very briefly (Figure 13.27), and this acetylation is apparently not required for recruitment of either TFIID or SWI/SNF. Finally, substitution of serine 10 with alanine blocked recruitment of TAF1, but not BRG1. Thus, loss of serine 10 has the same effect as loss of lysines 9 or 14. The effect of loss of serine 10 is consistent with the hypothesis that phosphorylation of serine 10 is required for acetylation of lysine 14.

All of these results can be summarized by a model like the one in Figure 13.29. The core idea of the model is that the enhancer has all the genetic information needed to assemble the enhanceosome, and the enhanceosome can then recruit the appropriate factors to remove the nucleosome blocking initiation of transcription. Thus, information flows from the enhancer to the nucleosome, and not in the reverse direction.

In particular, the model calls for the following sequence of events: On virus infection, activators appear and assemble the enhanceosome on the enhancer. The enhanceosome then recruits the HAT GCN5, which acetylates lysine 8 of histone H4 and lysine 9 of histone H3. The enhanceosome also recruits an unknown protein kinase that

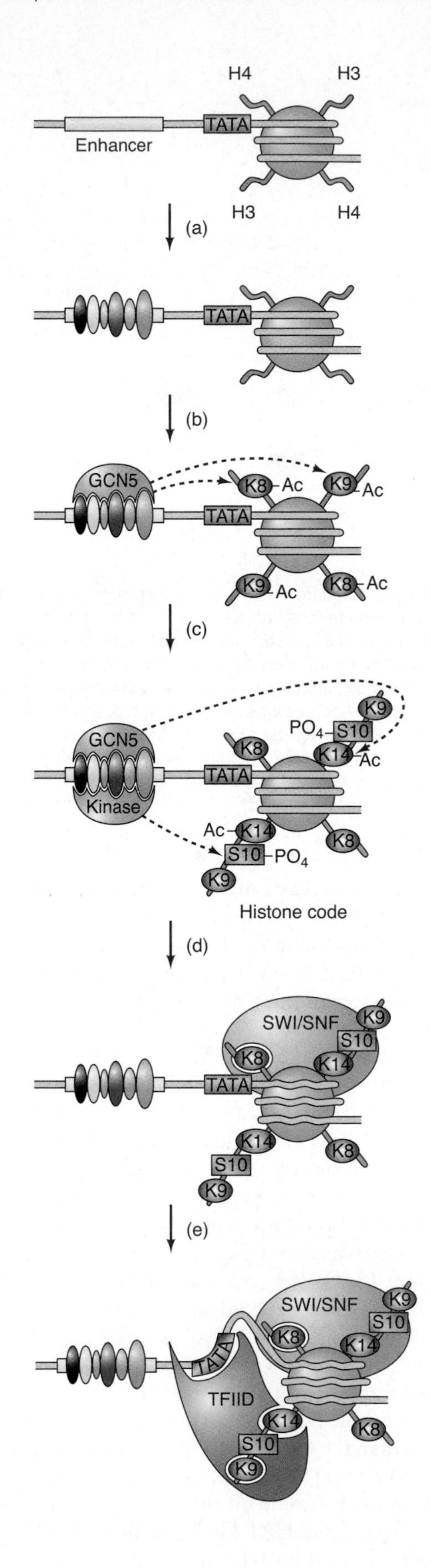

Figure 13.29 Model for the histone code at the human IFN-β promoter. **(a)** The enhanceosome assembles at the promoter according to the DNA code (the collection of enhancer elements). **(b)** The activators in the enhanceosome recruit GCN5, and this HAT acetylates lysine 8 (K8) on the tail of histone H4 and lysine 9 (K9) on the tail of histone H3. Arrows indicate acetylation only on the upper histone tails, but acetylation occurs on all four tails. **(c)** The enhanceosome also recruits a protein kinase that phosphorylates serine 10 (S10) of histone H3. Again, phosphorylation occurs on both H3 tails. This phosphorylation allows GCN5 to acetylate lysine 14 (K14) of histone H3. This completes the histone code, which is interpreted in the last two steps of the model. **(d)** Acetylated lysine 8 of histone H4 attracts the SWI/SNF complex, which remodels the nucleosome. This remodeling is represented by the wavy DNA lines in the nucleosome. **(e)** The remodeled nucleosome can now permit the binding of TFIID, which is attracted not only by the TATA box, but by the acetylated lysines 9 and 14 on the tail of histone H3. TFIID bends the DNA and moves the remodeled nucleosome 36 bp downstream. Now transcription can begin. (*Source:* Adapted from Agalioti, T., G. Chen, and D. Thanos, Deciphering the transcriptional histone acetylation code for a human gene. *Cell* 111 [2002] p. 389, f. 5.)

phosphorylates serine 10 of histone H3. Once that serine is phosphorylated, lysine 14 of histone H3 can be acetylated by GCN5. At this point, the histone code is complete.

Next, bromodomain-containing proteins interpret the histone code as follows: The single-bromodomain protein BRG1 binds to the acetylated lysine 8 of histone H4, bringing the whole SWI/SNF complex along with it. The rest of the polymerase II holoenzyme is presumably also recruited at this time but, for simplicity's sake, it is not shown. SWI/SNF then remodels the nucleosome in such a way that the double-bromodomain protein TAF1 can bind to histone H3, with its two acetylated lysines (9 and 14), and TAF1 brings the whole TFIID along with it. The binding of TFIID bends the DNA and causes the remodeled nucleosome to move out of the way downstream. The complex can now associate with the coactivator CBP, and transcription can begin.

In this context, it is worth mentioning another activity of TAF1 that has the potential to activate transcription, though it probably does not do so at the IFN-β promoter. That is, TAF1 has ubiquitin-conjugating activity, and one of its targets appears to be histone H1. Thus, when TAF1 is recruited to a promoter, possibly by binding to acetylated core histone tails, it can ubiquitylate a neighboring histone H1, targeting it for degradation by the 26S proteasome (Chapter 12). Because histone H1 helps repress transcription by cross-linking nucleosomes, the destruction of histone H1 would tend to activate neighboring genes.

SUMMARY The activators in the IFN-β enhanceosome can recruit a HAT (GCN5), which acetylates some of the lysines on histones H3 and H4 in a nucleosome at the promoter. A protein kinase also phosphorylates one of the serines on histone H3 of the same nucleosome, and this permits acetylation of one more lysine on histone H3, completing the histone code. One of the acetylated lysines then recruits the SWI/SNF complex, which remodels the nucleosome. This remodeling allows TFIID to bind to two acetylated lysines in the nucleosome through the dual bromodomain in TAF1. TFIID binding bends the DNA and moves the remodeled nucleosome aside, paving the way for transcription to begin.

Heterochromatin and Silencing

Most of the chromatin we have discussed in this chapter is in a class known as **euchromatin.** This chromatin is relatively extended and open and at least potentially active. By contrast, **heterochromatin** is very condensed and its DNA is inaccessible. In higher eukaryotes it even appears as clumps when viewed microscopically (Figure 13.30). In the yeast *Saccharomyces cerevisiae,* the chromosomes are too small to produce such clumps, but heterochromatin still exists, and it has the same repressive character as in higher eukaryotes. In fact, it can silence gene activity up to 3 kb away. Yeast heterochromatin is found at the telomeres, or tips of the chromosomes, and in the permanently repressed mating loci mentioned at the end of Chapter 10. Generally speaking, heterochromatin is found at the telomeres and the centromeres of chromosomes.

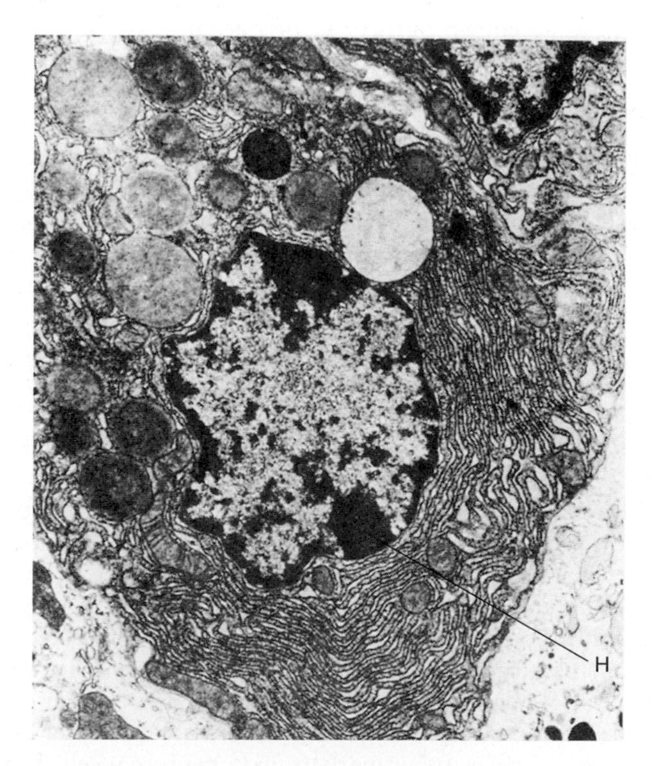

Figure 13.30 Interphase nucleus showing heterochromatin. Rat stomach lining cell with nucleus at center. Dark areas around periphery of nucleus are heterochromatin (H). (*Source:* Courtesy Dr. Keith Porter.)

It is particularly convenient to do genetic and biochemical experiments in yeast, so molecular biologists have exploited this organism to learn about the structure of heterochromatin and the way in which it silences genes, not only within the heterochromatin, but in neighboring regions of the chromosome. The silencing of genes near the telomere is called the **telomere position effect** (**TPE**) because the silencing of a gene is dependent on its position in the chromosome: If it is within about 3 kb of the telomere, it is silenced; if it is farther away, it is not.

Studies on yeast telomeric heterochromatin have shown that several proteins bind to the telomeres and are presumably involved in forming heterochromatin. These are **RAP1, SIR2, SIR3, SIR4,** and histones H3 and H4. (SIR stands for silencing information regulator.) Yeast telomeres consist of many repeats of this sequence: $C_{2-3}A(CA)_{1-5}$. (Of course, the opposite strand of the telomere has the complementary sequence.) This sequence, commonly called $C_{1-3}A$, is the binding site for the RAP1 protein, the only telomeric protein that binds to a specific site in DNA. RAP1 then recruits the SIR proteins to the telomere in this order: SIR3-SIR4-SIR2. As we have already seen, histones H3 and H4 are core histones of the nucleosome. Both SIR3 and SIR4 bind directly to the N-terminal tails of these two histones at residues 4–20 of histone H3 and residues 16–29 of histone H4.

Because RAP1 binds only to telomeric DNA, we might expect to find it associated only with the telomere, but we find it in the "subtelomeric" region adjacent to the telomere, along with the SIR proteins. To explain this finding, Michael Grunstein and his colleagues have proposed a model similar to the one in Figure 13.31: RAP1 binds to

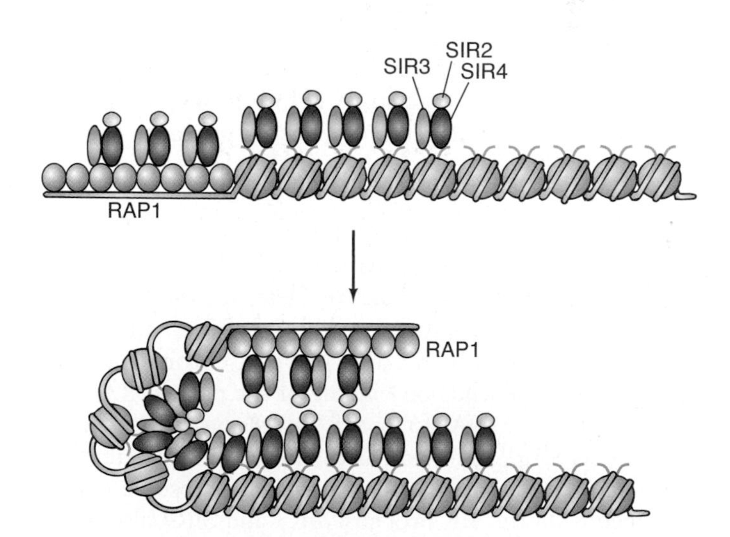

Figure 13.31 Model of telomere structure. RAP1 (red) binds to the telomere, and recruits SIR3 (green) and SIR4 (purple), which in turn attract SIR2 (yellow). SIR3 and SIR4 also bind to the N-terminal tails of histones H3 and H4 (thin blue lines). Interaction among the SIR proteins then causes the end of the chromosome to fold back on itself, so RAP1 is associated with the subtelomeric part of the chromosome. (*Source:* Adapted from Grunstein, M. 1998. Yeast heterochromatin: Regulation of its assembly and inheritance by histones. *Cell* 93: 325–28. Cell Press, Cambridge, MA.)

the telomeric DNA, the SIR proteins bind to RAP1 and to histones in the nucleosomes of the subtelomeric region. Then, protein–protein interactions cause the telomere to fold back on the subtelomeric region.

Earlier in this chapter, we learned that removing acetyl groups from core histones has a repressive effect on gene activity. Thus, we would predict that core histones in silenced chromatin would be poor in acetyl groups, or hypoacetylated. Indeed, whereas histone H4 in euchromatin is acetylated on lysines 5, 8, 12, and 16, histone H4 in yeast heterochromatin is acetylated only on lysine 12. What role might this hypoacetylation play in silencing? We know that lysine 16 of histone H4 is part of the domain (residues 16–29) that interacts with the SIR proteins (SIR3 in particular). Thus, acetylation of lysine 16 of histone H4 may block its interaction with SIR3, averting the formation of heterochromatin, and therefore preventing silencing.

Genetic experiments in yeast provide support for this hypothesis. Changing lysine 16 of histone H4 to a glutamine mimics the acetylation of this residue by removing its positive charge. This mutation also mimics acetylation in blocking the silencing of genes placed close to yeast telomeres and mating loci. On the other hand, changing lysine 16 to an arginine preserves the positive charge of the amino acid and thus mimics to some extent the deacetylated form of lysine. As expected, this mutation has less of an effect on silencing.

Because deacetylation of lysine 16 of histone H4 appears to attract the silencing complex, it is interesting that the SIR2 component of the yeast silencing complex has histone deacetylase activity (an NAD-dependent HDAC called N-HDAC). Thus, SIR2 is a good candidate for the enzyme that deacetylates lysine 16 of histone H4. If this hypothesis is valid, then SIR2 attracted to a nucleosome with a deacetylated lysine 16 of histone H4 could then deacetylate lysine 16 of histone H4 on a neighboring nucleosome and so propagate the silencing process.

SUMMARY Euchromatin is relatively extended and potentially active, whereas heterochromatin is condensed and genetically inactive. Heterochromatin can also silence genes as much as 3 kb away. Formation of heterochromatin at the tips of yeast chromosomes (telomeres) depends on binding of the protein RAP1 to telomeric DNA, followed by recruitment of the proteins SIR3, SIR4, and SIR2, in that order. Heterochromatin at other locations in the chromosome also depends on the SIR proteins. SIR3 and SIR4 also interact directly with histones H3 and H4 in nucleosomes. Acetylation of lysine 16 of histone H4 in nucleosomes prevents its interaction with SIR3 and therefore blocks heterochromatin formation. This is another way in which histone acetylation promotes gene activity.

Histone Methylation In addition to the other modifications we have seen, core histone tails are also subject to methylation, and methylation can have either an activating or a repressing effect. As we have seen, certain proteins, such as HATs, interact with specific acetylated lysines in core histone tails through acetyl-lysine-binding domains known as bromodomains. Thomas Jenuwein and colleagues noted that certain proteins involved in forming heterochromatin have conserved regions called **chromodomains.** One such protein is a **histone methyltransferase (HMTase)** whose human form is known as SUV39H HMTase. Another is a histone methyltransferase-associated protein called **HP1.**

Jenuwein and colleagues, and another group led by Tony Kouzarides, tested these and other proteins for binding to methylated and unmethylated peptides that included lysine 9 of histone H3, which is a target for methylation. Both groups found that HP1 binds to these peptides, but only if lysine 9 was methylated. This finding suggested a mechanism for spreading of methylated, and therefore repressive, chromatin: When lysine 9 of one histone H3 is methylated, it attracts HP1 through the latter's chromodomain. HP1 could then recruit SUV39H HMTase, which could methylate another nearby histone H3 on its lysine 9. In this way, the process could continue until many nucleosomes had become methylated. This methylation could lead to spreading of the heterochromatin state, as illustrated in Figure 13.32.

Lysine 9 of histone H3 is by no means the only histone target for methylation. All the core histones can be methylated on lysines and arginines, and the amino groups of lysines can accept up to three methyl groups each. Another favorite methylation site on histone H3 is lysine 4, and methylation of this site generally has an activating effect on transcription, owing to at least two mechanisms. First, it inhibits binding of the NuRD chromatin-remodeling and histone deacetylase complex to the histone H3 tail. This interferes with histone deacetylation, which would have a repressive effect. Second, methylation of lysine 4 of histone H3 blocks methylation of the nearby lysine 9, which would also be repressive. By inhibiting both of these repressive events, methylation of H3 lysine 4 has a net activating effect. Just as histone acetylation can be reversed by deacetylases, methylation of histone lysines and arginines can be reversed by demethylases, which reverse whatever repressive or stimulatory effect the methylation had.

Methylation of lysine 4 of histone H3 is generally trimethylation (designated H3K4Me3), and is usually associated with the 5′-end of an active gene. Thus, this modification appears to be a sign of transcription initiation. By contrast, trimethylation of lysine 36 of histone H3 (H3K36Me3) is usually associated with the 3’-end of an active gene, and therefore is taken as a marker for transcription elongation.

In a 2007 genome-wide ChIP-chip assay (Chapter 24) of these, as well as other markers, in human stem cell chromatin, Richard Young and colleagues made the following interesting discovery: Many protein-encoding genes are

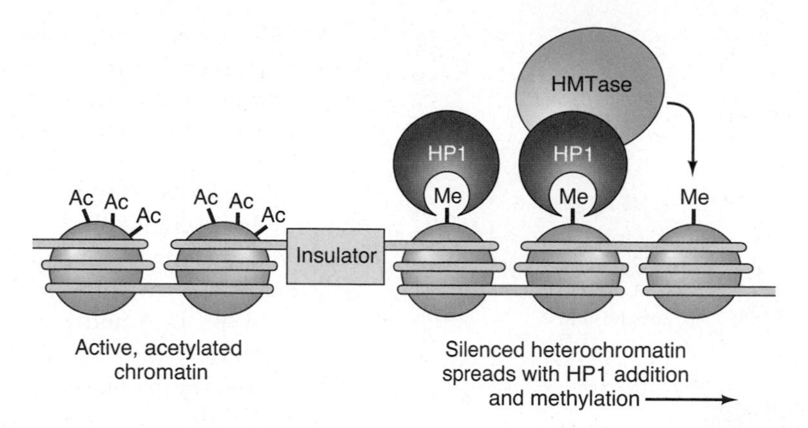

Figure 13.32 Model for involvement of histone methylation in chromatin repression. Nucleosomes to the right of the insulator have become methylated on lysine 9 of the histone H3 tails. This recruits HP1 (purple), which binds to a methylated lysine 9 on one nucleosome and recruits a histone methyltransferase (HMTase, green), to methylate lysine 9 on a neighboring nucleosome. Thus, the methylated, repressive state is propagated from one nucleosome to the next. (*Source:* Adapted from Bannister, S.D., P. Zegerman, J.F. Partridge, E.A. Miska, J.O. Thomas, R.C. Allshire, and T. Kouzarides, Selected recognation of methylated lysine 9 on histone H3 by the HP1 chromodomain. *Nature* 410 [2001] p. 123, f. 5.)

associated with nucleosomes having H3K4Me3, and therefore have presumably experienced transcription initiation, but they are not associated with nucleosomes having H3K36Me3, and therefore have probably not experienced transcription elongation. The simplest way to reconcile these two findings is to propose that many human genes contain RNA polymerase paused a short distance downstream of their promoters. This condition would open up a new potential means of controlling gene expression by controlling the restarting of paused RNA polymerase.

So far, we have dealt with individual methylations in isolation, but they do not really occur that way. Instead, many histone residues in a given nucleosome can be modified in various ways. Some will be acetylated, others will be methylated, others will be phosphorylated, and still others will be ubiquitylated. Figure 13.33 summarizes the modifications that can happen to the core histones.

As we have already seen, there is evidence for a histone code in which histone acetylation and phosphorylation can participate in a cascade of events leading to gene activation. Some investigators have wondered whether this histone code idea can be generalized to all histone modifications. A cell could read the different combinations of histone modifications in a given nucleosome as a combinatorial code that tells how much to express or silence genes in the neighborhood.

To address this question in the context of histone methylation, Frank Sauer and colleagues investigated the combined effects of methylations on three lysines in two histones: lysines 4 and 9 of histone H3 and lysine 20 of histone H4. They found that this combination of methylated lysines, created by a single HMTase called Ash1, had two effects in *Drosophila,* both of them positive. First, these methylations stimulated the binding of an activator called Brahma. Second, they inhibited the binding of the repressors HP1 and polycomb. Thus, the normal repressive effect of methylated histone H3 lysine 9 is masked in the context of the other two histone methylations. The cell must be able to read the whole combination of histone modifications, not just one.

Histone modifications not only mark chromatin for either activation or repression, they also affect other histone modifications. For example, methylation of histone H3 lysine 9 can be inhibited by several modifications on the same histone tail, including acetylation of lysine 9 (and perhaps lysine 14), methylation of lysine 4, and phosphorylation of serine 10.

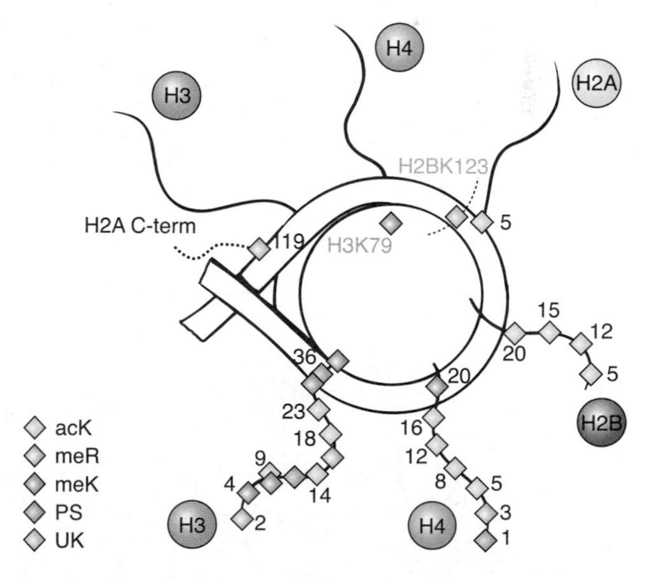

Figure 13.33 Summary of core histone modifications. Modifications are coded as shown at lower left: yellow, acetylated lysine (acK); gray, methylated arginine (meR); blue, methylated lysine (meK); pink, phosphorylated serine (PS); green, ubiquitylated lysine (UK). Modifications are shown on only one of the two histone H3 and H4 tails. Only one tail each is shown for histones H2A and H2B. The C-terminal tails of H2A and H2B are illustrated by dotted lines. The position of histone H3 lysine 79 (H3K79) is shown, though it is not on a histone tail. (*Source:* Adapted from Turner, B.M., Cellular memory and the histone code. *Cell* 111 [2002] p. 286, f. 1.)

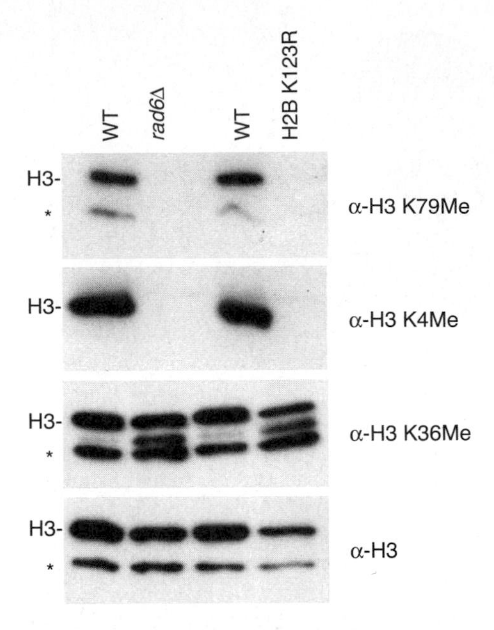

Figure 13.34 Effect of ubiquitylation of histone H2B on methylation of histone H3. Strahl and colleagues tested wild-type and mutant strains of yeast for the ability to methylate lysine 79 of histone H3. One mutant (*rad6Δ*) had the *rad6* gene deleted, so it could not ubiquitylate lysine 123 of histone H2B. In the other mutant (H2B K123R), lysine 123 of histone H2B was changed to arginine, so it could not be ubiquitylated, even with Rad6 functioning. Nuclear extracts from wild-type (lanes 1 and 3), and *rad6Δ* (lane 2) and H2B K123R (lane 4) were subjected to Western blotting by electrophoresis, followed by blotting and probing with antibodies against: methylated lysine 79 in histone H3 (top row); methylated lysine 4 in histone H3 (second row); lysine 36 in histone H3 (third row); and histone H3 (bottom row). The last row, with anti-H3 antibody, served as a positive control to make sure all lanes contained histone H3. The mutants did not support methylation of either lysine 4 or 79, but they did support methylation of lysine 36 of histone H3. The asterisk denotes a proteolytic product of H3 that removes the lysine 4 methylation site. (*Source:* Reprinted with permission from *Nature* 418: from Briggs et al., fig. 1, p. 498. © 2001 Macmillan Magazines Limited.)

Modifications in one histone can also affect modifications in another histone in the same nucleosome. For example, Brian Strahl and coworkers tested the effects of deleting the yeast gene *rad6,* which encodes the ubiquitin ligase **Rad6.** This enzyme is required for ubiquitylation of lysine 123 of histone H2B. This mutation blocked methylation of lysines 4 and 79 but had no effect on methylation of lysine 36 of histone H3 (Figure 13.34). Changing lysine 123 of histone H2B to arginine prevented ubiquitylation in cells with wild-type *rad6* and had the same negative effect on methylation of lysines 4 and 79 in histone H3. Thus, ubiquitylation of a lysine on one histone (H2B) can profoundly affect methylation of at least two sites on another (H3). By the way, lysine 79 is not on a histone tail. But it is on the surface of the nucleosome, as illustrated in Figure 13.33, and is accessible to the methylation machinery.

Finally, let us consider a regulatory interaction among modifications of three amino acids in the tail of histone H3: lysine 9, serine 10, and lysine 14. As we have seen, acetylation of lysine 14 is required for activation of some genes, including the human IFN-β gene. But, as we have also seen, this acetylation depends on phosphorylation of serine 10. Furthermore, phosphorylation of serine 10 is inhibited by methylation of lysine 9. Thus, methylation of lysine 9 can repress transcription by blocking phosphorylation of serine 10, thus blocking the needed acetylation of lysine 14. But the other side of the coin is that phosphorylation of serine 10, and probably acetylation of lysine 14, block methylation of lysine 9. Thus, once serine 10 and lysine 14 are appropriately modified, they tend to perpetuate the active state by preventing the repressive methylation of lysine 9. Moreover, acetylation of lysine 9 prevents methylation of the same residue, so that acetylation also works against repression. Figure 13.35 illustrates these interactions, interactions

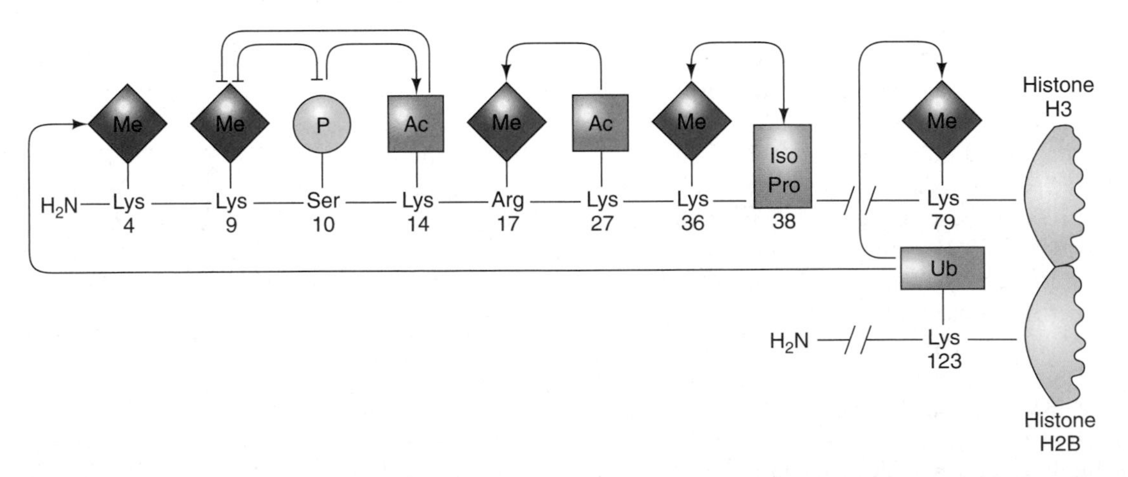

Figure 13.35 A model for the crosstalk among modifications on histone tails. The known interactions among modified residues on histones H3 and H2B are shown, but some crosstalk with at least histone H2A is also known. Activating interactions are shown with arrows, and inhibiting interactions are shown with a blocking symbol. For example, phosphorylation on serine 10 activates acetylation of lysine 14 and inhibits methylation of lysine 9. Me, methylation; Ac, acetylation; P, phosphorylation; Iso, proline isomeration; Ub, ubiquitylation.

with other histone H3 modifications, and crosstalk among modifications on histones H3 and H2A.

So far in this section we have learned that histone modifications can affect gene activity by two mechanisms: First, by altering the way histone tails interact with DNA and with histone tails in neighboring nucleosomes, and thereby altering nucleosome cross-linking. Second, by attracting proteins that can affect chromatin structure and activity. For example, acetylated lysines attract bromodomain proteins; methylated lysines attract proteins with chromodomains and chromo-like domains such as tudor and MBT, or other domains such as PHD fingers; and phosphorylated serines attract so-called **14-3-3 proteins** (this uninformative name derives from the electrophoretic mobilities of these proteins). These proteins frequently have catalytic activities of their own and can further modify histones or remodel chromatin. They can also recruit other proteins with their own activities.

For example, two of the subunits of the Rpd3C(S) histone deacetylase complex are the chromodomain protein Eaf3 and the PHD finger protein Rco1. Together, these proteins recognize histone H3 molecules methylated on lysine 36 downstream of promoters, and assure association of the Rpd3C(S) deacetylase with this downstream chromatin. The resulting histone deacetylation slows transcription elongation, which can be counteracted by one or more positive elongation factors. This deacetylation also prevents transcription initiation at any cryptic class II promoters that happen to lie within the body of the gene.

SUMMARY Methylation of lysine 9 in the N-terminal tail of histone H3 attracts the protein HP1, which in turn recruits a histone methyltransferase, which presumably methylates lysine 9 on a neighboring nucleosome, propagating the repressed, heterochromatic state. Methylation of other lysine and arginine side chains in the core histones can have either repressive or activating effects. These effects are achieved by proteins that recognize and bind to nucleosomes with specific patterns of histone methylation, and further modify the chromatin or directly affect transcription. Methylations occur in a given nucleosome in combination with other histone modifications, including acetylations, phosphorylations, and ubiquitylations. In principle, each particular combination can send a different message to the cell about activation or repression of transcription. A given histone modification can also influence other, nearby modifications.

Nucleosomes and Transcription Elongation

We have seen that nucleosomes must be absent from a gene's control region, or at least nudged aside as activators and general transcription factors bind to their respective DNA sites. But how does RNA polymerase deal with the nucleosomes that lie within the transcribed region of a gene?

The Role of FACT One important factor is a protein called FACT (facilitates chromatin transcription), which expedites elongation through nucleosomes by RNA polymerase II in vitro. Human FACT is composed of two polypeptides: the human homolog of the yeast Spt16 protein, and SSRP1, which is an HMG-1-like protein. FACT has been shown to interact strongly with histones H2A and H2B, which leads to the hypothesis that it can remove these two histones from nucleosomes, at least temporarily, and thereby destabilize the nucleosomes so RNA polymerase can transcribe through.

Several early lines of evidence supported this hypothesis. First, cross-linking the histones so none can be removed from the nucleosome blocks the action of FACT. Second, mutations in the yeast gene encoding histone H4 that alter histone–histone interactions have the same phenotype as mutations in the gene for the Spt16 subunit of FACT. Finally, actively transcribed chromatin is poor in histones H2A and H2B.

In 2003, Danny Reinberg and colleagues provided direct evidence that FACT facilitates chromatin transcription by RNA polymerase II by removing at least a histone H2A–H2B dimer from nucleosomes. They also showed that these proteins have a **histone chaperone** activity that can deposit histones back onto chromatin, reconstituting nucleosomes after the transcription machinery has passed through.

First, these workers used co-immunoprecipitation experiments to show that the Spt16 subunit of FACT binds to histone H2A–H2B dimers, and that the SSRP1 subunit binds to H3–H4 tetramers. The Spt16 subunit has a very acidic C-terminus, and Reinberg and colleagues demonstrated that recombinant FACT with an Spt16 subunit lacking this C-terminus (FACTΔC) can neither interact with histones in nucleosomes, nor facilitate transcription through chromatin.

Next, they labeled H2A–H2B dimers and H3–H4 tetramers with two different fluorescent tags. Then, after treatment with FACT or FACTΔC, they washed with buffer containing 350 mM KCl and detected the loss of dimers from nucleosomes by measuring the dimer/tetramer ratio by SDS-PAGE, followed by fluorimaging. (A fluorimager quantitatively measures the fluorescence of bands in a gel.) Figure 13.36 shows that FACT caused up to a 50% loss of H2A–H2B dimers from treated nucleosomes, but FACTΔC caused no more loss than washing with buffer alone (about 20%). Thus, FACT appears to weaken the association between H2A–H2B dimers and H3–H4 tetramers, and this effect depends on the C-terminus of the Spt16 subunit.

Reinberg and colleagues also demonstrated that FACT stimulated transcription through nucleosomes, and that the transcribed templates contained so-called hexasomes, which are nucleosomes lacking one H2A–H2B dimer.

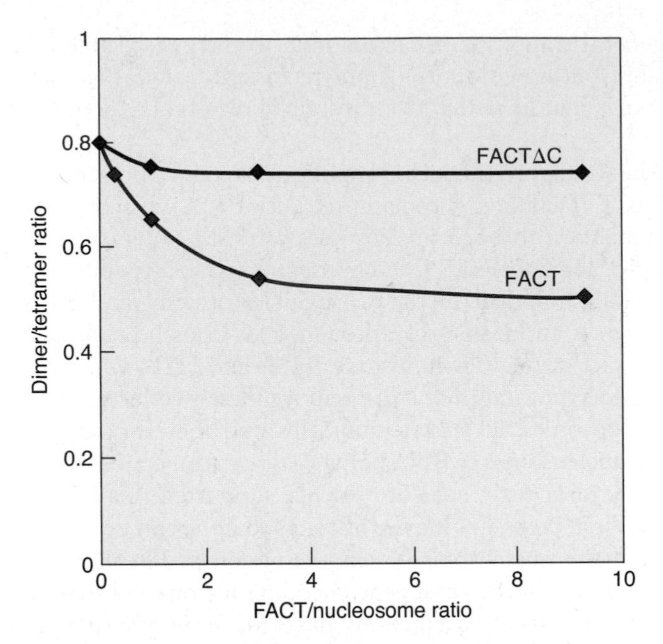

Figure 13.36 FACT stimulates loss of histone H2A–H2B dimers from nucleosomes. Reinberg and colleagues labeled H2A–H2B dimers and H3–H4 tetramers in nucleosomes with two different fluorescent tags, added FACT or FACTΔC for a one-hour incubation, then washed the nucleosomes to remove any loosely bound histones. Then they followed the loss of H2A–H2B dimers by measuring the ratio of dimers to tetramers using SDS-PAGE. The fluorescent tags were detected quantitatively in the SDS-PAGE gel with a fluorimager. (*Source:* Adapted from Belotserkovskaya, et al., *Science* 301, 2003, f. 3, p. 1092.)

To do this experiment, the investigators used a template with a single nucleosome positioned downstream of the transcription start site. They assembled transcription complexes on this template, and tethered the complexes to beads through a tag on RNA polymerase II. Then they carried out transcription with labeled nucleotides, in the presence of FACT or FACTΔC. When they electrophoresed the transcripts, they found that FACT, but not FACTΔC, stimulated transcription through the nucleosomes to form full-length run-off transcripts. That is, transcription with no FACT, or with FACTΔC, yielded a number of transcripts that stalled in the region of DNA involved in the nucleosome, but that FACT reduced such stalling, and yielded a higher percentage of full-length transcripts.

This experiment also allowed Reinberg and colleagues to examine the templates released, along with full-length run-off transcripts, from RNA polymerase. They labeled the DNA prior to transcription, and then electrophoresed the released templates, which had presumably been fully transcribed. These templates contained hexasomes if transcription was done in the presence of FACT, but not in the presence of FACTΔC. Furthermore, adding H2A and H2B back to the hexasomes converted them to full-size nucleosomes, indicating that the hexasomes really are nucleosomes lacking an H2A–H2B dimer. Thus, FACT appears to facilitate transcription through nucleosomes, at least in part, by loosening nucleosome structure enough to allow loss of at least one H2A–H2B dimer.

But, as we have mentioned, FACT is more than a nucleosome-disrupter. It can also deposit histones on DNA to reconstitute nucleosomes. Reinberg and colleagues demonstrated this histone chaperone effect of FACT with two experiments. First, they mixed core histones with labeled DNA with no FACT, FACT, or FACTΔC, and then electrophoresed the products. Without FACT, an aggregate formed that would not enter the electrophoretic gel. But with FACT, a well-behaved DNA-histone complex formed. Predictably, this complex did not form with FACTΔC. In the second experiment, Reinberg and colleagues labeled H2A–H2B dimers and H3–H4 tetramers with two different fluorescent tags, and then visualized the histone-DNA complexes on the electrophoretic gel with a fluorimager to see whether they contained the fluorescent tags associated with both sets of histones. Indeed they did, showing that FACT, but not FACTΔC, has histone chaperone activity. They also showed that neither of the FACT subunits alone has this activity.

If FACT really does play the role of a chromatin remodeler during transcription elongation, it should be found on chromatin along with RNA polymerase. Reinberg and John Lis and their colleagues demonstrated this behavior using the *Drosophila* heat shock gene *hsp70* as their experimental system. In the salivary gland cells of fruit fly larvae, the chromosomes replicate repeatedly without cell division, giving rise to large **polytene chromosomes,** with many sister chromatids packed side by side. These polytene chromosomes are visible with the aid of a light microscope, and active transcription sites are visible as swollen sites, or **chromosome puffs.** In particular, raising the temperature creates puffs at heat shock loci, such as *Hsp70.*

First, Reinberg and Lis isolated *Drosophila* polytene chromosomes before and after a 20-min heat shock and stained them with fluorescently labeled antibodies directed against RNA polymerase II and Spt16. After heat shock, the two antibodies co-localized over two chromosome puffs containing *hsp70* loci.

If FACT really does accompany RNA polymerase II, remodeling chromatin as transcription progesses, then FACT should be recruited to the heat shock gene as rapidly as polymerase II is, and it should be found downstream of promoter-associated transcription factors soon after transcription begins. To test this hypothesis, Reinberg, Lis, and colleagues examined chromatin stained with antibodies against the two subunits of FACT and against HSF, an activator that binds to the control region upstream of the *hsp70* gene. They looked before, and at 2.5 and 10 min after heat shock.

Figure 13.37 shows the results. Even at 2.5 min after heat shock, the two subunits of FACT are associated with the *hsp70* gene, just as HSF is. However, the FACT subunits

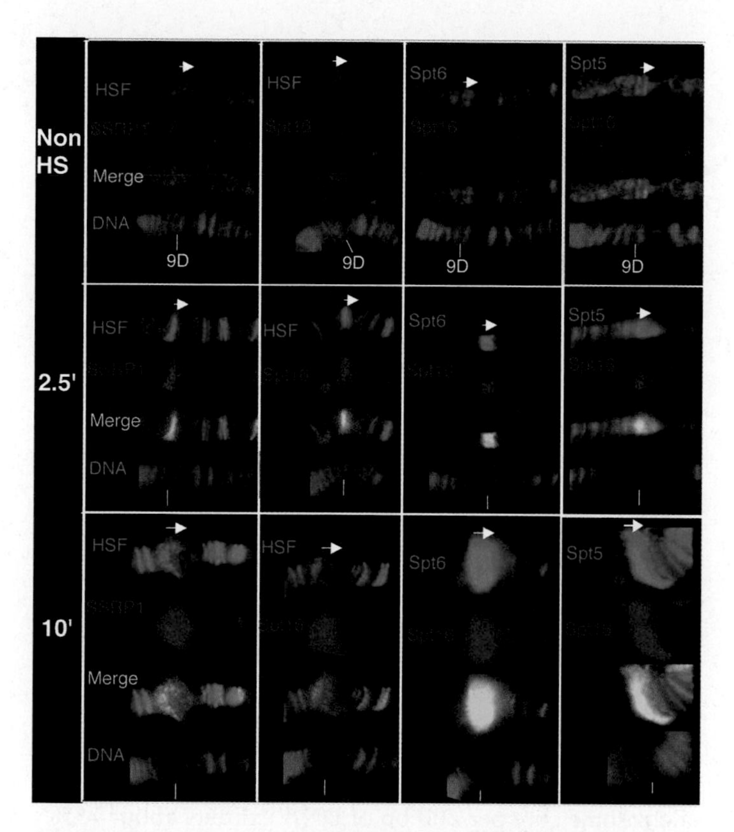

Figure 13.37 FACT is recruited rapidly to a transcribed gene and localizes downstream of an activator bound to the promoter. Reinberg, Lis, and colleagues stained *Drosophila* chromosomes with fluorescent antibodies in nonstimulated cells, and in cells 2.5 and 10 min after heat shock, as indicated at left. The antibodies used are indicated beside each stained chromosome, in the same color as the fluorescent antibody. Thus, the antibodies specific for HSF and Spt6 fluoresce green, and the antibody for SSRP1 and Spt16 fluoresce red. They also merged the two fluorescence images to check for overlap. Wherever the red and green fluorescence overlapped, it appeared yellow. Wherever there was not perfect overlap, some red fluorescence appeared to the right (downstream) of the yellow. This was especially evident in the merger of HSF and SSRP1 fluorescence at 10 min after heat shock (lower left panel). The chromosomes were also stained with Hoechst dye, which stains DNA violet (bottom of each panel.) *(Source:* Reprinted with permission from *Science,* Vol. 301, Abbie Saunders, Janis Werner, Erik D. Andrulis, Takahiro Nakayama, Susumu Hirose, Danny Reinberg, and John T. Lis, "Tracking FACT and the RNA Polymerase II Elongation Complex Through Chromatin in Vivo," Fig. 2, p. 1095. Copyright 2003, AAAS.)

are both located significantly further downstream than HSF. We can see this separation by comparing the red staining due to either SSRP1 or Spt16 and the green staining due to HSF. Separately, they are hard to distinguish, but when the two images are merged, we can see a leading edge of red (FACT fluorescence) downstream of the yellow, which corresponds to overlapping red (FACT) and green (HSF) fluorescence. This effect is also apparent 10 min after heat shock, especially with SSRP1.

By contrast, when another putative chromatin remodeler, Spt6, is stained with a green fluorescent tag, it colocalizes perfectly with FACT (see the yellow bands in the third panel either 2.5 or 10 min after heat shock). This behavior suggests that Spt6 and FACT both travel along with RNA polymerase II so they are in a position to help remodel chromatin to facilitate transcription.

SUMMARY FACT is a transcription elongation facilitator composed of two subunits, Spt16 and SSRP1. Spt16 binds to histone H2A–H2B dimers, and SSRP1 binds to H3–H4 tetramers. FACT can facilitate transcription through a nucleosome by promoting the loss of at least one H2A–H2B dimer from the nucleosome. It can also act as a histone chaperone by promoting the re-addition of an H2A–H2B dimer to a nucleosome that has lost such a dimer. The Spt16 subunit of FACT has an acid-rich C-terminus that is essential for both of these nucleosome remodeling activities.

The Role of PARP-1 The heat shock genes of *Drosophila* provide another example of removing nucleosomes to allow transcription. In 2008, Stephen Petesch and John Lis presented data elucidating the loss of nucleosomes from the *Hsp70* locus in *Drosophila* polytene chromosomes. They found that nucleosomes begin to disappear across the *Hsp70* locus only 30 s after heat shock, and this disappearance intensifies within two minutes. Thirty seconds is too short a time to allow for transcription of the whole locus, suggesting that loss of nucleosomes is not dependent on transcription. This hypothesis is supported by the finding that nucleosomes are lost even when transcription elongation is blocked by drugs. But nucleosome loss does require three proteins: heat shock factor (HSF), GAGA factor (discussed earlier in this chapter), and a **poly(ADP-ribose) polymerase (PARP)** known as **PARP1.**

PARP extracts ADP-ribose units from the substrate nicotinamide adenine dinucleotide (NAD) and links them together in a polymer [**poly(ADP-ribose), (PAR)**] attached through a glutamate carboxyl group to a protein, usually PARP itself (Figure 13.38). The polymer typically branches (by links between the ribose parts of the ADP-ribose units) every 40 to 50 units. The formation of PAR can be reversed by the enzyme **poly(ADP-ribose) glycohydrolase, (PARG)**, which breaks the bonds between ADP-ribose units.

How does PARP1 participate in nucleosome removal? First of all, PARP1 is able to bind to core nucleosomes much as histone H1 does, and that has a repressive effect. Activation of PARP1 causes it to poly(ADP-ribosyl)ate itself, which causes it to dissociate from nucleosomes, which should have an activating effect. Second, the PAR produced by PARP1 resembles a polynucleotide, particularly in its acidic nature. Thus, PAR can presumably compete with

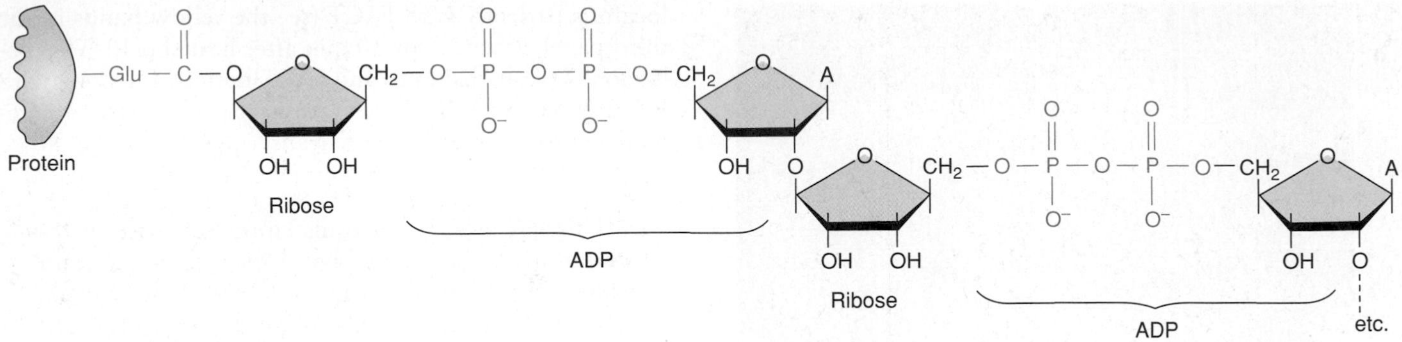

Figure 13.38 Poly(ADP-ribose). The first ADP-ribose unit is linked to a protein glutamate via an ester bond. The remaining ADP-ribose units are linked together via glycosidic bonds between the 2′-carbon of an ADP on one unit and the 1-carbon of the ribose on the next unit. The enzyme PARP forms these glycosidic linkages, and PARG breaks them.

DNA for association with the basic histones, thereby loosening the binding between histones and DNA and facilitating the breakup of nucleosomes.

SUMMARY Heat shock causes rapid loss of nucleosomes from chromatin in *Drosophila* polytene chromosome puffs. One of the agents required for this nucleosome loss is a poly(ADP-ribose) polymerase (PARP1). In response to heat shock, this enzyme poly(ADP-ribosyl)ates itself, removing it from its histone H1-like binding to core nucleosomes, thereby helping to destabilize the nucleosomes. Also, the poly(ADP-ribose), which is a polyanion, could bind directly to histones, further destabilizing the nucleosomes.

SUMMARY

Eukaryotic DNA combines with basic protein molecules called histones to form structures known as nucleosomes. These structures contain four pairs of histones (H2A, H2B, H3, and H4) in a wedge-shaped disc, around which is wrapped a stretch of 146 bp of DNA. Histone H1 is more easily removed from chromatin than the core histones and is not part of the core nucleosome.

In the second order of chromatin folding in vitro, and presumably also in vivo, a string of nucleosomes folds into a 30-nm fiber. Structural studies suggest that the 30-nm chromatin fiber in the nucleus exists in at least two forms: Inactive chromatin tends to have a high nucleosome repeat length (about 197 bp) and favors a solenoid folding structure. This kind of chromatin interacts with histone H1, which helps to stabilize its structure. Active chromatin tends to have a low nucleosome repeat length (about 167 bp) and folds according to the two-start double helical model.

The third order of chromatin condensation appears to involve formation of a radial loop structure in eukaryotic chromosomes. The 30-nm fiber seems to form loops between 35 and 85 kb long, anchored to the central matrix of the chromosome.

The core histones (H2A, H2B, H3, and H4) assemble nucleosome cores on naked DNA. Transcription of a class II gene in reconstituted chromatin with an average of one nucleosome core per 200 bp of DNA exhibits about 75% repression relative to naked DNA. The remaining 25% is due to promoter sites not covered by nucleosome cores. Histone H1 causes a further repression of template activity, in addition to that produced by core nucleosomes. This repression can be counteracted by transcription factors. Some, like Sp1 and GAL4, act as both antirepressors (preventing repression by histone H1) and as transcription activators. Others, like GAGA factor, are just antirepressors. The antirepressors presumably compete with histone H1 for binding sites on the DNA template.

Active genes tend to have DNase-hypersensitive control regions. At least part of this hypersensitivity is due to the absence of nucleosomes.

Histone acetylation occurs in both the cytoplasm and nucleus. Cytoplasmic acetylation is carried out by a HAT B and prepares histones for incorporation into nucleosomes. The acetyl groups are later removed in the nucleus. Nuclear acetylation is catalyzed by a HAT A and correlates with transcription activation. A variety of coactivators have HAT A activity, which may allow them to loosen the association of nucleosomes with each other and with a gene's control region. Acetylation of core histone tails also attracts bromodomain proteins such as TAF1, which are essential for transcription.

Transcription repressors such as unliganded nuclear receptors and Mad-Max bind to DNA sites and interact with corepressors such as NCoR/SMRT and SIN3, which in turn bind to histone deacetylases such as HDAC1 and 2. This assembly of ternary protein complexes brings the histone deacetylases close to nucleosomes in the neighborhood. The deacetylation of core histones allows the basic tails of the histones to bind strongly to DNA and to histones in neighboring nucleosomes, stabilizing the nucleosomes and inhibiting transcription.

Activation of many eukaryotic genes requires chromatin remodeling. Several different protein complexes carry out this remodeling, and all of them have an ATPase that harvests the energy from ATP hydrolysis to use for remodeling. The remodeling complexes are distinguished by their ATPase component, and two of the best-studied complexes are SWI/SNF and ISWI. The SWI/SNF complex in mammals has BRG1 as its ATPase, and 9-12 BRG1-associated factors (BAFs). One of the highly conserved BAFs is called BAF 155 or 170. It has a SANT domain that appears to be responsible for histone binding. This would help SWI/SNF bind to nucleosomes. Members of the ISWI class of remodeling complexes have a SANT domain, and another domain called SLIDE that appears to be involved in DNA binding.

The mechanism of chromatin remodeling is not understood in detail, but it does involve mobilization of nucleosomes, with loosening of the association between DNA and core histones. In contrast to uncatalyzed DNA exposure in nucleosomes, or simple sliding of nucleosomes along a stretch of DNA, catalyzed remodeling of nucleosomes involves the formation of distinct conformations of the nucleosomal DNA with respect to the core histones.

ChIP analysis can reveal the order of binding of factors to a gene during activation. As the yeast *HO* gene is activated, the first factor to bind is Swi5, followed by SWI/SNF and SAGA, which contains the HAT Gcn5p. Next, the general transcription factors and other proteins bind. Thus, chromatin remodeling is among the first steps in activation of this gene, but the order can be different in other genes.

The pattern of core histone modifications in a given nucleosome appears to constitute a histone code that can determine what happens to the nucleosome. For example, the activators in the IFN-β enhanceosome can recruit a histone acetyltransferase, which acetylates some of the lysines on the tails of histones H3 and H4 at the promoter. One of the serines on histone H3 also becomes phosphorylated, which allows acetylation of another lysine on histone H3, completing the histone code. One of the acetylated lysines on histone H4 then recruits the SWI/SNF complex, which remodels the nucleosome. Then TFIID can bind to two acetylated lysines on histone H3. TFIID binding bends the DNA and moves the remodeled nucleosome aside, paving the way for transcription to begin.

Euchromatin is relatively extended and potentially active, whereas heterochromatin is condensed and genetically inactive. Heterochromatin can also silence genes as much as 3 kb away. Formation of heterochromatin at the tips of yeast chromosomes (telomeres) depends on binding of the protein RAP1 to telomeric DNA, followed by recruitment of the proteins SIR3, SIR4, and SIR2, in that order. Heterochromatin at other locations in the chromosome also depends on the SIR proteins. SIR3 and SIR4 also interact directly with histones H3 and H4 in nucleosomes. Acetylation of lysine 16 of histone H4 in nucleosomes prevents its interaction with SIR3 and therefore prevents heterochromatin formation. This is another way in which histone acetylation promotes gene activity.

Methylation of lysine 9 in the N-terminal tail of histone H3 attracts the protein HP1, which in turn recruits a histone methyltransferase, which presumably methylates lysine 9 on a neighboring nucleosome, propagating the repressed, heterochromatic state. Methylation of other lysine and arginine side chains in the core histones can have either repressive or activating effects, and these methylations occur in a given nucleosome in combination with other histone modifications, including acetylations, phosphorylations, and ubiquitylations. In principle, each particular combination can send a different message to the cell about activation or repression of transcription. A given histone modification can also influence other, nearby modifications.

FACT is a transcription elongation facilitator composed of two subunits, Spt16 and SSRP1. Spt16 binds to histone H2A–H2B dimers, and SSRP1 binds to H3–H4 tetramers. FACT can facilitate transcription through a nucleosome by promoting the loss of at least one H2A–H2B dimer from the nucleosome. It can also act as a histone chaperone by promoting the re-addition of an H2A–H2B dimer to a nucleosome that has lost such a dimer. The Spt16 subunit of FACT has an acid-rich C-terminus that is essential for both of these nucleosome remodeling activities.

Heat shock causes rapid loss of nucleosomes from chromatin in *Drosophila* polytene chromosome puffs. One of the agents required for this nucleosome loss is a poly(ADP-ribose) polymerase (PARP1). In response to heat shock, this enzyme poly(ADP-ribosyl)ates itself, removing it from its histone H1-like binding to core nucleosomes, thereby helping to destabilize the nucleosomes. Also, the poly(ADP-ribose), which is a polyanion, could bind directly to histones, further destabilizing the nucleosomes.

REVIEW QUESTIONS

1. Diagram a nucleosome as follows: (a) On a drawing of the histones without the DNA, show the rough positions of all the histones. (b) On a separate drawing, show the path of DNA around the histones.
2. Cite electron microscopic evidence for a six- to sevenfold condensation of DNA in nucleosomes.

3. Cite electron microscopic evidence for formation of a condensed fiber (30-nm fiber) at high ionic strength.
4. Diagram the solenoid model of the 30-nm chromatin fiber.
5. Diagram the structure of a tetranucleosome revealed by x-ray crystallography. What structure for the 30-nm fiber does this tetranucleosome structure suggest?
6. How can single-molecule force spectroscopy shed light on the structure of the 30-nm chromatin fiber? What conclusions does it suggest?
7. Draw a model to explain the next order of chromatin folding after the 30-nm fiber. Cite biochemical and microscopic evidence to support the model.
8. Describe and give the results of an experiment that shows the competing effects of histone H1 and the activator GAL4-VP16 on transcription of the adenovirus E4 gene in reconstituted chromatin.
9. Present two models for antirepression by transcription activators, one in which the gene's control region is not blocked by a nucleosome, the other in which it is.
10. Describe and give the results of an experiment that shows that the nucleosome-free zone in active SV40 chromatin lies at the viral late gene control region.
11. Describe and give the results of an experiment that shows that the zone of DNase hypersensitivity in SV40 chromatin lies at the viral late gene control region.
12. Diagram and describe a general technique for detecting a DNase-hypersensitive DNA region.
13. Describe and give the results of an activity gel assay that shows the existence of a histone acetyltransferase (HAT) activity.
14. Present a model for the involvement of a corepressor and histone deacetylase in transcription repression.
15. Describe and give the results of an epitope-tagging experiment that shows interaction among the following three proteins: the repressor Mad1, the corepressor SIN3A, and the histone deacetylase HDAC2.
16. Present a model for activation and repression by the same protein, depending on the presence or absence of that protein's ligand.
17. Present models for uncatalyzed nucleosomal DNA exposure and for catalyzed nucleosome remodeling. Present evidence for the catalyzed model.
18. Describe how you could use a chromatin immunoprecipitation procedure to detect the proteins associated with a particular gene at various points in the cell cycle.
19. Describe and give the results of an experiment using chromatin immunoprecipitation to discover the timing of acetylation and phosphorylation of particular sites on core histones in a nucleosome at the IFN-β promoter.
20. Describe and give the results of an experiment to measure recruitment of SWI/SNF and TFIID to the IFN-β promoter with wild-type and mutant histones.
21. Present a model depicting the establishment and decoding of a histone code at the IFN-β promoter.
22. Present a model to explain why lysine 16 in histone H4 is thought to be critical for silencing. What evidence supports this hypothesis?
23. Present a model depicting the spread of chromatin repression via histone methylation.
24. Present a model of the interactions among the modifications of lysines 9 and 14, and serine 10 in the N-terminal tail of histone H3. Show both positive and negative interactions.
25. Present evidence that FACT causes a loss of histone H2A–H2B dimers from nucleosomes, and that this activity depends on the C-terminus of the Spt16 subunit of FACT.

ANALYTICAL QUESTIONS

1. If the globin locus did have the same DNase-hypersensitive sites in J6 cells as in HEL cells, approximately what size fragments would have been detected in Figure 13.20d? Which hypersensitive sites would not be detected?
2. Explain why brief digestion of eukaryotic chromatin with micrococcal nuclease gives DNA fragments about 200 bp long, but longer digestion yields 146-bp fragments.
3. The amino acid sequences of the core histones are highly conserved between plants and animals. Present a hypothesis to explain this finding.
4. Type A histone acetyltransferases (HAT A's) contain a bromodomain and HAT B's do not. What do you predict would occur if HAT A's were missing this bromodomain? What if HAT B's possessed this bromodomain? If the bromodomains were reversed so that all HAT B's gained bromodomains and all HAT A's lost them, would HAT A's take over the role of HAT B's and vice versa? Why or why not? How would you answer this question experimentally?

SUGGESTED READINGS

General References and Reviews

Brownell, J.E. and C.D. Allis. 1996. Special HATs for special occasions: Linking histone acetylation to chromatin assembly and gene activation. *Current Opinion in Genetics and Development* 6:176–84.

Felsenfeld, G. 1996. Chromatin unfolds. *Cell* 86:13–19.

Grunstein, M. 1998. Yeast heterochromatin: Regulation of its assembly and inheritance by histones. *Cell* 93:325–28.

Kouzarides, T. 2007. Chromatin modifications and their function. *Cell* 128:693–705.

Narlikar, G.J., H.-Y. Fan, and R.E. Kingston. 2002. Cooperation between complexes that regulate chromatin structure and transcription. *Cell* 108:475–87.

Pazin, M.J. and J.T. Kadonaga. 1997. What's up and down with histone deacetylation and transcription? *Cell* 89:325–28.

Pennisi, E. 1996. Linker histones, DNA's protein custodians, gain new respect. *Science* 274:503–4.

Pennisi, E. 1997. Opening the way to gene activity. *Science* 275:155–57.

Pennisi, E. 2000. Matching the transcription machinery to the right DNA. *Science* 288:1372–73.

Perlman, T. and B. Vennström. 1995. The sound of silence. *Nature* 377:387–88.

Rhodes, D. 1997. The nucleosome core all wrapped up. *Nature* 389:231–33.

Roth, S.Y. and C.D. Allis. 1996. Histone acetylation and chromatin assembly: A single escort, multiple dances? *Cell* 87:5–8.

Svejstrup, J.Q. 2003. Histones face the FACT. *Science* 301:1053–55.

Turner, B.M. 2002. Cellular memory and the histone code. *Cell* 111:285–91.

Wolffe, A.P. 1996. Histone deacetylase: A regulator of transcription. *Science* 272:371–72.

Wolffe, A.P. 1997. Sinful repression. *Nature* 387:16–17.

Wolffe, A.P. and D. Pruss. 1996. Targeting chromatin disruption: Transcription regulators that acetylate histones. *Cell* 84:817–19.

Woodcock, C.L. 2005. A milestone in the odyssey of higher-order chromatin structure. *Nature Structural & Molecular Biology* 12:639–40.

Xu, L., C.K. Glass, and M.G. Rosenfeld. 1999. Coactivator and corepressor complexes in nuclear receptor function. *Current Opinion in Genetics and Development* 9:140–47.

Research Articles

Agalioti, T., G. Chen, and D. Thanos. 2002. Deciphering the transcriptional histone acetylation code for a human gene. *Cell* 111:381–92.

Arents, G., R.W. Burlingame, B.-C. Wang, W.E. Love, and E.N. Moudrianakis. 1991. The nucleosomal core histone octamer at 3.1 Å resolution: A tripartite protein assembly and a left-handed superhelix. *Proceedings of the National Academy of Sciences USA* 88:10148–52.

Arents, G. and E.N. Moudrianakis. 1993. Topography of the histone octamer surface: Repeating structural motifs utilized in the docking of nucleosomal DNA. *Proceedings of the National Academy of Sciences USA* 90:10489–93.

Bannister, A.J., P. Zegerman, J.F. Partridge, E.A. Miska, J.O. Thomas, R.C. Allshire, and T. Kouzarides. 2001. Selective recognition of methylated lysine 9 on histone H3 by the HP1 chromodomain. *Nature* 410:120–24.

Belotserkovskaya, R., S. Oh, V.A. Bondarenko, G. Orphanides, V.M. Studitsky, and D. Reinberg. 2003. FACT facilitates transcription-dependent nucleosome alteration. *Science* 301:1090–93.

Benyajati, C. and A. Worcel. 1976. Isolation, characterization, and structure of the folded interphase genome of *Drosophila melanogaster. Cell* 9:393–407.

Briggs, S.D., T. Xiao, Z.-W. Sun, J.A. Caldwell, J. Shabanowitz, D.F. Hunt, C.D. Allis, and B.D. Strahl. 2002. *Trans*-histone regulatory pathway in chromatin. *Nature* 418:498.

Brownell, J.E. and C.D. Allis. 1996. An activity gel assay detects a single, catalytically active histone acetyltransferase subunit in *Tetrahymena* macronuclei. *Proceedings of the National Academy of Sciences USA* 92:6364–68.

Brownell, J.E., J. Zhou, T. Ranalli, R. Kobayashi, D.G. Edmondson, S.Y. Roth, and C.D. Allis. 1996. *Tetrahymena* histone acetyltransferase A: A homolog of yeast GCN5p linking histone acetylation to gene activation. *Cell* 84:843–51.

Cosma, M.P., T. Tanaka, and K. Nasmyth. 1999. Ordered recruitment of transcription and chromatin remodeling factors to a cell cycle- and developmentally regulated promoter. *Cell* 97:299–311.

Griffith, J. 1975. Chromatin structure: Deduced from a minichromosome. *Science* 187:1202–3.

Grosveld, F., G.B. van Assendelft, D.R. Greaves, and G. Kollias. 1987. Position-independent, high-level expression of the human β-globin gene in transgenic mice. *Cell* 51:975–85.

Jacobson, R.H., A.G. Ladurner, D.S. King, and R. Tjian. 2000. Structure and function of a human $TAF_{II}250$ double bromodomain module. *Science* 288:1422–28.

Kruithof, M., F.-T. Chien, A. Routh, C. Logie, D. Rhodes, and J. van Noort. 2009. Single-molecule force spectroscopy reveals highly compliant helical folding for the 30-nm chromatin fiber. *Nature Structural and Molecular Biology* 16:534–40.

Lachner, M., D. O'Carroll, S. Rea, K. Mechtier, and T. Jenuwein. 2001. Methylation of histone H3 lysine 9 creates a binding site for HP1 proteins. *Nature* 410:116–120.

Laherty, C.D., W.-M. Yang, J.-M. Sun, J.R. Davie, E. Seto, and R.N. Eisenman. 1997. Histone deacetylases associated with the mSin3 corepressor mediate Mad transcriptional repression. *Cell* 89:349–56.

Laybourn, P.J. and J.T. Kadonaga. 1991. Role of nucleosomal cores and histone H1 in regulation of transcription by RNA polymerase II. *Science* 254:238–45.

Luger, K., A.W. Mäder, R.K. Richmond, D.F. Sargent, and T.J. Richmond. 1997. Crystal structure of the nucleosome core particle at 2.8 Å resolution. *Nature* 389:251–60.

Marsden, M.P.F. and U.K. Laemmli. 1979. Metaphase chromosome structure: Evidence for a radial loop model. *Cell* 17:849–58.

Narlikar, G.J., M.L. Phelan, and R.E. Kingston. 2001. Generation and interconversion of multiple distinct nucleosomal states as a mechanism for catalyzing chromatin fluidity. *Molecular Cell* 8:1219–30.

Ogryzko, V.V., R.L. Schiltz, V. Russanova, B.H. Howard, and Y. Nakatani. 1996. The transcriptional coactivators p300 and CBP are histone acetyltransferases. *Cell* 87:953–59.

Saragosti, S., G. Moyne, and M. Yaniv. 1980. Absence of nucleosomes in a fraction of SV40 chromatin between the origin of replication and the region coding for the late leader RNA. *Cell* 20:65–73.

Saunders, A., J. Werner, E.D. Andrulis, T. Nakayama, S. Hirose, D. Reinberg, and J.T. Lis. 2003. Tracking FACT and RNA polymerase II elongation complex through chromatin in vivo. *Science* 301:1094–96.

Schalch, T., S. Duda, D.F. Sargent, and T.J. Richmond. 2005. X-ray structure of a tetranucleosome and its implications for the chromatin fibre. *Nature* 436:138–41.

Taunton, J., C.A. Hassig, and S.L. Schreiber. 1996. A mammalian histone deacetylase related to the yeast transcriptional regulator Rpd3p. *Science* 272:408–11.

Thoma, F., T. Koller, and A. Klug. 1979. Involvement of histone H1 in the organization of the nucleosome and of the salt-dependent superstructure of chromatin. *Journal of Cell Biology* 83:403–27.

CHAPTER 14

RNA Processing I: Splicing

A cowboy's lariat. © *Royalty-Free/Corbis*.

Bacterial gene expression can be summarized very briefly as follows: First, RNA polymerase transcribes a gene, or set of genes, in an operon. Then, even while transcription is still occurring, ribosomes bind to the mRNA and translate it to make protein. We have already studied the transcription part of this scheme in Chapters 6–8, and it may seem to be quite complex. However, the situation in eukaryotes is much more intricate. In Chapter 13 we discussed the complex chromatin structure that distinguishes eukaryotes, but the complexities do not stop there.

In eukaryotes, the compartments in which transcription and translation occur are different. Transcription takes place in the nucleus, whereas translation takes place in the cytoplasm. This means that transcription and translation cannot occur simultaneously as they do in bacteria. Instead, transcription has to finish, then the transcript has to make its way into the cytoplasm before translation can begin. This allows an interval between transcription and translation traditionally known as the posttranscriptional phase.

In this chapter we will see that most eukaryotic genes, in contrast to typical bacterial genes, are interrupted by noncoding DNA. RNA polymerase cannot distinguish the coding region of the gene from the noncoding regions, so it transcribes everything. Thus, the cell must remove the noncoding RNA from the original transcript, in a process called splicing. Eukaryotes also tack special structures onto the 5′- and 3′-ends of their mRNAs. The 5′-structure is called a cap, and the 3′-structure is a string of AMPs called poly(A). All three of these events occur in the nucleus before the mRNA emigrates to the cytoplasm, and it is becoming increasingly clear that all three occur before transcription is over. Thus, it might be more correct to refer to them as *cotranscripional,* rather than *posttranscriptional,* events. To avoid any confusion, we will refer to them as mRNA-processing events. It appears that all three of these events are coordinated. We will return to this theme at the end of Chapter 15, after we have studied splicing (this chapter) and capping and polyadenylation (Chapter 15) in detail.

14.1 Genes in Pieces

If we expressed the sequence of the human β-globin gene as a sentence, here is how it might look:

This is *bhgty* the human β-globin *qwtzptlrbn* gene.

Two regions (italicized) within the gene obviously make no sense: they contain sequences totally unrelated to the globin coding sequences surrounding them. These are sometimes called **intervening sequences,** or **IVSs,** but they usually go by the name Walter Gilbert gave them: **introns.** Similarly, the parts of the gene that make sense are sometimes called coding regions, or expressed regions, but Gilbert's name for them is more popular: **exons.** Some genes, especially in lower eukaryotes, have no introns at all; others have an abundance. The current record (362 introns) is held by the human titin gene, which codes for a huge muscle protein.

Evidence for Split Genes

Consider the major late locus of adenovirus—the first place introns were found, by Phillip Sharp and his colleagues in 1977. The adenovirus major late locus contains several genes that are transcribed late in infection. These genes encode structural proteins, such as hexon, one of the viral coat proteins. Several lines of evidence converged at that time to show that the genes of the adenovirus major late locus are interrupted, but perhaps the easiest to understand comes from studies using a technique called **R-looping.**

In R-looping experiments, RNA is hybridized to its DNA template. In other words, the DNA template strands are separated to allow a double-stranded hybrid to form between one of these strands and the RNA product. Such a hybrid double-stranded polynucleotide is actually a bit more stable than a double-stranded DNA under the conditions of the experiment. After the hybrid forms, it is examined by electron microscopy. These experiments can be done in two basic ways: (1) using DNA whose two strands are separated only enough to let the RNA hybridize or (2) completely separating the two DNA strands before hybridization. Sharp and colleagues used the latter method, hybridizing single-stranded adenovirus DNA to mature mRNA for one of the viral coat proteins: the hexon protein. Figure 14.1 shows the results. (Do not be confused by the similarity between the terms exon and hexon. They are not related.)

If the hexon gene had no introns, a smooth, linear hybrid would occur where the mRNA lined up with its DNA template. But what if introns *do* occur in this gene? Clearly, no introns are present in the mature mRNA, or they would code for nonsense that would appear in the protein product. Therefore, introns are sequences that occur in the DNA but are missing from mRNA. That means the hexon DNA and hexon mRNA will not be able to form a smooth hybrid. Instead, the intron regions of the DNA will not find counterparts in the mRNA and so will form unhybridized loops. That is exactly what happened in the experiment shown in Figure 14.1. The loops there are made of DNA, but we still call them R loops because hybridization with RNA caused them to form.

The electron micrograph shows an RNA–DNA hybrid interrupted by three single-stranded DNA loops (labeled A, B, and C). These loops represent the introns in the hexon gene. Each loop is preceded by a short hybrid region, and the last loop is followed by a long hybrid region. Thus, the gene has four exons: three short ones near the beginning, followed by one large one. The three short exons are transcribed into a leader region that appears at the 5′-end of the hexon mRNA before the coding region; the long exon contains the coding region of the gene. In fact, the major late genes have different coding regions, but all share the same leader region encoded in the same three short exons.

When we discover something as surprising as introns in a virus, we wonder whether it is just a bizarre viral phenomenon that has no relationship to eukaryotic cellular processes. Thus, it was important to determine whether eukaryotic cellular genes also have introns. One of the first such demonstrations was an R-looping experiment done by Pierre Chambon and colleagues, using the chicken ovalbumin gene. They observed six DNA loops of various sizes that could not hybridize to the mRNA, so this gene contains

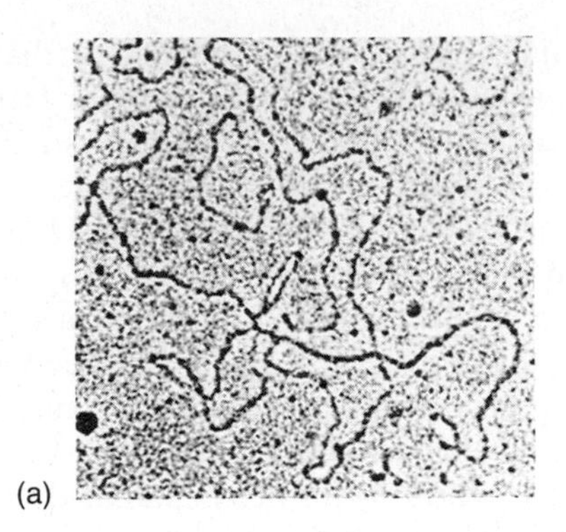

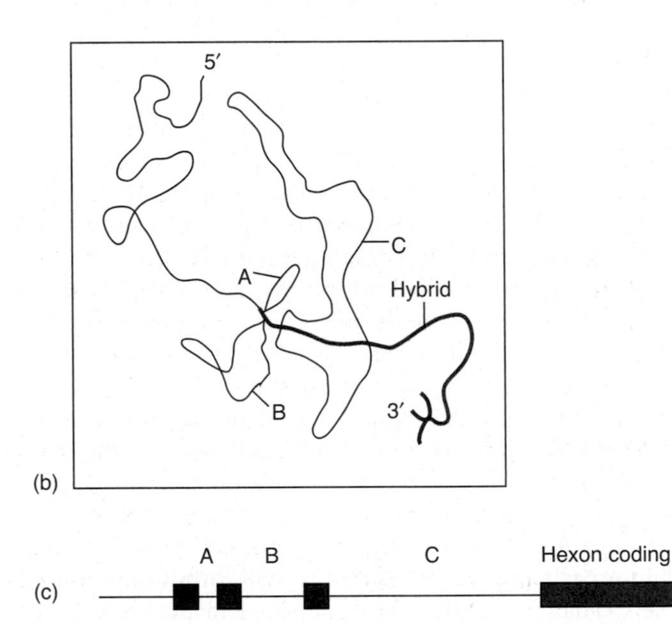

Figure 14.1 R-looping experiments reveal introns in adenovirus. (a) Electron micrograph of a cloned fragment of adenovirus DNA containing the 5′-part of the late hexon gene, hybridized to mature hexon mRNA. The loops represent introns in the gene that cannot hybridize to mRNA. **(b)** Interpretation of the electron micrograph, showing the three intron loops (labeled A, B, and C), the hybrid (heavy red line), and the unhybridized region of DNA upstream of the gene (upper left). The fork at the lower right is due to the 3′-end of the mRNA, which cannot hybridize because the 3′-end of the gene is not included. Therefore, the mRNA forms intramolecular double-stranded structures that have a forked appearance. **(c)** Linear arrangements of the hexon gene, showing the three short leader exons, the two introns separating them (A and B), and the long intron (C) separating the leaders from the coding exon of the hexon gene. All exons are represented by red boxes. (*Source:* (a) Berget, M., Moore, and Sharp, Spliced segments at the 3′ terminus of adenovirus 2 late mRNA. *Proceedings of the National Academy of Sciences USA* 74:3173, 1977.)

six introns spaced among seven exons. It is also interesting that most of the introns were considerably longer than most of the exons. This preponderance of introns is typical of higher eukaryotic genes. Introns in lower eukaryotes such as yeast tend to be shorter and much rarer.

So far we have discussed introns only in mRNA genes, but some tRNA genes also have introns, and even rRNA genes sometimes do. The introns in both these latter types of genes are a bit different from those in mRNA genes. For example, tRNA introns are relatively small, ranging in size from 4 to about 50 bp long. Not all tRNA genes have introns; those that do have only one, and it is adjacent to the DNA bases corresponding to the anticodon of the tRNA. Genes in mitochondria and chloroplasts can also have introns. Indeed, these introns are some of the most interesting, as we will see.

SUMMARY Most higher eukaryotic genes coding for mRNA and tRNA, and a few coding for rRNA, are interrupted by unrelated regions called introns. The other parts of the gene, surrounding the introns, are called exons; the exons contain the sequences that finally appear in the mature RNA product. Genes for mRNAs have been found with anywhere from zero to 362 introns. Transfer RNA genes have either zero or one.

RNA Splicing

Consider the problem introns pose. They are present in genes but not in mature RNA. How is it that the information in introns does not find its way into the mature RNA products of the genes? The two main possibilities are: (1) The introns are never transcribed; the polymerase somehow jumps from one exon to the next and ignores the introns in between. (2) The introns are transcribed, yielding a primary transcript, an overlarge gene product that is cut down to size by removing the introns. As wasteful as it seems, the latter possibility is the correct one. The process of cutting introns out of immature RNAs and stitching together the exons to form the final product is called **RNA splicing.** The splicing process is outlined in Figure 14.2, although, as we will see later in the chapter, this picture is considerably oversimplified.

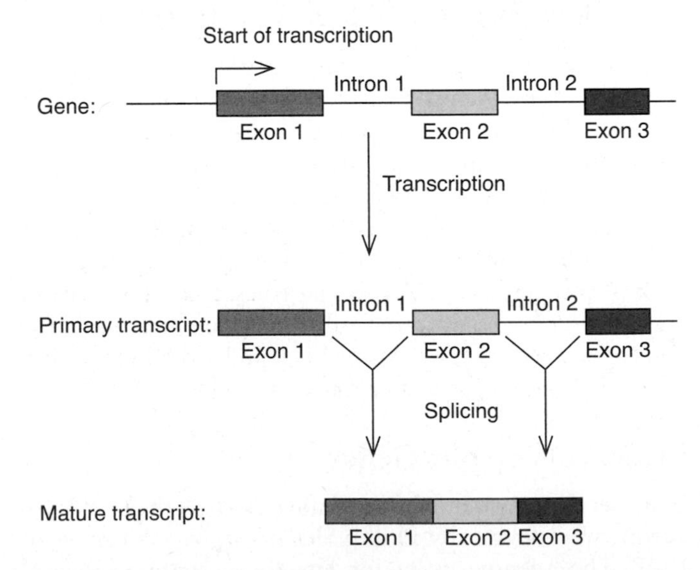

Figure 14.2 Outline of splicing. The introns in a gene are transcribed along with the exons (colored boxes) in the primary transcript. Then they are removed as the exons are spliced together.

How do we know splicing takes place? Actually, at the time introns were discovered, circumstantial evidence to support splicing already existed. A class of large nuclear RNAs called heterogeneous nuclear RNA (hnRNA), widely believed to be precursors to mRNA, had been found (Chapter 10). These hnRNAs are the right size (larger than mRNAs) and have the right location (nuclear) to be unspliced mRNA precursors. Furthermore, hnRNA turns over very rapidly, which means it is made and converted to smaller RNAs quickly. This too suggested that these RNAs are merely intermediates in the formation of more stable RNAs. However, no direct evidence existed to show that hnRNA could be spliced to yield mRNA.

The mouse β-globin mRNA and its precursor provided an ideal place to look for such evidence. The mouse globin mRNA precursor is a member of the hnRNA population. It is found only in the nucleus, turns over very rapidly, and is about twice as large (1500 bases) as mature globin mRNA (750 bases). Also, mouse immature red blood cells make so much globin (about 90% of their protein) that α- and β-globin mRNAs are abundant and can be purified relatively easily; even their precursors exist in appreciable quantities. This abundance made experiments feasible. Furthermore, the β-globin precursor is the right size to contain both exons and introns. Charles Weissmann and Philip Leder and their coworkers used R-looping to test the hypothesis that the precursor still contained the introns.

The experimental plan was to hybridize mature globin mRNA, or its precursor, to the cloned globin gene, then observe the resulting R loops (Figure 14.3). We know what the results with the mature mRNA should be. Because this RNA has no intron sequences, the introns in the gene will loop out. On the other hand, if the precursor RNA still has all the intron sequences, no such loops will form. That is what happened. You may have a little difficulty recognizing the structures in Figure 14.3 because this R-looping was done with double-, instead of single-stranded, DNA. Thus, the RNA hybridized to one of the DNA strands, displacing the other. The precursor RNA gave a smooth, uninterrupted R-loop; the mature mRNA gave an R-loop interrupted by an obvious loop of double stranded DNA, which represents the large intron. The small intron was not visible in this experiment. Notice that the term *intron* can be used for intervening sequences in either DNA or RNA.

SUMMARY Messenger RNA synthesis in eukaryotes occurs in stages. The first stage is synthesis of the primary transcription product, an mRNA precursor that still contains introns copied from the gene, if any were present. This precursor is part of a pool of large nuclear RNAs called hnRNAs. The second stage is mRNA maturation. Part of the maturation of an mRNA precursor is the removal of its introns in a process called splicing. This yields the mature-sized mRNA.

Splicing Signals

Consider the importance of accurate splicing. If too little RNA is removed from an mRNA precursor, the mature RNA will be interrupted by nonsense regions. If too much is removed, important sequences may be left out.

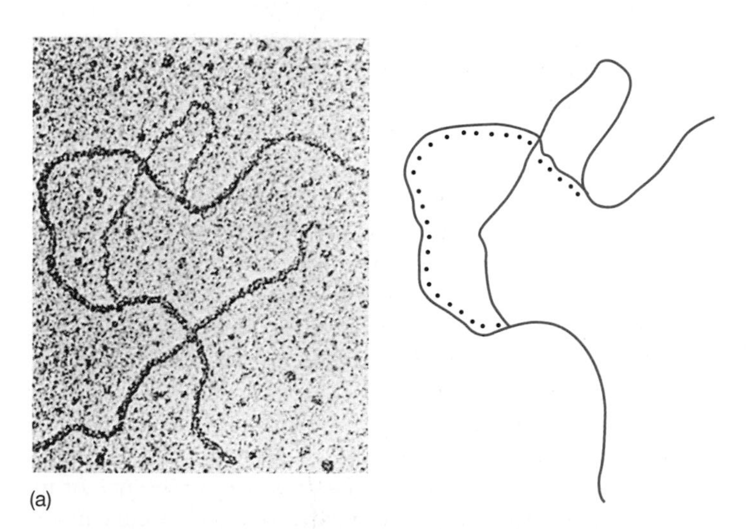

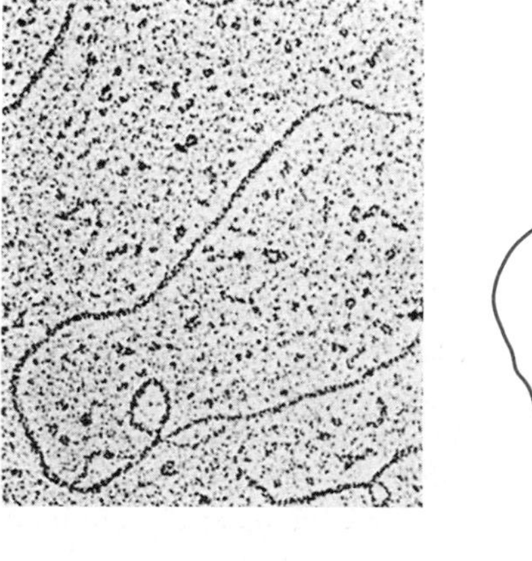

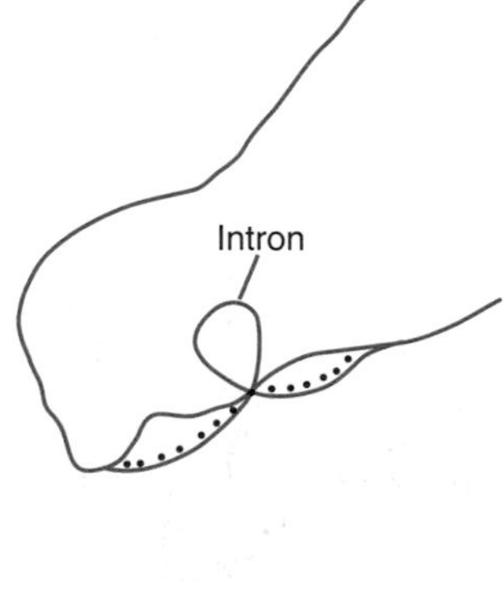

Figure 14.3 Introns are transcribed. (a) R-looping experiment in which the mouse globin mRNA precursor was hybridized to a cloned mouse β-globin gene. A smooth hybrid formed, demonstrating that the introns are represented in the mRNA precursor. **(b)** Similar R-looping experiment in which mature mouse globin mRNA was used. Here, the large intron in the gene looped out, showing that this intron was no longer present in the mRNA. The small intron was not detected in this experiment. In the interpretive drawings, the dotted black lines represent RNA and the solid red lines represent DNA. (*Source:* Tilghman, S., P. J. Curtis, D. C. Tiemeier, P. Leder, and C. Weissmann, The intervening sequence of a mouse β-globin gene is transcribed within the 15S β-globin mRNA precursor. *Proceedings of the National Academy of Sciences USA* 75:1312, 1978.)

Given the importance of accurate splicing, signals must occur in the mRNA precursor that tell the splicing machinery exactly where to "cut and paste." What are these signals? One way to find out is to look at the base sequences of a number of different genes, locate the intron boundaries, and see what sequences are common to all of them. In principle, these common sequences could be part of the signal for splicing. The most striking observation, first made by Chambon, is that almost all introns in nuclear mRNA precursors begin and end the same way:

exon/GU–intron–AG/exon

In other words, the first two bases in the intron of a transcript are GU and the last two are AG. This kind of conservation does not occur by accident; surely the GU–AG motif is part of the signal that says, "Splice here." However, a typical intron will contain several GU's and AG's within it. Why are these not used as splice sites? The answer is that splicing signals are more complex than that. They contain sequences at the exon-intron boundaries that extend beyond simply GU and AG, and they include a "branchpoint" sequence near the 3'-end of the intron, which we will discuss later in this chapter. Sequencing of many genes has revealed the following mammalian consensus sequences:

5′-AG/GUAAGU–intron–YNCUR$\underline{A}$C–Y_nNYAG/G-3′

where the slashes denote the exon–intron borders, Y is either pyrimidine (U or C), Y_n denotes a string of about nine pyrimidines, R is either purine (A or G), $\underline{A}$ is a special A in the "branchpoint" sequence within the intron, and N is any base. The consensus sequences in yeast mRNA precursors are also well studied, and a little different from those in mammals:

5′-/GUAUGU–intron–UACUA$\underline{A}$C–YAG/-3′

Finding consensus sequences is one thing; showing that they are really important is another. Several research groups have found ample evidence supporting the importance of these splice junction consensus sequences. Their experiments were of two basic types. In one, they mutated the consensus sequences at the splice junctions in cloned genes, then checked whether proper splicing still occurred. In the other, they collected defective genes from human patients with presumed splicing problems and examined the genes for mutations near the splice junctions. Both approaches gave the same answer: Disturbing the consensus sequences usually inhibits normal splicing.

Although the splice signals at the borders of an exon are necessary, they are not sufficient to define an exon. We will learn later in this chapter that the "branchpoint" sequence near the end of an intron is also required for the next exon to be recognized as such. Even all three consensus sequences are not always sufficient. That is because many introns in higher eukaryotes are enormous, ranging up to over 100 kb, and they can contain many exon-size sequences that are bounded by normal-looking splicing signals, including branchpoint sequences. Yet somehow, these "pseudoexons" rarely if ever get spliced into mature mRNAs. What sets the real exons apart from these pseudoexons? Part of the answer is that real exons tend to contain sequences known as exonic splicing enhancers (ESEs), which stimulate splicing, and pseudoexons tend to contain exonic splicing silencers (ESSs), which inhibit splicing. We will discuss these phenomena more fully later in this chapter.

SUMMARY The splicing signals in nuclear mRNA precursors are remarkably uniform. The first two bases of the intron are almost always GU, and the last two are almost always AG. The 5′- and 3′-splice sites have consensus sequences that extend beyond the GU and AG motifs, and there is also a branchpoint consensus sequence. All three consensus sequences are important to proper splicing; when they are mutated, abnormal splicing can occur.

Effect of Splicing on Gene Expression

It seems obvious that splicing introduces a degree of inefficiency into the gene expression process. Introns must be transcribed, only to be immediately removed from pre-mRNAs and degraded. Moreover, inaccurate splicing can disrupt an mRNA and lead to mistranslation. So it is fair to ask why evolution has not eliminated splicing from eukaryotes. Indeed, introns are relatively rare and small in simple eukaryotes like yeasts, but they are abundant and long—typically much longer than exons—in higher eukaryotes, including humans.

One reason that splicing may have evolved to become so prominent in higher eukaryotes is that it actually facilitates gene expression. In 2003, Shihua Lu and Bryan Cullen surveyed 10 human genes with and without introns in their 5′-untranslated regions and found that the introns improved gene expression in every case—from a relatively modest two-fold to about 35-fold in the case of the β-globin gene, which actually depends on introns for efficient expression. The advantage of introns comes from at least two sources: They stimulate efficient mRNA 3′-end formation, and they make translation more efficient.

It seems paradoxical that the presence or absence of introns could affect translation, as translation occurs in the cytoplasm, long after the introns have been removed. But we need to consider the fact that mRNAs do not exist as naked RNAs. Rather, they are complexed with a wide variety of proteins in the nucleus, and many of these proteins travel with the mRNA as a **messenger ribonucleoprotein (mRNP)** as it is transported to the cytoplasm. And some of the proteins are added to the mRNP at the exon junctions during splicing to form the **exon junction complex (EJC).** The presence of EJCs is necessary and sufficient for stimulation of gene expression by introns, probably by facilitating the association of mRNAs with ribosomes. Thus, it is the proteins added to the mRNP during splicing, rather than splicing itself, that causes the stimulation. In Chapter 18, we will see that the EJC also makes possible the destruction of faulty mRNAs that have premature stop codons. This also enhances efficiency by removing damaged mRNAs that would occupy ribosomes unproductively.

SUMMARY Splicing, by attracting the exon junction complex to mRNAs, enhances gene expression, primarily by making translation more efficient.

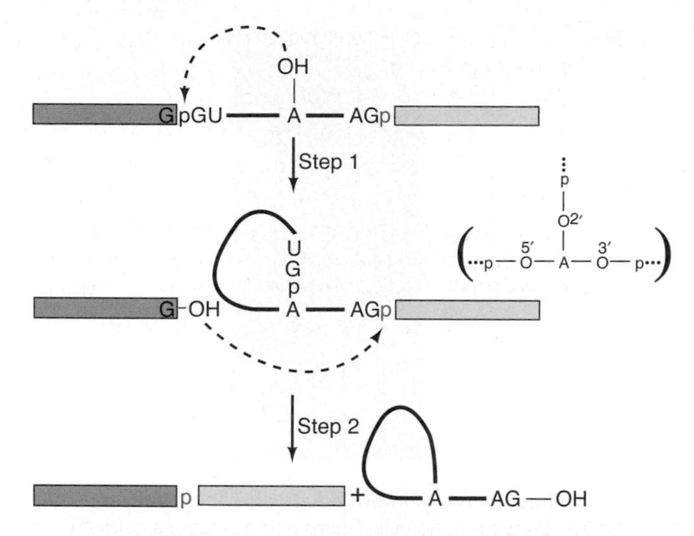

Figure 14.4 Simplified mechanism of nuclear mRNA precursor splicing. In step 1, the 2′-hydroxyl group of an adenine nucleotide within the intron attacks the phosphodiester bond linking the first exon (blue) to the intron. This attack, indicated by the dashed arrow at top, breaks the bond between exon 1 and intron, yielding the free exon 1 and the lariat-shaped intron–exon 2 intermediate, with the GU at the 5′-end of the intron linked through a phosphodiester bond to the branchpoint A. The lariat is a consequence of the internal attack of one part of the RNA precursor on another part of the same molecule. At right in parentheses is the branchpoint showing that the adenine nucleotide is involved in phosophdiester bonds through its 2′-, 3′-, and 5′-hydroxyl groups. In step 2, the free 3′-hydroxyl group on exon 1 attacks the phosphodiester bond between the intron and exon 2. This yields the spliced exon 1/exon 2 product and the lariat-shaped intron. Note that the phosphate (red) at the 5′-end of exon 2 becomes the phosphate linking the two exons in the spliced product.

14.2 The Mechanism of Splicing of Nuclear mRNA Precursors

The splicing scheme in Figure 14.2 gave only the precursor and the product, with no indication about the mechanism cells use to get from one to the other. Let us now explore the interesting and quite unexpected mechanism of nuclear mRNA precursor splicing.

A Branched Intermediate

One of the essential details missing from Figure 14.2 is that the intermediate in nuclear mRNA precursor splicing is branched, so it looks like a **lariat,** or cowboy's lasso. Figure 14.4 outlines the two-step lariat model of splicing. The first step is the formation of the lariat-shaped intermediate. This occurs when the 2′-hydroxyl group of an adenosine nucleotide in the middle of the intron attacks the phosphodiester bond between the first exon and the G at the beginning of the intron (the **5′-splice site**), forming the loop of the lariat and simultaneously separating the first exon from the intron. The second step completes the splicing process: The 3′-hydroxyl group left at the end of the first exon attacks the phosphodiester bond linking the intron to the second exon (the **3′-splice site**). This forms the exon–exon phosphodiester bond and releases the intron, in lariat form, at the same time.

This mechanism seemed unlikely enough that rigorous proof had to be presented for it to be accepted. In fact, very good evidence supports the existence of all the intermediates and products shown in Figure 14.4, much of it collected by Sharp and his research group.

First and foremost, what is the evidence for the branched intermediate? The first indication of a strangely shaped RNA created during splicing came in 1984, when Sharp and colleagues made a cell-free splicing extract and used it to splice an RNA with an intron. This splicing substrate was a radioactive transcript of the first few hundred base pairs of the adenovirus major late region. This transcript contained the first two leader exons, with a 231-nt intron in between. After allowing some time for splicing, these workers electrophoresed the RNAs and found the unspliced precursor plus a novel band with unusual behavior on gel electrophoresis. It migrated faster than the precursor on a 4% polyacrylamide gel, but slower than the precursor on a 10% polyacrylamide gel. This kind of behavior is characteristic of circular or branched RNAs, such as lariat-shaped RNAs.

Was this strange RNA a splicing product? Yes; its appearance was inhibited by an antiserum that blocks splicing, or by omitting ATP, which is required for splicing. Furthermore, another experiment by Sharp's group (Figure 14.5) showed that it accumulated more and more as splicing progressed. It turned out to be the lariat-shaped intron that had been removed from the precursor. This experiment also showed the existence of another RNA with anomalous electrophoretic behavior. Its concentration rose during the first part of the splicing process, then fell later

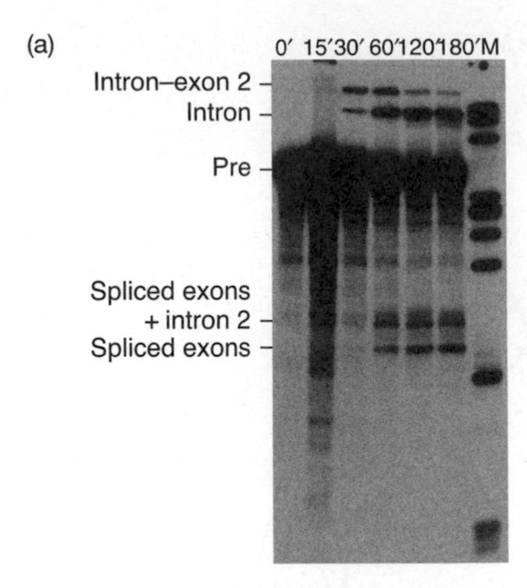

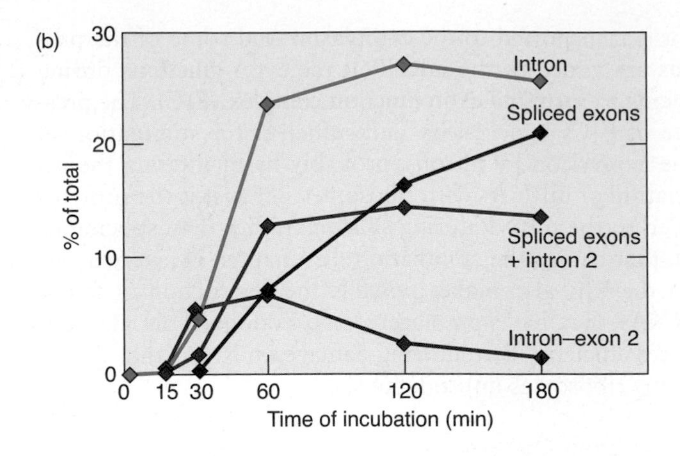

Figure 14.5 Time course of intermediate and liberated intron appearance. (a) Electrophoresis. Sharp and colleagues carried out splicing reactions in vitro and electrophoresed the products after various times, indicated at top, on a 10% polyacrylamide gel. The products are identified at left. The top band contained the intron–exon 2 intermediate. The next band contained the intron. Both these RNAs were lariat-shaped, as suggested by their anomalously low electrophoretic mobilities. The next band contained the precursor. The bottom two bands contained two forms of the spliced exons: the upper one was still attached to a piece of intron 2, and the lower one seemed to lack that extra RNA. **(b)** Graphic presentation. Sharp and colleagues plotted the intensities of each band from panel (a) to show the accumulation of each RNA species as a function of time. (*Source:* Grabowski P., R.A. Padgett, and P.A. Sharp, Messenger RNA splicing in vitro: An excised intervening sequence and a potential intermediate. *Cell* 37 (June 1984) f. 4, p. 419. Reprinted by permission of Elsevier Science.)

on, suggesting that it was a splicing intermediate. It is actually exon 2 with the lariat-shaped intron still attached. Both this RNA and the intron have anomalous electrophoretic behavior because they are lariat-shaped.

The two-step mechanism in Figure 14.4 allows the following predictions, each of which Sharp and colleagues verified.

1. The excised intron has a 3′-hydroxyl group. This is required if exon 1 attacks the phosphodiester bond as shown at the beginning of step 2, because this will remove the phosphate attached to the 3′-end of the intron, leaving just a hydroxyl group.
2. The phosphorus atom between the 2 exons in the spliced product comes from the 3′- (downstream) splice site.
3. The intermediate (exon 2 plus intron) and the spliced intron contain a branched nucleotide that has its 2′-, 3′- and 5′-hydroxyl groups bonded to other nucleotides.
4. The branch involves the 5′-end of the intron binding to a site within the intron.

Let us look at the evidence for the branched nucleotide. The intermediate (exon 2 plus intron) and the spliced intron contain a branched nucleotide that has its 2′-, 3′- and 5′-hydroxyl groups bonded to other nucleotides. Sharp and coworkers cut the splicing intermediate with either RNase T2 or RNase P1. Both enzymes cut after every nucleotide in an RNA, but RNase T2 leaves nucleoside-3′-phosphates just as RNase T1 does (Figure 14.6), whereas RNase P1 generates nucleoside-5′-phosphates. Both enzymes yielded novel products among the normal nucleoside monophosphates. Thin-layer chromatography allowed the charges of

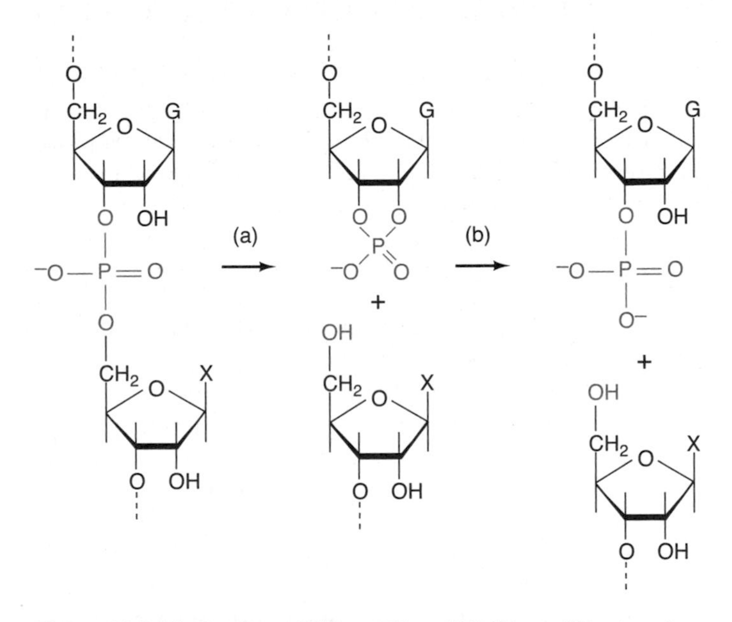

Figure 14.6 Mechanism of RNase T1 and T2. These RNases cut RNA as follows: **(a)** The RNase cleaves the bond between the phosphate attached to the 3′-hydroxyl group of a guanine nucleotide and the 5′-hydroxyl group of the next nucleotide, generating a cyclic 2′, 3′-phosphate intermediate. **(b)** The cyclic intermediate opens up, yielding an oligonucleotide ending in a guanosine 3′-phosphate.

these two products to be determined. The T2 product had a charge of −6, whereas the P1 product had a charge of −4. An ordinary mononucleotide would have a charge of −2.

What are these unusual products? Their charges are consistent with the structures shown in Figure 14.7, given that each phosphodiester bond has one negative charge and each terminal phosphate has two negative charges. To prove

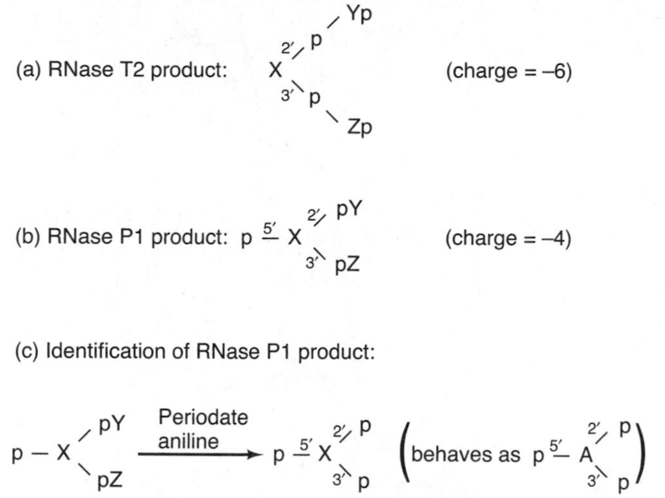

(a) RNase T2 product: X with 2′–p–Yp and 3′–p–Zp (charge = –6)

(b) RNase P1 product: p $\overset{5'}{-}$ X with 2′–pY and 3′–pZ (charge = –4)

(c) Identification of RNase P1 product:

p — X (pY, pZ) $\xrightarrow{\text{Periodate, aniline}}$ p $\overset{5'}{-}$ X (2′ p, 3′ p) (behaves as p $\overset{5'}{-}$ A (2′ p, 3′ p))

Figure 14.7 Direct evidence for a branched nucleotide. (a) Sharp and colleagues digested the splicing intermediate with RNase T2. This yielded a product with a charge of −6. This is consistent with the branched structure pictured here. **(b)** Digestion with RNase P1 gave a product with a charge of −4, consistent with this branched structure. **(c)** Sharp and colleagues treated the P1 product with periodate and aniline to eliminate the nucleosides bound to the 2′- and 3′-phosphates of the branched nucleotide. The resulting product copurified with adenosine-2′3′5′-trisphosphate, verifying the presence of a branch and demonstrating that the branch occurs at an adenine nucleotide.

that these structures were correct, Sharp and colleagues treated the RNase P1 product with periodate and aniline to remove the 2′- and 3′-nucleosides by β-elimination. The product of this reaction should be a nucleoside 2′, 3′, 5′-trisphosphate. To verify this assignment, these workers subjected the product to two-dimensional thin-layer chromatography and found that it comigrated with adenosine 2′,3′,5′-trisphosphate. Thus, a branched nucleotide occurs, and it is an adenine nucleotide.

SUMMARY Several lines of evidence demonstrate that nuclear mRNA precursors are spliced via a lariat-shaped, or branched, intermediate.

A Signal at the Branch

Is there something special about the adenine nucleotide that participates in the branch, or can any A in the intron serve this function? Study of many different introns has revealed the existence of a consensus sequence, and the fact that this sequence, and no other, can form the branch.

The first hint of a special region within the intron came from experiments with the yeast actin gene performed by Christopher Langford and Dieter Gallwitz in 1983. These workers cloned the actin gene, made numerous mutations in it, and reintroduced these mutant genes into normal yeast cells. Then they assayed for splicing by S1 mapping. Figure 14.8 shows the results: First, when they removed a region between 35 and 70 bp upstream of the intron's 3′-splice site (mutant #1), they blocked splicing. This suggested that this 35-bp region contains a sequence, represented in the figure by a small red box, that is important for splicing. When they inserted an extra DNA segment between this "special sequence" and the second exon (mutant #2), splicing occurred from the usual 5′-splice site, but not to the correct 3′-splice site. Instead, the aberrant 3′-splice site was

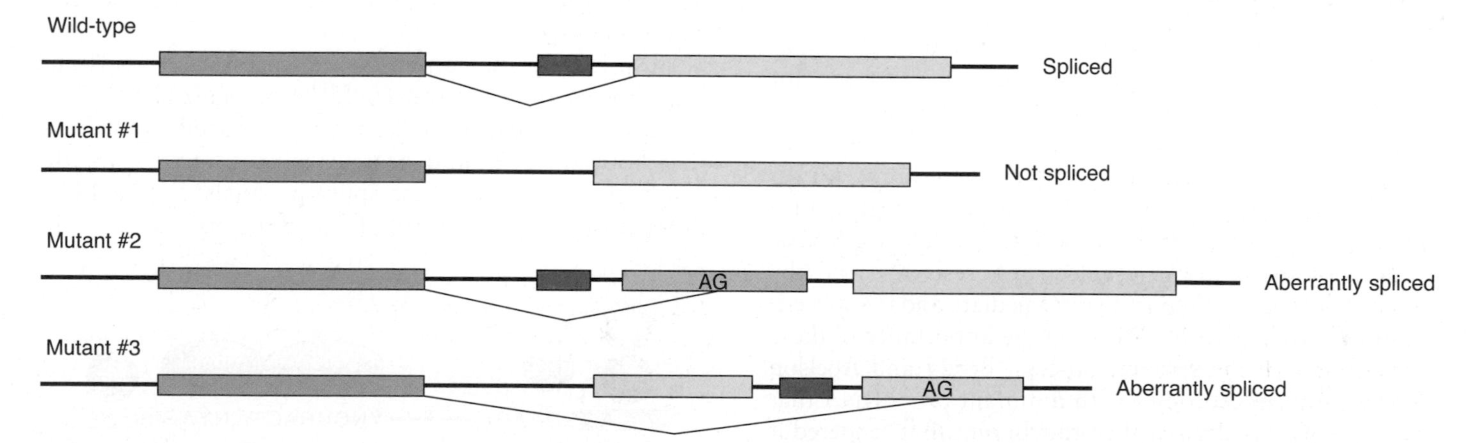

Figure 14.8 Demonstration of a critical signal within a yeast intron. Langford and Gallwitz made mutant yeast actin genes in vitro, reintroduced them into yeast cells, and tested them for splicing there. The wild-type gene contained two exons (blue and yellow). The intron contained a conserved sequence (red) found in all yeast introns. Yeast cells spliced this gene properly. To make mutant #1, Langford and Gallwitz deleted the conserved intron sequence, which destroyed the ability of this gene's transcript to be spliced. Mutant #2 had extra, nonintron DNA (pink) inserted into the intron downstream of the conserved intron sequence. The transcript of this gene was aberrantly spliced to the first AG within the insert. To construct mutant #3, Langford and Gallwitz moved the conserved intron sequence downstream into the second exon. The transcript of this gene was aberrantly spliced to the first AG downstream of the relocated conserved sequence. These experiments suggested that the conserved sequence is critical for splicing and that it designates a downstream AG as the 3′-splice site.

the first AG downstream of the special intron sequence. This AG lay within the inserted segment of DNA. This result suggested that the special intron sequence tells the splicing machinery to splice to an AG at some appropriate distance downstream. If one inserts a new AG in front of the usual one, splicing may go to the new site. Finally, mutant #3 contained the special intron sequence within the second exon. Again in this case, the 3′-splice site became the first AG downstream of the special sequence in its new location, which happened to be in the second exon.

The special intron sequence is so important because it contains the branchpoint adenine nucleotide: the final A in the sequence UACUAAC. In fact, this is the nearly invariant sequence around the branchpoint in all yeast nuclear introns. Higher eukaryotes have a more variable consensus sequence surrounding the branchpoint A: $U_{47}NC_{63}U_{53}R_{72}\underline{A}_{91}C_{47}$, where R is either purine (A or G), and N is any base. The subscripts indicate the frequency with which a base is found in that position. For example, the branchpoint A (underlined) is found in this position 91% of the time. The first U is frequently replaced by a C, so this position usually contains a pyrimidine.

SUMMARY In addition to the consensus sequences at the 5′- and 3′-ends of nuclear introns, branchpoint consensus sequences also occur. In yeast, this sequence is almost invariant: UACUAAC. In higher eukaryotes, the consensus sequence is more variable. In all cases, the branched nucleotide is the final A in the sequence.

Spliceosomes

Edward Brody and John Abelson discovered in 1985 that the lariat-shaped splicing intermediates in yeast are not free in solution, but bound to 40S particles they called **spliceosomes.** These workers added labeled pre-mRNAs to cell-free extracts and used a glycerol gradient ultracentrifugation procedure to purify the spliceosomes. Figure 14.9 shows a prominent 40S peak containing labeled RNAs. Analysis of these RNAs by electrophoresis revealed the presence of lariats: the splicing intermediate and the spliced-out intron. To further demonstrate the importance of these spliceosomes to the splicing process, Brody and Abelson tried to form spliceosomes with a mutant pre-mRNA that had an A→C mutation at the branchpoint that rendered it unspliceable. This RNA was severely impaired in its ability to form spliceosomes. Sharp and his colleagues isolated spliceosomes from human (HeLa) cells, also in 1985, and showed that they sedimented at 60S.

Spliceosomes contain the pre-mRNA, of course, but they also contain many RNAs and proteins. Some of these RNAs and proteins come in the form of **small nuclear ribonucleoproteins** (**snRNPs,** pronounced "snurps"), which

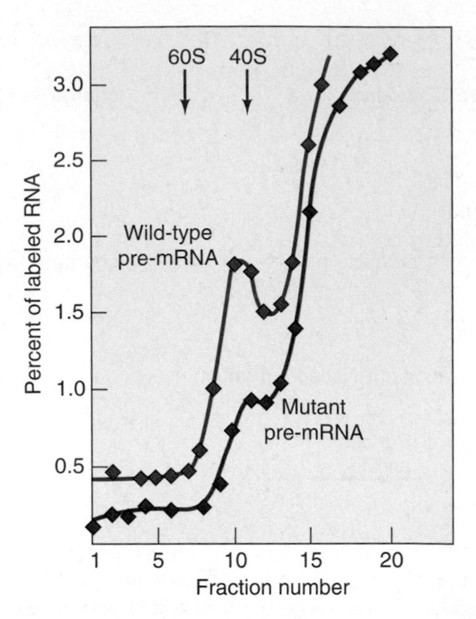

Figure 14.9 Yeast spliceosomes. Brody and Abelson incubated a labeled yeast pre-mRNA with a yeast splicing extract, then subjected the mixture to glycerol gradient ultracentrifugation. Finally, they determined the radioactivity in each gradient fraction by scintillation counting. Two different experiments with a wild-type pre-mRNA (red) and two different experiments with a mutant pre-mRNA with a base alteration at the 5′-splice site (blue) are shown. The wild-type pre-mRNA shows a clear association with a 40S aggregate. This association is much weaker with the mutant pre-mRNA. (*Source:* Adapted from Brody, E. and J. Abelson, The spliceosome: Yeast premessenger RNA associated with a 40S complex in a splicing-dependent reaction. *Science* 228:965, 1985.)

consist of **small nuclear RNAs (snRNAs)** coupled to proteins. The snRNAs can be resolved by gel electrophoresis into individual species designated **U1, U2, U4, U5,** and **U6.** All five of these RNAs join the spliceosome and play crucial roles in splicing.

In principle, the consensus sequences at the ends and branchpoint of an intron could be recognized by either proteins or nucleic acids. We now have excellent evidence that both snRNAs and protein splicing factors are the agents that recognize these splicing signals. Figure 14.10 illustrates a typical intron flanked by exons, and the

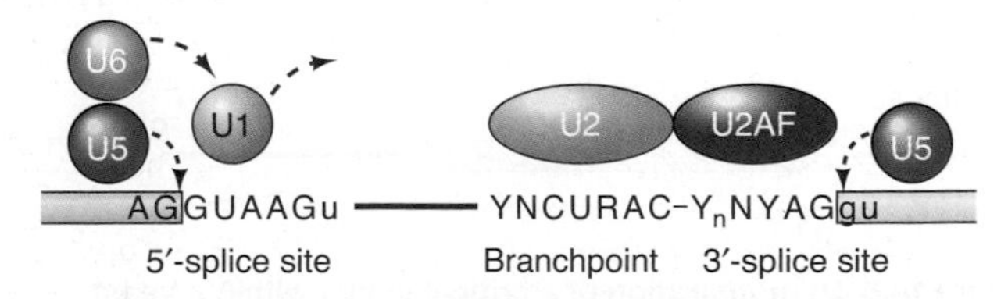

Figure 14.10 Recognition of a typical mammalian pre-mRNA intron by RNAs and proteins. The capital letters represent bases that are well conserved, and the lowercase letters represent less conserved bases. Y stands for both pyrimidines, R stands for both purines, and N is any base. U1 snRNP recognizes the 5′-splice site first, and then is replaced by U6 snRNP. U2 snRNP recognizes the branchpoint, and the protein U2AF (U2-associated factor) recognizes the 3′-splice site. U5 snRNP binds to the 5′- and 3′-splice sites after initial recognition by other factors.

molecular species that interact at the critical sites. We will examine the evidence for all these interactions in the following sections of this chapter.

SUMMARY Splicing takes place on a particle called a spliceosome. Yeast spliceosomes and mammalian spliceosomes have sedimentation coefficients of about 40S, and about 60S, respectively. Spliceosomes contain the pre-mRNA, as well as snRNPs and protein splicing factors that recognize key splicing signals and orchestrate the splicing process.

U1 snRNP Joan Steitz and, independently, J. Rogers and R. Wall, noticed in 1980 that U1 snRNA has a region whose sequence is almost perfectly complementary to both 5′- and 3′-splice site consensus sequences. They proposed that U1 snRNA base-paired with these splice sites, bringing them together for splicing. We now know that splicing involves a branch within the intron, which rules out such a simple mechanism. Nevertheless, base pairing between U1 snRNA and the 5′-splice site not only occurs, it is essential for splicing.

We know that this base pairing with U1 is essential because of genetic experiments performed by Yuan Zhuang and Alan Weiner in 1986. They introduced alterations into one of the three alternative 5′-splice sites of the adenovirus E1A gene. Splicing of this gene normally occurs from each of these 5′-sites to a common 3′-site to yield three different mature mRNAs, called 9S, 12S, and 13S (Figure 14.11). The mutations (at the 12S 5′-splice site) disturbed the potential base pairing with U1. To measure the effects of these mutations on splicing, Zhuang and Weiner performed an

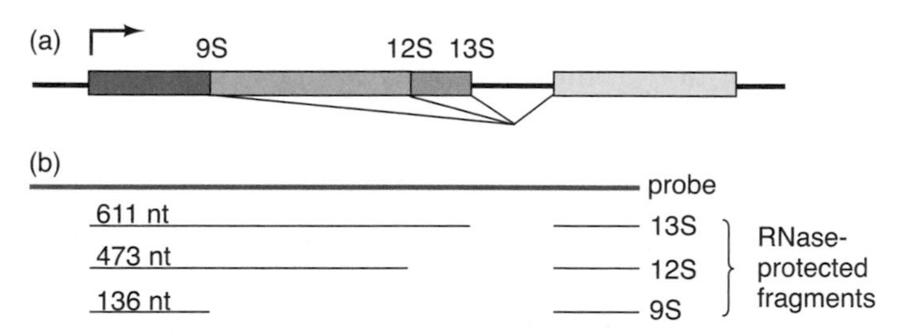

Figure 14.11 Splicing scheme of adenovirus E1A gene and RNase protection assay to detect each spliced product. **(a)** Splicing scheme. Three alternative 5′-splice sites (at the borders of the red, orange, and blue blocks and at the end of the blue block) combine with one 3′-splice site at the beginning of the yellow block to produce three different spliced mRNAs: the 9S, 12S, and 13S mRNAs, respectively. **(b)** RNase protection assay. The labeled riboprobe is represented by the purple line at top. Each alternative splicing product protects different-size fragments of this probe from digestion by RNase. (These sizes in nucleotides (nt) are given above each fragment. The three splicing products also produce identical protected fragments corresponding to the downstream exon.) (*Source:* Adapted from Zhuang, Y. and A.M. Weiner, A compensatory base change in U1 snRNA suppresses a 5′-splice site mutation. *Cell* 46:829, 1986.)

RNase protection assay (Chapter 5) on RNA from cells transfected with plasmids bearing the 5′-splice site mutations in the E1A gene. Figure 14.11 shows the length in nucleotides (nt) of the signals expected from splicing at each of the three sites.

The first mutation Zhuang and Weiner tested was actually a double mutation. The fifth and sixth bases (+5 and +6) of the intron were changed from GG to AU (Figure 14.12). This disrupted a GC base pair between the G(+5) of the intron and a C in U1, but introduced a new potential base pair between U(+6) of the intron and an A in U1. In spite of this new potential base pair, the overall base pairing between mutant splice site and U1 should have been weakened because the number of *contiguous* base pairs was lower. Was splicing affected? Figure 14.13 (lane 4) shows that the mutation essentially abolished splicing at the 12S site and caused a concomitant increase in splicing at the 13S and 9S sites. Next, these workers made a compensating mutation in the U1 gene that restored base pairing with the mutant splice site. They introduced the mutant U1 gene into HeLa cells on the same plasmid that bore the mutant E1A gene. Figure 14.13 (lane 5) shows that this mutant U1 not only restored base pairing, it also restored splicing at the 12S site.

Thus, base pairing between the splice site and U1 is required for splicing. But is it sufficient? If one could make a mutant splice site with weakened base pairing to U1 whose splicing could not be suppressed by a compensating mutation in U1, one could prove that this base pairing is not enough to ensure splicing. Figures 14.12 and 14.13 show how Zhuang and Weiner demonstrated just this. This time, they mutated the 13S 5′-splice site, changing an A to a U in the +3 position, which interrupted a string of six base pairs. This abolished 13S splicing, while stimulating 12S and, to a lesser degree, 9S splicing (Figure 14.13, lane 6). A compensating mutation in the U1 gene restored the six base pairs, but failed to restore splicing at the 13S site (lane 7). Thus, base pairing between the 5′-splice site and U1 is not sufficient for splicing.

SUMMARY Genetic experiments have shown that base pairing between U1 snRNA and the 5′-splice site of an mRNA precursor is necessary, but not sufficient, for splicing.

U6 snRNP Why do base changes in U1 sometimes fail to compensate for base changes in the 5′-splice site? We can imagine a variety of answers to this question, including the possibility that some protein or proteins must also recognize the sequence at the 5′-splice site. In that case, changes in U1 might not be enough to restore recognition of this site by the spliceosome. It is also possible that *another* snRNA must interact with the 5′-splice site. Altering the U1 sequence to match a

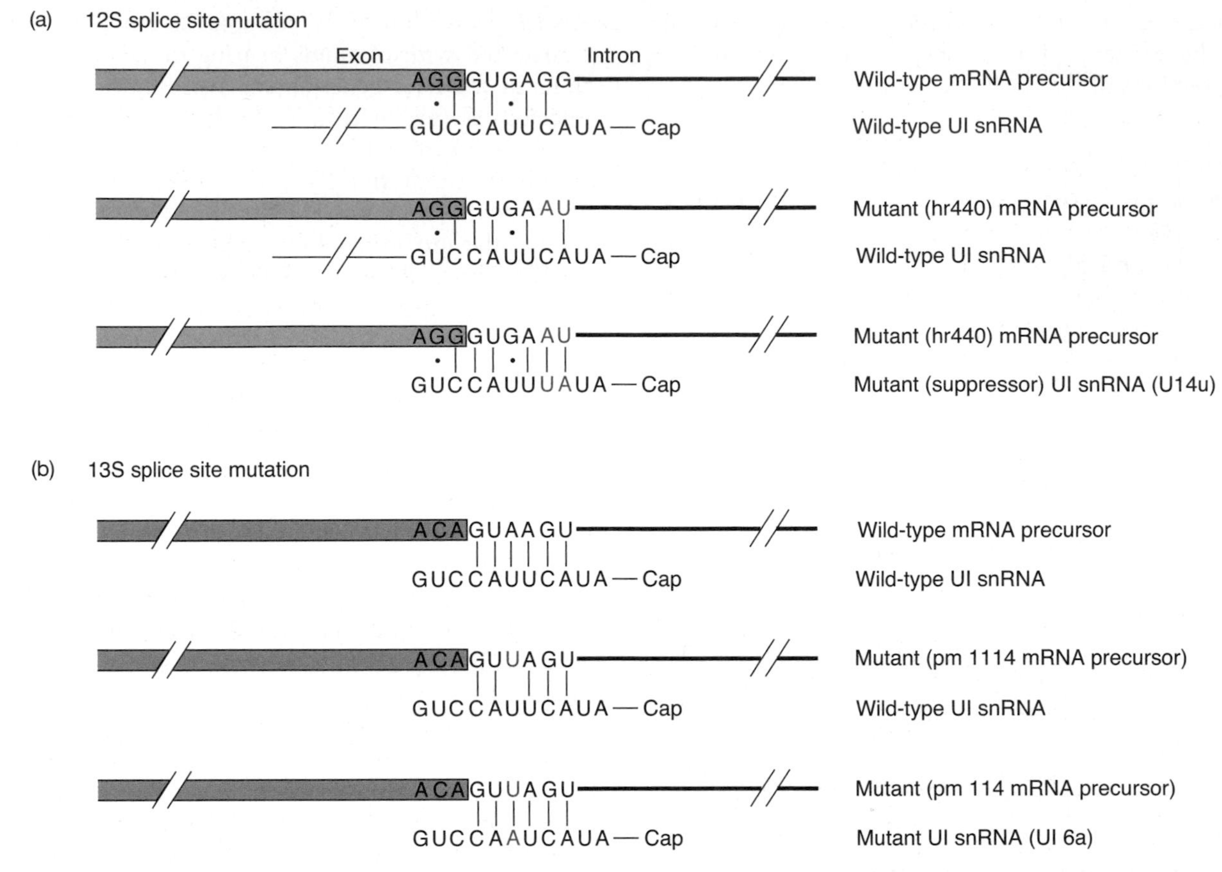

Figure 14.12 Alignment of wild-type and mutant 5′-splice sites with wild-type and mutant U1 snRNAs. (a) 12S splice site mutation. The wild-type and mutant sequences are identified at right. Watson–Crick base pairs between the mRNA precursor and U1 RNA are represented by vertical lines; wobble base pairs, by dots. Mutated bases are represented by red letters. The end of the exon is represented by an orange box as in Figure 14.11. **(b)** 13S splice site mutation. All symbols as in panel (a) except that the end of the exon is represented by a blue box as in Figure 14.11.

mutant splice site might not restore the splice site's interaction with this other snRNA, so splicing could still be prevented.

Two research groups, led by Christine Guthrie and Joan Steitz, have shown that another snRNA does indeed base-pair with the 5′-splice site. This is U6 snRNA. Steitz first demonstrated that U6 might be involved in events near the 5′-splice site when she showed that U6 could be chemically cross-linked to intron position +5. Based on this finding, she postulated that the ACA in the invariant sequence ACAGAG in U6 base-pairs with the conserved UGU in positions +4 to +6 of 5′-splice sites (Figure 14.14).

Erik Sontheimer and Joan Steitz also used cross-linking studies to show that U6 binds to a site very close to the 5′-end of the intron in the spliceosome. Their experimental strategy went like this: First they made a model splicing precursor with a single intron, flanked by two exons. Then they substituted 4-thiouridine (4-thioU) for the nucleotides at either of two positions: the last nucleotide in the first exon, or the second nucleotide of the intron. The 4-thioU residue is photosensitive; when it is activated by ultraviolet light, it forms covalent cross-links to other RNAs with which it is in contact. By isolating these cross-linked structures, the researchers could discover the RNAs that base-pair with the nucleotides at the 5′-splice site.

When Sontheimer and Steitz placed the 4-thioU in the second position of the intron they found a linkage to U6. Moreover, this and other cross-linking experiments showed that U6 binds to the splicing substrate both before and after the initial step in splicing, and that there is a U2–U6 complex, which can also be predicted based on sequence complementarity between these two RNAs. Later in this chapter we will see how base pairing between U2 and U6 helps to form a structure that constitutes the active site of the spliceosome.

SUMMARY The U6 snRNP associates with the 5′-end of the intron by base pairing through the U6 snRNA. This association first occurs prior to formation of the lariat intermediate, but it persists after this first step in splicing. The association between U6 and the splicing substrate is essential for the splicing process. U6 also associates with U2 during splicing.

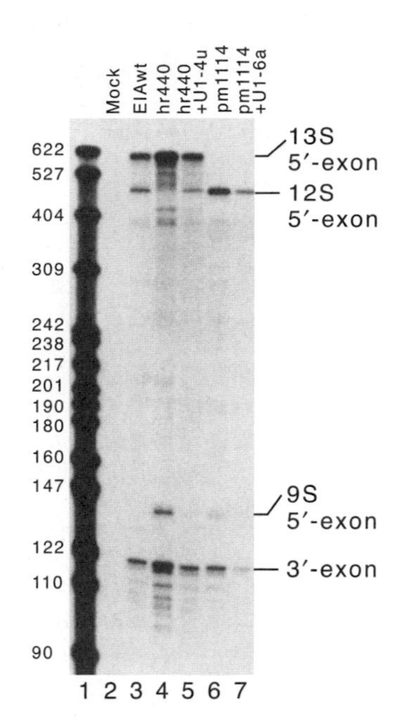

Figure 14.13 Results of RNase protection assay. Zhuang and Weiner tested the wild-type and mutant 5′-splice sites and wild-type and mutant U1 snRNAs pictured in Figure 14.12 by transfecting HeLa cells with plasmids containing these genes, then detected splicing by RNase protection as illustrated in Figure 14.11. Lane 1, size markers, with lengths in base pairs indicated at left. Lane 2, mock-transfected cells (negative control). Lane 3, wild-type E1A gene with wild-type U1 snRNA. Signals were visible for the 13S and 12S products, but not for the 9S product, which normally does not appear until late in infection. Lane 4, mutant hr440 with an altered 12S 5′-splice site. No 12S signal was apparent. Lane 5, mutant hr440 plus mutant U1 snRNA (U1–4u). Splicing at the 12S 5′-site was restored. Lane 6, mutant pm1114 with an altered 13S 5′-splice site. No 13S signal was apparent. Lane 7, mutant pm1114 plus mutant U1 snRNA (U1–6a). Even though base pairing between the 5′-splice site and U1 snRNA was restored, no 13S splicing occurred. (*Source:* Zhuang Y. and A.M. Weiner, A compensatory base change in U1 snRNA suppresses a 5′-splice site mutation. *Cell* 46 (12 Sept 1986) f. 1a, p. 829. Reprinted by permission of Elsevier Science.)

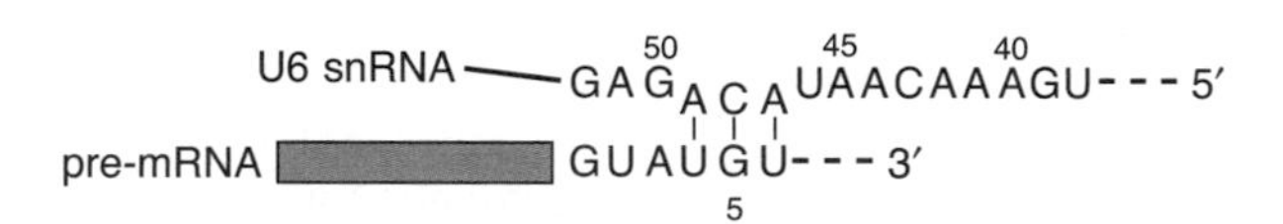

Figure 14.14 A model for interaction between a yeast 5′-splice site and U6 snRNA. The invariant ACA (nt 47–49) of yeast U6 base-pairs with the UGU (nt 4–6) of the intron. (*Source:* Adapted from Lesser, C.F. and C. Guthrie, Mutations in U6 snRNA that alter splice site specificity: Implications for the active site. *Science* 262:1983, 1993.)

U2 snRNP The consensus branchpoint sequence in yeast is complementary to a sequence in U2 snRNA, as shown in Figure 14.15, and genetic analysis has shown that base pairing between these two sequences is essential for splicing. Christine Guthrie and her colleagues provided such genetic evidence when they mutated the branchpoint sequence and showed that the defective splicing this caused could be reversed by a complementary mutation in the yeast U2 gene.

To do these experiments, these workers provided a histidine-dependent yeast mutant with a fused actin-*HIS4* gene containing an intron in the actin portion. If the transcript of this gene is spliced properly, the *HIS4* part of the fusion protein product will be active, and the cells can live on media containing the histidine precursor histidinol, because the *HIS4* product converts histidinol to histidine. Next, they introduced mutations into the splicing branchpoint. One of these, a U to A change in position 257, converted the nearly invariant sequence UAC<u>U</u>AAC to UAC<u>A</u>AAC and inhibited splicing by 95%. This also prevented growth on histidinol. Another mutation, a C to A transversion in position 256, converted the branch sequence UA<u>C</u>UAAC to UA<u>A</u>UAAC and inhibited splicing by 50%.

To test for suppression of these mutations by mutant U2s, Guthrie and colleagues introduced a plasmid bearing the mutant U2s into yeast. They made sure the plasmid was retained by endowing it with a selectable marker: the *LEU2* gene. (The host cells were *LEU*$^-$.) It was necessary to provide an *extra* copy of the U2 gene because making a mutation in the cell's only copy of the U2 gene could cause the splicing of all other genes to fail. Figure 14.16 shows that the U2s that restored complementary binding to the mutant branch sites really did restore splicing. This was especially apparent in the case of the A257 mutant, where no growth was observed with the wild-type U2, but abundant growth occurred with the U2 that had the mutation that restored base pairing with the mutant branch site.

Besides base-pairing with the branchpoint, U2 also base-pairs with U6. This association can be predicted on the basis of the sequences of the two RNAs, and genetic analysis by Guthrie and her colleagues provided direct evidence for the base pairing. First, Guthrie and colleagues discovered lethal mutations in the ACG sequence of yeast U6, which base-pairs to another snRNA, U4. These workers showed in two ways that the ability of these mutations to disrupt base pairing with U4 was not the problem. First, they introduced corresponding mutations into U4 that would cause the same disruption of the U4–U6 interaction and showed that these did not affect cell growth. Second, they introduced compensating mutations into U4 that would restore base pairing with the mutant U6 and showed that these did not suppress the lethal U6 mutations.

Apparently, U6 interacts with something else besides U4, and the lethal U6 mutations interfere with this interaction. Hiten Madhani and Christine Guthrie demonstrated that U2 is the other molecule with which U6 interacts. They introduced lethal mutations into residues 56–59 of U6 and found that these mutations could be suppressed by compensating mutations in residues 23 and 26–28 of U2, which restored base pairing with the mutant U6 molecules. This crucial base pairing between U2 and U6 forms a region called helix I, which will be summarized later in Figure 14.20.

Other workers (Jian Wu and James Manley, and Banshidar Datta and Alan Weiner) have used similar genetic

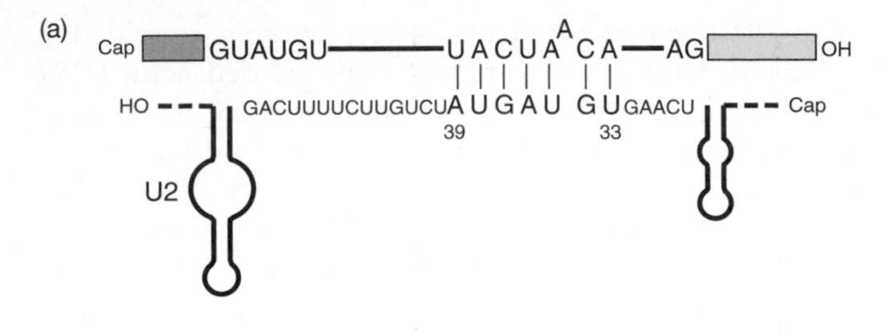

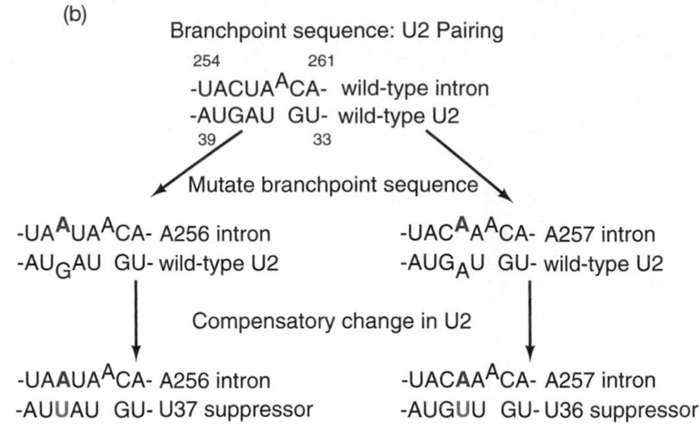

Figure 14.15 Base pairing between yeast U2 and yeast branchpoint sequences. (a) Proposed base pairing between wild-type yeast U2 and the invariant yeast branchpoint sequence. Note that the A at the branch site bulges out (top) and does not participate in the base pairing. **(b)** Proposed base pairing between wild-type and mutant yeast U2s and branchpoints. The red letters indicate mutations (A's) introduced into the branchpoint sequence at positions 256 and 257; the green letters represent compensating mutations (U's) introduced into U2. (*Source:* Adapted from Parker R., P.G. Sliciano, and C. Guthrie, Recognition of the TACTAAC box during mRNA splicing in yeast involves base pairing to the U_2-like snRNA. *Cell* 49:230, 1987.)

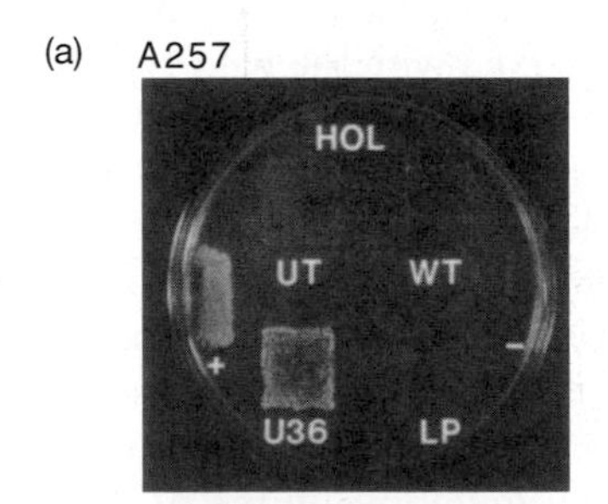

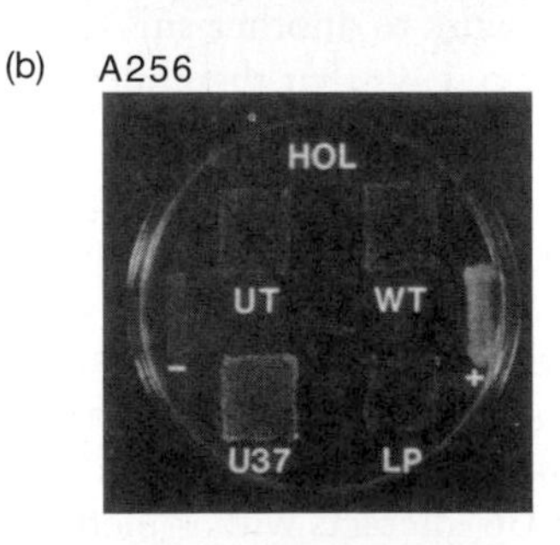

Figure 14.16 Demonstration of U2 snRNP-branchpoint base pairing by mutation suppression. Growth of A257 **(a)** and A256 **(b)** mutants on HOL medium was measured in the presence of wild-type and suppressor mutant U2. The abbreviations under each patch of cells denote the nature of the U2 added, if any: UT, untransformed (no U2 added); WT, wild-type U2; U36, U2 with mutation that restores base pairing with A257; U37, U2 with mutation that restores base pairing with A256; LP, a colony that lost its U2 plasmid. The positive control in each plate (+) contained a wild-type fusion gene and no extra U2. The negative control in each plate contained no fusion gene. (*Source:* Parker R., P.G. Siciliano, and C. Guthrie, Recognition of the TACTAAC box during mRNA splicing in yeast involves base pairing to the U2-like snRNA. *Cell* 49 (24 Apr 1987) f. 3, p. 232. Reprinted by permission of Elsevier Science.)

analysis of splicing efficiency in mammalian cells to demonstrate interaction between the 5′-end of U2 and the 3′-end of U6, to form another base-paired domain called helix II. Mutations in U2 could be suppressed by compensating mutations in U6 that restored base pairing. This interaction is nonessential in yeast, but necessary in mammals, at least for high splicing efficiency.

SUMMARY The U2 snRNA base-pairs with the conserved sequence at the splicing branchpoint. This base pairing is essential for splicing. U2 also forms vital base pairs with U6, forming a region called helix I, that apparently helps orient these snRNPs for splicing. In addition, the 5′-end of U2 interacts with the 3′-end of U6, forming a region called helix II, that is important for splicing in mammalian cells, but not in yeast cells.

U5 snRNP We have now seen evidence for the participation of U1, U2, and U6 snRNPs in splicing. What about U5? It has no obvious complementarity with any snRNA or conserved region of a splicing substrate, yet it does seem to associate with both exons, perhaps positioning them for the second splicing step.

Sontheimer and Steitz provided evidence for the involvement of U5 with the ends of the exons during splicing, again using 4-thioU-substituted splicing substrates. In one such experiment, they substituted 4-thioU for the normal C in the first position of the second exon of an adenovirus major late splicing substrate. This change still allowed normal splicing to occur. When they cross-linked the 4-thioU to whatever snRNA was near the 5′-end of the second exon, they created a doublet complex (U5/intron–E2) that appeared at 30 min after the onset of splicing

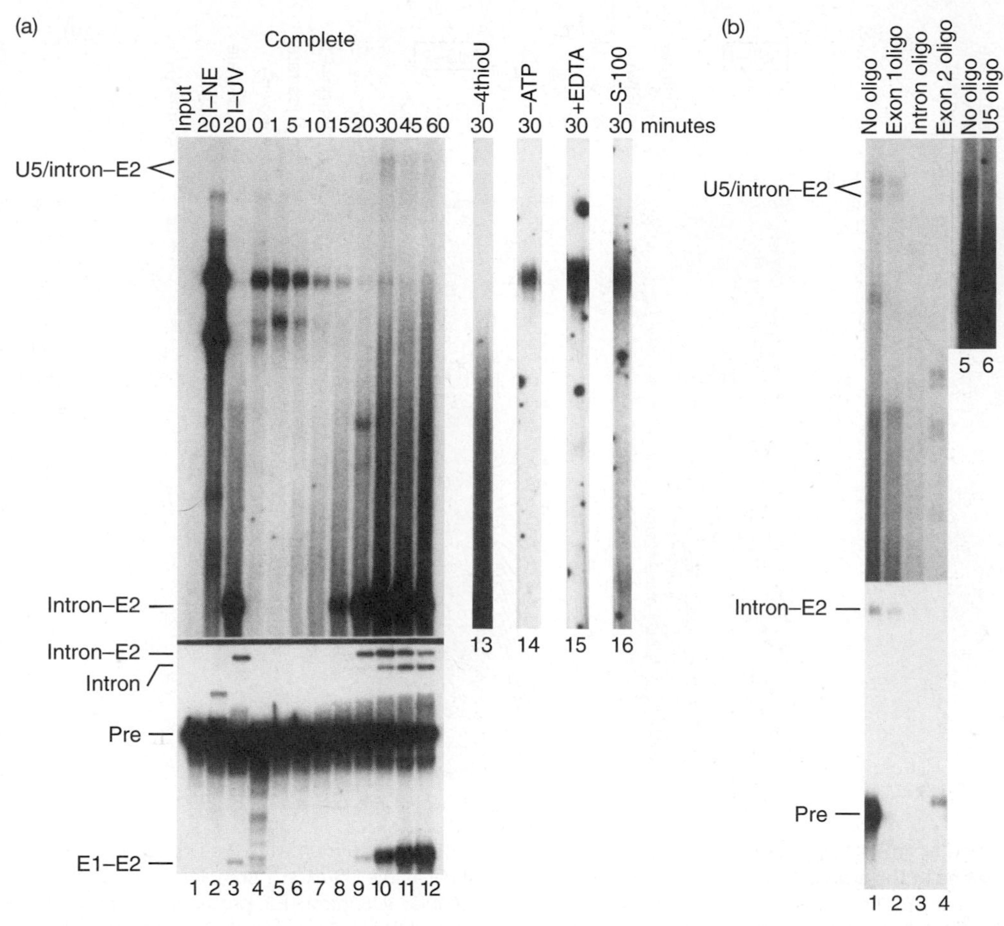

Figure 14.17 Detection of a complex between U5 and the 5′-end of the second exon. (a) Forming the complex. Sontheimer and Steitz placed 4-thioU in the first position of the second exon of a labeled splicing substrate and cross-linked it to whatever RNAs were nearby at various times during splicing. Then they electrophoresed the products and detected them by autoradiography. The U5/intron–E2 doublet appears near the top, late in the splicing process (after 30 min). Lane 1, input RNA with no incubation; lane 2, 20-min incubation with no nuclear extract (NE); lane 3, 20-min incubation followed by no UV irradiation; lanes 4–12, incubation for the times indicated at top; lane 13, no 4-thioU labeling; lane 14, no ATP; lane 15, EDTA added to chelate magnesium and block splicing; lane 16, a fraction clarified by high-speed ultracentrifugation was used instead of nuclear extract. **(b)** Identification of the RNAs in the complex. Sontheimer and Steitz irradiated the splicing mix after 30 min of splicing to form cross-links, then incubated it with DNA oligonucleotides complementary to U5 and other RNAs, then added RNase H to degrade any RNAs hybridized to the oligonucleotides. Finally, they electrophoresed and autoradiographed the products. The oligonucleotides (oligos) used were as follows: lanes 1 and 5, no oligo; lane 2, anti-exon-1 oligo; lane 3, anti-intron oligo; Lane 4, anti-exon-2 oligo; lane 6, anti-U5 oligo. The anti-intron, anti-exon-2, and anti-U5 oligos all helped destroy the complex, indicating that the complex is composed of the intron, second exon, and U5. *(Source:* Sontheimer E.J. and J.A. Steitz, The U5 and U6 small nuclear RNAs as active site components of the spliceosome. *Science* 262 (24 Dec 1993) f. 4, p. 1992. Copyright © American Association for the Advancement of Science.)

(Figure 14.17). This was late enough that the first splicing step had already occurred. Many other complexes also formed, but we will not discuss them here.

To show that this doublet complex really does include U5, the intron, and exon 2, Sontheimer and Steitz hybridized the complex to DNA oligonucleotides complementary to these RNAs, then treated the complex with RNase H, which degrades the RNA strand of an RNA–DNA hybrid. Figure 14.17 shows that oligonucleotides complementary to U5, the intron, and the second exon, but not the first exon, cooperated with RNase H to degrade the complex. Thus, the complex appears to include U5 and the intron–exon-2 splicing intermediate. The interaction between U5 and the second exon is position-specific because substitution of 4-thioU for the second base in the second exon did not result in formation of any bimolecular RNA complexes.

To identify the bases in U5 or U6 involved in the 4-thioU cross-links to the splicing intermediates, Sontheimer and Steitz exploited primer extension blockage. They used oligonucleotides complementary to sequences in the snRNAs as primers for reverse transcription of the snRNAs in the complexes. Wherever reverse transcriptase encounters a cross-link, it will stop, yielding a DNA of defined length. This length corresponds to the distance between the primer binding site and the cross-link, and therefore the exact postion of the cross-link. Figure 14.18 shows the results. Panels (a) and (b) demonstrate that two adjacent U's in U5 cross-link to the last base in the first exon, when either the intact splicing

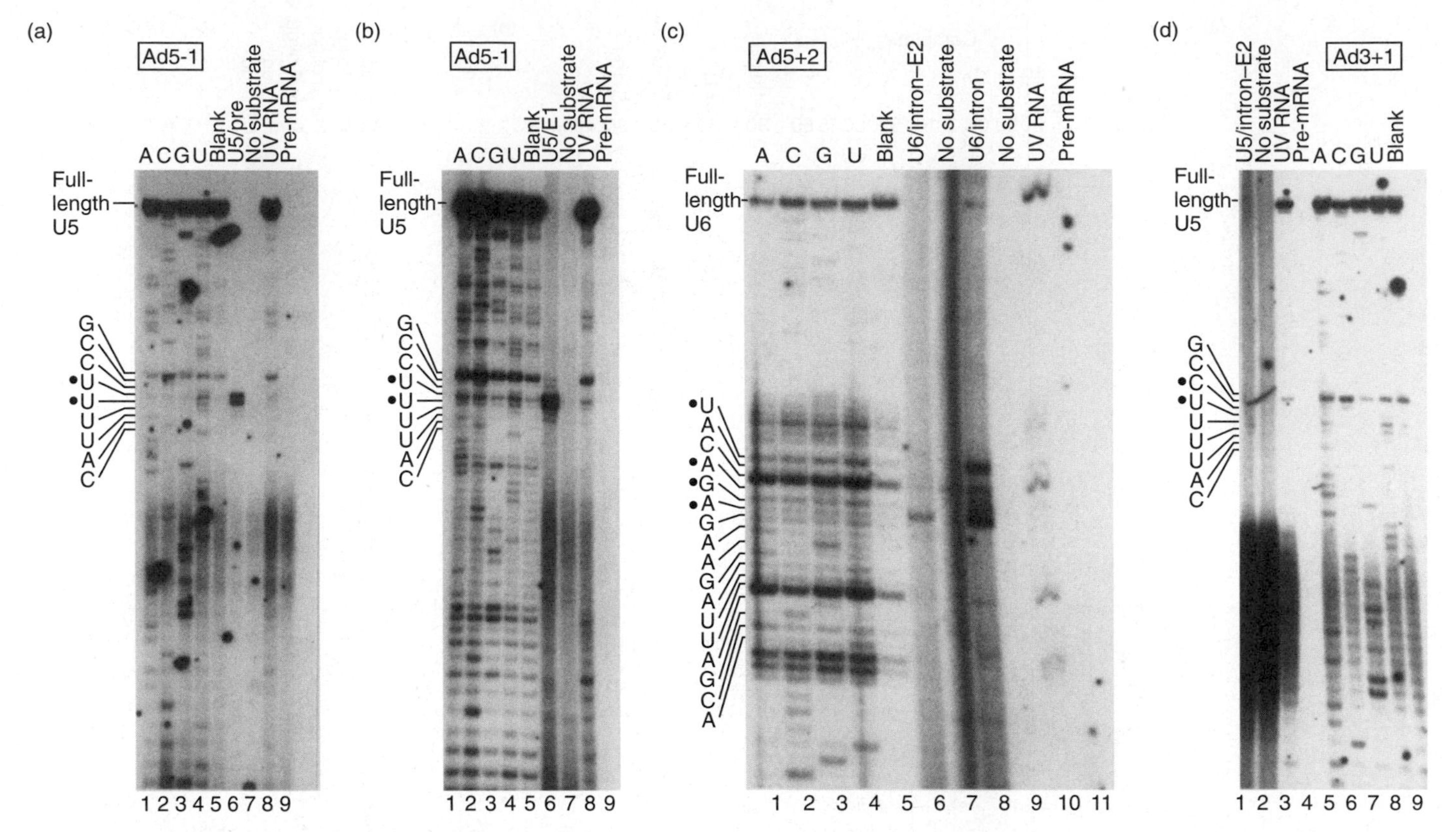

Figure 14.18 Identification of snRNP bases cross-linked to 4-thioU in various positions in the splicing substrate. Sontheimer and Steitz used primer extension to map the bases in U5 and U6 cross-linked to 4-thioU in the following positions: the last base in the first exon (Ad5-1, panels **a** and **b**); the second base in the intron (Ad5+2, panel **c**); or the first base in the second exon (Ad3+1, panel **d**). They formed cross-linked complexes with these RNAs, then excised the complexes from the electrophoresis gels and added primers specific for either U5 or U6, and performed primer extension analysis. The first four lanes in panels (a–c) and lanes 5–8 in panel (d) are sequencing lanes using the same primer as in the primer extension assays. The lanes marked "blank" are control sequencing lanes with no template. The experimental lanes are lanes 6 in panels (a and b), lanes 6 and 8 in panel (c), and lane 1 in panel (d). These are the results of primer extension with: the U5/splicing precursor complex (U5/pre, panel a); the U5/exon 1 complex (U5/E1, panel b); the U6/intron–exon-2 complex (U6/intron–E2, panel c), and the U6/intron complex, panel (c); and the U5/intron–exon-2 complex (U5/intron–E2, panel d). The other lanes are controls as follows: "no substrate," substrate was omitted from the reaction mix, then a slice of gel was cut out from the position where complex would be if substrate were included; "UV RNA," total RNA from an extract lacking substrate; "pre-mRNA," uncross-linked substrate. The cross-linked bases in the snRNPs are marked with dots at the left of each panel. (*Source:* Sontheimer, E.J. and J.A. Steitz, The U5 and U6 small nuclear RNAs as active site components of the spliceosome. *Science* 262 (24 Dec 1993) f. 5, p. 1993. Copyright © American Association for the Advancement of Science.)

substrate or just the first exon was used. Skipping panel (c) for a moment, panel (d) demonstrates that one of the same U's that were involved in cross-links to the end of the first exon is also involved, along with an adjacent C, in cross-links to the first base in the second exon. Panel (c) shows that four bases in U6 cross-link to the second base in the intron. The sum of the results with U5 suggest that this snRNP is involved in binding to the 3′-end of the first exon and the 5′-end of the second exon, as illustrated in Figure 14.19. This would allow it to position the two exons for splicing.

SUMMARY The U5 snRNA associates with the last nucleotide in one exon and the first nucleotide of the next. This presumably lines up the two exons for splicing.

U4 snRNP Most of what we know about U4 concerns its association with U6. We have known for some time that the sequences of U4 and U6 snRNAs suggest an association to form two base-paired stems, called stem I and stem II. Cross-linking experiments have also indicated an association between U4 and U6. Does U4 have any direct role to play in splicing? Apparently not. U4 dissociates from U6 after splicing is underway and can then be removed from the spliceosome using gentle procedures. Thus, its role may be to bind and sequester U6 until it is time for U6 to participate in splicing. It is worth noting that some U6 bases that participate in base pairing with U4 to form stem I are also involved in the essential base pairing to U2 that we discussed earlier in this chapter. This underscores the importance of removing U4, so U6 can base-pair to U2 and help form an active spliceosome.

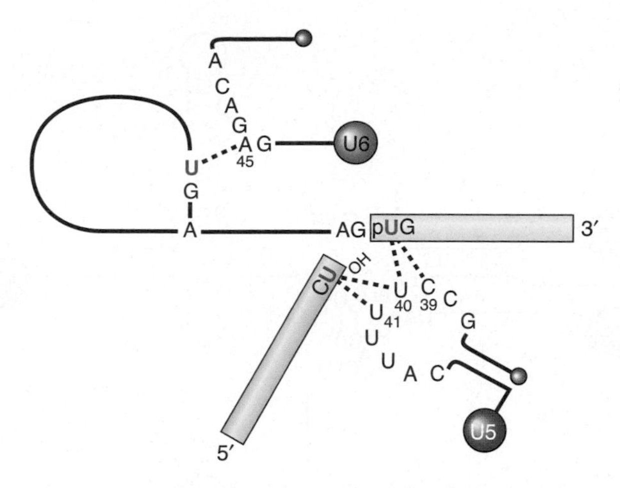

Figure 14.19 Summary of U5 and U6 interactions with the splicing substrates revealed by 4-thioU cross-linking. The red, boldfaced U's represent 4-thioUs introduced into the splicing substrate. The dotted lines illustrate cross-links between snRNP bases and 4-thioUs in the splicing substrates. Exon 1 is blue and exon 2 is yellow. The small purple dots are caps at the 5′-ends of the snRNAs. Note the role U5 can play in positioning the two exons for the second step in splicing. (*Source:* Adapted from Sontheimer, E.J. and J.A. Steitz, The U5 and U6 small nuclear RNAs as active site components of the spliceosome. *Science* 262:1995, 1993.)

SUMMARY U4 base-pairs with U6, and its role seems to be to bind U6 until U6 is needed in the splicing reaction.

snRNP Involvement in mRNA Splicing We will see later in this chapter that some other types of introns are self-splicing. That is, they do not rely on a spliceosome, but have all the catalytic activity they need to splice themselves. These self-splicing introns fall into two classes. One class, group II introns, use a lariat intermediate just like the lariat intermediates in spliceosomal mRNA splicing. Thus, it is tempting to speculate that the spliceosomal snRNPs substitute for parts of the group II intron in forming a similar structure that juxtaposes exons 1 and 2 for splicing.

Figure 14.20 depicts models for splicing both spliceosomal and group II introns. Panel (a) shows a variation on the model for the second step in nuclear mRNA splicing presented in Figure 14.19; panel (b) shows an equivalent model for a group II intron. Several features are noteworthy. First, the U5 loop, by contacting exons 1 and 2 and positioning them for splicing, substitutes for domain ID of a group II intron. Such RNA regions are called **internal guide sequences** because of their function in guiding other RNA regions into the proper position for catalysis. Second, the U6 region that base-pairs with the 5′-splice site substitutes for domain IC of a group II intron. Third, the U2–U6 helix I resembles domain V of a group II intron. Finally, the U2–branchpoint helix substitutes for domain VI of a group II intron. In both cases, base pairing around the branchpoint A causes this key nucleotide to bulge out, presumably helping it in its task of forming the branch. Because group II introns are catalytic RNAs (ribozymes), the similarities presented in Figure 14.20 suggest that the snRNPs, which substitute for group II intron elements at the center of splicing activity, also catalyze the splicing reactions.

Ren-Jang Lin and colleagues provided evidence in 2000 that U6 snRNA is indeed involved in catalysis. Their argument begins as follows: Each of the two splicing steps (recall Figure 14.4) is a transesterification reaction, in which one phosphodiester bond is broken and another is

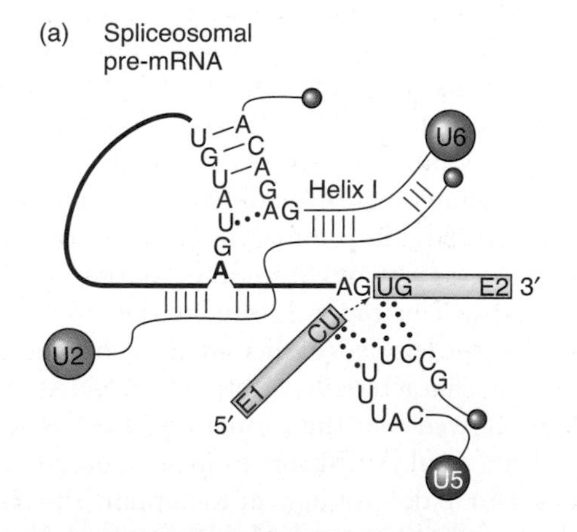

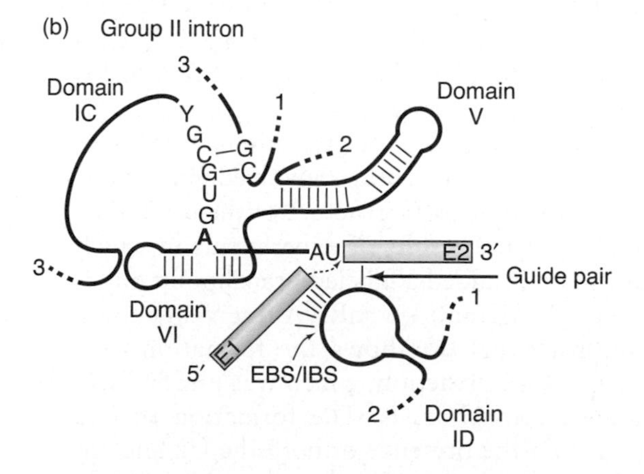

Figure 14.20 A model to compare the active center of a spliceosome to the active center of a group II intron. **(a)** Spliceosome. This is a variation on Figure 14.19, but including U2 (reddish brown). All other colors have the same significance as in Figure 14.19. The branchpoint A is bold, and the intron is rendered with a thick line. Dashed arrow represents the attack by exon 1 on the intron–exon-2 bond that is about to occur. **(b)** Group II intron. The intron is drawn in the same shape as the proposed spliceosomal structure in panel (a), to illustrate the similarities. Only parts of the intron are shown; the missing parts are suggested by dotted lines with numbers to indicate connections between parts. The exons are colored and the branchpoint A is bold. Dashed arrow represents the attack by exon 1 on the intron–exon-2 bond that is about to occur. (*Sources:* (a) Adapted from Wise, J.A., Guides to the heart of the spliceosome. *Science* 262:1978, 1993. (b) Adapted from Sontheimer, E.J. and J.A. Steitz, The U5 and U6 small nuclear RNAs as active site components of the spliceosome. *Science* 262:1995, 1993.)

formed. In the first step, for example, the bond between the first exon and the intron is broken and a new bond between the branchpoint A and the 5′-end of the intron forms, generating the lariat intermediate. Catalysts in reactions like this must do two things: activate the nucleophile (the 2′-OH of the branchpoint A) and stabilize the leaving group (the oxygen that will become the 3′-OH at the end of the first exon). Metal ions such as magnesium can perform both of these functions. Indeed, self-splicing group II introns use magnesium in this way.

Lin and colleagues found that replacing one of the oxygens of U6 snRNA with sulfur completely blocks splicing. This substitution would also be expected to hinder the ability of U6 to bind to magnesium. And if this is the critical magnesium at the catalytic site, it would mean that U6 also plays a direct role in catalysis. If this is so, then adding manganese might reverse the effects of substituting sulfur for oxygen in U6. That is because manganese can perform like magnesium in catalysis but, unlike magnesium, it can bind to RNA in which a key oxygen is replaced by sulfur.

Lin and colleagues found that manganese can indeed reverse the effect of the sulfur substitution in U6 snRNA. This suggests that U6 binds to the magnesium ion at the catalytic center of the spliceosome, but it does not prove the case because metal ions can be essential for catalysis without being at the catalytic center.

In 2001, Saba Valadkhan and James Manley added more support to the RNA catalysis hypothesis by showing that a mixture of in-vitro-synthesized U2 and U6 snRNA fragments, plus a yeast intron oligonucleotide containing a branchpoint consensus sequence, can catalyze a transesterification reaction related to the first reaction in splicing. In a normal first splicing step, the branchpoint A attacks the phosphodiester bond linking the first exon to the intron (the 5′-splice site). In the reaction catalyzed by the U2, U6, and intron fragments in vitro, there was no 5′-splice site, so the branchpoint A attacked a phosphodiester bond in U6 itself, forming a branched oligonucleotide Figure 14.21 illustrates the base pairing that occurs among the three RNAs in this reaction, the nucleotides involved in the catalytic reaction, and the proposed structure of the product.

Figure 14.22 gives the results of experiments in which Valadkhan and Manley added a labeled branchpoint oligonucleotide (Br) to the U2 and U6 snRNA fragments under various conditions. Panel (a) shows the formation of a product (X) after 24 h of reaction, which was purified and displayed by gel electrophoresis. The formation of this product depended on the presence of both the U2 and U6 fragments and was blocked by heating to a temperature near the melting temperature of the U2–U6 complex. Thus, both the U2 and U6 fragments appeared to be required for the reaction. Panels (b) and (c) show that the reaction that formed X was linear for about 2 h, continued for almost 20 h, and was stimulated by adding more U2 and U6 fragments, up to a saturation level at about 2 μM.

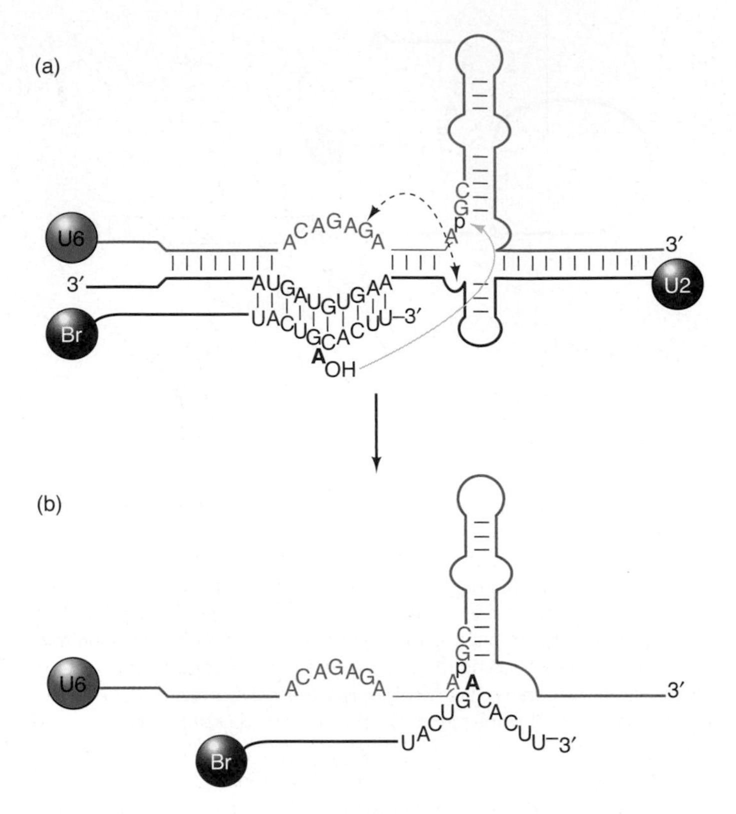

Figure 14.21 In vitro reaction resembling the first step in spliceosomal splicing. (a) Base-pairing among the three RNAs in the complex assembled in vitro. The U6 fragment (red) is on top, the U2 fragment (blue) in the middle, and the branchpoint fragment (Br, black) is on the bottom, with the bulged branchpoint A in boldface. The gray arrow points to the A52–G53 phosphodiester bond (black) that is the target for attack by the branchpoint A. The dashed arrow connects bases in U6 and U2 that can be cross-linked with UV light. **(b)** Proposed chemical structure of the product.

This same series of experiments also demonstrated that RNA X probably contains a branched nucleotide. Figure 14.22d shows that RNA X is not formed by unusually strong base pairing between two RNAs, because it withstood heating up to 90°C. Thus, RNA X appears to involve a covalent bond between RNAs, not just base pairing. In results not shown here, Valadkhan and Manley also showed that RNA X exhibits anomalous electrophoretic behavior. It electrophoreses just above an 87-nt marker in 8% polyacrylamide and just below a 236-nt marker in 16% polyacrylamide. As we learned earlier in this chapter, this kind of behavior is characteristic of branched RNAs. Finally, these workers showed that the formation of RNA X depends on Mg^{2+}. Ca^{2+} could substitute for Mg^{2+}, but not as efficiently, whereas Mn^{2+} did not appear to support the reaction at all.

Next, Valadkhan and Manley reacted 5′- and 3′-end-labeled Br and U2 and U6 fragments and found that label from both ends of U6 and Br, but no label from U2, appeared in RNA X. Thus, RNA X includes *all* of both U6 and Br, but does not include U2. And, because the linkage between U6 and Br is not mere base pairing, the two RNAs are probably covalently linked. Valadkhan and Manley

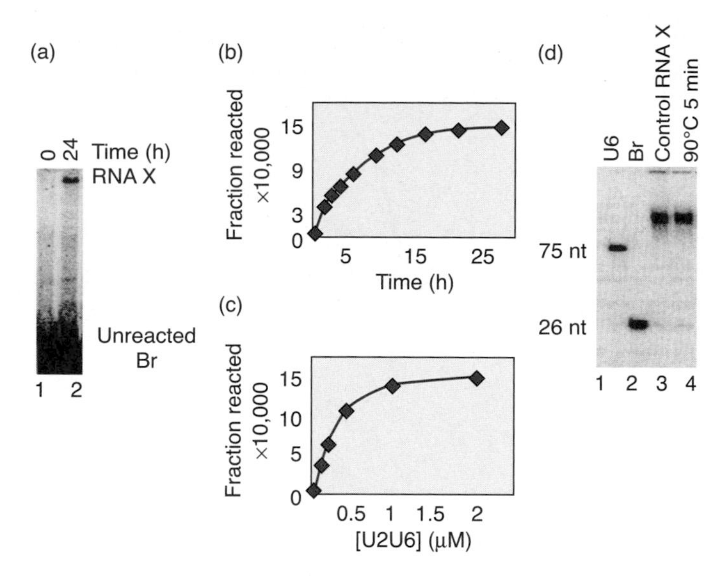

Figure 14.22 Formation of RNA X. **(a)** Detection of RNA X by SDS-PAGE. Valadkhan and Manley incubated in vitro-synthesized U2, U6, and Br fragments for 0 h or 24 h in the presence of Mg^{2+}, then electrophoresed the products. **(b)** Reaction time course. **(c)** Dependence of the reaction on U2 and U6. **(d)** Resistance of RNA X to heat-denaturation. Lane 3 shows the eletrophoretic mobility of unheated RNA X and lane 4 shows that this does not change upon heating RNA X to 90°C for 5 min. Lanes 1 and 2 are controls with the U6 and Br fragments, respectively. (*Source:* Reprinted with permission from *Nature* 413: from Valadkhan and Manley fig. 2, p. 702. © 2001 Macmillan Magazines Limited.)

also showed that blocking the 5′-ends of Br and the U6 fragment (by dephosphorylation and introduction of a cyclic phosphate, respectively) did not inhibit the formation of RNA X. Thus, the ends of the two RNAs are not involved in the linkage, so the linkage must be somewhere within each of the RNAs, which would produce an X-shaped product.

Finally, Valadkhan and Manley mapped the link between the two RNAs to the branchpoint A in Br and the phosphate between A53 and G54 of the invariant AGC triad in U6 (see Figure 14.21). To do this mapping, they employed the same kind of primer extension analysis used to map the 4-thioU cross-links between U5 and U6 and the splicing substrate (recall Figure 14.18). They also used chemical cleavage of end-labeled RNA X to detect nucleotides where RNA–RNA interactions prevented cleavage.

The result of this line of experimentation is that Mg^{2+} U2, U6, and Br, with no help from proteins, can catalyze a reaction similar to the first step in splicing. Of course, this reaction is not the same as the first step in splicing because there is no 5′-splice site for the branchpoint A to attack. However, this kind of attack on U6 is not unprecedented: Sometimes abnormal splicing in vivo involves the same kind of attack on the U6 backbone. Indeed, a yeast U6 gene has been found with an intron inserted adjacent to the conserved AGC triad, and this insertion presumably resulted from just this sort of abnormal attack by the branchpoint A on U6, rather than on the 5′-splice site.

Taken together, these results strongly suggest that the catalytic center of the spliceosome involves Mg^{2+} and three base-paired RNAs: U2 and U6 snRNAs, and the branchpoint part of the intron. Proteins may be involved in vivo, but they appear not to be required at the catalytic center, at least under these experimental conditions in vitro.

SUMMARY The spliceosomal complex (substrate, U2, U5, and U6) poised for the second step in splicing can be drawn in the same way as a group II intron at the same stage of splicing. Thus, the spliceosomal snRNPs seem to substitute for elements at the center of catalytic activity of the group II introns and probably have the spliceosome's catalytic activity. The catalytic center of the spliceosome appears to include Mg^{2+} and a base-paired complex of three RNAs: U2 and U6 snRNAs, and the branchpoint region of the intron. Protein-free fragments of these three RNAs can catalyze a reaction related to the first splicing step.

Spliceosome Assembly and Function

The spliceosome is composed of many components, proteins as well as RNAs. The components of the spliceosome assemble in a stepwise manner, and part of the order of assembly has been discovered. We call the assembly, function, and disassembly of the spliceosome the **spliceosome cycle.** In this section, we will discuss this cycle. We will see that by controlling the assembly of the spliceosome, a cell can regulate the quality and quantity of splicing and thereby regulate gene expression.

The Spliceosome Cycle When various research groups first isolated spliceosomes, they did not find U1 snRNP. This was surprising because U1 is clearly involved in base pairing to the 5′-splice site and is essential for splicing. The fact is that U1 *is* part of the spliceosome, but the methods used in the first spliceosome purifications were probably too harsh to retain U1. To emphasize the importance of this snRNP, Stephanie Ruby and John Abelson discovered in 1988 that U1 is the first snRNP to bind to the splicing precursor. These workers used a clever technique to measure spliceosome assembly. They immobilized a yeast pre-mRNA on agarose beads by hybridizing it to an "anchor RNA" joined to the beads through a biotin–avidin linkage. Then they added yeast nuclear extract for varying periods of time. They washed away unbound material, then extracted the RNAs, which they electrophoresed, blotted, and probed with radioactive probes for all spliceosomal snRNAs.

Figure 14.23 contains the results, which show that U1 was the first snRNP to bind to the splicing substrate. At the 2-min time point, it was the only snRNP whose association

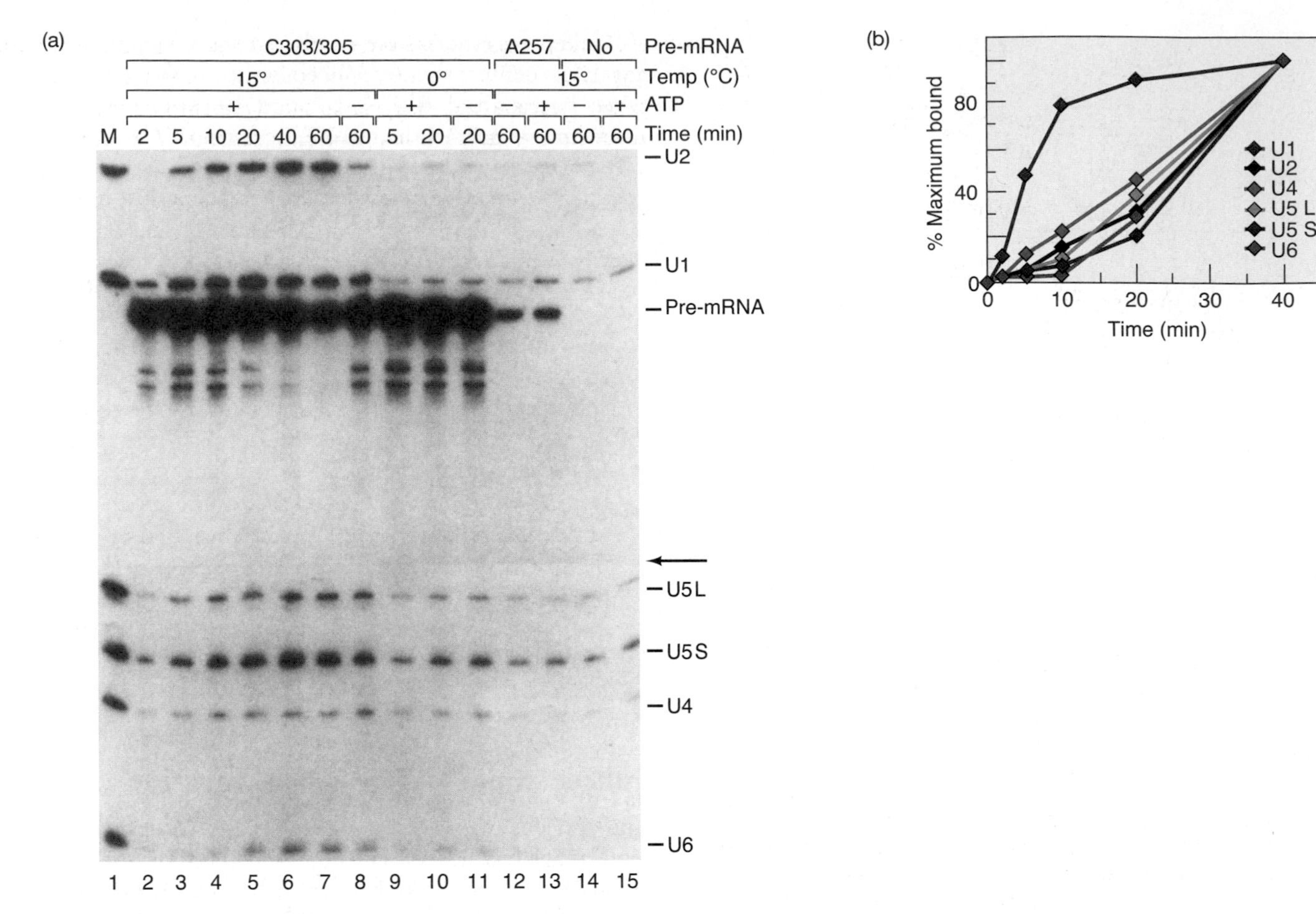

Figure 14.23 Kinetics of association of spliceosomal snRNPs with pre-mRNA. (a) Northern blot. Ruby and Abelson immobilized a yeast actin pre-mRNA to agarose beads by hybridizing it to an RNA (the anchor RNA) tethered through biotin–avidin links to the beads. They incubated this RNA–bead construct with yeast nuclear extract at either 15 or 0°C, in the presence or absence of ATP, and for 2–60 min, as indicated at top. The pre-mRNA was mutated in the 3′-splice site (C303/305), or in the conserved branchpoint (A257). The former would assemble a spliceosome, but the latter would not. The lanes marked "No" contained no pre-mRNA, only anchor RNA. After the incubation step, these workers washed away unbound material, extracted RNAs from the complexes, electrophoresed and blotted the RNAs, and hybridized the blots to probes for U1, U2, U4, U5, and U6. Two forms of U5 (U5 L and U5 S) were recognized. Lane 15, with no pre-mRNA, showed background binding of most snRNAs and served as a control for the other lanes. U1 bound first, then the other snRNPs bound. None of the snRNPs bound in significant amounts to the A257 mutant RNA. All snRNPs, including U1 and U4, remained bound after 60 min. **(b)** Graphic representation of amount of each snRNA bound to the complex as a function of time. U1 (red) clearly bound first, with all the others following later. (*Source:* Ruby, S.W. and J. Abelson, An early hierarchic role of U1 small nuclear ribonucleoprotein in spliceosome assembly. *Science* 242 (18 Nov 1988) f. 6a, p. 1032. Copyright © American Association for the Advancement of Science.)

with the pre-mRNA was above background; compare lane 2 with lane 15 in panel (a). Panel (a) also demonstrates that ATP was required for optimum binding of all snRNPs except U1. Figure 14.23b is a graph of the time course of association of all spliceosomal snRNPs with the substrate. U1 stands out from all the others as the first snRNP to join the spliceosome.

To probe more deeply into the order of spliceosome assembly, these workers inactivated either U1 or U2 by incubating extracts with DNA oligonucleotides complementary to key parts of these two snRNAs plus RNase H, then used the same spliceosome assembly assay as before. As we have seen, RNase H degrades the RNA part of an RNA–DNA hybrid, so the parts of the snRNAs in a hybrid with the DNA oligomers were degraded. The parts that hybridized to the pre-mRNA (the 5′-splice site and the branchpoint, respectively) were selected for degradation. The results in Figure 14.24 make two main points: (1) Inactivating U1 prevented U1 binding, as expected, and also prevented binding of all other snRNPs (compare lanes 2 and 4). (2) Inactivating U2 prevented U2 binding, as expected, and also prevented U5 binding. However, it did not prevent U1 binding (compare lanes 2 and 6). Taken together, these results indicate that U1 binds first, then U2 binds with the help of ATP, and then the rest of the snRNPs join the spliceosome.

As we will discuss later in this chapter, U6, once freed from association with U4, displaces U1 from its binding site at the 5′-splice site. We know from other experiments that, when U1 is displaced, it exits the spliceosome along with U4. This leaves an active spliceosome containing only U2, U5, and U6. Indeed, the replacement of U1 by U6

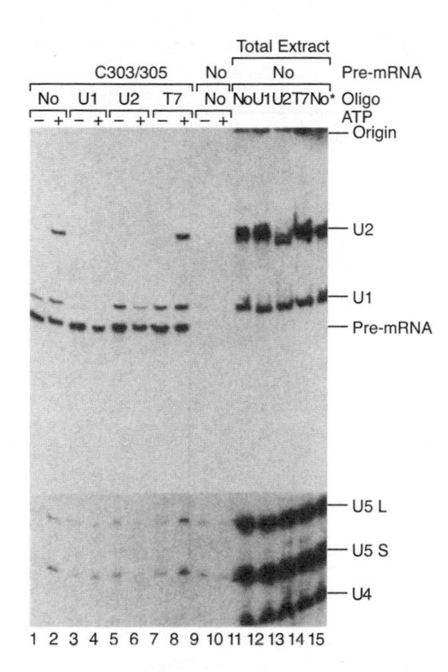

Figure 14.24 Effect of inactivation of U1 or U2 on assembly of the spliceosome. Ruby and Abelson inactivated either U1 or U2 by incubation with RNase H and a DNA oligonucleotide complementary to a key part of either snRNA. Lanes 11–15 show the patterns of labeled snRNAs in an extract after treating with RNase H and: no oligonucleotide (No); an anti-U1 oligonucleotide (U1); an anti-U2 oligonucleotide (U2); or an anti-phage T7 oligonucleotide (T7). The latter served as a second negative control. Treatment with RNase H and anti-U1 led to essentially complete conversion to a truncated form that electrophoresed slightly faster than the parent RNA. Treatment with RNase H and anti-U2 led to near-elimination of full-size U2, and appearance of a small amount of truncated U2. Lanes 1–10 show the results of spliceosome assembly experiments, as described in Figure 14.23, under the following conditions, as indicated at top: C303/305 pre-mRNA, or no pre-mRNA; extracts treated with RNase H and no oligonucleotide, anti-U1, anti-U2, or anti-T7 oligonucleotides; and with or without ATP. Inactivating U1 prevented binding of U1, U2, and U5. Inactivating U2 prevented binding of U2 and U5. (*Source:* Ruby, S.W. and J. Abelson, An early hierarchic role of U1 small nuclear ribonucleicprotein in spliceosome assembly. *Science* 242 (18 Nov 1988) f. 7, p. 1032. Copyright © American Association for the Advancement of Science.)

seems to be the event that activates the spliceosome to carry out the splicing reaction. Jonathan Staley and Christine Guthrie demonstrated in 1999 that activation can be blocked by changing the base sequence of the 5′-splice site so that it base-pairs even better with U1. This presumably made it harder for U6 to compete with U1 for binding to the 5′-splice site, and as a result, release of U1 and U4, as well as splicing, was inhibited. Conversely, with binding between U1 and the 5′-splice site held constant, enhancing the base pairing between U6 and the 5′-splice site allowed more activation (release of U1 and U4) and therefore more splicing. Staley and Guthrie went on to show that a protein known as **Prp28**, one of the proteins in U5 snRNP, appears to be required, along with ATP, for exchange of U1 for U6 at the 5′-splice site.

Figure 14.25 illustrates the yeast spliceosome cycle. The first complex to form, composed of splicing substrate plus U1 and perhaps other substances, is called the **commitment complex** (CC). As its name implies, the commitment complex is committed to splicing out the intron at which it assembles. Next, U2 joins, with help from ATP, to form the *A complex*. Next, U4–U6 and U5 join to form the *B1 complex*. U4 then dissociates from U6 to allow: (1) U6 to displace U1 from the 5′-splice site in an ATP-dependent reaction that activates the spliceosome, (2) U1 and U4 to exit the spliceosome, and (3) U6 to base-pair with U2. The activated spliceosome is also known as the *B2 complex*. ATP then provides the energy for the first splicing step, which separates the two exons and forms the lariat splicing intermediate, both held in the *C1 complex*. With energy from a second molecule of ATP, the second splicing step occurs, joining the two exons and removing the lariat-shaped intron, all held in the *C2 complex*. In the next step, the spliced, mature mRNA exits the complex, leaving the intron bound to the *I complex*. Finally, the I complex dissociates into its component snRNPs, which can be recycled into another splicing complex, and the lariat intermediate, which is debranched and degraded.

SUMMARY The spliceosome cycle includes the assembly, splicing activity, and disassembly of the spliceosome. Assembly begins with the binding of U1 to the splicing substrate to form a commitment complex. U2 is the next snRNP to join the complex, followed by the others. The binding of U2 requires ATP. When U6 dissociates from U4, it displaces U1 at the 5′-splice site. This ATP-dependent step activates the spliceosome and allows U1 and U4 to be released.

snRNP Structure All snRNPs have the same set of seven **Sm proteins.** These proteins are common targets of antibodies that appear in patients with systemic autoimmune diseases such as systemic lupus erythematosis, in which the body attacks its own tissues. Indeed, the Sm proteins were named in honor of the SLE patient in which they were discovered, Stephanie Smith. The Sm proteins bind to a common **Sm site.** (AAUUUGUGG) on the snRNAs. In addition to the Sm proteins, each snRNP has its own set of specific proteins. For example, U1 snRNP has three specific proteins, 70K, A, and C, with M_r's of 52, 31, and 17.5 kD, respectively.

Holger Stark and colleagues used single-particle electron cryomicroscopy to obtain a structure of the U1 snRNP at 10-Å resolution. This structure (Figure 14.26) shows that the Sm proteins form a doughnut-shaped structure with a hole through the middle, rather like a flattened funnel. The two largest U1-specific proteins, 70K and A, are attached to the Sm "doughnut" and also bind to stem-loop structures in the U1 snRNA. These protrusions were identified by performing electron microscopy on negative-stained U1 snRNPs lacking either the 70K or the A protein,

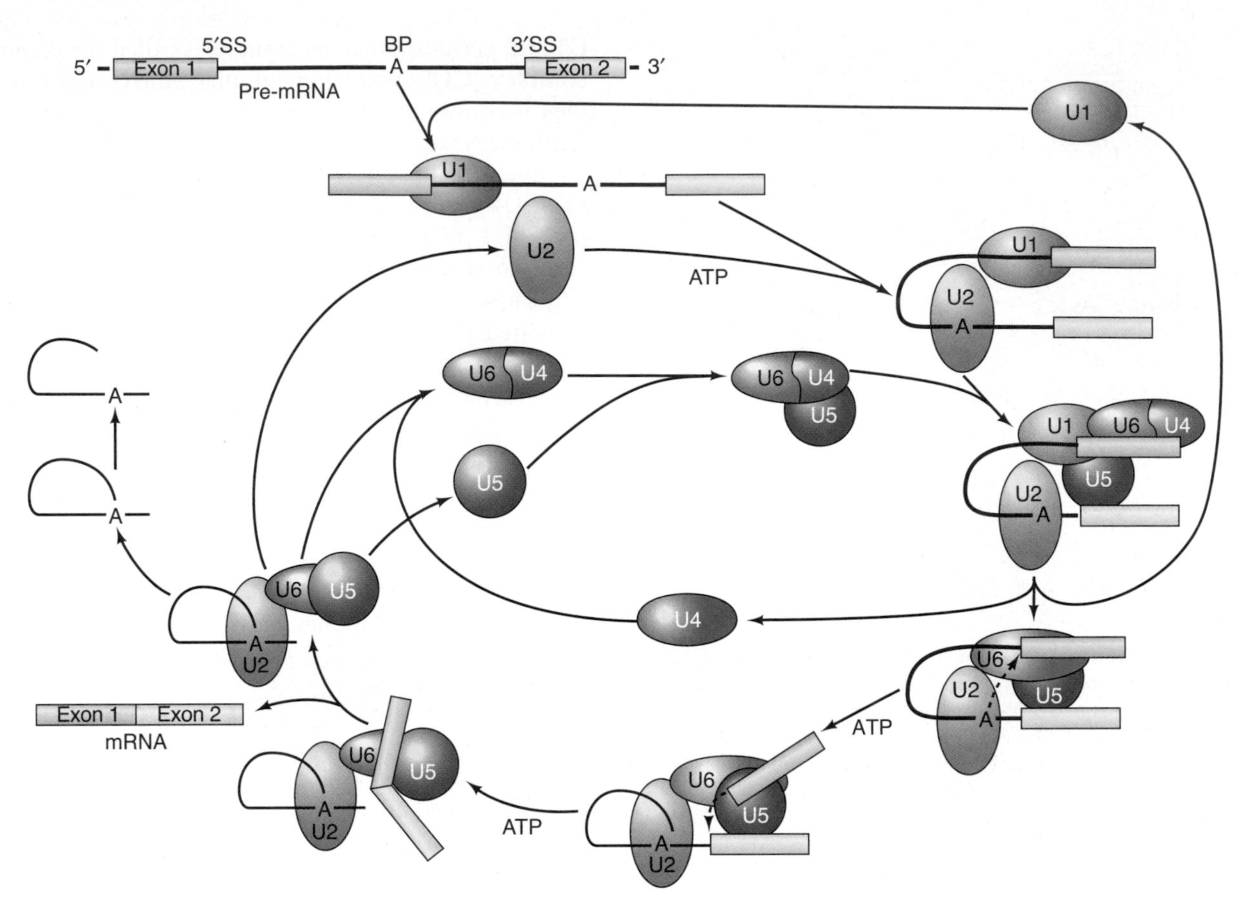

Figure 14.25 The spliceosome cycle. The text gives a description of the events in the cycle. (*Source:* Adapted from Sharp, P.A. Split genes and RNA splicing. *Cell* 77:811, 1994.)

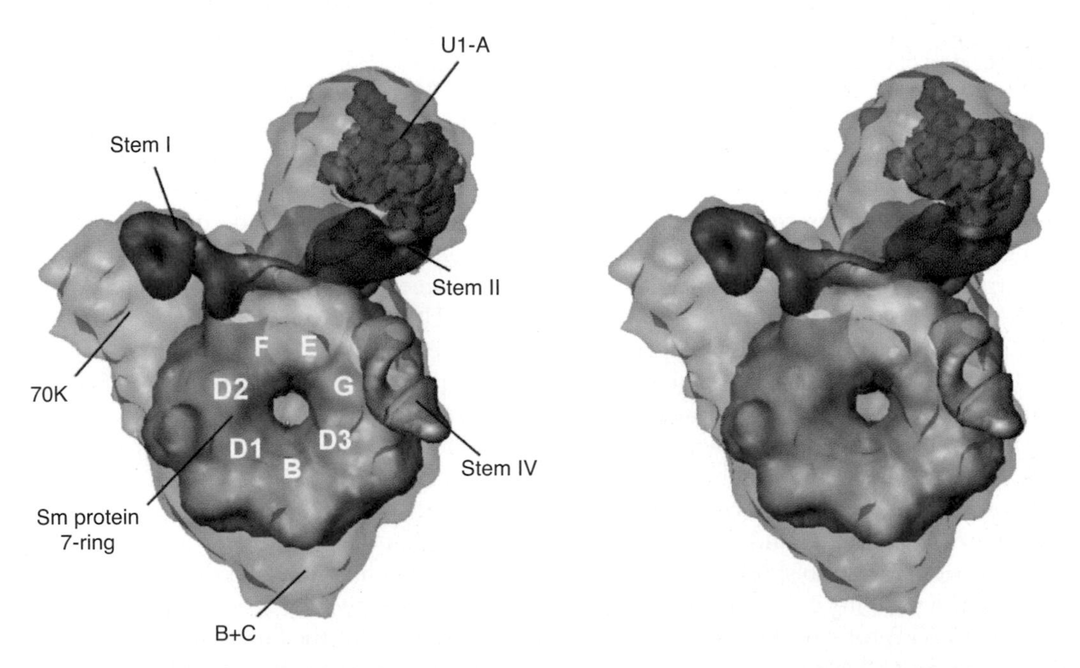

Figure 14.26 Structure of U1 snRNP. Stark and colleagues used single-particle electron cryomicroscopy to obtain this stereo model of the snRNP structure. The major protrusions, including the U1-A and 70K proteins, from the central Sm "doughnut" are labeled. Stems I, II, and IV are regions of the U1 snRNA. (*Source:* Reprinted with permission from *Nature* 409: from Stark et al., fig. 2, p. 540. © 2001 Macmillan Magazines Limited.)

and showing which protrusions were missing in each case, and therefore which protrusion corresponds to which protein.

The RNA with the Sm site is in a single-stranded region, and it could pass through the hole in the "doughnut." In fact, previous x-ray crystallography studies on subassemblies of the Sm proteins had predicted a ring-shaped structure with a hole lined with basic amino acid side chains. This basic character of the hole would facilitate binding to the Sm site in the U1 snRNA.

SUMMARY The five snRNPs that participate in splicing all contain a common set of seven Sm proteins and several other proteins that are specific to each snRNP. The structure of U1 snRNP reveals that the Sm proteins form a doughnut-shaped structure to which the other proteins are attached.

A Minor Spliceosome In the mid-1990s, a rare variant type of intron was discovered in metazoans (animals with distinct organs). The 5′-splice sites and branchpoint sequences in these variant introns are highly conserved and quite different from their relatively weakly conserved counterparts in the major introns. This finding raised the question: How can transcripts of these genes with variant introns be spliced if their sequences do not match those of the known snRNAs, U1 and U2, in particular? The answer is that metazoan cells contain a **minor spliceosome** with minor snRNAs known as: **U11,** which performs the same function as U1; **U12,** which performs the same function as U2; and **U4atac** and **U6atac,** which perform like U4 and U6, respectively. The minor spliceosome uses the same U5 snRNA as the major spliceosome.

The existence of this alternative splicing system serves as a check on the importance of base pairing between snRNAs and key sites in pre-mRNAs. In fact, the variant U11 snRNA base-pairs with the 5′-splice site and U12 snRNA can base-pair with the branchpoint in the variant pre-mRNAs. Furthermore, U4atac and U6atac can base-pair with each other in the same way that U4 and U6 do.

What about the proteins that associate with the minor snRNAs to make snRNPs? The first thing to notice is that U11 and U12 bind together in a single U11/U12 snRNP, in addition to individual U11 and U12 snRNPs. Some of the proteins associated with U11 and U12 in snRNPs are shared with the major snRNPs, but some are distinct. Among the shared proteins are the seven Sm proteins that are found in all the major snRNPs.

In 2007, Ferenc Müller and colleagues demonstrated that the major and minor spliceosomes are spatially separated: The major spliceosome resides in the nucleus, as we have seen, and the minor spliceosome is found, at least primarily, in the cytoplasm. Certain transcripts have some introns that are recognized by the major spliceosome, and others recognized by the minor spliceosome. Together, these findings give rise to the hypothesis that the major spliceosomal introns are removed in the nucleus, then the partially spliced pre-mRNA leaves the nucleus and its minor introns are removed in the cytoplasm. The physiological significance of this division of labor is not yet clear.

SUMMARY A minor class of introns with variant but highly conserved 5′-splice sites and branchpoints can be spliced with the help of a minor spliceosome containing a variant class of snRNAs, including U11, U12, U4atac, and U6atac. The minor spliceosomes are found at least primarily in the cytoplasm. Some pre-mRNAs appear to have some introns removed by the major spliceosome in the nucleus, and others removed by the minor spliceosome in the cytoplasm.

Commitment, Splice Site Selection, and Alternative Splicing

The snRNPs by themselves do not have enough specificity and affinity to bind exclusively and tightly at exon–intron boundaries and thus set the exons in a transcript off from the introns. Therefore, additional splicing factors are needed to help the snRNPs bind. Furthermore, some splicing factors are needed to bridge across introns and exons and thus define these RNA elements for splicing. In this section, we will see some examples of splicing factors and how they participate in commitment to splice at certain sites. Then we will see how other factors can shift splicing from one site to another.

Exon and Intron Definition In principle, the spliceosome can recognize either exons or introns in the splicing commitment process, presumably by assembling splicing factors to bridge across exons or introns, respectively. If exons are recognized, we call it **exon definition,** while if introns are recognized, it is **intron definition.** One can distinguish between the two possibilities by mutating an exon–intron boundary (splice site) and observing what happens to splicing (Figure 14.27). If exon definition operates, then mutating a splice site at the 3′-end of an exon should result in loss of recognition of that exon, and therefore splicing will skip that exon. That is, it will be spliced out along with the introns on either side (Figure 14.27a). On the other hand, if intron definition operates, then mutating a splice site at the end of an exon should result in loss of recognition of the intron that follows, so that intron will not be spliced out and will be included in the mature RNA along with the exons on either side (Figure 14.27b).

Applying this test, many investigators have shown that spliceosomes in higher eukaryotes, including vertebrates,

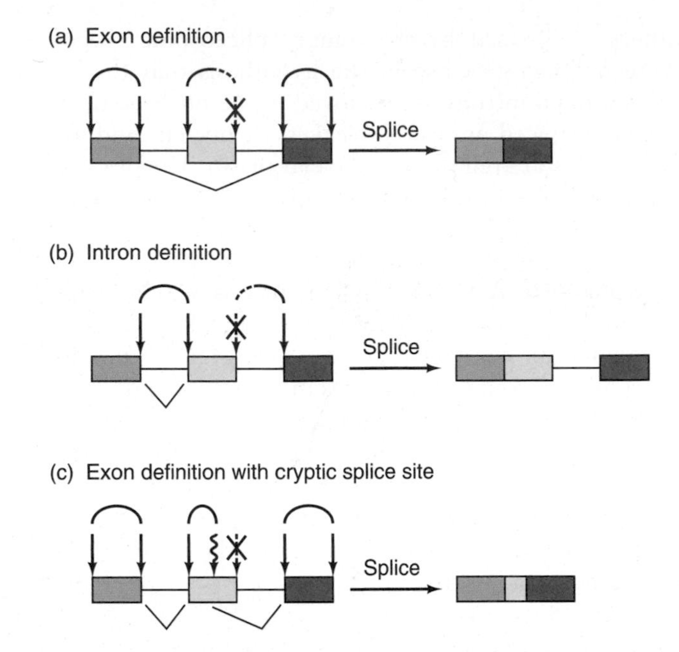

Figure 14.27 Analysis of exon vs. intron definition. (a) Exon definition. Exons are defined by factors bridging across the three exons, as indicated by the arcs above the arrows denoting the borders of the exons. The splice site at the 3′-end of the middle exon (yellow) is mutated, as indicated by the X, resulting in loss of recognition of this exon, indicated by the dashed arrow and dashed right end of the arc representing the definition of this exon. As a result, splicing skips this exon, and it is spliced out. **(b)** Intron definition. Introns are defined by factors as indicated by arcs as in (a). Again, the splice site at the 3′-end of the middle exon (5′-end of the second intron) is mutated. As a result the second intron is included in the mature RNA. **(c)** Exon definition with cryptic splice site. Again, the splice site at the 3′-end of the middle exon is mutated. This time, the spliceosome finds a cryptic splice site upstream in the middle exon and splices from there.

primarily use the exon definition scheme. Other lines of evidence also point in this direction. Sometimes, instead of skipping the exon that has a mutation in the splice site at its 3′-end, the spliceosome will splice from a cryptic (previously hidden) splice site, and this cryptic splice site is almost always within that exon (Figure 14.27c). This behavior is most easily explained if the exon is the unit that is being recognized: The spliceosome searches for a splice site in an exon, not in an intron. Moreover, we find that exons in higher eukaryotes tend to be small (usually less than 300 nt), while introns can be enormous—many thousands of nucleotides long. This makes sense if exon definition requires splicing factors to bridge across the exon: The exon cannot be too long for the factors to reach across. Indeed, if exons are artificially expanded beyond about 300 nt, they are usually skipped.

In contrast to higher eukaryotes, the fission yeast *Schistosaccharomyces pombe* appears to use intron definition in splicing. This hypothesis seems plausible in light of the fact that small introns are the rule in both fission and budding yeasts, while there seems to be no limit to exon size. This is just the opposite of the situation in higher eukaryotes, where exon definition predominates. Jo Ann Wise and her colleagues applied the tests outlined in Figure 14.27 to fission yeast and found that mutating one or both splice sites surrounding an intron resulted in intron retention, as in Figure 14.27b, rather than exon skipping, as in Figure 14.27a. Furthermore, when cryptic 5′-splice sites were used, they were in the intron, rather than the exon, arguing that the intron is the unit being recognized by the yeast spliceosome. Moreover, when the size of an intron was expanded, these cryptic sites could even compete with the normal 5′-splice site if they were closer to the 3′-splice site, even if they deviated strongly from the consensus sequence. This is consistent with a spliceosome searching for splice sites across an intron, and favoring those that are reasonably close together. Finally, there is a tiny exon within the *S. pombe cdc2* gene. This microexon would be skipped in vertebrates because it would be too small to be recognized by exon definition, but it was never skipped in *S. pombe*.

SUMMARY Splicing in a given organism typically uses either exon definition or intron definition. In exon definition, splicing factors appear to bridge across exons, while in intron definition, the factors bridge across introns.

Commitment Several splicing factors play critical roles in commitment, but Xiang-Dong Fu discovered in 1993 that, at least in certain circumstances, a single splicing factor can cause a committed complex to form. The splicing substrate he used was the human β-globin pre-mRNA; the splicing factor is called **SC35.** Fu's commitment assay worked as follows: He preincubated a labeled splicing substrate with purified SC35, then added a nuclear extract for 2 h to allow splicing to occur. Finally, he electrophoresed the labeled RNAs to see if spliced mRNA appeared.

Figure 14.28 shows the results. First, Fu determined that a 40-fold excess of an unlabeled RNA with a 5′-splice site could prevent splicing of the labeled β-globin pre-mRNA, presumably by competing for some splicing factor (compare lanes 1 and 4). An RNA containing a 3′-splice site was not as good a competitor (compare lanes 1 and 5). To show that SC35 was the limiting factor, Fu preincubated the labeled RNA with SC35, then added the nuclear extract plus competitor RNA. A comparison of lanes 4 and 6 shows that a preincubation with SC35 allowed splicing to occur even in the face of a challenge by competitor RNA. Therefore, SC35 can cause commitment. A similar experiment demonstrated that this commitment even survived a challenge by full-length human β-globin pre-mRNA as competitor. The SC35 used in these experiments was a cloned gene product made in insect cells, so it was unlikely to contain contaminating splicing factors. Thus, it seems that SC35 alone is sufficient to cause commitment. Further

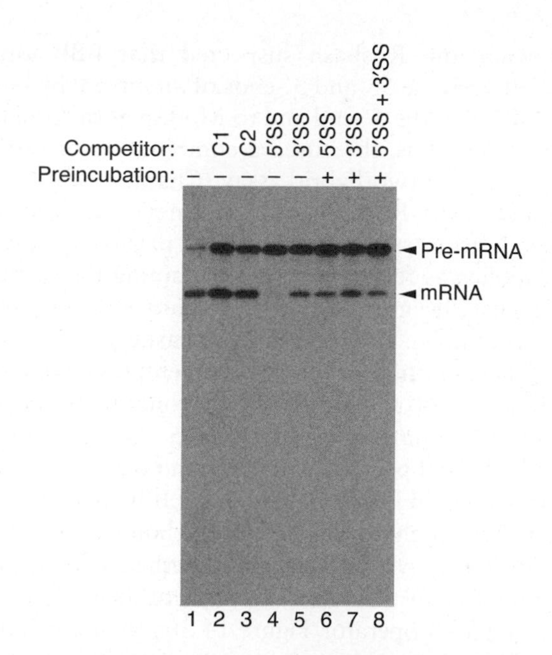

Figure 14.28 Commitment of the human β-globin pre-mRNA. Xiang-Dong Fu used a competition assay for commitment as follows: He incubated a labeled human β-globin pre-mRNA with or without SC35, as indicated at top (+ and −, respectively). Then he added a nuclear extract with or without a competitor RNA, as indicated at top. No competitor is indicated by (−). C1 and C2 are nonspecific RNAs that should not interfere with splicing. RNAs containing 5′- and 3′-splice sites are indicated as 5′SS and 3′SS, respectively. After allowing 2 h for splicing, Fu electrophoresed the labeled RNAs and autoradiographed the gel. The positions of pre-mRNA and mature mRNA are indicated at right. SC35 caused commitment. (*Source:* Fu, X.-D. Specific commitment of different pre-mRNAs to splicing by single SR proteins. *Nature* 365 (2 Sept 1993) f. 1, p. 83. Copyright © Macmillan Magazines Ltd.)

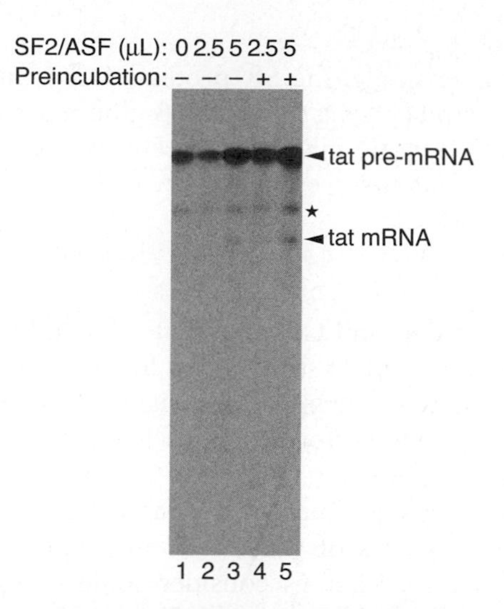

Figure 14.29 Commitment activities of several RNA-binding proteins: Effect of SF2/ASF on commitment with tat pre-mRNA. Fu ran the commitment assay with the concentrations of the SF2/ASF shown at top, and either without (lanes 1–3) or with (lanes 4 and 5) preincubation with the splicing factor. Comparing lanes 5 and 3 gives the clearest view of the effect of SF2/ASF. The star denotes a band resulting from artifactual tat pre-mRNA degradation. (*Source:* Fu, X.-D., Specific commitment of different pre-mRNA to splicing by single SR proteins. *Nature* 365 (2 Sept 1993) f. 3, p. 84. Copyright © Macmillan Magazines Ltd.)

experiments showed the conditions necessary for this commitment. It occured very rapidly (within 1 min) and even occurred at a reasonable level on ice or in the absence of ATP and Mg^{2+}.

SC35 is a member of a group of RNA-binding proteins called **SR proteins** because they contain domains that are rich in serine (S) and arginine (R). Therefore, Fu tested several other SR proteins and other RNA-binding proteins (hnRNP proteins) in the same commitment assay. SC35 worked best, followed by SF2 (which is also called ASF), then SRp55. SRp20 and hnRNP A1 showed no detectable activity, and hnRNP C1 and PTB (also called hnRNP 1) actually inhibited splicing activity. Thus, the commitment activity of SC35 is specific and does not derive from a general RNA-binding capability.

As further proof of the specificity of commitment, Fu tried a different splicing substrate, the tat pre-mRNA from human immunodeficiency virus (HIV), whose splicing had been reported to be stimulated by SF2/ASF. Figure 14.29 shows that SF2/ASF caused splicing commitment with this pre-mRNA. Fu also compared the commitment activities toward tat pre-mRNA of the same panel of RNA-binding proteins tested with the β-globin pre-mRNA. Only SF2/ASF could cause commitment with the tat pre-mRNA splicing substrate. Even SC35 had no effect. Thus, commitment with different pre-mRNAs requires different splicing factors.

We do not know yet exactly how commitment works, although it seems clear that one facet is the attraction of U1 to the commitment complex. James Manley and colleagues demonstrated this point with a gel mobility shift assay to measure formation of a stable complex between U1 snRNP and a labeled pre-mRNA. When they added U1 or SF2/ASF to the pre-mRNA separately, they got no complex formation. But when they added the two proteins together, they did get a complex. Furthermore, SF2/ASF appears to bind first: When they added the two proteins in sequence with a wash in between, they had to add SF2/ASF first in order to get a complex to form.

But if U1 snRNP binding to the 5′-splice site of a pre-mRNA depends on SF2/ASF, why did U1 appear to bind on its own to pre-mRNA in previous experiments? The reason is probably that these earlier experiments used crude nuclear extracts that naturally contained splicing factors. Complexes between these factors and the splicing substrates might have been detected if that is what the experimenters were looking for, but they were focusing on binding of snRNPs, not simple proteins.

SUMMARY Commitment to splice at a given site can be determined by an RNA-binding protein, which presumably binds to the splicing substrate and recruits other spliceosomal components, starting with

U1. For example, the SR proteins SC35 and SF2/ASF commit splicing on human β-globin pre-mRNA and HIV tat pre-mRNA, respectively. Part of this commitment involves attraction of U1, at least in some cases.

Bridging Proteins and Commitment An additional wrinkle to the commitment story is that SR proteins do not exist in yeast, in which many of the original spliceosome cycle experiments were performed. This finding suggested that commitment may work differently in yeast than in mammals. However, subsequent work has shown that the commitment complexes of yeast and mammals share many common features. Let us consider some of the proteins involved in bridging between the 5′- and 3′-ends of the intron in a yeast commitment complex, and compare these with their mammalian counterparts.

In 1993, Michael Rosbash and colleagues presented studies designed to find genes that encode proteins involved in the yeast commitment complex. Because U1 snRNA is a prominent and early participant in commitment, they decided to look for genes encoding proteins that interacted with U1 snRNA. To find these genes, they employed a **synthetic lethal screen** as follows: First, they introduced a temperature-sensitive mutation into the gene encoding U1 snRNA. The mutant U1 snRNA functioned at low temperature (30°C) but not at high temperature (37°C). They reasoned that the strain carrying this altered U1 snRNA would be especially sensitive to mutations in proteins that interact with snRNA. These second mutations could render the yeast strain inviable, even at the low temperature, so such mutations were called "Mutant-u-die," abbreviated Mud. Thus, the second mutations were not lethal in wild-type cells, but they became lethal in cells bearing the first mutation. In this sense, their lethality was "synthetic"—it depended on a conditional lethal mutation already created in the cell. One mutation discovered this way mapped to the *MUD2* gene, which encodes the protein **Mud2p.**

Subsequent work showed that the function of Mud2p depended on a natural sequence at the lariat branchpoint, near the 3′-end of the intron. This suggested that Mud2p interacted not only with U1 snRNA at the 5′-end of the intron, but with some other substance near the 3′-end of the intron. A major question remained: Does Mud2p by itself make these interactions with the 5′- and 3′-ends of the intron, or does it rely on other factors? In 1997, Nadja Abovich and Rosbash used another synthetic lethal screen to answer this question. They introduced a mutation into the *MUD2* gene, then looked for second mutations that would kill the *MUD2* mutant cells, but not wild-type cells. One gene identified by this screen is called *MSL-5* (Mud synthetic lethal-5). It encodes a protein originally named Msl5p, but renamed **BBP (branchpoint bridging protein)** once its binding properties were clarified.

Abovich and Rosbash suspected that BBP forms a bridge between the 5′- and 3′-ends of an intron, by binding to U1 snRNP at the 5′-end and to Mud2p at the 3′-end. To test this hypothesis, they used a combination of methods, including a yeast two-hybrid assay (Chapter 5).

Abovich and Rosbash already knew which proteins were likely to interact, so they made plasmids expressing these proteins as fusion proteins containing the protein of interest plus either a DNA-binding domain or a transcription-activating domain. They transfected yeast cells with various pairs of these plasmids. In one experiment, for example, one plasmid encoded a hybrid protein containing the LexA DNA-binding domain linked to BBP; the other plasmid encoded a hybrid protein containing the B42 transcription-activating domain linked to Mud2p. If BBP and Mud2p interact in the cell, that brings the DNA-binding domain and transcription activating domain together, constituting a transcription activator that can activate the *lacZ* reporter gene near a *lexA* operator. Figure 14.30a (first column, first

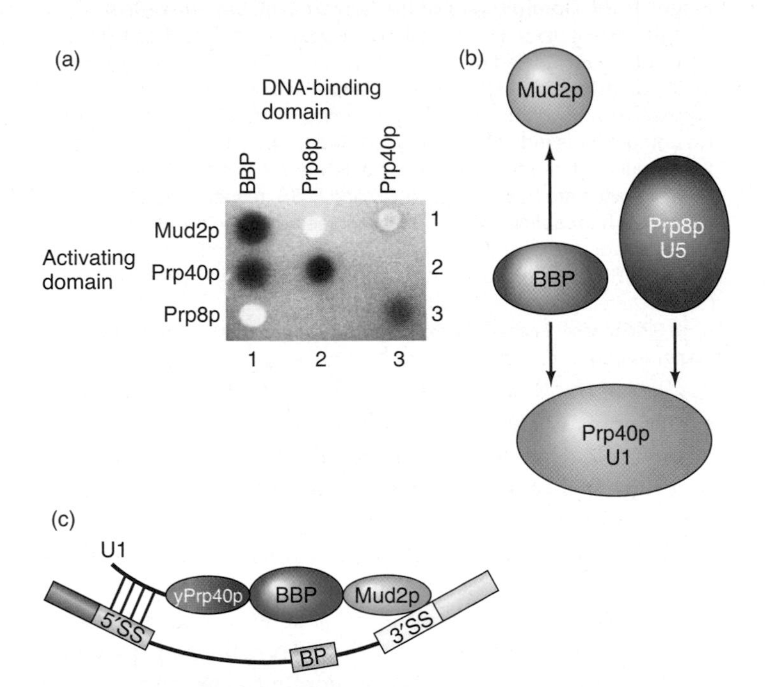

Figure 14.30 Yeast two-hybrid assays for interactions between BBP and other proteins. (a) Results of the assays. The proteins linked to the DNA-binding domain are listed at top, and the proteins linked to the transcription-activating domain are listed at left. Abovich and Rosbash spotted cells bearing the indicated pairs of plasmids on an indicator plate containing X-gal to measure the activation of the *lacZ* reporter gene. A dark stain indicates activation. For example, the darkly stained yeast cells in column 1, rows 1 and 2, indicated interaction between BBP and Mud2p, and between BBP and Prp40p (a component of U1 snRNP). The other positive reactions indicated interactions between Prp40p and Prp8p (a component of U5 snRNP). **(b)** Summary of results. This schematic shows the protein–protein interactions revealed by the yeast two-hybrid assay results in panel (a). **(c)** Summary of intron-bridging protein–protein interactions in yeast. 5′SS is the 5′-splicing signal; BP is the branchpoint, and 3′SS is the 3′-splicing signal. (*Source:* Abovich N. and M. Rosbash, Cross-intron bridging interactions in the yeast commitment complex are conserved in mammals. *Cell* 89 (2 May 1997) f. 5 and 8, pp. 406 and 409. Reprinted by permission of Elsevier Science.)

row) shows that cells bearing these two plasmids experienced activation of the *lacZ* gene, as demonstrated by the dark stain on the X-gal indicator plate. Thus, BBP bound to Mud2p in this assay. Figure 14.30a (first column, second row) shows that BBP also bound to Prp40p, a polypeptide component of U1 snRNP. On the other hand, Mud2p did not bind to Prp40p. Thus, BBP serves as a bridge between Mud2p, presumably bound at the branchpoint near the 3′-end of the intron, and to U1 snRNP at the 5′-end of the intron. In this way, BBP could help define the intron and help bring the two ends of the intron together for splicing. Abovich and Rosbash included Prp8p in this experiment as a positive control because they already knew it bound to Prp40p. Figure 14.30b summarizes the protein–protein interactions suggested by this yeast two-hybrid assay, and Figure 14.30c illustrates the bridging function of BBP. Abovich and Rosbash confirmed these interactions by showing that BBP tethered to Sepharose beads coprecipitated both Prp40p and Mud2p.

Abovich and Rosbash noted that the yeast Mud2p and BBP proteins resemble two mammalian proteins called $U2AF^{65}$ and **SF1**, respectively. If these two mammalian proteins behave like their yeast counterparts, they should bind to each other. To test this hypothesis, these workers used the same yeast two-hybrid assay and coprecipitation procedure and found that $U2AF^{65}$ and SF1 do indeed interact. Figure 14.30c SF1, the mammalian counterpart of yeast BBP, by interacting with $U2AF^{65}$, presumably forms bridges. However, because mammals primarily use exon definition, this bridging is likely to be across exons, rather than introns. **$U2AF^{65}$** is a 65-kD protein that is part of the splicing factor **U2AF (U2-associated factor)**, which also contains a 35-kD protein known as **$U2AF^{35}$**. The large subunit, $U2AF^{65}$, binds to the pyrimidine tract near the 3′-splice site, and Michael Green and colleagues have shown by cross-linking experiments that the small subunit binds to the AG at the 3′-splice site.

Further work by Rosbash's group demonstrated that BBP also recognizes the branchpoint UACUACC sequence and binds at (or very close to) this sequence in the commitment complex. Thus, BBP is also an RNA-binding protein, and the BBP now also stands for "branchpoint binding protein."

SUMMARY In the yeast commitment complex, the branchpoint bridging protein (BBP) binds to a U1 snRNP protein at the 5′-end of the intron, and to Mud2p near the 3′-end of the intron. It also binds to the RNA near the 3′-end of the intron. Thus, it bridges the intron and could play a role in defining the intron prior to splicing. The mammalian BBP counterpart, SF1, might serve a similar bridging function in the mammalian commitment complex, but its role is probably in exon definition.

3′-Splice Site Selection During step 2 of the splicing process, the 3′-hydroxyl group of exon 1 attacks the phosphodiester bond linking an AG at the end of the intron to the first nucleotide of exon 2. This AG is ideally between 18 and 40 nt downstream of the branchpoint. AG's that are closer to the branchpoint are usually skipped. What determines which AG is used? We have already seen that $U2AF^{35}$ recognizes the AG at the 3′-splice site. In addition, Robin Reed and colleagues have found that a **splicing factor** known as **Slu7** is required for selection of the proper AG. Without Slu7, the correct AG is not used, but an incorrect AG may come into play.

Katrin Chua and Reed immunodepleted a HeLa cell extract of Slu7 by treating the extract with an anti-Slu7 antiserum linked to Sepharose beads. Separation of the extract from the beads leaves an extract depleted of Slu7. They also prepared a mock-depleted extract by treating the extract with Sepharose beads linked to preimmune serum, which contained no anti-Slu7 antibodies. Then they tested these extracts for ability to splice a labeled model pre-mRNA made from part of the adenovirus major late transcript that was modified so it contained a single AG located 23 nt downstream of the branchpoint sequence (Figure 14.31). After incubating the model splicing substrate

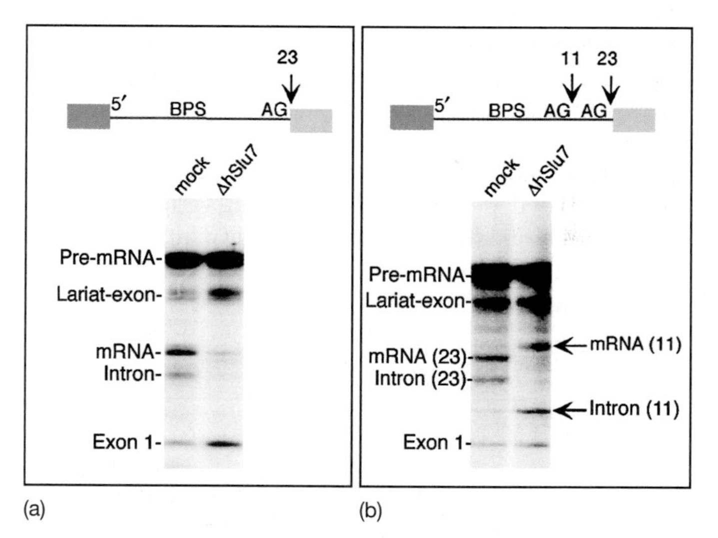

Figure 14.31 Slu7 is required for splicing to the correct AG at the 3′-splice site. Chua and Reed tested HeLa cell extracts that had been mock-depleted (mock) or immunodepleted with an anti-Slu7 antiserum (ΔhSlu7) for selection of the AG at the 3′-splice site. The labeled splicing substrate was modeled on the first two exons and first intron from the adenovirus major late pre-mRNA. After the splicing reaction, Chua and Reed electrophoresed the products and detected them by autoradiography. The positions of the substrates and products are indicated at left in each panel. **(a)** The splicing substrate contained a single AG 23 nt downstream of the branchpoint sequence (BPS). Splicing to the normal AG was suppressed in the extract lacking Slu7. **(b)** The splicing substrate contained two AG sequences downstream of the branchpoint sequence, one 11 nt downstream and the other 23 nt downstream. Splicing shifted to the AG 11 nt downstream in the extract lacking Slu7, a splice site that was scarcely used in the mock-depleted extract. (*Source:* (*photos*) Chua, K., and Reed, R. The RNA splicing factor hSlu7 is required for correct 3′-splice-site choice. *Nature* 402 (11 Nov 1999) f. 1, p. 208. © Macmillan Magazines Ltd.)

with an extract, Chua and Reed tested for splicing by electrophoresing the products.

Figure 14.31a shows the results with the "natural" substrate. The mock-depleted extract completed steps 1 and 2 of the splicing reaction, yielding mature mRNA, the intron, and relatively little of the unspliced exons. On the other hand, the extract depleted of human Slu7 (ΔhSlu7) yielded almost no mature mRNA or intron, but abundant exon 1 and lariat-exon 2. Thus, step 2 of splicing was blocked. This could mean that Slu7 is necessary for recognizing the normal AG at the 3′-splice site.

Chua and Reed next asked what would happen if they inserted an extra AG only 11 nt downstream of the branchpoint sequence. Figure 14.31b shows that the mock-depleted extract yielded mRNA spliced at the natural AG 23 nt downstream of the branchpoint sequence, but very little mRNA spliced at the AG unnaturally close to the branchpoint. By contrast, the extract depleted in Slu7 spliced most of the mRNA at the unnatural AG and very little at the natural AG. In further experiments, the depleted extract exhibited the same aberrant behavior when the two AGs were at 11 and 18 nt or 9 and 23 nt downstream of the branchpoint. Furthermore, it spliced to an incorrect AG placed downstream, as well as upstream, of the proper one, but not to the proper one itself. (In all cases, the incorrect AG had to be within about 30 nt of the branchpoint to be a target for aberrant splicing.) Thus, not only is Slu7 needed to recognize the correct splice site AG, but splicing to the correct splice site AG seems to be specifically suppressed in the absence of Slu7.

What accounts for this aberrant 3′-splice site selection? Chua and Reed purified spliceosomes at various stages of splicing and found that spliceosomes formed in Slu7-depleted extracts lacked exon 1, at least under certain conditions. Therefore, they concluded that exon 1 was held only loosely in spliceosomes from Slu7-depleted extracts. The loosely bound exon 1 was incapable of splicing to the correct AG, possibly because that AG was sequestered somehow in the active site of the spliceosome. Because it could not access the correct AG, this loosely bound exon 1 spliced to another nearby AG.

SUMMARY The splicing factor Slu7 is required for correct 3′-splice site selection. In its absence, splicing to the correct 3′-splice site AG is specifically suppressed and splicing to aberrant AG's within about 30 nt of the branchpoint is activated. U2AF is also required for 3′-splice site recognition.

Role of the RNA Polymerase II CTD As mentioned at the beginning of this chapter, splicing, as well as capping and polyadenylation, appear to be coordinated by the CTD of Rpb1, the largest subunit of RNA polymerase II. How do we know that the CTD plays a role in splicing? In 2000, Changqing Zeng and Susan Berget performed an in vitro splicing reaction using the labeled splicing substrate illustrated at the top of Figure 14.32b. This substrate contained two complete exons separated by an intron. To this reaction, Zeng and Berget added a recombinant CTD linked to glutathione-*S*-transferase (GST), or simply recombinant GST.

Figure 14.32a shows that the CTD–GST fusion protein stimulated splicing, as measured by production of the lariat

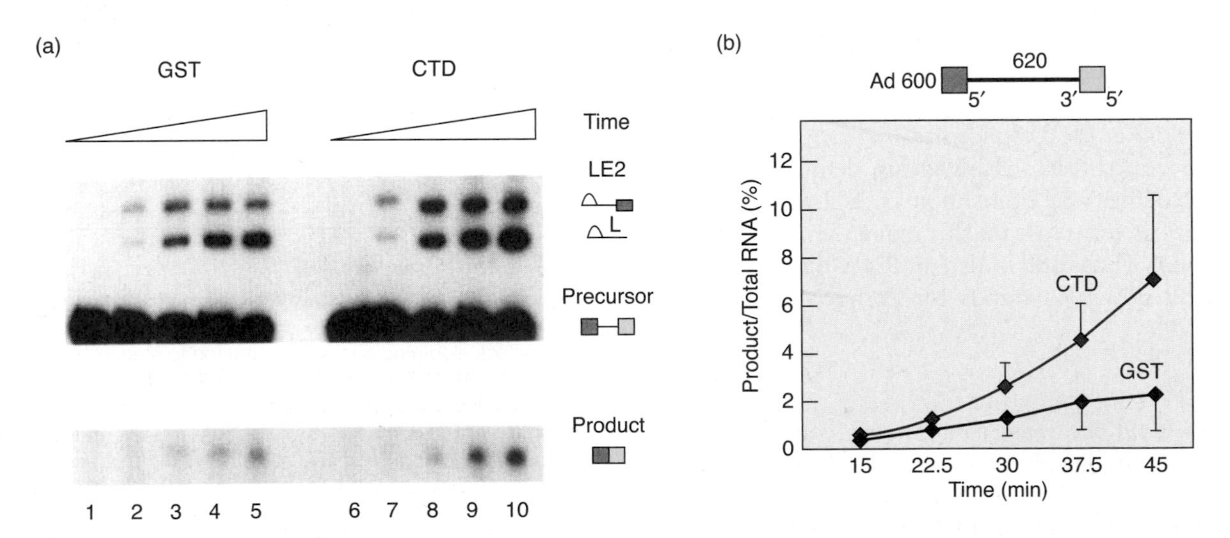

Figure 14.32 CTD–GST stimulates splicing in vitro. (a) Splicing reactions. Zeng and Berget incubated a ^{32}P-labeled splicing substrate (Ad600), illustrated at the top of panel **(b)** with a splicing extract supplemented with GST (left), or CTD–GST (right). The wedges at top indicate increasing time of incubation. Then they electrophoresed the extracts to separate the precursor, intermediate, and products. The positions of these RNA species are indicated at left, with drawings to aid in identification. The CTD stimulated the reaction three- to fivefold. **(b)** Graphical representation of results. The amount of product as a percent of total RNA is plotted against time in min. Blue, reaction with GST alone added; red, reaction with CTD–GST added. *(Source:* Copyright © American Society for Microbiology, *Molecular and Cellular Biology* vol. 20, No. 21, p. 8294, fig. 1, 2000.)

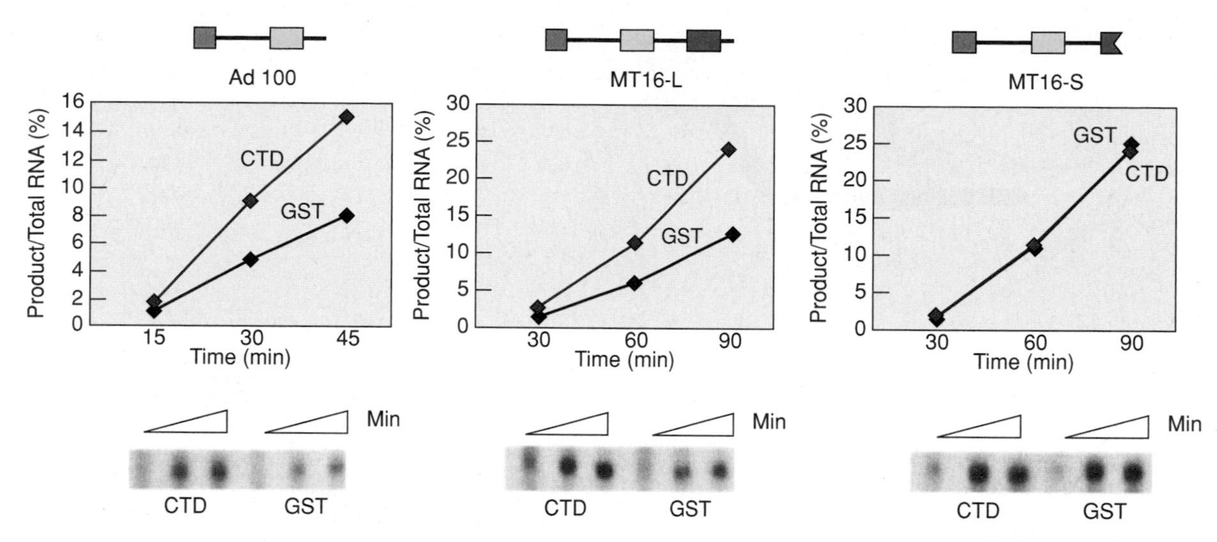

Figure 14.33 Effect of CTD–GST on splicing using exon or intron definition. Zeng and Berget carried out splicing assays as described in Figure 14.32 with the three labeled substrates illustrated at top. The first two contain complete exons and can be spliced by the exon definition pathway. The last, MT16-S, has an incomplete exon and can be spliced by the intron definition pathway. The gel electrophoresis results are presented at bottom, and these results are graphed above. Blue, reactions with GST alone added; red, reactions with CTD–GST added. (*Source:* Copyright © American Society for Microbiology, *Molecular and Cellular Biology* vol. 20, No. 21, p. 8294, fig. 4, 2000.)

exon intermediate, the lariat intron and spliced exon products. The degree of stimulation by CTD–GST was about 3- to 5-fold, compared with GST alone, which should have no effect. Note that the timing of appearance of the splicing intermediate and products was not accelerated, but the amount of intermediate and products appearing at each time was increased. Thus, the CTD appears to help recruit the splicing substrate to active spliceosomes.

It is interesting that CTD–GST did not stimulate splicing of a substrate containing an incomplete exon. Figure 14.33 illustrates this phenomenon. The substrates Ad 100 and MT16-L contain only complete exons, and CTD–GST stimulated their splicing. But the substrate MT16-S has two complete exons and one incomplete exon, and CTD–GST had no effect on its splicing. In a similar experiment, splicing of a substrate with one complete and one incomplete exon was not stimulated by CTD.

Previous experiments had shown that the CTD could bind to snRNPs and SR proteins, so Zeng and Berget proposed that the CTD facilitates splicing by assembling splicing factors on exons as the latter are synthesized by RNA polymerase (Figure 14.34). But why does this work only in a substrate with all complete exons? Zeng and Berget interpreted these results in terms of exon definition, which we discussed earlier in this chapter. For exon definition to work, all the exons must be complete; that way, there is no ambiguity about what is an exon and what is not. If there is ambiguity about one or more exons, intron definition can still work. If this hypothesis is correct, splicing by intron definition is apparently not facilitated by the CTD.

Further support for the hypothesis that the CTD plays a role in exon definition came from an immunodepletion experiment. Zeng and Berget immunodepleted an extract of RNA polymerase II and found that partial removal of the polymerase depressed splicing of a substrate that depended on exon definition, but had little effect on a substrate that could use intron definition. Adding CTD back to the depleted extract restored splicing activity with the exon definition-dependent substrate.

SUMMARY The CTD of the Rpb1 subunit of RNA polymerase II stimulates splicing of substrates that use exon definition, but not those that use intron definition, to prepare the substrate for splicing. The CTD binds to splicing factors and could therefore assemble the factors at the ends of exons to set them off for splicing.

Alternative Splicing Our previous discussion of commitment leads naturally to another important topic: **alternative splicing.** Many eukaryotic pre-mRNAs can be spliced in more than one way, leading to two or more alternative mRNAs that encode different proteins. In humans, about 75% of transcripts are subject to alternative splicing. The switch from one alternative splicing pattern to another undoubtedly involves commitment, and we will return to this theme at the end of this section.

Leroy Hood and colleagues discovered the first example of alternative splicing, the mouse immunoglobulin μ heavy-chain gene, in 1980. The μ heavy chain exists in two forms, a secreted form (μ_s), and a membrane-bound form (μ_m). The difference in the two proteins lies at the carboxyl terminus, where the membrane-bound form has a hydrophobic region that anchors it to the membrane, and the

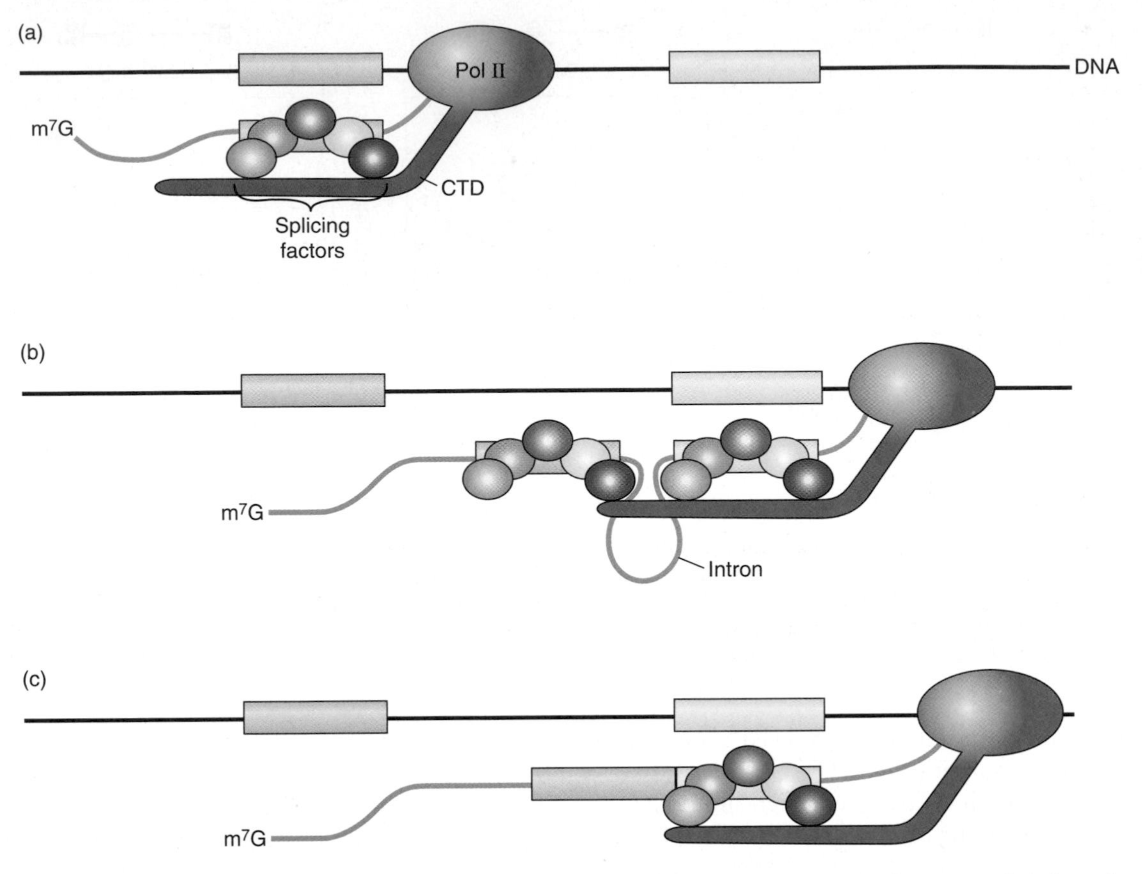

Figure 14.34 Model for participation of CTD in exon definition. **(a)** The polymerase has transcribed the first exon, and the CTD mediates the assembly of splicing factors at either end of the exon in the pre-mRNA, thus defining the exon. **(b)** The polymerase has transcribed the second exon, and the CTD mediates the definition of this exon in the same way as the first. The CTD also positions the two exons close to each other so they are ready to be spliced together. **(c)** The two exons have been spliced together, as the polymerase continues to transcribe the gene. (*Source:* Adapted from Zeng, C. and S. Berget, Participation of the C-terminal domain of RNA polymerase II in exon definition during pre-mRNA splicing. *Molecular and Cellular Biology* 20 (2000) p. 8299, F.9.)

secreted form lacks this membrane anchor. Using hybridization, Hood and colleagues found that the two proteins are encoded in two separate mRNAs that are identical at their 5′-ends, but differ at their 3′-ends. When these workers cloned the germline gene for the constant region of the μ heavy chain (the C_μ gene), they noticed that it encoded both the secreted and membrane-bound 3′-regions, and each of these was contained in a separate exon. Thus, two different modes of splicing of a common pre-mRNA could give two alternative mature mRNAs encoding μ_s and μ_m, as illustrated in Figure 14.35. In this way, alternative splicing can determine the nature of the protein product of a gene and therefore control gene expression.

Alternative splicing can have profound biological effects. One good example is the sex determination system in *Drosophila*. Sex in the fruit fly is determined by a pathway that includes alternative splicing of the pre-mRNAs from three different genes: *Sex lethal (Sxl)*; *transformer (tra)*; and *doublesex (dsx)*. Figure 14.36 illustrates this alternative splicing pattern. Males splice the transcripts of these genes in one way, which leads to male development; females splice them in a different way, which leads to development of a female.

Moreover, these genes function in a cascade as follows: Female-specific splicing of *Sxl* transcripts gives an active product that reinforces female-specific splicing of *Sxl* transcripts and also causes female-specific splicing of *tra* transcripts, which leads to an active *tra* product. (Actually, about half the *tra* transcripts are spliced according to the male pattern even in females, but this simply yields inactive product, so the female pattern is dominant.) The active *tra* product, together with the product of another gene, *tra-2*, causes female-specific splicing of transcripts of the *dsx* gene. This female-specific *dsx* product inactivates male-specific genes and therefore leads to female development.

By contrast, male-specific splicing of *Sxl* transcripts gives an inactive product because it includes an exon with a stop codon. This permits default (male-specific) splicing of *tra* transcripts, which again leads to an inactive product because of the inclusion of an exon with a stop codon. With no *tra* product, the developing cells splice the *dsx* transcripts according to the default, male-specific pattern, yielding a product that inactivates female-specific genes and therefore leads to development of a male.

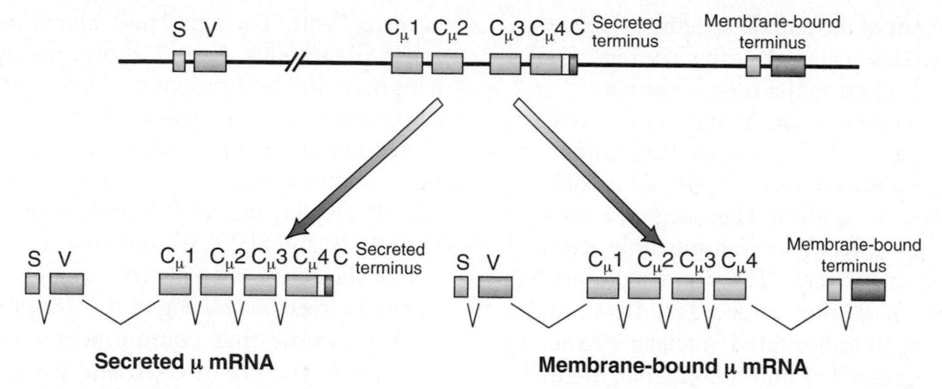

Figure 14.35 Alternative splicing pattern in the mouse immunoglobulin μ heavy-chain gene. The structure of the gene is shown at top. The boxes represent exons: The S exon (pink) encodes the signal peptide that allows the protein product to be exported to the plasma membrane, or secreted from the cell. The V exons (orange) encode the variable region of the protein. The C exons (blue) encode the constant region of the protein. Near the end of the fourth constant exon ($C_\mu 4$) lies the coding region (yellow) for the secreted terminus of the μ_s protein. This is followed by a short untranslated region (red), then by a long intron, then by two exons. The first of these (green) encodes the membrane anchor region of the μ_m mRNA. The second (red) is the untranslated region found at the end of the μ_m mRNA. The arrows pointing left and right indicate the splicing patterns that produce the secreted and membrane versions of the μ heavy chain (μ_s and μ_m, respectively). (*Source:* Adapted from Early P., J. Rogers, M. Davis, K. Calame, M. Bond, R. Wall, and L. Hood, Two mRNAs can be produced from a single immunoglobulin γ gene by alternative RNA processing pathways. *Cell* 20:318, 1980.)

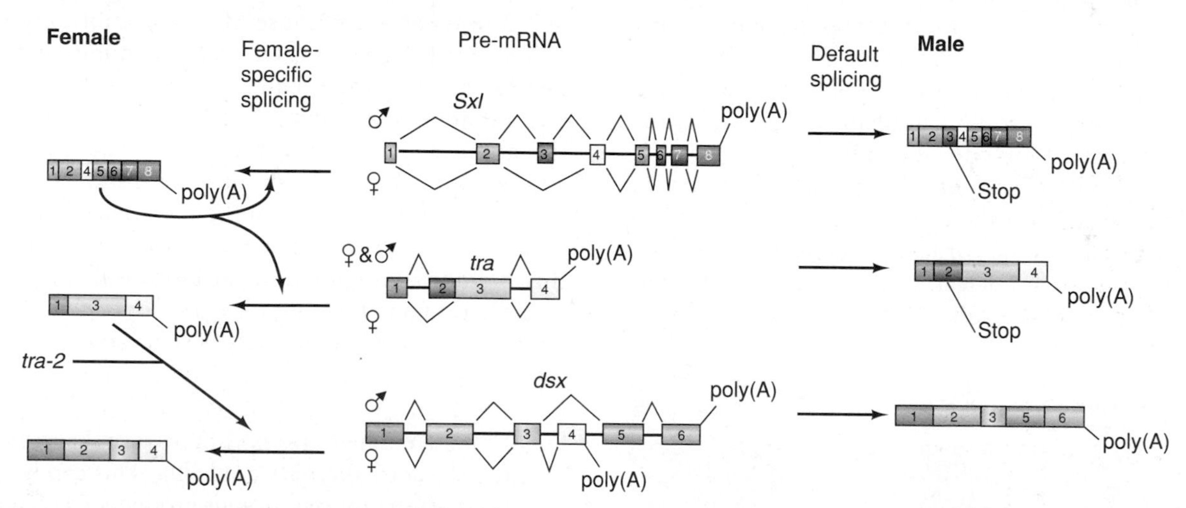

Figure 14.36 Alternative splicing cascade in *Drosophila* sex determination. The structures of the *Sxl, tra,* and *dsx* pre-mRNAs common to both males and females are shown at center, with the female-specific splicing pattern indicated below each, and the male-specific pattern above. Thus, female-specific splicing of the *Sxl* pre-mRNA includes exons 1, 2, and 4–8, whereas male-specific (default) splicing of the same transcript includes all exons (1–8), including exon 3, which has a stop codon. This means that male-specific splicing of this transcript gives a shortened, inactive protein product. Similarly, female-specific splicing of the *tra* pre-mRNA includes exons 1, 3, and 4, leading to an active protein product, whereas male-specific splicing of the same transcript includes all four exons, including exon 2 with a stop codon. Again, the male protein is inactive. The long arrows at far left indicate the positive effects of gene products on splicing. That is, the female *Sxl* product causes female-specific splicing of both *Sxl* and *tra* pre-mRNAs, and the female *tra* product, together with the *tra-2* product, causes female-specific splicing of *dsx* transcripts. (*Source:* Adapted from Baker, B.S. Sex in flies: The spice of life. *Nature* 340:523, 1989.)

How is this alternative splicing controlled? Knowing what we do about splicing commitment, we might guess that RNA-binding splicing factors would be involved. Indeed, because the products of *Sxl* and *tra* can determine which splice sites will be used in *tra* and *dsx* transcripts, respectively, we would predict that these proteins are splicing factors that cause commitment to the female-specific pattern of splicing. In accord with this hypothesis, the products of both *Sxl* and *tra* are SR proteins.

To further elucidate the mechanism of splice site selection, Tom Maniatis and his colleagues focused on the female-specific splicing of *dsx* pre-mRNA by Tra and Tra-2 (the products of *tra* and *tra-2,* respectively). They discovered that these two proteins act by binding to a regulatory region about 300 nt downstream of the female-specific 3′-splice site in the *dsx* pre-mRNA. This region contains six repeats of a 13-nt sequence, so it is known as the *repeat element.*

Tra and Tra-2 are necessary for commitment to female-specific splicing of *dsx* pre-mRNA, but are they sufficient? To find out, Ming Tian and Maniatis developed a commitment assay that worked as follows: They began with a labeled, shortened *dsx* pre-mRNA containing only exons 3 and 4, with the intron in between. This model pre-mRNA can be spliced in vitro. Then they added Tra, Tra-2, and a **micrococcal nuclease (MNase)**-treated nuclear extract to supply any proteins, besides Tra and Tra-2, that might be needed for commitment. The MNase degrades snRNAs, but leaves proteins intact. Then the experimenters added an untreated nuclear extract, along with an excess of competitor RNA. If commitment occurred during the preincubation, the labeled pre-mRNA would be spliced. If not, the competitor RNA would block splicing. To assay for splicing, Tian and Maniatis electrophoresed the RNAs and detected RNA species by autoradiography. They found that Tra and Tra-2 alone, without the MNase-treated extract, were not enough to cause commitment. However, something in the extract could complement these proteins, resulting in commitment.

To identify the other required factors, Tian and Maniatis first did a bulk purification of SR proteins and found that this SR protein mixture could complement Tra and Tra-2. Next, they obtained four pure recombinant SR proteins, and highly purified, nonrecombinant preparations of two others and tested them in the commitment assay with Tra and Tra-2. In this assay, the purified proteins took the place of the MNase-treated nuclear extract in the previous experiment. Figure 14.37, lane 1, shows that no splicing occurred with Tra and Tra-2 alone, in the absence of any other SR proteins. Lane 2 shows that a mixture of SR proteins prepared by ammonium sulfate (AS) precipitation could complement Tra and Tra-2. The other lanes show the effects of recombinant and highly purified SR proteins. Among these, some worked, and some did not. In particular, SC35, SRp40, SRp55, and SRp75 could complement Tra and Tra-2, but SRp20 and SF2/ASF could not. Thus, Tra, Tra-2, plus any one of the active proteins was enough to cause commitment to female-specific splicing of the *dsx* pre-mRNA.

We assume that commitment involves binding of SR proteins to the pre-mRNA, and we already know that Tra and Tra-2 bind to the repeat element, but do the other SR proteins also bind there? To find out, Tian and Maniatis performed affinity chromatography with a resin linked to an RNA containing the repeat element. After eluting the proteins from this RNA, they electrophoresed and immunoblotted (Western blotted) them. Finally, they probed the immunoblot in three separate experiments with antibodies against Tra, Tra-2, and SR proteins in general. They detected Tra and Tra-2 as expected, and also found large amounts of SRp40 and a band that could contain either SF2/ASF or SC35. Because SC35, but not SF2/ASF, could complement Tra and Tra-2 in the commitment assay, we assume that this latter band corresponds to SC35. No significant amounts of any SR proteins bound to the RNA in the absence of Tra and Tra-2. This experiment demonstrated only that two SR proteins bind well to repeat-element-containing RNA in the presence of Tra and Tra-2. It does not necessarily mean a relationship exists between this binding and commitment. However, the fact that the two SR proteins that bind are also ones that complement Tra and Tra-2 in commitment is suggestive.

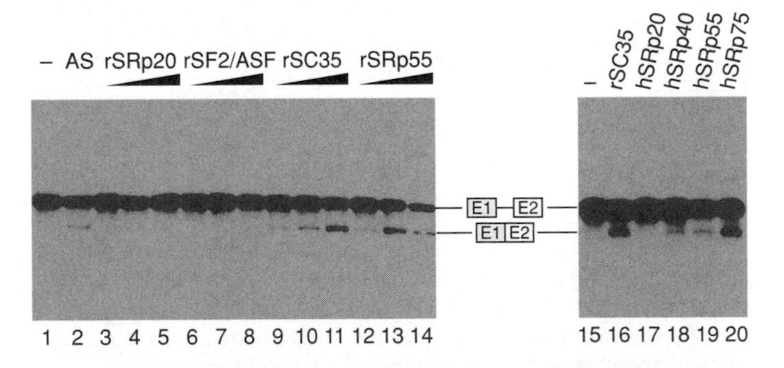

Figure 14.37 Commitment assay for female-specific splicing of *dsx* pre-mRNA. Tian and Maniatis assayed for the ability of various SR proteins to complement Tra and Tra-2 in an in vitro *dsx* splicing assay. Lane 1 contained no complementing protein. Lane 2 contained a mixture of SR proteins precipitated by ammonium sulfate (AS). Lanes 3–14 contained various amounts of the SR proteins indicated at the top of the lanes. Lane 15 is another negative control identical to lane 1. Lane 16 contained the highest amount of recombinant SC35, as in lane 11. Lanes 17–20 contained the purified nonrecombinant SR proteins indicated at the top of each lane. The electrophoretic mobilities of the splicing substrate (top band) and the spliced product (bottom band) are indicated between the two autoradiographs. (*Source:* Tian, and M. Maniatis, A splicing enhancer complex controls alternative splicing of *doublesex* pre-mRNA. *Cell* 74 (16 July 1993) f. 5, p. 108. Reprinted by permission of Elsevier Science.)

SUMMARY The transcripts of many eukaryotic genes are subject to alternative splicing. This can have profound effects on the protein products of a gene. For example, it can make the difference between a secreted or a membrane-bound protein; it can even make the difference between activity and inactivity. In the fruit fly, the products of three genes in the sex determination pathway are subject to alternative splicing. Female-specific splicing of the *tra* transcript gives an active product that causes female-specific splicing of the *dsx* pre-mRNA, which produces a female fly. Male-specific splicing of the *tra* transcript gives an inactive product that allows default, or male-specific, splicing of the *dsx* pre-mRNA, producing a male fly. Tra and its partner Tra-2 act in conjuction with one or more other SR proteins to commit splicing at the female-specific splice site on the *dsx* pre-mRNA. Such commitment is probably the basis of most, if not all, alternative splicing schemes.

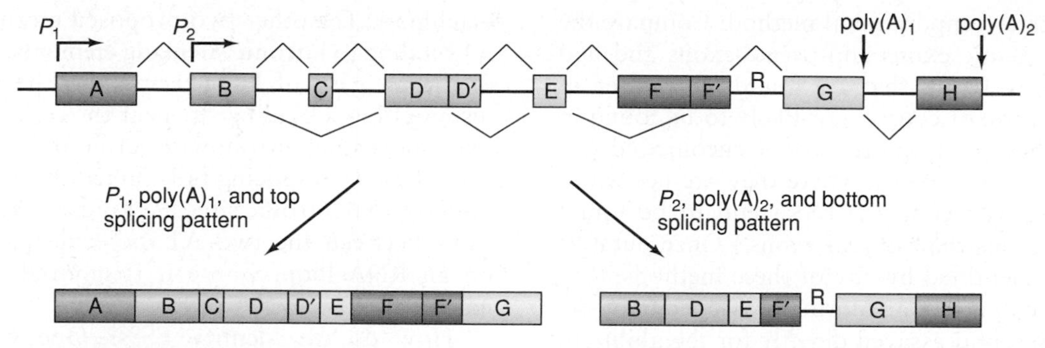

Figure 14.38 Alternative splicing patterns, coupled with alternative promoters and polyadenylation sites. Only two of 64 possible mRNAs are shown. The six different decision points are, from left to right: 1. Use of the first of two different promoters includes exon A, whereas use of the second promoter deletes that exon. 2. Failure to recognize exon C causes that exon to be omitted in the lower splicing pattern. 3. Recognition of an alternative 5′-splice site within exon D (between D and D′) causes deletion of D′ in the lower splicing pattern. 4. Recognition of an alternative 3′-splice site within exon F (between F and F′) causes deletion of F in the lower splicing pattern. 5. failure to recognize the retained intron (R) causes retention of that intron in the lower splicing pattern. 6. Polyadenylation, with cleavage of the pre-mRNA after poly(A) site 1 deletes exon H in the upper pattern.

Control of Splicing

We have seen two examples of systems in which alternative splicing of the same pre-mRNA gives rise to two very different products. But alternative splicing is not a rare curiosity. It has been estimated to occur in well over half the genes in humans. Many genes have more than two splicing patterns, and some have thousands.

Figure 14.38 illustrates several different kinds of alternative splicing. First, transcripts can begin at alternative promoters. In this example, transcripts beginning at the first promoter will include the first exon (A), but those starting at the second promoter will not. Second, some exons, such as exon C here, can simply be ignored, resulting in the deletion of that exon from the mRNA. Third, alternative 5′-splice sites can lead to inclusion or deletion of part of an exon (the D′ part, in this case). Fourth, alternative 3′-splice sites can lead to inclusion or deletion of part of an exon (the F part, in this case). Fifth, a so-called **retained intron** can be retained in the mRNA if it is not recognized as an intron, as in the lower splicing pattern. Sixth, polyadenylation, which we will study in Chapter 15, causes cleavage of the pre-mRNA, and loss of any downstream exons. For example, cleavage at poly(A) site 1 deletes exon H. So we have six sites at which two different things can happen, yielding $2^6 = 64$ different outcomes.

Alternative splicing is obviously carefully controlled by cells. It would not do, for example, to have female-specific splicing of the *dsx* pre-mRNA in male fruit flies. All of this implies that something that is recognized as an exon in one context is simply part of an intron in another context.

But what stimulates recognition of these signals under certain circumstances and inhibits such recognition in another context? Part of the answer, as we have just seen, is splicing factors that stimulate commitment at certain splice sites. Another part of the answer is that exons can contain sequences known as **exonic splicing enhancers (ESEs),** which stimulate splicing, and **exonic splicing silencers (ESSs),** which inhibit splicing. (Intronic splicing enhancers and silencers also exist.) These sequences presumably bind protein factors that are produced in certain cell types, or at certain stages in a cell's life, or in response to external agents, such as hormones. Such binding can then presumably either activate or repress splicing at nearby splice sites.

The *Drosophila* sex-determination gene *dsx* provides a good example of an exonic splicing enhancer. Exon 4 of this gene (Figure 14.36) has a very weak 3′-splice site that U2AF has a difficult time recognizing. Thus, in male flies, exon 4 is not recognized and is omitted from the mature mRNA. But in female flies, the *tra* gene product (Tra), along with two SR proteins, binds to an ESE in exon 4, and this activates recognition of the 3′-splice site preceding exon 4, presumably by attracting U2AF; therefore, exon 4 is included in the mature mRNA.

Many ESEs have now been identified. One way of finding them is to knock them out and observe the loss of splicing at a particular site. Another way of identifying ESEs is by a functional SELEX procedure (Chapter 5) that depends on the ability to stimulate splicing, rather than binding to particular molecules. Adrian Krainer and his colleagues started with a cloned DNA containing an exon-intron-exon, in which the second exon bore an ESE. They replaced this ESE with a large random set of DNA 20-mers by PCR. Then they transcribed these 1.2×10^{10} DNA sequences and selected the RNAs that could be spliced in a cell-free extract. The selection relied on gel electrophoresis, which separated spliced from unspliced RNAs.

The disadvantage of this functional SELEX procedure is that you have to know in advance what SR proteins to put in the cell-free extract, so ESEs that work with unknown proteins can be missed. One way around that problem is to use a computational method: Compare the sequences of authentic exons and pseudoexons and find short sequences (6–10 nt) that are found more often in real exons. ESEs are, of course, not likely to be found in pseudoexons, where splicing need not be encouraged, but they are present in real exons, where they are needed to promote splicing. (By contrast, ESSs tend to be found more in pseudoexons than in real exons.) Once putative ESEs have been identified by any of these methods, they can be placed in exons that are normally skipped in model splicing substrates, and assayed directly for the ability to stimulate splicing.

ESEs tend to interact with SR proteins, while ESSs interact with **hnRNP proteins,** which are the proteins that bind to hnRNAs, most of which are pre-mRNAs. An hnRNP protein commonly associated with ESS activity is **hnRNP A1.** Molecular biologists have found evidence for at least three different mechanisms for A1 action (Figure 14.39), and all three are probably valid, with different mechanisms applying to repression of splicing at different exons.

The first mechanism involves an ESS: A1 binding to an ESS within an exon nucleates binding of additional A1 molecules, such that bound A1 spreads throughout the exon and hides the splicing signals from the splicing machinery. The other two proposed mechanisms involve A1 binding to **intronic silencing elements.** The third exon of the *tat* gene of HIV exemplifies the second mechanism: A1 has a binding site near the splicing branchpoint in the preceding intron; with A1 bound there, U2 snRNP cannot bind, so splicing fails. In the third mechanism, A1 binds to two intronic sites flanking an exon, and interactions between the two A1 molecules isolate the exon on an RNA loop, where it is ignored by the splicing machinery.

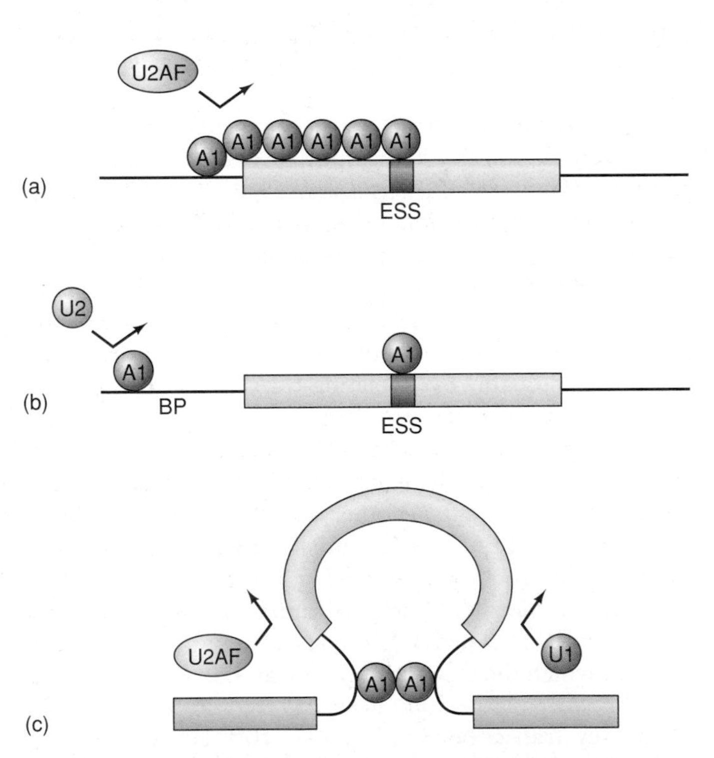

Figure 14.39 Models for hnRNP A1 silencing of splicing. (a) A1 binds first at an ESS and nucleates spreading of A1 binding, in this case toward the 3′-splice site at the end of the previous intron. This prevents U2AF from binding. **(b)** A1 binds to an intronic silencing element near the branchpoint (BP) in the intron. This prevents U2 from binding. **(c)** A1 binds to two intronic silencing elements in the introns flanking the yellow exon. Interactions between these two A1 molecules create an RNA loop, which isolates the exon, hiding it from the splicing machinery.

How do we identify ESSs? One way, as already suggested, is to apply a computational method and look for sequences that are enriched in pseudoexons, compared to real exons. Another is to look directly for sequences that inhibit splicing. Christopher Burge and colleagues have designed a reporter construct (Figure 14.40) to do just that. Their construct is a plasmid containing the two exons of the gene that encodes green fluorescent protein (GFP). Between these two exons is another exon, which, if included with the other two in the mature mRNA, interrupts the GFP mRNA and prevents production of GFP protein. So Burge and colleagues introduced random 10-bp sequences into this central exon, placed the constructs into cells, and then looked for green cells under fluorescent light.

Green cells indicated the production of GFP, which indicated that the central exon had not been included in the mRNA, which in turn indicated that the 10-mer in the central exon in that cell was acting as an ESS. Using this method, Burge and colleagues identified 141 10-mers with ESS activity, 133 of which were unique.

The concept of retained intron raises a question: How does a partially spliced transcript make it into the cytoplasm? Ordinarily, transcripts are retained in the nucleus until they are fully spliced. This retention is governed in part by the **exon junction complex** (EJC), a group of proteins that assemble at the junction of newly joined exons and facilitate export of the RNA from the nucleus. But there are many examples of transcripts that are exported even though they are incompletely spliced, and they rely on specific factors to guide them out of the nucleus and protect them from degradation once in the cytoplasm.

SUMMARY Alternative splicing is a very common phenomenon in higher eukaryotes. It represents a way to get more than one protein product out of the same gene, and a way to control gene expression in cells. Such control is exerted by splicing factors that bind to the splice sites and branchpoint, and also by proteins that interact with exonic splicing enhancers (ESEs), exonic splicing silencers (ESSs), and intronic silencing elements. SR proteins tend to bind to ESEs, while hnRNP proteins, such as hnRNP A1, bind to ESSs and intronic silencing elements.

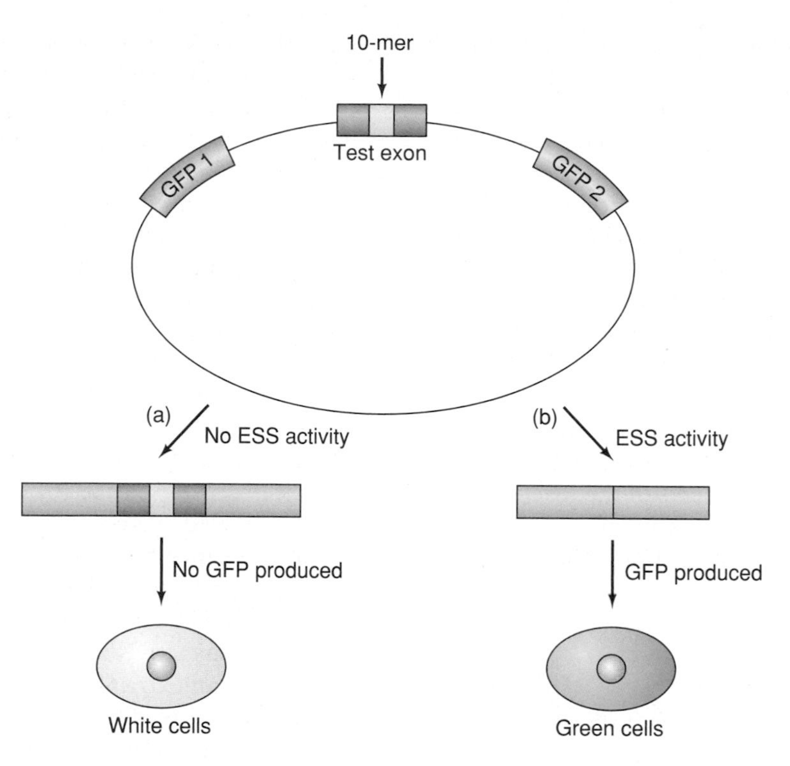

Figure 14.40 A reporter construct to detect ESS activity. Burge and colleagues constructed a plasmid containing the two exons of the GFP gene, separated by an intron that held a test exon (red) into which random 10-mers (yellow) had been placed. They transfected cells with collections of these plasmids, and then screened for green color. **(a)** If the 10-mer has no ESS activity, splicing of the test exon will not be silenced, so it will be included in the middle of the GFP mRNA, disrupting its activity, and producing white cells. **(b)** If the 10-mer does have ESS activity, the test exon will not be recognized, so it will be spliced out along with the surrounding intron. Thus, a normal GFP mRNA will be produced and the cells will be green.

14.3 Self-Splicing RNAs

One of the most stunning discoveries in molecular biology in the 1980s was that some RNAs could splice themselves without aid from a spliceosome or any other proteins. Thomas Cech (pronounced "Check") and his coworkers made this discovery in their study of the 26S rRNA gene of the ciliated protozoan, *Tetrahymena*. This rRNA gene is a bit unusual in that it has an intron, but the thing that really attracted attention when this work was published in 1982 was that the purified 26S rRNA precursor spliced itself in vitro. In fact, this was just the first example of self-splicing RNAs containing introns called **group I introns.** Subsequent work revealed another class of RNAs containing introns called **group II introns,** some of whose members are also self-splicing.

Group I Introns

To make the self-splicing RNA, Cech and coworkers cloned part of the 26S rRNA gene containing the intron, and transcribed it in vitro with *E. coli* RNA polymerase. When they electrophoresed the labeled products of these transcription reactions, they observed four large RNA products, plus three smaller RNAs corresponding in size to the linear and circular introns (plus a linear intron missing 15 nt), which they had observed in previous studies. This suggested that the RNA was being spliced, and that the excised intron was circularizing.

Was this splicing carried out by the RNA itself, or was the RNA polymerase somehow involved? To answer this question, Cech and coworkers ran the RNA polymerase reaction in the presence of polyamines (spermine, spermidine, and putrescine) that inhibit splicing. Then they electrophoresed the products, excised all four RNA bands plus the material that remained at the origin, and purified the RNAs. Next they incubated these RNAs under splicing conditions (no polyamines) and reelectrophoresed them. When they autoradiographed the electrophoretic gel, they could see the intron in the lanes containing RNA from three of the bands. Thus, these bands appear to be 26S rRNA precursors that can splice themselves without any protein, even RNA polymerase.

The band we are calling the intron is the right size, but is it really what we think it is? Cech and coworkers sequenced the first 39 nt of this RNA and showed that they corresponded exactly to the first 39 nt of the intron. Therefore, it seemed clear that this RNA really was the intron.

Cech's group also discovered that the linear intron—the RNA we have been discussing so far—can cyclize by itself.

All they had to do was raise the temperature, and the Mg^{2+} and salt concentrations, and at least some of the purified linear intron would convert to circular intron.

So far, we have seen that the rRNA precursor can remove its intron, but can it splice its exons together? Cech and coworkers used a model splicing reaction to show that it can (Figure 14.41a). They began by cloning a part of the *Tetrahymena* 26S rRNA gene including 303 bp of the first exon, the whole intron, and 624 bp of the second exon into a vector with a promoter for phage SP6 polymerase. To generate the labeled splicing substrate, they transcribed this DNA in vitro with SP6 polymerase in the presence of [α-^{32}P]ATP. Then they incubated this RNA under splicing conditions with and without GTP and electrophoresed the products. Lane 1 displays the products of the reaction with GTP. The familiar linear intron is present, as well as a small amount of circular intron. In addition, a prominent band representing the ligated exons appeared. By contrast, lane 2 shows that no such products appeared in the absence of GTP; only the substrate was present. This is what we expect because splicing of group I introns is dependent on GTP, and it reinforces the conclusion that these products are all the result of splicing. In summary, these data argue strongly for true splicing, including the joining of exons.

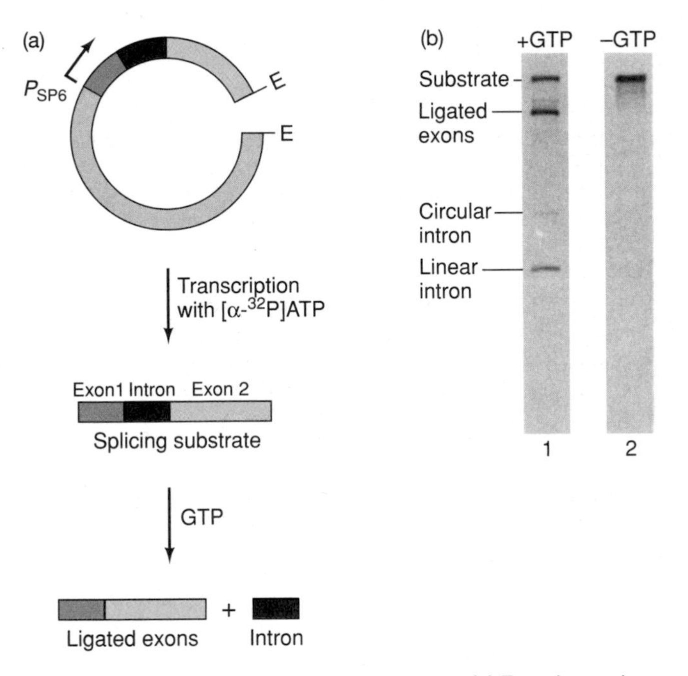

Figure 14.41 Demonstration of exon ligation. (a) Experimental scheme. Cech and coworkers constructed a plasmid containing part of the *Tetrahymena* 26S rRNA gene: 303 bp of exon 1 (blue); the 413-bp intron (red); and 624 bp of exon 2 (yellow). They linearized the plasmid by cutting it with *Eco*RI, creating *Eco*RI ends (E), then transcribed the plasmid in vitro with phage SP6 RNA polymerase and [α-^{32}P]ATP. This yielded the labeled splicing substrate. They incubated this substrate under splicing conditions in the presence or absence of GTP, then electrophoresed the splicing reactions and detected the labeled RNAs by autoradiography. **(b)** Experimental results. In the presence of GTP (lane 1), a prominent band representing the ligated exons appeared, in addition to bands representing the linear and circular intron. In the absence of GTP (lane 2), only the substrate band appeared. Thus, exon ligation appears to be a part of the self-splicing reaction catalyzed by this RNA. (*Source:* (*b*) Inane, T., F.X. Sullivan, and T.R. Cech, Intermolecular exon ligation of the rRNA precursor of *Tetrahymena:* Oligonucleotides can function as 5′-exons. *Cell* 43 (Dec 1985) f. 1a, p. 432. Reprinted by permission by Elsevier Science.)

Cech's group had already shown that splicing of the 26S rRNA precursor involved addition of a guanine nucleotide at the 5′-end of the intron. To verify that self-splicing in the absence of protein used the same mechanism, they performed a two-part experiment. In the first part, they incubated the splicing precursor with [α-^{32}P]GTP under splicing and nonsplicing conditions, then electrophoresed the products to see if the intron had become labeled. Figure 14.42a shows that it had, and a similar experiment with [γ-^{32}P]GTP gave the same results. In the second part, these workers 5′-end-labeled the intron with [α-^{32}P]GTP in the same way and sequenced the product. It gave exactly the sequence expected for the linear intron, with an extra G at the 5′-end (Figure 14.42b). This G could be removed by RNase T1, demonstrating that it is attached to the end of the intron by a normal 5′-3′-phosphodiester bond. Figure 14.43 presents a model for the splicing of the *Tetrahymena* 26S rRNA precursor, up to the point of ligating the two exons together and formation of the linear intron.

We have seen that the excised intron can cyclize itself. Cech and his coworkers showed that this cyclization actually involves the loss of 15 nt from the 5′-end of the linear intron. Three lines of evidence led to this conclusion: (1) When the 5′-end of the linear intron is labeled, none of this label appears in the circularized intron. (2) At least two RNase T1 products (actually three) found at the 5′-end of the linear intron are missing from the circular intron. (3) Cyclization of the intron is accompanied by the accumulation of an RNA 15-mer that contains the missing RNase T1 products.

But this is not the end of the process. After cyclization, the circular intron opens up again at the very same phosphodiester bond that formed the circle in the first place. Then the intron recyclizes by removing four more nucleotides from the 5′-end. Finally, the intron opens up at the same bond that just formed, yielding a shortened linear intron.

Figure 14.44 presents a detailed mechanism of the cyclization and relinearization of the excised intron. Notice that throughout the splicing process, for every phosphodiester bond that breaks a new one forms. Thus, the free energy change of each step is near zero, so no exogenous source of energy, such as ATP, is required. Another general feature of the process is that the bonds that form to make the circular introns are the same ones that break when the circle opens up again. This tells us that these bonds are special; the three-dimensional shape of the RNA must strain these bonds to

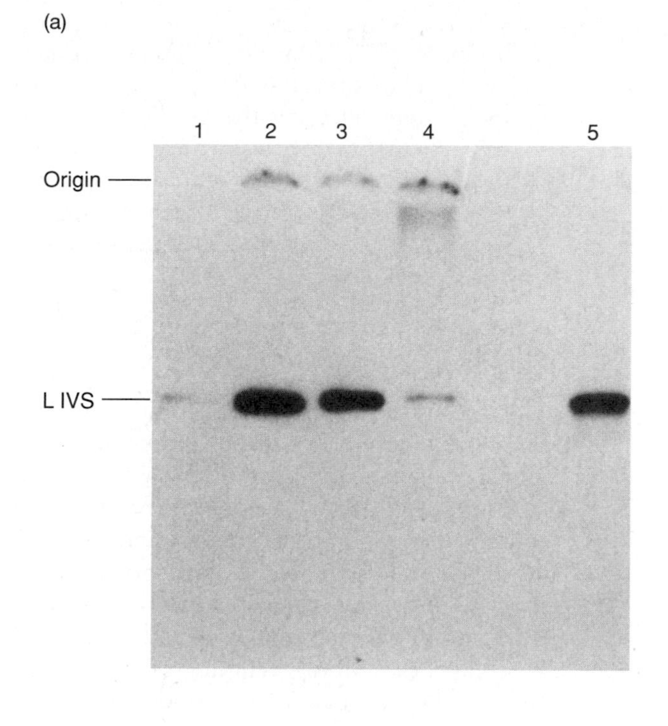

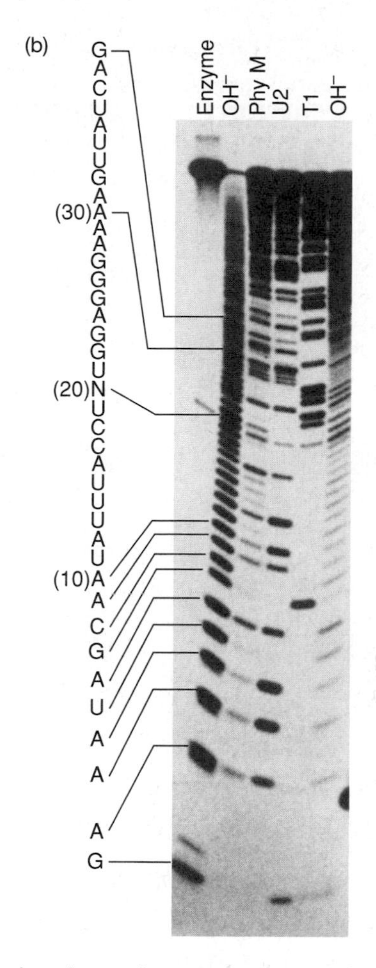

Figure 14.42 Addition of GMP to the 5′-end of the excised intron. (a) Radioactive GTP labels the intron during splicing. Cech and coworkers transcribed plasmid pIVS11 under nonsplicing conditions with no labeled nucleotides. They isolated this unlabeled 26S rRNA precursor and incubated it under splicing conditions in the presence of [α-^{32}P]GTP. Then they chromatographed the products on Sephadex G-50, electrophoresed the column fractions, and autoradiographed the gel. Lanes 1–4 are successive fractions from the Sephadex column. Lane 5 is a linear intron marker. Lanes 2 and 3 contain the bulk of the linear intron, and it is labeled, indicating that it had incorporated a labeled guanine nucleotide. **(b)** Sequence of the labeled intron. Cech and coworkers used an enzymatic method to sequence the 5′-end of the RNA. They cut it with base (OH$^-$), which cuts after every nucleotide; RNase Phy M, which cuts after A and U; RNase U2, which cuts after A; and RNase T1, which cuts after G. Treatment of each RNA sample is indicated at top. The deduced sequence is given at left. Note the 5′-G at bottom. (*Source:* Kruger K., P.J. Grabowski, A.J. Zaug, J. Sands, D.E. Gottschling and T.R. Cech, Self-splicing RNA: Autoexcision and autocyclization of the ribosomal RNA intervening sequence of *Tetrahymena. Cell* 31 (Nov 1982) f. 4, p. 151. Reprinted by permission of Elsevier Science.)

make them easiest to break during relinearization. This strain would help to explain the catalytic power of the RNA.

At first glance, there appears to be a major difference between the splicing mechanisms of spliceosomal introns and group I introns: Whereas the group I introns use an exogenous nucleotide in the first step of splicing, spliceosomal introns use a nucleotide that is integral to the intron itself. However, on closer examination we see that the difference might not be as great as it seems. Michael Yarus and his colleagues used molecular modeling techniques to predict the lowest energy conformation of the *Tetrahymena* 26S rRNA intron as it associates with GMP. They proposed that part of the intron folds into a double helix with a pocket that holds the guanine nucleotide through hydrogen bonds (Figure 14.45). This guanine, held fast to the intron, behaves in essentially the same way as the adenine in spliceosomal introns. Of course, it cannot form a lariat because it is not covalently linked to the intron.

Until the discovery of self-splicing RNAs, biochemists thought that the catalytic parts of enzymes were made only of protein. Sidney Altman had shown a few years earlier that **RNase P,** which cleaves extra nucleotides off the 5′-ends of tRNA precursors, has an RNA component called **M1.** But RNase P also has a protein component, which could have held the catalytic activity of the enzyme. In 1983, Altman confirmed that the M1 RNA is the catalytic component of RNase P (Chapter 16). This enzyme and self-splicing RNAs are examples of catalytic RNAs, which we call **ribozymes.**

Actually, the reactions we have seen so far, in which group I introns participate, are not enzymatic in the strict

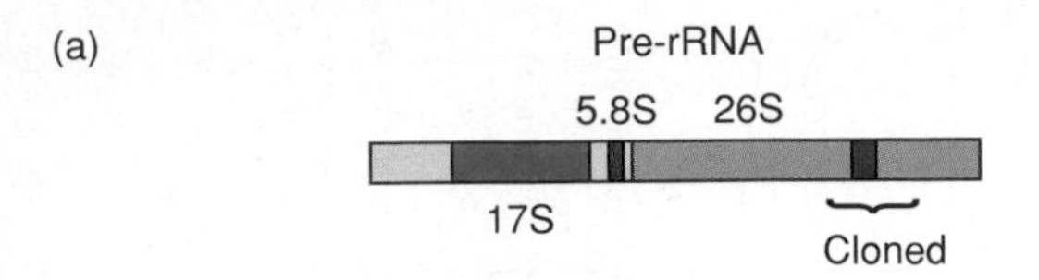

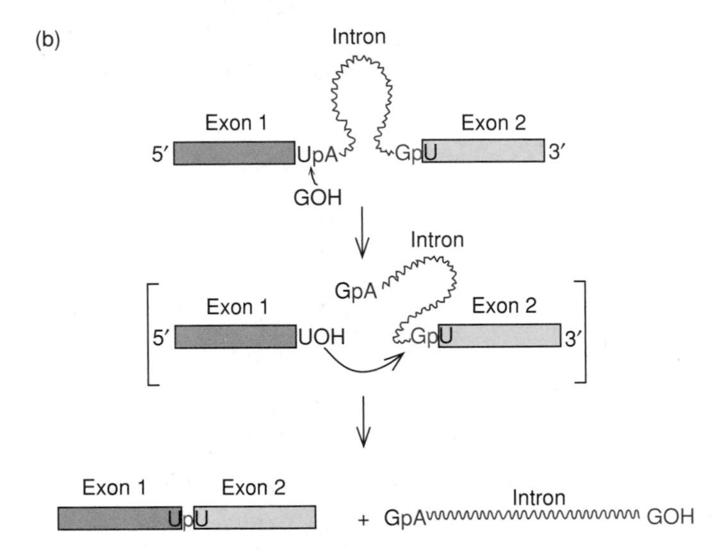

Figure 14.43 Self-splicing of *Tetrahymena* rRNA precursor. **(a)** Structure of the rRNA precursor, containing the 17S, 5.8S, and 26S sequences. Note the intron within the 26S region (red). The cloned segment used in subsequent experiments is indicated by a bracket. **(b)** Self-splicing scheme. In the first step (top), a guanine nucleotide attacks the adenine nucleotide at the 5′-end of the intron, releasing exon 1 (blue) from the rest of the molecule and generating the hypothetical intermediates shown in brackets. In the second step, exon 1 (blue) attacks exon 2 (yellow), performing the splicing reaction that releases a linear intron (red), and joins the two exons together. Finally, in a series of reactions not shown here, the linear intron loses 19 nt from its 5′-end.

sense, because the RNA itself changes. A true enzyme is supposed to emerge unchanged at the end of the reaction. But the final linearized group I intron from the *Tetrahymena* 26S rRNA precursor can act as a true enzyme by adding nucleotides to, and subtracting them from, an oligonucleotide. We should also make another qualification about ribozymes. They can operate on their own in vitro. But many, including many group I introns, are aided by proteins in vivo. These proteins have no catalytic activity of their own, but they can stabilize the catalytically active structure of the ribozyme. As such, these ribonucleoprotein complexes can be called **RNPzymes.**

SUMMARY Group I introns, such as the one in the *Tetrahymena* 26S rRNA precursor, can be removed in vitro with no help from protein. The reaction begins with an attack by a guanine nucleotide on the 5′-splice site, adding the G to the 5′-end of the intron, and releasing the first exon. In the second step, the first exon attacks the 3′-splice site, ligating the two exons together, and releasing the linear intron. The intron cyclizes twice, losing nucleotides each time, then linearizes for the last time.

Group II Introns

The introns of fungal mitochondrial genes were originally classified as group I or group II according to certain conserved sequences they contained. Later, it became clear that mitochondrial and chloroplast genes from many species contained group I and II introns, and that RNAs containing both classes of intron have members that are self-splicing. However, the mechanisms of splicing used by RNAs with group I and group II introns are different. Whereas the initiating event in group I splicing is attack by an independent guanine nucleotide, the initiating event in group II splicing involves intramolecular attack by an A residue in the intron to form a lariat.

The lariat formation by group II introns sounds very similar to the situation in spliceosomal splicing of nuclear mRNA precursors, and the similarity extends to the overall shapes of the RNAs in the spliceosomal complex and of the group II introns, as we saw in Figure 14.20. This implies a similarity in function between the spliceosomal snRNPs and the catalytic part of the group II introns. It may even point to a common evolutionary origin of these RNA species. In fact, it has been proposed that nuclear pre-mRNA introns descended from bacterial group II introns. These bacterial introns presumably got into eukaryotic cells because they inhabited the bacteria that invaded the precursors of modern eukaryotic cells and evolved into mitochondria. This hypothesis has become even more attractive since the discovery of group II introns in archaea, as well as in two classes of bacteria: cyanobacteria and purple bacteria. If we assume that the group II introns are older than the common ancestor of these two bacterial lineages, then they are old enough to have inhabited the bacteria that were the ancestors of modern eukaryotic organelles. Nevertheless, convergent evolution to a common mechanism also remains a possibility.

SUMMARY RNAs containing group II introns self-splice by a pathway that uses an A-branched lariat intermediate, just like the spliceosomal lariats. The secondary structures of the splicing complexes involving spliceosomal systems and group II introns are also strikingly similar.

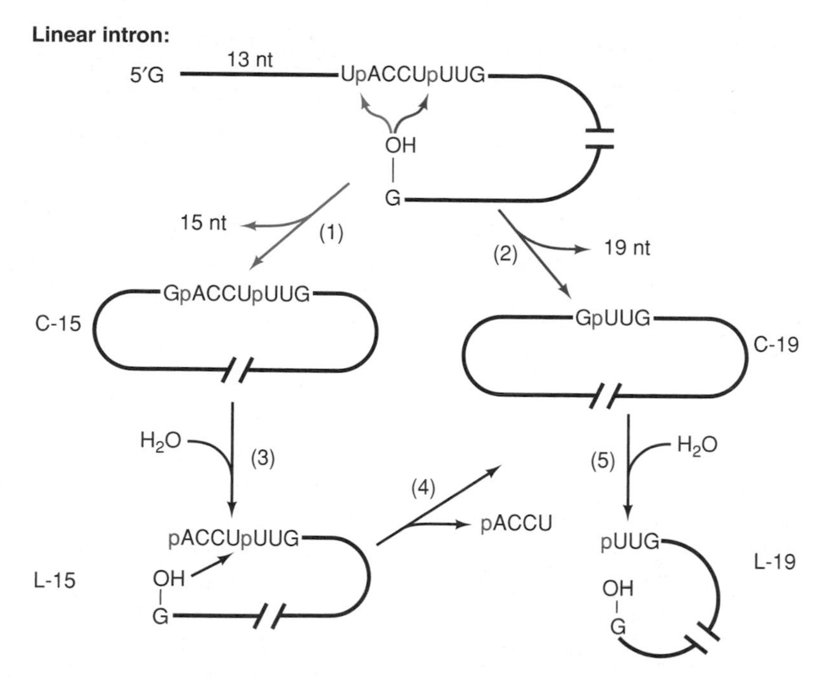

Figure 14.44 Fate of the linear intron. We begin with the linear intron originally excised from the 26S rRNA precursor. This can be cyclized in two ways: In reaction 1 (green arrows), the 3′-terminal G attacks the bond between U-15 and A-16, removing a 15-nt fragment and giving a circular intron (C-15). In the alternative reaction (2, blue arrows), the terminal G attacks 4 nt farther into the intron, removing a 19-nt fragment and leaving a smaller circular intron (C-19). Reaction 3, C-15 can open up at the same bond that closed the circle, yielding a linear intron (L-15). Reaction 4, the terminal G of L-15 can attack the bond between the first two U's, yielding the circular intron C-19. Reaction 5, C-19 opens up to yield the linear intron L-19.

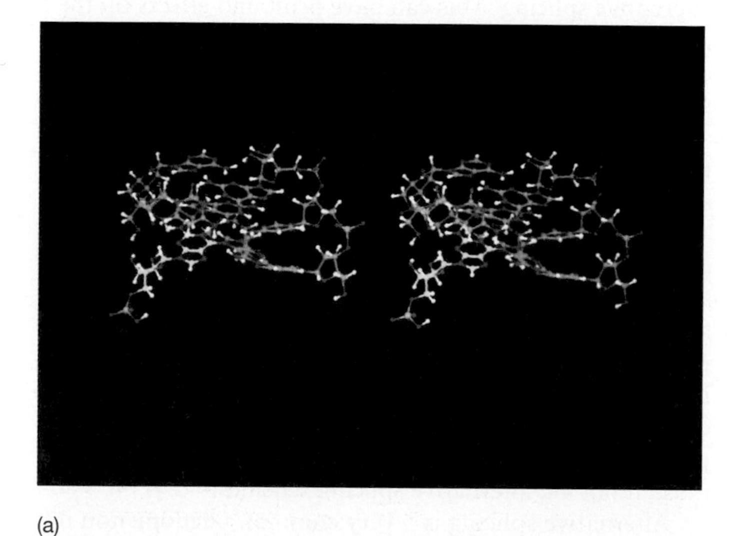

(a)

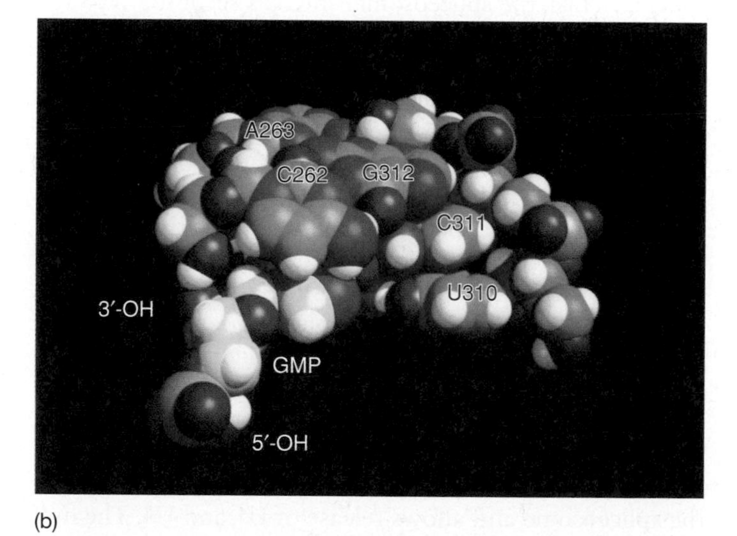

(b)

Figure 14.45 Two views of GMP held in a pocket of the 26S rRNA intron. (a) A cross-eyed stereogram that can be viewed in three dimensions by crossing the eyes until the two images merge. Carbon atoms of RNA, green; carbon atoms of G, yellow; phosphorus, lavender. Other atoms are standard colors. The GMP is at lower left. **(b)** Space-filling model. Colors are as in part (a). (*Source:* Yarus, M., I. Illangesekare, and E. Christian, An axial binding site in the *Tetrahymena* precursor RNA. *Journal of Molecular Biology.* 222 (1991) f. 7c–d, p. 1005, by permission of Elsevier.)

SUMMARY

Nuclear mRNA precursors are spliced via a lariat-shaped, or branched, intermediate. In addition to the consensus sequences at the 5′- and 3′-ends of nuclear introns, branchpoint consensus sequences also occur. In yeast, this sequence is nearly invariant: UACUAAC. In higher eukaryotes, the consensus sequence is more variable: YNCURAC. In all cases, the branched nucleotide is the final A in the sequence. The yeast branchpoint sequence also determines which downstream AG is the 3′-splice site.

Splicing appears to take place on a particle called a spliceosome. Yeast and mammalian spliceosomes have sedimentation coefficients of about 40S and 60S, respectively. Genetic experiments have shown that base pairing between U1 snRNA and the 5′-splice site of an mRNA precursor is necessary, but not sufficient, for splicing. The U6 snRNP also associates with the 5′-end of the intron by base pairing. This association first occurs prior to formation of the lariat intermediate, but its character may change after this first step in splicing. The association between U6 and the splicing substrate is essential for the splicing process. U6 also associates with U2 during splicing.

The U2 snRNA base-pairs with the conserved sequence at the splicing branchpoint. This base pairing is essential for splicing. U2 also forms vital base pairs with U6, forming a region called helix I, which apparently helps orient these snRNPs for splicing. The U4 snRNA base-pairs with U6, and its role seems to be to bind U6 until U6 is needed in the splicing reaction. The U5 snRNP associates with the last nucleotide in one exon and the first nucleotide of the next. This presumably lines the 5′- and 3′-splice sites up for splicing.

The spliceosomal complex (substrate, U2, U5, and U6) poised for the second step in splicing can be drawn in the same way as a group II intron at the same stage of splicing. Thus, the spliceosomal snRNPs seem to substitute for elements at the center of catalytic activity of the group II introns, and probably have the spliceosome's catalytic activity.

Indeed, the catalytic center of the spliceosome appears to include Mg^{2+} and a base-paired complex of three RNAs: U2 and U6 snRNAs, and the branchpoint region of the intron. Protein-free fragments of these three RNAs can catalyze a reaction related to the first splicing step.

The spliceosome cycle includes the assembly, splicing activity, and disassembly of the spliceosome. Assembly begins with the binding of U1 to the splicing substrate to form a commitment complex. U2 is the next snRNP to join the complex, followed by the others. The binding of U2 requires ATP. U6 dissociates from U4, then displaces U1 at the 5′-splice site. This ATP-dependent step activates the spliceosome and allows release of U1 and U4. The five snRNPs that participate in splicing all contain a common set of seven Sm proteins and several other proteins that are specific to each snRNP. The structure of U1 snRNP reveals that the Sm proteins form a doughnut-shaped structure to which the other proteins are attached. A minor class of introns with 5′-splice sites and branchpoints can be spliced with the help of a minor spliceosome containing a variant class of snRNAs, including U11, U12, U4atac, and U6atac.

The splicing factor Slu7 is required for correct 3′-splice site selection. In its absence, splicing to the correct 3′-splice site AG is specifically suppressed and splicing to aberrant AG's within about 30 nt of the branchpoint is activated. U2AF is also required for 3′-splice site recognition. The 65-kD U2AF subunit binds to the polypyrimidine tract upstream of the 3′-splice site, and the 35-kD subunit binds to the 3′-splice site AG.

Commitment to splice at a given site is determined by an RNA-binding protein, which presumably binds to the splicing substrate and recruits other spliceosomal components, starting with U1. For example, the SR proteins SC35 and SF2/ASF commit splicing on human β-globin pre-mRNA and HIV tat pre-mRNA, respectively. In the yeast commitment complex, the branchpoint bridging protein (BBP) binds to a U1 snRNP protein at the 5′-end of the intron, and to Mud2p near the 3′-end of the intron. It also binds to the RNA near the 3′-end of the intron. Thus, it bridges the intron and could play a role in defining the intron prior to splicing. The mammalian counterpart of BBP, SF1, may serve the a similar function, but in exon definition, in the mammalian commitment complex.

The CTD of the Rpb1 subunit of RNA polymerase II stimulates splicing of substrates that use exon definition, but not those that use intron definition, to prepare the substrate for splicing. The CTD binds to splicing factors and could therefore assemble the factors at the ends of exons to set them off for splicing.

The transcripts of many eukaryotic genes are subject to alternative splicing. This can have profound effects on the protein products of a gene. For example, it can make the difference between a secreted or a membrane-bound protein; it can even make the difference between activity and inactivity. In the fruit fly, the products of three genes in the sex determination pathway are subject to alternative splicing. Female-specific splicing of the *tra* transcript gives an active product that causes female-specific splicing of the *dsx* pre-mRNA, which produces a female fly. Male-specific splicing of the *tra* transcript gives an inactive product that allows default, or male-specific, splicing of the *dsx* pre-mRNA, producing a male fly. Tra and its partner Tra-2 act in conjuction with one or more other SR proteins to commit splicing at the female-specific splice site on the *dsx* pre-mRNA. Such commitment is undoubtedly the basis of most, if not all, alternative splicing schemes.

Alternative splicing is a very common phenomenon in higher eukaryotes. It represents a way to get more than one protein product out of the same gene, and a way to control gene expression in cells. Such control is exerted by splicing factors that bind to the splice sites and branchpoint, and also by proteins that interact with exonic splicing enhancers (ESEs), exonic splicing silencers (ESSs), and intronic silencing elements. SR proteins tend to bind to ESEs, while hnRNP proteins, such as hnRNP A1, bind to ESSs and intronic silencing elements.

Group I introns, such as the one in the *Tetrahymena* 26S rRNA precursor, can be removed with no help from protein in vitro. The reaction begins with an attack by a guanine nucleotide on the 5′-splice site, adding the G to the 5′-end of the intron and releasing the first exon. In the second step, the first exon attacks the 3′-splice site, ligating the two exons together and releasing the linear intron. The intron cyclizes twice, losing nucleotides each time, then linearizes for the last time.

RNAs containing group II introns self-splice by a pathway that uses an A-branched lariat intermediate, just like the spliceosomal lariats. The secondary structures of the splicing complexes involving spliceosomal systems and group II introns are also strikingly similar.

REVIEW QUESTIONS

1. Describe and show the results of an R-looping experiment that demonstrates that an intron is transcribed.
2. Diagram the lariat mechanism of splicing.
3. Present gel electrophoretic data that suggest that the excised intron is circular, or lariat-shaped.
4. Present gel electrophoretic data that distinguish between a lariat-shaped splicing intermediate (the intron—exon-2 intermediate) and a lariat-shaped product (the excised intron).
5. The lariat model predicts an intermediate with a branched nucleotide. Describe and show the results of an experiment that confirms this prediction.
6. Describe and give the results of an experiment that shows that a sequence (UACUAAC) within a yeast intron is required for splicing.
7. Describe and show the results of an experiment that demonstrates that the UACUAAC sequence within a yeast intron dictates splicing to an AG downstream.
8. What role does the UACUAAC sequence play in the lariat model of splicing?
9. Describe and show the results of an experiment that demonstrates that yeast spliceosomes have a sedimentation coefficient of 40S.
10. Describe and show the results of an experiment that demonstrates that base pairing between U1 snRNA and the 5′-splice site is required for splicing.
11. Describe and show the results of an experiment that demonstrates that base pairing between U1 and the 5′-splice site is not sufficient for splicing.
12. What snRNP besides U1 and U5 must bind near the 5′-splice site in order for splicing to occur? Present cross-linking data to support this conclusion.
13. Describe and show the results of an experiment that demonstrates that base pairing between U2 snRNA and the branchpoint sequence is required for splicing. In this experiment, why was it not possible to mutate the cell's only copy of the U2 gene?
14. Besides base-pairing with the pre-mRNA, U6 base-pairs with two snRNAs. Which ones are they?
15. Describe and show the results of an experiment that demonstrates that U5 contacts the 3′-end of the upstream exon and the 5′-end of the downstream exon during splicing. Make sure your experiment(s) provide positive identification of the RNA species involved, not just electrophoretic mobilities.
16. Describe and show the results of an experiment that demonstrates which bases in U5 can be cross-linked to bases in the pre-mRNA.
17. Summarize the evidence for a catalytic Mg^{2+} in spliceosomal splicing.
18. Summarize the evidence that a mixture of spliceosonal RNA fragments can catalyze a reaction related to the first splicing step.
19. Draw a diagram of a pre-mRNA as it exists in a spliceosome just before the second step in splicing. Show the interactions with U2, U5, and U6 snRNPs. This scheme resembles the intermediate stage for splicing of what kind of self-splicing RNA?
20. Describe and show the results of an experiment that demonstrates that U1 is the first snRNP to bind to the splicing substrate.
21. Describe and show the results of an experiment that demonstrates that binding of all other snRNPs to the spliceosome depends on U1, and that binding of U2 requires ATP.
22. What are Sm proteins?
23. How do the characteristics of minor spliceosomes help show the importance of base-pairing between snRNAs and pre-mRNA sites?
24. Describe and show the results of an experiment that demonstrates that Slu7 is required for selection of the proper AG at the 3′-splice site.
25. Describe a splicing commitment assay to screen for splicing factors involved in commitment. Show sample results.
26. Describe and give the results of a yeast two-hybrid assay that shows interaction between yeast branchpoint bridging protein (BBP) and two other proteins. What are the two other proteins, and where are they found with respect to the ends of the intron in the commitment complex?
27. Describe and give the results of an experiment that shows that the RNA polymerase II CTD stimulates splicing of pre-mRNAs that use exon definition.
28. Diagram the alternative splicing of the immunoglobulin μ heavy-chain transcript. Focus on the exons that are involved in one or the other of the alternative pathways, rather than the ones that are involved in both. What difference in the protein products is caused by the two pathways of splicing?
29. Describe a computational and an experimental method to identify sequences that act as exonic splicing silencers (ESSs).
30. Describe and show the results of an experiment that demonstrates self-splicing by a group I intron.
31. Describe and show the results of an experiment that demonstrates that a guanine nucleotide is added to the end of a spliced-out group I intron.

32. Draw a diagram of the steps involved in autosplicing of an RNA containing a group I intron. You do not need to show cyclization of the intron.
33. Diagram the steps involved in forming the L-19 intron from the original excised linear intron product of the *Tetrahymena* 26S pre-rRNA. Do not go through the C-15 intermediate.

ANALYTICAL QUESTIONS

1. You are investigating a gene with one large intron and two short exons. Show the results of R-looping experiments performed with:
   a. mRNA and single-stranded DNA
   b. mRNA and double-stranded DNA
   c. mRNA precursor and single-stranded DNA
   d. mRNA precursor and double-stranded DNA
2. You have discovered a new class of introns that do not require any proteins for splicing, but do require several small RNAs. One of these small RNAs, V3, has a sequence of 7 nt (CCUUGAG) complementary to the 3′-splice site. You suspect that base-pairing between V3 and the 3′-splice site is required for splicing. Design an experiment to test this hypothesis and show sample positive results.
3. Diagram the mechanism of RNase T1 (or T2) action. Because this is the same mechanism used in base hydrolysis, how does this explain why DNA is not subject to base hydrolysis?
4. You are studying a grave human disease called β-thalassemia in which no β-globin protein is produced. You find that the β-globin gene's coding region in people with this disease is normal, but the mRNA is over a hundred nucleotides longer than normal. You sequence the β-globin gene in these people and find a single base change within the gene's first intron. Present a hypothesis to explain the absence of β-globin in these patients.
5. Consider the gene illustrated in Figure 14.38, but remove P_2 and poly$(A)_1$, so there is only one promoter (P_1) and one polyadenylation site [poly$(A)_2$]. How many different spliced mRNAs can now be produced by this gene?
6. Consider the RNA sequencing results in Figure 14.42b. Knowing the cutting specificities of each enzyme, how do we know (a) that the band at the bottom in the first lane represents G? (b) that the next band represents A? (c) that the eighth band from the bottom represents C? (d) that the 13th, 14th, and 15th bands from the bottom represent U's? (Hint: PhyM cut inefficiently after U's in this experiment.)

SUGGESTED READINGS

General References and Reviews

Baker, B.S. 1989. Sex in flies: The splice of life. *Nature* 340:521–24.

Black, D.L. 2003. Mechanisms of alternative pre-messenger RNA splicing. *Annual Review of Biochemistry* 72:291–336.

Fu, X.-D. 2004. Towards a splicing code. *Cell* 119:736–38.

Guthrie, C. 1991. Messenger RNA splicing in yeast: Clues to why the spliceosome is a ribonucleoprotein. *Science* 253:157–63.

Hirose, Y. and J.L. Manley. 2000. RNA polymerase II and the integration of nuclear events. *Genes and Development* 14:1415–29.

Lamm, G.M. and A.I. Lamond. 1993. Non-snRNP protein splicing factors. *Biochimica et Biophysica Acta* 1173:247–65.

Manley, J. 1993. Question of commitment. *Nature* 365:14.

Murray, H.L. and K.A. Jarrell. 1999. Flipping the switch to an active spliceosome. *Cell* 96:599–602.

Newman, A. 2002. RNA enzymes for RNA splicing. *Nature* 413:695–96.

Nilsen, T.W. 2000. The case for an RNA enzyme. *Nature* 408:782–83.

Orphanides, G. and D. Reinberg. 2002. A unified theory of gene expression. *Cell* 108:439–51.

Proudfoot, N., A. Furger, and M.J. Dye. 2002. Integrating mRNA processing with transcription. *Cell* 108:501–12.

Sharp, P.A. 1994. Split genes and RNA splicing. (Nobel Lecture.) *Cell* 77:805–15.

Villa, T., J.A. Pleiss, and C. Guthrie. 2002. Spliceosomal snRNAs: Mg^{2+}-dependent chemistry at the catalytic core? *Cell* 109:149–52.

Weiner, A.M. 1993. mRNA splicing and autocatalytic introns: Distant cousins or the products of chemical determinism? *Cell* 72:161–64.

Wise, J.A. 1993. Guides to the heart of the spliceosome. *Science* 262:1978–79.

Research Articles

Abovich, N. and M. Rosbash. 1997. Cross-intron bridging interactions in the yeast commitment complex are conserved in mammals. *Cell* 89:403–12.

Berget, S.M., C. Moore, and P. Sharp. 1977. Spliced segments at the 5′ terminus of adenovirus 2 late mRNA. *Proceedings of the National Academy of Sciences USA* 74:3171–75.

Berglund, J.A., K. Chua, N. Abovich, R. Reed, and M. Rosbash. 1997. The splicing factor BBP interacts specifically with the pre-mRNA branchpoint sequence UACUAAC. *Cell* 89:781–87.

Brody, E. and J. Abelson. 1985. The spliceosome: Yeast pre-messenger RNA associated with a 40S complex in a splicing-dependent reaction. *Science* 228:963–67.

Chua, K. and R. Reed. 1999. The RNA splicing factor hSlu7 is required for correct 3′ splice-site choice. *Nature* 402:207–10.

Early, P., J. Rogers, M. Davis, K. Calame, M. Bond, R. Wall, and L. Hood. 1980. Two mRNAs can be produced from a single immunoglobulin γ gene by alternative RNA processing pathways. *Cell* 20:313–19.

Fu, X.-D. 1993. Specific commitment of different pre-mRNAs to splicing by single SR proteins. *Nature* 365:82–85.

Grabowski, P., R.A. Padgett, and P.A. Sharp. 1984. Messenger RNA splicing in vitro: An excised intervening sequence and a potential intermediate. *Cell* 37:415–27.

Grabowski, P., S.R. Seiler, and P.A. Sharp. 1985. A multicomponent complex is involved in the splicing of messenger RNA precursors. *Cell* 42:345–53.

Kruger, K., P.J. Grabowski, A.J. Zaug, J. Sands, D.E. Gottschling, and T.R. Cech. 1982. Self-splicing RNA: Autoexcision and autocylization of the ribosomal RNA intervening sequence of *Tetrahymena*. *Cell* 31:147–57.

Langford, C. and D. Gallwitz. 1983. Evidence for an intron-contained sequence required for the splicing of yeast RNA polymerase II transcripts. *Cell* 33:519–27.

Lesser, C.F. and C. Guthrie. 1993. Mutations in U6 snRNA that alter splice site specificity: Implications for the active site. *Science* 262:1982–88.

Liao, X.C., J. Tang, and M. Rosbash. 1993. An enhancer screen identifies a gene that encodes the yeast U1 snRNP A protein: Implications for snRNP protein function in pre-mRNA splicing. *Genes and Development* 7:419–28.

Parker, R., P.G. Siliciano, and C. Guthrie. 1987. Recognition of the TACTAAC box during mRNA splicing in yeast involves base pairing to the U2-like snRNA. *Cell* 49:229–39.

Peebles, C.L., P. Gegenheimer, and J. Abelson. 1983. Precise excision of intervening sequences from precursor tRNAs by a membrane-associated yeast endonuclease. *Cell* 32:525–36.

Ruby, S.W. and J. Abelson. 1988. An early hierarchic role of U1 small nuclear ribonucleoprotein in spliceosome assembly. *Science* 242:1028–35.

Sontheimer, E.J. and J.A. Steitz. 1993. The U5 and U6 small nuclear RNAs as active site components of the spliceosome. *Science* 262:1989–96.

Staley, J.P. and C. Guthrie. 1999. An RNA switch at the 5′ splice site requires ATP and the DEAD box protein Prp28p. *Molecular Cell* 3:55–64.

Stark, H., P. Dube, R. Lührmann and B. Kastner. 2001. Arrangement of RNA and proteins in the spliceosomal U1 small nuclear ribonucleoprotein particle. *Nature* 409:539–42.

Tian, M. and T. Maniatis. 1993. A splicing enhancer complex controls alternative splicing of *doublesex* pre-mRNA. *Cell* 74:105–14.

Tilghman, S.M., P. Curtis, D. Tiemeier, P. Leder, and C. Weissmann. 1978. The intervening sequence of a mouse β-globin gene is transcribed within the 15S β-globin mRNA precursor. *Proceedings of the National Academy of Sciences USA* 75:1309–13.

Valadkhan, S. and J.L. Manley. 2001. Splicing-related catalysis by protein-free snRNAs. *Nature* 413:701–07.

Wang, Z., M.E. Rolish, G. Yeo, V. Tung, M. Mawson, and C.B. Burge. 2004. Systematic identification and analysis of exonic splicing silencers. *Cell* 119:831–45.

Yarus, M., I. Illangesekare, and E. Christian. 1991. An axial binding site in the *Tetrahymena* precursor RNA. *Journal of Molecular Biology* 222:995–1012.

Yean, S.-L., G. Wuenschell, J. Termini, and R.-J. Lin. 2001. Metal-ion coordination by U6 small nuclear RNA contributes to catalysis in the spliceosome. *Nature* 408:881–84.

Zeng, C. and S.M. Berget. 2000. Participation of the C-terminal domain of RNA polymerase II in exon definition during pre-mRNA splicing. *Molecular and Cellular Biology* 20:8290–8301.

Zhuang, Y. and A.M. Weiner. 1986. A compensatory base change in U1 snRNA suppresses a 5′-splice site mutation. *Cell* 46:827–35.

CHAPTER 15

RNA Processing II: Capping and Polyadenylation

Caps in an Egyptian market. © *Iconotec.com.*

Besides splicing, eukaryotic cells perform several other kinds of processing on their RNAs. Messenger RNAs are subject to two kinds of processing, known as capping and polyadenylation. In capping, a special blocking nucleotide (a cap) is added to the 5′-end of a pre-mRNA. In polyadenylation, a string of AMPs (poly[A]) is added to the 3′-end of the pre-mRNA. These steps are essential for the proper function of mRNAs and will be our topics in this chapter.

15.1 Capping

By 1974, several investigators had discovered that mRNA from a variety of eukaryotic species and viruses was methylated. Moreover, a significant amount of this methylation was clustered at the 5′-end of mRNAs, in structures we call **caps.** In this section we will examine the structure and synthesis of these caps.

Cap Structure

Before gene cloning became routine, viral mRNAs were much easier to purify and investigate than cellular mRNAs. Thus, the first caps to be characterized came from viral RNAs. Bernard Moss and his colleagues produced vaccinia virus mRNAs in vitro and isolated their caps as follows: They labeled the methyl groups in the RNA with [^{3}H] S-adenosylmethionine (AdoMet, a methyl donor), or with ^{32}P-nucleotides, then subjected the labeled RNA to base hydrolysis. The major products of this hydrolysis were mononucleotides, but the cap could be separated from these by DEAE-cellulose chromatography. Figure 15.1 shows the chromatographic behavior of the vaccinia virus caps. They behaved as a substance with a net charge near −5. Furthermore, the red and blue curves in Figure 15.1b show that the ^{3}H(methyl) and ^{32}P labels essentially coincided, demonstrating that the caps were methylated. Aaron Shatkin and his coworkers obtained very similar results with reovirus caps.

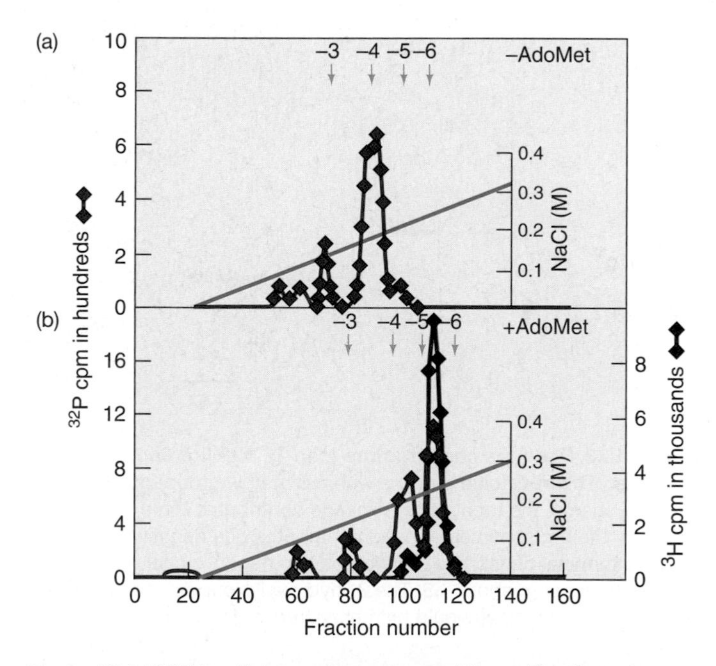

Figure 15.1 DEAE-cellulose chromatographic purification of vaccinia virus caps. Wei and Moss allowed vaccinia virus particles to synthesize caps in the presence of [β, γ-^{32}P]GTP and in the **(a)** absence and **(b)** presence of *S*-adenosyl[methyl-^{3}H]methionine. Then they digested the labeled, capped RNAs with KOH and separated the products by DEAE-cellulose column chromatography. ^{3}H (blue) and ^{32}P (red) radioactivities (in counts per minute) are plotted versus column fraction number. Salt concentrations (green) of each fraction are also plotted. The positions and net charges of markers are shown at the top of each panel. (*Source:* Adapted from Wei, C.M. and B. Moss, Methylated nucleotides block 5′-terminus of vaccinia virus messenger RNA, *Proceedings of the National Academy of Sciences USA* 72(1):318–322, January 1975.)

To determine the exact structure of the reovirus cap, Yasuhiro Furuichi and Kin-Ichiro Miura performed the following series of experiments. They found that they could label the cap with [β,γ-^{32}P]ATP (but not with [γ-^{32}P]ATP). This result indicated that the β-phosphate, but not the γ-phosphate, was retained in the cap. Because the β-phosphate of a nucleoside triphosphate remains only in the first nucleotide in an RNA, this finding reinforced the notion that the cap was at the 5′-terminus of the RNA. But the β-phosphate must be protected, or blocked, by some substance (X), because it cannot be removed with alkaline phosphatase.

This raised the next question: What is X? The blocking agent could be removed with phosphodiesterase, which cuts both phosphodiester and phosphoanhydride bonds (e.g., the bond between the α- and β-phosphates in a nucleotide). This enzyme released a charged substance likely to be Xp. Next, Furuichi and Miura removed the phosphate from Xp with phosphomonoesterase, leaving just X, and subjected this substance to paper electrophoresis, followed by paper chromatography. Figure 15.2 shows that X coelectrophoresed with **7-methylguanosine (m^{7}G).** Thus, the capping substance is m^{7}G.

Another product of phosphodiesterase cleavage of the cap was pAm (2′-O-methyl-AMP). Thus, m^{7}G is linked to pAm in the cap. What is the nature of the linkage? The following two considerations tell us that it is a triphosphate: (1) The α-phosphate, but not the β- or γ-phosphate, of GTP was retained in the cap. (2) The β- and α-phosphates of ATP are retained in the cap. Thus, because one phosphate comes from the capping GTP, and two come from the nucleotide (ATP) that initiated RNA synthesis, there are three phosphates (a triphosphate linkage) between the capping nucleotide (m^{7}G) and the next nucleotide. Furthermore, because both ATP and GTP have their phosphates in the 5′-position, the linkage is very likely to be 5′ to 5′.

How do we explain the charge of the reovirus cap, about −5? Figure 15.3 provides a rationale. Three negative charges come from the triphosphate linkage between the m^{7}G and the penultimate (next-to-end) nucleotide. One negative charge comes from the phosphodiester bond between the penultimate nucleotide and the next nucleotide. (This bond is not broken by alkali because the 2′-hydroxyl group, which is needed for cleavage, is methylated.) Two more negative charges come from the terminal phosphate in the cap. This makes a total of six negative charges, but the m^{7}G provides a positive charge, which gives the purified reovirus cap a charge of about −5.

Other viral and cellular mRNAs have similar caps, although the extent of 2′-O-methylation can vary to produce

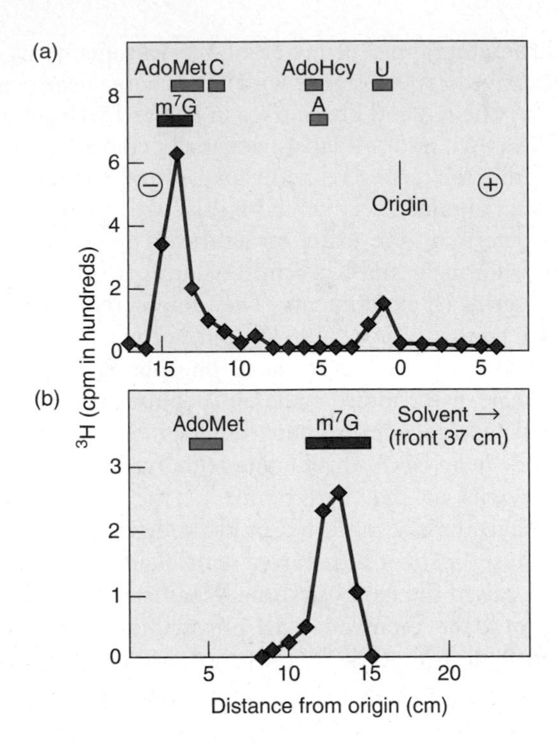

Figure 15.2 Identification of the capping substance (X) as 7-methylguanosine. Miura and Furuichi used phosphomonoesterase to digest the ^3H-labeled capping substance (Xp) to yield X. They electrophoresed this digest **(a)** along with a series of markers (*S*-adenosylmethionine, AdoMet; m^7G; *S*-adenosylhomocysteine, AdoHcy; adenosine, A; and uridine, U). Because electrophoresis did not resolve AdoMet and m^7G, these workers subjected the digest to paper chromatography **(b)** along with markers for AdoMet and m^7G. The radioactivity in X cochromatographed with the m^7G marker. (*Source:* Data from Furuichi, Y. and K. -I. Miura, A blocked structure at the 5′ terminus of mRNA from cytoplasmic polyhedrosis virus. *Nature* 253:375, 1975.)

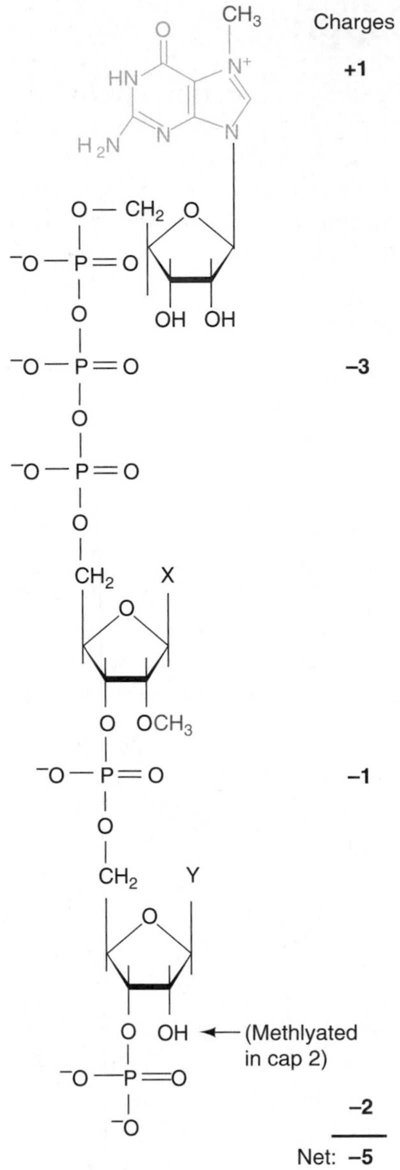

Figure 15.3 Reovirus cap structure (cap 1), highlighting the charges. The m^7G (blue guanine with red methyl group) contributes a positive charge, the triphosphate linkage contributes three negative charges, the phosphodiester bond contributes one negative charge, and the terminal phosphate contributes two negative charges. The net charge is therefore about −5. The 2′-hydroxyl group on the ribose attached to the Y base would be methylated in cap 2.

three forms of cap. **Cap 1** is the same as the cap shown in Figure 15.3. **Cap 2** has another 2′-O-methylated nucleotide (two in a row). And **cap 0** has no 2′-O-methylated nucleotides. Cap 2 is found only in eukaryotic cells, cap 1 is found in both cellular and viral RNAs, and cap 0 is found only in certain viral RNAs. Most of the snRNAs (Chapter 14) have another kind of cap, which contains a trimethylated guanosine. We will discuss these caps later in this chapter.

Cap Synthesis

To determine how caps are made, Moss and his colleagues, and Furuichi and Shatkin and their colleagues, studied capping of model substrates in vitro. These investigators used cores from vaccinia virus and reovirus, respectively, to provide the capping enzymes. Both these human viruses replicate in the cytoplasm of their host cells, so they do not have access to the host nuclear machinery. Therefore, they must carry their own transcription and capping systems right in their virus cores. In both viruses, we observe the same sequence of events, as illustrated in Figure 15.4. (a) A nucleotide phosphohydrolase (also called **RNA triphosphatase**) clips the γ-phosphate off the triphosphate at the 5′-end of the growing RNA (or model substrate), leaving a diphosphate. (b) A guanylyl transferase attaches GMP from GTP to the diphosphate at the end of the RNA, forming the 5′–5′-triphosphate linkage. (c) A methyltransferase transfers the methyl group from *S*-adenosylmethionine (AdoMet) to the 7-nitrogen of the capping guanine. (d) Another methyltransferase uses another molecule of AdoMet to methylate the 2′-hydroxyl of the penultimate nucleotide.

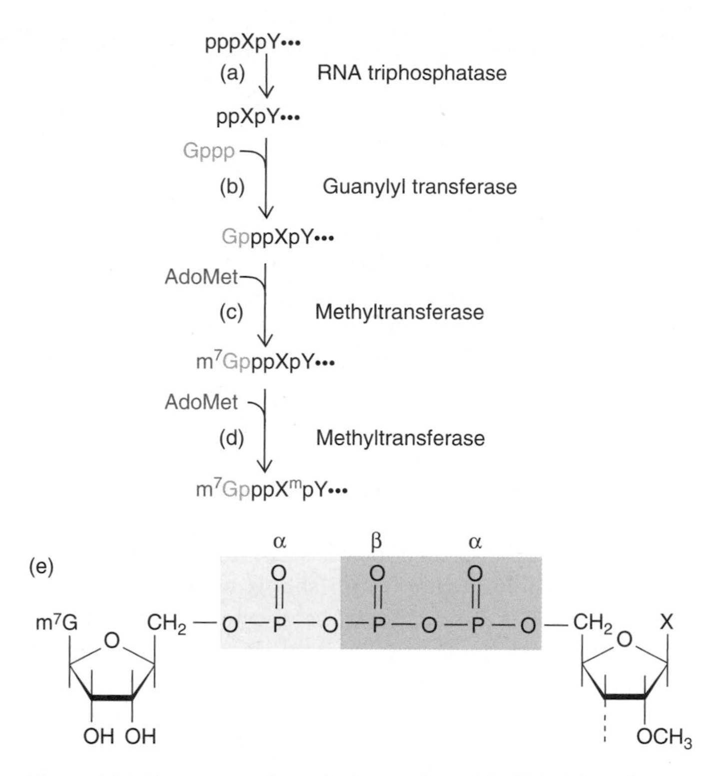

Figure 15.4 Sequence of events in capping. (a) RNA triphosphatase cleaves the γ-phosphate from the 5′-end of the growing RNA. **(b)** Guanylyl transferase adds the GMP part of GTP (blue) to form a triphosphate linkage, blocking the 5′-end of the RNA. **(c)** A methyltransferase adds a methyl group (red) from AdoMet to the N^7 of the blocking guanine. **(d)** Another methyltransferase adds a methyl group (red) from AdoMet to the 2′-hydroxyl group of the penultimate nucleotide. The product is cap 1. To form a cap 2, the next nucleotide (Y) would be methylated in a repeat of step (d). **(e)** The origin of the phosphates in the triphosphate linkage. The α- and β-phosphates from the initiating nucleotide (XTP) are highlighted in green, and the α-phosphate from the capping GTP is highlighted in yellow.

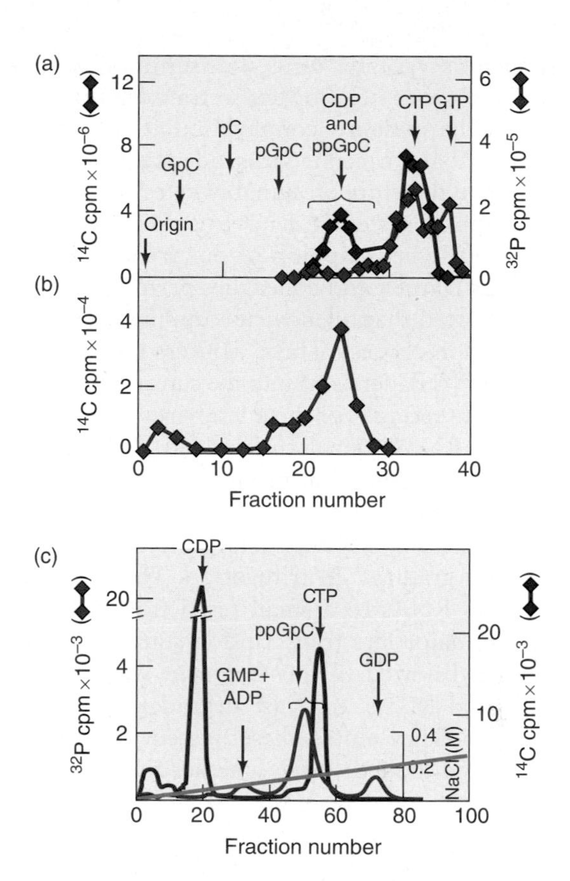

Figure 15.5 Identification of ppGpC as an intermediate in reovirus cap synthesis. (a) First purification step. Furuichi and colleagues added [^{14}C]CTP and [^{32}P]GTP to reovirus cores to label caps and capping intermediates. Then they analyzed the mixture by paper electrophoresis with the markers listed at top. One radioactive intermediate (bracket) coelectrophoresed with the ppGpC and CDP markers. **(b)** Conversion of ppGpC to GpC. Furuichi and colleagues treated the bracketed radioactive material from panel (a) with alkaline phosphatase, which should convert ppGpC to GpC, then electrophoresed the products. This time, a significant peak (though not the main peak) coelectrophoresed with the GpC marker. **(c)** Positive identification of ppGpC. Furuichi and colleagues subjected the bracketed material in (a) to ion-exchange chromatography on Dowex resin with the markers indicated at top. The major ^{32}P peak (red) coincided with the ppGpC marker. (*Source:* Adapted from Furuichi Y., S. Muthukrishnan, J. Tomasz, and A.J. Shatkin, Mechanism of formation of reovirus mRNA 5′-terminal blocked and methylated sequence m^{7}GpppG^{m}pC. *Journal of Biological Chemistry* 251:5051, 1976.)

To verify that this really is the correct pathway, the investigators isolated each of the enzymes we have listed and all of the intermediates. For example, Furuichi and colleagues started with the labeled model substrate pppGpC, which resembles the 5′-end of a newly initiated reovirus mRNA. How do we know that the virus cores can remove a terminal phosphate and convert this starting material to ppGpC? These workers blocked the guanylyl transferase reaction with an excess of by-product (PP_i), which should cause ppGpC to build up, if it exists. They looked directly for this intermediate by the scheme in Figure 15.5. First, they performed paper electrophoresis with markers and showed that a significant labeled product coelectrophoresed with the ppGpC marker. Unfortunately, CDP also electrophoresed to this position, so the product could not be clearly identified. Next, they treated the product with alkaline phosphatase to convert any ppGpC to GpC and reelectrophoresed it. Now a peak of radioactivity appeared in the GpC position. This was encouraging, but to positively identify ppGpC, these workers subjected the putative ppGpC peak from panel (a) to ion-exchange chromatography on a Dowex resin and obtained a radioactive peak that comigrated uniquely with the ppGpC marker. Thus, ppGpC is a real intermediate in the capping scheme. Relatively little ^{14}C radioactivity appeared in the ppGpC peak because of the lower radioactivity of the ^{14}C label.

When is the cap added? In some viruses, such as cytoplasmic polyhedrosis virus (CPV), lack of AdoMet completely inhibits transcription, suggesting that transcription depends on capping. This implies that capping in this virus is a very early event and presumably occurs soon after the first phosphodiester bond forms in the pre-mRNA. In other viruses, such as vaccinia virus, transcription occurs normally in the absence of AdoMet, so transcription and capping may not be so tightly coupled in that virus.

Unlike CPV and vaccinia virus, adenovirus replicates in the nucleus and therefore presumably takes advantage of the host cell's capping system. Adenovirus should therefore tell us more about when capping of eukaryotic pre-mRNAs occurs. James Darnell and colleagues performed an experiment that showed that adenovirus capping occurs early in the transcription process. These workers measured the incorporation of [^{3}H]adenosine into the cap and the first dozen or so adenylate residues of the adenovirus major late transcripts (pre-mRNAs). First, they added [^{3}H]adenosine to label the cap (the bold A in m^{7}Gppp**A**) and other adenosines in adenovirus pre-mRNAs during the late phase of infection. Then they separated large from small mRNA precursors by gradient centrifugation. Then they hybridized the small RNAs to a small restriction fragment that included the major late transcription start site. Any short RNAs that hybridized to this fragment were likely to be newly initiated RNAs, not just degradation products of mature RNAs. They eluted these nascent fragments from the hybrids and looked to see whether they were capped. Indeed they were, and no pppA, which would have been present on uncapped RNA, could be detected. This experiment demonstrated that caps are added to adenovirus major late pre-mRNA before the chain length reaches about 70 nt. It is now generally accepted that capping in eukaryotic cells occurs even earlier than that: before the pre-mRNA chain length reaches 30 nt.

SUMMARY Caps are made in steps: First, an RNA triphosphatase removes the terminal phosphate from a pre-mRNA; next, a guanylyl transferase adds the capping GMP (from GTP). Next, two methyltransferases methylate the N^7 of the capping guanosine and the 2′-O-methyl group of the penultimate nucleotide. These events occur early in the transcription process, before the chain length reaches 30 nt.

Functions of Caps

Caps appear to serve at least four functions. (1) They protect mRNAs from degradation. (2) They enhance the translatability of mRNAs. (3) They enhance the transport of mRNAs from the nucleus into the cytoplasm. (4) They enhance the efficiency of splicing of mRNAs. In this section we will discuss the first three of these functions, then deal with the fourth later in the chapter.

Protection The cap is joined to the rest of the mRNA through a triphosphate linkage found nowhere else in the RNA. The cap might therefore be expected to protect the mRNA from attack by RNases that begin at the 5′-end of their substrates and that cannot cleave triphosphate linkages. In fact, good evidence supports the notion that caps protect mRNAs from degradation.

Furuichi, Shatkin, and colleagues showed in 1977 that capped reovirus RNAs are much more stable than uncapped RNAs. They synthesized newly labeled reovirus RNA that was either capped with m^{7}GpppG, "blocked" with GpppG, or uncapped. Then they injected each of the three kinds of RNA into *Xenopus* oocytes, left them there for 8 h, then purified them and analyzed them by glycerol gradient ultracentrifugation. Reovirus RNAs exist in three size classes, termed large (l), medium (m), and small (s). Figure 15.6a shows a glycerol gradient ultracentrifugation separation of these three RNA classes. Furuichi and colleagues included RNAs with all three kinds of 5′-ends in this experiment, and no significant differences could be seen. All three size classes are clearly visible. Figure 15.6b shows what happened to these RNAs after 8 h in *Xenopus* oocytes. RNAs with all three kinds of 5′-ends had suffered degradation, but this degradation was much more pronounced for the uncapped RNAs. Thus, the *Xenopus* oocytes contain nucleases that

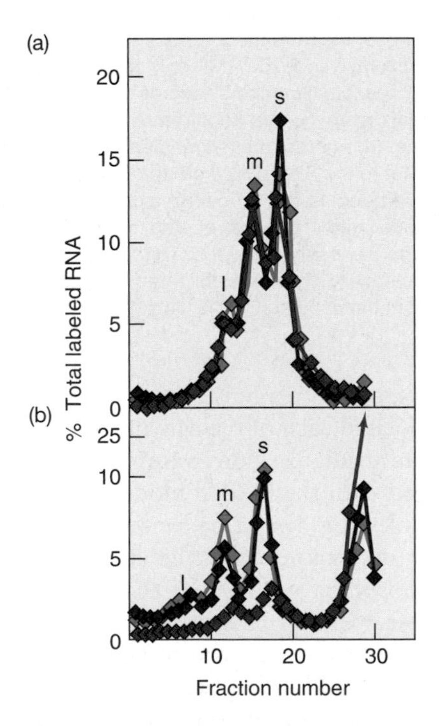

Figure 15.6 Effect of cap on reovirus RNA stability. (a) Appearance of newly synthesized RNAs. Furuichi and colleagues made labeled reovirus RNAs with capped (green), blocked (blue), or uncapped (red) 5′-ends, then subjected these RNAs to glycerol gradient ultracentrifugation. The three size classes of RNA are labeled l, m, and s. **(b)** Effect of incubation in *Xenopus* oocytes. Furuichi and colleagues injected the RNAs with the three different 5′-ends into *Xenopus* oocytes. After 8 h they purified the RNAs and performed the same sedimentation analysis as in panel (a). Colors have the same meaning as in panel (a). (*Source:* Adapted from Furuichi, Y., A. LaFiandra, and A.J. Shatkin, 5′-terminal structure and mRNA stability. *Nature* 266:236, 1977.)

Table 15.1 Synergism Between Poly(A) and Cap during Translation of Luciferase mRNA in Tobacco Protoplasts

mRNA	Luciferase mRNA Half-Life (min)	Luciferase Activity (light units/mg protein)	Relative Effect of Poly(A) on Activity	Relative Effect of Cap on Activity
Uncapped				
Poly(A)$^-$	31	2941	1	1
Poly(A)$^+$	44	4480	1.5	1
Capped				
Poly(A)$^-$	53	62,595	1	21
Poly(A)$^+$	100	1,331,917	21	297

Source: Gallie, D.R., The cap and poly(A) tail function synergistically to regulate mRNA translational efficiency, *Genes & Development* 5:2108–2116, 1991. Copyright © Cold Spring Harbor, NY. Reprinted by permission.

degrade the viral RNAs, but the caps appear to provide some protection from these nucleases.

Translatability Another important function of the cap is to provide translatability. We will see in Chapter 17 that a eukaryotic mRNA gains access to the ribosome for translation via a cap-binding protein that recognizes the cap. If there is no cap, the cap-binding protein cannot bind and the mRNA is very poorly translated. Using an in vivo assay, Daniel Gallie documented the stimulatory effect of the cap on translation. In this procedure, Gallie introduced the firefly luciferase mRNA, with and without a cap, and with and without poly(A), into tobacco cells. Luciferase is an easy product to detect because of the light it generates in the presence of luciferin and ATP. Table 15.1 illustrates that the poly(A) at the 3′-end and the cap at the 5′-end act synergistically to stabilize and, especially, to enhance the translation of luciferase mRNA. Poly(A) provided a 21-fold boost in translation of a capped mRNA, but that was a minor effect compared with the 297-fold stimulation of translation that the cap conferred on a polyadenylated mRNA. Of course, mRNA stability also figured into these numbers, but its effect was not great.

Transport of RNA The cap also appears to facilitate the transport of at least some mature RNAs out of the nucleus. Jörg Hamm and Iain Mattaj studied the behavior of U1 snRNA to reach this conclusion. Most of the snRNA genes, including the U1 snRNA gene, are normally transcribed by RNA polymerase II, and the transcripts receive monomethylated (m^7G) caps in the nucleus. They migrate briefly to the cytoplasm, where they bind to proteins to form snRNPs, and their caps are modified to trimethylated ($m^{2,2,7}G$) structures. Then they reenter the nucleus, where they participate in splicing and other activities. The U6 snRNA is exceptional. It is made by polymerase III and is not capped. It retains its terminal triphosphate and remains in the nucleus. Hamm and Mattaj wondered what would happen if they arranged for the U1 snRNA gene to be transcribed by polymerase III instead of polymerase II. If it failed to be capped and remained in the nucleus, that would suggest that capping is important for transporting an RNA out of the nucleus.

Thus, Hamm and Mattaj placed the *Xenopus* U1 snRNA gene under the control of the human U6 snRNA promoter, so it would be transcribed by polymerase III. Then they injected this construct into *Xenopus* oocyte nuclei, along with a labeled nucleotide and a *Xenopus* 5S rRNA gene, which acted as an internal control. They also included 1 μg/mL of α-amanitin to inhibit RNA polymerase II and therefore ensure that no transcripts of the U1 gene would be made by polymerase II. In addition to the wild-type U1 gene, these workers also used several mutant U1 genes, with lesions in the regions coding for protein-binding sites. Loss of ability to associate with the proper proteins in the cytoplasm rendered the products of these mutant genes unable to return to the nucleus once they had been transported to the cytoplasm. Twelve hours after injection, Hamm and Mattaj dissected the oocytes into nuclear and cytoplasmic fractions and electrophoresed the labeled products in each. They compared the cellular locations of capped U1 snRNAs made by RNA polymerase II and uncapped U1 snRNA made by polymerase III.

Virtually all the uncapped U1 snRNA made by polymerase III remained in the nucleus. On the other hand, the U1 snRNAs made by polymerase II were transported to the cytoplasm. These results are consistent with the hypothesis that capping is required for U1 snRNA to be transported out of the nucleus.

Finally, as we will see later in this chapter, the cap is essential for proper splicing of a pre-mRNA.

SUMMARY The cap provides: (1) protection of the mRNA from degradation; (2) enhancement of the mRNA's translatability; (3) transport of at least some RNAs out of the nucleus; and (4) proper splicing of the pre-mRNA.

15.2 Polyadenylation

We have already seen that hnRNA is a precursor to mRNA. One finding that suggested such a relationship between these two types of RNA was that they shared a unique structure at their 3′-ends: a long chain of AMP residues called **poly(A).** Neither rRNA nor tRNA has a poly(A) tail. The process of adding poly(A) to RNA is called **polyadenylation.** Let us examine first the nature of poly(A) and then the polyadenylation process.

Poly(A)

James Darnell and his coworkers performed much of the early work on poly(A) and polyadenylation. To purify HeLa cell poly(A) from the rest of the mRNA molecule, Diana Sheiness and Darnell released it with two enzymes: RNase A, which cuts after the pyrimidine nucleotides C and U, and RNase T1, which cuts after G nucleotides. In other words, they cut the RNA after every nucleotide except the A's, preserving only pure runs of A's. Next, Sheiness and Darnell electrophoresed the poly(A)s from nuclei and from cytoplasm to determine their sizes. Figure 15.7 shows the results, which demonstrate that both poly(A)s have major peaks that electrophoresed more slowly than 5S rRNA, at about 7S. Sheiness and Darnell estimated that this corresponded to about 150–200 nt. The poly(A) species observed in this experiment were labeled for only 12 min, so they were newly synthesized. Little difference in size between these fresh nuclear and cytoplasmic poly(A)s is noticeable. However, cytoplasmic poly(A) is subject to shortening, as we will see later in this chapter. Now that poly(A)s from many different organisms have been analyzed, we see an average size of fresh poly(A) of about 250 nt.

It is apparent that the poly(A) goes on the 3′-end of the mRNA or hnRNA because it can be released very quickly with an enzyme that degrades RNAs from the 3′-end inward. Furthermore, complete RNase digestion of poly(A) yielded one molecule of adenosine and about 200 molecules of AMP. Figure 15.8 demonstrates that this requires poly(A) to be at the 3′-end of the molecule. This experiment also reinforced the conclusion that poly(A) is about 200 nt long.

We also know that poly(A) is not made by transcribing DNA because genomes contain no runs of T's long enough to encode it. In particular, we find no runs of T's at the ends of any of the thousands of eukaryotic genes that have been sequenced. Furthermore, actinomycin D, which inhibits DNA-directed transcription, does not inhibit polyadenylation. Thus, poly(A) must be added posttranscriptionally. In fact, there is an enzyme in nuclei called **poly(A) polymerase (PAP)** that adds AMP residues one at a time to mRNA precursors.

We know that poly(A) is added to mRNA precursors because it is found on hnRNA. Even specific unspliced mRNA precursors (the 15S mouse globin mRNA precursor, for example) contain poly(A). However, as we will see later in this chapter, splicing of some introns in a pre-mRNA can occur before polyadenylation. Once an mRNA enters the cytoplasm, its poly(A) turns over; in other words, it is constantly being broken down by RNases and rebuilt by a cytoplasmic poly(A) polymerase.

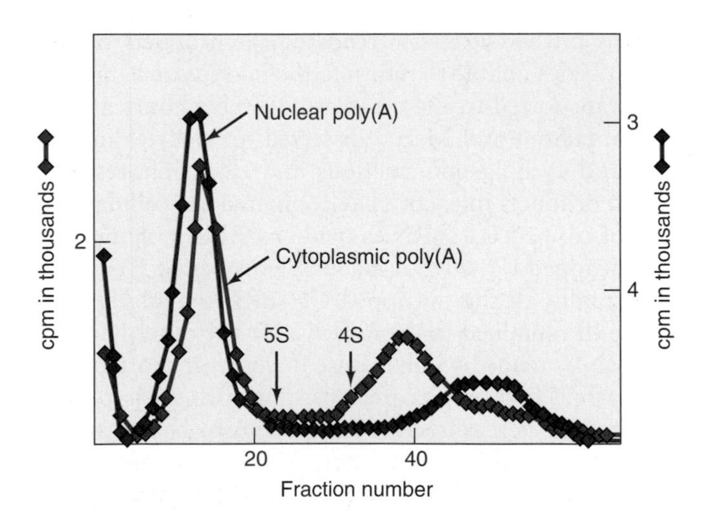

Figure 15.7 Size of poly(A). Sheiness and Darnell isolated radioactively labeled hnRNA from the nuclei (blue), and mRNA from the cytoplasm (red) of HeLa cells, then released poly(A) from these RNAs by RNase A and RNase T1 treatment. They electrophoresed the poly(A)s, collected fractions, and determined their radioactivities by scintillation counting (Chapter 5). They included 4S tRNA and 5S rRNA as size markers. Both poly(A)s electrophoresed more slowly than the 5S marker, corresponding to molecules about 200 nt long. (*Source:* Adapted from Sheiness, D. and J.E. Darnell, Polyadenylic acid segment in mRNA becomes shorter with age. *Nature New Biology* 241:267, 1973.)

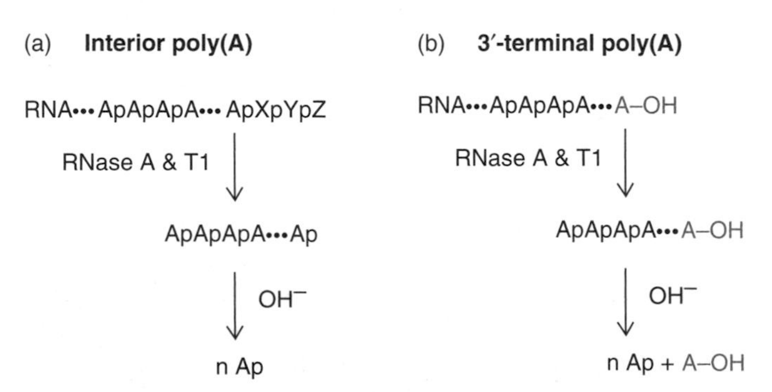

Figure 15.8 Finding poly(A) at the 3′-end of hnRNA and mRNA. (**a**) Interior poly(A). If poly(A) were located in the interior of an RNA molecule, RNase A and RNase T1 digestion would yield poly(A) with a phosphate at the 3′-end, then base hydrolysis would give only AMP. (**b**) Poly(A) at the 3′-end of hnRNA and mRNA. Because poly(A) is located at the 3′-end of these RNA molecules, RNase A and T1 digestion yields poly(A) with an unphosphorylated adenosine at the 3′-end. Base hydrolysis gives AMP plus one molecule of adenosine. In fact, the ratio of AMP to adenosine was 200, suggesting a poly(A) length of about 200 nt.

SUMMARY Most eukaryotic mRNAs and their precursors have a chain of AMP residues about 250 nt long at their 3′-ends. This poly(A) is added posttranscriptionally by poly(A) polymerase.

Functions of Poly(A)

Most mRNAs contain poly(A). One noteworthy exception is the histone mRNAs, which manage to perform their functions without detectable poly(A) tails. This exception notwithstanding, the near universality of poly(A) in eukaryotes raises the question: What is the purpose of poly(A)? One line of evidence suggests that it helps protect mRNAs from degradation. Another indicates that it stimulates translation of mRNAs to which it is attached. Still others show that poly(A) plays a role in splicing and transport of mRNA out of the nucleus. Here we will consider evidence for the effect of poly(A) on mRNA stability and translatability. We will return to the themes of splicing and transport at the end of this chapter.

Protection of mRNA To examine the stabilizing effect of poly(A), Michel Revel and colleagues injected globin mRNA, with and without poly(A) attached, into *Xenopus* oocytes and measured the rate of globin synthesis at various intervals over a 2-day period. They found that there was little difference at first. However, after only 6 h, the mRNA without poly(A) [**poly(A)$^-$ RNA**] could no longer support translation, while the mRNA with poly(A) [**poly(A)$^+$ RNA**] was still quite actively translated (Figure 15.9). The simplest explanation for this behavior is that the poly(A)$^+$ RNA has a longer lifetime than the poly(A)$^-$ RNA, and that poly(A) is therefore the protective agent. However, as we will see, other experiments have shown no protective effect of poly(A) on certain other mRNAs. Regardless, it is clear that poly(A) plays an even bigger role in efficiency of translation of mRNA.

Translatability of mRNA Several lines of evidence indicate that poly(A) also enhances the translatability of an mRNA. One of the proteins that binds to a eukaryotic mRNA during translation is **poly(A)-binding protein I, (PAB I).** Binding to this protein seems to boost the efficiency with which an mRNA is translated. One line of evidence in favor of this hypothesis is that excess poly(A) added to an in vitro reaction inhibited translation of a capped, polyadenylated mRNA. This finding suggested that the excess poly(A) was competing with the poly(A) on the mRNA for an essential factor, presumably for PAB I. Without this factor, the mRNA could not be translated well. Carrying this argument one step further leads to the conclusion that poly(A)$^-$ RNA, because it cannot bind PAB I, cannot be translated efficiently.

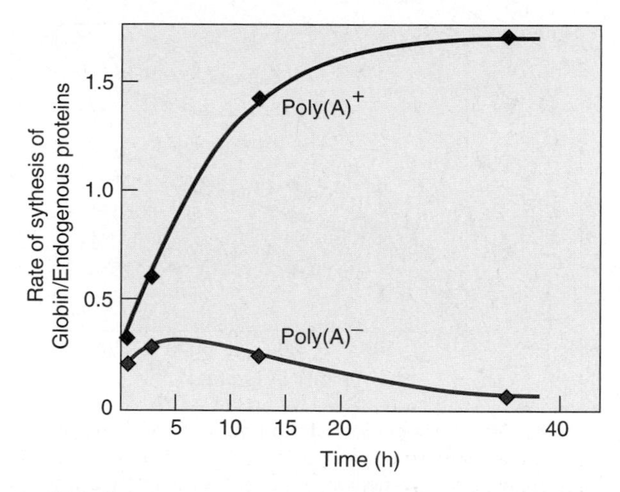

Figure 15.9 Time course of translation of poly(A)$^+$ (blue) and poly(A)$^-$ (red) globin mRNA. Revel and colleagues plotted the ratio of radioactivity incorporated into globin and endogenous protein versus the midpoint of the labeling time. (*Source:* Adapted from Huez, G., G. Marbaix, E. Hubert, M. Leclereq, U. Nudel, H. Soreq, R. Solomon, B. Lebleu, M. Revel, and U.Z. Littauer, Role of the polyadenylate segment in the translation of globin messenger RNA in *Xenopus* oocytes. *Proceedings of the National Academy of Sciences USA* 71(8):3143–3146, August 1974.)

To test the hypothesis that poly(A)$^-$ RNA is not translated efficiently, David Munroe and Allan Jacobson compared the rates of translation of two synthetic mRNAs, with and without poly(A), in rabbit reticulocyte extracts. They made the mRNAs (rabbit β-globin [RBG] mRNA and vesicular stomatitis virus N gene [VSV.N] mRNA) by cloning their respective genes into plasmids under the control of the phage SP6 promoter, then transcribing these genes in vitro with SP6 RNA polymerase. They endowed the synthetic mRNAs with various length poly(A) tails by adding poly(T) to their respective genes with terminal transferase and dTTP for varying lengths of time before cloning and transcription.

Munroe and Jacobson tested the poly(A)$^+$ and poly(A)$^-$ mRNAs for both translatability and stability in the reticulocyte extract. Figure 15.10 shows the effects of both capping and polyadenylation on translatability of the VSV.N mRNA. Both capped and uncapped mRNAs were translated better with poly(A) than without. Further experiments showed that polyadenylation made no difference in the stability of either mRNA. Munroe and Jacobson interpreted these results to mean that the extra translatability conferred by poly(A) was not due to stabilization of the mRNAs, but to enhanced translation per se. If so, what aspect of translation is enhanced by poly(A)? These studies suggested that it is a step at the very beginning of the translation process: association between mRNA and ribosomes. We will see in Chapter 17 that many ribosomes bind sequentially at the beginning of eukaryotic mRNAs and read the message in tandem. An mRNA with more than one ribosome translating it at once is called a polysome. Munroe

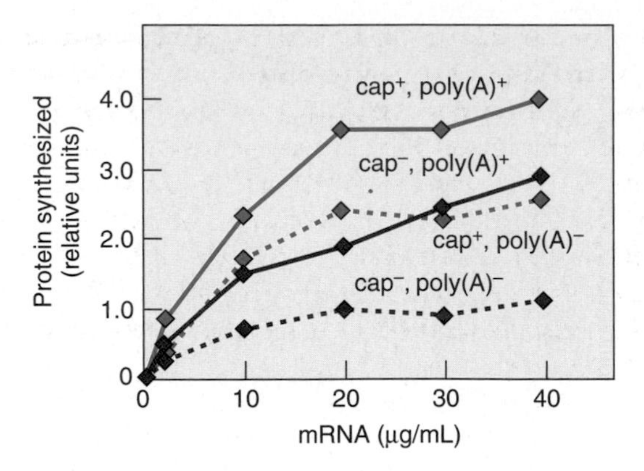

Figure 15.10 Effect of polyadenylation on translatability of mRNAs. Munroe and Jacobson incubated VSV.N mRNAs with [^{35}S]methionine in rabbit reticulocyte extracts. The mRNAs were capped (green) or uncapped (red), and poly(A)$^+$ (68 As; solid lines) or poly(A)$^-$ (dashed lines). After allowing 30 min for protein synthesis, these workers electrophoresed the labeled products and measured the radioactivity of the newly made protein by quantitative fluorography. **Poly(A) enhanced the translatability of both capped and uncapped mRNAs.** (*Source:* Adapted from Munroe, D. and A. Jacobson, mRNA poly(A) tail, a 3′ enhancer of a translational initiation. *Molecular and Cellular Biology* 10:3445, 1990.)

and Jacobson contended that poly(A)$^+$ mRNA forms polysomes more successfully than poly(A)$^-$ mRNA.

These workers measured the incorporation of labeled mRNAs into polysomes as follows: They labeled poly(A)$^+$ mRNA with ^{32}P and poly(A)$^-$ mRNA with ^{3}H, then incubated these RNAs together in a reticulocyte extract. Then they separated polysomes from monosomes by sucrose gradient ultracentrifugation. Figure 15.11a indicates that the poly(A)$^+$ VSV.N mRNA was significantly more associated with polysomes than was poly(A)$^-$ mRNA. In parallel experiments, the RBG mRNA exhibited the same behavior. Figure 15.11b shows the effect of *length* of poly(A) attached to RBG mRNA on the extent of polysome formation. We see the greatest increase as the poly(A) grows from 5 to 30 nt, and a more gradual increase as more A residues are added.

Munroe and Jacobson's finding that poly(A) did not affect the stability of mRNAs seems to contradict the earlier work by Revel and colleagues. Perhaps the discrepancy arises from the fact that the early work was done in intact frog eggs, whereas the later work used a cell-free system. Earlier in this chapter, Table 15.1 showed that poly(A) stimulated transcription of luciferase mRNA. The stabilizing effect of poly(A) on this mRNA was twofold at most, whereas the overall increase in luciferase production caused by poly(A) was up to 20-fold. Thus, this system also suggested that enhancement of translatability by poly(A) seems to be more important than mRNA stabilization.

In Chapter 17, we will see how poly(A) can both protect and stimulate the translation of an mRNA. Briefly, poly(A) can bind to cytoplasmic poly(A)-binding proteins. These in turn can bind to a translation initiation factor (eIF4G), which binds to a cap-binding protein, bound to the cap. In this way, the poly(A) at the 3′-end, and the cap at the 5′-end of the mRNA are brought together, effectively circularizing the mRNA. The mRNA in this closed loop

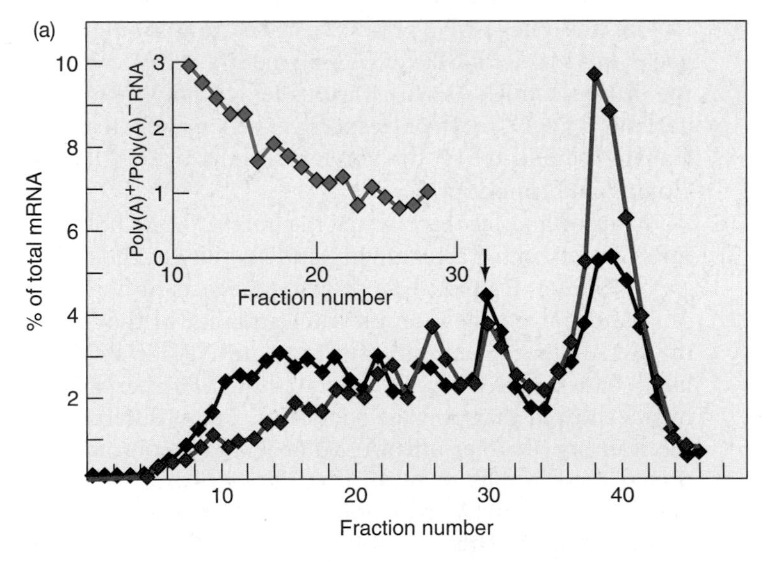

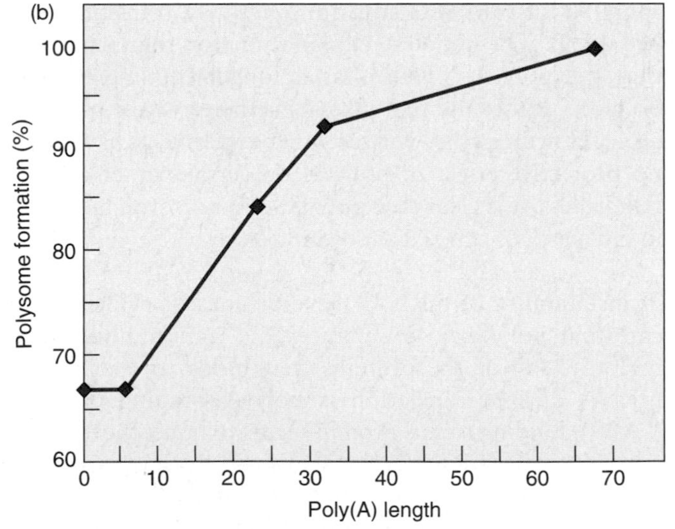

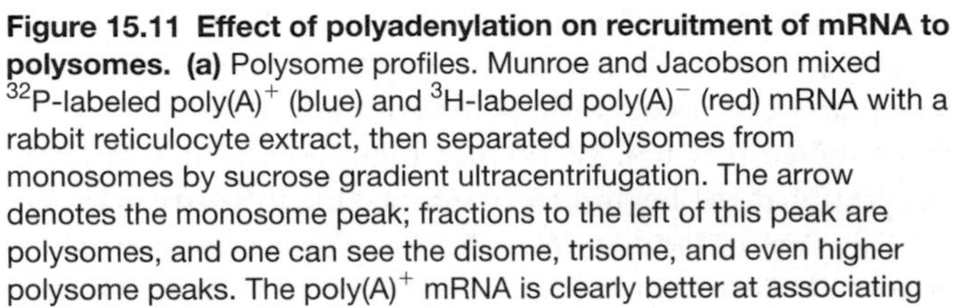

Figure 15.11 Effect of polyadenylation on recruitment of mRNA to polysomes. (a) Polysome profiles. Munroe and Jacobson mixed ^{32}P-labeled poly(A)$^+$ (blue) and ^{3}H-labeled poly(A)$^-$ (red) mRNA with a rabbit reticulocyte extract, then separated polysomes from monosomes by sucrose gradient ultracentrifugation. The arrow denotes the monosome peak; fractions to the left of this peak are polysomes, and one can see the disome, trisome, and even higher polysome peaks. The poly(A)$^+$ mRNA is clearly better at associating with polysomes, especially the higher polysomes. The inset shows the ratio of poly(A)$^+$ to poly(A)$^-$ RNA in fractions 11–28. Again, this demonstrates a preferential association of poly(A)$^+$ mRNA with polysomes (the lower fraction numbers). **(b)** Efficiency of polysome formation as a function of poly(A) length on VSV.N mRNA. The efficiency at a tail length of 68 is taken as 100%. (*Source:* Adapted from Munroe, D. and A. Jacobson, mRNA poly(A) tail, a 3′ enhancer of a translational initiation. *Molecular and Cellular Biology* 10:3447–8, 1990.)

form, with proteins binding to both its ends, is more stable than linear, naked mRNA would be. The mRNA is also more readily translated in this loop form, partly because the eIF4G, which ties the loop together, can help recruit the ribosomes to the mRNA.

SUMMARY Poly(A) enhances both the lifetime and translatability of mRNA. The relative importance of these two effects seems to vary from one system to another. At least in rabbit reticulocyte extracts, poly(A) seems to enhance translatability by helping to recruit mRNA to polysomes.

Basic Mechanism of Polyadenylation

It would be logical to assume that poly(A) polymerase simply waits for a transcript to be finished, then adds poly(A) to the 3′-end of the RNA. However, this is not what ordinarily happens. Instead, the mechanism of polyadenylation usually involves clipping an mRNA precursor, even before transcription has terminated, and then adding poly(A) to the newly exposed 3′-end (Figure 15.12). Thus, contrary to expectations, RNA polymerase can still be elongating an RNA chain, while the polyadenylation apparatus has already located a polyadenylation signal somewhere upstream, cut the growing RNA, and polyadenylated it.

Joseph Nevins and James Darnell provided some of the first evidence for this model of polyadenylation. They chose to study the adenovirus major late transcription unit because it serves as the template for several different overlapping mRNAs, each of which is polyadenylated at one of five separate sites. Recall from Chapter 14 that each of these mature mRNAs has the same three leader exons spliced to a different coding region. The poly(A) of each is attached

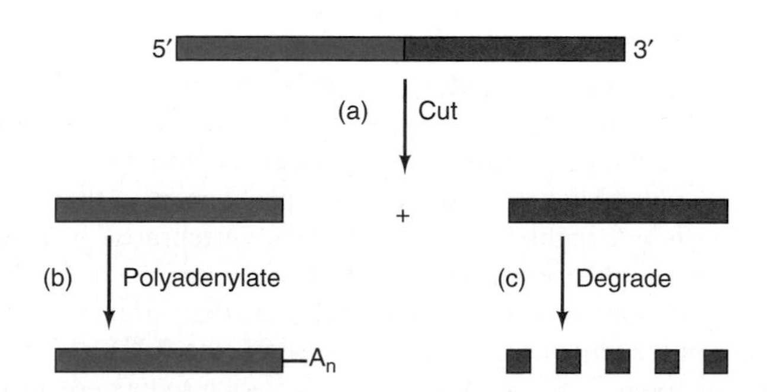

Figure 15.12 Overview of the polyadenylation process. (a) Cutting. The first step is cleaving the transcript, which may actually still be in the process of being made. The cut occurs at the end of the RNA region (green) that will be included in the mature mRNA. **(b)** Polyadenylation. The poly(A) polymerase adds poly(A) to the 3′-end of the mRNA. **(c)** Degradation of the extra RNA. All RNA (red) lying beyond the polyadenylation site is superfluous and is destroyed.

to the 3′-end of the coding region. There are two alternative hypotheses for the relationship between transcription termination and polyadenylation in this system. (1) Transcription terminates immediately downstream of a polyadenylation site, and then polyadenylation occurs. For example, if gene A is being expressed, transcription will proceed only to the end of coding region A, then terminate, and then polyadenylation will occur at the 3′-end left by that termination event. (2) Transcription goes at least to the end of the last coding exon, and polyadenylation can occur at any polyadenylation site, presumably even before transcription of the whole major late region is complete.

The first hypothesis, that transcription does not always go clear to the end, was easy to eliminate. Nevins and Darnell hybridized radioactive RNA made in cells late in infection to DNA fragments from various positions throughout the major late region. If primary transcripts of the first gene stopped after the first polyadenylation site, and only transcripts of the last gene made it all the way to the end, then much more RNA would hybridize to fragments near the 5′-end of the major late region than to fragments near the 3′-end. But RNA hybridized to all the fragments equally well—to fragments near the 3′-end of the region just as well as to fragments near the 5′-end. Therefore, once a transcript of the major late region is begun, it is elongated all the way to the end of the region before it terminates. In other words, the major late region contains only one transcription terminator, and it lies at the end of the region. Thus, this whole region can be called a **transcription unit** to denote the fact that it is transcribed as a whole, even though it contains multiple genes. Nevins and Darnell went on to show that clipping and polyadenylation usually occurred before transcription had terminated.

This behavior of transcribing far past a polyadenylation site before clipping and polyadenylating the transcript seems wasteful because all the RNA past the polyadenylation site will be destroyed without being used. So the question naturally arises: Is this method of polyadenylation unique to viruses, or does it also occur in ordinary cellular transcripts? To find out, Erhard Hofer and James Darnell isolated labeled RNA from Friend mouse erythroleukemia cells that had been induced with dimethyl sulfoxide (DMSO) to synthesize large quantities of globin, and therefore to transcribe the globin genes at a high rate. They hybridized the labeled transcripts to cloned fragments representing various parts of the mouse β-globin gene, and regions downstream of the gene (Figure 15.13). They observed just as much hybridization to fragments lying over 500 bp downstream of the polyadenylation site as to fragments within the globin gene. This demonstrated that transcription continues at least 500 bp downstream of this polyadenylation site. In further studies, these workers found that transcription finally terminated in regions lying even farther downstream. Thus, transcription significantly beyond the polyadenylation site occurs in cellular, as well as viral, transcripts.

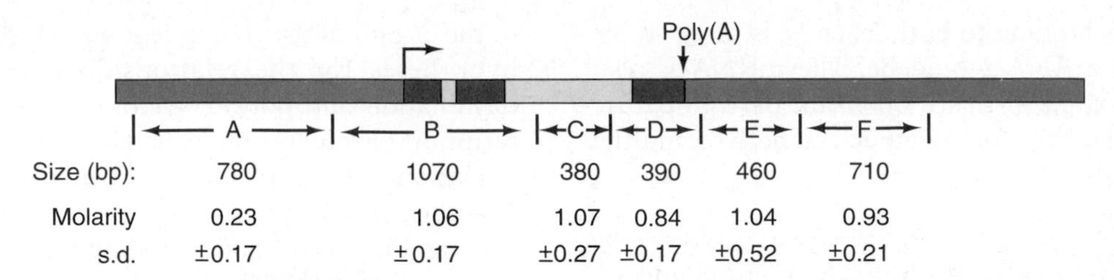

Figure 15.13 Transcription beyond the polyadenylation site. Hofer and Darnell isolated nuclei from DMSO-stimulated Friend erythroleukemia cells and incubated them with [^{32}P]UTP to label run-on RNA—mostly globin pre-mRNA. Then they hybridized this labeled RNA to DNA fragments A–F, whose locations and sizes are given in the diagram at top. The molarities of RNA hybridization to each fragment are given beneath each, with their standard deviations (s.d.). In the physical map at top, the exons are in red and the introns are in yellow. (*Source:* Adapted from E. Hofer and J.E. Darnell, The primary transcription unit of the mouse β-major globin gene. *Cell* 23:586, 1981.)

SUMMARY Transcription of eukaryotic genes extends beyond the polyadenylation site. Then the transcript is cleaved and polyadenylated at the 3′-end created by the cleavage.

Polyadenylation Signals

If the polyadenylation apparatus does not recognize the ends of transcripts, but binds somewhere in the middle to cleave and polyadenylate, what is it about a polyadenylation site that attracts this apparatus? The answer to this question depends on what kind of eukaryote or virus we are discussing. Let us first consider mammalian **polyadenylation signals.** By 1981, molecular biologists had examined the sequences of dozens of mammalian genes and had found that the most obvious common feature they had was the sequence AATAAA about 20 bp before the polyadenylation site. At the RNA level, the sequence AAUAAA occurs in most mammalian mRNAs about 20 nt upstream of their poly(A). Molly Fitzgerald and Thomas Shenk tested the importance of the **AAUAAA** sequence in two ways. First, they deleted nucleotides between this sequence and the polyadenylation site and sequenced the 3′-ends of the resulting RNAs. They found that the deletions simply shifted the polyadenylation site downstream by roughly the number of nucleotides deleted.

This result suggested that the AAUAAA sequence is at least part of a signal that causes polyadenylation approximately 20 nt downstream. If so, then deleting this sequence should abolish polyadenylation altogether. These workers used an S1 assay as follows to show that it did. They created a recombinant SV40 virus (mutant 1471) with duplicate polyadenylation signals 240 bp apart, at the end of the late region. S1 analysis of the 3′-ends of the late transcripts (Chapter 5) revealed two signals 240 bp apart (Figure 15.14). [We can ignore the poly(A) in this kind of experiment because it does not hybridize to the probe.] Thus, both polyadenylation sites worked, implying some readthrough of the first site. Then Fitzgerald and Shenk deleted either the first AATAAA (mutant 1474) or the second AATAAA (mutant 1475) and reran the S1 assay. This time, the polyadenylation site just downstream of the deleted AATAAA did not function, demonstrating that AAUAAA in the pre-mRNA is necessary for polyadenylation. We shall see shortly, however, that this is only part of the mammalian polyadenylation signal.

Is the AAUAAA invariant, or is some variation tolerated? Early experiments with manipulated signals (AAUACA, AAUUAA, AACAAA, and AAUGAA) suggested that no deviation from AAUAAA could occur without destroying polyadenylation. But by 1990, a compilation of polyadenylation signals from 269 vertebrate cDNAs showed some variation in these natural signals, especially in the second nucleotide. Marvin Wickens compiled these data, which defined a consensus sequence (Figure 15.15). The most common sequence, at the RNA level, is AAUAAA, and it is the most efficient in promoting polyadenylation. The most common variant is AUUAAA, and it is about 80% as efficient as AAUAAA. The other variants are much less common, and also much less efficient.

By now it has also become clear that AAUAAA by itself is not sufficient for polyadenylation. If it were, then polyadenylation would occur downstream of the many AAUAAA sequences found in introns, but it does not. Several investigators found that polyadenylation can be disrupted by deleting sequences immediately downstream of the polyadenylation site. This raised the suspicion that the region just downstream of the polyadenylation site contains another element of the polyadenylation signal. The problem was that that region is not highly conserved among vertebrates. Instead, there is simply a tendency for it to be GU- or U-rich.

These considerations suggested that the minimum efficient polyadenylation signal is the sequence AAUAAA followed about 20 bp later by a GU- or U-rich sequence. Anna Gil and Nicholas Proudfoot tested this hypothesis by examining the very efficient rabbit β-globin polyadenylation signal, which contains an AAUAAA, followed 24 bp later by a GU-rich region, immediately followed by a U-rich region. Throughout this discussion, we will refer to the sequences of the RNA (e.g., AAUAAA), even though the

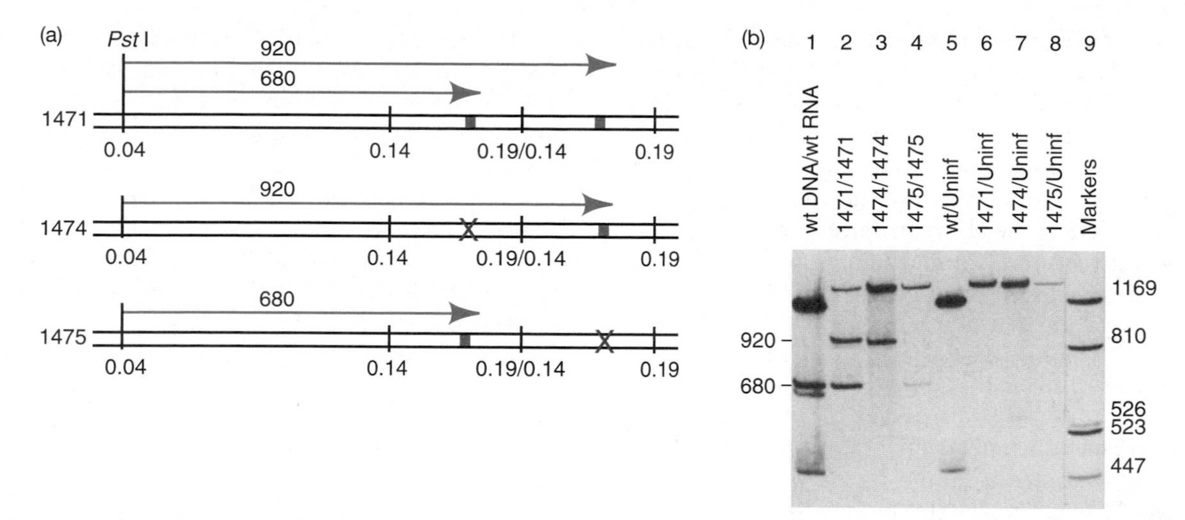

Figure 15.14 Importance of the AAUAAA sequence to polyadenylation. Fitzgerald and Shenk created recombinant SV40 viruses with the following characteristics **(a)** Mutant 1471 contained duplicate late polyadenylation sites (green) 240 bp apart within the duplicated region, which extends from 0.14 to 0.19 map units. Mutant 1474 contained a 16-bp deletion (red) at the AAUAAA in the upstream site, and mutant 1475 contained the same kind of 16-bp deletion (red) in the downstream site, resulting in the loss of the corresponding AAUAAA sequences in the pre-mRNAs produced by these mutant genes. Then they performed S1 analysis with a probe that should yield a 680-nt signal if the upstream polyadenylation signal works, and a 920-nt signal if the downstream polyadenylation signal works (blue arrows). **(b)** Experimental results. The lanes are marked at the top with the probe designation, followed by the RNA (or template) designation. Lane 1, using only wild-type probe and template, showed the wild-type signal at 680 nt, as well as an artifactual signal not usually seen. Lanes 5–8 are uninfected negative controls. The top band in each lane represents reannealed S1 probe and can be ignored. The results, also diagrammed in panel (a), show that deletion of an AAUAAA prevents polyadenylation at that site. (*Source:* Adapted from Fitzgerald, M. and T. Shenk, The sequence 5′-AAUAAA-3′ forms part of the recognition site for polyadenylation of late SV40 mRNAs. *Cell* 24 (April 1981) p. 257, f. 7.)

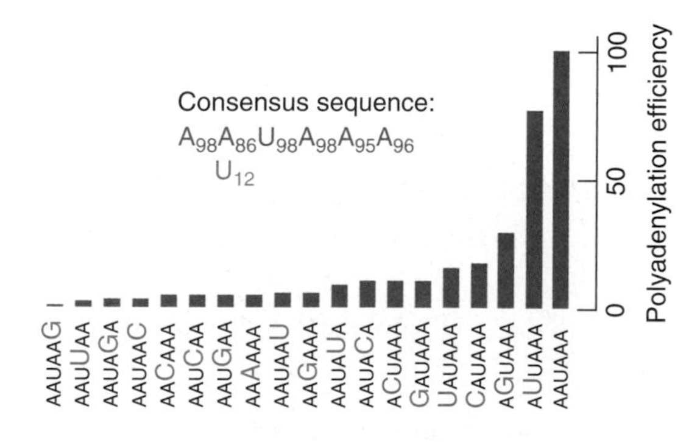

Figure 15.15 Summary of data on 369 vertebrate polyadenylation signals. The consensus sequence (in RNA form) appears at top, with the frequency of appearance of each base. The substitution of U for A in the second position is frequent enough (12%) that it is listed separately, below the main consensus sequence. Below, the polyadenylation efficiency is plotted for each variant polyadenylation signal. The base that deviates from normal is printed larger than the others in blue. The standard AAUAAA is given at the bottom, with the next most frequent (and active) variant (AUUAAA) just above it. (*Source:* Adapted from Wickens, M., How the messenger got its tail: addition of poly(A) in the nucleus. *Trends in Biochemical Sciences,* 15:278, 1990.)

mutations were of course made in the DNA. They began by inserting an extra copy of the whole polyadenylation signal upstream of the natural one, then testing for polyadenylation at the two sites of this mutant clone (clone 3) by S1 analysis. This DNA supported polyadenylation at the new site at a rate 90% as high as at the original site. Thus, the inserted polyadenylation site is active. Next, they created a new mutant clone [clone 2(v)] by deleting a 35-bp fragment containing the GU- and U-rich region (GU/U) in the new (upstream) polyadenylation signal. This abolished polyadenylation at the new site, reaffirming that this 35-bp fragment is a vital part of the polyadenylation signal.

To define the minimum efficient polyadenylation site, these workers added back various sequences to clone 2(v) and tested for polyadenylation. They showed that neither the GU-rich nor the U-rich sequence by itself could reconstitute an efficient polyadenylation signal: Clone GT had the GU-rich region, but was only 30% as active as the wild-type signal; clone A–T had the U-rich region, but had only 30% of the normal activity. Furthermore, the position of the GU/U region was important. In clone C–GT/T it was shifted 16 bp further downstream of the AAUAAA element, and this clone had less than 10% of normal activity. Moreover, the spacing between the GU-rich and U-rich sequences was important. Clone GT–T had both, but they were separated by an extra 5 bp, and this mutant signal had only 30% of the normal activity. Thus, an efficient polyadenylation signal has an AAUAAA motif followed 23–24 bp later by a GU-rich motif, followed immediately by a U-rich motif.

Plants and yeast mRNAs are also polyadenylated, but their polyadenylation signals are different from those of mammals. Yeast genes usually lack an AAUAAA sequence near their polyadenylation sites. In fact, it is difficult to discern a pattern in the yeast polyadenylation signals, other than a general AU-richness upstream of the polyadenylation site. Plant genes may have an AAUAAA in the appropriate position, and deletion of this sequence prevents polyadenylation. But plant and animal polyadenylation signals are not the same: Single-base substitutions within the AAUAAA of the cauliflower mosaic virus do not have near the negative effect they have in vertebrate polyadenylation signals. Furthermore, animal signals do not function when placed at the ends of plant genes in plant cells.

SUMMARY An efficient mammalian polyadenylation signal consists of an AAUAAA motif about 20 nt upstream of a polyadenylation site in a pre-mRNA, followed 23 or 24 bp later by a GU-rich motif, followed immediately by a U-rich motif. Many variations on this theme occur in nature, which results in variations in efficiency of polyadenylation. Plant polyadenylation signals also usually contain an AAUAAA motif, but more variation is allowed in this region than in an animal AAUAAA. Yeast polyadenylation signals differ even more, and rarely contain an AAUAAA motif.

Cleavage and Polyadenylation of a Pre-mRNA

The process commonly known as polyadenylation really involves both RNA cleavage and polyadenylation. In this section we will briefly discuss the factors involved in the cleavage reaction, then discuss the polyadenylation reaction in more detail.

Pre-mRNA Cleavage Several proteins are necessary for cleavage of mammalian pre-mRNAs prior to polyadenylation. One of these proteins is also required for polyadenylation, so it was initially called "cleavage and polyadenylation factor," or "CPF," but it is now known as **cleavage and polyadenylation specificity factor (CPSF).** Cross-linking experiments have demonstrated that this protein binds to the AAUAAA signal. Shenk and colleagues reported in 1994 that another factor participates in recognizing the polyadenylation site. This is the **cleavage stimulation factor (CstF),** which, according to cross-linking data, binds to the G/U-rich region. Thus, CPSF and CstF bind to sites flanking the cleavage and polyadenylation site. Binding of either CPSF or CstF alone is unstable, but together the two factors bind cooperatively and stably.

Still another pair of RNA-binding proteins required for cleavage are the **cleavage factors I** and **II (CF I** and **CF II).** It is also likely that poly(A) polymerase itself is required for cleavage because cleavage is followed immediately by polyadenylation. In fact, the coupling between cleavage and polyadenylation is so strong that no cleaved, unpolyadenylated RNAs can be detected.

Another protein that is intimately involved in cleavage is RNA polymerase II. The first hint of this involvement was the discovery that RNAs made in vitro by RNA polymerase II were capped, spliced, and polyadenylated properly, but those made by polymerases I and III were not. In fact, even RNAs made by RNA polymerase II lacking the carboxyl-terminal domain (CTD) of the largest subunit were not efficiently spliced and polyadenylated. These data suggested that the CTD was involved somehow in splicing and polyadenylation.

In light of these data, Yutaka Hirose and James Manley performed experiments to test the role of the CTD, including its phosphorylation status, in polyadenylation. In 1998 they reported that the CTD stimulates the cleavage reaction, and this stimulation is not dependent on transcription. First, these workers tested the phosphorylated and unphosphorylated forms of polymerase II (IIO and IIA, Chapter 10) for ability to stimulate cleavage in the presence of all the other cleavage and polyadenylation factors. They incubated ^{32}P-labeled adenovirus L3 pre-mRNA with CPSF, CstF, CF I, CF II, poly(A) polymerase, and either RNA polymerase IIA or IIO. After the incubation period, they electrophoresed the products and autoradiographed the gel to see if the pre-mRNA had been cleaved in the right place. Figure 15.16 depicts the results. Both polymerases IIA and IIO stimulated correct cleavage

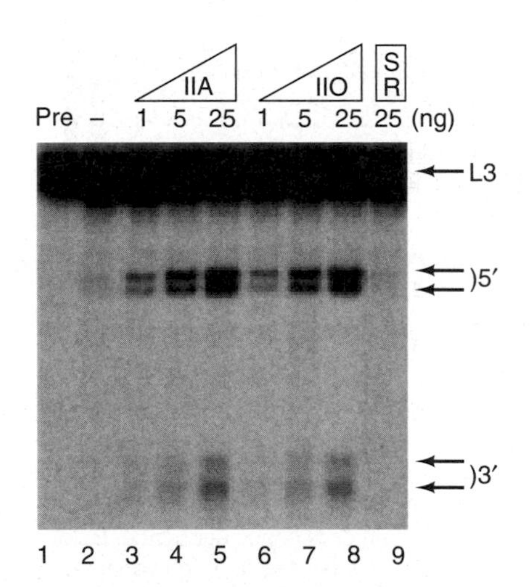

Figure 15.16 Effect of RNA polymerases IIA and IIO on prepolyadenylation mRNA cleavage in vitro. Hirose and Manley prepared a ^{32}P-labeled adenovirus L3 pre-mRNA and incubated it with all the cleavage and polyadenylation factors [CPSF, CstF, CF I, CF II, and poly(A) polymerase] plus polymerase IIA, IIO, no protein (–), or purified HeLa cell SR proteins, as indicated at top. (The amounts of the various proteins are given in nanograms.) Then the investigators electrophoresed the RNA products and detected them by autoradiography. The positions of the 5′- and 3′-cleavage fragments, and the pre-mRNA are indicated at right. Lane 1 contained precursor alone. Both IIA and IIO stimulated cleavage of the pre-mRNA to the appropriate 5′- and 3′-fragments. (*Source:* Hirose, Y. and Manley, J. RNA polymerase II is an essential mRNA polyadenylation factor. *Nature* 395 (3 Sep 1998) f. 2, p. 94. Copyright © Macmillan Magazines Ltd.)

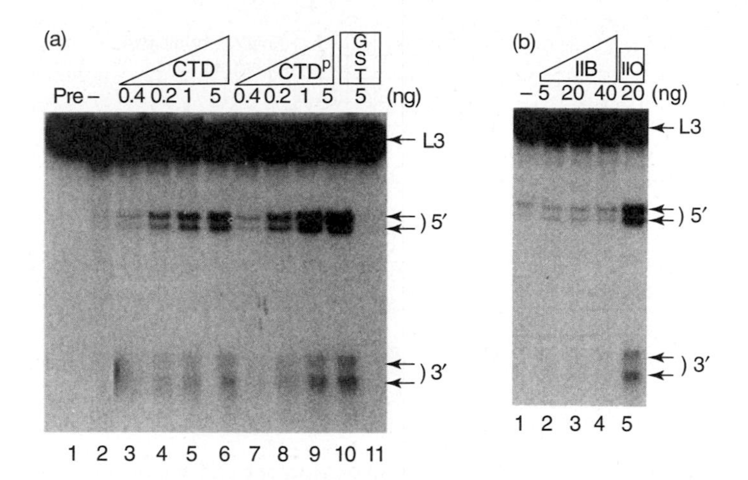

Figure 15.17 Effect of the Rpb1 CTD on prepolyadenylation mRNA cleavage in vitro. Hirose and Manley incubated a labeled pre-mRNA with cleavage and polyadenylation factors and assayed for cleavage as in Figure 15.16. **(a)** They included phosphorylated or unphosphorylated GST–CTD fusion proteins or GST alone, as indicated at top, in the cleavage reaction. **(b)** They included RNA polymerase IIB or IIO, as indicated at top, in the cleavage reaction. The phosphorylated CTD stimulated cleavage more than the unphosphorylated CTD; polymerase IIB, which lacks the CTD, did not stimulate cleavage at all. (*Source:* Hirose, Y. and Manley, J. RNA polymerase II is an essential mRNA polyadenylation factor. *Nature* 395 (3 Sep 1998) f. 3, p. 94. Copyright © Macmillan Magazines Ltd.)

of the pre-mRNA, yielding 5′- and 3′-fragments of the expected sizes.

To verify that the CTD is the important part of polymerase II in stimulating cleavage, Hirose and Manley expressed the CTD as a fusion protein with glutathione-*S*-transferase (Chapter 4), then purified the fusion protein by glutathione affinity chromatography. They phosphorylated part of the fusion protein preparation on its CTD component and tested the phosphorylated and unphosphorylated fusion proteins in the cleavage assay with the adenovirus L3 pre-mRNA. Figure 15.17a shows that both the phosphorylated and unphosphorylated CTDs stimulated cleavage, but the phosphorylated form worked about five times better than the unphosphorylated one. That makes sense because the CTD is phosphorylated in polymerase IIO, which is the form that carries out transcription. It is unclear why phosphorylation made no difference when whole polymerase II was used in Figure 15.16.

If the CTD is the key to stimulating cleavage of the pre-mRNA, then polymerase IIB, the proteolytic product of polymerase IIA that lacks the CTD, should not stimulate, and Figure 15.17b shows that it does not. Thus, RNA polymerase II, and the CTD in particular, appears to be required for efficient cleavage of a pre-mRNA prior to polyadenylation. Figure 15.18 summarizes our knowledge about the complex of proteins that assembles on a pre-mRNA just before cleavage.

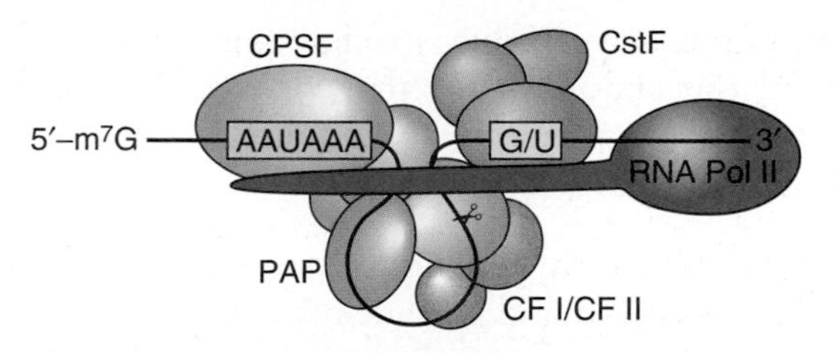

Figure 15.18 A model for the precleavage complex. This partly hypothetical model shows the apparent positions of all the proteins presumed to be involved in cleavage, with respect to the two parts of the polyadenylation signal (green and yellow). The scissors symbol denotes the active site of CPSF-73. (*Source:* Adapted from Wahle, E. and W. Keller, The biochemistry of polyadenylation, *Trends in Biochemical Sciences* 21 [1996] pp. 247–250, 1996.)

We have seen that an array of multisubunit complexes are required for cleavage at the polyadenylation site, but what protein carries out the cleavage itself? That question remained open until 2003, when Masayuki Nashimoto and colleagues discovered that one of the subunits of CPSF (**CPSF-73**) is related to the enzyme (ELAC2) that cleaves pre-tRNAs to generate their 3′-ends (Chapter 16). This finding led to the suggestion that CPSF-73 is the cleavage enzyme. This is an attractive notion because of the symmetry between ELAC2, which cleaves off the 3′-ends of pre-tRNAs prior to the untemplated addition of CCA, and CPSF-73, which cleaves off the 3′-ends of pre-mRNAs prior to the untemplated addition of poly(A). Both ELAC2 and CPSF-73 are unusual RNases that contain two zinc ions at their active sites. They belong to a family of hydrolases (enzymes that carry out hydrolytic reactions, such as hydrolyzing RNA phosphodiester bonds) known as the β-lactamase superfamily of zinc-dependent hydrolases.

Now James Manley and Liang Tong have provided strong evidence that CPSF-73 really is the enzyme that cleaves pre-mRNAs prior to polyadenylation. First, they obtained the crystal structure of human CPSF-73 (amino acids 1–460) in complex with a sulfate group, which mimics the scissile phosphodiester group (the one where the break will occur) in the pre-mRNA at the active site of the enzyme. They found that CPSF-73 contains a Zn-binding motif that coordinates two zinc ions that are essential for its RNase activity. These two zinc ions coordinate a hydroxide ion that is in perfect position to attack the scissile phosphodiester bond (represented by the sulfate) in the active site of the enzyme.

To demonstrate that CPSF-73 has endonuclease activity, Manley and Tong expressed the human CPSF-73 gene in bacteria and tested the product for the ability to cleave an SV40 late pre-mRNA. It did have weak endonuclease activity, producing a variety of cleavage products. By contrast, a mutant CPSF-73, which was missing two of the ligands for the zinc ions, was inactive. Although these data were not as clean as one might hope, taken together with the structural studies on the enzyme, they strongly suggest that CPSF-73 is indeed the endonuclease that cleaves the pre-mRNA prior to polyadenylation.

SUMMARY Polyadenylation requires both cleavage of the pre-mRNA and polyadenylation at the cleavage site. Cleavage in mammals requires several proteins: CPSF, CstF, CF I, CF II, poly(A) polymerase, and RNA polymerase II (in particular, the CTD of Rpb1). One of the subunits of CPSF (CPSF-73) appears to cleave the pre-mRNA prior to polyadenylation.

Initiation of Polyadenylation Once a pre-mRNA has been cleaved downstream of its AAUAAA motif, it is ready to be polyadenylated. The polyadenylation of a cleaved RNA occurs in two phases. The first, initiation, depends on the AAUAAA signal and involves slow addition of at least 10 A's to the pre-mRNA. The second phase, elongation, is independent of the AAUAAA motif, but depends on the oligo(A) added in the first phase. This second phase involves the rapid addition of 200 or more A's to the RNA. Let us begin with the initiation phase.

Strictly speaking, the entity we have been calling "the polyadenylation signal" is really the cleavage signal. It is what attracts the cleavage enzyme to cut the RNA about 20 nt downstream of the AAUAAA motif. Polyadenylation itself, that is, the addition of poly(A) to the 3′-end created by the cleavage enzyme, cannot use the same signal. This must be true because the cleavage enzyme has already removed the downstream part of the signal (the GU-rich and U-rich elements).

What is the signal that causes polyadenylation itself? It seems to be AAUAAA, followed by at least 8 nt at the end of the RNA. We know this because short synthetic oligonucleotides (as short as 11 nt) containing AAUAAA can be polyadenylated in vitro. The optimal length between the AAUAAA and the end of the RNA is 8 nt.

To study the process of polyadenylation by itself in vitro, it is necessary to divorce it from the cleavage reaction. Molecular biologists accomplish this by using labeled, short RNAs that have an AAUAAA sequence at least 8 nt from the 3′-end. These substrates mimic pre-mRNAs that have just been cleaved and are ready to be polyadenylated. The assay for polyadenylation is electrophoresis of the labeled RNA. If poly(A) has been added, the RNA will be much bigger and will therefore electrophorese much more slowly. It will also be less discrete in size, because the poly(A) tail varies somewhat in length from molecule to molecule. In this section, we will use the term *polyadenylation* to refer to the addition of poly(A) to the 3′-end of such a model RNA substrate.

Figure 15.19 shows how Marvin Wickens and his colleagues used this assay to demonstrate that two fractions are needed for polyadenylation: poly(A) polymerase and a

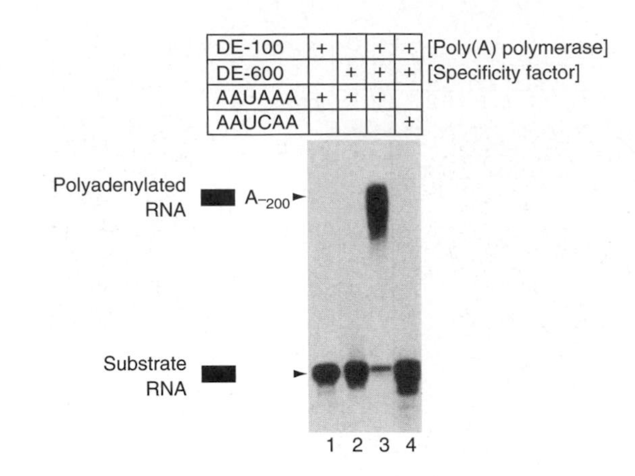

Figure 15.19 Separation of poly(A) polymerase and specificity factor activities. Wickens and colleagues separated HeLa cell poly(A) polymerase and specificity factor activities by DEAE-Sepharose chromatography. The polymerase eluted at 100 mM salt, so it is called the DE-100 fraction; the specificity factor eluted at 600 mM salt, so it is designated the DE-600 fraction. These workers tested the separated activities on a labeled synthetic substrate consisting of nucleotides −58 to +1 of SV40 late mRNA, whose 3′-end is at the normal polyadenylation site. After they incubated the two fractions, separately or together, with the substrate and ATP, they electrophoresed the labeled RNA and autoradiographed the gel. The components in the reactions in each lane are listed at top. The positions of substrate and polyadenylated product are listed at left. (*Source:* Bardwell, V.J., D. Zarkower, M. Edmonds, and M. Wickens, The enzyme that adds poly(A) to mRNAs is a classical poly(A) polymerase. *Molecular and Cellular Biology* 10 (Feb 1990) p. 847, f. 1. American Society for Microbiology.)

specificity factor. We now know that this specificity factor is CPSF. At high substrate concentrations, the poly(A) polymerase can catalyze the addition of poly(A) to the 3′-end of any RNA, but at low substrate concentrations it cannot polyadenylate by itself (lane 1). Neither can CPSF, which recognizes the AAUAAA signal (lane 2). But together, these two substances can polyadenylate the synthetic substrate (lane 3). Lane 4 demonstrates that both fractions together will not polyadenylate a substrate with an aberrant signal (AAUCAA).

Michael Sheets and Wickens questioned whether polyadenylation is carried out in phases, and they used several different model RNA substrates to answer this question. The first substrate is simply the same terminal 58 nt of the SV40 late mRNA, including the AAUAAA, used in Figure 15.19. The second is the same RNA with 40 A's [a short poly(A)] at the 3′-end. The third is the same RNA with 40 nt from the vector instead of a short poly(A) at the 3′-end. They also used an analogous set of three substrates that had an AAGAAA signal instead of AAUAAA.

Sheets and Wickens used each of these substrates in standard polyadenylation reactions with HeLa cell nuclear extracts. Figure 15.20, lanes 1–4, shows that the extract could polyadenylate the usual model substrate with an AAUAAA signal. Lanes 5–8 show that polyadenylation also occurred with the model substrate that already had 40 A's at

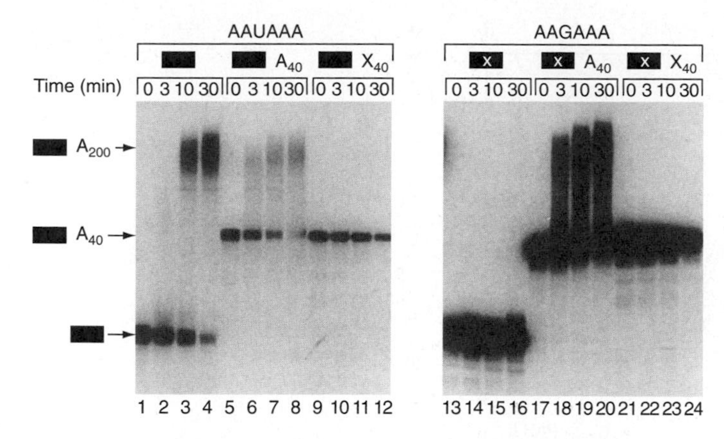

Figure 15.20 Demonstration of two phases in polyadenylation. Sheets and Wickens performed polyadenylation reactions in HeLa nuclear extracts with the following labeled substrates: 1. The standard 58-nt substrate containing the 3′-end of an SV40 late mRNA, represented by a black box; 2. The same RNA with a 40-nt poly(A), represented by a black box followed by A_{40}; 3. The same RNA with a 40-nt 3′-tag containing vector sequence, represented by a black box followed by X_{40}; substrates 1–3 containing an aberrant AAGAAA instead of AAUAAA are represented with white X's within the black boxes. Sheets and Wickens used four different reaction times with each substrate, and the substrate in each set of lanes is indicated by its symbol at top. The electrophoretic mobility of substrates and products are indicated at left. (*Source:* Sheets and Wickens, Two phases in the addition of a poly(A) tail. *Genes & Development* 3 (1989) p. 1402, f. 1. Cold Spring Harbor Laboratory Press.)

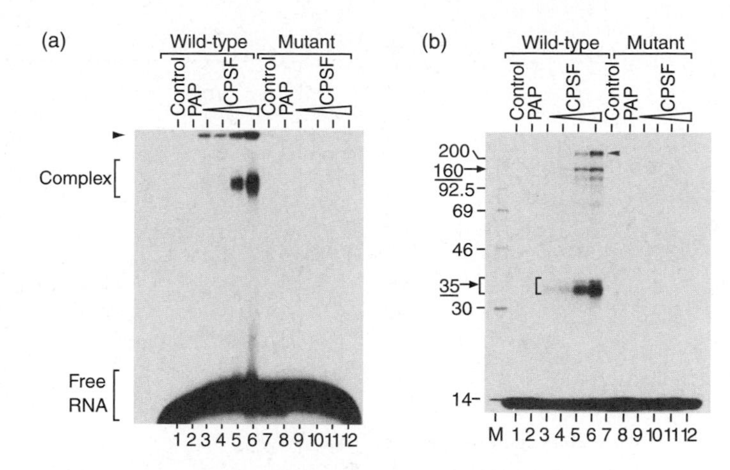

Figure 15.21 CPSF binds to the AAUAAA motif. (a) Gel mobility shift assay. Keller and colleagues mixed a labeled oligoribonucleotide with poly(A) polymerase (PAP), or CPSF in various concentrations, then electrophoresed the mixture. The wild-type oligo contained the AAUAAA motif, and the mutant oligo contained an AAGAAA motif. The controls contained no added proteins. CPSF could form a complex with the wild-type but not the mutant oligo. The band at the top in both panels (arrowheads) is material that remained at the top of the gel, rather than a specific band. **(b)** SDS-PAGE of proteins cross-linked to oligoribonucleotides. Keller and colleagues illuminated each of the mixtures from panel (a) with ultraviolet light to cross-link proteins to the oligo. Then they electrophoresed the complexes on an SDS polyacrylamide gel. Major bands appeared at about 35 and 160 kD (arrows). (*Source:* Keller, W., S. Bienroth, K.M. Lang, and G. Christofori, Cleavage and polyadenylation factor CPF specifically interacts with the pre-mRNA 3′ processing signal AAUAAA. *EMBO Journal* 10 (1991) p. 4243, f. 2.)

its end (A_{40}). The polyadenylated signal was weaker in this case, but the radioactivity of the substrate was also lower. On the other hand, the extract could not polyadenylate the model substrate with 40 non-poly(A) nucleotides at its end (X_{40}). Lanes 13–16 demonstrate that the extract could not polyadenylate the substrate with an aberrant AAGAAA signal and no poly(A) pre-added. However, lanes 17–20 make the most telling point: The extract is able to polyadenylate the substrate with an aberrant AAGAAA signal and 40 A's already added to the end. Thus, by the time 40 A's have been added, polyadenylation is independent of the AAUAAA signal. But these extra nucleotides must be A's; the X_{40} substrate with an aberrant AAGAAA signal could not be polyadenylated (lanes 21–24).

Sheets and Wickens went on to show that the shortest poly(A) that could override the effect of a mutation in AAUAAA is 9 A's, but 10 A's work even better. These findings suggest the following hypothesis: After cleavage of the pre-mRNA, the first phase of polyadenylation, initiation, begins. It depends on the AAUAAA signal and CPSF until the poly(A) reaches about 10 A's in length. At that point, polyadenylation enters the elongation phase and is independent of the AAUAAA and CPSF, but dependent on the poly(A) at the 3′-end of the RNA.

If CPSF recognizes the poladenylation signal AAUAAA, we would predict that CPSF binds to this signal in the pre-mRNA. Walter Keller and colleagues have demonstrated this directly, using gel mobility shift and RNA–protein cross-linking procedures. Figure 15.21 illustrates the results of both kinds of experiments. Panel (a) shows that CPSF binds to a labeled RNA containing an AAUAAA signal, but not to the same RNA with a U→G mutation in the AAUAAA motif. Panel (b) demonstrates that an oligonucleotide bearing an AAUAAA motif, but not an AAGAAA motif, can be cross-linked to two polypeptides (about 35 and 160 kD) in a CPSF preparation. Furthermore, these complexes will not form in the presence of unlabeled competitor RNAs containing AAUAAA; competitor RNAs containing AAGAAA cannot compete. All of these findings bolster the conclusion that CPSF binds directly to the AAUAAA motif.

SUMMARY Short RNAs that mimic a newly created mRNA 3′-end can be polyadenylated. The optimal signal for initiation of such polyadenylation of a cleaved substrate is AAUAAA followed by at least 8 nt. Once the poly(A) reaches about 10 nt in length, further polyadenylation becomes independent of the AAUAAA signal and depends on the poly(A) itself. Two proteins participate in the initiation process: poly(A) polymerase and CPSF, which binds to the AAUAAA motif.

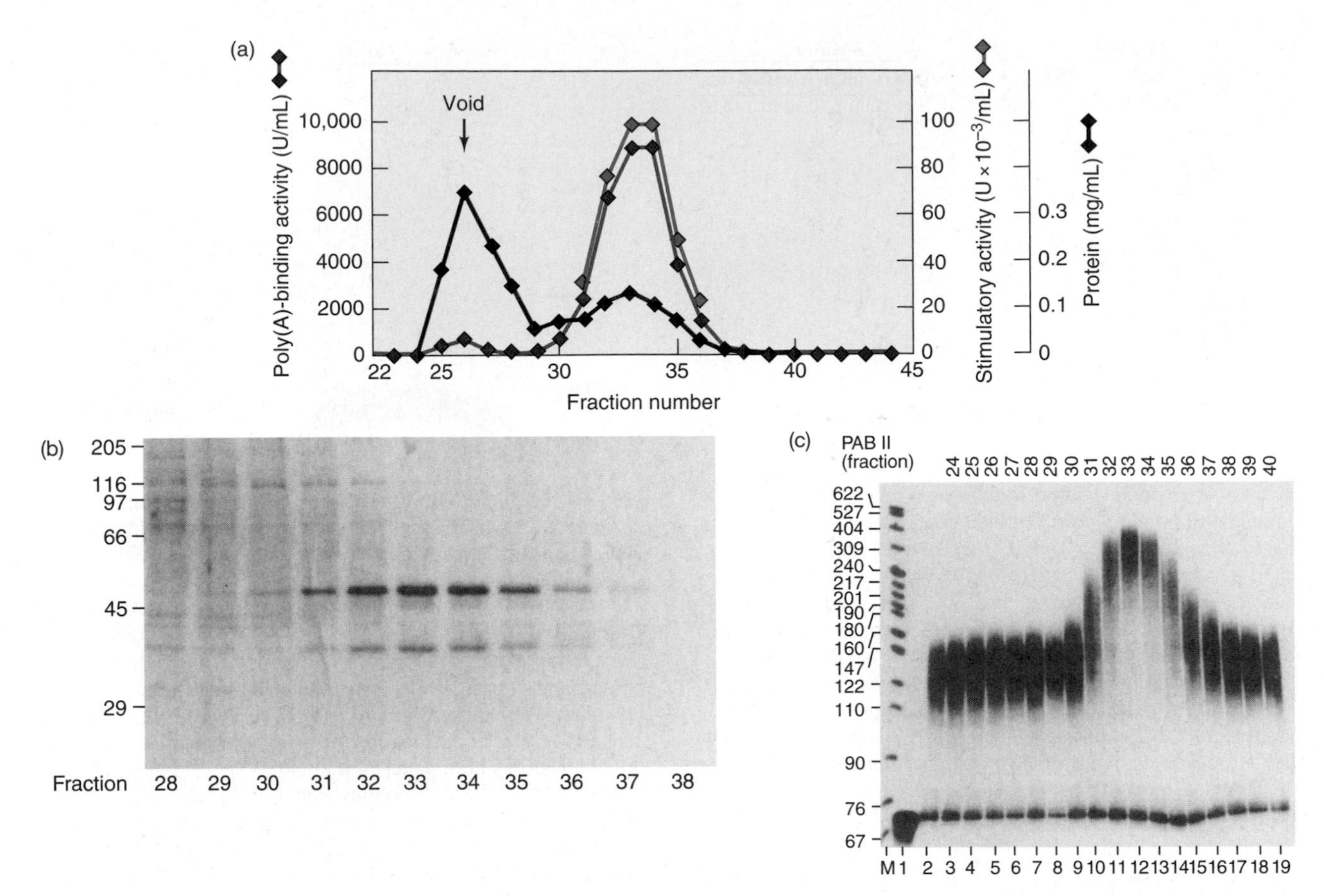

Figure 15.22 Purification of a poly(A)-binding protein. (a) Summary of results. Wahle subjected the poly(A)-binding protein to a final gel filtration chromatographic purification step on Sephadex G-100. In this panel, he plotted three parameters against fraction number from the G-100 column. Red, poly(A)-binding activity determined by a filter binding assay; green, polyadenylation-stimulating activity [see panel (c)]; blue, protein concentration. "Void" indicates proteins that eluted in the void volume. These large proteins were not included in the gel spaces on the column. **(b)** SDS-PAGE analysis. Wahle subjected aliquots of fractions from the G-100 column in panel (a) to SDS-PAGE and stained the proteins in the gel with Coomassie Blue. Sizes of marker polypeptides are given at left. A 49-kD polypeptide reached maximum concentration in the fractions (32–35) that had peak poly(A)-binding activity and polyadenylation-stimulatory activity. **(c)** Assay for polyadenylation stimulatory activity. Wahle added aliquots of each fraction from the G-100 column to standard polyadenylation reactions containing labeled L3pre RNA substrate. Lane 1 contained only substrate, with no poly(A) polymerase. The increase in size of poly(A) indicates stimulatory activity, which peaked in fractions 32–35. (*Source:* Wahle, E., A novel poly(A)-binding protein acts as a specificity factor in the second phase of messenger RNA polyadenylation. *Cell* 66 (23 Aug 1991) p. 761, f. 1. Reprinted by permission of Elsevier Science.)

Elongation of Poly(A) We have seen that elongation of an initiated poly(A) chain 10 nt or more in length is independent of CPSF. However, purified poly(A) polymerase binds to and elongates poly(A) only very poorly by itself. This implies that another specificity factor can recognize an initiated poly(A) and direct poly(A) polymerase to elongate it. Elmar Wahle has purified a poly(A)-binding protein that has these characteristics.

Figure 15.22b shows the results of PAGE on fractions from the last step in purification of the poly(A)-binding protein. A major 49-kD polypeptide is visible, as well as a minor polypeptide with a lower molecular mass. Because the latter band varied in abundance, and was even invisible in some preparations, Wahle concluded that it was not related to the poly(A)-binding protein. Wahle tested the fractions containing the 49-kD protein for poly(A) binding by a nitrocellulose filter binding assay [panel (a)], and found that the peak of poly(A)-binding activity coincided with the peak of abundance of the 49-kD polypeptide. Next, he tested the same fractions for ability to stimulate polyadenylation of a model RNA substrate in the presence of poly(A) polymerase and CPSF [panel (c)]. Again, he found that the peak of activity coincided with the abundance of the 49-kD polypeptide. Thus, the 49-kD polypeptide is a poly(A)-binding protein, but differs from the major, 70-kD poly(A)-binding protein, (**PAB I**) found earlier in the cytoplasm, so Wahle named it **poly(A)-binding protein II (PAB II).**

PAB II can stimulate polyadenylation of a model substrate, just as CPSF can, but it binds to poly(A) rather than to the AAUAAA motif. This suggests that PAB II is active in elongation, rather than initiation, of polyadenylation. If so,

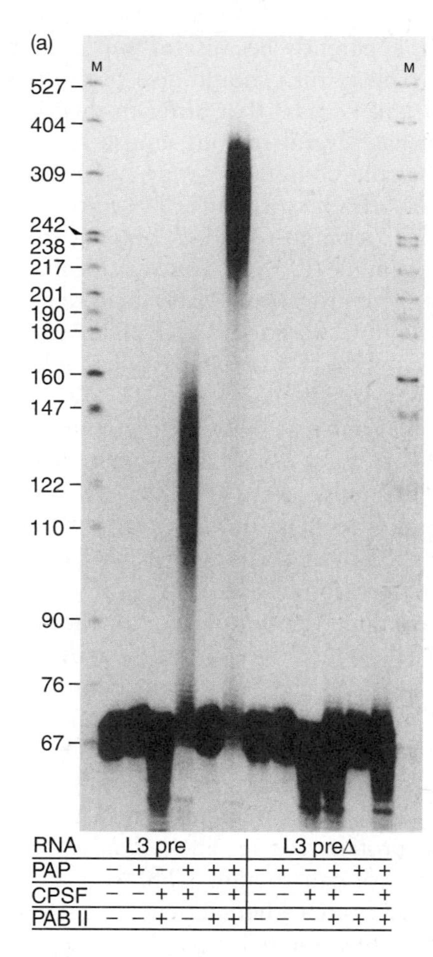

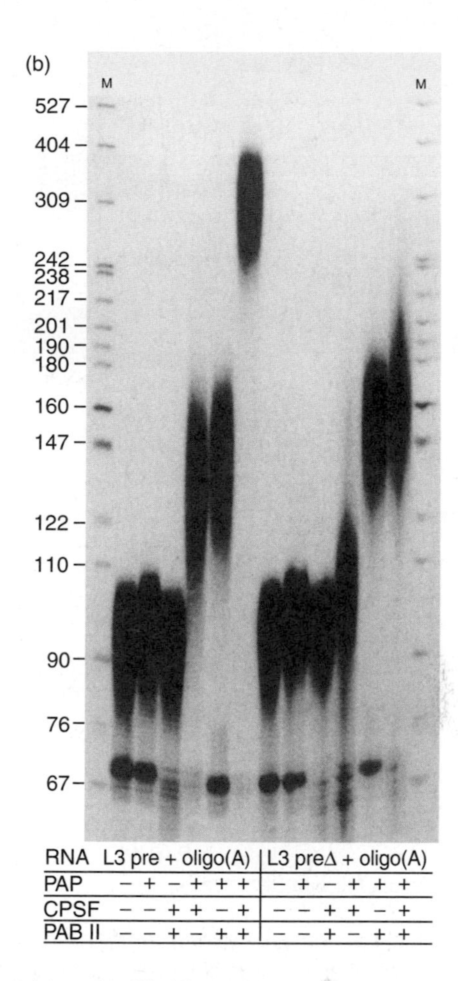

Figure 15.23 Effect of CPSF and PAB II on polyadenylation of model substrates. (a) Polyadenylation of RNAs lacking oligo(A). Wahle carried out polyadenylation reactions in the presence of the RNAs and proteins listed at bottom. L3 pre was the standard substrate RNA with an AAUAAA motif; L3 preΔ was the same, except that AAUAAA was mutated to AAGAAA. PAB II could not direct polyadenylation of L3 pre without help from CPSF. **(b)** Polyadenylation of RNAs containing oligo(A). All conditions were the same as in panel (a) except that the substrates contained oligo(A) at their 3′-ends. This allowed PAB II to work in the absence of CPSF and to work on the substrate with a mutant AAUAAA motif. The first and last lanes in both panels contained markers. (*Source:* Wahle, E., A novel poly(A)-binding protein acts as a specificity factor in the second phase of messenger RNA polyadenylation. *Cell* 66 (23 Aug 1991) p. 764, f. 5. Reprinted by permission of Elsevier Science.)

then its substrate preference should be different from that of CPSF. In particular, it should stimulate polyadenylation of RNAs that already have an oligo(A) attached, but not RNAs with no oligo(A). The results in Figure 15.23 confirm this prediction. Panel (a) shows that an RNA lacking oligo(A) (L3 pre) could be polyadenylated by poly(A) polymerase (PAP) plus CPSF, but not by PAP plus PAB II. However, PAP plus CPSF plus PAB II polyadenylated this substrate best of all. Presumably, CPSF serves as the initiation factor, then PAB II directs the polyadenylation of the substrate once an oligo(A) has been added, and does this better than CPSF can. Predictably, an L3 pre substrate with a mutant AAUAAA signal (AAGAAA) could not be polyadenylated by any combination of factors, because it depends on CPSF for initiation, and CPSF depends on an AAUAAA signal.

Figure 15.23b shows that the same RNA with an oligo(A) at the end behaved differently. It could be polyadenylated by PAP in conjunction with *either* CPSF or PAB II. This makes sense because this substrate has an oligo(A) that PAB II can recognize. It is interesting that both factors together produced even better polyadenylation of this substrate. This suggests that PAP might interact with both factors, directly or indirectly, during the elongation phase. Finally, panel (b) demonstrates that PAB II, in the absence of CPSF, could direct efficient polyadenylation of the mutant RNA with an AAGAAA motif, as long as the RNA had an oligo(A) to begin with. Again, this makes sense because the oligo(A) provides a recognition site for PAB II and therefore makes it independent of CPSF and the AAUAAA motif.

Figure 15.24 presents a model of initiation and elongation of polyadenylation. Optimal activity during the initiation phase requires PAP, CPSF, CstF, CF I, CF II and the two-part polyadenylation signal (the AAUAAA and G/U motifs flanking the polyadenylation site). The elongation phase requires PAP, PAB II, and an oligo(A) at least 10 nt long.

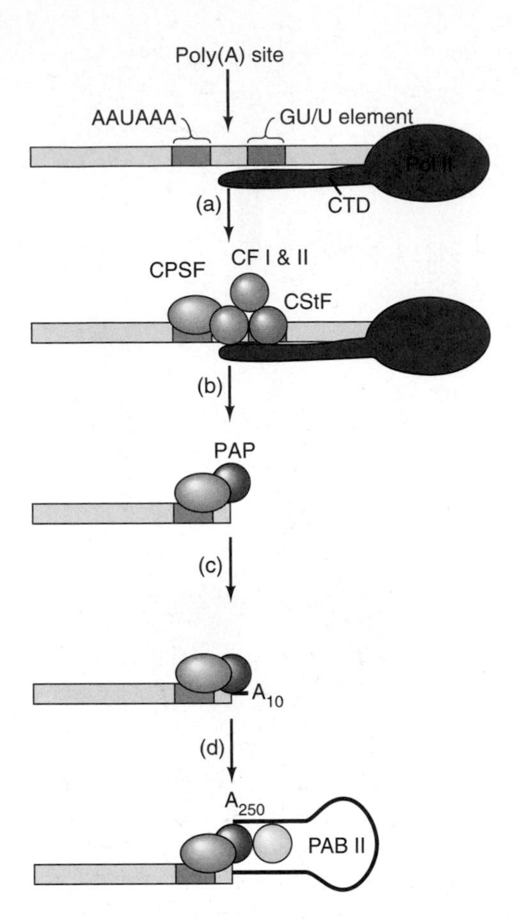

Figure 15.24 Model for polyadenylation. **(a)** CPSF (blue), CstF (brown), and CF I and II (gray) assemble on the pre-mRNA, guided by the AAUAAA and GU/U motifs. **(b)** Cleavage occurs, stimulated by the CTD of RNA polymerase II; CstF and CF I and II leave the complex; and poly(A) polymerase (PAP, purple) enters. **(c)** poly(A) polymerase, aided by CPSF, initiates poly(A) synthesis, yielding an oligo(A) at least 10 nt long. **(d)** PAB II (yellow) enters the complex and allows the rapid extension of the oligo(A) to a full-length poly(A). At this point, the complex presumably dissociates.

It is enhanced by CPSF. Table 15.2 lists all these protein factors, their structures, and their roles.

SUMMARY Elongation of poly(A) in mammals requires a specificity factor called poly(A)-binding protein II (PAB II). This protein binds to a preinitiated oligo(A) and aids poly(A) polymerase in elongating the poly(A) to 250 nt or more. PAB II acts independently of the AAUAAA motif. It depends only on poly(A), but its activity is enhanced by CPSF.

Poly(A) Polymerase

In 1991, James Manley and colleagues cloned cDNAs encoding bovine poly(A) polymerase (PAP). Sequencing of these clones revealed two different cDNAs that differed at their 3′-ends, apparently because of two alternative splicing schemes. This in turn should give rise to two different PAPs (*PAP I* and *PAP II*) that differ in their carboxyl termini. PAP II has several regions whose sequences match (more or less) the consensus sequences of known functional domains of other proteins. These are, in order from N-terminus to C-terminus: an RNA-binding domain (RBD); a polymerase module (PM); two nuclear localization signals (NLS 1 and 2); and several serine/threonine-rich regions (S/T). By 1996, four additional PAP cDNAs had been discovered. Two of these were short and could arise from polyadenylation within the pre-mRNA. Another was long and could come from a pseudogene (Chapter 23). The most important PAP in most tissues is probably PAP II.

Because the polymerase module, which presumably catalyzes the polyadenylation reaction, lies near the amino terminus of the protein, it would be interesting to know how much of the carboxyl end of the protein is required for activity. To examine the importance of the carboxyl end, Manley and colleagues expressed full-length and 3′-deleted versions of the PAP I cDNA by transcribing them in vitro with SP6 RNA polymerase, then translating these transcripts in cell-free reticulocyte extracts. This generated a full-length protein of 689 amino acids, and truncated proteins of 538, 379, and 308 amino acids. Then they tested each of these proteins for specific polyadenylation activity in the presence of calf thymus CPSF. The full-length and 538-amino-acid proteins had activity, but the smaller proteins did not. Thus, the S/T domain is not necessary for activity, but sequences extending at least 150 amino acids toward the carboxyl terminus from the polymerase module are essential, at least in vitro.

SUMMARY Cloning and sequencing cDNAs encoding calf thymus poly(A) polymerase reveal a mixture of 5 cDNAs derived from alternative splicing and alternative polyadenylation. The structures of the enzymes predicted from the longest of these sequences include an RNA-binding domain, a polymerase module, two nuclear localization signals, and a serine/threonine-rich region. The latter region, but none of the rest, is dispensable for activity in vitro.

Turnover of Poly(A)

Figure 15.7 showed some evidence of a slight difference in size between nuclear and cytoplasmic poly(A). However, that experiment involved newly labeled RNA, so the poly(A) had not had much time to break down. Sheiness and Darnell performed another study on RNA from cells that were continuously labeled with RNA precursors for 48 h. This procedure gave a population of poly(A)s at their "steady-state" sizes; that is, the natural sizes one would

Table 15.2 Mammalian Factors Required for 3′-Cleavage and Polyadenylation

Factor	Polypeptides (kD)	Properties
Poly(A) polymerase (PAP)	82	Required for cleavage and polyadenylation; catalyzes poly(A) synthesis
Cleavage and polyadenylation specificity factor (CPSF)	160 100 73 30	Required for cleavage and polyadenylation; binds AAUAAA and interacts with PAP and CstF; CPSF-73 cleaves RNA
Cleavage stimulation factor (CstF)	77 64 50	Required only for cleavage; binds the downstream element and interacts with CPSF
Cleavage factor I (CF I)	68 59 25	Required only for cleavage; binds RNA
Cleavage factor II (CF II)	Unknown	Required only for cleavage
RNA polymerase II (especially CTD)	Many	Required only for cleavage
Poly(A)-binding protein II (PAB II)	49	Stimulates poly(A) elongation; binds growing poly(A) tail; essential for poly(A) tail length control

Source: Adapted from Wahle, E. and W. Keller, The biochemistry of polyadenylation, *Trends in Biochemical Sciences* 21: 247–250. Copyright © 1996 with permission of Elseiver Science.

observe by peeking into a cell at any given time. Figure 15.25 shows an apparent difference in the sizes of nuclear and cytoplasmic poly(A)s. The major peak of nuclear poly(A) was 210 ± 20 nt, whereas the major peak of cytoplasmic poly(A) was 190 ± 20 nt. Furthermore, the cytoplasmic poly(A) peak showed a much broader skew toward smaller species than the nuclear poly(A) peak. This broad peak encompassed RNAs at least as small as 50 nt. Thus, poly(A) seems to undergo considerable shortening in the cytoplasm.

In 1970, Maurice Sussman proposed a "ticketing" hypothesis that held that each mRNA has a "ticket" that allows it entry to the ribosome for translation. Each time it is translated, the mRNA gets its "ticket punched." When it accumulates enough "punches," it can no longer be translated. Poly(A) would make an ideal ticket; the punches would then be progressive shortening of the poly(A) every time it is translated. To test this idea, Sheiness and Darnell tested the rate of shortening of poly(A) in the cytoplasm under normal conditions, and in the presence of emetine, which inhibits translation. They observed no difference in the size of cytoplasmic poly(A), whether or not translation was occurring. Thus, the shortening of poly(A) does not depend on translation, and the ticket, if it exists at all, seems not to be poly(A).

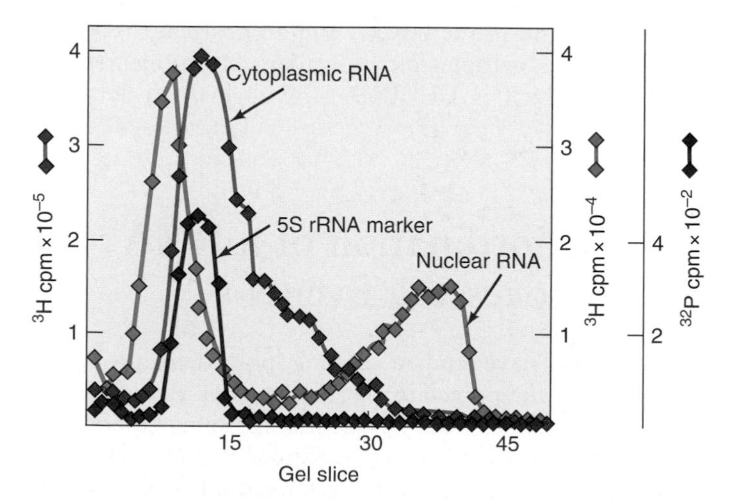

Figure 15.25 Shortening of cytoplasmic poly(A). Sheiness and Damell labeled HeLa cells with ^{3}H-adenine for 48 h, then isolated nuclear (green) and cytoplasmic (red) poly(A)$^+$ RNA and analyzed it by gel electrophoresis. They also included a [^{32}P]5S rRNA as a marker (blue). (*Source:* Adapted from Sheiness, D. and J.E. Darnell, Polyadenylic acid segment in mRNA becomes shorter with age. *Nature New Biology* 241:266, 1973.)

Poly(A) is not just shortened in the cytoplasm; it turns over. That is, it is constantly being shortened by RNases and lengthened by a cytoplasmic poly(A) polymerase. The general trend, however, is toward shortening, and ultimately an mRNA will lose all or almost all of its poly(A). By that time, its demise is near.

SUMMARY Poly(A) turns over in the cytoplasm. RNases tear it down, and poly(A) polymerase builds it back up. When the poly(A) is gone, the mRNA is slated for destruction.

Cytoplasmic Polyadenylation The best studied cases of cytoplasmic polyadenylation are those that occur during oocyte maturation. Maturation of *Xenopus* oocytes, for example, occurs in vitro on stimulation by progesterone. The immature oocyte cytoplasm contains a large store of mRNAs called **maternal messages, or maternal mRNAs,** many of which are almost fully deadenylated and are

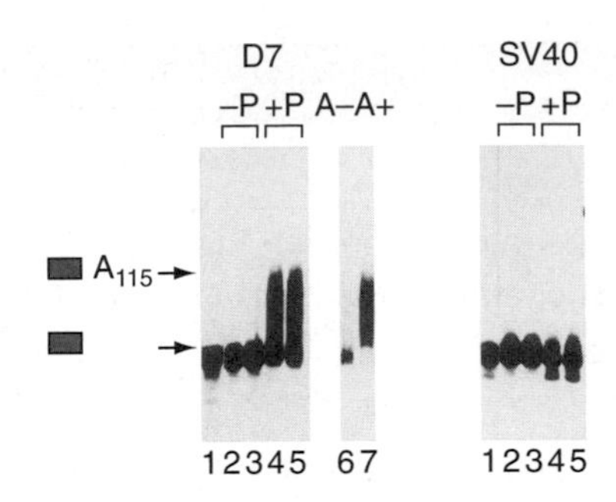

Figure 15.26 Maturation-specific polyadenylation of two RNAs. Wickens and colleagues injected labeled RNAs into *Xenopus* oocyte cytoplasm and stimulated maturation-specific polyadenylation with progesterone. After a 12-h incubation, they isolated the labeled RNA products, electrophoresed them, and visualized them by autoradiography. The two RNAs, as indicated at top, were synthetic 3′-fragments of either the *Xenopus* mRNA (D7), which normally undergoes maturation-specific polyadenylation, or an SV40 mRNA, which does not. The mobilities of unpolyadenylated RNA and RNA with a 115-nt poly(A) are indicated by the red boxes at left. The presence or absence of progesterone during the incubation is indicated at top by +P and −P, respectively. Lanes 6 and 7 contained RNA that was fractionated by oligo(dT)-cellulose chromatography. RNA that did not bind to the resin is designated A−, and RNA that did bind is designated A+. (*Source:* Fox et al., Poly(A) addition during maturation of frog oocytes: Distinct nuclear and cytoplasmic activities and regulation by the sequence UUUUUAU. *Genes & Development* 3 (1989) p. 2154, f. 3. Cold Spring Harbor Laboratory Press.)

UAA**UUUUUAU**AAGCUGC**AAUAAA**CAAGUUAACAACCUCUAG$_{OH}$

UAACCAUUAUAAGCUGC**AAUAAA**CAAGUUAACAACCUCUAG$_{OH}$

(a)

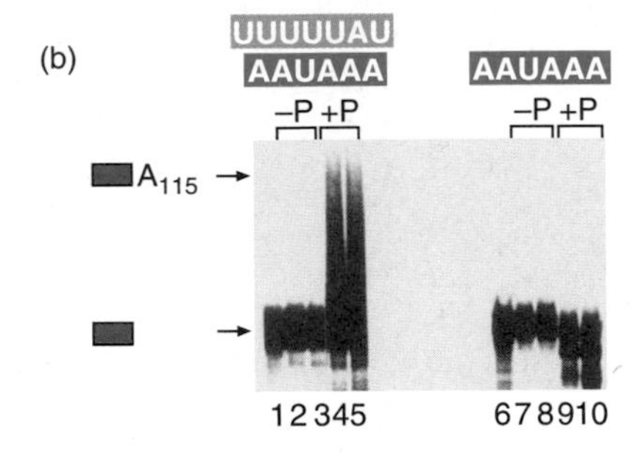

Figure 15.27 Demonstration that UUUUUAU confers maturation-specific polyadenylation. Wickens and colleagues performed the same experiment as described in Figure 15.26, using the same SV40 3′-mRNA fragment with and without an added UUUUUAU motif upstream of the AAUAAA motif. **(a)** Sequences of the two injected RNAs, with the UUUUUAU and AAUAAA motifs highlighted. **(b)** Results. Lanes 2–5 contained RNA from oocytes injected with the RNA having both a UUUUUAU and an AAUAAA sequence, as shown at top. Lanes 7–10 contained RNA from oocytes injected with the RNA having only an AAUAAA sequence. Presence or absence of progesterone during the incubation is indicated at top as in Figure 15.26. Lanes 1 and 6 had uninjected RNA. Markers at left as in Figure 15.26. The UUUUUAU motif was essential for polyadenylation. (*Source:* Fox et al., *Genes & Development* 3 (1989) p. 2155, f. 5. Cold Spring Harbor Laboratory Press.)

not translated. During maturation, some maternal mRNAs are polyadenylated, and others are deadenylated.

To find out what controls this maturation-specific cytoplasmic polyadenylation, Wickens and colleagues injected two mRNAs into *Xenopus* oocyte cytoplasm. The first was a synthetic 3′-fragment of D7 mRNA, a *Xenopus* mRNA known to undergo maturation-specific polyadenylation. The second was a synthetic 3′-fragment of an SV40 mRNA. As Figure 15.26 shows, the D7 RNA was polyadenylated, but the SV40 RNA was not. This implied that the D7 RNA contained a sequence or sequences that are required for maturation-specific polyadenylation, and that these are lacking in the SV40 RNA.

Wickens and colleagues noted that *Xenopus* RNAs that were known to undergo polyadenylation during oocyte maturation all contained the sequence UUUUUAU, or a close relative, upstream of the AAUAAA signal. Is this the key? To find out, these workers inserted this sequence upstream of the AAUAAA in the SV40 RNA and retested it. Figure 15.27 demonstrates that addition of this sequence caused polyadenylation of the SV40 RNA. In light of this character, the UUUUUAU sequence has been dubbed the **cytoplasmic polyadenylation element (CPE).**

Is the AAUAAA also required for cytoplasmic polyadenylation? To answer this question, Wickens and colleagues made point mutations in the AAUAAA motif and injected the mutated RNAs into oocyte cytoplasm. They found that alteration of AAUAAA to either AAUAUA or AAGAAA completely abolished polyadenylation. Thus, this motif is required for both nuclear and cytoplasmic polyadenylation.

SUMMARY Maturation-specific polyadenylation of *Xenopus* maternal mRNAs in the cytoplasm depends on two sequence motifs: the AAUAAA motif near the end of the mRNA and an upstream motif called the cytoplasmic polyadenylation element (CPE), which is UUUUUAU or a closely related sequence.

15.3 Coordination of mRNA Processing Events

Now that we have studied capping, polyadenylation, and splicing, we can appreciate that these processes are related. In particular, the cap can be essential for splicing, but only for splicing out the first intron. Similarly, the poly(A) can be essential for splicing out the last intron. Let us first consider the role of the CTD of the Rpb1 subunit of RNA polymerase II in coordinating capping, splicing, and polyadenylation. Then we will discuss the mechanism of termination of transcription of class II genes and its relationship to polyadenylation.

Binding of the CTD of Rpb1 to mRNA-Processing Proteins

In this chapter and in Chapter 14, we have seen evidence that all three of the mRNA-processing events—splicing, capping, and polyadenylation—take place during transcription. Capping occurs when the nascent mRNA is less than 30 nt long, when the 5′-end of the RNA first emerges from the polymerase. Polyadenylation occurs when the still-growing mRNA is cut at the polyadenylation site. And splicing at least begins when transcription is still underway. We have also just learned that capping and polyadenylation both stimulate splicing, at least of the first and last introns, respectively.

The unifying element for all these processing activities is the CTD of the Rpb1 subunit of RNA polymerase II. We have seen evidence in this chapter for the involvement of the CTD in polyadenylation, but it also plays a part in splicing and capping. In fact, direct evidence shows that the capping, polyadenylating, and splicing enzymes bind directly to the CTD, which provides a platform for all three activities.

For example, consider the evidence for interaction between the capping enzymes and the CTD, presented in 1997 by David Bentley and colleagues. They made affinity columns containing glutathione-*S*-transferase (GST) coupled to: wild-type CTD; wild-type phosphorylated CTD; mutant CTD; or just GST with no CTD attached. Then they subjected HeLa cell extracts to affinity chromatography on each of these columns and tested the eluates for guanylyl transferase activity. The guanylyl transferase assay was done by mixing an eluate with [^{32}P]GTP and observing the transfer of [^{32}P]GMP to form a covalent adduct with the enzyme. This labeled enzyme was then detected by SDS-PAGE and autoradiography. Figure 15.28 shows that the guanylyl transferase bound to the CTD, but only to its phosphorylated form.

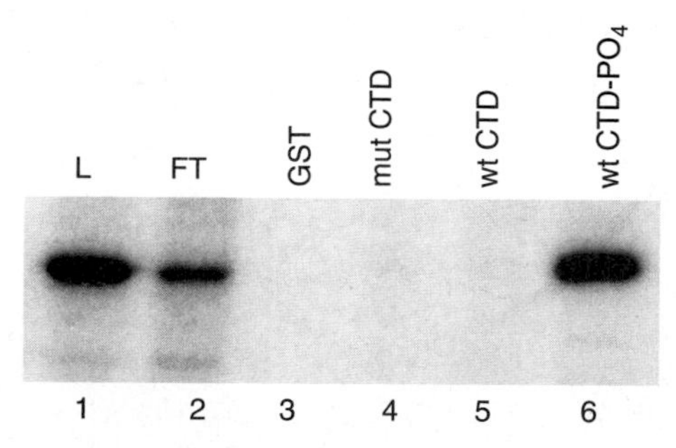

Figure 15.28 A mammalian capping guanylyl transferase binds to the phosphorylated CTD. Bentley and colleagues subjected HeLa cell nuclear extracts to affinity chromatography on resins containing the substances indicated at top, then tested the eluates for guanylyl transferase by observing the formation of a [^{32}P]GMP adduct with the enzyme, which could be identified by SDS-PAGE and autoradiography. L (lane 1) refers to the whole extract loaded onto the column; FT (lane 2) refers to the material that flowed through the column. Lanes 3–6 contain the results of guanylyl transferase assays on material subjected to affinity chromatography on resins containing GST (lane 3), and GST coupled to mutated CTD (lane 4); wild-type CTD (lane 5); and phosphorylated wild-type CTD (lane 6). The guanylyl transferase bound only to the phosphorylated CTD. (*Source:* McCracken et al., *Genes and Development* v. 11, p. 3310.)

Using a very similar experimental approach, Nick Proudfoot and colleagues demonstrated in 2001 that several subunits of the yeast cleavage/polyadenylation factor 1A (CF 1A) bind to the CTD in its phosphorylated form. Other components of the cleavage and polyadenylation complex appeared not to bind directly to the CTD, but they are tightly bound in the complex with other proteins that do bind to the CTD. Other, more indirect evidence also points to the association between the polyadenylation complex and the CTD: Polyadenylation does not function very well when RNA polymerase is lacking its CTD; and the CTD, particularly in its phosphorylated form, stimulates polyadenylation in vitro.

Strong evidence also exists for interactions between the CTD and proteins involved in splicing pre-mRNAs. For example, Daniel Morris and Arno Greenleaf showed in 2000 that a yeast splicing factor, Prp40 (a component of U1 snRNP) binds to the phosphorylated CTD. Morris and Greenleaf used a "**Far Western blot**" to demonstrate binding between Prp40 and the CTD. A Far Western blot is similar to a Western blot in that it begins with electrophoresis of a protein or proteins by SDS-PAGE and blotting of the electrophoresed proteins to a membrane such as nitrocellulose. However, whereas a Western blot would be probed with an antibody, a Far Western blot is probed with another protein suspected of binding to a protein on the blot. In this case, Prp40 (and other so-called WW proteins) were electrophoresed and blotted, then probed with [^{32}P]β-galactosidase-CTD. (The CTD was expressed as a fusion protein with β-galactosidase, for ease of purification, then labeled by phosphorylation in vitro.) WW proteins are characterized by a domain including two tryptophan (W) residues and are frequently involved in RNA synthesis and processing.

Figure 15.29 shows the results of this analysis. Panel (a) depicts a gel stained with Coomassie Blue, a dye that binds to all proteins; so this panel shows the spectrum of polypeptides contained in all the protein preparations, including Prp40, loaded on the gel. The largest polypeptide in each lane is the parent; the smaller polypeptides are likely to be degradation products of the parent. Panel (b) depicts the same gel subjected to Far Western blotting and probed with [^{32}P]β-galactosidase-CTD. Clearly, Ess1, Prp40, and Rsp5 bind to the CTD. However, simply having a WW domain does not guarantee CTD-binding activity, as the other two WW proteins failed to bind the CTD probe.

SUMMARY Capping, polyadenylation, and splicing proteins all associate with the CTD during transcription.

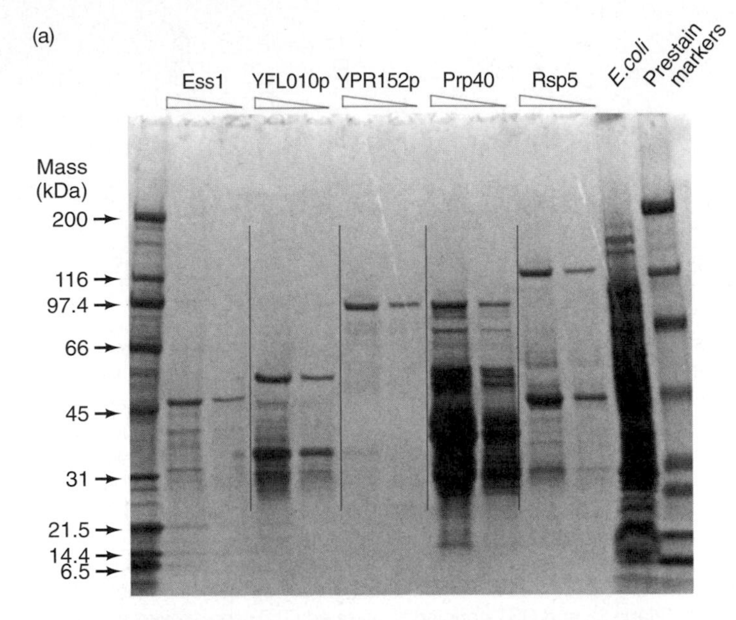

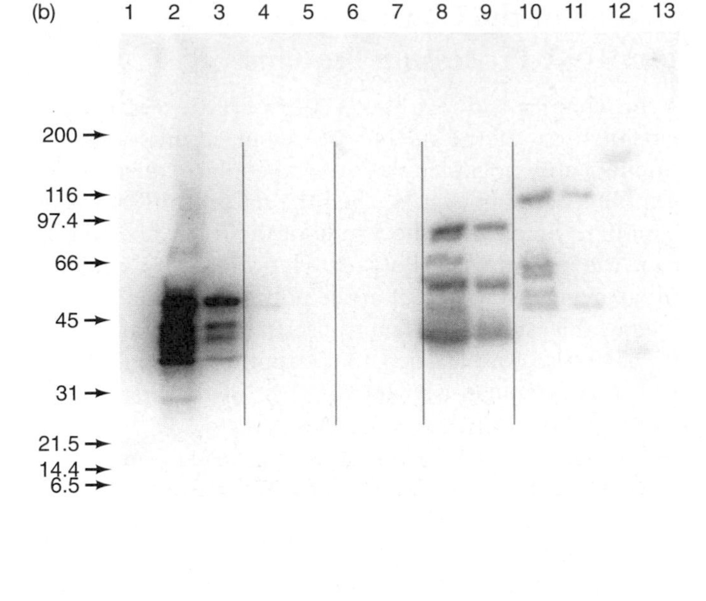

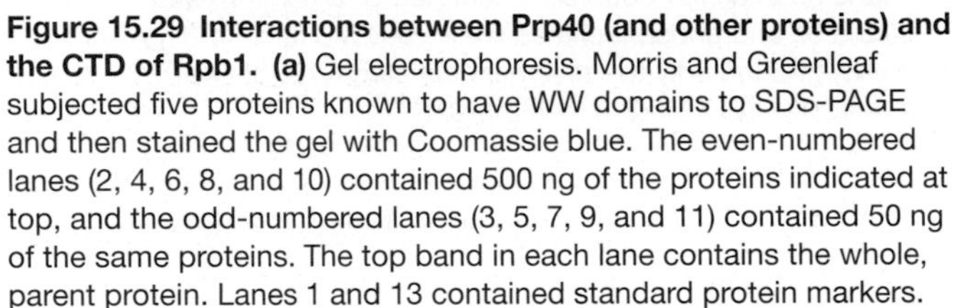

Figure 15.29 Interactions between Prp40 (and other proteins) and the CTD of Rpb1. (a) Gel electrophoresis. Morris and Greenleaf subjected five proteins known to have WW domains to SDS-PAGE and then stained the gel with Coomassie blue. The even-numbered lanes (2, 4, 6, 8, and 10) contained 500 ng of the proteins indicated at top, and the odd-numbered lanes (3, 5, 7, 9, and 11) contained 50 ng of the same proteins. The top band in each lane contains the whole, parent protein. Lanes 1 and 13 contained standard protein markers. Lane 12 contained *E. coli* proteins. **(b)** Far Western blot analysis. A gel electrophoresed in duplicate with the stained gel in panel (a) was blotted to a nitrocellulose membrane, and probed with [^{32}P] β-galactosidase-CTD, then subjected to phosphorimaging.

(Source: *Journal of Biological Chemistry* by Morris and Greenleaf. Copyright 2000 by Am. Soc. For Biochemistry & Molecular Biol. Reproduced with permission of Am. Soc. For Biochemistry & Molecular Biol. in the format Textbook via Copyright Clearance Center.)

Changes in Association of RNA-Processing Proteins with the CTD Correlate with Changes in CTD Phosphorylation

The fact that all three classes of major mRNA-processing proteins bind to the CTD raises a question: We know that the CTD is long and could bind to many proteins at once, but does it associate simultaneously with all the proteins and RNAs involved in all three processing events?

The answer is that proteins come to and go from the CTD as they are needed for the task at hand. Moreover, these comings and goings are correlated with changes in CTD phosphorylation during transcription. Steven Buratowski and coworkers investigated the association of capping and polyadenylation enzymes with yeast polymerase II near the promoter (shortly after initiation) and remote from the promoter (during elongation, long after initiation). They also examined the state of phosphorylation of the CTD near promoters or remote from promoters.

They discovered that the capping enzyme (the guanylyl transferase) associates with the CTD near the promoter (shortly after initiation), but not in the interior of the gene. By contrast, the cap methyl transferase and the polyadenylation factor Hrp1/CFIB associate with the CTD both near and remote from the promoter. Thus, these factors are present on the transcription complex during both initiation and elongation. Moreover, these workers discovered that serine 5 of the CTD heptads is phosphorylated when the complex is near promoters, but not later during elongation, while serine 2 of the CTD heptads has a complementary pattern of phosphorylation: It is phosphorylated during elongation (remote from promoters) but not earlier, when the polymerase is still near the promoter.

To reach these conclusions, Buratowski and coworkers exploited the chromatin immunoprecipitation (ChIP) technique described in Chapter 5. They immunoprecipitated chromatin with antibodies against the capping and polyadenylation proteins to catch chromatin being transcribed by polymerase that is interacting with these proteins. Then they probed the precipitated chromatin by PCR with primers that would amplify DNA regions close to promoters or remote from promoters of several different genes.

What can we learn from such an assay? One possible outcome is the following: Chromatin immunoprecipitated with an antibody directed against a particular protein gives a strong PCR signal with primers that hybridize near a promoter, but only a weak signal with primers that hybridize to the interior of a gene. This would indicate that this protein is associated with the transcribing complex at or shortly after initiation of transcription, but not later during the elongation phase.

Figure 15.30 shows the results of the ChIP assay with antibodies against: the yeast capping enzyme guanylyl transferase (α-Ceg1); yeast polyadenylation factor (α-Hrp 1); and the Rpb3 subunit of yeast RNA polymerase II (α-HA-Rpb3).

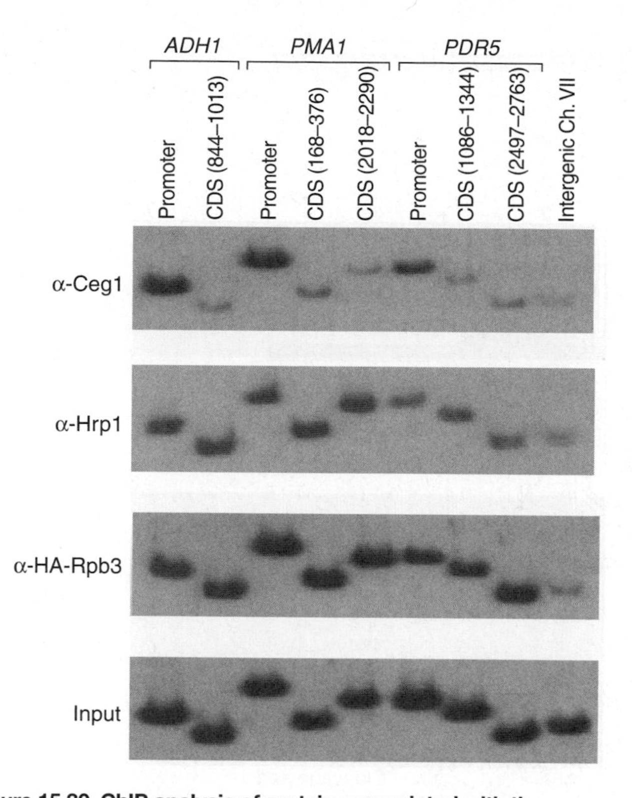

Figure 15.30 ChIP analysis of proteins associated with the transcription complex on three yeast genes. Buratowski and coworkers performed ChIP analysis of the association of three proteins (the capping guanylyl transferase, a polyadenylation factor, and the Rpb3 subunit of RNA polymerase II) with the transcription complex when it is near the promoter or remote from the promoter of three different genes (*ADH1, PMA1,* and *PDR5*). They used the following antibodies to immunoprecipitate chromatin: an antibody against the capping guanylyl transferase (α-Cegl); an antibody against a polyadenylation factor (α-Hrp1); and an antibody against the Rpb3 subunit of RNA polymerase II (α-HA-Rpb3). The antibodies used in each experiment are listed at left. Then they performed PCR on the precipitated chromatin with primers specific for promoter regions or coding sequences (CDS) of the three genes to determine whether the transcription complex was near the promoters of the genes or not. Strong signals, with abundant PCR product, indicate that the corresponding DNA, near or remote from the promoter, was present in the precipitated chromatin. The bottom panel contains PCR results on the input chromatin, showing that all areas of the genes were equally represented before immunoprecipitation. The last lane in each panel is a negative control, with the results of PCR with primers specific for an intergenic, untranscribed region of chromosome VII. This region was present in the input chromatin, but not immunoprecipitated by any of the antibodies. (*Source:* Reprinted by permission of S. Buratowski from "Komarnitsky, Cho, and Buratowski (2000) *Genes and Development* v. 14, pp. 2452–2460" © Cold Spring Harbor Laboratory Press.)

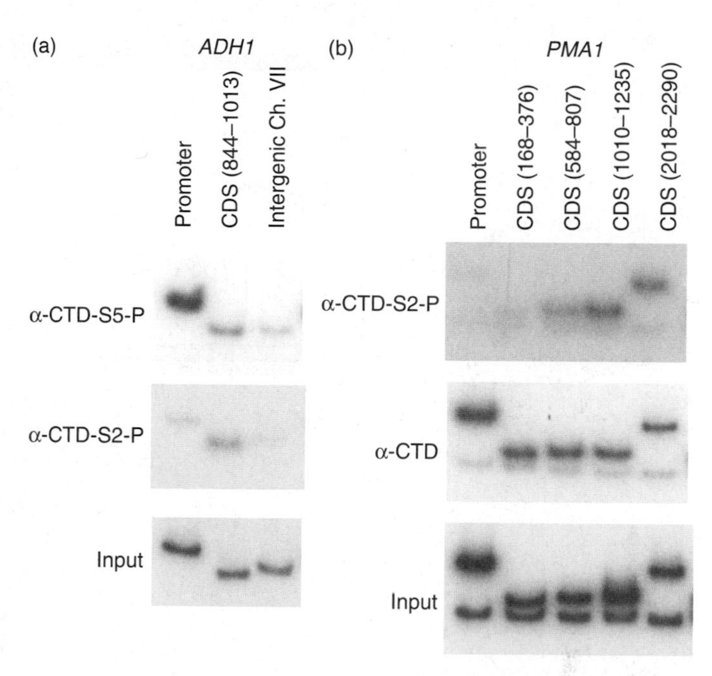

Figure 15.31 ChIP analysis of the phosphorylation state of the CTD of RNA polymerase II at various stages of transcription. Buratowski and coworkers performed ChIP analysis of the association of two phosphorylated forms of the CTD of the Rpb1 subunit of RNA polymerase II with chromatin near or remote from the promoters of two genes. **(a)** Transcription of the *ADH1* gene. Chromatin was immunoprecipitated with antibodies against the CTD phosphorylated on either the serine 2 or serine 5 of the heptad, as indicated at left (α-CTD-S2-P and α-CTD-S5-P, respectively). Then the precipitated chromatin was subjected to PCR with primers specific for regions near the promoter, or remote from the promoter, or an intergenic region, as indicated at top. **(b)** Transcription of the *PMA1* gene. Chromatin was immunoprecipitated with antibodies against the CTD phosphorylated on serine 2 or the unphosphorylated CTD, as indicated at left. PCR primers, indicated at top, were specific for the promoter, or regions progressively more remote from the promoter (CDS = coding sequences). Input chromatin controls are at bottom in both panels. (*Source:* Reprinted by permission of S. Buratowski from "Komarnitsky, Cho, and Buratowski (2000) *Genes and Development* v. 14, pp. 2452–2460" © Cold Spring Harbor Laboratory Press.)

The chromatin immunoprecipitated with each of these antibodies was subjected to PCR with primers specific for promoter regions and interiors of three yeast genes: alcohol dehydrogenase (*ADH1*); cytoplasmic H^+ ATPase (*PMA1*); and a multidrug resistance factor (*PDR5*). The results with all three genes were consistent and demonstrated that: (1) the guanylyl transferase (capping enzyme) associates with the transcription complex only when it is near the promoter; (2) the polyadenylation factor associates with the transcription complex both near and remote from the promoter; and, as expected, the Rpb3 subunit of RNA polymerase is present in the transcription complex both near and remote from the promoter.

Thus, there is a dynamic shift of proteins associating with the transcription complex through the CTD of Rpb1. Some are present only early during the transcription process; others are present for much longer. What causes these changes in the spectrum of proteins associated with the CTD? It is known that the phosphorylation state of the CTD changes during transcription, so perhaps this plays a role.

To investigate this possibility, Buratowski and coworkers performed ChIP assays using antibodies directed against specific phosphorylated amino acids (serine 2 and serine 5) within the heptad repeats of the CTD. The ChIP assays in Figure 15.31 reveal that serine 5 phosphorylation is found primarily in transcription complexes close to the promoter, while serine 2

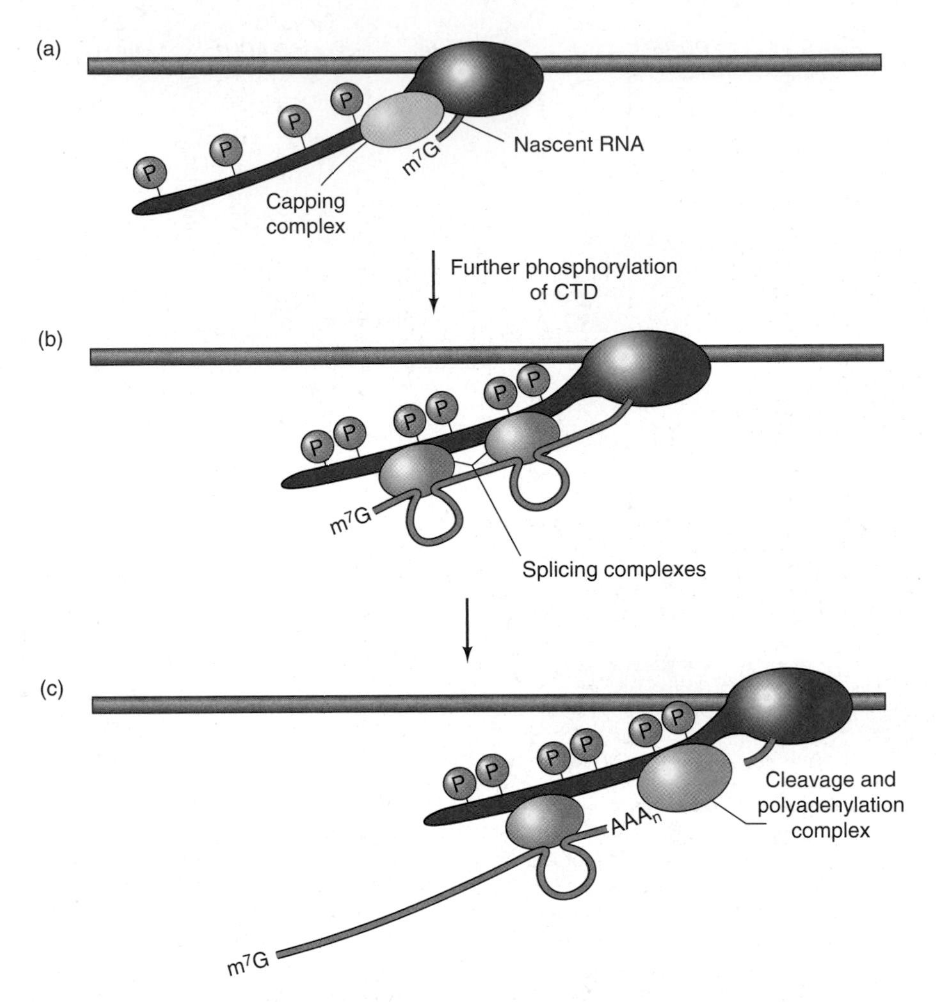

Figure 15.32 Hypothesis of RNA processing organized by CTD. **(a)** RNA polymerase (red) has begun synthesizing a nascent RNA (green). The partially phosphorylated CTD has attracted the capping complex (yellow), which adds a cap to the new RNA as soon as it is available. **(b)** The CTD has become further phosphorylated (presumably including a shift from serine 5 to serine 2 phosphorylation) and has attracted the splicing complex (blue), which defines exons as they are transcribed and splices out the introns in between. **(c)** The CTD is associated with the cleavage and polyadenylation complex (orange), which may have been present since initiation, and this complex has cleaved and begun polyadenylating the transcript. (*Source:* Adapted from Orphanides, G. and D. Reinberg, A unified theory of gene expression. *Cell* 108 [2000] p. 446, f. 3.)

phosphorylation occurs chiefly in transcription complexes remote from the promoter. Thus, it is not surprising that phosphorylation of serine 5 of the CTD helps recruit the capping complex, which needs to operate shortly after elongation begins. It is also quite possible that the shift in CTD phosphorylation from serine 5 to serine 2, as the transcription complex moves away from the promoter, causes some RNA-processing proteins (e.g., the capping complex) to leave the transcription complex and may even attract a new class of proteins. Figure 15.32 summarizes this hypothesis.

SUMMARY The phosphorylation state of the CTD of Rpb1 in transcription complexes in yeast changes as transcription progresses. Transcription complexes close to the promoter contain phosphorylated serine 5, while complexes farther from the promoter contain phosphorylated serine 2. The spectrum of proteins associated with the CTD also changes. For example, the capping guanylyl transferase is present early in the transcription process, when the complex is close to the promoter, but not later. And this enzyme, along with the rest of the capping complex, is recruited by phosphorylation of serine 5 of the heptad in the polymerase II CTD. By contrast, the polyadenylation factor Hrp1 is present in transcription complexes both near and remote from the promoter.

A CTD Code?

In 2007, Shona Murphy and colleagues showed that serine 7 of the CTD can also be phosphorylated. This raises the number of different phosphorylation states in a given repeat within the CTD to eight (ranging from no phosphates to three phosphates per repeat). It is also possible that the phosphorylation varies from repeat to repeat, opening up many more variations in CTD phosphorylation state.

Even the potential for eight different states in a given repeat raises the possibility of a "CTD code" that signals for transcription of different gene sets and for different RNA modifications. Indeed, there is evidence for such a CTD code. Murphy and colleagues showed in 2007 that phosphorylation of serine 7 is required for expression of the U2 snRNA gene in human cells. On the other hand, Dirk Eick and colleagues demonstrated that phosphorylation of serine 7 is not required for expression of protein-encoding genes.

Human snRNAs synthesized by polymerase II, including U1 and U2 snRNAs, are not polyadenylated. Instead, their genes contain a conserved **3′ box** element that is essential for proper 3′-end processing. Transcription termination occurs downstream of the 3′ box, and this 3′ box is required for the subsequent clipping that yields the primary 3′-ends that can then be processed in the cytoplasm to mature 3′-ends.

Murphy and colleagues started with an α-amanitin-resistant human polymerase II with an Rpb1 CTD containing only the first 25 heptads. These are the ones with canonical sequences ending in serine 7; most of the last 27 heptads have lysine or threonine instead of serine in the seventh position. The α-amanitin-resistance of this polymerase allowed it to be assayed in cells that also carried an endogenous wild-type polymerase II. Next, Murphy and colleagues mutated the α-amanitin-resistant polymerase to change all 25 serine 7's to alanines, and assayed for proper 3′-end processing by RNase protection analysis. They found that the mutant polymerase was deficient in U2 snRNA processing, but was normal in processing a protein-encoding pre-mRNA.

Note that this transcription control does not occur at the initiation level; the mutant polymerase still initiates at a normal level. Instead, control occurs at the termination or 3′-end processing level. Murphy and colleagues investigated the binding of the **Integrator complex,** a group of 12 polypeptides that are required for U1 and U2 snRNA 3′-end processing, to the mutant polymerase with all its serine 7's changed to alanines. They tagged one of the subunits of the Integrator complex with a TAP epitope and used ChIP to detect binding of the Integrator complex to the mutant RNA polymerase II. Whereas the Integrator complex binds well to the CTD of normal polymerase II, Murphy and colleagues found that it does not bind to the mutant polymerase lacking serine 7 in its CTD. This suggested that serine 7 phosphorylation is required for Integrator complex binding, and thus for proper 3′-end processing of U1 and U2 snRNA transcripts. This is the best evidence to date for a CTD code that affects gene expression.

SUMMARY In addition to serines 2 and 5, serine 7 of the heptad repeat in the Rpb1 CTD is phosphorylated during transcription. This raises the number of combinations of phosphorylated and unphosphorylated serines in each repeat to eight, and raises the possibility of a CTD code that governs which genes are expressed. One piece of evidence for such a code is the fact that loss of serine 7 from the repeats prevents 3′-end processing of U2 snRNA transcripts, and therefore prevents expression of the U2 snRNA gene.

Coupling Transcription Termination with mRNA 3′-end Processing

Termination of transcription of class II genes has been notoriously difficult to study, largely because the mature 3′-end of the mRNA is not the same as the termination site. Instead, as we have already learned, a longer, pre-mRNA must be cleaved at the polyadenylation site and then polyadenylated. This leaves a relatively stable mRNA and an unstable 3′-fragment that is rapidly degraded. It is the 3′-end of this unstable part of the RNA that is the true termination site. Despite this difficulty, several investigators have successfully studied termination in class II genes and have discovered that termination is coupled to cleavage at the polyadenylation site, in that each process depends on the other. Indeed, cleavage of the nascent RNA at the termination site may even precede cleavage at the polyadenylation site.

First of all, how do we know that termination is coupled to mRNA processing? Proudfoot and colleagues made this connection in their studies of yeast class II transcription termination. In particular, they examined the *CYC1* gene of the yeast *Saccharomyces cerevisiae* and found that mutations in proteins involved in cleavage at the polyadenylation site inhibited termination, whereas mutations in proteins involved in polyadenylation per se had little effect on termination.

Proudfoot and colleagues cloned the yeast *CYC1* gene into a plasmid (p*GCYC1*) in which it would be expressed under the control of the strong *GAL1/10* promoter. They made a similar construct (p*Gcyc1-512*), which lacked the normal polyadenylation signal at the end of the *CYC1* gene. Next, they transfected yeast cells with these plasmids and assayed first for the expression level of the gene by Northern blotting. Figure 15.33a shows the results: The loss of the polyadenylation site greatly reduced expression from the gene. The control showed that expression of another gene (*ACT1*) was not affected, so the loss of the *CYC1* signal was not due to differences in loading or blotting of the two lanes.

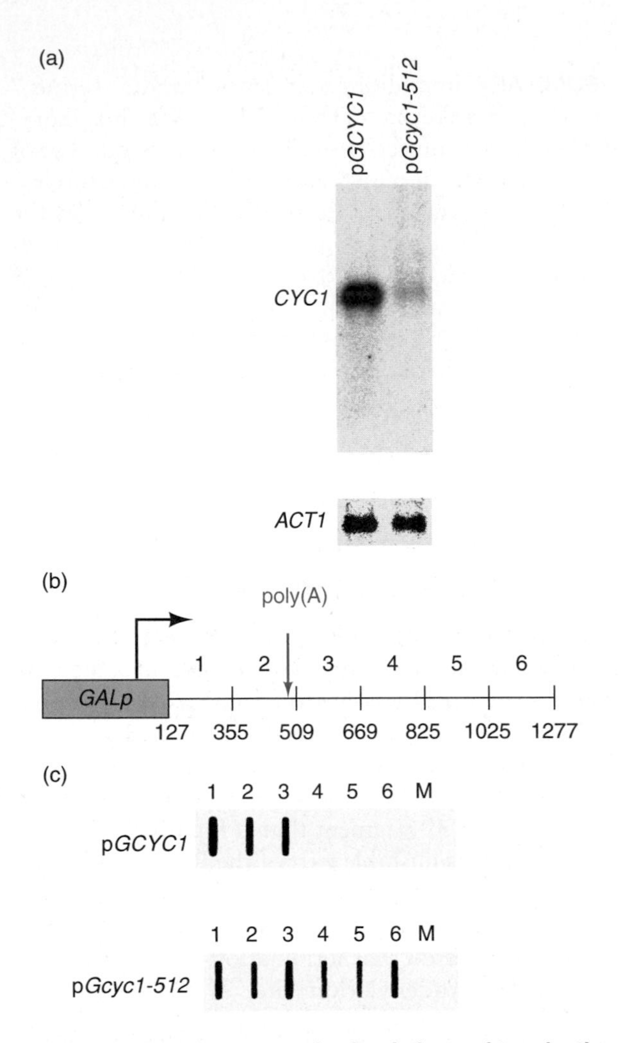

Figure 15.33 Linkage between polyadenylation and termination of transcription. **(a)** Northern blot analysis. Proudfoot and colleagues Northern blotted transcripts from cells bearing the wild-type gene (p*GCYC1*) or a gene lacking the *CYC1* polyadenylation site (p*Gcyc1-512*). Then they hybridized the blot with a labeled *CYC1* probe. After the first hybridization, they stripped the blot and reprobed with an actin gene probe (*ACT1*) as a control for blotting efficiency. **(b)** Map of the region used in nuclear run-on transcription analysis. Proudfoot and colleagues cloned the yeast *CYC1* gene under the control of the strong *GAL1/10* promoter (*GALp,* green) into a plasmid and placed this construct into yeast cells for analysis. For nuclear run-on analysis, they dot-blotted fragments 1–6, whose relative positions are given. The location of the polyadenylation site (red) in fragment 2 is indicated. **(c)** Results of run-on analysis. Proudfoot and colleagues hybridized dot blots of fragments 1–6, (panel b) to labeled nuclear run-on transcripts from cells carrying the wild-type or mutant *CYC1* gene, as indicated at left. M designates a negative control with M13 DNA on the dot blot. (*Source:* Birse et al *Science* 280: p. 299. © 1988 by the AAAS.)

One reason for the poor expression could be failure to terminate transcription properly. To see if termination really did fail, Proudfoot and colleagues performed a nuclear run-on analysis as follows: They dot-blotted fragments of the *CYC1* gene, including fragments encompassing about 800 bp downstream of the polyadenylation site, as illustrated in Figure 15.33b. Then they hybridized labeled nuclear run-on RNA from cells transfected with either the wild-type *CYC1* gene or the mutant gene lacking the polyadenylation site. Figure 15.33c shows the results. Transcription of the wild-type gene terminated in fragment 3, just downstream of the polyadenylation site. We know that termination occurred in fragment 3 because no transcripts hybridized to fragment 4. But transcription of the mutant gene extended far past the normal termination site, at least into fragment 6, showing that normal termination had failed.

As we have learned, polyadenylation really consists of two steps: RNA cleavage and then polyadenylation. In principle, one of these steps, and not the other, could be coupled to termination. To explore this issue, Proudfoot and colleagues performed a new run-on transcription assay with yeast strains bearing temperature-sensitive mutations in the genes encoding cleavage and polyadenylation factors. Again, they did Northern blots first and discovered that all of the mutants showed depressed levels of *CYC1* mRNA at the nonpermissive temperature. Again, failure to polyadenylate the transcript and failure to terminate the transcript could both have led to its instability.

The run-on transcription assay gave a more complete answer. Some of the mutations caused a failure of termination, but others did not. Is there a pattern here? Indeed, there is. The former set of genes encode proteins involved in cleavage prior to polyadenylation, while the latter set encode proteins involved in polyadenylation after cleavage. Thus, it appears that cleavage at the polyadenylation site, not polyadenylation per se, is coupled to termination of transcription.

We know that the cleavage and polyadenylation factors associate with the CTD of the Rpb1 subunit of RNA polymerase II. The fact that active cleavage factors are required for termination implicates the CTD in termination as well as in other aspects of mRNA maturation. We will return to this theme in the next section.

SUMMARY Transcription termination and mRNA 3′-end processing are coupled in the following way: An intact polyadenylation site and active factors that cleave at the polyadenylation site are required for transcription termination, at least in yeast. Active factors that polyadenylate a cleaved pre-mRNA are not required for termination.

Mechanism of Termination

Michael Dye and Proudfoot performed a detailed analysis of termination in the human β- and ε-globin genes in 2001. They made the following discoveries: (1) The region downstream of the polyadenylation site is essential for termination. (2) Cleavage of the nascent transcript at multiple sites downstream of the polyadenylation site is required for termination. (3) This transcript cleavage occurs cotranscriptionally and, presumably, precedes cleavage at the polyadenylation site. Then, in 2004, they discovered that the cleavage of the nascent transcript is an autocatalytic event: The RNA cleaves itself.

In their 2001 study, Dye and Proudfoot put the human β-globin gene, including 1.7 kb of its 3′-flanking region, into a plasmid under control of a strong enhancer–promoter combination from the human immunodeficiency virus (HIV). Then they placed this construct into HeLa cells where the β-globin gene could be expressed. The HIV enhancer–promoter has the advantage that the transcription it directs depends on a viral transactivating factor called Tat, so transcription can be turned on and off easily by adding or removing Tat.

Next, these workers performed nuclear run-on analysis of the cloned gene and compared the results to those from the β-globin gene in its natural chromosomal context, under control of its own promoter. Figure 15.34a shows a map of the β-globin gene, including the downstream region, with its own promoter, and the results of the nuclear run-on experiment. Transcription continued through region 10, which lies 1.7 kb downstream of the polyadenylation site. Figure 15.34b shows a map of the cloned β-globin gene under control of the HIV enhancer–promoter, and the results of the nuclear run-on experiment. Again, transcription continued through region 10, but fell off significantly after region 10. The DNA beyond region 10 encompassed regions A and B of the vector, and region U3 of the HIV enhancer–promoter. Thus, termination had occurred at least by region 10, and transcription and termination appeared to be working normally in this cloned construct.

Next, Dye and Proudfoot narrowed down the part of the 3′-flanking region that was important for termination of transcription. They did this by deleting parts of the region and testing by nuclear run-on analysis to see whether termination still occurred. They discovered that deleting regions 8–10 prevented termination. Thus, regions 4–7 were not sufficient for termination. On the other hand, they discovered that deleting regions 5–8, but retaining 9 and 10, or even deleting regions 5–9, but retaining 10, maintained termination. Most strikingly, deleting all regions downstream of 4, except region 8, maintained termination. Thus, regions 8, 9, and 10, individually or together, all could direct termination.

Because region 8 (as well as 9 and 10) appeared to have a termination sequence that operated by causing cleavage of the growing transcript during transcription, Proudfoot and colleagues named it the cotranscriptional cleavage element (**CoTC element**). Then, in 2004, Proudfoot and Alexander Akoulitchev and their colleagues discovered an important secret of the CoTC element: It encodes an autocatalytic domain that can cleave the growing RNA. When they incubated a transcript containing the full-length CoTC

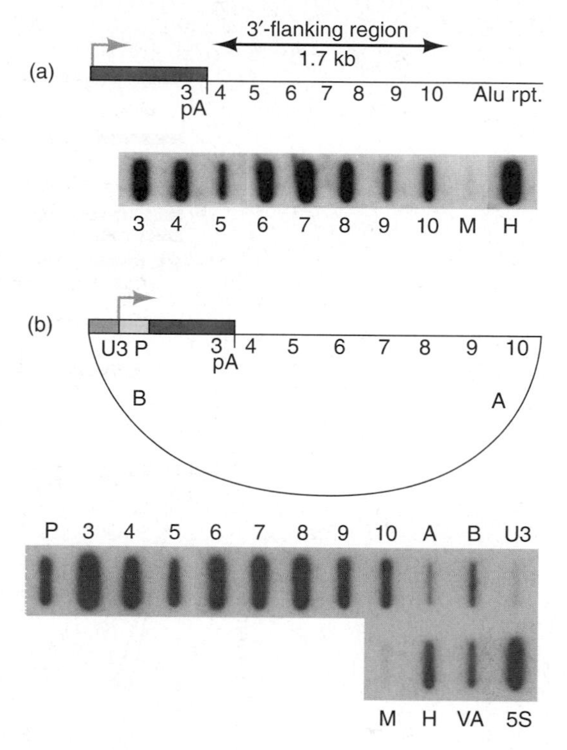

Figure 15.34 Nuclear run-on analysis of natural and cloned β-globin genes. **(a)** Gene in its chromosomal context. A map of the human gene is shown, including the promoter (purple arrow denotes transcription start site), the coding region (red), the polyadenylation site (pA), and 1.7 kb of downstream sequence (regions 4–10). The results of nuclear run-on analysis are shown below the map, including regions 3–10 and two controls, M and H. M is a negative control containing phage M13 DNA. H is a positive control containing human histone DNA. The histone gene will be transcribed by RNA polymerase II in the cell. **(b)** Gene under control of the HIV enhancer/promoter. The map shows the HIV enhancer region (blue), the HIV promoter region (yellow), the start of transcription (purple arrow), and the coding region (red). Regions A and B lie within the plasmid cloning vector. The results of nuclear run-on analysis are shown below the map. M and H have the same meaning as in panel (a). VA represents an adenovirus VA1 gene, cotransfected along with the β-globin plasmid. This gene is transcribed by RNA polymerase III. 5S denotes hybridization to a 5S rRNA probe, which detects in vivo transcription of the human 5S rRNA gene by RNA polymerase III. (*Source:* Reprinted from *Cell* v. 105, Dye and Proudfoot, p. 670 © 2001, with permission from Elsevier Science.)

element with Mg^{2+} and GTP, but no proteins, the RNA decayed much faster than a control RNA, with a half-life of just 38 min. By making deletions within the CoTC element, these workers were able to narrow the autocatalytic site's location down to a 200-nt sequence [**CoTC(r)**] at the 5′-end of the CoTC element (Figure 15.35). This 200-nt sequence decayed with a half-life of just 15 min in vitro. By contrast, the mutant sequence (*mut*Δ) containing nucleotides 50–150 had no autocatalytic activity.

Is the CoTC element important in transcription termination? To find out, the investigators inserted the β-globin gene into a plasmid and placed the plasmid into HeLa cells. They also replaced the CoTC element at the end of the β-globin

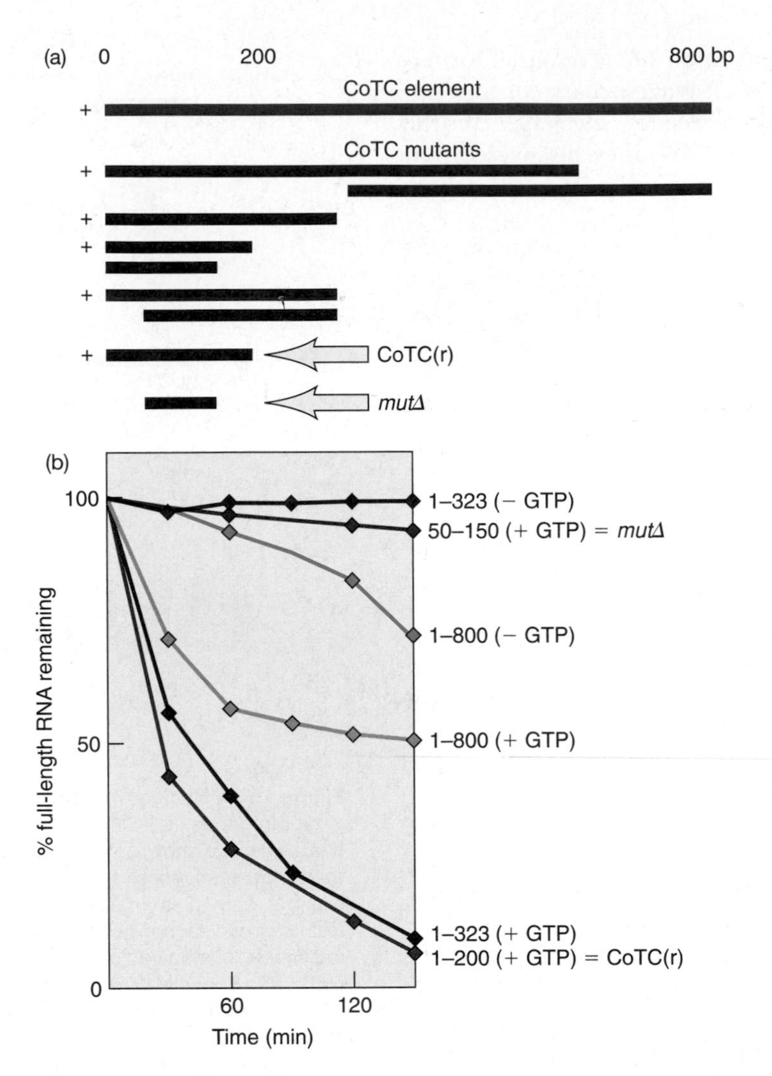

Figure 15.35 Finding the catalytic site in the CoTC element. **(a)** The mutants. Proudfoot, Akoulitchev, and colleagues started with the 800-bp CoTC element at top (red bar) and made deletion mutants that were transcribed to yield the RNAs illustrated below (blue bars). Deletions are denoted by gaps in the bars. Mutant RNAs that retained catalytic activity are marked with plus signs at left. The arrows point to: CoTC(r), the RNA containing nucleotides 1–200, which retained activity; and *mut*Δ, the RNA containing nucleotides 50–150, which lacked activity. **(b)** Experimental results. The fraction of full-length RNA remaining is plotted versus reaction time. We see that the reaction depends on GTP, and that the CoTC(r) RNA that includes nucleotides 1–200 retains full catalytic activity. (*Source:* Adapted from A. Teixeira et al., Autocatalytic RNA cleavage in the human beta-globin pre-mRNA promotes transcription termination. *Nature* 432:526, 2006.)

gene with its mutant forms, including CoTC(r) (the minimal autocatalytic element) and *mut*Δ (the element lacking autocatalytic activity). Then they performed nuclear run-on analysis to see whether transcription termination occurred normally. They found that the gene with the CoTC(r) element at its end terminated transcription almost as well as wild-type, while the gene with the *mut*Δ element at its end allowed transcription to continue past the normal termination site. In experiments with other mutant CoTC elements, they found that the autocatalytic activity of CoTC correlated very well with termination activity. Thus, the autocatalytic activity appears to be required for proper termination.

Is an autocatalytic CoTC-like element a general requirement for transcription termination in eukaryotes? The β-globin genes of primates do contain a conserved CoTC element, with the highest level of conservation in the catalytic core. Such elements are not detected in less related organisms, presumably because of greater sequence divergence. However, the CoTC element itself could not have been identified as a self-cleaving ribozyme on the basis of sequence alone, so there may be CoTC-like elements downstream of the poly(A) sites of many more eukaryotic genes.

Is simple cleavage of a growing RNA at a CoTC or other site sufficient to cause termination? Perhaps not, as we now have evidence for another phenomenon that operates on RNA polymerases that are extending transcripts beyond their poly(A) sites: The polymerases are "torpedoed." Figure 15.36 illustrates this torpedo mechanism, which resembles the rho-dependent mechanism of termination we studied in Chapter 6. First the RNA is cleaved

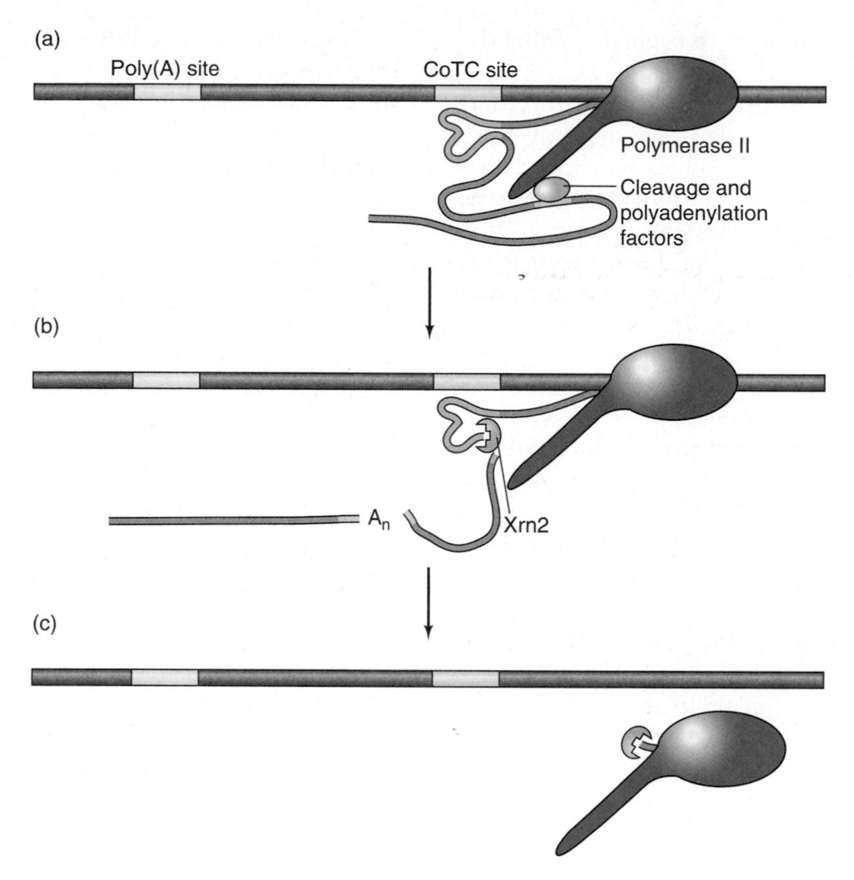

Figure 15.36 A torpedo model for transcription termination in the human β-globin gene. (a) The RNA polymerase (red) has transcribed both the poly(A) site (yellow) and the CoTC site (blue). Cleavage and polyadenylation factors (green) have assembled at the poly(A) site and are also attached to the CTD of the polymerase. **(b)** The cleavage and polyadenylation process is complete, and the mRNA has its poly(A) tail. Also, the CoTC sequence in the transcript has undergone self-cleavage, and the Xrn2 exonuclease (orange) has loaded onto the newly-created RNA 5′-end. **(c)** Xrn2 has degraded the growing RNA nucleotide by nucleotide, has caught the RNA polymerase, and has somehow torpedoed it, causing the polymerase to dissociate from the template and terminate transcription.

downstream of the poly(A) site at a CoTC or other site, then an exonuclease binds to the newly generated RNA free end and begins degrading the RNA, "chasing" the polymerase that is elongating the RNA. When the exonuclease catches the polymerase, it "torpedoes" it, terminating transcription.

In the context of the human β-globin gene, the torpedo model implies that cleavage of the growing transcript at the CoTC site provides an entry site for a 5′→3′ exonuclease that will ultimately torpedo the polymerase. If so, then depleting cells of the relevant 5′→3′ exonuclease should interfere with proper termination. Proudfoot and colleagues tested this notion by using RNAi (Chapter 16) to "knock down" the level of the major human nuclear 5′→3′ exonuclease, **Xrn2.** Using this technique, they depleted the Xrn2 activity to about 25% of its normal value, then tested these cells for proper termination by nuclear run-on assay. They discovered that depletion of Xrn2 activity resulted in a two- to three-fold decrease in normal termination. That is, transcription was two- to three-fold more likely to continue beyond the normal termination site.

Proudfoot and colleagues considered the possibility that cleavage at the poly(A) site, and not the CoTC site, is the entry site for Xrn2. If this were the case, then RNA derived from the region between the poly(A) site and the CoTC site should be less depleted in Xrn2 knock-down cells than in untreated cells. But an RNase protection assay with a probe to measure the steady-state level of transcript from the region between the poly(A) site and the CoTC site showed no difference between Xrn2 knock-down and untreated cells.

Will any 5′-end in the CoTC region provide an entry site for Xrn2? Proudfoot and colleagues addressed this question by substituting a hammerhead ribozyme sequence for the normal CoTC sequence. Hammerhead ribozymes are self-cleaving RNAs, but they produce 5′-hydroxyl groups instead of the 5′-phosphates produced by CoTC. And nuclear run-on analysis showed that although the hammerhead ribozyme did cleave the growing β-globin transcript cotranscriptionally, the downstream RNA was not degraded, as it is in cells with the normal CoTC sequence. Thus, Xrn2 at least appears to require a 5′-phosphate group, such as provided by CoTC, in order to begin degrading the downstream RNA.

How widespread is the torpedo mechanism for transcription termination? Jack Greenblatt, Steven Buratowski and their colleagues have found a 5′→3′ exonuclease called Rat1 that promotes transcription termination in yeast. There is no evidence for a CoTC element in yeast, so it is assumed that Rat1 gains access to the downstream RNA following cleavage at the poly(A) site, then chases the polymerase until it catches and torpedoes it.

SUMMARY Termination of transcription by RNA polymerase II occurs in two steps. First, the transcript experiences a cotranscriptional cleavage (CoTC) within the termination region downstream of the polyadenylation site. This step occurs before cleavage and polyadenylation at the poly(A) site and is independent of that process. Second, cleavage and polyadenylation occur at the poly(A) site, signaling the polymerase, which is still elongating RNA, to dissociate from the template. In certain genes, at least, this signal could be delivered by a "torpedo," as follows: The CoTC element downstream of the polyadenylation site in the human β-globin mRNA is a ribozyme that cleaves itself, generating a free RNA 5′-end. This cleavage is required for normal transcription termination, apparently because it provides an entry site for Xrn2, a 5′→3′ exonuclease that loads onto the RNA and "chases" the RNA polymerase by degrading the RNA. When it catches up to the polymerase, Xrn2 presumably "torpedoes" it, terminating transcription. A similar torpedo mechanism appears to operate in yeast.

Role of Polyadenylation in mRNA Transport

We have known since 1991 that polyadenylation plays a role in transport of mature mRNA out of the nucleus. That is when Max Birnstiel and colleagues demonstrated that transcripts of a bacterial neomycin gene transplanted into monkey COS1 cells remained in the nucleus. They reasoned that the lack of a polyadenylation signal in the bacterial gene would have left the transcripts without a mature 3′-end, and that might be the reason for defective transport to the cytoplasm.

To test this hypothesis, they provided the neomycin gene with the strong polyadenylation signal from a mammalian β-globin gene. This allowed for polyadenylation of the neomycin transcripts, which were then efficiently transported out of the nucleus into the cytoplasm.

In 2001, Patricia Hilleren and colleagues studied a strain of yeast carrying a temperature-sensitive mutation in the poly(A) polymerase gene. These cells could be shifted to the nonpermissive temperature to shut off polyadenylation of newly made transcripts. These workers focused their attention on transcripts of the *SSA4* gene, a heat-shock gene whose transcripts begin to accumulate at the time of the shift to the nonpermissive temperature. Then they showed by fluorescence in situ hybridization (FISH, Chapter 5) that the *SSA4* transcripts remained in small foci within the nucleus, presumably at or close to the site of their transcription. In wild-type cells, or in mutant cells at the permissive temperature, these transcripts could not be detected in the nucleus and had presumably been polyadenylated and transported to the cytoplasm. Again, it appeared that polyadenylation is required for active transport of mRNAs out of the nucleus. Without polyadenylation, transcripts didn't even seem to move far from their transcription site.

SUMMARY Polyadenylation is required for efficient transport of mRNAs from their point of origin in the nucleus to the cytoplasm.

SUMMARY

Caps are made in steps: First, an RNA triphosphatase removes the terminal phosphate from a pre-mRNA. Next, a guanylyl transferase adds the capping GMP (from GTP). Next, two methyl transferases methylate the N^7 of the capping guanosine and the 2′-O-methyl group of the penultimate nucleotide. These events occur early in the transcription process, before the chain length reaches 30. The cap ensures proper splicing of at least some pre-mRNAs, facilitates transport of at least some mature mRNAs out of the nucleus, protects the mRNA from degradation, and enhances the mRNA's translatability.

Most eukaryotic mRNAs and their precursors have a poly(A) about 250 nt long at their 3′-ends. This poly(A) is added posttranscriptionally by poly(A) polymerase. Poly(A) enhances both the lifetime and translatability of mRNA. The relative importance of these two effects seems to vary from one system to another.

Transcription of eukaryotic genes extends beyond the polyadenylation site. Then the transcript is cleaved and polyadenylated at the 3′-end created by the cleavage. An efficient mammalian polyadenylation signal consists of an AAUAAA motif about 20 nt upstream of a polyadenylation site in a pre-mRNA, followed 23 or 24 bp later by a GU-rich motif, followed immediately by a

U-rich motif. Many variations on this theme occur in nature, which results in variations in efficiency of polyadenylation. Plant polyadenylation signals also usually contain an AAUAAA motif, but more variation is allowed in this region than in an animal AAUAAA. Yeast polyadenylation signals are more different yet and rarely contain an AAUAAA motif.

Polyadenylation requires both cleavage of the pre-mRNA and polyadenylation at the cleavage site. Cleavage requires several proteins: CPSF, CstF, CF I, CF II, poly(A) polymerase, and the CTD of the RNA polymerase II largest subunit. One of the subunits of CPSF (CPSF-73) cleaves the pre-mRNA prior to polyadenylation. Short RNAs that mimic a newly created mRNA 3′-end can be polyadenylated. The optimal signal for initiation of such polyadenylation of a cleaved substrate is AAUAAA, followed by at least 8 nt. Once the poly(A) reaches about 10 nt in length, further polyadenylation becomes independent of the AAUAAA signal, and depends on the poly(A) itself. Two proteins participate in the initiation process: poly(A) polymerase and CPSF, which binds to the AAUAAA motif.

Elongation requires a specificity factor called poly(A)-binding protein II (PAB II). This protein binds to a preinitiated oligo(A) and aids poly(A) polymerase in elongating the poly(A) up to 250 nt or more. PAB II acts independently of the AAUAAA motif. It depends only on poly(A), but its activity is enhanced by CPSF.

Calf thymus poly(A) polymerase is probably a mixture of at least three proteins derived from alternative RNA processing. The structures of the enzymes predicted from these sequences include an RNA-binding domain, a polymerase module, two nuclear localization signals, and a serine/threonine-rich region. The latter region, but none of the rest, is dispensable for activity in vitro.

Poly(A) turns over in the cytoplasm. RNases tear it down, and poly(A) polymerase builds it back up. When the poly(A) is gone, the mRNA is slated for destruction. Maturation-specific polyadenylation of maternal mRNAs in the cytoplasm depends on two sequence motifs: the AAUAAA motif near the end of the mRNA, and an upstream motif called the cytoplasmic polyadenylation element (CPE), which is UUUUUAU or a closely related sequence.

Caps and poly(A) play a role in splicing, at least in removal of the introns closest to the 5′ and 3′ ends, respectively, of the pre-mRNA. Capping, polyadenylation, and splicing proteins all associate with the CTD during transcription.

The phosphorylation state of the CTD of Rpb1 in transcription complexes in yeast changes as transcription progresses. Transcription complexes close to the promoter contain phosphorylated serine 5, while complexes farther from the promoter contain phosphorylated serine 2. The spectrum of proteins associated with the CTD also changes. For example, the capping guanylyl transferase is present early in the transcription process, when the complex is close to the promoter, but not later. By contrast, the polyadenylation factor Hrp1 is present in transcription complexes both near and remote from the promoter. In addition to serines 2 and 5, serine 7 of the heptad repeat in the Rpb1 CTD is phosphorylated during transcription. This raises the number of combinations of phosphorylated and unphosphorylated serines in each repeat to eight, and raises the possibility of a CTD code that governs which genes are expressed. One piece of evidence for such a code is the fact that loss of serine 7 from the repeats prevents 3′-end processing of U2 snRNA transcripts, and therefore prevents expression of the U2 snRNA gene.

An intact polyadenylation site and active factors that cleave at the polyadenylation site are required for transcription termination, at least in yeast. Active factors that polyadenylate a cleaved pre-mRNA are not required for termination. Termination of transcription by RNA polymerase II occurs in two steps. First, the transcript experiences a cotranscriptional cleavage (CoTC) within the termination region downstream of the polyadenylation site. This step occurs before cleavage and polyadenylation at the poly(A) site and is independent of that process. Second, cleavage and polyadenylation occur at the poly(A) site, signaling the polymerase, which is still elongating RNA, to dissociate from the template. The CoTC element downstream of the polyadenylation site in the human β-globin mRNA is a ribozyme that cleaves itself, generating a free RNA 5′-end. This cleavage is required for normal transcription termination, apparently because it provides an entry site for Xrn2, a 5′→3′ exonuclease that loads onto the RNA and "chases" the RNA polymerase by degrading the RNA. When it catches up to the polymerase, Xrn2 presumably "torpedoes" it, terminating transcription. A similar torpedo mechanism appears to operate in yeast.

REVIEW QUESTIONS

1. You label a capped eukaryotic mRNA with ^{3}H-AdoMet and ^{32}P, then digest it with base and subject the products to DEAE-cellulose chromatography. Show the elution of cap 1 with respect to oligonucleotide markers of known charge. Draw the structure of cap 1 and account for its apparent charge.
2. How do we know that the cap contains 7-methylguanosine?
3. Outline the steps in capping.

4. Describe and show the results of an experiment that demonstrates the effect of capping on RNA stability.
5. Describe and give the results of an experiment that shows the synergistic effects of capping and polyadenylation on translation.
6. Describe and give the results of an experiment that shows the effect of capping on mRNA transport into the cytoplasm.
7. Describe and give the results of an experiment that shows the size of poly(A).
8. How do we know that poly(A) is at the 3′-end of mRNAs?
9. How do we know that poly(A) is added posttranscriptionally?
10. Describe and give the results of experiments that show the effects of poly(A) on mRNA translatability, mRNA stability, and recruitment of mRNA into polysomes.
11. With a simple sketch, summarize the polyadenylation process, beginning with an RNA that is being elongated past the polyadenylation site.
12. Describe and give the results of an experiment that shows that transcription does not stop at the polyadenylation site.
13. Describe and give the results of an experiment that shows the importance of the AAUAAA polyadenylation motif. What other motif is frequently found in place of AAUAAA? Where are these motifs found with respect to the polyadenylation site?
14. Describe and give the results of an experiment that shows the importance of the GU-rich and U-rich polyadenylation motifs. Where are these motifs with respect to the polyadenylation site?
15. Describe and give the results of an experiment that shows the effect of the Rpb1 CTD on pre-mRNA cleavage prior to polyadenylation.
16. Describe and give the results of an experiment that shows the importance to polyadenylation of poly(A) polymerase and the specificity factor CPSF.
17. Describe and give the results of an experiment that shows the effect on polyadenylation of adding 40 A's to the end of a polyadenylation substrate that has an altered AAUAAA motif.
18. Describe and give the results of an experiment that shows that CPSF binds to AAUAAA, but not AAGAAA.
19. Describe and give the results of an experiment that shows the effects of CPSF and PAB II on polyadenylation of substrates with AAUAAA or AAGAAA motifs, with and without oligo(A) added. How do you interpret these results?
20. Present a diagram of polyadenylation that illustrates the roles of CPSF, CStF, poly(A) polymerase (PAP), RNA polymerase II, and PAB II.
21. What part of the poly(A) polymerase PAP I is required for polyadenylation activity? Cite evidence.
22. Describe and give the results of an experiment that identifies the cytoplasmic polyadenylation element (CPE) that is necessary for cytoplasmic polyadenylation.
23. Describe and give the results of an experiment that shows that a capping enzyme binds to the RNA polymerase II CTD.
24. Describe and give the results of a Far Western blotting experiment that shows that a component of the U1 snRNP binds to the RNA polymerase II CTD.
25. Describe and give the results of ChIP analysis that shows: (a) that a capping enzyme associates with the RNA polymerase II CTD when it is close to the promoter but not when it is far from the promoter; and (b) that the phosphorylation state of the CTD changes as the RNA polymerase moves away from the promoter.
26. Describe and give the results of an experiment that shows that failure of polyadenylation results in failure of proper transcription termination. Is this behavior due to failure of polyadenylation per se, or is it due to failure of cleavage of the transcript at the polyadenylation site?
27. Describe and give the results of an experiment that indicates that transcription termination requires autocatalytic cleavage of the transcript, even as it is being elongated (cotranscriptional cleavage).
28. Present a torpedo model for transcription termination in eukaryotes.

ANALYTICAL QUESTIONS

1. You are studying a virus that produces mRNAs with extraordinary caps having a net charge of −4 instead of −5. You find these caps have the usual methylations of cap 1: the m^7G and the 2′-O-methyl on the penultimate nucleotide, but no additional methylations. Propose a hypothesis to explain the reduced negative charge and describe experiments to test your hypothesis. Describe sample positive results.
2. Design an experiment to demonstrate that CstF binds to the GU/U element of the cleavage and polyadenylation signal. How would you determine whether one or the other (GU-rich or U-rich) or both parts of this element are required for CstF binding?
3. You are working in a research laboratory that studies the biochemisty of mRNA processing. You have developed an in vitro assay for both splicing and polyadenylation. You produce in vitro the following radioactive mRNA substrates (see table, next page) that either include a 5′-cap or lack the 5′-cap. You incubate these radioactive mRNA substrates with HeLa nuclear extract for 20 min at 30°C and electrophorese the products on a high resolution gel. You then distinguish the splicing products based on their relative sizes in the gel. You count the amount of radioactivity found in the unprocessed mRNA (pre-mRNA), the amount with intron 1 removed (splice 1), the amount with intron 2 removed (splice 2), both introns

removed, and the amount of polyadenylated (poly A). You get the following results, where the number of pluses is related to the relative amount of radioactivity found in that band on the gel:

	Pre-mRNA	*Splice 1 only*	*Splice 2 only*	*Splice 1 and 2*	*Poly (A)*
RNA A uncapped	++	+	+++	+	+++
RNA A capped	+	+	+	+++	+++
RNA B uncapped	++++	+	+	+	+
RNA B capped	++	+++	+	+	+

Propose a hypothesis that explains all these results.

4. In yeast transcription complexes, the phosphorylation state of the CTD of Rpb1, as well as the spectrum of proteins associated with it, changes as transcription progresses. Currently the thought is that the shift in CTD phosphorylation from serine 5 to serine 2 may cause some RNA-processing proteins to leave the complex and possibly attract new proteins to the CTD (as depicted in Figure 15.32). Design and outline the experiments you would perform to demonstrate that the shift in CTD phosphorylation does indeed result in the release (or removal) of RNA-processing proteins as well as the addition of new RNA-processing proteins. Be sure to thoroughly explain your hypotheses to back up your experimental plans.

SUGGESTED READINGS

General References and Reviews

Barabino, S.M.L. and W. Keller. 1999. Last but not least: Regulated poly(A) tail formation. *Cell* 99:9–11.

Bentley, D. 1998. A tale of two tails. *Nature* 395:21–22.

Colgan, D.F. and Manley, J.L. 1997. Mechanism and regulation of mRNA polyadenylation. *Genes and Development* 11:2755–66.

Corden, J.L. 2007. Seven ups the code. *Science* 318:1735–36.

Manley, J.L. and Y. Takagaki. 1996. The end of the message—Another link between yeast and mammals. *Science* 274:1481–82.

Orphanides, G. and D. Reinberg. 2002. A unified theory of gene expression. *Cell* 108:439–51.

Proudfoot, N.J. 1996. Ending the message is not so simple. *Cell* 87:779–81.

Proudfoot, N.J., A. Furger, and M.J. Dye. 2002. Integrating mRNA processing with transcription. *Cell* 108:501–12.

Tollervey, D. 2004. Molecular biology: Termination by torpedo. *Nature* 432:456–57.

Wahle, E. and W. Keller. 1996. The biochemistry of polyadenylation. *Trends in Biochemical Sciences* 21:247–51.

Wickens, M. 1990. How the messenger got its tail: Addition of poly(A) in the nucleus. *Trends in Biochemical Sciences* 15:277–81.

Wickens, M. and T.N. Gonzalez. 2004. Knives, accomplices, and RNA. *Science* 306:1299–1300.

Research Articles

Bardwell, V.J., D. Zarkower, M. Edmonds, and M. Wickens. 1990. The enzyme that adds poly(A) to mRNA is a classical poly(A) polymerase. *Molecular and Cellular Biology* 10:846–49.

Barillà, D., B.A. Lee, and N.J. Proudfoot. 2001. Cleavage/polyadenylation factor IA associates with the carboxyl-terminal domain of RNA polymerase II in *Saccharomyces cerevisiae. Proceedings of the National Academy of Sciences USA* 98:445–50.

Birse, C.E., L. Minvielle-Sebastia, B.A. Lee, W. Keller, and N.J. Proudfoot. 1998. Coupling termination of transcription to messenger RNA maturation in yeast. *Science* 280:298–301.

Dye, M.J. and N.J. Proudfoot. 2001. Multiple transcript cleavage precedes polymerase release in termination by RNA polymerase II. *Cell* 105:669–81.

Egloff, S., D. O'Reilly, R.D. Chapman, A. Taylor, K. Tanzhaus, L. Pitts, D. Eick, and S. Murphy. 2007. Serine 7 of the RNA polymerase II CTD is specifically required for snRNA Gene Expression. *Science* 318:1777–79.

Fitzgerald, M. and T. Shenk. 1981. The sequence 5′-AAUAAA-3′ forms part of the recognition site for polyadenylation of late SV40 mRNAs. *Cell* 24:251–60.

Fox, C.A., M.D. Sheets, and M.P. Wickens. 1989. Poly(A) addition during maturation of frog oocytes: Distinct nuclear and cytoplasmic activities and regulation by the sequence UUUUUAU. *Genes and Development* 3:2151–56.

Furuichi, Y., A. LaFiandra, and A.J. Shatkin. 1977. 5′-terminal structure and mRNA stability. *Nature* 266:235–39.

Furuichi, Y. and K.-I. Miura. 1975. A blocked structure at the 5′-terminus of mRNA from cytoplasmic polyhedrosis virus. *Nature* 253:374–75.

Furuichi, Y., S. Muthukrishnan, J. Tomasz, and A.J. Shatkin. 1976. Mechanism of formation of reovirus mRNA 5′-terminal blocked and methylated sequence m^7GpppG^mpC. *Journal of Biological Chemistry* 251:5043–53.

Gallie, D.R. 1991. The cap and poly(A) tail function synergistically to regulate mRNA translational efficiency. *Genes and Development* 5:2108–16.

Gil, A. and N.J. Proudfoot. 1987. Position-dependent sequence elements downstream of AAUAAA are required for efficient rabbit β-globin mRNA 3′-end formation. *Cell* 49:399–406.

Hamm, J. and I.W. Mattaj. 1990. Monomethylated cap structures facilitate RNA export from the nucleus. *Cell* 63:109–18.

Hirose, Y. and J.L. Manley. 1998. RNA polymerase II is an essential mRNA polyadenylation factor. *Nature* 395:93–96.

Hofer, E. and J.E. Darnell. 1981. The primary transcription unit of the mouse β-major globin gene. *Cell* 23:585–93.

Huez, G., G. Marbaix, E. Hubert, M. Leclereq, U. Nudel, H. Soreq, R. Salomon, B. Lebleu, M. Revel, and U.Z.

Littauer. 1974. Role of the polyadenylate segment in the translation of globin messenger RNA in *Xenopus* oocytes. *Proceedings of the National Academy of Sciences USA* 71:3143–46.

Izaurralde, E., J. Lewis, C. McGuigan, M. Jankowska, E. Darzynkiewicz, and I.W. Mattaj. 1994. A nuclear cap binding protein complex involved in pre-mRNA splicing. *Cell* 78:657–68.

Keller, W., S. Bienroth, K.M. Lang, and G. Christofori. 1991. Cleavage and polyadenylation factor CPF specifically interacts with the pre-mRNA 3′ processing signal AAUAAA. *EMBO Journal* 10:4241–49.

Kim, M., N.J. Krogan, L. Vasiljeva, O.J. Rando, E. Nedea, J.F. Greenblatt, and S. Buratowski. 2004. The yeast Rat1 exonuclease promotes transcription termination by RNA polymerase II. *Nature* 432:517–22.

Komarnitsky, P., E.-J. Cho, and S. Buratowski. 2000. Different phosphorylated forms of RNA polymerase II and associated mRNA processing factors during transcription. *Genes and Development* 14:2452–60.

Mandel, C.R., S. Kaneko, H. Zhang, D. Gebauer, V. Vethantham, J.L. Manley, and L. Tong. 2006. Polyadenylation factor CPSF-73 is the pre-mRNA 3′-end-processing endonuclease. *Nature* 444:953–56.

McCracken, S., N. Fong, E. Rosonina, K. Yankulov, G. Brothers, D. Siderovski, A. Hessel, S. Foster, Amgen EST Program, S. Shuman, and D.L. Bentley. 1997. 5′-capping enzymes are targeted to pre-mRNA by binding to the phosphorylated carboxy-terminal domain of RNA polymerase II. *Genes and Development* 11:3306–18.

McDonald, C.C., J. Wilusz, and T. Shenk. 1994. The 64-kilodalton subunit of the CstF polyadenylation factor binds to pre-mRNAs downstream of the cleavage site and influences cleavage site location. *Molecular and Cellular Biology* 14:6647–54.

Morris, D.P. and A.L. Greenleaf. 2000. The splicing factor, Prp40, binds the phosphorylated carboxy-terminal domain of RNA polymerase II. *Journal of Biological Chemistry* 275:39935–43.

Munroe, D. and A. Jacobson. 1990. mRNA poly(A) tail, a 3′ enhancer of a translational initiation. *Molecular and Cellular Biology* 10:3441–55.

Sheets, M.D. and M. Wickens. 1989. Two phases in the addition of a poly(A) tail. *Genes and Development* 3:1401–12.

Sheiness, D. and J.E. Darnell. 1973. Polyadenylic acid segment in mRNA becomes shorter with age. *Nature New Biology* 241:265–68.

Teixeira, A., A. Tahirir-Alaoui, S. West, B. Thomas, A. Ramadass, I. Martianov, M. Dye, W. James, N.J. Proudfoot, and A. Akoulitchev. 2004. Autocatalytic RNA cleavage in the human β-globin pre-mRNA promotes transcription termination. *Nature* 432:526–30.

Wahle, E. 1991. A novel poly(A)-binding protein acts as a specificity factor in the second phase of messenger RNA polyadenylation. *Cell* 66:759–68.

Wei, C.M. and Moss, B. 1975. Methylated nucleotides block 5′-terminus of vaccinia virus mRNA. *Proceedings of the National Academy of Sciences USA* 72:318–22.

West, S., N. Gromak, and N.J. Proudfoot. 2004. Human 5′→3′ exonuclease Xrn2 promotes transcription termination at cotranscriptional cleavage sites. *Nature* 432:522–25.

CHAPTER 16

Other RNA Processing Events and Post-Transcriptional Control of Gene Expression

A petunia flower showing the effects of silencing by adding extra copies of the purple color gene. *© Courtesy of Dr. Richard A. Jorgensen,* The Plant Cell.

In the previous two chapters, we examined splicing, capping, and polyadenylation, which covers most of what happens to pre-mRNAs in eukaryotic cells. However, in a few organisms, other specialized pre-mRNA processing events occur. For example, parasitic protozoa called trypanosomes, as well as some parasitic worms and the free-living protist *Euglena,* carry out *trans*-splicing of pre-mRNAs. This involves splicing together two independent transcripts. Trypanosomes also have mitochondria, called kinetoplasts, that edit their mRNAs by adding or deleting nucleotides after transcription. In contrast to these rather esoteric processing events, most organisms process their rRNAs and tRNAs by more conventional mechanisms. Also eukaryotes control some of their gene expression by regulating posttranscriptional processes, primarily mRNA degradation. Finally, eukaryotes can react to foreign genes or double-stranded RNA by destroying the corresponding mRNA. All of these posttranscriptional events will be our subjects in this chapter.

16.1 Ribosomal RNA Processing

The rRNA genes of both eukaryotes and bacteria are transcribed as larger precursors that must be processed (cut into pieces) to yield rRNAs of mature size. However, this is not just a matter of removing unwanted material at either end of an overly long molecule. Instead, several different rRNA molecules are embedded in a long precursor, and each of these must be cut out. Let us consider rRNA processing, first in eukaryotes, then in bacteria.

Eukaryotic rRNA Processing

The rRNA genes in eukaryotes are repeated several hundred times and clustered together in the nucleolus of the cell. Their arrangement in amphibians has been especially well studied, and, as Figure 16.1a shows, they are separated by regions called **nontranscribed spacers (NTSs).** NTSs are distinguished from **transcribed spacers,** regions of the gene that are transcribed as part of the rRNA precursor and then removed in the processing of the precursor to mature rRNA species.

This clustering of the reiterated rRNA genes in the nucleolus made them easy to find and therefore provided Oscar Miller and his colleagues with an excellent opportunity to observe genes in action. These workers looked at amphibian nuclei with the electron microscope and uncovered a visually appealing phenomenon, shown in Figure 16.1b. The DNA containing the rRNA genes can be seen winding through the picture, but the most obvious feature of the micrograph is a series of "tree" structures. These include the rRNA genes (the trunk of the tree) and growing rRNA transcripts (the branches of the tree). We will see shortly that these transcripts are actually rRNA precursors, not mature rRNA molecules. The spaces between "trees" are the nontranscribed spacers. You can even tell the direction of transcription from the lengths of the transcripts within a given gene; the shorter RNAs are at the beginning of the gene and the longer ones are at the end.

We have seen that mRNA precursors frequently require splicing but no other trimming. On the other hand, rRNAs and tRNAs first appear as precursors that sometimes need splicing, but they also have excess nucleotides at their ends, or even between regions that will become separate mature

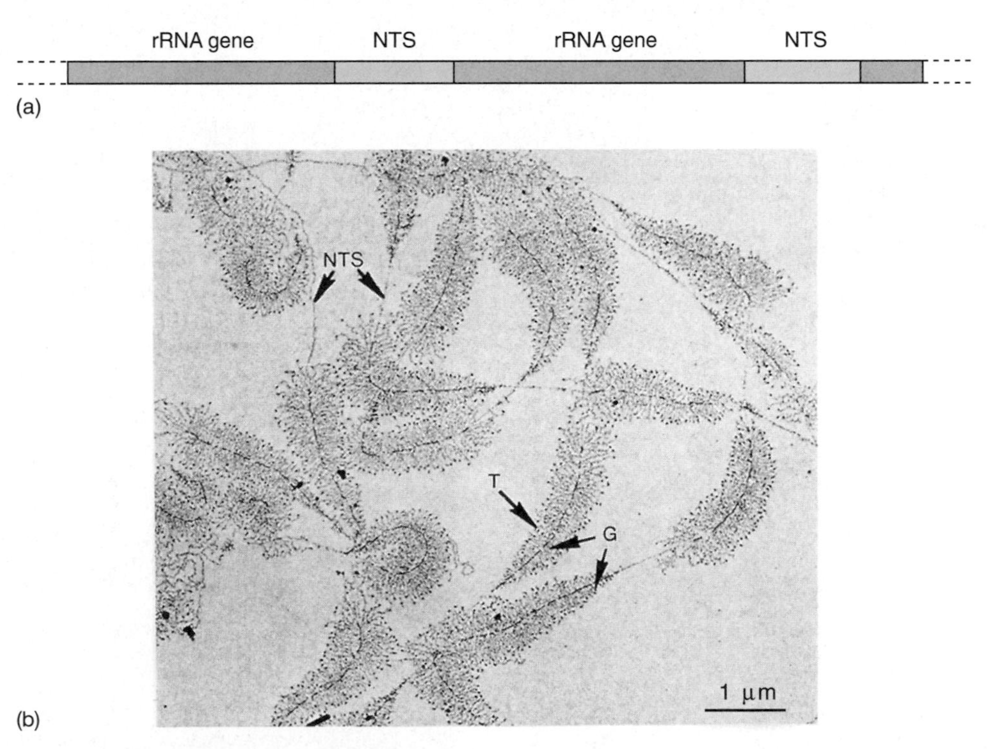

Figure 16.1 Transcription of rRNA precursor genes. (a) Map of a portion of the newt (amphibian) rRNA precursor gene cluster, showing the alternating rRNA genes (orange) and nontranscribed spacers (NTS, green). **(b)** Electron micrograph of part of a newt nucleolus, showing rRNA precursor transcripts (T) being synthesized in a "tree" pattern on the tandemly duplicated rRNA precursor genes (G). At the base of each transcript is an RNA polymerase I, not visible in this picture. The genes are separated by nontranscribed spacer DNA (NTS). (*Source:* (b) O.L. Miller, Jr., B.R. Beatty, B.A. Hamkalo, and C.A. Thomas, Electron microscopic visualization of transcription. *Cold Spring Harbor. Symposia on Quantitative Biology* 35 (1970) p. 506.)

RNA sequences. These excess regions must also be removed. This trimming of excess regions from an RNA precursor is another kind of **processing.** It is similar to splicing in that unnecessary RNA is removed, but it differs from splicing in that no RNAs are stitched together.

For example, mammalian RNA polymerase I makes a **45S rRNA precursor,** which contains the **28S, 18S,** and **5.8S rRNAs,** embedded between transcribed spacer RNA regions. The processing of the precursor (Figure 16.2) takes place in the **nucleolus,** the nuclear compartment where rRNAs are made and ribosomes are assembled. The first step is to cut off the spacer at the 5′-end, leaving a 41S intermediate. The next step involves cleaving the 41S RNA into two pieces, 32S and 20S, that contain the 28S and 18S sequences, respectively. The 32S precursor also retains the 5.8S sequence. Finally, the 32S intermediate is split to yield the mature 28S and 5.8S RNAs, which base-pair with each other, and the 20S intermediate is trimmed to mature 18S size.

What is the evidence for this sequence of events? As long ago as 1964, Robert Perry used a **pulse-chase** procedure to establish a precursor–product relationship between the 45S precursor and the 18S and 28S mature rRNAs. He labeled mouse L cells for a short time (a short pulse) with [^{3}H]uridine and found that the labeled RNA sedimented as a broad peak centered at about 45S. Then he "chased" the label in this RNA into 18S and 28S rRNAs. That is, he added excess unlabeled uridine to dilute the labeled nucleoside and observed that the amount of label in the 45S precursor decreased as the amount of label in the mature 18S and 28S rRNAs increased. This suggested that one or more RNA species in the 45S peak was a precursor to 18S and 28S rRNAs. In 1970, Robert Weinberg and Sheldon Penman found the key intermediates by labeling poliovirus-infected HeLa cells with [^{3}H]methionine and [^{32}P]phosphate and separating the labeled RNAs by gel electrophoresis. Ordinarily, processing intermediates are too short-lived to accumulate to detectable levels, but poliovirus infection slowed processing down enough that the intermediates could be seen. The major species observed were 45S, 41S, 32S, 28S, 20S, and 18S (Figure 16.3). Dual labeling was possible because rRNA precursors in eukaryotes are methylated.

In 1973, Peter Wellauer and Igor Dawid visualized the precursor, intermediates, and products of human rRNA processing by electron microscopy. Each RNA species had its own capacity for intramolecular base pairing, so each had its own secondary structure. Once David and Wellauer

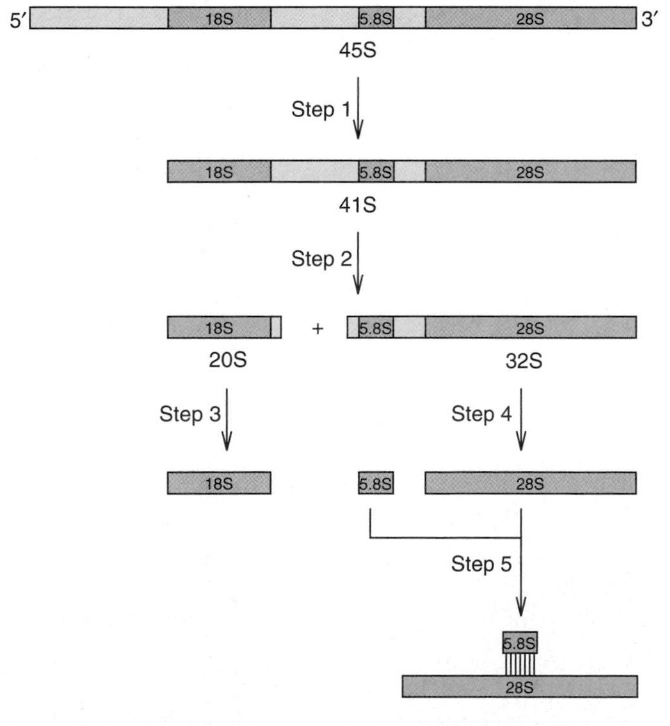

Figure 16.2 Processing scheme of 45S human (HeLa) rRNA precursor. Step 1: The 5′-end of the 45S precursor RNA is removed, yielding the 41S precursor. Step 2: The 41S precursor is cut into two parts, the 20S precursor of the 18S rRNA, and the 32S precursor of the 5.8S and 28S rRNAs. Step 3: The 3′-end of the 20S precursor is removed, yielding the mature 18S rRNA. Step 4: The 32S precursor is cut to liberate the 5.8S and 28S rRNAs. Step 5: The 5.8S and 28S rRNAs associate by base-pairing.

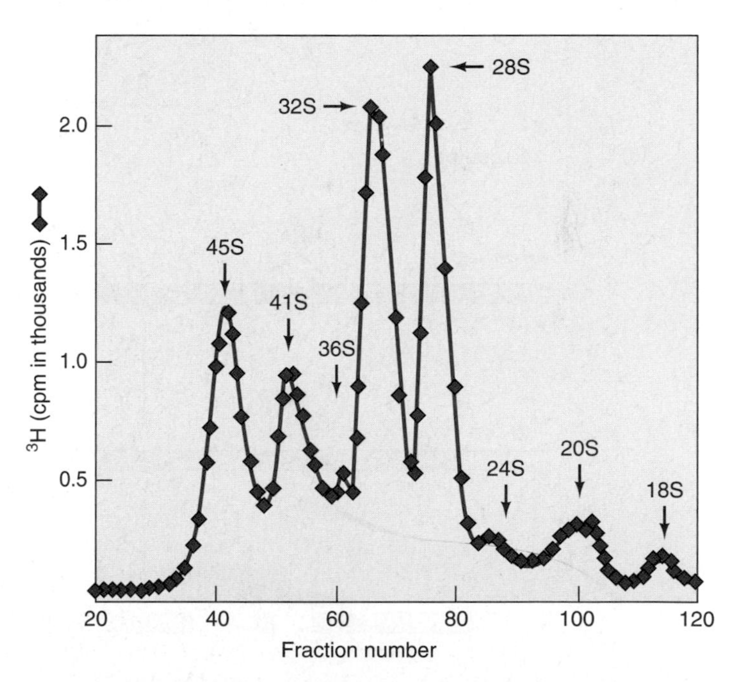

Figure 16.3 Isolation of 45S rRNA-processing intermediates from poliovirus-infected HeLa cells. Penman and colleagues labeled RNA in virus-infected cells with [^{3}H]methionine, which labeled the many methyl groups in rRNAs and their precursors. They isolated nucleolar RNA (mostly rRNA) from these cells, subjected it to gel electrophoresis, sliced the gel, determined the radioactivity in each slice, then plotted these radioactivity values in cpm versus slice, or fraction number. The mobilities of the RNA species were compared with those of markers of known sedimentation coefficients. (*Source:* Adapted from Weinberg, R.A. and S. Penman, Processing of 45S nucleolar RNA. *Journal of Molecular Biology* 47:169 (1970).)

had identified these "signatures" of all the RNA species, they could recognize them in the 45S precursor and thereby locate the 28S and 18S species in the precursor. Although they originally got the order backwards, we now know that the arrangement is: 5′-18S-5.8S-28S-3′. The details of this processing scheme are not universal; even the mouse does things a little differently, and the frog precursor is only 40S, which is quite a bit smaller than 45S. Still, the basic mechanism of rRNA processing, including the order of mature sequences in the precursor, is preserved throughout the eukaryotic kingdom.

The rRNA-processing steps are orchestrated in the nucleolus by a class of **small nucleolar RNAs (snoRNAs)**, associated with proteins in **small nucleolar ribonucleoproteins, (snoRNPs)**. There are many hundreds of snoRNPs, and quite a few of them participate in rRNA processing by modifying nucleotides within the rRNA precursor. The rRNA precursor contains about 110 2′-*O*-methyl groups and about 100 pseudouridines. (In pseudouridine, the ribose joins to the 5-carbon of the uracil, rather than the 1-nitrogen; Chapter 19). Because these modified nucleotides persist in the mature rRNAs, it appears that they help define what regions of the precursor to remove and what regions to preserve. The RNA parts (**guide snoRNAs**) of the snoRNPs base-pair to specific sites within the rRNA precursor and dictate either methylation or pseudouridylation at those sites.

SUMMARY Ribosomal RNAs are made in eukaryotic nucleoli as precursors that must be processed to release the mature rRNAs. The order of RNAs in the precursor is 18S, 5.8S, 28S in all eukaryotes, although the exact sizes of the mature rRNAs vary from one species to another. In human cells, the precursor is 45S, and the processing scheme creates 41S, 32S, and 20S intermediates. snoRNPs play vital roles in these processing steps by methylating and pseudouridylating specific sites within the rRNA precursor.

Bacterial rRNA Processing

The bacterium *E. coli* has seven *rrn* operons that contain rRNA genes. Figure 16.4a presents an example, *rrnD,* which has three tRNA genes in addition to the three rRNA genes. Transcription of the operon yields a 30S precursor, which must be cut up to release the three rRNAs and three tRNAs.

RNase III is the enzyme that performs at least the initial cleavages that separate the individual large rRNAs. One type of evidence leading to this conclusion is genetic: A mutant with a defective RNase III gene accumulates 30S rRNA precursors. In 1980, Joan Steitz and her colleagues compared the sequences of the spacers between the rRNAs

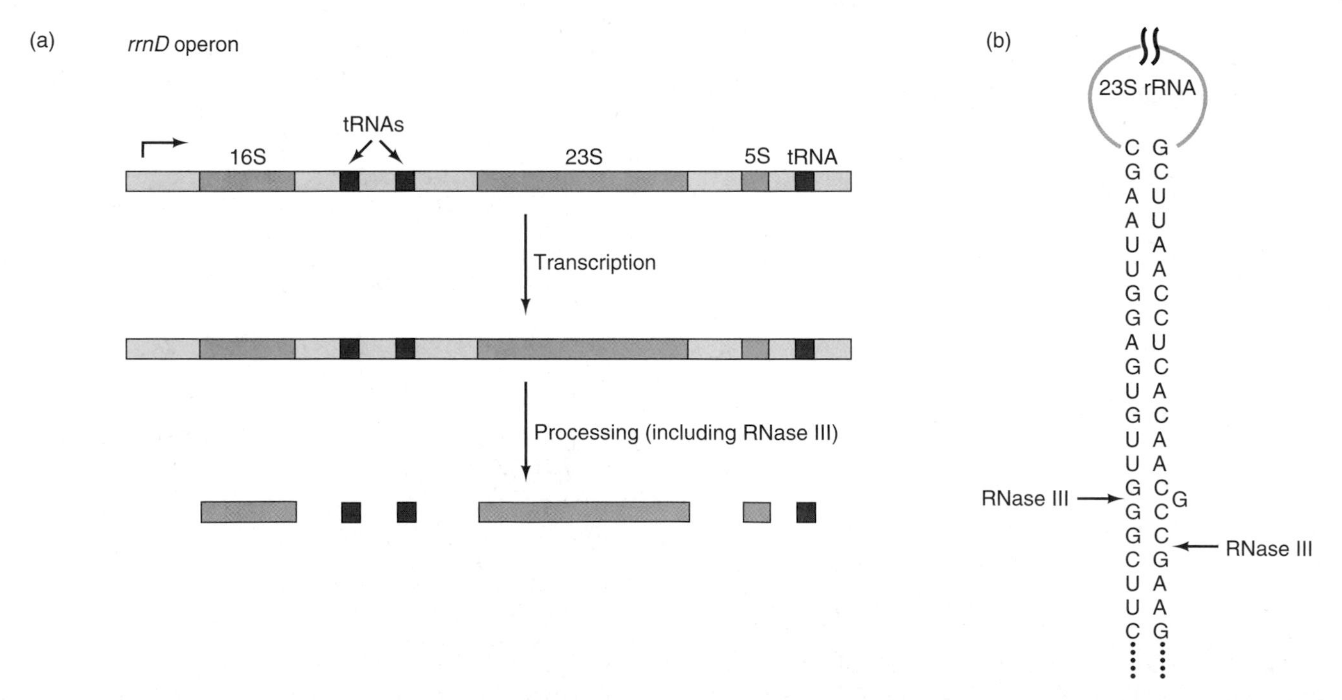

Figure 16.4 Processing bacterial rRNA precursors. (a) Structure of the *E. coli rrnD* operon. This operon is typical of the rRNA-encoding operons of *E. coli* in that it includes regions that code for tRNAs (red), as well as rRNA-coding regions (orange), embedded in transcribed spacers (yellow). As usual with bacterial operons, this one is transcribed to produce a long composite RNA. This RNA is then processed by enzymes, including RNase III, to yield mature products. **(b)** Sequence analysis has shown that the spacers surrounding the 23S rRNA gene are complementary, so they can form an extended hairpin with the 23S rRNA region at the top. The observed cleavage sites for RNase III are in the stem, offset by 2 bp. The regions surrounding the 16S rRNA gene can also form a hairpin stem, with a somewhat more complex structure.

in two different precursors (from the *rrnX* and *rrnD* operons) and found considerable similarity. These sequences revealed complementary sequences flanking both 16S and 23S rRNA regions of the precursors. This complementarity predicts two extended hairpins (Figure 16.4b) involving stems created by base pairing between two spacers, with the rRNA regions looping out in between. The RNase III cleavage sites in this model are in the stems. Another ribonuclease, **RNase E**, is responsible for removing the 5S rRNA from the precursor.

SUMMARY Bacterial rRNA precursors contain tRNAs as well as all three rRNAs. The rRNAs are released from their precursors by RNase III and RNase E.

16.2 Transfer RNA Processing

Transfer RNAs are made in all cells as overly long precursors that must be processed by removing RNA at both ends. In the nuclei of eukaryotes, these precursors contain a single tRNA; in bacteria, a precursor may contain one or more tRNAs, and sometimes a mixture of rRNAs and tRNAs, as we saw in Figure 16.4. Because the tRNA processing schemes in eukaryotes and bacteria are so similar, we will consider them together.

Cutting Apart Polycistronic Precursors

The first step in processing bacterial RNAs that contain more than one tRNA is to cut the precursor up into fragments with just one tRNA each. This means cutting between tRNAs in precursors that have two or more tRNAs, or cutting between tRNAs and rRNAs in precursors, such as the one in Figure 16.4, that have both tRNAs and rRNAs. The enzyme that performs both these chores seems to be RNase III.

Forming Mature 5′-Ends

After RNase III has cut the tRNA precursor into pieces, the tRNA still contains extra nucleotides at both 5′- and 3′-ends. As such, it resembles the primary transcripts of eukaryotic tRNA genes, which are monocistronic (single-gene) precursors with extended 5′- and 3′-ends. Maturation of the 5′-end of a bacterial or eukaryotic tRNA involves a single cut just at the point that will be the 5′-end of the mature tRNA, as shown in Figure 16.5. The enzyme that catalyzes this cleavage is **RNase P.**

RNase P from both bacteria and eukaryotic nuclei is a fascinating enzyme. It contains two subunits, but unlike other dimeric enzymes we have studied, one of these

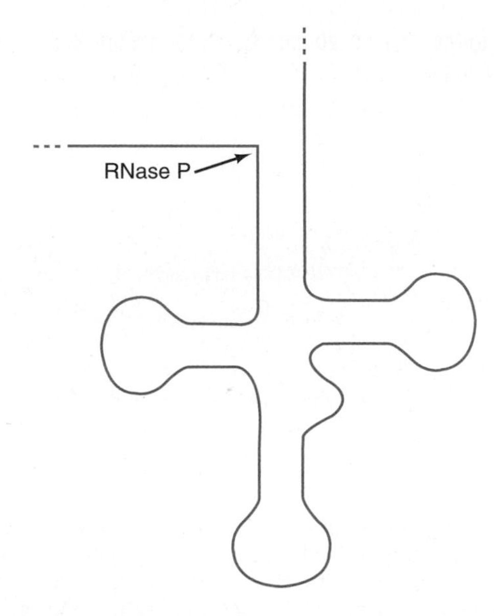

Figure 16.5 RNase P action. RNase P makes a cut at the site that will become the mature 5′-end of a tRNA. Thus, this enzyme is all that is needed to form mature 5′-ends.

subunits is made of RNA, not protein. In fact, the majority of the enzyme is RNA because the RNA (the **M1 RNA**) has a molecular mass of about 125 kD, and the protein has a mass of only about 14 kD. When Sidney Altman and his colleagues first isolated this enzyme and discovered that it is a ribonucleoprotein, they faced a critical question: Which part has the catalytic activity, the RNA or the protein? The heavy betting at that time was on the protein because all enzymes that had ever been studied were made of protein, not RNA. In fact, early studies on RNase P showed that the enzyme lost all activity when the RNA and protein parts were separated.

Then, in 1982, Thomas Cech and colleagues found autocatalytic activity in a self-splicing intron (Chapter 14). Shortly thereafter, Altman and Norman Pace and their colleagues demonstrated the catalytic activity of the M1 part of RNase P in 1983. As Figure 16.6 illustrates, the trick was magnesium concentration. The early studies had been performed with 5–10 mM Mg^{2+}, under these conditions, both the protein and RNA parts of RNase P are required for activity. Figure 16.6 shows the effect of Mg^{2+} concentration over the range 5 mM to 50 mM using M1 RNA alone. Altman, Pace, and colleagues used two different substrates: pre-$tRNA^{Tyr}$ and pre-4.5S RNA from *E. coli.* Figure 16.6, lanes 1–3 show the differences among 5, 10 and 20 mM Mg^{2+}, respectively. At 5 mM Mg^{2+}, neither substrate showed any maturation by cleavage of the extra nucleotides from the 5′-end. Even at 10 mM Mg^{2+}, the cleavage of pre-tRNA was barely detectable. By contrast, at 20 mM Mg^{2+}, approximately half the pre-tRNA was cleaved to mature form, releasing the extra nucleotides as a single fragment, labeled "5′-Tyr" in the figure. Increasing the Mg^{2+} concentration to 30, 40, and 50 mM

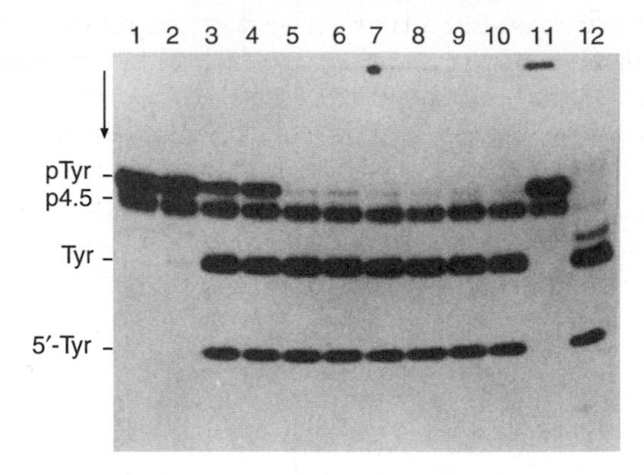

Figure 16.6 The M1 RNA of *E. coli* RNase P has enzymatic activity. Altman and Pace and colleagues purified the M1 RNA from RNase P and incubated it with ^{32}P-labeled pre-tRNA^{Tyr} (pTyr) and p4.5S RNA from *E. coli* (p4.5) for 15 min at the Mg^{2+} and NH_4Cl concentrations indicated at top. Then they electrophoresed the RNAs and visualized them by autoradiography. Lane 11, no additions; lane 12, crude *E. coli* RNase P. At the higher Mg^{2+} concentrations, the M1 RNA by itself cleaved the pTyr to form mature 5′-ends, but had no effect on the p4.5 substrate under any of the conditions used. (*Source:* Guerrier-Takada, C., K. Gardiner, T. Marsh, N. Pace, and S. Altman, The RNA moiety of ribonuclease P is the catalytic subunit of the enzyme. Cell 35 (Dec 1983) p. 851, f. 4A. Reprinted by permission of Elsevier Science.)

Mg^{2+} (lanes 5, 7, and 9, respectively) further enhanced 5′-processing of the pre-tRNA, but did not cause any pre-4.5S processing. Lane 12 demonstrates that crude RNase P (the dimeric form of the enzyme that contains both the RNA and protein subunits) can cleave both substrates at 10 mM Mg^{2+}.

Eukaryotic nuclear RNase P is very much like the bacterial enzyme. For example, the yeast nuclear RNase P contains a protein and an RNA part, and the RNA has the catalytic activity. However, Peter Gegenheimer and his colleagues, in papers beginning in 1988, showed that spinach chloroplast RNase P appears not to have an RNA at all. This enzyme is not inhibited by micrococcal nuclease, as it should be if it contains a catalytic RNA, and it has the density expected of pure protein, not a ribonucleoprotein that is mostly RNA. In 2008, Walter Rossmanith demonstrated that human mitochondrial RNase P also lacks an RNA component.

The archaeon *Nanoarchaeum equitans* gets along without RNase P. It synthesizes its tRNAs without 5′-leaders, so no RNase P is required to remove them.

SUMMARY Extra nucleotides are removed from the 5′-ends of pre-tRNAs in one step by an endonucleolytic cleavage catalyzed by RNase P. RNase P's from bacteria and eukaryotic nuclei have a catalytic RNA subunit called M1 RNA. Spinach chloroplast RNase P appears to lack an RNA subunit.

Forming Mature 3′-Ends

Transfer RNA 3′-end maturation is considerably more complex than 5′-maturation because not one, but six RNases take part. Murray Deutscher and other investigators have shown that the following RNases can remove nucleotides from the 3′-ends of tRNAs in vitro: *RNase D, RNase BN, RNase T, RNase PH, RNase II,* and *polynucleotide phosphorylase (PNPase).* Genetic experiments by Deutscher and colleagues have also demonstrated that each of these enzymes is necessary for the most efficient 3′-end processing. If the genes encoding any of these enzymes were inactivated, the efficiency of tRNA processing suffered. Inactivation of all of the genes at once was lethal to bacterial cells. On the other hand, the presence of any one of the enzymes was sufficient to ensure viability and tRNA maturation, although the efficiency varied depending on the active RNase.

A combination of genetic and biochemical experiments has shown that RNase II and PNPase cooperate to remove the bulk of the 3′-trailer from pre-tRNA. This opens the way for RNases PH and T to complete the job by removing the last two nucleotides. RNase T is the most active in removing the last nucleotide.

The situation in eukaryotes seems a bit simpler. A single enzyme, **tRNA 3′-processing endoribonuclease (3′-tRNase)** cleaves the excess nucleotides from the 3′-end of a tRNA precursor. In 2003, Masayuki Nashimoto and colleagues purified a 3′-tRNase from pig liver. Comparison of a partial sequence of the purified protein to the human genomic database revealed a close similarity to a poorly characterized human protein (**ELAC2**), mutations in which are risk factors for prostate cancer. Nashimoto and colleagues cloned and expressed the human ELAC2 gene in bacteria and tested the protein product for 3′-tRNase activity in vitro. It was able to efficiently remove the excess nucleotides from the end of human tRNA^{Arg}, showing that ELAC2 is at least one of the 3′-tRNase enzymes in humans.

SUMMARY RNase II and polynucleotide phosphorylase cooperate to remove most of the extra nucleotides at the end of an *E. coli* tRNA precursor, but stop at the +2 stage, with two extra nucleotides remaining. RNases PH and T are most active in removing the last two nucleotides from the RNA, with RNase T being the major participant in removing the very last nucleotide. In eukaryotes, a single enzyme, tRNA 3′-processing endoribonuclease (3′-tRNase), processes the 3′-end of a pre-tRNA.

16.3 *Trans*-Splicing

In Chapter 14 we considered the sort of splicing that occurs in almost all eukaryotic species. This splicing can be called ***cis*-splicing,** because it involves two or more exons that exist together in the same gene. As unlikely as it may seem, in another alternative, ***trans*-splicing,** the exons are not part of the same gene at all and may not even be found on the same chromosome.

The Mechanism of *Trans*-Splicing

Trans-splicing occurs in several oganisms, including parasitic and free-living worms (e.g., *Caenorhabditis elegans*), but it was first discovered in **trypanosomes,** a group of parasitic flagellated protozoa, one species of which causes African sleeping sickness. The genes of trypanosomes are expressed in a manner we would never have predicted based on what we have discussed in this book so far. Piet Borst and his colleagues laid the groundwork for these surprising discoveries in 1982 when they sequenced the 5′-end of an mRNA encoding a trypanosome surface coat protein and the 5′-end of the gene that encodes this same protein and discovered that they did not match. The mRNA had 35 extra nucleotides that were missing from the gene. As molecular biologists sequenced more and more trypanosome mRNAs, they discovered that they all had the same 35-nt leader, called the **spliced leader (SL),** but none of the genes encoded the SL. Instead, the SL is encoded by a separate gene that is repeated about 200 times in the trypanosome genome. This gene encodes only the SL, plus a 100-nt sequence that is joined to the leader through a consensus 5′-splice sequence. Thus, this minigene is composed of a short SL exon, followed by what looks like the 5′-part of an intron.

How can we explain the production of an mRNA derived from two widely separated DNA regions that are sometimes even found on separate chromosomes? Two classes of explanations are plausible. First (Figure 16.7a), the SL (with or without its intron) could be transcribed, and this transcript could then serve as a primer for transcription of any one of the coding regions elsewhere in the genome. Alternatively (Figure 16.7b), RNA polymerases could transcribe an SL and a coding region separately, and these two independent transcripts could then be spliced together.

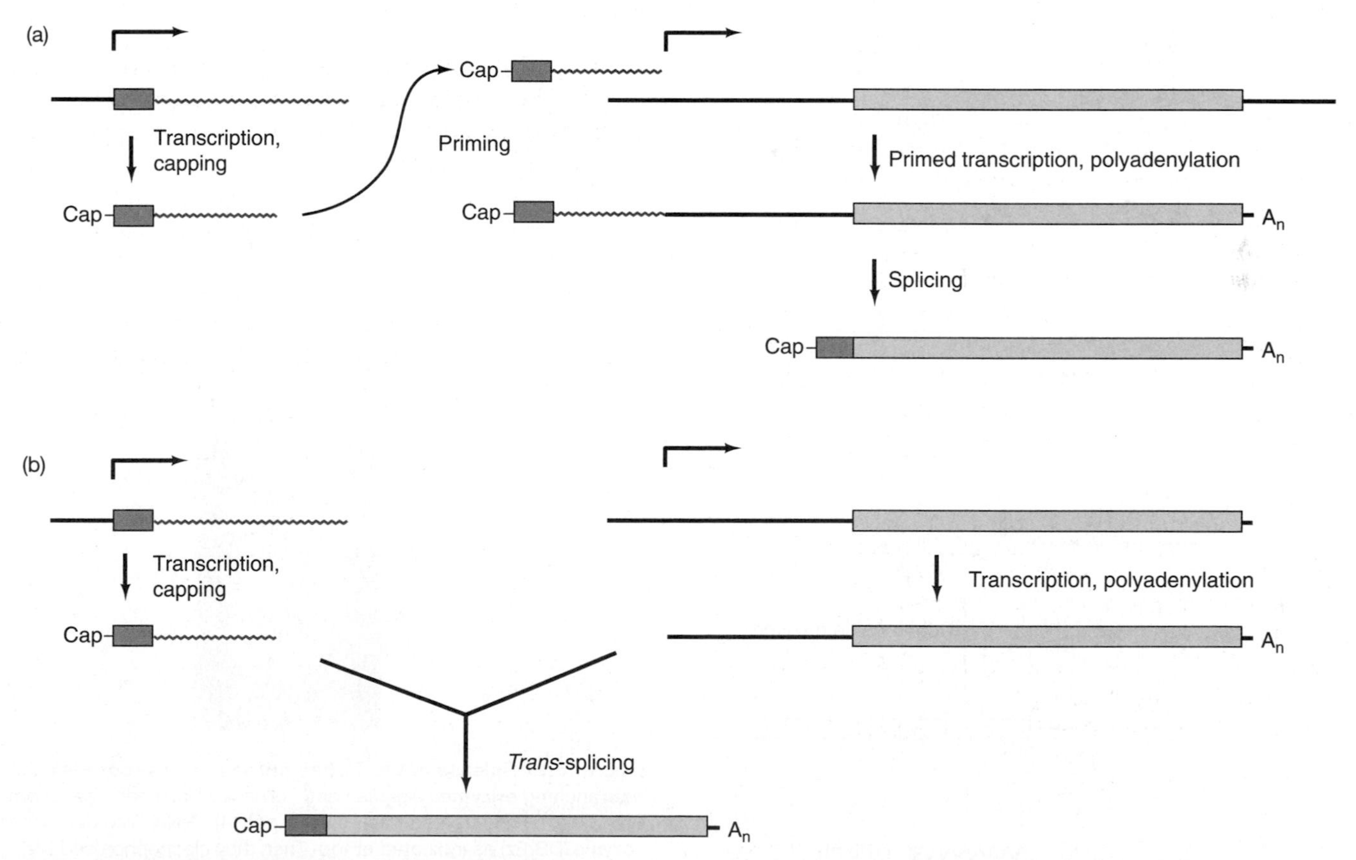

Figure 16.7 Two hypotheses for joining the SL to the coding region of an mRNA. (a) Priming by the SL intron. The SL (blue), with its attached half-intron (red), is transcribed to yield a 135-nt RNA. This RNA then serves as a primer for transcription of a coding region (yellow), including its attached half-intron (black). This produces a transcript including the SL plus the coding region, with a whole intron in between. The intron can then be spliced out to yield the mature mRNA. **(b)** *Trans*-splicing. The SL with its attached half-intron is transcribed; independently, the coding region with its half-intron is transcribed. Then these two separate RNAs undergo *trans*-splicing to produce the mature mRNA.

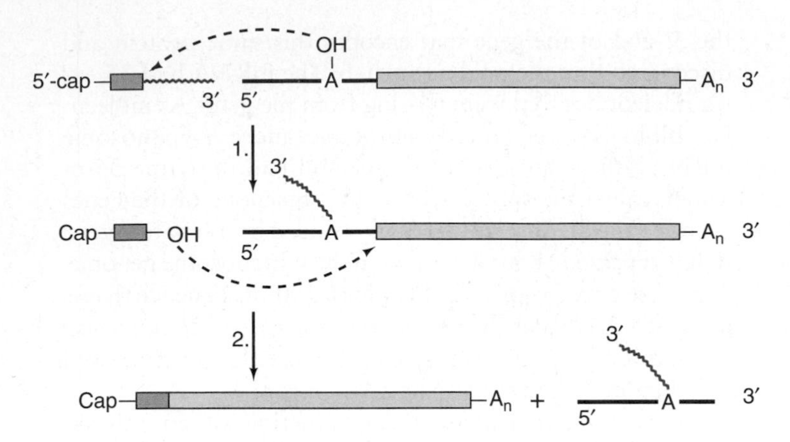

Figure 16.8 Detailed *trans*-splicing scheme for a trypanosome mRNA. Step 1: The branchpoint adenosine within the half-intron (black) attached to the coding exon (yellow) attacks the junction between the leader exon (blue) and its half-intron (red). This creates a Y-shaped intron–exon intermediate analogous to the lariat intermediate created by *cis*-splicing. Step 2: The leader exon attacks the splice site between the branched intron and the coding exon. This produces the spliced, mature mRNA plus the Y-shaped intron.

If such *trans*-splicing really occurs, then we would not expect to see lariat-shaped intermediates. Instead, we should find Y-shaped intermediates that form when the branchpoint in the intron attacks the 5′-end of the intron attached to the short leader exon, as illustrated in Figure 16.8. Finding the Y-shaped intermediate would go a long way toward proving that *trans*-splicing really takes place. Nina Agabian and colleagues reported evidence for the intermediate in 1986.

The unique feature of the Y-shaped structure, which distinguishes it from a normal, lariat intermediate, is that the 3′-end of the SL intron in the Y-shaped structure is free (see Figure 16.8). This means that treatment of the Y-shaped splicing intermediate with debranching enzyme, which breaks the 2′–5′-phosphodiester bond at the branchpoint, should yield a 100-nt fragment as a by-product (Figure 16.9). This contrasts with the results we expect from a lariat-shaped intermediate, which would simply be linearized. Figure 16.10 shows the results of a Northern blot of total RNA and poly(A)$^+$ RNA after treatment with debranching enzyme probed with an oligonucleotide specific for the 100-nt fragment. In both cases, the expected 100-nt fragment appeared, thus corroborating the *trans*-splicing hypothesis.

Trans-splicing is very widespread in some organisms. In *C. elegans*, for example, all or nearly all mRNAs are *trans*-spliced to a small group of spliced leaders. And more than 15% of these *trans*-spliced mRNAs are encoded in groups of two to eight genes that can be considered a kind of operon. Such a group of genes resembles a prokaryotic operon in that they belong to a transcription unit controlled by a single

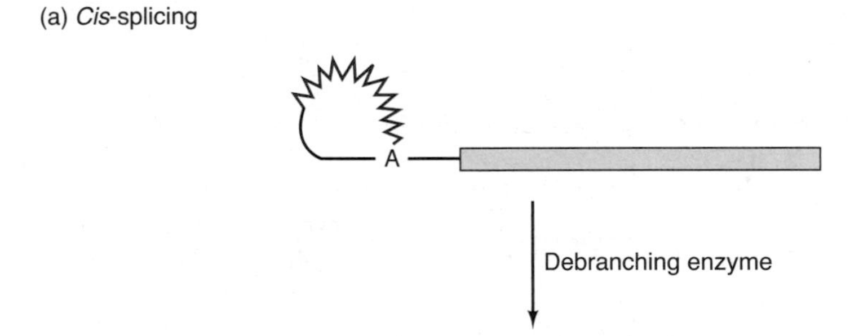

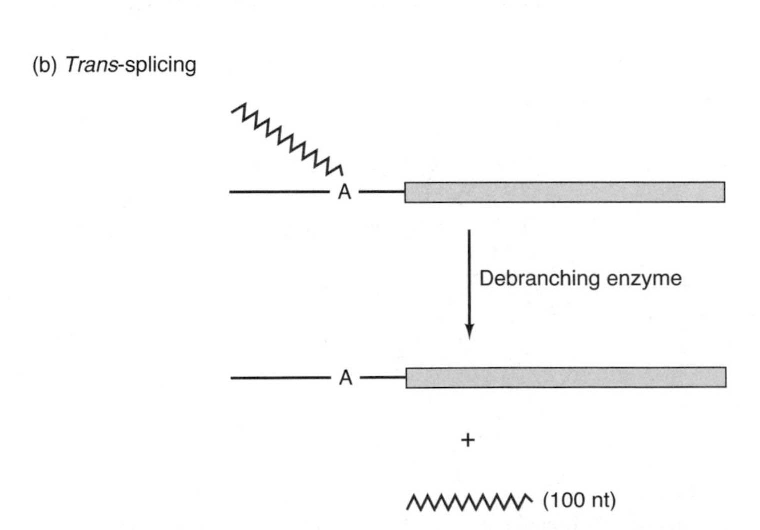

Figure 16.9 Treating hypothetical splicing intermediates with debranching enzyme. (a) *Cis*-splicing. The debranching enzyme simply opens the lariat up to a linear form. **(b)** *Trans*-splicing. Because the 100-nt half-intron (red) is open at its 3′-end instead of being involved in a lariat, debranching enzyme releases it as an independent RNA.

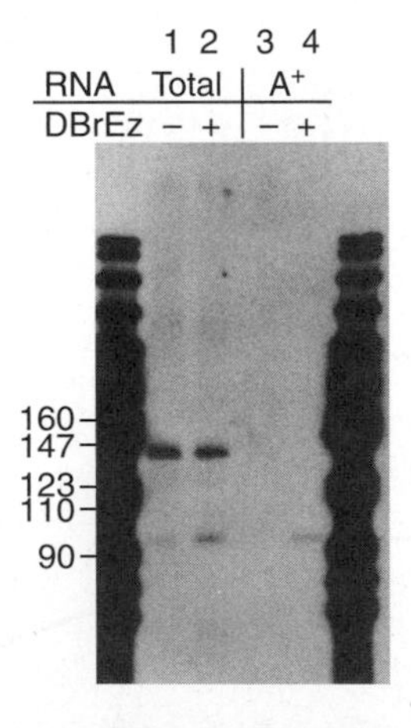

Figure 16.10 Release of the SL half-intron from a larger RNA by debranching enzyme. Agabian and colleagues labeled trypanosome RNA with ^{32}P and treated total RNA, or poly(A)$^+$ RNA, with debranching enzyme (DBrEz) as indicated at top. Then they electrophoresed the products, blotted them, and probed the blot with an oligonucleotide specific for the 100-nt SL half-intron, which is clearly detectable in both enzyme-treated RNA samples. (*Source:* Murphy W.J., K.P. Watkins, and N. Agabian, Identification of a novel Y branch structure as an intermediate in trypanosome mRNA processing. Evidence of *trans*-splicing. *Cell* 47 (21 Nov 1986) p. 521, f. 5. Reprinted by permission of Elsevier Science.)

promoter. But it differs from a true operon in that the primary transcript is ultimately broken into pieces by *trans*-splicing, with each coding region being supplied with its own spliced leader. Indeed, *trans*-splicing makes such eukaryotic "operons" possible by providing each of the internal coding regions with its own cap. Otherwise, only the first coding region would receive a cap upon transcription, and therefore would be the only one to be efficiently translated. This is not a problem in bacteria, which have unique translation start sites for each gene within a polycistronic mRNA (Chapter 7), but it would be in eukaryotes, whose mRNAs generally do not have internal translation start sites and instead depend on caps to recruit ribosomes (Chapter 17).

SUMMARY Trypanosome mRNAs are formed by *trans*-splicing between a short leader exon and any one of many independent coding exons. *Trans*-splicing is common in organisms such as *C. elegans,* in which polycistronic pre-mRNAs are broken up into their individual gene transcripts by *trans*-splicing each of those parts of the pre-mRNA to a common spliced leader.

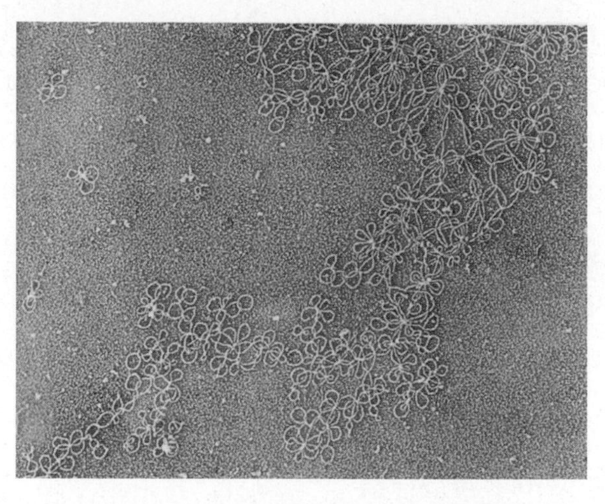

Figure 16.11 Part of the network of kinetoplast minicircles and maxicircles from *Leishmania tarentolae*. (Source: Cell 61 (1 June 1990) cover (acc. Sturm & Simpson, pp. 871–84). Reprinted by permission of Elsevier Science.)

COX II DNA: ···GTATAAAAGTAGA G A ACCTGG···

COX II RNA: ···GUAUAAAAGUAGAUUGUAUACCUGG···

Figure 16.12 Comparison of the sequence of part of the *COX II* gene of a trypanosome with its mRNA product. Four U's in the mRNA are not represented by T's in the gene. These four U's are presumably added to the RNA by editing.

16.4 RNA Editing

Trans-splicing is not the only bizarre occurrence in trypanosomatids. These organisms also have unusual mitochondria called **kinetoplasts,** which contain two types of circular DNA linked together into large networks (Figure 16.11). There are 25–50 identical **maxicircles,** 20–40 kb in size, which contain the mitochondrial genes, and about 10,000 1–3-kb **minicircles,** which have a role in mitochondrial gene expression. In 1986, Rob Benne and his colleagues discovered that the sequence of the cytochrome oxidase (*COX II*) mRNA from trypanosomes does not match the sequence of the *COX II* gene; the mRNA contains four nucleotides that are missing from the gene (Figure 16.12). Furthermore, these missing nucleotides cause a frameshift (a shift in the frame in which a ribosome reads the mRNA; see Chapter 18) that should seemingly inactivate the gene. But somehow the mRNA has been supplied with these four nucleotides, averting the frameshift.

Of course, one possibility is that the gene Benne and colleagues sequenced did not actually code for the mRNA, but was a **pseudogene,** a duplicate copy of a gene that has been mutated so it does not function and is no longer used. The active gene could reside elsewhere, and these workers could have missed it. The problem with this explanation is that, try as they might, Benne and his coworkers could find no other *COX II* gene in either the kinetoplast or the nucleus. Furthermore, they found the same missing nucleotides in the *COX II* genes of two other trypanosomatids. For these and other reasons, Benne and coworkers concluded that the mRNAs of trypanosomatids are copied from incomplete genes called **cryptogenes** and then edited by adding the missing nucleotides, which are all UMPs.

By 1988, a number of trypanosomatid kinetoplast genes and corresponding mRNAs had been sequenced, revealing editing as a common phenomenon in these organisms. In fact, some RNAs are very extensively edited (**panedited**). For example, a 731-nt stretch of the *COIII* mRNA of *Trypanosoma brucei* contains 407 UMPs added by editing; editing also deletes 19 encoded UMPs from this stretch of the *COIII* mRNA. Part of this sequence is presented in Figure 16.13.

Mechanism of Editing

We have been assuming that editing is a posttranscriptional event. This seems like a good bet because unedited transcripts can be found along with edited versions of the same mRNAs. Moreover, editing occurs in the poly(A) tails of mRNAs, which are added posttranscriptionally.

One important clue about the mechanism of editing is that partially edited transcripts have been isolated, and these are always edited at their 3′-ends but not at their 5′-ends. This suggests strongly that editing proceeds in a $3' \rightarrow 5'$ direction. Kenneth Stuart and colleagues first

UAUAUGUUUUGUUGUUUAUUAUGUGAUUAUGGUUUUGUUUUUUA

UUGGUAUUUUUUAGAUUUAUUUAAUUUGUUGAUAAAUACAUUUU

AUUUGUUUGUUAGUGGUUUAUUUGUUAAUUUUUUUGUUUUGUGU

UUUUGGUUUAGGUUUUUUUGUUGUUGUUGUUUUGUAUUAUGAUU

GAGUUUGUUGUUUGGUUUUUUGUUUUUGUGAAACCAGUUAUGAG

AGUUUGCAUUGUUAUUUAUUACAUUAAGUUG GGUGUUUUUGGU

UCUAUUUUAUUUUUAUUGGAUUUAUUACAUUUUAUGCAUGUUUU

UUUAGGUGUUUUGUUGUUGUUUAUUUGUUUUAGCGUUUGUUUA

AUUUUUUGUGUAUGGAUACACGUUUUGUUUUUUUGUAUUGUGUU

UGUUUAUAUUGACAUUUUGUUGAUUUAGUUUGAUUUUUUUUAUU

GCGAUUUGUUUAUUUUGAUGUUUUAUGGUUAUGU UUUGUGU

GUGUAAUUUUAUUGGUGUUUUUUUAGUUGUUGAAGUUA

Figure 16.13 Part of the edited sequence of the *COIII* mRNA of *T. brucei*. The U's added by editing are shown in gray; the T's present in the gene, but absent (as U's) in the mRNA are shown in blue above the sequence. (*Source:* Adapted from *Cell* 53:cover, 1988.)

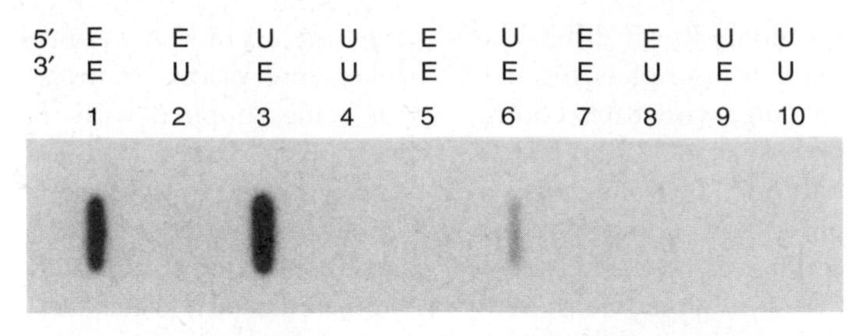

Figure 16.14 PCR analysis of direction of editing. Stuart and colleagues performed RT-PCR with kinetoplast RNA and edited (E) or unedited (U) 5′- and 3′-primers for the cytochrome c oxidase III transcript, as indicated at top. Then they slot-blotted the PCR products and hybridized them to a labeled probe and detected hybridization by autoradiography. PCR templates: lanes 1–4, RNA from wild-type cells; lanes 5–6, a 3′-edited cDNA (positive control); lanes 7–10, RNA from a mutant that lacks mitochondrial DNA (negative control). (*Source:* Abraham, J.M., J.E. Feagin, and K. Stuart, Characterization of cytochrome c oxidase III transcripts that are edited only in the 39 region. Cell 55 (21 Oct 1988) p. 269, f. 2a. Reprinted by permission of Elsevier Science.)

reported this phenomenon in 1988. Their experimental tool was RT-PCR, starting with reverse transcriptase to make the first DNA strand from an RNA template, followed by standard PCR (see Chapter 4).

In one experiment, Stuart and coworkers used pairs of PCR primers in which both were edited primers, both unedited primers, or one of each. A completely edited RNA will hybridize only to edited primers and give a PCR signal, whereas it will not hybridize to unedited primers, so any PCR protocol including at least one unedited primer will not give a signal from this RNA. By contrast, a completely unedited RNA will react only with unedited primers. But the real test is to use an unedited 5′-primer and an edited 3′-primer to detect 3′-edited transcripts, or an edited 5′-primer and an unedited 3′-primer to detect 5′-edited transcripts. If editing goes from 3′ to 5′ in the transcript, then 3′-edited transcripts, but not 5′-edited transcripts, should be detected. The advantage of the PCR method is that it amplifies very small amounts of RNA, such as partially edited RNAs, to easily detectable bands of DNA.

Figure 16.14 depicts the results of this analysis. Lanes 1–4 show the PCR products of *Trypanosoma brucei* kinetoplast RNA with different combinations of primers. We see signals only when both primers were edited, or the 3′-primer was edited. We see no signal when only the 5′-primer was edited. Thus, 3′-editing occurred in the absence of 5′-editing, but 5′-editing did not occur without 3′-editing. This is consistent with editing in the 3′→5′ direction. Lanes 5–6 and 7–10 are positive and negative controls, respectively.

This experiment is valuable, but it has a flaw: None of the lanes involving the unedited 3′-primer shows a signal. We might have expected to see a signal in lane 4, which used unedited 5′- and 3′-primers, but none was observed. This could (and probably does) mean that the concentration of totally unedited RNA is so small that it is undetectable using this method. But it could also mean that there is something wrong with the 3′-unedited primer. Thus, this experiment could have been improved by including a positive control for the 3′-unedited primer—some RNA, such as an in vitro transcript of the gene, which would be totally unedited and should therefore give a signal. If it did, it would remove any doubt about the quality of the 3′-unedited primer. Such controls are especially important in PCR experiments, which have enormous power to amplify tiny quantities of nucleic acids, including contaminants.

What determines where the editing system should add or delete UMPs? Larry Simpson and colleagues found the answer in 1990 when they discovered **guide RNAs (gRNAs)** encoded in *Leishmania* maxicircles. They began with a computer search of the 21-kb part of the maxicircle DNA sequence that was known at that time. This search revealed seven short sequences that could produce short RNAs (gRNAs) complementary to parts of five different edited mitochondrial mRNAs. In principle, such gRNAs could direct the insertion and deletion of UMPs over a stretch of several dozen nucleotides in the mRNA, as illustrated in Figure 16.15a and b. Once that editing is done, another gRNA could hybridize near the 5′-end of the newly edited region and direct editing of a new segment, as Figure 16.15c and d demonstrate. Working in this way from the 3′-end of the mRNA toward the 5′-end, successive gRNAs bind to regions edited by their predecessor gRNAs and direct further editing until they have finished the whole editing job. The sequences of the gRNAs reinforce the conclusion that editing proceeds in the 3′→5′direction: Only the gRNAs at

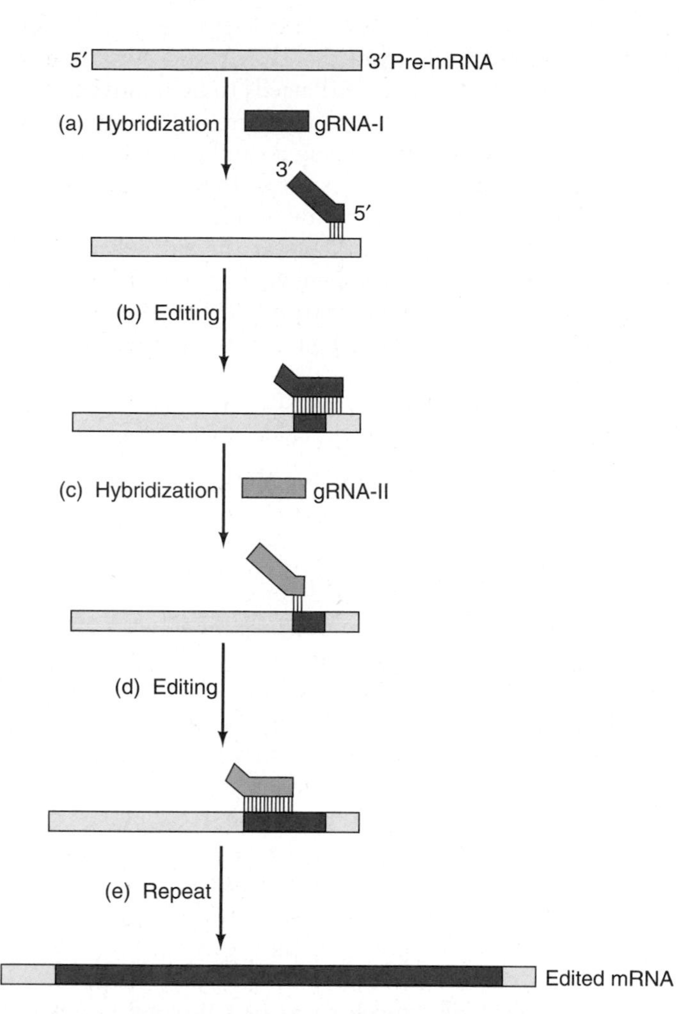

Figure 16.15 Model for the role of gRNAs in editing. (a) In the first step, gRNA-I (dark blue) hybridizes through its 5′-end to a region of the pre-mRNA that requires no editing. Its 3′-end also hybridizes through an oligo(U) region, but that is not illustrated here. **(b)** Most of the rest of the gRNA-I directs editing of part of the pre-mRNA. The edited portion is shown in red, and the pre-mRNA has grown in length, due to the inserted UMPs. **(c)** A new gRNA, gRNA-II (light blue), displaces gRNA-I by hybridizing to the 5′-end of the newly edited region of the pre-mRNA. **(d)** gRNA-II directs editing of a new part of the pre-mRNA. **(e)** The previous steps are repeated with additional gRNAs until the RNA is completely edited.

the 3′-border of editing can hybridize to unedited sequences. All the other gRNAs hybridize to edited sequences. This makes sense only if editing goes 3′→5′.

One notable feature of the base-pairing between gRNAs and mRNA is the existence of G–U base pairs, as well as standard Watson–Crick base pairs. In Chapter 18 we will learn that G–U base pairs are also common during codon–anticodon pairing in translation, and one of the two bases can accommodate these nonstandard base pairs by wobbling slightly from the position it would occupy in Watson–Crick base pairs. The importance of these G–U base pairs in editing probably derives from the fact that they are weaker than Watson–Crick base pairs. This means that the

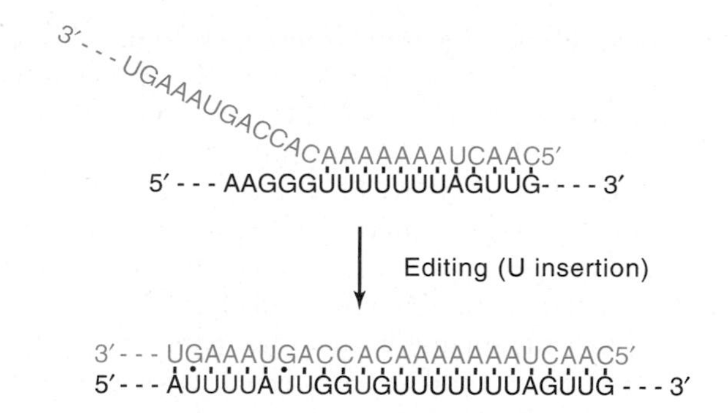

Figure 16.16 Editing of part of a hypothetical RNA. The gRNA (blue) binds via Watson–Crick base pairs to an edited portion of a pre-mRNA. The 3′-end of the gRNA then serves as the template for insertion of U residues (pink). Most of the base pairs between newly inserted U's and the gRNA are Watson–Crick A–U pairs, but two are wobble G–U pairs, denoted by dots.

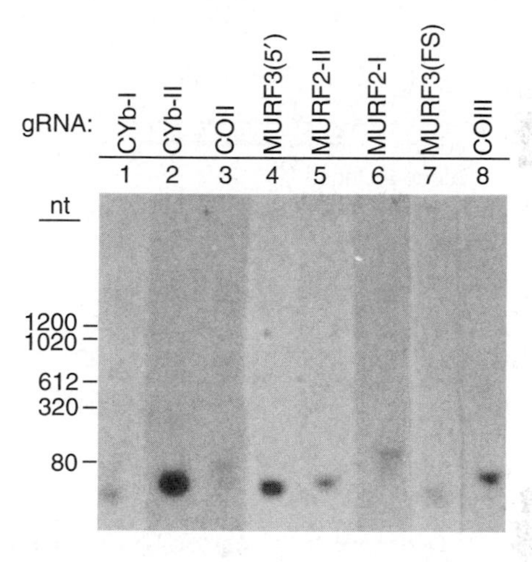

Figure 16.17 Evidence for gRNAs. Simpson and colleagues Northern blotted RNA from the mitochondria of *Leishmania tarentolae* and probed the blots with labeled oligonucleotides that would hybridize to gRNAs. The gRNAs are identified at top. (*Source:* Blum, B., N. Bakalara, and L. Simpson, A model for RNA editing in kinetoplastid mitochondria: "Guide" RNA molecules transcribed from maxicircle DNA provide the edited information. *Cell* 60 (26 Jan 1990) p. 191, f. 3a. Reprinted by permission of Elsevier Science.)

5′-end of a new gRNA, by forming Watson–Crick base pairs with the newly edited region of an mRNA, can displace the 3′-end of the base-paired region of an old gRNA, whose base-pairing with the mRNA includes weak G–U pairs (Figure 16.16).

Later in 1990, Nancy Sturm and Larry Simpson found that minicircles also encode gRNAs. But besides the coding potential, Simpson and colleagues found direct evidence for the existence of gRNAs. They electrophoresed kinetoplastid RNA, Northern blotted it, and hybridized it to labeled oligonucleotide probes designed to detect gRNAs, according to the sequences of putative gRNA genes in maxicircles. Figure 16.17 shows that this procedure detected small RNAs, most of which appeared to be shorter than 80 nt.

The precise mechanism of editing, the cutting and pasting required to insert and delete UMPs, remained unclear for several years, but the enzyme activities found in kinetoplasts provided some hints. For example, kinetoplasts have a **terminal uridylyl transferase (TUTase)** that could add extra UMPs (uridylates) to the mRNA during editing. Because the mRNA has to be cut to accept these new UMPs, it must also be ligated together again, and kinetoplasts also contain an **RNA ligase.** The major remaining question concerned the source of uridylates for editing. UTP could provide them. On the other hand, uridylates at the ends of gRNAs could be transferred to the pre-mRNA by transesterification. That is, the uridylates could be plucked off of the ends of gRNAs and transferred directly to the pre-mRNA.

Then, in 1994, Scott Seiwert and Stuart used a mitochondrial extract and a gRNA to edit a synthetic pre-mRNA. They found that deletion of UMPs required three enzymatic activities (Figure 16.18a): (1) an endonuclease that follows directions from the gRNA and cuts the pre-mRNA at the site where a UMP needs to be removed; (2) a 3′-exonuclease that is specific for terminal uridines; and (3) an RNA ligase. In 1996, using a similar in vitro system, Stuart and colleagues demonstrated that UMP insertion follows a similar three-step pathway (Figure 16.18b): (1) a gRNA-directed endonuclease cuts at the site where UMP insertion is required; (2) an enzyme (probably TUTase) transfers UMPs from UTP (not from gRNA), as directed by the gRNA; and (3) an RNA ligase puts the two pieces of RNA back together.

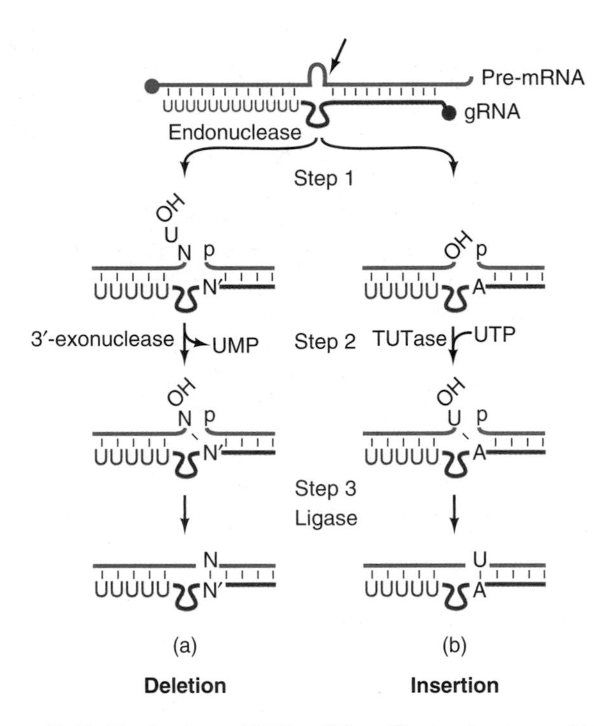

Figure 16.18 Mechanism of RNA editing. The mechanisms of **(a)** U deletion, and **(b)** U insertion are shown, starting with a hybrid between a pre-mRNA (pink) and a gRNA (dark blue) at top. The bulge in the gRNA denotes a stretch of bases that do not match those found in the pre-mRNA, and will be used as a template for editing. The arrow indicates the position at which the nuclease cuts the pre-mRNA for editing. **(a)** U deletion. Step 1: An endonuclease clips the pre-mRNA just to the 3′-side of the U to be deleted. Step 2: An exonuclease removes the UMP at the end of the left-hand RNA fragment. Base pairing occurs between base N in the pre-mRNA and base N′ in the gRNA. Step 3: RNA ligase puts the two halves of the pre-mRNA back together. **(b)** U insertion. Step 1: An endonuclease clips the pre-mRNA at the site where the gRNA dictates that a U should be inserted. Step 2: TUTase transfers a UMP from UTP to the 3′-end of the left-hand RNA fragment. This U base-pairs with an A in the gRNA. Step 3: RNA ligase puts the pre-mRNA back together. (*Source:* Adapted from Seiwert, S.D., Pharmacia Biotech in Science Prize. 1996 grand prize winner. RNA editing hints of a remarkable diversity in gene expression pathways. Science 274:1637, 1996.)

It is interesting that the gRNAs are encoded in the mitochondrial DNAs, while the proteins required for editing are encoded in the nucleus and imported into the mitochondria.

SUMMARY Trypanosomatid mitochondria encode incomplete mRNAs that must be edited before they can be translated. Editing occurs in the 3′→5′ direction by successive action of one or more guide RNAs. The 5′-end of the first gRNA hybridizes to an unedited region at the 3′-border of editing in the pre-mRNA; the 5′-ends of the rest of the gRNAs hybridize to edited regions progressively closer to the 5′-end of the region to be edited in the pre-mRNA. All of these gRNAs provide A's and G's as templates for the incorporation of U's missing from the mRNA. Sometimes the gRNA is missing an A or G to pair with a U in the mRNA, in which case the U is removed. The mechanism of removing U's involves: (1) cutting the pre-mRNA just beyond the U to be removed; (2) removal of the U by an exonuclease; and (3) ligating the two pieces of pre-mRNA together. The mechanism of adding U's uses the same first and last step, but the middle step (step 2) involves addition of one or more U's from UTP by TUTase instead of removing U's.

Editing by Nucleotide Deamination

RNA editing is not just something strange that happens in weird organisms, it also plays a vital role in higher organisms—even mammals. As yet, there has been no indication that mammals carry out the type of uridine addition and deletion that occurs in trypanosomes, but abundant evidence has been found for another kind of editing: deamination of adenosine, which converts adenosine to inosine, which has an oxygen in place of adenine's amino group. Because inosine forms base pairs with cytidine in the same way as guanosine, the deamination of adenosine changes the meaning of a codon. For example, an ACG (threonine) codon becomes an ICG codon, which would be read by the ribosome as GCG (alanine).

This kind of RNA editing is directed by an enzyme called **adenosine deaminase acting on RNA (ADAR)**. Humans and mice contain three ADAR genes: *ADAR1, ADAR2,* and *ADAR3*. The products of the first two are ubiquitous in the body, but the third gene product is found only in the brain. These enzymes are very specific. It would be disastrous if they deaminated every adenosine in an mRNA, so they select only certain adenosines in certain mRNAs. For example, ADAR2 deaminates one adenosine in the glutamate-sensitive ion-channel receptor subunit B (GluR-B) mRNA, with greater than 99% efficiency. This alteration in the mRNA changes a glutamine codon to an arginine codon. Is this an important change? We know it is because an ion channel containing the GluR-B protein with a glutamine instead of an arginine is too permeable to calcium ions. We would therefore predict that mice with a defective *ADAR2* gene would have serious problems. Indeed, mice homozygous for a defective *ADAR2* gene do not carry out the appropriate *GluR-B* mRNA editing. They seem to develop normally, but die shortly after weaning.

Peter Seeburg and colleagues wondered what would happen if the mouse *GluR-B* gene were simply changed so that it encoded arginine at the edited position; then, no editing of this gene's transcript would be necessary. When they performed this experiment, they found that their mice were viable, even if they had a homozygous-defective *ADAR2* gene. Thus, this experiment also demonstrated that the only critical target of ADAR2 is the *GluR-B* transcript.

The *Drosophila* genome contains only one *ADAR* gene. When this gene is mutated so the flies lack all ADAR activity, they do not carry out any mRNA editing at known editing sites. These mutant flies are viable, but they have difficulty walking, cannot fly, and suffer progressive neural degeneration, particularly in the brain. Thus, the phenotype of this mutation is similar to the phenotype of mutations in the gene for ADAR2 in mammals. The *Drosophila* work bolsters the hypothesis that mRNA editing by ADAR is essential for normal central nervous system development.

ADAR1 also appears to be essential for mammalian life. Kazuko Nishikura and coworkers mutated mouse stem cells to heterozygous mutant ($ADAR1^{+/-}$), then injected these cells into normal mouse blastocysts in an attempt to create chimeric mice (see Chapter 5). But they found it impossible even to generate chimeric mice with a sizeable proportion of mutant cells. No embryo with more than a limited complement of mutant cells survived to birth. Thus, even heterozygous mutations in *ADAR1* appear to be embryonic lethal.

Why do embryos with a low ADAR1 activity die? Most tissues in the affected embryos appeared normal, but red blood cells (erythrocytes) did not. They remained nucleated, like erythrocytes derived from the yolk sac, long after erythropoiesis (creation of erythrocytes) would normally have shifted from the yolk sac to the liver, which generates erythrocytes that lose their nuclei. Thus, some aspect of erythropoiesis depends on a full complement of ADAR1 in the embryo.

Interestingly, certain tumors lose ADAR activity. In particular, a very malignant human brain tumor called glioblastoma multiforme (GBM) has very low ADAR2 activity, and a corresponding underediting in the *GluR-B* mRNA. Some epileptics also have this underedited mRNA, and GBM patients often are afflicted with epileptic seizures.

Another kind of editing is carried out by **cytidine deaminase acting on RNA (CDAR)**, which converts cytidine to uridine. This C→U editing is defective in about 25% of the benign peripheral nerve sheath tumors found in neurofibromatosis type I patients. C→U editing also appears to occur in HIV transcripts in human cells. Still another kind of editing that occurs in HIV-infected human cells is G→A editing. But this kind of editing cannot be explained by a single-step deamination, and it is unclear how it is accomplished.

SUMMARY Some adenosines in mRNAs of higher eukaryotes, including fruit flies and mammals, must be deaminated to inosine posttranscriptionally for the mRNAs to code for the proper proteins. Enzymes known as adenosine deaminases active on RNAs (ADARs) carry out this kind of RNA editing. In addition, some cytidines must be deaminated to uridine for an mRNA to code properly.

16.5 Post-Transcriptional Control of Gene Expression: mRNA Stability

In our discussions of the mechanisms of prokaryotic and eukaryotic transcription, we saw many examples of transcriptional control. It makes sense to control gene expression by blocking the first step—transcription. That is the least wasteful method because the cell expends no energy making an mRNA for a protein that is not needed.

Although transcriptional control is the most prevalent form of control of gene expression, it is by no means the only way. We have already seen in Chapter 15 that poly (A) stabilizes and confers translatability on an mRNA, and special sequences in the 3′-untranslated region of an mRNA, called cytoplasmic polyadenylation elements (CPEs), govern the efficiency of polyadenylation of maternal messages during oocyte maturation. In this way, these CPEs serve as controllers of gene expression.

But an even more important posttranscriptional control of gene expression is control of mRNA stability. In fact, Joe Harford has pointed out that "cellular mRNA levels often correlate more closely with transcript stability than with transcription rate."

Casein mRNA Stability

The response of mammary gland tissue to the hormone prolactin provides a good example of control of mRNA stability. When cultured mammary gland tissue is stimulated with prolactin, it responds by producing the milk protein casein. One would expect an increase in casein mRNA concentration to accompany this casein buildup, and it does. The number of casein mRNA molecules increases about 20-fold in 24 h following the hormone treatment. But this does not mean the rate of casein mRNA synthesis has increased 20-fold. In fact it only increases about two- to threefold. The rest of the increase in casein mRNA level depends on an approximately 20-fold increase in stability of the casein mRNA.

Jeffrey Rosen and his colleagues performed a **pulse-chase** experiment to measure the **half-life** of casein mRNA. The half-life is the time it takes for half the RNA molecules to be degraded. Rosen and colleagues radioactively labeled casein mRNA for a short time in vivo in the presence or absence of prolactin. In other words, they gave the cells a pulse of radioactive nucleotides, which the cells incorporated into their RNAs. Then they transferred the cells to medium lacking radioactivity. This chased the radioactivity out of the RNA, as labeled RNAs broke down and were replaced by unlabeled ones. After various chase times, the experimenters measured the level of labeled casein mRNA by hybridizing it to a cloned casein gene. The faster the labeled casein mRNA disappeared, the shorter its half-life. The conclusion, shown in Table 16.1, was that the half-life of casein mRNA increased dramatically, from 1.1 h to 28.5 h, in the presence of prolactin. At the same time, the half-life of total polyadenylated mRNA increased only 1.3- to 4-fold in response to the hormone. It appears prolactin causes a selective stabilization of casein mRNA that is largely responsible for the enhanced expression of the casein gene. Note that pulse-chase experiments can do more than measure the half-life of a molecule. They can also show precursor-product relationships, as a labeled precursor is chased into labeled products. We saw a good example—rRNA precursor and products—earlier in this chapter.

Table 16.1 Effect of Prolactin on Half-Life of Casein mRNA

	RNA Half-life (h)	
Species of RNA	**– Prolactin**	**+ Prolactin**
rRNA	>790	>790
Poly(A)$^+$ RNA (short-lived)	3.3	12.8
Poly(A)$^+$ RNA (long-lived)	29	39
Casein mRNA	1.1	28.5

Source: Reprinted from Guyette, W.A., R.J. Matusik, and J.M. Rosen, Prolactin-mediated transcriptional and post-transcriptional control of casein gene expression. *Cell* 17:1013, 1979. Copyright © 1979, with permission from Elsevier Science.

SUMMARY A common form of posttranscriptional control of gene expression is control of mRNA stability. For example, when mammary gland tissue is stimulated by prolactin, the synthesis of casein protein increases dramatically. However, most of this increase in casein is not due to an increase in the rate of transcription of the casein gene. Instead, it is caused by an increase in the half-life of casein mRNA.

Transferrin Receptor mRNA Stability

One of the best studied examples of posttranscriptional control concerns iron homeostasis (control of iron concentration) in mammalian cells. Iron is an essential mineral for all eukaryotic cells, yet it is toxic in high concentrations. Consequently, cells have to regulate the intracellular iron concentration carefully. Mammalian cells do this by regulating the amounts of two proteins: an iron import protein called the **transferrin receptor (TfR)**, and an iron storage protein called **ferritin.** Transferrin is an iron-bearing protein that can get into a cell via the transferrin receptor on the cell surface. Once the cell imports transferrin, it passes the iron to cellular proteins, such as cytochromes, that need iron. Alternatively, if the cell receives too much iron, it stores the iron in the form of ferritin.

Thus, when a cell needs more iron, it increases the concentration of transferrin receptors to get more iron into the cell and decreases the concentration of ferritin, so not as much iron will be stored and more will be available. On the other hand, if a cell has too much iron, it decreases the concentration of transferrin receptors and increases the concentration of ferritin. It employs posttranscriptional strategies to do both these things: It regulates the rate of translation of ferritin mRNA, and it regulates the stability of the transferrin receptor mRNA. We will deal with the regulation of ferritin mRNA translation in Chapter 17. Here we are concerned with the latter process: controlling the stability of the mRNA encoding the transferrin receptor.

Joe Harford and his colleagues reported in 1986 that depleting intracellular iron by chelation resulted in an increase in transferrin receptor (TfR) mRNA concentration. On the other hand, increasing the intracellular iron concentration by adding hemin or iron salts decreased the TfR mRNA concentration. The changes in TfR mRNA concentrations with fluctuating intracellular iron concentration are not caused primarily by changes in the rate of synthesis of TfR mRNA. Instead, these alterations in TfR mRNA concentration largely depend on changes in the TfR mRNA half-life. In particular, the TfR mRNA half-life increases from about 45 min when iron is plentiful to many hours when iron is in short supply. We will examine the data on mRNA half-life but first we need to inspect the structure of the mRNA, which makes possible the modulation in its lifetime.

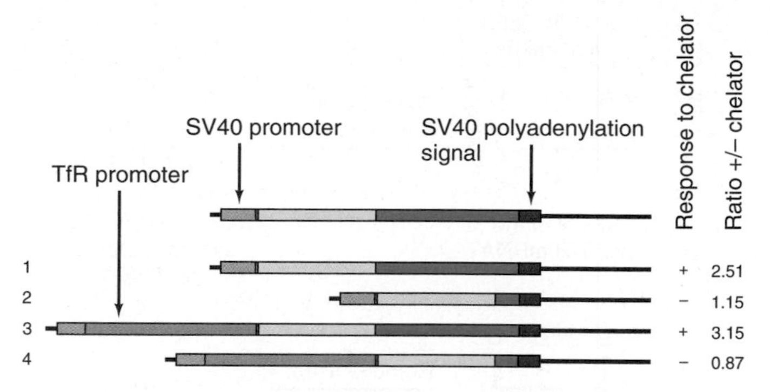

Figure 16.19 Effect of the 3′-UTR on the iron-responsiveness of cell surface concentration of TfR. Owen and Kühn made the TfR gene constructs diagrammed here. The DNA regions within the boxes are color-coded as follows: SV40 promoter, orange; TfR promoter, blue; TfR 5′-UTR, black; TfR-coding region, yellow; TfR 3′-UTR, green; SV40 polyadenylation signal, purple. These workers then transfected cells with each construct and assayed for concentration of TfR on the cell surface, using fluorescent antibodies. The ratio of cell surface TfR in the presence and absence of the iron chelator (desferrioxamine) is given at right, along with a qualitative index of response to chelator (+ or –). (*Source:* Adapted from Owen, D. and L.C. Kühn, Noncode 39 sequences of the transferrin receptor gene are required for mRNA regulation by iron. *The EMBO Journal* 6:1288, 1987.)

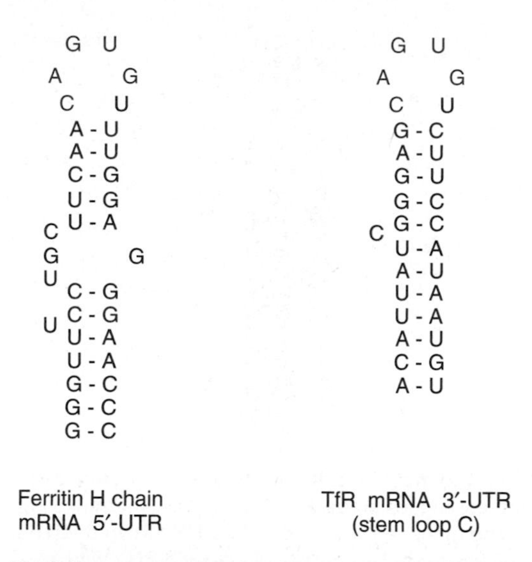

Figure 16.20 Comparison of stem loop structures in the 3′-UTR of the human TfR mRNA with the IRE in the 5′-UTR of the human ferritin mRNA. Only one (stem-loop C) of the five TfR mRNA stem-loops is shown. The conserved, looped-out C and the conserved bases in the loop are in blue and red, respectively. (*Source*: Adapted from Casey, J.L., M.W. Hentze, D.M. Koeller, S.W. Caughman. T.A. Rovault, R.D. Klausner, and J.B. Harford, Iron-responsive elements: Regulatory RNA sequences that control mRNA levels and translation. Science 240:926, 1988.)

Iron Response Elements Lukas Kühn and his colleagues cloned a human TfR cDNA in 1985 and found that it encoded an mRNA with a 96-nt 5′-untranslated region (**5′-UTR**), a 2280-nt coding region, and a 2.6-kb 3′-untranslated region (**3′-UTR**). To test the effect of this long 3′-UTR, Dianne Owen and Kühn deleted 2.3 kb of the 3′-UTR and transfected mouse L cells with this shortened construct. They also made similar constructs with the normal TfR promoter replaced by an SV40 viral promoter. Then they used a monoclonal antibody specific for the human TfR and a fluorescent secondary antibody to detect TfR on the cell surfaces. Figure 16.19 summarizes the results. With the wild-type gene, the cells responded to an iron chelator by increasing the surface concentration of TfR about threefold. Owen and Kühn observed the same behavior when the TfR gene was controlled by the SV40 promoter, demonstrating that the TfR promoter was not responsible for iron responsiveness. On the other hand, the gene with the deleted 3′-UTR did not respond to iron; the same concentration of TfR appeared on the cell surface in the presence or in the absence of the iron chelator. Thus, the part of the 3′-UTR deleted in this experiment apparently included the iron response element.

Of course, the appearance of TfR receptor on the cell surface does not necessarily reflect the concentration of TfR mRNA. To check directly for an effect of iron on TfR mRNA concentration, Owen and Kühn performed S1 analysis (Chapter 5) of TfR mRNA in cells treated and untreated with iron chelator. As expected, the iron chelator increased the concentration of TfR mRNA considerably. But this response to iron disappeared when the gene had a deleted 3′-UTR.

What part of the 3′-UTR confers responsiveness to iron? Harford and colleagues narrowed the search when they discovered that deletion of just 678 nt from the middle of the 3′-UTR eliminated most of the iron responsiveness.

Computer analysis of the critical 678-nt region of the 3′-UTR revealed that its most probable structure includes five hairpins, or stem-loops, as illustrated in Figure 16.20. Even more interesting is the fact that the overall structures of these stem-loops, including the base sequences in the loops, bear a strong resemblance to a stem loop found in the 5′-UTR of the ferritin mRNA. This stem-loop, called an **iron response element (IRE),** is responsible for the ability of iron to stimulate translation of the ferritin mRNA. The implication is that these TfR IREs are the mediators of the responsiveness of TfR expression to iron.

Harford and colleagues went on to show by gel mobility shift assays (Chapter 5) that human cells contain a protein or proteins that bind specifically to the human TfR IREs (Figure 16.21). This binding could be competed with excess TfR mRNA or ferritin mRNA, which also has an IRE, but it could not be competed by β-globin mRNA, which has no IRE. Thus, the binding is IRE-specific. This finding underscores the similarity between the ferritin and TfR IREs and suggests that they may even bind the same protein(s). However, binding of the protein(s) to the two mRNAs has different effects, as we have seen.

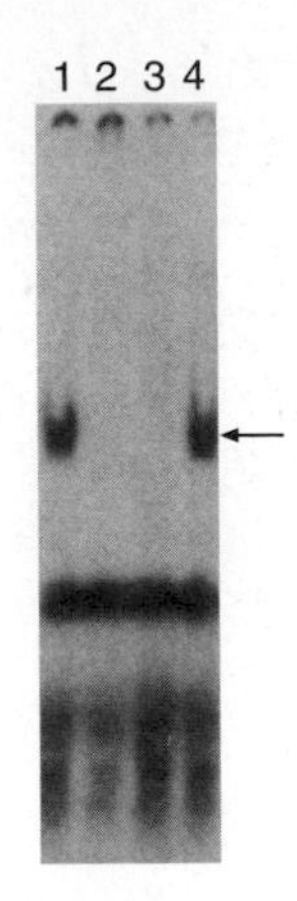

Figure 16.21 Gel mobility shift assay for IRE-binding proteins. Harford and colleagues prepared a labeled 1059-nt transcript corresponding to the region of the human TfR mRNA 3′-UTR that contains the five IREs. They mixed this labeled RNA with a cytoplasmic extract from human cells (with or without competitor RNA), electrophoresed the complexes, and visualized them by autoradiography. Lane 1, no competitor; lane 2, TfR mRNA competitor; lane 3, ferritin mRNA competitor; lane 4, β-globin mRNA competitor. The arrow points to a specific protein–RNA complex, presumably involving one or more IRE-binding proteins. (*Source*: Koeller, D.M., J.L. Casey, M.W. Hentze, E.M. Gerhardt, L.-N.L. Chan, R.D. Klausner, and J.B. Harford, A cytosolic protein binds to structural elements within the nonregulatory region of the transferrin receptor mRNA. *Proceedings of the National Academy of Sciences USA* 86 (1989) p. 3576, f. 3.)

SUMMARY The transferrin receptor-TfR concentration is low when iron concentration is high, and this loss of TfR is largely due to decreased stability of the TfR mRNA. This response to iron depends on the 3′-UTR of the mRNA, which contains five stem loops called iron response elements (IREs).

The Rapid Turnover Determinant Knowing that iron regulates the TfR gene by controlling mRNA stability, and knowing that a protein binds to one or more IREs in the 3′-UTR of TfR mRNA, we assume that the IRE-binding protein protects the mRNA from degradation. This kind of regulation demands that the TfR mRNA be inherently unstable. If it were a stable mRNA, relatively little would be gained by stabilizing it further. In fact, the mRNA *is* unstable, and Harford and coworkers have demonstrated that this instability is caused by a **rapid turnover determinant** that also lies in the 3′-UTR.

What is this rapid turnover determinant? Because the human and chicken TfR genes are controlled in the same manner, they probably have the same kind of rapid turnover determinant. Therefore, a comparison of the 3′-UTRs of these two mRNAs might reveal common features that would suggest where to start the search. Harford and colleagues compared the 678-nt region of the TfR mRNA from human with the corresponding region of the chicken TfR

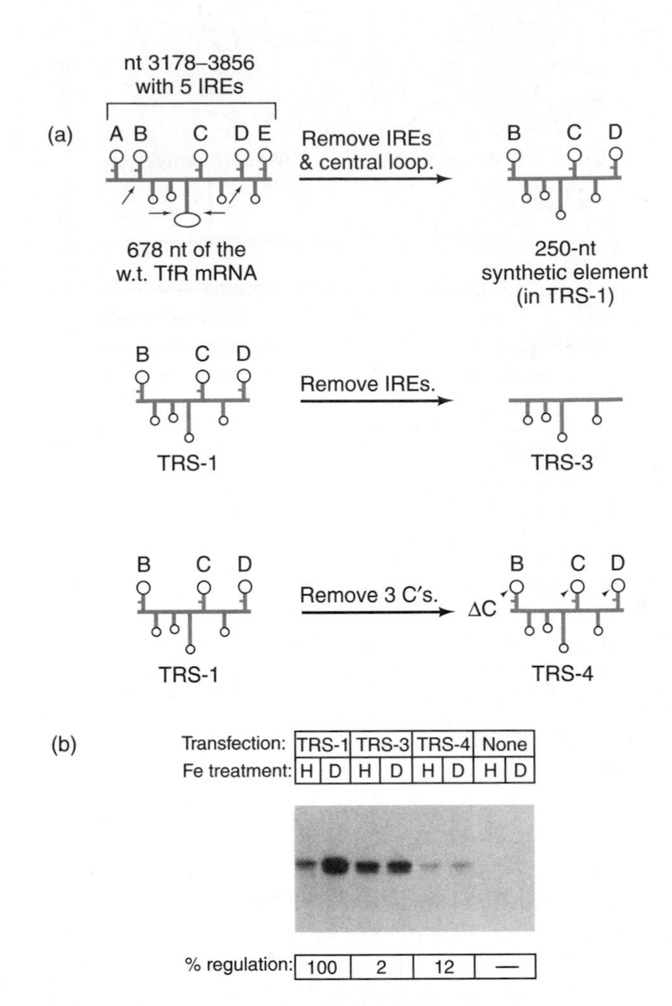

Figure 16.22 Effects of deletions in the IRE region of the TfR 3′-UTR on iron responsiveness. (a) Creation of deletion mutants. Harford and colleagues generated the TRS-1 mutant by removing IREs A and E, and the large central loop, as shown by the arrows. From TRS-1, they generated TRS-3 by removing the remaining three IREs, and TRS-4 by deleting a single C at the 5′-end of each IRE loop. **(b)** Testing mutants for iron response. These workers transfected cells with each construct, treated half the cells with hemin (H) and the other half with desferrioxamine (D), and assayed for TfR biosynthesis by immunoprecipitation. The autoradiograph is shown, with transfected construct and iron treatment shown at top. A summary of the percentage regulation by iron is given at bottom. This is the fold induction by iron chelator vs. hemin (D/H) compared with wild-type, which is defined as 100% regulation. TRS-3 shows essentially no regulation and a constitutively high level of TfR synthesis, suggesting a stable mRNA. TRS-4 shows little regulation and a low level of TfR synthesis, suggesting an unstable mRNA. (*Source:* Casey, J.L., D.M. Koeller, V.C. Ramin, R.D. Klausner, and J.B. Harford, Iron regulation of transferrin receptor mRNA levels requires iron-responsive elements and a rapid turnover determinant in the 39 untranslated region of the mRNA. *EMBO Journal* 8 (8 Jul 1989) p. 3695, f. 3B.)

mRNA and found a great deal of similarity in the region containing the IREs. Figure 16.22a (left) depicts the human structure. Both have two IREs in the 5′-part of the region, then a stem with a large loop (250 nt in human and 332 nt in chicken), then the other three IREs. The 5′-and 3′-IRE-containing regions in the human mRNA are very similar to the corresponding regions in the chicken mRNA, but the loop region in between and the regions farther upstream and downstream have no detectable similarity. This suggested that the rapid turnover determinant should be somewhere among the IREs. Harford and coworkers identified some of its elements by mutagenizing the TfR mRNA 3′-UTR and observing which mutations stabilized the mRNA.

The first mutants they looked at were simple 5′- or 3′-deletions. They transfected cells with these constructs and assayed for iron regulation by comparing the TfR mRNA and protein levels after treatment with either hemin or the iron chelator desferrioxamine. They measured mRNA levels by Northern blotting and protein levels by immunoprecipitation. They found that deletion of the 250-nt central loop or deletion of IRE A had no effect on iron regulation. However, deletion of both IREs A and B eliminated iron regulation: The levels of TfR mRNA and protein were the same (and high) with both treatments. Thus, the TfR mRNA is stable when IRE B is removed, so this IRE seems to be part of the rapid turnover determinant. The 3′-deletions gave a similar result. Deletion of IRE E had little effect on iron regulation, but deletion of both IREs D and E stabilized the TfR mRNA, even in the presence of hemin. Thus, IRE D appears to be part of the rapid turnover determinant.

Based on these findings, we would predict that IRE A, IRE E, and the central loop could be deleted without altering iron regulation. Accordingly, Harford and colleagues made a synthetic element they called TRS-1 that was missing these three parts as illustrated in Figure 16.22a. As expected, mRNAs containing this element retained full iron responsiveness. Next, these workers made two alterations to TRS-1 (Figure 16.22a). The first, TRS-3, had lost all three of its IREs. All that remained were the other stem-loops, pictured pointing downward in Figure 16.22. The other, TRS-4, had lost only three bases, the C's at the 5′-end of the loop in each IRE. Figure 16.22b shows the effects of these two alterations. TRS-3, with no IREs, had lost virtually all iron responsiveness, and the TfR RNA appeared to be much more stable than the wild-type mRNA. That is, there was abundant TfR even in the presence of hemin. TRS-4, with a C missing from each IRE, had lost most of its iron responsiveness, but the mRNA remained unstable. That is, there was not much TfR even in the presence of the iron chelator. Thus, this mRNA retained its rapid turnover determinant, but had lost the ability to be stabilized by the IRE-binding protein. In fact, as we would expect, gel mobility shift assays showed that TRS-4 could not bind the IRE-binding protein.

To pin down the rapid turnover determinant still further, Harford and colleagues made two new constructs in which they deleted one or the other of the two (downward-pointing) non-IRE stem-loops on either side of the large central stem-loop. Then they tested these constructs by transfection and immunoprecipitation as before. Both constructs showed almost total loss of iron responsiveness and a constitutively high level of TfR expression (the same pattern shown by TRS-3 in Figure 16.22). Thus, both of the deleted stem-loops appear to be essential to confer rapid turnover of the mRNA. To demonstrate that this effect was not due to an inability of the mRNAs to interact with the IRE-binding protein, these workers assayed protein–RNA binding as before by gel mobility shift. Both constructs were just as capable of binding to the IRE-binding protein as was the wild-type mRNA, and excess unlabeled IRE successfully competed with the labeled constructs for binding.

SUMMARY IREs A and E, and the large central loop of the TfR 3′-UTR can be deleted without altering the response to iron. However, removing IREs A and B, or IREs D and E, or all five IREs renders the TfR mRNA constitutively stable. Thus, IREs B and D, at least, are part of the rapid turnover determinant. Removing a C from IREs B–D renders the TfR mRNA constitutively unstable and unable to bind the IRE-binding protein.

TfR mRNA Stability and Degradation Pathway The data presented so far strongly suggest that iron regulates the TfR mRNA half-life, rather than the rate of mRNA synthesis. To provide direct evidence for this hypothesis, Ernst Müllner and Lukas Kühn measured the rate of TfR mRNA decay in the presence and absence of the iron chelator desferrioxamine. They found that the TfR mRNA was very stable when the iron concentration was low. On the other hand, at high iron concentration the TfR mRNA decayed much faster. These two half-lives were 30 and 1.5 h, respectively, so iron appears to destabilize the TfR mRNA by approximately 30/1.5, or 20-fold.

Harford and colleagues investigated the mechanism by which TfR mRNA is degraded and found that the first event appears to be an endonucleolytic cut within the IRE region. Unlike the degradation of many other mRNAs, there seems to be no requirement for deadenylation (removal of poly[A]) before TfR degradation can begin.

These workers began their study by treating human plasmacytoma cells (ARH-77 cells) with hemin and showing by Northern blotting that the level of TfR mRNA dropped precipitously in 8 h. When they exposed the blot for a longer time, they found that a new RNA species, about 1000–1500 nt shorter than full-length TfR mRNA, appeared during the period in which the TfR mRNA was breaking down. This RNA was also found in the poly(A)$^-$ fraction, suggesting that it had lost its poly(A). But the size of this shortened RNA suggested that it had lost much more than just its poly(A). The simplest explanation was that it had been cut by an endonuclease within its 3′-UTR, which removed over 1000 3′-terminal nucleotides, including the poly(A).

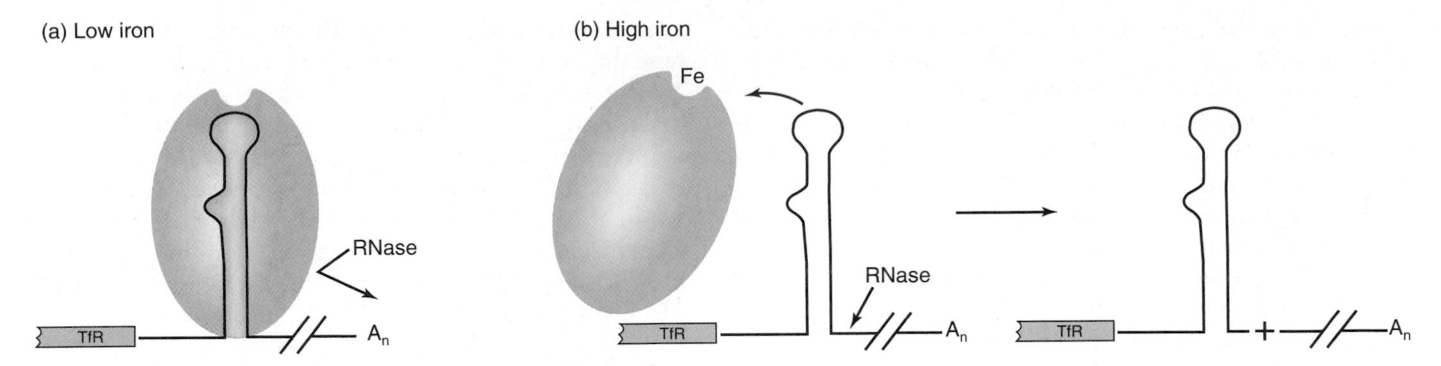

Figure 16.23 Model for destabilization of TfR mRNA by iron. **(a)** Under low-iron conditions, the aconitase apoprotein (orange) binds to the IREs in the 3′-UTR of the TfR mRNA. This protects the RNA from degradation by RNases. **(b)** Under high-iron conditions, iron binds to the aconitase apoprotein, removing it from the IREs, and opening the IREs up to attack by RNase. The RNase clips the mRNA at least once, exposing its 3′-end to further degradation.

All the data we have considered are consistent with the following hypothesis (Figure 16.23): When iron concentrations are low, an IRE-binding protein, or **iron regulatory protein (IRP)**, binds to the rapid turnover determinant in the 3′-UTR of the TfR mRNA. This protein protects the mRNA from degradation. When iron concentrations are high, iron binds to the IRE-binding protein, causing it to dissociate from the rapid turnover determinant, opening it up to attack by a specific endonuclease that clips off a 1-kb fragment from the 3′-end of the TfR mRNA. This destabilizes the mRNA and leads to its rapid degradation.

One of the proteins (IRP1) that bind to the IREs in both the transferrin receptor mRNA and the ferritin mRNA (Chapter 17) has now been identified as a form of **aconitase,** an enzyme that converts citrate to isocitrate in the citric acid cycle. The enzymatically active form of aconitase is an iron-containing protein that does not bind to the IREs. However, the apoprotein form of aconitase, which lacks iron, binds to the IREs in mRNAs.

SUMMARY When the iron concentration is high, the TfR mRNA decays rapidly. When the iron concentration is low, the TfR mRNA decays much more slowly. This difference in mRNA stability is about 20-fold and plays a major role in control of the gene's expression. The initiating event in TfR mRNA degradation seems to be an endonucleolytic cleavage of the mRNA more than 1000 nt from its 3′-end, within the IRE region. This cleavage does not require prior deadenylation of the mRNA. Iron controls TfR mRNA stability as follows: When iron concentration is low, aconitase exists at least partly in an apoprotein form that lacks iron. This protein binds to the IREs in the TfR mRNA and protects the RNA against attack by RNases. But when iron concentration is high, the aconitase apoprotein binds to iron and therefore cannot bind to the TfR mRNA IREs. This leaves the RNA vulnerable to degradation.

16.6 Post-Transcriptional Control of Gene Expression: RNA Interference

For years, molecular biologists have been using antisense RNA to inhibit expression of selected genes in living cells. At first, the rationale was that the antisense RNA, which is complementary to mRNA, would base-pair to the mRNA and inhibit its translation. The strategy usually worked, but the rationale was incomplete. As Su Guo and Kenneth Kemphues established in 1995, injecting sense RNA into cells worked just as well as antisense RNA in blocking expression of a particular gene. Then, in 1998, Andrew Fire and Craig Mello and their colleagues showed that double-stranded RNA (**dsRNA**) worked much better than either sense or antisense RNA. In fact, the main reason sense and antisense RNAs worked appears to be that they were contaminated with (or produced) small amounts of dsRNA, and the dsRNA actually did the most to block gene expression.

Also, beginning in 1990, molecular biologists began noticing that placing transgenes into various organisms sometimes had the opposite of the desired effect. Instead of turning on the transgene, organisms sometimes turned off, not only the transgene, but the normal cellular copy of the gene as well. One of the first examples was an attempt to intensify the purple color of a petunia by supplying extra copies of the pigment-producing genes. But in up to 25% of the transformed plants, blossoms were white or patchy purple and white—the opposite of the intended

Figure 16.24 Silencing of a purple color gene in petunia by adding extra copies of the color gene. The central white stripe in each petal shows where silencing occurred. (*Source*: Courtesy of Dr. Richard A. Jorgensen, The Plant Cell.)

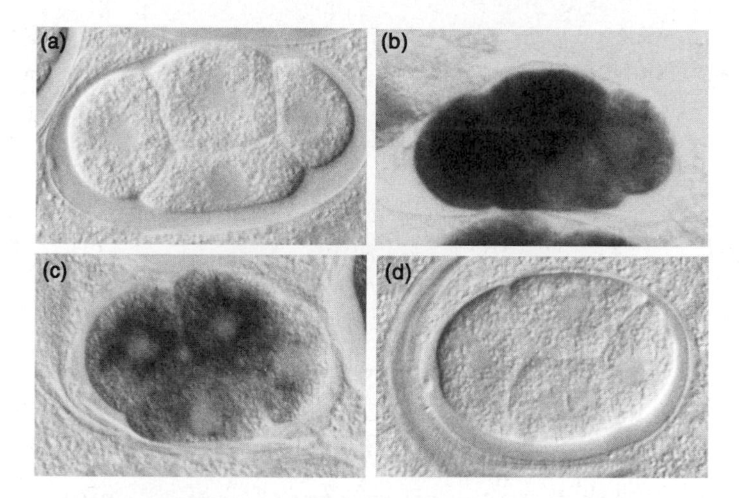

Figure 16.25 Double-stranded RNA-induced RNA interference causes destruction of a specific mRNA. Fire and colleagues injected antisense or dsRNA corresponding to the *C. elegans mex-3* mRNA into *C. elegans* ovaries. After 24 h, they fixed the embryos in the treated ovaries and subjected them to in situ hybridization (Chapter 5) with a probe for *mex-3* mRNA. **(a)** Embryo from a negative control parent with no hybridization probe. **(b)** Embryo from a positive control parent that was not injected with RNA. **(c)** Embryo from a parent that was injected with *mex-3* antisense RNA. A considerable amount of *mex-3* mRNA remained. **(d)** Embryo from a parent that was injected with dsRNA corresponding to part of the *mex-3* mRNA. No detectable *mex-3* mRNA remained. (*Source*: Fire, A., S. Xu, M.K. Montgomery, S.A. Kostas, S.E. Driver, and C.C. Mello, Potent and specific genetic interference by double-stranded RNA in *Caenorhabditis elegans*. *Nature* 391 (1998) f. 3, p. 809. Copyright © Macmillan Magazines Ltd.)

effect (Figure 16.24). This phenomenon was called by several names: cosuppression and **post-transcriptional gene silencing (PTGS)** in plants, **RNA interference (RNAi)** in animals such as nematodes *(Caenorhabditis elegans)* and fruit flies, and **quelling** in fungi. To avoid confusion, we will refer to this phenomenon as RNAi from now on, regardless of the species under study.

Mechanism of RNAi

Fire and colleagues showed that injecting *C. elegans* gonads with dsRNA (the **trigger dsRNA**) caused RNAi in the resulting embryos. Furthermore, they detected a loss of the corresponding mRNA (the **target mRNA**) in embryos undergoing RNAi (Figure 16.25). However, the dsRNA had to include exon regions; dsRNA corresponding to introns and promoter sequences did not cause RNAi. Finally, these workers demonstrated that the effect of the dsRNA crossed cell boundaries, at least in *C. elegans*. That is, the effect spread throughout the whole organism.

Is this loss of a particular mRNA in response to the corresponding dsRNA caused by repression of transcription of the gene or destruction of the mRNA? In 1998, Fire and colleagues, as well as others, demonstrated that RNAi is a post-transcriptional process that involves mRNA degradation. Several investigators reported the presence of short pieces of dsRNA called **short interfering RNA (siRNA)** in cells undergoing RNAi. In 2000, Scott Hammond and collaborators purified a nuclease from *Drosophila* embryos undergoing RNAi that digests the targeted mRNA. The partially purified preparation that contained this nuclease activity also contained a 25-nt RNA fraction that could be detected on Northern blots with probes for either the sense or antisense strand of the targeted mRNA. Degradation of the 25-nt RNA with micrococcal nuclease destroyed the ability of the preparation to digest the mRNA. These data suggested that a nuclease digests the trigger dsRNA into fragments about 25 nt long, and these fragments then associate with a nuclease and provide guide sequences that allow the nuclease to target the corresponding mRNA.

Phillip Zamore and collaborators developed a system based on *Drosophila* embryo lysates that carried out RNAi in vitro. This system allowed these workers to look at individual steps in the RNAi process. The embryos had been injected with trigger dsRNA corresponding to luciferase mRNA, so they targeted that mRNA for destruction. First, Zamore and collaborators showed that RNAi requires ATP. They depleted their extract of ATP by incubating it with hexokinase and glucose, which converts ATP to ADP and transfers the lost phosphate group to glucose. The ATP-depleted extract no longer carried out the degradation of the target, luciferase mRNA.

Next, these workers performed experiments in which they labeled one strand of the dsRNA at a time (or both) and showed that labeled short siRNAs of 21–23 nt appeared, no matter which strand was labeled (Figure 16.26). The appearance of the siRNAs did not require the presence of mRNA (e.g., compare lanes 2 and 3), so these short RNAs apparently derived from dsRNA, not mRNA. When capped antisense luciferase RNA was labeled (lanes 11 and 12),

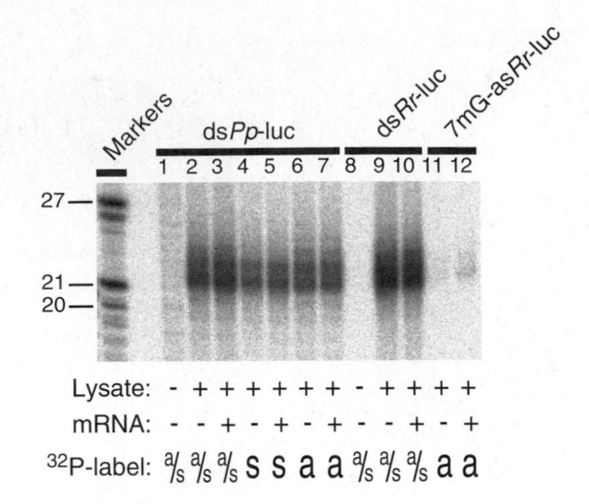

Figure 16.26 Generation of 21–23-nt RNA fragments in an RNAi-competent *Drosophila* embryo extract. Zamore and collaborators added ds luciferase RNA from *Photinus pyralis* (*Pp*-luc RNA) or from *Renilla reniformis* (*Rr*-luc RNA), as indicated at top, to lysates in the presence or absence of the corresponding mRNA, as indicated at bottom. The dsRNAs were labeled in the sense strand (s), in the antisense strand (a), or in both strands (a/s), as indicated at bottom. RNA markers from 17–27 nt long were included in the lane at left. Lanes 11 and 12 contained labeled, capped antisense *Rr*-luc RNA in the absence and presence of mRNA, respectively. (*Source:* Zamore, P.D., T. Tuschl, P.A. Sharp, and D.P. Bartel, RNAi: Double-stranded RNA directs the ATP-dependent cleavage of mRNA at 21 to 23 nucleotide intervals. *Cell* 101 (2000) f. 3, p. 28. Reprinted by permission of Elsevier Science.)

a small amount of siRNAs appeared, and that amount increased in the presence of mRNA (lane 12). This result suggested that the labeled antisense RNA was hybridizing to the added mRNA to generate a dsRNA that could be degraded to the short RNA pieces. In summary, all these results suggest that a nuclease degrades the trigger dsRNA into short pieces. Further work has shown that these siRNAs are about 21–23 nt long.

Next, Zamore and collaborators showed that the trigger dsRNA dictated where the corresponding mRNA would be cleaved. They added three different trigger dsRNAs, whose ends differed by about 100 nt, to their RNAi extracts, then added 5′-labeled mRNA, allowed RNA cleavage to occur, and electrophoresed the products. Figure 16.27 shows the results: The dsRNA (C) whose 5′-end was closest to the 5′-end of the mRNA yielded the shortest fragments; the next dsRNA(B), whose 5′-end was about 100 nt farther downstream, yielded mRNA fragments about 100 nt longer; and the third dsRNA, whose 5′-end was about another 100 nt farther downstream, yielded mRNA fragments about another 100 nt longer. This close relationship between the position of the trigger dsRNA relative to the mRNA, and the position at which cleavage began, strongly suggests that the dsRNA determined the sites of cleavage of the mRNA.

Next, Zamore and collaborators performed high-resolution gel electrophoresis of the mRNA degradation

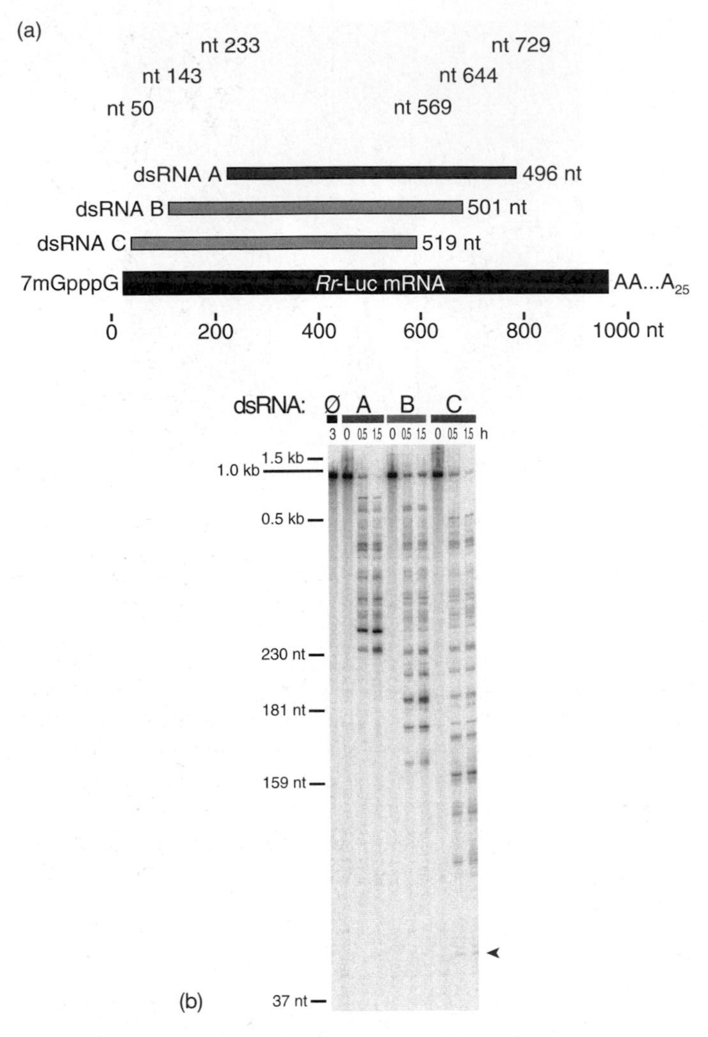

Figure 16.27 The trigger dsRNA dictates the boundaries of cleavage of mRNA in RNAi. Zamore and collaborators added the three dsRNAs pictured in panel **(a)** to an embryo extract along with an *Rr*-luc mRNA, 5′-labeled in one of the phosphates of the cap. **(b)** Experimental results. The 5′-end-labeled mRNA degradation products were electrophoresed. The dsRNAs included in the reactions are indicated and color-coded at top. The first lane, marked 0, contained no dsRNA. Reactions were incubated for the times (in h) indicated at top. The arrowhead indicates a faint cleavage site that lies outside the position of RNA C. Otherwise, the sites cleaved lie within the positions of the three dsRNAs on the mRNA. (*Source:* Zamore, P.D., T. Tuschl, P.A. Sharp, and D.P. Bartel, 2000. RNAi: Double-stranded RNA directs the ATP-dependent cleavage of mRNA at 21 to 23 nucleotide intervals. *Cell* 101 (2000) f. 5, p. 30. Reprinted by permission of Elsevier Science.)

products from Figure 16.27. The results, presented in Figure 16.28, are striking. The major cleavage sites in the mRNA are mostly at 21–23-nt intervals, producing a set of RNA fragments whose lengths differ by multiples of 21–23 nt. The one obvious exception is the site marked by an arrowhead, which lies only 9 nt from the previous cleavage site. This exceptional site lies within a run of seven uracil residues, which is interesting in light of the fact that 14 of 16 cleavage sites mapped were at uracils. After this exceptional site, the 21–23-nt interval resumed

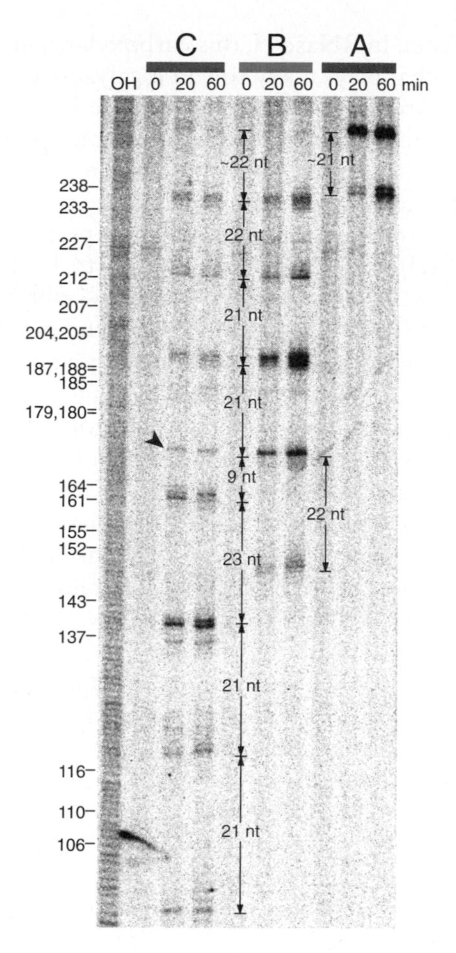

Figure 16.28 Cleavages of target mRNA in RNAi occur at 21–23-nt intervals. Zamore and collaborators performed high-resolution denaturing polyacrylamide gel electrophoresis on the products of RNAi in the presence of all three of the trigger dsRNAs from Figure 16.27. The cleavages, with one notable exception (arrowhead), occurred at 21–23-nt intervals. The exceptional band indicates a cleavage at only a 9-nt interval, but cleavages thereafter were at 21–23-nt intervals. (*Source:* Zamore, P.D., T. Tuschl, P.A. Sharp, and D.P. Bartel, RNAi: Double-stranded RNA directs the ATP-dependent cleavage of mRNA at 21 to 23 nucleotide intervals. *Cell* 101 (2000) f. 6, p. 31. Reprinted by permission of Elsevier Science.)

for the rest of the mapped cleavage sites. These results support the hypothesis that the 21–23-nt siRNAs determine where the mRNA will be cut and suggest that cleavage takes place preferentially at uracils.

In 2001, Hammond and colleagues reported that they had purified from *Drosophila* the enzyme that cleaves the trigger double-stranded RNA into short pieces. They named it **Dicer,** because it dices double-stranded RNA up into uniform-sized pieces. Dicer is a member of the RNase III family discussed earlier in this chapter. In fact, Hammond and colleagues narrowed their search for Dicer by looking for enzymes in this family because RNase III was the only known nuclease specific for dsRNA. Like RNase III, Dicer leaves 2-nt 3′-overhangs (protruding 3′-ends) at the ends of the double-stranded siRNAs, and phosphorylated 5′-ends.

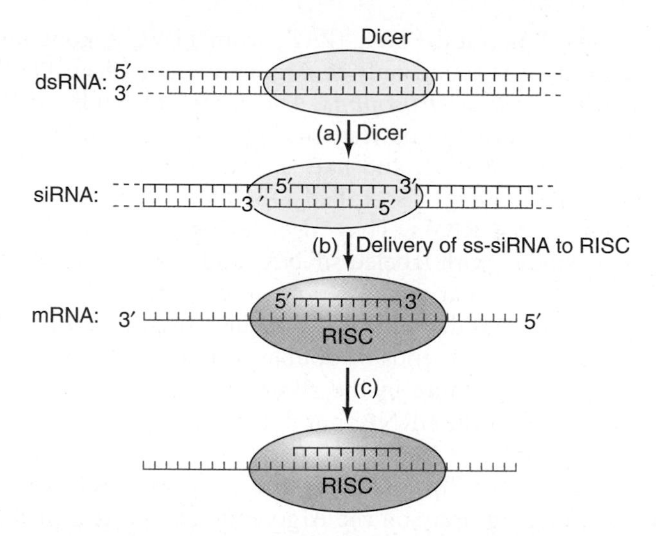

Figure 16.29 A simplified model for RNAi. (a) Dicer (yellow) recognizes and binds to a double-stranded RNA (red and blue), then cleaves the RNA into siRNAs about 21–23 nt long (depicted here as 10 nt long, for simplicity), with 2-nt 3′-overhangs. The ends of the central siRNA are labeled to illustrate the 3′-overhangs. **(b)** One of the siRNA strands (red) associates with RISC (orange) and base-pairs to a target mRNA (blue). **(c)** The siRNA strand in the RISC complex serves as a guide RNA to direct the cleavage of the target mRNA in the middle of the sequence opposite the siRNA.

Three early lines of evidence implicated Dicer in RNA cleavage in RNAi. First, *dicer,* the gene that encodes Dicer, produces a protein that can cut dsRNA into 22-nt pieces. Second, antibodies against this protein bind to an enzyme in *Drosophila* extracts that cuts dsRNA into short pieces. Finally, when *dicer* dsRNA is introduced into *Drosophila* cells, it partially blocks RNAi. It is ironic that Hammond and colleagues could use RNAi to block RNAi! But, of course, if you think about it, the blockage could never be complete.

Dicer also has RNA helicase activity, so it can separate the two strands of the siRNAs it creates, at least in principle. However, Dicer does not carry out the second step in RNAi, cleavage of the target mRNA. That appears to be the job of another enzyme, called **slicer,** which resides in a complex called the **RNA-induced silencing complex** (**RISC**). Figure 16.29 summarizes what we have learned so far about the mechanism of RNAi.

Hammond and others have implicated another *Drosophila* protein, **Argonaute,** known from genetic experiments to be required for RNAi, in the second (slicer) step. Argonaute does not have an RNase III motif, so molecular biologists discounted it at first as a slicer candidate. However, structural, biochemical, and genetic studies of Argonaute carried out by Leemor Joshua-Tor, Gregory Hannon, and their colleagues in 2004 showed that Argonaute almost certainly has slicer activity.

These workers had shown in structural studies in 2003 that Argonaute2 of *Drosophila* contains two characteristic domains, **PAZ**, and **PIWI**. (PAZ, from PIWI, Argonaute, and Zwili, was found only in Argonaute and Dicer; PIWI was discovered in *Drosophila*. The acronym stands for P-element-induced wimpy testis.) They had also determined the structure of PAZ, and had shown that it contained a module resembling a so-called OB fold, which can bind single-stranded RNAs. They also demonstrated by cross-linking studies with labeled siRNAs and cloned GST–PAZ fusion proteins that the PAZ domain was capable of binding to single-stranded siRNAs, or to the 2-nt single-stranded overhangs at the 3′-ends of double-stranded siRNAs. This implicated Argonaute in the slicer reaction, at least as a docking site for the siRNA, but not necessarily as the slicer enzyme itself.

Next, Joshua-Tor, Hannon, and colleagues performed x-ray crystallography on the Argonaute-like protein of the archaeon *Pyrococcus furiosus*. (No full-length eukaryotic Argonaute structure could be obtained.) They found that three domains of the protein (the middle domain, PIWI, and the N-terminal domain) form a crescent shape at the bottom of the structure, with the PIWI domain in the middle. The PAZ domain lies above the crescent and is connected to it by a stalk domain. Figure 16.30 depicts this structure, and illustrates that the crescent forms a groove, capped by the PAZ domain. This groove is big enough to accommodate a double-stranded RNA, and it is lined with basic residues, which could form electrostatic bridges to an RNA substrate.

However, the most telling part of the structure is that the PIWI domain resembles a similar domain in RNase H, which cleaves the RNA strand in an RNA–DNA hybrid. Thus, RNase H can recognize a double-stranded polynucleotide and cleave one of its strands (the RNA). In addition to their overall architectural similarities, both proteins have a cluster of three acidic residues (two aspartates and

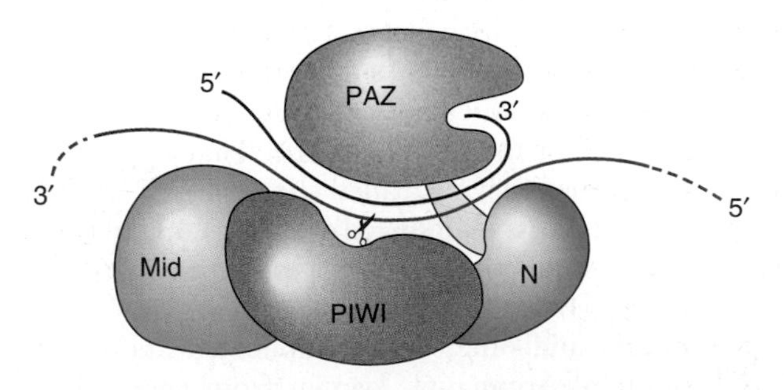

Figure 16.30 Model for slicer activity of Argonaute. The hybrid involving an siRNA and a target mRNA is held in the active site, at least partly due to the interaction between the 3′-end of the siRNA and the PAZ domain of Argonaute. This places the target mRNA in position to be cut by the slicer active site, represented by the scissors. Cleavage occurs opposite the middle of the siRNA, which serves as a guide RNA. The PAZ, middle, PIWI, and N-terminal domains of Argonaute are labeled. (*Source:* Adapted from Science, Vol. 305, Ji-Joon Song, Stephanie K. Smith, Gregory J. Hannon, and Leemor Joshua-Tor, "Crystal Structure of Argonaute and Its Implications for RISC Slicer Activity," Fig. 4, p. 1436, AAAS.)

one glutamate). In RNase H, this carboxylate cluster binds a Mg^{2+} ion that plays a key role in catalyzing the cleavage of the RNA strand. These similarities are very interesting because slicer has an analogous activity: It must also recognize a double-stranded polynucleotide (an siRNA–mRNA hybrid) and cleave one of its strands (the mRNA). Thus, Argonaute has all the attributes we expect of slicer: a domain (PIWI) with a site that appears to be capable of cleaving one strand of an siRNA–mRNA hybrid, and another domain (PAZ) that can bind to the end of the siRNA.

To investigate further the role of Argonaute in mammals, Hannon, Joshua-Tor, and colleagues performed genetic and biochemical studies on the Argonaute genes and proteins in the mouse. Mammals have four Argonaute proteins, designated Argonaute 1–4. The investigators transfected cells with genes encoding Argonautes 1–3, along with an siRNA that targets firefly luciferase mRNA. Then they immunoprecipitated the RISC complexes and tested them for ability to cleave luciferase mRNA in vitro. Only **Argonaute2 (Ago2)** had this capability.

Next, these workers knocked out the Ago2 gene in mice and observed that all such animals died in the embryonic stage of development, with severe developmental defects and delay. The reason for this profound phenotype is that Ago2 participates, not only in RNAi, but in a normal (and critical) developmental process involving microRNAs, which we will discuss later in this chapter. Furthermore, mouse embryo fibroblasts (MEFs) from wild-type cells showed normal RNAi, but MEFs from Ago2 knockout mice were defective in RNAi, as expected if Ago2 is important in RNAi.

All of the studies cited so far are consistent with the hypothesis that Ago2 has slicer activity, but none addressed this question directly. However, if Argonaute really has slicer activity, then mutating any of the three acidic amino acids at the putative active site should block cleavage of mRNA by RISC. Hannon, Joshua-Tor, and colleagues mutated each of the two key aspartate residues and found that either mutation abolished the RNAi-mRNA cleavage step both in vitro and in vivo. Taken together, all this evidence strongly implicates Ago2 as the slicer enzyme.

In 2005, Joshua-Tor and colleagues demonstrated definitively that human Ago2 really does have slicer activity. They reconstituted a minimal RISC with human recombinant Ago2 and an siRNA, which could accurately cleave a substrate RNA complementary to the siRNA. Figure 16.31 shows the results. The first siRNA (siRNA1) caused cleavage of the substrate RNA (S500) about 180 nt from its 3′-end, yielding a 3′-product about 180 nt long and a 5′-product about 320 nt long. The second siRNA (siRNA2) caused cleavage of the S500 about 140 nt from its 5′-end, yielding a 5′-product about 140 nt long and a 3′-product about 360 nt long. As expected, no products were produced in the absence of siRNA. Nor did products appear in the absence of Mg^{2+}, showing that a divalent metal ion is required for slicer activity.

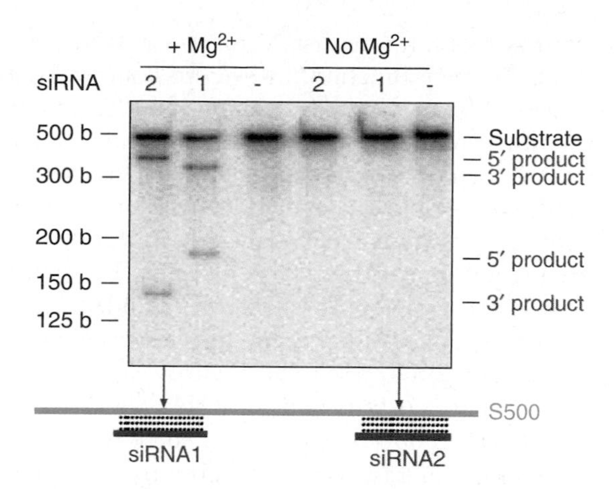

Figure 16.31 Ago2 plus an siRNA form a minimal RISC with slicer activity in vitro. Joshua-Tor and colleagues mixed recombinant human Ago2 (produced in bacteria) with either of two siRNAs that were specific for two different sites on a target 500-nt RNA, as shown at bottom. Then they added the labeled target RNA in the presence or absence of Mg^{2+} ions, as indicated at top. The siRNA used (either #1, or #2, or neither) is also indicated at top. Finally, they displayed the labeled RNA products by gel electrophoresis. Cleavage depended on Mg^{2+} and on an siRNA. The two siRNAs yielded different products, whose sizes were predicted from the known sites on the target RNA to which they hybridized. (*Source:* Reprinted from *Nature Structural & Molecular Biology*, vol 12, Fabiola V Rivas, Niraj H Tolia, Ji-Joon Song, Juan P Aragon, Jidong Liu, Gregory J. Hannon, Leemor Joshua-Tor, "Purified Argonaute2 and an siRNA form recombinant human RISC," fig. 1d, p. 341, Copyright 2005, reprinted by permission from Macmillan Publishers Ltd)

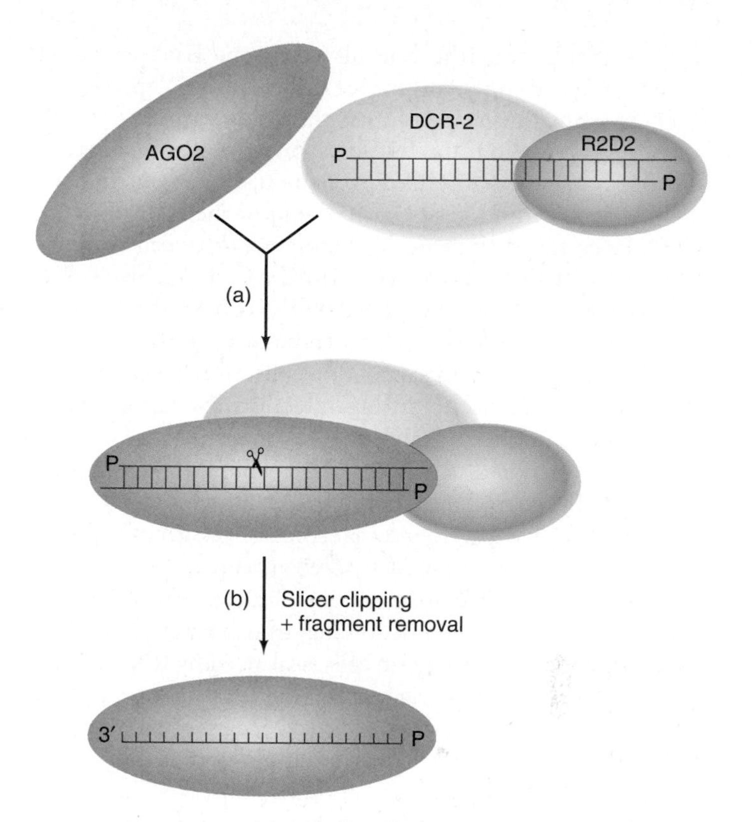

Figure 16.32 Delivery of single-stranded siRNA to RISC. The names of the proteins are from *Drosophila*, in which this process has been well studied. **(a)** Ago2 is attracted to a Dicer (DCR-2)-R2D2-dsRNA, forming a pre-RISC complex. The ds siRNA has already been created by DCR-2, leaving phosphorylated 5′-ends and 2-nt 3′-overhangs. **(b)** The slicer activity of Ago2 cuts the passenger strand (top) in half, weakening its base-pairing to the guide strand. The passenger strand fragments are lost, leaving the guide strand bound to Ago2, which is the catalytic center of the mature RISC. Other proteins besides Ago2 are part of mature RISC, though they are not shown here.

For mRNA cleavage to occur, a catalytically active RISC must form (Figure 16.32). We have seen that an Argonaute protein contains the slicer active site in a RISC, and we also know that a single-stranded siRNA must be present to serve as a guide to select mRNAs to degrade. So Ago2 plus siRNA constitutes a minimal RISC, at least in mammalian cells. But this complex does not form directly. Instead, siRNA must be delivered to Ago2 by a **RISC loading complex (RLC).** The composition of the RLC is presumed to include at least Dicer and a Dicer-associated protein, cutely-named **R2D2,** in addition to siRNA, and it could also include **Armitage,** which is essential for converting an RLC to a RISC in *Drosophila*.

What is the role of R2D2? It is not required for double-stranded siRNA formation, as Dicer can carry out this process efficiently without R2D2 in vitro. However, gel mobility shift and protein–RNA cross-linking experiments have shown that Dicer alone cannot retain contact with siRNAs once it has made them, but Dicer plus R2D2 can. Furthermore, R2D2 contains two double-stranded RNA-binding domains, and mutations in these domains render the Dicer–R2D2 complex incapable of binding double-stranded siRNAs. Thus, it appears that R2D2 is an essential part of the RLC because it can shepherd the siRNA between the time it is formed by Dicer and the time it is delivered to the RISC.

How are the two strands of the ds-siRNA separated to yield the ss-siRNA that ultimately associates with the RISC? An early hypothesis was that Armitage, which has RNA helicase activity, separated the two strands. However, that would require ATP, and the two RNA strands can be separated without ATP, at least in *Drosophila*. Figure 16.32 presents a model that incorporates that fact and other data. A complex composed of double-stranded siRNA plus Dicer (DCR-2 in *Drosophila*) and R2D2 attracts an Argonaute protein (Ago2 in *Drosophila*). Then Ago2 cleaves the **passenger strand** (the discarded strand) of the siRNA in the middle, weakening its grip on the **guide strand** (the strand that will associate with the RISC), so the passenger strand fragments are lost. This leaves a RISC active center composed of Ago2 and the siRNA guide strand.

What determines which strand is the guide strand, and which is the discarded passenger strand of the siRNA? This distinction is made in a complex that forms before the RLC, and contains Dicer and R2D2, each of which binds to an end of the double-stranded siRNA. The two proteins appear to bind asymmetrically, with Dicer associated with the less stable end (the one in which the base pairs are easiest to dissociate). And the strand with its 5′-end bound to Dicer is the one that becomes the guide strand.

X-ray crystallography studies on complexes between siRNAs and Argonaute-like proteins have shown that the siRNA guide strand binds with 3′-end in the PAZ domain. This places the active site of Argonaute between residues 10 and 11 of the siRNA, so the mRNA would be cleaved right in the middle of the siRNA–mRNA hybrid.

What is the physiological significance of RNAi? True double-stranded RNA does not normally occur in eukaryotic cells, but it does occur during infection by certain RNA viruses that replicate through dsRNA intermediates. So one important function of RNAi may be to inhibit the replication of viruses by degrading their mRNAs. But Fire and other investigators have also found that some of the genes required for RNAi are also required to prevent certain transposons from transposing within the genome. Indeed, Titia Sijen and Ronald Plasterk showed in 2003 that transposition of the Tc1 transposon in *C. elegans* germ cells is silenced by RNAi. What double-stranded RNA triggers this RNAi? It appears that transcription of the terminal inverted repeats of the transposon yields an RNA that can form a stem-loop structure, which is double-stranded in the stem portion. Thus, RNAi can protect cells not only against viruses, but also against transposition that can threaten the genomic integrity of germ cells.

RNAi can also silence transgenes and their genomic homologs. How is double-stranded RNA made from transgenes? It seems that some transcription of both strands of transgenes occurs, in contrast to the behavior of normal genes. This symmetric transcription yields enough double-stranded RNA to trigger RNAi.

Aside from its natural functions, RNAi has been a terrific boon to molecular biologists because it enables them to inactivate genes at will, simply by introducing double-stranded RNAs corresponding to the target genes. This process, known as **knockdown**, is usually much more convenient than the laborious process of producing knockout organisms, as described in Chapter 5. Also, it has not escaped the notice of the biotechnology industry that RNAi represents a potential bonanza. We know of many genes which, when overactive, can have devastating effects. For example, many **oncogenes** become hyperactive in various cancer cells, and that hyperactivity is what drives the cancer cells to lose control over their growth. RNAi directed against these oncogenes could control their activities, and thereby restore growth control to the cancer cells.

In spite of all this optimism, some caution is warranted because data began accumulating in 2004 that RNAi is not as exquisitely specific as had been thought. Genes that do not match the trigger double-stranded RNA perfectly are still targeted for repression to some extent. We do not know yet whether this nonspecificity will seriously compromise the effectiveness of RNAi in research and medicine.

Furthermore, if scientists want to use RNAi to investigate human gene function, or even to combat human disease, they will have to take account of another fact: Unlike in roundworms and fruit flies, the RNAi induced by adding dsRNA to mammalian cells is transient. But there is a way around this problem: Lasting RNAi can be induced by transforming mammalian cells with genes encoding RNAs with inverted repeats that form hairpins. These genes provide a continuous supply of double-stranded RNA in the form of hairpins, and that is enough to keep the RNAi process going. By 2004, researchers had already built libraries of genes encoding **short hairpin RNAs (shRNAs)** that targeted almost 10,000 human genes. These represent a valuable resource for research, and perhaps even intervention in human disease.

SUMMARY RNA interference (RNAi) occurs when a cell encounters dsRNA from a virus, a transposon, or a transgene (or experimentally added dsRNA), and results in destruction of the mRNA corresponding to the trigger dsRNA. The mechanism of RNAi in *Drosophila* is as follows: The trigger dsRNA is degraded into 21–23-nt fragments (siRNAs) by an RNase III-like enzyme called Dicer. The double-stranded siRNA, with Dicer and the Dicer-associated protein R2D2 recruit Ago2 to form a pre-RISC complex that can separate the siRNA into its two component strands: the guide strand, which will base-pair with the target mRNA in the RNA-induced silencing complex (RISC) and guide cleavage of the mRNA, and the passenger strand, which will be discarded. Ago2 cleaves the passenger strand, which then falls off the pre-RISC complex. The guide strand of the siRNA then base-pairs with the target mRNA in the active site in the PIWI domain of Ago2, which is an RNase H-like enzyme, also known as slicer. Slicer cleaves the target mRNA in the middle of the region of its base-pairing with the siRNA. In an ATP-dependent step, the cleaved mRNA is ejected from the RISC, which can then accept a new molecule of mRNA to be degraded.

Amplification of siRNA

One aspect of RNAi in some organisms, including plants and nematodes, has been difficult to explain: its great sensitivity. Just a few molecules of dsRNA can set in motion a process that totally silences a gene, not only in one cell, but in a whole organism—and even the descendants of that organism. This phenomenon led to the proposal that the process is catalytic. Indeed, Dicer does create many molecules of siRNA out of the trigger dsRNA and the target mRNA, but that seems insufficient to explain the power of

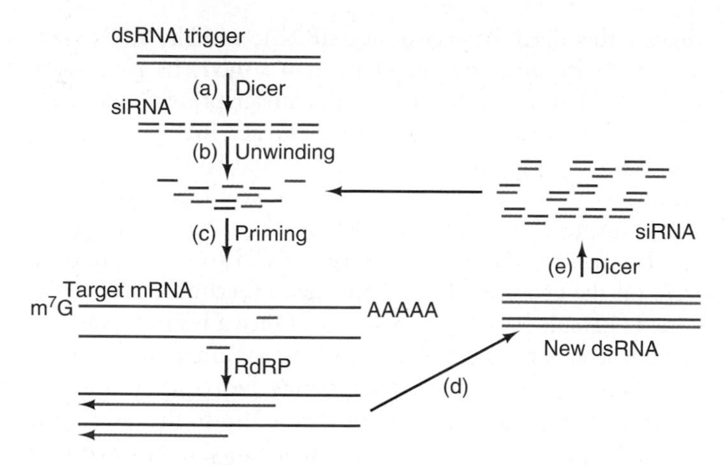

Figure 16.33 Amplification of siRNA. (a) Dicer chops up trigger dsRNA to make siRNA. **(b)** The antisense strands of siRNA hybridize to target mRNA. **(c)** RdRP uses the siRNA antisense strands as primers and target mRNA as template to make long antisense strands. **(d)** The product of step (c) is new trigger dsRNA. **(e)** Dicer chops up the new trigger dsRNA to make more siRNA, which can start a new round of priming and siRNA amplification. (*Source:* Adapted from Nishikura. *Cell* 107 (2001) f. 1, p. 416.)

RNAi in organisms like *C. elegans*. Fire and colleagues solved this riddle by showing that *C. elegans* cells employ an enzyme: **RNA-directed RNA polymerase** (**RdRP**) that uses antisense siRNAs as primers to make many copies of siRNA, as shown in Figure 16.33.

To test this hypothesis, Fire and colleagues used an RNase protection assay with a labeled sense strand probe to detect antisense siRNA in *C. elegans* fed on bacteria expressing trigger dsRNA at high levels. They used two different triggers and found large amounts of new siRNA produced in both cases. In addition, they discovered some **secondary siRNAs** outside the bounds of the trigger RNA. It is significant that these secondary siRNAs always corresponded only to the mRNA region upstream of the trigger sequence. This finding makes sense in the context of RdRP activity, because the trigger siRNA should prime synthesis toward the 5′ (upstream)-end of the mRNA. Thus, the discovery of secondary siRNAs also supports the hypothesis that an RdRP amplifies the siRNA, using the target mRNA as the template.

Thus, a mechanism does exist for amplifying the input dsRNA, and this could explain the great power of RNAi. The first round of this mechanism depends on priming by antisense siRNA on an mRNA template. This model can explain the earlier finding of Fire and colleagues that modification of the antisense, but not the sense, strand of the trigger dsRNA blocks RNAi. The model is also compatible with the earlier discovery of an RdRP in tomato cells, and the presence of homologous genes in fungi, and other plants, that are required for efficiency of RNAi.

SUMMARY In certain organisms, including *C. elegans*, siRNA is amplified during RNAi. This happens when antisense siRNAs hybridize to target mRNA and prime synthesis of full-length antisense RNA by an RNA-dependent RNA polymerase. This new dsRNA is then digested by Dicer into new pieces of siRNA.

Role of the RNAi Machinery in Heterochromatin Formation and Gene Silencing

In 2002, evidence began accumulating that implicated the RNAi machinery in heterochromatin formation and gene silencing, known as **transcriptional gene silencing** (**TGS**), as well as in RNAi itself. Then investigators found that siRNA-induced gene silencing can target a gene's control region through DNA and histone methylation.

RNAi and Heterochromatization Shiv Grewal, Robert Martienssen, and their colleagues deleted the RNAi genes encoding Dicer, Argonaute, and RdRP (*dcr1*, *ago1*, and *rdp1*, respectively) in the fission yeast *Schizosaccharomyces pombe* and found that all of these mutants were defective in the silencing that normally affects transgenes inserted near the centromere. That is, these transgenes became active in the RNAi mutants. Note that no trigger dsRNAs for the transgenes had been added, so RNAi was not directly involved in silencing the transgenes.

The investigators also looked to see whether the repeated DNA sequences (*cen3* sequences) at the centromere were transcribed in wild-type cells and in the mutants. Using Northern blots, they found no trace of such transcripts in wild-type cells, but they found three abundant transcripts in the RNAi mutants. A more detailed investigation using RNA dot blots showed that the reverse transcript of the *cen3* sequences appeared in wild-type and mutant cells, but the forward transcript appeared only in the mutants. Furthermore, nuclear run-on analysis demonstrated the same pattern: forward transcripts only in the mutants. Thus, the concentration of *cen3* transcripts is controlled at the transcriptional, rather than the post-transcriptional, level.

Next, the investigators examined specific core histone methylation in centromeric repeats using ChIP with antibodies against methylated histone H3 lysine 4 and lysine 9. As we learned in Chapter 13, methylated lysine 4 of histone H3 is associated with active genes, whereas methylated lysine 9 correlates with heterochromatin and gene inactivity. As expected from the activities we have already discussed, wild-type cells had lysines 4 and 9 that were both methylated in the centromeric region, but all three RNAi mutants showed an aberrant pattern of centromeric histone H3 methylation:

a high level of lysine 4 methylation, but a very low level of lysine 9 methylation. The same pattern was found in a *ura4+* transgene placed in the outermost centromere region (*otr*): a high level of lysine 9 methylation in wild-type cells, but a greatly depressed level in all three RNAi mutants.

Is RNAi responsible for histone methylation, and the resulting heterochromatization at the centromere? If so, we would expect at least some RNAi proteins to interact with centromeric chromatin, and we would also expect to find siRNAs corresponding to centromeric RNA. Martienssen and colleagues did indeed find that the Rdp1 part of the RNAi machinery binds to centromeric chromatin. And B.J. Reinhard and David Bartel had already found evidence to support the second prediction of the hypothesis when they cloned apparent Dicer products from wild-type cells and showed that all 12 clones came from transcripts of the centromeric region.

Thus, at least one component of the RNAi machinery is found at the centromere, and siRNAs are made from centromeric transcripts. All these data, and more, led Martienssen and colleagues to propose that RNAi is involved in heterochromatic silencing at the centromere (Figure 16.34). In particular, they proposed that the abundant reverse transcripts of the *otr* region base-pair with forward transcripts produced occasionally by RNA polymerase II, or perhaps by RdRP, to form trigger dsRNA. Dicer then digests this dsRNA to produce siRNA, and the siRNA associates with an Argonaute1 protein (**Ago1**) in a complex called **RITS** (for RNA-induced transcriptional silencing complex). This complex can then attract RdRP in a complex known as RDRC (for RNA-directed RNA polymerase complex) which amplifies the double-stranded siRNA. By base-pairing either to the DNA directly or to transcripts of the DNA, the siRNA then escorts RITS to corresponding sites on the genome. RITS then causes recruitment of a histone H3 lysine 9 methyltransferase. Once a lysine 9 is methylated, it can recruit Swi6, which is required for forming heterochromatin. Other proteins may be required, but the end result is spreading of heterochromatin to the *otr* region of the centromere. Whatever the mechanism, it is likely to be highly conserved, because mammalian pericentromeric heterochromatin structure also involves histone H3 lysine 9 modification and some RNase-sensitive substance, which could be one or more of the RNAi intermediates.

Does the RITS complex associate directly with DNA, or is it attracted by transcripts of chromatin regions that are targeted for silencing? In 2006, Danesh Moazed and colleagues provided evidence for the importance of transcripts in this process by showing that artificially tethering RITS to a nascent transcript of the $ura4^+$ gene resulted in silencing of this normally active gene.

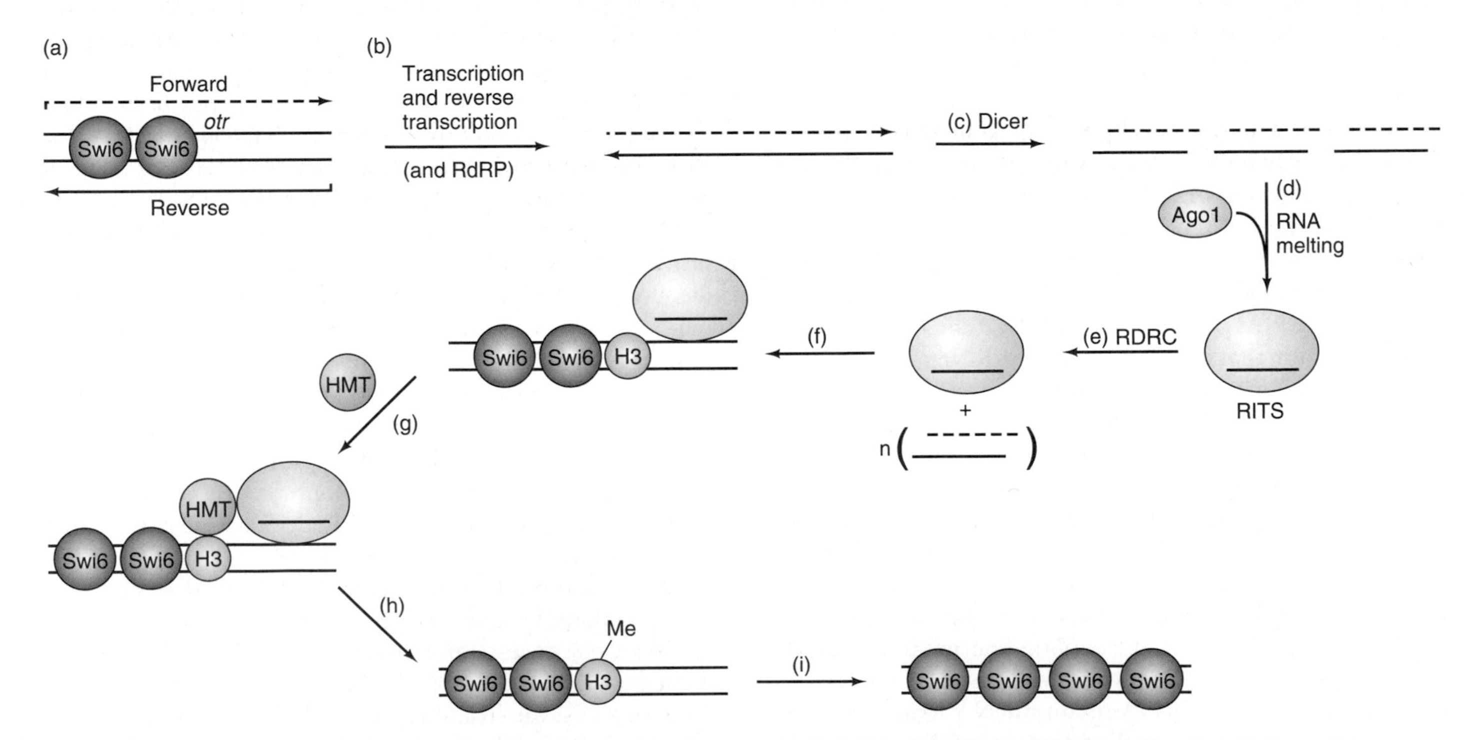

Figure 16.34 A model for the involvement of the RNAi machinery in the heterochromatization at the *S. pombe* centromere. (a) The outermost region *(otr)* of the centromere is constantly being transcribed to produce reverse transcripts, and production of forward transcripts probably also occurs at a low (undetectable) level. **(b)** After transcription and reverse transcription (or after reverse transcription and RdRP action), we have double-stranded RNA (dsRNA). **(c)** Dicer cuts the dsRNA into siRNAs. **(d)** Ago1 (yellow, perhaps along with other proteins) associates with single-stranded siRNAs to produce the RITS. **(e)** The RdRP in the RDRC amplifies the siRNA, producing double-stranded siRNAs. **(f)** The RITS, through its siRNA, associates with the *otr,* either through direct interaction with the DNA, or through interaction with transcripts in this region. **(g)** The RITS attracts a histone methyltransferase (HMT, green) to the *otr.* **(h)** The HMT methylates the lysine 9 of a histone H3 (blue). Of course, this histone is part of a nucleosome, which is not shown here, for simplicity. **(i)** This methylation in turn attracts more Swi6 (red), which helps to spread heterochromatization.

It seems paradoxical that, in order for a region like a centromere to be silenced, it has to be expressed. How, then, does expression occur after mitosis to preserve heterochromatization in the genomes of both progeny cells? A solution to this paradox was proposed by Rob Martienssen and colleagues and Grewal and colleagues in 2008. Together, the work of these two groups showed that serine 10 of histone H3 in centromeric heterochromatin in *S. pombe* becomes phosphorylated during mitosis, and that this results in the loss of methylation of lysine 9 of histone H3, and therefore in the loss of the Swi6 protein that is necessary for heterochromatization. As a result, the chromatin opens up enough that it is transcribed during the S phase. This produces centromere transcripts, presumably in both directions, that attract the RNAi machinery, so the centromere can be heterochromatized again during the ensuing long G_2 phase.

This hypothesis views heterochromatin as more dynamic than the traditional view of a static, condensed, inactive structure. Does it also open up the possibility of real expression of centromeric DNA? Apparently not. For one thing, centromeric transcription is confined to the S phase, in which gene expression is very restricted. For another, the centromeric transcripts are rapidly degraded, either by the RNAi machinery, or by other RNA-degrading systems that recognize aberrant transcripts.

Grewal and colleagues noted that centromere-like sequences are also found at sites such as the silent mating-type region, which lies far from the centromere but is also silenced by heterochromatization. In separate experiments, these workers showed that the RNAi machinery is required for initiating heterochromatization at the silent mating-type region, but is expendable for maintaining and inheriting the silencing. Swi6 is apparently sufficient for such heterochromatin maintenance.

The role of the RNAi machinery in centromeric events is not confined to lower organisms. In 2004, Tatsuo Fukagawa and colleagues reported tests on a chicken–human hybrid cell line whose only human chromosome was chromosome 21. These workers then made the Dicer gene tetracycline-repressible in these hybrid cells and observed what happened, particularly to human chromosome 21, when Dicer expression was blocked by tetracycline. The most obvious effect of the loss of Dicer was that the cells died after about five days.

Moreover, the specific pathologies of these cells point to problems with the centromere: The cells showed abnormal mitoses with evidence of premature sister chromatid separation. As in yeast cells with defective RNAi, these vertebrate cells exhibited abnormal buildup of transcripts of the centromeric repeat region of human chromosome 21. They also showed abnormal localization of some, but not all, centromeric proteins. The problems at the centromere were presumably caused by the loss of Dicer, and this in turn led to the failure of cell division and to cell death.

We assume that the events that occur in the centromeric region in fission yeast, illustrated in Figure 16.34, help to explain these results in cells from higher organisms. However, one caveat to bear in mind is that mammals appear to lack an RdRP. So any dsRNA that appears at the centromere in mammals must be made by bidirectional transcription of this region, or of a homologous region elsewhere in the genome.

Another major difference between heterochromatization in fission yeast and in plants and mammals is that the latter organisms experience DNA methylation in addition to histone methylation. The methyl groups are added to the C's of **CpG sequences** in both strands, and these help to attract the proteins that induce heterochromatization. Again, the presence of double-stranded RNA appears to play a key role by recruiting the RNAi machinery, which stimulates DNA methylation.

One significant advantage of this mechanism is that it is permanent. Once the DNA is methylated on the C's of both strands of a CpG sequence, this methylation is inherited from one cell generation to the next, as the methylated C on one strand ensures that the new C on the opposite strand will also be methylated after DNA replication. Although this methylation is permanent, it is not a true genetic change, which would be a change of one base to another (e.g., a C changed to a T). Instead, we call it an **epigenetic** modification of the DNA. It is every bit as important as a genetic change because it can cause the silencing of a gene or even heterochromatization of a whole region of a chromosome.

RNAi may also play a role in X chromosome inactivation in mammals. In each cell of a female mammal, one of the X chromosomes is inactivated by heterochromatization. This prevents the very deleterious consequences of elevated levels of X chromosome products. One of the first steps in X chromosome inactivation is histone H3 lysine 9 methylation. And this methylation occurs immediately after the appearance of a noncoding transcript of the *Xist* locus. We also know that *Xist* is controlled by the antisense RNA, *Tsix,* and by *Xist* promoter methylation. The presence of *Tsix* and *Xist* transcripts in the same cell would of course invoke the RNAi system, and that could recruit the histone methylase that kicks off the formation of heterochromatin.

SUMMARY The RNAi machinery is involved in heterochromatization at yeast centromeres and silent mating-type regions and is also involved in heterochromatization in other organisms. At the outermost regions of centromeres of fission yeast, active transcription of the reverse strand occurs. Occasional forward transcripts, or forward transcripts made by RdRP, base-pair with the reverse transcripts to kick off RNAi, which in turn recruits a histone methyltransferase, which methylates lysine 9 of histone H3, which recruits Swi6, which causes heterochromatization. In plants and mammals, this process is abetted by DNA methylation, which can also attract the heterochromatization machinery.

Transcriptional Gene Silencing Induced by siRNA Directed at a Gene's Control Region Kevin Morris and colleagues found in 2004 that mammalian genes can also be silenced by the RNAi machinery and, as we have seen with heterochromatization in plants and mammals, this silencing involves DNA methylation. Furthermore, in contrast to normal RNAi, this silencing involves an siRNA directed at the control region, rather than the coding region, of a gene.

Morris and colleagues targeted a green fluorescent protein reporter gene driven by the human elongation factor 1α gene (EF1A) promoter-enhancer region. They transduced human cells with feline immunodeficiency virus (FIV) containing this reporter construct, which caused integration of the reporter gene and its control region into the human genome. The FIV vector also made the nuclear membrane permeable to the siRNA, which otherwise would not have been taken up by the mammalian nuclei.

Because the siRNA in this case was directed against the gene's control region, and not its coding region, we would predict that it could not cause mRNA destruction or block translation. Indeed, we would predict that it would block transcription, and indeed that is what Morris and colleagues showed. Using real-time RT-PCR (Chapter 4), they demonstrated almost total disappearance of the GFP transcript upon transducing cells with the EF52 siRNA, which targets the control region of the fusion gene. By contrast, an siRNA that targets the coding region of the GFP mRNA caused a relatively modest 78% reduction in the concentration of the GFP transcript (Figure 16.35a).

Because a common feature of transcriptional silencing in mammals is histone and DNA (cytosine) methylation, Morris and colleagues tested the effect of trichostatin (TSA) and 5-azacytidine (5-azaC), which inhibit histone and DNA methylation, respectively. These drugs completely reversed the silencing caused by the EF52 siRNA, but had no effect on silencing caused by the GFP coding region siRNA. These results supported the hypothesis that DNA and/or histone methylation are involved in silencing caused by the EF52 siRNA.

To check whether the silencing by the EF52 siRNA was at the transcription level, Morris and colleagues performed nuclear run-on assays (Chapter 5). Figure 16.35b shows that EF52 did indeed dramatically reduce the number of initiated GFP transcripts, while it had no effect on irrelevant glyceraldehyde-phosphate dehydrogenase (GAPDH) transcripts.

To see whether DNA in the gene's control region was really methylated during transcriptional silencing, Morris and colleagues used *Hin*P1I, a restriction enzyme that cuts at a site that includes a CpG. If the C in this sequence is unmethylated, *Hin*P1I will cut, but if it is methylated it will not. There is a *Hin*P1I site in the control region of the EF1A gene. Thus, if this site is methylated, it will be protected from *Hin*P1I cleavage, and PCR using primers on opposite sides of the site will produce a product. On the other hand, if the site is unmethylated, *Hin*P1I will cut it, and no PCR product will appear.

Figure 16.36 shows the results of this experiment. The control in lane 1 shows that a plasmid with a *Hin*P1I site methylated in vitro really does yield a PCR product, even after attempted cleavage with *Hin*P1I. Lanes 2 and 3 are controls with DNA from cells that had been transduced with an irrelevant siRNA or a GFP coding region siRNA,

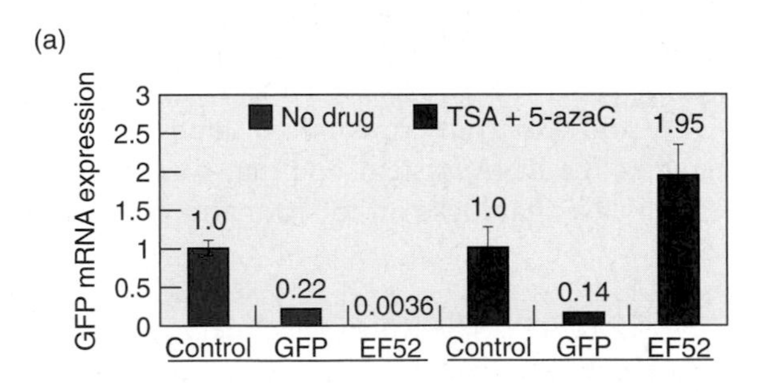

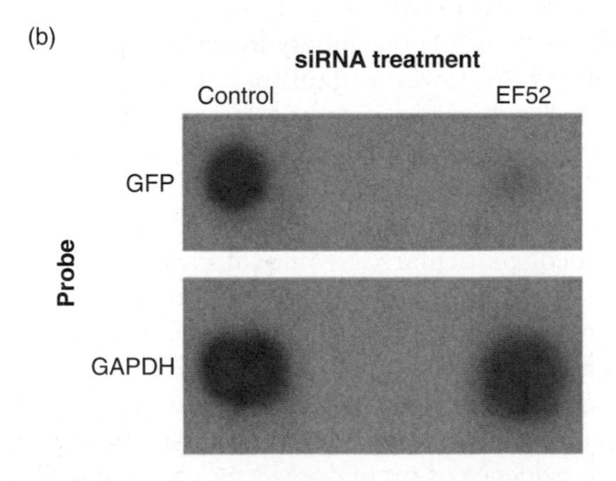

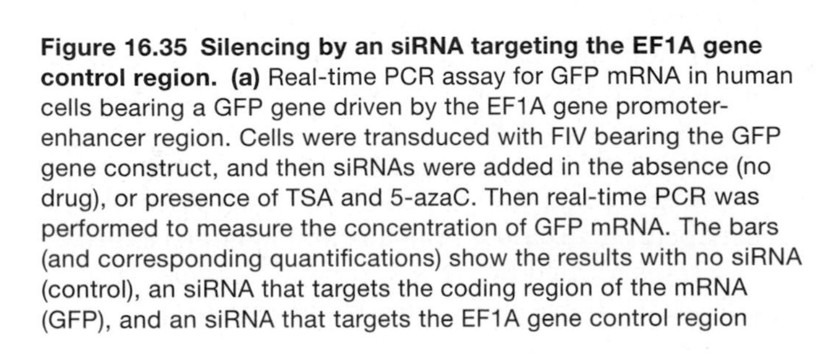

Figure 16.35 Silencing by an siRNA targeting the EF1A gene control region. **(a)** Real-time PCR assay for GFP mRNA in human cells bearing a GFP gene driven by the EF1A gene promoter-enhancer region. Cells were transduced with FIV bearing the GFP gene construct, and then siRNAs were added in the absence (no drug), or presence of TSA and 5-azaC. Then real-time PCR was performed to measure the concentration of GFP mRNA. The bars (and corresponding quantifications) show the results with no siRNA (control), an siRNA that targets the coding region of the mRNA (GFP), and an siRNA that targets the EF1A gene control region (EF52). **(b)** Nuclear run-on assay for transcription. Nuclei were isolated from cells transduced with the EF1A-GFP construct, plus either the EF52 siRNA or no siRNA (control). Labeled nuclear run-on mRNA was synthesized and hybridized to blots of GFP DNA, or GAPDH DNA, as indicated at left. The EF52 siRNA silenced the GFP gene, but not the GAPDH gene, at the transcriptional level. (*Source:* Reprinted with permission from *Science,* Vol. 305, Kevin V. Morris, Simon W.-L. Chan, Steven E. Jacobsen, and David J. Looney, "Small Interfering RNA-Induced Transcriptional Gene Silencing in Human Cells," Fig. 1, p. 1290, Copyright 2004, AAAS.)

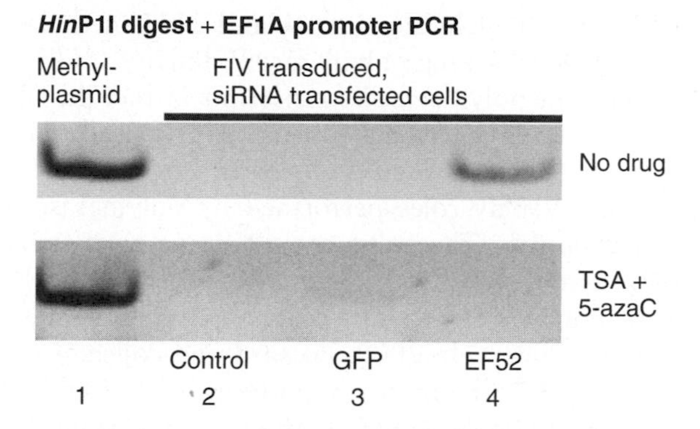

Figure 16.36 Demonstration of methylation of the EF1A gene control region in response to siRNA. Morris and colleagues tested for methylation of a CpG sequence in the EF1A control region by cleavage with *Hin*P1I, which cleaves unmethylated, but not methylated sites including CpG sequences. They performed the cleavage on DNA from cells either untreated (top row, "No drug") or treated (bottom row) with TSA plus 5-azaC to block methylation of CpG sequences. After treatment with *Hin*P1I, they performed PCR with primers flanking the CpG site. Only uncut (methylated) DNA should yield a signal. Lane 1, positive control with synthetically methylated site. Lane 2, negative control with irrelevant siRNA. Lane 3, negative control with an siRNA directed against the GFP coding region, rather than the control region. Lane 4, experimental result with an siRNA that targets the control region. With this siRNA, the CpG is methylated (uncut, and therefore yields a PCR signal) in the absence of drug, but is not methylated when the methylation blocker was included. (*Source:* Reprinted with permission from *Science,* Vol. 305, Kevin V. Morris, Simon W.-L. Chan, Steven E. Jacobsen, and David J. Looney, "Small Interfering RNA-Induced Transcriptional Gene Silencing in Human Cells," Fig. 1, p. 1290, Copyright 2004, AAAS.)

respectively. Lane 4 shows the results with cells transduced with the EF52 siRNA. The top row shows that the DNA must have been methylated, because it was protected from *Hin*P1I cleavage, and a PCR product appeared. However, the bottom row shows that the methylation-blocking drugs TSA and 5-azaC, blocked methylation, rendering the *Hin*P1I site cleavable, so no PCR product appeared.

All of the experiments described so far used cells that were transduced with FIV, which inserted the EF1A gene into the human genome, but not in its natural location. To check for siRNA silencing of the endogenous human gene, Morris and colleagues performed the same kinds of experiments as in Figures 16.35 and 16.36, but with cells rendered permeable to siRNAs with MPG, a fusion peptide that contains an HIV-1 transmembrane peptide linked to the nuclear localization signal from SV40 virus. In these experiments, no EF1A gene was introduced into the cells, so only the endogenous gene was present, and it was silenced (though not as dramatically as in the previous experiments) by the EF52 siRNA. As before, this silencing was accompanied by DNA methylation, and could be blocked by methylation inhibitors.

Where does the siRNA in these experiments come from? After all, it is directed at the control region, not the coding region, of the gene, so it cannot come from a normal gene transcript. Morris and colleagues showed that the sense strand part of the siRNA probably came from a 5′-extended transcript of the EF1a gene—that is, a transcript that started in the promoter, upstream of the normal transcription start site. They detected this extended transcript with an **RNA pull-down** procedure that used a 5′-biotin-labeled promoter antisense RNA and avidin bound to magnetic beads. The biotin-labeled promoter antisense RNA hybridized in vivo to the RNA transcribed through the promoter region, and the avidin-tagged beads bound to the biotin, allowing the whole RNA-RNA-bead complex to be isolated ("pulled down") magnetically.

Quantification of the promoter-associated RNA and the normal EF1a transcripts by real-time RT-PCR yielded a ratio of about 1:570. Thus, about one in 570 transcripts of the EF1a gene begins within the promoter. A 5′-RACE procedure (Chapter 5) showed that these promoter-associated transcripts begin about 230 bp upstream of the normal transcription start site, and a 3′-RACE procedure showed that these transcripts extend as far in the 3′-direction as the normal transcripts and are spliced and polyadenylated.

Does the promoter-associated RNA play a role in transcriptional gene silencing (TGS)? To answer this question, Morris and colleagues targeted the promoter-associated RNA for destruction by RNase H (Chapter 14), by transfecting cells with a promoter-associated RNA-specific phosphorothioate oligonucleotide, which acts like a deoxyribo-oligonucleotide in this procedure. The destruction of the EF1a promoter-associated RNA abolished transcriptional silencing by added promoter-associated siRNA. By contrast, RNase H-mediated destruction of a promoter-associated RNA from another gene (CCR5) had no effect on TGS of the EF1a gene. Thus, a promoter-associated RNA appears to be essential for TGS.

One of the epigenetic changes that occurs in the EF1a control region during gene silencing is a trimethylation of lysine 27 of histone H3 (H3K27me3) in a nucleosome at that site. Does the promoter-associated RNA play a role in this epigenetic change? A pull-down assay showed that it does. When the EF1a promoter-associated RNA was destroyed by oligonucleotide and RNase treatment, the chromatin could no longer be precipitated with an anti-H3K27me3 antibody. On the other hand, treatment with the irrelevant oligonucleotide directed at the CCR5 control region did not block precipitation of the EF1a promoter-associated nucleosome with an anti-H3K27me3 antibody.

Thus, the presence of the promoter-associated RNA is required for the silencing methylation of H3K27. The exact nature of that requirement is still unclear, but one can imagine that the promoter-associated RNA would hybridize to an antisense RNA (perhaps the antisense strand of an siRNA). This hybrid would in turn recruit a chromatin remodeling complex, including the H3K27 methyltransferase, which would trimethylate H3K27, helping to silence the gene.

All of the silencing we have discussed so far is due to epigenetic modification (usually methylation) of chromatin.

Another silencing mechanism targets nuclear RNA: Endogenous double-stranded siRNAs can enter the nucleus and cause degradation of nuclear RNAs by the familiar RNAi mechanism. Scott Kennedy and colleagues showed in 2008 that siRNAs bind to an Argonaute protein (NRDE-3 in *C. elegans*) in the cytoplasm. NRDE-3 has a **nuclear localization signal** that targets it to the nucleus, so the siRNA-NRDE-3 complex can enter the nucleus and collaborate in the destruction of cognate nuclear pre-mRNAs. Note that the nuclear location distinguishes this mechanism from ordinary RNAi, which occurs in the cytoplasm.

SUMMARY Individual genes in mammals can also be silenced by an RNAi mechanism that targets the control region, rather than the coding region, of the gene. This silencing process involves DNA and histone methylation, rather than mRNA destruction. One requirement for such histone methylation in siRNA-induced gene silencing, at least in some genes, is production of a 5′-extended transcript that begins within the gene's control region (a promoter-associated transcript). This transcript presumably associates with an antisense RNA, and then recruits a chromatin remodeling complex, including a histone methyltransferase, which methylates H3K27 on a nearby nucleosome, helping to silence the gene. Genes can also be silenced by a nuclear RNAi process that involves Argonaute proteins that are targeted to the nucleus by a nuclear localization signal.

Transcriptional Gene Silencing in Plants The short RNAs required for TGS in fission yeast and animals are made by RNA polymerase II. But in TGS in flowering plants, two other polymerases, **RNA polymerase IV** and **RNA polymerase V,** which are evolutionarily derived from polymerase II, play the key roles. Polymerase IV produces the 24-nt heterochromatic siRNAs whose yeast and animal counterparts are made by polymerase II. The role of polymerase V is more subtle, and was therefore more difficult to unravel.

Polymerase V produces transcripts of non-coding regions that are more than 200 nt long, have either caps or triphosphates at their 5′-ends, and are not polyadenylated. Transcripts in a given region have multiple 5′-ends, which suggests they are made in a promoter-independent manner. In 2008, Craig Pikaard and colleagues demonstrated the involvement of polymerase V in transcriptional gene silencing by mutating the largest subunit of the enzyme. They observed, in addition to loss of polymerase V activity, loss of transcripts of certain non-coding regions, and defective silencing in overlapping and adjacent chromatin regions. Furthermore, they found that some of the hallmarks of heterochromatin, including histone and DNA methylation, were lost in cells lacking polymerase V activity.

How do the polymerase V transcripts attract the silencing machinery? Pikaard and colleagues proposed a model very similar to that in Figure 16.34, except that polymerases IV and V play roles performed by polymerase II in fungi and animals. The polymerase V transcripts attract a complex composed of Argonaute 4 (Ago4) and siRNA (made by polymerase IV). This complex in turn attracts the silencing machinery. In 2009, Pikaard and colleagues provided more support for this hypothesis, as follows. First, they performed ChIP analysis with chromatin from *Arabidopsis* plants that produce mutant Ago4 and polymerase V. They found that both wild-type Ago4 and polymerase V bound to transposon genes that are normally silenced, but mutations in either the *Ago4* gene or the *nrpe1* gene, which encodes the largest polymerase V subunit, abolished this association. Thus, Ago4 and polymerase V are necessary for Ago4 to associate with chromatin that is to be silenced.

To test whether polymerase V transcripts are required to recruit Ago4 to chromatin, Pikaard and colleagues performed ChiP analysis in wild-type plants, and in plants bearing a mutation at the active site of the largest subunit of polymerase V. The mutant polypeptide is stable and can still bind normally to the second-largest subunit, but it is utterly incapable of making transcripts. ChIP analysis showed no binding of Ago4 to target chromatin sites in the mutant plants. This binding could be restored by transforming plants with the wild-type *nrpe1* gene, but not with the mutant gene. Thus, transcription by polymerase V is required to recruit Ago4, in accord with the hypothesis.

It is important to note that polymerase V transcripts are found throughout the genome of *Arabidopsis thaliana,* a member of the mustard family, in heterochromatic and euchromatic regions alike. How then do the euchromatic regions avoid silencing? Pikaard and colleagues proposed that polymerase V transcripts are necessary, but not sufficient, for silencing. The silencing process also requires siRNAs. Therefore, because euchromatic regions do not give rise to siRNAs, they are not silenced.

Earlier in this chapter, we discussed the paradox that silenced chromatin must be transcribed in order to be silenced. The existence of polymerases IV and V gives flowering plants a way to deal with this problem: These polymerases appear not to initiate at promoters, and they are not subject to the same rules as polymerase II. Thus, they can presumably initiate transcription even in chromatin regions that are silenced with respect to polymerase II.

SUMMARY Flowering plants have two nuclear RNA polymerases, polymerase IV and polymerase V, that are not found in animals and fungi. Polymerase IV makes siRNAs corresponding to chromatin regions to be silenced. Polymerase V makes longer RNAs from regions throughout the plant genome. These longer RNAs attract siRNA-Ago4 complexes, but only to regions that are targets for silencing, from which these siRNAs were made. These complexes in turn attract the enzymes required to methylate both DNA and histones, which in turn leads to heterochromatization.

16.7 Piwi-Interacting RNAs and Transposon Control

In Chapter 23 we will learn that DNA elements known as transposons can transpose, or jump from place to place in a genome. In doing so, they can interrupt and inactivate genes, or even break chromosomes. Thus, transposition is a dangerous process that can lead to cell death or disease, such as cancer. Accordingly, it is important that cells be able to control transposition. This is particularly true in germ cells, which give rise to gametes that will pass genes on to the next generation. The serious mutations or cell death caused by transposition in germ cells reduce reproductive success and therefore threaten a species's survival.

It is not surprising, therefore, that organisms have evolved mechanisms for dealing with transposons, and that these can be targeted to germ cells. In fact, germ cells produce another class of small RNAs, 24 to 30 nt long, called **Piwi-interacting RNAs (piRNAs)**. Like siRNAs and miRNAs, piRNAs associate with Argonaute proteins, but these proteins belong to a different branch, or clade, of the Argonaute superfamily than the Ago proteins we have been discussing. The piRNAs bind to members of the Piwi clade, while siRNAs and miRNAs bind to members of the Ago clade.

The piRNAs of fruit flies and mammals tend to be complementary to either the sense or antisense strand of transposons from the same organism. These piRNAs derive from clusters of piRNA genes, apparently via transcription of a long cluster and subsequent processing of the precursor RNA into mature piRNAs. Some, if not most, of this processing may actually occur simultaneously with inactivation of transposons, by a so-called ping-pong amplification loop, as follows (Figure 16.37):

In *Drosophila,* Piwi proteins such as **Piwi** and **Aubergine** tend to associate with piRNAs that are complementary to transposon mRNAs; these piRNAs usually have a U in the first position. This piRNA-Piwi or -Aubergine complex can associate through base-pairing with a transposon mRNA, which triggers slicer cutting 10 nt upstream of an A that is base-paired to the U at the 5′-end of the piRNA. This cut, together with processing at the 3′-end of the transposon mRNA, creates a short RNA that can associate with another protein, Ago3, which preferentially binds to RNAs that represent parts of transposon mRNAs. The RNA-Ago3 complex can then bind to a piRNA precursor RNA by base-pairing, and the slicer activity of Ago3 cuts just upstream of the U of the A–U base pair. This cut, together with end processing of the piRNA precursor, creates a mature piRNA that can bind to Piwi or Aubergine to start the cycle over.

Note that this mechanism accomplishes two things: It slices up transposon mRNA, thereby blocking transposition, and it amplifies the amount of piRNA available, thus stimulating the process. Because the transcription of piRNA clusters is confined to germ cells, and somatic cells immediately

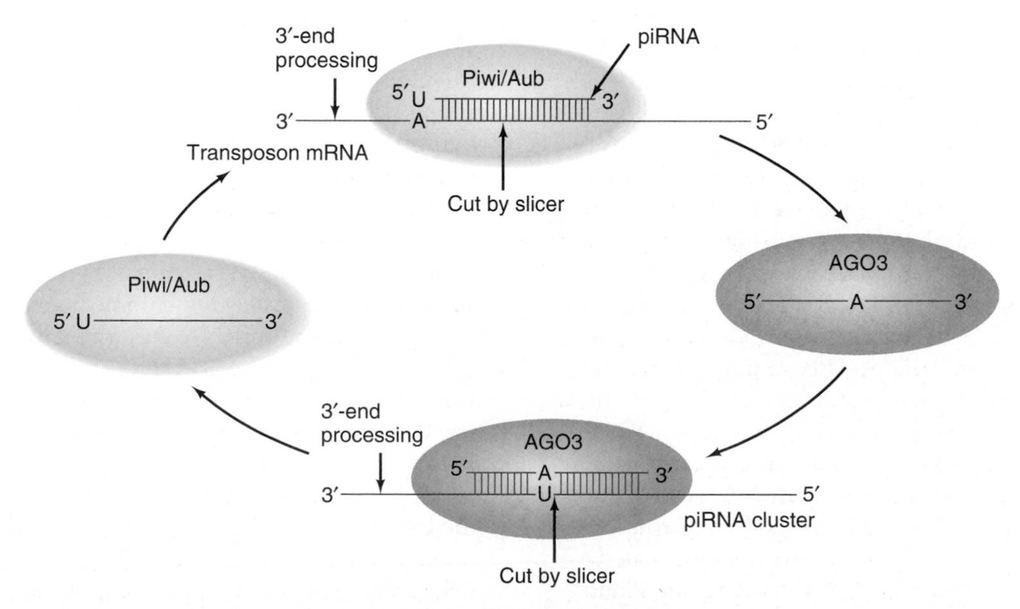

Figure 16.37 Model for a ping-pong amplification loop for piRNAs. Details are in the text.

surrounding the germ cells, transposition is specifically blocked in germ cells, where it would be especially dangerous.

Animal somatic cells do not produce piRNAs, so transposons must be inactivated by another mechanism in these cells. Phillip Zamore and colleagues showed in 2008 that *Drosophila* somatic cells produce endogenous siRNAs complementary to transposon mRNAs (and to some normal cellular mRNAs). These endogenous siRNAs are distinguished from miRNAs, which we will discuss later in this chapter, by two features: They contain a 2′-O-methylation at their 3′-ends; and they have a very narrow size distribution centered on 21 nt. Furthermore, they are not derived from stable stem-loop precursors, as miRNAs are. These endogenous siRNAs are also unlike piRNAs in that they have no tendency to begin with U or to have an A at position 10. Thus, *Drosophila* somatic cells use an endogenous RNAi mechanism, rather than a piRNA-based mechanism, to control transposition. Furthermore, although animal germ cells have the piRNA pathway to inactivate transposons, they also appear to produce endogenous siRNAs directed against at least some transposons, so they can bring at least two different mechanisms to bear on the transposon problem.

Plants lack Piwi proteins, so they must use a different pathway to produce and amplify RNAs complementary to transposon mRNAs. *Arabidopsis* cells produce short RNAs from transposons by an unknown mechanism, and these RNAs bind to the Ago protein Ago4. Without Piwi proteins to produce complementary RNAs by an amplification loop, these complementary RNAs are made by RNA-dependent RNA polymerases (see previous section). The short RNAs complementary to both strands of a transposon can anneal to form a trigger dsRNA that initiates destruction of transposon mRNA by RNAi.

SUMMARY Transposition of transposons is blocked in animal germ cells by a ping-pong amplification and mRNA destruction mechanism involving piRNAs. A piRNA complementary to a transposon mRNA binds to Piwi or Aubergine, and then base-pairs to a transposon mRNA. This initiates cleavage of the transposon mRNA by a slicer activity in the Piwi protein, and the 3′-end of the transposon mRNA is also processed. The resulting small RNA binds to Ago3, where it can base-pair to a piRNA precursor RNA. This initiates cleavage of the precursor RNA at a specific A–U base pair 10 nt from the 5′-end of the transposon mRNA fragment. Together with 3′-end processing of the precursor RNA, this generates a mature piRNA that can participate in a new round of transposon mRNA destruction and piRNA amplification. No piRNAs are produced in animal somatic cells, but transposition can be blocked by an endogenous RNAi mechanism. Plants lack Piwi proteins, so they must rely on an RNAi mechanism to control transposition in somatic and germ cells alike. Plants do have RNA-dependent RNA polymerases, so they can readily amplify siRNAs directed at transposon mRNAs.

16.8 Post-Transcriptional Control of Gene Expression: MicroRNAs

The siRNAs and piRNAs are not the only small RNAs that participate in gene silencing. Another class of small RNAs called **microRNAs (miRNAs)** are 22-nt RNAs produced naturally in plant and animal cells by cleavage from a larger, stem-loop precursor. In animals, these miRNAs then base-pair (though imperfectly) with the 3′-untranslated regions of specific mRNAs and silence gene expression primarily by blocking translation of those mRNAs. In plants, miRNAs base-pair perfectly (or almost so) with the interiors of mRNAs and direct the cleavage of those mRNAs. Let us consider the actions of miRNAs, and then their biogenesis.

Silencing of Translation by miRNAs

The first inkling of the importance of miRNAs came from work that began in 1981, which showed that mutations in the *lin-4* gene of the roundworm *(Caenorhabditis elegans)* caused developmental abnormalities. Subsequent genetic work suggested that the *lin-4* gene product acted by suppressing the level of LIN-14, the protein product of the *lin-14* gene. Interestingly, Gary Ruvkun and his colleagues showed that *lin-4* needed the **3′-untranslated region (3′-UTR)** of the *lin-14* mRNA in order to exert its LIN-14 suppression. Finally, in 1993, Victor Ambros and colleagues mapped the *lin-4* mutation, and found that it did not map to a protein-encoding gene. Instead, it mapped to the gene encoding the precursor of an miRNA. This suggested that an miRNA played an important role in *C. elegans* development, by reducing the expression of the *lin-14* gene. The sequence of the *C. elegans* genome bolstered this suggestion, showing that the miRNA was partially complementary to sequences within the 3′-UTR of the *lin-14* mRNA—the very sequences that are required for *lin-4* function.

We now know that miRNAs play crucial roles in the regulation of plant and animal genes. There are hundreds of miRNA genes in most plant and animal species examined so far, and each miRNA potentially controls many other genes. Mutations in miRNA genes typically have very deleterious effects, especially on development, underscoring the importance of these mRNAs, and suggesting that many disease states may be caused by mutations in, or improper regulation of, miRNA genes.

Indeed, miRNAs are so important in regulating genes in normal and diseased cells that they have enormous potential as drug targets in treating diseases such as cancer. Typically, cancer cells have abnormal spectra of miRNA expression, with some miRNAs unusually scarce and others unusually abundant. The trick will be to find which of these are important to the disease state, and then try to use drugs, possibly including the miRNA precursors themselves, to adjust the concentrations of those key miRNAs. However, macromolecules like miRNA precursors are notoriously difficult to use as drugs, and it is not clear how to selectively control the genes that encode miRNAs.

Given the importance of miRNAs, it is important to understand the mechanism by which they control genes. We will examine some of the evidence leading to different conclusions, but we will see that no one mechanism can explain all the data at hand.

In 1999, Philip Olsen and Ambros first demonstrated that the *lin-4* miRNA acts by limiting translation of the *lin-14* mRNA. The LIN-14 protein plays an important role in *C. elegans* development. During the first larval stage (L1), LIN-14 levels are high because this protein helps to specify the fates of cells that develop in that stage. However, at the end of L1, LIN-14 levels must drop so that other proteins can determine cell fate in the second larval stage, L2. This suppression of LIN-14 level depends on the *lin-4* RNA, a 22-nt miRNA that base-pairs to seven imperfect repeats of a sequence partially complementary to *lin-4* in the 3′-UTR of the *lin-14* mRNA.

Olsen and Ambros performed Western blots (Chapter 5) that showed at least a 10-fold decrease in LIN-14 protein between the L1 and L2 stages. On the other hand, their nuclear run-on analysis (Chapter 5) showed that the steady-state level of *lin-14* mRNA decreased less than two-fold between L1 and L2. Thus, control of *lin-14* appears to be at the translational level, not the transcriptional level.

Next, Olsen and Ambros used RT-PCR (Chapter 4) to amplify the 3′-ends, and thereby measure the sizes of the poly(A) tails, of *lin-14* mRNAs from the L1 and L2 stages. This analysis showed that the poly(A) tails of the mRNAs from the two stages were unchanged. Thus, the *lin-14* mRNA is not destabilized by shrinking its poly(A) tail in the L2 stage. In fact, Olsen and Ambros showed that *lin-14* mRNA was associated with polysomes (ribosomes in the act of translating an mRNA [Chapter 19]) just as much in L2 as in L1. Thus, translation initiation on *lin-14* mRNA appeared to be working just as well in stage L2 as in L1.

If appearance of LIN-14 protein is blocked in L2, but initiation of translation of its mRNA is normal, a reasonable conclusion would be that elongation or termination of translation on this mRNA is somehow blocked. Indeed, if *lin-4* miRNA really does bind to its target sites in the 3′-UTR of the *lin-14* mRNA, it would be well positioned to interfere with termination of translation. If so, both *lin-4* miRNA and *lin-14* mRNA should be found together on polysomes.

To test this hypothesis, Olsen and Ambros purified polysomes from L1 and L2 larvae by sucrose gradient ultracentrifugation (Chapter 17), and checked them for the presence of *lin-14* mRNA and *lin-4* miRNA by RNase protection assay (Chapter 5). Figure 16.38 shows the results. The "hump" to the right in each diagram (top) contains the fast-sedimenting polysomes. The polysomes are also contained in the middle two lanes in the electropherograms

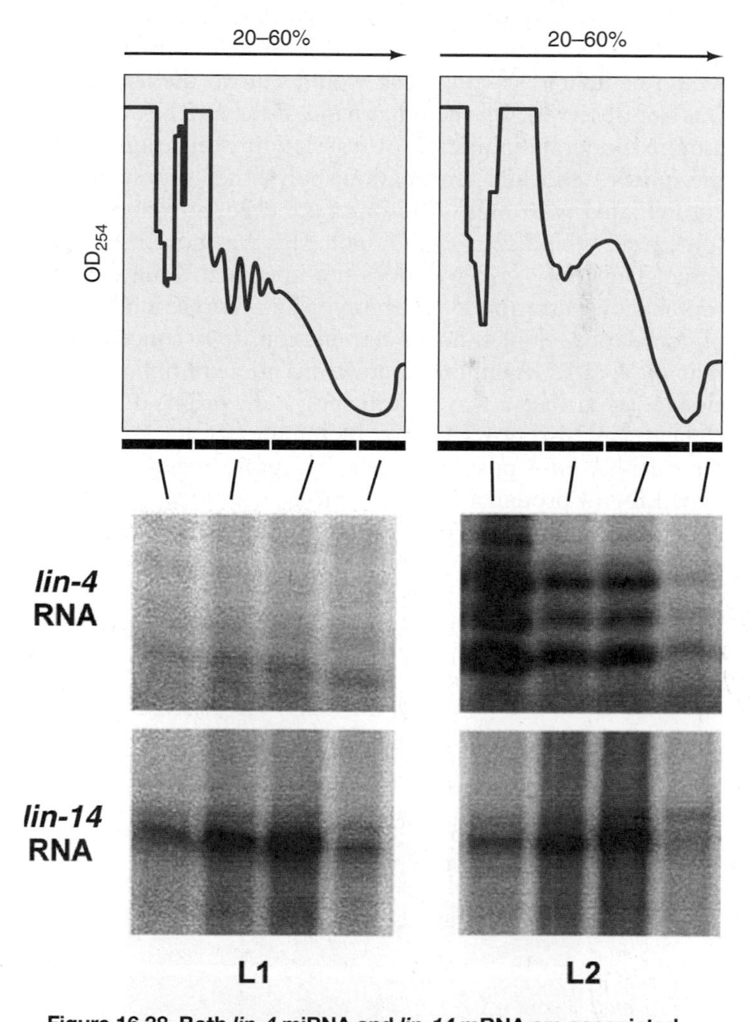

Figure 16.38 Both *lin-4* miRNA and *lin-14* mRNA are associated with polysomes in L1 and L2 larvae. Olsen and Ambros used sucrose gradient ultracentrifugation to display polysomes from *C. elegans* L1 (left) and L2 (right) larvae. They collected four fractions from the gradients, the middle two containing polysomes, and hybridized the RNAs from these fractions to labeled RNA probes for *lin-4* and *lin-14* RNAs. After they treated the RNA hybrids with RNase, they electrophoresed the protected probes on polyacrylamide gels. The results with *lin-4* and *lin-14* probes are at middle and bottom, respectively. The multiple bands represent protected probes differing by one nucleotide, and are presumably caused by "nibbling" at the ends of the hybrids by RNase. (*Source: Developmental Biology,* Volume 216, Philip H. Olsen and Victor Ambros, "The lin-4 Regulatory RNA Controls Developmental Timing in Caenorhabditis elegans by Blocking LIN-14 Protein Synthesis after the Initiation of Translation." fig. 8, p. 671–680, Copyright 1999, with permission from Elsevier.)

below the diagrams, which show the results of the RNase protection assays. We can see that the polysomes from both L1 and L2 larvae appear identical and contain approximately equal amounts of both *lin-4* miRNA (middle) and *lin-14* mRNA (bottom), presumably because the two RNAs are base-paired together.

These results present a difficulty: It is true that *lin-4* miRNA and *lin-14* mRNA are found together on polysomes, suggesting that they are base-paired together. But the polysome profile looks identical in L1 and L2 larvae. If the miRNA blocked translation elongation completely, or nearly completely, polysomes should have accumulated with very few ribosomes attached to the mRNA, so the polysomes would be lighter, and the peak would shift to the left. This was not observed. On the other hand, if the miRNA caused a more moderate inhibition of translation elongation, or if the miRNA blocked termination, polysomes should have accumulated with more ribosomes attached, and the polysome peak would shift to the right. This was not observed, either. Thus, *lin-4* miRNA does not appear to limit *lin-14* protein concentration in L2 embryos by a simple inhibition of translation elongation or termination. It is conceivable that *lin-4* miRNA inhibits both translation initiation and elongation in such a way that the polysome profile does not change. It is also possible that, by binding to the 3′-end of the mRNA, *lin-4* positions itself to capture newly synthesized LIN-14 protein and causes it to be degraded.

At least part of this question about *lin-4* miRNA activity could be explained by work by Amy Pasquinelli and her colleagues, reported in 2005. These workers used Northern blotting of *C. elegans* RNA (Figure 16.39) to show that

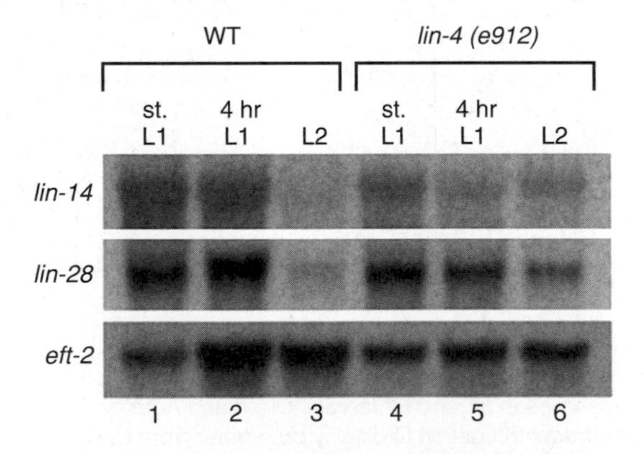

Figure 16.39 Concentrations of various mRNAs during development in *C. elegans*. Pasquinelli and colleagues Northern blotted RNAs from the following time points during *C. elegans* development, as indicated at top: starved L1; 4h L1; and L2. Then they hybridized the blot to probes for *lin-14* and *lin-28* mRNAs, as well as *eft-2* mRNA as a control (an mRNA known not to be influenced by *lin-4*). The concentrations of *lin-14* and *lin-28* mRNAs fell significantly between phases L1 and L2 in wild-type cells, but not in *lin-4(e912)* cells. (*Source*: Reprinted from *Cell*, Vol 122, Shveta Bagga, John Bracht, Shaun Hunter, Katlin Massirer, Janette Holtz, Rachel Eachus, and Amy E. Pasquinelli, "Regulation by let-7 and *lin-4* miRNAs Results in Target mRNA Degradation," p. 553–563, fig. 6a, Copyright 2005, with permission from Elsevier.)

lin-14 (and *lin-28*) mRNA levels actually do decrease about four-fold between stages L1 and L2. This figure also shows that this decrease depends on *lin-4* miRNA: Only modest decreases, at most, occurred in the *lin-4 e912* mutant. Thus, *lin-4* miRNA may exert its control via more than one mechanism.

Another approach to understanding the mechanism of miRNA action has been to use synthetic reporter mRNAs with one or more target sites for a particular miRNA, and then examine the effect of the miRNA (strictly speaking, a transfected siRNA that mimics the miRNA) on the behavior of the reporter mRNA. Phillip Sharp and colleagues tried one such strategy in 2006 and found that, when they inhibited translation initiation, the association of the reporter mRNA with ribosomes decayed more rapidly in the presence of the miRNA than in its absence. This suggested that the miRNA causes premature release of ribosomes from the mRNA (**ribosome drop-off**). These investigators also found that a reporter mRNA lacking a cap, but containing an internal ribosome initiation site (IRES), was also responsive to silencing by an miRNA. As we will learn in Chapter 17, cap recognition is the initiating step in eukaryotic translation, so this again indicated that the miRNA was acting downstream of the initiation step. Thus, the data were consistent with the ribosome drop-off model.

On the other hand, Filipowicz and colleagues presented evidence in 2005 for miRNA action at the translation initiation stage. They performed sucrose gradient ultracentrifugation to separate polysomes (actively translating ribosomes, Chapter 19) from mRNPs (proteins coupled to mRNAs that are not being translated). They found miRNAs and their target mRNAs associated with the mRNPs, rather than with polysomes. This suggested that the target mRNAs were not being translated, and therefore that the miRNAs were preventing translation initiation. Furthermore, if miRNAs act at the initiation step, which we will learn in Chapter 17 involves recognition of the cap at the 5′-end of the mRNA, allowing cap-independent initiation at an IRES should avoid silencing by miRNAs. That is exactly what Filipowicz and colleagues found, thereby reinforcing the hypothesis that miRNAs can block initiation of translation. There is also evidence that miRNAs team up with Argonaute proteins to compete with translation initiation factors for binding to mRNA caps, thereby blocking initiation.

Later in this chapter, we will see evidence that miRNAs can act by helping to degrade mRNAs. Thus, there are at least three major hypotheses for miRNA action: Blocking translation initiation; blocking translation elongation; and degradation of mRNAs. How do we reconcile all these ideas? It is possible that the differences we see reflect the different experimental approaches and the different organisms studied. But there is clear evidence for multiple mechanisms even within the same organism. It is also possible that different miRNAs act in different ways, or that the same miRNA can act in different ways, depending on the cellular context. Finally, Elisa Izaurralde and her colleagues have suggested that the different mechanisms that have been observed are different manifestations of the same unknown underlying mechanism. We will have to wait for more studies to fully answer this fascinating question.

In animals, at least, it appears that the degree of base-pairing between a small RNA and the target mRNA, not the origin of the small RNA, determines the kind of silencing that occurs. If the base-pairing is perfect, the mRNA tends to be degraded, even if the small RNA is an miRNA, rather than an siRNA. And if the base-pairing is imperfect, translation of the mRNA tends to be blocked, even if the small RNA is an siRNA, rather than an miRNA.

A good example of perfect base-pairing between an miRNA and mRNA, leading to mRNA destruction, is the miR-196 miRNA and the *HOXB8* mRNA in mice. Mammals and other animals possess clusters of **homeobox (HOX)** genes, which encode transcription factors that contain homeodomains (Chapter 12). These transcription factors tend to play critical roles in embryonic development. The HOX genes are down-regulated by miRNAs transcribed from genes that reside within the HOX clusters. One of these miRNAs, miR-196, base-pairs perfectly with the *HOXB8* mRNA, except for a single G–U wobble base pair (Chapter 18). In 2004, David Bartel and colleagues used rapid amplification of cDNA ends (RACE, Chapter 4) to detect the 5′-ends of fragments of *HOXB8* mRNA that were cut within the region that base-pairs with miR-196. They focused on mRNA fragments between days 15 and 17 of mouse embryogenesis because they knew that miR-196 miRNA was present during that time period. The RACE assay did indeed produce eight cDNA clones corresponding to broken *HOXB8* mRNA, and seven of these ended within the region of base-pairing with miR-196 miRNA.

These results suggested that the miRNA was causing breakage of the mRNA within the region of base-pairing between the two RNAs. To check this hypothesis, Bartel and colleagues placed the miR-196 complementary sequence into a firefly luciferase reporter gene and transfected this gene into HeLa (human) cells, along with either miR-196 miRNA, or a noncognate miRNA. Then they used their RACE assay to detect cleavage of the reporter gene's mRNA. They found that the miR-196 miRNA, but not the noncognate miRNA, caused cleavage of the luciferase mRNA. Thus, mammalian miRNAs, if they match their target mRNAs perfectly or nearly perfectly, can cause cleavage of the target mRNAs.

Note three important distinctions between the actions of siRNAs and miRNAs in animals:

1. The siRNAs silence genes by inducing degradation of the target mRNAs, while the miRNAs tend to silence genes by interfering with accumulation of the protein products of the target mRNAs. However, if base-pairing between an animal miRNA and its target mRNA is perfect or near perfect, the miRNA can cause cleavage of the target mRNA.
2. The siRNAs are formed by Dicer action on double-stranded RNAs that usually contain at least one strand that is foreign to the cell, or derive from transposons. On the other hand, the miRNAs are formed by Dicer action on the double-stranded part of a stem-loop RNA that is a normal cellular product.
3. The siRNAs base-pair perfectly with the target mRNAs, whereas the miRNAs usually base-pair imperfectly with their target mRNAs.

Silencing with both kinds of small RNA, siRNA and miRNA, depends on a RISC complex. In *Drosophila,* there are two Dicers (Dicer-1 and Dicer-2) and two RISCs, **siRISC** and **miRISC**, but there is no simple one-to-one correspondence. Silencing by siRNAs requires siRISC, and both Dicers, but Dicer-2 is more important in producing siRNAs. Silencing by miRNAs requires miRISC, and only Dicer-1 is required for producing miRNAs. However, this division of labor cannot be a general mechanism because other organisms, including yeast and mammals, have only one RISC. In spite of these complexities, it is becoming increasingly clear that the basic mechanisms of mRNA degradation mediated by siRNAs and miRNAs, at least in plants, are very similar, if not identical. They both require a Dicer to create the double-stranded siRNA or miRNA, and these double-stranded RNAs give rise to single-stranded RNAs that bind to an Argonaute-containing RISC. The single-stranded siRNAs or miRNAs then attract mRNAs with complementary sequences, which are broken by the RISC.

It is important to emphasize that not all animal miRNAs act at the translational level. They can also decrease mRNA concentrations, presumably by destabilizing the mRNAs. We have already seen two examples, including *lin-4,* the founding member of the miRNA class, which can decrease mRNA concentration, as well as inhibit translation. However, such decreases in mRNA concentration caused by miRNAs like *lin-4* cannot operate by an RNAi-like mechanism because RNAi requires perfect complementarity between miRNA and mRNA.

In Chapter 25, we will learn that transfection of human (HeLa) cells with either of two miRNAs caused a reduction in the levels of about 100 mRNAs. In fact, one miRNA, normally expressed in the brain, shifted the HeLa cell mRNA profile to something resembling the profile of mRNAs in the brain. By contrast, the other miRNA, normally expressed in muscle, shifted the mRNA profile closer to that of muscle cells. Moreover, the 3′-untranslated regions (3′-UTRs) of the destabilized mRNAs tended to contain sequences complementary to sequences near the 5′-ends of the respective miRNAs, the miRNA **seed regions** (usually residues 1-7 or 2-8). Thus, base-pairing between the miRNA and target mRNAs appeared to be important to the mRNA destabilization. The fact that each miRNA seemed to affect, directly or indirectly, the levels of about 100 mRNAs, also suggests that the miRNAs play a very widespread role in controlling gene expression in animals—a role whose importance may even rival that of the protein transcription factors.

The discovery of miRNAs and their function in destabilizing mRNAs has elucidated the role of **AU-rich elements** (**AREs**), which have been known since 1986 to exist in the 3′-UTRs of certain unstable mRNAs. In 2005, Jiahuai Han and colleagues reported that the instability of the *Drosophila* tumor necrosis factor-α mRNA depends on Dicer-1, Ago1 and Ago2, which are all involved in miRNA-mediated mRNA degradation. They went on to show that the instability of human ARE-containing mRNAs also depends on Dicer. Furthermore, a specific human miRNA (mi-R16), which is complementary to the ARE sequence (AAUAUUUA), is required for mRNA instability.

In contrast to the translation blockage model in animals, miRNAs in plants appear to silence by base-pairing perfectly or nearly perfectly with their target mRNAs and sponsoring degradation of those mRNAs. For example, James Carrington and colleagues showed in 2002 that a 21-nt RNA, known as miRNA 39, from *Arabidopsis thaliana* accumulates in flowering tissues and base-pairs to target sites in the middle of the mRNAs from several members of a family of transcription factors known as *Scarecrow-like* (*SCL*). This base pairing results in cleavage of the mRNAs within the region of base-pairing with the miRNA. Relatively little miRNA 39 accumulates in leaf and stem tissues, and no dectectable *SCL* mRNA cleavage occurs in those tissues.

To demonstrate miRNA-directed cleavage of mRNAs, Carrington and colleagues introduced the gene encoding the precursor to miRNA 39 into leaf tissue. They observed a high level of miRNA 39, suggesting that leaf tissue contains a Dicer-like enzyme that can produce miRNA from its precursor. More significantly, they observed active cleavage of *SCL* mRNA to a smaller, inactive product, in the leaf tissue expressing miRNA 39.

On the other hand, some plant miRNAs, although they base-pair very well with their target mRNAs, silence gene expression by interfering with translation. Xuemei Chen presented an example in 2004: miRNA172 of *Arabidopsis* base-pairs almost perfectly with the mRNA from a floral homeotic gene called *APETALA2,* yet it silences that gene by blocking translation, not by mRNA degradation. Thus, plant miRNAs, regardless of the degree of base-pairing with their target mRNAs, can use either mRNA degradation or translation blocking to silence genes.

Figure 16.40 summarizes the actions of miRNAs when base-pairing is imperfect (the typical situation in animals) and when it is perfect or near-perfect (the typical situation in plants; also observed in animals). In the former situation, translation, or at least appearance of protein product, is blocked. In the latter situation, the mRNA is cleaved. However, one should keep in mind that each of these canonical pathways has exceptions. That is, animal miRNAs, though they may base-pair imperfectly with their targets, can cause mRNA degradation, and plant miRNAs, though they may base-pair perfectly with their targets, can cause blockage of translation.

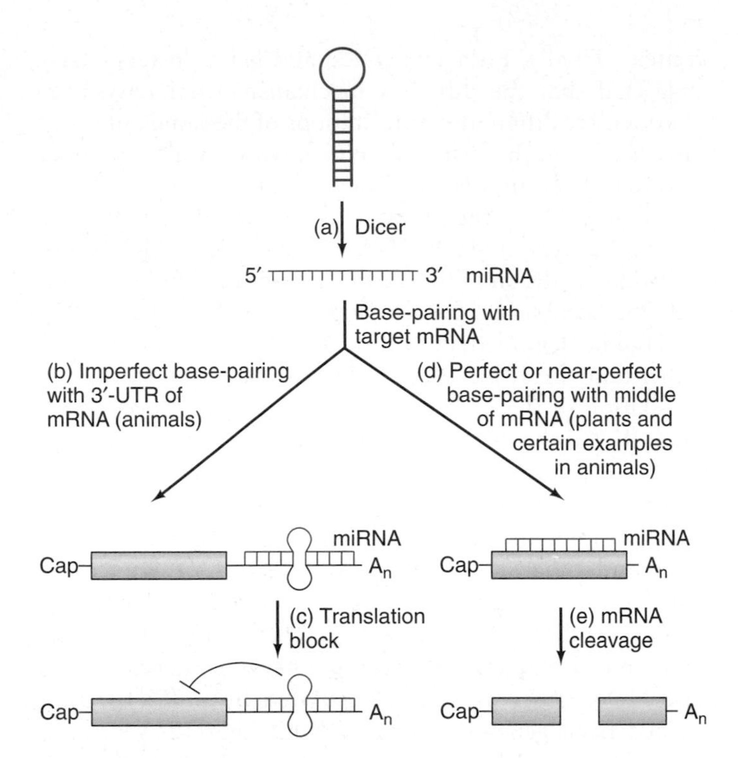

Figure 16.40 Two pathways to gene silencing by miRNAs. (a) A stem-loop miRNA precursor is cleaved by Dicer to yield a short miRNA about 21 nt long. **(b)** If the base-pairing between the miRNA and the 3′-UTR of its target mRNA is imperfect, as usually occurs in animals, the miRNA causes blockage of translation, or at least accumulation of the mRNA's protein product **(c). (d)** If the base-pairing between the miRNA and the middle of its target mRNA is perfect, or nearly so, as usually occurs in plants, and sometimes in animals, the mRNA is cleaved **(e),** which inactivates the mRNA.

MicroRNAs do not serve solely as modulators of cellular gene activity. There is also good evidence that they act as antiviral agents in plants and invertebrates by targeting viral mRNAs. It was widely assumed that vertebrates relied on their potent interferon systems, rather than on miRNAs, to combat viral infections. However, Michael David and colleagues showed in 2007 that miRNAs can also target viral mRNAs, and that these miRNAs are themselves a product of the interferon system.

In particular, David and colleagues demonstrated that interferon-β (IFN-β) stimulates the production of many miRNAs. Among these are eight miRNAs that are complementary to parts of the hepatitis C virus (HCV). These miRNAs appear to be effective in combating HCV because introduction of corresponding synthetic miRNAs mimics the effects of IFN-β on HCV infection and replication.

SUMMARY MicroRNAs (miRNAs) are 18–25-nt RNAs produced from a cellular RNA with a stem-loop structure. In the last step in miRNA synthesis, Dicer cleaves the double-stranded stem part of the precursor to yield the miRNA in double-stranded form. The single-stranded forms of these miRNAs can team up with an Argonaute protein in a RISC to control the expression of other genes by base-pairing to their mRNAs. In animals, miRNAs tend to base-pair imperfectly to the 3′-UTRs of their target mRNAs and inhibit accumulation of the protein products of these mRNAs. However, perfect or perhaps even imperfect base-pairing between an animal miRNA and its target mRNA can result in mRNA cleavage. In plants, miRNAs tend to base-pair perfectly or near-perfectly with their target mRNAs and cause cleavage of these mRNAs, although there are exceptions in which translation blockage can occur.

Stimulation of Translation by miRNAs

MicroRNAs do not always inhibit translation. Joan Steitz and her colleagues first noticed indications of positive action by miRNAs when they found that the ARE of the human tumor necrosis factor-α (TNFα) mRNA activates translation during serum starvation, which arrests the cell cycle in the G_1 phase. They also found that Ago2 and fragile X mental retardation-related protein (FXR1) associate with the ARE during translation activation, and are required for the activation.

This work suggested that miRNAs, which bind along with proteins to AREs, might be capable of directing activation, rather than inactivation, of translation under certain conditions. To test this hypothesis, Steitz and colleagues first used bioinformatics techniques (Chapter 25) to search the human genome for miRNAs with seed sequences complementary to the TNFα ARE. They identified five miRNA candidates, not counting miR16, which is known to reduce TNFα mRNA levels by binding outside the ARE region.

To screen the five miRNAs for effects on TNFα mRNA translation, they attached the TNFα ARE to the firefly luciferase reporter gene and tested this construct for translation efficiency in transfected cells under a variety of conditions. Only one miRNA, miR369-3, had an effect. It stimulated translation, but only in serum-starved cells.

First, Steitz and colleagues tested the effect of serum on miR369-3 levels using an RNase protection assay. Figure 16.41b shows that the level of the miRNA rose under serum starvation conditions, but that this rise was blocked by treatment with an siRNA that targets the loop of the pre-miR369-3. By contrast, serum had no effect on the levels of three control RNAs: miR369-5, which is essentially the complementary strand of miR369-3 in the stem of the pre-miRNA; miR16; or U6 snRNA. As expected, the siRNA also knocked down the level of miR369-5.

Next, Steitz and colleagues tested the effect of serum on reporter mRNA translation in the presence and absence of serum, and in the presence and absence of the siRNA that blocks accumulation of miR369-3. Figure 16.41c shows that translation efficiency increased about five-fold under serum-starved conditions. However, when the siRNA targeting pre-miR369-3 was included, the stimulation of translation disappeared. On the other hand, when the investigators rescued miR369-3 by adding a synthetic miR369-3 immune to the siRNA, translation again rose about five-fold upon serum starvation. Furthermore, serum had no effect on translation when the ARE did not match the seed sequence of the miRNA.

To test the importance of base-pairing between miR369-3 and the ARE, Steitz and colleagues used an intergenic suppression approach. They mutated the ARE to the sequence they called mtARE (Figure 16.41a) and tested the altered gene for activation with the wild-type miR369-3. As Figure 16.41d shows, no activation occurred upon serum starvation. Next, they added a mutant miR369-3 (miRmt369-3, Figure 16.41a) with a sequence complementary to that of mtARE, and re-tested for activation. This time, serum starvation caused activation. As expected, a control miRNA (miRcxcr4) caused no activation. Thus, complementarity between the ARE and the miRNA appears to be important.

To probe the importance of the seed regions in particular, Steitz and colleagues mutated each of the identical regions (seed1 and seed2) in the ARE of the mRNA that are complementary to the seed regions in miR369-3, and then made compensating mutations in the seed region of the miRNA. The mutant AREs are called mtAREseed1 and mtAREseed2, and the compensating mutant miRNA is called miRseedmt369-3. These sequences are all given in Figure 16.41a, and Figure 16.41e shows the results. As predicted, changing the sequences of each of the anti-seed regions in the mRNA eliminated activation by serum starvation, and making compensating mutations in the seed region of the miRNA restored activation. Thus, miR369-3 really is responsible for the activation, and base-pairing between the seed region of the miRNA and the ARE in the mRNA is critical for this activation.

Finally, Steitz and colleagues looked directly for miR369-3 associated with the reporter mRNA. They tagged the reporter mRNA with an S1 aptamer that allowed it to be affinity purified by binding to streptavidin. Then they cross-linked any associated RNAs with formaldehyde, performed streptavidin affinity purification of the reporter mRNA, and detected any miR369-3 associated with it by RNase protection assay. Figure 16.41f shows the results. The miR369-3 was associated with the reporter mRNA in serum-starved cells, but not in cells grown in serum. No association was detected in cells treated with the siRNA that targets the pre-miR369-3, but it was detected when these cells were rescued with miR369-3 and

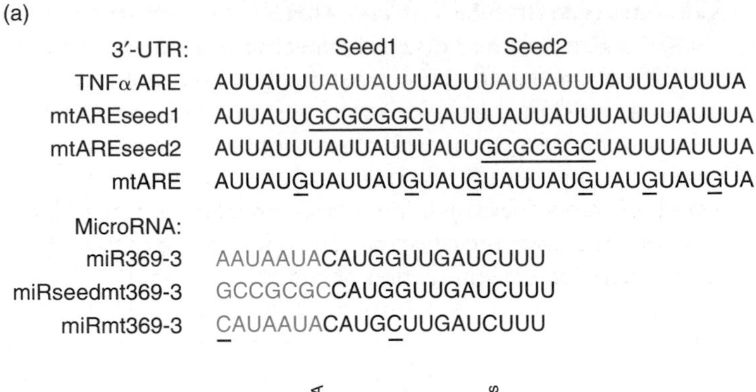

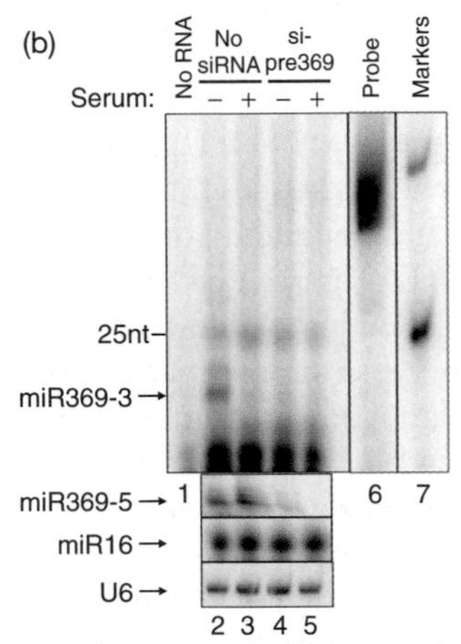

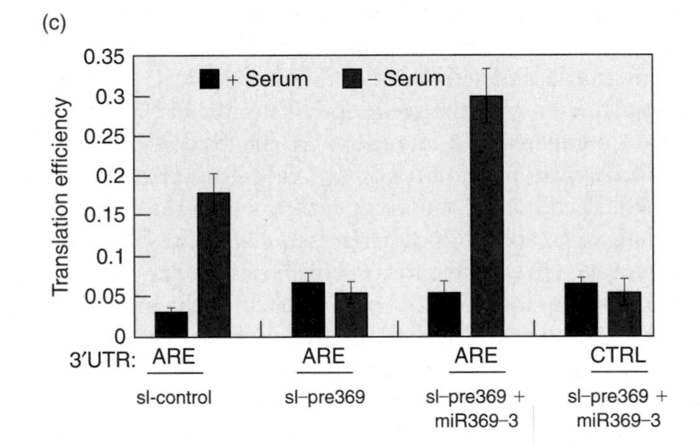

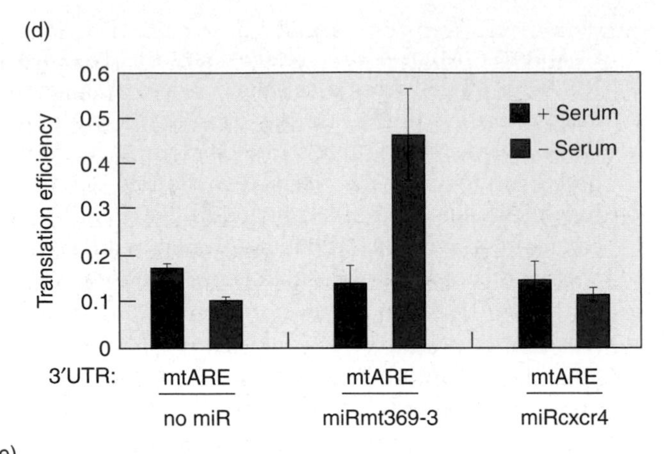

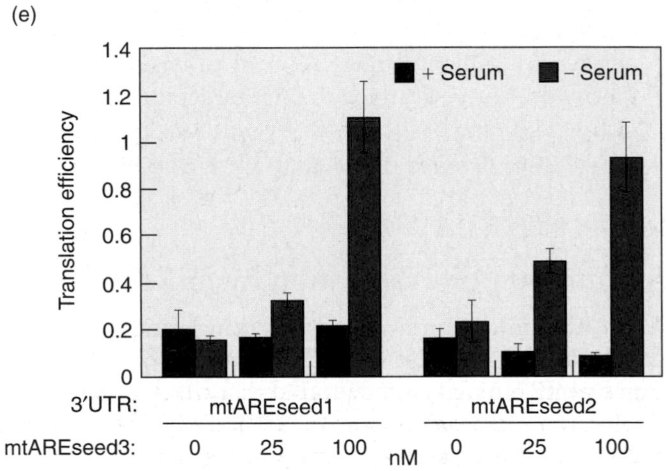

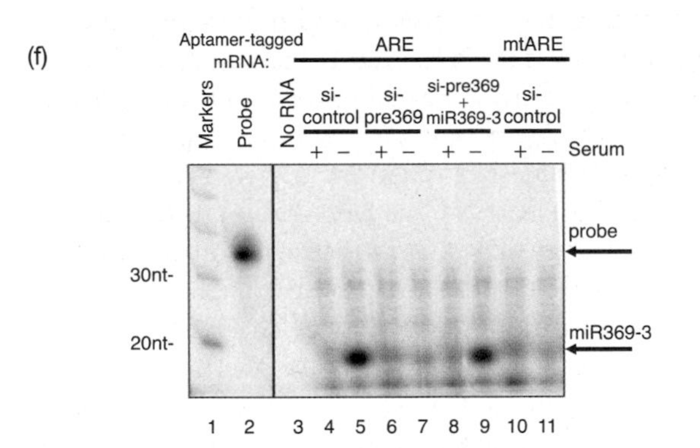

Figure 16.41 Role of MiR369-3 activation of reporter mRNA translation. **(a)** Sequences of wild-type and mutant TNFα 3′-UTRs linked to the luciferase reporter mRNA, and wild-type and mutant miRNAs. All sequences are written 5′→3′, so one must be inverted for complementarity with the other to be obvious. Note that the wild-type ARE has two regions (pink) that are complementary to the seed region (5′-AAUAAUA-3′, blue) in miR369-3. **(b)** Concentration of miR369-3, measured by RNase protection assay. RNA levels were measured with and without serum, as indicated at top, and with without an siRNA that targets the pre-miR369-3. At bottom, concentrations of miR369-5 (the passenger starand of miR369-3), as well as two control RNAs (miR16 and U6 snRNA) were measured. The position of miR369-3 is indicated at left, along with the position of a 25-nt marker RNA. **(c)** Translation efficiencies of mRNAs bearing the wild-type ARE, or a control ARE (CTRL) are shown with and without serum (blue and red, respectively). The experiments were run with no siRNA (si-control), with an siRNA targeting the pre-miR369-3 (si-pre369), or with the siRNA plus a rescuing miR369-3 (si-pre369 + miR369-3), as indicated at bottom. **(d)** Translation efficiencies of mRNAs bearing the mutated ARE (mtARE) are shown with and without a complementary mutated miR369-3 (miR369-3) or with a control miRNA (miRcxcr4). **(e)** Translation efficiencies of mRNAs bearing AREs with mutated anti-seed 1 or anti-seed 2 regions (mtAREseed 1 and mtAREseed 2, respectively indicated at bottom) are shown with and without serum (blue and red, respectively) and with three concentrations of an miRNA with a seed region complementary to the mutated anti-seed region (miRseedmt369-3), as indicated at bottom. **(f)** Detection of association between reporter mRNA and miR369-3. Formaldehyde-cross-linked RNAs were affinity-purified via an S1 aptamer tag on the reporter mRNA, and miR369-3 was delected by RNase protection assay. The experiments were run with no siRNA (si-control), with an siRNA targeting the pre-miR369-3 (si-pre369), or with the siRNA plus a rescuing miR369-3 (si-pre369 + miR369-3), as indicated at top. Also, a tagged control mRNA (mtARE) with a mutated ARE was used (lanes 10 and 11). (*Source:* Reprinted with permission of *Science*, 21 December 2007, Vol. 318, no. 5858, pp. 1931–1934, Vasudevan et al, "Switching from Repression to Activation: MicroRNAs Can Up-Regulate Translation." © 2007 AAAS.)

serum-starved. Also, no miR369-3 associated with a reporter mRNA with a mutated ARE (mtARE). Taken together, the results in Figure 16.41 show that the activation of reporter mRNA translation by serum starvation depends on an association between miR369-3 and the ARE of the mRNA.

Steitz and colleagues extended these studies to two other reporter mRNAs. One (CX) contained four synthetic miRNA (miRcxcr4) target sites; the other (Let-7) contained seven target sites for the endogenous Let-7 miRNA. Translation of both reporter mRNAs was activated by serum starvation in two different cell lines. Thus, all three of the miRNAs in this study can respond to serum starvation by activating translation.

Steitz and colleagues knew from previous experiments that translation activation was cell cycle-dependent, so they reasoned that synchronized cells might show more dramatic effects of serum than the nonsynchronized cells used in Figure 16.41. Accordingly, they synchronized cells by starving them of serum, and then released them to reenter the cell cycle by adding serum. When they measured translation efficiency, they found that synchronized cells growing in serum actually had about a five-fold lower translation efficiency than unsynchronized serum-grown cells. Furthermore, this translation repression depended on miR369-3. Thus, this miRNA can activate translation under some conditions, and repress it under other conditions.

Previous studies had shown that Ago2 and FXR1 are both required for translation activation upon serum starvation, so Steitz and colleagues measured the recruitment of these two proteins to ribonucleoprotein (RNP) complexes on aptamer-tagged mRNAs. They found both Ago2 and FXR1 in the RNP complex associated with the reporter mRNA under serum-starved conditions. However, when miR369-3 was depleted with the siRNA directed against pre-miR369-3, the amount of Ago2 in the RNP complex fell, but it was restored by adding miR369-3. In RNP complexes isolated from synchronized cells growing in serum, Ago2 was prominent, but FXR1 was not, and the amount of Ago2 in the complex dropped when miR369-3 was depleted. Steitz and colleagues concluded that miR369-3 recruits both proteins to the mRNA under serum-starved conditions, and these proteins participate in translation activation. On the other hand, miR369-3 recruits Ago2, but not FXR1, to the mRNA in synchronized proliferating cells, so Ago2, but not FXR1 appears to be involved in translation repression.

SUMMARY MicroRNAs can activate, as well as repress translation. In particular, miR369-3, with the help of AGO2 and FXR1, activates translation of the TNFα mRNA in serum-starved cells. On the other hand, miR369-3, with the help of Ago2, represses translation of the mRNA in synchronized cells growing in serum.

Biogenesis of miRNAs MicroRNAs are synthesized by RNA polymerase II as longer precursors known as **primary miRNAs (pri-miRNAs)**. We know that RNA polymerase II transcribes the pri-miRNA genes because the pri-miRNAs are capped and polyadenylated, which is characteristic of class II transcripts, because low concentrations of α-amanitin inhibit pri-miRNA synthesis, and because ChIP analysis shows association between polymerase II and chromatin containing pre-miRNA promoters.

A well-studied human pri-miRNA gene contains the coding regions for three miRNAs (miR23a, miR27a, and miR24-2). The pri-miRNA is about 2.2 kb long, including its poly(A) tail, which lies about 1.8 kb downstream of the last miRNA coding region. Although this gene is clearly transcribed by polymerase II, its promoter, which extends as much as 600 nt upstream of the transcription start site, has none of the typical class II core promoter elements we studied in Chapter 10, nor the PSE element characteristic of the class II snRNA promoters.

The pri-miRNAs contain each miRNA coding region as part of a stable stem-loop. The first step in processing this precursor to a mature miRNA occurs in the nucleus and requires an RNase III known as **Drosha**, which cleaves near the base of the stem, releasing a **pre-miRNA** consisting of a 60-70-nt stem-loop with a 5′-phosphate and a 2-nt 3′-overhang. However, Drosha cannot recognize and cleave a pri-miRNA on its own. It needs a double-stranded RNA-binding protein partner. In humans, this partner is called **DGCR8**; in *C. elegans* and *Drosophila* it is called **Pasha**. Together, Drosha and Pasha make up an RNA processing complex called **Microprocessor**. The final processing of a pre-miRNA to a mature miRNA is carried out in the cytoplasm by Dicer, the same RNase III responsible for siRNA production in RNAi. Figure 16.42a illustrates the two-step process of miRNA biogenesis.

Another mode of miRNA biogenesis bypasses the Drosha cleavage step. Many miRNAs are encoded in introns, and some of these, known as **mirtrons** ("mir" from miRNA, and "trons" from introns), take advantage of the splicing mechanism, rather than Drosha, to generate the pre-miRNA. As Figure 16.42b shows, the whole intron is a pre-miRNA. Therefore, the normal splicing machinery will cut it out of the primary transcript as a lariat-shaped intron, which will then be linearized by the debranching enzyme, whereupon it can fold into the stem-loop shape of a pre-miRNA.

Some miRNAs require A → I editing, which we discussed earlier in this chapter. For example, all but one member of the miR-376 RNA cluster in mice and humans undergo A → I editing in certain tissues, including the brain, at specific sites in the pri-miRNA. One of the most commonly edited sites is four bases from the 5′-end of the miRNA, within the seed region that base-pairs to the complementary site in the 3′-UTR of the target mRNA. Thus, this change in base sequence of the miRNAs changes the identity of their targets, with important implications for brain function.

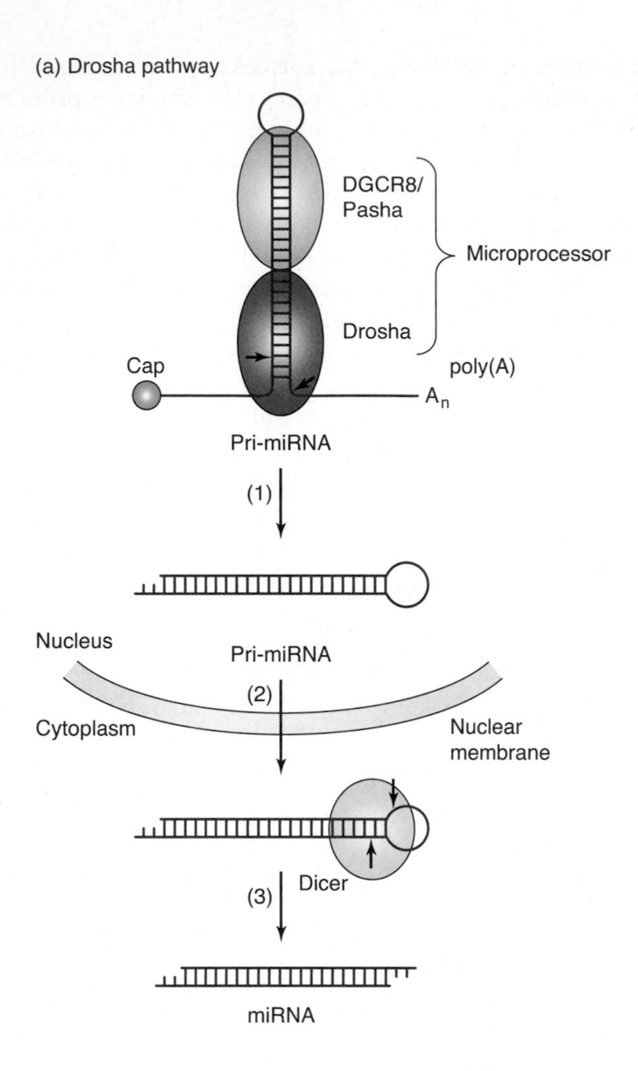

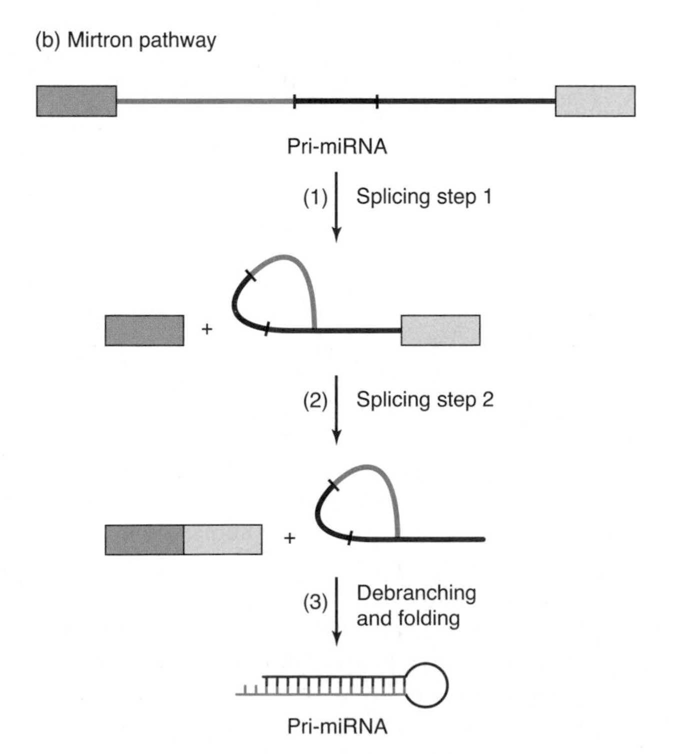

Figure 16.42 Maturation of a human miRNA. The primary transcription product of an miRNA gene is a pri-miRNA. It is made by RNA polymerase II and it may contain more than one miRNA sequence. For simplicity, this one contains just one. **(a)** The Drosha pathway. (1) Microprocessor, which consists of a double-stranded RNA-binding protein (DGCR8, or Pasha) and an RNase III (Drosha), binds to the pri-miRNA and cleaves it at the base of the stem, releasing a 60–70-nt stem-loop pre-miRNA. (2) The pre-miRNA is transported from the nucleus to the cytoplasm. (3) Dicer binds to the pre-miRNA in the cytoplasm and cuts 22 nt from the cut made by Drosha, yielding the mature miRNA. **(b)** The mirtron pathway. (1) The mirtron is color coded cyan, black, and magenta, corresponding to the three parts of the pre-miRNA it will become: the top strand of the stem; the loop; and the bottom strand of the stem, respectively. The first step of splicing separates the mirtron from the first exon and forms it into a lariat that is still attached to the second exon. (2) The second splicing step separates the mirtron from the second exon, still in lariat shape. (3) Debranching of the lariat, and folding (which occurs naturally) yields the mirtron as a pre-miRNA. It has the usual approximately 22 base pairs, but fewer are shown here for simplicity.

SUMMARY RNA polymerase II transcribes the miRNA precursor genes, to produce pri-miRNAs, which may encode more than one miRNA. Processing a pri-mRNA to a mature miRNA is a two-step process. In the first step, a nuclear RNase III known as Drosha cleaves the pri-miRNA to release a 60–70-nt stem-loop RNA known as a pre-miRNA. In the second step, which occurs in the cytoplasm, Dicer cuts the pre-miRNA within the stem to release a mature double-stranded miRNA. A mirtron is an intron that consists of a pre-miRNA. Thus, the spliceosome cuts it out of its pre-mRNA, then it is debranched and folded into a stem-loop pre-miRNA, without any participation by Drosha. Some miRNAs require A → I editing at the pri-miRNA stage, and some of this editing changes the targeting of the miRNAs to different mRNAs.

16.9 Translation Repression, mRNA Degradation, and P-Bodies

Processing bodies

(**P-bodies,** also known as **PBs**) are discrete cytoplasmic collections of RNAs and proteins that are involved in mRNA decay and translational repression. These cellular foci are enriched in enzymes that deadenylate mRNAs (deadenylases); decap mRNAs (the decapping enzyme, which, in *Drospophila,* contains two subunits, Dcp1 and Dcp2); and catalyze 5′→3′ degradation of mRNAs (exonuclease Xrn1). Thus, P-bodies appear to be involved in translational repression and also in degradation of mRNAs by a non-RNAi-like mechanism that entails deadenylation and decapping prior to 5′→3′ exonucleolytic destruction.

Degradation of mRNAs in P-bodies

One of the important partners for the miRNAs in mRNA silencing in P-bodies, at least in higher eukaryotes, is **GW182.** The "GW" in the name refers to repeats of glycine (G) and tryptophan (W) in the protein. GW182 is required for P-body integrity, but its role extends far beyond a simple structural one: This protein appears to be an essential part of the mRNA silencing machinery. One clue to the importance of GW182 is that it associates with DCP1, Ago1, and Ago2—all key players in mRNA silencing—in human cell P-bodies. Another indication of the importance of GW182 is that RNAi-mediated knockdown experiments in human cells showed that reducing the levels of GW182 impaired both miRNA function and the mRNA decay that is an essential part of RNAi. In *Drosophila* cells, by contrast, knockdown of GW182 impaired miRNA function, which depends on Ago1, but not RNAi, which depends on Ago2.

In 2006, Elisa Izaurralde and colleagues presented the results of their inquiry into the exact role of GW182 in miRNA-mediated silencing of mRNA function in *Drosophila.* Because GW182 and Ago1 both appear to be involved in miRNA-mediated mRNA silencing in *Drosophila* cells, these workers employed high-density oligonucleotide arrays (Chapter 24) to investigate the profiles of RNAs in cells depleted of GW182, Ago1, or Ago2 by knockdown using dsRNAs specific for each of the three genes. They found that there was a high correlation between the mRNAs up-regulated in response to knockdown of GW182 and Ago1 (a rank correlation coefficient r of 0.92). Rank correlation coefficients are computed by arranging two groups of values by rank and then calculating how closely the two ranks compare with each other. In this case, the mRNAs were ranked according to the degree to which they were up-regulated (or down-regulated) in response to knockdown of GW182 (first ranking) or Ago1 (second ranking). So an r of 0.92 indicates that mRNAs strongly up-regulated by a GW182 knockdown are also usually strongly up-regulated by an Ago1 knockdown. By contrast, there was much less correlation between the mRNAs up-regulated in response to knockdown of GW182 and Ago2 ($r = 0.64$).

Figure 16.43a shows the impressive similarity between the profiles of mRNAs regulated in the same way by both GW182 and Ago1. In this figure, 6345 transcripts were analyzed to see if they were up-regulated or down-regulated in response to a given knockdown. Red represents transcripts that are up-regulated at least two-fold, blue represents transcripts down-regulated at least two-fold, and yellow represents all the other transcripts, which were up- or down-regulated less than two-fold. Next, Izaurralde and colleagues focused on the mRNAs that were at least two-fold up- or down-regulated in response to GW182 or Ago1 knockdowns. Figure 16.43b illustrates the very high degree of concordance.

If GW182 and Ago1 knockdowns are up-regulating certain mRNAs because these mRNAs would otherwise be silenced by miRNA-mediated degradation, one should observe that known miRNA target mRNAs are up-regulated by knocking down either GW182 or Ago1. Indeed, when Izaurralde and colleagues did that experiment, they got exactly the predicted results. Figure 16.43c shows that all nine of the known miRNA targets were up-regulated at least two-fold by knockdowns of either GW182 or Ago1. In fact, even the degree of up-regulation of each mRNA correlated well between the two knockdowns. Izaurralde and colleagues also checked the oligonucleotide array data by performing classical Northern blots with selected mRNAs. Figure 16.43d shows that the Northern blot and array data match very well. Thus, GW182 and Ago1 seem to have the same effect: silencing genes by reducing mRNA concentration.

Izaurralde and colleagues wondered if GW182 by itself could silence the expression of target mRNAs. To find out, they physically tethered GW182 to a firefly luciferase reporter mRNA by the following strategy (further illustrated in Chapter 17): They added five λ phage box B coding sequences to the 3′-UTR of the reporter gene. As we learned in Chapter 8, box B sequences in an RNA are binding sites for the λN protein. Accordingly, these workers fused the GW182 gene to a gene fragment encoding the part of λN (the N-peptide) that binds to box B. Then they transfected *Drosophila* cells with the λN-GW182 construct, the reporter gene, and a control plasmid containing the *Renilla* (sea pansy) luciferase gene, whose protein product they could assay as a control for transfection efficiency.

Note that this combination of constructs yields a reporter mRNA containing box B sequences in its 3′-UTR, and a λN-GW182 protein with a natural affinity for box B. Thus, the λN-GW182 protein becomes tethered to the reporter mRNA. When Izaurralde and colleagues assayed for firefly luciferase activity (corrected for transfection efficiency), they found a 16-fold reduction in expression of the reporter mRNA with tethered λN-GW182, compared to a reporter mRNA tethered to λN protein by itself. Thus, GW182 alone is capable of strongly silencing expression of a bound mRNA. Is this silencing due to reduction of mRNA level alone? To answer this question, Izaurralde and colleagues performed Northern blots on RNA from cells expressing λN-GW182, or λN alone. They found only a four-fold decrease in reporter mRNA concentration when it was tethered to λN-GW182. This four-fold loss of mRNA clearly cannot fully explain the 16-fold decrease in expression, so it appears that GW182 also controls translation of at least some mRNAs to which it binds.

Is the silencing observed with tethered λN-GW182 independent of Ago1? To find out, Izaurralde and colleagues repeated the tethering experiment in ordinary cells, and in Ago1 knockdown cells. They found no difference, so

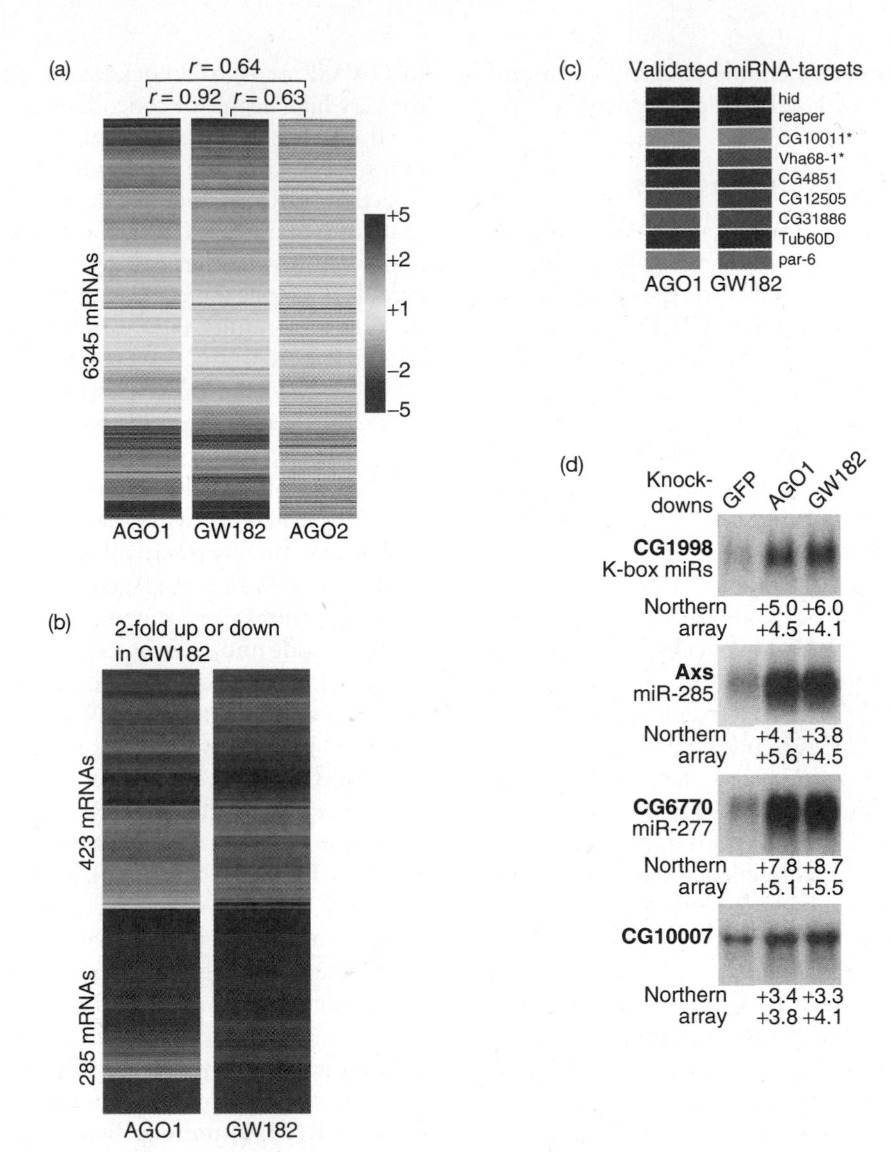

Figure 16.43 Effect of knockdowns of Ago1, GW182, ad Ago2 on abundance of other transcripts. **(a)** Izaurralde and colleagues isolated transcripts from untreated *Drosophila* cells, and from cells treated with dsRNAs to knock down Ago1, GW182, and Ago2 by RNAi. They hybridized transcripts from each of the three groups of treated cells, and untreated cells, to oligonucleotide arrays and determined the abundance of each of 6345 miRNAs before and after treatment. They coded up-regulation by at least two-fold as red, down-regulation by at least two-fold as blue, and less than two-fold change in either direction as yellow, according to the key at right. Note the similarity between the mRNA profiles form Ago1 and GW182 knockdowns, and the relative dissimilarity between either Ago1 or GW182 and Ago2. **(b)** Results of the same study, but only mRNAs up- or down-regulated by at least two-fold in Ago1 or GW182 knockdowns are presented. **(c)** The results from nine mRNAs that are known miRNA targets are shown for Ago1 and GW182 knockdowns. Note again the great similarity in the effects of knocking down Ago1 and GW182. **(d)** Northern blots of four different mRNAs, identified at left, are shown for Ago1 and GW182 knockdowns, along with a control green fluorescent protein (GFP) knockdown, which should not have any effect on the abundance of any of these mRNAs. The degrees of up-regulation of each mRNA in the Ago1 and GW182 knockdowns were calculated from these Northern blots and from the microarry analysis in panel (a), and are given below the respective blots. Note the similarity in degree of up-regulation determined by Northern blots and microarrays. (*Source*: Reprinted by permission of E. Izaurralde from Behm-Ansmant et al, mRNA degradation by miRNAs and GW182 requires both CCR4: NOT deadenylase and DCP1: DCP2 decapping complexes, *Genes and Development*, V. 20, pp. 1885–1898. Copyright © 2006 Cold Spring Harbor Laboratory Press.)

silencing appeared to work just as well without Ago1. Thus, binding GW182 to an mRNA appears to sidestep the requirement for Ago1, which may mean that Ago1 helps recruit GW182 to mRNAs targeted for silencing.

We have seen that tethering λN-GW182 to a reporter mRNA causes about a 75% degradation of the mRNA. In addition, Izaurralde and colleagues noticed that the remaining mRNA was a little shorter than the same reporter mRNA in cells without λN-GW182. They wondered whether this shortening was due to deadenylation, and whether this deadenylation would occur under normal circumstances. To find out, they isolated RNA from cells at time zero and 15 min after stopping transcription with actinomycin D. Then they deadenylated the mRNAs by oligo(dT)-targeted RNase H degradation (Chapter 14). Finally, they subjected these RNAs to Northern blot analysis with probes specific for the reporter mRNA and for rp49, an endogenous mRNA (not an miRNA target) that encodes the ribosomal protein L32. They found that the control RNA contained poly(A) at both time points, as it could be shortened by oligo(dT)-directed RNase H destruction of poly(A). On the other hand, the luciferase reporter mRNA contained poly(A) immediately after transcription, at time zero, but it appeared to be deadenylated by 15 min after transcription was halted, as it could not be further shortened by oligo(dT)-directed RNase H treatment. Thus, deadenylation appears to be part of the silencing caused by GW182. Furthermore, knockdown experiments showed that silencing by GW182 depends on the CCR4/NOT deadenylase in *Drosophila*.

Decapping of mRNA is also part of the miRNA-mediated mRNA degradation pathway, so Izaurralde and colleagues examined the effects of knocking down DCP1 and DCP2 in the λN-GW182 reporter mRNA tethering assay. They found that depleting cells of the DCP1/DCP2 decapping complex restores reporter mRNA levels to normal. However, loss of DCP1 and 2 had little effect on the strong silencing of luciferase activity by tethering λN-GW182 to its mRNA. A probable explanation comes from the finding that the reporter mRNA was still deadenylated in the DCP1/DCP2-depleted cells—and deadenylated mRNAs are expected to be poorly translated.

The GW182-mRNA tethering studies not only bypassed the need for Ago1, they also bypassed miRNAs. So we are left with the impression that GW182, along with Ago1, is an important player in miRNA-mediated silencing, but we have so far seen no direct evidence for this hypothesis. Accordingly, Izaurralde and colleagues examined the mechanism of miRNA-mediated mRNA decay and found that it depends on deadenylation by CCR4/NOT, decapping by DCP1/DCP2, as well as on GW182 and Ago1. These workers constructed three luciferase reporter mRNAs that were silenced by two miRNAs. The first contained the 3′-UTR from the *Drosophila* gene CG10011, including a binding site for miR-12. The second contained the 3′-UTR from the Nerfin gene, including a binding site for miR-9b. The third contained the 3′-UTr from the Vha68-1 gene, also including a miR-9b binding site. When these workers measured mRNA levels and luciferase activities in cells co-transfected with each of the reporter genes and their cognate miRNAs, they found the following: (1) Silencing of the luciferase-CG10011 reporter by miR-12 appeared to operate exclusively by reducing the level of the transcript. (2) Silencing of the luciferase-Nerfin reporter by miR-9 involved primarily a reduction in translation efficiency. (3) Silencing of the luciferase-Vha68-1 reporter used a combination of the two mechanisms, mRNA level reduction and translation inhibition.

Next, Izaurralde and colleagues measured luciferase activities and mRNA levels in *Drosophila* S2 cells transfected with each of the reporters and the miRNAs, and also depleted of CAF1, NOT1, DCP1/DCP2, or GW182 by knockdown. Control knockdowns were depleted of the essential Ago1 or the irrelevant green fluorescent protein (GFP). As expected, knockdown of Ago1 or GW182 resulted in normal luciferase activities and mRNA levels from all reporters, even in the presence of cognate miRNAs. That is because silencing by miRNAs depends on both Ago1 and GW182. And because silencing of these reporter mRNAs depends on both translation inhibition and mRNA decay, it appears that both Ago1 and GW182 are involved in both silencing mechanisms.

In miRNA-treated, NOT1-depleted cells, CG10011 and Vha68-1 mRNAs were restored to non-miRNA-treated levels, and luciferase activities were partially restored. Silencing of these two reporters depends wholly or principally on mRNA decay and deadenylation is a key part of that decay. Thus, it is not surprising that removing the deadenylation enzyme NOT1 prevents such mRNA decay. On the other hand, depleting NOT1 in miRNA-treated cells had no effect on the loss of luciferase activity from the luciferase-Nerfin reporter. Because the luciferase-Nerfin reporter responds to miRNA by decreasing translation efficiency, rather than by mRNA decay, this result suggests that, while deadenylation is an essential part of mRNA decay, it is not required for miR-9a-mediated translation silencing of the luciferase-Nerfin reporter.

Depletion of DCP1/DCP2 in miRNA-treated cells restored the levels of all three reporter mRNAs to normal. Although none of the mRNAs presumably suffered decapping in these cells, they all were deadenylated. Taken together, these two findings suggest that deadenylation alone cannot initiate mRNA decay, for example by a $3' \rightarrow 5'$ exonuclease. Thus, it is more likely that deadenylation and decapping are followed by mRNA degradation by a $5' \rightarrow 3'$ exonuclease. Also, the fact that all three reporter mRNAs were deadenylated helps explain why the luciferase activities from all three reporter mRNAs remained low: Deadenylation presumably inhibited translation of these mRNAs.

SUMMARY P-bodies are cellular foci where mRNAs are destroyed or translationally repressed. GW182 is an essential part of the *Drosophila* miRNA silencing mechanism in P-bodies, whether this mechanism involves translation inhibition or mRNA decay. Ago1 probably recruits GW182 to an mRNA within a P-body, and this marks that mRNA for silencing. GW182 and Ago1-mediated mRNA decay in P-bodies appears to involve both deadenylation and decapping, followed by mRNA degradation by a $5' \rightarrow 3'$ exonuclease.

Relief of Repression in P-Bodies

There is a flow of mRNAs back and forth between polysomes and P-bodies. Therefore, the more an mRNA is associated with polysomes, and is therefore being actively translated, the less that mRNA will be found in P-bodies. And conversely, mRNAs that are enriched in P-bodies are poorly represented in polysomes. Although many mRNAs are degraded in P-bodies, many others are merely held and repressed there, and may rejoin polysomes once cellular conditions change.

Witold Filipowicz and colleagues provided good evidence for this dynamic association between repressed mRNAs and P-bodies in their studies on the human cationic amino acid transporter (CAT-1), which transports lysine and arginine into cells. CAT-1 is normally kept at low levels in liver cells to prevent loss of arginine from serum. That loss would occur because liver cells have a high concentration of arginase, which rapidly degrades imported arginine. But, under certain stress conditions, including amino acid starvation, liver cells need to import more arginine, and the CAT-1 level is up-regulated. Filipowicz and colleagues showed that the reason CAT-1 levels are low in liver cells is that a miRNA represses CAT-1 mRNA translation in those cells. Furthermore, the relief of repression of CAT-1 mRNA translation under stress conditions is accompanied by a loss of CAT-1 mRNA from P-bodies.

Filipowicz and colleagues chose to study Huh7 hepatoma cells because evidence suggested that CAT-1 expression in these cells was controlled by an miRNA known as miR-122. First, these workers used a Western blot to show that the CAT-1 concentration was significantly lower in Huh7 cells than in three other human cell lines (Figure 16.44a). Then they used a Northern blot to establish that the CAT-1 mRNA levels were essentially the same in all four human cell lines (Figure 16.44b). Thus, control of CAT-1 levels in Huh7 cells does not occur at the transcriptional level, or even at the level of mRNA stability, but probably at the translational level.

Is this control dependent on miR-122? Possibly, because the Northern blot in Figure 16.44c reveals that, of the four cell lines, only Huh7 expresses miR-122. Furthermore, if miR-122 is really responsible, we would expect that treatment of cells with an anti-miR-122 oligonucleotide would abolish the control, and CAT-1 levels would rise in cells treated with the antisense oligonucleotide. Figure 16.44d shows that this is indeed what happened, whereas irrelevant oligonucleotides had no effect. This increase in CAT-1 protein was not reflected in an increase in CAT-1 mRNA, suggesting again that the regulation was occurring at the translational level.

To investigate further the role of miR-122 in control of CAT-1 production, Filipowicz and colleagues made a series

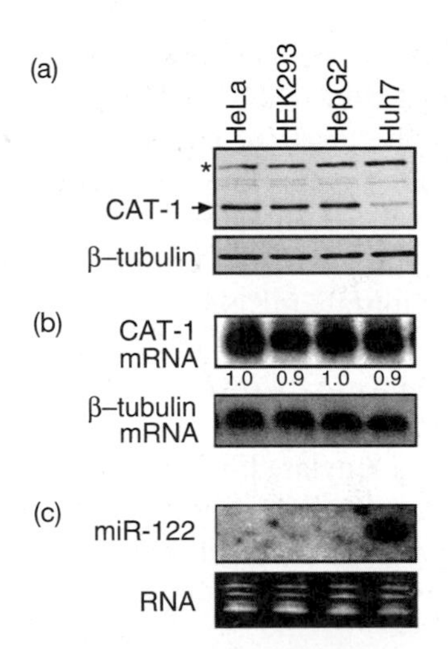

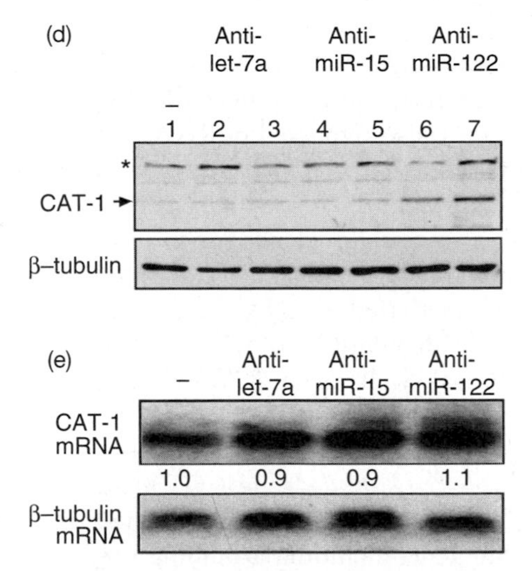

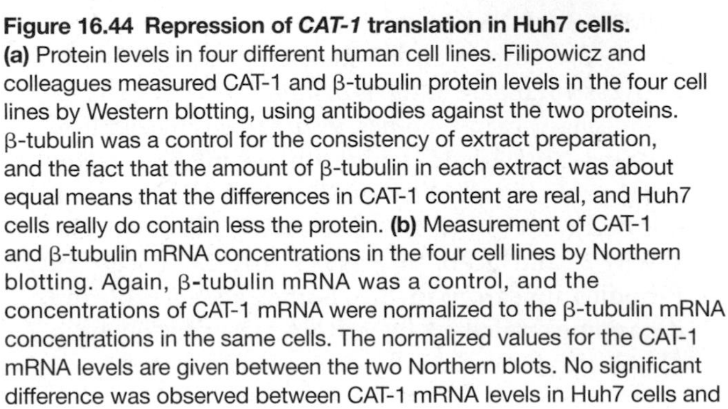

Figure 16.44 Repression of *CAT-1* translation in Huh7 cells. **(a)** Protein levels in four different human cell lines. Filipowicz and colleagues measured CAT-1 and β-tubulin protein levels in the four cell lines by Western blotting, using antibodies against the two proteins. β-tubulin was a control for the consistency of extract preparation, and the fact that the amount of β-tubulin in each extract was about equal means that the differences in CAT-1 content are real, and Huh7 cells really do contain less the protein. **(b)** Measurement of CAT-1 and β-tubulin mRNA concentrations in the four cell lines by Northern blotting. Again, β-tubulin mRNA was a control, and the concentrations of CAT-1 mRNA were normalized to the β-tubulin mRNA concentrations in the same cells. The normalized values for the CAT-1 mRNA levels are given between the two Northern blots. No significant difference was observed between CAT-1 mRNA levels in Huh7 cells and in the other three cell lines. **(c)** Upper panel: Northern blot analysis of miR-122 concentration in the four cells lines. Lower panel: Ethidium bromide staining of the gel used for the Northern blot, showing roughly equal amounts of RNA in all lanes. **(d)** Western blot analysis of the effects of miRNA antisense oligonucleotides on CAT-1 levels in Huh7 cells. Only the anti-miR-122 had a stimulatory effect. **(e)** Northern blot analysis of the effects of miRNA antisense oligonucleotides on CAT-1 and β-tubulin mRNA levels in Huh7 cells. CAT-1 mRNA levels were normalized to β-tubulin levels in the same extracts and the normalized values are presented between the two Northern blots. The anti-miR-122 oligonucleotides had no significant effect on CAT-1 mRNA level. (*Source:* Reprinted from CELL, Vol. 125, Bhattacharyya et al, Relief of microRNA-Mediated Translational Repression in Human Cells Subjected to Stress, Issue 6, 13 June 2006, pages 1111–1124, © 2006, with permission from Elsevier.)

of reporter constructs containing the *Renilla* luciferase coding region fused to various versions of the CAT-1 mRNA 3′-UTR. Then they tested these constructs in Huh7 and HepG2 cells. In HepG2 cells, in which the CAT-1 gene is not regulated, they found that constructs containing the miR-122 binding sites produced the same amount of luciferase as constructs lacking these sites. However, in Huh7 cells, in which the CAT-1 gene is regulated, reporter constructs lacking the miR-122 binding sites produced about three times more luciferase than constructs that contained these sites. Again, Northern blot analysis showed that mRNA levels did not vary, even though luciferase levels did. These findings support the hypothesis that CAT-1 production is controlled negatively by miR-122.

Based on what we know so far, we would predict that starvation for amino acids should derepress CAT-1 production in Huh7 cells, and this stimulatory effect should depend on miR-122. Accordingly, Filipowicz and colleagues starved Huh7 and HepG2 cells for amino acids and used Western blots to assay the effects on CAT-1 expression. As predicted, they observed a four-fold increase in CAT-1 level upon starvation of Huh7 cells, but not HepG2 cells, and this effect occurred within one hour. On the other hand, Northern blots showed that, while there was a 1.8-fold increase in CAT-1 mRNA level, this effect was undetectable until after three h of starvation. These results indicate that the stimulatory effect of starvation on Huh7 cells occurs via enhanced translation of preexisting CAT-1 mRNA.

The use of luciferase reporter constructs with and without miR-122 binding sites showed that the stimulatory response to starvation in Huh7 cells occurred only with constructs containing these sites. Thus, the derepression appeared to be dependent on miR-122. To check this conclusion, Filipowicz and colleagues turned to HepG2 cells, which do not normally express miR-122, and in which CAT-1 production is not inducible by starvation. To these cells, they added a miR-122 gene construct that would be expressed constitutively. In these engineered cells, a luciferase reporter construct with the CAT-1 mRNA 3′-UTR was activated by starvation, indicating that miR-122 is really involved in the repression observed in Huh7 cells.

Another interesting finding came from these studies in HepG2 cells: A luciferase reporter construct containing just the miR-122 binding sites from the CAT-1 mRNA 3′-UTR was *not* responsive to starvation. This result spurred Filipowicz and colleagues to look more closely at the CAT-1 mRNA 3′-UTR. They focused on a part of the 3′-UTR known as region D, which contains an ARE, which they named ARD. This is not a binding site for miR-122, or any other known miRNA, but it is a binding site for a protein known as HuR. This finding led to the hypothesis that HuR, in addition to miR-122, is required for regulation of CAT-1 production in starved Huh7 cells.

To test this hypothesis, Filipowicz and colleagues first demonstrated that knocking down the cellular level of HuR by RNAi abolished the responsiveness to starvation of luciferase reporters bearing the CAT-1 mRNA 3′-UTR in Huh7 cells. Thus, HuR does seem to be required for CAT-1 regulation. Second, they showed that HuR binds to the CAT-1 mRNA 3′-UTR by immunoprecipitating reporter constructs bearing the CAT-1 mRNA 3′-UTR with an anti-HuR antibody. As expected, the construct containing only the miR-122 binding sites, but not the region D, could not be immunoprecipitated with this antibody. A second set of binding studies using a gel mobility shift assay showed that complexes formed between a labeled region D RNA fragment and a GST-HuR fusion protein. It is significant that reporter constructs containing only a region D, with no miR-122 binding sites, were not subject to regulation in Huh7 cells. Thus, HuR and miR-122 act together to regulate expression of the CAT-1 gene.

Because it was known that repressed mRNAs could be found in P-bodies, while actively translated mRNAs are found in polysomes, Filipowicz and colleagues looked in these compartments for CAT-1 mRNA and luciferase reporters under starved and unstarved conditions. Figure 16.45a shows immunofluorescence data for CAT-1 mRNA (detected by in situ hybridization with a red-fluorescent-tagged CAT-1 antisense probe). In fed cells, the red CAT-1 mRNA was found in discrete cytoplasmic bodies. We know they are P-bodies because a marker for P-bodies, GFP-Dcp1a, which fluoresces green, co-localizes with the red fluorescing CAT-1 mRNA. Together, the red and green fluorescence produce the yellow color seen in the right hand panel. Transfecting the cells with an anti-miR-122 antisense RNA abolished the P-body location of the CAT-1 mRNA in fed cells (Figure 16.45b), demonstrating that this localization is miR-122-dependent.

On the other hand, in starved cells, CAT-1 mRNA was no longer detectable in P-bodies (Figure 16.45a). Was all miR-122 lost from the P-bodies along with the CAT-1 mRNA? Figure 16.45c, in which miR-122 was detected by in situ hybridization with a red-fluorescing probe, shows that it was not. Thus, miR-122 presumably regulates the translation of a large number of mRNAs in liver cell P-bodies, so the loss of one (or perhaps a few) regulated mRNAs during starvation did not significantly lower the miR-122 concentration in these P-bodies.

Did the CAT-1 mRNA in starved cells move from the P-bodies to polysomes? To find out, Filipowicz and colleagues displayed polysomes by sucrose gradient ultracentrifugation and assayed each sample for CAT-1 mRNA by Northern blotting. Figure 16.45d shows a big increase in CAT-1 mRNA in polysomes upon starvation of Huh7 cells, and Figure 16.45e quantifies this effect. This effect is specific to CAT-1 mRNAs. Most mRNAs react to starvation as the control β-tubulin mRNA did in Figure 16.45d and e: They move out of polysomes.

Filipowicz and colleagues also showed that the migration of CAT-1 mRNA from P-bodies to polysomes in

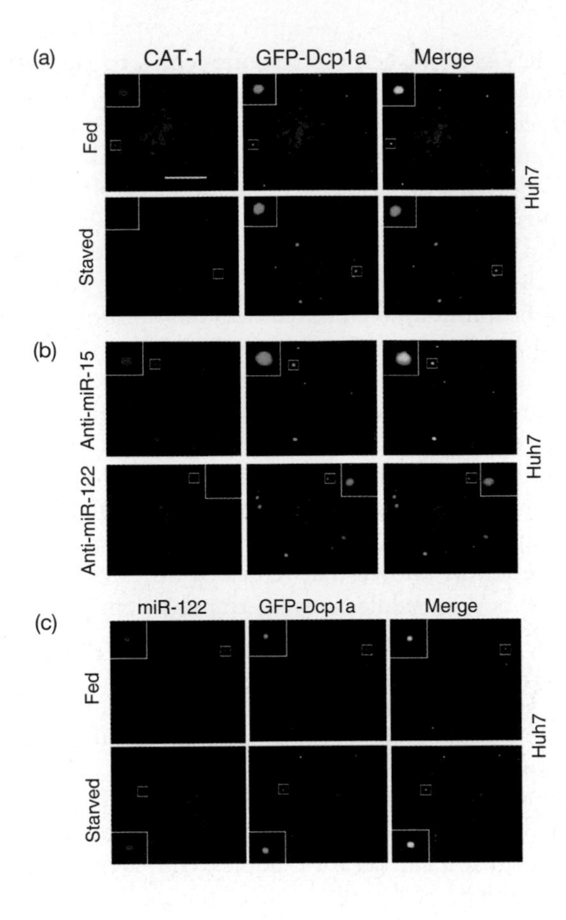

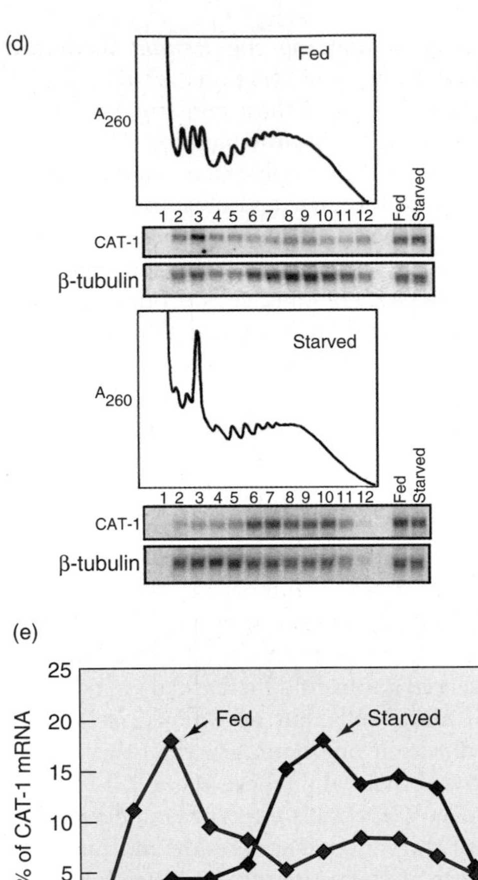

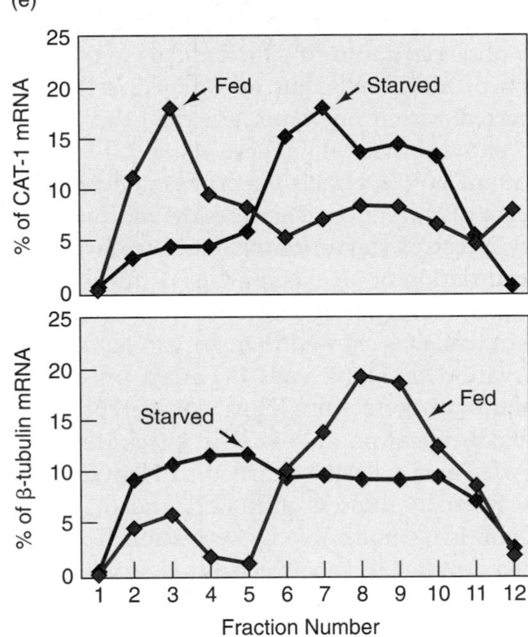

Figure 16.45 Starvation-induced relocation of CAT-1 mRNA from P-bodies to polysomes. (a) Loss of CAT-1 mRNA from P-bodies upon starvation in Huh7 cells. CAT-1 mRNA (left column) was detected by in situ hybridization with a red-fluorescent-tagged probe. The P-body marker, GFP-Dcp1a (middle column) fluoresces green. The right column is a merged view of the other two columns. In each micrograph, a P-body (small square) was selected, enlarged and presented in the large square at the upper left corner. The top row contains fed cells, and the bottom row, starved cells, as indicated at left. In fed cells, the merged view is yellow, reflecting the co-localization of the CAT-1 mRNA (red) and GFP-Dcp1a (green). In starved cells, there is essentially no red fluorescence in the P-bodies, so the merged view is green. **(b)** Effect of two antisense miRNAs on P-body localization of CAT-1 mRNA in fed cells. The irrelevant anti-miR-15 had no effect, but the anti-miR-122 blocked the localization of CAT-1 mRNA to P-bodies. Staining of the cells in the three columns was as in panel (a). **(c)** Presence of miR-122 in P-bodies in fed and starved Huh7 cells. Staining of the cells in the three columns was as in panel (a) except that a red-fluorescing anti-miR-122 oligonucleotide was used in the left-hand coumn. **(d)** Polysome analysis. Polysomes from fed and starved cells were displayed by sucrose gradient ultracentrifugation, and gradient fractions were subjected to Northern blotting and probed for either CAT-1 mRNA or β-tubulin mRNA, as indicated at left. Input RNA from fed and starved cells is probed at right. Starvation caused an increase in CAT-1 mRNA, but a decrease in β-tubulin mRNA, in heavy polysomes. **(e)** Graphic representation of the data from panel (d). The amount of CAT-1 (top) and β-tubulin (bottom) mRNAs are plotted vs. gradient fraction number in polysome profiles from fed (red) and starved (blue) cells. (*Source:* Reprinted from CELL, Vol. 125, Bhattacharyya et al, Relief of microRNA-Mediated Translational Repression in Human Cells Subjected to Stress, Issue 6, 13 June 2006, pages 1111–1124, © 2006, with permission from Elsevier.)

starved cells depended on HuR and region D of the CAT-1 mRNA 3′-UTR. They demonstrated that HuR moved with CAT-1 mRNA from P-bodies to polysomes upon amino acid starvation. Furthermore, when they knocked down HuR in starved Huh7 cells, they found that CAT-1 mRNA no longer relocated from P-bodies to polysomes.

If HuR helps move CAT-1 mRNA out of P-bodies upon starvation, then perhaps endowing another mRNA with the HuR binding site (region D) would enable it to move out of P-bodies under the same conditions. Filipowicz and colleagues tested this prediction by placing region D into another luciferase reporter mRNA (RL-3XBulge) that is responsive to the miRNA *let-7*. Ordinarily, this reporter mRNA is directed to P-bodies in cells, such as HeLa cells, that express *let-7*, and does not move out of P-bodies upon starvation. However, with region D added, the mRNA responded to starvation in HeLa cells by exiting the P-bodies. All this evidence points to an important role for HuR in transporting CAT-1 mRNA out of P-bodies in starved cells. It also suggests that the stress-related reactivation of mRNAs undergoing miRNA-mediated repression may be a general phenomenon that applies to a variety of mRNAs in a variety of cell types.

SUMMARY In a liver cell line (Huh7), translation of the CAT-1 mRNA is repressed by the miRNA miR-122, and the mRNA is sequestered in P-bodies. Upon starvation, the translation repression of the CAT-1 mRNA is relieved and the mRNA migrates from P-bodies to polysomes. This derepression and translocation of the mRNA depends on the mRNA-binding protein HuR, and on its binding site (region D) in the 3′-UTR of the mRNA. Such derepression and translocation in response to stress may be a common response of miRNA-repressed mRNAs.

Other Small RNAs

Since the discoveries of siRNAs, miRNAs, and piRNAs, other small RNAs have been found, although the functions of these RNAs are largely still unknown. One example is the **endo-siRNAs** of *Drosophila*. Like miRNAs, these are made from *Drosophila* genes as double-stranded RNA precursors. However, like siRNAs, these RNA precursors are processed by the Dicer-2 (DCR-2) pathway, and are loaded onto a RISC that contains Ago2. Thus, even though these RNAs are produced endogenously, their processing pathway suggests that they should be called siRNAs, rather than miRNAs. Accordingly, we call them endo-siRNAs, even as we acknowledge that these RNAs blur the line between siRNAs and miRNAs.

It is interesting that fruit flies with defective DCR-2 or Ago2 experience an increased level of transposon expression in somatic cells. This finding suggests that endo-siRNAs may help protect somatic cells against transposition, just as piRNAs protect germ cells.

SUMMARY Endo-siRNAs of *Drosophila* are encoded in the cellular genome, yet they are processed like siRNAs, rather than miRNAs. They may help protect somatic cells against transposons.

SUMMARY

Ribosomal RNAs are made in eukaryotic nucleoli as precursors that must be processed to release the mature rRNAs. The order of RNAs in the precursor is 18S, 5.8S, 28S in all eukaryotes, although the exact sizes of the mature rRNAs vary from one species to another. In human cells, the precursor is 45S, and the processing scheme creates 41S, 32S, and 20S intermediates. The snoRNAs play vital roles in these processing steps.

Extra nucleotides are removed from the 5′-ends of pre-tRNAs in one step by an endonucleolytic cleavage catalyzed by RNase P. RNase P's from bacteria and eukaryotic nuclei have a catalytic RNA subunit called M1 RNA. RNase II and polynucleotide phosphorylase cooperate to remove most of the extra nucleotides at the 3′-end of an *E. coli* tRNA precursor, but stop at the +2 stage. RNases PH and T are most active in removing the last two nucleotides from the RNA. In eukaryotes, a single enzyme, tRNA 3′-processing endoribonuclease (3′-tRNase), processes the 3′-end of a pre-tRNA.

Trypanosome mRNAs are formed by *trans*-splicing between a short leader exon and any one of many independent coding exons.

Trypanosomatid mitochondria (kinetoplastids) encode incomplete mRNAs that must be edited before they can be translated. Editing occurs in the 3′→5′ direction by successive action of one or more guide RNAs. These gRNAs hybridize to the unedited region of the mRNA and provide A's and G's as templates for the incorporation of U's missing from the mRNA or deletion of extra U's.

Some adenosines in mRNAs of higher eukaryotes, including fruit flies and mammals, must be deaminated to inosine post-transcriptionally for the mRNAs to code for the proper proteins. Enzymes known as adenosine deaminases active on RNAs (ADARs) carry out this kind of RNA editing. In addition, some cytidines must be deaminated to uridine for an mRNA to code properly.

A common form of post-transcriptional control of gene expression is control of mRNA stability. For example, the mammalian casein and transferrin receptor (Tfr) genes are controlled primarily by altering the stabilities of their mRNAs. When cells have abundant iron, the level of tranferrin receptor is reduced to avoid accumulation of too much iron in cells. Conversely, when cells are starved for iron, they increase the concentration of transferrin receptor to transport as much iron as possible into the cells. The transferrin receptor (TfR) mRNA stability is controlled as follows: The 3′-UTR of the TfR mRNA contains five stem-loops called iron response elements (IREs), which render the mRNA susceptible to degradation by RNase. When iron concentration is low, aconitase exists as an apoprotein that lacks iron. This protein binds to the IREs in the TfR mRNA and protects the RNA against attack by RNases. But when iron concentration is high, the aconitase apoprotein binds to iron and therefore cannot bind to the mRNA IREs. This leaves the RNA vulnerable to degradation.

RNA interference occurs when a cell encounters dsRNA from a virus, a transposon, or a transgene (or experimentally added dsRNA). This trigger dsRNA is degraded into 21–23-nt fragments (siRNAs) by an RNase III-like enzyme called Dicer. The double-stranded siRNA, with Dicer and the Dicer-associated protein R2D2, recruit Ago2 to form a pre-RISC complex that can separate the siRNA into its two component strands: the guide strand, which will base-pair with the target mRNA in the RNA-induced silencing complex (RISC) and guide cleavage of the mRNA, and the passenger strand, which will be discarded. Ago2 cleaves the passenger strand, which then falls off the pre-RISC complex. The guide strand of the siRNA then base-pairs with the target mRNA in the active site in the PIWI domain of Ago2, which is an RNase H-like enzyme, also known as slicer. Slicer cleaves the target mRNA in the middle of the region of its base-pairing with the siRNA. In an ATP-dependent step, the cleaved mRNA is ejected from the RISC, which can then accept a new molecule of mRNA to be degraded. In certain species, the siRNA is amplified during RNAi when antisense siRNAs hybridize to target mRNA and prime synthesis of full-length antisense RNA by an RNA-dependent RNA polymerase. This new dsRNA is then digested by Dicer into new pieces of siRNA.

The RNAi machinery is involved in heterochromatization at yeast centromeres and silent mating-type regions, and is also involved in heterochromatization in other organisms. At the outermost regions of centromeres of fission yeast, active transcription of the reverse strand occurs. Occasional forward transcripts, or forward transcripts made by RdRP, base-pair with the reverse transcripts to kick off RNAi, which in turn recruits a histone methyltransferase, which methylates lysine 9 of histone H3, which recruits Swi6, which causes heterochromatization. In plants and mammals, this process is abetted by DNA methylation, which can also attract the heterochromatization machinery. Individual genes in mammals can also be silenced by RNAi, which targets the control region, rather than the coding region, of the gene. This silencing process involves DNA methylation, rather than mRNA destruction.

MicroRNAs (miRNAs) are 18–25-nt RNAs produced from a cellular RNA with a stem-loop structure. In the last step in miRNA synthesis, Dicer cleaves the double-stranded stem part of the precursor to yield the miRNA in double-stranded form. The single-stranded forms of these miRNAs can team up with an Argonaute protein in a RISC to control the expression of other genes by base-pairing to their mRNAs. In animals, miRNAs tend to base-pair imperfectly to the 3′-UTRs of their target mRNAs and inhibit accumulation of the protein products of these mRNAs. However, perfect or perhaps even imperfect base-pairing between an animal miRNA and its target mRNA can result in mRNA cleavage. In plants, miRNAs tend to base-pair perfectly or near-perfectly with their target mRNAs and cause cleavage of these mRNAs, although there are exceptions in which translation blockage can occur.

MicroRNAs can activate, as well as repress translation. In particular, miR369-3, with the help of Ago2 and FXR1, activates translation of the TNFα mRNA in serum-starved cells. On the other hand, miR369-3, with the help of Ago2, represses translation of the mRNA in synchronized cells growing in serum.

RNA polymerase II transcribes the miRNA precursor genes, to produce pri-miRNAs, which may encode more than one miRNA. Processing a pri-mRNA to a mature miRNA is a two-step process. In the first step, a nuclear RNase III known as Drosha cleaves the pri-miRNA to release a 60–70-nt stem-loop RNA known as a pre-miRNA. In the second step, which occurs in the cytoplasm, Dicer cuts the pre-miRNA within the stem to release a mature double-stranded miRNA. A mirtron is an intron that consists of a pre-miRNA. Thus, the spliceosome cuts it out of its pre-mRNA, then it is debranched and folded into a stem-loop pre-miRNA, without any participation by Drosha.

P-bodies are cellular foci where mRNAs are stored, destroyed, and translationally repressed. GW182 is an essential part of the *Drosophila* miRNA silencing mechanism in P-bodies, whether this mechanism involves translation inhibition or mRNA decay. AGO1 probably recruits GW182 to an mRNA within a P-body, and this marks that mRNA for silencing. GW182 and AGO1-mediated mRNA decay in P-bodies appears to involve both deadenylation and decapping, followed by mRNA degradation by a 5′→3′ exonuclease.

In a liver cell line (Huh7), translation of the CAT-1 mRNA is repressed by the miRNA miR-122, and the mRNA is sequestered in P-bodies. Upon starvation, the translation repression of the CAT-1 mRNA is relieved and the mRNA migrates from P-bodies to polysomes. This derepression and translocation of the mRNA depends on the mRNA-binding protein HuR, and on its binding site (region D) in the 3′-UTR of the mRNA. Such derepression and translocation in response to stress may be a common response of miRNA-repressed mRNAs.

Endo-siRNAs of *Drosophila* are encoded in the cellular genome, yet they are processed like siRNAs, rather than miRNAs. They may help protect somatic cells against transposons.

REVIEW QUESTIONS

1. Draw the structure of a mammalian rRNA precursor, showing the locations of all three mature rRNAs.
2. What is the function of RNase P? What is unusual about this enzyme (at least the bacterial and eukaryotic nuclear forms of the enzyme)?
3. Illustrate the difference between *cis*- and *trans*-splicing.
4. Describe and give the results of an experiment that shows that a Y-shaped intermediate exists in the splicing of a trypanosome pre-mRNA. Show how this result is compatible with *trans*-splicing, but not with *cis*-splicing.
5. Describe what we mean by RNA editing. What is a cryptogene?
6. Describe and give the results of an experiment that shows that editing of kinetoplast mRNA goes in the 3′→5′ direction.
7. Draw a diagram of a model of RNA editing that fits the data at hand. What enzymes are involved?
8. Present direct evidence for guide RNAs.
9. Outline the evidence that shows that editing of the mouse *GluR-B* transcript by ADAR2 is essential, and that this transcript is the only critical target of ADAR2.
10. Describe and give the results of an experiment that shows that prolactin controls the casein gene primarily at the post-transcriptional level.
11. What two proteins are most directly involved in iron homeostasis in mammalian cells? How do their levels respond to changes in iron concentration?
12. How do we know that a protein binds to the iron response elements (IREs) of the TfR mRNA?
13. Describe and give the results of an experiment that shows that one kind of mutation in the TfR IRE region results in an iron-unresponsive and stable mRNA, and another kind of mutation results in an iron-unresponsive and unstable mRNA. Interpret these results in terms of the rapid turnover determinant and interaction with IRE-binding protein(s).
14. Present a model for the involvement of aconitase in determining the stability of TfR mRNA.
15. What evidence suggests that RNA interference depends on mRNA degradation?
16. Present a model for the mechanism of RNA interference.
17. Describe and give the results of an experiment that shows that Argonaute2 has slicer activity.
18. What roles do R2D2 and Ago2 play in formation of the RISC? What happens if R2D2 is absent?
19. Diagram the ping-pong mechanism whereby piRNAs are thought to amplify themselves and inactivate transposons at the same time.
20. Present a model for the involvement of the RNAi machinery in heterochromatization in fission yeast. How would this model have to be modified to describe the situation in mammals?
21. Present a model for gene silencing and heterochromatization in flowering plants. In what major ways does this differ from the model in fission yeast?
22. What is the evidence for the importance of non-siRNA transcripts in gene silencing in fission yeast and in flowering plants?
23. Chromatin targets for heterochromatization in dividing cells must be transcribed in order to be silenced. How is this problem resolved in fission yeast and in flowering plants?
24. Describe and give the results of experiments showing: (1) that a mammalian gene can be silenced by a mechanism involving an siRNA directed at the gene's control region; and (2) that DNA methylation is involved in the silencing.
25. Outline the processes by which siRNAs and miRNAs are produced. List the key players in these processes. Be sure to include two different ways to produce pre-miRNAs.
26. How can siRNAs that target the promoter region of a gene be made? Present evidence to support your hypothesis.
27. Compare and contrast the typical actions of siRNAs and miRNAs in animals.
28. MicroRNAs in animals typically base-pair imperfectly to their targets in the 3′-UTRs of mRNAs. How does their activity change if they base-pair perfectly, or near-perfectly? Present evidence.
29. Describe an example in which an miRNA activates translation of a gene. How was this activation assayed? Present evidence that base-pairing between this miRNA and the mRNA's ARE is important in activation.
30. Describe and present the results of an experiment that shows that the protein GW182 can reduce translation of an mRNA in P-bodies. Include a description of how the protein can be physically tethered to the mRNA. How much of the loss of protein product is due to mRNA destruction, and how much is due to translation repression? How can these two effects be experimentally separated?
31. Describe and give the results of experiments that show that:
    (a) translation of an mRNA is repressed by an miRNA in P-bodies.
    (b) this repression can be overcome in stressed cells.
    (c) an mRNA-binding protein is also required for relief of repression.
    (d) relief of repression is accompanied by the translocation of the mRNA from P-bodies to polysomes.

ANALYTICAL QUESTIONS

1. Why can *dicer* dsRNA never completely block RNAi?
2. Predict the effects of the following mutations on the abundance of the TfR mRNA. That is, would the mutations result in a constitutively low or high level of the TfR mRNA regardless of iron concentration, or would they have no effect on the mRNA level?
   a. A mutation that blocks the production of aconitase.
   b. A mutation that prevents aconitase from binding iron.
   c. A mutation that prevents aconitase from binding to the IREs.
3. Discuss the conflicting evidence about the effect of *lin-4* miRNA on expression of the *lin-14* gene in *C. elegans.*

SUGGESTED READINGS

General References and Reviews

Aravin, A.A., G.J. Hannon, and J. Brennecke. 2007. The Piwi-piRNA pathway provides an adaptive defense in the transposon arms race. *Science* 318:761–64.

Bass, B.L. 2000. Double-stranded RNA as a template for gene silencing. *Cell* 101:235–38.

Carrington, J.C. and V. Ambros. 2003. Role of microRNAs in plant and animal development. *Science* 301:336–38.

Daxinger, L., T. Kanno, and M. Matzke. 2008. Pol V transcribes to silence. *Cell* 135:592–94.

Dernburg, A.F. 2002. A Chromosome RNAissance. *Cell* 11:159–62.

Eulalio, A., E. Huntzinger, and E. Izaurralde. 2008. Getting to the root of miRNA-mediated gene silencing. *Cell* 132:9–14.

Filipowicz, W. 2005. RNAi: The nuts and bolts of the RISC machine. *Cell* 122:17–20.

Keegan, L.P., A. Gallo, and M.A. O'Connell. 2000. Survival is impossible without an editor. *Science* 290:1707–09.

Nilsen, T.W. 1994. Unusual strategies of gene expression and control in parasites. *Science* 264:1868–69.

Pillai, R.S., S.N. Bhattacharyya, and W. Filipowicz. 2007. Repression of protein synthesis by miRNAs: How many mechanisms? *Trends in Cell Biology* 17:118–26.

Rouault, T.A. 2006. If the RNA fits, use it. *Science* 314:1886–87.

Seiwert, S.D. 1996. RNA editing hints of a remarkable diversity in gene expression pathways. *Science* 274:1636–37.

Simpson, L. and D.A. Maslov. 1994. RNA editing and the evolution of parasites. *Science* 264:1870–71.

Solner-Webb, B. 1996. Trypanosome RNA editing: Resolved. *Science* 273:1182–83.

Sontheimer, E.J. and R.W. Carthew. 2004. Argonaute journeys into the heart of RISC. *Science* 305:1409–10.

Research Articles

Abraham, J.M., J.E. Feagin, and K. Stuart. 1988. Characterization of cytochrome *c* oxidase III transcripts that are edited only in the 3′ region. *Cell* 55:267–72.

Bagga, S., J. Bracht, S. Hunter, K. Massirer, J. Holtz, R. Eachus, and A.E. Pasquinelli. 2005. Regulation by *let-7* and *lin-4* miRNAs results in target mRNA degradation. *Cell* 122:553–63.

Behm-Ansmant, I., J. Rehwinkel, T. Doerks, A. Stark, P. Bork, and E. Izaurralde. 2006. mRNA degradation by miRNAs and GW182 requires both CCR4:NOT deadenylase and CDP1:DCP2 decapping complexes. *Genes and Development* 20:1885–98.

Bhattacharyya, S., R. Habermacher, U. Martine, E.I. Closs, and W. Filipowicz. 2006. Relief of microRNA-mediated translational repression in human cells subjected to stress. *Cell* 125:1111–24.

Blum, B., N. Bakalara, and L. Simpson. 1990. A model for RNA editing in kinetoplastid mitochondria: "Guide" RNA molecules transcribed from maxicircle DNA provide the edited information. *Cell* 60:189–98.

Casey, J.L., M.W. Hentze, D.M. Koeller, S.W. Caughman, T.A. Rovault, R.D. Klausner, and J.B. Harford. 1988. Iron-responsive elements: Regulatory RNA sequences that control mRNA levels and translation. *Science* 240:924–28.

Casey, J.L., D.M. Koeller, V.C. Ramin, R.D. Klausner, and J.B. Harford. 1989. Iron regulation of transferrin receptor mRNA levels requires iron-responsive elements and a rapid turnover determinant in the 3′ untranslated region of the mRNA. *EMBO Journal* 8:3693–99.

Feagin, J.E., J.M. Abraham, and K. Stuart. 1988. Extensive editing of the cytochrome c oxidase III transcript in *Trypanosoma brucei. Cell* 53:413–22.

Fire, A., S. Xu, M.K. Montgomery, S.A. Kostas, S.E. Driver, and C.C. Mello. 1998. Potent and specific genetic interference by double-stranded RNA in *Caenorhabditis elegans. Nature* 391:806–11.

Fukagawa, T., M. Nogami, M. Yoshikawa, M. Ikeno, T. Okazaki, Y. Takami, T. Nakayama, and M. Oshimura. 2004. Dicer is essential for formation of the heterochromatin structure in vertebrate cells. *Nature Cell Biology* 6:784–91.

Guerrier-Takada, C., K. Gardiner, T. Marsh, N. Pace, and S. Altman. 1983. The RNA moiety of ribonuclease P is the catalytic subunit of the enzyme. *Cell* 35:849–57.

Guyette, W.A., R.J. Matusik, and J.M. Rosen. 1979. Prolactin-mediated transcriptional and post-transcriptional control of casein gene expression. *Cell* 17:1013–23.

Hall, I.M., G.D. Shankaranarayana, K.-i. Noma, N. Ayoub, A. Cohen, and S.I.S. Grewal. 2002. Establishment and maintenance of a heterochromatin domain. *Science* 297:2232–37.

Hammond, S.M., E. Bernstein, D. Beach, and G.J. Hannon. 2000. An RNA-directed nuclease mediates post-transcriptional gene silencing in *Drosophila* cells. *Nature* 404:293–96.

Han, J., D. Kim, and K.V. Morris. 2007. Promoter-associated RNA is required for RNA-directed transcriptional gene silencing in human cells. *Proceedings of the National Academy of Sciences* 104:12422–27.

Johnson, P.J., J.M. Kooter, and P. Borst. 1987. Inactivation of transcription by UV irradiation of *T. brucei* provides evidence for a multicistronic transcription unit including a VSG gene. *Cell* 51:273–81.

Kable, M.L., S.D. Seiwart, S. Heidmann, and K. Stuart. 1996. RNA editing: A mechanism for gRNA-specified uridylate insertion into precursor mRNA. *Science* 273:1189–95.

Koeller, D.M., J.L. Casey, M.W. Hentze, E.M. Gerhardt, L.-N.L. Chan, R.D. Klausner, and J.B. Harford. 1989. A cytosolic protein binds to structural elements within the iron regulatory region of the transferrin receptor mRNA. *Proceedings of the National Academy of Sciences USA* 86:3574–78.

Koeller, D.M., J.A. Horowitz, J.L. Casey, R.D. Klausner, and J.B. Harford. 1991. Translation and the stability of mRNAs encoding the transferrin receptor and *c-fos*. *Proceedings of the National Academy of Sciences USA* 88:7778–82.

Lee, R.C., R.L. Feinbaum, and V. Ambros. 1993. The C. *elegans* heterochronic gene *lin-4* encodes small RNAs with antisense complementarity to *lin-14*. *Cell* 75:843–54.

Li, Z. and M.P. Deutscher. 1994. The role of individual exoribonucleases in processing at the 3′ end of *Escherichia coli* tRNA precursors. *Journal of Biological Chemistry* 269:6064–71.

Lipardi, C., Q. Wei, and B.M. Paterson. 2001. RNAi as random degradative PCR: siRNA primers convert mRNA into dsRNAs that are degraded to generate new siRNAs. *Cell* 107:297–307.

Liu, J., M.A. Carmell, F.V. Rivas, C.G. Marsden, J.M. Thomson, J.-J. Song, S.M. Hammond, L. Joshua-Tor, and G.J. Hannon. 2004. Argonaute2 is the catalytic engine of mammalian RNAi. *Science* 305:1437–41.

Miller, O.L., Jr., B.R. Beatty, B.A. Hamkalo, and C.A. Thomas, Jr. 1970. Electron microscopic visualization of transcription. *Cold Spring Harbor Symposia on Quantitative Biology*. 35:505–12.

Morris, K.V., S.W.-L. Chan, S.E. Jacobsen, and D.J. Looney. 2004. Small interfering RNA-induced transcriptional gene silencing in human cells. *Science* 305:1289–92.

Müllner, E.W. and L.C. Kühn. 1988. A stem-loop in the 3′ untranslated region mediates iron-dependent regulation of transferrin receptor mRNA stability in the cytoplasm. *Cell* 53:815–25.

Murphy, W.J., K.P. Watkins, and N. Agabian. 1986. Identification of a novel Y branch structure as an intermediate in trypanosome mRNA processing: Evidence for *trans* splicing. *Cell* 47:517–25.

Olsen, P.H. and V. Ambros. 1999. The *lin-4* regulatory RNA controls developmental timing in *Caenorhabditis elegans* by blocking LIN-14 protein synthesis after the initiation of translation. *Developmental Biology* 216:671–80.

Owen, D. and L.C. Kühn. 1987. Noncoding 3′ sequences of the transferrin receptor gene are required for mRNA regulation by iron. *EMBO Journal* 6:1287–93.

Seiwert, S.D. and K. Stuart. 1994. RNA editing: Transfer of genetic information from gRNA to precursor mRNA in vitro. *Science* 266:114–17.

Sijen, T. J., Fleenor, F., Simmer, K.L., Thijssen, S., Parrish, L., Timmons, R.H.A., Plasterk, and A. Fire. 2001. On the role of RNA amplification in dsRNA-triggered gene silencing. *Cell* 107:465–76.

Song, J.-J., J. Liu, N.H. Tolia, J. Schneiderman, S.K. Smith, R.A. Martienssen, G.J. Hannon, and L. Joshua-Tor. 2003. The crystal structure of the Argonaute2 PAZ domain reveals an RNA binding motif in RNAi effector complexes. *Nature Structural Biology* 10:1026–32.

Song, J.-J., S.K. Smith, G.J. Hannon, and L. Joshua-Tor. 2004. Crystal structure of Argonaute and its implications for RISC slicer activity. *Science* 305:1434–37.

Vasudevan, S., Y. Tong, and J.A. Steitz. 2007. Switching from repression to activation: MicroRNAs can up-regulate translation. *Science* 318:1931–34.

Volpe, T.A., Kidner, I.M. Hall, G. Teng, S.I.S. Grewal, and R.A. Martienssen. 2002. Regulation of heterochromatic silencing and histone H3 lysine-9 methylation by RNAi. *Science* 297:1833–37.

Wang, Q., J. Khillan, P. Gaude, and K. Nishikura. 2000. Requirement of the RNA editing deaminase ADAR1 gene for embryonic erythropoiesis. *Science* 290:1765–8.

Weinberg, R.A., and S. Penman. 1970. Processing of 45S nucleolar RNA. *Journal of Molecular Biology* 47:169–78.

Wierzbicki, A.T., T.S. Ream, J.R. Haag, and C.S. Pikaard. 2009. RNA polymerase V transcription guides ARGONAUTE4 to chromatin. *Nature Genetics* 41:630–34.

Yekta, S., I.-h. Shih, and D.P. Bartel. 2004. MicroRNA-directed cleavage of *HOXB8* mRNA. *Science* 304:594–96.

Zamore, P.D., T. Tuschl, P.A. Sharp, and D.P. Bartel. 2000. RNAi: Double-stranded RNA directs the ATP-dependent cleavage of mRNA at 21 to 23 nucleotide intervals. *Cell* 101:25–33.

CHAPTER 17

The Mechanism of Translation I: Initiation

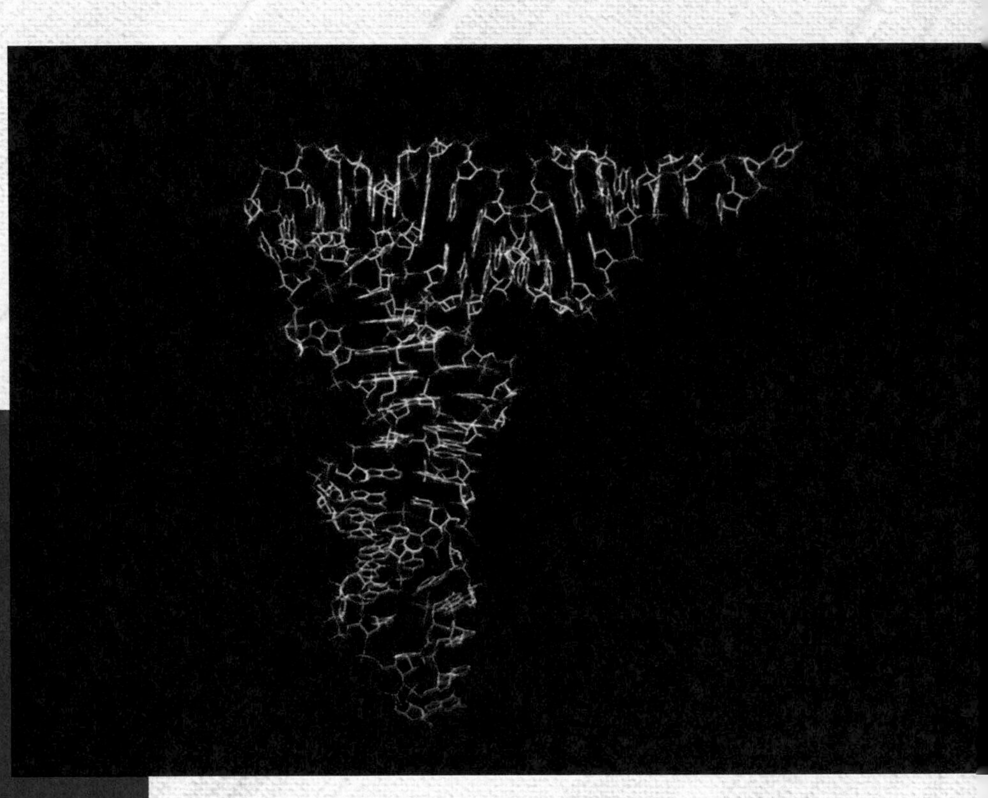

Cryo-electron microscopy model of the eIF3-mRNA-40S ribosomal particle complex. Yellow-green, ribosomal particle; magenta, eIF3; red, mRNA, with purple internal ribosomal entry site (IRES); e1, site of attachment of eIF1. *(© Tripos Associates/Peter Arnold/ PhotoLibrary Group)*

Translation is the process by which ribosomes read the genetic message in mRNA and produce a protein product according to the message's instructions. Ribosomes therefore serve as protein factories. Transfer RNAs (tRNAs) play an equally important role as adapters that can bind an amino acid at one end and interact with the mRNA at the other. Chapter 3 presented an outline of the translation process. In this chapter we will begin to fill in some of the details.

We can conveniently divide the mechanism of translation into three phases: initiation, elongation, and termination. In the initiation phase, the ribosome binds to the mRNA, and the first amino acid, attached to its tRNA, also binds. During the elongation phase, the ribosome adds one amino acid at a time to the growing polypeptide chain. Finally, in the termination phase, the ribosome releases the mRNA and the finished polypeptide. The overall scheme is similar in bacteria and eukaryotes, but there are significant differences, especially in the added complexity of the eukaryotic translation initiation system.

This chapter concerns the initiation of translation in eukaryotes and bacteria. Because the nomenclatures of the two systems are different, it is easier to consider them separately. Therefore, let us begin with a discussion of the simpler system, initiation in bacteria. Then we will move on to the more complex eukaryotic scheme.

17.1 Initiation of Translation in Bacteria

Two important events must occur even before translation initiation can take place. One of these prerequisites is to generate a supply of **aminoacyl-tRNAs** (tRNAs with their cognate amino acids attached). In other words, amino acids must be covalently bound to tRNAs. This process is called **tRNA charging;** the tRNA is said to be "charged" with an amino acid. Another preinitiation event is the dissociation of ribosomes into their two subunits. This is necessary because the cell assembles the initiation complex on the small ribosomal subunit, so the two subunits must separate to make this assembly possible.

tRNA Charging

All tRNAs have the same three bases (CCA) at their 3′-ends, and the terminal adenosine is the target for charging. An amino acid is attached by an ester bond between its carboxyl group and the 2′- or 3′-hydroxyl group of the terminal adenosine of the tRNA, as shown in Figure 17.1. Charging takes place in two steps (Figure 17.2), both catalyzed by the enzyme **aminoacyl-tRNA synthetase.** In the first reaction (1), the amino acid is activated, using energy from ATP; the product of the reaction is aminoacyl-AMP. The pyrophosphate by-product is simply the two end phosphate groups (the β- and γ-phosphates), which the ATP lost in forming AMP.

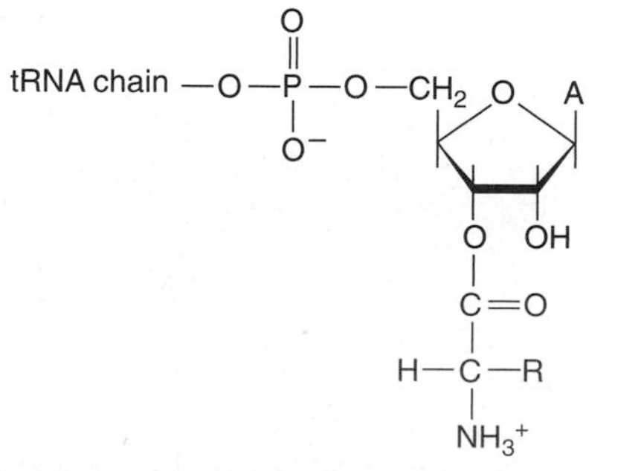

(1) amino acid + ATP → aminoacyl-AMP + pyrophosphate (PP_i)

The bonds between phosphate groups in ATP (and the other nucleoside triphosphates) are high-energy bonds. When they are broken, this energy is released. In this case, the energy is trapped in the **aminoacyl-AMP,** which is why we call this an *activated amino acid.* In the second reaction of charging, the energy in the aminoacyl-AMP is used to transfer the amino acid to a tRNA, forming *aminoacyl-tRNA.*

(2) aminoacyl-AMP + tRNA → aminoacyl-tRNA + AMP

The sum of reactions 1 and 2 is this:

(3) amino acid + ATP + tRNA → aminoacyl-tRNA + AMP + PP_i

Just like other enzymes, an aminoacyl-tRNA synthetase plays a dual role. Not only does it catalyze the reaction leading to an aminoacyl-tRNA, but it determines the specificity of this reaction. Only 20 synthetases exist, one for each amino acid, and they are very specific. Each will almost always place an amino acid on the right kind of tRNA. This is essential to life: If the aminoacyl-tRNA synthetases made many mistakes, proteins would be put together with a correspondingly large number of incorrect amino acids and could not function properly. We will return to this theme and see how the synthetases select the proper tRNAs and amino acids in Chapter 19.

SUMMARY Aminoacyl-tRNA synthetases join amino acids to their cognate tRNAs. They do this very specifically in a two-step reaction that begins with activation of the amino acid with AMP, derived from ATP.

Figure 17.1 Linkage between tRNA and an amino acid. Some amino acids are bound initially by an ester linkage to the 3′-hydroxyl group of the terminal adenosine of the tRNA as shown, but some bind initially to the 2′-hydroxyl group. In any event, the amino acid is transferred to the 3′-hydroxyl group before it is incorporated into a protein.

Dissociation of Ribosomes

We learned in Chapter 3 that ribosomes consist of two subunits. The 70S ribosomes of *E. coli,* for example, contain one 30S and one 50S subunit. Each subunit has one or two ribosomal RNAs and a large collection of ribosomal proteins. The 30S subunit binds the mRNA and the anticodon ends of the tRNAs. Thus, it is the decoding agent of the ribosome that reads the genetic code in the mRNA and allows binding with the appropriate aminoacyl-tRNAs. The 50S subunit binds the ends of the tRNAs that are charged with amino acids and has the peptidyl transferase activity that links amino acids together through peptide bonds.

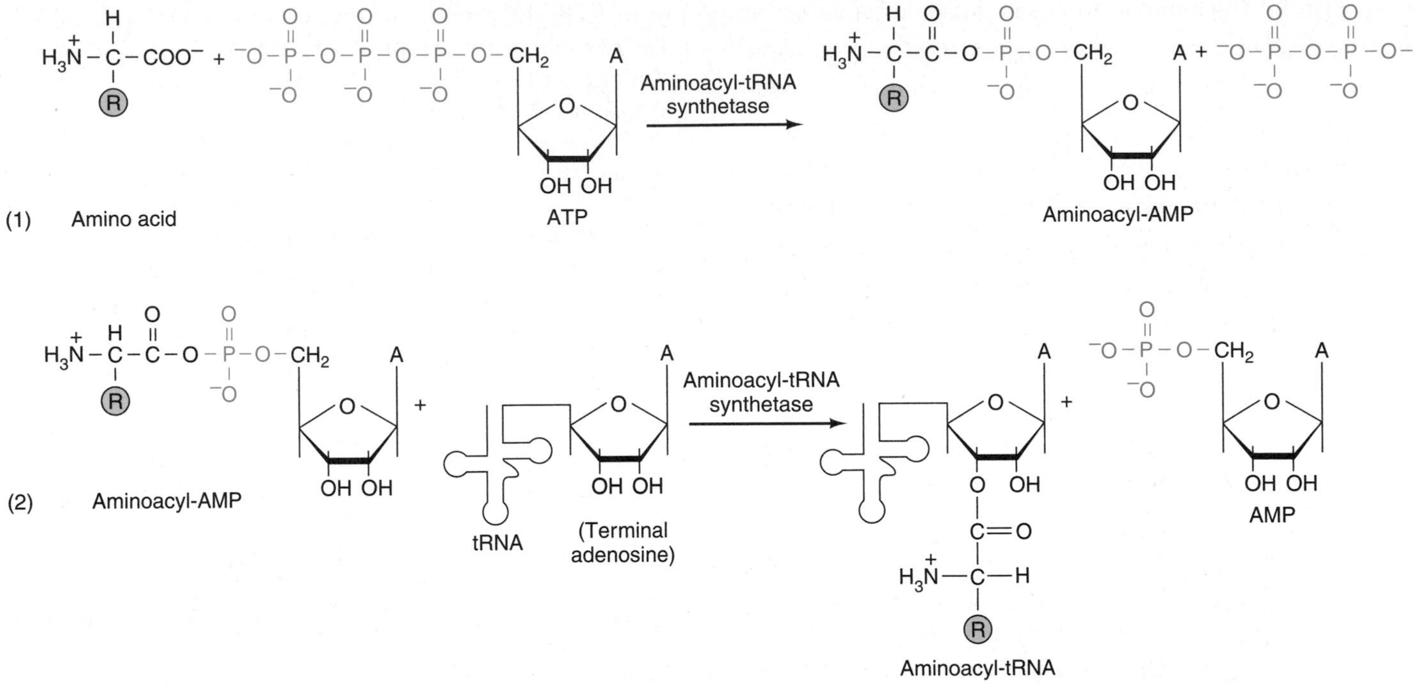

Figure 17.2 Aminoacyl-tRNA synthetase activity. Reaction 1: The aminoacyl-tRNA synthetase couples an amino acid to AMP, derived from ATP, to form an aminoacyl-AMP, with pyrophosphate (P-P) as a by-product. Reaction 2: The synthetase replaces the AMP in the aminoacyl-AMP with tRNA, to form an aminoacyl-tRNA, with AMP as a by-product. The amino acid is joined to the 3′-hydroxyl group of the terminal adenosine of the tRNA.

We will see shortly that both bacterial and eukaryotic cells build translation initiation complexes on the small ribosomal subunit. This implies that the two ribosomal subunits must dissociate after each round of translation for a new initiation complex to form. And as early as 1968, Matthew Meselson and colleagues provided direct evidence for the dissociation of ribosomes, using an experiment outlined in Figure 17.3. These workers labeled *E. coli* ribosomes with heavy isotopes of nitrogen (^{15}N), carbon (^{13}C), and hydrogen (^{2}H, deuterium), plus a little ^{3}H

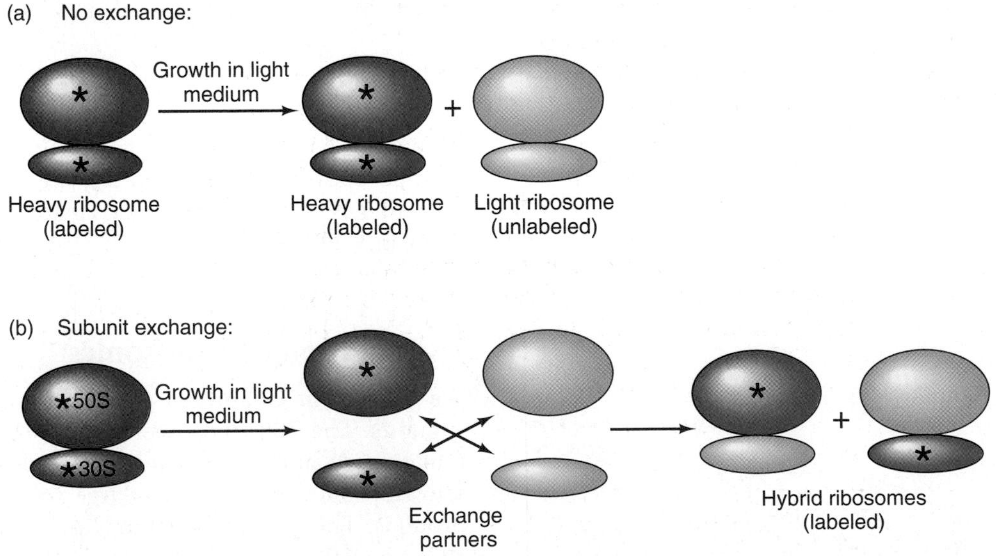

Figure 17.3 Experimental plan to demonstrate ribosomal subunit exchange. Meselson and colleagues made ribosomes heavy (red) by growing *E. coli* in the presence of heavy isotopes of nitrogen, carbon, and hydrogen, and made them radioactive (asterisks) by including some ^{3}H. Then they shifted the cells with labeled, heavy ribosomes to light medium containing the standard isotopes of nitrogen, carbon, and hydrogen. **(a)** No exchange. If no ribosome subunit exchange occurs, the heavy ribosomal subunits will stay together, and the only labeled ribosomes observed will be heavy. The light ribosomes made in the light medium will not be detected because they are not radioactive. **(b)** Subunit exchange. If the ribosomes dissociate into 50S and 30S subunits, heavy subunits can associate with light ones to form labeled hybrid ribosomes.

as a radioactive tracer. The ribosomes so labeled became much denser than their normal counterparts grown in ^{14}N, ^{12}C, and hydrogen, as illustrated in Figure 17.4a. Next, the investigators placed cells with labeled, heavy ribosomes in medium with ordinary light isotopes of nitrogen, carbon, and hydrogen. After 3.5 generations, they isolated the ribosomes and measured their masses by sucrose density gradient centrifugation with ^{14}C-labeled light ribosomes for comparison. Figure 17.4b shows the results. As expected, they observed heavy radioactively labeled ribosomal subunits (38S and 61S instead of the standard 30S and 50S). But the labeled whole ribosomes had a hybrid sedimentation coefficient, in between the standard 70S and the 86S they would have had if both subunits were heavy. This indicated that subunit exchange had occurred. Heavy ribosomes had dissociated into subunits and taken new, light partners.

More precise resolution of the ribosomes on CsCl gradients demonstrated two species: one with a heavy large subunit and a light small subunit, and one with a light large subunit and a heavy small subunit, as predicted in Figure 17.3. Meselson and colleagues performed the same experiments on yeast cells and obtained the same results, so eukaryotic ribosomes also cycle between intact ribosomes (80S) and ribosomal subunits (40S and 60S). What causes the ribosomal subunits to dissociate? We will learn in Chapter 18 that bacteria have a ribosome release factor (RRF) that acts in conjunction with an elongation factor (EF-G) to separate the subunits. In addition, an initiation factor, IF3 binds to the small subunit and keeps it from reassociating with the large subunit.

SUMMARY *E. coli* ribosomes dissociate into subunits at the end of each round of translation. RRF and EF-G actively promote this dissociation, and IF3 binds to the free 30S subunit and prevents its reassociation with a 50S subunit to form a whole ribosome.

Formation of the 30S Initiation Complex

Once the ribosomal subunits have dissociated, the cell builds a complex on the 30S ribosomal subunit, including mRNA, aminoacyl-tRNA, and **initiation factors.** This is known as the **30S initiation complex.** The three initiation factors are **IF1, IF2,** and **IF3.** IF3 is capable of binding by itself to 30S subunits, and IF1 and IF2 stabilize this binding.

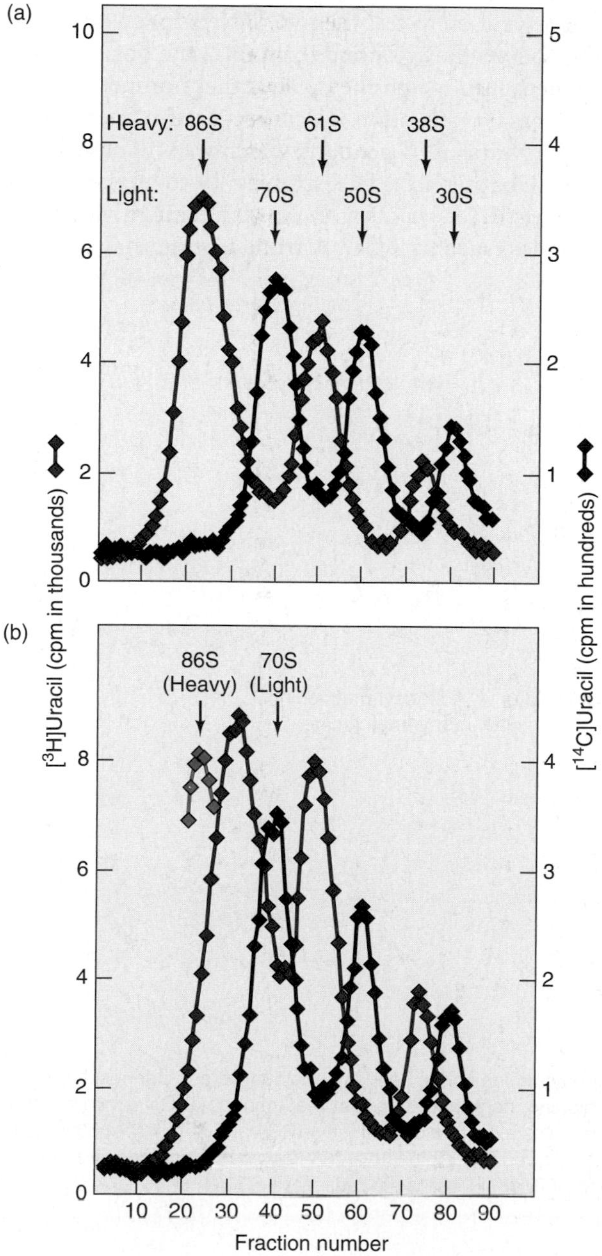

Figure 17.4 Demonstration of ribosomal subunit exchange. **(a)** Sedimentation behavior of heavy and light ribosomes. Meselson and coworkers made heavy ribosomes labeled with [^3H]uracil as described in Figure 17.3, and light (ordinary) ribosomes labeled with [^{14}C]uracil. Then they subjected these ribosomes to sucrose gradient centrifugation, collected fractions from the gradient, and detected the two radioisotopes by liquid scintillation counting. The positions of the light ribosomes and subunits (70S, 50S, and 30S; blue) and of the heavy ribosomes and subunits (86S, 61S, and 38S; red) are indicated at top. **(b)** Experimental results. Meselson and colleagues cultured *E. coli* cells with ^3H-labeled heavy ribosomes as in panel (a) and shifted these cells to light medium for 3.5 generations. Then they extracted the ribosomes, added ^{14}C-labeled light ribosomes as a reference, and subjected the mixture of ribosomes to sucrose gradient ultracentrifugation. They collected fractions and determined their radioactivity as in panel (a): ^3H, red; ^{14}C, blue. The position of the 86S heavy ribosomes (green) was determined from heavy ribosomes centrifuged in a parallel tube. The ^3H-labeled ribosomes (leftmost red peak) were hybrids that sedimented midway between the light (70S) and heavy (86S) ribosomes. (*Source:* Adapted from Kaempfer, R.O.R., M. Meselson, and H.J. Raskas, Cyclic dissociation into stable subunits and reformation of ribosomes during bacterial growth, *Journal of Molecular Biology* 31:277–89, 1968.)

Similarly, IF2 can bind to 30S particles, but achieves much more stable binding with the help of IF1 and IF3. IF1 does not bind by itself, but does so with the assistance of the other two factors. In other words, the three initiation factors bind cooperatively to the 30S ribosomal subunit. Therefore, it is not surprising that all three factors bind close together at a site on the 30S subunit near the 3′-end of the 16S rRNA. Once the three initiation factors have bound, they attract two other key players to the complex: mRNA and the first aminoacyl-tRNA. The order of binding of these two substances appears to be random. We will return to the roles of the initiation factors later in this section. First, let us consider the initiation codon and the aminoacyl-tRNA that responds to it.

The First Codon and the First Aminoacyl-tRNA In 1964, Fritz Lipmann showed that digestion of leucyl-tRNA from *E. coli* with RNase yielded the adenosyl ester of leucine (Figure 17.5a). This is what we expect, because we know that the amino acid is bound to the 3′-hydroxyl group of the terminal adenosine of the tRNA. However, when K.A. Marcker and Frederick Sanger tried the same procedure with methionyl-tRNA from *E. coli,* they found not only the expected adenosyl-methionine ester, but also an adenosyl-*N*-formyl-methionine ester (Figure 17.5b). This demonstrated that the tRNA with which they started was esterified, not only to methionine, but also to a methionine derivative, ***N*-formyl-methionine,** which is abbreviated **fMet.** Figure 17.5c compares the structures of methionine and *N*-formyl-methionine.

Next, B.F.C. Clark and Marcker showed that *E. coli* cells contain two different tRNAs that can be charged with methionine. They separated these two tRNAs by an old purification method called countercurrent distribution. The faster moving tRNA, now called $\mathbf{tRNA_m^{Met}}$ could be charged with methionine, but the methionine could not be formylated. That is, it could not accept a formyl group onto its amino group. The slower moving tRNA was called $\mathbf{tRNA_f^{Met}}$, to denote the fact that the methionine attached to it could be formylated. Notice that the methionine formylation takes place *on* the tRNA. The tRNA cannot be charged directly with formyl-methionine. Clark and Marcker went on to test the two tRNAs for two properties: (1) the codons they respond to, and (2) the positions within the protein into which they placed methionine.

The assay for codon specificity used a method introduced by Marshall Nirenberg, which we will describe more fully in Chapter 18. The strategy is to make a labeled aminoacyl-tRNA, mix it with ribosomes and a variety of trinucleotides, such as AUG. A trinucleotide that codes for a

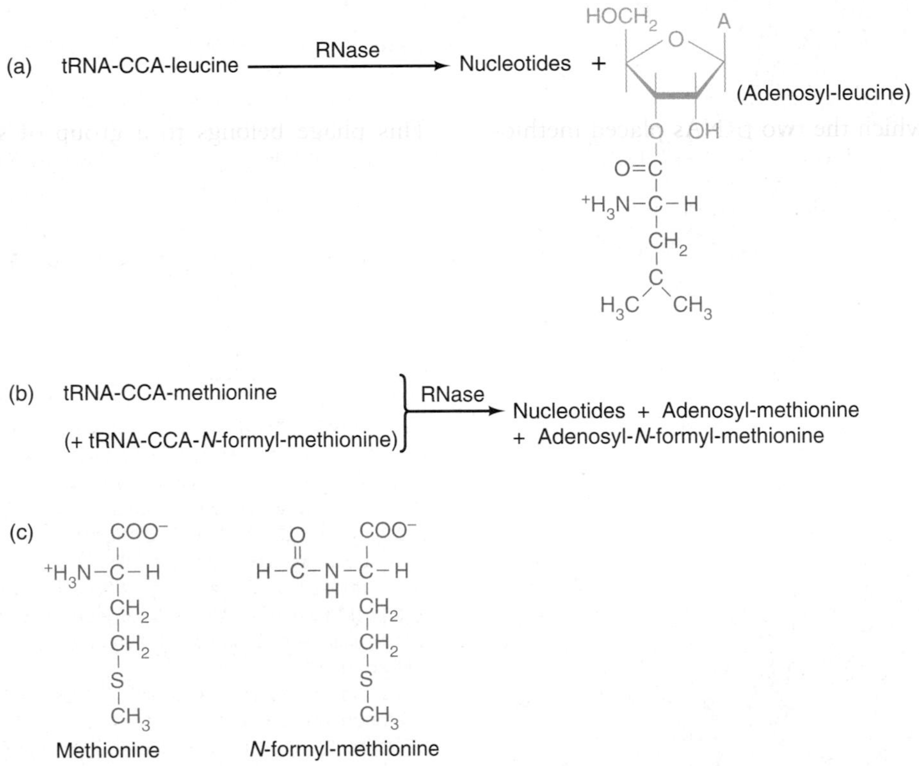

(a) tRNA-CCA-leucine —RNase→ Nucleotides + (Adenosyl-leucine)

(b) tRNA-CCA-methionine (+ tRNA-CCA-*N*-formyl-methionine) —RNase→ Nucleotides + Adenosyl-methionine + Adenosyl-*N*-formyl-methionine

(c) Methionine   *N*-formyl-methionine

Figure 17.5 Discovery of *N*-formyl-methionine. (a) Lipmann and colleagues degraded leucyl-tRNA with RNase to yield nucleotides plus adenosyl-leucine. The leucine was attached to the terminal A of the ubiquitous CCA sequence at the 3′-end of the tRNA. **(b)** Marcker and Sanger performed the same experiment with what they assumed was pure methionyl-tRNA. However, they obtained a mixture of adenosyl-amino acids: adenosyl-methionine and adenosyl-*N*-formyl-methionine, demonstrating that the aminoacyl-tRNA with which they started was a mixture of methionyl-tRNA and *N*-formyl-methionyl-tRNA. **(c)** Structures of methionine and *N*-formyl-methionine, with the formyl group of fMet highlighted in red.

given amino acid will usually cause the appropriate aminoacyl-tRNA to bind to the ribosomes. In the case at hand, $tRNA_m^{Met}$ responded to the codon AUG, whereas $tRNA_f^{Met}$ responded to AUG, GUG, and UUG. As we have already indicated, $tRNA_f^{Met}$ is involved in initiation, which suggests that all three of these codons, AUG, GUG, and UUG, can serve as initiation codons. Indeed, sequencing of many *E. coli* genes has confirmed that AUG is the initiating codon in about 83% of the genes, whereas GUG and UUG are initiating codons in about 14% and 3% of the genes, respectively.

By the way, in addition to the three well-recognized initiation codons (AUG, GUG, and UUG), AUU can serve as an initiation codon, but only two genes in *E. coli* use it. One of these genes encodes a toxic protein, which makes sense because AUU is an inefficient start codon and it would be dangerous to translate this gene too actively. The other gene encodes IF3, which is interesting because one of the roles of IF3 is to help ribosomes bind to the standard initiation codons and avoid the inefficient nonstandard initiation codons such as AUU. In other words, IF3 works against recognition of its own start codon. This provides a neat autoregulation mechanism: When the level of IF3 is high and there is little need for more, this protein inhibits translation of the IF3 mRNA. But when the level of IF3 drops and more IF3 is needed, there is little IF3 to prevent access to the AUU initiation codon, so more IF3 is produced.

Next, Clark and Marcker determined the positions in the protein chain in which the two tRNAs placed methionines. To do this, they used an in vitro translation system with a synthetic mRNA that had AUG codons scattered throughout it. When they used $tRNA_m^{Met}$, methionines were incorporated primarily into the interior of the protein product. By contrast, when they used $tRNA_f^{Met}$, methionines (actually, formyl-methionines) went only into the first position of the polypeptide. Thus, $tRNA_f^{Met}$ appears to serve as the initiating aminoacyl-tRNA. Is this due to the formylation of the amino acid, or to some characteristic of the tRNA? To find out, Clark and Marcker tried their experiment with formylated and unformylatated methionyl-$tRNA_f^{Met}$. They found that formylation made no difference; in both cases, this tRNA directed incorporation of the first amino acid. Thus, the tRNA part of formyl-methionyl-$tRNA_f^{Met}$ is what makes it the initiating aminoacyl-tRNA.

Martin Weigert and Alan Garen reinforced the conclusion that $tRNA_f^{Met}$ is the initiating aminoacyl-tRNA with an in vivo experiment. When they infected *E. coli* with R17 phage and isolated newly synthesized phage coat protein, they found fMet in the N-terminal position, as it should be if it is the initiating amino acid. Alanine was the second amino acid in this new coat protein. On the other hand, mature phage R17 coat protein has alanine in the N-terminal position, so maturation of this protein must involve removal of the N-terminal fMet. Examination of many different bacterial and phage proteins has shown that the fMet is frequently removed. In some cases the methionine remains, but the formyl group is always removed.

SUMMARY The initiation codon in bacteria is usually AUG, but it can also be GUG, or more rarely, UUG. The initiating aminoacyl-tRNA in bacteria is *N*-formyl-methionyl-$tRNA_f^{Met}$. *N*-formyl-methionine (fMet) is therefore the first amino acid incorporated into a polypeptide, but it is frequently removed from the protein during maturation.

Binding mRNA to the 30S Ribosomal Subunit We have seen that the initiating codon is AUG, or sometimes GUG or UUG. But these codons also occur in the interior of a message. An interior AUG codes for ordinary methionine, and GUG and UUG code for valine and leucine, respectively. How does the cell detect the difference between an initiation codon and an ordinary codon with the same sequence? Two explanations come readily to mind: Either a special primary structure (RNA sequence) or a special secondary RNA structure (e.g., a base-paired stem-loop) occurs near the initiation codon that identifies it as an initiation codon and allows the ribosome to bind there. In 1969, Joan Steitz searched for such distinguishing characteristics in the mRNA from an *E. coli* phage called R17. This phage belongs to a group of small spherical RNA phages, which also includes phages f2 and MS2. These are **positive strand phages,** which means that their genomes are also their mRNAs. Thus, these phages provide a convenient source of pure mRNA. These phages are also very simple; for example, each has only three genes, which encode the A protein (or maturation protein), the coat protein, and the replicase. Steitz searched the neighborhoods of the three initiation codons in phage R17 mRNA for distinguishing primary or secondary structures. She began by binding ribosomes to R17 mRNA under conditions in which the ribosomes would remain at the initiation sites. Then she used RNase A to digest the RNA not protected by ribosomes. Finally, she sequenced the initiation regions protected by the ribosomes. She found no obvious sequence or secondary structure similarities around the start sites.

In fact, subsequent work on phage MS2 has shown that the secondary structures at all three start sites are inhibitory; relaxing these secondary structures actually enhances initiation. This is particularly true of the A protein gene, where the base-pairing around the initiation codon is so strong that the gene can be translated only in a short period just after the RNA has replicated. This brief window of opportunity occurs because the RNA has not yet had a chance to form the base pairs that hide the initiation codon. In the replicase gene, the initiation codon is buried in a double-stranded structure that also involves part of the coat gene, as illustrated in Figure 17.6a. This base-pairing is not strong enough on its own to block translation, but a repressor protein stabilizes the base-paired stem enough that translation of the replicase gene cannot occur. This explains why the replicase gene of these phages cannot be translated until the coat gene is translated: The ribosomes moving through the coat gene open up the secondary structure hiding the initiation codon of the replicase gene (Figure 17.6b).

We have seen that secondary structure does not identify the start codons, and the first start site sequences did not reveal any obvious similarities, so what does constitute a ribosome binding site? The answer is that it *is* a special sequence, but sometimes, as in the case of the R17 coat protein gene, it diverges so far from the consensus sequence that it is hard to recognize. Richard Lodish and his colleagues laid some of the groundwork for the discovery of this sequence in their work on the translation of the f2 coat mRNA by ribosomes from different bacteria. They found that *E. coli* ribosomes could translate all three f2 genes in vitro, but that ribosomes from the bacterium *Bacillus stearothermophilus* could translate only the A protein gene. The real problem was in translating the coat gene; as we have seen, the translation of the replicase gene depends on translating the coat gene, so the inability of *B. stearothermophilus* ribosomes to translate the f2 replicase gene was simply an indirect effect of their inability to translate the coat gene. With mixing experiments, Lodish and coworkers demonstrated that the *B. stearothermophilus* ribosomes, not the initiation factors, were at fault.

Next, Nomura and his colleagues performed more detailed mixing experiments using R17 phage RNA. They found that the important element lay in the 30S ribosomal subunit. If the 30S subunit came from *E. coli,* the R17 coat gene could be translated. If it came from *B. stearothermophilus,* this gene could not be translated. Finally, they dissociated the 30S subunit into its RNA and protein components and tried them in mixing experiments. This time, two components stood out: one of the ribosomal proteins, called S12, and the 16S ribosomal RNA. If either of these components came from *E. coli,* translation of the coat gene was active. If either came from *B. stearothermophilus,* translation was depressed (though not as much as if the whole ribosomal subunit came from *B. stearothermophilus*).

These findings stimulated John Shine and Lynn Dalgarno to look for possible interactions between the 16S rRNA and sequences around the start sites of the R17 genes. They noted that all binding sites contained, just upstream of the initiation codon, all or part of this sequence: AGGAGGU, which is complementary to the underlined part of the following sequence, found at the very 3′-end of *E. coli* 16S rRNA: 3′HO-AU<u>UCCUCCA</u>C5′. Note that the hydroxyl group denotes the 3′-end of the 16S rRNA, and that this sequence is written 3′→5′, so its complementarity to the AGGAGGU sequence is obvious. This relationship is very

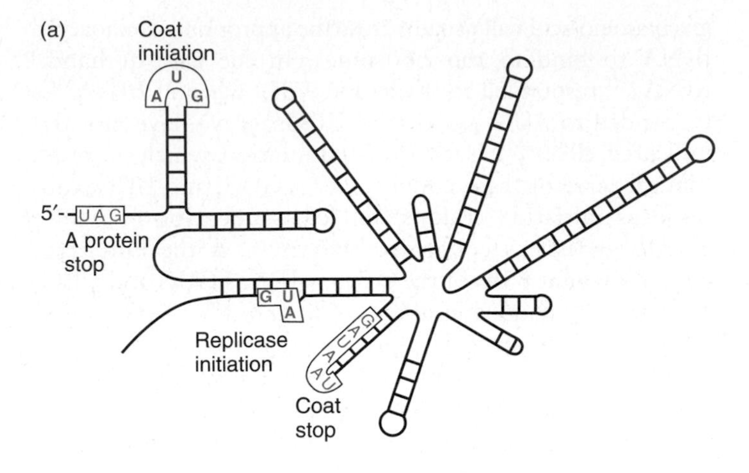

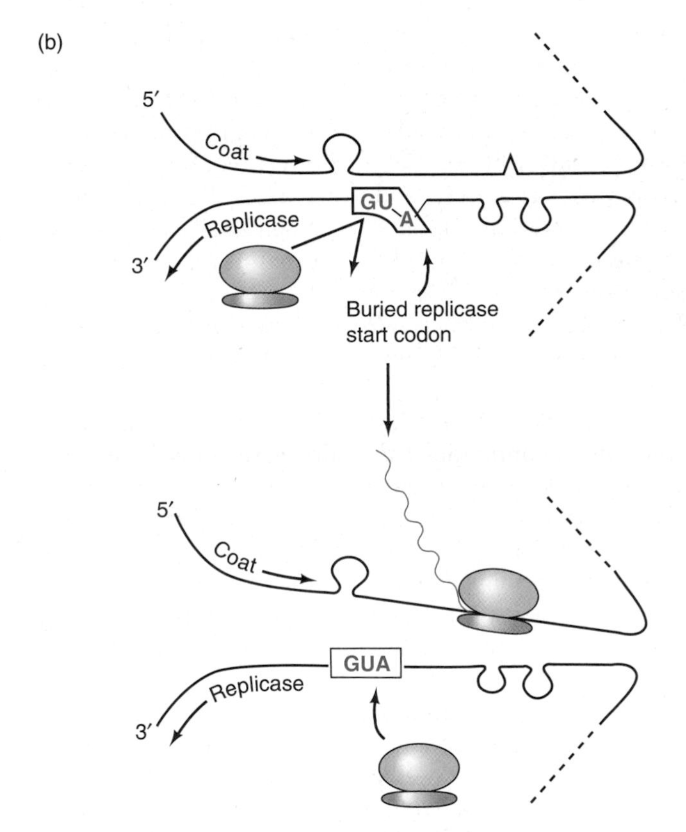

Figure 17.6 Potential secondary structure in MS2 phage RNA and its effect on translation. (a) The simplified secondary structure of the coat gene and surrounding regions in the MS2 RNA. Initiation and termination codons are boxed and labeled. **(b)** Effect of translation of coat gene on replicase translation. At top, the coat gene is not being translated, and the replicase initiation codon (AUG, green, written right to left here) is buried in a stem that is base-paired to part of the coat gene. Thus, the replicase gene cannot be translated. At bottom, a ribosome is translating the coat gene. This disrupts the base pairing around the replicase initiation codon and opens it up to ribosomes that can now translate the replicase gene. (*Source:* (a) Adapted from Min Jou, W., G. Haegeman, M. Ysebaert, and W. Fiers, Nucleotide sequence of the gene coding for the bacteriophage MS2 coat protein. *Nature* 237:84, 1972.)

suggestive, especially considering that the complementarity between the coat protein sequence and the 16S rRNA is the weakest of the three genes, and therefore would be likely to be the most sensitive to alterations in the sequence of the 16S rRNA.

The story gets even more intriguing when we compare the sequences of the *E. coli* and *B. stearothermophilus* 16S rRNAs and find an even poorer match between the R17 coat ribosome binding site and the *Bacillus* 16S rRNA. The *Bacillus* 16S rRNA can make four Watson–Crick base pairs with the A protein and replicase ribosome-binding sites, but only two such base pairs with the coat protein gene. The *E. coli* 16S rRNA can make at least three base pairs with the ribosome-binding sites of all three genes. Could the base pairing between 16S rRNA and the region upstream of the translation initiation site be vital to ribosome binding? If so, it would explain the inability of the *Bacillus* ribosomes to bind to the R17 coat protein initiation site, and it would also identify the AGGAGGU sequence as the ribosome-binding site. As we will see, other evidence shows that this really is the ribosome-binding site, and it has come to be called the **Shine–Dalgarno sequence,** or **SD sequence,** in honor of its discoverers.

To bolster their hypothesis, Shine and Dalgarno isolated ribosomes from two other bacterial species, *Pseudomonas aeruginosa* and *Caulobacter crescentus,* sequenced the 3′-ends of their 16S rRNAs, and tested the ribosomes for the ability to bind to the three R17 initiation sites. In accord with their other results, they found that whenever three or more contiguous base pairs were possible between the 16S rRNA and the sequence upstream of the initiation codon, ribosome binding occurred. Whenever fewer than 3 bp were possible, no ribosome binding occurred. It has since been shown that SD sequences as short as 3 nt must allow at least two G-C pairs with the 16S rRNA in order to support ribosome binding.

Steitz and Karen Jakes added strong evidence in favor of the Shine–Dalgarno hypothesis. They bound *E. coli* ribosomes to the R17 A protein gene's initiation region, then treated the complexes with a sequence-specific RNase called colicin E3, which cuts near the 3′-end of the 16S rRNA of *E. coli.* Next, they fingerprinted the RNA and found a double-stranded RNA fragment, as pictured in Figure 17.7. One strand of this RNA was an oligonucleotide from the A protein gene initiation site, including the Shine–Dalgarno sequence. Base-paired to it was an oligonucleotide from the 3′-end of the 16S rRNA. This demonstrated directly that the Shine–Dalgarno sequence base-paired to the 3′-end of the 16S rRNA and left little doubt that this was indeed the ribosome binding site. It is also important to remember that prokaryotic mRNAs are usually polycistronic. That is, they contain information from more than one cistron, or gene. Each cistron represented in the mRNA has its own initiation codon and its own ribosome-binding site. Thus, ribosomes bind independently to each initiation site, and this provides a means for controlling gene expression, by making some initiation sites more attractive to ribosomes than others.

Anna Hui and Herman De Boer produced excellent evidence for the importance of base pairing between the Shine–Dalgarno sequence and the 3′-end of the 16S rRNA in 1987. They cloned a mutant human growth hormone gene into an *E. coli* expression vector bearing a wild-type Shine–Dalgarno (SD) sequence (GGAGG), which is

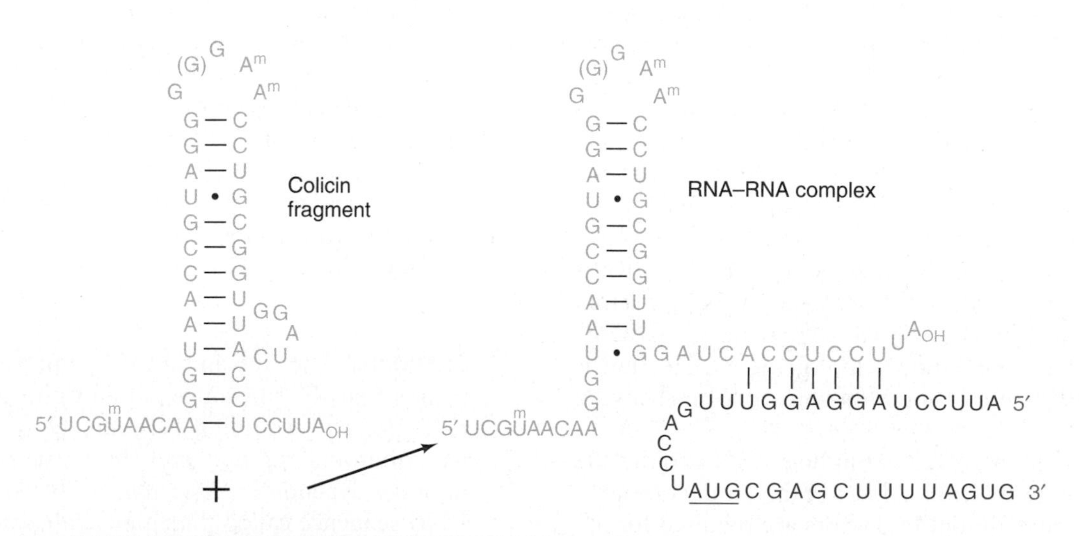

Figure 17.7 Potential structure of the colicin fragment from the 3′-end of *E. coli* 16S rRNA and the initiator region of the R17 phage A protein cistron. The initiation codon (AUG) is underlined. An "m" on the colicin fragment denotes a methylated base. G • U wobble base pairs are denoted by dots. (*Source:* Adapted from Steitz, J.A. and K. Jakes, How ribosomes select initiator regions in mRNA, *Proceedings of the National Academy of Sciences USA* 72(12):4734–38, December 1975.)

Table 17.1 Roles of Initiation Factors in Formation of the 30S Initiation Complex with Natural mRNAs

				Ribosomal binding (pmol)	
Experiment	Ribosomes	mRNA	Factor additions	mRNA	fMet-tRNA$_f^{Met}$
1	30S + 50S	R17	IF1 + IF2	0.4	0.4
			IF2	0.3	0.3
			IF3	2.7	0.1
			IF1 + IF3	4.8	0.2
			IF2 + IF3	2.5	1.3
			IF1 + IF2 + IF3	6.2	6.6
2	30S	MS2	IF1 + IF3		0.0
			IF2		1.8
			IF1 + IF2		3.7
			IF2 + IF3		2.7
			IF1 + IF2 + IF3		7.3
3	30S + 50S	TMV	IF1 + IF3		0.5
			IF2		1.7
			IF1 + IF2		3.1
			IF2 + IF3		8.3
			IF1 + IF2 + IF3		16.9

Source: Role of Initiation Factors in Formation of the 30S Initiation Complex with Natural mRNA from A.J. Wahba, K. Iwasaki, M.J. Miller, S. Sabol, M.A.G. Sillero, & C. Vasquez, "Initiation of Protein Synthesis in *Escherichia Coli* II," *Cold Spring Harbor Symposia in Quantitative Biology,* 34:292.

complementary to the wild-type 16S rRNA anti-SD sequence (CCUCC). This gave high levels of human growth hormone protein. Then they mutated the SD sequence to either CCUCC or GUGUG, which would not base-pair with the anti-SD sequence on the 16S rRNA. Neither of these constructs produced very much human growth hormone. But the clincher came when they mutated the anti-SD sequence in a 16S rRNA gene (on the same vector) to either GGAGG or CACAC, which restored the base pairing with CCUCC and GUGUG, respectively. Now the mRNA with the mutant CCUCC SD sequence was translated very well by the mutant cells with the 16S rRNA having the GGAGG anti-SD sequence, and the mRNA with the mutant GUGUG SD sequence was translated very well in cells with the 16S rRNA having the CACAC anti-SD sequence. This kind of intergenic suppression is strong evidence that important base-pairing occurs between these sequences.

What factors are involved in binding mRNA to the 30S ribosomal subunit? In 1969, Albert Wahba and colleagues showed that all three initiation factors are required for optimum binding, but that IF3 is the most important of the three. They mixed ^{32}P-labeled mRNAs from two *E. coli* phages, R17 and MS2, and from tobacco mosaic virus (TMV), with ribosomal subunits and initiation factors, either singly or in combinations. These viruses all have RNA genomes that serve as mRNAs, so they are convenient sources of mRNAs for experiments like this. Table 17.1, experiment 1, shows the results. IF2 or IF2 + IF1 showed little ability to cause R17 mRNA to bind to ribosomes, but IF3 by itself could cause significant binding. IF1 stimulated this binding further, and all three factors worked best of all. Thus, IF3 seems to be the primary factor involved in mRNA binding to ribosomes, but the other two factors also assist in this task. We have seen that IF3 is already bound to the 30S subunit, by virtue of its role in keeping 50S subunits from associating with the free 30S particles. The other two initiation factors also bind near the IF3 binding site on the 30S subunit, where they can participate in assembling the 30S initiation complex.

SUMMARY The 30S initiation complex is formed from a free 30S ribosomal subunit plus mRNA and fMet-tRNA$_f^{Met}$. Binding between the 30S prokaryotic ribosomal subunit and the initiation site of a message depends on base pairing between a short RNA sequence called the Shine–Dalgarno sequence just upstream of the initiation codon, and a complementary sequence at the 3′-end of the 16S rRNA. This binding is mediated by IF3, with help from IF1 and IF2. All three initiation factors have bound to the 30S subunit by this time.

Binding fMet-tRNA$_f^{Met}$ to the 30S Initiation Complex If IF3 bears the primary responsibility for binding mRNA to the 30S ribosome, which initiation factor plays this role for fMet-tRNA$_f^{Met}$? Table 17.1 shows that the answer is IF2. IF1 and IF3 together yielded little or no fMet-tRNA$_f^{Met}$ binding, whereas IF2 by itself could cause significant binding. Again, as is the case with mRNA binding, all three factors together yielded optimum fMet-tRNA$_f^{Met}$ binding.

In 1971, Sigrid and Robert Thach showed that one mole of GTP binds to the 30S ribosomal subunit along with every mole of fMet-tRNA$_f^{Met}$, but the GTP is not hydrolyzed until the 50S ribosomal subunit joins the complex and IF2 departs. We will discuss this matter further later in this chapter.

In 1973, John Fakunding and John Hershey performed in vitro experiments with labeled IF2 and fMet-tRNA$_f^{Met}$ to show the binding of both to the 30S ribosomal subunit, and the lack of necessity for GTP hydrolysis for such binding to occur. They labeled fMet-tRNA$_f^{Met}$ with ^3H, and IF2 by phosphorylating it with [^{32}P]ATP. This phosphorylated IF2 retained full activity. Then they mixed these components with 30S ribosomal subunits in the presence of either GTP or an unhydrolyzable analog of GTP, **GDPCP.** This analog has a methylene linkage (-CH_2-) between the β- and γ-phosphates where ordinary GTP would have an oxygen atom, which explains why it cannot be hydrolyzed to GDP and phosphate. After mixing all these components together, Fakunding and Hershey displayed the initiation complexes by sucrose gradient ultracentrifugation. Figure 17.8 shows the results. All of the labeled IF2 and a significant amount of the fMet-tRNA$_f^{Met}$ comigrated with the 30S ribosomal subunit, indicating the formation of an initiation complex. The same results were seen in the presence of either authentic GTP or GDPCP, demonstrating that GTP hydrolysis is not required for binding of either IF2 or fMet-tRNA$_f^{Met}$ to the complex. Indeed, IF2 can bind to 30S subunits in the absence of GTP, but only at unnaturally high concentrations of IF2.

This kind of experiment also allowed Fakunding and Hershey to estimate the stoichiometry of binding between the 30S subunit, IF2, and fMet-tRNA$_f^{Met}$. They added more and more IF2 to generate a saturation curve. The curve leveled off at 0.7 molecule of IF2 bound per 30S subunit. Because some of the 30S subunits were probably not competent to bind IF2, this number seems close enough to 1.0 to conclude that the real stoichiometry is 1:1. Furthermore, at saturating IF2 concentration, 0.69 molecule of fMet-tRNA$_f^{Met}$ bound to the 30S subunits. This is almost exactly the amount of IF2 that bound, so the stoichiometry of fMet-tRNA$_f^{Met}$ also appears to be 1:1. However, as we will see, IF2 is ultimately released from the initiation complex, so it can recycle and bind another fMet-tRNA$_f^{Met}$ to another complex. In this way, it really acts catalytically.

As we learned earlier in this chapter, all three factors can bind cooperatively to the 30S subunit. Indeed, the binding of all three factors seems to be the first step in formation of the 30S initiation complex. Once bound, the factors can direct the binding of mRNA and fMet-tRNA$_f^{Met}$, yielding a complete 30S initiation complex, which consists of a 30S ribosomal subunit plus one molecule each of mRNA, fMet-tRNA$_f^{Met}$, GTP, IF1, IF2, and IF3.

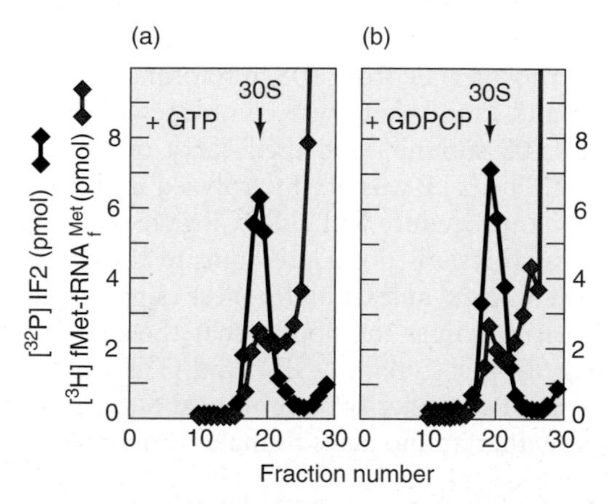

Figure 17.8 Formation of 30S initiation complex with GTP or GDPCP. Fakunding and Hershey mixed [^{32}P]IF2, [^3H]fMet-tRNA$_f^{Met}$ and AUG, an mRNA substitute, with 30S ribosomal subunits and either **(a)** GTP or **(b)** the unhydrolyzable GTP analog GDPCP. Then they centrifuged the mixtures in sucrose gradients and assayed each gradient fraction for radioactive IF2 (blue) and fMet-tRNA$_f^{Met}$ (red). Both substances bound to 30S ribosomes equally well with GTP and GDPCP. (*Source:* Adapted from Fakunding, J.L. and J.W.B., Hershey, The interaction of radioactive initiation factor IF2 with ribosomes during initiation of protein synthesis. *Journal of Biological Chemistry* 248:4208, 1973.)

SUMMARY IF2 is the major factor promoting binding of fMet-tRNA$_f^{Met}$ to the 30S initiation complex. The other two initiation factors play important supporting roles. GTP is also required for IF2 binding at physiological IF2 concentrations, but it is not hydrolyzed in the process. The complete 30S initiation complex contains one 30S ribosomal subunit plus one molecule each of mRNA, fMet-tRNA$_f^{Met}$, GTP, IF1, IF2, and IF3.

Formation of the 70S Initiation Complex

For elongation to occur, the 50S ribosomal subunit must join the 30S initiation complex to form the **70S initiation complex.** In this process, IF1 and IF3 dissociate from the complex. Then GTP is hydrolyzed to GDP and inorganic phosphate, as IF2 leaves the complex. We will see that GTP hydrolysis does not drive the binding of the 50S ribosomal subunit. Instead, it drives the release of IF2, which would otherwise interfere with formation of an active 70S initiation complex.

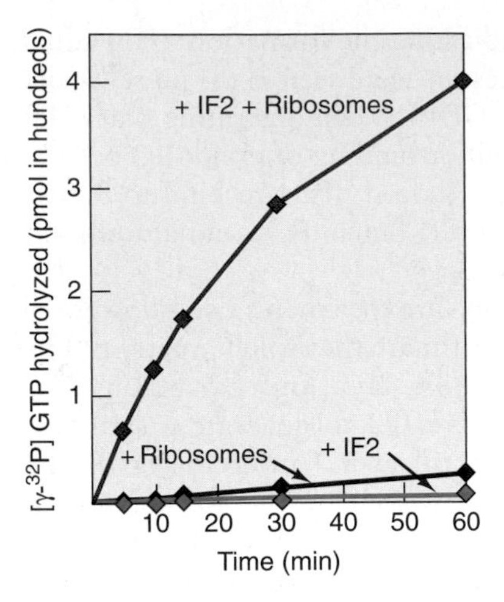

Figure 17.9 Ribosome-dependent GTPase activity of IF2. Dubnoff and Maitra measured the release of labeled inorganic phosphate from [γ-^{32}P]GTP in the presence of IF2 (green), ribosomes (blue), and IF2 plus ribosomes (red). Together, ribosomes and IF2 could hydrolyze the GTP. (*Source:* Adapted from Dubhoff, J.S., A.H. Lockwood, and U. Maitra, Studies on the role of guanosine triphosphate in polypeptide chain initiation in *Escherichia coli. Journal of Biological Chemistry* 247:2878, 1972.)

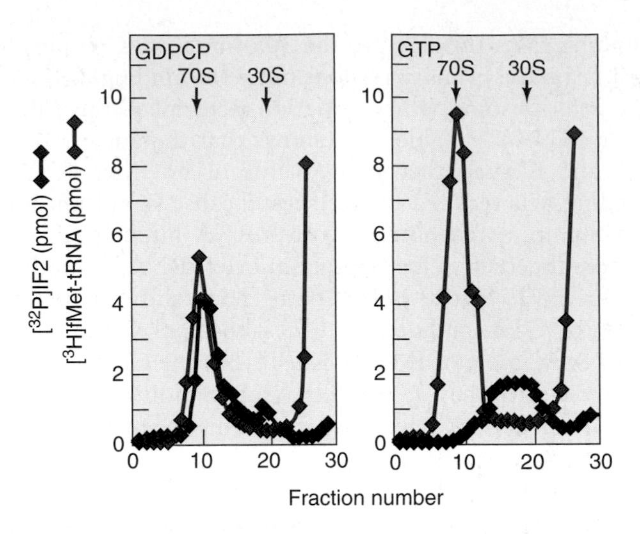

Figure 17.10 Effect of GTP hydrolysis on release of IF2 from the ribosome. Fakunding and Hershey mixed [^{32}P]IF2 (blue) and [^{3}H] fMet-tRNA$_f^{Met}$ (red) with 30S ribosomal subunits to form 30S initiation complexes. Then they added 50S ribosomal subunits in the presence of either **(a)** GDPCP, or **(b)** GTP, and then analyzed the complexes by sucrose gradient ultracentrifugation as in Figure 17.8. (*Source:* Adapted from Fakunding, J.L. and J.W.B. Hershey, The interaction of radioactive initiation factor IF2 with ribosomes during initiation of protein synthesis. *Journal of Biological Chemistry* 248:4210, 1973.)

We have already seen that GTP is part of the 30S initiation complex, and that it is removed when the 50S ribosomal subunit joins the complex. But how is it removed? Jerry Dubnoff and Umadas Maitra demonstrated in 1972 that IF2 contains a ribosome-dependent GTPase activity that hydrolyzes the GTP to GDP and inorganic phosphate (P_i). They mixed [γ-^{32}P]GTP with salt-washed ribosomes (devoid of initiation factors), or with IF2, or with both, and plotted the $^{32}P_i$ released. Figure 17.9 shows that ribosomes or IF2 separately could not hydrolyze the GTP, but together they could. Thus, IF2 and ribosomes together constitute a GTPase. Our examination of the 30S initiation complex in the previous section showed that the 30S ribosomal subunit cannot complement IF2 this way because GTP is not hydrolyzed until the 50S particle joins the complex.

What is the function of GTP hydrolysis? Fakunding and Hershey's experiments with labeled IF2 also shed light on this question: They showed that GTP hydrolysis is necessary for removal of IF2 from the ribosome. These workers formed 30S initiation complexes with labeled IF2 and fMet-tRNA$_f^{Met}$ and either GDPCP or GTP, added 50S subunits and then ultracentrifuged the mixtures to see which components remained associated with the 70S initiation complexes. Figure 17.10 shows the results. With GDPCP, both IF2 and fMet-tRNA$_f^{Met}$ remained associated with the 70S complex. By contrast, GTP allowed IF2 to dissociate, while fMet-tRNA$_f^{Met}$ remained with the 70S complex. This demonstrated that GTP hydrolysis is required for IF2 to leave the ribosome.

Another feature of Figure 17.10 is that much *more* fMet-tRNA$_f^{Met}$ bound to the 70S initiation complex in the presence of GTP than in the presence of GDPCP. This hints at the catalytic function of IF2: Hydrolysis of GTP is necessary to release IF2 from the 70S initiation complex so it can bind another molecule of fMet-tRNA$_f^{Met}$ to another 30S initiation complex. This recycling constitutes catalytic activity. However, if the factor remains stuck to the 70S complex because of failure of GTP to be hydrolyzed, it cannot recycle and therefore acts only stoichiometrically.

Is GTP hydrolysis also required to prime the ribosome for translation? Apparently not, since Maitra and colleagues removed GTP from 30S initiation complexes by gel filtration and found that these complexes were competent to accept 50S subunits and then carry out peptide bond formation. The GTP was not hydrolyzed in this procedure, and a similar procedure with GDPCP gave the same results, so GTP hydrolysis is not a prerequisite for an active 70S initiation complex, at least under these experimental conditions. This reinforces the notion that the real function of GTP hydrolysis is to remove IF2 (and GTP itself) from the 70S initiation complex so it can go about its business of linking together amino acids to make proteins.

SUMMARY GTP is hydrolyzed after the 50S subunit joins the 30S complex to form the 70S initiation complex. This GTP hydrolysis is carried out by IF2 in conjunction with the 50S ribosomal subunit. The purpose of this hydrolysis is to release IF2 and GTP from the complex so polypeptide chain elongation can begin.

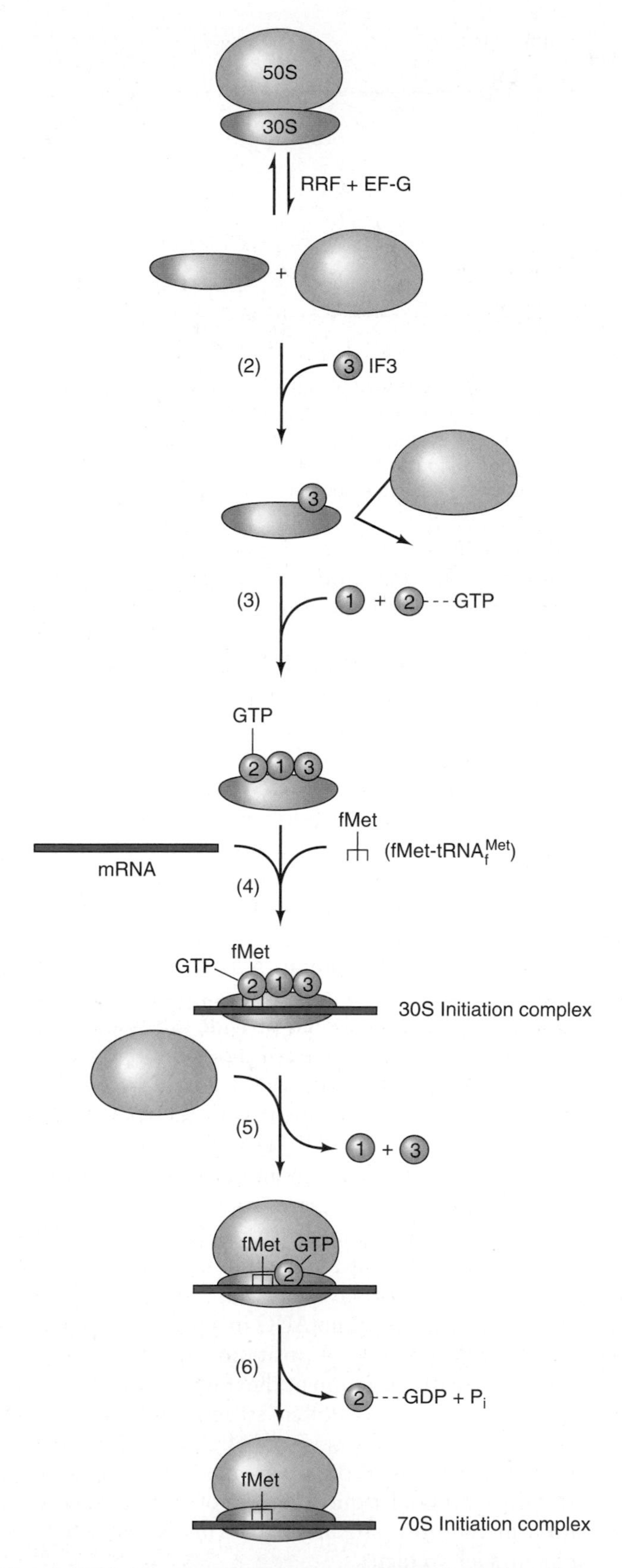

Figure 17.11 Summary of bacterial translation initiation. See the text for a description of steps 1–6. Steps 2 and 3 may be combined in vivo.

Summary of Initiation in Bacteria

Figure 17.11 summarizes what we have learned about translation initiation in bacteria. It includes the following features:

1. Dissociation of the 70S ribosome into 50S and 30S subunits, under the influence of RRF and EF-G.
2. Binding of IF3 to the 30S subunit, which prevents reassociation between the ribosomal subunits.
3. Binding of IF1 and IF2–GTP alongside IF3. This step probably occurs simultaneously with step 2.
4. Binding of mRNA and fMet-tRNA$_f^{Met}$ to form the 30S initiation complex. These two components can apparently bind in either order, but IF2 sponsors fMet-tRNA$_f^{Met}$ binding, and IF3 sponsors mRNA binding. In each case, the other initiation factors also help.
5. Binding of the 50S subunit, with loss of IF1 and IF3.
6. Dissociation of IF2 from the complex, with simultaneous hydrolysis of GTP. The product is the 70S initiation complex, ready to begin elongation.

17.2 Initiation in Eukaryotes

Several features distinguish eukaryotic translation initiation from bacterial. First, eukaryotic initiation begins with methionine, not *N*-formyl-methionine. But the initiating tRNA is different from the one that adds methionines to the interiors of polypeptides (tRNA$_m^{Met}$). The initiating tRNA bears an unformylated methionine, so it seems improper to call it tRNA$_f^{Met}$. Accordingly, it is frequently called **tRNA$_i^{Met}$**, or just **tRNA$_i$**. A second major difference distinguishing eukaryotic translation initiation from bacterial is that eukaryotic mRNAs contain no Shine–Dalgarno sequence to show the ribosomes where to start translating. Instead, most eukaryotic mRNAs have caps (Chapter 15) at their 5′-ends, which direct initiation factors to bind and begin searching for an initiation codon. This less direct recognition of the proper translation start site requires at least 12 factors, in contrast to the three that bacteria use. The eukaryotic mechanism of initiation and the initiation factors it requires will be our topics in this section.

The Scanning Model of Initiation

Most bacterial mRNAs are polycistronic. They contain information from multiple genes, or cistrons, and each cistron has its own initiation codon and ribosome-binding site. But polycistronic mRNAs that are translated intact are rare in eukaryotes, except for the transcripts of certain viruses. Thus, eukaryotic cells are usually faced with the task of finding a start codon near the 5′-end of a transcript. They accomplish this task by recognizing the cap at the 5′-end,

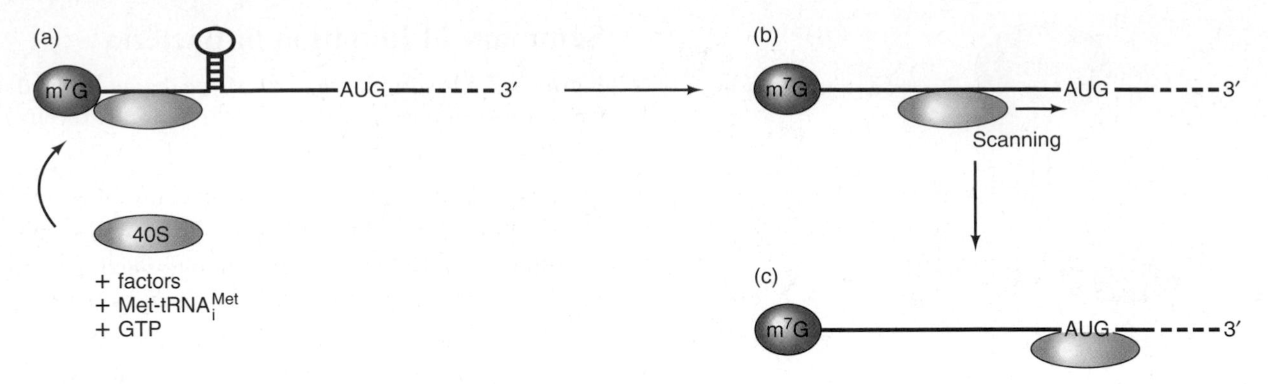

Figure 17.12 A simplified version of the scanning model for translation initiation. (a) The 40S ribosomal subunit, along with initiation factors, Met-tRNA$_i^{Met}$, and GTP, recognize the m^{7}G cap (red) at the 5′-end of an mRNA and allow the ribosomal subunit to bind at the end of the mRNA. All the other components (factors, etc.) are omitted for simplicity. **(b)** The 40S subunit is scanning the mRNA toward the 3′-end, searching for an initiation codon. It has melted a stem-loop structure in its way. **(c)** The ribosomal subunit has located an AUG initiation codon and has stopped scanning. Now the 60S ribosomal subunit can join the complex and initiation can occur.

then **scanning** the mRNA in the 5′→3′ direction until they encounter a start codon, as illustrated in Figure 17.12.

Marilyn Kozak first developed this scanning model in 1978, based on four considerations: (1) In no known instance was eukaryotic translation initiated at an internal AUG, as in a polycistronic mRNA. (2) Initiation did not occur at a fixed distance from the 5′-end of an mRNA. (3) In all of the first 22 eukaryotic mRNAs examined, the first AUG downstream of the cap was used for initiation. (4) As we saw in Chapter 15, the cap at the 5′-end of the mRNA facilitates initiation. We will see more definitive evidence for the scanning model later in this chapter.

The simplest version of the scanning model has the ribosome recognizing the first AUG it encounters and initiating translation there. However, a survey of 699 eukaryotic mRNAs revealed that the first AUG is not the primary initiation site in 5–10% of the cases. Instead, in those cases, most ribosomes skip over one or more AUGs before encountering the right one and initiating translation, a process Kozak called "leaky scanning." This raises the question: What sets the right AUG apart from the wrong ones? To find out, Kozak examined the sequences surrounding initiating AUGs and found that the consensus sequence in mammals was CCRCC<u>AUG</u>G, where R is a purine (A or G), and the initiation codon is underlined.

If this is really the optimum sequence, then mutations should reduce its efficiency. To check this hypothesis, Kozak systematically mutated nucleotides around the initiation codon in a cloned rat preproinsulin gene. She substituted a synthetic ATG-containing oligonucleotide for the normal initiating ATG, then introduced mutations into this initiation region, placed the mutated genes under control of the SV40 virus promoter, introduced them into monkey (COS) cells, then labeled newly synthesized proteins with [^{35}S]methionine, immunoprecipitated the proinsulin, electrophoresed it, and detected it by fluorography, a technique akin to autoradiography (Chapter 5). Finally, she scanned the fluorograph with a densitometer to quantify the production of proinsulin. The better the translation initiation, the more proinsulin was made. Throughout this discussion we will refer to the initiation codon as AUG, even though the mutations were done at the DNA level.

Figure 17.13 shows some of the results, which include alterations in positions −3 and +4, where the A in AUG is position +1. The best initiation occurred with a G or an A in position −3 and a G in position +4. Similar experiments showed that the best initiation of all occurred with the sequence ACC<u>AUG</u>G, and the −3 and +4 positions are the most important. These requirements are sometimes called **Kozak's rules.**

If this really is the optimum sequence for translation initiation, introducing it out of frame and upstream of the normal initiation codon should provide a barrier to scanning ribosomes and force them to initiate out of frame. The more this occurs, the less proinsulin should be produced. Kozak performed this experiment with the A's of the two AUGs 8 nt apart as follows: <u>AUG</u>NCACC<u>AUG</u>G. Note that the downstream AUG is in an optimal neighborhood, so initiation should start there readily if the ribosome can reach it without initiating upstream first. Figure 17.14 shows the results. Mutant F10 had no upstream AUG, and initiation from the normal AUG was predictably strong. Mutant F9 had the upstream AUG in a very weak context, with U's in both −3 and +4 positions. Again, this did not interfere much with initiation at the downstream AUG. But all the other mutants exhibited strong interference with normal initiation, and the strength of this interference was related to the context of the upstream AUG. The closer it resembled the optimal sequence, the more it interfered with initiation at the downstream AUG. This is just what the scanning model predicts.

What about natural mRNAs that have an upstream AUG in a favorable context, yet still manage to initiate from a downstream AUG? Kozak noted that these mRNAs

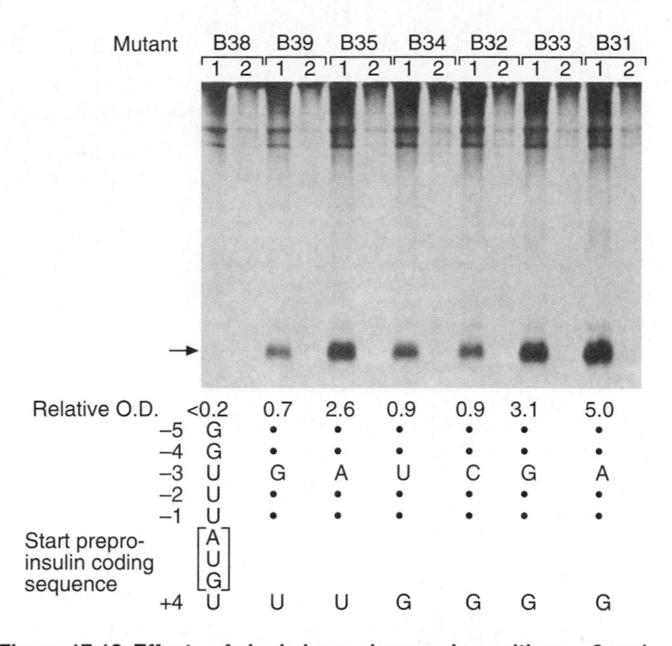

Figure 17.13 Effects of single base changes in positions −3 and +4 surrounding the initiating AUG. Starting with a cloned rat preproinsulin gene under the control of an SV40 viral promoter, Kozak replaced the natural initiation codon with a synthetic oligonucleotide containing an ATG, which was transcribed to AUG in the mRNA. She then mutagenized the nucleotides at positions −3 and +4 as shown at bottom, introduced the manipulated genes into COS cells growing in medium containing [^{35}S]methionine to label any proinsulin produced. She purified the proinsulin by immunoprecipitation, then electrophoresed it and detected the labeled protein by fluorography. This is a technique similar to autoradiography in which the electrophoresis gel is impregnated with a fluorescent compound to amplify the relatively weak radioactive emissions from an isotope such as ^{35}S. The arrow at left indicates the position of the proinsulin product. Kozak subjected the proinsulin bands in the fluorograph to densitometry to quantify their intensities. These are listed as relative O. D., or optical density, beneath each band. Optimal initiation occurred with a purine in position −3 and a G in position +4. Proinsulin is the product of the preproinsulin gene because the "signal peptide" at the amino terminus of preproinsulin is removed during translation, yielding proinsulin. The signal peptide directs the growing polypeptide, along with the ribosome and mRNA, to the endoplasmic reticulum (ER). This ensures that the polypeptide enters the ER and can therefore be secreted from the cell. All sequences are shown as they appear in mRNA. (*Source*: Kozak, M. Point mutations define a sequence flanking the AUG initiator codon that modulates translation by eukaryotic ribosomes. Cell 44 (31 Jan 1986) p. 286, f. 2. Reprinted by permission of Elsevier Science.)

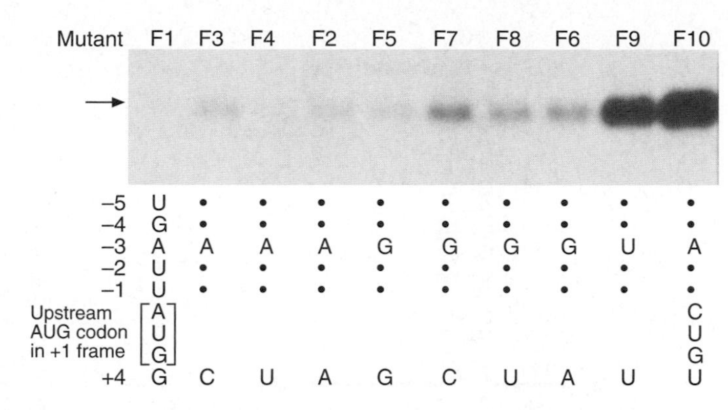

Figure 17.14 Influence of the context of an upstream "barrier" AUG. Kozak made a construct having the normal AUG initiation codon of the rat preproinsulin transcript preceded by an out-of-frame AUG, then made mutations in the −3 and +4 positions surrounding the upstream AUG (shown at bottom) and assayed the effect on proinsulin synthesis as in Figure 17.13. The arrow at left indicates the position of correctly initiated proinsulin. The more favorable the context of the upstream AUG, the better it serves as a barrier to correct downstream initiation. All sequences are presented as they appear in mRNA. (*Source:* Kozak, M., Point mutations define a sequence flanking the AUG initiation codon that modulates translation by eukaryotic ribosomes. *Cell* 44 (31 Jan 1986) p. 288, f. 6. Reprinted by permission of Elsevier Science.)

have in-frame stop codons between the two AUGs, and she argued that initiation at the downstream AUG actually represents reinitiation by ribosomes that have initiated at the upstream start codon, terminated at the stop codon, then continued scanning for another start codon. To illustrate the effect of a stop codon between the two AUGs, Kozak made another set of constructs with such a stop codon and tested them by the same assay. Abundant initiation occurred at the downstream AUG in this case, as long as the downstream AUG was in a good environment.

Note that an initiation codon and a downstream termination codon in the same reading frame define the boundaries of an **open reading frame (ORF).** Such an ORF potentially encodes a protein; whether it is actually translated in vivo is another matter. Further experiments have revealed another requirement for efficient reinitiation at a downstream ORF: The upstream ORF must be short. In every case in which a dicistronic mRNA with a full-sized upstream ORF has been examined, reinitiation at the downstream ORF has been extremely inefficient. Perhaps by the time a ribosome finishes translating a long ORF, the initiation factors needed for reinitiation have diffused away, so it ignores the second ORF.

To check rigorously the hypothesis that an upstream AUG is favored over downstream AUGs, Kozak created mRNAs with exact repeats of the initiation region of the rat preproinsulin cistron. She then tested these for the actual translation initiation site by isolating the resulting proteins and electrophoresing them to determine their sizes, which tell us which initiation site the ribosomes used in making them. In each case, the farthest upstream AUG was used, which is again consistent with the scanning model.

What is the effect of mRNA secondary structure on efficiency of initiation? Hairpins in the mRNA can affect initiation both positively and negatively. Kozak showed that a stem loop 12–15 nt downstream of an AUG in a weak context could act positively by preventing 40S ribosomal subunits from skipping that initiation site. The hairpin presumably stalled the ribosomal subunit at the AUG long enough for initiation to occur. Secondary structure can also have a negative effect. Kozak tested the effects of two different stem-loop structures in the leader of an

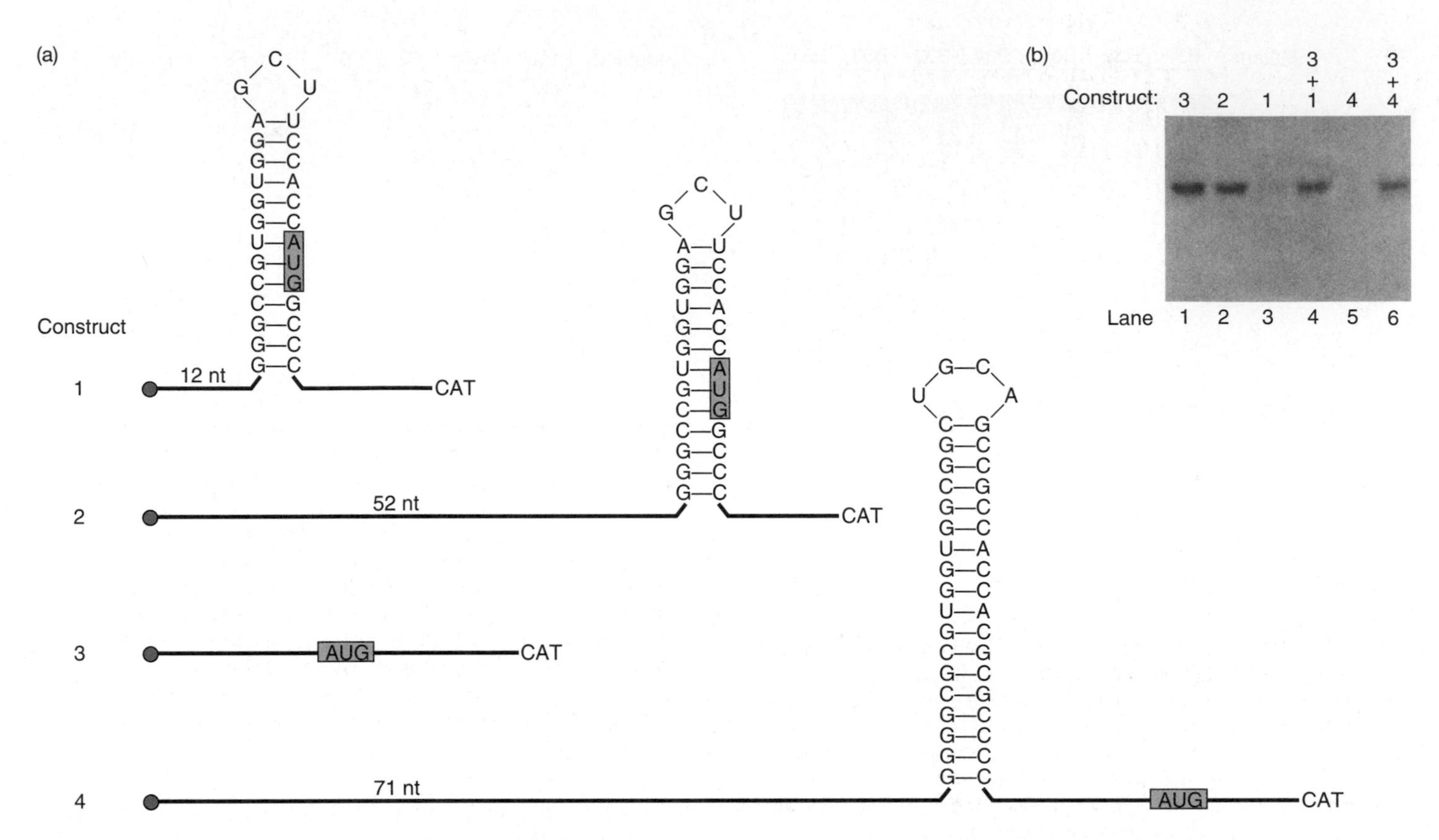

Figure 17.15 Effect of secondary structure in an mRNA leader on translation efficiency. (a) mRNA constructs. Kozak made the synthetic leader constructs pictured here, with the cap in red and the initiation codon highlighted in green, with the CAT ORF attached to the 3′-end of each. **(b)** Results of in vitro translation. Kozak translated each mRNA in vitro in a rabbit reticulocyte extract with [^{35}S]methionine. She electrophoresed the labeled proteins and detected them by fluorography. The short hairpin near the cap (construct 1) interfered, as did the long hairpin between the cap and the initiation codon (construct 4). (*Source:* Kozak, M., Circumstances and mechanisms of inhibition of translation by secondary structure in eukaryotic mRNAs. *Molecular and Cellular Biology* 9 (1989) p. 5136, f. 3. American Society for Microbiology.)

mRNA (Figure 17.15a). One was relatively short and had a free energy of formation (or stability) of −30 kcal/mol; the other was much longer, with a higher stability of −62 kcal/mol. She introduced these stem loops into various positions in the leader of the chloramphenicol acetyl transferase (CAT) gene, then transcribed the altered genes and translated their transcripts in vitro in the presence of [^{35}S] methionine. Finally, she electrophoresed the CAT proteins and detected them by fluorography. The results in Figure 17.15b show that a −30-kcal stem loop 52 nt downstream of the cap does not interfere with translation, even if it includes the initiating AUG. However, a −30-kcal stem loop only 12 nt downstream of the cap strongly inhibits translation, presumably because it interferes with binding of the 40S ribosomal subunit and factors at the cap. Furthermore, a −62-kcal stem loop placed 71 nt downstream of the cap completely blocked appearance of the CAT protein.

Why was the construct with the stable hairpin not translated? The simplest explanation is that the very stable stem loop blocked the scanning 40S ribosomal subunit and would not let it through to the initiation codon. This effect was observed only *in cis* (on the same molecule). When construct 3 and 4 (or 3 and 1) were tested together, translation occurred on the linear mRNA made from construct 3 (lanes 4 and 6). This indicates that the untranslatable constructs were not poisoning the translation system somehow.

The fact that construct 2 is translated well, even though its initiation codon lies buried in a hairpin, suggests that the scanning ribosomal subunit and initiation factors can unwind a certain amount of double-stranded RNA, as predicted by Kozak in her original scanning model (see Figure 17.12). However, as we have just seen, this unwinding ability has limits; the long hairpin in construct 4 effectively blocks the ribosomal subunits from reaching the initiation codon.

How do 40S ribosomal subunits recognize an AUG start codon? Thomas Donahue and colleagues have shown that the initiator tRNA ($tRNA_i^{Met}$) plays a critical role. They changed the anticodon of one of the four yeast $tRNA_i^{Met}$s to 3′-UCC-5′ so it would recognize the codon AGG instead of AUG. Then they placed *his4* genes with various mutant initiation codons into a *his4*$^-$ yeast strain. Figure 17.16a shows that the *his4* gene bearing an AGG

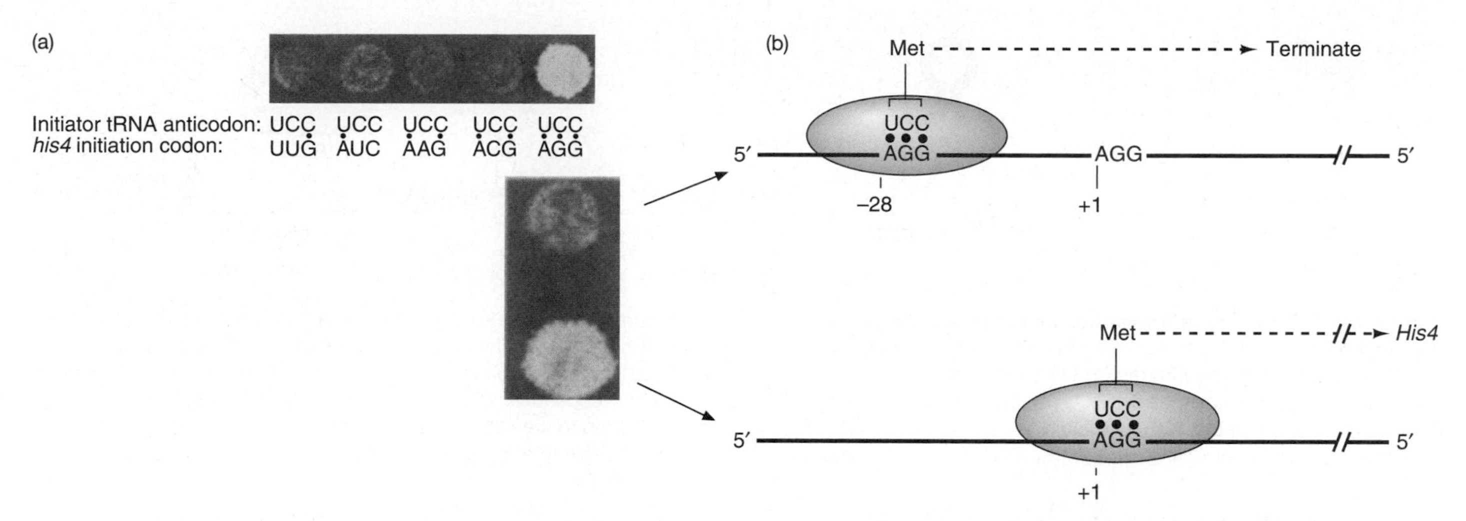

Figure 17.16 Role of initiator tRNA in scanning. **(a)** An initiator tRNA with an altered anticodon can recognize a complementary initiation codon. Donahue and colleagues mutated the anticodon of one of the initiator tRNAs in the yeast *Saccharomyces cerevisiae* to 3′-UCC-5′- and introduced the gene encoding this altered tRNA into *his4*$^-$ cells, using a high-copy yeast vector. Then they changed the initiation codon of the *his4* gene to any of the five versions listed at the bottom and tested the mutant yeast cells for growth in the absence of histidine. When the initiation codon was AGG, it could base-pair with the UCC anticodon on the initiator tRNA, so the mutant mRNA could be translated and growth occurred. **(b)** Effect of an extra AGG upstream and out of frame. Donahue and colleagues made a *his4* construct with an extra AGG in good context beginning at position −28 (top), placed it in cells bearing the initiator tRNA with the UCC anticodon, and tested these cells for ability to grow in the absence of histidine. Growth was much reduced compared with cells with no upstream AGG (bottom). The scanning 40S ribosomal subunit, together with the mutant $tRNA_i^{Met}$, apparently encountered the first AGG and initiated there, producing a shortened *his4* product. (*Source:* (a) Cigan, A.M., L. Feng, and T.F. Donohne, $tRNA_i^{Met}$ functions in directing the scanning ribosomes to the start site of translation. *Science* 242 (7 Oct 1988) p. 94, f. 1B & C (left). Copyright © AAAS.)

codon in place of the initiation codon could support yeast growth. None of the other substitute initiation codons worked, presumably because they could not pair with the UCC anticodon in the altered initiator tRNA. In another experiment, these workers placed a second AGG 28 nt upstream of the AGG in the initiation site and out of frame with it. This construct could not support growth. This result supports the scanning model, as illustrated in Figure 17.16b. The initiator tRNA, with a UCC anticodon in this case, binds to the 40S ribosomal subunit and the complex scans the mRNA searching for the first initiation codon (AGG in this case). Since the first AGG is out-of-frame with the *his4* coding region, translation will occur in the wrong reading frame and will soon encounter a stop codon and terminate prematurely.

The scanning model has some apparent exceptions. The best documented of these concern the polycistronic mRNAs of the picornaviruses such as poliovirus, which lack caps. In these cases, ribosomes can apparently enter at internal initiation codons using **internal ribosome entry sequences (IRESs)** that can attract ribosomes directly without help from the cap. We will discuss this phenomenon in more detail later in this chapter.

SUMMARY Eukaryotic 40S ribosomal subunits, together with the initiator Met-tRNA (Met-$tRNA_i^{Met}$), generally locate the appropriate start codon by binding to the 5′-cap of an mRNA and scanning downstream until they find the first AUG in a favorable context. The best context is a purine in the –3 position and a G in the +4 position where the A of the AUG is +1. In 5–10% of genes, most ribosomal subunits will bypass the first AUG and continue to scan for a more favorable one. Sometimes ribosomes apparently initiate at an upstream AUG, translate a short ORF, then continue scanning and reinitiate at a downstream AUG. This mechanism works only with short upstream ORFs. Secondary structure near the 5′-end of an mRNA can have positive or negative effects. A hairpin just past an AUG can force a ribosomal subunit to pause at the AUG and thus stimulate initiation. A very stable stem loop between the cap and an initiation site can block ribosomal subunit scanning and thus inhibit initiation. Some viral mRNAs that lack caps contain IRESs that attract ribosomes directly to the mRNAs.

Eukaryotic Initiation Factors

We have seen that bacterial translation initiation requires initiation factors and so does initiation in eukaryotes. As you might expect, though, the eukaryotic system is more complex than the bacterial. One level of extra complexity

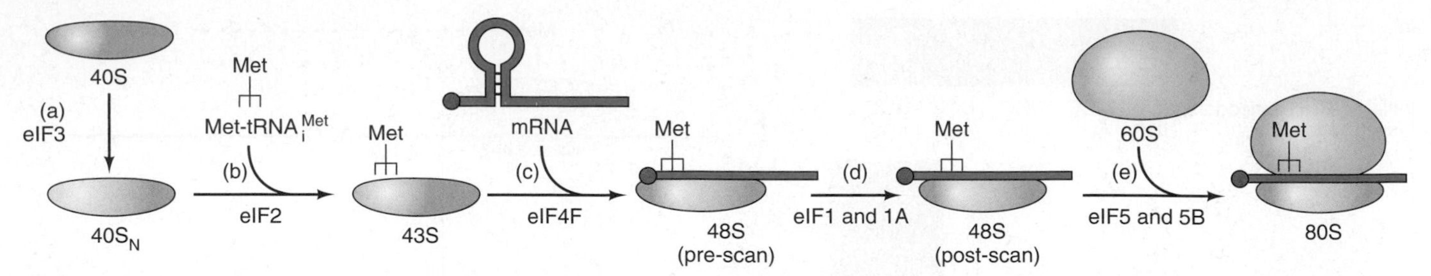

Figure 17.17 Summary of translation initiation in eukaryotes. **(a)** The eIF3 factor converts the 40S ribosomal subunit to 40S$_N$, which resists association with the 60S ribosomal particle and is ready to accept the initiator aminoacyl-tRNA. **(b)** With the help of eIF2, Met-tRNA$_i^{Met}$ binds to the 40S$_N$ particle, forming the 43S complex. **(c)** Aided by eIF4F the mRNA binds to the 43S complex, forming the 48S complex. **(d)** The eIF1 and 1A factors promote scanning to the initiation codon. **(e)** The eIF5 factor promotes hydrolysis of eIF2-bound GTP, which is a precondition for ribosomal subunit joining. eIF5B has a ribosome-dependent GTPase activity that helps the 60S ribosomal particle bind to the 48S complex, yielding the 80S complex that is ready to begin translating the mRNA.

we have already seen is the scanning process. Factors are needed to recognize the cap at the 5′-end of an mRNA and bind the 40S ribosomal subunit nearby. In this section we will examine the factors involved at the various stages of initiation in eukaryotes. We will also see that some of these steps are natural sites for regulation of the translation process.

Overview of Translation Initiation in Eukaryotes Figure 17.17 provides an outline of the initiation process in eukaryotes, showing the major classes of initiation factors involved. Notice that the eukaryotic initiation factor names all begin with e, which stands for "eukaryotic." An example is **eIF2,** which, like bacterial IF2 is responsible for binding the initiating aminoacyl-tRNA (Met-tRNA$_i^{Met}$) to the ribosome.

Another way in which eIF2 resembles IF2 is that it requires GTP to do its job, and this GTP is hydrolyzed to GDP when the factor dissociates from the ribosome. Then GTP must replace GDP on the factor for it to function again. This requires an exchange factor, **eIF2B,** which exchanges GTP for GDP on eIF2. This factor is also called **GEF,** for guanine nucleotide exchange factor. Notice that all of the factors acting at a given step are given the same number. For example, we have seen that at least two factors (eIF2 and eIF2B) are required for initiator aminoacyl-tRNA binding, and both of these share the number 2. Despite all the functional similarities between IF2 and eIF2, the two proteins are not homologous. Instead, IF2 is homologous to eIF5B, which we will discuss later in this chapter.

Another eukaryotic factor whose function bears at least some resemblance to that of a bacterial factor is **eIF3,** which binds to the 40S (small) ribosomal subunit and discourages its reassociation with the 60S (large) subunit. In this way, it resembles IF3. **eIF4F** is a complex cap-binding protein that allows the 40S ribosomal particle to bind to the 5′-end of an mRNA. This binding is mediated by eIF3, which binds to both eIF4F and the 40S ribosomal particle. Once the 40S particle has bound at the cap, it requires **eIF1** (and **eIF1A**) to scan to the initiation codon. **eIF5** is a factor with no known bacterial counterpart. It stimulates association between the 60S ribosomal subunit and the 40S initiation complex, which is actually called the 48S complex because it includes mRNA and many factors in addition to the 40S ribosomal subunit, and these raise the sedimentation coefficient. **eIF6** is another antiassociation factor, like eIF3. It binds to the 60S ribosomal subunit and discourages premature association with the 40S subunit.

SUMMARY The eukaryotic initiation factors have the following general functions: eIF2 is involved in binding Met-tRNA$_i^{Met}$ to the ribosome. eIF2B activates eIF2 by replacing its GDP with GTP. eIF1 and eIF1A aid in scanning to the initiation codon. eIF3 binds to the 40S ribosomal subunit and inhibits its reassociation with the 60S subunit. eIF4F is a cap-binding protein that allows the 40S ribosomal subunit to bind (through eIF3) to the 5′-end of an mRNA. eIF5 encourages association between the 60S ribosomal subunit and the 48S complex (40S subunit plus mRNA and Met-tRNA$_i^{Met}$). eIF6 binds to the 60S subunit and blocks its reassociation with the 40S subunit.

Function of eIF4F Now we come to a major novelty of eukaryotic translation initiation: the role of the cap. We have seen in Chapter 15 that the cap greatly stimulates the efficiency of translation of an mRNA. That implies that some factor can recognize the cap at the 5′-end of an mRNA and aid in the translation of that mRNA. Nahum Sonenberg, William Merrick, Aaron Shatkin, and colleagues identified a cap-binding protein in 1978 by crosslinking it to a modified cap as follows: First they oxidized the ribose of the capping nucleotide on a ^{3}H-reovirus mRNA to convert its 2′- and 3′-hydroxyl groups to a reactive dialdehyde. Then they incubated this altered mRNA with initiation factors. Free amino groups of any factor that binds to the modified cap should bind covalently to

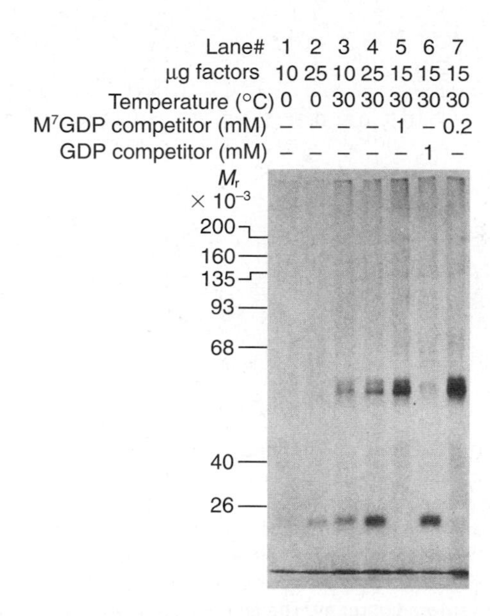

Figure 17.18 Identifying a cap-binding protein by chemical cross-linking. Sonenberg and colleagues placed a reactive dialdehyde in the ribose of the capping nucleotide of a ^{3}H-reovirus mRNA. Then they mixed initiation factors with this mRNA to cross-link any cap-binding protein via a Schiff base between an aldehyde on the cap and a free amino group on the protein. They made this covalent bond permanent by reduction with $NaBH_3CN$. Then they digested these complexes with RNase to remove everything but the cap, and electrophoresed the labeled cap–protein complexes to detect the sizes of any polypeptides that bound to the cap. The conditions in each lane were as listed at top. Note that m^{7}GDP competed with the 24-kD band for binding, but that the 50–55-kD bands did not. (*Source:* Sonenberg, N., M.A., Morgan, W.C. Merrick, and A.J. Shatkin, A polypeptide in eukaryotic initiation factors that crosslinks specifically to the 59-terminal cap in mRNA. *Proceedings of the National Academy of Science USA* 75 (1978) p. 4844, f. 1.)

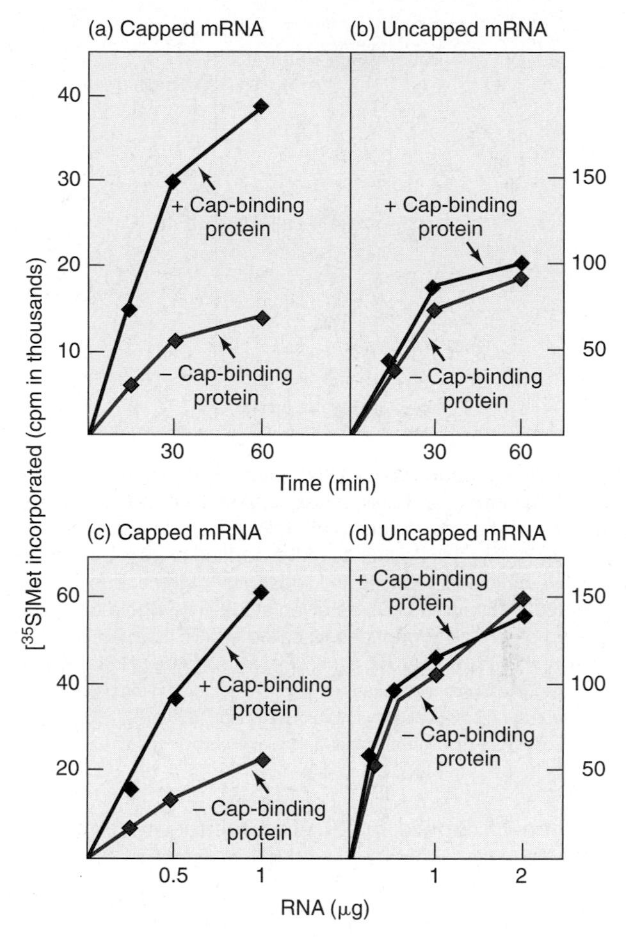

Figure 17.19 Cap-binding protein stimulates translation of capped, but not uncapped, mRNA. Shatkin and collaborators used HeLa cell-free extracts to translate capped and uncapped mRNAs in the presence of [^{35}S]methionine. Panels **(a)** and **(c):** translation of capped Sindbis virus mRNA with (blue) or without (red) cap-binding protein. Panels **(b)** and **(d):** translation of uncapped encephalomyocarditis virus (EMC) with (blue) or without (red) cap-binding protein. (*Source:* Adapted from Sonenberg, N., H. Trachsel, S. Hecht, and A.J. Shatkin, Differential stimulation of capped mRNA translation in vitro by cap-binding protein. *Nature* 285:331, 1980.)

one of the reactive aldehydes. This bond can be made permanent by reduction. After cross-linking, the investigators digested all of the RNA but the cap with RNase, then electrophoresed the products to measure the sizes of any proteins cross-linked to the labeled cap. Figure 17.18 shows that a polypeptide with a M_r of about 24 kD bound, even at low temperature. At higher temperature, another pair of polypeptides of higher molecular mass (50–55 kD) bound. However, unlabeled m^{7}GDP did not compete with these high M_r polypeptides for binding to the mRNA, whereas the unlabeled cap analog did compete with the 24-kD polypeptide for binding. This suggested that the 24-kD polypeptide bound specifically to the cap, but the 50–55-kD-polypeptides did not. On the other hand, GDP competed with the 50–55-kD polypeptides for binding to the mRNA, but it did not compete with the 24-kD polypeptide. This may mean that the larger polypeptides are GDP-binding proteins, rather than cap-binding proteins.

Sonnenberg, Shatkin, and colleagues followed up their discovery of the cap-binding protein by purifying it by affinity chromatography on an m^{7}GDP-Sepharose column. Then they added this protein to HeLa cell-free extracts and demonstrated that it stimulated transcription of capped, but not uncapped, mRNAs (Figure 17.19). They used viral mRNAs in both experiments: Sindbis virus mRNA for capped mRNA, and encephalomyocarditis virus mRNA for uncapped mRNA. (Encephalomyocarditis virus is a picornavirus similar to poliovirus.)

As we have seen, picornavirus mRNAs are not capped. Nevertheless, these viruses have mechanisms for ensuring that their mRNAs are translated. In fact, they take advantage of the cap-free nature of their mRNAs to eliminate competition from capped host mRNAs. They do this by inactivating the host cap-binding protein, thus blocking translation of capped host mRNAs, at least in certain cells. Molecular biologists have taken advantage of this situation by using poliovirus-infected cell extracts as an assay system for the cap-binding protein. Any protein that can restore

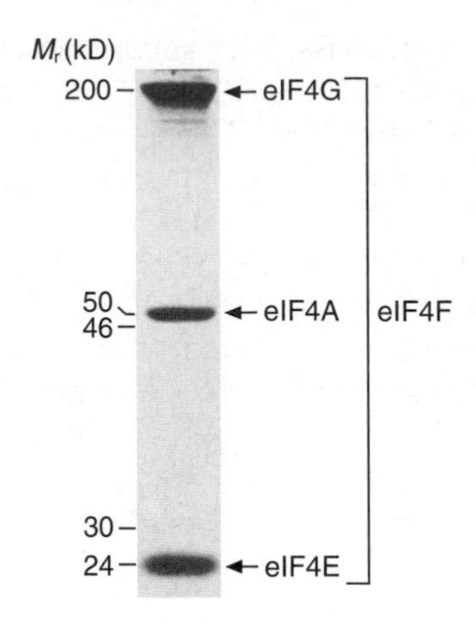

Figure 17.20 Components of eIF4F (complete cap-binding protein). Sonenberg and colleagues purified the cap-binding protein using a series of steps, including m^7GTP affinity chromatography. Then they displayed the subunits of the purified protein by SDS-PAGE. The relative molecular masses (in kilodaltons) of the subunits and markers (200, 46, and 30 kD) are given at left. The whole complex, composed of three polypeptides, is called eIF4F. (*Source:* Edery, I., M. Hümbelin, A. Darveau, K.A.W. Lee, S. Milburn, J.W.B. Hershey, H. Trachsel, and N. Sonenberg, Involvement of eukaryotic initiation factor 4A in the cap recognition process. *Journal of Biological Chemistry* 258 (25 Sept 1983) p. 11400, f. 2. American Society for Biochemistry and Molecular Biology.)

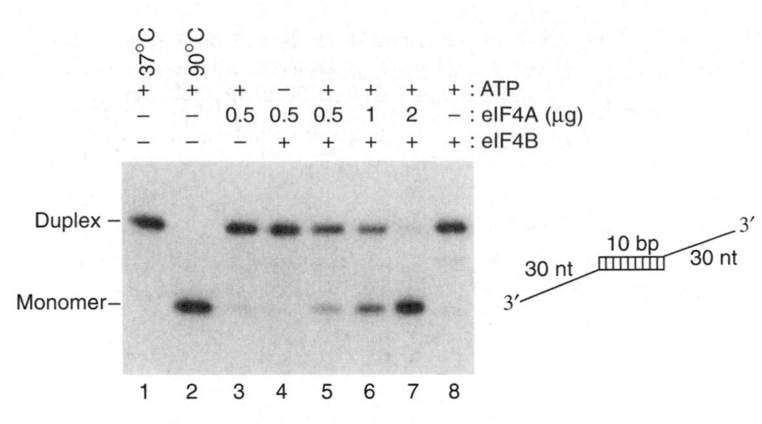

Figure 17.21 RNA helicase activity of eIF4A. Pause and Sonenberg tested combinations of ATP, eIF4A, and eIF4B (as indicated at top) on the radioactive helicase substrate shown at right. RNA helicase unwinds the 10-bp double-stranded region of the substrate, converting the dimer to two monomers. The dimer and monomers are then easily separated by gel electrophoresis, as indicated at left, and detected by autoradiography. The first two lanes are just substrate at low and high temperatures. The high temperature melts the double-stranded region of the substrate, yielding monomers. Lanes 3–8 show that ATP and eIF4A are required for helicase activity, and eIF4B stimulates this activity. (*Source:* Pause A. and N. Sonenberg, Mutational analysis of a DEAD box RNA helicase: The mammalian initiation translation factor eIF-4A. *EMBO Journal* 11 (1992) p. 2644, f. 1.)

translation of capped mRNAs to such extracts must contain the cap-binding protein. This assay revealed that the 24-kD protein by itself was quite labile, but a higher molecular mass complex was much more stable. Sonenberg and collaborators have refined this analysis to demonstrate that the active purified complex contains three polypeptides: the original 24-kD cap-binding protein, and two other polypeptides with M_rs of 50 kD and 220 kD (Figure 17.20). These polypeptides were then given new names: The 24-kD cap-binding protein is **eIF4E;** the 50-kD polypeptide is **eIF4A,** and the 220-kDa polypeptide is **eIF4G.** The whole three-polypeptide complex is called **eIF4F.**

SUMMARY eIF4F is a cap-binding protein composed of three parts: eIF4E has the actual cap-binding activity; it is accompanied by the two other subunits: eIF4A and eIF4G.

Functions of eIF4A and eIF4B The eIF4A polypeptide is a subunit of eIF4F, but it also has an independent function: It is a member of the so-called **DEAD protein** family, which has the consensus amino acid sequence Asp (D), Glu (E), Ala (A), Asp (D), and has **RNA helicase** activity. It can therefore unwind the hairpins that are frequently found in the 5′-leaders of eukaryotic mRNAs. To do this job effectively, eIF4A needs the help of **eIF4B,** which has an RNA-binding domain and can stimulate the binding of eIF4A to mRNA. Arnim Pause and Sonenberg used a well-defined in vitro system to demonstrate the activities of both eIF4A and 4B. They started with the products of the eIF4A and 4B genes cloned in bacteria, so there was no possibility of contamination by other eukaryotic proteins. Then they added the labeled RNA helicase substrate pictured on the right in Figure 17.21. This is actually two 40-nt RNAs with complementary 5′-ends, which form a 10-bp RNA double helix. If an RNA helicase unwinds this 10-bp structure, it separates the two 40-nt monomers. Electrophoresis then easily discriminates between monomers and dimer. The more monomers form, the greater is the RNA helicase activity.

Figure 17.21 depicts the results. A small amount of eIF4A (with ATP) caused a very modest amount of unwinding (lane 3), suggesting that this factor has some RNA helicase activity of its own. However, this helicase activity was stimulated by eIF4B (lane 5), and this activity depended on ATP (compare lanes 4 and 5). Greater amounts of eIF4A produced even greater RNA helicase activity (lanes 6 and 7). To show that eIF4B has no helicase activity of its own, Pause and Sonenberg added eIF4B and ATP without eIF4A and observed no helicase activity (lane 8). Thus, these two factors cooperate to unwind RNA helices, including hairpins, and this activity depends on ATP.

SUMMARY eIF4A has RNA helicase activity that can unwind hairpins found in the 5′-leaders of eukaryotic mRNAs. It is aided in this task by another factor, eIF4B, and requires ATP for activity.

Functions of eIF4G We have seen that most eukaryotic mRNAs are capped, and the cap serves to help the ribosome bind. But some viral mRNAs are uncapped; these mRNAs, and perhaps a few cellular mRNAs, have IRESs that can help ribosomes bind. Furthermore, we know that the poly(A) tail at the 3′-end of mRNAs stimulates translation. This latter process involves recruitment of ribosomes to the mRNA via a poly(A)-binding protein called **Pab1p** (yeast) or **PABP1** (human). The eIF4G protein participates in all of these kinds of initiations by serving as an adapter, or "scaffold" protein, that can interact with a variety of different proteins.

Figure 17.22 illustrates three different ways in which eIF4G can participate in translation initiation. In panel (a) we see the function eIF4G performs in initiating on ordinary, capped mRNAs. The amino terminus of eIF4G binds to eIF4E, which in turn binds to the cap. The central portion of eIF4G binds to eIF3, which in turn binds to the 40S ribosomal particle. Thus, by tethering together eIF4E and eIF3, eIF4G can bring the 40S subunit close to the 5′-end of the mRNA, where it can begin scanning.

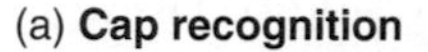

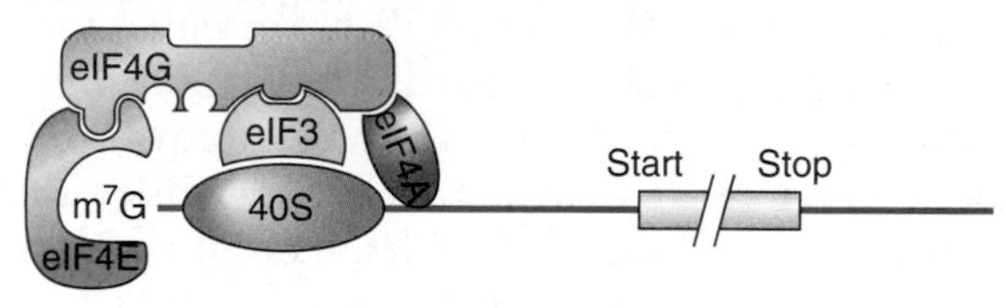

(b) IRES recognition (poliovirus mRNA)

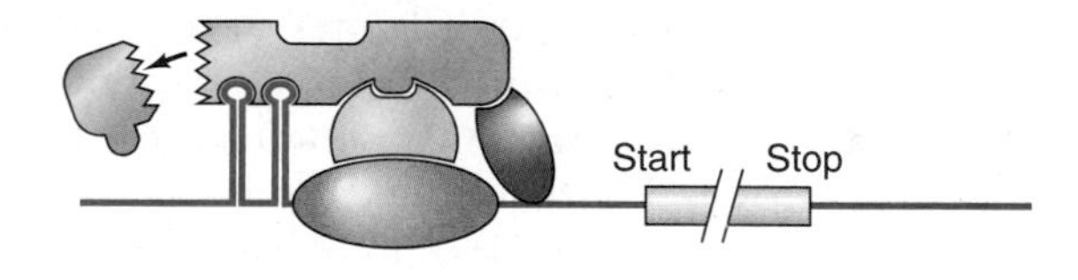

(c) Cap + poly(A) recognition

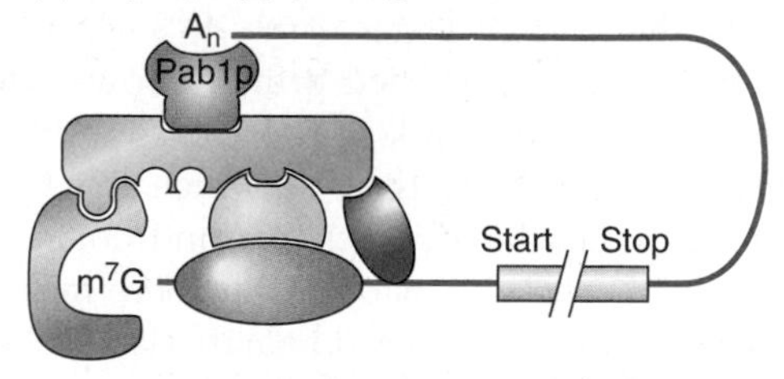

Figure 17.22 The adapter role of eIF4G in recruiting the 40S ribosomal particle in four different situations. (a) Capped mRNA. eIF4G (orange) serves as an adapter between eIF4E (green), bound to the cap, and eIF3 (yellow), bound to the 40S ribosomal particle (blue). The formation of this chain of molecules recruits the 40S particle to a site on the mRNA (dark green) near the cap, where it can begin scanning. eIF4A (red) is also bound to eIF4G, but does not play a role in the interactions illustrated here. **(b)** An mRNA, such as poliovirus mRNA, with an IRES. The IRES interacts directly with the remnant (p100) of eIF4G after a viral protease has cleaved it, ensuring recruitment of the 40S particle. This interaction happens even after removal of the N-terminal part of eIF4G, which blocks binding to capped cellular mRNAs, at least in certain cells. **(c)** Synergism between cap and poly(A). eIF4E bound to the cap and Pab1p (purple) bound to the poly(A) both bind to eIF4G and act synergistically in recruiting the 40S particle. (*Source:* Adapted from Hentze, M.W., eIF4: A multipurpose ribosome adapter? *Science* 275:501, 1997.)

Panel (b) depicts the corruption of translation initiation by a picornavirus such as poliovirus. A viral protease cleaves off the amino terminal domain from eIF4G, impairing its ability to interact with eIF4E in recognizing caps. Thus, capped cellular mRNAs go untranslated. However, the remaining part of eIF4G is still capable of binding to the poliovirus IRES, so 40S subunits are still recruited to the viral mRNA. In fact, the famous Sabin vaccine, which has helped in the ongoing effort to eradicate polio, contains three attenuated strains of the virus. In each strain, an important attenuating event was an alteration in the viral IRES that reduced the affinity for eIF4G, thus impairing translation of the viral mRNA.

When the viral protease cleaves off the N-terminal domain of eIF4G, it leaves a C-terminal domain called p100. Although the poliovirus IRES binds directly to p100, it depends on several cellular proteins (not pictured in Figure 17.22b) for optimum binding. Other viruses, including hepatitis C virus (HCV, another picornavirus), contain IRESs that bind directly to eIF3, without any need for p100 or intact eIF4G. Still other viruses, including hepatitis A virus (HVA, a flavivirus), have IRESs that bind directly to the 40S ribosomal subunit, bypassing the need for all the subunits of eIF4F, and even for eIF3.

It has been commonly assumed that p100 is ineffective in binding to eIF4E, and therefore that cleavage of eIF4G blocks cap-dependent host protein synthesis. On the other hand, Richard Jackson and colleagues demonstrated in 2001 that p100 can stimulate translation of capped mRNAs in a cell-free reticulocyte extract depleted of its own eIF4G, suggesting that p100 is indeed capable of supporting cap-dependent translation. However, maximum levels of cap-dependent translation required a concentration of p100 that is about four times higher than the natural concentration of eIF4G in reticulocyte lysates, leading Jackson and colleagues to suggest the following hypothesis: The loss of cap-dependent host protein synthesis in poliovirus-infected cells is due to competition by viral RNA for the limiting amount of p100, not to an inherent inability of p100 to support the translation of host mRNAs.

A further qualification of the model in Figure 17.22b is also necessary. Although the model appears to describe the situation in HeLa cells accurately, it should not be taken to imply that cleavage of eIF4G blocks host protein synthesis in all kinds of cells. Indeed, Akio Nomoto and colleagues have shown that, although eIF4G cleavage appears to be complete by about 5 h post-infection in human neural cells, host protein synthesis continues unabated. These workers suggested that another factor in neural cells can compensate for the loss of eIF4G, but no direct evidence for such a factor has been presented.

Finally, panel (c) illustrates the simultaneous interactions between eIF4G and eIF4E bound to the cap and between eIF4G and Pab1p bound to the poly(A) tail of the mRNA. This dual binding of eIF4G to proteins at both ends of the mRNA effectively circularizes the mRNA, which appears to aid translation in at least three ways: First, regulatory proteins and miRNAs bound to the 3′-UTR are close to the cap, which could help them influence initiation of translation. Second, ribosomes completing one round of translation are close to the cap, which may facilitate re-initiation. Finally, the two ends of the mRNA are sequestered and therefore relatively unavailable to RNases that would otherwise degrade the mRNA.

It is important to note that the cap-binding initiation factors we have just studied are the ones used *after* the so-called **pioneer round** of translation, in which the first ribosome binds to the mRNA and translates it. For the pioneer round, the ribosome uses a different set of proteins known as the **cap-binding complex** (**CBC**), which binds to the cap in the nucleus and is exported to the cytoplasm along with the mRNA, as part of an mRNA–protein complex known as the **mRNP** (messenger ribonucleoprotein). The cap-binding protein within the CBC in humans is a heterodimeric cap-binding protein, **CBP80/20**, named for the molecular masses (in kD) of its two subunits. After the pioneer round, the cytoplasmic eIF4F complex replaces the nuclear CBC.

CBP80 is important not only in cap binding, but also in the export of the mRNP out of the nucleus. This export requires a complex of proteins called the **TREX** (transcription export) complex. Mammalian TREX is composed of a seven-subunit complex known as THO, and two other proteins, UAP56 and Aly. Robin Reed and colleagues showed in 2006 that the CBP80 subunit of the cap-binding complex associates with Aly, recruiting TREX to a position near the cap of the growing mRNA. This association with TREX will allow the mature mRNP to be exported 5′-end first, from the nucleus to the cytoplasm, where it can be translated.

TREX is not recruited to pre-mRNAs before they are spliced, nor to the transcripts of synthetic cDNAs, which lack introns, leading to the hypothesis that splicing is necessary for recruitment of TREX to an mRNP. However, TREX does appear to be involved in the export of mRNPs derived from natural genes that lack introns, suggesting that splicing is not always required to attract TREX.

SUMMARY eIF4G is a scaffold protein that is capable of binding to a variety of other proteins, including eIF4E (the cap-binding protein), eIF3 (the 40S ribosomal subunit-binding protein), and Pab1p (a poly[A]-binding protein). By interacting with these proteins, eIF4G can recruit 40S ribosomal subunits to the mRNA and thereby stimulate translation initiation. In the pioneer round of translation, the cap-binding role of eIF4F is played by the CBC, which binds to the cap before export of the mRNP out of the nucleus. A subunit of the CBC also attracts TREX, which guides the mRNP, 5′-end first, out of the nucleus.

Functions of eIF1 and eIF1A eIF1 causes only a modest (about 20%) stimulation of translation activity in vitro. Thus, it was long thought to be dispensable. However, the genes encoding both eIF1 and eIF1A are essential for yeast viability, so their products are hardly dispensable. But what roles do they play? In 1998, Tatyana Pestova and colleagues found the answer: Without eIF1 and eIF1A, the 40S subunit scans only a few nucleotides, if at all, and remains only loosely bound to the mRNA. With these factors, the 40S particle scans to the initiation codon and forms a stable 48S complex.

Pestova and coworkers used a **toeprint assay** based on the primer extension technique (Chapter 5) to locate the leading edge of the 40S ribosomal subunit as it bound to an mRNA. They isolated complexes between the 40S subunit and a mammalian β-globin mRNA, then mixed them with a primer that binds downstream of the initiation codon on the mRNA. Then they extended the primer with nucleotides and reverse transcriptase. When the reverse transcriptase hits the leading edge of the 40S subunit, it stops, so the length of the extended primer shows where that leading edge lies. If you think of the 40S subunit as a foot, its leading edge would be the toe, which is why we call this a toeprint assay. Finally, Pestova and colleagues electrophoresed the primer extension products to measure their sizes. Figure 17.23 presents a schematic view of this procedure.

The actual results are presented in Figure 17.24. Lanes 1 and 2 contained only mRNA or mRNA and 40S subunits, with no factors, so it is not surprising that no complex formed. Lane 3 contained mRNA, 40S subunits, and eIF2, 3, 4A, 4B, and 4F. These factors promoted formation of complex I (the pre-scan complex) only, with no trace of complex II (the post-scan complex). The leading edge of the 40S particle under these circumstances was between positions +21 and +24 relative to the cap of the mRNA, about where we would expect it if the 40S subunit bound at the cap and did not begin scanning or scanned at most a short distance. Lane 4 contains all the factors in lane 3, plus a mixture of initiation factors obtained by washing ribosomes with a saline solution, then

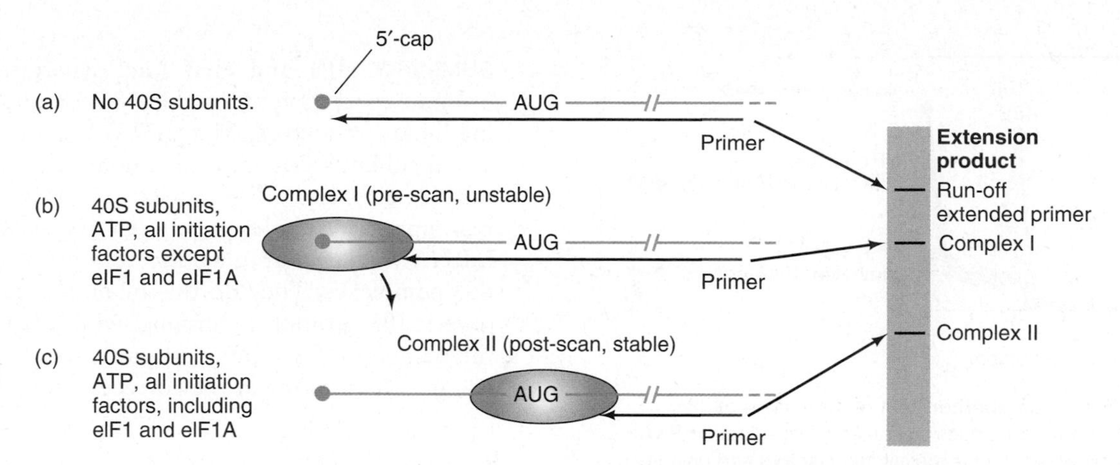

Figure 17.23 Principle of toeprint assay. (a) Negative control. Leave out an essential ingredient, such as 40S subunits, so no complex can form between 40S ribosomal subunits and mRNA. With no 40S particle to block the reverse transcriptase, the primer is extended to the 5′-end of the mRNA. This yields a run-off extended primer corresponding to naked mRNA. **(b)** Complex formed in the absence of eIF1 and eIF1A. Add all the components listed at left, but omit eIF1 and 1A. Complex I forms at the cap, but does not progress far, if at all. Thus, the primer is extended a long distance to the leading edge of the 40S particle. **(c)** Complex formed in the presence of eIF1 and eIF1A. The 40S ribosomal particle has scanned downstream to the initiation codon (AUG) and formed a stable complex (complex II). Thus, the primer is extended only a short distance before it is blocked by the leading edge of the 40S particle in the 48S complex. (*Source:* Adapted from Jackson, R.J., Cinderella factors have a ball. *Nature* 394:830, 1998.)

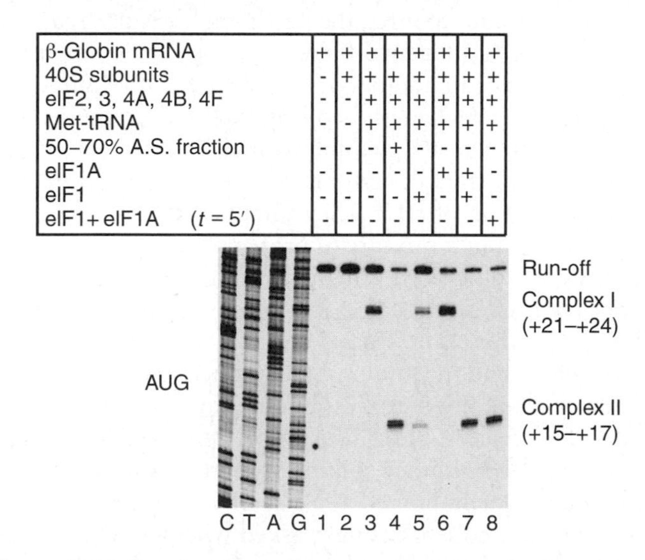

Figure 17.24 Results of toeprint assay. Pestova and colleagues carried out a toeprint assay as described in Figure 17.23, using mammalian β-globin mRNA. The components added to each assay are listed at the top of lanes 1–8. "50–70% A.S. fraction" (lane 4) refers to the factors obtained by precipitating proteins from a ribosome salt wash with ammomium sulfate concentrations between 50 and 70% saturated. "eIF1 + eIF1A (t = 5′)" refers to eIF1 and eIF1A added 5 min after adding the other components of the assay. Lanes C, T, A, and G were the results of sequencing a DNA corresponding to the β-globin mRNA. These sequencing lanes were included as markers to determine the exact positions of the leading edges (toeprints) of the 40S ribosomal particle in the complexes. The position of the initiation codon (AUG) is given at left. The bands corresponding to full-length run-off extended primer and complexes I and II are given at right, with the leading edge of the 40S particle relative to the cap and the initiation codon, respectively. eIF1 and eIF1A were required for complex II formation. (*Source:* Pestova, T.V., S.I. Borukhov, and C.V.T. Hellen, Eukaryotic ribosomes require initiation factors 1 and 1A to locate initiation codons. *Nature* 394 (27 Aug 1998) f. 2, p. 855. Copyright © Macmillan Magazines Ltd.)

collecting those proteins that could be precipitated by ammonium sulfate concentrations between 50 and 70%. Clearly, this mixture of factors, along with others, could promote the formation of complex II, whose leading edge was between positions +15 and +17 relative to the A of the AUG initiation codon, about where we would expect it if the 40S particle was centered on the initiation codon.

Next, Pestova and colleagues purified the important proteins in the 50–70% ammonium sulfate fraction to homogeneity and obtained partial amino acid sequences to identify them. They turned out to be eIF1 and eIF1A. Figure 17.24, lanes 5 and 6 show that each of these factors individually had little or no ability to stimulate complex II formation. On the other hand, lane 7 demonstrates that these two factors together caused complex II to be formed almost exclusively. Thus, these two factors act synergistically to promote complex II formation. In lane 8, complex I was allowed to form for 5 min, then eIF1 and eIF1A were added. Under these conditions, only complex II formed. Thus, complex I was not a dead end; initiation factors could convert it to complex II.

Did eIF1 and eIF1A convert complex I to complex II by simply causing the 40S subunit to scan farther on the same mRNA, or did these factors cause the 40S particle to dissociate from the mRNA and bind again to scan to the initiation codon? To find out, Pestova and colleagues formed complex I on a radiolabeled mRNA, then added eIF1 and eIF1A with and without a 15-fold excess of unlabeled competitor mRNA. They purified 48S complexes (presumably equivalent to complex II) by sucrose gradient ultracentrifugation and checked these complexes for radioactivity by scintillation counting (Chapter 5).

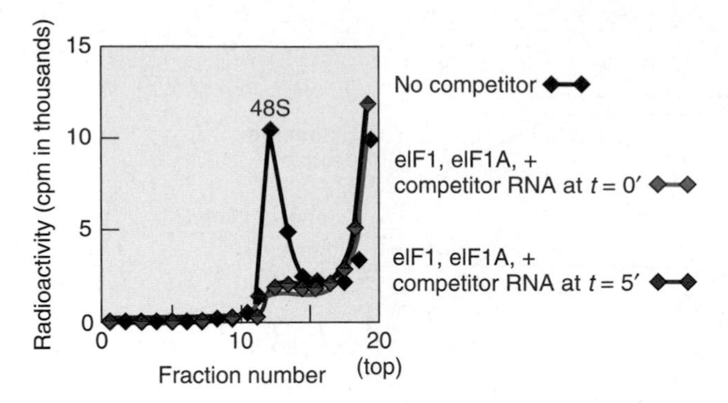

Figure 17.25 Effect of competitor RNA on formation of 48S complex. Pestova and colleagues incubated $[^{32}P]\beta$-globin mRNA with 40S ribosomal particles plus the initiation factors and unlabeled competitor RNA combinations indicated at right: blue, no competitor; green, competitor, along with eIF1 and eIF1A, added at time zero; red, competitor, along with eIF1 and eIF1A, added after 5 min of incubation (by which time complex I had formed). After the incubations, the investigators subjected the mixtures to sucrose gradient ultracentrifugation to detect the formation of stable 48S complexes involving 40S particles, $[^{32}P]$mRNA, and Met-tRNA$_i^{Met}$. They plotted the radioactivity in counts per minute (cpm) detected in each fraction by scintillation counting. The top of the gradient was in fraction 19, as indicated at bottom right. (*Source:* Adapted from Pestova, T.V., S.I. Borukhov, and C.V.T. Hellen, Eukaryotic ribosomes require initiation factors 1 and 1A to locate initiation codons. *Nature* 394:856, 1998.)

As expected (Figure 17.25), they found a clear radioactive peak of 48S complexes in the absence of competitor mRNA. However, they found no radioactive peak of 48S complexes when they added the competitor mRNA at the beginning of the incubation *or* when they added the competitor mRNA after complex I had formed for 5 min. Thus, eIF1 and eIF1A did not simply allow 40S subunits in complex I to scan downstream and form complex II on the same, labeled mRNA. If they did, labeled 48S complexes would have been seen when these factors and the competitor mRNA were added after 5 min, when complex I had already formed on the labeled mRNA. Instead, these factors disrupted complex I on the labeled mRNA and forced a new complex to form on the excess, unlabeled mRNA. Presumably, the 40S subunits abandoned the labeled mRNA, bound to the caps of (mostly) unlabeled mRNAs, and scanned to the initiation codons of these unlabeled mRNAs, forming complex II.

Thus, eIF1 and eIF1A are not only essential for proper 48S complex formation, they also appear to disrupt improper complexes between 40S ribosomal subunits and mRNA.

In fact, later work has shown that the interaction between eIF1 and eIF1A is antagonistic: eIF1 tends to prevent the scanning 40S subunit from committing to initiate at a given start codon, and this helps to ensure that the wrong codon will not be chosen. In other words, eIF1 promotes scanning. On the other hand, eIF1A slows scanning down. It helps the scanning complex pause long enough at the right start codon to facilitate commitment to initiate there.

SUMMARY eIF1 and eIF1A act synergistically to promote formation of a stable 48S complex, involving initiation factors, Met-tRNA$_i^{Met}$, and 40S ribosomal subunits bound at the initiation codon of an mRNA. eIF1 and eIF1A appear to act by dissociating improper complexes between 40S subunits and mRNA and encouraging the formation of stable 48S complexes. They do this by antagonizing each other: eIF1 promotes scanning, while eIF1A causes the scanning 40S subunit to pause long enough to commit to initiating at the correct start codon.

Functions of eIF5 and eIF5B Once eIF2 has delivered Met-tRNA to the 40S ribosomal subunit and mRNA has also bound to complete the 48S initiation complex, eIF2 needs to dissociate from the complex. To accomplish this dissociation, GTP hydrolysis is required. However, unlike IF2, eIF2 needs the help of another factor—eIF5—to hydrolyze its bound GTP. Even after the eIF5-induced hydrolysis of the GTP bound to eIF2, the 48S complex is not ready to accept the 60S ribosomal subunit to finish the initiation process. Instead, an additional factor, **eIF5B,** is required.

Christopher Hellen and colleagues discovered eIF5B in 2000 when they tested recombinant eIF5 for the ability to induce 60S ribosomal subunits to bind to 48S complexes after dissociation of eIF2. They found that eIF5 alone was not sufficient, but a mixture of proteins released from ribosomes by washing with a high-ionic-strength buffer could complement eIF5 and cause joining of the ribosomal subunits. From this "salt wash," these investigators purified eIF5B, which had the joining-inducing activity. The purified eIF5B (or a modified eIF5B obtained by cloning its gene) could not induce subunit joining on its own. However, it could stimulate subunit joining in a reaction containing other factors, including eIF1, eIF2, eIF3, and eIF5.

Hellen and colleagues next asked whether GTP hydrolysis is required for the subunit-joining reaction. For this experiment, they mixed preformed 48S complexes with eIF5, eIF5B, 60S subunits, and either GTP or the unhydrolyzable analog, GDPNP. No subunit joining took place without either GTP or GDPNP. Thus, we know that GTP is required. Furthermore, GDPNP could support subunit joining, but it required stoichiomentric quantities of eIF5B. On the other hand, eIF5B acted catalytically with GTP in stimulating subunit joining. Thus, because GDPNP will suffice, GTP hydrolysis is not required for subunit joining.

Hellen and colleagues also showed that eIF5B was not released from 80S complexes formed in the presence of GDPNP, but it was released from complexes formed with GTP. Thus, GTP hydrolysis appears to be required for release of eIF5B from the ribosome. In this respect, eIF5B resembles bacterial IF2, which also requires GTP hydrolysis in order to be released from the ribosome. The two factors are also similar in having a ribosome-stimulated GTPase, and they both play a similar role in ribosomal subunit joining. In fact, the two factors are homologous, so their similarity of functions is not surprising. On the other hand, eIF5B is quite different from IF2 in that it cannot stimulate binding of Met-tRNA$_i^{Met}$, whereas IF2 can carry out the equivalent reaction in bacteria. Instead of eIF5B, eIF2 is responsible for this reaction in eukaryotes.

SUMMARY eIF5B is homologous to the prokaryotic factor IF2. It resembles IF2 in binding GTP and stimulating association of the two ribosomal subunits. eIF5B works with eIF5 in this reaction. eIF5B also resembles IF2 in using GTP hydrolysis to promote its own dissociation from the ribosome so protein synthesis can begin. But it differs from IF2 in that it cannot stimulate the binding of the initiating aminoacyl-tRNA to the small ribosomal subunit. That task is performed by eIF2 in eukaryotes.

17.3 Control of Initiation

We have already examined control of gene expression at the transcriptional and post-transcriptional levels. But control also occurs at the translational level. Given the extensive control we see at the transcriptional and post-transcriptional levels, it is fair to ask why organisms have also evolved mechanisms to control gene expression at the translational level. The major advantage of translational control is speed. New gene products can be produced quickly, simply by turning on translation of preexisting mRNAs. This is especially valuable in eukaryotes, where transcripts are relatively long and take a correspondingly long time to make. Naturally enough, most of this translational control happens at the initiation step.

Bacterial Translational Control

We have learned that most of the control of bacterial gene expression occurs at the transcription level. The very short lifetime (only 1–3 min) of the great majority of bacterial mRNAs is consistent with this scheme, because it allows bacteria to respond quickly to changing circumstances. It is true that different cistrons on a polycistronic transcript can be translated better than others. For example, the *lacZ, Y,* and *A* cistrons yield protein products in a molar ratio of 10:5:2. However, this ratio is constant under a variety of conditions, so it seems to reflect the relative efficiencies of the ribosome-binding sites of the three cistrons as well as differential degradation of parts of the polycistronic mRNA. However, some examples of real control of bacterial translation do occur. Let us consider several of them.

Shifts in mRNA Secondary Structure RNA secondary structure can play a role in translation efficiency, as we observed in Figure 17.6 earlier in this chapter. We learned that the initiation codon of the replicase cistron of the MS2 family of RNA phages is buried in a double-stranded structure that also involves part of the coat gene. This explains why the replicase gene of these phages cannot be translated until the coat protein is translated: The ribosomes moving through the coat gene open up the secondary structure that hides the initiation codon of the replicase gene.

Another example of control via mRNA structure comes from the induction of σ^{32} synthesis during heat shock in *E. coli,* which we mentioned in Chapter 8. When *E. coli* cells experience a rise in temperature from the normal 37°C to 42°C, they switch on a set of heat shock genes that help them cope with the higher temperature. These new, heat shock genes respond to σ^{32}, rather than the normal σ^{70}. But σ^{32} begins accumulating in less than a minute after heat shock, which is too little time for transcription of the σ^{32} gene *(rpoH)* and translation of the corresponding mRNA. So how can we account for such rapid accumulation of σ^{32}?

The data support two answers. First, preexisting σ^{32}, which is normally unstable, becomes stabilized. Second, and more relevant to our discussion here, the σ^{32} gene is controlled at the level of translation initiation. The mRNA encoding σ^{32} is normally folded in such a way that its initiation codon is hidden in secondary structure. That is, the initiation codon is base-paired to another, downstream region of the mRNA. But when the temperature rises, the base pairs causing this secondary structure melt, unmasking the initiation codon so the mRNA can be translated. Thus, there is always plenty of mRNA for this special σ-factor, but it is untranslatable until the temperature rises to dangerous levels. In other words, the built-in thermosensor in the mRNA allows for heating to stimulate gene expression at the translation level.

Takashi Yura and colleagues provided strong support for this hypothesis in 1999 using a derivative of the *rpoH* gene that produced an mRNA with the secondary structure shown in Figure 17.26. This mRNA showed the same regulation characteristics as the wild-type mRNA. Note the base pairing between the initiation codon (boxed) and a region near the 3′-end of the mRNA, forming "stem I," which would presumably prevent translation of this mRNA under physiological conditions. Next, Yura and colleagues made mutations in the stem I region that made the base pairing either stronger or weaker and measured the effects of these mutations on induction by heat.

When the mutations made the base-pairing in stem I stronger, induction was weakened. For example, the C in position +5 with respect to the A of the AUG codon is normally not paired with the U in the opposite strand.

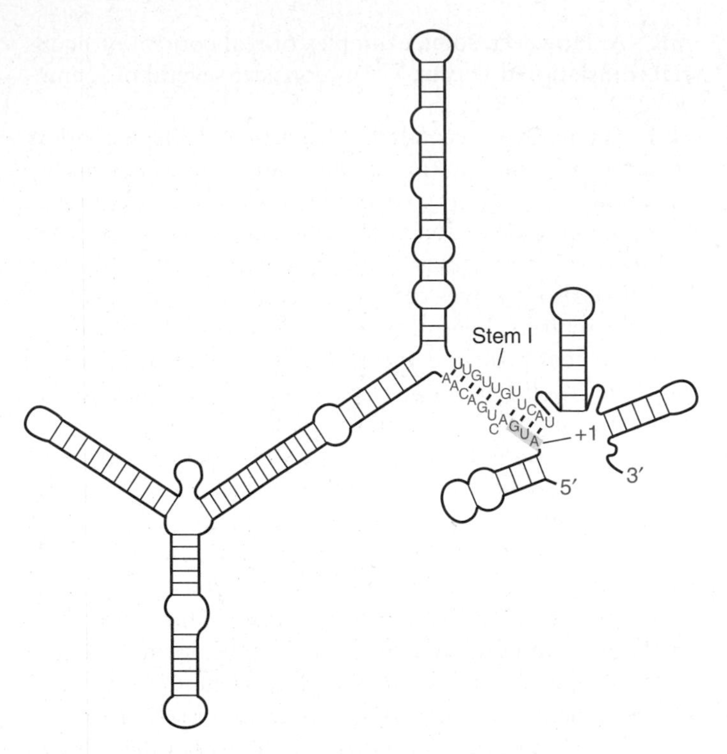

Figure 17.26 Secondary structure of a portion of the *rpoH* mRNA. The sequence in the base-paired region of stem I is shown, including the AUG initiation codon, which is shaded gray. (*Source:* Adapted from Morita, M.T., Y. Tanaka, T.S. Kodama, Y. Kyogoku, K. Yanagi, and T. Yura, Translational induction of heat shock transcription Factor σ^{32}. Evidence for a built-in RNA thermosensor. *Genes and Development* 13 [1999] p. 656, f. 1b.)

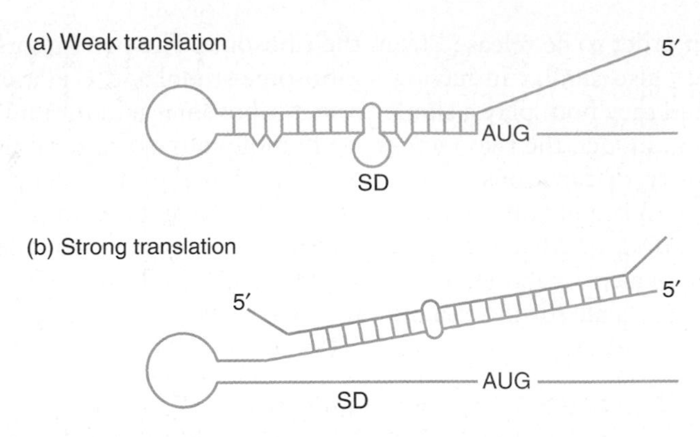

Figure 17.27 Model for activation of *rpoS* mRNA translation by an sRNA. (a) Base-pairing within the 5′-UTR of the *rpoS* mRNA creates a stem loop that hides the Shine–Dalgarno sequence (SD) and the initiation codon (AUG, pink). **(b)** The DsrA sRNA binds to the RNA-binding protein Hfq and base-pairs with part of the 5′-UTR, opening up the SD sequence and initiation codon for binding to the ribosome.

However, when this C was changed to A, it could pair to the U and increase the stability of stem I by 2.9 kcal/mol. This reduced induction from the normal 3.5-fold to only 1.4-fold. This makes sense because stronger base pairing is more difficult to disrupt by heating. On the other hand, most mutations that weakened base pairing also increased gene expression at both high and low temperatures. Again, this makes sense because weaker base pairing would be easier to disrupt even at lower temperatures.

SUMMARY The fact that bacterial mRNAs are very short-lived means that transcriptional control is a very efficient way to control gene expression in these organisms. However, translational control also occurs. Messenger RNA secondary structure can govern translation initiation, as in the replicase gene of the MS2 class of phages, whose initiation codon is buried in secondary structure until ribosomes translating the coat gene open up this structure. In another example, the initiation codon in the mRNA for the *E. coli* heat shock σ-factor, σ^{32}, is repressed by secondary structure that is relaxed by heating. Thus, heat can cause an immediate unmasking of σ^{32} mRNA initiation codons, and a burst of σ^{32} synthesis.

Shifts in mRNA Secondary Structure Induced by Proteins and RNAs In Chapter 16, we learned that small RNAs called microRNAs can control mRNA stability and translation in eukaryotes. Translation in bacteria can also be controlled by a class of short RNAs known simply as **small RNAs (sRNAs),** and these can act on mRNA secondary structure. For example, the initiation codon of the mRNA (*rpoS*) for the stress sigma factor (σ^S, or σ^{38}) is normally buried in secondary structure, so little if any protein is made. However, as shown in Figure 17.27, the DsrA sRNA, in concert with the chaperone protein Hfq, can base-pair with the upstream region of the mRNA, unmasking the *rpoS* initiation codon, and allowing translation to occur.

As we learned in Chapter 7, **riboswitches** are regions within mRNAs that can bind to small molecules, change conformation, and thereby switch gene expression on or off—for example, by shifting from an antiterminator to a terminator to cause attenuation of transcription. The region of the RNA that binds to the small molecule is known as an **aptamer.**

One of the first examples of a riboswitch was discovered by Ronald Breaker and colleagues in 2002. They showed that the *E. coli* mRNAs that encode the enzymes required to synthesize thiamine (vitamin B_1) can assume at least two different conformations. When thiamine or thiamine pyrophosphate binds to an aptamer in the mRNA, the mRNA assumes a conformation that hides the ribosome binding site, so the mRNA cannot be translated. Of course, this is helpful because the presence of thiamine indicates that the cell does not need to waste energy making more enzymes to make this vitamin. Notice that no proteins are involved in this riboswitch. The small molecule thiamine can change the conformation of the mRNA by itself.

Breaker and colleagues had already demonstrated that the leader of the mRNA encoding one of the enzymes in coenzyme B_{12} synthesis could bind to the coenzyme, and this caused a structural change in the mRNA that was important in control of coenzyme synthesis. They wondered if a similar mechanism applied to the thiamine biosynthesis pathway because two of the genes (*thiM* and *thiC*) encoding enzymes in this pathway contained *thi boxes* with conserved sequences and secondary structures.

Accordingly, they linked the thi boxes to a *lacZ* reporter gene, and tested these constructs for ability to produce β-galactosidase in the presence and absence of thiamine. They found that thiamine suppressed the production of β-galactosidase by 18- and 110-fold, respectively. Thus, the thi boxes were indeed involved in suppression of gene activity. Much of the suppression by the thi box in the *thiC* construct turned out to be at the transcriptional level, whereas all of the suppression by the *thiM* thi box was at the translational level. Since we are concerned with translational control in this chapter, let us focus on the *thiM* gene.

Breaker and colleagues next applied an **in-line probing** technique (Chapter 7) to see if thiamine or its derivatives could cause a structural change in the mRNA leader. This strategy is based on the fact that an unstructured RNA is more susceptible to spontaneous cleavage than one with lots of secondary structure (intramolecular base pairs) or tertiary structure (three-dimensional structure). So the investigators incubated a 165-nt fragment of the mRNA containing the thi box (165 *thiM* RNA) for 40 h in the presence or absence of thiamine pyrophosphate (TPP) and then electrophoresed the products to see where cleavage had occurred. Figure 17.28a reveals that plenty of cleavage occurred with or without TPP, but there were significant

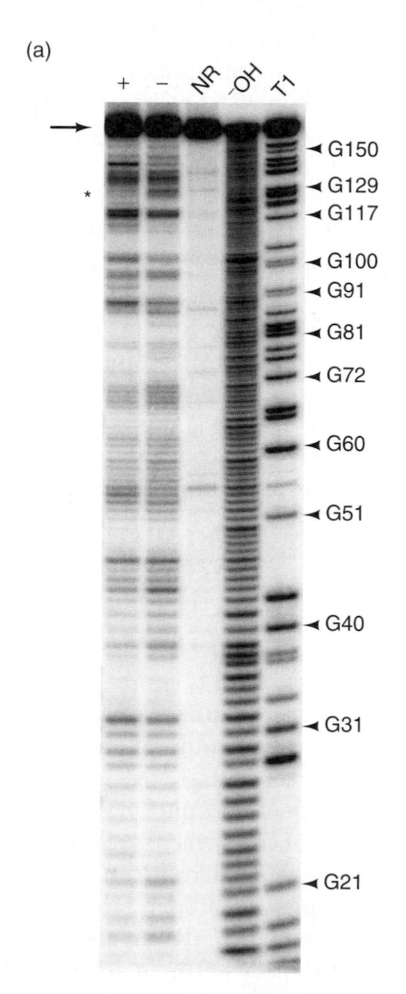

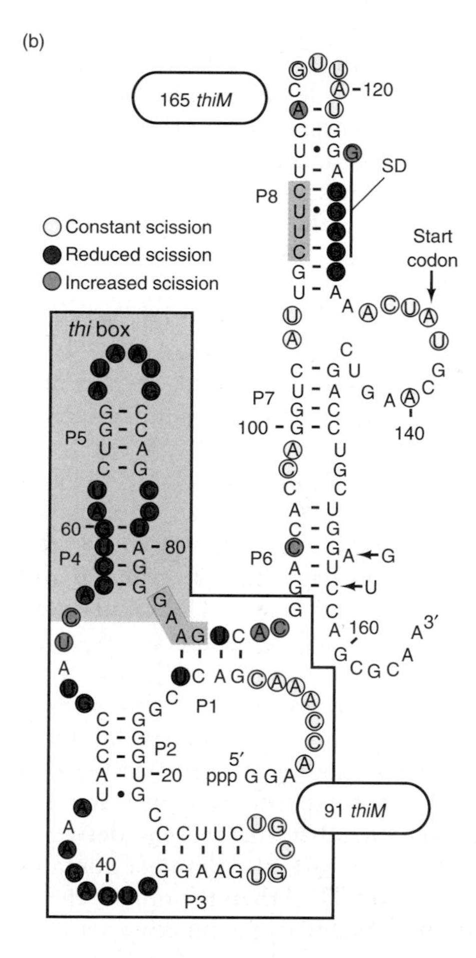

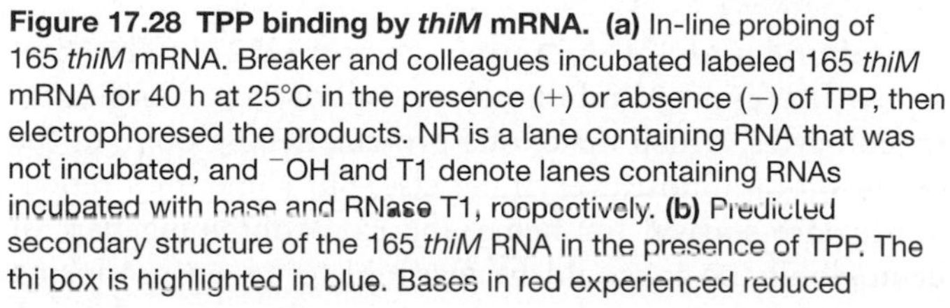

Figure 17.28 TPP binding by *thiM* mRNA. (a) In-line probing of 165 *thiM* mRNA. Breaker and colleagues incubated labeled 165 *thiM* mRNA for 40 h at 25°C in the presence (+) or absence (−) of TPP, then electrophoresed the products. NR is a lane containing RNA that was not incubated, and ⁻OH and T1 denote lanes containing RNAs incubated with base and RNase T1, respectively. **(b)** Predicted secondary structure of the 165 *thiM* RNA in the presence of TPP. The thi box is highlighted in blue. Bases in red experienced reduced cleavage in the presence of TPP, while those in green experienced increased cleavage. Unpaired bases in yellow experienced no change in cleavage. The bases in orange are the CUUC that is shown here paired with GGAG in the Shine–Dalgarno sequence (SG), and an AGGA that is another potential partner for the CUUC. (*Source: Nature*, 419, Wade Winkler, Ali Nahvi, Ronald R. Breaker, "Thiamine derivatives bind messenger RNAs directly to regulate bacterial gene expression," fig. 1 a&b, p. 953, Copyright 2002, reprinted by permission from Macmillan Publishers Ltd.)

differences. In particular, less cleavage in the region spanning positions 39–80 (including the thi box) occurred in the presence of TPP.

Notice also the region (bases 126–130) denoted by the asterisk. This is the only region that is more ordered (less cleavage) in the presence of TPP, aside from the thi box and nucleotides on the immediate 5′-side of the thi box. And this region encompasses the Shine–Dalgarno sequence, where the ribosome binds. Thus, these results suggest that TPP causes a shift in conformation of the *thiM* mRNA that hides the Shine–Dalgarno sequence in a base-paired stem. This would impede ribosome binding and lower the efficiency of translation of the mRNA.

Breaker and colleagues identified a GAAG sequence, highlighted in orange in Figure 17.28b just at the end of the thi box, that could base-pair with the CUUC at position 108–111 (also highlighted in orange) across from the Shine–Dalgarno sequence in stem P8. This suggested a model in which the CUUC (positions 108–111) normally base-pairs with the GAAG at the end of the thi box, leaving the Shine–Dalgarno sequence available for ribosome binding. This mRNA structure allows active translation. However, TPP, by binding to an aptamer in the thi box, changes the mRNA secondary structure such that the CUUC at position 108–111 base-pairs to the GGAG in the Shine–Dalgarno sequence, hiding it from the ribosomes, and slowing down translation.

This hypothesis makes several predictions. First, a piece of the mRNA containing the thi box should respond to low concentrations of TPP. Indeed, Breaker and colleagues showed that the structural modification of 165 *thiM* RNA was half-complete at a TPP concentration of only 600 nM. Second, TPP should be able to bind tightly to 165 *thiM* RNA, and Breaker and colleagues used a technique called equilibrium dialysis to demonstrate that it does indeed bind tightly. Equilibrium dialysis uses a labeled ligand (tritium-labeled TPP in this case) placed in one chamber, and a large molecule (a *thiM* RNA fragment) in a second chamber, separated from the first by a dialysis membrane which allows small molecules like TPP to pass through, but retains large molecules like RNA. After equilibrium between the two chambers is established, the experimenter measures the amount of label in each chamber and thereby derives a dissociation constant. In this case, the chamber containing the RNA had much more label than the other, reflecting a low dissociation constant (tight binding between TPP and the RNA).

A third prediction is that the binding between thiamine family members and *thiM* mRNA should be specific. Indeed, thiamine, thiamine phosphate (TP), and TPP bound well to the RNA, but oxythiamine and other thiamine derivatives did not. Finally, RNAs with alterations that would disrupt the important structural elements of the *thiM* leader sequence should block both TPP binding and control of *thiM* expression. Breaker and colleagues tested this prediction by making alterations in bases that participate in the predicted stems P3, P5, and P8. These mutant RNAs all failed to bind TPP, and failed to show reduced *thiM* expression in the presence of TPP. However, compensating mutations that restored base-pairing in stems P3, P5, and P8, all restored TPP binding and *thiM* control. For example, changing bases 106 and 107 from U and G, respectively, to A and C, respectively, blocked base-pairing with A and C, respectively at positions 130 and 131. This weakened stem P8, and blocked TPP binding and control. However, if the A and C at positions 130 and 131 were changed to G and U, respectively, TPP binding and control were restored. Thus, base-pairing in all three of these stems appears to be essential for control, as the hypothesis predicts.

SUMMARY Small RNAs, in concert with proteins, can affect mRNA secondary structure to control translation initiation. Riboswitches can also be used to control translation initiation via mRNA secondary structure. The 5′-untranslated region of the *E. coli thiM* mRNA contains a riboswitch, including an aptamer that binds thiamine and its metabolites, thiamine phosphate and, especially, thiamine pyrophosphate (TPP). When TPP is abundant, it binds to this aptamer, causing a conformational shift in the mRNA that ties up the Shine–Dalgarno sequence in secondary structure. This shift hides the SD sequence from ribosomes, and inhibits translation of the mRNA. This saves energy because the *thiM* mRNA encodes an enzyme that is needed to produce more thiamine and, thus, TPP.

Eukaryotic Translational Control

Eukaryotic mRNAs are much longer-lived than bacterial ones, so there is more opportunity for translational control. The rate-limiting factor in translation is usually initiation, so we would expect to find most control exerted at this level. In fact, the most common mechanism of such control is phosphorylation of initiation factors, and we know of cases where such phosphorylation can be inhibitory, and others where it can be stimulatory. Finally, there is an example of a protein binding directly to the 5′-untranslated region of an mRNA and preventing its translation. Removal of this protein activates translation.

Phosphorylation of Initiation Factor eIF2α The best known example of inhibitory phosphorylation occurs in reticulocytes, which make one protein, hemoglobin, to the exclusion of almost everything else. But sometimes reticulocytes are starved for heme, the iron-containing part of hemoglobin, so it would be wasteful to go on producing α- and β-globins, the protein parts. Instead of stopping the production of the globin mRNAs, reticulocytes block their translation as follows (Figure 17.29): The absence of heme unmasks the activity of a protein kinase called the **heme-controlled repressor,** or **HCR.** This enzyme phosphorylates one of the subunits of eIF2, known as **eIF2α.** The phosphorylated form of eIF2 binds more tightly than usual to eIF2B, which is an initiation factor whose job is to exchange GTP for GDP on eIF2. When eIF2B is stuck fast to phosphorylated eIF2, it cannot get free to exchange GTP for GDP on other molecules of eIF2, so eIF2 remains in the inactive GDP-bound form and cannot attach Met-$tRNA_i^{Met}$ to 40S ribosomes. Thus, translation initiation grinds to a halt.

The antiviral proteins known as **interferons** follow this same pathway. In the presence of interferon and double-stranded RNA, which appears in many viral infections, but not in normal cellular life, another eIF2α kinase is activated. This one is called **DAI,** for **double-stranded RNA-activated inhibitor of protein synthesis.** The effect of DAI is the same as that of HCR—blocking translation initiation. This is useful in a virus-infected cell because the virus has taken over the cell, and blocking translation will block production of progeny viruses, thus short-circuiting the infection.

SUMMARY Eukaryotic mRNA lifetimes are relatively long, so there is more opportunity for translation control than in bacteria. The α-subunit of eIF2 is a favorite target for translation control. In heme-starved reticulocytes, HCR is activated, so it can phosphorylate eIF2α and inhibit initiation. In virus-infected cells, another kinase, DAI, is activated; it also phosphorylates eIF2α and inhibits translation initiation.

Phosphorylation of an eIF4E-Binding Protein The rate-limiting step in translation initiation is cap binding by the cap-binding factor eIF4E. Thus, it is intriguing that eIF4E is also subject to phosphorylation, which stimulates, rather than represses, translation initiation. Phosphorylated eIF4E binds the cap with about four times the affinity of unphosphorylated eIF4E, which explains the stimulation of translation. We saw that the conditions that favor eIF2α phosphorylation and translation repression are unfavorable for cell growth, (e.g., heme starvation and virus infection). This suggests that the conditions that favor eIF4E phosphorylation and translation stimulation should be favorable for cell growth, and this is generally true. Indeed, stimulation of cell division with insulin or mitogens leads to an increase in eIF4E phosphorylation.

Insulin and various growth factors, such as platelet-derived growth factor (PDGF), also stimulate translation in

(a) **Heme abundance: No repression**

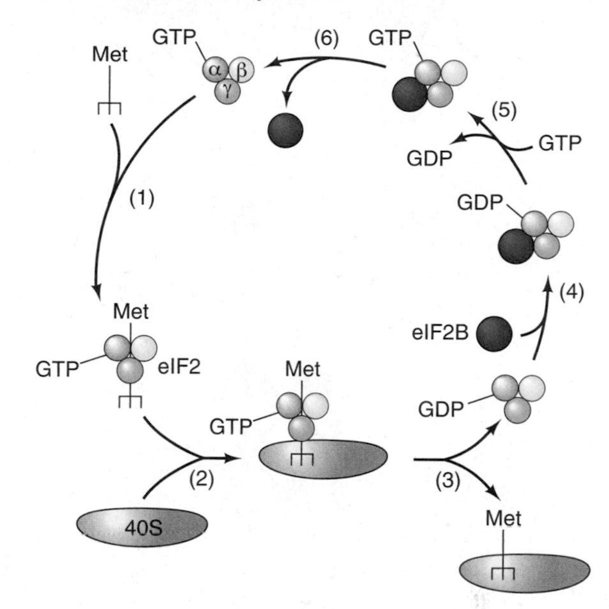

(b) **Heme starvation: Translation repression**

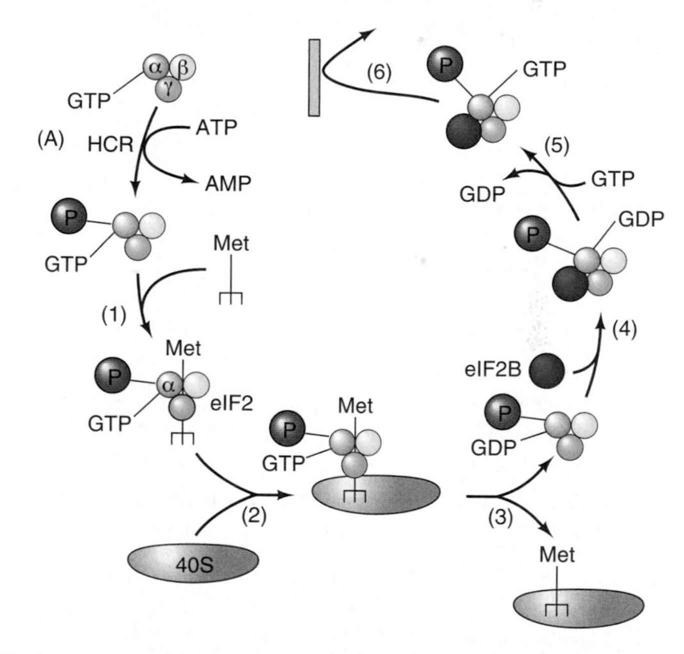

Figure 17.29 Repression of translation by phosphorylation of eIF2α (a) Heme abundance, no repression. Step 1, Met-$tRNA_i^{Met}$ binds to the eIF2-GTP complex, forming the ternary Met-$tRNA_i^{Met}$ GTP-eIF2 complex. The eIF2 factor is a trimer of nonidentical subunits (α [green], β [yellow], and γ [orange]). Step 2, the ternary complex binds to the 40S ribosomal subunit (blue). Step 3, GTP is hydrolyzed to GDP and phosphate, allowing the GDP–eIF2 complex to dissociate from the 40S ribosome, leaving Met-$tRNA_i^{Met}$ attached. Step 4, eIF2B (red) binds to the eIF2–GDP complex. Step 5, eIF2B exchanges GTP for GDP on the complex. Step 6, eIF2B dissociates from the complex. Now eIF2–GTP and Met-$tRNA_i^{Met}$ can get together to form a new complex to start a new round of initiation. **(b)** Heme starvation leads to translational repression. Step A, HCR (activated by heme starvation) attaches a phosphate group (purple) to the α-subunit of eIF2. Then, steps 1–5 are identical to those in panel (a), but step 6 is blocked because the high affinity of eIF2B for the phosphorylated eIF2α prevents its dissociation. Now eIF2B will be tied up in such complexes, and translation initiation will be repressed.

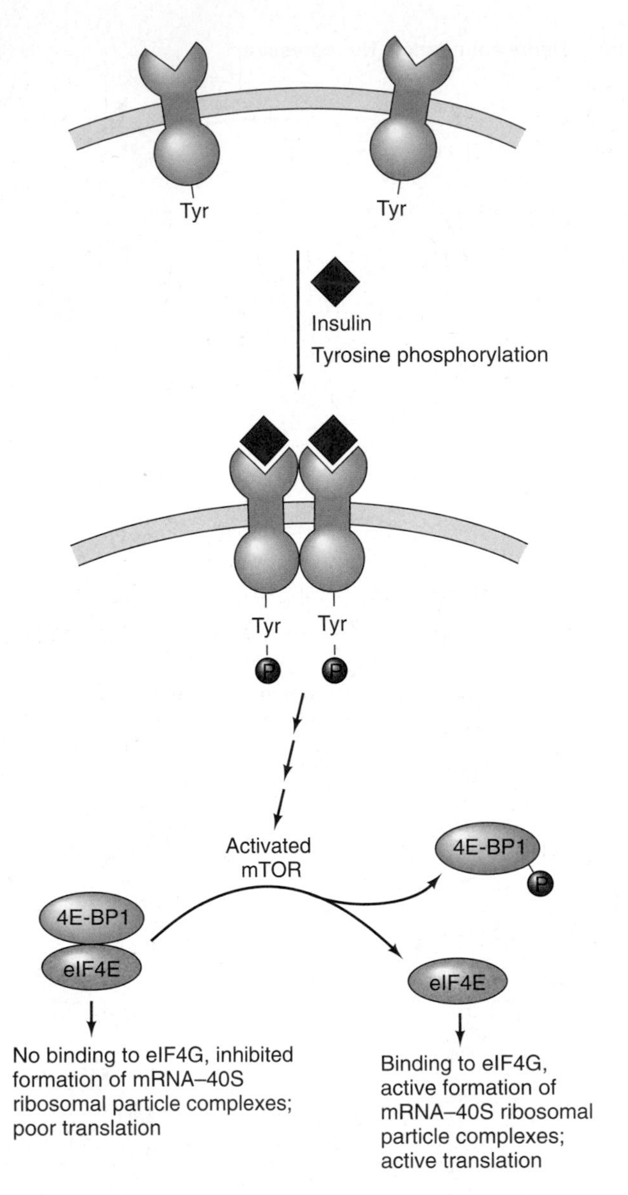

Figure 17.30 Stimulation of translation by phosphorylation of PHAS-I. Insulin, or a growth factor such as EGF, binds to its receptor at the cell surface. Through a series of steps, this activates the protein kinase mTOR. One of the targets of mTOR is 4E-BP1. When 4E-BP1 is phosphorylated by mTOR, it dissociates from eIF4E, releasing it to bind to eIF4G and therefore to participate in active translation initiation.

mammals by an alternative signal transduction pathway that involves eIF4E. We have known for many years that insulin and many growth factors interact with specific receptors at the cell surface (Figure 17.30). These receptors have intracellular domains with protein tyrosine kinase activity. When they interact with their ligands, these receptors can dimerize and autophosphorylate. In other words, the tyrosine kinase domain of one monomer phosphorylates a tyrosine on the other monomer. This triggers several signal transduction pathways (Chapter 12). One of these activates a protein called **mTOR** (**target of rapamycin**, where rapamycin is an antibiotic that inhibits translation initiation). mTor is a protein kinase, and is part of a complex called **mTOR complex 1** (**mTORC1**), which binds to eIF3 in the translation preinitiation complex. From that vantage point, mTOR can stimulate translation initiation by phosphorylating at least two other proteins in the preinitiation complex.

One of the targets of mTORC1 is a protein called **4E-BP1** (**eIF4E-binding protein**). In rats, the same protein is called PHAS-1. 4E-BP1 binds to eIF4E and inhibits its activity. In particular, 4E-BP1 inhibits binding between eIF4E and eIF4G. But once phosphorylated by mTOR, 4E-BP1 dissociates from eIF4E, which is then free to bind eIF4G and promote formation of active complexes between mRNA and 40S ribosomal subunits (Figures 17.30 and 17.22). Thus, translation is stimulated.

Sonenberg and John Lawrence and colleagues discovered human 4E-BP1 in 1994 in a Far Western screen for proteins that bind to eIF4E. A **Far Western screen** is similar to a screen of an expression library with an antibody (Chapter 4), except that the probe is a labeled ordinary protein instead of an antibody. Thus, one is looking for the interaction between two non-antibody proteins instead of the recognition of a protein by an antibody. In this case, the investigators probed a human expression library (in λgt11) with a derivative of eIF4E, looking for eIF4E-binding proteins. The probe was eIF4E, coupled to the phosphorylation site of heart muscle kinase (HMK), which was then phosphorylated with [γ-^{32}P]ATP to label it. Of about one million plaques screened, nine contained genes encoding proteins that bound the eIF4E probe. Three of these contained at least part of the gene that codes for the eIF4G subunit of eIF4F, so it is not surprising that these bound to eIF4E. The other six positive clones coded for two related proteins, 4E-BP1 and 4E-BP2.

The binding of mTORC1 to eIF3 activates translation in other ways besides removing 4E-BP1. It also causes phosphorylation of another eIF3-bound protein, **S6K1** (**S6 kinase-1**), one of whose functions is to phosphorylate the ribosomal protein S6 (Chapter 19). But S6K1 has two more important roles in the present context. First, once phosphorylated and dissociated from the eIF3 complex, S6K1 phosphorylates eIF4B, which facilitates its association with eIF4A. Second, S6K1 phosphorylates an inhibitor of eIF4A known as PDCD4. This phosphorylation leads to ubiquitylation and destruction of PDCD4, which relieves the inhibition of eIF4A. As we learned earlier in this chapter, eIF4A and eIF4B collaborate to unwind mRNA leaders and expedite scanning for the initiation codon. By encouraging the association between eIF4A and eIF4B, and removing an inhibitor of eIF4A, S6K1 stimulates scanning, thereby accelerating translation.

We have seen that mTORC1 responds to insulin and growth factors by stimulating translation. We also know from Chapter 14 that splicing stimulates translation. John

Blenis and colleagues proposed that there was a connection between these two phenomena, and this hypothesis gained support from their finding that rapamycin, which inhibits mTOR, blocks the stimulation of translation by splicing. In 2008, Blenis and colleagues showed that the connection between splicing and mTOR is mediated by a protein known as **SKAR** (**S6K1 Aly/REF-like substrate**). SKAR is recruited to the **exon junction complex** (**EJC**), a collection of proteins placed on mRNAs as they are spliced. Once in the cytoplasm, SKAR, now a part of the messenger ribonucleoprotein (mRNP), can recruit S6K1, activated by mTOR, to the mRNA. And activated S6K1, as we have seen, stimulates translation.

It is important to note that this model of translation stimulation can apply only to the first ribosome translating the newly made mRNA—the so-called pioneer round of translation. That is so because the first ribosome to translate an mRNA removes the EJC, including SKAR, so it can no longer recruit S6K1. We can only speculate about how splicing stimulates the overall rate of translation. Perhaps the efficiency of the pioneer round of translation somehow affects the efficiency of subsequent rounds. Another possibility is based on the fact that recruitment of eIF4E to the cap is rate limiting in translation. Blenis and colleagues speculated that, during remodeling of the mRNP during the pioneer round, mTOR and S6K1 help with the replacement of CBP80/20 by eIF4E and thereby enhance the efficiency of translation.

SUMMARY Insulin and a number of growth factors stimulate a pathway involving a protein kinase complex known as mTORC1, which binds to eIF3 and then phosphorylates its target proteins in the preinitiation complex. One of the targets for mTOR kinase is a protein called 4E-BP1. Upon phosphorylation by mTOR, this protein dissociates from eIF4E and releases it to participate in more active translation initiation. Another target of mTOR is S6K1. Once phosphorylated, activated S6K1, itself a protein kinase, phosphorylates eIF4B, which facilitates that protein's association with eIF4A, stimulating translation initiation. It also phosphorylates PDCD4, which leads to that protein's destruction. Because PDCD4 is an eIF4A inhibitor, its removal also stimulates initiation. Splicing stimulates translation via SKAR, a component of the EJC. SKAR recruits activated S6K1 for the pioneering round of translation.

Control of Translation Initiation via Maskin, an eIF4E-Binding Protein Eukaryotic cells can also use other proteins to target eIF4E, thereby inhibiting translation initiation. One of these proteins, discovered in the frog *Xenopus laevis,* is called **Maskin.** Figure 17.31 illustrates the current hypothesis for how Maskin acts to inhibit translation of the cyclin B mRNA in *Xenopus* oocytes. As we learned in Chapter 15, many mRNAs in *Xenopus* oocytes have very short poly(A) tails and are not well translated. One reason for this situation may be that the cytoplasmic polyadenylation element (CPE) is occupied by a binding protein, **CPEB.** This protein in turn binds to Maskin, which binds to eIF4E. In this interaction, Maskin behaves like 4E-BP1 in blocking the interaction between eIF4E and eIF4G, thereby inhibiting initiation of translation.

When the *Xenopus* oocyte is activated, CPEB is phosphorylated by an enzyme called Eg2. This phosphorylation appears to have two major effects. First, it attracts the cleavage and polyadenylation specificity factor (CPSF) to the polyadenylation signal in the mRNA (AAUAAA), and this stimulates polyadenylation of the dormant mRNA.

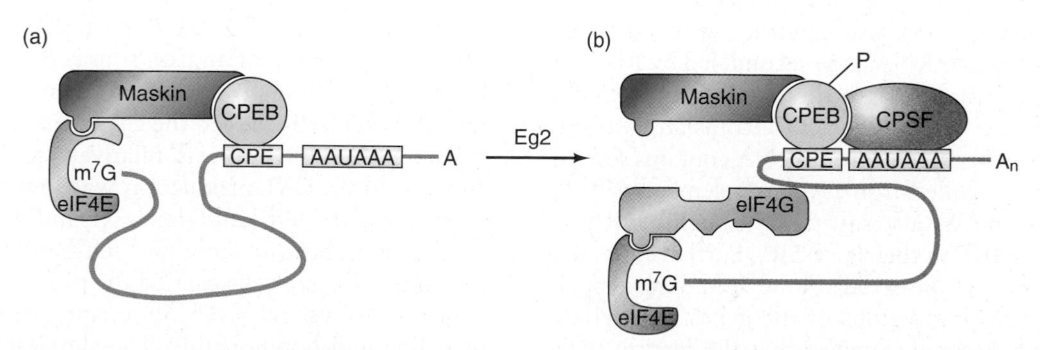

Figure 17.31 Model for control of translation initiation by Maskin. (a) In dormant *Xenopus* oocytes, CPEB is bound to CPE on cyclin B mRNA, Maskin is bound to CPEB, and eIF4E is bound to Maskin. The last interaction interferes with the ability of eIF4E to bind to eIF4G, which is necessary for translation initiation. As a result, the cyclin B mRNAs are dormant. **(b)** Upon activation, Eg2 phosphorylates CPEB, allowing recruitment of CPSF and polyadenylation of the mRNA. This event also apparently causes Maskin to dissociate from eIF4E, which enables eIF4E to bind to eIF4G, stimulating translation initiation. (*Source:* Adapted from Richter, J.D. and W.E. Theurkauf, The message is in the translation. *Science* 293 [2001] p. 61, f. 1.)

Second, phosphorylation of CPEB (or perhaps the polyadenylation resulting from this phosphorylation) apparently causes Maskin to lose its grip on eIF4E, allowing eIF4E to bind to eIF4G, stimulating initiation of translation.

It is important to note that cyclin B, one of the genes controlled by Maskin, is a key activator of the cell cycle. Thus, a process as fundamental as cell division is subject to control at the level of translation.

SUMMARY In *Xenopus* oocytes, Maskin binds to eIF4E and to CPEB bound to dormant cyclin B mRNAs. With Maskin bound to it, eIF4E cannot bind to eIF4G, so translation is inhibited. Upon activation of the oocytes, CPEB is phosphorylated, which stimulates polyadenylation and causes Maskin to dissociate from eIF4E. With Maskin no longer attached, eIF4E is free to associate with eIF4G, and translation can initiate.

Repression by an mRNA-Binding Protein We have seen that mRNA secondary structure can influence translation of bacterial genes. This is also true in eukaryotes. Let us consider a well-studied example of repression of translation of an mRNA by interaction between an RNA secondary structure element (a stem loop) and an RNA-binding protein. In Chapter 16 we learned that the concentrations of two iron-associated proteins, the transferrin receptor and ferritin, are regulated by iron concentration. When the serum concentration of iron is high, the synthesis of the transferrin receptor slows down due to destabilization of the mRNA encoding this protein. At the same time, the synthesis of **ferritin,** an intracellular iron storage protein, increases. Ferritin consists of two polypeptide chains, L and H. Iron causes an increased level of translation of the mRNAs encoding both ferritin chains.

What causes this increased efficiency of translation? Two groups arrived at the same conclusion almost simultaneously. The first, led by Hamish Munro, examined translation of the rat ferritin mRNAs; the second, led by Richard Klausner, studied translation of the human ferritin mRNAs. Recall from Chapter 16 that the 3′-untranslated region (3′-UTR) of the transferrin receptor mRNA contains several stem-loop structures called iron response elements (IREs) that can bind proteins. We also saw that the ferritin mRNAs have a very similar IRE in their 5′-UTRs. Furthermore, the ferritin IREs are highly conserved among vertebrates, much more so than the coding regions of the genes themselves. These observations strongly suggest that the ferritin IREs play a role in ferritin mRNA translation.

To test this prediction, Munro and colleagues made DNA constructs containing the CAT reporter gene flanked by the 5′- and 3′-UTRs from the rat ferritin L gene. In one construct (pLJ5CAT3), CAT transcription was driven by a very strong retroviral promoter–enhancer. In the other (pWE5CAT3), CAT transcription was under the control of the weak β-actin promoter. Next, they introduced these DNAs into mammalian cells and tested for CAT production in the presence of an iron source (hemin), an iron chelator (desferal), or no additions. Figure 17.32 shows the results. When cells carried the CAT gene in the pWE5CAT3 plasmid, CAT mRNA was relatively scarce. Under these circumstances, CAT production was low, but inducible by iron (compare left-hand lanes C and H) and inhibited by the iron chelator (compare left-hand lanes C and D). By contrast, when cells carried the pLJ5CAT3 plasmid, the CAT mRNA was relatively abundant, and CAT production was high and noninducible. The simplest explanation for these results is that a repressor binds to the IRE in the ferritin 5′-UTR and blocks translation of the associated CAT cistron. Iron somehow removes the repressor and allows translation to occur. CAT production was not inducible when the CAT mRNA was abundant because the

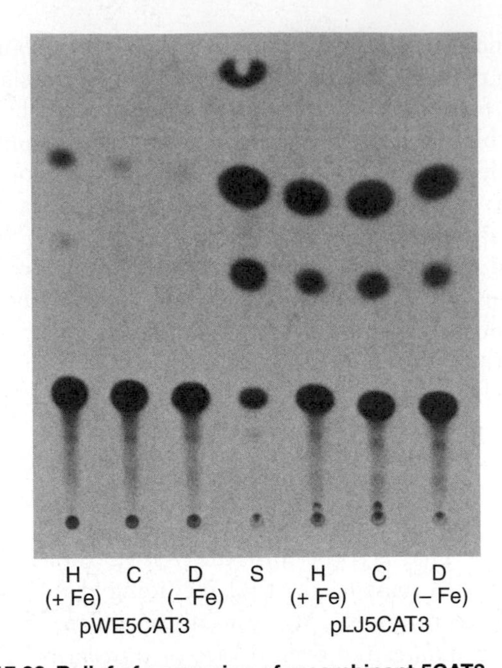

Figure 17.32 Relief of repression of recombinant 5CAT3 translation by iron. Munro and colleagues prepared two recombinant genes with the CAT reporter gene flanked by the 5′-and 3′-UTRs of the rat ferritin L gene. They introduced this construct into cells under control of a weak promoter (the β-actin promoter in the plasmid pWE5CAT3) or a strong promoter (a retrovirus promoter–enhancer in the plasmid pLJ5CAT3). They treated the cells in lanes H with hemin, and those in lanes D with the iron chelator desferal to remove iron. The cells in lanes C were untreated. They assayed CAT activity in each group of cells as described in Chapter 5. Lane S was a standard CAT reaction showing the positions of the chloramphenicol substrate and the acetylated forms of the antibiotic. The lanes on the left show that when the CAT mRNA is not abundant, its translation is inducible by iron. By contrast, the lanes on the right show that when the mRNA is abundant, its translation is not inducible by iron. (*Source:* Adapted from Aziz, N. and H.N. Munro, Iron regulates ferritin mRNA translation through a segment of its 5′ untranslated region. *Proceedings of the National Academy of Sciences USA* 84 (1997) p. 8481, f. 6.)

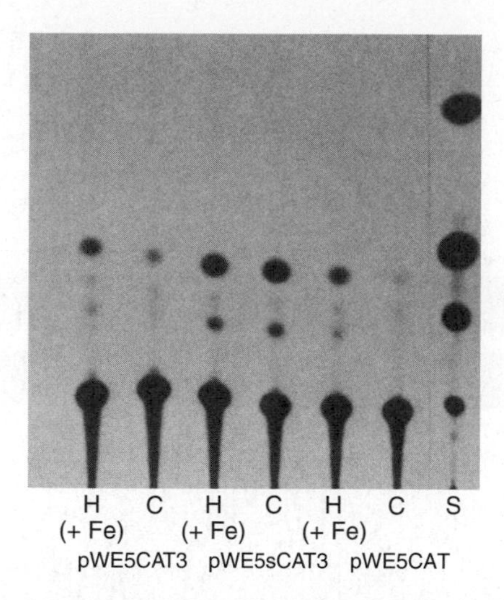

Figure 17.33 Importance of the IRE in the 5′-UTR of pWE5CAT3 for iron inducibility. Munro and colleagues transfected cells with the parent plasmid pWE5CAT3, as described in Figure 17.32, and with two derivatives: pWE5sCAT3, which lacked the first 67 nt of the ferritin 5′-UTR, including the IRE; and pWE5CAT, which lacked the ferritin 3′-UTR. These cells were either treated (H) or not treated (C) with hemin. Then the experimenters assayed each batch of cells for CAT activity. Loss of the IRE caused a loss of iron inducibility. (*Source:* Adapted from Aziz, N. and H.N. Munro, Iron regulates ferritin mRNA translation through a segment of its 5′-untranslated region. *Proceedings of the National Academy of Sciences USA* 84 (1987) p. 8482, f. 7.)

mRNA molecules greatly outnumbered the repressor molecules. With little repression happening, induction cannot be observed.

How do we know that the IRE is involved in repression? In fact, how do we even know that the 5′-UTR, and not the 3′-UTR, is important? Munro and colleagues answered these questions by preparing two new constructs, one containing the 5′-UTR, but lacking the 3′-UTR, and one containing both UTRs, but lacking the first 67 nt, including the IRE in the 5′-UTR. Figure 17.33 shows that pWE5CAT, the plasmid lacking the ferritin mRNA's 3′-UTR, still supported iron induction of CAT. On the other hand, pWE5sCAT3, which lacked the IRE, was expressed at a high level with or without added iron. This result not only indicates that the IRE is responsible for induction, it also reinforces the conclusion that the IRE mediates repression because loss of the IRE leads to high CAT production even without iron.

We can conclude that some repressor protein(s) must bind to the IRE in the ferritin mRNA 5′-UTR and cause repression until removed somehow by iron. Because such great conservation of the IREs occurs in the ferritin mRNAs and the transferrin receptor mRNAs, we suspect that at least some of these proteins might operate in both cases. In fact, as we learned in Chapter 16, the aconitase apoprotein is the IRE-binding protein. When it binds to iron, it dissociates from the IRE. In this case, that would relieve repression.

SUMMARY Ferritin mRNA translation is subject to induction by iron. This induction seems to work as follows: A repressor protein (aconitase apoprotein), binds to a stem-loop iron response element (IRE) near the 5′-end of the 5′-UTR of the ferritin mRNA. Iron removes this repressor and allows translation of the mRNA to proceed.

Blockage of Translation Initiation by an miRNA We have seen in Chapter 16 that miRNAs can control gene expression in two ways: They can cause degradation of mRNAs when base-paired perfectly to their target mRNAs, or, if base-pairing is not perfect, they can inhibit protein production by an unexplained mechanism. Witold Filipowicz and colleagues set out to elucidate that mysterious mechanism, and presented results in 2005 that indicated that imperfectly-paired mammalian *let-7* miRNA can inhibit initiation of translation, probably by interfering with cap recognition.

These workers used reporter genes as probes. In particular, they used the *Renilla reniformis* (sea pansy) luciferase (RL) and firefly luciferase (FL) genes, because the gene products (luciferase) are easily assayed: When mixed with luciferin and ATP, they generate light. The 3′-UTRs of these reporter genes were engineered to have a region that aligns perfectly with *let-7* miRNA (Perf), or to have one or three mismatched regions of complementarity that cause bulges in the miRNA–mRNA duplex. These altered genes were named 1xBulge and 3xBulge, respectively. The wild-type control gene (Con) had no complementarity to *let-7* miRNA.

When they transfected human cells with the reporter genes, Filipowicz and colleagues found that the expression of the RL-Perf and the RL-3xBulge genes decreased dramatically (up to 10-fold) compared to the control gene. Furthermore, this decrease was blocked by co-transfection with a competitor RNA that was complementary to *let-7* miRNA, suggesting that this miRNA was involved in the decrease, as we would expect.

According to the paradigm presented in Chapter 16, we would predict that the amount of RL-Perf mRNA would decrease, because the perfect alignment between the mRNA and miRNA would lead to mRNA degradation. Indeed, Filipowicz and colleagues observed a five-fold reduction in the amount of this mRNA. Furthermore, we would predict that the amount of RL-3xBulge mRNA would not decrease significantly, because the imperfect alignment between the mRNA and miRNA would lead to interference with translation, rather than to mRNA destruction. And, in fact, the amount of this mRNA decreased only 20%.

These data are consistent with the hypothesis that the decline in RL-3xBulge expression is explained by blocking translation, rather than by degradation of mRNA. But it is also possible that the miRNA somehow targets the nascent protein for degradation by proteolysis. If that were true, then hiding the nascent protein in the endoplasmic reticulum (ER) should shield it from destruction, and little or no drop in expression should be observed. To test this hypothesis, Filipowicz and colleagues coupled the RL-3xBulge gene to the hemaglutinin gene, which contained a signal sequence expressed at the N-terminus of the fusion protein. This signal sequence directed the nascent protein to the lumen of the ER. The protein product of this construct suffered the same decrease compared to the control as the RL-3xBulge product itself did. Thus, protein synthesis, rather than the protein product itself, appears to be the target of the *let-7* miRNA.

What part of the translation process is inhibited by *let-7* miRNA? To begin to answer this question, Filipowicz and colleagues collected polysomes (mRNAs being translated by multiple ribosomes, Chapter 18) from cells transfected with the RL-3xBulge gene. To detect the RL-3xBulge mRNA in the polysome profile, they performed Northern blots on polysome fractions (Figure 17.34). The more active the translation initiation on a given mRNA, the more ribosomes will be attached to the mRNA, and therefore the heavier the polysomes will be. The heaviest polysomes are found toward the right in Figure 17.34, and it is clear that the control RL mRNAs were in much larger polysomes (farther to the right, panel [a]) than the RL-3xBulge mRNAs (panel [b]). These results are depicted graphically in Figure 17.34c. The shift in polysome profile was mostly eliminated by co-transfection with an anti-*let-7* miRNA, which would block miRNA–mRNA interaction (results not shown). The shift was also eliminated when the RL-3xBulge mRNA was mutated to remove the 3′-UTR region that hybridizes to the miRNA. Taken together, these data indicate that translation initiation on RL-3xBulge mRNA is significantly inhibited compared to initiation on the control mRNA. Thus, initiation (binding of ribosomes to mRNA) seems to be the part of translation that is the target of the *let-7* miRNA.

Further study showed that the poly(A) tail on the mRNA played no role in *let-7* miRNA inhibition of translation: Translation of poly(A)$^+$ and poly(A)$^-$ mRNAs were equally inhibited by *let-7* miRNA. But the cap did play a big role. As we have seen, translation of uncapped mRNAs is very poor, so Filipowicz and colleagues endowed either the RL or FL mRNA with the internal ribosome entry site (IRES) from the encephalomyocarditis virus (EMCV), which allows cap-independent translation. Then they compared the effect of *let-7* miRNA on cap-dependent and -independent translation. As usual, *let-7* inhibited cap-dependent translation of FL-3xBulge mRNA, but it had no effect on the cap-independent translation of FL-3xBulge mRNA with an EMCV IRES. Thus, *let-7* miRNA appears to target cap-dependent initiation of translation.

To pin down the part of cap-dependent initiation that is affected by *let-7* miRNA, Filipowicz and colleagues built a

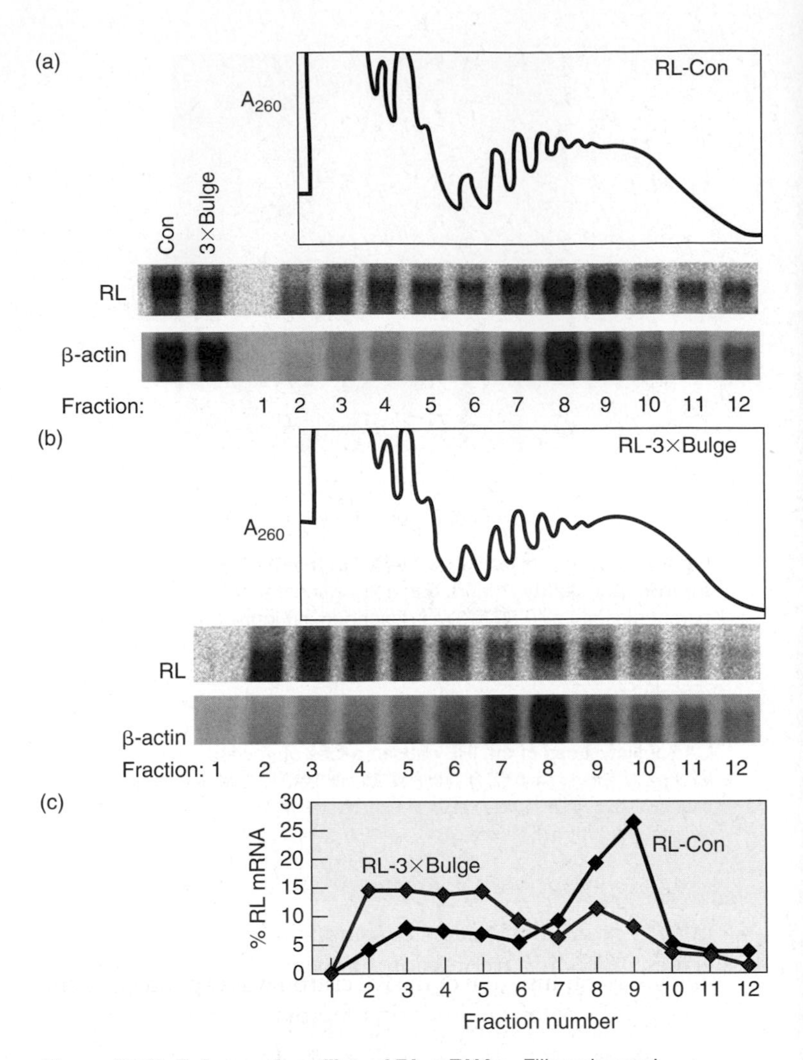

Figure 17.34 Polysomal profiles of RL mRNAs. Filipowicz and colleagues transfected human cells with genes that encoded either **(a)** the control RL mRNA (RL-Con) or **(b)** RL-3xBulge mRNA. Then they displayed the polysomes by sucrose gradient ultracentrifugation, subjected RNAs from fractions from the polysome profile to Northern blotting, and hybridized the blots to radioactive probes for RL or β-actin mRNA. The latter is an ordinary cellular mRNA, used as a positive control. The two lanes on the far left of the Northern blots in panel (a) contain RNAs from the inputs into the ultracentrifugation step. **(c)** The percentages of total radioactivity in each fraction from the control and RL-3xBulge polysome profiles are presented. (*Source:* (*a–c*) Reprinted with permission from *Science,* Vol. 309, Ramesh S. Pillai, Suvendra N. Bhattacharyya, Caroline G. Artus, Tabea Zoller, Nicolas Cougot, Eugenia Basyuk, Edouard Bertrand, and Witold Filipowicz, "Inhibition of Translational Initiation by Let-7 MicroRNA in Human Cells" Fig. 1 c&e, p. 1574, Copyright 2004, AAAS.)

DNA construct encoding a dicistronic mRNA with either eIF4E or eIF4G tethered in the intercistronic region just before the RL cistron. They performed the tethering as follows (Figure 17.35a): In the intercistronic region, they placed so-called BoxB stem-loops that have affinity for a peptide called the N peptide. Then they engineered genes for eIF4E and eIF4G, adding N peptide-hemagglutinin coding regions, so the initiation factors were each produced as fusion proteins tagged with the N peptide. These fusion proteins in turn bound to the BoxB stem-loops, so they

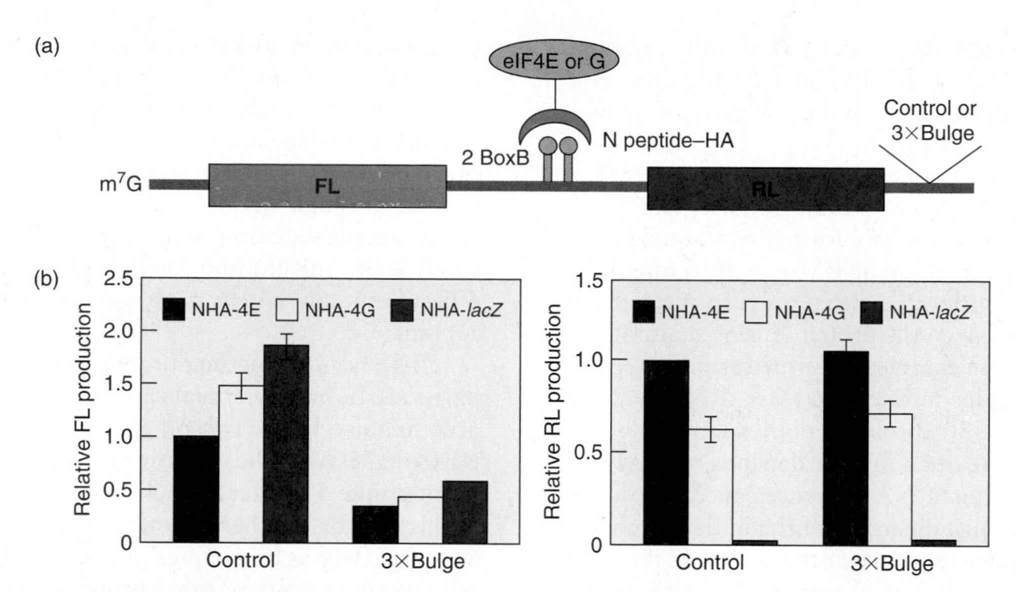

Figure 17.35 Effect of tethering translation initiation factors to the intercistronic region of a dicistronic mRNA. (a) Diagram of the construct with two BoxB stem loops (purple), between the two cistrons, bound to the N peptide part (green) of a fusion protein that also contained either eIF4E or eIF4G (orange). The 3′-UTR contained either the control RL sequence (Con) or the 3xBulge sequence. **(b)** Production of FL (left) and RL (right) from the control and 3xBulge mRNAs, as indicated at bottom, with various proteins tethered to the intercistronic region. The N peptide-hemaglutinin (NHA)-tagged protein tethered to the intercistronic region is indicated by color in the bar graphs: eIF4E, blue; eIF4G, yellow; *lacZ* product, red. (*Source:* Adapted from Ramesh, S., et al., 2004 Inhibition of translational initiation by let-7 microRNA in human cells. *Science* 309:1575, fig. 2.)

could stimulate translation of the RL cistron on the dicistronic mRNA. The translation of the FL cistron was cap-dependent, since this cistron came first in the capped mRNA. But translation of the RL cistron was cap-independent as long as one of the initiation factors was tethered to the intercistronic region. This protein apparently attracted all the other factors needed for initiation.

So Filipowicz and colleagues tested expression of the FL and RL parts of the fusion gene with either a control 3′-UTR or the 3xBulge 3′-UTR, and either of the initiation factors (or, as a negative control, the *lacZ* product, β-galactosidase) tethered to the intercistronic region. Figure 17.35b shows the results. As expected, translation of the FL cistron was cap-dependent, and the *let-7* miRNA inhibited translation of the FL cistron of the 3xBulge mRNA compared to the control mRNA. But, when either eIF4E or eIF4G was tethered to the intercistronic region, *let-7* miRNA did not inhibit translation of the RL cistron in the 3xBulge mRNA. (With the *lacZ* product, rather than an initiation factor, tethered in the intercistronic region, almost no translation occurred, even with the control mRNA.) Thus, having either eIF4E or eIF4G available (in this case by tethering) circumvents the *let-7*-mediated inhibition of translation initiation. This suggests that *let-7* blocks some step before eIF4E recruits eIF4G to the cap. One obvious candidate for this *let-7*-sensitive step is eIF4E binding to the cap.

These results in mammalian cells, showing that *let-7* miRNA interferes with translation initiation, differ from some of the results presented in Chapter 16, which indicated that *lin-4* miRNA does not alter the polysome profile of its target mRNA in *C. elegans* cells, and therefore does not appear to block translation initiation. As pointed out in Chapter 16, this discrepancy can be explained if different miRNAs have different modes of action, or if miRNAs work differently in different organisms, or both.

SUMMARY The *let-7* miRNA shifts the polysomal profile of target mRNAs in human cells toward smaller polysomes, indicating that this miRNA blocks translation initiation in human cells. Translation initiation that is cap-independent because of the presence of an IRES, or tethered initiation factors, is not affected by *let-7* miRNA, suggesting that this miRNA blocks binding of eIF4E to the cap of target mRNAs in human cells.

SUMMARY

Two events must occur as a prelude to protein synthesis: First, aminoacyl-tRNA synthetases join amino acids to their cognate tRNAs. They do this very specifically in a two-step reaction that begins with activation of the amino acid with AMP, derived from ATP. Second, ribosomes must dissociate into subunits at the end of each round of translation. In bacteria, RRF and EF-G actively promote this dissociation, whereas IF3 binds to the free 30S subunit and prevents its reassociation with a 50S subunit to form a whole ribosome.

The initiation codon in prokaryotes is usually AUG, but it can also be GUG, or more rarely, UUG. The initiating aminoacyl-tRNA is *N*-formyl-methionyl-$tRNA_f^{Met}$. *N*-formyl-methionine (fMet) is therefore the first amino acid incorporated into a polypeptide, but it is frequently removed from the protein during maturation.

The 30S initiation complex is formed from a free 30S ribosomal subunit plus mRNA and fMet-$tRNA_f^{Met}$. Binding between the 30S prokaryotic ribosomal subunit and the initiation site of an mRNA depends on base pairing between a short RNA sequence called the Shine–Dalgarno sequence just upstream of the initiation codon, and a complementary sequence at the 3′-end of the 16S rRNA. This binding is mediated by IF3, with help from IF1 and IF2. All three initiation factors have bound to the 30S subunit by this time.

IF2 is the major factor promoting binding of fMet-$tRNA_f^{Met}$ to the 30S initiation complex. The other two initiation factors play important supporting roles. GTP is also required for IF2 binding at physiological IF2 concentrations, but it is not hydrolyzed in the process. The complete 30S initiation complex contains one 30S ribosomal subunit plus one molecule each of mRNA, fMet-$tRNA_f^{Met}$, GTP, IF1, IF2, and IF3. GTP is hydrolyzed after the 50S subunit joins the 30S complex to form the 70S initiation complex. This GTP hydrolysis is carried out by IF2 in conjunction with the 50S ribosomal subunit. The purpose of this hydrolysis is to release IF2 and GTP from the complex so polypeptide chain elongation can begin.

Eukaryotic 40S ribosomal subunits, together with the initiating Met-tRNA (Met-$tRNA_i^{Met}$), generally locate the appropriate start codon by binding to the 5′-cap of an mRNA and scanning downstream until they find the first AUG in a favorable context. The best context contains a purine at position −3 and a G at position +4. In 5–10% of the cases, most ribosomal subunits will bypass the first AUG and continue to scan for a more favorable one. Sometimes ribosomes apparently initiate at an upstream AUG, translate a short ORF, then continue scanning and reinitiate at a downstream AUG. This mechanism works only with short upstream ORFs. Some viral mRNAs that lack caps have IRESs that attract ribosomes directly to the mRNAs.

Secondary structure near the 5′-end of an mRNA can have positive or negative effects. A hairpin just past an AUG can force a ribosomal subunit to pause at the AUG and thus stimulate initiation. A very stable stem loop between the cap and an initiation site can block ribosomal subunit scanning and thus inhibit initiation.

The eukaryotic initiation factors have the following general functions: eIF1 and eIF1A aid in scanning to the initiation codon. eIF2 is involved in binding Met-$tRNA_i^{Met}$ to the ribosome. eIF2B activates eIF2 by replacing its GDP with GTP. eIF3 binds to the 40S ribosomal subunit and inhibits its reassociation with the 60S subunit. eIF4F is a cap-binding protein that allows the 40S ribosomal subunit to bind (through eIF3) to the 5′-end of an mRNA. eIF5 encourages association between the 43S complex (40S subunit plus mRNA and Met-$tRNA_i^{Met}$). eIF6 binds to the 60S subunit and blocks its reassociation with the 40S subunit.

eIF4F is a cap-binding protein composed of three parts: eIF4E has the actual cap-binding activity; it is accompanied by the two other subunits, eIF4A and eIF4G. eIF4A has RNA helicase activity that can unwind hairpins found in the 5′-leaders of eukaryotic mRNAs. It is aided in this task by another factor, eIF4B, and requires ATP for activity. eIF4G is an adapter protein that is capable of binding to a variety of other proteins, including eIF4E (the cap-binding protein), eIF3 (the 40S ribosomal subunit-binding protein), and Pab1p (a poly[A]-binding protein). By interacting with these proteins, eIF4G can recruit 40S ribosomal subunits to the mRNA and thereby stimulate translation initiation.

eIF1 and eIF1A act synergistically to promote formation of a stable 48S complex, involving initiation factors, Met-$tRNA_i^{Met}$, and a 40S ribosomal subunit that has scanned to the initiation codon of an mRNA. eIF1 and eIF1A appear to act by dissociating improper complexes between 40S subunits and mRNA and encouraging the formation of stable 48S complexes.

eIF5B is homologous to the prokaryotic factor IF2. It resembles IF2 in binding GTP and stimulating association of the two ribosomal subunits. eIF5B works with eIF5 in this reaction. eIF5B also resembles IF2 in using GTP hydrolysis to promote its own dissociation from the ribosome so protein synthesis can begin. But it differs from IF2 in that it cannot stimulate the binding of the initiating aminoacyl-tRNA to the small ribosomal subunit. That task is performed by eIF2 in eukaryotes.

Prokaryotic mRNAs are very short-lived, so control of translation is not common in these organisms. However, some translational control does occur. Messenger RNA secondary structure can govern translation initiation, as in the replicase gene of the MS2 class of phages, or in the mRNA for *E. coli* σ^{32}, whose translation is repressed by secondary structure that is relaxed by heating.

Small RNAs, in concert with proteins, can also affect mRNA secondary structure to control translation initiation, and riboswitches are one way this control can be exercised. The 5′-untranslated region of the *E. coli thiM* mRNA contains a riboswitch, including an aptamer that binds thiamine and its metabolites, including thiamine pyrophosphate (TPP). When TPP is abundant, it binds to this aptamer, causing a conformational shift in the mRNA that ties up the Shine–Dalgarno sequence in secondary structure. This shift hides the SD sequence from ribosomes, and inhibits translation of the mRNA.

Eukaryotic mRNA lifetimes are relatively long, so there is more opportunity for translation control than in prokaryotes. The α-subunit of eIF2 is a favorite target for translation control. In heme-starved reticulocytes, HCR is activated, so it can phosphorylate eIF2α and inhibit initiation. In virus-infected cells, another kinase, DAI is activated; it also phosphorylates eIF2α and inhibits translation initiation.

Insulin and a number of growth factors stimulate a pathway involving a protein kinase called mTOR. One of the targets for mTOR is a protein called 4E-BP1. On phosphorylation by mTOR, this protein dissociates from eIF4E and releases it to participate in more active translation initiation. Another target of mTOR is S6K1. Once phosphorylated, activated S6K1, itself a protein kinase, phosphorylates targets that enhance translation. Splicing stimulates translation via SKAR, a component of the EJC. SKAR recruits activated S6K1 for the pioneering round of translation.

In *Xenopus* oocytes, Maskin binds to eIF4E and to CPEB bound to dormant cyclin B mRNAs. With Maskin bound to it, eIF4E cannot bind to eIF4G, so translation is inhibited. Upon activation of the oocytes, CPEB is phosphorylated, which stimulates polyadenylation and causes Maskin to dissociate from eIF4E. With Maskin no longer attached, eIF4E is free to associate with eIF4G, and translation can initiate.

Ferritin mRNA translation is subject to induction by iron. This induction seems to work as follows: A repressor protein (aconitase apoprotein), binds to a stem-loop iron response element (IRE) near the 5′-end of the 5′-UTR of the ferritin mRNA. Iron removes this repressor and allows translation of the mRNA to proceed.

The *let-7* miRNA shifts the polysomal profile of target mRNAs in human cells toward smaller polysomes, indicating that this miRNA blocks translation initiation in human cells. Translation initiation that is cap-independent because of the presence of an IRES, or tethered initiation factors, is not affected by *let-7* miRNA, suggesting that this miRNA blocks binding of eIF4E to the cap of target mRNAs in human cells.

REVIEW QUESTIONS

1. Describe and give the results of an experiment that shows that ribosomes dissociate and reassociate.
2. How does IF3 participate in ribosome dissociation?
3. What are the two bacterial methionyl-tRNAs called? What are their roles?
4. Why does translation of the MS2 phage replicase cistron depend on translation of the coat cistron?
5. Present data (exact base sequences are not necessary) to support the importance of base-pairing between the Shine–Dalgarno sequence and the 16S rRNA in translation initiation. Select the most convincing data.
6. Present data to show the effects of the three initiation factors in mRNA-ribosome binding.
7. Describe and give the results of an experiment that shows the role (if any) of GTP hydrolysis in forming the 30S initiation complex.
8. Describe and give the results of an experiment that shows the role of GTP hydrolysis in release of IF2 from the ribosome.
9. Present data to show the effects of the three initiation factors in fMet-$tRNA_f^{Met}$ binding to the ribosome.
10. Draw a diagram to summarize the initiation process in *E. coli*.
11. Explain what the Shine–Dalgarno sequence and the Kozak consensus sequence are and compare and contrast their roles.
12. Write the sequence of an ideal eukaryotic translation initiation site. Aside from the AUG, what are the most important positions?
13. Draw a diagram of the scanning model of translation initiation.
14. Present evidence that a scanning ribosome can bypass an AUG and initiate at a downstream AUG.
15. Under what circumstances is an upstream AUG in good context not a barrier to initiation at a downstream AUG? Present evidence.
16. Describe and give the results of an experiment that shows the effects of secondary structure in an mRNA leader on scanning.
17. Draw a diagram of the steps in translation initiation in eukaryotes, showing the effects of each class of initiation factor.
18. Describe and give the results of an experiment that identified the cap-binding protein.
19. Describe and give the results of an experiment that shows that cap-binding protein stimulates translation of capped, but not uncapped, mRNAs.
20. What is the subunit structure of eIF4F? Molecular masses are not required.
21. Describe and give the results of an experiment that shows the roles of eIF4A and eIF4B in translation.
22. How does the poliovirus genetic material resemble a typical cellular mRNA? How it is different? How does the virus take advantage of this difference? Compare and contrast this behavior with that of the hepatitis C virus.
23. How do we know that eIF1 and eIF1A do not cause conversion of complex I to complex II by stimulating scanning on the same mRNA?
24. Compare the initiation factors IF2 and eIF5B. What functions do they have in common? What function can IF2 perform that eIF5B cannot? What factor performs this function in eukaryotes?

25. Describe the mechanism by which the *rpoH* mRNA senses high temperature and turns on its own translation. What is the evidence for this model?
26. Describe the mechanism by which the riboswitch in the *E. coli thiM* gene controls translation.
27. Present a model for repression of translation by phosphorylation of eIF2α.
28. Present a model to explain the effect of 4E-BP1 phosphorylation on translation efficiency.
29. Describe and give the results of an experiment that shows the importance of the IRE in the ferritin mRNA to iron inducibility of ferritin production.
30. Present a hypothesis for iron inducibility of ferritin production in mammalian cells. Make sure your hypothesis explains why ferritin production is not inducible in cells in which the ferritin gene is driven by a strong promoter.
31. How is the human *let-7* miRNA thought to control expression of its target genes? Summarize the evidence for this model.

ANALYTICAL QUESTIONS

1. Describe a toeprint assay involving *E. coli* ribosomal subunits and a fictious mRNA in a cell-free extract that contains all the factors necessary for translation. What results would you expect to see with 30S ribosomal subunits alone? With 50S subunits alone? With both subunits and all amino acids except leucine, which is required in the 20th position of the polypeptide?
2. Predict the effects of the following mutations on phage R17 coat gene and replicase gene translation:
   a. An amber mutation (premature stop codon) six codons downstream of the coat gene initiation codon.
   b. Mutations in the stem loop around the coat gene initiation codon that weaken the base-pairing in the stem loop.
   c. Mutations in the interior of the replicase gene that cause it to base-pair with the coat gene initiation codon.
3. You are studying a eukaryotic gene in which translation normally begins with the second AUG in the mRNA. The sequence surrounding the two AUG codons is:

   CGG<u>AUG</u>CACAGGACAUCCUAUGGAG<u>AUG</u>A

   where the two AUG codons are underlined. Predict the effects of the following mutations on translation of this mRNA.
   a. Changing the first and second C's to G's.
   b. Changing the first and second C's to G's, and also changing the UAU codon before the second AUG codon to UAG.
   c. Changing the GAG<u>AUG</u>A sequence at the end to CAG<u>AUG</u>U
4. You are studying a eukaryotic mRNA that you believe exhibits control at the level of translation, particularly the initiation of translation. You think that the 5′-UTR plays a role in the control of translation. To definitively determine the role of the 5′-UTR, describe in detail experiments that you could perform to prove this. Be sure to include how you would experimentally determine if a protein binds to the 5′-UTR to prevent translation and the possible effects a mutation in the 5′-UTR might have on gene expression at the RNA level.

SUGGESTED READINGS

General References and Reviews

Cech, T.R. 2004. RNA finds a simpler way. *Nature* 428:263–64.

Gottesman, S. 2004. The small RNA regulators of *Escherichia coli:* Roles and mechanisms. *Annual Review of Microbiology* 58:303–28.

Hentze, M.W. 1997. eIF4G: A multipurpose ribosome adapter? *Science* 275:500–1.

Jackson, R.J. 1998. Cinderella factors have a ball. *Nature* 394:829–31.

Kozak, M. 1989. The scanning model for translation: An update. *Journal of Cell Biology* 108:229–41.

Kozak, M. 1991. Structural features in eukaryotic mRNAs that modulate the initiation of translation. *Journal of Biological Chemistry* 266:19867–70.

Kozak, M. 2005. Regulation of translation via mRNA structure in prokaryotes and eukaryotes. *Gene* 361:13–37.

Lawrence, J.C. and Abraham, R.T. 1997. PHAS/4E-BPs as regulators of mRNA translation and cell proliferation. *Trends in Biochemical Sciences.* 22:345–49.

Proud, C.G. 1994. Turned on by insulin. *Nature* 371:747–48.

Rhoads, R.E. 1993. Regulation of eukaryotic protein synthesis by initiation factors. *Journal of Biological Chemistry* 268:3017–20.

Richter, J.D. and W.E. Theurkauf. 2001. The message is in the translation. *Science* 293:60–62.

Roll-Mecak, A., B.-S. Shin, T.E. Dever, and S.K. Burley. 2001. Engaging the ribosome: Universal IFs of translation. *Trends in Biochemical Sciences* 26:705–9.

Sachs, A.B. 1997. Starting at the beginning, middle, and end: Translation initiation in eukaryotes. *Cell* 89:831–38.

Thach, R.E. 1992. Cap recap: The involvement of eIF4F in regulating gene expression. *Cell* 68:177–80.

Research Articles

Aziz, N. and H.N. Munro. 1987. Iron regulates ferritin mRNA translation through a segment of its 5′-untranslated region. *Proceedings of the National Academy of Sciences USA* 84:8478–82.

Brown, L. and T. Elliott. 1997. Mutations that increase expression of the *rpoS* gene and decrease its dependence on *hfq* function in *Salmonella typhimurium. Journal of Bacteriology* 179:656–62.

Cigan, A.M., L. Feng, and T.F. Donahue. 1988. $tRNA_f^{Met}$ functions in directing the scanning ribosome to the start site of translation. *Science* 242:93–96.

Dubnoff, J.S., A.H. Lockwood, and U. Maitra. 1972. Studies on the role of guanosine triphosphate in polypeptide chain initiation in *Escherichia coli*. *Journal of Biological Chemistry* 247:2884–94.

Edery, I., M. Hümbelin, A. Darveau, K.A.W. Lee, S. Milburn, J.W.B. Hershey, H. Trachsel, and N. Sonenberg. 1983. Involvement of eukaryotic initiation factor 4A in the cap recognition process. *Journal of Biological Chemistry* 258:11398–403.

Fakunding, J.L. and J.W.B. Hershey. 1973. The interaction of radioactive initiation factor IF2 with ribosomes during initiation of protein synthesis. *Journal of Biological Chemistry* 248:4206–12.

Guthrie, C. and M. Nomura. 1968. Initiation of protein synthesis: A critical test of the 30S subunit model. *Nature* 219:232–35.

Hui, A. and H.A. De Boer. 1987. Specialized ribosome system: Preferential translation of a single mRNA species by a subpopulation of mutated ribosomes in *Escherichia coli*. *Proceedings of the National Academy of Sciences USA* 84:4762–66.

Kaempfer, R.O.R., M. Meselson, and H.J. Raskas. 1968. Cyclic dissociation into stable subunits and reformation of ribosomes during bacterial growth. *Journal of Molecular Biology* 31:277–89.

Kozak, M. 1986. Point mutations define a sequence flanking the AUG initiator codon that modulates translation by eukaryotic ribosomes. *Cell* 44:283–92.

Kozak, M. 1989. Circumstances and mechanisms of inhibition of translation by secondary structure in eucaryotic mRNAs. *Molecular and Cellular Biology* 9:5134–42.

Min Jou, W., G. Haegeman, M. Ysebaert, and W. Fiers. 1972. Nucleotide sequence of the gene coding for the bacteriophage MS2 coat protein. *Nature* 237:82–88.

Morita, M.T., Y. Tanaka, T.S. Kodama, Y. Kyogoku, K. Yanagi, and T. Yura. 1999. Translational induction of heat shock transcription factor σ^{32}. Evidence for a built-in RNA thermosensor. *Genes and Development* 13:655–65.

Noll, M. and H. Noll. 1972. Mechanism and control of initiation in the translation of R17 RNA. *Nature New Biology* 238:225–28.

Pause, A. and N. Sonenberg. 1992. Mutational analysis of a DEAD box RNA helicase: The mammalian translation initiation factor eIF4A. *EMBO Journal* 11:2643–54.

Pestova, T.V., S.I. Borukhov, and C.V.T. Hellen. 1998. Eukaryotic ribosomes require initiation factors 1 and 1A to locate initiation codons. *Nature* 394:854–59.

Pestova, T.V., I.B. Lomakin, J.H. Lee, S.K. Choi, T.E. Dever, and C.U.T. Hellen. 2000. The joining of ribosomal subunits in eukaryotes requires eIF5B. *Nature* 403:332–35.

Pillai, R.S., S.N. Bhattacharyya, C.G. Artus, T. Zoller, N. Cougot, E. Basyuk, E. Bertrand, and W. Filipowicz. 2005. Inhibition of translational initiation by Let-7 microRNA in human cells. *Science* 309:1573–76.

Sonenberg, N., M.A. Morgan, W.C. Merrick, and A.J. Shatkin. 1978. A polypeptide in eukaryotic initiation factors that crosslinks specifically to the 5′-terminal cap in mRNA. *Proceedings of the National Academy of Sciences USA* 75:4843–47.

Sonenberg, N., H. Trachsel, S. Hecht, and A.J. Shatkin. 1980. Differential stimulation of capped mRNA translation in vitro by cap binding protein. *Nature* 285:331–33.

Steitz, J.A. and K. Jakes. 1975. How ribosomes select initiator regions in mRNA: Base pair formation between the 3′-terminus of 16S rRNA and the mRNA during initiation of protein synthesis in *Escherichia coli*. *Proceedings of the National Academy of Sciences USA* 72:4734–38.

Wahba, A.J., K. Iwasaki, M. J. Miller, S. Sabol, M.A.G. Sillero, and C. Vasquez. 1969. Initiation of protein synthesis in *Escherichia coli*, II. Role of the initiation factors in polypeptide synthesis. *Cold Spring Harbor Symposia* 34:291–99.

Winkler, W., A. Nahvi, and R.R. Breaker. 2002. Thiamine derivatives bind messenger RNAs directly to regulate bacterial gene expression. *Nature* 419:952–56.

CHAPTER 18

The Mechanism of Translation II: Elongation and Termination

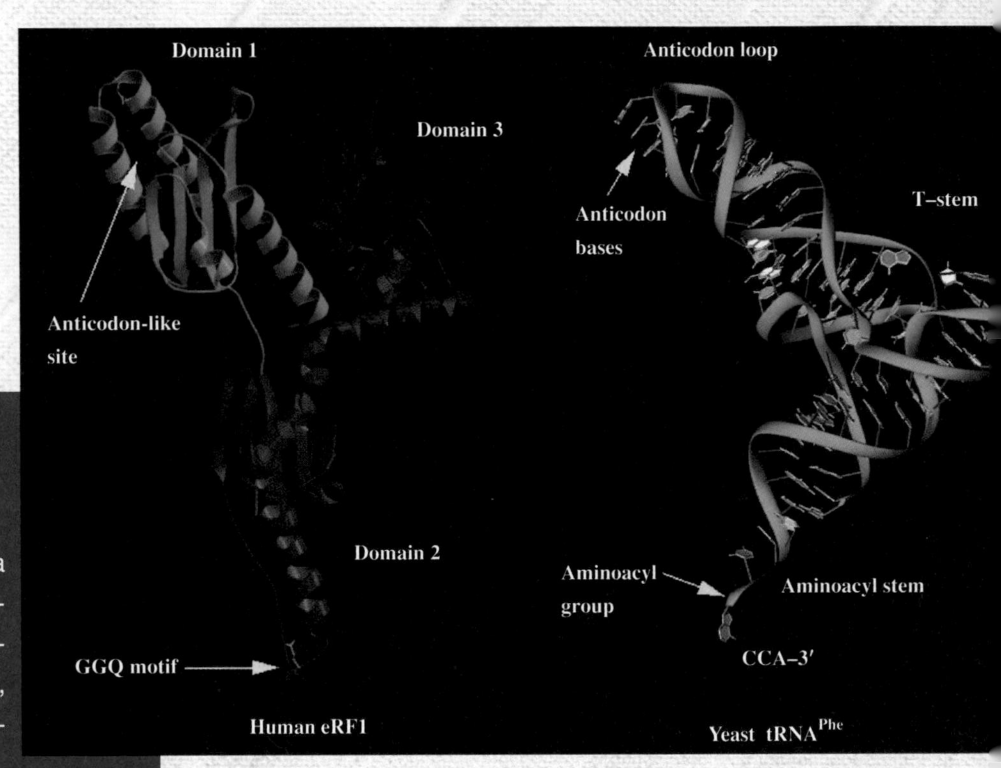

Comparison of crystal structures of human eRF1 and yeast tRNA^{Phe}.

The elongation processes in bacteria and eukaryotes are very similar. Accordingly, we will consider the processes together, discussing the bacterial system first, then noting some differences in the eukaryotic system.

As we learned in Chapter 17, the initiation process in bacteria creates a ribosome primed with an mRNA and the initiating aminoacyl-tRNA, fMet-tRNA$_f^{Met}$, ready to begin elongating a polypeptide chain. Before we look at the steps involved in this elongation process, let us consider some fundamental questions about the nature of elongation: (1) In what direction is a polypeptide synthesized? (2) In what direction does the ribosome read the mRNA? (3) What is the nature of the genetic code that dictates which amino acids will be incorporated in response to the mRNA?

18.1 The Direction of Polypeptide Synthesis and of mRNA Translation

Proteins are made one amino acid at a time, but where does synthesis begin? Do protein chains grow in the amino-to-carboxyl direction, or the reverse? In other words, which amino acid is inserted first into a growing polypeptide—the amino-terminal amino acid, or the carboxyl-terminal one? Howard Dintzis provided definitive proof of the amino → carboxyl direction in 1961 with a study of α- and β-globin synthesis in isolated rabbit reticulocytes (immature red blood cells). He labeled the growing globin chains for various short lengths of time with [^{3}H]leucine, and for a long time with [^{14}C]leucine. Then he separated the α- and β-globins, cut them into peptides with trypsin, and separated the peptides. He then plotted the relative amounts of [^{3}H]leucine incorporated into the peptides versus the positions of the peptides, from N-terminus to C-terminus, in the proteins. The long labeling with [^{14}C]leucine should have labeled all peptides equally, so it could be used as an internal control for losses of certain peptides during purification, and for differences in leucine content from one peptide to another.

Figure 18.1 shows how this procedure can tell us the direction of translation. It is important to notice that the protein chains are in all stages of completion when the ^{3}H- labeled amino acid is added. Thus, some are just starting, some are partly finished, and some are almost finished. This means that label will be incorporated into the first peptide only in those proteins whose synthesis had just begun when the label was added. The others will be labeled in downstream peptides, but not in the first one. By contrast, the end of the protein where protein synthesis ends will be

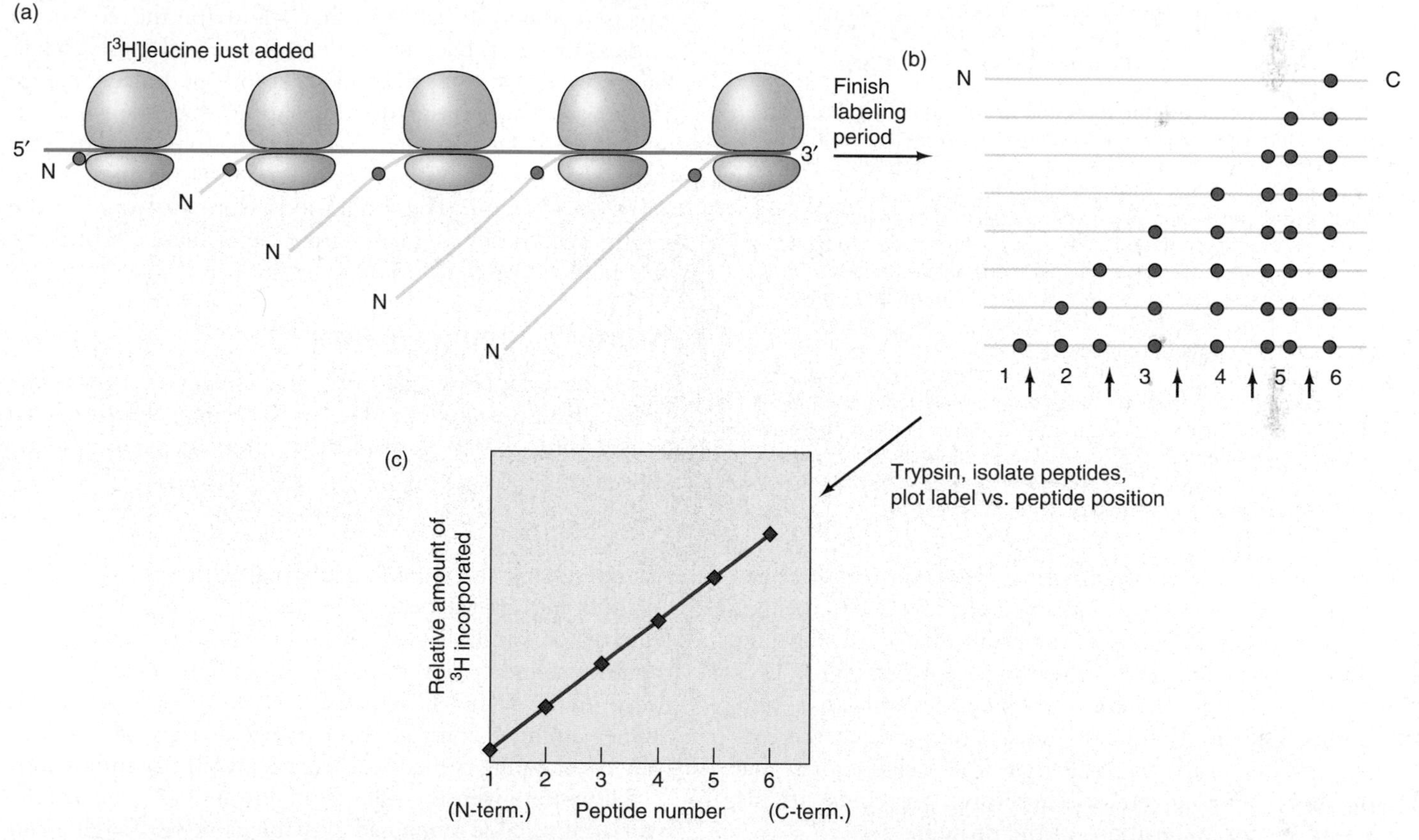

Figure 18.1 Experimental strategy to determine the direction of translation. (a) Labeling the protein. Consider an mRNA (green) being translated by several ribosomes (pink and blue), assuming that the mRNA is translated in the 5′→3′ direction and the proteins are made in the amino (N) to carboxyl (C) direction. A labeled amino acid ([^{3}H]leucine) has just been added to the system, so it has begun to be incorporated into the growing protein chains (blue), as indicated by the red dots. It is incorporated near the N-terminus in the polypeptides on the left, where protein synthesis has just begun, but only near the C-terminus in the polypeptides on the right, which are almost completed. **(b)** Distribution of label in completed proteins after a moderate labeling period. The proteins near the top, with label only near the C-terminus correspond to the nearly completed proteins near the right in panel (a). Those near the bottom, with label distributed toward the N-terminus, correspond to the growing proteins near the left in panel (a). These have had time to incorporate label throughout a greater length of the protein. Cutting sites for trypsin within the protein are indicated by arrows at bottom, and the resulting peptides are numbered 1–6 according to their positions in the protein. **(c)** Model experimental results. One plots the relative amount of ^{3}H labeling in each of the peptides, 1–6, and finds that the C-terminal peptides are the most highly labeled. This is what we expect if translation started at the N-terminus. If it had started at the C-terminus (opposite to the picture in panel [a]), then the N-terminal peptides would be the most highly labeled.

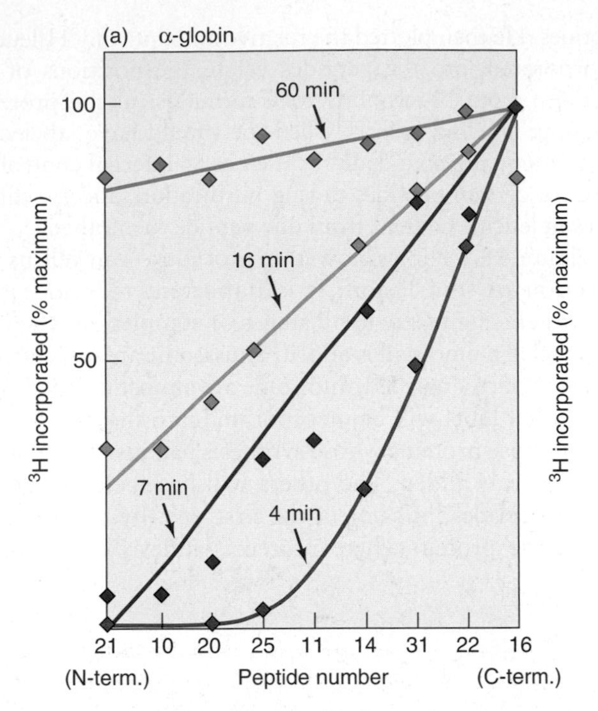

Figure 18.2 Determining the direction of translation. Dintzis carried out the experimental plan outlined in Figure 18.1 with rabbit reticulocytes, which make almost nothing but α- and β-globins. He labeled the reticulocytes with [^{3}H]leucine for various lengths of time, then separated the α- and β-globins, cut each protein into peptides with trypsin, and determined the label in each peptide. He plotted the relative amount of ^{3}H label against the peptide number, with the N-terminal peptide on the left, and the C-terminal peptide on the right. The curves for α- and β-globin showed the most label in the C-terminal peptides, especially after short labeling times. (Only the α-globin results are shown here.) This is what we expect if translation starts at the N-terminus of a protein. Note that the peptide numbers are not related to their position in the protein, as they are in the example in Figure 18.1. (*Source:* Adapted from Dintzis, H.M., Assembly of the peptide chains of hemoglobin. *Proceedings of the National Academy of Sciences USA* 47:255, 1961.)

relatively rich in label after a short labeling time. Intermediate peptides will show intermediate levels of labeling. Thus, if translation starts at the amino terminus, labeling will be strongest in carboxyl-terminal peptides. Figure 18.2 shows the results. Labeling of the peptides of both α- and β-globins increased from the amino terminus to the carboxyl terminus, and this disparity was especially noticeable with short labeling times. Therefore, protein synthesis starts at the amino terminus of the protein.

Is the mRNA read in the 5′→3′ direction or the reverse? Knowing that proteins grow in the amino→carboxyl direction, it is easy to show that mRNAs are read in the 5′→3′ direction. When molecular biologists first started using synthetic mRNAs as templates for protein synthesis in the 1960s, some of these messages held the answer to our question. For example, when Ochoa and his colleagues translated the mRNA: 5′-AUGUUU$_n$-3′, they obtained fMet-Phe$_n$, where the fMet was at the amino terminus. We know that AUG codes for fMet and UUU codes for phenylalanine (Phe). We see that fMet is incorporated into the amino terminal position of the protein, which means it was added first, before any of the phenylalanines. Therefore the mRNA must have been read from the 5′-end, because that is where the fMet codon is.

SUMMARY Messenger RNAs are read in the 5′→3′ direction, the same direction in which they are synthesized. Proteins are made in the amino→carboxyl direction, which means that the amino terminal amino acid is added first.

18.2 The Genetic Code

The term **genetic code** refers to the set of three-base code words (**codons**) in mRNAs that stand for the 20 amino acids in proteins. Like any code, this one had to be broken before we knew what the codons stood for. Indeed, before 1960, other more basic questions about the code were still unanswered. These included: Do the codons overlap? Are there gaps, or "commas," in the code? How many bases make up a codon? These questions were answered in the 1960s by a series of imaginative experiments, which we will examine here.

Nonoverlapping Codons

In a nonoverlapping code, each base is part of at most one codon. In an overlapping code, one base may be part of two or even three codons. Consider the following micromessage:

AUGUUC

Assuming that the code is triplet (three bases per codon) and this message is read from the beginning, the codons will be AUG and UUC if the code is nonoverlapping. On the other hand, an overlapping code might yield four codons: AUG, UGU, GUU, and UUC. As early as 1957, Sydney Brenner concluded on theoretical grounds that a fully overlapping triplet code like this would be impossible.

However, given the data available in 1957, a *partially* overlapping code remained possible, but A. Tsugita and H. Frankel-Conrat laid it to rest with the following line of reasoning: If the code is nonoverlapping, a change of one base in an mRNA (a missense mutation) would change no more than one amino acid in the resulting protein. For example, consider another micromessage:

AUGCUA

Assuming that the code is triplet (three bases per codon) and this message is read from the beginning, the codons will be AUG and CUA if the code is nonoverlapping. A change in the fourth base (C) would change only one codon (CUA) and therefore at most only one amino acid. On the other hand, if the code were overlapping, base C could be part of three adjacent codons (UGC, GCU, and CUA). Therefore, if the C were changed, up to three adjacent amino acids could be changed in the resulting protein. But when the investigators introduced one-base alterations into mRNA from tobacco mosaic virus (TMV), they found that these never caused changes in more than one amino acid. Hence, the code must be nonoverlapping.

No Gaps in the Code

If the code contained untranslated gaps, or "commas," mutations that add or subtract a base from a message might change a few codons, but we would expect the ribosome to get back on track after the next comma. In other words, these mutations might frequently be lethal, but in many cases the mutation should occur just before a comma in the message and therefore have little, if any, effect. If no commas were present to get the ribosome back on track, these mutations would be lethal except when they occur right at the end of a message.

Such mutations do occur, and they are called **frameshift mutations;** they work as follows. Consider another tiny message:

AUGCAGCCAACG

If translation starts at the beginning, the codons will be AUG, CAG, CCA, and ACG. If we insert an extra base (X) right after base U, we get:

AUXGCAGCCAACG

Now this would be translated from the beginning as AUX, GCA, GCC, AAC. Notice that the extra base changes not only the codon (AUX) in which it appears, but every codon from that point on. The **reading frame** has shifted one base to the left; whereas C was originally the first base of the second codon, G is now in that position.

On the other hand, a code with commas would be one in which each codon is flanked by one or more untranslated bases, represented by Z's in the following message. The commas would serve to set off each codon so the ribosome could recognize it:

AUGZCAGZCCAZACGZ

Deletion or insertion of a base anywhere in this message would change only a single codon. The comma (Z) at the end of the damaged codon would then put the ribosome back on the right track. Thus, addition of an extra base (X) to the first codon would give the message:

AUXGZCAGZCCAZACGZ

The first codon (AUXG) is now wrong, but all the others, still neatly set off by Z's, would be translated normally.

When Francis Crick and his colleagues treated bacteria with acridine dyes that usually cause single-base insertions or deletions, they found that such mutations were very severe; the mutant genes gave no functional product. This is what we would expect of a "comma-less" code with no gaps; base insertions or deletions cause a shift in the reading frame of the message that persists until the end of the message.

Moreover, Crick found that adding a base could cancel the effect of deleting a base, and vice versa. This phenomenon is illustrated in Figure 18.3, where we start with an artificial gene composed of the same codon, CAT, repeated over and over. When we add a base, G, in the third position, we change the reading frame so that all codons thereafter read TCA. When we start with the wild-type gene and delete the fifth base, A, we change the reading frame in the other direction, so that all subsequent codons read ATC. Crossing these two mutants sometimes gives a recombined "pseudo-wild-type" gene like the one on line 4 of the figure. Its first two codons, CAG and TCT, are wrong, but thereafter the insertion and deletion cancel, and the original reading frame is restored. All codons from that point on read CAT.

The Triplet Code

Francis Crick and Leslie Barnett discovered that a presumed set of three insertions or deletions could produce a

1. Wild-type: CAT | CAT | CAT | CAT | CAT
2. Add a base: CAG | TCA | TCA | TCA | TCA
3. Delete a base: CAT | CTC | ATC | ATC | ATC
4. Cross #2 and #3: CAG | TCT | CAT | CAT | CAT
5. Add 3 bases: CAG | GGT | CAT | CAT | CAT

Figure 18.3 Frameshift mutations. Line 1: An imaginary gene has the same codon, CAT, repeated over and over. The vertical dashed lines show the reading frame, starting from the beginning. Line 2: Adding a base, G (pink), in the third position changes the first codon to CAG and shifts the reading frame one base to the left so that every subsequent codon reads TCA. Line 3: Deleting the fifth base, A (marked by the triangle), from the wild-type gene changes the second codon to CTC and shifts the reading frame one base to the right so that every subsequent codon reads ATC. Line 4: Crossing the mutants in lines 2 and 3 occasionally gives a recombined "pseudo-wild-type" revertant with an insertion and a deletion close together. The end result is a DNA with its first two codons altered, but all the other ones put back into the correct reading frame. Line 5: Adding three bases, GGG (pink), after the first two bases disrupts the first two codons, but leaves the reading frame unchanged. The same would be true of deleting three bases.

pseudo-wild-type gene (Figure 18.3, line 5). This of course demands that a codon consist of three bases. As Crick remarked to Barnett when he saw the experimental result, "We're the only two [who] know it's a triplet code!" Actually, Crick and Bartlett were inferring that their pseudo-wild-type genes contained three insertions or deletions. They had no way of sequencing the genes to make sure, so more experiments were needed.

In 1961, Marshall Nirenberg and Johann Heinrich Matthaei performed a groundbreaking experiment that laid the foundation for confirming the triplet nature of the code and for breaking the genetic code itself. The experiment was deceptively simple; it showed that synthetic RNA could be translated in vitro. In particular, when Nirenberg and Matthaei translated poly(U), a synthetic RNA composed only of U's, they made polyphenylalanine. Of course, that told them that a codon for phenylalanine contains only U's. This finding by itself was important, but the long-range implication was that one could design synthetic mRNAs of defined sequence and analyze the protein products to shed light on the nature of the code. Gobind Khorana and his colleagues were the chief practitioners of this strategy.

Here is how Khorana's synthetic messenger experiments confirmed that the codons contain three bases: First, if the codons contain an odd number of bases, then a repeating dinucleotide poly(UC) or UCUCUCUC . . . should contain two alternating codons (UCU and CUC, in this case), no matter where translation starts. The resulting protein would be a repeating dipeptide—two amino acids alternating with each other. If codons have an even number of bases, only one codon (UCUC, for example) should be repeated over and over. Of course, if translation started at the second base, the single repeated codon would be different (CUCU). In either case, the resulting protein would be a homopolypeptide, containing only one amino acid repeated over and over. Khorana found that poly(UC) translated to a repeating dipeptide, poly(serine-leucine) (Figure 18.4a), proving that the codons contained an odd number of bases.

Repeating triplets were translated to homopolypeptides, as had been expected if the number of bases in a codon was three or a multiple of three. For example, poly(UUC) translated to polyphenylalanine plus polyserine plus polyleucine (Figure 18.4b). The reason for three different products is that translation can start at any point in the synthetic message. Therefore, poly(UUC) can be read as UUC, UUC, and so on, UCU, UCU, and so on, or CUU, CUU, and so on, depending on where translation starts. In all cases, once translation begins, only one codon is encountered, as long as the number of bases in a codon is divisible by 3.

Repeating tetranucleotides were translated to repeating tetrapeptides. For example, poly(UAUC) yielded poly(tyrosine-leucine-serine-isoleucine) (Figure 18.4c). As an exercise, you can write out the sequence of such a message

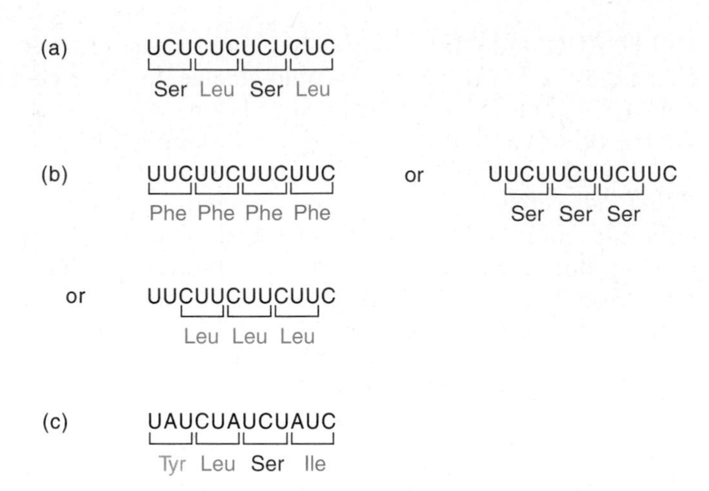

Figure 18.4 Coding properties of several synthetic mRNAs. **(a)** Poly(UC) contains two alternating codons, UCU and CUC, which code for serine (Ser) and leucine (Leu), respectively. Thus, the product is poly(Ser-Leu). **(b)** Poly(UUC) contains three codons, UUC, UCU, and CUU, which code for phenylalanine (Phe), serine (Ser), and leucine (Leu), respectively. The product is therefore poly(Phe), or poly(Ser), or poly(Leu), depending on which of the three reading frames the ribosome uses. **(c)** Poly(UAUC) contains four codons in a repeating sequence: UAU, CUA, UCU, and AUC, which code for tyrosine (Tyr), leucine (Leu), serine (Ser), and isoleucine (Ile), respectively. The product is therefore poly(Tyr-Leu-Ser-Ile).

and satisfy yourself that it is compatible with codons having three bases, or nine, or even more, but not six. (We already know six cannot be right because it is not an odd number.) Because codons are not likely to be as cumbersome as nine bases long, three is the best choice. Look at the problem another way: Three is the lowest number that gives enough different codons to specify all 20 amino acids. (The number of permutations of four different bases taken 3 at a time is 4^3, or 64.) There would be only 16 two-base codons ($4^2 = 16$), not quite enough. But there would be over 200,000 ($4^9 = 262{,}144$) nine-base codons. Nature is usually more economical than that.

SUMMARY The genetic code is a set of three-base code words, or codons, in mRNA that instruct the ribosome to incorporate specific amino acids into a polypeptide. The code is nonoverlapping: that is, each base is part of only one codon. It is also devoid of gaps, or commas; that is, each base in the coding region of an mRNA is part of a codon.

Breaking the Code

Obviously, Khorana's synthetic mRNAs gave strong hints about some of the codons. For example, because poly(UC) yields poly(serine-leucine), we know that one of the codons (UCU or CUC) codes for serine and the other codes for leucine. The question remains: Which is which? Nirenberg developed a powerful assay to answer this question. He found that a trinucleotide was usually enough like an mRNA to cause a specific aminoacyl-tRNA to bind to ribosomes. For example, the triplet UUU will cause phenylalanyl-tRNA to bind, but not lysyl-tRNA or any other aminoacyl-tRNA. Therefore, UUU is a codon for phenylalanine. This method was not perfect; some codons did not cause any aminoacyl-tRNA to bind, even though they were authentic codons for amino acids. But it provided a nice complement to Khorana's method, which by itself would not have given all the answers either, at least not easily.

Here is an example of how the two methods could be used together: Translation of the polynucleotide poly(AAG) yielded polylysine plus polyglutamate plus polyarginine. There are three different codons in that synthetic message: AAG, AGA, and GAA. Which one codes for lysine? All three were tested by Nirenberg's assay, yielding the results shown in Figure 18.5. Clearly, AGA and GAA caused no binding of [^{14}C]lysyl-tRNA to ribosomes, but AAG did. Therefore, AAG is the lysine codon in poly(AAG). Something else to notice about this experiment is that the triplet AAA also caused lysyl-tRNA to bind. Therefore, AAA is another lysine codon. This illustrates a general feature of the code: In most cases, more than one triplet codes for a given amino acid. In other words, the code is **degenerate.**

Figure 18.6 shows the entire genetic code. As predicted, there are 64 different codons and only 20 different amino acids, yet all of the codons are used. Three are "stop" codons found at the ends of messages, but all the others specify amino acids, which means that the code is highly degenerate. Leucine, serine, and arginine have six different codons; several others, including proline, threonine, and alanine, have four; isoleucine has three; and many others have two. Just two amino acids, methionine and tryptophan, have only one codon.

SUMMARY The genetic code was broken by using either synthetic messengers or synthetic trinucleotides and observing the polypeptides synthesized or aminoacyl-tRNAs bound to ribosomes, respectively. There are 64 codons in all. Three are stop signals, and the rest code for amino acids. This means that the code is highly degenerate.

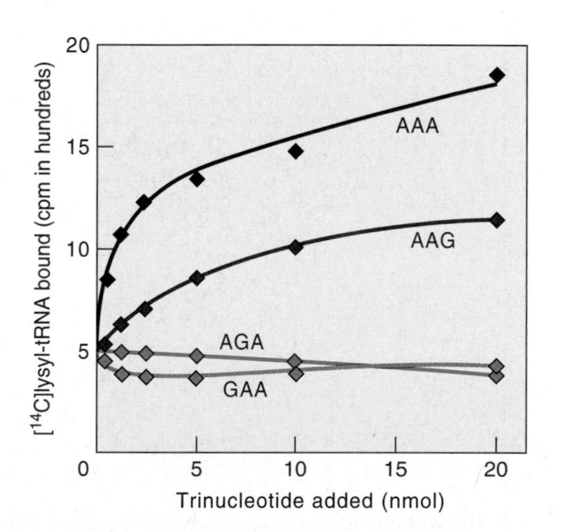

Figure 18.5 Binding of lysyl-tRNA to ribosomes in response to various codons. Lysyl-tRNA was labeled with radioactive carbon (^{14}C) and mixed with *E. coli* ribosomes in the presence of the following trinucleotides: AAA, AAG, AGA, and GAA. Lysyl-tRNA-ribosome complex formation was measured by binding to nitrocellulose filters. (Unbound lysyl-tRNA does not stick to these filters, but a lysyl-tRNA–ribosome complex does.) AAA was a known lysine codon, so binding was expected with this trinucleotide. (*Source:* Adapted from Khorana, H.G., Synthesis in the study of nucleic acids, *Biochemical Journal* 109:715, 1968.)

First position (5′-end)	Second position: U	C	A	G	Third position (3′-end)
U	UUU Phe	UCU Ser	UAU Tyr	UGU Cys	U
U	UUC Phe	UCC Ser	UAC Tyr	UGC Cys	C
U	UUA Leu	UCA Ser	UAA STOP	UGA STOP	A
U	UUG Leu	UCG Ser	UAG STOP	UGG Trp	G
C	CUU Leu	CCU Pro	CAU His	CGU Arg	U
C	CUC Leu	CCC Pro	CAC His	CGC Arg	C
C	CUA Leu	CCA Pro	CAA Gln	CGA Arg	A
C	CUG Leu	CCG Pro	CAG Gln	CGG Arg	G
A	AUU Ile	ACU Thr	AAU Asn	AGU Ser	U
A	AUC Ile	ACC Thr	AAC Asn	AGC Ser	C
A	AUA Ile	ACA Thr	AAA Lys	AGA Arg	A
A	AUG Met	ACG Thr	AAG Lys	AGG Arg	G
G	GUU Val	GCU Ala	GAU Asp	GGU Gly	U
G	GUC Val	GCC Ala	GAC Asp	GGC Gly	C
G	GUA Val	GCA Ala	GAA Glu	GGA Gly	A
G	GUG Val	GCG Ala	GAG Glu	GGG Gly	G

Figure 18.6 The genetic code. All 64 codons are listed, along with the amino acid for which each codes. To find a given codon—ACU, for example—we start with the wide horizontal row labeled with the name of the first base of the codon (A) on the left border. Then we move across to the vertical column corresponding to the second base (C). This brings us to a box containing all four codons beginning with AC. It is now a simple matter to find the one among these four we are seeking, ACU. We see that this triplet codes for threonine (Thr), as do all the other codons in the box: ACC, ACA, and ACG. This is an example of the degeneracy of the code. Notice that three codons (pink) do not code for amino acids; instead, they are stop signals.

Unusual Base Pairs Between Codon and Anticodon

How does an organism cope with multiple codons for the same amino acid? One way would be to have multiple tRNAs (**isoaccepting species**) for the same amino acid, each one specific for a different codon. This is part of the answer, and indeed a given organism contains about 60 different tRNAs. But, in principle, we can get along with considerably fewer tRNAs than that simple hypothesis would predict. Again Francis Crick anticipated experimental results with insightful theory. In this case, Crick hypothesized that the first two bases of a codon must pair correctly with the anticodon according to Watson–Crick base-pairing rules (Figure 18.7a), but the last base of the codon can "wobble" from its normal position to form unusual base pairs with the anticodon. This proposal was called the **wobble hypothesis.** In particular, Crick proposed that a G in an anticodon can pair not only with a C in the third position of a codon (the **wobble position**), but also with a U. This would give the **wobble base pair** shown in Figure 18.7b. Notice how the U has moved, or wobbled from its normal position to form this base pair.

Furthermore, Crick noted that one of the unusual nucleosides found in tRNA is **inosine (I)**, which has a structure similar to that of guanosine. This nucleoside can ordinarily pair like G, so we would expect it to pair with C (Watson–Crick base pair) or U (wobble base pair) in the third position (the wobble position) of a codon. But Crick proposed that inosine could form still another kind of wobble pair, this time with A in the third position of a codon (Figure 18.7c). That means an anticodon with I in the first position can potentially pair with three different codons ending with C, U, or A.

The wobble phenomenon reduces the number of tRNAs required to translate the genetic code. For example, consider the two codons for phenylalanine, UUU and UUC, listed at the top left of Figure 18.6. According to the wobble hypothesis, they can both be recognized by an anticodon that reads 3′-AAG-5′ (Figure 18.8a). The G in the 5′-position of the anticodon could form a Watson–Crick G–C base pair with the C in the UUC, or a G–U wobble base pair with the U in UUU. Similarly, the two leucine codons in the same box, UUA and UUG, can both be recognized by the anticodon 3′-AAU-5′ (Figure 18.8b). The U can form a Watson–Crick pair with the A in UUA, or a wobble pair with the G in UUG.

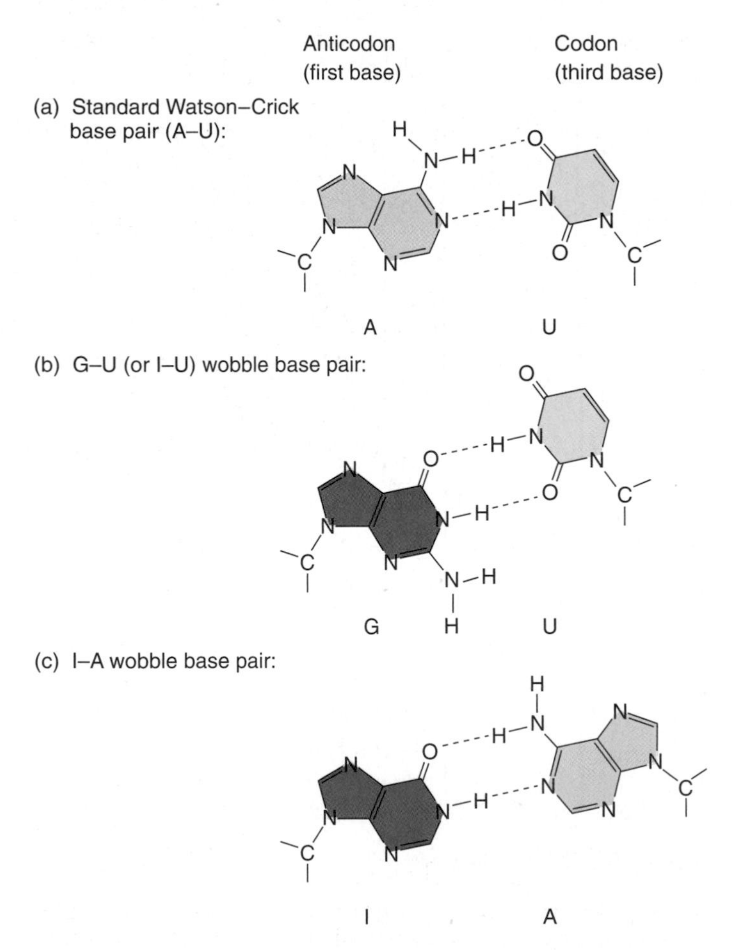

Figure 18.7 Wobble base pairs. **(a)** Relative positions of bases in a standard (A–U) base pair. The base on the left here and in the wobble base pairs (b) and (c) is the first base in the anticodon. The base on the right is the third base in the codon. **(b)** Relative positions of bases in a G–U (or I–U) wobble base pair. Notice that U has to "wobble" upward to pair with the G (or I). **(c)** Relative positions of bases in an I–A wobble base pair. The A has to "wobble" upward in order to form this pair.

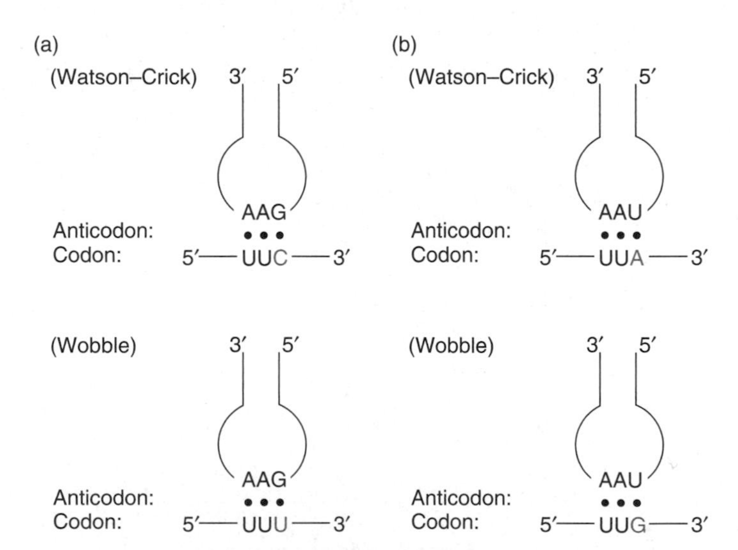

Figure 18.8 The wobble position. **(a)** An abbreviated tRNA with anticodon 3′-AAG-5′ is shown base-pairing with two different codons for phenylalanine: UUC and UUU. The wobble position (the third base of the codon) is highlighted in red. The base-pairing with the UUC codon (top) uses only Watson–Crick pairs; the base-pairing with the UUU codon (bottom) uses two Watson–Crick pairs in the first two positions of the codon, but requires a wobble pair (G–U) in the wobble position. **(b)** A similar situation, in which a tRNA with anticodon AAU base-pairs with two different codons for leucine: UUA and UUG. Pairing with the UUG codon requires a G–U wobble pair in the wobble position.

According to the wobble hypothesis, a cell should be able to get by with only 31 tRNAs to read all 64 codons, assuming no tRNA is needed to read the UAA and UAG stop codons. But human mitochondria and plant plastids contain fewer than 31 tRNAs, so something besides wobble appears to be in play. This has led to the **superwobble** hypothesis, which holds that a single tRNA with a U in its wobble position (the first base in its anticodon) can, at least in certain circumstances, recognize codons ending in any of the four bases.

Ralph Bock and colleagues put the superwobble hypothesis to the test in 2008 when they knocked out both $tRNA^{Gly}$ genes in tobacco plastids, then added back only $tRNA^{Gly}$(UCC), which, using superwobble, should be able to translate all four glycine codons. The resulting tobacco cells were indeed viable, though translation efficiency was reduced. Thus, superwobble appears to work, but not perfectly, which probably explains why it has not evolved very often.

SUMMARY Part of the degeneracy of the genetic code is accommodated by isoaccepting species of tRNA that bind the same amino acid but recognize different codons. The rest is handled by wobble, in which the third base of a codon is allowed to move slightly from its normal position to form a non-Watson–Crick base pair with the anticodon. This allows the same aminoacyl-tRNA to pair with more than one codon. The wobble pairs are G–U (or I–U) and I–A. Some organelles have evolved with fewer tRNAs than are required to translate all the sense codons. In these cases, codons with U in the wobble position can apparently translate codons with all four bases in the last position by superwobble.

The (Almost) Universal Code

In the years after the genetic code was broken, all organisms examined, from bacteria to humans, were shown to share the same code. Therefore it was generally assumed (incorrectly, as we will see) that the code was universal, with no deviations whatsoever. This apparent universality led in turn to the notion of a single origin of present life on earth.

The reasoning for this idea goes like this: Nothing is inherently advantageous about each specific codon assignment we see. There is no obvious reason, for example, why UUC should make a good codon for phenylalanine, whereas AAG is a good one for lysine. Rather, the genetic code may be an "accident"; it just happened to evolve that way. However, once these codons were established, there was a very good reason why they did not change: A change that fundamental would almost certainly be lethal.

Consider, for instance, a tRNA for the amino acid cysteine and the codon it recognizes, UGU. For that relationship to change, the anticodon of the cysteinyl-tRNA would have to change so it can recognize a different codon, say UCU, which is a serine codon. At the same time, all the UCU codons in that organism's genome that code for important serines would have to change to alternate serine codons so they would not be recognized as cysteine codons. The chances of all these things happening together, even over vast evolutionary time, are negligible. That is why the genetic code is sometimes called a "frozen accident"; once it was established, for whatever reasons, it had to stay that way. So a universal code would be powerful evidence for a single origin of life. After all, if life started independently in two places, we would hardly expect the two lines to evolve the same genetic code by accident!

In light of all this, it is remarkable that the genetic code is not absolutely universal; there are some exceptions to the rule. The first of these to be discovered were in the genomes of mitochondria. In mitochondria of the fruit fly *D. melanogaster,* UGA is a codon for tryptophan rather than for "stop." Even more remarkably, AGA in these mitochondria codes for serine, whereas it is an arginine codon in the standard code. Mammalian mitochondria show some deviations, too. Both AGA and AGG, though they are arginine codons in the standard code, have a different meaning in human and bovine mitochondria; there they code for "stop." Furthermore, AUA, ordinarily an isoleucine codon, codes for methionine in these mitochondria.

These aberrations might be dismissed as relatively unimportant, occurring as they do in mitochondria, which have very small genomes coding for only a few proteins and therefore more latitude to change than nuclear genomes. But exceptional codons also occur in nuclear genomes and bacterial genomes. In at least three ciliated protozoa, including *Paramecium,* UAA and UAG, which are normally stop codons, code for glutamine. In the prokaryote *Mycoplasma capricolum,* UGA, normally a stop codon, codes for tryptophan. In the pathogenic yeast, *Candida albicans,* CTG, usually a leucine codon, codes for serine. Deviations from the standard genetic code are summarized in Table 18.1.

Clearly, the so-called universal code is not really universal. Does this mean that the evidence now favors more than one origin of present life on earth? If the deviant codes were radically different from the standard one, this might be an attractive possibility, but they are not. In many cases, the novel codons are stop codons that have been recruited to code for an amino acid: glutamine or tryptophan. There is a well-established mechanism for this sort of occurrence, as we will see later in this chapter. The vast majority of known examples of codons that have switched their meaning from one amino acid to another occur in mitochondria. Again, mitochondrial genomes, because they code for far fewer proteins than nuclear genomes or even bacterial genomes, might be expected to change a codon safely every now and then. In summary, even if the code is not universal, a standard code does exist from which the deviant ones almost certainly evolved. Therefore, the evidence still strongly favors a single origin of life.

Table 18.1 Deviations from the "Universal" Genetic Code

Source	Codon	Usual meaning	New meaning
Fruit fly mitochondria	UGA	Stop	Tryptophan
	AGA & AGG	Arginine	Serine
	AUA	Isoleucine	Methionine
Mammalian mitochondria	AGA & AGG	Arginine	Stop
	AUA	Isoleucine	Methionine
	UGA	Stop	Tryptophan
Yeast mitochondria	CUN*	Leucine	Threonine
	AUA	Isoleucine	Methionine
	UGA	Stop	Tryptophan
Higher plant mitochondria	UGA	Stop	Tryptophan
	CGG	Arginine	Tryptophan
Candida albicans nuclei	CTG	Leucine	Serine
Protozoa nuclei	UAA & UAG	Stop	Glutamine
Mycoplasma	UGA	Stop	Tryptophan

*N = Any base.

What about the argument that the code is random: that the existing codons have no inherent advantage? Actually, when we consider the code's effectiveness in dealing with mutations, we find that it is an excellent code indeed. First, consider the fact that single-base changes in the code are likely to result in a shift to a chemically similar amino acid. For example, leucine, isoleucine, and valine all have very similar hydrophobic side chains. And their codons are also very similar, differing only in the first base. So, to pick a particularly advantageous example, a mutation in the first base of the isoleucine codon AUA, could yield UUA, CUA, or GUA. The first two are leucine codons, and the last is a valine codon. Thus, none of these mutations would cause much change in the corresponding amino acid, which minimizes the chance of causing serious damage to the protein product of the mutated gene.

When we consider two other factors, the code looks even better: First, **transitions** (the change of one purine to another, or one pyrimidine to another), are much more common mutations than **transversions**, the change of a purine to a pyrimidine, or vice versa. Second, the ribosome is much more likely to misread the first and third bases in a codon than the second. Considering these things, we can calculate the probability that a single base change will result in no change or just a modest change in the encoded amino acid, for all the possible three-base codes. Then we can see how our natural code stacks up against the others. Figure 18.9 presents a result of this mathematical analysis, which shows that our code is literally one in a million. Only one in a million other possible codes would work better than ours in minimizing the effects of mutations. Given those odds, it seems less likely that our code is just an accident, and not the result of honing by evolution.

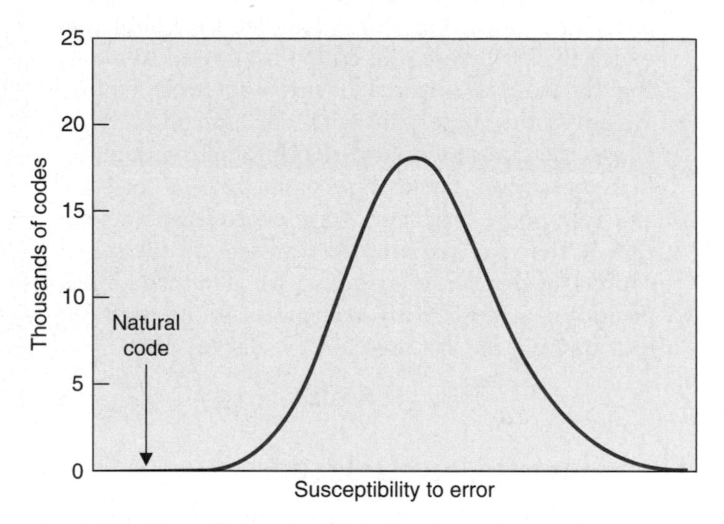

Figure 18.9 Susceptibility of genetic codes to error. The susceptibility to error of all possible triplet genetic codes with four bases is plotted against the number of codes (in thousands) having each susceptibility value. Our own natural code lies far outside the normal distribution, with a very low susceptibility to error. In fact, only one code in a million has a lower susceptibility. (*Source*: Adapted from Vogel, G. Tracking the history of the genetic code. *Science* 281 (17 Jul 1998) 329–331.)

SUMMARY The genetic code is not strictly universal. In certain eukaryotic nuclei and mitochondria and in at least one bacterium, codons that cause termination in the standard genetic code can code for amino acids such as tryptophan and glutamine. In several mitochondrial genomes, and in the nuclei of at least one yeast, the sense of a codon is changed from one amino acid to another. These deviant codes are still closely related to the standard one from which they probably evolved. It is not clear whether the genetic code is a frozen accident or the product of evolution, but its ability to cope with mutations suggests that it has been subject to evolution.

18.3 The Elongation Cycle

Elongation of a polypeptide chain occurs in a three-step cycle (the **elongation cycle**) that is repeated over and over. We will survey these steps first, then come back and fill in the details, along with experimental evidence.

Overview of Elongation

Figure 18.10 schematically depicts the elongation cycle through two rounds (adding two amino acids to a growing polypeptide chain) in *E. coli.* We start with mRNA and fMet-$tRNA_f^{Met}$ bound to a ribosome. There are three binding sites for aminoacyl-tRNAs on the ribosome. Two of these are called the **P (peptidyl) site** and the **A (aminoacyl) site.** In our schematic diagram, the P site is on the left and the A site is on the right. The fMet-$tRNA_f^{Met}$ is in the P site. A binding site for deacylated tRNA called the **E (exit) site** is empty because the translation process has just begun. Detailed below are the elongation events as shown in Figure 18.10:

a. To begin elongation, we need another amino acid to join with the first. This second amino acid arrives

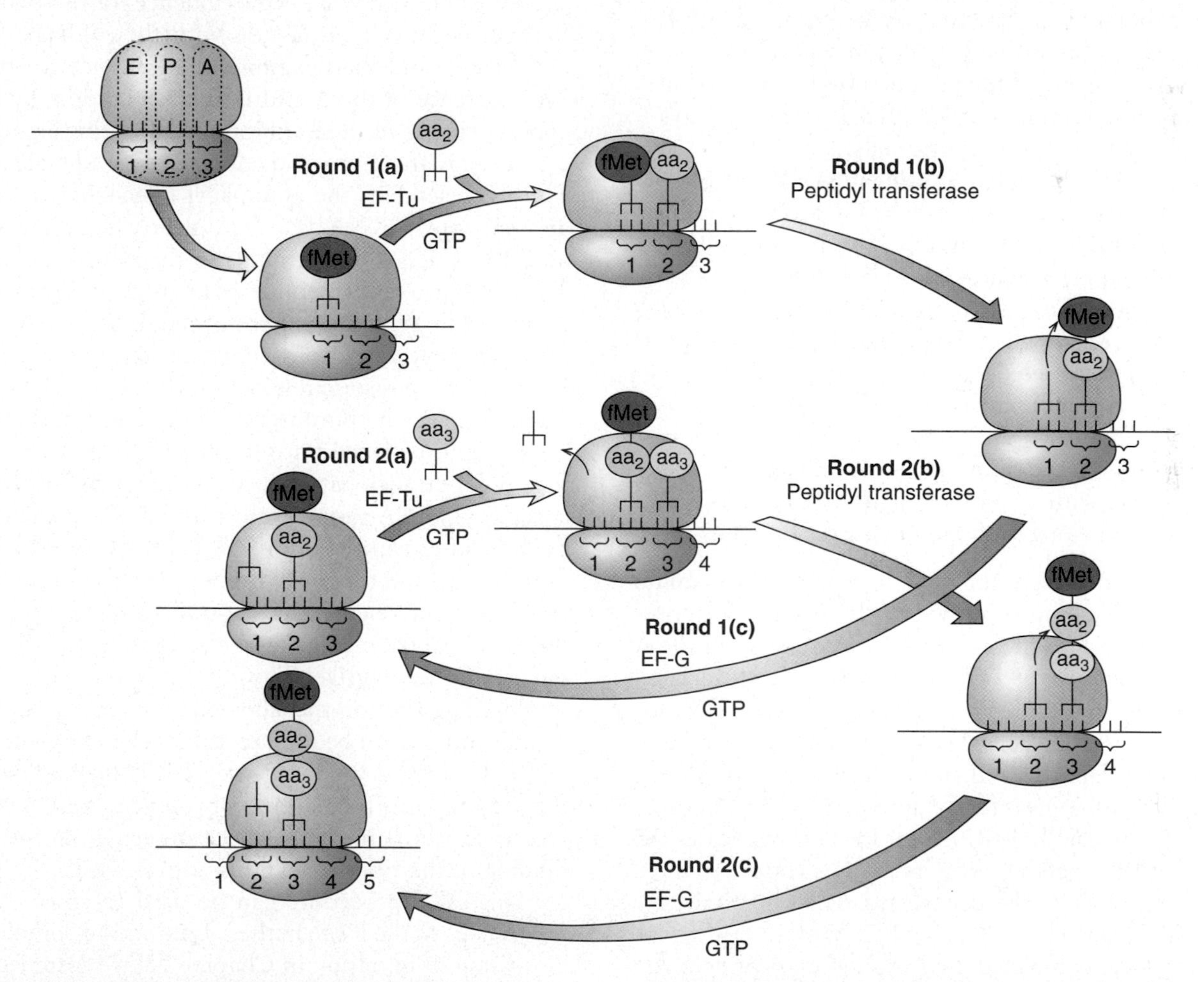

Figure 18.10 Elongation in translation. Note first of all that this is a highly schematic view of protein synthesis. For example, tRNAs are represented by fork-like structures that merely suggest the two business ends of the molecule. Upper left: A ribosome with an mRNA attached is shown to illustrate three sites, E, P and A, indicated with dotted lines. Round I: **(a)** EF-Tu brings in the second aminoacyl-tRNA (yellow) to the A site on the ribosome. The P site is already occupied by fMet-tRNA (magenta). **(b)** Peptidyl transferase forms a peptide bond between fMet and the second aminoacyl-tRNA. **(c)** In the translocation step, EF-G shifts the message and the tRNAs one codon's width to the left. This moves the dipeptidyl-tRNA into the P site, moves the deacylated tRNA in the P site into the E site, and opens up the A site for a new aminoacyl-tRNA. In round 2, these steps are repeated to add one more amino acid (green) to the growing polypeptide. This time, there is a deacylated tRNA in the E site. When EF-Tu brings in the third aminoacyl-tRNA, hydrolysis of the bound GTP allows release of the tRNA from the E site. This opens up the E site for the next translocation step.

bound to a tRNA, and the nature of this aminoacyl-tRNA is dictated by the second codon in the message. The second codon is in the A site, which is otherwise empty, so our second aminoacyl-tRNA will bind to this site. Such binding requires a protein **elongation factor** known as **EF-Tu** (where EF stands for elongation factor) and GTP.

b. Next, the first peptide bond forms. An enzyme called **peptidyl transferase**—an integral part of the large ribosomal subunit—transfers the fMet from its tRNA in the P site to the aminoacyl-tRNA in the A site. This forms a two-amino acid unit called a dipeptide linked to the tRNA in the A site. This whole assembly in the A site is a dipeptidyl-tRNA. What remains in the P site is a deacylated tRNA—a tRNA without its amino acid.

   The formation of the first peptide bond in bacteria is aided by an essential factor known as **EF-P.** Its role appears to be to position the fMet-$tRNA_f^{Met}$ properly for peptide bond formation. A eukaryotic homolog called **eIF5A** probably plays the same role in eukaryotic cells.

c. In the next step, called **translocation,** the mRNA with its peptidyl-tRNA attached in the A site moves one codon's length to the left. This has the following results: (1) The deacylated tRNA in the P site (the one that lost its amino acid during the peptidyl transferase step when the peptide bond formed) moves to the E site. (2) The dipeptidyl-tRNA in the A site, along with its corresponding codon, moves into the P site. (3) The codon that was "waiting in the wings" to the right moves into the A site, ready to interact with an aminoacyl-tRNA. Translocation requires an elongation factor called **EF-G** plus GTP.

The process then repeats itself to add another amino acid: (a) EF-Tu, in conjunction with GTP, brings the appropriate aminoacyl-tRNA to match the new codon in the A site. Upon hydrolysis of GTP by EF-Tu, the deacylated tRNA is ejected from the E site, which makes room for another deacylated tRNA at the end of the second round of elongation. (b) Peptidyl transferase brings the dipeptide from the P site and joins it to the aminoacyl-tRNA in the A site, forming a tripeptidyl-tRNA. (c) EF-G translocates the tripeptidyl-tRNA, together with its mRNA codon, to the P site. At the same time, the deacylated tRNA in the P site moves to the E site.

We have now completed two rounds of peptide chain elongation. We started with an aminoacyl-tRNA (fMet-$tRNA_f^{Met}$) in the P site, and we have lengthened the chain by two amino acids to a tripeptidyl-tRNA. This process continues over and over until the ribosome reaches the last codon in the message. The polypeptide is now complete; it is time for chain termination. The elongation process has been greatly simplified in this brief presentation. It will be fleshed out later in this chapter, and even more in Chapter 19.

SUMMARY Elongation takes place in three steps: (1) EF-Tu, with GTP, binds an aminoacyl-tRNA to the ribosomal A site. (2) Peptidyl transferase forms a peptide bond between the peptide in the P site and the newly arrived aminoacyl-tRNA in the A site. This lengthens the peptide by one amino acid and shifts it to the A site. (3) EF-G, with GTP, translocates the growing peptidyl-tRNA, with its mRNA codon, to the P site, and moves the deacylated tRNA in the P site to the E site.

A Three-Site Model of the Ribosome

The previous section introduced the concept of the three-site ribosome. But what is the evidence for these three sites? We will begin our discussion with the evidence for the A and P sites, and then examine the evidence for the E site. The existence of the A and P sites was originally based on experiments with the antibiotic **puromycin** (Figure 18.11). This drug is an amino acid coupled to an adenosine analog. Thus, it resembles the aminoacyl adenosine at the end of an aminoacyl-tRNA. In fact, it looks enough like an aminoacyl-tRNA that it binds to the A site of a ribosome. Then it can form a peptide bond with the peptide in the P site, yielding a peptidyl puromycin. At this point the ruse is over. The peptidyl puromycin is not tightly bound to the ribosome and so is soon released, aborting translation prematurely. This is why puromycin kills bacteria and other cells.

The link between puromycin and the two-site model is this: Before translocation, because the A site is occupied by a peptidyl-tRNA, puromycin cannot bind and release the peptide; after translocation, the peptidyl-tRNA has moved to the P site, and the A site is open. At this point puromycin can bind and release the peptide. We therefore see two states the ribosome can assume: puromycin reactive and puromycin unreactive. Those two states require at least two binding sites on the ribosome for the peptidyl-tRNA.

Puromycin can be used to show whether an aminoacyl-tRNA is in the A or the P site. If it is in the P site, it can form a peptide bond with puromycin and be released. However, if it is in the A site, it prevents puromycin from binding to the ribosome and is not released.

This same procedure can be used to show that fMet-tRNA goes to the P site in the 70s initiation complex. In our discussion of initiation in Chapter 17, we assumed that the fMet-$tRNA_f^{Met}$ goes to the P site. This certainly makes sense, because it would leave the A site open for the second aminoacyl-tRNA. Using the puromycin assay, M.S. Bretscher and Marcker showed in 1966 that it does indeed go to the P site. They mixed [^{35}S]fMet-$tRNA_f^{Met}$ with ribosomes, the trinucleotide AUG, and puromycin. If AUG attracted fMet-$tRNA_f^{Met}$ to the P site, then the labeled fMet should have been able to react with puromycin, releasing

(a)

Tyrosyl-tRNA

Puromycin

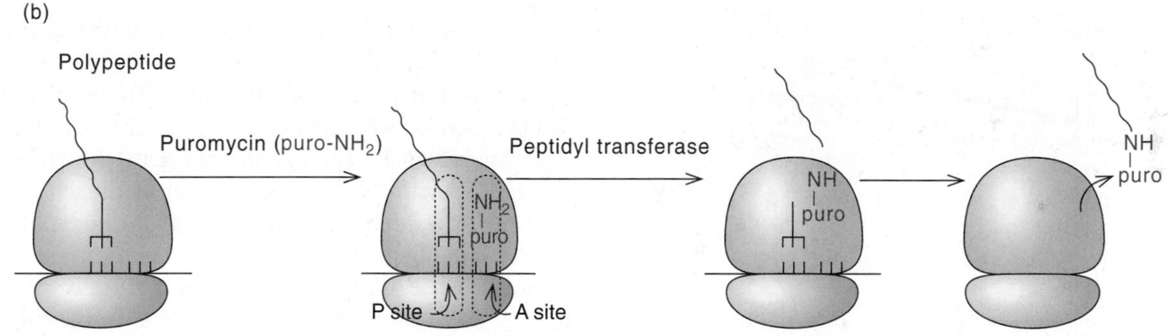

Figure 18.11 Puromycin structure and activity. (a) Comparison of structures of tyrosyl-tRNA and puromycin. Note the rest of the tRNA attached to the 5′-carbon in the aminoacyl-tRNA, where there is only a hydroxyl group in puromycin. The differences between puromycin and tyrosyl-tRNA are highlighted in magenta. **(b)** Mode of action of puromycin. First, puromycin (puro-NH_2) binds to the open A site on the ribosome. (The A site must be open for puromycin to bind.) Next, peptidyl transferase joins the peptide in the P site to the amino group of puromycin in the A site. Finally, the peptidyl-puromycin dissociates from the ribosome, terminating translation prematurely.

labeled fMet-puromycin. On the other hand, if the fMet-$tRNA_f^{Met}$ went to the A site, puromycin should not have been able to bind, so no release of labeled amino acid should have occurred. Figure 18.12 shows that the fMet attached to $tRNA_f^{Met}$ was indeed released by puromycin, whereas the methionine attached to $tRNA_m^{Met}$ was not. Thus, fMet-$tRNA_f^{Met}$ goes to the P site, but methionyl-$tRNA_m^{Met}$ goes to the A site. One could argue that it was the fMet, not the $tRNA_f^{Met}$ that made the difference in this experiment. To eliminate that possibility, Bretscher and Marcker performed the same experiment with Met-$tRNA_f^{Met}$ and found that its methionine was also released by puromycin (Figure 18.12c). Thus, the tRNA, not the formyl group on the methionine, is what targets the aminoacyl-tRNA to the P site.

Actually, x-ray crystallography studies in 2009 showed that fMet-$tRNA_f^{Met}$ does not automatically go to the P site. Instead, on its own, it goes first into a hybrid state called the **P/I state** in which the anticodon of the tRNA is in the P site of the 30S subunit, but the fMet and acceptor stem of the tRNA are not in the P site of the 50S subunit, which encompasses the peptidyl transferase center. Instead, the fMet and acceptor stem are in an "initiator" site to the left of the P site (toward the E site) as the ribosome is conventionally depicted (recall Figure 18.10). It is the job of a protein factor called EF-P to bind to the left of fMet-$tRNA_f^{Met}$ and nudge the fMet and acceptor stem to the right into the peptidyl transferase center. That action puts the fMet-$tRNA_f^{Met}$ fully in the P site.

In 1981, Knud Nierhaus and coworkers presented evidence for a third ribosomal site called the E site. Their experimental strategy was to bind radioactive deacylated $tRNA^{Phe}$ ($tRNA^{Phe}$ lacking phenylalanine), or Phe-$tRNA^{Phe}$, or acetyl-Phe-$tRNA^{Phe}$ to *E. coli* ribosomes and to measure the number of molecules bound per 70S ribosome. Table 18.2 shows the results of binding experiments carried out in the presence or absence of poly(U) mRNA. Only one molecule of acetyl-Phe-$tRNA^{Phe}$ could bind at a time to a ribosome, and the binding site could be either the A site or P site. On the other hand, two molecules of Phe-$tRNA^{Phe}$ could bind, one to the A site, and the other to the P site. Finally, three molecules of deacylated $tRNA^{Phe}$ could bind. We can explain these results most easily by postulating a third site that presumably binds deacylated tRNA on its way out of the ribosome. Hence the E, for exit. In the absence of mRNA, only one tRNA can bind. This can be either deacylated $tRNA^{Phe}$

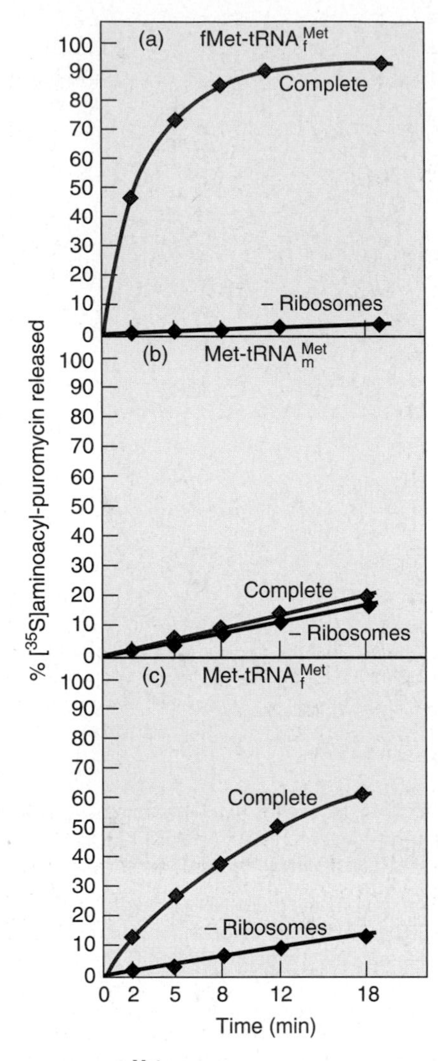

Figure 18.12 fMet-tRNA$_f^{Met}$ occupies the ribosomal P site. Bretscher and Marcker used a puromycin-release assay to determine the location of fMet-tRNA$_f^{Met}$ on the ribosome. They mixed ^{35}S-labeled fMet-tRNA$_f^{Met}$ **(a),** Met-tRNA$_m^{Met}$ **(b),** or Met-tRNA$_f^{Met}$ **(c)** with ribosomes, AUG, and puromycin, and tested for release of labeled fMet- or Met-puromycin by precipitating tRNA and protein with perchloric acid. Aminoacyl-puromycin released from the ribosome is acid-soluble, whereas aminoacyl-tRNA bound to the ribosome is acid-insoluble. The complete reactions contained all ingredients; control reactions lacked one ingredient, as indicated beside each curve. Met or fMet attached to tRNA$_f^{Met}$ went to the P site and was released. Met attached to tRNA$_m^{Met}$ stayed in the A site and was not released by puromycin. (*Source*: Adapted from Bretscher, M.S. and K.A. Marcker, Peptidyl-sRibonucleic acid and amino-acyl-sRibonucleic acid binding sites on ribosomes. *Nature* 211:382–83, 1966.)

or acetyl-Phe-tRNA^{Phe}. Nierhaus and colleagues speculated that the binding site was the P site, and subsequent work has confirmed this suspicion.

We will discuss the E site in greater detail in Chapter 19, but we should note at this point that the E site is not just a way station for deacylated tRNA on its way out of the ribosome. It plays a critical role in maintaining the reading frame of an mRNA. Ordinarily, reading frame shifts occur only about once in 30,000 codons, which is a good thing because such shifts generally give rise to meaningless proteins. But proper translation of some mRNAs actually depends on frameshifting.

Table 18.2 Binding of tRNAs and Aminoacyl-tRNAs to *E. coli* Ribosomes

	tRNA	Binding sites	
mRNA	**Species**	**No.**	**Location**
Poly(U)	Acetyl-Phe-tRNA^{Phe}	1	P or A
Poly(U)	Phe-tRNA^{Phe}	2	P and A
Poly(U)	tRNA^{Phe}	3	P, E, and A
None	tRNA^{Phe}	1	P
None	Phe-tRNA^{Phe}	0	—
None	Acetyl-Phe-tRNA^{Phe}	1	P

Source: Rheinberger, H.-J., H. Sternbach, and K.H. Nierhaus, Three tRNA binding sites on *Escherichia coli* ribosomes, *Proceedings of the National Academy of Sciences USA* 78(9):5310–14, September 1981. Reprinted with permission.

An example is the *E. coli prfB* gene, which encodes RF2, a release factor we will study later in this chapter. In order for the *prfB* mRNA to be translated correctly, a frameshift to the +1 reading frame must occur within the mRNA. Thus, the sequence CUUUGAC would normally be read: CUU UGA (Leu, Stop). But, with the +1 frameshift, it is read CUU*U*GAC (Leu, Asp). The italicized U is skipped, and the next codon is the underlined GAC, which encodes aspartate.

In 2004, Knud Nierhaus and colleagues examined translation of the *prfB* mRNA in vitro and found that the presence of a deacylated tRNA in the E site prevented this frameshift. When they removed the deacylated tRNA from the E site, the frameshift occurred with high frequency. Thus, they concluded that deacylated tRNA in the E site is normally required for the vital purpose of maintaining the proper reading frame. When frameshifting is required for proper translation of a particular mRNA, the cell must remove the deacylated tRNA from the E site.

SUMMARY Puromycin resembles an aminoacyl-tRNA and so can bind to the A site, couple with the peptide in the P site, and release it as peptidyl puromycin. On the other hand, if the peptidyl-tRNA is in the A site, puromycin will not bind to the ribosome, and the peptide will not be released. This defines two sites on the ribosome: a puromycin-reactive site (P), and a puromycin unreactive site (A). fMet-tRNA$_f^{Met}$ is puromycin reactive in the 70S initiation complex, so it is in the P site. Other studies have identified a third binding site (the E site) for deacylated tRNA. Such tRNAs presumably bind to the E site as they exit the ribosome, and this binding helps maintain the reading frame of the mRNA.

Elongation Step 1: Binding an Aminoacyl-tRNA to the A Site of the Ribosome

Our detailed understanding of the elongation process began in 1965 when Yasutomi Nishizuka and Fritz Lipmann used anion exchange chromatography to separate two protein factors required for peptide bond formation in *E. coli.* They named one factor T, for transfer, because it transfers aminoacyl-tRNAs to the ribosome. The second factor they called G because of its GTPase activity. (T also has GTPase activity, as we will see.) Then Jean Lucas-Lenard and Lipmann showed that T is actually composed of two different proteins, which they called Tu (where the **u** stands for unstable) and Ts (where the **s** stands for stable). These three factors, which we now call EF-Tu (or EF1A), **EF-Ts** (or EF1B), and EF-G (or EF2), participate in the first and third steps in elongation. (In eukaryotes, the roles of EF-Tu and EF-Ts are played by a three-subunit protein known as **EF1.** The EF1 α subunit performs the EF-Tu role, and the β and γ subunits perform the EF-Ts role. The EF-G role in eukaryotes is played by **EF2.**) Let us consider first the activities of EF-Tu and -Ts because they are involved in the first elongation step.

Joanne Ravel showed in 1967 that unfractionated EF-T (Tu plus Ts) had GTPase activity, and that EF-T required GTP to bind an aminoacyl-tRNA to the ribosome. To demonstrate this phenomenon, she made [^{14}C]Phe-tRNA^{Phe} and added it to washed ribosomes along with EF-T and an increasing concentration of GTP. Then she filtered the ribosomes through nitrocellulose. Labeled Phe-tRNA^{Phe} that bound to ribosomes stuck to the filter, but unbound Phe-tRNA^{Phe} washed through. Figure 18.13a depicts the results. Background nonenzymatic binding of the Phe-tRNA^{Phe} to the ribosomes was rather high in the absence of EF-T and GTP, but this was not physiologically significant. Ignoring that background, we can see that GTP was necessary for EF-T-dependent binding of Phe-tRNA^{Phe} to the ribosomes.

When Ravel added both EF-T and EF-G to washed ribosomes in the presence of poly(U) and labeled Phe-tRNA^{Phe} she found that the ribosomes made labeled polyphenylalanine. And this polymerization of amino acids required an even higher concentration of GTP than the aminoacyl-tRNA-binding reaction did.

When we examined initiation of translation, we learned that IF-2-mediated binding of fMet-tRNA$_f^{Met}$ to ribosomes also required GTP, but that GTP hydrolysis was not required. Could the same be true of EF-T and binding of ordinary aminoacyl-tRNAs to ribosomes? Anne-Lise Haenni and Lucas-Lenard showed in 1968 that this is indeed the case. They labeled *N*-acetyl-Phe-tRNA with ^{14}C and Phe-tRNA with ^{3}H. Then they mixed these labeled aminoacyl-tRNAs with EF-T and either GTP or the unhydrolyzable analog, GDPCP. Under the non-physiological conditions of this experiment, the *N*-acetyl-Phe-tRNA^{Phe} went to the P site. These workers measured binding of aminoacyl-tRNAs to ribosomes by filter binding, as described in Figure 18.13. They also measured

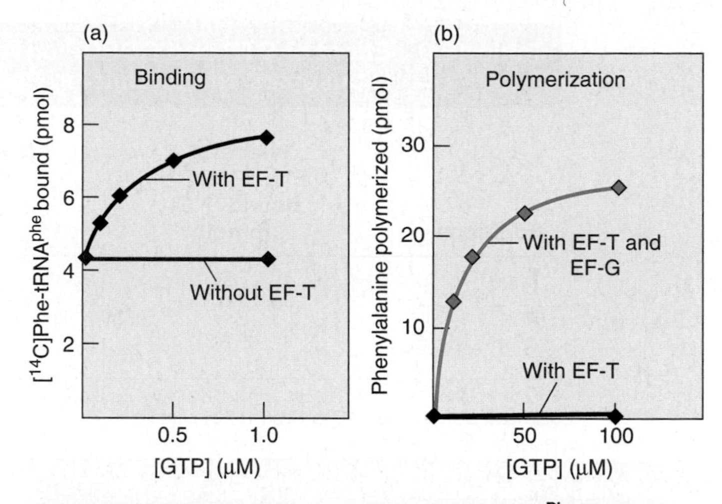

Figure 18.13 Effects of EF-T and GTP on Phe-tRNA^{Phe} binding to ribosomes and on poly-Phe synthesis. (a) Binding Phe-tRNA^{Phe} to ribosomes. Ravel mixed ^{14}C-Phe-tRNA^{Phe} with washed ribosomes and various concentrations of GTP in the presence or absence of EF-T. She measured Phe-tRNA^{Phe}–ribosome binding by filtering the mixture and determining the labeled Phe bound to the ribosomes on the filter. Considerable nonenzymatic binding occurred in the absence of EF-T and GTP, but the EF-T-dependent binding required GTP. **(b)** Polymerization of phenylalanine. Ravel mixed labeled Phe-tRNA^{Phe} with ribosomes, EF-T, and various concentrations of GTP in the presence and absence of EF-G. She measured polymerization of Phe by acid precipitation as follows: She precipitated the poly(Phe) with trichloroacetic acid (TCA), heated the precipitate in the presence of TCA to hydrolyze any phe-tRNA^{Phe}, and trapped the precipitated poly(Phe) on filters. Polymerization required both EF-T and EF-G and a high concentration of GTP. (*Source:* Adapted from Ravel, J.M., Demonstration of a guanosine triphosphate-dependent enzymatic binding of aminoacyl-ribonucleic acid to *Escherichia coli* ribosomes. *Proceedings of the National Academy of Sciences USA* 57:1815, 1967.)

peptide bond formation between the *N*-acetyl-Phe in the P site and the Phe-tRNA^{Phe} in the A site by extracting the dipeptide product and identifying it by paper electrophoresis. Table 18.3 shows that *N*-acetyl-Phe-tRNA^{Phe} could bind to the P site, and that Phe-tRNA^{Phe} could bind to the A site, with the help of EF-T and either GTP or GDPCP. (In fact, *N*-acetyl-Phe-tRNA^{Phe} did not even need EF-T to bind to the P site.) Thus, GTP hydrolysis is not needed for EF-T to promote aminoacyl-tRNA binding to the ribosomal A site. In marked contrast, formation of the peptide bond between *N*-acetyl-Phe and Phe-tRNA^{Phe} required GTP hydrolysis. This is analogous to the situation in initiation, where IF-2 can bind fMet-tRNA$_f^{Met}$ to the P site without GTP hydrolysis, but subsequent events are blocked until GTP is hydrolyzed.

These same scientists also demonstrated that *both* EF-Tu and EF-Ts are required for Phe-tRNA^{Phe} binding to the ribosome. The assay was the same as in Table 18.3, except that no GDPCP was used and that EF-Tu and EF-Ts were separated from each other (except for some residual contamination of the EF-Tu fraction with EF-Ts), and added separately. Table 18.4 shows that both EF-Tu and -Ts are required for Phe-tRNA^{Phe}-ribosome binding. The small amount of binding seen with EF-Tu alone resulted from contamination of the factor by EF-Ts.

Table 18.3 Effect of GTP and GDPCP on Aminoacyl-tRNA Binding to Ribosomes and on Binding Plus Peptide Bond Formation

Additions	*N*-acetyl-Phe-$tRNA^{Phe}$ bound (^{14}C) (pmol)	*N*-acetyl diPhe-tRNA formed (^{14}C or ^{3}H) (pmol)	Phe-tRNA bound (^{3}H) (pmol)
None	7.6	0.4	0.1
EF-T + GTP	3.0	4.5	2.8
EF-T + GDPCP	7.0	0.5	4.8

Source: Haenni, A.L. and J. Lucas-Lenard, Stepwise synthesis of a tripeptide, *Proceedings of the National Academy of Sciences, USA* 61:1365, 1968. Reprinted by permission.

Table 18.4 Requirement for Both EF-Ts and EF-Tu to Bind [^{3}H]Phe-tRNA to Ribosomes Carrying Prebound *N*-acetyl-[^{14}C]Phe-tRNA

Additions	[^{3}H]Phe-tRNA bound (pmol)
None	2.8
EF-Ts + GTP	2.8
EF-Tu + GTP	5.2
EF-Ts + EF-Tu + GTP	11.6

Source: Naenni, A.L., and J. Lucas-Lenard, Stepwise synthesis of a tripeptide, *Proceedings of the National Academy of Sciences USA* 61:1365, 1968. Reprinted with permission.

Figure 18.14 presents a model for the detailed mechanism by which EF-Tu and EF-Ts cooperate to cause transfer of aminoacyl-tRNAs to the ribosome. First, EF-Tu and GTP form a binary (two-part) complex. Then aminoacyl-tRNA joins the complex, forming a ternary (three-part) complex composed of EF-Tu, GTP, and aminoacyl-tRNA. This ternary complex then delivers its aminoacyl-tRNA to the ribosome's A site. EF-Tu and GTP remain bound to the ribosome. Next, GTP is hydrolyzed and an EF-Tu–GDP complex dissociates from the ribosome. Finally, EF-Ts exchanges GTP for GDP on the complex, yielding an EF-Tu–GTP complex.

What is the evidence for this scheme? Herbert Weissbach and colleagues found in 1967 that an EF-T preparation and GTP could form a complex that was retained by a nitrocellulose filter. They labeled GTP, mixed it with EF-T, and found that the labeled nucleotide bound to the filter. This meant that GTP had bound to a protein in the EF-T preparation,

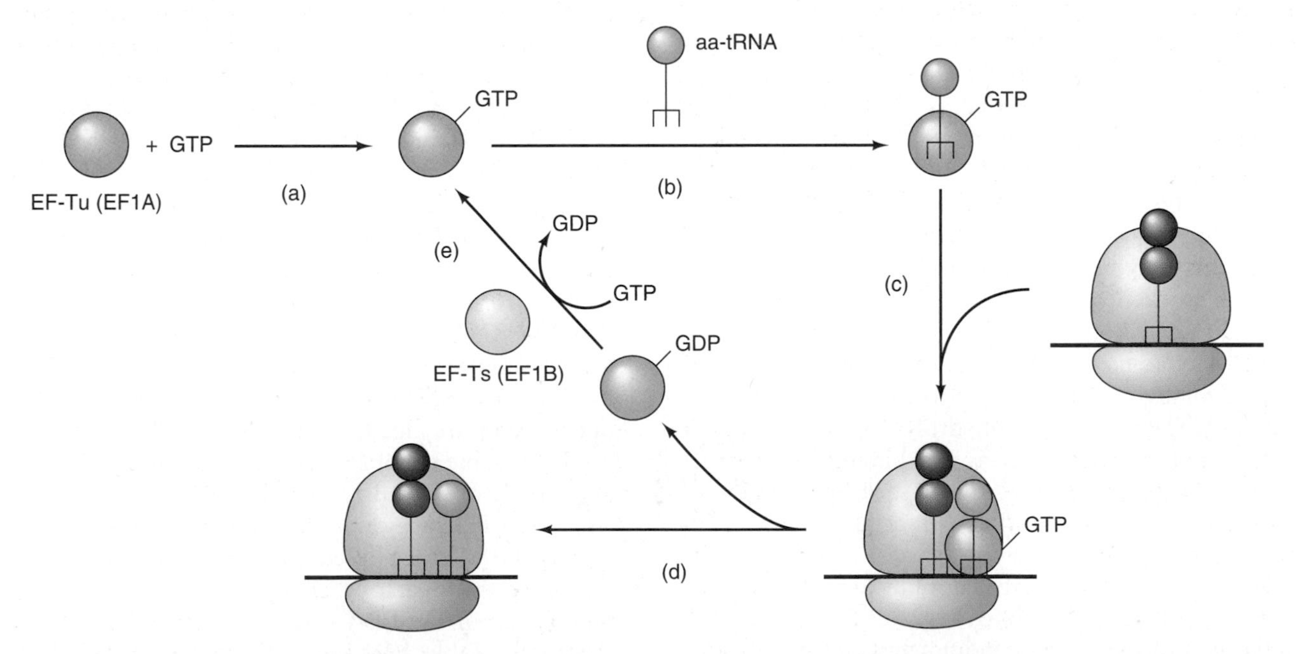

Figure 18.14 Model of binding aminoacyl-tRNAs to the ribosome A site. **(a)** EF-Tu couples with GTP to form a binary complex. **(b)** This complex associates with an aminoacyl-tRNA to form a ternary complex. **(c)** The ternary complex binds to a ribosome with a peptidyl-tRNA in its P site and an empty A site. **(d)** GTP is hydrolyzed and the resulting EF-Tu–GDP complex dissociates from the ribosome, leaving the new aminoacyl-tRNA in the A site. **(e)** EF-Ts exchanges GTP for GDP on EF-Tu, regenerating the EF-Tu–GTP complex.

presumably EF-T itself, to form a complex. Julian Gordon then discovered that adding an aminoacyl-tRNA to the EF-Tu–GTP complex caused the complex to be released from the filter. One interpretation of this behavior is that the aminoacyl-tRNA joined the EF-Tu–GTP complex to form a ternary complex that could no longer bind to the filter.

Ravel and her collaborators gave us additional evidence for the formation of the ternary complex with the following experiment. They labeled GTP with ^{3}H and ^{32}P, and Phe-$tRNA^{Phe}$ with ^{14}C, and mixed them with EF-T, then subjected the mixture to gel filtration on Sephadex G100 (Chapter 5). This gel filtration resin excludes relatively large proteins, such as EF-T, so they flow through rapidly in a fraction called the void volume. By contrast, relatively small substances like GTP, and even Phe-$tRNA^{Phe}$, enter the pores in the resin and are thereby retarded; they emerge later from the column, after the void volume. In fact, the smaller the molecule, the longer it takes to elute from the column. Figure 18.15 shows the results of this gel filtration experiment. A fraction of both labeled substances, GTP and Phe-$tRNA^{Phe}$, emerged relatively late, in their usual positions. These fractions represented free GTP and Phe-$tRNA^{Phe}$, although very little free GTP was observed. However, significant fractions of both substances eluted much earlier, around fraction 20, demonstrating that they must be complexed to something larger. The predominant larger substance in this experiment was EF-T, and the experiments we have already discussed implicate EF-T in this complex, so we infer that a ternary complex, involving Phe-$tRNA^{Phe}$, GTP, and EF-T has formed.

So far, we have not distinguished between EF-Ts and EF-Tu in these experiments. Herbert Weissbach and his collaborators did this by separating the two proteins and testing them separately. They found that EF-Tu is the factor that binds GTP in the binary complex. What then is the role of EF-Ts? These investigators demonstrated that this factor is essential for conversion of the EF-Tu–GDP complex to the EF-Tu–GTP complex. However, EF-Ts has little, if any, effect when it is presented with the pre-formed EF-Tu–GTP complex or with EF-Tu itself (Figure 18.16).

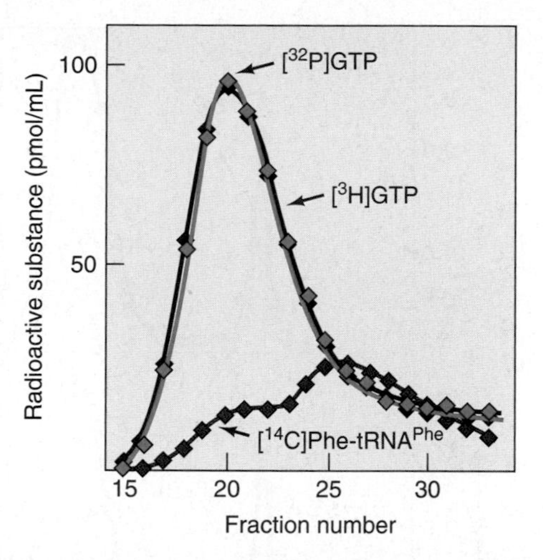

Figure 18.15 Formation of a ternary complex among EF-T, aminoacyl-tRNA, and GTP. Ravel and colleagues mixed [^{14}C]Phe-$tRNA^{Phe}$ with GTP (labeled in the guanine part with ^{3}H and in the γ-phosphate with ^{32}P), and EF-T. Then they passed the mixture through a Sephadex G100 gel filtration column to separate large molecules, such as EF-T, from relatively small molecules such as GTP and Phe-$tRNA^{Phe}$. They assayed each fraction for the three radioisotopes to detect GTP and Phe-$tRNA^{Phe}$. Both of these substances were found at least partly in a large-molecule fraction (around fraction 20), so they were bound to the EF-T in a complex. (*Source:* Adapted from Ravel, J.M., R.L. Shorey, and W. Shive, The composition of the active intermediate in the transfer of aminoacyl-RNA to ribosomes. *Biochemical and Biophysical Research Communications* 32:12, 1968.)

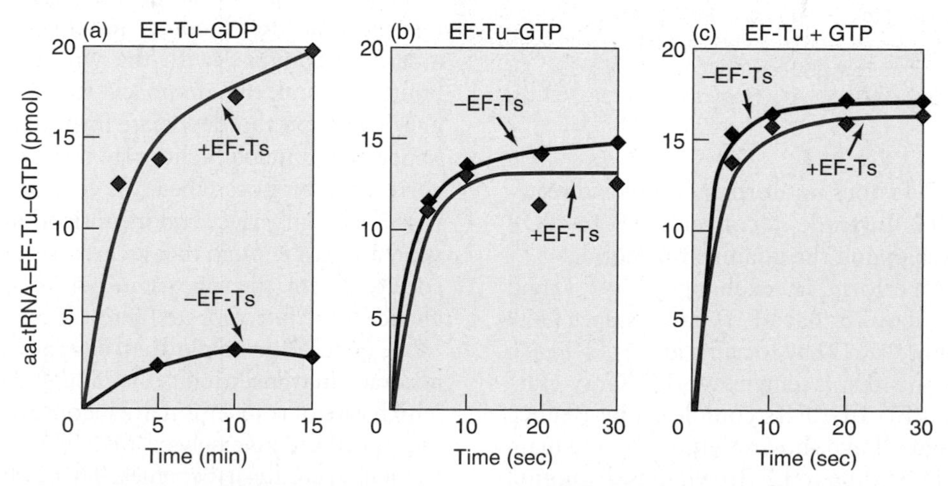

Figure 18.16 Effect of EF-Ts on ternary complex formation. Weissbach and colleagues attempted to form the ternary complex with [^{14}C]Phe-tRNA, [^{3}H]GTP, and the EF-Tu preparations listed at top, with (red) and without (blue) EF-Ts. They measured ternary complex formation by loss of radioactivity trapped by nitrocellulose filtration. EF-Ts stimulated complex formation only when EF-Tu–GDP was the substrate (panel **a**). EF-Tu–GTP (panel **b**) or EF-Tu+GTP (panel **c**) could form the complex spontaneously, with no help from EF-Ts. (aa-tRNA = aminoacyl-tRNA). (*Source:* Adapted from Weissbach, H., D.L. Miller, and J. Hachmann, Studies on the role of factor Ts in polypeptide synthesis. *Archives of Biochemistry and Biophysics*, 137:267, 1970.)

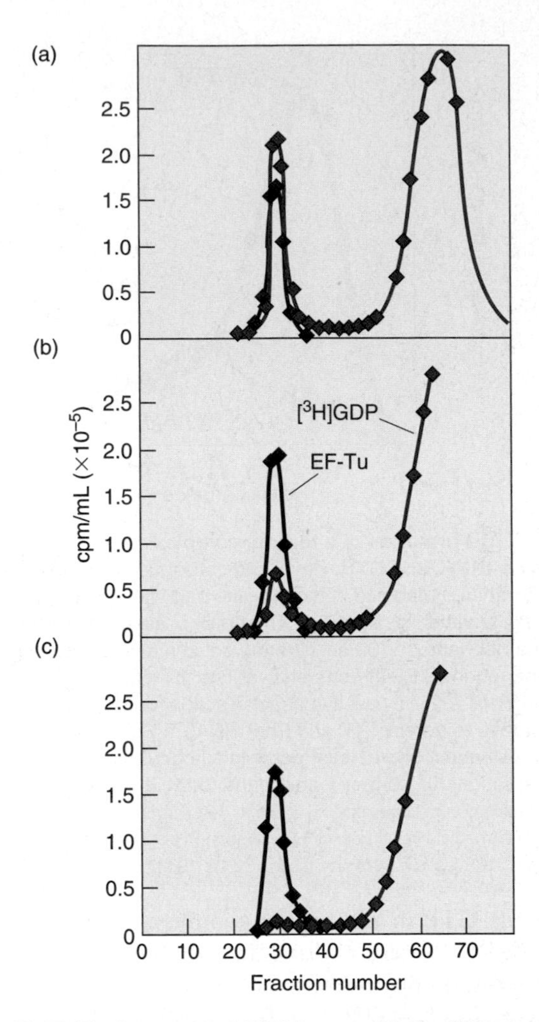

Figure 18.17 Displacement of GDP from an EF-Tu–GDP complex by EF-Ts. Miller and Weissbach mixed an EF-Tu–[^{3}H]GDP complex with three different amounts of EF-Ts, then detected the amount of GDP remaining in the complex by gel filtration through Sephadex G-25. The three panels contained the following amounts of EF-Ts: **(a),** 500 units; **(b),** 14,000 units; **(c),** 25,000 units. Red, [^{3}H]GDP; blue, EF-Tu. (*Source*: Adapted from Miller, D.L. and H. Weissbach, Interactions between the elongation factors: The displacement of GDP from the Tu-GDP complex by factor Ts. *Biochemical and Biophysical Research Communications* 38:1019, 1970.)

Thus, it seems that EF-Ts does not form a complex directly from EF-Tu and GTP. Instead, it converts EF-Tu–GDP to EF-Tu–GTP by exchanging the guanine nucleotide.

How does EF-Ts perform its exchange duty? David Miller and Weissbach showed that EF-Ts can displace GDP from EF-Tu–GDP (Figure 18.17) by forming an EF-Ts–EF-Tu complex. How does this displacement work? X-ray crystallography studies on EF-Tu–EF-Ts complexes by Reuben Leberman and colleagues have shown that one of the main consequences of EF-Ts binding to EF-Tu–GDP is disruption of the Mg^{2+}-binding center of EF-Tu. The weakened binding between EF-Tu and Mg^{2+} leads to dissociation of GDP, which opens the way for binding of GTP to EF-Tu.

Why is EF-Tu needed to escort aminoacyl-tRNAs to the ribosome? The ester bond joining the amino acid to its cognate tRNA is easily broken, and sequestering the aminoacyl-tRNA within the EF-Tu protein protects this labile compound from hydrolysis. But the concentration of aminoacyl-tRNAs in the cell is quite high. Is there enough EF-Tu to go around? Yes, because EF-Tu is one of the most abundant proteins in the cell. For example, EF-Tu constitutes 5% of the total protein in *E. coli* cells, and the reason for this abundance appears to be the important protective role that EF-Tu plays.

SUMMARY A ternary complex formed from EF-Tu, aminoacyl-tRNA, and GTP delivers an aminoacyl-tRNA to the ribosome's A site, without hydrolysis of the GTP. In the next step, EF-Tu hydrolyzes GTP with its ribosome-dependent GTPase activity, and an EF-Tu–GDP complex dissociates from the ribosome. EF-Ts regenerates an EF-Tu–GTP complex by exchanging GTP for GDP attached to EF-Tu. Addition of aminoacyl-tRNA then reconstitutes the ternary complex for another round of translation elongation.

Proofreading As we will see in Chapter 19, part of the accuracy of protein synthesis comes from charging of tRNAs with the correct amino acids. But part also comes in elongation step 1: The ribosome usually binds the aminoacyl-tRNA called for by the codon in the A site. However, if it makes a mistake in this initial recognition step, it still has a chance to correct it by rejecting an incorrect aminoacyl-tRNA before it can donate its amino acid to the growing polypeptide. This process is called **proofreading.**

Proofreading can occur at two steps within step 1 of elongation: First, the ternary complex can dissociate from the ribosome after binding, and this happens more readily if a ternary complex with the wrong aminoacyl-tRNA has bound. Second, the aminoacyl-tRNA (derived from the ternary complex) can dissociate from the ribosome. Again, this happens at a much higher rate if the aminoacyl-tRNA is incorrect than it does when it is correct, because of the weakness of the imperfect codon–anticodon base pairing. This is generally fast enough that an incorrect aminoacyl-tRNA dissociates from the ribosome before its amino acid has a chance to be incorporated into the nascent polypeptide.

A general principle that emerges from the analysis of accuracy in translation is that a high degree of accuracy and a high rate of translation are incompatible. In fact, accuracy and speed are inversely related: The faster translation goes, the less accurate it becomes. This is because the ribosome must allow enough time for incorrect ternary complexes and aminoacyl-tRNAs to leave before the incorrect amino acid is irreversibly incorporated into the growing polypeptide. If translation goes faster, more incorrect amino acids will be incorporated. Conversely, if translation goes more slowly, accuracy will be higher, but then proteins may not be made fast enough to sustain life. So there is a delicate balance between speed and accuracy of translation.

One of the most important factors in this balance is the rate of hydrolysis of GTP by EF-Tu. If the rate were higher, less time would be available for the first proofreading step: EF-Tu would hydrolyze GTP to GDP quickly without giving sufficient time for ternary complexes bearing improper aminoacyl-tRNAs to dissociate from the ribosome. On the other hand, if the rate were lower, there would be ample time for proofreading, but translation would be too slow. What is the proper rate? In *E. coli,* the average time between binding of the ternary complex and hydrolysis of GTP is several milliseconds. Then it takes several milliseconds more for EF-Tu–GDP to dissociate from the ribosome. Proofreading takes place during both of these pauses, and shortening one or both could be devastating to accuracy of translation.

How large an error rate in translation can a cell tolerate? What if it were 1%, for example? Ninety-nine percent accuracy sounds pretty good until you consider that the lengths of most polypeptides are much more than 100 amino acids. They average about 300 amino acids long, and some are more than 1000 amino acids long. The probability p of producing an error-free polypeptide, given an error rate per amino acid (ε) and a polypeptide length (n) is given by the following expression:

$$p = (1 - \varepsilon)^n$$

For example, with an error rate of 1%, an average-size polypeptide would be produced error-free only about 5% of the time, and a 1000-amino-acid polypeptide would almost never be error-free. With a 10-fold better error rate, 0.1%, an average-size polypeptide would be produced error-free about 74% of the time, but a 1000-amino-acid polypeptide would be made error-free only about 37% of the time. This would still pose a problem for large polypeptides. What if the error rate were only 0.01%? At that rate, about 97% of average-size polypeptides, and about 91% of 1000-amino-acid polypeptides would be produced error-free. That seems like an acceptable rate of production of defective proteins, and the observed error rate per amino acid added, at least in *E. coli,* is in fact close to 0.01%.

An important antibiotic known as **streptomycin** interferes with proofreading so the ribosome makes more mistakes. For example, normal ribosomes incorporate phenylalanine almost exclusively in response to the synthetic message poly(U). But streptomycin greatly stimulates the incorporation of isoleucine and, to a lesser extent, serine and leucine in response to poly(U).

Certain natural conditions allow us to see what happens when the rate of translation is either faster or slower than normal. For example, mutants in ribosomal proteins, such as *ram,* or in EF-Tu, such as *tufAr,* double the rate of peptide bond formation. In these mutants, accuracy of translation suffers because not enough time is available for incorrect aminoacyl-tRNAs to dissociate from the ribosome.

By contrast, in streptomycin-resistant mutants such as *strA,* the rate of peptide bond formation is only half the normal value. This allows extra time for incorrect aminoacyl-tRNAs to leave the ribosome, so translation is extra accurate.

SUMMARY The protein-synthesizing machinery achieves accuracy during elongation in a two-step process. First, it gets rid of ternary complexes bearing the wrong aminoacyl-tRNA before GTP hydrolysis occurs. If this screen fails, it can still eliminate the incorrect aminoacyl-tRNA in the proofreading step before the wrong amino acid can be incorporated into the growing protein chain. Presumably, both these screens rely on the weakness of incorrect codon–anticodon base pairing to ensure that dissociation will occur more rapidly than either GTP hydrolysis or peptide bond formation. The balance between speed and accuracy of translation is delicate. If peptide bond formation goes too fast, incorrect aminoacyl-tRNAs do not have enough time to leave the ribosome, so their amino acids are incorporated into protein. But if translation goes too slowly, proteins are not made fast enough for the organism to grow successfully. The actual error rate, about 0.01% per amino acid added, strikes a good balance between speed and accuracy.

Elongation Step 2: Peptide Bond Formation

After the initiation factors and EF-Tu have done their jobs, the ribosome has fMet-$tRNA_f^{Met}$ in the P site and an aminoacyl-tRNA in the A site. Now it is time to form the first peptide bond. You might be expecting a new group of elongation factors to participate in this event, but there are none. Instead, the ribosome itself contains the enzymatic activity, called **peptidyl transferase,** that forms peptide bonds. No soluble factors are needed.

The peptidyl transferase step in prokaryotes is inhibited by an important antibiotic called **chloramphenicol.** This drug has no effect on most eukaryotic ribosomes, which makes it selective for bacterial invaders in higher organisms. However, the mitochondria of eukaryotes have their own ribosomes, and chloramphenicol does inhibit their peptidyl transferase. Thus, chloramphenicol's selectivity for bacteria is not absolute.

The classic assay for peptidyl transferase was invented by Robert Traut and Robert Monro and uses a labeled aminoacyl-tRNA or peptidyl-tRNA bound to the ribosomal P site, and puromycin. The release of labeled aminoacyl- or

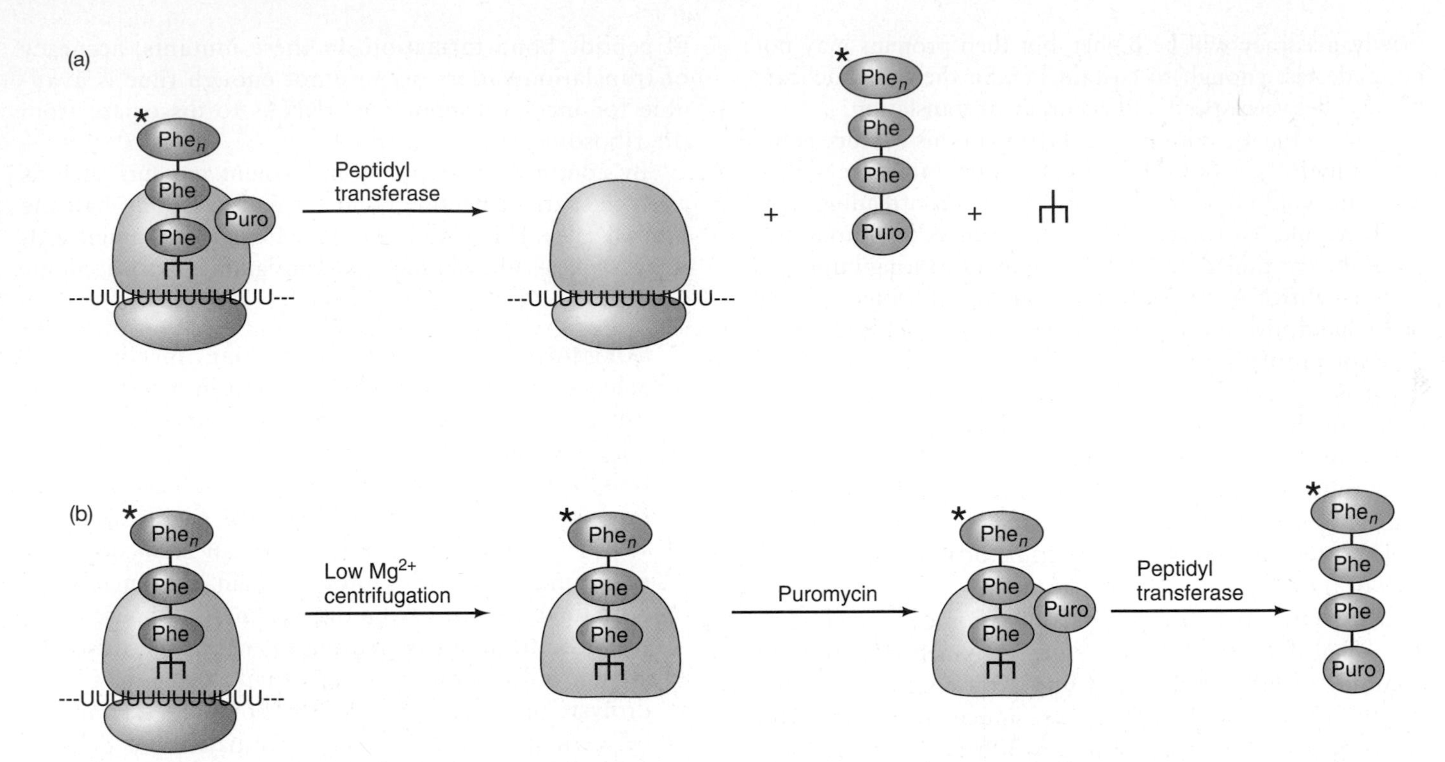

Figure 18.18 The puromycin reaction as an assay for peptidyl transferase. (a) Standard puromycin reaction. Add labeled poly(Phe)–tRNA into the P site by running a translation reaction with poly(U) as messenger. Then add puromycin. When a peptide bond forms between the labeled poly(Phe) and puromycin, the labeled peptidyl-puromycin is released. **(b)** Reaction with 50S subunits only. Again, add labeled poly(Phe)-tRNA into the ribosome's P site, incubate in a low Mg^{2+} buffer, and then centrifuge to separate the 50S-poly(Phe)–tRNA complex from the 30S subunit and the mRNA. Then add puromycin and detect peptidyl transferase by the release of labeled peptidyl-puromycin. The by-products of the reaction (50S subunits and tRNA) are not pictured. The asterisks denote the label in the poly(Phe).

peptidyl-puromycin depends on forming a peptide bond between the amino acid or peptide in the P site, and puromycin in the A site, as depicted in Figure 18.18a. Traut and Monro also discovered that this system could be modified somewhat to show that the 50S ribosomal subunit, without any help from the 30S subunit or soluble factors, could carry out the peptidyl transferase reaction (Figure 18.18b). First, they allowed ribosomes to carry out poly(Phe) synthesis, using poly(U) as mRNA. This placed labeled poly(Phe)-tRNA in the P site. Then they removed the 30S subunits by incubation with buffer having a low Mg^{2+} concentration, followed by ultracentrifugation. Then they washed away any remaining initiation or elongation factors with salt solutions, leaving the 50S subunits bound to poly(Phe)-tRNA. Ordinarily, such primed 50S subunits would be unreactive with puromycin, but these workers found that they could elicit puromycin reactivity with 33% methanol (ethanol also worked). In both assays, one must distinguish the released peptidyl-tRNA from the peptidyl-tRNA still bound to ribosomes. Traut and Monro originally accomplished this by sucrose gradient centrifugation, as shown in Figure 18.19. Later, a more convenient filter-binding assay was developed. Figure 18.19a is a negative control with no puromycin, and the poly(Phe) remained bound to the 50S subunit, as expected. Figure 18.19b is a positive control in which the poly(Phe) was released by destroying the ribosomes with urea and RNase. Figure 18.19c and 18.19d show the experimental results with puromycin plus and minus GTP, respectively. Peptidyl transferase appeared to be working, since puromycin could release the poly(Phe). This reaction occurred even in the absence of GTP, as the peptidyl transferase reaction should.

The puromycin reaction with 50S subunits seems to demonstrate that the 50S subunit contains the peptidyl transferase activity, but could the rather unphysiological conditions (33% methanol, puromycin) be distorting the picture? One encouraging sign is that the reaction of a peptide with puromycin seems to follow the same mechanism as normal peptide synthesis. Also, M.A. Gottesmann substituted poly(A) for poly(U), and therefore poly(Lys) for poly(Phe), and also substituted lysyl-tRNA for puromycin, and found the same kind of reaction, demonstrating that the puromycin reaction is a valid model for peptide bond formation. Furthermore, these reactions are all blocked by chloramphenicol and other antibiotics that inhibit the normal peptidyl transferase reaction, suggesting that the model reactions use the same pathway as the normal one.

For decades, no one knew what part of the 50S subunit had the peptidyl transferase activity. However, as soon as Thomas Cech and coworkers demonstrated in the early

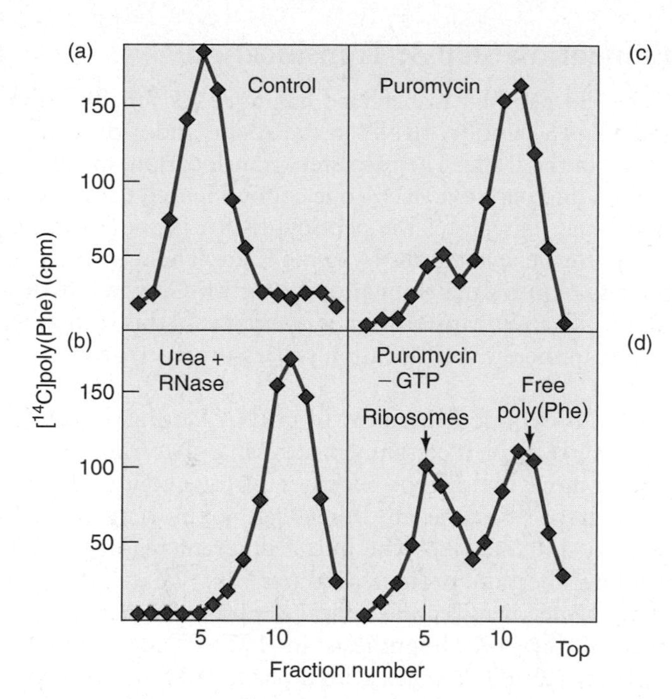

Figure 18.19 Puromycin assay for peptide bond formation. Traut and Monro loaded ribosomes with [^{14}C]polyphenylalanine, incubated them with or without puromycin, then subjected the products to sucrose gradient centrifugation to separate ribosome-bound poly(Phe) from free poly(Phe) that had been released from the ribosomes. The poly(Phe)-loaded ribosomes were treated as follows: **(a)** no treatment; **(b)** treated with urea and RNase; **(c)** treated with puromycin; **(d)** treated with puromycin in the absence of GTP. The positions of ribosomes and free poly(Phe) are indicated in (d). (*Source*: Adapted from Traut, R.R. and R.E. Monro, The puromycin reaction and its relation to protein synthesis. *Journal of Molecular Biology,* 10:63–72, 1964.)

1980s that some RNAs have catalytic activity, some molecular biologists began to suspect that the 23S rRNA might actually catalyze the peptidyl transferase reaction. In 1992, Harry Noller and his coworkers presented evidence that this is so. As their assay for peptidyl transferase, they used a modification of the puromycin reaction called the **fragment reaction.** This procedure, pioneered by Monro in the 1960s, uses a fragment of labeled fMet-tRNA$_f^{Met}$ in the P site and puromycin in the A site. The fragment can be CCA-fMet, or CAACCA-fMet. Either one resembles the whole fMet-tRNA$_f^{Met}$ enough that it can bind to the P site. Then the labeled fMet can react with puromycin to release labeled fMet-puromycin.

The task facing Noller and collaborators was to show that they could remove all the protein from 50S particles, leaving only the rRNA, and that this rRNA could catalyze the fragment reaction. To remove the protein from the rRNA, these workers treated 50S subunits with three harsh agents known for their ability to denature or degrade protein: phenol, SDS, and proteinase K (PK). Figure 18.20, lanes 1–4, shows that the peptidyl transferase activity of *E. coli* 50S subunits survived SDS and proteinase K treatment, but not extraction with phenol. The ability to withstand SDS and PK was impressive, but it leaves us wondering

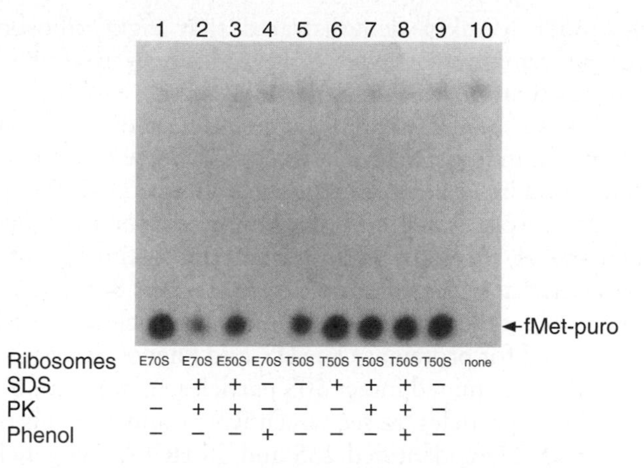

Figure 18.20 Effects of protein-removing reagents on peptidyl transferase activities of *E. coli* and *Thermus aquaticus* ribosomes. Noller and collaborators treated ribosomes with SDS, proteinase K (PK), or phenol, or combinations of these treatments, as indicated at bottom. Then they tested the treated ribosomes for peptidyl transferase by the fragment reaction using CAACCA-f[^{35}S] Met. They isolated f[^{35}S]Met-puromycin by high-voltage paper electrophoresis and detected it by autoradiography. The ribosome source is listed at bottom: E70S and E50S are 70S ribosomes and 50S ribosomal subunits, respectively, from *E. coli;* T50S refers to 50S ribosomal subunits from *Thermus aquaticus.* The position of fMet-puromycin is indicated at right. (*Source:* Adapted from Noller, H.F., V. Hoffarth, and L. Zimniak, Unusual resistance of peptidyl transferase to protein extraction procedures. *Science* 256 (1992) p. 1417, f. 2.)

why phenol extraction would disrupt the peptidyl transferase any more than the other two agents.

Noller and colleagues reasoned that phenol might be disrupting some higher-order RNA structure that is essential for peptidyl transferase activity. If so, they postulated that the rRNA from a thermophilic bacterium might be more sturdy and therefore might keep its native structure even after phenol extraction. To test this hypothesis, they tried the same experiment with 50S subunits from a thermophilic bacterium, *Thermus aquaticus,* that inhabits scalding hot springs. Lanes 5–9 of Figure 18.20 demonstrate that the peptidyl transferase activity of *T. aquaticus* 50S subunits survives treatment with all three of these agents.

If the fragment activity really represents peptidyl transferase, it should be blocked by peptidyl transferase inhibitors like chloramphenicol and carbomycin. Furthermore, if rRNA is a key factor in peptidyl transferase, then the fragment reaction should be inhibited by RNase. Noller and colleagues verified both of these predictions. The fragment reactions carried out by either intact or treated *T. aquaticus* 50S subunits are inhibited by carbomycin, chloramphenicol, and RNase, just as they should be.

Do these experiments show that ribosomal RNA is the only component of peptidyl transferase? Noller and coworkers stopped short of that conclusion, in part because they could not eliminate all protein from their preparations, even after vigorous treatment with protein-destroying agents. In fact, their subsequent work in collaboration with

Alexander Mankin demonstrated that eight ribosomal proteins remained associated with rRNA even after such vigorous treatment.

Mankin, Noller, and colleagues subjected *T. aquaticus* 50S ribosomal particles to the same protein-destroying agents used in Noller's original experiments. Then they performed sucrose gradient ultracentrifugation on the remaining material, and found that the material retaining peptidyl transferase activity sedimented as 50S and 80S particles, which they called *KSP50* and *KSP80* particles. The K, S, and P stand for proteinase K, SDS, and Phenol, respectively. Next, they examined intact 50S particles, as well as KSP50 and KSP80 particles, to see which RNAs and proteins they contained. They identified 23S and 5S rRNAs by gel electrophoresis. To separate and identify the protein in these particles, they used two-dimensional electrophoresis (Chapter 5). Amazingly, eight proteins remained more-or-less intact, and four of them (L2, L3, L13, and L22) were present in near-stoichiometric quantities. The other four (L15, L17, L18, and L21) were reduced in quantity. Mankin, Noller, and colleagues double-checked the identities of these eight proteins by sequencing N-terminal peptides derived from each one. Because identical proteins and RNAs appeared in both particles, it is likely that the KSP80 particles are simply dimers of KSP50 particles.

Earlier studies on reconstitution of peptidyl transferase from purified components had shown that peptidyl transferase activity could be reconstituted from just 23S rRNA and proteins L2, L3, and L4. Of these, only L4 was missing from the KSP particles. Thus, considering the reconstitution data together with the KSP particle data, Mankin, Noller, and colleagues concluded that the minimum components necessary for peptidyl transferase activity are 23S rRNA and proteins L2 and L3.

What role does 23S rRNA play in the peptidyl transferase activity? It is tempting to speculate that it has a catalytic role, but we cannot reach that conclusion based on the data presented so far. However, in 2000 Thomas Steitz and his colleagues performed x-ray crystallography studies on 50S ribosomal particles and they found no proteins—only 23S rRNA—near the peptidyl transferase active center. So it appears that 23S rRNA really does have the peptidyl transferase catalytic activity. We will examine this subject in detail in Chapter 19.

SUMMARY Peptide bonds are formed by a ribosomal enzyme called peptidyl transferase, which resides on the 50S ribosomal particle. The minimum components necessary for peptidyl transferase activity in vitro are 23S rRNA and proteins L2 and L3. X-ray crystallography studies show that 23S rRNA is at the catalytic center of peptidyl transferase and therefore appears to have peptidyl transferase activity in vivo.

Elongation Step 3: Translocation

Once the peptidyl transferase has done its job, the ribosome has a peptidyl-tRNA in the A site and a deacylated tRNA in the P site. The next step, translocation, moves the mRNA and peptidyl-tRNA one codon's length through the ribosome. This places the peptidyl-tRNA in the P site and moves the deacylated tRNA to the E site. The translocation process requires the elongation factor EF-G, which hydrolyzes GTP after translocation is complete. In this section we will examine the translocation process in more detail.

Three-Nucleotide Movement of mRNA During Translocation First of all, it certainly makes sense that translocation should move the mRNA exactly 3 nt (one codon's length) through the ribosome; any other length of movement would tend to shift the ribosome into a different reading frame, yielding aberrant protein products. But what is the evidence? Peter Lengyel and colleagues provided data in support of the 3-nt hypothesis in 1971. They created a pretranslocation complex with a phage mRNA, ribosomes, and aminoacyl-tRNAs, but left out EF-G and GTP to prevent translocation. Then they made a posttranslocation complex by adding EF-G and GTP. They treated each of these complexes with pancreatic ribonuclease to digest any mRNA not protected by the ribosome, then released the protected RNA fragment and sequenced it. They found that the sequence of the 3′-end of the fragment was UUU in the pretranslocation complex, and UUUACU in the posttranslocation complex. This indicated that translocation moved the mRNA 3 nt to the left, so three additional nucleotides (ACU) entered the ribosome and became protected. As an added check on the 3′-terminal sequences of the protected RNAs, these workers finished translating them before they released them for sequencing. They found that the protected mRNA fragment in the pretranslocation complex produced a peptide ending in phenylalanine, encoded by UUU, but the protected mRNA fragment in the posttranslocation complex produced a peptide ending in threonine, encoded by ACU. Thus, translocation had moved the mRNA exactly 3 nt, one codon's worth, through the ribosome.

SUMMARY Each translocation event moves the mRNA one codon's length, 3 nt, through the ribosome.

Role of GTP and EF-G Translocation in *E. coli* depends on GTP and a GTP-binding protein called EF-G, as we learned earlier in this chapter. In eukaryotes, a homologous protein known as **EF-2** carries out the same process. Yoshito Kaziro and colleagues demonstrated this dependence on GTP and EF-G in 1970. Then, in 1974, they amplified their findings by showing *when* during the translocation process GTP is required. First, they created the translocation substrate

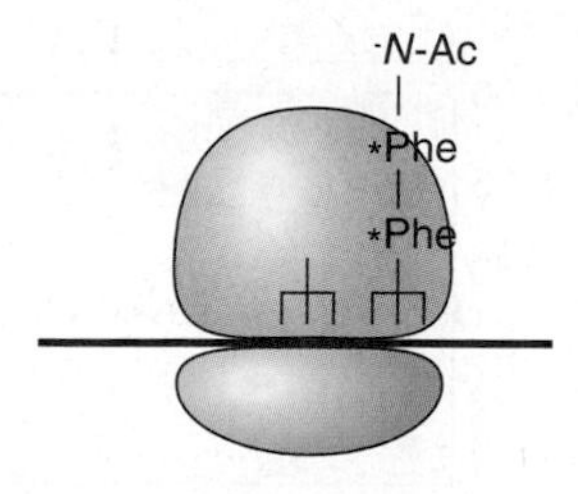

Figure 18.21 Translocation substrate used to measure dependence of translocation on EF-G and GTP. Kaziro and colleagues created a translocation substrate by loading ribosomes with *N*-acetyl-di-Phe-tRNA in the A site and a deacylated tRNA in the P site as follows: First, they mixed ribosomes and poly(U) RNA with *N*-acetyl-Phe-tRNA, which went to the P site. Then they added ordinary Phe-tRNA, which went to the A site. Peptidyl transferase then formed a peptide bond, yielding *N*-acetyl-diPhe-tRNA in the A site, and a deacylated tRNA in the P site.

pictured in Figure 18.21, with ^{14}C-labeled *N*-acetyl-diPhe-tRNA in the A site and a deacylated tRNA in the P site. This substrate is poised to undergo translocation, which can be measured in two ways: The first assay was the release of the deacylated tRNA from the ribosome. This is a nonphysiological reaction. In vivo, the deacylated tRNA would simply go to the E site. The second assay for translocation was puromycin reactivity. As soon as translocation occurs, the labeled dipeptide in the P site can join with puromycin and be released. Table 18.5 shows that neither GTP nor EF-G alone caused significant translocation, but that both together did promote translocation, measured by the release of deacylated tRNA.

Table 18.5 Roles of EF-G and GTP in Translocation

Additions	tRNA released	
	(pmol)	Δ
Experiment 1		
None	0.8	
GTP	1.8	1.0
EF-G	2.4	1.6
EF-G, GTP	12.6	11.8
EF-G, GDPCP	7.5	6.7
Experiment 2		
None	1.6	
EF-G	1.5	0
EF-G, GTP	5.1	3.5
EF-G, GTP, fusidic acid	6.7	5.1
EF-G, GDPCP	4.3	2.7
EF-G, GDPCP, fusidic acid	4.7	3.1

Source: Inove-Yokosawa, N., C. Ishikawa, and Y. Kaziro, The role of guanosine triphosphate in translocation reaction catalyzed by elongation factor G. *Journal of Biological Chemistry* 249:4322, 1974. Copyright © 1974. The American Society for Biochemistry & Molecular Biology, Bethesda, MD. Reprinted by permission.

At what point in this process is GTP hydrolyzed? There are two main possibilities: Model I calls for GTP hydrolysis pretranslocation. Model II allows for GTP hydrolysis after translocation has occurred. As unlikely as it may sound, model II was once the preferred hypothesis, based on the following experiments.

Kaziro and colleagues performed experiments with an unhydrolyzable analog of GTP, GDPCP. If GTP is not needed until after translocation, then this GTP analog should promote translocation, as natural GTP does. Table 18.5 shows that GDPCP does yield a significant amount of translocation, though not quite as much as GTP does. However, when they used GDPCP, the investigators found that they had to add stoichiometric quantities of EF-G (equimolar with the ribosomes). Ordinarily, translation requires only catalytic amounts of EF-G, because EF-G can be recycled over and over. But when GTP hydrolysis is not possible, as with GDPCP, recycling cannot occur. This suggested a function for GTP hydrolysis: release of EF-G from the ribosome, so both EF-G and ribosome can participate in another round of elongation.

Experiment 2, reported in the bottom part of Table 18.5, includes data on the effect of the antibiotic **fusidic acid.** This substance blocks the release of EF-G from the ribosome after GTP hydrolysis. This would normally greatly inhibit translation because it would halt the process after only one round of translocation. In this experiment, however, one round of translocation was all that could occur in any event, so fusidic acid had no effect. Kaziro and colleagues repeated these same experiments, using puromycin reactivity as their assay for translocation, and obtained essentially the same results. They also tried GDP in place of GTP and found that it could not support translocation.

Kaziro and colleagues concluded that GTP hydrolysis is not absolutely required for translocation (although it did help). Therefore, they reasoned, GTP hydrolysis must follow translocation. But their assays took several minutes, much longer than the millisecond time scale at which translation reactions take place. So they could not measure GTP hydrolysis and translocation and really tell which one happened first. To answer the question rigorously, we need a **kinetic experiment** that can measure events from one millisecond to the next. In 1997, Wolfgang Wintermeyer and colleagues performed such kinetic experiments and showed conclusively that GTP hydrolysis is very rapid and occurs before translocation.

Part of these workers' experimental plan was to load pretranslocation ribosomes in vitro with a fluorescent peptidyl-tRNA in the A site and a deacylated tRNA in the P site. Then they added EF-G–GTP and instantly began measuring the fluorescence of the complex. Such kinetic experiments on millisecond (ms) time scales are possible using a **stopped-flow apparatus** in which two or more

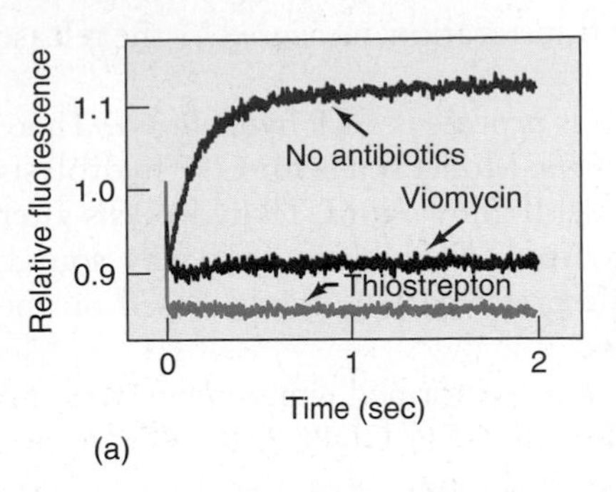

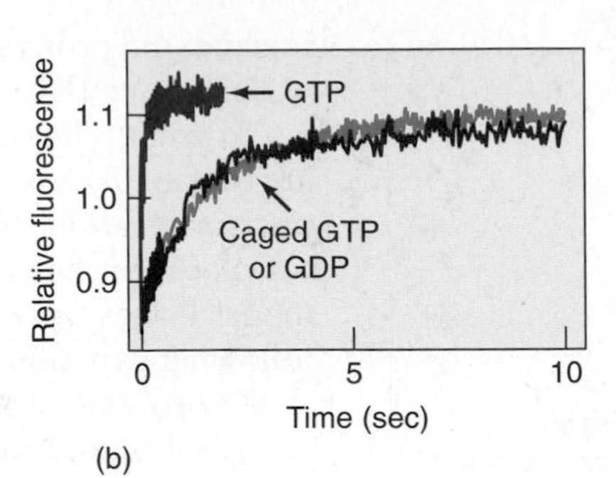

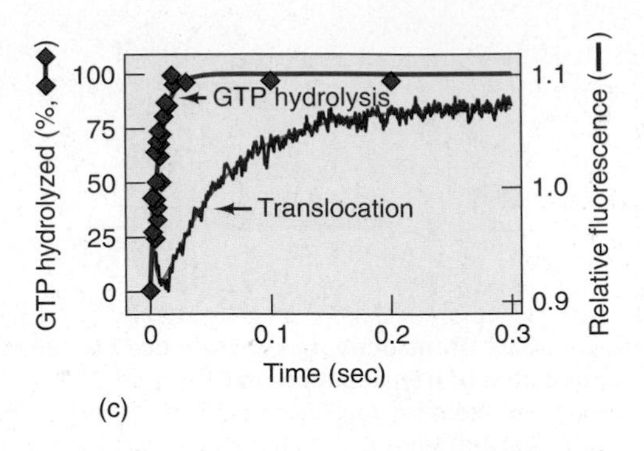

Figure 18.22 Kinetics of translocation. Wintermeyer and colleagues used stopped-flow kinetic experiments to measure translocation. They plotted the relative fluorescence of a fluorescent derivative of fMet-Phe-$tRNA^{Phe}$ bound to the A site as a function of time in seconds. The rise in fluorescence was taken as a measure of translocation. **(a)** Effect of antitranslocation antibiotics as follows: red, no antibiotics; blue, viomycin; green, thiostrepton. **(b)** Effect of GTP analogs. The following GTP analogs were added to the translocation reaction: red, GTP; blue, an unhydrolyzable GTP analog (caged GTP); green, GDP. **(c)** Timing of GTP hydrolysis and translocation. Wintermeyer and colleagues measured translocation by stopped-flow kinetics as in panels (a) and (b) and GTP hydrolysis by release of $^{32}P_i$ from [^{32}P]GTP in a stopped-flow device. GTP hydrolysis occurs first, and about five times faster than translocation. (*Source:* Adapted from (a) Rodnina, M.V., A. Savelsbergh, V.I. Katunin, and W. Wintermeyer, Hydrolysis of GTP by elongation factor G drives tRNA movement on the ribosome. *Nature* 385 (2 Jan 1997) f. 1, p. 37. (b) f. 1, p. 37. (c) f. 2, p. 38.)

solutions are forced simultaneously into a mixing chamber, and then immediately into another chamber for analysis. The mixing time in these experiments is of the order of only 2 ms. After an initial drop, the fluorescence increased significantly, as shown in Figure 18.22a, red trace. This increase in fluorescence appears to be related to translocation, because it is prevented by two antibiotics that block translocation, viomycin and thiostrepton (Figure 18.22a, blue and green traces, respectively). Translocation worked much better with GTP (Figure 18.22b, red trace) than with an unhydrolyzable GTP analog (a "caged" GTP, Figure 18.22b, blue trace) or with GDP (Figure 18.22b, green trace).

Next, Wintermeyer and colleagues compared the timing and speed of GTP hydrolysis and translocation. They measured GTP hydrolysis with [γ-^{32}P]GTP, again with a stopped-flow device. This time, they rapidly mixed the radioactive GTP with the other components and then, after only milliseconds, forced the mixture into another chamber where the reaction was stopped with perchlorate solution. They measured $^{32}P_i$ released by liquid scintillation counting. Again, they assayed translocation by fluorescence increase. Figure 18.22c shows that GTP hydrolysis occurred first, and about five times faster than translocation. Thus, Wintermeyer and colleagues concluded that GTP hydrolysis precedes and drives translocation.

It is clear that EF-G, using energy from GTP, catalyzes the translocation process. Does that mean that no translocation can occur in the absence of EF-G? Actually, certain in vitro conditions have been found to allow some translocation even in the absence of EF-G. In 2003, Kurt Fredrick and Harry Noller performed the most convincing study to date on this topic, demonstrating that the antibiotic sparsomycin can catalyze translocation in the absence of EF-G and GTP. This finding suggests that the ribosome itself has the ability to perform translocation even without help from EF-G, and that the energy required for translocation is stored in the complex of ribosome, tRNAs, and mRNA after each peptide bond forms.

SUMMARY GTP and EF-G are necessary for translocation, although translocation activity appears to be inherent in the ribosome and can be expressed without EF-G and GTP in vitro. GTP hydrolysis precedes translocation and significantly accelerates it. For a new round of elongation to occur, EF-G must be released from the ribosome, and that release depends on GTP hydrolysis.

G Proteins and Translation

We have now seen two examples of proteins that use hydrolysis of GTP to drive important steps in the elongation phase of translation: EF-Tu and EF-G. Recall from Chapter 17 that IF2 plays a similar role in the initiation phase. Finally, at the end of this chapter we will discover that another factor (RF3) plays the same role in translation termination.

What do all of these processes have in common? All use energy from GTP to drive molecular movements essential for translation. IF2 and EF-Tu both bring aminoacyl-tRNAs to the ribosome (IF2 transports the initiating aminoacyl-tRNA (fMet-$tRNA_f^{Met}$) to the P site of the ribosome, while EF-Tu transports the elongating aminoacyl-tRNAs to the A site of the ribosome). EF-G sponsors translocation, in which the mRNA and the peptidyl-tRNA move from the A site to the P site and the deacylated tRNA moves from the P site to the E site of the ribosome. And RF3 helps catalyze termination, in which the bond linking the finished polypeptide to the tRNA is broken and the polypeptide exits the ribosome.

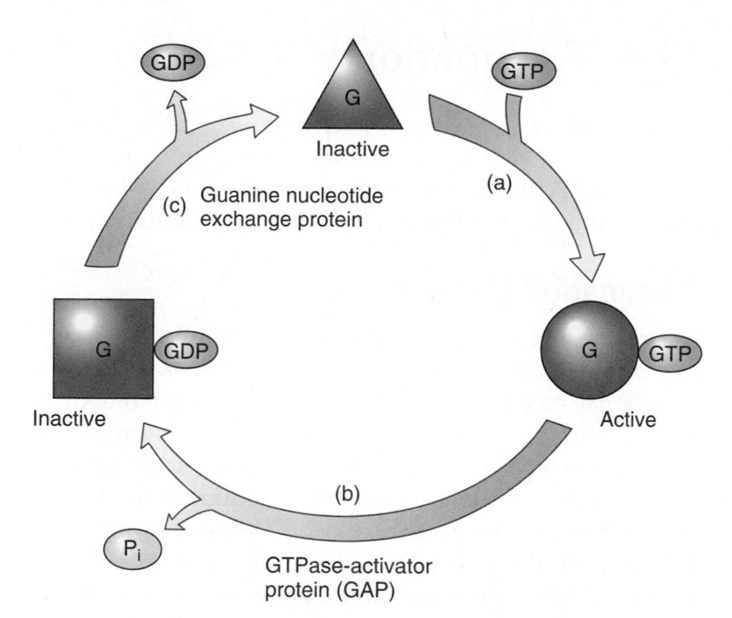

Figure 18.23 Generalized G protein cycle. The G protein at top (red triangle) is in the unbound state with neither GDP nor GTP bound. This state is normally short-lived. **(a)** GTP binds to the unbound G protein, changing its conformation (represented by the change from triangular to circular shape), and thereby activating it. **(b)** A GTPase-activator protein (GAP) stimulates the intrinsic GTPase activity of the G protein, causing it to hydrolyze its GTP to GDP. This results in another conformational change, represented by the change to square shape, which inactivates the G protein. **(c)** A guanine nucleotide exchange protein removes the GDP from the G protein, changing it back to the original unbound state, which is ready to accept another GTP.

All of these factors belong to a large class of proteins known as **G proteins** that perform a wide variety of cellular functions. Most of the G proteins share the following features, illustrated in Figure 18.23:

1. They are GDP- and GTP-binding proteins. In fact the "G" in "G protein" comes from "guanine nucleotide."
2. They cycle among three conformational states, depending on whether they are bound to GDP, GTP, or neither nucleotide, and these conformational states determine their activities.
3. When they are bound to GTP they are activated to carry out their functions.
4. They have intrinsic GTPase activity.
5. Their GTPase activity is stimulated by another agent called a **GTPase activator protein (GAP).**
6. When a GAP stimulates their GTPase activity, they cleave their bound GTP to GDP, inactivating themselves.
7. They are reactivated by another protein called a **guanine nucleotide exchange protein.** This factor removes GDP from the inactive G protein and allows another molecule of GTP to bind. One guanine nucleotide exchange protein comes immediately to mind: EF-Ts. We have seen that EF-Ts is essential for replacing GDP with GTP on EF-Tu.

The GTPases of all the G proteins involved in translation are stimulated by the ribosome. Thus, we might predict that the GAP for all these G proteins would be a protein or proteins at some site(s) on the ribosome. In fact, a set of ribosomal proteins and parts of a ribosomal RNA, collectively known as the **GTPase-associated site** or **GTPase center** has been discovered on the ribosome. It consists of the ribosomal protein L11, a complex of the ribosomal proteins L10 and L12, and the 23S rRNA. Note that the GTPase center merely stimulates the GTPase activity of the associated G protein; it does not have GTPase activity of its own.

The GTPase center is located on a stalk of the 50S subunit conventionally shown on the right of the ribosome, and called either the **L7/L12 stalk**, or the **L10–L12 stalk.** L7 and L12 are 50S ribosomal proteins that have identical amino acid sequences, but L7 is acetylated on its N-terminal amino group. One molecule each of L7 and L12 form a dimer that binds to the rest of the 50S particle via protein L10. *E. coli* ribosomes have two dimers of L7/L12. *Thermus thermophilus* and some other thermophilic bacterial ribosomes have three dimers of L12. The L12 molecules in these bacteria are not acetylated, but some of them are phosphorylated.

SUMMARY Several translation factors harness the energy of GTP to catalyze molecular motions. These factors belong to a large class of G proteins that are activated by GTP, have intrinsic GTPase activity that is activated by an external factor (GAP), are inactivated when they cleave their own GTP to GDP, and are reactivated by another external factor (a guanine nucleotide exchange protein) that replaces GDP with GTP.

The Structures of EF-Tu and EF-G

If EF-Tu and EF-G really bind to the same ribosomal GTPase center, then the two factors should have similar structures, just as two keys that fit the same lock must have similar shapes. X-ray crystallography studies on the two proteins have shown that this is true, with one qualification: It is actually the EF-Tu–tRNA–GTP ternary complex that has a shape very similar to that of the EF-G–GTP binary complex. This makes sense because EF-Tu binds to the ribosome as a ternary complex with tRNA and GTP, whereas EF-G binds as a binary complex with GTP only. To avoid GTP hydrolysis, the experimenters used unhydrolyzable GTP analogs, GDP in the case of EF-G, and GDPNP in the case of EF-Tu–tRNA.

Figure 18.24 depicts the three-dimensional structures of the two complexes. We can see that the lower part of the

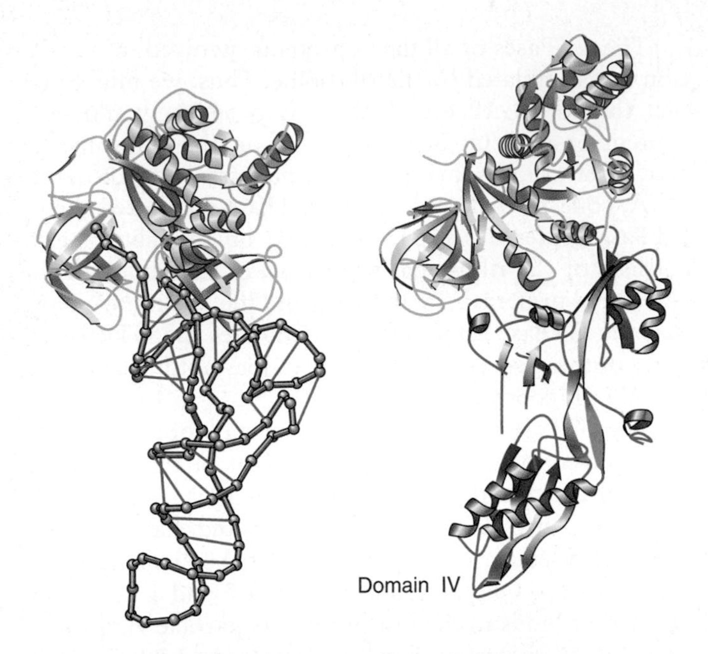

Figure 18.24 Comparison of the three-dimensional shapes of the EF-Tu–tRNA–GDPNP ternary complex (left) and the EF-G–GDP binary complex (right). The tRNA part of the ternary complex and the corresponding protein part of the binary complex are highlighted in red.

EF-G protein (**domain IV**) mimics the shape of the anticodon stem loop portion of the tRNA (red, left) in the EF-Tu ternary complex. This presumably allows both complexes to bind at or close to the same site on the ribosome.

Two other translation factors also have ribosome-dependent GTPase activities: the prokaryotic initiation factor IF2 (Chapter 17) and the termination factor RF3 (see later in this chapter). Because they also seem to rely on the same GTPase-activating center on the ribosome, it is reasonable to predict that they are structurally similar to at least parts of the two complexes depicted in Figure 18.24. Later in this chapter, we will learn that the structure of *E. coli* RF3-GDP is indeed very similar to that of EF-Tu–GTP.

Furthermore, if EF-G and IF2 bind to the same GTPase center of the ribosome, we would expect the two to compete for binding there. In fact, Albert Dahlberg and colleagues demonstrated in 2002 that IF2 does indeed compete with EF-G for ribosome binding. Moreover, they showed that two antibiotics, thiostrepton and micrococcin, that were known to bind to the GTPase center, also interfere with binding of both EF-G and IF2 at that site. Thus, IF2, EF-G, EF-Tu, and, quite probably, RF3 all bind to at least overlapping GTPase centers on the ribosome.

SUMMARY The three-dimensional shapes of the EF-Tu–tRNA–GDPNP ternary complex and the EF-G–GDP binary complex have been determined by x-ray crystallography. As predicted, they are very similar.

18.4 Termination

The elongation cycle repeats over and over, adding amino acids one at a time to the growing polypeptide product. Finally, the ribosome encounters a stop codon, signaling that it is time for the last step in translation: termination.

Termination Codons

The first termination codon (the **amber codon**) was discovered by Seymour Benzer and Sewell Champe in 1962 as a conditional mutation in a T4 phage. The amber mutation was conditional in that the mutant phage was unable to replicate in wild-type *E. coli* cells, but could replicate in a mutant, **suppressor** strain. Certain mutations in the *E. coli* alkaline phosphatase gene were also suppressed by the same suppressor strain, so it appeared that they were also **amber mutations.** We now know that amber mutations create termination codons that cause translation to stop prematurely in the middle of an mRNA, and therefore give rise to incomplete proteins. What was the evidence for this conclusion?

First of all, amber mutations have severe effects. Ordinary missense mutations change at most one amino acid in a protein, which may or may not affect the function of the protein, but even if the protein is inactive, it can usually be detected with an antibody. By contrast, *E. coli* strains with amber mutations in the alkaline phosphatase gene produce no detectable alkaline phosphatase activity or protein. This fits the hypothesis that the amber mutations caused premature termination of the alkaline phosphatase, so no full-size protein could be found.

A genetic experiment by Benzer and Champe further strengthened this hypothesis. They introduced a deletion into the adjacent *rIIA* and *B* genes of phage T4 that fused the two genes together, as shown in Figure 18.25. The fused gene gave a fusion protein with B activity, but no A activity. Then they introduced an amber mutation into the *rIIA* part of the fused gene. This mutation blocked *rIIB* activity, and this block was removed by an amber suppressor. How could a mutation in the *A* cistron block the expression of the *B* cistron, which lies downstream? Translation termination at the amber mutation is an obvious explanation. If translation stops at the amber codon, it would never reach the *B* cistron. Moreover, according to this logic, the amber suppressor overrides the translation termination at the amber codon and allows translation to continue on into the *B* cistron.

More direct evidence for the amber mutation as a translation terminator came from studies by Brenner and colleagues on the head protein gene of phage T4. When this phage infects *E. coli* B, head protein accounts for more than 50% of the protein made late in infection, which makes it easy to purify. When these investigators introduced amber mutations into the head protein gene, they were unable to isolate intact head protein from infected cells, but they could isolate fragments of head protein. And tryptic digestion of these fragments yielded peptides that could be

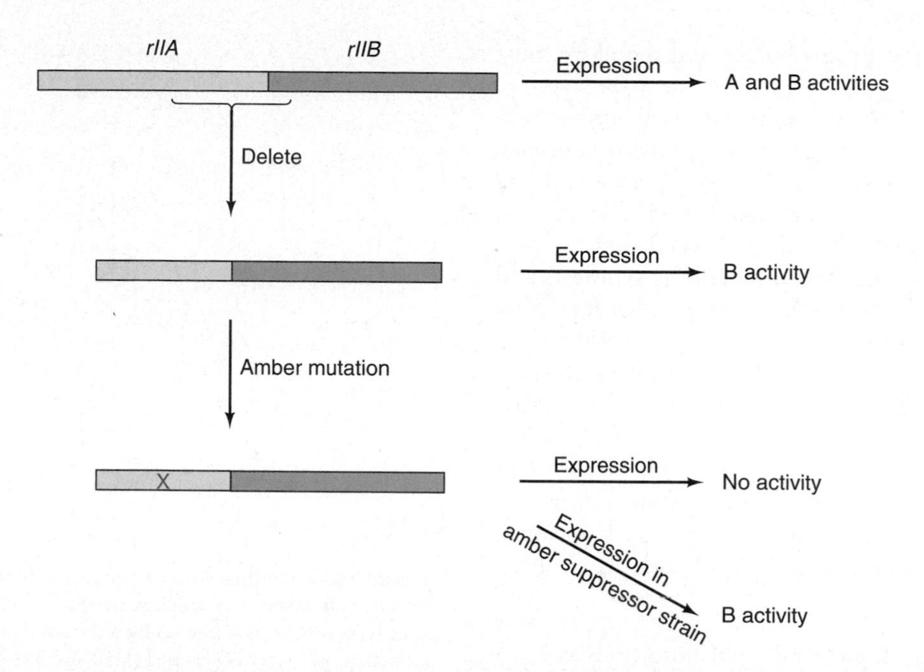

Figure 18.25 Effects of an amber mutation in a fused gene. Benzer and Champe deleted the DNA shown by the bracket, fusing the *rIIA* and *B* cistrons together. Expression of this fused gene yielded *B* activity, but no *A* activity. An amber mutation in the *A* cistron inactivated *B* activity, which could be restored by transferring the gene to an amber suppressor strain (*E. coli* CR63). The amber mutation caused premature translation termination in the *A* cistron, and the amber suppressor prevented this termination, allowing production of the *B* part of the fusion protein.

identified as amino-terminal peptides. Thus, the products of head protein genes with amber mutations were all amino-terminal protein fragments. Because translation starts at a protein's amino terminus, this experiment demonstrated that the amber mutations caused termination of translation before it had a chance to reach the carboxyl terminus.

The amber mutation defined one translation stop codon, but the two others have similarly colorful names, **ochre** and **opal.** Ochre mutations were originally distinguished by the fact that they were not suppressed by **amber suppressors.** Instead, they have their own class of **ochre suppressors.** Similarly, opal mutations are suppressed by **opal suppressors.**

How did the amber mutation get its name? In was named in honor of the mother of a graduate student named Harris Bernstein to settle a bet he made with two fellow students about the mutant they were making. He accurately predicted the properties of the mutant, so it now bears his mother's (and his) name—translated into English (German: *bernstein* = amber). Mutants that create the other two stop codons were named in the same colorful style.

Since amber mutations are caused by mutagens that give rise to missense mutations, we suspect that these mutations come from the conversion of an ordinary codon to a stop codon by a one-base change. We know that only three unassigned "nonsense" codons occur in the genetic code: UAG, UAA, and UGA. We assume these are stop codons, so the simplest explanation for the results we have seen so far is that one of these is the **amber codon,** one is the **ochre codon,** and one is the **opal codon.** But which is which?

Martin Weigert and Alan Garen answered this question in 1965, not by sequencing DNA or RNA, but by sequencing protein. They studied an amber mutation at one position in the alkaline phosphatase gene of *E. coli.* The amino acid at this position in wild-type cells was tryptophan, whose sole codon is UGG. Because the amber mutation originated with a one-base change, we already know that the amber codon is related to UGG by a one-base change. To find out what that change was, Weigert and Garen determined the amino acids inserted in this position by several different revertants. The revertants presumably arose by one-base changes from the amber codon. Some of these had tryptophan in the key position, but most had other amino acids: serine, tyrosine, leucine, glutamate, glutamine, and lysine. These other amino acids could substitute for tryptophan well enough to give at least some alkaline phosphatase activity. The puzzle is to deduce the one codon that is related by one-base changes to at least one codon for each of these amino acids, including tryptophan. Figure 18.26 demonstrates

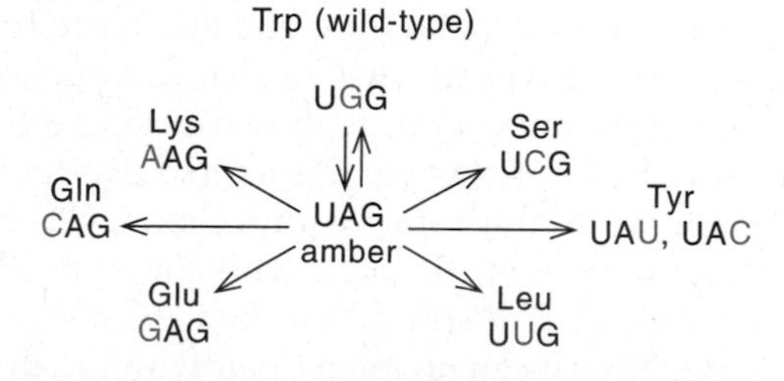

Figure 18.26 The amber codon is UAG. The amber codon (middle) came via a one-base change from the tryptophan codon (UGG), and the gene reverts to a functional condition in which one of the following amino acids replaces tryptophan: serine, tyrosine, leucine, glutamate, glutamine, or lysine. The pink color represents the single base that is changed in all these revertants, including the wild-type revertant that codes for tryptophan.

that UAG is the solution to this puzzle and therefore must be the amber codon.

By the same logic, including the fact that amber mutants can mutate by single-base changes to ochre mutants, Sydney Brenner and collaborators reasoned that the ochre codon must be UAA. Severo Ochoa and colleagues verified that UAA is a stop signal when they showed that the synthetic message AUGUUUUAAA_n directed the synthesis and release of the dipeptide fMet-Phe. (AUG codes for fMet; UUU codes for Phe; and UAA codes for stop.) With UAG and UAA assigned to the amber and ochre codons, respectively, UGA must be the opal codon, by elimination. Now that we have the base sequences of thousands of genes, it is abundantly clear that these three codons really do serve as stop signals. Sometimes we even find two stop codons in a row (e.g., UAAUAG), which provides a fail-safe stop signal even if termination at one codon is suppressed.

SUMMARY Amber, ochre, and opal mutations create termination codons (UAG, UAA, and UGA, respectively) within an mRNA and thereby cause premature termination of translation. These three codons are also the natural stop signals at the ends of coding regions in mRNAs.

Stop Codon Suppression

How do suppressors overcome the lethal effects of premature termination signals? Mario Capecchi and Gary Gussin showed in 1965 that tRNA from a suppressor strain of *E. coli* could suppress an amber mutation in the coat cistron of phage R17 mRNA. This identified tRNA as the suppressor molecule, but how does it work? Brenner and collaborators found the answer when they sequenced a suppressor tRNA. They placed the gene for an amber suppressor tRNA on a φ80 phage and used this recombinant phage to infect *E. coli* cells bearing an amber mutation in the *lacZ* gene. Because of this suppressor tRNA, infected cells were able to suppress the amber mutation by inserting a tyrosine instead of terminating. When Brenner and colleagues sequenced this suppressor tRNA they found only one difference from the sequence of the wild-type $tRNA^{Tyr}$: a change from C to G in the first base of the anticodon, as shown in Figure 18.27.

Figure 18.28 illustrates how this altered tRNA can suppress an amber codon. We start with a codon, CAG, which encodes glutamine (Gln). It pairs with the anticodon 3′-GUC-5′ on a $tRNA^{Gln}$. Assume that the CAG codon is mutated to UAG. Now it can no longer pair with the $tRNA^{Gln}$; instead, it attracts the termination machinery to stop translation. Now a second mutation occurs in the anticodon of a $tRNA^{Tyr}$, changing it from AUG to AUC (again reading 3′→5′). This new tRNA is a suppressor tRNA because it has an anticodon complementary to the amber codon

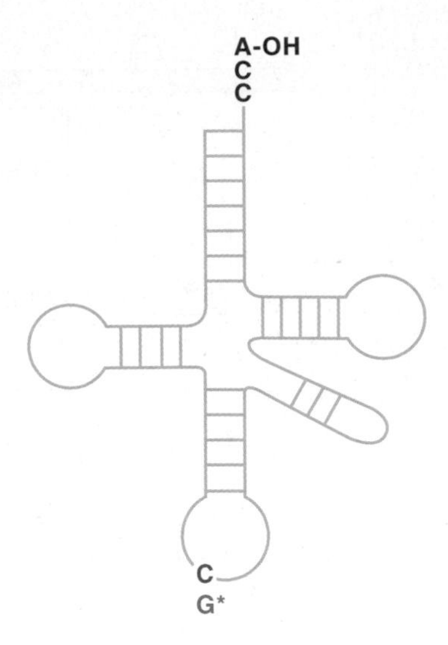

Figure 18.27 Comparison of sequence of wild-type *E. coli* $tRNA^{Tyr}$ and *E. coli* amber suppressor tRNA. The G* (green) present in the wild-type $tRNA^{Tyr}$ is replaced by a C (red) in the suppressor tRNA. (*Source:* Adapted from Goodman, H.M., J. Abelson, A. Landy, S. Brenner, and J.D. Smith, Amber suppression: A nucleotide change in the anticodon of a tyrosine transfer RNA. *Nature* 217:1021, 1968.)

UAG. Thus, it can pair with the UAG stop codon and insert tyrosine into the growing polypeptide, allowing the ribosome to get past the stop codon without terminating translation.

SUMMARY Most suppressor tRNAs have altered anticodons that can recognize stop codons and prevent termination by inserting an amino acid and allowing the ribosome to move on to the next codon.

Release Factors

Because the stop codons are triplets, just like ordinary codons, one might expect that these stop codons would be decoded by tRNAs, just as other codons are. However, work begun by Capecchi in 1967 proved that tRNAs do not ordinarily recognize stop codons. Instead, proteins called **release factors (RFs)** do. Capecchi devised the following scheme to identify the release factors: He began with *E. coli* ribosomes plus an R17 phage mRNA that was mutated in the seventh codon of the coat cistron to UAG (amber). The codon preceding this amber codon was ACC, which codes for threonine. He incubated the ribosomes with this mRNA in the absence of threonine so they would make a pentapeptide and then stall at the threonine codon. Then he isolated the ribosomes with the pentapeptide attached and placed them in a system containing only EF-Tu, EF-G (attached to the ribosomes) and [^{14}C]threonyl-tRNA. The ribosomes incorporated the labeled threonine into the peptide, producing a labeled hexapeptide in the P site, poised on the brink of release. To find the release factor, Capecchi added

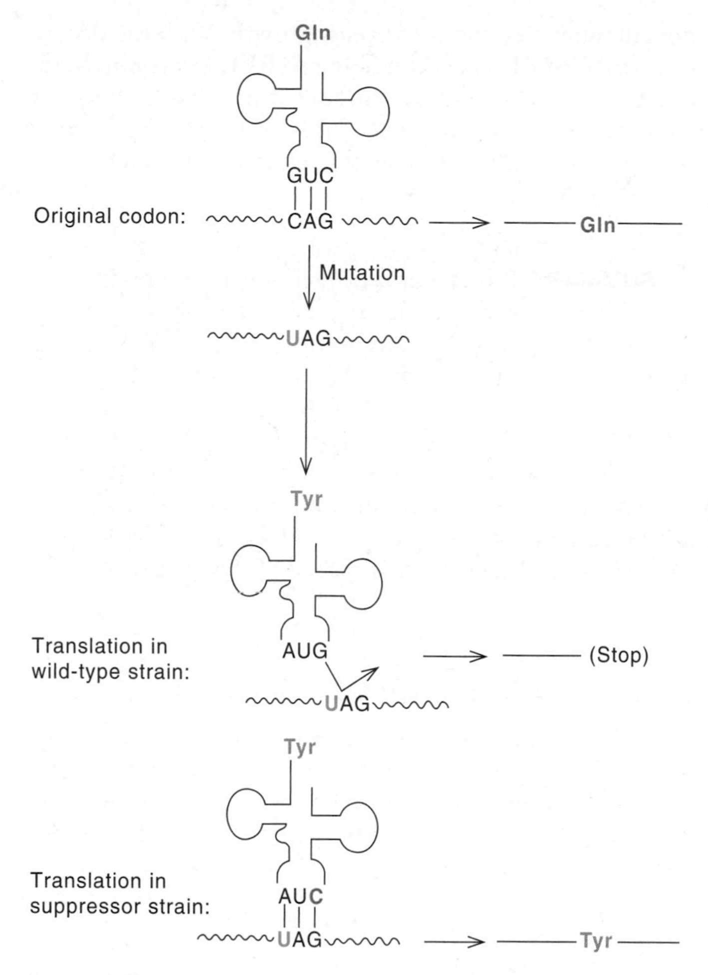

Figure 18.28 Mechanism of suppression. Top: The original codon in the wild-type *E. coli* gene was CAG, which was recognized by a glutamine tRNA. Middle: This codon mutated to UAG, which was translated as a stop codon by a wild-type strain of *E. coli.* Notice the tyrosine tRNA, whose anticodon (AUG) cannot translate the amber codon. Bottom: A suppressor strain contains a mutant tyrosine tRNA with the anticodon AUC instead of AUG. This altered anticodon recognizes the amber codon and causes the insertion of tyrosine (gray) instead of allowing termination.

ribosomal supernatant fractions until one released the labeled peptide. He discovered that this factor, which he called release factor (RF), was not a tRNA, but a protein.

Nirenberg and colleagues devised a simpler technique (Figure 18.29), which was a takeoff on their assay for identifying codons, examined earlier in this chapter. They formed a ternary complex with ribosomes, the triplet AUG, and $[^3H]$fMet-tRNA$_f^{Met}$. The initiation codon and aminoacyl-tRNA went to the P site in the complex, and the labeled amino acid was therefore eligible for release. Incubation of this complex with a crude release factor preparation and any of the three termination codons (UAG, UAA, or UGA) caused release of the labeled fMet. In this assay, the termination trinucleotide went to the A site and dictated release if the appropriate release factor was present. Table 18.6 shows that one factor (**RF1**)

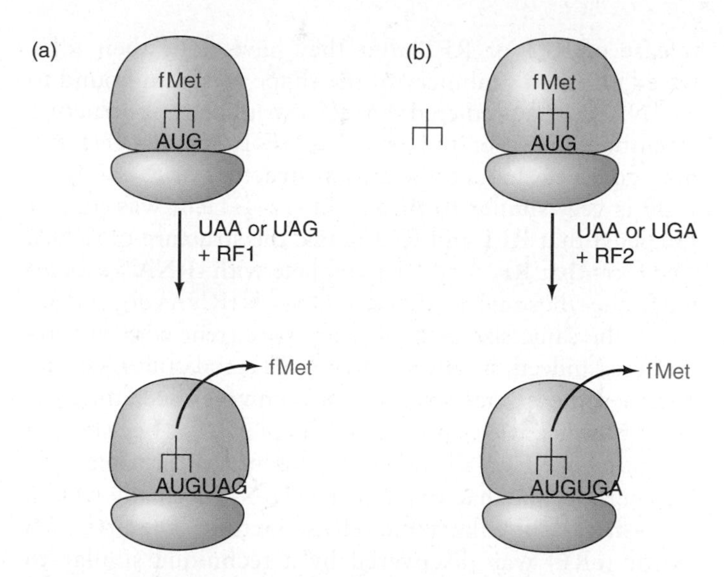

Figure 18.29 Nirenberg's assay for release factors. Nirenberg loaded the P site of ribosomes with the initiation codon AUG and $[^3H]$ fMet-tRNA$_f^{Met}$. Then he added one of the termination codons plus a release factor, which released the labeled fMet. **(a)** RF1 is active with UAA or UAG. **(b)** RF2 is active with UAA or UGA.

Table 18.6 Response of RF1 and RF2 to Stop Codons

Additions		pmol $[^3H]$fMet released in presence of:	
Release factor	**Stop codon**	**0.012 M Mg^{2+}**	**0.030 M Mg^{2+}**
RF1	None	0.12	0.15
RF1	UAA	0.47	0.86
RF1	UAG	0.53	1.20
RF1	UGA	0.08	0.10
RF2	None	0.02	0.14
RF2	UAA	0.22	0.77
RF2	UAG	0.02	0.14
RF2	UGA	0.33	1.08

Source: From "Release Factors Differing in Specificity for Terminator codons," by W. Scolnick, R. Tompkins, T. Caskey, and M. Nirenberg, *Proceedings of the National Academy of Sciences, USA,* 61:772, 1968. Reprinted with permission of the authors.

cooperated with the stop codons UAA and UAG to cause release of the fMet, while another factor (**RF2**) cooperated with UAA and UGA. Subsequent studies showed that UAA or UAG could direct the binding of purified RF1 to the ribosome, while UAA or UGA could direct RF2 binding. This reinforced the idea that the RFs could recognize specific translation stop signals. A third release factor, (**RF3**), a ribosome-dependent GTPase, binds GTP, then binds to the ribosome and induces a large conformational change in the ribosome that apparently facilitates the release of RF1 or RF2 after they have done their jobs. Based on EF-G's mimicry of the shape of EF-Tu bound to a tRNA, it was predicted that RF3 would have a structure resembling the protein part of the EF-Tu–tRNA–GTP ternary complex. In fact, the crystal structure of *E. coli* RF3-GDP is very similar to that of EF-Tu–GTP. It was further predicted that RF1 and RF2 mimic the structure of tRNA. The facts that RF1 and RF2 compete with tRNA for binding to the ribosome, recognize codons as tRNAs do, and are about the same size as tRNAs are consistent with this hypothesis. Indeed, in 2008 Harry Noller and colleagues determined the crystal structure of a complex including the 70S ribosome, RF1, and tRNA (Chapter 19). They showed that parts of RF1 really do occupy essentially the same position in the A site that an aminoacyl-tRNA normally would.

What about eukaryotic release factors? The first such factor (eRF) was discovered by a technique similar to Nirenberg's in 1971. Then, in 1994, a collaborative group led by Lev Kisselev finally purified eRF, still using an assay based on Nirenberg's, and succeeded in cloning and sequencing the eRF gene. Their approach to cloning and sequencing the gene was a widely used one: Using an fMet release assay similar to Nirenberg's to detect eRF, they purified the eRF activity until it gave one major band on SDS-PAGE, then subjected this protein to two-dimensional gel electrophoresis to purify it away from all other proteins. They cut out the eRF spot from this electrophoresis step, cleaved the protein with trypsin, and subjected four of the tryptic peptides to microsequencing. The sequences strongly resembled those of proteins from humans, *Xenopus laevis,* yeast, and the small flowering plant *Arabidopsis thaliana*. Thus, they were able to use the *Xenopus* gene (C11), which had already been cloned, as a probe to find the corresponding human gene in a human cDNA library. To verify that the products of the cloned *Xenopus* and human genes (C11 and TB3-1, respectively) had eRF activity, Kisselev and colleagues expressed these genes in bacteria or yeast, respectively, and tested them in the fMet release assay with tetranucleotides, some of which contained stop codons. Both proteins released fMet from loaded ribosomes, but only in the presence of a stop codon. The *Xenopus* protein was expressed with an oligohistidine (His) tag, so Kisselev and colleagues included unrelated His-tagged proteins as negative controls. They also showed that an antibody against C11 blocked its release factor activity, but an irrelevant antibody (anti-Eg5) did not.

Furthermore, eRF can recognize all three stop codons, unlike either of the two prokaryotic release factors, which can recognize only two. Does eRF collaborate with a G protein as prokaryotic RF1 and RF2 do? Michel Philippe and colleagues found that the answer is yes when they discovered a protein factor, now called **eRF3,** in *X. laevis* cells in 1995. Another member of the eRF3 family, a yeast protein known as Sup35, has a guanine nucleotide-binding domain and is essential for yeast growth. With the discovery of eRF3, eRF has been renamed **eRF1.** Interestingly, the function of eRF3 is much different from that of bacterial RF3. It collaborates with eRF1 both in recognizing the three stop codons, and in releasing the finished polypeptide from the ribosome.

SUMMARY Prokaryotic translation termination is mediated by three factors: RF1, RF2, and RF3. RF1 recognizes the termination codons UAA and UAG; RF2 recognizes UAA and UGA. RF3 is a GTP-binding protein that facilitates release of RF1 and RF2 from the ribosome. Eukaryotes have two release factors: eRF1, which recognizes all three termination codons, and eRF3, a ribosome-dependent GTPase that helps eRF1 recognize stop codons and release the finished polypeptide.

Dealing with Aberrant Termination

Two kinds of aberrant mRNAs can lead to aberrant termination. First, as we have seen, "nonsense" mutations can occur that cause premature termination. Second, some mRNAs (**non-stop mRNAs**) lack termination codons, sometimes because the synthesis of the mRNA was aborted upstream of the termination codon. Ribosomes translate through these non-stop mRNAs and then stall. Both of these events cause problems for the cell. Either premature termination or a stalled ribosome yields incomplete proteins that might have adverse effects on the cell. Stalled ribosomes present a cell with the additional problem that the stalled ribosome is out of action and unable to participate in any further protein synthesis. Let us first examine the ways that cells deal with non-stop mRNAs, then we will look at mechanisms for degrading the products of premature termination.

Non-Stop mRNAs To deal with non-stop mRNAs, cells need to degrade the aberrant protein product and release the ribosomal subunits so they can participate in productive translation instead of remaining stalled forever. The mechanisms of this process differ between bacteria and eukaryotes. Bacteria use so called **transfer-messenger RNAs (tmRNAs)** to rescue stalled ribosomes and tag the non-stop mRNAs for destruction (**tmRNA-mediated ribosome rescue**). The tmRNAs are about 300 nt long, and their 5′- and 3′-ends come together to form a **tRNA-like domain (TLD)** that resembles a tRNA (Figure 18.30). In fact, the resemblance is so strong that a tmRNA can be charged with alanine. Once charged, the alanyl-tmRNA can bind to the ribosome's A site and, via the ribosome's peptidyl transferase, can donate its alanine to the stalled polypeptide.

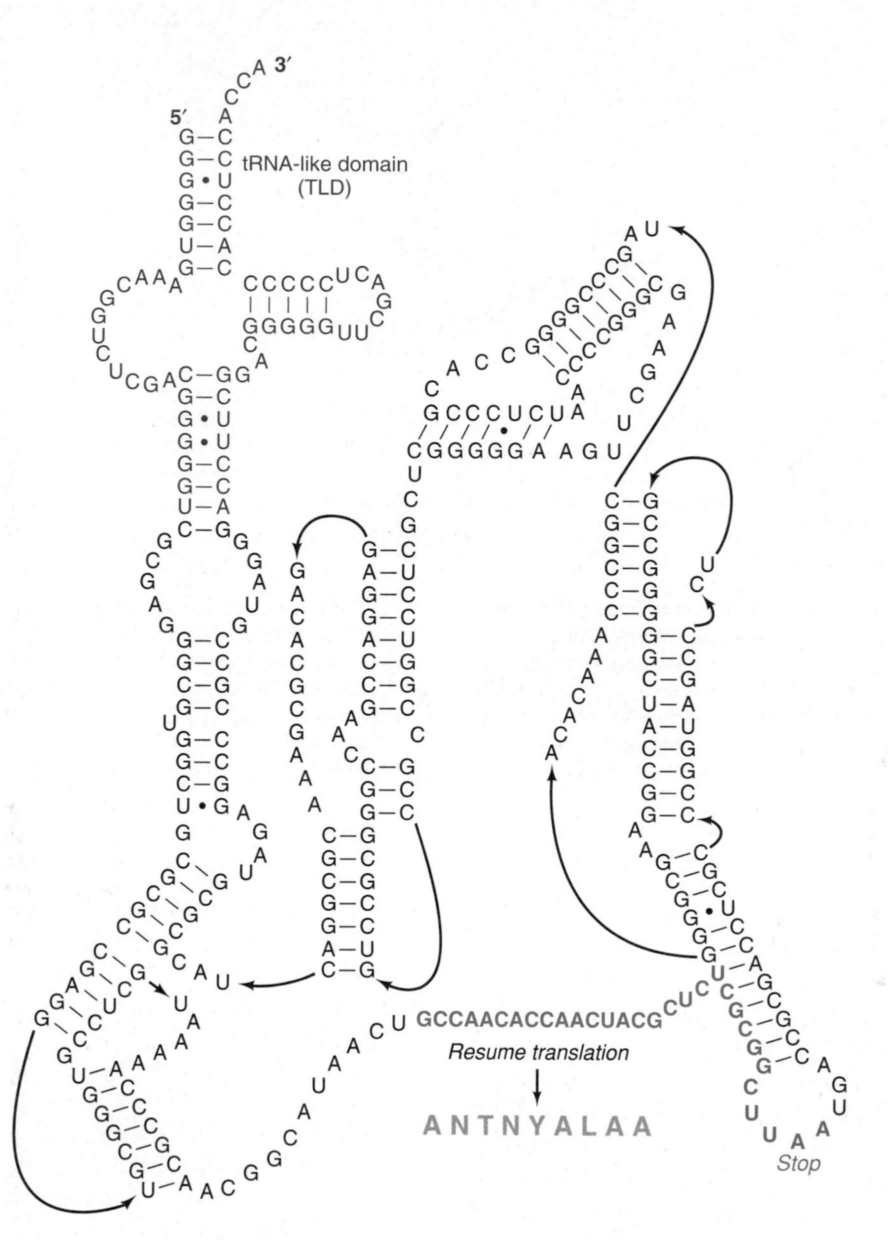

Figure 18.30 Structure of the *Thermus thermophilus* tmRNA. The TLD is at upper left in pink, and the ORF is at bottom in blue. The peptide encoded by the ORF is in orange. (*Source*: Adapted from Valle et al., Visualizing tmRNA entry into a stalled ribosome, *Science* 300:128, fig.1, 2003.)

After this peptidyl transferase reaction, the central part of the tmRNA comes into play (Figure 18.31). This part of the tmRNA contains a short open reading frame (ORF) that is positioned in the A site such that the ribosome switches from translating the non-stop mRNA to translating the tmRNA, a process called ***trans*-translation.** The ORF of the tmRNA encodes a short, hydrophobic peptide that is added to the carboxyl terminus of the stalled polypeptide. This peptide targets the whole polypeptide for destruction, minimizing its ability to harm the cell.

Obviously, a tmRNA is not just like a tRNA. For one thing, it lacks an anticodon, so there can be no codon–anticodon pairing. And, as we have seen, codon–anticodon pairing is essential to avoid dissociation of an aminoacyl-tRNA during proofreading. A second difference between a tmRNA and a real tRNA is that the tmRNA does not have a standard D loop. But the tmRNA systems gets around these problems using a protein known as **SmpB.** In 2003, Joachim Frank and V. Ramakrishnan obtained cryo-electron microscopy images of a complex of EF-Tu, tmRNA, and SmpB bound to ribosomes from *Thermus thermophilus*. This study showed that SmpB binds to tmRNA and EF-Tu and makes contacts with the ribosome that would normally come from the D loop of an RNA. Thus, SmpB helps to hold the tmRNA to the ribosome even though the tmRNA lacks some of the elements it needs to bind tightly by itself.

What happens to the non-stop mRNA once the ribosome has been released by tmRNA? We do not know the answer for sure, but tmRNAs do copurify with a 3′→5′ exonuclease known as **RNase R.** It is an attractive hypothesis

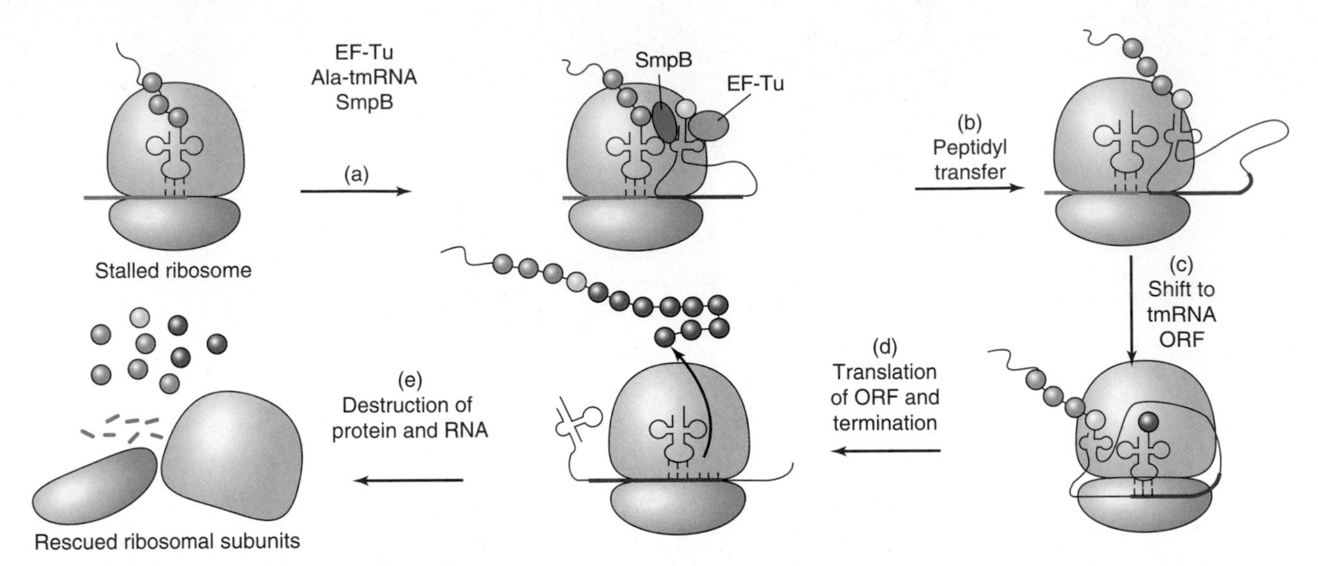

Figure 18.31 Mechanism of tmRNA-mediated release of non-stop mRNA and polypeptide. (a) EF-Tu, alanyl-tmRNA and SmpB (turquoise) bind to the A site of the ribosome stalled on a non-stop mRNA (brown). SmpB helps the tRNA-like domain of the tmRNA bind to the ribosome. **(b)** The ribosome's peptidyl transferase transfers the alanine (yellow) from the tmRNA to the stalled polypeptide (green). **(c)** The ribosome shifts to reading the ORF (purple) of the tmRNA. **(d)** The ribosome completes translating the ORF of the tmRNA, adding nine more amino acids (red) to the end of the stalled polypeptide and releasing it. **(e)** Together, these extra amino acids target the whole polypeptide for destruction. At the same time, the non-stop mRNA is destroyed, perhaps by RNase R, which associates with tmRNA. (*Source:* Adapted from Moore, S.D., K.E. McGinness, and R.T. Sauer, A glimpse into tmRNA-mediated ribosome rescue. *Science* 300 [2003] p. 73, f. 1.)

that RNase R degrades the non-stop mRNA before it can complex with a new ribosome.

Eukaryotes do not have tmRNAs, so how do they deal with non-stop mRNAs? Figure 18.32 illustrates the current hypothesis. The A site of a ribosome stalled at the end of a nonstop mRNA will contain zero to three nucleotides of the terminal poly(A). This state is recognized by the carboxyl-terminal domain of a protein called **Ski7p.** This protein domain resembles the GTPase domains of the elongation and termination factors EF1A and eRF3, respectively. These domains normally associate with the ribosomal A site, and so does Ski7p. In addition, Ski7p associates tightly with the cytoplasmic **exosome,** a complex of 9-11 proteins, including a $3' \rightarrow 5'$ exonuclease that degrades RNA. The Ski7p–exosome complex then recruits the **Ski complex** to the ribosomal A site, adajacent to the end of the non-stop mRNA. Finally, the exosome degrades the non-stop mRNA. (in a process known as **non-stop decay** [**NSD**])

SUMMARY Prokaryotes deal with non-stop mRNAs by tmRNA-mediated ribosome rescue. An alanyl-tmRNA, which resembles an alanyl-tRNA, binds to the vacant A site of a ribosome stalled on a non-stop mRNA, and donates its alanine to the stalled polypeptide. Then the ribosome shifts to translating an ORF on the tmRNA, adding another nine amino acids to the polypeptide before terminating. These extra amino acids target the polypeptide for destruction, and a nuclease destroys the non-stop mRNA. Eukaryotic ribosomes at the end of the poly(A) tail of a non-stop mRNA recruit the Ski7p–exosome complex to the vacant A site. Next, the Ski complex is recruited to the A site, and the exosome, positioned just at the end of the non-stop mRNA, degrades that RNA. The aberrant polypeptide is presumably also destroyed.

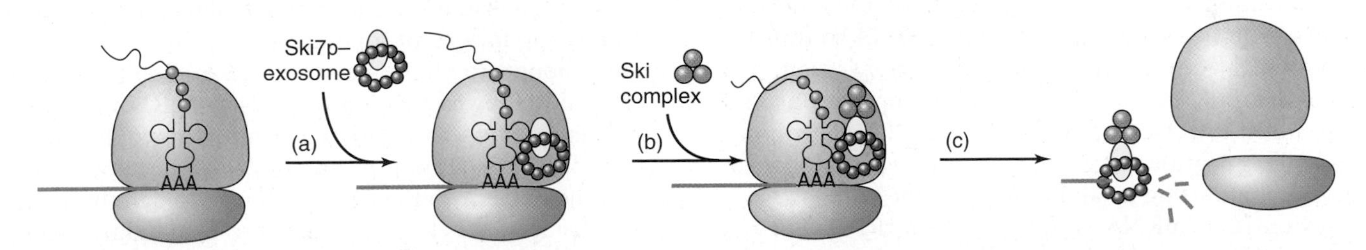

Figure 18.32 Model for exosome-mediated degradation of eukaryotic non-stop mRNA. (a) The A site of a ribosome stalled at the end of a non-stop mRNA (brown) contains zero to three nucleotides of the mRNA's poly(A) tail. Here, no A's are in the A site. This state of the ribosome is attractive to the Ski7p–exosome complex (yellow and red), which binds to the vacant A site. **(b)** Next, the Ski complex (purple) binds to the A site, and **(c)** this triggers degradation of the non-stop mRNA and release of the ribosomal subunits.

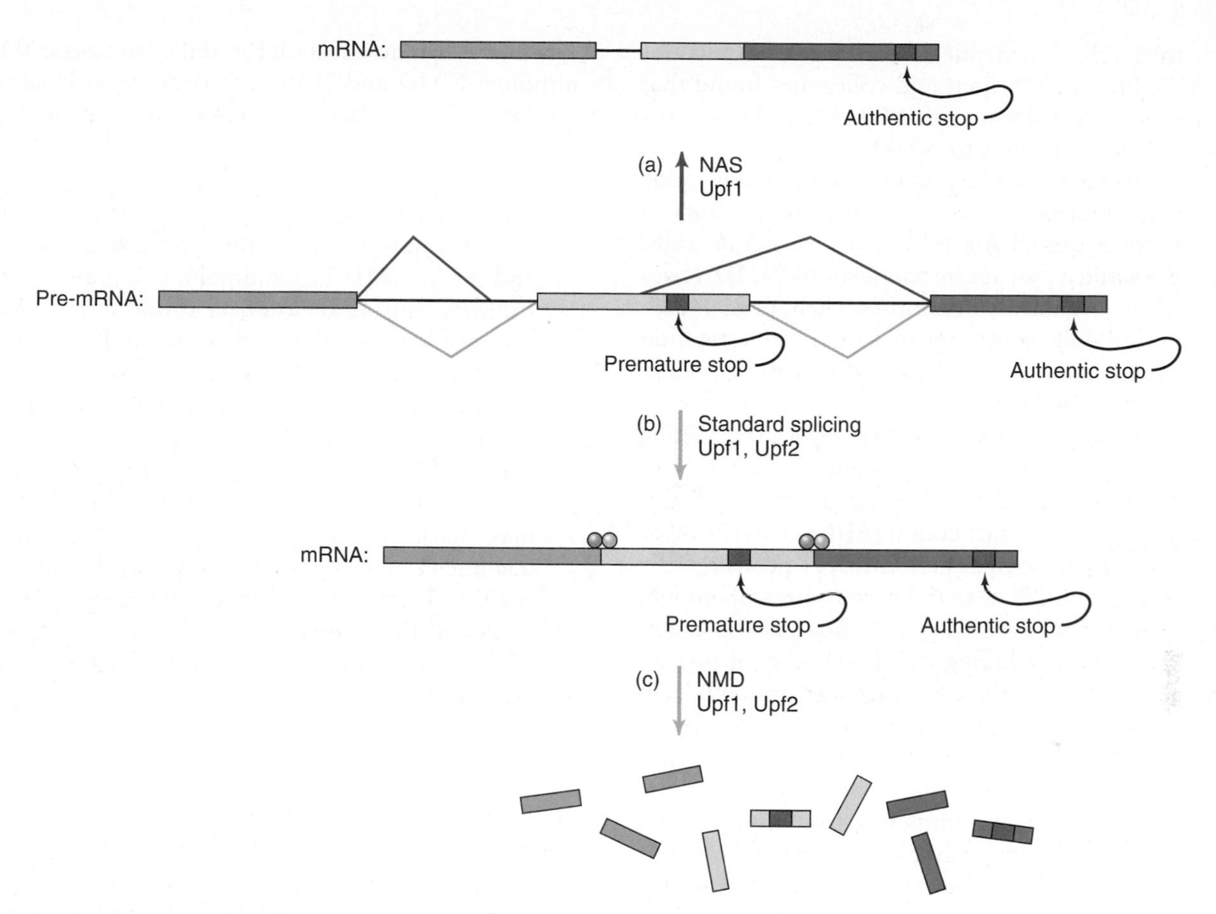

Figure 18.33 Models for NAS and NMD. (a) NAS. Upf1, perhaps in conjunction with other proteins, senses a premature stop codon in the reading frame of the future mRNA and induces an alternative splicing pattern (purple) to produce the mature mRNA at top, which lacks the premature stop codon. **(b)** Standard splicing (orange) produces a mature mRNA with a premature stop codon, and Upf1 and Upf2 bound at the exon/exon boundaries. **(c)** NMD. Upf1 and Upf2 (brown and gray), perhaps in conjunction with other proteins, sense the in-frame premature stop codon too close to the second exon/exon boundary and induce destruction of the mRNA.

Premature Termination Messenger RNAs with premature termination codons (nonsense codons) also give rise to aberrant, truncated protein products that are potentially harmful to the cell. Eukaryotic cells have evolved two ways of dealing with this problem (Figure 18.33): **nonsense-mediated mRNA decay (NMD) and nonsense-associated altered splicing (NAS).**

NMD depends on indentifying a stop codon as premature (a **premature termination codon [PTC]**). Obviously, there is an authentic stop codon at the end of every mRNA, and the cell must somehow discriminate between authentic and premature stop codons. Mammalian cells do this by measuring the distance between the stop codon and the **exon junction complex (EJC)** during the pioneer round of translation. (The EJC is a collection of proteins deposited about 20 to 25 nt upstream of exon-exon junctions at the time of splicing. If the distance between the stop codon and the EJC is short (less than about 55 nt), the stop codon is likely to be authentic, but if it is longer than about 55 nt, the stop codon is likely to be premature.

Two of the EJC proteins that are active in mammalian T cells are **Upf1** and **Upf2.** If either of these proteins is removed from a cell by RNAi (Chapter 16), NMD is inhibited. When these proteins are bound to an mRNA at a sufficiently long distance downstream of a stop codon, they recognize the stop codon as premature and activate the NMD process. On the other hand, if these proteins are relatively close to the stop codon, they are simply removed by the ribosome translating the mRNA in the pioneer round.

Lynne Macquat and colleagues presented data in 2008 that further illuminated the role of Upf1 in human NMD. They found that when translation terminates prematurely at a PTC, Upf1 binds to the downstream EJC and becomes phosphorylated. Phospho-Upf1 then binds to eIF3 and prevents the eIF3-dependent conversion of the 48S initiation complex to the 80S initiation complex that is competent to begin translation. Thus, translation is repressed, and the PTC-bearing mRNA is degraded, probably in P bodies (Chapter 16). If this model, which critically involves eIF3, is correct, then eIF3-independent translation should not exhibit NMD. Indeed, Macquat and colleagues found that eIF3-independent translation of cricket paralysis virus (CrPV) mRNA is not subject to NMD.

In contrast to the model just described, Elisa Izaurralde and colleagues reported in 2003 that the components of the EJC are not required for NMD in *Drosophila* cells, raising the possibility that the mechanism of NMD varies from one class of organisms to another. Then, in 2004, Allan Jacobson and colleagues reported on an investigation of NMD in yeast, showing that the mechanism of premature termination is itself aberrant.

In particular, Jacobson and colleagues used a toeprinting assay (Chapter 17) to show that ribosomes, once they had terminated prematurely, did not dissociate from the mRNA, but moved upstream to a start codon (AUG). This behavior could be blocked by removing the yeast Upf1 protein, or by placing a normal 3′-UTR near the premature stop codon. Furthermore, an mRNA containing a premature stop codon could be stabilized by tethering a poly(A)-binding protein (Pab1p) to the mRNA. All these findings support a model in which the ribosome recognizes a normal stop codon by its context near a 3′-UTR, or near a poly(A), and terminates normally. By contrast, the ribosome recognizes a premature stop codon as aberrant by its remoteness from these normal cues, and terminates abnormally by going back to an upstream AUG. In principle, any eukaryotic cell should be able to recognize this unusual termination and degrade the associated mRNA, but it is not yet clear how uniform the NMD mechanism is in eukaryotes.

NAS is more mysterious than NMD. When the NAS machinery detects an in-frame (but not an out-of-frame) premature stop codon, it causes the splicing apparatus to splice the pre-mRNA in an alternative way that eliminates the premature stop codon from the mature mRNA. But that scheme raises a very intriguing question: How does the NAS machinery detect the future reading frame before the pre-mRNA is even spliced?

So far, we have no answer to that question, but we do know that one of the essential players in NAS is also one of the key agents in NMD: Upf1. Harry Dietz and colleagues used RNAi to show that Upf1, but not Upf2, is required for NAS. Then they refined their technique to ask whether the same parts of Upf1 are required for both NMD and NAS. To do this, they used **allele-specific RNAi** as follows: They made an altered *Upf1* gene that was not subject to RNAi caused by the double-stranded RNA that blocks expression of the endogenous gene. Then they introduced this altered gene, on a plasmid, into cells experiencing RNAi directed at the endogenous *Upf1* gene. The altered gene could rescue both NAS and NMD, which would otherwise have been blocked due to loss of *Upf1* expression.

Next, Dietz and colleagues made mutations to conserved regions of the altered *Upf1* gene. One of these mutations knocked out the ability of the altered gene to rescue NMD, but had no effect on the ability to rescue NAS. Thus, although NMD and NAS both depend on Upf1, they apparently rely on different functions of the protein.

SUMMARY Eukaryotes deal with premature termination codons by two different mechanisms: NMD and NAS. NMD in mammalian cells relies on the ribosome during the pioneer round to measure the distance between the stop codon and the EJC. If it is too long, the mRNA is destroyed. In yeast, the cell appears to recognize a premature stop codon by the absence of a normal 3′-UTR or poly(A) nearby. When a ribosome stops at a premature stop codon, it moves to an upstream AUG, and this may mark the mRNA for destruction. The NAS machinery senses a stop codon in the middle of a reading frame and changes the splicing pattern such that the premature stop codon is spliced out of the mature mRNA. Like NMD, this process also requires Upf1.

No-go Decay In 2006, Meenakshi Doma and Roy Parker identified another kind of mRNA decay, which they dubbed **“no-go decay (NGD).”** They artificially induced a ribosome stall by creating an mRNA with a very stable stem-loop that the ribosome was incapable of traversing. Yeast cells degraded this mRNA faster than they did the wild-type mRNA lacking the stem-loop.

Doma and Parker found that this accelerated decay occurred in cells that were deficient in either decapping or 3′→5′ exonucleases, which are key elements of the usual 5′→3′ and 3′→5′ decay, respectively, in yeast. And they found that decay is also accelerated in cells defective in NMD because of a mutation in *Upf1*.

If decay is not happening by the usual pathways, how is it accomplished? Doma and Parker showed that the no-go mRNA was cleaved by an endonuclease at a site near the stable stem-loop that had stalled the ribosome. This cut within the mRNA created new 3′- and 5′-ends that are substrates for degradation by the usual 3′- and 5′-endonucleases.

Natural mRNAs are not likely to contain stable stem-loops that arrest ribosomes, so no-go decay probably acts on ribosomes that are stalled because of natural causes such as defective mRNAs or ribosomes. It also provides another potential means of post-transcriptional control by selective degradation of mRNAs.

SUMMARY Stalled ribosomes can trigger no-go decay of mRNA, which begins with an endonucleolytic cleavage near the stalled ribosome.

Use of Stop Codons to Insert Unusual Amino Acids

Most proteins contain only the 20 amino acids pictured in Figure 3.2. However, a few proteins require unusual amino acids. The first unusual amino acids to be discovered, such as hydroxyproline, were found to arise through posttranslational modification of proteins made from the standard 20 amino acids. More recently, other unusual amino acids, such as selenocysteine and pyrrolysine, have been shown to be incorporated directly into growing polypeptides. In these cases, mechanisms have evolved to take advantage of stop codons in the middle of coding regions. Cells interpret these stop codons, not as termination signals, but as codons for unusual amino acids.

The first unusual amino acid discovered in proteins (the "21st amino acid") was **selenocysteine,** which looks just like cysteine except that it has a selenium atom in place of the sulfur atom. Some enzymes, such as glutathione peroxidase and formate dehydrogenase, do not work without selenocysteine. Each requires a single selenocysteine residue as part of its active site. But how can this unusual amino acid be incorporated into proteins? The genes that encode these enzymes produce mRNAs with UGA stop codons in the positions where selenocysteine is needed. Furthermore, in the absence of selenium, translation stops prematurely at these stop codons. These findings suggest that the cell somehow interprets these UGA codons as selenocysteine codons. But how?

A special tRNA with an anticodon that recognizes the UGA stop codon can be charged with serine by a normal seryl-tRNA synthetase. Then, the serine in this special seryl-tRNA is converted to selenocysteine. A special EF-Tu can then deliver this altered aminoacyl-tRNA to the ribosome in response to the UGA codon in the middle of the mRNA—but not to UGA codons at the ends of coding regions. If the latter were the case, selenocysteines would be incorporated in response to authentic stop codons, hindering termination.

Thus, the UGA codons within an mRNA are only part of the signal that recruits the selenocysteinyl-tRNA. Other parts of the mRNA must also play a role. In the case of the formate dehydrogenase mRNA, this is a region about 40 nt downstream of the internal UGA, and in another mRNA, it is a region about 1000 nt downstream, in the 3′-untranslated region of the mRNA. Such an mRNA region, which dictates that a UGA codon should be recognized as a selenocysteine (Sec) codon, is called a **Sec insertion sequence,** or **SECIS.** A SECIS is a stem-loop in the mRNA with three short conserved motifs. These conserved sequences are clearly important, because mutations within them prevent selenocysteine incorporation.

The "22nd amino acid" is **pyrrolysine,** which has the structure shown in Figure 18.34. Unlike selenocysteine, which is widespread, pyrrolysine has so far been found only in certain methanogenic (methane-producing) archaea. Also unlike selenocysteine, which is built from a normal amino acid (serine) on seryl-tRNA, pyrrolysine is first synthesized and then added to a special tRNA by a special pyrrolysyl-tRNA synthetase. This is the 21st aminoacyl-tRNA synthetase ever found—the only one aside from the 20 that charge normal tRNAs with the 20 normal amino acids.

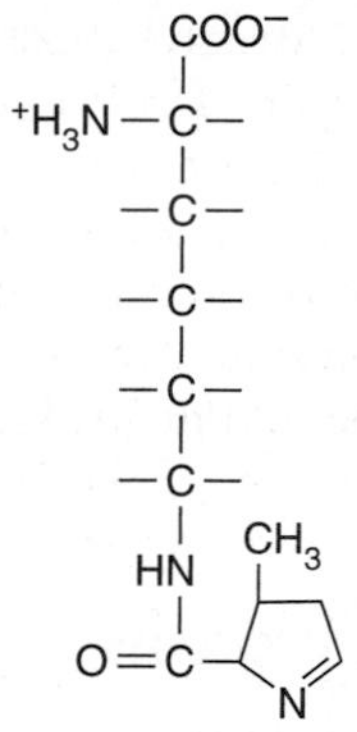

Figure 18.34 Pyrrolysine.

E. coli cells cannot normally incorporate pyrrolysine into their proteins. But Joseph Krzycki and colleagues showed in 2004 that they could endow *E. coli* cells with the ability to do this incorporation if they added three things to the cells: a gene for the special tRNA, a gene for the special pyrrolysyl-tRNA synthetase, and pyrrolysine itself. Furthermore, they showed that the tRNA can accept preformed pyrrolysine in vitro, strongly suggesting that this is the way it works in vivo.

As is the case with selenocysteine, pyrrolysine is incorporated into growing polypeptides in response to a stop codon, but it is the UAG codon instead of UGA. This implies that the anticodon of the special tRNA is 5′-CUA-3′, and that is indeed the case.

SUMMARY The unusual amino acids selenocysteine and pyrrolysine are incorporated into growing polypeptides in response to the termination codons UGA and UAG, respectively, as follows: (1) Selenocysteine: A special tRNA (with an anticodon that recognizes the UGA codon) is charged with serine, which is then converted to selenocysteine, and the selenocysteyl-tRNA is escorted to the ribosome by a special EF-Tu. (2) Pyrrolysine: A special pyrrolysyl-tRNA synthetase joins preformed pyrrolysine with a special tRNA that has an anticodon that recognizes the codon UAG.

18.5 Posttranslation

The story of translation does not end with termination. Proteins must fold properly and ribosomes need to be released from the mRNA so they can engage in further rounds of translation. Strictly speaking, the first of these processes does not occur after translation; rather, it is a cotranslational event that occurs as the nascent polypeptide is being made. However it is convenient to deal with it separately, as it has no direct relationship to the initiation, elongation, and termination events we have been discussing. Let us consider the folding problem first, then the ribosomal release problem.

Folding Nascent Proteins

Native proteins are folded so that any hydrophobic (Greek: "water-fearing") regions are buried in the interiors of the proteins, away from the aqueous environment in the cell. But most proteins do not fold into their proper shapes by themselves. They need help from molecular chaperones, just as proteins that have been unfolded by heat shock do (Chapter 8). The problem is that any exposed hydrophobic sections of a nascent polypeptide would try to interact with any other exposed hydrophobic regions they could find, to hide from the water surrounding them. But the nearest hydrophobic region is likely to be the wrong partner, so that interaction would lead to a misfolded and therefore inactive protein. In fact, some misfolded proteins, such as the one involved in bovine spongiform encephalopathy (BSE, or "mad cow disease") can be deadly toxic to a cell.

Here is another example of the importance of proper protein folding: Silent mutations occur when a codon for an amino acid is changed into another codon for that same amino acid. Ordinarily, such mutations have no effect, which is why we call them silent. Occasionally, however, "silent" mutations can actually cause problems. This has been documented to occur in several ways: The change of one codon for an amino acid to another codon for the same amino acid sounds harmless, but if the new codon is much rarer for that organism (a phenomenon known as **codon bias**), the corresponding tRNA is probably also rare, so the ribosome slows down at that codon waiting for the rare aminoacyl-tRNA to appear. Some proteins fold differently depending on their rate of synthesis, so slowing down translation while waiting for a rare aminoacyl-tRNA can cause misfolding, and perhaps inactivation, of the protein product. Michael Gottesman and colleagues demonstrated in 2007 that a mutation in the human *multidrug resistant* 1 (*MDR*1) gene, though it is a "silent" mutation, creates a rare codon and yields a product with altered, and less effective, activity, presumably because of misfolding.

On the other hand, ribosomal pausing between domains (independently folded parts) of a protein can be beneficial because it allows these domains to fold without interference from irrelevant other parts of the protein. Thus, it was intriguing that Joseph Watts, Kevin Weeks, and their colleagues showed in 2009 that the HIV (human immunodeficiency virus) RNA, which serves as both genome and mRNA, has its highest levels of secondary structure in the regions of the mRNAs that encode loops between protein domains. These regions of secondary structure (intramolecular base-pairing) would presumably impede the progress of the ribosome and allow the recently completed protein domain to fold before beginning the synthesis of the next domain.

To probe the secondary structure of the HIV RNA, Watts and Weeks and their colleagues used a technique known as **selective 2′-hydroxyl acylation analyzed by primer extension (SHAPE)**. This method relies on the fact that certain reagents, such as 1-methyl-7-nitroisatoic anhydride (1M7), selectively acylate the 2′-hydroxyl groups of RNA nucleotides that are conformationally flexible. Nucleotides that are base-paired are rigid and relatively protected from acylation. After reacting the RNA with 1M7, the investigators subjected it to primer extension (Chapter 5) with reverse transcriptase and fluorescent primers. Then they analyzed the lengths of the extended primers to locate regions of base-pairing, where the primer extension tends to stop.

Combining this direct analysis of secondary structure with computational analysis of likely secondary structure allowed Watts, Weeks and colleagues to build a low-resolution model of secondary structure encompassing the entire RNA. The HIV RNA encodes 15 mature proteins. Three of its nine open reading frames encode polyproteins that must be cleaved by a protease to yield the mature proteins. For example, the Gag-Pol polyprotein contains the protease, the reverse transcriptase, and the integrase. In Chapter 23 we will discuss HIV and other retroviruses in more detail. The secondary structure model showed a striking correspondence between likely secondary structure and the coding regions for the loops between protein domains, and between mature protein sequences in the polyproteins. Thus, the RNA appears to have a regulatory code written into its sequence that would cause ribosomes to encounter RNA secondary structure and pause between coding regions for protein domains. And this pausing should help with protein folding during translation.

Joshua Plotkin and colleagues enriched this discussion in 2009 when they created a library of 154 genes encoding green fluorescent protein (GFP), all containing "silent" mutations that did not change the coding of the gene. But, when these genes were expressed in *E. coli*, they yielded protein levels that differed by a factor of 250. Codon bias played little or no role in this variation; instead, the stability of mRNA folding, particularly around the Shine-Dalgarno sequence, was the most important factor.

To minimize misfolding, the cell needs a mechanism to hide hydrophobic sections of a nascent polypeptide until the right partner is made. Ordinary molecular chaperones do this by enveloping exposed hydrophobic protein regions in a hydrophobic pocket of their own, and preventing inappropriate associations with other exposed hydrophobic regions. But *E. coli* has a special chaperone called **trigger factor** that associates with the large ribosomal subunit and catches newly-synthesized hydrophobic regions in a hydrophobic basket to protect them from water.

To see how trigger factor does its job, it would be ideal to have the crystal structure of the chaperone bound to its ribosomal docking site. But that presents a problem: The only large ribosomal subunit that has been crystallized is from the archaeon *Haloarcula marismortui* (Chapter 19), but archaea do not have trigger factor. So Nenad Ban and colleagues crystallized the whole *E. coli* trigger factor to see its shape, and then crystallized the ribosome-binding part of *E. coli* trigger factor together with the archaeal large ribosomal subunit, in hopes that the ribosomal binding site was conserved well enough between archaea and bacteria that such a cross-kingdom complex would form.

And the strategy worked! The binding site for trigger factor (on ribosomal protein L23) is highly conserved between bacteria and archaea, so the ribosomal subunit for an archaeon can bind to a bacterial trigger factor. The crystal structure of trigger factor alone suggested to Ban and colleagues a "crouching dragon" with a head, back, arms, and tail, as illustrated in Figure 18.35. Based on the cocrystal structure of the 50S ribosomal subunit with the tail domain of trigger factor, Ban and colleagues positioned trigger factor as shown in Figure 18.35, with the "dragon crouching" upside down. This places the hydrophobic surface of the tail and arm domains in perfect position to catch the nascent polypeptide as it exits through the ribosomal exit tunnel. This would effectively sequester any exposed hydrophobic regions of the nascent polypeptide until they can associate with the appropriate partner hydrophobic regions.

Trigger factor is not essential for *E. coli* life, because bacteria have a backup system: a chaperone called DnaK. It is freestanding protein, rather than a ribosome-associated protein like trigger factor. Instead of a basket to catch nascent proteins, DnaK has a hydrophobic arch that protects exposed hydrophobic regions of nascent proteins until they can fold properly. Archaea and eukaryotes lack trigger-factor-like proteins entirely, so they rely exclusively on freestanding chaperones for proper folding of nascent proteins.

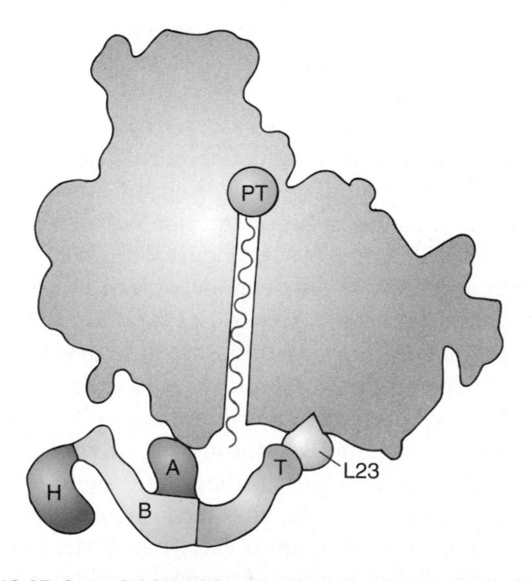

Figure 18.35 A model for trigger factor bound to a ribosome. The chaperone protein, trigger factor, is bound like an upside-down crouching dragon to the bottom of the ribosome, covering the exit tunnel. In this position, the hydrophobic domains of trigger factor (arm [A] and tail [T], purple and blue, respectively) can catch hydrophobic regions of a nascent polypeptide as they emerge from the exit tunnel, and keep them in a hydrophobic environment until they can pair with other hydrophobic regions of the nascent polypeptide, promoting proper folding. The other domains of trigger factor are the head (H, red), and the back (B, yellow). L23 (green) is one of the proteins of the large ribosomal subunit, and is the site of major contacts with trigger factor. PT (orange) is the peptidyl transferase site at the beginning of the exit tunnel. (*Source:* Adapted from Ferbitz, L., T. Maier, H. Patzelt, B. Bukau, E. Deverling, and N. Ban, Trigger factor in complex with the ribosome forms a molecular cradle for nascent proteins, *Nature* 431:593, 2004.)

SUMMARY Most newly-made polypeptides do not fold properly by themselves, but require help from molecular chaperones. *E. coli* cells have a protein called trigger factor that associates with the ribosome in such a way as to catch the nascent polypeptide as it emerges from the ribosome's exit tunnel. Thus, hydrophobic regions of the nascent polypeptide are protected from inappropriate associations until the appropriate partner is available. Archaea and eukaryotes lack trigger factor, so they must use freestanding chaperones, which are also present in bacteria. "Silent" mutations can affect translation rates, even though they do not change the sequence of the protein product.

Release of Ribosomes from mRNA

Early studies on termination used model systems, including just AUG and UAG as mRNA analogs, and these studies did not detect a need for ribosome release, in part because some of the model mRNAs dissociated from ribosomes spontaneously.

Then A. Kaji and colleagues discovered a protein factor that could release ribosomes from natural mRNAs in **post-termination complexes (post-TCs)**. They named it **ribosomal recycling factor (RRF).** Then in 1994, Kaji and colleagues demonstrated that RRF is essential for bacterial life. In temperature-sensitive mutants in the gene for RRF, shift to the nonpermissive temperature killed bacteria in lag phase and arrested the growth of bacteria in log phase. Thus, release of ribosomes from mRNAs after termination of translation is essential.

Kaji and colleagues purified RRF from the bacterium *Thermotoga maritima* using the following assay to detect RRF: They treated bacterial polysomes with puromycin to

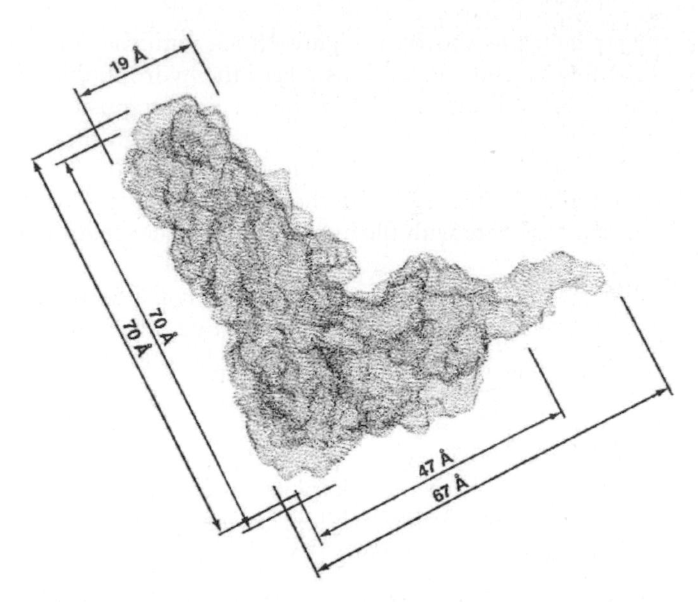

Figure 18.36 Superimposition of the structures of RRF and a tRNA. The surfaces of the *Thermotoga maritima* RRF (blue) and yeast $tRNA^{Phe}$ (red) are superimposed to show their great similarity. (*Source:* From Selmer M., Al-Karadaghi S., Hirokawa G., Kaji A., and Liljas A. 1999. Crystal structure of Thermotoga maritima ribosome recycling factor: A tRNA mimic. Science 286:2349. © 1999 AAAS.)

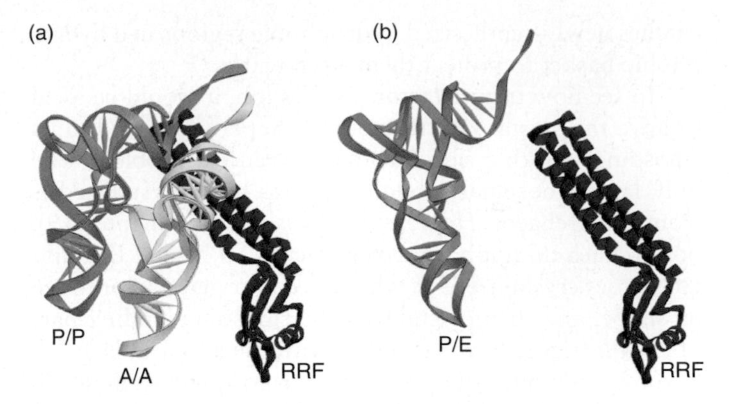

Figure 18.37 Model for the position of RRF in the ribosome. **(a)** Position of RRF (red) relative to tRNAs bound in the pure A site (A/A, yellow) and the pure P site (P/P, orange). **(b)** Position of RRF (red) relative to a tRNA in the hybrid P/E site (orange). (*Source:* Reprinted from Cell v. III, Lancaster et al., p. 444 © 2002, with permission from Elsevier Science.)

release the nascent polypeptide. This left each of the ribosomes with two deacylated tRNAs, one in the P site and one in the E site. Thus, each of the ribosomes in these polysomes resembled a ribosome that had just experienced termination, except that there was no termination codon in the A site. To these puromycin-treated polysomes, these workers added RRF, which converted the polysomes to monosomes. Once it was purified, Kaji and colleagues, in collaboration with Anders Liljas and colleagues, determined the crystal structure of RRF.

The crystal structure was striking—an almost perfect mimic of a tRNA. Figure 18.36 shows the structure of the *T. maritima* RRF superimposed on the structure of $tRNA^{Phe}$. The fit is nearly perfect; the only things missing from RRF are amino acids to fill in the space normally occupied by the terminal CCA of the tRNA, and a small piece of the anticodon. Based on this structure and other information, Kaji and colleagues proposed that RRF binds to the A site, just like an aminoacyl-tRNA would, thereby allowing translocation to occur in the presence of EF-G, and then somehow releases the ribosome from the mRNA.

Then in 2002, Kaji and colleagues, in collaboration with Noller and colleagues, performed structural studies on RRF–ribosome complexes using **hydroxyl radical probing.** They employed this method as follows: First, they used site-directed mutagenesis to replace the single cysteine in the RRF molecule with serine. Then they mutagenized this cysteine-free RRF, which still retained activity, to place cysteine at each of 10 different locations throughout the RRF molecule. Each of these RRF molecules with a single cysteine could be coupled to a molecule bearing Fe^{2+}, and then the RRF-Fe^{2+} could be bound to ribosomes. The Fe^{2+} creates hydroxyl radicals that break nearby segments of rRNA, and these breaks can be detected by primer extension (Chapter 5). Because we know exactly where each part of the 16S and 23S rRNAs are located in the ribosome (Chapter 19), different parts of RRF could be mapped to specific locations on the ribosome.

This experiment demonstrated that, despite its near-perfect structural resemblance to tRNA, RRF does not behave just like a tRNA in binding to the ribosome. It binds to the A site of the ribosome in an orientation very different from that of a tRNA in the A site (Figure 18.37a). This result called into question the simple model of Kaji and colleagues. In fact, it even raised the question of how RRF could bind to the ribosome in the way it does because the end of RRF would overlap with the acceptor stem of a deacylated tRNA bound in the P site. But Kaji, Noller, and colleagues noted that a tRNA deacylated by puromycin, or presumably by RF1 or RF2, does not exist in the pure P site-bound state. Instead, as Noller and colleagues have shown, it is in a hybrid P/E state, with its acceptor end in the E site and its anticodon in the P site. In this position, it would not interfere with RRF's binding, as illustrated in Figure 18.37b.

What happens after RRF binds to the A site? That is still poorly understood, though we know it acts with EF-G to release the ribosome from the mRNA. Some of the time, it could release just the 50S subunit, leaving the 30S subunit to be released by another mechanism, perhaps by binding to IF3.

Eukaryotes do not encode an RRF, so how do they dissociate post-TCs? Tatyana Pestova and colleages showed in 2007 that eIF3 is the most important factor in eukaryotic ribosome release, and it gets help from eIF1, eIF1A, and eIF3j, which is a loosely bound subunit of eIF3.

SUMMARY Ribosomes do not release from the mRNA spontaneously after termination. Bacterial ribosomes need help from ribosome recycling factor (RRF) and EF-G. RRF strongly resembles a tRNA and can bind to the ribosome's A site, but in a position not normally taken by a tRNA. Then it collaborates with EF-G in releasing either the 50S ribosomal subunit, or the whole ribosome. Eukaryotic ribosomes are released from post-TCs by eIF3, aided by eIF1, eIF1A, and eIF3j.

SUMMARY

Messenger RNAs are read in the $5' \rightarrow 3'$ direction, the same direction in which they are synthesized. Proteins are made in the amino to carboxyl direction, which means that the amino-terminal amino acid is added first.

The genetic code is a set of three-base code words, or codons, in mRNA that instruct the ribosome to incorporate specific amino acids into a polypeptide. The code is nonoverlapping: that is, each base is part of only one codon. It is also devoid of gaps, or commas; that is, each base in the coding region of an mRNA is part of a codon. There are 64 codons in all. Three are stop signals, and the rest code for amino acids. This means that the code is highly degenerate.

Part of the degeneracy of the genetic code is accommodated by isoaccepting species of tRNA that bind the same amino acid but recognize different codons. The rest is handled by wobble, in which the third base of a codon is allowed to move slightly from its normal position to form a non-Watson–Crick base pair with the anticodon. This allows the same aminoacyl-tRNA to pair with more than one codon. The wobble pairs are G–U (or I–U) and I–A.

The genetic code is not strictly universal. In certain eukaryotic nuclei and mitochondria and in at least one bacterium, codons that cause termination in the standard genetic code can code for amino acids such as tryptophan and glutamine. In several mitochondrial genomes, the sense of a codon is changed from one amino acid to another. These deviant codes are still closely related to the standard one from which they probably evolved.

Elongation takes place in three steps: (1) EF-Tu, with GTP, binds an aminoacyl-tRNA to the ribosomal A site. (2) Peptidyl transferase forms a peptide bond between the peptide in the P site and the newly arrived aminoacyl-tRNA in the A site. This lengthens the peptide by one amino acid and shifts it to the A site. (3) EF-G, with GTP, translocates the growing peptidyl-tRNA, with its mRNA codon, to the P site, and moves the deacylated tRNA in the P site to the E site.

Puromycin resembles an aminoacyl-tRNA, and so can bind to the A site, couple with the peptide in the P site, and release it as peptidyl puromycin. On the other hand, if the peptidyl-tRNA is in the A site, puromycin will not bind to the ribosome, and the peptide will not be released. This defines two sites on the ribosome: the P site, in which the peptide in a peptidyl-tRNA is puromycin reactive, and the A site, in which the peptide in a peptidyl-tRNA is puromycin unreactive. fMet-$tRNA_f^{Met}$ is puromycin reactive in the 70S initiation complex, so it is in the P site. Binding and structural studies have identified a third binding site (the E site) for deacylated tRNA. Such tRNAs bind to the E site as they exit the ribosome, and this binding helps maintain the reading frame of the mRNA.

A ternary complex formed from EF-Tu, aminoacyl-tRNA, and GTP delivers an aminoacyl-tRNA to the ribosome's A site, without hydrolysis of the GTP. In the next step, GTP is hydrolyzed by a ribosome-dependent GTPase activity of EF-Tu, and an EF-Tu–GDP complex dissociates from the ribosome. EF-Ts regenerates an EF-Tu–GTP complex by exchanging GTP for GDP attached to EF-Tu. Addition of aminoacyl-tRNA then reconstitutes the ternary complex for another round of translation elongation.

The protein-synthesizing machinery achieves accuracy during elongation in a two-step process. First, it gets rid of ternary complexes bearing the wrong aminoacyl-tRNA before GTP hydrolysis occurs. If this screen fails, it can still eliminate the incorrect aminoacyl-tRNA in the proofreading step before the wrong amino acid can be incorporated into the growing protein chain. Both these screens may rely on the weakness of incorrect codon–anticodon base pairing to ensure that dissociation will occur more rapidly than either GTP hydrolysis or peptide bond formation. The balance between speed and accuracy of translation is delicate. If peptide bond formation goes too fast, incorrect aminoacyl-tRNAs do not have enough time to leave the ribosome, so their amino acids are incorporated into protein. But if translation goes too slowly, proteins are not made fast enough for the organism to grow successfully.

Peptide bonds are formed by a ribosomal enzyme called peptidyl transferase. This activity resides on the 50S subunit. The 23S rRNA contains the catalytic center of the peptidyl transferase.

Each translocation event moves the mRNA one codon's length, 3 nt, through the ribosome. GTP and EF-G are necessary for translocation, although translocation activity can be expressed without EF-G and GTP in vitro. For a new round of elongation to occur, GTP hydrolysis releases EF-G from the ribosome. The three-dimensional shapes of the EF-Tu–tRNA–GDPNP ternary complex and the EF-G–GDP binary complex have been determined by x-ray crystallography. As predicted, they are very similar.

Amber, ochre, and opal mutations create termination codons (UAG, UAA, and UGA, respectively) in the middle of a message and thereby cause premature termination of translation. These three codons are also the natural stop signals at the ends of coding regions in mRNAs. Most suppressor tRNAs have altered anticodons that can recognize stop codons and prevent termination by inserting an amino acid and allowing the ribosome to move on to the next codon.

Prokaryotic translation termination is mediated by three factors: RF1, RF2, and RF3. RF1 recognizes the termination codons UAA and UAG; RF2 recognizes UAA and UGA. RF3 is a GTP-binding protein that facilitates release of RF1 and RF2 from the ribosome. Eukaryotes have two release factors: eRF1, which recognizes all three termination codons, and eRF3, a ribosome-dependent GTPase that helps eRF1 recognize stop codons and release the finished polypeptide.

Prokaryotes deal with non-stop mRNAs by tmRNA-mediated ribosome rescue. An alanyl-tmRNA, which resembles an alanyl-tRNA, binds to the vacant A site of a ribosome stalled on a non-stop mRNA and donates its alanine to the stalled polypeptide. Then the ribosome shifts to translating an ORF on the tmRNA, adding another nine amino acids to the polypeptide before terminating. These extra amino acids target the polypeptide for destruction, and a nuclease destroys the non-stop mRNA. Eukaryotic ribosomes at the end of the poly(A) tail of a non-stop mRNA recruit the Ski7p–exosome complex to the vacant A site. Next, the Ski complex is recruited to the A site, and the exosome, positioned just at the end of the non-stop mRNA, degrades that RNA. The aberrant polypeptide is presumably also destroyed.

Eukaryotes deal with premature termination codons by two different mechanisms: NMD and NAS. NMD in mammalian cells involves a downstream destabilizing element, including Upf1 and Upf2 bound to an mRNA at exon–exon junctions that measures the distance to a stop codon. If the codon is far enough upstream, it looks like a premature stop codon and activates the downstream destabilizing element to degrade the mRNA. In yeast, the absence of a normal 3′-UTR or poly(A) near a stop codon may identify it as abnormal. The NAS machinery senses a stop codon in the middle of a reading frame and changes the splicing pattern such that the premature stop codon is spliced out of the mature mRNA. Like NMD, this process also requires Upf1.

The unusual amino acids selenocysteine and pyrrolysine are incorporated into growing polypeptides in response to the termination codons UGA and UAG, respectively, as follows: (1) Selenocysteine: A special tRNA (with an anticodon that recognizes the UGA codon) is charged with serine, which is then converted to selenocysteine, and the selenocysteyl-tRNA is escorted to the ribosome by a special EF-Tu. (2) Pyrrolysine: A special pyrrolysyl-tRNA synthetase joins preformed pyrrolysine with a special tRNA that has an anticodon that recognizes the codon UAG.

Most newly-made polypeptides do not fold properly by themselves, but require help from molecular chaperones. *E. coli* cells have a protein called trigger factor that associates with the ribosome in such a way as to catch the nascent polypeptide as it emerges from the ribosome's exit tunnel. Thus, hydrophobic regions of the nascent polypeptide are protected from inappropriate associations until the appropriate partner is available. Archaea and eukaryotes lack trigger factor, so they must use freestanding chaperones, which are also present in bacteria.

Ribosomes do not release from the mRNA spontaneously after termination; they need help from ribosome recycling factor (RRF) and EFG. RRF strongly resembles a tRNA and can bind to the ribosome's A site, but in a position not normally taken by a tRNA. Then it collaborates with EFG in releasing either the 50S ribosomal subunit, or the whole ribosome, by an unknown mechanism.

REVIEW QUESTIONS

1. Describe and give the results of an experiment that shows that translation starts at the amino terminus of a protein.
2. How do we know that mRNAs are read in the 5′→3′ direction?
3. How do we know that the genetic code is: (a) nonoverlapping; (b) commaless; (c) triplet; (d) degenerate?
4. Describe and give the results of an experiment that reveals two of the codons for an amino acid.
5. Diagram a wobble base pair. You do not have to show the positions of all the atoms, just the shape of the base pair. Contrast this with the shape of a Watson–Crick base pair. What is the importance of wobble in translation?
6. Diagram the translation elongation process in prokaryotes.
7. Diagram the mode of action of puromycin.
8. Describe and give the results of an experiment that shows that fMet-$tRNA_f^{Met}$ occupies the P site of the ribosome.
9. Describe and give the results of an experiment that shows that EF-Ts releases GDP from EF-Tu.
10. What step in translation does chloramphenicol block?
11. Diagram the roles of EF-Tu and EF-Ts in translation.
12. Present evidence for the formation of a ternary complex among EF-Tu, GTP, and aminoacyl-tRNA.
13. Describe and give the results of an experiment that shows that ribosomal RNA is likely to be the catalytic agent in peptidyl transferase.
14. What are the initial recognition and proofreading steps in protein synthesis?

15. Describe and give the results of an experiment that shows that the mRNA moves in 3-nt units in the translocation step.
16. Describe and give the results of an experiment that shows that EF-G and GTP are both required for translocation. What are the effects of (a) substituting GDPCP for GTP, and (b) adding fusidic acid in this single-translocation event assay?
17. Describe an experiment that shows that GTP hydrolysis precedes translocation.
18. Present direct evidence that the amber codon is a translation terminator.
19. Present evidence that the amber codon is UAG.
20. Explain how an amber suppressor works.
21. Present evidence that the amber suppressor is a tRNA.
22. Describe an assay for a release factor.
23. What are the roles of RF1, RF2, and RF3?
24. How do we know which termination codons RF1 and RF2 recognize?
25. What are the roles of eRF1 and eRF3?
26. Diagram the mechanism by which prokaryotes deal with non-stop mRNAs.
27. What differences between tmRNAs and tRNAs limit the ability of tmRNAs to bind tightly to the ribosome? How does the cell deal with these deficiencies?
28. Diagram the mechanism by which mammalian cells deal with non-stop mRNAs.
29. Diagram two mechanisms by which eukaryotic cells deal with premature termination codons.
30. Describe the mechanisms by which selenocysteine and pyrrolysine are incorporated into proteins.
31. How does trigger factor's cellular location help it in its chaperone function?

ANALYTICAL QUESTIONS

1. What would be the effect on a G protein's activity if:
   a. its GAP were inhibited?
   b. its guanine nucleotide exchange protein were inhibited?
2. You have isolated an *E. coli* mutant with an aminoacyl-tRNA synthetase that causes a tRNA with the anticodon 3′-UUC-5′ to be charged with asparagine at the elevated temperature of 42°C. What effect would you expect this to have on protein synthesis in these cells at 42°C, and why? You then isolate another mutant that suppresses the first mutation, and you trace the second mutation to a tRNA gene. What tRNA would you expect to be altered in the second mutant, and where? Predict the nature of this alteration.
3. Consider this short mRNA: 5′-AUGGCAGUGCCA-3′. Answer the following questions, assuming first that the code is fully overlapping and then that it is nonoverlapping.
   a. How many codons would be represented in this oligonucleotide?
   b. If the second G were changed to a C, how many codons would be changed?
4. What would be the effect on reading frame and gene function if
   a. two bases were inserted into the middle of an mRNA?
   b. three bases were inserted into the middle of an mRNA?
   c. one base were inserted into one codon and one subtracted from the next?
5. If codons were six bases long, what kind of product would you expect from a repeating tetranucleotide such as poly (UUCG)?
6. How many codons would exist in a genetic code that had codons that were four bases long?
7. A certain ochre suppressor inserts glutamine in response to the ochre codon. What is the likeliest change in the anticodon of a $tRNA^{Gln}$ that created this suppressor strain?
8. Describe the evolutionary changes that had to occur to give an organism the ability to incorporate pyrrolysine into its proteins. In what order do you think these changes occurred? Why? *Hint:* See Wang, L. (2003). Expanding the genetic code. *Science* 302:584–85.
9. Each of the 20 amino acids can be found in natural proteins adjacent to each of the other amino acids. How does this prove that the genetic code is nonoverlapping?

SUGGESTED READINGS

General References and Reviews

Horwich, A. 2004. Sight at the end of the tunnel. *Nature* 431:520–22.

Kaji, A., M.C. Kiel, G. Hirokawa, A.R. Muto, Y. Inokuchi, and H. Kaji. 2001. The fourth step of protein synthesis: Disassembly of the posttermination complex is catalyzed by elongation factor G and ribosome recycling factor, a near-perfect mimic of tRNA. *Cold Spring Harbor Symposia on Quantitative Biology* 66:515–29.

Kaziro, Y. 1978. The role of guanosine 5′-triphosphate in polypeptide chain elongation. *Biochimica et Biophysica Acta* 505:95–127.

Khorana, H.G. 1968. Synthesis in the study of nucleic acids. *Biochemistry Journal* 109:709–25.

Maquat, L.E. 2002. Skiing toward nonstop mRNA decay. *Science* 295:2221–22.

Moore, M.J. 2002. No end to nonsense. *Science* 298:370–71.

Moore, S.D., K.E. McGinness, and R.T. Sauer. 2003. A glimpse into tmRNA-mediated ribosome rescue, *Science* 300:72–73.

Nakamura, Y., K. Ito, and M. Ehrenberg. 2000. Mimicry grasps reality in translation termination. *Cell* 101:349–52.

Nakamura, Y., K. Ito, and L.A. Isaksson. 1996. Emerging understanding of translation termination. *Cell* 87:147–50.

Nierhaus, K.H. 1996. An elongation factor turn-on. *Nature* 379:491–92.

Ramakrishnan, V. 2002. Ribosome structure and the mechanism of translation. *Cell* 108:557–72.

Schimmel, P. and K. Beebe. 2004. Genetic code seizes pyrrolysine. *Nature* 431:257–58.

Schmeing, T.M. and V. Ramakrishnan. 2009. What recent ribosome structures have revealed about the mechanism of translation. *Nature* 461:1234–42.

Thompson, R.C. 1988. EFTu provides an internal kinetic standard for translational accuracy. *Trends in Biochemical Sciences* 13:91–93.

Tuite, M.F. and I. Stansfield. 1994. Knowing when to stop. *Nature* 372:614–15.

Research Articles

Benzer, S. and S.P. Champe. 1962. A change from nonsense to sense in the genetic code. *Proceedings of the National Academy of Sciences USA* 48:1114–21.

Brenner, S., A.O.W. Stretton, and S. Kaplan. 1965. Genetic code: The "nonsense" triplets for chain termination and their suppression. *Nature* 206:994–98.

Bretscher, M.S., and K.A. Marcker. 1966. Peptidyl-sRibonucleic acid and amino-acyl-sRibonucleic acid binding sites on ribosomes. *Nature* 211:380–84.

Crick, F.H.C., L. Barnett, S. Brenner, and R.J. Watts-Tobin. 1961. General nature of the genetic code for proteins. *Nature* 192:1227–32.

Dintzis, H.M. 1961. Assembly of the peptide chains of hemoglobin. *Proceedings of the National Academy of Sciences USA* 47:247–61.

Ferbitz, L., T. Maier, H. Patzelt, B. Bukau, E. Deuerling, and N. Ban. 2004. Trigger factor in complex with the ribosome forms a molecular cradle for nascent proteins. *Nature* 431:590–96.

Fredrick, K. and H.F. Noller. 2003. Catalysis of ribosomal translocation by sparsomycin. *Science* 300:1159–62.

Goodman, H.M., J. Abelson, A. Landy, S. Brenner, and J.D. Smith. 1968. Amber suppression: A nucleotide change in the anticodon of a tyrosine transfer RNA. *Nature* 217:1019–24.

Haenni, A.-L. and J. Lucas-Lenard. 1968. Stepwise synthesis of a tripeptide. *Proceedings of the National Academy of Sciences USA* 61:1363–69.

Inoue-Yokosawa, N., C. Ishikawa, and Y. Kaziro. 1974. The role of guanosine triphosphate in translocation reaction catalyzed by elongation factor G. *Journal of Biological Chemistry* 249:4321–23.

Ito, K., M. Uno, and Y. Nakamura. 2000. A tripeptide "anticodon" deciphers stop codons in messenger RNA. *Nature* 403:680–84.

Khaitovich, P., A.S. Mankin, R. Green, L. Lancaster, and H.F. Noller. 1999. Characterization of functionally active subribosomal particles from *Thermus aquaticus*. *Proceedings of the National Academy of Sciences USA* 96:85–90.

Lancaster, L., M.C. Kiel, A. Kaji, and H.F. Noller. 2002. Orientation of ribosome recycling factor in the ribosome from directed hydroxyl radical probing. *Cell* 111:129–40.

Last, J.A., W.M. Stanley, Jr., M. Salas, M.B. Hille, A.J. Wahba, and S. Ochoa. 1967. Translation of the genetic message, IV. UAA as a chain termination codon. *Proceedings of the National Academy of Sciences USA* 57:1062–67.

Miller, D.L. and H. Weissbach. 1970. Interactions between the elongation factors: The displacement of GDP from the Tu-GDP complex by factor Ts. *Biochemical and Biophysical Research Communications* 38:1016–22.

Nirenberg, M. and P. Leder. 1964. RNA codewords and protein synthesis: The effect of trinucleotides upon binding of sRNA to ribosomes. *Science* 145:1399–1407.

Nissen, P., M. Kjeldgaard, S. Thirup, G. Polekhina, L. Reshetnikova, B.F.C. Clark, and J. Nyborg. 1995. Crystal structure of the ternary complex of Phe-$tRNA^{Phe}$, EF-Tu, and a GTP analog. *Science* 270:1464–71.

Noller, H.F., V. Hoffarth, and L. Zimniak. 1992. Unusual resistance of peptidyl transferase to protein extraction procedures. *Science* 256:1416–19.

Ravel, J.M. 1967. Demonstration of a guanine triphosphate-dependent enzymatic binding of aminoacyl-ribonucleic acid to *Escherichia coli* ribosomes. *Proceedings of the National Academy of Sciences USA* 57:1811–16.

Ravel, J.M., R.L. Shorey, and W. Shire. 1968. The composition of the active intermediate in the transfer of aminoacyl-RNA to ribosomes. *Biochemical and Biophysical Research Communications* 32:9–14.

Rheinberger, H.-J., H. Sternbach, and K.H. Nierhaus. 1981. Three tRNA binding sites on *Escherichia coli* ribosomes. *Proceedings of the National Academy of Sciences USA* 78:5310–14.

Rodnina, M.V., A. Savelsbergh, V.I. Katunin, and W. Wintermeyer. 1997. Hydrolysis of GTP by elongation factor G drives tRNA movement on the ribosome. *Nature* 385:37–41.

Sarabhai, A.S., A.O.W. Stretton, S. Brenner, and A. Bolle. 1964. Co-linearity of the gene with the polypeptide chain. *Nature* 201:13–17.

Scolnick, E., R. Tompkins, T. Caskey, and M. Nirenberg. 1968. Release factors differing in specificity for terminator codons. *Proceedings of the National Academy of Sciences USA* 61:768–74.

Thach, S.S. and R.E. Thach. 1971. Translocation of messenger RNA and "accommodation" of fMet-tRNA. *Proceedings of the National Academy of Sciences USA* 68:1791–95.

Traut, R.R. and R.E. Monro. 1964. The puromycin reaction and its relation to protein synthesis. *Journal of Molecular Biology* 10:63–72.

Valle, M., R. Gillet, S. Kaur, A. Henne, V. Ramakrishnan, and J. Frank. 2003 Visualizing tmRNA entry into a stalled ribosome. *Science* 300:127–30.

Weigert, M.G. and A. Garen. 1965. Base composition of nonsense codons in *E. coli*. *Nature* 206:992–94.

Weissbach, H., D.L. Miller, and J. Hachmann. 1970. Studies on the role of factor Ts in polypeptide synthesis. *Archives of Biochemistry and Biophysics* 137:262–69.

Zhouravleva, G., L. Frolova, X. Le Goff, R. Le Guellec, S. Inge-Vechtomov, L. Kisselev, and M. Philippe. 1995. Termination of translation in eukaryotes is governed by two interacting polypeptide chain release factors, eRF1 and eRF3. *EMBO Journal* 14:4065–72.

CHAPTER 19

Ribosomes and Transfer RNA

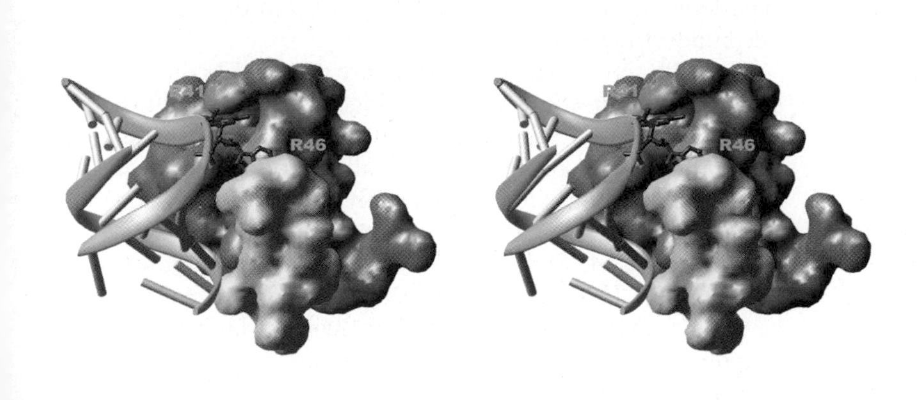

Stereo view of intimate association between 16S rRNA bases A1492 and A1493 (red stick structures) and pockets formed from IF1 (magenta) and S12 (yellow). *From Carter et al., Science 291: p. 500. © 2001 AAAS.*

In Chapter 3 we examined a few aspects of translation. We learned that ribosomes are the cell's protein factories and that transfer RNA plays a crucial adapter role, binding an amino acid at one end and an mRNA codon at the other. Chapters 17 and 18 expanded on the mechanisms of translation initiation, elongation, and termination, without dealing in depth with ribosomes and tRNA. Let us continue our discussion of translation with a closer look at these two essential agents.

19.1 Ribosomes

Chapter 3 introduced the *E. coli* ribosome as a two-part structure with a sedimentation coefficient of 70S. The two subunits of this structure are the **30S** and **50S** ribosomal subunits. We also learned in chapter 3 that the small subunit decodes the mRNA and the large subunit links amino acids together through peptide bonds. In this section we will focus on the bacterial ribosome, its overall structure, composition, assembly, and function.

Fine Structure of the 70S Ribosome

X-ray crystallography provides the best structural information but that is a difficult task with an asymmetric object as large as a ribosome. Despite the difficulty, Harry Noller and colleagues succeeded in obtaining crystals of ribosomes from the bacterium *Thermus thermophilus* that were suitable for x-ray crystallography. By 1999, they had obtained crystal structures of these ribosomes. These studies provided the most detailed structure to that time of the intact ribosome, at a resolution as great as 7.8 Å.

Then, in 2001 Noller and colleagues crystallized a complex of *T. thermophilus* 70S ribosomes plus an mRNA analog, and tRNAs bound to the P and E sites of the ribosome. These crystals yielded a structure at 5.5 Å resolution, a considerable improvement over the previous structure. These workers also crystallized these same complexes with and without tRNA bound to the A site and obtained the structure of the tRNA in the A site by difference, to a resolution of 7 Å.

Figure 19.1 shows the crystal structure of the 70S ribosome. Panels (a–d) show the ribosome in four different orientations: front, right side, back, and left side. The 16S rRNA of the 30S subunit is in cyan and the 30S proteins are in blue. The 23S rRNA of the 50S subunit is in gray, the 5S rRNA is in dark blue, and the 50S proteins are in purple. The tRNAs in the A, P, and E sites are in gold, orange, and red, respectively, although they are difficult to see in panels a–d because they lie in a cleft between the two ribosomal subunits. Most of the ribosomal proteins are identified. Notice L9 sticking out far to the side of the main body of the ribosome (to the left in panel [a]). Figure 19.1e shows a top view of the ribosome, in which the three tRNAs are clearly visible. Notice the anticodon stem-loops of all three pointing down into the 30S subunit at the bottom.

Panels f and g show the two subunits separated to reveal the positions of the tRNAs. The 30S particle has been rotated 180 degrees around its vertical axis so we can see the three tRNAs. Notice that the cleft where the tRNAs bind is lined mostly with rRNA in both subunits; the proteins are mostly peripheral in these views. This finding suggests that rRNAs, not proteins, dominate in the crucial interactions with tRNAs in decoding in the 30S subunit and peptide bond synthesis in the 50S subunit. Furthermore, the ribosome interacts with the conserved portions of all three t-RNAs, allowing it to bind in exactly the same way to all the different tRNAs it encounters.

Notice again in panel (g) the anticodon stem-loops pointing down into the 30S subunit. The anticodons of the tRNAs in the A and P sites approach each other within 10 Å, which does not seem close enough to allow them to bind to adjacent codons. The ribosome solves this problem by kinking the mRNA by 45 degrees between the codons in the A and P sites (Figure 19.2). This points the two codons in the proper directions to be decoded by the tRNAs. Figure 19.1f shows that the tRNAs in the A and P sites also approach each other closely in the 50S subunit. Although it is difficult to see in this view, the acceptor stems of these two tRNAs insert into the peptidyl transferase pocket in the 50S subunit and approach each other within 5 Å. This close approach is necessary because the amino acid and the peptide bound to these two tRNAs must join during peptide bond formation.

The 70S ribosomal crystal structure reveals 12 contacts between subunits (intersubunit bridges), which are illustrated in Figure 19.3. Most of these bridges consist of RNA, rather than protein. Indeed, all of the bridges near the tRNA-binding sites involve only RNA. Notice that bridges B2a, B3, B5, and B6, all involve a single helical domain (helix 44) of the 16S rRNA in the 30S subunit (see Figure 19.2). This helix is a major contributor to contact between the two subunits, and, as we will see later in this chapter, it also plays a role in codon–anticodon recognition. Because the translocation of tRNAs from A to P to E sites requires movement of 20–50 Å, it is very likely that at least some of the intersubunit bridges are dynamic, breaking and reforming to allow translocation to occur.

Figure 19.4 is a more schematic view of the ribosome that emphasizes three important points: First, a large cavity exists between the two ribosomal subunits that can accommodate the three tRNAs. Second, the tRNAs interact with the 30S subunit through their anticodon ends, which bind to the mRNA that is also bound to the 30S subunit. Third, the tRNAs interact with the 50S subunit through their acceptor stems. This makes sense because the acceptor stems must come together during the peptidyl transferase reaction, which takes place on the 50S subunit. During this reaction, the peptide, linked to the acceptor stem of the peptidyl-tRNA in the P site, joins the amino acid, linked to the acceptor stem of the aminoacyl-tRNA in the A site.

In 2005, Jamie Doudna Cate and colleagues achieved a major coup: They obtained the crystal structure of the *E. coli* 70S ribosome at 3.5-Å resolution. Not only was this the best resolution to date of any 70S ribosome, it was the long-sought structure of the *E. coli* ribosome, which is complemented by decades of biochemical and genetic data. Before this structure was available, scientists had to try to fit these biochemical and genetic data on the *E. coli*

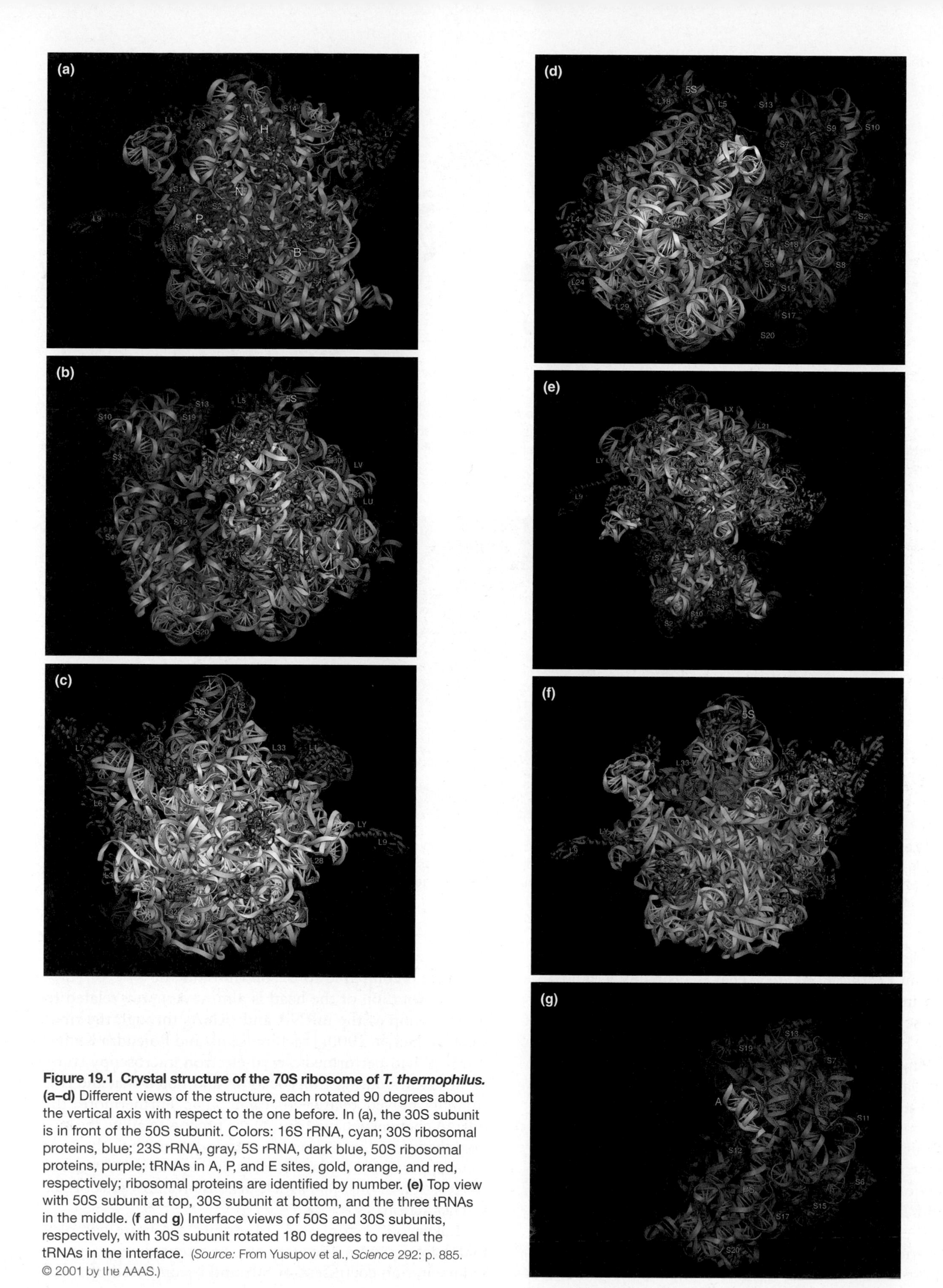

Figure 19.1 Crystal structure of the 70S ribosome of *T. thermophilus*. **(a–d)** Different views of the structure, each rotated 90 degrees about the vertical axis with respect to the one before. In (a), the 30S subunit is in front of the 50S subunit. Colors: 16S rRNA, cyan; 30S ribosomal proteins, blue; 23S rRNA, gray, 5S rRNA, dark blue, 50S ribosomal proteins, purple; tRNAs in A, P, and E sites, gold, orange, and red, respectively; ribosomal proteins are identified by number. **(e)** Top view with 50S subunit at top, 30S subunit at bottom, and the three tRNAs in the middle. (**f** and **g**) Interface views of 50S and 30S subunits, respectively, with 30S subunit rotated 180 degrees to reveal the tRNAs in the interface. (*Source:* From Yusupov et al., *Science* 292: p. 885. © 2001 by the AAAS.)

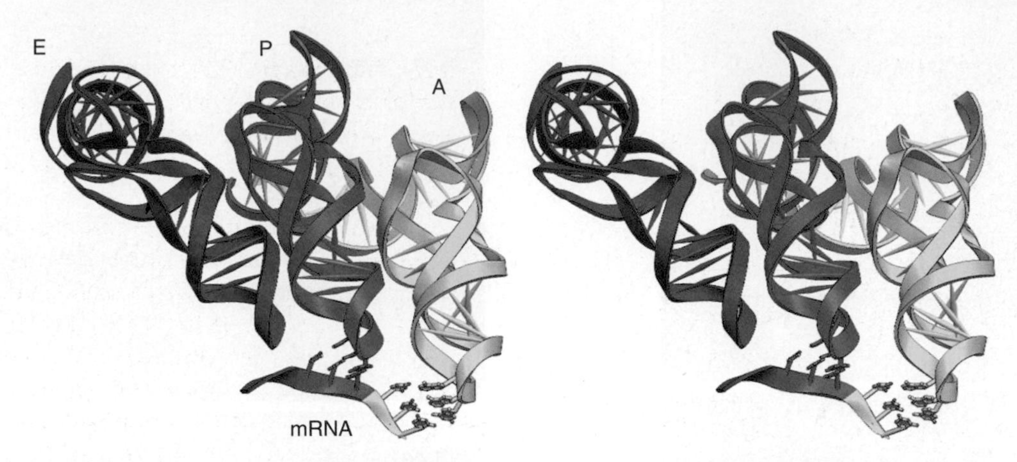

Figure 19.2 Stereo view of the codon–anticodon base-pairing in the A and P sites. All three tRNAs are shown, color-coded as in Figure 19.1 (A, gold; P, orange; and E, red). The bases of the codons and anticodons are shown as stick figures at bottom. Note the 45-degree kink in the mRNA between codons. The anticodon of the tRNA in the E site is not shown because it is not base-paired to mRNA. (*Source:* From Yusupov et al., *Science* 292: p. 893. © 2001 by the AAAS.)

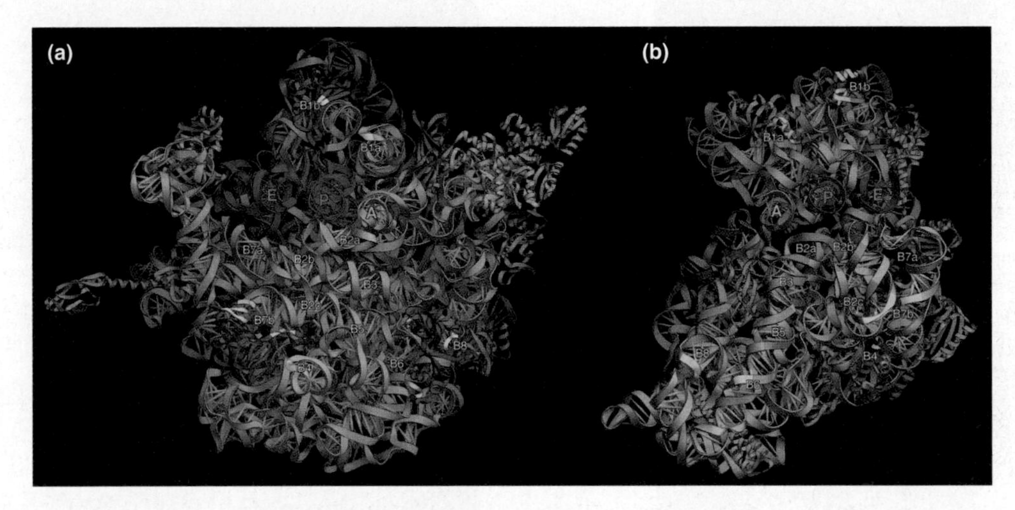

Figure 19.3 Interface view showing intersubunit bridges. (**a** and **b**) 50S and 30S subunits, respectively. In both subunits, large rRNAs are in gray, 5S rRNA is in dark blue at top of 50S particle, and proteins are in light blue. tRNAs are colored as in Figure 19.1 in gold, orange, and red. RNA–RNA bridges between subunits are in pink, and protein–protein bridges are in yellow. All bridges are numbered (B1a, B1b, B2a, etc.) (*Source:* From Yusupov et al., *Science* 292: p. 890. © 2001 by the AAAS.)

ribosome to the structure of a ribosome from another bacterium (*T. thermophilus*). That is probably a valid approach in most cases, but there are always doubts, especially because of the very different environments in which the two bacteria grow: mammalian intestines and boiling hot springs, respectively.

The latest structure contains a massive amount of data, and these data are not yet fully analyzed. Nevertheless, several interesting findings have emerged. Most strikingly, each unit cell of the crystal contained two different ribosomal structures, termed "ribosome I" and "ribosome II." The major differences between the two structures were due to rigid body motions of ribosomal domains. The most obvious of these motions was a rotation of the head of the 30S particle, 6 degrees toward the E site, from ribosome I to ribosome II. This rotation is even more pronounced (12 degrees toward the E site) when the *T. thermophilus* structure is compared to *E. coli* ribosome II.

This rotation of the head is almost certainly related to translocation of the mRNA and tRNAs through the ribosome. In fact, in 2000, Joachim Frank and Rajendra Kumar Agrawal had performed a cryo-electron microscopy study of ribosomes during translocation, and noted that the two subunits moved relative to each other. Furthermore, the mRNA channel widened during the process to allow the motion, then closed up again after translocation. Thus, the ribosome appears to act like a ratchet during translocation, and the rotation of the 30S particle head is probably part of this ratchet action.

Eukaryotic cytoplasmic ribosomes are more complex than bacterial ones. In mammals, the whole ribosome has a sedimentation coefficient of 80S and is composed of a 40S

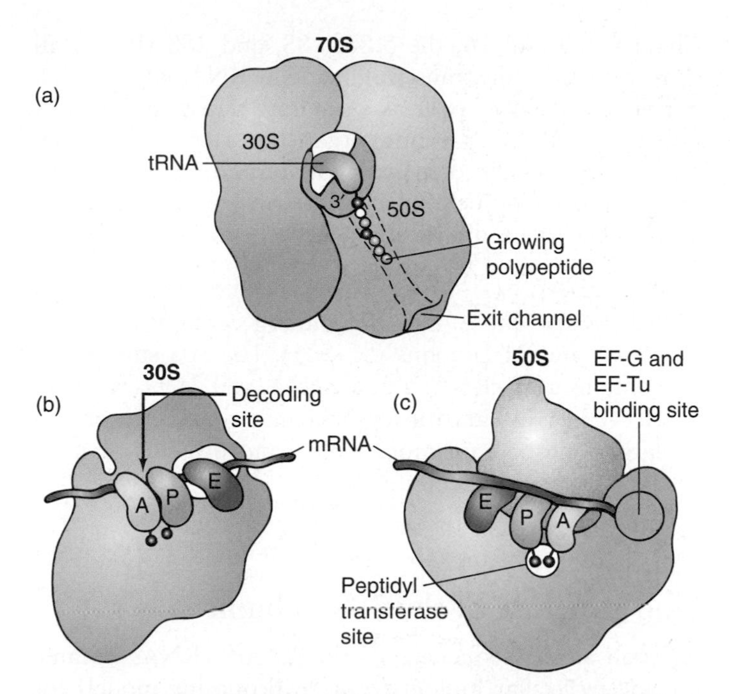

Figure 19.4 Schematic representation of the ribosome. (a) 70S ribosome, showing the large cavern between subunits, which can accommodate three tRNAs at a time. The peptidyl tRNA in the P site is shown, with the nascent polypeptide feeding through an exit tunnel in the 50S subunit. Notice that the interaction between the tRNA and the 30S subunit is through the tRNA's anticodon end, but the interaction between the tRNA and the 50S subunit is through the tRNA's acceptor stem. **(b)** The 30S subunit, with an mRNA and all three tRNAs bound. **(c)** The 50S subunit with an mRNA and all three tRNAs bound. (*Source:* Adapted from Liljas, A., Function is structure. *Science* 285:2078, 1999.)

and a 60S subunit. The 40S subunit contains one (18S) rRNA, and the 60S subunit contains three (28S, 5.8S, and 5S) rRNAs. Budding yeast ribosomes contain 79 ribosomal proteins, compared to 55 in *E. coli*. Eukaryotic organelles also have their own ribosomes, but these are less complex. In fact, they are even simpler than bacterial ribosomes.

SUMMARY The crystal structure of the *T. thermophilus* 70S ribosome in a complex with an mRNA analog and three tRNAs reveals the following: The positions and tertiary structures of all three rRNAs and most of the proteins can be determined. The shapes and locations of tRNAs in the A, P, and E sites are evident. The binding sites for the tRNAs in the ribosome are composed primarily of rRNA, rather than protein. The anticodons of the tRNAs in the A and P sites approach each other closely enough to base-pair with adjacent codons bound to the 30S subunit, given that the mRNA kinks 45 degrees between the two codons. The acceptor stems of the tRNAs in the A and P sites also approach each other closely—within just 5 Å—in the peptidyl transferase pocket in the 50S subunit. This is consistent with the need for the two stems to interact during peptide bond formation. Twelve contacts between subunits can be seen, and most of these are mediated by RNA–RNA interactions.

The crystal structure of the *E. coli* ribosome contains two structures that differ from each other by rigid body motions of domains of the ribosome, relative to each other. In particular, the head of the 30S particle rotates by 6 degrees, and by 12 degrees compared to the *T. thermophilus* ribosome. This rotation is probably part of the ratchet action of the ribosome that occurs during translocation.

Eukaryotic cytoplasmic ribosomes are larger and more complex than their prokaryotic counterparts, but eukaryotic organellar ribosomes are smaller than prokaryotic ones.

Ribosome Composition

We learned in Chapter 3 that the *E. coli* 30S ribosomal subunit is composed of a molecule of 16S rRNA and 21 ribosomal proteins, whereas the 50S particle contains two rRNAs (5S and 23S) and 34 ribosomal proteins. The rRNAs were relatively easy to purify by phenol extracting ribosomes to remove the proteins, leaving rRNA in solution. Then the sizes of the rRNAs could be determined by ultracentrifugation.

On the other hand, the ribosomal proteins are much more complex mixtures and had to be resolved by finer methods. The 30S ribosomal proteins can be displayed by one-dimensional SDS-PAGE to reveal a number of different bands ranging in mass from about 60 down to about 8 kD, but some of the proteins are incompletely resolved by this method. In 1970, E. Kaldschmidt and H.G. Wittmann used two-dimensional gel electrophoresis to give almost complete resolution of the ribosomal proteins from both subunits. In this version of the technique, the two steps were simply native PAGE (no SDS) performed at two different pH values and acrylamide concentrations.

Figure 19.5 depicts the results of two-dimensional electrophoresis on *E. coli* 30S and 50S proteins. Each spot contains a protein, identified as S1–S21 for the 30S proteins, and L1–L33 (L34 is not visible) for the 50S proteins. The S and L stand for small and large ribosomal subunits. The numbering starts with the largest protein and ends with the smallest. Thus, S1 is about 60 kD and S21 is about 8 kD. You can see almost all the proteins, and almost all of them are resolved from their neighbors.

Eukaryotic ribosomes are more complex. The mammalian 40S subunit contains an 18S rRNA and about 30 proteins. The mammalian 60S subunit holds three rRNAs (5S, 5.8S, and 28S) and about 40 proteins. As we learned in

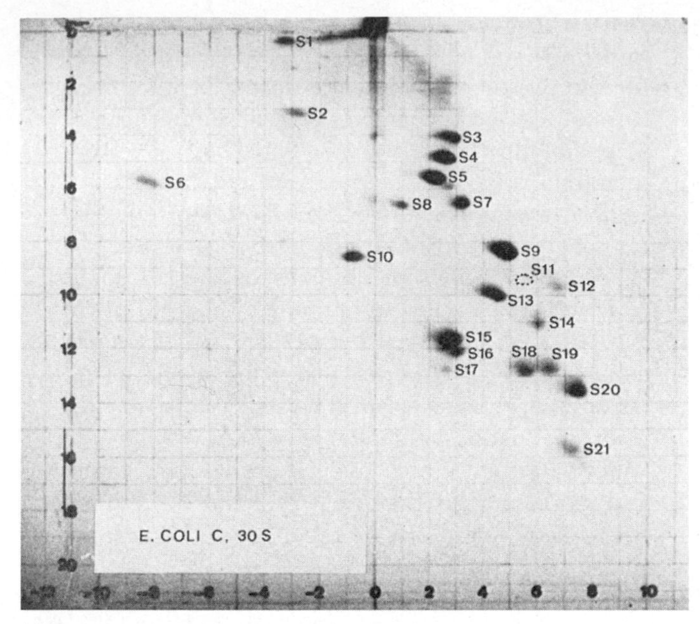

(a)

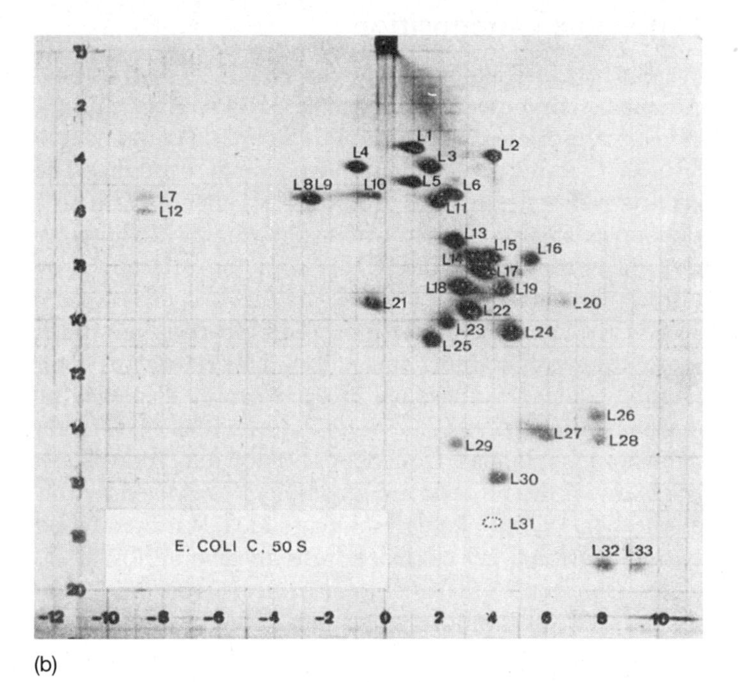

(b)

Figure 19.5 Two-dimensional gel electrophoresis of proteins from (a) *E. coli* 30S subunits and (b) *E. coli* 50S subunits. Proteins are identified by number, with S designating the small ribosomal subunit, and L, the large subunit. Electrophoresis in the first dimension (horizontal) was run at pH 8.6 and 8% acrylamide; electrophoresis in the second dimension (vertical) was run at pH 4.6 and 18% acrylamide. Proteins S11 and L31 were not visible on these gels, but their positions from other experiments are marked with dotted circles. (*Source:* Kaltschmidt, E. and H.G. Wittmann, Ribosomal proteins XII: Number of proteins in small and large ribosomal subunits of *Escherichia coli* as determined by two-dimensional gel electrophoresis. *Proceedings of the National Academy of Sciences USA* 67 (1970) f. 1–2, pp. 1277–78.)

Chapters 10 and 16, the 5.8S, 18S, and 28S rRNAs all come from the same transcript, made by RNA polymerase I, but the 5S rRNA is made as a separate transcript by RNA polymerase III. Eukaryotic organellar rRNAs are even smaller than their prokaryotic counterparts. For example, the mammalian mitochondrial small ribosomal subunit has an rRNA with a sedimentation coefficient of only 12S.

SUMMARY The *E. coli* 30S subunit contains a 16S rRNA and 21 proteins (S1–S21). The 50S subunit contains a 5S rRNA, a 23S rRNA, and 34 proteins (L1–L34). Eukaryotic cytoplasmic ribosomes are larger and contain more RNAs and proteins than their prokaryotic counterparts.

Fine Structure of the 30S Subunit

As soon as the sequences of the *E. coli* rRNAs became known, molecular biologists began proposing models for their secondary structures. The idea is to find the most stable molecule—the one with the most intramolecular base pairing. Figure 19.6 depicts a consensus secondary structure for the 16S rRNA that has been verified by x-ray crystallography of 30S ribosomal subunits. Note the extensive base pairing proposed for this molecule. Note also how the molecule can be divided into three almost independently folded domains (one of which has two subdomains), highlighted in different colors.

How does the three-dimensional arrangement of the 16S rRNA relate to the positions of the ribosomal proteins in the intact ribosomal subunit? The best way to obtain such information is to perform x-ray crystallography, and V. Ramakrishnan and colleagues succeeded in 2000 in solving the crystal structure for the *T. thermophilus* 30S subunit to a resolution of 3.0 Å. At almost the same time, a group led by François Franceschi determined the same structure to 3.3 Å resolution. The structure of Ramakrishnan and colleagues contained all of the ordered regions of the 16S rRNA (over 99% of the RNA molecule) and of 20 ribosomal proteins (95% of the protein). The parts of the proteins missing from the structure were only at their disordered ends.

Figure 19.7a is a stereo diagram of the 16S rRNA alone, and the RNA clearly outlines all of the important parts of the ribosome, including the head, platform, and body. In addition, we can see a neck joining the head to the body, a beak (sometimes called a nose) protruding to the left from the head, and a spur at the lower left of the body. The color coding is the same as in Figure 19.6, emphasizing the fact that the 16S rRNA secondary structural elements correspond to independent three-dimensional elements. Figure 19.7b shows front and back views of the 30S subunit with proteins added to the RNA. The proteins do not cause major changes in the overall shape of the subunit. In other words,

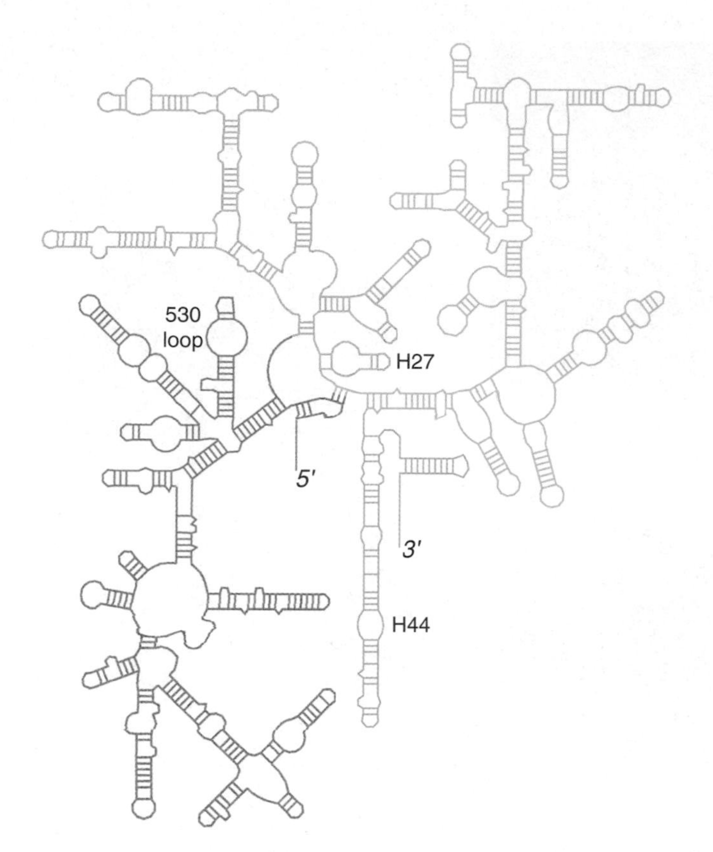

Figure 19.6 Secondary structure of 16S rRNA. This structure is based on optimal base-pairing and on x-ray crystallography of 30S ribosomal subunits from *T. thermophilus.* Two helices (H27 and H44) and the 530 loop, discussed later in the chapter, are labeled. Red, 5′-domain; green, central domain; yellow, 3′-major domain; turquoise, 3′-minor domain. (*Source:* Adapted from Wimberly, B.T., D.E. Brodersen, W.M. Clemons Jr., R.J. Morgan-Warren, A.P. Carter, C. Vonrhein, T. Hartsch, and V. Ramakrishnan, Structure of the 30S ribosomal subunit. *Nature* 407 (21 Sep 2000) f. 2a, p. 329.)

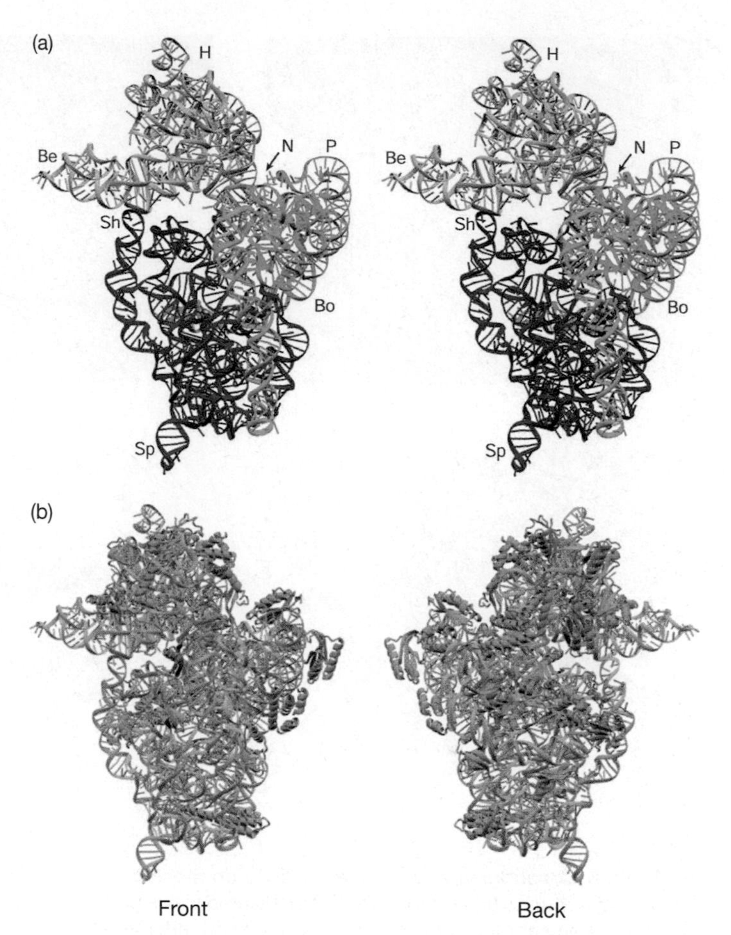

Figure 19.7 Crystal structure of the 30S ribosomal subunit. **(a)** Stereo diagram of the 16S rRNA portion of the 30S subunit from *T. thermophilus.* The major features are identified as follows: H, head; Be, beak; Sh, shoulder; N, neck; P, platform; Bo, body; and Sp, spur. Colors have the same meaning as in Figure 19.6. **(b)** Front and back views of the 30S subunit with the proteins (purple) added to the RNA (gray). The front is conventionally recognized as the side of the 30S subunit that interacts with the 50S subunit. Note that these are two different views of the ribosome, not a stereo diagram. (*Source:* Wimberly, B.T., D.E. Brodersen, W.M. Clemons Jr., R.J. Morgan-Warren, A.P. Carter, C. Vonrhein, T. Hartsch, and V. Ramakrishnan, Structure of the 30S ribosomal subunit. *Nature* 407 (21 Sep 2000) f. 2b, p. 329. Copyright © Macmillan Magazines Ltd.)

the proteins do not contribute exclusively to any of the major parts of the subunit. These statements do not mean that the 16S rRNA would take the shape shown here in the absence of proteins, just that the rRNA is such a major part of the 30S subunit that its shape in the intact subunit resembles a skeleton of the subunit itself. The locations of most of the proteins agree well with the locations determined earlier by other methods.

SUMMARY Sequence studies of 16S rRNA led to a proposal for the secondary structure (intramolecular base pairing) of this molecule. X-ray crystallography studies have confirmed the conclusions of these studies. They show a 30S subunit with an extensively base-paired 16S rRNA whose shape essentially outlines that of the whole particle. The x-ray crystallography studies have also confirmed the locations of most of the 30S ribosomal proteins.

Interaction of the 30S Subunit with Antibiotics Ramakrishnan and colleagues also obtained the crystal structures of the 30S subunit bound to three different antibiotics: spectinomycin, which inhibits translocation; streptomycin, which causes errors in translation; and paromomycin, which increases the error rate by another mechanism. These data, together with the structure of the 30S subunit by itself, gave further insights about the mechanism of translation.

First, Ramakrishnan and coworkers superimposed on their 30S subunit structure the positions of the three aminoacyl-tRNAs from the structure of the whole 70S ribosome (recall Figure 19.1). Figure 19.8a and b show two different views of the positions of the anticodon stem loops

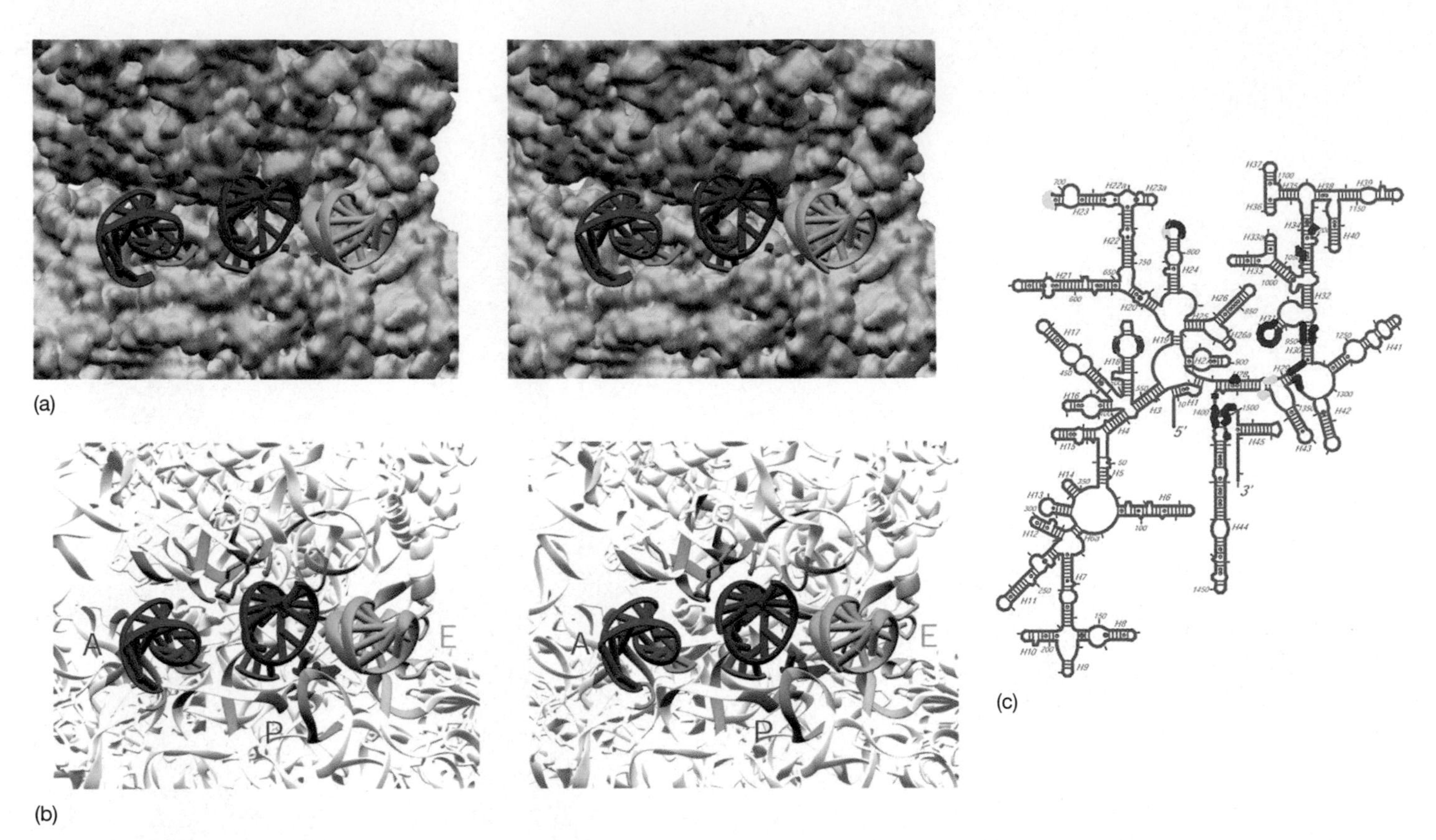

Figure 19.8 Locations of the A, P, and E sites on the 30S ribosomal subunit. (a) and **(b)** Two different stereo views of the inferred placement of the anticodon stem-loops and mRNA codons on the 30S ribosomal subunit. The anticodon stem-loops are colored magenta (A site), red (P site), and gold (E site). The mRNA codons are colored green (A site) blue, (P site), and dotted magenta (E site). **(c)** Secondary structure of the 16S rRNA showing the regions involved in each of the three sites, color-coded the same as the anticodon stem-loops in parts (a) and (b): magenta, A site; red, P site; and gold, E site. (*Source:* Carter, A.P., W.M. Clemons Jr., D.E. Brodersen, R.J. Morgan-Warren, B.T. Wimberly, and V. Ramakrishnan, Functional insights from the structure of the 30S ribosomal subunit and its interactions with antibiotics. *Nature* 407 (21 Sep 2000) f. 1, p. 341. Copyright © MacMillan Magazines Ltd.)

of the aminoacyl-tRNAs, and codons of a hypothetical mRNA, bound to the A, P, and E sites on the 30S subunit. It is striking that the codons and anticodons in the A and P sites lie in a region near the neck of the 30S subunit that is almost devoid of protein. Thus, codon–anticodon recognition occurs in an environment that is surrounded by segments of the 16S rRNA, and very little protein. Figure 19.8c shows which parts of the 16S rRNA are involved at each of the three sites.

The positions of the three antibiotics on the 30S subunit help elucidate the two activities of the 30S subunit: translocation and **decoding** (codon–anticodon recognition). The geometry of the 30S subunit suggests that translocation must involve movement of the head relative to the body. **Spectinomycin** is a rigid three-ring molecule that inhibits translocation. Its binding site on the 30S subunit lies near the point around which the head presumably pivots during translocation. Thus, it is in position to block the turning of the head that is necessary for translocation.

Streptomycin increases the error rate of translation by interfering with initial codon–anticodon recognition and with proofreading. The position of streptomycin on the 30S subunit (Figure 19.9) provides some clues about how this antibiotic works. Streptomycin lies very close to the A site, where decoding occurs. In particular, it makes a close contact with A913 in helix H27 of the 16S rRNA.

This placement of streptomycin is significant because the H27 helix is thought to have two alternative base-pairing patterns during translation, and these patterns affect accuracy. The first is called the ***ram*** state (from **ribosome ambiguity**). As its name implies, this base-pairing scheme for H27 stabilizes interactions between codons and anticodons, even noncognate anticodons, so accuracy is low in the *ram* state. (The crystal structures obtained by Ramakrishnan and colleagues contain the H27 helix in the *ram* state.) The alternative base-pairing pattern is **restrictive,** and it demands accurate pairing between codon and anticodon. If the ribosome is locked into the *ram* state it accepts noncognate aminoacyl-tRNAs too readily and cannot switch to the restrictive state required for proofreading. As a result, translation is inaccurate. On the other hand, if the ribosome is locked into the restrictive state,

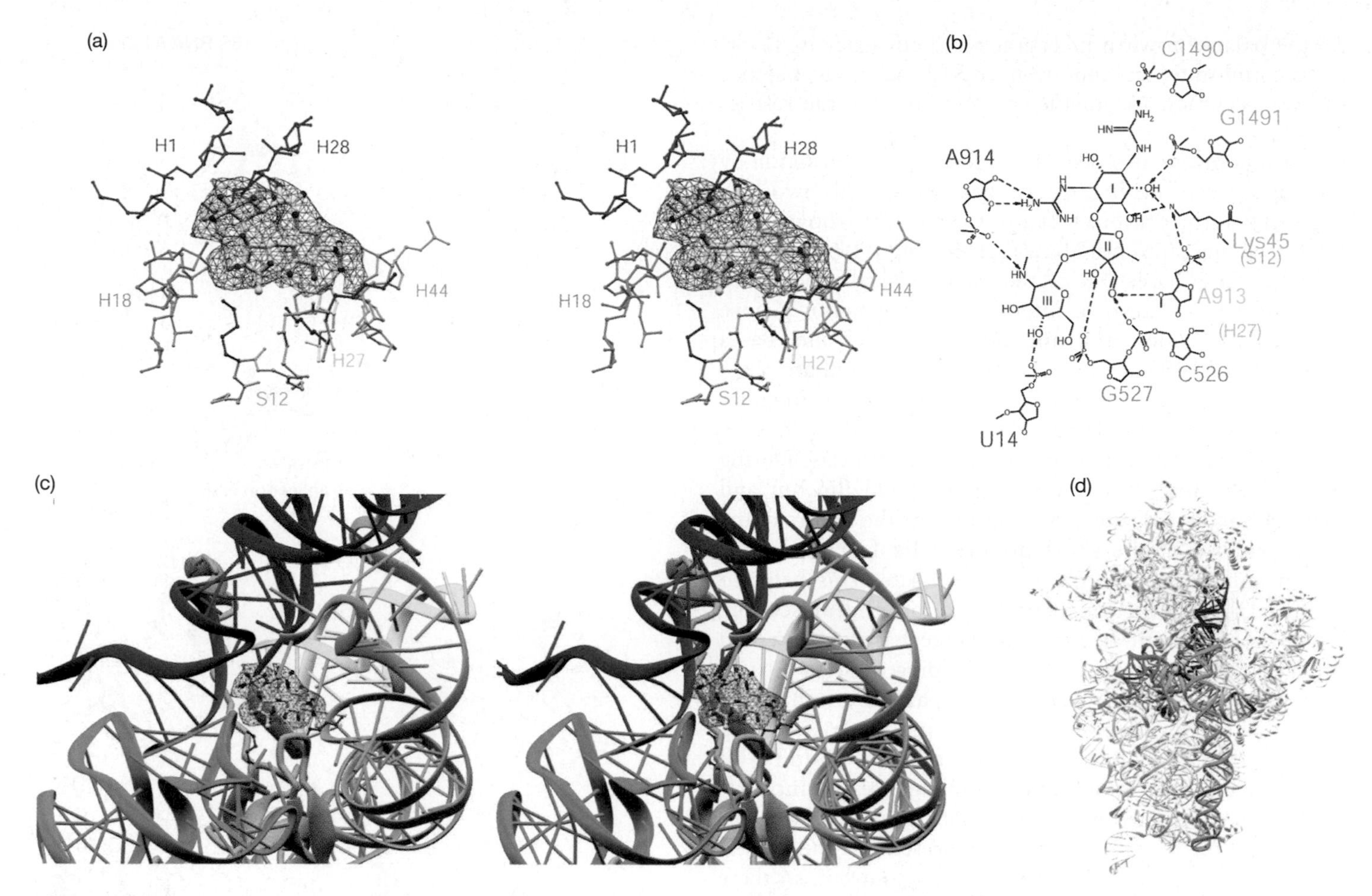

Figure 19.9 Interaction of streptomycin with the 30S ribosomal subunit. (a) Stereo diagram of streptomycin and its nearest neighbors in the 30S subunit. The streptomycin molecule is shown as a ball-and-stick model within a cage of electron density (actually the difference in density between 30S subunits with and without the antibiotic). The nearby helices of the 16S rRNA are shown. Notice especially the H27 helix (yellow), which is crucial for the activity of this antibiotic. Notice also the position of the only protein near the A site—S12 (tan and red), which is also important in streptomycin activity. Amino acids of S12 that are altered in streptomycin-resistant cells are shown in red. **(b)** Interactions of specific groups of streptomycin (containing rings numbered I, II, and III) with neighboring atoms on the 30S subunit. Notice the interactions with A913 of H27 and Lys45 of S12. **(c)** Another stereo view of streptomycin and its nearest neighbors. Color coding is the same as in panel (a). Notice again H27 (yellow) and S12 (tan). **(d)** Location of the streptomycin-binding site on the whole 30S subunit. Streptomycin is shown as a small, red space-filling model at the point where all the colored 16S rRNA helices converge. (*Source:* Carter, A.P., W.M. Clemons Jr., D.E. Brodersen, R.J. Morgan-Warren, B.T. Wimberly, and V. Ramakrishnan, Functional insights from the structure of the 30S ribosomal subunit and its interactions with antibiotics. *Nature* 407 (21 Sep 2000) f. 5, p. 345. Copyright © Macmillan Magazines Ltd.)

it is hyperaccurate—it rarely makes mistakes, but aminoacyl-tRNAs have a difficult time binding to the A site, so translation is inefficient.

The interactions between streptomycin and the 30S subunit indicate the antibiotic stabilizes the *ram* state. This would reduce accuracy in two ways. First, it would favor the *ram* state during decoding and thereby encourage pairing between a codon and noncognate aminoacyl-tRNAs. Second, it would inhibit the switching to the restrictive state that is necessary for proofreading.

Mutations in the ribosomal protein S12 can confer streptomycin resistance or even streptomycin dependence. Almost all of these S12 mutations are in regions of the protein that stabilize the 908–915 part of H27 and the 524–527 part of H18. These are also parts of the 16S rRNA that stabilize the *ram* state. These considerations led Ramakrishnan and colleagues to propose the following two-part hypothesis: First, S12 mutations that cause streptomycin resistance destabilize the *ram* state enough to counteract the *ram* state stabilization produced by the antibiotic. The result is a ribosome that works properly even in the presence of streptomycin. Second, S12 mutations that cause streptomycin dependence destabilize the *ram* state so much that the mutant ribosomes need the antibiotic to confer normal stability to the *ram* state. The result is a ribosome that cannot carry out normal translation without streptomycin.

In other words, translation that is both accurate and efficient depends on a balance between the *ram* state and the restrictive state of the ribosome. Streptomycin can tip the balance toward inaccuracy and efficiency by favoring the *ram* state, and mutations in S12 can tip the balance toward accuracy and inefficiency by favoring the restrictive state.

Paromomycin also decreases accuracy of translation by binding to the A site. In 2000, Ramakrishnan and coworkers showed that this antibiotic binds in the major groove of the H44 helix and "flips out" bases **A1492** and **A1493.** That is, it forces these bases out of the major helical groove and puts them in position to interact with the minor groove between the codon and anticodon in the A site. Bases A1492 and A1493 are universally conserved and are absolutely required for translation activity. Mutations in either of these two bases are lethal.

These factors led to the following hypothesis: During normal decoding, bases A1492 and A1493 flip out and form H bonds with the 2′-OH groups of the sugars in the minor groove of the short double helix formed by the codon–anticodon base pairs in the A site. This helps to stabilize the interaction between codon and anticodon, which is important because the three base pairs would otherwise provide little stability. Flipping these two bases out ordinarily requires energy but paromomycin eliminates this energy requirement by forcing the bases to flip out. In this way, paromomycin stabilizes binding of aminoacyl-tRNAs, including noncognate aminoacyl-tRNAs, to the A site and thereby increases the error rate.

No codon or anticodon were present in the crystal structure of the 30S subunit with paromomycin, so there was no direct evidence for the proposed interactions between bases A1492 and A1493 on the one hand, and the minor groove of the codon–anticodon duplex on the other.

In 2001, Ramakrishnan and coworkers provided direct evidence for their hypothesis. They soaked crystals of *T. thermophilus* 30S ribosomal subunits in a solution containing a 17-nt oligonucleotide corresponding to the anticodon stem-loop of $tRNA^{Phe}$, plus a U_6 oligonucleotide that codes for diphenylalanine. These molecules were both small enough to insert into their proper locations on the 30S subunit, mimicking the anticodon and codon of a full aminoacyl-tRNA and an mRNA, respectively.

Figure 19.10 shows stereo views of selected parts of the crystal structure of this complex. Panel (a) shows clearly that A1493 of helix H44 contacts the 2′-hydroxyl groups of the sugars of both nucleotides in the minor groove of the first codon–anticodon base pair (U1–A36). Panel (b) shows the less favorable interactions with A1493 if A36 of the anticodon is replaced by G. In panel (c), A1492 of helix H44 and G530 of the 530 loop of the 16S rRNA contact the 2′-hydroxyl groups of the sugars of both nucleotides in the second codon–anticodon base pair (U2–A35). These are the two most important base pairs in decoding, and both are stabilized by the flipped-out bases A1492 and A1493, in addition to some other ribosomal elements.

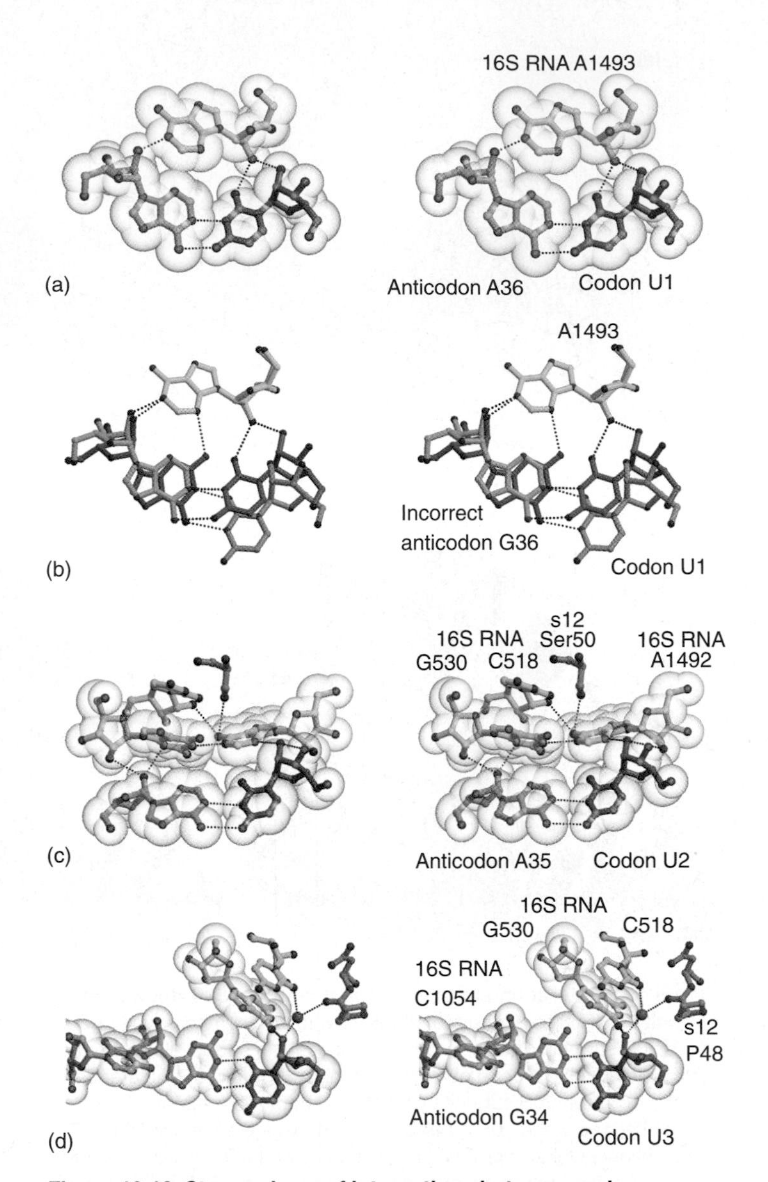

Figure 19.10 Stereo views of interactions between codon-anticodon base pairs and elements of the 30S ribosomal subunit. **(a)** A1493 of helix H44 binding in the minor groove of the U1-A36 base pair. **(b)** Same as in panel (a), but also showing the result of replacing A36 in the anticodon with G, so a wobble G–U pair forms between G36 and U1. Now the positions of G36 (red) and U1 (lavender) can be contrasted with the normal positions of A36 (gold) and U1 (purple). Notice that U1 has been displaced such that it loses its normal interactions with A1493 (represented by a black dotted line). This destabilizes the interaction and helps the ribosome discriminate between a cognate A-U anticodon-codon base pair and a noncognate G–U anticodon-codon base pair involving the first base in the codon. **(c)** A1492 and G530 binding in the minor groove of the U2-A35 base pair. **(d)** The wobble base pair U3-G34 interacts through U3 with G530, and, through a Mg^{2+} ion (magenta sphere), with C518 and proline 48 of protein S12. Base C1054 of the 16S rRNA stacks next to G34. (*Source:* From Ogle et al., *Science* 292: p. 900. © 2001 by the AAAS.)

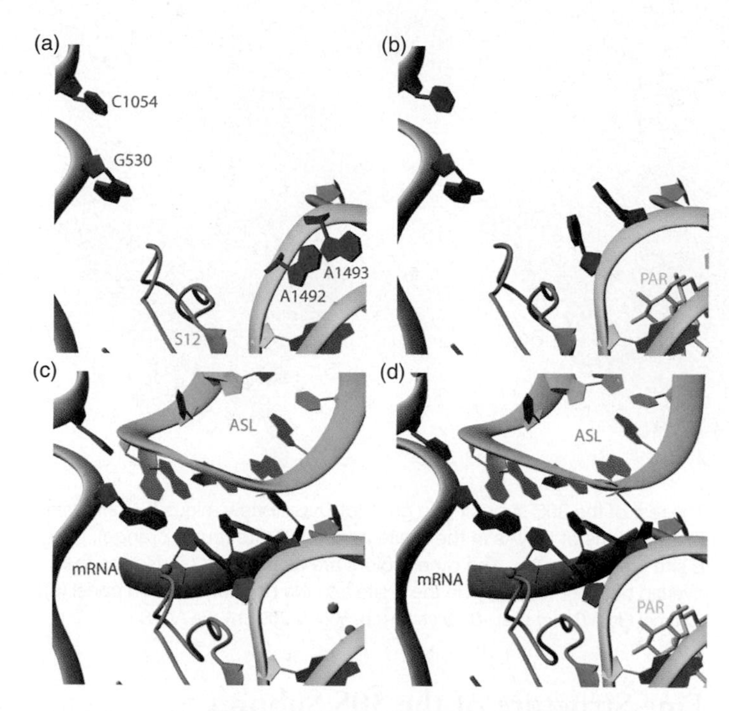

Figure 19.11 Structure of the decoding center in the presence and absence of tRNA, mRNA, and paromomycin. (a) The decoding center by itself. Note the positions of A1492 and A1493 in the H44 helix. The positions of these bases are very flexible. **(b)** The decoding center in the presence of paromomycin. Binding of the antibiotic inside helix H44 has forced A1492 and A1493 to positions outside the helix and into the decoding center. **(c)** The decoding center in the presence of mRNA and the anticodon stem loop (ASL) of the decoding center tRNA. A1492 and A1493 assume the same position in the decoding center that they would in the presence of paromomycin alone. **(d)** Same as in panel (c) except that paromomycin is present. The antibiotic makes little difference because A1492 and A1493 are already interacting in the decoding center. (*Source:* From Ogle et al., *Science* 292: p. 900. © 2001 by the AAAS.)

The third codon–anticodon base pair (wobble pair U3–G34, panel d) is also stabilized by ribosomal elements, including P48 of ribosomal protein S12 and G530 of 16S rRNA, but not by A1492 and 1493.

Figure 19.11 summarizes what these crystal structures tell us about the roles of A1492, A1493, and paromomycin in decoding. Comparing panels (a) and (b), we can see that paromomycin binds inside helix H44 and forces A1492 and A1493 out of the helix into the **decoding center** of the A site. Panel (c) illustrates decoding in the absence of paromomycin, and shows that A1492 and A1493 occupy the same positions as with paromomycin, and that these two rRNA bases are in perfect position to sense the fit between the bases in the first and second base pairs by feeling the positions of the ribose sugars in the minor groove of the codon–anticodon double helix. Indeed, A1492 and A1493, together with G530, are the key components of the decoding center of the ribosome. Panel (d) illustrates the same structure in the presence of paromomycin and again shows little change from the structure without the antibiotic.

All of these findings are consistent with the hypothesis that paromomycin, by nudging A1492 and A1493 out of helix H44, pays part of the energy cost of the induced fit between codon and anticodon at the decoding center. By so doing, the antibiotic makes base pairing between noncognate codons and anticodons easier, thereby increasing the frequency of mRNA misreading.

SUMMARY The 30S ribosomal subunit plays two roles. It facilitates proper decoding between codons and aminoacyl-tRNA anticodons, including proofreading. It also participates in translocation. Crystal structures of the 30S subunit with three antibiotics that interfere with these two roles shed light on translocation and decoding. Spectinomycin binds to the 30S subunit near the neck, where it can interfere with the movement of the head that is required for translocation. Streptomycin binds near the decoding center of the 30S subunit and stabilizes the *ram* state of the ribosome. This reduces fidelity of translation by allowing noncognate aminoacyl-tRNAs to bind relatively easily to the decoding center and by preventing the shift to the restrictive state that is necessary for proofreading. Paromomycin binds in the major groove of the 16S rRNA H44 helix near the decoding center. This flips out bases A1492 and A1493, so they can stabilize base pairing between codon and anticodon. This flipping-out process normally requires energy, but paromomycin forces it to occur and keeps the stabilizing bases in place. This state of the decoding center stabilizes codon–anticodon interaction, including interaction between noncognate codons and anticodons, so fidelity declines.

Interaction of the 30S Subunit with Initiation Factors We have seen in Chapter 17 that IF1 helps the other initiation factors do their jobs. Another postulated role of IF1 is to prevent aminoacyl-tRNAs from binding to the ribosomal A site until the initiation phase is over. This blockage of the A site presumably plays two roles. First, until the 50S particle joins the initiation complex, EF-Tu-directed proofreading of the aminoacyl-tRNA in the A site cannot occur. Thus, blockage of the A site prevents such inaccurate binding of aminoacyl-tRNAs and thereby promotes fidelity of translation. Second, it ensures that the initiator aminoacyl-tRNA binds to the P site, not the A site.

Ramakrishnan and coworkers have determined the crystal structure of IF1 bound to *T. thermophilus* 30S ribosomal subunits. The structure, presented in Figure 19.12b and c shows clearly that IF1 binds to and occludes the A site of the 30S subunit. It occupies much of the spot to which the tRNA would bind in the A site.

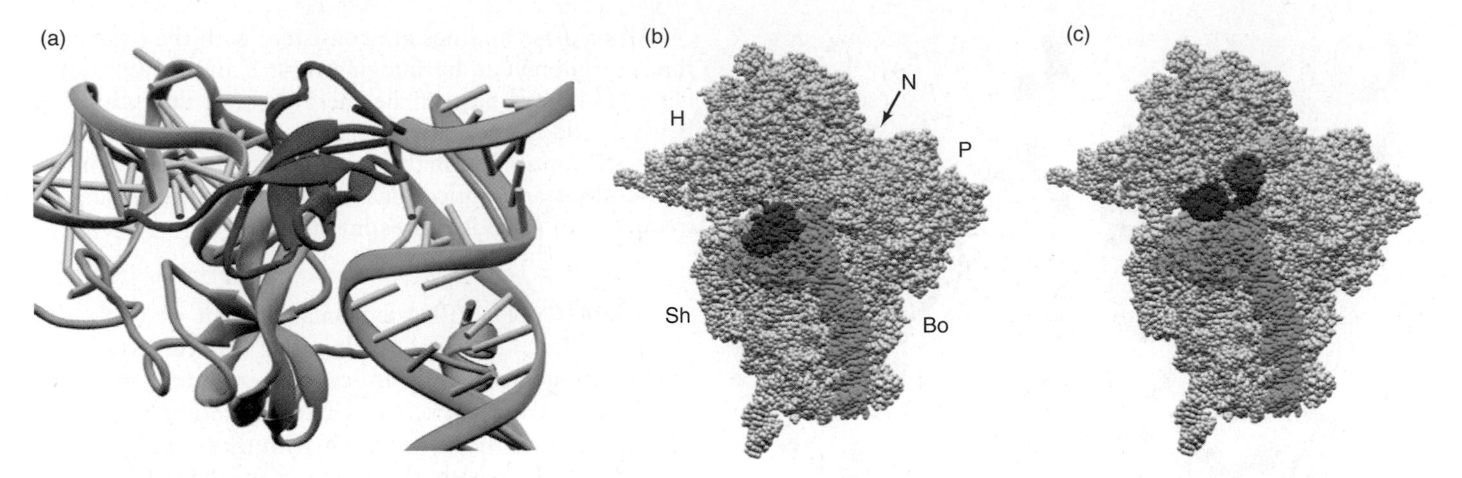

Figure 19.12 Crystal structure of the IF1–30S ribosomal subunit complex. **(a)** Close-up view showing IF1 in magenta, helix H44 of the 16S rRNA in turquoise (with A1492 and A1493 as red sticks), the 530 loop of the 16S rRNA in green, and the S12 protein in orange. **(b)** Overall view of the complex, with the same colors as in panel (a). The rest of the 30S subunit is in gray. **(c)** Overall view minus IF1, showing the positions of tRNAs in the A site (purple), P site (burnt orange), and E site (yellow-green). The other colors are as in panel (a). Notice the overlap between the tRNA in the A site and the position of IF1 in panel (a). (*Source:* From Carter et al., *Science* 291: p. 500. © 2001 by the AAAS.)

The crystals in this study did not include IF2, but we know from Chapter 17 that IF1 aids IF2 in binding fMet-tRNA to the P site, and it is also known that IF1 and IF2 interact. Thus, it is quite possible that binding of IF1 to the A site allows IF1 to help IF2 bind to the 30S subunit in such a way as to facilitate the binding of fMet-tRNA to the P site.

Experiments in the early 1970s appeared to show that IF1 facilitates the dissociation of the two ribosomal subunits. Actually, it also helps the two subunits reassociate, so it does not change the equilibrium between the two. It is only with the help of IF3, which prevents reassociation, that IF1 appears to be an agent of ribosomal dissociation. The structures in Figure 19.12 all show intimate contact between IF1 and helix H44 of the 16S rRNA in the 30S subunit. Helix H44 is also known to make extensive contact with the 50S ribosomal subunit. Ramakrishnan and coworkers speculated that the contact between IF1 and helix H44 perturbs the structure of helix H44 so as to resemble its structure in the transition state between association and dissociation of the ribosomal subunits. This would explain how IF1 accelerates both ribosomal association and dissociation.

SUMMARY The x-ray crystal structure of IF1 bound to the 30S ribosomal subunit shows that IF1 binds to the A site. In that position, it clearly blocks fMet-tRNA from binding to the A site, and may also actively promote fMet-tRNA binding to the P site through a presumed interaction between IF1 and IF2. IF1 also interacts intimately with helix H44 of the 30S subunit, and this may explain how IF1 accelerates both association and dissociation of the ribosomal subunits.

Fine Structure of the 50S Subunit

In 2000, Peter Moore and Thomas Steitz and their colleagues achieved a milestone in the study of ribosomal structure, and in the field of x-ray crystallography, by determining the crystal structure of a 50S ribosomal subunit at 2.4 Å resolution. They performed these studies on 50S subunits from the archaeon *Haloarcula marismortui*, because crystals of 50S subunits suitable for x-ray diffraction could be prepared from this organism. The structure, shown in Figure 19.13, includes 2833 of 3045 nucleotides in the rRNAs of the subunit (all 122 of the 5S rRNA nucleotides), and 27 of the subunit's proteins. The other proteins were not well ordered and could not be located accurately.

One clear difference between the two subunits lies in the tertiary structures of their rRNAs. Whereas the 16S rRNA in the 30S subunit assumed a three-domain structure, the 23S rRNA of the 50S subunit is a monolithic structure with no clear boundaries between domains. Moore, Steitz, and colleagues speculated that the reason for this difference is that the structural domains of the 30S subunit have to move relative to one another, whereas most of those in the 50S subunit do not.

The smaller structures in Figure 19.13 show the locations of the proteins in the 50S subunit. As we saw earlier in this chapter, the proteins in the 50S subunit are generally missing from the interface between the two subunits, particularly in the center, where the peptidyl transferase active site is thought to lie. This was a provocative finding because some uncertainty (Chapter 18) surrounded the question whether the peptidyl transferase activity lies in the RNA or protein of the 50S subunit.

To determine whether proteins are present at the peptidyl transferase active site, one needs to identify the active

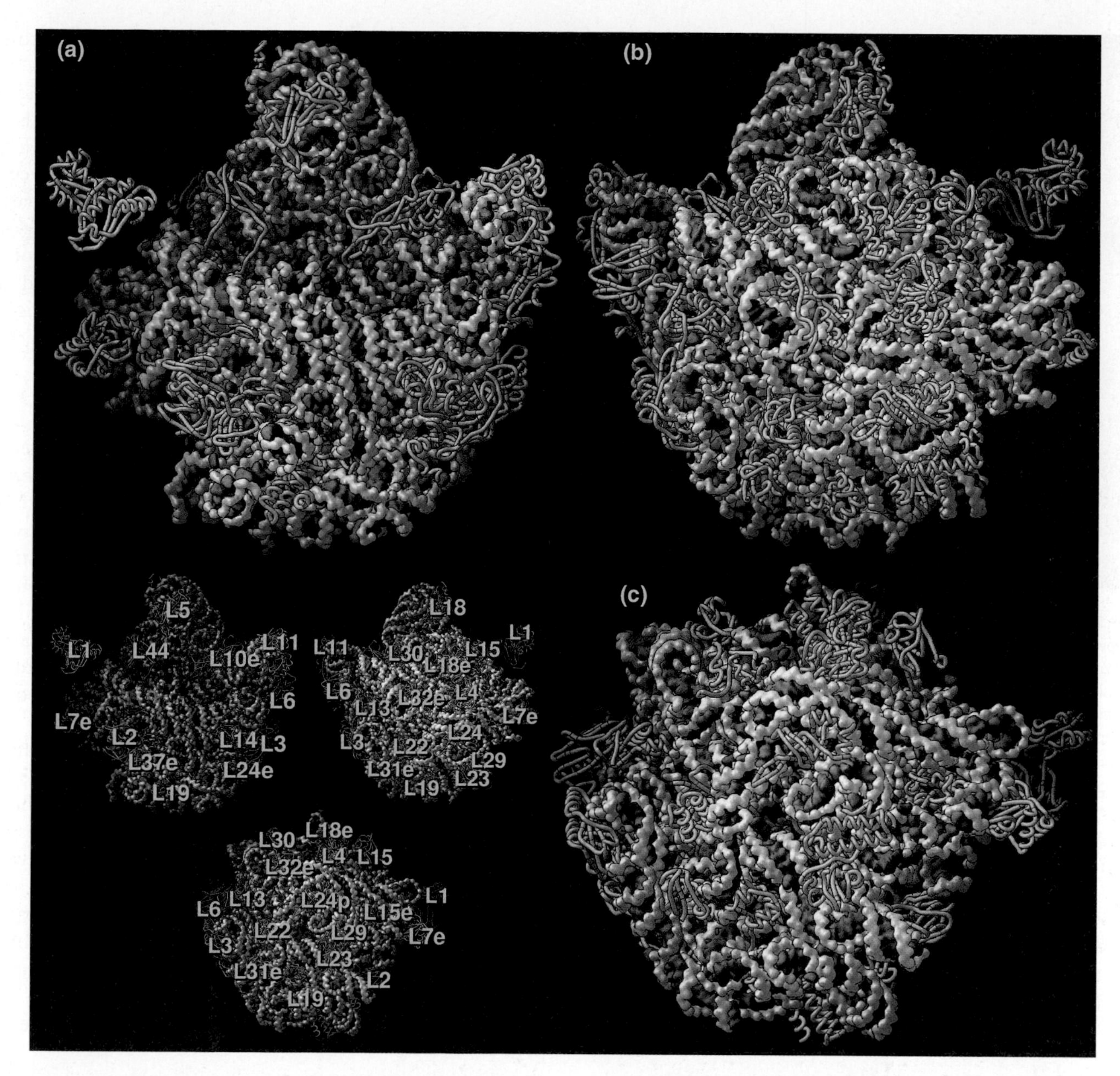

Figure 19.13 Crystal structure of the 50S ribosomal subunit from *Haloarcula marismortui*. The three large structures show the subunit in three different orientations: **(a)** front, or "crown" view (so named because of the resemblance to a three-pointed crown); **(b)** back view (crown view rotated 180 degrees); **(c)** bottom view, showing the end of the polypeptide exit tunnel at center. The RNA is gray and the proteins are gold. The three small structures at lower left are the same three orientations, with the proteins identified. The letter "e" after some numbers designates archaeal proteins that have only eukaryotic (not bacterial) homologs. (*Source:* Ban, N., P. Nissen, J. Hansen, P.B. Moore, and T.A. Steitz, The complete atomic structure of the large ribosomal subunit at 2.4 Å resolution. *Science* 289 (11 Aug 2000) f. 7, p. 917. Copyright © AAAS.)

site in a crystal structure. To accomplish this goal, Moore, Steitz, and coworkers soaked crystals of 50S subunits with two different peptidyl transferase substrate analogs, then performed x-ray crystallography and calculated electron difference maps. This located the electron densities corresponding to the substrate analogs, and therefore to the active site. One analog (CCdAp-puromycin) was designed by Michael Yarus to resemble the transition state, or intermediate, during the peptidyl transferase reaction. Thus, it is called the "Yarus analog."

Figure 19.14 shows that the Yarus analog lies in the cleft in the face of the 50S subunit, right where the active site was predicted to be. And no proteins are around, only RNA. The same behavior was observed for the other analog. Figure 19.15 is a model of the active site with all RNA removed, so we can see just how far the proteins are from the phosphate of the Yarus analog, which corresponds to the tetrahedral carbon atom at the very center of the transition state in the active site. The nearest protein is L3, which is more than 18 Å away from this

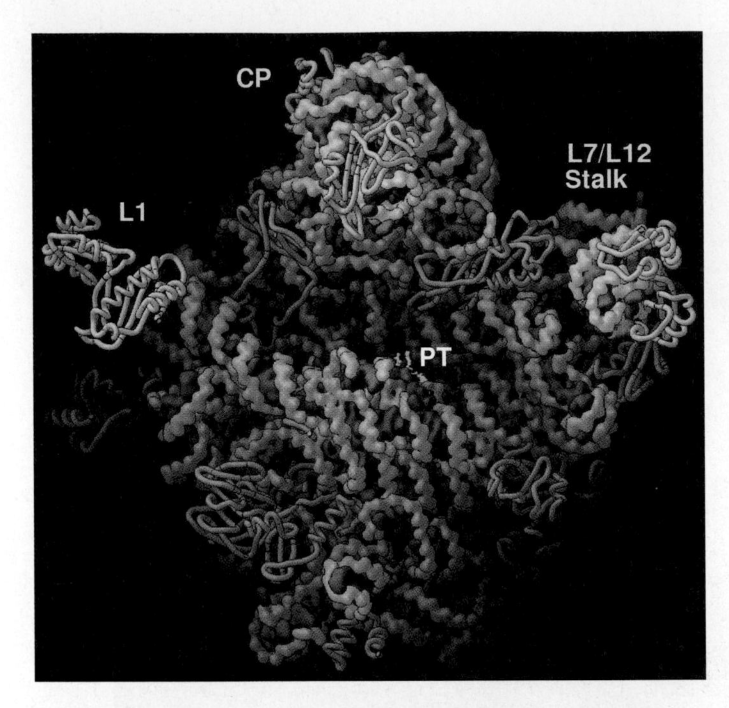

Figure 19.14 Location of the peptidyl transferase active site. This is a crown view of the 50S subunit as in Figure 19.13, with the location of the Yarus analog, which should be at the peptidyl transferase (PT) active site, in green. Notice the absence of proteins (gold) close to the active site. (*Source:* Ban, N., P. Nissen, J. Hansen, P.B. Moore, and T.A. Steitz, The complete atomic structure of the large ribosomal subunit at 2.4 Å resolution. *Science* 289 (11 Aug 2000) f. 2, p. 907. Copyright © AAAS.)

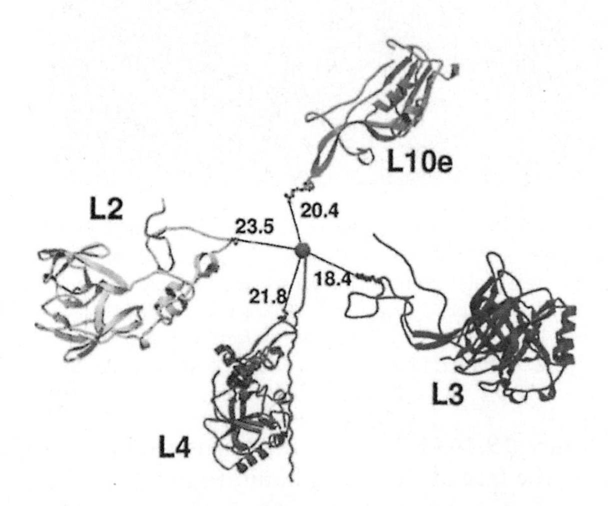

Figure 19.15 Peptidyl transferase active site with all RNA removed. The phosphate of the Yarus analog, at the center of the active site, is rendered in magenta (dark pink), with a long magenta tail representing a growing polypeptide. The four proteins closest to the active site are pictured, along with measurements of the closest approach (in Å) of each protein to the active site. (*Source:* Nissen, P., J. Hansen, N. Ban, P.B. Moore, and T.A. Steitz, The structural basis of ribosome activity in peptide bond synthesis. *Science* 289 (11 Aug 2000) f. 6b, p. 924. Copyright © AAAS.)

active site—much too far to play any direct role in catalysis.

If protein is absent from the active site, RNA must have the enzymatic activity. The crystal structure reveals that adenine 2486 (A2486), which corresponds to A2451 in *E. coli*, is closest to the tetrahedral carbon at the active center. This base is conserved in ribosomes from every species examined from all three kingdoms of life, which suggests it plays a crucial role. Furthermore, chloramphenicol and carbomycin, which inhibit peptidyl transferase, bind at or near A2451 in *E. coli*. And *E. coli* cells with mutations in A2451 are chloramphenicol-resistant, further implicating this base in the reaction.

If this model is correct, then mutations in A2486 would be expected to reduce peptidyl transferase activity by orders of magnitude. Alexander Mankin and colleagues tested this prediction in 2001 by reassembling a *T. aquaticus* 50S subunit from isolated proteins and 23S rRNAs with all three possible mutations in A2451, the base equivalent to A2486 in *H. marismortui*, then testing the reconstituted 50S subunits for peptidyl transferase activity by four different assays, including the fragment reaction described in Chapter 18. None of the mutations caused a dramatic decrease in activity; each mutated 23S rRNA could support at least 44% of wild-type activity in at least one of the assays.

If the adenine of A2486 does not play a major catalytic role in the peptidyl transferase reaction, what does? Scott Strobel and colleagues presented evidence in 2004 that implicates the 2′-hydroxyl group of the terminal adenosine of the peptidyl-tRNA in the P site. Figure 19.16 shows the position of this 2′-OH group with respect to the amino acid in the A site, which is making a nucleophilic attack on the carbonyl carbon that links the peptide to the tRNA in the P site. This attack will result in the joining of the peptide in the P site to the aminoacyl-tRNA in the A site, which is **transpeptidation**, the reaction catalyzed by peptidyl

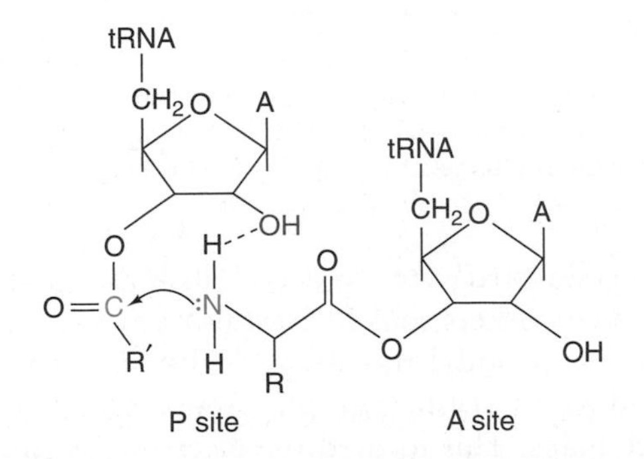

Figure 19.16 Positions of the tRNAs in the A and P sites during the peptidyl transferase reaction. The 2′-OH of the P site tRNA is in red; the amino nitrogen of the aminoacyl-tRNA in the A site is in green, and the carbonyl carbon of the peptidyl tRNA in the P site is in blue. Note the proximity of the 2′-OH of the P site tRNA to the attacking amino nitrogen in the A site.

transferase. It is clear that the 2′-OH group is very well positioned to play a role in this reaction by forming a hydrogen bond with one of the protons on the amino group, thus making the amino nitrogen a better nucleophile.

If this hypothesis is correct, removing the oxygen from the 2′-position of the terminal adenosine (A76) of the peptidyl-tRNA should impair the peptidyl transferase activity. Strobel and colleagues tested this idea in two ways: by replacing the 2′-hydroxyl group with a hydrogen atom (2′-deoxyadenosine, dA) or a fluorine atom (2′-deoxy, 2′-fluoroadenosine, fA). When they made either of these changes to the terminal adenosine of the tRNA in the P site, peptidyl transferase activity was severely inhibited.

To do their assay, Strobel and colleagues loaded [^{35}S] fMet-tRNA into the P site, then Lys-tRNA into the A site. This Lys-tRNA was added in separate experiments in three forms with respect to the terminal adenosine: normal, dA, and fA. Then they allowed peptidyl transferase and one round of translocation, placing [^{35}S]fMet-Lys-tRNA in the P site. This set the stage for adding puromycin and observing the rate of labeled peptidyl-puromycin release from the ribosome. Because puromycin binds very rapidly to the A site, peptidyl transferase is rate-limiting in peptidyl-puromycin release, so the release rate can be taken as a measure of the rate of peptidyl transferase. Strobel and colleagues separated the released labeled peptidyl-puromycin from other labeled substances using thin-layer electrophoresis, and determined the radioactivity in the product by phosphorimaging.

Figure 19.17 shows the results. With the normal tRNA substrate, the peptidyl transferase reaction was complete by the first time point (10 s). However, with either modified substrate, essentially no reaction occurred, even after 24 h. Thus, substituting either a hydrogen atom or a fluorine atom for the 2′-hydroxyl group of the tRNA in the P site completely blocked the peptidyl transferase reaction, strongly suggesting that this 2′-hydroxyl group is required for the reaction. The same behavior was observed with the three substrates and ordinary Phe-tRNA, rather than puromycin, in the A site, further supporting the importance of the 2′-hydroxyl group.

This study still left in question the role of the highly conserved A2451 (using the *E. coli* numbering) of the 23S rRNA. To probe that question, Norbert Polacek and colleagues devised a method to change the nature, not only of the base, but also of the sugar of A2451. When they removed the adenine base from A2451, creating an abasic site, little change occurred in peptidyl transferase activity, as measured by the familiar fMet-puromycin release assay. However, when they removed the 2′-hydroxyl group of A2451, they reduced activity almost 10-fold. Furthermore, when they removed the base as well as the 2′-OH group, they almost completely abolished activity. By contrast, performing the same changes in the adjoining nucleoside, A2450, had only modest effects on activity, emphasizing again the special importance of A2451.

The loss of activity in the ribosomes lacking the 2′-OH at position 2451 of the 23S rRNA could be due to lowered affinity for tRNA at the P site. If so, raising the concentration of fMet-tRNA should have enhanced activity, but it did not. So what is the role of this hydroxyl group? The evidence we just examined for the participation of the 2′-hydroxyl group

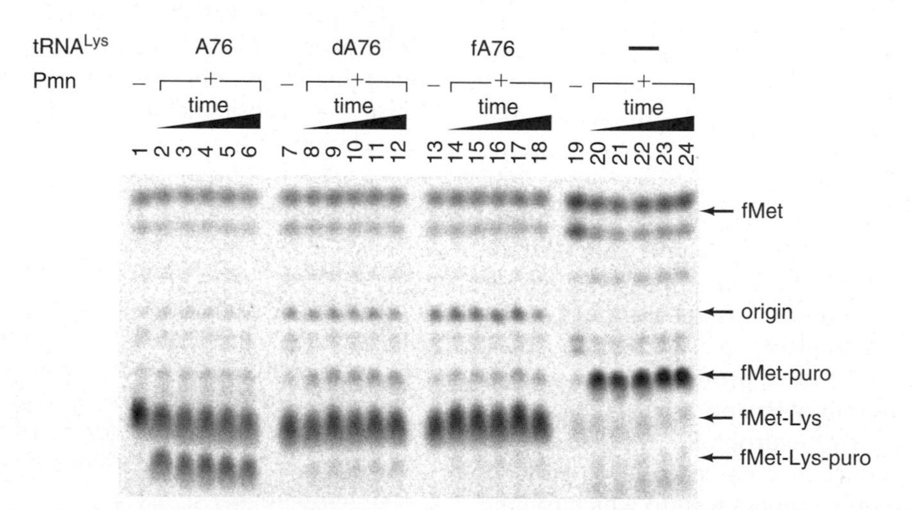

Figure 19.17 Peptidyl transferase activities with modified tRNAs. Strobel and colleagues carried out the peptidyl transferase reaction using a labeled dipeptidyl-tRNA in the P site and puromycin added to the A site. The tRNA in the P site contained a normal A76, dA76, or fA76, or simply fMet-tRNA with no modification (–), as indicated at top. They carried out the reactions for various times (10 s, 1 min, 6 min, 1 h, and 24 h in the presence of puromycin, or with no puromycin (–), also indicated at top. They separated labeled dipeptidyl-puromycin (fMet-Lys-puro) from other reactants and products by thin-layer electrophoresis, and subjected the electropherogram to phosphorimaging. Only the normal A76 in the P site tRNA was able to support measurable peptidyl transferase activity. (*Source:* Reprinted from *Nature Structural & Molecular Biology,* vol 11, Joshua S. Weinger, K. Mark Parnell, Silke Dorner, Rachel Green & Scott A. Strobel, "Substrate-assisted catalysis of peptide bond formation by the ribosome," Fig. 3a, p. 1103. Copyright 2004, reprinted by permission from Macmillan Publishers Ltd.)

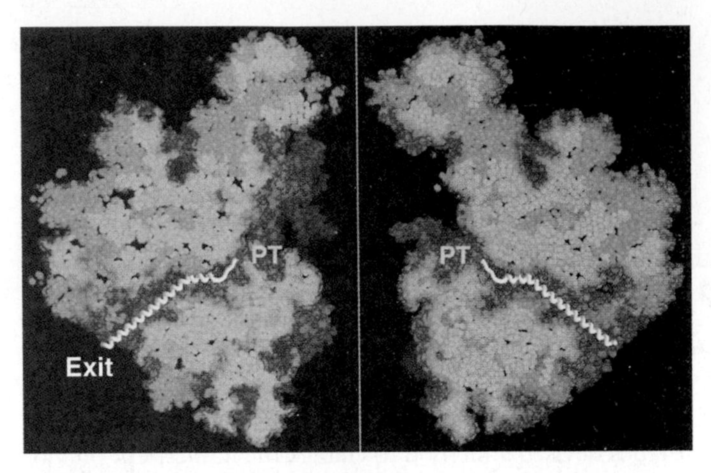

Figure 19.18 The polypeptide exit tunnel. The 50S subunit is pictured as if it were a fruit cut through the middle and opened up. This view reveals the exit channel leading away from the peptidyl transferase site (PT). A white α-helix is placed in the channel to represent an exiting polypeptide. (*Source:* Ban, N., P. Nissen, J. Hansen, P.B. Moore, and T.A. Steitz, The structural basis of ribosome activity in peptide bond synthesis. *Science* 289 (11 Aug 2000) f. 11a, p. 927. Copyright © AAAS.)

of the P site tRNA in the chemistry of transpeptidation is strong, but it remains possible that the 2′-hydroxyl group of A2451 also participates in this way. Alternatively, one or both of these hydroxyl groups could contribute to catalysis by helping to position the reactants properly in the active site. In contrast to the *Haloarcula* ribosome structure, a protein (the N-terminus of L27) in the *E. coli* ribosome is close enough to the peptidyl transferase center to be cross-linked to the 3′-end of the P site tRNA. However, given the strong evidence for RNA as the catalytic agent in one bacterium, it is unlikely that RNA does not play this role in another. Perhaps the N-terminus of L27 helps stabilize the peptidyl tRNA in the P site in the *E. coli* ribosome.

As the polypeptide product grows, it is thought to exit the ribosome through a tunnel in the 50S subunit. Moore, Steitz, and coworkers' studies also shed considerable light on this issue. Figure 19.18 shows a model of the 50S subunit cleaved in half to reveal the exit tunnel. The peptidyl transferase center has been marked, and a polypeptide modeled in the tunnel. The tunnel has an average diameter of 15 Å and narrows in two places to as little as 10 Å, just wide enough to accommodate a protein α-helix, so any further folding of the nascent polypeptide is unlikely. Much of the tunnel wall is made of hydrophilic RNA, so the exposed hydrophobic residues in a nascent polypeptide are not likely to find much in the tunnel wall to which to bind and retard the exit process.

SUMMARY The crystal structure of the 50S ribosomal subunit from *H. marismortui* has been determined to 2.4 Å resolution. This structure reveals relatively few proteins at the interface between ribosomal subunits, and no protein within 18 Å of the peptidyl transferase active center tagged with a transition state analog. The 2′-OH group of the tRNA in the P site is very well positioned to form a hydrogen bond to the amino group of the aminoacyl-tRNA in the A site, and therefore to help catalyze the peptidyl transferase reaction. In accord with this hypothesis, removal of this hydroxyl group eliminates almost all peptidyl transferase activity. Similarly, removal of the 2′-OH group of A2451 of the 23S rRNA strongly inhibits peptidyl transferase activity. This group may also participate in catalysis by hydrogen bonding, or it may help position the reactants properly for catalysis. The exit tunnel through the 50S subunit is just wide enough to allow a protein α-helix to pass through. Its walls are made of RNA, whose hydrophilicity is likely to allow exposed hydrophobic side chains of the nascent polypeptide to slide through easily.

Ribosome Structure and the Mechanism of Translation

As suggested in Chapter 18, the mechanism of translation presented there, including the three-site (A, P, E) model of the ribosome, was oversimplified. We have already seen that aminoacyl-tRNAs can exist in hybrid states that do not conform to the three-site model. The example we saw in Chapter 18 was the P/I state, which fMet-$tRNA_f^{Met}$ assumes without help from EF-P. But other hybrid states also exist. In this section we will examine structural studies that have shed considerably more light on the mechanism of translation.

Binding an Aminoacyl-tRNA to the A Site Single-particle cryo-electron microscopy (cryo-EM) studies as early as 1997 detected that an incoming aminoacyl-tRNA was first bent into the **A/T state**, in which the anticodon is interacting with the codon in the A site, but the amino acid and acceptor stem are still interacting with EF-Tu–GTP, rather than with the A site of the 50S subunit. Only upon GTP hydrolysis does the aminoacyl-tRNA unbend and fully enter the A site of the ribosome—a process known as **accommodation.**

In 2009, Ramakrishnan and colleagues used the higher-resolution x-ray crystallography method to clarify the details of the process by which EF-Tu brings a new aminoacyl-tRNA into the A site. They made crystals of the *T. thermophilus* ribosome complexed with mRNA, $tRNA^{Phe}$ in the P and E sites, and the ternary complex of EF-Tu–Thr-$tRNA^{Thr}$–GDP. They also included the antibiotic kirromycin,

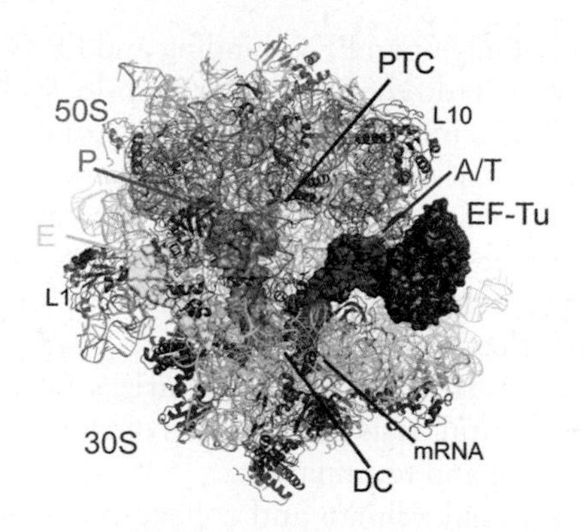

Figure 19.19 Crystal structure of the ribosome with deacylated tRNAs in the P and E sites and an aminoacyl-tRNA in the A/T state. EF-Tu and tRNAs are represented as surfaces, and the rRNA and proteins as cartoons. The 30S particle is depicted in cyan (RNA) and purple (proteins), and the 50S particle in orange (RNA) and brown (proteins). The tRNA in the E site is in yellow, the tRNA in the P site in green, and the aminoacyl-tRNA in the A/T state in magenta, bound to EF-Tu in red. DC, decoding center; PTC, peptidyl transferase center; L1, the L1 stalk of the 50S particle, which contains the L1 ribosomal protein. Note the empty A site in the 50S particle, into which the amino acid and acceptor stem of the aminoacyl-tRNA will move upon GTP hydrolysis. (*Source:* Reprinted with permission of *Science,* 30 October 2009, Vol. 326, no. 5953, pp. 688–694, Schmeing et al, The Crystal Structure of the Ribosome Bound to EF-Tu and Aminoacyl-tRNA. © 2009 AAAS.)

which prevents rearrangement of EF-Tu after GTP hydrolysis. The intent was to catch the aminoacyl-tRNA in the A/T state. Finally, they included paromomycin, which we have already learned stabilizes the binding between codon and anticodon.

As hoped, the aminoacyl-tRNA was in the A/T state, as shown in Figure 19.19. One can see that the +anticodon end of the aminoacyl-tRNA (magenta) is in the decoding center of the 30S ribosomal particle next to the mRNA, but the aminoacyl-tRNA is bent to the right by about 30° so its acceptor stem contacts EF-Tu, rather than inserting into the A site next to the peptidyl transferase center (PTC). Closer inspection showed that this bend is smooth and does not involve a kink in the tRNA.

What is the advantage of this tRNA bending? It requires energy, and this energy is provided by the correct interaction of a codon and its cognate anticodon. But binding a noncognate tRNA does not release as much energy, so the tRNA bend required to achieve the A/T state does not occur as readily. Thus, the requirement for the tRNA bend serves the purpose of translational fidelity by selecting against noncognate aminoacyl-tRNAs. This hypothesis is supported by the existence of several tRNA mutations that facilitate the bending required for the A/T state. These mutations result in lower translational fidelity because they make it easier to accommodate noncognate aminoacyl-tRNAs.

We know the bent aminoacyl-tRNA must straighten up to enter the A site, and this is relatively easy because the aminoacyl-tRNA makes contacts mostly with the decoding center and EF-Tu, with few contacts with the ribosome in between. The energy stored in the bent tRNA is more than enough to break these few contacts and cause the aminoacyl-tRNA to enter fully into the A site.

How does the ribosome collaborate with the GTPase of EF-Tu to cleave the GTP in the ternary complex, but only when a cognate aminoacyl-tRNA is in the decoding center? The GTPase center of EF-Tu is presumed to include elements called the P loop, switch I, and switch II. Switch II includes the putative catalytic residues Gly 83 and His 84. GTP cannot be hydrolyzed by the ternary complex itself because, in the absence of the ribosome, Gly 83 and His 84 are kept out of the GTPase active center by a hydrophobic gate composed of Ile 60 of switch I and Val 20 of the P loop. When this gate is opened, the catalytic residues can reach the catalytic center and activate a water molecule that hydrolyzes the GTP.

The present structure represents the post-GTP hydrolysis state, so we would expect the catalytic His 84 to be remote from the GDP, and it is. In addition, the P loop and switch II elements are well-ordered, but the region of switch I that contains the Ile 60 gate is not. This means that this part of switch I can move in the crystal structure, which gives rise to the hypothesis that this is the gate that swings open to allow the catalytic residues access to the GTP.

But what opens the gate? Figure 19.20 presents Ramakrishnan and colleagues' hypothesis, with the numbers in black circles representing the following events in order: (1) The process begins with the interaction of a codon and its cognate anticodon in the decoding center (16S rRNA residues A1492, A1493, and G530). (2) When the decoding center senses the proper fit between codon and anticodon, it causes the 30S subunit to undergo "domain closure," which shifts the 16S rRNA shoulder region into contact with EF-Tu. (3) This contact shifts the position of the β-turn of EF-Tu domain 2. (4) This shift in the β-turn changes the conformation of the acceptor stem of the aminoacyl-tRNA

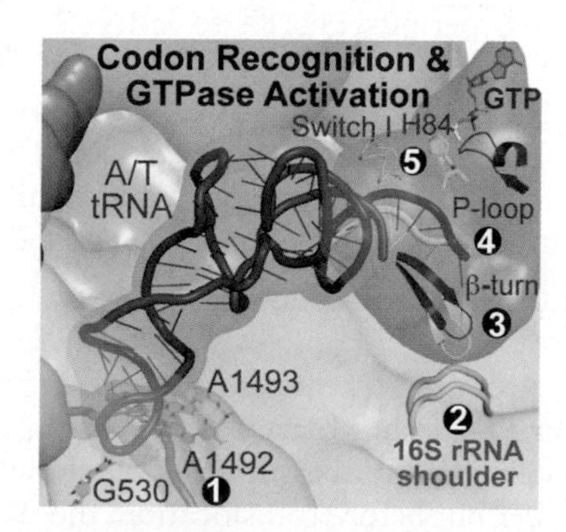

Figure 19.20 Codon recognition and GTPase activation. The aminoacyl-tRNA (magenta) is shown in the A/T state with its anticodon in the decoding center, and its accepter stem bound to EF-Tu. Only relevant parts of EF-Tu (β-turn [or loop], P-loop, switch I, and His 84 [H84]) are shown. The steps denoted by the white numbers in black circles are described in the text. (*Source:* Reprinted with permission of *Science,* 30 October 2009, Vol. 326, no. 5953, pp. 688–694, Schmeing et al, The Crystal Structure of the Ribosome Bound to EF-Tu and Aminoacyl-tRNA. © 2009 AAAS.)

to help bend the tRNA into the A/T state. (5) The change in conformation of the acceptor stem of the tRNA breaks its contacts with switch I, which allows the latter to move, opening the gate and allowing His 84 to move into the GTPase catalytic center and hydrolyze the GTP. One feature not illuminated by this study is the role of the L10–L12 stalk of the 50S particle, which is known to stimulate the GTPase activity of EF-Tu. The L10–L12 stalk was disordered in this crystal structure, and was therefore not seen.

The molecular interactions described in this section, including the bending of the aminoacyl-tRNA in the A/T state, the activation of the GTPase of EF-Tu, and the unbending of the aminoacyl-tRNA are shown in a movie (movie s1) at www.sciencemag.org/cgi/content/full/1179700/DC1. The three-dimensional effect of the movie shows these events much more clearly than a static, two-dimensional picture can. In addition, the movie shows what happens after GTP hydrolysis: EF-Tu–GDP leaves the A site, which allows the aminoacyl-tRNA to unbend into the full A/A state. This "accommodation" of the aminoacyl-tRNA by the A site causes a shift in the conformations of both the 30S and 50S ribosomal subunits. In particular, the mobile L1 stalk of the 50S particle moves, opening the E site and allowing the deacylated tRNA to leave the ribosome. Other studies had previously implicated the L1 stalk in release of the E site tRNA.

SUMMARY An aminoacyl-tRNA, upon binding to a ribosome, first enters the A/T state with its anticodon in the decoding site of the 30S particle, and its acceptor stem still bound to EF-Tu. This forces a bend in the tRNA, which occurs most readily with a perfect match between codon and anticodon, thus enhancing accuracy. Upon bending, the tRNA loses contact with switch I of EF-Tu, allowing switch I to move, which permits His 84 to enter the GTPase active center and hydrolyze GTP. Upon GTP hydrolysis, EF-Tu–GDP leaves the ribosome, allowing the aminoacyl-tRNA to enter the A/A state. This rearrangement in turn causes a conformational shift in the ribosome that releases the deacylated tRNA from the E site.

Translocation Danesh Moazed and Harry Noller used chemical footprinting studies in 1989 to show that, after peptidyl transfer but before translocation, the tRNAs in the A and P sites spontaneously shift their acceptor stems to the P and E sites, respectively, of the 50S subunit. This shift occurs even before EF-G binds to the ribosome and is driven by a ratcheting motion of the 30S and 50S subunits by 6° relative to each other. However, the anticodons remain paired with codons in the A and P sites, respectively, of the 30S subunit. Thus, these tRNAs have assumed hybrid A/P and P/E states. Only upon EF-G binding and EF-G-dependent hydrolysis of GTP do the anticodon stem-loops shift, along with the mRNA, in the 30S subunit to bring the tRNAs fully into the P and E sites. These events are shown in Figure 19.21, and in a movie at www.mrc-lmb.cam.ac.uk/ribo/homepage/movies/translation_bacterial.mov. The movie shows things much more clearly because of the three-dimensional effect, and the ability to show changes smoothly through time. Furthermore, it summarizes what we know about the structural basis of all phases of translation: initiation, elongation, and termination.

In 2009, Ramakrishnan and colleagues determined the crystal structure of the *T. thermophilus* ribosome complexed with mRNA, EFG-GDP, and the antibiotic fusidic acid, which allows translocation and GTP hydrolysis, but blocks EFG-GDP release from the ribosome. This structure was predicted to be in the post-translocation state, with the tRNAs in the classic P and E states, rather than in pre-translocation hybrid A/P and P/E states, and indeed that was what Ramakrishnan and colleagues found. Also, as predicted, EF-G interacts with the ribosome via its domain IV in much the same way that the EF-Tu—aminoacyl-tRNA—GTP complex does.

A novel feature of this crystal structure is that it stabilized the mobile L1 and L10–L12 stalks of the 50S particle so they could be visualized. In the present context, the shape and position of the L10–L12 stalk is particularly important because it is known to participate in the GTPase reaction catalyzed by EF-G. Indeed, this structure shows that the carboxyl terminal domain (CTD) of L12 contacts the G′ domain of EF-G. However, Ramakrishnan and colleagues noted that mutations that would disrupt this contact inhibit only the release of the inorganic phosphate byproduct of the GTPase reaction, not the reaction itself. This led them to speculate that the spatial relationship of L12 and EF-G is somewhat different at the time of GTP hydrolysis, and that it then converts to the shape they observed, which is important for phosphate release. It is also likely that L12 behaves in the same way with respect to the GTPase center of EF-Tu.

SUMMARY Translocation begins with a spontaneous ratcheting of the 30S particle with respect to the 50S particle, which brings the tRNAs into hybrid A/P and P/E states. Upon EF-G–GTP binding and hydrolysis of GTP, the tRNAs and mRNA translocate on the 30S particle to enter the classical P and E sites, and the ratchet has reset. Structural studies on a complex containing the 70S ribosome, EFG–GDP, mRNA and fusidic acid have revealed that EF-G binds to the ribosome in much the same way that EF-Tu–aminoacyl-tRNA–GDP does. These studies have also shown how the L10-L12 stalk may stimulate the GTPase of EF-G (and EF-Tu).

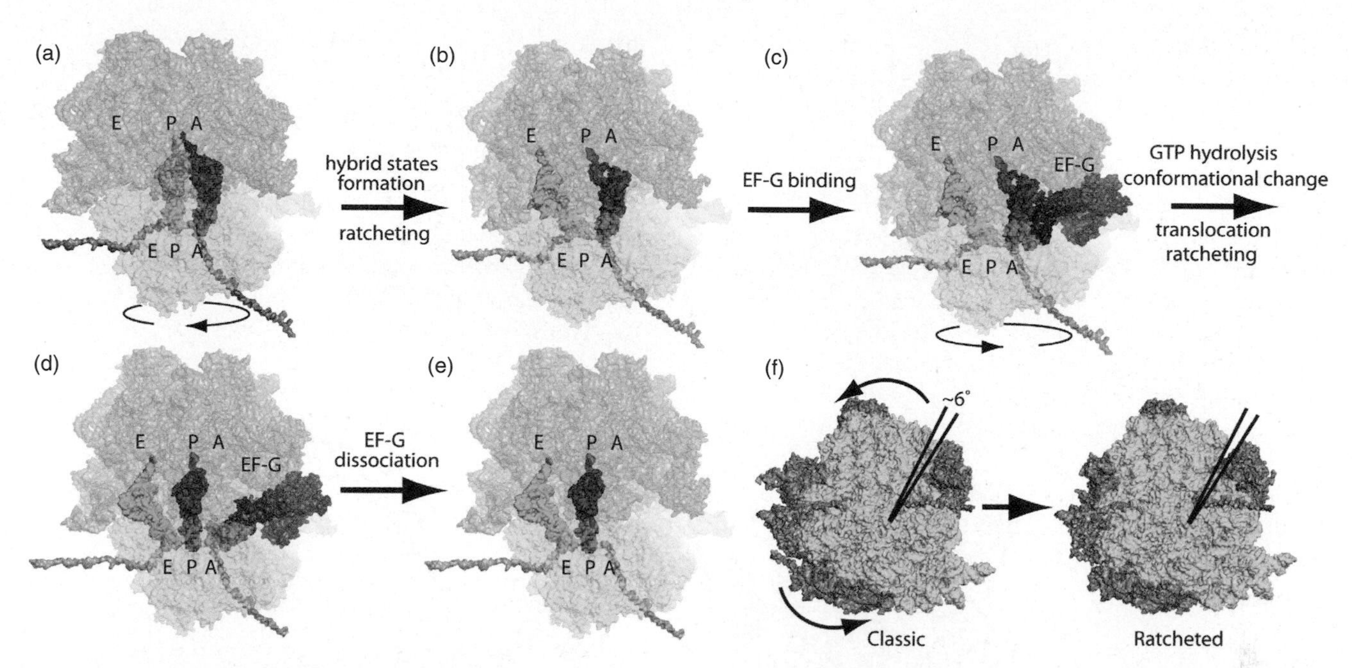

Figure 19.21 Structural basis of the translocation process. **(a)** The pretranslocation state with tRNAs in the classic A and P sites. The P site tRNA is deacylated. **(b)** Spontaneous ratcheting of the two subunits of the ribosome brings the two tRNAs into hybrid A/P and P/E states. **(c)** EF-G–GTP binds to the ribosome, with its domain IV closest to the A site. **(d)** GTP is hydrolyzed, which allows the mRNA and anticodon ends of the tRNAs to translocate on the 30S particle. This brings the two tRNAs into the classic P and E sites, and also allows relaxation of the ratchet back to its initial, pretranslocation state. **(e)** EF-G–GTP dissociates from the ribosome. **(f)** The ratchet. The 30S particle (cyan) rotates about 6° counter-clockwise relative to the 50S particle (brown) in going from the classic (left) to ratcheted (right) state. (*Source:* Reprinted by permission from Macmillan Publishers Ltd: *Nature* 461, 1234–1242 (29 October 2009) Schmeing & Ramakrishnan, What recent ribosome structures have revealed about the mechanism of translation. © 2009.)

Interaction of the 70S Ribosome with RF1 and RF2 Several structural studies have shown that the release factors, both prokaryotic and eukaryotic, resemble tRNAs and that certain amino acids at one end of the release factor molecule may act like an anticodon in interacting with the stop codon. In particular, a string of three amino acids in RF1 (PXT, where P is proline, T is threonine, and X is any amino acid) was predicted to recognize two stop codons, UAA and UAG. In 2008, Harry Noller and colleagues shed more light on this and other issues when they presented the x-ray crystal structure of a complex containing the *T. thermophilus* 70S ribosome, RF1, tRNA, and an mRNA that included a UAA stop codon.

Figure 19.22a and b compare the positions of RF1 and an aminoacyl-tRNA in the A site of the ribosome. These panels, as well as the details shown in panels c and d, make it clear that parts of RF1, including domains 2 and 3, occupy essentially the same position in the A site that an aminoacyl-tRNA would normally fill. In particular, panels c and d suggest that a part of domain 2 (yellow), including the PXT motif (in this case, PVT, red), constitute a kind of "reading head" that closely approaches the stop codon in the mRNA and has the potential to make specific contacts to "read" the stop codon. Panels c and d also show that the other end of RF1 in the A site, the tip of domain 3 (purple), including the universally conserved GGQ motif (red) closely approaches the peptidyl transferase center (PTC) and therefore is in position to participate in the conversion of the peptidyl transferase activity to an esterase activity that cleaves the polypeptide from the tRNA, terminating translation. Below, we will examine the role of the codon recognition end (the reading head) of RF1 in more detail.

Figure 19.23 depicts the codon recognition site of the complex, and demonstrates that the previously suggested simple recognition of UAA by the PXT motif was far too simple. The PXT motif does indeed play an important role, but it discriminates the first two bases of the UAA codon, rather than the last two, as previously proposed, and it is aided by other conserved parts of RF1 and the 16S rRNA. Specifically, Figure 19.23b shows that T186 of the PXT motif helps to recognize U1 and A2 of the UAA codon by forming hydrogen bonds with both bases. In addition, the protein backbone at glycine 116 and glutamate 119 makes two hydrogen bonds with U1 of the UAA codon. Also, A2 of the stop codon stacks between stop codon base A1 and histidine 193 of RF1. Finally, the 2′-hydroxyl groups of the ribose moieties of U1 and A2 make hydrogen bonds to phosphate 1493 and the ribose of A1492, respectively, of the 16S rRNA (using the *E. coli* numbering system). All of these interactions work best with the U and A in the first two positions of the stop codon. It is interesting that A1492 and A1493 participate in binding normal codons (see earlier in this chapter) and the stop codon, but their roles are much different with the two types of codon.

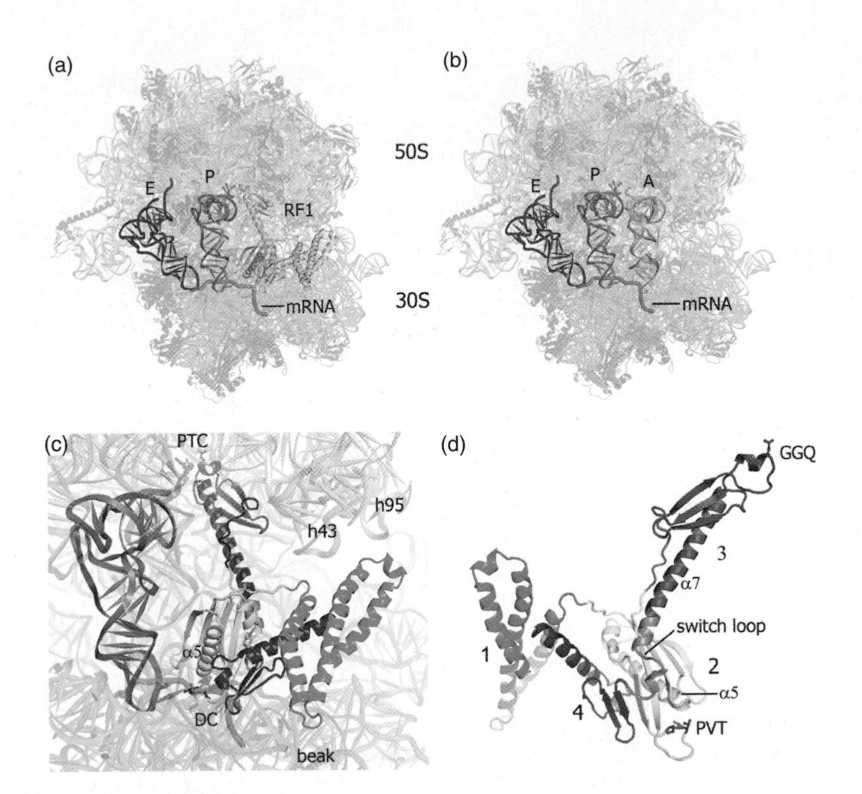

Figure 19.22 Structure of the RF1-ribosome complex. **(a)** Positions of RF1, P site tRNA, E site tRNA, and mRNA in the 70S-ribosome. **(b)** Positions of A site tRNA, P site tRNA, E site tRNA, and mRNA in the 70S ribosome. **(c)** Detail of the positions of RF1 and P site tRNA (orange) in the ribosome. PTC, peptidyltransferase center; DC, decoding center; h43 and h95, helices of 23S rRNA. **(d)** RF1 rotated 180° relative to panel (c). The domains of RF1 are denoted by the same colors as in panel (c): domain 1, green; domain 2, yellow; domain 3, purple; domain 4, magenta; PVT and GGQ motifs, red; switch loop, orange. (*Source:* Reprinted by permission from Macmillan Publishers Ltd: *Nature,* 454, 852–857, 14 August 2008. Laurberg et al, Structural basis for translation termination on the 70S ribosome. © 2008.)

An amino acid-encoding codon has all three bases stacked together, so they can base-pair with the three stacked bases of the corresponding anticodon. However, the crystal structure in Figure 19.23a and c shows that the third base (A3) of the stop codon UAA is widely separated from the others. This separation is caused by several factors. For one thing, His193 of RF1 inserts roughly where the third base of a normal codon would be, and stacks with A2. This pushes A3 away from A2 (to the right in Figure 19.20a), where it can interact with the following residues of IF1: Thr 194, Q 181, and the backbone carbonyl of I 192. In addition, G530 of the 16S rRNA stacks with A3, helping to stabilize its separation from A2.

Later in 2008, Ramakrishnan and colleagues published the crystal structure of RF2 bound to the *T. thermophilus* ribosome, including the UGA stop codon, which is specific for RF2. This structure confirmed that the anticodon-like tripeptide corresponding to PXT in RF1, which is (SPF; Ser-Pro-Phe) in RF2, acts like PTX in RF1 by closely approaching the decoding center, where it helps recognize the stop codon. In addition, just as the PXT motif in RF1 gets help from other residues in RF1 and 16S rRNA, the SPF motif in RF2 is important, but by no means acts alone in recognizing the UGA stop codon.

Ramakrishnan and colleagues also showed that the invariant GGQ motif in RF2, just like the same motif in RF1, is positioned very close to the peptidyl transferase center, where it presumably takes part in release of the polypeptide from the tRNA. Their structure showed that the two glycines in the motif assume conformations that would be impossible for any other amino acid, which explains why these two amino acids are universally conserved. The conformation of the GGQ places the Q in position to participate in the hydrolysis of the ester bond linking the polypeptide to the tRNA. This is also the way RF1 presumably works, which explains why the glutamine in the motif is universally conserved.

SUMMARY RF1 Domains 2 and 3 fill the codon recognition site and the peptidyl transferase site, respectively, of the ribosome's A site, in recognizing the UAA stop codon. The "reading head" portion of domain 2 of RF1, including its conserved PXT motif, occupies the decoding center within the A site and collaborates with A1493 and A1492 of the 16S rRNA to recognize the stop codon. The universally conserved GGQ motif at the tip of domain 3 of RF1 closely approaches the peptidyl transferase center and participates in cleavage of the ester

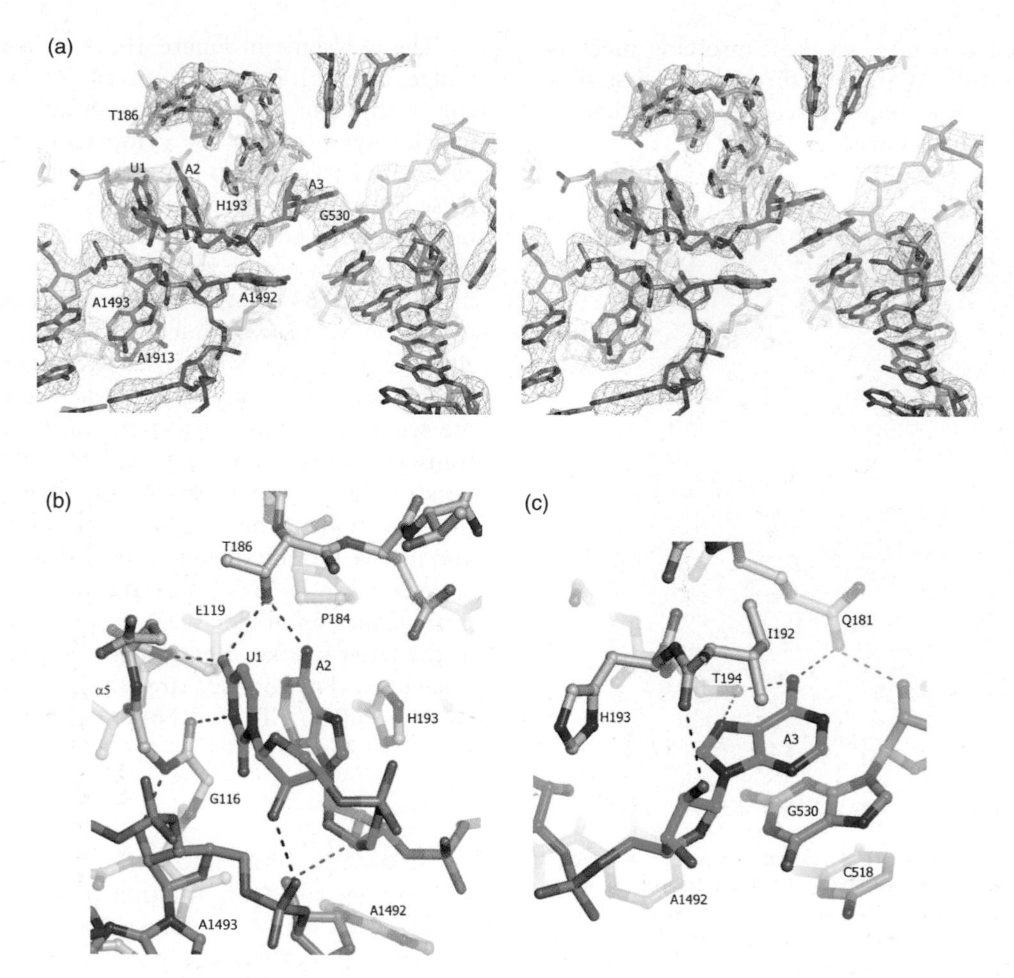

Figure 19.23 Detail of interactions between UAA stop codon and the decoding center. (a) Stereo diagram of the stop codon (green), the RFI reading head (yellow), 16S rRNA (cyan), and one base of 23S rRNA (A1913, gray). U1, A2, and A3 of the stop codon are labeled, as are key amino acids of RFI, and key bases of 16S rRNA. **(b and c)** Detail of interactions between the first two bases **(b)** and the last base **(c)** of the stop codon and the decoding center. Hydrogen bonds between key parts of the RFI protein and the 16S rRNA are shown as dashed lines. (*Source:* Reprinted by permission from Macmillan Publishers Ltd: *Nature,* 454, 852–857, 14 August 2008. Laurberg et al, Structural basis for translation termination on the 70S ribosome. © 2008.)

bond linking the completed polypeptide to the tRNA. RF2 binds to the ribosome in much the same way in response to the UGA stop codon. Its SPF motif, which corresponds to the PXT motif in RF1, is in position to recognize the stop codon, in collaboration with other residues in RF2 and the 16S rRNA. Its GGQ motif is at the peptidyl transferase center, where it can participate in cleavage of the polypeptide–tRNA bond, which terminates translation.

Polysomes

We have seen in previous chapters that more than one RNA polymerase can transcribe a gene at a time. The same is true of ribosomes and mRNA. In fact, it is common for many ribosomes to be traversing the same mRNA in tandem at any given time. The result is a polyribosome, or **polysome,** such as the one pictured in Figure 19.24. In this polysome we can count 74 ribosomes translating the mRNA simultaneously. We can also tell which end of the polysome is which by looking at the nascent polypeptide chains. These grow longer as the ribosome moves from the 5′-end (where translation begins) to the 3′-end (where translation ends). Therefore, the 5′-end is at lower left, and the 3′-end is at lower right.

Consider the process of forming a eukaryotic polysome. The first ribosome to load onto the mRNA faces the most difficult task in its "pioneer round" of translation. The mRNA comes from the nucleus loaded with proteins: Some of these are left over from the processes of splicing and polyadenylation; other mRNA-bound proteins help guide the mRNA out of the nucleus and protect it from destruction. But there is barely room for the mRNA itself between the two ribosomal subunits, so these proteins must be stripped off as the mRNA threads through the first ribosome. These proteins are soon replaced by others that are required for the translation process.

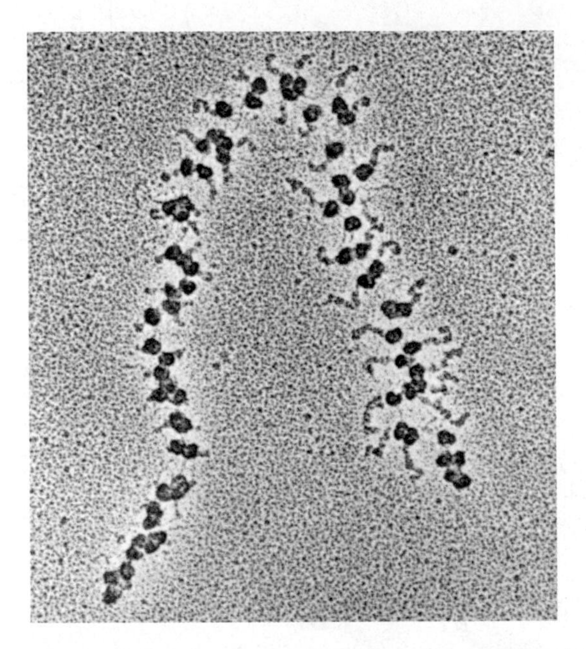

Figure 19.24 Electron micrograph of a polysome from the midge *Chironomus.* The 5′-end on the mRNA is at lower left, and the mRNA bends up and then down to the 3′-end at lower right. The dark blobs attached to the mRNA are ribosomes. The fact that many (about 74) of them are present is the reason for the name *polysome.* Nascent polypeptides extend away from each ribosome and grow longer as the ribosomes approach the end of the mRNA. The faint blobs on the nascent polypeptides are not individual amino acids but domains containing groups of amino acids. (*Source:* Francke et al., Electron microscopic visualization of a discreet class of giant translation units in salivary glands of *Chironomus tetans. EMBO Journal* 1, 1982, pp. 59–62. European Molecular Biology Organization.)

The polysome in Figure 19.24 is from a eukaryote (a midge, or gnat). Because transcription and translation occur in different compartments in eukaryotes, polysomes will always occur in the cytoplasm, independent of the genes. Prokaryotes also have polysomes, but the picture in these organisms is complicated by the fact that transcription and translation of a given gene and its mRNA occur simultaneously and in the same location. Thus, we can see nascent mRNAs being synthesized and being translated by ribosomes at the same time. Figure 19.25 shows just such a situation in *E. coli.* We can see two segments of the bacterial chromosome running parallel from left to right. Only the segment on top is being transcribed. We can tell that transcription is occurring from left to right in this picture because the polysomes are getting longer as they move in that direction; as they get longer, they have room for more and more ribosomes. Do not be misled by the difference in scale between Figures 19.24 and 19.25; the ribosomes appear smaller, and the nascent protein chains are not visible in the latter picture. Remember also that the strands running across Figure 19.25 are DNA, whereas that in Figure 19.24 is mRNA. The mRNAs are more or less vertical in Figure 19.25.

SUMMARY Most mRNAs are translated by more than one ribosome at a time; the result, a structure in which many ribosomes translate an mRNA in tandem, is called a polysome. In eukaryotes, polysomes are found in the cytoplasm. In prokaryotes, transcription of a gene and translation of the resulting mRNA occur simultaneously. Therefore, many polysomes are found associated with an active gene.

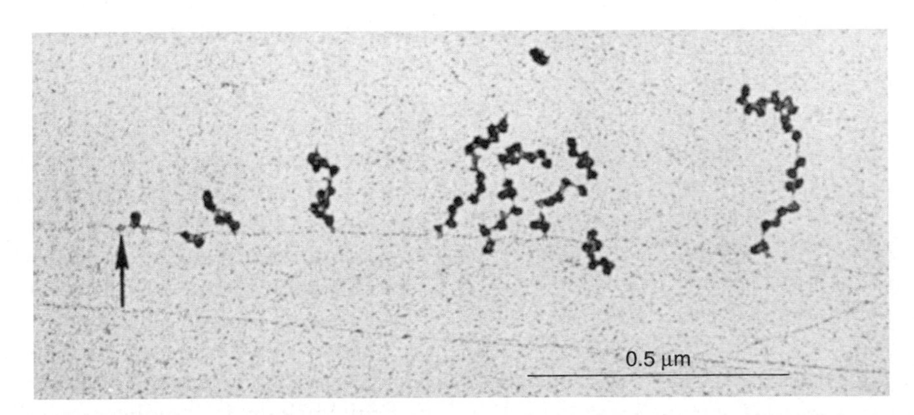

Figure 19.25 Simultaneous transcription and translation in *E. coli.* Two DNA segments stretch horizontally across the picture. The top segment is being transcribed from left to right. As the mRNAs grow, more and more ribosomes attach and carry out translation. This gives rise to polysomes, which are arrayed more or less perpendicular to the DNA. The nascent polypeptides are not visible in this picture. The arrow at left points to a faint spot, which may be an RNA polymerase just starting to transcribe the gene. Other such spots denoting RNA polymerase appear at the bases of some of the polysomes, where the mRNAs join the DNA. (*Source:* O.L. Miller, B.A. Hamkalo, and C.A. Thomas Jr., Visualization of bacterial genes in action. *Science* 169 (July 1970) p. 394. Copyright © AAAS.)

19.2 Transfer RNA

In 1958, Francis Crick postulated the existence of an adaptor molecule, presumably RNA, that could serve as a mediator between the string of nucleotides in DNA (actually in mRNA) and the string of amino acids in the corresponding protein. Crick favored the idea that the adapter contained two or three nucleotides that could pair with nucleotides in codons, although no one knew the nature of codons, or even of the existence of mRNA, at that time. Transfer RNA had already been discovered by Paul Zamecnik and coworkers a year earlier, although they did not realize that it played an adapter role.

The Discovery of tRNA

By 1957, Zamecnik and colleagues had worked out a cell-free protein synthesis system from the rat. One of the components of the system was a so-called pH 5 enzyme fraction that contained the soluble factors that worked with ribosomes to direct translation of added mRNAs. Most of the components in the pH 5 enzyme fraction were proteins, but Zamecnik's group discovered that this mixture also included a small RNA. Of even more interest was their finding that this RNA could be coupled to amino acids. To demonstrate this, they mixed the RNA with the pH 5 enzymes, ATP, and [^{14}C]leucine. Figure 19.26a shows that the more labeled leucine these workers added to the mixture, the more was attached to the RNA, which they separated from protein by phenol extraction. Furthermore, when they left out ATP, no reaction occurred. We now know that this reaction was the charging of tRNA with an amino acid.

Not only did Zamecnik and his coworkers show that the small RNA could be charged with an amino acid, they also demonstrated that it could pass its amino acid to a growing protein. They performed this experiment by mixing the [^{14}C]leucine-charged pH 5 RNA with microsomes—small sections of endoplasmic reticulum containing ribosomes. Figure 19.26b shows a near-perfect correspondence between the loss of radioactive leucine from the pH 5 RNA and gain of the leucine by the protein in the microsomes. This represented the incorporation of leucine from leucyl-tRNA into nascent polypeptides on ribosomes.

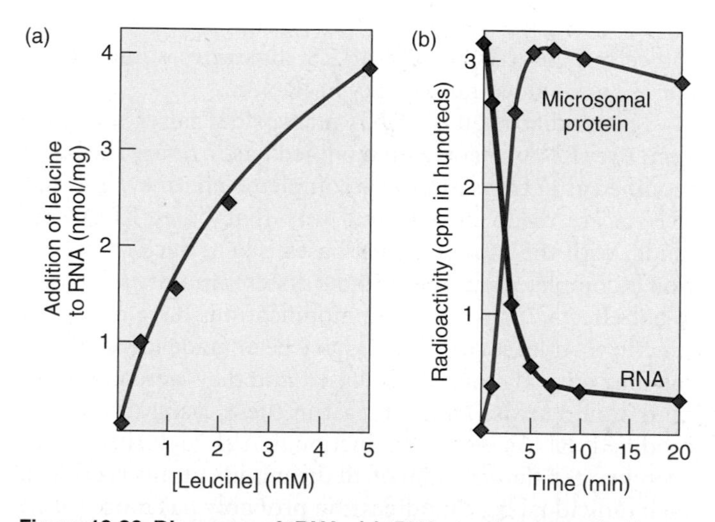

Figure 19.26 Discovery of tRNA. (a) tRNAs can be charged with leucine. Zamecnik and colleagues added labeled leucine to the tRNA-containing fraction and plotted the binding of leucine to the RNA as a function of labeled leucine added. **(b)** The charged tRNA can donate its amino acid to nascent protein. Zamecnik and colleagues followed the radioactivity (cpm) lost from the RNA (blue) and gained by the nascent proteins (red) in the microsomes, which contained the ribosomes. The reciprocal relationship between these curves suggested that the RNA was donating its amino acid to the growing protein. (*Source:* Adapted from Hoagland, M. B., et al., *Journal of Biological Chemistry* 231:244 & 252, 1958.)

SUMMARY Transfer RNA was discovered as a small RNA species independent of ribosomes that could be charged with an amino acid and could then pass the amino acid to a growing polypeptide.

tRNA Structure

To understand how a tRNA carries out its functions, we need to know the structure of the molecule, and tRNAs have a surprisingly complex structure considering their small size. Just as a protein has primary, secondary, and tertiary structure, so does a tRNA. The primary structure is the linear sequence of bases in the RNA; the secondary structure is the way different regions of the tRNA base-pair with each other to form stem-loops; and the tertiary structure is the overall three-dimensional shape of the molecule. In this section, we will survey tRNA structure and its relationship to tRNA function.

In 1965, Robert Holley and his colleagues completed the first determination ever of the base sequence of a natural nucleic acid, an alanine tRNA from yeast. This primary sequence suggested at least three attractive secondary structures, including one that had a cloverleaf shape. By 1969, 14 tRNA sequences had been determined, and it became clear that, despite considerable differences in primary structure, all could assume essentially the same "cloverleaf" secondary structure, as illustrated in Figure 19.27a. As we study this structure we should bear in mind that the real three-dimensional structure of a tRNA is not cloverleaf-shaped at all; the cloverleaf merely describes the base-pairing pattern in the molecule.

The cloverleaf has four base-paired stems that define the four major regions of the molecule (Figure 19.27b). The first, seen at the top of the diagram, is the **acceptor stem,** which includes the two ends of the tRNA, which are base-paired to each other. The 3′-end, bearing the invariant sequence CCA, protrudes beyond the 5′-end. On the left is

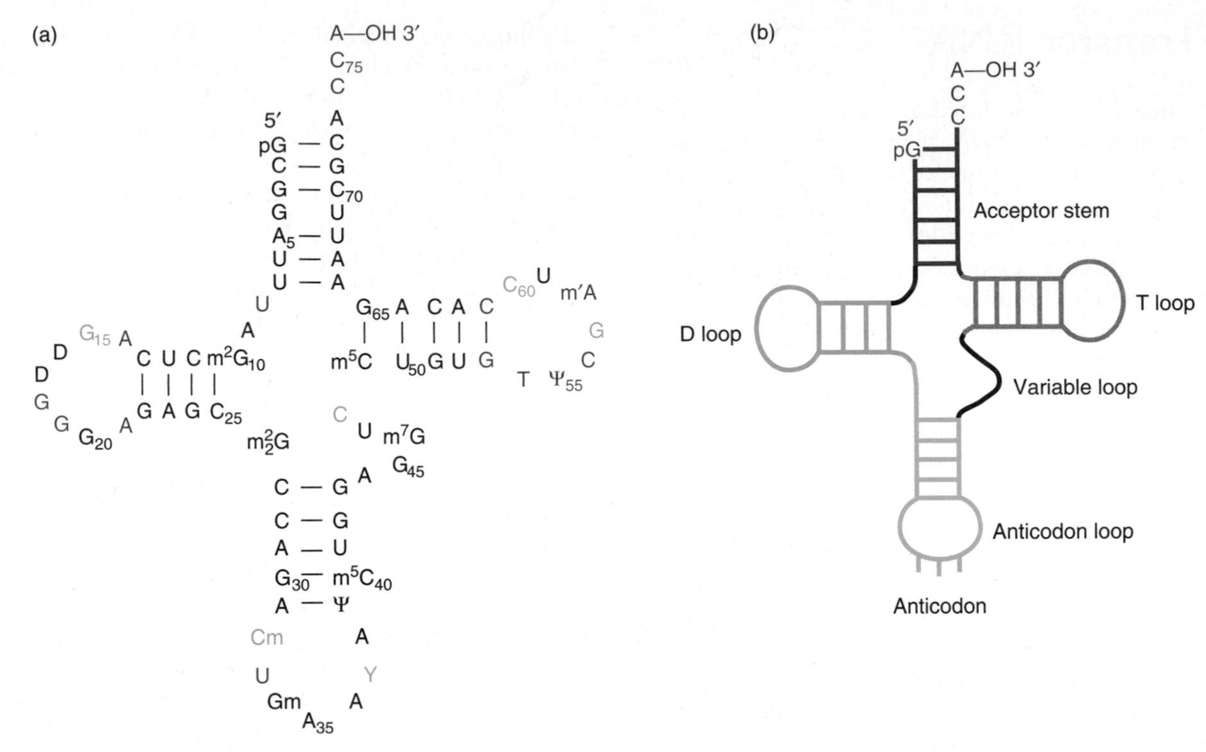

Figure 19.27 Two views of the cloverleaf structure of tRNA. **(a)** Base sequence of yeast $tRNA^{Phe}$, shown in cloverleaf form. Invariant nucleotides are in red. Bases that are always purines or always pyrimidines are in blue. **(b)** Cloverleaf structure of yeast $tRNA^{Phe}$. At top is the acceptor stem (red), where the amino acid binds to the 3′-terminal adenosine. At left is the dihydro U loop (D loop, blue), which contains at least one dihydrouracil base. At bottom is the anticodon loop (green), containing the anticodon. The T loop (right, gray) contains the virtually invariant sequence TΨC. Each loop is defined by a base-paired stem of the same color. (*Source:* (*a*) Adapted from Kim, S.H., F.L. Suddath, G.J. Quigley, A. McPherson, J.L. Sussman, A.H.J. Wang, N.C. Seeman, and A. Rich, Three-dimensional tertiary structure of yeast phenylalanine transfer RNA, *Science* 185:435, 1974.)

the **dihydrouracil loop (D loop)**, named for the modified uracil bases this region always contains. At the bottom is the **anticodon loop**, named for the all-important anticodon at its apex. As we learned in Chapter 3, the anticodon base-pairs with an mRNA codon and therefore allows decoding of the mRNA. At right is the T loop, which takes its name from a nearly invariant sequence of three bases: TΨC. The Ψ stands for a modified nucleoside in tRNA, **pseudouridine.** It is the same as normal uridine, except that the base is linked to the ribose through the 5-carbon of the base instead of the 1-nitrogen. The region between the anticodon loop and the T loop in Figure 19.27 is called the **variable loop** because it varies in length from 4 to 13 nt; some of the longer variable loops contain base-paired stems.

Transfer RNAs contain many modified nucleosides in addition to dihydrouridine and pseudouridine. Some of the modifications are simple methylations. Others are more elaborate, such as the conversion of guanosine to a nucleoside called **wyosine,** which contains a complex three-ring base called the Y base (Figure 19.28). Some tRNA modifications are general. For example, virtually all tRNAs have a pseudouridine in the same position in the T loop, and most tRNAs have a hypermodified nucleoside such as wyosine next to the anticodon. Other modifications are specific for certain tRNAs. Figure 19.28 illustrates some of the common modified nucleosides in tRNAs.

The modification of tRNA nucleosides raises the question: Are tRNAs made with modified bases, or are the bases modified after transcription is complete? The answer is that tRNAs are made in the same way that other RNAs are made, with the four standard bases. Then, once transcription is complete, multiple enzyme systems modify the bases. What effects, if any, do these modifications have on tRNA function? At least two tRNAs have been made in vitro with the four normal, unmodified bases, and they were unable to bind amino acids. Thus, at least in these cases, totally unmodified tRNAs were nonfunctional. Although these studies suggested that the sum of all the modifications is critical, each individual base modification probably has more subtle effects on the efficiency of charging and tRNA usage.

In the 1970s, Alexander Rich and his colleagues used x-ray diffraction techniques to reveal the tertiary structure of tRNAs. Because all tRNAs have essentially the same secondary structure, represented by the cloverleaf model, it is perhaps not too surprising that they all have essentially the same tertiary structure as well. Figure 19.29 illustrates this inverted L-shaped structure for yeast $tRNA^{Phe}$. Perhaps the most important aspect of this structure is that it

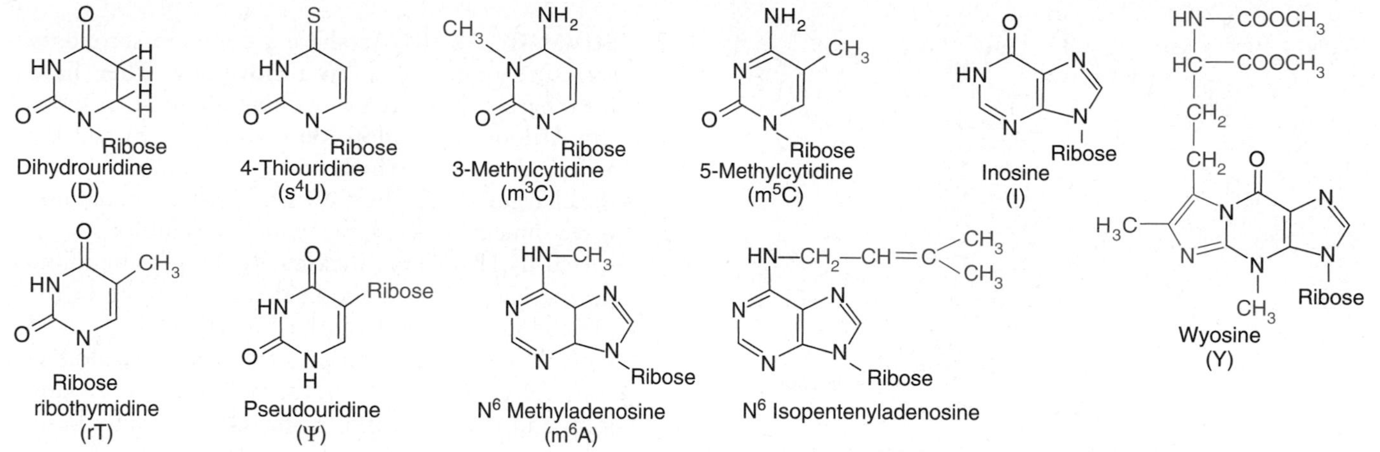

Dihydrouridine
(D)
4-Thiouridine
(s4U)
3-Methylcytidine
(m3C)
5-Methylcytidine
(m5C)
Inosine
(I)
Wyosine
(Y)
ribothymidine
(rT)
Pseudouridine
(Ψ)
N6 Methyladenosine
(m6A)
N6 Isopentenyladenosine
Ribose

Figure 19.28 Some modified nucleosides in tRNA. Red indicates the variation from one of the four normal RNA nucleosides. Inosine is a special case; it is a normal precursor to both adenosine and guanosine.

T loop
T stem
Acceptor stem
D loop
D stem
Variable loop
Anticodon stem
Anticodon
Anticodon loop
(a)
(b)
(c)
54
64
1
56
72
20
7
69
12
44
26
38
32
2
3
4

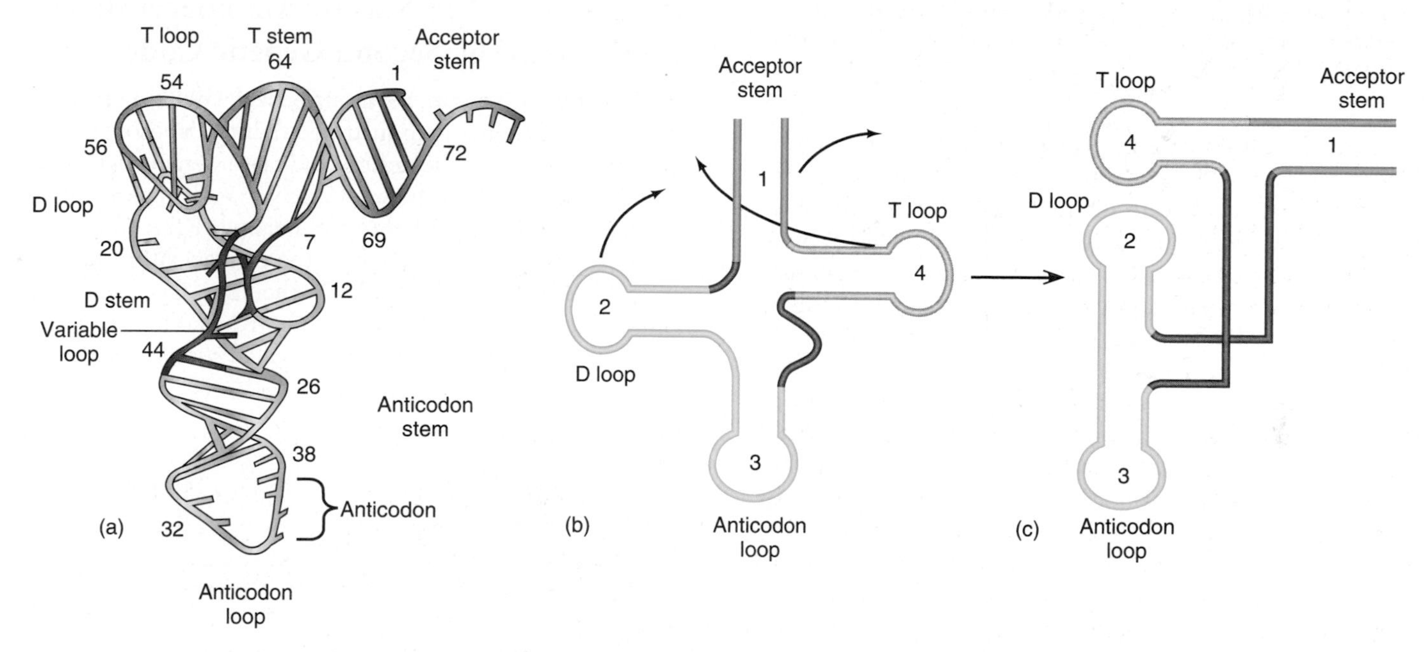

Figure 19.29 Three-dimensional structure of tRNA. (a) A planar projection of the three-dimensional structure of yeast tRNA^{Phe}. The various parts of the molecule are color-coded to correspond to (b) and (c). **(b)** Familiar cloverleaf structure of tRNA with same color scheme as part (a). Arrows indicate the contortions this cloverleaf would have to go through to achieve the approximate shape of a real tRNA, shown in part **(c).** (*Source:* Adapted from Quigley, G.J. and A. Rich, Structural domains of transfer RNA molecules, *Science* 194:197, Fig. 1b, 1976.)

maximizes the lengths of its base-paired stems by stacking them in sets of two to form relatively long extended base-paired regions. One of these regions lies horizontally at the top of the molecule and encompasses the acceptor stem and the T stem; the other forms the vertical axis of the molecule and includes the D stem and the anticodon stem. Even though the two parts of each stem are not aligned perfectly and the stems therefore bend slightly, the alignment allows the base pairs to stack on each other, and therefore confers stability. The base-paired stems of the molecule are RNA–RNA double helices. As we learned in Chapter 2, such RNA helices should assume an A-helix form with about 11 bp per helical turn, and the x-ray diffraction studies verified this prediction.

Figure 19.30 is a stereo diagram of the yeast tRNA^{Phe} molecule. The base-paired regions are particularly easy to see in three dimensions, but you can even visualize them in two dimensions in the T stem-acceptor region because they are depicted almost perpendicular to the plane of the page, so they appear as almost parallel lines.

As we have seen, a tRNA is stabilized primarily by the secondary interactions that form the base-paired regions,

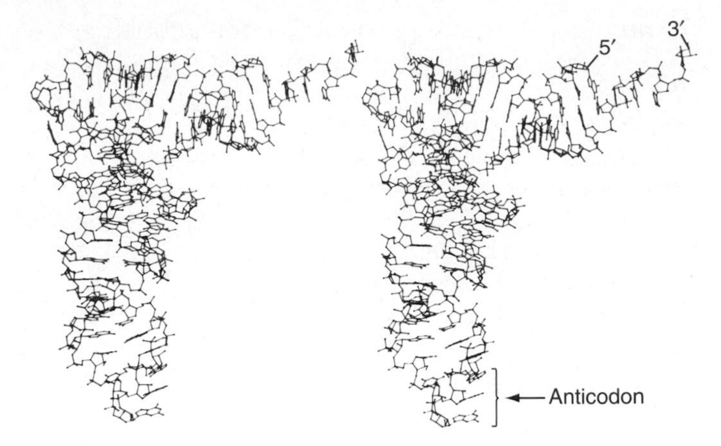

Figure 19.30 Stereo view of tRNA. To see the molecule in three dimensions, use a stereo viewer, or force the two images to merge either by relaxing your eyes as if focusing on something in the distance (the "magic eye" technique) or by crossing your eyes slightly. It may take a little time for the three-dimensional effect to develop. (*Source:* From Quigley, G.J. and A. Rich, Structural domains of transfer RNA molecules. *Science* 194 (19 Nov 1976) f. 2, p. 798. Copyright © AAAS. Reprinted with permission from AAAS.)

but it is also stabilized by dozens of tertiary interactions between regions. These include base–base, base–backbone, and backbone–backbone interactions. Most of the base–base tertiary interactions that involve hydrogen bonds occur between invariant or semi-invariant bases (the semi-invariant bases are always purines or always pyrimidines). Because these interactions allow the tRNA to fold into the proper shape, it makes sense that the bases involved tend not to vary; any variance would hinder the proper folding and hence the proper functioning of the tRNA. Only one of the base–base interactions is a normal Watson–Crick base pair (G19–C56). All the others are extraordinary. The G15–C48 pair, for example, which joins the D loop to the variable loop, cannot be a Watson–Crick base pair because the two strands are parallel here, rather than antiparallel. We call this a *trans*-pair. Several examples also occur of one base interacting with two other bases. One of these involves U8, A14, and A21. Now that the tertiary interactions have been discussed, you can look again at Figure 19.29a and see them in a more realistic form. Note for example the interactions between bases 18 and 55, and between bases 19 and 56. At first glance, these look like base pairs within the T loop; on closer inspection we can now see that they link the T loop and the D loop.

One other striking aspect of tRNA tertiary structure is the structure of the anticodon. Figure 19.30 demonstrates that the anticodon bases are stacked, but this stacking occurs with the bases projecting out to the right, away from the backbone of the tRNA. This places them in position to interact with the bases of the codon in an mRNA. In fact, the anticodon backbone is already twisted into a partial helix shape, which presumably facilitates base-pairing with the corresponding codon (recall Figure 19.2)

SUMMARY All tRNAs share a common secondary structure represented by a cloverleaf. They have four base-paired stems defining three stem-loops (the D loop, anticodon loop, and T loop) and the acceptor stem, to which amino acids are added in the charging step. The tRNAs also share a common three-dimensional shape, which resembles an inverted L. This shape maximizes stability by lining up the base pairs in the D stem with those in the anticodon stem, and the base pairs in the T stem with those in the acceptor stem. The anticodon of the tRNA protrudes from the side of the anticodon loop and is twisted into a shape that readily base-pairs with the corresponding codon in mRNA.

Recognition of tRNAs by Aminoacyl-tRNA Synthetase: The Second Genetic Code

In 1962, Fritz Lipmann, Seymour Benzer, Günter von Ehrenstein, and colleagues demonstrated that the ribosome recognizes the tRNA, not the amino acid, in an aminoacyl-tRNA. They did this by forming cysteyl-$tRNA^{Cys}$, then reducing the cysteine with Raney nickel to yield alanyl-$tRNA^{Cys}$, as illustrated in Figure 19.31. (Notice the nomenclature here. In cysteyl-$tRNA^{Cys}$ [Cys-$tRNA^{Cys}$] the first Cys tells what amino acid is actually attached to the tRNA. The second Cys [in the superscript] tells what amino acid *should be* attached

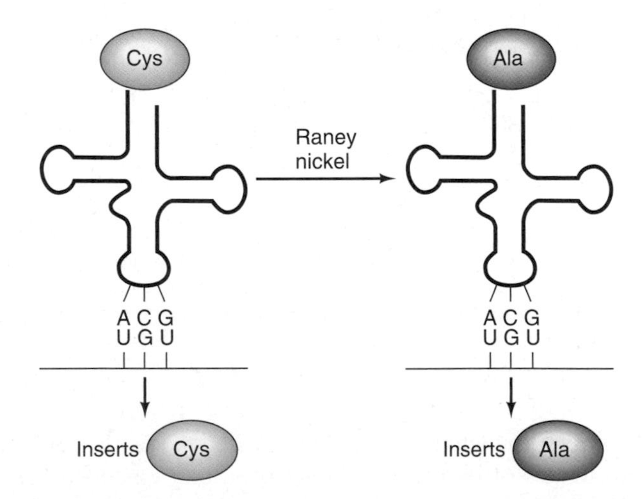

Figure 19.31 The ribosome responds to the tRNA, not the amino acid of an aminoacyl-tRNA. Lipmann, Ehrenstein, Benzer, and colleagues started with a cysteyl-$tRNA^{Cys}$, which inserted cysteine (Cys, blue) into a protein chain, as shown at left. They treated this aminoacyl-tRNA with Raney nickel, which reduced the cysteine to alanine (Ala, red), but had no effect on the tRNA. This alanyl-$tRNA^{Cys}$ inserted alanine into a protein chain at a position normally occupied by cysteine, as depicted at right. Thus, the nature of the amino acid attached to the tRNA does not matter; it is the nature of the tRNA that matters, because its anticodon has to match the mRNA codon.

to this tRNA. Thus, alanyl-$tRNA^{Cys}$ is a tRNA that *should* bind cysteine, but in this case is bound to alanine.) Then Lipmann and colleagues added this altered aminoacyl-tRNA to an in vitro translation system, along with a synthetic mRNA that was a random polymer of U and G, in a 5:1 ratio. This mRNA had many UGU codons, which encode cysteine, so it normally caused incorporation of cysteine. It should not cause incorporation of alanine because the codons for alanine are GCN, where N is any base, and the UG polymer contained no C's. However, in this case alanine was incorporated because it was attached to a $tRNA^{Cys}$. This showed that ribosomes do not discriminate among amino acids attached to tRNAs; they recognize only the tRNA part of an aminoacyl-tRNA.

This experiment pointed to the importance of fidelity in the aminoacyl-tRNA synthetase step. The fact that ribosomes recognize only the tRNA part of an aminoacyl-tRNA means that if the synthetases make mistakes and put the wrong amino acids on tRNAs, then these amino acids will be inserted into proteins in the wrong places. That could be very damaging because a protein with the wrong amino acid sequence is likely not to function properly. Thus, it is not surprising that aminoacyl-tRNA synthetases are very specific for the tRNAs and amino acids they bring together. This raises a major question related to the structure of tRNAs: Given that the secondary and tertiary structures of all tRNAs are essentially the same, what base sequences in tRNAs do the synthetases recognize when they are selecting one tRNA out of a pool of over 20? This set of sequences has even been dubbed the "second genetic code" to highlight its importance. This question is complicated by the fact that some **isoaccepting species** of tRNA can be charged with the same amino acid by the same synthetase, yet they have different sequences, and even different anticodons.

If we were to guess about the locations of the tRNA elements that an aminoacyl-tRNA synthetase recognizes, two sites would probably occur to us. First, the acceptor stem seems a logical choice, because that is the locus on the tRNA that accepts the amino acid and is therefore likely to lie at or near the enzyme's active site as it is being charged. Because the enzyme presumably makes such intimate contact with the acceptor stem, it should be able to discriminate among tRNAs with different base sequences in the acceptor stem. Of course, the last three bases are irrelevant for this purpose because they are the same, CCA, in all tRNAs. Second, the anticodon is a reasonable selection, because it is different in each tRNA, and it has a direct relationship to the amino acid with which the tRNA should be charged. We will see that both these predictions are correct in most cases, and some other areas of certain tRNAs also play a role in recognition by aminoacyl-tRNA synthetases.

The Acceptor Stem In 1972, Dieter Söll and his colleagues noticed a pattern in the nature of the fourth base from the 3′-end, position 73 in most tRNAs. That is, this base tended to be the same in tRNAs specific for a certain class of amino acids. For example, virtually all the hydrophobic amino acids are coupled to tRNAs with A in position 73, regardless of the species in which we find the tRNA. However, this obviously cannot be the whole story because one base does not provide enough variation to account for specific charging of 20 different classes of tRNAs. At best, it fills the role of a rough discriminator.

Bruce Roe and Bernard Dudock used another approach. They examined the base sequence of all the tRNAs from several species that could be charged by a single synthetase. This included some tRNAs that were charged with the wrong amino acid, in a process called *heterologous mischarging.* This term refers to the ability of a synthetase from one species to charge an incorrect tRNA from another species, although this mischarging is always slower and requires a higher enzyme concentration than normal. For example, yeast phenylalanyl-tRNA synthetase (PheRS) can charge $tRNA^{Phe}$ from *E. coli,* yeast, and wheat germ correctly, but it can also charge *E. coli* $tRNA^{Val}$ with phenylalanine.

Because all these tRNAs can be charged by the same synthetase, they should all have the elements that the synthetase uses to tell it which tRNAs to charge. So Roe and Dudock compared the sequences of all these tRNAs, looking for things they have in common, but are not common to all tRNAs. Two features stood out: base 73, and nine nucleotides in the D stem.

In 1973, J.D. Smith and Julio Celis studied a mutant suppressor tRNA that inserted Gln instead of Tyr. In other words, the wild-type suppressor tRNA was charged by the GlnRS, but some change in its sequence caused it to be charged by the TyrRS instead. The only difference between the mutant and wild-type tRNAs was a change in base 73 from G to A.

In 1988, Ya-Ming Hou and Paul Schimmel used genetic means to demonstrate the importance of a single base pair in the acceptor stem to charging specificity. They started with a $tRNA^{Ala}$ that had its anticodon mutated to 5′-CUA-3′ so it became an amber suppressor capable of inserting alanine in response to the amber codon UAG. Then they looked for mutations in the tRNA that changed its charging specificity. Their assay was a convenient one they could run in vivo. They built a *trpA* gene with an amber mutation in codon 10. This mutation could be suppressed only by a tRNA that could insert an alanine (or glycine) in response to the amber codon. Any other amino acid in position 10 yielded an inactive protein. Finally, they challenged their mutants by growing them in the absence of tryptophan. If the mutant could suppress the amber mutation in the *trpA* gene, it had a suppressor tRNA that could still be correctly charged with alanine (or glycine). If not, the suppressor tRNA was altered so it was charged with another amino acid. They found that all the cells that grew in the absence of tryptophan had a G in position 3 of the suppressor tRNA and a U in position 70, so a G3-U70 wobble base pair could form in the acceptor stem three bases from the end of the stem.

This experiment suggested that the G3–U70 base pair is a key determinant of charging by AlaRS. If so, these workers reasoned, they might be able to take another suppressor tRNA that inserted another amino acid, change its bases at positions 3 and 70 to G and U, respectively, and convert the charging specificity of the suppressor tRNA to alanine. They did this with two different suppressor tRNAs: $tRNA^{Cys/CUA}$ and $tRNA^{Phe/CUA}$, where the CUA designation refers to the anticodon, which recognizes the UAG amber codon. Both of the tRNAs originally had a C3–G70 base pair in their acceptor stems. However, when Hou and Schimmel changed this one base pair to G3–U70, they converted the tRNAs to $tRNA^{Ala/CUA}$, as indicated by their ability to suppress the amber mutation in codon 10 of the *trpA* gene.

Did these altered amber suppressor tRNAs really insert alanine into the TrpA protein? Amino acid sequencing revealed that they did. Furthermore, these altered tRNAs could be charged with alanine in vitro. Thus, even though these two tRNAs differed from natural $tRNA^{Ala/CUA}$ in 38 and 31 bases, respectively, changing just one base pair from C–G to G–U changed the charging specificity from Cys or Phe to Ala.

In 1989, Christopher Francklyn and Schimmel presented another line of evidence that implicates the acceptor stem, and the G3–U70 base pair in particular, in AlaRS charging specificity. They showed that a synthetic 35-nt "minihelix" resembling the top part of the inverted L-shaped $tRNA^{Ala}$, including the acceptor stem and the TΨC loop, can be efficiently charged with alanine. In fact, as long as the G3–U70 base pair was present, charging with alanine occurred even when many other bases were changed.

It is also interesting that the Ala-minihelix binds to the P site of the ribosome, and participates just as well as intact Ala-$tRNA^{Ala}$ in the peptidyl transferase reaction with puromycin. These observations have led to the speculation that the top part of the tRNA molecule evolved first, and could have participated, along with an ancestor of 23S rRNA, in a crude version of protein synthesis in the "RNA world" before ribosomes evolved.

SUMMARY Biochemical and genetic experiments have demonstrated the importance of the acceptor stem in recognition of a tRNA by its cognate aminoacyl-tRNA synthetase. In certain cases, changing one base pair in the acceptor stem can change the charging specificity.

The Anticodon In 1973, LaDonne Schulman pioneered a technique in which she treated $tRNA_f^{Met}$ with bisulfite, which converts cytosines to uracils. She and her colleagues found that many of these base alterations had no effect, but some destroyed the ability of the tRNA to be charged with methionine. One such change was a C→U change in base 73; another was a C→U change in the anticodon. Since then, Schulman and her colleagues have amassed a large body of evidence that shows the importance of the anticodon in charging specificity.

In 1983, Schulman and Heike Pelka developed a method to change specifically one or more bases at a time in the anticodon of the initiator tRNA, $tRNA_f^{Met}$. First, they cut the wild-type tRNA in two with a limited digestion with pancreatic RNase. This removed the anticodon from the tRNA 5′-fragment, and also cut off the last two nucleotides of the CCA terminus of the 3′-fragment. Then they used T4 RNA ligase to attach a small oligonucleotide to the 5′-fragment that would replace the lost anticodon, with one or more bases altered, ligated the two halves of the molecule back together, and then added back the lost terminal CA with tRNA nucleotidyltransferase. Finally, they tested the tRNAs with altered anticodons in charging reactions in vitro. Table 19.1 shows that changing one base in the anticodon of $tRNA_f^{Met}$ was sufficient to lower the rate of charging with Met by at least a factor of 10^5. The first base in the anticodon (the "wobble" position) was the most sensitive; changing this one base always had

Table 19.1 Initial Rates of Aminoacylation of $tRNA_f^{Met}$ Derivatives

tRNA*	Mol Met-tRNA/mol Met-tRNA synthetase per min	Relative rate, CAU/other
$tRNA_f^{Met}$	28.45	0.8
$tRNA_f^{Met}$ (gel)†	22.80	1
CAU	22.15	1
CAUA	1.59	14
CCU	4.0×10^{-1}	55
CUU	2.6×10^{-2}	850
CUA	2.0×10^{-2}	1100
CAG	1.7×10^{-2}	1300
CAC	1.2×10^{-3}	18,500
CA	0.5×10^{-3}	44,000
C	$<10^{-4}$	$>10^5$
ACU	$<10^{-4}$	$>10^5$
UAU	$<10^{-4}$	$>10^5$
AAU	$<10^{-4}$	$>10^5$
GAU	$<10^{-4}$	$>10^5$

*The oligonucleotide inserted in the anticodon loop of synthesized $tRNA_f^{Met}$ derivatives is indicated.
†Control sample isolated from a denaturing polyacrylamide gel in parallel with the synthesized $tRNA_f^{Met}$ derivatives.
Source: L.H. Schulman and H. Pelka, "Anticodon Loop Size and Sequence Requirements for Recognition of Formylmehionine tRNA by Methionyl-tRNA Synthetase," *Proceedings of the National Academy of Sciences,* November 1983. Reprinted with permission of the author.

a drastic effect on charging. Thus, the anticodon seems to be required for charging of this tRNA in vitro.

In 1991, Schulman and Leo Pallanck followed up the earlier in vitro studies with an in vivo study of the effects of altering the anticodon. Again, they changed the anticodon of the $tRNA_f^{Met}$, but this time they tested the ability of the altered tRNA to be mischarged with the amino acid corresponding to the new anticodon. They tested mischarging with a reporter gene encoding dihydrofolate reductase (DHFR), which is easy to isolate in highly purified form. Here is an example of how the assay worked: They altered the gene for $tRNA_f^{Met}$ so its anticodon was changed from CAU to GAU, which is an isoleucine (Ile) anticodon. Then they placed this mutant gene into *E. coli* cells, along with a mutant DHFR gene bearing an AUC initiation codon.

Ordinarily, AUC would not work well as an initiation codon, but in the presence of a $tRNA_f^{Met}$ with a complementary anticodon, it did. Sequencing of the resulting DHFR protein demonstrated that the amino acid in the first position was primarily Ile. Some Met occurred in the first position, showing that the endogenous wild-type $tRNA_f^{Met}$ could recognize the AUC initiation codon to some extent.

Pallanck and Schulman used the same procedure to change the $tRNA_f^{Met}$ anticodon to GUC (valine, Val) or UUC (phenylalanine, Phe). In each case, they made a corresponding change in the DHFR initiation codon so it was complementary to the anticodon in the altered $tRNA_f^{Met}$. In both cases, the gene functioned significantly better in the presence than in the absence of the complementary $tRNA_f^{Met}$. More importantly, this experiment showed that the nature of the initiating amino acid can change with the alteration in the tRNA anticodon. In fact, with the $tRNA_f^{Met}$ bearing the valine anticodon, valine was the only amino acid found at the amino terminus of the DHFR protein. This means that a change of the $tRNA_f^{Met}$ anticodon from CAU to GAC altered the charging specificity of this tRNA from methionine to valine. Thus, in this case, the anticodon seems to be the crucial factor in determining the charging specificity of the tRNA.

On the other hand, changing the anticodon of the $tRNA_f^{Met}$ always reduced its efficiency. In fact, most such alterations yielded $tRNA_f^{Met}$ molecules whose efficiency was too low to analyze further, even in the presence of complementary initiation codons. Thus, some aminoacyl-tRNA synthetases could charge a noncognate tRNA with an altered anticodon, but others could not. These latter enzymes apparently required more cues than just the anticodon.

SUMMARY Biochemical and genetic experiments have shown that the anticodon, like the acceptor stem, is an important element in charging specificity. Sometimes the anticodon can be the absolute determinant of specificity.

Structures of Synthetase–tRNA Complexes X-ray crystallography studies of complexes between tRNAs and their cognate aminoacyl-tRNA synthetases have shown that both the acceptor stem and the anticodon have docking sites on the synthetases. Thus, these findings underline the importance of the acceptor stem and anticodon in synthetase recognition. In 1989, Dieter Söll and Thomas Steitz and their colleagues used x-ray crystallography to determine the first three-dimensional structure of an aminoacyl-tRNA synthetase (*E. coli* GlnRS) bound to its cognate tRNA. Figure 19.32 presents this structure. Near the top, we see a deep cleft in the enzyme that enfolds the acceptor stem, including base 73 and the 3–70 base pair. At lower left, we observe a smaller cleft in the enzyme into which the anticodon of the tRNA protrudes. This would allow for

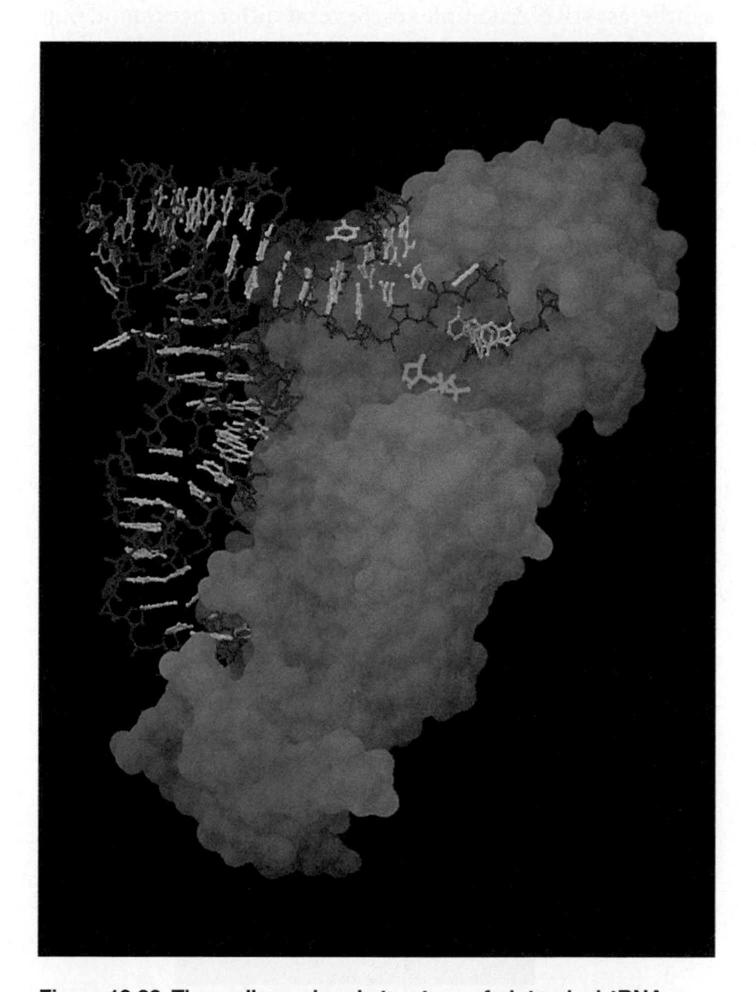

Figure 19.32 Three-dimensional structure of glutaminyl-tRNA synthetase complexed with tRNA and ATP. The synthetase is shown in blue, the tRNA in brown and yellow, and the ATP in green. Note the three areas of contact between enzyme and tRNA: (1) the deep cleft at top that holds the acceptor stem of the tRNA, and the ATP; (2) the smaller pocket at lower left into which the tRNA's anticodon inserts; and (3) the area in between these two clefts, which contacts much of the inside of the L of the tRNA. (*Source:* Courtesy T.A. Steitz; from Rould, Perona, Vogt, and Steitz, *Science* 246 (1 Dec 1989) cover. Copyright © AAAS.)

specific recognition of the anticodon by the synthetase. In addition, we see that most of the left side of the enzyme is in intimate contact with the inside of the L of the tRNA, which includes the D loop side and the minor groove of the acceptor stem.

About half the synthetases, including GlnRS, are in a group called **class I.** These are all structurally similar and initially aminoacylate the 2′-hydroxyl group of the terminal adenosine of the tRNA. The other half of the synthetases are in **class II;** they are structurally similar to other members of their group, but quite different from the members of class I, and they initially aminoacylate the 3′-hydroxyl group of their cognate tRNAs. In 1991, D. Moras and colleagues obtained the x-ray crystal structure of a member of this group, yeast AspRS, together with $tRNA^{Asp}$. Figure 19.33 contrasts the structures of the class I and class II synthetase–tRNA complexes. Several differences stand out. First, although the synthetase still contacts the inside of the L, it does so on the tRNA's opposite face, including the variable loop and the major groove of the acceptor stem.

(a)

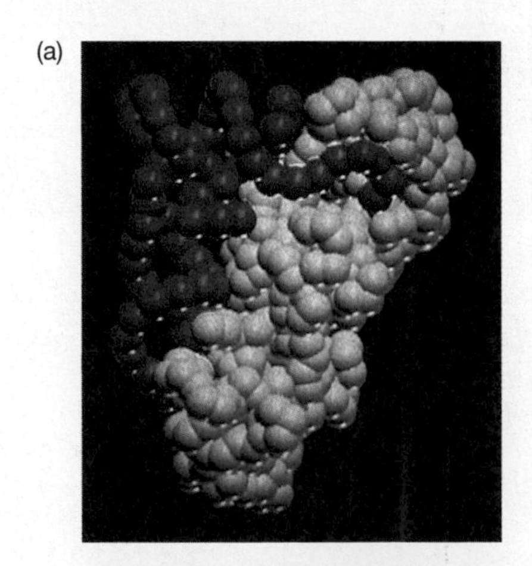

(b)

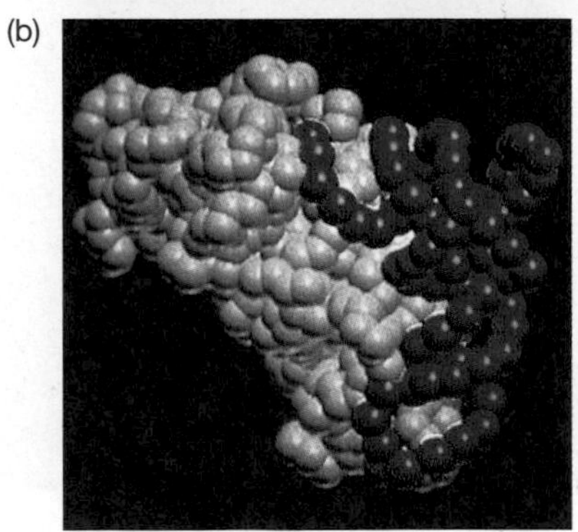

Figure 19.33 Models of (a) a class I complex: *E. coli* GlnRS-$tRNA^{Gln}$, and (b) a class II complex: yeast AspRS-$tRNA^{Asp}$. For simplicity, only the phosphate backbones of the tRNAs (red) and the α-carbon backbones of the synthetases (blue) are shown. Notice the approach of the two synthetases to the opposite sides of their cognate tRNAs. (*Source:* Ruff, M., S. Krishnaswamy, M. Boeglin, A. Poterszman, A. Mitschler, A. Podjarny, B. Rees, J.C. Thierry, and D. Moras, Class II aminoacyl transfer RNA synthetases: Crystal structure of yeast aspartyl-tRNA synthetase complexed with $tRNA^{Asp}$. *Science* 252 (21 June 1991) f. 3, p. 1686. Copyright © AAAS.)

Also, the acceptor stem, including the terminal CCA, is in a regular helical conformation. This contrasts with the class I structure, in which the first base pair is broken and the 3′-end of the molecule makes a hairpin turn. Thus, x-ray crystallography has corroborated the major conclusions of biochemical and genetic studies on synthetase–tRNA interactions: Both the anticodon and acceptor stem are in intimate contact with the enzyme and are therefore in a position to determine specificity of enzyme–tRNA interactions.

SUMMARY X-ray crystallography has shown that synthetase–tRNA interactions differ between the two classes of aminoacyl-tRNA synthetases. Class I synthetases have pockets for the acceptor stem and anticodon of their cognate tRNAs and approach the tRNAs from the D loop and acceptor stem minor groove side. Class II synthetases also have pockets for the acceptor stem and anticodon, but approach their tRNAs from the opposite side, which includes the variable arm and major groove of the acceptor stem.

Proofreading and Editing by Aminoacyl-tRNA Synthetases

As good as aminoacyl-tRNA synthetases are at recognizing the correct (cognate) tRNAs, they have a more difficult job recognizing the cognate amino acids. The reason is clear: tRNAs are large, complex molecules that vary from one another in nucleotide sequence and in nucleoside modifications, but amino acids are simple molecules that resemble one another fairly closely—sometimes very closely. Consider isoleucine and valine, for example. The two amino acids are identical except for an extra methylene (CH_2) group in isoleucine. In 1958, Linus Pauling used thermodynamic considerations to calculate that isoleucyl-tRNA synthetase (IleRS) should make about one-fifth as much incorrect Val-$tRNA^{Ile}$ couples as correct Ile-$tRNA^{Ile}$ couples. In fact, however, only one in 150 amino acids activated by IleRS is valine, and only one in 3000 aminoacyl-tRNAs produced by this enzyme is Val-$tRNA^{Ile}$. How does isoleucyl-tRNA synthetase prevent formation of Val-$tRNA^{Ile}$?

As first proposed by Alan Fersht in 1977, the enzyme uses a *double-sieve* mechanism to avoid producing tRNAs with the wrong amino acid attached. Figure 19.34 illustrates this concept. The first sieve is accomplished by the **activation site** of the enzyme, which rejects substrates that are too large. However, substrates such as valine that are too small can fit into the activation site and so get activated to the aminoacyl adenylate form and sometimes make it all the way to the aminoacyl-tRNA form. That is where the second sieve comes into play. Activated amino acids or, less

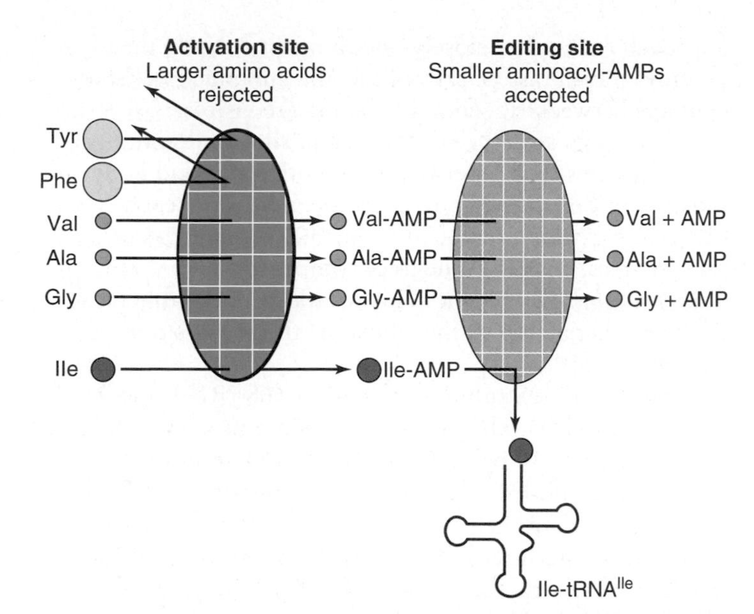

Figure 19.34 The double sieve of isoleucine-tRNA synthetase. The activation site is the coarse sieve in which large amino acids, such as Tyr and Phe, are excluded because they don't fit. The editing (hydrolytic) site is the fine sieve, which accepts activated amino acids smaller than Ile-AMP, such as Val-AMP, Ala-AMP, and Gly-AMP, but rejects Ile-AMP because it is too large. As a result, the smaller activated amino acids are hydrolyzed to AMP and amino acids, whereas Ile-AMP is converted to Ile-tRNA^{Ile}. (*Source:* Adapted from Fersht, A.R., Sieves in sequence. *Science* 280:541, 1998.)

commonly, aminoacyl-tRNAs that are too small are hydrolyzed by another site on the enzyme: the **editing site.**

For example, IleRS uses the first sieve to exclude amino acids that are too large, or the wrong shape. Thus, the enzyme excludes phenylalanine because it is too large and leucine because it is the wrong shape. (One of the terminal methyl groups of leucine cannot fit into the activation site.) But what about smaller amino acids such as valine? In fact, they do fit into the activation site of IleRS, and so they become activated. But then they are transported to the editing site, where they are recognized as incorrect and deactivated. This second sieve is called either **proofreading** or **editing.**

Shigeyuki Yokoyama and colleagues have obtained the crystal structure of the *T. thermophilus* IleRS alone, coupled to its cognate amino acid, isoleucine, and to the noncognate amino acid valine. These structures have amply verified Fersht's elegant hypothesis. Figure 19.35 shows the structure of the activation site, with either (a) isoleucine, or (b) valine bound. We can see that both amino acids fit well into this site, although valine makes slightly weaker contact with two of the hydrophobic amino acid side chains (Pro46 and Trp558) that surround the site. On the other hand, it is clear that this site is too small to admit large amino acids such as phenylalanine, and even leucine would be sterically hindered from binding by one of its two terminal methyl groups. This picture is fully consistent with the coarse sieve part of the double-sieve hypothesis.

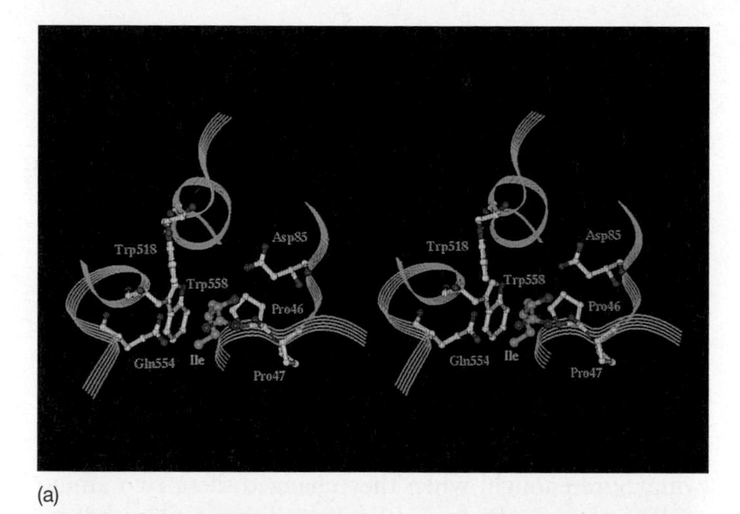

(a)

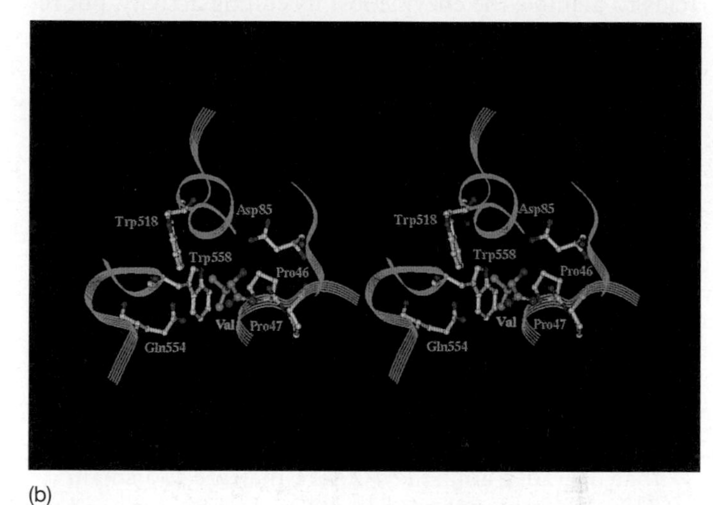

(b)

Figure 19.35 Stereo views of isoleucine and valine in the activation site of IleRS. The backbone of the enzyme is represented by turquoise ribbons, with the carbons of amino acid side chains in yellow. The carbons of the substrates [isoleucine **(a)**, valine **(b)**] are rendered in green. Oxygens of all amino acids are in red and nitrogens are in blue. Note that both isoleucine and valine fit into the activation site. (*Source:* Nureki, O., D.G. Vassylyev, M. Tateno, A. Shimada, T. Nakama, S. Fukai, M. Konno, T.L. Henrickson, P. Schimmel, and S. Yokoyama, Enzyme structure with two catalytic sites for double-sieve selection of substrate. *Science* 280 (24 Apr 1998) f. 2, p. 579. Copyright © AAAS.)

The enzyme has a second deep cleft comparable in size to the cleft of the activation site, but 34 Å away. This second cleft is thought to be the editing site, based in part on the fact that a fragment of the enzyme containing this cleft still retains editing activity. The crystal structure confirms this hypothesis: When Yokoyama and colleagues prepared crystals of the IleRS with valine, they found a molecule of valine at the bottom of the deep cleft. However, when they prepared crystals with isoleucine, no amino acid was found in the cleft. Thus, because the cleft seems to be specific for valine, it appears to be the editing site. Furthermore, inspection of the pocket in which valine is found, shows that the space in between the side chains of Trp232 and Tyr386 is just big enough to accommodate valine, but too small to admit isoleucine.

If this really is the editing site, we would expect that its removal would abolish editing. Indeed, when Yokoyama and colleagues removed 47 amino acids from this region, including Trp232, they abolished editing activity while retaining full activation activity. Thus, the second cleft really does appear to be the editing site. Several amino acid side chains are particularly close to the valine in the cleft, and Thr230 and Asn237 are well-positioned to take part in the hydrolysis reaction that is the essence of editing. To test this hypothesis, Yokoyama and coworkers changed the amino acids in the *E. coli* IleRS (Thr243 and Asn250) that correspond to Thr230 and Asn237 in the *T. thermophilus* enzyme. Sure enough, when they changed these two amino acids to alanine, the enzyme lost its editing activity, but retained its activation activity. All these data are consistent with the hypothesis that the second cleft is the editing site, and that hydrolysis of noncognate aminoacyl-AMPs such as Val-AMP occurs there.

SUMMARY The amino acid selectivity of at least some aminoacyl-tRNA synthetases is controlled by a double-sieve mechanism. The first sieve is a coarse one that excludes amino acids that are too big. The enzyme accomplishes this task with an active site for activation of amino acids that is just big enough to accommodate the cognate amino acid, but not larger amino acids. The second sieve is a fine one that degrades aminoacyl-AMPs that are too small. The enzyme accomplishes this task with a second active site (the editing site) that admits small aminoacyl-AMPs and hydrolyzes them. The cognate aminoacyl-AMP is too big to fit into the editing site, so it escapes being hydrolyzed. Instead, the enzyme transfers the activated amino acid to its cognate tRNA.

SUMMARY

X-ray crystallography studies on bacterial ribosomes with and without tRNAs have shown that the tRNAs occupy the cleft between the two subunits. They interact with the 30S subunit through their anticodon ends, and with the 50S subunit through their acceptor stems. The binding sites for the tRNAs are composed primarily of rRNA. The anticodons of the tRNAs in the A and P sites approach each other closely enough to base-pair with adjacent codons in the mRNA bound to the 30S subunit, given that the mRNA kinks 45 degrees between the two codons. The acceptor stems of the tRNAs in the A and P sites also approach each other closely—within just 5 Å—in the peptidyl transferase pocket of the 50S subunit. Twelve contacts between ribosomal subunits are visible.

The crystal structure of the *E. coli* ribosome contains two structures that differ from each other by rigid body motions of domains of the ribosome, relative to each other. In particular, the head of the 30S particle rotates by 6 degrees, and by 12 degrees compared to the *T. thermophilus* ribosome. This rotation is probably part of the ratchet action of the ribosome that occurs during translocation.

The *E. coli* 30S subunit contains a 16S rRNA and 21 proteins (S1–S21). The 50S subunit contains a 5S rRNA, a 23S rRNA, and 34 proteins (L1–L34). Eukaryotic cytoplasmic ribosomes are larger and contain more RNAs and proteins than their prokaryotic counterparts.

Sequence studies of 16S rRNA led to a proposal for the secondary structure (intramolecular base pairing) of this molecule. X-ray crystallography studies have confirmed the conclusions of these studies. They show a 30S subunit with an extensively base-paired 16S rRNA whose shape essentially outlines that of the whole particle. The x-ray crystallography studies have also confirmed the locations of most of the 30S ribosomal proteins.

The 30S ribosomal subunit plays two roles. It facilitates proper decoding between codons and aminoacyl-tRNA anticodons, including proofreading. It also participates in translocation. Crystal structures of the 30S subunit with three antibiotics that interfere with these two roles shed light on translocation and decoding. Spectinomycin binds to the 30S subunit near the neck, where it can interfere with the movement of the head that is required for translocation. Streptomycin binds near the A site of the 30S subunit and stabilizes the *ram* state of the ribosome. This reduces fidelity of translation by allowing noncognate aminoacyl-tRNAs to bind relatively easily to the A site and by preventing the shift to the restrictive state that is necessary for proofreading. Paromomycin binds in the major groove of the 16S rRNA H44 helix near the A site. This flips out bases A1492 and A1493, so they can stabilize base-pairing between codon and anticodon, including anticodons on noncognate aminoacyl-tRNAs, so fidelity declines.

The x-ray crystal structure of IF1 bound to the 30S ribosomal subunit shows that IF1 binds to the A site. In that position, it clearly blocks fMet-tRNA from binding to the A site, and may also actively promote fMet-tRNA binding to the P site through a presumed interaction between IF1 and IF2. IF1 also interacts intimately with helix H44 of the 30S subunit, and this may explain how IF1 accelerates both association and dissociation of the ribosomal subunits.

The crystal structure of the 50S ribosomal subunit has been determined to 2.4 Å resolution. This structure reveals relatively few proteins at the interface between ribosomal subunits, and no protein within 18 Å of the peptidyl transferase active center tagged with a transition state analog. The 2′-OH group of the tRNA in the P site is very well positioned to form a hydrogen bond to the amino group of the aminoacyl-tRNA in the A site, and therefore to help catalyze the peptidyl transferase reaction. In accord with this hypothesis, removal of this hydroxyl group eliminates almost all peptidyl transferase activity. Similarly, removal of the 2′-OH group of A2451 of the 23S rRNA strongly inhibits peptidyl transferase activity. This group may also participate in catalysis by hydrogen bonding, or it may help position the reactants properly for catalysis. The exit tunnel through the 50S subunit is just wide enough to allow a protein α-helix to pass through. Its walls are made of RNA, whose hydrophilicity is likely to allow exposed hydrophobic side chains of the nascent polypeptide to slide through easily. RF1 domains 2 and 3 fill the codon recognition site and the peptidyl transferase site, respectively, of the ribosome's A site, in recognizing the UAA stop codon. The "reading head" portion of domain 2 of RF1, including its conserved PXT motif, occupies the decoding center within the A site and collaborates with A1493 and A1492 of the 16S rRNA to recognize the stop codon. The universally conserved GGQ motif at the tip of domain 3 of RF1 closely approaches the peptidyl transferase center and participates in cleavage of the ester bond linking the completed polypeptide to the tRNA. RF2 binds to the ribosome and operates in much the same way in response to the UGA stop codon.

Most mRNAs are translated by more than one ribosome at a time; the result, a structure in which many ribosomes translate an mRNA in tandem, is called a polysome. In eukaryotes, polysomes are found in the cytoplasm. In prokaryotes, transcription of a gene and translation of the resulting mRNA occur simultaneously. Therefore, many polysomes are found associated with an active gene.

Transfer RNA was discovered as a small RNA species independent of ribosomes that could be charged with an amino acid and could then pass the amino acid to a growing polypeptide. All tRNAs share a common secondary structure represented by a cloverleaf. They have four base-paired stems defining three stem loops (the D loop, anticodon loop, and T loop) and the acceptor stem, to which amino acids are added in the charging step. The tRNAs also share a common three-dimensional shape that resembles an inverted L. This shape maximizes stability by lining up the base pairs in the D stem with those in the anticodon stem, and the base pairs in the T stem with those in the acceptor stem. The anticodon of the tRNA protrudes from the side of the anticodon loop and is twisted into a shape that readily base-pairs with the corresponding codon in mRNA.

The acceptor stem and anticodon are important cues in recognition of a tRNA by its cognate aminoacyl-tRNA synthetase. In certain cases, each of these elements can be the absolute determinant of charging specificity. X-ray crystallography has shown that synthetase–tRNA interactions differ between the two classes of aminoacyl-tRNA synthetases. Class I synthetases have pockets for the acceptor stem and anticodon of their cognate tRNAs and approach the tRNAs from the D loop and acceptor stem minor groove side. Class II synthetases also have pockets for the acceptor stem and anticodon, but approach their tRNAs from the opposite side, which includes the variable arm and major groove of the acceptor stem.

The amino acid selectivity of at least some aminoacyl-tRNA synthetases is controlled by a double-sieve mechanism. The first sieve is a coarse one that excludes amino acids that are too big. The enzyme accomplishes this task with an active site for activation of amino acids that is just big enough to accommodate the cognate amino acid, but not larger amino acids. The second sieve is a fine one that degrades aminoacyl-AMPs that are too small. The enzyme accomplishes this task with a second active site (the editing site) that admits small aminoacyl-AMPs and hydrolyzes them. The cognate aminoacyl-AMP is too big to fit into the editing site, so it escapes being hydrolyzed.

REVIEW QUESTIONS

1. Draw rough sketches of the *E. coli* 30S and 50S ribosomal subunits and show how they fit together to form a 70S ribosome.
2. Draw rough sketches of interface views of both 50S and 30S ribosomal subunits. Point out the rough positions of tRNAs in the A, P, and E sites.
3. What parts of the tRNAs interact with the 30S subunit? With the 50S subunit?
4. Why is it important that the anticodons of the tRNAs in the A and P sites approach each other closely?
5. Why is it important that the acceptor stems of the tRNAs in the A and P sites approach each other closely?
6. Describe the process of two-dimensional gel electrophoresis described in this chapter. In what way is two-dimensional superior to one-dimensional electrophoresis?
7. Present plausible hypotheses to explain how the following antibiotics interfere with translation. Present evidence for each hypothesis.
   a. Streptomycin
   b. Paromomycin
8. How can x-ray diffraction data rule out ribosomal proteins as the active site in peptidyl transferase?
9. Outline the evidence for the importance of the 2′-OH of the terminal adenosine of the peptidyl-tRNA in the P site in transpeptidation. How is this hydroxyl group likely to participate in transpeptidation?

10. Outline the evidence for the importance of the 2′-OH of A2451 of the 23S rRNA in transpeptidation. How is this hydroxyl group likely to participate in transpeptidation?
11. How do we know the base of A2451 (A2486 in *H. marismortui*) is not important in transpeptidation?
12. What part of RF1 recognizes the stop codon UAA? What ribosomal elements participate in this recognition? What part of RF1 participates in cleavage of the bond between the tRNA and the peptide?
13. Explain how the bending of the tRNA in an aminoacyl-tRNA as it first binds to the A site (actually the A/T site), and the unbending of the tRNA during accommodation in the A site, contribute to accuracy of translation.
14. Describe the experiments that led to the discovery of tRNA.
15. How was the "cloverleaf" secondary structure of tRNA discovered?
16. Draw the cloverleaf tRNA structure and point out the important structural elements.
17. Describe and give the results of an experiment that shows that the ribosome responds to the tRNA part, not the amino acid part, of an aminoacyl-tRNA.
18. Describe and give the results of an experiment that shows that the G3–U70 base pair in a tRNA acceptor stem is a key determinant in the charging of the tRNA with alanine.
19. Present at least one line of evidence for the importance of the anticodon in the recognition of a tRNA by an aminoacyl-tRNA synthetase.
20. Based on x-ray crystallographic studies, what parts of a tRNA are in contact with the cognate aminoacyl-tRNA synthetase?
21. Diagram a double-sieve mechanism that ensures amino acid selectivity in aminoacyl-tRNA synthetases.
22. Outline the evidence for the double sieve in the isoleucine–tRNA synthetase that excludes larger and smaller amino acids.

ANALYTICAL QUESTIONS

1. Draw a diagram of a hypothetical eukaryotic polysome in which nascent protein chains are visible. Identify the 5′- and 3′-ends of the mRNA and use an arrow to indicate the direction the ribosomes are moving along the mRNA. Use N and C to indicate the amino and carboxyl ends of one of the growing polypeptides.
2. Draw a diagram of a hypothetical prokaryotic gene being transcribed and translated simultaneously. Show the nascent mRNAs with ribosomes attached, but do not show nascent proteins. With an arrow, indicate the direction of transcription.
3. You are investigating a $tRNA^{Phe}$ whose charging specificity appears to be affected by a C11–G24 base pair in the D stem. Design two experiments to show that changing this base pair changes the charging specificity of the tRNA. The first experiment should be a biochemical one using an in vitro reaction. The second should be a genetic one performed in vivo.
4. Consider the process of bringing a new aminoacyl-tRNA to the A site, as revealed by x-ray crystallography. Describe the probable effects of each of the following mutations on speed and fidelity of translation:
   a. A mutation in the 16S rRNA that facilitates "domain closure" in the 30S subunit.
   b. A mutation in the acceptor stem of the tRNA that inhibits the change in conformation that normally helps the tRNA bend into the A/T state.
   c. A mutation in switch I of EF-Tu that strengthens its binding to the acceptor stem of tRNA.
   d. Mutating His 84 of EF-Tu to Alanine.

SUGGESTED READINGS

General References and Reviews

Cech, T.R. 2000. The ribosome is a ribozyme. *Science* 289:878–79.

Dahlberg, A.E. 2001. The ribosome in action. *Science* 292:868–69.

Fersht, A.R. 1998. Sieves in sequence. *Science* 280:541.

Liljas, A. 2009. Leaps in translation elongation. *Science* 326:677–78.

Moore, P.B. 2005. A ribosomal coup: *E. coli* at last! *Science* 310:793–95.

Noller, H.F. 1990. Structure of rRNA and its functional interactions in translation. In Hill, W.E., et al., eds. *The Ribosome: Structure, Function and Evolution*. Washington, D.C.: American Society for Microbiology, chapter 3, pp. 73–92.

Pennisi, E. 2001. Ribosome's inner workings come into sharper view. *Science* 291:2526–27.

Saks, M.E., J.R. Sampson, and J.N. Abelson. 1994. The transfer RNA identity problem: A search for rules. *Science* 263:191–97.

Schmeing, T.M. and V. Ramakrishnan. 2009. What recent ribosome structures have revealed about the mechanism of translation. *Nature* 461:1234–42.

Waldrop, M.M. 1990. The structure of the "second genetic code." *Science* 246:1122.

Research Articles

Ban, N., P. Nissen, J. Hansen, P.B. Moore, and T.A. Steitz. 2000. The complete atomic structure of the large ribosomal subunit at 2.4 Å resolution. *Science* 289:905–20.

Carter, A.P., W.M. Clemons, Jr., D.E. Brodersen, R.J. Morgan-Warren, T. Hartsch, B.T. Wimberly, and V. Ramakrishnan. 2000. Crystal structure of an initiation factor bound to the 30S ribosomal subunit. *Science* 291:498–501.

Carter, A.P., W.M. Clemons, Jr., D.E. Brodersen, R.J. Morgan-Warren, T. Hartsch, B.T. Wimberly, and V. Ramakrishnan. 2000. Functional insights from the structure of the 30S

ribosomal subunit and its interactions with antibiotics. *Nature* 407:340–48.
Gao, Y.-G., M. Selmer, C.M. Dunham, A. Weixlbaumer, A.C. Kelley, and V. Ramakrishnan. 2009. The structure of the ribosome with elongation factor G trapped in the posttranslocation state. *Science* 326:694–99.
Hoagland, M.B., M.L. Stephenson, J.F. Scott, L.I. Hecht, and P.C. Zamecnik. 1958. A soluble ribonucleic acid intermediate in protein synthesis. *Journal of Biological Chemistry* 231:241–57.
Holley, R.W., J. Apgar, G.A. Everett, J.T. Madison, M. Marquisee, S.H. Merrill, J.R. Penswick, and A. Zamir. 1965. Structure of a ribonucleic acid. *Science* 147:1462–65.
Kaltschmidt, E. and H.G. Wittmann. 1970. Ribosomal proteins XII: Number of proteins in small and large ribosomal subunits of *Escherichia coli* as determined by two-dimensional gel electrophoresis. *Proceedings of the National Academy of Sciences USA* 67:1276–82.
Kim, S.H., F.L. Suddath, G.J. Quigley, A. McPherson, J.L. Sussman, A.H.J. Wang, N. C. Seeman, and A. Rich. 1974. Three-dimensional tertiary structure of yeast phenylalanine transfer RNA. *Science* 185:435–40.
Lake, J.A. 1976. Ribosome structure determined by electron microscopy of *Escherichia coli* small subunits, large subunits and monomeric ribosomes. *Journal of Molecular Biology* 105:131–59.
Laurberg, M., H. Asahara, A. Korostelev, J. Zhu, S. Trakhanov, and H.F. Noller. 2008. Structural basis for translation termination on the 70S ribosome. *Nature* 454:852–57.
Miller, O., B.A. Hamkalo, and C.A. Thomas, Jr. 1970. Visualization of bacterial genes in action. *Science* 169:392–95.
Mizushima, S. and M. Nomura. 1970. Assembly mapping of 30S ribosomal proteins from *E. coli*. *Nature* 226:1214–18.
Muth, G.W., L. Ortoleva-Donnelly, and S.A. Strobel. 2000. A single adenosine with a neutral pK_a in the ribosomal peptidyl transferase center. *Science* 289:947–50.
Nissen, P., J. Hansen, N. Ban, P.B. Moore, and T.A. Steitz. 2000. The structural basis of ribosome activity in peptide bond synthesis. *Science* 289:920–30.
Nureki, O., D.G. Vassylyev, M. Tateno, A. Shimada, T. Nakama, S. Fukai, M. Konno, T.L. Henrickson. P. Schimmel, and S. Yokoyama. 1998. Enzyme structure with two catalytic sites for double-sieve selection of substrates. *Science* 280:578–82.
Ogle, J.M., D.E Brodersen, W.M.Clemons Jr., M.J. Tarry, A.P. Carter, and V. Ramakrishnan. 2001. Recognition of cognate transfer RNA by the 30S ribosomal subunit. *Science* 292:897–902.
Polacek, N., M. Gaynor, A. Yassin, and A.S Mankin. 2001. Ribosomal peptidyl transferase can withstand mutations at the putative catalytic nucleotide. *Nature* 411:498–501.
Quigley, G.J. and A. Rich. 1976. Structural domains of transfer RNA molecules. *Science* 194:796–806.
Rould, M.A., J.J. Perona, D. Söll, and T.A. Steitz. 1989. Structure of *E. coli* glutaminyl-tRNA synthetase complexed with $tRNA^{Gln}$ and ATP at 2.8 Å resolution. *Science* 246:1135–42.
Ruff, M., S. Krishnaswamy, M. Boeglin, A. Poterszman, A. Mitschler, A. Podjarny, B. Rees, J.C. Thierry, and D. Moras. 1991. Class II aminoacyl transfer RNA synthetases: Crystal structure of yeast aspartyl-tRNA synthetase complexed with $tRNA^{Asp}$. *Science* 252:1682–89.
Schluenzen, F., A. Tocilj, R. Zarivach, J. Harms, M. Gluehmann, D. Janell, A. Bashan, H. Bartels, I. Agmon, F. Franceschi, and A. Yonath. 2000. Structure of functionally activated small ribosomal subunit at 3.3 Å resolution. *Cell* 102:615–23.
Schmeing, T.M., R.M. Voorhees, A.C. Kelley, Y.-G. Gao, F.V. Murphy IV, J.R. Weir, and V. Ramakrishnan. 2009. The crystal structure of the ribosome bound to EF-Tu and aminoacyl-tRNA. *Science* 326:688–94.
Schulman, L.H. and H. Pelka. 1983. Anticodon loop size and sequence requirements for recognition of formylmethionine tRNA by methionyl-tRNA synthetase. *Proceedings of the National Academy of Sciences USA* 80:6755–59.
Schuwirth, B.S., M.A. Borovinskaya, C.W. Hau, W. Zhang, A. Vila-Sanjurjo, J.M. Holton, and J.H. Doundna Cate. 2005. Structures of the bacterial ribosome at 3.5 Å resolution. *Science* 310:827–34.
Stern, S., B. Weiser, and H.F. Noller. 1988. Model for the three-dimensional folding of 16S ribosomal RNA. *Journal of Molecular Biology* 204:447–81.
Weinger, J.S., K.M. Parnell, S. Dorner, R. Green, and S.A. Strobel. 2004. Substrate-assisted catalysis of peptide bond formation by the ribosome. *Nature Structural and Molecular Biology* 11:1101–06.
Wimberly, B.T., D.E. Brodersen, W.M. Clemons Jr., R.J. Morgan-Warren, A.P. Carter, C. Vonrhein, T. Hartsch, and V. Ramakrishnan. 2000. Structure of the 30S ribosomal subunit. *Nature* 407:327–39.
Yusupov, M.M., G. Zh. Yusupova, A. Baucom, K. Lieberman, T.N. Earnest, J.H.D. Cate, and H.F Noller. 2001. Crystal structure of the ribosome at 5.5 Å resolution. *Science* 292:883–96.

CHAPTER 20

DNA Replication, Damage, and Repair

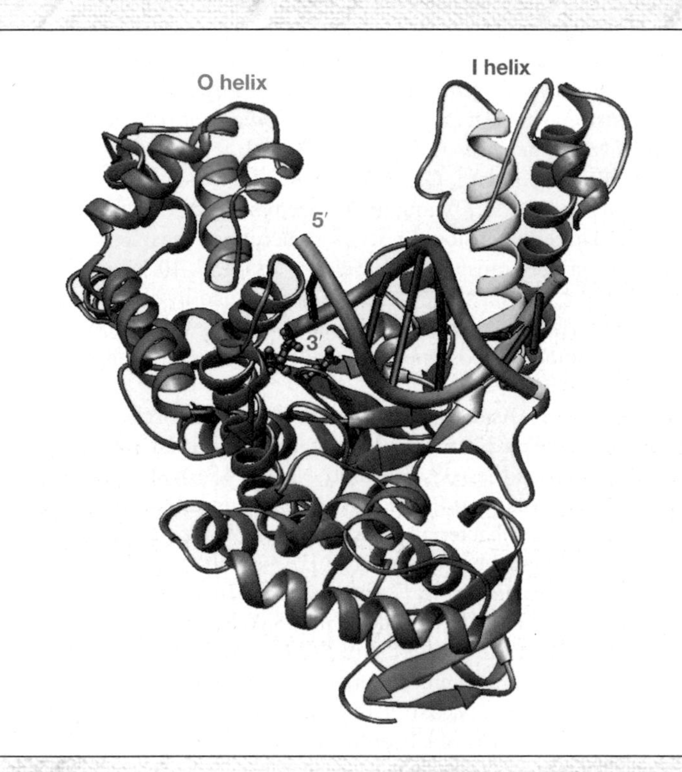

Cocrystal structure of *Taq* DNA polymerase with a double-stranded model DNA template (orange). *From Eom, S.H., Wang, J., and Steitz, T.A. Structure of Taq polymerase with DNA at the polymerase active site. Nature 382 (18 July 1996) f. 2a, p. 280. Copyright © Macmillan Magazines, Ltd.*

In Chapter 3 we learned that genes have three main activities. One is to carry information, and we have spent most of the intervening chapters examining how cells decode this information through transcription and translation. Another activity of genes is to participate in replication. The next two chapters will examine this process in detail. In this chapter we will also consider DNA damage and repair.

20.1 General Features of DNA Replication

Let us first consider the general mechanism of DNA replication. The double-helical model for DNA includes the concept that the two strands are complementary. Thus, each strand can in principle serve as the template for making its own partner. As we will see, this semiconservative model for DNA replication is the correct one. In addition, molecular biologists have uncovered the following interesting general features of DNA replication: It is half discontinuous (made in short pieces that are later stitched together); it requires RNA primers; and it is usually bidirectional. Let us look at each of these features in turn.

Semiconservative Replication

The Watson–Crick model for DNA replication (introduced in Chapter 2) assumed that as new strands of DNA are made, they follow the usual base-pairing rules of A with T and G with C. The model also proposed that the two parental strands separate and that each then serves as a template for a new progeny strand. This is called **semiconservative replication** because each daughter duplex has one parental strand and one new strand (Figure 20.1a). In other words, one of the parental strands is "conserved" in each daughter duplex. However, this is not the only possibility. Another potential mechanism (Figure 20.1b) is **conservative replication,** in which the two parental strands stay together and somehow produce another daughter helix with two completely new strands. Yet another possibility is **dispersive replication,** in which the DNA becomes fragmented so that new and old DNAs coexist in the same strand after replication (Figure 20.1c). This mechanism was envisioned to avoid the formidable problem of unwinding the two DNA strands.

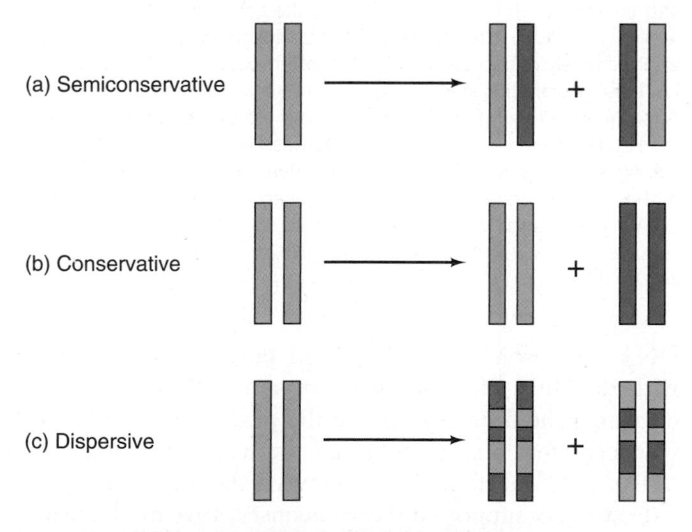

Figure 20.1 Three hypotheses for DNA replication. **(a)** Semiconservative replication gives two daughter duplex DNAs, each of which contains one old strand (blue) and one new strand (red). **(b)** Conservative replication yields two daughter duplexes, one of which has two old strands (blue) and one of which has two new strands (red). **(c)** Dispersive replication gives two daughter duplexes, each of which contains strands that are a mixture of old and new.

In 1958, Matthew Meselson and Franklin Stahl performed a classic experiment to distinguish among these three possibilities. They labeled *E. coli* DNA with heavy nitrogen (^{15}N) by growing cells in a medium enriched in this nitrogen isotope. This made the DNA denser than normal. Then they switched the cells to an ordinary medium containing primarily ^{14}N, for various lengths of time. Finally, they subjected the DNA to CsCl gradient ultracentrifugation to determine the density of the DNA. Figure 20.2 depicts the results of a control experiment that shows that ^{15}N- and ^{14}N-DNAs are clearly separated by this method.

What outcomes would we expect after one round of replication according to the three different mechanisms? If replication is conservative, the two heavy parental strands will stay together, and another, newly made DNA duplex will appear. Because this second duplex will be made in the presence of light nitrogen, both its strands will be light. The heavy/heavy (H/H) parental duplex and light/light (L/L) progeny duplex will separate readily in the CsCl gradient (Figure 20.3a). On the other hand, if replication is semiconservative, the two heavy parental strands will separate and each will be supplied with a new, light partner. These H/L hybrid duplexes will have a density halfway between the H/H parental duplexes and L/L ordinary DNA (Figure 20.3b). Figure 20.4 shows that this is exactly what happened; after the first DNA doubling, a single band appeared midway between the labeled H/H DNA and a normal L/L DNA. This ruled out conservative replication, but was still consistent with either semiconservative or dispersive replication.

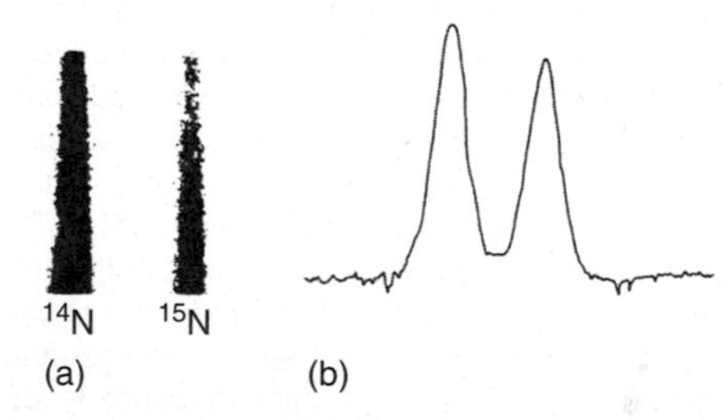

Figure 20.2 Separation of DNAs by cesium chloride density gradient centrifugation. DNA containing the normal isotope of nitrogen (^{14}N) was mixed with DNA labeled with a heavy isotope of nitrogen (^{15}N) and subjected to cesium chloride density gradient centrifugation. The two bands had different densities, so they separated cleanly. **(a)** A photograph of the spinning rotor under ultraviolet illumination. Note that this is a photograph through a window in the rotor as it spins. The ultracentrifuge rotor was designed to allow the experimenter to check its contents without stopping the centrifuge. The two dark bands correspond to the two different DNAs that absorb ultraviolet light. **(b)** A graph of the darkness of each band, which gives an idea of the relative amounts of the two kinds of DNA. (*Source:* Adapted from Meselson, M. and F. Stahl, The replication of DNA in *Escherichia coli. Proceedings of the National Academy of Sciences USA* 44 (1958) p. 673, f. 2.)

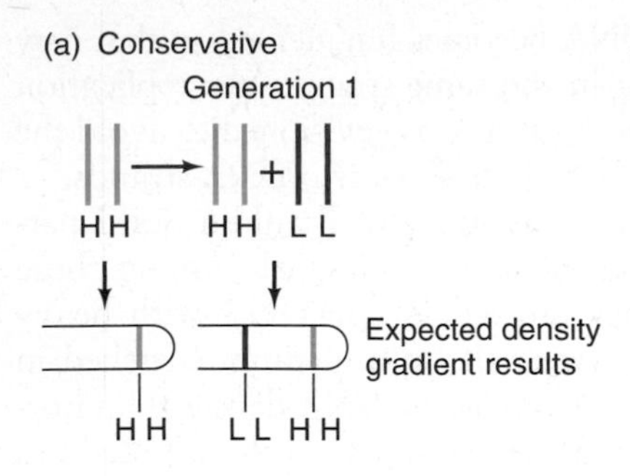

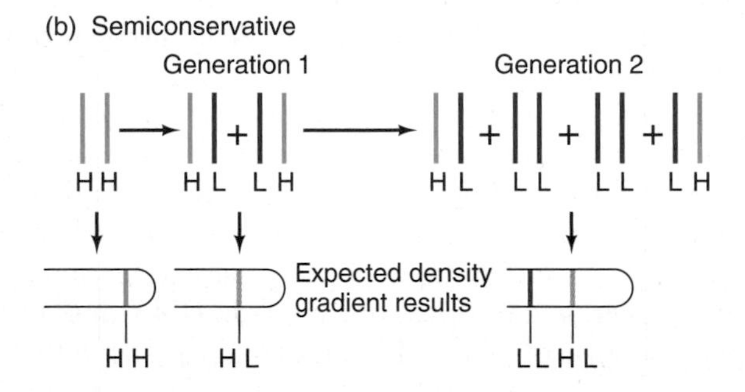

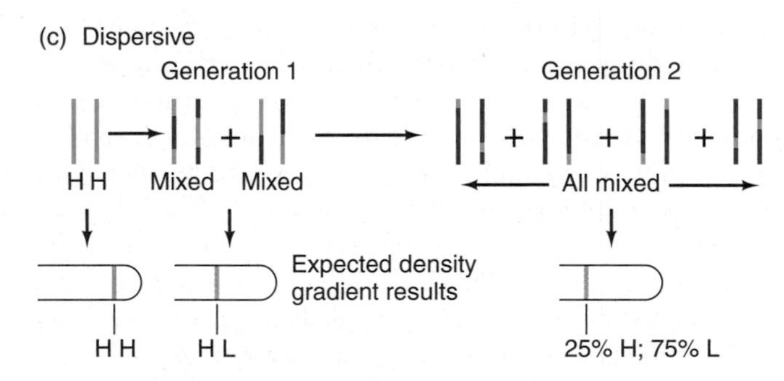

Figure 20.3 Three replication hypotheses. The conservative model **(a)** predicts that after one generation equal amounts of two different DNAs (heavy/heavy [H/H] and light/light [L/L]) will occur. Both the semiconservative **(b)** and dispersive **(c)** models predict a single band of DNA with a density halfway between the H/H and L/L densities. Meselson and Stahl's results confirmed the latter prediction, so the conservative mechanism was ruled out. The dispersive model predicts that the DNA after the second generation will have a single density, corresponding to molecules that are 25% H and 75% L. This should give one band of DNA halfway between the L/L and the H/L band. The semiconservative model predicts that equal amounts of two different DNAs (L/L and H/L) will be present after the second generation. Again, the latter prediction matched the experimental results, supporting the semiconservative model.

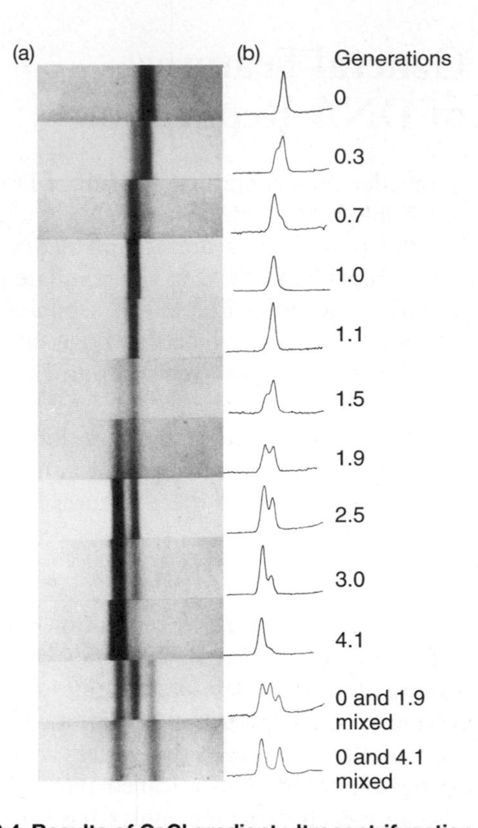

Figure 20.4 Results of CsCl gradient ultracentrifugation experiment that demonstrates semiconservative DNA replication. Meselson and Stahl shifted ^{15}N-labeled *E. coli* cells to a ^{14}N medium for the number of generations given at right, then subjected the bacterial DNA to CsCl gradient ultracentrifugation. **(a)** Photographs of the spinning centrifuge tubes under ultraviolet illumination. The dark bands correspond to heavy DNA (right) and light DNA (left). A band of intermediate density was also observed between these two and is virtually the only band observed at 1.0 and 1.1 generations. This band corresponds to duplex DNAs in which one strand is labeled with ^{15}N, and the other with ^{14}N, as predicted by the semiconservative replication model. After 1.9 generations, Meselson and Stahl observed approximately equal quantities of the intermediate band (H/L) and the L/L band. Again, this is what the semiconservative model predicts. After three and four generations, they saw a progressive depletion of the H/L band, and a corresponding increase in the L/L band, again as we expect if replication is semiconservative. **(b)** Densitometer tracings of the bands in panel (a), which can be used to quantify the amount of DNA in each band. (*Source:* Meselson, M. and F.W. Stahl, The replication of DNA in *Escherichia coli, Proceedings of the National Academy of Sciences USA* 44:675, 1958.)

The results of one more round of DNA replication ruled out the dispersive hypothesis. Dispersive replication would give a product with one-fourth ^{15}N and three-fourths ^{14}N after two rounds of replication in a ^{14}N medium. Semiconservative replication would yield half of the products as H/L and half as L/L (see Figure 20.3b). In other words, the hybrid H/L products of the first round of replication would each split and be supplied with new, light partners, giving the 1:1 ratio of H/L to L/L DNAs. Again, this is precisely what occurred (see Figure 20.4). To make sure that the intermediate-density peak was really a 1:1 mixture of the heavy and light DNA, Meselson and Stahl mixed pure ^{15}N-labeled DNA with the DNA after 1.9 generations in ^{14}N medium, then measured the distances among the peaks. The middle peak was centered almost perfectly between the other two (50% ± 2% of the distance between them). Therefore, the data strongly supported the semiconservative mechanism.

SUMMARY DNA replicates in a semiconservative manner. When the parental strands separate, each serves as the template for making a new, complementary strand.

At Least Semidiscontinuous Replication

If we were charged with the task of designing a DNA-replicating machine, we might come up with a system such as the one pictured in Figure 20.5a. DNA would unwind to create a fork, and two new DNA strands would be synthesized continuously in the same direction as the moving fork. However, this scheme has a fatal flaw. It demands that the replicating machine be able to make DNA in both the 5′→3′ and 3′→5′ directions. That is because of the antiparallel nature of the two strands of DNA; if one runs 5′→3′ left to right, the other must run 3′→5′ left to right. But the DNA synthesizing part (**DNA polymerase**) of all natural replicating machines can make DNA in only one direction: 5′→3′. That is, it inserts the 5′-most nucleotide first and extends the chain toward the 3′-end by adding nucleotides to the 3′-end of the growing chain.

Following this line of reasoning, Reiji Okazaki concluded that both strands could not replicate continuously. DNA polymerase could theoretically make one strand (the **leading strand**) *continuously* in the 5′→3′ direction, but the other strand (the **lagging strand**) would have to be made *discontinuously* as shown in Figure 20.5b and c. The

(a) **Continuous:**

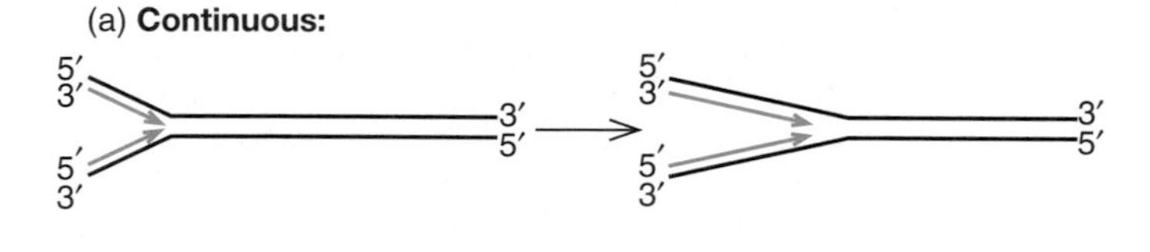

(b) **Semidiscontinuous:**

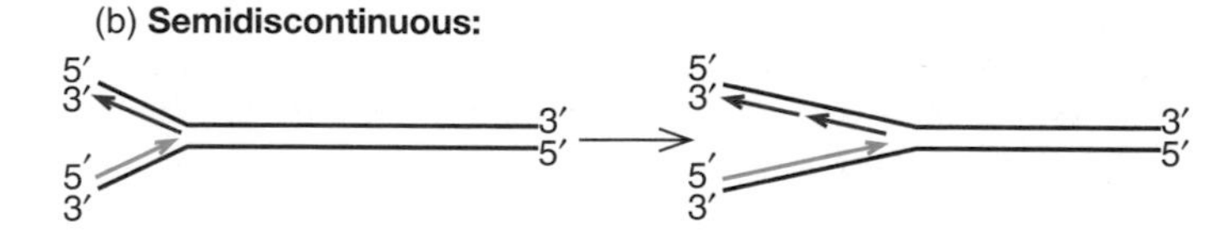

(c) **Discontinuous:**

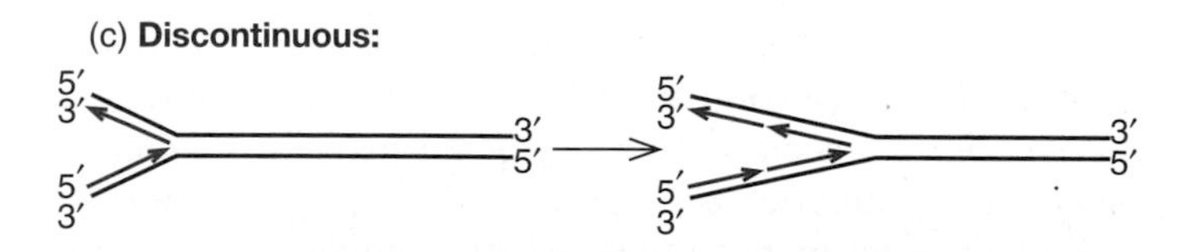

Figure 20.5 Continuous, semidiscontinuous, and discontinuous models of DNA replication. (a) Continuous model. As the replicating fork moves to the right, both strands are replicated continuously in the same direction, left to right (blue arrows). The top strand grows in the 3′→5′ direction, the bottom strand in the 5′→3′ direction. **(b)** Semidiscontinuous model. Synthesis of one of the new strands (the leading strand, bottom) is continuous (blue arrow), as in the model in panel (a); synthesis of the other (the lagging strand, top) is discontinuous (pink arrows), with the DNA being made in short pieces. Both strands grow in the 5′→3′ direction. **(c)** Discontinuous model. Both leading and lagging strands are made in short pieces (i.e., discontinuously; pink arrows). Both strands grow in the 5′→3′ direction.

discontinuity of synthesis of the lagging strand comes about because its direction of synthesis is opposite to the direction in which the replicating fork is moving. Therefore, as the fork opens up and exposes a new region of DNA to replicate, the lagging strand is growing in the "wrong" direction, away from the fork. The only way to replicate this newly exposed region is to restart DNA synthesis at the fork, behind the piece of DNA that has already been made. This starting and restarting of DNA synthesis occurs over and over again. The short pieces of DNA thus created would of course have to be joined together somehow to produce the continuous strand that is the final product of DNA replication.

The model of semidiscontinuous replication makes two predictions that Okazaki's team tested experimentally: (1) Because at least half of the newly synthesized DNA appears first as short pieces, one ought to be able to label and catch these before they are stitched together by allowing only very short periods (pulses) of labeling with a radioactive DNA precursor. (2) If one eliminates the enzyme (**DNA ligase**) responsible for stitching together the short pieces of DNA, these short pieces ought to be detectable even with relatively long pulses of DNA precursor.

For his model system, Okazaki chose replication of phage T4 DNA. This had the advantage of simplicity, as well as the availability of T4 ligase mutants. To test the first prediction, Okazaki and colleagues gave shorter and shorter pulses of ^{3}H-labeled thymidine to *E. coli* cells that were replicating T4 DNA. To be sure of catching short pieces of DNA before they could be joined together, they even administered pulses as short as 2 sec. Finally, they measured the approximate sizes of the newly synthesized DNAs by ultracentrifugation.

Figure 20.6a shows the results. Already at 2 sec, some labeled DNA was visible in the gradient; within the limits of detection, it appeared that all of the label was in very small DNA pieces, 1000–2000 nt long, which remained near the top of the centrifuge tube. With increasing pulse time, another peak of labeled DNA appeared much nearer the bottom of the tube. This was the result of attaching the small, newly formed pieces of labeled DNA to much larger, preformed pieces of DNA that were made before labeling began. These large pieces, because they were unlabeled before the experiment began, did not show up until enough time had elapsed for DNA ligase to join the smaller, labeled pieces to them; this took only a few seconds. The small pieces of DNA that are the initial products of replication have come to be known as **Okazaki fragments.**

The discovery of Okazaki fragments provided evidence for at least partially discontinuous replication of T4 DNA. This hypothesis was supported by the demonstration that these small DNA fragments accumulated to very high levels when the stitching enzyme, DNA ligase, did not operate. Okazaki's group performed this experiment with the T4 mutant containing a defective DNA ligase gene. Figure 20.6b

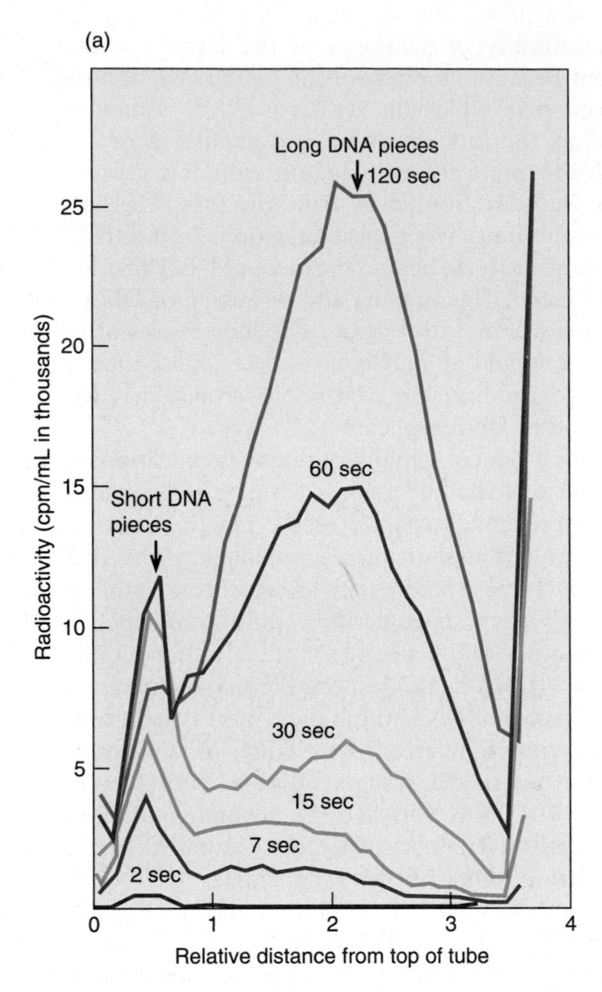

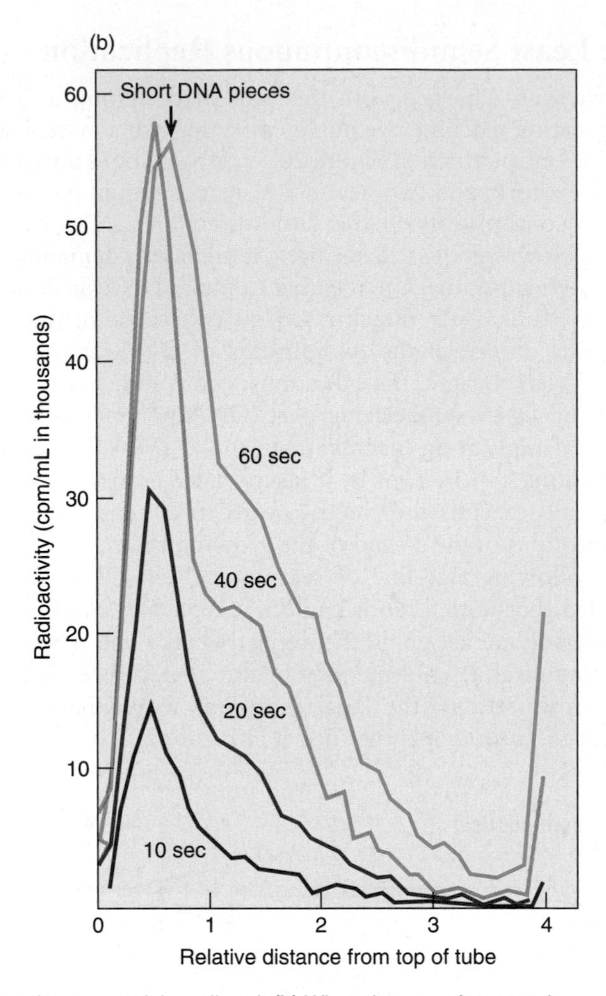

Figure 20.6 Experimental demonstration of at least semidiscontinuous DNA replication. (a) Okazaki and his colleagues labeled replicating phage T4 DNA with very short pulses of radioactive DNA precursor and separated the product DNAs according to size by ultracentrifugation. At the shortest times, the label went primarily into short DNA pieces (found near the top of the tube), as the discontinuous model predicted. **(b)** When these workers used a mutant phage with a defective DNA ligase gene, short DNA pieces accumulated even after relatively long labeling times (1 min in the results shown here). (*Source:* Adapted from R. Okazaki et al., In vivo mechanism of DNA chain growth, *Cold Spring Harbor Symposia on Quantitative Biology,* 33:129–143, 1968.)

shows that the peak of Okazaki fragments predominated in this mutant. Even after a full minute of labeling, this was still the major species of labeled DNA, suggesting that Okazaki fragments are not just an artifact of very short labeling times.

The predominant accumulation of small pieces of labeled DNA could be interpreted to mean that replication proceeded discontinuously on *both* strands, as pictured in Figure 20.5c. Indeed, this was Okazaki and colleagues' interpretation. But a commonly invoked alternative explanation is that some of the small DNA pieces are created by a DNA repair system that removes dUMP residues incorporated into DNA. UTP is an essential precursor of RNA, but the cell also makes dUTP, which can be accidentally incorporated into DNA (as dUMP) in place of dTMP. Two enzymes help to minimize this problem. One of these, **dUTPase**—the product of the *dut* gene, degrades dUTP. The other, **uracil N-glycosylase**—the product of the *ung* gene, removes uracil bases from DNA, creating "abasic sites" that are subject to breakage as part of the repair process. Thus, this repair process generates a certain component of short DNA pieces regardless of whether the replicating DNA is made continuously or discontinuously. The question is, what proportion of the Okazaki fragments observed in experiments such as those depicted in Figure 20.6 are due to discontinuous replication and what proportion to the repair of misincorporated dUMP residues?

One way to answer this question would be to look at the sizes of newly labeled DNA fragments in dut^+ ung^- cells. These cells minimize dUMP incorporation (because of the presence of dUTPase) and cannot create abasic sites (because of the absence of uracil N-glycosylase). Therefore, strand breakage due to dUMP incorporation should be minimized. In fact, this experiment has been done, and most newly labeled DNAs are still small—Okazaki fragment size. Indeed, it appears that, even in wild-type cells, the amount of dUMP incorporation is quite low—far too low to explain the preponderance of Okazaki fragments observed at short labeling times in Figure 20.6. These data suggest a conclusion, though it is not generally accepted, that replication on *both* strands occurs discontinuously, at least in *E. coli*.

SUMMARY DNA replication in *E. coli* (and in other organisms) is at least semidiscontinuous. One strand (the leading strand) is replicated in the direction of the movement of the replicating fork. This strand is commonly thought to replicate continuously, though there is evidence that it replicates discontinuously. The other strand (the lagging strand) is replicated discontinuously as 1–2 kb Okazaki fragments in the opposite direction. This allows both strands to be replicated in the 5′→3′ direction.

Priming of DNA Synthesis

We have seen in previous chapters that RNA polymerase initiates transcription simply by starting a new RNA chain; it puts the first nucleotide in place and then joins the next to it. But DNA polymerases cannot perform the same trick with initiation of DNA synthesis. If we supply a DNA polymerase with all the nucleotides and other small molecules it needs to make DNA, then add either single-stranded or double-stranded DNA with no strand breaks, the polymerase will make no new DNA. What is missing?

We now know that the missing component is a **primer**, a piece of nucleic acid that the polymerase can "grab onto" and extend by adding nucleotides to its 3′-end. This primer is not DNA, but a short piece of RNA. Figure 20.7 shows a simplified version of this process. First, a replicating fork opens up; next, short RNA primers are made; next, DNA polymerase adds deoxyribonucleotides to these primers, forming DNA, as indicated by the arrows.

The first line of evidence supporting RNA priming was the finding that replication of M13 phage DNA by an *E. coli* extract is inhibited by the antibiotic rifampicin. This was a surprise because rifampicin inhibits *E. coli* RNA polymerase, not DNA polymerase. The explanation is that M13 uses the *E. coli* RNA polymerase to make RNA primers for its DNA synthesis. However, this is not a general phenomenon. Even *E. coli* does not use its own RNA polymerase for priming; it has a special enzyme system for that purpose.

Perhaps the best evidence for RNA priming was the discovery that DNase cannot completely destroy Okazaki fragments. It leaves little pieces of RNA 10–12 bases long. Most of this work was carried out by Tuneko Okazaki, Reiji Okazaki's wife and scientific colleague. She and her coworkers' first estimate of the primer size was too low only 1–3 nt. Two problems contributed to this underestimation: (1) Nucleases had already reduced the size of the

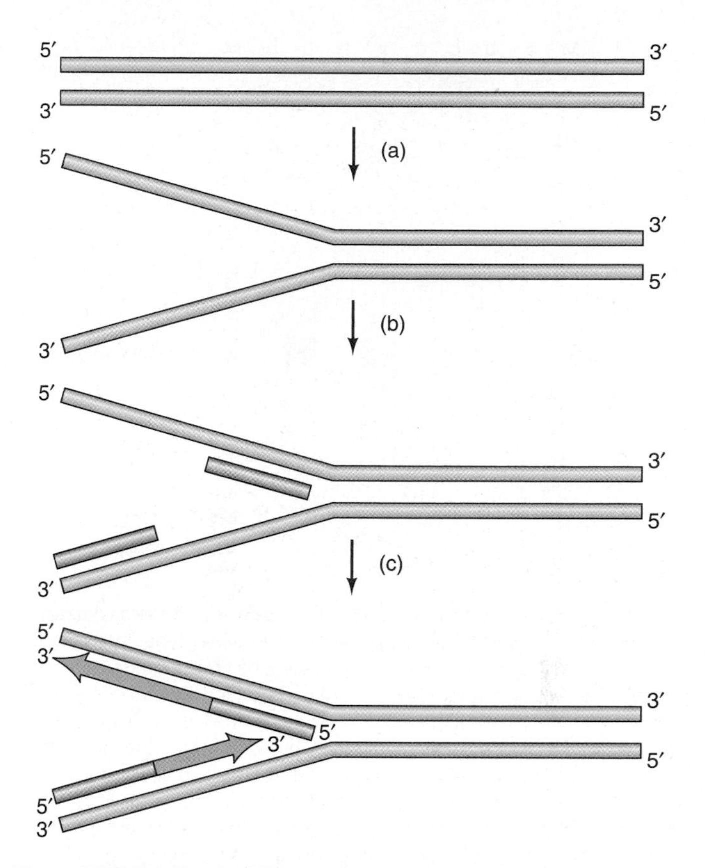

Figure 20.7 Priming in DNA synthesis. (a) The two parental strands (blue) separate. **(b)** Short RNA primers (pink) are made. **(c)** DNA polymerase uses the primers as starting points to synthesize progeny DNA strands (green arrows).

primers by the time they could be purified, and (2) the investigators had no way of distinguishing degraded from intact primers. In a second set of experiments, completed in 1985, Okazaki's group solved both of these problems and found that intact primers are really about 10–12 nt long.

To reduce nuclease activity, these workers used mutant bacteria that lacked ribonuclease H or the nuclease activity of DNA polymerase I, or both. This greatly enhanced the yield of the intact primer. To label only intact primer, they used the capping enzyme, guanylyl transferase, and [α-^{32}P]GTP, to label the 5′-ends of these RNAs. Recall from Chapter 15 that guanylyl transferase adds GMP to RNAs with 5′-terminal phosphates (ideally, a terminal diphosphate). If the primer were degraded at its 5′-end, it would no longer have these phosphates and would therefore not become labeled.

After radiolabeling the primers in this way, these investigators removed the DNA parts of the Okazaki fragments with DNase, then subjected the surviving labeled primers to gel electrophoresis. Figure 20.8 depicts the result. The primers from all the mutant bacteria produced clearly visible bands that corresponded to an RNA with a length of 11 ± 1 nt. The wild-type bacteria did not yield a detectable

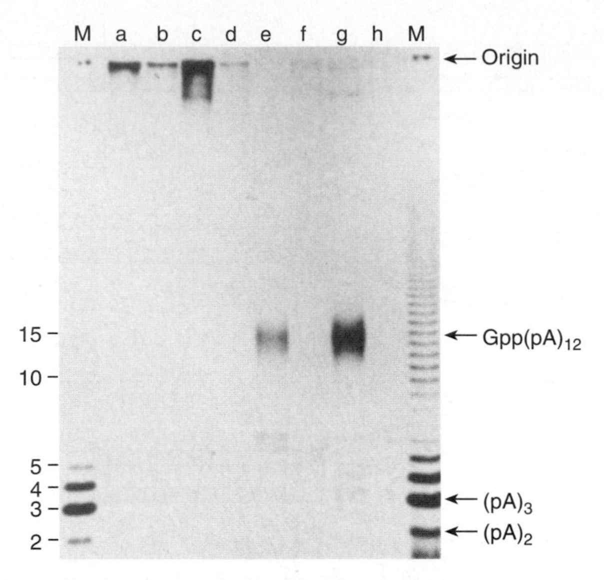

Figure 20.8 Finding and measuring RNA primers. Tuneko Okazaki and colleagues isolated Okazaki fragments from wild-type and mutant *E. coli* cells lacking one or both of the nucleases that degrade RNA primers. Next, they labeled the intact primers on the Okazaki fragments with [^{32}P]GTP and a capping enzyme. They destroyed the DNA in the fragments with DNase, leaving only the labeled primers. They subjected these primers to electrophoresis and detected their positions by autoradiography. Lanes M are markers. Lanes a–d, before DNase digestion; lanes e–h, after digestion. Lanes a and e, cells were defective in RNase H; lanes b and f, cells were defective in the nuclease activity of DNA polymerase I; lanes c and g, cells were defective in both RNase H and the nuclease activity of DNA polymerase I; lanes d and h, cells were wild-type. The best yield of primers occurred when both nucleases were defective (lane g), and the primers in all cases were 11 ± 1 nt long. The position of the 13-mer Gpp(pA)$_{12}$ marker is indicated at right. (*Source:* Kitani, T., K.-Y. Yoda, T. Ogawa, and T. Okazaki, Evidence that discontinuous DNA replication in *Escherichia coli* is primed by approximately 10 to 12 residues of RNA starting with a purine. *Journal of Molecular Biology* 184 (1985) p. 49, f. 2, by permission of Elsevier.)

band; nucleases had apparently degraded most or all of their intact primers. Further experiments actually resolved the broad band in Figure 20.8 into three discrete bands with lengths of 10, 11, and 12 nt.

SUMMARY Okazaki fragments in *E. coli* are initiated with RNA primers 10–12 nt long. Intact primers are difficult to detect in wild-type cells because of enzymes that attack RNAs.

Bidirectional Replication

In the early 1960s, John Cairns labeled replicating *E. coli* DNA with a radioactive DNA precursor, then subjected the labeled DNA to autoradiography. Figure 20.9a shows the results, along with Cairns's interpretation. The structure represented in Figure 20.9a is a so-called theta structure because of its resemblance to the Greek letter θ (theta). Because it may not be immediately obvious that the DNA in Figure 20.9a looks like a theta, Figure 20.9b provides a schematic diagram of the events in the second round of replication that led to the autoradiograph. This drawing shows that DNA replication begins with the creation of a "bubble"—a small region where the parental strands have separated and progeny DNA has been synthesized. As the bubble expands, the replicating DNA begins to take on the theta shape. We can now recognize the autoradiograph as representing a structure shown in the middle of Figure 20.9b, where the crossbar of the theta has grown long enough to extend above the circular part.

The θ structure contains two **replicating forks,** marked X and Y in Figure 20.9. This raises an important question: Does one of these forks, or do both, represent sites of active DNA replication? In other words, is DNA replication **unidirectional,** with one fork moving away from the other, which remains fixed at the origin of replication? Or is it

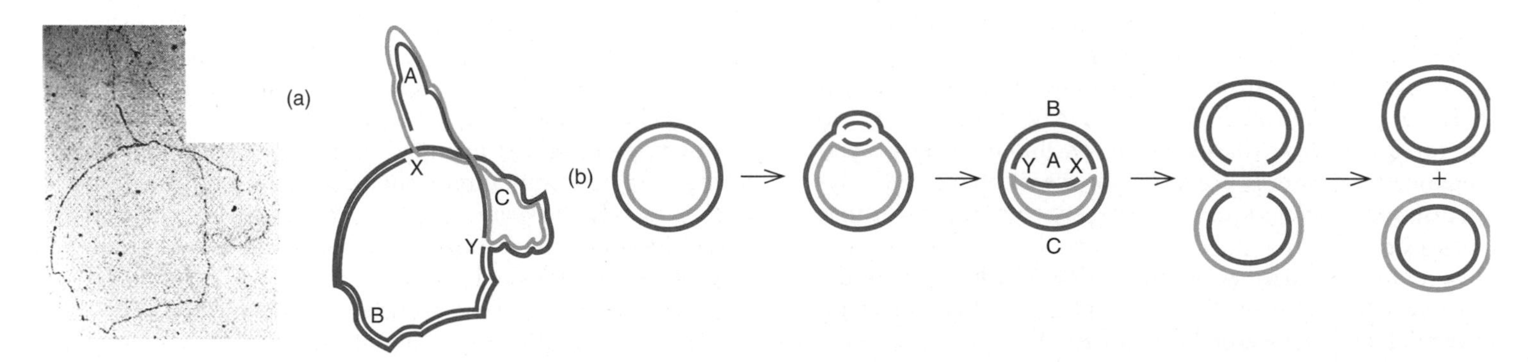

Figure 20.9 The theta mode of DNA replication in *Escherichia coli.* **(a)** An autoradiograph of replicating *E. coli* DNA with an interpretive diagram. The DNA was allowed to replicate for one whole generation and part of a second in the presence of radioactive nucleotides to label the DNA. The interpretive diagram to the right uses red to represent labeled DNA and blue to represent unlabeled parental DNA. **(b)** Detailed description of the theta mode of DNA replication. The colors have the same meaning as in panel (a) (*Source:* (a) Cairns, J., The chromosome of *Escherichia coli. Cold Spring Harbor Symposia on Quantitative Biology* 28 (1963) p. 44.)

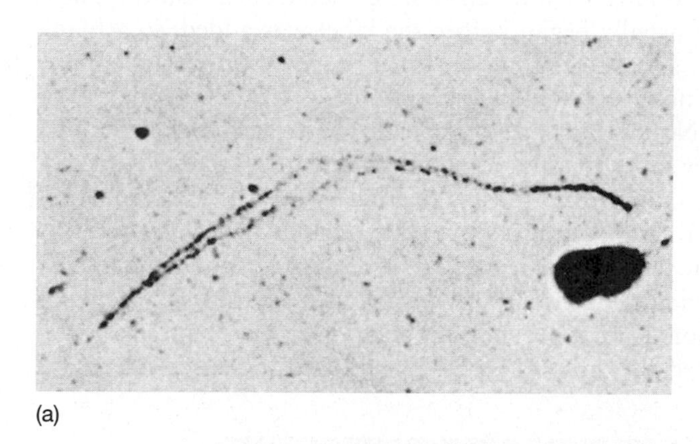

(a)

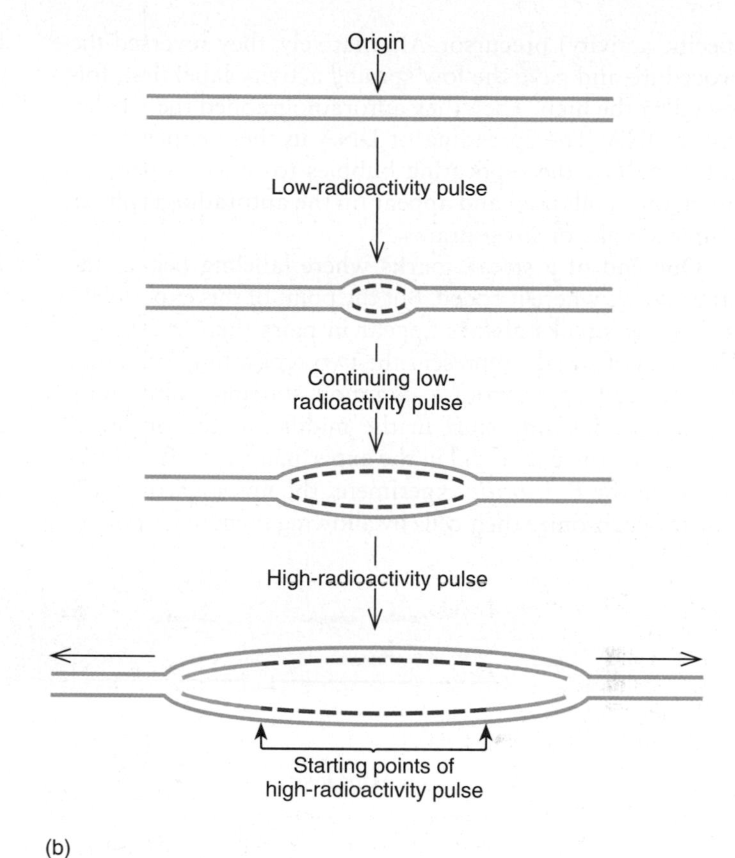

(b)

Figure 20.10 Experimental demonstration of bidirectional DNA replication. (a) Autoradiograph of replicating *Bacillus subtilis* DNA. Dormant bacterial spores were germinated in the presence of low-radioactivity DNA precursor, so the newly formed replicating bubbles immediately became slightly labeled. After the bubbles had grown somewhat, a more radioactive DNA precursor was added to label the DNA for a short period. **(b)** Interpretation of the autoradiograph. The purple color represents the slightly labeled DNA strands produced during the low-radioactivity pulse. The orange color represents the more highly labeled DNA strands produced during the later, high-radioactivity pulse. Because both forks picked up the high-radioactivity label, both must have been functioning during the high-radioactivity pulse. DNA replication in *B. subtilis* is therefore bidirectional. (*Source:* (a) Gyurasits, E.B. and R.J. Wake, Bidirectional chromosome replication in *Bacillus subtilis. Journal of Molecular Biology* 73 (1973) p. 58, by permission of Elsevier.)

bidirectional, with two replicating forks moving in opposite directions away from the origin? Cairns's autoradiographs were not designed to answer this question, but a subsequent study on *Bacillus subtilis* replication performed by Elizabeth Gyurasits and R.B. Wake showed clearly that DNA replication in that bacterium is bidirectional.

These investigators' strategy was to allow *B. subtilis* cells to grow for a short time in the presence of a weakly radioactive DNA precursor, then for a short time with a more strongly radioactive precursor. The labeled precursor was the same in both cases: [^{3}H]thymidine. Tritium (^{3}H) is especially useful for this type of autoradiography because its radioactive emissions are so weak that they do not travel far from their point of origin before they stop in the photographic emulsion and create silver grains. This means that the pattern of silver grains in the autoradiograph will bear a close relationship to the shape of the radioactive DNA. It is important to note that unlabeled DNA does not show up in the autoradiograph. The pulses of label in this experiment were short enough that only the replicating bubbles are visible (Figure 20.10a). You should not mistake these for whole bacterial chromosomes such as in Figure 20.9.

If you look carefully at Figure 20.10a, you will notice that the pattern of silver grains is not uniform. They are concentrated near both forks in the bubble. This extra labeling identifies the regions of DNA that were replicating during the "hot," or high-radioactivity, pulse period. Both forks incorporated extra label, showing that they were both active during the hot pulse. Therefore, DNA replication in *B. subtilis* is bidirectional; two forks arise at a fixed starting point—the **origin of replication**—and move in opposite directions around the circle until they meet on the other side. Later experiments employing this and other techniques have shown that the *E. coli* chromosome also replicates bidirectionally.

J. Huberman and A. Tsai have performed the same kind of autoradiography experiments in a eukaryote, the fruit fly *Drosophila melanogaster.* Here, the experimenters gave a pulse of strongly radioactive (high specific activity) DNA precursor, followed by a pulse of weakly radioactive (low specific activity) precursor. Alternatively, they reversed the procedure and gave the low specific activity label first, followed by the high. Then they autoradiographed the labeled insect DNA. The spreading of DNA in these experiments did not allow the replicating bubbles to remain open; instead, they collapsed and appear on the autoradiographs as simple streaks of silver grains.

One end of a streak marks where labeling began; the other shows where it ended. But the point of this experiment is that the streaks always appear in pairs (Figure 20.11a). The pairs of streaks represent the two replicating forks that have moved apart from a common starting point. Why doesn't the labeling start in the middle, at the origin of replication, the way it did in the experiment with *B. subtilis* DNA? In the *B. subtilis* experiment, the investigators were able to synchronize their cells by allowing them to germinate from spores, all starting at the same time. That way they could get label into the cells before any of them had started making DNA (i.e., before germination). Such synchronization was not tried in the *Drosophila* experiments, where it would have been much more difficult. As a result, replication usually began before the label was added, so a blank area arises in the middle where replication was occurring but no label could be incorporated.

Notice the shape of the pairs of streaks in Figure 20.11a. They taper to a point, moving outward, rather like an old-fashioned waxed mustache. That means the DNA incorporated highly radioactive label first, then more weakly radioactive label, leading to a tapering off of radioactivity moving outward in both directions from the origin of replication. The opposite experiment—"cooler" label first, followed by "hotter" label—would give a reverse mustache,

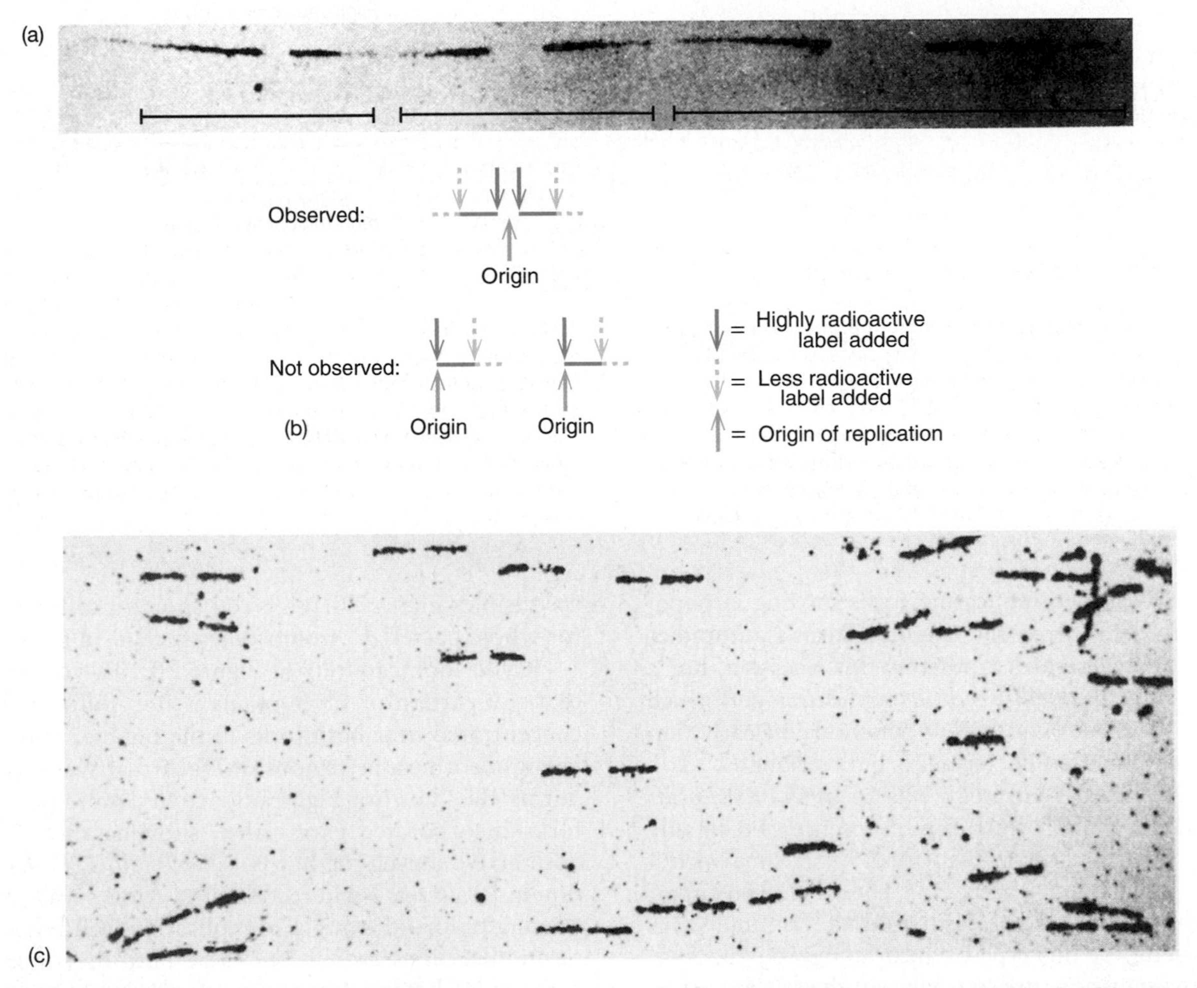

Figure 20.11 Bidirectional DNA replication in eukaryotes. **(a)** Autoradiograph of replicating *Drosophila melanogaster* DNA, pulse-labeled first with high-radioactivity DNA precursor, then with low. Note the pairs of streaks (denoted by brackets) tapering away from the middle. This reflects the pattern of labeling of a replicon with a central origin and two replicating forks. **(b)** Idealized diagram showing the patterns observed with high-, then low-radioactivity labeling, and the pattern expected if the pairs of streaks represent two independent unidirectional replicons whose replicating forks move in the same direction. The latter pattern was not observed. **(c)** Autoradiograph of replicating embryonic *Triturus vulgaris* DNA. Note the constant size and shape of the pairs of streaks, suggesting that all the corresponding replicons began replicating at the same time. (*Sources:* (a) Huberman, J.A. and A. Tsai, Direction of DNA replication in mammalian cells. *Journal of Molecular Biology* 75 (1973) p. 8, by permission of Elsevier. (c) Callan, H.G., DNA replication in the chromosomes of eukaryotes. *Cold Spring Harbor Symposia on Quantitative Biology* 38 (1973) f. 4c, p. 195.)

with points on the inside. It is possible, of course, that closely spaced, independent origins of replication gave rise to these pairs of streaks. But we would not expect that such origins would always give replication in opposite directions. Surely some would lead to replication in the same direction, producing asymmetric autoradiographs such as the hypothetical one in Figure 20.11b. But these were not seen. Thus, these autoradiography experiments confirm that each pair of streaks we see really represents one origin of replication, rather than two that are close together. It therefore appears that replication of *Drosophila* DNA is bidirectional.

These experiments were done with *Drosophila* cells originally derived from mature fruit flies and then cultured in vitro. H.G. Callan and his colleagues performed the same type of experiment using highly radioactive label and embryonic amphibian cells. These experiments (with embryonic cells of the newt) gave the striking results shown in Figure 20.11c. In contrast to the pattern in adult insect cells, the pairs of streaks here are all the same. They are all approximately the same length and they all have the same size space in the middle. This tells us that replication at all these origins began simultaneously. This must be so, because the addition of label caught all the forks at the same point—the same distance away from their respective origins of replication. This phenomenon probably helps explain how embryonic newt cells complete their DNA replication so rapidly (in as little as an hour, compared to 40 h in adult cells): Replication at all origins begins simultaneously, rather than in a staggered fashion.

This discussion of origins of replication helps us define an important term: **replicon.** The DNA under the control of one origin of replication is called a replicon. The *E. coli* chromosome is a single replicon because it replicates from a single origin. Obviously, eukaryotic chromosomes have many replicons; otherwise, it would take far too long to replicate a whole chromosome.

Not all DNAs replicate bidirectionally. Michael Lovett used electron microscopic evidence to show that the replication of the plasmid ColE1 in *E. coli* occurs unidirectionally, with only one replicating fork.

SUMMARY Most eukaryotic and bacterial DNAs replicate bidirectionally. ColE1 is an example of a DNA that replicates unidirectionally.

Rolling Circle Replication

Certain circular DNAs replicate, not by the θ mode we have already discussed, but by a mechanism called **rolling circle** replication. The *E. coli* phages with single-stranded circular DNA genomes, such as ϕX174, use a relatively simple form of rolling circle replication in which a double-stranded **replicative form** (RFI) gives rise to many copies of

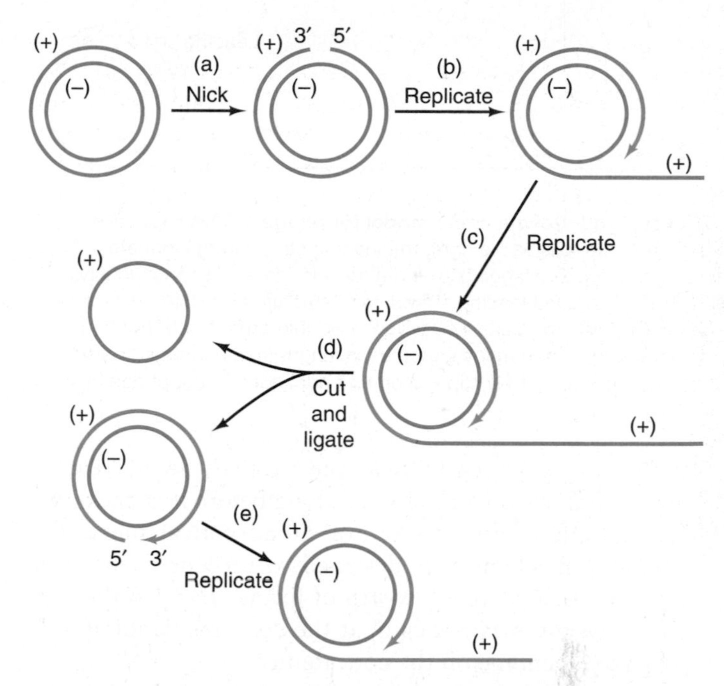

Figure 20.12 Schematic representation of rolling circle replication that produces single-stranded circular progeny DNAs. (a) An endonuclease creates a nick in the positive strand of the double-stranded replicative form. **(b)** The free 3′-end created by the nick serves as the primer for positive strand elongation, as the other end of the positive strand is displaced. The negative strand is the template. Red denotes newly-synthesized DNA. **(c)** Further replication occurs, as the positive strand approaches double length. The circle can be considered to be rolling counterclockwise. **(d)** The unit length of positive strand DNA that has been displaced is cleaved off by an endonuclease and circularized. **(e)** Replication continues, producing another new positive strand, using the negative strand as template. This process repeats over and over to yield many copies of the circular positive strand.

a single-stranded progeny DNA, as illustrated in simplified form in Figure 20.12. The intermediates (steps b and c in Figure 20.12) give this mechanism the rolling circle name because the double-stranded part of the replicating DNA can be considered to be rolling counterclockwise and trailing out the progeny single-stranded DNA, rather like a roll of toilet paper unrolling as it speeds across the floor. This intermediate also somewhat resembles an upside-down Greek letter σ (sigma), so this mechanism is sometimes called the σ mode, to distinguish it from the θ mode.

The rolling circle mechanism is not confined to production of single-stranded DNA. Some phages (e.g., λ) use this mechanism to replicate double-stranded DNA. During the early phase of λ DNA replication, the phage follows the θ mode of replication to produce several copies of circular DNA. These circular DNAs are not packaged into phage particles; they serve as templates for rolling circle synthesis of linear λ DNA molecules that *are* packaged. Figure 20.13 shows how this rolling circle operates. Here, the replicating fork looks much more like that in *E. coli* DNA replication, with (perhaps) continuous synthesis on the leading strand (the one going around the circle) and discontinuous synthesis

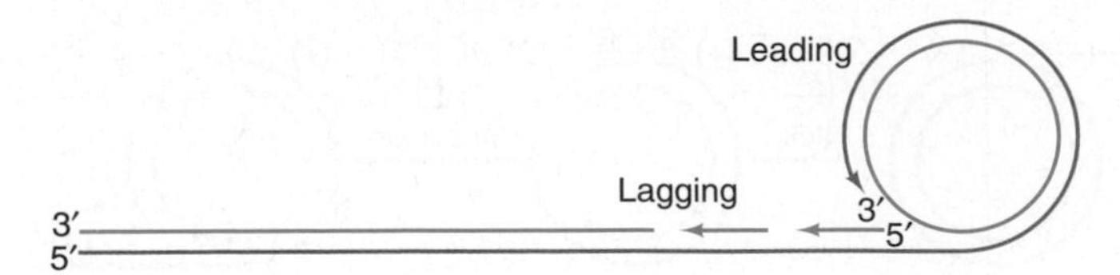

Figure 20.13 Rolling circle model for phage λ DNA replication. As the circle rolls to the right, the leading strand (red) elongates continuously. The lagging strand (blue) elongates discontinuously, using the unrolled leading strand as a template and RNA primers for each Okazaki fragment. The progeny double-stranded DNA thus produced grows to many genomes in length (a concatemer) before one genome's worth is clipped off and packaged into a phage head.

on the lagging strand. In λ, the progeny DNA reaches lengths that are several genomes long before it is packaged. The multiple-length DNAs are called **concatemers.** The packaging mechanism is designed to provide each phage head with one genome's worth of linear DNA, so the concatemer is cut enzymatically at the *cos* sites flanking each complete λ genome on the concatemer.

SUMMARY Circular DNAs can replicate by a rolling circle mechanism. One strand of a double-stranded DNA is nicked and the 3′-end is extended, using the intact DNA strand as template. This displaces the 5′-end. In phage ϕX174 replication, when one round of replication is complete, a full-length, single-stranded circle of DNA is released. In phage λ, the displaced strand serves as the template for discontinuous, lagging strand synthesis.

20.2 Enzymology of DNA Replication

Over 30 different polypeptides cooperate in replicating the *E. coli* DNA. Let us begin by examining the activities of some of these proteins and their homologs in other organisms, starting with the DNA polymerases—the enzymes that make DNA.

Three DNA Polymerases in *E. coli*

Arthur Kornberg discovered the first *E. coli* DNA polymerase in 1958. Because we now know that it is only one of three DNA polymerases, we call it **DNA polymerase I (pol I).** In the absence of evidence for other cellular DNA polymerases, many molecular biologists assumed that pol I was the polymerase responsible for replicating the bacterial genome. As we will see, this assumption was incorrect. Nevertheless, we begin our discussion of DNA polymerases with pol I because it is relatively simple and well understood, yet it exhibits the essential characteristics of a DNA synthesizing enzyme.

Pol I Although pol I is a single 102-kD polypeptide chain, it is remarkably versatile. It catalyzes three quite distinct reactions. It has a DNA polymerase activity, of course, but it also has two different exonuclease activities: a 3′→5′, and a 5′→3′ exonuclease activity. Why does a DNA polymerase also need two exonuclease activities? The 3′→5′ activity is important in **proofreading** newly synthesized DNA (Figure 20.14). If pol I has just added the wrong nucleotide to a growing DNA chain, this nucleotide will not base-pair properly with its partner in the parental strand and should be removed. Accordingly, pol I pauses and the 3′→5′ exonuclease removes the mispaired nucleotide, allowing replication to continue. This greatly increases the fidelity, or accuracy, of DNA synthesis.

The 5′→3′ exonuclease activity allows pol I to degrade a strand ahead of the advancing polymerase, so it can remove and replace a strand all in one pass of the polymerase, at least in vitro. This DNA degradation function is useful because pol I seems to be involved primarily in DNA repair (including removal and replacement of RNA primers), for which destruction of damaged or mispaired DNA (or RNA primers) and its replacement by good DNA is required. Figure 20.15 illustrates this process for primer removal and replacement.

Another important feature of pol I is that it can be cleaved by mild proteolytic treatment into two polypeptides: a large fragment (the **Klenow fragment**), which has the polymerase and proofreading (3′→5′ exonuclease) activities; and a small fragment with the 5′→3′ exonuclease activity. The Klenow fragment is frequently used in molecular biology when DNA synthesis is required and destruction

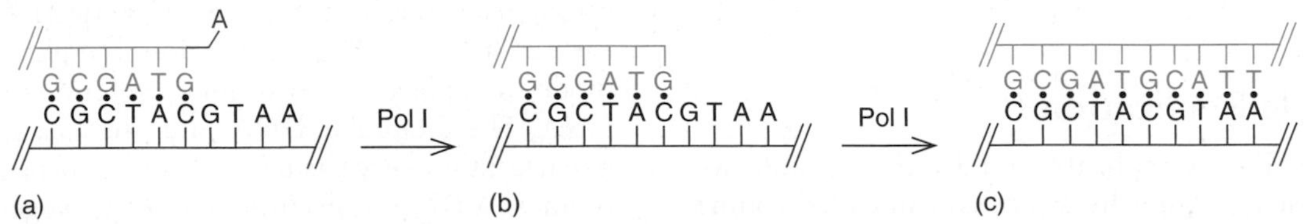

A
G C G A T G
C G C T A C G T A A
Pol I
G C G A T G
C G C T A C G T A A
Pol I
G C G A T G C A T T
C G C T A C G T A A
(a) (b) (c)

Figure 20.14 Proofreading in DNA synthesis. (a) An adenine nucleotide (pink) has been mistakenly incorporated across from a guanine. This destroys the perfect base pairing required at the 3′-end of the primer, so the replicating machinery stalls. **(b)** This pause then allows Pol I to use its 3′→5′ exonuclease function to remove the mispaired nucleotide. **(c)** With the appropriate base-pairing restored, Pol I is free to continue DNA synthesis.

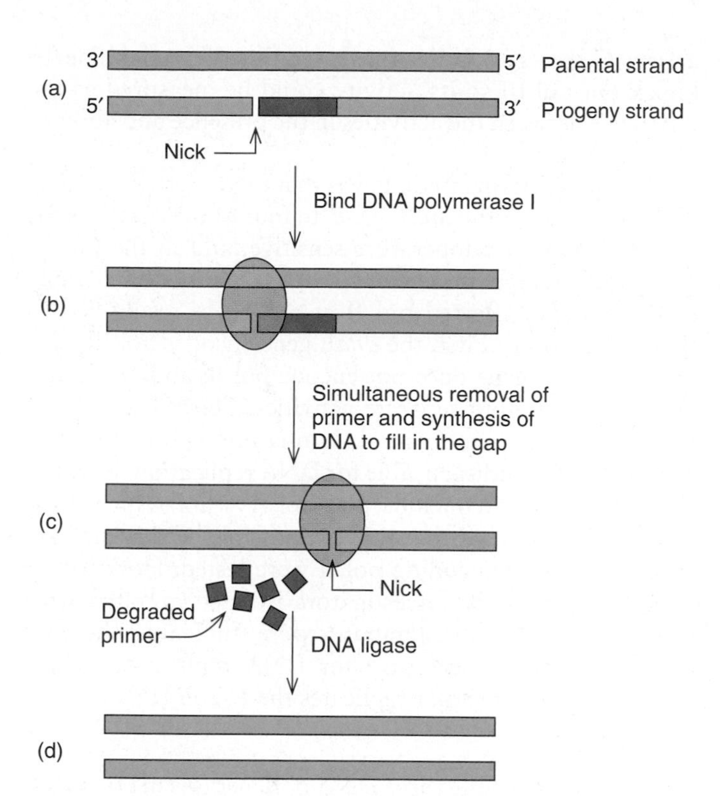

Figure 20.15 Removing primers and joining nascent DNA fragments. (a) There are two adjacent progeny DNA fragments, the right-hand one containing an RNA primer (red) at its 5′-end. The two fragments are separated by a single-stranded break called a nick. **(b)** DNA polymerase I binds to the double-stranded DNA at the nick. **(c)** The 5′→3′ exonuclease and polymerase activities of DNA polymerase I simultaneously remove the primer and fill in the resulting gap by extending the left-hand DNA fragment rightward. The polymerase leaves degraded primer in its wake. **(d)** DNA ligase seals the remaining nick by forming a phosphodiester bond between the left-hand and right-hand progeny DNA fragments.

of one of the parental DNA strands, or the primer, is undesirable. For example, the Klenow fragment is often used to perform DNA end-filling (Chapter 5) and can also be used to sequence a DNA. On the other hand, the whole pol I is used to perform nick translation (Chapter 4) to label a probe in vitro, because nick translation depends on 5′→3′ degradation of DNA ahead of the moving fork.

Thomas Steitz and colleagues determined the crystal structure of the Klenow fragment in 1987, giving us our first look at the fine structure of a DNA-synthesizing machine. The most obvious feature of the structure is a great cleft between two α-helices. This is the presumed binding site for the DNA that is being replicated. In fact, all of the known polymerase structures, including that of T7 RNA polymerase, are very similar, and have been likened to a hand. In the Klenow fragment, one α-helix is part of the "fingers" domain, the other is part of the "thumb" domain, and the β-pleated sheet between them is part of the "palm" domain. The palm domain contains three conserved aspartate residues that are essential for catalysis. They are thought to coordinate magnesium ions that catalyze the polymerase reaction.

Is the cleft in the polymerase structure really the DNA binding site? To find out, Steitz and colleagues turned to another DNA polymerase, the *Taq* polymerase. They made a cocrystal of *Taq* polymerase and a model double-stranded DNA template containing 8 bp and a blunt end at the 3′-end of the nontemplate (primer) strand. *Taq* polymerase is the polymerase from the thermophilic bacterium *Thermus aquaticus* that is widely used in PCR (Chapter 4). Its polymerase domain is very similar to that of the Klenow fragment—so much so that it is called the "KF portion," for "Klenow fragment" portion, of the enzyme. Figure 20.16 shows the results of x-ray crystallography studies on the *Taq* polymerase–DNA complex. The primer strand (red) has its 3′-end close to the three essential aspartate residues in the palm domain, but not quite close enough for magnesium ions to bridge between the carboxyl groups of the aspartates and the 3′-hydroxyl group of the primer strand. Thus, this structure is not exactly like a catalytically productive one, perhaps in part because the magnesium ions are missing.

In 1969, Paula DeLucia and John Cairns isolated a mutant with a defect in the *polA* gene, which encodes pol I. This mutant (*polA1*) lacked pol I activity, yet it was viable,

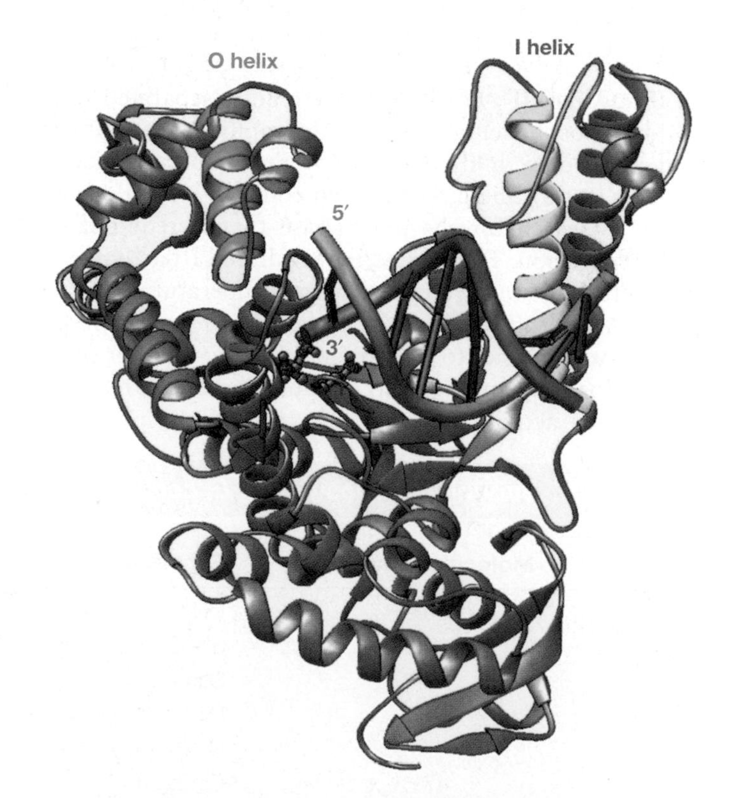

Figure 20.16 Cocrystal structure of *Taq* DNA polymerase with a double-stranded model DNA template. The O helix and I helix of the "fingers" and "thumb" of the polymerase "hand" are in green and yellow, respectively. The template and primer strands of the model DNA are in orange and red, respectively. The three essential aspartate side chains in the "palm" are represented by small red balls near the 3′-end of the primer strand. (*Source:* Eom, S.H., J. Wang and T.A. Steitz, Structure of *Taq* polymerase with DNA at the polymerase active site. *Nature* 382 (18 July 1996) f. 2a, p. 280. Copyright © Macmillan Magazines, Ltd.)

strongly suggesting that pol I was not really the DNA-replicating enzyme. Instead, pol I seems to play a dominant role in repair of DNA damage. It fills in the gaps left when damaged DNA is removed. The finding that pol I is not essential spurred a renewed search for the real DNA replicase, and in 1971, Thomas Kornberg and Malcolm Gefter discovered two new polymerase activities: **DNA polymerases II** and **III** (**pol II** and **pol III**). We will see that pol III is the actual replicating enzyme.

SUMMARY Pol I is a versatile enzyme with three distinct activities: DNA polymerase; 3′→5′ exonuclease; and 5′→3′ exonuclease. The first two activities are found on a large domain of the enzyme, and the last is on a separate, small domain. The large domain (the Klenow fragment) can be separated from the small by mild protease treatment, yielding two protein fragments with all three activities intact. The structure of the Klenow fragment (and all other known DNA polymerases) shows a wide cleft for binding to DNA. This polymerase active site is remote from the 3′→5′ exonuclease active site on the Klenow fragment.

Pol II and Pol III Pol II could be readily separated from pol I by phosphocellulose chromatography, but pol III had been masked in wild-type cells by the preponderance of pol I. Next, Kornberg, Gefter, and colleagues used genetic means to search for the polymerase that is required for DNA replication. They tested the pol II and III activities in 15 different *E. coli* strains that were temperature-sensitive for DNA replication. Most of these strains were *polA1*⁻, which made it easier to measure pol III activity after phosphocellulose chromatography because there was no competing pol I activity. In those few cases where pol I was active, Gefter and colleagues used *N*-ethylmaleimide to knock out pol III so its activity could be measured as the difference between the activities in the presence and absence of the inhibitor.

The most striking finding was that there were five strains with mutations in the *dnaE* gene. In four of these, the pol III activity was very temperature-sensitive, and in the fifth it was slightly temperature-sensitive. On the other hand, none of the mutants affected pol II at all. These results led to three conclusions: First, the *dnaE* gene encodes pol III. Second, the *dnaE* gene does not encode pol II, and pol II and pol III are therefore separate activities. Third, because defects in the gene encoding pol III interfere with DNA replication, pol III is indispensable for DNA replication. It would have been nice to conclude that pol II is *not* required for DNA replication, but that was not possible because no mutants in the gene encoding pol II were tested. However, in separate work, these investigators isolated mutants with inactive pol II, and these mutants were still viable, showing that pol II is not necessary for DNA replication. Thus, pol III is the enzyme that replicates the *E. coli* DNA.

SUMMARY Of the three DNA polymerases in *E. coli* cells, pol I, pol II, and pol III, only pol III is required for DNA replication. Thus, this polymerase is the enzyme that replicates the bacterial DNA.

The Pol III Holoenzyme The enzyme that carries out the elongation of primers to make both the leading and lagging strands of DNA is called **DNA polymerase III holoenzyme** (**pol III holoenzyme**). The "holoenzyme" designation indicates that this is a multisubunit enzyme, and indeed it is: As Table 20.1 illustrates, the holoenzyme contains 10 different polypeptides. On dilution, this holoenzyme dissociates into

Table 20.1 Subunit Composition of *E. coli* DNA Polymerase III Holoenzyme

Subunit	Molecular mass (kD)	Function	Subassemblies			
α	129.9	DNA polymerase	Core	Pol III′	Pol III*	Pol III holoenzyme
ε	27.5	3′→5′ exonuclease	Core	Pol III′	Pol III*	Pol III holoenzyme
θ	8.6	Stimulates ε exonuclease	Core	Pol III′	Pol III*	Pol III holoenzyme
τ	71.1	Dimerizes core Binds γ complex		Pol III′	Pol III*	Pol III holoenzyme
γ	47.5	Binds ATP	γ complex (DNA-dependent ATPase)		Pol III*	Pol III holoenzyme
δ	38.7	Binds to β	γ complex (DNA-dependent ATPase)		Pol III*	Pol III holoenzyme
δ'	36.9	Binds to γ and δ	γ complex (DNA-dependent ATPase)		Pol III*	Pol III holoenzyme
χ	16.6	Binds to SSB	γ complex (DNA-dependent ATPase)		Pol III*	Pol III holoenzyme
ψ	15.2	Binds to χ and γ	γ complex (DNA-dependent ATPase)		Pol III*	Pol III holoenzyme
β	40.6	Sliding clamp				Pol III holoenzyme

*Pol III holoenzyme minus the β-subunit.
Source: Reprinted from Herendee, D.R. and T.T. Kelly, DNA Polymerase III: Running rings around the fork *Cell* 84:6, 1996. Copyright © 1996, with permission from Elsevier.

several different subassemblies, also as indicated in Table 20.1. Each pol III subassembly is capable of DNA polymerization, but only very slowly. This suggested that something important is missing from the subassemblies because DNA replication in vivo is extremely rapid. The replicating fork in *E. coli* moves at the amazing rate of 1000 nt/sec. (Imagine the sheer mechanics involved in unwinding parental DNA, correctly pairing 1000 nt with partners in the parental DNA strands, and forming 1000 phosphodiester bonds every second!) In vitro, the holoenzyme goes almost that fast: about 700 nt/sec, suggesting that this is the entity that replicates DNA in vivo. The other two DNA polymerases in the cell, pol I and pol II, are not ordinarily found in holoenzyme forms, and they replicate DNA much more slowly than the pol III holoenzyme does.

Charles McHenry and Weldon Crow purified DNA polymerase III to near-homogeneity and found that three polypeptides compose the core of pol III: the α-, ε-, and θ-subunits. These have molecular masses of 130, 27.5, and 10 kD, respectively. The rest of the subunits of the holoenzyme dissociated during purification, but the core subunits were bound tightly together. In this section, we will examine the pol III core more thoroughly, but we will save our discussion of the other polypeptides in the pol III holoenzyme for Chapter 21 because they play important roles in initiation and elongation of DNA synthesis.

The α-subunit of the pol III core has the DNA polymerase activity, but this was not easy to determine because the α-subunit is so difficult to separate from the other core subunits. When Hisaji Maki and Arthur Kornberg cloned and overexpressed the gene for the α-subunit, they finally paved the way for purifying the polymerase activity because the overproduced α-subunit was in great excess over the other two subunits. When they tested this purified α-subunit for DNA polymerase activity, they found that it had activity similar to the same amount of core. Thus, the α-subunit contributes the DNA polymerase activity to the core.

The pol III core has a 3′→5′ exonuclease activity that removes mispaired bases as soon as they are incorporated, allowing the polymerase to proofread its work. This is similar to the 3′→5′ exonuclease activity of the pol I Klenow fragment. Scheuermann and Echols used the overexpression strategy to demonstrate that the core ε-subunit has this exonuclease activity. They overexpressed the ε-subunit (the product of the *dnaQ* gene) and purified it through various steps. After the last step, DEAE-Sephacel chromatography, the ε-subunit was essentially pure. Next, Richard Scheuermann and Harrison Echols tested this purified ε-subunit, as well as core pol III, for exonuclease activity. Figure 20.17 shows that the core and the ε-subunit both have exonuclease activity, and they are both specific for mispaired DNA substrates, having no measurable activity on perfectly paired DNAs. This is what we expect for the proofreading activity. This activity also explains why *dnaQ* mutants are subject to excess mutations (10^3–10^5 more

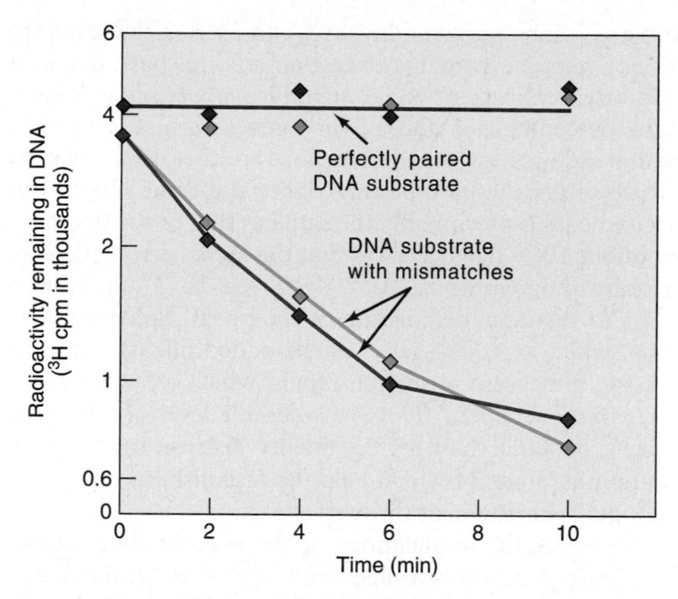

Figure 20.17 Exonuclease activity of ε-subunit and pol III core with substrates that are perfectly base-paired or that have mismatches. Scheuermann and Echols incubated the purified ε-subunit with ^{3}H-labeled synthetic DNAs and measured the amount of radioactivity remaining in the DNAs after increasing lengths of time. Symbols: blue and green, pol III core; orange and red, ε-subunit. (*Source*: Adapted from Scheuermann, R.H. and H. Echols, A separate editing exonuclease for DNA replication: The ε subunit of *Escherichia coli* DNA polymerase III holoenzyme. *Proceedings of the National Academy of Sciences USA* 81:7747–51, December 1984.)

than in wild-type cells). Without adequate proofreading, many more mismatched bases fail to be removed and persist as mutations. Thus, we call *dnaQ* mutants **mutator mutants,** and the gene has even been referred to as the *mutD* gene because of this mutator phenotype.

Relatively little work has been performed on the θ-subunit of the core. Its function, other than a stimulation of ε exonuclease activity, is unknown. However, it is clear that the α- and ε-subunits cooperate to boost each other's activity in the core polymerase. The DNA polymerase activity of the α-subunit increases by about two-fold in the core, compared with the free subunit, and the activity of the ε-subunit increases by about 10–80-fold when it joins the core.

SUMMARY The pol III core is composed of three subunits, α, ε, and θ. The α-subunit has the DNA polymerase activity. The ε-subunit has the 3′→5′ exonuclease activity that carries out proofreading. The role of the θ-subunit is not yet clear.

Fidelity of Replication

The proofreading mechanism of pol III (and pol I) greatly increases the fidelity of DNA replication. The pol III core makes about one pairing mistake in one hundred thousand in vitro—not a very good record, considering that even the *E. coli* genome contains over four million base pairs. At this rate, replication would introduce errors into a significant percentage of genes every generation. Fortunately, proofreading allows the polymerase another mechanism by which to get the base pairing right. The error rate of this second pass is presumably the same as that of the first pass, or about 10^{-5}. This predicts that the actual error rate with proofreading would be $10^{-5} \times 10^{-5} = 10^{-10}$, and that is close to the actual error rate of the pol III holoenzyme in vivo, which is 10^{-10}–10^{-11}. (The added fidelity comes at least in part from mismatch repair, which we will discuss later in this chapter.) This is a tolerable level of fidelity. In fact, it is better than perfect fidelity because it allows for mutations, some of which help the organism to adapt to a changing environment through evolution.

Consider the implications of the proofreading mechanism, which removes a mispaired nucleotide at the 3′-end of a DNA progeny strand (recall Figure 20.14). DNA polymerase cannot operate without a base-paired nucleotide to add to, which means that it cannot start a new DNA chain unless a primer is already there. That explains the need for primers, but why primers made of RNA? The reason seems to be the following: Primers are made with more errors, because their synthesis is not subject to proofreading. Making primers out of RNA guarantees that they will be recognized, removed, and replaced with DNA by extending the neighboring Okazaki fragment. The latter process is, of course, relatively error-free, because it is catalyzed by pol I, which has a proofreading function.

SUMMARY Faithful DNA replication is essential to life. To help provide this fidelity, the *E. coli* DNA replication machinery has a built-in proofreading system that requires priming. Only a base-paired nucleotide can serve as a primer for the pol III holoenzyme. Therefore, if the wrong nucleotide is incorporated by accident, replication stalls until the 3′→5′ exonuclease of the pol III holoenzyme removes it. The fact that the primers are made of RNA may help mark them for degradation.

Multiple Eukaryotic DNA Polymerases

Much less is known about the proteins involved in eukaryotic DNA replication, but we do know that multiple DNA polymerases take part in the process, and we also have a good idea of the roles these enzymes play. Table 20.2 lists the major mammalian DNA polymerases and their probable roles.

Table 20.2 Probable Roles of Some Eukaryotic DNA Polymerases

Enzyme	Probable role
DNA polymerase α	Priming of replication of both strands
DNA polymerase δ	Elongation of lagging strand
DNA polymerase ε	Elongation of leading strand
DNA polymerase β	DNA repair
DNA polymerase γ	Replication of mitochondrial DNA

It had been thought that polymerase α synthesized the lagging strand because of the low processivity of this enzyme. **Processivity** is the tendency of a polymerase to stick with the replicating job once it starts. The *E. coli* polymerase III holoenzyme is highly processive. Once it starts on a DNA chain, it remains bound to the template, making DNA for a long time. Because it does not fall off the template very often, which would require a pause as a new polymerase bound and took over, the overall speed of *E. coli* DNA replication is very rapid. Polymerase δ is much more processive than polymerase α. Thus, it was proposed that the less processive DNA polymerase α synthesized the lagging strand, which is made in short pieces. However, it now appears that polymerase α, the only eukaryotic DNA polymerase with primase activity, makes the primers for both strands. Then DNA polymerase epsilon ε elongates the leading strand and DNA polymerase δ elongates the lagging strand.

Actually, much of the processivity of polymerases δ and ε comes, not from the polymerase itself, but from an associated protein called **proliferating cell nuclear antigen,** or **PCNA.** This protein, which is enriched in proliferating cells that are actively replicating their DNA, enhances the processivity of polymerase δ by a factor of 40. That is, PCNA causes the polymerase to travel 40 times farther elongating a DNA chain before falling off the template. PCNA works by physically clamping the polymerase onto the template. We will examine this clamping phenomenon more fully when we consider the detailed mechanism of DNA replication in *E. coli* in Chapter 21.

In marked contrast, polymerase β is not processive at all. It usually adds only one nucleotide to a growing DNA chain and then falls off, requiring a new polymerase to bind and add the next nucleotide. This fits with its postulated role as a repair enzyme that needs to make only short stretches of DNA to fill in gaps created when primers or mismatched bases are excised. In addition, the level of polymerase β in a cell is not affected by the rate of division of the cell, which suggests that this enzyme is not involved in DNA replication. If it were, we would expect it to be more prevalent in rapidly dividing cells, as polymerases δ and α are.

Polymerase γ is found in mitochondria, not in the nucleus. Therefore, we conclude that this enzyme is responsible for replicating mitochondrial DNA.

SUMMARY Mammalian cells contain five different DNA polymerases. Polymerases ε, δ, and α appear to participate in replicating both DNA strands: α by priming DNA synthesis ε by elongating the leading strand, and δ by elongating the lagging strand. Polymerase β seems to function in DNA repair. Polymerase γ probably replicates mitochondrial DNA.

Strand Separation

In our discussion of the general features of DNA replication, we have been assuming that the two DNA strands at the fork somehow unwind. This does not happen automatically as DNA polymerase does its job; the two parental strands hold tightly to each other, and it takes energy and enzyme action to separate them.

Helicase The enzyme that harnesses the chemical energy of ATP to separate the two parental DNA strands at the replicating fork is called a **helicase.** We have already seen an example of helicase action in Chapter 11, in our discussion of the DNA helicase activity of TFIIH, which unwinds a short region of DNA to help create the transcription bubble in eukaryotes. That DNA melting is transient, in contrast to the permanent strand separation needed to advance a replicating fork.

Many DNA helicases have been identified in *E. coli* cells. The problem is finding which of these is involved in DNA replication. The first three to be investigated—the *rep* helicase, and DNA helicases II and III—could be mutated without inhibiting cellular multiplication. This made it unlikely that any of these three enzymes could participate in something as vital to cell survival as DNA replication; we would anticipate that defects in the helicase that participates in DNA replication would be lethal.

One way to generate mutants with defects in essential genes is to make the mutations conditional, usually temperature-sensitive. That way, one can grow the mutant cells at a low temperature at which the mutation is not expressed, then shift the temperature up to observe the mutant phenotype. As early as 1968, François Jacob and his colleagues discovered two classes of temperature-sensitive mutants in *E. coli* DNA replication. Type 1 mutants showed an immediate shut-off of DNA synthesis on raising the temperature from 30°C to 40°C, whereas type 2 mutants showed only a gradual decrease in the rate of DNA synthesis at elevated temperature.

One of the type 1 mutants was the *dnaB* mutant; DNA synthesis in *E. coli* cells carrying temperature-sensitive mutations in the *dnaB* gene stopped short as soon as the temperature rose to the nonpermissive level. This is what we would expect if *dnaB* encodes the DNA helicase required for replication. Without a functional helicase, the fork cannot move, and DNA synthesis must halt immediately. Furthermore, the *dnaB* product (DnaB) was known to be an ATPase, which we also expect of a DNA helicase, and the DnaB protein was found associated with the primase, which makes primers for DNA replication.

All of these findings suggested that DnaB is the DNA helicase that unwinds the DNA double helix during *E. coli* DNA replication. All that remained was to show that DnaB has DNA helicase activity. Jonathan LeBowitz and Roger McMacken did this in 1986. They used the helicase substrate shown in Figure 20.18a, which is a circular M13 phage DNA, annealed to a shorter piece of linear DNA, which was labeled at its 5′-end. Figure 20.18a also shows how the helicase assay worked. LeBowitz and McMacken incubated the labeled substrate with DnaB, or other proteins, and then electrophoresed the products. If the protein had helicase activity, it would unwind the double-helical DNA and separate the two strands. Then the short, labeled DNA would migrate independently of the larger, unlabeled DNA, and would have a much higher electrophoretic mobility.

Figure 20.18b shows the results of the assay. DnaB alone had helicase activity, and this was stimulated by DnaG (which we will see in Chapter 21 is a primase), and by SSB, a single-stranded DNA-binding protein that we will introduce next. Neither DnaG nor SSB, by themselves or together, had any DNA helicase activity. Thus, DnaB is the helicase that unwinds the DNA at the replicating fork.

SUMMARY The helicase that unwinds double-stranded DNA at the replicating fork is encoded by the *E. coli dnaB* gene.

Single-Strand DNA-Binding Proteins

Another class of proteins, called **single-strand DNA-binding proteins (SSBs),** also participate in DNA strand separation during replication. These proteins do not catalyze strand separation, as helicases do. Instead, they bind selectively to single-stranded DNA as soon as it forms and coat it so it cannot anneal to re-form a double helix. The single-stranded DNA can form by natural "breathing" (transient local separation of strands, especially in A–T-rich regions) or as a result of helicase action, then SSB catches it and keeps it in single-stranded form.

The best-studied SSBs are bacterial. The *E. coli* protein is called SSB and is the product of the *ssb* gene. The T4 phage protein is **gp32,** which stands for "gene product *32*" (the product of gene *32* of phage T4). The M13 phage protein is **gp5** (the product of the phage gene *5*). All of these proteins act cooperatively: The binding of one protein facilitates the binding of the next. For example, the binding of the first molecule of gp32 to single-stranded DNA raises the affinity for the next molecule a thousandfold.

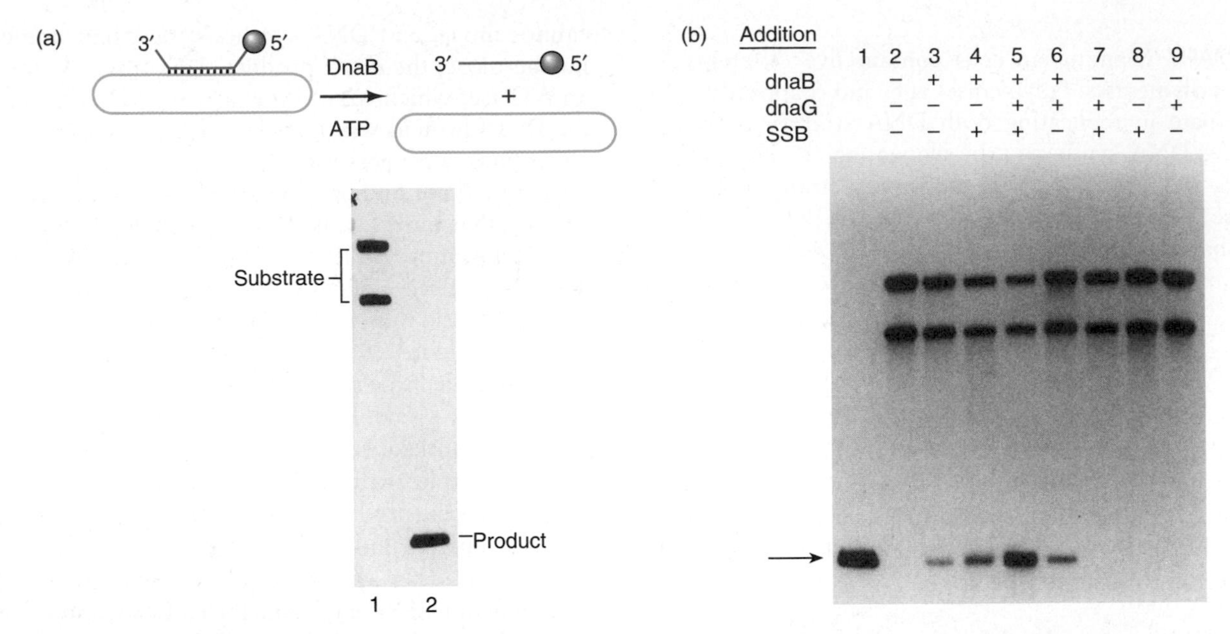

Figure 20.18 DNA helicase assay. (a) Principle of assay. LeBowitz and McMacken made a helicase substrate (top) by ^{32}P-labeling a single-stranded 1.06-kb DNA fragment (red) at its 5′-end and annealing the fragment to an unlabeled single-stranded recombinant M13 DNA bearing a complementary 1.06-kb region. The dnaB protein, or any DNA helicase, can unwind the double-stranded region of the substrate and liberate the labeled short piece of DNA (red) from its longer, circular partner. Bottom: Electrophoresis of the substrate (lane 1) yields two bands, which probably correspond to linear and circular versions of the long DNA annealed to the labeled, short DNA. Electrophoresis of the short DNA by itself (lane 2) shows that it has a much higher mobility than the substrate (see band labeled "Product"). **(b)** Helicase assay results. LeBowitz and McMacken performed the assay outlined in (a) with the additions (DnaB, DnaG, and SSB) indicated at top. The electrophoresis results are given at bottom. Lane 1 is a control with the unannealed, labeled short DNA to show its electrophoretic behavior (arrow). Lane 3 shows that DnaB has helicase activity on its own, but lanes 4 and 5 demonstrate that the other proteins stimulate this activity. On the other hand, lanes 7–9 show that the other two proteins have no helicase activity without DnaB. (*Source:* LeBowitz, J.H. and R. McMacken, The *Escherichia coli* dnaB replication protein is a DNA helicase. *Journal of Biological Chemistry* 261 (5 April 1986) figs. 2, 3, pp. 4740–41. American Society for Biochemistry and Molecular Biology.)

Thus, once the first molecule of gp32 binds, the second binds easily, and so does the third, and so forth. This results in a chain of gp32 molecules coating a single-stranded DNA region. The chain will even extend into a double-stranded hairpin, melting it, as long as the free energy released in cooperative gp32 binding through the hairpin exceeds the free energy released by forming the hairpin. In practice, this means that relatively small, or poorly base-paired hairpins will be melted, but long, or well base-paired ones will remain intact. The gp32 protein binds to DNA as a chain of monomers, whereas gp5 binds as a string of dimers, and *E. coli* SSB binds as a chain of tetramers, with about 65 nt of single-stranded DNA wound around each SSB tetramer.

By now we have had some hints that the name "single-strand DNA-binding protein" is a little misleading. These proteins do indeed bind to single-stranded DNA, but so do many other proteins we have studied in previous chapters, including RNA polymerase. But the SSBs do much more. We have already seen that they trap DNA in single-stranded form, but they also specifically stimulate their homologous DNA polymerases. For example, gp32 stimulates the T4 DNA polymerase, but it does not stimulate phage T7 polymerase or *E. coli* DNA polymerase I.

Are the activities of the SSBs important? In fact, they are essential. Temperature-sensitive mutations in the *ssb* gene of *E. coli* render the cell inviable at the nonpermissive temperature. In cells infected by the tsP7 mutant of phage T4, with a temperature-sensitive gp32, phage DNA replication stops within 2 min after shifting to the nonpermissive temperature (Figure 20.19). Furthermore, the phage DNA begins to be degraded. This behavior suggests that one function of gp32 is to protect from degradation the single-stranded DNA created during phage DNA replication.

Based on the importance of the SSBs in prokaryotes, it is surprising that SSBs with similar importance have not yet been found in eukaryotes. However, a host SSB has been found to be essential for replication of SV40 DNA in human cells. This protein, called **RF-A,** or human SSB, binds selectively to single-stranded DNA and stimulates the DNA helicase activity of the viral large T antigen. Because this is a host protein, we assume that it plays a role in the uninfected human cell as well, but we do not know yet what that role is. We also know that virus-encoded SSBs play a major role in replication of certain eukaryotic viral DNAs, including adenovirus and herpesvirus DNAs.

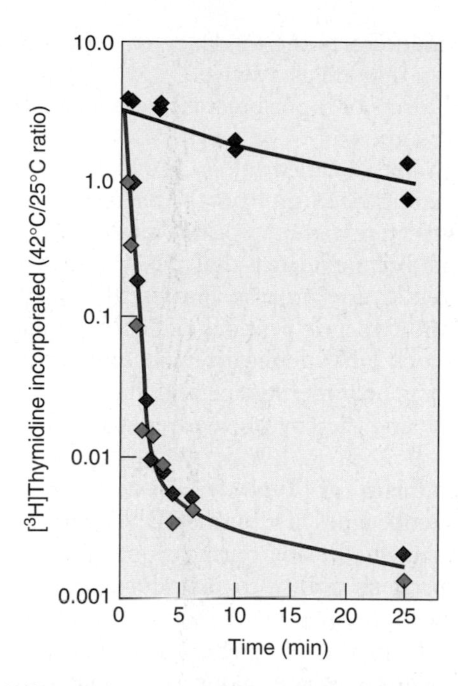

Figure 20.19 Temperature-sensitivity of DNA synthesis in cells infected by T4 phage with a temperature-sensitive mutation in the SSB (gp32) gene. Curtis and Alberts measured the relative incorporation of [^{3}H]thymidine after 1 min pulses at 42° and 25°C in cells infected with T4 phage mutants having mutations in the following genes: gene *23*, blue; gene *32* plus gene *23*, red; and gene *32* plus gene *49*, green. The amber mutations in genes *23* and *49* have no effect on DNA synthesis. Thus, the observed drop in DNA synthesis is due to the *ts* mutation in gene *32*. (*Source:* Adapted from Curtis, M.J. and B. Alberts, Studies on the structure of intracellular bacteriophage T4 DNA, *Journal of Molecular Biology,* 102: 793–816, 1976.)

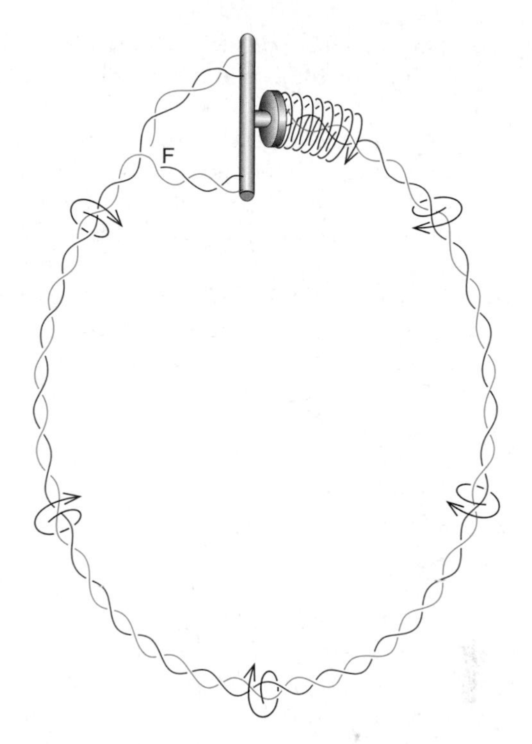

Figure 20.20 Cairns's swivel concept. As the closed circular DNA replicates, the two strands must separate at the fork (F). The strain of this unwinding would be released by a swivel mechanism. Cairns actually envisioned the swivel as a machine that rotated actively and thus drove the unwinding of DNA at the fork.

SUMMARY The prokaryotic single-stranded DNA-binding proteins bind much more strongly to single-stranded than to double-stranded DNA. They aid helicase action by binding tightly and cooperatively to newly formed single-stranded DNA and keeping it from annealing with its partner. By coating the single-stranded DNA, SSBs also protect it from degradation. They also stimulate their homologous DNA polymerases. These activities make SSBs essential for prokaryotic DNA replication.

Topoisomerases

Sometimes we refer to the separation of DNA strands as "unzipping." We should not forget, when using this term, that DNA is not like a zipper with straight, parallel sides. It is a double helix. Therefore, when the two strands of DNA separate, they must rotate around each other. Helicase could handle this task alone if the DNA were linear and unnaturally short, but closed circular DNAs, such as the *E. coli* chromosome, present a special problem. As the DNA unwinds at the replicating fork, a compensating winding up of DNA will occur elsewhere in the circle. This tightening of the helix will create intolerable strain unless it is relieved. Cairns recognized this problem in 1963 when he first observed circular DNA molecules in *E. coli,* and he proposed a "swivel" in the DNA duplex that would allow the DNA strands on either side to rotate to relieve the strain (Figure 20.20). We now know that an enzyme known as DNA gyrase serves the swivel function. **DNA gyrase** belongs to a class of enzymes called **topoisomerases** that introduce transient single- or double-stranded breaks into DNA and thereby allow it to change its shape, or topology.

To understand how the topoisomerases work, we need to look more closely at the phenomenon of supercoiled, or superhelical, DNA mentioned in Chapters 2 and 6. All naturally occurring, closed circular, double-stranded DNAs studied so far exist as supercoils. Closed circular DNAs are those with no single-strand breaks, or nicks. When a cell makes such a DNA, it causes some unwinding of the double helix; the DNA is then said to be "underwound." As long as both strands are intact, no free rotation can occur around the bonds in either strand's backbone, so the DNA cannot relieve the strain of underwinding except by supercoiling. The supercoils introduced by underwinding are called "negative," by convention. This is the kind of supercoiling found in most organisms; however, positive supercoils do exist in extreme thermophiles, which have a reverse

DNA gyrase that introduces positive supercoils, thus stabilizing the DNA against the boiling temperatures in which these organisms live.

You can visualize the supercoiling process as follows: Take a medium to large rubber band, and hold it at the top with one hand. With your other hand, twist the side of the rubber band through one full turn. You should notice that the rubber band resists the turning as strain is introduced, then relieves the strain by forming a supercoil (a figure 8). The more you twist, the more supercoiling you will observe: one superhelical turn for every full twist you introduce. Reverse the twist and you will see supercoiling of the opposite handedness or sign.

If you release your grip on the side of the rubber band, of course the superhelix will relax. In DNA, it is only necessary to cut one strand to relax a supercoil because the other strand can rotate freely.

Unwinding DNA at the replicating fork would form positive rather than negative supercoils if no other way for relaxing the strain existed. That is because replication permanently unwinds one region of the DNA without nicking it, forcing the rest of the DNA to become overwound, and therefore positively supercoiled, to compensate. To visualize this, look at the circular arrow ahead of the replicating fork (F) in Figure 20.20. Notice how twisting the DNA in the direction of the arrow causes unwinding behind the arrow but overwinding ahead of it. Imagine inserting your finger into the DNA just behind the fork and moving it in the direction of the moving fork to force the DNA strands apart. You can imagine how this would force the DNA to rotate in the direction of the circular arrow, which overwinds the DNA helix. This overwinding strain would resist your finger more and more as it moved around the circle. Therefore, unwinding the DNA at the replicating fork introduces positive superhelical strain that must be constantly relaxed so replication will not be retarded. You can appreciate this when you think of how the rubber band increasingly resisted your twisting as it became more tightly wound. In principle, any enzyme that is able to relax this strain could serve as a swivel. In fact, of all the topoisomerases in an *E. coli* cell, only one, DNA gyrase, appears to perform this function.

Topoisomerases are classified according to whether they operate by causing single- or double-stranded breaks in DNA. Those in the first class (**type I topoisomerases,** e.g., topoisomerase I of *E. coli*) introduce temporary single-stranded breaks. Enzymes in the second class (**type II topoisomerases,** e.g., DNA gyrase of *E. coli*) break and reseal both DNA strands. Why is *E. coli* topoisomerase I incapable of providing the swivel function needed in DNA replication? Because it can relax only negative supercoils, not the positive ones that form in replicating DNA ahead of the fork. Obviously, the nicks created by these enzymes do not allow free rotation in either direction. But DNA gyrase pumps negative supercoils into closed circular DNA and therefore counteracts the tendency to form positive ones. Hence, it can operate as a swivel.

Not all forms of topoisomerase I are incapable of relaxing positive supercoils. Topoisomerases I from eukaryotes and archaea (the so-called eukaryotic-like topoisomerases I) use a different mechanism from the bacterial-like topoisomerases I, and can relax both positive and negative supercoils.

There is direct evidence that DNA gyrase is crucial to the DNA replication process. First of all, mutations in the genes for the two polypeptides of DNA gyrase are lethal and they block DNA replication. Second, antibiotics such as novobiocin, coumermycin, and nalidixic acid inhibit DNA gyrase and thereby prevent replication.

The Mechanism of Type II Topoisomerases Martin Gellert and colleagues first purified DNA gyrase in 1976. To detect the enzyme during purification, they used an assay that measured its ability to introduce superhelical turns into a relaxed circular DNA (the colE1 plasmid we discussed earlier in this chapter). Then they added varying amounts of DNA gyrase, along with ATP. After an hour, they electrophoresed the DNA and stained it with ethidium bromide so it would fluoresce under UV light.

Figure 20.21 depicts the results of one such assay. In the absence of gyrase (lane 2) or in the absence of ATP (lane 11)

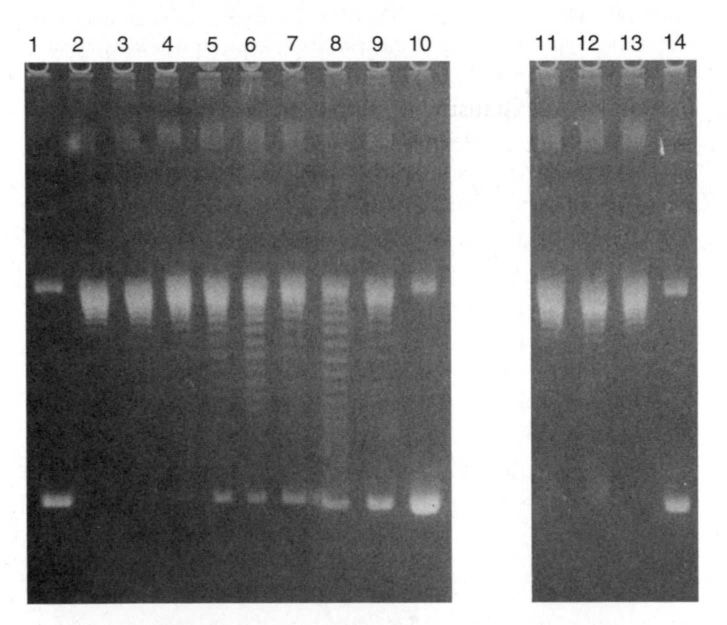

Figure 20.21 Assay for a DNA topoisomerase. Gellert and colleagues incubated relaxed circular ColE1 DNA with varying amounts of *E. coli* DNA gyrase, plus ATP, spermidine, and $MgCl_2$, except where indicated. Lane 1, supercoiled ColE1 DNA as isolated from cells; lane 2, no DNA gyrase; lanes 3–10, DNA gyrase increasing as follows: 24 ng, 48 ng, 72 ng, 96 ng, 120 ng, 120 ng, 240 ng, and 360 ng. Lane 11, ATP omitted; lane 12, spermidine omitted; lane 13, $MgCl_2$ omitted; lane 14, supercoiled ColE1 DNA incubated with 240 ng of gyrase in the absence of ATP. (*Source:* Gellert, M., K. Mizuuchi, M.H. O'Dea, and H.A. Nash, DNA gyrase: An enzyme that introduces superhelical turns into DNA. *Proceedings of the National Academy of Sciences USA* 73 (1976) fig. 1, p. 3873.)

we see essentially only the low-mobility relaxed circular form of the plasmid. On the other hand, as the experimenters added more and more DNA gyrase (lanes 3–10), they observed more and more of the high-mobility form of the plasmid with many superhelical turns. At intermediate levels of gyrase, intermediate forms of the plasmid appeared as distinct bands, with each band representing a plasmid with a different, integral number of superhelical turns.

This experiment demonstrates the dependence of DNA gyrase on ATP, but the enzyme does not use as much ATP as you might predict based on all the breaking and reforming of phosphodiester bonds. The reason for this modest energy requirement is that the gyrase itself (not a water molecule) is the agent that breaks the DNA bonds, so it forms a covalent enzyme–DNA intermediate. This intermediate conserves the energy in the DNA phosphodiester bond so it can be reused when the DNA ends are rejoined and the enzyme is released in its original form.

What is the evidence for the enzyme–DNA bond? James Wang and colleagues trapped DNA–gyrase complexes by denaturing the enzyme midway through the breaking–rejoining cycle and found DNA with nicks in both strands, staggered by four bases, with the gyrase covalently linked to each protruding DNA end. In 1980, Wang and colleagues went on to show that the covalent bond between enzyme and DNA is through a tyrosine on the enzyme. They incubated [^{32}P]DNA with DNA gyrase, trapped the DNA–gyrase complex as before by denaturing the enzyme, then isolated the complex. They digested the DNA in the complex exhaustively with nuclease, and finally isolated [^{32}P]enzyme, with the label in the A subunits. (DNA gyrase, like all forms of bacterial DNA topoisomerase II, is a tetramer of two different subunits: A_2B_2).

The fact that the enzyme's A subunits became labeled with ^{32}P strongly suggested that these subunits had been linked through one of their amino acids to the ^{32}P[DNA]. Which amino acid in the enzyme was linked to the DNA? Wang and colleagues digested the labeled A subunit in boiling HCl to break it down into its component amino acids. Then they purified the labeled amino acid, which copurified with phosphotyrosine. Thus, the enzyme is linked covalently through a tyrosine residue in each A subunit to the DNA.

How do DNA gyrase and the other DNA topoisomerase IIs perform their task of introducing negative superhelical turns into DNA? The simplest explanation is that they allow one part of the double helix to pass through another part. Figure 20.22 shows a representation of the structure of yeast topoisomerase II, based on x-ray crystallography. Like all eukaryotic forms of topoisomerase II, it is a dimer of identical subunits, and each monomer has domains corresponding to the A and B subunits of the bacterial topoisomerase IIs. Yeast topoisomerase II is a heart-shaped protein made out of two crescent-shaped monomers. The protein can be considered as a double-jawed structure, with one jaw at the top and the other at the bottom.

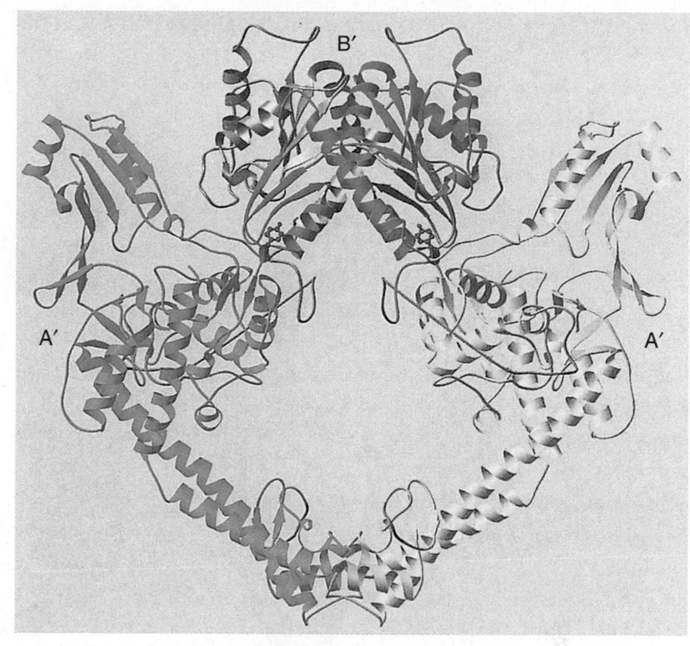

Figure 20.22 Crystal structure of yeast topoisomerase II. The monomer on the left is represented in green and orange, and the monomer on the right is in yellow and blue. The domains of each monomer corresponding to prokaryotic A subunits are in green and yellow (and labeled A′), and the domains corresponding to prokaryotic B subunits are in orange and blue (and labeled B′). The B′ domains, with ATPase activity, form an upper "jaw" of the enzyme, and A′ domains form a lower jaw. The jaws are closed in this representation. The active-site tyrosines that become linked to DNA during the reaction are represented by purple hexagons near the interfaces between the A′ and B′ domains. The primary contact between the monomers is indicated at bottom. (*Source:* Adapted from Berger, J.M., S.J. Gamblin, S.C. Harrison, and J.C. Wang, Structure and mechanism of DNA topoisomerase II. *Nature* 379:231, 1996.)

Figure 20.23 presents a model for how these two jaws could cooperate in the DNA segment-passing process. The upper jaw binds one DNA segment, called the **G-segment** because it will contain the gate through which the other segment will pass. Then, after activation by ATP, the upper jaws bind the other DNA segment, called the **T-segment** because it will be transported through the G-segment. The two segments are perpendicular to each other. The enzyme breaks the G-segment to form a gate, and the T-segment passes through into the lower gate, from which it is ejected.

SUMMARY One or more enzymes called helicases use ATP energy to separate the two parental DNA strands at the replicating fork. As helicase unwinds the two parental strands of a closed circular DNA, it introduces a compensating positive supercoiling

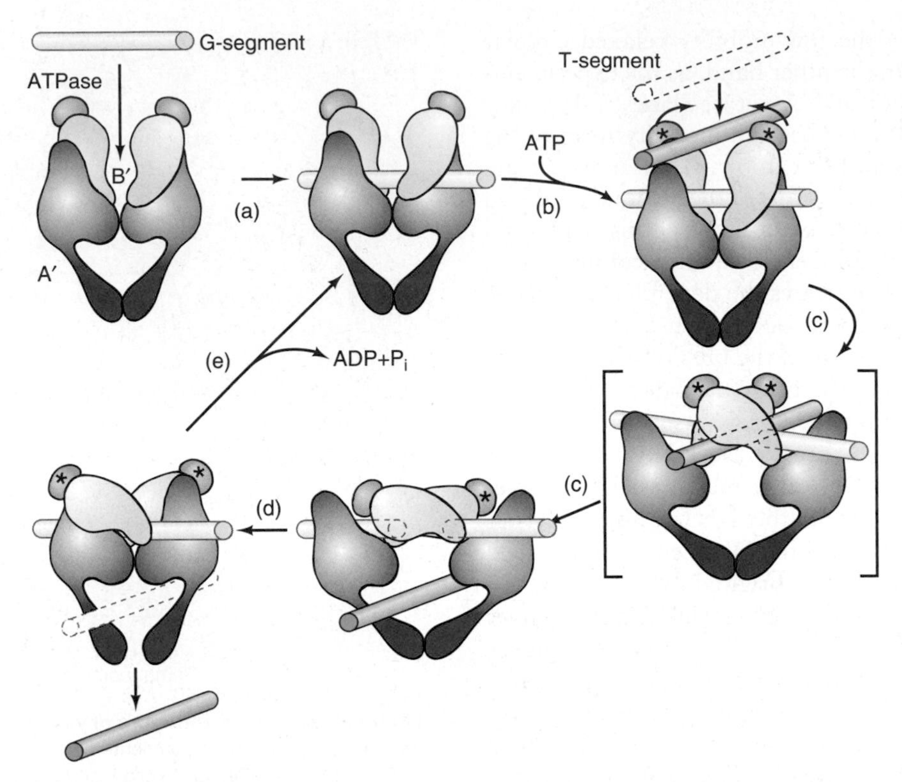

Figure 20.23 Model of the segment-passing step in the topoisomerase II reaction. Based on the crystal structure of the enzyme, and other evidence, Wang and colleagues proposed the following model: **(a)** The upper jaws of the enzyme open to bind the DNA G-segment (a double-stranded DNA), which is the one that will break to form a gate that will allow the other DNA segment to pass through. This binding of DNA induces a conformational change in the enzyme that brings the active-site tyrosines on the B′ domain into position to attack the DNA. **(b)** The ATPase domain of each upper jaw binds ATP (represented by an asterisk), and the upper jaw also binds the double-stranded DNA T-segment, which will be passed through the G-segment. **(c)** In a series of conformational changes, including a hypothetical intermediate (in brackets), the active site breaks the DNA G-segment, and allows the T-segment to pass through into the lower jaws. The front B′ domain during step (c) is transparent so the DNA behind it can be seen. **(d)** The lower jaws open to release the T-segment and the G-segment fragments are rejoined. **(e)** The enzyme hydrolyzes the bound ATP, returning the enzyme to a state in which it can accept another T-segment and repeat the segment-passing process. (*Source:* Adapted from Berger, J.M., S.J. Gamblin, S.C. Harrison, and J.C. Wang, Structure and mechanism of DNA topoisomerase II. *Nature* 379:231, 1996.)

force into the DNA. The stress of this force must be overcome or it will resist progression of the replicating fork. The name given to this stress-release mechanism is the swivel. DNA gyrase is the leading candidate for this role in *E. coli.* By pumping negative supercoils into the replicating DNA, DNA gyrase neutralizes the positive supercoils that would otherwise halt replication.

20.3 DNA Damage and Repair

DNA can be damaged in many different ways, and this damage, if left unrepaired, can lead to mutations: changes in the base sequence of a DNA. This distinction is worth emphasizing at the outset: *DNA damage is not the same as mutation, although it can lead to mutation.* DNA damage is simply a chemical alteration to DNA. A mutation is a change in a base pair. For example, the change from a G–C pair to an ethyl-G–C pair is DNA damage; the change from a G–C pair to any other natural base pair (A–T or T–A or C–G) is a mutation. If a particular kind of DNA damage is likely to lead to a mutation, we call it **genotoxic.** Indeed, we will see in the next section that the ethyl-G in our example is genotoxic because it is likely to mispair with T instead of C during DNA replication. If this happens, then another round of replication will place an A across from the mispaired T, and conversion of the normal G–C pair to an A–T pair (a true mutation) will be complete. Notice that this example illustrates the importance of DNA replication in conversion of DNA damage to mutation.

Let us look at two common examples of DNA damage: base modifications caused by alkylating agents and pyrimidine dimers caused by ultraviolet radiation. Then we will examine the mechanisms that bacterial and eukaryotic cells use to deal with such damage. Most of these mechanisms involve DNA replication.

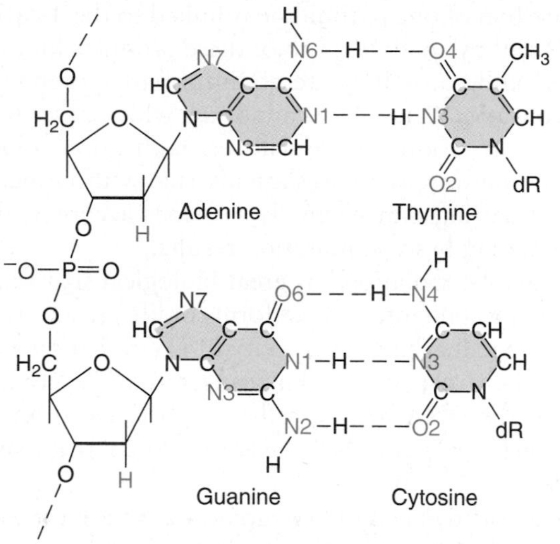

Figure 20.24 Electron-rich centers in DNA. The targets most commonly attacked by electrophiles are the phosphodiester bonds, N7 of guanine, and N3 of adenine (red); other targets are in blue.

Damage Caused by Alkylation of Bases

Some substances in our environment, both natural and synthetic, are *electrophilic,* meaning electron- (or negative charge-) loving. Thus, **electrophiles** seek centers of negative charge in other molecules and bind to them. Many other environmental substances are metabolized in the body to electrophilic compounds. One of the most obvious centers of negative charge in biology is the DNA molecule. Every nucleotide contains one full negative charge on the phosphodiester bond and partial negative charges on the bases. When electrophiles encounter these negative centers, they attack them, usually adding carbon-containing groups called *alkyl groups.* Thus, we refer to this process as **alkylation.**

Figure 20.24 shows the centers of negative charge in DNA. Aside from the phosphodiester bonds, the favorite sites of attack by alkylating agents are the N7 of guanine and the N3 of adenine, but many other targets are available, and different alkylating agents have different preferences for these targets.

What are the consequences of alkylations at these DNA sites? Consider the two predominant sites of alkylation, the N7 of guanine and the N3 of adenine. N7 alkylation of guanine does not change the base-pairing properties of the target base and is generally harmless. Alkylation of the N3 of adenine is more serious because it creates a base (e.g., 3-methyl adenine [3mA]) that cannot base-pair properly with any other base—a so-called **noncoding base.** Because a DNA polymerase does not recognize any base pair involving 3mA as correct, it stops at the 3mA damage, stalling DNA replication. Such blockage of DNA replication can kill a cell, so we say it is **cytotoxic.** On the other hand, as we will see later in this chapter, such stalled replication can be resumed without repairing the damage, but the mechanism of such resumption is error-prone and therefore leads to mutations.

Moreover, all of the nitrogen and oxygen atoms involved in base pairing (see Figure 20.24) are also subject to alkylation, which can directly disrupt base pairing and lead to mutation. The alkylation target that leads to most mutations is the O6 of guanine. Even though this atom is relatively rarely attacked by alkylating agents, such alkylations are very mutagenic because they allow the product to base-pair with thymine rather than cytosine. For example, consider the alkylation of the O6 of guanine by the common laboratory mutagen ethylmethane sulfonate (EMS), which transfers ethyl (CH_3CH_2) groups to DNA (Figure 20.25). The alkylation of the guanine O6 changes the tautomeric form (the pattern of double bonds) of the guanine so it base-pairs naturally with thymine. This leads to the replacement of a G–C base pair by an A–T base pair.

Many environmental **carcinogens,** or cancer-causing agents, are electrophiles that act by attacking DNA and alkylating it. As we have just seen, this can lead to mutations. If the mutations occur in genes that control or otherwise influence cell division, they can cause a cell to lose control over its replication and therefore change into a cancer cell.

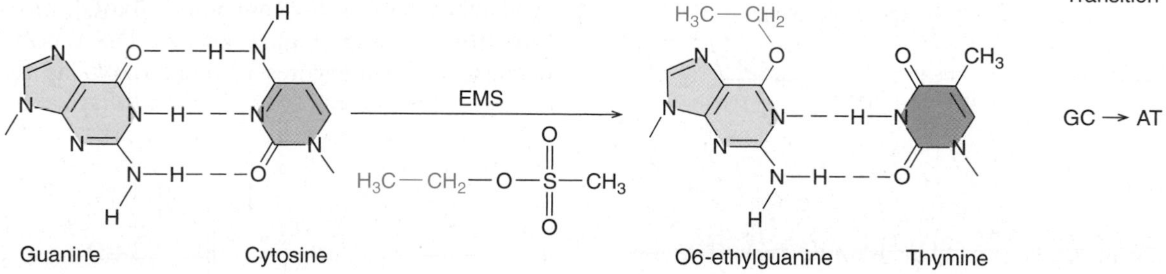

Figure 20.25 Alkylation of guanine by EMS. At the left is a normal guanine–cytosine base pair. Note the free O6 oxygen (red) on the guanine. Ethylmethane sulfonate (EMS) donates an ethyl group (blue) to the O6 oxygen, creating O6-ethylguanine (right), which base-pairs with thymine instead of cystosine. After one more round of replication, an A–T base pair will have replaced a G–C pair.

SUMMARY Alkylating agents like ethylmethane sulfonate add alkyl groups to bases. Some of these alkylations do not change base-pairing, so they are innocuous. Others cause DNA replication to stall, so they are cytotoxic, and can lead to mutations if the cell attempts to replicate its DNA without repairing the damage. Other alkylations change the base-pairing properties of a base, so they are mutagenic, and thus genotoxic.

Damage Caused by Ultraviolet Radiation

Ultraviolet (UV) radiation cross-links adjacent pyrimidines on the same DNA strand, forming two major lesions. Eighty to 90 percent of these are **pyrimidine dimers** (see Figure 20.26), which are also called **cyclobutane pyrimidine dimers (CPDs)** because of the four-member cyclobutane ring that forms between the two bases. Ten to 20 percent of the lesions are **(6-4) photoproducts**, in which the 6-carbon of one pyrimidine is linked to the 4-carbon of an adjacent pyrimidine. Both of these products block DNA replication because they are noninformative (non-coding): The replication machinery cannot tell which bases to insert opposite the lesion. As we will see, replication sometimes proceeds anyway, and bases are inserted without benefit of the base pairing that normally provides accuracy. If these are the wrong bases, a mutation results.

(a)
ACTTGC UV → ACT=TGC
TGAACG TGAACG

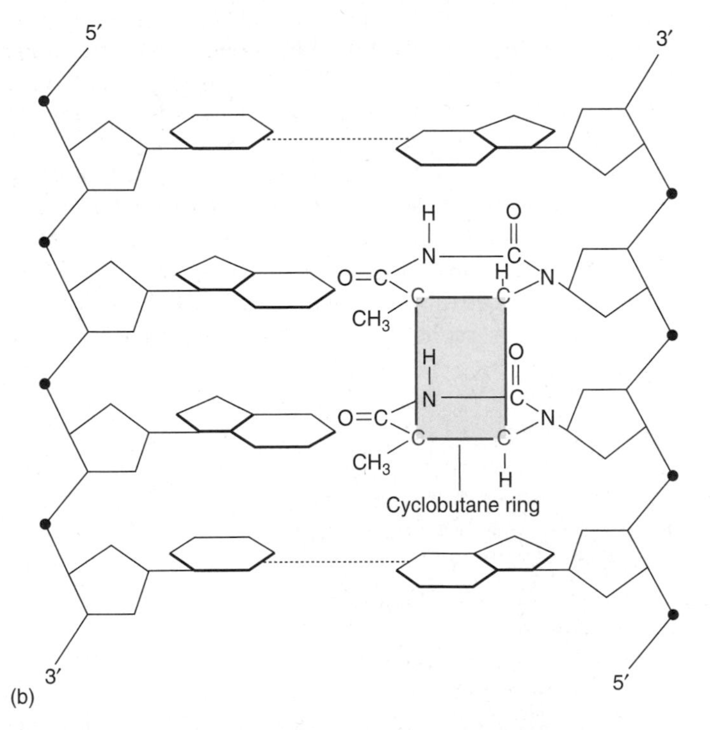

Figure 20.26 Pyrimidine dimers. (a) Ultraviolet light cross-links two pyrimidine bases (thymines in this case) on the top strand. This distorts the DNA so that these two noncoding bases no longer pair with their adenine partners. **(b)** The two bonds joining the two pyrimidines form a four-member cyclobutane ring (pink).

Ultraviolet radiation has great biological significance; it is present in sunlight, so most forms of life are exposed to it to some extent. The mutagenicity of UV radiation explains why sunlight can cause skin cancer: Its UV component damages the DNA in skin cells, which leads to mutations that sometimes cause those cells to lose control over their division.

Given the dangers of UV radiation, we are fortunate to have a shield—the ozone layer—in the earth's upper atmosphere to absorb the bulk of such radiation. However, scientists have noticed alarming holes in this protective shield—the most prominent one located over Antarctica. The causes of this ozone depletion are somewhat controversial, but they probably include the release of compounds traditionally used in air conditioners and in plastics into the atmosphere. Unless we can arrest the destruction of the ozone layer, we are destined to suffer more of the effects of UV radiation, including skin cancer.

Damage Caused by Gamma and X-Rays

The much more energetic **gamma rays** and **x-rays,** like UV rays, can interact directly with the DNA molecule. However, they cause most of their damage by ionizing the molecules, especially water, surrounding the DNA. This forms **free radicals,** chemical substances with an unpaired electron. These free radicals, especially those containing oxygen (e.g., $OH^{\cdot}$), are extremely reactive, and they immediately attack neighboring molecules. When such a free radical attacks a DNA molecule, it can change a base, or it can cause a single- or double-stranded break.

DNA bases are subject to at least 20 kinds of oxidative damage, and these can be caused by reactive oxygen species derived from ionizing radiation, or simply from normal oxidative metabolism. The best-studied oxidatively damaged DNA base is **8-oxoguanine (oxoG)**, also known as **8-hydroxyguanine** (Figure 20.27). DNA polymerases in bacteria and eukaryotes misread oxoG as thymine and

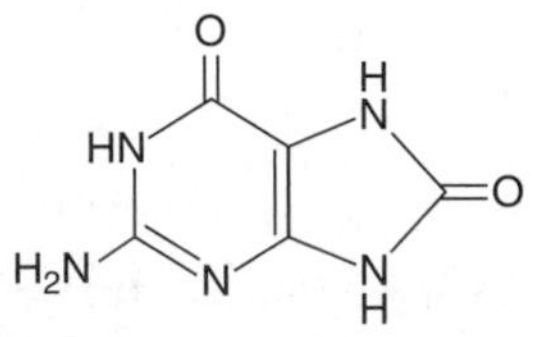

Figure 20.27 8-oxoguanine.

insert adenine instead of cytosine, resulting in an oxoG–A pair. Both bases in this pair are genotoxic because they both will probably lead to mutations if they are not removed before the DNA replicates again.

Single-stranded breaks are ordinarily not serious because they are easily repaired, just by rejoining the ends of the severed strand, but double-stranded breaks are very difficult to repair properly, so they frequently cause a lasting mutation. Because ionizing radiation can break chromosomes, it is referred to not only as a mutagen, or mutation-causing substance, but also as a **clastogen,** which means "breaker."

SUMMARY Different kinds of radiation cause different kinds of damage. Ultraviolet rays have comparatively low energy, and they cause a moderate type of damage: pyrimidine dimers. Gamma and x-rays are much more energetic. They ionize the molecules around DNA and form highly reactive free radicals that can attack DNA, altering bases or breaking strands.

Directly Undoing DNA Damage

One way to cope with DNA damage is to repair it, or restore it to its original, undamaged state. There are two basic ways to do this: (1) Directly undo the damage, or (2) remove the damaged section of DNA and fill it in with new, undamaged DNA. Let us begin by looking at two methods *E. coli* cells use to directly undo DNA damage.

In the late 1940s, Albert Kelner was trying to measure the effect of temperature on repair of ultraviolet damage to DNA in the bacterium *Streptomyces*. However, he noticed that damage was repaired much faster in some bacterial spores than in others kept at the same temperature. Obviously, some factor other than temperature was operating. Finally, Kelner noticed that the spores whose damage was repaired fastest were the ones kept most directly exposed to light from a laboratory window. When he performed control experiments with spores kept in the dark, he could detect no repair at all. Renato Dulbecco soon observed the same effect in bacteria infected with UV radiation-damaged phages. It now appears that most forms of life share this important mechanism of repair, which is termed **photoreactivation,** or **light repair.** However, placental mammals, including humans, do not have a photoreactivation pathway.

It was discovered in the late 1950s that photoreactivation is catalyzed by an enzyme called **photoreactivating enzyme** or **photolyase.** Actually, two separate enzymes catalyze the repair of CPDs and (6-4) photoproducts. The former is called **CPD photolyase,** or simply photolyase; the latter is known as **(6-4) photolyase.** The CPD photolyase

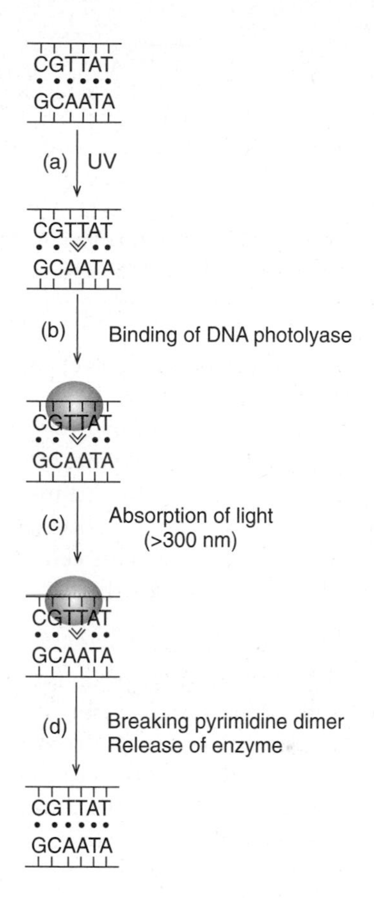

Figure 20.28 Model for photoreactivation. (a) Ultraviolet radiation causes a pyrimidine dimer to form. **(b)** The DNA photolyase enzyme (red) binds to this region of the DNA. **(c)** The enzyme absorbs near-UV to visible light. **(d)** The enzyme breaks the dimer and finally dissociates from the repaired DNA.

operates by the mechanism sketched in Figure 20.28. First, the enzyme detects and binds to the damaged DNA site (a pyrimidine dimer). Then the enzyme absorbs light in the UV-A to blue region of the spectrum, which activates it so it can break the bonds holding the pyrimidine dimer together. This restores the pyrimidines to their original independent state. Finally, the enzyme dissociates from the DNA; the damage is repaired.

Organisms ranging from *E. coli* to human beings can directly reverse another kind of damage, alkylation of the O6 of guanine. After DNA is methylated or ethylated, an enzyme called **O6-methylguanine methyltransferase** comes on the scene to repair the damage. It does this by accepting the methyl or ethyl group itself, as outlined in Figure 20.29.

The acceptor site on the enzyme for the alkyl group is the sulfur atom of a cysteine residue. Strictly speaking, this means that the methyltransferase does not fulfill one part of the definition of an enzyme—that it be regenerated unchanged after the reaction. Instead, this protein seems to be irreversibly inactivated, so we call it a "suicide enzyme" to denote the fact that it "dies" in performing its function.

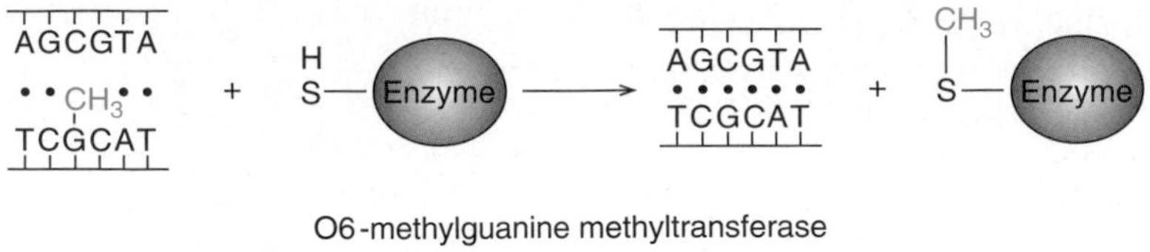

O6-methylguanine methyltransferase

Figure 20.29 Mechanism of O6-methylguanine methyltransferase. A sulfhydryl group of the enzyme accepts the methyl group (blue) from a guanine on the DNA, thus inactivating the enzyme.

The repair process is therefore expensive; each repair event costs one protein molecule.

One more property of the O6-methylguanine methyltransferase is worth noting. The enzyme, at least in *E. coli,* is induced by DNA alkylation. This means bacterial cells that have already been exposed to alkylating agents are more resistant to DNA damage than cells that have just been exposed to such mutagens for the first time.

SUMMARY Ultraviolet radiation damage to DNA (pyrimidine dimers) can be directly repaired by a DNA photolyase that uses energy from near-UV to blue light to break the bonds holding the two pyrimidines together. O6 alkylations on guanine residues can be directly reversed by the suicide enzyme O6-methylguanine methyltransferase, which accepts the alkyl group onto one of its amino acids.

Excision Repair

The percentage of DNA damage products that can be handled by direct reversal is necessarily small. Most such damage products involve neither pyrimidine dimers nor O6-alkylguanine, so they must be handled by a different mechanism. Most are removed by a process called **excision repair.** The damaged DNA is first removed, then replaced with fresh DNA, by one of two mechanisms: base excision repair or nucleotide excision repair. Base excision repair is more prevalent and usually works on common, relatively subtle changes to DNA bases, such as chemical modifications caused by cellular agents. Nucleotide excision repair generally deals with more drastic changes to bases, many of which distort the DNA double helix. These changes tend to be caused by mutagenic agents from outside of the cell. A good example of such damage is a pyrimidine dimer caused by UV light.

Base Excision Repair In **base excision repair (BER),** a damaged base is recognized by an enzyme called **DNA glycosylase,** which distorts the DNA in such a way as to extrude the damaged base out of its association with its base-paired partner, then breaks the **glycosidic bond** between the damaged base and its sugar (Figure 20.30). This leaves an **apurinic** or **apyrimidinic site (AP site),** which is a sugar without its purine or pyrimidine base. Once the AP site is created, it is recognized by an **AP endonuclease** that cuts, or **nicks,** the DNA strand on the 5′-side of the AP site. (The "endo" in endonuclease means the enzyme cuts inside a DNA strand, not at a free end; Greek *endo,* meaning within.) In *E. coli,* DNA phosphodiesterase removes the AP

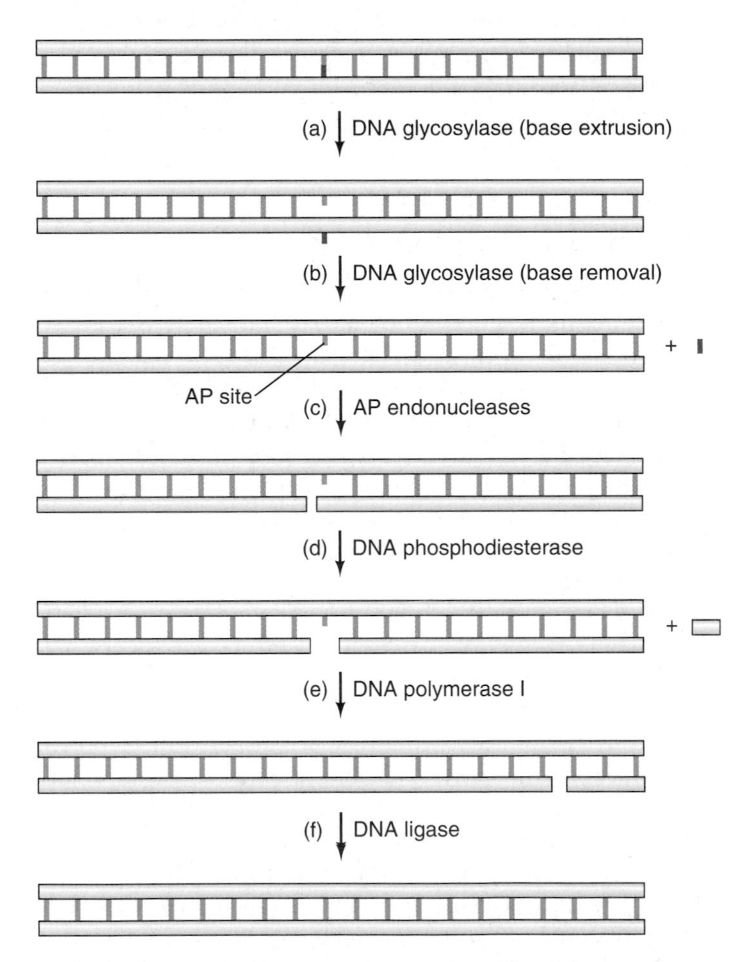

Figure 20.30 Base excision repair in *E. coli.* **(a)** DNA glycosylase extrudes the damaged base (red). **(b)** DNA glycosylase removes the extruded base, leaving an apurinic or apyrimidinic site on the bottom DNA strand. **(c)** An AP endonuclease cuts the DNA on the 5′-side of the AP site. **(d)** DNA phosphodiesterase removes the AP-deoxyribose phosphate (yellow block at right) that was left by the DNA glycosylase. **(e)** DNA polymerase I fills in the gap and continues repair synthesis for a few nucleotides downstream, degrading DNA and simultaneously replacing it. **(f)** DNA ligase seals the nick left by the DNA polymerase.

sugar phosphate, then DNA polymerase I performs repair synthesis by degrading DNA in the 5′→3′ direction, while filling in with new DNA. But DNA polymerase cannot repair nicks, so DNA ligase seals the remaining nick to complete the job. Many different DNA glycosylases have evolved to recognize different kinds of damaged bases. Humans have at least eight of these enzymes. Because subtle chemical modifications of bases frequently allow DNA replication, but still cause miscoding, BER is important in preventing mutations.

Most BER in eukaryotes proceeds by a pathway (Figure 20.31a–e), that is similar to BER in bacteria, except that there is no participation by a DNA phosphodiesterase. Instead, DNA polymerase β fills in the gap left after AP-site cleavage, and simultaneously removes the hanging sugar-phosphate flap (blue). But this scheme has a fundamental problem: Whereas DNA polymerase I in bacteria has a built-in editing activity, DNA polymerase β does not. It tends to make mistakes—about one every 4000 nt—and

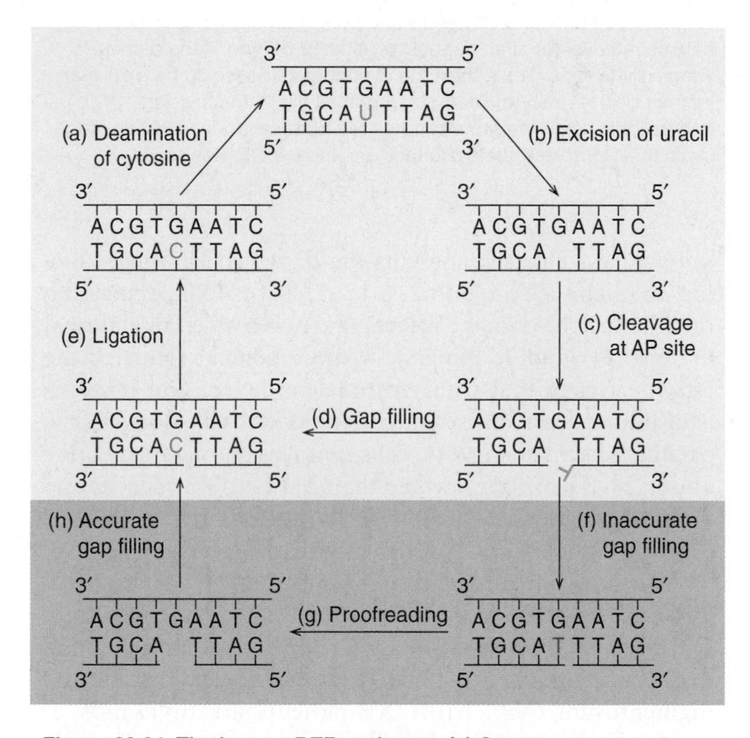

Figure 20.31 The human BER pathway. **(a)** Spontaneous cytosine deamination has converted a C (blue) to a U (orange) in the lower strand of the DNA. **(b)** A glycosylase removes the uracil. **(c)** APE1 cleaves on the 5′-side of the apyrimidinic site. **(d)** DNA polymerase β correctly fills in the gap with a C (blue) and simultaneously removes the hanging sugar-phosphate tag (green). **(e)** DNA ligase I seals the nick, returning the DNA to normal. **(f)** Occasionally, the DNA polymerase makes a mistake. This time it has incorporated a T (red) rather than a C, leaving a mismatch at the 3′-end of the fragment to the left of the nick. **(g)** APE1 uses its 3′-exonuclease to remove the mispaired T, again leaving a gap. **(h)** This time, DNA polymerase β correctly places a C (blue) across from the G. Now the mismatch is repaired, and the DNA just needs to be ligated to be back to normal. (*Source:* Adapted from Jiricny, J., An APE that proofreads. *Nature* 415 [2002] p. 593, f. 1.)

cannot repair them by itself. That may not sound so bad, but considering that between 20,000 and 80,000 damaged bases occur in our genomes every day, that error rate means that the BER system would introduce about 5–20 mutations into our genome daily.

Fortunately, eukaryotic cells have a solution for that problem. In 2002, Kai-Ming Chou and Yung-Chi Cheng showed that the human apurinic/apyrimidinic (AP) endonuclease (**APE1**) works in conjunction with the DNA polymerase β to edit the latter enzyme's mistakes. It had been known for years that APE1 had a 3′→5′ exonuclease in addition to its dominant endonuclease activity, but the exonuclease activity appeared to be too weak to be significant. Chou and Cheng showed that, although the 3′→5′ exonuclease activity is indeed weak on properly base-paired nucleotides, it is 50–150-fold stronger when faced with a terminal mispair, such as would occur after DNA polymerase β has performed inaccurate gap-filling (Figure 20.31f).

DNA ligase I is relatively inefficient at ligating two adjacent DNA strands when one of them has a mispair at the end, as in the structure after step f in Figure 20.31. In fact, its efficiency in ligating such substrates is less than 10%. If APE1 really does participate in repairing mispaired DNA created by DNA polymerase β, one would expect it to work with DNA ligase by repairing the mismatches and stimulating the efficiency of the ligase. Chou and Cheng used a reconstituted system with purified DNA ligase I, DNA polymerase β, and APE1 to demonstrate that APE1 stimulated the efficiency of ligation in a concentration-dependent manner from <10–95%. Thus, APE1 really does appear to be the enzyme that repairs mismatches introduced by DNA polmerase β.

A special case of base excision repair occurs when cells deal with 8-oxoguanine, which we encountered earlier in this chapter as a consequence of oxidative damage to DNA. Recall that oxoG tends to pair with A, forming oxoG–A base pairs, and that both bases in this pair are genotoxic because they both will probably take the wrong partner in the next round of replication, causing mutations. In humans, these mutations lead to cancer. But aerobic organisms have evolved mechanisms for dealing with both of these bases.

Gregory Verdine and colleagues elucidated the mechanism for dealing with the mispaired A in 2004. The enzyme responsible is an **adenine DNA glycosylase** called MutY in bacteria and hMYH in humans. It can remove an A that is mispaired with oxoG, but it leaves a correctly paired C alone. Moreover, it ignores all the A's that are correctly paired with T's. How does it make those distinctions? X-ray crystallography of a complex between MutY and model DNAs containing oxoG would shed considerable light on this problem, but those complexes were apparently too unstable to crystallize. So Verdine and colleagues formed a covalent disulfide bond between oxoG-containing oligonucleotides and MutY, and the complexes held together and formed crystals.

The crystal structure revealed close and specific contacts between the oxoG–A pair and the enzyme. Furthermore the adenine base is extruded, or "flipped out" such that it loses contact with its oxoG partner, and enters the active site of the enzyme. There, the glycosidic bond linking the adenine to the deoxyribose sugar is severed, and the adenine is thus removed from the DNA. By contrast, an ordinary T–A base pair does not make these close and specific contacts, so those base pairs are left alone. Furthermore, an oxoG–C pair makes the same contacts between the enzyme and the oxoG base as an oxoG–A pair does, but the cytosine base is not extruded, so it does not enter the enzyme's active site, and therefore is not removed.

What about removing the oxoG itself? That BER process is initiated by another DNA glycosylase, known as the **oxoG repair enzyme,** which cleaves the glycosidic bond linking oxoG to its deoxyribose. In humans, this enzyme is called hOGG1, and it can distinguish an oxoG–C pair from a normal G–C pair, extrude the oxoG out of its association with its C partner, and excise it.

SUMMARY Base excision repair (BER) typically acts on subtle base damage. This process begins with a DNA glycosylase, which extrudes a base in a damaged base pair, then clips out the damaged base, leaving an apurinic or apyrimidinic site that attracts the DNA repair enzymes that remove the remaining deoxyribose phosphate and replace it with a normal nucleotide. In bacteria, DNA polymerase I is the enzyme that fills in the missing nucleotide in BER; in eukaryotes, DNA polymerase β plays this role. However, this enzyme makes mistakes, and has no proofreading activity, so APE1 carries out the necessary proofreading. Repair of 8-oxoguanine sites in DNA is a special case of BER, that can happen in two ways. Since oxoG mispairs with A, the A can be removed after DNA replication by a specialized adenine DNA glycosylase. However, if replication has not yet occurred, the oxoG will still be paired with C, and the oxoG can be removed by another DNA glycosylase, the oxoG repair enzyme.

Nucleotide Excision Repair Bulky base damage, including pyrimidine dimers, can be removed directly, without help from a DNA glycosylase. In this **nucleotide excision repair** (**NER**) pathway (Figure 20.32), the incising enzyme system recognizes the strand with the bulky damage and makes cuts on either side of the damage, removing an oligonucleotide with the damage. The key enzyme *E. coli* cells use in this process is called the uvrABC endonuclease because it contains three polypeptides, the products of the *uvrA, uvrB,* and

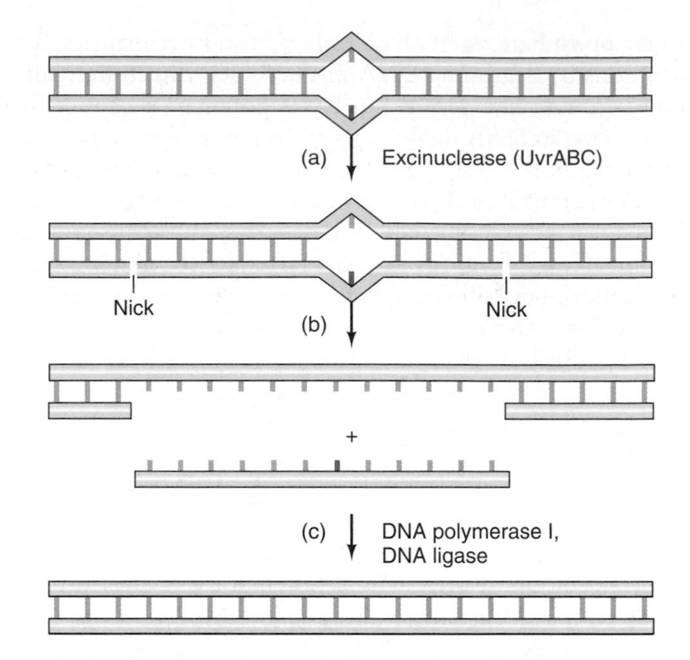

Figure 20.32 Nucleotide excision repair in *E. coli* (a) The UvrABC excinuclease cuts on either side of a bulky damaged base (red). This causes removal **(b)** of an oligonucleotide 12 nt long. If the damage were a pyrimidine dimer, then the oligonucleotide would be a 13-mer instead of a 12-mer. **(c)** DNA polymerase I fills in the missing nucleotides, using the top strand as template, and then DNA ligase seals the nick to complete the task, as in base excision repair.

uvrC genes. This enzyme cuts the damaged DNA, producing an oligonucleotide that is 12–13 bases long, depending on whether the damage affects one nucleotide (alkylations) or two (pyrimidine dimers). A more general term for the enzyme system that catalyzes nucleotide excision repair is excision nuclease, or **excinuclease.** As we will soon see, the excinuclease in eukaryotic cells removes an oligonucleotide about 24–32 nt long, rather than a 12- to 13-mer. In any case, DNA polymerase fills in the gap left by the excised oligonucleotide and DNA ligase seals the final nick.

Much of our information about repair mechanisms in humans has come from the study of congenital defects in DNA repair. These repair disorders cause a group of human diseases, including Cockayne's syndrome and **xeroderma pigmentosum (XP).** Most XP patients are thousands of times more likely to develop skin cancer than normal people if they are exposed to the sun. In fact, their skin can become literally freckled with skin cancers. However, if XP patients are kept out of sunlight, they suffer only normal incidence of skin cancer. Even if XP patients are exposed to sunlight, the parts of their skin that are shielded from light have essentially no cancers. These findings underscore the potency of sunlight as a mutating agent.

Why are XP patients so extraordinarily sensitive to sunlight? XP cells are defective in NER and therefore cannot repair helix-distorting DNA damage, including pyrimidine dimers, effectively. Thus, the damage persists and ultimately leads to mutations, which ultimately lead to cancer. Because NER is also responsible for repairing chemically induced DNA damage that is helix-distorting, we would expect XP patients to have a somewhat higher than average incidence of internal cancers caused by chemical mutagens, and they do. However, the incidence of such cancers in XP patients is only marginally higher than that in normal people. This suggests that most internal DNA damage in humans is not helix-distorting and we have an alternative pathway for correcting that milder kind of damage: the BER pathway. But we have no alternative pathway for correcting UV damage because we do not have a photoreactivation system.

Nucleotide excision repair takes two forms in eukaryotes. It can involve all lesions throughout the genome (**global genome NER**, or **GG-NER**), or it can be confined to the transcribed strands in genetically active regions of the genome (**transcription-coupled NER**, or **TC-NER**). The mechanisms of these two forms of NER share many aspects in common, but the method of recognition of the damage differs, as we will see. Let us examine both processes as they occur in humans.

Global Genome NER What repair steps are defective in XP cells? There are at least eight answers to this question. The problem has been investigated by fusing cells from different patients to see if the fused cells still show the defect. Frequently they do not; instead, the genes from two different patients complement each other. This probably means that a different gene was defective in each patient. So far, seven different complementation groups affecting excision repair have been identified this way. In addition, some patients have a variant form of XP (**XP-V**) in which excision repair is normal, and the patients' cells are only slightly more sensitive to UV light than normal cells are. We will discuss the gene responsible for XP-V later in this chapter. Taken together, these studies suggest that the defect can lie in any of at least eight different genes. Seven of these genes are responsible for excision repair, and they are named ***XPA–XPG.*** Most often, the first step in excision repair, incision, or cutting the affected DNA strand, seems to be defective.

The first step in human global genome NER (Figure 20.33) is the recognition of a distortion in the double helix caused by DNA damage. This is where the first XP protein (**XPC**) gets involved. XPC, together with another protein called hHR23B, recognizes a lesion in the DNA, binds to it, and causes melting of a small DNA region around the damage. This role in melting DNA is supported by in vitro studies performed in 1997 with templates that contain lesions surrounded by or adjacent to a small "bubble" of melted DNA. These templates do not require XPC, suggesting that this protein's job had already been performed when the DNA was melted. Also, Jan Hoeijmakers and colleagues used DNase footprinting in 1998 to show that XPC binds

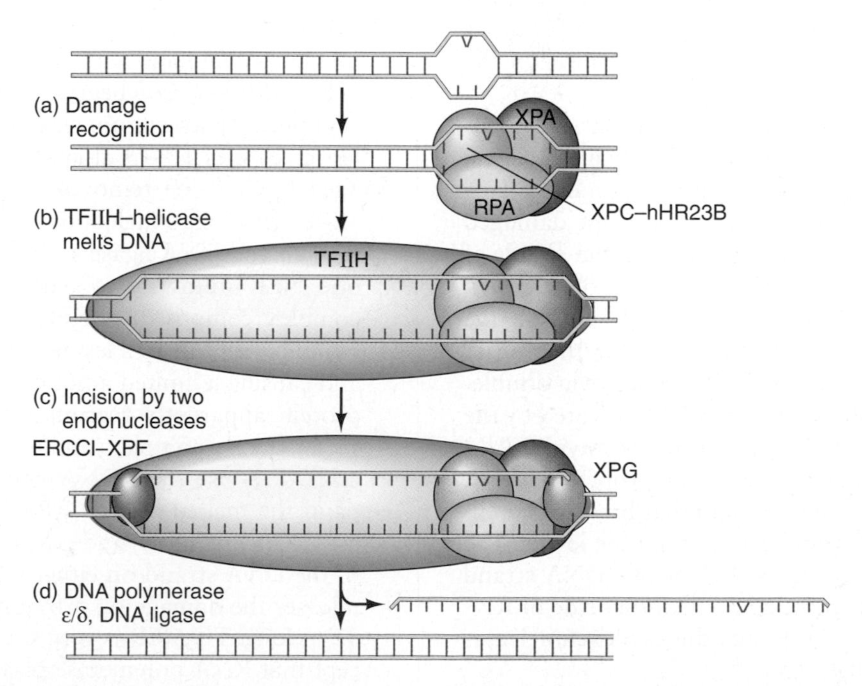

Figure 20.33 Human global genome NER. (a) In the damage recognition step, the XPC–hHR23B complex recognizes the damage (a pyrimidine dimer in this case), binds to it, and causes localized DNA melting. XPA also aids this process. RPA binds to the undamaged DNA strand across from the damage. **(b)** The DNA helicase activity of TFIIH causes increased DNA melting. **(c)** RPA helps position two endonucleases (the ERCC1–XPF complex and XPG) on either side of the damage, and these endonucleases clip the DNA. **(d)** With the damaged DNA removed on a fragment 24–32 nt long, DNA polymerase fills in the gap with good DNA and DNA ligase seals the final nick.

directly to a site of helix distortion in DNA and causes a change in the DNA's conformation (presumably a strand separation).

XPA, which has an affinity for damaged DNA, is also involved in an early stage of damage recognition. Because both XPC and XPA can bind to damaged DNA, why do we believe that XPC is the first factor on the scene? Competition studies performed by Hoeijmakers and colleagues, with different sized templates, support this hypothesis. These workers incubated XPC with one damaged template, and all the other factors *except* XPC with the other damaged template. Then they mixed the two together. Repair began first on the template that was originally incubated with XPC alone, suggesting that XPC binds first to the damaged DNA. Then what is the role of XPA? It can bind to many of the other factors involved in NER, so it may verify the presence of a DNA lesion in DNA that is already denatured (by XPC or by other means), and help to recruit the other NER factors.

At first, it may seem surprising to learn that two of the other XP genes—***XPB*** and ***XPD***—code for two subunits of the general transcription factor TFIIH, implicating this general transcription factor in NER. However, we now know that these two polypeptides have the DNA helicase activity inherent in TFIIH (Chapter 11). So one role of TFIIH is to enlarge the region of melted DNA around the damage. But TFIIH is required for NER in vitro even with damaged DNAs that have large melted regions, so this protein must have a function beyond providing DNA helicases. The fact that TFIIH interacts with a number of other NER factors suggests that it serves as an organizer of the NER complex.

The melting of the DNA by TFIIH attracts nucleases that nick one strand on either side of the damage, excising a 24–32-nt oligonucleotide that contains the damage. Two excinucleases make the cuts on either side of the damaged DNA. One is the ***XPG*** product, which cuts on the 3′-side of the damage. The other is a complex composed of a protein called *ERCC1* plus the ***XPF*** product, which cuts on the 5′-side. These nucleases are ideally suited for their task: They specifically cut DNA at the junction between double-stranded DNA and the single-stranded DNA created by the TFIIH around the damage. Another protein known as *RPA* helps position the two excinucleases for proper cleavage. RPA is a single-strand-binding protein that binds preferentially to the undamaged strand across from the lesion. The side of RPA facing toward the 3′-end of this DNA strand binds the ERCC1–XPF complex, and the other side of RPA binds XPG. This automatically puts the two excinucleases on the correct sides of the lesion.

Once the defective DNA is removed, DNA polymerase ε or δ fills in the gap, and DNA ligase seals the remaining nick. The role of **XPE** is not clear yet. It appears not to participate in NER, but it does bind to damaged DNA, so it is presumably involved somehow in DNA repair.

Transcription-Coupled NER Transcription-coupled NER uses all of the same factors as does global genome NER, except for XPC. Because XPC appears to be responsible for initial damage recognition and limited DNA melting in GG-NER, what plays these roles in TC-NER? The answer is RNA polymerase. When RNA polymerase encounters a distortion of the double helix caused by DNA damage, it stalls. This places the bubble of melted DNA, which is created by the polymerase, at the site of the lesion. At that point, XPA could recognize the lesion in the denatured DNA and recruit the other factors. From that point on, these factors would behave much as they do in GG-NER, enlarging the melted region, clipping the DNA in two places, and removing the piece of DNA containing the lesion.

Consider the usefulness of RNA polymerase as a DNA damage detector. It is constantly scanning the genome as it transcribes, and lesions block its passage, demanding attention. Lesions in parts of the DNA that are not transcribed (or even on the nontranscribed strand in a transcribed region) would not be detected this way, but they can wait longer to be repaired because they are not blocking gene expression. Thus, the fact that noncoding lesions such as pyrimidine dimers and 3mA block transcription as well as DNA replication is useful to the cell in that these lesions stall the transcribing polymerase, which recruits the repair machinery.

SUMMARY Nucleotide excision repair typically handles bulky damage that distorts the DNA double helix. NER in *E. coli* begins when the damaged DNA is clipped by an endonuclease on either side of the lesion, at sites 12–13 nt apart. This allows the damaged DNA to be removed as part of the resulting 12–13-base oligonucleotide. DNA polymerase I fills the gap and DNA ligase seals the final nick. Eukaryotic NER follows two pathways. In GG-NER, a complex composed of XPC and hHR23B initiates repair by binding to a lesion anywhere in the genome and causing a limited amount of DNA melting. This protein apparently recruits XPA and RPA. TFIIH then joins the complex, and two of its subunits (XPB and XPD) use their DNA helicase activities to expand the melted region. RPA binds two excinucleases (XPF and XPG) and positions them for cleavage of the DNA strand on either side of the lesion. This releases the damage on a fragment between 24 and 32 nt long. TC-NER is very similar to GG-NER, except that RNA polymerase plays the role of XPC in damage sensing and initial DNA melting. In either kind of NER, DNA polymerase ε or δ fills in the gap left by the removal of the damaged fragment, and DNA ligase seals the DNA.

Double-Strand Break Repair in Eukaryotes

Double-strand breaks in eukaryotes are probably the most dangerous form of DNA damage. They are really broken chromosomes, and if they are not repaired, they can lead to cell death or, in vertebrates, to cancer. Eukaryotic cells deal with double-strand breaks in DNA (**DSBs**) in two ways: First, they can use homologous recombination, with the unbroken sister chromatid as the recombining partner. This mechanism is similar to recombination repair in bacteria, discussed later in this chapter, except that both strands must participate in recombination. Second, eukaryotic cells can use **nonhomologous end-joining (NHEJ)**. In replicating cells in S and G2 phases, homologous recombination is the dominant mechanism, because only one DNA copy is broken and the other is available to align the breaks properly. Yeast cells, which divide frequently, rely primarily on homologous recombination to repair their double-strand breaks. On the other hand, mammalian cells in G1 phase preferentially use nonhomologous end-joining because the DNA has not replicated and no second, homologous chromosome is yet available to serve as a template for repair. In this section, we will focus on the latter mechanism.

Nonhomologous End-Joining J. Phillips and W. Morgan investigated nonhomologous end-joining in 1994 by introducing a restriction endonuclease into Chinese hamster ovary cells. This enzyme made double-stranded cuts in chromosomes, including a site within the adenine phosphoribosyltransferase (*APRT*) gene, which was present in only one copy in these cells. Then these workers looked for viable cells with mutations in the *APRT* gene and sequenced the mutated genes to see what had happened during the rejoining process. They found mostly short insertions and deletions of DNA around the cleavage site. Furthermore, these insertions and deletions appeared to have been directed by microhomology—small areas of homology (1–6 bp)—in the DNA ends. Figure 20.34 shows a model for nonhomologous end-joining that explains these and other findings.

First, the DNA ends attract **Ku,** a dimer of two polypeptides (Ku70 [M_r = 69 kD] and Ku80 [M_r = 83 kD]). One of the important functions of this protein is to protect the DNA ends from degradation until end-joining is complete. Ku has DNA-dependent ATPase activity and is the regulatory subunit for **DNA protein kinase (DNA-PK)**, whose catalytic subunit is known as **DNA-PK$_{cs}$**. X-ray crystallography studies have shown that Ku binds to DNA ends like a ring on a finger. Its two subunits form a ring that is lined with basic amino acids, which help it bind to acidic DNA.

Once Ku has bound to a DNA end, it can recruit the DNA-PK$_{cs}$ and perhaps other proteins, completing the DNA-PK complex. The protein complexes on each DNA end have binding sites, not only for the DNA ends, but also for double-stranded DNA adjacent to the ends. Thus, these DNA-PK complexes, by binding to the other DNA

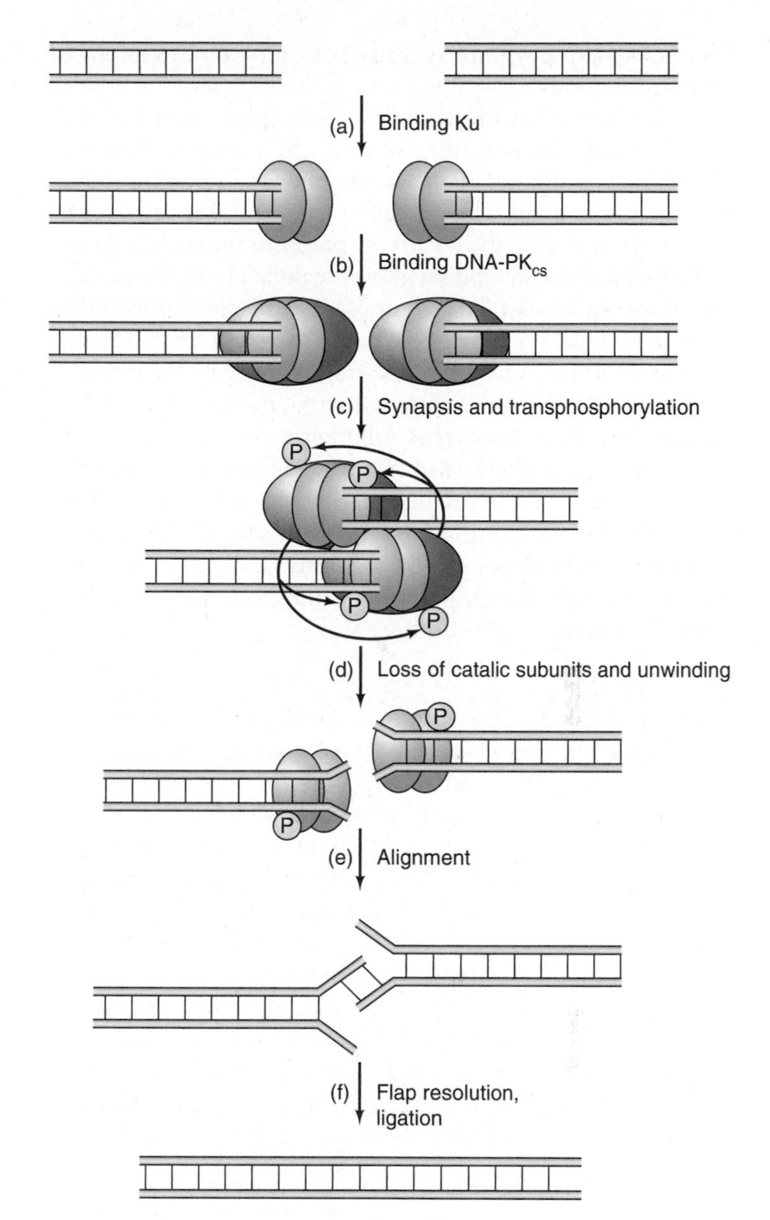

Figure 20.34 Model for nonhomologous end-joining. (a) Free DNA ends attract Ku (blue), which protects them from degradation. **(b)** Ku attracts DNA-PK$_{cs}$ (red), constituting the full DNA-PK complex. **(c)** The DNA-PK complexes promote synapsis, or lining up of regions of microhomology near the DNA ends. The two DNA-PK complexes phosphorylate each other on both the regulatory (Ku) and catalytic subunits. **(d)** The phosphorylation from step (c) has two effects: (1) The phosphorylated catalytic subunits dissociate from the complex. (2) Phosphorylation activates the DNA helicase activity of Ku, which unwinds the two DNA ends. The phosophorylation of Ku activates its DNA helicase activity, which unwinds the DNA of the two ends. **(e)** Regions of microhomology in the two ends base-pair with each other in the alignment step. **(f)** Flap resolution removes extra flaps of DNA, and fills in gaps. Finally, DNA ligase joins the ends of the DNA strands together permanently.

(*Source:* Adapted from Chu, G., Double strand break repair. *Journal of Biological Chemistry* 272 [1997] p. 24099, f. 4.)

fragment, can promote synapsis, or lining up of regions of microhomology.

The two DNA-PK complexes also phosphorylate each other, which has two effects: First, the phosphorylation of DNA-PK_{cs} promotes dissociation of that catalytic subunit, whose job is done. The phosphorylation of Ku activates its DNA helicase activity, so it can promote unwinding of the DNA ends. This unwinding allows regions of microhomology to base-pair, leaving flaps composed of the ends of the other, nonpairing strands. Finally, the flaps are removed by nucleases, gaps are filled in, and the DNA strands are ligated together.

When the flaps are removed a few nucleotides of DNA are lost, but this process is inherently inaccurate, and nucleotides can also be added. We will encounter nonhomologous end-joining again in Chapter 23 when we discuss recombination of antibody genes. This scheme deliberately introduces double-strand breaks into DNA and then rearranges the DNA fragments by joining selected free DNA ends in a process that requires Ku.

SUMMARY Double-strand DNA breaks in mammals can be repaired by homologous recombination or by nonhomologous end joining. The latter process requires Ku and DNA-PK_{cs}, which bind together at the DNA ends, constituting active DNA-PK complexes that allow the ends to find regions of microhomology with each other. Once the regions of microhomology line up, the two DNA-PK complexes phosphorylate each other. This phosphorylation activates the catalytic subunit (DNA-PK_{cs}) to dissociate, and it also activates the DNA helicase activity of Ku to unwind the DNA ends so the microhomology regions can base-pair. Finally, extra flaps of DNA are removed, gaps are filled, and the DNA ends are ligated permanently together.

The Role of Chromatin Remodeling in Double-Stranded Break Repair We learned in Chapter 13 that nucleosomes can block association of gene control regions with transcription factors, and therefore that chromatin remodeling is required for activation of eukaryotic genes. By the same token, it seems reasonable to expect that nucleosomes would block association between damaged DNA and repair factors, and therefore that chromatin remodeling would be required for DNA repair. Indeed, work in 2004 by Susan Gasser and colleagues and by Xuetong Shen and colleagues showed that double-stranded chromosome break (DSB) repair in yeast, which is accomplished primarily by homologous recombination, depends on a chromatin remodeling complex known as **INO80.**

INO80, a member of the SWI/SNF family of chromatin remodelers (Chapter 13), is composed of 12 polypeptides, including the *ino80* gene product **Ino80.** This polypeptide has the ATPase/translocase domain characteristic of chromatin remodeling proteins. Mutations in *ino80* block both transcription and DSB repair, presumably because of chromatin remodeling defects in both cases.

Both groups of investigators induced a unique double-stranded break at a defined site at the *MAT* locus in yeast chromatin, then used chromatin immunoprecipitation (ChIP, Chapter 13) to measure recruitment of proteins to the break. INO80 appeared at the break within 30–60 min, suggesting that it is involved in DSB repair. The next question concerns the other proteins that are required to recruit INO80. One clue to the answer is that two yeast protein kinases, Mec1 and Tel1, were already known to phosphorylate serine 129 of histone H2A on nucleosomes near DSBs, and that replacement of serine 129 with alanine renders yeast cells sensitive to radiation and chemicals that damage DNA. Because alanine, unlike serine, cannot be phosphorylated, this finding indicates that phosphorylation of serine 129 on histone H2A promotes DSB repair.

Moreover, both groups showed that mutations in the genes encoding Mec1 and Tel1, or mutations that changed serine 129 to alanine, inhibited recruitment of INO80 to DSBs. These findings suggested a direct interaction between phosphorylated H2A and INO80. Indeed, Shen and colleagues showed that INO80 co-purified with phosphorylated H2A and other histones, but not with unphosphorylated H2A.

What roles does INO80 play in DSB repair? Gasser and colleagues showed that yeast strains with mutations in genes encoding the subunits of INO80, or mutations that changed serine 129 of histone H2A, do not form the 3′-single-stranded overhangs at the broken ends of chromosomes with DSBs. Thus, formation of these essential overhangs appears to be one of the functions of INO80, and it could help in this process by sliding nucleosomes away from the broken ends.

A suggestion for how INO80 could perform this remodeling comes from the finding that INO80 contains two ATPases, Rvb1 and Rvb2, that are similar to RuvB, a protein involved in recombination and DSB repair in *E. coli.* RuvB is composed of two cyclic hexamers of identical subunits (Chapter 22) and it uses its DNA helicase activity to drive "branch migration," the sliding of a branch connecting two recombining DNA duplexes. Similarly, Rvb1/Rvb2 has DNA helicase activity, and the human homolog has been proposed to have a double hexamer structure, although the yeast protein appears to be a single heterohexamer. Because a DNA helicase tracks along a DNA duplex as it unwinds the DNA, it is possible to imagine that INO80 uses its DNA helicase activity to nudge aside nucleosomes as it tracks along the DNA, pushing the nucleosomes away from a DSB.

Another chromatin remodeler, **SWR1**, is also recruited to DSBs. Like INO80, SWR1 contains Rvb1/Rvb2, but it has an additional intriguing activity: the ability to replace histone H2A with the H2A variant Htz1. Thus, SWR1 might replace phospho-H2A with Htz1, which cannot be phosphorylated. In this way, SWR1 would return the histone phosphorylation in nucleosomes near DSBs to the pre-broken state once DSB repair is at least underway. In support of this hypothesis, Jerry Workman and colleagues have shown that Domino/p400, the *Drosophila* homolog of SWR1, replaces phospho-H2A with unphosphorylated H2A in vitro.

Another chromatin remodeler recruited to double-strand breaks, and other sites of DNA damage, is ALC1 (amplified in liver cancer). This protein contains a **macrodomain**, which binds specifically to **poly(ADP-ribose)** (Chapter 13) that is formed at the sites of DNA damage by **poly(ADP-ribose) polymerase** (**PARP-1**). This binding to poly(ADP-ribose) also stimulates the remodeling activity of ALC1. A histone H2A variant known as macroH2A1.1 also has a macrodomain, and is also attracted to poly(ADP-ribose) at sites of damaged DNA. The substitution of macroH2A1.1 for ordinary H2A may facilitate the remodeling catalyzed by ALC1, or other chromatin remodelers. Assuming that this remodeling aids in DNA repair, it appears that PARP-1 plays a role in DNA repair. The fact that PARP-1 inhibitors are highly toxic to cells defective in homologous recombination repair supports this hypothesis. So does the fact that cells with excessive DNA damage have hyperactive PARP-1.

Both of these findings have important clinical implications. Cancer cells, especially breast cancer cells with impaired homologous recombination repair due to faulty BRCA1 and BRCA2 genes, are readily killed by PARP-1 inhibitors. And heart and brain cells can have their DNA damaged by the oxidative stress of a cut-off blood supply (ischemia) due to heart attack or stroke, respectively; the sudden return of oxygen-rich blood (reperfusion) can result in hyperactive PARP-1 in these cells. This is good for repairing the DNA, but making so much poly(ADP-ribose) depletes the ATP stores of the cells, which can rapidly kill them. PARP-1 inhibitors could protect such cells.

SUMMARY Two protein kinases, Mec1 and Tel1, are recruited to DSBs, where they phosphorylate serine 129 of histone H2A in nearby nucleosomes. This phosphorylation recruits the chromatin remodeler INO80 to the DSB, where it appears to use its DNA helicase activity to push nucleosomes away from the ends of the DSB, enabling formation of single-stranded 3′-DNA overhangs, which are essential for both nonhomologous end-joining and homologous recombination. Another chromatin remodeler known as SWR1, which shares many components with INO80, also appears at DSBs, and replaces phospho-H2A with the H2A variant Htz1, which cannot be phosphorylated. This returns the phosphorylation state of H2A on nucleosomes near DSBs to normal. PARP-1 is recruited to DSBs and other damaged DNA sites. It poly(ADP-ribosyl)ates itself and other proteins at the damage site, which recruits chromatin remodelers such as ALC1 and the histone variant macroH2A1.1, both via their macrodomains.

Mismatch Repair

So far, we have been discussing repair of DNA damage caused by mutagenic agents. What about DNA that simply has a mismatch due to incorporation of the wrong base and failure of the proofreading system? At first, it would seem tricky to repair such a mistake because of the apparent difficulty in determining which strand is the newly synthesized one that has the mistake and which is the parental one that should be left alone. At least in *E. coli* this is not a problem because the parental strand has identification tags that distinguish it from the progeny strand. These tags are methylated adenines, created by a methylating enzyme that recognizes the sequence GATC and places a methyl group on the A. Because this 4-base sequence occurs approximately every 250 bp, one is usually not far from a newly created mismatch.

Moreover, GATC is a palindrome, so the opposite strand also reads GATC in its 5′→3′ direction. This means that a newly synthesized strand across from a methylated GATC is also destined to become methylated, but a little time elapses before that can happen. The **mismatch repair** system (Figure 20.35) takes advantage of this delay; it uses the methylation on the parental strand as a signal to leave that strand alone and correct the nearby mismatch in the unmethylated progeny strand. This process must occur fairly soon after the mismatch is created, or both strands will be methylated and no distinction between them will be possible. Eukaryotic mismatch repair is not as well understood as that in *E. coli.* The genes encoding the mismatch recognition and excision enzymes (MutS and MutL) are very well conserved, so the mechanisms that depend on these enzymes are likely to be similar in eukaryotes and bacteria. However, the gene encoding the strand recognition protein (MutH) is not found in eukaryotes, so eukaryotes appear not to use the methylation recognition trick. It is not clear yet how eukaryotic cells distinguish the progeny strand from the parental strand at a mismatch.

SUMMARY The *E. coli* mismatch repair system recognizes the parental strand by its methylated adenines in GATC sequences. Then it corrects the mismatch in the complementary (progeny) strand. Eukaryotes use part of this repair system, but they rely on a different, uncharacterized method for distinguishing the strands at a mismatch.

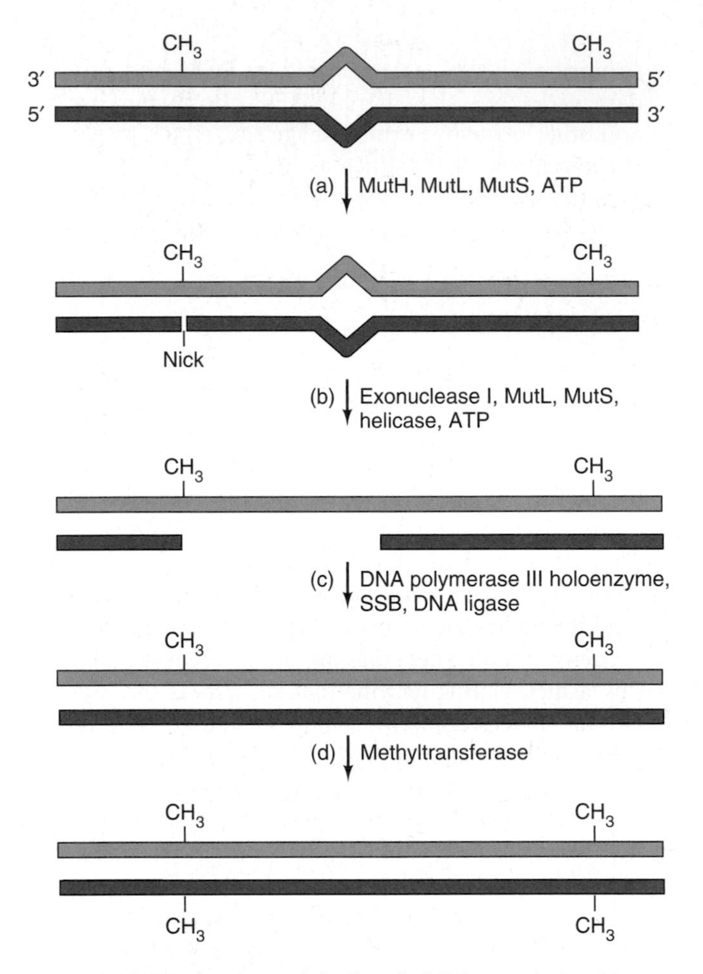

Figure 20.35 Mismatch repair in *E. coli.* **(a)** The products of the *mutH, L,* and *S* genes along with ATP, recognize a base mismatch (center), identify the newly synthesized strand by the absence of methyl groups on GATC sequences, and introduce a nick into that new strand, across from a methylated GATC and upstream of the incorrect nucleotide. **(b)** Exonuclease I, along with MutL, MutS, DNA helicase, and ATP, removes DNA downstream of the nick, including the incorrect nucleotide. **(c)** DNA polymerase III holoenzyme, with help from single-stranded binding protein (SSB), fills in the gap left by the exonuclease, and DNA ligase seals the remaining nick. **(d)** A methyltransferase methylates GATC sequences in the progeny strand across from methylated GATC sequences in the parental strand. Once this happens, mismatch repair nearby cannot occur because the progeny and parental strands are indistinguishable.

Failure of Mismatch Repair in Humans

Failure of human mismatch repair has serious consequences, including cancer. One of the most common forms of hereditary cancer is **hereditary nonpolyposis colon cancer (HNPCC)**, also known as Lynch syndrome. Approximately 1 American in 200 is affected by this disease, and it accounts for about 15% of all colon cancers. One of the characteristics of HNPCC patients is microsatellite instability, which means that DNA **microsatellites,** tandem repeats of 1–4-bp sequences, change in size (number of repeats) during the patient's lifetime. This is unusual; the number of repeats in a given microsatellite may differ from one normal individual to another, but it should be the same in all tissues and remain constant throughout an individual's lifetime. The relationship between microsatellite instability and mismatch repair is that the mismatch repair system is responsible for recognizing and repairing the "bubble" created by the inaccurate insertion of too many or too few copies of a short repeat because of "slippage" during DNA replication. When this system breaks down, such slippage goes unrepaired, leading to mutations in many genes whenever DNA replicates in preparation for cell division. This kind of genetic instability presumably leads to cancer, by mechanisms involving mutated genes (oncogenes and tumor suppressor genes) that are responsible for control of cell division.

SUMMARY The failure of human mismatch repair leads to microsatellite instability, and ultimately to cancer.

Coping with DNA Damage Without Repairing It

The direct reversal and excision repair mechanisms described so far are all true repair processes. They eliminate the defective DNA entirely. However, cells have other means of coping with damage that do not remove it but simply skirt around it. These are sometimes called repair mechanisms, even though they really are not. A better term might be *damage bypass mechanism.* These mechanisms come into play when a cell has not performed true repair of a lesion, but has either replicated its DNA or both replicated its DNA and divided before repairing the lesion. At each of these steps (DNA replication and cell division), the cell loses attractive options for dealing with DNA damage and is increasingly faced with more dangerous options.

Recombination Repair **Recombination repair** is the most important of these mechanisms. It is also sometimes called **postreplication repair** because replication past a pyrimidine dimer can leave a problem: a gap opposite the dimer that must be repaired. Excision repair will not work any longer because there is no undamaged DNA opposite the dimer—only a gap—so recombination repair is one of the few alternatives left. Figure 20.36 shows how recombination repair works. First, the DNA is replicated. This creates a problem for DNA with pyrimidine dimers because the dimers stop the replication machinery. Nevertheless, after a pause, replication continues, leaving a gap (a **daughter strand gap**) across from the dimer. (A new primer is presumably required to restart DNA synthesis.) Next, recombination occurs between the gapped strand and its homolog

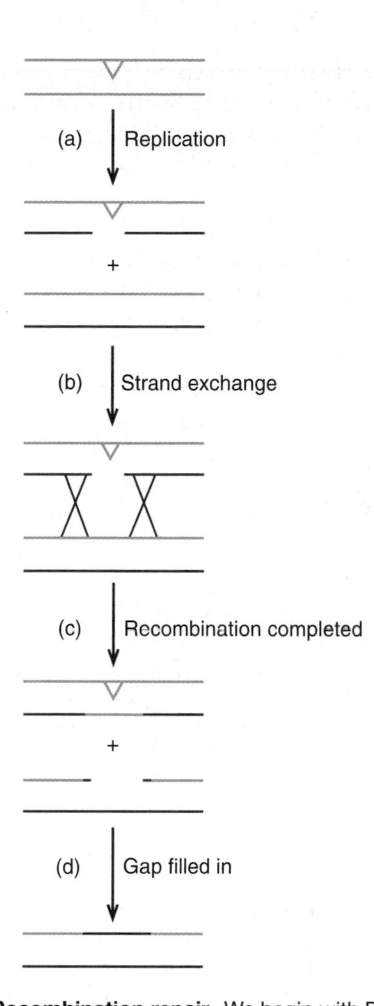

Figure 20.36 Recombination repair. We begin with DNA with a pyrimidine dimer, represented by a V shape. **(a)** During replication, the replication machinery skips over the region with the dimer, leaving a gap; the complementary strand is replicated normally. The two newly synthesized strands are shown in pink. **(b)** Strand exchange between homologous strands occurs. **(c)** Recombination is completed, filling in the gap opposite the pyrimidine dimer, but leaving a gap in the other daughter duplex. The duplex with the pyrimidine dimer has not been repaired, but it has replicated successfully and may be repaired properly in the next generation. **(d)** This last gap is easily filled, using the normal complementary strand as the template.

on the other daughter DNA duplex. This recombination depends on the *recA* gene product, which exchanges the homologous DNA strands. We have encountered *recA* before in our discussion of the induction of a λ prophage during the SOS response (Chapter 8)—and we will discuss it more fully in our consideration of recombination in Chapter 22. The net effect of this recombination is to fill in the gap across from the pyrimidine dimer and to create a new gap in the other DNA duplex. However, because the other duplex has no dimer, the gap can easily be filled in by DNA polymerase and ligase. Note that the DNA damage still exists, but the cell has at least managed to replicate its DNA. Sooner or later, true DNA repair could presumably occur.

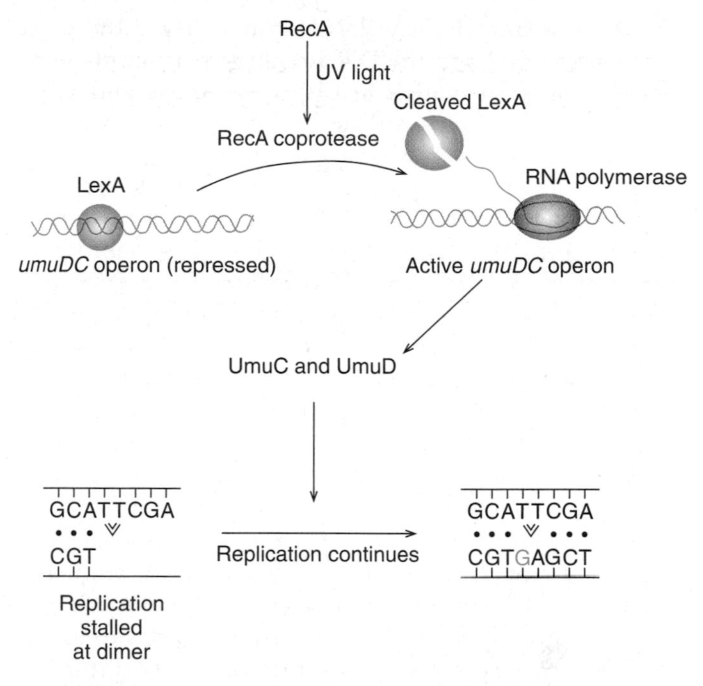

Figure 20.37 Error-prone (SOS) bypass. Ultraviolet light activates the RecA coprotease, which stimulates the LexA protein (purple) to cleave itself, releasing it from the *umuDC* operon. This results in synthesis of UmuC and UmuD proteins, which allow DNA synthesis across from a pyrimidine dimer, even though mistakes (blue) will frequently be made.

Error-Prone Bypass So-called **error-prone bypass** is another way of dealing with damage without really repairing it. In *E. coli,* this pathway is induced as part of the SOS response by DNA damage, including UV damage, and depends on the product of the *recA* gene. The chain of events seems to be as follows (Figure 20.37): UV light or another mutagenic treatment somehow activates the RecA coprotease activity. This coprotease has several targets. One we have studied already is the λ repressor, but its main target is the product of the *lexA* gene. This product, **LexA,** is a repressor for many genes, including repair genes; when it is stimulated by RecA coprotease to cleave itself, all these genes are induced.

Two of the newly induced genes are ***umuC*** and ***umuD,*** which make up a single operon (***umuDC***). The product of the *umuD* gene (**UmuD**) is clipped by a protease to form **UmuD′,** which associates with the *umuC* product, **UmuC,** to form a complex **UmuD′$_2$C.** This complex has DNA polymerase activity, so it is also referred to as **DNA pol V.** Pol V can cause error-prone bypass of DNA lesions in vitro on its own, but it is activated by RecA-ATP. This RecA-ATP comes from the 3′-end of a nucleoprotein filament of RecA and DNA (RecA*), which may have assembled at a site remote from the site of error-prone bypass. Such bypass involves replication of DNA across from the DNA lesion even though correct "reading" of the lesion itself is impossible. This avoids leaving a gap, but it frequently puts the wrong bases into the new DNA strand (hence the name "error-prone"). When the DNA replicates again, these errors will be perpetuated. Error-prone bypass and other, more error-free bypass mechanisms found in eukaryotes, are also called **translesion synthesis (TLS).**

DNA polymerase V can efficiently bypass the three most common types of DNA lesion: pyrimidine dimers, related lesions also caused by UV light—(6-4) photoproducts, and abasic (AP) sites. However, this enzyme performs this translesion synthesis with varying degrees of fidelity. In 2000, Myron Goodman and colleagues measured the incorporation of A and G across from the two T's of a thymine dimer, or of a (6-4) photoproduct, and across from an AP site. Opposite a pyrimidine dimer, DNA polymerase V tended to incorporate A's in both positions, which is fine for thymine dimers, but not if the dimer contains cytosines. Opposite a (6-4) photoproduct containing two thymines, DNA polymerase V tended to incorporate a G in the first position and an A in the second—obviously not very faithful replication. Opposite an AP site, DNA polymerase V incorporated about two-thirds A and about one-third G. All of these ratios, and the fact that pyrimidines were not detectably incorporated, agree with in vivo observations, suggesting that DNA polymerase V is indeed the enzyme that performs translesion synthesis in vivo.

If the *umu* genes are really responsible for error-prone bypass, we might expect mutations in one of these genes to make *E. coli* cells *less* susceptible to mutation. These mutant cells would be just as prone to DNA damage, but the damage would not be as readily converted into mutations. In 1981, Graham Walker and colleagues verified this expectation by creating a **null allele** of the *umuC* gene (a version of the gene with no activity), and showing that bacteria harboring this gene were essentially unmutable. In fact, "umu" stands for "unmutable."

These workers established an *E. coli* strain carrying the *umuC* mutant, and a his^- mutation that is ordinarily revertable by UV radiation. Then they challenged this bacterial strain with UV radiation and counted the his^+ revertants. The more revertants, the more mutation was allowed because a reversion is just a back-mutation. Figure 20.38 shows the results. A reasonable number of revertants occurred in wild-type cells (about 200 at the highest UV dose). By stark contrast, in $umuC^-$ cells almost no revertants occur. Furthermore, addition of a plasmid bearing the *muc* gene, which can suppress the unmutable phenotype of $umuC^-$ cells, caused a dramatic increase in the number of revertants (about 500, even at a relatively low UV dose).

The null allele in this experiment was created by insertion of the *lac* structural genes, without the *lac* promoter, into the *umuC* gene, then screening for lac^+ cells. The cells were originally lac^-, so the appearance of lac^+ cells indicated that the *lac* genes had inserted downstream of a promoter—the *umuDC* promoter, in this case. The fact that the *lac* genes fell under control of the *umuDC* promoter allowed Walker and colleagues to test the inducibility of this promoter by UV radiation, simply by measuring β-galactosidase activity. Figure 20.39 shows that the promoter was indeed inducible by UV radiation at a dose of 10 J/m^2 (blue curve). But the promoter was not inducible in *lexA* mutant or $recA^-$ cells (green and red curves). The *lexA* mutant cells used in this experiment encoded a LexA protein that was not cleavable and therefore could not be removed from the *umuDC* operator.

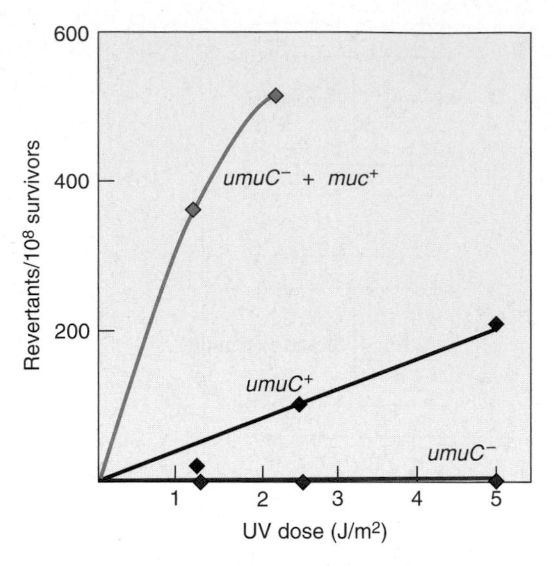

Figure 20.38 An *umuC* strain of *E. coli* is unmutable. Walker and colleagues tested three his^- strains of bacteria for the ability to generate his^+ revertants after UV irradiation. The strains were: wild-type with respect to *umuC* (blue), a $umuC^-$ strain (red), and a $umuC^-$ strain supplemented with a plasmid containing the *muc* gene (green). (*Source:* Adapted from Bagg, A., C.J. Kenyon, and G.C. Walker, Inducibility of a gene product required for UV and chemical mutagenesis in *Escherichia. coli. Proceedings of the National Academy of Sciences USA* 78:5750, 1981.)

Wild-type *E. coli* cells can tolerate as many as 50 pyrimidine dimers in their genome without ill effect because of their active repair mechanisms. Bacteria lacking one of the *uvr* genes cannot carry out excision repair, so their susceptibility to UV damage is greater. However, they are still somewhat resistant to DNA damage. On the other hand, double mutants in *uvr* and *recA* can perform neither excision repair nor recombination repair, and they are very sensitive to UV damage, perhaps because they have to rely on error-prone bypass. Under these conditions, only one to two pyrimidine dimers per genome is a lethal dose.

Obviously, if bacterial cells had evolved without the error-prone bypass, they would be subject to many fewer mutations. If that is the case, then why have they retained this mutation-causing mechanism? It is likely that the error-prone bypass system does more good than harm by allowing an organism to replicate its damaged genome even at the risk of mutation. This is especially obvious if the

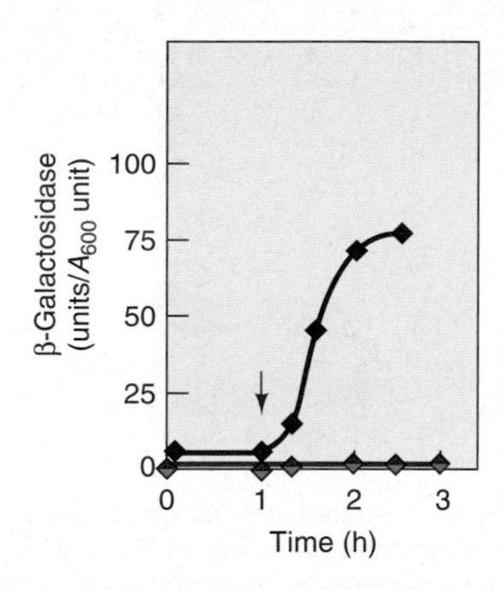

Figure 20.39 The *umuDC* promoter is UV-inducible. Walker and colleagues irradiated cells with the *lac* genes under control of the *umuDC* promoter with a UV dose of 10 J/m^2. They performed the irradiation at 1 h, as indicated by the arrow. Then they measured the accumulation of β-galactosidase activity (blue) per OD_{600} unit (an index of turbidity and therefore of cell density). They also performed the same experiment in *lexA* mutant (green) and *recA*$^-$ cells (red). The *lexA* mutant was an "uninducible" one encoding a LexA protein that cannot be cleaved and therefore cannot be removed from the *umuDC* operator. (*Source:* Adapted from Bagg, A., C.J. Kenyon, and G.C. Walker, Inducibility of a gene product required for UV and chemical mutagenesis in *Escherichia coli. Proceedings of the National Academy of Sciences USA* 78:5751, 1981.)

price for failure to replicate is death, as would be the case after a cell replicates its damaged DNA and then divides without repairing the damage. This chain of events would produce one daughter cell with a DNA gap across from a lesion. By this time, excision repair and even recombination repair are no longer possible. So the last resort is error-prone bypass to stave off cell death.

It is also true that a certain level of mutation is good for a species because it allows the genomes of a group of organisms to diverge so they do not all have equal susceptibility to disease and other insults. That way, when a new challenge arises, some of the members of a population have evolved resistance and can survive to perpetuate the species.

SUMMARY Cells can employ nonrepair methods to circumvent DNA damage. One of these is recombination repair, in which the gapped DNA strand across from a damaged strand recombines with a normal strand in the other daughter DNA duplex after replication. This solves the gap problem but leaves the original damage unrepaired. Another mechanism to deal with DNA damage, at least in *E. coli,* is to induce the SOS response, which causes the DNA to replicate even though the damaged region cannot be read correctly. This results in errors in the newly made DNA, so the process is called error-prone bypass.

Error-Prone and Error-Free Bypass in Humans All of the DNA repair processes are well conserved throughout all kingdoms of life, probably because DNA damage has been part of life from the very beginning, so damage repair had to evolve early, before the three kingdoms diverged. Error-prone bypass is no exception: Human cells have systems similar to those in prokaryotes to deal with lesions like pyrimidine dimers. These bypass systems depend on specialized DNA polymerases, including **DNA polymerases ζ** (zeta), **η** (eta), **θ** (theta), **ι** (iota), and **κ** (kappa). These specialized polymerases take over from polymerases δ and ε, which synthesize the lagging and leading strands, respectively, but stall at uninstructive DNA lesions like pyrimidine dimers.

Some of these enzymes insert bases at random to get past the lesion, which is obviously an error-prone strategy. But some of them have specificities that minimize errors and are therefore relatively error-free. For example, DNA polymerase η automatically inserts two dAMPs into the DNA strand across from a pyrimidine dimmer. Thus, even though the bases in the dimer cannot base-pair, this system is able to make the correct choice if both bases in the dimer are thymines—which is often the case. DNA polymerase η can also bypass adjacent guanines (Pt-GGs) that have been cross-linked via platinum by the anti-cancer drug cisplatin. It does a good job of replicating the 3′-dG, usually inserting a dC in the opposite strand, but it randomly inserts either dC or dA opposite the 5′-dG.

In 1999, Fumio Hanaoka and colleagues discovered that the defective gene in patients with the variant form of XP (XP-V) is the gene that codes for DNA polymerase η. Thus, these patients cannot carry out the comparatively error-free bypass of pyrimidine dimers catalyzed by DNA polymerase η and must therefore rely on the error-prone bypass catalyzed by other specialized DNA polymerases, including DNA polymerase ζ. This error-prone system introduces mutations during replication of pyrimidine dimers not removed by the excision repair system. However, because these patients have normal excision repair, few dimers are left for the error-prone system to deal with. This argument accounts for the relatively low sensitivity of XP-V cells to ultraviolet radiation.

Polymerase η cannot carry out error-free bypass by itself. After it inserts two A's across from a pyrimidine dimer, the 3′-end of the newly synthesized strand is not base-paired to a T because the T's in the template strand are locked up in the pyrimidine dimer. Without a base-paired nucleotide to add to, the replicative DNA polymerases (ε and δ) cannot resume DNA synthesis. Thus, another polymerase, perhaps polymerase ζ, must do the job.

Why doesn't polymerase η simply continue synthesizing enough DNA for one of the replicative polymerases to get started again? The answer is that this would be a very error-prone process. Although the term "error-free" for

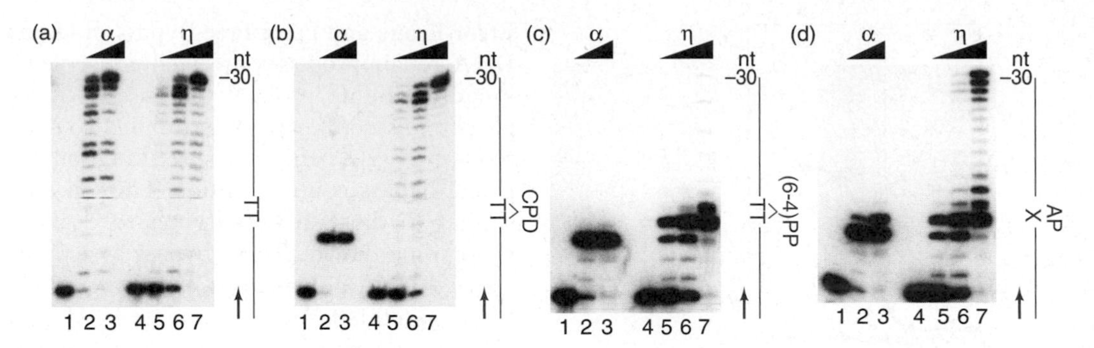

Figure 20.40 Activities of DNA polymerases α and η on undamaged and damaged templates. Hanaoka and colleagues prepared double-stranded DNAs containing on the template strand: **(a)** no damage; **(b)** a cyclobutane pyrimidine dimer (CPD); **(c)** a (6-4) photoproduct [(6-4)PP]; or **(d)** an AP site. The nontemplate strand of these DNAs was a ^{32}P-labeled primer that was poised to be extended through the damage (or normal pair of thymines) on the template strand. The DNAs are illustrated with cartoons adjacent to each panel. The workers added increasing amounts of either DNA polymerase α or η, along with nucleotides, and electrophoresed the products on polyacrylamide gels. If translesion synthesis was successful, the primer was extended to the full length of the template strand, 30 nt. If not, synthesis stalled at the lesion. (*Source:* From Masutani et al., *Cold spring Harbor Symposia* p. 76. © 2000.)

DNA polymerase η is justified in terms of its ability to deal with thymine dimers, this enzyme is remarkably error-prone when replicating ordinary DNA. When Hanaoka, Thomas Kunkel, and colleagues tested the fidelity of this enzyme in vitro, using a double-stranded DNA with a gap in it, they found that DNA polymerase η had a lower fidelity than any other template-dependent DNA polymerase ever studied until that time: one mistake per 18–380 nt incorporated. By contrast; DNA polymerase ζ is about 20 times more accurate. Thus, it is a good thing that cells normally have the NER system. Without it, DNA polymerase η would be a very poor backstop for dealing with anything but thymine dimers—as typical XP patients can attest.

DNA polymerase η is specific for translesion synthesis at certain kinds of DNA damage. This enzyme can perform TLS at a pyrimidine dimer, but not at a (6-4) photoproduct. DNA polymerase η can also bypass an abasic (AP) site. Hanaoka and colleagues performed an assay to measure TLS in vitro at each of these kinds of DNA damage, using either polymerase α or polymerase η. They used templates that contained one damaged strand and one ^{32}P-labeled primer strand that had its 3′-end just upstream of the damage. Then they added nucleotides to allow TLS and electrophoresed the products.

Figure 20.40 depicts the results. Panel (a) shows that polymerases α and η could both extend the primer on an undamaged template, but polymerase α was ineffective in extending the primer past any of the DNA lesions. This failure of polymerase α is not surprising because it is designed for accurate copying of normal DNA to make primers, not for dealing with the noninformative DNA in these lesions. Panels (b–d) show that polymerase η could extend the primer past a cyclic pyrimidine dimer (CPD) and an AP site, but not past a (6-4) photoproduct.

SUMMARY Humans have a relatively error-free bypass system that inserts dAMPs across from a pyrimidine dimer, thus replicating thymine dimers (but not dimers involving cytosines) correctly. This system uses DNA polymerase η plus another enzyme to replicate a few bases beyond the lesion. When the gene for DNA polymerase η is defective, DNA polymerase ζ and perhaps other DNA polymerases take over. But these polymerases insert random nucleotides across from a pyrimidine dimer, so they are error-prone. These errors in correcting UV damage lead to a variant form of XP known as XP-V. DNA polymerase η is active on templates with thymine dimers and AP sites, but not on (6-4) photoproducts. This polymersase is not really error-free. With a gapped template it is one of the least accurate template-dependent polymerases known.

SUMMARY

Several principles apply to all (or most) DNA replication: (1) Double-stranded DNA replicates in a semiconservative manner. When the parental strands separate, each serves as the template for making a new, complementary strand. (2) DNA replication in *E. coli* (and in other organisms) is at least semidiscontinuous. One strand is replicated in the direction of the movement of the replicating fork; This strand is commonly thought to replicate continuously, though there is evidence that it replicates discontinuously. the other is replicated discontinuously, forming 1–2 kb Okazaki fragments in the opposite direction. This allows both strands to be replicated in the 5′→3′ direction. (3) Initiation of DNA replication requires a primer. Okazaki fragments in *E. coli* are initiated with RNA primers 10–12 nt long. (4) Most eukaryotic and bacterial DNAs replicate bidirectionally. ColE1 is an example of a DNA that replicates unidirectionally.

Circular DNAs can replicate by a rolling circle mechanism. One strand of a double-stranded DNA is nicked and the 3′-end is extended, using the intact DNA strand as template. This displaces the 5′-end. In phage λ, the displaced strand serves as the template for discontinuous, lagging strand synthesis.

Pol I is a versatile enzyme with three distinct activities: DNA polymerase; 3′→5′ exonuclease; and 5′→3′ exonuclease. The first two activities are found on a large domain of the enzyme, and the last is on a separate, small domain. The large domain (the Klenow fragment) can be separated from the small by mild protease treatment, yielding two protein fragments with all three activities intact. The structure of the Klenow fragment shows a wide cleft for binding to DNA. This polymerase active site is remote from the 3′→5′ exonuclease active site on the Klenow fragment.

Of the three DNA polymerases in *E. coli* cells, pol I, pol II, and pol III, only pol III is required for DNA replication. Thus, this polymerase is the enzyme that replicates the bacterial DNA. The pol III core is composed of three subunits, α, ε, and θ. The α-subunit has the DNA polymerase activity. The ε-subunit has the 3′→5′ activity that carries out proofreading.

Faithful DNA replication is essential to life. To help provide this fidelity, the *E. coli* DNA replication machinery has a built-in proofreading system that requires priming. Only a base-paired nucleotide can serve as a primer for the pol III holenzyme. Therefore, if the wrong nucleotide is incorporated by accident, replication stalls until the 3′→5′ exonuclease of the pol III holoenzyme removes it. The fact that the primers are made of RNA may help mark them for degradation.

Mammalian cells contain five different DNA polymerases. Polymerases ε, δ, and α appear to participate in replicating both DNA strands. Polymerase α makes the primers for both strands, polymerase ε elongates the leading strand, and polymerase δ elongates the lagging strand. Polymerase β seems to function in DNA repair. Polymerase γ probably replicates mitochondrial DNA.

The helicase that unwinds double-stranded DNA at the replicating fork is encoded by the *E. coli dnaB* gene. The bacterial single-strand DNA-binding proteins bind much more strongly to single-stranded than to double-stranded DNA. They aid helicase action by binding tightly and cooperatively to newly formed single-stranded DNA and keeping it from annealing with its partner. By coating the single-stranded DNA, SSBs also protect it from degradation. They also stimulate their homologous DNA polymerases. These activities make SSBs essential for bacterial DNA replication.

As a helicase unwinds the two parental strands of a closed circular DNA, it introduces a compensating positive supercoiling force into the DNA. The stress of this force must be overcome or it will resist progression of the replicating fork. A name given to this stress-release mechanism is the swivel. DNA gyrase, a bacterial topoisomerase, is the leading candidate for this role in *E. coli*.

Alkylating agents like ethylmethane sulfonate add bulky alkyl groups to bases, either disrupting base pairing directly or causing loss of bases, either of which can lead to faulty DNA replication or repair.

Different kinds of radiation cause different kinds of damage. Ultraviolet rays have comparatively low energy, and they cause a moderate type of damage: pyrimidine dimers. Gamma and x-rays are much more energetic. They ionize the molecules around DNA and form highly reactive free radicals that can attack DNA, altering bases or breaking strands.

Ultraviolet radiation damage to DNA (pyrimidine dimers) can be directly repaired by a DNA photolyase that uses energy from visible light to break the bonds holding the two pyrimidines together. O6 alkylations on guanine residues can be directly reversed by the suicide enzyme O6-methylguanine methyltransferase, which accepts the alkyl group onto one of its amino acids.

Base excision repair (BER) typically acts on subtle base damage. This process begins with a DNA glycosylase, which extrudes a base in a damaged base pair, then clips out the damaged base, leaving an apurinic or apyrimidinic site that attracts the DNA repair enzymes that remove the remaining deoxyribose phosphate and replace it with a normal nucleotide. In bacteria, DNA polymerase I is the enzyme that fills in the missing nucleotide in BER, in eukaryotes, DNA polymerase β plays this role. However, this enzyme makes mistakes, and has no proofreading activity, so APE1 carries out the necessary proofreading. Repair of 8-oxoguanine sites in DNA is a special case of BER that can happen in two ways. Since oxoG mispairs with A, the A can be removed after DNA replication by a specialized adenine DNA glycosylase. However, if replication has not yet occurred, the oxoG will still be paired with C, and the oxoG can be removed by another DNA glycosylase, the oxoG repair enzyme.

Nucleotide excision repair (NER) generally deals with drastic, helix-distorting base changes. In bacterial NER, the damaged DNA is clipped out directly by cutting on both sides of the lesion with an endonuclease to remove the damaged DNA as part of an oligonucleotide. DNA polymerase I fills in the gap and DNA ligase seals the final nick.

Eukaryotic NER follows two pathways. In global genome NER (GG-NER), a complex composed of XPC and hHR23B initiates repair by binding to a lesion anywhere in the genome and causing a limited amount of DNA melting. This protein apparently recruits XPA and RPA. TFIIH then joins the complex, and two of its subunits (XPB and XPD) use their DNA helicase activities to expand the melted region. RPA binds two excinucleases (XPF and XPG) and positions them for cleavage of the DNA strand on either side of the lesion. This releases the damage on a fragment between 24 and 32 nt long. Transcription-coupled NER (TC-NER) is very similar to global genome NER, except that RNA polymerase plays the role of XPC in damage sensing and initial DNA melting. In either kind of NER, DNA polymerase ε or δ fills in the gap left by the removal of the damaged fragment, and DNA ligase seals the DNA.

Double-strand DNA breaks can be repaired by homologous recombination or by nonhomologous end joining. The latter process requires Ku and DNA–PK_{cs}, which bind together at the DNA ends, constituting active DNA–PK complexes that allow the ends to find regions of microhomology with each other. Once the regions of microhomology line up, the two DNA–PK complexes phosphorylate each other. This phosphorylation activates the catalytic subunit (DNA–PK_{cs}) to dissociate, and it also activates the DNA helicase activity of Ku to unwind the DNA ends so the microhomology regions can base-pair. Finally, extra flaps of DNA are removed, gaps are filled, and the DNA ends are ligated permanently together.

Chromatin remodeling is required for both nonhomologous end-joining and homologous recombination. In yeast, two protein kinases, Mec1 and Tel1, are recruited to DSBs, where they phosphorylate serine 129 of histone H2A in nearby nucleosomes. This phosphorylation recruits the chromatin remodeler INO80 to the DSB, where it appears to use its DNA helicase activity to push nucleosomes away from the ends of the DSB, enabling formation of single-stranded 3′-DNA overhangs, which are essential for both NHEJ and homologous recombination. Another chromatin remodeler known as SWR1, which shares many components with INO80, also appears at DSBs, and replaces phospho-H2A with the H2A variant Htz1, which cannot be phosphorylated. This returns the phosphorylation state of H2A on nucleosomes near DSBs to normal.

Errors in DNA replication leave mismatches that can be detected and repaired. The *E. coli* mismatch repair system recognizes the parental strand by its methylated adenines in GATC sequences. Then it corrects the mismatch in the complementary (progeny) strand. The failure of human mismatch repair leads to microsatellite instability, and ultimately to cancer.

Cells can employ nonrepair methods to circumvent DNA damage. One of these is recombination repair, in which the gapped DNA strand across from a damaged strand recombines with a normal strand in the other daughter DNA duplex after replication. This solves the gap problem but leaves the original damage unrepaired. Another mechanism to deal with DNA damage, at least in *E. coli,* is to induce the SOS response, which causes the DNA to replicate even though the damaged region cannot be read correctly. This results in errors in the newly made DNA, so the process is called error-prone bypass.

Humans have a relatively error-free bypass system that inserts dAMPs across from a pryimidine dimer, thus replicating thymine dimers (but not dimers involving cytosines) correctly. This system uses DNA polymerase η plus another enzyme to replicate a few bases beyond the lesion. When the gene for DNA polymerase η is defective, DNA polymerase ζ, and perhaps other DNA polmerases, take over. But these polymerases insert random nucleotides across from a pryimidine dimer, so they are error-prone. These errors in correcting UV damage lead to a variant form of XP known as XP-V.

REVIEW QUESTIONS

1. Compare and contrast the conservative, semiconservative, and dispersive mechanisms of DNA replication.
2. Describe and give the results of an experiment that shows that DNA replication is semiconservative.
3. Compare and contrast the continuous, discontinuous, and semidiscontinuous modes of DNA replication.
4. Describe and give the results of an experiment that shows that DNA replication is at least semidiscontinuous.
5. What is the evidence for fully discontinuous DNA replication in *E. coli* cells?
6. Describe and give the results of an experiment that measures the size of the primers on Okazaki fragments.
7. Present electron microscopic evidence that DNA replication of the *B. subtilis* chromosome is bidirectional, whereas replication of the colE1 plasmid is unidirectional.
8. Diagram the rolling circle replication mechanism used by the λ phage.
9. Diagram the proofreading process used by *E. coli* DNA polymerases.
10. What activities are contained in *E. coli* DNA polymerase I? What is the role of each in DNA replication?
11. How does the Klenow fragment differ from the intact *E. coli* DNA polymerase I? Which enzyme would you use in nick translation? DNA end-filling? Why?
12. Of the three DNA polymerases in *E. coli,* which is essential for DNA replication? Present evidence.
13. Which pol III core subunit has the DNA polymerase activity? How do we know?

14. Which pol III core subunit has the proofreading activity? How do we know?
15. Explain how the necessity for proofreading rationalizes the existence of priming in DNA replication.
16. List the eukaryotic DNA polymerases and their roles. Outline evidence for these roles.
17. Compare and contrast the activity of a helicase with that of a topoisomerase in the context of DNA replication.
18. What roles do SSBs play in DNA replication?
19. Explain why nicking one strand of a supercoiled DNA removes the supercoiling.
20. How do we know that DNA gyrase forms a covalent bond between an enzyme tyrosine and DNA? What is the advantage of forming this bond?
21. Present a model, based on the structure of yeast DNA topoisomerase II, for the DNA segment-passing step.
22. Compare and contrast the DNA damage done by UV rays and x-rays or gamma rays.
23. What two enzymes catalyze direct reversal of DNA damage? Diagram the mechanisms they use.
24. Compare and contrast base excision repair and nucleotide excision repair. Diagram both processes. For what types of damage is each primarily responsible?
25. What enzyme performs proofreading in human base excision repair? Outline the evidence supporting your answer.
26. Briefly describe the crystal structures of complexes between the human oxoG repair enzyme (hOGG1) and an oxoG–C pair, or a normal G–C pair. How do these structures explain why oxoG is removed, while ordinary G is not.
27. How does transcription-coupled NER differ from global genome NER?
28. Outline the nonhomologous end-joining mechanism mammals use to repair double-stand DNA breaks. Show how this process can lead to loss of nucleotides at the repair site.
29. What DNA repair system is missing in most cases of xeroderma pigmentosum? Why does that make XP patients so sensitive to UV light? What is the primary backup system for these patients?
30. What DNA repair system is missing in XP-V patients? Why is the incidence of skin cancer lower in these people than in typical XP patients? What is the backup system for lesions missed by the NER system in XP-V patients?
31. Why is chromatin remodeling needed for double-strand break repair in eukaryotes?
32. Diagram the mismatch repair mechanism in *E. coli*.
33. Diagram the recombination repair mechanism in *E. coli*.
34. Diagram the error-prone bypass system in *E. coli*.
35. Explain why recombination repair and error-prone bypass are not real repair systems.
36. Present evidence that shows that DNA polymerase η can bypass a thymine dimer and an AP site but not a (6-4) photoproduct, and that DNA polymerase α cannot bypass any of these lesions.

ANALYTICAL QUESTIONS

1. Why is it improbable that we will ever observe continuous DNA replication of both strands in nature?
2. You are studying a protein that you suspect has DNA helicase activity. Describe how you would assay the protein for this activity and show sample positive results.
3. You are studying a protein that you suspect has DNA topoisomerase activity. Describe how you would assay the protein for the activity and show sample positive results.
4. Explain the difference between DNA damage and mutation. How do mutations in *E. coli* DNA polymerase V illustrate this difference?
5. Recently, as a post-doc in a highly reputable laboratory, you designed a new single-celled organism only capable of three DNA repair mechanisms. You have been asked to present your research at a prestigious Molecular Biology conference Describe how you will support your reason for choosing the three repair mechanisms and discuss if there are overlaps or gaps between the chosen mechanisms. Additionally, explain the types of mutations your cell can overcome and the types of damage that may potentially destroy your new organism. You may assume that your organism already has a homologous recombination system.

SUGGESTED READINGS

General References and Reviews

Cairns, B.R. 2004. Around the world of DNA damage INO80 days. *Cell* 119:733–34.

Chu, G. 1997. Double strand break repair. *Journal of Biological Chemistry.* 272:24097–100.

Citterio, E., W. Vermeulen, and J.H.J. Hoeijmakers. 2000. Transcriptional healing. *Cell* 101:447–50.

David, S.S. 2005. DNA search and rescue. *Nature* 434:569–70.

de Latt, W.L., N.G.J. Jaspers, and J.H.J. Hoeijmakers. 1999. Molecular mechanism of nucleotide excision repair. *Genes and Development* 13:768–85.

Friedberg, E.C., R. Wagner, and M. Radman. 2002. Specialized DNA polymerases, cellular survival, and the genesis of mutations. *Science* 296:1627–30.

Herendeen, D.R. and T.J. Kelly. 1996. DNA polymerase III: Running rings around the fork. *Cell* 84:5–8.

Jiricny, J. 2002. An APE that proofreads. *Nature* 415:593–94.

Joyce, C.M. and T.A. Steitz. 1987. DNA polymerase I: From crystal structure to function via genetics. *Trends in Biochemical Sciences* 12:288–92.

Kornberg, A. and T. Baker. 1992. *DNA Replication.* New York: W.H. Freeman and Company.

Lindahl, T. 2004. Molecular biology: Ensuring error-free DNA repair. *Nature* 427:598.

Lindahl, T. and R.D. Wood. 1999. Quality control by DNA repair. *Science* 286:1897–1905.

Maxwell, A. 1996. Protein gates in DNA topoisomerase II. *Nature Structural Biology.* 3:109–12.

Sharma, A. and A. Mondragón. 1995. DNA topoisomerases. *Current Opinion in Structural Biology* 5:39–47.
Wood, R.D. 1997. Nucleotide excision repair in mammalian cells. *Journal of Biological Chemistry* 272:23465–68.
Wood, R.D. 1999. Variants on a theme. *Nature* 399:639–70.

Research Articles

Bagg, A., C.J. Kenyon, and G.C. Walker. 1981. Inducibility of a gene product required for UV and chemical mutagenesis in *Escherichia coli. Proceedings of the National Academy of Sciences USA* 78:5749–53.
Banerjee, A., W. Yang, M. Karplus, and G.L. Verdine. 2005. Structure of a repair enzyme interrogating undamaged DNA elucidates recognition of damaged DNA. *Nature* 434:612–18.
Berger, J.M., S.J. Gamblin, S.C. Harrison, and J.C. Wang. 1996. Structure and mechanism of DNA topoisomerase II. *Nature* 379:225–32.
Cairns, J. 1963. The chromosome of *Escherichia coli. Cold Spring Harbor Symposia on Quantitative Biology* 28:43–46.
Chou, K.-M. and Y.-C. Cheng. 2002. An exonucleolytic activity of human apurinic/apyrimidinic endonuclease on 3′ mispaired DNA. *Nature* 415:655–59.
Curtis, M.J. and B. Alberts. 1976. Studies on the structure of intracellular bacteriophage T4 DNA. *Journal of Molecular Biology* 102:793–816.
Drapkin, R., J.T. Reardon, A. Ansari, J.-C. Huang, L. Zawel, K. Ahn, A. Sancar, and D. Reinberg. 1994. Dual role of TFIIH in DNA excision repair and in transcription by RNA polymerase II. *Nature* 368:769–72.
Eom, S.H., T. Wang, and T.A. Steitz. 1996. Structure of *Taq* polymerase with DNA at the active site. *Nature* 382: 278–281.
Gefter, M.L., Y. Hirota, T. Kornberg, J.A. Wechster, and C. Barnoux. 1971. Analysis of DNA polymerases II and III in mutants of *Escherichia coli* thermosensitive for DNA synthesis. *Proceedings of the National Academy of Sciences USA* 68:3150–53.
Gellert, M., K. Mizuuchi, M.H. O'Dea, and H.A. Nash. 1976. DNA gyrase: An enzyme that introduces superhelical turns into DNA. *Proceedings of the National Academy of Sciences USA* 73:3872–76.
Gyurasits, E.B. and R.J. Wake. 1973. Bidirectional chromosome replication in *Bacillus subtilis. Journal of Molecular Biology* 73:55–63.
Hirota, G.H., A. Ryter, and F. Jacob. 1968. Thermosensitive mutants in *E. coli* affected in the processes of DNA synthesis and cellular division. *Cold Spring Harbor Symposia on Quantitative Biology* 33:677–93.
Huberman, J.A., A. Kornberg, and B.M. Alberts. 1971. Stimulation of T4 bacteriophage DNA polymerase by the protein product of T4 gene 32. *Journal of Molecular Biology* 62:39–52.
Kitani, T., K.-Y. Yoda, T. Ogawa, and T. Okazaki. 1985. Evidence that discontinuous DNA replication in *Escherichia coli* is primed by approximately 10 to 12 residues of RNA starting with a purine. *Journal of Molecular Biology* 184:45–52.
LeBowitz, J.H. and R. McMacken. 1986. The *Escherichia coli* dnaB replication protein is a DNA helicase. *Journal of Biological Chemistry* 261:4738–48.
Maki, H. and A. Kornberg. 1985. The polymerase subunit of DNA polymerase III of *Escherichia coli. Journal of Biological Chemistry* 260:12987–92.
Masutani, C., R. Kusumoto, A. Yamada, N. Dohmae, M. Yokoi, M. Yuasa, M. Araki, S. Iwai, K. Takio, and F. Hanaoka. 1999. The *XPV* (xeroderma pigmentosum variant) gene encodes human DNA polymerase η. *Nature* 399: 700–04.
Masutani, C., R. Kusumoto, A. Yamada, M. Yuasa, M. Araki, T. Nogimori, M. Yokoi, T. Eki, S. Iwai, and F. Hanaoka. 2000. Xeroderma pigmentosum variant: From a human genetic disorder to a novel DNA polymerase. *Cold Spring Harbor Symposia on Quantitative Biology.* 65:71–80.
Matsuda, T., K. Bebenek, C. Masutani, F. Hanaoka, and T.A. Kunkel. 2000. Low fidelity DNA synthesis by human DNA polymerase-η. *Nature* 404:1011–13.
Meselson, M. and F. Stahl. 1958. The replication of DNA in *Escherichia coli. Proceedings of the National Academy of Sciences USA* 44:671–82.
Okazaki, R., T. Okazaki, K. Sakabe, K. Sugimoto, R. Kainuma, A. Sugino, and N. Iwatsuki. 1968. In vivo mechanism of DNA chain growth. *Cold Spring Harbor Symposia on Quantitative Biology* 33:129–43.
Scheuermann, R.H. and H. Echols. 1984. A separate editing exonuclease for DNA replication: The ε subunit of *Escherichia coli* DNA polymerase III holoenzyme. *Proceedings of the National Academy of Sciences USA* 81:7747–57.
Sugasawa, K., J.M.Y. Ng, C. Masutani, S. Iwai, P.J. van der Spek, A.P.M. Eker, F. Hanaoka, D. Bootsma, and J.H.J. Hoeijmakers. 1998. Xeroderma pigmentosum group C protein complex is the initiator of global genome nucleotide excision repair. *Molecular Cell* 2:223–32.
Tse, Y.-C., K. Kirkegaard, and J.C. Wang. 1980. Covalent bonds between protein and DNA. *Journal of Biological Chemistry* 255:5560–65.
Wakasugi, M. and A. Sancar. 1999. Order of assembly of human DNA repair excision nuclease. *Journal of Biological Chemistry* 274:18759–68.

CHAPTER 21

DNA Replication II: Detailed Mechanism

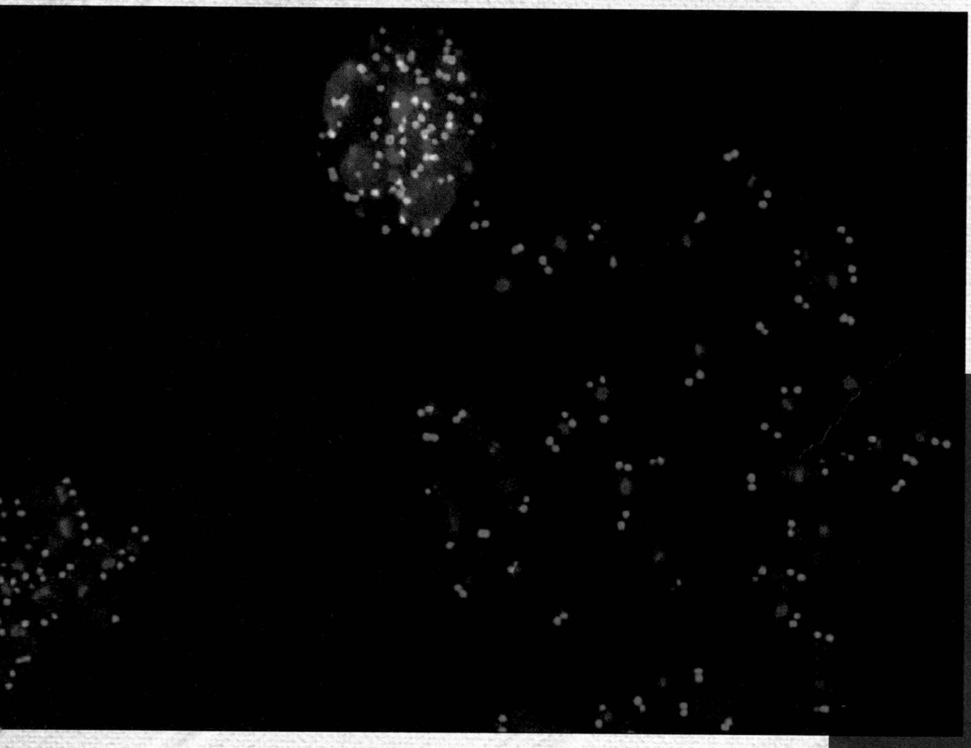

Telomeres in human chromosomes. The telomeres are stained green and the centromeres are stained pink. *Cal Harley/Geron Corporation & Peter Rabinovitch, Univ. of Washington.*

We learned in Chapter 20 that DNA replication is at least semidiscontinuous and requires the synthesis of primers before DNA synthesis can begin. We have also learned about some of the major proteins involved in DNA replication in *E. coli*. Thus, we know that DNA replication is complex and involves more than just a DNA polymerase. This chapter presents a close look at the mechanism of this process in *E. coli* and in eukaryotes. We will look at the three stages of replication—initiation, elongation, and termination—in a variety of systems.

21.1 Initiation

As we have seen, initiation of DNA replication means primer synthesis. Different organisms use different mechanisms to make primers; even different phages that infect *E. coli* (coliphages) use quite different primer synthesis strategies. The coliphages were convenient tools to probe *E. coli* DNA replication because they are so simple that they have to rely primarily on host proteins to replicate their DNAs.

Priming in *E. coli*

As mentioned in Chapter 20, the first example of coliphage primer synthesis was found by accident in M13 phage, when this phage was discovered to use the host RNA polymerase as its **primase** (primer-synthesizing enzyme). But *E. coli* and its other phages do not use the host RNA polymerase as a primase. Instead, they employ a primase called **DnaG,** which is the product of the *E. coli dnaG* gene. Arthur Kornberg noted that *E. coli* and most of its phages need at least one more protein (**DnaB,** a DNA helicase introduced in Chapter 20) to form primers, at least on the lagging strand.

Arthur Kornberg and colleagues discovered the importance of DnaB with an assay in which single-stranded ϕX174 phage DNA (without SSB) is converted to double-stranded form. Synthesis of the second strand of phage DNA required primer synthesis, then DNA replication. The DNA replication part used pol III holoenzyme, so the other required proteins should be the ones needed for primer synthesis. Kornberg and colleagues found that three proteins: DnaG (the primase), DnaB, and pol III holoenzyme were required in this assay. Thus, DnaG and DnaB were apparently needed for primer synthesis. Kornberg coined the term **primosome** to refer to the collection of proteins needed to make primers for a given replicating DNA. Usually this is just two proteins, DnaG and DnaB, although other proteins may be needed to assemble the primosome.

The *E. coli* primosome is mobile and can repeatedly synthesize primers as it moves around the uncoated circular ϕX174 phage DNA. As such, it is also well suited for the repetitious task of priming Okazaki fragments on at least the lagging strand of *E. coli* DNA. This contrasts with the activity of RNA polymerase or primase alone, which prime DNA synthesis at only one spot—the origin of replication.

Two different general approaches were used to identify the important components of the *E. coli* DNA replication system, with DNA from phages ϕX174 and G4 as model substrates. The first approach was a combination genetic–biochemical one, the strategy of which was to isolate mutants with defects in their ability to replicate phage DNA, then to complement extracts from these mutants with proteins from wild-type cells. The mutant extracts were incapable of replicating the phage DNA in vitro unless the right wild-type protein was added. Using this system as an assay, the protein can be highly purified and then characterized. The second approach was the classical biochemical one: Purify all of the components needed and then add them all back together to reconstitute the replication system in vitro.

The Origin of Replication in *E. coli* Before we discuss priming further, let us consider the unique site at which DNA replication begins in ***E. coli: oriC*.** An origin of replication is a DNA site at which DNA replication begins and which is essential for proper replication to occur. We can locate the place where replication begins by several means, but how do we know how much of the DNA around the initiation site is essential for replication to begin? One way is to clone a DNA fragment, including the initiation site, into a plasmid that lacks its own origin of replication but has an antibiotic resistance gene. Then we can use the antibiotic to select for autonomously replicating plasmids. Any cell that replicates in the presence of the antibiotic must have a plasmid with a functional origin. Once we have such an *oriC* plasmid, we can begin trimming and mutating the DNA fragment containing *oriC* to find the minimal effective DNA sequence. The minimal origin in *E. coli* is 245 bp long. Some features of the origins are conserved in bacteria, and the spacing between them is also conserved.

Figure 21.1 illustrates the steps in initiation at *oriC*. The origin includes four 9-mers with the consensus sequence TTATCCACA. Two of these are in one orientation, and two are in the opposite orientation. DNase foot-printing shows that these 9-mers are binding sites for the ***dnaA* product (DnaA).** These 9-mers are therefore sometimes called ***dnaA* boxes.** DnaA appears to facilitate the binding of DnaB to the origin.

DnaA helps DnaB bind at the origin by stimulating the melting of three 13-mer repeats at the left end of *oriC* to form an **open complex.** This is analogous to the open promoter complex we discussed in Chapter 6. DnaB can then bind to the melted DNA region. Another protein, DnaC, binds to DnaB and helps deliver it to the origin.

The evidence also strongly suggests that DnaA directly assists the binding of DnaB. Here is one line of evidence that points in this direction. A *dnaA* box resides in the stem of a hairpin stem loop in a plasmid called R6K. When DnaA binds to this DNA, DnaB (with the help of DnaC) can also bind. Here, no DNA melting appears to occur, so we infer that DnaA directly affects binding between DNA and DnaB.

At least two other factors participate in open complex formation at *oriC*. The first of these is RNA polymerase. This enzyme does not serve as a primase, as it does in M13 phage replication, but it still serves an essential function. We know RNA polymerase action is required, because rifampicin blocks primosome assembly. The role of RNA polymerase seems to be to synthesize a short piece of RNA that creates an R loop (Chapter 14). The R loop can be adjacent to *oriC,* rather than within it. The second factor is

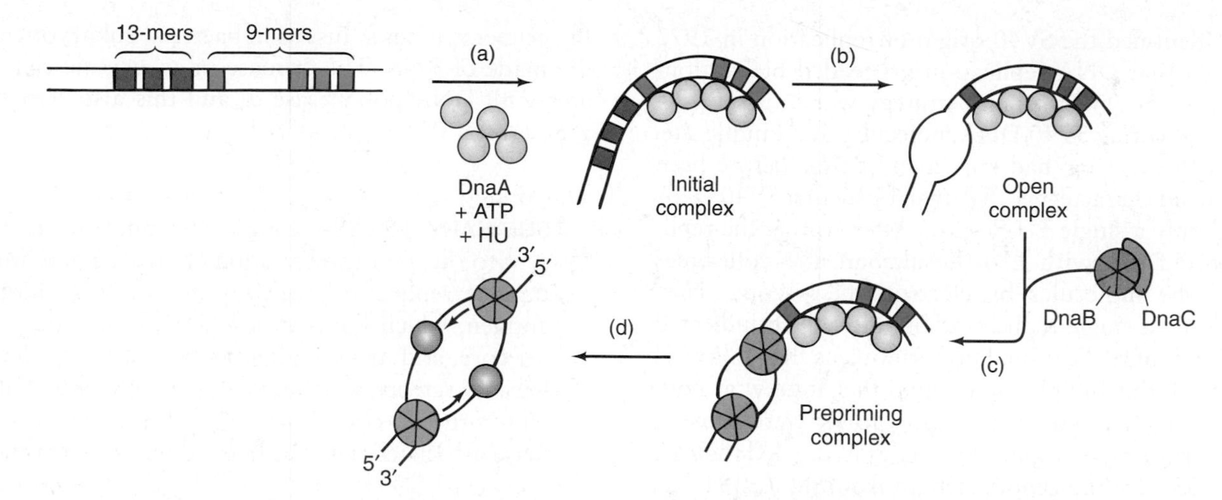

Figure 21.1 Priming at *oriC*. **(a)** Formation of the initial complex. First, DnaA (yellow) binds ATP and forms a multimer. Along with the HU protein, the DnaA/ATP complex binds to the DNA, encompassing the four 9-mers. In all, this complex covers about 200 bp. HU protein probably induces the bend in the DNA pictured here. **(b)** Formation of the open complex. The binding of DnaA, along with the bending induced by HU protein, apparently destabilizes the adjacent 13-mer repeats and causes local DNA melting there. This allows the binding of DnaB protein to the melted region. **(c)** Formation of the prepriming complex. DnaC binds to the DnaB protein and helps deliver it to the DNA. **(d)** Priming. Finally, primase (purple) binds to the prepriming complex and converts it to the primosome, which can make primers to initiate DNA replication. Primers are represented by arrows. (*Source:* Adapted from *DNA Replication*, 2/e, (plate 15) by Arthur Kornberg and Tania Baker.)

HU protein. This is a small basic DNA-binding protein that can induce bending in double-stranded DNA. This bending, together with the R loop, presumably destabilizes the DNA double helix and facilitates melting of the DNA to form the open complex.

Finally, DnaB stimulates the binding of the primase (DnaG), completing the primosome. Priming can now occur, so DNA replication can get started. The primosome remains with the replication machinery, or **replisome,** as it carries out elongation, and serves at least two functions. First, it must operate repeatedly in priming Okazaki fragment synthesis to build the lagging strand. Second, DnaB serves as the helicase that unwinds DNA to provide templates for both the leading and lagging strands. To accomplish this task, DnaB moves in the 5′→3′ direction on the lagging strand template—the same direction in which the replicating fork is moving. This anchors the primosome to the lagging strand template, where it is needed for priming Okazaki fragment synthesis.

SUMMARY Primer synthesis in *E. coli* requires a primosome composed of the DNA helicase, DnaB, and the primase, DnaG. Primosome assembly at the origin of replication, *oriC*, occurs as follows: DnaA binds to *oriC* at sites called *dnaA* boxes and cooperates with RNA polymerase and HU protein in melting a DNA region adjacent to the leftmost *dnaA* box. DnaB then binds to the open complex and facilitates binding of the primase to complete the primosome. The primosome remains with the replisome, repeatedly priming Okazaki fragment synthesis, at least on the lagging strand. DnaB also has a helicase activity that unwinds the DNA as the replisome progresses.

Priming in Eukaryotes

Eukaryotic replication is considerably more complex than the bacterial replication we have just studied. One complicating factor is the much bigger size of eukaryotic genomes. This, coupled with the slower movement of eukaryotic replicating forks, means that each chromosome must have multiple origins. Otherwise, replication would not finish within the time allotted—the S phase of the cell cycle—which can be as short as a few minutes. Because of this multiplicity and other factors, identification of eukaryotic origins of replication has lagged considerably behind similar work in prokaryotes. However, when molecular biologists face a complex problem, they frequently resort to simpler systems such as viruses to give them clues about the viruses' more complex hosts. Scientists followed this strategy to identify the origin of replication in the simple monkey virus SV40 as early as 1972. Let us begin our study of eukaryotic origins of replication there, then move on to origins in yeast.

The Origin of Replication in SV40 Two research groups, one headed by Norman Salzman, the other by Daniel

Nathans, identified the SV40 origin of replication in 1972 and showed that DNA replication proceeded bidirectionally from this origin. Salzman's strategy was to use *Eco*RI to cleave replicating SV40 DNA molecules at a unique site. (Although this enzyme had only a short time before been discovered and characterized, Salzman knew that SV40 DNA contained only a single *Eco*RI site.) After cutting the replicating SV40 DNA with *Eco*RI, Salzman and colleagues visualized the molecules by electron microscopy. They observed only a single replicating bubble, which indicated a single origin of replication. Furthermore, as they followed the growth of this bubble, they found that it grew at both ends, showing that both replicating forks were moving away from the single origin. This analysis revealed that the origin lies 33% of the genome length from the *Eco*RI site. But which direction from the *Eco*RI site? Because the SV40 DNA is circular, and these pictures contain no other markers besides the single *Eco*RI site, we cannot tell. But Nathans used another restriction enzyme (*Hin*dII), and his results, combined with these, placed the origin at a site overlapping the SV40 control region, adjacent to the GC boxes and the 72-bp repeat enhancer we discussed in Chapters 10 and 12 (Figure 21.2).

The minimal *ori* sequence (the *ori* core) is 64 bp long and includes several essential elements (1) four pentamers (5′-GAGGC-3′), which are the binding site for **large T antigen,** the major product of the viral early region; (2) a 15-bp palindrome, which is the earliest region melted during DNA replication; and (3) a 17-bp region consisting only of A–T pairs, which probably facilitates melting of the nearby palindrome region.

Other elements surrounding the *ori* core also participate in initiation. These include two additional large T antigen-binding sites, and the GC boxes to the left of the *ori* core. The GC boxes provide about a 10-fold stimulation of initiation of replication. If the number of GC boxes is reduced, or if they are moved only 180 bp away from *ori,* this stimulation is reduced or eliminated. This effect is somewhat akin to the participation of RNA polymerase in initiation at *ori*C in *E. coli.* One difference: At the SV40 *ori,* no transcription need occur; binding of the transcription factor Sp1 to the GC boxes is sufficient to stimulate initiation of replication.

Once large T antigen binds at the SV40 *ori,* its DNA helicase activity unwinds the DNA and prepares the way

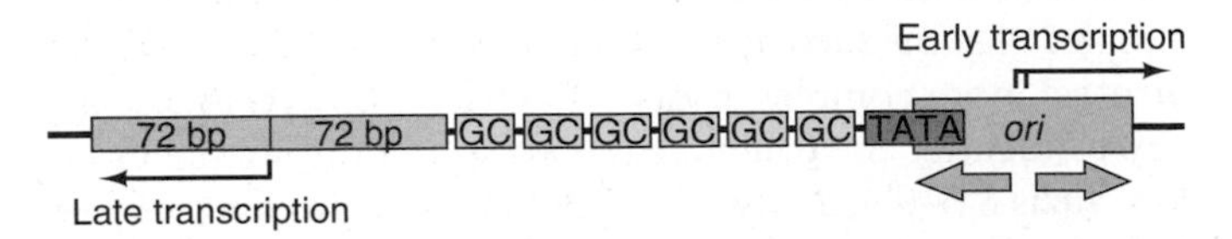

Figure 21.2 Location of the SV40 *ori* in the transcription control region. The core *ori* sequence (green) encompasses part of the early region TATA box and the cluster of early transcription initiation sites. Pink arrows denote bidirectional replication away from the replication initiation site. Black arrows denote transcription initiation sites.

for primer synthesis. Just as in bacteria, eukaryotic primers are made of RNA. The primase in eukaryotic cells associates with DNA polymerase α, and this also serves as the primase for SV40 replication.

SUMMARY The SV40 origin of replication is adjacent to the viral transcription control region. Initiation of replication depends on the viral large T antigen, which binds to a region within the 64-bp *ori* core, and at two adjacent sites, and exercises a helicase activity, which opens up a replication bubble within the *ori* core. Priming is carried out by a primase associated with the host DNA polymerase α.

The Origin of Replication in Yeast So far, yeast has provided most of our information about eukaryotic origins of replication. This is not surprising, because yeasts are among the simplest eukaryotes, and they lend themselves well to genetic analysis. As a result, yeast genetics are well understood. As early as 1979, C.L. Hsiao and J. Carbon discovered a yeast DNA sequence that could replicate independently of the yeast chromosomes, suggesting that it contains an origin of replication. This DNA fragment contained the yeast $ARG4^+$ gene. Cloned into a plasmid, it transformed $arg4^-$ yeast cells to $ARG4^+$, as demonstrated by their growth on medium lacking arginine. Any yeast cells that grew must have incorporated the $ARG4^+$ gene of the plasmid and, furthermore, must be propagating that gene somehow. One way to propagate the gene would be by incorporating it into the host chromosomes by recombination, but that was known to occur with a low frequency—about 10^{-6}–10^{-7}. Hsiao and Carbon obtained $ARG4^+$ cells at a much higher frequency—about 10^{-4}. Furthermore, shuttling the plasmid back and forth between yeast and *E. coli* caused no change in the plasmid structure, whereas recombination with the yeast genome would have changed it noticeably. Thus, these investigators concluded that the yeast DNA fragment they had cloned in the plasmid probably contained an origin of replication. Also in 1979, R.W. Davis and colleagues performed a similar study with a plasmid containing a yeast DNA fragment that converted trp^- yeast cells to TRP^+. They named the 850-bp yeast fragment **autonomously replicating sequence 1, or ARS1.**

Although these early studies were suggestive, they failed to establish that DNA replication actually begins in the ARS sequences. To demonstrate that ARS1 really does have this key characteristic of an origin of replication, Bonita Brewer and Walton Fangman used two-dimensional electrophoresis to detect the site of replication initiation in a plasmid bearing ARS1. This technique depends on the fact that circular and branched DNAs migrate more slowly than linear DNAs of the same size during gel electrophoresis, especially at high voltage or high agarose concentration.

Brewer and Fangman prepared a yeast plasmid bearing ARS1 as the only origin of replication. They allowed this plasmid to replicate in synchronized yeast cells and then isolated replication intermediates (RIs). They linearized these RIs with a restriction endonuclease, then electrophoresed them in the first dimension under conditions (low voltage and low agarose concentration) that separate DNA molecules roughly according to their sizes. Then they electrophoresed the DNAs in the second dimension using higher voltage and agarose concentrations that cause retardation of branched and circular molecules. Finally, they Southern blotted the DNAs in the gel and probed the blot with a labeled plasmid-specific DNA.

Figure 21.3 shows an idealized version of the behavior of various branched and circular RIs of a hypothetical 1-kb fragment. Simple Y's (panel a) begin as essentially linear 1-kb fragments with a tiny Y at their right ends; these would behave almost like linear 1-kb fragments. As the fork moves from right to left, the Y grows larger and the mobility of the fragment in the second (vertical) dimension slows. Then, as the Y grows even larger, the fragment begins to look more and more like a linear 2-kb fragment, with just a short stem on the Y. This is represented by the horizontal linear form with a short vertical stem in panel (a). Because these forms resemble linear shapes more and more as the fork moves, their mobility increases correspondingly, until the fork has nearly reached the end of the fragment. At this point, they have a shape and mobility that is almost like a true linear 2-kb fragment. This behavior gives rise to an arc-shaped pattern, where the apex of the arc corresponds to a Y that is half-replicated, at which point it is least like a linear molecule.

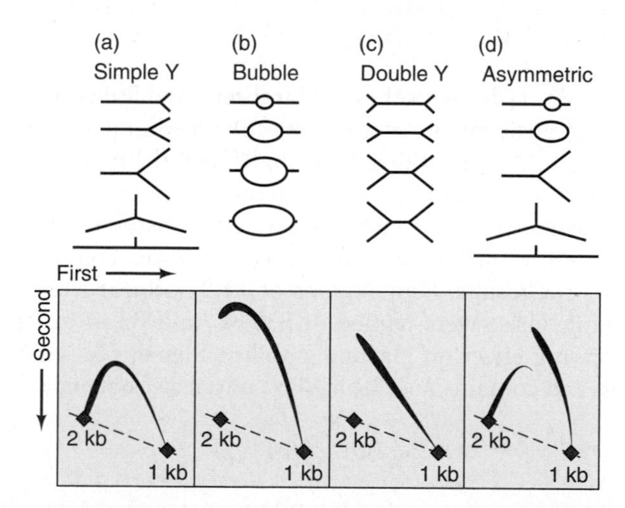

Figure 21.3 Theoretical behaviors of various types of replication intermediates on two-dimensional gel electrophoresis. The top parts of panels **a–d** are cartoons showing the shapes of growing simple Y's, bubbles, double Y's, and asymmetric bubbles that convert to simple Y's as replication progresses. The bottom parts of each panel are cartoons that depict the expected deviation of the changing mobilities of each type of growing RI from the mobilities of linear forms growing progressively from 1 to 2 kb (dashed lines). (*Source*: Adapted from Brewer, B.J. and W.L. Fangman, The localization of replication origins on ARS plasmids in *S. cerevisiae*. *Cell* 51:464, 1987.)

Figure 21.3b shows what to expect for a bubble-shaped fragment. Again, we begin with a 1-kb linear fragment, but this time with a tiny bubble right in the middle. As the bubble grows larger, the mobility of the fragment slows more and more, yielding the arc shown at the bottom of the panel. Panel (c) shows the behavior of a double Y, where the RI becomes progressively more branched as the two forks approach the center of the fragment. Accordingly, the mobility of the RI decreases almost linearly. Finally, panel (d) shows what happens to a bubble that is asymmetrically placed in the fragment. It begins as a bubble, but then, when one fork passes the restriction site at the right end of the fragment, it converts to a Y. The mobilities of the RIs reflect this discontinuity: The curve begins like that of a bubble, then abruptly changes to that of a Y, with an obvious discontinuity showing exactly when the fork passed the restriction site and converted the bubble to a Y.

This kind of behavior is especially valuable in mapping the origin of replication. In panel (d), for example, we can see that the discontinuity occurs in the middle of the curve, when the mobility in the first dimension was that of a 1.5-kb fragment. This tells us that the arms of the Y are each 500 bp long. Assuming that the two forks are moving at an equal rate, we can conclude that the origin of replication was 250 bp from the right end of the fragment.

Now let us see how this works in practice. Brewer and Fangman chose restriction enzymes that would cleave the plasmid with its ARS1 just once, but in locations that would be especially informative if the origin of replication really lies within ARS1. Figure 21.4 shows the locations of the two restriction sites, at top, and the experimental results, at bottom. The first thing to notice about the autoradiographs is that they are simple and correspond to the patterns we have seen in Figure 21.3. This means that there is a single origin of replication; otherwise, there would have been a mixture of different kinds of RIs, and the results would have been more complex.

The predicted origin within ARS1 lies adjacent to a *Bgl*II site (B, in panel a). Thus, if the RI is cleaved with this enzyme, it should yield double-Y RIs. Indeed, as we see in the lower part of panel (a), the autoradiograph is nearly linear—just as we expect for a double-Y RI. Panel (b) shows that a *Pvu*I site (P) lies almost halfway around the plasmid from the predicted origin. Therefore, cleaving with *Pvu*I should yield the bubble-shaped RI shown at the top of panel (b). The autoradiograph at the bottom of panel (b) shows that Brewer and Fangman observed the discontinuity expected for a bubble-shaped RI that converts at the very end to a very large single Y, as one fork reaches the *Pvu*I site, then perhaps to a very asymmetric double Y as the fork passes that site. Both of these results place the origin of replication adjacent to the *Bgl*II site, just where we expect it if ARS1 contains the origin.

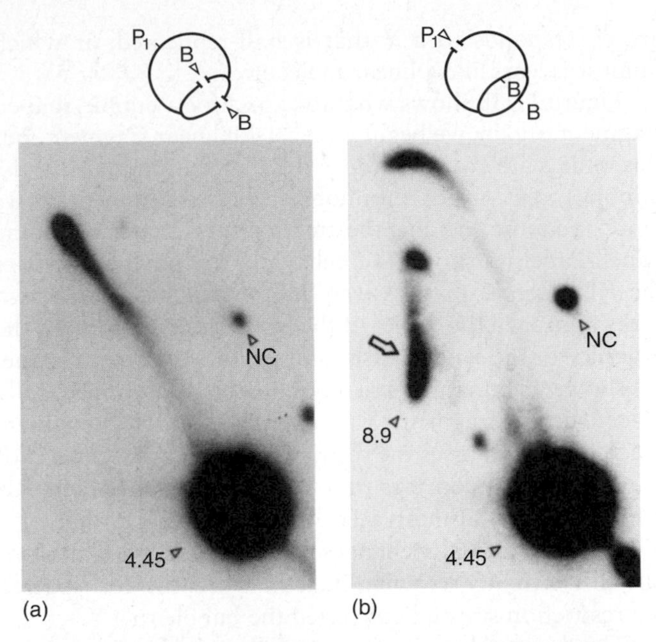

Figure 21.4 Locating the origin of replication in ARS1. (a) Results of cleaving 2-μm plasmid with *Bgl*II. Top: cartoon showing the shape expected when an RI is cut with *Bgl*II, assuming the origin lies adjacent to the *Bgl*II site within ARS1. The bubble contains DNA that has already replicated, so there are two copies of the *Bgl*II site (arrowheads labeled B), both of which are cut to yield the double-Y intermediate depicted. Bottom: experimental results showing the straight curve expected of double-Y intermediates. **(b)** Results of cleaving the plasmid with *Pvu*I. Top: cartoon showing the shape expected when an RI is cut with *Pvu*I, assuming the origin lies almost across the circle from the *Pvu*I site within ARS1. Bottom: experimental results showing the rising arc, with a discontinuity near the end. This is what we expect for a bubble-shaped RI that converts to a nearly linear Y as one of the replication forks passes a *Pvu*I site. Both of these results confirm the expectations for an origin of replication within ARS1. NC denotes nicked circles. The large open arrow points to large Y's or very asymmetric double Y's that result when a replicating fork passes a *Pvu*I site. Numbers refer to sizes in kb. (*Source:* Brewer, B.J. and W.L. Fangman, The localization of replication origins on ARS plasmids in *S. cerevisiae*. *Cell* 51 (6 Nov 1987) f. 8, p. 469. Reprinted by permission of Elsevier Science.)

York Marahrens and Bruce Stillman performed linker scanning experiments to define the important regions within ARS1. They constructed a plasmid very similar to the one used by Brewer and Fangman, containing (1) ARS1 in a 185-bp DNA sequence; (2) a yeast centromere; and (3) a selectable marker—*URA3*—which confers on *ura3-52* yeast cells the ability to grow in uracil-free medium. Then they performed linker scanning (Chapter 10) by systematically substituting an 8-bp *Xho*I linker for the normal DNA at sites spanning the ARS1 region. They transformed yeast cells with each of the linker scanning mutants and selected for transformed cells with uracil-free medium. Some of the transformants containing mutant ARS1 sequences grew more slowly than those containing wild-type ARS1 sequences. Because the centromere in each plasmid ensured proper segregation of the plasmid, the most likely explanation for poor growth was poor replication due to mutation of ARS1.

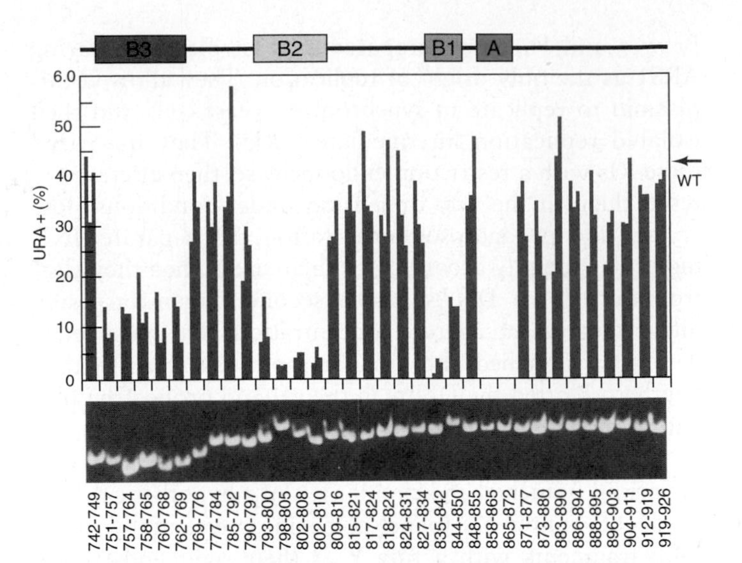

Figure 21.5 Linker scanning analysis of ARS1. Marahrens and Stillman substituted linkers throughout an ARS1 sequence within a plasmid bearing a yeast centromere and the *URA3* selectable marker. To test for replication efficiency of the mutants, they grew them for 14 generations in nonselective medium, then tested them for growth on selective (uracil-free) medium. The vertical bars show the results of three independent determinations for each mutant plasmid. Results are presented as a percentage of the yeast cells that retained the plasmid (as assayed by their ability to grow). Note that even the wild-type plasmid was retained with only 43% efficiency in nonselective medium (arrow at right). Four important regions (A, B1, B2, and B3) were identified. The regions that were mutated are identified by base number at bottom. The stained gel at bottom shows the electrophoretic mobility of each mutant plasmid. Note the altered mobility of the B3 mutant plasmids, which suggests altered bending. (*Source:* From Marahrens, Y. and B. Stillman, A yeast chromosomal origin of DNA replication defined by multiple functional elements. *Science* 255 (14 Feb 1992) f. 2, p. 819. Copyright © AAAS. Reprinted with permission from AAAS.)

To check this hypothesis, Marahrens and Stillman grew all the transformants in a nonselective medium containing uracil for 14 generations, then challenged them again with a uracil-free medium to see which ones had not maintained the plasmid well. The mutations in these unstable plasmids presumably interfered with ARS1 function. Figure 21.5 shows the results. Four regions of ARS1 appear to be important. These were named A, B1, B2, and B3 in order of decreasing effect on plasmid stability. Element A is 15 bp long, and contains an 11-bp ARS consensus sequence:

$$5'\text{-}{}^{T}_{A}TTTA^{TA}_{CG}TTT^{T}_{A}\text{-}3'$$

When it was mutated, *all* ARS1 activity was lost. The other regions had a less drastic effect, especially in selective medium. However, mutations in B3 had an apparent effect on the bending of the plasmid, as assayed by gel electrophoresis. The stained gel below the bar graph shows increased electrophoretic mobility of the mutants in the B3 region. Marahrens and Stillman interpreted this as altered bending of the ARS1 in the presence of the replicating machinery.

The existence of four important regions within ARS1 raises the question whether these are also *sufficient* for ARS function. To find out, Marahrens and Stillman constructed a synthetic ARS1 with wild-type versions of all four regions, spaced just as in the wild-type ARS1, but with random sequences in between. A plasmid bearing this synthetic ARS1 was almost as stable under nonselective conditions as one bearing a wild-type ARS1. Thus, the four DNA elements defined by linker scanning are sufficient for ARS1 activity. Finally, these workers replaced the wild-type 15-bp region A with the 11-bp ARS consensus sequence. This reduced plasmid stability dramatically, suggesting that the other 4 bp in region A are also important for ARS activity.

SUMMARY The yeast origins of replication are contained within autonomously replicating sequences (ARSs) that are composed of four important regions (A, B1, B2, and B3). Region A is 15 bp long and contains an 11-bp consensus sequence that is highly conserved in ARSs. Region B3 may allow for an important DNA bend within ARS1.

21.2 Elongation

Once a primer is in place, real DNA synthesis (elongation) can begin. We have already identified the pol III holoenzyme as the enzyme that carries out elongation in *E. coli,* and DNA polymerases δ and ε as the enzymes that elongate the lagging and leading strands, respectively, in eukaryotes. The *E. coli* system is especially well characterized, and the data point to an elegant method of coordinating the synthesis of lagging and leading strands in a way that keeps the pol III holoenzyme engaged with the template so replication can be highly processive, and therefore very rapid. Let us focus on this *E. coli* elongation mechanism, beginning with a discussion of the speed of elongation.

Speed of Replication

Minsen Mok and Kenneth Marians performed one of the studies that measured the rate of fork movement in vitro with the pol III holoenzyme. They created a synthetic circular template for rolling circle replication, illustrated in Figure 21.6. This template contained a ^{32}P-labeled, tailed, full-length strand with a free 3′-hydroxyl group for priming. Mok and Marians incubated this template with either holoenzyme plus preprimosomal proteins and SSB, or plus DnaB helicase alone. At 10-sec intervals, they removed the labeled product DNAs and measured their lengths by electrophoresis. Panels (a) and (b) in Figure 21.7 depict the results with the two reactions, and Figure 21.7c shows plots of the rates of fork movement with the two reactions.

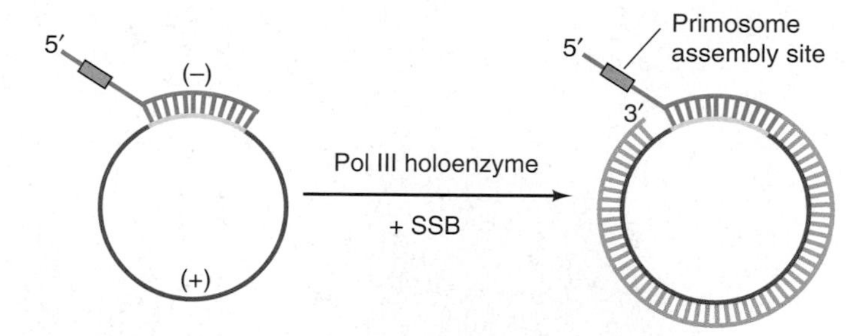

Figure 21.6 Synthesis of template used to measure fork velocity in vitro. Mok and Marians started with the 6702-nt positive strand (red) from the f1 phage and annealed it to a primer (green) that hybridized over a 282-nt region (yellow). This primer contained a primosome assembly site (orange). Mok and Marians elongated the primer with pol III holoenzyme and single-strand binding protein (SSB) to create the negative strand (blue). The product was a double-stranded template for multiple rounds of rolling circle replication, in which the free 3′-end could serve as the primer. (*Source:* Adapted from Mok, M. and K.J. Marians, The *Escherichia coli* preprimosome and DNA B helicase can form replication forks that move at the same rate. *Journal of Biological Chemistry* 262:16645, 1987.)

Both plots yielded rates of 730 nt/sec, close to the in vivo rate of almost 1000 nt/sec.

Furthermore, the elongation in these reactions with holoenzyme was highly processive. As we have mentioned, processivity is the ability of the enzyme to stick to its job a long time without falling off and having to reinitiate. This is essential because reinitiation is a time-consuming process, and little time can be wasted in DNA replication. To measure processivity, Mok and Marians performed the same elongation assay as described in Figure 21.7, but included either of two substances that would prevent reinitiation if the holoenzyme dissociated from the template. These substances were a competing DNA, poly(dA), and an antibody directed against the β-subunit of the holoenzyme. In the presence of either of these competitors, the elongation rate was just as fast as in their absence, indicating that the holoenzyme did not dissociate from the template throughout the process of elongation of the primer by at least 30 kb. Thus, the holoenzyme is highly processive in vitro, just as it is in vivo.

SUMMARY The pol III holoenzyme synthesizes DNA at the rate of about 730 nt/sec in vitro, just a little slower than the rate of almost 1000 nt/sec observed in vivo. This enzyme is also highly processive, both in vitro and in vivo.

The Pol III Holoenzyme and Processivity of Replication

The pol III core by itself is a very poor polymerase. It puts together about 10 nt and then falls off the template. Then it has to spend about a minute reassociating with the

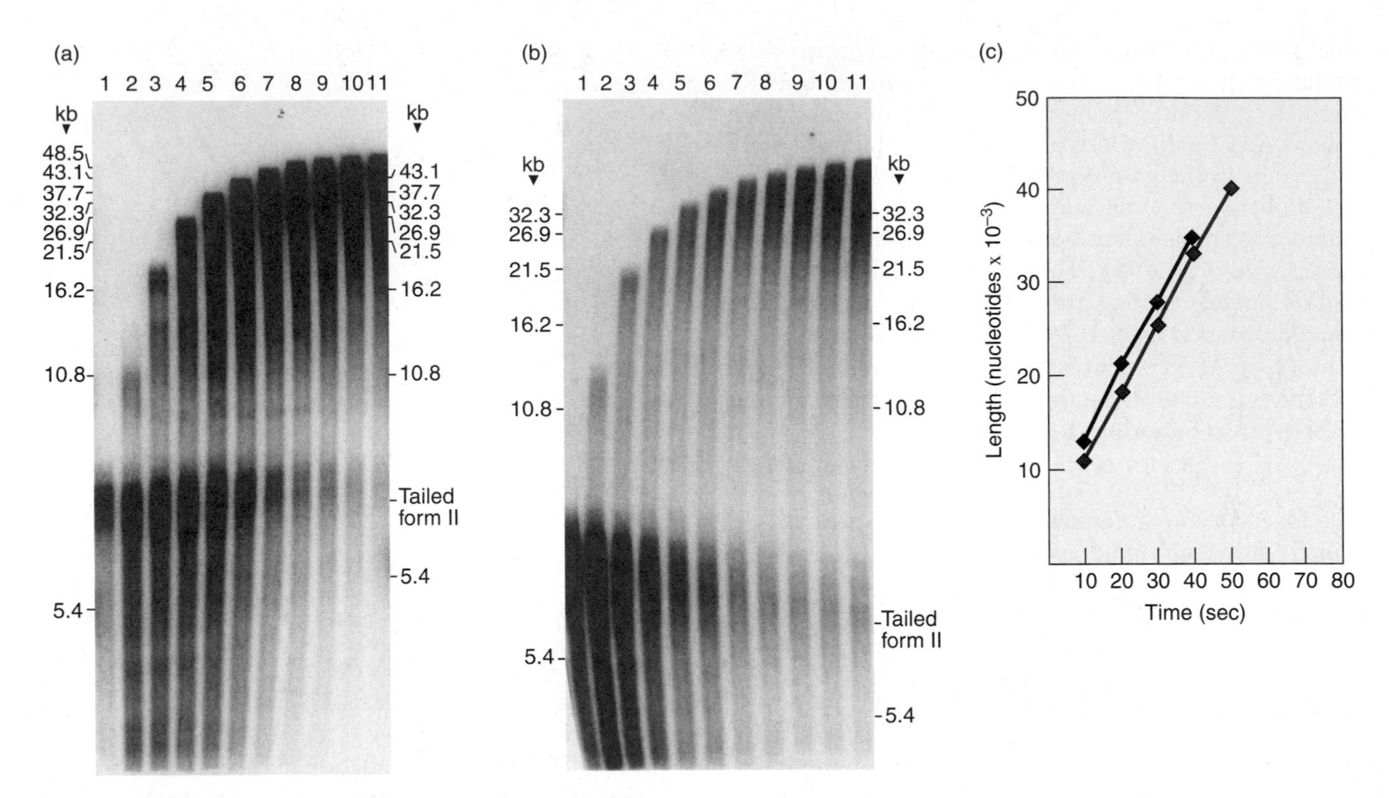

Figure 21.7 Measuring the rate of fork movement in vitro. Mok and Marians labeled the negative strand of the tailed template in Figure 21.6 and used it in in vitro reactions with pol III holoenzyme plus: **(a)** the preprimosomal proteins (the primosomal proteins minus DnaG); or **(b)** DnaB alone. They took samples from the reactions at 10-sec intervals, beginning with lanes 1 at zero time and lanes 2 at 10 sec, electrophoresed them, and then autoradiographed the gel. Recall that electrophoretic mobilities are a log function, not a linear function, of mass. The numbers on the left in each panel are marker sizes, not the sizes of DNA products. Panel **(c)** shows a plot of the results from the first five and four time points from panels (a) (red) and (b) (blue), respectively. (*Source:* Mok M. and K.J. Marians, The *Escherichia coli* preprimosome and DNA B helicase can form replication forks that move at the same rate. *Journal of Biological Chemistry* 262 no. 34 (5 Dec 1987) f. 6a–b, p. 16650. Copyright © American Society for Biochemistry and Molecular Biology.)

template and the nascent DNA strand. This contrasts sharply with the situation in the cell, where the replicating fork moves at the rate of almost 1000 nt/sec. Obviously, something important is missing from the core.

That "something" is an agent that confers processivity on the holoenzyme, allowing it to remain engaged with the template while polymerizing at least 50,000 nt before stopping—quite a contrast to the 10 nt polymerized by the core before it stops. Why such a drastic difference? The holoenzyme owes its processivity to a **"sliding clamp"** that holds the enzyme on the template for a long time. The β-subunit of the holoenzyme performs this sliding clamp function, but it cannot associate by itself with the preinitiation complex (core plus DNA template). It needs a **clamp loader** to help it join the complex, and a group of subunits called the **γ complex** provides this help. The γ complex includes the γ-, δ-, δ′-, χ-, and ψ-subunits. In this section, we will examine the activities of the β clamp and the clamp loader.

The β clamp One way we can imagine the β-subunit conferring processivity on the pol III core is by binding both the core complex and DNA. That way, it would tie the core to the DNA and keep it there—hence the term **β clamp.** In the course of probing this possibility, Mike O'Donnell and colleagues demonstrated direct interaction between the β- and α-subunits. They mixed various combinations of subunits, then separated subunit complexes from individual subunits by gel filtration. They detected subunits by gel electrophoresis, and activity by adding the missing subunits and measuring DNA synthesis. Figure 21.8 depicts the electrophoresis results. It is clear that α and ε bind to each other, as we would expect, because they are both part of the core. Furthermore, α, ε, and β form a complex, but which subunit does β bind to, α or ε? Panels (d) and (e) show the answer: β binds to α alone (both subunits peak in fractions 60–64), but not to ε alone (β peaks in fractions 68–70, whereas ε peaks in fractions 76–78). Thus, α is the core subunit to which β binds.

This scheme demands that β be able to slide along the DNA as α and ε together replicate it. This in turn suggests that the β clamp would remain bound to a circular DNA, but could slide right off the ends of a linear DNA. To test this possibility, O'Donnell and colleagues performed the experiment reported in Figure 21.9. The general strategy of this experiment was to load ^{3}H-labeled β dimers onto circular, double-stranded phage DNA with the help of the γ complex, then to treat the DNA in various ways to see if

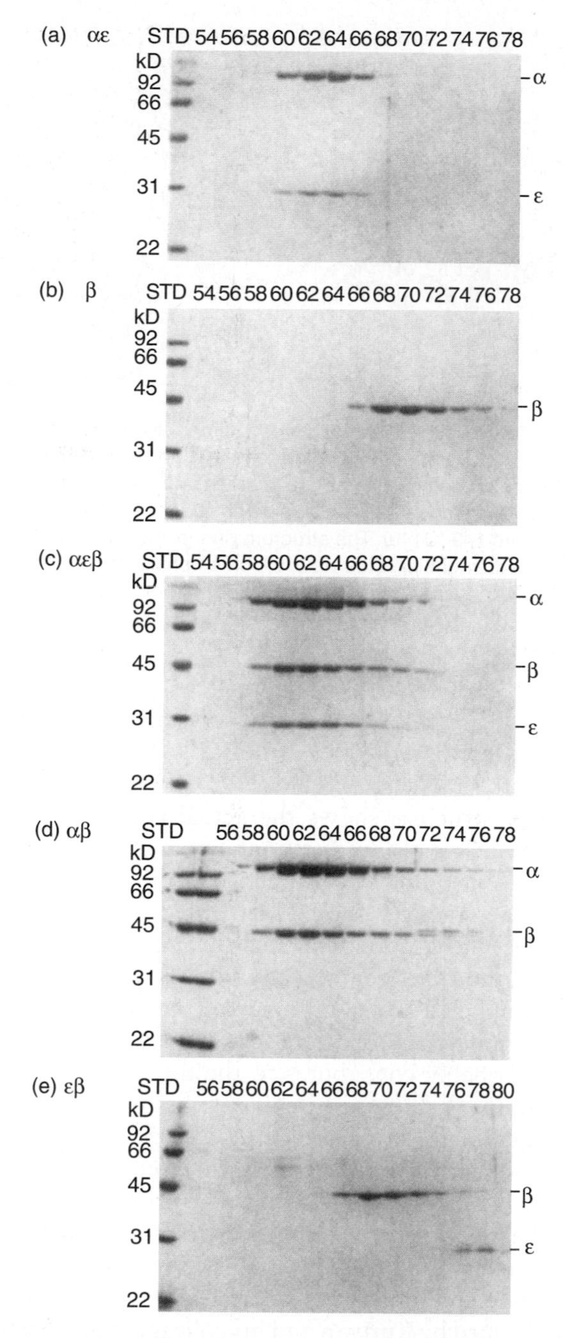

Figure 21.8 The Pol III subunits α and β bind to each other. O'Donnell and colleagues mixed various combinations of pol III subunits as follows: **(a)** α+ε; **(b)** β; **(c)** α+ε+β; **(d)** α+β; **(e)** ε+β. Then they subjected the mixtures to gel filtration to separate complexes from free subunits, then electrophoresed fractions from the gel filtration column to detect complexes. If a complex formed, the subunits in the complex should appear in the same fractions, as the α and ε fractions do in panel (a). (*Source:* Stukenberg, P.T., P.S. Studwell-Vaughn, and M. O'Donnell, Mechanism of the sliding β-clamp of DNA polymerase III holoenzyme. *Journal of Biological Chemistry* 266 no. 17(15 June 1991) figs. 2a–e, 3, pp. 11330–31. American Society for Biochemistry and Molecular Biology.)

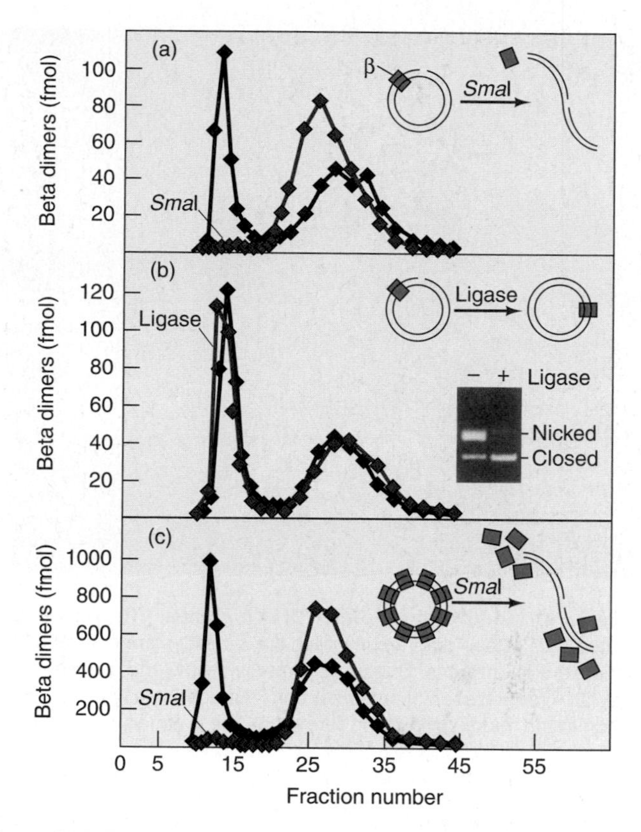

Figure 21.9 The β clamp can slide off the ends of a linear DNA. O'Donnell and colleagues loaded ^{3}H-labeled β dimers onto various DNAs, with the help of the γ complex, then treated the complexes in various ways as described. Finally, they subjected the mixtures to gel filtration to separate protein–DNA complexes (which were large and eluted quickly from the column, around fraction 15), from free protein (which was relatively small and eluted later, around fraction 28). **(a)** Effect of linearizing the DNA with *Sma*I. DNA was cut once with *Sma*I and then assayed (red). Uncut DNA was also assayed (blue). **(b)** Effect of removing a nick in the template. The nick in the template was removed with DNA ligase before assay (red), or left alone (blue). The inset shows the results of electrophoresis of DNAs before and after the ligase reaction. **(c)** Many β dimers can be loaded onto the DNA and then lost when it is linearized. The ratio of β dimers loaded onto DNA templates was increased by raising the concentration of β-subunits and lowering the concentration of DNA templates. Then the DNA was either cut with *Sma*I before assay (red) or not cut (blue). (*Source:* Stukenberg, P.T., P.S. Studwell-Vaughn, and M. O'Donnell, Mechanism of the sliding β-clamp of DNA polymerase III holoenzyme. *Journal of Biological Chemistry* 266 no. 17 (15 June 1991) fig. 3, p. 11331. American Society for Biochemistry and Molecular Biology.)

the β dimers could dissociate from the DNA. The assay for β-binding to DNA was gel filtration. Independent β dimers emerge from a gel filtration column much later than they do when they are bound to DNA.

In panel (a), the DNA was treated with *Sma*I to linearize the DNA, then examined to see whether the β clamp had slid off. It remained bound to circular DNA, but had dissociated from linearized DNA, apparently by sliding off the ends. Panel (b) demonstrates that the nick in the circular DNA is not what caused retention of the β dimer, because the nick can be removed with DNA ligase, and the

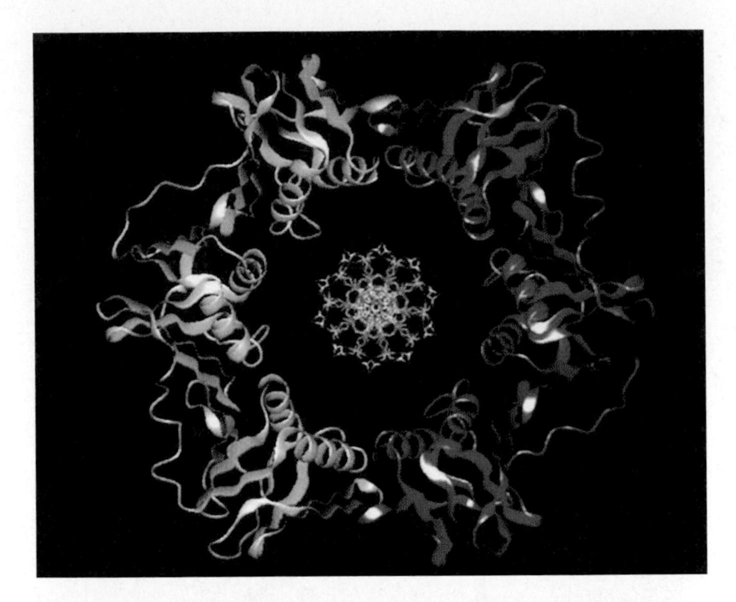

Figure 21.10 Model of the β dimer/DNA complex. The β dimer is depicted by a ribbon diagram in which the α-helices are coils and the β-sheets are flat ribbons. One β monomer is yellow and the other is red. A DNA model, seen in cross section, is placed in a hypothetical position in the middle of the ring formed by the β dimer. (*Source:* Kong, X.-P., R. Onrust, M. O'Donnell, and J. Kuriyan, Three-dimensional structure of the beta subunit of *E. coli* DNA polymerase III holoenzyme: A sliding DNA clamp. *Cell* 69 (1 May 1992) f. 1, p. 426. Reprinted by permission of Elsevier Science.)

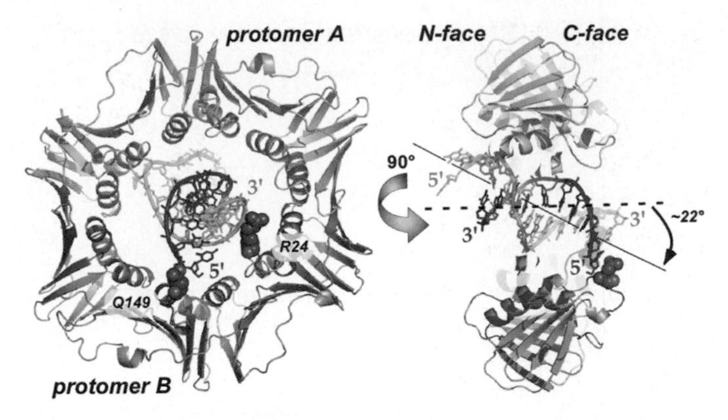

Figure 21.11 Co-crystal structure of β dimer and primed DNA template. The two β monomers (protomers A and B) are in gold and blue, with the primed DNA template in green and red. Magenta and blue space-filling models show the side chains of arginine 24 (R24) and glutamine 149 (Q149). The structure on the left is a front view; the structure on the right is a side view, which emphasizes the 22-degree tilt of the DNA. (*Source:* Georgescu et al., Structure of a sliding clamp on DNA. *Cell* 132 (11 January 2008) f. 3a, p. 48. Reprinted by permission of Elsevier Science.)

β dimer remains bound to the DNA. The inset shows electrophoretic evidence that the ligase really did remove the nick because the nicked form disappeared and the closed circular form was enhanced. Panel (c) shows that adding more β-subunit to the loading reaction increased the number of β dimers bound to the circular DNA. In fact, more than 20 molecules of β-subunit could be bound per molecule of circular DNA. This is what we would expect if many holoenzymes can replicate the DNA in tandem.

If the β dimers are lost from linear DNA by sliding off the ends, one ought to be able to prevent their loss by binding other proteins to the ends of the DNA. O'Donnell's group did this in experiments, not shown here, by binding two different proteins to the ends of the DNA and demonstrating that the β dimers no longer fell off. Indeed, single-stranded tails at the ends of the DNA, even without protein attached, proved to be an impediment to the β dimers sliding off.

Mike O'Donnell and John Kuriyan used x-ray crystallography to study the structure of the β clamp. The pictures they produced provided a perfect rationale for the ability of the β clamp to remain bound to a circular DNA but not to a linear one: The β dimer forms a ring that can fit around the DNA. Thus, like a ring on a string, it can readily fall off if the string is linear, but not if the string is circular. Figure 21.10 is one of the models O'Donnell and Kuriyan constructed; it shows the ring structure of the β dimer, with a scale model of B-form DNA placed in the middle.

In 2008, O'Donnell and colleagues obtained the structure of a co-crystal of a β dimer bound to a primed DNA template. Figure 21.11 shows this crystal structure, which demonstrates that the β clamp really does encircle the DNA, as the model in Figure 21.10 predicted. However, this newer structure shows the actual geometry of DNA within the β clamp, and it contains a bit of a surprise: Instead of extending straight through the β clamp, like a finger through a ring, the DNA is tilted about 22 degrees with respect to a horizontal line through the clamp. Furthermore, the DNA contacts the side chains of two amino acids, arginine 24 and glutamine 149, both of which lie on the C-terminal face of the β clamp. This protein–DNA contact probably contributes to the tilt of the DNA with respect to the β dimer.

As mentioned in Chapter 20, eukaryotes also have a processivity factor called PCNA, which performs the same function as the bacterial β clamp. The primary structure of PCNA bears no apparent similarity to that of the β clamp, and the eukaryotic protein is only two-thirds the size of its prokaryotic counterpart. Nevertheless, x-ray crystallography performed by Kuriyan and his colleagues demonstrated that yeast PCNA forms a trimer with a structure arrestingly similar to that of the β clamp dimer: a ring that can encircle a DNA molecule, as shown in Figure 21.12.

SUMMARY The Pol III core (αε or αεθ) does not function processively by itself, so it can replicate only a short stretch of DNA before falling off the template. By contrast, the core plus the β-subunit can replicate DNA processively at a rate approaching 1000 nt/sec. The β-subunit forms a dimer that is ring-shaped. This ring fits around a DNA template and interacts with the α-subunit of the core to tether

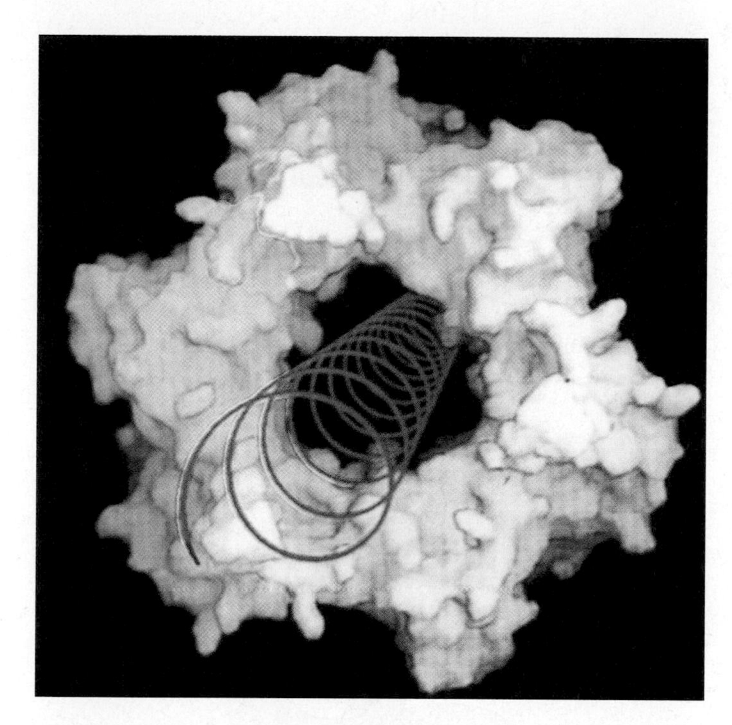

Figure 21.12 Model of PCNA–DNA complex. Each of the monomers of the PCNA trimer is represented by a different pastel color. The shape of the trimer is based on x-ray crystallography analysis. The red helix represents the probable location of the sugar–phosphate backbone of a DNA associated with the PCNA trimer. (*Source:* Krishna, T.S.R., X.-P. Kong, S. Gary, P.M. Burgers, and J. Kuriyan, Crystal structure of the eukaryotic DNA polymerase processivity factor PCNA. *Cell* 79 (30 Dec 1994) f. 3b, p. 1236. Reprinted by permission of Elsevier Science.)

the whole polymerase and template together. This is why the holoenzyme stays on its template so long and is therefore so processive. The eukaryotic processivity factor PCNA forms a trimer with a similar ring shape that can encircle DNA and hold DNA polymerase on the template.

The Clamp Loader O'Donnell and his colleagues demonstrated the function of the clamp loader in an experiment presented in Figure 21.13. These scientists used the α- and ε-subunits instead of the whole core, because the θ-subunit was not essential in their in vitro experiments. As template, they used a single-stranded M13 phage DNA annealed to a primer. They knew that highly processive holoenzyme could replicate this DNA in about 15 sec but that the αε core could not give a detectable amount of replication in that time. Thus, they reasoned that a 20-sec pulse of replication would allow all processive polymerase molecules the chance to complete one cycle of replication, and therefore the number of DNA circles replicated would equal the number of processive polymerases. Figure 21.13a shows that each femtomole (fmol, or 10^{-15} mol) of γ complex resulted in about 10 fmol of circles replicated in the presence of αε core and β-subunit. Thus, the γ complex acts

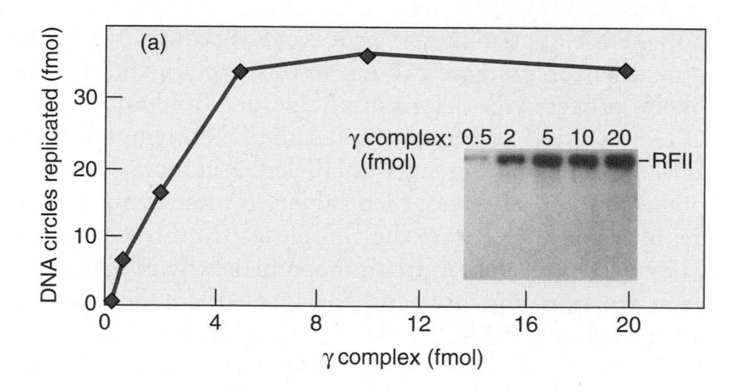

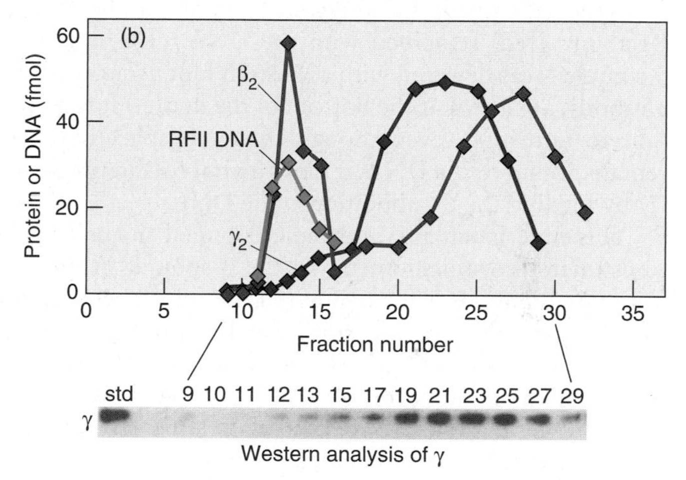

Figure 21.13 Involvement of β and γ complex in processivity. (a) The γ complex acts catalytically in forming a processive polymerase. O'Donnell and coworkers added increasing amounts of γ complex (indicated on the *x* axis) to a primed M13 phage DNA template coated with SSB, along with αε core, and the β-subunit of pol III holoenzyme. Then they allowed a 20-sec pulse of DNA synthesis in the presence of [α-^{32}P]ATP to label the DNA product. They determined the radioactivity of part of each reaction and converted this to fmol of DNA circles replicated. To check for full circle replication, they subjected another part of each reaction to gel electrophoresis. The inset shows the result: The great majority of each product is full-circle size (RFII). **(b)** The β-subunit, but not the γ complex associates with DNA in the preinitiation complex. O'Donnell and colleagues added ^{3}H-labeled β-subunit and unlabeled γ complex to primed DNA coated with SSB, along with ATP to form a preinitiation complex. Then they subjected the mixture to gel filtration to separate preinitiation complexes from free proteins. They detected the β-subunit in each fraction by radioactivity, and the γ complex by Western blotting, with an anti-γ antibody as probe (bottom). The plot shows that the β-subunit (as dimers) bound to the DNA in the preinitiation complex, but the γ complex did not. (*Source:* Stukenberg, P.T., P.S. Studwell-Vaughn, and M. O'Donnell, Mechanism of the sliding [beta]-clamp of DNA polymerase III holoenzyme. *Journal of Biological Chemistry* 266 (15 June 1991) f. 1a&c, p. 11329. American Society for Biochemistry and Molecular Biology.)

catalytically: One molecule of γ complex can sponsor the creation of many molecules of processive polymerase. The inset in this figure shows the results of gel electrophoresis of the replication products. As expected of processive replication, they are all full-length circles.

This experiment suggested that the γ complex itself is not the agent that provides processivity. Instead, the γ complex could act catalytically to add something else to the core polymerase that makes it processive. Because β was the only other polymerase subunit in this experiment, it is the likely processivity-determining factor. To confirm this, O'Donnell and colleagues mixed the DNA template with ^{3}H-labeled β-subunit and unlabeled γ complex to form preinitiation complexes, then subjected these complexes to gel filtration to separate the complexes from free proteins. They detected the preinitiation complexes by adding αε to each fraction and assaying for labeled double-stranded circles formed (RFII, green). Figure 21.13b demonstrates that only a trace of γ complex (blue) remained associated with the DNA, but a significant fraction of the labeled β-subunit (red) remained with the DNA. (The unlabeled γ complex was detected with a Western blot using an anti-γ antibody, as shown at the bottom of the figure.) It is important to note that, even though the γ complex does not remain bound to the DNA, it plays a vital role in processivity by loading the β-subunit onto the DNA.

This experiment also allowed O'Donnell and colleagues to estimate the stoichiometry of the β-subunit in the preinitiation complex. They compared the fmol of β with the fmol of complex, as measured by the fmol of double-stranded circles produced. This analysis yielded a value of about 2.8 β-subunits/complex, which would be close to one β *dimer*/complex, in accord with other studies that suggested that β acts as a dimer.

Implicit in the discussion so far is the fact that ATP is required to load the β clamp onto the template. Peter Burgers and Kornberg demonstrated the necessity for ATP (or dATP) with an assay that did not require dATP for replication. The template in this case was poly(dA) primed with oligo(dT). The results showed that ATP or dATP is required for high-activity elongation of the oligo(dT) primer with dTMP.

How does the clamp loader pry apart the β dimer to allow it to clamp around DNA? O'Donnell, Kuriyan, and colleagues have determined the crystal structures of two complexes that give strong hints about how the clamp loader works. One of these was the structure of the active part of the clamp loader (a γδδ′ complex). The other was the structure of a modified β–δ complex composed of: a monomer of a mutant form of β (β_{mt}) that is unable to dimerize; and a fragment of δ that can interact with β.

The crystal structure of this modified β–δ complex showed that the interaction between δ and a β monomer would be expected to weaken the binding at one interface between the two β monomers in two ways. First, δ acts as a molecular wrench by inducing a conformational change in the β dimer interface such that it no longer dimerizes as readily. Second, δ changes the curvature of one β-subunit so that it no longer naturally forms a ring with the other subunit. Instead, it forms a structure that resembles a lock washer. Figure 21.14 illustrates these concepts. Notice that δ binds to only one β monomer in the β clamp (there is only one δ per β dimer in the pol III holoenzyme), so it weakens only one dimer interface, and therefore forces ring opening. If δ bound to both β monomers, it would presumably cause the two monomers to dissociate entirely.

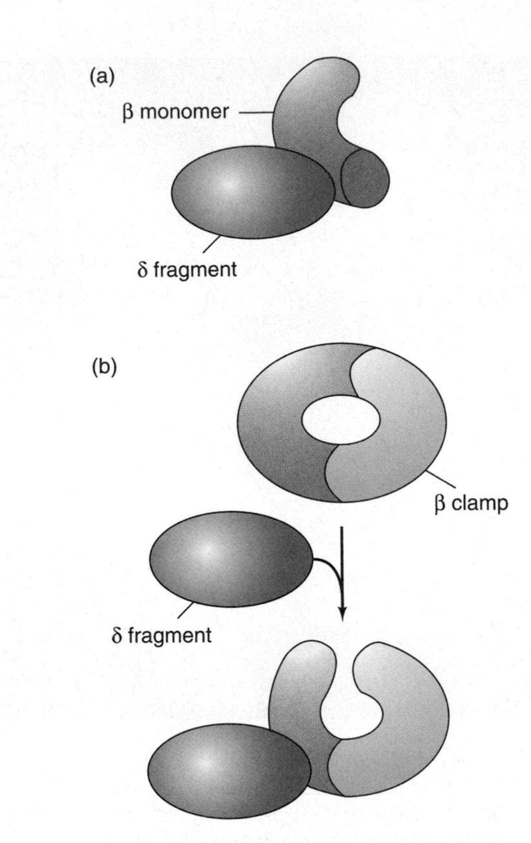

Figure 21.14 Model for the effect of δ binding on the β dimer. **(a)** Shape of the complex between the δ fragment and the β_{mt} monomer. **(b)** Effect of δ binding on the β clamp. The δ-subunit (or the δ fragment) causes the β dimer interface at the top to weaken and also changes the curvature of the β monomer on the left such that it can no longer form a complete circle with the other monomer. The result is an opening of the clamp. (*Source:* Adapted from Ellison, V. and B. Stillman, Opening of the clamp: An intimate view of an ATP-driven biological machine. *Cell* 106 [2001] p. 657, f. 3.)

These structural studies and earlier biochemical studies, some of which we will discuss later in this chapter, showed that δ on its own binds readily to a β monomer, but that δ in the context of the clamp loader complex cannot bind to the β clamp unless ATP is present. So the role of ATP appears to be to change the shape of the clamp loader to expose the δ-subunit so it can bind to one of the β-subunits and pry open the β clamp.

SUMMARY The β-subunit needs help from the γ complex (γ, δ, δ′, χ, and ψ) to load onto the DNA template. The γ complex acts catalytically in forming this processive αδβ complex, so it does not remain associated with the complex during processive replication. Clamp loading is an ATP-dependent process. The energy from ATP changes the conformation of the clamp loader such that the δ-subunit can bind to one of the β-subunits of the clamp. This binding opens the clamp and allows it to encircle DNA.

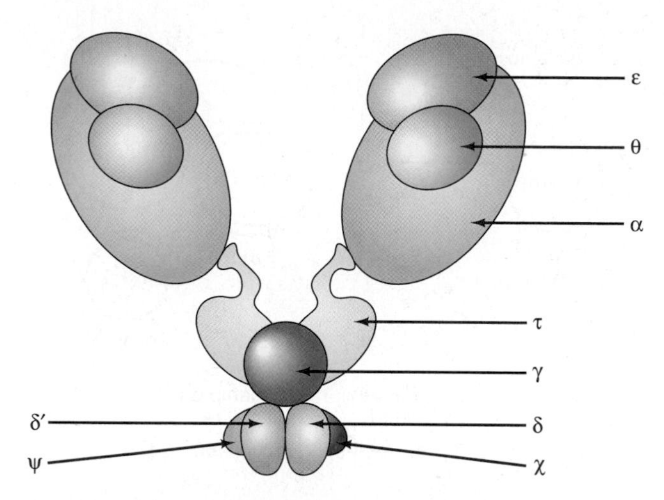

Figure 21.15 Model of the Pol III* subassembly. Note that two cores and two τ-subunits are present, but only one γ-complex (γ, δ, δ′, χ, and ψ). The τ-subunits are joined to the cores by their flexible C-terminal domains.

Lagging Strand Synthesis Structural studies on pol III* (holoenzyme minus the β clamp) have shown that the enzyme consists of two core polymerases, linked through a dimer of the τ-subunit to a clamp loader, as illustrated in Figure 21.15. The following reasoning suggests that the τ-subunit serves as a dimerizing agent for the core enzyme: The α-subunit is a monomer in its native state, but τ is a dimer. Furthermore, τ binds directly to α, so α is automatically dimerized by binding to the two τ-subunits. In turn, ε is dimerized by binding to the two α-subunits, and θ is dimerized by binding to the two ε-subunits. The two τ-subunits are products of the same gene that produces the γ-subunit. However, the γ-subunit lacks a 24-kDa domain (τ_c) at the C-terminus of the τ-subunits because of a programmed frameshift during translation. The two τ_c domains provide flexible linkers between the core polymerases and the γ complex.

The fact that the holoenzyme contains two core polymerases fits very nicely with the fact that two DNA strands need to be replicated. This leads directly to the suggestion that each of the core polymerases replicates one of the strands as the holoenzyme follows the moving fork. This is straightforward for the core polymerase replicating the leading strand, as that replication moves in the same direction as the fork. But it is more complicated for the core polymerase replicating the lagging strand, because that replication occurs in the direction opposite to that of the moving fork. This means that the lagging strand must form a loop, as pictured in Figure 21.16. Because this loop extends as an Okazaki fragment grows and then retracts to begin synthesis of a new Okazaki fragment, the loop resembles the slide of a trombone, and this model is sometimes called the "trombone model."

Because discontinuous synthesis of the lagging strand must involve repeated dissociation and reassociation of the

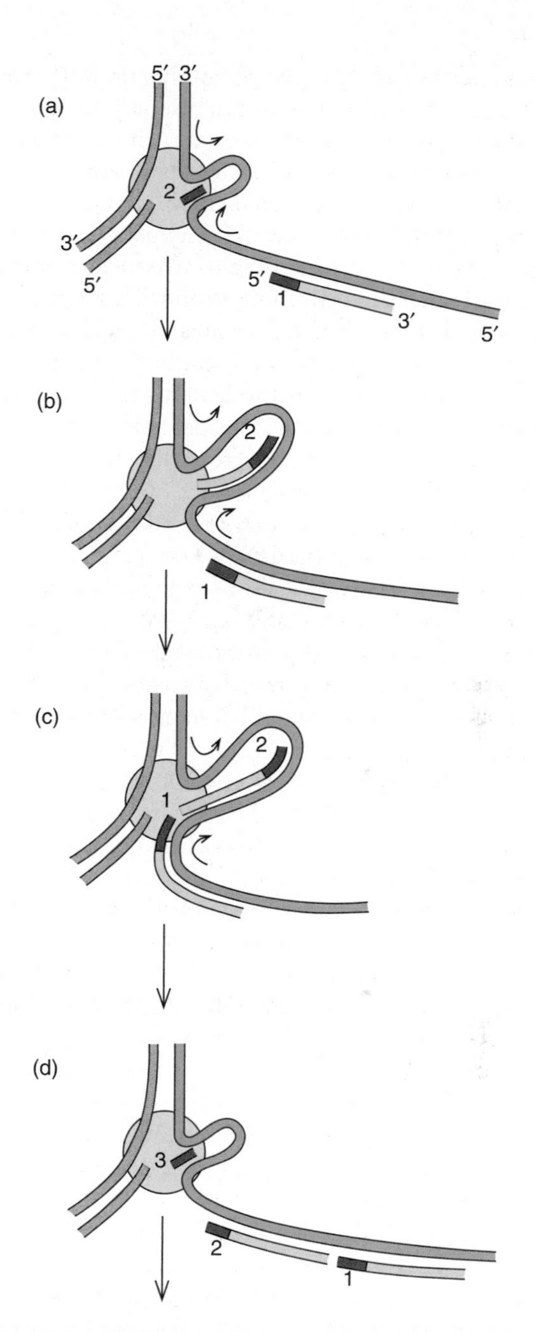

Figure 21.16 A model for simultaneous synthesis of both DNA strands. **(a)** The lagging template strand (blue) has formed a loop through the replisome (gold), and a new primer, labeled 2 (red), has been formed by the primase. A previously synthesized Okazaki fragment (green, with red primer labeled 1) is also visible. The leading strand template and its progeny strand are shown at left (gray), but the growth of the leading strand is not considered here. **(b)** The lagging strand template has formed a bigger loop by feeding through the replisome from the top and bottom, as shown by the arrows. The motion of the lower part of the loop (lower arrow) allows the second Okazaki fragment to be elongated. **(c)** Further elongation of the second Okazaki fragment brings its end to a position adjacent to the primer of the first Okazaki fragment. **(d)** The replisome releases the loop, which permits the primase to form a new primer (number 3). The process can now begin anew.

core polymerase from the template, this model raises two important questions: First, how can discontinuous synthesis of the lagging strand possibly keep up with continuous (or perhaps discontinuous) synthesis of the leading strand? If the pol III core really dissociated completely from the template after making each Okazaki fragment of the lagging strand, it would take a long time to reassociate and would fall hopelessly behind the leading strand. This would be true even if the leading strand replicated discontinuously, because no dissociation and reassociation of the pol III core is necessary in synthesizing the leading strand. A second, related question is this: How is repeated dissociation and reassociation of the pol III core from the template compatible with the highly processive nature of DNA replication? After all, the β clamp is essential for processive replication, but once it clamps onto the DNA, how can the core polymerase dissociate every 1–2 kb as it finishes one Okazaki fragment and jumps forward to begin elongating the next?

The answer to the first question seems to be that the pol III core making the lagging strand does not really dissociate completely from the template. It remains tethered to it by its association with the core that is making the leading strand. Thus, it can release its grip on its template strand without straying far from the DNA. This enables it to find the next primer and reassociate with its template within a fraction of a second, instead of the many seconds that would be required if it completely left the DNA.

The second question requires us to look more carefully at the way the β clamp interacts with the clamp loader and with the core polymerase. We will see that these two proteins compete for the same binding site on the β clamp, and that the relative affinities of the clamp for one or the other of them shifts back and forth to allow dissociation and reassociation of the core from the DNA. We will also see that the clamp loader can act as a clamp unloader to facilitate this cycling process.

Theory predicts that the pol III* synthesizing the lagging strand must dissociate from one β clamp as it finishes one Okazaki fragment and reassociate with another β clamp to begin making the next Okazaki fragment. But does dissociation of pol III* from its β clamp actually occur? To find out, O'Donnell and his colleagues prepared a primed M13 phage template (M13mp18) and loaded a β clamp and pol III* onto it. Then they added two more primed phage DNA templates, one (M13Gori) preloaded with a β clamp and the other (ϕX174) lacking a β clamp. Then they incubated the templates together under replication conditions long enough for the original template and secondary template to be replicated. They knew they would see replicated M13mp18 DNA, but the interesting question is this: Which secondary template will be replicated, the one with, or the one without, the β clamp? Figure 21.17 (lanes 1–4) demonstrates that replication occurred preferentially on the M13Gori template—the one with the β clamp. What if they put the β clamp on the other template instead? Lanes 5–8 show that in that case, the other template (ϕX174) was preferentially replicated. If the pol III* kept its original β clamp, it could have begun replicating either secondary template, regardless of which was preloaded with a β clamp. Thus, the results of this experiment imply that dissociation of pol III* from the template, and its β clamp, really does happen, and the enzyme can bind to another template (or another part of the same template), if another β clamp is present.

To check this conclusion, these workers labeled the β clamp with ^{32}P by phosphorylating it with [γ-^{32}P]ATP, then labeled pol III* with ^{3}H in either the θ- or τ-subunits, or in the γ complex. Then they allowed these labeled complexes to either idle on a gapped template in the presence of only dGTP and dCTP or to fill in the whole gap with all four dNTPs and thus terminate. Finally, they subjected the

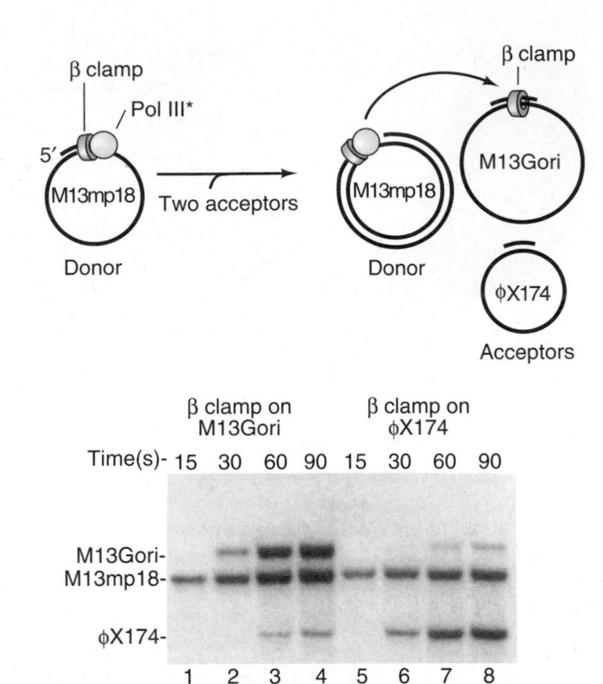

Figure 21.17 Test of the cycling model. If one assembles a pol III* complex with a β clamp on one primed template (M13mp18, top left) and presents it with two acceptor primed templates, one with a β clamp (M13Gori) and one without (ϕX174), the pol III* complex should choose the template with the clamp (M13Gori, in this case) to replicate when it has finished replicating the original template. O'Donnell and colleagues carried out this experiment, allowing enough time to replicate both the donor and acceptor templates. They also included labeled nucleotides so the replicated DNA would be labeled. Then they electrophoresed the DNAs and detected the labeled DNA products by gel electrophoresis. The electrophoresis of the replicated DNA products (bottom) show that the acceptor template with the β clamp was the one that was replicated. When the β clamp was on the M13Gori acceptor template, replication of this template predominated. On the other hand, when the β clamp was on the ϕX174 template, this was the one that was favored for replication. The positions of the replicated templates are indicated at left. (*Source:* Stukenberg, P.T., J. Turner, and M. O'Donnell, An explanation for lagging strand replication: Polymerase hopping among DNA sliding clamps. *Cell* 78 (9 Sept 1994) f. 2, p. 878. Reprinted by permission of Elsevier Science.)

reaction mixtures to gel filtration and determined whether the two labels had separated. When the polymerase merely idled, the labeled β clamp and pol III* stayed together on the DNA template. By contrast, when termination occurred, the pol III* separated from its β clamp, leaving it behind on the DNA. O'Donnell and coworkers observed the same behavior regardless of which subunit of pol III* was labeled, so this whole entity, not just the core enzyme, must separate from the β clamp and DNA template upon termination of replication.

The *E. coli* genome is 4.6 Mb long, and its lagging strand, at least, is replicated in Okazaki fragments only 1–2 kb long. This means that over 2000 priming events are required on each template, so at least 2000 β clamps are needed. Because an *E. coli* cell holds only about 300 β dimers, the supply of β clamps would be rapidly exhausted if they could not recycle somehow. This would require that they dissociate from the DNA template. Does this happen? To find out, O'Donnell and colleagues assembled several β clamps onto a gapped template, then removed all other protein by gel filtration. Then they added pol III* and reran the gel filtration step. Figure 21.18a shows that, sure enough, the β clamps dissociated in the presence of pol III*, but not without the enzyme. Figure 21.18b demonstrates that these liberated β clamps were also competent to be loaded onto an acceptor template.

It is clear from what we have learned so far that the β clamp can interact with both the core polymerase and the γ complex (the clamp loader). It must associate with the core during synthesis of DNA to keep the polymerase on the template. Then it must dissociate from the template so it can move to a new site on the DNA where it can interact with another core to make a new Okazaki fragment. This movement to a new DNA site, of course, requires the β clamp to interact with a clamp loader again. One crucial question remains: How does the cell orchestrate the shifting back and forth of the β clamp's association with core and with clamp loader?

To begin to answer this question, it would help to show how and when the core and the clamp loader interact with the β clamp. O'Donnell and associates first answered the "how" question, demonstrating that the α-subunit of the core contacts β, and the δ-subunit of the clamp loader also contacts β. One assay these workers used to reveal these interactions was **protein footprinting.** This method works on the same principle as DNase footprinting, except the starting material is a labeled protein instead of a DNA, and protein-cleaving reagents are used instead of DNase. In this case, O'Donnell and colleagues introduced a six-amino acid protein kinase recognition sequence into the C-terminus of the β-subunit by manipulating its gene. They named the altered product β^{PK}. Then they phosphorylated this protein in vitro using protein kinase and labeled ATP (an ATP derivative with an oxygen in the γ-phosphate replaced by ^{35}S); this procedure labeled the protein at its C-terminus.

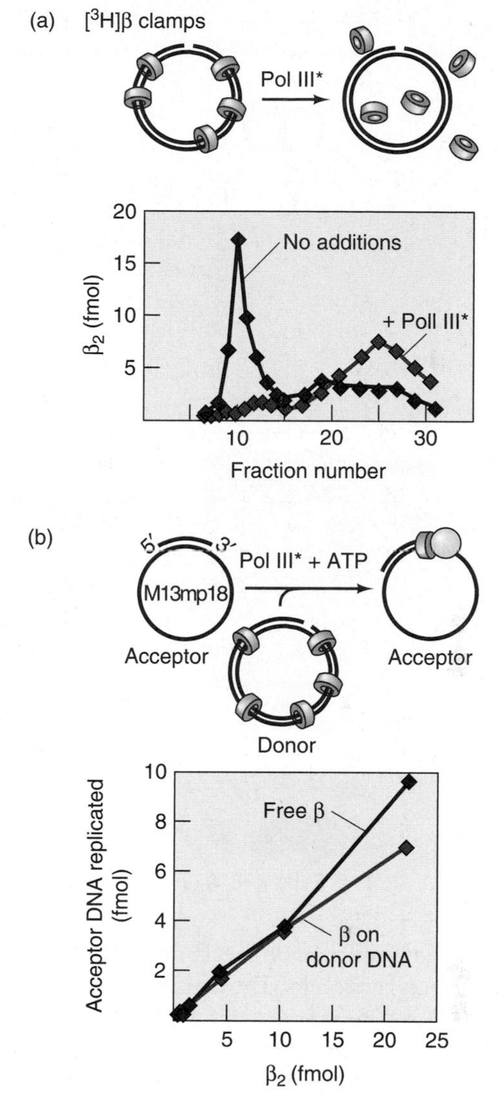

Figure 21.18 Pol III* has clamp unloading activity. (a) Clamp unloading. O'Donnell and colleagues used the γ complex to load β clamps (blue, top) onto a gapped circular template, then removed the γ complex by gel filtration. Then they added pol III* and performed gel filtration again. The graph of the results (bottom) shows β clamps that were treated with pol III* (red) were released from the template, whereas those that were not treated with pol III* (blue) remained associated with the template. **(b)** Recycling of β clamps. The β clamps from a donor β clamp–template complex treated with pol III* (red) were just as good at rebinding to an acceptor template as were β clamps that were free in solution (blue). (*Source:* Adapted from Stukenberg, P.T., J. Turner, and M. O'Donnell, An explanation for lagging strand replication: Polymerase hopping among DNA sliding clamps. *Cell* 78:883, 1994.)

(Note that this is similar to labeling a DNA at one of its ends for DNase footprinting.) First they showed that the δ-subunit of the clamp loader and the α-subunit of the core could each protect β^{PK} from phosphorylation, suggesting that both of these proteins contact β^{PK}.

Protein footprinting reinforced these conclusions. O'Donnell and colleagues mixed labeled β^{PK} with various proteins, then cleaved the protein mixture with two

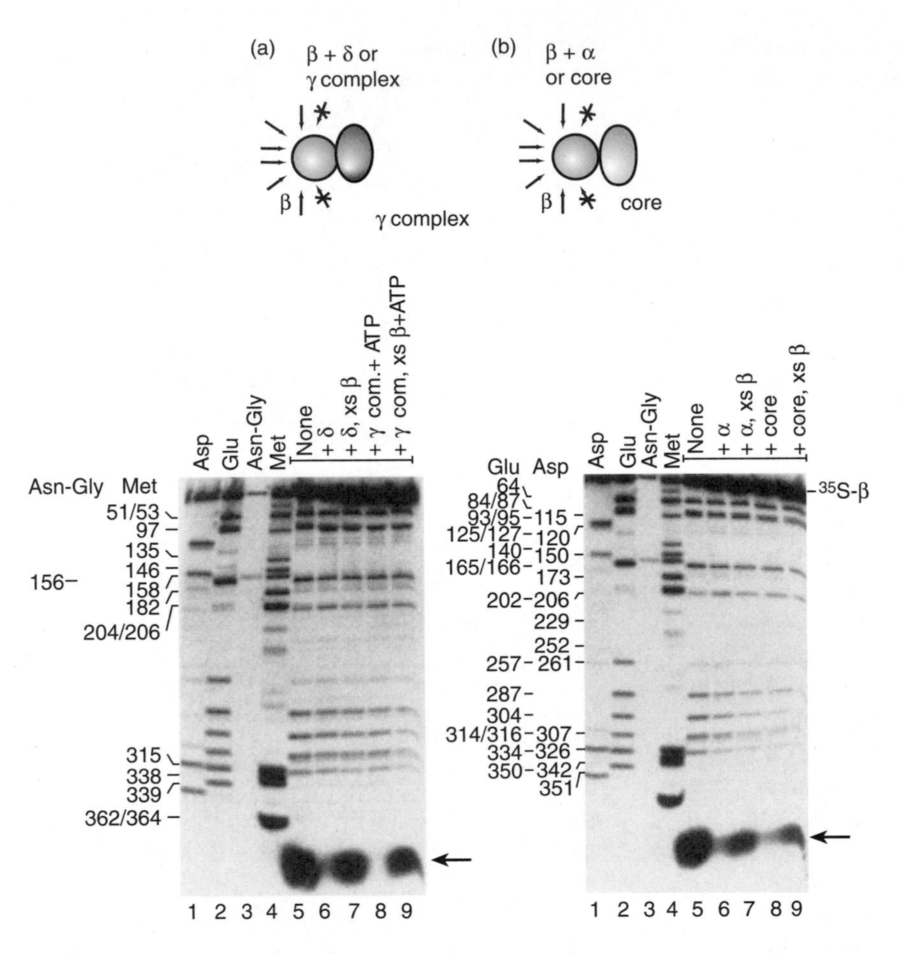

Figure 21.19 Protein footprinting of β with the γ complex and core polymerase. O'Donnell and colleagues labeled β^{PK} at its C-terminus by phosphorylation with protein kinase and [^{35}S]ATP. Then they mixed this end-labeled β with either δ or the whole γ complex (panel **a**) or with either α or the whole core (panel **b**). Then they subjected the protein complexes to mild cleavage with a mixture of pronase E and V8 protease to generate a series of end-labeled digestion products. Finally, they electrophoresed these products and autoradiographed the gel to detect them. The first four lanes in each panel are digestion products that serve as markers. The amino acid specificity of each treatment is given at top. Thus, in lane 1, the protein was treated with a protease that cleaves after aspartate (Asp) residues. Lane 5 in both panels represents β^{PK} cleaved in the absence of other proteins. Lanes 6–9 in both panels represent β^{PK} cleaved in the presence of the proteins listed at the top of each lane. The δ- and α-subunits and the γ and core complexes all protect the same site from digestion. Thus, they reduce the yield of the fragment indicated by the arrow at the bottom of the gel. The drawings at top illustrate the binding between the β clamp and either the γ complex (a) or the core (b), emphasizing that both contact the β clamp at the same places near the C-terminus of each β monomer and prevent cleavage there (arrows with Xs). (*Source:* Naktinis, V., J. Turner, and M. O'Donnell, A molecular switch in a replicating machine defined by an internal competition for protein rings. *Cell* 84 (12 June 1996) f. 3ab bottoms, p. 138. Reprinted by permission of Elsevier Science.)

proteolytic enzymes: pronase E and V8 protease. Figure 21.19 depicts the results. The first four lanes at the bottom of each panel are markers formed by cleaving the labeled β-subunit with four different reagents that cleave at known positions. Lane 5 in both panels shows the end-labeled peptides created by cleaving β in the absence of another protein. We observe a typical ladder of end-labeled products. Lane 6 in panel (a) shows what happens in the presence of δ. We see the same ladder as in lane 5, with the exception of the smallest fragment (arrow), which is either missing or greatly reduced in abundance. This suggests that the δ-subunit binds to β near its C-terminus and blocks a protease from cleaving there. If this δ–β interaction is specific, one should be able to restore cleavage of the labeled β^{PK} by adding an abundance of unlabeled β to bind to δ and prevent its binding to the labeled β^{PK}. Lane 7 shows that this is what happened. Lanes 8 and 9 in panel (a) are similar to 6 and 7, except that O'Donnell and coworkers used whole γ complex instead of purified δ. Again, the γ complex protected a site near the C-terminus of β^{PK} from cleavage, and unlabeled β prevented this protection.

Panel (b) of Figure 21.19 is just like panel (a), except that the investigators used the α-subunit and whole core instead of the δ-subunit and whole γ complex to footprint labeled β^{PK}. They observed exactly the same results: α or whole core protected the same site from cleavage as did δ or whole γ complex. This suggests that the core and the clamp loader both contact β at the same site, and that the α- and δ-subunits, respectively, mediate these contacts. In a further experiment, these workers used whole pol III* to footprint β^{PK}. Because pol III* contains both the core and the clamp loader, one might have expected it to yield a larger footprint than either subassembly separately. But it did not. This is consistent with the hypothesis that pol III* contacts β through either the core or the clamp loader, but not both at the same time.

If the β clamp can bind to the core or the clamp loader, but not both simultaneously, which does it prefer? O'Donnell and colleagues used gel filtration to show that when the proteins are free in solution, β prefers to bind to the clamp loader, rather than the core polymerase. This is satisfying because free β needs to be loaded onto DNA by the γ complex before it can interact with the core polymerase. However, that situation should change once the β clamp is loaded onto a primed DNA template; once that happens, β needs to associate with the core polymerase and begin making DNA. To test this prediction, O'Donnell and colleagues loaded ^{35}S-labeled β clamps onto primed M13 phage DNA and then added either ^{3}H-labeled clamp loader (γ complex) and unlabeled core, or ^{3}H-labeled core and unlabeled γ complex. Then they subjected these mixtures to gel filtration to separate DNA–protein complexes from free proteins.

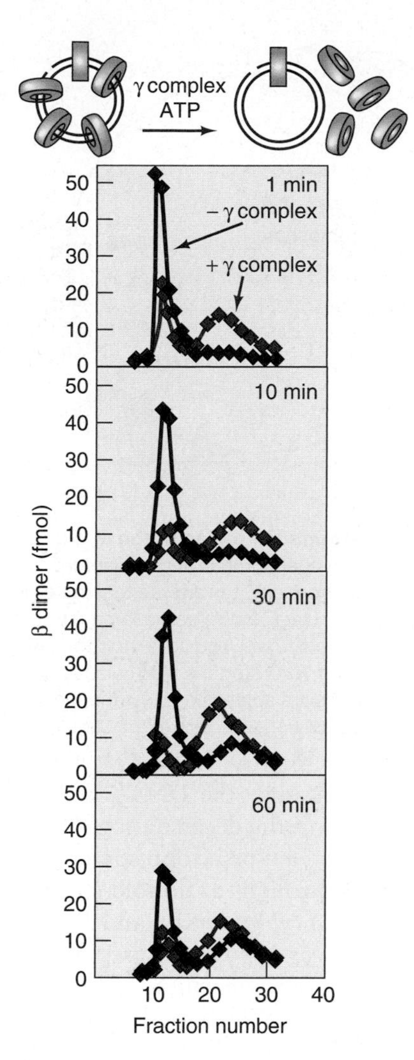

Figure 21.20 Clamp-unloading activity of the γ complex. O'Donnell and coworkers loaded β clamps onto a nicked circular DNA template, as shown at top, then incubated these complexes in the presence (red) or absence (blue) of the γ complex and ATP for the times indicated. Finally, they subjected the mixtures to gel filtration to determine how much β clamp remained associated with the DNA and how much had dissociated. The cartoon at top interprets the results: The γ complex and ATP served to accelerate the unloading of β clamps from the nicked DNA. (*Source:* Adapted from Naktinis, V., J. Turner, and M. O'Donnell, A molecular switch in a replication machine defined by an internal competition for protein rings. *Cell* 84:141, 1996.)

Under these conditions, it was clear that the β clamp on the DNA preferred to associate with the core polymerase. Almost no γ complex bound to the β clamp–DNA complex.

Once the holoenzyme has completed an Okazaki fragment, it must dissociate from the β clamp and move to a new one. Then the original β clamp must be removed from the template so it can participate in the synthesis of another Okazaki fragment. We have already seen that pol III* has clamp-unloading activity, but we have not seen what part of pol III* has this activity. O'Donnell and associates performed gel filtration assays that showed that the γ complex has clamp-unloading activity. Figure 21.20 illustrates this experiment. The investigators loaded β clamps onto a

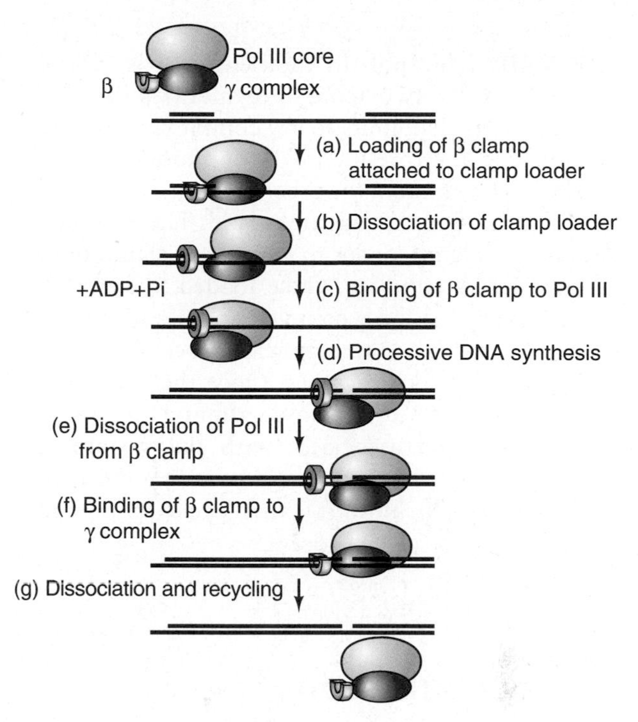

Figure 21.21 Summary of lagging strand replication. We begin with a β clamp associated with the γ complex part (red) of a pol III*. **(a)** The γ complex loads the β clamp (blue) onto a primed DNA template. **(b)** The γ complex, or clamp loader, dissociates from the β clamp. **(c)** The core (green) associates with the clamp. **(d)** The core and clamp cooperate to processively synthesize an Okazaki fragment, leaving just a nick between two Okazaki fragments. **(e)** The polymerase core dissociates from the clamp. **(f)** The γ complex reassociates with the β clamp. **(g)** The γ complex acts as a clamp unloader, removing the β clamp from the template. Now it is free to repeat the process, recycling to another primer on the template. (*Source:* Adapted from Herendeen, D.R. and T.J. Kelly, DNA polymerase III: Running rings around the fork. *Cell* 84:7, 1996.)

nicked DNA template, then removed all other proteins. Then they incubated these DNA–protein complexes in the presence and absence of the γ complex. We can see that the β clamps are unloaded from the nicked DNA much faster in the presence of the γ complex and ATP than in their absence.

Thus the γ complex is both a clamp loader and a clamp unloader. But what determines when it will load clamps and when it will unload them? The state of the DNA seems to throw this switch, as illustrated in Figure 21.21. Thus, when β clamps are free in solution and there is a primed template available, the clamps associate preferentially with the γ complex, which serves as a clamp loader to bind the β clamp to the DNA. Once associated with the DNA, the clamp binds preferentially to the core polymerase and sponsors processive synthesis of an Okazaki fragment. When the fragment has been synthesized, and only a nick remains, the core loses its affinity for the β clamp. The clamp reassociates with the γ complex, which now acts as a clamp unloader, removing the clamp from the template so it can recycle to the next primer and begin the cycle anew.

SUMMARY The pol III holoenzyme is double-headed, with two core polymerases attached through two τ-subunits to a γ complex. One core is responsible for (presumably) continuous synthesis of the leading strand, the other performs discontinuous synthesis of the lagging strand. The γ complex serves as a clamp loader to load the β clamp onto a primed DNA template. Once loaded, the β clamp loses affinity for the γ complex and associates with the core polymerase to help with processive synthesis of an Okazaki fragment. Once the fragment is completed, the β clamp loses affinity for the core polymerase and associates with the γ complex, which acts as a clamp unloader, removing the clamp from the DNA. Now it can recycle to the next primer and repeat the process.

21.3 Termination

Termination of replication is relatively straightforward for λ and other phages that produce a long, linear concatemer. The concatemer simply continues to grow as genome-sized parts of it are snipped off and packaged into phage heads. But for bacteria and eukaryotes, where replication has a definite end as well as a beginning, the mechanisms of termination are more complex and more interesting. In bacterial DNA replication, the two replication forks approach each other in the terminus region, which contains 22-bp terminator sites that bind specific proteins. In *E. coli,* the terminator (*Ter*) sites are *TerA–TerF,* and they are arranged as pictured in Figure 21.22. The *Ter* sites bind proteins called **Tus** (for terminus utilization substance). Replicating forks enter the terminus region and pause before quite completing the replication process. This leaves the two daughter duplexes entangled. They must become disentangled before cell division occurs, or they cannot separate to the two daughter cells. Instead, they would remain caught in the middle of the cell, cell division would fail, and the cell would probably die. These considerations raise the question: How do the daughter duplexes become disentangled? For eukaryotes, we would like to know how cells fill in the gaps left by removing primers at the 5′-ends of the linear chromosomes. Let us examine each of these problems.

Decatenation: Disentangling Daughter DNAs

Bacteria face a problem near the end of DNA replication. Because of their circular nature, the two daughter duplexes remain entwined as two interlocking rings, a type of **catenane.** For these interlocked DNAs to move to the two daughter cells, they must be unlinked, or **decatenated.** If decatenation occurs before repair synthesis, a single nick

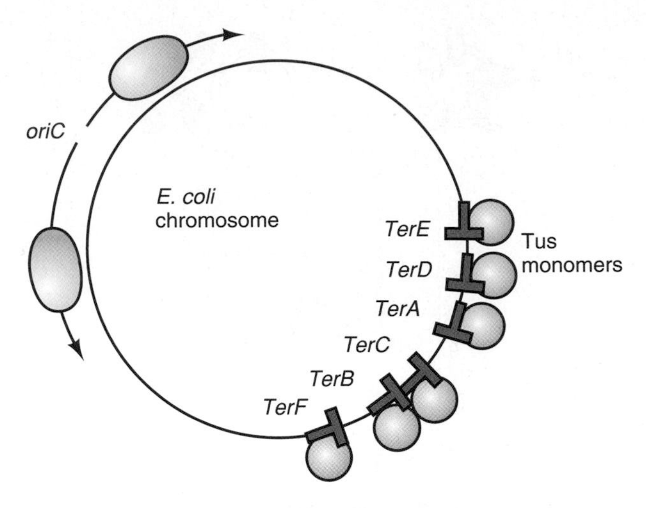

Figure 21.22 The termination region of the *E. coli* genome. Two replicating forks with their accompanying replisomes (green) are pictured moving away from *oriC* toward the terminator region on the opposite side of the circular *E. coli* chromosome. Three terminator sites operate for each fork: *TerE, TerD, and TerA* stop the counterclockwise fork; and *TerF, TerB,* and *TerC* stop the clockwise fork. The Tus protein binds to the terminator sites and helps arrest the moving forks. (*Source:* Adapted from Baker, T.A., Replication arrest. *Cell* 80:521, 1995.)

will suffice to disentangle the DNAs, and a type I topoisomerase can perform the decatenation. However, if repair synthesis occurs first, a type II topoisomerase, which passes a DNA duplex through a double-stranded break, is required. *Salmonella typhimurium* and *E. coli* cells contain four topoisomerases: topoisomerases I–IV (topo I–IV). Topo I and III are type I enzymes, and topo II and IV are type II. The question is: Which topoisomerase is involved in decatenation?

Because DNA gyrase (topo II) acts as the swivel during DNA replication, many molecular biologists assumed that it also decatenates the daughter duplexes. But Nicholas Cozzarelli and his colleages demonstrated that **topo IV** is really the decatenating enzyme. They tested various temperature-sensitive mutant strains of *S. typhimurium,* a close relative of *E. coli,* for ability to decatenate dimers of the plasmid pBR322 in vivo at the permissive and nonpermissive temperatures. They showed that bacteria with mutations in the genes encoding the subunits of topo IV failed to decatenate the plasmid at the nonpermissive temperature (44°C) in the absence of norfloxacin. This suggests that topo IV is important in decatenation. Norfloxacin, by blocking DNA gyrase, halted DNA replication and presumably allowed subsequent decatenation by the small amount of residual topo IV, or by another topoisomerase. By contrast, the strain with the mutant DNA gyrase did not accumulate catenanes at the nonpermissive temperature, either in the presence or absence of norfloxacin, suggesting that this enzyme does not participate in decatenation. When they tested temperature-sensitive mutants of *E. coli,* Cozzarelli and colleagues observed similar behavior, indicating that topo IV also participates in decatenation in *E. coli.*

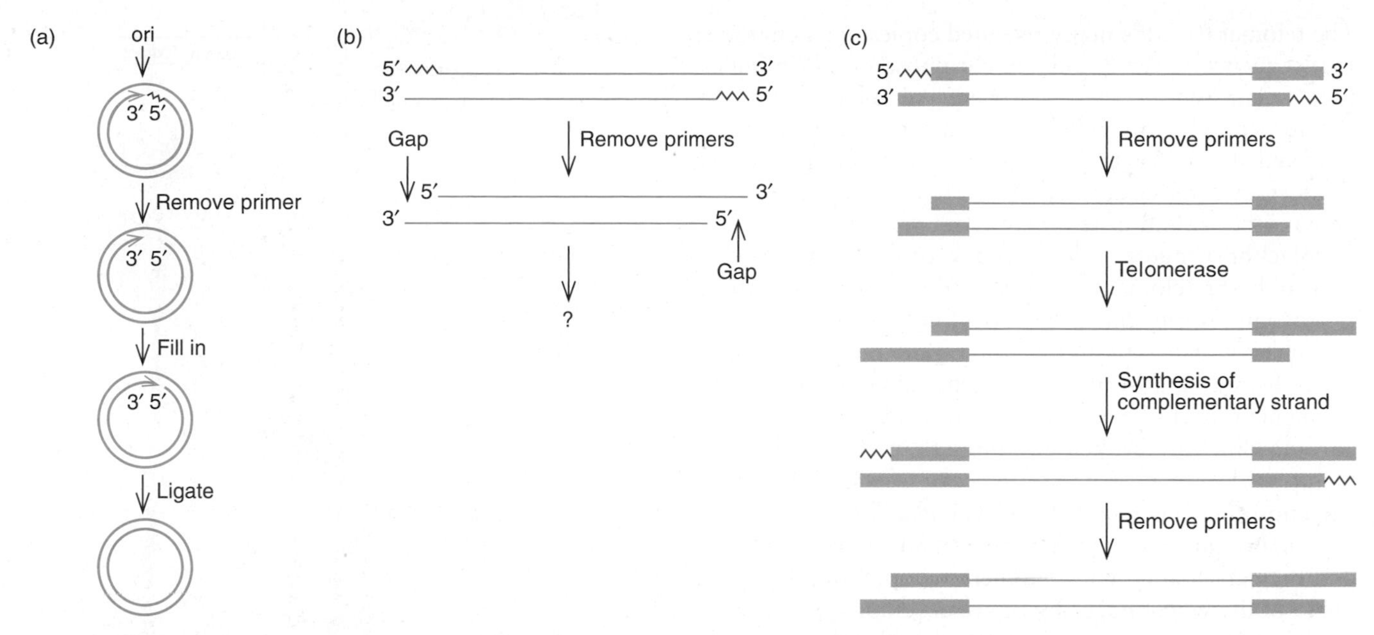

Figure 21.23 Coping with the gaps left by primer removal. (a) In bacteria, the 3′-end of a circular DNA strand can prime the synthesis of DNA to fill in the gap left by the first primer (pink). For simplicity, only one replicating strand is shown. **(b)** Hypothetical model to show what would happen if primers were simply removed from the 5′-end of linear DNA strands with no telomerase action. The gaps at the ends of chromosomes would grow longer each time the DNA replicated. **(c)** How telomerase can solve the problem. In the first step, the primers (pink) are removed from the 5′-ends of the daughter strands, leaving gaps. In the second step, telomerase adds extra telomeric DNA (green boxes) to the 3′-ends of the other daughter strands. In the third step, DNA synthesis occurs, using the newly made telomeric DNA as a template. In the fourth step, the primers used in step three are removed. This leaves gaps, but the telomerase action has ensured that no net loss of DNA has occurred. The telomeres represented here are not drawn to scale with the primers. In reality, human telomeres are thousands of nucleotides long. (*Source: (c)* Adapted from Greider, C.W. and E.H. Blackburn, Identification of a specific telomere terminal transferase activity in tetramere extracts. *Cell* 43 (Dec Pt1 1985) f. 1A, p. 406.)

Eukaryotic chromosomes are not circular, but they have multiple replicons, so replication forks from neighboring replicons approach one another just as the two replication forks of a bacterial chromosome approach each other near the termination point opposite the origin of replication. Apparently, this inhibits completion of DNA replication, so eukaryotic chromosomes also form catenanes that must be disentangled. Eukaryotic topo II resembles bacterial topo IV more than it does DNA gyrase, and it is a strong candidate for the decatenating enzyme.

SUMMARY At the end of replication, circular bacterial chromosomes form catenanes that must be decatenated for the two daughter duplexes to separate. In *E. coli* and related bacteria, topoisomerase IV performs this decatenation. Linear eukaryotic chromosomes also require decatenation during DNA replication.

Termination in Eukaryotes

Eukaryotes face a difficulty at the end of DNA replication that prokaryotes do not: filling in the gaps left when RNA primers are removed. With circular DNAs, such as those in bacteria, there is no problem filling all the gaps because another DNA 3′-end is always upstream to serve as primer (Figure 21.23a). But consider the problem faced by eukaryotes, with their linear chromosomes. Once the first primer on each strand is removed (Figure 21.23b), there is no way to fill in the gaps because DNA cannot be extended in the 3′→5′ direction, and no 3′-end is upstream, as there would be in a circle. If this were actually the situation, the DNA strands would get shorter every time they replicated. This is a termination problem in that it deals with the formation of the ends of the DNA strands, but how do cells solve this problem?

Telomere Maintenance Elizabeth Blackburn and her colleagues provided the answer, which is summarized in Figure 21.23c. The **telomeres,** or ends of eukaryotic chromosomes, are composed of repeats of short, GC-rich sequences. The G-rich strand of a telomere is added at the very 3′-ends of DNA strands, not by semiconservative replication, but by an enzyme called **telomerase.** The exact sequence of the repeat in a telomere is species-specific. In *Tetrahymena,* it is TTGGGG/AACCCC; in vertebrates, including humans, it is TTAGGG/AATCCC. Blackburn showed that this specificity resides in the telomerase itself and is due to a small RNA in the enzyme that serves as the template for telomere synthesis. This solves the problem:

The telomerase adds many repeated copies of its characteristic sequence to the 3′-ends of chromosomes. Priming can then occur within these telomeres to make the C-rich strand. There is no problem when terminal primers are removed and not replaced, because only telomere sequences are lost, and these can always be replaced by telomerase and another round of telomere synthesis.

Blackburn made a clever choice of organism in which to search for telomerase activity: *Tetrahymena,* a ciliated protozoan. *Tetrahymena* has two kinds of nuclei: (1) micronuclei, which contain the whole genome in five pairs of chromosomes that serve to pass genes from one generation to the next; and (2) macronuclei, in which the five pairs of chromosomes are broken into more than 200 smaller fragments used for gene expression. Because each of these minichromosomes has telomeres at its ends, *Tetrahymena* cells have many more telomeres than human cells, for example, and they are loaded with telomerase, especially during the phase of life when macronuclei are developing and the new minichromosomes must be supplied with telomeres. This made isolation of the telomerase enzyme from *Tetrahymena* relatively easy.

In 1985, Carol Greider and Blackburn succeeded in identifying a telomerase activity in extracts from synchronized *Tetrahymena* cells that were undergoing macronuclear development. They assayed for telomerase activity in vitro using a synthetic primer with four repeats of the TTGGGG telomere sequence and included a radioactive nucleotide to label the extended telomere-like DNA. Figure 21.24 shows the results. Lanes 1–4 each contained a different labeled nucleotide (dATP, dCTP, dGTP, and dTTP, respectively), plus all three of the other, unlabeled nucleotides. Lane 1, with labeled dATP showed only a smear, and lanes 2 and 4 showed no extension of the synthetic telomere. But lane 3, with labeled dGTP, exhibited an obvious periodic extension of the telomere. Each of the clusters of bands represents an addition of one more TTGGGG sequence (with some variation in the degree of completion), which accounts for the fact that we see *clusters* of bands, rather than *single* bands. Of course, we should observe telomere extension with labeled dTTP, as well as with dGTP. Further investigation showed that the concentration of dTTP was too low in this experiment, and that dTTP could be incorporated into telomeres at higher concentration. Lanes 5–8 show the results of an experiment with one labeled, and only one unlabeled nucleotide. This experiment verifed that dGTP could be incorporated into the telomere, but only if unlabeled dTTP was also present. This is what we expect because this strand of the telomere contains only G and T. Controls in lanes 9–12 showed that an ordinary DNA polymerase, Klenow fragment, cannot extend the telomere. Further controls in lanes 13–16 demonstrated that telomerase activity depends on the telomere-like primer.

How does telomerase add the correct sequence of bases to the ends of telomeres without a complementary DNA strand to read? It uses its own RNA constituent as a template. (Note that this is a template, not a primer.) Greider and Blackburn demonstrated in 1987 that telomerase is a ribonucleoprotein with essential RNA and protein subunits. Then in 1989 they cloned and sequenced the gene that encodes the 159-nt RNA subunit of the *Tetrahymena* telomerase and found that it contains the sequence CAACCCCAA. In principle, this sequence can serve as template for repeated additions of TTGGGG sequences to the ends of *Tetrahymena* telomeres as illustrated in Figure 21.25.

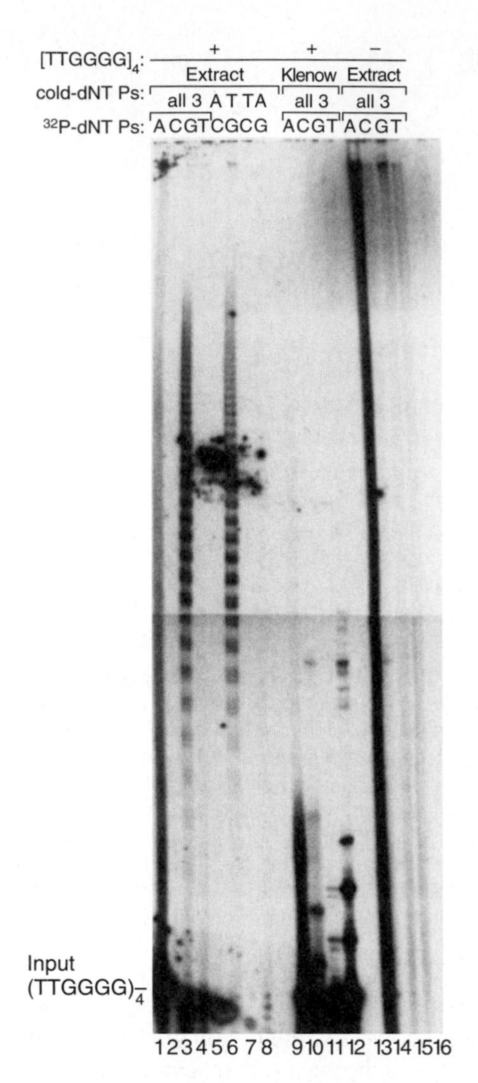

Figure 21.24 Identification of telomerase activity. Greider and Blackburn synchronized mating of *Tetrahymena* cells and let the offspring develop to the macronucleus development stage. They prepared cell-free extracts and incubated them for 90 min with a synthetic oligomer having four repeats of the TTGGGG telomere repeat sequence, plus the labeled and unlabeled nucleotides indicated at top. After incubation, they electrophoresed the products and detected them by autoradiography. Lanes 9–12 contained the Klenow fragment of *E. coli* DNA polymerase I instead of *Tetrahymena* extract. Lanes 13–16 contained extract, but no primer. Telomerase activity is apparent only when both dGTP and dTTP are present. (*Source:* Greider, C.W., and E.H. Blackburn, Identification of a specific telomere terminal transferase activity in tetramere extracts. *Cell* 43 (Dec Pt1 1985) f. 1A, p. 406. Reprinted by permission of Elsevier Science.)

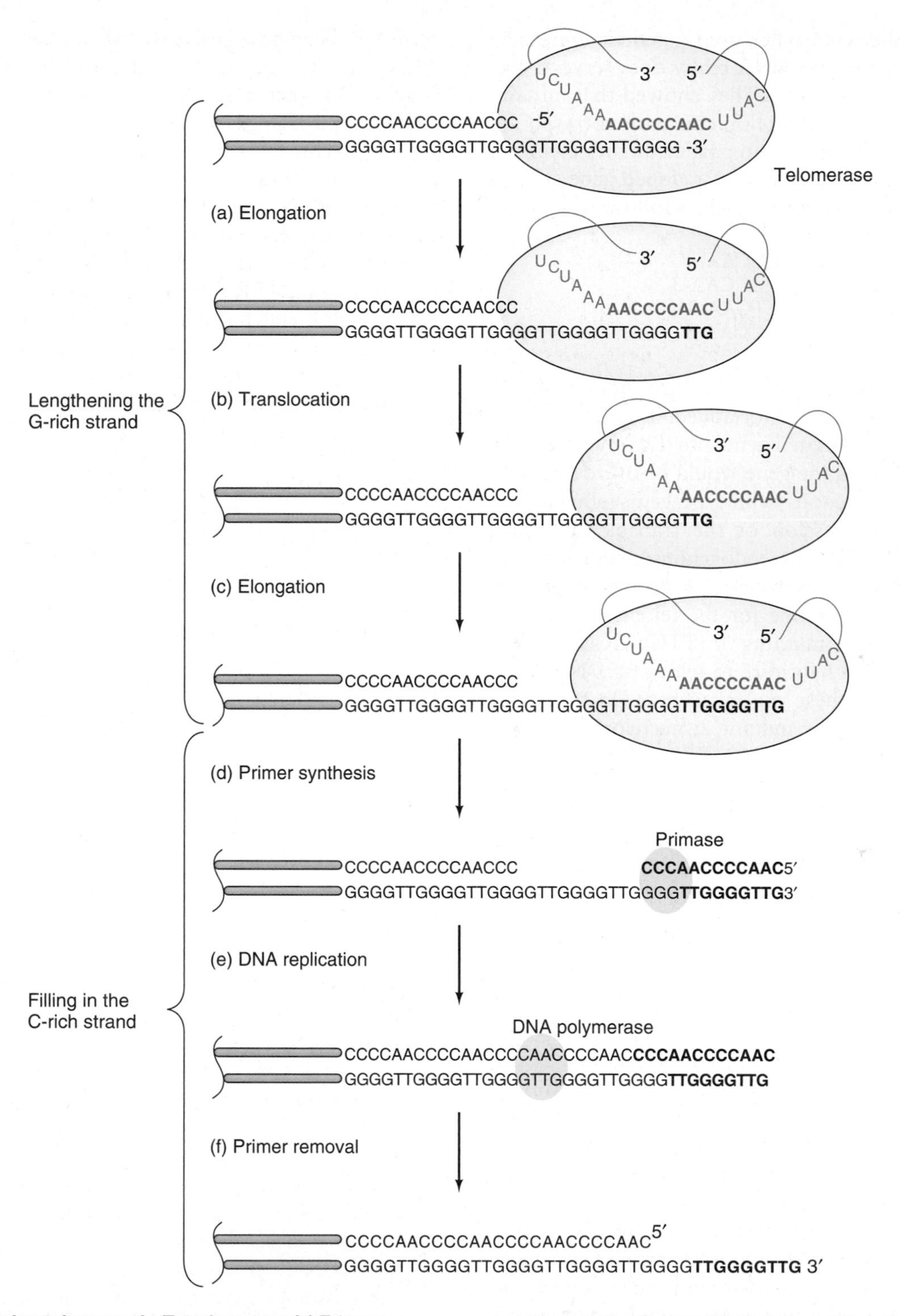

Figure 21.25 Forming telomeres in *Tetrahymena*. **(a)** Telomerase (yellow) promotes hybridization between the 3′-end of the G-rich telomere strand and the template RNA (red) of the telomerase. The telomerase uses three bases (AAC) of its RNA as a template for the addition of three bases (TTG, boldface) to the 3′-end of the telomere. **(b)** The telomerase translocates to the new 3′-end of the telomere, pairing the left-hand AAC sequence of its template RNA with the newly incorporated TTG in the telomere. **(c)** The telomerase uses the template RNA to add six more nucleotides (GGGTTG, boldface) to the 3′-end of the telomere. Steps (a) through (c) can repeat indefinitely to lengthen the G-rich strand of the telomere. **(d)** When the G-rich strand is sufficiently long (probably longer than shown here), primase (orange) can make an RNA primer (boldface), complementary to the 3′-end of the telomere's G-rich strand. **(e)** DNA polymerase (green) uses the newly made primer to prime synthesis of DNA to fill in the remaining gap on the C-rich telomere strand and DNA ligase seals the nick. **(f)** The primer is removed, leaving a 12–16-nt overhang on the G-rich strand.

Blackburn and her colleagues used a genetic approach to prove that the telomerase RNA really does serve as the template for telomere synthesis. They showed that mutant telomerase RNAs gave rise to telomeres with corresponding alterations in their sequence. In particular, they changed the sequence 5′-CAACCCCAA-3′ of a cloned gene encoding the *Tetrahymena* telomerase RNA as follows:

wt: 5′-CAACCCCAA-3′
1: 5′-CAACCCCCAA-3′
2: 5′-CAACCTCAA-3′
3: 5′-CGACCCCAA-3′

The underlined bases in each of the three mutants (1, 2, and 3) denote the base changed (or added, in 1). They introduced the wild-type or mutated gene into *Tetrahymena* cells in a plasmid that ensured the gene would be overexpressed. Even though the endogenous wild-type gene remained in each case, the overexpression of the transplanted gene swamped out the effect of the endogenous gene. Southern blotting of telomeric DNA from cells transformed with each construct showed that a probe for the telomere sequence expected to result from mutants 1 (TTGGGGG) and 3 (GGGGTC) actually did hybridize to telomeric DNA from cells transformed with these mutant genes. On the other hand, this did not work for mutant 2; no telomeric DNA that hybridized to a probe for GAGGTT was observed.

These results suggested that mutant telomerase RNAs 1 and 3, but not 2, served as templates for telomere elongation. To confirm this suggestion, Blackburn and colleagues sequenced a telomere fragment from cells transformed with mutant telomerase RNA 3. They found the following sequence:

5′-CTTTTACTCAATGTCAAAGAAATTATTAAATT(GGGGTT)$_{30}$
(GGGGTC)$_{2}$GGGGTT(GGGGTC)$_{8}$GGGGTTGGGGTC(GGGGTT)$_{N}$-3′

where the underlined bases must have been encoded by the mutant telomerase RNA. This nonuniform sequence differs stikingly from the normal, very uniform telomeric sequence in this species. The first 30 repeats appear to have been encoded by the wild-type telomerase RNA before transformation. These are followed by 11 mutant repeats interspersed with 2 wild-type repeats, then by all wild-type repeats. The terminal wild-type sequences may have resulted from recombination with a wild-type telomere, or from telomere synthesis after loss of the mutant telomerase RNA gene from the cell. Nevertheless, the fact remains that a significant number of repeats have exactly the sequence we would expect if they were encoded by the mutant telomerase RNA. Thus, we can conclude that the telomerase RNA does serve as the template for telomere synthesis, as Figure 21.25 suggests.

The fact that telomerase uses an RNA template to make a DNA strand implies that telomerase acts as a reverse transcriptase. Thus, Blackburn and others set about to purify the enzyme to prove that this is indeed how it works. Although the enzyme eluded purification for 10 years, Joachim Lingner and Thomas Cech finally succeeded in 1996 in purifying it from another ciliated protozoan, *Euplotes*. This telomerase contains two proteins, p43 and p123, in addition to the RNA subunit that serves as the template for extending telomeres. The p123 protein has the signature sequence of a reverse transcriptase, indicating that it provides the catalytic activity of the enzyme. We therefore call it **TERT,** for **telomerase reverse transcriptase.** Because this enzyme was discovered when the Human Genome Project was well along, it did not take long to find a complementary sequence in the human genome and use it to clone the human TERT gene, *hTERT,* in 1997.

Structural analysis has shown that the C-terminal part of the TERT protein contains the reverse transcriptase activity, and the N-terminal part binds to the RNA. In fact, the RNA appears to be tethered to the protein so as to give the RNA, which is hundreds of nucleotides long, considerable flexibility. This allows the RNA to fulfill its template role by moving with respect to the active site of the enzyme as each nucleotide is added to the growing telomere.

Until 2003, it appeared that the somatic cells of higher eukaryotes, including humans, lack telomerase activity, whereas germ cells retain this activity. Then, William Hahn and colleagues showed that cultured normal human cells do express telomerase at a low level, but only transiently, during S phase, when DNA is replicated. On the other hand, cancer cells have much higher telomerase activity, which is expressed constitutively—all the time. These findings have profound implications for the characteristics of cancer cells, and perhaps even for their control (see Box 21.1).

SUMMARY Eukaryotic chromosomes have special structures known as telomeres at their ends. One strand of these telomeres is composed of many tandem repeats of short, G-rich regions whose sequence varies from one species to another. The G-rich telomere strand is made by an enzyme called telomerase, which contains a short RNA that serves as the template for telomere synthesis. The C-rich telomere strand is synthesized by ordinary RNA-primed DNA synthesis, like the lagging strand in conventional DNA replication. This mechanism ensures that chromosome ends can be rebuilt and therefore do not suffer shortening with each round of replication.

Telomere Structure Besides protecting the ends of chromosomes from degradation, telomeres play another critical role: They prevent the DNA repair machinery from recognizing the ends of chromosomes as chromosome breaks and sticking chromosomes together. This inapproriate joining of chromosomes would be potentially lethal to the cell. Furthermore, cells have a DNA damage checkpoint that

BOX 21.1

Telomeres, the Hayflick Limit, and Cancer

Everyone knows that organisms, including humans, are mortal. But biologists used to assume that cells cultured from humans were immortal. Each individual cell would ultimately die, of course, but the cell *line* would go on dividing indefinitely. Then in the 1960s Leonard Hayflick discovered that ordinary human cells are not immortal. They can be grown in culture for a finite period—about 50 generations (or cycles of subculturing). Then they enter a period of senescence, when they slow down and then stop dividing, and finally they reach a crisis stage and die. This ceiling on the lifetime of normal cells is known as the Hayflick limit. But cancer cells do not obey any such limit. They *do* go on dividing generation after generation, indefinitely.

Investigators have discovered a significant difference between normal cells and cancer cells that may explain why cancer cells are immortal and normal cells are not: Human cancer cells contain abundant telomerase that is expressed constitutively, whereas normal somatic cells generally produce this enzyme only weakly and transiently. (Germ cells must retain telomerase, of course, to safeguard the ends of the chromosomes handed down to the next generation.) Thus, we see that cancer cells can repair their telomeres after every cell replication, but most normal cells cannot. Therefore, cancer cells can go on dividing without degrading their chromosomes, whereas normal cells' chromosomes grow shorter with each cell division. Sooner or later the telomeres are lost, and the ends of chromosomes that lack telomeres look like the ends of broken chromosomes. Most cells react to this apparent assault by halting their replication and ultimately by dying. But this does not happen to cancer cells; telomerase saves them from that fate.

One of the typical changes that occurs in a cell to make it cancerous is the reactivation of the telomerase gene. This leads to the immortality that is the hallmark of cancer cells. This discussion also suggests a potential treatment for cancer: Turn off the telomerase gene in cancer cells or, more simply, administer a drug that inhibits telomerase. Such a drug may not harm most normal cells because they have very little telomerase to begin with. Cancer researchers are hard at work on this strategy, but the discovery in 2003 that human fibroblasts in culture express low levels of *hTERT* and have a little telomerase activity casts some doubt on this idea. Further reservations come from the findings that expression of an inactive form of *hTERT,* or inhibiting the expression of normal *hTERT* by RNAi, causes premature senescence in human fibroblasts. The trick will be to kill cancer cells without dooming the patient's normal cells to an early death.

Some signs indicate that simply inhibiting the telomerase of cancer cells may not cause the cells to die. For one thing, knockout mice totally lacking telomerase activity survive and reproduce for at least six generations, though eventually the loss of telomeres leads to sterility. However, cells from these telomerase knockout mice can be immortalized, they can be transformed by tumor viruses, and these transformed cells can give rise to tumors when transplanted to immunodeficient mice. Thus, the presence of telomerase is not an absolute requirement for the development of a cancer cell. It may be that mouse cells have a way of preserving their telomeres without telomerase. We will have to see whether human cells behave differently.

Finally, immortalizing human cells in culture leads to the idea of immortalizing human beings themselves. Could it be that reactivating telomerase activity in human somatic cells would lengthen human lifetimes? Or would it just make us more susceptible to cancer? To begin answering this question, Serge Lichtsteiner, Woodring Wright, and their colleagues transplanted the *hTERT* gene into human somatic cells in culture, so these cells were forced to express telomerase activity. The results were striking: The telomeres in these cells grew longer and the cells went on dividing far past their normal lifetimes. They remained youthful in appearance and in their chromosome content. Furthermore, they did not show any signs of becoming cancerous. These findings were certainly encouraging, but they do not herald a fountain of youth. For now, that remains in the realm of science fiction.

detects damage and stops cell division until the damage can be repaired. Because chromosome ends without telomeres look like broken chromosomes, they invoke the checkpoint, so cells stop dividing and eventually die. If telomeres really looked the way they are pictured in Figures 21.23 and 21.25, little would distinguish them from real chromosome breaks. In fact, the critical telomere length in humans is 12.8 repeats of the core 6-bp sequence. Below that threshold, human chromosomes began to fuse. How do telomeres allow the cell to recognize the difference between a real chromosome end and a broken chromosome?

For years, molecular biologists pondered this question and, as telomere-binding proteins were discovered, they theorized that these proteins bind to the ends of chromosomes and in that way identify the ends. Indeed, eukaryotes from yeasts to mammals have a suite of telomere-binding proteins that protect the telomeres from degradation, and also hide the telomere ends from the DNA damage factors that would otherwise recognize them as chromosome breaks. We will discuss the telomere-binding proteins from three groups of eukaryotes and see how they solve the telomere protection problem.

The Mammalian Telomere-Binding Proteins: Shelterin In mammals, the group of telomere-binding proteins is appropriately known as **shelterin,** because it "shelters" the telomere. There are six known mammalian shelterin proteins: **TRF1, TRF2, TIN2, POT1, TPP1,** and **RAP1.** TRF1 was the first of these proteins to be discovered. Because it bound to double-stranded telomere DNA, which includes repeats of the sequence TTAGGG, it was named **TTAGGG repeat-binding factor-1** (TRF1). TRF2 is a product of a paralog of the TRF1 gene (paralogs are homologous genes in the same organism), and it also binds to the double-stranded parts of telomeres. POT1 (**protection of telomeres -1**) binds to the single-stranded 3′-tails of telomeres, beginning at a position just 2 nt away from the 5′-end of the other strand. In this way it is positioned to protect the single-stranded telomeric DNA from endonucleases, and the 5′-end of the other strand within the double-stranded telomeric DNA from 5′-exonucleases. TPP1 is a POT1-binding protein. Indeed, it appears to be a partner of POT1 in a heterodimer. TIN2 (**TRF1-interacting factor-2**) plays an organizing role in shelterin. It connects TRF1 and TRF2 together, and connects the dimer TPP1/POT1 to TRF1 and TRF2. Finally, RAP1, with the uninformative name "repressor activator protein-1," binds to the telomere by interacting with TRF2.

Other proteins besides shelterin bind to telomeres, but shelterin proteins can be distinguished from the others in three ways: They are found only at telomeres; they associate with telomeres throughout the cell cycle; and they are known to function nowhere else in the cell. Other proteins may fulfill one of these criteria, but not two or all three.

Shelterin can affect the structure of telomeres in three ways. First, it can remodel the telomere into a loop called a **t-loop** (for "telomere-loop"). In 1999, Jack Griffith and Titia de Lange and their colleagues discovered that mammalian telomeres are not linear, as had been assumed, but form a DNA loop they called a t-loop. These loops are unique in the chromosome and therefore quite readily set the ends of chromosomes apart from breaks that occur in the middle and would yield linear ends to the chromosome fragments.

What is the evidence for t-loops? Griffith, de Lange and colleagues started by making a model mammalian telomeric DNA with about 2 kb of repeating TTAGGG sequences, and a 150–200-nt single-stranded 3′-overhang at the end. They added one of the telomere-binding proteins, TRF2, then subjected the complex to electron microscopy. Figure 21.26a shows that a loop really did form, with a ball of TRF2 protein right at the loop–tail junction. Such structures appeared about 20% of the time. By contrast, when these workers cut off the single-stranded 3′-overhang, or left out TRF2, they found a drastic reduction in loop formation.

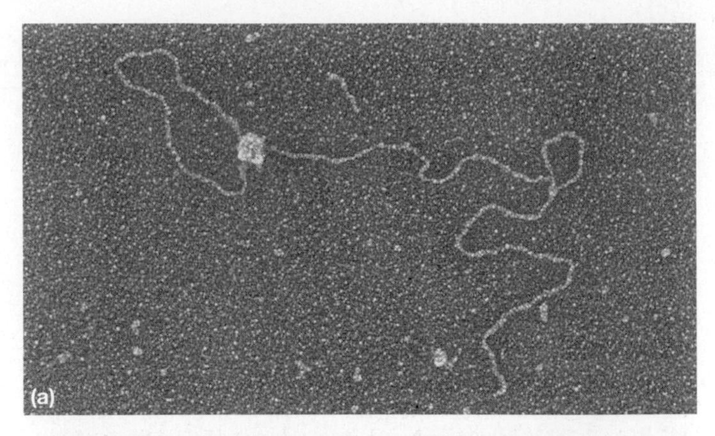

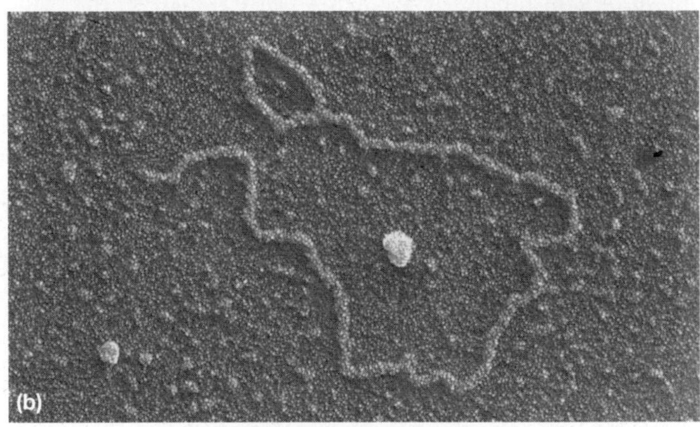

Figure 21.26 Formation of t-loops in vitro. (a) Direct detection of loops. Griffith and colleagues mixed a model DNA having a telomere-like structure with TRF2, then spread the mixture on an EM grid, shadowed the DNA and protein with tungsten, and observed the shadowed molecules with an electron microscope. An obvious loop appeared, with a blob of TRF2 at the junction between the loop and the tail. **(b)** Stabilization of the loop by cross-linking. Griffith and coworkers formed the t-loop as in panel (a), then cross-linked double-stranded DNA with psoralen and UV radiation, then removed the protein, spread the cross-linked DNA on an EM grid, shadowed with platinum and paladium, and visualized the shadowed DNA with an electron microscope. Again, an obvious loop appeared. The bar represents 1 kb. (*Source:* Griffith, J.D., L. Comeau, S. Rosenfield, R.M Stansel, A. Bianchi, H. Moss, and T. de Lange, Mammalian telomeres end in a large duplex loop. *Cell* 97 (14 May 1999) f. 1, p. 504. Reprinted by permission of Elsevier Science.)

One way for a telomere to form such a loop would be for the single-stranded 3′-overhang to invade the double-stranded telomeric DNA upstream, as depicted in Figure 21.27. If this hypothesis is correct, one should be able to stabilize the loop with psoralen and UV radiation, which cross-link thymines on opposite strands of a double-stranded DNA. Because the invading strand base-pairs with one of the strands in the invaded DNA, this creates double-stranded DNA that is subject to cross-linking and therefore stabilization. Figure 21.26b shows the results of an experiment in which Griffith, de Lange, and coworkers cross-linked the model DNA with psoralen and UV, then

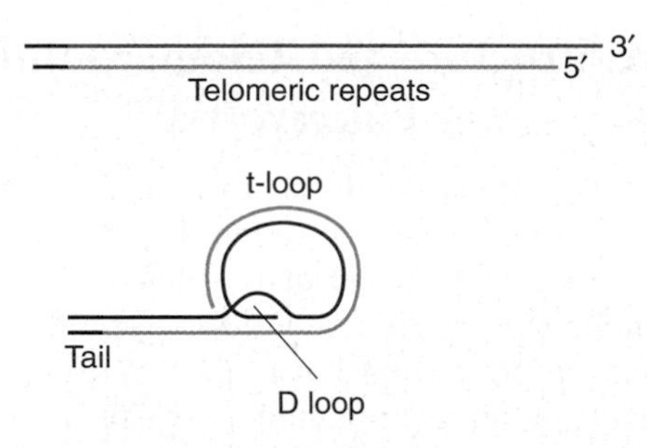

Figure 21.27 A model of a mammalian t-loop. The single-stranded 3′-end of the G-rich strand (red) invades the double-stranded telomeric DNA upstream, forming a long t-loop and a 75–200-nt displacement loop at the junction between the loop and the tail. A short subtelomeric region (black) is pictured adjoining the telomere (blue and red). (*Source:* Adapted from Griffith, D., L. Comeau, S. Rosenfield, R.M. Stansel, A. Bianchi, H. Moss, and T. de Lange, Mammalian telomeres end in a large duplex loop. *Cell,* 97:511, 1999).

deproteinized the complex, then subjected it to electron microscopy. The loop is still clearly visible, even in the absence of TRF2, showing that the DNA itself has been cross-linked, stabilizing the t-loop.

Next, these workers purified natural telomeres from several human cell lines and from mouse cells and subjected them to psoralen–UV treatment and electron microscopy. They obtained the same result as in Figure 21.26b, showing that t-loops appear to form in vivo. Furthermore, the sizes of these putative t-loops correlated well with the known lengths of the telomeres in the human or mouse cells, reinforcing the hypothesis that these loops really do represent telomeres.

To test further the notion that the loops they observed contain telomeric DNA, Griffith, de Lange and colleagues added TRF1, which is known to bind very specifically to double-stranded telomeric DNA, to their looped DNA. They observed loops coated with TRF1, as shown in Figure 21.28a.

If the strand invasion hypothesis in Figure 21.27 is valid, the single-stranded DNA displaced by the invading DNA (the displacement loop, or D-loop) should be able to bind *E. coli* single-strand-binding protein (SSB, recall Chapter 20) if the displaced DNA is long enough. Figure 21.28b demonstrates that SSB is indeed visible, right at the tail–loop junction. That is just where the hypothesis predicts we should find the displaced DNA.

Shelterin is essential for t-loop formation. In particular, TRF2 can form t-loops in a model DNA substrate. However, this remodeling reaction is weak in the absence of the other shelterin subunits. TRF1, the other telomere repeat-binding protein, is especially helpful, as it can bend, loop, and pair telomeric repeats. It is striking that this remodeling reaction can occur in vitro even in the absence of ATP. Based on all we know about shelterin proteins, de Lange proposed the model for t-loop formation depicted in Figure 21.29. Figure 21.29a shows the members of the shelterin complex bound to an unlooped telomere. Figure 21.29b is a model for the interaction of shelterin with a t-loop.

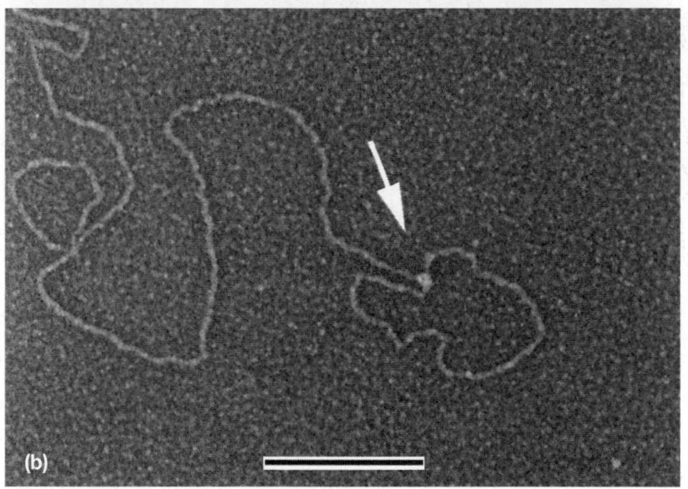

Figure 21.28 Binding of TRF1 and SSB to t-loops. (a) TRF1. Griffith, de Lange, and colleagues purified natural HeLa cell t-loops, cross-linked them with psoralen and UV radiation, and added TRF1, which binds specifically to double-stranded telomeric DNA. Then they shadowed the loop with platinum and paladium and performed electron microscopy. The t-loop, but not the tail, is coated uniformly with TRF1. **(b)** SSB. These workers followed the same procedure as in panel (a), but substituted *E. coli* SSB for TRF1. SSB should bind to single-stranded DNA, and it was observed at the loop–tail junction (arrow), where the single-stranded displacement loop was predicted to be. The bar represents 1 kb. (*Source:* Griffith, J.D., L. Comeau, S. Rosenfield, R.M. Stansel, A. Bianchi, H. Moss, and T. de Lange, Mammalian telomeres end in a large duplex loop. *Cell* 97 (14 May 1999) f. 5, p. 510. Reprinted by permission of Elsevier Science.)

Figure 21.29b also hints at an explanation for the paradox that POT1 is a single-stranded telomere-binding protein and yet the single-stranded telomeric DNA is hidden in the t-loop. But the figure shows that formation of the t-loop also creates a D-loop, and the displaced single-stranded region is a potential binding site for POT1. There is also the possibility that not all mammalian telomeres form t-loops. Any telomeres that remain linear would provide obvious binding sites for POT1.

The second way shelterin affects the structure of telomeres is by determining the structure of the end of the

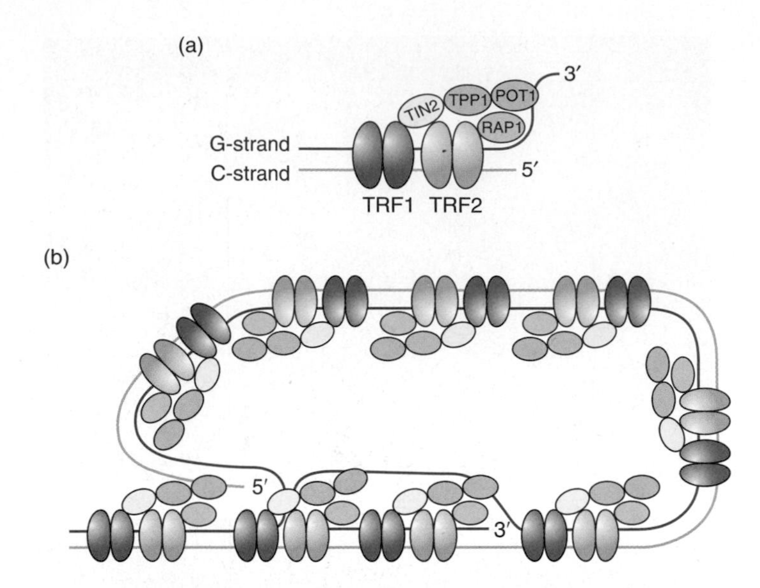

Figure 21.29 The shelterin-telomere complex. (**a**) Interaction with shelterin proteins and a linear telomere. TRF1 and TRF2 are shown interacting as dimers with the double-stranded part of the telomere, as POT1 interacts with the single-stranded part. The known interactions among shelterin proteins are also shown. (**b**) Model for the interaction of shelterin complexes with a t-loop. Colors are as in panel (a). Note the binding of POT1 (orange) to the single-stranded telomeric DNA in the D-loop, and the binding of TRF1 and TRF2 to the double-stranded telomeric DNA elsewhere in the t-loop.

telomere. It does this in two ways: by promoting 3′-end elongation, and protecting both the 5′- and 3′-ends from degradation. Finally, the third effect of shelterin on the structure of telomeres is to maintain telomere length within close tolerances. When the telomere gets too long, shelterin inhibits further telomerase action, limiting the growth of the telomere. POT1 plays a critical role in this process: When POT1 activity is eliminated, mammalian telomeres grow to abnormal lengths.

SUMMARY In mammals, telomeres are protected by a group of six proteins collectively known as shelterin. Two of the shelterin proteins, TRF1 and TRF2, bind to the double-stranded telomeric repeats. A third protein, POT1, binds to the single-stranded 3′-tail of the telomere. A fourth protein, TIN2, organizes shelterin by facilitating interaction between TRF1 and TRF2, and tethering POT1, via its partner, TPP1, to TRF2. Shelterin affects telomere structure in three ways: First, it remodels telomeres into t-loops, wherein the single-stranded 3′-tail invades the double-stranded telomeric DNA, creating a D-loop. In this way, the 3′-tail is protected. Second, it determines the structure of the telomeric end by promoting 3′-end elongation and protecting both 3′- and 5′-telomeric ends from degradation. Third, it maintains the telomere length within close tolerances.

Telomere Structure and Telomere-Binding Proteins in Lower Eukaryotes

Yeasts also have telomere-binding proteins, but they appear not to form t-loops. Thus, the proteins themselves must protect the telomere ends, without the benefit of hiding the single-stranded end within a D-loop. The fission yeast, *Schizosaccharomyces pombe,* has a group of telomere-binding proteins that resemble mammalian shelterin proteins. A protein called Taz1 plays the double-stranded telomere-binding role of mammalian TRF in fission yeast, and binds through Rap1 and Poz1 to a dimer of Tpz1 and Pot1. That resembles the TPP1-POT1 dimer in mammals, not only in structure, but in ability to bind to single-stranded telomeric DNA. These proteins can bind to a linear telomere, and they may also bend the telomere by 180 degrees by protein-protein interactions between proteins bound to the double-stranded telomere, and those bound to its single-stranded tail. This bending does not seem to form t-loops, however.

The budding yeast *Saccharomyces cerevisiae* also has telomere-binding proteins, but their evolutionary relationship to mammalian shelterin proteins is limited to one protein: Rap1. However, unlike mammalian RAP1, yeast Rap1 binds directly to double-stranded DNA, as the mammalian TRF proteins do. RAP1 has two partners, Rif1 and Rif2. In addition, a second protein complex, composed of Cdc13, Stn1, and Ten1, binds to the single-stranded telomeric tail.

Telomere-binding proteins were first discovered in the ciliated protozoan *Oxytricha*. This organism makes do with just two such proteins, TEBPα and TEBPβ, which are evolutionarily related to POT1 and TPP1 in mammals. These proteins bind to the single-stranded 3′-end of the organism's telomeres and protect them from degradation. By covering the ends of the telomeres, these proteins also prevent the telomeres from appearing like the ends of broken chromosomes—and all the negative consequences that would have.

SUMMARY Yeasts and ciliated protozoa do not form t-loops, but their telomeres are still associated with proteins that protect them. Fission yeasts have shelterin-like telomere-binding proteins, while budding yeasts have only one shelterin relative, Rap1, which binds to the double-stranded part of the telomere, plus two Rap1-binding proteins and three proteins that protect the single-stranded 3′-end of the telomere. The ciliated protozoan *Oxytricha* has only two telomere-binding proteins, which bind to the single-stranded 3′-ends of telomeres.

The Role of Pot1 in Protecting Telomeres In *S. pombe,* Pot1, instead of limiting the growth of telomeres, as mammalian POT1 does, plays a critical role in maintaining their integrity. Indeed loss of Pot1 can cause the loss of telomeres from this organism.

In 2001, Peter Baumann and Thomas Cech reported that they had found a protein in *S. pombe* that binds the single-stranded tails of telomeres. They named the *S. pombe* gene *pot1*, for protection of telomeres, and its product is now known as Pot1.

To test their hypothesis that *pot1* encodes a protein that protects telomeres, Baumann and Cech generated a $pot1^+$/ $pot1^-$ diploid strain and germinated the spores from this strain. The $pot1^-$ spores gave rise to very small colonies compared with the colonies from $pot1^+$ spores. And the $pot1^-$ cells tended to be elongated, to show defects in chromosome segregation, and to stop dividing. All of these effects are consistent with loss of telomere function.

To test directly the effect of *pot1* on telomeres, Baumann and Cech looked for the presence of telomeres in $pot1^-$ strains by Southern blotting DNA from these strains and probing with a telomere-specific probe. Figure 21.30 shows the results. DNA from the $pot1^+$ strains, and from the diploid strains containing at least one $pot1^+$ allele, reacted strongly with the telomere probe, indicating the presence of telomeres. But DNA from the $pot1^-$ strains did not react with the probe, indicating that their telomeres had disappeared. Thus, the *pot1* gene product, Pot1p (or Pot1), really does seem to protect telomeres.

If Pot1 really protects telomeres, we would expect it to bind to telomeres. To check this prediction, Baumann and Cech cloned the *pot1* gene into an *E. coli* vector so it could be expressed as a fusion protein with a tag of six histidines (Chapter 4). They purified this fusion protein and used a gel mobility shift assay (Chapter 5) to detect its binding to either the C-rich or G-rich strand of the telomere, or a double-stranded telomeric DNA. Figure 21.31a shows that Pot1 bound to the G-rich strand, but not to the C-rich or duplex DNA. Furthermore, an N-terminal fragment of Pot1 was even more effective in binding to the G-rich strand of the telomere (Figure 21.31b).

It is interesting that the phenotype of the $pot1^-$ strains, though it was originally quite aberrant, returned to normal after about 75 generations. The same effect had previously

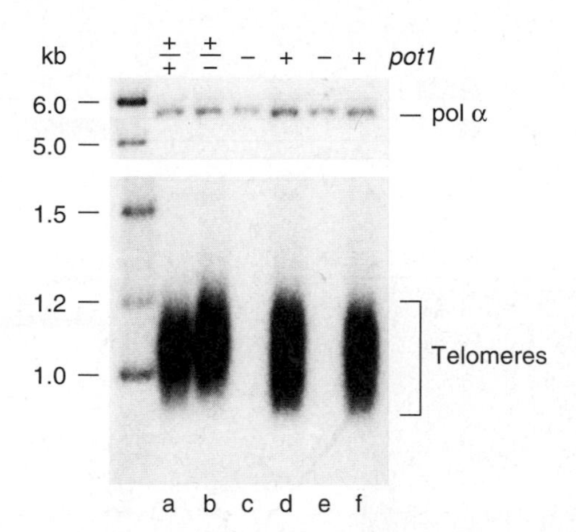

Figure 21.30 Fission yeast strains defective in *pot1* lose their telomeres. Baumann and Cech generated homozygous and heterozygous diploid, and $pot1^-$ and $pot1^+$ haploid strains of *S. pombe*, as indicated at top, then isolated DNA from these strains, digested the DNA with *Eco*RI, electrophoresed and Southern blotted the fragments, then probed the blot with a telomere-specific probe. As a control for uniform loading of the blot, the blot was also probed for DNA polymerase α, as indicated at top right. (*Source:* From Baumann and Cech, *Science* 292: p. 1172. © 2001 by the AAAS.)

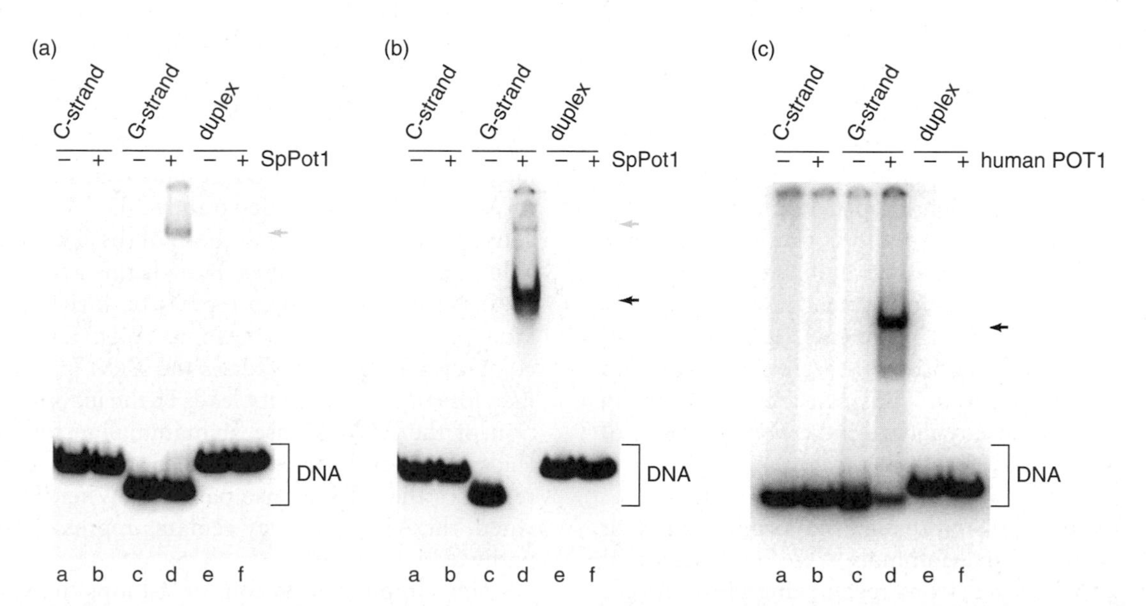

Figure 21.31 Pot1 binding to telomeric DNA. Baumann and Cech performed gel mobility shift experiments with *S. pombe* Pot1 and labeled *S. pombe* telomeric DNA **(a** and **b)** and human hPot1 and labeled human telomeric DNA **(c).** The telomeric DNA was either from the C-rich strand, the G-rich strand, or duplex DNA, as indicated at top. Panel **(a)** contained full-length Pot1. Panel **(b)** contained mostly an N-terminal fragment of Pot1, with slight contamination from full-length Pot1. Panel **(c)** contained an N-terminal fragment of human POT1. Arrows indicate the positions of shifted bands containing full-length Pot1 (yellow arrows) or N-terminal fragments of Pot1 or human POT1 (blue arrows). (*Source:* From Baumann and Cech, *Science* 292: p. 1172. © 2001 by the AAAS.)

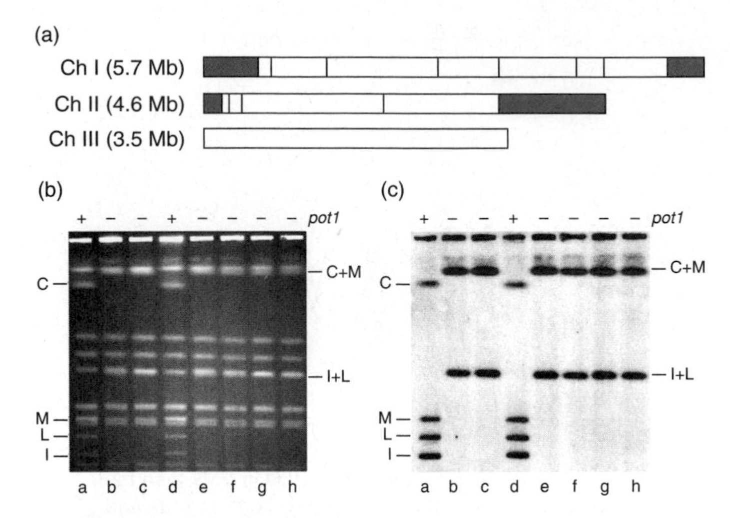

Figure 21.32 Surviving *Pot1*$^-$ stains have circularized chromosomes. (a) Maps of the three chromosomes of *S. pombe* showing the restriction sites for *Not*I as vertical lines. The terminal *Not*I fragments in chromosomes I and II are in red. Chromosome III is not cut by *Not*I. **(b)** Stained gel after pulsed-field gel electrophoresis of *Not*I DNA fragments from *pot1*$^+$ and *pot1*$^-$ cells, as indicated at top. The positions of terminal fragments (C, M, L, and I) of chromosomes I and II are indicated at left, and the positions of fused C+M and I+L fragments are indicated at right. **(c)** Baumann and Cech Southern blotted the gel from panel (b) and probed it with labeled DNA fragments C, M, L and I, representing the ends of chromosomes I and II. (*Source:* From Baumann and Cech. *Science* 292: p. 1172. © 2001 by the AAAS.)

been observed in strains lacking telomerase. This behavior can be explained if yeast chromosomes lacking telomeres can protect their ends by circularizing. To test this hypothesis, Baumann and Cech cleaved DNA from surviving *pot1*$^-$ strains with the rare cutter *Not*I (Chapter 4) and subjected the resulting DNA fragments to pulsed-field gel electrophoresis. If the chromosomes really had circularized, the *Not*I fragments at the ends of chromosomes should be missing and new fragments composed of the fused terminal fragments should appear. Figure 21.32 shows that this is exactly what happened for the two chromosomes tested, chromosomes I and II. The two fragments (I and L) normally at the ends of chromosome I were missing, and a new band (I+L), not present in *pot1*$^+$ strains, appeared. Similarly, the two fragments (C and M) normally at the ends of chromosome II were missing, and a new band (C+M) appeared. Thus, the chromosomes in *pot1*$^-$ strains really do circularize in response to loss of their telomeres.

The Role of Shelterin in Suppressing Inappropriate Repair and Cell Cycle Arrest in Mammals We have seen that telomeres prevent the cell from recognizing chromosome ends as chromosome breaks and invoking two processes that would threaten the life of the cell and even the organism. These processes are **homology-directed repair (HDR)** and nonhomologous end-joining (NHEJ, Chapter 20). HDR would promote homologous recombination between telomeres on separate chromosomes, or between telomeres and other chromosomal regions, resulting in potentially drastic shortening or lengthening of telomeres. The shortening would be especially dangerous because it could lead to loss of the whole telomere. NHEJ would lead to chromosome fusion, which is often lethal to the cell because the chromosomes do not separate properly during mitosis. If the cell doesn't die, the results could be even worse for the organism because they can lead to cancer.

In addition to HR and NHEJ, broken chromosomes also activate a checkpoint whereby the cell cycle can be arrested until the damage is repaired. If it is not repaired, the cells irreversibly enter a senescence phase and ultimately die, or they undergo a process called **apoptosis,** or programmed cell death, that results in rapid, controlled death of the cell. If normal chromosome ends invoked such a checkpoint, cells could not grow and life would cease. This is another reason that telomeres must prevent the cell from recognizing the normal ends of chromosomes as breaks.

Chromosome breaks do not by themselves activate cell cycle arrest. Instead, they are recognized by two protein kinases that autophosphorylate (phosphorylate themselves) and thereby initiate signal transduction pathways that lead to cell cycle arrest. One of these kinases is the ataxia telangiectasia mutated kinase (**ATM kinase**), which responds directly to unprotected DNA ends. Ataxia telangiectasia is an inherited disease caused by mutations in the ATM kinase gene. It is characterized by poor coordination (ataxia), prominent blood vessels in the whites of the eyes (telangiectasias), and susceptibility to cancer, among other symptoms.

The second kinase that senses chromosome breaks is the ataxia telangiectasia and Rad3 related kinase (**ATR kinase**), which responds to the single-stranded DNA end that appears when one DNA strand at a chromosome break is nibbled back by nucleases. As we have seen, mammalian telomeres have DNA ends that could activate the ATM kinase, and single-stranded DNA ends that could activate the ATR kinase, so both of these kinases need to be held in check at telomeres. How is this accomplished?

It is shelterin's job to repress both the ATM and ATR kinase at normal chromosome ends. One of shelterin's components, TRF2, represses the ATM kinase pathway. In fact, loss of TRF2 activity leads to the inappropriate activation of the ATM kinase at mammalian telomeres, which leads to cell cycle arrest. Another shelterin subunit, POT1, represses the ATR kinase pathway. When POT1 is inactivated, the ATM pathway remains repressed, but the ATR pathway is activated.

The simple formation of t-loops may explain the repression of the ATM pathway because the t-loops hide the DNA ends. However, t-loops cannot explain the repression of the ATR pathway, which is actually initiated by replication protein A (RPA), which binds directly to single-stranded DNA—and single-stranded DNA persists in the

D-loop part of a t-loop. Presumably, POT1 blocks binding of RPA to this single-stranded DNA simply by out-competing it for those binding sites. POT1 has an advantage over RPA in that it is automatically concentrated at telomeres by being part of the shelterin complex.

Shelterin also blocks the two DNA repair pathways that threaten telomeres: NHEJ and HDR. TRF2 represses NHEJ at telomeres during the G_1 phase of the cell cycle, before DNA replication, while POT1 and TRF2 team up to repress NHEJ at telomeres in the G_2 phase, after DNA replication. POT1 and TRF2 also collaborate to block HDR at telomeres. Ku (Chapter 20) can also block HDR at telomeres. This is interesting, because Ku's other role is to promote NHEJ when chromosomes are broken. Thus, telomeres must take advantage of Ku's ability to suppress HDR, while keeping in check its ability to promote NHEJ.

SUMMARY Unprotected chromosome ends would look like broken chromosomes and cause two potentially dangerous DNA repair activities, HDR and NHEJ. They would also stimulate two dangerous pathways (the ATM kinase and ATR kinase pathways) leading to cell cycle arrest. Two subunits of shelterin, TRF2 and POT1, block HDR and NHEJ. These two shelterin subunits also repress the two cell cycle arrest pathways. TRF2 represses the ATM kinase pathway, and POT1 represses the ATR kinase pathway.

SUMMARY

Primer synthesis in *E. coli* requires a primosome composed of the DNA helicase, DnaB, and the primase, DnaG. Primosome assembly at the origin of replication, *oriC,* occurs as follows: DnaA binds to *oriC* at sites called *dnaA* boxes and cooperates with RNA polymerase and HU protein in melting a DNA region adjacent to the leftmost *dnaA* box. DnaB then binds to the open complex and facilitates binding of the primase to complete the primosome. The primosome remains with the replisome, repeatedly priming Okazaki fragment synthesis, at least on the lagging strand. DnaB also has a helicase activity that unwinds the DNA as the replisome progresses.

The SV40 origin of replication is adjacent to the viral transcription control region. Initiation of replication depends on the viral large T antigen, which binds to a region within the 64-bp minimal *ori,* and at two adjacent sites, and exercises a helicase activity, which opens up a replication bubble within the minimal *ori.* Priming is carried out by a primase associated with the host DNA polymerase α.

The yeast origins of replication are contained within autonomously replicating sequences (ARSs) that are composed of four important regions (A, B1, B2, and B3). Region A is 15 bp long and contains an 11-bp consensus sequence that is highly conserved in ARSs. Region B3 may allow for an important DNA bend within ARS1.

The pol III holoenzyme synthesizes DNA at the rate of about 730 nt/sec in vitro, just a little slower than the rate of almost 1000 nt/sec observed in vivo. This enzyme is also highly processive, both in vitro and in vivo.

The pol III core ($\alpha\varepsilon$ or $\alpha\varepsilon\theta$) does not function processively by itself, so it can replicate only a short stretch of DNA before falling off the template. By contrast, the core plus the β-subunit can replicate DNA processively at a rate approaching 1000 nt/sec. The β-subunit forms a dimer that is ring-shaped. This ring fits around a DNA template and interacts with the α-subunit of the core to tether the whole polymerase and template together. This is why the holoenzyme stays on its template so long and is therefore so processive. The eukaryotic processivity factor PCNA forms a trimer with a similar ring shape that can encircle DNA and hold DNA polymerase on the template.

The β-subunit needs help from the γ complex (γ, δ, δ', χ, and ψ) to load onto the complex. The γ complex acts catalytically in forming this processive $\alpha\varepsilon\beta$ complex, so it does not remain associated with the complex during processive replication. Clamp loading is an ATP-dependent process.

The pol III holoenzyme is double-headed, with two core polymerases attached through two τ-subunits to a γ complex. One core is responsible for (presumably) continuous synthesis of the leading strand, the other performs discontinuous synthesis of the lagging strand. The γ complex serves as a clamp loader to load the β clamp onto a primed DNA template. Once loaded, the β clamp loses affinity for the γ complex and associates with the core polymerase to help with processive synthesis of an Okazaki fragment. Once the fragment is completed, the β clamp loses affinity for the core polymerase and associates with the γ complex, which acts as a clamp unloader, removing the clamp from the DNA. Then it can recycle to the next primer and repeat the process.

At the end of replication, circular bacterial chromosomes form catenanes that must be decatenated for the two daughter duplexes to separate. In *E. coli* and related bacteria, topoisomerase IV performs this decatenation. Linear eukaryotic chromosomes also require decatenation during DNA replication.

Eukaryotic chromosomes have special structures known as telomeres at their ends. One strand of these telomeres is composed of many tandem repeats of short, G-rich regions whose sequence varies from one species to another. The G-rich telomere strand is made by an enzyme called telomerase, which contains a short RNA that serves as the template for telomere synthesis. The C-rich telomere strand is synthesized by ordinary RNA-primed DNA synthesis, like the lagging strand in conventional DNA replication. This mechanism ensures that chromosome ends can be rebuilt and therefore do not suffer shortening with each round of replication.

In mammals, telomeres are protected by a group of six proteins collectively known as shelterin. Two of the shelterin proteins, TRF1 and TRF2, bind to the double-stranded telomeric repeats. A third protein, POT1, binds to the single-stranded 3′-tail of the telomere. A fourth protein, TIN2, organizes shelterin by facilitating interaction between TRF1 and TRF2, and tethering POT1, via its partner, TPP1, to TRF2. Shelterin affects telomere structure in three ways: First, it remodels telomeres into t-loops, wherein the single-stranded 3′-tail invades the double-stranded telomeric DNA, creating a D-loop. In this way, the 3′-tail is protected. Second, it determines the structure of the telomeric end by promoting 3′-end elongation and protecting both 3′- and 5′-telomeric ends from degradation. Third, it maintains the telomere length within close tolerances.

Yeasts and ciliated protozoa do not form t-loops, but their telomeres are still associated with proteins that protect them. Fission yeasts have shelterin-like telomere-binding proteins, while budding yeasts have only one shelterin relative, Rap1, which binds to the double-stranded part of the telomere, plus two Rap1-binding proteins and three proteins that protect the single-stranded 3′-end of the telomere. The ciliated protozoan *Oxytricha* has only two telomere-binding proteins, which bind to the single-stranded 3′-ends of telomeres.

Unprotected chromosome ends would look like broken chromosomes and cause two potentially dangerous DNA repair activities, HDR and NHEJ. They would also stimulate two dangerous pathways (the ATM kinase and ATR kinase pathways) leading to cell cycle arrest. Two subunits of shelterin, TRF2 and POT1, block HDR and NHEJ. These two shelterin subunits also repress the two cell cycle arrest pathways. TRF2 represses the ATM kinase pathway, and POT1 represses the ATR kinase pathway.

REVIEW QUESTIONS

1. Describe an assay to locate and determine the minimal length of an origin of replication.
2. List the components of the *E. coli* primosome and their roles in primer synthesis.
3. Outline a strategy for locating the SV40 origin of replication.
4. Outline a strategy for identifying an autonomously replicating sequence (ARS1) in yeast.
5. Outline a strategy to show that DNA replication begins in ARS1 in yeast.
6. Describe and give the results of an experiment that shows the rate of elongation of a DNA strand in vitro.
7. Describe a procedure to check the processivity of DNA synthesis in vitro.
8. Which subunit of the pol III holoenzyme provides processivity? What proteins load this subunit (the clamp) onto the DNA? To which core subunit does this clamp bind?
9. Describe and give the results of an experiment that shows the different behavior of the β clamp on circular and linear DNA. What does this behavior suggest about the mode of interaction between the clamp and the DNA?
10. What mode of interaction between the β clamp and DNA do x-ray crystallography studies suggest?
11. What mode of interaction between PCNA and DNA do x-ray crystallography studies suggest?
12. Describe and give the results of an experiment that shows that the clamp loader acts catalytically. What is the composition of the clamp loader?
13. Outline a hypothesis to explain how the clamp loader uses ATP energy to open the β clamp to allow entry to DNA.
14. How can discontinuous synthesis of the lagging strand keep up with synthesis of the leading strand?
15. Describe and give the results of an experiment that shows that pol III* can dissociate from its β clamp.
16. Describe a protein footprinting procedure. Show how such a procedure can be used to demonstrate that the pol III core and the clamp loader both interact with the same site on the β clamp.
17. Describe and give the results of an experiment that shows that the γ complex has clamp-unloading activity.
18. Describe how the β clamp cycles between binding to the core pol III and to the clamp unloader during discontinuous DNA replication.
19. Why is decatenation required after replication of circular DNAs?
20. Outline the evidence that topoisomerase IV is required for decatenation of plasmids in *Salmonella typhimurium* and *E. coli.*
21. Why do eukaryotes need telomeres, but prokaryotes do not?
22. Diagram the process of telomere synthesis.
23. Why was *Tetrahymena* a good choice of organism in which to study telomerase?
24. Describe an assay for telomerase activity and show sample results.
25. Describe and give the results of an experiment that shows that the telomerase RNA serves as the template for telomere synthesis.
26. Diagram the t-loop model of telomere structure.
27. What evidence supports the existence of t-loops?

28. What evidence supports the strand-invasion hypothesis of t-loop formation?
29. Present a model for the structure of mammalian shelterin, showing each of the subunits, and how they participate in t-loop formation.
30. How does mammalian shelterin protect chromosome ends from HDR and NHEJ and block the two pathways leading to cell cycle arrest? What would be the consequences of failure to block each of these pathways?

ANALYTICAL QUESTIONS

1. Starting with the nucleotide sequence of the *hpot1* gene (or the amino acid sequence of hPot1) from humans, describe how you would search for a homologous gene (or protein) in another organism whose genome has been sequenced, such as the nematode *Caenorhabditis elegans*. Then describe how you would obtain the protein and test it for Pot1 activity.
2. You are investigating the *pot1* gene of a newly-discovered protozoan species. You find that cells with a defective *pot1* gene return to normal after 50 generations. Wild-type cells have only two chromosomes with the following restriction maps with respect to the restriction enzyme *Zap*I:

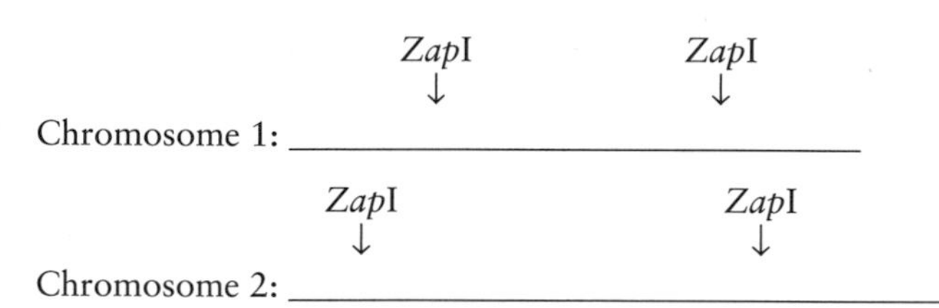

Propose a hypothesis to explain how the mutant cells returned to normal, and describe an experiment you would perform to test it. Show the results you would obtain if your hypothesis is correct.

3. You are studying a eukaryotic virus with a 130-kb double-stranded DNA genome. You suspect that it has more than one origin of replication. Propose an experiment to test your hypothesis and find all of the origins.
4. You are investigating DNA replication in a new species of bacteria. You discover that this organism has a β clamp and pol III*, similar to their counterparts in *E. coli*. You want to know whether this β clamp and pol III* separate during idling and after termination on a model template. Describe the experiment you would use to answer this question. Include the assay for separation you would use, and present sample results.
5. You are investigating the elongation rate during replication of the DNA from a new extreme thermophile, *Rapidus royi*. Here are the results of electrophoresis on DNA elongated in vitro for various times. What is the elongation rate? Does it set a new world record?

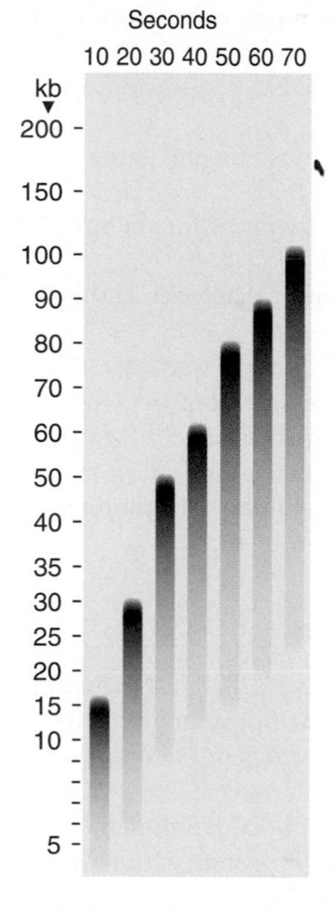

6. Assuming they could be made in eukaryotes, what would be the advantages and disadvantages of primers made of DNA, rather than RNA? Would such primers eliminate the need for telomeres?

SUGGESTED READINGS

General References and Reviews

Baker, T.A. 1995. Replication arrest. *Cell* 80:521–24.

Blackburn, E.H. 1990. Telomeres: Structure and synthesis. *Journal of Biological Chemistry* 265:5919–21.

Blackburn, E.H. 1994. Telomeres: No end in sight. *Cell* 77:621–23.

Cech, T. R. 2004. Beginning to understand the end of the chromosome. *Cell* 116:273–79.

de Lange, T. 2001. Telomere capping—one strand fits all. *Science* 292:1075–76.

de Lange, T. 2005. Shelterin, the protein complex that shapes and safeguards human telomeres. *Genes and Development* 19:2100–10.

de Lange, T. 2009. How telomeres solve the end-protection problem. *Science* 326:948–52.

Ellison, V. and B. Stillman. 2001. Opening of the clamp: An intimate view of an ATP-driven biological machine. *Cell* 106:655–60.

Greider, C.W. 1999. Telomeres do D-loop-T-loop. *Cell* 97:419–22.

Herendeen, D.R. and T.J. Kelly. 1996. DNA polymerase III: Running rings around the fork. *Cell* 84:5–8.

Kornberg, A. and T.A. Baker. 1992. *DNA Replication*, 2nd ed. New York: W.H. Freeman.

Marx, J. 1994. DNA repair comes into its own. *Science* 266:728–30.
Marx, J. 1995. How DNA replication originates. *Science* 270:1585–86.
Marx, J. 2002. Chromosome end game draws a crowd. *Science* 295:2348–51.
Newlon, C.S. 1993. Two jobs for the origin replication complex. *Science* 262:1830–31.
Stillman, B. 1994. Smart machines at the DNA replication fork. *Cell* 78:725–28.
Wang, J.C. 1991. DNA topoisomerases: Why so many? *Journal of Biological Chemistry* 266:6659–62.
West, S.C. 1996. DNA helicases: New breeds of translocating motors and molecular pumps. *Cell* 86:177–80.
Zakian, V.A. 1995. Telomeres: Beginning to understand the end. *Science* 270:1601–6.

Research Articles

Arai, K. and A. Kornberg. 1979. A general priming system employing only *dnaB* protein and primase for DNA replication. *Proceedings of the National Academy of Sciences USA* 76:4309–13.
Arai, K., R. Low, J. Kobori, J. Shlomai, and A. Kornberg. 1981. Mechanism of *dnaB* protein action V. Association of *dnaB* protein, protein n′, and other prepriming proteins in the primosome of DNA replication. *Journal of Biological Chemistry* 256:5273–80.
Baumann, P. and T. Cech. 2001. Pot 1, the putative telomere end-binding protein in fission yeast and humans. *Science* 292:1171–75.
Blackburn, E.H. 1990. Functional evidence for an RNA template in telomerase. *Science* 247:546–52.
Blackburn, E.H. 2001. Switching and signaling at the telomere. *Cell* 106:661–73.
Bouché, J.-P., L. Rowen, and A. Kornberg. 1978. The RNA primer synthesized by primase to initiate phage G4 DNA replication. *Journal of Biological Chemistry* 253:765–69.
Brewer, B.J. and W.L. Fangman. 1987. The localization of replication origins on ARS plasmids in *S. cerevisiae*. *Cell* 51:463–71.
Georgescu, R.E., S.-S. Kim, O. Yuryieva, J. Kuriyan, X.-P. Kong, and M. O'Donnell. 2008. Structure of a sliding clamp on DNA. *Cell* 132:43–54.
Greider, C.W. and E.H. Blackburn. 1985. Identification of a specific telomere terminal transferase activity in *Tetrahymena* extracts. *Cell* 43:405–13.
Greider, C.W. and E.H. Blackburn. 1989. A telomeric sequence in the RNA of *Tetrahymena* telomerase required for telomere repeat synthesis. *Nature* 337:331–37.
Griffith, J.D., L. Comeau, S. Rosenfield, R.M. Stansel, A. Bianchi, H. Moss, and T. de Lange. 1999. Mammalian telomeres end in a large duplex loop. *Cell* 97:503–19.
Jeruzalmi, D., M. O'Donnell, and J. Kuriyan. 2001. Crystal structure of the processivity clamp loader gamma (γ) complex of *E. coli* DNA polymerase III. *Cell* 106:429–41.
Jeruzalmi, D., O. Yurieva, Y. Zhao, M. Young, J. Stewart, M. Hingorani, M. O'Donnell, and J. Kuriyan. 2001. Mechanism of processivity clamp opening by the delta subunit wrench of the clamp loader complex of *E. coli* DNA polymerase III. *Cell* 106:417–28.
Kong, X.-P., R. Onrust, M. O'Donnell, and J. Kuriyan. 1992. Three-dimensional structure of the β subunit of *E. coli* DNA polymerase III holoenzyme: A sliding DNA clamp. *Cell* 69:425–37.
Krishna, T.S.R., X.-P. Kong, S. Gary, P.M. Burgers, and J. Kuriyan. 1994. Crystal structure of the eukaryotic DNA polymerase processivity factor PCNA. *Cell* 79:1233–43.
Marahrens, Y. and B. Stillman. 1992. A yeast chromosomal origin of DNA replication defined by multiple functional elements. *Science* 255:817–23.
Mok, M. and K.J. Marians. 1987. The *Escherichia coli* preprimosome and DNA B helicase can form replication forks that move at the same rate. *Journal of Biological Chemistry* 262:16644–54.
Naktinis, V., J. Turner, and M. O'Donnell. 1996. A molecular switch in a replication machine defined by an internal competition for protein rings. *Cell* 84:137–45.
Stukenberg, P.T., P.S. Studwell-Vaughan, and M. O'Donnell. 1991. Mechanism of the sliding β-clamp of DNA polymerase III holoenzyme. *Journal of Biological Chemistry* 266:11328–34.

CHAPTER 22

Homologous Recombination

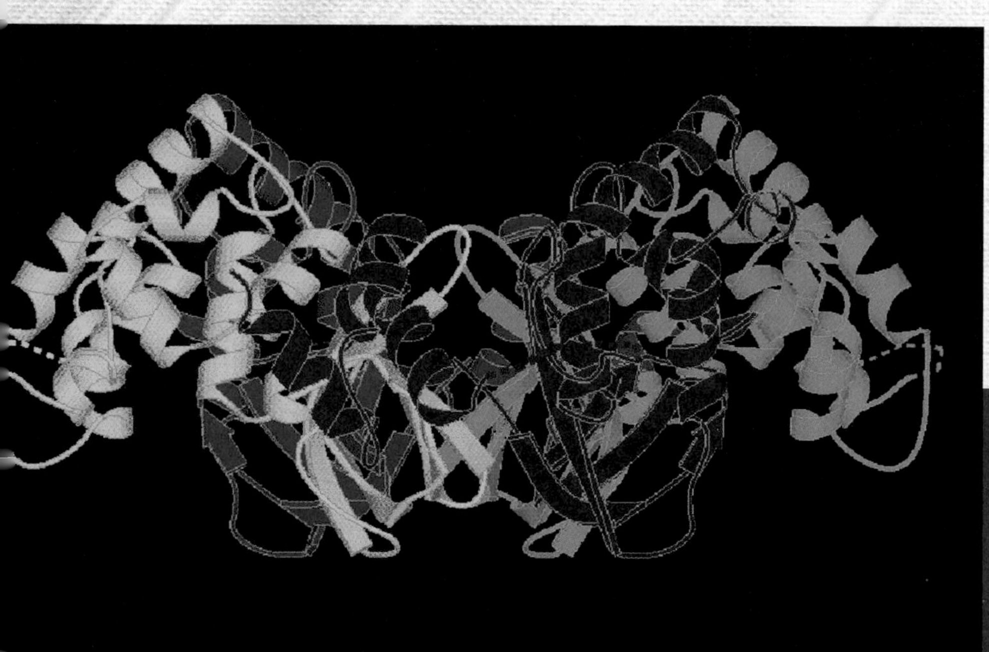

Crystal structure of the RuvA tetramer, with each monomer represented by a different color. RuvA binds to Holliday junctions and facilitates branch migration during recombination in *E. coli.* Rafferty, J.B., S.E. Sedelnikova, D. Hargreaves, P.J. Artymink, P.J. Baker, G.J. Sharples, A.A. Mahdi, R.G. Lloyd, and D.W. Rice, Crystal structure of DNA recombination protein RuvA and a model for its binding to the Holliday junction. *Science* 274 (18 Oct 1996) f. 2e, p. 417. Copyright © AAAS.

Geneticists have known for a long time that sexual reproduction gives offspring a different genetic makeup from their parents. Some of this variation comes from independent assortment of parental chromosomes. Most of the rest results from homologous recombination, which occurs between homologous chromosomes during meiosis. This process scrambles the genes of maternal and paternal chromosomes, so nonparental combinations occur in the offspring. This scrambling is valuable because the new combinations sometimes allow the progeny organisms a better chance of survival than their parents. Furthermore, meiotic recombination forms physical links between homologous chromosomes that allow the chromosomes to align properly during meiotic prophase so they separate properly during meiotic metaphase. These links are vital: It is estimated that 10 to 30%

of fertilized human eggs are aneuploid; that is, they contain an abnormal number of chromosomes, which is usually a lethal problem. And one of the leading causes of this aneuploidy is a reduction in the number, or abnormal placement, of recombination events during meiosis. Also, as we saw in Chapter 20, homologous recombination plays an important role in allowing cells to deal with DNA damage by so-called recombination repair.

Figure 22.1 illustrates several variations on the theme of homologous recombination. Each variation is characterized by a crossover event that joins DNA segments that were previously separated. This does not mean the two segments must start out on separate DNA molecules. Recombination can be intramolecular, in which case crossover between two sites on the same chromosome either removes or inverts the DNA segment in between. On the other hand, bimolecular recombination involves crossover between two independent DNA molecules. Ordinarily, recombination is reciprocal—a two-way street in which the two participants trade DNA segments. DNA molecules can undergo one crossover event, or two, or more, and the number of events strongly influences the nature of the final products.

22.1 The RecBCD Pathway for Homologous Recombination

To illustrate the principles of homologous recombination, let us consider the well-studied **RecBCD pathway,** one of the homologous recombination pathways used by *E. coli.*

This recombination process (Figure 22.2) begins with the induction of a double-stranded break in one of the recombining DNAs. The **RecBCD protein,** the product of the *recB*, –*C*, and –*D* genes, binds to a DNA double-stranded break and uses its DNA helicase activity to unwind the DNA toward a so-called **Chi site** or **χ** (Chi = crossover hotspot instigator), which has the sequence 5′-GCTGGTGG-3′. Chi sites are found on average every 5000 bp in the *E. coli* genome. RecBCD also has double-stranded and single-stranded exonuclease and single-stranded endonuclease activities. These allow RecBCD to produce a single-stranded tail, which can then be coated by **RecA protein** (the product of the ***recA* gene**). RecBCD also helps load RecA onto the 3′-DNA tail.

RecA allows the tail to invade a double-stranded DNA duplex and search for a region of homology. This creates a displacement loop (**D-loop**), defined by the displaced DNA strand. Once the tail finds a homologous region, a nick occurs in the D-looped DNA, possibly with the aid of RecBCD. This nick allows RecA and SSB to create a new tail that can pair with the gap in the other DNA. DNA

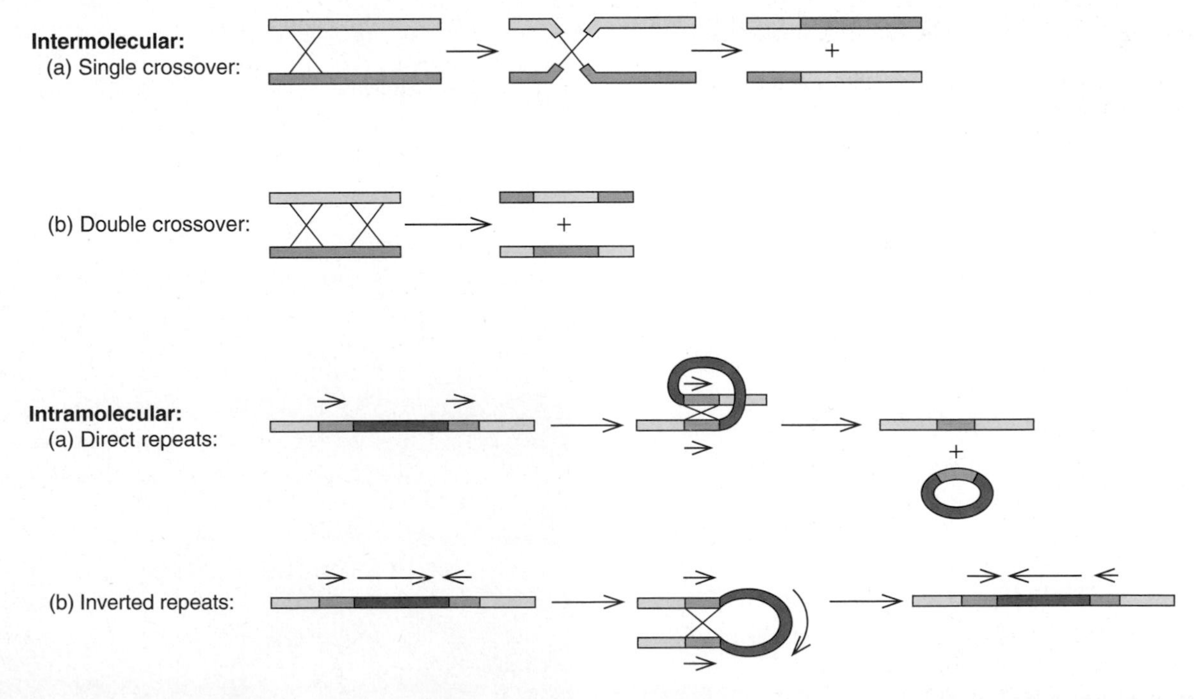

Figure 22.1 Examples of recombination. The X's represent crossover events between the two chromosomes or parts of the same chromosome. To visualize how these work, look at the intermediate form of the reciprocal recombination on the top line. Imagine the DNAs breaking and forming new, interstrand bonds as indicated by the arms of the X. This same principle applies to all the examples shown.

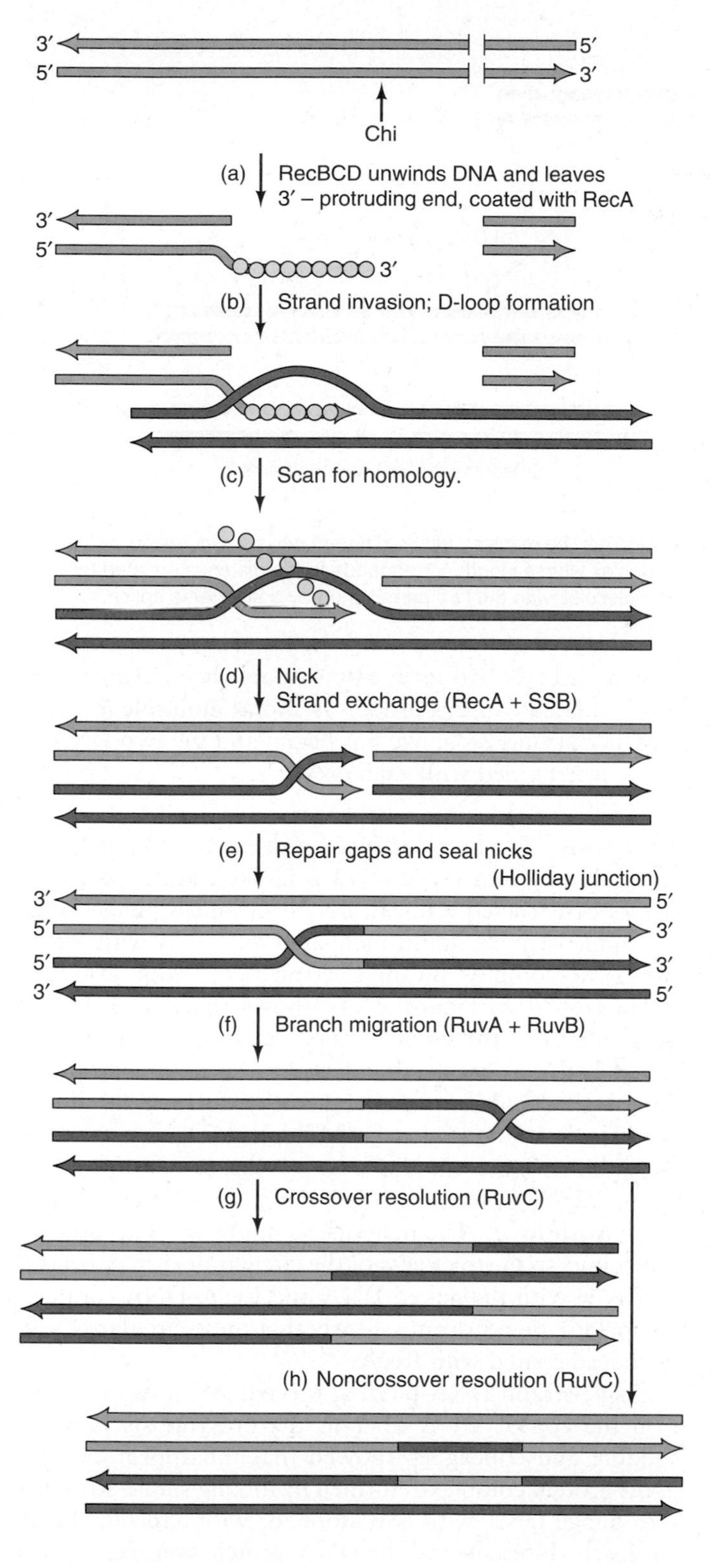

Figure 22.2 The RecBCD pathway of homologous recombination. (a) The RecBCD protein (omitted for the sake of clarity) binds at a double-stranded DNA break, and the DNA helicase activity of RecBCD then unwinds the DNA toward a Chi site, ultimately creating a 3′-terminal, single-stranded DNA that is coated with RecA protein (yellow spheres) **(b)** RecA promotes invasion of another DNA duplex, forming a D-loop. **(c)** RecA helps the invading strand scan for a region of homology in the recipient DNA duplex. Here, the invading strand has base-paired with a homologous region, releasing RecA. **(d)** Once a homologous region is found, a nick in the looped-out DNA appears, perhaps caused by RecBCD. This permits the tail of the newly nicked DNA to base-pair with the single-stranded region in the other DNA, probably aided by RecA. **(e)** The remaining gaps are filled in and nicks are sealed by DNA ligase, yielding a four-stranded complex with a Holliday junction. **(f)** Branch migration occurs, sponsored by RuvA and RuvB. Notice that the branch has migrated to the right. (**g** and **h**) Nicking by RuvC resolves the structure into two molecules, crossover recombinants or heteroduplexes, respectively.

ligase seals both nicks to generate a **Holliday junction,** named for Robin Holliday, who first proposed them in 1964. Holiday junctions are also known as half chiasmas and Chi structures. The branch in the Holliday junction can migrate in either direction simply by breaking old base pairs and forming new ones in a process called **branch migration.**

Branch migration does not occur at a useful rate spontaneously. Just as in DNA replication, DNA unwinding is required, and this in turn requires helicase activity and energy from ATP. Two proteins, **RuvA** and **RuvB,** collaborate in this function. Both have DNA helicase activity, and RuvB is an ATPase, so it can harvest energy from ATP for the branch migration process. Finally, two DNA strands must be nicked to resolve each Holliday junction into heteroduplexes or recombinant products. The **RuvC** protein carries out this function. Two alternative products can be produced, depending on which strands are nicked by RuvC. If the inner strands of the Holliday junction are nicked (Figure 22.3a), the structure resolves into a **noncrossover recombinant,** also known as a patch recombinant, or heteroduplex. If the outer strands of the Holliday junction are nicked (Figure 22.3b), the structure resolves into a **crossover recombinant,** also known as a splice recombinant, in which the DNA duplex changes from one genotype (represented by blue) at one end to another (represented by red) at the other.

SUMMARY RecBCD-sponsored homologous recombination in *E. coli* begins with invasion of a duplex DNA by a RecA-coated single-stranded DNA from another duplex that has suffered a double-stranded break. The invading strand forms a D-loop. Subsequent degradation of the D-loop strand leads to the formation of a branched intermediate. Branch migration in this intermediate yields a Holliday junction with two strands exchanging between homologous chromosomes. Finally, the Holliday junction can be resolved by nicking two of its strands. This can yield two noncrossover recombinant DNAs with patches of heteroduplex, or two crossover recombinant DNAs that have traded flanking DNA regions.

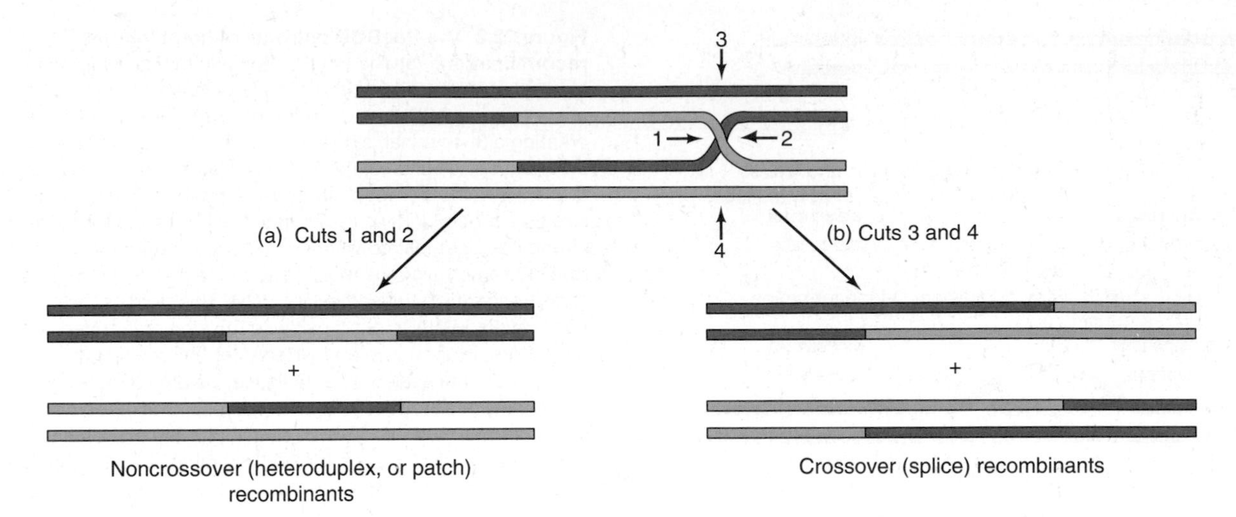

Figure 22.3 Resolution of a Holliday junction. The Holliday junction pictured at top can be resolved in two different ways, as indicated by the numbered arrows. **(a)** Cuts 1 and 2 yield two duplex DNAs with patches of heteroduplex whose length corresponds to the distance covered by branch migration before resolution. **(b)** Cuts 3 and 4 yield crossover recombinant molecules with the two parts joined by a staggered splice.

22.2 Experimental Support for the RecBCD Pathway

Now that we have seen a brief overview of the RecBCD pathway, let us look at the experimental evidence that supports this important bacterial recombination mechanism.

RecA

We have encountered RecA before, in our discussion of induction of the λ phage (Chapter 8). Indeed, this is a protein of many functions, but it was first discovered in the context of recombination, and that is how it got its name. In 1965, Alvin Clark and Ann Dee Margulies isolated two *E. coli* mutants that could accept F plasmids, but could not integrate their DNA permanently by recombination. These mutants were also highly sensitive to ultraviolet light, presumably because they were defective in recombination repair of UV damage (Chapter 20). Characterizing these mutants led ultimately to the discovery of two key proteins in the RecBCD pathway: RecA and RecBCD.

The *recA* gene had been cloned and overexpressed, so abundant RecA protein was available for study. It is a 38-kD protein that can promote a variety of strand exchange reactions in vitro. Using such in vitro assays, Charles Radding and colleagues discerned the three stages of participation of RecA in strand exchange:

1. **Presynapsis,** in which RecA coats the single-stranded DNA.
2. **Synapsis,** or alignment of complementary sequences in the single-stranded and double-stranded DNAs that will participate in strand exchange.
3. **Postsynapsis,** or **strand exchange,** in which the single-stranded DNA replaces the (+) strand in the double-stranded DNA to form a new double helix. An intermediate in this process is a **joint molecule** in which strand exchange has begun and the two DNAs are intertwined with each other.

Presynapsis The best evidence for the association between RecA and single-stranded DNA is visual. Radding and colleagues constructed a linear, double-stranded phage DNA with single-stranded tails, incubated this DNA with RecA, spread the complex on an electron microscope grid and photographed it. Figure 22.4a shows that RecA bound preferentially to the single-stranded ends, forming protein-coated DNA filaments, but leaving the double-stranded DNA in the middle uncoated. These workers also incubated single-stranded circular M13 phage DNA with RecA and subjected these complexes to the same procedure. Figure 22.4b shows extended DNA circles coated uniformly with RecA. The magnification in panels (a) and (b) is the same, so the thickness of the circular fiber in panel (b), compared with the naked DNA and RecA–DNA complex in panel (a), demonstrates clearly that these circular DNAs are indeed coated with RecA.

Single-strand DNA-binding protein (SSB) also helps to form the coated DNA fiber in the presynapsis process. Radding and colleagues showed that the appearances of DNA–protein complexes formed by mixing single-stranded M13 phage DNA with SSB alone, or with SSB plus RecA are clearly different, and the DNA–protein complexes with both SSB and RecA strongly resemble those with RecA alone in Figure 22.4. Furthermore, Radding and colleagues showed that SSB accelerates the formation of the coated DNA. In the presence of SSB plus RecA, formation of extended circular filaments was complete after only 10 min. By contrast, the process had barely begun after 10 min when SSB was absent.

(a)

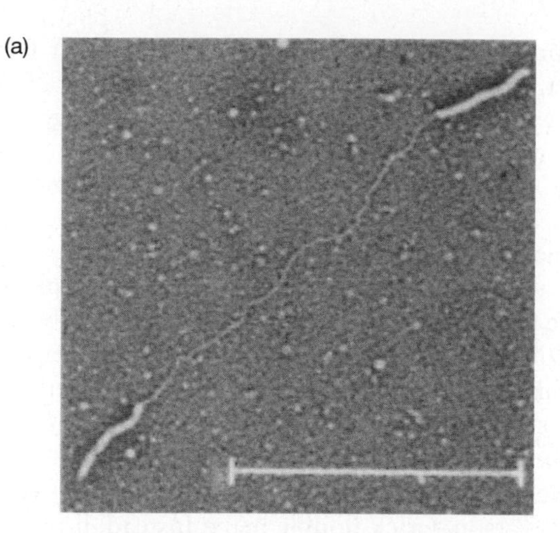

(b)

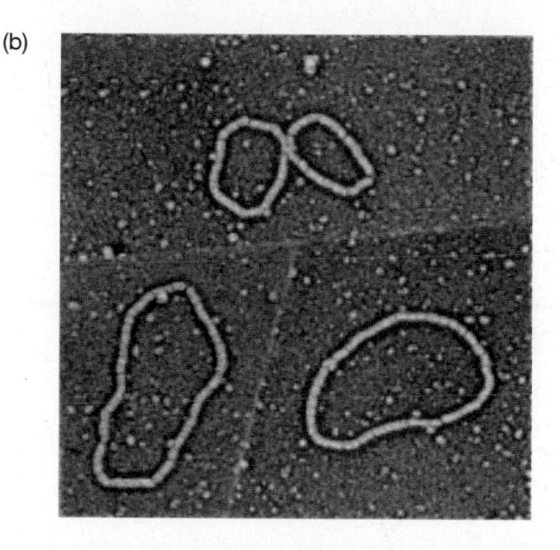

Figure 22.4 Binding of RecA to single-stranded DNA. Radding and colleagues prepared **(a)** a linear double-stranded DNA with single-stranded ends and **(b)** a circular single-stranded phage DNA. Then they added RecA, allowed time for a complex to form, spread the complexes on coated electron microscope grids, and photographed them. The bar in panel (a) represents 500 nm for both panels. (*Source:* Radding, C.M., J. Flory, A. Wu, R. Kahn, C. DasGupta, D. Gonda, M. Bianchi, and S.S. Tsang, Three phases in homologous pairing: Polymerization of *recA* protein on single-stranded DNA, synapsis, and polar strand exchange. *Cold Spring Harbor Symposia of Quantitative Biology* 47 (1982) f. 3 f&j, p. 823.)

Because RecA can coat a single-stranded DNA by itself, what is the role of SSB? It seems to be required to melt secondary structure (hairpins) in the single-stranded DNA that would otherwise impede the expansion of the RecA-coated fiber. Evidence for this notion comes from several sources. Radding and colleagues assayed for strand exchange when RecA was incubated with single-stranded DNA at low and high concentrations of $MgCl_2$. Low $MgCl_2$ concentrations destabilize DNA secondary structure, but high $MgCl_2$ concentrations stabilize it. In this experiment, SSB was required under high, but not low, $MgCl_2$ concentration conditions, which is the result we expect if SSB is needed to relax DNA secondary structure.

Later in this section, we will see that ATP hydrolysis is normally required for strand exchange, but I.R. Lehman and coworkers showed that ATPγS, an unhydrolyzable analog of ATP, will support a limited amount of strand exchange if SSB is present. Because ATPγS causes RecA to bind essentially irreversibly to both single- and double-stranded DNA, Radding and colleagues posed the hypothesis that ATPγS causes RecA to trap DNA in secondary structure that is unfavorable for strand exchange. If this is true, then SSB should be able to override this difficulty by removing secondary structure if it is added to DNA before RecA. As expected, SSB did indeed accelerate strand exchange if it was added before RecA. This provided more evidence that the role of SSB is to unwind secondary structure in a single-stranded DNA participating in recombination.

SUMMARY In the presynapsis step of recombination, RecA coats a single-stranded DNA that is participating in recombination. SSB accelerates the recombination process, apparently by melting secondary structure and preventing RecA from trapping any secondary structure that would inhibit strand exchange later in the recombination process.

Synapsis: Alignment of Complementary Sequences We will see later in this section that RecA stimulates strand exchange, which involves invasion of a duplex DNA by a single strand from another DNA. In this process, the invading strand forms a new double helix with one of the strands of the other duplex. But the step that precedes strand exchange, synapsis, entails a simple alignment of complementary sequences, without the formation of an intertwined double helix. This process yields a less stable product and is therefore more difficult to detect than strand exchange. Nevertheless, Radding and colleagues presented good evidence for synapsis as early as 1980.

As in the presynapsis experiments, electron microscopy was a key technique in the first demonstration of synapsis. Radding and coworkers used a favorite pair of substrates for their synapsis experiments: a single-stranded circular phage DNA and a double-stranded linear DNA. However, in this case, the single-stranded circular DNA was G4 phage DNA, and the double-stranded linear DNA was M13 phage DNA with a 274-bp G4 phage DNA insert near the middle. Because this target for the single-stranded G4 phage DNA lay thousands of base pairs from either end, and nicks were very rare in this DNA, it was unlikely that true strand exchange could happen. Instead, simple synapsis of complementary sequences could occur, as illustrated in Figure 22.5.

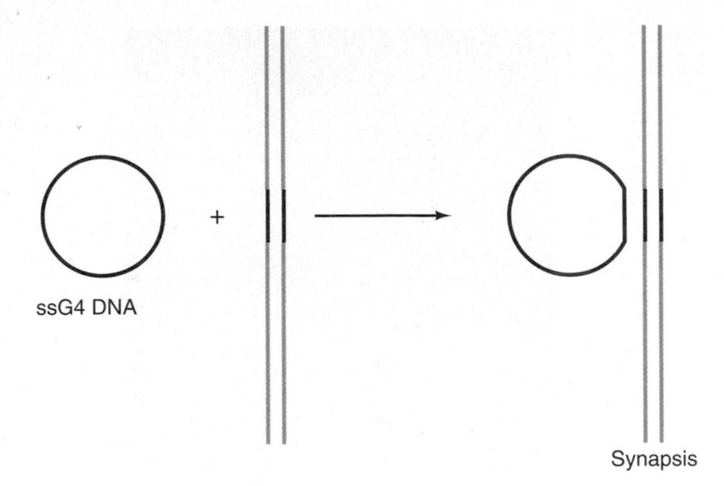

Figure 22.5 Synapsis. Synapsis is shown between circular single-stranded G4 phage DNA (red) and linear double-stranded M13(G4) DNA (M13 phage DNA [blue] with a 274-bp insert of G4 phage DNA [red]). The synapsis does not involve any intertwining of the linear and circular DNAs.

To measure synapsis between the two DNAs, Radding and colleagues mixed the two DNAs in the presence and absence of RecA and subjected the mixture to electron microscopy. In the presence of RecA, they found a significant proportion of the DNA molecules undergoing synapsis, as depicted in Figure 22.6. In most of the aligned molecules, the length of the aligned regions was appropriate, and the position of the aligned region within the linear DNA was correct. Furthermore, synapsis was reduced by 20–40-fold with two DNAs that did not share a homologous region. Synapsis also failed in the absence of RecA.

Although the percentage of nicked double-stranded DNAs was very low in this experiment, it was conceivable that nicks could create free ends in the linear DNA that could allow formation of a true, intertwined **plectonemic double helix.** If this had happened, the linkage between the two DNAs would have been stable to temperatures approaching the melting point of DNA. However, the alignments shown in Figure 22.6 were destroyed by heating at 20°C below the melting point for 5 min. Thus, the synapsis observed here does not involve the formation of a base-paired Watson–Crick double helix. Instead, it probably involves a **paranemic double helix** in which the two aligned DNA strands are side by side, but not intertwined. Further support for the notion that nicks are not required for synapsis comes from the finding that supercoiled DNA (unnicked by definition) works just as well as linear double-stranded DNA in these experiments.

How much homology is necessary for synapsis to occur? David Gonda and Radding provided an estimate by showing that a 151-bp homologous region gave just as efficient synapsis as did 274 bp of homology. But DNAs with just 30 bp of homology gave only background levels of aligned DNA molecules. Thus, the minimum degree of homology for efficient synapsis is somewhere between 30 and 151 bp.

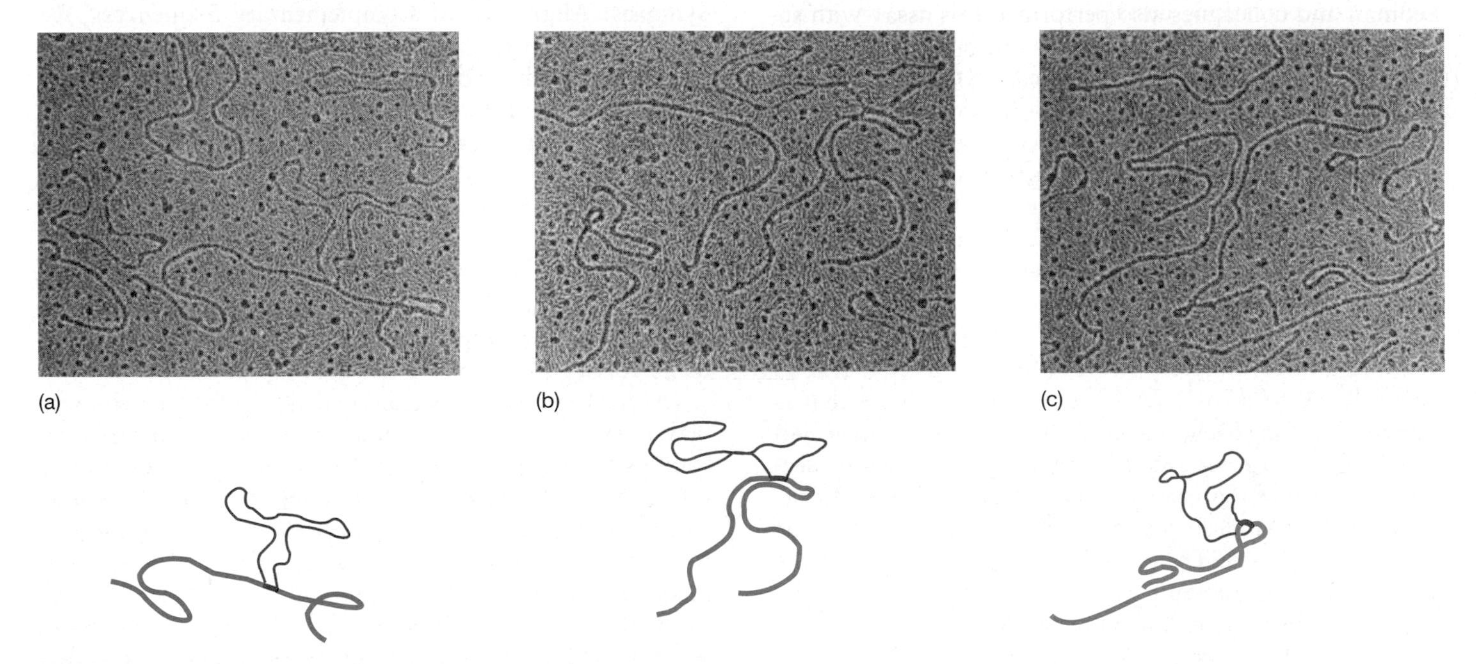

Figure 22.6 Demonstration of RecA-dependent synapsis in vitro. Radding and colleagues mixed the circular single-stranded and linear double-stranded DNAs described in Figure 22.5 with RecA and then examined the products by electron microscopy. Panels **(a–c)** show three different examples of aligned DNA molecules. Below each electron micrograph is an interpretive diagram, showing the linear double-stranded DNA in blue and the circular single-stranded DNA in red. Thick red lines denote the zones of synapsis between the two DNAs in each case. (*Source:* DasGupta C., T. Shibata, R.P. Cunningham, and C.M. Radding, The topology of homologous pairing promoted by *recA* protein. *Cell* 22 (Nov 1980 Pt2) f. 9 d–f, p. 443. Reprinted by permission of Elsevier Science.)

SUMMARY Synapsis occurs when a single-stranded DNA finds a homologous region in a double-stranded DNA and aligns with it. No intertwining of the two DNAs occurs at this point.

Postsynapsis: Strand Exchange We have learned that RecA is required for the first two steps in strand exchange: presynapsis and synapsis. Now we will see that it is also required for the last: postsynapsis, or strand exchange itself. Lehman and colleagues measured strand exchange between double-stranded and single-stranded phage DNA by a filter-binding assay for D-loop formation as follows: They incubated ^{3}H-labeled duplex P22 phage DNA with unlabeled single-stranded P22 DNA in the presence or absence of RecA. They used high-salt and low-temperature conditions that restricted branch migration, which would have completely assimilated the single-stranded DNA and eliminated the D-loops. Then they removed protein from the DNA with a detergent (either Sarkosyl or sodium dodecyl sulfate). Finally, they filtered the mixture through a nitrocellulose filter. If D-loops formed in the duplex DNA, the single-stranded D-loop would cause the complex to bind to the filter, so labeled DNA would be retained. If no D-loop formed, the unlabeled single-stranded DNA would stick to the filter, but the labeled duplex DNA would flow through. The detergent prevented DNA from sticking to the filter simply because of its association with RecA. Lehman and colleagues also performed this assay with supercoiled duplex M13 phage DNA and linear M13 DNA. In both cases, about 50% of the DNA duplexes formed D-loops, but only in the presence of RecA. Without RecA, retention of D-looped DNA was less than 1%. Also, with a nonhomologous single-stranded DNA, retention of D-looped DNA was only 2%.

To verify that D-loops had actually formed, these workers treated the complexes with S1 nuclease to remove single-stranded DNA, then filtered the product. This greatly reduced the retention of the labeled DNA on the filter, suggesting that a D-loop was really involved. To be sure, they directly visualized the D-loops by electron microscopy. D-loops were clearly visible in both kinds of DNA—linear and supercoiled. This experiment thus had the added benefit of demonstrating that supercoiled DNA is not required for strand exchange.

Table 22.1 shows the effects of nucleotides on D-loop formation, as measured by retention by nitrocellulose filtration. We see that ATP is required for D-loop formation, and its role cannot be performed by GTP, UTP, or ATPγS. The fact that ATPγS cannot substitute for ATP in D-loop formation indicates that ATP hydrolysis is required. In fact, it appears that ATP hydrolysis permits RecA to dissociate from DNA, which permits the new base pairing that must occur in strand exchange.

Table 22.1 Requirements for D-loop Formation

Duplex DNA	Reaction components	D-loops formed (%)
P22 phage	Complete	100
	−RecA	<1
	−ATP	<1
	−ATP + GTP	<1
	−ATP + UTP	<1
	−ATP + ATPγS	<1
M13 phage	Complete	100
	−RecA	1
	−ATP	1
	−ATP + GTP	2

Thus, this experiment demonstrated that ATP hydrolysis is essential for D-loop formation. RecA, an incredibly versatile protein, has ATPase activity, which cleaves ATP as RecA falls off the DNAs to allow the D-loops to form.

SUMMARY RecA and ATP collaborate to promote strand exchange between a single-stranded and double-stranded DNA. ATP is necessary to clear RecA off the synapsing DNAs to make way for formation of double-stranded DNA involving the single strand and one of the strands of the DNA duplex.

RecBCD

In our discussion of RecA, we have been considering model reactions involving a single-stranded and a duplex DNA. The reason, of course, is that RecA requires a single-stranded DNA to initiate strand exchange. But naturally recombining DNAs are usually both double-stranded, so how does RecA get the single strand it needs? We have already learned that RecBCD provides it. Two elements intimately involved in this process are Chi sites on the DNA and the DNA helicase activity of RecBCD. Let us consider the evidence for these two things.

Chi sites were discovered in genetic experiments with bacteriophage λ. Lambda *red gam* phages lacked Chi sites, but their efficient replication depended on recombination by the RecBCD pathway. Because, as we will see, the RecBCD pathway depends on Chi sites, these mutants made small plaques. Franklin Stahl and colleagues showed that certain λ *red gam* mutants made large plaques, suggesting that these mutants had more active RecBCD recombination. Stahl and colleagues then discovered that recombination was enhanced near the point of the mutation and named these mutated sites Chi, for crossover hotspot instigator. The fact that the mutations promoted recombination nearby suggested that these mutations did not behave like ordinary gene mutations that occur in coding regions and change the structures of gene products. Instead, it appeared that these mutations created new Chi sites that stimulated recombination nearby.

We know that Chi sites stimulate the RecBCD pathway, but not the λ Red (homologous recombination) pathway, the λ Int (site-specific recombination) pathway, or the *E. coli* RecE and RecF pathways (both homologous). This points strongly to the participation of the RecBCD protein at Chi sites, because this is the only component of the RecBCD pathway not found in any of the others. In fact, because RecBCD has endonuclease activity, one attractive hypothesis is that it nicks DNA near Chi sites to initiate recombination.

Gerald Smith and his colleagues found evidence for this hypothesis. They generated a 3′-end-labeled, double-stranded fragment of plasmid pBR322 with a Chi site near its end. The labeled 3′-end lay only about 80 bp from the Chi site, as shown in Figure 22.7a. Then they added purified RecBCD protein. After heat-denaturing the DNAs in some of the reactions, they electrophoresed the DNA products and looked for the 80-nt fragment that would be generated by nicking at the Chi site. (They had to run the reaction briefly to avoid general degradation of the DNA by the nonspecific RecBCD nuclease activity.) Figure 22.7b shows that the 80-nt product was indeed observed and that its appearance depended on the presence of the RecBCD protein. Heat-denaturation of the DNA was not necessary to yield the 80-nt single-stranded DNA, which suggested that the RecBCD protein was not only nicking, but also unwinding the DNA beyond the nick. Smith and colleagues also mapped the exact cleavage sites by running the labeled ≈ 80-nt fragment alongside sequencing lanes generated by chemical cleavage of the same labeled substrate. They observed two bands, one nucleotide apart, that showed that RecBCD cut this substrate in two places, as indicated by the asterisks in this sequence:

5′-<u>GCTGGTGG</u>GTT*G*CCT-3′

Thus, RecBCD cut this substrate 3 and 4 nt to the 3′-side of the GCTGGTGG Chi site (underlined). Another substrate could be cut in three places, 4, 5, and 6 nt to the 3′-side of the Chi site. Thus, the exact cleavage sites depend on the substrate.

These findings supported the idea that RecBCD nicks DNA near a Chi site and also suggested that RecBCD can unwind the DNA, starting at the nick. Further support for the role of RecBCD in unwinding DNA came from work by Stephen Kowalczykowski and colleagues. Figure 22.8

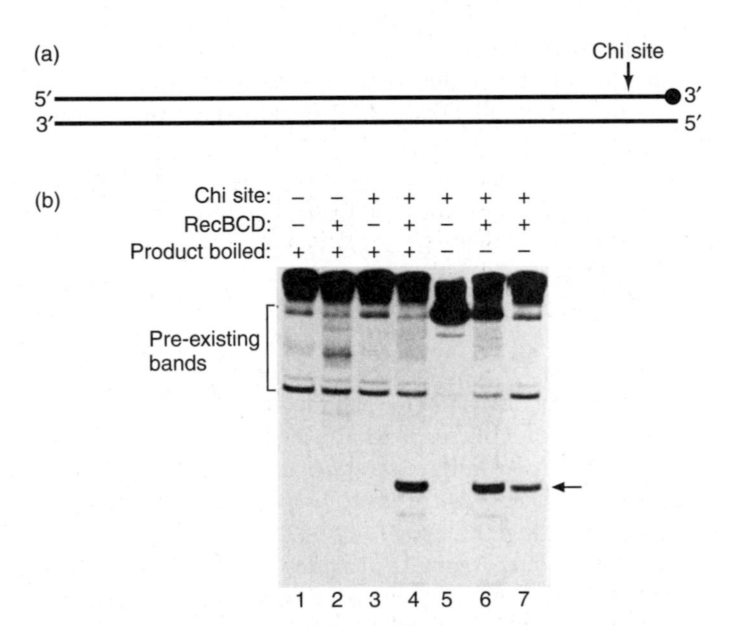

Figure 22.7 Chi-specific nicking of DNA by RecBCD. (a) Substrate for nicking assay. Smith and colleagues prepared a 1.58-kb *Eco*RI-*Dde*I restriction fragment with a Chi site about 80 bp from the *Dde*I end. They 3′-end-labeled the *Dde*I end by end-filling with [^{32}P] nucleotide (red). **(b)** Nicking assay. Smith and colleagues incubated the end-labeled DNA fragment in panel (a), designated "+" in the top line, or a similar fragment lacking the Chi site, designated "−" in the top line, with or without RecBCD (designated "+" or "−" in the middle line) for 30 sec. Then they terminated the reactions and electrophoresed the products. Some of the reaction products were boiled for 3 min as indicated at top. The arrow at right denotes the 80-nt labeled fragment released by nicking at the Chi site. The appearance of this product depended on RecBCD and a Chi site, but not on boiling the product. (*Source:* (*b*) Ponticelli, A.S., D.W. Schultz, A.F. Taylor, and G.R. Smith, Chi-dependent DNA strand cleavage by *rec*BC enzyme. *Cell* 41 (May 1985) f. 2, p. 146. Reprinted by permission of Elsevier Science.)

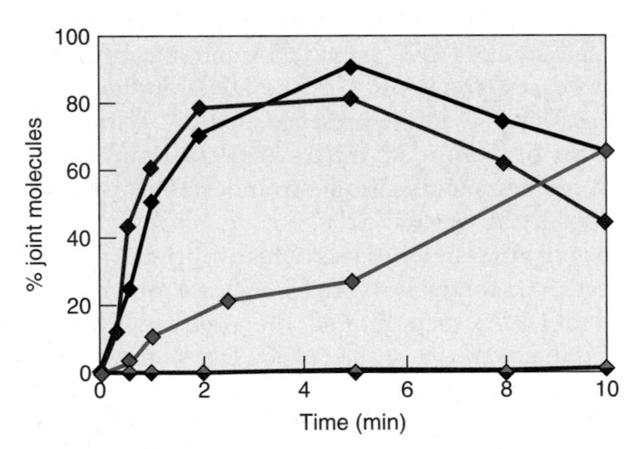

Figure 22.8 RecBCD-dependence of strand exchange between two duplex DNAs. Kowalczykowski and coworkers incubated two duplex DNAs with RecA, RecBCD, and SSB (red) and assayed for joint molecules (strand exchange) by filter binding or by gel electrophoresis. (The joint molecules have a lower electrophoretic mobility than the nonrecombining DNAs.) They also assayed for joint molecules without RecA or without RecBCD (orange and purple symbols at bottom). The blue line shows the results when RecBCD was omitted, but one of the DNAs was heat-denatured. The green line shows the results when all components except RecA were preincubated together, then RecA was added to start the reaction. RecBCD was added last in all other reactions. (*Source:* Adapted from Roman, L.J., D.A. Dixon, and S.C. Kowalczykowski, "RecBCD-dependent joint molecule formation promoted by the *Escherichia coli* RecA and SSB proteins," *Proceedings of the National Academy of Sciences USA* 88:3367–71, April 1991.)

presents the results of one of their experiments, which demonstrated that: (1) RecA alone, or even RecA plus SSB, could not cause pairing between two homologous double-stranded DNAs. (2) However, with RecBCD in addition to RecA and SSB, strand exchange, which depends on DNA unwinding, occurred rapidly, as long as the two DNAs were homologous. (3) RecBCD was dispensible if one of the DNAs was heat-denatured. This last finding implied that one function of RecBCD is to unwind one of the DNAs to provide a free DNA end; RecA and SSB can coat and then use this free DNA end to initiate strand invasion.

Stuart Linn and colleagues provided direct evidence for the DNA helicase activity of RecBCD using electron microscopy to detect the unwound T7 phage DNA products. When the experimenters added SSB and RecBCD together, they observed forked DNAs with duplex DNA adjoining two single strands. This implied that RecBCD began unwinding at the end of the duplex, and SSB trapped the two single-stranded DNAs that were generated. As expected, the forks grew longer with time.

SUMMARY RecBCD has a DNA endonuclease activity that can nick double-stranded DNA, especially near Chi sites, and a DNA helicase activity that can unwind double-stranded DNAs from their ends. These activities help RecBCD to provide the single-stranded DNA ends that RecA needs to initiate strand exchange.

RuvA and RuvB

RuvA and RuvB form a DNA helicase that catalyzes the branch migration of a Holliday junction. We have seen that Holliday junctions can be created in vitro; in fact they are a by-product of experiments that measure the effect of RecA on strand exchange. Early work on RuvA and RuvB used such RecA products as the Holliday junctions that could interact with RuvA and RuvB. Later, Stephen West and his colleagues devised a method for using four synthetic oligonucleotides whose sequences required that they base-pair in such a way as to form a Holliday junction, as illustrated in Figure 22.9.

Carol Parsons and West end-labeled such a synthetic Holliday junction and used a gel mobility shift assay to measure binding of RuvA and RuvB to the Holliday junction. Because branch migration was known to require ATP, they used the unhydrolyzable ATP analog, ATPγS. In principle, this should allow assembly of RuvA and RuvB on the DNA, but should prevent the branch migration that would dissociate the Holliday junction. They were successful in demonstrating a complex between RuvA and the Holliday junction, but did not see a supershift with RuvB, which would have indicated a RuvA–RuvB–Holliday junction complex. This suggested that this ternary complex was too unstable under these experimental conditions. To stabilize the putative complex, they added glutaraldehyde, which should cross-link the proteins in the complex and prevent them from dissociating during gel electrophoresis.

Figure 22.10 demonstrates cooperative binding between RuvA and RuvB. At low RuvA concentration (lane b), little, if any, binding to the Holliday junction occurred. On the

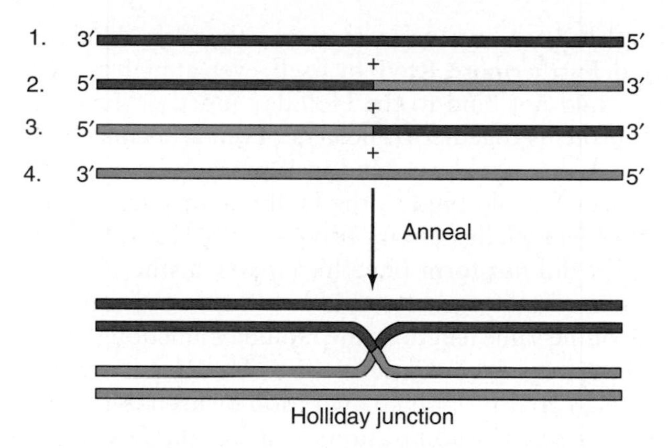

Figure 22.9 Forming a synthetic Holliday junction. Oligonucleotides 1–4 are mixed under annealing conditions so the complementary parts of each can base-pair. The 5′-end of oligo 2 (red) is complementary to the 3′-end of oligo 1 (red), so those two half-molecules can base-pair, but the 3′-end of oligo 2 (blue) is complementary to the 5′-end of oligo 4 (blue), so those two half-molecules form base pairs. Similarly, the two ends of oligo 3 are complementary to the other ends of oligos 1 and 4, so oligo 3 crosses over in its base pairing, in a manner complementary to that of oligo 2. The result is a synthetic Holliday junction.

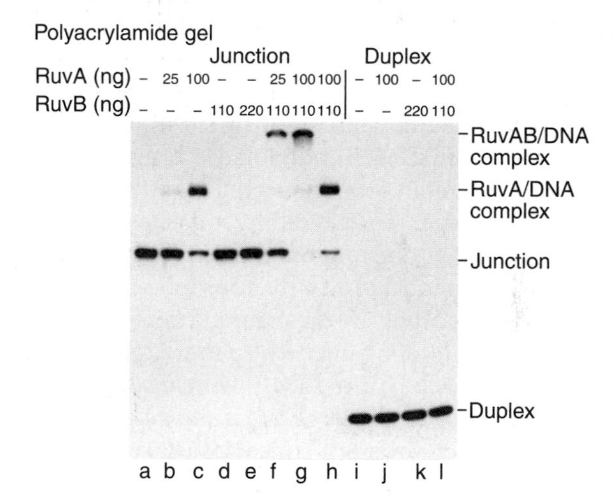

Figure 22.10 Detecting a RuvA–RuvB–Holliday junction complex. Parsons and West constructed a labeled synthetic Holliday junction and mixed it with varying amounts of RuvA and RuvB, as indicated at top. All mixtures contained ATPγS except the one in lane h. Parsons and West then treated the mixtures with glutaraldehyde to cross-link proteins in the same complex and prevent their dissociation. Finally, they subjected the complexes to polyacrylamide gel electrophoresis and autoradiography to detect the labeled complexes. (*Source:* Parsons, C.A. and S.C. West, Formation of a RuvAB–Holliday junction complex *in vitro*. *Journal of Molecular Biology* 232 (1993) f. 2, p. 400, by permission of Elsevier.)

other hand, at high concentration, abundant binding occurred. Furthermore, RuvB by itself, even at high concentration, could not bind to the Holliday junction (lane e), but both proteins together could bind, even at a concentration of RuvA that could not bind well by itself (lanes f and g). Only RuvA could bind to the Holliday junction in the absence of ATPγS; the ternary RuvA–RuvB–Holliday junction complex did not form (lane h). Finally, neither RuvA, nor RuvB, nor both together could bind to an ordinary duplex DNA of the same length as the Holliday junction (lanes j–l), so these proteins bind specifically to Holliday junctions.

RuvB can drive branch migration by itself if it is present in high enough concentration, so it has the DNA helicase and attendant ATPase activity. What, then, is the role of RuvA? It binds to the center of a Holliday junction and facilitates binding of RuvB, so branch migration can take place at a much lower RuvB concentration. Furthermore, as we will soon see, it appears to hold the Holliday junction in a square planar conformation that is favorable for rapid branch migration.

What is the nature of the binding between RuvA and the Holliday junction? David Rice and colleagues have bolstered the hypothesis of a square planar conformation for the RuvA–Holliday junction complex by performing x-ray crystallography on RuvA tetramers and showing that they have a square planar shape. This is illustrated in Figure 22.11a, in which we see that each monomer is roughly L-shaped, containing a leg and a foot connected by a flexible loop of indeterminate shape, represented by a dashed colored line. The foot of the L of one monomer interacts with the leg of the next to form a lobe; one of the four lobes is surrounded by a white dashed line in the figure. These lobes are arranged in a four-fold symmetrical pattern, with a natural groove between each pair of lobes; white dashed lines lie in two of these grooves. Figure 22.11b shows a side view of the tetramer, which reveals a concave surface on top and a convex surface on the bottom.

Molecular modeling showed that this RuvA tetramer could mate naturally with a Holliday junction in a corresponding square planar conformation, as shown in Figure 22.12. Note the neat fit between the DNA and the concave surface of the protein. The four branches of the Holliday junction could lie in the four grooves on the surface of the protein. Four β turns, one on each monomer, form a hollow-looking pin that protrudes through the center of the Holliday junction. This square planar shape would allow rapid branch migration. Any deviation from this shape would slow branch migration, which emphasizes the importance of the square planar shape of RuvA. What is the relationship between the square planar Holliday junction and the familiar, branched Holliday junction we have seen so far? They are really just two representations of the same structure, as we will soon see.

Stephen West and Edward Egelman, along with Xiong Yu, performed electron microscopy of the RuvAB–Holliday junction complex. They made 100 micrographs of the complex, scanned them, and combined them to create an average image. Figure 22.13a presents a model based on this image, with color-coded DNA strands added. As expected, the RuvA tetramer is in the center at the junction, with two RuvB hexameric rings flanking it. Panel (b) shows what happens on bending two of the arms of the complex

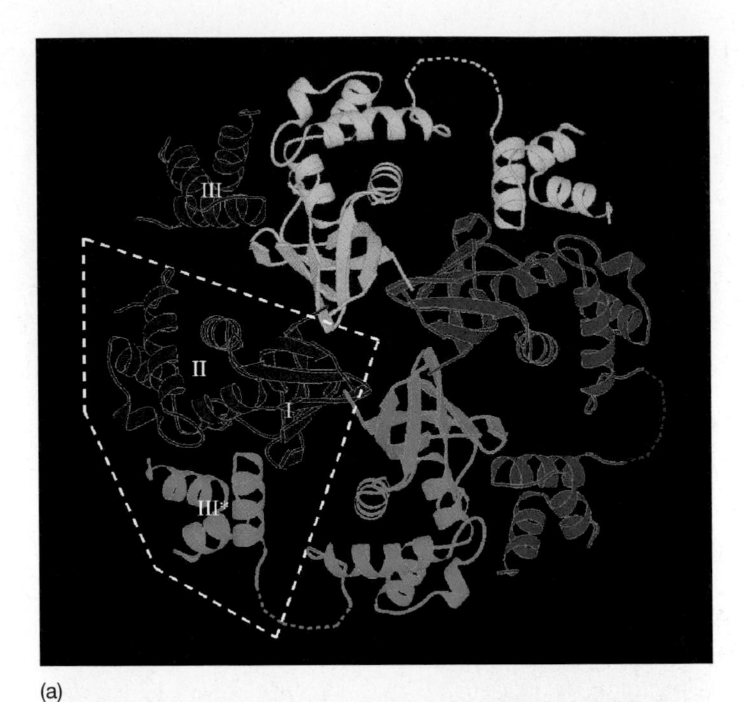

(a)

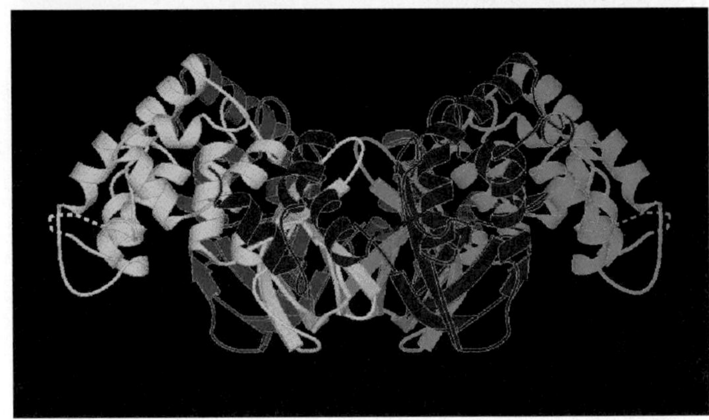

(b)

Figure 22.11 Structure of RuvA tetramer as revealed by x-ray crystallography. (a) Top view. The four monomers are represented by different-colored ribbons, and one of the four lobes in the square planar structure is outlined with a dashed white line. The three domains of the blue monomer are numbered, as is the third domain (the "foot" of the L) of the green monomer. **(b)** Side view. The same structure, represented by the same colored ribbons, is shown from the side. The concave and convex surfaces at top and bottom, respectively, are evident. (*Source:* Rafferty J.B., S.E. Sedelnikova, D. Hargreaves, P.J. Artymiuk, P.J. Baker, G.J. Sharples, A.A. Mahdi, R.G. Lloyd, and D.W. Rice, Crystal structure of DNA recombination protein RuvA and a model for its binding to the Holliday junction. *Science* 274 (18 Oct 1996) f. 2 d–e, p. 417. Copyright © AAAS.)

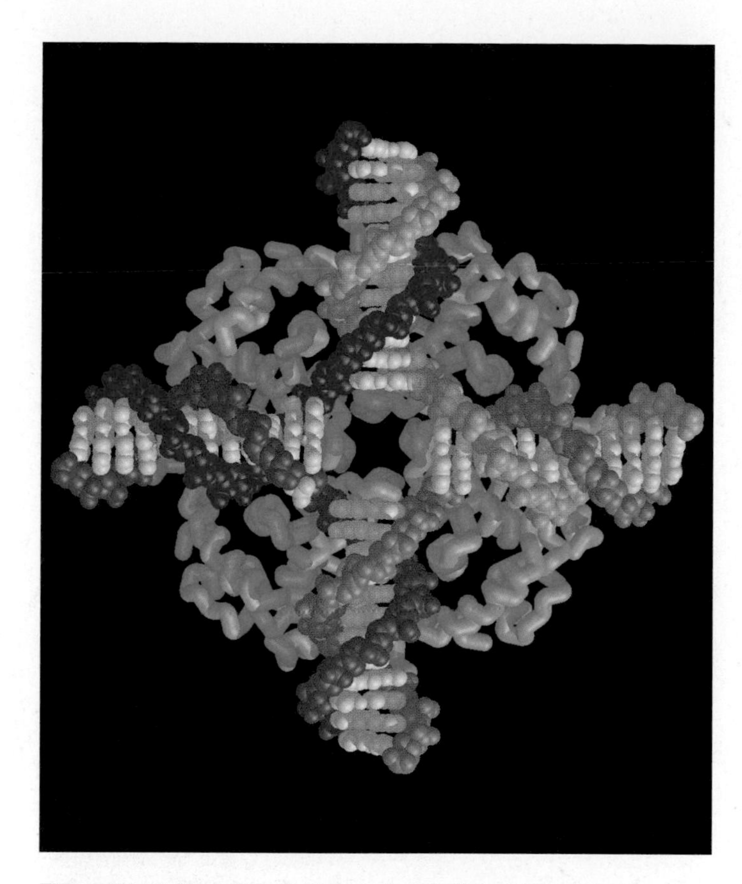

Figure 22.12 Model for the interaction between RuvA and a Holliday junction. The RuvA monomers are represented by green tubes that trace the α-carbon backbones of the polypeptides. The DNAs in the Holliday junction are represented by space-filling models containing dark and light pink and blue backbones and silver base pairs. The yellow balls denote the phosphate groups of one of the two pairs of sites that can be cut by RuvC to resolve the Holliday junction. (*Source:* Rafferty, J.B., S.E. Sedelnikova, D. Hargreaves, P.J. Artymiuk, P.J. Baker, G.J. Sharples, A.A. Mahdi, R.G. Lloyd, and D.W. Rice, Crystal structure of DNA recombination protein RuvA and a model for its binding to the Holliday junction. *Science* 274 (18 Oct 1996) f. 3d, p. 418. Copyright © AAAS.)

(with the RuvA tetramer removed for clarity), and panel (c) shows the result of rotating the bottom DNA duplex through 180 degrees out of the plane of the paper (again the RuvA tetramer is removed for clarity). The RuvB rings on this familiar Holliday junction are poised to catalyze branch migration by moving in the direction of the arrows. Though it is harder to visualize, they can do the same with the DNA in the shape shown in panel (a).

SUMMARY RuvA and RuvB form a DNA helicase that can drive branch migration. A RuvA tetramer with square planar symmetry recognizes the center of a Holliday junction and binds to it. This presumably induces the Holliday junction itself to adopt a square planar conformation, and promotes binding of hexamer rings of RuvB to two diametrically opposed branches of the Holliday junction. Then RuvB uses its ATPase to drive the DNA unwinding and rewinding that is necessary for branch migration.

RuvC

What nuclease is responsible for making the cuts that resolve Holliday junctions? West and colleagues showed in 1991 that it is RuvC. They built the ^{32}P-labeled synthetic Holliday junction pictured in Figure 22.14a, with a short (12 bp) homologous region (J) at the joint, but the rest of the structure composed of nonhomologous regions. Next, they used a gel mobility shift assay to test the ability of RuvC to bind to the Holliday junction and to a linear duplex DNA. Figure 22.14b shows the results: As they added more and more RuvC to the Holliday junction, West and colleagues observed more and more DNA–protein complex, indicating RuvC–Holliday junction binding. But the

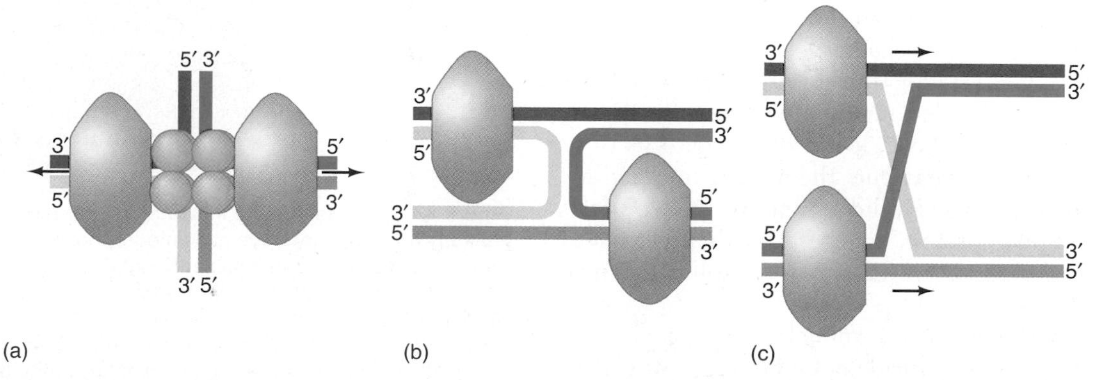

Figure 22.13 Model for RuvAB–Holliday junction complex based on combining EM images of the complex. (a) Complex with DNA branches perpendicular to each other. The DNA moves through the complex in the directions indicated by the arrows. **(b)** The blue-yellow and red-green branches from panel (a) have been rotated 90 degrees in the plane of the paper, and the RuvA tetramer has been removed so we can see the center of the junction. **(c)** The blue-green and blue-yellow branches from panel (b) have been interchanged by rotating the lower limb of the junction 180 degrees out of the plane of the paper. This produces a familiar Holliday junction with RuvB hexamer rings in position to catalyze branch migration by moving in the direction of the arrows. (*Source:* Adapted from Yu, X., S.C. West, and E.H. Egelman, Structure and subunit composition of the RuvAB–Holliday junction complex. *Journal of Molecular Biology* 266:217–222, 1997.)

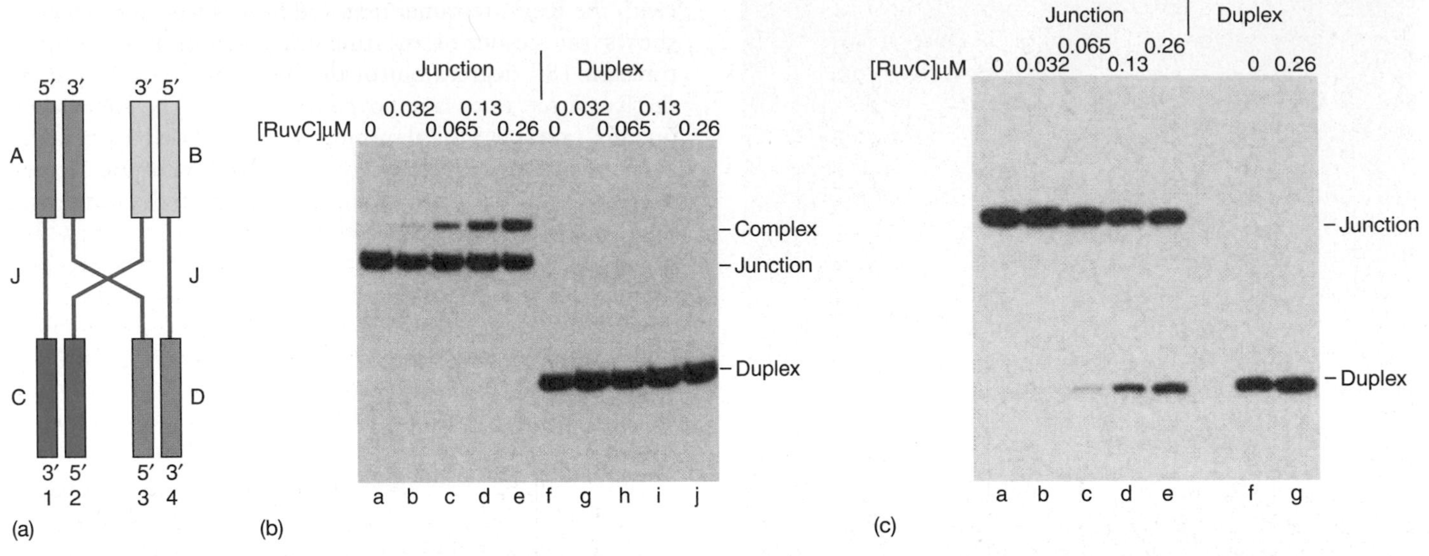

Figure 22.14 RuvC can resolve a synthetic Holliday junction. **(a)** Structure of the synthetic Holliday junction. Only the 12-bp central region (J, red) is formed from homologous DNA. The other parts of the Holliday junction (A, B, C, and D) are nonhomologous, as indicated by the different colors. **(b)** Binding of RuvC to the synthetic Holliday junction. West and colleagues end-labeled the Holliday junction (and a linear duplex DNA) and bound them to increasing amounts of pure RuvC under noncleavage conditions (low temperature and absence of $MgCl_2$). Then they electrophoresed the products. RuvC binds to the Holliday junction, but not to ordinary duplex DNA. **(c)** Resolution of the Holliday junction by RuvC. West and coworkers mixed the labeled Holliday junction (or linear duplex DNA) with increasing concentrations of RuvC under cleavage conditions (37°C and 5 mM $MgCl_2$). Then they electrophoresed the products. RuvC resolved some of the Holliday junction to a linear duplex form. (*Source:* Dunderdale, H.J., F.E. Benson, C A. Parsons, G.J. Sharples, R.G. Lloyd, and S.C. West, Formation resolution of recombination intermediates by *E. coli recA* and RuvC proteins. *Nature* 354 (19–26 Dec 1991) f. 5b–c. p. 509. Copyright © Macmillan Magazines Ltd.)

same assay showed no binding between RuvC and a linear duplex DNA made from strand 1 and its complement.

Thus, RuvC binds specifically to a Holliday junction, but can it resolve the junction? Figure 22.14c shows that it can. West and coworkers added increasing amounts of RuvC to the labeled Holliday junction, or to the duplex DNA. They found that RuvC caused resolution of the Holliday junction to a labeled species with the same mobility as the duplex DNA, which is what we expect for resolution. More complex experiments that can distinguish between patch or splice resolution showed that the splice products predominate, at least in vitro.

Thanks to x-ray crystallography studies performed by Kosuke Morikawa and colleagues, we now know the three-dimensional structure of RuvC. It is a dimer, with its two active sites 30 Å apart. That puts them right in position to cleave the square planar Holliday junction at two sites, as shown in Figure 22.15a. Figure 22.15b presents a more detailed representation of a RuvC–Holliday junction complex.

Does RuvC act alone, as this model implies, or does it act on a Holliday junction already bound to RuvB or RuvA plus RuvB? The evidence strongly suggests the latter. West and colleagues have reconstituted a system that carries out the intermediate to late stages of recombination in vitro and have shown that monoclonal antibodies against RuvA, RuvB, and RuvC each block resolution of Holliday junctions.

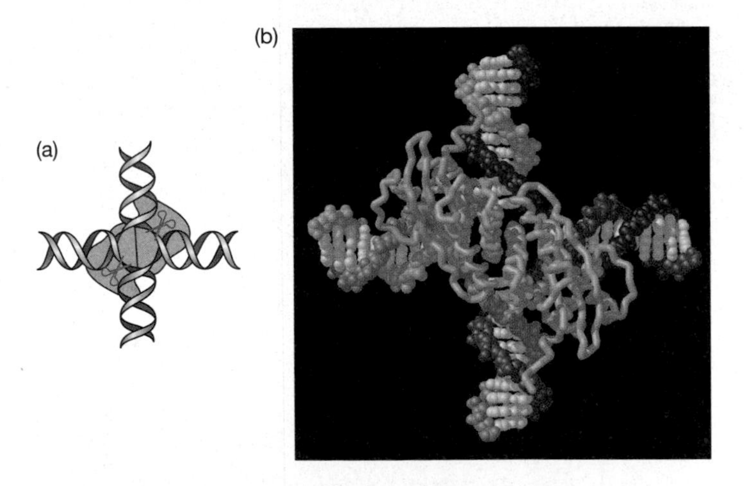

Figure 22.15 Model for the interaction between RuvC and a Holliday junction. **(a)** Schematic model showing the RuvC dimer (gray) bound to the square planar Holliday junction. Scissors symbols (green) denote the active sites on the two RuvC monomers. Note how the location of these active sites fits with the positioning of the DNA strands to be cleaved in resolving the complex. **(b)** Detailed model. Gray tubes represent the carbon backbone of the RuvC dimer. The Holliday junction is represented by the same blue and pink backbone and silver base pairs as in Figure 22.12. (*Source:* From Rafferty, J.B., S.E. Sedelnikova, D. Hargreaves, P.J. Artymiuk, P.J. Baker, G.J. Sharples, A.A. Mahdi, R.G. Lloyd, and D.W. Rice, Crystal structure of DNA recombination protein RuvA and a model for its binding to the Holliday junction. *Science* 274 (18 Oct 1996) f. 3e, p. 418. Copyright © AAAS. Reprinted with permission from AAAS.)

One way to explain this result is that RuvA, RuvB, and RuvC work together. If that is true, then the proteins probably naturally associate with one another, and one should be able to cross-link them. So West and colleagues prepared various mixtures of the two proteins, added glutaraldehyde to cross-link them, and electrophoresed them to detect cross-links. They found that RuvA and RuvB could be cross-linked, as expected, and RuvB and RuvC could also be cross-linked, but RuvA and RuvC could not. Thus, RuvB can bind to both RuvA and RuvC, suggesting that all three proteins can bind together to a Holliday junction.

This hypothesis of concerted action by RuvA, RuvB, and RuvC is consistent with the notion that branch migration is necessary during resolution to help RuvC find its preferred sites of cleavage. It is also consistent with x-ray crystallography data showing that RuvA can associate with a Holliday junction as a tetramer, as we have already seen, or as an octamer, with tetramers on either side of the DNA. West hypothesized that the complex involving the RuvA octamer is specific for efficient branch migration (Figure 22.16a). Later, RuvC could replace one of the RuvA tetramers to form the putative RuvABC–junction complex, or "resolvasome" (Figure 22.16b) that is specific for resolution of the Holliday junction.

Mutations in *ruvA, ruvB,* and *ruvC* all produce the same phenotype: heightened sensitivity to UV light, ionizing radiation, and the antibiotic mitomycin C because of defective recombination repair. But RuvA and B promote branch migration, whereas RuvC catalyzes resolution of Holliday junctions. Why should defects in all three of these proteins have the same end result? One way to answer this question would be to show that resolution depends on branch migration. Then defective RuvA or RuvB would block resolution indirectly by blocking branch migration.

West and colleagues did not exactly do that, but they did show that hotspots for RuvC resolution occur, which implies that branch migration is needed to reach those hotspots. To determine the sequences at the RuvC cutting sites, West and coworkers performed primer extension analysis on the RuvC products, using the same primers for DNA sequencing of the same DNAs. In all, they identified 19 cutting sites, and they observed a clear consensus sequence: 5′-(A/T)TT ↓ (G/C)-3′. RuvA and B are presumably needed to catalyze branch migration in vivo to reach such consensus sites. This hypothesis also implies that resolution to patch or splice products depends on the frequencies of the RuvC resolution sequence in the two DNA strands. Overall, this should be a 50/50 mix.

(a) **Alternate RuvAB–junction complex**

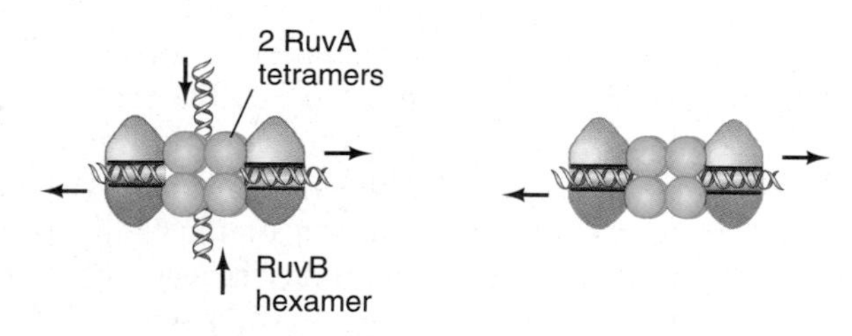

(b) **Putative RuvABC–junction complex**

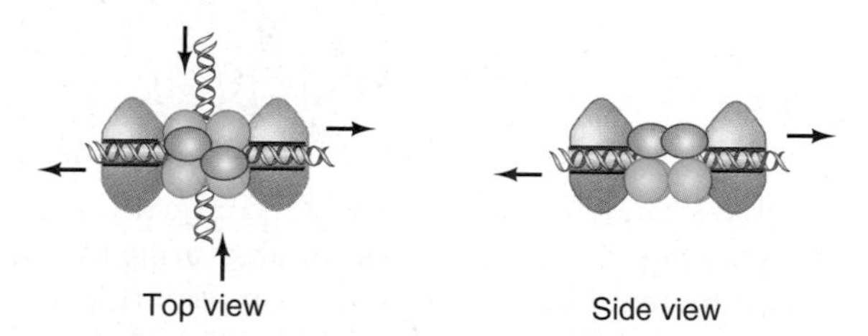

Figure 22.16 Models of Ruv protein–junction complexes. (a) A RuvAB–junction complex discovered by Pearl and colleagues. In contrast to the complex with a RuvA tetramer described by other workers, this one contains a RuvA octamer at the Holliday junction. This could be the form the complex takes during active migration. **(b)** West's model of the RuvABC–junction complex, with RuvC in purple. This could be the form the complex takes during resolution. Top and side views are on the left and right, respectively, in both panels. (*Source:* Adapted from West, S.C., RuvA gets x-rayed on Holliday. *Cell* 94:700, 1998.)

SUMMARY Resolution of Holliday junctions in *E. coli* is catalyzed by the RuvC resolvase. This protein acts as a dimer to clip two DNA strands to yield either patch or splice recombinant products. This clipping occurs preferentially at the consensus sequence 5′-(A/T)TT ↓ (G/C)-3′. Branch migration is essential for efficient resolution of Holliday junctions, presumably because it is essential to reach the preferred cutting sites. Accordingly, RuvA, B, and C appear to work together in a complex to locate and cut those sites.

22.3 Meiotic Recombination

As mentioned early in this chapter, meiosis in most eukaryotes is accompanied by recombination. This process shares many characteristics in common with homologous recombination in bacteria. In this section, we will examine the mechanism of meiotic recombination in yeast.

The Mechanism of Meiotic Recombination: Overview

Figure 22.17 presents a hypothesis for meiotic recombination in budding yeast (*Saccharomyces cerevisiae*), where it has been most thoroughly studied. The process starts with a chromosomal lesion: a double-stranded break. Next, an exonuclease recognizes the break and digests the 5′-ends of two of the strands, creating 3′-single-stranded overhangs. One of these single-stranded ends can then invade the other DNA duplex,

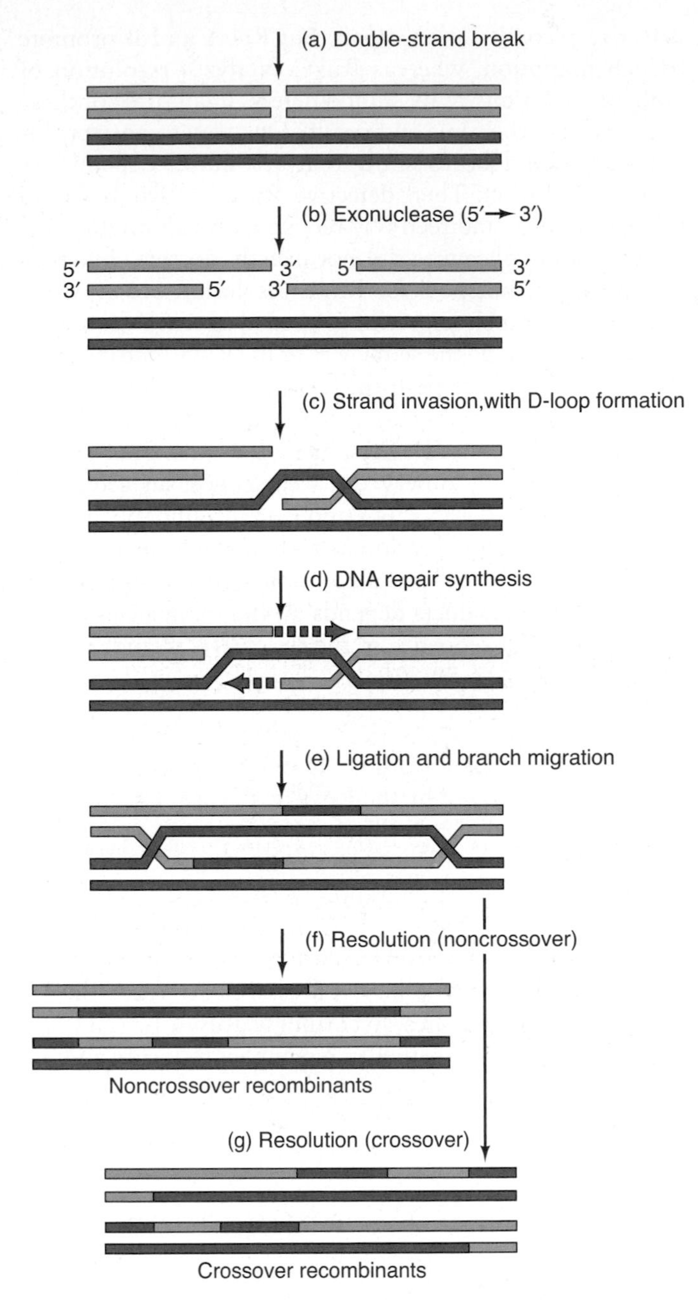

Figure 22.17 Model for meiotic recombination in yeast. (a) A double-stranded break occurs in one DNA duplex (blue), which is paired with another DNA duplex (red). **(b)** An exonuclease digests the DNA 5′-ends at the newly created break. **(c)** A single-stranded 3′-end of the top duplex invades the bottom duplex, creating a D-loop. **(d)** DNA repair synthesis extends the free 3′-ends, with enlargement of the D-loop. **(e)** Branch migration occurs both leftward and rightward to yield two Holliday junctions. **(f)** The Holliday junctions are resolved by cleaving the inside strands at both Holliday junctions, yielding noncrossover recombinant DNAs with patches of heteroduplex, but no exchange of DNA arms beyond the Holliday junctions. **(g)** The Holliday junctions are resolved by cleaving the inside strands at the left Holliday junction and the outside strands at the right Holliday junction. This yields crossover recombinant DNAs with exchange of DNA arms to the right of the right Holliday junction.

forming a D-loop as we observed in bacterial homologous recombination. Next, DNA repair synthesis fills in the gaps in the top duplex, expanding the D-loop in the process. Next, branch migration can occur in both directions, leading to two Holliday junctions. Finally, the Holliday junctions can be resolved to yield either a noncrossover recombinant with two sections of heteroduplex, or a crossover recombinant that has exchanged flanking DNA regions.

There is good evidence for most of these steps, but a few features of the hypothesis are contradicted by experiment. In particular, the model predicts that hybrid DNA will be produced on both sides of the double-stranded break. However, when this prediction was tested genetically, hybrid DNA was usually found only on one side of the break. In the few cases where hybrid DNA was found on both sides of the break, it was in the same chromatid, not in both chromatids as the model predicts. Thus, more data are needed to resolve this discrepancy and perhaps amend the hypothesis.

It is also worth noting that different organisms may do things somewhat differently. The classical work on meiotic recombination was performed in budding yeast, where the double Holliday junction model seems to predominate. However, Gerald Smith and colleagues reported in 2006 that the fission yeast *Schizosaccharomyces pombe* carries out meiotic recombination through a single Holliday junction intermediate, such as that pictured in Figures 22.2 and 22.3. Furthermore, these authors suggested that this organism may initiate some meiotic recombination by single-strand nicking, rather than double-strand breaks.

In another departure from the canonical model, Thorsten Allers and Michael Licten reported in 2001 that noncrossover recombinants appeared at the same time as Holliday junctions in budding yeast. Only later did crossover recombinants appear, through resolution of the Holliday junctions. This finding suggests that noncrossover recombinants in this organism do not result primarily from resolution of Holliday junctions, but from another mechanism that does not involve Holliday junctions.

The Double-Stranded DNA Break

How do we know that recombination in yeast initiates with a double-stranded DNA break (**DSB**)? Jack Szostak and colleagues laid the groundwork for answering this question in 1989 by mapping a recombination initiation site in the *ARG4* gene of the yeast *Saccharomyces cerevisiae*. They did not look at recombination per se, but at meiotic gene conversion, which depends on meiotic recombination in yeast. Because both gene conversion and recombination initiate at the same site, these workers could use gene conversion as a surrogate for recombination. We will examine the mechanism of gene conversion later in this chapter.

Szostak and coworkers verified earlier work that had shown that meiotic gene conversion in the *ARG4* locus was polar: It was common (about 9% of total meioses) near the 5′-end of the gene, and relatively rare (about 0.4% of total meioses) at the 3′-end. This behavior suggested that the initiation site of recombination lies near the 5′-end of the gene. So Szostak and colleagues made deletions in this region to try to remove the initiation site and therefore to block gene conversion. They found that deletions having their 3′-ends in the −316 to +1 region all greatly decreased gene conversion rates, suggesting that the initiation site for recombination lies in the promoter region of the *ARG4* gene.

This information allowed these workers to look for a DNA break, either single- or double-stranded, in a very restricted region of the yeast genome. Accordingly, they cloned a 15-kb fragment of DNA, including the *ARG4* gene, into a yeast plasmid, and introduced it into a strain of yeast that carries out synchronous meiosis immediately on transfer to sporulation medium. They extracted plasmid DNA at various times after induction of sporulation and subjected it to electrophoresis.

Figure 22.18 depicts the results of the electrophoresis. At time zero, we can see mostly supercoiled monomers, with some supercoiled dimers, relaxed circular monomers, and a band of lower mobility that is probably some form of dimer. These same bands appeared throughout the time course after induction of sporulation. The one novelty during sporulation was a relatively faint band of linear monomer that first appeared at 3 h, peaked at 4 h, and decreased after that time. This linear DNA must have been created by a double-stranded break in the plasmid. The timing of this DSB coincided with the timing of commitment to meiotic recombination in these cells (2.5–5 h), and with the appearance of recombination products (4 h). These findings are all consistent with the hypothesis that the first step in meiotic recombination is the formation of a DSB.

Szostak and colleagues used restriction mapping to demonstrate that DSBs occurred in three different locations in the plasmid, indicated by arrows in Figure 22.18a. One of these breaks (site 2) lies within a 216-bp restriction fragment in the control region just 5′ of the *ARG4* gene. Szostak and colleagues' previous work had shown that a 142-bp deletion in this same region depressed the level of meiotic gene conversion in the *ARG4* gene, so they tested the same deletion for effect on the DSB. They found that this deletion did indeed eliminate the DSB at site 2, but had no effect on the DSBs at sites 1 and 3. Thus, the ability to form the DSB at site 2 is correlated with the efficiency of meiotic gene conversion just downstream in the *ARG4* gene.

If a DSB is the initiating event in meiotic recombination, then it should occur in a yeast chromosome, not just in a plasmid. Therefore, Szostak and colleagues used restriction mapping in cells lacking the plasmid to search for these same DSBs. They found that a DSB at site 2 (and at site 1) also occurred in yeast chromosomal DNA, and that the timing of appearance of these DSBs was the same as in the plasmid.

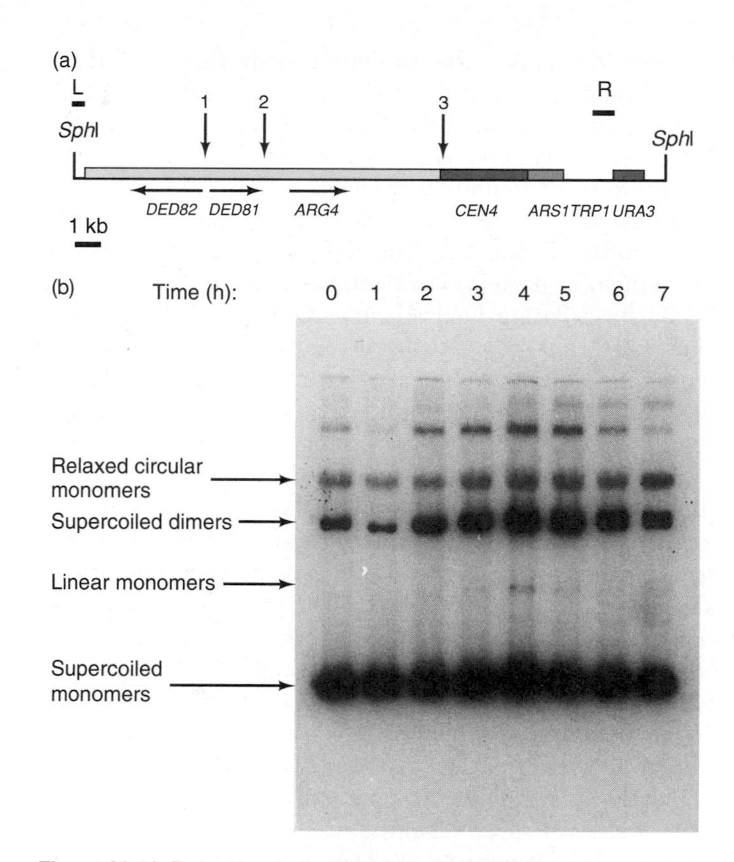

Figure 22.18 Detecting a double-stranded DNA break in a plasmid bearing a recombination initiation site. (a) Map of the plasmid used to detect the DSB. The yellow bar represents the 15-kb insert of yeast DNA containing the recombination initiation site. The other colored bars represent loci within the vector, including the centromere (*CEN4*). The locations of genes are indicated below the bars. L and R are the locations of probes used to hybridize to blots. The arrows marked 1, 2, and 3, are the locations of DSBs mapped in this experiment. **(b)** Electrophoresis results. Szostak and colleagues transformed yeast cells with the plasmid depicted in panel (a), induced sporulation, then electrophoresed samples of plasmid collected at the indicated times after induction. Finally, they Southern blotted the DNAs and probed them with a ^{32}P-labeled probe that hybridized to the vector. The identities of the bands are indicated at left. Note the appearance of linear monomers around 4 h, which indicates a double-stranded break. (*Source:* Sun, H., D. Treco, N.P. Schultes, and J.W. Szostak, Double-strand breaks at an initiation site for meiotic gene conversion. *Nature* 338 (2 Mar 1989) f. 1, p. 88. Copyright © Macmillan Magazines Ltd.)

Nancy Kleckner and colleagues demonstrated similar double-stranded breaks in a yeast chromosome into which they had inserted a *LEU2* gene next to a *HIS4* gene to create a hotspot for meiotic recombination. Actually, two DSBs occurred close together at this hotspot. They also discovered a nonnull mutation (a mutation that does not totally inactivate a gene) in the *RAD50* gene *(rad50S)* that caused a buildup of the fragments caused by the DSBs. This mutation apparently blocked a step downstream of DSB formation, so it allowed the DSBs to accumulate.

In 1995, Scott Keeney and Kleckner discovered that the 5′-ends created by a DSB are covalently bound to a protein in *rad50S* mutants. One attractive hypothesis to explain this behavior is that the catalytic protein that created the DSB would normally dissociate from the DNA immediately, but in this mutant the protein remained bound to the DNA ends it had created. If that is the case, then identifying the protein bound to the DSB ends would identify a prime suspect for the endonuclease that created the DSB.

Accordingly, Kleckner and colleagues set out to identify the protein or proteins covalently bound to the DSB. They began by isolating nuclei from meiotic *rad50S* cells. Because of their accumulation of the protein–DSB complex, these cells should provide a rich source of the DSB-bound protein. To purify the bound protein, Kleckner and coworkers used a two-stage screening procedure: First, they extracted the nuclei and denatured the protein with guanidine and detergent, then purified DNA and DNA–protein complexes by $CsCl_2$ gradient ultracentrifugation. Any proteins bound to DNA under these denaturing conditions should be covalently attached. Then, they passed this mixture through a glass fiber filter that bound the DNA–protein complexes but allowed pure DNA to flow through. The material bound to the filter should be highly enriched in covalent DNA–protein complexes, so Kleckner and colleagues digested the DNA in these complexes with a nuclease and subjected the liberated proteins to SDS-PAGE. They observed several bands, two of which appeared in *rad50S* cells, but not in a *spo11Δ* mutant that was blocked in DSB formation. These same two bands appeared in preparative-scale, as well as pilot-scale preparations, and their appearance depended on nuclease treatment of DNA–protein complexes.

Next, Kleckner and coworkers excised the two candidate bands (M_r = 34 kD and 45 kD) from a preparative gel, subjected them to trypic digestion, then sequenced some of the trypic peptides. The short protein sequences obtained from these peptides yielded the corresponding DNA sequences. Then, because the sequence of the entire yeast genome was already known, these short DNA sequences allowed for easy identification of the corresponding genes. The 45-kD protein corresponded to (coincidentally) the *spo11* gene product, **Spo11,** and the 34-kD protein corresponded to a mixture of five different proteins, including two ribosomal proteins.

Because Spo11 was already known to be required for meiosis, it was an attractive candidate for the DSB-bound protein. To reinforce this hypothesis, Kleckner and colleagues demonstrated that Spo11 is bound specifically to DSBs, and not to bulk DNA. To do this, they used an epitope tagging approach. They created a *Spo11* gene fused to the coding region for an epitope of the protein hemagglutinin. The protein product of this gene (Spo11-HA), and any DNA attached, could therefore be immunoprecipitated with an antihemagglutinin antibody.

In a preliminary experiment, these workers isolated DNA (with any covalently attached proteins) from meiotic *rad50S* cells containing the same *HIS4LEU2* recombination

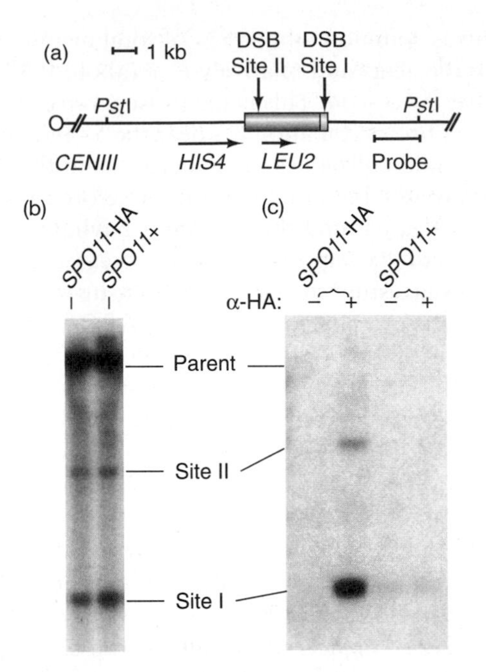

Figure 22.19 Association of Spo11 with DSB fragments. (a) Map of the hotspot region. A fragment (red and blue) with the *LEU2* gene (red) has been inserted adjacent to the *HIS4* gene in yeast chromosome III. The centromere (*CENIII*), the locations of the DSB sites I and II, and the site to which the Southern blot probe hybridizes are shown, along with the locations of two *Pst*I sites flanking the DSBs. **(b)** Southern blot of total DNA, cut with *Pst*I, electrophoresed, blotted, and hybridized to the probe. The parent fragment, as well as the subfragments generated by DSBs, are present. **(c)** Southern blot of DNA, cut with *Pst*I, then immunoprecipitated with an anti-HA antibody, then blotted and probed as in panel (b). The subfragments generated by DSBs are greatly enriched relative to the parent fragment. (*Source:* Keeney, S., C. Giroux, and N. Kleckner, Meiosis-specific DNA double-strand breaks are catalyzed by Spo11, a member of a widely conserved protein family. *Cell* 88 (Feb 1997) f. 3, p. 378. Reprinted by permission of Elsevier Science.)

hotspot as in their previous studies. They cut the DNA with *Pst*I, electrophoresed it, blotted the fragments, and probed the blot with a DNA probe for the hotspot region. Figure 22.19a depicts a map of the hotspot region, showing the two sites where DSBs occur during meiotic recombination, two *Pst*I sites flanking the DSB sites, and the location to which the probe hybridizes downstream of *HIS4LEU2*. Thus, if no DSBs occur, only the parent *Pst*I fragment should be observed. On the other hand, if the DSBs do occur, two additional smaller fragments, corresponding to DSB sites I and II, should appear. Figure 22.19b demonstrates that both these smaller fragments do indeed appear, both in wild-type cells ($SPO11^+$) and in SPO11-HA cells.

Next, Kleckner and colleagues checked to see whether Spo11-HA was specifically bound to these fragments created by DSBs. They repeated the experiment we have just discussed, but this time they immunoprecipitated Spo11-HA–DNA complexes after cutting the DNA with *Pst*I. Figure 22.19c presents the results. It is clear that the two DNA fragments created by DSBs (but very little of the parental fragment) were immunoprecipitated along with Spo11-HA. However, they were not immunoprecipitated in the absence of the anti-HA antibody, nor were they precipitated from a yeast strain with the wild-type *SPO11* gene with no HA tag attached. Further analysis showed that these fragments were not immunoprecipitated from a wild-type *RAD50* strain that did not accumulate DSBs, nor from a mutant strain that did not form DSBs at all.

If Spo11-HA merely bound nonspecifically to DNA, it should have been attached to the parental DNA fragment as well as the two subfragments created by DSBs, but the subfragments were enriched over 600-fold relative to the parental DNA in the immunoprecipitates. Thus, Spo11 appears to bind specifically to DSBs and is likely to be the catalytic part of the enzyme that created the DSBs. Furthermore, Spo11 was known at that time to be homologous to proteins in other organisms, including an archaeon, a fission yeast, and a roundworm. All four proteins have only one conserved tyrosine, which is likely to be the catalytic amino acid that becomes covalently attached to the DSB. In this way, it would resemble the active site tyrosine of a topoisomerase (Chapter 21). The conserved tyrosine in Spo11 is Tyr-135 and, as expected, it is essential for activity. Accordingly, one can propose a model such as the one in Figure 22.20, which calls for the participation of two molecules of Spo11, one to attack each strand of the DNA at slightly offset positions. This process creates the DSB, and leaves a transient intermediate with a molecule of SPO11 covalently attached through its active site tyrosine to the newly created 5′-phosphate on each strand. Thus, the creation of DSBs appears not to occur by a simple hydrolysis, but by a **transesterification**, in which the attacking group is a tyrosine residue of the enzyme, rather than a water molecule.

The covalent association between Spo11 and the DSB ends is only transient, so the two molecules of Spo11 must be removed somehow. This process could occur by direct hydrolysis of the protein–DNA bonds, or by the action of an endonuclease that would remove the proteins along with a short stretch of DNA from each end.

In 2005, Scott Keeney and colleagues showed that the latter mechanism, illustrated in Figure 22.20, is correct. Like Kleckner and colleagues, they engineered a yeast strain to express Spo11 tagged with the hemagglutinin (HA) epitope. Then they immunoprecipitated Spo11 from meiotic cells using an anti-HA antibody. To detect oligonucleotides bound to Spo11, they treated the immunoprecipitates with terminal deoxynucleotidyl transferase (TdT) and ^{32}P-labeled cordycepin triphosphate. TdT adds nucleotides nonspecifically to the 3′-ends of DNAs, and cordycepin triphosphate (3′-deoxyadenosine triphosphate) terminates the reaction because it provides no 3′-hydroxyl group to link to the next nucleotide.

Figure 22.21 shows the results. The asterisks mark two bands that must not have anything to do with DSBs or Spo11, because they are seen in lane 1, in which no cell extract or antibody was used. The arrows point to two bands (in lane 3) that were Spo11-specific. This claim of specificity comes from the fact that these bands were not seen when mock-immunoprecipitation was performed without the HA antibody (lane 2), or when Spo11 was not tagged with the HA epitope (lane 4). We know that the appearance of these bands also depended on the formation of DSBs, because they did not appear when the catalytic tyrosine of Spo11 was changed to phenylalanine (lane 5), or when DSBs were blocked by the *mei4* mutation (lane 6).

The fact that the oligonucleotide-tagged Spo11-HA appeared in two bands suggested that Spo11 is associated

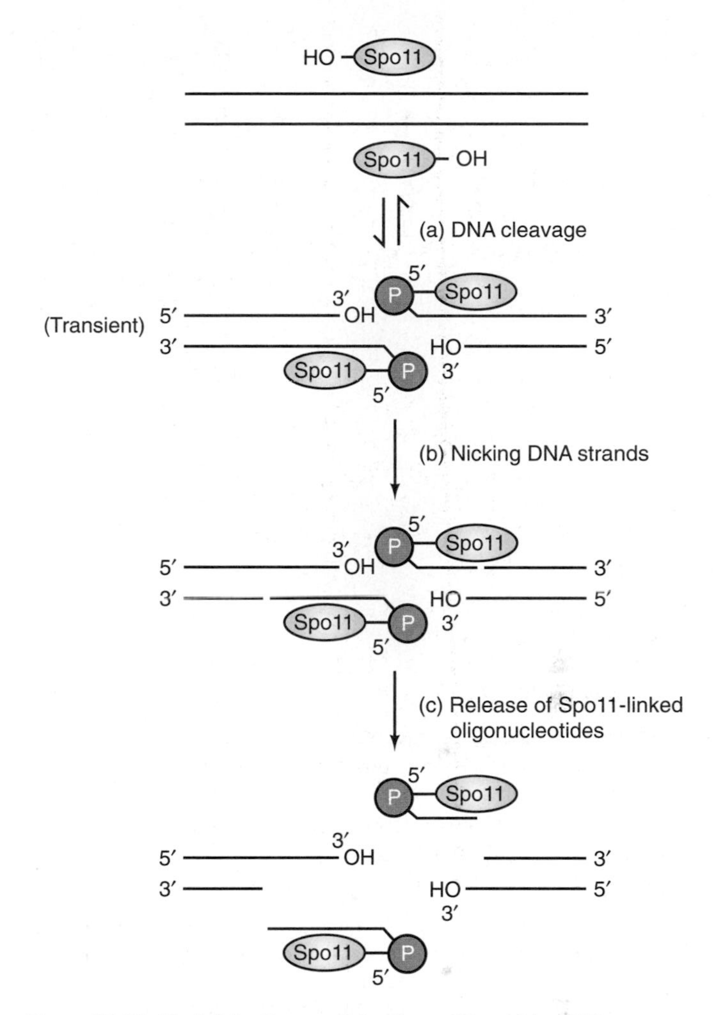

Figure 22.20 Model for the participation of Spo11 in DSB formation. (a) DNA cleavage. Two molecules of Spo11, with active site tyrosines represented by their OH groups, attack the two DNA strands at slightly offset positions. This transesterification reaction breaks phosphodiester bonds within the DNA strands and creates new phosphodiester bonds between the new DNA 5′-ends and the Spo11 tyrosines. **(b)** Nicking DNA strands. The nicking is asymmetric, yielding two sizes of Spo11-linked oligonucleotides. **(c)** Release of Spo11-linked oligonucleotides. The release could occur before DNA end resection, as shown here, but there is evidence for a later release.

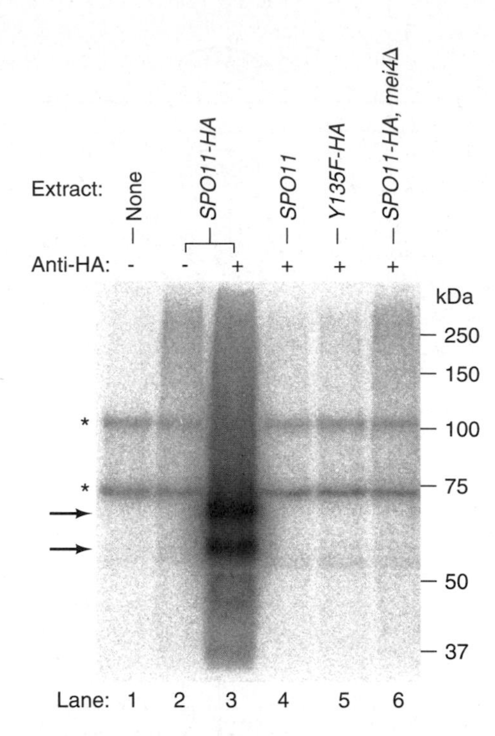

Figure 22.21 Evidence for Spo11-linked oligonucleotides. Keeney and colleagues immunoprecipitated yeast proteins from cellular extracts with no antibody (lanes 1 and 2), or with anti-HA antibody, as indicated at top. The genotypes of the cells from which extracts were prepared, also indicated at top, were: lanes 2 and 3, *SPO11,* fused to a coding region for an HA epitope; lane 4, wild-type *SPO11,* including no HA epitope; lane 5, the *SPO11* mutant *Y135F,* in which the active site tyrosine (Y) is changed to phenylalanine (F), fused to a coding region for an HA epitope; lane 6, the meiosis mutant *mei4*Δ, in which DSBs are not formed, with *SPO11* fused to a coding region for an HA epitope. Keeney and colleagues terminally labeled any oligonucleotides attached to the immunoprecipitates with TdT and [α-^{32}P]cordycepin triphosphate. Finally, they subjected the proteins to SDS-PAGE and detected labeled proteins by autoradiography. Asterisks indicate nonspecific labeled bands that appear even without extracts or antibody. Arrows indicate Spo11-specific proteins that are labeled only when DSBs form. (*Source:* Reprinted by permission from Macmillan Publisher Ltd: *Nature* 436, 1053-1057, Thomas Schalch, Sylwia Duda, David F. Sargent and Timothy J. Richmond, "Endonucleolytic processing of covalent protein-linked DNA double-strand breaks," fig. 16, p. 1054 copyright 2005.)

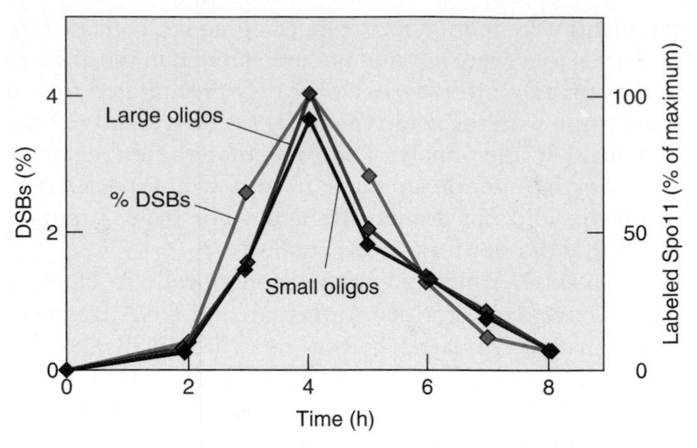

Figure 22.22 Time course of DSB and Spo11-oligonucleotide formation and disappearance. DSBs as a percentage of total DNA (green) were measured at the yeast *HIS4LEU2* recombination hotspot. The large (red) and small (blue) oligonucleotide-linked Spo11 species are plotted as a percent of the maximum for the large species. (*Source:* Adapted from Neale, M. J., et al., Endonucleolytic processing of covalent protein-linked DNA double strand breaks. *Nature* 436: 1054, fig. 1f, 2005.)

with oligonucleotides of two sizes. Accordingly, Keeney and colleagues digested the protein from each band with protease and electrophoresed the remaining oligonucleotides to determine their sizes. The upper band yielded a smeared oligonucleotide band centered around 24–40 nt long, and the lower band yielded a smeared oligonucleotide band centered around 10–15 nt long. This result confirmed that oligonucleotides of two sizes were bound to Spo11, and the smearing suggested that the oligonucleotides were of varying lengths, or that there was heterogeneity in the cutting of Spo11 by protease. Allowing for an average of three amino acids remaining attached to the oligonucleotides, and knowing that oligonucleotides less than 10 nt long were not retained well by the gel, Keeney and colleagues estimated that the two oligonucleotides were about 21–37 and ≤ 12 nt long. These workers also obtained similar results in their studies on mouse DSB processing, but the sizes of the two classes of oligonucleotides linked to the mouse Spo11 homolog were somewhat different from those discovered in yeast.

Figure 22.22 shows that the timing of appearance of the Spo11-oligonucleotides exactly corresponded to the timing of appearance of Spo11-free resected DSBs, which is what we expect if Spo11-oligonucleotides are natural products of DSB processing. Furthermore, the larger and smaller bands were always in a strict 1:1 ratio, indicating that they are produced simultaneously by the same process. These findings lead to some intriguing tentative conclusions.

First, the accumulation and disappearance of Spo11-oligonucleotides closely mirrors the accumulation and disappearance of Spo11-free resected DSBs. As already noted, the correspondence of the accumulation of the species is expected, but one would not have predicted that the disappearance of Spo11-oligonucleotides would coincide with disappearance of Spo11-free DSBs. Instead, the simplest model would be that Spo11 (with or without attached oligonucleotides) would be released from the DSB before resection (Figure 22.20c), which in turn would occur before the DSBs disappeared due to formation of Holliday junctions. This model would therefore predict that destruction of Spo11-oligonucleotides would begin before the loss of Spo11-free DSBs. The fact that the two phenomena occur simultaneously can be explained in two ways. First, it could just be a coincidence that the destruction of Spo11-oligonucleotides is slow enough that it occurs at the same time that resected DSBs are forming Holliday junctions. But the more interesting possibility is that Spo11-oligonucleotides do not begin to be degraded until after resection because they are not released until that time. This concept is illustrated in Figure 22.23.

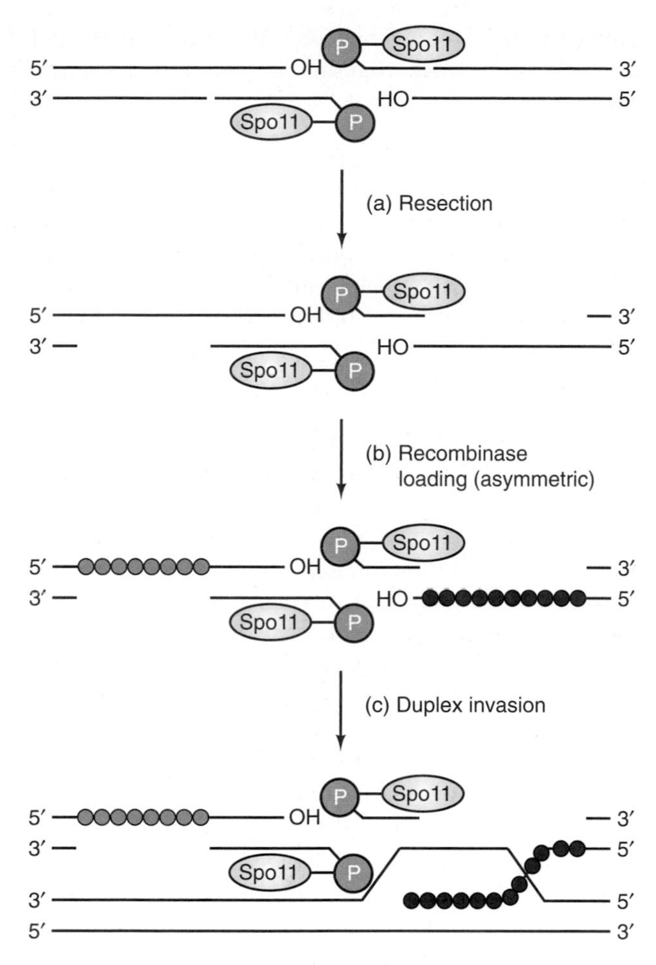

Figure 22.23 A model for DSB end resection prior to release of Spo11-oligonucleotides. (a) Resection occurs on both strands, using the nicks created at a previous step (Figure 22.20). **(b)** Both recombinases (Rad51 and Dmc1) load asymmetrically onto the newly created single-stranded regions, with one protein (blue) coating one strand, and the other (orange) coating the other strand. At this point, we do not know which protein promotes duplex invasion, so the colors are arbitrary. **(c)** One of the proteins (blue) tags the coated free 3′-end for invasion into a homologous duplex, initiating Holliday complex formation. At this point, the Spo11-linked oligonucleotides would dissociate and be degraded.

The second intriguing feature of Figure 22.22 is that the two size classes of Spo11-linked oligonucleotides are produced in equal quantities. This suggests that the larger oligonucleotides come from one of the DSB ends, and the smaller ones come from the other. This could mean that there is an inherent asymmetry in the DSB that predetermines which free 3′-end will invade the other DNA duplex to initiate Holliday junction formation. Keeney and colleagues envisioned a model similar to the one illustrated in Figure 22.23. The asymmetry of cutting of the two strands leads to an asymmetry in the lengths of the free 3′-ends base-paired to Spo11-linked oligonucleotides after resection. The one on the right in this illustration is less tied up in base pairing, and it could attract one of the two recombinases (Rad51 or Dmc1), while the other would bind to the other recombinase. This asymmetry could then dictate which free end invades the homologous duplex to initiate Holliday junction formation. In this case, it is the one on the right.

We will see in the next section that a complex including Rad50 and Mre11 is involved in resecting the DSB ends, and the following evidence suggests that this complex also contains the endonuclease that cuts the DNA near DSBs, leading to the release of the Spo11-oligonucleotides: First, we know that Mre11 has the required endonuclease activity; second, mutations in both the *RAD50* and *MRE11* genes block removal of Spo11 from DSB ends; and third, the oligonucleotides attached to Spo11 have 3′-hydroxyl groups, which is consistent with the mechanism used by the Mre11 endonuclease.

Now we know that the *Spo11* gene is highly conserved throughout the eukaryotic kingdom, including yeasts, plants, and animals. Thus, it is very likely that the double-stranded break model for initiation of recombination is also conserved. In one study that supports this conclusion, Kim McKim and Aki Hayashi-Hagihara performed experiments similar to those of Kleckner and colleagues, looking for mutations in *Drosophila* that blocked gene conversion. In 1998, they reported that mutations in the *mei-W68* gene had this phenotype, and that *mei-W68* is a *Drosophila* homolog of the yeast *Spo11* gene. Interestingly, *mei-W68* mutations affect gene conversion in somatic, as well as meiotic, cells. Thus, in contrast to yeast, where *Spo11* is required only for meiotic recombination, *mei-W68* is required for both meiotic and somatic recombination. Thus, the double-stranded breaks that *mei-W68* presumably induces appear to be required for both meiotic and somatic recombination in *Drosophila*.

SUMMARY The DSB model of meiotic recombination in yeast begins with a double-strand break. DSBs can be directly observed in a plasmid DNA or in chromosomal DNA. DSBs accumulate in *rad50S* mutants, where Spo11 can be found covalently bound to the DSB ends. Two molecules of Spo11 collaborate to create DSBs by cleaving both strands at closely spaced sites. The cleavage operates through transesterification reactions involving active site tyrosines on the two molecules of Spo11, which leads to covalent bonds between the two molecules of Spo11 and the newly formed DSBs. Spo11 then could be released from the DSBs in a complex with oligonucleotides ranging from about 12–37 nt long. The Spo11-oligonucleotides may even be released after resection of the DSBs. The cleavage by Spo11 appears to be asymmetric, yielding a longer free 3′-end on one side of the DSB than on the other. This may set up one free end for invasion of a homologous DNA duplex, which initiates Holliday junction formation.

Creation of Single-Stranded Ends at DSBs

Once Spo11 has created a DSB, the new 5′-ends are digested to yield free 3′-ends that can invade another DNA duplex. Szostak and coworkers first discovered the single-stranded ends in 1989 when they examined the structure of the DNA termini created by DSBs. They digested the DNAs with S1 nuclease, which specifically degrades single-stranded DNA, but leaves double-stranded DNA intact. The S1 nuclease had no effect on the intensities of the bands, but it reduced the lengths of all three DNAs. This result is exactly what the model in Figure 22.17 predicts: After the DSB occurs, an exonuclease digests the two 5′-ends at the break, creating single-stranded DNA that would then be susceptible to S1 nuclease digestion. This S1 nuclease digestion would reduce the length of the DNA fragment.

Kleckner and colleagues also found evidence for digestion, or resection, of one strand at DSBs. They discovered that the fragments that accumulated in wild-type cells produced diffuse bands on gel electrophoresis, as if the ends had been nibbled to varying extents. By contrast, the bands were discrete in *rad50S* mutants that blocked resection of the DSB ends.

Mutations in *RAD50, MRE11,* and *COM1/SAE2* obstruct DSB resection. Actually, null alleles of *RAD50* and *MRE11* block DSB formation altogether, so only certain nonnull alleles of these genes permit DSB formation but impede DSB resection. Thus, the products of both of these genes are required for both DSB formation and resection. Evidence for the role of these two gene products comes from a comparison with *E. coli.* The *E. coli* proteins SbcC and SbcD are homologous to yeast Rad50 and Mre11, respectively. Furthermore, the two bacterial proteins work as an SbcC/SbcD dimer, which has exonuclease activity on double-stranded DNA. This finding suggests that Rad50 and Mre11 also work together to resect the 5′-ends of DSBs in yeast.

SUMMARY Formation of the DSB in meiotic recombination is followed by 5′→3′ exonuclease digestion of the 5′-ends at the break, yielding overhanging 3′-ends that can invade another DNA duplex. Rad50 and Mre11 probably collaborate to carry out this resection.

22.4 Gene Conversion

When fungi such as pink bread mold (*Neurospora crassa*) sporulate, two haploid nuclei fuse, producing a diploid nucleus that undergoes meiosis to give four haploid nuclei. These nuclei then experience mitosis to produce eight haploid nuclei, each appearing in a separate spore. In principle, if one of the original nuclei contained one allele (*A*) at a given locus, and the other contained another allele (*a*) at the same locus, then the mixture of alleles in the spores should be equal: four *A*'s and four *a*'s. It is difficult to imagine any other outcome (five *A*'s and three *a*'s, for example), because that would require conversion of one *a* to an *A*. In fact, aberrant ratios *are* observed about 0.1% of the time, depending on the fungal species. This phenomenon is called **gene conversion.** We discuss this topic under the heading of recombination because the two processes are related.

The mechanism of meiotic recombination discussed in this chapter, and illustrated in Figure 22.17, suggests a mechanism for gene conversion during meiosis. Figure 22.24

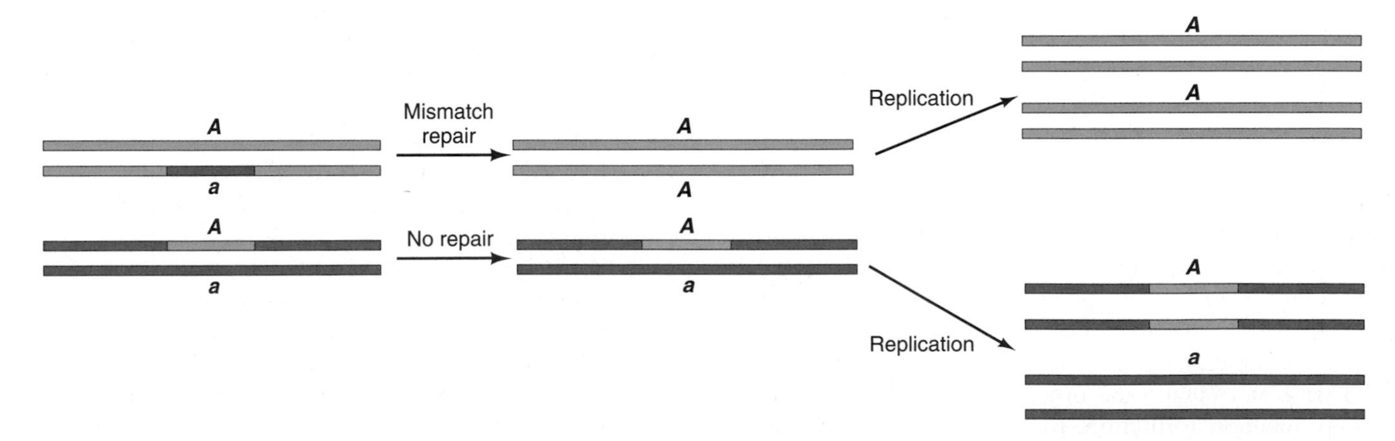

Figure 22.24 A model for gene conversion in sporulating *Neurospora*. A strand exchange event with branch migration during sporulation has resolved to yield two duplex DNAs with patches of heteroduplex in a region where a one-base difference occurs between allele *A* (blue) and allele *a* (red). The other two daughter chromosomes are homoduplexes, one pure *A* and one pure *a* (not shown). The top heteroduplex undergoes mismatch repair to convert the a strand to *A;* the bottom heteroduplex is not repaired. Replication of the repaired DNA yields two *A* duplexes; replication of the unrepaired DNA yields one *A* and one *a*. Thus, the sum of the daughter duplexes pictured at right is three *A*'s and one *a*. Replication of the two DNAs not pictured yields two *A*'s and two *a*'s. The sum of all daughter duplexes is therefore five *A*'s and three *a*'s, instead of the normal four of each.

depicts this hypothesis in the case of *N. crassa*. We start with a nucleus in which DNA duplication has already occurred, so it contains four chromatids. In principle, two chromosomes should bear the *A* allele and two the *a* allele. But in this case, strand exchange and branch migration have occurred, followed by resolution yielding two chromosomes with patches of heteroduplex, as illustrated in Figure 22.24. These heteroduplex regions just happen to be in the region where alleles *A* and *a* differ at one base, so each chromosome has one strand with one allele and the other strand with the other allele. If DNA replication occurred immediately, this situation would resolve itself simply, yielding two *A* duplexes and two *a* duplexes. However, before replication, one or both of the heteroduplexes may attract the enzymes that repair base mismatches. In the example shown here, only the top heteroduplex is repaired, with the *a* being converted to *A*. This leaves three strands with the *A* allele, and only one with the *a* allele. Now DNA replication will produce three *A* duplexes and only one *a* duplex. When we add the two *A* and two *a* duplexes resulting from the chromosomes that did not undergo heteroduplex formation, a final ratio of five *A*'s and only three *a*'s results.

Figure 22.25 presents another pathway by which gene conversion can occur, this time without mismatch repair. We start with the situation after step (c) in Figure 22.17, just after strand invasion and D-loop formation. The region in which allele *A* differs from allele *a* is indicated at the top. Blue indicates *A* and red indicates *a*. This scheme differs from Figure 22.17 in that the invading strand is subject to partial resection, with shrinkage of the D-loop, before repair synthesis begins. That resection allows a longer stretch of repair synthesis to occur, which converts more of the invading strand from *A* to *a*. And the conversion occurs in the exact region where alleles *A* and *a* differ. After branch migration and resolution (either crossover or noncrossover resolution), we have all four DNA strands representing the *a* allele in the significant region, whereas we started with two of each. Gene conversion has occurred.

Gene conversion is not confined to meiotic events. It is also the mechanism that switches the mating type of baker's yeast (*S. cerevisiae*). This gene conversion event involves the transient interaction of two versions of the *MAT* locus, followed by conversion of one gene sequence to that of the other.

SUMMARY When two similar but not identical DNA sequences interact, the possibility exists for gene conversion—the conversion of one DNA sequence to that of the other. The sequences participating in gene conversion can be alleles, as in meiosis, on nonallelic genes, such as the MAT genes that determine mating type in yeast.

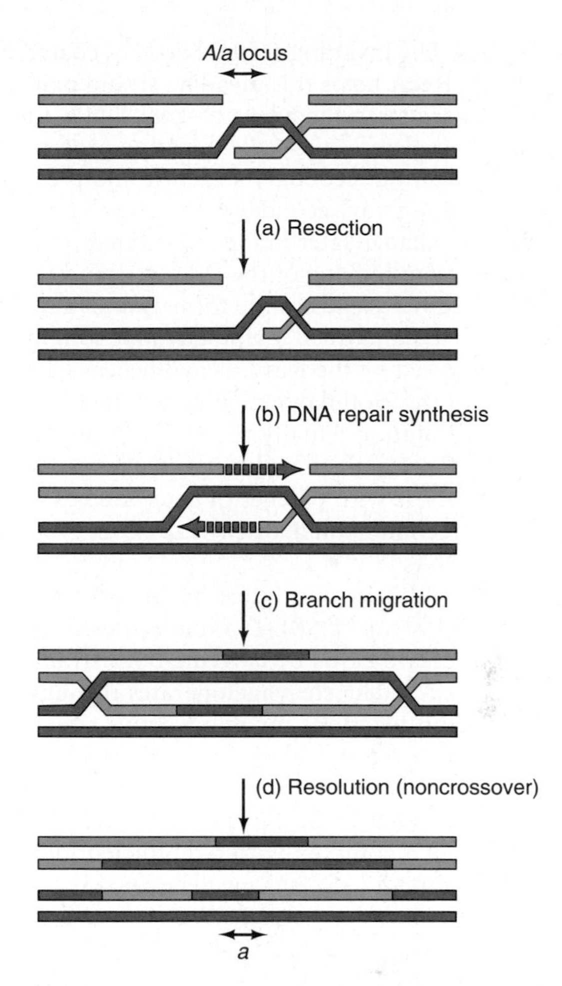

Figure 22.25 A model for gene conversion without mismatch repair. This figure begins in the middle of the DSB recombination scheme illustrated in Figure 22.17, just after strand invasion. **(a)** This time, the invading strand is partially resected, which causes partial collapse of the D-loop. **(b)** DNA repair synthesis is more extensive because of the resection, which produces a region (indicated at top and bottom) in which all four DNA strands are allele *a* (red). **(c)** and **(d)** Branch migration and resolution do not change the nature of the four DNA strands in the region in which alleles *A* and *a* differ: All are allele *a*. Thus, this process has converted a DNA duplex that was allele *A* to allele *a*.

SUMMARY

Homologous recombination is essential to life: In eukaryotic meiosis, it locks homologous chromosomes together so they separate properly. It also scrambles parental genes in offspring. In all forms of life, homologous recombination helps to cope with DNA damage.

Homologous recombination by the RecBCD pathway in *E. coli* begins with invasion of a duplex DNA by a single-stranded DNA from another duplex that has experienced a double-stranded break. A free end can be generated by the nuclease and helicase activities of RecBCD, which prefers to nick DNA at special sequences called Chi sites. The invading strand becomes coated with RecA and SSB. RecA helps this invading strand pair with its complementary strand in a homologous DNA, forming a D-loop. SSB accelerates the recombination process, apparently by melting secondary structure and preventing RecA from trapping any secondary structure that would inhibit strand exchange later in the recombination process. Subsequent nicking of the D-loop strand, probably by RecBCD, leads to the formation of a branched intermediate called a Holliday junction. Branch migration, catalyzed by the RuvA–RuvB helicase, brings the crossover of the Holliday junction to a site that is favorable for resolution. Finally, the Holliday junction can be resolved by RuvC, which nicks two of its strands. This can yield two DNAs with patches of heteroduplex (noncrossover recombinants), or two crossover recombinant DNAs.

Meiotic recombination in yeast begins with a double-stranded break (DSB). Two molecules of Spo11 collaborate to create DSBs by cleaving both strands at closely spaced sites. The cleavage operates through transesterification reactions involving active site tyrosines on the two molecules of Spo11, which leads to covalent bonds between the two molecules of Spo11, and the newly formed DSBs. Spo11 is then released from the DSBs in a complex with oligonucleotides ranging from about 12–37 nt long. The Spo11-oligonucleotides may even be released after resection of the DSBs. Formation of the DSB in meiotic recombination is followed by $5' \rightarrow 3'$ exonuclease digestion of the 5′-ends at the break. Rad50 and Mre11 probably collaborate to carry out this resection. Next, the newly generated 3′-overhang invades the other DNA duplex, creating a D-loop. DNA repair synthesis and branch migration yield two Holliday junctions that can be resolved to produce either noncrossover or crossover recombinants.

When two similar but not identical DNA sequences interact, the possibility exists for gene conversion—the conversion of one DNA sequence to that of the other. The sequences participating in gene conversion can be alleles, as in meiosis, or nonallelic genes, such as the *MAT* genes that determine mating type in yeast.

REVIEW QUESTIONS

1. List the three steps in homologous recombination in which RecA participates, with a short explanation of each.
2. What evidence indicates that RecA coats single-stranded DNA? What role does SSB play in the interaction between RecA and single-stranded DNA?
3. Describe and give the results of an experiment that shows that RecA is required for synapsis at the beginning of recombination.
4. How would you show that the apparent synapsis you observe by electron microscopy is really synapsis, rather than true base pairing?
5. Describe and give the results of an experiment that shows that RecBCD nicks DNA near a Chi site. How could you demonstrate that it is RecBCD, and not a contaminant, that causes the nicking?
6. Show how you could use a gel mobility shift assay to demonstrate that RuvA can bind to a Holliday junction by itself at high concentration, that RuvB cannot bind by itself at all, and that RuvA and RuvB can bind cooperatively at relatively low concentrations. What is the function of glutaraldehyde in this experiment?
7. Draw a diagram of the RuvAB–Holliday junction complex, with the Holliday junction in its familiar cross-shaped form, ready for branch migration. Include the RuvB rings, but not the RuvA tetramer.
8. Describe and give the results of an experiment that shows that RuvC can resolve a Holliday junction.
9. What evidence suggests that RuvA, B, and C are all together in a complex with a Holliday junction?
10. Present a model for meiotic recombination in yeast.
11. Describe and give the results of an experiment that shows that a DSB forms during meiotic recombination in yeast.
12. Describe and give the results of an experiment that shows that Spo11 is covalently attached to a DSB during meiotic recombination in yeast.
13. Describe and give the results of an experiment that demonstrates the formation of Spo11-linked oligonucleotides of two size classes.
14. What do the two size classes of Spo11-linked oligonucleotides, and the timing of their appearance and disappearance, suggest about the mechanism of homologous recombination in yeast? Illustrate your answer with a drawing.
15. Present a model for meiotic gene conversion.

ANALYTICAL QUESTIONS

1. Draw a diagram of a Holliday junction. Starting with that diagram, illustrate:
   a. Branch migration to the right, then resolution to yield a short heteroduplex, or resolution to yield crossover recombinant DNAs
   b. Branch migration to the left, then resolution to yield a short heteroduplex, or resolution to yield crossover recombinant DNAs
2. Describe or diagram the products of RecBCD pathway recombination in *E. coli* cells with mutations in the following genes:
   a. *recB*
   b. *recA*
   c. *ruvA*
   d. *ruvB*
   e. *ruvC*

3. Show how you could use DNase footprinting to demonstrate that RuvA binds to the center of the Holliday junction, and that RuvB binds to the upstream side relative to the direction of branch migration.
4. What would be the final mixture of alleles in the gene conversion depicted in Figure 22.24 if both heteroduplexes were converted to *A*/*A* by mismatch repair? What if one heteroduplex were converted to *A*/*A* and the other to *a*/*a*?
5. One of the two elements intimately involved in the process of strand exchange is the Chi sites on the DNA. It is believed that Chi sites stimulate the RecBCD pathway but not several other homologous pathways (λ Red, *E.coli* RecE and RecF). Describe an experiment that would test this hypothesis.

SUGGESTED READINGS

General References and Reviews

Fincham, J.R.S. and P. Oliver. 1989. Initiation of recombination. *Nature* 338:14–15.

McEntee, K. 1992. RecA: From locus to lattice. *Nature* 355:302–3.

Meselson, M. and C.M. Radding. 1975. A general model for genetic recombination. *Proceedings of the National Academy of Sciences USA* 72:358–61.

Roeder, G.S. 1997. Meiotic chromosomes: It takes two to tango. *Genes and Development* 11:2600–21.

Smith, G.R. 1991. Conjugational recombination in *E. coli:* Myths and mechanisms. *Cell* 64:19–27.

West, S.C. 1998. RuvA gets x-rayed on Holliday. *Cell* 94:699–701.

Research Articles

Cao, L., E. Alani, and N. Kleckner. 1990. A pathway for generation and processing of double-strand breaks during meiotic recombination in *S. cerevisiae. Cell* 61:1089–1101.

DasGupta, C., T. Shibata, R.P. Cunningham, and C.M. Radding. 1980. The topology of homologous pairing promoted by RecA protein. *Cell* 22:437–46.

Dunderdale, H.J., F.E. Benson, C.A. Parsons, G.J. Sharples, R.G. Lloyd, and S.C. West. 1991. Formation and resolution of recombination intermediates by *E. coli* RecA and RuvC proteins. *Nature* 354:506–10.

Eggleston, A.K., A.H. Mitchell, and S.C. West. 1997. In vitro reconstitution of the late steps of genetic recombination in *E. coli. Cell* 89:607–17.

Honigberg, S.M., D.K. Gonda, J. Flory, and C.M. Radding. 1985. The pairing activity of stable nucleoprotein filaments made from RecA protein, single-stranded DNA, and adenosine 5′-(γ-thio)triphosphate. *Journal of Biological Chemistry* 260:11845–51.

Keeney, S, C.N. Giroux, and N. Kleckner. 1997. Meiosis-specific DNA double-strand breaks are catalyzed by Spo11, a member of a widely conserved protein family. *Cell* 88:375–84.

Neale, M.J., J. Pan, and S. Keeney. 2005. Endonucleolytic processing of covalent protein-linked DNA double-strand breaks. *Nature* 436:1053–57.

Parsons, C.A. and S.C. West. 1993. Formation of a RuvAB–Holliday junction complex in vitro. *Journal of Molecular Biology* 232:397–405.

Ponticelli, A.S., D.W. Schultz, A.F. Taylor, and G.R. Smith. 1985. Chi-dependent DNA strand cleavage by RecBC enzyme. *Cell* 41:145–51.

Radding, C.M. 1991. Helical interactions in homologous pairing and strand exchange driven by RecA protein. *Journal of Biological Chemistry* 266:5355–58.

Radding, C.M., J. Flory, A. Wu, R. Kahn, C. DasGupta, D. Gonda, M. Bianchi, and S.S. Tsang. 1982. Three phases in homologous pairing: Polymerization of *recA* protein on single-stranded DNA, synapsis, and polar strand exchange. *Cold Spring Harbor Symposia on Quantitative Biology* 47:821–28.

Rafferty, J.B., S.E. Sedelnikova, D. Hargreaves, P.J. Artymiuk, P.J. Baker, G.J. Sharples, A.A. Mahdi, R.G. Lloyd, and D.W. Rice. 1996. Crystal structure of DNA recombination protein RuvA and a model for its binding to the Holliday junction. *Science* 274:415–21.

Roman, L. J., D.A. Dixon, and S.C. Kowalczykowski. 1991. RecBCD-dependent joint molecule formation promoted by the *Escherichia coli* RecA and SSB proteins. *Proceedings of the National Academy of Sciences USA* 88:3367–71.

Shah, R., R.J. Bennett, and S.C. West. 1994. Genetic recombination in *E. coli:* RuvC protein cleaves Holliday junctions at resolution hotspots in vitro. *Cell* 79:853–64.

Sun, H., D. Treco, N.P. Schultes, and J.W. Szostak. 1989. Double-strand breaks at an initiation site for meiotic gene conversion. *Nature* 338:87–90.

Yu, X., S.C. West, and E.H. Egelman. 1997. Structure and subunit composition of the RuvAB–Holliday junction complex. *Journal of Molecular Biology* 266:217–22.

CHAPTER 23

Transposition

Three ears of Indian corn. © *Creatas/PunchStock.*

We have already learned that an organism's DNA does not remain absolutely unchanged from the beginning of its life to the end. In Chapter 20, we learned that DNA can be damaged and then repaired, and can even be mutated beyond repair. In Chapter 22 we saw that DNAs can recombine by homologous recombination to bring together new combinations of genes. DNAs can also undergo site-specific recombination, which requires much less sequence homology than homologous recombination does. This kind of recombination almost always involves defined DNA sequences—hence the name "site-specific." A favorite example is the insertion of λ phage DNA into the *E. coli* host DNA, and excision of the phage DNA back out again. By contrast, transposition has little if any requirement for homology between recombining DNAs, and is therefore not site-specific. Transposition will be our subject in this chapter.

23.1 Bacterial Transposons

In transposition, a **transposable element,** or **transposon,** moves from one DNA address to another. Barbara McClintock discovered transposons in the 1940s in her studies on the genetics of maize. Since then, transposons have been found in all kinds of organisms, from bacteria to humans. We will begin with a discussion of the bacterial transposons.

Discovery of Bacterial Transposons

James Shapiro and others laid the groundwork for the discovery of bacterial transposons with their discovery in the late 1960s of phage mutations that did not behave normally. For example, they did not revert readily the way point mutations do, and the mutant genes contained long stretches of extra DNA. Shapiro demonstrated this by taking advantage of the fact that a λ phage will sometimes pick up a piece of host DNA during lytic infection of *E. coli* cells, incorporating the "passenger" DNA into its own genome. He allowed λ phages to pick up either a wild-type *E. coli* galactose utilization gene (gal^+) or its mutant counterpart (gal^-), then measured the sizes of the recombinant DNAs, which contained λ DNA plus host DNA. He measured the DNA sizes by measuring the densities of the two types of phage using cesium chloride gradient centrifugation (Chapter 20). Because the phage coat is made of protein and always has the same volume, and because DNA is much denser than protein, the more DNA the phage contains the denser it will be. It turned out that the phages harboring the gal^- gene were denser than the phages with the wild-type gene and therefore held more DNA. The simplest explanation is that foreign DNA had inserted into the *gal* gene and thereby inactivated it. Indeed, later experiments revealed 800–1400-bp inserts in the mutant *gal* gene, which were not found in the wild-type gene. In the rare cases when such mutants did revert, they lost the extra DNA. These extra DNAs that could inactivate a gene by inserting into it were the first transposons discovered in bacteria. They are called **insertion sequences (ISs).**

Insertion Sequences: The Simplest Bacterial Transposons

Bacterial insertion sequences contain only the elements necessary for transposition. The first of these elements is a set of special sequences at a transposon's ends, one of which is the inverted repeat of the other. The second element is the set of genes that code for the enzymes that catalyze transposition.

Because the ends of an insertion sequence are inverted repeats, if one end of an insertion sequence is 5′-ACCG TAG, the other end of that strand will be the reverse complement: CTACGGT-3′. The inverted repeats given here are hypothetical and are presented to illustrate the point. Typical insertion sequences have somewhat longer inverted repeats, from 15 to 25 bp long. IS1, for example, has inverted repeats 23 bp long. Larger transposons can have inverted repeats hundreds of base pairs long.

Stanley Cohen provided one graphic demonstration of inverted repeats at the ends of a transposon with the experiment illustrated in Figure 23.1. He started with a plasmid containing a transposon with the structure shown on the left in Figure 23.1a. The original plasmid was linked to the ends of the transposon, which were inverted repeats. Cohen reasoned that if the transposon really had inverted repeats at its ends, he could separate the two strands of the recombinant plasmid, and get the inverted repeats on one strand to base-pair with each other, forming a stem-loop structure as shown on the right in Figure 23.1a. The stems would be double-stranded DNA composed of the two inverted repeats: the loops would be the rest of the DNA in single-stranded form. The electron micrograph in Figure 23.1b shows the expected stem-loop structure.

The main body of an insertion sequence codes for at least two proteins that catalyze transposition. These proteins are collectively known as **transposase;** we will discuss their mechanism of action later in this chapter. We know that these proteins are necessary for transposition because mutations in the body of an insertion sequence can render that transposon immobile.

One other feature of an insertion sequence, shared with more complex transposons, is found just outside the transposon itself. This is a pair of short direct repeats in the DNA immediately surrounding the transposon. These repeats did not exist before the transposon inserted; they result from the insertion process itself and tell us that the transposase cuts the target DNA in a staggered fashion rather than with two cuts right across from each other. Figure 23.2 shows how staggered cuts in the two strands of the target DNA at the site of insertion lead automatically to direct repeats. The length of these direct repeats depends on the distance between the two cuts in the target DNA strands. This distance depends in turn on the nature of the insertion sequence. The transposase of IS1 makes cuts 9 bp apart and therefore generates direct repeats that are 9 bp long.

SUMMARY Insertion sequences are the simplest of the bacterial transposons. They contain only the elements necessary for their own transposition; short inverted repeats at their ends and at least two genes coding for an enzyme called transposase that carries out transposition. Transposition involves duplication of a short sequence in the target DNA; one copy of this short sequence flanks the insertion sequence on each side after transposition.

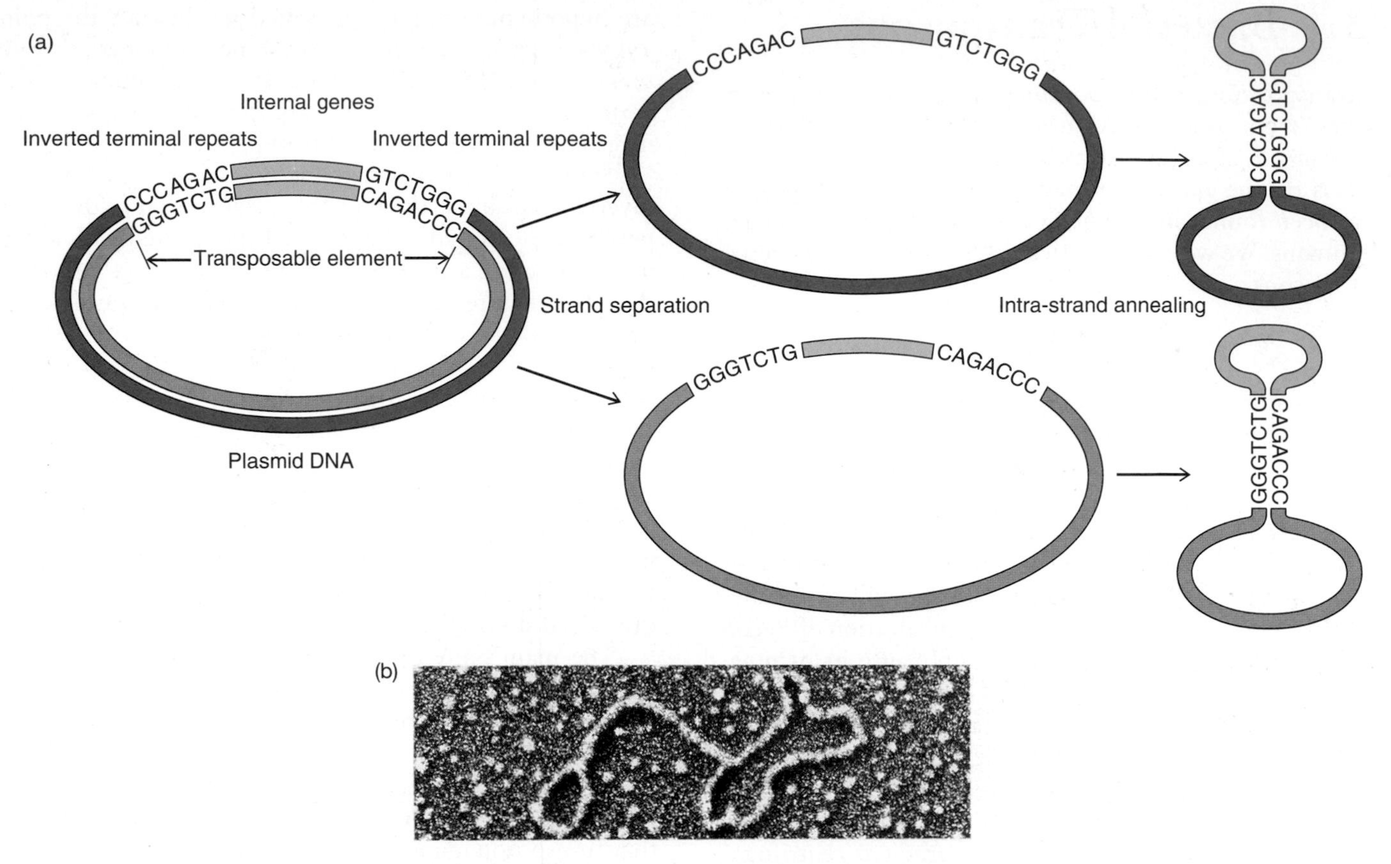

Figure 23.1 Transposons contain inverted terminal repeats. **(a)** Schematic diagram of experiment. The two strands of a transposon-bearing plasmid were separated and allowed to anneal with themselves separately. The inverted terminal repeats will form a base-paired stem between two single-stranded loops corresponding to the internal genes of the transposon (small loop, green) and host plasmid (large loop, purple and pink). **(b)** Experimental results. The DNA was shadowed with heavy metal and subjected to electron microscopy. The loop-stem-loop structure is obvious. The stem is hundreds of base pairs long, demonstrating that the inverted terminal repeats in this transposon are much longer than the 7 bp shown for convenience in part (a). (*Source:* (*b*) Courtesy Stanley N. Cohen, Stanford University.)

More Complex Transposons

Insertion sequences and other transposons are sometimes called "selfish DNA," implying that they replicate at the expense of their hosts and apparently provide nothing useful in return. However, some transposons do carry genes that are valuable to their hosts, the most familiar being genes for antibiotic resistance. Not only is this a clear benefit to the bacterial host, it is also valuable to molecular biologists, because it makes the transposon much easier to track.

For example, consider the situation in Figure 23.3, in which we start with a donor plasmid containing a gene for kanamycin resistance (Kan^r) and harboring a transposon (**Tn3**) with a gene for ampicillin resistance (Amp^r); in addition, we have a target plasmid with a gene for tetracycline resistance (Tet^r). After transposition, Tn3 has replicated and a copy has moved to the target plasmid. Now the target plasmid confers both tetracycline and ampicillin resistance, properties that we can easily monitor by transforming antibiotic-sensitive bacteria with the target plasmid and growing these host bacteria in medium containing both antibiotics. If the bacteria survive, they must have taken up both antibiotic resistance genes; therefore, Tn3 must have transposed to the target plasmid.

Mechanisms of Transposition

Because of their ability to move from one place to another, transposons are sometimes called "jumping genes." However, the term is a little misleading because it implies that the DNA always leaves one place and jumps to the other. This mode of transposition does occur and is called **nonreplicative transposition** (or "cut and paste") because both strands of the original DNA move together from one place to the other without replicating. However, transposition frequently involves DNA replication, so one copy of the transposon remains at its original site as another copy inserts at the new site. This is called **replicative transposition** (or "copy and paste") because a transposon moving by this route also replicates itself. Let us discuss how both kinds of transposition take place.

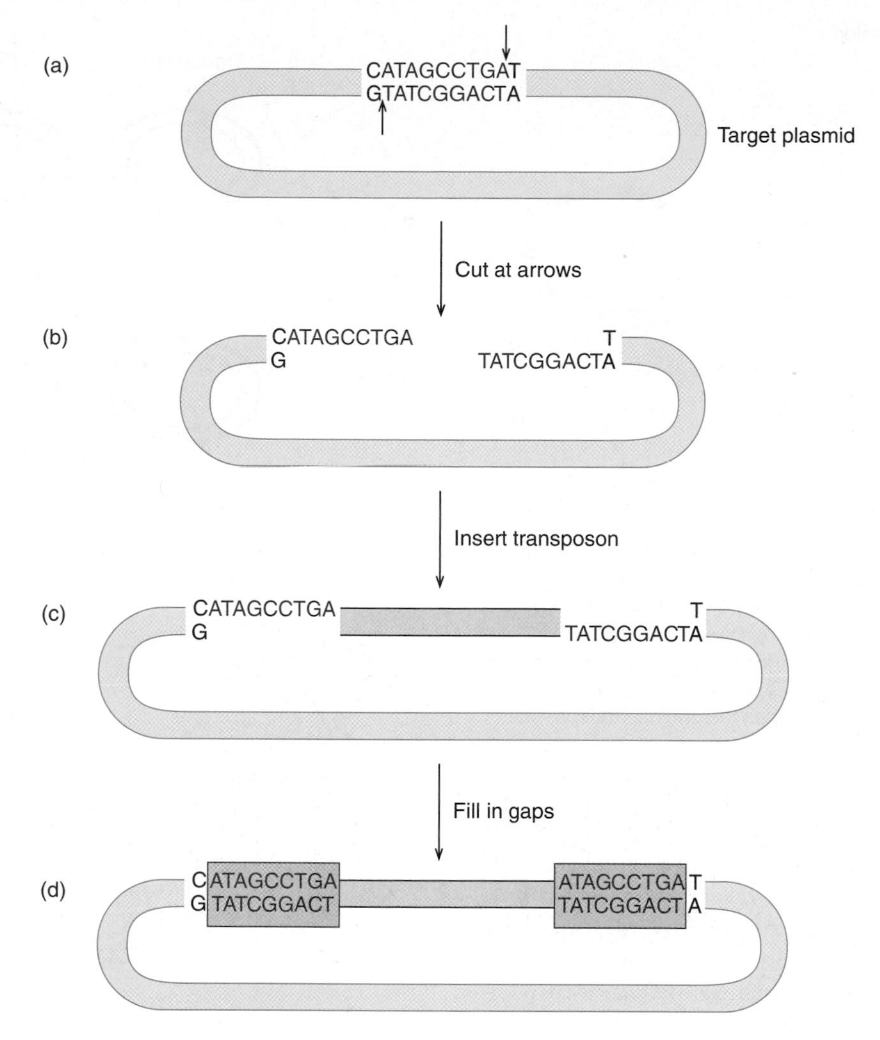

Figure 23.2 Generation of direct repeats in host DNA flanking a transposon. (a) The arrows indicate where the two strands of host DNA will be cut in a staggered fashion, 9 bp apart. **(b)** After cutting. **(c)** The transposon (yellow) has been ligated to one strand of host DNA at each end, leaving two 9-bp base gaps. **(d)** After the gaps are filled in, 9-bp repeats of host DNA (pink boxes) are apparent at each end of the transposon.

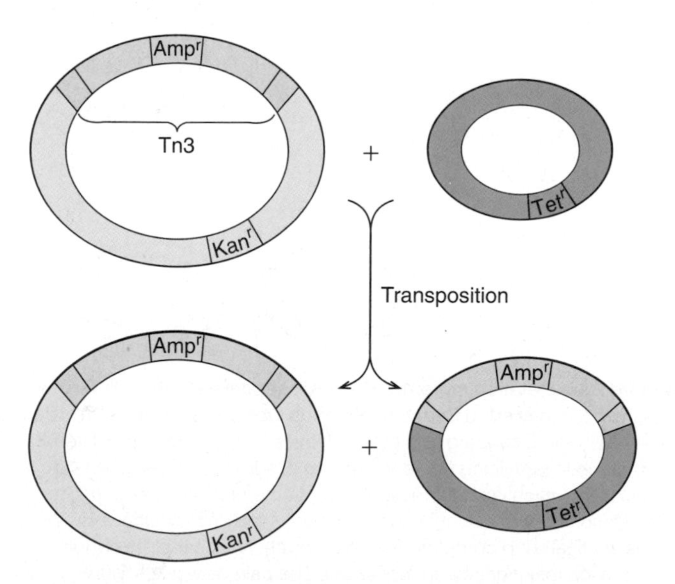

Figure 23.3 Tracking transposition with antibiotic resistance genes. We begin with two plasmids: The larger (blue) encodes kanamycin resistance (Kan^r) and bears the transposon Tn3 (yellow), which codes for ampicillin resistance (Amp^r); the smaller (green) encodes tetracycline resistance (Tet^r). After transposition, the smaller plasmid bears both the Tet^r and Amp^r genes.

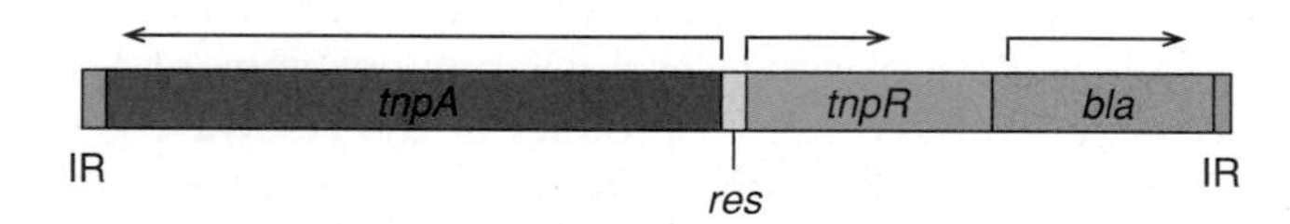

Figure 23.4 Structure of Tn3. The *tnpA* and *tnpR* genes are necessary for transposition; *res* is the site of the recombination that occurs during the resolution step in transposition; the *bla* gene encodes β-lactamase, which protects bacteria against the antibiotic ampicillin. This gene is also called Amp^r. Inverted repeats (IR) are found on each end. The arrows indicate the direction of transcription of each gene.

Replicative Transposition of Tn3 Tn3, whose structure is shown in Figure 23.4, illustrates one well-studied mechanism of transposition. In addition to the *bla* gene, which encodes ampicillin-inactivating β-lactamase, Tn3 contains two genes that are instrumental in transposition. Tn3 transposes by a two-step process, each step of which requires one of the Tn3 gene products. Figure 23.5 shows a simplified version of the sequence of events. We begin with two plasmids; the donor, which harbors Tn3, and the target. In the first step, the two plasmids fuse, with Tn3 replication, to form a **cointegrate** in which they are

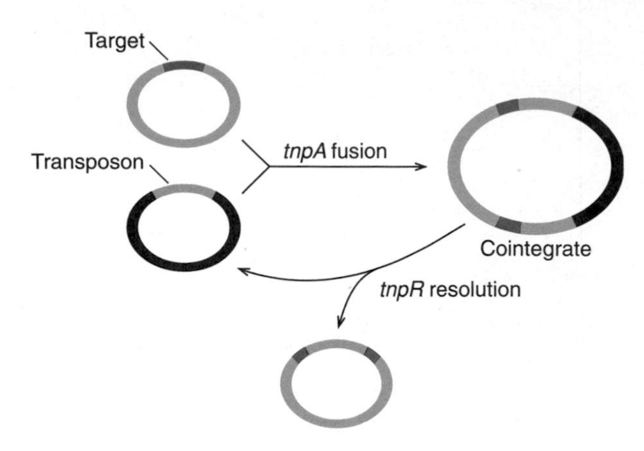

Figure 23.5 Simplified scheme of the two-step Tn3 transposition. In the first step, catalyzed by the *tnpA* gene product, the plasmid (black) bearing the transposon (blue) fuses with the target plasmid (green, target in red) to form a cointegrate. During cointegrate formation, the transposon replicates. In the second step, catalyzed by the *tnpR* gene product, the cointegrate resolves into the target plasmid, with the transposon inserted, plus the original transposon-bearing plasmid.

coupled through a pair of Tn3 copies. This step requires recombination between the two plasmids, which is catalyzed by the product of the Tn3 transposase gene *tnpA*. Figure 23.6 shows a detailed picture of how all four DNA strands involved in transposition might interact to form the cointegrate. Figures 23.5 and 23.6 illustrate transposition between two plasmids, but the donor and target DNAs can be other kinds of DNA, including phage DNAs or the bacterial chromosome itself.

The second step in Tn3 transposition is a **resolution** of the cointegrate, in which the cointegrate breaks down into two independent plasmids, each bearing one copy of Tn3. This step, catalyzed by the product of the **resolvase** gene *tnpR*, is a recombination between homologous sites on Tn3 itself, called ***res* sites.** Several lines of evidence show that Tn3 transposition is a two-step process. First, mutants in the *tnpR* gene cannot resolve cointegrates, so they cause formation of cointegrates as the final product of transposition. This demonstrates that the cointegrate is normally an intermediate in the reaction. Second, even if the *tnpR* gene is defective, cointegrates can be resolved if a functional *tnpR* gene is provided by another DNA molecule—the host chromosome or another plasmid, for example.

Nonreplicative Transposition Figures 23.5 and 23.6 illustrate the replicative transposition mechanism, but transposition does not always work this way. Some transposons (e.g., Tn10) move without replicating, leaving the donor DNA and appearing in the target DNA. How does this occur? It may be that nonreplicative transposition starts out in the same way as replicative transposition, by nicking and joining strands of the donor and target DNAs, but then something different happens (Figure 23.7). Instead of replication occurring through the transposon, new nicks appear

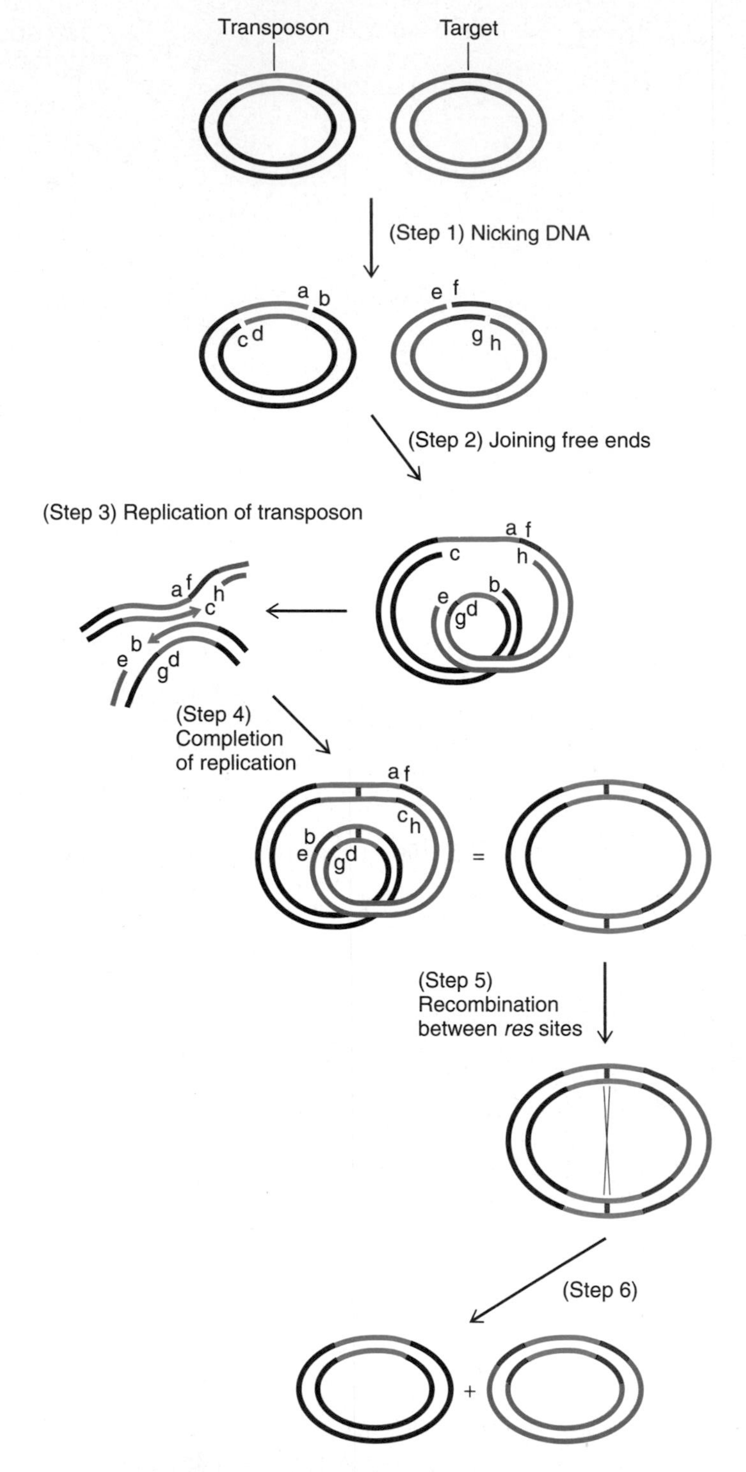

Figure 23.6 Detailed scheme of Tn3 transposition. Step 1: The two plasmids are nicked to form the free ends labeled a–h. Step 2: Ends a and f are joined, as are g and d. This leaves b, c, e, and h free. Step 3: Two of these remaining free ends (b and c) serve as primers for DNA replication, which is shown in a blowup of the replicating region. Step 4: Replication continues until end b reaches e and end c reaches h. These ends are ligated to complete the cointegrate. Notice that the whole transposon (blue) has been replicated. The paired *res* sites (purple) are shown for the first time here, even though one *res* site existed in the previous steps. The cointegrate is drawn with a loop in it, so its derivation from the previous drawing is clearer; however, if the loop were opened up, the cointegrate would look just like the one in Figure 23.5 (shown here at right). Steps 5 and 6 (resolution): A crossover occurs between the two *res* sites in the two copies of the transposon, leaving two independent plasmids, each bearing a copy of the transposon.

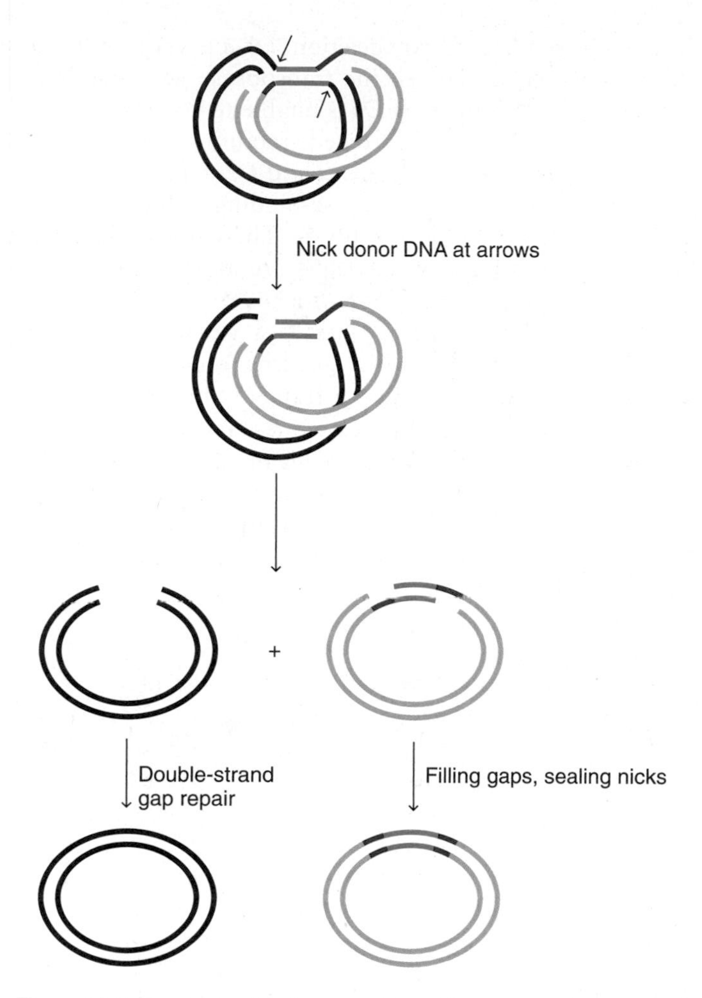

Figure 23.7 Nonreplicative transposition. The first two steps are just like those in replicative transposition, and the structure at the top is the same as that between steps 2 and 3 in Figure 23.6. Next, however, new nicks occur at the positions indicated by the arrows. This liberates the donor plasmid minus the transposon, which remains attached to the target DNA. Filling gaps and sealing nicks completes the target plasmid with its new transposon. The free ends of the donor plasmid may or may not join. In any event, this plasmid has lost its transposon.

in the donor DNA on either side of the transposon. This releases the gapped donor DNA but leaves the transposon still bound to the target DNA. The remaining nicks in the target DNA can be sealed, yielding a recombinant DNA with the transposon integrated into the target DNA. The donor DNA has a double-stranded gap, so it may be lost or, as shown in Figure 23.7, here, the gap may be repaired.

SUMMARY Many transposons contain genes aside from the ones necessary for transposition. These are commonly antibiotic resistance genes. For example, Tn3 contains a gene that confers ampicillin resistance. Tn3 and its relatives transpose by a two-step process (replicative transposition). First the transposon replicates and the donor DNA fuses to the target DNA, forming a cointegrate. In the second step, the cointegrate is resolved into two DNA circles, each of which bears a copy of the transposon. An alternative pathway, not used by Tn3, is nonreplicative transposition, in which no replication of the transposon occurs.

23.2 Eukaryotic Transposons

It would be surprising if bacteria were the only organisms to harbor transposable elements, especially because these elements have powerful selective forces on their side. First, many transposons carry genes that are an advantage to their hosts. Therefore, their host can multiply at the expense of competing organisms and can multiply the transposons along with the rest of their DNA. Second, even if transposons are not advantageous to their hosts, they can replicate themselves within their hosts in a "selfish" way. Indeed, transposable elements *are* also present in eukaryotes. In fact, they were first identified in eukaryotes.

The First Examples of Transposable Elements: *Ds* and *Ac* of Maize

Barbara McClintock discovered the first transposable elements in a study of maize (corn) in the late 1940s. It had been known for some time that the variegation in color observed in the kernels of so-called Indian corn was caused by an unstable mutation. In Figure 23.8a, for example, we see a kernel that is colored. This color is due to a factor encoded by the maize *C* locus. Figure 23.8b shows what happens when the *C* gene is mutated; no purple pigment is made, and the kernel appears almost white. The spotted kernel in Figure 23.8c shows the results of reversion in some of the kernel's cells. Wherever the mutation has reverted, the revertant cell and its progeny will be able to make pigment, giving rise to a dark spot on the kernel. It is striking that so many spots occur in this kernel. That means the mutation is very unstable: It reverts at a rate much higher than we would expect of an ordinary mutation.

In this case, McClintock discovered that the original mutation resulted from an insertion of a transposable element, called ***Ds*** for "dissociation," into the *C* gene (Figure 23.9a and b). Another transposable element, ***Ac*** for "activator," could induce *Ds* to transpose by a nonreplicative mechanism out of *C*, causing reversion (Figure 23.9c). In other words, *Ds* can transpose, but only with the help of *Ac*. *Ac* on the other hand, is an autonomous transposon. It can transpose itself and therefore inactivate other genes without help from other elements.

Now that molecular biological tools are available, we can isolate and characterize these genetic elements decades after McClintock found them. Nina Fedoroff and her collaborators obtained the structures of *Ac* and three different

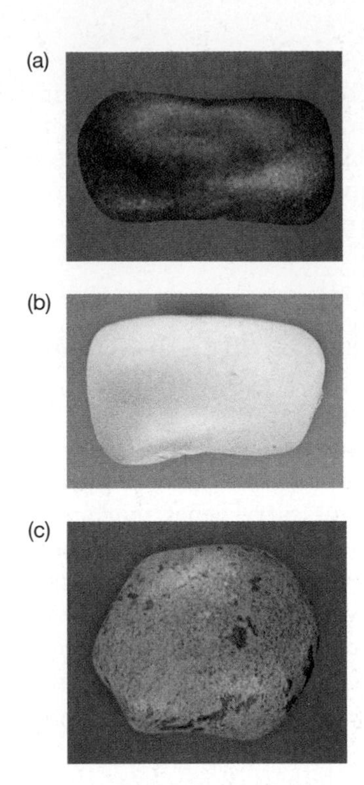

Figure 23.8 Effects of mutations and reversions on maize kernel color. **(a)** Wild-type kernel has an active *C* locus that causes synthesis of purple pigment. **(b)** The *C* locus has mutated, preventing pigment synthesis, so the kernel is colorless. **(c)** The spots correspond to patches of cells in which the mutation in *C* has reverted, again allowing pigment synthesis. (*Source:* F.W. Goro, from Fedoroff, N., Transposable genetic elements in maize. *Scientific American* 86 (June 1984).)

forms of *Ds*. *Ac* resembles the bacterial transposons we have already studied (Figure 23.10). It is about 4500 bp long, contains a transposase gene, and is bounded by short, imperfect inverted repeats, and adjacent *subterminal repetitive regions* that bind transposase. The various forms of *Ds* are derived from *Ac* by deletion. *Ds-a* is very similar to *Ac*, except that a piece of the transposase gene has been deleted. This explains why *Ds* is unable to transpose itself. *Ds-b* is more severely shortened, retaining only a small fragment of the transposase gene, and *Ds-c* retains only the inverted repeats and transposase-binding subterminal repetitive regions in common with *Ac*. These inverted repeats and transposase-binding sequences are all that *Ds-c* needs to be a target for transposition directed by *Ac*.

It is interesting that the first pea gene described by Mendel himself (*R* or *r*), which governs round versus wrinkled seeds, seems to involve a transposable element. We now know that the *R* locus encodes an enzyme (starch branching enzyme) that participates in starch metabolism. The wrinkled phenotype results from a malfunction of this gene; this mutation is in turn caused by an insertion of an 800-bp piece of DNA that seems to be a member of the *Ac*/*Ds* family.

SUMMARY The variegation in the color of maize kernels is caused by multiple reversions of an unstable mutation in the *C* locus, which is responsible for the kernel's color. The mutation and its reversion result from a *Ds* (dissociation) element, which transposes into the *C* gene, mutating it, and then transposes out again, causing it to revert to wild-type. *Ds* cannot transpose on its own; it must have help from an autonomous transposon called *Ac* (for activator), which supplies the transposase. *Ds* is an *Ac* element with more or less of its middle removed. All *Ds* needs in order to be transposed is a pair of inverted terminal repeats and adjacent short sequences that the *Ac* transposase can recognize.

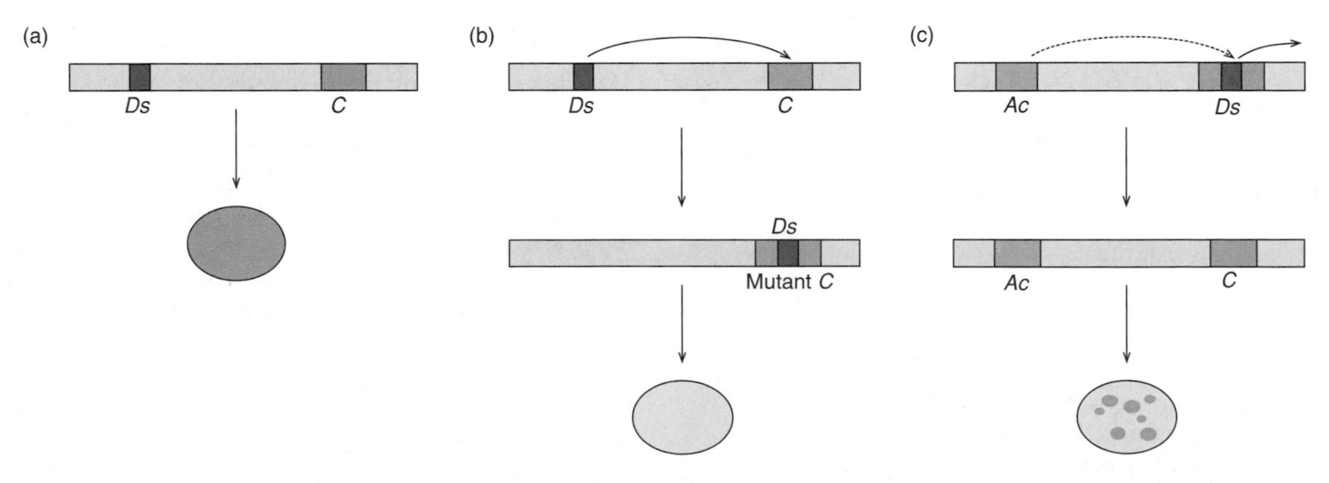

Figure 23.9 Transposable elements cause mutations and reversions in maize. **(a)** A wild-type maize kernel has an uninterrupted, active *C* locus (blue) that causes synthesis of purple pigment. **(b)** A *Ds* element (red) inserts into *C*, inactivating it and preventing pigment synthesis. The kernel is therefore colorless. **(c)** *Ac* (green) is present, as well as *Ds*. This allows *Ds* to transpose out of *C* in many cells, giving rise to groups of cells that make pigment. Such groups of pigmented cells account for the purple spots on the kernel. Of course, *Ds* must have transposed into *C* before it (*Ds*) became defective, or else it had help from an *Ac* element.

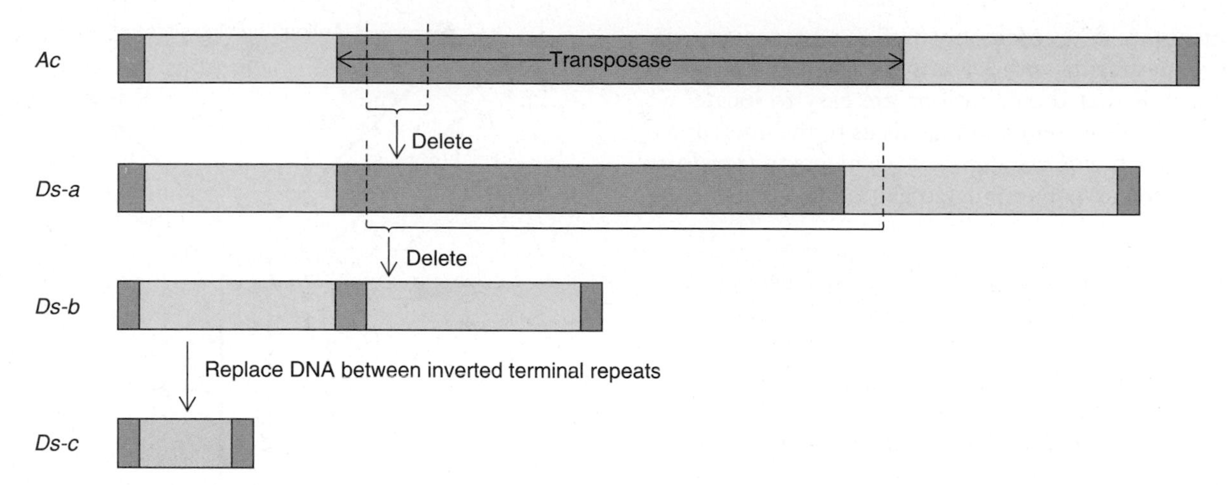

Figure 23.10 Structures of *Ac* and *Ds*. *Ac* contains the transposase gene (purple) and two imperfect inverted terminal repeats (blue), including the subterminal repetitive regions. *Ds-a* is missing a 194-bp region from the transposase gene (dashed lines); otherwise, it is almost identical to *Ac*. *Ds-b* is missing a much larger segment of *Ac*. *Ds-c* has no similarity to *Ac* except for the inverted terminal repeats and subterminal repetitive regions.

P Elements

The phenomenon called **hybrid dysgenesis** illustrates another obvious kind of mutation enhancement caused by a eukaryotic transposon. In hybrid dysgenesis, one strain of *Drosophila* mates with another to produce hybrid offspring that suffer so much chromosomal damage that they are dysgenic, or sterile. Hybrid dysgenesis requires a contribution from both parents; for example, in the *P-M system* the father must be from strain P (paternal contributing) and the mother must be from strain *M* (maternal contributing). The reverse cross, with an M father and a P mother, produces normal offspring, as do crosses within a strain (P × P or M × M).

What makes us suspect that a transposon is involved in this phenomenon? First of all, any P male chromosome can cause dysgenesis in a cross with an M female. Moreover, recombinant male chromosomes derived in part from P males and in part from M males usually can cause dysgenesis, showing that the P trait is carried on multiple sites on the chromosomes.

One possible explanation for this behavior is that the P trait is governed by a transposable element, and that is why we find it in so many different sites. In fact, this is the correct explanation. The transposon responsible for the P trait is called the **P element** and it is found only in wild flies, not in laboratory strains—unless a biologist puts it there. Margaret Kidwell and her colleagues investigated the P elements that inserted into the *white* locus of dysgenic flies. They found that these elements had great similarities in base sequence but differed considerably in size (from about 500 to about 2900 bp). Furthermore, the P elements had direct terminal repeats and were flanked by short direct repeats of host DNA, both signatures of transposons. Finally, the *white* mutations reverted at a high rate by losing the entire P element—again a property of a transposon.

If P elements act like transposons, why do they transpose and cause dysgenesis only in hybrids? The answer is that the P element also encodes a suppressor of transposition, which accumulates in the cytoplasm of the developing germ cells. Thus, in a cross of a P or M male with a P female, the female cytoplasm contains the suppressor, which binds to any P elements and prevents their transposition. But in a cross of a P male with an M female, the early embryo contains no suppressor, and none is made at first, because the P element becomes active only in developing germ cells. When the P element is finally activated, the transposase and suppressor are both made, but the transposase alone goes to the nucleus, where it freely stimulates transposition.

In 2009, Gregory Hannon and colleagues identified a good candidate for the suppressor: a group of anti-P element piRNAs. Recall from Chapter 16 that piRNAs target transposons in germ cells and suppress their transposition. Hannon and colleagues found that P females had abundant P element-specific piRNAs in their germ cells, while M females did not. The same principle applies to the similar I element: Crosses between "inducer" (I) males, carrying the I element and "reactive" (R) females lacking an I suppressor gave sterile progeny, whereas crosses involving I females, which had the suppressor, gave fertile offspring. And, in accord with the hypothesis, Hannon and colleagues showed that I females contained piRNAs targeting the I element, while R females did not.

Hybrid dysgenesis may have important consequences for speciation—the formation of new species that cannot interbreed. Two strains of the same species (such as P and M) that frequently produce sterile offspring will tend to become genetically isolated—their genes no longer mix as often—and eventually will be so different that they will not be able to interbreed at all. When this happens, they have become separate species.

P elements are now commonly used as mutagenic agents in genetic experiments with *Drosophila*. One advantage of this approach is that the mutations are easy to locate; we just look for the P element and it leads us to the interrupted gene. Molecular biologists also use P elements to *transform* flies—that is, to carry manipulated genes into flies.

SUMMARY The P-M system of hybrid dysgenesis in *Drosophila* is caused by the conjunction of two factors: (1) a transposable element (P) contributed by the male, and (2) M cytoplasm contributed by the female, which allows transposition of the P element. Hybrid offspring of P males and M females therefore suffer multiple transpositions of the P element. This causes damaging chromosomal mutations that render the hybrids sterile. On the other hand, P females contain a suppressor of transposition (a group of piRNAs targeting the P element), so offspring of either P or M males and P females are fertile. P elements have practical value as mutagenic and transforming agents in genetic experiments with *Drosophila*.

23.3 Rearrangement of Immunoglobulin Genes

Rearrangements of the mammalian genes in B cells that produce **antibodies,** or **immunoglobulins,** and in T cells that produce **T-cell receptors,** use a process that closely resembles transposition. Even the recombinases involved in antibody and T-cell receptor gene rearrangements resemble transposases. Because of these similarities, we include these rearrangements in this chapter.

As mentioned in Chapter 3, an antibody is composed of four polypeptides: two heavy chains and two light chains. (Similarly, T-cell receptors contain one large β-chain and one smaller α-chain.) Figure 23.11 illustrates an antibody schematically and shows the sites that combine with an invading antigen. These sites, called **variable regions,** vary from one antibody to the next and give these proteins their specificities; the rest of the protein (the **constant region**) does not vary from one antibody to another within an antibody class, though some variation occurs between the few classes of antibodies. Any given immune cell can make antibody with only one kind of specificity. Remarkably enough, humans have immune cells capable of producing antibodies to react with virtually any foreign substance we would ever encounter. That means we can make many millions of different antibodies.

Does this imply that we have millions of different antibody genes? That is an untenable hypothesis; it would place an impossible burden on our genomes to carry all the

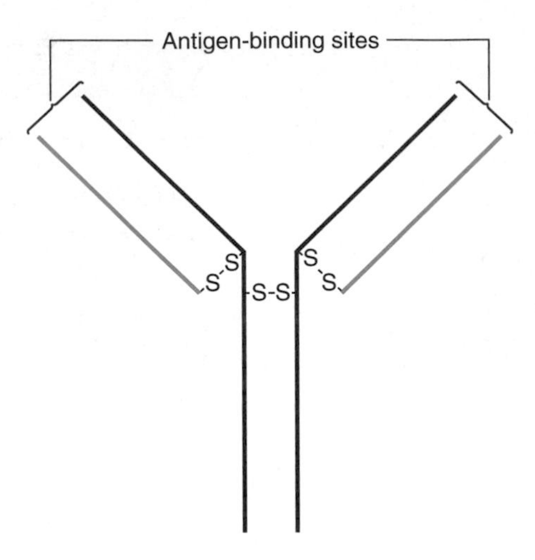

Figure 23.11 Structure of an antibody. The antibody is composed of two light chains (blue) bound through disulfide bridges to two heavy chains (pink), which are themselves held together by a disulfide bridge. The antigen-binding sites are at the amino termini of the protein chains, where the variable regions lie.

necessary genes. So how do we solve the antibody diversity problem? As unlikely as it may seem, a maturing B cell, a cell that is destined to make an antibody, rearranges its genome to bring together separate parts of its antibody genes. The machinery that puts together the gene selects these parts at random from heterogeneous groups of parts, rather like ordering from a luncheon menu ("Choose one from column A and one from column B"). This arrangement greatly increases the variability of the genes. For instance, if 41 possibilities are present in "column A" and 5 in "column B," the total number of combinations of A + B is 41 × 5 or 205. Thus, from 46 gene fragments, we can assemble 205 genes. And this is just for one of the antibody polypeptides. If a similar situation exists for the other, the total number of antibodies will be the product of the numbers of the two polypeptides. This description, though correct in principle, is actually an oversimplification of the situation in the antibody genes; as we will see, they have somewhat more complex mechanisms for introducing diversity, which lead to an even greater number of possible antibody products.

Studies on mammalian antibodies have revealed two families of antibody light chains called kappa (κ) and lambda (λ). Figure 23.12 illustrates the arrangement of the gene parts for a human κ light chain. "Column A" of this "menu" contains 41 variable region parts (V); "Column B" contains 5 **joining region** parts (J). The J segments actually encode the last 12 amino acids of the variable region, but they are located far away from the rest of the V region and close to a single constant region part. This is the situation in the germ cells, before the antibody-producing cells differentiate and before rearrangement brings the two unlinked regions together. The rearrangement and expression events are depicted in Figure 23.12.

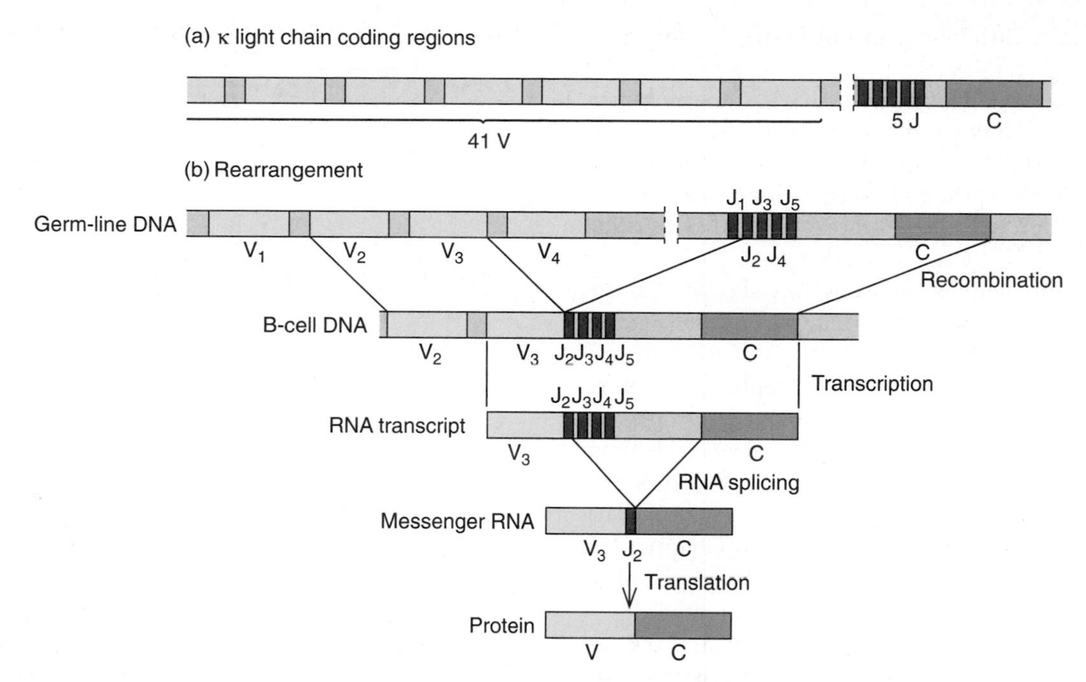

Figure 23.12 Rearrangement of an antibody light chain gene. **(a)** The human κ-antibody light chain is encoded in 41 variable gene segments (V; light green), five joining segments (J; red), and one constant segment (C; blue). **(b)** During maturation of an antibody-producing cell, a DNA segment is deleted, bringing a V segment (V_3, in this case) together with a J segment (J_2 in this case). The gene can now be transcribed to produce the mRNA precursor shown here, with extra J segments and intervening sequences. The material between J_2 and C is then spliced out, yielding the mature mRNA, which is translated to the antibody protein shown at the bottom. The J segment of the mRNA is translated into part of the variable region of the antibody.

First, a recombination event brings one of the V regions together with one of the J regions. In this case, V_3 and J_2 fuse together, but it could just as easily have been V_1 and J_4; the selection is random. After the two parts of the gene assemble, transcription occurs, starting at the beginning of V_3 and continuing until the end of C. Next, the splicing machinery joins the J_2 region of the transcript to C, removing the extra J regions and the intervening sequence between the J regions and C. It is important to remember that the rearrangement step takes place at the DNA level, but this splicing step occurs at the RNA level by mechanisms we studied in Chapter 14. The messenger RNA thus assembled moves into the cytoplasm to be translated into an antibody light chain with a variable region (encoded in both V and J) and a constant region (encoded in C).

Why does transcription begin at the beginning of V_3 and not farther upstream? The answer seems to be that an enhancer in the intron between the J regions and the C region activates the promoter closest to it: the V_3 promoter in this case. This also provides a convenient way of activating the gene after it rearranges; only then is the enhancer close enough to turn on the promoter.

The rearrangement of the heavy chain gene is even more complex, because there is an extra set of gene parts in between the V's and J's. These gene fragments are called D, for "diversity," and they represent a third column on our menu. Figure 23.13 shows that the heavy chain is assembled from 48 V regions, 23 D regions, and 6 J regions. On this basis alone, the cell can put together $48 \times 23 \times 6$, or 6624 different heavy chain genes. Furthermore, 6624 different heavy chains combined with 205 κ light chains and 170 λ light chains yield almost 2.5 million different antibodies or, strictly speaking, 2.5 million different combinations of variable regions.

But there are even more sources of diversity. The first derives from the fact that the mechanism joining V, D, and J segments, which we call **V(D)J joining,** is not precise. It can add or delete bases on either side of the joining site.

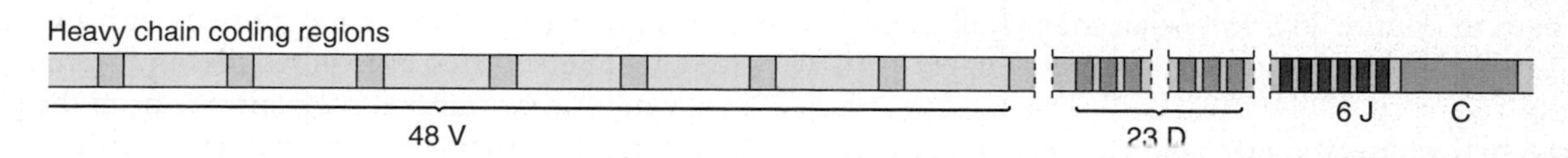

Figure 23.13 Structure of antibody heavy chain coding regions. The human heavy chain is encoded in 48 variable segments (V; light green), 23 diversity segments (D; purple), 6 joining segments (J; red), and 1 constant segment (C; blue).

This leads to extra differences in antibodies' amino acid sequences.

Another source of antibody diversity is somatic hypermutation, or rapid mutation in an organism's somatic (nonsex) cells. In this case, the mutations occur in antibody genes, probably at the time that a clone of antibody-producing B cells proliferates to meet the challenge of an invader.

Genetic and biochemical analysis has shown that somatic hypermutation occurs in two steps. First, a cytidine deaminase that is induced during B cell activation deaminates cytosines to uracils during DNA replication. Next, the uracils attract either the mismatch repair process or uracil-N-glycosylase, which removes the uracils, leaving abasic sites. In either case, a single-strand break occurs, and the cell then "repairs" the break with the same auxiliary DNA polymerases used in translesion bypass (Chapter 20): DNA polymerases ζ, η, θ, and possibly ι. These polymerases are error prone, so many mutations are created.

Together, imprecise joining of gene segments and somatic hypermutation magnify the number of possible antibodies tremendously. In fact, it has been estimated that the total number of antibodies one can make in a lifetime is as high as 100 billion. This surely seems enough to match any attacker.

SUMMARY The immune systems of vertebrates can produce billions of different antibodies to react with virtually any foreign substance. These immune systems generate such enormous diversity by three basic mechanisms: (1) assembling genes for antibody light chains and heavy chains from two or three component parts, respectively, each part selected from heterogeneous pools of parts; (2) joining the gene parts by an imprecise mechanism that can delete bases or even add extra bases, thus changing the gene; and (3) causing a high rate of somatic mutations, probably during proliferation of a clone of immune cells, thus creating slightly different genes.

Recombination Signals

How does the recombination machinery determine where to cut and paste to bring together the disparate parts of an immunoglobulin gene? Susumu Tonegawa examined the sequences of many mouse immunoglobulin genes (encoding κ and λ light chains, and heavy chains) and noticed a consistent pattern (Figure 23.14a): Adjacent to each coding region lies a conserved palindromic heptamer (7-mer), with the consensus sequence 5′-CACAGTG-3′. This heptamer is accompanied by a conserved nonamer (9-mer) whose consensus sequence is 5′-ACAAAAACC-3′. The heptamer and nonamer are separated by a nonconserved spacer containing either 12 bp (a **12 signal**) or 23 (±1) bp (a **23 signal**). The arrangement of these **recombination signal sequences (RSSs,** Figure 23.14b) is such that recombination always joins a 12 signal to a 23 signal. This **12/23 rule** stipulates that 12 signals are never joined to each other, nor are 23 signals joined to each other, and thus ensures that one, and only one, of each coding region is incorporated into the mature immunoglobulin gene.

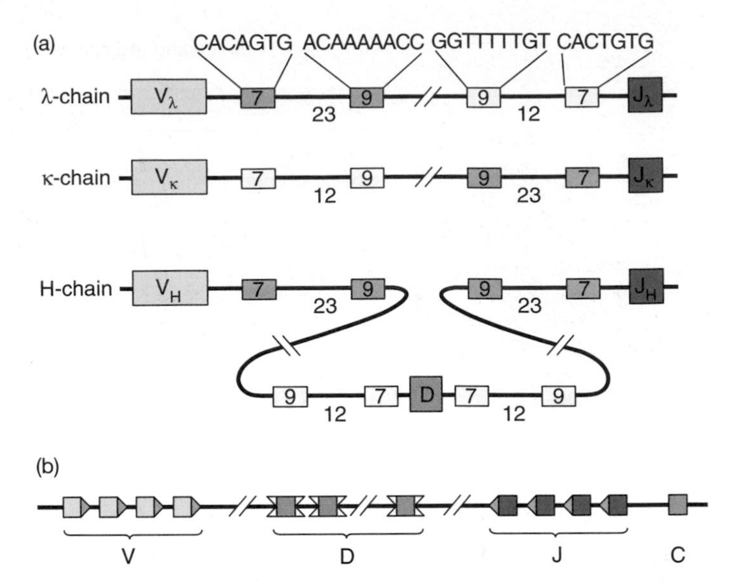

Figure 23.14 Signals for V(D)J joining. (a) Arrangement of signals around coding regions for immunoglobulin κ and λ light chain genes and heavy chain gene. Boxes labeled "7" or "9" are conserved heptamers or nonamers, respectively. Their consensus sequences are given at top. The 12-mer and 23-mer spacers are also labeled. Notice the arrangement of the 12 signals and 23 signals such that joining one kind to the other naturally allows assembly of a complete gene. **(b)** Schematic illustration of the arrangement of the 12 and 23 signals in an immunoglobulin heavy chain gene. The yellow symbols represent 12 signals, and the orange triangles represent 23 signals. Notice again how the 12/23 rule guarantees inclusion of one of each coding region (V, D, and J) in the rearranged gene. (*Source:* (a) Adapted from Tonegawa, S., Somatic generation of antibody diversity. *Nature* 302:577, 1983.)

Aside from the existence of consensus RSSs, what is the evidence for their importance? Martin Gellert and colleagues have systematically mutated the heptamer and nonamer by substituting bases, and the spacer regions by adding or subtracting bases, and observed the effects of these alterations on recombination. They measured recombination efficiency in the following way: They built a recombinant plasmid with the construct shown in Figure 23.15. The first element in this construct is a *lac* promoter. This is followed by a 12 signal, then a prokaryotic transcription terminator, then a 23 signal, and finally a *cat* reporter gene. They made mutations throughout these RSSs, then introduced the altered plasmids into a pre-B cell line. Finally, they purified the plasmids from the pre-B cells and introduced them into chloramphenicol-sensitive *E. coli* cells and tested them for chloramphenicol resistance. If no recombination took place, the transcription terminator

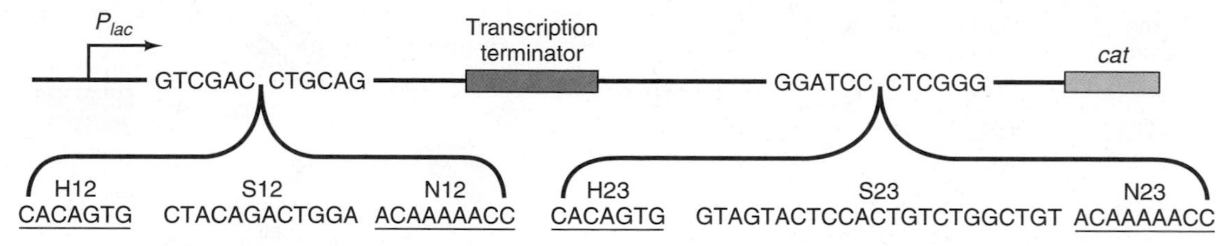

Figure 23.15 Structure of reporter construct used to measure effects of mutations in RSSs on recombination efficiency. Gellert and coworkers made a recombination reporter plasmid containing a *lac* promoter and *cat* gene separated by an insert containing a transcription terminator flanked by a 12 signal and a 23 signal. Recombination between the two RSSs either inverts or deletes the terminator, allowing expression of *cat.* Transformation of bacterial cells with the rearranged plasmid yields many CAT-producing colonies that are chloramphenicol-resistant. On the other hand, transformation of bacteria with the unrearranged plasmid yields almost no chloramphenicol-resistant colonies. (*Source:* Adapted from Hesse, J., M. R. Lieber, K. Mizuuchi, and M. Gellert, V(D)J recombination: a functional definition of the joining signals. *Genes and Development* 3:1053–61, 1989.)

prevented *cat* expression, and therefore chloramphenicol resistance was almost nonexistent. On the other hand, if recombination between the 12 signal and the 23 signal occurred, the terminator was either inverted or deleted, and therefore inactivated. In that case, *cat* expression occurred under control of the *lac* promoter, and many chloramphenicol-resistant colonies formed. This experiment showed that many alterations in bases in the heptamer or nonamer reduced recombination efficiency to background level. The same was true of insertions and deletions of bases in the spacer regions. Thus, all these elements of the RSSs are important in V(D)J recombination.

SUMMARY The recombination signal sequences (RSSs) in V(D)J recombination consist of a heptamer and a nonamer separated by either 12-bp or 23-bp spacers. Recombination occurs only between a 12 signal and a 23 signal, which guarantees that only one of each coding region is incorporated into the rearranged gene.

The Recombinase

David Baltimore and his colleagues searched for the gene(s) encoding the V(D)J recombinase using a recombination reporter plasmid similar to the one we just discussed, but designed to operate in eukaryotic cells by conferring resistance to the drug mycophenolic acid. They introduced this plasmid, along with fragments of mouse genomic DNA, into NIH 3T3 cells, which lack V(D)J recombination activity, and tested for recombination by assaying for drug-resistant 3T3 cells. This led to the identification of a recombination-activating gene (*RAG-1*) that stimulated V(D)J joining activity in vivo.

However, the degree of stimulation by a genomic clone containing most of *RAG-1* was modest—no more than that obtained with whole genomic DNA. Furthermore, cDNA clones containing the whole *RAG-1* sequence did no better, so something seemed to be missing. Baltimore's group sequenced the whole genomic fragment containing most of *RAG-1* and found another whole gene tightly linked to it. They wondered whether this other gene might also have something to do with V(D)J joining, so they tested this genomic fragment plus a *RAG-1* cDNA in the same transfection experiment. When they introduced the two DNAs together into the same cell, they found many more drug-resistant cells. In this way, they discovered that two genes are responsible for V(D)J recombination, and they named the second *RAG-2.*

RAG-1 and *RAG-2* are expressed only in pre-B and pre-T cells, where V(D)J joining of immunoglobulin and T-cell receptor gene segments, respectively, are occurring. The T-cell receptors are membrane-bound antigen-binding proteins with an architecture similar to that of the immunoglobulins. The genes encoding the T-cell receptors rearrange according to the same rules that apply to the immunoglobulin genes, complete with RSSs containing 12 signals and 23 signals. Thus, *RAG-1* and *RAG-2* are apparently involved in both immunoglobulin and T-cell receptor V(D)J joining.

Mechanism of V(D)J Recombination

V(D)J joining is imprecise, which contributes to the diversity of products from the process. Both loss of bases and addition of extra bases at the joints are frequently observed. This is good for immunoglobulin and T-cell receptor production, because it adds to the variety of proteins that can be made from a limited repertoire of gene segments.

How do we explain this imprecision? Figure 23.16 illustrates the mechanism of cleavage at the RSSs that flank an intervening segment between two coding segments. We see that the products of the *RAG-1* and *RAG-2* genes, **Rag-1** and **Rag-2**, respectively, first nick the DNAs at the joints. Then the new 3′-hydroxyl groups attack phosphodiester bonds on the complementary strands, liberating the intervening segment and forming hairpins at the ends of the coding segments. These hairpins are the key to the

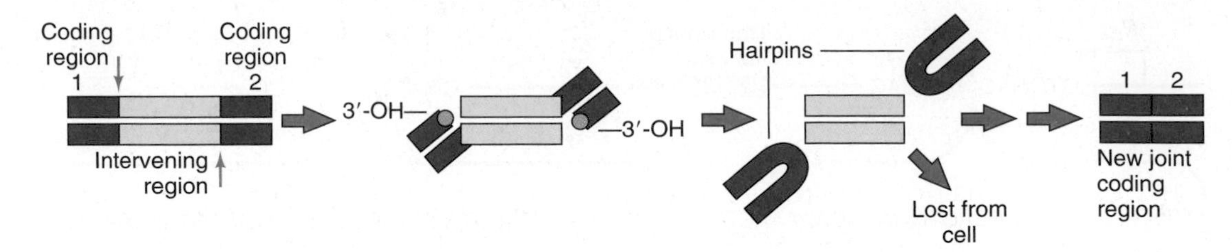

Figure 23.16 Mechanism of cleavage at RSSs. Nicking of opposite strands (vertical arrows) occurs at RSSs at the junctions between coding regions (red) and the intervening region (yellow). The new 3′-hydroxyl groups (blue) attack and break the opposite strands, forming hairpins and releasing the intervening segment, which is lost. Finally, the hairpins open, and the two coding regions are joined by an imprecise mechanism. (*Source:* Adapted from Craig, N.L., V(D)J recombination and transposition: closer than expected. *Science* 271:1512, 1996.)

imprecision of joining; they can open up on either side of the apex of the hairpin, and bases can then be added or subtracted to make the DNA ends blunt for joining. The Rag-1 and Rag-2 proteins hold both hairpins together in a complex so they can join covalently with each other.

How do we know hairpins form? They were first found in vivo, but in very low concentration. Gellert and his colleagues later developed an in vitro system in which they could be readily observed. Figure 23.17a illustrates one of the labeled substrates these workers used. It was a 50-mer labeled at one 5′-end with ^{32}P. It contained a 12 signal, represented by a yellow symbol, flanked by a 16-bp segment on the left; the right-hand end of the fragment was, therefore, a 34-bp segment, which included the 12 signal. A similar substrate contained the same flanking segments, but had a 23 signal instead of a 12 signal. Thus, it was 61 bp long.

Gellert and colleagues incubated these substrates with RAG1 and RAG2, the human homologs of mouse Rag1 and Rag2, respectively, then electrophoresed the products under nondenaturing conditions to see if any DNA cleavages had occurred (Figure 23.17b). They found a 16-mer, demonstrating that a double-stranded cleavage had occurred. However, nondenaturing gel electrophoresis could not distinguish between a true double-stranded 16-mer and a 16-mer with a hairpin end, so these workers subjected the same products to denaturing polyacrylamide gel electrophoresis in the presence of urea and at an elevated temperature (Figure 23.17c). Under these conditions, a double-stranded 16-mer would give rise to two single-stranded 16-mers. On the other hand, a 16-mer with a hairpin at the end would give rise to a single-stranded 32-mer. This is what Gellert and coworkers observed whenever the DNA contained either a 12 signal or a 23 signal and both RAG1 and RAG2 proteins were present. A DNA with no 12 or 23 signal gave no product, hairpin or otherwise, and reactions lacking either RAG1 or RAG2 protein gave no product (Figure 23.17d). Thus, RAG1 and RAG2 recognize both the 12 signal and the 23 signal and cleave the DNA adjacent to the signal, forming a hairpin at the end of the coding segment.

Moreover, the 16-mer product from the nondenaturing gel yielded only hairpin product on the denaturing gel, demonstrating that no simple double-stranded 16-mer formed. But labeled DNA migrating with the substrate in the nondenaturing gel yielded a small amount of 16-mer in the denaturing gel. This cannot have come from a double-stranded break, or it would not have remained with the substrate in the nondenaturing gel. Thus, it must have come from a nick in the labeled strand. The 16-mer created by the nick would have remained base-paired to its partner during nondenaturing electrophoresis, but would have migrated independently as a 16-mer during denaturing electrophoresis. Thus, single-stranded nicking is apparently also part of the action of RAG1 and RAG2 proteins.

To investigate further the relationship between nicking and hairpin formation, Gellert and colleagues ran a time-course study in which they incubated the substrate for increasing lengths of time with RAG1 and RAG2 proteins and then subjected the products to denaturing gel electrophoresis. They found that the nicked species appeared first, followed by the hairpin species. This suggested that the nicked species is a precursor of the hairpin species. To test this hypothesis, they created nicked intermediates and incubated them with RAG1 and RAG2. Sure enough, the RAG1 and RAG2 converted the nicked DNAs to hairpins. Subsequent work by Gellert's group has shown the sequence of events seems to be: RAG1 and RAG2 nick one DNA strand adjacent to a 12 signal or a 23 signal; then the newly formed hydroxyl group attacks the other strand in a transesterification reaction, forming the hairpin, as was illustrated in Figure 23.16.

What enzyme opens up the hairpins created by RAG1 and RAG2? Michael Lieber and colleagues demonstrated in 2002 that an enzyme called **Artemis** carries out this function. On its own, Artemis has exonuclease activity. However, in conjuction with DNA-PK$_{cs}$, Artemis gains endonuclease activity that can cleave hairpins. You may recongnize DNA-PK$_{cs}$ from our discussion in Chapter 20 of nonhomologous DNA end-joining (NHEJ) for repair of double-strand DNA breaks. In fact, joining of the opened hairpins resembles NHEJ and relies on the NHEJ machinery.

Artemis is also required to cleave the hairpins created during the rearrangement of T cell receptor genes, which

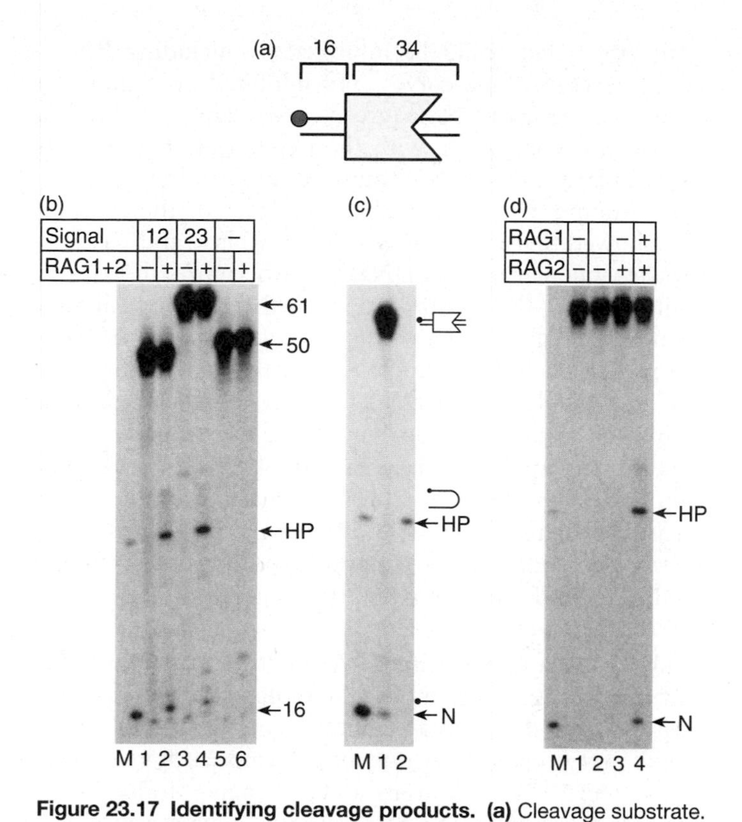

Figure 23.17 Identifying cleavage products. (a) Cleavage substrate. Gellert and colleagues constructed this labeled 50-mer, which included 16 bp of DNA on the left, then a 12 signal (yellow), included in a 34-bp segment on the right. The single 5′-end label is indicated by the red dot. These workers also made an analogous 61-mer substrate with a 23 signal. **(b)** Identifying the hairpin product. Gellert and coworkers incubated RAG1 and RAG2 proteins, as indicated at top, with either the labeled 12-signal or 23-signal substrate, also as indicated at top. After the incubation, they subjected the products to nondenaturing gel electrophoresis and autoradiographed the gel to detect the labeled products. The positions of the 61-mer and 50-mer substrates, the hairpin (HP), and the 16-mer are indicated at right. **(c)** Identifying the products from a nondenaturing gel. Gellert and colleagues recovered the labeled products (apparently uncleaved 50-mer substrate and 16-mer fragment) from the bands of a nondenaturing gel. They then electrophoresed these DNAs again in lanes 1 and 2, respectively, of a denaturing gel, along with markers (identified with diagrams at right) corresponding to the uncleaved substrate, the 16-bp hairpin (HP), and the single-stranded 16-mer released by denaturing the nicked substrate. **(d)** Requirement for RAG1 and RAG2. This experiment was very similar to the one in panel (b) except that the presence of RAG1 and RAG2 proteins (indicated at top) were the only variables. "N" denotes the position of the 16-mer released from the nicked species. (*Source:* McBlane, J.F., D.C. Van Gent, D.A. Ramsden, C. Romeo, C.A. Cuomo, M. Gellert, and M.A. Oettinger, Cleavage at a V(D)J recombination signal requires only RAG1 and RAG2 proteins and occurs in two steps. *Cell* 83 (3 Nov 1995) f. 4 a–c, p. 390. Reprinted by permission of Elsevier Science.)

closely resembles rearrangement of immunoglobulin genes. Without antibodies, B cells are useless, and without T cell receptors, T cells are useless. Thus, loss of Artemis function means loss of both B cell and T cell function. Indeed, people with defective Artemis genes have a very serious condition known as **severe combined immunodeficiency (SCID,** "bubble boy" syndrome) and cannot mount an immune response against any pathogen. They must be isolated from the rest of the world in order to survive.

SUMMARY RAG1 and RAG2 introduce single-strand nicks into DNA adjacent to either a 12 signal or a 23 signal. This leads to a transesterification in which the newly created 3′-hydroxyl group attacks the opposite strand, breaking it, and forming a hairpin at the end of the coding segment. The hairpins then break in an imprecise way, allowing joining of coding regions with loss of bases or gain of extra bases.

23.4 Retrotransposons

McClintock's maize transposons are examples of so-called cut-and-paste or copy-and-paste transposons, similar to the bacterial transposons we discussed earlier in this chapter. If DNA replication is involved, it is direct replication. Humans also carry transposons in this class, which constitute about 1.6% of the human genome. The most prevalent example is called **mariner,** but all of the mariner elements studied so far have been defective in transposition. Eukaryotes also carry many more transposons of another kind: **retrotransposons,** which replicate through an RNA intermediate. In this respect, the retrotransposons resemble **retroviruses,** some of which cause tumors in vertebrates, and some of which (the human immunodeficiency viruses, or HIVs) cause AIDS. As an introduction to the replication scheme of the retrotransposons, let us first examine the replication of the retroviruses.

Retroviruses

The most salient feature of a retrovirus, indeed the feature that gives this class of viruses its name, is its ability to make a DNA copy of its RNA genome. This reaction, RNA→DNA, is the reverse of the transcription reaction, so it is commonly called **reverse transcription.** In 1970, Howard Temin and, simultaneously, David Baltimore convinced a skeptical scientific community that this reaction takes place. They did so by finding that the virus particles contain an enzyme that catalyzes the reverse transcription reaction. Inevitably, this enzyme has been dubbed **reverse transcriptase.** A more proper name is **RNA-dependent DNA polymerase.**

Figure 23.18 illustrates the retrovirus replication cycle. We start with a virus infecting a cell. The virus contains two copies of its RNA genome, linked together by base pairing at their 5′-ends (for simplicity, only one copy is

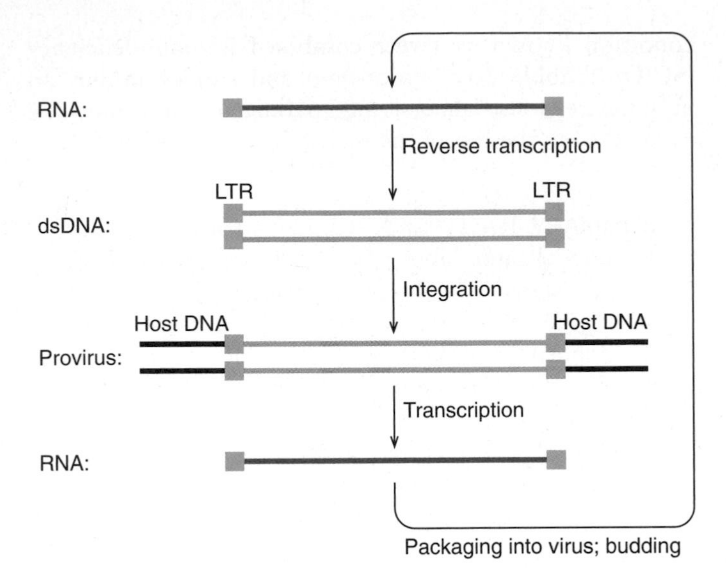

Figure 23.18 Retrovirus replication cycle. The viral genome is an RNA, with long terminal repeats (LTRs, green) at each end. Reverse transcriptase makes a linear, double-stranded DNA copy of the RNA, which then integrates into the host DNA (black), creating the provirus form. The host RNA polymerase II transcribes the provirus, forming genomic RNA. The viral RNA is packaged into a virus particle, which buds out of the cell and infects another cell, starting the cycle over again.

shown). When the virus enters a cell, its reverse transcriptase (a product of the viral ***pol*** gene) makes a double-stranded DNA copy of the viral RNA, with **long terminal repeats (LTRs)** at each end. This DNA recombines with the host genome to yield an integrated form of the viral genome called the **provirus.** The host RNA polymerase II transcribes the provirus, yielding viral mRNAs, which are then translated to viral proteins. To complete the replication cycle, polymerase II also makes RNA copies of the provirus, which are new viral genomes. These genomic RNAs are packaged into virus particles (Figure 23.19) that bud out of the infected cell and go on to infect other cells.

Evidence for Reverse Transcriptase The skepticism about the reverse transcription reaction arose from the fact that no one had ever observed it, and the notion that it violated the "central dogma of molecular biology" promulgated by Watson and Crick, which said that the flow of genetic information is from DNA to RNA to protein, not the reverse. Crick later stated that the DNA→RNA arrow was intended to be double-headed, but that was clearly not the popular perception at the time. What evidence did Baltimore and Temin bring to bear to dispel this skepticism?

Figure 23.20 shows the result of one of Baltimore's experiments. He incubated purified retrovirus particles (Raucher mouse leukemia virus, or R-MLV) with all four dNTPs, including [^{3}H]dTTP, then measured the incorporation of the labeled TTP into a polymer (DNA) that could be precipitated with acid. He observed a clear incorporation (red curve) that could be inhibited by including RNase in the reaction (blue curve), and inhibited even more by preincubating with RNase (green curve). This sensitivity to RNase was compatible with the hypothesis that RNA is the template in the reverse transcription reaction.

Baltimore also examined the product of the reaction and showed that it was insensitive to RNase and base hydrolysis, but sensitive to DNase. Furthermore, the virions could support the incorporation of dNTPs only. Ribonucleotides, including ATP, could not be incorporated. Thus, the product behaved like DNA, and the enzyme behaved like an RNA-dependent DNA polymerase—a reverse transcriptase. Baltimore and Temin both performed similar experiments on Rous sarcoma virus particles, with very similar results. Thus, it appeared that all RNA tumor viruses probably contained reverse transcriptase and behaved according to the provirus hypothesis illustrated in Figure 23.18. This has proven to be true.

Evidence for a tRNA Primer As molecular biologists began to investigate the molecular biology of reverse transcription, they discovered that the viral reverse transcriptase is like every other DNA polymerase known: It requires a primer. In 1971, Baltimore and colleagues found RNA primers attached to the 5′-ends of nascent reverse transcripts using the following strategy: They labeled the nascent reverse transcripts in avian myeloblastosis virus (AMV) by the same method Baltimore and Temin had used—incubating virus particles with labeled dNTPs. Then they subjected the products to Cs_2SO_4 gradient ultracentrifugation to separate RNA from DNA based on their densities (RNA being denser than DNA).

In the first experiment, Baltimore and colleagues isolated the nucleic acids from the virus particles and subjected them immediately to ultracentrifugation. Figure 23.21a shows the results: a peak of labeled DNA that appeared to have the density of RNA. This finding is consistent with the hypothesis that the nascent DNA is still base-paired to the much bigger RNA template, so the whole complex behaves like RNA. If this hypothesis is true, then heating the RNA–DNA hybrid should denature it and release the DNA product as an independent molecule. When Baltimore and colleagues performed that experiment, they observed the behavior in Figure 23.21b: Now the nascent DNA product had a density much closer to that of DNA, but still a little too dense, as if there were still some RNA attached.

That behavior could be explained if the nascent DNA still had an RNA primer covalently attached to it. To check this possibility, Baltimore and coworkers treated the nascent DNA with RNase and again subjected it to ultracentrifugation. This time, the density of the product behaved exactly as expected for pure DNA (Figure 23.21c). Thus, the nascent reverse transcript appears to be primed by RNA. But what RNA?

Figure 23.19 AIDS virion internal structure, cutaway artwork. AIDS (acquired immune deficiency syndrome) is caused by the human immunodeficiency virus (HIV). The core of this HIV virus particle is a capsule (pink) containing RNA strands (ribonucleic acid, yellow). Around the core is an icosahedral shell of matrix proteins (blue). Over this is a membrane envelope (yellow bilayer) taken from the membrane of the host cell that made this virus particle. Anchored to the shell are viral knobs (yellow) that allow the virus particle to attach to cells. AIDS impairs the immune system and allows often fatal secondary infections. (*Source:* © Russell Kightley/Photo Researchers, Inc.)

In the process of making an inventory of all the molecules within the retrovirus particle, molecular biologists had discovered some tRNAs, one of which, host $tRNA^{Trp}$, appeared to be partially base-paired to the viral RNA. Could this be the primer? If so, it should bind to the reverse transcriptase. To see if it does, Baltimore, James Dahlberg, and colleagues labeled host $tRNA^{Trp}$, or the $tRNA^{Trp}$ from virus particles, with ^{32}P and mixed these labeled tRNAs with AMV reverse transcriptase. Then they subjected these mixtures to gel filtration on Sephadex G-100 (Chapter 5). By itself, $tRNA^{Trp}$ was included in the gel and eluted in a peak centered at about fraction #25. However, both host and virion tRNAs, when mixed with reverse transcriptase eluted with the enzyme in a peak centered at about fraction #20. Thus, this reverse transcriptase binds $tRNA^{Trp}$. Together with the data we have already discussed, the binding data strongly suggest that $tRNA^{Trp}$ serves as the primer for this enzyme. The virus does not encode a tRNA, so the primer must be picked up from the host cell.

The Mechanism of Retrovirus Replication The initial product of reverse transcription in vitro is a short piece of DNA called **strong-stop DNA.** The reason for the strong-stop is obvious when we consider the site on the viral RNA to which the tRNA primer hybridizes (the **primer-binding site, or PBS**). It is only about 150 nt (depending on the retrovirus) from the 5′-end of the viral RNA. This means that the reverse transcriptase will synthesize DNA for just 150 nt or so before reaching the end of the RNA template and stopping. This raises the interesting question: What happens next?

That question is related to another paradox of retrovirus replication, illustrated in Figure 23.22. The provirus is longer than the viral RNA, yet the viral RNA serves as the template for making the provirus. In particular, the LTRs

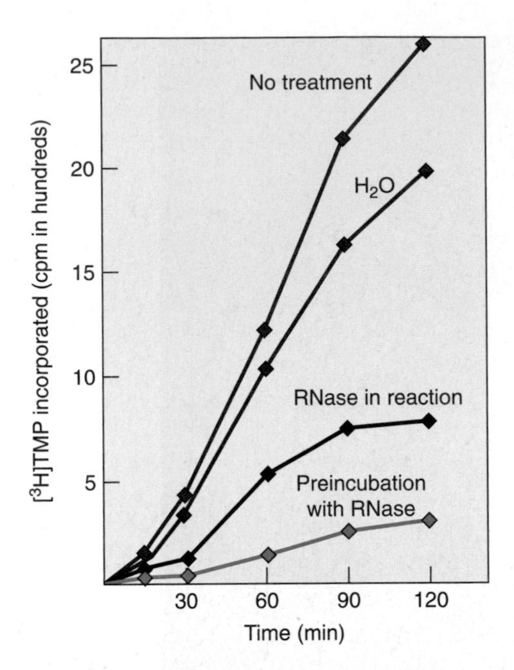

Figure 23.20 Effect of RNase on reverse transcriptase activity. Baltimore incubated R-MLV particles with the four dNTPs, including [^{3}H]dTTP, under various conditions, then acid-precipitated the product and measured the radioactivity of the product by liquid scintillation counting. Treatments: red, no extra treatment; purple, preincubation for 20 min with water; blue, RNase included in the reaction; green, preincubated with RNase. (*Source:* Adapted from Baltimore, D., Viral RNA-dependent DNA polymerase. *Nature* 226:1210, 1970.)

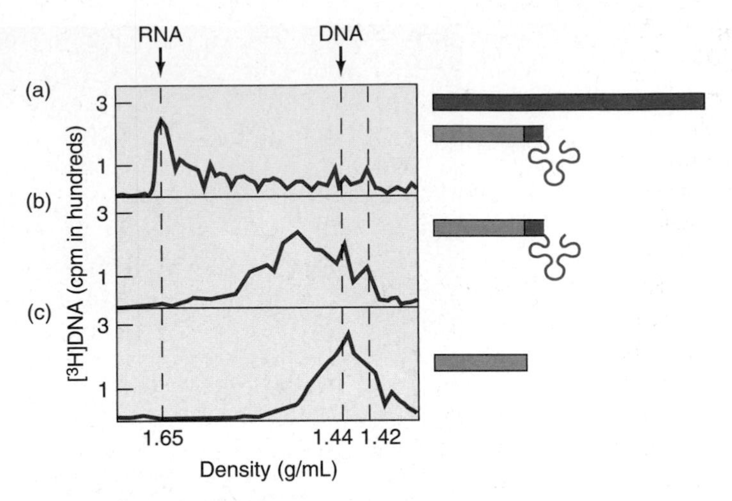

Figure 23.21 Reverse transcripts contain an RNA primer. Baltimore and colleagues labeled reverse transcripts in AMV particles with [^{3}H]dTTP, then subjected them to Cs_2SO_4 gradient ultracentrifugation after the following treatments: **(a)** no treatment; **(b)** heating to denature double-stranded polynucleotides; and **(c)** heating and RNase to remove any primers attached to the reverse transcripts. Interpretive drawings at right provide an explanation for the results: (a) The untreated material has a high density like RNA because the reverse transcript is short and is base-paired to a much longer viral RNA template. (b) The heated material has a density closer to that of DNA because the RNA template has been removed, but it is still denser than pure DNA because of an RNA primer that is covalently attached. (c) The heated and RNase-treated material has the density of a pure DNA because the RNase has removed the RNA primer. The approximate densities of pure RNA and DNA are indicated at top. (*Source*: Adapted from Verma, I.M., N.L. Menth, E. Bromfeld, K.F. Manly, and D. Baltimore, Covalently linked RNA–DNA molecules as initial product of RNA tumor virus DNA polymerase. *Nature New Biology* 233:133, 1971.)

in the viral RNA are incomplete. The left LTR contains a redundant region (**R**) plus a 5′-untranslated region (**U5**), whereas the right LTR contains an R region plus a 3′-untranslated region (**U3**). How can the provirus have complete LTRs on each end while its template is missing a U3 region at its left end and a U5 region at its right end? Harold Varmus proposed an answer based on the important fact that reverse transcriptase has another distinct activity: an RNase activity. The RNase inherent in reverse transcriptase is **RNase H,** which specifically degrades the RNA part of an RNA–DNA hybrid.

Varmus's hypothesis is illustrated in Figure 23.23. First, (a) the reverse transcriptase uses the tRNA to prime synthesis of strong-stop DNA. This appears at first to be the end of the line, but then (b) RNase H recognizes a stretch of RNA hybrid between the strong-stop DNA and the RNA template, and degrades the R and U5 parts of the RNA. The removal of this RNA leaves a tail of DNA (blue) that can hybridize through its R region with the RNA at the other end of the RNA template, or with another RNA template (c). This hybridization to another R region is called the "first jump." In principle, the DNA could jump to the other end of the same RNA, and this could be facilitated by looping the RNA around so the strong-stop DNA does not even need to leave the left end of the RNA to pair with the right end. But the DNA can also jump to another viral RNA, and this seems likely because each virus particle contains two copies of the RNA genome.

After the first jump, the strong-stop DNA is at the right end of the template and can serve as a primer for the reverse transcriptase to copy the rest of the viral RNA (d).

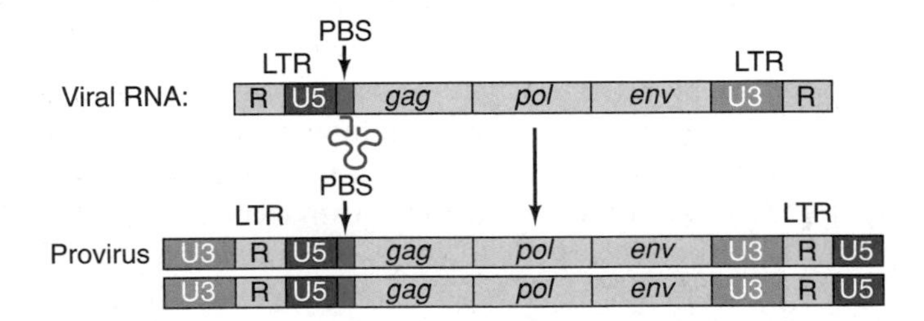

Figure 23.22 Structures of retroviral RNA and provirus DNA. This is a nondefective retroviral RNA that contains all the genes necessary for replication: a coat protein gene (*gag*), a reverse transcriptase gene (*pol*), and an envelope protein gene (*env*). In addition, it contains long terminal repeats (LTRs) at both ends, but these repeats are not identical. The left LTR contains an R and a U5 region, including a primer-binding site (PBS), shown here bound to a tRNA primer, but the right LTR contains a U3 and an R region. On the other hand, the proviral DNA, made using the viral RNA as a template, contains full LTRs (U3, R, and U5) at each end.

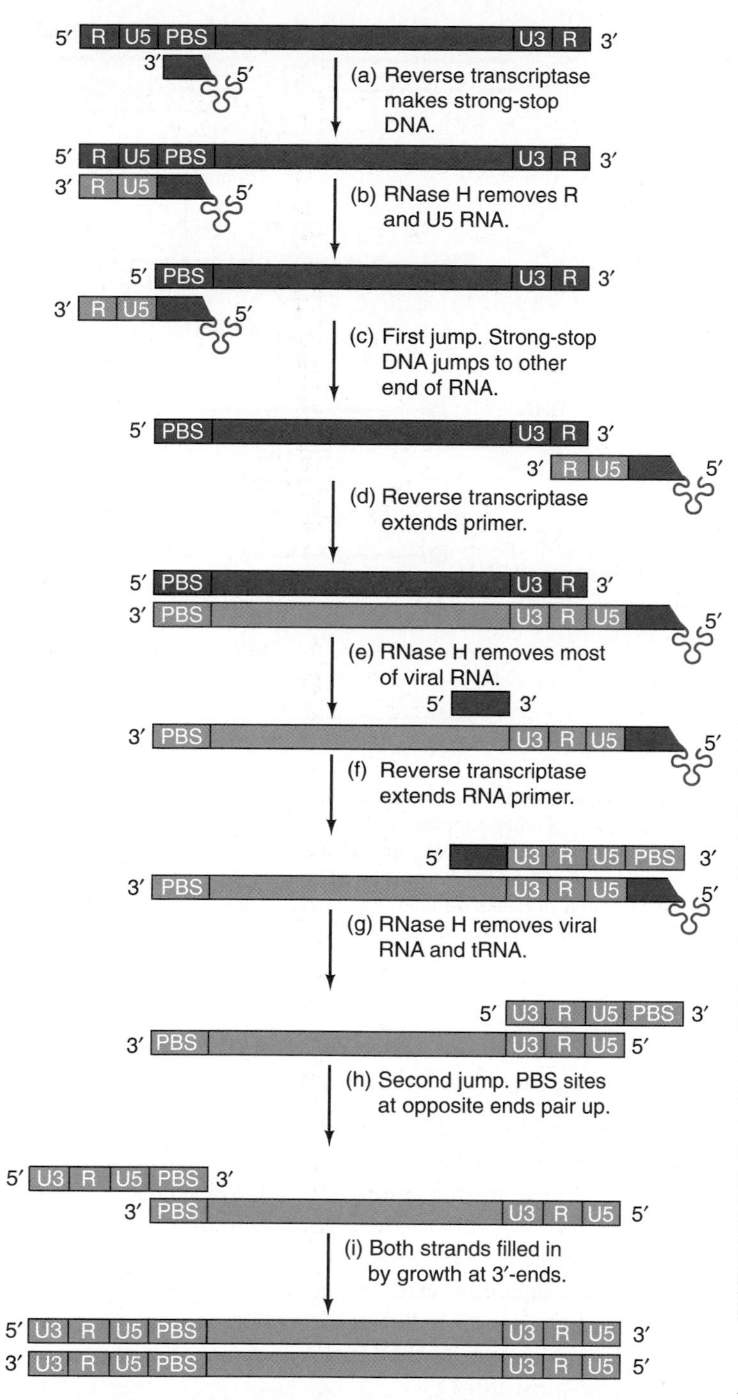

Figure 23.23 A model for the synthesis of the provirus DNA from a retroviral RNA template. RNA is in red and DNA is in blue, throughout. The tRNA primer is represented by a cloverleaf with a 3′-tag that hybridizes to the primer-binding site (PBS) in the viral RNA. The steps are described more fully in the text.

Notice that the first jump has allowed the right LTR to be completed. The U5 and R regions were copied from the left LTR of the viral RNA and the U3 region was copied from the right LTR. In step (e), the RNase H removes most of the viral RNA, but it leaves a small piece of RNA adjacent to the right LTR to serve as a primer for second strand synthesis (f). After the reverse transcriptase extends this primer to the end, including the PBS region, RNase H removes the remaining RNA (g)—the second strand primer and the tRNA—both of which were paired to DNA. This sets up the second jump (h), in which the PBS region on the right pairs with the one on the left. Like the first jump, the second jump can be visualized as a jump to another molecule, or the other end of the same molecule. If the same molecule is involved in the jump, the DNA can loop around to allow the two PBS regions to base-pair. After the second jump, the stage is set for reverse transcriptase, which can use DNA as a template, or another DNA polymerase to complete both strands (i), using the long single-stranded overhangs at each end as templates.

Once the provirus is synthesized, it can be inserted into the host genome by an **integrase.** This enzyme is originally part of a **polyprotein** derived from the *pol* gene, which we have seen also encodes reverse transcriptase and RNase H. The integrase is cut from the polyprotein by a **protease,** which also starts out as part of the same polyprotein. The protease also cuts itself out of the polyprotein. (It is worth noting that some of the most promising drugs for combatting AIDS are protease inhibitors that target the HIV version of this enzyme.) Once the provirus is integrated into the host genome, it is transcribed by host RNA polymerase II to yield viral RNAs.

SUMMARY Retroviruses replicate through an RNA intermediate. When a retrovirus infects a cell, it makes a DNA copy of itself, using a virus-encoded reverse transcriptase to carry out the RNA→DNA reaction, and an RNase H to degrade the RNA parts of RNA–DNA hybrids created during the replication process. A host tRNA serves as the primer for the reverse transcriptase. The finished double-stranded DNA copy of the viral RNA is then inserted into the host genome, where it can be transcribed by host polymerase II.

Retrotransposons

All eukaryotic organisms appear to harbor transposons that replicate through an RNA intermediate and therefore depend on reverse transcriptase. These **retrotransposons** fall into two groups with different modes of replication. The first group includes the retrotransposons with LTRs, which replicate in a manner very similar to retroviruses, except that they do not pass from cell to cell in virus particles. Not surprisingly, these are called **LTR-containing retrotransposons.** The second group includes the retrotransposons that lack LTRs (the **non-LTR retrotransposons**).

LTR-Containing Retrotransposons The first examples of retrotransposons were discovered in the fruit fly (*Drosophila melanogaster*) and yeast (*Saccharomyces cerevisiae*). The prototype *Drosophila* transposon is called ***copia*** because it is present in the genome in copious quantity. In fact, *copia* and related transposons called *copia*-like elements account for about 1% of the total fruit fly genome. Similar transposable elements in yeast are called **Ty**, for "transposon yeast." These transposons have LTRs that are very similar to the LTRs in retroviruses, which suggests that their transposition resembles the replication of a retrovirus. Indeed, several lines of evidence indicate that this is true. Here is a summary of the evidence that the Ty1 elements replicate through an RNA intermediate, just as retroviruses do:

1. Ty1 encodes a reverse transcriptase. The *tyb* gene in Ty codes for a protein with an amino acid sequence closely resembling that of the reverse transcriptases encoded in the *pol* genes of retroviruses. If the Ty1 element really codes for a reverse transcriptase, then this enzyme should appear when Ty1 is induced to transpose; moreover, mutations in *tyb* should block the appearance of reverse transcriptase. Gerald Fink and his colleagues have performed experiments that bear out both of these predictions.
2. Full-length Ty1 RNA and reverse transcriptase activity are both associated with particles that closely resemble retrovirus particles. These particles appear only in yeast cells that are induced for Ty1 transposition.
3. In a clever experiment, Fink and colleagues inserted an intron into a Ty1 element and then analyzed the element again after transposition. The intron was gone! This finding is incompatible with the kind of transposition bacteria employ, in which the transposed DNA looks just like its parent. But it is consistent with the following mechanism (Figure 23.24): The Ty element is first transcribed, intron and all; then the RNA is spliced to remove the intron; and finally, the spliced RNA is reverse transcribed, perhaps within a virus-like particle, and the resulting DNA is inserted back into the yeast genome at a new location.
4. Jef Boeke and colleagues demonstrated that the host $tRNA_i^{Met}$ serves as the primer for Ty1 reverse transcription. First, they mutated 5 of 10 nucleotides in the Ty1 element's PBS that are complementary to the host $tRNA_i^{Met}$. These changes abolished transposition, presumably because they made it impossible for the tRNA primer to bind to its PBS. Then Boeke and coworkers made five compensating mutations in a copy of the host $tRNA_i^{Met}$ gene that restored binding to the mutated PBS. These mutations restored transposition activity to the mutant Ty1 element. As we have seen many times throughout this book, this kind of mutation suppression is powerful evidence for the importance of interaction between two molecules: in this case, interaction between the $tRNA_i^{Met}$ primer and its binding site in the Ty1 element.

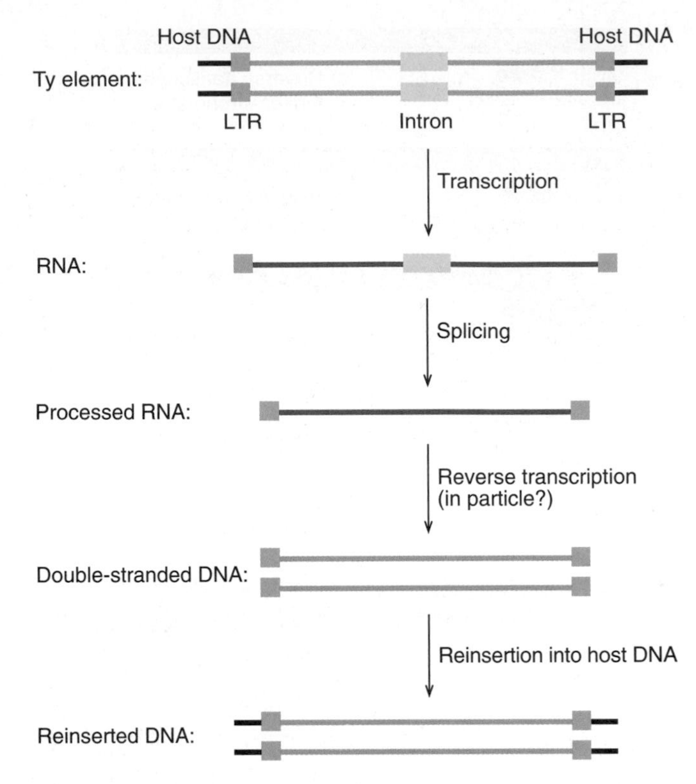

Figure 23.24 Model for transposition of Ty. The Ty element has been experimentally supplied with an intron (yellow). The Ty element is transcribed to yield an RNA copy containing the intron. This transcript is spliced, and then the processed RNA is reverse-transcribed, possibly in a virus-like particle. The resulting double-stranded DNA then reinserts into the yeast genome. *Abbreviation:* LTR = long terminal repeat.

Copia and its relatives share many of the characteristics we have described for Ty, and it is clear that they also transpose in the same way as Ty. Humans also have LTR-containing retrotransposons, but they lack a functional *env* gene. The most prominent examples are the **human endogenous retroviruses (HERVs)**, which make up 1–2% of the genome. So far, no transposition-competent HERVs are known, so the HERVs may be relics of previous retrotransposition.

SUMMARY Several eukaryotic transposons, including Ty of yeast and *copia* of *Drosophila,* apparently transpose by a mechanism similar to that of retrovirus replication. They start with DNA in the host genome, make an RNA copy, then reverse transcribe it—probably within a virus-like particle—to DNA that can insert in a new location. HERVs probably transposed in the same way until most or all of them lost the ability to transpose.

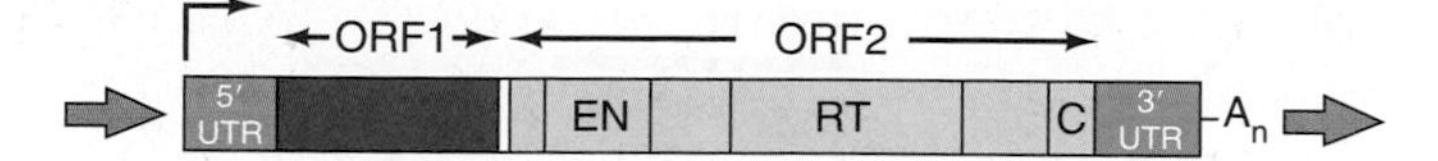

Figure 23.25 Map of the L1 element. The subregions within ORF2 (yellow) are designated EN (endonuclease), RT (reverse transcriptase), and C (cysteine-rich). The purple arrows at each end indicate direct repeats of host DNA, and the A_n on the right indicates the poly(A).

Non-LTR Retrotransposons Retrotransposons that lack LTRs are much more abundant than those with LTRs, at least in mammals. The most abundant of all are the **long interspersed elements (LINEs),** one of which (**L1**) is present in at least 100,000 copies and makes up about 17% of the human genome, although about 97% of the copies of L1 are missing parts of their 5′-ends and the great majority (all but ~60–100 copies) have mutations that prevent their transposition. The prevalence of L1 elements means that this retrotransposon, which has been traditionally classified as "junk DNA," occupies about five times as much of the genome as all the human exons do. Figure 23.25 is a map of an intact L1 element, showing its two ORFs. ORF1 encodes an RNA-binding protein (p40), and ORF2 encodes a protein with two activities: an endonuclease and a reverse transcriptase. L1, like all retrotransposons in this class, is polyadenylated.

We have just seen that the LTR is crucial for replication of most retrotransposons with LTRs, so how do non-LTR retrotransposons replicate? In particular, what do they use for a primer? The answer is that their endonuclease creates a single-stranded break in the target DNA and their reverse transcriptase uses the newly formed DNA 3′-end as a primer. Our best information on this mechanism comes from Thomas Eickbush and colleagues' studies on **R2Bm,** a LINE-like element from the silkworm *Bombyx mori.* This element resembles the mammalian LINEs in that it encodes a reverse transcriptase, but no RNase H, protease, or integrase, and it lacks LTRs. But it differs from the LINEs in that it has a specific target site—in the 28S rRNA gene of the host. This latter property made the insertion mechanism easier to investigate.

Eickbush and colleagues first showed that the single ORF of R2Bm encodes an endonuclease that specifically cleaves the 28S rDNA target site. Next, they purified the endonuclease (and an RNA cofactor that was required for activity) and added it to a supercoiled plasmid containing the target site. If a single strand of the plasmid is cut, the supercoiled plasmid will be converted to a relaxed circle. If both strands are cut, a linear DNA should appear. Figure 23.26a and b show the rapid appearance of relaxed (open) circles, followed by the slower conversion of open circles to linear DNA. Thus, the R2Bm endonuclease rapidly cleaves one of the DNA strands at the target site, then much more slowly cleaves the other strand. This cleavage is specific: The nuclease cannot cut even one strand of a plasmid that lacks the target site.

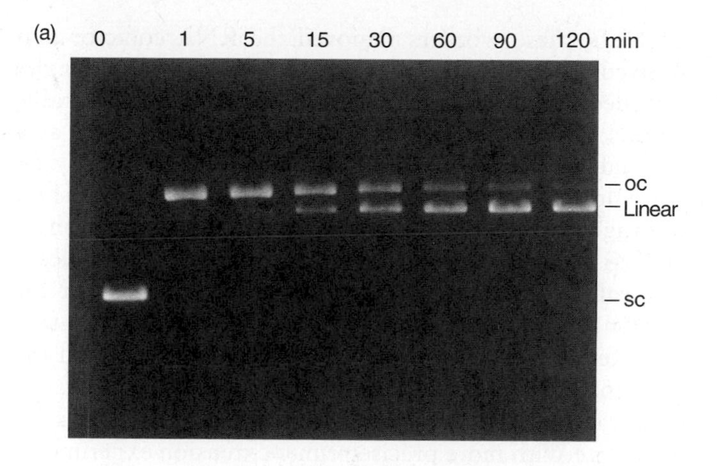

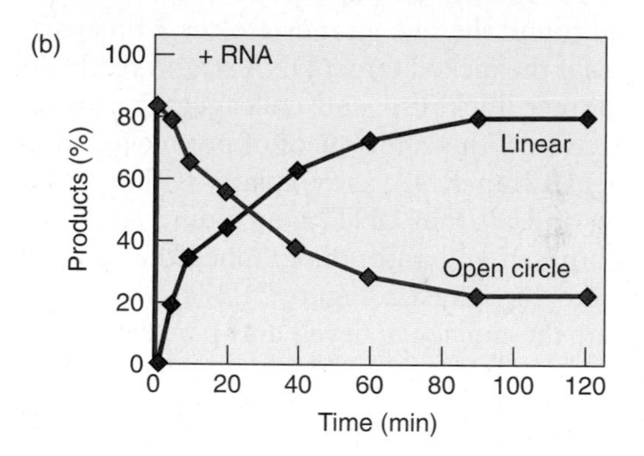

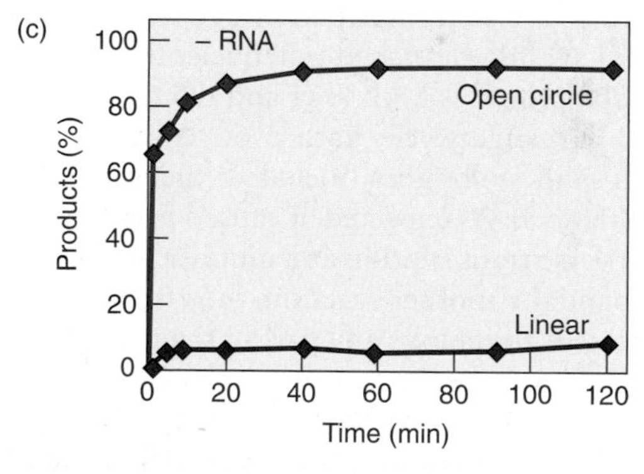

Figure 23.26 DNA nicking and cleavage activity of the R2Bm endonuclease. Eickbush and colleagues mixed a supercoiled plasmid bearing the target site for the R2Bm retrotransposon with the purified R2Bm endonuclease, with or without its RNA cofactor, then electrophoresed the plasmid to see if it had been nicked (relaxed to an open circular form) or cut in both strands to yield a linear DNA. **(a)** Electrophoretic gel stained with ethidium bromide. The positions of the supercoiled plasmid (sc), the open circular plasmid (oc), and the linear plasmid (linear) are indicated at right. **(b)** Graphical representation of the results from panel (a). **(c)** Results from a similar experiment in which the RNA cofactor was omitted. (*Source:* Adapted from Luan, D.D., M.H. Korman, J.L. Jakubczak, and T.H. Eickbush, Reverse transcription of R2Bm RNA is primed by a nick at the chromosomal target sige: a mechanism for non-LTR retrotransposition. *Cell* 72 (Feb 1993) f. 2, p. 597. Reprinted by permission of Elsevier.)

Next, these workers removed the RNA cofactor and showed that the protein by itself still caused rapid single-stranded nicking of the target site, but barely detectable cutting of the other strand (Figure 23.26c). They also showed that the linear DNA could be recircularized by T4 DNA ligase, which requires a 5′-phosphate group. Thus, cleavage by the R2Bm endonuclease leaves a 5′-phosphate and a 3′-hydroxyl group. Next, they used the endonuclease to create single-stranded nicks and showed by primer extension analysis that the transcribed strand is the one that is nicked. (The nick in the transcribed strand stopped the DNA polymerase in the primer extension experiment, but primer extension on the other strand proceeded unimpeded by nicks.) With more precise primer extension experiments on DNA cut in both strands, they located the cut sites exactly and found the two strands are cut 2 bp apart.

To see if the nicked target DNA strand really does serve as the primer, Eickbush and colleagues performed an in vitro reaction with a short piece of pre-nicked target DNA as primer, R2Bm RNA as template, R2Bm reverse transcriptase, and all four dNTPs, including [^{32}P]dATP. They electrophoresed and autoradiographed the products to see if they were the right size. Figure 23.27a shows what should happen at the molecular level, and panel b shows the results. When a nonspecific RNA was added as template, no product was made (lane 1), but when the R2Bm RNA was added, a strong band at 1.9 kb appeared. Is this what we expect? It is hard to know because we do not know exactly how far the reverse transcriptase traveled and we are dealing with a slightly branched polynucleotide, but it is close because the primer is 1 kb long and the template is 802 nt long. To investigate the nature of the product further, Eickbush and colleagues included dideoxy-CTP in the reaction (lane 3). As expected, it caused premature termination of reverse transcription at a number of sites, leading to a fuzzy band. In another reaction, they treated the product with RNase A to remove any part of the template not base-paired to the reverse transcription product before electrophoresis. Lane 4 shows that this sharpened the product to a 1.8-kb band, suggesting that about 100 nt had been removed from the 5′-end of the RNA template, so the reverse transcriptase had apparently not completed its task in the majority of cases. These workers also treated the product with RNase H prior to electrophoresis (lane 5) and obtained a diffuse band of about 1.5 kb. This procedure should remove the RNA template because it is in a hybrid with the product. The fact that the band is still longer than 1 kb indicates that a strand of DNA has been extended. Lane 6 is another negative control in which a nonspecific DNA was used instead of the target DNA.

Similar experiments with a target DNA that extended farther to the left (with the target site in the middle) showed a predominance of large, Y-shaped products (as predicted in Figure 23.27), suggesting that reverse transcription occurred before second-strand cleavage. If second-strand cleavage had occurred first, the products would have been linear and smaller. To confirm that the target DNA was serving as the primer, Eickbush and coworkers performed PCR with primers that hybridized to the target DNA and to the reverse transcript, and obtained PCR products of the expected size and sequence.

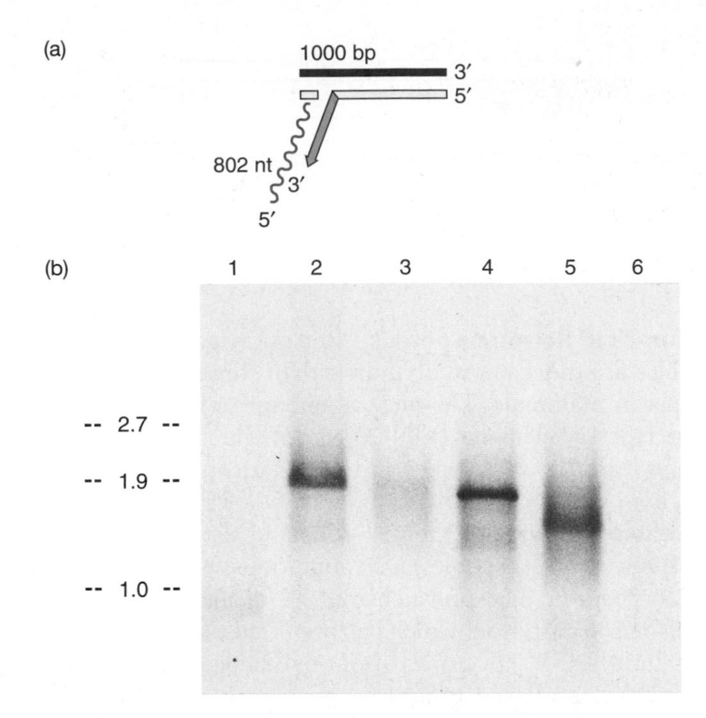

Figure 23.27 Evidence for target priming of reverse transcription of R2Bm. **(a)** Model of the product we expect if the R2Bm endonuclease makes a nick near the left end of a 1-kb target DNA and uses the new 3′-end to prime reverse transcription of an 802-nt transposon RNA. The reverse transcript (blue) is covalently attached to the primer (yellow). The rest of the lower DNA strand is also rendered in yellow at left. The opposite DNA strand is black. **(b)** Experimental results. Eickbush and colleagues started with a 1-kb target DNA with the target site close to the left end. They added R2Bm RNA and the ORF2 product and dNTPs, including [^{32}P]dATP to allow labeled reverse transcripts to be formed. Then they electrophoresed the products and autoradiographed them. Lane 1, a nonspecific RNA was used instead of R2Bm RNA; lanes 2–6, R2Bm RNA was used; lane 3, dideoxy-CTP was included in the reverse transcription reaction; lane 4, the product was treated with RNase A before electrophoresis; lane 5, the product was treated with RNase H before electrophoresis; lane 6, a nonspecific target DNA was used. (*Source:* Luan, D.D., M.H. Korman, J.L. Jakubczak, and T.H. Eickbush, Reverse transcription of R2Bm RNA is primed by a nick at the chromosomal target site: a mechanism for non-LTR retrotransposition *Cell* 72 (Feb 1993) f. 4, p. 599. Reprinted by permission of Elsevier Science.)

Based on these and other data, H.H. Kazazian and John Moran proposed the model of L1 transposition presented in Figure 23.28. First, the transposon is transcribed and the transcript is processed. The processed mRNA leaves the nucleus to be translated in the cytoplasm. It associates with its two products, p40 and the ORF2 product, and reenters the nucleus. There, the endonuclease activity of the ORF2 product nicks the target DNA. For L1, the target can be

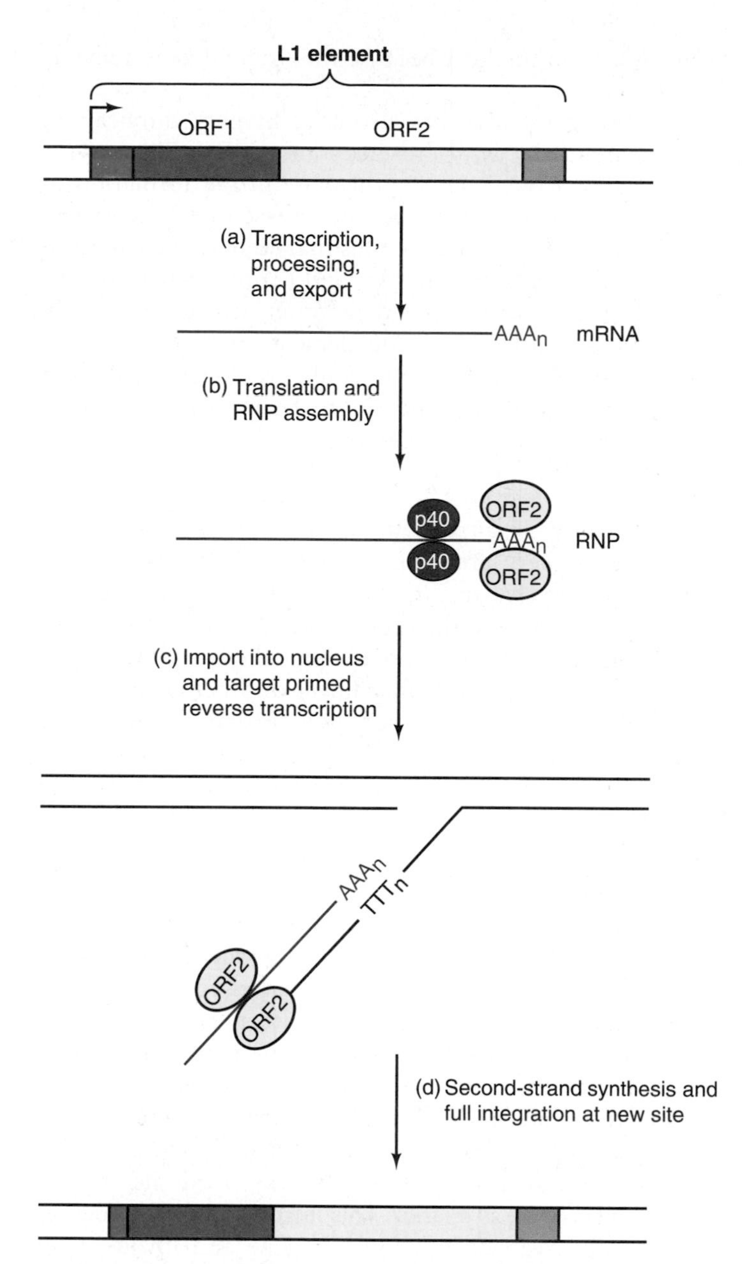

Figure 23.28 A model for L1 transposition. **(a)** The L1 element is transcribed, processed, and exported from the nucleus. **(b)** The mRNA is translated to yield the ORF1 product (p40), and the ORF2 product, with endonuclease and reverse transcriptase activities. These proteins associate with the mRNA to form an RNP **(c)** The ribonucleoprotein reenters the nucleus. The endonuclease nicks the target DNA (anywhere in the genome), and the reverse transcriptase uses the new DNA 3′-end to prime synthesis of the reverse transcript. **(d)** In a series of unspecified steps, the second L1 strand is made and the element, usually truncated at its 5′-end, is ligated into the target DNA.

any region of the DNA. Then the reverse transcriptase activity of the ORF2 product uses the target DNA 3′-end created by the endonuclease as a primer to copy the L1 RNA. Thus, this mechanism is called **target-primed retrotransposition.** Finally, in steps that are still poorly understood, the second strand of L1 is made, the second strand of the target is cleaved, and the L1 element is ligated into its new home.

At the beginning of this section, we learned that L1 elements comprise about 17% of the human genome. And, as we will soon see, these elements can carry pieces of genomic DNA with them as they transpose. Thus, one can estimate that, directly or indirectly, L1 elements have sculpted about 30% of the human genome. Furthermore, L1-like elements have been found in both plants and animals. Thus, these elements are ancient—at least 600 million years old. And, because identical DNA sequences can lose all resemblance to each other after about 200 million years of evolution, the true contribution of L1 elements to the human genome may actually be about 50%.

You would suspect that anything as prevalent as L1 is in the human genome must have some negative consequences, and indeed a number of L1-mediated mutations have been discovered that have led to human disease. In particular, copies of L1 have been found: in the blood clotting factor VIII gene, causing hemophilia; in the *DMD* gene, causing Duchenne muscular dystrophy; and in the *APC* gene, helping to cause adenomatous polyposis coli, a kind of colon cancer. In this last case, the patient's cancer cells had the L1 element in their *APC* gene, but the normal cells did not. Thus, this transposition had occurred during the patient's lifetime as a somatic mutation.

What is more surprising is that the L1 elements may actually have beneficial consequences as well. For example, significant homology occurs between the reverse transcriptase of L1 and human telomerase, suggesting that L1 may have been the origin of the enzyme that maintains the ends of our chromosomes (although the reverse may also have been true). But the most plausible beneficial aspect of L1 is that it may facilitate exon shuffling, the exchange of exons among genes. This happens because the polyadenylation signal of L1 is weak, so the polyadenylation machinery frequently bypasses it in favor of a polyadenylation site downstream in the host part of the transcript. RNAs polyadenylated in that way will include a piece of human RNA attached to the L1 RNA, and this human RNA will be incorporated as a reverse transcript wherever the L1 element goes next. This is bound to have deleterious consequences sometimes, but it also creates new genes out of parts of old genes, and that can give rise to proteins with new and useful characteristics.

Why are the polyadenylation signals of L1 elements weak? Moran offers the following explanation: If the polyadenylation signals were strong, insertion of these elements into the introns of human genes would cause premature polyadenylation of transcripts, so all the exons downstream would be lost. That would probably inactivate the gene and might well lead to the death of the host. And, unlike retroviruses, which can move from one individual to another, the L1 elements live and die with their hosts. On the other hand, weak polyadenylation signals allow these elements to insert into introns of human genes without disrupting a very high percentage of the transcripts of these genes. Thus, because the amount of DNA devoted to introns is much higher than that devoted to exons, the L1 elements have a large area of the human genome to colonize relatively safely.

SUMMARY LINEs and LINE-like elements are retrotransposons that lack LTRs. These elements encode an endonuclease that nicks the target DNA. Then the element takes advantage of the new DNA 3′-end to prime reverse transcription of element RNA. After second-strand synthesis, the element has become replicated at its target site. A new round of transposition begins when the LINE is transcribed. Because the LINE polyadenylation signal is weak, transcription of a LINE can include one or more downstream exons of host DNA.

Nonautonomous Retrotransposons Members of another class of non-LTR retrotransposons (**nonautonomous retrotransposons**) encode no proteins, so they are not autonomous like the transposition-competent LINEs. Instead, they depend on other elements, probably the LINEs due to their prevalence, to supply the proteins, including the reverse transcriptase they need to transpose. The best studied of these nonautonomous retrotransposons are the **Alu elements,** so-called because they contain the sequence AGCT that is recognized by the restriction enzyme *Alu*I. These are about 300 bp long and are present in up to a million copies in the human genome. Thus, they have been even more successful than the LINEs. One reason for this success may be that the transcripts of the Alu elements contain a domain that resembles the 7SL RNA that is normally part of the signal recognition particle that helps attach certain ribosomes to the endoplasmic reticulum. Two signal recognition particle proteins bind tightly to Alu element RNA and may carry it to the ribosomes, where the LINE RNA is being translated. This may put the Alu element RNA in a position to help itself to the proteins it needs to be reverse transcribed and inserted at a new site. Because of their small size, Alu elements and similar elements are called **short interspersed elements (SINEs).**

The LINEs have probably also played a role in shaping the human genome by facilitating the creation of **processed pseudogenes.** Ordinary **pseudogenes** are DNA sequences that resemble normal genes, but for one reason or another cannot function. Sometimes they have internal translation stop signals; sometimes they have inactive or missing splicing signals; sometimes they have inactive promoters; usually a combination of problems prevents their expression. They apparently arise by gene duplication and subsequently accumulate mutations. This process has no deleterious effect on the host because the original gene remains functional.

Processed pseudogenes also arise by gene duplication, but apparently by way of reverse transcription. We strongly suspect that RNA is an intermediate in the formation of processed pseudogenes because: (1) these pseudogenes frequently have short poly(dA) tails that seem to have derived from poly(A) tails on mRNAs; and (2) processed pseudogenes lack the introns that their progenitor genes usually have. As in the case of the Alu elements, which are not derived from mRNAs, the LINEs could provide the molecular machinery that allows mRNAs to be reverse transcribed and inserted into the host genome.

SUMMARY Nonautonomous retrotransposons include the very abundant Alu elements in humans and similar elements in other vertebrates. They cannot transpose by themselves because they do not encode any proteins. Instead they take advantage of the retrotransposition machinery of other elements, such as LINEs. Processed pseudogenes probably arose in the same way: mRNAs were reverse-transcribed by LINE machinery and then inserted into the genome.

Group II introns In Chapter 14 we learned that group II introns, which inhabit bacterial, mitochondrial, and chloroplast genomes, are self-splicing introns that form a lariat intermediate. In 1998, Marlene Belfort and colleagues discovered that a group II intron in a particular gene could insert into an intronless version of the same gene somewhere else in the genome. This process, called **retrohoming,** appears to occur by the mechanism outlined in Figure 23.29. The gene bearing the intron is first transcribed, then the intron is spliced out as a lariat. This intron can then recognize an intronless version of the same gene and invade it by reverse-splicing. Reverse transcription creates a cDNA copy of the intron, and second-strand synthesis replaces the RNA intron with a second strand of DNA.

In 1991, Phillip Sharp proposed that group II introns could be the ancestors of modern spliceosomal introns, in part because of their very similar mechanisms of splicing. In 2002, Belfort and colleagues showed how this could have happened. They detected true retrotransposition, not just retrohoming, of a bacterial group II intron. Thus, the intron moved to a variety of new sites, not just to an intronless copy of the intron's home gene.

To detect retrotransposition, Belfort and colleagues built a plasmid with a modified version of the group II *Lactococcus lactis* L1. LtrB intron, containing a kanamycin resistance gene in reverse orientation, interrupted by a self-splicing group I intron. In order for kanamycin resistance to be expressed, this group II intron would first have to be

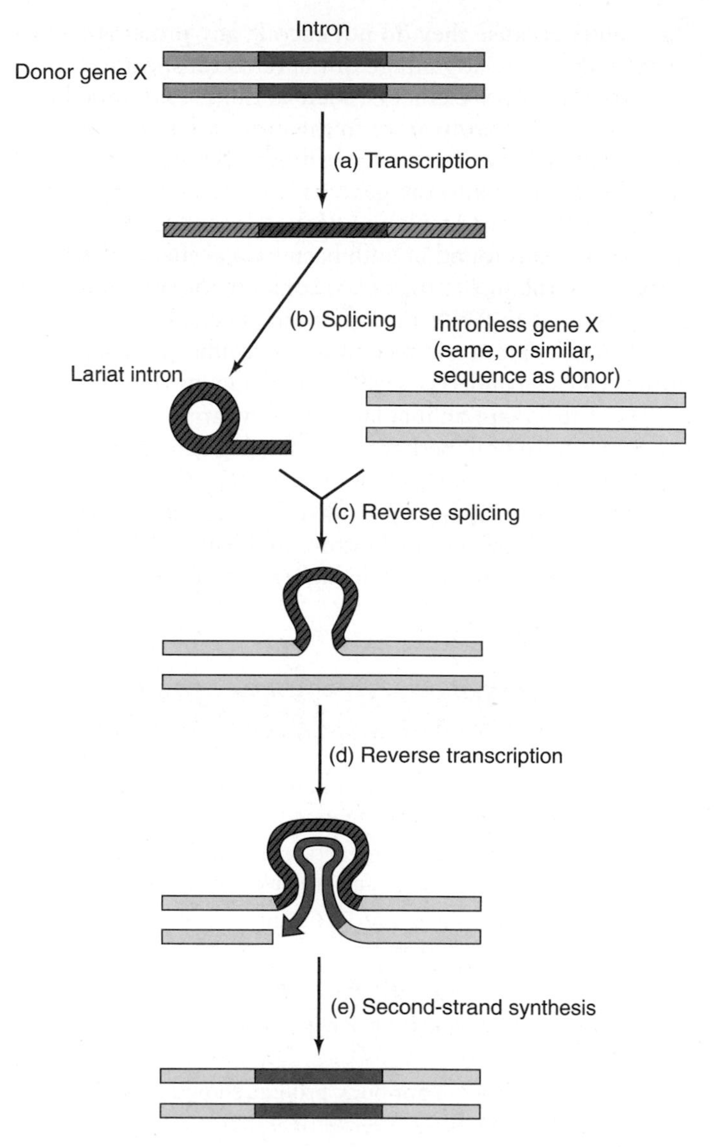

Figure 23.29 Retrohoming. (a) The donor gene X (blue) bearing a group II intron (red) is transcribed to yield an RNA (RNAs are shaded throughout). **(b)** The transcript is spliced, yielding a lariat-shaped intron. **(c)** The intron reverse-splices itself into another copy of gene X that has the same or similar sequence as the first except that it lacks the intron. **(d)** The intron-encoded reverse transcriptase makes a DNA copy of the intron, using a nick in the bottom DNA strand as primer. The arrowhead marks the 3′-end of the growing reverse transcript. **(e)** The second strand (DNA version) of the intron is made, replacing the RNA intron in the top strand. This completes the retrohoming process.

transcribed, so the interrupting group I intron could be removed. Then, the transcript would have to be reverse-transcribed to yield a DNA that could insert into the host DNA, where it could be transcribed in the forward, rather than the reverse direction. As long as the group II intron remained in RNA form, it could not code for kanamycin resistance because its resistance gene had been transcribed in the reverse direction, yielding an antisense RNA.

When Belfort and colleagues selected for kanamycin-resistant cells, they found that transposition was relatively rare, but did occur at a measurable rate. An interesting feature of this transposition was that most of it occurred into the DNA replication lagging strand. This finding suggested that transposition happened during replication and used the short DNA fragments created in the lagging strand (Chapter 20) as primers for the kind of target-primed reverse transcription we saw in the L1 transposition scheme in Figure 23.28. Notice that no homology between the transposon and the target DNA is required for this mechanism, as nicks in replicating lagging strands occur everywhere in the genome.

Once a group II intron has retrotransposed, it retains its ability to splice itself out, so the target gene should usually continue to function. Thus, the proliferation of group II introns may have occurred readily and with relative safety in the precursors to modern eukaryotes. Ultimately, eukaryotes appear to have developed spliceosomes to make the splicing process more efficient.

SUMMARY Group II introns can retrohome to intronless copies of the same gene by insertion of an RNA intron into the gene, followed by reverse transcription and second-strand synthesis. Group II introns can also undergo retrotransposition by insertion of an RNA intron into an unrelated gene by target-primed reverse transcription, using lagging strand DNA fragments as primers. This kind of retrotransposition of group II introns may have provided the ancestors of modern-day eukaryotic spliceosomal introns and may account for their widespread appearance in higher eukaryotes.

SUMMARY

Transposable elements, or transposons, are pieces of DNA that can move from one site to another. Some transposable elements replicate, leaving one copy at the original location and placing one copy at a new site; others transpose without replication, leaving the original location altogether. Bacterial transposons include the following types: (1) insertion sequences such as IS1 that contain only the genes necessary for transposition, flanked by inverted terminal repeats; (2) transposons such as Tn3 that are like insertion sequences but contain at least one extra gene, usually a gene that confers antibiotic resistance.

Eukaryotic transposons use a wide variety of replication strategies. The DNA transposons, such as *Ds* and *Ac* of maize and the P elements of *Drosophila* behave like the DNA transposons, such as Tn3, of bacteria.

The immunoglobulin genes of mammals rearrange using a mechanism that resembles transposition. Vertebrate immune systems create enormous diversity in the kinds of immunoglobulins they can make. The primary source of this diversity is the assembly of genes from two or three component parts, each selected from a heterogeneous pool of parts. This assembly of gene segments is known as V(D)J recombination. The recombination signal sequences (RSSs) in V(D)J recombination consist of a heptamer and a nonamer separated by either 12-bp or 23-bp spacers. Recombination occurs only between a 12 signal and a 23 signal, which ensures that only one of each kind of coding region is incorporated into the rearranged gene. RAG1 and RAG2 are the principal players in human V(D)J recombination. They introduce single-strand nicks into DNA adjacent to either a 12 signal or a 23 signal. This leads to a transesterification in which the newly created 3′-hydroxyl group attacks the opposite strand, breaking it, and forming a hairpin at the end of the coding segment. The hairpins then break and join with each other in an imprecise way, allowing joining of coding regions with loss of bases or gain of extra bases.

The retrotransposons come in two different types. The LTR-containing retrotransposons replicate like retroviruses, which replicate through an RNA intermediate as follows: When a retrovirus infects a cell, it makes a DNA copy of itself, using a virus-encoded reverse transcriptase to carry out the RNA→DNA reaction, and an RNase H to degrade the RNA parts of RNA–DNA hybrids created during the replication process. A host tRNA serves as the primer for the reverse transcriptase. The finished double-stranded DNA copy of the viral RNA is then inserted into the host genome, where it can be transcribed by host polymerase II. The retrotransposons Ty of yeast and *copia* of *Drosophila* replicate in much the same way. They start with DNA in the host genome, make an RNA copy, then reverse-transcribe it—probably within a virus-like particle—to DNA that can insert in a new location.

The other class of eukaryotic retrotransposons are the non-LTR retrotransposons, and they use different methods of priming reverse transcription. For example, LINEs and LINE-like elements encode an endonuclease that nicks the target DNA. Then the element takes advantage of the new DNA 3′-end to prime reverse transcription of element RNA. After second-strand synthesis, the element has become replicated at its target site. A new round of transposition begins when the LINE is transcribed. Because the LINE polyadenylation signal is weak, transcription of a LINE frequently includes one or more downstream exons of host DNA and this can transport host exons to new locations in the genome.

Nonautonomous, non-LTR retrotransposons include the very abundant Alu elements in humans and similar elements in other vertebrates. They cannot transpose by themselves because they do not encode any proteins. Instead, they take advantage of the retrotransposition machinery of other elements, such as LINEs. Processed pseudogenes probably arose in the same way: mRNAs were probably reverse-transcribed by LINE machinery and then inserted into the genome.

Group II introns represent another class of non-LTR retrotransposons found in both bacteria and eukaryotes. They can retrohome to intronless copies of the same gene by insertion of an RNA intron into the gene, followed by reverse transcription and second-strand synthesis. Group II introns can also undergo retrotransposition by insertion of an RNA intron into an unrelated gene by target-primed reverse transcription, perhaps using lagging strand DNA fragments as primers. This kind of retrotransposition of group II introns may have provided the ancestors of modern-day eukaryotic spliceosomal introns and may account for their widespread appearance in higher eukaryotes.

REVIEW QUESTIONS

1. Describe and give the results of an experiment that shows that bacterial transposons contain inverted terminal repeats.
2. Compare and contrast the genetic maps of the bacterial transposons IS1 and Tn3, and the eukaryotic transposon *Ac*.
3. Diagram the mechanism of Tn3 transposition, first in simplified form, then in detail.
4. Diagram a mechanism for nonreplicative transposition.
5. Explain how transposition can give rise to speckled maize kernels.
6. Draw a sketch of an antibody protein, showing the light and heavy chains.
7. Explain how thousands of immunoglobulin genes can give rise to many millions of antibody proteins.
8. Diagram the rearrangement of immunoglobulin light- and heavy-chain genes that occurs during B-lymphocyte maturation.
9. Explain how the signals for V(D)J joining ensure that one and only one of each of the parts of an immunoglobulin gene will be included in the mature, rearranged gene.
10. Diagram a reporter plasmid designed to test the importance of the heptamer, nonamer, and spacer in a recombination signal sequence. Explain how this plasmid detects recombination.
11. Present a model for cleavage and rejoining of DNA strands at immunoglobulin gene recombination signal sequences. How does this mechanism contribute to antibody diversity?
12. Describe and give the results of an experiment that shows that cleavage at an immunoglobulin recombination signal sequence leads to formation of a hairpin in vitro.
13. Present evidence for a reverse transcriptase activity in retrovirus particles and the effects of RNase on this activity.

14. Describe and show the results of an experiment that demonstrates that strong-stop reverse transcripts in retroviruses are base-paired to the RNA genome and covalently attached to an RNA primer.
15. Illustrate the difference between the structures of the LTRs in genomic retroviral RNAs and retroviral proviruses.
16. Diagram the conversion of a retrovirus RNA to a provirus. Show how this explains the difference in the previous question.
17. Compare and contrast the mechanisms of retrovirus replication and retrotransposon transposition.
18. Summarize the evidence that retrotransposons transpose via an RNA intermediate.
19. Describe and show the results of an experiment that demonstrates that the endonuclease of a LINE-like element can specifically nick one strand of the element's target DNA.
20. Describe and show the results of an experiment that demonstrates that a LINE-like element can use a nicked strand of its target DNA as a primer for reverse transcription of the element.
21. Present a model for retrotransposition of a LINE-like element.

ANALYTICAL QUESTIONS

1. A certain transposon's transposase creates staggered cuts in the host DNA five base pairs apart. What consequence does this have for the host DNA surrounding the inserted transposon? Draw a diagram to explain how the staggered cuts affect the host DNA.
2. You are interested in measuring the rate of transfer of a hypothetical transposon, *Stealth,* from one plasmid, carrying two antibiotic resistance genes of its own, to another plasmid, which carries the gene for chloramphenicol resistance. (*Stealth* carries an ampicillin resistance gene.) Describe an experiment you would perform to assay for this transposition.
3. Identify the end product of abortive transposition carried out by Tn3 transposons with mutations in the following genes.
   a. Transposase
   b. Resolvase
4. Transposon TnT in plasmid A transposes to plasmid B. How many copies of TnT are in the cointegrate? Where are they with respect to the two plasmids in the cointegrate?
5. If the transposable element *Ds* of maize transposed by the same mechanism as Tn3, would we see the speckled kernels with the same high frequency? Why, or why not?
6. Assume you have two cell-free transposition systems that have all the enzymes necessary for transposition of Tn3 and Ty, respectively. What effect would the following inhibitors have on these two systems, and why?
   a. Inhibitors of double-stranded DNA replication
   b. Inhibitors of transcription
   c. Inhibitors of reverse transcription
   d. Inhibitors of translation
7. You have identified a new transposon you call Rover. You want to determine whether Rover transposes by a retrotransposon mechanism or by a standard replication transposition mechanism such as that used by Tn3. Describe an experiment you would use to answer this question, and tell what the results would be in each case.
8. You are a molecular biologist interested in learning more about the fascinating process of V(D)J recombination. Assuming that you are capable of generating all of the following possible variants, explain what effect (from a molecular process standpoint as well as a physiological and/or immunological standpoint) you would expect to observe if the following were created in your laboratory:
   a. the removal of all of the D gene segments from the section of the genome encoding the heavy chain of antibodies.
   b. the removal of all of the D gene segments from the section of the genome encoding the beta chain of T-cell receptors.
   c. the genetic alteration of the RSS flanking the D gene segments from a 12 signal to a 23 signal.
   d. the elimination of expression of the RAG gene products.

SUGGESTED READINGS

General References and Reviews

Baltimore, D. 1985. Retroviruses and retrotransposons: The role of reverse transcription in shaping the eukaryotic genome. *Cell* 40:481–82.

Cohen, S.N. and J.A. Shapiro. 1980. Transposable genetic elements. *Scientific American* 242 (February):40–49.

Craig, N.L. 1996 V(D)J recombination and transposition: Closer than expected. *Science* 271:1512.

Doerling, H.-P. and P. Starlinger. 1984. Barbara McClintock's controlling elements: Now at the DNA level. *Cell* 39:253–59.

Eickbush, T.H. 2000. Introns gain ground. *Nature* 404:940–41.

Engels, W.R. 1983. The P family of transposable elements in *Drosophila. Annual Review of Genetics* 17:315–44.

Federoff, N.V. 1984. Transposable genetic elements in maize. *Scientific American* 250(June):84–99.

Grindley, N.G.F. and A.E. Leschziner. 1995. DNA transposition: From a black box to a color monitor. *Cell* 83:1063–66.

Kazazian, H.H., Jr. and J.V. Moran. 1998. The impact of L1 retrotransposons on the human genome. *Nature Genetics* 19:19–24.

Lambowitz, A.M. and S. Zimmerly. 2004. Mobile group II introns. *Annual Review of Genetics* 38:1–35.

Levin, K.L. 1997. It's prime time for reverse transcriptase. *Cell* 88:5–8.

Lewis, S.M. 1994. The mechanism of V(D)J joining: Lessons from molecular, immunological, and comparative analyses. *Advances in Immunology* 56:27–50.

Tonegawa, S. 1983. Somatic generation of antibody diversity. *Nature* 302:575–81.
Voytas, D.F. 1996. Retroelements in genome organization. *Science* 274:737–38.

Research Articles

Baltimore, D. 1970. Viral RNA-dependent DNA polymerase. *Nature* 226:1209–11.
Boland, S. and N. Kleckner, 1996. The three chemical steps of Tn10/IS10 transposition involve repeated utilization of a single active site. *Cell* 84:223–33.
Chapman, K.B., A.S. Byström, and J.D. Boeke. 1992. Initiator methionine tRNA is essential for Ty1 transcription. *Proceedings of the National Academy of Sciences USA* 89:3236–40.
Cousineau, B., S. Lawrence, D. Smith, and M. Belfort. 2002. Retrotransposition of a bacterial group II intron. *Nature* 404:1018–21.
Davies, D.R., I.Y. Goryshin, W.S. Reznikoff, and I. Rayment. 2000. Three-dimensional structure of the Tn5 synaptic complex transposition intermediate. *Science* 289:77–85.
Difilippantonio, M.J., C.J. McMahan, Q.M. Eastman, E. Spanopoulou, and D.G. Schatz. 1996. RAG1 mediates signal sequence recognition and recruitment of RAG2 in V(D)J recombination. *Cell* 87:253–62.
Garfinkel, D.J., J.F. Boeke, and G.R. Fink. 1985. Ty element transposition: Reverse transcription and virus-like particles. *Cell* 42:507–17.
Hesse, J.E., M.R. Lieber, K. Mizuuchi, and M. Gellert. 1989. V(D)J recombination: A functional definition of the joining signals. *Genes and Development* 3:1053–61.
Luan, D.D., M.H. Korman, J.L. Jakubczak, and T.H. Eickbush. 1993. Reverse transcription of R2Bm RNA is primed by a nick at the chromosomal target site: A mechanism for non-LTR retrotransposition. *Cell* 72:595–605.
Oettinger, M.A., D.G. Schatz, C. Gorka, and D. Baltimore. 1990. RAG-1 and RAG-2, adjacent genes that synergistically activate V(D)J recombination. *Science* 248:1517–22.
Panet, A., W.A. Haseltine, D. Baltimore, G. Peters, F. Harada, and J. E. Dahlberg. 1975. Specific binding of tryptophan transfer RNA to avian myeloblastosis virus RNA-dependent DNA polymerase (reverse transcriptase). *Proceedings of the National Academy of Sciences USA* 72:2535–39.
Temin, H.M. and Mizutani, S. 1970. RNA-dependent DNA polymerase in virions of Rous sarcoma virus. *Nature* 226:1211–13.
Wessler, S.R. 1988. Phenotypic diversity mediated by the maize transposable elements *Ac* and *Spm*. *Science* 242:399–405.

CHAPTER 24

Introduction to Genomics: DNA Sequencing on a Genomic Scale

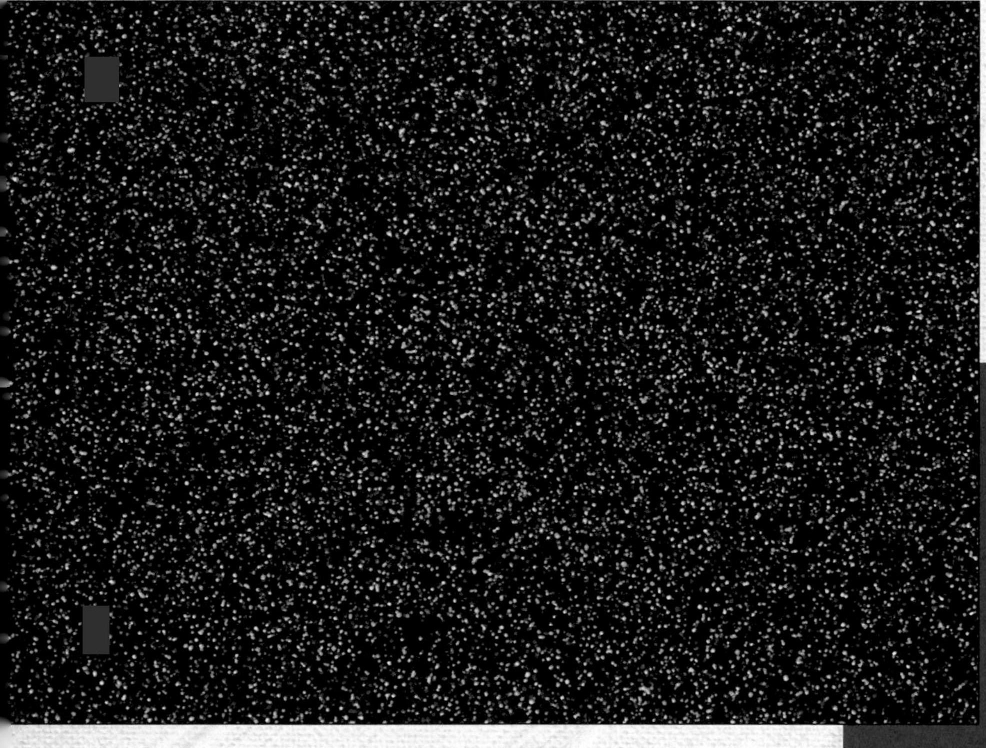

A composite image of human DNA clusters from a next-generation DNA sequencing run. This artificially generated image is composed by adding four images, each of which detects a different base, and assigning a color to depict each base: blue for G; green for T; red for C, and yellow for A.

Throughout most of this book, we have been concerned with the activities of genes—taken one gene at a time. But, with the advent of rapid and relatively inexpensive sequencing methods, molecular biologists have been able to obtain the base sequences of entire genomes, and a new subdiscipline has been born: genomics, the study of the structure and function of whole genomes.

We will begin this chapter with a discussion of positional cloning, a technique for identifying a gene responsible for a given trait, and see how much easier this process is when the sequence of the organism's genome is known. Then we will examine the techniques scientists use to sequence DNA on a massive scale. We will also discuss some of the lessons we have learned from the sequences of genomes, especially the evolutionary insights to be gained by comparing the genomic sequences of different organisms.

24.1 Positional Cloning: An Introduction to Genomics

Before we examine the techniques of genomic research, let us consider one of the important uses of genomic information: **positional cloning**, which is one method for the discovery of the genes involved in genetic traits. In humans, this frequently involves the identification of genes that govern genetic diseases. We will begin by considering an example of positional cloning that was done before the genomic era: finding the gene whose malfunction causes Huntington disease in humans. We will see that much of the effort went into narrowing down the region in which to look for the faulty gene. One reason for all this effort was to avoid having to sequence a huge chunk of DNA. Nowadays, that is not a problem because the sequencing has already been done. Nevertheless, this example serves as a good introduction to genomics for several reasons: It illustrates the principle of positional cloning, which is still a major use of genomic information; it shows how difficult positional cloning was in the absence of genomic information; and it is a heroic story that still deserves to be told.

Classical Tools of Positional Cloning

Geneticists seeking the genes responsible for human genetic disorders frequently face a problem: They do not know the identity of the defective protein, so they are looking for a gene without knowing its function. Thus, they have to identify the gene by finding its position on the human genetic map, and this process therefore has come to be called positional cloning.

The strategy of positional cloning begins with the study of a family or families afflicted with the disorder, with the goal of finding one or more markers that are tightly linked to the "disease gene," that is, the gene which, when mutated, causes the disease. Frequently, these markers are not genes, but stretches of DNA whose pattern of cleavage by restriction enzymes or other physical attributes vary from one individual to another.

Because the position of the marker is known, the disease gene can be pinned down to a relatively small region of the genome. However, that "relatively small" region usually contains about a million base pairs, so the job is not over. The next step is to search through the million or so base pairs to find a gene that is the likely culprit. Several tools have traditionally been used in the search, and we will describe two here. These are: (1) finding exons with exon traps; and (2) locating the CpG islands that tend to be associated with genes. We will see how these tools have been used as we discuss our example in the next section of this chapter. First, let us examine a favorite method to map a gene to a fairly small region of the genome.

Restriction Fragment Length Polymorphisms In the late twentieth century, we knew the locations of relatively few human genes, so the likelihood of finding one of these close to a new gene we were trying to map was small. Another approach, which does not depend on finding linkage with a known gene, is to establish linkage with an "anonymous" stretch of DNA that may not even contain any genes. We can recognize such a piece of DNA by its pattern of cleavage by restriction enzymes.

Because each person differs genetically from every other, the sequences of their DNAs will differ a little bit, as will the pattern of cutting by restriction enzymes. Consider the restriction enzyme *Hin*dIII, which recognizes the sequence AAGCTT. One individual may have three such sites separated by 4 and 2 kb, respectively, in a given region of a chromosome (Figure 24.1). Another individual may lack the middle site but have the other two, which are 6 kb apart. This means that if we cut the first person's DNA with *Hin*dIII, we will produce two fragments, 2 kb and 4 kb long, respectively. The second person's DNA will yield a 6-kb fragment instead. In other words, we are dealing with a **restriction fragment length polymorphism (RFLP)**. Polymorphism means that a genetic locus has different forms, or alleles (Chapter 1), so this clumsy term simply means that cutting the DNA from any two individuals with a restriction enzyme may yield fragments of different lengths. The abbreviated term, RFLP, is usually pronounced "rifflip."

How do we go about looking for a RFLP? Clearly, we cannot analyze the whole human genome at once. It contains approximately a million cleavage sites for a typical restriction enzyme, so each time we cut the whole genome with such an enzyme, we release about a million fragments. No one would relish sorting through that morass for subtle differences between individuals.

Fortunately, there is an easier way. With a Southern blot (Chapter 5) one can highlight small portions of the total genome with various probes, so any differences are easy to see. However, there is a catch. Because each labeled probe hybridizes only to a small fraction of the total human DNA, the chances are very poor that any given one will reveal a RFLP linked to the gene of interest. We may have to screen many thousands of probes before we find the right one. As laborious as it is, this procedure at least provides a starting point, and it has been a key to finding the genes responsible for several genetic diseases.

Exon Traps Once a gene has been pinned down to a region stretching over hundreds of kilobases, how does one sort out the genes from the other DNA? If that DNA region has not yet been sequenced, one can sequence it and look for **open reading frames (ORFs)**. An ORF is a sequence of bases that, if translated in one reading frame, contains no stop codons for a relatively long distance. But

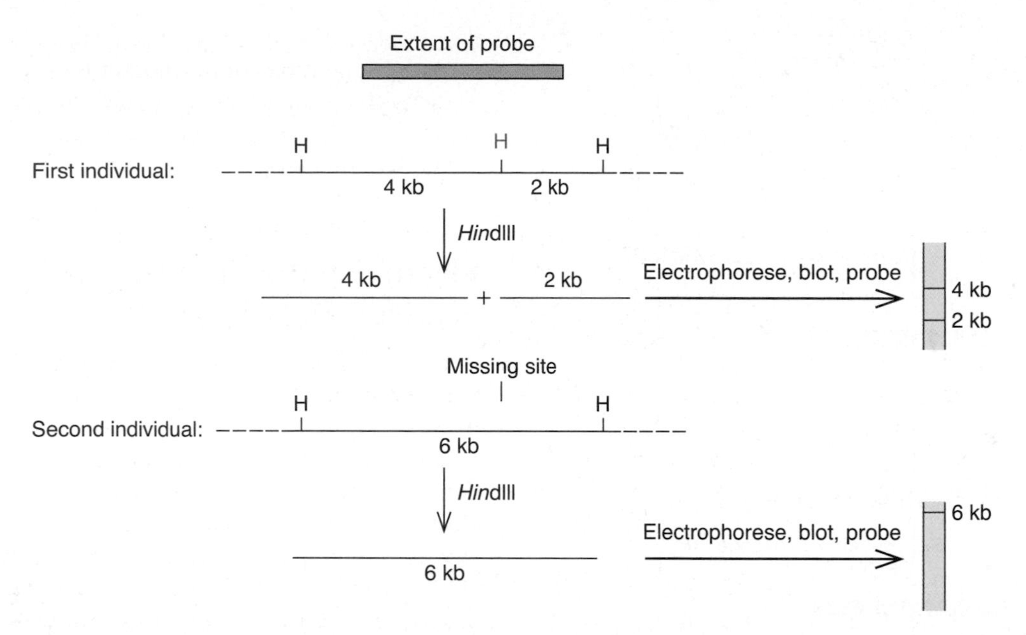

Figure 24.1 Detecting a RFLP. Two individuals are polymorphic with respect to a *Hin*dIII restriction site (red).The first individual contains the site, so cutting the DNA with *Hin*dIII yields two fragments, 2 and 4 kb long, that can hybridize with the probe, whose extent is shown at top. The second individual lacks this site, so cutting that DNA with *Hin*dIII yields only one fragment, 6 kb long, which can hybridize with the probe. The results from electrophoresis of these fragments, followed by blotting, hybridization to the radioactive probe, and autoradiography, are shown at right. The fragments at either end, represented by dashed lines, do not show up because they cannot hybridize to the probe.

searching for ORFs is very laborious. Several more efficient methods are available, including a procedure invented by Alan Buckler called exon amplification or **exon trapping.** Figure 24.2 shows how an exon trap works. We begin with a plasmid vector such as pSPL1, which Buckler designed for this purpose. This vector contains a chimeric gene under the control of the SV40 early promoter. The gene was derived from the rabbit β-globin gene by removing its second intron and substituting a foreign intron from the human immunodeficiency virus (HIV), with its own 5′- and 3′-splice sites. We insert human genomic DNA fragments into a restriction site within the intron of this plasmid, then place the recombinant vector into monkey cells (COS-7 cells) that can transcribe the gene from the SV40 promoter. Now if any of the genomic DNA fragments we put into the intron are complete exons, with their own 5′- and 3′-splice sites, this exon will become part of the processed transcript in the COS cells. We purify the RNA made by the COS cells, reverse transcribe it to make cDNA, then subject this cDNA to amplification by PCR, using primers that are specific for the regions surrounding the insert. Thus, any new exon inserted between the primer-binding sites will be amplified. Finally, we clone the PCR products, which should represent only exons. Any other piece of DNA inserted into the intron will not have splicing signals; thus, after being transcribed, it will be spliced out along with the surrounding intron and will be lost.

CpG Islands Another gene-finding technique takes advantage of the fact that the control regions of active human genes tend to be associated with unmethylated CpG sequences, whereas the CpGs in inactive regions are almost always methylated. Moreover, many methylated CpG sites have been lost over evolutionary time because of the following phenomenon, known as **CpG suppression:** Methyldeoxycytidine (methylC) in a methylCpG site can be deaminated spontaneously to methylU, which is the same as T. Thus, once a methylC is deaminated, it becomes a T. If this change is not immediately recognized and repaired, the T will take an A partner in the next round of DNA replication, and the mutation will be permanent. By contrast, in an ordinary, unmethylated CpG sequence, deamination yields a U, which is subject to immediate recognition and removal by a uracil-*N*-glycosylase (Chapter 20) and replacement by an ordinary C. So unmethylated CpG sequences have been retained in the genome.

Furthermore, the restriction enzyme *Hpa*II cuts at the sequence CCGG, but only if the second C is unmethylated. In other words, it will cut active genes that have unmethylated CpGs within CCGG sites, but it will leave inactive sequences (with methylated CCGGs) alone. Thus, geneticists can scan large regions of DNA for "islands" of sites that could be cut with *Hpa*II in a "sea" of other DNA sequences that could not be cut. Such a site is called a **CpG island,** or an **HTF island** because it yields *Hpa*II tiny fragments.

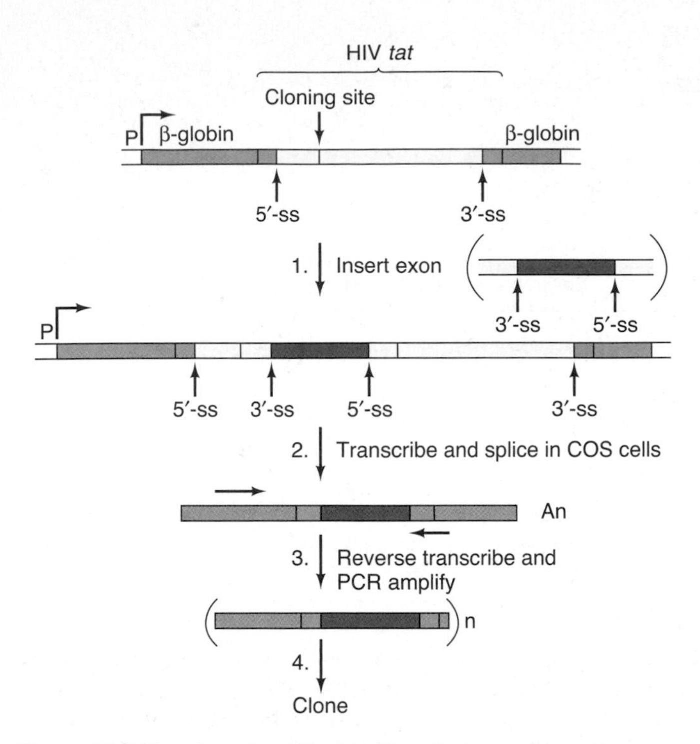

Figure 24.2 Exon trapping. Begin with a cloning vector, such as pSPL1, shown here in slightly simplified form. This vector has an SV40 promoter (P), which drives expression of a hybrid gene containing the rabbit β-globin gene (orange), interrupted by part of the HIV *tat* gene, which includes two exon fragments (blue) surrounding an intron (yellow). The exon–intron borders contain 5′- and 3′-splice sites (ss). The *tat* intron contains a cloning site, into which random DNA fragments can be inserted. In step 1, an exon (red) has been inserted, flanked by parts of its own introns, and its own 5′- and 3′-splice sites. In step 2, insert this construct into COS cells, where it can be transcribed and then the transcript can be spliced. Note that the foreign exon (red) has been retained in the spliced transcript, because it had its own splice sites. Finally (steps 3 and 4), subject the transcripts to reverse transcription and PCR amplification, with primers indicated by the arrows. This gives many copies of a DNA fragment containing the foreign exon, which can now be cloned and examined. Note that a non-exon will not have splice sites and will therefore be spliced out of the transcript along with the intron. It will not survive to be amplified in step 3, so one does not waste time studying it.

SUMMARY Positional cloning begins with mapping studies (Chapter 1) to pin down the location of the gene of interest to a reasonably small region of DNA. Mapping depends on a set of landmarks to which the position of a gene can be related. Sometimes such landmarks are genes, but more often they are RFLPs—sites at which the lengths of restriction fragments generated by a given restriction enzyme vary from one individual to another. Several methods are available for identifying the genes in a large region of unsequenced DNA. One of these is the exon trap, which uses a special vector to help clone exons only. Another is to use methylation-sensitive restriction enzymes to search for CpG islands—DNA regions containing unmethylated CpG sequences.

Identifying the Gene Mutated in a Human Disease

Let us conclude this section with a classic example of positional cloning: pinpointing the gene for Huntington disease.

Huntington disease (HD) is a progressive nerve disorder. It begins almost imperceptibly with small tics and clumsiness. Over a period of years, these symptoms intensify and are accompanied by emotional disturbances. Nancy Wexler, an HD researcher, describes the advanced disease as follows: "The entire body is encompassed by adventitious movements. The trunk is writhing and the face is twisting. The full-fledged Huntington patient is very dramatic to look at." Finally, after 10–20 years, the patient dies.

Huntington disease is controlled by a single dominant gene. Therefore, a child of an HD patient has a 50:50 chance of being affected. People who have the disease could avoid passing it on by not having children, except that the first symptoms usually do not appear until after the child-bearing years.

Because they did not know the nature of the product of the HD gene (*HD*), geneticists could not look for the gene directly. The next best approach was to look for a gene or other marker that is tightly linked to *HD*. Michael Conneally and his colleagues spent more than a decade trying to find such a linked gene, but with no success.

In their attempt to find a genetic marker linked to *HD*, Wexler, Conneally, and James Gusella turned next to RFLPs. They were fortunate to have a very large family to study. Living around Lake Maracaibo in Venezuela is a family whose members have suffered from HD since the early nineteenth century. The first member of the family to be so afflicted was a woman whose father, presumably a European, carried the defective gene. So the pedigree of this family can be traced through seven generations, and the number of individuals is unusually large: It is not uncommon for a family to have 15–18 children.

Gusella and colleagues knew they might have to test hundreds of probes to detect a RFLP linked to *HD*, but they were amazingly lucky. Among the first dozen probes they tried, they found one (called G8) that detected a RFLP that is very tightly linked to *HD* in the Venezuelan family. Figure 24.3 shows the locations of *Hin*dIII sites in the stretch of DNA that hybridizes to the probe. We can see seven sites in all, but only five of these are found in all family members. The other two, marked with asterisks and

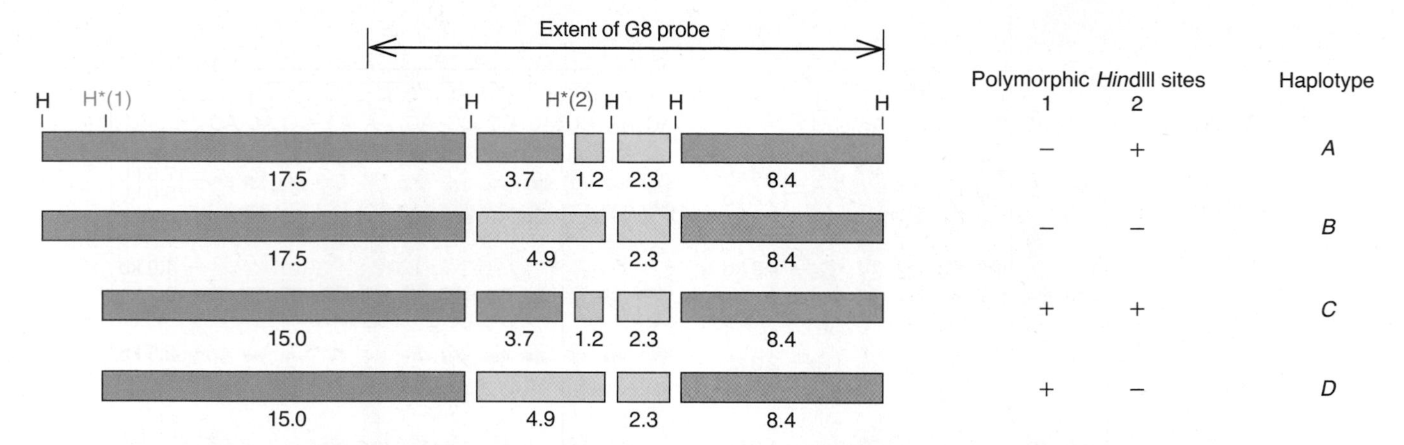

Figure 24.3 The RFLP associated with the Huntington disease gene. The *Hin*dIII sites in the region that hybridizes to the G8 probe are shown. The families studied show polymorphisms in two of these sites, marked with an asterisk and numbered 1 (blue) and 2 (red). Presence of site 1 results in a 15-kb fragment plus a 2.5-kb fragment that is not detected because it lies outside the region that hybridizes to the G8 probe. Absence of this site results in a 17.5-kb fragment. Presence of site 2 results in two fragments of 3.7 and 1.2 kb. Absence of this site results in a 4.9-kb fragment. Four haplotypes (*A–D*) result from the four combinations of presence or absence of these two sites. These are listed at right, beside a list of polymorphic *Hin*dIII sites and a diagram of the *Hin*dIII restriction fragments detected by the G8 probe for each haplotype. For example, haplotype *A* lacks site 1 but has site 2. As a result, *Hin*dIII fragments of 17.5, 3.7, and 1.2 are produced. The 2.3- and 8.4-kb fragments are also detected by the probe, but we ignore them because they are common to all four haplotypes.

numbered 1 and 2, may or may not be present. These latter two sites are therefore polymorphic, or variable.

Let us see how the presence or absence of these two restriction sites gives rise to a RFLP. If site 1 is absent, a single fragment 17.5 kb long will be produced. However, if site 1 is present, the 17.5-kb fragment will be cut into two pieces having lengths of 15 kb and 2.5 kb, respectively. Only the 15-kb band will show up on the autoradiograph because the 2.5-kb fragment lies outside the region that hybridizes to the G8 probe. If site 2 is absent, a 4.9-kb fragment will be produced. On the other hand, if site 2 is present, the 4.9-kb fragment will be subdivided into a 3.7-kb fragment and a 1.2-kb fragment.

There are four possible **haplotypes** (clusters of alleles on a single chromosome) with respect to these two polymorphic *Hin*dIII sites, and they have been labeled *A–D:*

Haplotype	Site 1	Site 2	Fragments Observed
A	Absent	Present	17.5; 3.7; 1.2
B	Absent	Absent	17.5; 4.9
C	Present	Present	15.0; 3.7; 1.2
D	Present	Absent	15.0; 4.9

The term *haplotype* is a contraction of *haploid genotype,* which emphasizes that each member of the family will inherit two haplotypes, one from each parent. For example, an individual might inherit the *A* haplotype from one parent and the *D* haplotype from the other. This person would have the *AD* genotype. Sometimes different genotypes (pairs of haplotypes) can be indistinguishable. For example, a person with the *AD* genotype will have the same RFLP pattern as one with the *BC* genotype because all five fragments will be present in both cases. However, the true genotype can be deduced by examining the parents' genotypes. Figure 24.4 shows autoradiographs of Southern blots of two families, using the radioactive G8 probe. The 17.5- and 15-kb fragments migrate very close together, so they are difficult to distinguish when both are present, as in the *AC* genotype; nevertheless, the *AA* genotype with only the 17.5-kb fragment is relatively easy to distinguish from the *CC* genotype with only the 15-kb fragment. The *B* haplotype in the first family is obvious because of the presence of the 4.9-kb fragment.

Which haplotype is associated with the disease in the Venezuelan family? Figure 24.5 demonstrates that it is *C*. Nearly all individuals with this haplotype have the disease. Those who do not have the disease yet will almost certainly develop it later. Equally telling is the fact that no individual lacking the *C* haplotype has the disease. Thus, this is a very accurate way of predicting whether a member of this family is carrying the Huntington disease gene. A similar study of an American family showed that, in this family, the *A* haplotype was linked with the disease. Therefore, each family varies in the haplotype associated with the disease, but within a family, the linkage between the RFLP site and *HD* is so close that recombination between these sites is very rare. Thus we see that a RFLP can be used as a genetic marker for mapping, just as if it were a gene.

Finding linkage between *HD* and the DNA region that hybridizes to the G8 probe also allowed Gusella and colleagues to locate *HD* to chromosome 4. They did this by making mouse–human hybrid cell lines, each containing only a few human chromosomes. They then prepared DNA from each of these lines and hybridized it to the

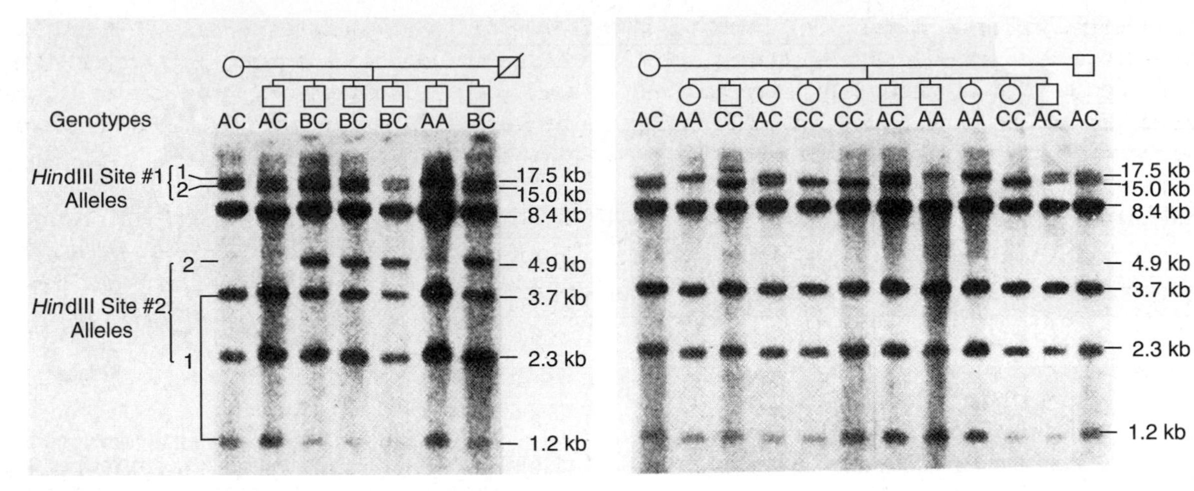

Figure 24.4 Southern blots of *Hin*dIII fragments from members of two families, hybridized to the G8 probe. The bands in the autoradiographs represent DNA fragments whose sizes are listed at right. The genotypes of all the children and three of the parents are shown at top. The fourth parent was deceased, so his genotype could not be determined. (*Source:* Gusella, J.F., N.S. Wexler, P.M. Conneally, S.L. Naylor, M.A. Anderson, R.E. Tauzi, et al., A polymorphic DNA marker genetically linked to Huntington's disease. *Nature* 306:236. Copyright © 1983 Macmillan Magazines Limited.)

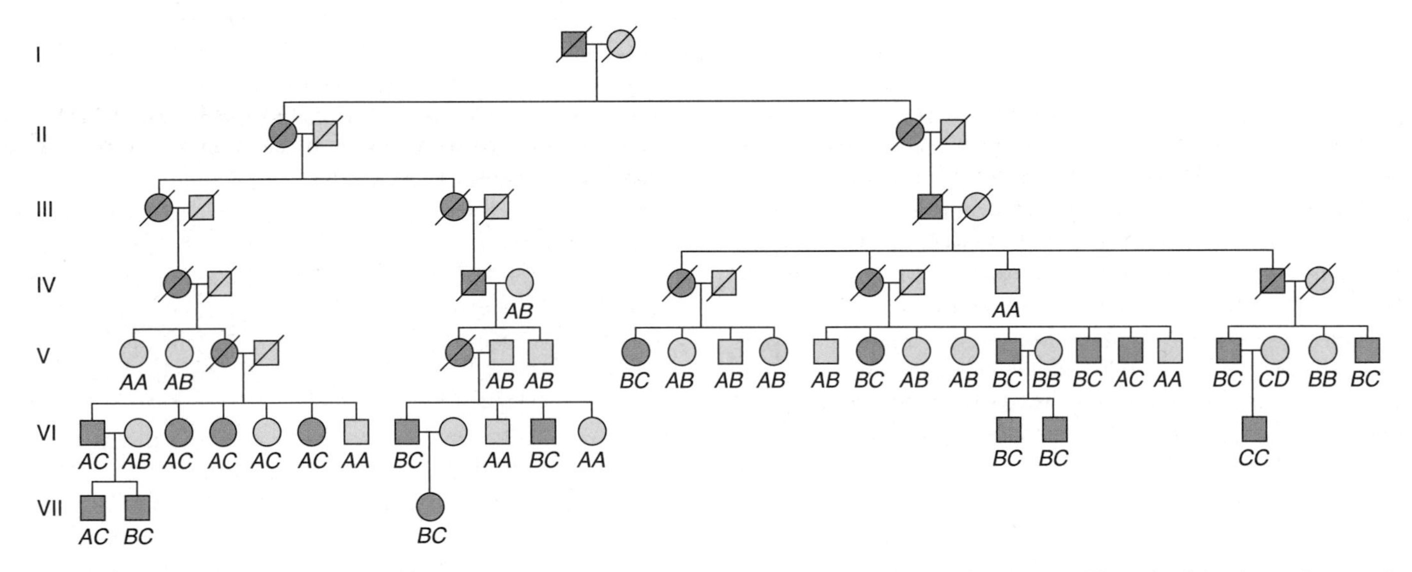

Figure 24.5 Pedigree of the large Venezuelan family with Huntington disease. Family members with confirmed disease are represented by purple symbols. Notice that most of the individuals with the *C* haplotype already have the disease, and that no sufferers of the disease lack the *C* haplotype. Thus, the *C* haplotype is strongly associated with the disease, and the corresponding RFLP is tightly linked to the Huntington disease gene.

radioactive G8 probe. Only the cell lines having chromosome 4 hybridized; the presence or absence of all other chromosomes did not matter. Therefore, human chromosome 4 carries *HD*.

At this point, the *HD* mapping team's luck ran out. One long detour arose from a mapping study that indicated the gene lay far out at the end of chromosome 4. This made the search much more difficult because the tip of the chromosome is a genetic wasteland, full of repetitive sequences, and apparently devoid of genes. Finally, after wandering for years in what he called a genetic "junkyard," Gusella and his group turned their attention to a more promising region. Some mapping work suggested that *HD* resided, not at the tip of the chromosome, but in a 2.2-Mb region several megabases removed from the tip. Unless you know the DNA sequence, over 2 Mb is a tremendous amount of DNA to sift through to find a gene, so Gusella decided to focus on a 500-kb region that was highly conserved among about one-third of *HD* patients, who seemed to have a common ancestor.

On average, a 500-kb region of the human genome contains about five genes. To find them, Gusella and colleagues used an exon-trapping strategy and identified a handful of exon clones. They then used these exons to probe a cDNA library to identify the DNA copies of mRNAs transcribed from the target region. One of the clones, called IT15, for "interesting transcript number 15," hybridized to cDNAs that identified a large (10,366 nt) transcript that codes for a large (3144 amino acid) protein called huntingtin. The presumed protein product did not resemble any known proteins, so that did not provide any evidence that this is indeed *HD*. However, the gene had an intriguing repeat of 23 copies of the triplet CAG (one copy is actually CAA), encoding a stretch of 23 glutamines.

Is this really *HD*? Gusella's team's comparison of the gene in affected and unaffected individuals in 75 HD families demonstrated that it is. In all unaffected individuals, the number of CAG repeats ranged from 11 to 34, and 98% of these unaffected people had 24 or fewer CAG repeats. In all affected individuals, the number of CAG repeats had expanded to at least 42, up to a high of about 100. Thus, we can predict whether an individual will be affected by the disease by looking at the number of CAG repeats in this gene.

Furthermore, the severity, or age of onset of the disease correlates at least roughly with the number of CAG repeats. People with a number of repeats at the low end of the affected range (now known to be 36–40) generally survive well into adulthood before symptoms appear, whereas people with a number of repeats at the high end of the range tend to show symptoms in childhood. In one extreme example, an individual with the highest number of repeats detected (about 100) started showing disease symptoms at the extraordinarily early age of 2.

Finally, two people were affected, even though their parents were not. In both cases, the affected individuals had expanded CAG repeats, whereas their parents did not. New mutations (expanded CAG repeats), although a rare occurrence in HD, apparently caused both these cases of disease.

Another way of demonstrating that this gene is really *HD* would be to deliberately mutate it and show that the mutation has neurological effects. Obviously, one cannot perform such an experiment in humans, but it would be feasible in mice, if the gene corresponding to *HD* is known. Fortunately, *HD* is conserved in many species, including the mouse, where the gene is known as *Hdh*. In 1995, a team of geneticists led by Michael Hayden created knockout mice (Chapter 5) with a targeted disruption in exon 5 of *Hdh*. Mice that are homozygous for this mutation die in utero. Heterozygotes are viable, but they show loss of neurons with corresponding lowering of intelligence. This reinforces the notion that *Hdh,* and therefore *HD,* plays an important role in the brain—exactly what we would expect of the gene that causes *HD*.

How can we put this new knowledge to work? One obvious way is to perform accurate genetic screening to detect people who will be affected by the disease. In fact, by counting the CAG repeats, we may even be able to predict the age of onset of the disease. However, that kind of information is a mixed blessing, as it can be psychologically devastating. What we really need, of course, is a cure, but that may be a long way off.

The Advantage of Genomic Data The positional cloning study we have just examined took years, and much of that time was spent sequencing DNA in the suspected regions and trying to determine which gene in the sequence was the most likely culprit. With the human genome now finished, that job has become much easier. Just how much easier is indicated by Neal Copeland, a mouse geneticist who has been doing positional cloning in mice for years. He says, "It took us 15 years to get 10 possible cancer genes before we had the sequence. And it took us a few months to get 130 genes once we had the sequence." He was talking about the mouse sequence, of course, but the same principle applies to humans, and mouse positional-cloning studies very often identify genes that cause similar problems in humans. So one of the biggest anticipated payoffs of genomics research will be the acceleration of discovery of disease genes in humans. You should not conclude from this discussion that positional cloning is obsolete. It will be important as long as we are curious about finding genes responsible for traits in any organism. Sequenced genomes simply make positional cloning much easier.

SUMMARY Using RFLPs, geneticists mapped the Huntington disease gene (*HD*) to a region near the end of chromosome 4. Then they used an exon trap to identify the gene itself. The mutation that causes the disease is an expansion of a CAG repeat from the normal range of 11–34 copies, to the abnormal range of at least 38 copies. The extra CAG repeats cause extra glutamines to be inserted into huntingtin.

24.2 Techniques in Genomic Sequencing

The first genome to be sequenced, as you might expect, was a very simple one: The small DNA genome of an *E. coli* phage called ϕX174. Frederick Sanger, the inventor of the dideoxy chain termination method of DNA sequencing, obtained the sequence of this 5375-nt genome in 1977.

What kind of information can we glean from this sequence? First, we can locate exactly the coding regions for all the genes. This tells us the spatial relationships among genes and the distances between them to the exact nucleotide. How do we recognize a coding region? It contains an ORF that is long enough to code for one of the phage proteins. Furthermore, the ORF must start with an ATG (or occasionally a GTG) triplet, corresponding to an AUG (or GUG) translation initiation codon, and end with the DNA equivalent of a stop codon (UAG, UAA, or UGA). In other words, an ORF in a bacterium or phage is the same as a gene's coding region.

The base sequence of the phage DNA also tells us the amino acid sequences of all the phage proteins. All we have to do is use the genetic code to translate the DNA base sequence of each open reading frame into the corresponding amino acid sequence. This may sound like a laborious process, but a personal computer can do it in a split second.

Sanger's analysis of the open reading frames of the ϕX174 DNA revealed something unexpected and fascinating: Some of the phage genes overlap. Figure 24.6a shows that the coding region for gene B lies within gene A and the coding region for gene E lies within gene D. Furthermore, genes D and J overlap by 1 bp. How can two genes occupy the same space and code for different proteins? The answer is that the two genes are translated in different reading frames (Figure 24.6b). Because entirely different sets of codons will be encountered in these two frames, the two protein products will also be quite different.

This was certainly an interesting finding, and it raised the question of how common this phenomenon would be. So far, major overlaps seem to be confined almost exclusively to viruses, which is not surprising because these simple infectious agents have small genomes in which the premium is on efficient use of the genetic material. Moreover, viruses have prodigious power to replicate, so enormous numbers of generations have passed during which evolution has honed the viral genomes.

With the advent of automated sequencing, geneticists have added much larger genomes to the list of total known sequences. In 1988, D.J. McGeoch and colleagues published the sequence of an important human virus (herpes simplex virus I) with a relatively large genome: 152,260 bp. In 1995, Craig Venter and Hamilton Smith and colleagues determined the entire base sequences of the genomes of two bacteria: *Haemophilus influenzae* and *Mycoplasma genitalium*. The *H. influenzae* (strain Rd) genome contains 1,830,137 bp and it was the first genome from a free-living organism to be completely sequenced. The *M. genitalium* genome, at only 580,000 bp, is the smallest of any known free-living organism and contains only about 470 genes.

In April 1996, the leaders of an international consortium of laboratories announced another milestone: The 12-million-bp genome of baker's yeast (*Saccharomyces cerevisiae*) had been sequenced. This was the first eukaryotic genome to be entirely sequenced. Later in 1996, the first genome of an organism (*Methanococcus jannaschii*) from the third domain of life, the archaea, was sequenced.

Then, in 1997, the long-awaited sequence of the 4.6 million-bp *E. coli* genome was reported. This is only about one-third the size of the yeast genome, but the importance of *E. coli* as a genetic tool made this a milestone as well.

In 1998, the sequence of the first animal genome, from the roundworm *Caenorhabditis elegans,* was reported. The first plant genome (from the mustard family member *Arabidopsis thaliana*) was completed in 2000. *C. elegans* and *A. thaliana* are both **model organisms** chosen for study because of their small genome size, short generation time, and their ease of manipulation in genetic experiments. *C. elegans* has the additional advantages of having fewer than 1000 cells, and being transparent, so the development of each of its cells can be tracked visually. Two other famous model organisms are the fruit fly *Drosophila melanogaster* and the house mouse *Mus musculus.* The sequences of the genomes of these two organisms were reported in 2000

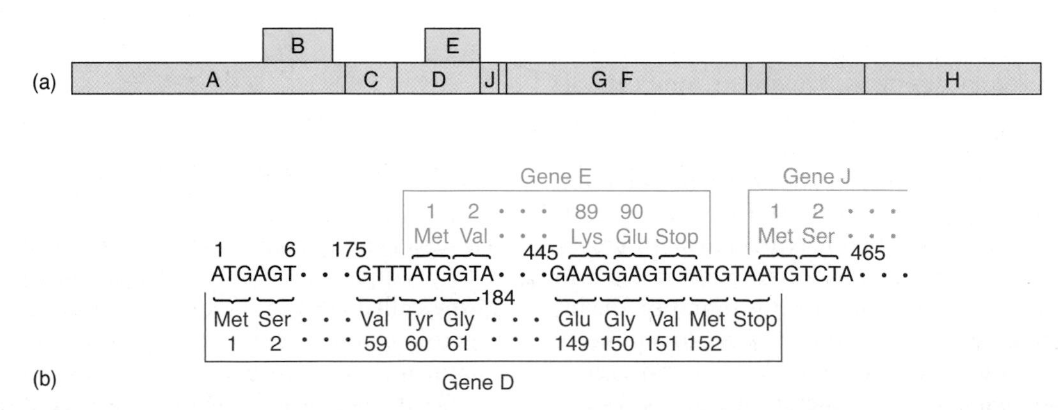

Figure 24.6 The genetic map of phage ϕX174. (a) Each letter stands for a phage gene. **(b)** Overlapping reading frames of ϕX174. Gene D (pink) begins with the base numbered 1 in this diagram and continues through base number 459. This corresponds to amino acids 1–152 plus the stop codon TAA. Dots represent bases or amino acids not shown. Only the nontemplate strand is shown. Gene E (blue) begins at base number 179 and continues through base number 454, corresponding to amino acids 1–90 plus the stop codon TGA. This gene uses the reading frame one base to the right, relative to the reading frame of gene D. Gene J (gray) begins at the base number 459 and uses the reading frame one base to the left, relative to gene D.

and 2002, respectively. Also in 2000, the eagerly awaited rough draft of the human genome sequence was announced. By 2001, this "working draft" of the human genome was published.

In 2002, several important genomes were reported, in at least draft form. These included the genomes of the single-celled parasite *Plasmodium falciparum,* which causes malaria, and the mosquito *Anopheles gambiae,* which is the major carrier of the parasite. Together, these genomes promise to help in designing better ways of combating the terrible scourge of malaria. The year 2002 also saw the publication of draft sequences of the genomes of two common varieties of rice (*Oryza sativa*). This is the first cereal plant genome to be sequenced, and it has enormous potential significance for human nutrition. Much of the world's population relies on cereals, and rice in particular, for the bulk of their food.

The genomic sequences of two more vertebrates also appeared in 2002: The tiger pufferfish (*Fugu rubripes*), and the house mouse (*Mus musculus*). Comparison of these sequences to that of the human genome has already shed light on vertebrate evolution. Additional help on this evolutionary investigation has come from the sequence of the genome of the sea squirt, *Ciona intestinalis.* The adult of this species is a sessile marine organism that attaches itself to rocks and pier pilings. It bears scant resemblance to a vertebrate, but its larval form resembles a tadpole, complete with a dorsal column made of cartilage that bears some resemblance to a spine. Thus, the sea squirt is a chordate, in the same phylum with the vertebrates. Comparison of the genome of this organism with those of vertebrates and invertebrates, such as nematodes and fruit flies, will give us additional insight into vertebrate evolution.

Most molecular evolution studies depend on comparisons of base sequences of parts of genomes from different organisms. The guiding principle is that there is a relationship between the divergence of the genomic sequences between any two organisms and the evolutionary distance between those two organisms. Thus, the genomes of organisms that diverged relatively recently, such as the mouse and human, should be more similar than the genomes of organisms that diverged longer ago, such as the sea squirt and human. In general, this is certainly true, but genomic studies on these and other organisms have revealed some unexpected features. For example, the rate of evolution of the human genome is not constant throughout. Instead, there are regions of relatively rapid change interspersed with regions that have changed relatively slowly over time. It will be fascinating to discover the reasons for these differences.

Another lesson from the genomes sequenced so far is that the size of an organism's genome tends to correlate with the organism's complexity. (On the other hand, we discovered in Chapter 2 when we discussed the C-value paradox that there are many exceptions to this general rule.) In accord with the rule, prokaryotic genomes tend to be much smaller than eukaryotic ones. However, it is interesting that there is some overlap. For example, the smallest eukaryotic genome sequenced to date is that of the obligate intracellular parasite of humans and other mammals, *Encephalitozoon cuniculi.* This organism has a genome comprising only about 2.9 Mb, and has only 1997 ORFs that could potentially code for proteins. (Of course, a parasitic lifestyle enables an organism to survive with fewer genes because it can rely on its host for many of its needs.) By contrast, the largest bacterial genome, as of 2008, is that of the social bacterium *Sorangium cellulosum.* It has a genome composed of about 13 Mb, which is even larger than the genome of budding yeast.

On April 14, 2003, the International Human Genome Sequencing Consortium announced that it had produced a "finished" human genome sequence—two years ahead of schedule. That is, it had done 99% of the sequencing that was possible with 2003 technology, the sequence was subject to an error rate of only one in 100,000, and all sequences were in the proper order. This was a significant improvement over the rough draft announced two years earlier. Several hundred gaps remained to be filled, but they were mostly very challenging repetitive regions and centromeres.

As of December 6, 2010, more than 1440 complete genomes had been sequenced, of which 1372 were from microbes, according to the NCBI website (www.ncbi.nlm.nih.gov/genome). Table 24.1 presents a time line of some of the most important achievements in genome sequencing. In the following sections we will discuss the lessons we have learned from these sequences.

SUMMARY The base sequences of viruses and organisms ranging from phages to bacteria to animals and plants have been obtained. A rough draft and finished version of the human genome have also been obtained. Comparison of the genomes of closely related and more distantly related organisms can shed light on the evolution of these species.

The Human Genome Project

In 1990, American geneticists embarked on an ambitious quest: to map and ultimately sequence the entire human genome. This effort, which quickly became an international program, was somewhat controversial at first, partly because of the enormous effort and cost of carrying it through to its ultimate goal: knowing the entire base sequence of every one of the human chromosomes. The reason for the high cost, of course, is that the human genome is huge—more than 3 billion bp. To get an idea of the magnitude of this task, consider that if all 3 billion bases were written down, it would take about 500,000 pages of the journal *Nature* to contain all the information. If you could

Table 24.1 Milestones in Genomic Sequencing

Genome (Importance)	Size (bp)	Year
Phage ϕX174 (first genome)	5375	1977
Phage λ (large-DNA phage)	48,513	1983
Herpes simplex virus I (large-DNA eukaryotic virus)	152,260	1988
Haemophilus influenzae (bacterium, first organism)	1,830,000	1995
Mycoplasma genitalium (smallest bacterial genome)	580,000	1995
Saccharomyces cerevisiae (yeast, first eukaryote)	12,068,000	1996
Methanococcus jannaschii (first archaeon)	1,660,000	1996
Escherichia coli (best studied bacterium)	4,639,221	1997
Caenorhabditis elegans (first animal, roundworm)	97,000,000	1998
Human chromosome 22 (first human chromosome)	53,000,000	1999
Arabidopsis thaliana (first plant, mustard family)	120,000,000	2000
Drosophila melanogaster (a favorite genetic model)	180,000,000	2000
Human (working draft of the "holy grail" of genomics)	3,200,000,000	2001
Plasmodium falciparum (the malaria parasite)	23,000,000	2002
Anopheles gambiae (the major mosquito malaria carrier)	278,000,000	2002
Fugu rubripes (tiger pufferfish)	365,000,000	2002
Mus musculus (house mouse)	2,500,000,000	2002
Ciona intestinalis (sea squirt, a primitive chordate)	117,000,000	2002
Canis lupus familiaris (dog, working draft)	~2,400,000,000	2003
Gallus gallus (chicken, first farm animal)	1,050,000,000	2004
Human (finished sequence)	3,200,000,000	2004
Oryza sativa (rice, first cereal grain)	489,000,000	2005
Pan troglodytes (chimpanzee, our closest relative, working draft)	~3,000,000,000	2005
Three trypanosomatids (*Trypanosoma cruzi, T. brucei,* and *Leishmania major,* parasites that cause severe human illness)	25–55,000,000	2005
Populus trichocarpa (black cottonwood, first tree)	~485,000,000	2006
First individual humans (two Caucasians, one African, and one Han Chinese)	3,200,000,000	2007 and 2008
Homo Neanderthalensis (our closest evolutionary relative, working draft)	~3,000,000,000	2010

stand the boredom, it would take you about 60 years, working 8 h/day, every day, at 5 bases a second, to read it all. Assuming a 1990 cost of about a dollar a base, the project would consume more than $3 billion, vastly more than we are used to devoting to a single biological project. In the end, more efficient sequencing methods allowed the project to be completed much sooner and at a lower cost than originally estimated.

The original plan for the Human Genome Project was systematic and conservative: First, geneticists would prepare genetic and physical maps of the genome. These would contain the markers, or signposts, that would allow DNA sequences to be pieced together in the proper order. The bulk of the sequencing would be done only after the mapping was complete and clones representing all points on the map were in hand—systematically stored in freezers around the world. The original target date for completion of the sequence was 2005.

Then, in May of 1998, Craig Venter, who had established a private, for-profit company, Celera, to sequence the human genome (and other genomes), shocked the genomics community by announcing that Celera would complete a rough draft of the human genome by the end of 2000. That timetable was astonishing enough, but the method by which he proposed to do the sequencing was even more arresting. Instead of relying on a map, with the ordered clones used to build it, Venter proposed a **shotgun sequencing** approach in which the whole human genome would be chopped up and cloned, then the clones would be sequenced at random, and finally the sequences would be pieced together using powerful computer programs that find overlapping sequences. It was not long before Francis

Collins, director of the publicly financed Human Genome Project, rose to Venter's challenge and promised that he and his colleagues would also produce a rough draft by the end of 2000, and a polished final draft by 2003, using the map-then-sequence strategy.

The upshot of this race was a tie of sorts. Venter and Collins appeared with President Clinton and other dignitaries at a ceremony in the East Room of the White House on June 26, 2000, to announce the completion of a rough draft of the human genome. We will examine the two approaches to sequencing large genomes: mapping, then sequencing (clone by clone); and shotgun sequencing. But first, let us examine the cloning vectors that have been developed for massive projects like the Human Genome Project.

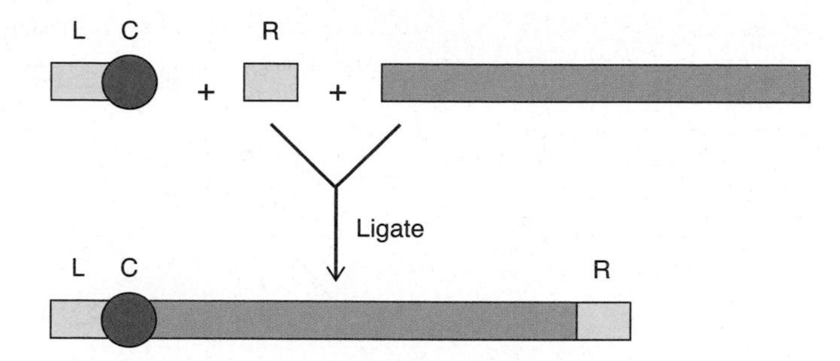

Figure 24.7 Cloning in yeast artificial chromosomes. We begin with two tiny pieces of DNA from the two ends of a yeast chromosome. One of these, the left arm, contains the left telomere (yellow, labeled L) plus the centromere (red, labeled C). The right arm contains the right telomere (yellow, labeled R). These two arms are ligated to a large piece of foreign DNA (blue)—several hundred kilobases of human DNA, for example—to form the YAC, which can replicate in yeast cells along with the real chromosomes.

Vectors for Large-Scale Genome Projects

No matter which sequencing strategy is used, one must first clone fragments of the genome in appropriate vectors, and large fragments are particularly valuable. We will describe two of the most popular here: yeast artificial chromosomes and bacterial artificial chromosomes. The early mapping work relied on yeast artificial chromosomes, so we will begin with those.

Yeast Artificial Chromosomes The main problem with the cloning tools described in Chapter 4 is that they do not hold enough DNA for large-scale physical mapping of the human genome. Even the cosmids accommodate DNA inserts up to only about 50 kb, which is too small for efficient mapping of regions spanning more than a million bases.

Vectors called **yeast artificial chromosomes,** or **YACs,** were very useful in mapping the human genome because they could accommodate hundreds of thousands of kilobases each. YACs containing a megabase or more are known as "megaYACs." A YAC contains a left and right yeast chromosomal telomere (Chapter 21), which are both necessary to protect the chromosome's ends, and a yeast centromere, which is necessary for segregation of sister chromatids to opposite poles of the dividing yeast cell. The centromere is placed adjacent to the left telomere, and a huge piece of human (or any other) DNA can be placed in between the centromere and the right telomere, as shown in Figure 24.7. The large DNA inserts are prepared by slightly digesting long pieces of human DNA with a restriction enzyme. The YACs, with their huge DNA inserts, can then be introduced into yeast cells, where they will replicate just as if they were normal yeast chromosomes.

Using YACs, geneticists made great strides in the mapping phase of the Human Genome Project. They produced a genetic map of the whole genome that provided an average resolution of 0.7 centimorgan. A **centimorgan (cM)** is the distance that yields a 1% recombination frequency between two markers and corresponds to an average of about 1 Mb in humans. These researchers also produced relatively high-resolution physical maps of two of the smallest chromosomes, 21 and Y. These maps were especially useful in that they represented long stretches of overlapping DNA segments cloned in YACs. Thus, in the days before the human genome was sequenced, if you were interested in a disease gene that mapped to one of these chromosomes, you had a much simplified task. You needed only to discover two markers flanking the gene of interest, look on the map to find which YAC or YACs contained these markers, obtain the YACs, and begin your final search for the gene.

Bacterial Artificial Chromosomes Despite all the success they made possible in human genome mapping, YACs suffer from several serious drawbacks: They are inefficient (not many clones are obtained per microgram of DNA); they are hard to isolate from yeast cells; they are unstable; and they tend to contain scrambled inserts that are really composites of DNA fragments from more than one site. **Bacterial artificial chromosomes (BACs)** solve all of these problems and were therefore the vector of choice for much of the sequencing phase of the Human Genome Project.

BACs are based on a well-known natural plasmid that inhabits *E. coli* cells: the **F plasmid.** This plasmid allows conjugation between bacterial cells. In some conjugation events, the F plasmid itself is transferred from a donor **F^+ cell** to a recipient **F^- cell,** converting the latter to an F^+ cell. In other events, a small piece of host DNA is transferred as an insert in the F plasmid (which is called an **F′ plasmid** if it has an insert of foreign DNA). And in still other events, the F plasmid inserts into the host chromosome and mobilizes the whole chromosome to pass from the donor cell to the recipient cell. Thus, because the *E. coli* chromosome contains over 4 million bp, the F plasmid can obviously accommodate a large insert of DNA. In practice, BACs usually have inserts less than 300,000 bp (average about 150,000 bp), and these plasmids are stable

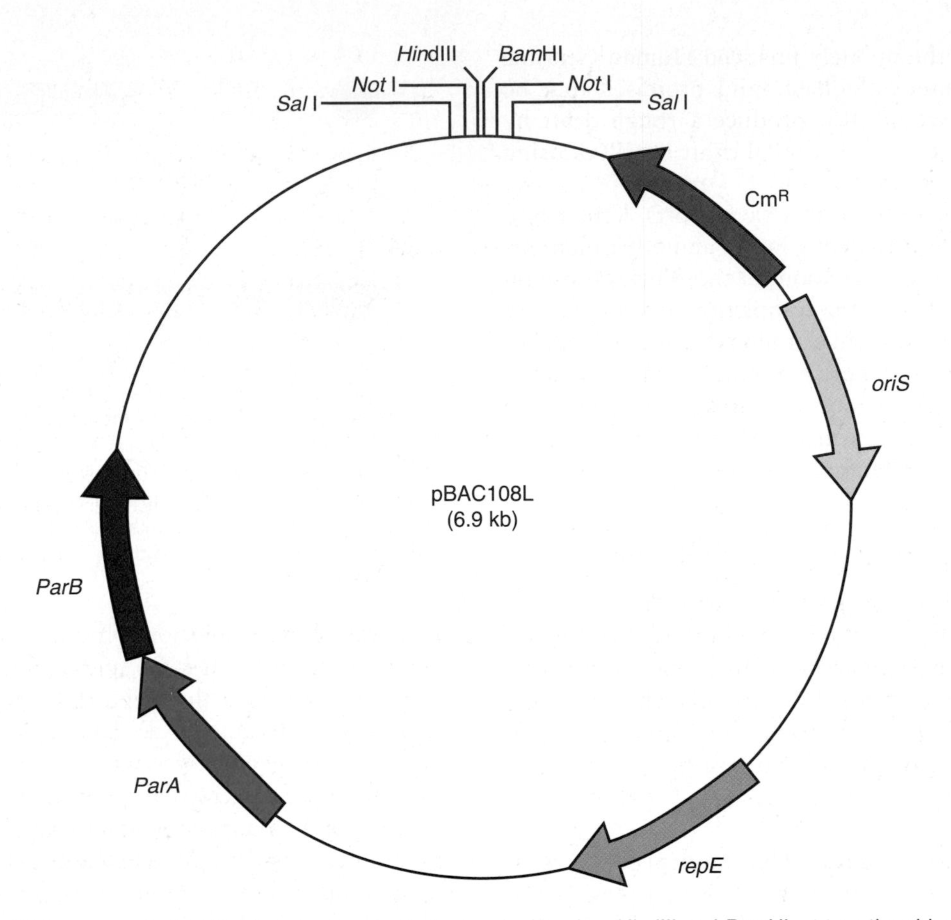

Figure 24.8 Map of the BAC vector, pBAC108L. Key features include the cloning sites *Hin*dIII and *Bam*HI, at top; the chloramphenicol resistance gene (Cm^R), used as a selection tool; the origin of replication (*oriS*); and the genes governing partition of plasmids to daughter cells (*ParA* and *ParB*).

in vivo and in vitro. Unlike the linear YACs, which tend to break under shearing forces, the circular, supercoiled BACs resist breakage.

Figure 24.8 shows the map of one of the first BACs, which was developed by Melvin Simon and colleagues in 1992. It has an origin of replication, a cloning site with two restriction sites (for *Hin*dIII and *Bam*HI) into which large DNA fragments may be inserted. It also has genes (the Par genes) that govern plasmid partition to the daughter cells that keep the plasmid copy number at about two per cell. This contributes to the stability of the plasmid, and it has a chloramphenicol-resistance gene to enable selection of cells that have the plasmid.

SUMMARY Two high-capacity vectors have been used extensively in the Human Genome Project. Much of the mapping work was done with yeast artificial chromosomes (YACs), which can accept inserts of a million or more base pairs. Most of the sequencing work was performed with bacterial artificial chromosomes (BACs) which can accept up to about 300,000 bp. The BACs are more stable and easier to work with than the YACs.

The Clone-by-Clone Strategy

This strategy has inherent appeal because it is so systematic. First, the whole genome is mapped by finding markers regularly spaced along each chromosome. A by-product of the mapping is a collection of clones corresponding to the markers. Because we already know the order of these clones, we can sequence each one and put that sequence in its proper place in the whole genome. Thus, this method is commonly called the **clone-by-clone sequencing** strategy. Aside from their usefulness in cloning, genetic and physical maps have another important benefit: They give us signposts to use when searching for the genes responsible for diseases. In the next section, we will consider some of the most powerful methods used in mapping large genomes in preparation for sequencing. As you read this section, bear in mind that these techniques are designed to map markers that are not genes but simply stretches of DNA that vary from one individual to another. We have already seen one example of such markers: restriction fragment length polymorphisms (RFLPs).

Variable Number of Tandem Repeats The greater the degree of polymorphism of a RFLP, the more useful it will be. If only 1 person in 100 has one form of the RFLP (the 6-kb fragment in Figure 24.1, for example), and the other 99 have the other form (the 4-kb and 2-kb fragments), one must screen many individuals before finding the one rare variant. This makes mapping very tedious. However, some RFLPs, called **variable number tandem repeats,** or **VNTRs,** are more useful. These derive from **minisatellites** (Chapter 5), stretches of DNA that contain a short core sequence repeated over and over in tandem (head to tail). Because the number of repeats of the core sequence in a VNTR is likely to be different from one individual to another, VNTRs are highly polymorphic, and therefore relatively easy to map. However, VNTRs have a disadvantage as genetic markers: They tend to bunch together at the ends of chromosomes, leaving the interiors of the chromosomes relatively devoid of markers.

Sequence-Tagged Sites Another kind of anonymous marker, which is very useful to genome mappers, is the **sequence-tagged site (STS).** STSs are short sequences, about 60–1000 bp long, that can be detected by PCR. Figure 24.9 illustrates how to use PCR to detect an STS. One must first know enough about the DNA sequence in the region being mapped to design short primers that will hybridize a few hundred base pairs apart and cause amplification of a predictable length of DNA in between. One can then apply PCR with these two primers to any unknown DNA; if the proper size amplified DNA fragment appears, then the unknown DNA has the STS of interest. Notice that hybridization of the primers to the unknown DNA is not enough; they must hybridize a specific number of base pairs apart to give the right size PCR fragment. This provides a check on the specificity of hybridization. One great advantage of STSs as a mapping tool is that no DNA must be cloned and examined and kept in someone's freezer. Instead, the sequences of the primers used to generate an STS are published and then anyone in the world can order those same primers and find the same STS in an experiment that takes just a few hours. Another big advantage is that it takes much less DNA to perform PCR than to do a Southern blot.

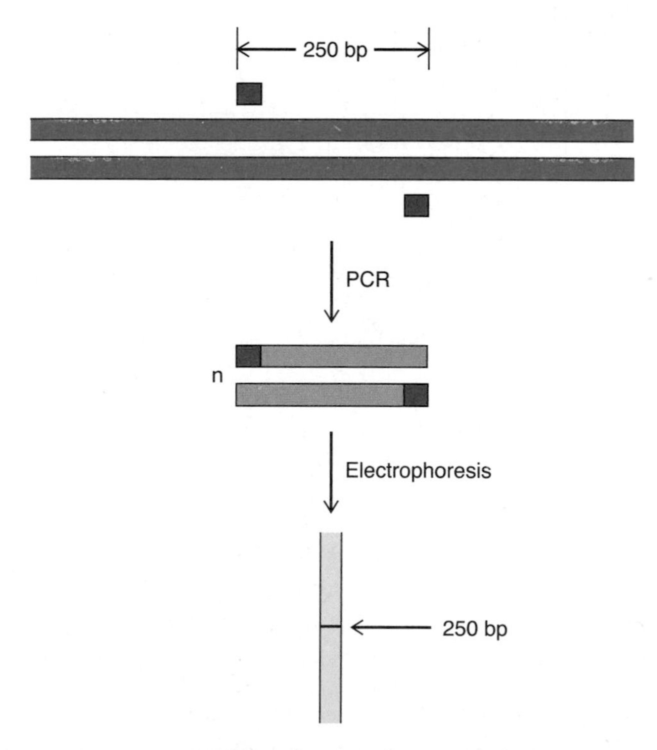

Figure 24.9 Sequence-tagged sites. We start with a large cloned piece of DNA, extending indefinitely in either direction. The sequences of small areas of this DNA are known, so one can design primers that will hybridize to these regions and allow PCR to produce double-stranded fragments of predictable lengths. In this example, two PCR primers (red) spaced 250 bp apart have been used. Several cycles of PCR generate many copies of a double-stranded PCR product that is precisely 250 bp long. Electrophoresis of this product allows one to measure its size exactly and confirm that it is the correct one.

Microsatellites STSs are very useful in physical mapping or locating specific sequences in the genome. But they are worthless as markers in traditional genetic mapping unless they are polymorphic. Only then can we use them to determine genetic linkage. Fortunately, geneticists have discovered a class of STSs called **microsatellites** that are highly polymorphic. Microsatellites are similar to minisatellites in that they consist of a core sequence repeated over and over many times in a row. However, whereas the core sequence in typical minisatellites is a dozen or more base pairs long, the core in microsatellites is much smaller—usually only 2–4 bp long. In 1992, Jean Weissenbach and his colleagues produced a linkage map of the entire human genome based on 814 microsatellites containing a C–A dinucleotide repeat. They isolated cloned DNAs containing these microsatellites and used their sequences to design PCR primers that flank the repeats at each locus. A given pair of primers yielded a PCR product whose size depended on the number of C–A repeats in a given individual's microsatellite at that locus. Happily, the number of repeats varied quite a bit from one individual to another. Besides the fact that microsatellites are highly polymorphic, they are also widespread and relatively uniformly distributed in the human genome. Thus, they are ideal as markers for both linkage and physical mapping.

Genetic (linkage) mapping with microsatellites is done by the same technique outlined in Chapter 1 for traditional genetic markers in fruit flies. Instead of determining the recombination frequency between, say, wing shape and eye color, geneticists would determine the recombination frequency between two microsatellites. For example, consider an example in which a man's DNA yields a microsatellite at one locus that is 78 bp long and a microsatellite at a nearby locus that is 42 bp long. His wife has a microsatellite at the first locus that is 102 bp long and a microsatellite at the second locus that is 36 bp long. Within limits, the more their children show nonparental combinations of these two markers in their gametes (e.g., grandchild with microsatellites that are 78 and 36 bp long, respectively), the more recombination has occurred between the markers, and the farther apart the markers are on the chromosome.

Geneticists interested in physically mapping or sequencing a given region of a genome aim to assemble a set of clones called a **contig,** which contains contiguous (actually overlapping) DNAs spanning long distances. This is rather like putting together a jigsaw puzzle; the bigger the pieces, the easier the puzzle. Thus, it is essential to have vectors like BACs and YACs that hold big chunks of DNA. Assuming we have a BAC library of the human genome, we need some way to identify the clones that contain the region we want to map. This can be done in several ways. We could hybridize BAC DNA to a labeled DNA probe corresponding to the region of interest, but this is subject to some uncertainty due to possible nonspecific hybridization. A more reliable method is to look for STSs in the BACs. It is best to screen the BAC library for at least two STSs, spaced hundreds of kilobases apart, so BACs spanning a long distance are selected.

After we have found a number of positive BACs, we begin mapping by screening them for several additional STSs, so we can line them up in an overlapping fashion as shown in Figure 24.10. This set of overlapping BACs is our new contig. We can now begin finer mapping, and even sequencing, of the contig.

Radiation Hybrid Mapping Mapping with BACs sounds straightforward, but it presents difficulties. One of the most important is that BACs are so small relative to a whole human chromosome that creating a BAC contig of a whole chromosome would be unbearably laborious. So we need a method to find linkage between STSs that are even farther apart than those that could fit into a single BAC. **Radiation hybrid mapping** provides a way. We begin by irradiating

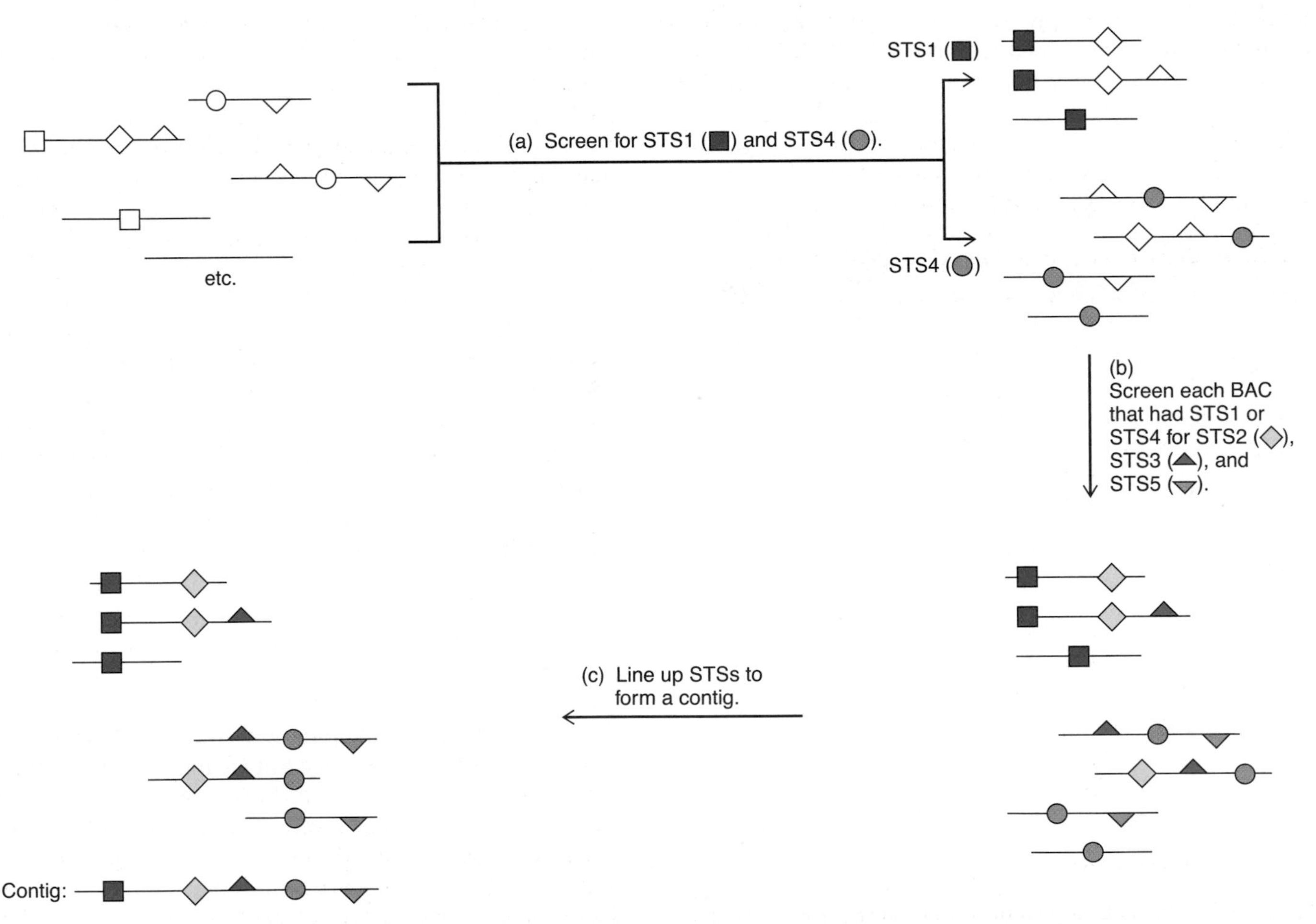

Figure 24.10 Mapping with STSs. At top left, several representative BACs are shown, with different symbols representing different STSs placed at specific intervals. In step **(a)** of the mapping procedure, screen for two or more widely spaced STSs. In this case screen for STS1 and STS4. All those BACs with either STS1 or 4 are shown at top right. The identified STSs are shown in color. In step **(b),** each of these positive BACs is further screened for the presence of STS2, STS3, and STS5.The colored symbols on the BACs at bottom right denote the STSs detected in each BAC. In step **(c),** align the STSs in each BAC to form the contig. Measuring the lengths of the BACs by pulsed-field gel electrophoresis helps to pin down the spacing between pairs of BACs.

human cells with lethal doses of ionizing radiation, such as x-rays or gamma rays, which break the human chromosomes into pieces. Next, we fuse these doomed human cells with hamster cells to form hybrid cells that contain only some of the human chromosome fragments. Then, we form clones of identical hybrid cells by growing groups of cells—each group deriving from a single progenitor cell. Finally, we examine clones of hybrid cells to see which STSs tend to be found together in the hybrid cells. The more often they are together, the closer together they are likely to be on a human chromosome.

In 1996, an international consortium of geneticists, including G.D. Schuler, published a human map based on STSs mapped by this technique. It contained more than 16,000 STS markers, plus about a thousand genetic markers mapped by classical linkage methods (family studies), which provided an overall framework for the map. The STS markers used in this study were a special class called **expressed sequence tags (ESTs).** These are STSs that are generated by starting with mRNAs and using the enzyme reverse transcriptase to make corresponding cDNAs. These cDNAs can then be amplified by PCR and cloned. Finally, both ends of the cDNAs are sequenced, yielding two "sequence tags" that are usually less than 500 bases long. Thus, ESTs represent genes that are expressed in the cell from which the mRNAs were isolated. Because the STS (or EST) method yields the sequence of only a small part of a gene, a given gene may be represented by many different ESTs in an EST database. To minimize such duplications, the mapping consortium confined their mapping to ESTs that represented the 3′-untranslated regions (3′-UTRs) of genes. This strategy also has the advantage of avoiding most introns, which tend not to be found in 3′-UTRs. By 1998, the international consortium (P. Deloukas et al.) had refined and extended the map to include over 30,000 genes.

SUMMARY Mapping the human genome requires a set of landmarks to which we can relate the positions of genes. Some of these markers are genes, but many more are nameless stretches of DNA, such as RFLPs, VNTRs, STSs (including ESTs and microsatellites). The latter two are regions of DNA that can be identified by formation of a predictable length of amplified DNA by PCR with pairs of primers.

Shotgun Sequencing

The shotgun-sequencing strategy, first proposed by Craig Venter, Hamilton Smith, and Leroy Hood in 1996, bypasses the mapping stage and goes right to the sequencing stage. The sequencing starts with a set of BAC clones containing large DNA inserts, averaging about 150 kb. The insert in each BAC is sequenced on both ends using an automated sequencer that can usually read about 500 bases at a time, so 500 bases at each end of the clone will be determined. Assuming that 300,000 clones of human DNA are sequenced this way, that would generate 300 million bases of sequence, or about 10% of the total human genome, and the 500-base sequenced regions would therefore occur on average every 5 kb in the genome. These 500-base sequences serve as an identity tag, called a **sequence-tagged connector (STC)**, for each BAC clone. On average, assuming an average clone size of 150 kb, and an STC every 5 kb, 30 clones (150 kb/5 kb = 30) should share a given STC somewhere within their span. This is the origin of the term *connector*—each clone should be "connected" via its STCs to about 30 other clones.

The next step is to **fingerprint** each clone by digesting it with a restriction enzyme. This serves two important purposes. First, it tells the insert size (the sum of the sizes of all the fragments generated by the restriction enzyme). Second, it allows one to eliminate aberrant clones whose fragmentation patterns do not fit the consensus of the overlapping clones. Note that this clone fingerprinting is not the same as mapping; it is just a simple check before sequencing begins.

The next step is to obtain the entire sequence of a BAC that looks interesting (a **seed BAC**). This is done by subdividing the BAC into smaller clones, frequently in a pUC-type vector with inserts averaging only about 2 kb. This whole BAC sequence allows the identification of the 30 or so other BACs that overlap with the seed: They are the ones with STCs that occur somewhere in the seed BAC.

Next, one selects other BACs with minimal overlap with the original one and proceeds to sequence them. Then this process is repeated with other BACs with minimal overlap with the second set, and so forth. This strategy, called **BAC walking,** would in principle allow one laboratory to sequence the whole human genome—given enough time.

But they did not have that much time, so Venter and colleagues modified the procedure by sequencing BACs at random until they had about 35 billion nt of sequence. In principle that should cover the human genome ten times over, giving a high degree of coverage and accuracy. Then they fed all the sequence into a computer with a powerful program that found areas of overlap between clones and fit their sequences together, building the sequence of the whole genome.

As mentioned a little earlier, the bulk of the sequencing is done with pUC clones with relatively small inserts—only about 2 kb each. But these small inserts would not provide enough overlaps to piece together the whole genome. This drawback is especially apparent in regions of repeated DNA. A 2-kb cloned sequence from a 10-kb region of tandem DNA repeats would give no clues about where the cloned sequence fit within the larger repeat region—one part looks the same as another. That is one way the BAC clones come in handy: They are large enough to cover almost any repeated region. They also provide overlaps spanning large DNA regions, so they can help to organize the smaller cloned fragments. This job was also facilitated by the physical maps, especially the STS maps, that were already available. So the shotgun strategy for sequencing the human genome was in practice a hybrid of a pure shotgun and a map-then-sequence strategy.

Any strategy to sequence over 3 billion bp depends on a high-volume, low-cost sequencing method. We now have sequencing devices that perform electrophoresis of DNA fragments in capillary tubes instead of the traditional thin gel slabs. These instruments are fully automated and each can handle about 1000 samples per day with only 15 min of human attention. Another of Venter's companies, The Institute for Genomic Research (TIGR), had 230 such instruments; together, they could produce about 100 Mb of DNA sequence every day, with a relatively low labor cost.

SUMMARY Massive sequencing projects can take two forms: (1) In the map-then-sequence strategy, one produces a physical map of the genome including STSs, then sequences the clones (mostly BACs) used in the mapping. This places the sequences in order so they can be pieced together. (2) In the shotgun approach, one assembles libraries of clones with different size inserts, then sequences the inserts at random. This method relies on a computer program to find areas of overlap among the sequences and piece them together. In practice, a combination of these methods was used to sequence the human genome.

Sequencing Standards

What do we mean by "rough draft," or "working draft," and "final draft" of a genome? That depends on whom you ask. Most investigators agree that a working draft may be only 90% complete and may have an error rate of up to 1%. Although there is less agreement about what qualifies as a final draft, there is consensus that it should have an error rate of less than 1/10,000 (0.01%) and should have as few gaps as possible. Some molecular biologists insist that a genome is not completely sequenced until every last gap is filled, but it would be very difficult to eliminate all gaps in the human genome. As we will see in the next section, some regions of DNA, for mostly unexplained reasons, resist cloning. The cost of overcoming the obstacles to cloning these regions will likely be prohibitively high, so the task of filling in the last few million bases of the human genome may never be done. As detailed in the next section, the consortium that sequenced human chromosome 22 decided that their sequence was "functionally complete" when they had obtained all the sequence possible with the cloning and sequencing tools currently available, even though significant gaps remained.

24.3 Studying and Comparing Genomic Sequences

Once a genomic sequence is in hand, scientists can mine it for the wealth of information it contains. They can also compare it to the sequences of other genomes to shed light on the evolution of these species. We will begin this section with a discussion of the human genome, and then compare it with the genomes of closely related, and then more distantly related organisms.

The Human Genome

At the end of 1999, we tasted the first fruit of the Human Genome Project: The final draft of human chromosome 22. In February 2001, the Venter group and the public consortium each published their versions of a working draft of the whole human genome. In 2004, the international consortium announced the finished sequence of the euchromatic part of the human genome.

In this section we will look at the lessons we learned from the finished sequence of chromosome 22 (the first human chromosome to be sequenced), and the working draft and finished sequence of the whole genome. Before we begin, one lesson worth noting is that the finished sequences came from the more orderly clone-by-clone approach. This strategy yields the final draft sequences of whole chromosomes as soon as the groups sequencing each chromosome complete their work. On the other hand, the raw sequence in the shotgun sequencing approach is not pieced together until the very end, when the computer finds the overlaps necessary to build contigs. Thus, this strategy may not yield the final draft sequence of any chromosome until the whole genome is finished.

Chromosome 22 In reality, only the long arm (22q) of the chromosome was sequenced; the short arm (22p) is composed of pure heterochromatin and is thought to be devoid of genes. Also, 11 gaps remained in the sequence. Ten of these were gaps between contigs that could not be filled with clones—presumably due to "unclonable" DNA. The other corresponded to a 1.5-kb region of cloned DNA that resisted sequencing. The reasons that some DNAs, sometimes called "poison regions," are unclonable are not completely clear, but it is known that DNAs with unusual secondary structure or repetitive sequences are frequently lost from bacterial cells. This is one reason that heterochromatin (Chapter 13) is very poorly represented, even in the final draft of the human genome. It is found primarily at the centromeres and near the telomeres of chromosomes and is rich in repetitive sequences. By failing to sequence the heterochromatin in the genome, scientists are not missing very many, if any, genes, because genes are not thought to reside there. But there could be other interesting aspects of these heterochromatin regions that will be missed.

What did we learn from the first completed sequence of chromosome 22? Several findings were interesting. First, we are going to have to learn to live with gaps in our sequence of the human genome, although perhaps not as many as first appeared in the sequence of this chromosome. Already by the summer of 2000, one of the gaps had been filled, and by December of 2010, only four gaps remained, not counting the short arm of the chromosome. Still, the same problems encountered in spanning the gaps in chromosome 22 bedeviled investigators sequencing the other chromosomes. Table 24.2 lists the sequenced contigs in chromosome 22 and the gaps between them as of 1999. The contigs accounted for 33,464 kb, or about 97% of the long arm of the chromosome, and they were sequenced with very high accuracy—estimated at less than one error per 50,000 bases. It is interesting that all of the gaps occurred in the regions of the chromosome close to the centromere and telomeres. Between gaps 4 and 5 was an enormous contig composed of 23,006 kb that covered more than two-thirds of chromosome 22q. By December of 2010, 34,894,566 bases of chromosome 22q had been sequenced.

Table 24.2 Chromosome 22 Contigs and Gaps as of 1999

Contig	Gap	Size (kb)	
1		234	
	1		1.9
2		406	
	2		~150
3		1394	
	3		~150
4		1790	
	4		~100
5		23,006	
	5		~50
6		767	
	6		~50–100
7		1528	
	7		~150
8		2485	
	8		~50
9		190	
	9		~100
10		993	
	10		~100
11		291	
	11		~100
12		380	
Total sequence length		33,464	
Total length of 22q		34,491	

(*Source:* Adapted from Dunham, I., N. Shimizu, B.A. Roe, S. Chissoe, A.R. Hunt, J.E. Collins, et al., The DNA sequence of human chromosome 22. *Nature* 402:491, 1999.)

The second major finding was that chromosome 22 was estimated to contain 679 **annotated genes** (genes or gene-like sequences that were at least partially identified). These can be categorized as follows: **known genes,** whose sequences are identical to known human genes or to the sequences deduced from known human proteins; **related genes,** whose sequences are homologous to known genes of human or other species or which have regions of similarity to known genes; **predicted genes,** which contain sequences homologous to ESTs (so we are fairly sure they are expressed); and **pseudogenes,** whose sequences are homologous to known genes, but they contain defects that preclude proper expression. There were 247 known genes, 150 related genes, 148 predicted genes, and 134 pseudogenes in chromosome 22q. Thus, not counting the pseudogenes, there were 545 annotated genes. Computer analysis of the sequence predicted another 325 genes, but such analyses are still very inaccurate because the algorithms depend on finding exons, and the many long introns in human genes make exons hard to spot. As of December, 2010, 855 genes had been found in chromosome 22q, including pseudogenes.

The third major finding was that the coding regions of genes accounted for only a tiny fraction of the length of the chromosome. Even counting introns, the annotated genes accounted for only 39% of the total length of 22q, and the exons accounted for only 3%. By contrast, fully 41% of 22q is devoted to repeat sequences, especially Alu sequences and LINEs (Chapter 23). Table 24.3 lists the interspersed repeat elements found in chromosome 22 and their prevalences.

A fourth major finding was that the rate of recombination varied across the chromosome, with long regions in which recombination is relatively low interspersed with short regions of relatively high rates of recombination (Figure 24.11). As we have seen earlier in this chapter, geneticists had already made a genetic map of the human genome, including chromosome 22, based on microsatellites. This map was based on recombination frequencies between microsatellites and was therefore calibrated in centimorgans. The chromosome 22 sequencing team was able to find these microsatellites in the sequence and measure the real physical distance between them. Figure 24.11 shows that a plot of the genetic distance between markers versus the physical distance between the same markers is not linear. The numbers indicate regions of high rates of recombination, and therefore high apparent genetic distance, separated by longer regions of relatively low rates of recombination. The average ratio of genetic distance to physical distance in this chromosome is 1.87 cM/Mb. Of course, we should remember that the y axis represents *cumulative* genetic distance, that is, the sum of the distances between closely spaced markers. The actual genetic distance between widely separated markers is not the same as the sum

Table 24.3 Repetitive DNA Content of Human Chromosome 22

Type	Number	Total base pairs	% of chromosome
Alu	20,188	5,621,998	16.80
HERV	255	160,697	0.48
LINE 1	8043	3,256,913	9.73
LINE 2	6381	1,273,571	3.81
LTR	848	256,412	0.77
MER	3757	763,390	2.28
MIR	8426	1,063,419	3.18
MLT	2483	605,813	1.81
THE	304	93,159	0.28
Other	2313	625,562	1.87
Dinucleotide	1775	133,765	0.40
Trinucleotide	166	18,410	0.06
Tetranucleotide	404	47,691	0.14
Pentanucleotide	16	1612	0.0048
Other tandem repeats	305	102,245	0.31
Total	55,664	14,024,657	41.91

(*Source:* Adapted from Dunham, I., N. Shimizu, B.A. Roe, S. Chissoe, A.R. Hunt, J.E. Collins, et al., The DNA sequence of human chromosome 22. *Nature* 402:491, 1999.)

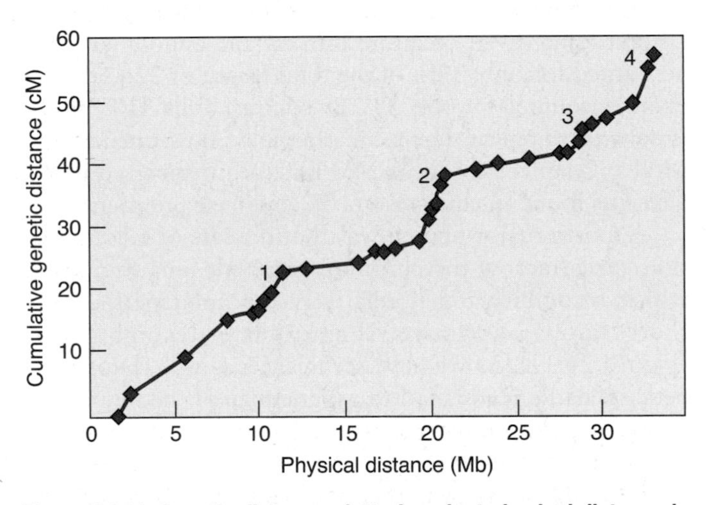

Figure 24.11 Genetic distance plotted against physical distance in chromosome 22q. The cumulative genetic distance between markers (in cM) is graphed versus the physical distance between the same markers (in Mb). The numbers denote four areas of relatively high rates of recombination (as reflected in the steeply rising curves). (*Source:* Adapted from Dunham, I., N. Shimizu, B.A. Roe, S. Chissoe, A.R. Hunt, J.E. Collins, et al. (The Chromosome 22 Sequencing Consortium), The DNA sequence of human chromosome 22. *Nature* 402:492, 1999.)

of the distances between intervening markers. That is because *multiple* recombination events are more probable between distant markers, which makes them appear closer together than they really are (Chapter 1).

A fifth major finding was that chromosome 22q had several local and long-range duplications. The most obvious involved the immunoglobulin λ locus. Clustered together at this locus are 36 gene segments that are at least potentially able to encode λ variable regions (V-λ gene segments), as well as 56 V-λ pseudogenes and 27 partial V-λ pseudo-genes known as "relics." Other duplications are separated by long distances. In one striking example, a 60-kb region is duplicated with greater than 90% fidelity almost 12 Mb away. Compared with the interspersed repeats, such as Alu sequences and LINEs, these duplications are found in few copies, so they are known as **low-copy repeats** or **LCRs.** Seven of the eight previously described LCR22s in the centromeric end of 22q were sequenced; the eighth (LCR22-1) probably lies in the sequence gap closest to the centromere.

The sixth major finding was that large chunks of human chromosome 22q are conserved in several different mouse chromosomes. The sequencing team found 113 human genes whose mouse orthologs had been mapped to mouse chromosomes. (**Orthologs** are homologous genes in different species that have evolved from a common ancestral gene. **Paralogs,** by contrast, are homologous genes that have evolved by gene duplication within a species. **Homologs** are any kind of homologous genes—orthologs or paralogs.) These mouse orthologs clustered into eight regions on seven different mouse chromosomes, as shown in Figure 24.12. The mouse chromosomes represented in human 22q are chromosomes 5, 6, 8, 10, 11, 15, and 16. Mouse chromosome 10 is represented in two regions of human 22q. As the two species have diverged, their chromosomes have rearranged, but linkage among many markers

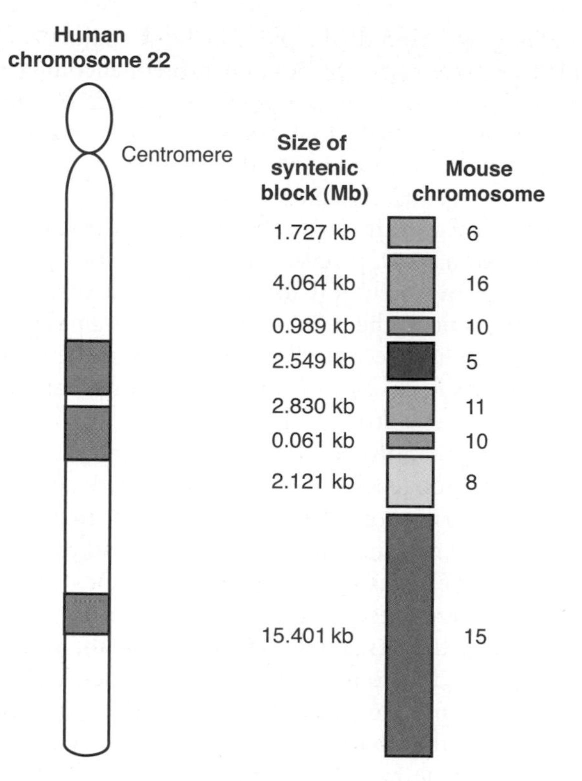

Figure 24.12 Regions of conservation between human and mouse chromosomes. Human chromosome 22 is depicted on the left, with the centromere near the top, and prominent bands in white and brown. Seven different mouse chromosomes contain syntenic blocks (orthologs in conserved order) and these are shown on the right. Colors correspond to the mouse chromosomes listed at far right. (*Source:* Adapted from Dunham, I., N. Shimizu, B.A. Roe, S. Chissoe, A.R. Hunt, J.E. Collins, et al. (The Chromosome 22 Consortium), The DNA sequence of human chromosome 22. *Nature* 402:494, 1999.)

has been preserved in **syntenic blocks.** (The preservation of gene order between two species is known as **synteny.**) Clearly, our knowledge of the sequence of the human genome has sped the sequencing of the mouse genome.

SUMMARY Human chromosome 22q has been sequenced to high accuracy, but the sequence still has 10 gaps that cannot be filled with available methods. There are 679 annotated genes, but the great bulk of the chromosome is made up of noncoding DNA, over 40% of it in interspersed repeats such as Alu sequences and LINEs. The rate of recombination varies across the chromosome, with long regions of low rates of recombination punctuated by short regions with relatively high rates. The chromosome contains several examples of local and long-range duplications. The human chromosome contains large regions where linkage among genes has been conserved with that in seven different mouse chromosomes.

Working Draft and Finished Version of the Human Genome In February 2001, the Venter group and the public consortium published separately their own versions of the working draft of the whole human genome. The drafts of the human genome presented by the two groups were by no means complete. They had many gaps and inaccuracies, but they also contained a wealth of information that kept scientists busy for years analyzing and extending it. Furthermore, the public draft continued to improve as groups working on its separate parts completed the laborious finishing phase that eliminates gaps and corrects errors.

The most striking discovery from both groups was the low number of genes in the genome. The Venter group found 26,588 genes for which there were at least two independent lines of evidence, and about 12,000 more potential genes. These potential genes were identified computationally, but there were no other supporting data. Venter and colleagues assumed that most of these latter sequences were false-positives. The public consortium estimated that the human genome contains 30,000–40,000 genes. As we will see later in this section, the estimate from the finished human genome sequence is even lower—fewer than 23,000 genes.

Thus, contrary to earlier estimates, the number of human genes seems to be scarcely larger than the number of genes in a lowly roundworm or a fruit fly. Clearly, the complexity of an organism is not directly proportional to the number of genes it contains. How then can we explain human complexity? One emerging explanation is that the *expression* of genes in humans is more complex than it is in simpler organisms. For example, it is estimated that at least 40% of human transcripts experience alternative splicing (Chapter 14). Thus, a relatively small number of gene regions encoding domains and motifs of proteins can be shuffled in different ways to give a rich variety of proteins with different functions. Moreover, posttranslational modification of proteins in humans seems more complex than that in simpler organisms, and this also gives rise to a greater variety of protein functions.

Another important finding is that about half of the human genome appears to have come from transposable elements duplicating themselves and carrying human DNA from place to place within the genome (Chapter 23). However, even though transposons have contributed so greatly to the genome, the vast majority of them are now inactive. In fact, all of the non-retrotransposons are inactive, and all of the LTR-containing retrotransposons seem to be. On the other hand, as we learned in Chapter 23, a few L1 transposons are still active in the human genome and continue to contribute to human disease.

Dozens of human genes appear to have come via horizontal transmission from bacteria, and some others came from new transposons entering human cells. Thus, the human genome has been shaped not entirely by internal mutations and rearrangements, but also by importation of genes from the outside world.

The total size of the human genome appears to be close to the 3 billion bp (3 Gb) predicted for many years. The Venter group sequenced about 2.9 Gb, and the public consortium predicted that the total size of the genome is about 3.2 Gb.

As mentioned earlier in this chapter, the international consortium of labs sequencing the human genome announced in the spring of 2003 that they had produced a finished draft of the human genome, two years earlier than originally planned. They published their work in 2004. The major advantages of this version over the rough draft were:

1. It was more complete. Ninety-nine percent of the sequence that was possible to obtain had been obtained—2,851,330,913 base pairs, or about 2.85 **gigabase pairs (Gb)** worth.
2. It was more accurate. The inaccuracy rate was a tiny 0.001%, and all the sequences were in the proper order.

However, there were still 341 gaps, though 33 of those were in the heterochromatic regions of the genome, which were not a target of this project. Still, biologists generally concede that we will have to live with many of these gaps, perhaps forever. Also, in spite of the polish on the finished product, annotation is still difficult, and we still do not know the real number of genes in the human genome. The international consortium found 22,287 protein-encoding genes (19,438 known genes and 2188 predicted genes), considerably fewer than estimated by both rough drafts of the human genome. The difference appears to be largely due to earlier double counting of apparent genes that actually map to the same true gene.

The estimated number of human genes has mostly decreased with time, at least if one includes only protein-encoding genes. In 2007, Michele Clamp reported an estimate of only 20,488 human genes, and allowed that a hundred or so remained undiscovered. She approached the question from a bioinformatics angle—using only computational tools. For example, she looked in a database called Ensembl for human genes and then compared those with counterparts in the dog and mouse genomes. This check of the presumed human genes showed that 19,209 really do code for proteins, while 3009 were on the list by mistake. Another 1177 putative genes remained in doubt, so Clamp analyzed them by comparing them to random DNA sequences for qualities of "geneness," such as genelike GC contents. All but 10 failed this test, yielding an estimate of 19,219 genes. Combining this with similar analyses of two other databases yielded a final estimate of 20,488.

What else have we learned from the finished draft? Here are a few examples: The estimated 22,289 genes appear to give rise to 34,214 transcripts, or about 1.5 per gene. These genes are represented by 231,667 exons, or about 10.4 per gene. The amount of DNA included in all these exons is just 34 Mb, which is only 1.2% of the euchromatic part of the human genome. This confirms something we already knew: The vast majority of the human genome does not contain protein-encoding genes. Some of it codes for useful RNAs, such as rRNAs, tRNAs, snRNAs, and miRNAs that are, of course, not translated. But the bulk of it appears not to be transcribed at all, and its functions, if any, remain a mystery.

The finished draft is also a great aid in the study of human evolution. First, it reveals newly duplicated genes that provide the raw material for new genes with new functions: One gene in the pair can retain its original function, but the other is free to collect mutations and evolve new activities, without compromising the original activity, which may be essential to life.

Second, the finished draft reveals newly inactivated genes, or pseudogenes. The search for pseudogenes began with a comparison of the rat, mouse, and human genomes to find strings of genes that were found in all three organisms. Then the investigators looked for genes within this string that were present in the rodents, but not in the human. Finally, they examined the region in the human genome predicted to contain these missing genes. They found 37 candidate pseudogenes that were still clearly recognizable, though they had all been inactivated. On average, each pseudogene had 0.8 premature stop codons and 1.6 frameshifts. Either of these types of mutation would have rendered the gene inactive. It is clear that these genes must not be essential to human life, though they presumably were to the common ancestor of humans, rats, and mice, and may still be to the rodents.

To verify that these apparent pseudogenes were really what they appeared to be, the investigators went back and sequenced 34 of them. In 33 cases, the inactivations were real, and in one case the apparent inactivation was due to a sequencing error. Then they compared these 33 sequences to the corresponding sequences in the chimpanzee genome. Nineteen of these pseudogenes had two or more inactivating mutations, and these were all pseudogenes in the chimpanzee as well. The other 14, with just one inactivating mutation, were more interesting. Eight of these were pseudogenes in the chimpanzee, but five were functional genes in the chimpanzee, and one is a polymorphism (present as a pseudogene in a fraction of the human population, but as a functional gene in the others). Thus, we can see the traces of gene inactivation through evolutionary time—since the rodent and human lineages diverged, and since the chimpanzee and human lineages diverged.

SUMMARY The working draft of the human genome reported by two separate groups allowed estimates that the genome contains fewer genes than anticipated. About half of the genome has derived from the action of transposons, and transposons themselves have contributed dozens of genes to the genome. In addition, bacteria appear to have donated at least dozens of genes. The finished draft of the human genome is much more accurate and complete than the working draft, but it still contains some gaps. On the basis of the finished draft, geneticists estimate that the genome contains about 20,000–25,000 genes. The finished draft also gives valuable information about gene birth and death during human evolution.

Personal Genomics

By 2007, two groups had used traditional sequencing techniques to sequence the genomes of two major players in the human genome project, James Watson and Craig Venter. By 2008, two different groups used high throughput sequencing to sequence the genomes of two non-Caucasian individuals, one of Nigerian descent, and one of Han Chinese descent. The addition of the genomes of these two individuals to the two previously sequenced genomes of individuals of European descent added more diversity to the growing pool of human genomes. One can detect millions of SNPs, hundreds of thousands of insertions and deletions, and thousands of structural variants among the four genomes. By 2010, several more individual genomes had been sequenced, including a European (French), a Southern African (San) and a Papua New Guinean.

As the speed and economy of DNA sequencing have improved, it has become possible to envision sequencing the genome of anyone who wants it and who is willing to pay the cost. The goal (with a significant cash prize) is to sequence a whole human genome for $1000. No one has claimed the prize yet, but high-throughput sequencing techniques (Chapter 5) are making it seem feasible that millions of people will one day have their whole genomic sequence on a flash drive, or whatever data storage medium is popular at that time. That wealth of information is bound to be valuable, but it also will create ethical problems.

Other Vertebrate Genomes

The complete sequences of the mouse and a pufferfish (the tiger pufferfish, *Fugu rubripes*) have been published. What lessons have these genomes taught us? Here are some of the most important:

The *Fugu* genome was chosen for sequencing because it is a vertebrate with a much smaller genome than human—only one-ninth the size. But despite the difference in size, the two genomes have about the same number of genes (31,059 predicted genes in *Fugu*). The difference lies, not in gene content, but in the size of introns and amount of repetitive DNA. The *Fugu* genome has much smaller introns than the human, and much less repetitive DNA. Comparing the *Fugu* and human genomes has allowed genomics researchers to identify 1000 human genes.

Because genetic mutations that cause human diseases are more likely to occur at important sites in genes, and because these important sites are especially well conserved, comparing two relatively distantly related vertebrate genomes, such as human and *Fugu,* should help identify these important sites. The mouse genome is not as useful for this purpose because it is relatively similar to the human genome. There simply has not been enough time for the mouse and human genomes to diverge very far, and many sites, not just important ones, have been conserved.

The mouse genome is a little smaller than the human, about 2.5 Gb compared with about 3 Gb, but both organisms have about the same number of genes, and a high percentage of these are the same in the two organisms: 99% of mouse genes have a counterpart we can identify in humans. This 1% difference is obviously much too little to account for the biological differences between humans and mice, so something besides sheer DNA sequence must be at work. Preliminary studies suggest that it is the control of the genes, not the genes themselves, that plays the biggest role in distinguishing humans from mice. Knowing the great similarity in genomic structure between mice and humans, scientists can use the mouse as a human surrogate in which to do experiments they could not do in humans. For example, they can knock out genes in mice and observe the effects. The results give us clues about what the homologous genes do in humans. Molecular biologists can also examine the expression patterns of mouse genes to learn when and where these genes are expressed during development and in adults. Again, these results give information about the expression of homologous genes in humans.

By the beginning of 2003, some of the best studies comparing the human and mouse genomes focused on chromosomes whose sequences were finished, including human chromosome 21 and mouse chromosome 16. Let us consider some results from each of these studies.

A comparison of the DNA in human chromosome 21 and equivalent DNA in the mouse has revealed about 3000 conserved sequences. Surprisingly, only half of these conserved sequences contain genes. However, the fact that they are so well conserved suggests that they are important, and we need to find out why. Perhaps they play a role in gene expression. Humans have 234 so-called "gene deserts" that are poor in genes. Again, it is surprising that 178 of these deserts are conserved in the mouse. And again, this degree of conservation of seemingly useless DNA demands an explanation. Accordingly, geneticists are knocking out some of those gene deserts in the mouse to see what effect their loss will have.

In 2002, Venter and colleagues reported a detailed comparison of the sequence of mouse chromosome 16 with sequences in the human genome. They found many regions

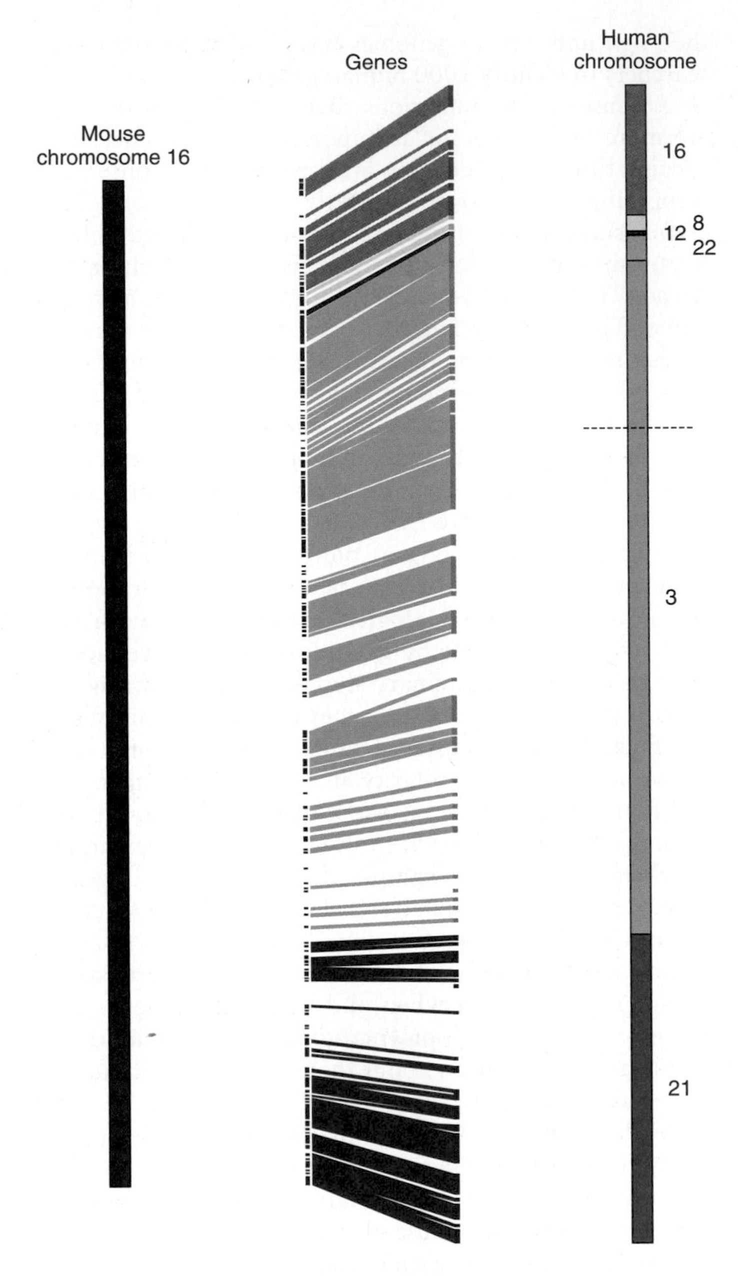

Figure 24.13 Regions of conserved synteny between mouse chromosome 16 and the human genome. Homologous genes were detected by analysis at the protein level. Mouse chromosome 16 is depicted at left, with the syntenic regions on six different human chromosomes illustrated at right (different colors indicate different human chromosomes). Orthologous genes in mouse and human are connected by colored lines in the middle of the diagram and indicated by tiny horizontal lines (purple, mouse; various colors, human). Genes homologous to mouse chromosome 16 in human chromosome 3 are found in two distinct syntenic blocks, separated by the dotted line. Above that line are human chromosome regions 3q27–29; below the line are regions 3q11.1–13.3. (*Source:* Adapted from Mural et al., *Science* 296 (2002) Fig. 3, p. 1666.)

of synteny, that is, regions with conserved gene order that appear to have derived from an ancestral mammalian chromosome. Figure 24.13 illustrates these syntenic regions, analyzed at the protein level. In all, mouse chromosome 16 has homologs on six human chromosomes (represented by different colors); the homologous genes on human chromosome 3 are found in two syntenic blocks, separated by the dotted line. Thus, all told, the genes on mouse chromosome 16 are represented in seven syntenic blocks in the human genome.

The degree of homology between syntenic regions in the two species is striking. Of 731 mouse genes that could be predicted with high confidence on the mouse chromosome, 717 (98%) have homologs in the human genome. This great homology far overshadows the fact that the mouse chromosome is represented in six separate human chromosomes and seven different syntenic blocks. Chromosomes frequently become scrambled during evolution without changing much if anything about gene expression, and without changing gene orders within large, syntenic blocks of genes. This can happen by chromosome breakage and translocation. For example, two closely related species of muntjac deer have experienced so much chromosome breakage (or joining, or both) since the two species diverged that one has 3 pairs of chromosomes, and the other has 23 pairs! Nevertheless, the two species can interbreed to produce healthy, albeit infertile, hybrids.

The degree of similarity of mice and humans at the genomic level is clearly out of proportion to the obvious differences in appearance and behavior between these two species. How do we explain this discrepancy? If we cannot find the answer in the genes themselves, it must lie in the way the genes are expressed. But some answers are already determined. We know that human genes are subject to an extraordinary amount of alternative splicing. In fact, it has been estimated that about 75% of human genes are spliced in at least two different ways in vivo (Chapter 14). This makes the human proteome (the total complement of human proteins) much more complex than the genome suggests. We also have evidence that the pattern of expression of human genes varies considerably from the expression of the almost identical set of genes in our closest relative, the chimpanzee, and varies even more from the pattern in mice. This could derive from control by miRNAs, which, in contrast to protein-encoding genes, seem to be much different in mice and humans.

Another source of variation in gene expression between two closely related species could come from the interaction between transcription factors and their binding sites on the DNA. As we have learned, eukaryotic genes have *cis*-control elements known as promoters and enhancers, and these are the targets of many transcription factors. We might predict that closely related species with highly conserved gene sets would also have highly conserved *cis*-control elements, but that seems not necessarily to be true. For example, Michael Snyder and colleagues reported in 2007 on ChIP analysis coupled with DNA microchip assays on the DNA targets for two transcription factors from three closely related species of yeast, which showed that these factors bound in the same places relative to the genes they control only 20% of the time in all three species. (This kind of experimentation is called ChIP-chip analysis and is described in more detail in Chapter 25.)

The great variation in transcription factor binding observed among these three yeast species is partly due to elements missing in one or two of the genomes, but it also sometimes occurs because a factor fails to bind, even when the element is still present. A similar phenomenon has been observed in a comparison of factor binding in human and mouse genomes.

How do we relate this rapid evolution of *cis*-regulatory elements to changes in phenotype between organisms? At this point, it is very difficult, because of uncertainty about how much each element contributes to expression of a particular gene. It is possible that most of the differences Snyder and colleagues observed play no role in the phenotypic differences among the three species, especially because of the redundancy that appears to be built into many *cis*-regulatory elements. On the other hand, it seems likely that some of these differences really are important to phenotype.

In 2005, scientists presented a working draft of the chimpanzee genome. Because the chimpanzee is our closest living relative, this sequence has special significance for evolutionary studies. Everyone wants to know what sets us apart from the chimpanzee. What genes give us the intelligence to build a city or write a symphony—or, for that matter, to wonder what makes us human? But a comparison of the chimpanzee and human genomes shows that we share almost all our protein-encoding genes in common, and our genomes differ by only 1.23% at the nucleotide level. Three hypotheses have been put forward to explain these data: (1) The important differences are changes in protein-encoding genes. (2) The "less is more" hypothesis, which holds that inactivation of certain genes in the human can explain the differences. (3) The differences are found in changes in gene control regions.

Each hypothesis has some data to support it. Despite the paucity of differences between chimpanzees and humans in protein-encoding genes, geneticists have noticed some differences that could make a big difference. For example, the *FOXP2* gene is highly conserved. It experienced only one change in amino acid coding in the approximately 130 million years between the divergence of the human and mouse lineages and the divergence of the human and chimpanzee lineages. But in the approximately 5 million years since the human and chimpanzee lineages diverged, two amino acid changes occurred. Why might the *FOXP2* gene be important? It encodes a forkhead class transcription factor, and mutations in this gene cause severe speech impairment in humans. And, of course, speech is one of the key traits that sets humans and chimpanzees apart.

The "less is more" hypothesis also has some support. For example, it is easy to imagine that the relative lack of hair in humans is due to the loss or inactivation of a gene responsible for hairiness. And a comparison of the human and chimpanzee genomes has uncovered 53 examples of human genes that have been disrupted by insertions or deletions (**indels**). These genes are functional in chimpanzees, but inactive in humans.

There is less direct experimental support for the third hypothesis—differences in gene control—because of the difficulty in identifying the genetic elements responsible. But the great similarity in the protein-coding regions of the two species suggests we look elsewhere, and genetic control is an attractive place to look. Indeed, as we will see, the most rapidly changing DNA sequences that distinguish the human and chimpanzee genomes are in apparently noncoding DNA regions. The easiest way to make sense of this finding is to say that these DNA regions are involved in controlling the protein-encoding genes.

David Haussler and colleagues took the following approach to finding important differences—coding or noncoding—between the human and chimpanzee genomes. They used computational techniques to identify genome regions that are strongly conserved among vertebrates. Then they looked in these regions to find regions of DNA that had experienced a high rate of change since the divergence of humans and chimpanzees. They found 49 such regions, which they named HAR1–HAR49 (HAR = human accelerated regions). HAR1, a 118-bp DNA region, stood out most of all. In the 310 million years since the chicken and chimpanzee lineages diverged, only two changes occurred. However, in the 5 million years since the human and chimpanzee lineages diverged, fully 18 changes have occurred.

Haussler and colleagues then used in situ hybridization on brain slices and found that one of the two RNAs (HAR1F) that includes the HAR1 region is expressed in the developing cerebral neocortex of humans and other primates. The neocortex is thought to be central to higher cognitive function—perhaps the most salient difference between chimpanzees and humans.

Thus, we know that HAR1 gives rise to two RNAs, but these RNAs appear not to encode any proteins. However, the base sequence of HAR1F allows a prediction of a stable secondary structure (intramolecular base-pairing). And the changes between the chimpanzee and human forms of HAR1F are predicted to cause a significant difference in secondary structure, including a strengthening of base-pairing. We do not know yet what HAR1F and HAR1R do, but a reasonable hypothesis is that one or both of these RNAs influence the expression of protein-encoding genes in the developing human brain and give it some of its cognitive power.

One striking finding from the work of Haussler and colleagues, as well as other workers in this field, is that the most rapid changes in the genomes of humans and chimpanzees has not been in protein-encoding genes, but in noncoding regions of the genome.

Although the chimpanzee is our closest living relative, our closest evolutionary relative is the Neanderthal (*Homo neanderthalensis*), which has been extinct for about 30,000 years. In 2010, a group led by Svante Pääbo succeeded in a task many people assumed was impossible—they reported a draft sequence of the Neanderthal genome. The problem with sequencing the genome of a fossil organism is that the DNA is badly degraded, and therefore commonly thought to be unfit for sequencing. But Pääbo and colleagues solved this problem by using next-generation sequencing techniques, in which DNAs are intentionally fragmented to begin with, so DNAs that are already fragmented pose less of a problem. Another difficulty was that the bone samples from which the Neanderthal DNA came were massively contaminated with bacterial DNA, but Pääbo and colleagues minimized that problem by cutting the DNA with restriction enzymes whose recognition sites include CG sequences, which are rare in mammals, but common in microbes. This reduced the size of most microbial DNA fragments to the point that they did not interfere with the sequencing.

One limitation of next-generation sequencing is that the DNA fragments are frequently too short to exhibit obvious overlaps, so they cannot be pieced together to form a whole genome. But that is not a problem if a closely related species has already had its genome sequenced, so the fragments can be compared to that sequence and placed in the proper order. Because the human genome was already available, Pääbo and colleagues could use it as a framework for their Neanderthal sequence, which they obtained from DNA extracted from well-preserved fossil remains.

It is fascinating to have the Neanderthal sequence for many reasons. For example, it appears to be able to answer the question whether modern humans and Neanderthals interbred. The two species coexisted for at least 10,000 years in Europe and Asia, until the Neanderthals disappeared, so interbreeding was certainly possible. If interbreeding occurred, and the offspring were fertile, it should be possible to find traces of the Neanderthal genome in the present human genome. Indeed, Pääbo and colleagues found similarities between the Neanderthal genome and the genomes of a modern European (French), a modern East Asian (Han Chinese), and a modern Papua New Guinean, but these similarities did not extend to the genomes of two modern sub-Saharan Africans (a San from Southern Africa and a Yoruba from West Africa). Thus, Neanderthals did apparently interbreed with the ancestors of modern Eurasians, but this happened after the Eurasian and African lineages diverged. Also, because the Neanderthal genome resembles the Papua New Guinean, Chinese, and European genomes equally closely, the interbreeding appears to have happened before those lineages diverged.

Pääbo and colleagues also reported the full Neanderthal mitochondrial DNA sequence, in 2008. They eliminated errors and minimized the effect of contamination by sequencing so thoroughly that each base was represented in at least 35 independent reads. Gaps and ambiguities were resolved by traditional sequencing. The modern human and Neanderthal mitochondrial sequences differ in an average of 206 bases. This contrasts with differences between modern human mitochondrial sequences that vary between 2 and 118 bases. These data allowed Pääbo and colleagues to estimate the time of divergence between the modern human and Neanderthal lineages at about 660,000 years ago.

SUMMARY Comparing the human genome with that of other vertebrates has already taught us much about the similarities and differences among genomes. Such comparisons have also helped to identify many human genes. In the future, such comparisons will help find the genes that are defective in human genetic diseases. One can also use closely related species like the mouse to find when and where their genes are expressed and therefore to estimate when and where the corresponding human genes are expressed. Detailed comparison of mouse and human chromosomes has revealed a high degree of synteny between the two species. Comparisons of the human genome with that of our closest living relative, the chimpanzee, have identified a few DNA regions that have changed rapidly since the two species diverged. These are good candidates for the DNA sequences that set humans and chimps apart, yet very few of them are in protein-encoding genes. Thus, the thing that really sets us apart may be control of genes, rather than the genes themselves. Studies in yeasts have shown that even closely related species have great variation in the *cis*-regulatory elements that control their genes, though the genes themselves are highly conserved. Thus, *cis*-regulatory elements are subject to relatively rapid evolution, and that may help to explain differences in gene control, and therefore in phenotype. More insight into what makes us human will come from the genome of the Neanderthal. A working draft of this genome, as well as a finished version of the mitochondrial DNA, have already been published.

The Minimal Genome

By early 2002, over 50 bacterial genomes had been sequenced. The smallest of these genomes belong to intracellular parasites, such as mycoplasmas, *Rickettsia* (one of whose members causes Rocky Mountain spotted fever), and parasitic spirochetes like *Borrelia burgdorferi,* which causes Lyme disease. The record for smallest bacterial genome is held by *Mycoplasma genitalium,* at only 530 kb. This kind of analysis has led some geneticists to ask, "What is the smallest genome that is still compatible with life?"

One way to answer this question would be to compare the genomes of bacteria and find the lowest common denominator: the set of genes they all have in common. But that yields a set of only about 80 genes, which is clearly too few to sustain life. Thus, different bacteria have followed different paths to streamlining their genomes, and it is therefore not useful simply to find where the endpoints of these different paths overlap.

In 1999, Craig Venter and colleagues reported the results of another approach to finding the minimal genome. They systematically mutagenized the genes in *Mycoplasma genitalium* and the related species *M. Pneumoniae,* using transposons to interrupt the genes. Then they looked to see which genes were essential, and which were not. They discovered that 265–350 of the 480 protein-encoding genes in these organisms are essential. Surprisingly, 111 of these genes had unknown functions, suggesting that we still have a lot to learn about what it takes to sustain life.

This experiment identified the **essential gene set,** that is the set of genes whose loss is incompatible with life. But that is not the same as the **minimal genome,** the collection of genes that would sustain life in a real organism. The distinction comes from the fact that an organism can afford to lose certain genes by themselves, but loss of two or more of these same genes together is not compatible with life. Thus, these genes are not part of the essential gene set, but they are part of the minimal genome.

The next task was to discover which genes need to be added to the essential gene set to produce a minimal genome. Venter and colleagues proposed to perform this task in a spectacularly ambitious way. They aimed to synthesize DNA from scratch, building DNA cassettes carrying several genes. Then they would place these cassettes into *Mycoplasma* cells whose own genes had been disabled so they would not confuse the issue. They would experiment with different combinations of genes until they found the combination with the smallest number of genes that could still support life.

This plan had to deal with a difficult hurdle to get the genes to function appropriately in a new cell without any genes of its own. It is true that one can place one or a few foreign genes into a normal bacterial cell and get them to turn on very well. But what about an entirely new gene set? There was a significant chance that the genes would not turn on, but would just sit there. Bernhard Palsson has stated the problem this way: "How do you boot up a new genome?"

However, by 2007, Venter and colleagues had reported progress that showed that booting up a genome really does work. They transplanted the genome of *Mycoplasma mycoides* to another bacterium, *Mycoplasma capricolum,* and the resulting cell thrived with its new genome. However, they had to use some creative manipulations to make the transplant work. First, they added an antibiotic resistance gene to the donor bacteria (*M. mycoides*), and embedded these cells in an agarose gel. Then they broke open the cells and digested their proteins with proteolytic enzymes. (*Mycoplasma* cells lack a cell wall, which makes it easier to break them.) With the released circular genome protected from physical stress by the agarose, the recipient bacteria (*M. capricolum*) were added, along with the membrane-fusing agent polyethylene glycol. Apparently, some of these recipient bacterial membranes opened up and then fused around the naked donor genomes.

Instead of destroying the recipient cell's genome, Venter and colleagues played a clever trick involving the antibiotic resistance gene they had placed in the donor cell genome. After fusion, the recipient cell found itself with two genomes: one it had always had, and one from the donor. With two genomes, the cell was ready to divide, and proceeded to do so. One daughter cell got the donor genome, and the other got the recipient genome. But only the daughter cell with the donor genome had the antibiotic resistance gene, so growing the cells in the presence of the antibiotic automatically removed the cells with the recipient genome. The result was that all of the cells that formed early in the experiment were *M. capricolum* cells with a *M. mycoides* genome.

In 2010, Venter and colleagues used a similar technique to introduce an entirely synthetic *M. mycoides* genome into *M. capricolum* cells. The success of this experiment ushered in a new era of "synthetic biology." Of course, the engineered organisms are not truly synthetic—only their genome is—but they represent a milestone nonetheless. A potential ethical question might remain: Is it ethical to create life from nonliving ingredients? Recognizing this issue, Venter and colleagues submitted their plan to a panel of ethicists, who decided in 1999 that it presented no serious ethical problems. But they did see some safety issues, and recommended that public officials should examine the possibility that the artificial life forms Venter and his colleagues would create could pose an environmental hazard, or that they might lend themselves to being modified for use as agents of bioterrorism or biowarfare.

To at least partially address the safety issue, Venter and colleagues have endowed their synthetic genome with a watermark—a DNA sequence not found in nature—that will enable the engineered organisms to be identified. Use of these organisms by terrorists seems very unlikely because a great deal of sophistication will be needed to create them, and there is no indication that they would be any more dangerous than highly toxic natural organisms that are already available. Ethical questions remain, however, and President Obama convened an ethics panel to study the issues and issue a report by the end of 2010.

Why build an organism with a minimal genome? On a purely scientific level, it will be important to show that there is such a thing as a minimal genome, and then to investigate why these particular genes are required. But practical applications are also possible. Indeed, Venter and colleagues plan to supplement the minimal genome with genes that will enable the bacteria to create fuels such as hydrogen, or to clean up industrial waste, including CO_2 from power plants.

This does not mean that traditional organisms will lose out to synthetic ones with minimal genomes as the microbial workhorses of the future. Frederick Blattner and his colleagues have been trimming away the genome of *E. coli* to build an organism with a reduced genome that is hospitable to new genes. Their strategy is to identify genes that differ from one strain to another, and are therefore probably dispensable. They have found that these genes tend to cluster in "islands" that can be conveniently deleted. As of late 2005, they had made 43 deletions, cutting the genome's size down by more than 10%. Already, this altered bacterium was ten times better at accepting new genes than typical laboratory strains. By late 2007, Blattner's group had pared away 14% of the *E. coli* genome without harming the ability of the cells to grow and express foreign genes, and more trimming remains to be done.

Finally, it is worth noting that *M. genitalium* and other intracellular parasites can get away with such a small genome because of their parasitic lifestyle. They get many of their nutrients from their hosts, so they can safely shed the genes that produce those nutrients. In fact, *M. genitalium* may already have honed its genome to something close to the minimum required for life in its human host. But scientists may be able to hone it even further—to the minimum required to live under rigidly controlled laboratory conditions.

SUMMARY It is possible to define the essential gene set of a simple organism by mutating one gene at a time to see which genes are required for life. In principle, it is also possible to define the minimal genome—the set of genes that is the minimum required for life. It is likely that this minimal genome is larger than the essential gene set. It is also possible to place this minimal genome into a cell lacking genes of its own and thereby create a new form of life that can live and reproduce under laboratory conditions. With selected genes added, such a life form could be modified to perform many useful tasks.

The Barcode of Life

Taxonomists are in the business of classifying organisms and understanding their differences and relationships. Traditionally, they have relied on simple appearances, or morphological characteristics, to distinguish among different species. Now, in the era of DNA sequencing, they have gained another tool, because different species have different DNA sequences, as well as different appearances. Moreover, the degree of difference in the DNA sequences between two species is a good measure of their evolutionary distance, or the time since the two diverged, assuming a constant rate of mutation.

However, with millions of species to study, there is no hope with present technology of sequencing the whole genomes of even a significant fraction of all these species. Instead, taxonomists focus on small regions of the genome that show a significant amount of variation among the species they are studying. Now, a group of scientists called the Consortium for the Barcode of Life (CBOL) is proposing to obtain a relatively short DNA sequence, or **barcode,** from the genome of every species on earth. In principle, this would allow the rapid identification of any known species, including agents of bioterrorism, and it would help to place new species on the proper branch of the tree of life. The work would start with the 1.7 known species of animals and plants and then move to the rest of the 10 million or more unknown species (not counting microbes).

CBOL scientists settled on a 648-bp region from the mitochondrial cytochrome c oxidase subunit I (COI) gene as the barcode, at least for animals. This gene is present in all organisms. And, at least in animals, it shows a good degree of difference between closely related species, but little difference between members of the same species. For example, the barcodes in different human beings differ from one another by only one or two base pairs out of 648, while those in humans and chimpanzees, our closest living relatives in the tree of life, differ by 60 bp. Moreover, a sequence of 648 bp is easy and cheap to obtain in one run of a traditional automated sequencer, and mitochondrial DNA is relatively easy to purify because each cell contains 100–10,000 copies instead of just two copies in nuclear DNA.

One drawback to the COI barcode is that plant mitochondrial DNA sequences show much less variation than animal sequences do, so the COI barcode will not work well for plants. Instead, a consortium of plant systematists, known as the Plant Working Group of COBOL, has proposed using sequences from two chloroplast genes (*matK* and *rbcL*) for the plant barcode. This is not a perfect solution, as this barcode works better for some plant species than for others. But it has correctly identified 72% of all plant species, and has a perfect record in placing plants in the proper genus.

In Richard Preston's novel, *The Cobra Event,* a deranged man creates a very nasty virus and releases it in New York City. But scientists in the book have an invaluable tool for detecting such agents—a handheld device that almost instantly identifies microbes. We are clearly not at that point yet. However, someday it may be possible to miniaturize DNA sequencers to the point that they could be used as field devices for quick identification, via barcodes, of unknown organisms.

SUMMARY A movement has begun to create a barcode to identify any species of life on earth. The first "barcode of life" will consist of the sequence of a 648-bp piece of the mitochondrial COI gene from each organism. This sequence is sufficient to uniquely identify almost any animal. Other sequences, or barcodes, are being worked out for plants.

SUMMARY

Several methods are available for identifying the genes in a large, unsequenced DNA region. One of these is the exon trap, which uses a special vector to help clone exons only. Another is to use methylation-sensitive restriction enzymes to search for CpG islands—DNA regions containing unmethylated CpG sequences. Before the genomics era, geneticists mapped the Huntington disease gene (*HD*) to a region near the end of chromosome 4. Then they used an exon trap to identify the gene itself.

Rapid, automated DNA sequencing methods have allowed molecular biologists to obtain the base sequences of viruses and organisms ranging from simple phages to bacteria to yeast, simple animals, plants, mice, and humans. Much of the mapping work in the Human Genome Project was done with yeast artificial chromosomes (YACs), vectors that contain a yeast origin of replication, a centromere, and two telomeres. Foreign DNA up to 1 million bp long can be inserted between the centromere and one of the telomeres. It will then replicate along with the YAC. On the other hand, because of their superior stability and ease of use, most of the sequencing work in the Human Genome Project was done with bacterial artificial chromosomes (BACs). BACs are vectors based on the F plasmid of *E. coli*. They can accept inserts up to about 300 kb, but their inserts average about 150 kb.

Mapping the human genome, or any large genome, requires a set of landmarks (markers) to which one can relate the positions of genes. Genes can be used as markers in mapping, but markers are usually anonymous stretches of DNA such as RFLPs, VNTRs, STSs (including ESTs), and microsatellites. RFLPs (restriction fragment length polymorphisms) are differences in the lengths of restriction fragments generated by cutting the DNA of two or more different individuals with a restriction endonuclease. RFLPs can be caused by the presence or absence of a restriction site in a particular place or insertions and deletions between restriction sites. They can also be caused by a variable number of tandem (head-to-tail) repeats (VNTRs) between two restriction sites. STSs (sequence-tagged sites) are regions of DNA that can be identified by formation of a predictable length of amplified DNA by PCR with pairs of primers. ESTs (expressed sequence tags) are a subset of STSs generated from cDNAs, so they represent expressed genes. Microsatellites are a subset of STSs generated by PCR with pairs of primers flanking tandem repeats of just a few nucleotides (usually 2–4 nt).

Radiation hybrid mapping allows mapping of STSs and other markers that are too far apart to fit on one BAC. In radiation hybrid mapping, human cells are irradiated to break chromosomes, then these dying cells are fused with hamster cells. Each hybrid cell has a different subset of human chromosome fragments. The closer together two markers are, the more likely they are to be found in the same hybrid cell.

Massive sequencing projects can take two forms: The map-then-sequence (clone-by-clone) approach or the shotgun approach. Actually, a combination of these methods was used to sequence the human genome. The clone-by-clone strategy calls for production of a physical map of the genome including STSs, then sequencing the overlapping clones (mostly BACs) used in the mapping. This places the sequences in order so they can be pieced together. The shotgun strategy calls for the assembly of libraries of clones with different size inserts, then sequencing the inserts at random. This method relies on a computer program to find areas of overlap among the sequences and piece them together.

Sequencing of human chromosome 22q has revealed: (1) gaps that cannot be filled with available methods; (2) 855 annotated genes; (3) the great bulk (about 97%) of the chromosome is made up of noncoding DNA; (4) over 40% of the chromosome is in interspersed repeats such as Alu sequences and LINEs; (5) the rate of recombination varies across the chromosome, with long regions of low rates of recombination punctuated by short regions with relatively high rates; (6) several examples of local and long-range duplications; (7) large regions where linkage among genes has been conserved with that in seven different mouse chromosomes.

The working draft of the human genome reported by two separate groups allowed estimates that the genome probably contains fewer genes than anticipated. About half of the genome has derived from the action of transposons, and transposons themselves have contributed dozens of genes to the genome. In addition, bacteria appear to have donated at least dozens of genes. The finished draft of the human genome is much more accurate and complete than the working drafts, but it still contains some gaps. On the basis of the finished draft, geneticists estimate that the genome contains about 20,000–25,000 genes. The finished draft also gives valuable information about human evolution.

Comparing the human genome to that of other vertebrates has already taught us much about the similarities and differences among genomes. Such comparisons have also helped to identify many human genes. In the future, such comparisons will help find the genes that are defective in human genetic diseases. One can also use closely related species like the mouse to find when and where their genes are expressed and therefore to estimate when and where the corresponding human genes are expressed. Detailed comparison of mouse and human chromosomes has revealed a high degree of synteny between the two species.

It is possible to define the essential gene set of a simple organism by mutating one gene at a time to see which genes are required for life. It is also possible to define the minimal genome—the set of genes that is the minimum required for life. It is likely that this minimal genome is larger than the essential gene set. In principle, it is also possible to place this minimal genome into a cell lacking genes of its own and thereby create a new form of life that can live and reproduce under laboratory conditions. With selected genes added, such a life form could be modified to perform many useful tasks.

A movement has begun to create a barcode to identify any species of life on earth. The first "barcode of life" will consist of the sequence of a 648-bp piece of the mitochondrial COI gene from each organism. This sequence is sufficient to uniquely identify almost any animal. Other sequences, or barcodes, are being worked out for plants.

REVIEW QUESTIONS

1. What is a CpG island? Why have CpG sequences tended to disappear from the human genome?
2. a. What kind of mutation gave rise to Huntington disease?
   b. What is the evidence that the gene identified as *HD* is really the gene that causes HD?
3. What is an open reading frame (ORF)? Write a DNA sequence containing a short ORF.
4. What are the essential elements of a YAC vector?
5. On what plasmid are the BAC vectors based? What essential elements do they contain?
6. Describe the procedure for finding an STS in a genome.
7. Describe microsatellites and minisatellites. Why are microsatellites better tools for linkage mapping than minisatellites?
8. Show how to use STSs in a set of BAC clones to form a contig. Illustrate with a diagram different from the one given in the text.
9. Describe the use of radiation hybrid mapping to map STSs.
10. How does an expressed sequence tag (EST) differ from an ordinary STS?
11. Compare and contrast the clone-by-clone sequencing strategy and the shotgun sequencing strategy for large genomes.
12. What major conclusions can we draw from the sequence of human chromosome 22?
13. What is a pseudogene?
14. What is the difference between an ortholog and a paralog?
15. How do scientists estimate the number of genes in complex eukaryotes like humans?
16. The tiger pufferfish (*Fugu rubripes*) genome is nine times smaller than the human genome, but it contains just as many genes. How can that be?
17. What do we mean by "syntenic regions" in the mouse and human genomes?
18. Humans appear to have about as many protein-encoding genes as roundworms. How do you explain the lack of correspondence between the apparent numbers of genes and the complexities of these two organisms?
19. What is the difference between an organism's "essential gene set" and its "minimal genome?"

ANALYTICAL QUESTIONS

1. Will the following DNA fragments be detected by an exon trap? Why or why not?
   a. An intron
   b. Part of an exon
   c. A whole exon with parts of introns on both sides
   d. A whole exon with part of an intron on one side
2. The following is a physical map of a region you are mapping by RFLP analysis.

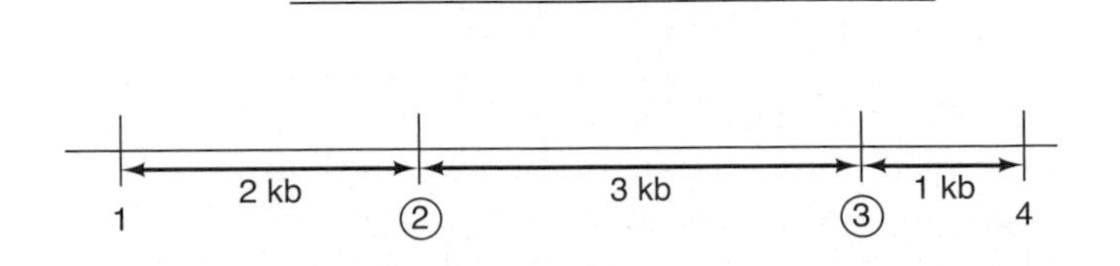

The numbered vertical lines represent restriction sites recognized by *Sma*I. The circled sites (2 and 3) are polymorphic, the others are not. You cut the DNA with *Sma*I, electrophorese the fragments, blot them to a membrane, and probe with a DNA whose extent is shown at top. Give the sizes of fragments you will detect in individuals homozygous for the following haplotypes with respect to sites 2 and 3.

Haplotype	Site 2	Site 3	Fragment sizes
A	Present	Present	
B	Present	Absent	
C	Absent	Present	
D	Absent	Absent	

3. You are mapping the gene responsible for a human genetic disease. You find that the gene is linked to a RFLP detected with a probe called X-21. You hybridize labeled X-21 DNA to DNAs from a panel of mouse–human hybrid cells. The following shows the human chromosomes present in each hybrid cell line, and whether the probe hybridized to DNA from each. Which human chromosome carries the disease gene?

Cell Line	Human chromosome content	Hybridization to X-21
A	1, 5, 21	+
B	6, 7	–
C	1, 22, Y	–
D	4, 5, 18, 21	+
E	8, 21, Y	–
F	2, 5, 6	+

4. You have just obtained the sequence of the genome of an organism that has been the subject of considerable genetic study. Describe how you would identify genomic regions that have experienced high rates of recombination. Explain the reasoning behind your approach.

SUGGESTED READINGS

General References and Reviews

Ball, P. 2007. Designs for life. *Nature* 448:32–33.

Collins, F.S., M.S. Guyer, and A. Chakravarti. 1997. Variations on a theme: Cataloging human DNA sequence variation. *Science* 278:1580–81.

Fields, S. 2007. Site-seeing by sequencing. *Science* 316:1441–42.

Goffeau, A. 1995. Life with 482 genes. *Science* 270:445–46.

Goffeau, A., B.G. Barrell, H. Bussey, R.W. Davis, B. Dujon, H. Feldmann, et al. 1996. Life with 6000 genes. *Science* 274:546–67.

Levy, S., and R.L. Strausberg. 2008. Individual genomes diversify. *Nature* 456:49–51.

Morell, V. 1996. Life's last domain. *Science* 273:1043–45.

Murray, T.H. 1991. Ethical issues in human genome research. *FASEB Journal* 5:55–60.

Ponting, C.P. and G. Lunter. 2006. Human brain gene wins genome race. *Nature* 443:149–50.

Reeves, R.H. 2000. Recounting a genetic story. *Nature* 405:283–34.

Venter, J.C., H.O. Smith, and L. Hood. 1996. A new strategy for genome sequencing. *Nature* 381:364–66.

Zimmer, C. 2003. Tinker, tailor: Can Venter stitch together a genome from scratch? *Science* 299:1006–07.

Research Articles

Bentley, D.R. et al. 2008. Accurate whole human genome sequencing using reversible terminator chemistry. *Nature* 456:53–59.

Blattner, F.R., G. Plunkett 3rd, C.A. Bloch, N.T. Perna, V. Burland, M. Riley, et al. 1997. The complete genomic sequence of *Escherichia coli* K12. *Science* 277: 1453–62.

Bult, C.J., O. White, G.J. Olsen, L. Zhou, R.D. Fleischmann, G.G. Sutton, et al. 1996. Complete genome sequence of the methanogenic archaeon, *Methanococcus jannaschii*. *Science* 273:1058–73.

C. elegans Sequencing Consortium. 1998. Genome sequence of the nematode *C. elegans:* A platform for investigating biology. *Science* 282:2013–18.

Deloukas, P., G.D. Schuler, G. Gyapay, E.M. Beasley, C. Soderlund, P. Rodriguez-Tome, et al. 1998. A physical map of 30,000 human genes. *Science* 282:744–46.

Dunham, I., N. Shimizu, B.A. Roe, S. Chissoe, A.R. Hunt, J.E. Collins, (The Chromosome 22 Sequencing Consortium). 1999. The DNA sequence of human chromosome 22. *Nature* 402:489–95.

Grimson, A., M. Srivastava, B. Fahey, B.J. Woodcroft, H.R. Chiang, N. King, B.M. Degnan, D.S. Rokhsar, and D.P. Bartel. 2008. Early origins and evolution of microRNAs and Piwi-interacting RNAs in animals. *Nature* 455:1193–97.

Gusella, J.F., N.S. Wexler, P.M. Conneally, S.L. Naylor, M.A. Anderson, R.E. Tauzi, et al. 1983. A polymorphic DNA marker genetically linked to Huntington's disease. *Nature* 306:234–38.

Hudson, T.J., L.D. Stein, S.S. Gerety, J. Ma, A.B. Castle, J. Silva, et al. 1995. An STS-based map of the human genome. *Science* 270:1945–54.

Hutchinson, C.A. III, S.N. Peterson, S.R. Gill, R.T. Cline, O. White, C.M. Fraser, H.O. Smith, and J.C. Venter. 1999. Global transposon mutagenesis and a minimal mycoplasma genome. *Science* 286:2165–69.

International HapMap Consortium. 2005. A haplotype map of the human genome. *Nature* 437:1299–1320.

International Human Genome Sequencing Consortium. 2001. Initial sequencing and analysis of the human genome. *Nature* 409:860–921.

Mural, R.J., M.D. Adams, E.W. Myers, H.O. Smith, G.L. Miklos, R. Wides, et al. 2002. A comparison of whole-genome shotgun-derived mouse chromosome 16 and the human genome. *Science* 296:1661–71.

Pääbo, S. and many other authors. 2008. A complete Neandertal mitochondrial genome sequence determined by high-throughput sequencing. *Cell* 134:416–26.

Pääbo, S. and many other authors. 2010. A draft sequence of the Neandertal genome. *Science* 328:710–22.

Schuler, G.D., M.S. Boguski, E.A. Stewart, L.D. Stein, G. Gyapay, K. Rice, et al. 1996. A gene map of the human genome. *Science* 274:540–46.

Shizuya, H., B. Birren, U.-J. Kim, V. Mancino, T. Slepak, Y. Tachiiri, and M. Simon. 1992. Cloning and stable maintenance of 300-kilobase-pair fragments of human DNA in *Escherichia coli* using an F-factor-based vector. *Proceedings of the National Academy of Sciences USA* 89:8794–97.

Venter, J.C., M.D. Adams, E.W. Myers, P.W. Li, R.J. Mural, G.G. Sutton, et al. 2001. The sequence of the human genome. *Science* 291:1304–51.

CHAPTER 25

Genomics II: Functional Genomics, Proteomics, and Bioinformatics

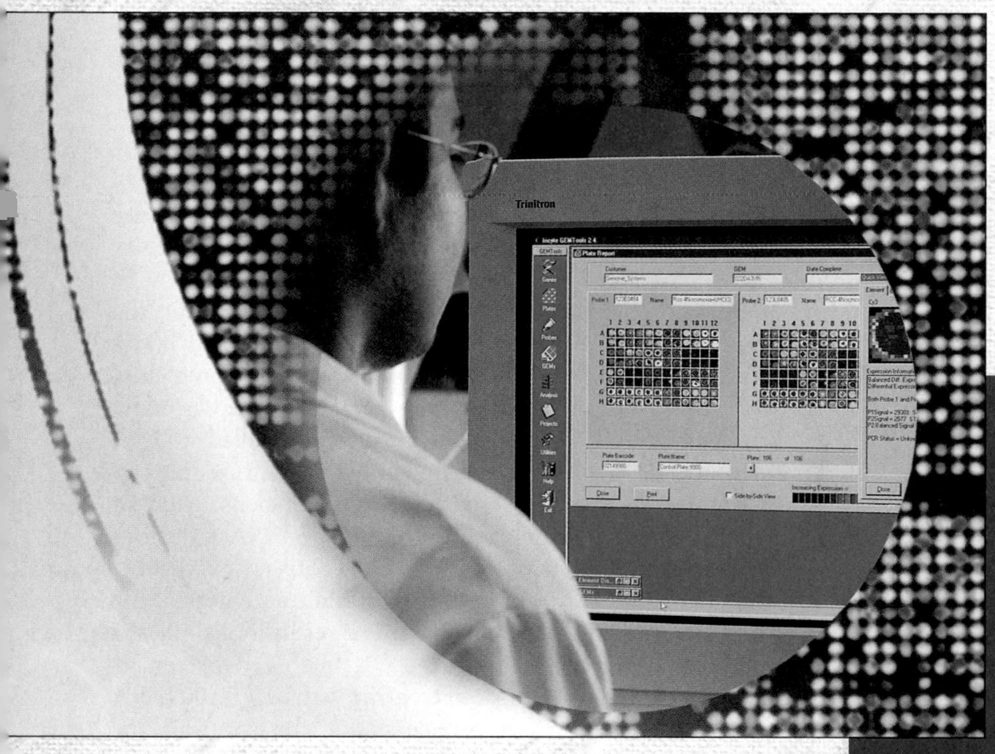

A DNA microarray used to measure expression of thousands of genes at a time. *Inset:* A technician analyzes expression of particular genes. *(Copyright © IncyteGenomics.)*

In Chapter 24, we dealt mostly with the process of finding out the sequences of genomes, and the lessons to be learned simply by looking at those sequences and comparing them with others. There are many other applications, all of which could be labeled "postgenomic" because they depend on the preexistence of genomic information. One major class of applications can be called **functional genomics** because they deal with the function, or expression, of genomes.

We will begin this chapter with an examination of functional genomics. Then we will consider a quest that is even more complex than genomics: **proteomics,** the study of an organism's **proteome**—the properties and activities of all the proteins an organism makes in its lifetime. Finally, we will introduce **bioinformatics,** the discipline concerned with managing and using the vast stores of data that come from genomic, proteomic, and other massive biological studies.

25.1 Functional Genomics: Gene Expression on a Genomic Scale

First of all, one can focus on expression of genomes at the RNA level. If we consider all the transcripts an organism makes at any given time, we call that the organism's **transcriptome,** by analogy with the term "genome," which refers to all the genes in an organism. And functional genomics studies that measure the levels of RNAs produced from many genes at a time are part of a field called **transcriptomics.** Second, one can use genomic information to try to determine the pattern of expression of all the genes in an organism at all stages of the organism's life. This kind of analysis is called **genomic functional profiling.**

Third, one can compare many individuals' genomes to find significant differences. For example, differences in single nucleotides are called **single-nucleotide polymorphisms (SNPs).** Sometimes these SNPs are associated with genetic disorders or other, less dramatic characteristics, such as susceptibilities to drugs. But SNPs are not the only common differences among human genomes. The more geneticists look, the more they find major chromosomal structural variations, such as inversions, duplications, and deletions. Moreover, at least some of these variations appear to have important consequences. For example, one long inversion has been found to be common in Europeans, but not in Africans and Asians, and women with this inversion have more children than those without it. Thus, the inversion seems to provide an evolutionary advantage.

Finally, one can study the structures and functions of the protein products of genomes. To the extent that it focuses on protein structure, this latter enterprise can be called **structural genomics,** but the whole endeavor is called **proteomics,** and will be the subject of a later section of this chapter. In this section, we will consider transcriptomics, genomic functional profiling, and SNPs.

Transcriptomics

To discover the pattern of expression of a gene in a given tissue over time, one can perform a **dot blot** analysis as described in Chapter 5. In a classical dot blot, one makes spots a few mm in diameter containing a single-stranded DNA from the gene in question on filters and then hybridizes these dot blots to labeled RNAs made in the tissue in question at different times. But suppose one wants to know the pattern of expression of all the genes in that tissue over time. In principle, one could make a large dot blot with tens of thousands of single-stranded DNAs corresponding to all the potential mRNAs in a cell and hybridize labeled cellular RNAs to that monster dot blot. But the sheer size of that blot would present a serious problem. Fortunately, molecular biologists have devised some methods to miniaturize such

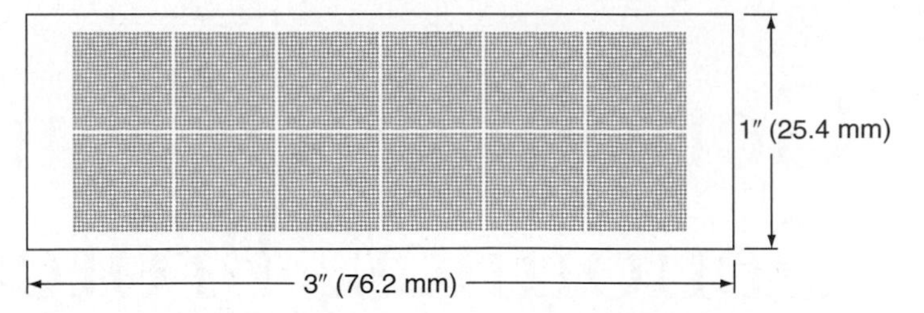

Figure 25.1 Schematic diagram of a DNA microarray. This drawing represents a standard, 1″ × 3″ glass microscope slide with an array of 7500 tiny spots of DNA. Each dot is 200 μm in diameter, and the distance between the dot centers is 400 μm. This is by no means the highest density of spots presently attainable. It is actually possible to place more than 50,000 spots on a slide of this size. (*Source:* Adapted from Cheung, V.G., M. Morley, F. Aguilar, A. Massimi, R. Kucherlapati, and G. Childs, Making and reading microarrays. *Nature Genetics Supplement* Vol. 21 (1999) f. 2, p. 17.)

blots, and some novel methods to analyze the expression of whole genomes. We will look first at DNA arrays and gene microchips, and then at a more exotic method.

DNA Microarrays and Microchips To circumvent the problem of size, molecular biologists have adapted inkjet printer technology to spot tiny volumes of DNA on a chip, so the dots of the dot blot are very small. This allows many different DNAs to be spotted on one chip, called a **DNA microarray.** One system, developed by Vivian Cheung and colleagues, uses a robot with 12 parallel pens, each of which can squirt out a tiny volume of DNA solution: 0.25–1.0 nL (billionth of a liter). The spots are exquisitely small, only 100–150 μm in diameter, and the centers of the spots are only 200–250 μm apart. The result looks like the schematic diagram in Figure 25.1, but even better, as the figure represents a DNA microarray with only 7500 DNA spots on a common microscope slide. After spotting, the DNAs are air dried, and covalently attached by ultraviolet radiation to a thin silane layer on top of the glass.

Another strategy for reducing the size of a blot has been to synthesize many oligonucleotides simultaneously, right on the surface of a chip. Steven Fodor and his colleagues pioneered this method in 1991, using the same kind of photolithographic techniques employed in computer chip manufacture, to build short DNAs (oligonucleotides) on tiny, closely spaced spots on a small glass microchip. In a 1999 version of this technique (Figure 25.2), these workers started with a small glass slide coated with a synthetic linker that was blocked with a photoreactive group that can be removed by light. They masked some of the areas of the slide and illuminated it, so the blocking agent was removed only from the unmasked areas. Then they added a nucleotide (also blocked with a photoreactive group) and chemically coupled it to all the areas of the slide that had been unblocked in the previous step. The result: A nucleotide was attached to a subset of the tiny spots on the chip. Next, they masked a different

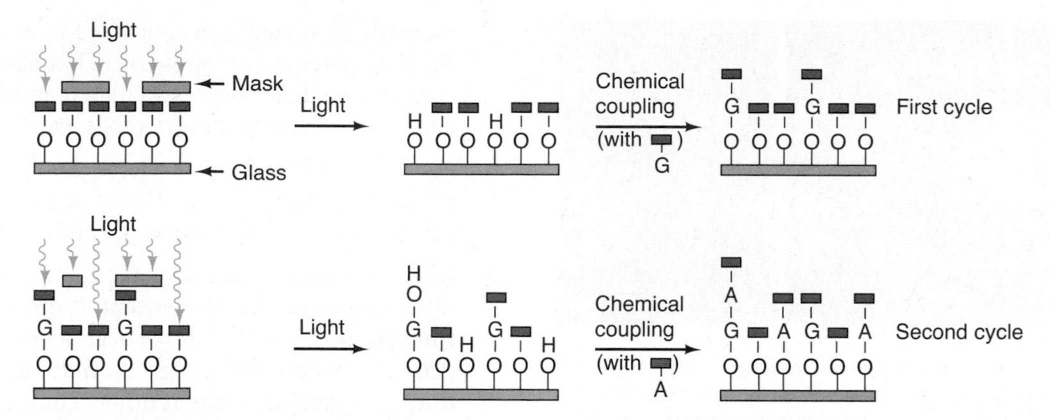

Figure 25.2 Growing oligonucleotides on a glass substrate. The glass is coated with a reactive group that is blocked with a photosensitive agent (red). This blocking agent can be removed with light, but parts of the plate are masked (blue) so the light cannot get through. In the first cycle, four of the six spots pictured are masked, so the light reaches only two unmasked spots and removes the blocking agent. Then a blocked guanosine nucleotide is chemically coupled to the unblocked spots. In the second cycle, three spots are masked, and the other three are therefore exposed to the light. This removes the blocking agent from three spots, including the first one, which already has a G attached. Thus, after a blocked adenosine nucleotide is chemically coupled to the three unblocked spots, the first spot has a G–A dinucleotide, the third and sixth spots have an A mononucleotide, the fourth has a G mononucleotide, and the second and fifth spots, which were masked in both cycles, have no nucleotides attached yet. As the cycle is repeated over and over with different masking patterns and different nucleotides, unique oligonucleotides are built up in each spot.

subset of spots, illuminated the others to remove the blocking groups, and attached another nucleotide. On the spots that were unmasked in both steps, dinucleotides were formed. By repeating this process, they could build up different oligonucleotides on each spot.

The resulting chip is known as a **DNA microchip** or **oligonucleotide array,** although these terms and "DNA microarray" are often used interchangeably. In fact, the generic term "microarray" can be used to refer to any kind of DNA or oligonucleotide microarray. The technology is so miniaturized that about 300,000 oligonucleotides can be built on a chip only 1.28 × 1.28 cm (about ½″ square). And the process is so efficient that a set of 4^n different oligonucleotides can be built in only $4 \times n$ cycles. So if our goal is to generate all the possible 9-mers (4^9, or about 250,000 different oligonucleotides), we can do it in only $4 \times 9 = 36$ cycles. How long must an oligonucleotide be to uniquely identify one human gene product in a mixture of all the others? Knowing the sequence of the human genome helps us answer this question with great accuracy. However, even without that information, we can do a calculation to give us a minimum estimate. A given sequence of n bases will occur in a DNA about every 4^n bases. In other words, a DNA sequence needs to be n bases long to occur about once in a DNA 4^n bases long. Thus, we need to solve the following equation for n to find the minimum size of an oligonucleotide we would expect to find only once in the whole human genome, which may be as much as 3.5×10^9 bases long:

$$4^n = 3.5 \times 10^9$$

The answer is that if $n = 16$, $4^n > 3.5 \times 10^9$. So our oligonucleotides need to be at least 16 bases long, and that would require $4 \times 16 = 64$ cycles to build them all on an oligonucleotide array. Again, however, this is a minimum estimate, so it would be a good idea to start with longer oligonucleotides to be reasonably sure that they occur only once in the human genome and therefore uniquely identify human genes.

Even before the publication of the sequence of the first human chromosome, scientists at Affymetrix, Inc. were already producing microchips containing 25-mers designed to recognize single genes. They based their design on the sequence that was available, including the many ESTs already in the database. To enhance the reliability of their chips, they included multiple oligonucleotides designed to hybridize to single transcripts, so the results obtained with each of these oligonucleotides could be checked against one another.

The oligonucleotides on a microchip or the cDNAs on a microarray can be hybridized to labeled RNA isolated from cells (or to corresponding cDNAs) to see which genes in the cell were being transcribed. For example, consider a study by Patrick Brown and colleagues in which they used the DNA microarray technique to examine the effect of serum on the RNAs made by a human cell. They isolated RNA from cells grown in the presence and absence of serum, then reverse transcribed the two RNA samples in the presence of nucleotides tagged with fluorescent dyes, so the cDNA products would be labeled with the fluorescent tags. They used a green-fluorescing nucleotide to label the cDNA from serum-deprived cells, and a red-fluorescing nucleotide to label the cDNA from serum-stimulated human cells. Then they mixed the cDNAs, hybridized them to DNA microarrays containing unlabeled cDNAs corresponding to 8613 different human genes, and detected the resulting

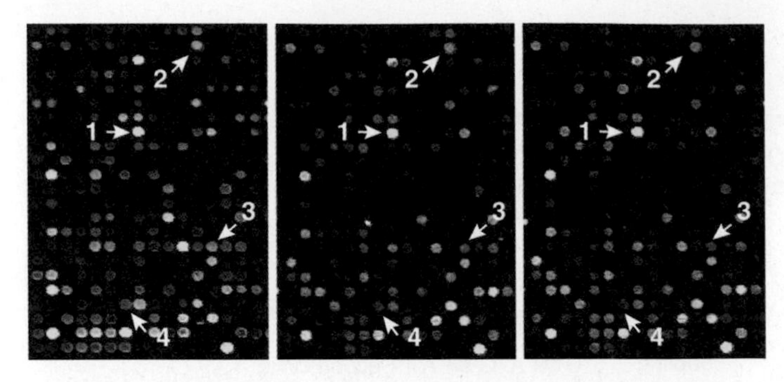

Figure 25.3 Using a DNA chip. Brown and colleagues made cDNAs from RNAs from serum-starved and serum-stimulated human cells. They labeled the cDNAs corresponding to RNAs from serum-starved cells with a green fluorescent nucleotide; they labeled the cDNAs corresponding to RNAs from serum-stimulated cells with a red fluorescent nucleotide. Then they hybridized these fluorescent cDNAs together to DNA chips containing cDNAs corresponding to over 8600 human genes. The figure shows the same part of the DNA chip from three different hybridizations. The red spots (e.g., spots 2 and 4) correspond to genes that are more active in the presence of serum. The green spots (e.g., spot 3) correspond to genes that are more active in the absence of serum. The yellow spots (e.g., spot 1) correspond to genes that are roughly equally active in the presence or absence of serum. (*Source:* Lyer, V.R., M.B. Eisen, D.T. Ross, G. Schuler, T. Moore, J.C. Lee, et al., The transcriptional program in the response of human fibroblasts to serum. *Science* 283 (1 Jan 1999) f. 1, p. 83. Copyright © AAAS.)

fluorescence. Figure 25.3 shows the same region of the microarray from triplicate hybridizations. The red spots correspond to genes that are turned on by serum, and the green spots represent genes that are active in serum-deprived cells. The yellow spots result from hybridization of both probes to the same spot (the green and red fluorescence together produce a yellow color). Thus, the yellow spots correspond to genes that are active in both the presence and absence of serum.

Microarrays allow one to examine changes in gene expression in systems much more complex than the one we have just described. For example, our knowledge of the complete yeast genome sequence has enabled molecular biologists to use DNA chips to analyze the expression of every yeast gene at once, under a variety of conditions.

In another example, Kevin White and colleagues used DNA chips in 2002 to follow the expression of 4028 *Drosophila* genes during 66 distinct periods throughout the fly's life cycle. Figure 25.4a shows the 66 developmental stages at which RNAs were collected for gene expression analysis. Notice that almost half (30) of these time points were in the embryonic phase of development, in which gene expression changes most rapidly. In fact, early in the embryonic phase, when gene expression is most dynamic, RNAs were collected every half-hour. This analysis yielded several conclusions:

- A large number of genes (3219) experienced a substantial change in expression (four-fold or more) during the fly's life cycle. Figure 25.4b shows all of these developmentally regulated genes, ordered by time of onset of the first increase in expression. That is, the topmost genes in the figure were stimulated earliest in the life cycle, and the bottommost genes were stimulated last.
- More than 88% of the developmentally regulated genes are active during the first 20 h of development, which is before the end of the embryonic phase (see Figure 25.4c).
- RNAs from about 33% of the developmentally regulated genes are already present at the very earliest time point (Figure 25.4c). These represent **maternal genes,** or **maternal effect genes,** those that are expressed during oogenesis in the mother. Thus, the maturing oocyte either transcribes these genes or receives their transcripts from surrounding nurse cells so the mRNAs are already present in the egg and are available for translation as soon as fertilization occurs.
- As illustrated in Figure 25.4d, expression of some genes is maintained throughout the life cycle, whereas expression of others peaks and declines. In particular, as further illustrated in Figure 25.4e, genes that reach peak expression during early embryonic life tend to peak again in early pupal development, whereas genes that peak in the late embryonic phase tend to achieve another peak in late pupal development. A related phenomenon, not illustrated here, is that genes that peak in larval development tend to reach another peak of expression during adult life.
- Genes encoding components of a given supramolecular complex tended to be coexpressed. Thus, the genes encoding the ribosomal proteins tended to be regulated coordinately, as did the genes encoding the proteins in the mitochondrion.
- Genes encoding proteins with related functions tended to be coexpressed, even if the proteins did not form complexes. Thus, genes encoding transcription factors, or cell cycle regulators, tended to be expressed together.
- Coexpression of some genes was tissue-specific. For example, one cluster of 23 coregulated genes included eight genes that were already known to be expressed in muscle cells. Upon further examination, the control regions of 15 of the genes in this cluster had pairs of binding sites for the transcription factor dMEF2, which is known to activate genes in differentiating muscle cells. Seven of the genes in the cluster had unknown function, and six of these had dMEF2-binding sites and were expressed in differentiating muscle. Thus, this analysis allowed White and colleagues to assign a function in muscle differentiation to these six unknown genes. This is important because it is very difficult to determine the function of genes based solely on their sequences. The additional clues about timing and location of expression are a tremendous help. Indeed, they allowed White and colleagues to assign functions to 53% of the genes they analyzed.

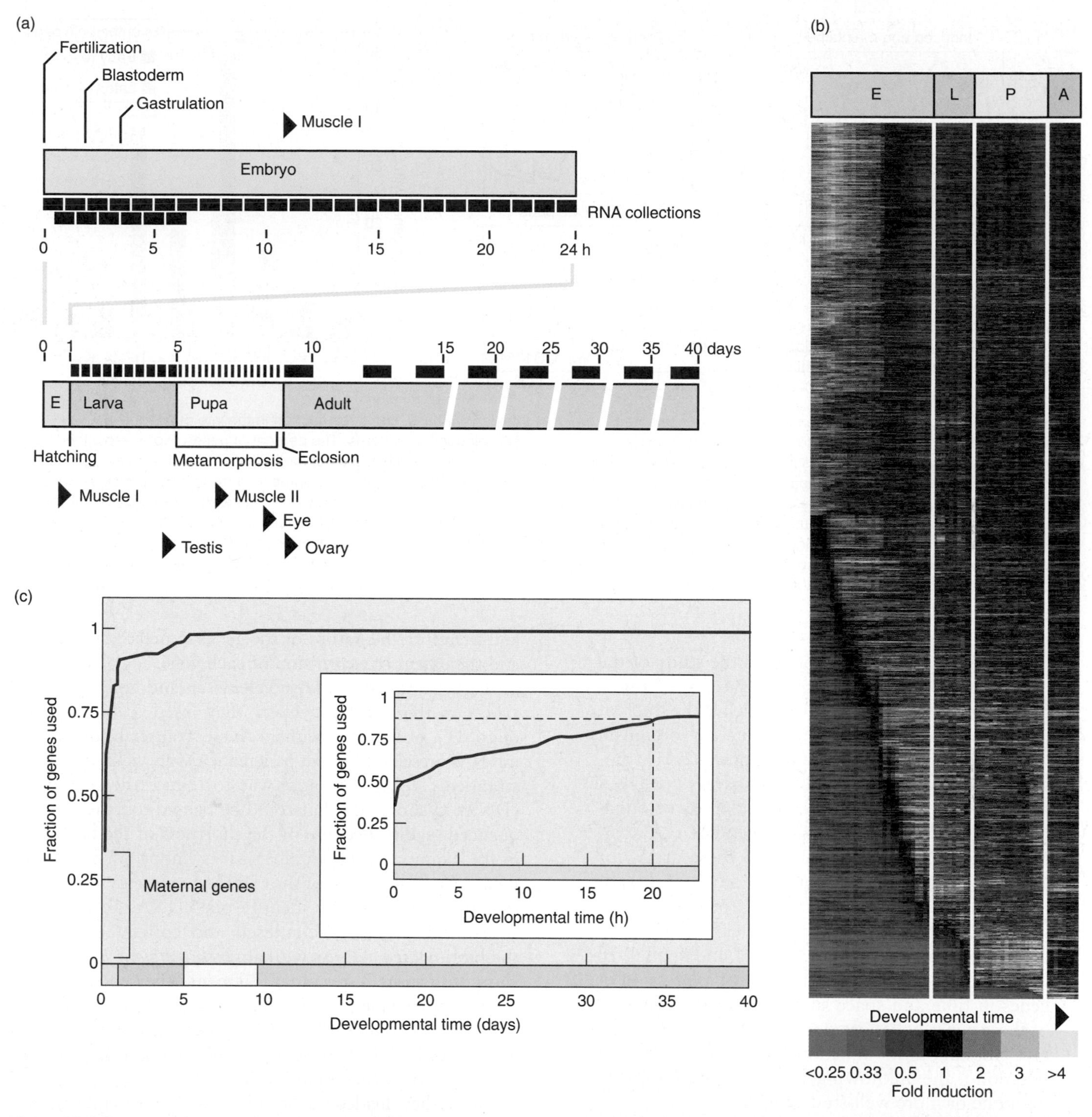

Figure 25.4 Patterns of expression of *Drosophila* genes during development. (a) Outline of RNA collection periods. White and colleagues collected RNAs from whole animals at the indicated times during development (E, embryonic; L; larval; P, pupal; A, the first 40 days of the adult phase). The embryonic period is expanded to show all of the overlapping collection periods. They purified Poly(A)$^+$ RNA by oligo(dT)-cellulose chromatography and made fluorescent cDNAs by reverse transcribing the poly(A)$^+$ RNAs in the presence of a fluorescent nucleotide. Then they hybridized the fluorescent cDNA from a given time point to a microarray and measured the extent of hybridization. They normalized all such hybridization values against the extent of hybridization of a reference standard cDNA prepared from a mixture of RNAs from all phases of the life cycle. **(b)** Gene expression profiles. The profiles of 3219 genes whose expression levels changed by more than four-fold during the fly life cycle are arranged in order of the onset of the first increase in abundance of transcript. The developmental phase is indicated at top, with the same abbreviations and color coding as in (a). The expression level is indicated by color, as indicated at bottom, blue stands for low expression and yellow stands for high expression. **(c)** Graphic representation of the cumulative fraction of genes that have shown a strong increase in expression. Note that a large fraction (about 33%) of genes are already represented by a large amount of RNA at the earliest time point. These are labeled maternal genes. The inset is an expansion of the first 20 h of the embryonic phase, which also shows the large proportion of transcripts already present in the first hour of development.

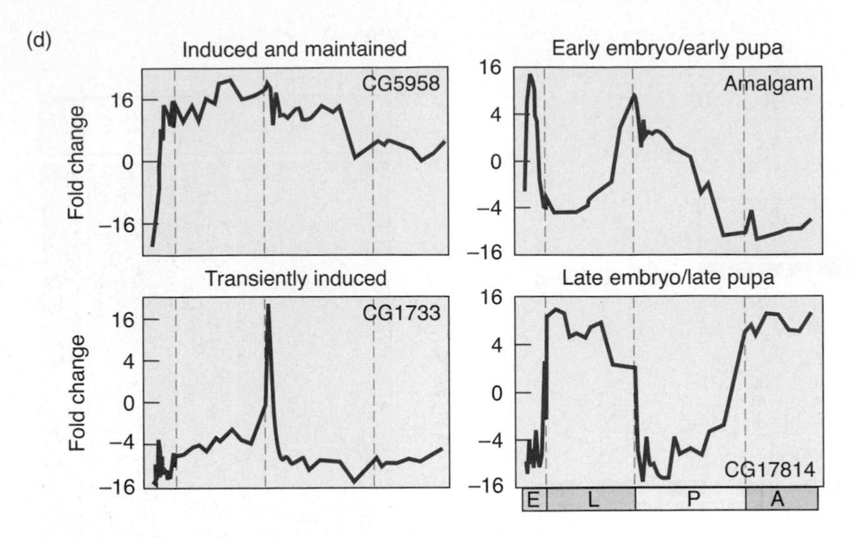

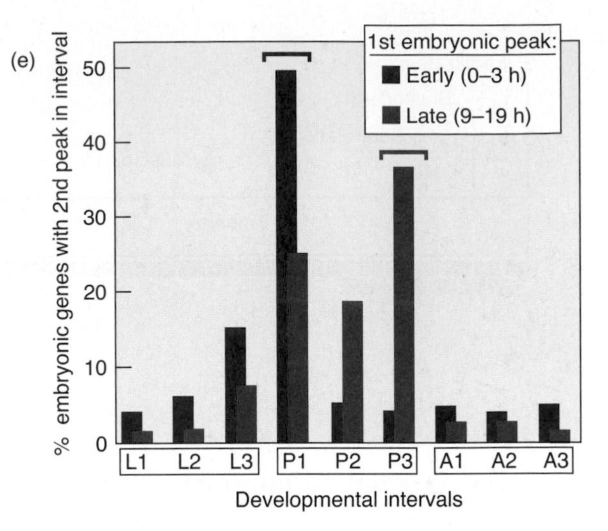

Figure 25.4 *Continued* (d) Expression patterns of four selected genes. At upper left, gene CG5958 shows an induction in early embryonic phase to a high level that is largely maintained throughout the life cycle. At upper right, the *Amalgam* gene shows an induction in the early embryonic phase, a decrease in the larval phase, and a reinduction at the boundary between the larval and pupal stages. At lower left, gene CG1733 shows a distinct peak of expression at the larval–pupal boundary. At lower right, gene CG17814 shows one burst of induction that begins in the late embryonic phase and lasts through the larval phase, and a reinduction in the late pupal phase. **(e)** Reinduction patterns. The percent of genes expressed either early (blue) or late (red) in the embryonic phase that show a reinduction at the given times later in development. Note that the genes expressed in early embryogenesis tend to be reinduced in the early pupal stage (P1, bracket over blue bar), whereas the genes expressed in late embryogenesis tend to be reinduced in the late pupal stage (P3, bracket over red bar). (*Source:* Adapted from Arbeitman et al., *Science* 297, 2002. Fig. 1, p. 2271. © 2002 by the AAAs.)

SUMMARY Functional genomics is the study of the expression of large numbers of genes. One branch of this study is transcriptomics, which is the study of transcriptomes—all the transcripts an organism makes at any given time. One approach to transcriptomics is to create DNA microarrays or DNA microchips, holding thousands of cDNAs or oligonucleotides, then to hybridize labeled RNAs (or corresponding cDNAs) from cells to these arrays or chips. The intensity of hybridization to each spot reveals the extent of expression of the corresponding gene. With a microarray one can canvass the expression patterns (both temporal and spatial) of many genes at once. The clustering of expression of genes in time and space suggests that the products of these genes collaborate in some process. This can give clues about the functions of genes of unknown function if the unknown gene is expressed together with one or more well-studied genes.

Serial Analysis of Gene Expression In 1995, Victor Velculescu, working with Kenneth Kinzler and colleagues, developed a novel method of analyzing the range of genes expressed in a given cell. They called this method **serial analysis of gene expression (SAGE).** The underlying strategy of SAGE is to synthesize short cDNAs, or **tags,** from all the mRNAs in a cell, and then link these tags together in clones that can be sequenced to learn the nature of the tags, and therefore the nature of the genes expressed in the cell, and the extent of expression of each gene.

Figure 25.5 shows how Velculescu and colleagues carried out this strategy. First, they used a biotinylated oligo(dT) primer to prime reverse transcription of the mRNAs present in human pancreatic tissue, yielding double-stranded cDNAs. The goal was to reduce the size of the cDNAs to short tags that could be ligated together and sequenced readily. Because of the shortness of the tags (9 bp in the example in Figure 25.5), it is important to confine them to a small region of the cDNAs to increase the chance that they will uniquely identify one cDNA. To begin the shortening process, Velculescu and colleagues cleaved the biotinylated cDNAs with an *anchoring enzyme (AE)* to chop off a short 3′-terminal fragment. They chose as their anchoring enzyme *Nla*III, which recognizes 4-base restriction sites and therefore yields fragments averaging 250 bp long. They bound these biotinylated 3′-fragments to streptavidin beads, which bind biotin.

Next, they divided the bead-bound cDNA fragments into two pools and ligated one pool to a linker (Y) and the other pool to a second linker (Z). Both linkers contained the recognition site for a type IIS restriction endonuclease (the *tagging enzyme* [TE]) that cuts 20 bp downstream of this recognition site. The result of cleavage of the cDNA fragments with the tagging enzyme *Fok*I was a set of short fragments, each containing the linker (Y or Z) followed by the 4-bp anchoring enzyme site, followed by 9 bp from the cDNA. That 9-bp piece of cDNA is the tag. If the tagging enzyme leaves overhangs, these can be filled in to yield blunt ends.

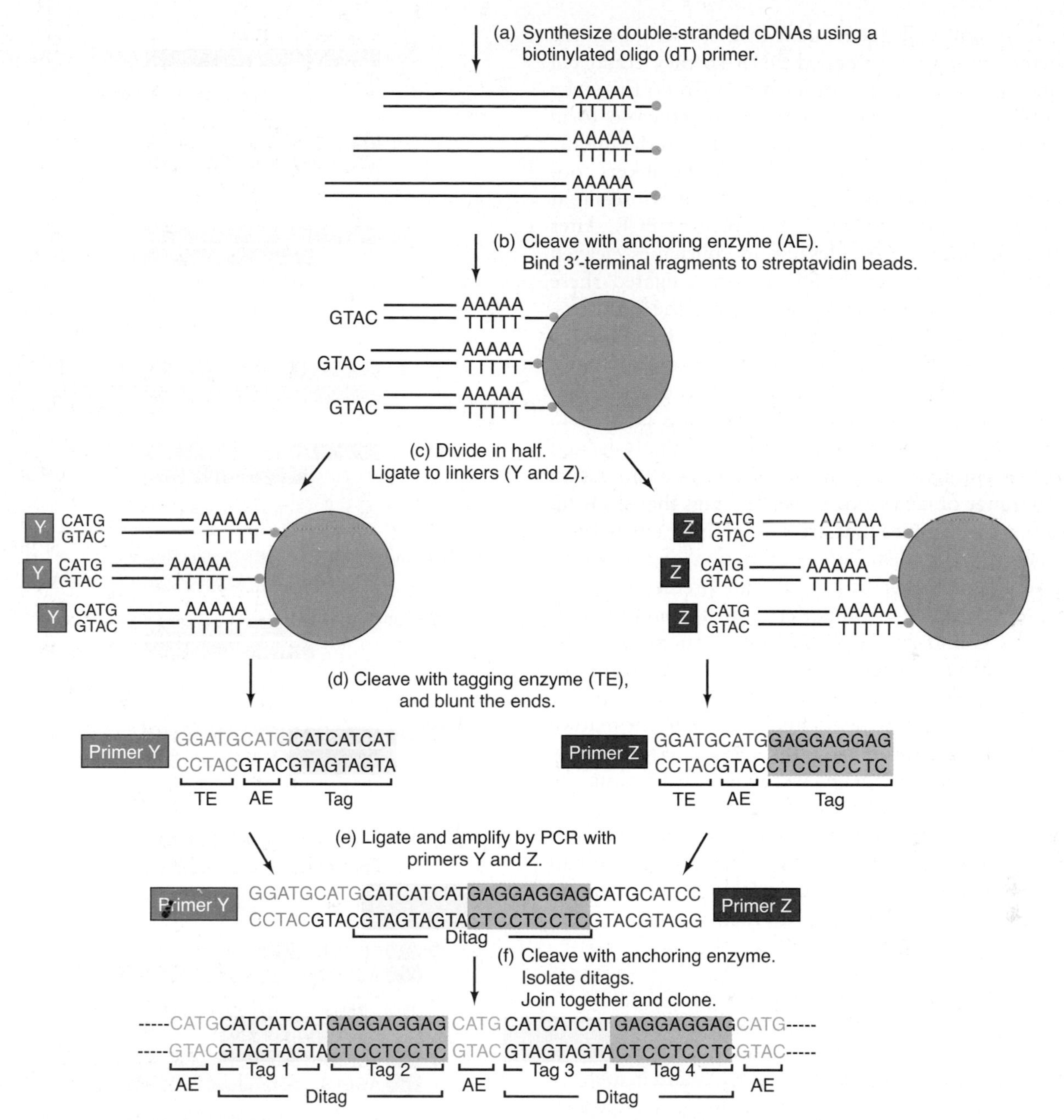

Figure 25.5 Serial analysis of gene expression (SAGE). **(a)** Double-stranded cDNAs are formed from cellular mRNAs, using biotinylated oligo(dT) to prime first-strand cDNA synthesis. Orange balls represent biotin. **(b)** Biotinylated cDNAs are cleaved with an anchoring enzyme (AE, *Nla*III in this case), and the biotinylated 3′-end fragments are bound to streptavidin beads (blue). **(c)** The bead-bound fragments are divided into two pools; the fragments in one pool are ligated to linker Y (blue) and the fragments in the other pool are ligated to linker Z (pink). **(d)** The fragments are cleaved with the tagging enzyme (TE), and ends are filled in if necessary to create blunt ends. In this case, the tagging enzyme is *Fok*I, which leaves 9-bp tags attached to the linkers. The tag attached to linker Y is represented by the arbitrary sequence CATCATCAT and its complement highlighted in yellow, and the tag attached to linker Z is represented by the arbitrary sequence GAGGAGGAG and its complement (light purple highlight). **(e)** Tag-containing fragments are blunt-end-ligated together and amplified by PCR with primers that hybridize to primer Y and primer Z regions in each linker. Only fragments ligated with tags joined tail to tail (ditags) will be amplified by PCR. **(f)** The amplified ditag-containing fragments are cleaved with the anchoring enzyme to yield ditags with sticky ends. The ditags are ligated together to form concatemers, which are cloned. Part of a concatemer of ditags is shown, with the 4-base recognition sites for the anchoring enzyme shown in green. Note that these 4-base sites set off each ditag so it can be recognized easily. The clones are then sequenced to discover which tags are represented, and in what quantity. This tells which genes are expressed, and how actively. (*Source:* Adapted from Velculescu, V.E., L. Zhang, B. Vogelstein, and K.W. Kinsler, Serial analysis of gene expression. *Science* 270:484, 1995.)

Velculescu and colleagues' next task was to ligate the tags together, along with defined DNA so they could tell where one tag left off and another began. To do this, they blunt-end-ligated the tagged fragments together to form fragments with two tags abutting each other in the middle (forming a *ditag*) and linkers on each end. The linkers contain sites that are complementary to a pair of primers that can be used to amplify the whole fragment by PCR. After the PCR amplification, Velculescu and colleagues cleaved the products with the anchoring enzyme, ligated these restriction fragments together, and cloned the products. Now the ditags can be easily identified because each one is flanked by the 4-bp anchoring enzyme recognition sites. And, of course, half of each ditag belongs to one tag, and half to the other. Clones with at least 10 tags (some had more than 50) can be identified by PCR analysis and sequenced. If enough clones are sequenced, we can get an idea of the range of genes expressed, and tags that show up repeatedly indicate genes that are very actively expressed.

Velculescu and colleagues' examination of expression in the human pancreas by SAGE had predictable, and therefore encouraging, results. The most common tags (GAGCACACC and TTCTGTGTG) corresponded to the genes for procarboxypeptidase A1 and pancreatic trypsinogen 2, respectively. These are two abundantly expressed pancreatic proenzymes, which, after cleavage to the mature enzyme forms, digest proteins in the small intestine. Many other familiar pancreatic genes were identified among the plentiful tags, but many of the tags did not match any gene sequences in the database, so their identities were unknown. As the database expands to include all human genes, all tags should at least be correlated to genes, even if the functions of some of those genes remain obscure.

SUMMARY SAGE allows us to determine which genes are expressed in a given tissue and the extent of that expression. Short tags, characteristic of particular genes, are generated from cDNAs and ligated together between linkers. The ligated tags are then sequenced to determine which genes are expressed and how abundantly.

Cap Analysis of Gene Expression (CAGE) SAGE is a useful method for global analysis of gene expression, but it focuses on the 3′-ends of transcripts. Sometimes it is necessary to identify the 5′-ends of transcripts—for example, if one is interested in identifying promoters on a genomic scale. In that case, a related method known as **cap analysis of gene expression** (**CAGE,** Figure 25.6) is available.

The CAGE procedure starts with reverse transcription (RT), as SAGE does, but with two important differences that ensure production of full-length cDNAs that copy the mRNA all the way to the 5′-end. First, the RT reaction includes a disaccharide known as trehalose. This substance

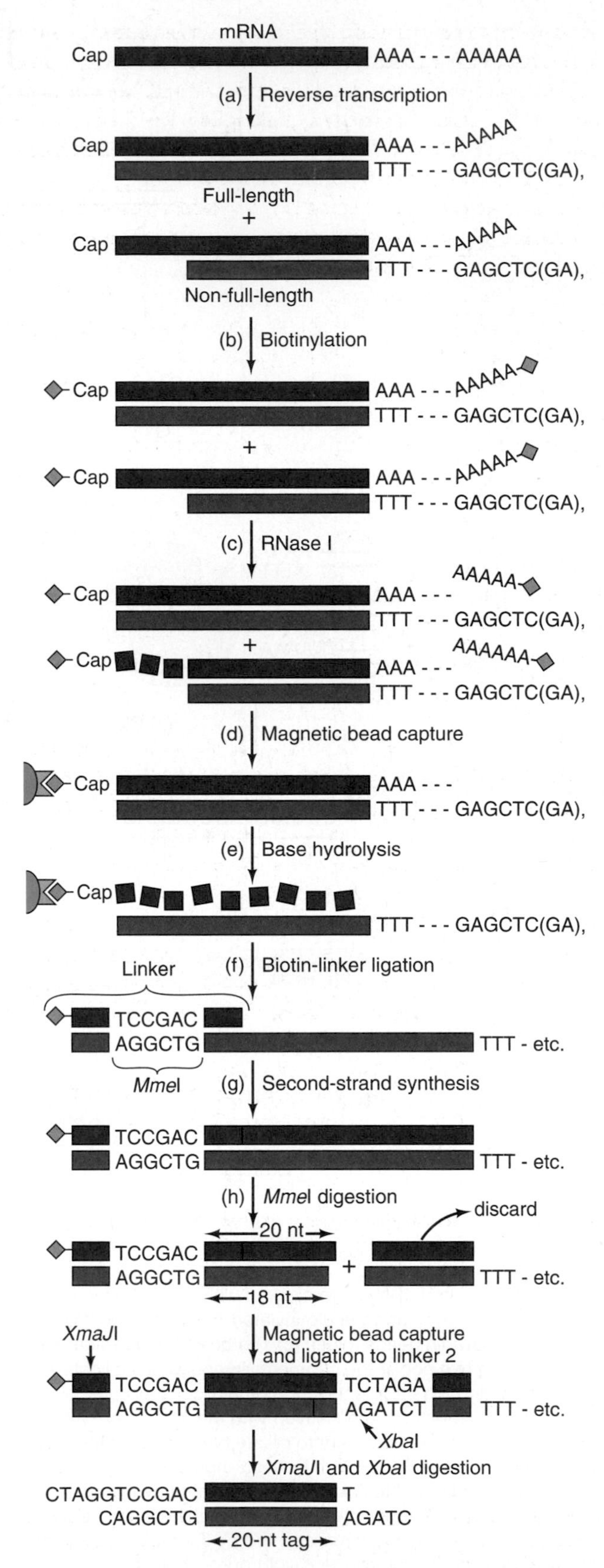

Figure 25.6 Use of CAGE to produce 20-nt tags representing the 5′-ends of mRNAs. The procedure is described in the text. After the tags are produced as shown here, they can be ligated together via their identical sticky ends to form concatemers, cloned, and sequenced.

stabilizes reverse transcriptase at high temperature, so the RT reaction can be run at 60°C. This elevated temperature weakens mRNA secondary structure that otherwise would stop the RT reaction before it reached the 5′-end of the mRNA. Second, a **cap trapper** method is used: The caps of the mRNAs in the mRNA–cDNA hybrids are tagged with biotin. As we will see, this allows hybrids with full-length cDNAs to be purified away from hybrids containing less-than-full-length cDNAs.

Figure 25.6 shows how the tagging works. First, the RT priming is done, not with oligo(dT), but with oligo(dT), preceded by a stretch of random nucleotides that do not hybridize with the poly(A) tail. The importance of this feature will become apparent shortly. After first strand cDNA synthesis, both ends of the mRNA are tagged with biotin by reacting the RNA–DNA hybrid with a biotin-containing reagent that attaches to diols. There are only two diols (adjacent hydroxyl groups) in a capped mRNA: the free 2′- and 3′-hydroxyl groups in the cap and the 3′-terminal nucleotide.

One would like to tag just the cap, but the 3′-terminal nucleotide is unavoidably tagged in the same step. But that problem is resolved in the next step, in which the hybrids are treated with RNase I. The RNase degrades any single-stranded RNA that is not hybridized to the cDNA. Thus, it not only removes the biotin tag from any hybrids that contain incomplete cDNAs, it also removes the biotin tag from the 3′-hydroxyl group at the end of every mRNA's poly(A) tail, which cannot hybridize to the random tail at the beginning of the primer. After the RNase treatment, the only remaining biotin-tagged hybrids are those containing full-length cDNAs, and these are collected using magnetic beads coated with the biotin-binding protein streptavidin. After the hybrids are purified, their mRNA parts, including the biotin-tagged caps, are destroyed by base hydrolysis, leaving just the single-stranded cDNAs.

Next, the full-length, single-stranded cDNAs are ligated to biotin-tagged linkers that contain a recognition site for the tagging enzyme *Mme*I, which dictates cleavage 20 and 18 nt away. Thus, after second-strand cDNA synthesis, the tagged cDNAs can be cut with *Mme*I to yield 20-nt tags that can be purified via their biotin parts, and ligated to a second linker (linker 2) via their 2-nt overhangs. Linker 1 also contains a recognition site for *XmaJ*I and linker 2 contains a recognition site for *Xba*I, so the tags can be cut with those two enzymes, ligated together into concatemers, cloned, and sequenced as in the SAGE procedure.

The 20-nt tags would be expected to be found every 4^{20}, or about 1.1×10^{12} base-pairs. Thus, since the human genome contains only about 3×10^9 bp, most of the 20-nt tags should identify a unique sequence in even the large human genome, which can be found by consulting the known human genome sequence. This sequence should begin with the transcription start site, so the promoter should be in the immediate neighborhood. When Piero Carninci and colleagues performed this kind of CAGE analysis on mouse mRNAs from whole brain and three distinct brain regions, they found many CAGE tags that mapped close to previously mapped start sites, but many more that did not. This could help identify a number of new promoters and alternative start sites.

SUMMARY Cap analysis of gene expression (CAGE) gives the same information as SAGE about which genes are expressed, and how abundantly, in a given tissue. Because it focuses on the 5′-ends of mRNAs, it also allows the identification of transcription start sites and, therefore, helps locate promoters.

Whole Chromosome Transcriptional Mapping Transcriptomics studies have become sophisticated enough that they can map transcripts with great accuracy to sites in whole chromosomes. This kind of study, called **transcriptional mapping,** is shedding light on a paradox mentioned earlier in this chapter: The number of protein-encoding genes in humans is scarcely larger than the number of such genes in a lowly roundworm! How can we reconcile that fact with the vastly greater complexity of human beings? One emerging answer is that transcripts of protein-encoding genes make up only a small fraction of the whole human transcriptome. And the closer we look at this problem, the more complex the human transcriptome becomes.

If we consider only exons in protein-coding genes, we would predict that only 1–2% of the whole human genome would be expressed in RNAs found in the cytoplasm of cells. However, as early as 2002, Thomas Gingeras and colleagues, using microarrays to study expression of human chromosomes 21 and 22, discovered that polyadenylated RNAs in the cytoplasm of human cells covered about an order of magnitude more of those two chromosomes than could be accounted for by protein-encoding exons. This excess of unexpected transcripts has been dubbed **transcripts of unknown function,** or **TUFs.** All of the transcribed regions (exons and TUFs alike) detected by such arrays are called **transcribed fragments,** or **transfrags.**

Furthermore, approximately two-thirds of the transcripts in human cells and hamster cells have been reported to be nonpolyadenylated [poly(A)$^-$]. These poly(A)$^-$ transcripts therefore represent another chunk of the human genome, whose extent is unknown, but apparently large. Taken together, these findings suggest that protein-encoding exons make up only a small fraction of the total genomic sequences represented by cytoplasmic RNAs.

To investigate this intriguing conclusion further, Gingeras and colleagues used high-density oligonucleotide arrays with 25-mers spaced on average only 5 bp apart, thus providing an average of a 20-bp overlap. Why use such a high density? For one thing, it allows one to detect shorter exons, and, for another, hybridizations to overlapping oligonucleotides give greater confidence that transcription in that region really occurs. The oligonucleotide on the arrays came from the sequences of ten human chromosomes (6, 7, 13, 14, 19, 20, 21, 22, X, and Y), representing 30% of the total length of the human genome. To the arrays, Gingeras and colleagues hybridized double-stranded cDNAs representing cytoplasmic poly(A)$^+$ RNAs from eight different human cell lines, or cytoplasmic and nuclear poly(A)$^+$ and poly(A)$^-$ RNAs from a single cell line (HepG2). In all cases, transfrags that overlapped pseudogenes or repetitive DNA regions were dropped from consideration.

About 9% of more than 74 million probe pairs (both strands) hybridized to cDNAs from poly(A)$^+$ RNA, per cell line. Applying a "1 of 8" rule, in which a probe pair needs to hybridize to a cDNA from only one of the eight cell lines, the percentage of positive probes rose to 16.5%. This is the "1 of 8 map." An average of 4.9% of the nucleotides in the 10 chromosomes were expressed as cytoplasmic RNA in each cell line. In the 1 of 8 map, this figure rose to 10.1%. These findings suggest that about 10.1% of the sequences in the 10 human chromosomes are expressed as polyadenylated RNA in the cytoplasm in at least one cell line. Furthermore, the difference between 4.9% and 10.1% indicates that considerable cell-line-specific transcription occurs.

Figure 25.7 shows the proportions of each of the 10 chromosomes from which cytoplasmic polyadenylated transcripts are made. Such transcripts from intergenic regions and introns are, by definition, unannotated. And these regions make up the majority (57%) of the transcripts from the 10 chromosomes as a whole (central pie chart). The annotated transcripts overlap with one of three annotations: Known, which is a combination of two exon databases; mRNA, which contains the mRNAs from a third database that do not overlap with the Known exons; and EST, which contains all publicly available ESTs that do not overlap with either the Known or mRNA databases.

What about poly(A)$^-$ transcripts? For this analysis, Gingeras and colleagues focused on a single cell line, HepG2. They looked for stable poly(A)$^+$, poly(A)$^-$, and bimorphic transcripts in both the nucleus and cytoplasm of these cells. (Bimorphic transcripts start out polyadenylated,

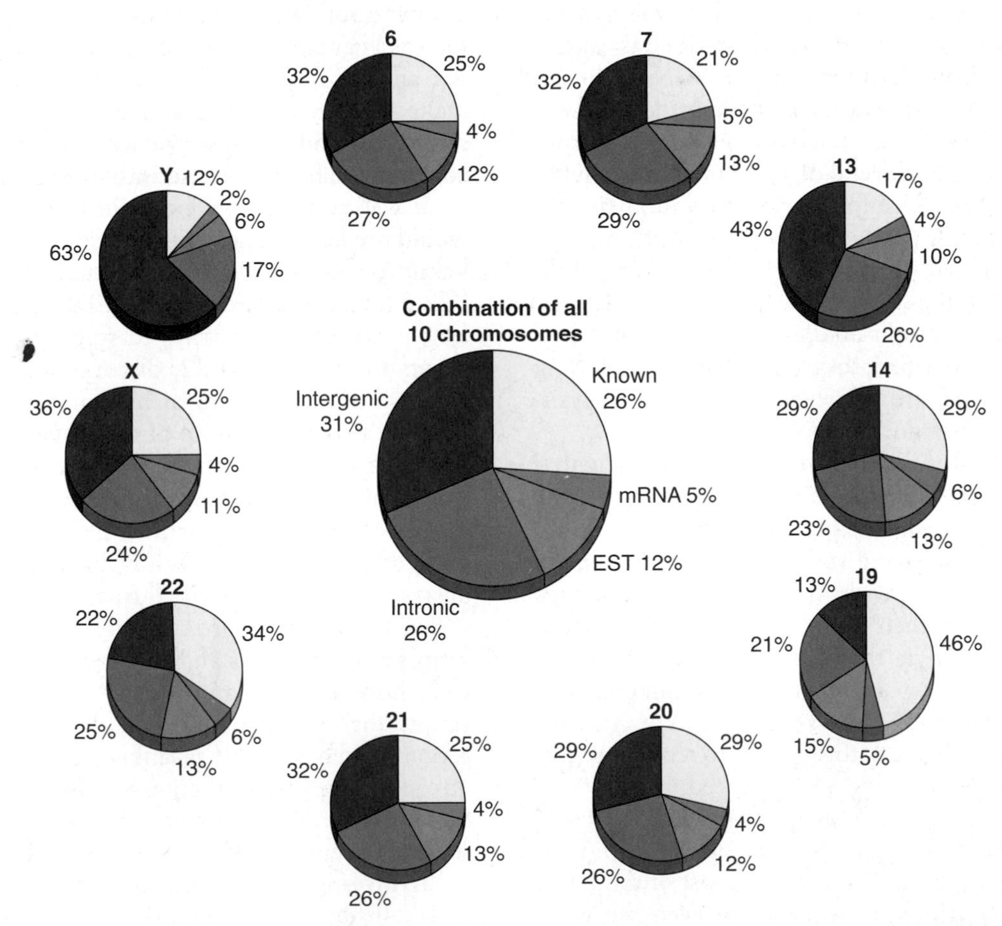

Figure 25.7 Transcription maps of 10 human chromosomes. The percentages of different categories of sequences found in polyadenylated cytoplasmic transcripts in the 1 in 8 map are represented by the wedges of each pie chart. Each of the chromosomes represented by the small pie charts is identified in boldface, as is the collective of all 10 chromosomes (large pie chart in the middle). Sequence categories are given in the collective pie chart, and the same color coding is used throughout. The unannotated sequences are intergenic and intronic. The annotated sequences are designated Known, mRNAs, and ESTs. (*Source:* Cheng, J., T.R. Gingeras, et al. 2005. Transcriptional maps of 10 human chromosomes at 5-nucleotide resolution. *Science* 308:1149–54.)

but then lose their poly[A] tail.) They found that fully 15.4% of nucleotides in the 10 chromosomes are represented in one of these classes of transcripts (almost half of which are poly[A]$^-$). Thus, about 10 times as much of the genome is represented in stable transcripts than we would expect on the basis of exons alone. Of course, the majority of most human genes is in introns, so this result may not sound surprising at first. But if spliced-out introns have no function, we would expect them to be degraded rapidly and not contribute so heavily to the cDNAs made from presumably stable nuclear RNAs.

Another conclusion from this study is that about half of the human transcriptome appears to be overlapping. There are two kinds of overlaps: those on the same strand, and those on opposite strands. Of course, transcripts that overlap on opposite strands represent sense/antisense pairs, which should invoke an RNAi response. Thus, this may represent a kind of gene expression control mechanism.

Studies like this that show abundant cytoplasmic poly(A)$^+$ and poly(A)$^-$ transcripts of non-exon regions may help to explain the differences between organisms. Although the exons of humans and chimpanzees are extremely similar, the non-exon regions have diverged considerably more. And transcription of those regions may give rise to some of the differences we see in the two species.

SUMMARY High-density whole chromosome transcriptional mapping studies have shown that the majority of sequences in cytoplasmic polyadenylated RNAs derive from non-exon regions of 10 human chromosomes. Furthermore, almost half of the transcription from these same 10 chromosomes is nonpolyadenylated. Taken together, these results indicate that the great majority of stable nuclear and cytoplasmic transcripts of these chromosomes comes from regions outside the exons. This may help to explain the great differences between species, such as humans and chimpanzees, whose exons are almost identical.

Genomic Functional Profiling

The ultimate goal of **genomic functional profiling** is to determine the pattern of expression of all the genes in an organism at all stages of the organism's life. That is a daunting task even in the simplest of eukaryotes, but it is even more difficult in complex multicellular organisms. So far, the puzzle for each organism is being put together piece by piece, with each research group contributing its own piece. Let us consider some general techniques for attacking the problem.

Deletion Analysis Once all the genes in a genome have been identified, one can investigate what happens when each of them is removed. That kind of experiment is ethically impossible in humans, of course, but it can be done in other vertebrates as their genomes are completely sequenced—at least in principle. Logistical problems may delay this kind of analysis of a genome as large as that of a vertebrate, but the yeast genome has already been profiled in this way.

In 2002, a large consortium of investigators led by Ronald Davis reported that they had generated a set of yeast mutants, in each of which one gene had been replaced with an antibiotic resistance gene flanked by 20-mer sequences that were different for each replaced gene. Thus, each gene replacement has a "molecular barcode" so it can be uniquely identified. In all, these investigators replaced over 96% of the annotated ORFs in *Saccharomyces cerevisiae*. Next, they examined the mutants for ability to grow in a mixed culture under six different conditions: high salt; sorbitol; galactose; pH 8; minimal medium; and the antifungal agent nystatin. They also examined gene expression under each of these conditions by hybridization of RNA to oligonucleotide microarrays.

To do this genomic functional profile, Davis and colleagues grew a mixed culture of all 5916 mutants under each of the conditions and collected cells at various times and tested for each barcode by hybridization to an oligonucleotide array containing sequences complementary to the barcodes. If a gene is important for dealing with a given condition, such as the presence of galactose, then mutants lacking that gene should disappear rapidly from the mixture when that condition is imposed. In fact, the rate at which the mutant disappears should correlate with the importance of the deleted gene in dealing with the condition.

When the investigators applied this kind of profiling to yeast mutants responding to the presence of galactose, they found several genes that were already known through years of study to be involved in yeast metabolism of galactose. But they also found 10 new genes that had previously not been implicated in galactose metabolism. Wild-type yeast and 11 of the mutants identified by the profiling as important in galactose metabolism were tested individually, and the results are presented in Figure 25.8. As predicted, all 11 mutant strains grew more slowly in galactose than the wild-type strain did. Their growth rates varied from 44% to 91% of wild-type.

SUMMARY Genomic functional profiling can be performed in several ways. In one kind of mutation analysis, called deletion analysis, mutants are created by replacing genes one at a time with an antibiotic resistance gene flanked by oligomers that serve as a barcode to identity each mutant. Then, a functional profile can be obtained by growing the whole group of mutants together under various conditions to see which mutants disappear most rapidly.

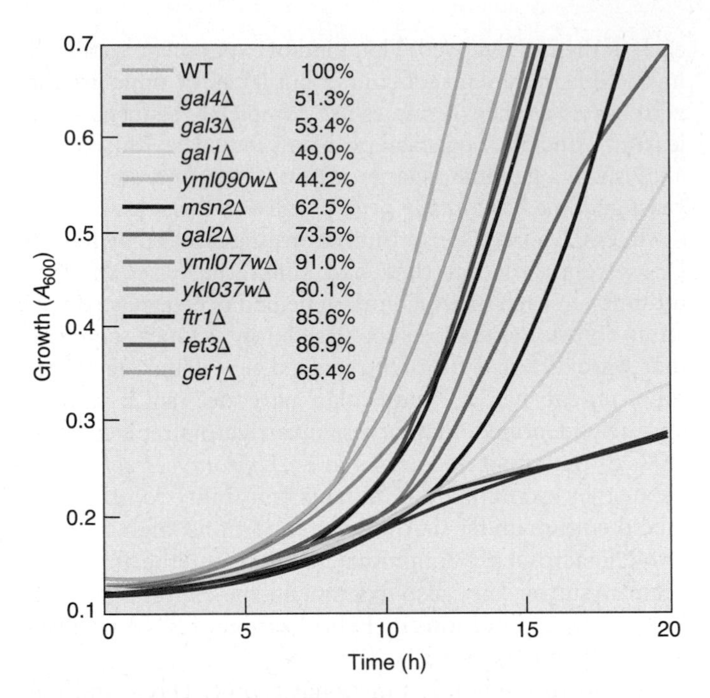

Figure 25.8 Growth curves of various mutants discovered by profiling to be deficient in response to galactose. Davis and colleagues tested wild-type yeast cells and 11 deletion mutants individually for growth in galactose-containing medium. All of the mutants had been identified by profiling in a mixture of strains as defective in growth with galactose. A_{600} (absorbance of 600-nm light) is a measure of turbidity, which in turn is a measure of yeast growth. (*Source:* Adapted from Giaever, G., A.M. Chu, L. Ni, C. Connelly, L. Riles, S. Veronneau, et al., Functional profiling of the *Saccharomyces cerevisiae* genome. *Nature* 418, 2002, p. 388, f. 2.)

RNAi Analysis "Knocking out" genes by mutagenesis is laborious, and has so far been accomplished on a genome-wide scale only in yeast. But some more complex organisms are amenable to a simpler alternative: "knocking down" genes by RNA interference (RNAi, Chapter 16). The nematode worm *Caenorhabditis elegans* is particularly susceptible to RNAi, which even affects the progeny of treated worms; it can reproduce parthenogenically, which means that only one parent is required; it contains fewer than 1000 cells, and its whole genome has been sequenced. Thus, this organism is an obvious target for genomic functional profiling by RNAi analysis.

Birte Sönnichsen and colleagues have exploited this technique to inactivate 19,075 of the worm's genes, over 98% of the total, and observe the effects on early embryogenesis—the first two cell divisions after fertilization. They injected 25-bp double-stranded RNAs into worms and then followed the first two cell divisions in the progeny of the injected worms by time-lapse microscopy. They also checked for the viability of the embryos beyond the two-cell stage and for gross phenotypic alterations in the larval and adult stages.

In all, inactivation of 1668 genes by RNAi produced detectable phenotypic defects. Of these 1668, inactivation of 661 genes gave reproducible defects in the first two cell divisions; the rest gave defects at later stages of development (Figure 25.9). (It is not surprising that inactivating virtually all of the 661 genes that gave defects in early embryogenesis also produced embryonic lethality.)

One problem with RNAi is that it sometimes fails to inactivate genes (false-negatives), so negative results are difficult to interpret. As a check on their procedure, Sönnichsen and colleagues evaluated the 65 genes that had previously been shown by mutagenesis to affect the first cell division. Of these genes, 62 (95%) had been detected by the RNAi analysis. The three genes that had been missed the first time were rechecked by RNAi analysis, and two were detected the second time, increasing the success rate to 98%.

It is also true that mutations are detected only if they give clear phenotypes, so the mutagenesis strategy also produces false-negatives. Thus, as another check on their procedure, the researchers compared their data to other RNAi analyses that targeted early embryogenesis, and found that

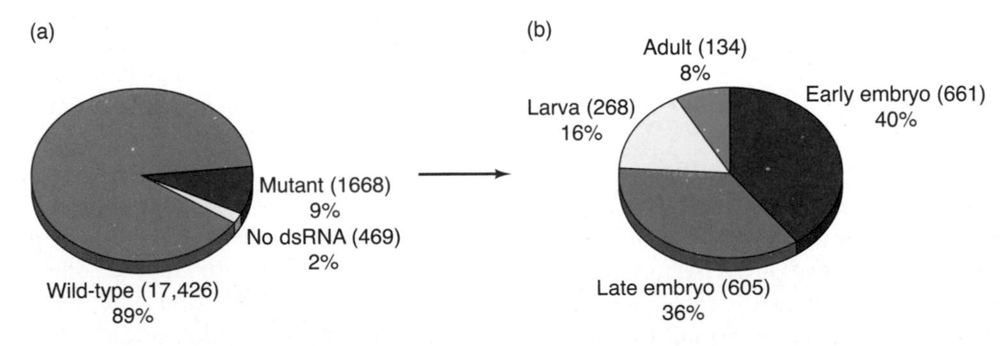

Figure 25.9 Distribution of phenotypes from a genomic functional profile of *C. elegans* using RNAi. (a) Initial screen. Sönnichsen and colleagues targeted 19,075 genes with dsRNAs. Of these, 17,426 ("wild-type," blue) caused no change in phenotype in the screens the authors used, and 1,668 ("Mutant," red) showed an alteration in phenotype. Four hundred sixty-nine genes ("No dsRNA," yellow) were not targeted in this experiment. **(b)** Distribution of mutant phenotypes. Starting with the 1668 genes whose inactivation yielded mutant phenotypes, Sönnichsen and colleagues sorted the developmental stages at which defects were seen. For example, 661 of these (red) exhibited defects in the early embryo stage (first two cell divisions). (*Source:* Adapted from Sönnichsen, et al., Full-genome RNAi profiling of early embryogenesis in *Caenorhabditis elegans*. *Nature*. Vol. 434 (2005) f. 2, p. 465.)

they had detected 75% of the genes that others had found. Accordingly, Sönnichsen and colleagues concluded conservatively that their RNAi analysis could detect 75–90% of genes involved in early embryogenesis.

Next, the researchers grouped the 661 genes according to their specific phenotypes. They found that inactivation of about half (326) of the genes produced defects in embryogenesis per se, while the remainder (335) simply affected the general cell metabolism required to keep the embryo alive long enough to divide twice. By careful annotation of the specific defects, the researchers were able to group the former 326 genes into defects in 23 aspects of embryogenesis, such as spindle assembly (9 genes) and sister chromatid separation (64 genes).

SUMMARY Genomic functional analysis on complex organisms can be done by inactivating genes via RNAi. An application of this approach targeting the genes involved in early embryogenesis in *C. elegans* has identified 661 important genes, 326 of which are involved in embryogenesis per se.

Tissue-Specific Functional Profiling Another approach to genomic functional profiling is to observe the tissue-specificity of the genes that are inactivated by mutation or other means. In one notable study, Lee Lim and colleagues used two miRNAs to knock down expression of genes in human (HeLa) cells in culture, and then looked at the profile of genes whose expression was significantly reduced. Remarkably, miR-124, an miRNA expressed in brain, knocked down expression of genes that are expressed at low levels in brain, while miR-1, an miRNA expressed in muscle, knocked down expression of genes that are expressed at low levels in muscle. In other words, these two miRNAs shifted the expression of genes in HeLa cells towards that seen in the tissues in which the respective miRNAs are prominent. This is exactly what we would expect if these two miRNAs play a major role in turning down the expression of these same genes in vivo.

A further striking feature of this study is that the miRNAs reduced the concentrations of the mRNAs in question, even though, as we learned in Chapter 16, animal miRNAs generally affect mRNA translation, not mRNA concentrations. Thus, Lim and colleagues introduced double-stranded miRNAs into HeLa cells and then used microarrays to measure the levels of mRNAs purified from the treated cells. The result was clear reduction in the concentrations of 100 or more mRNAs with each miRNA.

Here is how Lim and colleagues did their analysis, considering miR-124 first. They began by plotting the expression levels of 10,000 human genes in each of 46 tissues, using data from a previous genome-wide survey. The histogram in Figure 25.10a contains the data for gene expression

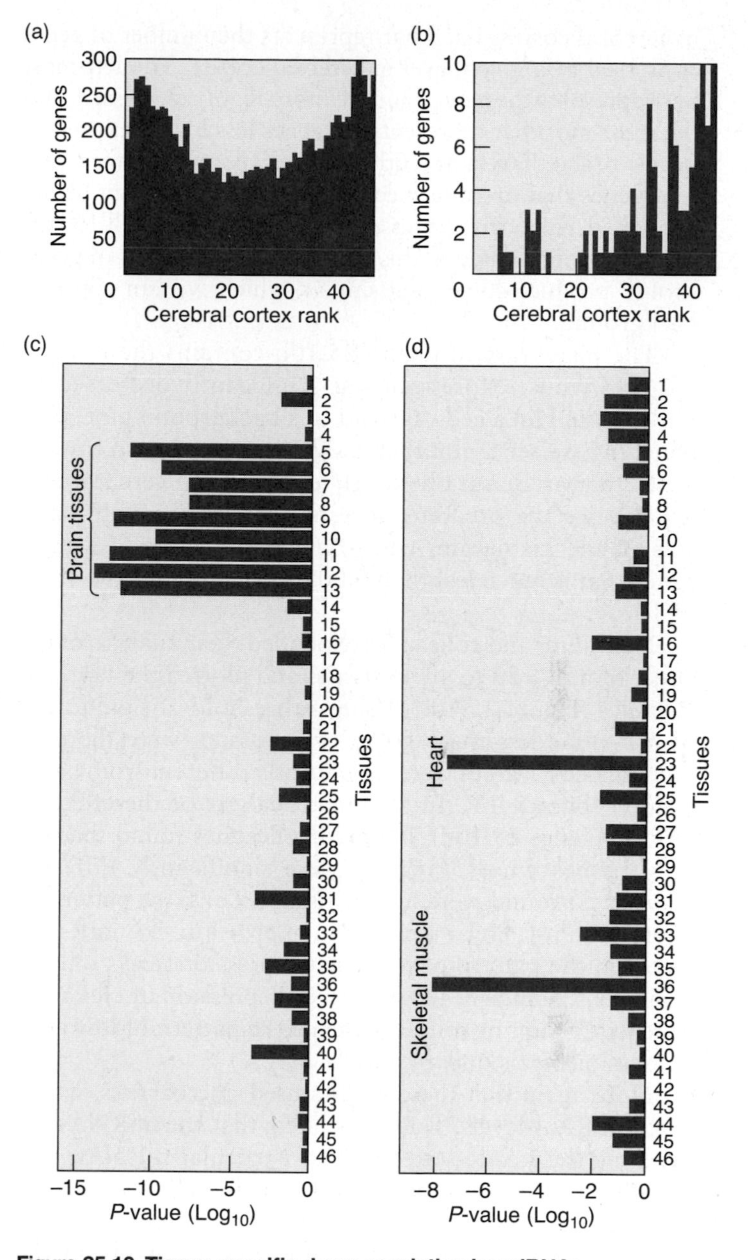

Figure 25.10 Tissue-specific down-regulation by miRNAs. **(a)** Ranking of expression of genes in cerebral cortex. The rankings of all 10,000 genes in each of 46 tissues are plotted as follows: The left-most bar (rank 1) represents the genes that are expressed at a higher level in cerebral cortex than in any other tissue; the next bar (rank 2) represents genes that are expressed at a higher level in cerebral cortex than in any other tissue except one, and the last bar (rank 46) represents the genes that are expressed at a lower level in cerebral cortex than in any other tissue. **(b)** Ranking of genes whose mRNA levels are significantly decreased by miR-124. Note the skew toward genes that are poorly expressed in cerebral cortex compared to the background in panel (a), which gives a *P*-value of significance of about 10^{-12}. **(c)** Plot of the Log_{10} of *P*-values derived from plots like that in panel (b) for all 46 tissues. The only tissues with significant *P*-values (<0.001) are brain tissues: 5, whole brain; 6, amygdala; 7, caudate nucleus; 8, cerebellum; 9, cerebral cortex; 10, fetal brain; 11, hippocampus; 12, postcentral gyrus; and 13, thalamus. **(d)** Similar to (c), except that the analysis was performed on cells to which miR-1, instead of miR-124, had been added. (*Source:* Adapted from Lim et al., Microarray analysis shows that some microRNAs downregulate large numbers of target mRNAs. *Nature*. Vol. 433 (2005) f. 1, p. 770.)

in cerebral cortex. Each bar represents the number of genes expressed at a given level in cerebral cortex. The left-most bar represents the genes that are more highly expressed, and the right-most bar represents the genes less highly expressed in this tissue than in any other tissue. The other bars represent genes that are intermediate in expression, from highly expressed, to poorly expressed in cerebral cortex. All 10,000 genes are represented in this panel, so a random set of genes should produce something similar, which we can consider background.

The histogram in Figure 25.10b contains the ranking of genes whose expression was significantly decreased by miR-124 in HeLa cells. Instead of a background plot, as in panel (a), we see a plot that is significantly skewed toward genes that are naturally poorly expressed in cerebral cortex. Notice the predominance of bars on the right-hand side of the histogram, which yields a *P*-value of significance that is much less than 0.001. In fact, it is of the order of 10^{-12}.

Next, Lim and colleagues expanded their analysis of the effect of miR-124 to all 46 tissues and plotted the Log_{10} of *P*-values (Figure 25.10c). Using a threshold of significance of a *P*-value less than 0.001, brain tissues were the only ones whose *P*-values were significantly different from background (bars 5–13). In a similar analysis of the effect of miR-1 (Figure 25.10d), Lim and colleagues found that the only tissues whose *P*-values were significantly different from background were muscle tissues. Thus, the pattern of depression of HeLa cell gene expression by miR-124 matched the pattern of low gene expression levels only in brain cells. Similarly, the pattern of depression of HeLa cell gene expression by miR-1 matched the pattern of low gene expression levels only in muscle cells.

Note again that these studies used microarrays, which detect mRNA levels. Thus, it is likely that the miRNAs are affecting the steady-state levels of particular mRNAs, presumably by destabilizing them. If this is so, we would expect to see evidence of complementarity between the miRNAs and the destabilized mRNAs, probably in the 3′-UTRs of the mRNAs, where such complementarity has typically been found.

So Lim and colleagues compared the sequences of the miRNAs to the sequences of the 3′-UTRs of the mRNAs whose levels were significantly depressed. They used a "motif discovery tool" called MEME to do the matching, and obtained striking results. Fully 88% of the mRNAs down-regulated by miR-1 had strings of at least six bases, with the consensus sequence CAUUCC, that is complementary to a string of bases in miR-1. And 76% of the mRNAs down-regulated by miR-124 had strings of at least six bases, with the consensus sequence GUGCCU, that is complementary to a string of bases in miR-124. This is strong evidence that the miRNAs really do interact with the 3′-UTRs of their target mRNAs, and presumably destabilize them.

An attractive hypothesis emerges from these studies: miRNAs play an important role in cell differentiation by inhibiting the expression of **gene batteries,** or sets of functionally related effector genes. For example, miR124 inhibits the expression of a battery of hundreds of non-neuronal genes that help to keep a human cell in an undifferentiated state. Presumably, suppression of these non-neuronal genes is a key to differentiation of neuronal cells.

Gail Mandel and her colleagues have provided support for this hypothesis by identifying a protein factor, RE1 silencing transcription factor (REST) that inhibits the expression of a battery of neuron-specific genes, including miR-124 and a number of other miRNAs. REST inhibits miR-124 expression in non-neuronal and pre-neuronal cells. However, during differentiation of neuronal cells, REST dissociates from the miR-124 gene and allows its expression. The newly made miR-124 then inhibits the expression of non-neuronal genes, helping the cell develop into a neuronal cell. Indeed, one of the mRNAs targeted by miR-124 encodes one of the subunits of REST. Thus, miR-124 and REST antagonize each other's expression, as we might expect of two factors that lead to different developmental fates.

SUMMARY Tissue-specific expression profiling can be done by examining the spectrum of mRNAs whose levels are decreased by an exogenous miRNA, and comparing that to the spectrum of expression of genes at the mRNA level in various tissues. If the miRNA in question causes a decrease in the levels of the mRNAs that are naturally low in cells in which the miRNA is expressed, it suggests that the miRNA is at least part of the cause of those natural low levels. This kind of analysis has implicated miR-124 in destabilizing mRNAs in brain tissue, and miR-1 in destabilizing mRNAs in muscle tissue. By inhibiting the expression of batteries of genes, miRNAs can influence the differentiation of cells. For example, miR-124 inhibits the expression of non-neuronal genes. Thus, expression of miR-124 in a pre-neuronal cell pushes the cell toward neuronal differentiation.

Locating Target Sites for Transcription Factors As we learned in Chapter 12, genes are stimulated by activators, which bind to enhancers. Many activators have many enhancer targets in a genome and therefore activate many genes. Such a set of genes that tend to be regulated together is sometimes called a **regulon.** To understand fully the effects of a given activator, it is important to identify all the genes that respond to that activator, and several methods have been developed to accomplish this task.

The most straightforward method is to compare the microarray hybridization patterns of RNAs from organisms that do not express, express at a low level, or overexpress the gene for a given activator. This analysis reveals the genes that are turned on by high expression of the activator and has been useful for that purpose. But two problems limit the utility of this sort of experiment. First, the genes that are turned on may not be direct targets of the activator, but may be targets of other activators whose genes were stimulated by the first activator. Second, the genes that are turned on when the activator is overexpressed may not be turned on in vivo by physiological levels of the activator. Still, there are ways to get around these problems by examining directly the interaction of an activator with the control regions of specific genes.

One such strategy, employed by Richard Young and colleagues (Ren et al., 2000), melds two different techniques: chromatin immunoprecipitation (ChIP, Chapter 13) and DNA microarray hybridization on a DNA microarray, or chip. The technique is therefore called **ChIP-chip** or, sometimes, **ChIP on chip.** Figure 25.11 shows the general plan of the method, which Young and colleagues adapted to identify the binding sites for the activator GAL4 throughout the yeast genome. First, they chemically cross-linked proteins to DNA in chromatin so they could not separate. Then they broke open the cells and sheared the chromatin into small segments. Next, they immunoprecipitated the sheared yeast chromatin with an antibody against GAL4 to precipitate DNA bound to GAL4. Then they reversed the cross-links between the protein and DNA, and labeled copies of this DNA with a red fluorescent dye (Cy5) by PCR. By a parallel procedure, they labeled copies of DNA that was *not* immunoprecipitated by the anti-GAL4 antibody with a green fluorescent dye (Cy3). Then they probed DNA microarrays representing all the intergenic regions of the yeast genome with the two labeled DNAs. Figure 25.12 shows the results of a small section of the array. One spot, denoted by the arrow, clearly shows a preponderance of red fluorescence, suggesting that it hybridized preferentially to the DNA that was associated with GAL4. Using this technique, Young and colleagues identified DNA sequences associated with 10 genes, all of which are known to be activated by GAL4. Thus, the method worked well in this trial.

This method is well suited for yeast because of the limited size of the yeast genome and the fact that the yeast genome has been completely sequenced. But could one perform a similar experiment with the human genome? There would be a serious problem, because the whole intergenic fraction of the human genome is almost as large as the genome itself, so a microarray containing all those sequences would be very complex and difficult to produce. But there are some ways to narrow the field of DNA sequences to make the experiment practical. Two of these were reported in work on the same activator, human E2F4, in 2002.

In their approach to narrowing the field, Peggy Farnham and coworkers used a microarray containing only CpG islands (7776 of them). As we learned in Chapter 24, such CpG islands are associated with gene control regions and therefore should be highly enriched in the activator-binding sequences being sought by this technique. Using that strategy, Farnham and coworkers identified 68 target

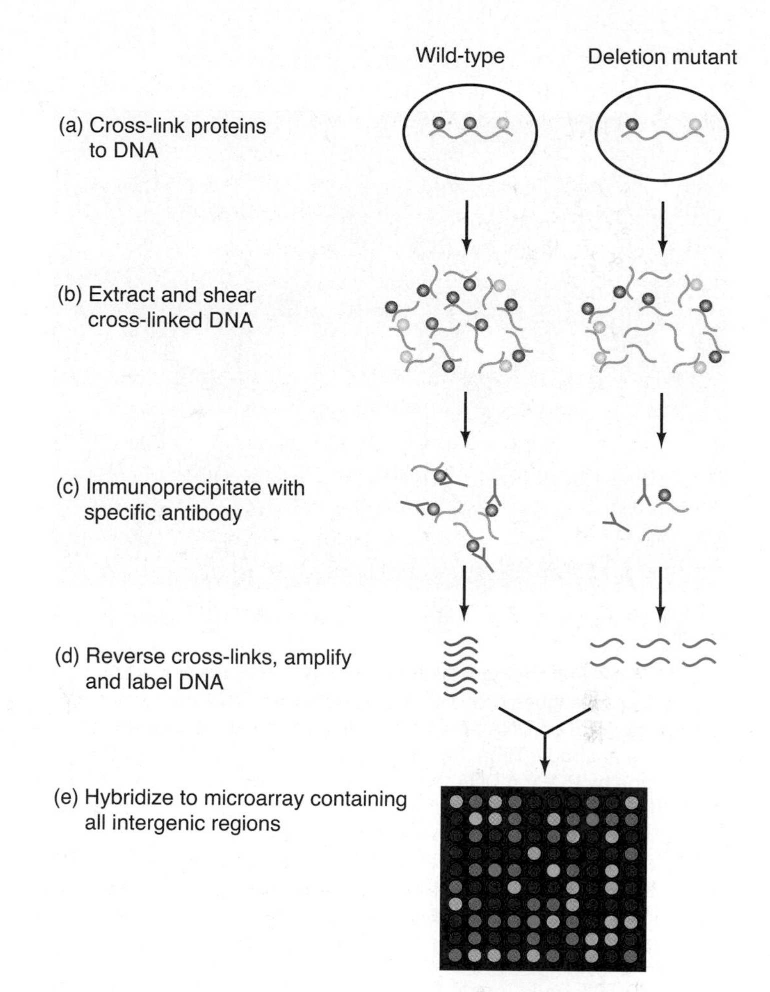

Figure 25.11 Genome-wide search for DNA–protein interactions in yeast by ChIP-chip analysis. (a) First, proteins are chemically cross-linked to DNA in yeast cells. This is done in wild-type cells and in reference cells missing the gene encoding the protein of interest (red). **(b)** The protein–DNA complexes (cross-linked chromatin) are extracted from the cells and sheared by sonication. **(c)** Sheared chromatin is immunoprecipitated with an antibody directed against the protein of interest. **(d)** After precipitation, the cross-links are reversed, and the precipitated DNA is amplified and labeled by PCR. **(e)** The labeled DNA from both kinds of cells is hybridized to a microarray containing DNA representing all intergenic regions in the yeast genome. The precipitated DNA from the wild-type cells is labeled with a red fluorescent dye, and the precipitated DNA from the mutant cells lacking the protein of interest is labeled with a green fluorescent dye. Thus, if a DNA spot on the microarray hybridizes to DNA that binds to the protein of interest more than to other proteins, that spot will fluoresce red. If the DNA hybridizes to DNA that binds to other proteins preferentially, the spot will fluoresce green. If it hybridizes to both DNA probes, it will fluoresce yellow. Careful normalization of the relative intensities of fluorescence of the two DNA probes allows one to determine the ratio of red and green fluorescence at each spot and therefore the significance of the preference a given DNA region has for binding to the protein of interest. (*Source:* Adapted from *Nature* 409: from Lyer et al., 2001, Fig. 1, p. 534).

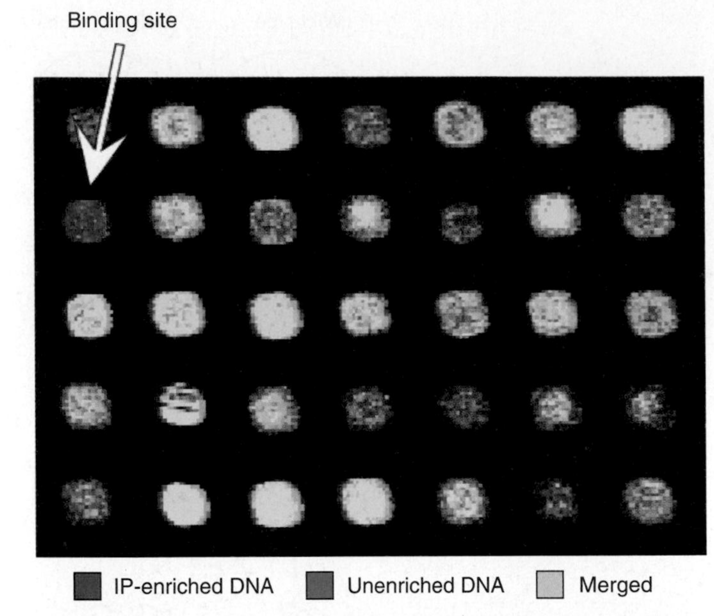

Figure 25.12 Identifying a DNA sequence that binds to GAL4. Young and colleagues prepared a red fluorescent DNA probe by performing PCR on DNA from chromatin immunoprecipitated by an anti-GAL4 antibody. Then they prepared a similar, green fluorescent DNA probe by PCR on DNA that was not immunoprecipitated by the antibody. Then they hybridized these two probes to a DNA microarray with DNAs representing all the intergenic regions in the yeast genome. This is a small section of that array, showing one red spot (arrow) that indicates a putative GAL4-binding DNA, several green spots, indicating DNA that does not bind GAL4, and several yellow spots (binding both red and green probes) that do not show significant preferential binding of GAL4. (*Source:* Adapted from Ren et al., *Science* 290 (2000) Fig. 1A. p. 2306.)

sites for their activator. Instead of CpG islands, David Dynlacht and colleagues chose the control regions of approximately 1200 genes that were known to be activated as cells entered the cell cycle (a time when E2F4 is active). From this panel of DNAs on the microarray, they found that 127 bound to E2F4 in human fibroblasts. Thus, some foreknowledge of the timing and selectivity of an activator can be very useful in designing a microarray to seek out more target genes.

One problem with the ChIP-chip technique for finding transcription factor binding sites is that it is limited to the sequences placed on the chip. In order to contain all the possible sequences in the euchromatic part of the human genome, such a chip (or chips) would have to contain of the order of a billion spots—beyond the reach of current technology. Even when chips with **tiling arrays** (DNAs with overlapping sequences) approach the resolution of just a few nucleotides, they are predicted to be quite expensive, at least at first. Another problem is that hybridization efficiency to spots on a chip is different for different DNAs, so some binding sites will be missed because their hybridization conditions are not met. Also, it is an unfortunate fact of life that hybridization specificity is not perfect: Sometimes one DNA will hybridize to more than one spot, or will fail to hybridize where it should because of DNA secondary structure. Finally, excellent coverage of the genome by ChIP-chip will be realized in the near future only for the human genome, in which high-resolution tiling arrays will be available. Investigators studying other genomes will not have that advantage.

An alternative that solves these problems is a technique called **tag sequencing,** in which the amplified pieces of DNA precipitated in the ChIP procedure are not hybridized to a chip, but repeatedly sequenced using one of the new high-throughput, next-generation techniques described in Chapter 5. With 2007 technology, one instrument could do about 400,000 200-nt reads, or 40 million 25-nt reads at a time. Barbara Wold and colleagues tested such a method, which they dubbed ChipSeq (more commonly known as **ChIP-seq**) in 2007. They performed millions of 25-nt reads on DNAs isolated by ChIP with an antibody specific for a transcription factor called neuron-restrictive silencing factor (NRSF), which represses neuronal genes in non-neuronal cells and in neuronal precursor cells. Then they used a computer program to show where these 25-nt reads mapped to the human genome. They counted as significant any site where 13 or more reads clustered, and where this clustering was at least five-fold enriched over a control in which no antibody was used during the ChIP procedure. Figure 25.13 depicts a cluster of reads that defines a binding site for a hypothetical protein.

NRSF binding sites were attractive subjects because they had already been carefully studied by other techniques, and a canonical binding site sequence had been recognized. The ChIPSeq procedure identified almost all of the canonical binding sites, and found new binding sites as well. Some of these had canonical half-sites separated by noncanonical spacers. Others had only one half-site. Thus, this technique appears to be comprehensive in its ability to identify binding sites.

Mathieu Blanchette, François Robert, and their colleagues adopted a different approach to finding transcription

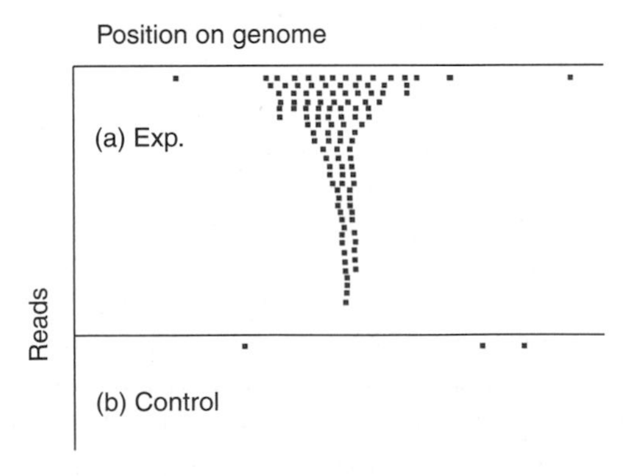

Figure 25.13 Mapping a transcription factor binding site by ChIPSeq. (**a**) Short (25-nt) reads of sequence of DNAs precipitated by ChIP using an antibody specific for a transcription factor are plotted vs. genome position at one particular place in the genome. Each red block represents one read. The peak defines the binding site for the transcription factor. (**b**) A control is run without an antibody in the ChIP step, and this shows only background binding.

factor binding sites in the human genome. Instead of searching for binding sites for a single protein, they looked for clusters of such binding sites (***cis* regulatory modules [CRMs]**, Chapter 12). Whereas each individual transcription factor binding site can be quite variable in sequence, and thus escape notice, clusters of such sites are relatively easy to find.

Blanchette, Robert, and colleagues took advantage of the Transfac database, containing binding site sequence information for 229 different transcription factors. They also realized that CRMs are well conserved, relative to surrounding DNA sequences. Accordingly, they focused on nonrepetitive, noncoding DNA regions that are conserved in the human, mouse, and rat genomes and searched in those regions for transcription factor binding sites from the Transfac database.

This scan, encompassing the 34% of the human genome that can be aligned with both the mouse and rat genomes, yielded 118,402 predicted CRMs (pCRMs). This number surely includes some false positives, but it represents only about one-third of the human genome. While that part of the genome is likeliest to be enriched in CRMS, we can still conclude that the human genome probably contains at least two hundred thousand CRMs. That number may seem surprisingly large, but the authors have validated their data in several ways. For example, they found a strong enrichment of their pCRMs in known promoter regions (defined as DNA regions within 1 kb upstream of the transcription start site), particularly promoters within CpG islands. They also found good correspondence between the pCRMs and DNase hypersensitive regions, which, as we learned in Chapter 13, tend to contain gene regulatory elements.

One somewhat surprising result of this work was the large number of pCRMs that lie in regions thought to be devoid of genes. This finding could be explained in several ways: (1) It may reflect our inability to identify all the genes in the human genome. (2) It may indicate that some genes have cryptic transcription start sites that lie far upstream of the canonical start sites. (3) The pCRMs may be regulating the production of noncoding RNAs. (4) The pCRMs may be regulating the transcription of genes a great distance away.

Figure 25.14 depicts the frequency of pCRMs within and surrounding known genes. As expected, there is a strong preference for pCRMs in the immediate 5′-flanking region of a gene, where enhancers are classically found. But there is also a preponderance of pCRMs in regions where we would not expect them, beginning with the region just downstream of the transcription start site. This could reflect alternative, downstream transcription start sites, or it could be the first indication of widespread regulatory elements within genes. A second surprise in Figure 25.14 is the abundance of pCRMs in the region surrounding the transcription termination site. Again, this has at least two possible explanations. It could indicate a large class of enhancers just downstream of the genes they control, or it could represent antisense transcripts that could play a negative role in gene expression. There is a poverty of pCRMs in the regions 10–50 kb upstream and 10–30 kb downstream of genes, and at the edges of introns (except the first and last ones). Some of this may be only apparent. For example, there could be a selection in these regions for pCRMs with few enough factor binding sites that they escaped notice in this study.

SUMMARY ChIP-chip analysis can be used to identify DNA-binding sites for activators and other proteins. In organisms with small genomes, such as yeast, all of the intergenic regions can be included in the microarray. But with large genomes, such as the human genome, that is now impractical. To narrow the field, CpG islands can be used, since they are associated with gene control regions. Also, if the timing or conditions of an activator's activity are known, the control regions of genes known to be activated at those times, or under those conditions, can be used.

Tag sequencing, or ChIP-seq, in which the chromatin pieces precipitated by ChIP are repeatedly sequenced, can also be used to identify transcription factor binding sites. Knowledge of the sequences of multiple mammalian genomes also allows one to narrow the search for human transcription factor binding sites by beginning with conserved regions of the genome. In addition, it is easier to search for CRMs, which contain several transcription factor binding sites. There are more than 100,000 CRMs in the human genome. They tend to cluster in the regions surrounding the transcription start and termination sites, but a surprising number are found in gene deserts far from any known genes.

Locating Enhancers that Bind Unknown Proteins The "gene-centric" strategy we have just studied is applicable only to enhancers that bind known proteins. But there are still many enhancers whose protein partners are unknown. In order to identify such enhancers, Len Pennacchio and colleagues reasoned that they needed a genomic approach, and they described a very effective one in 2006. They started their search for vertebrate enhancers by looking for highly conserved noncoding DNA regions. These DNA regions could meet their definition of "highly conserved" in two ways: They were either conserved in distantly related species (say, human and pufferfish), or 100% conserved over at least 200 base pairs in more closely related species (e.g., human and mouse).

Pennacchio and colleagues found 167 such enhancer candidates. To test these DNA sequences for enhancer

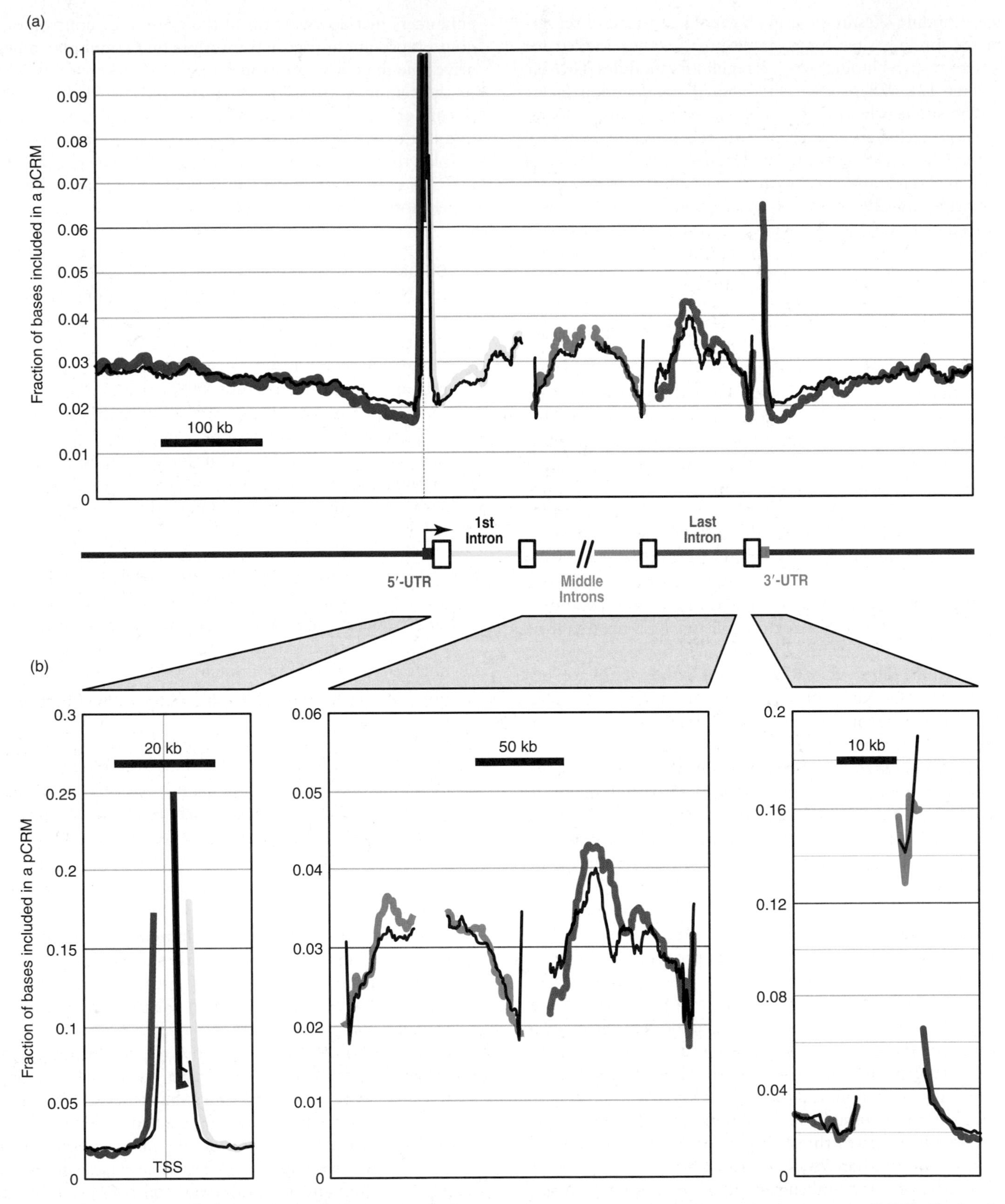

Figure 25.14 Distribution of pCRMs within and surrounding genes. **(a)** The fraction of bases included in a pCRM is plotted vs. position within or outside a gene. Colors in the graph, and in the gene diagram below, represent various gene regions as follows: Dark blue, upstream and downstream flanking regions; red, 5′-UTR; yellow, first intron; light blue, middle introns; brown, last intron; aqua, 3′-UTR. (The fraction of bases in a pCRM is off scale for the 3′-UTR, so no aqua line is visible. **(b)** Same as in (a), except that the horizontal scale has been lengthened to show the individual regions more clearly.

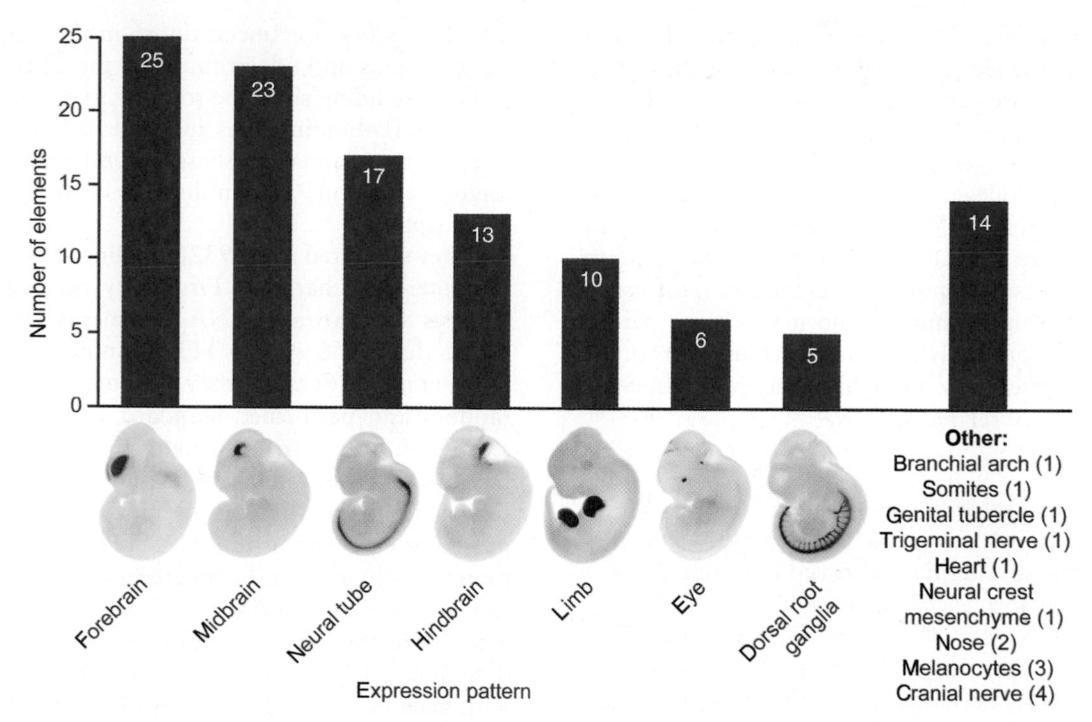

Figure 25.15 Expression patterns driven by enhancers discovered by transgenic mouse enhancer assay. The expression patterns are pictured in typical X-gal-stained mouse 11.5-day embryonic whole mounts, below the bar graph. The number of DNA elements giving rise to each expression pattern is shown. Some enhancers produced more than one expression pattern, which explains why the number of elements is higher than the total number (75) of enhancers tested. (*Source:* Reprinted by permission from Macmilllan Publishers Ltd: *Nature,* 444, 499–502, 23 November 2006. Pennacchio et al, *In vivo* enhancer anylysis of human conserved non-coding sequences. © 2006.)

activity, they hooked them up to *lacZ* reporter genes under the control of a mouse minimal promoter. Then they placed these constructs into mouse zygotes, creating transgenic mice. They allowed the transgenic embryos to grow to embryonic day 11.5, then stained whole embryo mounts with X-gal to detect β-galactosidase. Strong blue staining with X-gal indicates abundant β-galactosidase, and therefore strong transcription stimulated by proteins binding to an enhancer. Pennacchio and colleagues chose day 11.5 embryos for several reasons: First, they can be stained and visualized as whole embryo mounts. Furthermore, major organ systems are visible by this stage. Finally, highly conserved enhancers are known to be clustered near genes that are expressed during embryonic life.

Of the 167 enhancer candidates tested in this way, Pennacchio and colleagues found that 75 (45%) were positive in this transgenic mouse enhancer assay. Figure 25.15 shows the number of enhancers that operated in each of several different tissues, and the pattern of staining that demonstrates each of the tissue-specificities. The numbers add up to more than 75, because many of the enhancers are active in more than one tissue. It is striking that nervous tissue is by far the most common locus of enhancer activity in this experiment, but that is not surprising, considering that a large percentage of vertebrate genes are expressed in nervous tissue, and that the development of the nervous system is complex and requires the function of many genes.

Thus, this strategy has a remarkably high success rate: 45%, achieved by sampling only one stage of embryonic development. One expects that many of the sequences that gave negative results in this experiment would be positive if other stages of life were sampled. Also, it is already known that some of the negative sequences are in fact silencers, so they are also interesting gene control elements. Pennacchio and colleagues reported that there are 5500 more noncoding sequences in the human genome that are conserved between humans and pufferfish, and are thus good candidates for additional enhancers. This strategy therefore shows great promise for locating enhancers in the human and in other genomes.

As successful as this method may be for locating gene control regions, it suffers from the drawback that it only detects highly conserved sequences. And there is reason to believe that not all important gene control regions are conserved. We have already seen examples of poorly conserved control regions in different species of yeast earlier in this chapter, and the same phenomenon is also found in vertebrates. In 2008, Duncan Odom and colleagues reported their studies on gene expression in mouse cells carrying a copy of human chromosome 21. They found that the levels of transcription of human chromosome 21 genes in mouse cells more closely resembles their transcription levels in human cells than the levels of transcription of homologous mouse genes in mouse cells. This implies that mouse transcription factors recognize human gene control regions and homologous mouse gene control regions differently. Indeed, Odom and colleagues also showed by ChIP analysis that mouse transcription factors bind to human chromosome 21 in a more human-like than mouse-like pattern. The most likely reason for these differences is a difference in sequence between the human and mouse gene control regions. Thus, one probably misses important gene control regions if one focuses only on highly conserved sequences, even between closely related species.

SUMMARY To find enhancers whose protein partners are unknown, one can look for noncoding sequences that are highly conserved between moderately related species, or absolutely conserved between closely related species. These putative enhancers can then be verified by linking them to a reporter gene, such as *lacZ*, and looking for reporter gene activity in embryos, in which many genes are active. In the case of the *lacZ* reporter gene, one looks for blue tissue in the presence of the indicator X-gal. One limitation to this kind of study is that some important gene control regions are not well conserved, even between closely related species.

Locating Promoters In principle, class II promoters should be easier to locate than enhancers, as they lie at or very near the transcription start sites of genes, which are usually known. Nevertheless, when Bing Ren and colleagues performed a genome-wide search for human promoters, they got a surprise: Many genes have alternative promoters that are located hundreds of base pairs away from the primary ones.

Ren and colleagues searched for promoters in human fibroblasts using a ChIP-chip strategy. As mentioned earlier in this chapter, the ChIP-chip technique seeks to identify regions in the genome that bind a particular protein. Ren and colleagues performed ChIP using a monoclonal antibody against the TAF1 subunit of TFIID, reasoning that preinitiation complexes forming at promoters should contain this key general transcription factor. Then they amplified the DNA precipitated by ChIP and used it to probe DNA microarrays containing about 14.5 million 50-mers representing all the nonrepetitive DNA in the human genome. Figure 24.16 summarizes the method and presents some of their findings.

They found 12,150 TFIID-binding sites, of which 10,553 (87%) mapped within 2.5 kb of a known transcription start site. They had to use the fairly large window of 2.5 kb to allow for uncertainties in the mapping of transcript 5′-ends and uncertainties in the ChIP-chip mapping of TFIID-binding sites due to noise in the microarray data. Some TFIID-binding sites mapped to the same transcript 5′-ends; by eliminating these redundancies, Ren and colleagues settled on 9328 binding sites that mapped to unique transcripts.

They subjected these 9328 binding sites to four tests for promoter-like character. First, they performed ChIP-chip analysis with an anti-RNA polymerase II antibody and found that 97% of the TFIID-binding sites also bound polymerase II. Second, they selected 28 of these sites at random and performed standard ChIP analysis with an anti-RNA polymerase antibody to verify polymerase II binding. All but one site passed this test. Third, they searched for CpG islands and Inr, DPE, and TATA box core promoter elements in the 9328 TFIID-binding sites. They found enrichment for the first three but not for TATA boxes (Figure 24.17c). Fourth, they used ChIP-chip analysis to look for histone modifications (acetylated histone H3 and dimethylated lysine 4 on histone H3) that are associated with gene activity. Again, 97% of the TFIID-binding sites were associated with these modifications. In summary, the ChIP-chip method appears to have selected promoters very accurately, and most of these promoters lack TATA boxes, in accord with other data showing a paucity of TATA boxes in yeast and *Drosophila*.

Ren and colleagues discovered that over 1600 of the genes they identified had multiple promoters. In most cases, these promoters gave rise to transcripts that differed only in the lengths of their 5′-UTRs, or in having a distinct first exon, but did not affect the protein products of the genes. In other cases, they gave rise to transcripts that were spliced, polyadenylated, or translated differently. These latter cases could provide another layer of control over gene expression, if cells can select which promoter to use at a given time.

SUMMARY Class II promoters can be identified using ChIP-chip analysis with an anti-TAF1 antibody. In one such study with human fibroblasts, over 9000 promoters were identified, and over 1600 genes had multiple promoters.

In Situ Expression Analysis Consider the following opportunity: As is well known, human chromosome 21 is involved in Down syndrome. To discover which gene(s) on this chromosome are responsible for the disorder, it would be useful to know the pattern of expression during embryonic life of all the genes on this chromosome.

Such studies are routinely done in lower organisms, typically by performing in situ hybridization (Chapter 5)

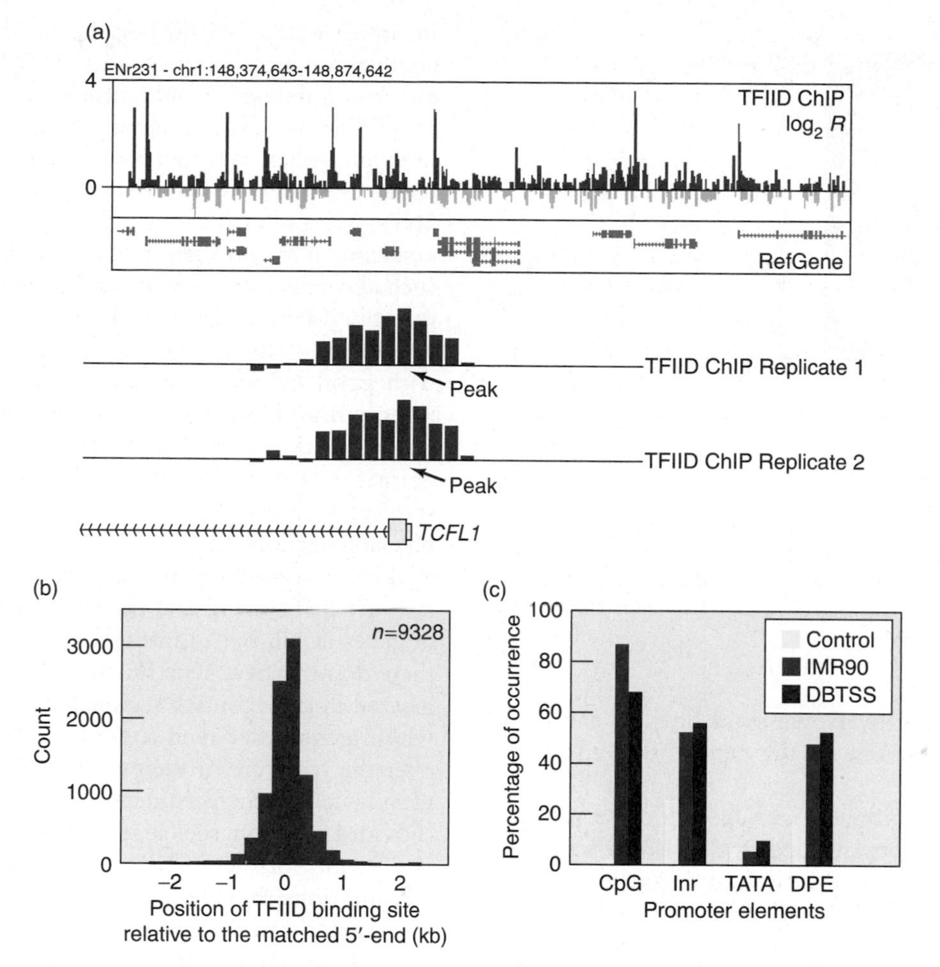

Figure 25.16 Finding promoters. Ren and colleagues performed ChIP-chip analysis using an anti-TAF1 antibody to identify TFIID-binding sites in human fibroblasts. **(a)** Representative results from a relatively small region of human chromosome 1. The top panel presents the logarithmic ratio ($\log_2 R$) of hybridization of DNA precipitated by TAF1-ChIP to hybridization of a control DNA. Peaks show putative TFIID-binding sites. The middle panel shows a gene annotation of this DNA region from the RefSeq database. Note that the peaks in the top panel generally align with the 5′-ends of the annotated genes. The bottom panel presents a blow-up of two replicate ChIP analyses of the *TCFL1* gene. Arrows show the peak of hybridization, determined by a peak-finding algorithm, and the position of the gene is given below, with the 5′-end on the right. **(b)** Alignment of TFIID-binding sites with 5′-ends of genes. The bulk of the binding sites (83%) fall within 500 bp of the 5′-ends of genes. **(c)** Association of CpG islands and three core promoter elements with promoters. Red, TFIID-binding sites identified in this study; blue, promoters from the DBTSS database; yellow, control DNA.

with cDNA probes in embryonic sections. But that presents a serious problem: Such studies are ethically problematic when performed on human embryos. Fortunately, now that we have the sequence of the mouse genome, there is a way around this problem. The mouse genome harbors orthologs for 161 of the 178 confirmed genes on human chromosome 21. So the expression of these genes can be followed through time and space during development of mouse embryos, and we can assume a similar pattern of expression applies to the homologous genes in the human embryo.

Two research groups applied this strategy to the mouse orthologs of the genes on human chromosome 21. In one, Gregor Eichele, Stylianos Antonarakis, and Andrea Ballabio and their colleagues looked at expression of 158 of the mouse orthologs at three times during gestation by in situ hybridization. They also checked the expression patterns of all 161 orthologs in adults by RT-PCR (Chapter 5). They found patterned expression (expression confined to specific sites at specific times) of several genes. Moreover, some of this patterned expression was in sites (central nervous system, heart, gastrointestinal tract, and limbs) that are consistent with the pathology of Down syndrome.

For example, Figure 25.17 shows the expression of the *Pcp4* gene in day 10.5 mouse embryos (by in situ hybridization to whole mount sections) and in day 14.5 embryos (by in situ hybridization to embryonic sections). At day 10.5, the gene is expressed in the eye (black arrow), brain, and dorsal root ganglia (white arrow). At day 14.5, the gene is expressed in many tissues, including the cortical plate (red arrow) in the brain, the midbrain, cerebellum, spinal cord, intestine, heart, and dorsal root ganglia. All of

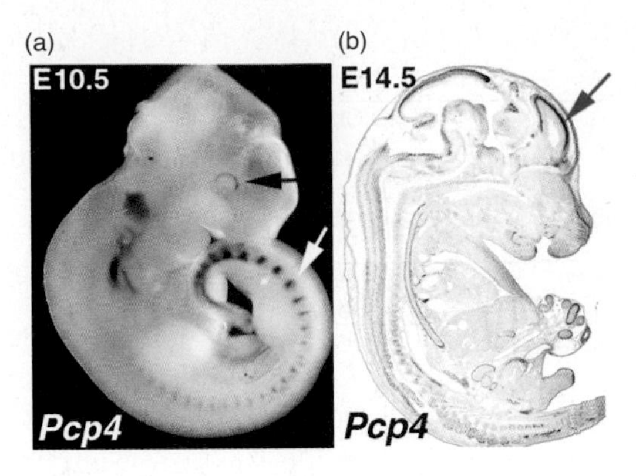

Figure 25.17 Expression of two genes in mouse embryos. Gene expression was assayed by in situ hybridization (Chapter 5), using either a whole mount embryo (panel a), or a sectioned embryo (panel b). **(a)** Expression of *Pcp4* in a whole mount of a day 10.5 embryo. The black arrow indicates the eye, and the white arrow indicates a dorsal root ganglion. **(b)** Expression of *Pcp4* in a section of a day 14.5 embryo. The red arrow indicates the cortical plate of the brain. Dark staining denotes expression of the gene. (*Source:* Adapted from *Nature* 420: from Reymond et al., fig. 2, p. 583, 2002.)

these are areas affected by Down syndrome, so the *Pcp4* gene is a candidate for one of the genes involved in the disorder.

Another example combines work from Eichele and colleagues and another group headed by Ariel Ruiz i Altaba, Bernhard Herrmann, and Marie-Laure Yaspo on the expression of the mouse *SH3BGR* gene in days 9.5, 10.5, and 14.5 of gestation. These studies show that this gene is prominently expressed in the heart at all three stages of development. Because the heart is one of the organs affected by Down syndrome, the *SH3BGR* gene is another candidate for involvement in the disorder.

SUMMARY The mouse can be used as a human surrogate in large-scale expression studies that would be impermissible to perform on humans. For example, scientists have studied the expression of almost all the mouse orthologs of the genes on human chromosome 21. They have followed the expression of these genes through various stages of embryonic development and have catalogued the embryonic tissues in which the genes are expressed.

Single-Nucleotide Polymorphisms: Pharmacogenomics

Now that we have a finished draft of the human genome sequence, we can look for differences among individuals. So far, most of these are differences in single nucleotides, and we classify them as **single-nucleotide polymorphisms,** or **SNPs** (pronounced "snips") if the minor variant is present in at least 1% of the population. The human genome contains at least 10 million such SNPs, and on average, any two unrelated people differ in millions of SNPs. If we can link these SNPs to human diseases governed by defects in single genes, we could then screen individuals for the tendency to develop those diseases simply by screening for SNPs. We might also be able to find sets of SNPs that associate with polygenic traits, such as susceptibility to such disorders as cardiovascular disease and cancer and thus pin down the genes responsible for these traits.

We may also be able to identify SNPs that correlate with good or poor response to certain drugs. Using this information, physicians should be able to screen a patient for key SNPs, then custom design a drug treatment program for that patient based on his or her predicted responses to a range of drugs. This field of study is called **pharmacogenomics.**

However, these tasks will not be easy. Already, geneticists are discovering that the vast majority of SNPs are not in genes at all, but in intergenic regions of DNA. Most of these do not affect gene function, but a few will if they are located in gene control regions. Even when they are found within genes, they tend to be silent mutations that do not alter the structure of the protein product, and thus do not usually cause any malfunction that could lead to a disease. (For an exception, see Chapter 18.) The reason for the preponderance of silent SNPs is clear: Polymorphisms caused by mutations that change the products of genes are generally deleterious, and are therefore selected against. That is, the individuals with these damaging mutations generally die before they can reproduce and thus the mutations are lost. Finally, if history is a guide, even knowing which SNPs correlate with diseases may not be of immediate benefit. It will take time to figure out how to use this information.

One can detect SNPs correlating with disease or other traits in any given individual by a variety of genotyping techniques. One of these is to hybridize a primer adjacent to a SNP and then perform primer extension with fluorescent nucleotides and observe which nucleotide is incorporated in the SNP position. Another is to hybridize a person's DNA to DNA microarrays containing oligonucleotides with the wild-type and mutated sequences. Still another is sequencing: either shotgun sequencing, or amplifying a region surrounding a SNP by PCR and then sequencing it. Such knowledge can be useful in helping to prevent or treat disease.

How do SNPs differ from RFLPs? RFLPs are identical to SNPs if the single-nucleotide difference between two individuals lies in a restriction site, as we observed in Chapter 24 in the RFLPs involving *Hin*dIII sites in Huntington disease patients. In such a case, a single-nucleotide difference makes a difference in the pattern of restriction fragments. However, RFLPs can also result from insertion of a chunk of DNA between two restriction sites in one individual, but not another—VNTRs, for example. That would not be a SNP because it involves more than just a single-nucleotide difference.

For those who are enthusiastic about the potential of SNPs to help identify the causes of common diseases, 2005 was a banner year. The International HapMap Consortium published a haplotype map including over 1 million human SNPs, discovered by genotyping 269 DNA samples from four distinct human populations (one in Nigeria; one in Utah, USA; one in China; and one in Japan). A **haplotype map** shows the locations of haplotypes, blocks of DNA that tend to be inherited intact, because of the low rate of recombination within the block. We have already seen in our discussion of the human genome that the rate of recombination varies considerably from spot to spot, and regions of high recombination rate alternate with regions in which recombination is rare. The latter regions are likely to contain genetic markers that are inherited together and therefore make up a haplotype.

By focusing on certain well-chosen SNPs (**tag SNPs**), the International HapMap Consortium was able to identify other SNPs in the same region, thus cutting down on the total amount of genotyping they had to do. They did this genotyping largely by hybridizing labeled human DNA fragments to DNA microarrays designed to detect tag SNPs. This procedure is highly automated, allowing one worker to scan 500,000 SNPs covering the whole genome in only two days.

One immediate payoff of the project was the identification of millions of new SNPs (only 1.7 million were known at the beginning of the project). Another was new insight into recombination and natural selection in human evolution.

But the potential payoff that attracts the most attention is the identification of genes that are involved in human diseases. This process was straightforward in the case of HD and other diseases caused by a mutation in a single gene, because people with particular mutations are all but certain to have the disease. But it is vastly more difficult when many genes contribute to a disease, because each mutation may contribute only a little bit, and so each is difficult to spot. Unfortunately, the diseases that kill and disable most people (cancer, heart disease, and dementia, for example) are of the latter kind. In principle, the HapMap should make this job easier.

Indeed, in 2005, Josephine Hoh, Margaret Pericak-Vance, and Albert Edwards and their colleagues reported their work on age-related macular degeneration (AMD), a common cause of blindness in elderly people. They scanned 116,204 human SNPs looking for linkage to AMD and found one with a high degree of correlation. That is, one allele is found significantly more frequently in AMD patients than in normal controls. These workers traced this SNP to a gene called *CFH,* which encodes complement factor H. This protein regulates the complement cascade, which governs inflammation. Later in 2005, Gregory Hageman, Rando Alikmets, Bert God, Michael Dean, and their colleagues confirmed the linkage between *CFH* and AMD, finding a high-risk variant of the gene and also several variants of the gene that appeared to be protective.

These results led Hageman, Alikmets, God, Dean, and their colleagues to look for participation of other components of the complement cascade. Sure enough, they found a strong association between AMD and the factor B gene, and both high-risk and protective variants. These findings validated Hageman's earlier hypothesis that inflammation is central to the disease process in AMD, and suggest that controlling inflammation may be a way to help prevent or control the disease. But genes in the complement cascade are not the only ones linked to AMD. Another group has linked a gene (*LOC387715*), with a product of unknown function, in AMD, and there are sure to be others.

Other workers have looked beyond SNPs in comparing the genomes of different people, and have been surprised by what they found. The genomes of seemingly normal people frequently contain not just SNPs, but deletions, insertions, inversions, and other rearrangements of whole chunks of DNA. Geneticists are now calling such differences in genomes **structural variation.** For example, Michael Wigler and his colleagues examined the genomes of 20 healthy individuals and found 221 places where these people had different numbers of copies of particular chunks of DNA. While these variations in copy number had no apparent effect on health in these people, it is possible that, in combination with certain environmental factors, they could predispose other people to disease.

On the other hand, some structural variants appear to be beneficial. Sunil Ahuja and colleagues have shown that extra copies of a particular immune system gene help protect people against AIDS. And a team of Icelandic scientists has discovered a large inversion that is carried by 20% of Europeans. Strikingly, women carrying this inversion have more children than those who do not, suggesting that the inversion confers some kind of reproductive advantage, and that it is therefore probably spreading.

The complete sequences of genomes of simpler organisms can also be important in understanding and treating human diseases. For example, as soon as the complete yeast genome had been sequenced, molecular biologists began systematically mutating every one of the 6000 yeast genes to see what effects those mutations would have. They also began systematically screening all 18 million possible protein–protein interactions using a yeast two-hybrid screen (Chapter 5, and see later in this chapter). The results of such experiments can tell us much about the activities of gene products that are still uncharacterized. And knowing the activities of all the proteins in an organism, and the other proteins with which they interact, should lead to greater understanding of biochemical pathways, such as the ones that metabolize drugs, or signal transduction pathways that control gene expression. This understanding, in turn, should give us important clues about how these pathways work in humans. Moreover, yeast cells can be used as human surrogates to test the effects of knocking out the yeast ortholog of a known human disease gene.

SUMMARY Single-nucleotide polymorphisms can probably account for many genetic conditions caused by single genes, and even multiple genes. They might also be able to predict a person's response to drugs. A haplotype map with over 1 million SNPs will make it easier to sort out the important SNPs from those with no effect. Structural variation (insertions, deletions, inversions, and other rearrangements of chunks of DNA) is also a surprisingly prominent source of variation in human genomes. Some structural variation can in principle predispose certain people to contract diseases, but some is presumably benign, and some is demonstrably beneficial.

25.2 Proteomics

Earlier in this chapter, we learned that studies of an organism's **proteome,** that is, the properties and activities of all the proteins that organism makes in its lifetime, is called **proteomics.** Whereas the task of analyzing an organism's genome, or even its transcriptome, is relatively straightforward, the task of analyzing an organism's proteome is anything but simple, in large part because of the complexity of proteins relative to nucleic acids. Indeed, with current techniques, proteomics studies on complex organisms can examine only a fraction of the total proteome.

Given this difficulty, why are scientists even interested in studying gene expression at the protein level, when they already have transcriptomics, in which they can probe the expression of vast numbers of genes simultaneously by looking at the levels of their transcripts? Part of the answer is that we now know that a large fraction, perhaps 50% or more, of polyadenylated RNAs in human cells do not code for proteins. These are called **noncoding RNAs (ncRNAs)**, and, as we have seen, they are also known as transcripts of unknown function (TUFs). They are interesting in their own right, but their level of expression tells us nothing about protein expression levels. Another part of the answer is that the sequence of a protein-encoding gene and the level of its expression may give little or no information about the activity of its protein product.

Another part of the answer is that the level of transcription of a gene gives only a rough idea of the real level of expression of that gene. For one thing, an mRNA may be produced in abundance, but degraded rapidly, or translated inefficiently, so the amount of protein produced is minimal. For another, many proteins experience posttranslational modifications that have a profound effect on their activities. For example, some proteins are not active until they are phosphorylated. Thus, if the cell is not phosphorylating such a protein at a given time, production of a large amount of mRNA for that protein would give a misleading picture of the true level of expression of the corresponding gene. Furthermore, many transcripts give rise to more than one protein—through alternative splicing, or alternative posttranslational modification. So measuring the level of a gene transcript doesn't necessarily tell what protein products will be made. Finally, many polypeptides form large complexes with other polypeptides, and the true expression of each polypeptide's function occurs only in the context of the complex.

Therefore, if we want to measure real gene expression, we must look at the protein level. To analyze all the proteins in an organism, we need to do two things: First, we need to separate all those proteins from one another. Second, we have to analyze each protein by identifying it and measuring its activity. In the next two sections we will introduce some of the ways molecular biologists do these things.

SUMMARY The sum of all proteins produced by an organism is its proteome, and the study of these proteins, even smaller sets of them, is called proteomics. Such studies give a more accurate picture of gene expression than transcriptomics studies do.

Protein Separations

One of the best separation tools available is two-dimensional gel electrophoresis, which was invented in the 1970s (Chapter 5). As powerful as that technique is, it is not up to the job of resolving all the tens of thousands of proteins in the human proteome. An average 2-D gel can resolve only about 2000 proteins, and even the best gel in the best hands can resolve only about 11,000 proteins. This problem is compounded by the fact that the performance of 2-D electrophoresis is unpredictable, and it frequently seems to be more art than science. Another problem is that many very interesting membrane proteins are too hydrophobic to dissolve in the buffers used in 2-D electrophoresis, so they cannot be seen at all. Finally, many proteins are present in such tiny quantities in the cell that a 2-D gel cannot detect them.

Most of these problems are presently intractable, but scientists have dealt with the 2-D gel resolution problem by analyzing different cellular compartments separately. For example, they can start with just the nucleus, or even a subcompartment like the nucleolus or a protein assembly like the nuclear pore complex. With many fewer proteins to separate, resolution is not such a serious problem.

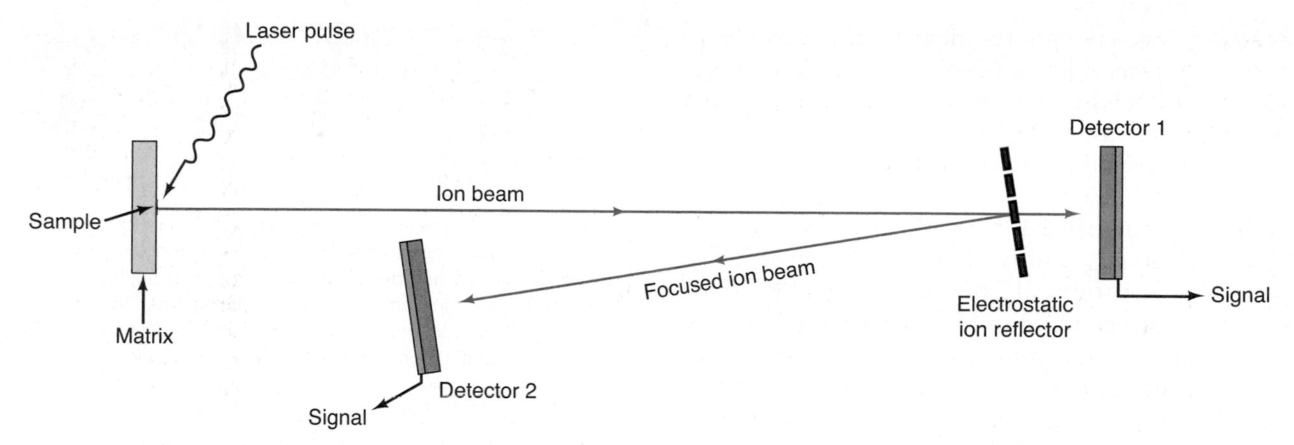

Figure 25.18 Principle behind MALDI-TOF mass spectrometry. Place a sample (a peptide in this case) on the matrix at left and ionize it with a laser pulse. An electrical potential difference between the matrix and the sample then accelerates the ionized sample toward detector 1. The time it takes the ions to reach detector 1 depends on their masses, so one can learn much about their masses by analyzing the time of flight to detector 1. Alternatively, one can turn on an electrostatic ion reflector in front of detector 1 to focus the ions and reflect them toward detector 2. This detector gives even more precise data about the masses of the ions, according to their times of flight.

Protein Analysis

Once the proteins are separated and quantified, how are they analyzed? First, they have to be identified, and the best method now available works like this: Individual spots are cut out of the gel and cleaved into peptides with proteolytic enzymes. These peptides can then be identified by **mass spectrometry.** Figure 25.18 illustrates a popular technique known by the cumbersome title matrix-assisted laser desorption-ionization time-of-flight (MALDI-TOF) mass spectrometry. In this procedure, a peptide is placed on a matrix, which causes the peptide to form crystals. Then the peptide on the matrix is ionized with a laser beam (the matrix helps the peptide ionize), and an increase in voltage at the matrix is used to shoot the ions toward a detector. Assuming all the ions have just one charge (and almost all do), the time it takes an ion to reach the detector depends on its mass. The higher the mass, the longer the time of flight of the ion. In a MALDI-TOF mass spectrometer, the ions can also be deflected with an electrostatic reflector that also focuses the ion beam. Thus, we can determine the masses of the ions reaching the second detector with high precision, and these masses can reveal the exact chemical compositions of the peptides.

Then these ions can be broken at their peptide bonds by a process known as **collision-induced dissociation (CID).** Experimenters do this by accelerating the ions and colliding them with a neutral gas to break them, mostly at their peptide bonds, then sending the new peptide ions to another analyzer to determine their molecular makeup. Because this involves two mass spectrometry steps in a row, it is called **MS/MS.** By comparing the masses of ions differing by just one amino acid, the nature of the lost amino acids can be determined one by one, which leads to a sequence, as illustrated in Figure 25.19.

If the sequence of the whole genome is known, we know what proteins to expect, so a computer can use the information from the mass spectrometer to match each spot on the 2-D gel with one of the genes in the genome, and therefore predict the sequence of the whole protein. For example, the sequence information determined in

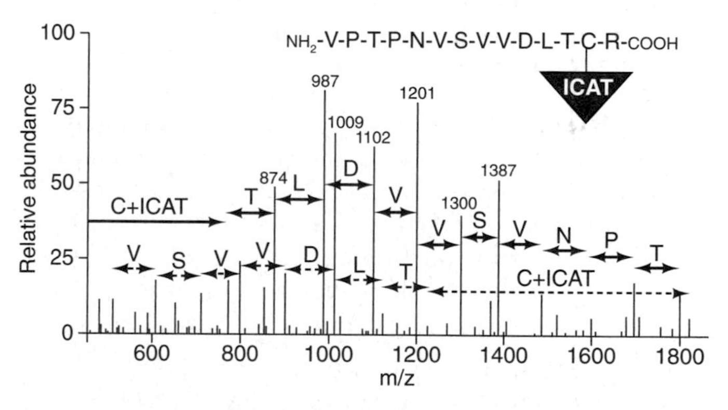

Figure 25.19 Sequencing a peptide by mass spectronomy (MS/MS). The molecular ion is the ionized peptide at top, linked through its cysteine residue (C) to an adduct known as ICAT, which we will discuss in the next section. Its nature is not important here. The molecular ion was fragmented by CID, and the fragment ions were then subjected to a second round of MS, yielding the spectrum shown below the sequence. The relative abundance of each ion is plotted against its mass/charge ratio (m/z). The charge of each ion is assumed to be +1 in this experiment. Starting at the right, measuring the exact mass differences between the most prominent ions, one can deduce the amino acid that was lost to generate the next ion to the left. For example, the difference between the masses of the last two ions on the right shows that a threonine (T) was lost. Continuing in this way, and following the top (solid) arrows, one can read the sequence TPNVSVVDLTC-ICAT. The ion also fragmented from the other end, giving the sequence shown on the bottom with dashed arrows between major ions.

Figure 25.19 was enough to identify the peptide as belonging to glyceraldehyde-3-phosphate dehydrogenase. However, knowing the sequence of a protein does not necessarily tell us that protein's activity, so further research will be necessary to determine the activities of many proteins.

You may be thinking that it would be nice to make a microchip that could identify thousands of proteins at once, as DNA microarrays identify thousands of RNAs at once in functional genomics studies. That would remove the need to separate the proteins because a mixture of many proteins could simply be incubated with the chip to see what binds. One such strategy would be to produce antibodies that can recognize proteins specifically and quantitatively and place them on microchips. But many obstacles stand in the way of realizing that dream. To begin with, antibodies are much more expensive and time-consuming to produce than oligonucleotides. In fact, the task of generating antibodies for every human protein is unthinkably vast at present. Moreover, the task of detecting low-abundance proteins, already impossible for many proteins using 2-D gels, would only be exacerbated by the miniaturization of microchip technology. On the other hand, the technology to complete the human genome in a reasonable period of time was not available when that project was first proposed in the mid-1980s, but the project stimulated the development of the technology. Perhaps we will experience a similar phenomenon if a full-scale human proteome project is initiated.

SUMMARY Current research in proteomics requires first that proteins be resolved, sometimes on a massive scale. The best tool available for separation of many proteins at once is 2-D gel electrophoresis. After they are separated, proteins must be identified, and the best method for doing that involves digestion of the proteins one by one with proteases, and identifying the resulting peptides by mass spectrometry. Someday microchips with antibodies attached may allow analysis of proteins in complex mixtures without separation.

Quantitative Proteomics

Mass spectrometry is now able to identify proteins as they emerge from high-performance separation procedures, such as capillary chromatography, or even in mixtures without separation. But mass spectrometry is not a quantitative method, so it has been difficult to use it to analyze the expression levels of proteins. However, beginning at the end of the 1990s, analytical chemists developed methods that can tell us how much of a given protein is present in cells under one set of conditions, compared to the concentration of that same protein in cells under a different set of conditions. For example, it can measure the increase in concentration of a protein when the gene for that protein is turned on by an inducer.

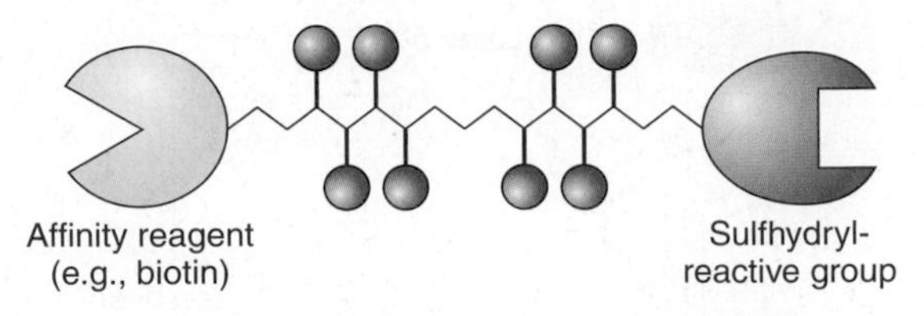

Figure 25.20 A generic ICAT tag. One end (blue) contains a sulfhydryl-reactive group that binds to cysteine side chains. The middle contains a number of positions (red) that can be either all light isotopes (e.g., hydrogen) or all heavy isotopes (e.g., deuterium). The left end (yellow) contains an affinity reagent such as biotin, which allows easy purification of tagged proteins or peptides.

Here is how one such method, using **isotope coded affinity tags (ICATs)**, works. Experimenters couple affinity tags to proteins through the sulfhydryl groups of their cysteine side chains. These affinity tags typically contain three parts, illustrated generically in Figure 25.20: a sulfhydryl-reactive group at one end that can link to a protein's cysteine side chains; a linker in the middle that contains several atoms of either a normal isotope (e.g., hydrogen), or a heavy isotope (e.g., deuterium); and an affinity reagent such as biotin at the other end, which allows convenient purification of a protein or peptide bearing the tag. In the example in Figure 25.20, the heavy tag would be 8 Daltons heavier than the light tag, by virtue of its eight deuteriums. This permits tagged peptides and their untagged counterparts to be identified easily in mass spectra because they appear as a pair of peaks exactly 8 Da apart.

How does this help in quantification? Consider cells grown under two conditions: with and without serum, for example. The question is how much change we see in the concentrations of proteins when serum is added to the medium in which cells are growing. Figure 25.21 shows one approach to this question. In this case, the investigator could add light ICATs to proteins from cells grown in the absence of serum (condition 1), and heavy ICATs to proteins from cells grown in the presence of serum (condition 2). Then the proteins could be mixed, hydrolyzed with a protease such as trypsin, affinity-purified using the affinity reagent, and subjected to **liquid chromatography-mass spectrometry (LC-MS)**, in which the peptides are separated by liquid chromatography in a fine capillary, then fed into a mass spectrometer, in which each peptide appears as a pair of peaks, separated by a molecular mass defined by the ICATs in use (e.g., 8 Da).

The heavier of the peaks in each pair comes from the cells grown in the presence of serum, and the lighter of the two comes from cells grown without serum. Their relative areas, which can be determined by expanding the spectrum to reveal true peaks instead of lines, tell us the change in the

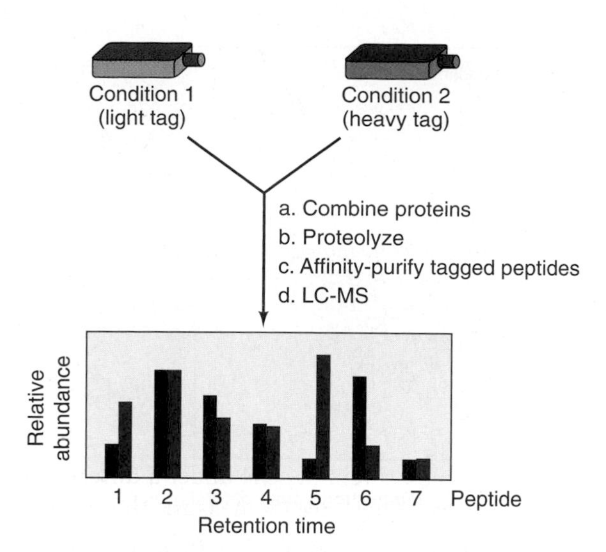

Figure 25.21 Using ICATs to measure the change in protein concentrations upon shift in growth conditions. Cells are grown under two different conditions (e.g., without [condition 1] and with [condition 2] serum). Proteins are extracted from cells grown under both conditions and tagged with either a light ICAT (condition 1, blue) or a heavy ICAT (condition 2, red). The tagged proteins are combined and proteolyzed, and the resulting tagged peptides are subjected to LC-MS. MS resolves the peptides derived from condition 1 and condition 2 because of their small difference in mass (8 Da, in the example in Figure 25.20). Thus, each peptide appears as a pair of peaks, and the relative areas under these peaks corresponds to the change in concentration of the protein to which each peptide belongs. That protein can frequently be identified by sequencing the peptide by MS/MS.

amount of each peptide upon addition of serum to the medium. Even without expanding the spectrum, we can estimate that the concentration of peptide #1 appears to double, peptide #2 to remain the same, and peptide #3 to fall about 25%, upon addition of serum. Of course, these peptides represent proteins, and many of those proteins can be indentified by sequencing the peptides by MS/MS as described earlier in this chapter. In this way, the change in the concentration of a large number of proteins can be quantified relatively quickly and easily.

Since the introduction of the ICAT labeling method, other methods have been developed. For example, proteins can be labeled in vivo by including heavy-isotope-tagged amino acids in the growth medium. This is called **stable isotope labeling by amino acids in cell culture** (SILAC), and it has the advantage of labeling a wider range of peptides—not just those that contain cysteines. It also eliminates all variation in sample preparation because the two cell cultures are mixed prior to protein preparation.

The power of these techniques led Jürgen Cox and Matthias Mann to ask: "Is proteomics the new genomics?" In other words, can we hope to examine massive numbers of proteins simultaneously, in the same way that a DNA microchip allows us to examine massive numbers of RNAs? Clearly, the proteomic method is more time-consuming than the genomic method, and only a subset of proteins can be identified at one time, because of the time limitations of the MS/MS technique. But, with some readily imaginable improvements, these proteomic techniques will become even more powerful.

Note that the methods described here quantify the change in proteins' concentrations, rather than the absolute concentrations of proteins. Fortunately, the former is frequently the more useful information. However, if one wants to quantify a particular protein's absolute cellular concentration, one can take a protein mixture labeled with a light tag and spike it with a known amount of that protein, labeled with a heavy tag. MS on peptides derived from the tagged protein will reveal the ratio of the known, heavy peak to the unknown, light peak, and therefore the concentration of the protein.

SUMMARY To determine the changes in protein levels upon perturbation of a cell culture, one can label the cells under the first condition with a light isotopic tag, and under the second condition with a heavy isotopic tag. If the proteins are labeled in vivo, the cell cultures can be mixed, proteins can be extracted and fragmented by proteolysis. Then the peptides can be separated and subjected to mass spectronomy. Peptides will appear as pairs of peaks separated by the mass difference in the tags. The ratio of heavy to light peak area tells us the change in protein concentration as the growth conditions change.

Comparative Proteomics What makes a worm a worm and a fly a fly? As stated in Chapter 3, it is the proteins produced in these organisms that set them apart. And, presumably, not just the sum total of proteins produced, but when and where they are made. Quantitative proteomics techniques such as those described in the previous section can shed a good deal of light on these questions.

For example, in 2009, Michael Hengartner and colleagues examined the *C. elegans* (roundworm) proteome using these techniques, and compared it to the *D. melanogaster* proteome that had been reported in 2007. They looked at proteins in eggs and worms at various stages of development, and identified 10,977 different proteins, representing 10,631 different genes, which is 54% of the 19,735 predicted genes in the *C. elegans* genome. When they compared the proteins they identified with the proteins predicted from the genome, they found certain classes of proteins underrepresented. These missing proteins tended to be short (less than 400 amino acids) and to have high hydrophobicity (presumably membrane proteins with many fatty transmembrane domains).

Hengartner and colleagues estimated protein concentrations in *C. elegans* from their mass spectrometry with ICAT data, and compared them with similar protein concentration data from the previous *Drosophila* study. They focused on 2695 pairs of orthologs present in both organisms, for which there was also transcript concentration data from microarray and SAGE experiments. The earlier transcript concentration data had shown only a modest correlation between the concentration of a given worm mRNA and its ortholog in the fly. But the protein concentrations of orthologs in worm and fly showed a much better correlation. Indeed, the correlation between orthologous protein concentrations in the two organisms is even better than the correlation between mRNA and corresponding protein concentrations within either organism. Apparently, orthologous proteins are needed in similar concentrations in the two organisms, so differences in mRNA concentrations between the two organisms are compensated by mechanisms affecting protein abundance. To make these comparisons, Hengartner and colleagues used **Spearman's rank correlation.** In this statistical technique, two data sets are arranged in rank order. In this case, the concentrations of the 2695 worm proteins were arranged in rank order from highest to lowest concentration, and the orthologous fly proteins were arranged in the same way. Then the correlation between the two ranks is expressed as Spearman's rank correlation, R_S. A perfect correlation would have an R_S of 1.0, and two totally unrelated data sets would have an R_S of 0.0, though random similarities in large data sets will raise this number above zero, even if there is no correlation.

Figure 25.22 shows the statistical data. Figure 25.22a shows a graphical representation of the protein data. If the two data sets were perfectly correlated, all the dots, each representing a comparison of the abundance of a single orthologous protein in the two organisms, would fall on a line with a slope of 1.0. In this case, there is considerable scatter in the data points, but they cluster around a line with a slope of 1.0. In fact, as shown in Figure 25.22b, the R_S for the protein data is high: 0.79, showing a clear correlation between protein concentrations of orthologous proteins in the two organisms. By contrast, the concentrations of orthologous mRNAs in the two organisms have an R_S of only 0.47 if measured by microarrays, and only 0.22 if measured by SAGE. Thus, the protein concentrations are much more highly conserved than their corresponding mRNA concentrations. In fact, the protein concentrations in the two organisms are even more highly correlated than the protein and mRNA concentrations in the same organism. The R_S values for protein–mRNA correlations in *C. elegans* are 0.59 with the microarray data and 0.44 with the SAGE data. The R_S values in *Drosophila* are 0.66 and 0.36 with the two data sets.

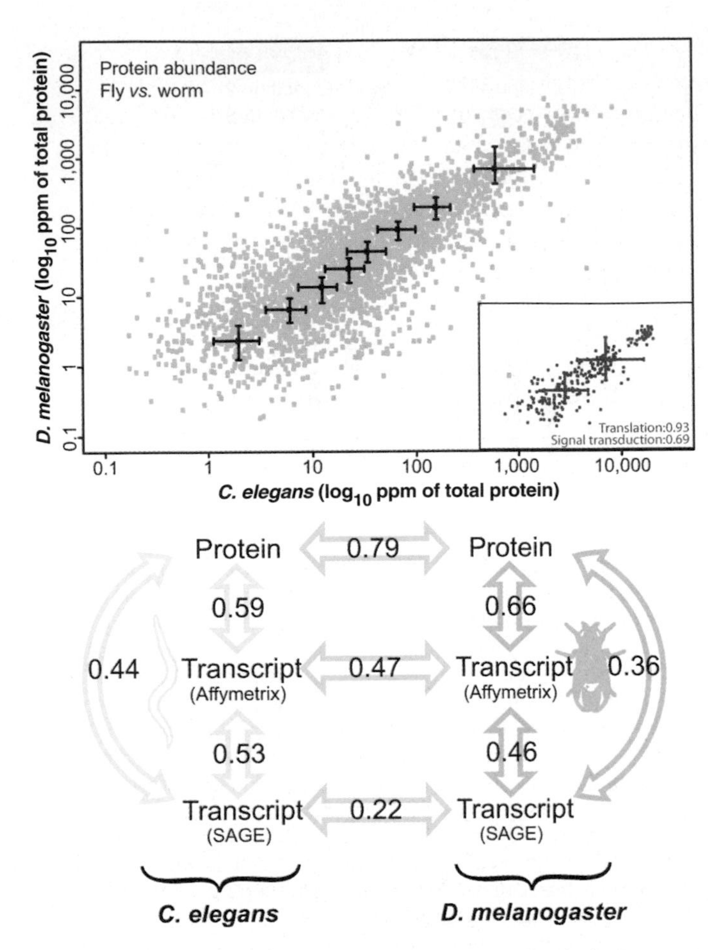

Figure 25.22 Correlation between abundances of orthologous proteins and transcripts in *C. elegans* and *D. melanogaster*. **(a)** Abundances (in parts per million [ppm]) of orthologous proteins in the two organisms, determined by mass spectrometry, are plotted against each other. Each dot represents one orthologous pair of proteins. Crosses represent medians of equal sized bins of values. The "whiskers" at the ends of the crosses represent the range from 25% to 75% of values (where the median, of course, is 50%). The inset contains a similar analysis of the subsets of proteins involved in signal transduction (blue) and translation (red). **(b)** Correlation coefficients (R_S) between proteins and transcripts (measured by microarray [Affymetrix] or SAGE, as noted) in the two species, and between proteins and transcripts within the two species. (*Source:* Figure 5 from, Schrimpf SP, Weiss M, Reiter L, Ahrens CH, Jovanovic M, et al. (2009). Comparative Functional Analysis of the *Caenorhabditis elegans* and *Drosophila melanogaster* Proteomes. PLoS Biol 7(3): e1000048. doi:10.1371/journal.pbio.100048. © 2009 Schrimpf et al.)

SUMMARY Mass spectrometry data can be used to compare protein concentrations in two different organisms. This kind of analysis, applied to *C. elegans* and *Drosophila,* showed that the concentrations of orthologous proteins in the two organisms are correlated much better than the orthologous mRNAs in the two organisms, and even better than the proteins and corresponding mRNAs in the same organism.

Protein Interactions

Most proteins do not function in isolation, but collaborate with other proteins, by participating in such things as biochemical or developmental pathways. Signal transduction pathways (Chapter 12) are good examples. Many other proteins form large multiprotein complexes dedicated to a specific task, such as the ribosome (protein synthesis) or the proteasome (protein degradation). So one goal of proteomics is to identify the proteins that interact with one another. This frequently can give important clues about the functions of newly discovered proteins.

Traditionally, protein–protein interactions have been detected by yeast two-hybrid analysis (Chapter 5), and some proteome-wide studies of protein–protein interactions have been performed using this technique. But two-hybrid analysis is indirect, using reporter gene activation to observe interaction between two parts of a chimeric transcription activator, and it suffers from both false-positives and false-negatives. Nevertheless, in conjunction with validation by an independent technique, yeast two-hybrid screens can be very powerful. In 2005, Erich Wanker and colleagues used a yeast two-hybrid screen, with partial independent validation, to detect over 3000 interactions between human proteins—a start down the arduous path toward elucidating the human **interactome,** the total set of interactions among human proteins.

Investigators have also used ultrasensitive protein mass spectrometry to do a better job of detecting protein–protein interactions. In one such study in 2002, Daniel Figeys and colleagues employed the following procedure (Figure 25.23) to screen protein–protein interactions in yeast: First, they chose a set of 725 "bait" proteins that were likely to interact with other, "fish" proteins. The bait proteins represented several different classes, including protein kinases, protein phosphatases, and proteins that participate in the response to DNA damage. The investigators engineered the genes for each of these proteins to include the coding region for the Flag epitope and then introduced the chimeric genes into yeast cells where they were expressed. (The word "Flag" simply refers to the fact that the epitope serves as a "flag" to make the proteins easy for a single antibody to recognize.)

Then the investigators used immunoaffinity chromatography with an anti-Flag antibody to purify protein complexes containing the bait protein from a cell extract. They separated the proteins from the complexes by SDS-PAGE, cut each band out of the gel, digested the protein in each band with trypsin, and subjected the resulting tryptic peptides to mass spectrometry. Because we know the sequence of the whole yeast genome, a computer can predict all of the proteins encoded in the genome, and the masses of the tryptic peptides that should be obtained from each of them. Thus, this kind of bioinformatic analysis (see next section) can use the mass spectrometer data to identify the tryptic peptides and therefore the proteins.

Using 10% of the predicted yeast proteins as bait, Figeys and colleagues fished out and identified 3617 associated proteins, which is about 25% of the predicted yeast proteome. This is about three-fold higher than the success rate for yeast two-hybrid analysis. Figure 25.24 shows the results obtained with two bait proteins that are protein kinases, Kss1 and Cdc28. Some known interactions (red arrows) were rediscovered, but many new interactions (green arrows) were also found.

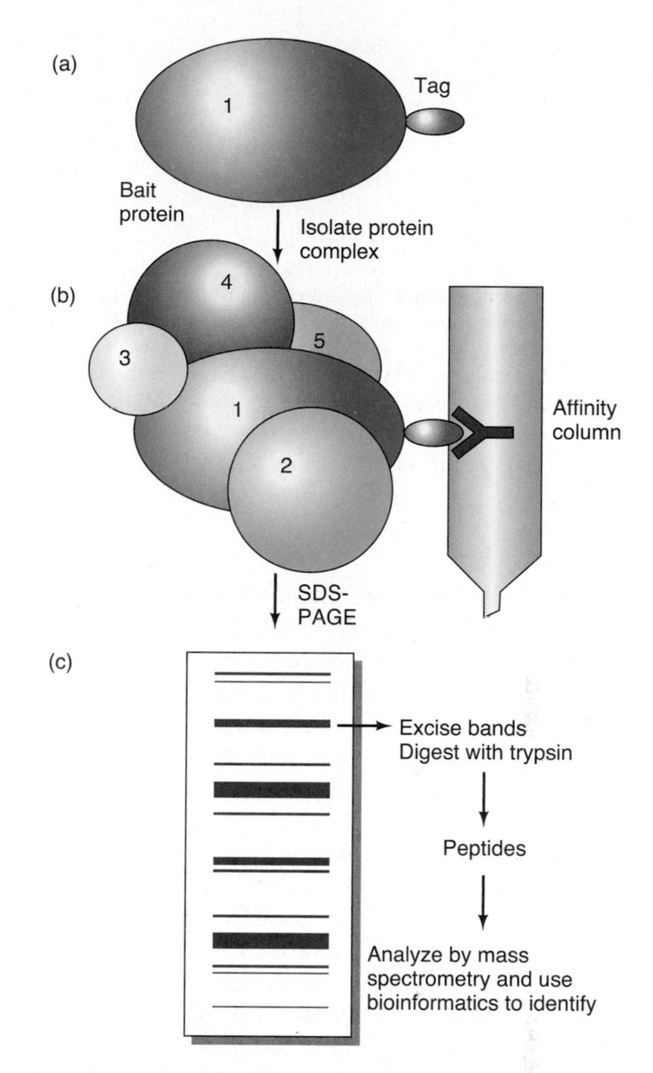

Figure 25.23 Using mass spectrometry to detect protein–protein interactions. (a) Generating the tagged bait protein. A yeast gene encoding a bait protein is engineered to include the coding region for a tag, such as the Flag epitope, then placed in yeast cells and expressed to yield the tagged bait protein. **(b)** Isolating complexes with the bait protein. Immunoaffinity chromatography is performed with a resin containing an antibody directed against the tag on the bait protein. This "fishes out" not only the bait protein, but any "fish" proteins that interact with it. In this case, there are four such proteins, numbered 2–5. **(c)** Purifying and identifying the proteins. SDS-PAGE is used to separate and purify the proteins in the complex. The proteins are excised from the gel and digested with trypsin, and the resulting peptides are analyzed by mass spectrometry. A computer compares the masses of the tryptic peptides with the predicted masses of peptides from all the proteins encoded in the yeast genome to identify the proteins. (*Source:* Adapted from Kumar, A. and M. Snyder, Protein complexes take the bait. *Nature* 415, 2002, p. 123, f. 1.)

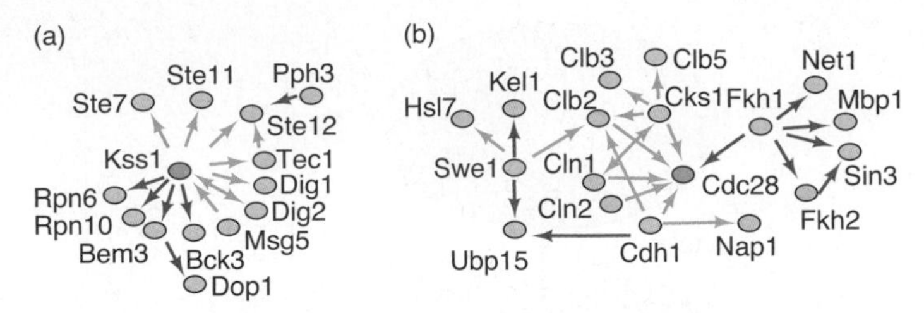

Figure 25.24 Examples of protein–protein interactions discovered by Figeys and colleagues. (a) Interactions discovered with Kss1 as bait. **(b)** Interactions discovered with Cdc28 as bait. In both panels, red arrows represent known interactions, and green arrows represent new interactions discovered in this study. (*Source:* Adapted from Ho, Y., A. Grahler, A. Heilbut, G.D. Bader, L. Moore, S.L. Adams, et al., Systematic identification of protein complexes in *Saccharomyces carevisiae* by mass spectrometry. *Nature* 415, 2002, p. 180, f. 1.)

In a similar study, Anne-Claude Gavin and colleagues discovered 589 yeast protein assemblies in 232 distinct multiprotein complexes. Most interesting was the fact that these associations could predict new roles for 344 proteins, including 231 proteins for which no function was previously known. This "guilt by association" technique is a powerful way to assign functions to unknown proteins.

Michael Snyder and colleagues have approached the problem from a different angle. They have used protein microarrays representing most of the yeast proteome to determine which yeast proteins (or lipids) bind to each protein in the array. Each tiny spot on the array contained a yeast protein coupled to glutathione-*S*-transferase and an oligohistidine tag. In fact, the proteins were tethered to the nickel-coated chip through their oligohistidine tags. In one test of the method (Figure 25.25), Snyder and colleagues probed the array with a protein or lipid coupled to biotin, then probed with streptavidin bound to a fluorescent tag. The streptavidin binds tightly to biotin, and its tag fluoresces green, indicating a positive interaction. The proteins on the microarray were spotted in duplicate, so true positives should appear as pairs of green spots. Figure 25.25 shows at least one positive interaction in each field. Calmodulin is a calcium-binding protein that interacts with many other proteins that require calcium for activity. The other five probes were liposomes containing biotinylated lipids, most of which are active in intracellular signaling. The arrays were also probed with anti-GST antibody and a secondary antibody that gave red fluorescence. This was a control for protein loading; all the proteins were tagged with GST, so they should all "light up" with the α-GST antibody.

Some proteins have binding modules for particular peptide sequences in other proteins. For example, SH3 and WW domains bind to proline-rich peptides, and SH2 domains bind to peptides containing a phosphotyrosine. Based on this knowledge, Stanley Fields, Charles Boone,

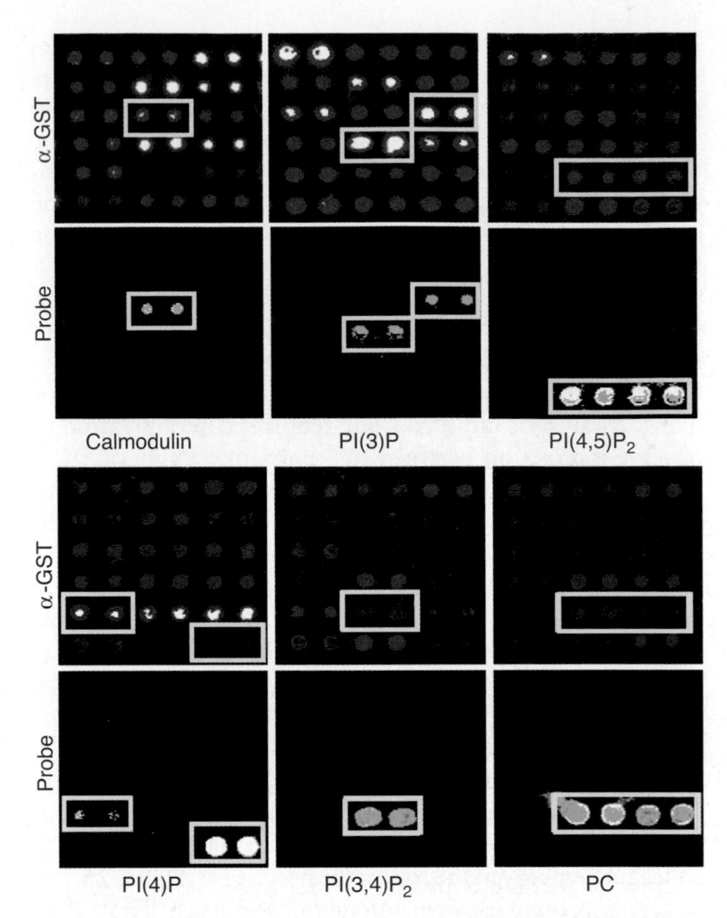

Figure 25.25 Using a protein microchip to detect protein–protein and protein–lipid interactions. Snyder and colleagues made protein microarrays with proteins spotted in duplicate side-by-side and probed them first with an α-GST antibody (first and third rows) or the probes listed beneath the second and fourth rows. The α-GST antibody was in turn detected with a fluorescent probe to yield the red spots. The intensity of the red fluorescence indicated the amount of protein in each spot. The probes in the second and fourth rows were coupled to biotin, which could be detected with streptavidin coupled to a green flourescent tag. The probes were calmodulin, a protein involved in many processes that require calcium, and liposomes containing the following signalling lipids: phosphatidylinositol(3)-phosphate [PI(3)P]; phosphatidylinositol(4,5)bisphosphate [PI(4,5)P_2]; phosphatidylinositol(4)phosphate [PI(4)P]; phosphatidylinositol(3,4)-bisphosphate [PI(3,4)P_2]; and phosphatidylcholine [PC]. Each pair of green spots corresponds to a protein on the microarray, spotted in duplicate, that binds to the protein or lipid probe. The red spots corresponding to the positive (green) spots in rows 2 and 4 are boxed. (*Source:* Adapted from Zhu et al., *Science* 293 (2001) Fig. 2A, p. 2102.)

and Gianni Cesareni and colleagues (Tong et al., 2002) have developed a procedure that meshes experimental and computational strategies to identify the specific partners of proteins having these and other peptide-binding domains.

The procedure employs the following four steps: First, the investigators used a technique called **phage display** to discover the consensus sequences recognized by a given peptide-binding domain. In phage display, the gene or gene

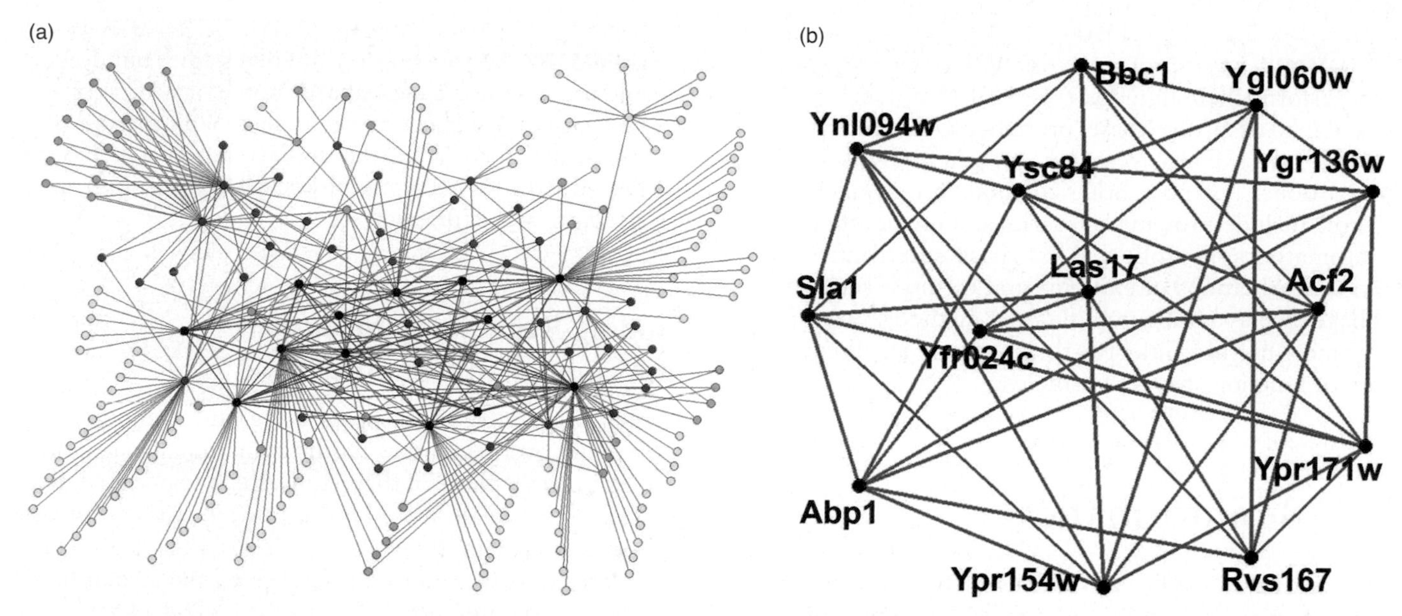

Figure 25.26 Predicted network of protein–protein interactions involving yeast SH3 domains and their targets. **(a)** All proteins and interactions predicted by phage display and searching the yeast proteome. Proteins are grouped into *k*-cores in which each protein makes *k* interactions. For example, a 3-core contains proteins that make 3 interactions. Each protein is color-coded by its *k*-core value as follows: 6-cores, black; 5-cores, cyan; 4-cores, blue; 3-cores, red; 2-cores, green; and 1-cores, yellow. The interactions of the 6-core proteins are represented by red lines. **(b)** Expansion of the 6-core network to show interactions with specific proteins. (*Source:* Adapted from Tong et al., *Science* 295 (2002) Fig. 2, p. 322.)

fragment encoding a protein or peptide is cloned into a phage vector coupled to a phage coat protein gene such that the protein or peptide will be displayed on the surface of the recombinant phage. The phages displaying a protein or peptide that interacts with a second protein can be fished out with the second protein linked to a resin bead. These positive phage clones can then be analyzed to see what protein or peptide they are displaying. These are putative targets for the second protein.

In this study, Tong and colleagues identified 24 different SH3 domains in yeast by a ψ-BLAST analysis (see next section) with the oncoprotein Src, which has an SH3 domain, as the query sequence. Twenty of these SH3 domains could be expressed as GST-fusion proteins in *E. coli,* and Tong and colleagues linked these fusion proteins to resin beads and screened them against a library of random nonapeptides (peptides of 9 amino acids) displayed on phage surfaces. Each SH3 domain bound preferentially to a subset of nonapeptides, which yielded a consensus sequence for the peptide target of each SH3 domain.

Second, Tong and colleagues used computational methods to find the consensus peptide target sequences in the yeast proteome. This process yielded the protein network shown in Figure 25.26a. It is a network because many target proteins have SH3 domains of their own that bind in turn to other targets. The proteins are grouped in "*k*-cores," where each protein has *k* interactions with other proteins. For example, the 6 core is a group of proteins, each of which is predicted to interact with at least six other proteins. The 6-core is shown in black, with red connecting lines, in Figure 25.26a, and is expanded in Figure 25.26b.

In the third step, Tong and colleagues detected interactions between SH3 domains and target proteins in a different way, using a yeast two-hybrid analysis. Finally, in the fourth step, they compared the results of the two methods to find interactions common to both. Of all the interactions, 59 were detected by both methods and, because they were independently identified by two methods, it is very likely that the great majority of them are authentic. As a test, Tong and colleagues chose one protein (Las17) with five different proline-rich domains, which is predicted to interact with nine different SH3 proteins. They then verified all of these interactions with direct in vitro assays. Indeed, the phage display experiments predicted which of the five proline-rich domains on Las17 would be the favorite target of each of the nine proteins. With one exception, the in vitro assays proved these predictions correct.

Each of these techniques for measuring protein–protein interaction is useful, but each has its own problems. All are subject to false-negatives (failure to discover an authentic interaction) and false-positives (detecting an apparent interaction that does not occur in vivo). The best data will probably come from a combination of different techniques.

SUMMARY Most proteins work with other proteins to perform their functions. Several techniques are available to probe these protein–protein interactions. Traditionally, yeast two-hybrid analysis has been done, but now other methods are available. These include protein microarrays, immunoaffinity chromatography followed by mass spectrometry, and combinations of experimental methods such as phage display with computational methods. One of the most useful fruits of such analyses is the discovery of functions for new proteins.

25.3 Bioinformatics

As our databases swell with billions of bases of sequence from the human and other genomes, and countless protein structures and protein–protein interactions, one crucial problem will be to access and manipulate all those data. Accordingly, a new specialty has arisen, known as **bioinformatics.** Practitioners of bioinformatics must understand both biology and computerized data processing, so they can manage the data collecting during genomic and proteomic studies and write programs that allow scientists to use the data. For example: **BLAST** is a program that searches a database for a DNA or protein sequence similar to a sequence of interest and shows how the two sequences line up; and **GRAIL** is a program that identifies genes in a database.

Two types of databases are already established. First, we have generalized databases that include DNA and protein sequences from all organisms. Two generalized databases for DNA sequences are GenBank (*http://www.ncbi.nlm.nih.gov/*) and EMBL (*http://www.ebi.ac.uk/embl/*). Swissprot (*http://www.ebi.ac.uk/swissprot/*) is a generalized protein sequence database. Second, we have specialized databases that deal with a particular organism. For example, FlyBase is a database of the genome of the fruitfly *Drosophila melanogaster.* You can access it online at *http://flybase.bio.indiana.edu:82,* and search it for genetic maps, genes, DNA sequences, and other information. A similar site, WormBase, provides the same kind of data for the nematode *C. elegans*.

The problem, as William Gelbart has pointed out, is that we are functional illiterates in understanding the genomic sequence. He uses a language analogy: We know a few of the "nouns," or polypeptide coding regions of the genome, but we don't know the "verbs," "adjectives," and "adverbs" that tell when and how much of each gene to express. And we don't know the "grammar" that tells how polypeptides assemble into complexes to do their jobs, such as catalyzing biochemical pathways. Bioinformatics will supply the databases and annotation that will be needed to understand genomic grammar fully.

SUMMARY Bioinformatics involves the building and use of biological databases, some of which contain the DNA sequences of genomes. Bioinformatics is essential for mining the massive amount of biological data for meaningful knowledge about gene structure and expression.

Finding Regulatory Motifs in Mammalian Genomes

Here is an example of scientists using pure computational tools to discover regulatory motifs in mammalian genomes. In the discovery phase of their study, these scientists did not use test tubes (which would be an in vitro study) or whole cells or animals (an in vivo study). Thus, their work could be described as **in silico,** in reference to the silicon-based chips in their computers.

Earlier in this chapter, we saw an example of an experimental approach to identifying target sites for some known transcription factors. But what about regulatory sites that interact with molecules nobody has identified yet? In 2005, Eric Lander and Manolis Kellis and their colleagues reported the results of a bioinformatic approach to this question. They reasoned that regulatory motifs (6–10 bp long) are most likely to be found in the upstream regulatory regions of genes, where transcription factors are likely to bind, and in the 3′-untranslated regions (UTRs) of genes, where miRNAs and other regulatory molecules bind and regulate mRNA stability and translatability. They further reasoned that regulatory motifs are likely to be conserved among related organisms. So they compared the human, mouse, rat, and dog genomes to find conserved sequences in the 5′-flanking regions, and in the 3′-UTRs of genes.

These researchers focused on about 17,000 genes in the four species that were well annotated, so there was little doubt that they were real genes. They defined the promoter region of each gene as the noncoding sequence within a 4-kb region centered on the transcription start site, and they defined the 3′-UTR as the region between the translation stop codon and the polyadenylation signal as annotated for each mRNA. As a control, they looked at approximately 123 Mb of sequence from the last two introns in many genes. The terminal introns are thought to be poor in regulatory motifs and so should provide a good negative control.

The authors defined conservation as follows: A "conserved occurrence" is a motif that is absolutely conserved in all four species. The "conservation rate" is the ratio of conserved occurrences of a motif to total occurrences of that motif in the part of the human genome under study (promoter regions, for example). Finally, the "motif conservation score," or "MCS," is the number of standard deviations by which the conservation rate of a motif exceeds the conservation rate for random motifs of the same size.

To illustrate conservation, the authors chose the 8-mer TGACCTTG, which is a binding site for the Err-α transcription factor. This motif occurs 434 times in human promoter regions, of which 162 are conserved occurrences. Thus, the conservation rate is 162/434, or 37%. On the other hand, a random 8-mer has a conservation rate of only 6.8% in promoter regions. Furthermore, the conservation of the 8-mer is specific to promoter regions. In introns, the conservation of this sequence is only 6.2%. Statistical analysis of these and other data allowed the authors to calculate a conservation score. The MCS for the Err-α motif is 25.2 standard deviations, which reflects the very small probability of finding a conservation rate of 37% against a background rate of 6.8%.

To get a more general idea of conservation of regulatory motifs, the authors calculated the MCSs of known transcription factor binding sites from the TRANSFAC database. They found that 63% of these motifs had MCS >3 and nearly 50% had MCS <6. So they defined "highly conserved motifs" as those having MCS >6. The authors listed three reasons why so many known regulatory motifs failed to achieve an MCS as high as 3: They may be erroneously identified; they may not be conserved in the four species studied; or they may not be common enough.

The authors identified 174 highly conserved motifs in promoter regions, of which 59 strongly matched, and 10 weakly matched, previously-identified regulatory motifs in TRANSFAC. The other 105 motifs are likely to represent new regulatory elements. If these new motifs are authentic regulatory elements, the genes with which they are associated are likely to show some tissue specificity of expression. That is because genes that are controlled by a common factor are likely to be active in the same tissues. The authors consulted databases listing gene expression data from 75 tissues and found that 86% of the known motifs and 50% of the new motifs were associated with genes whose activity was significantly enriched in one or more tissues.

Another check on authenticity is to see if the elements show positional bias with respect to the transcription start site. In fact, the highly conserved elements showed a strong tendency to cluster within 100 bp of the start site, while random elements were randomly distributed across the 4-kb region analyzed. Taken together, these data demonstrated that most of the identified motifs are likely to be part of authentic regulatory elements.

The authors found 106 highly conserved motifs in the 3′-UTRs of genes. However, because there was no database of 3′-UTR elements similar to TRANSFAC, they had to use other means to check their authenticity. Fortunately, two characteristics stood out. First, the 3′-UTR motifs, unlike those in the promoter region, showed a strong directional bias—they tended to be found on one strand and not the other. This is consistent with the hypothesis that the 3′-UTR motifs act in mRNAs, where they bind to miRNAs and other molecules to regulate mRNA stability or translatability. That is because the motif must be on the correct strand to be transcribed into mRNA. On the other hand, motifs in promoter regions act at the DNA level. They typically bind to activators, which can work in either orientation, so there is no strand bias.

The second characteristic of the highly conserved motifs in the 3′-UTRs of genes is that they had a strong preference for an 8-base length, and for A in the last position. The motifs in the promoter regions showed no such biases. These characteristics are consistent with the hypothesis that the 8-mers are sites for hybridization to miRNAs, which tend to begin with a U, followed by seven bases that are complementary to sites in the mRNAs they regulate.

The authors were interested in the apparent relationship of the highly conserved 8-mers and miRNAs, so they searched the miRNA registry, which contained 207 different human miRNAs, for matches with the 8-mers, and found 43.5% of the known human miRNAs matched one of the 8-mers perfectly, while only 2% matched an equal number of control 8-mers. The 8-mers that did not match a known miRNA were evolving faster than those that did, suggesting that the matching 8-mers cannot alter their sequences without impairing hybridization to miRNAs, which is important to gene regulation.

Finally, the authors used the conserved 8-mer motifs to find new miRNA genes. They searched the four genomes for conserved sequences complementary to the highly conserved 8-mer motifs. Then they examined the sequences surrounding the conserved sequences for ability to form the stable stem-loop structures that are characteristic of miRNAs. They found 242 such stable stem-loop structures, which presumably encode miRNAs. Of these, 113 encode known miRNAs, leaving 129 more that encode predicted miRNAs. The authors chose 12 of these at random and checked for expression in pooled adult tissues (the only in vitro experimental part of their work). They found that six were expressed. Thus, many of the 129 predicted miRNA genes probably really do encode miRNAs. This means that many miRNA genes probably remain to be discovered, and the control of gene expression by miRNAs is probably even more widespread than had been believed.

SUMMARY Using computational biology techniques, Lander and Kellis have discovered highly conserved sequence motifs in the promoter regions and 3′-UTRs of four mammalian species, including humans. The motifs in the promoter regions probably represent binding sites for transcription factors. Most of the motifs in the 3′-UTRs probably represent binding sites for miRNAs.

Using the Databases Yourself

Some very useful databases are kept at the National Center for Biological Information (NCBI). In this section we will see a few simple examples of how to access and use the data. To see how a search works, let us imagine you are a physician treating a patient with warts. Suspecting a viral cause, and even having a candidate virus (papilloma virus) in mind, you excise the warts, homogenize them in a buffer containing a detergent to break open the virus particles, purify the DNA, and perform PCR with primers specific for the candidate virus. You obtain a band, which confirms that the suspected viral DNA is present in the warts. To investigate further the exact strain of the virus, you sequence part of the DNA that you have amplified by PCR and obtain the sequencing gel pictured in Figure 5.19.

If you need practice reading a sequencing gel, ignore the sequence below and write down the sequence of the first 21 bases from the gel, beginning with the C at the extreme bottom of the gel. If you want to skip that step, here is the sequence, written in lowercase, which is less ambiguous than capitals because it is harder to confuse a **g** with a **c** than a **G** with a **C**:

caaaaaacggaccgggtgtac

To begin a search, go to the NCBI home page and click BLAST, or just start at the NCBI BLAST home page: *http://www.ncbi.nlm.nih.gov/BLAST/*. To search in a nucleotide database, look under the Nucleotide heading and click Nucleotide-nucleotide BLAST [blastn]. The large box near the top asks for the query sequence (the sequence you want to compare to the database). You can type in a sequence, but it is easier to copy and paste a sequence from another document.

When you have finished entering your sequence, you can choose just part of it to search for using the Query subrange boxes. For example, if you wanted to search for residues 10–21 of your sequence, enter 10 in the "From" window and 21 in the "To" window (but we will use all 21 nucleotides in our search). You can also select the database in which you want to search using the top Choose Search Set box. Because we think this is a viral sequence we ignore the human and mouse database options and select Others. The default database under Others is Nucleotide collection (nr/nt), where nr stands for "nonredundant." This includes all the nucleotide sequences in several different databases and is the most comprehensive of all. To start searching, click the BLAST button near the bottom of the page. You should receive a search status message, including a request ID (RID) number.

If you receive your results promptly, you can proceed with your analysis. However, you may have to wait. In that case, remember the RID, so you can log onto the NCBI website later and retrieve the results for that ID number.

You will receive your results in several forms. First, you will see a colored bar graph indicating the rough extent of match between your query sequence and various sequences in the database. In this case, blue indicates the best match. You can mouse-over each bar to see the identity of the DNA sequence that matches your query sequence. If you click on a bar you will get more details about the matching sequence. Below the bar graph are the sequences that match the query sequence, within certain limits, which we will discuss later. Each of the matching sequences is identified, with the best match given first and then the others in descending order of closeness of match.

Two scores are assigned to each sequence. The first is a **bit score** *(S)*, which is related to the number of matches between the query sequence and the sequence from the database. The larger the bit score, the better the match. For the best match in this case, it is 42.1, which is good. The second score is the **expect value (E value).** This is the number of matches yielding the corresponding bit score that we would expect to see by chance. Thus, the lower the E value, the better the match. Really good matches give E scores much less than 1.0. For the best match in this case it is 0.021.

What is the identity of the database sequence with the best match to your query sequence? The mouse-over on the top bar says it is the human papilloma virus (HPV) type 31, which suggests that this is the strain of virus that caused the warts in your patient. You can get the same information from the short list of matching sequences below the bars. The top black bar corresponds to a *Mus musculus* (house mouse) gene, which also gives a fairly good match.

Moving down the page in the results, we come to the alignments between the query sequence and the database sequences. You can see that your query sequence matches the HPV 31 sequence perfectly, but matches the mouse gene sequence in only 19 out of 21 positions. However, that apparently minor difference makes a big difference in the E value. The mouse gene has an E value of 0.33, which is significantly less exciting than 0.011.

The query sequence we have entered is only 21 bases long, which is unusually short. To see the effect of increasing the length of the query sequence, either read the sequencing gel further (to 42 bases), or enter the following sequence into a new search:

caaaaaacggaccgggtgtacaacttttactatggcgtgaca.

The new E value is $5e^{-14}$, which is a very small number and indicates that the perfect match in all 42 positions has a high degree of significance.

What if you do not have a DNA sequence of your own to investigate, but you want to use the NCBI database for general information? For example, you may be interested in finding genes that are associated with certain human diseases, such as colon cancer. To start such a search, you could go to the NCBI website and click on Genes and Expression in the menu on the left. Then enter "colon cancer" in the box at top, leave the database option on the default, "all databases," and click "Search." The next page asks you to limit your search, so click on "Gene: gene-centered information." The next page gives you a list of genes associated with colon cancer. The fifth entry is *MLH1* (at least it was in December, 2010, but the order will change with time).

Click on the *MLH1* link to receive data on this gene. The summary reveals that *MLH1* is the human homolog of the *mutL* gene in *E. coli,* which encodes a protein involved in mismatch repair (Chapter 20). The *MLH1* gene in humans is also involved in mismatch repair, and *MLH1* mutations cause mismatches to build up and, therefore, mutations to accumulate. This presumably predisposes people to develop cancer, and colon cancer in particular, because genes that normally keep cancer in check (**tumor suppressor genes**) can be inactivated by mutation, and genes that predispose cells to lose control of their growth (**oncogenes**) can be activated by mutation.

We can also use the NCBI site as a source of information about protein structure. For example, suppose we wanted to see the structure of the p53 protein (a tumor suppressor gene product), whose inactivation is a feature of the majority of human cancers. Go to the NCBI website. In the box at the top, type "p5 complexed with DNA" and click "Go." You will be back at the Entrez page, from which you selected "Gene" when looking for information about colon cancer. This time, select "Structure." You will be presented with a list of entries. Scroll down to the structure named "1TUP." In December, 2010, this was entry number 18, but that will change with time. Click on the structure. This will bring up a page of information about this structure. To see it in 3-D, you will need the appropriate free software. If you already have the Cn3D software on your computer, simply click "Structure view in Cn3D." If not, click "Download Cn3D."

Once you have installed Cn3D, click the "Structure view in Cn3D" button. You will see a structure based on an x-ray crystallography study of p53 complexed with DNA. The Cn3D software allows you to rotate the structure any way you wish with your mouse. Start with the mouse pointer on the left of the structure. Left click and hold the button down and move the mouse to the right. The structure will rotate from left to right. You can also rotate it from top to bottom, or through any angle in between horizontal and vertical. Rotate it so you can clearly see the interaction between the zinc module and the major groove of the DNA.

You can also look at the 3D structures of some of the proteins we have studied in previous chapters. For example, look for the structures of GAL4 and the glucocorticoid receptors. In both cases, the rotation will make the structures even clearer than they were in this book.

SUMMARY The NCBI website contains a vast store of biological information, including genomic and proteomic data. You can start with a sequence and discover the gene it belongs to, and compare that sequence with that of similar genes. You can also start with a topic you want to study and query the database for information on that topic. Or you can look up a protein of interest and view the structure of that protein in three dimensions by rotating the structure on your computer screen.

SUMMARY

Functional genomics is the study of the expression of large numbers of genes. One branch of this study is transcriptomics, which is the study of transcriptomes—all the transcripts an organism makes at any given time. One approach to transcriptomics is to create DNA microarrays or DNA microchips, holding thousands of cDNAs or oligonucleotides, then to hybridize labeled RNAs (or corresponding cDNAs) from cells to these arrays or chips. The intensity of hybridization to each spot reveals the extent of expression of the corresponding gene. Such arrays can be used to analyze the timing and location of expression of many genes at once.

SAGE (serial analysis of gene expression) allows one to determine which genes are expressed in a given tissue and the extent of that expression. Short tags, characteristic of particular genes, are generated from cDNAs and ligated together between linkers. The ligated tags are then sequenced to determine which genes are expressed and how abundantly. Cap analysis of gene expression (CAGE) gives the same information as SAGE about which genes are expressed, and how abundantly, in a given tissue. However, because it focuses on the 5′-ends of mRNAs, it also allows the identification of transcription start sites and, therefore, helps locate promoters.

High-density whole chromosome transcriptional mapping studies have shown that the majority of sequences in cytoplasmic polyadenylated RNAs derive from non-exon regions of 10 human chromosomes. Furthermore, almost half of the transcription from these same 10 chromosomes is nonpolyadenylated. Taken together, these results indicate that the great majority of stable nuclear and cytoplasmic transcripts of these chromosomes comes from regions outside the exons. This may help to explain the great differences between species, such as humans and chimpanzees, whose exons are almost identical.

Genomic functional profiling can be performed by creating mutants in an organism by replacing genes one at a time with an antibiotic resistance gene flanked by oligomers that serve as a barcode to identify each mutant.

Then the whole group of mutants can be grown together under various conditions to see which mutants disappear most rapidly. Functional profiling can also be done by inactivating genes via RNAi.

Tissue-specific expression profiling can be done by examining the spectrum of mRNAs whose levels are decreased by an exogenous miRNA, and comparing that to the spectrum of expression of genes at the mRNA level in various tissues. If the miRNA in question causes a decrease in the levels of the mRNAs that are naturally low in cells in which the miRNA is expressed, it suggests that the miRNA is at least part of the cause of those natural low levels. This kind of analysis has implicated miR-124 in destabilizing mRNAs in brain tissue, and miR-1 in destabilizing mRNAs in muscle tissue.

Chromatin immunoprecipitation followed by DNA microarray analysis (ChIP-chip analysis) can be used to identify DNA-binding sites for activators and other proteins. In organisms with small genomes, such as yeast, all of the intergenic regions can be included in the microarray. But with large genomes, such as the human genome, that is now impractical. To narrow the field, CpG islands can be used, since they are associated with gene control regions. Also, if the timing or conditions of an activator's activity are known, the control regions of genes known to be activated at those times, or under those conditions, can be used.

The mouse can be used as a human surrogate in large-scale expression studies that would be ethically impossible to perform on humans. For example, scientists have studied the expression of almost all the mouse orthologs of the genes on human chromosome 21. They have followed the expression of these genes through various stages of embryonic development and have catalogued the embryonic tissues in which the genes are expressed.

Single-nucleotide polymorphisms can probably account for many genetic conditions caused by single genes, and even multiple genes. They might also be able to predict a person's response to drugs. A haplotype map with over 10 million SNPs will make it easier to sort out the important SNPs from those with no effect. Structural variation (insertions, deletions, inversions, and other rearrangements of chunks of DNA) is also a surprisingly prominent source of variation in human genomes. Some structural variation can in principle predispose certain people to contract diseases, but some is presumably benign, and some is demonstrably beneficial.

The sum of all proteins produced by an organism is its proteome, and the study of these proteins, even smaller sets of them, is called proteomics. Current research in proteomics requires first that proteins be resolved, sometimes on a massive scale. One of the best tools available for separation of many proteins at once is 2-D gel electrophoresis. After they are separated, proteins must be identified, and the best method for doing that involves digestion of the proteins one by one with proteases and identifying the resulting peptides by mass spectrometry. Someday microchips with antibodies attached may allow analysis of proteins in complex mixtures without separation.

Most proteins work with other proteins to perform their functions. Several techniques are available to probe these protein–protein interactions. Traditionally, yeast two-hybrid analysis has been done, but now other methods are available. These include protein microarrays, immunoaffinity chromatography followed by mass spectrometry, and combinations of experimental methods such as phage display and computational methods. One of the most useful fruits of such analyses is the discovery of functions for new proteins.

Bioinformatics involves the building and use of biological databases, some of which contain the DNA sequences of genomes. Bioinformatics is essential for mining the massive amount of genomic information for meaningful knowledge about gene structure and expression.

Using computational biology techniques, Lander and Kellis have discovered highly conserved sequence motifs in the promoter regions and 3′-UTRs of four mammalian species, including humans. The motifs in the promoter regions probably represent binding sites for transcription factors. Most of the motifs in the 3′-UTRs probably represent binding sites for miRNAs.

The NCBI website contains a vast store of biological information, including genomic and proteomic data. You can start with a sequence and discover the gene it belongs to, and compare that sequence with that of similar genes. You can also start with a topic you want to study and query the database for information on that topic. Or you can look up a protein of interest and view the structure of that protein in three dimensions by rotating the structure on your computer screen.

REVIEW QUESTIONS

1. Describe the process of making a DNA microchip (oligonucleotide array).
2. Describe a SAGE experiment to measure transcription in cancer cells of a certain type. Show how the production of a ditag works, with actual sequences of your own invention.
3. Explain how the cap-trapper in a CAGE experiment ensures that only full-length cDNAs are captured.
4. Explain the roles of the *Mme*I, *Xma*JI, and *Xba*I restriction sites in the CAGE procedure.
5. Describe how genomic functional profiling can be performed by gene knockout in yeast.
6. Describe how genomic functional profiling can be performed by RNAi in higher eukaryotes.

7. Describe a tissue-specific functional profiling method that shows the effects of miRNAs on gene expression. Give a hypothetical example of positive results.
8. Explain how ChIP-chip analysis works. Show how it can be used to find DNA regions (enhancers) that bind to a particular activator.
9. Explain how tag sequencing (ChIP-seq) works. What problems in ChIP-chip analysis are solved by ChIP-seq?
10. What are *cis*-regulatory modules (CRMs)? Why are they easier to find than single enhancers?
11. Outline a genomic strategy for finding enhancers that bind to unknown proteins. Describe at least one drawback to this strategy.
12. Ren and colleagues employed ChIP-chip analysis using an anti-TAF1 antibody to locate promoters in human cells. They found that these promoters were not enriched in TATA boxes. Given that TAF1 is part of a transcription factor (TFIID) that binds to TATA boxes, why is it not surprising that many of the promoters identified lack TATA boxes? You will need information from Chapter 11 to answer this question.
13. Describe how in situ expression analysis works. Give a hypothetical example of a positive result.
14. What are SNPs? Why are most of them unimportant? How can some of them be useful? How can they be abused?
15. Compare and contrast in a general way the techniques used in, and information obtained in, transcriptomics and proteomics.
16. Describe a bioinformatic approach to identifying human gene control motifs.
17. Explain how MS/MS analysis can yield the sequence of a protein. Present hypothetical results.
18. Explain how isotope coded affinity tags (ICATs) can enable you to quantify the changes in protein concentration in cells grown under two different conditions.
19. In Figure 25.21, estimate what has happened to the concentrations of peptides 4–7 when cells are shifted from condition 1 (no serum) to condition 2 (+serum)?
20. Explain how you would use stable isotope labeling by amino acids in cell culture (SILAC) to quantify the changes in protein concentration in cells grown under two different conditions. Show sample results.
21. How would you measure the absolute concentration of a particular protein in a cell?
22. What do the data in Figure 25.22 tell us about the accuracy of estimating a protein's concentration from its mRNA's concentration?
23. What would happen to the gray data points in Figure 25.22 if there were a lower correlation between the abundances of orthologous proteins in the two organisms? What would happen if there were a higher correlation?
24. Describe how affinity tagging and mass spectrometry can be used to examine an organism's interactome.
25. Explain how a protein microarray could be used to examine an organism's interactome.
26. Describe an experiment in which you would use phage display to investigate an organism's interactome.

ANALYTICAL QUESTIONS

1. List in order the steps you would perform to make an oligonucleotide array in which two of the spots contain the dinucleotides AC and AT. You may ignore all of the other spots.
2. Describe a hypothetical experiment using a DNA microarray to measure the transcription from viral genes at two stages of infection of cells by the virus. Present sample results.
3. Perform a BLAST search on the first 20 nt of this sequence (the sequence is divided into blocks of 10 nt): ttaagtgaaa taaagagtga atgaaaaaat aatatcctta. What gene did you identify? What was the best E value you obtained? Now try again with all 40 nt. Did you still retrieve the same gene? What is the best E value you obtained this time? Why is the E value different this time? On what chromosome is this gene located? Is there any relationship between this gene and prostate cancer in men? If so, what is the relationship?
4. You are an MD/PhD developmental biologist (highly trained in techniques of molecular biology) studying the pathogenesis of Type I insulin-dependent diabetes mellitus (IDDM). You have several patients who are predisposed to becoming diabetic, and control subjects who have no family history or predisposition to becoming diabetic, enrolled in a clinical study that will involve the removal of a small section of pancreatic beta cells. You want to analyze the differences in gene expression between cells from these two groups of subjects. Describe the experimental method(s) you would use, and what information you hope to obtain from this study.

SUGGESTED READINGS

General References and Reviews

Abbott, A. 1999. A post-genomic challenge: Learning to read patterns of protein synthesis. *Nature* 402:715–20.

Cheung, V.G., M. Morley, F. Aguilar, A. Massimi, R. Kucherlapati, and G. Childs. 1999. Making and reading microarrays. *Nature Genetics Supplement* 21:15–19.

Cox, J. and M. Mann. 2007. Is proteomics the new genomics? *Cell* 130:395–98.

Hieter, P. and Boguski, M. 1997. Functional genomics: It's all how you read it. *Science* 278:601–02.

Kruglyak, L. and D.L. Stern. 2007. An embarrassment of switches. *Science* 317:758–59.

Kumar, A. and M. Snyder. 2002. Protein complexes take the bait. *Nature* 415:123–24.

Lipshutz, R.J., S.P.A. Fodor, T.R. Gingeras, and D.J. Lockhart. 1999. High density synthetic oligonucleotide arrays. *Nature Genetics Supplement* 21:20–24.

Marx, J. 2006. A clearer view of macular degeneration. *Science* 311:1704–05.

Service, R.F. 1998. Microchip arrays put DNA on the spot. *Science* 282:396–99.

Young, R.A. 2000. Biomedical discovery with DNA arrays. *Cell* 102:9–15.

Research Articles

Arbeitman, M.N., E.E.M. Furlong, F. Imam, E. Johnson, B.H. Null, B.S. Baker, M.A. Krasnow, M.P. Scott, R.W. Davis, and K.P. White. 2002. Gene expression during the life cycle of *Drosophila melanogaster. Science* 297:2270–75.

Blanchette, M., A.R. Bataille, X. Chen, C. Poitras, J. LaganiPre, G. Debois, V. GiguPre, V. Ferretti, D. Bergeron, B. Coulombe, and F. Robert. 2006. Genome-wide computational prediction of transcriptional regulatory modules reveals new insights into human gene expression. *Genome Research* 16:656–68.

Cheng, J., T.R. Gingeras, et al. 2005. Transcriptional maps of 10 human chromosomes at 5-nucleotide resolution. *Science* 308:1149–54.

Gavin, A.-C., M. Bosche, R. Krause, P. Grandi, M. Marzioch, A. Bauer, et al. 2002. Functional organization of the yeast proteome by systematic analysis of protein complexes. *Nature* 415:141–47.

Glaever, G., A.M. Chu, L. Ni, C. Connelly, L. Riles, S. Veronneau, et al. 2002. Functional profiling of the *Saccharomyces cerevisiae* genome. *Nature* 418:387–91.

Gygi, S.P., B. Rist, S.A. Gerber, F. Turecek, M.H. Gelb, and R. Aebersoldgene. 1999. Quantitative analysis of complex protein mixtures using isotope-coded affinity tags. *Nature Biotechnology* 17:994–99.

Ho, Y., A. Gruhler, A. Heilbut, G.D. Bader, L. Moore, S.L. Adams, D. Figeys, and many other authors. 2002. Systematic identification of protein complexes in *Saccharomyces cerevisiae* by mass spectrometry. *Nature* 415:180–83.

Iyer, V.R., M.B. Eisen, D.T. Ross, G. Schuler, T. Moore, J. Lee, et al. 1999. The transcriptional program in the response to human fibroblasts to serum. *Science* 283:83–87.

Lim, L.P., N.C. Lau, P. Garrett-Engele, A. Grimson, J.M. Schelter, J. Castle, D.P. Bartel, P.S. Linsley, and J.M. Johnson. 2005. Microarray analysis shows that some microRNAs downregulate large numbers of target mRNAs. *Nature* 433:769–73.

Pennacchio, L.A., et al. 2006. *In vivo* enhancer analysis of human conserved non-coding sequences. *Nature* 444:499–502.

Ren, B., F. Robert, J.J. Wyrick, O. Aparicio, E.G. Jennings, I. Simon, et al. 2000. Genome-wide location and function of DNA binding proteins. *Science* 290:2306–09.

Sönnichsen, B., et al. 2005. Full-genome RNAi profiling of early embryogenesis in *Caenorhabditis elegans. Nature* 434:462–69.

Stelzl, U., et al. 2005. A human protein–protein interaction network: A resource for annotating the proteome. *Cell* 122:957–68.

Tong, A.H., B. Drees, G. Nardelli, G.D. Bader, B. Brannetti, L. Castagnoli, et al. 2002. A combined experimental and computational strategy to define protein interaction networks for peptide recognition modules. *Science* 295:321–4.

Velculescu, V.E., L. Zhang, B. Vogelstein, and K.W. Kinsler. 1995. Serial analysis of gene expression. *Science* 270:484–87.

Wang, D.G., J.B. Fan, C.J. Sino, A. Berno, P. Young, R. Sapolsky, et al. 1998. Large-scale identification, mapping, and genotyping of single-nucleotide polymorphisms in the human genome. *Science* 280:1077–82.

Xie, X., J. Lu, E.J. Kulbokas, T.R. Golub, V. Mootha, K. Lindblad-Toh, E.S. Lander, and M. Kellis. 2005. Systematic discovery of regulatory motifs in human promoters and 3′ UTRs by comparison of several mammals. *Nature* 434:338–45.

Zhu, H., M. Bilgin, R. Bangham, D. Hall, A. Casamayor, P. Bertone, et al. 2001. Global analysis of protein activities using proteome chips. *Science* 293:2101–05.

GLOSSARY

AAUAAA An important part of the animal polyadenylation signal that dictates cleavage and polyadenylation about 20 nt downstream.

A site (ribosomal) The ribosomal site to which new aminoacyl-tRNAs (except the first one) bind.

A site (RNA polymerase) The site the incoming nucleotide occupies during phosphodiester formation.

abortive transcripts Very short transcripts (about 6 nt long) synthesized at prokaryotic promoters before promoter clearance occurs.

Ac A maize transposon ("activator") that can activate transposition of an inactive transposon like *Ds* by providing the necessary transposase.

acceptor stem The part of a tRNA molecule formed by base-pairing between the 5′- and 3′-ends of the molecule. The 3′-end can "accept" an amino acid by charging, hence the name.

accommodation Unbending of the aminoacyl-tRNA in the A/T state, to allow the aminoacyl-tRNA to bind fully in the A site.

acidic domain A transcription-activating domain rich in acidic amino acids.

aconitase An enzyme whose apoprotein form (lacking iron) binds to iron response elements in mRNAs and controls their translation or degradation.

activation region I (ARI) The region of CAP that is thought to bind to the carboxyl-terminal domain (CTD) of the α-subunit of *E. coli* RNA polymerase.

activation region II (ARII) The region of CAP that is thought to bind to the amino-terminal domain (NTD) of the α-subunit of *E. coli* RNA polymerase.

activation site The site on an aminoacyl-tRNA synthetase that activates the amino acid, forming an aminoacyl adenylate.

activator A protein that binds to an enhancer (or activator-binding region) and activates transcription from a nearby promoter. In eukaryotes, stimulates formation of a preinitiation complex.

activator interference *See* squelching.

activator-binding site The DNA site to which a prokaryotic transcription activator binds (e.g., the CAP–cAMP-binding site in a catabolite repressible operon).

ADAR *See* adenosine deaminase acting on RNA.

adenine DNA glycosylase Called MutY in bacteria and hMYH in humans. It can remove an A that is mispaired with oxoG.

adenosine deaminase acting on RNA (ADAR) An RNA-editing enzyme that deaminates certain adenosines in RNAs, converting them to inosines.

A-DNA A form of DNA found at low relative humidity, with 11 bp per helical turn. The form assumed in solution by an RNA–DNA hybrid.

affinity chromatography A chromatography method that purifies molecules based on their affinity for a bait molecule bound to a resin. The bait molecule can be an antibody, an enzyme substrate, or any other molecule with a known specific affinity for the molecule to be purified.

affinity labeling Labeling one substance by covalently attaching a reactive compound with specific affinity for the first substance (e.g., labeling the active site of an enzyme by linking a labeled substrate analog covalently to the enzyme).

A1492 and A1493 Two universally conserved bases in 16S rRNA that play a critical role in codon–anticodon recognition by inserting into the minor groove between codon and anticodon and stabilizing the interaction.

Ago1 *See* Argonaute1.

Ago2 *See* Argonaute2.

Ago3 An Argonaute protein that associates with transposon mRNAs, and then, in conjuction with piRNAs, cleaves the transposon mRNAs.

alarmone A compound produced in an organism in response to stress; initiates actions to cope with the stress.

α-amanitin A toxin produced by poisonous species of mushrooms in the *Amanita* genus. Inhibits RNA polymerase II at very low concentration, and RNA polymerase III at higher concentration. Usually does not inhibit RNA polymerase I at all.

alkylation The addition of carbon-containing groups to other molecules. Alkylation of DNA bases constitutes DNA damage that can lead to mutations.

allele-specific RNAi Introduction of an altered copy of a gene whose mRNA will not be destroyed by RNAi targeted to the endogenous copy of the gene. This allows one to manipulate the altered gene and examine the effects of these alterations in a background in which the endogenous gene is not expressed.

α-complementation In vivo complementation between the α-peptide and the ω-peptide of β-galactosidase to yield an active enzyme. Cloning vectors usually take advantage of α-complementation by having the vector encode the α-peptide and the host encode the ω-peptide. Thus, the vector alone will enable α-complementation, but a vector with an insert will not.

allolactose A rearranged version of lactose with a β-1, 6-galactosidic bond; the inducer of the *lac* operon.

allosteric protein A protein in which binding of a molecule to one site changes the conformation of a remote site and alters its interaction with a second molecule.

alternative splicing Splicing the same pre-RNA in two or more ways to yield two or more different mRNAs that produce two or more different protein products.

Alu element A human nonautonomous retrotransposon that contains the AGCT sequence recognized by the restriction enzyme *AluI*. Present in about 1 million copies in the human genome.

amber codon UAG, coding for termination.

amber mutation *See* nonsense mutation.

amber suppressor A tRNA bearing an anticondon that can recognize the amber codon (UAG) and thereby suppress amber mutations.

amino acid The building block of proteins.

aminoacyl-AMP An activated amino acid linked through a high-energy anhydride bond to the phosphate group of AMP. Created by aminoacyl-tRNA synthetase in the first step in tRNA charging.

aminoacyl-tRNA A tRNA with its cognate amino acid esterified to its 3′-hydroxyl group.

aminoacyl-tRNA synthetase The enzyme that links a tRNA to its cognate amino acid.

a. class I An aminoacyl-tRNA synthetase that charges the 2′-OH of the tRNA.

b. class II An aminoacyl-tRNA synthetase that charges the 3′-OH of the tRNA.

amino tautomer The normal tautomer of adenine or cytosine found in nucleic acids.

amino terminus The end of a polypeptide with a free amino group. The end at which protein synthesis begins.

amplification Selective replication of a gene to produce more than the normal single copy in a haploid genome.

anabolic metabolism The process of building up substances from relatively simple precursors. The *trp* operon encodes anabolic enzymes that build the amino acid tryprophan.

annealing of DNA The process of bringing back together the two separate strands of denatured DNA to re-form a double helix.

annotated genes Genes or gene-like sequences from a genomic sequencing project that are at least partially identified.

antibody A protein with the ability to recognize and bind to a substance, usually another protein, with great specificity. Helps the body's immune system recognize and trigger an attack on invading agents.

anticodon A 3-base sequence in a tRNA that base-pairs with a specific codon.

anticodon loop The loop, conventionally drawn at the bottom of tRNA molecule, that contains the anticodon.

antigen A substance recognized and bound by an antibody.

antiparallel The relative polarities of the two strands in a DNA double helix; if one strand goes 5′→3′, top to bottom, the other goes 3′→5′. The same antiparallel relationship applies to any double-stranded polynucleotide or oligonucleotide, including the RNAs in a codon–anticodon pair.

antirepression Prevention of repression by histones or other transcription-inhibiting factors. Antirepression is part of a typical activator's function.

antisense RNA An RNA complementary to an mRNA.

antiserum Serum containing an antibody or antibodies directed against a particular substance.

anti-σ-factor A protein that binds to a σ-factor and inhibits its activity.

anti-anti-σ-factor A protein that binds to an anti-σ-factor–σ-factor complex and releases the σ-factor.

anti-anti-anti-σ-factor A protein that phosphorylates and inactivates an anti anti-σ-factor.

antiterminator A protein, such as the λ N and Q proteins, that causes transcription to continue through terminators.

AP-1 A transcription activator composed of one molecule each of Fos and Jun. (Jun–Jun homodimers also have AP1 activity.) Mediates response to the mitogenic phorbol esters.

AP endonuclease An enzyme that cuts a strand of DNA on the 5′-side of an AP site.

APE1 (AP endonuclease 1) A mammalian enzyme that uses its 3′→5′ exonuclease activity to edit the errors made by DNA polymerase β during base excision repair.

AP site An apurinic or apyrimidinic site in a DNA strand.

aporepressor A repressor in an inactive form, without its corepressor.

aptamer A nucleic acid, or region of a nucleic acid, usually an RNA, that has a function. Such functions include specific binding to another molecule, ribozyme catalysis, and other activities.

apurinic site (AP site) A deoxyribose in a DNA strand that has lost its purine base.

apyrimidinic site (AP site) A deoxyribose in a DNA strand that has lost its pyrimidine base.

AraC The negative regulator of the *ara* operon.

archaea The kingdom of prokaryotic organisms whose biochemistry and molecular biology resemble those of eukaryotes as well as those of bacteria. The archaea typically live in extreme environments that are very hot or very salty. Some of the archaea are strict anaerobes that generate methane.

architectural transcription factor A protein that does not activate transcription by itself, but helps DNA bend so other activators can stimulate transcription.

ARE *See* AU-rich elements.

Argonaute1 (Ago1) An Argonaute protein that associates with siRNA in a RITS complex.

Argonaute2 (Ago2) The mammalian Argonaute protein with RISC (slicer) activity.

Armitage A possible RLC component. Known to be needed to convert the RLC to a RISC.

arrest (of transcription) A state in which RNA polymerase is permanently paused with the end of the transcript extruded from the enzyme. Transcription cannot resume unless the extruded end of the RNA is removed.

Artemis An enzyme that opens up DNA hairpins created by Rag-1 and Rag-2 during V(D)J recombination.

assembly factor A transcription factor that binds to DNA early in the formation of a preinitiation complex and helps the other transcription factors assemble the complex.

assembly map A scheme showing the order of addition of ribosomal proteins during self-assembly of a ribosomal particle in vitro.

asymmetrical transcription Transcription of only one strand of a given region of a double-stranded polynucleotide.

A/T state The state in which an aminoacyl-tRNA first binds to the bacterial ribosome. Its anticodon is paired with the codon in the A site, but the tRNA is bent to allow the amino acid and acceptor stem to remain bound to EF-Tu and to ribosomal elements to the right of the A site, as the 50S particle is conventionally depicted.

ATPase An enzyme that cleaves ATP, releasing energy for other cellular activities.

attachment sites *See att* sites.

attB The *att* site on the *E. coli* genome.

attenuation A mechanism of transcription control that involves premature transcription termination.

attenuator A region of DNA upstream from one or more structural genes, where premature transcription termination (attenuation) can occur.

attP The *att* site on the λ phage genome.

***att* sites** Sites on phage and host DNA where recombination occurs, allowing integration of the phage DNA into the host genome as a prophage.

AU-rich elements (AREs) Sequences in the 3′-UTRs of mRNAs that serve as targets for miRNAs that destabilize the mRNAs.

Aubergine A Piwi protein that associates with piRNAs.

autonomously replicating sequence 1 (ARS1) A yeast origin of replication.

autoradiography A technique in which a radioactive sample is allowed to expose a photographic emulsion, thus "taking a picture of itself."

autoregulation The control of a gene by its own product.

BAC *See* bacterial artificial chromosome.

BAC walking Sequencing a BAC with minimal overlap with a seed BAC, then sequencing another BAC with minimal overlap to the second, and so forth until all BACs in a contig are sequenced.

back mutation *See* reversion.

bacterial artificial chromosome (BAC) A vector based on the *E. coli* F plasmid, capable of holding inserts up to 300,000 bp (average insert size, about 150,000 bp).

bacteriophage *See* phage.

baculoviruses A class of rod-shaped viruses that contain a large, circular DNA genome. These viruses, most of which infect caterpillars, have been used to develop powerful expression vectors for eukaryotic genes.

barcode A relatively short DNA sequence from the genome of every species on earth. In principle, reading this barcode would allow the rapid identification of any known species.

barrier The negative activity an insulator exerts on a silencer by blocking the encroachment of condensed chromatin into an active region of a chromosome, thereby maintaining the activity of genes in that region.

basal level transcription A very low level of class II gene transcription achieved with general transcription factors and polymerase II alone.

base A cyclic, nitrogen-containing compound linked to deoxyribose in DNA and to ribose in RNA.

base excision repair An excision repair pathway that begins by removing a damaged base by a DNA glycosylase and continues by cleaving the 5′-side of the resulting AP site by an AP endonuclease, then removing the AP sugar phosphate and concludes with removing downstream bases and filling in the gap by DNA polymerase and DNA ligase.

base pair (bp) A pair of bases (A–T or G–C), one in each strand, that occur opposite each other in a double-stranded DNA.

β clamp A dimer of β-subunits of the DNA pol III holoenzyme that clamps around DNA, tethering the holoenzyme to the DNA and thereby conferring processivity.

B-DNA The standard Watson–Crick model of DNA favored at high relative humidity and in solution.

BER *See* base excision repair.

β-galactosidase An enzyme that breaks the bond between the two constituent sugars of lactose.

β-galactoside A complex sugar in which the 1-carbon of galactose is linked through a β-bond to another compound, usually another sugar.

β-galactosidic bond The bond linking the 1-carbon of galactose to the other compound in a β-galactoside.

bHLH domain An HLH motif coupled to a basic motif. When two bHLH proteins dimerize through their HLH motifs, the basic motifs are in position to interact with a specific region of DNA. The bHLH protein grasps the major groove of a DNA, like a pair of tongs.

bHLH–ZIP domain A dimerization and DNA binding domain. A basic region is coupled with both an HLH and a leucine zipper (ZIP) domain.

bidirectional DNA replication Replication that occurs in both directions at the same time from a common starting point, or origin of replication. Requires two active replicating forks.

bioinformatics A field that involves the building and manipulation of biological databases. In the context of genomics, this means managing massive amounts of sequencing data and providing useful access to and interpretation of the data.

biolistic transformation (or transfection) A method in which tiny metal pellets are coated with DNA and shot into cells.

bit score (S) A measure of the number of matches between a query sequence and a sequence in the database in a BLAST search.

β-lactamase An enzyme that breaks down ampicillin and related antibiotics and renders a bacterium resistant to the antibiotic.

BLAST A program that searches a database for a DNA or protein sequence and displays how the query sequence lines up with the database sequences.

branch migration Lateral motion of the branch of a Holliday junction during recombination.

branchpoint-bridging protein (BBP) A protein essential for splicing that binds to U1 snRNP at the 5′-end of the intron, and to Mud2p at the 3′-end of the intron.

bridge helix An α-helix near the active center of bacterial RNA polymerases that flexes to promote translocation during the transcription process.

BRG1 The catalytic subunit of SWI/SNF that has ATPase and chromatin remodeling activities.

BRG1-associated factors (BAFs) The 9–12 polypeptides, which, together with BRG1, make up SWI/SNF.

bromodomain A protein domain that binds specifically to acetylated lysine residues on other proteins, such as histones.

bZIP domain A leucine zipper motif coupled to a basic motif. When two bZIP proteins dimerize through their leucine zippers, the basic motifs are in position to interact with a specific region of DNA. The bZIP protein grasps the major groove of a DNA, like a pair of tongs.

C-value The amount of DNA in picograms (trillionths of a gram) in a haploid genome of a given species.

C-value paradox The fact that the C-value of a given species is not necessarily related to the genetic complexity of that species.

cAMP response element (CRE) The enhancer that responds to cAMP.

cap A methylated guanosine bound through a 5′-5′ triphosphate linkage to the 5′-end of an mRNA, an hnRNA, or an snRNA.

cap 0 A cap lacking any 2′-O-methylations. Found only on certain viral mRNAs.

cap 1 A typical cap, with a methyl group on the 2′-hydroxyl group of the penultimate nucleotide.

cap 2 A cap found in a minority of mRNAs with methyl groups on the 2′-hydroxyl groups of the first two nucleotides.

CAP (catabolite activator protein) A protein which, together with cAMP, activates operons that are subject to catabolite repression. Also known as CRP.

cap analysis of gene expression (CAGE) A gene expression monitoring technique similar to SAGE, but emphasizing detection of the 5′-ends of mRNAs.

cap-binding complex (CBC) The complex of proteins that binds to the cap of an mRNA during transcription, goes with the mRNP to the cytoplasm, and substitutes for eIF4F during the pioneer round of translation.

cap-binding protein (CBP) *See* eIF4F.

cap trapper A technique for ensuring that cDNA-mRNA hybrids contain mRNAs with caps. The caps are selectively tagged with a diol reactive biotin derivative and the hybrids are then purified by avidin affinity chromatography.

carboxyl-terminal domain (CTD, of Rpb1) The carboxyl-terminal region of the largest subunit of RNA polymerase II. Consists of dozens of repeats of a heptamer rich in serines and threonines.

carboxyl terminus The end of a polypeptide with a free carboxyl group.

CARM1 *See* coactivator-associated arginine methyltransferase.

catabolic metabolism The process of breaking substances down into simpler components. The *lac* operon encodes catabolic enzymes that break down lactose into its component parts, galactose and glucose.

catabolite activator protein *See* CAP.

catabolite repression The repression of a gene or operon by glucose or, more likely, by a catabolite, or breakdown product of glucose.

catalytic center The active site of an enzyme, where catalysis takes place.

catenane A structure composed of two or more circles linked in a chain.

CBC *See* cap-binding complex.

CBP *See* CREB-binding protein.

CCAAT-binding transcription factor (CTF) A transcription activator that binds to the CCAAT box.

CCAAT box An upstream motif, having the sequence CCAAT, found in many eukaryotic promoters recognized by RNA polymerase II.

Cdc13p A yeast protein that binds to single-stranded telomere ends and recruits Stn1p, which recruits Ten1p to the telomere ends. Together, all three proteins protect the telemore ends from degradation and DNA repair enzymes.

CDK1/CDK9 kinase The kinase that phosphorylates serine 2 in the CTD of Rpb1 in yeast and metazoans, respectively.

cDNA A DNA copy of an RNA, made by reverse transcription.

cDNA library A set of clones representing as many as possible of the mRNAs in a given cell type at a given time.

centimorgan (cM) The genetic distance that yields a 1% recombination frequency between two markers.

centromere Constricted region on the chromosome where spindle fibers are attached during cell division.

CF I and CF II *See* cleavage factors I and II.

chaperone proteins *See* chaperones.

chaperones Proteins that bind to unfolded proteins and help them fold properly.

charging Coupling a tRNA with its cognate amino acid.

Charon phages A set of cloning vectors based on λ phage.

Chi site An *E. coli* DNA site with the consensus sequence 5′-GCTGGTGG-3′. RecBCD cuts at the 3′-end of a Chi site during homologous recombination.

Chi structure *See* Holliday junction.

ChIP *See* chromatin immunoprecipitation.

ChIP-chip Chromatin immunoprecipitation followed by identification of the precipitated DNAs by hybridizing them to a DNA microarray (microchip).

ChIP-seq Chromatin immunoprecipitation followed by identification of the precipitated DNAs by repetitive sequencing.

chloramphenicol An antibiotic that kills bacteria by inhibiting the peptidyl transferase reaction catalyzed by SOS ribosomes.

chloramphenicol acetyl transferase (CAT) An enzyme whose bacterial gene is frequently used as a reporter gene in eukaryotic transcription and translation experiments. This enzyme adds acetyl groups to the antibiotic chloramphenicol.

chromatids Copies of a chromosomes produced in cell division.

chromatin The material of chromosomes, composed of DNA and chromosomal proteins.

chromatin immunoprecipitation (ChIP) A method for purifying chromatin containing a protein of interest by immunoprecipitating the chromatin with an antibody directed against that protein or against an epitope tag attached to the protein.

chromatin remodeling ATP-dependent alterations in the structure of nucleosomes that either move the nucleosomes or enable them to be moved by other proteins.

chromatography A group of techniques for separating molecules based on their relative affinities for a mobile and a stationary phase. In ion-exchange chromatography, the charged resin is the stationary phase, and the buffer of increasing ionic strength is the mobile phase.

chromodomain A conserved region found in proteins involved in heterochromatin formation. Probably binds to methylated histones.

chromogenic substrate A substrate that produces a colored product when acted on by an enzyme.

chromosome The physical structure, composed largely of DNA and protein, the contains the genes of an organism.

chromosome conformation capture (3C) A method for determining whether two remote chromosome sites are brought together in vivo by looping.

chromosome puff A physically enlarged site of active transcription on a polytene chromosome.

chromosome theory of inheritance The theory that genes are contained in chromosomes.

cI The gene that encodes the λ repressor.

CI The *cI* gene product. *See* λ repressor.

***cis*-acting** A term that describes a genetic element, such as an enhancer, a promoter, or an operator, that must be on the same chromosome in order to influence a gene's activity.

***cis*-dominant** Dominant, but only with respect to genes on the same piece of DNA. For example, an operator constitutive mutation in one copy of the *lac* operon of a merodiploid *E. coli* cell is dominant with respect to that copy of the *lac* operon, but not to the other. That is because the operator controls the operon that is directly attached to it, but it cannot control an unattached operon.

***cis*-splicing** Ordinary splicing, in which the exons are on the same precursor RNA molecule.

cistron A genetic unit defined by the *cis–trans* test. For all practical purposes, it is synonymous with the word gene.

clamp The part of RNA polymerase II that clamps down on the DNA (at least the template strand) and holds it in place during elongation.

clamp loader The γ complex portion of the DNA pol III holoenzyme that facilitates binding of the β clamp to the DNA.

clamp module The part of an RNA polymerase that opens to allow entry to the DNA template, then closes to hold the polymerase onto the template.

class I, II, and III promoters Promoters recognized by polymerases I, II, and III, respectively.

clastogen An agent that causes DNA strand breaks.

cleavage factors I and II (CF I and CF II) RNA-binding proteins that are important in cleavage of a pre-mRNA at the polyadenylation site.

cleavage and poly(A) specificity factor (CPSF) A protein that recognizes the AAUAAA part of the polyadenylation signal in a pre-mRNA and stimulates cleavage and polyadenylation.

cleavage stimulation factor (CstF) A protein that recognizes the GU-rich part of the polyadenylation signal in a pre-mRNA and stimulates cleavage.

CLIM (cofactor of LIM) A coactivator of the LIM-HD activator. Proteolytically destroyed after ubiquitination by RLIM.

clone-by-clone sequencing A systematic method of sequencing large genomes. First the whole genome is mapped, then clones corresponding to known regions of the genome are sequenced.

clones Individuals formed by an asexual process so that they are genetically identical to the original individual. Also, colonies of cells of groups of viruses that are genetically identical.

closed promoter complex The complex formed by relatively loose binding between RNA polymerase and a prokaryotic promoter. It is "closed" in the sense that the DNA duplex remains intact, with no "opening up," or melting of base pairs.

coactivator-associated arginine methyltransferase (CARM1) A eukaryotic protein that methylates proteins in the vicinity of a promoter, activating transcription.

coactivators Factors that have no transcription-activation ability of their own, but help other proteins stimulate transcription.

codon A 3-base sequence in mRNA that causes the insertion of a specific amino acid into protein or causes termination of translation.

codon bias Differences in synonymous codon usage in different organisms.

coiled coil A protein motif in which two α-helices (coils) wind around each other. When the two α-helices are on separate proteins, the formation of a coiled coil causes dimerization.

cointegrate An intermediate in transposition of a transposon such as Tn3 from one replicon to another. The transposon replicates, and the cointegrate contains the two replicons joined through the two transposon copies.

colE1 A plasmid found in certain strains of *E. coli,* which codes for a bacteriocidal toxin known as a colicin. The colE1 DNA replicates unidirectionally.

collision-induced dissociation (CID) A mass spectrometry (MS) technique in which a polypeptide ion is accelerated and collided with a neutral gas to fragment the polypeptide at some of its peptide bonds. The newly formed peptide ions are then subjected to a second round of MS to identify them.

colony hybridization A procedure for selecting a bacterial clone containing a gene of interest. DNAs from a large number of clones are simultaneously tested with a labeled probe that hybridizes to the gene of interest.

combinatorial code A metaphor to describe the actions of multiple enhancers and their activators on an associated promoter. Different combinations of activators will have different effects on the activity of the promoter because the battery of enhancers can sense the concentration of each activator and integrate the signals from all activators.

commitment complex (CC) A complex containing at least nuclear pre-mRNA and U1 snRNP that is committed to splicing out the intron to which the U1 snRNP has bound.

complementary polynucleotide strands Two strands of DNA or RNA that have complementary sequences; that is, wherever one has an adenine the other has a thymine, and wherever one has a guanine the other has a cytosine.

complex B The precursor to the RLC containing only Dicer and R2D2, plus the double-stranded siRNA.

composite transposon A bacterial transposon composed of two types of parts: two arms containing IS or IS-like elements, and a central region comprised of the genes for transposition and one or more antibiotic resistance genes.

concatemers DNAs of multiple genome length.

conditional lethal A mutation that is lethal under certain circumstances, but not under others (e.g., a temperature-sensitive mutation).

consensus sequence The average of several similar sequences. For example, the consensus sequence of the –10 box of an *E. coli* promoter is TATAAT. This means that if you examine a number of such sequences, T is most likely to be found in the first position, A in the second, and so forth.

conservative replication DNA (or RNA) replication in which both parental strands remain together, producing a progeny duplex both of whose strands are new.

conservative transposition "Cut and paste" transposition in which both strands of the transposon DNA are conserved as they leave their original location and move to the new site.

constant region The region of an antibody that is more or less the same from one antibody to the next.

constitutive Always turned on.

constitutive mutant An organism containing a constitutive mutation that causes a gene to be expressed at all times, regardless of normal controls.

contig A group of cloned DNAs containing contiguous or overlapping sequences.

copia A transposable element found in *Drosophila* cells.

core element An element of the eukaryotic promoter recognized by RNA polymerase I. Includes the bases surrounding the transcription start site.

core histones All the nucleosomal histones except H1. The histones inside the DNA coils of a nucleosome.

core polymerase *See* RNA polymerase core.

corepressor A substance that associates with an aporepressor to form an active repressor (e.g., tryprophan is the corepressor of the *trp* operon). A protein that works in conjunction with other proteins to repress gene transcription. Histone deacetylases can act as corepressors.

core promoter (class II) Whatever promoter elements are near (within about 37 bp of) the transcription initiation site.

core promoter elements (bacterial) The minimal elements of a promoter (e.g., the –10 and –35 boxes).

core promoter elements (eukaryotic, class II) Include TBE, TATA box, Inr, DPE, DCE, and MTE.

core TAFs The set of 13 TAFs conserved in a wide variety of eukaryotes.

cos The cohesive ends of the linear λ phage DNA.

cosmid A vector designed for cloning large DNA fragments. A cosmid contains the cos sites of λ phage, so it can be packaged into λ heads, and a plasmid origin of replication, so it can replicate as a plasmid.

CoTC element cotranscriptional cleavage (CoTC) The cleavage of a growing transcript downstream of the polyadenylational site; part of the transcription termination process.

counts per minute (cpm) The average number of scintillations detected per minute by a liquid scintillation counter. Generally, this is dpm times the efficiency of the counter.

CPD (cyclobutane pyrimidine dimer) *See* pyrimidine dimers.

CPEB Cytoplasmic polyadenylation element (CPE)-binding protein.

CpG island A region of DNA containing many unmethylated CpG sequences. Usually associated with active genes.

CpG sequences Motifs in mammalian DNAs that are targets of methylation (at the 5-position of the C).

CpG suppression Loss of CpG sequences from genomes over evolutionary time because of methylation of the C and subsequent deamination to T.

CPSF *See* cleavage and poly(A) specificity factor.

CPSF-73 The subunit of CPSF that has the endonuclease activity that cleaves a pre-mRNA prior to polyadenylation.

CRE *See* cAMP response element.

CREB *See* CRE-binding protein.

CREB-binding protein (CBP) A coactivator that binds to phosphorylated CREB at the CRE and then contacts one or more general transcription factors to stimulate assembly of a preintiation complex.

CRE-binding protein (CREB) The activator that is phosphorylated and thus activated by cAMP-stimulated protein kinase A. Binds to CRE and, together with CBP, stimulates transcription of associated genes.

Cro The product of the λ *cro* gene. A repressor that binds preferentially to O_R3 and turns off the λ repressor gene *(cI)*.

cross-linking A technique for probing the interaction between two species (e.g., a protein and a DNA). The two species are chemically cross-linked as they form a complex, then the nature of the cross-linked species is examined.

crossing over Physical exchange between DNAs that occurs during recombination.

cross talk Interaction between members of different signal transduction pathways.

crown galls Tumorous growths on plants caused by bacterial infection.

CRP Cyclic-AMP receptor protein. *See* CAP.

CRSP A coactivator that collaborates with Spl to activate transcription.

cryptogene A gene coding for the unedited version of an RNA that requires editing.

CstF *See* cleavage stimulation factor.

CTCF CCCTC-binding factor. A common vertebrate insulator-binding protein.

CTD *See* carboxyl-terminal domain.

α-CTD The carboxyl-terminal domain of the α-subunit of bacterial RNA polymerase.

cyanobacteria (blue-green algae) Photosynthetic bacteria. The ancestors of modern cyanobacteria are thought to have invaded eukaryotic cells and evolved into chloroplasts.

cyclic-AMP (cAMP) An adenine nucleotide with a cyclic phosphodiester linkage between the 3′ and 5′ carbons. Implicated in a variety of control mechanisms in prokaryotes and eukaryotes.

cytidine A nucleoside containing the base cytosine.

cytidine deaminase acting on RNA (CDAR) An RNA-editing enzyme that deaminates certain cytidines in RNAs, converting them to uridines.

cytoplasmic polyadenylation element (CPE) A sequence in the 3′-UTR of an mRNA (consensus, UUUUUAU) that is important in cytoplasmic polyadenylation.

cytosine (C) The pyrimidine base that pairs with guanosine in DNA.

cytotoxic Having the ability to kill cells.

DAI *See* double-stranded RNA-activated inhibitor of protein synthesis.

***dam* methylase** The deoxyadenosine methylase that adds methyl groups to the A in the sequence GATC in the DNA of *E. coli* cells. The mismatch repair system inspects the methylation of GATC sequences to determine which strand is newly synthesized, and therefore unmethylated.

daughter strand gap The gap left by the DNA replication machinery after it skips over a noncoding base or a pyrimidine dimer.

deadenylation The removal of AMP residues from poly(A) in the cytoplasm.

DEAD protein A member of a family of proteins containing the sequence Asp-Glu-Ala-Asp and having RNA helicase activity.

deamination of DNA The removal of an amino group (NH_2) from a cytosine or adenine in DNA, in which the amino group is replaced by a carbonyl group (C–O). This converts cytosine to uracil and adenine to hypoxanthine.

decatenation The process of unlinking the circles in a catenane.

decoding Interactions between codons and anticodons on the ribosome that lead to binding of the correct aminoacyl-tRNA.

defective virus A virus that is unable to replicate without a helper virus.

degenerate code A genetic code, such as the one employed by all life on earth, in which more than one codon can stand for a single amino acid.

deletion A mutation involving a loss of one or more base pairs.

denaturation (DNA) Separation of the two strands of DNA.

denaturation (protein) Disruption of the three-dimensional structure of a protein without breaking any covalent bonds.

densitometer An instrument that measures the darkness of a spot on a transparent film (e.g., an autoradiograph).

deoxyribose The sugar in DNA.

DGCR8 *See* Pasha/DGCR8.

Dicer The member of the RNase III family that chops the trigger RNA into pieces about 21 bp long during the RNAi process. Also part of the RISC complex that degrades the target mRNA.

dideoxyribonucleotide A nucleotide, deoxy at both 2′- and 3′-positions, used to stop DNA chain elongation in DNA sequencing.

dihydrouracil loop *See* D loop.

dimer (protein) A complex of two polypeptides. These can be the same (in a homodimer), or different (in a heterodimer).

dimerization domain The part of a protein that interacts with another protein to form a dimer (or higher multimer).

dimethyl sulfate (DMS) An agent used for methylating DNA. After methylation, the DNA can be chemically cleaved at the methylated sites.

diploid The chromosomal number of the human zygote and other cells (except the gametes). Symbolized as $2n$.

directional cloning Insertion of foreign DNA into two different restriction sites of a vector, such that the orientation of the insert can be predetermined.

disintegrations per minute (dpm) The average number of radioactive emissions produced each minute by a sample.

dispersive replication A hypothetical mechanism in which the DNA becomes fragmented so that new and old DNA coexist in the same strand after replication.

distal sequence element The nonessential part of a class II snRNA promoter that increases efficiency.

distributive Opposite of processive. Unable to continue a task without repeatedly dissociating and reassociating with the substrate or template.

D loop A loop formed when a free DNA or RNA end "invades" a double helix, base-pairing with one of the strands and forcing the other to "loop out."

DMS footprinting A technique similar to DNase footprinting that uses DMS methylation and chemical cleavage rather than DNase cleavage of DNA.

DNA (deoxyribonucleic acid) A polymer composed of deoxyribonucleotides linked together by phosphodiester bonds. The material of which most genes are made.

DnaA The first protein to bind to *oriC* in forming an *E. coli* primosome.

***dnaA* box** A 9-mer within *oriC* to which DnaA binds in forming an *E. coli* primosome.

DnaB A key component of the *E. coli* primosome. Helps assemble the primosome by facilitating primase binding. Also has DNA helicase activity to unwind the parental DNA strands prior to primer synthesis.

DNA-binding domain The part of a DNA-binding protein that makes specific contacts with a target site on the DNA.

DNA fingerprints The use of highly variable regions of DNA to identify particular individuals.

DnaG The *E. coli* primase.

DNA glycosylase An enzyme that breaks the glycosidic bond between a damaged base and its sugar.

DNA gyrase A topoisomerase that pumps negative superhelical turns into DNA. Relaxes the positive superhelical strain created by unwinding the *E. coli* DNA during replication.

DNA ligase An enzyme that joins two double-stranded DNAs end to end.

DNA melting *See* denaturation (DNA).

DNA microarray A chip containing many tiny spots of DNA or oligonucleotides. Used as a dot blot to measure expression of many genes at once.

DNA microchip *See* DNA microarray.

DNA photolyase The enzyme that catalyzes photoreactivation by breaking pyrimidine dimers.

DNA-PK (DNA protein kinase) The key enzyme in eukaryotic double-strand break repair.

DNA-PK$_{CS}$ The catalytic subunit of DNA-PK.

DNA polymerase An enzyme that synthesizes DNA by linking together deoxyribonucleoside monophosphates (dNMPs) in the order dictated by the complementary sequence of nucleotides in a template DNA strand.

DNA polymerase η A specialized eukaryotic polymerase that carries out translesion synthesis by inserting two dAMPs across from a pyrimidine dimer.

DNA polymerase θ A specialized eukaryotic DNA polymerase involved in translesion synthesis.

DNA polymerase I (pol I) One of three different DNA-synthesizing enzymes in *E. coli;* used primarily in DNA repair.

DNA polymerase II (pol II) Another DNA polymerase in *E. coli.*

DNA polymerase III holoenzyme The enzyme within the *E. coli* replisome, which actually makes DNA during replication.

DNA polymerase ζ A specialized DNA polymerase involved in extending a nascent DNA strand after translesion synthesis in eukaryotic cells.

DNA polymerase V *See* UmuD′$_2$C.

DNA protein kinase *See* DNA-PK.

DNA sequencing *See* sequencing.

DNase Deoxyribonuclease, an enzyme that degrades DNA.

DNase footprinting A method of detecting the binding site for a protein on DNA by observing the DNA region this protein protects from degradation by DNase.

DNase-hypersensitive sites Regions of chromatin that are about a hundred times more susceptible to attack by DNase I than bulk chromatin. These usually lie in the 5′-flanking regions of active or potentially active genes.

DNase-sensitive sites Regions of chromatin that are about ten times more sensitive to DNase I than bulk chromatin. Whole active genes tend to be DNase-sensitive.

DNA typing The use of molecular techniques, especially Southern blotting, to identify a particular individual.

domain (protein) An independently folded part of a protein.

domains of life The three distinct forms of life: bacteria, archaea, and eukarya (or eukaryotes), originally distinguished by their rRNA sequences.

dominant An allele or trait that expresses its phenotype when heterozygous with a recessive allele; for example, *A* is dominant over *a* because the phenotypes of *AA* and *Aa* are the same.

dominant-negative mutation A mutation that yields a protein that is not only inactive but spoils the activity of wild-type protein made in the same cell, by forming a mixed multimer, for example.

double helix The shape two complementary DNA strands assume in a chromosome.

double-stranded RNA-activated inhibitor of protein synthesis (DAI) A protein kinase that responds to interferon and double-stranded RNA by phosphorylating eIF-2α, strengthening its binding to eIF-2B and thereby blocking translation initiation. This prevents viral protein synthesis in an infected cell.

down mutation A mutation, usually in a promoter, that results in less expression of a gene.

downstream core element (DCE) A three-part class two core promoter element lying approximately between positions +6 and +33.

downstream destabilizing element A collection of proteins that bind to exon–exon junctions at the time of splicing during mRNA maturation. The cell uses these proteins as a point of reference to determine whether a nonsense codon is a real stop codon or a premature stop codon.

downstream promoter element (DPE) A class II core promote element centered on position +30.

Drosha The RNase III that converts pri-miRNAs to pre-miRNAs.

Drosophila melanogaster A species of fruit fly used widely by geneticists.

Ds A defective transposable element found in corn, which relies on an *Ac* element for transposition.

DSB Double-strand break in DNA. Required for initiation of meiotic recombination.

DskA A bacterial protein that collaborates with the alarmone ppGpp in decreasing rRNA production in response to starvation.

dsRNA Double-stranded RNA.

dUTPase An enzyme that degrades dUTP and thereby prevents its incorporation into DNA.

editing The process, which occurs at the editing site of an aminoacyl-tRNA synthetase, of breaking down noncognate aminoacyl-AMPs (or even aminoacyl-tRNAs) that have been activated in error by the synthetase. Also called proofreading.

editing site The site on an aminoacyl-tRNA synthetase that examines aminoacyl adenylates and, sometimes, aminoacyl-tRNAs and hydrolyzes those whose amino acids are too small.

EF-2 The eukaryotic homolog of EF-G.

EF-G The bacterial translation elongation factor that, along with GTP, fosters translocation.

EF-Ts An exchange factor that exchanges GDP for GTP on EF-Tu.

EF-Tu The bacterial translation elongation factor that, along with GTP, carries aminoacyl-tRNAs (except fMet-$tRNA_f^{Met}$ to the ribosomal A site.

EGF *See* epidermal growth factor.

eIF1 A eukaryotic initiation factor that stimulates scanning to locate the proper initiation codon.

eIF1A Acts synergistically with eIF1 to cause 40S ribosomal subunits to scan to the initiation codon.

eIF2 A eukaryotic initiation factor that is responsible for binding Met-$tRNA_i^{Met}$ to the 40S ribosomal subunit.

eIF2α One of the subunits of eIF2. It is subject to a phosphorylation that can inhibit translation initiation.

eIF2B A eukaryotic exchange factor that exchanges GTP for GDP on eIF2.

eIF3 A eukaryotic initiation factor that binds to 40S ribosomal subunits and prevents their reuniting prematurely with 60S subunits.

eIF4A One of the subunits of eIF4F; a DEAD family RNA-binding protein with RNA helicase activity. In conjunction with eIF4B, eIF4A can bind to the leader region of an mRNA and remove hairpins in advance of the scanning ribosomal subunit.

eIF4B An RNA-binding protein that helps eIF4A bind to mRNA during translation initiation.

eIF4E The cap-binding component of eIF4F.

eIF4F The cap-binding complex that participates in transcription initiation in eukaryotes.

eIF4G One of the subunits of eIF4F. Serves as an adapter by binding to two different proteins: Binds to eIF4E, which is bound to the cap; also binds to eIF3, which is bound to the 40S ribosomal particle. In this way, it brings the 40S particle together with the 5′-end of an mRNA, where it can begin scanning. Also binds to PAB I, which binds to poly(A).

eIF5 A eukaryotic initiation factor that promotes association between a 40S initiation complex and a 60S ribosomal subunit.

eIF5B The eukaryotic homolog of prokaryotic IF2. Helps eIF5 recruit the 60S ribosomal subunit to the initiation complex. Requires GTP hydrolysis to be released from the ribosome.

eIF6 A eukaryotic initiation factor with activity similar to that of eIF3.

8-oxoguanine (8-hydroxyguanine; oxoG) An oxidation product in DNA having a hydroxyl group in the 8 position of guanine.

EJC *See* exon junction complex.

ELAC2 A candidate for a human 3′ tRNA processing endoribonuclease.

electron-density map A three-dimensional representation of the electron density in a molecule or complex of molecules. Usually determined by x-ray crystallography.

electrophile A molecule that seeks centers of negative charge in other molecules and attacks them there.

electrophoresis A procedure in which voltage is applied to charged molecules, inducing them to migrate. This technique can be used to separate DNA fragments, RNAs, or proteins.

electrophoretic mobility shift assay (EMSA) *See* gel mobility shift assay.

electroporation The use of a strong electric current to introduce DNA into cells.

Elk-1 An activator that is a target of the signal transduction serine/threonine kinase ERK.

elongation factor A protein that is necessary for either the aminoacyl-tRNA binding or the translocation step in the elongation phase of translation.

embryonic stem (ES) cells Cells that can differentiate into any kind of cell in an organism.

encode To contain the information for making an RNA or polypeptide. A gene can encode an RNA or a polypeptide.

end-filling Filling in the recessed 3′-end of a double-stranded DNA using deoxynucleoside triphosphates and a DNA polymerase. This technique can be used to label the 3′-end of a DNA strand.

endonuclease An enzyme that makes cuts within a polynucleotide strand.

endoplasmic reticulum (ER) Literally, "cellular network"; a network of membranes in the cell on which proteins destined for export from the cell are synthesized.

endo-siRNAs siRNAs that are encoded by cells, at least in *Drosophila*.

endospores Dormant spores formed within a cell, as in *Bacillus subtilis*.

enhanceosome A collection of proteins bound to an enhancer, or group of enhancers, required for the complex to adopt a specific shape that can activate transcription efficiently.

enhancer A DNA element that binds one or more activators and stimulates transcription of a gene or genes. Enhancers are usually found upstream of the genes they influence, but they can also function if inverted or moved hundreds or even thousands of base pairs away.

enhancer-binding protein *See* activator.

enhancer-blocking The negative action an insulator exerts on an enhancer.

enzyme A molecule—usually a protein, but sometimes an RNA—that catalyzes, or accelerates and directs, a biochemical reaction.

E1, E2, and E3 snoRNAs *See* small nucleolar RNAs.

epidermal growth factor (EGF) A protein that binds to a transmembrane receptor, signaling the cell to divide.

epigenetic Not affecting the base sequence of DNA.

epitope tagging Using genetic means to attach a small group of amino acids (an epitopic tag) to a protein. This enables the protein to be purified readily by immunoprecipitation with the antibody that recognizes the epitope tag.

ERCC1 Together with XPF, cuts on the 5′-side of DNA damage during human NER.

eRF1 The eukaryotic release factor that recognizes all three stop codons and releases the finished polypeptide from the ribosome.

eRF3 The eukaryotic release factor with ribosome-dependent GTPase activity that collaborates with eRF1 in releasing finished polypeptides from the ribosome.

ERK (extracellular signal-regulated kinase) A signal transduction serine/threonine protein kinase that is activated by MEK and in turn activates activators such as Elk-1 in the nucleus.

error-prone bypass A mechanism cells use to replicate DNA with pyrimidine dimers or noncoding bases. An error-prone DNA polymerase is recruited to insert nucleotides across from the lesion.

Escherichia coli (E. coli) An intestinal bacterium; the favorite subject for bacterial molecular biology.

E site (ribosomal) The exit site to which deacylated tRNAs bind on their way out of the ribosome.

E site (RNA polymerase) The site the incoming nucleotide occupies just before moving (including rotation) to the A site.

essential gene set The set of genes whose loss is incompatible with the life of an organism.

EST *See* expressed sequence tag.

euchromatin Chromatin that is extended, accessible to RNA polymerase, and at least potentially active. These regions stain either lightly or normally with dyes and are thought to contain most of the genes.

eukaryote An organism whose cells have nuclei.

evolutionarily conserved regions (ECRs) Sequences of DNA found in a wide variety of organisms; likely to reside in exons.

excinuclease An endonuclease that participates in cutting out the oligonucleotide containing the damage in human NER.

excision repair Repair of damaged DNA that involves removing the damage and replacing it with normal DNA.

exon A region of a gene that is ultimately represented in that gene's mature transcript. The word refers to both the DNA and its RNA product.

exon definition A splicing scheme in which splicing factors recognize the ends of exons.

exon junction complex The complex of an mRNA with proteins added just upstream of the exon-exon junctions at the time of splicing. These proteins facilitate mRNP transport out of the nucleus.

exonic splicing enhancer (ESE) A region of an exon that stimulates splicing.

exonic splicing silencer (ESS) A region of an exon that inhibits splicing.

exon trapping A method for cloning exons by placing random DNA fragments into a vector that will express them only if they are complete exons.

exonuclease An enzyme that degrades a polynucleotide from the end inward.

exosome A protein complex that degrades RNAs. Different exosomes are found in the nucleus and cytoplasm.

expect value (E value) The number of matches yielding the corresponding bit score that we would expect to see by chance. The lower the E value, the better the match.

expressed sequence tag (EST) An STS generated by amplifying cellular mRNA by RT-PCR.

expression vector A cloning vector that allows expression of a cloned gene.

F plasmid An *E. coli* plasmid that allows conjugation between bacterial cells.

F′ plasmid An F plasmid that has picked up a piece of host DNA.

F_1 The progeny of a cross between two parental types that differ at one or more genes; the first filial generation.

F_2 The progeny of a cross between two F1 individuals or the progeny of a self-fertilized F1; the second filial generation.

FACT (facilitates chromatin transcription) A protein that expedites transcription through nucleosomes in vitro. Interacts strongly with histones H2A and H2B and may destabilize nucleosomes by removing these two core histones.

Far Western blot A blot similar to a Western blot, but probed with a labeled protein (not an antibody) that is suspected of binding to a protein on the blot.

50S ribosomal subunit The large bacterial ribosomal subunit, involved in peptide bond formation.

ferritin An intracellular iron storage protein.

fingerprint (protein) The specific pattern of peptide spots formed when a protein is cut into pieces (peptides) with an enzyme (e.g., trypsin), and then the peptides are separated by chromatography.

FISH Fluorescence in situ hybridization. A means of hybridizing a fluorescent probe to whole chromosomes to determine the location of a gene or other DNA sequence within a chromosome.

5′-end The end of a polynucleotide with a free (or phosphorylated or capped) 5′-hydroxyl group.

flap-tip helix An α-helical region at the tip of the flap of the β-subunit of *E. coli* RNA polymerase. Interacts with the pause helix loop in the transcript of a transcription terminator.

fluor A substance that emits photons when excited by a radioactive emission.

fluorescence resonance energy transfer (FRET) An analytical technique that enables one to measure the distance between two molecules or between two parts of the same macromolecule. It takes advantage of the fact that two fluorescent molecules close to each other will engage in transfer of resonance energy, but that the efficiency of this transfer will decrease with distance between the molecules.

fluorescent probe One of the fluorescent molecules in a FRET experiment.

fluorography A method for visualizing weak radioactive emissions in a medium, such as a gel, by soaking the medium in a fluor that can convert the radioactive emissions into light.

fMet *See* *N*-formyl methionine.

Fos One of the two subunits of the activator AP-1 (the other is Jun).

14-3-3 protein A member of a signal transduction pathway that binds to phospho-serines on other signaling proteins.

fragment reaction A substitute for the peptidyl transferase reaction using simpler substrates. A 6-nt fragment of $tRNA_f^{Met}$ linked to fMet substitutes for a peptidyl-tRNA, and puromycin substitutes for an aminoacyl tRNA. The product is fMet-puromycin released from the ribosome.

frameshift mutation An insertion or deletion of one or two bases in the coding region of a gene, which changes the reading frame of the corresponding mRNA.

free radicals Very reactive chemical substances with an unpaired electron. Can attack and damage DNA.

FRET *See* fluorescence resonance energy transfer.

FRET-ALEX (FRET with alternating pulsed excitation) An adaptation of FRET that corrects for the changing spectrum of a donor fluorophore due to changing protein environment.

functional genomics The study of the pattern of genome-wide gene expression at various times or under various conditions.

functional SELEX A SELEX method that enriches a nucleic acid based on its function (e.g., ability to be spliced).

fusidic acid An antibiotic that blocks the release of EF-G from the ribosome after GTP hydrolysis and thereby blocks translation after the translocation step.

fusion protein A protein resulting from the expression of a recombinant DNA containing two open reading frames (ORFs) fused together. One or both of the ORFs can be incomplete.

G protein A protein that is activated by binding to GTP and inactivated by hydrolysis of the bound GTP to GDP by an inherent GTPase activity.

G-segment The segment of DNA that breaks to form a gate through which the T segment passes during topoisomerase II activity.

galactoside permease An enzyme encoded in the *E. coli lac* operon that transports lactose into the cell.

galactoside transacetylase One of the three enzymes encoded by the *lac* operon. It can acetylate a galactoside such as lactose, but its importance to the *lac* operon is unclear.

GAGA box Element of certain *Drosophila* insulators.

GAL4 A transcription factor that activates the galactose utilization (GAL) genes of yeast by binding to an upstream control element (UAS_G)

gamete A haploid sex cell.

γ complex The complex of the γ, δ, δ', χ, and ψ subunits of the pol III holoenzyme. Has clamp loader activity.

gamma rays Very high energy radiation that ionizes cellular components. The ions then can cause chromosome breaks.

GAP *See* GTPase activator protein.

GC box A hexamer having the sequence GGGCGG on one strand, which occurs in a number of mammalian structural gene promoters. The binding site for the transcription factor Sp1.

GDPCP An unhydrolyzable analog of GTP with a methylene linkage between the β- and γ-phosphate groups. Also called GMPPCP.

gel electrophoresis An electrophoretic method in which substances (usually nucleic acids or proteins) are separated on a gel of agarose or polyacrylamide.

gel filtration A column chromatographic method for separating substances according to their sizes. Small molecules enter the beads of the gel and so take longer to move through the column than larger molecules, which cannot enter the beads.

gel mobility shift assay An assay for DNA–protein binding. A short labeled DNA is mixed with a protein and electrophoresed. If the DNA binds to the protein, its electrophoretic mobility is greatly decreased.

gene The basic unit of heredity. Contains the information for making one RNA and, in most cases, one polypeptide.

gene battery A set of functionally related effector genes controlled by a common agent (e.g., an miRNA).

gene cloning Generating many copies of a gene by inserting it into an organism, such as a bacterium, where it can replicate along with the host.

gene cluster A group of related genes grouped together on a eukaryotic chromosome.

gene conversion The conversion of one gene's sequence to that of another.

gene expression The process by which gene products are made.

general transcription factors Eukaryotic proteins that participate, along with one of the RNA polymerases, in forming a preinitiation complex.

genetic code The set of 64 codons and the amino acids (or terminations) they stand for.

genetic linkage The physical association of genes on the same chromosome.

genetic mapping Determining the linear order of genes and the distances between them.

genetic marker A mutant gene or other peculiarity in a genome that can be used to "mark" a spot in a genome for mapping purposes.

genome One complete set of genetic information from a genetic system; e.g., the single, circular chromosome of a bacterium is its genome.

genomic functional profiling Determining the pattern of expression of all the genes in an organism at all stages of the organism's life.

genomic library A set of clones containing DNA fragments derived directly from a genome, rather than from mRNA.

genomics The study of the structure and function of whole genomes.

genotype The allelic constitution of a given individual. The genotypes at locus *A* in a diploid individual may be *AA*, *Aa*, or *aa*.

GG-NER *See* global genome NER.

gigabase pairs (Gb) Billion base pairs.

G-less cassette A double-stranded piece of DNA lacking a G in the nontemplate strand. A G-less cassette can be placed under the control of a promoter and used to test transcription in vitro in the absence of GTP. Transcripts of the G-less cassette will be produced because GTP is not needed, but nonspecific transcripts originating elsewhere will nor grow beyond very small size due to the lack of GTP.

global genome NER (GG-NER) NER that can remove lesions throughout the genome.

glucose A simple, six-carbon sugar used as an energy source by many forms of life.

glutamine-rich domain A transcription-activating domain rich in glutamines.

glycosidic bond (in a nucleoside) The bond linking the base to the sugar (ribose or deoxyribose) in RNA or DNA.

Golgi apparatus A membranous organelle that packages newly synthesized proteins for export from the cell.

gp5 The product of the phage M13 gene *5*. The phage single-strand DNA-binding protein.

gp28 The product of phage SP01 gene *28*. A phage middle gene-specific σ-factor.

gp32 The product of the phage T4 gene *32*. The phage single-strand DNA-binding protein.

gp33 and gp34 The products of the phage SP01 genes *33* and *34*. Together they constitute a phage late gene-specific σ-factor.

gpA The product of the phage ϕX174 *A* gene. Plays a central role in phage DNA replication as a nuclease to nick one strand of the RF and as a helicase to unwind the double-stranded parental DNA.

GRAIL A program that identifies genes in a database.

GRB2 An adapter protein with an SH2 domain that recognizes phosphotyrosines on signal transduction proteins, and an SH3 domain that binds proline-rich helices in other signal transduction proteins, thus passing on the signal.

GreA and GreB Bacterial auxiliary proteins that bind to RNA polymerase and stimulate a latent RNase activity to cleave off the end of a nascent RNA that contains a misincorporated nucleotide.

g-RNAs *See* guide RNAs (editing).

group I introns Self-splicing introns in which splicing is initiated by a free guanosine or guanosine nucleotide.

group II introns Self-splicing introns in which splicing is initiated by formation of a lariat-shaped intermediate.

GTPase activator protein (GAP) A protein that activates the inherent GTPase of a G protein and thereby inactivates the G protein.

GTPase-associated site A site (or sites) on the ribosome that interacts with the G protein initiation, elongation, and termination factors and stimulates their GTPase activities.

guanine (G) The purine base that pairs with cytosine in DNA.

guanine nucleotide exchange protein A protein that replaces GDP with GTP on a G protein, thereby activating the G protein.

guanosine A nucleoside containing the base guanine.

guide RNAs (editing) Small RNAs that bind to regions of an mRNA precursor and serve as templates for editing a region upstream.

guide sequences (splicing) Regions of an RNA that bind to other RNA regions to help position them for splicing.

guide strand (of siRNA) The strand that associates with RISC to degrade cognate mRNA.

GW182 A protein required for P-body integrity, and also for mRNA silencing in P-bodies, at least in higher eukaryotes.

hairpin A structure resembling a hairpin (bobby pin), formed by intramolecular base-pairing in an inverted repeat of a single-stranded DNA or RNA.

half-life The time it takes for half of a population of molecules to disappear.

hammerhead ribozyme An RNA whose secondary structure resembles a hammer, and which has an RNase activity that can cleave the RNA itself.

haploid The chromosomal number in the gamete *(n)*.

haplotype A cluster of alleles on a single chromosome.

haplotype map A genomic map showing the locations of haplotypes.

HAT *See* histone acetyltransferase.

HAT-A A HAT that acetylates core histones and plays a role in gene regulation.

HAT-B A HAT that acetylates histone H3 and H4 prior to their assembly into nucleosomes.

HCR *See* heme-controlled repressor.

HDAC1 and HDAC2 Two histone deacetylases.

heat shock genes Genes that are switched on in response to environmental insults, including heat.

heat shock response The response of cells to heat and other environmental insults. Cells turn on heat shock genes that encode molecular chaperones that help unfolded proteins refold properly and proteases that degrade hopelessly unfolded proteins.

helicase An enzyme that unwinds a polynucleotide double helix.

helix-loop-helix domain (HLH domain) A protein domain that can cause dimerization by forming a coiled coil with another HLH domain.

helix-turn-helix A structural motif in certain DNA-binding proteins, especially those from prokaryotes, that fits into the DNA major groove and gives the protein its binding capacity and specificity.

helper virus (or phage) A virus that supplies functions lacking in a defective virus, allowing the latter to replicate.

heme-controlled repressor (HCR) A protein kinase that phosphorylates eIF-2α, strengthening its binding to eIF-2B and thereby blocking translation initiation.

hemoglobin The red, oxygen-carrying protein in the red blood cells.

hereditary nonpolyposis colon cancer (HNPCC) A common form of human hereditary colon cancer, caused by failure of mismatch repair.

heterochromatin Chromatin that is condensed and inactive.

heteroduplex A double-stranded polynucleotide whose two strands are not completely complementary.

heterogeneous nuclear RNA (hnRNA) A class of large, heterogeneous-sized RNAs found in the nucleus, including unspliced mRNA precursors.

heteroschizomers Restriction endonucleases that recognize the same restriction site, but cut at different places within the site.

heterozygote A diploid genotype in which the two alleles for a given gene are different; for example, A_1A_2.

high-throughput DNA sequencing. *See* sequencing (high throughput, or next generation).

histone acetyltransferase (HAT) An enzyme that transfers acetyl groups from acetyl CoA to histones.

histone chaperones Proteins that load histones onto naked DNA to form nucleosomes.

histone code A set of histone modifications in a nucleosome near a gene's control region that have a specific effect on transcription of that gene.

histone fold A structural motif found in histones, consisting of three helices linked by two loops.

histone methyltransferase (HMTase) A chromodomain-containing enzyme that transfers methyl groups to core histones.

histones A class of five small, basic proteins intimately associated with DNA in most eukaryotic chromosomes.

HLH domain *See* helix-loop-helix domain.

HMG domain A domain resembling a domain common to the HMG proteins and found in some architectural transcription factors.

HMG protein A nuclear protein with a high electrophoretic mobility (high-mobility group). Some of the HMG proteins have been implicated in control of transcription.

HMGA1a An architectural transcription factor that modulates the tendency of A-T-rich DNA regions to bend. Essential for activation of the IFN-β gene.

HNPCC *See* hereditary nonpolyposis colon cancer.

hnRNA *See* heterogeneous nuclear RNA.

hnRNP A1 An hnRNP that associates with ESSs and helps inhibit splicing.

hnRNP proteins Proteins that bind to hnRNAs.

Holliday junction The branched DNA structure formed by strand exchange during recombination.

homeobox (HOX) A sequence of about 180 bp found in homeotic genes and other development-controlling genes in eukaryotes. Encodes a homeodomain.

homeodomain (HD) A 60-amino acid domain of a DNA-binding protein that allows the protein to bind tightly to a specific DNA region. Resembles a helix-turn-helix domain in structure and mode of interaction with DNA.

homeotic gene A gene in which a mutation causes the transformation of one body part into another.

homologous chromosomes Chromosomes that are identical in size, shape, and, except for allelic differences, genetic composition.

homologous (genes or proteins) Similar because of evolutionary relatedness.

homologous recombination Recombination that requires extensive sequence similarity between the recombining DNAs.

homologs Genes that have evolved from a common ancestral gene. Includes orthologs and paralogs.

homology-directed repair (HDR) *See* recombination repair.

homozygote A diploid genotype in which both alleles for a given gene are identical; for example, A_1A_1 or *aa*.

hormone response elements Enhancers that respond to nuclear receptors bound to their ligands.

housekeeping genes Genes that code for proteins needed for basic processes in all kinds of cells.

HP1 A chromodomain-containing protein associated with histone methyltransferase.

HTF island *See* CpG island.

human endogenous retroviruses Transposition-defective LTR-containing retrotransposons in human cells.

human immunodeficiency virus (HIV) The retrovirus that causes acquired immune deficiency syndrome (AIDS).

HU protein A small, DNA-binding protein that induces bending of *oriC,* thereby encouraging formation of the open complex.

hybrid dysgenesis A phenomenon observed in *Drosophila* in which the hybrid offspring of two certain parental strains suffer so much chromosomal damage that they are sterile, or dysgenic.

hybridization (of polynucleotides) Forming a double-stranded structure from two polynucleotide strands (either DNA or RNA) from different sources.

hybrid polynucleotide The product of polynucleotide hybridization.

hydrogen bond network A network of multiple hydrogen bonds among two or more molecules.

hydroxyl radicals OH units with an unpaired electron. They are highly reactive and can attack DNA and break it. They are therefore useful reagents for footprining.

hydroxyl radical probing A technique in which an iron (Fe^{+})-containing reagent is attached to a cysteine in a protein, and then the protein is bound to an RNA or an RNA-containing complex. The iron ion creates hydroxyl radicals that break nearby sites in the RNA, and these breaks can be detected by primer extension. This reveals where the protein is bound with respect to the RNA.

hyperchromic shift The increase in a DNA solution's absorbance of 260-nm light on denaturation.

identity elements Bases or other DNA elements recognized by a DNA-binding domain.

IF1 The prokaryotic initiation factor that promotes dissociation of the ribosome after a round of translation. Also augments the activities of the other two initiation factors.

IF2 The prokaryotic initiation factor responsible for binding fMet-$tRNA_f^{Met}$ to the ribosome.

IF3 The prokaryotic initiation factor responsible for binding mRNA to the ribosome and for keeping ribosomal subunits apart once they have separated after a round of translation.

immune (λ phage) Lysogens of one lambdoid phage are immune to superinfection by another lambdoid phage if they cannot be infected by the second phage.

immunity region The control region of a λ or λ-like phage, containing the gene for the repressor as well as the operators recognized by this repressor.

immunoblot *See* Western blotting.

immunoglobulin (antibody) A protein that binds very specifically to an invading substance and alerts the body's immune defenses to destroy the invader.

immunoprecipitation A technique in which labeled proteins are reacted with a specific antibody or antiserum, then cross-linked and precipitated by centrifugation. The precipitated proteins are usually detected by electrophoresis and autoradiography.

imprinting Sex-specific silencing of genes by epigenetic means (methylation) during gametogenesis.

imprinting control region (ICR) The locus that controls imprinting of the *Igf2/H19* locus in mammals.

in cis A condition in which two genes are located on the same chromosome.

in silico Performed by computation only (using the silicon chips of a computer).

in trans A condition in which two genes are located on separate chromosomes.

incision Nicking a DNA strand with an endonuclease.

inclusion bodies Insoluble aggregates of protein frequently formed on high-level expression of a foreign gene in *E. coli.* The protein in these inclusion bodies is usually inactive, but can frequently be reactivated by controlled denaturation and renaturation.

indel An insertion or deletion in the genome of one individual or species, relative to another.

independent assortment A principle discovered by Mendel, which states that genes on different chromosomes are inherited independently.

inducer A substance that releases negative control of an operon.

initiation factor A protein that helps catalyze the initiation of translation.

initiator (Inr) A site surrounding the transcription start site that is important in the efficiency of transcription from some class II promoters, especially those lacking TATA boxes.

in-line probing A method that detects secondary structure in RNAs by measuring the ease with which the RNA is cleaved. Unstructured RNA can more easily assume the "in-line" arrangement of reactants, so it is more readily cleaved.

INO80 A yeast nucleosome remodeling factor homologous to SW12/SNF2.

inosine (I) A nucleoside containing the base hypoxanthine, which base-pairs with cytosine.

insertion sequence (IS) A simple type of transposon found in bacteria, containing only inverted terminal repeats and the genes needed for transposition.

insertion state A hypothetical second state, after the preinitiation state, in transcription elongation in which the incoming nucleotide can be examined again for proper fit with the template base.

in situ hybridization Hybridizing labeled probes directly to biological specimens such as sectioned embryos or even chromosome spreads to find genes or gene transcripts.

insulator A DNA element that shields a gene from the positive effect of an enhancer or the negative effect of a silencer.

insulator bodies Conglomerations of two or more insulators and their binding proteins.

integrase An enzyme that integrates one nucleic acid into another; for example, the provirus of a retrovirus into the host genome.

integrator complex A group of 12 polypeptides required for proper 3′-end processing of class II snRNA transcripts.

intensifying screen A screen that intensifies the autoradiographic signal produced by a radioactive substance. The screen contains a fluor that emits photons when excited by radioactive emissions.

interactome The total set of interactions among proteins of an organism.

intercalate To insert between two base pairs in DNA.

interferon A double-stranded RNA-activated antiviral protein with various effects on the cell.

interferon-like growth factor 2 (IGF2) A protein whose gene (*Igf2*) is subject to imprinting in mammals.

intergenic suppression Suppression of a mutation in one gene by a mutation in another.

intermediate A substrate–product in a biochemical pathway.

internal guide sequence A region within a ribozyme, such as a self-splicing intron, that positions other parts of the RNA for catalysis.

internal ribosome entry sequence (IRES) A sequence to which a ribosome can bind and begin translating in the middle of a transcript, without having to scan from the 5′-end.

intervening sequence (IVS) *See* intron.

intracistronic complementation Complementation of two mutations in the same gene. Can occur by cooperation among different defective monomers to form an active oligomeric protein.

intrinsic terminator A bacterial terminator that does not require help from termination factors such as rho.

intron A region that interrupts the transcribed part of a gene. An intron is transcribed, but is removed by splicing during maturation of the transcript. The word refers to the intervening sequence in both the DNA and its RNA product.

intron definition A splicing scheme in which splicing factors recognize the ends of introns.

intronic silencing element A region of an intron that inhibits splicing.

inverted repeat A symmetrical sequence of DNA, reading the same forward on one strand and backward on the opposite strand. For example:

GGATCC
CCTAGG

IRE *See* iron response element.

iron response element (IRE) A stem-loop structure in an untranslated region of an mRNA that binds an iron regulatory protein that in turn influences the lifetime or translatability of the mRNA.

iron regulatory protein (IRP) A protein that binds to an IRE. See also aconitase.

IRP *See* iron regulatory protein.

isoaccepting species (of tRNA) Two or more species of tRNA that can be charged with the same amino acid.

isoelectric focusing Electrophoresing a mixture of proteins through a pH gradient until each protein stops at the pH that matches its isoelectric point. Because the proteins have no net charge at their isoelectric points, they can no longer move toward the anode or cathode.

isoelectric point The pH at which a protein has no net charge.

isoschizomers Two or more restriction endonucleases that recognize and cut the same place in the same restriction site.

isotope-coded affinity tag A tag that can be attached to a protein and give it a higher molecular mass because of a known number of deuteriums in the tag, relative to the same tag that contains hydrogen instead of deuterium.

ISWI A family of coactivators that help remodel chromatin.

joining region (J) The segment of an immunoglobulin gene encoding the last 13 amino acids of the variable region. One of several joining regions is joined by a chromosomal rearrangement to the rest of the variable region, introducing extra variability into the gene.

joint molecule an intermediate in the postsynapsis phase of homologous recombination in *E. coli* in which strand exchange has begun and the two DNAs are intertwined with each other.

Jun One of the two subunits of the activator AP-1 (the other is Fos).

keto tautomer The normal tautomer of uracil, thymine, or guanine found in nucleic acids.

kilobase pair (kb) One thousand base pairs.

kinetic experiment An experiment that measures the kinetics (speed) of a reaction. Because chemical reactions take place on a very short time scale, such experiments require rapid measurements.

kinetoplasts The mitochondria of trypanosomes. Genome consists of many minicircles and maxicircles.

Klenow fragment A fragment of DNA polymerase I, created by cleaving with a protease, that lacks the 5′→3′ exonuclease activity of the parent enzyme.

knockouts Organisms, typically mice, with a gene inactivated by inserting engineered cells into embryos.

known genes Genes from a genomic sequencing project whose sequences are identical to previously characterized genes.

Kozak's rules The set of requirements for an optimal context for a eukaryotic translation initiation signal. The most important of Kozak's rules are that the base at position –3 relative to the AUG initiation codon should be a purine, preferably an A, and that the base at position +4 should be a G.

Ku The ATPase-containing regulatory subunit of DNA-PK. Binds to double-stranded DNA ends created by chromosome breaks and protects them until end-joining can occur.

L1 An abundant human LINE, present in at least 100,000 copies, which occupies about 15% of the human genome.

lacA The *E. coli* gene that encodes galactoside transacetylase.

lacI The *E. coli* gene that encodes the *lac* repressor.

***lac* operon** The operon that encodes enzymes that permit a cell to metabolize the milk sugar lactose.

***lac* repressor** A protein, the product of the *E. coli lacI* gene, that forms a retramer that binds to the *lac* operator and thereby represses the *lac* operon.

lactose A two-part sugar, or disaccharide, composed of two simple sugars, galactose and glucose.

lacY The *E. coli* gene that encodes galactoside permease.

lacZ The *E. coli* gene that encodes β-galactosidase.

lagging strand The strand that is made discontinuously in semidiscontinuous DNA replication.

λ gt11 An insertion cloning vector that accepts a foreign DNA into the *lacZ* gene engineered into a λ phage.

λ phage (lambda phage) A temperate phage of *E. coli*. Can replicate lytically or lysogenically.

λ repressor The protein that forms a dimer and binds to the lambda operators O_R and O_L, thus repressing all other phage genes except the repressor gene itself.

large T antigen The major product of the SV40 viral early region. A DNA helicase that binds to the viral *ori* and unwinds DNA in preparation for primer synthesis. Also causes malignant transformation of mammalian cells.

lariat The name given the lasso-shaped intermediate in certain splicing reactions.

LC-MS *See* liquid chromatography-mass spectronomy.

leader A sequence of untranslated bases at the 5′-end of an mRNA (the 5′-UTR).

leading strand The strand that is made continuously in semidiscontinuous DNA replication.

LEF-1 Lymphoid enhancer-binding factor. An architectural transcription factor.

leucine zipper A domain in a DNA-binding protein that includes several leucines spaced at regular intervals. Appears to permit formation of a dimer with another leucine zipper protein. The dimer is then empowered to bind to DNA.

LexA The product of the *E. coli lexA* gene. A repressor that represses, among other things, the *umuDC* operon.

light repair *See* photoreactivation.

LIM homeodomain (LIM-HD) activators Activators that associate with CLIM coactivators and RLIM corepressors.

limited proteolysis Mild treatment with a protease that can break a protein down into its component domains.

LINEs *See* long interspersed elements.

linker scanning mutagenesis Creation of clustered mutations by replacing small segments (roughly 10 bp) of natural DNA with synthetic double-stranded oligonucleotides (linkers).

liposome A lipid-bounded vesicle. Can be used to introduce DNA into cells.

liquid chromatography-mass spectronomy (LC-MS) Separation of substances by liquid chromatography (LC) in a fine capillary, followed by MS analysis of each substance as it emerges from the LC.

liquid scintillation counting A technique for measuring the degree of radioactivity in a substance by surrounding it with scintillation fluid, a liquid containing a fluor that emits photons when excited by radioactive emissions.

locus (loci, pl.) The position of a gene on a chromosome, used synonymously with the term gene in many instances.

locus control region (LCR) A chromatin region, such as that associated with the globin genes, that ensures activity of the associated genes, regardless of chromatin location.

long interspersed elements (LINEs) The most abundant non-LTR retrotransposons in mammals.

long terminal repeats (LTRs) Regions of several hundred base pairs of DNA found at both ends of the provirus of a retrovirus or an LTR-containing retrotransposon.

looping out The process by which DNA-binding proteins can interact simultaneously with one another and with remote sites on a DNA, by causing the DNA in between the sites to form a loop.

LTRs *See* long terminal repeats.

L7/L12 stalk (L10/L12 stalk) The stalk conventionally shown on the right of the 50S ribosomal particle. Contains the protein L12 and its acetylated counterpart, L7 in most bacteria. Binds to the rest of the ribosome via the protein L10. In some thermophilic bacteria, L7 is missing, so the stalk is called L10/L12.

LTR-containing retrotransposon A retrotransposon with LTRs at both ends. Replicates in a manner identical to that of retroviruses except that no transmissible virus is involved.

luciferase An enzyme that converts luciferin to a chemiluminescent product that emits light and is therefore easily assayed. The firefly luciferase gene is used as a reporter gene in eukaryotic transcription and translation experiments.

luxury genes Genes that code for specialized cell products.

lysis Rupturing the membrane of a cell, as by a virulent phage.

lysogen A bacterium harboring a prophage.

Mad-Max A mammalian repressor.

MAPK *See* mitogen-activated protein kinase.

mariner A defective human transposon that transposed, presumably only in the past, by direct DNA replication.

marker A gene or mutation that serves as a signpost at a known location in the genome.

Maskin A *Xenopus laevis* protein that can inhibit translation of cyclin B mRNAs by binding to CPEB and eIF4E and preventing the latter protein's binding to eIF4G. When CPEB is phosphorylated, it leaves the complex, and Maskin releases eIF4E so translation initiation can occur.

mass spectrometry A high-resolution analytical technique that ionizes molecules and shoots them toward a target. Assuming they are all singly charged, their times of flight to the target are related to their masses. The masses give important information about the nature of the molecules.

maternal genes Genes that are expressed during oogenesis in the mother.

maternal message An mRNA produced in an oocyte before fertilization. Many maternal messages are kept in an untranslated state until after fertilization.

maternal mRNA *See* maternal message.

maxicircles 20–40-kb circular DNAs found in kinetoplasts. Contain the genes (and cryptogenes) and encode some of the gRNAs of the kinetoplast.

Mediator A yeast coactivator that binds to an activator and helps it stimulate assembly of a preinitiation complex.

megabase pair (Mb) One million base pairs.

meiosis Cell division that produces gametes (or spores) having half the number of chromosomes of the parental cell.

MEK (MAPK/ERK kinase) A signal transduction serine/threonine protein kinase that is activated by Raf and, in turn, activates ERK by phosphorylation.

Mendelian genetics *See* transmission genetics.

merodiploid A bacterium that is only partially diploid—that is, diploid with respect to only some of its genes.

message *See* mRNA.

messenger RNA *See* mRNA.

messenger RNP *See* mRNP.

methylation interference assay A means of detecting the sites on a DNA that are important for interacting with a particular protein. These are the sites whose methylation interferes with binding to the protein.

7-methyl guanosine (m^7G) The capping nucleoside at the beginning of a eukaryotic mRNA.

micrococcal nuclease (MNase) A nuclease that degrades the DNA between nucleosomes, leaving the nucleosomal DNA alone.

Microprocessor The complex of Drosha and Pasha (or its homolog).

microRNA (miRNA) A short (18–25-nt) RNA produced naturally in cells that can control the expression of cellular genes by causing destruction of specific mRNAs, or blocking their translation.

microsatellite A short DNA sequence (usually 2–4 bp) repeated many times in tandem. A given microsatellite is found in varying lengths, scattered around a eukaryotic genome.

minicircles 1–3-kb circular DNAs found in kinetoplasts. Encode some of the gRNAs of the kinetoplast.

minimal genome The smallest collection of genes that can sustain life in an organism.

minisatellite A short sequence of (usually) 12 or more bp repeated over and over in tandem.

minus ten box (–10 box) An *E. coli* promoter element centered about 10 bp upstream of the start of transcription.

minus thirty-five box (–35 box) An *E. coli* promoter element centered about 35 bp upstream of the start of transcription.

miRISC The *Drosophila* RISC that participates with miRNAs in control of gene expression.

miRNA *See* microRNA.

mirtron An miRNA that is encoded in an intron. Splicing cuts the mirtron out of the pre-mRNA in lariat form. After debranching this intron folds into a stem-loop that can be processed into an miRNA by Dicer.

mismatch repair The correction of a mismatched base incorporated by accident—in spite of the editing system—into a newly synthesized DNA.

missense mutation A change in a codon that results in an amino acid change in the corresponding protein.

mitogen A substance, such as a hormone or growth factor, that stimulates cell division.

mitogen-activated protein kinase (MAPK) A protein kinase that is activated by phosphorylation as a result of a signal transduction pathway initiated by a mitogen such as a growth factor.

mitosis Cell division that produces two daughter cells having nuclei identical to the parental cell.

model organism An organism chosen for study as a surrogate for study on humans because of any or all of the following factors: small genome size, short generation time, and ease of manipulation in genetic experiments.

molecular chaperones *See* chaperones.

M1 RNA The catalytic RNA subunit of an RNase P.

motif ten element (MTE) A class II core promoter element lying approximately between positions +18 and +27.

mRNA (messenger RNA) A transcript that bears the information for making one or more proteins.

mRNP (messenger RNP) The complex of an mRNA and all the proteins that bind to it.

MS/MS A two-step mass spectronomy technique in which the ions created by the first MS step are manipulated in some way and then subjected to a second MS step.

Mud2p A yeast splicing factor that binds to the 3′-splice site and the branchpoint-bridging protein (BBP), thereby defining the 3′-splice site at an exon–intron boundary.

multiple cloning site (MCS) A region in a cloning vector that contains several restriction sites in tandem. Any of these can be used for inserting foreign DNA.

mutagen A mutation-causing agent.

mutant An organism (or genetic system) that has suffered at least one mutation.

mutation The original source of genetic variation caused, for example, by a change in a DNA base or a chromosome. Spontaneous mutations are those that appear without explanation, while induced mutations are those attributed to a particular mutagenic agent.

mutator mutants Mutants that accumulate mutations more rapidly than wild-type cells.

N The λ phage gene that encodes the antiterminator N.

N The N gene product, an antiterminator that inhibits transcription termination after the λ immediate early genes.

N utilization site (*nut* site) A site in the immediate early genes of phage λ that allows N to serve as an antiterminator. Transcription of the *nut* sites gives rise to a site in the corresponding transcripts that binds N. N can then interact with several proteins bound to RNA polymerase, converting the polymerase to a juggernaut that ignores the terminators at the ends of the immediate early genes.

NAS *See* nonsense-associated altered splicing.

NC2 *See* negative cofactor 2.

NCoR/SMRT Mammalian corepressors that work in conjunction with nuclear receptors.

ncRNAs *See* noncoding RNAs.

negative cofactor 2 (NC2) A protein that stimulates transcription from DPE-containing promoters and inhibits transcription from TATA-box-containing promoters.

negative control A control system in which gene expression is turned off unless a controlling element (e.g., repressor) is removed.

neoschizomers *See* heteroschizomers.

NER *See* nucleotide excision repair.

Neurospora crassa A common bread mold, developed by Beadle and Tatum as a subject for genetic investigation.

next-generation sequencing *See* sequencing (high throughput, or next generation).

NF-κB *See* nuclear factor kappa B.

***N*-formyl-methionine (fMet)** The initiating amino acid in bacterial translation.

nick A single-stranded break in DNA.

nick translation The process by which a DNA polymerase simultaneously degrades the DNA ahead of a nick and elongates the DNA behind the nick. The result is the movement (translation) of the nick in the 3′-direction in a DNA strand.

19S particle The regulatory part of the proteasome, which can also stimulate transcription elongation in certain genes.

nitrocellulose A type of paper that has been changed chemically so it binds single-stranded DNA and proteins. Used for blotting DNA prior to hybridizing with a labeled probe. Also used for blotting proteins prior to probing with antibodies.

NMD *See* nonsense-mediated mRNA decay.

no-go decay (NGD) Decay of an mRNA on which a ribosome has stalled for whatever reason.

nodes Points where two circles cross each other in a catenane.

nonautonomous retrotransposon A non-LTR retrotransposon that encodes no proteins, so it depends on other retrotransposons for transposition activity.

noncoding base A DNA base (e.g., 3-methyl adenine) that cannot base-pair properly with any natural base.

noncoding RNAs (ncRNAs) Transcripts that do not code for proteins. *See also* transcripts of unknown function.

nonhomologous end-joining (NHEJ) A eukaryotic mechanism for repairing double-strand breaks in DNA (chromosome breaks).

non-LTR retrotransposon A retrotransposon that lacks LTRs and replicates by a mechanism different from that used by the LTR-containing retrotransposons.

nonpermissive conditions Those conditions under which a conditional mutant gene product cannot function.

nonreplicative transposition A mode of transposition ("cut and paste") in which the transposon moves from one DNA locus to another without replicating, so no copy is left behind at the original locus. *See also* conservative transposition.

nonsense codons UAG, UAA, and UGA These codons tell the ribosome to stop protein synthesis.

nonsense-associated altered splicing (NAS) A eukaryotic system for dealing with a premature stop codon. When a premature stop codon is detected in frame with the initiation codon, the NAS system invokes an alternative splicing scheme that removes the part of the pre-mRNA containing the premature stop codon.

nonsense-mediated mRNA decay (NMD) A eukaryotic system for destroying mRNAs with premature termination codons. If the distance between the downstream destabilizing element and the nonsense codon in an mRNA is too long, the cell recognizes the nonsense codon as premature and triggers degradation of the mRNA.

nonsense mutation A mutation that creates a premature stop codon within a gene's coding region. Includes amber mutations (UAG), ochre mutations (UAA), and opal mutations (UGA).

non-stop mRNA An mRNA with no stop codon.

nontemplate DNA strand The strand complementary to the template strand. Sometimes called the coding strand or sense strand.

nontranscribed spacer (NTS) A DNA region lying between two rRNA precursor genes in a cluster of such genes.

Northern blotting Transferring RNA fragments to a support medium (*see* Southern blotting).

nr The default (nonredundant) database accessed in a BLAST search.

nt *See* nucleotide.

nTAFs (neural TAFs) TAFs associated with TRF1.

α-NTD The amino-terminal domain of the α-subunit of bacterial RNA polymerase.

nuclear factor kappa B (NF-κB) A mammalian activator that collaborates with other factors, such as HMG I(Y), to activate the interferon-β gene, and other genes of the immune system.

nuclear localization signal A sequence of amino acids, typically rich in basic amino acids, in a protein that target the protein to the nucleus.

nuclear receptor A protein that interacts with hormones, such as the sex hormones, glucocorticoids, or thyroid hormone, or other substances, such as vitamin D or retinoic acid, and binds to an enhancer to stimulate transcription. Some nuclear receptors remain in the nucleus, but some meet their ligands in the cytoplasm, form a complex, and then move into the nucleus to stimulate transcription.

nucleic acid A chain-like molecule (DNA or RNA) composed of nucleotide links.

nucleocapsid A structure containing a viral genome (DNA or RNA) with a coat of protein.

nucleolus A cell organelle found in the nucleus that disappears during part of cell division. Contains the rRNA genes.

nucleoside A base bound to a sugar—either ribose or deoxyribose.

nucleosome A repeating structural element in eukaryotic chromosomes, composed of a core of eight histone molecules with about 200 bp of DNA wrapped around the outside and one molecule of histone H1, also bound outside the core histone octamer.

nucleosome core particle The part of the nucleosome remaining after nuclease has digested all but about 146 bp of nucleosomal DNA. Contains the core histone octamer, but lacks histone H1.

nucleosome positioning The establishment of specific positions of nucleosomes with respect to a gene's promoter.

nucleotide (nt) The subunit, or chain-link, in DNA and RNA, composed of a sugar, a base, and at least one phosphate group.

nucleotide excision repair (NER) An excision repair pathway in which enzymes cut the DNA strand on either side of a damaged base, removing an oligonucleotide that contains the damage. The gap is then filled in with DNA polymerase and DNA ligase.

NuRD A nucleosome remodeling factor with core histone deacetylase activity.

NusA A bacterial protein that stimulates transcription termination by promoting hairpin formation in the transcript at a terminator.

O6-methylguanine methyltransferase A suicide enzyme that accepts methyl or ethyl groups from alkylated DNA bases and thereby reverse the DNA damage.

obligate release A version of the σ cycle that requires σ to be released upon promoter clearance.

ochre codon UAA, coding for termination.

ochre mutation *See* nonsense mutation.

ochre suppressor A tRNA bearing an anticodon that can recognize the ochre codon (UAA) and thereby suppress ochre mutations.

Okazaki fragments Small DNA fragments, 1000–2000 bases long, created by discontinuous synthesis of the lagging strand.

oligo(dT) cellulose affinity chromatography A method for purifying poly(A)$^+$ RNA by binding it to oligo(dT) cellulose in buffer at relatively high ionic strength, and eluting it with water.

oligomeric protein A protein that contains more than one polypeptide subunit.

oligonucleotide A short piece of RNA or DNA.

oligonucleotide array *See* DNA microchip.

oncogene A gene whose product can contribute to the transformation of cells to a malignant phenotype.

one gene–one polypeptide hypothesis The hypothesis, now generally regarded as valid, that one gene codes for one polypeptide.

oocyte 5S rRNA genes The 5S rRNA genes (haploid number about 19,500 in *Xenopus laevis*) that are expressed only in oocytes.

opal codon UGA, coding for termination.

opal mutation *See* nonsense mutation.

opal suppressor A tRNA bearing an anticodon that can recognize the opal codon (UGA) and thereby suppress opal mutations.

open complex A complex of *dnaA* protein and *oriC* in which three 13-mers within *oriC* are melted.

open promoter complex The complex formed by tight binding between RNA polymerase and a prokaryotic promoter. It is "open" in the sense that at least 10 bp of the DNA duplex open up, or separate.

open reading frame (ORF) A reading frame that is uninterrupted by translation stop codons.

operator A DNA element found in prokaryotes that binds tightly to a specific repressor and thereby regulates the expression of adjoining genes.

operator constitutive mutation A mutation in an operator that renders it unable to bind effectively to its repressor. Thus, the operon is constitutive, or somewhat active all the time.

operon A group of genes coordinately controlled by an operator.

O region The small region of homology between *attP* and *attB*.

ORF *See* open reading frame.

oriC The *E. coli* origin of replication.

origin of replication The unique spot in a replicon where replication begins.

orthologs Homologous genes in different species that have evolved from a common ancestral gene.

oxoG *See* 8-oxoguanine.

oxoG repair enzyme Cleaves the glycosidic bond linking oxoG to its deoxyribose.

P_{RE} The λ promoter from which transcription of the repressor gene occurs during the establishment of lysogeny.

P_{RM} The lambda promoter from which transcription of the repressor gene occurs during maintenance of the lysogenic state.

P site (ribosomal) The ribosomal site to which a peptidyl tRNA is bound at the time a new aminoacyl-tRNA enters the ribosome.

PAB I *See* poly(A)-binding protein I.

PAB II *See* poly(A)-binding protein II.

palindrome *See* inverted repeat.

panediting Extensive editing of a pre-mRNA.

paper chromatography A chromatography method that separates molecules based on their relative affinities for paper and for a solvent flowing through the paper.

PAR *See* poly(ADP-ribose).

PARG *See* poly(ADP-ribose) glycohydrolase.

PARP *See* poly(ADP-ribose) polymerase.

PARP1 A PARP that binds to core nucleosomes much as a linker histone would. Upon activation, PARP1 poly(ADP-ribos)ylates itself, which causes dissociation from the nucleosome and gene activation.

paralogs Homologous genes that have evolved by gene duplication within a species.

paromomycin An antibiotic that binds to the A site of the ribosome and decreases accuracy of translation.

paranemic double helix A double helix in which the two strands are not intertwined, but simply laid side-by-side. They can be separated without unwinding.

Pasha/DGCR8 The RNA-binding protein that partners with Drosha in binding to pri-miRNAs. Called Pasha in *Drosophila* and *C. elegans,* and DGCR8 in humans.

passenger strand The strand of an siRNA that is discarded from the RISC, leaving the guide strand associated with the RISC.

pathway (biochemical) A series of biochemical reactions in which the product of one reaction (an intermediate) becomes the substrate for the following reaction.

pause sites DNA sites where an RNA polymerase pauses before continuing elongation.

PAZ The domain of an Argonaute protein that binds single-stranded siRNAs.

P-bodies Discrete cytoplasmic structures that carry out mRNA decay and translational repression.

PBS *See* primer-binding site.

PCNA Proliferating cell nuclear antigen. A eukaryotic protein that confers processivity on DNA polymerase δ during leading strand synthesis.

PCR *See* polymerase chain reaction.

P element A transposable element of *Drosophila,* responsible for hybrid dysgenesis. Can be used to mutagenize *Drosophila* deliberately.

peptide bond The bond linking amino acids in aprotein.

peptidyl transferase An enzyme that is an integral part of the large ribosomal subunit and catalyzes the formation of peptide bonds during protein synthesis.

permissive conditions Those conditions under which a conditional mutant gene product can function.

phage A bacterial virus.

phage 434 A lambdoid (λ-like) phage with its own distinct immunity region.

phage display Expression of a foreign gene as a fusion protein with a phage coat protein so the foreign gene product will be displayed on the surface of the phage.

phagemid A plasmid cloning vector with the origin of replication of a single stranded phage, which gives it the ability to produce single-stranded cloned DNA on phage infection.

phage P1 A lytic phage of *E. coli* used in cloning large pieces of DNA.

phage P22 A lambdoid (λ-like) phage with its own distinct immunity region.

phage T7 A relatively simple DNA phage of *E. coli,* in the same group as phage T3. These phages encode their own single-subunit RNA polymerases.

pharmacogenomics Using a patient's SNPs to predict his or her reaction to various drugs so that therapy can be custom designed.

phenotype The morphological, biochemical, behavioral, or other properties of an organism. Often only a particular trait of interest, such as weight, is considered.

phorbol ester A compound capable of stimulating cell division by a cascade of events, including the activation of AP1.

phosphodiester bond The sugar-phosphate bond that links the nucleotides in a nucleic acid.

phosphorimager An instrument that performs phosphorimaging.

phosphorimaging A technique for measuring the degree of radioactivity of a substance (e.g., on a blot) electronically, without using film.

photoreactivating enzyme *See* DNA photolyase.

photoreactivation Direct repair of a pyrimidine dimer by DNA photolyase.

physical map A genetic map based on physical characteristics of the DNA, such as restriction sites, rather than on locations of genes.

pioneer round (of translation) Translation in which the first ribosome binds to an mRNA and translates it.

piRNA *See* Piwi-interacting RNA.

PIWI The domain of an Argonaute protein that has the slicer activity that cleaves target mRNAs.

Piwi-interacting RNA (piRNA) An RNA complementary to a transposon RNA, which binds to Piwi proteins and degrades the transposon RNA by a ping-pong mechanism.

P/I state A hybrid state in which the initiator aminoacyl-tRNA initially binds, with its anticodon in the P site and its amino acid and acceptor stem in an "initiator" site on the E site side of the P site.

plaque A hole that a virus makes on a layer of host cells by infecting and either killing the cells or slowing their growth.

plaque assay An assay for virus (or phage) concentration in which the number of plaques produced by a given dilution of virus is determined.

plaque-forming unit (pfu) A virus capable of forming a plaque in a plaque assay.

plaque hybridization A procedure for selecting a phage clone that contains a gene of interest. DNAs from a large number of phage plaques are simultaneously tested with a labeled probe that hybridizes to the gene of interest.

plasmid A circular DNA that replicates independently of the cell's chromosome.

plectonemic double helix A double helix, such as the Watson–Crick double helix, in which the two strands are wound around each other and cannot be separated without unwinding.

point mutation An alternation of one, or a very small number, of contiguous bases.

poly(A) Polyadenylic acid. The string of about two hundred A's added to the end of a typical eukaryotic mRNA.

poly(A)-binding protein I (PAB I, Pab1p) A protein that binds the poly(A) tails on mRNAs and apparently helps confer translatability on the mRNAs.

poly(A)-binding protein II (PAB II) A protein that binds to nascent poly(A) at the end of a pre-mRNA and stimulates lengthening of the poly(A).

polyadenylation Addition of poly(A) to the 3′-end of an RNA.

polyadenylation signal The set of RNA sequences that govern the cleavage and polyadenylation of a transcript. An AAUAAA sequence followed 20–30 nt later by a GU-rich region, then a U-rich region is the canonical cleavage signal. After cleavage, the AAUAAA sequence is the polyadenylation signal.

poly(A) polymerase (PAP) The enzyme that adds poly(A) to an mRNA or to its precursor.

poly(A)$^+$ RNA RNA that contains a poly(A) at its 3′-end.

poly(A)$^-$ RNA RNA that contains no poly(A).

poly(ADP-ribose) (PAR) A polymer of ADP-ribose, that is attached to a nuclear protein by poly(ADP)-ribose polymerase. The polymer usually branches every 40–50 ADP-ribose units.

poly(ADP-ribose) glycohydrolase (PARG) An enzyme that breaks the glycosidic bonds between ribose moieties in poly(ADP-ribose), thus breaking down the polymer.

poly(ADP-ribose) polymerase (PARP) A nuclear enzyme that tranfers ADP-ribose units one by one from nicotinamide adenine dinucleotide (NAD) to a target protein, thus attaching poly(ADP-ribose) to the protein.

polycistronic message An mRNA bearing information from more than one gene.

polymerase chain reaction (PCR) Amplification of a region of DNA using primers that flank the region and repeated cycles of DNA polymerase action.

polynucleotide A polymer composed of nucleotide subunits; DNA or RNA.

polypeptide A single protein chain.

polyprotein A long polypeptide that is processed to yield two or more smaller, functional polypeptides; for example, the *pol* polyprotein of a retrovirus.

polyribosome *See* polysome.

polysome A messenger RNA attached to (and presumably being translated by) several ribosomes.

polytene chromosome An enlarged chromosome in certain cells in certain species such as the salivary gland cells in *Drosophila* larvae. The chromosome replicates repeatedly without cell division, so the sister chromatids stick together to form a large chromosome.

pore 1 A pore in RNA polymerase II that admits nucleotides to the active site.

positional cloning Locating genes involved in particular genetic traits.

positive control A control system in which gene expression depends on the presence of a positive effector such as CAP (and cAMP).

positive strand The strand of a viral genome with the same sense as the viral mRNAs.

positive strand phage (or virus) An RNA phage (or virus) whose genome also serves as an mRNA.

postsynapsis The phase of homologous recombination in *E. coli* in which the single-stranded DNA replaces a strand in the double-stranded DNA to form a new double helix.

postreplication repair *See* recombination repair.

posttranscriptional control Control of gene expression that occurs during the posttranscriptional phase when transcripts are processed by splicing, clipping, and modification.

posttranscriptional gene silencing (PTGS) *See* RNA interference.

posttranslational modification The set of changes that occur in a protein after it is synthesized.

POT1 (Protection of telomeres-1) *See* shelterin.

Pot1p The yeast homolog of mammalian POT1. Binds to telomere ends and protects them from degradation and from DNA repair enzymes.

ppGpp The alarmone guanosine 3′-diphosphate, 5′-diphosphate. Slows down translation when bacterial cells are stressed, as by starvation.

predicted genes Genes from a genomic sequencing project that contain sequences homologous to ESTs.

preinitiation complex The combination of RNA polymerase and general transcription factors assembled at a promoter just before transcription begins.

preinsertion state A hypothetical state in transcription elongation in which the incoming nucleotide can be examined for proper fit with the template base and for the presence of the appropriate sugar.

pre-miRNA The stem-loop pecursor to an miRNA, cleaved from a pri-miRNA by Drosha.

presynapsis The phase of homologous recombination in *E. coli* in which RecA (and SSB) coat the single-stranded DNA prior to invasion of the DNA duplex.

Pribnow box *See* minus ten box (−10 box).

primary miRNA (pri-miRNA) The initial transcript of an miRNA gene. Contains a stem-loop with extra material on each side.

primary structure The sequence of amino acids in a polypeptide, or of nucleotides in a DNA or RNA.

primary transcript The initial, unprocessed RNA product of a gene.

primase The enzyme within the primosome that actually makes the primer.

primer A small piece of RNA that provides the free end needed for DNA replication to begin.

primer-binding site (PBS) The site on a retroviral RNA to which the tRNA primer binds to start reverse transcription.

primer extension A method for quantifying the amount of a transcript in a sample, and also for locating the 5′-end of the transcript. A labeled DNA primer is hybridized to a particular mRNA in a mixture, extended to the 5′-end of the transcript with reverse transcriptase, and the DNA product electrophoresed to determine its size and abundance.

primosome A complex of about 20 polypeptides, which makes primers for *E. coli* DNA replication.

probe (nucleic acid) A piece of nucleic acid, labeled with a tracer (traditionally radioactive) that allows an experimenter to track the hybridization of the probe to an unknown DNA. For example, a radioactive probe can be used to identify an unknown DNA band after electrophoresis.

processed pseudogene A pseudogene that has apparently arisen by retrotransposon-like activity: transcription of a normal gene, processing of the transcript, reverse transcription, and reinsertion into the genome.

processing (of RNA) The group of cuts that occur in RNA precursors during maturation, including splicing, 5′- or 3′-end clipping, or cutting rRNAs out of a large precursor.

processing bodies *See* P-bodies.

processivity The tendency of an enzyme to remain bound to one or more of its substrates during repetitions of the catalytic process. Thus, the longer a DNA or RNA polymerase continues making its product without dissociating from its template, the more processive it is.

prokaryotes Microorganisms that lack nuclei. Comprising bacteria, including cyanobacteria (blue-green algae), and archaea.

proliferating cell nuclear antigen (PCNA) A protein that associates with eukaryotic DNA polymerase δ and enhances its processivity.

proline-rich domain A transcription activation domain rich in prolines.

promoter A DNA sequence to which RNA polymerase binds prior to initiation of transcription—usually found just upstream of the transcription start site of a gene.

promoter clearance The process by which an RNA polymerase moves away from a promoter after initiation of transcription.

proofreading (aminoacyl-tRNA synthetase) The process by which aminoacyl adenylates and, less commonly, aminoacyl-tRNAs are hydrolyzed if their amino acids are too small for the sythetase.

proofreading (DNA) The process a cell uses to check the accuracy of DNA replication as it occurs and to replace a mispaired base with the right one.

proofreading (protein synthesis) The process by which aminoacyl-tRNAs are double-checked on the ribosome for correctness before the amino acids are incorporated into the growing protein chain.

prophage A phage genome integrated into the host's chromosome.

protease An enzyme that cleaves proteins; for example, the protease that cleaves a retroviral polyprotein into its functional parts.

proteasome A collection of proteins (sedimentation coefficient 26S) that proteolytically degrades a ubiquitinated protein.

protein A polymer, or polypeptide, composed of amino acid subunits. Sometimes the term *protein* denotes a functional collection of more than one polypeptide (e.g., the hemoglobin protein consists of four polypeptide chains).

protein footprinting A method, analogous to DNase footprinting, for determining the site at which one protein contacts another. One protein is end-labeled, bound to the other protein, then mildly digested with a protease. If the bound protein protects part of the labeled protein from digestion, a band will be missing when the proteolytic fragments are electrophoresed.

protein kinase A (PKA) A serine-threonine-specific protein kinase whose activity is stimulated by cAMP.

protein sequencing Determining the sequence of amino acids in a protein.

proteolytic processing Cleavage of a protein into pieces.

proteome The structures and activities of all the proteins an organism can make in its lifetime.

proteomics The study of proteomes.

provirus A double-stranded DNA copy of a retroviral RNA, which inserts into the host genome.

proximal promoter Class II promoter elements lying approximately between positions −37 and −250.

proximal sequence element The essential element of a class II snRNA promoter.

Prp28 A protein component of the U5 snRNP; required for exchange of U6 for U1 snRNP at the 5′-splice site.

pseudogene A nonallelic copy of a normal gene, which is mutated so that it cannot function.

pseudouridine A nucleoside, found in tRNA, in which the ribose is joined to the 5-carbon instead of the 1-nitrogen of the uracil base.

p300 A homolog of CBP.

pUC vectors Plasmid vectors based on pBR322, containing an ampicillin resistance gene, and a multiple cloning site that interrupts the *lacZ′* gene, which enables blue/white screening for inserts.

pulse-chase The process of giving a short period, or "pulse," of radioactive precursor so that a substance such as RNA becomes radioactive, then adding an excess of unlabeled precursor to "chase" the radioactivity out of the substance.

pulsed-field gel electrophoresis (PFGE) An electrophoresis technique in which the electric field is repeatedly reversed. Allows separation of very large pieces of DNA, up to several Mb in size.

pulse labeling Providing a radioactive precursor for only a short time. For example, DNA can be pulse labeled by incubating cells for a short time in radioactive thymidine.

purine The parent base of guanine and adenine.

puromycin An antibiotic that resembles an aminoacyl-tRNA and kills bacteria by forming a peptide bond with a growing polypeptide and then releasing the incomplete polypeptide from the ribosome.

pyrimidine The parent base of cytosine, thymine, and uracil.

pyrimidine dimers Two adjacent pyrimidines in one DNA strand linked covalently via a cyclobutane ring, interrupting their base pairing with purines in the opposite strand. The main DNA damage caused by UV light.

pyrogram The output of a camera monitoring a pyrosequencing run. Consists of a series of peaks corresponding to the incorporation of nucleotides.

pyrosequencing A high-throughput DNA sequencing method that converts inorganic phosphate released at the incorporation of each new nucleotide to light that can be quantified.

pyrrolysine The "22nd amino acid." Added to growing polypeptides in certain archaea by a special tRNA.

Q The λ phage gene that encodes the antiterminator Q.

Q The *Q* gene product, an antiterminator that inhibits transcription termination a short distance downstream of the λ late promoter $P_{R'}$.

Q utilization site (*qut* site) A Q-binding site overlapping the λ late promoter. When Q binds to *qut* it allows RNA polymerase to ignore the nearby terminator and extend transcription into the late genes.

quaternary structure The way two or more polypeptides interact in a complex protein.

quenching Quickly chilling heat-denatured DNA to keep it denatured.

R A "redundant" region in the LTR of a retrovirus that lies between the U3 and U5 regions.

R2Bm A LINE-like element from the silkworm *Bombyx mori.*

RACE *See* rapid amplification of cDNA ends.

Rad6 A ubiquitin ligase that ubiquitylates histone H2B.

RAD25 The subunit of yeast TFIIH that has DNA helicase activity.

radiation hybrid mapping A mapping technique in which human cells are treated with ionizing radiation to fragment their chromosomes, then these cells are fused with hamster cells to form hybrids with varying contents of human chromosome fragments. Genetic markers that are close together on a chromosome tend to be found in the same hybrid cells.

Raf A serine/threonine protein kinase that is targeted to the inner surface of the cell membrane by the signal transduction protein Ras. Once at the cell membrane, Raf is activated by phosphorylation.

Rag-1 Product of the human *RAG-1* gene. Cooperates with Rag-2 in cleaving immature immunoglobulin and T cell receptor genes at RSSs so the various gene segments can recombine with each other.

Rag-2 Product of the human *RAG-2* gene. Cooperates with Rag-1 in cleaving immature immunoglobulin and T cell receptor genes at RSSs so the various gene segments can recombine with each other.

***ram* state** Ribosome ambiguity state of the H27 helix of the 16S rRNA of *E. coli* ribosomes. In this state, the base pairing in the H27 helix stabilizes pairing between codons and anticodons (even noncognate anticodons), so decoding accuracy is decreased.

RAP1 A yeast telomere-binding protein that binds to a specific telomeric DNA sequence and recruits other telomeric proteins, including the SIR proteins. *See* shelterin.

rapid amplification of cDNA ends (RACE) A method for extending a partial cDNA to its 5′- or 3′-end.

rapid turnover determinant The set of structures in the 3′-UTR of the TfR mRNA that ensure the mRNA will have a short lifetime unless iron starvation conditions cause stabilization of the mRNA.

rare cutter A restriction endonuclease that cuts only rarely, because its recognition site is uncommon.

Ras Product of the *ras* oncogene. In its CTP-bound, activated form, it activates Raf, which passes on the signal to turn on genes that stimulate cell division.

Ras exchanger A protein that replaces GDP with GTP on Ras, thereby activating Ras.

RdRP *See* RNA-directed RNA polymerase.

read (in DNA sequencing) A contiguous sequence obtained from a single run of a sequencing aparatus.

reading frame One of three possible ways the triplet codons in an mRNA can be translated. For example, the message CAGUGCUCGAC has three possible reading frames, depending on where translation begins: (1) CAG UGC UCG; (2) AGU GCU CGA; (3) GUC CUC GAC. A natural mRNA generally has only one correct reading frame.

real-time PCR A method that uses fluorescent tags to measure the amplification of a DNA as it happens during PCR.

RecA The product of the *E. coli recA* gene. Along with SSB, coats a single-stranded DNA tail and allows it to invade a DNA duplex to search for a region of homology in homologous recombination. Also functions as a coprotease during the SOS response.

recA The *E. coli* gene that encodes the RecA protein.

RecBCD pathway The major homologous recombination pathway in *E. coli,* initiated by the RecBCD protein.

RecBCD protein The protein that nicks one of the strands near a chi site and creates a 3′-single-stranded DNA end during homologous recombination in *E. coli.*

recessive An allele or trait that does not express its phenotype when heterozygous with a dominant allele; for example, *a* is recessive to *A* because the phenotype for *Aa* is like *AA* and not like *aa.*

recognition helix The α-helix in a DNA-binding motif of a DNA-binding protein that fits into the major groove of its DNA target and makes sequence specific contacts that define the specificity of the protein. In effect, the recognition helix recognizes the specific sequence of its DNA target.

recombinant DNA The product of recombination between two (or more) fragments of DNA. Can occur naturally in a cell, or be fashioned by molecular biologists in vitro.

recombination Reassortment of genes or alleles in new combinations. Occurs by crossing over between or within DNAs.

recombination-activating gene A gene encoding Rag-1 or Rag-2.

recombination repair A mechanism that cells use to replicate DNA containing uninformative lesions such as pyrimidine dimers. First, the two strands are replicated, leaving a gap across from the lesion. Next, recombination between the progeny duplexes places the gap across from normal DNA so it can be filled in.

recombination signal sequence (RSS) A specific sequence at a recombination junction, recognized by the recombination apparatus during immunoglobulin and T-cell receptor gene maturation.

recruitment Encouraging the binding of a substance to a complex. Usually refers to enhancing the binding of RNA polymerase or transcription factors to a promoter.

regulon A set of genes that are regulated together.

related genes Genes from a genomic sequencing project whose sequences are homologous to known genes, or parts of genes, of the same or other species.

release factor (RF) A protein that causes termination of translation at stop codons.

renaturation of DNA *See* annealing of DNA.

repetitive DNA DNA sequences that are repeated many times in a haploid genome.

replacement vector A cloning vector derived from λ phage, in which a significant part of the phage DNA is removed and must be replaced by a segment of foreign DNA of similar size.

replicating fork The point where the two parental DNA strands separate to allow replication.

replicative form The double-stranded version of a single-stranded RNA or DNA phage or virus that exists during genome replication.

replicative transposition "Copy and paste" transposition in which the transposon DNA replicates, so one copy remains in the original location as another copy moves to the new site.

replicon All the DNA replicated from one origin of replication.

replisome The large complex of polypeptides, including the primosome, which replicates DNA in *E. coli.*

reporter gene A gene attached to a promoter or translation start site, and used to measure the activity of the resulting transcription or translation. The reporter gene serves as an easily assayed surrogate for the gene it replaces.

repressed Turned off. When an operon is repressed, it is turned off, or inactive.

resolution The second step in transposition through a cointegrate intermediate; it involves separation of the cointegrate into its two component replicons, each with its own copy of the transposon. Also, the final step in recombination, in which the second pair of strands is broken.

resolvase The enzyme that catalyzes resolution of a cointegrate; an endonuclease that nicks two DNA strands to resolve a Holliday junction after branch migration.

***res* sites** Sites on the two copies of a transposon in a cointegrate, between which crossing over occurs to accomplish resolution.

restriction endonuclease An enzyme that recognizes specific base sequences in DNA and cuts at or near those sites.

restriction fragment A piece of DNA cut from a larger DNA by a restriction endonuclease.

restriction fragment length polymorphism (RFLP) A variation from one individual to another in the number of cutting sites for a given restriction endonuclease in a given genetic locus.

restriction map A map that shows the locations of restriction sites in a region of DNA.

restriction–modification system (R-M system) The combination of a restriction endonuclease and the DNA methylase that recognizes the same DNA site.

restriction site A sequence of nucleotides recognized and cut by a restriction endonuclease.

restrictive conditions *See* nonpermissive conditions.

restrictive state Alternative to the *ram* state of the H27 helix of the 16S rRNA of *E. coli* ribosomes. In this state, required for proofreading, the base pairing in the H27 helix demands accurate pairing between codons and anticodons, so decoding accuracy is increased.

retained intron An intron retained in a mature mRNA via an alternate splicing scheme.

retrohoming The process by which a group II intron in one gene can transpose into an intronless version of the same gene somewhere else in the genome.

retrotransposon A transposable element such as *copia* or Ty that transposes via a retrovirus-like mechanism.

retrovirus An RNA virus whose replication depends on formation of a provirus by reverse transcription.

reverse transcriptase RNA-dependent DNA polymerase; the enzyme, commonly found in retroviruses, that catalyzes reverse transcription.

reverse transcriptase PCR (RT-PCR) A PCR procedure that begins with the synthesis of cDNA from an mRNA template, using reverse transcriptase. The cDNA then serves as the template for conventional PCR.

reverse transcription Synthesis of a DNA using an RNA template.

reversion A mutation that cancels the effects of an earlier mutation in the same gene.

RF (replicative form) The circular double-stranded form of the genome of a single-stranded DNA phage such as ϕX174. The DNA assumes this form in preparation for rolling circle replication.

RF1 The prokaryotic release factor that recognizes UAA and UAG stop codons.

RF2 The prokaryotic release factor that recognizes UAA and UGA stop codons.

RF3 The prokaryotic release factor with ribosome dependent GTPase activity. Along with GTP, helps RF1 and RF2 bind to the ribosome.

RF-A A human single-strand DNA-binding protein that is essential for SV40 virus DNA replication.

***RFN* element** A riboswitch in the 5′-UTR of the *ribD* operon. Upon binding FMN, the base-pairing in the riboswitch changes to create a terminator that attenuates transcription.

RFLP *See* restriction fragment length polymorphism.

rho (ρ) A protein that is needed for transcription termination at certain terminators in *E. coli* and its phages.

rho-dependent terminator A terminator that requires rho for activity.

rho-independent terminator *See* intrinsic terminator.

rho loading site A site on a growing mRNA to which rho can bind and begin pursuing the RNA polymerase.

ribonuclease (RNase) An enzyme that degrades RNA.

ribonuclease H *See* RNase H.

ribonucleoside triphosphates The building blocks of RNA: ATP, CTP, GTP, and UTP.

ribose The sugar in RNA.

riboprobe A labeled RNA probe, commonly used in RNase protection assays.

ribosomal RNA *See* rRNA.

ribosome An RNA–protein particle that translates mRNAs to produce proteins.

ribosome drop-off Premature release of ribosomes from an mRNA.

ribosome recycling factor (RRF) A protein with a strong resemblance to a tRNA, which can bind to the A site of a ribosome after translation termination and, together with EF-G, release the ribosome from the mRNA.

riboswitch A region of an RNA that can bind to a small molecule that alters gene expression by influencing transcription or translation.

ribozyme A catalytic RNA (RNA enzyme).

rifampicin An antibiotic that blocks transcription initiation by *E. coli* RNA polymerase.

RISC loading complex (RLC) The protein complex that (presumably) separates the two siRNA strands, then delivers the guide strand to the RISC.

RITS (RNA-induced initiator of transcriptional gene silencing) A protein complex that attracts RdRP and then amplifies an siRNA.

RLC *See* RISC loading complex.

RLIM (RING finger LIM domain-binding protein) A corepressor of LIM-HD activators that represses by ubiquitinating the coactivator CLIM, thereby marking it for proteolytic destruction.

R-looping A technique for visualizing hybrids between DNA and RNA by electron microscopy. A classic R loop is formed when an RNA hybridizes to one strand of a DNA and displaces the other strand as a loop. R-looping can also be performed with single-stranded DNA and RNA; in this procedure, classic R loops do not form, but loops can still be observed if the DNA contains information not found in the RNA.

RNA (ribonucleic acid) A polymer composed of ribonucleotides linked together by phosphodiester bonds.

RNA-dependent DNA polymerase *See* reverse transcriptase.

RNA-directed RNA polymerase (RdRP) The enzyme that elongates the siRNA primers, using target mRNA as a template, thus providing more substrate for Dicer and amplifying the siRNA.

RNA helicase An enzyme that can unwind a double-stranded RNA, or a double-stranded region within an RNA.

RNA-induced silencing complex (RISC) The RNase complex that degrades the target mRNA during RNAi. Includes Dicer, Argonaute, and an unknown RNase that cleaves the target mRNA.

RNA interference (RNAi) Control of gene expression by specific mRNA degradation caused by insertion of a double-stranded RNA into a cell.

RNA ligase An enzyme that can join two pieces of RNA, such as the two pieces of a pre-tRNA created by cutting out an intron.

RNA polymerase The enzyme that directs transcription, or synthesis of RNA.

RNA polymerase I The eukaryotic RNA polymerase that makes large rRNA precursors.

RNA polymerase II The eukaryotic RNA polymerase that makes mRNA precursors and most snRNAs.

RNA polymerase III The eukaryotic RNA polymerase that makes 5S rRNA and tRNA precursors, as well as the precursors of several other small RNAs, including U6 snRNA.

RNA polymerase IV The polymerase II-like enzyme that makes 24-nt heterochromatic siRNAs in plants.

RNA polymerase V The polymerase II-like enzyme that makes the long RNAs that collaborate with heterochromatic siRNAs and Ago4 to silence heterochromatin in plants.

RNA polymerase core The collection of subunits of a prokaryotic RNA polymerase having basic RNA chain-elongation capacity, but no specificity of initiation; all the RNA polymerase subunits except the σ-factor.

RNA polymerase IIA A form of polymerase II with the CTD in an unphosphorylated or underphosphorylated condition.

RNA polymerase II holoenzyme The combination of polymerase II, transcription factors, and other proteins that can be purified as a unit using mild techniques.

RNA polymerase holenzyme (bacterial) The collection of polypeptides that make up the whole enzyme. Usually includes β, β', α_2, ω, and σ subunits.

RNA polymerase IIO A form of polymerase II with the CTD in a highly phosphorylated condition.

RNA processing Modifying an initial transcript to its mature form by cleavage, splicing, capping, polyadenylation, etc.

RNA pull-down A procedure in which a specific RNA is precipitated by reacting with a biotin-tagged complementary RNA, then reacting that complex with avidin which is attached to magnetic beads. By precipitating the beads with a magnet, the attached RNAs can also be precipitated. A similar procedure uses a nonmagnetic bead that is easily precipitated in a centrifuge.

RNase E The enzyme that removes the *E. coli* 5S rRNA from its precursor RNA.

RNase H An RNase that is specific for the RNA part of an RNA–DNA hybrid. One of the activities of a retroviral reverse transcriptase.

RNase mapping A variation on S1 mapping in which the probe is RNA; RNase, instead of S1 nuclease, is used to digest the single-stranded RNA species.

RNase P The enzyme that cleaves the extra nucleotides from the 5′-end of a tRNA precursor. Most forms of RNase P have a catalytic RNA subunit.

RNase protection assay *See* RNase mapping.

RNase R A ribonuclease that associates with tmRNAs and may degrade non-stop mRNAs after they have been released by tmRNAs.

RNase III The enzyme that performs at least the first cuts in processing of *E. coli* rRNA precursors.

RNA splicing The process of removing introns from a primary transcript and attaching the exons to one another.

RNA triphosphatase The enzyme that removes the γ-phosphate at the 5′-end of a pre-mRNA prior to capping.

rolling circle replication A mechanism of replication in which one strand of a double-stranded circular DNA remains intact and serves as the template for elongation of the other strand at a nick.

RRF *See* ribosome recycling factor.

rRNA (ribosomal RNA) The RNA molecules contained in ribosomes.

rRNA precursor (45S) The large rRNA precursor in mammals, which contains the 28S, 18S, and 5.8S rRNA sequences.

rRNA (5.8S) The smallest of the mammalian rRNAs derived from the 45S precursor. Found in the large (60S) ribosomal subunit, base-paired to the 28S rRNA.

rRNA (18S) The mammalian rRNA found in the small (40S) ribosomal subunit.

rRNA (28S) The largest mammalian rRNA. Found in the large (60S) ribosomal subunit, base-paired to the 5.8S rRNA.

***rrn* genes** Bacterial genes encoding rRNAs.

Rsd The anti-σ factor that inhibits the major vegetative σ factor in *E. coli*, σ^{70} (σ^D).

***rsd* gene** The gene that encodes Rsd.

RSS *See* recombination signal sequence.

R2D2 A Dicer-associated protein that is presumed to be part of the RLC.

run-off transcription assay A method for quantifying the extent of transcription of a particular gene in vitro. A double-stranded DNA containing a gene's control region and the 5′-region of the gene is transcribed in vitro with labeled ribonucleoside triphosphates to label the product. The RNA polymerase "runs off" the end of the truncated gene, giving a short RNA product of predictable length. The abundance of this run-off product is a measure of the extent of transcription of the gene in vitro.

run-on transcription assay A method for measuring the amount of transcription of a particular gene in vivo. Nuclei are isolated, with RNA polymerases caught in the act of elongating various RNA chains. These chains are elongated in vitro in the presence of labeled nucleotides to label the RNAs. Then the labeled RNAs are hybridized to Southern blots or dot blots containing unlabeled samples of DNAs representing the genes to be assayed. The extent of hybridization to each band or spot on the blot is a measure of the number of elongated RNA chains made from the corresponding genes, and therefore of the extent of transcription of these genes.

RuvA Together with RuvB, forms a DNA helicase that promotes branch migration during homologous recombination in *E. coli*.

RuvB Together with RuvA, forms a DNA helicase that promotes branch migration during homologous recombination in *E. coli*. Contains the ATPase that provides energy to the helicase.

RuvC The resolvase of the RecBCD pathway of homologous recombination.

Saccharomyces cerevisiae Baker's yeast.

SAGA A transcription adapter complex with histone acetyltransferase activity. Mediates the effects of certain activators.

SAGE *See* serial analysis of gene expression.

scanning A model of translation initiation in eukaryotes that invokes a 40S ribosomal subunit binding to the 5′-end of the mRNA and scanning, or sliding along, the mRNA until it finds the first start codon in a good context for initiation.

scintillation A burst of light created by a radioactive emission striking a fluor in a liquid-scintillation counter.

scintillation fluid *See* liquid scintillation counting.

screen A genetic sorting procedure that allows one to distinguish desired organisms from unwanted ones, but does not automatically remove the latter.

scrunching A hypothesis to explain abortive transcription that invokes an RNA polymerase that squeezes (scrunches) more DNA into itself without moving relative to the DNA.

SC35 A mammalian SR-class RNA-binding protein that can singlehandedly cause commitment to splice at a particular site in certain pre-mRNAs.

SDS PAGE *See* sodium dodecyl sulfate.

secondary siRNAs Short double-stranded RNAs that lie outside the limits of the original trigger RNA. Always upstream with respect to the sense strand of the trigger RNA.

secondary structure The local folding of a polypeptide or RNA. In the latter case, the secondary structure is defined by intramolecular base pairing.

sedimentation coefficient A measure of the rate at which a molecule or particle travels toward the bottom of a centrifuge tube under the influence of a centrifugal force.

seed BAC A BAC chosen for complete sequencing during shotgun sequencing of a large genome.

selection A genetic sorting procedure that eliminates unwanted organisms, usually by preventing their growth or by killing them.

selenocysteine The "21st amino acid." An unusual amino acid with selenium in place of sulfur in the cysteine structure.

SELEX (systematic evolution of ligands by exponential enrichment) A method for enriching nucleic acids (usually RNAs) containing a functional region, or aptamer. Functional molecules are selected by a method like affinity chromatography, then amplified by PCR, then selected and amplified again and again, effecting an exponential enrichment.

semiconservative replication DNA replication in which the two strands of a parental duplex separate completely and pair with new progeny strands. One parental strand is therefore conserved in each progeny duplex.

semidiscontinuous replication A mechanism of DNA replication in which one strand is made continuously and the other is made discontinuously.

sequenator An automated DNA sequencer.

sequence-tagged connector (STC) A sequence of about 500 bp obtained from the end of a large clone, such as a BAC, during large-scale genomic sequencing.

sequence-tagged site (STS) A short stretch of DNA that can be identified by amplifying it using PCR with defined primers.

sequencing Determining the amino acid sequence of a protein, or the base sequence of a DNA or RNA.

sequencing (high throughput, or next generation) DNA sequencing that uses very rapid automated methods, such as pyrosequencing, to produce relatively short reads of DNA in a massively parallel manner that yields a great deal of sequence in a short time.

serial analysis of gene expression (SAGE) A method for determining the levels of expression of many genes at once. Uses short cDNAs, or tags, from many mRNAs that are linked together, cloned, and sequenced. Tags that are most frequently found are expressed most actively.

7SL RNA A small RNA involved in recognizing the signal peptides on proteins destined for secretion.

70S initiation complex The complex of 70S ribosome, mRNA, and fMet-$tRNA_f^{Met}$ that is poised to initiate translation.

severe combined immunodeficiency (SCID) A disease in which the patient has no immune system. One cause is a defective Artemis gene. Also called "bubble boy" syndrome.

SH2 domain A phosphotyrosine-binding domain found in many signal transduction proteins.

SH3 domain A proline-rich-helix-binding domain that causes protein-protein interactions.

shelterin The collection of proteins that bind to a telomere and "shelter" it from degradation or inappropriate chromosome end joining. In mammals there are six shelterin proteins: TRF1, TRF2, TIN2, POT1, TPP1, and RAP1.

Shine–Dalgarno (SD) sequence A G-rich sequence (consensus = AGGAGGU) that is complementary to a sequence at the 3′-end of *E. coli* 16S rRNA. Base pairing between these two sequences helps the ribosome bind an mRNA.

short hairpin RNAs (shRNAs) RNAs engineered with inverted repeats so they form hairpins in vivo and initiate RNAi.

short interfering RNAs *See* siRNAs.

shotgun sequencing Genomic sequencing performed by chopping the genome up into small pieces that are cloned and sequenced at random. Later these sequences are pieced together to give the sequence of the whole genome.

shuttle vector A cloning vector that can replicate in two or more different hosts, allowing the recombinant DNA to shuttle back and forth between hosts.

sickle cell disease A genetic disease in which abnormal β-globin is produced. Because of a single amino acid change, this blood protein tends to aggregate under low-oxygen conditions, distorting red blood cells into a sickle shape.

sigma (σ) The bacterial RNA polymerase subunit that confers specificity of transcription—that is, ability to recognize specific promoters.

σ-cycle The pattern of initiation of transcription with an RNA polymerase holoenzyme, loss of the σ-factor at some point during transcription, and reassociation of the σ with a core polymerase to form a new holoenzyme capable of initiation.

σ^{43} The principal σ-factor of *B. subtilis.*

σ^{70} The principal σ-factor of *E coli.*

signal peptide A stretch of about 20 amino acids, usually at the amino terminus of a polypeptide, that helps to anchor the nascent polypeptide and its ribosome in the endoplasmic reticulum. Polypeptides with a signal peptide are destined for packaging in the Golgi apparatus and are usually exported from the cell.

signal transduction pathway A biochemical pathway that connects a signal, such as a growth factor binding to the cell surface, with an intracellular effect, usually gene activation (or repression).

silencer A DNA element that can act at a distance to decrease transcription from a eukaryotic gene.

silencing Repression of eukaryotic gene activity. Can occur by forming heterochromatin in the region of the gene, or by more localized mechanisms, including tightening the binding between particular nucleosomes and DNA.

silent mutation Mutations that cause no detectable change in an organism, even in a haploid organism or in homozygous condition.

SIN3 A yeast corepressor.

SIN3A and SIN3B Mammalian counterparts of the yeast corepressor SIN3.

single-nucleotide polymorphism (SNP) A single-nucleotide difference between two or more individuals at a particular genetic locus.

single-strand DNA-binding protein *See* SSB.

siRISC The *Drosophila* RISC that participates with siRNAs in RNAi.

SIR2, SIR3, and SIR4 Proteins associated with, and required for, the formation of yeast heterochromatin, including telomeric heterochromatin.

siRNAs (short interfering RNAs) The short pieces (21–28 nt) of double-stranded trigger RNA created by Dicer during the RNAi process.

site-directed mutagenesis A method for introducing specific, predetermined alterations into a cloned gene.

site-specific recombination Recombination that always occurs in the same place and depends on limited sequence similarity between the recombining DNAs.

[6-4] photoproducts DNA lesions caused by UV radiation, in which the 6-carbon of one pyrimidine is covalently linked to the 4-carbon of an adjacent pyrimidine.

SKAR (S6K1 Aly/REF-like substrate) A protein that is recruited to the EJC in the nucleus, and then recruits S6K1 to the mRNP in the cytoplasm.

Ski complex Part of the complex, containing Ski7p and the exosome, which causes degradation of eukaryotic non-stop mRNAs.

Ski7p A protein that recognizes a eukaryotic non-stop mRNA with 0–3 A's of the terminal poly(A) in the A site of the ribosome, and recruits an RNase complex to degrade the non-stop mRNA.

SL1 A class I transcription factor that contains TBP and three TAF_Is. Acts synergistically with UBF to stimulate polymerase I binding to DNA and transcription.

slicer The activity of Argonaute2 that cleaves a target mRNA in a RISC.

sliding clamp The clamp-like structure of RNA polymerase II between the tip of the arm and the underlying shelf that keeps the enzyme from dissociating from the DNA template, thus enhancing processivity.

Slu7 A splicing factor required for selection of the proper AG at the 3′-splice site.

small nuclear RNAs (snRNAs) A set of small RNAs found in the nucleus, associated with proteins to form small nuclear ribonucleoproteins (snRNPs), which participate in splicing of pre-mRNAs.

small nucleolar RNAs (snoRNAs) A set of hundreds of small RNAs found in the nucleolus. A small subset of the snoRNAs (E1, E2, and E3), associated with proteins in small nucleolar ribonucleoproteins, (snoRNPs), participate in processing the large rRNA precursor.

small RNAs (sRNAs) A class of short bacterial RNAs that control translation by binding to mRNAs.

SMCC/TRAP A human Mediator-like protein.

Sm proteins A set of seven proteins found in all snRNPs, including minor snRNPs.

Sm site A sequence (AAUUUGUGG) on an snRNA that interacts with the Sm proteins.

SmpB A protein that associates with tmRNAs and helps them bind to ribosomes, probably in part by compensating for the lack of a D loop on the tmRNA.

snoRNAs *See* small nucleolar RNAs.

snoRNPs *See* small nucleolar RNAs.

SNP *See* single-nucleotide polymorphism.

snRNAs *See* small nuclear RNAs.

snRNPs *See* small nuclear RNAs.

sodium dodecyl sulfate (SDS) A strong, negatively charged detergent used to denature proteins in SDS polyacrylamide gel electrophoresis (SDS-PAGE).

somatic cells Nonsex cells.

somatic 5S rRNA genes The 5S rRNA genes (haploid number about 400 in *Xenopus laevis*) that are expressed in both somatic cells and oocytes.

somatic mutation A mutation that affects only somatic cells, so it cannot be passed on to progeny.

S1 mapping A method for mapping the ends of a specific transcript (or quantifying a specific transcript). A labeled DNA probe is hybridized to transcripts made in vivo or in vitro, the hybrid is treated with S1 nuclease to remove unhybridized parts, and the protected part of the probe is then electrophoresed alongside size standards.

S1 nuclease A nuclease specific for single-stranded RNA and DNA. Used in S1 mapping.

Sos A Ras exchanger.

SOS response The activation of a group of genes, including recA, that helps *E. coli* cells respond to environmental insults such as chemical mutagens or radiation.

Southern blotting Transferring DNA fragments separated by gel electrophoresis to a suitable support medium such as nitrocellulose, in preparation for hybridization to a labeled probe.

spacer DNA DNA sequences found between, and sometimes within, repeated genes such as rRNA genes.

Spearman's rank correlation A statistical technique in which two data sets are arranged in rank order, and the correlation between the rank orders of the two data sets is expressed as an R_S, where a perfect correlation has an R_S of 1.0 and two totally unrelated sets have an R_S of 0.

specialized genes Genes that are active only in one (or a very few) types of cells; e.g., the insulin gene in pancreas β-islet cells, or the globin genes in red blood cells. Also called luxury genes.

spliced leader (SL) The independently synthesized 35-nt leader that is *trans*-spliced to surface antigen mRNA coding regions in trypanosomes.

spliceosome The large RNA–protein body on which splicing of nuclear mRNA precursors occurs.

spliceosome cycle The process of forming the spliceosome, splicing, then dissociation of the spliceosome.

splicing The process of linking together two RNA exons while removing the intron that lies between them.

splicing factors Proteins besides snRNP proteins that are essential for splicing nuclear pre-mRNAs.

Spo11 The endonuclease that creates the DSBs that initiate meiotic recombination in yeast.

spore (1) A specialized haploid cell formed sexually by plants or fungi, or asexually by fungi. The latter can either serve as a gamete, or germinate to produce a new haploid cell. (2) A specialized cell formed asexually by certain bacteria in response to adverse conditions. Such a spore is relatively inert and resistant to environmental stress.

sporulation Formation of spores.

squelching Inhibition of one activator by increasing the concentration of a second one. Presumably caused by competition for a scarce common factor.

SRB and MED-containing cofactor (SMCC) *See* SMCC/TRAP.

SRC *See* steroid receptor coactivator.

SRC-1, SRC-2, and SRC-3 *See* steroid receptor coactivator (SRC).

sRNAs *See* small RNAs.

SR proteins A group of RNA-binding proteins having an abundance of serine (S) and arginine (R).

SSB Single-strand DNA-binding protein, used during DNA replication and recombination. Binds to single-stranded DNA and keeps it from base-pairing with a complementary strand.

S6K1 *See* S6 kinase-1.

S6 kinase-1 (S6K1) A protein kinase that phosphorylates the ribosomal protein S6. This enzyme is also phosphorylated by mTOR, which causes it to dissociate from eIF3 and phosphorylate eIF4B. This in turn enhances association between eIF4B and eIF4A. Activated S6K1 also phosphorylates (and inhibits) an inhibitor of eIF4A.

stable isotope labeling by amino acids in cell culture (SILAC) Labeling a protein with heavy isotopes by including heavy-isotope-labeled amino acids in the growth medium.

STC *See* sequence-tagged connector.

steroid receptor coactivator (SRC) A member of a family of proteins that bind to liganded, but not unliganded steroid receptors, and help these activators to recruit CBP.

sticky ends Single-stranded ends of double-stranded DNAs that are complementary and can therefore base-pair and stick together.

Stn1p *See* Cdc13p.

stochastic release A version of the σ-cycle that allows for release of σ at random during the elongation phase.

stop codon One of three codons (UAG, UAA, and UGA) that code for termination of translation.

stopped-flow apparatus An apparatus for performing kinetic experiments, in which reagents are forced together very rapidly, enabling their reaction to be measured.

strand exchange *See* postsynapsis.

streptavidin A protein made by *Streptomyces* bacteria that binds avidly to biotin.

streptomycin An antibiotic that kills bacteria by causing their ribosomes to misread mRNAs.

stringency (of hybridization) The combination of factors (temperature, salt, and organic solvent concentration) that influence the ability of two polynucleotide strands to hybridize. At high stringency, only perfectly complementary strands will hybridize. At reduced stringency, some mismatches can be tolerated.

strong-stop DNA The initial product of reverse transcription of a retroviral RNA. It initiates at the primer-binding site and terminates about 150 nt later, at the 5′-end of the viral RNA.

structural genomics The study of the expression of large sets of genes.

structural variation (genomic) Variation in large chunks of DNA, as opposed to SNPs.

STS *See* sequence-tagged site.

SUMO (small ubiquitin-related modifier) A small polypeptide that can be attached to other proteins, such as activators, which targets them to a compartment of the nucleus from which they cannot activate genes.

sumoylation The attachment of SUMO proteins to other proteins.

supercoil *See* superhelix.

superhelix A form of circular double-stranded DNA in which the double helix coils around itself like a twisted rubber band.

superinfection (λ phage) Infection of a lysogen of one lambdoid phage with a second lambdoid phage.

supershift The extra gel mobility shift observed when a new protein joins a protein–DNA complex.

superwobble hypothesis The idea that a tRNA with a U in the first position of its anticodon can, at least under certain conditions, recognize a codon ending in any of the four bases.

suppression Compensation by one mutation for the effects of another.

suppressor mutation A mutation that reverses the effects of a mutation in the same or another gene.

SV40 Simian virus 40. A DNA tumor virus with a small circular genome, capable of causing tumors in certain rodents.

SWI/SNF A family of coactivators that help remodel chromatin by disrupting nucleosome cores.

synapsis Alignment of complementary sequences in the single-stranded and double-stranded DNAs that will participate in strand exchange during homologous recombination in *E. coli*.

syntenic blocks Blocks of DNA in which the order of genes has been preserved in different organisms.

synteny Preservation of gene order in different organisms. Also, existence of genes on the same chromosome in a given organism.

synthetic lethal screen A screen for interacting genes that uses cells with a nonlethal mutation (e.g., a conditional lethal mutation) in one gene to search for other genes in which ordinarily nonlethal mutations are lethal. Presumably, this means that the products of these two genes interact in some way; a defect in one or the other is nonlethal, but a defect in both is lethal.

TAF *See* TBP-associated factor.

tag sequencing *See* ChIP-seq

tags (in SAGE) Short cDNAs corresponding to the mRNAs in a cell.

tag SNPs Diagnostic SNPs that reveal the probable nature of other SNPs in the same region.

T antigen The major product of the early region of the DNA tumor virus SV40. A DNA-binding protein with DNA helicase activity; has the ability to transform cells and thereby cause tumors.

Taq polymerase A heat-resistant DNA polymerase obtained from the thermophilic bacterium *Thermus aquaticus*.

target mRNA The mRNA that is targeted and degraded during RNAi.

TATA box An element with the consensus sequence TATAAAA that begins about 25–30 bp upstream of the start of transcription in most eukaryotic promoters recognized by RNA polymerase II.

TATA-box-binding protein (TBP) A subunit of SL1, TFIID, and TFIIIB in class I, II, and III preinitiation complexes, respectively. Binds to the TATA box in class II promoters that have a TATA box.

t loop A loop formed in the telomere at the end of a eukaryotic chromosome.

T loop The loop in a tRNA molecule, conventionally drawn on the right, that contains the nearly invariant sequence TψC, where ψ is pseudouridine.

TBP *See* TATA-box-binding protein.

TBP-associated factor A protein associated with TBP in SL1, TFIID, or TFIIIB.

TBP-free TAF_{II}-containing complex (TFTC) An alternative TFIID that lacks TBP.

TBP-like factor (TLF) A homolog of TBP that lacks the intercalating phenylalanines that help bend the TATA box. Appears to substitute for TBP in binding to at least some TATA-less class II promoters.

TBP-related factor 1 (TRF1) An alternative TBP found in *Drosophila* and active during neural development.

TC-NER *See* transcription-coupled NER.

T-cell receptor (TCR) Antigen-binding proteins on the surfaces of T cells. Composed of two heavy (β) and two light (α) chains.

T-DNA The tumor-inducing part of the Ti plasmid.

TEBP *See* telomere end-binding protein.

telomerase An enzyme that can extend the ends of telomeres after DNA replication.

telomere A structure at the end of a eukaryotic chromosome, containing tandem repeats of a short DNA sequence.

telomere end-binding protein (TEBP) A dimeric protein in ciliated protozoa that binds to telomere ends and protects them from degradation and from DNA repair enzymes.

telomerase reverse transcriptase (TERT) The subunit of telomerase with the reverse transcriptase active site.

telomere position effect (TPE) Silencing of genes near a telomere.

temperate phage A phage that can enter a lysogenic phase in which a prophage is formed.

temperature-sensitive mutation A mutation that causes a product to be made that is defective at high temperature (the nonpermissive temperature) but functional at low temperature (the permissive temperature).

template A polynucleotide (RNA or DNA) that serves as the guide for making a complementary polynucleotide. For example, a DNA strand serves as the template for ordinary transcription.

template DNA strand The DNA strand of a gene that is complementary to the RNA product of that gene; that is, the strand that served as the template for making the RNA. Sometimes called the anticoding strand or antisense strand.

Ten1p *See* Cdc13p.

teratogen A substance that causes abnormal development of an organism.

terminal transferase An enzyme that adds deoxyribonucleotides, one at a time, to the 3′-end of a DNA.

terminal uridylyl transferase (TUTase) An enzyme that adds UMP residues to pre-mRNAs during RNA editing.

terminator *See* transcription terminator.

***Ter* sites** DNA sties within the *E. coli* DNA replication termination region. There are six sites, called *TerA, TerB, TerC, TerD, TerE,* and *TerF.*

TERT *See* telomerase reverse transcriptase.

tertiary structure The overall three-dimensional shape of a polypeptide or RNA.

tetramer (protein) A complex of four polypeptides.

TFIIA A class II general transcription factor that stabilizes binding of TFIID to the TATA box.

TFIIB A class II general transcription factor that binds to a promoter after TFIID in vitro. Helps TFIIF plus RNA polymerase II bind to the promoter.

$TFIIB_C$ The C-terminal domain of TFIIB, which binds to the TATA box after it is bent by TBP.

$TFIIB_N$ The N-terminal domain of TFIIB, which binds near the active site of RNA polymerase II, positioning it for initiation the correct distance from the TATA box.

TFIID A class II general transcription factor that binds first to TATA-box-containing promoters in vitro and serves as a nucleation site around which the preinitiation complex assembles. Contains TATA-box-binding protein (TBP) and TBP-associated factors (TAF_{II}s).

TFIIE A class II general transcription factor that binds to the preinitiation complex after TFIIF and RNA polymerase and before TFIIH in vitro.

TFIIF A class II general transcription factor that binds to the preinitiation complex cooperatively with RNA polymerase II in vitro after TFIIB has bound.

TFIIH The last class II general transcription factor to bind to the preinitiation complex in vitro. Has protein kinase and DNA helicase activities.

TFIIS A protein that stimulates transcription elongation by RNA polymerase II by limiting pausing at pause sites.

TFIIIA A general transcription factor that works with TFIIIC to help activate eukaryotic 5S rRNA genes by facilitating binding of TFIIIB.

TFIIIB A general transcription factor that activates genes transcribed by RNA polymerase III by binding to a region just upstream of the gene.

TFIIIC A general transcription factor that stimulates binding of TFIIIB to classical class III genes.

TfR *See* transferrin receptor.

TFTC *See* TBP-free TAF_{II}-containing complex.

thermal cycler An instrument that performs PCR reactions automatically by repeatedly cycling among the three temperatures required for primer annealing, DNA elongation, and DNA denaturation.

4-thioU (sU) A photoreactive nucleotide that can be introduced (as 4-thioUMP) into an RNA and then UV-cross-linked to any RNA-binding proteins bound at the 4-thioUMP site.

30S initiation complex The complex composed of a 30S ribosomal particle, an mRNA, an fMet-$tRNA_f^{Met}$, initiation factors, and GTP. This complex is ready to join with a 50S ribosomal subunit.

30S ribosomal subunit The small bacterial ribosomal subunit, involved in mRNA decoding.

3′ box A DNA element near the end of a class II snRNA gene that ensures proper 3′ processing of the primary transcript.

3′-end The end of a polynucleotide with a free (or phosphorylated) 3′-hydroxyl group.

3′ tRNase *See* tRNA 3′ processing endoribonuclease.

thymidine A nucleoside containing the base thymine.

thymine (T) The pyrimidine base that pairs with adenine in DNA.

thymine dimer Two adjacent thymines in one DNA strand linked covalently, whose base pairing with adenines in the opposite strand is interrupted.

thyroid hormone receptor (TR) A nuclear receptor. In the absence of thyroid hormone, it acts as a repressor; in the presence of thyroid hormone, it acts as an activator.

thyroid-hormone-receptor-associated protein (TRAP) *See* SMCC/TRAP.

thyroid hormone response element (TRE) An enhancer that responds to the thyroid hormone receptor plus thyroid hormone.

TIF-1B The homolog of human SL1 in certain lower eukaryotes.

Ti plasmid The tumor-inducing plasmid from *Agrobacterium tumefaciens*. Used as a vector to carry foreign genes into plant cells.

tiling array A DNA microarray containing DNAs with overlapping sequences.

TIN2 (TRF1-interacting factor-2) *See* shelterin.

TLF *See* TBP-like factor.

tmRNA *See* transfer-messenger RNA.

tmRNA-mediated ribosome rescue Rescue of a stalled ribosome on a non-stop mRNA by tmRNA. The tmRNA is charged with alanine, then enters the A site of the stalled ribosome and donates its alanine to the stalled nascent polypeptide. On translocation, a short ORF of the tmRNA shifts to the P site, displacing the non-stop mRNA.

Tn3 An *E. coli* transposon that encodes an ampicillin-resistance gene.

toeprint assay A primer extension assay that locates the edge of a protein bound to a DNA or RNA.

topo IV The topoisomerase that decatenates the daughter duplexes at the end of *E. coli* DNA replication.

topoisomerase An enzyme that changes a DNA's superhelical form, or topology.

1. **type I topoisomerase** A topoisomerase that introduces transient single-strand breaks into substrate DNAs.

2. **type II topoisomerase** A topoisomerase that introduces transient double-strand breaks into substrate DNAs.

torus A donut-shaped structure.

TPE *See* telomere position effect.

trailer The untranslated region of bases at the 3′-end of an mRNA between the termination codon and the poly(A). Also called the 3′-UTR.

***trans*-acting** A term that describes a genetic element, such as a repressor gene or transcription factor gene, that can be on a separate chromosome and still influence another gene. These *trans*-acting genes function by producing a diffusible substance that can act at a distance.

transcribed fragments (transfrags) Small regions of a genome that are represented by transcripts, as detected by microarrays.

transcribed spacer A region encoding a part of an rRNA precursor that is removed during processing to produce the mature rRNAs.

transcript An RNA copy of a gene.

transcription The process by which an RNA copy of a gene is made.

transcription-activating domain The part of a transcription activator that stimulates transcription.

transcription arrest A more or less permanent pause in transcription, which requires external agents to restart.

transcription bubble The region of locally melted DNA that follows the RNA polymerase as it synthesizes the RNA product.

transcription-coupled NER (TC-NER) NER that can remove lesions only in transcribed strands because the RNA polymerase detects the lesions and attracts the NER apparatus.

transcription factor A protein that stimulates (or sometimes represses) transcription of a eukaryotic gene by binding to a promoter or enhancer element.

transcription factor B (TFB) An archaeal factor homologous to eukaryotic TFIIB.

transcription factories Discrete nuclear sites where transcription of multiple genes occurs.

transcription pause A temporary stoppage of transcription, which can be reversed by the polymerase itself.

transcription terminator A specific DNA sequence that signals transcription to terminate.

transcription unit A region of DNA bounded by a promoter and a terminator that is transcribed as a single unit. May contain multiple coding regions, as in the major late transcription unit of adenovirus.

transcriptional mapping Mapping of transcripts (not just genes) to specific sites on genomes.

transcriptome The sum of all the different transcripts an organism can make in its lifetime.

transcriptomics The global study of an organism's transcripts.

transcripts of unknown function (TUFs) Transcripts that do not code for proteins, and whose functions are not known.

transesterification A reaction that simultaneously breaks one ester bond and creates another. For example, the formation of the lariat intermediate in nuclear pre-mRNA splicing is a transesterification reaction.

transfection Transformation of eukaryotic cells by incorporating foreign DNA into the cells.

transferrin An iron-carrier protein that imports iron into cells via the transferrin receptor.

transferrin receptor (TfR) A membrane protein that binds transferrin and allows it to enter the cell with its payload of iron.

transfer-messenger RNA (tmRNA) A 300-nt RNA that resembles a tRNA and can rescue a stalled ribosome on a non-stop mRNA.

transfer RNA *See* tRNA.

transformation (genetic) An alteration in a cell's genetic makeup caused by introducing exogenous DNA.

transgene A foreign gene transplanted into an organism, making the recipient a transgenic organism.

transgenic organism An organism into which a new gene or set of genes has been transferred.

transition A mutation in which a pyrimidine replaces a pyrimidine, or a purine replaces a purine.

translation The process by which ribosomes use the information in mRNAs to synthesize proteins.

translesion synthesis (TLS) A mechanism for bypassing DNA damage by replicating through it.

translocation The translation elongation step, following the peptidyl transferase reaction, in which an mRNA moves one codon's length through the ribosome and brings a new codon into the ribosome's A site.

transmission genetics The study of the transmission of genes from one generation to the next.

transpeptidation The reaction (formation of peptide bonds) catalyzed by peptidyl transferase.

transposable element A DNA element that can move from one genomic location to another.

transposase The name for the collection of proteins, encoded by a transposon, that catalyze transposition.

transposition The movement of a DNA element (transposon) from one DNA location to another.

transposon *See* transposable element.

***trans*-splicing** Splicing together two RNA fragments transcribed from separate transcription units.

***trans* translation** The shift from reading a non-stop mRNA to reading the ORF of a tmRNA that occurs during tmRNA-mediated ribosomal rescue.

transversion A mutation in which a pyrimidine replaces a purine, or vice versa.

***trc* promoter** A hybrid promoter used in many expression vectors, which contains the –35 box of the *trp* promoter to provide strength, and the –10 box and operator of the *lac* promoter to provide inducibility.

TRE *See* thyroid hormone response element.

TRF1 and TRF2 (TTAGGG repeat-binding factors) Telomere-binding proteins that bind specifically to double-stranded DNA within telomeres. *See* shelterin.

TRF1 *See* TBP-related factor 1.

trigger dsRNA The double-stranded RNA that initiates the RNAi process.

trigger loop A part of Rpb1 that promotes accuracy of transcription by making several important contacts with correctly paired nucleotide substrates in the polymerase active site.

Trl A GAGA-box-binding protein.

tRNA (transfer RNA) A relatively small RNA molecule that binds an amino acid at one end and "reads" an mRNA codon at the other, thus serving as an "adapter" that translates the mRNA code into a sequence of amino acids.

tRNA charging The process of coupling a tRNA with its cognate amino acid, catalyzed by aminoacyl-tRNA synthetase.

$tRNA_f^{Met}$ The tRNA responsible for initiating protein synthesis in bacteria.

$tRNA_i^{Met}$ The tRNA responsible for initiating protein synthesis in eukaryotes, analogous to $tRNA_f^{Met}$.

$tRNA_m^{Met}$ The tRNA that inserts methionines into the interiors of proteins.

tRNA A-like domain The base-paired 5′- and 3′-ends of a tmRNA, which together resemble the acceptor stem of a tRNA.

tRNA 3′ processing endoribonuclease (3′ tRNase) The enzyme that removes excess nucleotides from the 3′-ends of tRNA precursors in eukaryotes.

***trp* operon** The operon that encodes that enzymes needed to make the amino acid tryptophan.

***trp* repressor** The repressor of the *trp* operon. Composed of the *trp* aporepressor plus the corepressor tryptophan.

trypanosomes Protozoa that parasitize both mammals and tsetse flies; the latter spread the disease by biting mammals.

T-segment The segment of DNA that passes through the G segment gate during topoisomerase II activity.

tumor suppressor gene A gene whose product tends to keep cell division under control, and thereby suppresses the development of malignant tumors.

Tus *Ter* utilization substance. An *E. coli* protein that binds to *Ter* sites and participates in replication termination.

TUTase *See* terminal uridylyl transferase.

12 signal An RSS composed of a conserved heptamer and nonamer separated by a nonconserved 12-bp sequence.

12/23 rule The recombination scheme used in immunoglobulin and T-cell receptor gene maturation, in which a 12 signal is always joined to a 23 signal, but like signals are never joined to each other.

23 signal An RSS composed of a conserved heptamer and nonamer separated by a nonconserved 23 bp sequence.

two-dimensional gel electrophoresis A high-resolution method for separating proteins. First the proteins are separated in the first dimension by isoelectric focusing. Then they are separated in the second dimension by SDS-PAGE.

2 μm plasmid A yeast plasmid that serves as the basis for yeast cloning vectors.

Ty A yeast transposon that transposes via a retrovirus-like mechanism.

U1 snRNP The first snRNP that recognizes the 5-splice site in a nuclear pre-mRNA.

$U2AF^{35}$ and $U2AF^{65}$ The two subunits of U2AF.

U2-associated factor (U2AF) A splicing factor that helps recognize the correct AG at the 3′-splice site by binding to both the polypyrimidine tract in the 3′-splice signal and the AG.

U2 snRNP The snRNP that recognizes the branchpoint in a nuclear pre-mRNA.

U3 The 3′-untranslated region in the LTR of a retrovirus.

U4 snRNP The snRNP whose RNA base-pairs with the RNA in U6 snRNP until U6 snRNP is needed to splice a nuclear pre-mRNA.

U4atac A minor snRNA that participates in splicing variant introns and fulfills the same role as U4 snRNA.

U5 The 5′-untranslated region in the LTR of a retrovirus.

U5 snRNP The snRNP that associates with both 5′ and 3′ exon–intron junctions, thus helping to bring the two exons together for splicing.

U6 snRNP The snRNP whose RNA base pairs with both the 5′-splice site and with the RNA in U2 snRNP in the spliceosome.

U6atac A minor snRNA that participates in splicing variant introns and fulfills the same role as U6 snRNA.

U11 snRNA A minor snRNA that participates in splicing variant introns and fulfills the same role as U1snRNA.

U12 snRNA A minor snRNA that participates in splicing variant introns and fulfills the same role as U2 snRNA.

UAF *See* upstream activating factor.

ubiquitin A small polypeptide that can be attached to proteins, including activators, either singly or in chains. Single ubiquitylation frequently activates an activator, but polyubiquitylation marks it for destruction by the proteasome.

ubiquitylated protein A protein with at least one molecule of ubiquitin attached.

UAS_G *See* upstream activating sequence.

UBF *See* upstream-binding factor.

ultraviolet (UV) radiation Radiation found in sunlight. Causes pyrimidine dimers in DNA.

UmuC One of the components of the $UmuD'_2C$ complex.

umuC One of the genes of the *umuDC* operon, which is induced by the SOS response to DNA damage.

UmuD Clipped by a protease to form UmuD′, one of the components of the $UmuD'_2C$ complex.

umuD One of the genes of the *umuDC* operon, which is induced by the SOS response to DNA damage.

***umu*DC** The operon containing the *umuD* and *umuC* genes.

$UmuD'_2C$ Also known as DNA polymerase V, which can cause error-prone bypass of pyrimidine dimers.

undermethylated region A region of a gene or its flank that is relatively poor in, or devoid of, methyl groups.

unidirectional DNA replication Replication that occurs in one direction, with only one active replicating fork.

unit cell A small repeating unit in a crystal.

untranslated region (UTR) A region at the 5′- or 3′-end of an mRNA that lies outside the coding region, so is not translated.

UP element An extra promoter element found upstream of the -35 box in certain strong bacterial promoters. Allows extra strong interaction between polymerase and promoter.

Upf1 and Upf2 Part of the mammalian T-cell downstream destabilizing element.

up mutation A mutation, frequently in a promoter, that results in more expression of a gene.

UPE *See* upstream promoter element.

upstream activating sequence (UAS_G) An enhancer for yeast galactose-utilization genes that binds the activator GAL4.

upstream activating factor (UAF) The yeast homolog of UBF.

upstream-binding factor (UBF) A class I transcription factor that binds to the upstream promoter element (UPE). Acts synergistically with SL1 to stimulate polymerase I binding and transcription.

upstream-binding site (UBS) A site on the bacterial core RNA polymerase that binds to the upstream half of the RNA hairpin at a terminator and thereby slows down hairpin formation and inhibits transcription termination.

upstream promoter element (UPE; class I) A promoter element found upstream of the core promoter in eukaryotic class I promoters.

upstream promoter element (class II) An element of the proximal promoter, upstream of the core promoter.

uracil (U) The pyrimidine base that replaces thymine in RNA.

uracil N-glycosylase An enzyme that removes uracil from a DNA strand, leaving an abasic site (a sugar without a base).

uridine A nucleoside containing the base uracil.

UTR An untranslated region on an mRNA—either at the 5′- or 3′-end.

variable loop The loop, or stem, in a tRNA molecule, which lies between the anticodon and T stems.

variable number tandem repeats (VNTR) A type of RFLP that includes tandem repeats of a minisatellite between the restriction sites.

variable region The region of an antibody that binds specifically to a foreign substance, or antigen. As its name implies, it varies considerably from one kind of antibody to another.

V(D)J joining The assembly of active immunoglobulin or T-cell receptor genes by recombination involving separate V and J or V, D, and J segments in the embryonic genes.

vector A DNA (a plasmid or a phage DNA) that serves as a carrier in gene cloning experiments.

vegetative cell A cell that is reproducing by division, rather than sporulating or reproducing sexually.

virulent phage A phage that lyses its host.

VNTR *See* variable number tandem repeats.

void volume The fraction in a gel filtration experiment that contains the large molecules that cannot enter the pores in the gel at all.

VP16 A herpesvirus transcription factor with an acidic transcription-activating domain, but no DNA-binding domain.

Western blotting Electrophoresing proteins, then blotting them to a membrane and reacting them with a specific antibody or antiserum. The antibody is detected with a labeled secondary antibody or protein A.

wobble The ability of the third base of a codon to shift slightly to form a non-Watson–Crick base pair with the first base of an anticodon, thus allowing a tRNA to translate more than one codon.

wobble base pair A base pair formed by wobble (a G–U or A–I pair).

wobble hypothesis Francis Crick's hypothesis that invoked wobble to explain how one anticodon could decode more than one codon.

wobble position The third base of a codon, where wobble base pairing is permitted.

wyosine A highly modified guanine nucleoside found in tRNA.

xeroderma pigmentosum (XP) A disease characterized by extreme sensitivity to sunlight. Even mild exposure leads to many skin cancers. Caused by a defect in nucleotide excision repair.

Xis The product of the λ *xis* gene. Responsible for excision of λ DNA from host DNA.

XPA A protein that verifies damaged DNA that is already bound to XPC and helps assemble the other components of the human NER complex.

XPA–XPG The human genes involved in NER. Mutations in any of these genes can cause XP.

XPB One of the two subunits of the human TFIIH DNA helicase. Necessary for DNA melting during NER.

XPC Together with another protein, can recognize DNA lesions and initiate human GG-NER.

XPD One of the two subunits of the human TFIIH DNA helicase. Necessary for DNA melting during NER.

XPF Together with ERCC1, cuts on the 5′-side of DNA damage during human NER.

XPG An endonuclease that cuts on the 3′-side of DNA damage during human NER.

XP-V A variant form of xeroderma pigmentosum caused by mutations in the DNA polymerase η gene.

x-ray crystallography *See* x-ray diffraction analysis.

x-ray diffraction analysis A method for determining the three-dimensional structure of molecules by measuring the diffraction of x-rays by crystals of the molecule or molecules.

x-rays High-energy radiation that is diffracted by crystals. The pattern of x-ray diffraction can then be used to determine the shape of the molecule(s) in the crystal. X-rays also ionize cellular components and can therefore cause chromosome breaks.

Xrn2 A human $5' \rightarrow 3'$ exonuclease that degrades the downstream RNA product after co-transcriptional cleavage, and causes termination of transcription.

YAC See yeast artificial chromosome.

yeast artificial chromosome A high-capacity cloning vector consisting of yeast left and right telomeres and a centromere. DNA placed between the centromere and one telomere becomes part of the YAC and will replicate in yeast cells.

yeast two-hybrid assay An assay for interaction between two proteins. One protein (the bait) is produced as a fusion protein with a DNA-binding domain from another protein. The other protein (the target, or prey) is produced as a fusion protein with a transcription-activation domain. If the two fusion proteins interact in the cell, they form an activator that can activate one or more reporter genes.

Z-DNA A left-handed helical form of double-stranded DNA whose backbone has a zigzag appearance. This form is stabilized by stretches of alternating purines and pyrimidines.

zinc finger A DNA-binding motif that contains a zinc ion complexed to four amino acid side chains, usually the side chains of two cysteines and two histidines. The motif is roughly finger-shaped and inserts into the DNA major groove, where it makes specific protein–DNA contacts.

I N D E X